Table of Atomic Masses*

Element	Symbol	Atomic Number	Atomic Mass	Element	Symbol	Atomic Number	Atomic Mass	Element	Symbol	Atomic Number	Atomic Mass
Actinium	Ac	89	(227)†	Hafnium	Hf	72	178.5	Promethium	Pm	61	(145)
Aluminum	Al	13	26.98	Helium	He	2	4.003	Protactinium	Pa	91	(231)
Americium	Am	95	(243)	Holmium	Ho	67	164.9	Radium	Ra	88	226
Antimony	Sb	51	121.8	Hydrogen	H	1	1.008	Radon	Rn	86	(222)
Argon	Ar	18	39.95	Indium	In	49	114.8	Rhenium	Re	75	186.2
Arsenic	As	33	74.92	Iodine	I	53	126.9	Rhodium	Rh	45	102.9
Astatine	At	85	(210)	Iridium	Ir	77	192.2	Rubidium	Rb	37	85.47
Barium	Ba	56	137.3	Iron	Fe	26	55.85	Ruthenium	Ru	44	101.1
Berkelium	Bk	97	(247)	Krypton	Kr	36	83.80	Samarium	Sm	62	150.4
Beryllium	Be	4	9.012	Lanthanum	La	57	138.9	Scandium	Sc	21	44.96
Bismuth	Bi	83	209.0	Lawrencium	Lr	103	(260)	Selenium	Se	34	78.96
Boron	B	5	10.81	Lead	Pb	82	207.2	Silicon	Si	14	28.09
Bromine	Br	35	79.90	Lithium	Li	3	6.941	Silver	Ag	47	107.9
Cadmium	Cd	48	112.4	Lutetium	Lu	71	175.0	Sodium	Na	11	22.99
Calcium	Ca	20	40.08	Magnesium	Mg	12	24.31	Strontium	Sr	38	87.62
Californium	Cf	98	(251)	Manganese	Mn	25	54.94	Sulfur	S	16	32.07
Carbon	C	6	12.01	Mendelevium	Md	101	(258)	Tantalum	Ta	73	180.9
Cerium	Ce	58	140.1	Mercury	Hg	80	200.6	Technetium	Tc	43	(98)
Cesium	Cs	55	132.9	Molybdenum	Mo	42	95.94	Tellurium	Te	52	127.6
Chlorine	Cl	17	35.45	Neodymium	Nd	60	144.2	Terbium	Tb	65	158.9
Chromium	Cr	24	52.00	Neon	Ne	10	20.18	Thallium	Tl	81	204.4
Cobalt	Co	27	58.93	Neptunium	Np	93	(237)	Thorium	Th	90	232.0
Copper	Cu	29	63.55	Nickel	Ni	28	58.69	Thulium	Tm	69	168.9
Curium	Cm	96	(247)	Niobium	Nb	41	92.91	Tin	Sn	50	118.7
Dysprosium	Dy	66	162.5	Nitrogen	N	7	14.01	Titanium	Ti	22	47.88
Einsteinium	Es	99	(252)	Nobelium	No	102	(259)	Tungsten	W	74	183.9
Erbium	Er	68	167.3	Osmium	Os	76	190.2	Uranium	U	92	238.0
Europium	Eu	63	152.0	Oxygen	O	8	16.00	Vanadium	V	23	50.94
Fermium	Fm	100	(257)	Palladium	Pd	46	106.4	Xenon	Xe	54	131.3
Fluorine	F	9	19.00	Phosphorus	P	15	30.97	Ytterbium	Yb	70	173.0
Francium	Fr	87	(223)	Platinum	Pt	78	195.1	Yttrium	Y	39	88.91
Gadolinium	Gd	64	157.3	Plutonium	Pu	94	(244)	Zinc	Zn	30	65.38
Gallium	Ga	31	69.72	Polonium	Po	84	(209)	Zirconium	Zr	40	91.22
Germanium	Ge	32	72.59	Potassium	K	19	39.10				
Gold	Au	79	197.0	Praseodymium	Pr	59	140.9				

*The values given here are to four significant figures where possible.
†A value given in parentheses denotes the mass of the longest-lived isotope.

CHEMICAL PRINCIPLES

CHEMICAL PRINCIPLES

Second Edition

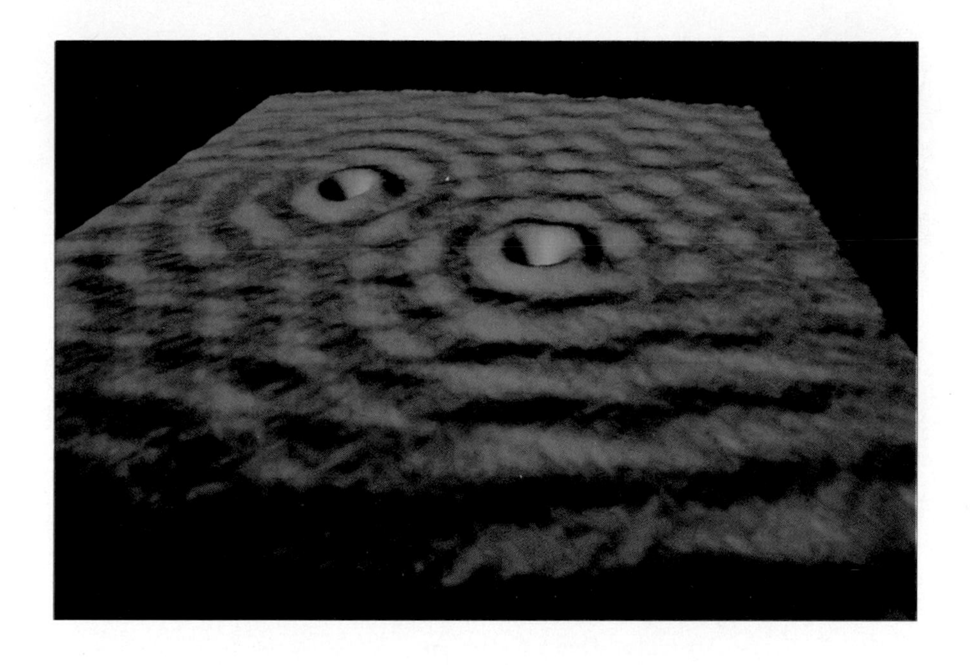

Steven S. Zumdahl

University of Illinois

D. C. Heath and Company

Lexington, Massachusetts Toronto

Address editorial correspondence to
D. C. Heath and Company
125 Spring Street
Lexington, MA 02173

Acquisitions Editor: Kent Porter Hamann
Developmental Editor: June Goldstein
Production Editor: Heather Garrison
Designer: Henry Rachlin
Photo Researcher: Sharon Donahue
Art Editor: Diane Grossman
Production Coordinator: Richard Tonachel
Permissions Editor: Margaret Roll

Cover photo: Courtesy, IBM Corporation, Research Division, Almaden Research Center.

For permission to use photographs, grateful acknowledgment is made to the copyright holders listed on pages A67–A68, which are hereby considered an extension of this copyright page.

Published simultaneously in Canada.

Printed in the United States of America.

International Standard Book Number: 0–669–39321–5

Library of Congress Catalog Number: 94–76934

10 9 8 7 6 5 4 3 2 1

PREFACE

In a time when there is great concern about the quality of chemical education, those of us who teach freshman chemistry must show that our discipline is accessible, fascinating, and important—one that offers very diverse, but always interesting and challenging opportunities. We must also let our own enthusiasm for our discipline show clearly to provide our students with the motivation they need to proceed successfully through their chemistry course. This is what I have tried to do in *Chemical Principles,* Second Edition.

This text is based on my experience at the University of Illinois teaching an accelerated general chemistry course for chemical sciences majors and other students who require a rigorous introductory course. These students typically have excellent credentials and an excellent aptitude for chemistry, but have had only limited exposure to the fundamental concepts of chemistry. Although they may know how to solve stoichiometry and gas problems when they arrive in my course, these students typically lack a thorough understanding of the chemical principles that underlie these applications. This is not because they have had inadequate preparation in high school; instead, I believe it results from the nature of chemistry itself—a subject that even chemists realize sometimes requires several passes before real mastery can take place.

So my point in writing this text is to produce a book that does not assume that students already know how to think like chemists. These students will do complicated and rigorous thinking eventually, but they must be brought to that point gradually. Thus this book covers the advanced topics (in gases, atomic theory, thermodynamics, and so on) that one expects in a course for chemical science majors, but it starts with the fundamentals, and then builds to the level required for more complete understanding. Chemistry is not the result of an inspired vision. It is the product of countless observations and many attempts, using logic and trial and error, to account for these observations. In this book I try to develop key chemical concepts in the same way—to show the observations first and then discuss the models that have been constructed to explain the observed behavior. I hope students will practice "thinking like a chemist" by carefully studying the observations to see if they can follow the thought process, rather than just jumping ahead to the equation or model that will follow.

In *Chemical Principles,* Second Edition, I take advantage of the excellent math skills that these students typically possess. As a result, there are fewer worked-out examples than would be found in most mainstream books. The end-of-chapter problems cover a wide range—from drill exercises to difficult problems, some of which would challenge the average senior chemistry major. Thus instructors can tailor the problem assignments to the level appropriate for their students.

Finally, this book maintains a student-friendly approach without being patronizing. In addition, to demonstrate the importance of chemistry in real life, I have incorporated throughout the book a number of applications and recent advances in chemistry in essay form.

I am pleased that the first edition of this text was well received and that the users did not recommend any major changes. Thus we have refined the discussions in areas where that was needed, and we have added several new "applications boxes," such as "Electrons as Waves" and "Chemicals to Protect the Ozone," that emphasize the environment and materials and engineering sciences. We have also updated and refined the end-of-chapter problems. In addition, many of the illustrations have been reconceived for additional visual clarification of chemical concepts and all have been rerendered to achieve a new standard of accuracy and clarity. Pedagogical consistency in the use of color has been a focus in the generation of this new artwork.

What's It Like to Be a Chemical Professional?

The most unusual chapter in this text is Chapter 1, which discusses what it means to be a chemical professional. I have included this material because students, especially freshmen, know very little about possible careers in the chemical sciences and tend not to think about these issues until it's time for them to start looking for jobs. In addition, they do not realize the incredible diversity of opportunities that exist in the chemical sciences or how often the typical person changes jobs. To inform students about these issues, Chapter 1 discusses some typical jobs, as well as some typical problems confronted by someone working in the chemical sciences. I have also included profiles of some real people—several young professionals who are pursuing diverse careers.

Organization

The early chapters in this book deal with chemical reactions. Stoichiometry is covered in Chapters 3 and 4, with special emphasis on reactions in aqueous solutions. The properties of gases are treated in Chapter 5, followed by coverage of gas phase equilibria in Chapter 6. Acid-base equilibria are covered in Chapter 7, and Chapter 8 deals with additional aqueous equilibria. Thermodynamics is covered in two chapters: Chapter 9 deals with thermochemistry and the first law of thermodynamics; Chapter 10 treats the topics associated with the second law of thermodynamics. The discussion of electrochemistry follows in Chapter 11. Atomic theory and quantum mechanics are covered in Chapter 12, followed by two chapters on chemical bonding (Chapters 13 and 14). Chemical kinetics is discussed in Chapter 15, followed by coverage of solids and liquids in Chapter 16, and the physical properties of solutions in Chapter 17. A systematic treatment of the descriptive chemistry of the representative elements is given in Chapters 18 and 19, and of the transition metals in Chapter 20. Chapters 21–23 cover topics in nuclear chemistry, organic chemistry, and biochemistry, respectively.

Flexibility of Topic Order

Instructors have several options for arranging the material to complement their syllabi. For example, the section on gas phase and aqueous equilibria (Chapters 6–8) could be moved to any point later in the course. The chapters on thermodynamics can be separated: Chapter 9 can be used early in the course, with Chapter 10 later. In addition, the chapters on atomic theory and bonding (Chapters 12–14) can be used near the beginning of the course. In summary, an instructor who wants to cover atomic theory early and equilibrium later might prefer the following order of chapters: 1–5, 9, 12, 13, 14, 10, 11, 6, 7, 8, 15–23. An alternative order might be: 1–5, 9, 12, 13, 14, 6, 7, 8, 10, 11, 15–23. The point is that the chapters on atomic theory and bonding (12–14), thermodynamics (9, 10), and equilibrium (6, 7, 8) can be moved around quite easily. In addition, the kinetics chapter (Chapter 15) can be covered at any time after bonding. It is also possible to use Chapter 21 (on nuclear chemistry) much earlier—after Chapter 12, for example—if desired.

Mathematical Level

This text assumes a solid background in algebra. All of the mathematical operations required are described in Appendix One or are illustrated in worked-out examples. A knowledge of calculus is not required for use of this text. Differential and integral notations are used only where absolutely necessary and are explained where they are used as in Section 10.2 and Section 15.1.

Supplements

A supplements package has been designed to make this book more useful to both student and instructor. These supplements include

- ***Chemistry in Motion Videotape and Videodisc Supplements.*** **D. C. Heath has an exciting and pedagogically valuable set of video demonstrations of chemical experiments. Available on videotape *and* videodisc is a series of 34 demonstrations done by Patricia L. Samuel of Boston University, with the assistance of Lorraine Kelly. The *Chemistry in Motion Videodisc* also includes illustrations from the second edition of *Chemical Principles.***
- ***Study Guide,*** **by Paul B. Kelter of the University of Nebraska, Lincoln. Written to be a self-study aid for students, this guide includes alternate strategies for solving various types of problems, supplemental explanations for the most difficult material, and self-tests. There are over 300 worked examples and approximately 900 practice problems (with answers) designed to give students mastery and confidence.**
- ***Selected Solutions Guide,*** **by Kenneth C. Brooks, of New Mexico State University, and Thomas J. Hummel and Steven S. Zumdahl, both of the University of Illinois, Urbana, provides detailed solutions for half of the end-of-chapter exercises (designated by blue question numbers or letters) using the strategies emphasized in the text.**

- ***Complete Solutions Guide,*** **by Kenneth C. Brooks, Thomas J. Hummel, and Steven S. Zumdahl, presents detailed solutions for all of the end-of-chapter exercises in the text for the convenience of faculty and staff involved in instruction and for instructors who wish their students to have solutions for all exercises. Departmental approval is required for the sale of the *Complete Solutions Guide* to students.**
- ***Test Item File,*** **by Susan Arena and Thomas J. Hummel (available to adopters), offers a printed version of more than 1000 exam questions referenced to the appropriate text section. Questions are available in multiple-choice, open-ended, and true-false formats.**
- ***Computerized Testing*** **presents the *Test Item File* questions in a computerized testing program by ESATest. Instructors can produce chapter tests, midterms, and final exams easily and with excellent graphics capability. The instructor can also edit existing questions or add new ones as desired, or preview questions on screen and add them to the test with a single keystroke. The testing program is available for Apple II, Macintosh, and IBM computers.**

Acknowledgments

The successful completion of this book is due to the efforts of many people. Kent Porter Hamann, Editorial Director, Science, has done a masterful job of spearheading this project. Kent's incredible knowledge of chemistry publishing, coupled with her high standards and endless supplies of energy and enthusiasm have made this project a real pleasure. In addition, I greatly appreciate the efforts of June Goldstein, Developmental Editor, who worked tirelessly gathering ideas and organizing and assimilating the information from the many reviewers of the first edition. June's creativity and powers of organization, her meticulous reading of the manuscript, and her cheerful and positive outlook contributed greatly to the second edition of this text. I am also grateful to Heather Garrison, Production Editor, for the excellent job she did in coordinating the production of a very complicated project.

I greatly appreciate the efforts of Tom Hummel from the University of Illinois who carefully read the end-of-chapter exercises, and provided many suggestions for improving their clarity. Tom also checked the answers that appear at the end of the book and provided a very thorough accuracy review of all the solutions in the Solutions Manual.

My thanks also go to Ken Brooks from New Mexico State University, who collaborated in writing the end-of-chapter exercises, wrote the Solutions Manual for these exercises in collaboration with Tom Hummel, provided many helpful suggestions concerning the content of the text, and who has been a good friend through many projects. I also thank Dr. Malcolm F. Nicol of the University of California, Los Angeles, and Dr. Christer Svensson of Lund University, Sweden, for the end-of-chapter problems they contributed.

Many other people at D. C. Heath contributed in very important ways to this project. Jim Porter Hamann, Senior Marketing Manager, who provided invaluable feedback from the marketplace and a great deal of enthusiastic general support; Sharon Donohue, Photo Researcher, who showed real creativity in securing photos for this text; Henry Rachlin, Senior Designer, who developed a functional and attractive design; Diane Grossman, Art Editor; and Daphne Zervoglos, Joanne Williams, and Mary Ned Fotis, Editorial Assistants, without whose help this book never would have been completed.

The sales reps play a vital role in the life of a text like this one. They furnish valuable information from the marketplace as the text is developed, and then they explain the goals and strengths of the finished text to the potential users. I extend my special thanks to these talented and energetic people.

Finally, I owe much to the following people who reviewed the manuscript for the first edition: J. Aaron Bertrand, Georgia Institute of Technology; Ernest R. Davidson, Indiana University; James E. Davis, Harvard University; Jack D. Graybeal, Virginia Polytechnic Institute and State University; John E. Harriman, University of Wisconsin—Madison; and John S. Hutchinson, Rice University. In addition, I want to extend special thanks to the following users of the first edition, who provided critical feedback: Roger E. Cramer, University of Hawaii; Morton Z. Hoffman, Boston University; R. Fehrmann, Technical University of Denmark; Joseph Lauher, SUNY at Stony Brook; Malcolm Nicol, University of California, Los Angeles; Bjørn Pedersen, University of Oslo; Norbert T. Porile, Purdue University; Rebecca Regan, Cornell University; Christer Svensson, Lund University; Jeff Tassin, University of the South. These people provided thoughtful, constructive criticism that helped immensely in producing a useful, accurate text. Their contribution to this book cannot be overestimated.

ABOUT THE AUTHOR

STEVEN S. ZUMDAHL received his B.S. degree in Chemistry from Wheaton College (Illinois) in 1964 and his Ph.D. in Chemistry from the University of Illinois (Urbana) in 1968.

In over 25 years of teaching he has been a faculty member at the University of Colorado (Boulder), Parkland College (Illinois), and the University of Illinois (Urbana). Currently he is Associate Head of the Chemistry Department and Professor of Chemistry as well as Director of Undergraduate Chemistry Programs at the University of Illinois. In 1994 Dr. Zumdahl received the National Catalyst Award from the Chemical Manufacturers Association in recognition of his contribution to chemical education in the United States.

Professor Zumdahl is known at the University of Illinois for his rapport with students and for his outstanding teaching ability. During his tenure at the University, he has received the University of Illinois Award for Excellence in Teaching, the Liberal Arts and Sciences College Award for Distinguished Teaching, and the School of Chemical Sciences Teaching Award (four times).

Dr. Z., as he is known to his students, is an avid collector of classic automobiles. He has recently restored a 1957 Corvette (which is parked in his study) and is now restoring a 1963 Corvette which was his daily transportation for more than 20 years.

CONTENTS

12 Quantum Mechanics and Atomic Theory 495

13 Bonding: General Concepts 567

14 Covalent Bonding: Orbitals 631

15 Chemical Kinetics 671

18 The Representative Elements: Groups 1A Through 4A 831

19 The Representative Elements: Groups 5A Through 8A 855

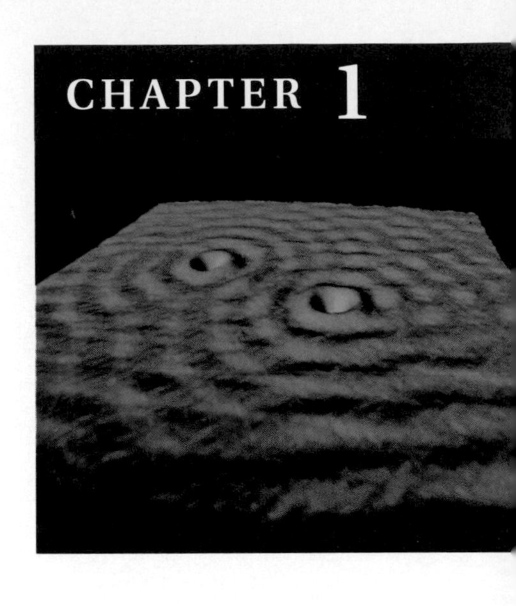

Chemists and Chemistry

Chemistry. It is a word that evokes various, and often dramatic, responses. It is a word that is impossible to define concisely, because the field is so diverse, and its practitioners perform such an incredible variety of jobs. Chemistry mainly deals with situations in which the nature of a substance is changed by altering its composition: entirely new substances are synthesized or the properties of existing substances are enhanced.

There are many misconceptions about the practitioners of chemistry. Many people picture a chemist as a solitary figure who works in a laboratory and does not talk to anyone else for days at a time. Nothing could be further from the truth. Many chemists do indeed work in laboratories but rarely by themselves. A typical day for a modern chemist would be spent as a member of a team solving a particular problem important to his/her company. This team might consist of chemists from various specialties, chemical engineers, development specialists, and possibly even lawyers. Figure 1.1 represents the people and organizations with which typical laboratory chemists might expect to interact in the course of their jobs.

On the other hand, many persons trained as chemists do not perform actual laboratory work but may work as patent lawyers, financial analysts, plant managers, salespeople, personnel managers, and so on. Also, it is quite common for a person trained as a chemist to have many different jobs during a career. This will become obvious as you read the profiles of chemists at the end of this chapter.

The goal of this chapter is to introduce some of the important aspects of chemistry not typically discussed in connection with learning chemistry. In the

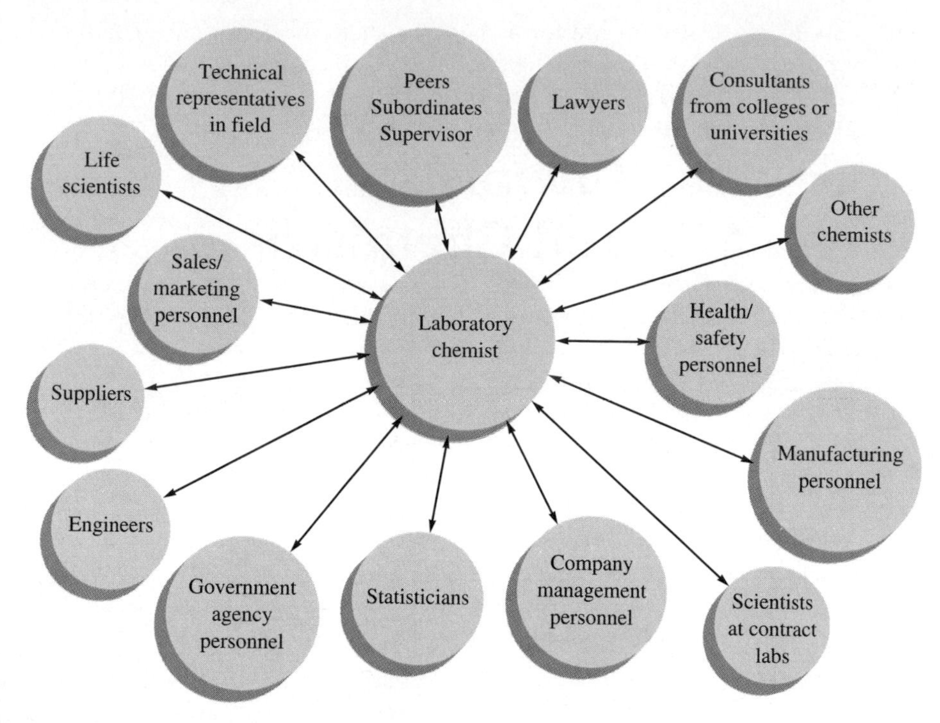

Figure 1.1
Typical chemists interact with a great variety of other people while doing their jobs.

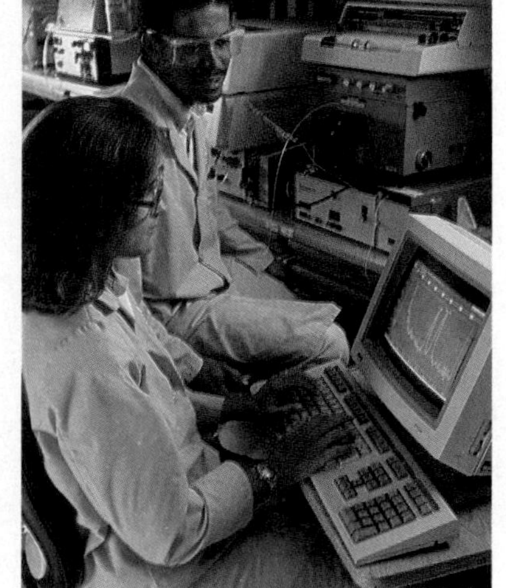

Chemists work together in an industrial setting to solve complex problems.

remainder of this text we will concentrate on the formal discipline of chemistry—its observations, theories, and applications. In this chapter we briefly discuss what it means to be a chemist, biochemist, or chemical engineer, and point out some of the opportunities that exist for someone trained in the chemical sciences. This discussion of opportunities is by no means exhaustive. Its purpose is to give some appreciation for the great variety of interesting jobs available to a person with a chemical sciences background. We include an introduction to the world of commercial chemistry and provide a couple of specific examples of the types of problems confronted by the practitioners of the "chemical arts." We begin by considering the chemical scientist as a problem solver.

1.1 Thinking Like a Chemist

Much of your life, both personal and professional, will involve problem solving. Most likely, the more creative you are at solving problems, the more effective and successful you will be. Chemists are usually excellent problem solvers because they get a lot of practice. Chemical problems are frequently very complicated—there is usually no neat and tidy solution. Often it is difficult to know where to begin. In response to this dilemma, a chemist makes an educated guess (formulates a hypothesis) and then tests it to see if the proposed solution correctly predicts the observed behavior of the system. This process of trial and error is virtually a way of life for a chemist. Chemists rarely solve a complex problem in a straightforward, elegant manner. More commonly, they poke and prod the problem and make progress only in fits and starts.

It's very important to keep this in mind as you study chemistry. Although "plug and chug" exercises are necessary to familiarize you with the relationships that govern chemical behavior, your ultimate goal should be to advance beyond this stage to true problem solving. Unfortunately, it is impossible to give a formula for becoming a successful problem solver. Creative problem solving is a rather mysterious activity that defies simple analysis. However, it is clear that practice helps. That's why we will make every attempt in this text to challenge you to be creative with the knowledge of chemistry you will be acquiring. Although this process can be frustrating at times, it is definitely worth the struggle—both because it is one of the most valuable skills you can develop and because it helps you test your understanding of chemical concepts. If your understanding of these concepts is not sufficient to allow you to solve problems involving "twists" that you have never encountered before, your knowledge is not very useful to you. The only way to develop your creativity is to expose yourself to new situations where you need to make new connections. A substantial part of creative problem solving involves developing the confidence necessary to think your way through unfamiliar situations. You must recognize that the entire solution to a complex problem is almost never visible in the beginning. Typically, one tries first to understand pieces of the problem and then puts these pieces together to form the solution.

1.2 A Real-World Chemistry Problem

As discussed above, the professional chemist is primarily a problem solver—one who daily confronts tough, but fascinating, situations that must be understood. To illustrate, we will consider an important current problem that requires chemical expertise to solve: the crumbling of the paper in many of the books published in the past century. The pages of many of these books are literally falling apart. To give some perspective on the magnitude of the problem, if the books in the New York Public Library were lined up, they would stretch for nearly 100 miles. Currently, about 40 miles of these books are quietly crumbling to dust.

Acid-damaged paper.

Because of the magnitude of this problem, the company that develops a successful process will reap considerable financial rewards, in addition to performing an important service to society. Assume that you work for a company that is interested in finding a method for saving the crumbling paper in books, and that you are put in charge of your company's efforts to develop such a process. What do you know about paper? Probably not much. So the first step is to go to the library to learn all you can about paper. Because paper manufacturing is a mature industry, there is a great deal of information available. Research at the library will show that paper is made of cellulose obtained from wood pulp and that the finished paper is "sized" to give it a smooth surface that prevents ink from "fuzzing." The agent typically used for sizing is alum, $Al_2(SO_4)_3$, which is the cause of the eventual decomposition of the paper. This happens as follows: in the presence of moisture, the Al^{3+} ions from alum

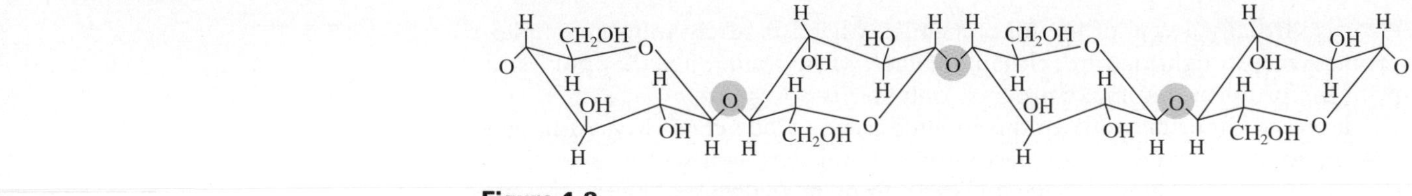

Figure 1.2
The polymer cellulose, which consists of β-D-glucose monomers.

become hydrated, forming $Al(H_2O)_6^{3+}$. The $Al(H_2O)_6^{3+}$ ion acts as an acid because the very strong $Al^{3+}—O$ bond causes changes in the O—H bonds of the attached water molecules, thus allowing H^+ ions to be produced by the following reaction:

$$Al(H_2O)_6^{3+} \rightleftharpoons [Al(OH)(H_2O)_5]^{2+} + H^+$$

Therefore, paper sized with alum contains significant numbers of H^+ ions. This is important because the H^+ assists in the breakdown of the polymeric cellulose structure of paper. Cellulose is composed of glucose molecules ($C_6H_{12}O_6$) bonded together to form long chains. A segment of cellulose is shown in Fig. 1.2. When the long chains of glucose units in cellulose are broken into shorter pieces, the structural integrity of the paper fails and it crumbles.

Although library research helps you to understand the fundamentals of the problem, now the tough part (and the most interesting part) begins. Can you find a creative solution to the problem? Can the paper in existing books be treated to stop the deterioration in a way that is economical, permanent, and safe?

The essence of the problem seems to be the H^+ present in the paper. How can it be removed or at least rendered harmless?

Your general knowledge of chemistry tells you that some sort of base (a substance that reacts with H^+) is needed. One of the most common and least expensive bases is sodium hydroxide. Why not dip the affected books in a solution of sodium hydroxide and remove the H^+ by the reaction: $H^+ + OH^- \rightarrow H_2O$? This seems to be a reasonable first idea; but as you consider it further and discuss it with your colleagues, several problems become apparent:

1. The $NaOH(aq)$ is a strong base and is therefore quite corrosive. It will destroy the paper by breaking down the cellulose just as acid does.
2. The binding of the books will be destroyed by dipping the books in water, and the pages will stick together after the book dries.
3. The process will be very labor-intensive, requiring the handling of individual books.

Some of these difficulties can be addressed. For example, a much weaker base than sodium hydroxide could be used. Also, the pages could be removed from the binding, soaked one at a time, dried and then rebound. In fact, this process is used for some very rare and valuable books, but the labor involved makes it very expensive—much too expensive for the miles of books in the New York Public Library. Obviously, this process is not what your company is seeking.

You need to find a way to treat large numbers of books without disassembling them. How about using a gaseous base? The books could be sealed in a chamber and the gaseous base allowed to permeate them. The first candidate that occurs to you is ammonia, a readily available gaseous base that reacts with H^+ to form NH_4^+:

$$NH_3 + H^+ \longrightarrow NH_4^+$$

This seems like a very promising idea, so you decide to construct a pilot treatment chamber. To construct this chamber, you need some help from coworkers. For example, you might consult a chemical engineer for help in the design of the plumbing and pumps needed to supply ammonia to the chamber. You might also consult a mechanical engineer about the appropriate material to use for the chamber and then discuss the actual construction of the chamber with machinists and other personnel from the company's machine shop. In addition, you probably would consult a safety specialist and possibly a toxicologist about the hazards associated with ammonia.

Before the chamber is built, you also have to think carefully about how to test the effectiveness of the process. How could you evaluate, in a relatively short time, how well the process protects paper from deterioration? At this stage, you would undoubtedly do more library research and consult with other experts, such as a paper chemist your company hires as an outside consultant.

Assume now that the chamber has been constructed and that the initial tests look encouraging. At first the H^+ level is greatly reduced in the treated paper. However, after a few days the H^+ level begins to rise again. Why? The fact that ammonia is a gas at room temperature (and pressure) is an advantage, because it allows you to treat many books simultaneously in a dry chamber. However, the volatility of ammonia works against you after the treatment. The process

$$NH_4^+ \longrightarrow NH_3 \uparrow + H^+$$

allows the ammonia to escape after a few days. Thus this treatment is too temporary. Even though this effort failed, it was still useful, because it provided an opportunity to understand what is required to solve this problem. You need a gaseous substance that *permanently* reacts with the paper and that also consumes H^+.

In discussing this problem over lunch, a colleague suggests the compound diethyl zinc $(C_2H_5)_2Zn$, which is quite volatile (boiling point = 117°C) and which reacts with water (moisture is present in paper) as follows:

$$(C_2H_5)_2Zn + H_2O \longrightarrow ZnO + 2C_2H_6$$

The C_2H_6 (ethane) is a gas that escapes, but the white solid, ZnO, becomes an integral part of the paper. The important part of ZnO is the oxide ion, O^{2-}, which reacts with H^+ to form water:

$$O^{2-} + 2H^+ \longrightarrow H_2O$$

Thus the ZnO is a nonvolatile base that can be placed in the paper by a gaseous substance. This process looks very promising and, in fact, is now being used for the mass treatment of books at the Library of Congress.

The major disadvantage of this process (there are always disadvantages) is that diethyl zinc is *very* flammable and great care must be exercised in its use.

This leads to another question: is the treatment effective enough to be worth the risks involved? Only through more experiments can this question be answered.

The type of problem solving illustrated by the investigation of the acid decomposition of paper is quite typical of that which a practicing chemist confronts daily. The first step in successful problem solving is to identify the exact nature of the problem. Although this may seem trivial, it is often the most difficult and most important part of the process. Poor problem solving often results from a fuzzy definition of the problem. You cannot efficiently solve a problem if you do not understand the essence of the problem. Once the problem is well-defined, then solutions can be advanced, usually by a process of intelligent trial and error. This process typically involves starting with the simplest potential solution and iterating to a final solution as the feedback from earlier attempts is used to refine the approach. Rarely, if ever, is the solution to a complex problem obvious immediately after the problem is defined. The best solution only becomes apparent as the results from various trial solutions are evaluated. A schematic summarizing the approach for dealing with the acid decomposition of paper described below is shown in Fig. 1.3.

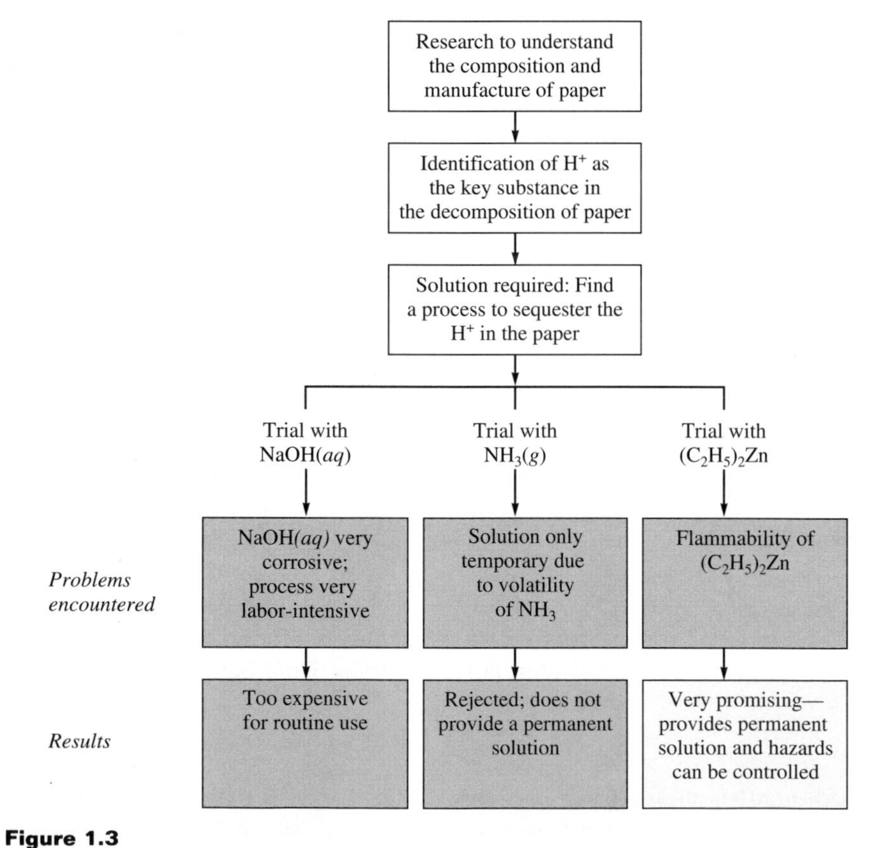

Figure 1.3
Schematic diagram of the strategy for solving the problem of the acid decomposition of paper.

1.3 The Scientific Method

Science is a framework for gaining and organizing knowledge. Science is not simply a set of facts but is also a plan of action—a *procedure* for processing and understanding certain types of information. Scientific thinking is useful in all aspects of life, but in this text we will use it to understand how the chemical world operates. The process that lies at the center of scientific inquiry is called the **scientific method.** There are actually many scientific methods, depending on the nature of the specific problem under study and on the particular investigator involved. However, it is useful to consider the following general framework for a generic scientific method:

Steps in the Scientific Method

1. *Making observations.* Observations may be *qualitative* (the sky is blue; water is a liquid) or *quantitative* (water boils at 100°C; a certain chemistry book weighs 2 kilograms). A qualitative observation does not involve a number. A quantitative observation (called a **measurement**) involves both a number and a unit.
2. *Formulating hypotheses.* A hypothesis is a *possible* explanation for the observation.
3. *Performing experiments.* An experiment is carried out to test the hypothesis. This involves gathering new information that enables a scientist to decide whether or not the hypothesis is correct—that is, whether it is supported by the new information learned from the experiment. Experiments always produce new observations, and this brings the process back to the beginning again.

See Appendix A1.6 for conventions regarding the use of significant figures in connection with measurements and the calculations involving measurements. Appendix 2 discusses methods for converting among various units.

To understand a given phenomenon, these steps are repeated many times, gradually accumulating the knowledge necessary to provide a possible explanation of the phenomenon.

Once a set of hypotheses that agree with the various observations is obtained, the hypotheses are assembled into a theory. A **theory**, which is often called a *model*, is a set of tested hypotheses that gives an overall explanation of some natural phenomenon.

It is very important to distinguish between observations and theories. An observation is something that is witnessed and can be recorded. A theory is an *interpretation*—a possible explanation of *why* nature behaves in a particular way. Theories inevitably change as more information becomes available. For example, the motions of the sun and stars have remained virtually the same over the thousands of years during which humans have been observing them, but our explanations—our theories—for these motions have changed greatly since ancient times.

The point is that scientists do not stop asking questions just because a given theory seems to account satisfactorily for some aspect of natural behavior. They continue doing experiments to refine or replace the existing theories. This is generally done by using the currently accepted theory to make a prediction and then performing an experiment (making a new observation) to see whether the results bear out this prediction.

Always remember that theories (models) are human inventions. They represent attempts to explain observed natural behavior in terms of human experiences. A theory is actually an educated guess. We must continue to do experiments and to refine our theories (making them consistent with new knowledge) if we hope to approach a more nearly complete understanding of nature.

As scientists observe nature, they often see that the same observation applies to many different systems. For example, studies of innumerable chemical changes have shown that the total observed mass of the materials involved is the same before and after the change. Such generally observed behavior is formulated into a statement called a **natural law.** For example, the observation that the total mass of materials is not affected by a chemical change in those materials is called the law of conservation of mass.

Note the difference between a natural law and a theory. A natural law is a summary of observed (measurable) behavior, whereas a theory is an explanation of behavior. *A law summarizes what happens; a theory (model) is an attempt to explain why it happens.*

In this section we have described the scientific method as it might ideally be applied (see Fig. 1.4). However, it is important to remember that science does not always progress smoothly and efficiently. For one thing, hypotheses and observations are not totally independent of each other, as we have assumed in the description of the idealized scientific method. The coupling of observations and hypotheses occurs because once we begin to proceed down a given theoretical path, our hypotheses are unavoidably couched in the language of those theoretical underpinnings. In other words, we tend to see what we expect to see and often fail to notice things that we do not expect. Thus the theory we are testing helps us because it focuses our questions. However, at the very same time, this focusing process may limit our ability to see other possible explanations.

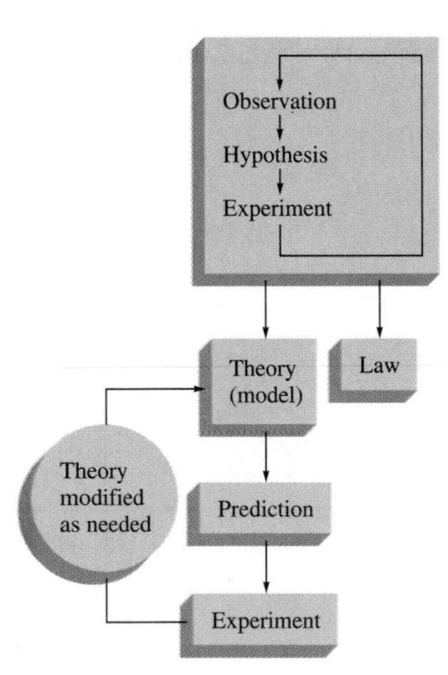

Figure 1.4
The various parts of the scientific method.

It is also important to keep in mind that scientists are human. They have prejudices; they misinterpret data; they become emotionally attached to their theories and thus lose objectivity; and they play politics. Science is affected by profit motives, budgets, fads, wars, and religious beliefs. Galileo, for example, was forced to recant his astronomical observations in the face of strong religious resistance. Lavoisier, the father of modern chemistry, was beheaded because of his political affiliations. And great progress in the chemistry of nitrogen fertilizers resulted from the desire to produce explosives to fight wars. The progress of science is often affected more by the frailties of humans and their institutions than by the limitations of scientific measuring devices. The scientific methods are only as effective as the humans using them. They do not automatically lead to progress.*

1.4 Industrial Chemistry

The impact of chemistry on our lives is due in no small measure to the many industries that process and manufacture chemicals to provide the fuels, fabrics,

* A very useful discussion of these issues is given in the pamphlet "On Being a Scientist" produced by the Committee on the Conduct of Science of the National Academy of Sciences (Washington, D.C.: National Academy Press, 1989).

fertilizers, food preservatives, detergents, and many other products that affect us daily. The chemical industry can be subdivided in terms of three basic types of activities:

1. The isolation of naturally occurring substances for use as raw materials
2. The processing of raw materials by chemical reactions to manufacture commercial products
3. The use of chemicals to provide services

A given industry may participate in one, two, or all of these activities.

Producing chemicals on a large industrial scale is very different from an academic laboratory experiment. Some of the important differences are described below.

In the academic laboratory practicality is typically the most important consideration. Because the amounts of substances used are usually small, hazardous materials can be handled by using fume hoods, safety shields, and so on; expense, although always a consideration, is not a primary factor. However, for any industrial process, economy and safety are critical.

In industry, containers and pipes are metal rather than glass, and corrosion is a constant problem. In addition, since the progress of reactions cannot be monitored visually, gauges must be used.

In the laboratory any by-products of a reaction are simply disposed of; in industry they are usually recycled or sold. If no current market exists for a given by-product, the manufacturer tries to develop such a market.

Industrial processes often run at very high temperatures and pressures and ideally are *continuous flow,* meaning that reactants are added and products are extracted continuously. In the laboratory, reactions are run in batches and typically at much lower temperatures and pressures.

The many criteria that must be satisfied to make a process feasible on the industrial scale require that great care be taken in the development of each process to ensure safe and economical operation. The development of an industrial chemical process typically involves the following steps:

STEP 1

A need for a particular product is identified.

STEP 2

The relevant chemistry is studied on a small scale in a laboratory. Various ways of producing the desired material are evaluated in terms of costs and potential hazards.

STEP 3

The data are evaluated by chemists, chemical engineers, business managers, safety engineers, and others to determine which possibility is most feasible.

STEP 4

A *pilot plant test* of the process is carried out. The scale of the pilot plant is between that of the laboratory and that of a manufacturing plant. The purpose

TABLE 1.1 The 20 Chemicals Produced in the Largest Quantities in the United States in 1992

Substance	Formula	Production in 1992 (billions of pounds)	Method of Preparation	Uses
1. Sulfuric acid	H_2SO_4	88.8	Oxidation of S to SO_3, which is reacted with H_2O	Manufacture of fertilizers, detergents, explosives, petroleum, plastics, pesticides, pharmaceuticals, dyes, storage batteries, and metals
2. Nitrogen	N_2	58.7	Distillation of liquid air	Manufacture of ammonia, nitric acid, and nitrates; for purging and protecting; as a low-temperature coolant
3. Oxygen	O_2	42.4	Distillation of liquid air	Manufacture of steel; rocket fuel; medical uses (respiration)
4. Ethylene	C_2H_4	40.4	Cracking of petroleum	Manufacture of ethylene glycol, ethanol, vinyl chloride, ethylene oxide, polyethylene, and other polymers; as a ripening agent for fruit
5. Ammonia	NH_3	36.0	Haber process (N_2 + H_2)	Fertilizers; in the manufacture of nitric acid, explosives, and fertilizers
6. Lime	CaO	34.7	Thermal decomposition of limestone ($CaCO_3$)	Manufacture of glass, cement, insecticides, and petroleum
7. Phosphoric acid	H_3PO_4	25.4	Treatment of phosphate rock with H_2SO_4	Manufacture of fertilizers, detergents, and pharmaceuticals; pickling and rustproofing of steel; water treatment; flavoring in beverages
8. Sodium hydroxide (caustic soda)	NaOH	24.0	Electrolysis of brine [NaCl(*aq*)]	Manufacture of rayon, cellophane, petroleum, pulp and paper, detergents, and aluminum
9. Propylene	C_3H_6	22.6	Cracking of petroleum	Manufacture of polypropylene, propylene oxide, isopropyl alcohol, and acetone
10. Chlorine	Cl_2	22.3	Electrolysis of brine	Manufacture of pesticides and bleach; for water purification
11. Sodium carbonate (soda ash)	Na_2CO_3	20.9	Reaction of CaO, NH_3, and NaCl	Manufacture of glass, pulp and paper, soaps and detergents, petroleum, and aluminum
12. Urea	$(NH_2)_2CO$	16.8	Reaction of NH_3 and CO_2	Manufacture of fertilizers, plastics, adhesives, animal feeds, and flameproofing agents
13. Nitric acid	HNO_3	16.1	Reaction of NH_3 and O_2	Manufacture of explosives, fertilizers, dyes, and pharmaceuticals
14. Ethylene dichloride (dichloroethane)	$C_2H_4Cl_2$	15.9	Reaction of C_2H_4 and Cl_2	Manufacture of vinyl chloride paint removers, solvents, and rubber
15. Ammonium nitrate	NH_4NO_3	15.3	Reaction of NH_3 and HNO_3	Manufacture of fertilizers, explosives, nitrous oxide, and insecticides
16. Vinyl chloride	CH_2CHCl	13.2	Thermal decomposition of ethylene dichloride	Manufacture of polyvinyl chloride and other polymers
17. Benzene	C_6H_6	12.0	Cracking and reforming petroleum	Manufacture of styrene, phenol, detergents, cyclohexane, dyes, paint removers, rubber cement, and gasoline; as a solvent
18. Ethylbenzene	$C_2H_5C_6H_5$	11.0	Reaction of C_6H_6 and C_2H_4	Manufacture of styrene; as a solvent
19. Carbon dioxide	CO_2	10.9	Recovered as a by-product in various industrial processes	Manufacture of various chemicals; for beverage carbonation, and fire extinguishing
20. Methyl *tert*-butyl ether	$(CH_3)_3COCH_3$	10.9	Reaction of CH_3Br and $(CH_3)CONa$	Fuel additive

of this test is to make sure that the reaction is efficient at a larger scale, to test reactor (reaction container) designs, to determine the costs of the process, to evaluate the hazards, and to gather information on environmental impact.

1.5 The Production of Chemicals

To give you some idea of what types of chemical products are the most important (in terms of the quantities produced) in the United States, the "top 20 chemicals" are listed in Table 1.1. Notice that most of the chemicals listed are relatively simple substances that are used in the manufacture of other materials. Do not be concerned if you are not familiar at this point with some of the compounds and processes listed. As the course progresses, you will encounter most of these substances and the reactions for making them.

1.6 Polyvinyl Chloride (PVC): Real-World Chemistry

To get a little better feel for how the world of industrial chemistry operates, we will now consider a particular product, polyvinyl chloride (PVC), to see what types of considerations have been important in making this a successful and important consumer product.

A landfill being constructed with a PVC liner.

When you put on a nylon jacket, use a polyethylene wash bottle in the lab, wear contact lenses, or accidentially drop your telephone (and it doesn't break), you are benefiting from the properties of polymers. Polymers are very large molecules that are assembled from small units (called monomers). Because of their many useful properties, polymers are manufactured in huge quantities. In fact, it has been estimated that over 50% of all industrial chemists have jobs that are directly related to polymers.

One particularly important polymer is polyvinyl chloride (PVC), which is made from the molecule commonly called vinyl chloride:

$$\begin{array}{ccc} H & & H \\ & \diagdown \; \diagup & \\ & C{=}C & \\ & \diagup \; \diagdown & \\ H & & Cl \end{array}$$

When many of these units are joined together, the polymer PVC results,

$$\begin{array}{c} \;\;H \;\;\; H \;\;\; H \;\;\; H \;\;\; H \;\;\; H \\ \sim\!\sim C-C-C-C-C-C \sim\!\sim \\ \;\;H \;\;\; Cl \;\;\; H \;\;\; Cl \;\;\; H \;\;\; Cl \end{array}$$

which can be represented as

$$\left(\begin{array}{cc} H & H \\ | & | \\ -C- & C- \\ | & | \\ H & Cl \end{array} \right)_n$$

where n is usually greater than 1000.

Because the development of PVC into a useful, important material is representative of the type of problem solving encountered in industrial chemistry, we will consider it in some detail.

In pure form PVC is a hard, brittle substance that decomposes easily at the high temperatures necessary to process it. This makes it almost useless. The fact that it has become a high-volume plastic ($\approx$10 billion pounds per year produced in the United States) is a tribute to chemical innovation. Depending on the additives used, PVC can be made rigid or highly flexible, and it can be tailored for use in inexpensive plastic novelty items or for use in precision engineering applications.

The development of PVC illustrates the interplay of logic and serendipity, as well as the importance of optimizing properties both for processing and for applications. PVC production has been beset with difficulties from the beginning, but solutions have been found for each problem through a combination of chemical deduction and trial and error. For example, many additives have been found that provide temperature stability so that PVC can be processed as a melt (liquid) and so that PVC products can be used at high temperatures. However, there is still controversy among chemists about exactly how PVC decomposes thermally, and thus the reason these stabilizers work is not well understood. Also, there are approximately one hundred different plasticizers (softeners) available for PVC, but the theory of its plasticization is too primitive to predict accurately which compounds might produce even better results.

PVC was discovered by a German chemical company in 1912, but its brittleness and thermal instability proved so problematical that in 1926 the company stopped paying the fees to maintain its patents. That same year Waldo Semon, a chemist at B. F. Goodrich, found that PVC could be made flexible by the addition of phosphate and phthalate esters. Semon also found that white lead, $Pb_3(OH)_2(CO_3)_2$, provided thermal stability to PVC. These advances led to the beginning of significant U.S. industrial production of PVC ($\approx$4 million pounds per year by 1936). In an attempt to further improve PVC, T. L. Gresham (also a chemist at B. F. Goodrich) tried approximately one thousand compounds, searching for a better plasticizer. The compound that he found (its identity is not important here) remains the most common plasticizer added to PVC. The types of additives commonly used in the production of PVC are listed in Table 1.2.

A *plasticizer* is a compound added to a polymer to soften it. The type and amount of plasticizer determine the pliability of the final product. One theory suggests that a plasticizer causes softening by its insertion between the polymer chains where they are held in place only by weak forces. This lessens polymer–polymer interactions and thus softens the material.

Although the exact mechanism of the thermal, heat-induced decomposition of PVC remains unknown, most chemists agree that the chlorine atoms present in the polymer play an important role. Lead salts are added to PVC both to provide anions less reactive than chloride and to provide lead ions to combine

TABLE 1.2 Types of Additives Commonly Used in the Production of PVC

Type of Additive	Effect
Plasticizer	Softens the material
Heat stabilizer	Increases resistance to thermal decomposition
Ultraviolet absorber	Prevents damage by sunlight
Flame retardant	Lowers flammability
Biocide	Prevents bacterial or fungal attack

with the released chloride ions. As a beneficial side effect, the lead chloride formed gives PVC enhanced electrical resistance, making lead stabilizers particularly useful in producing PVC for electrical wire insulation.

One major use of PVC is for pipes in plumbing systems. Here, even though the inexpensive lead stabilizers would be preferred from an economic standpoint, the possibility that the toxic lead could be leached from the pipes into the drinking water necessitates the use of more expensive tin and antimony compounds as thermal stabilizers. Because about half of the annual production of PVC is formed into piping, the PVC formulation used for pipes represents a huge market for companies that manufacture additives, and the competition is very intense. A recently developed low-cost thermal stabilizer for PVC is a mixture of antimony and calcium salts. This mixture has replaced stabilizers containing tin compounds that have become increasingly costly in recent years.

Outdoor applications of PVC often require that it contain ultraviolet light absorbers to protect against damage from sunlight. For pigmented applications such as vinyl siding, window frames, and building panels, titanium(IV) oxide (TiO_2) is usually used. For applications where the PVC must be transparent, other compounds are needed.

The additives used in PVC in the largest amounts are plasticizers, but one detrimental effect of these additives is an increase in flammability. Rigid PVC, which contains little plasticizer, is quite flame-resistant due to its high chloride content. However, as more plasticizer is added for flexibility, the flammability increases to the point where fire retardants must be added, the most common being antimony(III) oxide (Sb_2O_3). As the PVC is heated, this oxide forms antimony(III) chloride ($SbCl_3$), which migrates into the flame where it inhibits the burning process. Because antimony(III) oxide is a white salt, it cannot be used for transparent or darkly colored PVC. In these cases sodium antimonate (Na_3SbO_4), a transparent salt, is used.

Once the additives have been chosen for a particular PVC application, the materials must be blended. This is often done in a dry-blending process producing a powder that is then used for fabrication of the final product. The powdered mixture also can be melted and formed into pellets, which are easily shipped to manufacturers where they are remelted and formed into the desired products.

The production of PVC provides a good case study of an industrial process. It illustrates many of the factors that must be taken into account when any product is manufactured: effectiveness of the product, cost, ease of production, safety, and environmental impact. The last issue is becoming ever more important as our society struggles both to reduce the magnitude of the waste stream by recycling and to improve our waste disposal methods.

1.7 Profiles of People Working in the Chemical Sciences

As we mentioned earlier in this chapter, people trained in the chemical sciences hold an amazing variety of jobs. Because most positions in chemically based industries require a sound technical background, it is common for these industries to hire people with degrees in the chemical sciences and then to encourage these people to learn new skills that will allow them to assume the other responsibilities connected with the manufacture, sale, and use of chemicals. It is quite common for newly graduated chemical sciences majors to take entry-level positions as laboratory chemists who carry out experiments to help their companies develop new products (support of research and development). Other entry-level positions might involve ensuring that products that are to be sent out meet company standards (quality control or quality assurance), or assisting customers in the use of the company's products (support of technical sales and service). A person may make a career of laboratory work, or, as is commonly the case, in a few years may move out of the laboratory into areas such as marketing and sales, eventually becoming a manager and perhaps reaching the highest levels of management. The chief executive officer (CEO) of a chemically based industry is usually a person with a technical degree.

People who want to make a career of laboratory research in industry usually attend graduate school and earn a Ph.D. Graduate study provides a deeper understanding of a particular field and provides the training needed to carry out independent research. A person with a Ph.D. usually manages pure research activities that seek to establish new knowledge that will eventually lead to new products. A career in research often proceeds through a series of management positions, with increasing responsibility and scope, and can lead to the highest levels of management in the company.

Of course, someone interested in research may also pursue an academic career following completion of a Ph.D. and (often) postdoctoral studies. Academic research is pure research most often supported by government agencies such as the National Science Foundation (NSF) or the National Institutes of Health (NIH). Although an academic career is often not as lucrative as a career in private industry, an academic career can be extraordinarily satisfying and has the advantage of providing a great deal of personal freedom.

The following profiles are meant to give a sampling of careers open to persons trained in the chemical sciences. They illustrate the great variety of interesting and important jobs available in the chemical-based industries, and they also show how common it is to change jobs frequently during a career.

CHEMISTRY • CAREER • PROFILES

Kathleen M. Miller received her B.S. degree in chemistry from the University of Illinois in 1978 and then attended the University of California at Los Angeles (UCLA), receiving a Ph.D. degree in inorganic chemistry in 1983. After graduate school, Kathy accepted a position with TRW, Inc., near Los Angeles, working primarily on physics projects, such as plasma isotope separation. In 1987, she changed career paths, taking a position at Mallinckrodt Medical, Inc. (MMI), in St. Louis, Missouri to work in medical products R&D (research and development).

Kathy's initial position at MMI was as a Research Chemist in the Nuclear Medicine department. Her responsibilities included identifying, characterizing, and developing formulations for diagnostic medical products called radiopharmaceuticals, products which contain gamma-ray emitting radioactive isotopes. The location of the radioactivity in the body allows abnormalities, such as heart disease or cancer to

Kathleen M. Miller

be detected. In this position, Kathy was able to expand her knowledge of pharmaceutical R&D beyond chemistry to areas such as pharmacology and pharmacy.

Because of her success in this initial position, Kathy was appointed as the Project Team Leader for a product that contains the radioactive nuclide In-111, which binds to tumor tissue and is used to detect cancer. Her responsibilities in this job included developing a product that was acceptable in the US as well as in Europe and coordinating the efforts to file the New Drug Application with the FDA. In this Team Leader position, Kathy continued to learn about other aspects of pharmaceutical development, including conducting patient clinical trials and the challenges of dealing with governmental regulatory processes. Kathy says: "Working at MMI is very gratifying because I know the products I've worked on will be used to diagnose someone's disease or may affect a doctor's decision on how to treat a patient. The usefulness of my chemistry background is often reinforced; the analytical skills and problem solving training can be applied to various aspects of the business."

Recently, Kathy has taken a Position at MMI as Assistant Director of Special Projects, working with the R&D administration. This new position provides opportunities to move beyond radiopharmaceutical R&D and poses different challenges than those found at the lab bench, such as spearheading efforts to change R&D policies and procedures, evaluating outside product opportunities, and streamlining product development.

Betsy Hovey

Since graduating from the University of Illinois in May, 1986 with a Bachelor of Science degree in Chemical Engineering, **Betsy Hovey** has had experience in several aspects of manufacturing. She began her career as a Team Manager (first level manager) of a fatty alcohol product operation for Procter and Gamble (P&G). Betsy says, "This was the most 'chemical' of my experiences to date, involving not only daily supervision of the operation, but technical troubleshooting of the process. This was a 'hands-on' position; my wardrobe consisted of a hardhat, safety glasses, steel-toed shoes and clothes that could get dirty! This was also my first exposure to the relative rarity of women in manufacturing operations. I was one of only two women in an operation of 150 people."

From her position in alcohol production, Betsy moved to Production Planning Manager for a P&G operation that produces liquid dishwashing soap. In this position she was responsible for forecasting future needs for the product, coordinating production to meet these needs, and making sure the product was delivered to the customers.

Betsy was then promoted to Logistics Department Manager with responsibilities for raw material inventory management, production scheduling, finished goods inventory management and customer service. This assignment, while another step away for her chemical roots, gave her the broadest view to date of the complexities of managing a manufacturing operation.

In April 1992, Betsy decided to leave P&G to take a manufacturing management job at Hallmark Cards. In her current position, she is responsible for an injection molding operation. While her management-level responsibilities do not bring her into daily contact with the manufacturing operation, her understanding of plastics and the technical aspects of the production operation helps her as she makes decisions for developing the business.

Betsy says: "In general, I have not pursued a career path specific to my background in the chemical sciences. However, I have found an extremely rewarding career in manufacturing management. I have enjoyed close working relationships with a diverse collection of people, have found satisfaction in solving the problems that arise in daily operations, and am constantly challenged by the pace and variety of opportunities in manufacturing. More recently, I have begun teaching new managers how to be successful in manufacturing and I learn from this experience every day. My technical problem solving skills and understanding of manufacturing processes have allowed me to be successful in several different manufacturing roles. As I continue to pursue my interest in manufacturing management, I believe my background in chemical engineering will open the door for opportunities in many different manufacturing environments."

Keith Reese

Keith Reese combined his interests in chemistry, mathematics, and engineering by getting a degree in chemical engineering from the University of Illinois (Urbana). During summers while in college, Keith worked for oil companies in research labs and at pipeline facilities to get a feel for the world of industry and to gain confidence that he could succeed in the business world. Upon finishing his B.S. degree, Keith considered both graduate school and industry, deciding to take a position with Intel Corporation, an electronics company that makes computer chips.

"I interviewed with Intel in part because it seemed like a new and interesting business, and, in all honesty, because I wanted a trip to California. The most memorable part of that trip was the enthusiasm the people had for their jobs. That was what led me to take the job with them, in spite of the fact that I had almost no idea how a semiconductor was made or how it works.

"The opportunity for learning was enormous as I started in this field. Not only could I apply chemical principles I had learned in school, I could also learn other fields like mechanical and electrical engineering. Semiconductor manufacturing requires a broad spectrum of engineering skills. I spent the first seven years of my career in Santa Clara, California, the heart of Silicon Valley. In that time I worked in all four of the areas that comprise the manufacturing process, and gradually began to supervise more personnel. This happened with a technician, then an engineer, then more engineers working for me. Eventually, I was supervising a group of 12 people. It was after I began to supervise people that I appreciated the diversity of my education.

"The semiconductor field is especially challenging because it is constantly decreasing its cost per function. When you can lower the cost on such an important product, you really feel like you're helping the information age grow. That's one of the things that makes the field enjoyable. As an individual, you can make a great difference to your team in the company as well as to your customers.

Moving to a plant in Livermore, California, I became an engineering manager with a group of about 60 people working for me. I found I enjoyed management, and moved to Albuquerque, New Mexico, to build a group of 400 employees as manufacturing manager at Intel's first submicron manufacturing facility. I am currently the automation manager for Intel's largest factory, managing a group of 60 professionals. Again, the diversity of my education has helped me to take on this challenge.

"In looking back at my career, the times that I moved up were always prefaced by a time that I moved laterally. This required flexibility on my part but also provided reward to me in terms of seeing and learning new things. I think this has rounded my background and enabled me to see the different sides of a problem. I'd definitely do it all over again."

Kathy Wolfgram

Kathy Wolfgram graduated in 1976 from Carroll College (Waukesha, Wisconsin) with a BS degree in chemistry. During her college career she worked part-time as a chemical analyst for a local company and was a summer intern at Dow Chemical Company in Midland, Michigan. Upon graduation Kathy took a job at Dow Chemical where in 18 years she has had eight different jobs. After her initial 2½ years as a laboratory chemist in the Central Research organization, she shifted toward product development work. The project led to a 1980 transfer to Granville, Ohio to work in the Technical Service and Development function supporting Dow's Rigid Foam Products (sold primarily for building insulation).

After a couple of years servicing the residential construction market in the TS&D* function, Kathy was promoted through a series of supervisory positions in the same department. At the same time she served as a member of an interfunctional product/market management team responsible for developing marketing, production, and research strategies for some of the Rigid Foam Products. In 1990, she left supervision and assumed project management responsibilities within the same business, analyzing the economic impact of potential formulation changes, patent strategies, and recycling strategies. In 1992, she became a global project manager coordinating a major research project between laboratories in Europe, Japan, and the US.

Kathy writes: "What I love about my job is that it offers a dynamic mix of technical, business, and people issues . . . there isn't a day when I haven't learned new things. I'm also very excited by the world of opportunities a large corporation can offer . . . I've been able to consider research, technical service, marketing, and sales responsibilities within a company that markets over 2000 products around the world!

"What I would recommend to college freshmen . . . *seek out* opportunities to learn about different careers, industries, and companies. I was fortunate to be handed a job offer to work at Dow during college and to see what the life of a scientist in industry can be. I'm not sure I would have been as blessed careerwise if my job considerations and

*Technical Service and Development is a function in Dow which serves as the link between the research and marketing organization (translating market needs to research, translating research capabilities to marketing) and offers technical expertise in use of the product to the customer. It often involves traveling to meet with the customer to understand problems he or she might be having with your product, working within your organization to develop a solution to that problem, and finally testing that solution with the customer. Additionally, it involves introducing the technical features and benefits of new products and applications to the marketplace. This type of job provides great diversity and requires high-order problem-solving, planning, organization, and communication skills.

Paul Schiller

pursuit were left only to my own initiative—I was lucky!

"Don't assume you are not capable of certain job positions. Hard work, a curious mind, and a willingness to listen to others and learn from their ideas and methods can lead to tremendous opportunity.

"Communication skills are *critical!* I've seen too many intelligent scientists who are frustrated and frustrate others because they can't communicate simply, concisely, and logically to truly impact the thinking and direction within the company."

Paul Schiller has worked for Amoco Chemical Company since receiving a BS degree in Chemical Engineering from the University of Illinois in 1981. In the intervening years Paul has also earned an MBA from the University of Chicago.

Today Paul is a dynamic, articulate engineer/businessman who projects great enthusiasm about his career. In just over ten years Paul has held amazingly diverse job assignments in the areas of manufacturing, marketing, and sales. These jobs have included leading a companywide effort to improve the efficiency and safety of oxidation reactor operations and managing a project that significantly improved the containers that consumers use for soft drinks and many other purposes. In a less traditional role for a chemical engineer Paul has also been active in personnel issues which have included spearheading an effort to improve the appraisal system for employee performance, initiating a program to teach the use of statistics to all employees, and taking the lead in developing a new program for identifying and correcting work hazards.

Paul is currently working on a product that could revolutionize tire construction, a topic of keen interest to Paul, as he is an avid racer of high-performance automobiles. The marketing effort for this product has already led to safer airport lights and structures, advanced coatings for environmentally friendly paints and stains, high-tech inks for currency, and many other breakthroughs. Paul says, "The sales and marketing effort is truly exciting, with the possibility of a new breakthrough around every corner."

Paul has encountered several unique events during his career. In early 1991, he received a call from a desperate customer requiring a small emergency shipment of a product used to manufacture urgently needed radar antennae for the EF-111 radar jamming planes being used in the Gulf War. A sufficient quantity of the product was shipped within the hour to allow delivery of the antennae to Saudi Arabia within 48 hours. Paul also participated in a development project for a revolutionary sailboat showcasing Amoco materials. This project provided Paul with several enjoyable sails both in Lake Michigan and off Manhattan. Finally, Paul has taken advantage of Amoco's sponsorship of the Bill Elliot/Junior Johnson NASCAR race team to entertain several customers at races and get a behind-the-scenes explanation of NASCAR.

Paul says that two key things he has learned during his career are that everything is always negotiable, and that engineering, analytical skills, and the laws of physics are useful in many areas including auto racing, golf, and group dynamics.

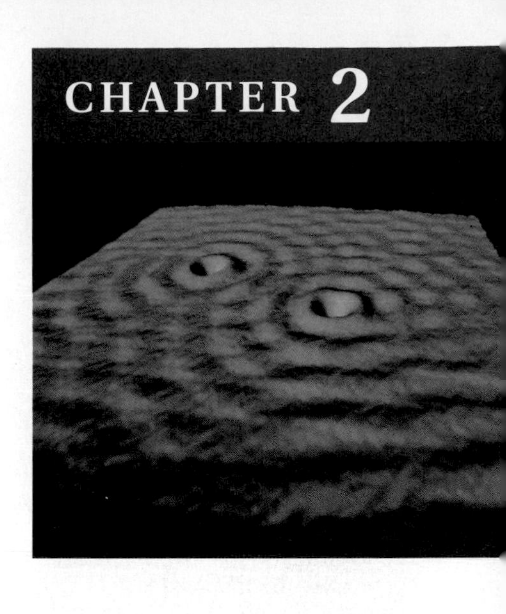

Atoms, Molecules, and Ions

In this chapter we present very briefly many of the fundamental concepts and some of the vocabulary of chemistry plus something about how the science developed. Depending on your specific background in chemistry, much of this material may be review. However, whatever your background, read this chapter carefully to be sure this material is fresh in your mind as we pursue the study of reaction chemistry in Chapters 3 and 4.

2.1 The Early History of Chemistry

Chemistry has been important since ancient times. The processing of natural ores to produce metals for ornaments and weapons and the use of embalming fluids are two applications of chemical phenomena that were utilized prior to 1000 B.C.

The Greeks were the first to try to explain why chemical changes occur. By about 400 B.C. they had proposed that all matter was composed of four fundamental substances: fire, earth, water, and air. The Greeks also considered the question of whether matter is continuous, and thus infinitely divisible into smaller pieces, or composed of small indivisible particles. One supporter of the latter position was Democritus, who used the term *atomos* (which later became *atoms*) to describe these ultimate particles. However, because the Greeks had no experiments to test their ideas, no definitive conclusion about the divisibility of matter was reached.

The next 2000 years of chemical history were dominated by a pseudoscience called alchemy. Alchemists were often mystics and fakes who were

The Priestley Medal is the highest honor given by the American Chemical Society. It is named for Joseph Priestley, who was born in England on March 13, 1733. He performed many important scientific experiments, among them the discovery that a gas later identified as carbon dioxide could be dissolved in water to produce *seltzer.* Also, as a result of meeting Benjamin Franklin in London in 1766, Priestley became interested in electricity and was the first to observe that graphite was an electrical conductor. However, his greatest discovery occurred in 1774 when he isolated oxygen by heating mercuric oxide.

Because of his nonconformist political views, he was forced to leave England. He died in the United States in 1804.

obsessed with the idea of turning cheap metals into gold. However, this period also saw important discoveries: elements such as mercury, sulfur, and antimony were discovered, and alchemists learned how to prepare the mineral acids.

The foundations of modern chemistry were laid in the sixteenth century with the development of systematic metallurgy (extraction of metals from ores) by a German, Georg Bauer, and the medicinal application of minerals by a Swiss alchemist called Paracelsus.

The first "chemist" to perform truly quantitative experiments was Robert Boyle (1627–1691), an Irish scientist, who carefully measured the relationship between the pressure and volume of gases. When Boyle published his book *The Sceptical Chemist* in 1661, the quantitative sciences of physics and chemistry were born. In addition to his results on the quantitative behavior of gases, Boyle's other major contribution to chemistry consisted of his ideas about the chemical elements. Boyle held no preconceived notion about the number of elements. In his view a substance was an element unless it could be broken down into two or more simpler substances. As Boyle's experimental definition of an element became generally accepted, the list of known elements began to grow, and the Greek system of four elements finally died. Although Boyle was an excellent scientist, he was not always right. For example, he clung to the alchemist's views that metals were not true elements and that a way would eventually be found to change one metal to another.

The phenomenon of combustion evoked intense interest in the seventeenth and eighteenth centuries. The German chemist Georg Stahl (1660–1734) suggested that a substance he called phlogiston flowed out of the burning material. Stahl postulated that a substance burning in a closed container eventually stopped burning because the air in the container became saturated with phlogiston. Oxygen gas, discovered by Joseph Priestley (1733–1804), an English clergyman and scientist, was found to support vigorous combustion and was thus supposed to be low in phlogiston. In fact, oxygen was originally called "dephlogisticated air."

2.2 Fundamental Chemical Laws

By the late eighteenth century, combustion had been studied extensively; the gases carbon dioxide, nitrogen, hydrogen, and oxygen had been discovered; and the list of elements continued to grow. However, it was Antoine Lavoisier (1743–1794), a French chemist (Fig. 2.1), who finally explained the true nature of combustion, thus clearing the way for the tremendous progress that was made near the end of the eighteenth century. Lavoisier, like Boyle, regarded measurement as the essential operation of chemistry. His experiments, in which he carefully weighed the reactants and products of various reactions, suggested that *mass is neither created nor destroyed.* Lavoisier's discovery of this **law of conservation of mass** was the basis for the developments in chemistry in the nineteenth century.

Lavoisier's quantitative experiments showed that combustion involved oxygen (which Lavoisier named), not phlogiston. He also discovered that life was supported by a process that also involved oxygen and was similar in many ways to combustion. In 1789 Lavoisier published the first modern chemistry textbook, *Elementary Treatise on Chemistry,* in which he presented a unified picture of the chemical knowledge assembled up to that time. Unfortunately, in

Figure 2.1
Antoine Lavoisier with his wife. Lavoisier was born in Paris on August 26, 1743. From the beginning of his scientific career, Lavoisier recognized the importance of accurate measurements. His careful weighings showed that mass is conserved in chemical reactions and that combustion involves reaction with oxygen. Also, he wrote the first modern chemistry textbook. He is often called the father of modern chemistry.

Because of his connection to a private tax-collecting firm, radical French revolutionaries demanded his execution, which occurred on the guillotine on May 8, 1794.

the same year the text was published, the French Revolution broke out. Lavoisier, who had been associated with collecting taxes for the government, was executed on the guillotine as an enemy of the people in 1794.

After 1800 chemistry was dominated by scientists who, following Lavoisier's lead, performed careful weighing experiments to study the course of chemical reactions and to determine the composition of various chemical compounds. One of these chemists, a Frenchman, Joseph Proust (1754–1826), showed that *a given compound always contains exactly the same proportion of elements by weight.* For example, Proust found that the substance copper carbonate is always 5.3 parts copper to 4 parts oxygen to 1 part carbon (by mass). The principle of the constant composition of compounds, originally called Proust's law, is now known as the **law of definite proportion.**

Proust's discovery stimulated John Dalton (1766–1844), an English schoolteacher (Fig. 2.2), to think about atoms. Dalton reasoned that if elements were composed of tiny individual particles, a given compound should always contain the same combination of these atoms. This concept explained why the same relative masses of elements were always found in a given compound.

But Dalton discovered another principle that convinced him even more of the existence of atoms. He noted, for example, that carbon and oxygen formed two different compounds that contained different relative amounts of carbon and oxygen, as shown by the following data:

	Mass of Oxygen That Combines with 1 g of Carbon
Compound I	1.33 g
Compound II	2.66 g

Dalton noted that compound II contained twice as much oxygen per gram of carbon as compound I, a fact that could be easily explained in terms of atoms.

Figure 2.2
John Dalton (1766–1844), an Englishman, began teaching at a Quaker school when he was 12. His fascination with science included an intense interest in meteorology (he kept careful daily weather records for 46 years), which led to an interest in the gases of the air and their ultimate components, atoms. Dalton is best known for his atomic theory, in which he postulated that the fundamental differences among atoms are their masses. He was the first to prepare a table of relative atomic weights.

Dalton was a humble man with several apparent handicaps: he was poor; he was not articulate; he was not a skilled experimentalist; and he was color-blind, a terrible problem for a chemist. In spite of these disadvantages, he helped to revolutionize the science of chemistry.

Compound I might be CO, and compound II might be CO_2. This principle, which was found to apply to compounds of other elements as well, became known as the **law of multiple proportions:** *when two elements form a series of compounds, the ratios of the masses of the second element that combine with 1 gram of the first element can always be reduced to small whole numbers.*

These ideas are also illustrated by the compounds of nitrogen and oxygen, as shown by the following data:

	Mass of Nitrogen That Combines with 1 g of Oxygen
Compound I	1.750 g
Compound II	0.8750 g
Compound III	0.4375 g

which yield the following ratios:

$$\frac{\text{I}}{\text{II}} = \frac{1.750}{0.8750} = \frac{2}{1}$$

$$\frac{\text{II}}{\text{III}} = \frac{0.8750}{0.4375} = \frac{2}{1}$$

$$\frac{\text{I}}{\text{III}} = \frac{1.750}{0.4375} = \frac{4}{1}$$

The significance of these data is that compound I contains twice as much nitrogen (N) per gram of oxygen (O) as does compound II and that compound II contains twice as much nitrogen per gram of oxygen as does compound III. In terms of the numbers of atoms combining, these data can be explained by any of the following sets of formulas:

Compound I	N_2O		NO		N_4O_2
Compound II	NO	or	NO_2	or	N_2O_2
Compound III	NO_2		NO_4		N_2O_4

In fact, an infinite number of other possibilities exists. Dalton could not deduce absolute formulas from the available data on relative masses. However, the data on the composition of compounds in terms of the relative masses of the elements supported his hypothesis that each element consisted of a certain type of atom and that compounds were formed from specific combinations of atoms.

2.3 Dalton's Atomic Theory

These statements are a modern paraphrase of Dalton's ideas.

In 1808 Dalton published *A New System of Chemical Philosophy*, in which he presented his theory of atoms:

1. Each element is made up of tiny particles called atoms.
2. The atoms of a given element are identical; the atoms of different elements are different in some fundamental way or ways.
3. Chemical compounds are formed when atoms combine with each other. A given compound always has the same relative numbers and types of atoms.

4. Chemical reactions involve reorganization of the atoms—changes in the way they are bound together. The atoms themselves are not changed in a chemical reaction.

It is instructive to consider Dalton's reasoning on the relative masses of the atoms of the various elements. In Dalton's time water was known to be composed of the elements hydrogen and oxygen, with 8 grams of oxygen present for every 1 gram of hydrogen. If the formula for water were OH, an oxygen atom would have to have 8 times the mass of a hydrogen atom. However, if the formula for water were H_2O (two atoms of hydrogen for every oxygen atom), this would mean that each atom of oxygen is 16 times as heavy as *each* atom of hydrogen (since the ratio of the mass of one oxygen to that of *two* hydrogens is 8 to 1). Because the formula for water was not then known, Dalton could not specify the relative masses of oxygen and hydrogen unambiguously. To solve the problem, Dalton made a fundamental assumption: he decided that nature would be as simple as possible. This assumption led him to conclude that the formula for water should be OH. He thus assigned hydrogen a mass of 1 and oxygen a mass of 8.

Joseph Louis Gay-Lussac (1778–1850), a French physicist and chemist, was remarkably versatile. Although he is now primarily known for his studies on the combining of volumes of gases, Gay-Lussac was instrumental in the studies of many of the other properties of gases. Some of Gay-Lussac's motivation to learn about gases arose from his passion for ballooning. In fact, he made ascents to heights of over 4 miles to collect air samples, setting altitude records that stood for about 50 years. Gay-Lussac also was the codiscoverer of boron and the developer of a process for manufacturing sulfuric acid. As chief assayer of the French mint, Gay-Lussac developed many techniques for chemical analysis and invented many types of glassware now used routinely in labs. Gay-Lussac spent his last 20 years as a lawmaker in the French government.

Using similar reasoning for other compounds, Dalton prepared the first table of **atomic masses** (often called **atomic weights** by chemists, since mass is usually determined by comparison to a standard mass—a process called *weighing**). Many of the masses were later proved to be wrong because of Dalton's incorrect assumptions about the formulas of certain compounds, but the construction of a table of masses was an important step forward.

Although not recognized as such for many years, the keys to determining absolute formulas for compounds were provided in the experimental work of the French chemist Joseph Gay-Lussac (1778–1850) and by the hypothesis of an Italian chemist named Amedeo Avogadro (1776–1856). In 1809 Gay-Lussac performed experiments in which he measured (under the same conditions of temperature and pressure) the volumes of gases that reacted with each other. For example, Gay-Lussac found that 2 volumes of hydrogen react with 1 volume of oxygen to form 2 volumes of gaseous water and that 1 volume of hydrogen reacts with 1 volume of chlorine to form 2 volumes of hydrogen chloride.

In 1811 Avogadro interpreted these results by proposing that, *at the same temperature and pressure, equal volumes of different gases contain the same number of particles.* This assumption (called **Avogadro's hypothesis**) makes sense if the distances between the particles in a gas are very great compared with the sizes of the particles. Under these conditions the volume of a gas is determined by the number of molecules present, not by the size of the individual particles.

If Avogadro's hypothesis is correct, Gay-Lussac's result,

2 volumes of hydrogen react with 1 volume of oxygen
$\longrightarrow$ 2 volumes of water vapor

can be expressed as follows:

2 molecules of hydrogen react with 1 molecule of oxygen
$\longrightarrow$ 2 molecules of water

*Technically, weight is the force exerted on an object by gravitational attraction to a body such as the earth. It is mass, not weight, that chemists use in their measurements, although the two terms are often used interchangeably.

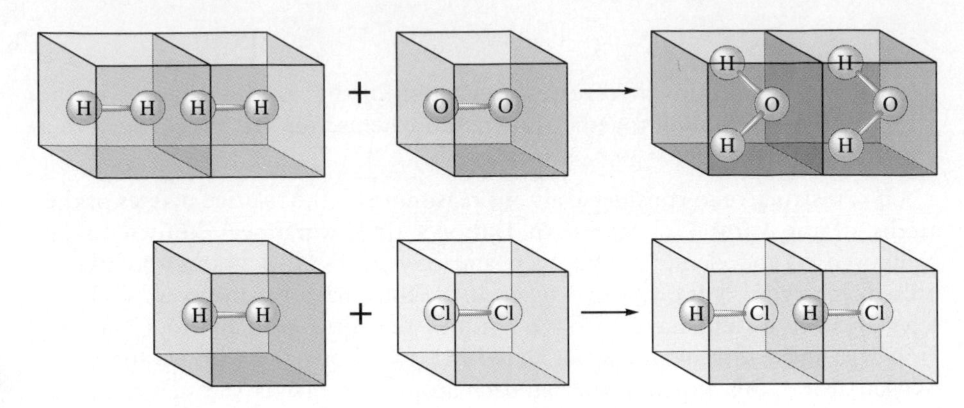

Figure 2.3
A representation of combining gases at the molecular level. The spheres represent atoms in the molecules and the boxes represent the relative volumes of the gases.

These observations can best be explained by assuming that gaseous hydrogen, oxygen, and chlorine are all composed of diatomic (two-atom) molecules: H_2, O_2, and Cl_2, respectively. Gay-Lussac's results can then be represented as shown in Fig. 2.3. (Note that this reasoning suggests that the formula for water is H_2O, not OH as Dalton believed.)

Unfortunately, Avogadro's interpretations were not accepted by most chemists. The main stumbling block seems to have been the prevailing belief that only atoms of different elements could attract each other to form molecules. Dalton and the other prominent chemists of the time assumed that identical atoms had no "affinity" for each other and thus would not form diatomic molecules.

Because no general agreement existed concerning the formulas for elements such as hydrogen, oxygen, and chlorine or for the compounds formed from these elements, chaos reigned in the first half of the nineteenth century. Although chemists, such as the Swedish chemist Jöns Jakob Berzelius (1779–1848), made painstaking measurements during this period of the masses of various elements that combined to form compounds, these results were interpreted in many different ways, depending on the assumptions about the formulas of the elements and compounds, and this led to many different tables of atomic masses. The situation was so confused that 19 different formulas for the compound acetic acid were given in a textbook written in 1861 by F. August Kekulé (1829–1896). In the next section we will see how this mess was finally cleaned up, largely due to the leadership of the Italian chemist Stanislao Cannizzaro.

2.4 Cannizzaro's Interpretation

Convinced that chemists had to find a way to agree on a common set of atomic masses, the German chemist F. August Kekulé organized the First International Chemical Congress held in 1860 at Karlsruhe, Germany. At this meeting the young Italian chemist Stanislao Cannizzaro (1826–1910) presented his ideas so clearly and forcefully, both in formal and informal talks, that a consensus about atomic masses began to develop in the chemical community. Cannizzaro was guided by two main beliefs:

1. Compounds contained whole numbers of atoms as Dalton postulated.
2. Avogadro's hypothesis was correct—equal volumes of gases under the same conditions contain the same number of molecules.

Applications of Avogadro's hypothesis to Gay-Lussac's results by combining volumes of gas convinced Cannizzaro that hydrogen gas consisted of H_2 molecules. Thus he arbitrarily assigned the relative molecular mass of hydrogen (H_2) to be 2. He then set out to measure the relative molecular masses for other gaseous substances. He did so by comparing the mass of 1 liter of a given gas with the mass of 1 liter of hydrogen gas (both gases at the same conditions of temperature and pressure). For example, the ratio of the masses of 1-liter samples of oxygen and hydrogen gas is 16:

Both gases are at the same temperature and pressure.

$$\frac{\text{Mass of 1.0 L oxygen gas}}{\text{Mass of 1.0 L hydrogen gas}} = \frac{16}{1} = \frac{32}{2}$$

Since by Avogadro's hypothesis both samples of gas contain the same number of molecules, the mass of an oxygen molecule (which he assumed to be O_2) must be 32 relative to a mass of 2 for the H_2 molecule. Since each molecule contains two atoms, the relative atomic masses for oxygen and hydrogen are then 16 and 1, respectively. Using this same method, Cannizzaro found the relative molecular mass of carbon dioxide to be 44 (relative to 2 for H_2). Chemical analysis of carbon dioxide had shown it to contain 27% carbon (by mass). This percentage corresponds to (0.27)(44 g), or 12 g, of carbon in 44 g of carbon dioxide, and 44 g − 12 g = 32 g of oxygen. Recall that the oxygen atom has a relative mass of 16. Thus if the formula of carbon dioxide is assumed to be CO_2, then the relative mass of carbon is 12, because [12 + 2(16) = 44]. However, if the formula of carbon dioxide is C_2O_2, then 12 represents the relative mass of two carbon atoms, giving carbon a relative mass of 6. Similarly, the formula C_3O_2 for carbon dioxide gives a relative mass of 4 for carbon. Thus the relative mass of the carbon atom cannot be determined from these data without knowing the formula for carbon dioxide. This is exactly the type of problem that had plagued chemists all along and was the reason for so many different mass tables.

Cannizzaro addressed this problem by obtaining the relative molecular masses of many other compounds containing carbon. For example, consider the data shown in Table 2.1. Notice from these data that the relative mass of carbon present in the compounds is always a multiple of 12. This observation strongly suggests that the relative mass of carbon is 12, which in turn would mean that the formula for carbon dioxide is CO_2.

TABLE 2.1 **Relative Mass Data for Several Gases Containing Carbon**

Compound	Relative Molecular Mass	Percent Carbon (by mass)	Relative Mass of Carbon Present
Methane	16	75	12
Ethane	30	80	24
Propane	44	82	36
Butane	58	83	48
Carbon dioxide	44	27	12

Berzelius, Selenium, and Silicon

Jöns Jakob Berzelius (see figure) was probably the best experimental chemist of his generation and, given the crudeness of his laboratory equipment, maybe the best of all time. Unlike Lavoisier, who could afford to buy the best laboratory equipment available, Berzelius worked with minimal equipment in very plain surroundings. One of Berzelius's students described the Swedish chemist's workplace: "The laboratory consisted of two ordinary rooms with the very simplest arrangements; there were neither furnaces nor hoods, neither water system nor gas. Against the walls stood some closets with the chemicals, in the middle the mercury trough and the blast lamp table. Beside this was the sink consisting of a stone water holder with a stopcock and a pot standing under it. [Next door in the kitchen] stood a small heating furnace."

In these simple facilities Berzelius performed more than 2000 experiments over a 10-year period to determine accurate atomic masses for the 50 elements then known. His success can be seen from the data in the table below. These remarkably accurate values attest to his experimental skills and patience.

Jöns Jakob Berzelius (1779–1848).

Comparison of Several of Berzelius's Atomic Masses with the Current Values

Element	Atomic Mass	
	Berzelius's Value	Current Value
Chlorine	35.41	35.45
Copper	63.00	63.55
Hydrogen	1.00	1.01
Lead	207.12	207.2
Nitrogen	14.05	14.01
Oxygen	16.00	16.00
Potassium	39.19	39.10
Silver	108.12	107.87
Sulfur	32.18	32.07

Besides his table of atomic masses, Berzelius made many other major contributions to chemistry. The most important of these was the invention of a simple set of symbols for the elements along with a system for writing the formulas of compounds to replace the awkward symbolic representations of the alchemists. Although some chemists, including Dalton, objected to the new system, it was gradually adopted and forms the basis of the system we use today.

In addition to these accomplishments, Berzelius also discovered the elements cerium, thorium, selenium, and silicon. Of these elements, selenium and silicon are particularly important in today's world. Berzelius discovered selenium in 1817 in connection with his studies of sulfuric acid. For years selenium's toxicity has been known, but only recently have we become aware that it may have a positive effect on human health. Studies have shown that trace amounts of selenium in the diet may protect people from heart disease and cancer. One study

The Alchemists' Symbols for Some Common Elements and Compounds

Substance	Alchemists' Symbol
Silver	☽
Lead	♄
Tin	♃
Platinum	☽☉
Sulfuric acid	+⊕
Alcohol	🝡
Sea salt	⊙

based on data from 27 countries showed an inverse relationship between the cancer death rate and the selenium content of soil in a particular region (low cancer death rate in areas with high selenium content). Another research paper reported an inverse relationship between the selenium content of the blood and the incidence of breast cancer in women. Selenium is also found in the heart muscle and may play an important role in proper heart function. Because of these and other studies, selenium's reputation has improved, and many scientists are now studying its function in the human body.

Silicon is the second most abundant element in the earth's crust, exceeded only by oxygen. As we will see in Chapter 16, compounds involving silicon bonded to oxygen make up most of the earth's sand, rock, and soil. Berzelius prepared silicon in its pure form in 1824 by heating silicon tetrafluoride (SiF_4) with potassium metal. Today, silicon forms the basis for the microelectronics industry centered near San Francisco in a place that has come to be known as Silicon Valley. The technology of the silicon chip (see figure) with its printed circuits has transformed computers from room-sized monsters with thousands of unreliable vacuum tubes to desktop and notebook-sized units with trouble-free "solid-state" circuitry.*

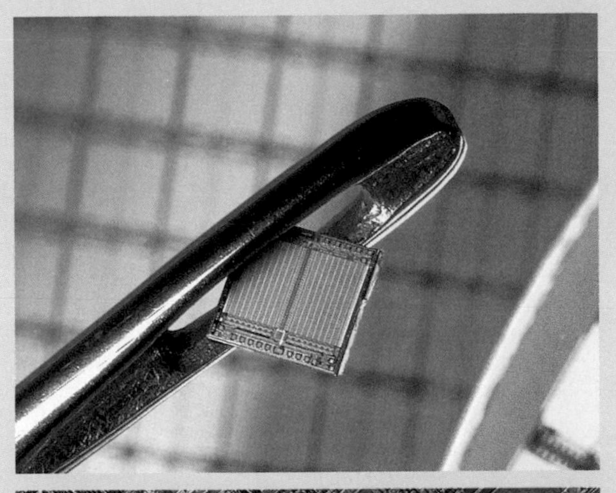

(top) A silicon chip microprocessor. (bottom) A highly magnified silicon chip microprocessor.

* For further reading, see Bernard Jaffe, *Crucibles: The Story of Chemistry* (Premiere Book, 1957).

EXAMPLE 2.1

The first four compounds listed in Table 2.1 contain only carbon and hydrogen. Predict the formulas for these compounds.

Solution

Since the compounds contain only carbon and hydrogen, the percent hydrogen in each compound (by mass) is 100 − % carbon. We can then find the relative mass of hydrogen present as follows:

$$\text{Relative mass of hydrogen} = \frac{\text{percent hydrogen}}{100} \times \text{relative molecular mass}$$

In tabular form the results are as follows:

Compound	Relative Molecular Mass	Percent Hydrogen	Relative Mass of Hydrogen
Methane	16	25	4
Ethane	30	20	6
Propane	44	18	8
Butane	58	17	10

Combining the above results with those from Table 2.1, we find that methane contains relative masses of carbon and hydrogen of 12 and 4, respectively. Using the relative atomic mass values of 12 and 1 for carbon and hydrogen gives a formula of CH_4 for methane. Similarly, the relative masses of carbon and hydrogen in ethane of 24 and 6, respectively, lead to a formula of C_2H_6 for ethane. Similar reasoning gives formulas for propane and butane of C_3H_8 and C_4H_{10}, respectively.

Cannizzaro's work was so convincing because he collected data on so many compounds. Although he couldn't absolutely prove that his atomic mass values were correct (because he had no way to verify absolutely the formulas of the compounds), the consistency of the large quantity of data he had collected eventually convinced virtually everyone that his interpretation made sense and that the relative values of atomic mass that he had determined were correct. The confusion was finally over. Chemistry had the universal (relative) mass standards that it needed.

It is worthwhile to note that Cannizzaro's work led to *approximate* values of the relative atomic masses. His goal was not to determine highly precise values for atomic masses but rather to pin down the approximate values (for example, to show that oxygen's relative mass was 16 rather than 8). The most precise values for atomic masses were determined by quantitative experiments in which the combining masses of elements were carefully measured, such as in the work of Berzelius.

In the next chapter we will have much more to say about atomic masses, including the origin of the very precise values used by today's chemists.

2.5 Early Experiments to Characterize the Atom

On the basis of the work of Dalton, Gay-Lussac, Avogadro, Cannizzaro, and others, chemistry was beginning to make sense. The concept of atoms was clearly a good idea. Inevitably, scientists began to wonder about the nature of the atom. What is an atom made of, and how do the atoms of the various elements differ?

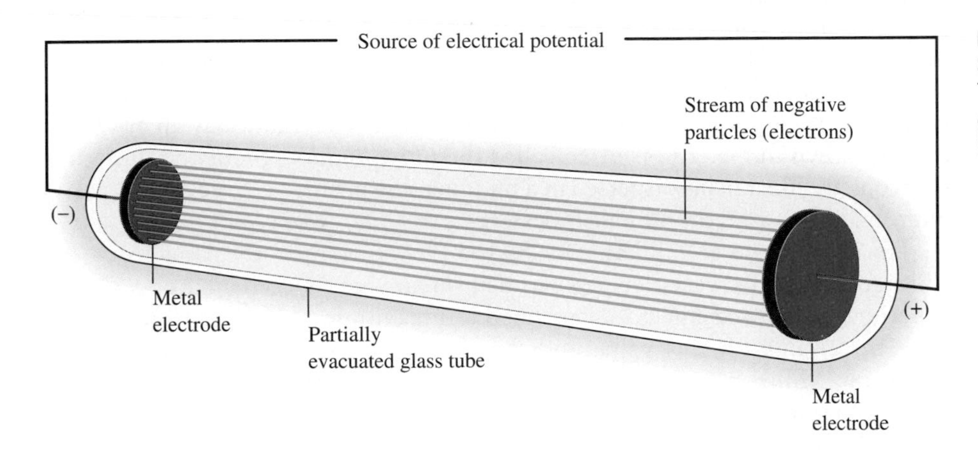

Figure 2.4
Schematic of a cathode-ray tube. The fast-moving electrons excite the gas in the tube, causing a glow between the electrodes.

The Electron

The first important experiments that led to an understanding of the composition of the atom were done by the English physicist J. J. Thomson (1856–1940), who studied electrical discharges in partially evacuated tubes called *cathode-ray tubes* (Fig. 2.4) during the period from 1898 to 1903. Thomson found that when high voltage was applied to the tube, a "ray" he called a **cathode ray** (because it emanated from the negative electrode, or cathode) was produced. Because this ray was produced at the negative electrode and was repelled by the negative pole of an applied electric field (see Fig. 2.5), Thomson

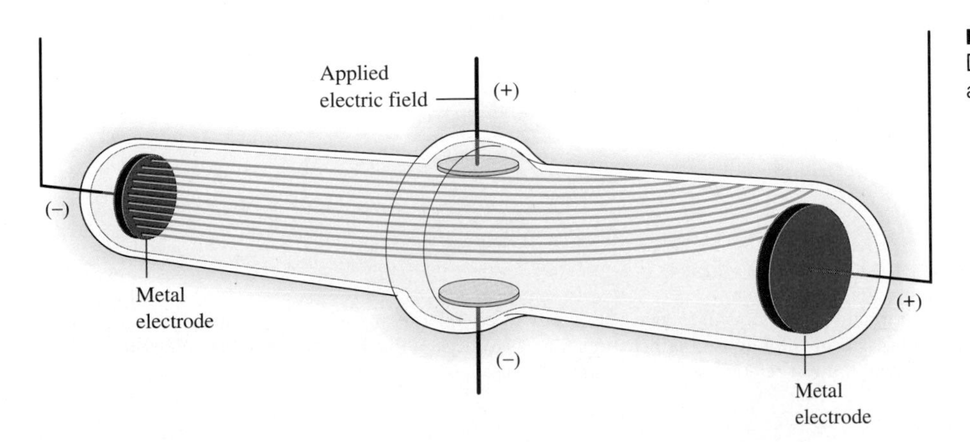

Figure 2.5
Deflection of cathode rays by an applied electric field.

postulated that the ray was a stream of negatively charged particles, now called **electrons.** From experiments in which he measured the deflection of the beam of electrons in a magnetic field, Thomson determined the *charge-to-mass ratio* of an electron:

$$\frac{e}{m} = -1.76 \times 10^{8} \text{ C/g}$$

where e represents the charge on the electron in coulombs and m represents the electron mass in grams.

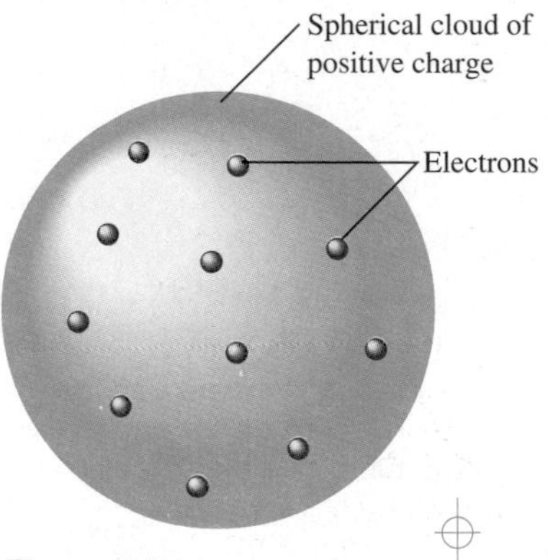

Figure 2.6
Thomson's plum pudding model.

One of Thomson's primary goals in his cathode-ray tube experiments was to gain an understanding of the structure of the atom. He reasoned that since electrons could be produced from electrodes made of various types of metals, *all* atoms must contain electrons. Since atoms were known to be electrically neutral, Thomson further assumed that atoms also must contain some positive charge. Thomson postulated that an atom consisted of a diffuse cloud of positive charge with the negative electrons embedded randomly in it. This model, shown in Fig. 2.6, is often called the *plum pudding model* because the electrons are like raisins dispersed in a pudding (the positive charge cloud), as in plum pudding, a favorite English dessert.*

In 1909 Robert Millikan (1868–1953), working at the University of Chicago, performed very clever experiments involving charged oil drops. These experiments allowed him to determine the magnitude of the electron charge (see Fig. 2.7). With this value and the charge-to-mass ratio determined by Thomson, Millikan was able to calculate the mass of the electron as 9.11×10^{-31} kilogram.

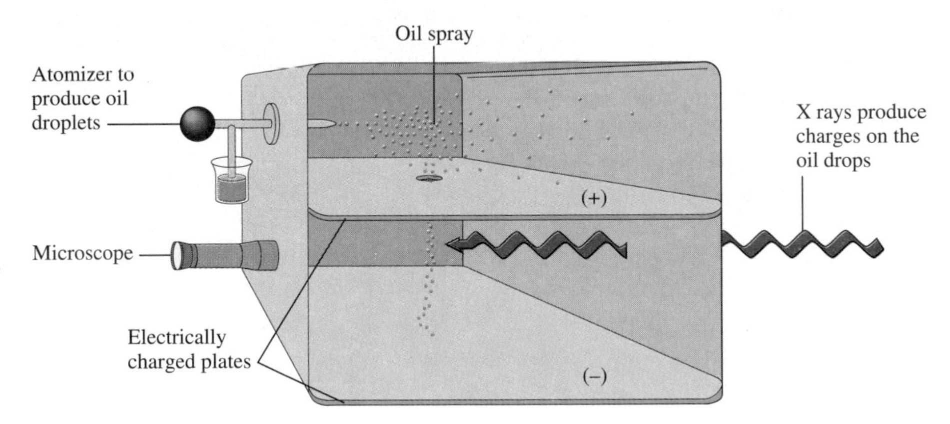

Figure 2.7
A schematic representation of the apparatus Millikan used to determine the charge on the electron. The fall of charged oil droplets due to gravity can be halted by adjusting the voltage across the two plates. The voltage and the mass of an oil drop can then be used to calculate the charge on the oil drop. Millikan's experiments showed that the charge on an oil drop is always a whole-number multiple of the electron charge.

*Although J. J. Thomson is generally given credit for this model, the idea was apparently first suggested by the English mathematician and physicist William Thomson (better known as Lord Kelvin and not related to J. J. Thomson).

Radioactivity

In the late nineteenth century scientists discovered that certain elements produce high-energy radiation. For example, in 1896 the French scientist Antoine Henri Becquerel found accidentally that the image of a piece of mineral containing uranium could be produced on a photographic plate in the absence of light. He attributed this phenomenon to a spontaneous emission of radiation by the uranium, which he called **radioactivity.** Studies in the early twentieth century demonstrated three types of radioactive emission: gamma (γ) rays, beta (β) particles, and alpha (α) particles. A γ ray is high-energy "light"; a β particle is a high-speed electron; and an α particle has a 2+ charge, that is, a charge twice that of the electron and with the opposite sign. The mass of an α particle is 7300 times that of the electron. More modes of radioactivity are now known, and we will discuss them in Chapter 21. Here we will consider only α particles because they were used in some crucial early experiments.

Figure 2.8
Ernest Rutherford (1871–1937) was born on a farm in New Zealand. In 1895 he placed second in a scholarship competition to attend Cambridge University but was awarded the scholarship when the winner decided to stay home and get married. As a scientist in England, Rutherford did much of the early work on characterizing radioactivity. He named the α and β particles and the γ ray and coined the term *half-life* to describe an important attribute of radioactive elements. His experiments on the behavior of α particles striking thin metal foils led him to postulate the nuclear atom. He also invented the name *proton* for the nucleus of the hydrogen atom. He received the Nobel prize in chemistry in 1908.

The Nuclear Atom

In 1911 Ernest Rutherford (Fig. 2.8), who performed many of the pioneering experiments to explore radioactivity, carried out an experiment to test Thomson's plum pudding model. The experiment involved directing α particles at a thin sheet of metal foil, as illustrated in Fig. 2.9. Rutherford reasoned that if Thomson's model were accurate, the massive α particles should crash through the thin foil like cannonballs through gauze, as shown in Fig. 2.10(a). He expected the α particles to travel through the foil with, at the most, very minor deflections in their paths. The results of the experiment were very different from those Rutherford anticipated. Although most of the α particles passed straight through, many of the particles were deflected at large angles, as shown in Fig. 2.10(b), and some were reflected, never hitting the detector. This outcome was a great surprise to Rutherford. (He wrote that this result was comparable to shooting a howitzer at a piece of paper and having the shell reflected back.)

Rutherford knew from these results that the plum pudding model for the atom could not be correct. The large deflections of the α particles could only be caused by a center of concentrated positive charge that contains most of the atom's mass, as illustrated in Fig. 2.10(b). Most of the α particles pass directly

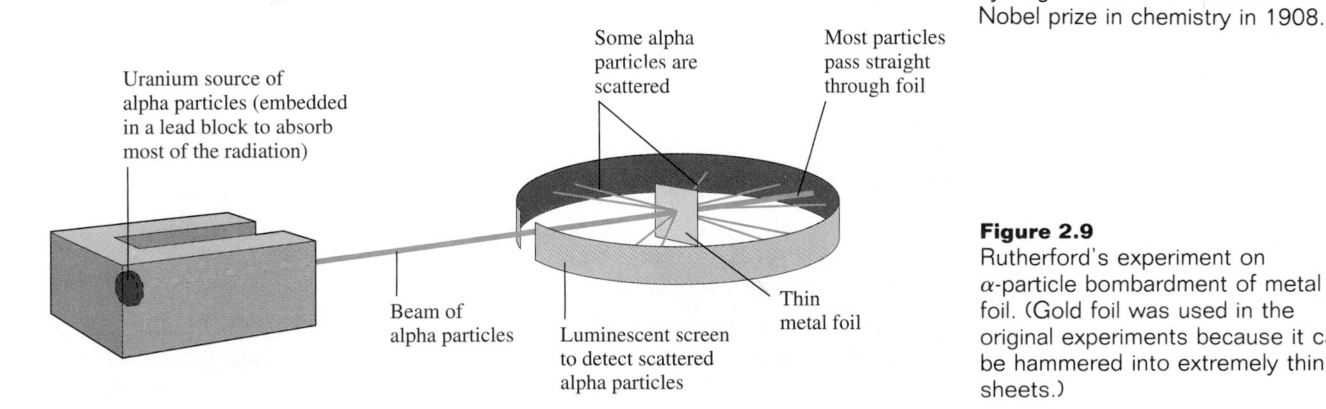

Figure 2.9
Rutherford's experiment on α-particle bombardment of metal foil. (Gold foil was used in the original experiments because it can be hammered into extremely thin sheets.)

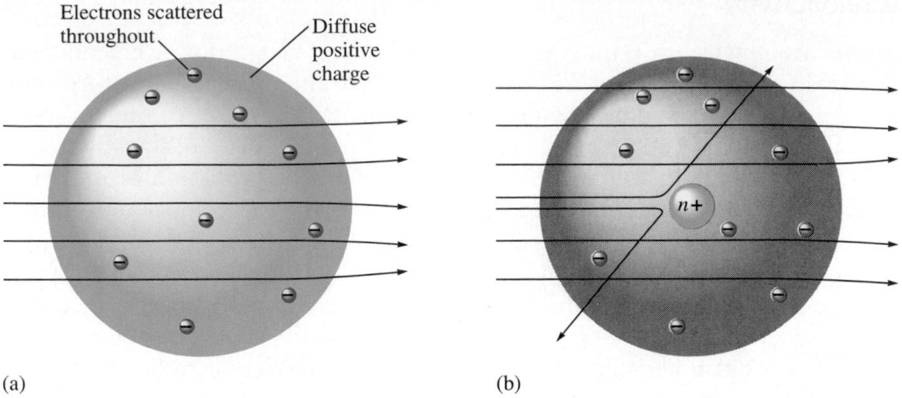

Electrons scattered throughout
Diffuse positive charge
$n+$
(a)
(b)

Figure 2.10
(a) The expected results of the metal foil experiment if Thomson's model were correct. (b) Actual results.

through the foil because the atom is mostly open space. The deflected α particles are those that had a "close encounter" with the massive positive center of the atom, and the few reflected α particles are those that made a "direct hit" on the much more massive positive center.

In Rutherford's mind these results could only be explained in terms of a **nuclear atom**—an atom with a dense center of positive charge (the **nucleus**) with electrons moving around the nucleus at a distance that is large relative to the nuclear radius.

2.6 The Modern View of Atomic Structure: An Introduction

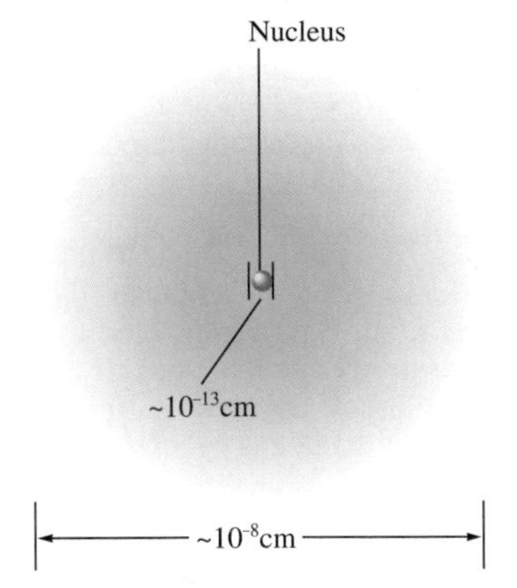

Figure 2.11
A nuclear atom viewed in cross section.

In the years since Thomson and Rutherford, a great deal has been learned about atomic structure. Because much of this material will be covered in detail in later chapters, only an introduction will be given here. The simplest view of the atom is that it consists of a tiny nucleus with a diameter of about 10^{-13} cm and electrons that move about the nucleus at an average distance of about 10^{-8} cm away from it (see Fig. 2.11).

As we will see later, the chemistry of an atom mainly results from its electrons. For this reason chemists can be satisfied with a relatively crude nuclear model. The nucleus is assumed to contain **protons,** which have a positive charge equal in magnitude to the electron's negative charge, and **neutrons,** which have virtually the same mass as a proton but no charge. The masses and charges of the electron, proton, and neutron are shown in Table 2.2.

Two striking things about the nucleus are its small size, compared with the overall size of the atom, and its extremely high density. The tiny nucleus accounts for almost all of the atom's mass. Its great density is dramatically demonstrated by the fact that a piece of nuclear material about the size of a pea would have a mass of 250 million tons!

An important question to consider at this point is, "*If all atoms are composed of these same components, why do different atoms have different chemical properties?*" The answer to this question lies in the number and the arrangement of the electrons. The electrons constitute most of the atomic vol-

TABLE 2.2 **The Mass and Charge of the Electron, Proton, and Neutron**

Particle	Mass	Charge*
Electron	9.11×10^{-31} kg	1−
Proton	1.67×10^{-27} kg	1+
Neutron	1.67×10^{-27} kg	none

* The magnitude of the charge of the electron and the proton is 1.60×10^{-19} C.

ume and thus are the parts that "intermingle" when atoms combine to form molecules. Therefore, the number of electrons possessed by a given atom greatly affects its ability to interact with other atoms. As a result, the atoms of different elements, which have different numbers of protons and electrons, show different chemical behavior.

The *chemistry* of an atom arises from its electrons.

A sodium atom has 11 protons in its nucleus. Since atoms have no net charge, the number of electrons must equal the number of protons. Therefore, a sodium atom has 11 electrons moving around its nucleus. It is *always* true that a sodium atom has 11 protons and 11 electrons. However, each sodium atom also has neutrons in its nucleus, and different types of sodium atoms exist that have different numbers of neutrons. For example, consider the sodium atoms represented in Fig. 2.12. These two atoms are **isotopes,** or *atoms with the same number of protons but different numbers of neutrons.* Note that the symbol for one particular type of sodium atom is written

Mass number → $^{23}_{11}Na$ ← Element symbol
Atomic number ↗

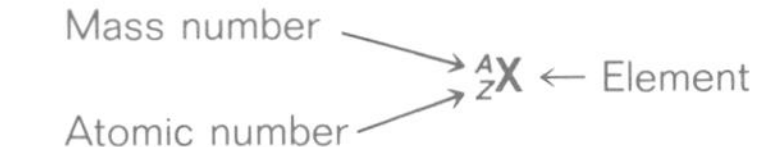

where the **atomic number** Z (number of protons) is written as a subscript and the **mass number** A (the total number of protons and neutrons) is written as a superscript. (The particular atom represented here is called "sodium twenty-three." It has 11 electrons, 11 protons, and 12 neutrons.) Because the chemistry of an atom is due to its electrons, isotopes show almost identical chemical properties. In nature most elements contain a mixture of isotopes.

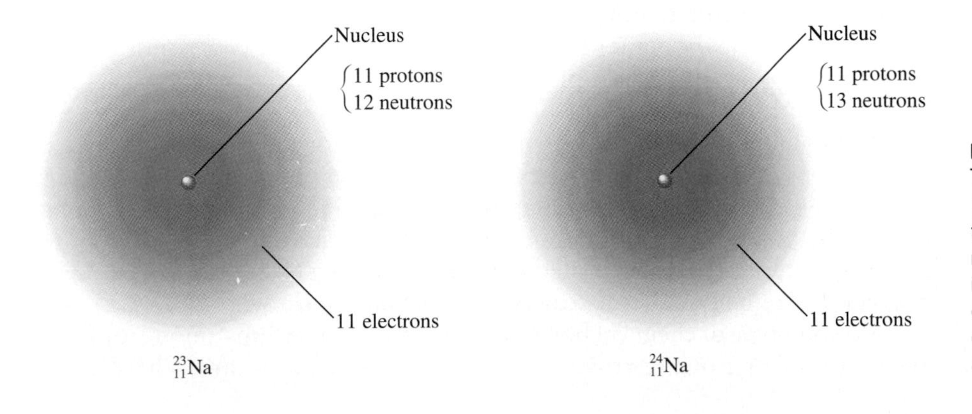

Figure 2.12
Two isotopes of sodium. Both have 11 protons and 11 electrons, but they differ in the number of neutrons in their nuclei. Sodium 23 is the only naturally occurring form of sodium. Sodium 24 does not occur naturally but can be made artificially.

2.7 Molecules and Ions

From a chemist's viewpoint the most interesting characteristic of an atom is its ability to combine with other atoms to form compounds. It was John Dalton who first recognized that chemical compounds were collections of atoms, but he could not determine the structure of atoms or their means for binding to each other. During the twentieth century scientists have learned that atoms have electrons and that these electrons participate in the bonding of one atom to another. We will discuss bonding thoroughly in Chapters 13 and 14; here we will consider some definitions that will be useful in the next few chapters.

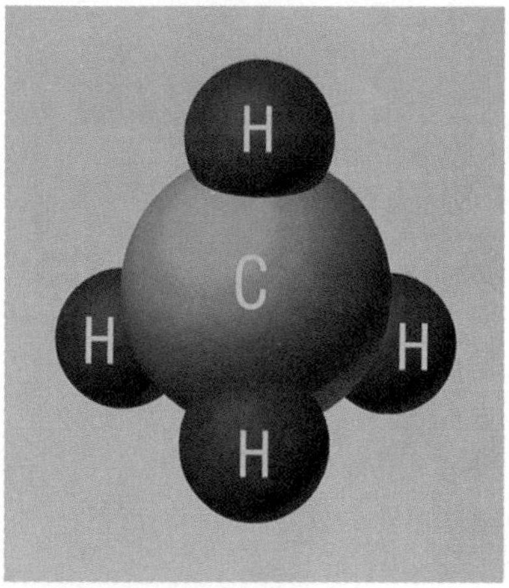

The forces that hold atoms together in compounds are called **chemical bonds.** One way that atoms can form bonds is by *sharing electrons.* These bonds are called **covalent bonds,** and the resulting collection of atoms is called a **molecule.** Molecules can be represented in several different ways. The simplest method is the **chemical formula,** in which the symbols for the elements are used to indicate the types of atoms present, and subscripts are used to indicate the relative numbers of atoms. For example, the formula for carbon dioxide is CO_2, meaning, of course, that each molecule contains 1 atom of carbon and 2 atoms of oxygen.

Figure 2.13
Space-filling model of the methane molecule. This type of model shows both the relative sizes of the atoms in the molecule and their spatial relationships.

Familiar examples of molecules that contain covalent bonds are hydrogen (H_2), water (H_2O), oxygen (O_2), ammonia (NH_3), and methane (CH_4). More information about a molecule is given by its **structural formula,** in which the individual bonds are shown (indicated by lines). Structural formulas may or may not indicate the actual shape of the molecule. For example, water might be represented as

H—O—H or H/O\H

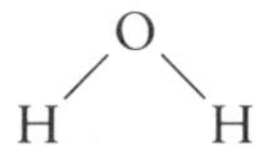

The structure on the right shows the actual shape of the water molecule, based on experimental evidence. Other examples of structural formulas are

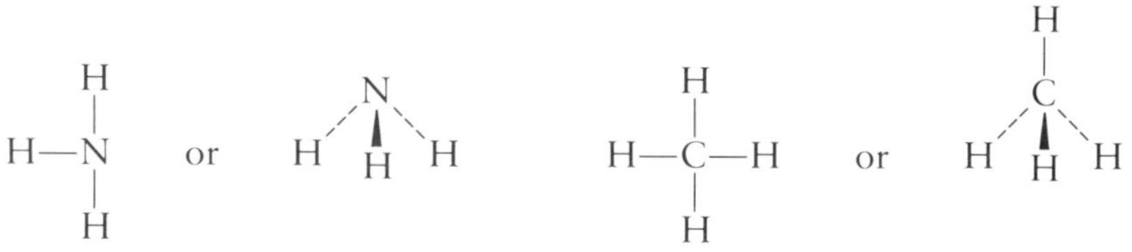

Ammonia: H—N(—H)—H or N with H, H, H (dashed and wedge bonds)

Methane: H—C(—H)(—H)—H or C with H above and H, H, H (dashed and wedge bonds)

In the actual structures on the right the central atom and the solid lines are understood to be in the plane of the paper. Atoms connected to the central atom by dashed lines are behind the plane of the paper, and atoms connected to the central atom by wedges are in front of the plane of the page.

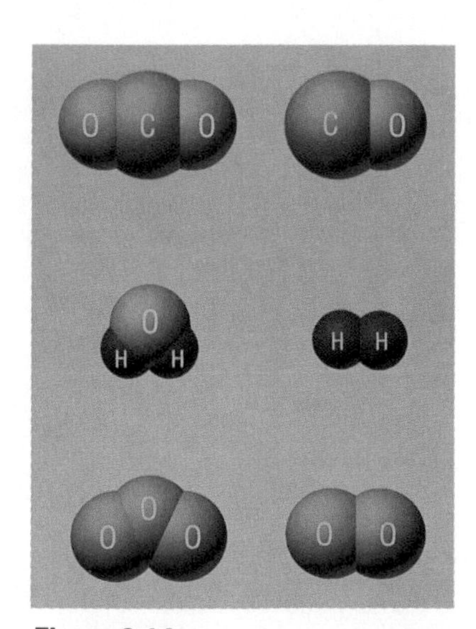

Figure 2.14
Space-filling models of various molecules.

In a compound composed of molecules, the individual molecules move around as independent units. For example, a sample of methane gas is represented in Fig. 2.13 using **space-filling models.** These models show the relative sizes of the atoms, as well as their relative orientation in the molecule. Figure 2.14 shows other examples. **Ball-and-stick models** are also used to represent molecules. The ball-and-stick structure of methane is shown in Fig. 2.15.

A second type of chemical bonding results from attractions among ions. An **ion** is an atom or group of atoms that has a net positive or negative charge. The

Figure 2.15
Ball-and-stick model of methane.

(upper left) Sodium metal is soft enough to be cut easily with a knife. (upper right) Chlorine is a green gas. (lower left) Sodium and chlorine react vigorously to form sodium chloride (lower right).

best-known ionic compound is common table salt, or sodium chloride, which forms when neutral chlorine and sodium react.

To see how ions are formed, consider what happens when an electron is transferred from sodium to chlorine (the neutrons in the nuclei will be ignored):

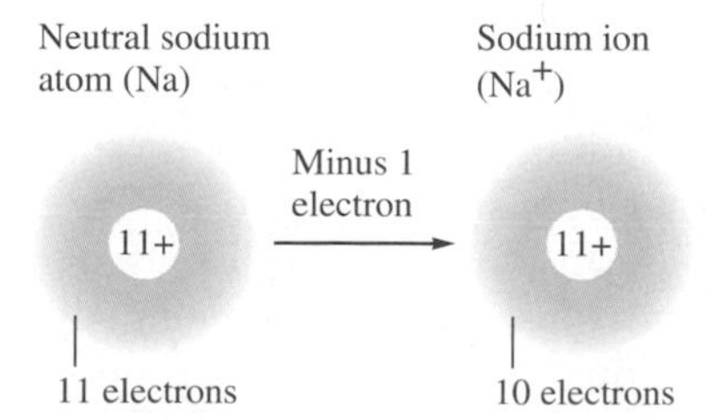

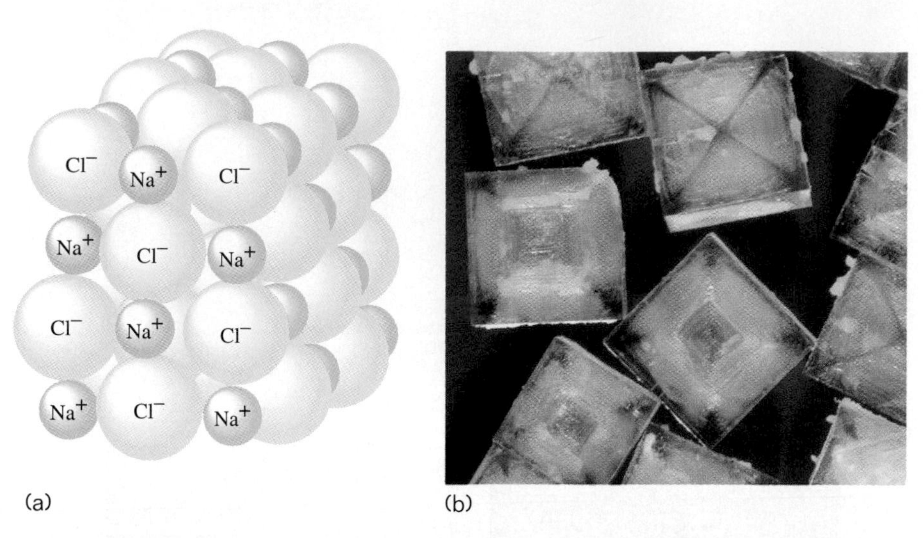

Figure 2.16
(a) The arrangement of sodium ions (Na^+) and chloride ions (Cl^-) in the ionic compound sodium chloride. (b) Sodium chloride forms cubic crystals.

With one electron stripped off, the sodium with its 11 protons and only 10 electrons has become a *positive ion* with a net 1+ charge. A positive ion is called a **cation.** The process can be represented in shorthand form as

$$Na \longrightarrow Na^+ + e^-$$

If an electron is added to chlorine,

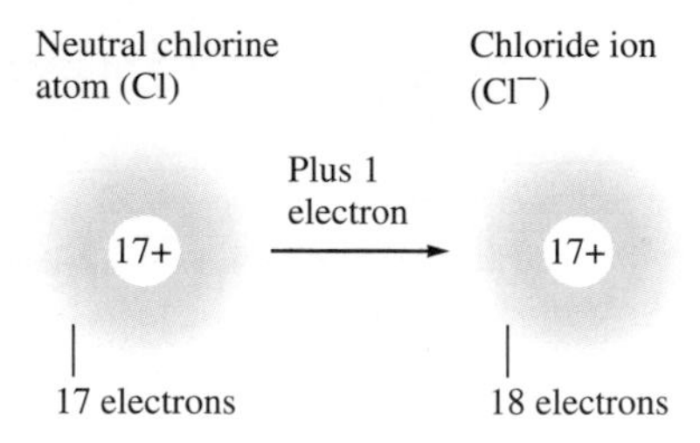

the 18 electrons produce a net 1− charge; the chlorine has become an *ion with a negative charge*—an **anion.** This process is represented as

$$Cl + e^- \longrightarrow Cl^-$$

Because anions and cations have opposite charges, they attract each other. This *force of attraction between oppositely charged ions* is called **ionic bonding.** As shown in the lower left photo on page 35, sodium metal and chlorine gas (a green gas composed of Cl_2 molecules) react to form solid sodium chloride, which contains many Na^+ and Cl^- ions packed together [Fig. 2.16(a)]. The solid forms the beautiful, colorless cubic crystals shown in Fig. 2.16(b).

A solid consisting of oppositely charged ions is called an *ionic solid,* or (often) a *salt.* Ionic solids can consist of simple ions, as in sodium chloride, or of **polyatomic** (many-atom) **ions,** as in ammonium nitrate (NH_4NO_3), which contains ammonium cations (NH_4^+) and nitrate anions (NO_3^-). The ball-and-stick models of these ions are shown in Fig. 2.17.

Figure 2.17
Ball-and-stick models of the ammonium ion (NH_4^+) and the nitrate ion (NO_3^-).

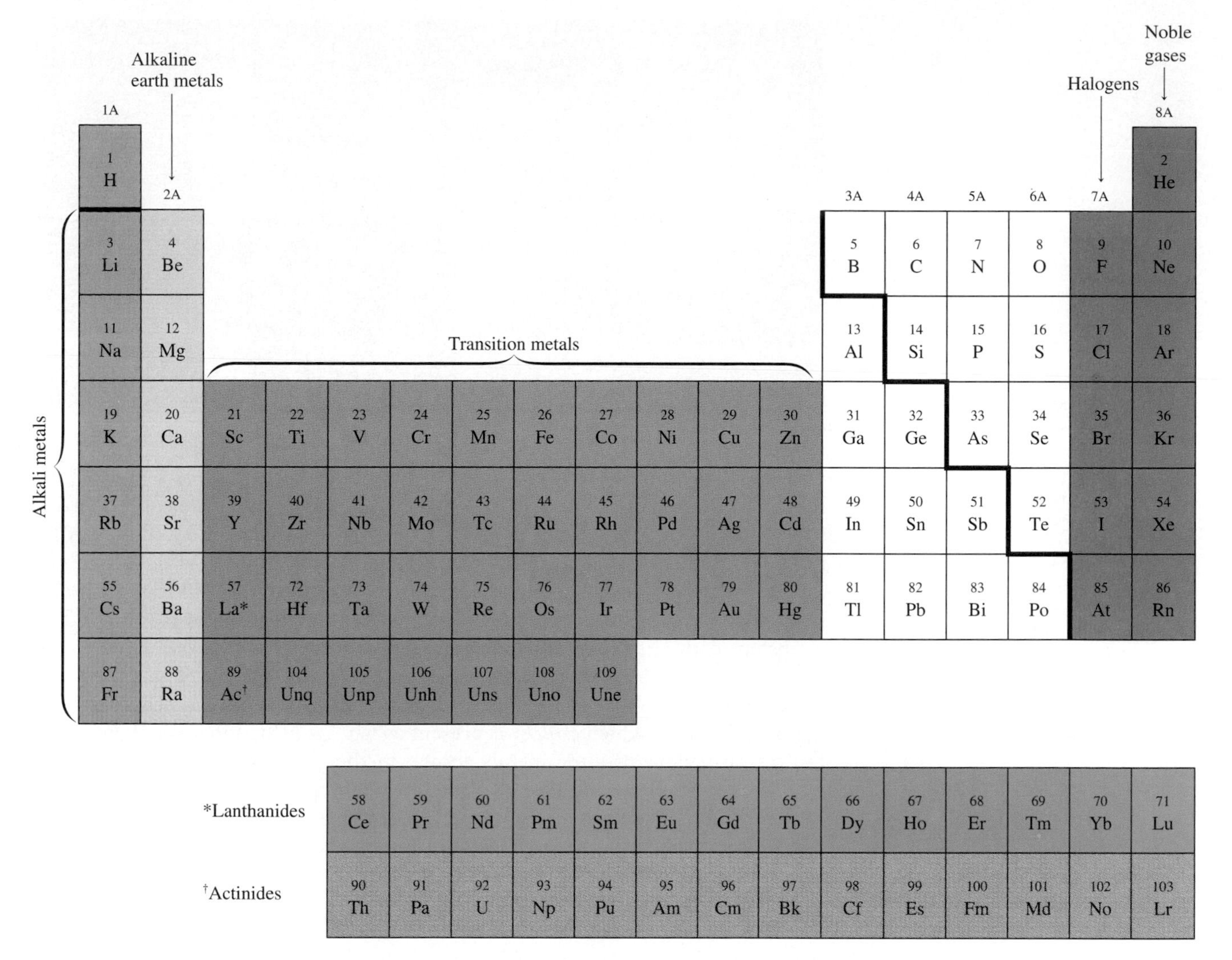

Figure 2.18
The periodic table. For elements 104 and beyond, a proposal has been made to name them systematically, using a letter abbreviation related to each atomic number. Another proposal is to name these elements after famous scientists. These names have not yet been officially approved by IUPAC and will not be used in this text.

2.8 An Introduction to the Periodic Table

In a room where chemistry is taught or practiced, a chart called the **periodic table** is almost certain to be found hanging on the wall. Recall that this chart shows all of the known elements and provides a good deal of information about each. As your knowledge of chemistry increases, the periodic table will become more and more useful to you. In this section we will remind you about its fundamental aspects.

A simple version of the periodic table is shown in Fig. 2.18. The letters given in the boxes are the symbols for the elements, and the number shown

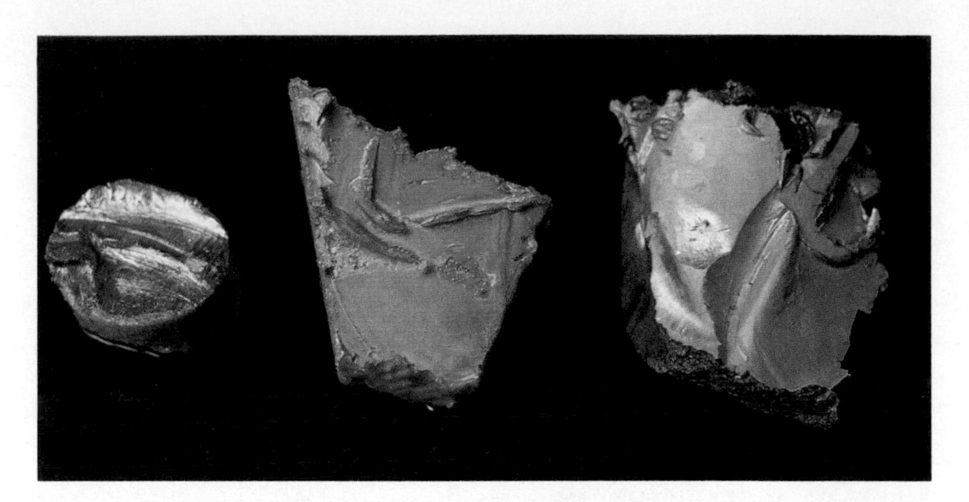

Samples of the alkali metals lithium, sodium, and potassium.

above each symbol is the atomic number (number of protons) for that element. Most of the elements are **metals.** Metals have characteristic physical properties such as efficient conduction of heat and electricity, malleability (they can be hammered into thin sheets), ductility (they can be pulled into wires), and (often) a lustrous appearance. Chemically, metal atoms tend to *lose* electrons to form positive ions. For example, copper is a typical metal. It is lustrous (although it tarnishes readily); it is an excellent conductor of electricity (it is widely used in electrical wires); and it is readily formed into various shapes such as pipes for water systems. Copper is also found in many salts, such as the beautiful blue copper sulfate, in which copper is present as Cu^{2+} ions. Copper is a member of the transition metals—the metals shown in the center of the periodic table.

Metals tend to form positive ions; nonmetals tend to form negative ions.

The relatively few **nonmetals** appear in the upper right-hand corner of the table (to the right of the heavy line in Fig. 2.18), except hydrogen, a nonmetal that is grouped with the metals. The nonmetals typically lack the physical properties that characterize the metals. Chemically, they tend to *gain* electrons to form anions in reactions with metals. Nonmetals often bond to each other by forming covalent bonds. For example, chlorine is a typical nonmetal. Under normal conditions it exists as Cl_2 molecules; it reacts with metals to form salts containing Cl^- ions (NaCl, for example); and it forms covalent bonds with nonmetals (for example, hydrogen chloride gas, or HCl).

The periodic table is arranged so that elements in the same vertical columns (called **groups** or **families**) have *similar chemical properties*. For example, all of the **alkali metals**, members of Group 1A—lithium (Li), sodium (Na), potassium (K), rubidium (Rb), cesium (Cs), and francium (Fr)—are very active elements that readily form ions with a 1+ charge when they react with nonmetals. The members of Group 2A—beryllium (Be), magnesium (Mg), calcium (Ca), strontium (Sr), barium (Ba), and radium (Ra)—are called the **alkaline earth metals.** They all form ions with a 2+ charge when they react with nonmetals. The **halogens,** the members of Group 7A—fluorine (F), chlorine (Cl), bromine (Br), iodine (I), and astatine (At)—all form diatomic molecules. Fluorine, chlorine, bromine, and iodine all react with metals to form salts containing ions with a 1− charge (F^-, Cl^-, Br^-, and I^-). The members of Group 8A—helium (He), neon (Ne), argon (Ar), krypton (Kr), xenon (Xe), and radon (Rn)—are known as the **noble gases.** They all exist under normal conditions as monatomic (single-atom) gases and have little chemical reactivity.

Three members of the halogen family: iodine (purple), bromine (reddish-brown), and chlorine (green).

The horizontal rows of elements in the periodic table are called **periods.** Horizontal row one is called the first period (it contains H and He); row two is called the second period (elements Li through Ne); and so on.

We will learn much more about the periodic table as we continue with our study of chemistry. Meanwhile, when an element is introduced in this text, you should always note its position on the periodic table.

2.9 Naming Simple Compounds

When chemistry was an infant science, there was no system for naming compounds. Names such as sugar of lead, blue vitrol, quicklime, Epsom salts, milk of magnesia, gypsum, and laughing gas were coined by early chemists. Such names are called *common names*. As chemistry grew, it became clear that using common names for compounds would lead to unacceptable chaos. More than 4 million chemical compounds are currently known. Memorizing common names for these compounds would be an impossible task.

The solution, of course, is to adopt a *system* for naming compounds in which the name tells something about the composition of the compound. After learning the system, a chemist given a formula should be able to name the compound, or given a name should be able to construct the compound's formula. In this section we will specify the most important rules for naming compounds other than organic compounds (those based on chains of carbon atoms).

The systematic naming of organic compounds will be discussed in Chapter 22.

We will begin with the systems for naming inorganic **binary compounds**—compounds composed of two elements—which we classify into various types for easier recognition. We will consider both ionic and covalent compounds.

Binary Compounds (Type I; Ionic)

Binary ionic compounds contain a positive ion (cation), always written first in the formula, and a negative ion (anion). In the naming of these compounds the following rules apply:

TABLE 2.3 **Common Monatomic Cations and Anions**

Cation	Name	Anion	Name
H^+	hydrogen	H^-	hydride
Li^+	lithium	F^-	fluoride
Na^+	sodium	Cl^-	chloride
K^+	potassium	Br^-	bromide
Cs^+	cesium	I^-	iodide
Be^{2+}	beryllium	O^{2-}	oxide
Mg^{2+}	magnesium	S^{2-}	sulfide
Ca^{2+}	calcium	N^{3-}	nitride
Ba^{2+}	barium	P^{3-}	phosphide
Al^{3+}	aluminum		
Ag^+	silver		

1. The cation is always named first and the anion second.
2. A monatomic (meaning from one-atom) cation takes its name from the name of the element. For example, Na^+ is called sodium in the names of compounds containing this ion.
3. A monatomic anion is named by taking the first part of the element name and adding *-ide*. Thus the Cl^- ion is called chloride.

A monatomic cation has the same name as its parent element.

Some common monatomic cations and anions and their names are given in Table 2.3.

A Type I binary ionic compound contains a metal that forms only one type of cation. The rules for naming Type I compounds are illustrated by the following examples:

In formulas of ionic compounds, simple ions are represented by the element symbol: Cl means Cl^-, Na means Na^+, and so on.

Compound	*Ions Present*	*Name*
NaCl	Na^+, Cl^-	sodium chloride
KI	K^+, I^-	potassium iodide
CaS	Ca^{2+}, S^{2-}	calcium sulfide
Li_3N	Li^+, N^{3-}	lithium nitride
CsBr	Cs^+, Br^-	cesium bromide
MgO	Mg^{2+}, O^{2-}	magnesium oxide

Binary Compounds (Type II; Ionic)

Type II binary ionic compounds contain a metal that can form more than one type of cation.

In the ionic compounds considered above (Type I), the metal involved forms only a single type of cation. That is, sodium forms only Na^+, calcium forms only Ca^{2+}, and so on. However, as we will see in more detail later in the text, there are many metals that can form more than one type of positive ion and thus form more than one type of ionic compound with a given anion. For example, the compound $FeCl_2$ contains Fe^{2+} ions, and the compound $FeCl_3$ contains Fe^{3+} ions. In a case such as this, the *charge on the metal ion must be specified.* The systematic names for these two iron compounds are iron(II) chloride and iron(III) chloride, respectively, where the *Roman numeral indicates the charge of the cation.*

Another system for naming these ionic compounds that is seen in the older literature was used for metals that form only two ions. *The ion with the higher*

(left) A dish of copper(II) sulfate. (right) A close-up photo of copper(II) sulfate crystals.

charge has a name ending in -ic, and the one with the lower charge has a name ending in -ous. In this system, for example, Fe^{3+} is called the ferric ion, and Fe^{2+} is called the ferrous ion. The names for $FeCl_3$ and $FeCl_2$ are then ferric chloride and ferrous chloride, respectively.

Table 2.4 gives both names for many common Type II cations. The system that uses Roman numerals will be used exclusively in this text.

Note that the use of a Roman numeral in a systematic name is required only in cases where more than one ionic compound forms between a given pair of elements. This case most commonly occurs for compounds containing transition metals, which often form more than one cation. *Elements that form only one cation do not need to be identified by a Roman numeral.* Common metals that do not require Roman numerals are the Group 1A elements, which form only 1+ ions; the Group 2A elements, which form only 2+ ions; and aluminum, which forms only Al^{3+}.

A compound containing a transition metal must have a Roman numeral in its name.

When a metal ion is present that forms more than one type of cation, the charge on the metal ion must be determined by balancing the positive and negative charges of the compound. To make this determination, you must be able to recognize the common cations and anions and know their charges (see Tables 2.3 and 2.5).

The following flow chart is useful when you are naming binary ionic compounds:

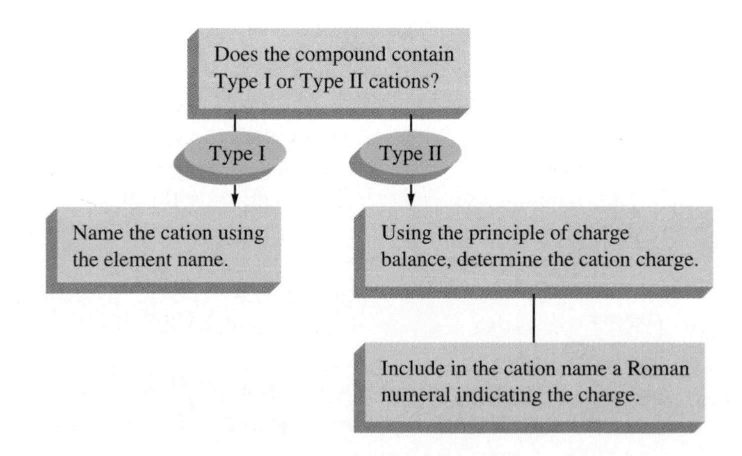

TABLE 2.4
Common Type II Cations

Ion	Systematic Name	Alternate Name
Fe^{3+}	iron(III)	ferric
Fe^{2+}	iron(II)	ferrous
Cu^{2+}	copper(II)	cupric
Cu^{+}	copper(I)	cuprous
Co^{3+}	cobalt(III)	cobaltic
Co^{2+}	cobalt(II)	cobaltous
Sn^{4+}	tin(IV)	stannic
Sn^{2+}	tin(II)	stannous
Pb^{4+}	lead(IV)	plumbic
Pb^{2+}	lead(II)	plumbous
Hg^{2+}	mercury(II)	mercuric
Hg_2^{2+}*	mercury(I)	mercurous

* Note that mercury(I) ions always occur bound together to form Hg_2^{2+}.

Buckminsterfullerene: A New Form of Carbon

A favorite chemistry exam question—What are the allotropes of elemental carbon?—now has a new answer: diamond, graphite, and buckminsterfullerene. Buckminsterfullerene? This fascinating new form of elemental carbon containing C_{60} molecules was first identified by chemists Richard E. Smalley, Howard F. Kroto, and their collaborators at the University of Sussex and has now been synthesized in large quantities using techniques developed by a team of physicists headed by Donald R. Huffman.

The structure of the C_{60} molecule resembles a soccer ball. It is a spherical molecule with the carbon atoms arranged at the 60 vertices of 32 interconnecting pentagons (12) and hexagons (20). It gets its name from the work of R. Buckminster Fuller, who proposed geodesic domes with frameworks similar to the arrangement of carbons in C_{60}.

Buckminsterfullerene was first synthesized in large quantities in the lab of Donald R. Huffman by vaporizing carbon from pure graphite electrodes in an atmosphere of helium gas at a pressure of 100 torr. The soot that forms in this environment is about 5% C_{60} by mass. The scientists have also found significant quantities of another "fullerene" in this soot that consists of C_{70} molecules. The C_{70} molecule is similar to the spherical C_{60} molecule except that the extra carbon atoms lead to an egg-shaped structure. Experiments also show that when the 60- and 70-carbon fullerenes are subjected to intense laser pulses they open their cages and combine to form huge spherical fullerenes with 400 or more carbon atoms.

The discovery of fullerenes has led to a flurry of activity in labs around the world as chemists rush to characterize this new substance. Scientists have produced the C_{60}^- and C_{60}^{2-} anions and have prepared C_{60} molecules with various substituents added. In addition, many molecules have been synthesized with metal atoms trapped inside the cages. Examples are La @ C_{60}, K @ C_{14}, and U @ C_{28}, where the @ symbol means that the metal is inside the cage. It's safe to say that a new branch of chemistry has been founded.

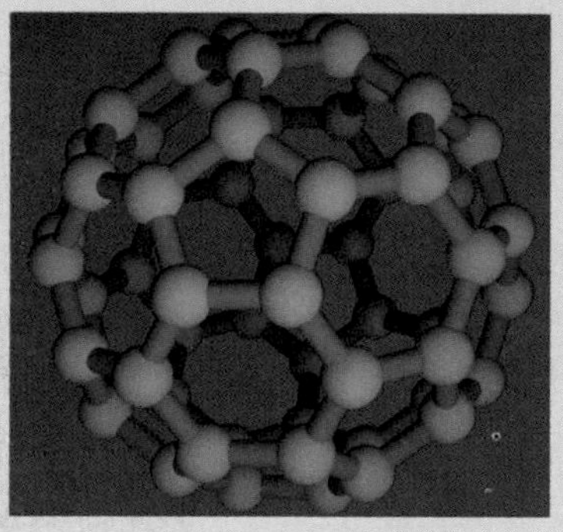

Computer-generated ball and stick model of buckminsterfullerene.

TABLE 2.5 Common Polyatomic Ions

Ion	Name	Ion	Name
NH_4^+	ammonium	CO_3^{2-}	carbonate
NO_2^-	nitrite	HCO_3^-	hydrogen carbonate (bicarbonate is a widely used common name)
NO_3^-	nitrate		
SO_3^{2-}	sulfite		
SO_4^{2-}	sulfate	ClO^-	hypochlorite
HSO_4^-	hydrogen sulfate (bisulfate is a widely used common name)	ClO_2^-	chlorite
		ClO_3^-	chlorate
		ClO_4^-	perchlorate
OH^-	hydroxide	$C_2H_3O_2^-$	acetate
CN^-	cyanide	MnO_4^-	permanganate
PO_4^{3-}	phosphate	$Cr_2O_7^{2-}$	dichromate
HPO_4^{2-}	hydrogen phosphate	CrO_4^{2-}	chromate
$H_2PO_4^-$	dihydrogen phosphate	O_2^{2-}	peroxide

EXAMPLE 2.2

Give the systematic name of each of the following compounds.

a. $CoBr_2$ b. $CaCl_2$ c. Al_2O_3 d. $CrCl_3$

Solution

Compound	*Name*	*Comment*
a. $CoBr_2$	cobalt(II) bromide	Cobalt is a transition metal; the compound name must have a Roman numeral. The two Br^- ions must be balanced by a Co^{2+} cation.
b. $CaCl_2$	calcium chloride	Calcium, an alkaline earth metal, forms only the Ca^{2+} ion. A Roman numeral is not necessary.
c. Al_2O_3	aluminum oxide	Aluminum forms only Al^{3+}. A Roman numeral is not necessary.
d. $CrCl_3$	chromium(III) chloride	Chromium is a transition metal. The compound name must have a Roman numeral. $CrCl_3$ contains Cr^{3+}.

Various chromium compounds in solid form.

Ionic Compounds with Polyatomic Ions

Ionic compounds that contain polyatomic ions are not binary compounds.

We have not yet considered ionic compounds that contain polyatomic ions. For example, the compound ammonium nitrate, NH_4NO_3, contains the polyatomic ions NH_4^+ and NO_3^-. Polyatomic ions are assigned special names that *must be memorized* in order to name the compounds containing them. The most important polyatomic ions and their names are listed in Table 2.5.

Note in Table 2.5 that several series of anions contain an atom of a given element and different numbers of oxygen atoms. These anions are called **oxyanions.** When there are two members in such a series, the name of the one with the smaller number of oxygen atoms ends in *-ite,* and the name of the one with the larger number ends in *-ate,* for example, sulfite (SO_3^{2-}) and sulfate (SO_4^{2-}). When more than two oxyanions make up a series, *hypo-* (less than) and *per-* (more than) are used as prefixes to name the members of the series with the fewest and the most oxygen atoms, respectively. The best example involves the oxyanions containing chlorine, as shown in Table 2.5.

Binary Compounds (Type III; Covalent—Contain Two Nonmetals)

In binary covalent compounds the element names follow the same rules as those for binary ionic compounds.

Binary covalent compounds are formed between *two nonmetals.* Although these compounds do not contain ions, they are named very similarly to binary ionic compounds.

In the naming of binary covalent compounds the following rules apply:

1. The first element in the formula is named first, using the full element name.
2. The second element is named as if it were an anion.
3. Prefixes are used to denote the numbers of atoms present. These prefixes are given in Table 2.6.
4. The prefix *mono-* is never used for naming the first element. For example, CO is called carbon monoxide, *not* monocarbon monoxide.

To see how these rules apply, we will now consider the names of the several covalent compounds formed by nitrogen and oxygen:

Compound	*Systematic Name*	*Common Name*
N_2O	dinitrogen monoxide	nitrous oxide
NO	nitrogen monoxide	nitric oxide
NO_2	nitrogen dioxide	
N_2O_3	dinitrogen trioxide	
N_2O_4	dinitrogen tetroxide	
N_2O_5	dinitrogen pentoxide	

Notice from the above examples that to avoid awkward pronunciations, we often drop the final *o* or *a* of the prefix when the element begins with a vowel. For example, N_2O_4 is called dinitrogen tetroxide, *not* dinitrogen tetr*a*oxide; and CO is called carbon monoxide, *not* carbon monooxide.

Some compounds are always referred to by their common names. The two best examples are water and ammonia. The systematic names for H_2O and NH_3 are never used.

TABLE 2.6
Prefixes Used to Indicate Number in Chemical Names

Prefix	Number Indicated
mono-	1
di-	2
tri-	3
tetra-	4
penta-	5
hexa-	6
hepta-	7
octa-	8

The rules for naming binary compounds are summarized in Fig. 2.19. Notice that prefixes to indicate the number of atoms are used only in Type III binary compounds (those containing two nonmetals). An overall strategy for naming compounds is summarized in Fig. 2.20.

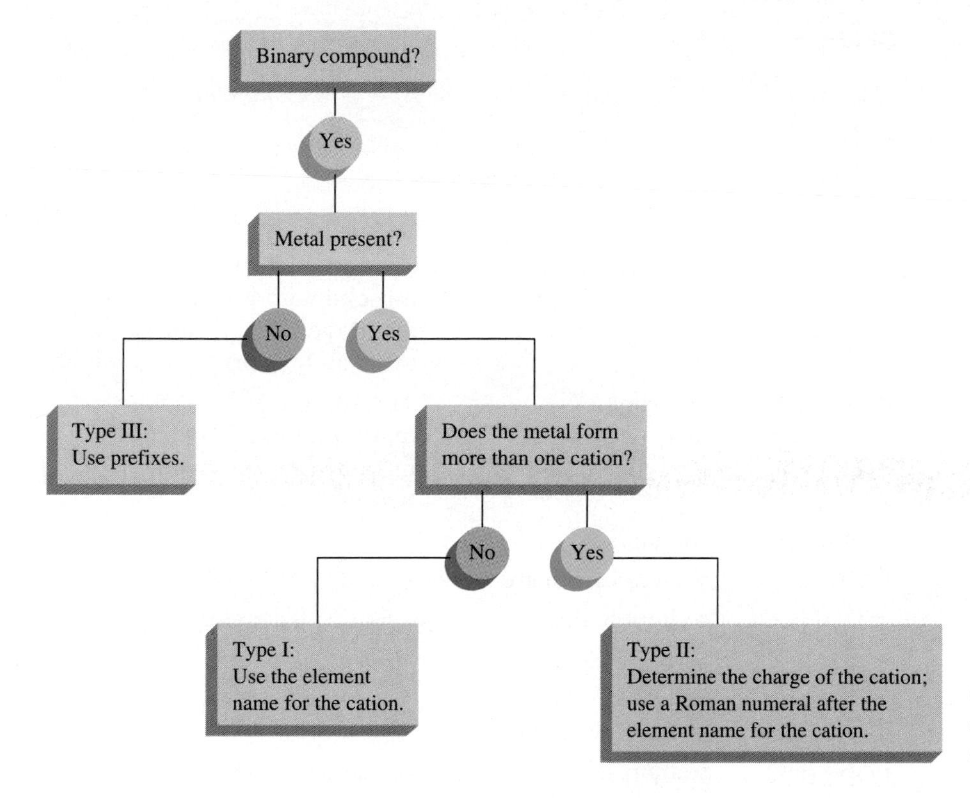

Figure 2.19
A flow chart for naming binary compounds.

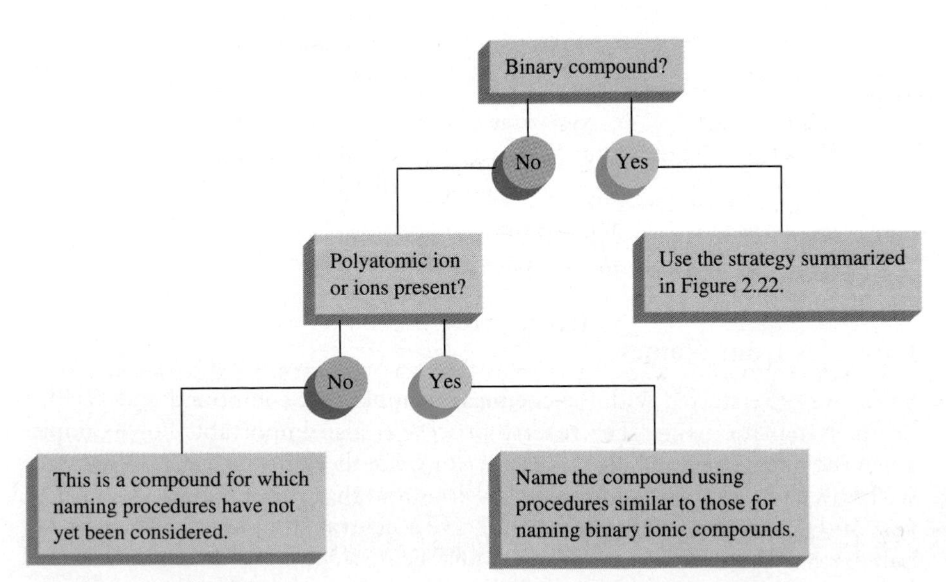

Figure 2.20
Overall strategy for naming chemical compounds.

EXAMPLE 2.3

Give the systematic name of each of the following compounds.

a. Na_2SO_4	d. $Mn(OH)_2$	g. $NaHCO_3$	j. Na_2SeO_4
b. KH_2PO_4	e. Na_2SO_3	h. $CsClO_4$	k. $KBrO_3$
c. $Fe(NO_3)_3$	f. Na_2CO_3	i. NaOCl	

Solution

Compound	*Name*	*Comment*
a. Na_2SO_4	sodium sulfate	
b. KH_2PO_4	potassium dihydrogen phosphate	
c. $Fe(NO_3)_3$	iron(III) nitrate	Transition metal—name must contain a Roman numeral. Fe^{3+} ion balances three NO_3^- ions.
d. $Mn(OH)_2$	manganese(II) hydroxide	Transition metal—name must contain a Roman numeral. Mn^{2+} ion balances two OH^- ions.
e. Na_2SO_3	sodium sulfite	
f. Na_2CO_3	sodium carbonate	
g. $NaHCO_3$	sodium hydrogen carbonate	Often called sodium bicarbonate.
h. $CsClO_4$	cesium perchlorate	
i. NaOCl	sodium hypochlorite	
j. Na_2SeO_4	sodium selenate	Atoms in the same group, like sulfur and selenium, often form similar ions that are named similarly. Thus SeO_4^{2-} is selenate, like SO_4^{2-} (sulfate).
k. $KBrO_3$	potassium bromate	As above, BrO_3^- is bromate, like ClO_3^- (chlorate).

Formulas from Names

So far we have started with the chemical formula of a compound and decided on its systematic name. The reverse process is also important. For example, given the name calcium hydroxide, we can write the formula as $Ca(OH)_2$ since we know that calcium forms only Ca^{2+} ions and that, since hydroxide is OH^-, two of these anions will be required to give a neutral compound. Similarly, the name iron(II) oxide implies the formula FeO, since the Roman numeral II indicates the presence of Fe^{2+} and since the oxide ion is O^{2-}.

EXAMPLE 2.4

Given the following systematic names, write the formula for each compound.

a. ammonium sulfate
b. vanadium(V) fluoride
c. dioxygen difluoride
d. rubidium peroxide
e. gallium oxide

Solution

Name	*Chemical Formula*	*Comment*
a. ammonium sulfate	$(NH_4)_2SO_4$	Two ammonium ions (NH_4^+) are required for each sulfate ion (SO_4^{2-}) to achieve charge balance.
b. vanadium(V) fluoride	VF_5	The compound contains V^{5+} ions and requires five F^- ions for charge balance.
c. dioxygen difluoride	O_2F_2	The prefix *di-* indicates two of each atom.
d. rubidium peroxide	Rb_2O_2	Since rubidium is in Group 1A, it forms only 1+ ions. Thus two Rb^+ ions are needed to balance the 2− charge on the peroxide ion (O_2^{2-}).
e. gallium oxide	Ga_2O_3	Since gallium is in Group 3A, like aluminum, it forms 3+ ions. Two Ga^{3+} ions are required to balance the charge on three O^{2-} ions.

Acids

When dissolved in water, certain molecules produce a solution containing free H^+ ions (protons). These substances, acids, will be discussed in detail in Chapters 4, 7, and 8. Here we will simply present the rules for naming acids.

An acid can be viewed as a molecule with one or more H^+ ions attached to an anion. The rules for naming acids depend on whether or not the anion contains oxygen. If the *anion does not contain oxygen,* the acid is named with the prefix *hydro-* and the suffix *-ic*. For example, when gaseous HCl is dissolved in water, it forms hydrochloric acid. Similarly, HCN and H_2S dissolved in water are called hydrocyanic and hydrosulfuric acids, respectively.

When the *anion contains oxygen,* the acid name is formed from the root name of the anion with a suffix of *-ic* or *-ous*. If the anion name ends in *-ate,* the acid name ends with *-ic* (or sometimes *-ric*). For example, H_2SO_4 contains the sulfate anion (SO_4^{2-}) and is called sulfuric acid; H_3PO_4 contains the phosphate anion (PO_4^{3-}) and is called phosphoric acid; and $HC_2H_3O_2$ contains the

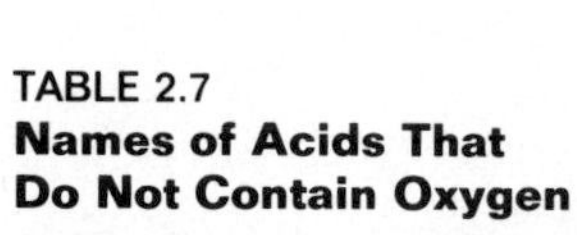

TABLE 2.7
Names of Acids That Do Not Contain Oxygen

Acid	Name
HF	hydrofluoric acid
HCl	hydrochloric acid
HBr	hydrobromic acid
HI	hydroiodic acid
HCN	hydrocyanic acid
H_2S	hydrosulfuric acid

Does the anion contain oxygen?
No → *hydro-* + anion root + *-ic* = *hydro*(anion root)*ic* acid
Yes → Check the ending of the anion.
-ite → anion or element root + *-ous* = (root)*ous* acid
-ate → anion or element root + *-ic* = (root)*ic* acid

Figure 2.21
A flow chart for naming acids. The acid is considered as one or more H^+ ions attached to an anion.

TABLE 2.8
Names of Some Oxygen-Containing Acids

Acid	Name
HNO_3	nitric acid
HNO_2	nitrous acid
H_2SO_4	sulfuric acid
H_2SO_3	sulfurous acid
H_3PO_4	phosphoric acid
$HC_2H_3O_2$	acetic acid

acetate ion ($C_2H_3O_2^-$) and is called acetic acid. If the anion has an *-ite* ending, the acid name ends with *-ous*. For example, H_2SO_3, which contains sulfite (SO_3^{2-}), is named sulfurous acid; HNO_2, which contains nitrite (NO_2^-), is named nitrous acid. The application of these rules can be seen in the names of the acids of the oxyanions of chlorine:

Acid	*Anion*	*Name*
$HClO_4$	perchlor*ate*	perchlor*ic* acid
$HClO_3$	chlor*ate*	chlor*ic* acid
$HClO_2$	chlor*ite*	chlor*ous* acid
HClO	hypochlor*ite*	hypochlor*ous* acid

The names of the most important acids are given in Tables 2.7 and 2.8. An overall strategy for naming acids is shown in Figure 2.21.

EXERCISES

A blue exercise number indicates that the answer to that exercise appears at the back of this book.

Development of the Atomic Theory

1. Several compounds containing only sulfur (S) and fluorine (F) are known. Three of them have the following compositions:
   i. 1.188 g of F for every 1.000 g of S
   ii. 2.375 g of F for every 1.000 g of S
   iii. 3.563 g of F for every 1.000 g of S

   How do these data illustrate the law of multiple proportions?
2. A reaction of 1 L of chlorine gas (Cl_2) with 3 L of fluorine gas (F_2) yields 2 L of a gaseous product. All gas volumes are at the same temperature and pressure. What is the formula of the gaseous product?
3. When mixtures of gaseous H_2 and gaseous Cl_2 react, a product forms that has the same properties regardless of the relative amounts of H_2 and Cl_2 used.
   a. How is this result interpreted in terms of the law of definite proportion?
   b. When a volume of H_2 reacts with an equal volume of Cl_2 at the same temperature and pressure, what volume of product having the formula HCl is formed?
4. Early tables of atomic weights (masses) were generated by measuring the mass of a substance that reacts with 1 g of oxygen. Given the following data and taking the atomic

mass of hydrogen as 1.00, generate a table of relative atomic masses for oxygen, sodium and magnesium.

Element	Mass That Combines with 1.00 g Oxygen	Assumed Formula
Hydrogen	0.126 g	HO
Sodium	2.875 g	NaO
Magnesium	1.500 g	MgO

How do your values compare with those in the periodic table? How do you account for any differences?

5. How does Dalton's atomic theory account for each of the following?
   a. the law of conservation of mass
   b. the law of definite proportion
   c. the law of multiple proportions
6. What refinements had to be made in Dalton's atomic theory to account for Gay-Lussac's results on the combining volumes of gases?
7. One of the best indications of a "good" theory is that it raises more questions for further experimentation than it originally answered. Does this apply to Dalton's atomic theory? If so, in what ways?
8. Dalton assumed that all atoms of the same element were identical in all of their properties. How have we had to modify this?

The Nature of the Atom

9. What evidence led to the conclusion that cathode rays had a negative charge?
10. Is there a difference between a cathode ray and a β particle?
11. From the information in this chapter on the mass of the proton, the mass of the electron, and the sizes of the nucleus and the atom, calculate the densities of a hydrogen nucleus and a hydrogen atom.
12. A chemist in a galaxy far, far away performed the Millikan oil drop experiment and got the following results for the charge on various drops. What is the charge of the electron in zirkombs?

    2.56×10^{-12} zirkombs 7.68×10^{-12} zirkombs
    3.84×10^{-12} zirkombs 6.40×10^{-13} zirkombs

13. What discoveries were made by J. J. Thomson, Antoine Henri Becquerel, and Lord Rutherford? How did Dalton's model of the atom have to be modified to account for these discoveries?

Elements and the Periodic Table

14. What is the distinction between atomic number and mass number? Between mass number and atomic mass?
15. Which lanthanide element and which transition element have only radioactive isotopes? (Elements in the periodic table with atomic masses in parentheses have only radioactive isotopes.)
16. How many elements are there in
    a. the second period of the periodic table?
    b. the third period?
    c. the fourth period?
    d. the iron group?
    e. the oxygen family?
    f. the nickel group?
17. Identify each of the following elements.
    a. $^{31}_{15}X$ b. $^{127}_{53}X$ c. $^{39}_{19}X$ d. $^{173}_{70}X$
18. How many protons, neutrons, and electrons are in each of the following atoms or ions?
    a. $^{24}_{12}Mg$ d. $^{59}_{27}Co^{3+}$ g. $^{79}_{34}Se^{2-}$
    b. $^{24}_{12}Mg^{2+}$ e. $^{59}_{27}Co$ h. $^{63}_{28}Ni$
    c. $^{59}_{27}Co^{2+}$ f. $^{79}_{34}Se$ i. $^{59}_{28}Ni^{2+}$
19. Complete the following table.

Symbol	Number of Protons	Number of Neutrons	Number of Electrons	Net Charge
	33	42		3+
	16	16	18	
	81	123		1+
$^{197}_{79}Au$				
$^{197}_{79}Au^{3+}$				

20. Would you expect each of the following atoms to gain or lose electrons when forming ions? What is the most likely ion each will form?
    a. Na d. I g. B
    b. Sr e. Al h. Cs
    c. Ba f. S i. Se
21. Consider the elements of the carbon family: C, Si, Ge, Sn, and Pb. What is the trend in metallic character as one goes down a group in the periodic table?
22. What is the trend in metallic character going from left to right across a period in the periodic table?

Nomenclature

23. Name the following compounds.
    a. $NaClO_4$ e. SF_6 i. NaOH
    b. $Mg_3(PO_4)_2$ f. Na_2HPO_4 j. $Mg(OH)_2$
    c. $Al_2(SO_4)_3$ g. NaH_2PO_4 k. $Al(OH)_3$
    d. SF_2 h. Li_3N l. GeO_2
24. Name the following compounds.
    a. NaBr b. $BaBr_2$ c. RbCl

d. $CsCl$ g. NO j. NF_3
e. AlF_3 h. NO_2 k. N_2F_4
f. HBr i. N_2O_4 l. $FeSO_4$

25. Name the following compounds.
 a. HNO_3 e. $NaHSO_4$ i. $Ru(NO_3)_3$
 b. HNO_2 f. $Ca(HSO_3)_2$ j. V_2O_5
 c. H_3PO_4 g. $NaBrO_3$ k. $PtCl_4$
 d. H_3PO_3 h. $Fe(IO_4)_3$ l. $PtCl_2$

26. Write formulas for the following compounds.
 a. sulfur dioxide
 b. sulfur trioxide
 c. sodium sulfite
 d. potassium hydrogen sulfite
 e. lithium nitride
 f. chromium(III) carbonate
 g. chromium(II) acetate
 h. tin(IV) fluoride
 i. ammonium hydrogen sulfate
 j. ammonium hydrogen phosphate
 k. potassium perchlorate
 l. sodium hydride

27. Write formulas for the following compounds.
 a. sodium oxide
 b. sodium peroxide
 c. potassium cyanide
 d. copper(II) nitrate
 e. silicon tetrachloride
 f. lead(II) oxide
 g. lead(IV) oxide (common name lead dioxide)
 h. copper(I) chloride
 i. gallium arsenide
 j. cadmium selenide
 k. zinc sulfide
 l. mercury(I) chloride

28. The common names and formulas for several substances are given below. What are the systematic names for these substances?
 a. sugar of lead $Pb(C_2H_3O_2)_2$
 b. blue vitrol $CuSO_4$
 c. quicklime CaO
 d. Epsom salts $MgSO_4$
 e. milk of magnesia $Mg(OH)_2$
 f. gypsum $CaSO_4$
 g. laughing gas N_2O

Additional Exercises

29. Insulin is a complex protein molecule produced by the pancreas in all vertebrates. It is a hormone that regulates carbohydrate metabolism. Inability to produce insulin results in diabetes mellitus. Diabetes is treated by injections of insulin. Given the law of definite proportion, would you expect any differences in chemical activity between human insulin extracted from pancreatic tissue and human insulin produced by genetically engineered bacteria? Why or why not?

30. Technetium (Tc) was the first synthetically produced element. Technetium comes from the Greek word for artificial. It was first produced by Perries and Segré in 1937 in Berkeley, California, by bombarding a molybdenum plate with ^2H nuclei. Elemental technetium is produced from ammonium pertechnetate. How many protons and neutrons are in the nuclei of ^{98}Tc and ^{99}Tc? What is the formula of ammonium pertechnetate?

31. The early alchemists used to do an experiment in which water was boiled for several days in a sealed glass container. Eventually, some solid residue would begin to appear in the bottom of the flask. This result was interpreted to mean that some of the water in the flask had been converted into earth. When Lavoisier repeated this experiment, he found that the water weighed the same before and after heating, and the weight of the flask plus the solid residue equaled the original weight of the flask. Were the alchemists correct? Explain what really happened. (This experiment is described in the article by A. F. Scott in *Scientific American*, January 1984.)

32. Elements in the same family often form oxyanions of the same general formula. The anions are named in a similar fashion. What are the names of the oxyanions of selenium and tellurium: SeO_4^{2-}, SeO_3^{2-}, TeO_4^{2-}, TeO_3^{2-}?

33. By analogy to phosphorus compounds, name the following: $Mg_3(AsO_4)_2$, H_3AsO_4, Na_3SbO_4, Na_3AsO_3, and Na_2HAsO_4.

34. The mass of beryllium that combines with 1.000 g of oxygen to form beryllium oxide is 0.5633 g. When atomic masses were first being measured, it was thought that the formula of beryllium oxide was Be_2O_3. What would be the atomic mass of beryllium if this were the case?

35. Predict the formula and name of a binary compound formed from the following elements.
 a. Ba and O d. Ca and N g. Al and H
 b. Li and H e. B and O h. In and P
 c. In and F f. O and F i. Mg and F

36. Hydrazine, ammonia, and hydrogen azide all contain only nitrogen and hydrogen. The mass of hydrogen that combines with 1.00 g of nitrogen for each compound is 1.44×10^{-1} g, 2.16×10^{-1} g, and 2.40×10^{-2} g, respectively. Show how these data illustrate the law of multiple proportions.

37. Identify each of the following elements.
 a. A noble gas with 36 protons in the nucleus.
 b. A member of the same family as oxygen. The anion with a 2− charge contains 54 electrons.
 c. A member of the alkaline earth metal family. The 2+ ion contains 18 electrons.
 d. A transition metal with 47 protons in the nucleus.
 e. A radioactive element with 94 protons and 94 electrons in the neutral atom.

38. Chlorine has two natural isotopes: $^{37}_{17}Cl$ and $^{35}_{17}Cl$. Hydrogen reacts with chlorine to form the compound HCl. Would a given amount of hydrogen react with different masses of the two chlorine isotopes? Does this conflict with the law of definite proportion? Why or why not?

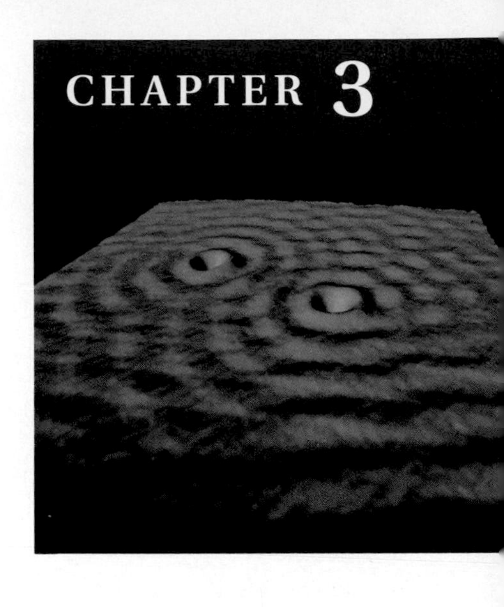

Stoichiometry

Chemical reactions have a profound effect on our lives. There are many examples: food is converted to energy in the human body; nitrogen and hydrogen are combined to form ammonia, which is used as a fertilizer; fuels and plastics are produced from petroleum; the starch in plants is synthesized from carbon dioxide and water using energy from sunlight; human insulin is produced in laboratories by bacteria; cancer is induced in humans by substances from our environment; and so on, in a seemingly endless list. The central activity of chemistry is to understand chemical changes such as these, and the study of reactions occupies a central place in this book. We will examine why reactions occur, how fast they occur, and the specific pathways they follow.

In this chapter we will consider the quantities of materials consumed and produced in chemical reactions. This area of study is called **chemical stoichiometry.** To understand chemical stoichiometry, you must first understand the concept of relative atomic masses.

3.1 Atomic Masses

As we saw in Chapter 2, the first quantitative information about atomic masses came from the work of Dalton, Gay-Lussac, Lavoisier, Avogadro, Cannizzaro, and Berzelius. By observing the proportions in which elements combine to form various compounds, nineteenth-century chemists calculated relative atomic masses. The modern system of atomic masses, instituted in 1961, is based on

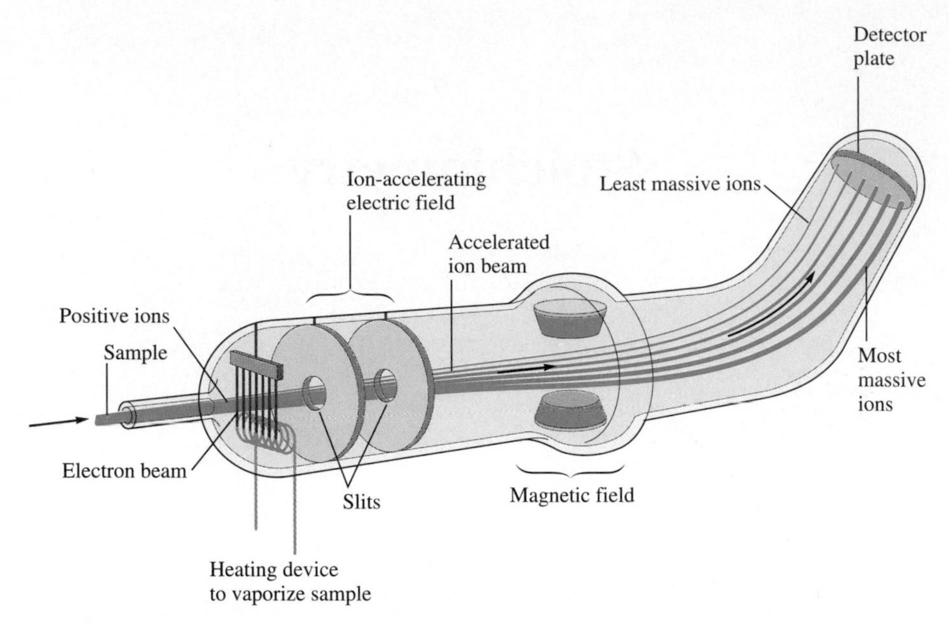

Figure 3.1
Schematic diagram of a mass spectrometer.

^{12}C (carbon twelve) as the standard. In this system *^{12}C is assigned a mass of exactly 12 atomic mass units* (amu), and the masses of all other atoms are given relative to this standard.

The most accurate method currently available for comparing the masses of atoms involves the use of the **mass spectrometer.** In this instrument, diagrammed in Fig. 3.1, atoms or molecules are passed into a beam of high-speed electrons. The high-speed electrons knock electrons off the atoms or molecules being analyzed and change them to positive ions. An applied electric field then accelerates these ions into a magnetic field. Because an accelerating ion produces its own magnetic field, an interaction with the applied magnetic field occurs, which tends to change the path of the ion. The amount of path deflection for each ion depends on its mass—the most massive ions are deflected the smallest amount—which causes the ions to separate, as shown in Fig. 3.1. A comparison of the positions where the ions hit the detector plate gives very accurate values of their relative masses. For example, when ^{12}C and ^{13}C are analyzed in a mass spectrometer, the ratio of their masses is found to be

$$\frac{\text{Mass } ^{13}\text{C}}{\text{Mass } ^{12}\text{C}} = 1.0836129$$

Since the atomic mass unit is defined such that the mass of ^{12}C is *exactly* 12 atomic mass units, then on this same scale,

$$\text{Mass of } ^{13}\text{C} = (1.0836129)(\underset{\substack{\uparrow \\ \text{Exact number,} \\ \text{by definition}}}{12 \text{ amu}}) = 13.003355 \text{ amu}$$

The masses of other atoms can be determined in a similar fashion.

Elemental Analysis Catches Elephant Poachers

In an effort to combat the poaching of elephants by controlling illegal exports of ivory, scientists are now using the isotopic composition of ivory trinkets and elephant tusks to identify the region of Africa where the elephant lived. Using a mass spectrometer, scientists analyze the ivory for the relative amounts of ^{12}C, ^{13}C, ^{14}N, ^{15}N, ^{86}Sr, and ^{87}Sr to determine the diet of the elephant and thus its place of origin. For example, because grasses use a different photosynthetic pathway to produce glucose than do trees, grasses have a slightly different $^{13}C/^{12}C$ ratio from that of trees. They have different ratios because each time a carbon is added in going from simpler to more complex compounds, the more massive ^{13}C is disfavored relative to ^{12}C since it reacts more slowly. Because trees use more steps to build up glucose, they end up with a smaller $^{13}C/^{12}C$ ratio in their leaves relative to grasses, and this difference is then reflected in the tissues of elephants. Thus scientists can tell whether a particular tusk came from a savanna-dwelling elephant (grass-eating) or from a tree-browsing elephant.

Similarly, because the ratios of $^{15}N/^{14}N$ and $^{87}Sr/^{86}Sr$ in elephant tusks also vary depending on the region of Africa the elephant inhabits, they can also be used to trace the elephant's origin. In fact, using these techniques, scientists have reported being able to discriminate between elephants living only about 100 miles apart.

There is now international concern about the dwindling elephant populations in Africa—their numbers have decreased by 40% over the past decade. This concern has led to bans in the export of ivory from many countries in Africa. However, a few nations still allow ivory to be exported. Thus to enforce the trade restrictions, the origin of a given piece of ivory must be established. It is hoped that the "isotope signature" of the ivory can be used for this purpose.

The mass for each element is given in the table inside the front cover of this book. This value, even though it is actually a mass, is often called (for historical reasons) the atomic weight for each element.

Look at the value of the atomic mass of carbon given in this table. You might expect to see 12 since we said the system of atomic masses is based on ^{12}C. However, the number given for carbon is 12.01, because the carbon found on earth (natural carbon) is a mixture of the isotopes ^{12}C, ^{13}C, and ^{14}C. All three isotopes have six protons, but they have six, seven, and eight neutrons, respectively. Because natural carbon is a mixture of isotopes, the atomic mass (weight) we use for carbon is an *average value* based on its isotopic composition.

Most elements occur in nature as mixtures of isotopes; thus atomic masses are usually average values.

The average atomic mass for carbon is computed as follows. Chemists know that natural carbon is composed of 98.89% ^{12}C atoms and 1.11% ^{13}C atoms. The amount of ^{14}C is negligibly small at this level of precision. Using the masses of ^{12}C (exactly 12 amu) and ^{13}C (13.003355 amu), the average atomic mass for natural carbon can be calculated.

See Appendix Section A1.5 for a discussion of significant figures.

$$98.89\% \text{ of } 12 \text{ amu} + 1.11\% \text{ of } 13.0034 \text{ amu}$$
$$= (0.9889)(12 \text{ amu}) + (0.0111)(13.0034 \text{ amu}) = 12.01 \text{ amu}$$

This average mass is often called the atomic weight of carbon.

Even though natural carbon does not contain a single atom with mass 12.01, for stoichiometric purposes we consider carbon to be composed of only

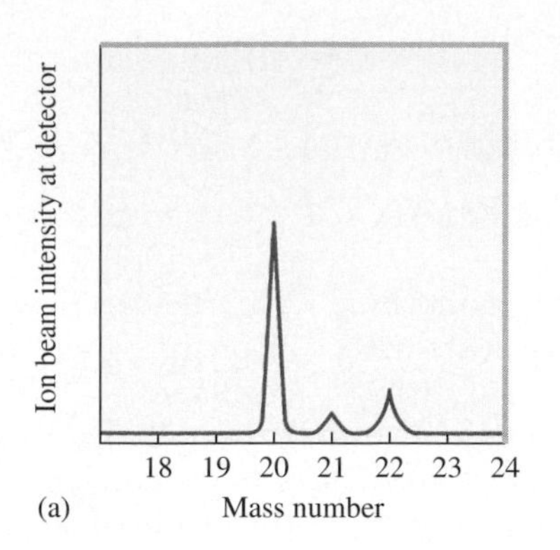

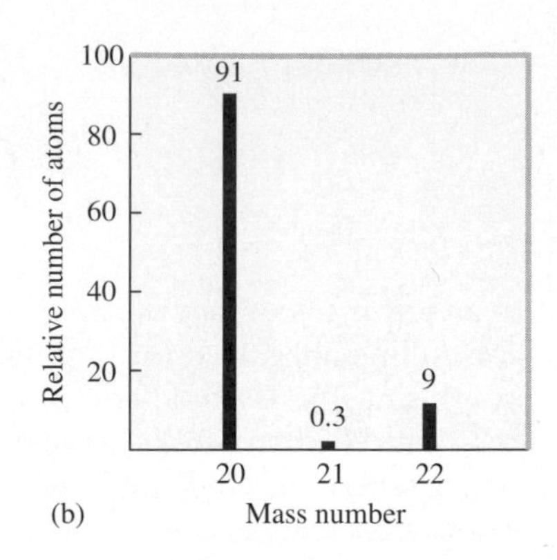

Figure 3.2
The relative intensities of the signals recorded when natural neon is injected into a mass spectrometer, represented in terms of (a) "peaks" and (b) a bar graph. The relative areas of the peaks are 0.9092 (^{20}Ne), 0.00257 (^{21}Ne), and 0.0882 (^{22}Ne); natural neon is therefore 90.92% ^{20}Ne, 0.257% ^{21}Ne, and 8.82% ^{22}Ne.

one type of atom with a mass of 12.01. We do this so that we can count atoms of natural carbon by weighing a sample of carbon. To cite a nonchemical example, it is much easier to weigh out 3000 grams of jelly beans (with an average mass of 3 grams per jelly bean) than to count out 1000 of them. Note that none of the jelly beans has to have a mass of 3 grams for this method to work; only the *average* mass must be 3 grams. We extend this same principle to counting atoms. For natural carbon with an average mass of 12.01 atomic mass units, to obtain 1000 atoms would require weighing out 12,010 atomic mass units of natural carbon (a mixture of ^{12}C and ^{13}C).

As in the case of carbon, the mass for each element given in the table inside the front cover of the book is an average value based on the isotopic composition of the naturally occurring element. For instance, the mass listed for hydrogen (1.008) is the average mass for natural hydrogen, which is a mixture of ^{1}H and ^{2}H (deuterium). *No* atom of hydrogen actually has the mass 1.008.

In addition to being used for determining accurate mass values for individual atoms, the mass spectrometer is also used to determine the isotopic composition of a natural element. For example, when a sample of natural neon is injected into a mass spectrometer, the results shown in Fig. 3.2 are obtained. The areas of the "peaks" or the heights of the bars indicate the relative numbers of $^{20}_{10}$Ne, $^{21}_{10}$Ne, and $^{22}_{10}$Ne atoms.

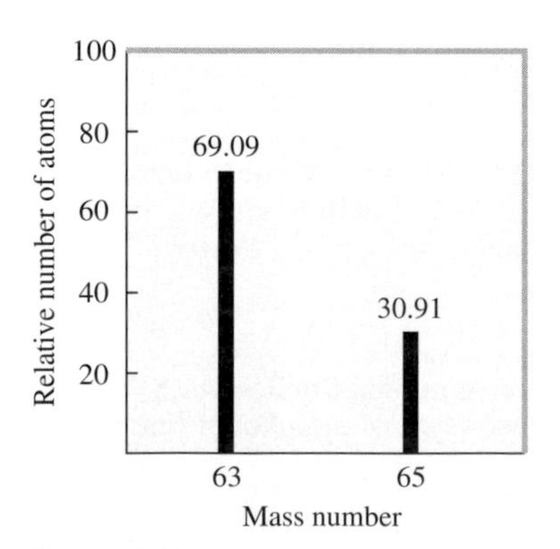

Figure 3.3
Mass spectrum of natural copper.

EXAMPLE 3.1

Copper is a very important metal used for water pipes, electrical wiring, roof coverings, and other materials. When a sample of natural copper is vaporized and injected into a mass spectrometer, the results shown in Fig. 3.3 are obtained. Use these data to compute the average mass of natural copper. (The mass values for ^{63}Cu and ^{65}Cu are 62.93 amu and 64.93 amu, respectively.)

Solution

As shown by the graph, of every 100 atoms of natural copper, on the average 69.09 are ^{63}Cu and 30.91 are ^{65}Cu. Thus the average mass of 100 atoms of natural copper is

$$(69.09 \text{ atoms})\left(62.93 \frac{\text{amu}}{\text{atom}}\right)+(30.91 \text{ atoms})\left(64.93 \frac{\text{amu}}{\text{atom}}\right) = 6355 \text{ amu}$$

The average mass per atom is

$$\frac{6355 \text{ amu}}{100 \text{ atoms}} = 63.55 \text{ amu/atom}$$

This mass value is used in calculations involving the reactions of copper and is the value given in the table inside the front cover of this book.

3.2 The Mole

Because samples of matter typically contain so many atoms, a unit of measure called the mole has been established for use in counting atoms. For our purposes it is most convenient to define the **mole** (abbreviated mol) as *the number equal to the number of carbon atoms in exactly 12 grams of pure* ^{12}C. Modern techniques that allow us to count atoms very precisely have been used to determine this number as 6.022137×10^{23} (6.022×10^{23} will be sufficient for our purposes). This number is called **Avogadro's number** to honor his contributions to chemistry. *One mole of something consists of* 6.022×10^{23} *units of that substance.* Just as a dozen eggs is 12 eggs, a mole of eggs is 6.022×10^{23} eggs.

The SI definition of the mole is the amount of a substance that contains as many entities as there are in exactly 0.012 kg (12 g) of carbon 12.

The magnitude of the number 6.022×10^{23} is very difficult to imagine. To give you some idea, 1 mole of seconds represents a span of time 4 million times as long as the earth has already existed, and 1 mole of marbles is enough to cover the entire earth to a depth of 50 miles! However, since atoms are so tiny, a mole of atoms or molecules is a perfectly manageable quantity to use in a reaction (see Fig. 3.4).

Figure 3.4
One-mole samples of sulfur, copper, mercury, and carbon.

TABLE 3.1 **Comparison of 1-Mole Samples of Various Elements**

Element	Number of Atoms Present	Mass of Sample (g)
Aluminum	6.022×10^{23}	26.98
Gold	6.022×10^{23}	196.97
Iron	6.022×10^{23}	55.85
Sulfur	6.022×10^{23}	32.06
Boron	6.022×10^{23}	10.81
Xenon	6.022×10^{23}	131.30

How do we use the mole in chemical calculations? Recall that Avogadro's number is defined as the number of atoms in exactly 12 grams of ^{12}C. Thus 12 grams of ^{12}C contain 6.022×10^{23} atoms. Also, a 12.01-gram sample of natural carbon contains 6.022×10^{23} atoms (a mixture of ^{12}C, ^{13}C, and ^{14}C atoms, with an average mass of 12.01). Since the ratio of the masses of the samples (12 g/12.01 g) is the same as the ratio of the masses of the individual components (12 amu/12.01 amu), the two samples contain the *same number* of components.

To be sure this point is clear, think of oranges with an average mass of 0.5 pound each and grapefruit with an average mass of 1.0 pound each. Any two sacks for which the sack of grapefruit weighs twice as much as the sack of oranges will contain the same number of pieces of fruit. The same idea extends to atoms. Compare natural carbon (average mass of 12.01) and natural helium (average mass of 4.003). A sample of 12.01 grams of natural carbon contains the same number of atoms as 4.003 grams of natural helium. Both samples contain 1 mole of atoms (6.022×10^{23}). Table 3.1 gives more examples that illustrate this basic idea.

The mass of 1 mole of an element is equal to its atomic mass in grams.

Thus the mole is defined such that a sample of a natural element with a mass equal to the element's atomic mass expressed in grams contains 1 mole of atoms. This definition also fixes the relationship between the atomic mass unit and the gram. Since 6.022×10^{23} atoms of carbon (each with a mass of 12 amu) have a mass of 12 grams, then

$$(6.022 \times 10^{23} \text{ atoms})\left(\frac{12 \text{ amu}}{\text{atom}}\right) = 12 \text{ g}$$

and

$$6.022 \times 10^{23} \text{ amu} = \underset{\substack{\uparrow \\ \text{Exact} \\ \text{number}}}{1 \text{ g}}$$

EXAMPLE 3.2

Americium is an element that does not occur naturally. It can be made in very small amounts in a device called a particle accelerator. Compute the mass in grams of a sample of americium containing six atoms.

Solution

From the table inside the front cover of the book, note that one americium atom has a mass of 243 amu. Thus the mass of six atoms is

$$6 \text{ atoms} \times 243 \frac{\text{amu}}{\text{atom}} = 1.46 \times 10^3 \text{ amu}$$

From the relationship 6.022×10^{23} amu = 1 g, the mass of six americium atoms in grams is

$$1.46 \times 10^3 \text{ amu} \times \frac{1 \text{ g}}{6.022 \times 10^{23} \text{ amu}} = 2.42 \times 10^{-21} \text{ g}$$

This relationship can be used to derive the factor needed to convert between atomic mass units and grams.

To do chemical calculations, you must understand what the mole means and how to determine the number of moles in a given mass of a substance. These procedures are illustrated in Example 3.3.

Refer to Appendix 2 for a discussion of units and the conversion from one unit to another.

EXAMPLE 3.3

A silicon chip used in an integrated circuit of a microcomputer has a mass of 5.68 mg. How many silicon (Si) atoms are present in this chip?

Solution

The strategy for doing this problem is to convert from milligrams of silicon to grams of silicon, then to moles of silicon, and finally to atoms of silicon:

$$5.68 \text{ mg Si} \times \frac{1 \text{g Si}}{1000 \text{ mg Si}} = 5.68 \times 10^{-3} \text{ g Si}$$

$$5.68 \times 10^{-3} \text{ g Si} \times \frac{1 \text{ mol Si}}{28.09 \text{ g Si}} = 2.02 \times 10^{-4} \text{ mol Si}$$

$$2.02 \times 10^{-4} \text{ mol Si} \times \frac{6.022 \times 10^{23} \text{ atoms}}{1 \text{ mol Si}} = 1.22 \times 10^{20} \text{ atoms}$$

It always makes sense to think about orders of magnitude as you do a calculation. In Example 3.3, the 5.68-milligram sample of silicon is clearly much less than 1 mole of silicon (which has a mass of 28.09 grams), so the final answer of 1.22×10^{20} atoms (compared with 6.022×10^{23} atoms) is at least in the right direction. Paying careful attention to units and making sure the answer is sensible can help you detect an inverted conversion factor or a number that was incorrectly entered in your calculator.

Always check to see if your answer is sensible.

Measuring the Masses of Large Molecules or Making Elephants Fly

When a chemist produces a new molecule, one crucial property for making a positive identification is the molecule's mass. There are many ways to determine the molar mass of a compound, but one of the fastest and most accurate methods involves mass spectrometry. This method requires that the substance be put into the gas phase and ionized. The deflection that the resulting ion exhibits as it is accelerated through a magnetic field can be used to obtain a very precise value of its mass. One drawback of this method is that it is difficult to use with large molecules because they are difficult to vaporize. That is, substances that contain large molecules typically have very high boiling points, and these molecules are often damaged when they are vaporized at such high temperatures. A case in point involves proteins, an extremely important class of large biological molecules that are quite fragile at high temperatures. Typical methods used to obtain the masses of protein molecules are slow and tedious.

Mass spectrometry has not previously been used to obtain protein masses, because proteins decompose at the temperatures necessary to vaporize them. However, a relatively new technique called "matrix-assisted laser desorption" has been developed that allows mass spectrometric determination of protein molar masses. In this technique the large "target" molecule is embedded in a matrix of smaller molecules. The matrix is then placed in a mass spectrometer and blasted with a laser beam, which causes its disintegration. Disintegration of the matrix frees the large target molecule, which is then swept into the mass spectrometer. One researcher involved in this project likened this method to an elephant on top of a tall building: "The elephant must fly if the building suddenly turns into fine grains of sand."

This technique allows scientists to determine the masses of huge molecules. So far researchers have measured proteins with masses up to 350,000 daltons (1 dalton is equal to one atomic mass unit). This method, which makes mass spectrometry a routine tool for the determination of protein masses, probably will be extended to even larger molecules such as DNA and could be a revolutionary development in the characterization of biomolecules.

3.3 Molar Mass

A chemical compound is, ultimately, a collection of atoms. For example, methane (the major component of natural gas) consists of molecules that each contain one carbon atom and four hydrogen atoms (CH_4). How can we calculate the mass of 1 mole of methane; that is, what is the mass of 6.022×10^{23} CH_4 molecules? Since each CH_4 molecule contains one carbon atom and four hydrogen atoms, 1 mole of CH_4 molecules consists of 1 mole of carbon atoms and 4 moles of hydrogen atoms. The mass of 1 mole of methane can be found by summing the masses of carbon and hydrogen present:

The atomic weight (mass) for carbon to five significant digits is 12.011.

$$\begin{aligned} \text{Mass of 1 mol of C} &= 12.011 \text{ g} \\ \text{Mass of 4 mol of H} &= \underline{4 \times 1.008 \text{ g}} \\ \text{Mass of 1 mol of } CH_4 &= 16.043 \text{ g} \end{aligned}$$

EXAMPLE 3.4

Isopentyl acetate ($C_7H_{14}O_2$), the compound responsible for the scent of bananas, can be produced commercially. Interestingly, bees release about 1 μg (1×10^{-6} g) of this compound when they sting, in order to attract other bees to join the attack. How many molecules of isopentyl acetate are released in a typical bee sting? How many atoms of carbon are present?

Solution

Since we are given a mass of isopentyl acetate and want the number of molecules, we must first compute the molar mass.

$$7 \text{ mol C} \times 12.011 \frac{\text{g}}{\text{mol}} = 84.077 \text{ g C}$$

$$14 \text{ mol H} \times 1.0079 \frac{\text{g}}{\text{mol}} = 14.111 \text{ g H}$$

$$2 \text{ mol O} \times 15.999 \frac{\text{g}}{\text{mol}} = 31.998 \text{ g O}$$

$$\text{Mass of 1 mol of } C_7H_{14}O_2 = 130.186 \text{ g}$$

The atomic weight (mass) for hydrogen to five significant digits is 1.0079.

Thus 1 mol of isopentyl acetate (6.022×10^{23} molecules) has a mass of 130.186 g.

To find the number of molecules released in a sting, we must first determine the number of moles of isopentyl acetate in 1×10^{-6} g:

$$1 \times 10^{-6} \text{ g } C_7H_{14}O_2 \times \frac{1 \text{ mol } C_7H_{14}O_2}{130.186 \text{ g } C_7H_{14}O_2} = 8 \times 10^{-9} \text{ mol } C_7H_{14}O_2$$

Since 1 mol is 6.022×10^{23} units, we can determine the number of molecules:

$$8 \times 10^{-9} \text{ mol } C_7H_{14}O_2 \times \frac{6.022 \times 10^{23} \text{ molecules}}{1 \text{ mol } C_7H_{14}O_2} = 5 \times 10^{15} \text{ molecules}$$

To determine the number of carbon atoms present, we must multiply the number of molecules by 7, since each molecule of isopentyl acetate contains seven carbon atoms:

$$5 \times 10^{15} \text{ molecules} \times \frac{7 \text{ carbon atoms}}{\text{molecule}} = 4 \times 10^{16} \text{ carbon atoms}$$

To show the correct number of significant figures in each calculation, we round off after each step. In your calculations always carry extra significant figures through to the end; then round off.

NOTE: In keeping with our practice of always showing the correct number of significant figures, we have rounded off after each step. However, if extra digits are carried throughout this problem, the final answer rounds to 3×10^{16}.

A substance's molecular weight or molar mass is the mass in grams of 1 mole of the substance.

Since the number 16.043 represents the mass of 1 mole of methane molecules, it makes sense to call it the *molar mass* for methane. However, traditionally, the term *molecular weight* has been used to describe the mass of 1 mole of a substance. Thus the terms **molar mass** and **molecular weight** mean exactly the same thing: *the mass in grams of 1 mole of a compound.* The molar mass of a known substance is obtained by summing the masses of the component atoms, as we did for methane.

Some substances exist as a collection of ions rather than as separate molecules. An example is ordinary table salt, sodium chloride (NaCl), which is composed of an array of Na^+ and Cl^- ions. There are no NaCl molecules present. However, in this book, for convenience, we will apply the term *molar mass* to both ionic and molecular substances. Thus we will refer to 58.44 (22.99 + 35.45) as the molar mass for NaCl. In some books the term *formula weight* is used for ionic compounds instead of molar mass or molecular weight.

3.4 Percent Composition of Compounds

So far we have discussed the composition of a compound in terms of the numbers of its constituent atoms. It is often useful to know a compound's composition in terms of the masses of its elements. We can obtain this information from the formula of the compound by comparing the mass of each element present in 1 mole of the compound with the total mass of 1 mole of the compound.

For example, consider ethanol, which has the formula C_2H_5OH. The mass of each element present and the molar mass are obtained through the following procedure.

$$\text{Mass of C} = 2\text{ mol} \times 12.011\,\frac{\text{g}}{\text{mol}} = 24.022\text{ g}$$

$$\text{Mass of H} = 6\text{ mol} \times 1.008\,\frac{\text{g}}{\text{mol}} = 6.048\text{ g}$$

$$\text{Mass of O} = 1\text{ mol} \times 15.999\,\frac{\text{g}}{\text{mol}} = \underline{15.999\text{ g}}$$

$$\text{Mass of 1 mol of } C_2H_5OH = 46.069\text{ g}$$

The *mass percent* (often called the weight percent) of carbon in ethanol can be computed by comparing the mass of carbon in 1 mole of ethanol with the total mass of 1 mole of ethanol and multiplying the result by 100:

$$\begin{aligned}\text{Mass percent of C} &= \frac{\text{mass of C in 1 mol } C_2H_5OH}{\text{mass of 1 mol } C_2H_5OH} \times 100 \\ &= \frac{24.022\text{ g}}{46.069\text{ g}} \times 100 \\ &= 52.144\%\end{aligned}$$

The mass percents of hydrogen and oxygen in ethanol are obtained in a similar manner:

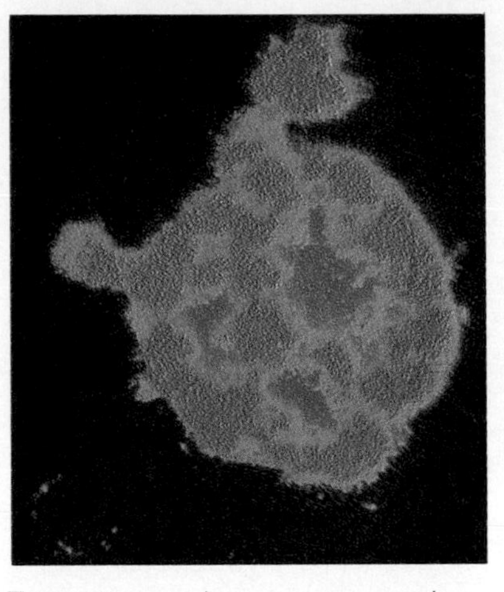
Transmission electron micrograph showing penicillin acting on a single bacterium. Magnification: ×40,000

EXAMPLE 3.5

Penicillin, the first of a now large number of antibiotics (antibacterial agents), was discovered accidentally by the Scottish bacteriologist Alexander Fleming in 1928, but he was never able to isolate it as a pure compound. This and similar antibiotics have saved millions of lives that might have been lost to infections. Penicillin F has the formula $C_{14}H_{20}N_2SO_4$. Compute the mass percent of each element.

Solution

The molar mass of penicillin F is computed as follows:

$$\text{C:}\quad 14\text{ mol} \times 12.011\ \frac{\text{g}}{\text{mol}} = 168.15\ \text{g}$$

$$\text{H:}\quad 20\text{ mol} \times 1.008\ \frac{\text{g}}{\text{mol}} = 20.16\ \text{g}$$

$$\text{N:}\quad 2\text{ mol} \times 14.007\ \frac{\text{g}}{\text{mol}} = 28.014\ \text{g}$$

$$\text{S:}\quad 1\text{ mol} \times 32.07\ \frac{\text{g}}{\text{mol}} = 32.07\ \text{g}$$

$$\text{O:}\quad 4\text{ mol} \times 15.999\ \frac{\text{g}}{\text{mol}} = \underline{63.996\ \text{g}}$$

$$\text{Mass of 1 mol of } C_{14}H_{20}N_2SO_4 = 312.39\ \text{g}$$

$$\text{Mass percent of C} = \frac{168.15\ \text{g C}}{312.39\ \text{g } C_{14}H_{20}N_2SO_4} \times 100 = 53.827\%$$

$$\text{Mass percent of H} = \frac{20.16\ \text{g H}}{312.39\ \text{g } C_{14}H_{20}N_2SO_4} \times 100 = 6.453\%$$

$$\text{Mass percent of N} = \frac{28.014\ \text{g N}}{312.39\ \text{g } C_{14}H_{20}N_2SO_4} \times 100 = 8.968\%$$

$$\text{Mass percent of S} = \frac{32.07\ \text{g S}}{312.39\ \text{g } C_{14}H_{20}N_2SO_4} \times 100 = 10.27\%$$

$$\text{Mass percent of O} = \frac{63.996\ \text{g O}}{312.39\ \text{g } C_{14}H_{20}N_2SO_4} \times 100 = 20.486\%$$

CHECK: The percentages add up to 100.00%.

$$\text{Mass percent of H} = \frac{\text{mass of H in 1 mol } C_2H_5OH}{\text{mass of 1 mol } C_2H_5OH} \times 100$$

$$= \frac{6.048\ \text{g}}{46.069\ \text{g}} \times 100$$

$$= 13.13\%$$

$$\text{Mass percent of O} = \frac{\text{mass of O in 1 mol C}_2\text{H}_5\text{OH}}{\text{mass of 1 mol C}_2\text{H}_5\text{OH}} \times 100$$
$$= \frac{15.999\ \text{g}}{46.069\ \text{g}} \times 100$$
$$= 34.728\%$$

Notice that the percentages add to 100% if rounded to two decimal places; this is the check of the calculations.

3.5 Determining the Formula of a Compound

When a new compound is prepared, one of the first items of interest is the formula of the compound. The formula is often determined by taking a weighed sample of the compound and either decomposing it into its component elements or reacting it with oxygen to produce substances such as CO_2, H_2O and N_2, which are then collected and weighed. A device for doing this type of analysis is shown in Fig. 3.5. The results of such analyses provide the mass of each type of element in the compound, which can be used to determine the mass percent of each element present.

We will see how information of this type can be used to compute the formula of a compound. Suppose a substance has been prepared that is composed of carbon, hydrogen, and nitrogen. When 0.1156 gram of this compound is reacted with oxygen, 0.1638 gram of carbon dioxide (CO_2) and 0.1676 gram of water (H_2O) are collected. Assuming that all of the carbon in the compound is converted to CO_2, we can determine the mass of carbon originally present in the 0.1156-gram sample. To do so, we must use the fraction (by mass) of carbon in CO_2. The molar mass of CO_2 is 12.011 g/mol plus 2(15.999) g/mol, or 44.009 g/mol. The fraction of carbon present by mass (12.011 grams C/44.009 grams CO_2) can now be used to determine the mass of carbon in 0.1638 gram of CO_2:

$$0.1638\ \text{g CO}_2 \times \frac{12.011\ \text{g C}}{44.009\ \text{g CO}_2} = 0.04470\ \text{g C}$$

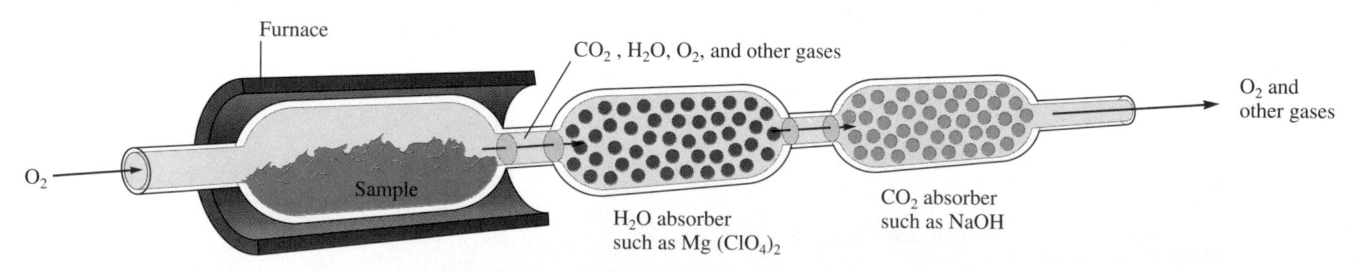

Figure 3.5
A schematic diagram of the combustion device used to analyze substances for carbon and hydrogen. The sample is burned in the presence of excess oxygen, which converts all of its carbon to carbon dioxide and all of its hydrogen to water. These compounds are collected by absorption using appropriate materials, and their amounts are determined by measuring the increase in weights of the absorbents.

Remember that this carbon originally came from the 0.1156-gram sample of the unknown compound. Thus the mass percent of carbon in this compound is

$$\frac{0.04470 \text{ g C}}{0.1156 \text{ g compound}} \times 100 = 38.67\% \text{ C}$$

The same procedure can be used to find the mass percent of hydrogen in the unknown compound. We assume that all of the hydrogen present in the original 0.1156 gram of compound was converted to H_2O. The molar mass of H_2O is 18.015 grams, and the fraction of hydrogen by mass in H_2O is 2.016 grams H/ 18.015 grams H_2O. Therefore, the mass of hydrogen in 0.1676 gram of H_2O is

$$0.1676 \text{ g } H_2O \times \frac{2.016 \text{ g H}}{18.015 \text{ g } H_2O} = 0.01876 \text{ g H}$$

And the mass percent of hydrogen in the compound is

$$\frac{0.01876 \text{ g H}}{0.1156 \text{ g compound}} \times 100 = 16.23\% \text{ H}$$

The unknown compound contains only carbon, hydrogen, and nitrogen. So far we have determined that it is 38.67% carbon and 16.23% hydrogen. The remainder must be nitrogen:

$$100.00\% - (\underset{\%\text{ C}}{\underset{\uparrow}{38.67\%}} + \underset{\%\text{ H}}{\underset{\uparrow}{16.23\%}}) = 45.10\% \text{ N}$$

We have determined that the compound contains 38.67% carbon, 16.23% hydrogen, and 45.10% nitrogen. Next, we use these data to obtain the formula.

Since the formula of a compound indicates the *numbers* of atoms in the compound, we must convert the masses of the elements to numbers of atoms. The easiest way to do this is to work with 100.00 grams of the compound. In the present case 38.67% carbon by mass means 38.67 grams of carbon per 100.00 grams of compound; 16.23% hydrogen means 16.23 grams of hydrogen per 100.00 grams of compound; and so on. To determine the formula, we must calculate the number of carbon atoms in 38.67 grams of carbon, the number of hydrogen atoms in 16.23 grams of hydrogen, and the number of nitrogen atoms in 45.10 grams of nitrogen. We can do this as follows:

$$38.67 \text{ g C} \times \frac{1 \text{ mol C}}{12.011 \text{ g C}} = 3.220 \text{ mol C}$$

$$16.23 \text{ g H} \times \frac{1 \text{ mol H}}{1.008 \text{ g H}} = 16.10 \text{ mol H}$$

$$45.10 \text{ g N} \times \frac{1 \text{ mol N}}{14.007 \text{ g N}} = 3.220 \text{ mol N}$$

Thus 100.00 grams of this compound contains 3.220 moles of carbon atoms, 16.10 moles of hydrogen atoms, and 3.220 moles of nitrogen atoms.

We can find the smallest *whole-number ratio* of atoms in this compound by dividing each of the mole values above by the smallest of the three:

$$\text{C:} \quad \frac{3.220}{3.220} = 1$$

$$\text{H:} \quad \frac{16.10}{3.220} = 5$$

$$\text{N:} \quad \frac{3.220}{3.220} = 1$$

Thus the formula of this compound can be written CH_5N. This formula is called the **empirical formula.** It represents the *simplest whole-number ratio of the various types of atoms in a compound.*

If this compound is molecular, then the formula might well be CH_5N. It might also be $C_2H_{10}N_2$, or $C_3H_{15}N_3$, and so on—that is, some multiple of the simplest whole-number ratio. Each of these alternatives also has the correct relative numbers of atoms. Any molecule that can be represented as $(CH_5N)_x$, where x is an integer, has the empirical formula CH_5N. To be able to specify the exact formula of the molecule involved, the **molecular formula,** we must know the molar mass.

Molecular formula = (empirical formula)$_x$, where x is an integer.

Suppose we know that this compound with empirical formula CH_5N has a molar mass of 31.06. How do we determine which of the possible choices represents the molecular formula? Since the molecular formula is always a whole-number multiple of the empirical formula, we must first find the empirical formula mass for CH_5N:

$$\begin{aligned}
\text{1 C:} &\quad 1 \times 12.011\text{ g} = 12.011\text{ g} \\
\text{5 H:} &\quad 5 \times 1.008\text{ g} = 5.040\text{ g} \\
\text{1 N:} &\quad 1 \times 14.007\text{ g} = \underline{14.007\text{ g}} \\
\text{Formula mass of } CH_5N &= 31.058\text{ g}
\end{aligned}$$

This value is the same as the known molar mass of the compound. Thus in this case the empirical formula and the molecular formula are the same; this substance consists of molecules with the formula CH_5N. It is quite common for the empirical and molecular formulas to be different; some examples where this is the case are shown in Fig. 3.6.

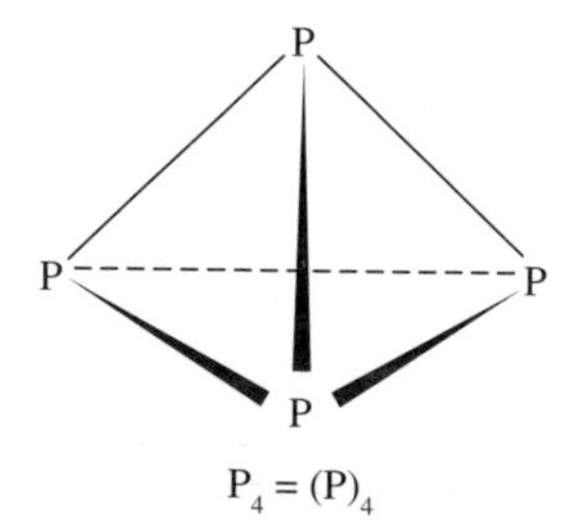

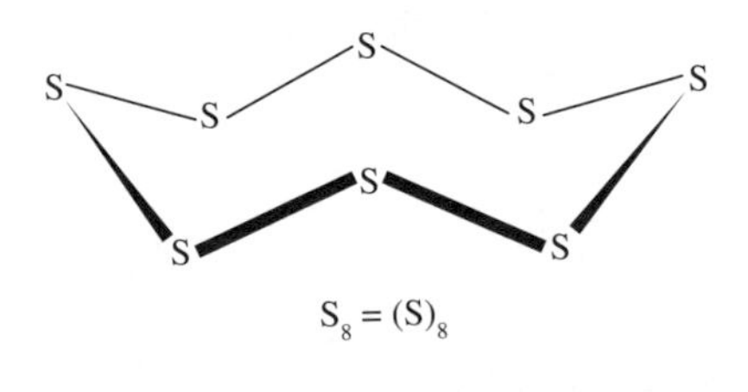

$C_6H_{12}O_6 = (CH_2O)_6$

Figure 3.6
Examples of substances whose empirical and molecular formulas differ. Notice that molecular formula = (empirical formula)$_x$, where x is an integer.

EXAMPLE 3.6

A white powder is analyzed and found to contain 43.64% phosphorus and 56.36% oxygen by mass. The compound has a molar mass of 283.88 g. What are the compound's empirical and molecular formulas?

Solution

In 100.00 g of this compound, there are 43.64 g of phosphorus and 56.36 g of oxygen. In terms of moles, in 100.00 g of compound we have

$$43.64 \text{ g P} \times \frac{1 \text{ mol P}}{30.97 \text{ g P}} = 1.409 \text{ mol P}$$

$$56.36 \text{ g O} \times \frac{1 \text{ mol O}}{15.999 \text{ g O}} = 3.523 \text{ mol O}$$

Dividing both mole values by the smaller one gives

$$\frac{1.409}{1.409} = 1 \text{ P} \quad \text{and} \quad \frac{3.523}{1.409} = 2.5 \text{ O}$$

This yields the formula $PO_{2.5}$. Since compounds must contain whole numbers of atoms, the empirical formula should contain only whole numbers. To obtain the simplest set of whole numbers, we multiply both numbers by 2 to give the empirical formula P_2O_5.

To obtain the molecular formula, we must compare the empirical formula mass with the molar mass. The empirical formula mass for P_2O_5 is 141.94.

$$\frac{\text{Molar mass}}{\text{Empirical formula mass}} = \frac{283.88}{141.94} = 2$$

The molecular formula is $(P_2O_5)_2$, or P_4O_{10}.

The structural formula of this interesting compound is given in Fig. 3.7.

Figure 3.7
The structural formula of P_4O_{10}. Note that the oxygen atoms act as "bridges" between the phosphorus atoms. This compound has a great affinity for water and is often used as a desiccant, or drying agent.

In Example 3.6 we found the molecular formula by comparing the empirical formula mass with the molar mass. There is an alternative way to obtain the molecular formula. The molar mass and the percentages (by mass) of each element present can be used to compute the moles of each element present in one mole of compound. These numbers of moles then represent directly the subscripts in the molecular formula. This procedure is illustrated in Example 3.7.

EXAMPLE 3.7

Caffeine, a stimulant found in coffee, tea, chocolate, and some medications, contains 49.48% carbon, 5.15% hydrogen, 28.87% nitrogen, and 16.49% oxygen by mass and has a molar mass of 194.2. Determine the molecular formula of caffeine.

Solution

We will first determine the mass of each element in 1 mol (194.2 g) of caffeine:

$$\frac{49.48 \text{ g C}}{100.0 \text{ g caffeine}} \times \frac{194.2 \text{ g}}{\text{mol}} = \frac{96.09 \text{ g C}}{\text{mol caffeine}}$$

$$\frac{5.15 \text{ g H}}{100.0 \text{ g caffeine}} \times \frac{194.2 \text{ g}}{\text{mol}} = \frac{10.0 \text{ g H}}{\text{mol caffeine}}$$

$$\frac{28.87 \text{ g N}}{100.0 \text{ g caffeine}} \times \frac{194.2 \text{ g}}{\text{mol}} = \frac{56.07 \text{ g N}}{\text{mol caffeine}}$$

$$\frac{16.49 \text{ g O}}{100.0 \text{ g caffeine}} \times \frac{194.2 \text{ g}}{\text{mol}} = \frac{32.02 \text{ g O}}{\text{mol caffeine}}$$

Now we will convert to moles:

$$\text{C:} \quad \frac{96.09 \text{ g C}}{\text{mol caffeine}} \times \frac{1 \text{ mol C}}{12.011 \text{ g C}} = \frac{8.000 \text{ mol C}}{\text{mol caffeine}}$$

$$\text{H:} \quad \frac{10.0 \text{ g H}}{\text{mol caffeine}} \times \frac{1 \text{ mol H}}{1.008 \text{ g H}} = \frac{9.92 \text{ mol H}}{\text{mol caffeine}}$$

$$\text{N:} \quad \frac{56.07 \text{ g N}}{\text{mol caffeine}} \times \frac{1 \text{ mol N}}{14.01 \text{ g N}} = \frac{4.002 \text{ mol N}}{\text{mol caffeine}}$$

$$\text{O:} \quad \frac{32.02 \text{ g O}}{\text{mol caffeine}} \times \frac{1 \text{ mol O}}{16.00 \text{ g O}} = \frac{2.001 \text{ mol O}}{\text{mol caffeine}}$$

Rounding the numbers to integers gives the molecular formula for caffeine: $C_8H_{10}N_4O_2$.

The methods for obtaining empirical and molecular formulas are summarized on the following page.

Summary: **Determination of the Empirical Formula**

- **Since mass percentage gives the number of grams of a particular element per 100 grams of compound, base the calculation on 100 grams of compound. Each percent will then represent the mass in grams of that element present in the compound.**
- **Determine the number of moles of each element present in 100 grams of compound, using the atomic weights (masses) of the elements present.**
- **Divide each value of the number of moles by the smallest of the values. If each resulting number is a whole number (after appropriate rounding), these numbers represent the subscripts of the elements in the empirical formula.**
- **If the numbers obtained in the previous step are not whole numbers, multiply each number by an integer so that the results are all whole numbers.**

Numbers very close to whole numbers, such as 9.92 and 1.08, should be rounded to whole numbers. Numbers such as 2.25, 4.33, and 2.72 should not be rounded to whole numbers.

Summary: **Determination of the Molecular Formula**

METHOD ONE

- **Obtain the empirical formula.**
- **Compute the empirical formula mass.**
- **Calculate the ratio**

$$\frac{\text{Molar mass}}{\text{Empirical formula mass}}$$

- **The integer from the previous step represents the number of empirical formula units in one molecule. When the empirical formula subscripts are multiplied by this integer, the molecular formula results.**

METHOD TWO

- **Using the mass percentages and the molar mass, determine the mass of each element present in 1 mole of compound.**
- **Determine the number of moles of each element present in 1 mole of compound.**
- **The integers from the previous step represent the subscripts in the molecular formula.**

3.6 Chemical Equations

Chemical Reactions

A chemical change involves reorganization of the atoms in one or more substances. For example, when the methane (CH_4) in natural gas combines with oxygen (O_2) in the air and burns, carbon dioxide (CO_2) and water (H_2O) are formed. This process is represented by a **chemical equation** with the

Carbon burning in oxygen.

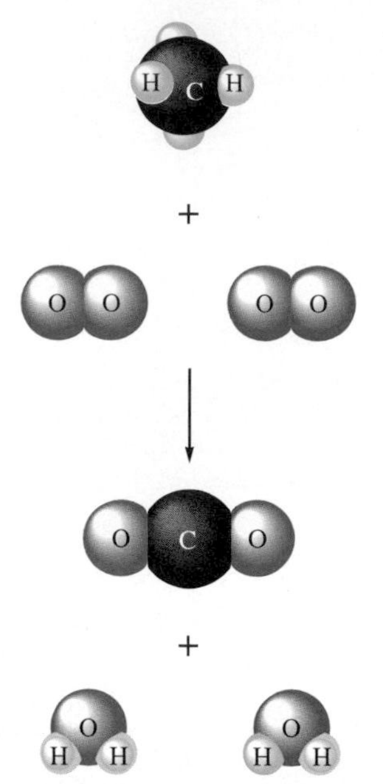

Figure 3.8
The reaction between methane and oxygen to give water and carbon dioxide. Note that no atoms have been gained or lost in the reaction. The reaction simply reorganizes the atoms.

reactants (here methane and oxygen) on the left side of an arrow and the **products** (carbon dioxide and water) on the right side:

$$\underbrace{CH_4 + O_2}_{\text{Reactants}} \longrightarrow \underbrace{CO_2 + H_2O}_{\text{Products}}$$

Notice that the atoms have been reorganized. *Bonds have been broken and new ones formed.* Remember that *in a chemical reaction atoms are neither created nor destroyed. All atoms present in the reactants must be accounted for among the products.* In other words, there must be the same number of each type of atom on the product side and on the reactant side of the arrow. Making sure that this rule is obeyed is called **balancing a chemical equation** for a reaction.

The equation (shown above) for the reaction between CH_4 and O_2 is not balanced. As we will see in the next section, the equation can be balanced to produce

$$CH_4 + 2O_2 \longrightarrow CO_2 + 2H_2O$$

This reaction is shown graphically in Fig. 3.8. We can check that the equation is balanced by comparing the number of each type of atom on both sides:

$$CH_4 + 2O_2 \longrightarrow CO_2 + 2H_2O$$

1 C, 4 H (from CH_4); 4 O (from $2O_2$); 1 C, 2 O (from CO_2); 4 H, 2 O (from $2H_2O$)

The Meaning of a Chemical Equation

The chemical equation for a reaction provides two important types of information: the nature of the reactants and products, and the relative numbers of each.

The reactants and products in a specific reaction must be identified by experiment. Besides specifying the compounds involved in the reaction, the equation often includes the *physical states* of the reactants and products:

State	Symbol
Solid	(*s*)
Liquid	(*l*)
Gas	(*g*)
Dissolved in water (in aqueous solution)	(*aq*)

For example, when hydrochloric acid in aqueous solution is added to solid sodium hydrogen carbonate, the products carbon dioxide gas, liquid water, and sodium chloride (which dissolves in the water) are formed:

$$HCl(aq) + NaHCO_3(s) \longrightarrow CO_2(g) + H_2O(l) + NaCl(aq)$$

Hydrochloric acid reacts with solid sodium hydrogen carbonate to produce gaseous carbon dioxide.

The relative numbers of reactants and products in a reaction are indicated by the *coefficients* in the balanced equation. (The coefficients can be determined since we know that the same number of each type of atom must occur on both sides of the equation.) For example, the balanced equation

$$CH_4(g) + 2O_2(g) \longrightarrow CO_2(g) + 2H_2O(g)$$

TABLE 3.2 Information Conveyed by the Balanced Equation for the Combustion of Methane

Reactants	$\longrightarrow$	Products
$CH_4(g) + 2O_2(g)$	$\longrightarrow$	$CO_2(g) + 2H_2O(g)$
1 molecule CH_4 + 2 molecules O_2	$\longrightarrow$	1 molecule CO_2 + 2 molecules H_2O
1 mol CH_4 molecules + 2 mol O_2 molecules	$\longrightarrow$	1 mol CO_2 molecules + 2 mol H_2O molecules
6.022×10^{23} CH_4 molecules + $2(6.022 \times 10^{23})$ O_2 molecules	$\longrightarrow$	6.022×10^{23} CO_2 molecules + $2(6.022 \times 10^{23})$ H_2O molecules
16 g CH_4 + 2(32 g) O_2	$\longrightarrow$	44 g CO_2 + 2(18 g) H_2O
80 g reactants	$\longrightarrow$	80 g products

can be interpreted in several equivalent ways, as shown in Table 3.2. Note that the total mass is 80 grams for both reactants and products. We should expect this result, since chemical reactions involve only a rearrangement of atoms. Atoms, and therefore mass, are conserved in a chemical reaction.

From this discussion you can see that a balanced chemical equation gives you a great deal of information.

3.7 Balancing Chemical Equations

An unbalanced chemical equation is of limited use. Whenever you see an equation, you should ask yourself whether or not it is balanced. The principle that lies at the heart of the balancing process is that atoms are conserved in a chemical reaction. The same number of each type of atom must be found among the reactants and products. Also, remember that the identities of the reactants and products of a reaction are determined by experimental observation. For example, when liquid ethanol is burned in the presence of sufficient oxygen gas, the products will always be carbon dioxide and water. When the equation for this reaction is balanced, the *identities* of the reactants and products must not be changed. *The formulas of the compounds must never be changed when balancing a chemical equation.* That is, the subscripts in a formula cannot be changed, nor can atoms be added or subtracted from a formula.

In balancing equations, start with the most complicated molecule.

Most chemical equations can be balanced by inspection, that is, by trial and error. It is always best to start with the most complicated molecules (those containing the greatest number of atoms). For example, consider the reaction of ethanol with oxygen, given by the unbalanced equation

$$C_2H_5OH(l) + O_2(g) \longrightarrow CO_2(g) + H_2O(g)$$

The most complicated molecule here is C_2H_5OH. We will begin by balancing the products that contain the atoms in C_2H_5OH. Since C_2H_5OH contains two carbon atoms, we place a 2 before the CO_2 to balance the carbon atoms:

$$\underset{\text{2 C atoms}}{C_2H_5OH(l)} + O_2(g) \longrightarrow \underset{\text{2 C atoms}}{2CO_2(g)} + H_2O(g)$$

Since C_2H_5OH contains six hydrogen atoms, the hydrogen atoms can be balanced by placing a 3 before the H_2O:

$$C_2H_5OH(l) + O_2(g) \longrightarrow 2CO_2(g) + 3H_2O(g)$$
$$(5 + 1)\ H \qquad\qquad\qquad\qquad (3 \times 2)\ H$$

Last, we balance the oxygen atoms. Note that the right side of the above equation contains seven oxygen atoms, while the left side has only three. We can correct this by putting a 3 before the O_2 to produce the balanced equation:

$$C_2H_5OH(l) + 3O_2(g) \longrightarrow 2CO_2(g) + 3H_2O(g)$$
$$1\ O \qquad 6\ O \qquad (2 \times 2)\ O \qquad 3\ O$$

Now we check:

$$C_2H_5OH(l) + 3O_2(g) \longrightarrow 2CO_2(g) + 3H_2O(g)$$

Reactants	Products
2 C atoms	2 C atoms
6 H atoms	6 H atoms
7 O atoms	7 O atoms

The equation is balanced.

Summary: Writing and Balancing the Equation for a Chemical Reaction

- **Determine what reaction is occurring. What are the reactants, the products, and the states involved?**
- **Write the *unbalanced* equation that summarizes the above information.**
- **Balance the equation by inspection, starting with the most complicated molecule(s). Determine what coefficients are necessary to ensure that the same number of each type of atom appears on both reactant and product sides. Do not change the identities (formulas) of any of the reactants or products.**

EXAMPLE 3.8

Chromate and dichromate compounds are suspected carcinogens (cancer-inducing agents) and should be handled carefully.

Chromium compounds exhibit a variety of bright colors. When solid ammonium dichromate, $(NH_4)_2Cr_2O_7$, a vivid orange compound, is ignited, a spectacular reaction occurs, as shown in the two photographs on this page. Although the reaction is somewhat more complex, let's assume here that the products are solid chromium(III) oxide, nitrogen gas (consisting of N_2 molecules), and water vapor. Balance the equation for this reaction.

Solution

From the description given, the reactant is solid ammonium dichromate, $(NH_4)_2Cr_2O_7(s)$, and the products are nitrogen gas, $N_2(g)$, water vapor, $H_2O(g)$, and solid chromium(III) oxide, $Cr_2O_3(s)$. The formula for chromium(III) oxide can be determined by recognizing that the Roman numeral III means that Cr^{3+} ions are present. For a neutral compound the formula must then be Cr_2O_3, since each oxide ion is O^{2-}.

The unbalanced equation is

$$(NH_4)_2Cr_2O_7(s) \longrightarrow Cr_2O_3(s) + N_2(g) + H_2O(g)$$

Note that nitrogen and chromium are balanced (two nitrogen atoms and two chromium atoms on each side), but hydrogen and oxygen are not. A coefficient of 4 for H_2O balances the hydrogen atoms:

$$\underset{(4 \times 2)\ H}{(NH_4)_2Cr_2O_7(s)} \longrightarrow Cr_2O_3(s) + N_2(g) + \underset{(4 \times 2)\ H}{4H_2O(g)}$$

Note that in balancing the hydrogen, we have also balanced the oxygen, since there are seven oxygen atoms in the reactants and in the products.

CHECK: $\underset{\text{Reactant atoms}}{2\ N,\ 8\ H,\ 2\ Cr,\ 7\ O} \longrightarrow \underset{\text{Product atoms}}{2\ N,\ 8\ H,\ 2\ Cr,\ 7\ O}$

The equation is balanced.

Decomposition of ammonium dichromate.

3.8 Stoichiometric Calculations: Amounts of Reactants and Products

Recall that the coefficients in chemical equations represent *numbers* of molecules, not masses of molecules. However, in the laboratory or chemical plant, when a reaction is to be run, the amounts of substances needed cannot be determined by counting molecules directly. Counting is always done by weighing. In this section we will see how chemical equations can be used to deal with *masses* of reacting chemicals.

To develop the principles involved in dealing with the stoichiometry of reactions, we will consider the combustion of propane (C_3H_8), a hydrocarbon used for gas barbecue grills and often used as a fuel in rural areas where natural gas pipelines are not available. Propane reacts with oxygen to produce carbon dioxide and water. We will consider the question: "What mass of oxygen will react with 96.1 grams of propane?" The first thing that must always be done when performing calculations involving chemical reactions is to *write the balanced chemical equation* for the reaction. In this case the balanced equation is

Before doing any calculations involving a chemical reaction, be sure the equation for the reaction is balanced.

$$C_3H_8(g) + 5O_2(g) \longrightarrow 3CO_2(g) + 4H_2O(g)$$

Recall that this equation means that 1 mole of C_3H_8 will react with 5 moles of O_2 to produce 3 moles of CO_2 and 4 moles of H_2O. To use this equation to find the masses of reactants and products, we must be able to convert between masses and moles of substances. Thus we must first ask: *"How many moles of propane are present in 96.1 grams of propane?"* The molar mass of propane to three significant figures is 44.1 grams per mole. The moles of propane can be calculated as follows:

$$96.1\ g\ C_3H_8 \times \frac{1\ mol\ C_3H_8}{44.1\ g\ C_3H_8} = 2.18\ mol\ C_3H_8$$

Sulfuric Acid: The Most Important Chemical

More sulfuric acid (H_2SO_4) is produced in the world than any other chemical. Over 40 million tons (4×10^{10} kilograms) of sulfuric acid is manufactured annually in the United States (see figure). Sulfuric acid is used in the production of fertilizers, explosives, petroleum products, detergents, dyes, insecticides, drugs, plastics, steel, storage batteries, and many other materials. The largest amount of sulfuric acid is used in the production of phosphate fertilizers. In this process calcium phosphate, $Ca_3(PO_4)_2$, in phosphate rock, which cannot be used by plants because of its insolubility in groundwater, is converted to forms that will dissolve in water, thus making the phosphate available to plants. This reaction can be represented as

$$Ca_3(PO_4)_2(s) + 3H_2SO_4(aq) \longrightarrow 3CaSO_4(s) + 2H_3PO_4(aq)$$

The mixture of $CaSO_4$ and H_3PO_4 (phosphoric acid) is dried, pulverized, and spread on fields, where the phosphate is dissolved by rainfall.

Sulfuric acid is produced by a sequence of three simple reactions.

1. The combustion of sulfur to form sulfur dioxide:

$$S(s) + O_2(g) \longrightarrow SO_2(g)$$

2. The conversion of sulfur dioxide to sulfur trioxide:

$$2SO_2(g) + O_2(g) \longrightarrow 2SO_3(g)$$

3. The combination of sulfur trioxide with water:

$$SO_3(g) + H_2O(l) \longrightarrow H_2SO_4(aq)$$

Over 90% of sulfuric acid is produced commercially by the *contact process.* This name was coined because the sulfur dioxide and oxygen molecules react in *contact* with the surface of solid vanadium(V) oxide (V_2O_5).

Because gaseous sulfur trioxide reacts so violently with water, it is absorbed during the production process by a sulfuric acid solution rather than by pure water. The sulfur trioxide is added to a flowing solution of sulfuric acid to which water is constantly added to keep the concentration at 98% sulfuric acid by mass. This is the substance sold as *concentrated sulfuric acid.*

One remarkable property of sulfuric acid is its great affinity for water. For example, when it is mixed with sugar, it dehydrates the sugar (takes the water out) and forms a column of black carbon (see Fig. 19.22). Because of this high affinity for water, sulfuric acid is often used as a drying agent in the production of explosives, dyes, detergents, and various anhydrous (water-free) materials.

The violence with which sulfuric acid and water combine makes the dilution of concentrated sulfuric acid potentially hazardous. The addition of water to the concentrated acid produces a vigorous reaction, which often causes acid droplets to spew in all directions. For obvious reasons, this must be avoided. *Always add the acid to water when diluting* so that any accidental splattering will involve dilute acid rather than concentrated acid.

A sulfuric acid plant.

Next, we must take into account that each mole of propane reacts with 5 moles of oxygen. The best way to do this is to use the balanced equation to construct a **mole ratio.** In this case we want to convert from moles of propane to moles of oxygen. From the balanced equation we see that 5 moles of O_2 are required for each mole of C_3H_8, so the appropriate ratio is

$$\frac{5 \text{ mol } O_2}{1 \text{ mol } C_3H_8}$$

which can be used to calculate the number of moles of O_2 required:

$$2.18 \text{ mol } C_3H_8 \times \frac{5 \text{ mol } O_2}{1 \text{ mol } C_3H_8} = 10.9 \text{ mol } O_2$$

Since the original question asked for the mass of oxygen needed to react with 96.1 grams of propane, the 10.9 moles of O_2 must be converted to *grams,* using the molar mass of O_2:

$$10.9 \text{ mol } O_2 \times \frac{32.0 \text{ g } O_2}{1 \text{ mol } O_2} = 349 \text{ g } O_2$$

Therefore, 349 grams of oxygen are required to burn 96.1 grams of propane.

This example can be extended by asking: "What mass of carbon dioxide is produced when 96.1 grams of propane are combusted with oxygen?" In this case we must convert between moles of propane and moles of carbon dioxide. This conversion can be done by looking at the balanced equation, which shows that 3 moles of CO_2 are produced for each mole of C_3H_8 reacted:

$$2.18 \text{ mol } C_3H_8 \times \frac{3 \text{ mol } CO_2}{1 \text{ mol } C_3H_8} = 6.54 \text{ mol } CO_2$$

Then we use the molar mass of CO_2 (44.0 grams per mole) to calculate the mass of CO_2 produced:

$$6.54 \text{ mol } CO_2 \times \frac{44.0 \text{ g } CO_2}{1 \text{ mol } CO_2} = 288 \text{ g } CO_2$$

The process for finding the mass of carbon dioxide produced from 96.1 grams of propane is summarized below:

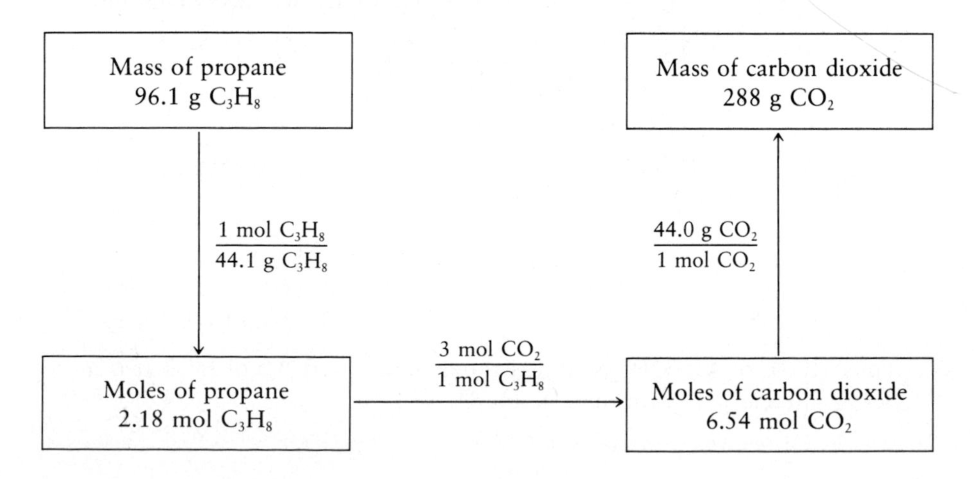

Summary: Calculation of Masses of Reactants and Products in Chemical Reactions

- **Balance the equation for the reaction.**
- **Convert the known masses of the substances to moles.**
- **Use the balanced equation to set up the appropriate mole ratios.**
- **Use the appropriate mole ratios to calculate the number of moles of the desired reactant or product.**
- **Convert from moles back to grams if required by the problem.**

EXAMPLE 3.9

Baking soda ($NaHCO_3$) is often used as an antacid. It neutralizes excess hydrochloric acid secreted by the stomach:

$$NaHCO_3(s) + HCl(aq) \longrightarrow NaCl(aq) + H_2O(l) + CO_2(aq)$$

Milk of magnesia, which is an aqueous suspension of magnesium hydroxide, is also used as an antacid:

$$Mg(OH)_2(s) + 2HCl(aq) \longrightarrow 2H_2O(l) + MgCl_2(aq)$$

Which is the more effective antacid per gram, $NaHCO_3$ or $Mg(OH)_2$?

Solution

To answer the question, we must determine the amount of HCl neutralized per gram of $NaHCO_3$ and per gram of $Mg(OH)_2$.

Using the molar mass of $NaHCO_3$, we determine the moles of $NaHCO_3$ in 1.00 g of $NaHCO_3$:

$$1.00 \text{ g NaHCO}_3 \times \frac{1 \text{ mol NaHCO}_3}{84.01 \text{ g NaHCO}_3} = 1.19 \times 10^{-2} \text{ mol NaHCO}_3$$

Because HCl and $NaHCO_3$ react 1:1, this answer also represents the moles of HCl required. Thus 1.00 g of $NaHCO_3$ will neutralize 1.19×10^{-2} mol HCl. Using the molar mass of $Mg(OH)_2$, we next determine the moles of $Mg(OH)_2$ in 1.00 g:

$$1.00 \text{ g Mg(OH)}_2 \times \frac{1 \text{ mol Mg(OH)}_2}{58.32 \text{ g Mg(OH)}_2} = 1.71 \times 10^{-2} \text{ mol Mg(OH)}_2$$

Using the balanced equation, we determine the moles of HCl that will react with this amount of $Mg(OH)_2$:

$$1.71 \times 10^{-2} \text{ mol Mg(OH)}_2 \times \frac{2 \text{ mol HCl}}{1 \text{ mol Mg(OH)}_2} = 3.42 \times 10^{-2} \text{ mol HCl}$$

Thus 1.00 g of $Mg(OH)_2$ will neutralize 3.42×10^{-2} mol HCl. It is a better antacid per gram than $NaHCO_3$.

3.9 Calculations Involving a Limiting Reactant

When chemicals are mixed together to undergo a reaction, they are often mixed in **stoichiometric quantities,** that is, in exactly the correct amounts so that all reactants "run out" (are used up) at the same time. To clarify this concept, let's consider the production of hydrogen for use in the manufacture of ammonia by the **Haber process.** Ammonia, a very important fertilizer itself and a starting material for other fertilizers, is made by combining nitrogen from the air with hydrogen according to the equation

$$N_2(g) + 3H_2(g) \longrightarrow 2NH_3(g)$$

The hydrogen for this process is produced from the reaction of methane with water:

$$CH_4(g) + H_2O(g) \longrightarrow 3H_2(g) + CO(g)$$

Now consider the following question: What mass of water is required to react *exactly* with 2.50×10^3 kilograms of methane? That is, how much water will just use up all of the 2.50×10^3 kilograms of methane, leaving no methane or water remaining? Using the principles developed in the previous section, we can calculate that if 2.50×10^3 kilograms of methane are mixed with 2.81×10^3 kilograms of water, both reactants will run out at the same time. The reactants have been mixed in stoichiometric quantities.

If, on the other hand, 2.50×10^3 kilograms of methane are mixed with 3.00×10^3 kilograms of water, the methane will be consumed before the water runs out. The water will be in *excess*. In this case the quantity of products formed will be determined by the quantity of methane present. Once the methane is consumed, no more products can be formed, even though some water still remains. In this situation, because the amount of methane *limits* the amount of products that can be formed, it is called the **limiting reactant,** or **limiting reagent.** In any stoichiometry problem it is essential to determine which reactant is the limiting one in order to calculate correctly the amounts of products that will be formed.

Suppose 25.0 kilograms of nitrogen and 5.00 kilograms of hydrogen are mixed and reacted to form ammonia. How do we calculate the mass of ammonia produced when this reaction is run to completion?

To solve this problem, we must use the balanced equation

$$N_2(g) + 3H_2(g) \longrightarrow 2NH_3(g)$$

to determine whether nitrogen or hydrogen is a limiting reactant and then to determine the amount of ammonia that is formed. We first calculate the moles of reactants present:

$$25.0 \text{ kg } N_2 \times \frac{1000 \text{ g } N_2}{1 \text{ kg } N_2} \times \frac{1 \text{ mol } N_2}{28.0 \text{ g } N_2} = 8.93 \times 10^2 \text{ mol } N_2$$

$$5.00 \text{ kg } H_2 \times \frac{1000 \text{ g } H_2}{1 \text{ kg } H_2} \times \frac{1 \text{ mol } H_2}{2.016 \text{ g } H_2} = 2.48 \times 10^3 \text{ mol } H_2$$

Since 1 mole of N_2 reacts with 3 moles of H_2, the number of moles of H_2 that will react exactly with 8.93×10^2 moles of N_2 is

$$8.93 \times 10^2 \text{ mol } N_2 \times \frac{3 \text{ mol } H_2}{1 \text{ mol } N_2} = 2.68 \times 10^3 \text{ mol } H_2$$

Always check to see which reactant is limiting.

Thus 8.93×10^2 moles of N_2 require 2.68×10^3 moles of H_2 to react completely. However, in this case only 2.48×10^3 moles of H_2 are present. Thus the hydrogen will be consumed before the nitrogen. Therefore, hydrogen is the *limiting reactant* in this particular situation, and we must use the amount of hydrogen to compute the quantity of ammonia formed:

$$2.48 \times 10^3 \text{ mol } H_2 \times \frac{2 \text{ mol } NH_3}{3 \text{ mol } H_2} = 1.65 \times 10^3 \text{ mol } NH_3$$

$$1.65 \times 10^3 \text{ mol } NH_3 \times \frac{17.0 \text{ g } NH_3}{1 \text{ mol } NH_3} = 2.80 \times 10^4 \text{ g } NH_3 = 28.0 \text{ kg } NH_3$$

Note that to determine the limiting reactant, we could have started instead with the given amount of hydrogen and calculated the moles of nitrogen required:

$$2.48 \times 10^3 \text{ mol } H_2 \times \frac{1 \text{ mol } N_2}{3 \text{ mol } H_2} = 8.27 \times 10^2 \text{ mol } N_2$$

Thus 2.48×10^3 moles of H_2 require 8.27×10^2 moles of N_2. Since 8.93×10^2 moles of N_2 are actually present, the nitrogen is in excess. The hydrogen will run out first, and thus again we find that hydrogen limits the amount of ammonia formed.

A related, but simpler, way to determine which reactant is limiting is to compare the mole ratio of the substances required by the balanced equation with the mole ratio of reactants actually present. For example, in this case the mole ratio of H_2 to N_2 required by the balanced equation is

$$\frac{3 \text{ mol } H_2}{1 \text{ mol } N_2}$$

That is,

$$\frac{\text{mol } H_2}{\text{mol } N_2} \text{ (required)} = \frac{3}{1} = 3$$

In this experiment we have 2.48×10^3 moles of H_2 and 8.93×10^2 moles of N_2. Thus the ratio

$$\frac{\text{mol } H_2}{\text{mol } N_2} \text{ (actual)} = \frac{2.48 \times 10^3}{8.93 \times 10^2} = 2.78$$

Since 2.78 is less than 3, the actual mole ratio of H_2 to N_2 is too small, and H_2 must be limiting. If the actual H_2-to-N_2 mole ratio had been greater than 3, then the H_2 would have been in excess and the N_2 would have been limiting.

EXAMPLE 3.10

Nitrogen gas can be prepared by passing gaseous ammonia over solid copper(II) oxide at high temperatures. The other products of the reaction are solid copper and water vapor. If 18.1 g of NH_3 is reacted with 90.4 g of CuO, which is the limiting reactant? How many grams of N_2 will be formed?

Solution

The description of the problem leads to the balanced equation

$$2NH_3(g) + 3CuO(s) \longrightarrow N_2(g) + 3Cu(s) + 3H_2O(g)$$

Next, we must compute the moles of NH_3 and of CuO:

$$18.1 \text{ g } NH_3 \times \frac{1 \text{ mol } NH_3}{17.0 \text{ g } NH_3} = 1.06 \text{ mol } NH_3$$

$$90.4 \text{ g CuO} \times \frac{1 \text{ mol CuO}}{79.5 \text{ g CuO}} = 1.14 \text{ mol CuO}$$

The limiting reactant is determined by using the mole ratio for CuO and NH_3:

$$1.06 \text{ mol } NH_3 \times \frac{3 \text{ mol CuO}}{2 \text{ mol } NH_3} = 1.59 \text{ mol CuO}$$

Thus 1.59 mol of CuO is required to react with 1.06 mol of NH_3. Since only 1.14 mol of CuO is actually present, the amount of CuO is limiting.

We can verify this conclusion by comparing the mole ratio of CuO and NH_3 required by the balanced equation,

$$\frac{\text{mol CuO}}{\text{mol } NH_3} \text{ (required)} = \frac{3}{2} = 1.5$$

with the mole ratio actually present,

$$\frac{\text{mol CuO}}{\text{mol } NH_3} \text{ (actual)} = \frac{1.14}{1.06} = 1.08$$

Since the actual ratio is too small (smaller than 1.5), CuO is the limiting reactant.

Since CuO is the limiting reactant, we must use the amount of CuO to calculate the amount of N_2 formed:

$$1.14 \text{ mol CuO} \times \frac{1 \text{ mol } N_2}{3 \text{ mol CuO}} = 0.380 \text{ mol } N_2$$

Using the molar mass of N_2, we can calculate the mass of N_2 produced:

$$0.380 \text{ mol } N_2 \times \frac{28.0 \text{ g } N_2}{1 \text{ mol } N_2} = 10.6 \text{ g } N_2$$

The amount of a given product formed when the limiting reactant is completely consumed is called the **theoretical yield** of that product. In Example 3.10, 10.6 grams of nitrogen is the theoretical yield. This is the *maximum amount* of nitrogen that can be produced from the quantities of reactants used. Actually, the amount of product predicted by the theoretical yield is seldom obtained because of side reactions (other reactions that involve one or more of the reactants or products) and other complications. The *actual yield* of product

Percent yield is important as an indicator of the efficiency of a particular laboratory or industrial reaction.

is often given as a percentage of the theoretical yield. This value is called the **percent yield:**

$$\frac{\text{Actual yield}}{\text{Theoretical yield}} \times 100 = \text{Percent yield}$$

For example, if the reaction considered in Example 3.10 actually produced 6.63 grams of nitrogen instead of the predicted 10.6 grams, the percent yield of nitrogen would be

$$\frac{6.63 \text{ g } N_2}{10.6 \text{ g } N_2} \times 100 = 62.5\%$$

Summary: **Solving a Stoichiometry Problem Involving Masses of Reactants and Products**

- **Write and balance the equation for the reaction.**
- **Convert the known masses of substances to moles.**
- **By comparing the mole ratio of reactants required by the balanced equation to the mole ratio of reactants actually present, determine which reactant is limiting.**
- **Using the amount of the limiting reactant and the appropriate mole ratios, compute the number of moles of the desired product.**
- **Convert from moles to grams using the molar mass.**

EXAMPLE 3.11

Potassium chromate, a bright yellow solid, is produced by the reaction of solid chromite ore ($FeCr_2O_4$) with solid potassium carbonate and gaseous oxygen at high temperatures. The other products of the reaction are solid iron(III) oxide and gaseous carbon dioxide. In a particular experiment 169 kg of chromite ore, 298 kg of potassium carbonate, and 75.0 kg of oxygen were sealed in a reaction vessel and reacted at a high temperature. The amount of potassium chromate obtained was 194 kg. Calculate the percent yield of potassium chromate.

Solution

The unbalanced equation, which can be written from the above description of the reaction, is

$$FeCr_2O_4(s) + K_2CO_3(s) + O_2(g) \longrightarrow K_2CrO_4(s) + Fe_2O_3(s) + CO_2(g)$$

The balanced equation is

$$4FeCr_2O_4(s) + 8K_2CO_3(s) + 7O_2(g) \longrightarrow 8K_2CrO_4(s) + 2Fe_2O_3(s) + 8CO_2(g)$$

The numbers of moles of the various reactants are obtained as follows:

$$169 \text{ kg } FeCr_2O_4 \times \frac{1000 \text{ g } FeCr_2O_4}{1 \text{ kg } FeCr_2O_4} \times \frac{1 \text{ mol } FeCr_2O_4}{223.84 \text{ g } FeCr_2O_4} = 7.55 \times 10^2 \text{ mol } FeCr_2O_4$$

$$298 \text{ kg } K_2CO_3 \times \frac{1000 \text{ g } K_2CO_3}{1 \text{ kg } K_2CO_3} \times \frac{1 \text{ mol } K_2CO_3}{138.21 \text{ g } K_2CO_3} = 2.16 \times 10^3 \text{ mol } K_2CO_3$$

$$75.0 \text{ kg } O_2 \times \frac{1000 \text{ g } O_2}{1 \text{ kg } O_2} \times \frac{1 \text{ mol } O_2}{32.00 \text{ g } O_2} = 2.34 \times 10^3 \text{ mol } O_2$$

Now we must determine which of the three reactants is limiting. To do so, we will compare the mole ratios of reactants required by the balanced equation with the actual mole ratios. For the reactants K_2CO_3 and $FeCr_2O_4$ the required mole ratio is

$$\frac{\text{mol } K_2CO_3}{\text{mol } FeCr_2O_4} \text{ (required)} = \frac{8}{4} = 2$$

The actual mole ratio is

$$\frac{\text{mol } K_2CO_3}{\text{mol } FeCr_2O_4} \text{ (actual)} = \frac{2.16 \times 10^3}{7.55 \times 10^2} = 2.86$$

Since the actual mole ratio is greater than that required, the K_2CO_3 is in excess compared with $FeCr_2O_4$. Thus either $FeCr_2O_4$ or O_2 must be limiting. To determine which of these will limit the amounts of products, we compare the required mole ratio,

$$\frac{\text{mol } O_2}{\text{mol } FeCr_2O_4} \text{ (required)} = \frac{7}{4} = 1.75$$

with the actual mole ratio,

$$\frac{\text{mol } O_2}{\text{mol } FeCr_2O_4} \text{ (actual)} = \frac{2.34 \times 10^3}{7.55 \times 10^2} = 3.10$$

Thus more K_2CO_3 and O_2 are present than required. These reactants are in excess, so $FeCr_2O_4$ is the limiting reactant.

We must use the amount of $FeCr_2O_4$ to calculate the maximum amount of K_2CrO_4 that can be formed:

$$7.55 \times 10^2 \text{ mol } FeCr_2O_4 \times \frac{8 \text{ mol } K_2CrO_4}{4 \text{ mol } FeCr_2O_4} = 1.51 \times 10^3 \text{ mol } K_2CrO_4$$

Using the molar mass of K_2CrO_4, we can determine the mass:

$$1.51 \times 10^3 \text{ mol } K_2CrO_4 \times \frac{194.19 \text{ g } K_2CrO_4}{1 \text{ mol } K_2CrO_4} = 2.93 \times 10^5 \text{ g } K_2CrO_4$$

This value represents the theoretical yield of K_2CrO_4. The actual yield was 194 kg, or 1.94×10^5 g. Thus the percent yield is

$$\frac{1.94 \times 10^5 \text{ g } K_2CrO_4}{2.93 \times 10^5 \text{ g } K_2CrO_4} \times 100 = 66.2\%$$

Exercises

A blue exercise number indicates that the answer to that exercise appears at the back of this book.

Atomic Masses and the Mass Spectrometer

1. The element magnesium (Mg) has three stable isotopes with the following masses and abundances:

Isotope	Mass (amu)	Abundance
^{24}Mg	23.9850	78.99%
^{25}Mg	24.9858	10.00%
^{26}Mg	25.9826	11.01%

Calculate the average atomic mass (the atomic weight) of magnesium from these data.

2. The element europium exists in nature as two isotopes: ^{151}Eu has a mass of 150.9196 amu, and ^{153}Eu has a mass of 152.9209 amu. The average atomic mass of europium is 151.96 amu. Calculate the relative abundance of the two europium isotopes.
3. The element rhenium (Re) has two naturally occurring isotopes, ^{185}Re and ^{187}Re, with an average atomic mass of 186.207 amu. Rhenium is 62.60% ^{187}Re and the atomic mass of ^{187}Re is 186.956 amu. Calculate the mass of ^{185}Re.
4. Assume that element Uus is synthesized and that it has the following isotopes:

^{284}Uus (283.9 amu), 21.00%

^{285}Uus (284.8 amu), 31.54%

^{288}Uus (287.8 amu), 47.46%

What is the value of the average atomic mass that would be listed on the periodic table?

5. An element consists of 1.40% of an isotope with mass 203.973 amu, 24.10% of an isotope with mass 205.9745 amu, 22.10% of an isotope with mass 206.9759 amu, and 52.40% of an isotope with mass 207.9766 amu. Calculate the average atomic mass and identify the element.
6. The mass spectrum of bromine (Br_2) consists of three peaks with the following relative sizes:

Mass (amu)	Relative size
157.84	0.2534
159.84	0.5000
161.84	0.2466

How do you interpret these data?

7. Gallium arsenide, GaAs, is gaining widespread use in semiconductor devices that interconvert light and electrical signals in fiber-optic communications systems. Gallium consists of 60.% ^{69}Ga and 40.% ^{71}Ga. Arsenic has only one naturally occurring isotope, ^{75}As. Gallium arsenide is a polymeric material but its mass spectrum shows fragments with formulas GaAs and Ga_2As_2. What would the distribution of peaks look like for these two fragments?

Moles and Molar Masses

8. Determine the mass in grams of the following.
   a. 3.00×10^{20} HF molecules
   b. 3.00×10^{-3} mol of HF
   c. 1.5×10^{2} mol of HF
   d. a single HF molecule
   e. 2.00×10^{-15} mol of HF
   f. 18.0 picomoles of HF
   g. 5.0 nanomoles of HF
9. How many moles are represented by each of these samples?
   a. 100 molecules (exactly) of H_2O
   b. 100.0 g of H_2O
   c. 500 atoms (exactly) of Fe
   d. 500.0 g of Fe
   e. 150 molecules (exactly) of N_2
   f. 150.0 g of Fe_2O_3
   g. 10.0 mg of NO_2
   h. 1.0 femtomoles of NO_2
   i. 1.5×10^{16} molecules of BF_3
   j. 2.6 mg of BF_3
10. Aspartame is an artificial sweetener that is 160 times sweeter than sucrose (table sugar) when dissolved in water. It is marketed as Nutra-Sweet. The molecular formula of aspartame is $C_{14}H_{18}N_2O_5$.
    a. Calculate the molar mass of aspartame.
    b. How many moles of molecules are in 10.0 g of aspartame?
    c. What is the mass in grams of 1.56 mol of aspartame?
    d. How many molecules are in 5.0 mg of aspartame?
    e. How many atoms of nitrogen are in 1.2 g of aspartame?
    f. What is the mass in grams of 1.0×10^{9} molecules of aspartame?
    g. What is the mass in grams of one molecule of aspartame?
11. Humulone, $C_{21}H_{30}O_5$, is one of the flavor components that gives a bitter taste to the hops used in making beer.
    a. What is the molar mass of humulone?
    b. How many moles of $C_{21}H_{30}O_5$ molecules are in 275 mg of humulone?
    c. What is the mass of 0.600 mol of humulone?
    d. How many atoms of hydrogen are in 1.00 pg of humulone?
    e. What is the mass of 1.00×10^{9} molecules of humulone?
    f. What is the mass of one molecule of humulone?

12. In the spring of 1984, concern arose over the presence of ethylene dibromide, or EDB, in grains and cereals. EDB has the molecular formula $C_2H_4Br_2$ and until 1984 was commonly used as a plant fumigant. The federal limit for EDB in finished cereal products is 30.0 parts per billion (ppb), where 1.0 ppb = 1.0×10^{-9} g of EDB for every 1.0 g of sample. How many molecules of EDB are in 1.0 lb of flour if 30.0 ppb of EDB are present?

Percent Composition

13. In 1987 the first substance to act as a superconductor at a temperature above that of liquid nitrogen (77 K) was discovered. The approximate formula of this substance is $YBa_2Cu_3O_7$. Calculate the percent composition by mass of this material.

14. Arrange the following compounds in order of increasing percent of phosphorus.
    a. PF_3
    b. P_4O_{10}
    c. $(NPCl_2)_3$
    d. InP

15. There are several important compounds that contain only nitrogen and oxygen. Calculate the mass percent of nitrogen in each of the following.
    a. NO, a gas formed by the reaction of N_2 and O_2 in internal combustion engines
    b. NO_2, a brown gas mainly responsible for the brownish color of photochemical smog
    c. N_2O_4, a colorless liquid used as a fuel in space shuttles
    d. N_2O, a colorless gas used as an anesthetic by dentists (known as laughing gas)

16. Vitamin B_{12}, cyanocobalamin, is essential for human nutrition. It is concentrated in animal tissue but not in higher plants. Although nutritional requirements for the vitamin are quite low, people who abstain completely from animal products may develop a deficiency anemia. Cyanocobalamin is the form used in vitamin supplements. It contains 4.34% cobalt by mass. Calculate the molar mass of cyanocobalamin, assuming there is one atom of cobalt in every molecule of cyanocobalamin.

17. Hemoglobin is the protein that transports oxygen in mammals. Hemoglobin is 0.342% Fe by mass, and each hemoglobin molecule contains four iron atoms. Calculate the molar mass of hemoglobin.

18. Portland cement acts as the binding agent in concrete. A typical Portland cement has the following composition:

Formula	Name	Mass Percent
Ca_3SiO_5	tricalcium silicate	50.
Ca_2SiO_4	dicalcium silicate	25
$Ca_3Al_2O_6$	tricalcium aluminate	12
Ca_2AlFeO_5	calcium aluminoferrite	8.0
$CaSO_4 \quad 2H_2O$	calcium sulfate dihydrate	3.5
other substances, mostly MgO		1.5

Assuming that the impurities contain no Ca, Al, or Fe, calculate the mass percent of these elements in this Portland cement.

Empirical and Molecular Formulas

19. A compound that contains only carbon, hydrogen, and oxygen is 48.64% C and 8.16% H by mass. What is the empirical formula of this substance?

20. A compound contains only carbon, hydrogen, nitrogen, and oxygen. Combustion of 0.157 g of the compound produced 0.213 g of CO_2 and 0.0310 g of H_2O. In another experiment, 0.103 g of the compound produced 0.0230 g of NH_3. What is the empirical formula of the compound? *Hint:* Combustion involves reacting with excess O_2. Assume that all of the carbon ends up in CO_2 and all of the hydrogen ends up in H_2O. Also assume that all of the nitrogen ends up in the NH_3 in the second experiment.

21. A confiscated white substance, suspected of being cocaine, was purified by a forensic chemist and subjected to elemental analysis. Combustion of a 50.86-mg sample yielded 150.0 mg of CO_2 and 46.05 mg of H_2O. Analysis for nitrogen showed that the compound contained 9.39% N by mass. The formula of cocaine is $C_{17}H_{21}NO_4$. Can the forensic chemist conclude that the suspected compound is cocaine?

22. The active ingredient in photographic fixer solution contains sodium, sulfur, and oxygen. Analysis of a sample shows that the sample contains 0.979 g Na, 1.365 g S, and 1.021 g O. What is the empirical formula of this substance?

23. One of the most commonly used white pigments in paint is a compound of titanium and oxygen that contains 59.9% Ti by mass. Calculate the empirical formula of the compound.

24. A compound contains only carbon, hydrogen, and oxygen. Combustion of 10.68 mg of the compound yields 16.01 mg of CO_2 and 4.37 mg of H_2O. The molar mass of the compound is 176.1 g. What are the empirical and molecular formulas of the compound?

25. Cumene is a hydrocarbon that is used in the production of acetone and phenol in the chemical industry. Combustion of 47.6 mg of cumene produces 156.8 mg of CO_2 and 42.8 mg of water. The molar mass is between 115 and 125 g. Determine the empirical and molecular formulas.

26. ABS plastic is a tough, hard plastic used in applications requiring shock resistance. (See Chapter 22.) The polymer consists of three monomer units: acrylonitrile (C_3H_3N), butadiene (C_4H_6), and styrene (C_8H_8).
    a. A sample of ABS plastic contains 8.80% N by mass. It took 0.605 g of Br_2 to react completely with a 1.20 g sample of ABS plastic. Bromine reacts 1:1 (by moles) with the butadiene molecules in the polymer and

nothing else. What is the percent by mass of acrylonitrile and butadiene in this polymer?

b. What are the relative numbers of each of the monomer units in this polymer?

Balancing Chemical Equations

27. Phosphorus occurs naturally in the form of fluorapatite, $CaF_2 \cdot 3Ca_3(PO_4)_2$, the dot indicating 1 part CaF_2 to 3 parts $Ca_3(PO_4)_2$. In the preparation of a fertilizer, this mineral is reacted with an aqueous solution of sulfuric acid. The products are phosphoric acid, hydrogen fluoride, and gypsum, $CaSO_4 \cdot 2H_2O$. Write and balance the chemical equation describing this process.

28. Write a balanced chemical equation that describes each of the following.
    a. Iron metal reacts with oxygen to form rust, iron(III) oxide.
    b. The fermentation of fruit juice to produce wine involves the conversion of glucose ($C_6H_{12}O_6$) to ethanol (C_2H_6O) and carbon dioxide (CO_2).
    c. Calcium metal reacts with water to produce aqueous calcium hydroxide and hydrogen gas.
    d. Aqueous barium hydroxide reacts with aqueous sulfuric acid to produce solid barium sulfate and water.

29. Lead hydrogen arsenate, an inorganic insecticide still used against the potato beetle, is usually produced by using the following reaction:

$$Pb(NO_3)_2(aq) + H_3AsO_4(aq) \longrightarrow PbHAsO_4(s) + HNO_3(aq)$$

Balance this equation.

30. The electrolysis of concentrated brine solutions is an important source of NaOH, H_2, and Cl_2 for the chemical industry. The reaction is

$$NaCl(aq) + H_2O(l) \xrightarrow{\text{Electricity}} Cl_2(g) + H_2(g) + NaOH(aq)$$

Balance this equation.

Reaction Stoichiometry

31. The reusable booster rockets of the U.S. space shuttle employ a mixture of aluminum and ammonium perchlorate for fuel. A possible equation for this reaction is

$$3Al(s) + 3NH_4ClO_4(s) \longrightarrow Al_2O_3(s) + AlCl_3(s) + 3NO(g) + 6H_2O(g)$$

What mass of NH_4ClO_4 should be used in the fuel mixture for every kilogram of Al?

32. Nitric acid is produced commercially by the Ostwald process. The three steps of the Ostwald process are shown in the following equations:

$$4NH_3(g) + 5O_2(g) \longrightarrow 4NO(g) + 6H_2O(g)$$

$$2NO(g) + O_2(g) \longrightarrow 2NO_2(g)$$

$$3NO_2(g) + H_2O(l) \longrightarrow 2HNO_3(aq) + NO(g)$$

What mass of NH_3 must be used to produce 1.0×10^6 kg of HNO_3 by the Ostwald process, assuming 100% yield in each reaction?

33. One of the major commercial uses for sulfuric acid is in the production of phosphoric acid and calcium sulfate. The phosphoric acid is used for fertilizer. The reaction is

$$Ca_3(PO_4)_2(s) + 3H_2SO_4(aq) \longrightarrow 3CaSO_4(s) + 2H_3PO_4(aq)$$

What mass of concentrated sulfuric acid (98% H_2SO_4 by mass) must be used to react completely with 1.0×10^2 g of $Ca_3(PO_4)_2$?

34. Elixirs such as Alka-Seltzer use the reaction of sodium bicarbonate with citric acid in aqueous solution to produce a fizz:

$$3NaHCO_3(aq) + C_6H_8O_7(aq) \longrightarrow 3CO_2(g) + 3H_2O(l) + Na_3C_6H_5O_7(aq)$$

a. What mass of $C_6H_8O_7$ should be used for every 1.0×10^2 mg of $NaHCO_3$?

b. What mass of $CO_2(g)$ would be produced from such a mixture?

35. a. Write the balanced equation for the combustion of isooctane (C_8H_{18}) to produce water vapor and carbon dioxide gas.
    b. Assuming gasoline is 100% isooctane, with a density of 0.692 g/mL, what mass of carbon dioxide is produced by the combustion of 1.2×10^{10} gal of gasoline (the approximate annual consumption of gasoline in the United States)?

36. In the production of printed circuit boards for the electronics industry, a 0.60 mm layer of copper is laminated onto an insulating plastic board. Next a circuit pattern made of a chemically resistant polymer is printed on the board. The unwanted copper is removed by chemical etching and the protective polymer is finally removed by solvents. One etching reaction is

$$Cu(NH_3)_4Cl_2(aq) + 4NH_3(aq) + Cu(s) \longrightarrow 2Cu(NH_3)_4Cl(aq)$$

A plant needs to manufacture 10,000 printed circuit boards, each 8.0×16.0 cm in area. An average of 80.% of the copper is removed from each board (density of copper = 8.96 g/cm³). What masses of $Cu(NH_3)_4Cl_2$ and NH_3 are needed to do this?

Limiting Reactants and Percent Yield

37. Consider the reaction

$$4Al(s) + 3O_2(g) \longrightarrow 2Al_2O_3(s)$$

Identify the limiting reagent in each of the following reaction mixtures.
a. 1.0 mol Al and 1.0 mol O_2
b. 2.0 mol Al and 4.0 mol O_2
c. 0.50 mol Al and 0.75 mol O_2
d. 64.75 g Al and 115.21 g O_2
e. 75.89 g Al and 112.25 g O_2
f. 51.28 g Al and 118.22 g O_2

38. When copper is heated with an excess of sulfur, copper(I) sulfide is the product. When 2.00 g of copper was heated with excess sulfur, 2.31 g of copper(I) sulfide was isolated. What is the theoretical yield? What was the percent yield?

39. Aluminum burns in bromine, producing aluminum bromide:

$$2Al(s) + 3Br_2(l) \longrightarrow 2AlBr_3(s)$$

When 10.0 g of aluminum was reacted with an excess of bromine, 79.8 g of aluminum bromide was isolated. Calculate the theoretical and percent yield of this reaction.

40. When a mixture of silver metal and sulfur is heated, silver sulfide is formed:

$$16Ag(s) + S_8(s) \xrightarrow{\text{Heat}} 8Ag_2S(s)$$

a. What mass of Ag_2S is produced from a mixture of 1.00 g of Ag and 2.00 g of S_8?
b. What mass of which reactant is left unreacted?

41. Hydrogen cyanide is produced industrially from the reaction of gaseous ammonia, oxygen, and methane:

$$2NH_3(g) + 3O_2(g) + 2CH_4(g) \longrightarrow 2HCN(g) + 6H_2O(g)$$

If 5.00×10^3 kg each of NH_3, O_2, and CH_4 are reacted, what mass of HCN and of H_2O will be produced, assuming 100% yield?

42. The production capacity for acrylonitrile (C_3H_3N) in the United States is over 2 million pounds per year. Acrylonitrile, the building block for polyacrylonitrile fibers and a variety of plastics, is produced from gaseous propylene, ammonia, and oxygen:

$$2C_3H_6(g) + 2NH_3(g) + 3O_2(g) \longrightarrow 2C_3H_3N(g) + 6H_2O(g)$$

a. What mass of acrylonitrile can be produced from a mixture of 1.00 kg of propylene, 1.50 kg of ammonia, and 2.00 kg of oxygen?
b. What mass of water is produced, and what masses of which starting materials are left in excess?

43. Hexamethylenediamine, $C_6H_{16}N_2$, is one of the starting materials for the production of nylon. It can be prepared from adipic acid, $C_6H_{10}O_4$, by the following overall reaction:

$$C_6H_{10}O_4(l) + 2NH_3(g) + 4H_2(g) \longrightarrow C_6H_{16}N_2(l) + 4H_2O(l)$$

a. What mass of hexamethylenediamine can be produced from 1.00×10^3 g of adipic acid?
b. What is the percent yield if 765 g of hexamethylenediamine is made from 1.00×10^3 g of adipic acid?

44. The aspirin substitute, acetaminophen ($C_8H_9O_2N$), is produced by the following three-step synthesis:

I. $C_6H_5O_3N(s) + 3H_2(g) + HCl(aq) \longrightarrow C_6H_8ONCl(s) + 2H_2O(l)$

II. $C_6H_8ONCl(s) + NaOH(aq) \longrightarrow C_6H_7ON(s) + H_2O(l) + NaCl(aq)$

III. $C_6H_7ON(s) + C_4H_6O_3(l) \longrightarrow C_8H_9O_2N(s) + HC_2H_3O_2(l)$

The first two reactions have percent yields of 87% and 98% by mass, respectively. The overall reaction yields 3 mol of acetaminophen product for every 4 mol of $C_6H_5O_3N$ reacted.
a. What is the percent yield by mass for the overall process?
b. What is the percent yield by mass of step III?

Additional Exercises

45. Only one isotope of this element occurs in nature. One atom of this isotope has a mass of 9.123×10^{-23} g. Identify the element and determine its atomic mass.

46. Automobile safety glass is made by laminating a sheet of polyvinylbutyral between two thin sheets of glass. Polyvinylbutyral has the general formula (the fragment shown is called a monomer unit)

$$\left[\begin{array}{c} -\!\!\left(CHCH_2CHCH_2\right)\!\!- \\ | \quad\quad\quad | \\ O \quad\quad O \\ \diagdown \;\; \diagup \\ CH \\ | \\ CH_2CH_2CH_3 \end{array} \right]_n$$

If the average molar mass of a sample of polyvinylbutyral is 100,000 g, what is the average number of monomer units present in the molecule?

47. The compound cisplatin, $Pt(NH_3)_2Cl_2$, has been extensively studied as an antitumor agent [see the *Journal of Chemical Education:* 54 (1977): 739].
a. Calculate the elemental percent composition by mass of cisplatin.
b. Cisplatin is synthesized as follows:

$$K_2PtCl_4(aq) + 2NH_3(aq) \longrightarrow Pt(NH_3)_2Cl_2(s) + 2KCl(aq)$$

What mass of cisplatin can be made from 100. g of K_2PtCl_4 and sufficient NH_3? What mass of KCl is also produced?

48. Indium oxide contains 4.784 g of indium for every gram of oxygen. In 1869, when Mendeleev first presented his version of the periodic table, he proposed the formula In_2O_3 indium oxide. Before that time it was thought that the formula was InO. What values for the atomic mass of indium are obtained using these two formulas?

49. A compound composed of only antimony and oxygen is 83.53% Sb by mass. The molar mass is between 550 and 600 g. Calculate the empirical and molecular formulas of the compound.

50. Many cereals are made with high moisture content so that the cereal can be formed into various shapes before it is dried. A cereal product containing 58% H_2O by mass is produced at the rate of 1000. kg/hr. How much water must be evaporated per hour if the final product contains only 20.% water?

51. When aluminum metal is heated with an element from Group 6A of the periodic table, an ionic compound forms. When the experiment is performed with an unknown Group 6A element, the product is 18.56% Al by mass. What is the formula of the compound?

52. A salt contains only barium and one of the halide ions. A 0.158-g sample of the salt was dissolved in water, and an excess of sulfuric acid was added to form barium sulfate ($BaSO_4$), which was filtered, dried, and weighed. Its mass was found to be 0.124 g. What is the formula of the barium halide?

53. Terephthalic acid is an important chemical used in the manufacture of polyesters and plasticizers. It contains only C, H, and O. Combustion of 19.81 mg of terephthalic acid produced 41.98 mg CO_2 and 6.45 mg of H_2O. The molar mass of terephthalic acid is 166 g. Calculate the empirical and molecular formulas for terephthalic acid.

54. Calcium oxide, or lime, is produced by the thermal decomposition of limestone:

$$CaCO_3(s) \xrightarrow{\text{Heat}} CaO(s) + CO_2(g)$$

What mass of CaO can be produced from 2.00×10^3 kg of limestone?

55. Consider the following unbalanced equation:

$$Ca_3(PO_4)_2(s) + H_2SO_4(aq) \longrightarrow CaSO_4(s) + H_3PO_4(aq)$$

What masses of calcium sulfate and phosphoric acid can be produced from the reaction of 1.0 kg of calcium phosphate with 1.0 kg of concentrated sulfuric acid (98% H_2SO_4 by mass)?

56. Impure nickel can be purified by first forming the compound $Ni(CO)_4$, which is then decomposed by heating to yield very pure nickel. Metallic nickel reacts directly with gaseous carbon monoxide as follows:

$$Ni(s) + 4CO(g) \longrightarrow Ni(CO)_4(g)$$

Other metals present do not react. If 94.2 g of a metal mixture produces 98.4 g of $Ni(CO)_4$, what is the mass percent of nickel in the original sample?

57. Mercury and bromine will react with each other to produce mercury(II) bromide:

$$Hg(l) + Br_2(l) \longrightarrow HgBr_2(s)$$

a. What mass of $HgBr_2$ is produced from the reaction of 10.0 g Hg and 10.0 Br_2? What mass of which reagent is left unreacted?
b. What mass of $HgBr_2$ is produced from the reaction of 5.00 mL of mercury (density = 13.5 g/mL) and 5.00 mL of bromine (density = 3.12 g/mL)?

58. Arrange the following samples in order according to the total number of atoms present.

4.0 g of hydrogen gas
4.0 g of helium gas
1.0 mol of fluorine gas
44.0 g of carbon dioxide gas
146 g of sulfur hexafluoride gas

59. In using a mass spectrometer, a chemist sees a peak at a mass of 30.0106. Of the choices $^{12}C_2{}^1H_6$, $^{12}C^1H_2{}^{16}O$, and $^{14}N^{16}O$, which is responsible for this peak? Pertinent masses are ^1H, 1.007825; ^{16}O, 15.994915; ^{14}N, 14.003074.

60. Natural rubidium has the average mass 85.4678 and is composed of isotopes ^{85}Rb (mass = 84.9117) and ^{87}Rb. The ratio of atoms $^{85}Rb/^{87}Rb$ in natural ribidium is 2.591. Calculate the mass of ^{87}Rb.

61. The compounds called chlorofluorocarbons, or Freons, have proved to be very valuable as refrigerants and as cleaning agents for circuit boards. Unfortunately, in the atmosphere these compounds produce chlorine atoms that catalyze the decomposition of the ozone that protects the earth from ultraviolet radiation. Two of these compounds have the following mass percentages:

	%C	%Cl	%F
I	9.93	58.6	31.4
II	11.5	33.9	54.6

Determine the empirical formulas of these compounds, and show how they obey the law of multiple proportions.

62. Consider the following data for three binary compounds of hydrogen and nitrogen:

	% H (by mass)	% N (by mass)
I	17.75	82.25
II	12.58	87.42
III	2.34	97.66

When 1.00 L of each gaseous compound is decomposed to its elements, the following volumes of $H_2(g)$ and $N_2(g)$ are obtained:

	H_2 (L)	N_2 (L)
I	1.50	0.50
II	2.00	1.00
III	0.50	1.50

Use these data to determine the molecular formulas of compounds I, II, and III and to determine the relative values for the atomic masses of hydrogen and nitrogen.

63. A component of cast iron called cementite has the composition (by mass) of 93.31% iron and 6.69% carbon. Determine the empirical formula of cementite.

64. A 0.200-g sample of protactinium(IV) oxide is converted to another oxide of protactinium by heating in the presence of oxygen to give 0.2081 g of the new oxide, Pa_xO_y. Determine the values of x and y.

65. An element X forms both a dichloride, XCl_2, and a tetrachloride, XCl_4. Treatment of 10.00 g of XCl_2 with excess chlorine forms 12.55 g of XCl_4. Calculate the atomic weight (mass) of X and identify X.

66. A 1.000-g sample of XI_2 is dissolved in water, and excess silver nitrate is added to precipitate all of the iodide as AgI. The mass of the dry AgI is found to be 1.375 g. Calculate the atomic weight (mass) of X.

67. An unknown binary compound containing hydrogen (XH_n) has a density as a gas that is 2.393 times that of oxygen gas under the same conditions. When 2.23×10^{-2} mol of this compound reacts with excess oxygen gas, 0.803 g of water is produced. Identify the element X in this compound.

68. In the 1920s the discovery of the element illinium (approximate mass of 150) was reported at the University of Illinois. However, later results indicated this report to be in error, and the element was renamed by the group that actually discovered the element. A sample of 6.05 g of illinium combines with 1.00 g of oxygen. What is the present name for illinium?

69. When $M_2S_3(s)$ is heated in air, it is converted to $MO_2(s)$. A 4.000-g sample of $M_2S_3(s)$ shows a decrease in mass of 0.277 g when it is heated in air. What is the atomic weight of M?

70. A substance X_2Z has the composition (by mass) of 40.0% X and 60.0% Z. What is the composition (by mass) of the compound XZ_2?

71. Pure carbon was burned in an excess of oxygen. The gaseous products were

CO_2	72.0 mol%
CO	16.0 mol%
O_2	12.0 mol%

How many moles of O_2 were present in the initial reaction mixture for every mole of carbon?

72. Bacterial digestion is an economical method of sewage treatment. The reaction

$$5CO_2(g) + 55NH_4^+(aq) + 76O_2(g) \xrightarrow{\text{bacteria}} \underset{\text{bacterial tissue}}{C_5H_7O_2N(s)} + 54NO_2^-(aq) + 52H_2O(l) + 109H^+(aq)$$

is an intermediate step in the conversion of the nitrogen in organic compounds into nitrate ions. How much bacterial tissue is produced in a treatment plant for every 1.0×10^4 kg of wastewater containing 3.0% NH_4^+ ions by mass? Assume 95% of the ammonium ions are consumed by the bacteria.

73. An electric furnace produces phosphorus by the following reaction:

$$Ca_3(PO_4)_2(s) + 5C(s) + 3SiO_2(s) \longrightarrow 3CaSiO_3(s) + 5CO(g) + 2P(l)$$

An initial reaction mixture contains 1500 kg calcium phosphate, 250 kg carbon, and 1.0×10^3 kg SiO_2.
a. What is the limiting reagent?
b. What is the theoretical yield of phosphorus?
c. After reaction the slag (solid residue) was analyzed. It contained 3.8% C, 5.8% P, and 26.6% Ca by mass. What was the actual yield of phosphorus in kg? What was the percent yield?

74. Lanthanum was reacted with hydrogen in a given experiment to produce the nonstoichiometric compound $LaH_{2.90}$. Assuming that the compound contains H^-, La^{2+}, and La^{3+}, calculate the fraction of La^{2+} and La^{3+} present.

75. A certain mixture contains only $SrCO_3$ and $BaCO_3$. When 1.60 g of this mixture is treated with hydrochloric acid, 0.421 g of carbon dioxide gas is liberated. Calculate the mass percentage of each component of the mixture.

76. A sample of a mixture containing only sodium chloride and potassium chloride has a mass of 4.000 g. When this sample is dissolved in water and excess silver nitrate is added, a white precipitate (silver chloride) forms. After filtration and drying, this precipitate has the mass 8.5904 g. Calculate the mass percentage of each mixture component.

77. A 1.500-g sample of a mixture containing only Cu_2O and CuO was treated with hydrogen to produce 1.297 g of pure copper metal. Calculate the percentage composition (by mass) of the mixture.

78. In a given experiment with a test engine, gasoline (assume the average composition is C_8H_{18}) and air undergo combustion according to the following equation:

$$aC_8H_{18}(l) + bO_2(g) \longrightarrow cCO_2(g) + dCO(g) + eCH_4(g) + fH_2(g) + gH_2O(g)$$

The exhaust gas (products) was found to have the following composition (by volume): 11.5% CO_2, 4.4% CO, 1.5% H_2, 0.5% CH_4, and 82.1% N_2. These measurements were made after all the water vapor had been removed. Assuming that the N_2 goes through the engine unchanged and that air consists of 21 mole percent O_2 and 78 mole percent N_2, determine the values of the coefficients as accurately as possible.

79. When 4.72 g of compound A (molar mass = 128.6 g) reacts with a few grams of compound B (molar mass unknown), 6.56 g of product C is produced. Analysis shows that C is actually a simple addition compound (that is, AB) and that the yield of product was 76.2% based on A. If A is the limiting reagent, what is the *least* amount of B that must be used to prepare the maximum amount of C obtainable from 4.72 g of compound A?

80. Boron consists of two isotopes, ^{10}B and ^{11}B. Chlorine also has two isotopes, ^{35}Cl and ^{37}Cl. How many peaks and at what approximate masses would you see them in the mass spectrum of BCl_3? Which would be the largest peak and which would be the smallest?

CHAPTER 4

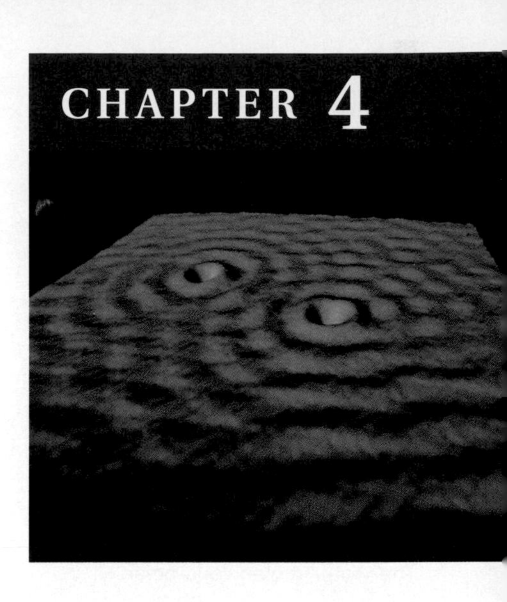

Types of Chemical Reactions and Solution Stoichiometry

Much of the chemistry that affects each of us occurs among substances dissolved in water. For example, virtually all of the chemistry that makes life possible occurs in an aqueous environment. Also, various tests for illnesses involve aqueous reactions. Modern medical practice depends heavily on analyses of blood and other body fluids. In addition to the common tests for sugar, cholesterol, and iron, analyses for specific chemical markers allow detection of many diseases before more obvious symptoms occur.

Aqueous chemistry is also important in our environment. In recent years contamination of the groundwater by substances such as chloroform and nitrates has been widely publicized. Water is essential for life, and the maintenance of an ample supply of clean water is crucial to all of civilization.

To understand the chemistry that occurs in such diverse places as the human body, the groundwater, the oceans, the local water treatment plant, your hair as you shampoo it, and so on, we must understand how substances dissolved in water react with each other.

However, before we can understand solution reactions, we need to discuss the nature of solutions in which water is the dissolving medium, or *solvent*. These solutions are called **aqueous solutions.** In this chapter we will study the nature of materials after they are dissolved in water and various types of reactions that occur among these substances. You will see that the procedures developed in Chapter 3 to deal with chemical reactions work very well for reactions that take place in aqueous solutions. To understand the types of reactions that occur in aqueous solutions, we must first explore the types of species present. This requires an understanding of the nature of water.

4.1 Water, the Common Solvent

Water is one of the most important substances on earth. It is, of course, crucial for sustaining the reactions that keep us alive, but it also affects our lives in many indirect ways. Water helps moderate the earth's temperature; it cools automobile engines, nuclear power plants, and many industrial processes; it provides a means of transportation on the earth's surface and a medium for the growth of a myriad of creatures we use as food; and much more.

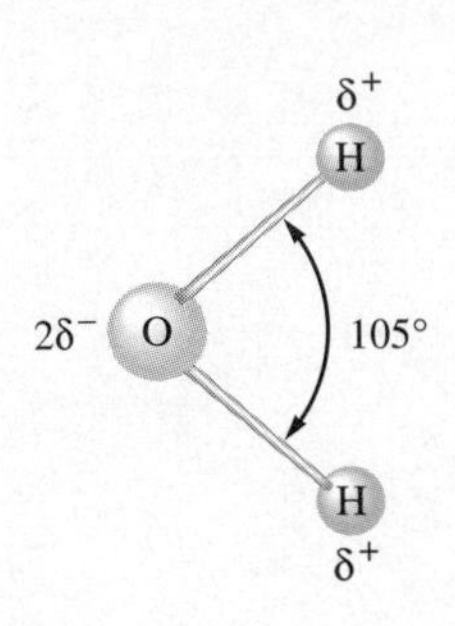

Figure 4.1
The water molecule is polar.

One of the most valuable functions of water involves its ability to dissolve many different substances. For example, salt disappears when you sprinkle it into the water used to cook vegetables, as does sugar when you add it to your iced tea. In each case the "disappearing" substance is obviously still present—you can taste it. What happens when a solid dissolves? To understand this process, we need to consider the nature of water. Liquid water consists of a collection of H_2O molecules. An individual H_2O molecule is "bent" or V-shaped, with an H—O—H angle of about 105°:

H 105° H
O

The O—H bonds in the water molecule are covalent bonds formed by electron sharing between the oxygen and hydrogen atoms. However, the electrons of the bond are not shared equally between these atoms. For reasons we will discuss in later chapters, oxygen has a greater attraction for electrons than does hydrogen. If the electrons were shared equally between the two atoms, both would be electrically neutral because, on the average, the number of electrons around each would equal the number of protons in that nucleus. However, because the oxygen atom has a greater attraction for electrons, the shared electrons tend to spend more time close to the oxygen than to either of the hydrogens. Thus the oxygen atom gains a slight excess of negative charge, and the hydrogen atoms become slightly positive. This is shown in Fig. 4.1, where δ (delta) indicates a *partial* charge *(less than one unit of charge)*. Because of this unequal charge distribution, water is said to be a **polar molecule.** It is this polarity that gives water its great ability to dissolve compounds.

A schematic of an ionic solid dissolving in water is shown in Fig. 4.2. Note that the "positive ends" of the water molecules are attracted to the negatively charged anions, and that the "negative ends" are attracted to the positively charged cations. This process is called **hydration.** The hydration of its ions tends to cause a salt to "fall apart" in the water, or to dissolve. The strong forces present among the positive and negative ions of the solid are replaced by strong water-ion interactions.

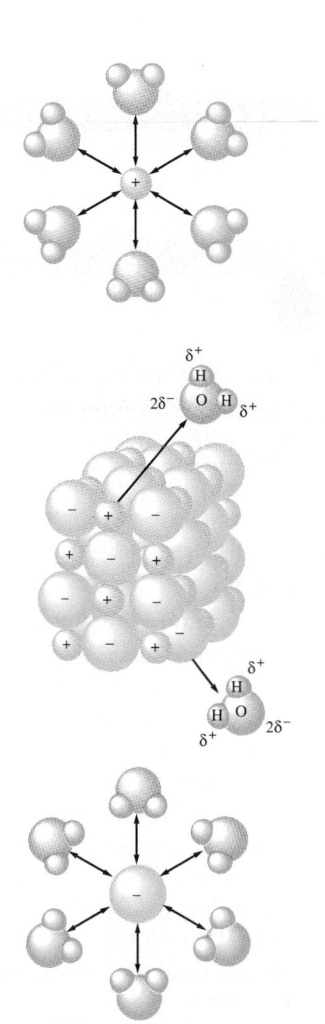

Figure 4.2
Polar water molecules interact with the positive and negative ions of a salt.

It is very important to recognize that when ionic substances (salts) dissolve in water, they break up into the *individual* cations and anions. For instance, when ammonium nitrate (NH_4NO_3) dissolves in water, the resulting solution contains NH_4^+ and NO_3^- ions floating around independently. This process can be represented as

$$NH_4NO_3(s) \xrightarrow{H_2O(l)} NH_4^+(aq) + NO_3^-(aq)$$

where (aq) designates that the ions are hydrated by unspecified numbers of water molecules.

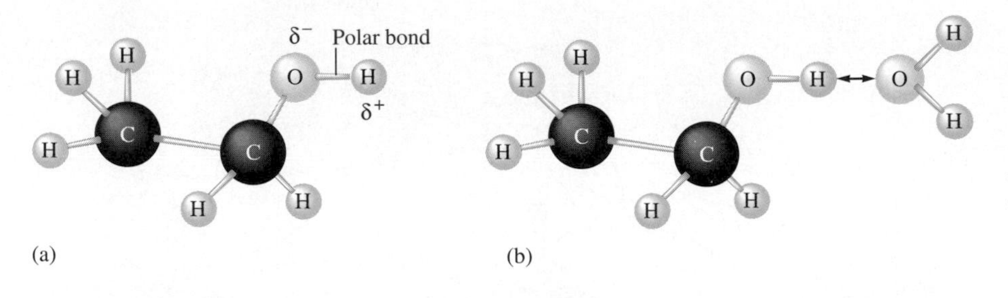

Figure 4.3
(a) The ethanol molecule contains a polar O—H bond similar to those in the water molecule. (b) The polar water molecule interacts strongly with the polar O—H bond in ethanol. This is a case of "like dissolving like."

The solubility of ionic substances in water varies greatly. For example, sodium chloride is quite soluble in water, whereas silver chloride (contains Ag^+ and Cl^- ions) is only very slightly soluble. The differences in the solubilities of ionic compounds in water typically depend on the relative affinities of the ions for each other (these forces hold the solid together) and the affinities of the ions for water molecules [which cause the solid to disperse (dissolve) in water]. Solubility is a complex issue that we will explore in much more detail in Chapter 17. However, the most important thing to remember at this point is that when an ionic solid does dissolve in water, the ions are dispersed and are assumed to move around independently.

Water also dissolves many nonionic substances. Ethanol (C_2H_5OH), for example, is very soluble in water. Wine, beer, and mixed drinks are aqueous solutions of alcohol and other substances. Why is ethanol so soluble in water? The answer lies in the structure of the alcohol molecule, which is shown in Fig. 4.3(a). The molecule contains a polar O—H bond like those in water, which makes it very compatible with water. The interaction of water with ethanol is represented in Fig. 4.3(b).

Many substances do not dissolve in water. Pure water will not, for example, dissolve animal fat, because fat molecules are nonpolar and do not interact effectively with polar water molecules. In general, polar and ionic substances are expected to be more soluble in water than nonpolar substances. "Like dissolves like" is a useful rule for predicting solubility. We will explore the basis for this generalization when we discuss the details of solution formation in Chapter 17.

4.2 The Nature of Aqueous Solutions: Strong and Weak Electrolytes

Recall that a solution is a homogeneous mixture. It is the same throughout (the first sip of a cup of coffee is the same as the last), but its composition can be varied by changing the amount of dissolved substances (one can make weak or strong coffee). In this section we will consider what happens when a substance, the **solute,** is dissolved in liquid water, the **solvent.**

One useful property for characterizing a solution is its **electrical conductivity,** its ability to conduct an electric current. This characteristic can be checked conveniently by using an apparatus like the one shown in Fig. 4.4. If the solution in the container conducts electricity, the bulb lights. Some solutions conduct current very efficiently, and the bulb shines very brightly; these solutions contain **strong electrolytes.** Other solutions conduct only a small current,

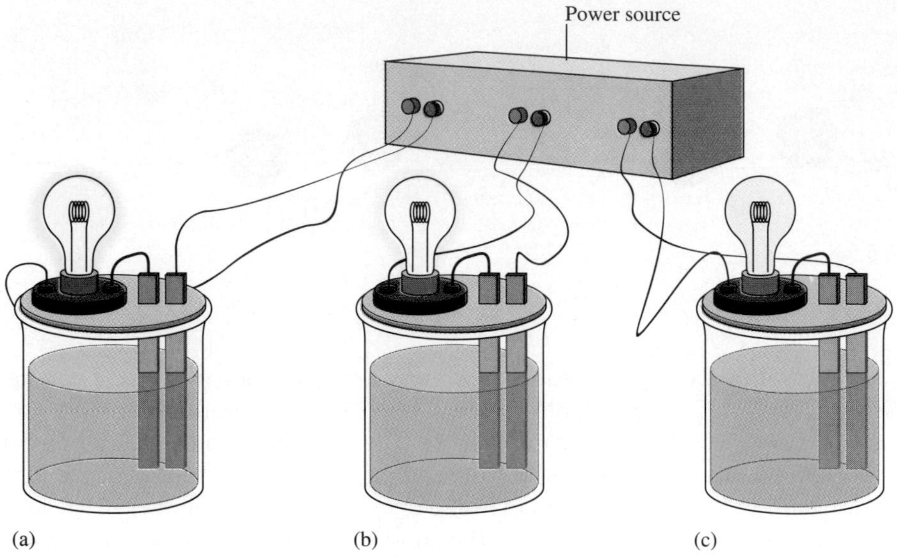

Power source

(a) (b) (c)

Figure 4.4
Electrical conductivity of aqueous solutions. The circuit will be completed and will allow current to flow only when there are charge carriers (ions) in the solution.
(a) Strong electrolyte in solution.
(b) Weak electrolyte in solution.
(c) Nonelectrolyte in solution.

and the bulb glows dimly; these solutions contain **weak electrolytes.** Some solutions permit no current to flow, and the bulb remains unlit; these solutions contain **nonelectrolytes.**

The basis for the conductivity properties of solutions was first correctly identified by Svante Arrhenius, then a Swedish graduate student in physics, who carried out research on the nature of solutions at the University of Uppsala in the early 1880s. Arrhenius came to believe that the conductivity of solutions arose from the presence of ions, an idea that was at first scorned by the majority of the scientific establishment. However, in the late 1890s when atoms were found to contain charged particles, the ionic theory suddenly made sense and became widely accepted.

As Arrhenius postulated, the extent to which a solution can conduct an electric current depends directly on the number of ions present. Some materials, such as sodium chloride, readily produce ions in aqueous solution and are thus strong electrolytes. Other substances, such as acetic acid, produce relatively few ions when dissolved in water and are weak electrolytes. A third class of materials, such as sugar, form virtually no ions when dissolved in water and are nonelectrolytes.

Strong Electrolytes

We will consider several classes of strong electrolytes: (1) soluble salts, (2) strong acids, and (3) strong bases.

As shown in Fig. 4.2, a salt consists of an array of cations and anions that separate and become hydrated when the salt dissolves. **Solubility** is usually measured in terms of the mass (grams) of solute that dissolves per given volume of solvent *or* in terms of the number of moles of solute that dissolve in a given volume of solution. Some salts, such as NaCl, KCl, and NH_4Cl, are very soluble in water. For example, approximately 357 grams of NaCl will dissolve in a liter of water at 25°C. On the other hand, many salts are only very slightly soluble in water; for example, silver chloride (AgCl) dissolves in water only to a slight extent (approximately 2×10^{-3} g/L at 25°C). We will consider only soluble salts at this point.

One of Arrhenius's most important discoveries concerned the nature of **acids.** Acidic behavior was first associated with the sour taste of citrus fruits. In fact, the word *acid* comes directly from the Latin word *acidus*, meaning "sour." The *mineral acids* sulfuric acid (H_2SO_4) and nitric acid (HNO_3), so named because they were originally obtained by the treatment of minerals, were discovered around 1300.

Acids were known to exist for hundreds of years before the time of Arrhenius, but no one had recognized their essential nature. In his studies of solutions, Arrhenius found that when the substances HCl, HNO_3, and H_2SO_4 were dissolved in water, they behaved as strong electrolytes. He postulated that this was the result of ionization reactions in water, for example:

$$\mathrm{HCl} \xrightarrow{\mathrm{H_2O}} \mathrm{H^+}(aq) + \mathrm{Cl^-}(aq)$$
$$\mathrm{HNO_3} \xrightarrow{\mathrm{H_2O}} \mathrm{H^+}(aq) + \mathrm{NO_3^-}(aq)$$
$$\mathrm{H_2SO_4} \xrightarrow{\mathrm{H_2O}} \mathrm{H^+}(aq) + \mathrm{HSO_4^-}(aq)$$

Thus Arrhenius proposed that an *acid is a substance that produces* H^+ *ions (protons) when it is dissolved in water.*

Studies of conductivity show that when HCl, HNO_3, and H_2SO_4 are placed in water, *virtually every molecule* dissociates to give ions. These substances are strong electrolytes and are thus called **strong acids.** All three are very important chemicals, and much more will be said about them as we proceed. However, at this point the following facts are important:

The Arrhenius definition of an acid: a substance that produces H^+ ions in solution.

Hydrochloric acid, nitric acid, and sulfuric acid are aqueous solutions and should be written in chemical equations as HCl(*aq*), HNO_3(*aq*), and H_2SO_4(*aq*), respectively, although they often appear without the (*aq*) symbol.

A strong acid is one that completely dissociates into its ions. Thus if 100 molecules of HCl are dissolved in water, 100 H^+ ions and 100 Cl^- ions are produced. Virtually no HCl molecules exist in aqueous solution.

Perchloric acid, $HClO_4$(*aq*), is another strong acid.

Sulfuric acid is a special case. The formula H_2SO_4 indicates that this acid can produce two H^+ ions per molecule when dissolved in water. However, only the first H^+ ion is completely dissociated. The second H^+ ion can be pulled off under certain conditions, which we will discuss later. Thus a solution of H_2SO_4 dissolved in water contains mostly H^+ ions and HSO_4^- ions.

Another important class of strong electrolytes is the **strong bases,** soluble compounds containing the *hydroxide ion* (OH^-) that completely dissociate when dissolved in water. Solutions containing bases have a bitter taste and a slippery feel. The most common basic solutions are those produced when solid sodium hydroxide (NaOH) or potassium hydroxide (KOH) is dissolved in water to produce ions, as follows:

Strong electrolytes dissociate completely in aqueous solution.

$$\mathrm{NaOH}(s) \xrightarrow{\mathrm{H_2O}} \mathrm{Na^+}(aq) + \mathrm{OH^-}(aq)$$
$$\mathrm{KOH}(s) \xrightarrow{\mathrm{H_2O}} \mathrm{K^+}(aq) + \mathrm{OH^-}(aq)$$

Weak Electrolytes

Weak electrolytes dissociate only to a small extent in aqueous solution.

Weak electrolytes are substances that produce relatively few ions when dissolved in water. The most common weak electrolytes are weak acids and weak bases.

The main acidic component of vinegar is acetic acid ($HC_2H_3O_2$). The formula is written to indicate that acetic acid has two chemically distinct types of hydrogen atoms. Formulas for acids are often written with the acidic hydrogen atom or atoms (any that will produce H^+ ions in solution) listed first. If any nonacidic hydrogens are present, they are written later in the formula. Thus the formula $HC_2H_3O_2$ indicates one acidic and three nonacidic hydrogen atoms. The dissociation reaction for acetic acid in water can be written as follows:

$$HC_2H_3O_2(aq) \xrightarrow{H_2O} H^+(aq) + C_2H_3O_2^-(aq)$$

Acetic acid is very different from the strong acids in that only about 1% of its molecules dissociate in aqueous solution. Thus when 100 molecules of $HC_2H_3O_2$ are dissolved in water, approximately 99 molecules of $HC_2H_3O_2$ remain intact, and only one H^+ ion and one $C_2H_3O_2^-$ ion are produced.

Because acetic acid is a weak electrolyte, it is called a **weak acid.** Any acid, such as acetic acid, that *dissociates only to a slight extent in aqueous solution is called a weak acid.* In Chapter 7 we will explore weak acids in detail.

The most common **weak base** is ammonia (NH_3). When ammonia is dissolved in water, it reacts as follows:

$$NH_3(aq) + H_2O(l) \longrightarrow NH_4^+(aq) + OH^-(aq)$$

The solution is *basic* since OH^- ions are produced. Ammonia is called a **weak base** because *the resulting solution is a weak electrolyte*—very few ions are present. In fact, for every 100 molecules of NH_3 that are dissolved, only one NH_4^+ ion and one OH^- ion are produced; 99 molecules of NH_3 remain unreacted.

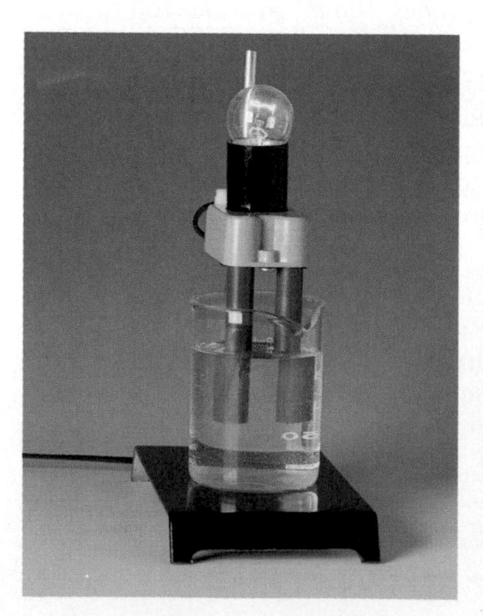

(left) Hydrochloric acid is a strong electrolyte—it readily conducts an electric current. (right) Acetic acid is a weak electrolyte as shown by the dimly lit bulb.

Nonelectrolytes

Nonelectrolytes are substances that dissolve in water but do not produce any ions. An example of a nonelectrolyte is ethanol (see Fig. 4.3 for the structural formula). When ethanol dissolves, entire C_2H_5OH molecules are dispersed in the water. Since the molecules do not break up into ions, the resulting solution does not conduct an electric current. Another common nonelectrolyte is table sugar (sucrose, $C_{12}H_{22}O_{11}$), which is very soluble in water but produces no ions when it dissolves.

4.3 The Composition of Solutions

Chemical reactions often take place when two solutions are mixed. To perform stoichiometric calculations in such cases, we must know two things: (1) the *nature of the reaction,* which depends on the exact forms the chemicals take when dissolved; and (2) the *amounts of chemicals* present in the solutions, that is, the composition of each solution.

The composition of a solution can be described in many different ways, as we will see in Chapter 17. At this point we will consider only the most commonly used expression of concentration, **molarity** (M), which is defined as *moles of solute per volume of solution (expressed in liters):*

$$M = \text{molarity} = \frac{\text{moles of solute}}{\text{liters of solution}}$$

A solution that is 1.0 molar (written as 1.0 M) contains 1.0 mole of solute per liter of solution.

EXAMPLE 4.1

Calculate the molarity of a solution prepared by bubbling 1.56 g of gaseous HCl into enough water to make 26.8 mL of solution.

Solution

First, we calculate the number of moles of HCl:

$$1.56 \text{ g HCl} \times \frac{1 \text{ mol HCl}}{36.5 \text{ g HCl}} = 4.27 \times 10^{-2} \text{ mol HCl}$$

Then we change the volume of the solution to liters,

$$26.8 \text{ mL} \times \frac{1 \text{ L}}{1000 \text{ mL}} = 2.68 \times 10^{-2} \text{ L}$$

and divide the moles of solute by the liters of solution:

$$\text{Molarity} = \frac{4.27 \times 10^{-2} \text{ mol HCl}}{2.68 \times 10^{-2} \text{ L solution}} = 1.59\ M \text{ HCl}$$

Note that the description of a solution's composition may not accurately reflect the true chemical nature of the solution. Solution concentration is always given in terms of the form of the solute *before* it dissolves. For example, when a solution is described as being 1.0 M NaCl, this means that the solution was prepared by dissolving 1.0 mole of solid NaCl in enough water to make 1.0 liter of solution; it does not mean that the solution contains 1.0 mole of NaCl units. Actually, the solution contains 1.0 mole of Na^+ ions and 1.0 mole of Cl^- ions.

Often we need to determine the number of moles of solute present in a given volume of a solution of known molarity. The procedure for doing so is easily derived from the definition of molarity:

$$M = \frac{\text{moles of solute}}{\text{liters of solution}}$$

$$\text{Liters of solution} \times \text{molarity} = \text{liters of solution} \times \frac{\text{moles of solute}}{\text{liters of solution}}$$
$$= \text{moles of solute}$$

EXAMPLE 4.2

Calculate the number of moles of Cl^- ions in 1.75 L of $1.0 \times 10^{-3}\,M$ $AlCl_3$.

Solution

When solid $AlCl_3$ dissolves, it produces ions as follows:

$$AlCl_3(s) \xrightarrow{H_2O} Al^{3+}(aq) + 3Cl^-(aq)$$

Thus a $1.0 \times 10^{-3}\,M$ $AlCl_3$ solution contains $1.0 \times 10^{-3}\,M$ Al^{3+} ions and $3.0 \times 10^{-3}\,M$ Cl^- ions.

To calculate the moles of Cl^- ions in 1.75 L of the $1.0 \times 10^{-3}\,M$ $AlCl_3$ solution, we must multiply the volume times the molarity:

$$1.75\text{ L solution} \times 3.0 \times 10^{-3}\,M\text{ Cl}^-$$
$$= 1.75\text{ L solution} \times \frac{3.0 \times 10^{-3}\text{ mol Cl}^-}{\text{L solution}}$$
$$= 5.3 \times 10^{-3}\text{ mol Cl}^-$$

EXAMPLE 4.3

Typical blood serum is about 0.14 M NaCl. What volume of blood contains 1.0 mg of NaCl?

Solution

We must first determine the number of moles represented by 1.0 mg of NaCl:

$$1.0\text{ mg NaCl} \times \frac{1\text{ g NaCl}}{1000\text{ mg NaCl}} \times \frac{1\text{ mol NaCl}}{58.45\text{ g NaCl}}$$
$$= 1.7 \times 10^{-5}\text{ mol NaCl}$$

Next, we must determine what volume of 0.14 M NaCl solution contains 1.7×10^{-5} mol of NaCl. There is some volume, call it V, that when multiplied by the molarity of this solution yields 1.7×10^{-5} mol of NaCl; that is:

$$V \times \frac{0.14 \text{ mol NaCl}}{\text{L solution}} = 1.7 \times 10^{-5} \text{ mol NaCl}$$

Solving for the volume gives

$$V = \frac{1.7 \times 10^{-5} \text{ mol NaCl}}{\dfrac{0.14 \text{ mol NaCl}}{\text{L solution}}}$$

$$= 1.2 \times 10^{-4} \text{ L solution}$$

Thus 0.12 mL of blood contains 1.7×10^{-5} mol of NaCl, or 1.0 mg of NaCl.

A **standard solution** is a solution *whose concentration is accurately known*. Standard solutions, often used in chemical analysis, can be prepared as shown in Fig. 4.5 and in Example 4.4.

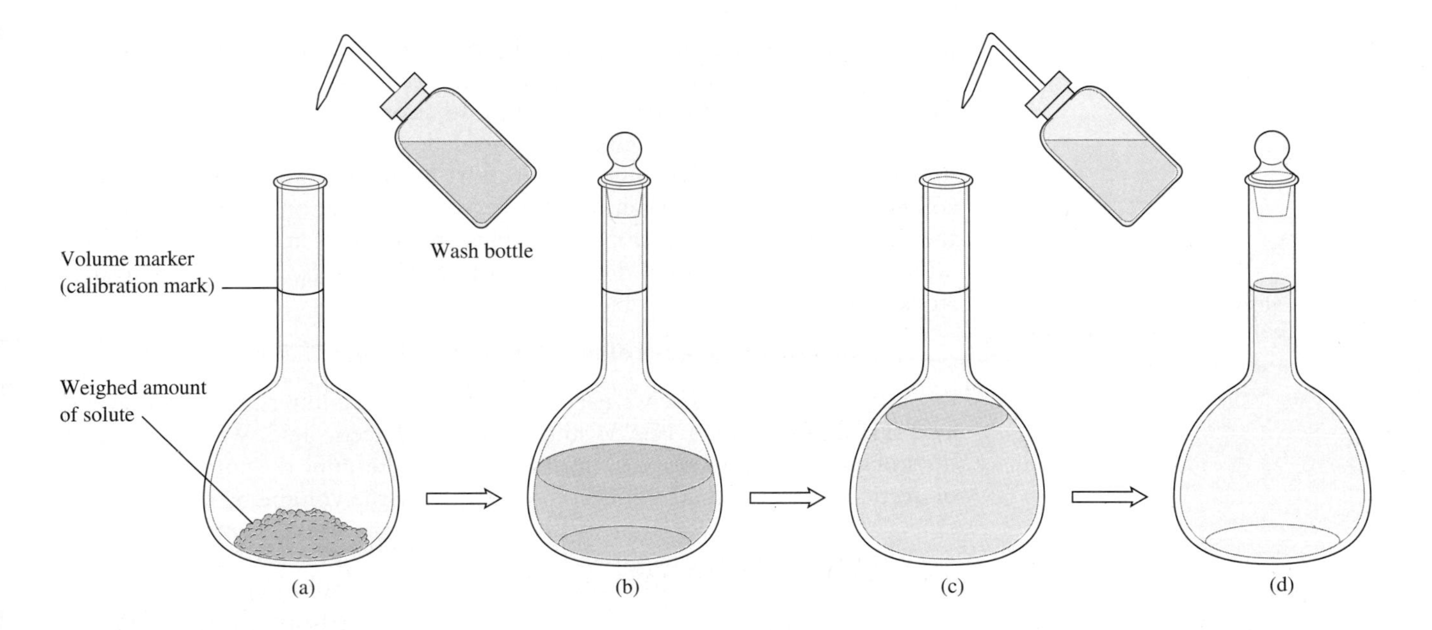

Figure 4.5
Steps involved in the preparation of a standard solution. (a) A weighed amount of a substance (the solute) is put into the volumetric flask, and (b) a small quantity of water is added. (c) The solid is dissolved in the water by gently swirling the flask (*with the stopper in place*). (d) More water is added, until the level of the solution just reaches the mark etched on the neck of the flask.

EXAMPLE 4.4

To analyze the alcohol content of a certain wine, a chemist needs 1.00 L of an aqueous 0.200 *M* $K_2Cr_2O_7$ (potassium dichromate) solution. How much solid $K_2Cr_2O_7$ must be weighed out to make this solution?

Solution

First, determine the moles of $K_2Cr_2O_7$ required:

$$1.00 \text{ L solution} \times \frac{0.200 \text{ mol } K_2Cr_2O_7}{\text{L solution}} = 0.200 \text{ mol } K_2Cr_2O_7$$

This amount can be converted to grams by using the molar mass of $K_2Cr_2O_7$:

$$0.200 \text{ mol } K_2Cr_2O_7 \times \frac{294.2 \text{ g } K_2Cr_2O_7}{\text{mol } K_2Cr_2O_7} = 58.8 \text{ g } K_2Cr_2O_7$$

Thus to make 1.00 L of 0.200 *M* $K_2Cr_2O_7$, the chemist must weigh out 58.8 g of $K_2Cr_2O_7$, put it in a 1.00-L volumetric flask, and add water up to the mark on the flask.

Dilution

To save time and space in the laboratory, routinely used solutions are often purchased or prepared in concentrated form (these are called *stock solutions*). Water is then added to achieve the molarity desired for a particular solution. This process is called **dilution.** For example, the common acids are purchased as concentrated solutions and diluted as needed. A typical dilution calculation involves determining how much water must be added to an amount of stock solution to achieve a solution of the desired concentration. The key to doing these calculations is to remember that since only water is added in the dilution, all of the solute in the final dilute solution must come from the concentrated stock solution. That is,

Dilution with water doesn't alter the number of moles of solute present.

$$\text{Moles of solute after dilution} = \text{moles of solute before dilution}$$

For example, suppose we need to prepare 500. milliliters of 1.00 *M* acetic acid ($HC_2H_3O_2$) from a 17.5 *M* stock solution of acetic acid. What volume of the stock solution is required? The first step is to determine the number of moles of acetic acid in the final solution by multiplying the volume by the molarity:

$$500. \text{ mL solution} \times \frac{1 \text{ L solution}}{1000 \text{ mL solution}} \times \frac{1.00 \text{ mol } HC_2H_3O_2}{\text{L solution}} = 0.500 \text{ mol } HC_2H_3O_2$$

Thus we need to use a volume of 17.5 *M* acetic acid that contains 0.500 mole of $HC_2H_3O_2$, that is,

$$V \times \frac{17.5 \text{ mol } HC_2H_3O_2}{\text{L solution}} = 0.500 \text{ mol } HC_2H_3O_2$$

Solving for V gives

$$V = \frac{0.500 \text{ mol } HC_2H_3O_2}{\dfrac{17.5 \text{ mol } HC_2H_3O_2}{\text{L solution}}} = 0.0287 \text{ L or } 28.7 \text{ mL solution}$$

Thus to make 500. milliliters of a 1.00 M acetic acid solution, we can take 28.7 milliliters of 17.5 M acetic acid and dilute it to a total volume of 500 milliliters.

A dilution procedure typically involves two types of glassware: a pipet and a volumetric flask. A pipet is a device for accurately measuring and transferring a given volume of solution. There are two common types of pipets: *measuring pipets* and *volumetric pipets,* as shown in Fig. 4.6. Measuring pipets are used to measure out volumes for which a volumetric pipet is not available. For example, we would use a measuring pipet as shown in Fig. 4.7 to deliver 28.7 mL of 17.5 M acetic acid into a 500-mL volumetric flask and then add water to the mark to perform the dilution described above.

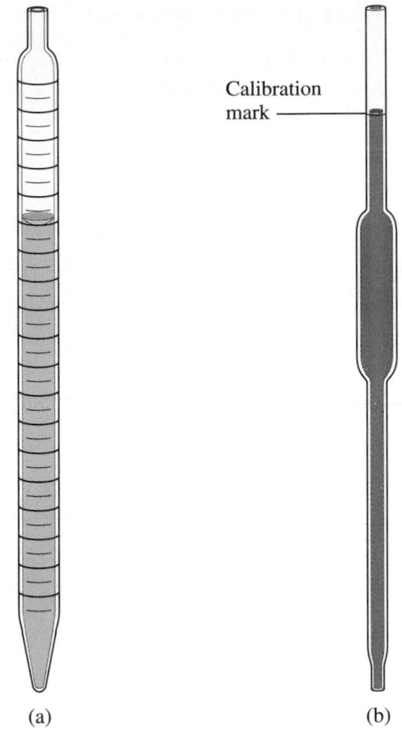

4.4 Types of Chemical Reactions

Although we have considered many reactions so far, we have examined only a tiny fraction of the millions of possible chemical reactions. In order to make sense of all these reactions, we need some system for grouping reactions into classes. Although there are many different ways to do this, we will use the system most commonly used by practicing chemists. They divide reactions into the following groups: **precipitation reactions, acid-base reactions,** and **oxidation-reduction reactions.**

Calibration mark

(a) (b)

Figure 4.6
(a) A measuring pipet is graduated and can be used to measure various volumes of liquid accurately. (b) A volumetric pipet is designed to measure *one* volume accurately. When filled to the calibration mark, it delivers the volume indicated on the pipet (in this case, 20 mL).

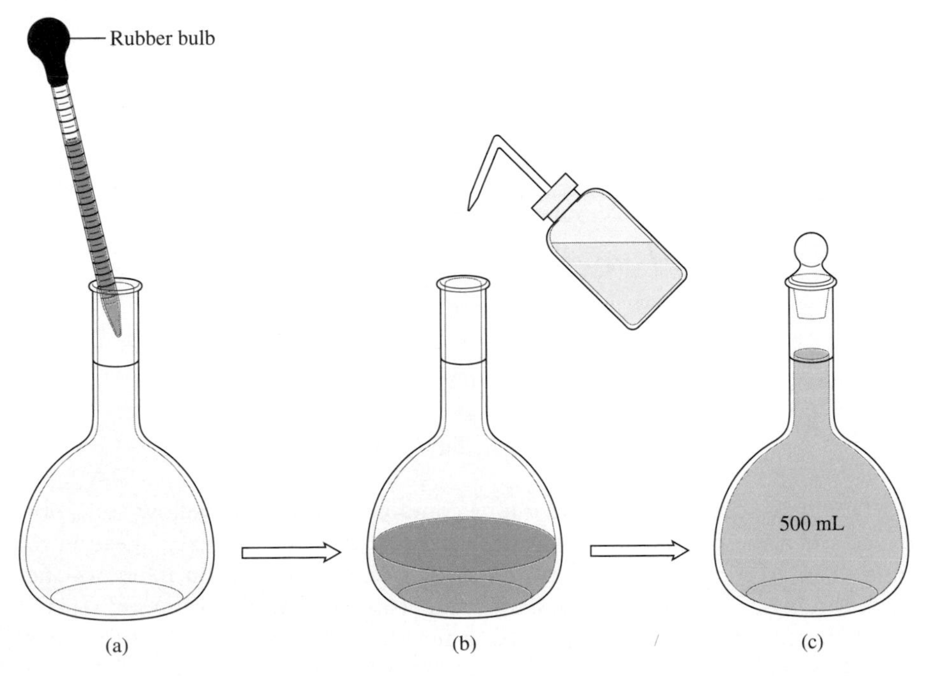

Figure 4.7
(a) A measuring pipet is used to transfer 28.6 mL of 17.5 M acetic acid solution to a volumetric flask. (b) Water is added to the flask to the calibration mark. (c) The resulting solution is 1 M acetic acid.

Types of Solution Reactions

Precipitation reactions
Acid-base reactions
Oxidation-reduction reactions

Virtually all reactions can be placed into one of these classes. We will define and illustrate each type in the following sections.

4.5 Precipitation Reactions

When two solutions are mixed, an insoluble substance sometimes forms; that is, a solid forms and separates from the solution. Such a reaction is called a **precipitation reaction** and the solid that forms is called a **precipitate.** For example, a precipitation reaction occurs when an aqueous solution of potassium chromate, $K_2CrO_4(aq)$, which is yellow, is mixed with a colorless aqueous solution containing barium nitrate, $Ba(NO_3)_2(aq)$. As shown in Fig. 4.8, when these solutions are mixed, a yellow solid forms. What is the equation that describes this chemical change? To write the equation we must know the identities of the reactants and products. The reactants have already been described: $K_2CrO_4(aq)$ and $Ba(NO_3)_2(aq)$. Is there some way we can predict the identities of the products? In particular, what is the yellow solid?

The best way to predict the identity of this solid is to think carefully about what products are possible. To do so, we need to know what species are present in the solution formed when the reactant solutions are mixed. First, let's think about the nature of each reactant solution. The designation $Ba(NO_3)_2(aq)$ means that barium nitrate (a white solid) has been dissolved in water. Notice that barium nitrate contains the Ba^{2+} and NO_3^- ions. ***Remember:*** **In virtually every case, when a solid containing ions dissolves in water, the ions separate** and move around independently. That is, $Ba(NO_3)_2(aq)$ does not contain $Ba(NO_3)_2$ units; it contains separated Ba^{2+} and NO_3^- ions.

When ionic compounds dissolve in water, the *resulting solution contains the separated ions.*

Similarly, since solid potassium chromate contains K^+ and CrO_4^{2-} ions, an aqueous solution of potassium chromate (which is prepared by dissolving solid K_2CrO_4 in water) contains these separated ions.

We can represent the mixing of $K_2CrO_4(aq)$ and $Ba(NO_3)_2(aq)$ in two ways. First, we can write

$$K_2CrO_4(aq) + Ba(NO_3)_2(aq) \longrightarrow \text{products}$$

However, a much more accurate representation is

$$\underbrace{2K^+(aq) + CrO_4^{2-}(aq)}_{\text{The ions in } K_2CrO_4(aq)} + \underbrace{Ba^{2+}(aq) + 2NO_3^-(aq)}_{\text{The ions in } Ba(NO_3)_2(aq)} \longrightarrow \text{products}$$

Thus the mixed solution contains the ions

$$K^+, CrO_4^{2-}, Ba^{2+}, NO_3^-$$

Figure 4.8
When aqueous barium nitrate is added to aqueous potassium chromate, yellow barium chromate precipitates.

How can some or all of these ions combine to form the yellow solid observed when the original solutions are mixed? This is not an easy question to answer. In fact, predicting the products of a chemical reaction is one of the hardest things a beginning chemistry student is asked to do. Even an experienced chemist, when confronted with a new reaction, is often not sure what will happen. The chemist tries to think of the various possibilities, considers the likelihood of each possibility, and then makes a prediction (an educated guess). Only after identifying each product *experimentally* is the chemist sure what reaction has taken place. However, an educated guess is useful because it provides a place to start. It tells us what kinds of products we are most likely to find.

We already know some things that will help us predict the products.

1. When ions form a solid compound, the compound must have a zero net charge. Thus the products of this reaction must contain *both anions and cations*. For example, K^+ and Ba^{2+} could not combine to form the solid, nor could CrO_4^{2-} and NO_3^-.

2. Most ionic materials contain only two types of ions: one type of cation, and one type of anion (for example, NaCl, KOH, Na_2SO_4, K_2CrO_4, $Co(NO_3)_2$, NH_4Cl, Na_2CO_3).

The possible combinations of a given cation and a given anion from the list of ions K^+, CrO_4^{2-}, Ba^{2+}, NO_3^- are

$$K_2CrO_4,\ KNO_3,\ BaCrO_4,\ Ba(NO_3)_2$$

Which of these possibilities is most likely to represent the yellow solid? We know it's not K_2CrO_4 or $Ba(NO_3)_2$. They are the reactants. They were present (dissolved) in the separate solutions that were mixed. The only real possibilities for the solid that formed are

$$KNO_3 \quad \text{and} \quad BaCrO_4$$

To decide which of these possibilities most likely represents the yellow solid, we need more facts. An experienced chemist knows that the K^+ ion and the NO_3^- ion are both colorless. Thus if the solid is KNO_3, it should be white, not yellow. On the other hand, the CrO_4^{2-} ion is yellow (note in Fig. 4.8 that $K_2CrO_4(aq)$ is yellow). Thus the yellow solid is almost certainly $BaCrO_4$. Further tests show that this is the case.

So far we have determined that one product of the reaction between $K_2CrO_4(aq)$ and $Ba(NO_3)_2(aq)$ is $BaCrO_4(s)$, but what happened to the K^+ and NO_3^- ions? The answer is that these ions are left dissolved in the solution. That is, KNO_3 does not form a solid when the K^+ and NO_3^- ions are present in this much water. In other words, if we took the white solid, $KNO_3(s)$, and put it in the same quantity of water as is present in the mixed solution, it would dissolve. Thus when we mix $K_2CrO_4(aq)$ and $Ba(NO_3)_2(aq)$, $BaCrO_4(s)$ forms, but KNO_3 is left behind in solution (we write it as $KNO_3(aq)$). Therefore the equation for this precipitation reaction is

$$K_2CrO_4(aq) + Ba(NO_3)_2(aq) \longrightarrow BaCrO_4(s) + KNO_3(aq)$$

If we removed the solid $BaCrO_4$ by filtration and then evaporated the water, the white solid, KNO_3, would be obtained.

Now let's consider another example. When an aqueous solution of silver nitrate is added to an aqueous solution of potassium chloride, a white precipitate forms. We can represent what we know so far as

$$AgNO_3(aq) + KCl(aq) \longrightarrow \text{unknown white solid}$$

Remembering that when ionic substances dissolve in water, the ions separate, we can write

$$\underbrace{Ag^+, NO_3^-}_{\text{In silver nitrate solution}} + \underbrace{K^+, Cl^-}_{\text{In potassium chloride solution}} \longrightarrow \underbrace{Ag^+, NO_3^-, K^+, Cl^-}_{\text{Combined solution, before reaction}} \longrightarrow \text{white solid}$$

Since we know the white solid must contain both positive and negative ions, the possible compounds that can be assembled from this collection of ions are

$$AgNO_3, KCl, AgCl, KNO_3$$

Since $AgNO_3$ and KCl are the substances dissolved in the reactant solutions, we know that they do not represent the white solid product. The only real possibilities are

$$AgCl \quad \text{and} \quad KNO_3$$

Doing chemistry requires both understanding ideas and remembering facts.

From the example considered above we know that KNO_3 is quite soluble in water. Thus solid KNO_3 will not form when the reactant solutions are mixed. The product must be AgCl(*s*) (which can be proved by experiment). The equation for the reaction now can be written:

$$AgNO_3(aq) + KCl(aq) \longrightarrow AgCl(s) + KNO_3(aq)$$

Notice that to do these two examples, we had to know both concepts (solids always have a zero net charge) and facts (KNO_3 is very soluble in water, the CrO_4^{2-} is yellow, and so on).

Predicting the identity of the solid product in a precipitation reaction requires knowledge of the solubilities of common ionic substances. As an aid in predicting the products of precipitation reactions, some simple solubility rules are given in Table 4.1. You should memorize these rules.

The phrase "slightly soluble" used in the solubility rules in Table 4.1 means that the tiny amount of solid that dissolves is not noticeable. The solid appears

Figure 4.9
Precipitation of silver chloride by mixing solutions of silver nitrate and potassium chloride. The K^+ and NO_3^- ions remain in solution.

TABLE 4.1 **Simple Rules for Solubility of Salts in Water**

1. Most nitrate (NO_3^-) salts are soluble.
2. Most salts of Na^+, K^+, and NH_4^+ are soluble.
3. Most chloride salts are soluble. Notable exceptions are AgCl, $PbCl_2$, and Hg_2Cl_2.
4. Most sulfate salts are soluble. Notable exceptions are $BaSO_4$, $PbSO_4$, and $CaSO_4$.
5. Most hydroxide salts are only slightly soluble. The important soluble hydroxides are NaOH, KOH, and $Ca(OH)_2$ (marginally soluble).
6. Most sulfide (S^{2-}), carbonate (CO_3^{2-}), and phosphate (PO_4^{3-}) salts are only slightly soluble.

to be insoluble to the naked eye. Thus the terms *insoluble* and *slightly soluble* are often used interchangeably.

Note that the information in Table 4.1 allows us to predict that AgCl is the white solid formed when solutions of $AgNO_3$ and KCl are mixed; Rules 1 and 2 indicate that KNO_3 is soluble and Rule 3 states that AgCl is (virtually) insoluble. Figure 4.9 shows the results of mixing silver nitrate and potassium chloride solutions.

When solutions containing ionic substances are mixed, it will be helpful in determining the products if you think in terms of *ion interchange*. For example, in the above discussion we considered the results of mixing $AgNO_3(aq)$ and $KCl(aq)$. In determining the products, we took the cation from one reactant and combined it with the anion of the other reactant:

$$Ag^+ \quad + \quad NO_3^- \quad + \quad K^+ \quad + \quad Cl^- \longrightarrow$$

Possible solid products

The solubility rules in Table 4.1 allow us to predict whether either product forms as a solid.

The key to dealing with the chemistry of an aqueous solution is first to *focus on the actual components of the solution before any reaction occurs* and then figure out how those components will react with each other. Example 4.5 illustrates this process for three different reactions.

Focus on the ions in solution before any reaction occurs.

EXAMPLE 4.5

Using the solubility rules in Table 4.1, predict what will happen when the following pairs of solutions are mixed.

a. $KNO_3(aq)$ and $BaCl_2(aq)$

b. $Na_2SO_4(aq)$ and $Pb(NO_3)_2(aq)$

c. $KOH(aq)$ and $Fe(NO_3)_3(aq)$

Solution

a. $KNO_3(aq)$ stands for an aqueous solution obtained by dissolving solid KNO_3 in water to form a solution containing the hydrated ions $K^+(aq)$ and $NO_3^-(aq)$. Likewise, $BaCl_2(aq)$ is a solution formed by dissolving solid $BaCl_2$ in water to produce $Ba^{2+}(aq)$ and $Cl^-(aq)$. When these two solutions are mixed, the resulting solution contains the ions K^+, NO_3^-, Ba^{2+}, and Cl^-. All will be hydrated, but (aq) is omitted for simplicity. To look for possible solid products, combine the cation from one reactant with the anion from the other:

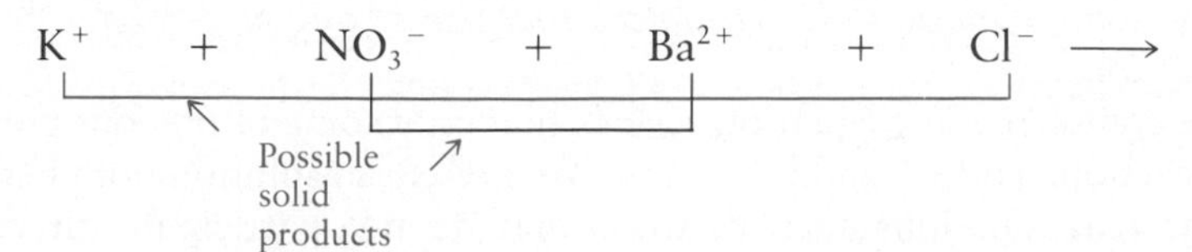

Lead sulfate is a white solid.

Solid $Fe(OH)_3$ forms when aqueous KOH and $Fe(NO_3)_3$ are mixed.

Note from Table 4.1 that the rules predict that both KCl and $Ba(NO_3)_2$ are soluble in water. Thus no precipitate will form when $KNO_3(aq)$ and $BaCl_2(aq)$ are mixed. All of the ions will remain dissolved in the solution. No reaction occurs.

b. Using the same procedures as in part a, we find that the ions present in the combined solution before any reaction occurs are Na^+, SO_4^{2-}, Pb^{2+}, and NO_3^-. The possible salts that could form precipitates are

$$Na^+ \quad + \quad SO_4^{2-} \quad + \quad Pb^{2+} \quad + \quad NO_3^- \longrightarrow$$

The compound $NaNO_3$ is soluble, but $PbSO_4$ is insoluble (see Rule 4 in Table 4.1). When these solutions are mixed, $PbSO_4$ will precipitate from the solution. The balanced equation is

$$Na_2SO_4(aq) + Pb(NO_3)_2(aq) \longrightarrow PbSO_4(s) + 2NaNO_3(aq)$$

c. The combined solution (before any reaction occurs) contains the ions K^+, OH^-, Fe^{3+}, and NO_3^-. The salts that might precipitate are KNO_3 and $Fe(OH)_3$. The solubility rules in Table 4.1 indicate that both K^+ and NO_3^- salts are soluble. However, $Fe(OH)_3$ is only slightly soluble (Rule 5) and hence will precipitate. The balanced equation is

$$3KOH(aq) + Fe(NO_3)_3(aq) \longrightarrow Fe(OH)_3(s) + 3KNO_3(aq)$$

4.6 Describing Reactions in Solution

In this section we will consider the types of equations used to represent reactions in solution. For example, when we mix aqueous potassium chromate with aqueous barium nitrate, a reaction occurs to form a precipitate ($BaCrO_4$) and dissolved potassium nitrate. So far we have written the **molecular equation** for this reaction:

$$K_2CrO_4(aq) + Ba(NO_3)_2(aq) \longrightarrow BaCrO_4(s) + 2KNO_3(aq)$$

Although this equation shows the reactants and products of the reaction, it does not give a very clear picture of what actually occurs in solution. As we have seen, aqueous solutions of potassium chromate, barium nitrate, and potassium nitrate contain the individual ions, not molecules, as is implied by the molecular equation. Thus the **complete ionic equation**

$$2K^+(aq) + CrO_4^{2-}(aq) + Ba^{2+}(aq) + 2NO_3^-(aq) \longrightarrow BaCrO_4(s) + 2K^+(aq) + 2NO_3^-(aq)$$

A strong electrolyte is a substance that completely breaks apart into ions when dissolved in water.

better represents the actual forms of the reactants and products in solution. *In a complete ionic equation all substances that are strong electrolytes are represented as ions.*

The complete ionic equation reveals that only some of the ions participate in the reaction. The K^+ and NO_3^- ions are present in solution both before and after the reaction. Ions such as these that do not participate directly in a

EXAMPLE 4.6

For each of the following reactions, write the molecular equation, the complete ionic equation, and the net ionic equation.

a. Aqueous potassium chloride is added to aqueous silver nitrate to form a silver chloride precipitate plus aqueous potassium nitrate.
b. Aqueous potassium hydroxide is mixed with aqueous iron(III) nitrate to form a precipitate of iron(III) hydroxide and aqueous potassium nitrate.

Solution

a. *Molecular:*

$$KCl(aq) + AgNO_3(aq) \longrightarrow AgCl(s) + KNO_3(aq)$$

Complete ionic (remember that any ionic compound dissolved in water will be present as the separated ions):

$$\underset{\substack{\uparrow \\ \text{Spectator} \\ \text{ion}}}{K^+(aq)} + Cl^-(aq) + Ag^+(aq) + \underset{\substack{\uparrow \\ \text{Spectator} \\ \text{ion}}}{NO_3^-(aq)} \longrightarrow$$

$$\underset{\substack{\uparrow \\ \text{Solid,} \\ \text{not} \\ \text{written} \\ \text{as separate ions}}}{AgCl(s)} + \underset{\substack{\uparrow \\ \text{Spectator} \\ \text{ion}}}{K^+(aq)} + \underset{\substack{\uparrow \\ \text{Spectator} \\ \text{ion}}}{NO_3^-(aq)}$$

Net ionic: Canceling the spectator ions,

$$\cancel{K^+}(aq) + Cl^-(aq) + Ag^+(aq) + \cancel{NO_3^-}(aq)$$
$$\longrightarrow AgCl(s) + \cancel{K^+}(aq) + \cancel{NO_3^-}(aq)$$

gives the following net ionic equation:

$$Cl^-(aq) + Ag^+(aq) \longrightarrow AgCl(s)$$

b. *Molecular:*

$$3KOH(aq) + Fe(NO_3)_3(aq) \longrightarrow Fe(OH)_3(s) + 3KNO_3(aq)$$

Complete ionic:

$$3K^+(aq) + 3OH^-(aq) + Fe^{3+}(aq) + 3NO_3^-(aq)$$
$$\longrightarrow Fe(OH)_3(s) + 3K^+(aq) + 3NO_3^-(aq)$$

Net ionic:

$$3OH^-(aq) + Fe^{3+}(aq) \longrightarrow Fe(OH)_3(s)$$

reaction in solution are called **spectator ions.** The ions that participate in this reaction are the Ba^{2+} and CrO_4^{2-} ions, which combine to form solid $BaCrO_4$:

$$Ba^{2+}(aq) + CrO_4^{2-}(aq) \longrightarrow BaCrO_4(s)$$

Net ionic equations include only those components that undergo changes in the reaction.

This equation, called the **net ionic equation,** includes only those solution components directly involved in the reaction. Chemists usually write the net ionic equation for a reaction in solution because it gives the actual forms of the reactants and products and includes only the species that undergo a change.

Summary: Three Types of Equations Used to Describe Reactions in Solution

- **The *molecular equation* gives the overall reaction stoichiometry but not necessarily the actual forms of the reactants and products in solution.**
- **The *complete ionic equation* represents as ions all reactants and products that are strong electrolytes.**
- **The *net ionic equation* includes only those solution components undergoing a change. Spectator ions are not included.**

4.7 Selective Precipitation

We can use the fact that salts have different solubilities to separate mixtures of ions. For example, suppose we have an aqueous solution containing the cations Ag^+, Ba^{2+} and Fe^{3+}, and the anion NO_3^-. We want to separate the cations by precipitating them one at a time, a process called **selective precipitation.**

How can the separation of these cations be accomplished? We can perform some preliminary tests and observe the reactivity of each cation toward the anions Cl^-, SO_4^{2-}, and OH^-. For example, to test the reactivity of Ag^+ toward Cl^-, we can mix the $AgNO_3$ solution with aqueous KCl or NaCl. As we have seen, this produces a precipitate. When we carry out tests of this type using all of the possible combinations, we obtain the results in Table 4.2.

TABLE 4.2 **Testing the Reactivity of the Cations Ag^+, Ba^{2+}, and Fe^{3+} with the Anions Cl^-, SO_4^{2-}, and OH^-**

	Test Solution (anion)		
Cation	NaCl(*aq*) (Cl^-)	Na_2SO_4(*aq*) (SO_4^{2-})	NaOH(*aq*) (OH^-)
Ag^+	White precipitate (AgCl)	No reaction	White precipitate that turns brown ($AgOH \rightarrow Ag_2O$, White → Brown)
Ba^{2+}	No reaction	White precipitate ($BaSO_4$)	No reaction
Fe^{3+}	Yellow color but no solid	No reaction	Reddish brown precipitate [$Fe(OH)_3$]

CHEMICAL ANALYSIS OF COCKROACHES

Cockroaches can be a big problem. Not only are these hardy pests unpleasant to live with, but they also consume significant quantities of the world's precious food and grain supplies. Because the many different species of cockroaches require different control measures, determining which species is causing a particular problem is important. Careful examination of a cockroach can reveal its species, but this is very time-consuming. However, a new method of cockroach identification, based on gas chromatography, has been developed at the U.S. Department of Agriculture by D. A. Carlson and R. J. Brenner. In gas chromatography the compounds to be separated are dispersed in a carrier gas that passes through a porous solid. Because different substances have differing tendencies to adhere to the solid, the components of the mixture travel at different rates through the system, causing them to spread out so they can be separated and identified.

In the cockroach identification study Carlson and Brenner found that the composition of the outer, waxy layer of a roach is distinct to a particular species. Thus by dissolving this waxy coating and injecting it into the gas stream of a gas chromatograph, scientists can identify the cockroach unambiguously in less than half an hour. This technique is particularly useful for identifying hybrid Asian-German cockroaches, which have become a major problem for the food industry.

Although biologists might argue that the gas chromatographic method takes the fun and the challenge out of identifying cockroaches, this technique should lead to significant advances in the control of these insects.

After studying these results, we might proceed to separate the cations as follows:

STEP 1

Add an aqueous solution of NaCl to the solution containing the Ag^+, Ba^{2+}, and Fe^{3+} ions. Solid AgCl will form and can be removed, leaving Ba^{2+} and Fe^{3+} ions in solution.

STEP 2

Add an aqueous solution of Na_2SO_4 to the solution containing the Ba^{2+} and Fe^{3+} ions. Solid $BaSO_4$ will form and can be removed, leaving only Fe^{3+} ions in solution.

STEP 3

Add an aqueous solution of NaOH to the solution containing the Fe^{3+} ions. Solid $Fe(OH)_3$ will form and can be removed.

Steps 1–3 are represented schematically in Fig. 4.10.

Note that adding the anions in this order precipitates the cations one at a time and thus separates them. The process whereby mixtures of ions are separated and identified is called **qualitative analysis.** In this example the qualitative analysis was carried out by selective precipitation, but it can also be accomplished by using other separation techniques that will not be discussed here.

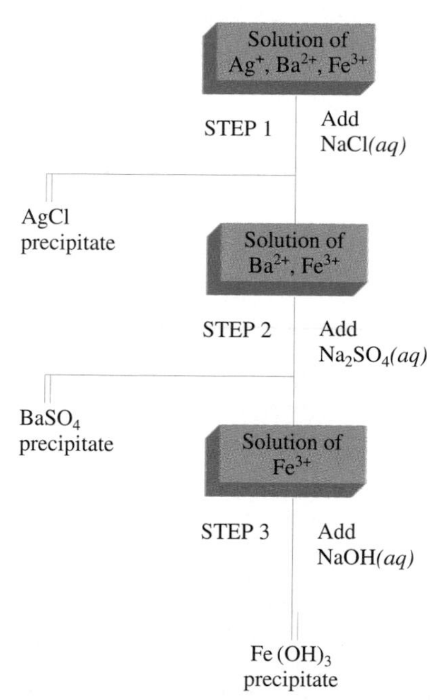

Figure 4.10
Selective precipitation of Ag^+, Ba^{2+} and Fe^{3+} ions. In this schematic representation a double line means that a solid forms, and a single line designates a solution.

4.8 Stoichiometry of Precipitation Reactions

In Chapter 3 we covered the principles of chemical stoichiometry: the procedures for calculating quantities of reactants and products involved in a chemical reaction. Recall that in performing these calculations, we first convert all quantities to moles and then use the coefficients of the balanced equation to assemble the appropriate molar ratios. In cases where reactants are mixed, we must determine which reactant is limiting, since the reactant that is consumed first will limit the amounts of products formed. *These same principles apply to reactions that take place in solutions.* However, there are two points about solution reactions that need special emphasis. The first is that it is sometimes difficult to tell immediately which reaction will occur when two solutions are mixed. Usually we must think about the various possibilities and then decide what will happen. The first step in this process *always* should be to write down the species that are actually present in the solution, as we did in Section 4.5.

The second special point about solution reactions is that to obtain the moles of reactants, we must use the volume of a particular solution and its molarity. This procedure was covered in Section 4.3.

We will introduce stoichiometric calculations for reactions in solution in Example 4.7.

Notice from Example 4.7 that the procedures for doing stoichiometric calculations for solution reactions are very similar to those for other types of

EXAMPLE 4.7

Calculate the mass of solid NaCl that must be added to 1.50 L of a 0.100 M $AgNO_3$ solution to precipitate all of the Ag^+ ions in the form of AgCl.

Solution

When added to the $AgNO_3$ solution (which contains Ag^+ and NO_3^- ions), the solid NaCl dissolves to yield Na^+ and Cl^- ions. Thus the mixed solution contains the ions

$$Ag^+, NO_3^-, Na^+, Cl^-$$

Since Table 4.1 indicates that $NaNO_3$ is soluble and AgCl is insoluble, solid AgCl forms according to the following net ionic reaction:

$$Ag^+(aq) + Cl^-(aq) \longrightarrow AgCl(s)$$

In this case enough Cl^- ions must be added to react with all of the Ag^+ ions present. Thus we must calculate the moles of Ag^+ ions present in 1.50 L of a 0.100 M $AgNO_3$ solution (remember that a 0.100 M $AgNO_3$ solution contains 0.100 M Ag^+ ions and 0.100 M NO_3^- ions):

$$1.50\ \text{L} \times \frac{0.100\ \text{mol Ag}^+}{\text{L}} = 0.150\ \text{mol Ag}^+$$

Since Ag^+ and Cl^- react in a 1:1 ratio, 0.150 mol of Cl^- ions and thus 0.150 mol of NaCl is required. We calculate the mass of NaCl required as follows:

$$0.150\ \text{mol NaCl} \times \frac{58.4\ \text{g NaCl}}{\text{mol NaCl}} = 8.76\ \text{g NaCl}$$

reactions. It is useful to think in terms of the following steps for reactions in solution.

Summary: Solving a Stoichiometry Problem Involving Reactions in Solution

- **Identify the species present in the combined solution and determine which reaction occurs.**
- **Write the balanced equation for the reaction.**
- **Calculate the moles of reactants.**
- **Determine which reactant is limiting.**
- **Calculate the moles of product or products, as required.**
- **Convert to grams or other units, as required.**

EXAMPLE 4.8

When aqueous solutions of Na_2SO_4 and $Pb(NO_3)_2$ are mixed, $PbSO_4$ precipitates. Calculate the mass of $PbSO_4$ formed when 1.25 L of 0.0500 *M* $Pb(NO_3)_2$ and 2.00 L of 0.0250 *M* Na_2SO_4 are mixed.

Solution

When the aqueous solutions of Na_2SO_4 (containing Na^+ and SO_4^{2-} ions) and $Pb(NO_3)_2$ (containing Pb^{2+} and NO_3^- ions) are mixed, the mixed solution contains the ions Na^+, SO_4^{2-}, Pb^{2+}, and NO_3^-. Since $NaNO_3$ is soluble and $PbSO_4$ is insoluble (see Table 4.1), solid $PbSO_4$ will form.

The net ionic equation is

$$Pb^{2+}(aq) + SO_4^{2-}(aq) \longrightarrow PbSO_4(s)$$

Since 0.0500 *M* $Pb(NO_3)_2$ contains 0.0500 *M* Pb^{2+} ions, we can calculate the moles of Pb^{2+} ions in 1.25 L of this solution as follows:

$$1.25 \text{ L} \times \frac{0.0500 \text{ mol Pb}^{2+}}{\text{L}} = 0.0625 \text{ mol Pb}^{2+}$$

The 0.0250 *M* Na_2SO_4 solution contains 0.0250 *M* SO_4^{2-} ions, and the number of moles of SO_4^{2-} ions in 2.00 L of this solution is

$$2.00 \text{ L} \times \frac{0.0250 \text{ mol SO}_4^{2-}}{\text{L}} = 0.0500 \text{ mol SO}_4^{2-}$$

Because Pb^{2+} and SO_4^{2-} react in a 1:1 ratio, the amount of SO_4^{2-} will be limiting.

Since the Pb^{2+} ions are present in excess, only 0.0500 mol of solid $PbSO_4$ will be formed. The mass of $PbSO_4$ formed can be calculated by using the molar mass of $PbSO_4$ (303.3):

$$0.0500 \text{ mol PbSO}_4 \times \frac{303.3 \text{ g PbSO}_4}{1 \text{ mol PbSO}_4} = 15.2 \text{ g PbSO}_4$$

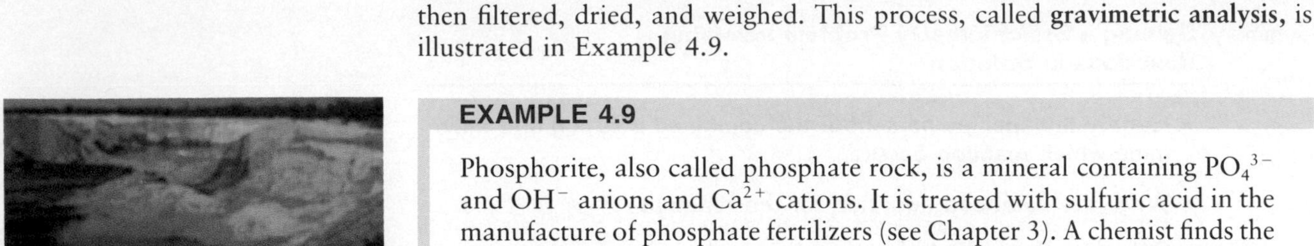

One method for determining the amount of a given substance present in a solution is to form a precipitate that includes the substance. The precipitate is then filtered, dried, and weighed. This process, called **gravimetric analysis,** is illustrated in Example 4.9.

Phosphate rock is used in the manufacturing of fertilizer. The mine pit appears in the background.

EXAMPLE 4.9

Phosphorite, also called phosphate rock, is a mineral containing PO_4^{3-} and OH^- anions and Ca^{2+} cations. It is treated with sulfuric acid in the manufacture of phosphate fertilizers (see Chapter 3). A chemist finds the calcium content in an impure sample of phosphate rock by weighing out a 0.4367-g sample, dissolving it in water, and precipitating the Ca^{2+} ions as the insoluble hydrated salt* $CaC_2O_4 \cdot H_2O$ ($C_2O_4^{2-}$ is called the oxalate ion). After being filtered and dried (at a temperature of about 100°C so that the extraneous water is driven off but not the water of hydration), the $CaC_2O_4 \cdot H_2O$ precipitate weighed 0.2920 g. Calculate the mass percent of calcium in the sample of phosphate rock.

Solution

This is a straightforward stoichiometry problem. The gravimetric procedure can be summarized as follows:

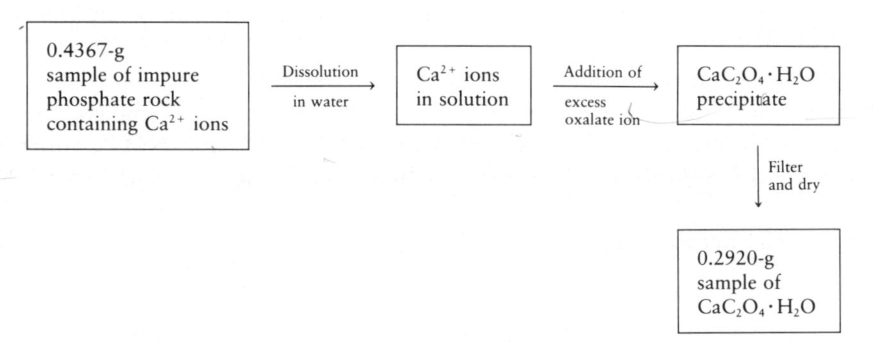

In this analysis excess $C_2O_4^{2-}$ ions are added to ensure that all Ca^{2+} ions are precipitated. Thus the number of moles of Ca^{2+} ions in the original sample determines the number of moles of $CaC_2O_4 \cdot H_2O$ formed. Using the molar mass of $CaC_2O_4 \cdot H_2O$, we can calculate the moles of $CaC_2O_4 \cdot H_2O$:

$$0.2920 \text{ g } CaC_2O_4{\cdot}H_2O \times \frac{1 \text{ mol } CaC_2O_4{\cdot}H_2O}{146.12 \text{ g } CaC_2O_4{\cdot}H_2O} = 1.998 \times 10^{-3} \text{ mol } CaC_2O_4{\cdot}H_2O$$

Thus the original sample of impure phosphate rock contained 1.998×10^{-3} mol of Ca^{2+} ions, which we convert to grams:

$$1.998 \times 10^{-3} \text{ mol } Ca^{2+} \times \frac{40.08 \text{ g } Ca^{2+}}{1 \text{ mol } Ca^{2+}} = 8.009 \times 10^{-2} \text{ g } Ca^{2+}$$

*Hydrated salts contain one or more H_2O molecules per formula unit in addition to the cations and anions. A dot is used in the formula of these salts.

The mass percent of calcium in the original sample is then

$$\frac{8.009 \times 10^{-2}\text{ g}}{0.4367\text{ g}} \times 100 = 18.34\%$$

4.9 Acid-Base Reactions

Earlier in this chapter we considered Arrhenius's concept of acids and bases: an acid is a substance that produces H^+ ions when dissolved in water, and a base is a substance that produces OH^- ions. Although these ideas are fundamentally correct, it is convenient to have a more general definition of a base, which covers substances that do not produce OH^- ions. Such a definition was provided by Brønsted and Lowry, who defined acids and bases as follows:

The Brønsted-Lowry concept of acids and bases will be discussed in detail in Chapter 7.

An acid is a proton donor.

A base is a proton acceptor.

How do we recognize acid-base reactions? One of the most difficult tasks for someone inexperienced in chemistry is to predict which reaction might occur when two solutions are mixed. With precipitation reactions we found that the best way to deal with this problem is to focus on the species actually present in the mixed solution. This also applies to acid-base reactions. For example, when an aqueous solution of hydrogen chloride (HCl) is mixed with an aqueous solution of sodium hydroxide (NaOH), the combined solution contains the ions H^+, Cl^-, Na^+ and OH^-, since HCl is a strong acid and NaOH is a strong base. How can we predict what reaction occurs, if any? First, will NaCl precipitate? From Table 4.1 we know that NaCl is soluble in water and thus will not precipitate. The Na^+ and Cl^- ions are spectator ions. On the other hand, because water is a nonelectrolyte, large quantities of H^+ and OH^- ions cannot coexist in solution. They will react to form H_2O molecules:

$$H^+(aq) + OH^-(aq) \longrightarrow H_2O(l)$$

This is the net ionic equation for the reaction that occurs when aqueous solutions of HCl and NaOH are mixed.

Next, consider mixing an aqueous solution of acetic acid ($HC_2H_3O_2$) with an aqueous solution of potassium hydroxide (KOH). In our earlier discussion of conductivity, we said that an aqueous solution of acetic acid is a weak electrolyte. Thus acetic acid does not dissociate into ions to any great extent. In fact, in aqueous solution 99% of the $HC_2H_3O_2$ molecules remain undissociated. However, when solid KOH is dissolved in water, it dissociates completely to produce K^+ and OH^- ions. So in the solution formed by mixing aqueous solutions of $HC_2H_3O_2$ and KOH, *before any reaction occurs* the principal species are H_2O, $HC_2H_3O_2$, K^+, and OH^-. Which reaction will occur? A possible precipitation reaction involves K^+ and OH^-, but we know that KOH is soluble. Another possibility is a reaction involving the hydroxide ion and a proton donor. Is there a source of protons in the solution? The answer is yes—the $HC_2H_3O_2$ molecules. The OH^- ion has such a strong affinity for protons that it can strip them from the $HC_2H_3O_2$ molecules. Thus the net ionic equation for the reaction is

$$OH^-(aq) + HC_2H_3O_2(aq) \longrightarrow H_2O(l) + C_2H_3O_2^-(aq)$$

This reaction illustrates a very important general principle: *the hydroxide ion is such a strong base that for purposes of stoichiometry it is assumed to react completely with any weak acid dissolved in water.* Of course, OH^- ions also react completely with the H^+ ions in the solutions of strong acids.

We will now deal with the stoichiometry of acid-base reactions in aqueous solutions. The procedure is fundamentally the same as that used previously.

Summary: Calculations for Acid-Base Reactions

- **List the species present in the combined solution *before reaction,* and decide which reaction will occur.**
- **Write the balanced net ionic equation for this reaction.**
- **Change the given quantities of reactants to moles. For reactions in solution, use the volumes of the original solutions and their molarities.**
- **Determine the limiting reactant where appropriate.**
- **Calculate the moles of the required reactant or product.**
- **Convert to grams or a volume of solution, as required by the problem.**

An acid-base reaction is often called a **neutralization reaction.** When just enough base is added to react exactly with all of the acid in a solution, we say the acid has been *neutralized.*

Acid-Base Titrations

Acid-base **titrations** are an example of **volumetric analysis,** a technique in which one solution is used to analyze another. The solution used to carry out the analysis is called the **titrant** and is delivered from a device called a **buret,** which measures the volume accurately. The point in the titration at which enough titrant has been added to react exactly with the substance being determined is called the **equivalence point** or the **stoichiometric point.** This point is often marked by the change in color of a chemical called an **indicator.** The titration procedure is illustrated in Fig. 4.11.

The following requirements must be met for a titration to be successful:

The concentration of the titrant must be known. Such a titrant is called a *standard solution.*

The exact reaction between titrant and substance being analyzed must be known.

The stoichiometric (equivalence) point must be known. An indicator that changes color at, or very near, the stoichiometric point is often used.

Ideally, the end point and stoichiometric point should coincide.

The point at which the indicator changes color is called the **end point.** The goal is to choose an indicator whose end point coincides with the stoichiometric point. An indicator very commonly used for acid-base titrations is *phenolphthalein,* which is colorless in acid and turns pink at the end point when an acid is titrated with a base.

The volume of titrant required to reach the stoichiometric point must be known as accurately as possible.

Figure 4.11
The titration of an acid with a base. (left) A beaker with an acidic solution containing the indicator phenolphthalein, ready to be titrated with aqueous sodium hydroxide. (center) As the sodium hydroxide is added to the acidic solution, the indicator turns pink near the surface because of the excess $[OH^-]$ in that region. When mixed, the solution again becomes colorless. (right) At the end point of the titration, the indicator turns the solution pink.

We will deal with acid-base titrations only briefly here but will return to the topic of titrations and indicators in more detail in Chapter 8. When a substance being analyzed contains an acid, the amount of acid present is usually determined by titration with a standard solution containing hydroxide ions.

EXAMPLE 4.10

What volume of a 0.100 *M* HCl solution is needed to neutralize 25.0 mL of a 0.350 *M* NaOH solution?

Solution

The species present in the mixed solutions before any reaction occurs are

$$\underbrace{H^+, Cl^-,}_{\text{From HCl}(aq)} \quad \underbrace{Na^+, OH^-,}_{\text{From NaOH}(aq)} \quad H_2O$$

Which reaction will occur? The two possibilities are

$$Na^+(aq) + Cl^-(aq) \longrightarrow NaCl(s)$$
$$H^+(aq) + OH^-(aq) \longrightarrow H_2O(l)$$

Since we know that NaCl is soluble, the first reaction does not take place (Na^+ and Cl^- are spectator ions). However, as we have seen before, the reaction of the H^+ and OH^- ions to form H_2O does occur.

The balanced net ionic equation for the reaction is

$$H^+(aq) + OH^-(aq) \longrightarrow H_2O(l)$$

Acid spills can be neutralized with a basic solution.

Next, we calculate the number of moles of OH^- ions in the 25.0-mL sample of 0.350 *M* NaOH:

$$25.0 \text{ mL NaOH} \times \frac{1 \text{ L}}{1000 \text{ mL}} \times \frac{0.350 \text{ mol } OH^-}{\text{L NaOH}} = 8.75 \times 10^{-3} \text{ mol } OH^-$$

This problem requires the addition of just enough H^+ ions to react exactly with the OH^- ions present. Thus we need not be concerned with determining a limiting reactant.

Since H^+ and OH^- ions react in a 1:1 ratio, 8.75×10^{-3} mol of H^+ ions is required to neutralize the OH^- ions present.

The volume (V) of 0.100 *M* HCl required to furnish this amount of H^+ ions can be calculated as follows:

$$V \times \frac{0.100 \text{ mol } H^+}{\text{L}} = 8.75 \times 10^{-3} \text{ mol } H^+$$

Solving for V gives

$$V = 8.75 \times 10^{-2} \text{ L}$$

Thus 8.75×10^{-2} L (87.5 mL) of 0.100 *M* HCl is required to neutralize 25.0 mL of 0.350 *M* NaOH.

EXAMPLE 4.11

In a certain experiment 28.0 mL of 0.250 *M* HNO_3 and 53.0 mL of 0.320 *M* KOH are mixed. Calculate the amount of water formed in the resulting reaction. What is the concentration of H^+ or OH^- ions in excess after the reaction goes to completion?

Solution

The ions available for reaction are

$$\underbrace{H^+, NO_3^-,}_{\text{From } HNO_3 \text{ solution}} \quad \underbrace{K^+, OH^-,}_{\text{From KOH solution}} \quad H_2O$$

Since KNO_3 is soluble, K^+ and NO_3^- are spectator ions, so the net ionic equation is

$$H^+(aq) + OH^-(aq) \longrightarrow H_2O(l)$$

We next compute the amounts of H^+ and OH^- ions present.

$$28.0 \text{ mL } HNO_3 \times \frac{1 \text{ L}}{1000 \text{ mL}} \times \frac{0.250 \text{ mol } H^+}{\text{L}} = 7.00 \times 10^{-3} \text{ mol } H^+$$

$$53.0 \text{ mL KOH} \times \frac{1 \text{ L}}{1000 \text{ mL}} \times \frac{0.320 \text{ mol } OH^-}{\text{L}} = 1.70 \times 10^{-2} \text{ mol } OH^-$$

Since H^+ and OH^- react in a 1:1 ratio, the limiting reactant is H^+. Thus 7.00×10^{-3} mol of H^+ ions will react with 7.00×10^{-3} mol of OH^- ions to form 7.00×10^{-3} mol of H_2O.

The amount of OH^- ions in excess is obtained from the following difference:

$$\text{Original amount} - \text{amount consumed} = \text{amount in excess}$$

$$1.70 \times 10^{-2}\text{ mol OH}^- - 7.00 \times 10^{-3}\text{ mol OH}^- = 1.00 \times 10^{-2}\text{ mol OH}^-$$

The volume of the combined solution is the sum of the individual volumes:

$$\text{Original volume of HNO}_3 + \text{original volume of KOH} = \text{total volume}$$

$$28.0\text{ mL} + 53.0\text{ mL} = 81.0\text{ mL} = 8.10 \times 10^{-2}\text{ L}$$

Thus the molarity of OH^- ions in excess is

$$M = \frac{\text{mol OH}^-}{\text{L solution}} = \frac{1.00 \times 10^{-2}\text{ mol OH}^-}{8.10 \times 10^{-2}\text{ L}} = 0.123\ M\text{ OH}^-$$

EXAMPLE 4.12

An environmental chemist analyzed the effluent (the released waste material) from an industrial process known to produce the compounds carbon tetrachloride (CCl_4) and benzoic acid ($HC_7H_5O_2$), a weak acid that has one acidic hydrogen atom per molecule. A sample of this effluent weighing 0.3518 g was placed in water and shaken vigorously to dissolve the benzoic acid. The resulting aqueous solution required 10.59 mL of 0.1546 *M* NaOH for neutralization. Calculate the mass percent of $HC_7H_5O_2$ in the original sample.

$c = \frac{m}{v}$

Solution

In this case the sample was a mixture containing CCl_4 and $HC_7H_5O_2$, and it was titrated with OH^- ions. Clearly, CCl_4 is not an acid (it contains no hydrogen atoms); so we can assume it does not react with OH^- ions. However, $HC_7H_5O_2$ is an acid. It donates one H^+ ion per molecule to react with an OH^- ion as follows:

$$HC_7H_5O_2(aq) + OH^-(aq) \longrightarrow H_2O(l) + C_7H_5O_2^-(aq)$$

Although $HC_7H_5O_2$ is a weak acid, the OH^- ion is such a strong base that we can assume that each OH^- ion added will react with a $HC_7H_5O_2$ molecule until all of the benzoic acid is consumed.

We must first determine the number of moles of OH^- ions required to react with all of the $HC_7H_5O_2$:

$$10.59 \text{ mL NaOH} \times \frac{1 \text{ L}}{1000 \text{ mL}} \times \frac{0.1546 \text{ mol OH}^-}{\text{L NaOH}} = 1.637 \times 10^{-3} \text{ mol OH}^-$$

This number is also the number of moles of $HC_7H_5O_2$ present. The number of grams of the acid is calculated by using its molar mass:

$$1.637 \times 10^{-3} \text{ mol HC}_7\text{H}_5\text{O}_2 \times \frac{122.125 \text{ g HC}_7\text{H}_5\text{O}_2}{1 \text{ mol HC}_7\text{H}_5\text{O}_2} = 0.1999 \text{ g HC}_7\text{H}_5\text{O}_2$$

The mass percent of $HC_7H_5O_2$ in the original sample is

$$\frac{0.1999 \text{ g}}{0.3518 \text{ g}} \times 100 = 56.82\%$$

The first step in the analysis of a complex solution is to write down the components present and to focus on the chemistry of each one.

Chemical systems often seem difficult to deal with simply because there are many components. Solving a problem involving a solution where several components are present is simplified if you *think* about the *chemistry* involved. *The key to success is to write down all the components in the solution and to focus on the chemistry of each one.* We have been emphasizing this approach in dealing with the reactions between ions in solution. Make it a habit to write down the components of solutions before trying to decide which reaction(s) might take place.

4.10 Oxidation-Reduction Reactions

As we have seen, many important substances are ionic. Sodium chloride, for example, can be formed by the reaction of elemental sodium and chlorine:

$$2Na(s) + Cl_2(g) \longrightarrow 2NaCl(s)$$

In this reaction an electron is transferred from a sodium atom to a chlorine atom, yielding a Na^+ ion and a Cl^- ion. *Reactions like this one, in which one or more electrons are transferred, are called* **oxidation-reduction reactions,** or **redox reactions.**

Oxidation of copper metal by nitric acid. The copper atoms lose two electrons to form Cu^{2+} ions, which produce a blue color in water.

Many important chemical reactions involve oxidation and reduction. In fact, most reactions used for energy production are redox reactions. In humans the oxidation of sugars, fats, and proteins provides the energy necessary for life. Combustion reactions, which provide most of the energy to power our civilization, also involve oxidation and reduction. An example is the reaction of methane with oxygen:

$$CH_4(g) + 2O_2(g) \longrightarrow CO_2(g) + 2H_2O(g) + \text{energy}$$

Even though none of the reactants or products in this reaction is ionic, the reaction is still assumed to involve a transfer of electrons from carbon to oxygen. To explain this, we must introduce the concept of oxidation states.

State-of-the-Art Analysis

The real world of chemical analysis is often quite different from what students do in the typical university laboratory. In the real world chemical analysis must be done quickly, accurately, economically, and often outside the laboratory setting. Analytical accuracy is crucial. A career can hinge on accuracy when a drug test is involved, and sometimes accuracy is truly a life-or-death matter, as in the screening of air travelers' luggage for explosives.

Chemical analysis can turn up in unexpected places. Modern engines in automobiles have been made much more fuel-efficient and less polluting by the inclusion of a sensor to analyze the oxygen (O_2) concentration in the exhaust gases. The signal from this sensor is sent to the computer that controls engine function so that instantaneous adjustments can be made in spark timing and air/fuel mixtures.

The automated screening of luggage for explosives is a very difficult and important analysis problem. The method being developed for luggage screening is called thermal neutron analysis (TNA), in which the substance to be analyzed is bombarded with neutrons. When nuclei in the sample absorb neutrons, they release gamma rays that are characteristic of a specific nucleus. For example, after the nucleus of a nitrogen atom absorbs a neutron, it emits a gamma ray that is unique to nitrogen, whereas an oxygen atom would produce a different gamma ray unique to oxygen, and so on. Thus when a sample is bombarded by neutrons and the resulting gamma rays are analyzed by a detector connected to a computer, the atoms present in the sample can be specified. In an airport the luggage passes through the TNA instrument on a conveyor belt, where it is bombarded by neutrons from californium-252. The detector is set up to look for unusually large quantities of nitrogen, because most chemical explosives are based on nitrogen compounds. Although this system is still under development, the Federal Aviation Administration is optimistic that it will work.

Analytical chemists have always admired the supersensitive natural detection devices built into organisms as part of elaborate control systems used to regulate the levels of various crucial chemicals, such as enzymes, hormones, and neurotransmitters. Because these "biosensors" are so sensitive, chemists are now attaching them to their instruments. For example, the sensory hairs from Hawaiian red swimming crabs can be connected to electrical analyzers and used to detect hormones at concentrations lower than 10^{-12} *M*. Also, slices from the tissues of pineapple cores can be used to detect hydrogen peroxide at levels of $\approx 10^{-6}$ *M*.

Another state-of-the-art detection system contains a surface acoustic wave (SAW) device, which is based on a piezoelectric crystal whose resonant frequency is sensitive to tiny changes in its mass—it can sense a change of 10^{-10} g/cm^2. In one use of this device as a detector, it was coated with a thin film of zeolite, a silicate mineral. Zeolite has intricate passages of a very uniform size. Thus it can act as a "molecular sieve," allowing only molecules of a certain size to pass through onto the detector, where their accumulation changes the mass and therefore alters the detector frequency. This sensor has been used to detect amounts of methyl alcohol (CH_3OH) as low as 10^{-9} g.

The face of chemical analysis is changing rapidly. In fact, although wet chemical analyses (titrations, for example) are still quite important in the chemical industry, increasingly these routine analyses are done by robots, which not only perform the analyses automatically but also send the results to a computer for interpretation.

Oxidation States

The concept of **oxidation states** (also called **oxidation numbers**) provides a way to keep track of electrons in oxidation-reduction reactions. Oxidation states are defined by a set of rules, most of which describe how to divide up the shared electrons in compounds containing covalent bonds. However, before we discuss these rules, we need to discuss the distribution of electrons in a bond.

Recall from the discussion of the water molecule in Section 4.1 that oxygen has a greater attraction for electrons than does hydrogen, causing the O—H

TABLE 4.3 **Rules for Assigning Oxidation States**

1. The oxidation state of an atom in an element is 0. For example, the oxidation state of each atom in the substances Na(*s*), $O_2(g)$, $O_3(g)$, and Hg(*l*) is 0.
2. The oxidation state of a monatomic ion is the same as its charge. For example, the oxidation state of the Na^+ ion is +1.
3. Oxygen is assigned an oxidation state of −2 in its covalent compounds, such as CO, CO_2, SO_2, and SO_3. The exception to this rule occurs in peroxides (compounds containing the O_2^{2-} group), where each oxygen is assigned an oxidation state of −1. The best-known example of a peroxide is hydrogen peroxide (H_2O_2).
4. In its covalent compounds with nonmetals, hydrogen is assigned an oxidation state of +1. For example, in the compounds HCl, NH_3, H_2O, and CH_4, hydrogen is assigned an oxidation state of +1.
5. In binary compounds the element with the greater attraction for the electrons in the bond is assigned a negative oxidation state equal to its charge in its ionic compounds. For example, fluorine is always assigned an oxidation state of −1. That is, for purposes of counting electrons, fluorine is assumed to be F^-. Nitrogen is usually assigned −3. For example, in NH_3, nitrogen is assigned an oxidation state of −3; in H_2S, sulfur is assigned an oxidation state of −2; in HI, iodine is assigned an oxidation state of −1; and so on.
6. The sum of the oxidation states must be zero for an electrically neutral compound and must be equal to the overall charge for an ionic species. For example, the sum of the oxidation states for the hydrogen and oxygen atoms in water is 0; the sum of the oxidation states for the carbon and oxygen atoms in CO_3^{2-} is −2; and the sum of oxidation states for the nitrogen and hydrogen atoms in NH_4^+ is +1.

bonds in the water molecule to be polar. This phenomenon occurs in other bonds as well, and we will discuss the topic of polarity in detail in Chapter 13. For now we will be satisfied with some general guidelines to help us keep track of electrons in oxidation-reduction reactions. The nonmetals with the highest attraction for shared electrons are in the upper right-hand corner of the periodic table. They are fluorine, oxygen, nitrogen, and chlorine. The relative ability of these atoms to attract shared electrons is

$$F > O > N \approx Cl$$

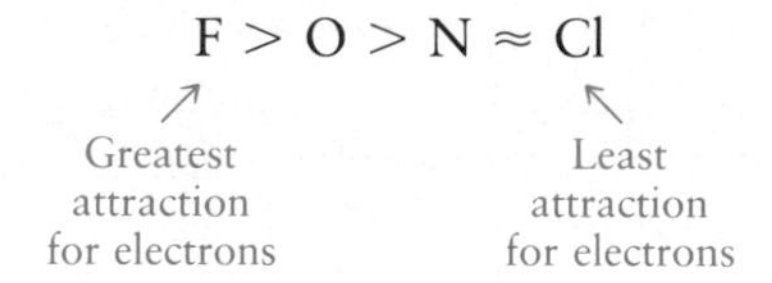

That is, fluorine attracts shared electrons to the greatest extent, followed by oxygen, then nitrogen and chlorine.

The rules for assigning oxidation states are given in Table 4.3. Application of these rules allows the assignment of oxidation states in most compounds. The principles are illustrated in Example 4.13.

EXAMPLE 4.13

Assign oxidation states to all atoms in the following:

a. CO_2 **b.** SF_6 **c.** NO_3^-

Solution

a. The rule that takes precedence here is that oxygen is assigned an oxidation state of −2. The oxidation state for carbon can be determined by recognizing that since CO_2 has no charge, the sum of the oxidation states for oxygen and carbon must be 0. Since each oxygen is −2 and there are two oxygen atoms, the carbon atom must be assigned an oxidation state of +4:

$$\underset{+4\quad -2 \text{ for each oxygen}}{CO_2}$$

b. Since fluorine has the greater attraction for the shared electrons, we assign its oxidation state first. Since its charge in ionic compounds is 1−, we assign −1 as the oxidation state of each fluorine atom. The sulfur must then be assigned an oxidation state of +6 to balance the total of −6 from the fluorine atoms:

$$\underset{+6\quad -1 \text{ for each fluorine}}{SF_6}$$

c. Since oxygen has a greater attraction than does nitrogen for the shared electrons, we assign its oxidation state of −2 first. Because the sum of the oxidation states of the three oxygens is −6 and the net charge on the NO_3^- ion is 1−, the nitrogen must have an oxidation state of +5:

$$\underset{+5\quad -2 \text{ for each oxygen}}{NO_3^-}$$

Actual charges are given as $n-$ or $n+$. Oxidation states (*not actual charges*) are given as $-n$ or $+n$. For example, for NO_3^-, the overall charge is 1−; the oxidation state for N is +5 and for O is −2.

Next, let us consider the oxidation states of the atoms in Fe_3O_4, which is the main component in magnetite, an iron ore that accounts for the reddish color of many types of rocks and soils. We assign each oxygen atom its usual oxidation state of −2. The three iron atoms must yield a total of +8 to balance the total of −8 from the four oxygens. Thus each iron atom has an oxidation state of $+\frac{8}{3}$. A noninteger value for oxidation state may seem strange since charge is expressed in whole numbers. However, although they are rare, noninteger oxidation states do occur because of the rather arbitrary way that the electrons are divided up by the rules in Table 4.3. In the compound Fe_3O_4, for example, the rules assume that all of the iron atoms are equal, when in fact this compound can best be viewed as containing four O^{2-} ions, two Fe^{3+} ions, and one Fe^{2+} ion per formula unit. (Note that the "average" charge on iron works out to be $\frac{8}{3}+$, which is equal to the oxidation state we determined above.) Noninteger oxidation states should not intimidate you. They serve the same purpose as integer oxidation states—for keeping track of electrons.

The Characteristics of Oxidation-Reduction Reactions

Oxidation-reduction reactions are characterized by a transfer of electrons. In some cases the transfer occurs in a literal sense to form ions, such as in this reaction:

$$2Na(s) + Cl_2(g) \longrightarrow 2NaCl(s)$$

However, sometimes the transfer occurs in a more formal sense, such as in the combustion of methane (the oxidation state for each atom is given):

$$\underset{\substack{-4 \quad +1 \\ \text{(each H)}}}{CH_4(g)} + \underset{0}{2O_2(g)} \longrightarrow \underset{\substack{+4 \quad -2 \\ \text{(each O)}}}{CO_2(g)} + \underset{\substack{+1 \quad -2 \\ \text{(each H)}}}{2H_2O(g)}$$

Oxidation state: -4 (C), $+1$ (each H); 0; $+4$, -2 (each O); $+1$ (each H), -2

Note that the oxidation state of oxygen in O_2 is 0 because it is in elemental form. In this reaction there are no ionic compounds, but we can still describe the process in terms of a transfer of electrons. Note that carbon undergoes a change in oxidation state from -4 in CH_4 to $+4$ in CO_2. Such a change can be accounted for by a loss of eight electrons (the symbol e^- stands for an electron):

$$\underset{-4}{CH_4} \longrightarrow \underset{+4}{CO_2} + 8e^-$$

On the other hand, each oxygen changes from an oxidation state of 0 in O_2 to -2 in H_2O and CO_2, signifying a gain of two electrons per atom. Since four oxygen atoms are involved, this is a gain of eight electrons:

$$\underset{0}{2O_2} + 8e^- \longrightarrow CO_2 + 2H_2O \qquad 4(-2) = -8$$

No change occurs in the oxidation state of hydrogen, so it is not formally involved in the electron transfer process.

Oxidation is an increase in oxidation state. Reduction is a decrease in oxidation state.

With this background, we can now define some important terms. **Oxidation** is an *increase* in oxidation state (a loss of electrons). **Reduction** is a *decrease* in oxidation state (a gain of electrons). Thus in the reaction

$$\underset{0}{2Na(s)} + \underset{0}{Cl_2(g)} \longrightarrow \underset{+1 \quad -1}{2NaCl(s)}$$

A helpful mnemonic device is OIL RIG (*O*xidation *I*nvolves *L*oss; *R*eduction *I*nvolves *G*ain).

sodium is oxidized and chlorine is reduced. In addition, Cl_2 is called the **oxidizing agent (electron acceptor)** and Na is called the **reducing agent (electron donor)**.

Concerning the reaction

$$\underset{-4 \quad +1}{CH_4(g)} + \underset{0}{2O_2(g)} \longrightarrow \underset{+4 \quad -2}{CO_2(g)} + \underset{+1 \quad -2}{2H_2O(g)}$$

we can say the following:

An oxidizing agent is reduced and a reducing agent is oxidized in a redox reaction.

Carbon is oxidized because there is an increase in its oxidation state (carbon has formally lost electrons).

Oxygen is reduced as shown by the decrease in its oxidation state (oxygen has formally gained electrons).

CH_4 is the reducing agent.

O_2 is the oxidizing agent.

Note that when the oxidizing or reducing agent is named, the *whole compound* is specified, not just the element that undergoes the change in oxidation state.

4.11 Balancing Oxidation-Reduction Equations

Oxidation-reduction reactions are often complicated, which means that it can be difficult to balance their equations by simple inspection. Two methods for balancing redox reactions will be considered here: (1) the oxidation states method, and (2) the half-reaction method.

The Oxidation States Method

Methanol (CH_3OH) is used as a fuel in high-performance engines such as those in the race cars in the Indianapolis 500. The unbalanced combustion reaction is

$$CH_3OH(l) + O_2(g) \longrightarrow CO_2(g) + H_2O(g)$$

We want to balance this equation by using the changes in oxidation state, so we must first specify all oxidation states. The only molecule here that we have not previously considered is CH_3OH. We assign oxidation states of $+1$ to each hydrogen and -2 to the oxygen, which means that the oxidation state of the carbon must be -2, since the compound is electrically neutral. Thus the oxidation states for the reaction participants are as follows:

$$\underset{\substack{\nearrow \\ -2}}{C}\underset{\substack{\uparrow \\ +1 \\ \text{(each H)}}}{H_3}\underset{\substack{\uparrow \\ -2}}{O}\underset{\substack{\nwarrow \\ +1}}{H}(l) + \underset{\substack{\uparrow \\ 0}}{O_2}(g) \longrightarrow \underset{\substack{\uparrow \\ +4}}{C}\underset{\substack{\uparrow \\ -2 \\ \text{(each O)}}}{O_2}(g) + \underset{\substack{\uparrow \\ +1 \\ \text{(each H)}}}{H_2}\underset{\substack{\uparrow \\ -2}}{O}(g)$$

Note that the oxidation state of carbon changes from -2 to $+4$, an increase of 6. On the other hand, the oxidation state of oxygen changes from 0 to -2, a decrease of 2. This means that three oxygen atoms are needed to balance the increase in the oxidation state of the single carbon atom. We can write this relationship as follows:

$$\underset{\substack{\uparrow \\ \text{1 carbon atom}}}{CH_3OH(l)} + \underset{\substack{\uparrow \\ \text{3 oxygen atoms}}}{\tfrac{3}{2}O_2(g)} \longrightarrow \text{products}$$

The rest of the equation can be balanced by inspection:

$$CH_3OH(l) + \tfrac{3}{2}O_2(g) \longrightarrow CO_2(g) + 2H_2O(g)$$

We then write it in conventional format (multiply through by 2):

$$2CH_3OH(l) + 3O_2(g) \longrightarrow 2CO_2(g) + 4H_2O(g)$$

In using the oxidation states method to balance an oxidation-reduction equation, we find the coefficients for the reactants that will make the total increase in oxidation state balance the total decrease. The remainder of the equation is then balanced by inspection.

EXAMPLE 4.14

Because metals are so reactive, very few are found in nature in pure form. Metallurgy involves reducing the metal ions in ores to the elemental form. The production of manganese from the ore pyrolusite, which contains MnO_2, employs aluminum as the reducing agent. Using oxidation states, balance the equation for this process.

$$MnO_2(s) + Al(s) \longrightarrow Mn(s) + Al_2O_3(s)$$

Solution

First, we assign oxidation states:

$$\underset{+4\quad -2\ (\text{each O})}{MnO_2(s)} + \underset{0}{Al(s)} \longrightarrow \underset{0}{Mn(s)} + \underset{+3\ (\text{each Al})\quad -2\ (\text{each O})}{Al_2O_3(s)}$$

Each Mn atom undergoes a decrease in oxidation state of 4 (from +4 to 0), while each Al atom undergoes an increase of 3 (from 0 to +3).

Thus we need three Mn atoms for every four Al atoms in order to balance the increase and decrease in oxidation states:

$$\text{Increase} = 4(3) = \text{decrease} = 3(4)$$

$$3MnO_2(s) + 4Al(s) \longrightarrow \text{products}$$

We balance the rest of the equation by inspection:

$$3MnO_2(s) + 4Al(s) \longrightarrow 3Mn(s) + 2Al_2O_3(s)$$

The procedures for balancing an oxidation-reduction equation by the oxidation states method are summarized below.

Summary: Balancing an Oxidation-Reduction Equation by the Oxidation States Method

- **Assign the oxidation states of all atoms.**
- **Decide which element is oxidized and determine the increase in oxidation state.**
- **Decide which element is reduced and determine the decrease in oxidation state.**
- **Choose coefficients for the species containing the atom oxidized and the atom reduced such that the total increase in oxidation state equals the total decrease in oxidation state.**
- **Balance the remainder of the equation by inspection.**

The Half-Reaction Method

For oxidation-reduction reactions that occur in aqueous solution, it is often useful to separate the reaction into two **half-reactions:** one involving oxidation and the other involving reduction. For example, consider the unbalanced equation for the oxidation-reduction reaction between cerium(IV) ion and tin(II) ion:

$$Ce^{4+}(aq) + Sn^{2+}(aq) \longrightarrow Ce^{3+}(aq) + Sn^{4+}(aq)$$

This reaction can be separated into a half-reaction involving the substance being *reduced,*

$$Ce^{4+}(aq) \longrightarrow Ce^{3+}(aq)$$

and one involving the substance being *oxidized,*

$$Sn^{2+}(aq) \longrightarrow Sn^{4+}(aq)$$

The general procedure is to balance the equations for the half-reactions separately and then to add them to obtain the overall balanced equation. The half-reaction method for balancing oxidation-reduction equations differs slightly depending on whether the reaction takes place in acidic or basic solution.

Summary: Balancing Oxidation-Reduction Equations Occurring in Acidic Solution by the Half-Reaction Method

- **Write the equations for the oxidation and reduction half-reactions.**
- **For each half-reaction:**
  - **a. Balance all of the elements except hydrogen and oxygen.**
  - **b. Balance oxygen using H_2O.**
  - **c. Balance hydrogen using H^+.**
  - **d. Balance the charge using electrons.**
- **If necessary, multiply one or both balanced half-reactions by integers to equalize the number of electrons transferred in the two half-reactions.**
- **Add the half-reactions, and cancel identical species.**
- **Check to be sure that the elements and charges balance.**

We will illustrate this method by balancing the equation for the reaction between permanganate and iron(II) ions in acidic solution:

$$MnO_4^-(aq) + Fe^{2+}(aq) \xrightarrow{\text{Acidic}} Fe^{3+}(aq) + Mn^{2+}(aq)$$

This reaction is used to analyze iron ore for its iron content.

- **Identify and write equations for the half-reactions.**

The oxidation states for the half-reaction involving the permanganate ion show that manganese is reduced:

$$\underset{+7\ \ -2\text{ (each O)}}{MnO_4^-} \longrightarrow \underset{+2}{Mn^{2+}}$$

This is the *reduction half-reaction*. The other half-reaction involves the oxidation of iron(II) to iron(III) ion and is the *oxidation half-reaction:*

$$\underset{+2}{\overset{\uparrow}{Fe^{2+}}} \longrightarrow \underset{+3}{\overset{\uparrow}{Fe^{3+}}}$$

- **Balance each half-reaction.**

For the reduction reaction, we have

$$MnO_4^-(aq) \longrightarrow Mn^{2+}(aq)$$

a. The manganese is balanced.

b. We balance oxygen by adding $4H_2O$ to the right side of the equation:

$$MnO_4^-(aq) \longrightarrow Mn^{2+}(aq) + 4H_2O(l)$$

c. Next, we balance hydrogen by adding $8H^+$ to the left side:

$$8H^+(aq) + MnO_4^-(aq) \longrightarrow Mn^{2+}(aq) + 4H_2O(l)$$

d. All of the elements have been balanced, but we need to balance the charge by using electrons. At this point we have the following charges for reactants and products in the reduction half-reaction:

$$8H^+(aq) + MnO_4^-(aq) \longrightarrow Mn^{2+}(aq) + 4H_2O(l)$$

$$\underbrace{8+ \quad + \quad 1-}_{7+} \qquad\qquad \underbrace{2+ \quad + \quad 0}_{2+}$$

We can equalize the charges by adding five electrons to the left side:

$$\underbrace{5e^- + 8H^+(aq) + MnO_4^-(aq)}_{2+} \longrightarrow \underbrace{Mn^{2+}(aq) + 4H_2O(l)}_{2+}$$

Both the *elements* and the *charges* are now balanced, so this represents the balanced reduction half-reaction. The fact that five electrons appear on the reactant side of the equation makes sense, since five electrons are required to reduce MnO_4^- (in which Mn has an oxidation state of $+7$) to Mn^{2+} (in which Mn has an oxidation state of $+2$).

For the oxidation reaction,

$$Fe^{2+}(aq) \longrightarrow Fe^{3+}(aq)$$

the elements are balanced, and we must simply balance the charge:

$$\underbrace{Fe^{2+}(aq)}_{2+} \longrightarrow \underbrace{Fe^{3+}(aq)}_{3+}$$

One electron is needed on the right side to give a net 2+ charge on both sides:

$$\underbrace{Fe^{2+}(aq)}_{2+} \longrightarrow \underbrace{Fe^{3+}(aq) + e^-}_{2+}$$

- **Equalize the electron transfer in the two half-reactions.**

  Since the reduction half-reaction involves a transfer of five electrons and the oxidation half-reaction involves a transfer of only one electron, the oxidation half-reaction must be multiplied by 5:

$$5Fe^{2+}(aq) \longrightarrow 5Fe^{3+}(aq) + 5e^-$$

The number of electrons gained in the reduction half-reaction must equal the number of electrons lost in the oxidation half-reaction.

- **Add the half-reactions.**

  The half-reactions are added to give

$$5e^- + 5Fe^{2+}(aq) + MnO_4^-(aq) + 8H^+(aq) \longrightarrow 5Fe^{3+}(aq) + Mn^{2+}(aq) + 4H_2O(l) + 5e^-$$

  Note that the electrons cancel (as they must) to give the final balanced equation:

$$5Fe^{2+}(aq) + MnO_4^-(aq) + 8H^+(aq) \longrightarrow 5Fe^{3+}(aq) + Mn^{2+}(aq) + 4H_2O(l)$$

- **Check that the elements and charges balance.**

  Elements balance: 5 Fe, 1 Mn, 4 O, 8 H $\longrightarrow$ 5 Fe, 1 Mn, 4 O, 8 H

  Charges balance: $5(2+) + (1-) + 8(1+) = 17+ \longrightarrow 5(3+) + (2+) + 0 = 17+$

  The equation is balanced.

Oxidation-reduction reactions can occur in basic as well as in acidic solutions. The half-reaction method for balancing equations is slightly different in such cases.

Summary: Balancing Oxidation-Reduction Equations Occurring in Basic Solution by the Half-Reaction Method

- **Use the half-reaction method as specified for acidic solutions to obtain the final balanced equation *as if H^+ ions were present.***
- **To both sides of the equation obtained above, add the number of OH^- ions that is equal to the number of H^+ ions. (We want to eliminate H^+ by forming H_2O.)**
- **Form H_2O on the side containing both H^+ and OH^- ions, and eliminate the number of H_2O molecules that appear on both sides of the equation.**
- **Check that the elements and charges balance.**

Silver ore in Gila County, Arizona.

EXAMPLE 4.15

Silver is sometimes found in nature as large nuggets; more often it is found mixed with other metals and their ores. Cyanide ion is often used to extract the silver by the following reaction that occurs in basic solution:

$$Ag(s) + CN^-(aq) + O_2(g) \xrightarrow{\text{Basic}} Ag(CN)_2^-(aq)$$

Balance this equation by using the half-reaction method.

Solution

- **Balance the equation as if H^+ ions were present.**

Balance the oxidation half-reaction:

$$CN^-(aq) + Ag(s) \longrightarrow Ag(CN)_2^-(aq)$$

Balance carbon and nitrogen:

$$2CN^-(aq) + Ag(s) \longrightarrow Ag(CN)_2^-(aq)$$

Balance the charge:

$$2CN^-(aq) + Ag(s) \longrightarrow Ag(CN)_2^-(aq) + e^-$$

Balance the reduction half-reaction:

$$O_2(g) \longrightarrow$$

Balance oxygen:

$$O_2(g) \longrightarrow 2H_2O(l)$$

Balance hydrogen:

$$O_2(g) + 4H^+(aq) \longrightarrow 2H_2O(l)$$

Balance the charge:

$$4e^- + O_2(g) + 4H^+(aq) \longrightarrow 2H_2O(l)$$

Multiply the balanced oxidation half-reaction by 4:

$$8CN^-(aq) + 4Ag(s) \longrightarrow 4Ag(CN)_2^-(aq) + 4e^-$$

Add the half-reactions, and cancel identical species:

Oxidation half-reaction:

$$8CN^-(aq) + 4Ag(s) \longrightarrow 4Ag(CN)_2^-(aq) + 4e^-$$

Reduction half-reaction:

$$4e^- + O_2(g) + 4H^+(aq) \longrightarrow 2H_2O(l)$$

$$8CN^-(aq) + 4Ag(s) + O_2(g) + 4H^+(aq) \longrightarrow 4Ag(CN)_2^-(aq) + 2H_2O(l)$$

- **Add OH^- ions to both sides of balanced equation.**

We need to add $4OH^-$ to each side:

$$8CN^-(aq) + 4Ag(s) + O_2(g) + \underbrace{4H^+(aq) + 4OH^-(aq)}_{4H_2O(l)} \longrightarrow 4Ag(CN)_2^-(aq) + 2H_2O(l) + 4OH^-(aq)$$

- **Eliminate as many H_2O molecules as possible.**

$$8CN^-(aq) + 4Ag(s) + O_2(g) + 2H_2O(l) \longrightarrow 4Ag(CN)_2^-(aq) + 4OH^-(aq)$$

- **Check that elements and charges balance.**

EXAMPLE 4.16

Cerium(IV) ion is a strong oxidizing agent that accepts one electron to produce cerium(III) ion:

$$Ce^{4+}(aq) + e^- \longrightarrow Ce^{3+}(aq)$$

A solution containing an unknown concentration of Sn^{2+} ions was titrated with a solution containing Ce^{4+} ions, which oxidize the Sn^{2+} ions to Sn^{4+} ions. In one titration 1.00 L of the unknown solution required 46.45 mL of a 0.1050 *M* Ce^{4+} solution to reach the stoichiometric point. Calculate the concentration of Sn^{2+} ions in the unknown solution.

Solution

The unbalanced equation for the titration reaction is

$$Ce^{4+}(aq) + Sn^{2+}(aq) \longrightarrow Ce^{3+}(aq) + Sn^{4+}(aq)$$

The balanced equation is

$$2Ce^{4+}(aq) + Sn^{2+}(aq) \longrightarrow 2Ce^{3+}(aq) + Sn^{4+}(aq)$$

We can obtain the number of moles of Ce^{4+} ions from the volume and molarity of the Ce^{4+} solution used as the titrant:

$$46.45\ \text{mL} \times \frac{1\ \text{L}}{1000\ \text{mL}} \times \frac{0.1050\ \text{mol Ce}^{4+}}{\text{L}} = 4.877 \times 10^{-3}\ \text{mol Ce}^{4+}$$

The number of moles of Sn^{2+} ions can be obtained by applying the appropriate mole ratio from the balanced equation:

$$4.877 \times 10^{-3}\ \text{mol Ce}^{4+} \times \frac{1\ \text{mol Sn}^{2+}}{2\ \text{mol Ce}^{4+}} = 2.439 \times 10^{-3}\ \text{mol Sn}^{2+}$$

This value represents the quantity of Sn^{2+} ions in 1.00 L of solution. Thus the concentration of Sn^{2+} in the unknown solution is

$$\text{Molarity} = \frac{\text{mol Sn}^{2+}}{\text{L solution}} = \frac{2.439 \times 10^{-3}\ \text{mol Sn}^{2+}}{1.00\ \text{L}} = 2.44 \times 10^{-3}\ M$$

Figure 4.12
Permanganate being introduced into a flask of reducing agent.

4.12 Simple Oxidation-Reduction Titrations

Oxidation-reduction reactions are commonly used as the basis for volumetric analytical procedures. For example, a reducing substance can be titrated with a solution of a strong oxidizing agent, or vice versa. Three of the most frequently used oxidizing agents are aqueous solutions of *potassium permanganate* ($KMnO_4$), *potassium dichromate* ($K_2Cr_2O_7$), and *cerium hydrogen sulfate* [$Ce(HSO_4)_4$].

The strong oxidizing agent, the permanganate ion (MnO_4^-), can undergo several different reactions. The reaction that occurs in acidic solution is the one most commonly used:

$$MnO_4^-(aq) + 8H^+(aq) + 5e^- \longrightarrow Mn^{2+}(aq) + 4H_2O(l)$$

Permanganate has the advantage of being its own indicator—the MnO_4^- ion is intensely purple, and the Mn^{2+} ion is almost colorless. A typical titration using permanganate is illustrated in Fig. 4.12. As long as some reducing agent remains in the solution being titrated, the solution remains colorless (assuming all other species present are colorless), since the purple MnO_4^- ion being added is converted to the essentially colorless Mn^{2+} ion. However, when all the reducing agent has been consumed, the next drop of permanganate titrant will turn the solution being titrated light purple (pink). Thus the end point (where the color change indicates the titration should stop) occurs approximately one drop beyond the stoichiometric point (the actual point at which all of the reducing agent has been consumed).

Example 4.17 describes a typical volumetric analysis using permanganate.

EXAMPLE 4.17

Iron ores often involve a mixture of oxides and contain both Fe^{2+} and Fe^{3+} ions. Such an ore can be analyzed for its iron content by dissolving it in acidic solution, reducing all the iron to Fe^{2+} ions, and then titrating with a standard solution of potassium permanganate. In the resulting solution, MnO_4^- is reduced to Mn^{2+}, and Fe^{2+} is oxidized to Fe^{3+}. A sample of iron ore weighing 0.3500 g was dissolved in acidic solution, and all of the iron was reduced to Fe^{2+}. Then the solution was titrated with a 1.621×10^{-2} M $KMnO_4$ solution. The titration required 41.56 mL of the permanganate solution to reach the light purple (pink) end point. Determine the mass percent of iron in the iron ore.

Solution

First, we write the unbalanced equation for the reaction:

$$H^+(aq) + MnO_4^-(aq) + Fe^{2+}(aq) \longrightarrow Fe^{3+}(aq) + Mn^{2+}(aq) + H_2O(l)$$

Using the half-reaction method, we balance the equation:

$$8H^+(aq) + MnO_4^-(aq) + 5Fe^{2+}(aq) \longrightarrow 5Fe^{3+}(aq) + Mn^{2+}(aq) + 4H_2O(l)$$

The number of moles of MnO_4^- ion required in the titration is found from the volume and concentration of permanganate solution used:

$$41.56 \text{ mL} \times \frac{1 \text{ L}}{1000 \text{ mL}} \times \frac{1.621 \times 10^{-2} \text{ mol } MnO_4^-}{\text{L}}$$
$$= 6.737 \times 10^{-4} \text{ mol } MnO_4^-$$

The balanced equation shows that five times as much Fe^{2+} as MnO_4^- is required:

$$6.737 \times 10^{-4} \text{ mol } MnO_4^- \times \frac{5 \text{ mol } Fe^{2+}}{1 \text{ mol } MnO_4^-}$$
$$= 3.368 \times 10^{-3} \text{ mol } Fe^{2+}$$

Thus the 0.3500-g sample of iron ore contained 3.368×10^{-3} mol of iron. The mass of iron present is

$$3.368 \times 10^{-3} \text{ mol Fe} \times \frac{55.85 \text{ g Fe}}{1 \text{ mol Fe}} = 0.1881 \text{ g Fe}$$

The mass percent of iron in the iron ore is

$$\frac{0.1881 \text{ g}}{0.3500 \text{ g}} \times 100 = 53.74\%$$

EXERCISES

A blue exercise number indicates that the answer to that exercise appears at the back of this book.

Aqueous Solutions: Strong and Weak Electrolytes

1. Distinguish between the terms *slightly soluble* and *weak electrolyte.*
2. How would you determine experimentally whether a substance is a strong or a weak electrolyte?
3. Commercial cold packs and hot packs are available for treating athletic injuries. Both types contain a pouch of water and a dry chemical. When the pack is struck, the pouch of water breaks, dissolving the chemical, and the solution becomes either hot or cold. Many hot packs use magnesium sulfate, and many cold packs use ammonium nitrate. Write reactions to show how these strong electrolytes break apart in water.

Solution Concentration: Molarity

4. How would you prepare 1.00 L of a 0.50 *M* solution of each of the following?
   a. H_2SO_4 from "concentrated" (18 *M*) sulfuric acid
   b. HCl from "concentrated" (12 *M*) reagent
   c. $NiCl_2$ from the salt $NiCl_2 \cdot 6H_2O$
   d. HNO_3 from "concentrated" (16 *M*) reagent
   e. sodium carbonate from the pure solid
5. Calculate the molarity of each of these solutions.
   a. 4.592 g of $NaHCO_3$ is dissolved in enough water to make 250.0 mL of solution.
   b. 275.9 mg of $K_2Cr_2O_7$ is dissolved in enough water to make 500.0 mL of solution.
   c. 0.1025 g of copper metal is dissolved in 35 mL of concentrated HNO_3 to form Cu^{2+} ions, and then water is added to make a total volume of 200.0 mL. (Calculate the molarity of Cu^{2+}.)
6. Disodium aurothiomalate ($Na_2C_4H_3O_4SAu$) has the trade name Myocrisin and is used in the treatment of rheumatoid arthritis. Patients receive weekly intramuscular injections of 50.0 mg of Myocrisin in 0.500 mL of solution. During treatment, serum levels of gold are as high as 300.0 μg of gold per 100.0 mL of serum. Calculate the above two concentrations of Myocrisin in units of molarity.
7. A solution is prepared by dissolving 25.0 g of ammonium sulfate in enough water to make 100.0 mL of stock solution. A 10.00-mL sample of this stock solution is added to

50.00 mL of water. Calculate the concentration of ammonium ions and sulfate ions in the final solution.

8. A solution of ethanol (C_2H_5OH) in water is prepared by dissolving 75.0 mL of ethanol (density = 0.79 g/cm^3) in enough water to make 250.0 mL of solution. What is the molarity of the ethanol in this solution?

9. The units of parts per million (ppm) and parts per billion (ppb) are commonly used by environmental chemists. In general, 1 ppm means 1 part of solute for every 10^6 parts of solution. (Both solute and solution are measured using the same units.) Mathematically, by mass:

$$\text{ppm} = \frac{\mu\text{g solute}}{\text{g solution}} = \frac{\text{mg solute}}{\text{kg solution}}$$

In the case of very dilute aqueous solutions, a concentration of 1.0 ppm is equal to 1.0 μg of solute per 1.0 mL, which equals 1.0 g of solution. Parts per billion is defined in a similar fashion. Calculate the molarity of each of the following aqueous solutions.
   a. 5.0 ppb Hg in H_2O
   b. 1.0 ppb $CHCl_3$ in H_2O
   c. 10.0 ppm As in H_2O
   d. 0.10 ppm DDT ($C_{14}H_9Cl_5$) in H_2O

10. a. A 150-mg sample of Na_2CO_3 is dissolved in H_2O to give 1.0 L of solution. What is the concentration of Na^+ in parts per million?
   b. A 2.5-mg sample of dioctyl phthalate ($C_{24}H_{38}O_4$), a plasticizer, is dissolved in H_2O to give 500.0 mL of solution. What is the concentration of dioctyl phthalate in parts per billion? (See Exercise 9 for definitions.)

11. A standard is prepared for the analysis of fluoxymesterone ($C_{20}H_{29}FO_3$), an anabolic steroid. A stock solution is prepared by dissolving 10.0 mg of fluoxymesterone in enough water to give a total volume of 500.0 mL. A 100.0-μL aliquot (portion) of this solution is diluted to a final volume of 100.0 mL. Calculate the concentration of the final solution in parts per billion and in terms of molarity. (See Exercise 9 for definitions.)

12. The World Health Organization lists the maximum desirable concentrations of Mg^{2+} and Ca^{2+} in drinking water as 30. and 75 mg/L, respectively. The maximum permissible concentrations are 1250 and 200. mg/L for the same ions. Calculate all four of these concentrations in moles per liter. (Recall that a decimal point makes trailing zeros count as significant digits.)

13. Although quite controversial in some quarters, low concentrations ($\leq$1 ppm) of fluoride ion in drinking water seem to furnish excellent protection against dental caries. However, at F^- concentrations of 2–3 ppm, a brown mottling of teeth can occur, and harmful toxic effects can occur at 50. ppm. Calculate these concentrations in moles per liter. (See Exercise 9 for definitions.)

14. In the spectroscopic analysis of many substances, a series of standard solutions of known concentration are measured in order to generate a calibration curve. How would you prepare standard solutions containing 10.0, 25.0, 50.0, 75.0, and 100. ppm of copper from a commercially produced 1000.0-ppm solution? Assume each solution has a final volume of 100.0 mL.

15. A stock solution containing Mn^{2+} ions is prepared by dissolving 1.584 g of pure manganese metal in nitric acid and diluting to a final volume of 1.000 L. The following solutions are prepared by dilution:

   For solution A, 50.00 mL of stock solution is diluted to 1000.0 mL.

   For solution B, 10.00 mL of A is diluted to 250.00 mL.

   For solution C, 10.00 mL of B is diluted to 500.0 mL.

   Calculate the concentrations of the stock solution and solutions A, B, and C.

16. You have a stock solution that is 0.200 *M* $FeSO_4$.
   a. Which of the following procedures would give a final solution that is most accurately 1.00×10^{-4} *M* $FeSO_4$? Why?
      i. A 10.00-mL portion (aliquot) of the stock solution is diluted to 200.0 mL using a volumetric pipet and a volumetric flask, followed by two 10:1 dilutions. Assume all volume measurements are accurate to $\pm$ 0.01 mL.
      ii. A 0.050-mL aliquot of the stock solution is diluted to a final volume of 100.0 mL. Assume the 0.050-mL aliquot can be measured to $\pm$1 μL.
   b. In units of mg Fe^{2+}/mL solution, calculate the Fe^{2+} concentration in the 0.200 *M* $FeSO_4$ solution.

Precipitation Reactions

17. Write net ionic equations for the reaction, if any, that occurs when aqueous solutions of the following are mixed.
   a. ammonium sulfate and barium nitrate
   b. lead(II) nitrate and sodium chloride
   c. sodium phosphate and potassium nitrate
   d. sodium bromide and hydrochloric acid
   e. copper(II) chloride and sodium hydroxide

18. How would you separate the following ions in aqueous solution by selective precipitation?
   a. Ag^+, Ba^{2+}, Cr^{3+}
   b. Ag^+, Pb^{2+}, Cu^{2+}
   c. Hg_2^{2+}, Ni^{2+}

19. A lake may be polluted with Pb^{2+} ions. How might you qualitatively test for Pb^{2+}?

20. A sample solution may contain any or all of the following ions: Hg_2^{2+}, Ba^{2+}, and Mn^{2+}. No precipitate formed when an aqueous solution of NaCl or Na_2SO_4 was added to the sample solution. A precipitate formed when the sample

solution was made basic with NaOH. Which ion or ions are present in the sample solution?

21. What volume of 0.100 *M* NaOH is required to precipitate all of the nickel(II) ion from 50.00 mL of 0.175 *M* $Ni(NO_3)_2$?

22. Aluminum can be determined gravimetrically by reaction with a solution of 8-hydroxyquinoline (C_9H_7NO). The net ionic equation is

$$Al^{3+}(aq) + 3C_9H_7NO(aq) \longrightarrow Al(C_9H_6NO)_3(s) + 3H^+(aq)$$

A mass of 0.1248 g of $Al(C_9H_6NO)_3$ was obtained by precipitating all of the Al^{3+} from a solution prepared by dissolving 1.8571 g of a mineral. What is the mass percent of aluminum in the mineral?

23. What mass of barium sulfate is produced when 100.0 mL of a 0.100 *M* solution of barium chloride is mixed with 100.0 mL of a 0.100 *M* solution of iron(III) sulfate?

24. During the developing process of black-and-white film, silver bromide is removed from photographic film by the fixer. The major component of the fixer is sodium thiosulfate. The net ionic equation for the reaction is

$$AgBr(s) + 2S_2O_3^{2-}(aq) \longrightarrow Ag(S_2O_3)_2^{3-}(aq) + Br^-(aq)$$

What mass of AgBr can be dissolved by 1.00 L of 0.200 *M* $Na_2S_2O_3$?

25. The thallium (present as Tl_2SO_4) in a 9.486-g pesticide sample was precipitated as thallium(I) iodide. Calculate the mass percent of Tl_2SO_4 in the sample if 0.1824 g of TlI was recovered.

26. Saccharin ($C_7H_5NO_3S$) is sometimes dispensed in tablet form. Ten tablets with a total mass of 0.5894 g were dissolved in water. This solution was then oxidized to convert all of the sulfur to sulfate ion, which was precipitated by adding an excess of barium chloride solution. The mass of $BaSO_4$ obtained was 0.5032 g. What is the average mass of saccharin per tablet? What is the average mass percent of saccharin in the tablets?

27. Douglasite is a mineral with the formula $2KCl \cdot FeCl_2 \cdot 2H_2O$. Calculate the mass percent of douglasite in a 455.0-mg sample if it took 37.20 mL of a 0.1000 *M* $AgNO_3$ solution to precipitate all of the Cl^- as AgCl. Assume the douglasite is the only source of chloride ion.

28. What mass of $BaSO_4(s)$ forms from the reaction of 100.0 mL of 0.0426 *M* $BaCl_2$ with 50.0 mL of 0.2000 *M* K_2SO_4?

Acid-Base Reactions

29. Write balanced equations (all three types) for the reactions that occur when the following aqueous solutions are mixed.
   a. ammonia and nitric acid
   b. barium hydroxide and hydrochloric acid
   c. perchloric acid [$HClO_4(aq)$] and solid iron(III) hydroxide
   d. solid silver hydroxide and hydrobromic acid

30. Some of the substances commonly used in stomach antacids are MgO, $Mg(OH)_2$, and $Al(OH)_3$.
   a. Write a balanced equation for the neutralization of hydrochloric acid by each of these substances.
   b. Which of these substances will neutralize the greatest amount of 0.10 M HCl per gram?

31. Carminic acid, a naturally occurring red pigment extracted from the cochineal insect, contains only carbon, hydrogen, and oxygen. It was commonly used as a dye in the first half of the nineteenth century. It is 53.66% C and 4.09% H by mass. A titration required 18.02 mL of 0.0406 *M* NaOH to neutralize 0.3602 g of carminic acid. Assuming that there is only one acidic hydrogen per molecule, what is the molecular formula of carminic acid?

32. What volume of each of the following acids will react completely with 50.00 mL of 0.100 *M* NaOH?
   a. 0.100 *M* HCl
   b. 0.100 *M* H_2SO_3 (2 acidic hydrogens)
   c. 0.200 *M* H_3PO_4 (3 acidic hydrogens)
   d. 0.150 *M* HNO_3
   e. 0.200 *M* $HC_2H_3O_2$ (1 acidic hydrogen)
   f. 0.300 *M* H_2SO_4 (2 acidic hydrogens)

33. Sodium hydroxide solution is usually standardized by titrating a pure sample of potassium hydrogen phthalate (KHP), an acid with one acidic hydrogen and a molar mass of 204.22 g. It takes 20.46 mL of a sodium hydroxide solution to titrate a 0.1082-g sample of KHP. What is the molarity of the sodium hydroxide?

34. A 0.500-L sample of H_2SO_4 solution was analyzed by taking a 100.0-mL aliquot and adding 50.0 mL of 0.213 *M* NaOH. After the reaction occurred, an excess of OH^- ions remained in the solution. The excess base required 13.21 mL of 0.103 *M* HCl for neutralization. Calculate the molarity of the original sample of H_2SO_4.

35. Calcium metal will react with water as follows:

$$Ca(s) + 2H_2O(l) \longrightarrow Ca(OH)_2(aq) + H_2(g)$$

What is the molarity of hydroxide ions in the solution formed when 2.58 g of calcium metal is dissolved in enough water to make a final volume of 100.0 mL?

36. A 10.00-mL sample of sulfuric acid from an automobile battery requires 35.08 mL of 2.12 *M* sodium hydroxide solution for complete neutralization. What is the molarity of the sulfuric acid? The reaction is

$$H_2SO_4(aq) + 2NaOH(aq) \longrightarrow 2H_2O(l) + Na_2SO_4(aq)$$

37. A 10.00-mL sample of vinegar, an aqueous solution of acetic acid ($HC_2H_3O_2$), is titrated with 0.5062 *M* NaOH, and 16.58 mL is required to reach the end point.
    a. What is the molarity of the acetic acid?
    b. If the density of the vinegar is 1.006 g/cm^3, what is the mass percent of acetic acid in the vinegar?

38. A 25.00-mL sample of hydrochloric acid solution requires 24.16 mL of 0.106 *M* sodium hydroxide for complete neutralization. What is the concentration of the hydrochloric acid solution?

39. A solution is prepared by dissolving 15.0 g of NaOH in 150.0 mL of 0.250 *M* nitric acid. Will the final solution be acidic, basic, or neutral? Calculate the concentrations of all of the ions present in the solution after the reaction has occurred.

40. A 50.00-mL sample of an ammonia solution is analyzed by titration with HCl. The reaction is

$$NH_3(aq) + H^+(aq) \longrightarrow NH_4^+(aq)$$

It took 39.47 mL of 0.0984 *M* HCl to titrate (react completely with) the ammonia. What is the concentration of the original ammonia solution?

41. Hydrochloric acid (75.0 mL of 0.250 *M*) is added to 225.0 mL of 0.0550 *M* $Ba(OH)_2$ solution. What is the concentration of the excess H^+ or OH^- left in this solution?

42. When organic compounds containing sulfur are burned, sulfur dioxide is produced. The amount of SO_2 formed can be determined by reaction with hydrogen peroxide:

$$H_2O_2(aq) + SO_2(g) \rightarrow H_2SO_4(aq)$$

The resulting sulfuric acid is then titrated with a standard NaOH solution. A 1.325-g sample of coal is burned and the SO_2 collected in a solution of hydrogen peroxide. It took 28.44 mL of 0.1000 *M* NaOH to titrate the resulting sulfuric acid. Calculate the mass percent of sulfur in the coal sample.

Oxidation-Reduction Reactions

43. Assign oxidation states to all atoms in each compound.
    a. $KMnO_4$
    b. NiO_2
    c. $K_4Fe(CN)_6$ (Fe only)
    d. $(NH_4)_2HPO_4$
    e. P_4O_6
    f. Fe_3O_4
    g. $XeOF_4$
    h. SF_4
    i. CO
    j. $Na_2C_2O_4$

44. Assign an oxidation state to chlorine in each of the following anions: OCl^-, ClO_2^-, ClO_3^-, and ClO_4^-.

45. Assign an oxidation state to nitrogen in the following.
    a. Li_3N
    b. NH_3
    c. N_2H_4
    d. NO
    e. N_2O
    f. NO_2
    g. NO_2^-
    h. NO_3^-
    i. N_2

46. Assign oxidation states to all atoms in the following.
    a. UO_2^{2+}
    b. As_2O_3
    c. $NaBiO_3$
    d. As_4
    e. $HAsO_2$
    f. $Mg_2P_2O_7$
    g. $Na_2S_2O_3$
    h. Hg_2Cl_2
    i. $Ca(NO_3)_2$

47. Tell which of the following are oxidation-reduction reactions. For those that are, identify the oxidizing agent, the reducing agent, the substance being oxidized, and the substance being reduced.
    a. $CH_4(g) + 2O_2(g) \rightarrow CO_2(g) + 2H_2O(g)$
    b. $Zn(s) + 2HCl(aq) \rightarrow ZnCl_2(aq) + H_2(g)$
    c. $Cr_2O_7^{2-}(aq) + 2OH^-(aq) \rightarrow 2CrO_4^{2-}(aq) + H_2O(l)$
    d. $O_3(g) + NO(g) \rightarrow O_2(g) + NO_2(g)$
    e. $2H_2O_2(l) \rightarrow 2H_2O(l) + O_2(g)$
    f. $2CuCl(aq) \rightarrow CuCl_2(aq) + Cu(s)$
    g. $HCl(g) + NH_3(g) \rightarrow NH_4Cl(s)$
    h. $SiCl_4(l) + 2H_2O(l) \rightarrow 4HCl(aq) + SiO_2(s)$
    i. $SiCl_4(l) + 2Mg(s) \rightarrow 2MgCl_2(s) + Si(s)$

48. Balance each of the following oxidation-reduction reactions by using the oxidation states method.
    a. $C_2H_6(g) + O_2(g) \rightarrow CO_2(g) + H_2O(g)$
    b. $Mg(s) + HCl(aq) \rightarrow Mg^{2+}(aq) + Cl^-(aq) + H_2(g)$
    c. $Cu(s) + Ag^+(aq) \rightarrow Cu^{2+}(aq) + Ag(s)$
    d. $Cu(s) + HNO_3(aq) \rightarrow Cu(NO_3)_2(aq) + NO(g)$
    e. $Zn(s) + H_2SO_4(aq) \rightarrow ZnSO_4(aq) + H_2(g)$

49. Balance the following oxidation-reduction reactions, which occur in acidic solution, using the half-reaction method.
    a. $Cu(s) + HNO_3(aq) \rightarrow Cu^{2+}(aq) + NO(g)$
    b. $Cr_2O_7^{2-}(aq) + Cl^-(aq) \rightarrow Cr^{3+}(aq) + Cl_2(g)$
    c. $Pb(s) + PbO_2(s) + H_2SO_4(aq) \rightarrow PbSO_4(s)$
    d. $Mn^{2+}(aq) + NaBiO_3(s) \rightarrow Bi^{3+}(aq) + MnO_4^-(aq)$
    e. $H_3AsO_4(aq) + Zn(s) \rightarrow AsH_3(g) + Zn^{2+}(aq)$
    f. $As_2O_3(s) + NO_3^-(aq) \rightarrow H_3AsO_4(aq) + NO(g)$
    g. $Br^-(aq) + MnO_4^-(aq) \rightarrow Br_2(l) + Mn^{2+}(aq)$
    h. $CH_3OH(aq) + Cr_2O_7^{2-}(aq) \rightarrow CH_2O(aq) + Cr^{3+}(aq)$

50. Balance the following oxidation-reduction reactions, which occur in basic solution, using the half-reaction method.
    a. $Al(s) + MnO_4^-(aq) \rightarrow MnO_2(s) + Al(OH)_4^-(aq)$
    b. $Cl_2(g) \rightarrow Cl^-(aq) + ClO^-(aq)$
    c. $NO_2^-(aq) + Al(s) \rightarrow NH_3(g) + AlO_2^-(aq)$
    d. $MnO_4^-(aq) + S^{2-}(aq) \rightarrow MnS(s) + S(s)$
    e. $CN^-(aq) + MnO_4^-(aq) \rightarrow CNO^-(aq) + MnO_2(s)$

51. Balance the following equations by the half-reaction method.
    a. $Fe(s) + HCl(aq) \longrightarrow HFeCl_4(aq) + H_2(g)$
    b. $IO_3^-(aq) + I^-(aq) \xrightarrow{\text{Acidic}} I_3^-(aq)$
    c. $Cr(NCS)_6^{4-}(aq) + Ce^{4+}(aq) \xrightarrow{\text{Acidic}} Cr^{3+}(aq) + Ce^{3+}(aq) + NO_3^-(aq) + CO_2(g) + SO_4^{2-}(aq)$

d. $CrI_3(s) + Cl_2(g) \xrightarrow{\text{Basic}} CrO_4^{2-}(aq) + IO_4^-(aq) + Cl^-(aq)$

e. $Fe(CN)_6^{4-}(aq) + Ce^{4+}(aq) \xrightarrow{\text{Basic}} Ce(OH)_3(s) + Fe(OH)_3(s) + CO_3^{2-}(aq) + NO_3^-(aq)$

f. $Fe(OH)_2(s) + H_2O_2(aq) \xrightarrow{\text{Basic}} Fe(OH)_3(s)$

52. The Ostwald process for the commercial production of nitric acid involves the following three steps:

$$4NH_3(g) + 5O_2(g) \longrightarrow 4NO(g) + 6H_2O(g)$$
$$2NO(g) + O_2(g) \longrightarrow 2NO_2(g)$$
$$3NO_2(g) + H_2O(l) \longrightarrow 2HNO_3(aq) + NO(g)$$

a. Which reactions in the Ostwald process are oxidation-reduction reactions?
b. Identify the oxidizing agent and the reducing agent in each redox reaction.
c. Consider the first step of the Ostwald process. How much nitric oxide, NO, can be produced from a mixture of 5.0×10^6 g of ammonia and 5.0×10^7 g of O_2?

53. Chlorine gas was first prepared in 1774 by C. W. Scheele by oxidizing sodium chloride with manganese(IV) oxide. The reaction is

$$NaCl(aq) + H_2SO_4(aq) + MnO_2(s) \longrightarrow Na_2SO_4(aq) + MnCl_2(aq) + H_2O(l) + Cl_2(g)$$

Balance this reaction by the oxidation states method.

54. One of the classical methods for the determination of the manganese content in steel involves converting all the manganese to the deeply colored permanganate ion and then measuring the absorption of light. The steel is first dissolved in nitric acid, producing the manganese(II) ion and nitrogen dioxide gas. This solution is then reacted with an acidic solution containing periodate ion; the products are the permanganate and iodate ions. Write balanced chemical equations for both of these steps.

55. Gold metal will not dissolve in either concentrated nitric acid or concentrated hydrochloric acid. It will dissolve, however, in *aqua regia,* a mixture of the two concentrated acids. The products of the reaction are the $AuCl_4^-$ ion and gaseous NO. Write a balanced equation for the dissolution of gold in aqua regia.

56. A solution of permanganate is standardized by titration with oxalic acid ($H_2C_2O_4$). It required 28.97 mL of the permanganate solution to react completely with 0.1058 g of oxalic acid. The unbalanced equation for the reaction is

$$MnO_4^-(aq) + H_2C_2O_4(aq) \xrightarrow{\text{Acidic}} Mn^{2+}(aq) + CO_2(g)$$

What is the molarity of the permanganate solution?

57. A 50.00-mL sample of solution containing Fe^{2+} ions is titrated with a 0.0216 *M* $KMnO_4$ solution. It required 20.62 mL of the $KMnO_4$ solution to oxidize all of the Fe^{2+} ions to Fe^{3+} ions by the reaction

$$MnO_4^-(aq) + Fe^{2+}(aq) \xrightarrow{\text{Acidic}} Mn^{2+}(aq) + Fe^{3+}(aq) \quad \text{(Unbalanced)}$$

a. What was the concentration of Fe^{2+} ions in the sample solution?
b. What volume of 0.0150 *M* $K_2Cr_2O_7$ solution would it take to do the same titration? The reaction is

$$Cr_2O_7^{2-}(aq) + Fe^{2+}(aq) \xrightarrow{\text{Acidic}} Cr^{3+}(aq) + Fe^{3+}(aq) \quad \text{(Unbalanced)}$$

58. The iron content of iron ore can be determined by titration with standard $KMnO_4$ solution. The iron ore is dissolved in HCl, and all of the iron is reduced to Fe^{2+} ions. This solution is then titrated with $KMnO_4$ solution, producing Fe^{3+} and Mn^{2+} ions in acidic solution. If it required 38.37 mL of 0.0198 *M* $KMnO_4$ to titrate a solution made from 0.6128 g of iron ore, what is the mass percent of iron in the iron ore?

59. What mass of CO_2 is produced from the reaction of 0.250 g of $Na_2C_2O_4$ with 50.00 mL of 0.0200 *M* $KMnO_4$ solution in the presence of acid? (See Exercise 56.)

60. A piece of copper metal (1.06 g) is placed in 250 mL of a 0.20 *M* $AgNO_3$ solution. Will all of the copper dissolve in this solution? The net ionic equation is

$$Cu(s) + 2Ag^+(aq) \longrightarrow 2Ag(s) + Cu^{2+}(aq)$$

61. Triiodide ions are generated in solution by the following (unbalanced) reaction in acidic solution:

$$IO_3^-(aq) + I^-(aq) \rightarrow I_3^-(aq)$$

Triiodide ion is determined by titration with a sodium thiosulfate ($Na_2S_2O_3$) solution. The products are iodide ion and tetrathionate ion ($S_4O_6^{2-}$).
a. Balance the equation for the reaction of IO_3^- with I^- ions.
b. A sample of 0.6013 g of potassium iodate was dissolved in water. Hydrochloric acid and solid potassium iodide were then added in excess. What is the minimum mass of solid KI and the minimum volume of 3.00 *M* HCl required to convert all of the IO_3^- ions to I_3^- ions?
c. Write and balance the equation for the reaction of $S_2O_3^{2-}$ with I_3^- in acidic solution.
d. A 25.00-mL sample of a 0.0100 *M* solution of KIO_3 is reacted with an excess of KI. It requires 32.04 mL of $Na_2S_2O_3$ solution to titrate the I_3^- ions present. What is the molarity of the $Na_2S_2O_3$ solution?
e. How would you prepare 500.0 mL of the KIO_3 solution in part d, using pure dry KIO_3?

Additional Exercises

62. The zinc in a 1.200-g sample of foot powder was precipitated as $ZnNH_4PO_4$. Strong heating of the precipitate yielded 0.4089 g of $Zn_2P_2O_7$. Calculate the mass percent of zinc in the sample of foot powder.

63. Lead chromate (commonly called chrome yellow) is used as a pigment in paints. What mass of chrome yellow is produced when 100.0 mL of 0.4100 *M* sodium chromate is mixed with 100.0 mL of 0.3200 *M* lead(II) nitrate?

64. Rust stains can be removed by washing the surface with a dilute solution of oxalic acid ($H_2C_2O_4$). The reaction is

$$Fe_2O_3(s) + 6H_2C_2O_4(aq) \longrightarrow 2Fe(C_2O_4)_3^{3-}(aq) + 3H_2O(l) + 6H^+(aq)$$

a. Is this an oxidation-reduction reaction?
b. What mass of rust can be removed by 1.0 L of a 0.14 M solution of oxalic acid?

65. A mixture contains only NaCl and $Fe(NO_3)_3$. A 0.456-g sample of the mixture is dissolved in water, and an excess of NaOH is added, producing a precipitate of $Fe(OH)_3$. The precipitate is filtered, dried, and weighed. Its mass is 0.107 g. Calculate the following.

a. the mass of iron in the sample
b. the mass of $Fe(NO_3)_3$ in the sample
c. the mass percent of $Fe(NO_3)_3$ in the sample

66. A sample is a mixture of KCl and KBr. When 0.1024 g of the sample is dissolved in water and reacted with excess silver nitrate, 0.1889 g of solid is obtained. What is the composition by mass percent of the mixture?

67. What mass of $Fe(OH)_3$ would be produced by reacting 75.0 mL of 0.105 *M* $Fe(NO_3)_3$ with 125 mL of 0.150 *M* NaOH?

68. A mixture contains only sodium chloride and potassium chloride. A 0.1586-g sample of the mixture was dissolved in water. It took 22.90 mL of 0.1000 *M* $AgNO_3$ to completely precipitate all of the chloride present. What is the composition (by mass) of the mixture?

69. The concentration of a potassium iodide solution can be determined by first adding excess silver nitrate:

$$I^-(aq) + Ag^+(aq) \rightarrow AgI(s)$$

The excess silver ion remaining in solution is then determined by reaction with a potassium thiocyanate (KSCN) solution of known concentration:

$$Ag^+(aq) + SCN^-(aq) \rightarrow AgSCN(s)$$

In an experiment, 50.00 mL of 0.0565 *M* $AgNO_3$ was added to 25.00 mL of a potassium iodide solution. It then took 8.32 mL of 0.0510 *M* KSCN solution to precipitate the unreacted silver ions. What is the concentration of the original KI solution?

70. Polychlorinated biphenyls (PCBs) have been used extensively as dielectric materials in electrical transformers. Because PCBs have been shown to be potentially harmful, analysis for their presence in the environment has become very important. PCBs are manufactured according to the following generic reaction:

$$C_{12}H_{10} + nCl_2 \xrightarrow[\text{catalyst}]{\text{Fe}} C_{12}H_{10-n}Cl_n + nHCl$$

This reaction results in a mixture of PCB products. The mixture is analyzed by decomposing the PCBs and then precipitating the resulting Cl^- as AgCl.

a. Develop a general equation that relates the average value of n to the mass of a given mixture of PCBs and the mass of AgCl produced.
b. A 0.1947-g sample of a commercial PCB yielded 0.4971 g of AgCl. What is the average value of n for this sample?

71. A stream flows at a rate of 5.00×10^4 liters per second (L/s) upstream of a manufacturing plant. The plant discharges 3.50×10^3 L/s of water that contains 65.0 ppm HCl into the stream. (See Exercise 9 for definitions.)

a. Calculate the stream's total flow rate downstream from this plant.
b. Calculate the concentration of HCl in ppm downstream from this plant.
c. Further downstream, another manufacturing plant diverts 1.80×10^4 L/s of water from the stream for its own use. This plant must first neutralize the acid and does so by adding lime:

$$CaO(s) + 2H^+(aq) \rightarrow Ca^{2+}(aq) + H_2O(l)$$

What mass of CaO is consumed in an 8.00-h work day by this plant?
d. The original stream water contained 10.2 ppm Ca^{2+}. Although no calcium was in the waste water from the first plant, the waste water of the second plant contains Ca^{2+} from the neutralization process. If 90.0% of the water used by the second plant is returned to the stream, calculate the concentration of Ca^{2+} in ppm downstream of the second plant.

72. Chromium has been investigated as a coating for steel cans. The thickness of the chromium film is determined by dissolving a sample of a can in acid and oxidizing the resulting Cr^{3+} to $Cr_2O_7^{2-}$ with the peroxydisulfate ion:

$$S_2O_8^{2-}(aq) + Cr^{3+}(aq) + H_2O(l) \rightarrow Cr_2O_7^{2-}(aq) + SO_4^{2-}(aq) + H^+(aq) \text{ (Unbalanced)}$$

After removal of unreacted $S_2O_8^{2-}$, an excess of ferrous ammonium sulfate [$Fe(NH_4)_2(SO_4)_2 \cdot 6H_2O$] is added, reacting with $Cr_2O_7^{2-}$ produced from the first reaction. The unreacted Fe^{2+} from the excess ferrous ammonium sulfate is titrated with a separate $K_2Cr_2O_7$ solution. The reaction is:

$$H^+(aq) + Fe^{2+}(aq) + Cr_2O_7^{2-}(aq) \rightarrow Fe^{3+}(aq) + Cr^{3+}(aq) + H_2O(l) \text{ (Unbalanced)}$$

a. Write balanced chemical equations for the two reactions.
b. In one analysis, a 40.0-cm^2 sample of a chromium plated can was treated according to this procedure. After dissolution and removal of excess $S_2O_8^{2-}$, 3.000 g of $Fe(NH_4)_2(SO_4)_2 \cdot 6H_2O$ was added. It took 8.58 mL of 0.0520 M $K_2Cr_2O_7$ solution to completely react with the excess Fe^{2+}. Calculate the thickness of the chromium film on the can. (The density of chromium is 7.19 g/cm^3.)

73. One high-temperature superconductor has the general formula $YBa_2Cu_3O_x$. The copper is a mixture of Cu(II) and Cu(III) oxidation states. This mixture of oxidation states appears vital for high-temperature superconductivity tooccur. A simple method for determining the average copper oxidation state has been reported [D. C. Harris, M. E. Hillis, and T. A. Hewston, *J. Chem. Educ.*64, 847(1987).] The described analysis takes place in two steps:

i. One superconductor sample is treated directly with I^-:

$$Cu^{2+}(aq) + I^-(aq) \rightarrow CuI(s) + I_3^-(aq) \text{ (Unbalanced)}$$

$$Cu^{3+}(aq) + I^-(aq) \rightarrow CuI(s) + I_3^-(aq) \text{ (Unbalanced)}$$

ii. A second superconductor sample is dissolved in acid, converting all copper to Cu(II). This solution is then treated with I^-:

$$Cu^{2+}(aq) + I^-(aq) \rightarrow CuI(s) + I_3^-(aq) \text{ (Unbalanced)}$$

In both steps the I_3^- is determined by titrating with a standard sodium thiosulfate ($Na_2S_2O_3$) solution:

$$I_3^-(aq) + S_2O_3^{2-}(aq) \rightarrow S_4O_6^{2-}(aq) + I^-(aq) \text{ (Unbalanced)}$$

a. Calculate the average copper oxidation states for materials with the formulas $YBa_2Cu_3O_{6,5}$, YBa_2Cu_3O, and $YBa_2Cu_3O_8$. Interpret your results in terms of a mixture of Cu(II) and Cu(III) ions, assuming that only Y^{3+}, Ba^{2+}, and O^{2-} are present in addition to the copper ions.
b. Balance the equations involved in the copper analysis.
c. A superconductor sample was analyzed by the above procedure. In step i it took 37.77 mL of 0.1000 M $Na_2S_2O_3$ to react completely with the I_3^- generated from a 562.5-mg sample. In step ii it took 22.57 mL of 0.1000 M $Na_2S_2O_3$ to react with the I_3^- generated by a 504.2-mg sample. Determine the formula of this superconductor sample (i.e., find the value of x in $YBa_2Cu_3O_x$). Calculate the average oxidation state of copper in this material.

74. You are given a compound that is either iron(II) sulfate or iron(III) sulfate. How could you determine the identity of the compound by using a dilute solution of potassium permanganate?

75. You are given a solid that is a mixture of Na_2SO_4 and K_2SO_4. A 0.205-g sample of the mixture is dissolved in water. An excess of an aqueous solution of $BaCl_2$ is added. The $BaSO_4$ that is formed is filtered, dried, and weighed. Its mass is 0.298 g. What mass of SO_4^{2-} ion is in the sample? What is the mass percent of SO_4^{2-} ion in the sample? What are the percent compositions by mass of Na_2SO_4 and K_2SO_4 in the sample?

76. A solution is prepared by dissolving 0.150 $\pm$ 0.003 g of NaCl in a volumetric flask. The volume of the flask is 100.0 $\pm$ 0.5 mL. What is the molarity of NaCl in the solution? What is the range of values for the molarity of NaCl? Express this as a value $\pm$ the uncertainty. (See Appendix Section A1.5.)

77. It took 25.06 $\pm$ 0.05 mL of a sodium hydroxide solution to titrate a 0.4016-g sample of KHP (see Exercise 33). Calculate the concentration and uncertainty in the concentration of the sodium hydroxide solution. (See Appendix Section A1.5.) Neglect any uncertainty in the mass.

78. You wish to prepare 1 L of a 0.02 M potassium iodate solution. You require that the final concentration be within 1% of 0.02 M and that the concentration must be known accurately to the fourth decimal place. How would you prepare this solution? Specify the glassware you would use, the accuracy needed for the balance, and the ranges of acceptable masses of KIO_3 that can be used.

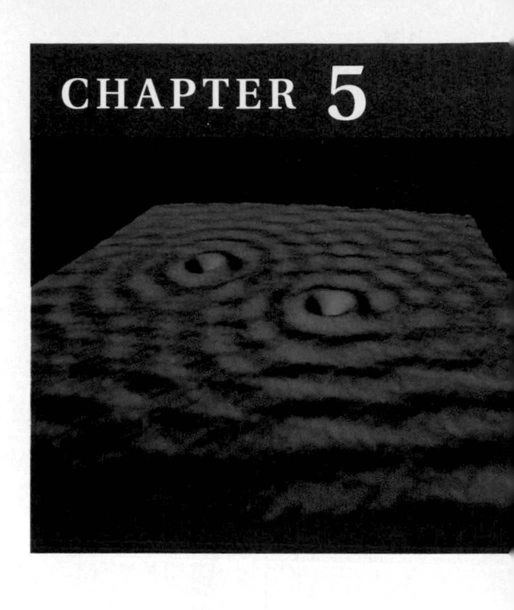

Gases

Matter exists in three distinct physical states: gas, liquid, and solid. Of these, the gaseous state is the easiest to describe both experimentally and theoretically. In particular, the study of gases provides an excellent example of the scientific method in action. It illustrates how observations lead to natural laws, which in turn can be accounted for by models. Then, as more accurate measurements become available, the models are modified.

In addition to providing a good illustration of the scientific method, gases are important in their own right. For example, gases are often produced in chemical reactions and thus must be dealt with in stoichiometric calculations. Also, the earth's atmosphere is a mixture of gases, mainly elemental nitrogen and oxygen; it both supports life and acts as a waste receptacle for the exhaust gases that accompany many industrial processes.

Therefore it is important to understand the behavior of gases. We will pursue this goal by considering the properties of gases, the laws and models that describe the behavior of gases, and finally the reactions that occur among the gases in the atmosphere.

5.1 Early Experiments

Even though the Greeks considered "air" to be one of the four fundamental elements and various alchemists obtained "airs," or "vapors," in their experiments, careful study of these elusive substances proved difficult. The first

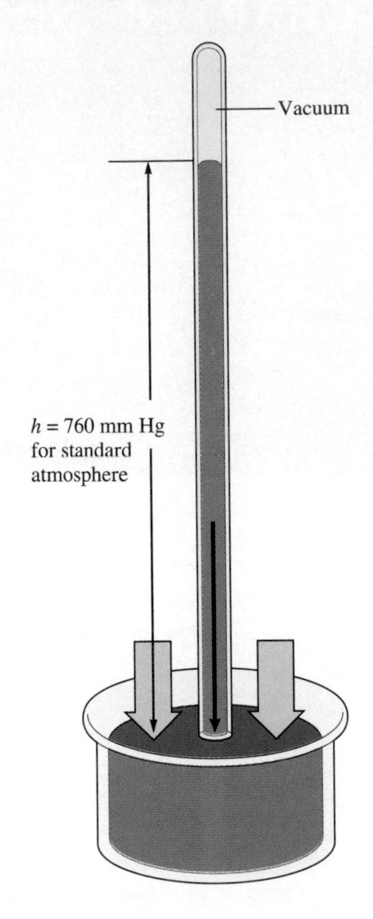

Figure 5.1
A torricellian barometer. The tube, completely filled with mercury, is inverted in a dish of mercury. Mercury flows out of the tube until the pressure of the column of mercury (shown by black arrow) "standing on the surface" of the mercury in the dish is equal to the pressure of the air (shown by yellow arrows) on the rest of the surface of the mercury in the dish.

person to attempt a scientific study of the "vapors" produced in chemical reactions was a Flemish physician named Jan Baptista Van Helmont (1577–1644). Thinking that air and similar substances must be akin to the "chaos" from which, according to Greek myth, the universe was created, Van Helmont described these substances using the Flemish word for *chaos,* which was *gas.*

Van Helmont extensively studied a gas he obtained from burning wood, which he called "gas sylvestre" and which we now know as carbon dioxide, and noted that this substance was similar in many ways but not identical to air. By the end of his life, the importance of gases, especially air, was becoming more apparent. In 1643 an Italian physicist named Evangelista Torricelli (1608–1647), who had been a student of Galileo, performed experiments that showed that *the air in the atmosphere exerts pressure.* (In fact, as we will see, all gases exert pressure.) Torricelli designed the first **barometer** by filling a tube that was closed at one end with mercury and then inverting it in a dish of mercury (see Fig. 5.1). He observed that a column of mercury approximately 760 millimeters long always remained in the tube as a result of the pressure of the atmosphere.

A few years later Otto von Guericke, a German physicist, invented an air pump, often called a vacuum pump, which he used in a famous demonstration for the king of Prussia in 1654. Guericke placed two hemispheres together and pumped the air out of the resulting sphere through a valve, which was subsequently closed. He then dramatically showed that teams of horses could not pull the hemispheres apart. However, after secretly opening the valve to let air in, Guericke was able to separate the hemispheres easily by hand. The king of Prussia was so impressed by Guericke's cleverness that he awarded him a lifetime pension.

Units of Pressure

Because instruments used for measuring pressure, such as the **manometer** (see Fig. 5.2), so often utilize columns of mercury because of its high density, the most commonly used units for pressure are based on the height of the mercury column (in millimeters) the gas pressure can support. The unit **mm Hg** (millimeters of mercury) is called the **torr** in honor of Torricelli. A related unit for pressure is the **standard atmosphere:**

$$1 \text{ standard atmosphere} = 1 \text{ atm} = 760 \text{ mm Hg} = 760 \text{ torr}$$

However, since pressure is defined as force per unit area,

$$\text{Pressure} = \frac{\text{force}}{\text{area}}$$

the fundamental units of pressure involve units of force divided by units of area. In the SI system the unit of force is the newton (N) and the unit of area is meters squared (m^2). (For a review of the SI system, see Appendix Two.) Thus the unit of pressure in the SI system is newtons per meter squared (N/m^2), called the **pascal (Pa).** In terms of pascals the standard atmosphere is

$$1 \text{ standard atmosphere} = 101{,}325 \text{ Pa}$$

$$\begin{aligned} 1 \text{ atm} &= 760 \text{ mm Hg} \\ &= 760 \text{ torr} \\ &= 101{,}325 \text{ Pa} \\ &= 29.92 \text{ in Hg} \\ &= 14.7 \text{ lb/in}^2 \end{aligned}$$

Thus 1 atmosphere is about 10^5 pascals. Since the pascal is so small, and since it

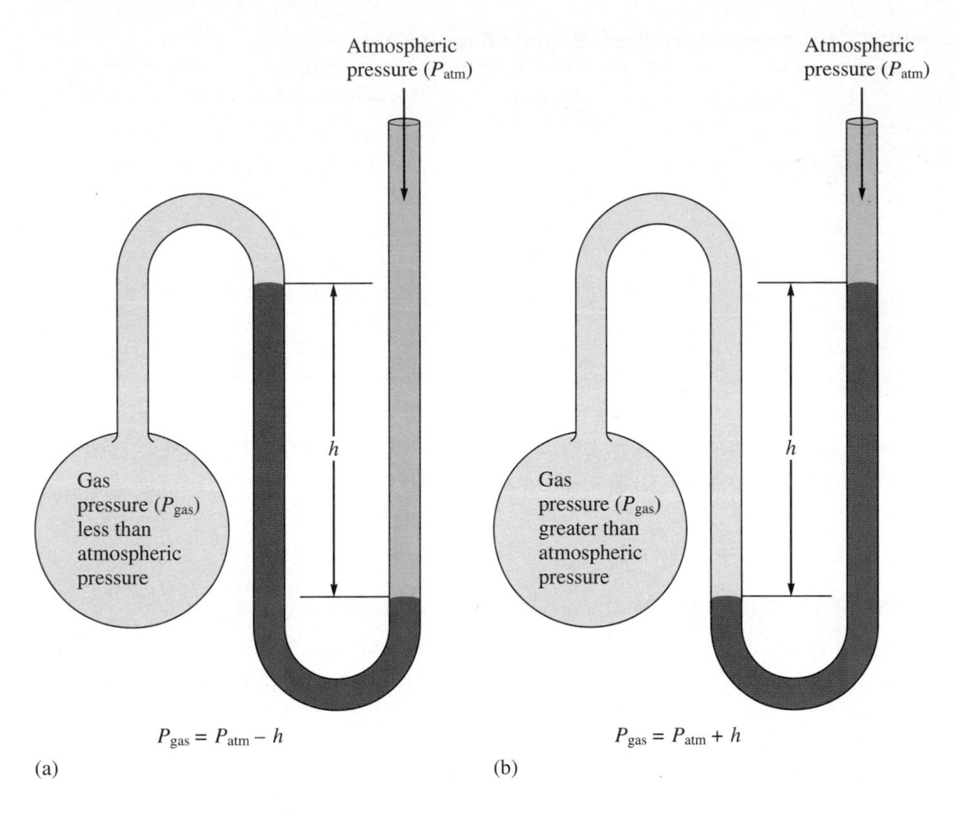

Figure 5.2
A simple manometer, a device for measuring the pressure of a gas in a container. The pressure of the gas is given by h (the difference in mercury levels) in units of torr (equivalent to mm Hg). (a) Gas pressure = atmospheric pressure − h. (b) Gas pressure = atmospheric pressure + h.

is not commonly used in the United States, we will use it sparingly in this book. However, converting from torrs or atmospheres to pascals is straightforward.

5.2 The Gas Laws of Boyle, Charles, and Avogadro

Boyle's Law

The first quantitative experiments on gases were performed by an Irish chemist, Robert Boyle (1627–1691). Using a J-shaped tube closed at one end (Fig. 5.3), which he reportedly set up in the multistory entryway of his house, Boyle studied the relationship between the pressure of the trapped gas and its volume. Representative values from Boyle's experiments are given in Table 5.1. These data show that the product of the pressure and volume for the trapped air sample is constant within the accuracies of Boyle's measurements (note the third column in Table 5.1). This behavior can be represented by the equation

$$PV = k$$

which is called **Boyle's law**, where k is a constant at a specific temperature for a given sample of air.

It is convenient to represent the data in Table 5.1 by using two different plots. Figure 5.4(a) shows a plot of P versus V, which produces a hyperbola. Notice that as the pressure drops by half, the volume doubles. Thus there is an

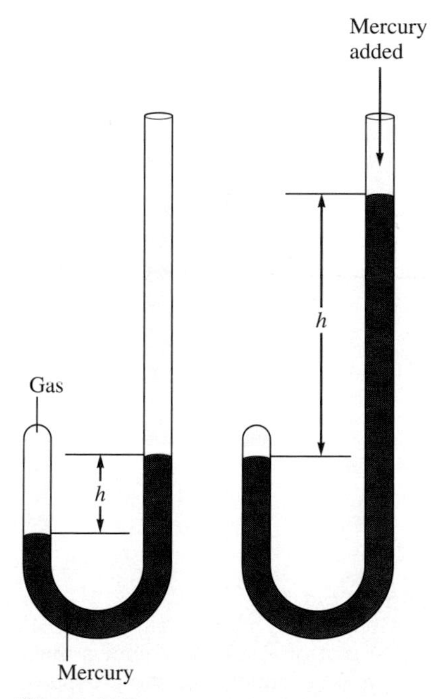

Figure 5.3
A J-tube similar to the one used by Boyle.

Boyle's law: $V \propto 1/P$ at constant temperature.

TABLE 5.1 **Actual Data from Boyle's Experiments**

Volume (in^3)	Pressure (in of Hg)	Pressure × Volume (in of Hg × in^3)
48.0	29.1	14.0×10^2
40.0	35.3	14.1×10^2
32.0	44.2	14.1×10^2
24.0	58.8	14.1×10^2
20.0	70.7	14.1×10^2
16.0	87.2	14.0×10^2
12.0	117.5	14.1×10^2

inverse relationship between pressure and volume. The second type of plot can be obtained by rearranging Boyle's law to give

$$V = \frac{k}{P}$$

which is the equation for a straight line of the type

$$y = mx + b$$

Graphing is reviewed in Appendix Section A1.3.

where m represents the slope and b the intercept of the straight line. In this case, $y = V$, $x = 1/P$, $m = k$, and $b = 0$. Thus a plot of V versus $1/P$ using Boyle's data gives a straight line with an intercept of zero, as shown in Fig. 5.4(b).

Boyle's law only approximately describes the relationship between pressure and volume for a gas. Highly accurate measurements on various gases at a constant temperature have shown that the product PV is not quite constant but changes with pressure. Results for several gases are shown in Fig. 5.5. Note the small changes that occur in the product PV as the pressure is varied. Such changes become very significant at pressures much higher than normal atmospheric pressure. We will discuss these deviations and the reasons for them in detail in Section 5.10. *A gas that obeys Boyle's law is called an* **ideal gas.** We will describe the characteristics of an ideal gas more completely in Section 5.3.

One common use of Boyle's law is to predict the new volume of a gas when the pressure is changed (at constant temperature), or vice versa.

We mentioned above that Boyle's law is only approximately followed for real gases. To determine the significance of the deviations, studies of the effect

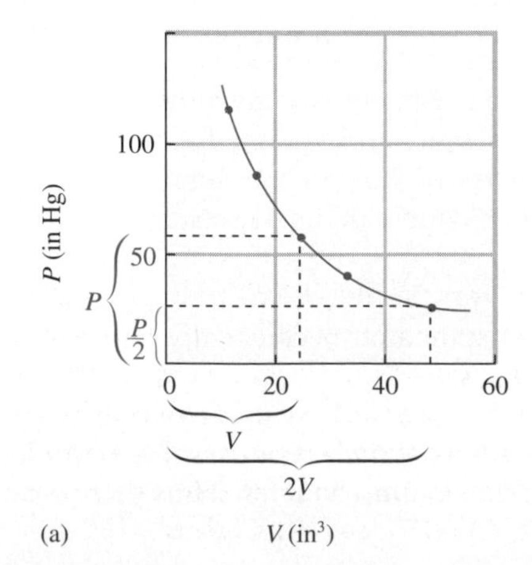

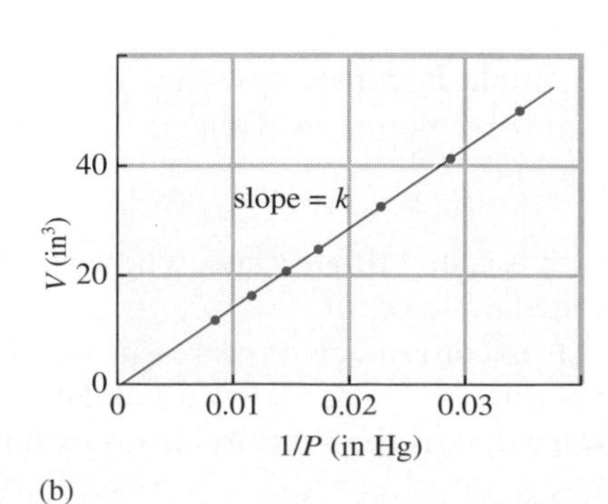

Figure 5.4
Plotting Boyle's data from Table 5.1. (a) A plot of P versus V shows that the volume doubles as the pressure is halved. (b) A plot of V versus $1/P$ gives a straight line. The slope of this line equals the value of the constant k.

of changing pressure on the volume of a gas are often carried out, as shown in Example 5.1.

EXAMPLE 5.1

In a study to see how closely gaseous ammonia obeys Boyle's law, several volume measurements were made at various pressures, using 1.0 mol of NH_3 gas at a temperature of 0°C. Using the results listed below, calculate the Boyle's law constant for NH_3 at the various pressures.

Experiment	Pressure (atm)	Volume (L)
1	0.13	172.1
2	0.25	89.28
3	0.30	74 35
4	0.50	44.49
5	0.75	29.55
6	1.00	22.08

Solution

To determine how closely NH_3 gas follows Boyle's law under these conditions, we calculate the value of k (in L atm) for each set of values:

Experiment	1	2	3	4	5	6
$k = PV$	22.37	22.32	22.31	22.25	22.16	22.08

Although the deviations from true Boyle's law behavior are quite small at these low pressures, note that the value of k changes regularly in one direction as the pressure is increased. Thus to calculate the "ideal" value of k for NH_3, we can plot PV versus P, as shown in Fig. 5.6, and extrapolate (extend the line beyond the experimental points) back to zero pressure, where, for reasons we will discuss later, a gas behaves most ideally. The value of k obtained by this extrapolation is 22.41 L atm. Notice that this is the same value obtained from similar plots for the gases CO_2, O_2, and Ne at 0°C, as shown in Fig. 5.5.

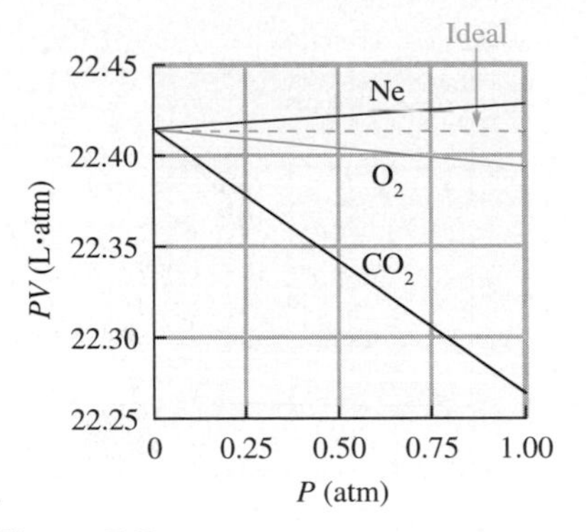

Figure 5.5
A plot of PV versus P for several gases. An ideal gas is expected to have a constant value of PV, as shown by the dotted line. Carbon dioxide shows the largest change in PV, and this change is actually quite small: PV changes from about 22.39 L atm at 0.25 atm to 22.26 L atm at 1.00 atm. Thus Boyle's law is a good approximation at these relatively low pressures.

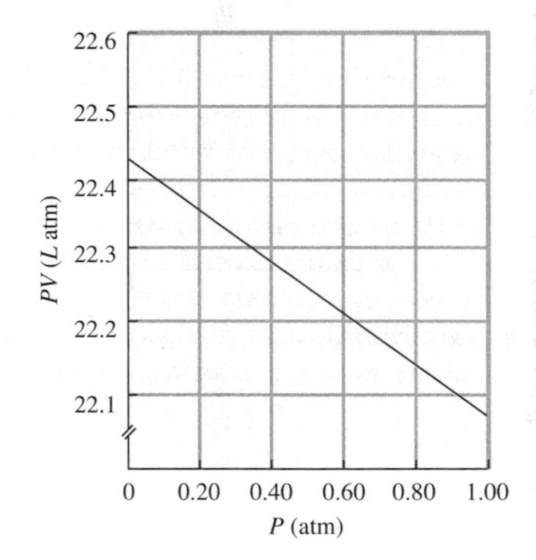

Figure 5.6
A plot of PV versus P for 1 mol of ammonia. The dashed line shows the extrapolation of the data to zero pressure to give the "ideal" value of PV of 22.41 L atm.

Charles's Law

As with Boyle's law, Charles's law is obeyed exactly only at relatively low pressures.

In the century following Boyle's findings, scientists continued to study the properties of gases. One of these scientists was a French physicist, Jacques Charles (1746–1823), who was the first person to fill a balloon with hydrogen gas and who made the first solo balloon flight. Charles found in 1787 that the volume of a gas at constant pressure increases *linearly* with the temperature of the gas. That is, a plot of the volume of a gas (at constant pressure) versus its temperature (°C) gives a straight line. This behavior is shown for several gases

Charles's law: $V \propto T$ (expressed in K) at constant pressure.

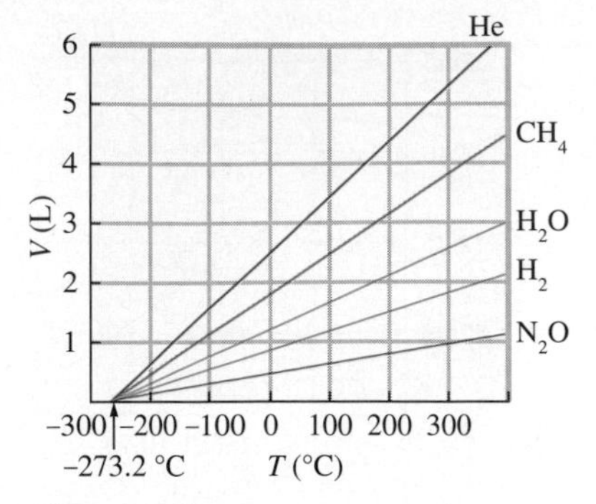

Figure 5.7
Plots of V versus T (°C) for several gases. The solid lines represent experimental measurements on gases. The dashed lines represent extrapolation of the data into regions where these gases would become liquids or solids.

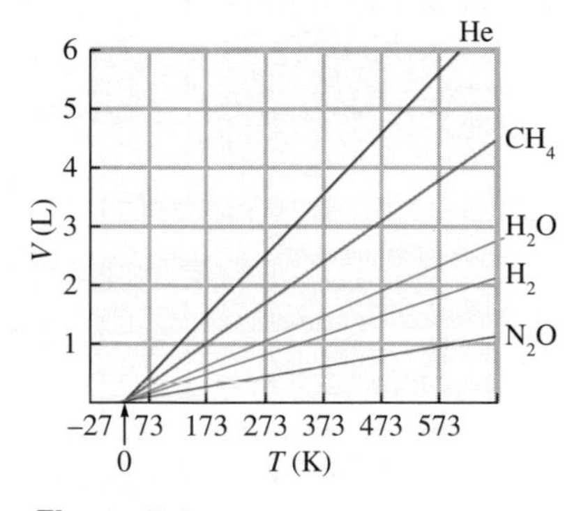

Figure 5.8
Plots of V versus T as in Fig. 5.7, except that here the Kelvin scale is used for temperature.

in Fig. 5.7. One very interesting feature of these plots is that the volumes of all the gases extrapolate to zero at the same temperature, $-273.2°C$. On the Kelvin temperature scale this point is defined as 0 K, which leads to the following relationship between the Kelvin and Celsius scales:

$$\text{Temperature (K)} = 0°\text{C} + 273$$

When the volumes of the gases shown in Fig. 5.7 are plotted versus temperature on the Kelvin scale, the plots in Fig. 5.8 result. In this case the volume of each gas is *directly proportional to temperature* and extrapolates to zero when the temperature is 0 K. This behavior is represented by the equation known as **Charles's law,**

$$V = bT$$

where T is the temperature (in Kelvins) and b is a proportionality constant.

Before we illustrate the uses of Charles's law, let us consider the importance of 0 K. At temperatures below this point, the extrapolated volumes would become negative. The fact that a gas cannot have a negative volume suggests that 0 K has a special significance. In fact, 0 K is called **absolute zero,** and there is much evidence to suggest that this temperature cannot be attained. Temperatures of approximately 10^{-6} K have been produced in laboratories, but 0 K has never been reached.

Avogadro's Law

In Chapter 2 we noted that in 1811 the Italian chemist Avogadro postulated that equal volumes of gases at the same temperature and pressure contain the same number of "particles." This observation is called **Avogadro's law,** which can be stated mathematically as

$$V = an$$

where V is the volume of the gas, n is the number of moles, and a is a proportionality constant. This equation states that *for a gas at constant temperature and pressure the volume is directly proportional to the number of moles of gas.* This relationship is obeyed closely by gases at low pressures.

5.3 The Ideal Gas Law

We have considered three laws that describe the behavior of gases as revealed by experimental observations:

Boyle's law:	$V = \dfrac{k}{P}$	(at constant T and n)
Charles's law:	$V = bT$	(at constant P and n)
Avogadro's law:	$V = an$	(at constant T and P)

These relationships showing how the volume of a gas depends on pressure, temperature, and number of moles of gas present, can be combined as follows:

$$V = R\left(\frac{Tn}{P}\right)$$

where R is the combined proportionality constant called the **universal gas constant.** When the pressure is expressed in atmospheres and the volume in liters, R has the value 0.08206 L atm K^{-1} mol^{-1}. The above equation can be rearranged to the more familiar form of the **ideal gas law:**

R = 0.08206 L atm K^{-1} mol^{-1}.

$$PV = nRT$$

The ideal gas law is an *equation of state* for a gas, where the state of the gas is its condition at a given time. A particular *state* of a gas is described by its pressure, volume, temperature, and number of moles. Knowledge of any three of these properties is enough to completely define the state of a gas, since the fourth property can then be determined from the equation for the ideal gas law.

It is important to recognize that the ideal gas law is an empirical equation—it is based on experimental measurements of the properties of gases. A gas that obeys this equation is said to behave *ideally*. That is, this equation defines the behavior of an ideal gas, which is a hypothetical substance. The ideal gas equation is best regarded as a limiting law—it expresses behavior that real gases *approach* at low pressures and high temperatures. Most gases obey this equation closely enough at pressures below 1 atmosphere that only minimal errors result from assuming ideal behavior. Unless you are given information to the contrary, you should assume ideal gas behavior when solving problems involving gases in this text.

The ideal gas law applies best at pressures smaller than 1 atm.

COLD ATOMS

Over two hundred years ago the work of Charles and Gay-Lussac led to the suspicion that an absolute low temperature exists for matter. In recent years scientists have come very close to cooling matter to 0 K. The latest low-temperature record was achieved at the University of Colorado in Boulder when a team of scientists led by Carl Wieman reported that they had cooled a sample containing 2×10^7 cesium atoms to 1.1×10^{-6} K, about one-millionth of a degree above absolute zero. This record-low temperature was achieved by a technique known as "laser cooling," in which a laser beam is directed against a beam of individual atoms, dramatically slowing the atoms. The atoms are then further cooled in an "optical molasses" produced by the intersection of six laser beams. By adding a trapping magnetic field, the Colorado scientists were able to hold the supercold cesium atoms for about 1 s, and the possibility exists that the cold atoms could be trapped for much longer periods of time with improvements in the apparatus.

Trapping atoms at these extremely low temperatures for several seconds should allow the study of low-energy collisions, of how atoms attract each other to form aggregates, and of other properties that would provide tests of the fundamental theories of matter.

EXAMPLE 5.2

A sample of hydrogen gas (H_2) has a volume of 8.56 L at a temperature of 0°C and a pressure of 1.5 atm. Calculate the moles of H_2 present in this gas sample.

Solution

Solving the ideal gas law for n gives

$$n = \frac{PV}{RT}$$

In this case $P = 1.5$ atm, $V = 8.56$ L, $T = 0°\text{C} + 273 = 273$ K, and $R = 0.08206 \text{ L atm K}^{-1} \text{ mol}^{-1}$. Thus

$$n = \frac{(1.5 \text{ atm})(8.56 \text{ L})}{\left(0.08206 \frac{\text{L atm}}{\text{K mol}}\right)(273 \text{ K})} = 0.57 \text{ mol}$$

The ideal gas law is often used to calculate the changes that will occur when the conditions of a gas are changed, as illustrated below.

EXAMPLE 5.3

Suppose we have a sample of ammonia gas with a volume of 3.5 L at a pressure of 1.68 atm. The gas is compressed to a volume of 1.35 L at a constant temperature. Use the ideal gas law to calculate the final pressure.

Solution

The basic assumption we make when using the ideal gas law to describe a change in state for a gas is that the equation applies equally well to both the initial and the final states. In dealing with a change in state, we always *place the variables on one side of the equals sign and the constants on the other.* In this case the pressure and volume change, while the temperature and the number of moles remain constant (as does R, by definition). Thus we write the ideal gas law as

$$PV = nRT$$

↗ Change ↖ Remain constant

Since n and T remain the same in this case, we can write $P_1V_1 = nRT$ and $P_2V_2 = nRT$. Combining these equations gives

$$P_1V_1 = nRT = P_2V_2 \quad \text{or} \quad P_1V_1 = P_2V_2$$

We are given $P_1 = 1.68$ atm, $V_1 = 3.5$ L, $V_2 = 1.35$ L. Solving for P_2 gives

$$P_2 = \left(\frac{V_1}{V_2}\right)P_1 = \left(\frac{3.5 \text{ L}}{1.35 \text{ L}}\right)1.68 \text{ atm} = 4.4 \text{ atm}$$

CHECK: Does this answer make sense? The volume decreased (at constant temperature), so the pressure should increase, as the result of the calculation indicates. Note that the calculated final pressure is 4.4 atm. Because most gases do not behave ideally above 1 atm, we might find that if we *measured* the pressure of this gas sample, the observed pressure would differ slightly from 4.4 atm.

EXAMPLE 5.4

A sample of methane gas that has a volume of 3.8 L at 5°C is heated to 86°C at constant pressure. Calculate its new volume.

Solution

To solve this problem, we take the ideal gas law and segregate the changing variables and the constants by placing them on opposite sides of the equation. In this case volume and temperature change, and number of moles and pressure (and of course R) remain constant. Thus $PV = nRT$ becomes

$$\frac{V}{T} = \frac{nR}{P}$$

which leads to

$$\frac{V_1}{T_1} = \frac{nR}{P} \quad \text{and} \quad \frac{V_2}{T_2} = \frac{nR}{P}$$

Combining these equations gives

$$\frac{V_1}{T_1} = \frac{nR}{P} = \frac{V_2}{T_2} \quad \text{or} \quad \frac{V_1}{T_1} = \frac{V_2}{T_2}$$

We are given

$T_1 = 5°\text{C} + 273 = 278\ \text{K}$ $\quad$ $T_2 = 86°\text{C} + 273 = 359\ \text{K}$

$V_1 = 3.8\ \text{L}$ $\quad$ $V_2 = ?$

Thus

$$V_2 = \frac{T_2V_1}{T_1} = \frac{(359\ \text{K})(3.8\ \text{L})}{278\ \text{K}} = 4.9\ \text{L}$$

CHECK: Is the answer sensible? In this case the temperature was increased (at constant pressure), so the volume should increase. Thus the answer makes sense.

The problem in Example 5.4 can be described as a Charles's law problem, while the problem in Example 5.3 can be said to be a Boyle's law problem. In both cases, however, we started with the ideal gas law. The real advantage of using the ideal gas law is that it applies to virtually any problem dealing with gases and is easy to remember.

EXAMPLE 5.5

A sample of diborane gas (B_2H_6), a substance that bursts into flames when exposed to air, has a pressure of 345 torr at a temperature of $-15°C$ and a volume of 3.48 L. If conditions are changed so that the temperature is 36°C and the pressure is 468 torr, what will be the volume of the sample?

Solution

Since, for this sample, pressure, temperature, and volume all change while the number of moles remains constant, we use the ideal gas law in the form

$$\frac{PV}{T} = nR$$

which leads to

$$\frac{P_1V_1}{T_1} = nR = \frac{P_2V_2}{T_2} \quad \text{or} \quad \frac{P_1V_1}{T_1} = \frac{P_2V_2}{T_2}$$

Then

$$V_2 = \frac{T_2P_1V_1}{T_1P_2}$$

We have

$$\begin{aligned} P_1 &= 345 \text{ torr} & P_2 &= 468 \text{ torr} \\ T_1 &= -15°\text{C} + 273 = 258 \text{ K} & T_2 &= 36°\text{C} + 273 = 309 \text{ K} \\ V_1 &= 3.48 \text{ L} & V_2 &= ? \end{aligned}$$

Thus $$V_2 = \frac{(309 \text{ K})(345 \text{ torr})(3.48 \text{ L})}{(258 \text{ K})(468 \text{ torr})} = 3.07 \text{ L}$$

Always convert the temperature to the Kelvin scale when applying the ideal gas law.

Since the equation used in Example 5.5 involved a *ratio* of pressures, it was unnecessary to convert pressures to units of atmospheres. The units of torrs cancel. (You will obtain the same answer by inserting $P_1 = \frac{345}{760}$ and $P_2 = \frac{468}{760}$ into the equation.) However, temperature *must always* be converted to the Kelvin scale; since this conversion involves *addition* of 273, the conversion factor does not cancel. Be careful.

5.4 Gas Stoichiometry

Suppose we have 1 mole of an ideal gas at 0°C (273.2 K) and 1 atmosphere. From the ideal gas law, the volume of the gas is given by

$$V = \frac{nRT}{P} = \frac{(1.000 \text{ mol})(0.08206 \text{ L atm K}^{-1} \text{ mol}^{-1})(273.2 \text{ K})}{1.000 \text{ atm}} = 22.42 \text{ L}$$

This volume of 22.42 liters is called the **molar volume** of an ideal gas. The measured molar volumes of several gases are listed in Table 5.2. Note that the

TABLE 5.2 **Molar Volumes for Various Gases at 0°C and 1 atm**

Gas	Molar Volume (L)
Oxygen (O_2)	22.397
Nitrogen (N_2)	22.402
Hydrogen (H_2)	22.433
Helium (He)	22.434
Argon (Ar)	22.397
Carbon dioxide (CO_2)	22.260
Ammonia (NH_3)	22.079

molar volumes of some of the gases are very close to the ideal value, but others deviate significantly. Later in this chapter we will discuss some of the reasons for the deviations.

The conditions 0°C and 1 atmosphere, called **standard temperature and pressure** (abbreviated **STP**), are common reference conditions for the properties of gases. For example, the molar volume of an ideal gas is 22.42 liters at STP.

STP: 0°C and 1 atm

Many chemical reactions involve gases. By assuming ideal behavior for these gases, we can carry out stoichiometric calculations if the pressure, volume, and temperature of the gases are known.

EXAMPLE 5.6

Quicklime (CaO) is produced by the thermal decomposition of calcium carbonate ($CaCO_3$). Calculate the volume of CO_2 produced at STP from the decomposition of 152 g of $CaCO_3$ according to the reaction

$$CaCO_3(s) \longrightarrow CaO(s) + CO_2(g)$$

Solution

We use the same strategy we used in the stoichiometry problems earlier in the book. That is, we compute the number of moles of $CaCO_3$ consumed and the number of moles of CO_2 produced. The moles of CO_2 can then be converted to volume by using the molar volume of an ideal gas.

Using the molar mass of $CaCO_3$, we can calculate the number of moles of $CaCO_3$:

$$152 \text{ g } CaCO_3 \times \frac{1 \text{ mol } CaCO_3}{100.1 \text{ g } CaCO_3} = 1.52 \text{ mol } CaCO_3$$

Since each mole of $CaCO_3$ produces a mole of CO_2, 1.52 mol of CO_2 will be formed. We can compute the volume of CO_2 at STP by using the molar volume:

$$1.52 \text{ mol } CO_2 \times \frac{22.42 \text{ L } CO_2}{1 \text{ mol } CO_2} = 34.1 \text{ L } CO_2$$

Thus the decomposition of 152 g of $CaCO_3$ will produce 34.1 L of CO_2 at STP.

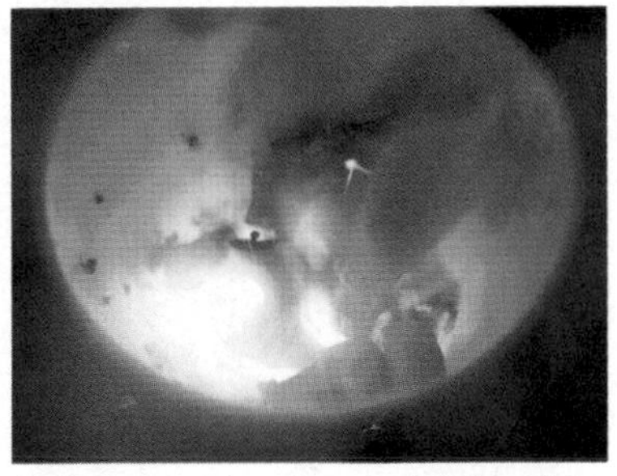

Calcium oxide gives a bright white flame when it burns.

Remember that the molar volume of an ideal gas is 22.42 L at STP.

Note that in Example 5.6 the final step involved calculation of the volume of gas from the number of moles. Since the conditions were specified as STP, we were able to use the molar volume of a gas at STP. If the conditions of a problem are different from STP, the ideal gas law must be used to compute the volume.

Molar Mass

One very important use of the ideal gas law is in the calculation of the molar mass (molecular weight) of a gas from its measured density. To understand the relationship between gas density and molar mass, note that the number of moles of gas n can be expressed as

$$n = \frac{\text{grams of gas}}{\text{molar mass}} = \frac{\text{mass}}{\text{molar mass}} = \frac{m}{\text{molar mass}}$$

Substitution into the ideal gas equation gives

$$P = \frac{nRT}{V} = \frac{(m/\text{molar mass})RT}{V} = \frac{m(RT)}{V(\text{molar mass})}$$

But m/V is the gas density, d, in units of grams per liter. Thus

$$P = \frac{dRT}{\text{molar mass}}$$

or

$$\text{Molar mass} = \frac{dRT}{P} \tag{5.1}$$

Thus if the density of a gas at a given temperature and pressure is known, its molar mass can be calculated.

You can memorize the equation involving gas density and molar mass, but it is better simply to remember the ideal gas equation, the definition of density, and the relationship between number of moles and molar mass. You can then derive this equation when you need it. This approach proves that you understand the concepts and means one less equation to memorize.

5.5 Dalton's Law of Partial Pressures

Among the experiments that led John Dalton to propose the atomic theory were his studies of mixtures of gases. In 1803 Dalton summarized his observations as follows: *For a mixture of gases in a container, the total pressure exerted is the sum of the pressures that each gas would exert if it were alone.* This statement, known as **Dalton's law of partial pressures,** can be expressed as follows:

$$P_{\text{TOTAL}} = P_1 + P_2 + P_3 + \cdots$$

where the subscripts refer to the individual gases (gas 1, gas 2, etc.). The pressures P_1, P_2, P_3, and so on, are called **partial pressures;** that is, each one is the pressure that gas would exert if it were alone in the container.

Assuming that each gas behaves ideally, the partial pressure of each gas can be calculated from the ideal gas law:

$$P_1 = \frac{n_1RT}{V}, \qquad P_2 = \frac{n_2RT}{V}, \qquad P_3 = \frac{n_3RT}{V}, \qquad \cdots$$

The total pressure of the mixture, P_{TOTAL}, can be represented as

$$P_{\text{TOTAL}} = P_1 + P_2 + P_3 + \cdots = \frac{n_1RT}{V} + \frac{n_2RT}{V} + \frac{n_3RT}{V} + \cdots$$

$$= (n_1 + n_2 + n_3 + \cdots)\left(\frac{RT}{V}\right) = n_{\text{TOTAL}}\left(\frac{RT}{V}\right)$$

where n_{TOTAL} is the sum of the numbers of moles of the various gases. Thus for a mixture of ideal gases, it is the *total number of moles of particles* that is important, not the identity or composition of the individual gas particles. This idea is illustrated in Fig. 5.9.

This important result indicates some fundamental characteristics of an ideal gas. The fact that the pressure exerted by an ideal gas is not affected by the identity (structure) of the gas particles reveals two things about ideal gases: (1) the volume of the individual gas particle must not be important; and (2) the forces among the particles must not be important. If these factors were important, the pressure exerted by the gas would depend on the nature of the individual particles. These observations will strongly influence the model that we will eventually construct to explain ideal gas behavior.

At this point we need to define the **mole fraction:** *the ratio of the number of moles of a given component in a mixture to the total number of moles in the mixture.* The Greek letter chi (χ) is used to symbolize the mole fraction. For a given component in a mixture, the mole fraction (χ_1) is

$$\chi_1 = \frac{n_1}{n_{\text{TOTAL}}} = \frac{n_1}{n_1 + n_2 + n_3 + \cdots}$$

From the ideal gas equation we know that the number of moles of a gas is directly proportional to the pressure of the gas, since

$$n = P\left(\frac{V}{RT}\right)$$

That is, for each component in the mixture,

$$n_1 = P_1\left(\frac{V}{RT}\right), \qquad n_2 = P_2\left(\frac{V}{RT}\right), \qquad \cdots$$

Therefore, we can represent the mole fraction in terms of pressures:

$$\chi_1 = \frac{n_1}{n_{\text{TOTAL}}} = \frac{\overbrace{P_1(V/RT)}^{n_1}}{\underbrace{P_1(V/RT)}_{n_1} + \underbrace{P_2(V/RT)}_{n_2} + \underbrace{P_3(V/RT)}_{n_3} + \cdots}$$

$$= \frac{(V/RT)P_1}{(V/RT)(P_1 + P_2 + P_3 + \cdots)}$$

$$= \frac{P_1}{P_1 + P_2 + P_3 + \cdots} = \frac{P_1}{P_{\text{TOTAL}}}$$

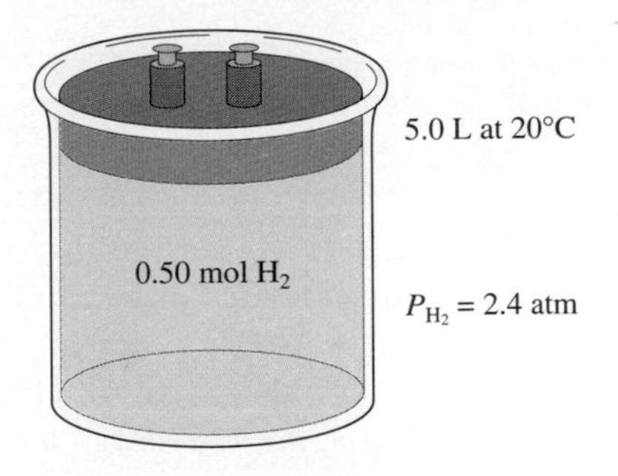

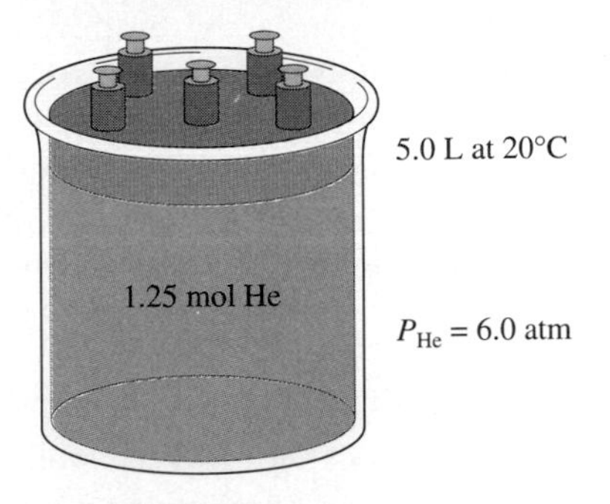

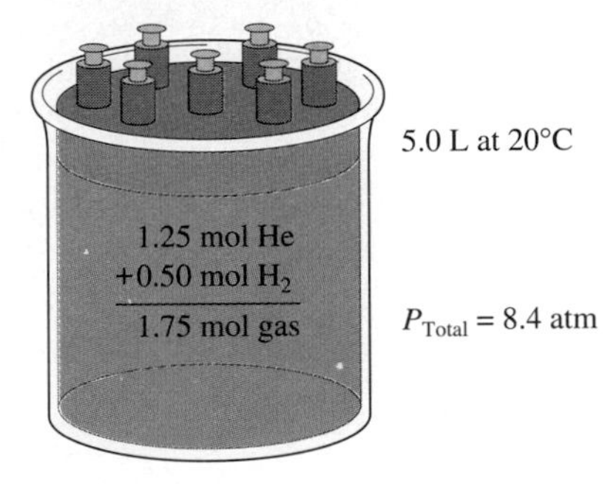

Figure 5.9
The partial pressure of each gas in a mixture of gases depends on the number of moles of that gas. The total pressure is the sum of the partial pressures and depends on the total moles of gas particles present, no matter what their identities.

THE CHEMISTRY OF AIR BAGS

Most experts agree that air bags represent a very important advance in automobile safety. These bags, which are stored in the auto's steering wheel or dash, are designed to inflate rapidly (within about 40 ms) in the event of a crash, cushioning the front seat occupants against impact. The bags then deflate immediately to allow vision and movement after the crash. Air bags are activated when a severe deceleration (an impact) causes a steel ball to compress a spring and electrically ignite a detonator cap, which, in turn, causes sodium azide, NaN_3, to decompose explosively, forming sodium and nitrogen gas:

$$2NaN_3(s) \longrightarrow 2Na(s) + 3N_2(g)$$

This system works very well and requires only a relatively small amount of sodium azide (100 g yields 33 L of $N_2(g)$ at 25°C).

When a vehicle containing air bags reaches the end of its useful life, the sodium azide present in the activators must be given proper disposal. Sodium azide, besides being explosive, has a toxicity roughly equal to that of sodium cyanide. It also forms hydrazoic acid (HN_3), a toxic and explosive liquid, when treated with acid.

The air bag represents an application of chemistry that will undoubtedly save thousands of lives in the years ahead.

Inflated dual airbags.

Similarly,
$$\chi_2 = \frac{n_2}{n_{TOTAL}} = \frac{P_2}{P_{TOTAL}}$$

and so on. Thus the mole fraction of a particular component in a mixture of ideal gases is directly related to its partial pressure.

The expression for the mole fraction,

$$\chi_1 = \frac{P_1}{P_{TOTAL}}$$

can be rearranged:
$$P_1 = \chi_1 \times P_{TOTAL}$$

That is, *the partial pressure of a particular component of a gaseous mixture is equal to the mole fraction of that component times the total pressure.*

A mixture of gases occurs whenever a gas is collected by displacement of water. For example, Fig. 5.10 shows the collection of oxygen gas produced by the decomposition of solid potassium chlorate. In this situation the gas in the bottle is a mixture of water vapor and the gas being collected. Water vapor is

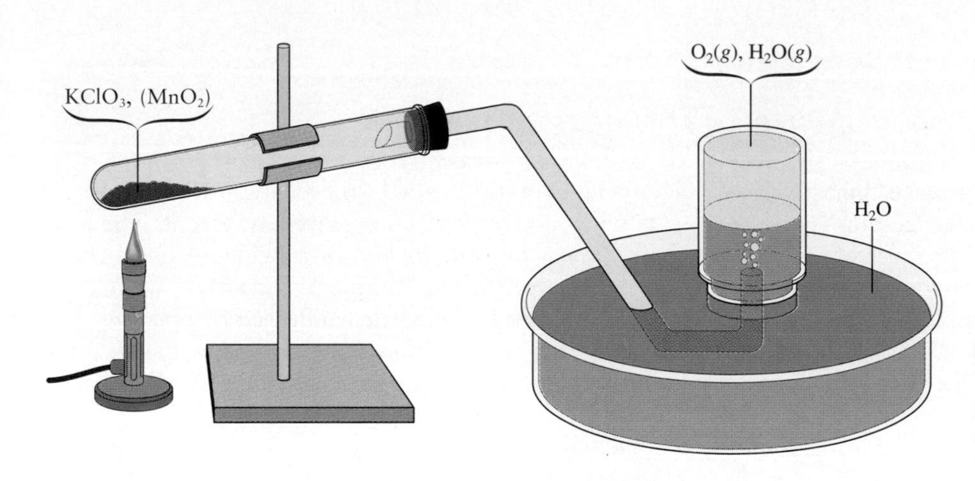

Figure 5.10
The production of oxygen by thermal decomposition of $KClO_3$. The MnO_2 catalyst is mixed with the $KClO_3$ to make the reaction faster.

present because molecules of water escape from the surface of the liquid and collect in the space above the liquid. Molecules of water also return to the liquid. When the rate of escape equals the rate of return, the number of water molecules in the vapor state remains constant, and thus the pressure of water vapor remains constant. This pressure, which depends on temperature, is called the *vapor pressure of water*.

Vapor pressure will be discussed in detail in Chapter 16.

EXAMPLE 5.7

The mole fraction of nitrogen in the air is 0.7808. Calculate the partial pressure of N_2 in air when the atmospheric pressure is 760. torr.

Solution

The partial pressure of N_2 can be calculated as follows:

$$P_{N_2} = \chi_{N_2} \times P_{TOTAL} = 0.7808 \times 760.\ \text{torr} = 593\ \text{torr}$$

5.6 The Kinetic Molecular Theory of Gases

We have so far considered the behavior of gases from an experimental point of view. On the basis of observations from different types of experiments, we know that at pressures less than 1 atmosphere, most gases closely approach the behavior described by the ideal gas law. Now we want to construct a model to explain this behavior.

Before we construct the model, we will briefly review the scientific method. Recall that a law is a way of generalizing behavior that has been observed in many experiments. Laws are very useful, since they allow us to predict the behavior of similar systems. For example, if a chemist prepares a new gaseous compound, a measurement of the gas density at known pressure and temperature can provide a reliable value for the compound's molar mass.

However, although laws summarize observed behavior, they do not tell us *why* nature behaves in the observed fashion. This is the central question for

Scuba Diving

Oxygen is essential to our existence, but surprisingly, it can be harmful under certain circumstances. At sea level the partial pressure of this life-sustaining gas is 0.21 atm,* and in each normal breath we inhale about 0.02 mol of O_2 molecules. Our bodies operate very effectively under these conditions, but the situation changes if we subject ourselves to greater pressures—for example, by diving in deep water. A scuba diver at a depth of 100 ft experiences approximately 4 atm of pressure, and at 300 ft the pressure is about 10 atm. This increased pressure affects the ear canals and squeezes the lungs, but a more serious effect involves the increased partial pressure of oxygen in air breathed at these pressures. Elevated concentrations of oxygen are actually quite harmful for reasons that are not well understood. The symptoms of oxygen poisoning include confusion, impaired vision and hearing, and nausea. Clearly, it is possible to get too much of a good thing.

At a depth of 300 ft under the ocean's surface, the partial pressure of oxygen is about 2 atm (0.21×10 atm) for a diver inhaling compressed air. This value is much too high, and the oxygen must be diluted by another gas. Although it might seem to be an ideal candidate, nitrogen cannot be used. At elevated pressures a large quantity of nitrogen dissolves in the blood, producing nitrogen narcosis, often called "rapture of the deep," which has an effect not unlike that from imbibing too many martinis. The increased solubility of nitrogen in blood at high pressures is also responsible for an agonizing condition called "the bends," which results when a diver makes too rapid an ascent. The diver's joints become locked in a bent position—hence the name. Just as a bottle of soda fizzes when the pressure inside the bottle is released by opening the cap, the excess dissolved nitrogen in the bloodstream forms bubbles that can stop blood flow and impair the nervous system.

Helium is the diluting agent most often used in scuba tanks. It is inert, and its solubility in blood is much lower than that of oxygen or nitrogen. However, it has the effect of raising the pitch of the human voice, producing a "Donald Duck" effect. This effect occurs because the pitch of the voice depends on the density of the gas surrounding the vocal cords: lower density yields a higher pitch. Because the mass of a He atom is much smaller than the masses of N_2 and O_2, the density of helium gas is much lower than that of air.

*This result is obtained from Dalton's law. Since the mole fraction of O_2 in air is 0.21, the partial pressure of O_2 is 0.21×1 atm $= 0.21$ atm.

Scuba divers off the coast of New Zealand.

scientists. To try to answer this question, we construct theories (build models). The models in chemistry consist of speculations about what the individual atoms or molecules (microscopic particles) might be doing to cause the observed behavior of the macroscopic systems (collections of very large numbers of atoms and molecules).

EXAMPLE 5.8

A sample of solid potassium chlorate ($KClO_3$) was heated in a test tube (see Fig. 5.10) and decomposed according to the following reaction:

$$2KClO_3(s) \longrightarrow 2KCl(s) + 3O_2(g)$$

The oxygen produced was collected by displacement of water at 22°C at a total pressure of 754 torr. The volume of the gas collected was 0.65 L, and the vapor pressure of water at 22°C is 21 torr. Calculate the partial pressure of O_2 in the gas collected and the mass of $KClO_3$ in the sample that was decomposed.

Solution

First, we find the partial pressure of O_2 from Dalton's law of partial pressures:

$$P_{\text{TOTAL}} = P_{O_2} + P_{H_2O} = P_{O_2} + 21 \text{ torr} = 754 \text{ torr}$$

Thus $$P_{O_2} = 754 \text{ torr} - 21 \text{ torr} = 733 \text{ torr}$$

Now we use the ideal gas law to find the number of moles of O_2:

$$n_{O_2} = \frac{P_{O_2}V}{RT}$$

In this case

$$P_{O_2} = 733 \text{ torr} = \frac{733 \text{ torr}}{760 \text{ torr/atm}} = 0.964 \text{ atm}$$

$$V = 0.650 \text{ L}$$

$$T = 22°\text{C} + 273 = 295 \text{ K}$$

$$R = 0.08206 \text{ L atm K}^{-1} \text{ mol}^{-1}$$

Thus

$$n_{O_2} = \frac{(0.964 \text{ atm})(0.650 \text{ L})}{(0.08206 \text{ L atm K}^{-1} \text{ mol}^{-1})(295 \text{ K})} = 2.59 \times 10^{-2} \text{ mol}$$

Next, we calculate the moles of $KClO_3$ needed to produce this quantity of O_2, using the mole ratio from the balanced equation for the decomposition of $KClO_3$:

$$2.59 \times 10^{-2} \text{ mol } O_2 \times \frac{2 \text{ mol } KClO_3}{3 \text{ mol } O_2} = 1.73 \times 10^{-2} \text{ mol } KClO_3$$

Using the molar mass of $KClO_3$, we calculate the grams of $KClO_3$:

$$1.73 \times 10^{-2} \text{ mol } KClO_3 \times \frac{122.6 \text{ g } KClO_3}{1 \text{ mol } KClO_3} = 2.12 \text{ g } KClO_3$$

Thus the original sample contained 2.12 g of $KClO_3$.

A model is considered successful if it explains the observed behavior in question and predicts correctly the results of future experiments. Note that a model can never be proved to be absolutely true. In fact, *any model is an approximation* by its very nature and is bound to fail at some point. Models range from the simple to the extraordinarily complex. We use simple models to predict approximate behavior and more complicated models to account very precisely for observed quantitative behavior. In this text we will stress simple models that provide an approximate picture of what might be happening and that fit the most important experimental results.

An example of this type of model is the **kinetic molecular theory,** a simple model that attempts to explain the properties of an ideal gas. This model is based on speculations about the behavior of the individual gas particles (atoms or molecules). The postulates of the kinetic molecular theory can be stated as follows:

An ideal gas consists of particles that have the following properties:

- The particles are so small compared with the distances between them that *the volume of the individual particles can be assumed to be negligible* (zero).
- *The particles are in constant motion. The collisions of the particles with the walls of the container are the cause of the pressure exerted by the gas.*
- *The particles are assumed to exert no forces on each other;* they are assumed neither to attract nor to repel each other.
- *The average kinetic energy of a collection of gas particles is assumed to be directly proportional to the Kelvin temperature of the gas.*

Of course, real gas particles do have a finite volume and do exert forces on each other. Thus they do not conform exactly to these assumptions. But we will see that these postulates do indeed explain *ideal* gas behavior.

The true test of a model is how well its predictions fit the experimental observations. The postulates of the kinetic molecular model picture an ideal gas as consisting of particles having no volume and no attraction for each other, and the model assumes that the gas produces pressure on its container by collisions with the walls. To test the validity of this model, we need to consider the question: "When we apply the principles of physics to a collection of these gas particles, can we derive an expression for pressure that agrees with the ideal gas law?" The answer is, "Yes, we can." We will now consider this derivation in detail.

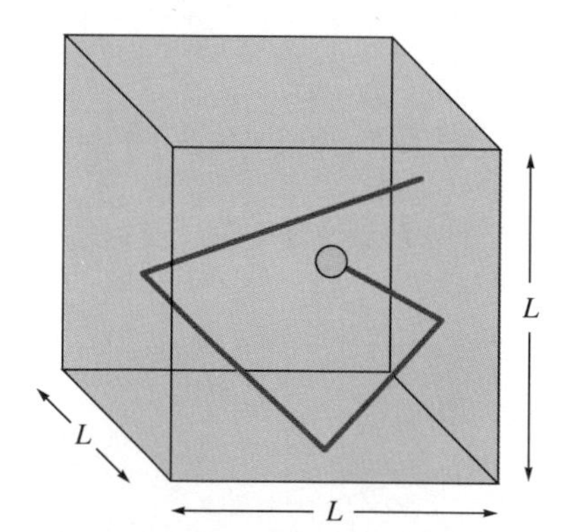

Figure 5.11
An ideal gas particle in a cube whose sides are of length L (in meters). The particle collides elastically with the walls in a random, straight-line motion.

The Quantitative Kinetic Molecular Model

Suppose there are n moles of an ideal gas in a cubical container with sides each of length L in meters. Assume each gas particle has a mass m and that it is in rapid, random, straight-line motion colliding with the walls, as shown in Fig. 5.11. The collisions will be assumed to be *elastic*—no loss of kinetic energy occurs. We want to compute the force on the walls from the colliding gas particles and then, since pressure is force per unit area, to obtain an expression for the pressure of the gas.

Before we can derive the expression for the pressure of a gas, we must first discuss some characteristics of velocity. Each particle in the gas has a particular

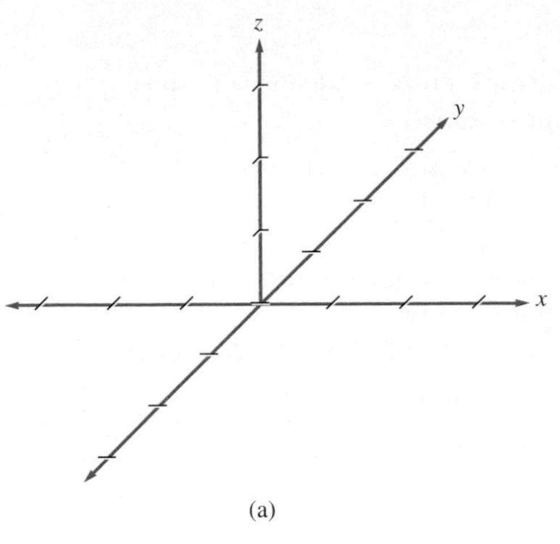

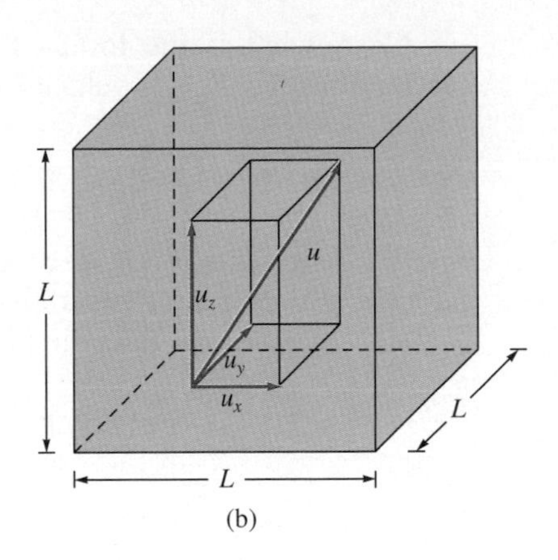

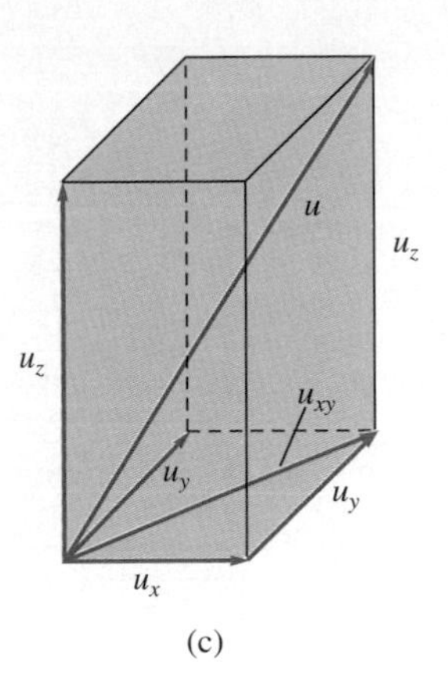

Figure 5.12
(a) The Cartesian coordinate axes.
(b) The velocity u of any gas particle can be broken down into three mutually perpendicular components, u_x, u_y, and u_z. This can be represented as a rectangular solid with sides u_x, u_y, and u_z and body diagonal u.
(c) In the xy plane,

$$u_x^2 + u_y^2 = u_{xy}^2$$

by the Pythagorean theorem. Since u_{xy} and u_z are also perpendicular,

$$u^2 = u_{xy}^2 + u_z^2 = u_x^2 + u_y^2 + u_z^2$$

velocity u, which can be broken into components u_x, u_y, and u_z, as shown in Fig. 5.12. First, using u_x and u_y and the Pythagorean theorem, we can obtain u_{xy} as shown in Fig. 5.12:

$$u_{xy}^2 = u_x^2 + u_y^2$$

Hypotenuse of right triangle (u_{xy}); Sides of right triangle (u_x, u_y)

Then, constructing another triangle as shown in Fig. 5.12, we find

$$u^2 = u_{xy}^2 + u_z^2$$

or

$$u^2 = u_x^2 + u_y^2 + u_z^2$$

Now let's consider how an individual gas particle moves. For example, how often does this particle strike the two walls of the box that are perpendicular to the x axis? Note that only the x component of the velocity affects the particle's impacts on these two walls, as shown in Fig. 5.13. The larger the x component of the velocity, the faster the particle travels between these two walls, thus producing more impacts per unit of time on these walls. Remember that the pressure of the gas is due to these collisions with the walls.

The collision frequency (collisions per unit of time) with the two walls that are perpendicular to the x axis is given by

$$(\text{Collision frequency})_x = \frac{\text{velocity in the } x \text{ direction}}{\text{distance between the walls}} = \frac{u_x}{L}$$

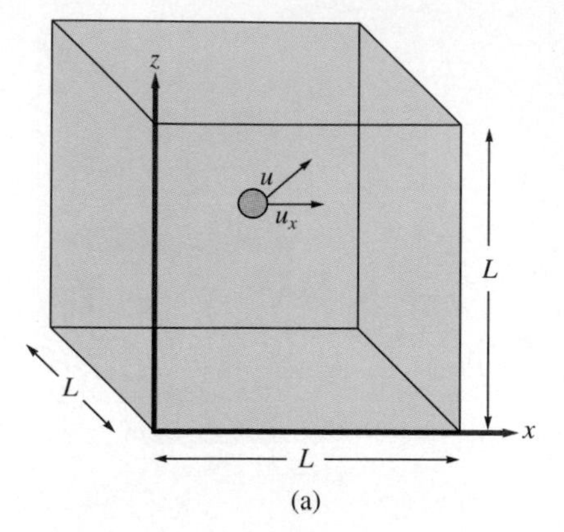

(a)

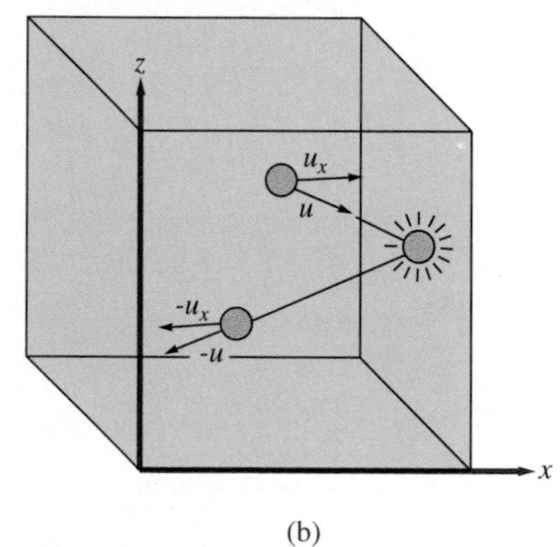

(b)

Figure 5.13
(a) Only the x component of the gas particle's velocity affects the frequency of impacts on the shaded walls, the walls that are perpendicular to the x axis. (b) For an elastic collision, there is an exact reversal of the x component of the velocity and of the total velocity. The change in momentum (final − initial) is then

$$-mu_x - mu_x = -2mu_x$$

Next, what is the force of a collision? Force is defined as mass times acceleration (change in velocity per unit of time):

$$F = ma = m\left(\frac{\Delta u}{\Delta t}\right)$$

where F represents force, a represents acceleration, Δu represents a change in velocity, and Δt represents a given length of time.

Since we assume that the particle has constant mass, we can write

$$F = \frac{m\Delta u}{\Delta t} = \frac{\Delta(mu)}{\Delta t}$$

The quantity mu is the momentum of the particle (momentum is the product of mass and velocity), and the expression $F = \Delta(mu)/\Delta t$ means that force is the change in momentum per unit of time. When a particle hits a wall perpendicular to the x axis, as shown in Fig. 5.13, an elastic collision occurs, resulting in an *exact reversal* of the x component of velocity. That is, the *sign,* or direction, of u_x reverses when the particle collides with one of the walls perpendicular to the x axis. Thus the final momentum is the *negative,* or opposite, of the initial momentum. Remember that an elastic collision means that there is no change in the *magnitude* of the velocity. The change in momentum in the x direction is:

$$\begin{aligned}\text{Change in momentum} &= \Delta(mu_x)\\ &= \text{final momentum} - \text{initial momentum}\\ &= -mu_x - mu_x\end{aligned}$$

(Final momentum in x direction) (Initial momentum in x direction)

$$= -2mu_x$$

We are interested in the force the gas particle exerts on the walls of the box. Since we know that every action produces an equal but opposite reaction, the change in momentum with respect to the wall on impact is $-(-2mu_x)$, or $2mu_x$.

Recall that since force is the change in momentum per unit of time, then

$$\text{Force}_x = \frac{\Delta(mu_x)}{\Delta t}$$

for the walls perpendicular to the x axis.

This expression can be obtained by multiplying the change in momentum per impact by the number of impacts per unit of time:

$$\text{Force}_x = (2mu_x)\left(\frac{u_x}{L}\right) = \text{change in momentum per unit of time}$$

(Change in momentum per impact) (Impacts per unit of time)

That is,

$$\text{Force}_x = \frac{2mu_x^{\,2}}{L}$$

So far, we have considered only the two walls of the box perpendicular to the x axis. We can assume that the force on the two walls perpendicular to the y axis is given by

$$\text{Force}_y = \frac{2mu_y^2}{L}$$

and that on the two walls perpendicular to the z axis by

$$\text{Force}_z = \frac{2mu_z^2}{L}$$

Since we have shown that

$$u^2 = u_x^2 + u_y^2 + u_z^2$$

the total force on the box is

$$\begin{aligned}\text{Force}_{\text{TOTAL}} &= \text{force}_x + \text{force}_y + \text{force}_z \\ &= \frac{2mu_x^2}{L} + \frac{2mu_y^2}{L} + \frac{2mu_z^2}{L} \\ &= \frac{2m}{L}(u_x^2 + u_y^2 + u_z^2) = \frac{2m}{L}(u^2)\end{aligned}$$

Now since we want the average force (the force created by an "average" particle), we use the average of the square of the velocity ($\overline{u^2}$) to obtain

$$\overline{\text{Force}_{\text{TOTAL}}} = \frac{2m}{L}(\overline{u^2})$$

Next, we need to compute the pressure (force per unit of area):

$$\begin{aligned}\text{Pressure due to "average" particle} &= \frac{\overline{\text{force}_{\text{TOTAL}}}}{\text{area}_{\text{TOTAL}}} \\ &= \frac{2m\overline{u^2}/L}{6L^2} = \frac{m\overline{u^2}}{3L^3}\end{aligned}$$

The 6 sides of the cube — Area of each side

Since the volume V of the cube is equal to L^3, we can write

$$\text{Pressure} = P = \frac{m\overline{u^2}}{3V}$$

So far we have considered the pressure on the walls due to a single, "average" particle. Of course we want the pressure due to the entire gas sample. The number of particles in a given gas sample can be expressed as follows:

$$\text{Number of gas particles} = nN_{\text{A}}$$

where n is the number of moles and N_{A} is Avogadro's number.

The total pressure on the box due to n moles of a gas is therefore

$$P = nN_{\text{A}}\frac{m\overline{u^2}}{3V}$$

Next, we want to express the pressure in terms of the kinetic energy of the gas molecules. Kinetic energy (the energy due to motion) is given by $\frac{1}{2}mu^2$, where m is the mass and u the velocity. Since we are using the average of the velocity squared ($\overline{u^2}$), and since $m\overline{u^2} = 2(\frac{1}{2}m\overline{u^2})$, we have

$$P = \frac{2}{3}\left[\frac{nN_A(\frac{1}{2}m\overline{u^2})}{V}\right]$$

Recall that P is the pressure of the gas, n is the number of moles of gas, N_A is Avogadro's number, m is the mass of each particle, $\overline{u^2}$ is the average of the squares of the velocities of the particles, and V is the volume of the container.

$KE = \frac{1}{2}mu^2$, the energy due to the motion of a particle.

The quantity $\frac{1}{2}m\overline{u^2}$ represents the average kinetic energy of a gas particle. If the average kinetic energy of an individual particle is multiplied by N_A, the number of particles in a mole, we get the average kinetic energy for a mole of gas particles:

$$(KE)_{avg} = N_A(\tfrac{1}{2}m\overline{u^2})$$

Using this definition, we can rewrite the expression for pressure as

$$P = \frac{2}{3}\left[\frac{n\,(KE)_{avg}}{V}\right] \quad \text{or} \quad \frac{PV}{n} = \frac{2}{3}(KE)_{avg}$$

The fourth postulate of the kinetic molecular theory is that the average kinetic energy of the particles in the gas sample is directly proportional to the temperature in Kelvins. Thus since $(KE)_{avg} \propto T$, we can write

$$\frac{PV}{n} = \frac{2}{3}(KE)_{avg} \propto T \quad \text{or} \quad \frac{PV}{n} \propto T$$

Note that this expression has been *derived* from the assumptions of the kinetic molecular theory. How does it compare with the ideal gas law—the equation obtained from experiment? Compare the ideal gas law,

$$\frac{PV}{n} = RT \quad \text{From experiment}$$

with the result from the kinetic molecular theory,

$$\frac{PV}{n} \propto T \quad \text{From theory}$$

(left) A balloon filled with air at room temperature. (center) Liquid nitrogen at 77 K is poured over the balloon. (right) The balloon collapses as the molecules inside slow down owing to the decreased temperature. Slower molecules produce a lower pressure.

These expressions have exactly the same form if R, the universal gas constant, is considered the proportionality constant in the second case.

The agreement between the ideal gas law and the kinetic molecular theory gives us confidence in the validity of the model. The characteristics we have assumed for ideal gas particles must agree, at least under certain conditions, with their actual behavior.

The Meaning of Temperature

We have seen from the kinetic molecular theory that the Kelvin temperature is a measure of the average kinetic energy of the gas particles. The exact relationship between temperature and average kinetic energy can be obtained by combining the equations

$$\frac{PV}{n} = RT = \frac{2}{3}(\mathrm{KE})_{\mathrm{avg}}$$

which yields the expression

$$(\mathrm{KE})_{\mathrm{avg}} = \tfrac{3}{2}RT$$

This is a very important relationship. It summarizes the meaning of the Kelvin temperature of a gas: the Kelvin temperature is an index of the random motions of the particles of a gas, with higher temperature meaning greater motion.

Root Mean Square Velocity

In the equation from the kinetic molecular theory, the average velocity of the gas particles is a special kind of average. The symbol $\overline{u^2}$ means the average of the *squares* of the particle velocities. The square root of $\overline{u^2}$ is called the **root mean square velocity** and is symbolized by u_{rms}:

$$u_{\mathrm{rms}} = \sqrt{\overline{u^2}}$$

We can obtain an expression for u_{rms} from the equations

$$(\mathrm{KE})_{\mathrm{avg}} = \mathrm{N_A}(\tfrac{1}{2}m\overline{u^2}) \quad \text{and} \quad (\mathrm{KE})_{\mathrm{avg}} = \tfrac{3}{2}RT$$

Combination of these equations gives

$$\mathrm{N_A}(\tfrac{1}{2}m\overline{u^2}) = \tfrac{3}{2}RT \quad \text{or} \quad \overline{u^2} = \frac{3RT}{\mathrm{N_A}m}$$

Taking the square root of both sides of the last equation produces

$$\sqrt{\overline{u^2}} = u_{\mathrm{rms}} = \sqrt{\frac{3RT}{\mathrm{N_A}m}}$$

$R = 0.08206$ L atm $\mathrm{K^{-1}\ mol^{-1}}$
$R = 8.3145$ J $\mathrm{K^{-1}\ mol^{-1}}$

In this expression m represents the mass in kilograms of a single gas particle. When $\mathrm{N_A}$, the number of particles in a mole, is multiplied by m, the product is the mass of a *mole* of gas particles in *kilograms*. We will call this quantity M. Substituting M for $\mathrm{N_A}m$ in the equation for u_{rms}, we obtain

$$u_{\mathrm{rms}} = \sqrt{\frac{3RT}{M}}$$

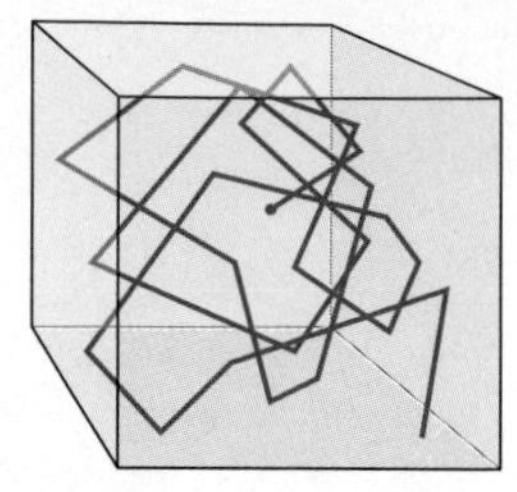

Figure 5.14
Path of one particle in a gas. Any given particle will continuously change its course as a result of collisions with other particles, as well as with the walls of the container.

Before we can use this equation, we need to consider the units for R. So far we have used 0.08206 L atm K^{-1} mol^{-1} as the value of R. But to obtain the desired units (meters per second) for u_{rms}, R must be expressed in different units. As we will see in more detail in Chapter 9, the energy unit most often used in the SI system is the joule (J). A **joule** is defined as a kilogram meter squared per second squared (kg m^2/s^2). When R is converted from liter atmospheres to joules, it has the value 8.3145 J K^{-1} mol^{-1}. When R with these units is used in the expression $\sqrt{3RT/M}$, u_{rms} has units of meters per second, as desired.

So far we have said nothing about the range of velocities actually found in a gas sample. In a real gas there are large numbers of collisions between particles. For example, when an odorous gas such as ammonia is released in a room, it takes some time for the odor to permeate the air, as we will see in the next section. This delay results from collisions between the NH_3 molecules and O_2 and N_2 molecules in the air, which greatly slow the mixing process.

If the path of a particular gas particle could be monitored, it would probably look very erratic, something like that shown in Fig. 5.14. The average distance a particle travels between collisions in a particular gas sample is called the **mean free path.** It is typically a very small distance (1×10^{-7} m for O_2 at STP). One effect of the many collisions among gas particles is to produce a large range of velocities as the particles collide and exchange kinetic energy. Although u_{rms} for oxygen gas at STP is approximately 500 meters per second, the majority of O_2 molecules do not have this velocity. The actual distribution of molecular velocities for oxygen gas at STP is shown in Fig. 5.15. This figure shows the relative number of gas molecules having each particular velocity.

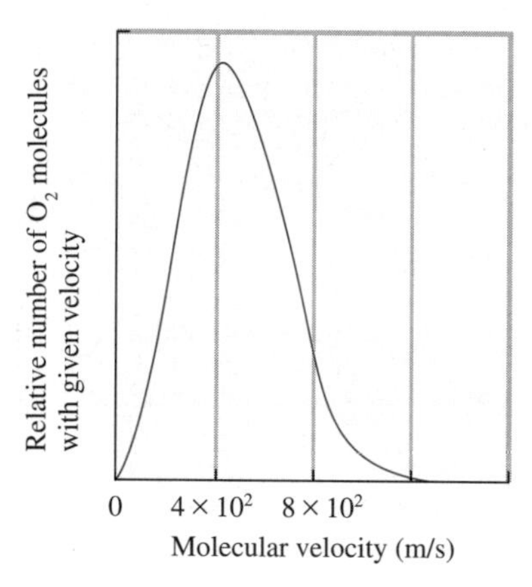

Figure 5.15
A plot of the relative number of O_2 molecules that have a given velocity at STP.

We are also interested in the effect of *temperature* on the velocity distribution in a gas. Figure 5.16 shows the velocity distribution for nitrogen gas at three temperatures. Note that as the temperature is increased, the curve maximum, which reflects the average velocity, moves toward higher values and the range of velocities becomes much larger.

The distribution of velocities of the particles in an ideal gas is described by the Maxwell-Boltzmann distribution law:

$$f(u) = 4\pi\left(\frac{m}{2\pi k_B T}\right)^{3/2} u^2 e^{-mu^2/2k_B T}$$

where

u = velocity in m/s
m = mass of a gas particle in kg
k_B = Boltzmann's constant = 1.38066×10^{-23} J/K
T = temperature in K

This equation was derived independently by James C. Maxwell, a Scottish physicist, and Ludwig E. Boltzmann, an Austrian physicist who did much of the fundamental theoretical work on the kinetic molecular description of an ideal gas. The meaning of $f(u)$ is best understood as the fraction of gas molecules with velocities between u and $u + du$, where du represents an infinitesimal velocity increment. This function is the one plotted in Figs. 5.15 and 5.16.

Analysis of the expression for $f(u)$ yields the following equation for the most probable velocity u_{mp} (the velocity possessed by the greatest number of gas particles):

$$u_{mp} = \sqrt{\frac{2k_BT}{m}} = \sqrt{\frac{2RT}{M}}$$

where

M = molar mass of the gas particles in kg = $6.022 \times 10^{23} \times m$
R = gas constant = $6.022 \times 10^{23} \times k_B$

Note that R and k_B are related by Avogadro's number. In fact, it is useful to think of k_B as the gas law constant per particle (per molecule).

Another type of velocity that can be obtained from $f(u)$ is the average velocity, or u_{avg} (sometimes written $\bar{u}$), which is given by the equation

$$u_{avg} = \bar{u} = \sqrt{\frac{8k_BT}{\pi m}} = \sqrt{\frac{8RT}{\pi M}}$$

Thus we have three ways to describe a "typical" velocity for the particles in an ideal gas: the root mean square velocity, the most probable velocity, and the average velocity. As can be seen from the equations for u_{rms}, u_{mp}, and u_{avg}, these velocities are not the same. In fact, they stand in the ratios

$$u_{mp}:u_{avg}:u_{rms} = 1.000:1.128:1.225$$

This relationship is shown for nitrogen gas at 0°C in Fig. 5.17.

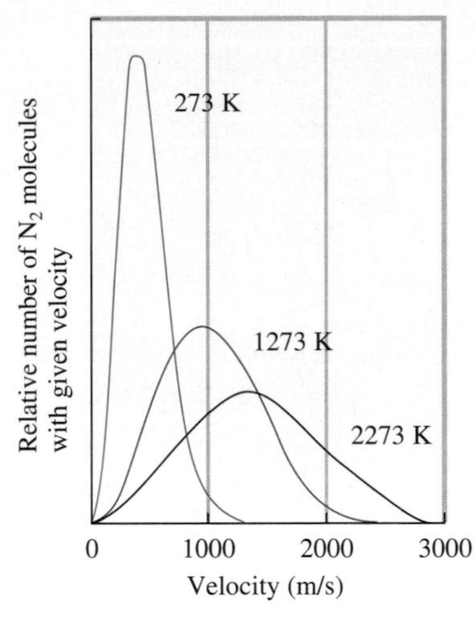

Figure 5.16
A plot of the relative number of N_2 molecules that have a given velocity at three temperatures. Note that as the temperature increases, both the average velocity (reflected by the curve's peak) and the spread of velocities increase.

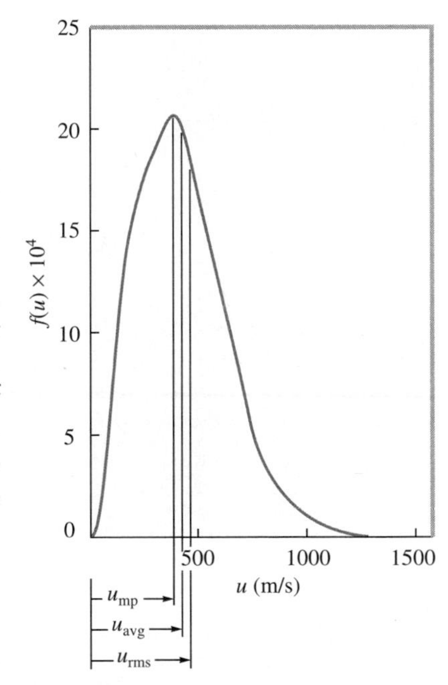

Figure 5.17
The velocity distribution for nitrogen gas at 273 K, with the values of most probable velocity (u_{mp}, the velocity at the curve maximum), the average velocity (u_{avg}), and the root mean square velocity (u_{rms}) indicated.

5.7 Effusion and Diffusion

We have seen that the postulates of the kinetic molecular theory, combined with the appropriate physical principles, produce an equation that successfully fits the experimentally observed properties of gases as they approach ideal behavior. Two phenomena involving gases provide further tests of this model.

Diffusion is the term used to describe the mixing of gases. When a small amount of pungent-smelling ammonia is released at the front of a classroom, it takes some time before everyone in the room can smell it, because time is required for the ammonia to mix with the air. The rate of diffusion is the rate of the mixing of gases. **Effusion** is the term used to describe the passage of a gas through a tiny orifice into an evacuated chamber, as shown in Fig. 5.18. The rate of effusion measures the rate at which the gas is transferred into the chamber.

Effusion

Thomas Graham (1805–1869), a Scottish chemist, found experimentally that the rate of effusion of a gas is inversely proportional to the square root of the mass of its particles. Stated in another way, the relative rates of effusion of two gases at the same temperature and pressure are given by the inverse ratio of the square roots of the masses of the gas particles:

$$\frac{\text{Rate of effusion for gas 1}}{\text{Rate of effusion for gas 2}} = \frac{\sqrt{M_2}}{\sqrt{M_1}}$$

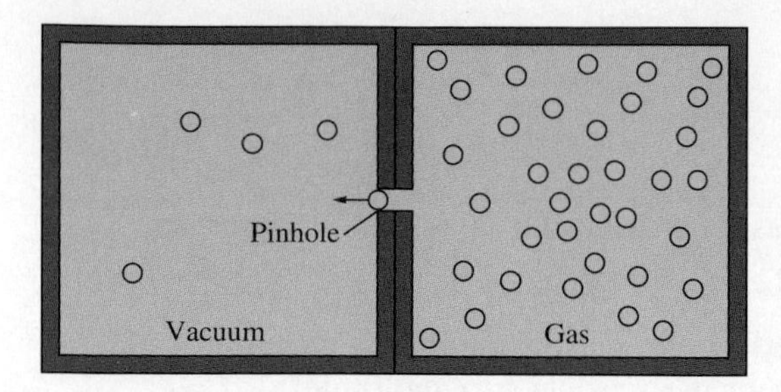

Figure 5.18
The effusion of a gas into an evacuated chamber. The rate of effusion (the rate at which the gas is transferred across the barrier through the pinhole) is inversely proportional to the square root of the molar mass of the gas.

In Graham's law the units for molar mass can be g/mol or kg/mol, since the units cancel in the ratio $\sqrt{M_2}/\sqrt{M_1}$.

where M_1 and M_2 represent the molar masses of the gases. This equation is called **Graham's law of effusion.**

Does the kinetic molecular model for gases correctly predict the relative effusion rates of gases summarized by Graham's law? To answer this question, we must recognize that the effusion rate for a gas depends directly on the average velocity of its particles. The faster the gas particles are moving, the more likely they are to pass through the effusion orifice. This reasoning leads to the following *prediction* for two gases at the same temperature (T):

$$\frac{\text{Effusion rate for gas 1}}{\text{Effusion rate for gas 2}} = \frac{u_{\text{avg}} \text{ for gas 1}}{u_{\text{avg}} \text{ for gas 2}} = \frac{\sqrt{8RT/\pi M_1}}{\sqrt{8RT/\pi M_2}} = \frac{\sqrt{M_2}}{\sqrt{M_1}}$$

This equation is Graham's law, and thus the kinetic molecular model fits the experimental results for the effusion of gases.

Diffusion

Diffusion is frequently illustrated by the lecture demonstration represented in Fig. 5.19, in which two cotton plugs, one soaked in ammonia and the other hydrochloric acid, are simultaneously placed at the ends of a long tube. A white ring of ammonium chloride (NH_4Cl) forms where the NH_3 and HCl molecules meet several minutes later:

$$NH_3(g) + HCl(g) \longrightarrow \underset{\text{White solid}}{NH_4Cl(s)}$$

The progress of the gases through the tube is surprisingly slow in light of the fact that the velocities of the HCl and NH_3 molecules at 25°C are about 450 and 660 meters per second, respectively. Why does it take several minutes for the NH_3 and HCl molecules to meet? The answer is that the tube contains air and thus the NH_3 and HCl molecules undergo many collisions with O_2 and N_2 molecules as they travel through the tube. Although these collisions greatly slow their progress through the tube, it still seems reasonable to expect the relative distances traveled by the NH_3 and HCl molecules to be related to their velocities:

If no air were present in the tube, the ratio of distances would be 1.5 as predicted from Graham's law.

$$\frac{d_{NH_3}}{d_{HCl}} = \frac{\text{distance traveled by } NH_3}{\text{distance traveled by HCl}} = \frac{u_{\text{avg}}(NH_3)}{u_{avg(HCl)}} = \sqrt{\frac{M_{HCl}}{M_{NH_3}}} = \sqrt{\frac{36.5}{17}} = 1.5$$

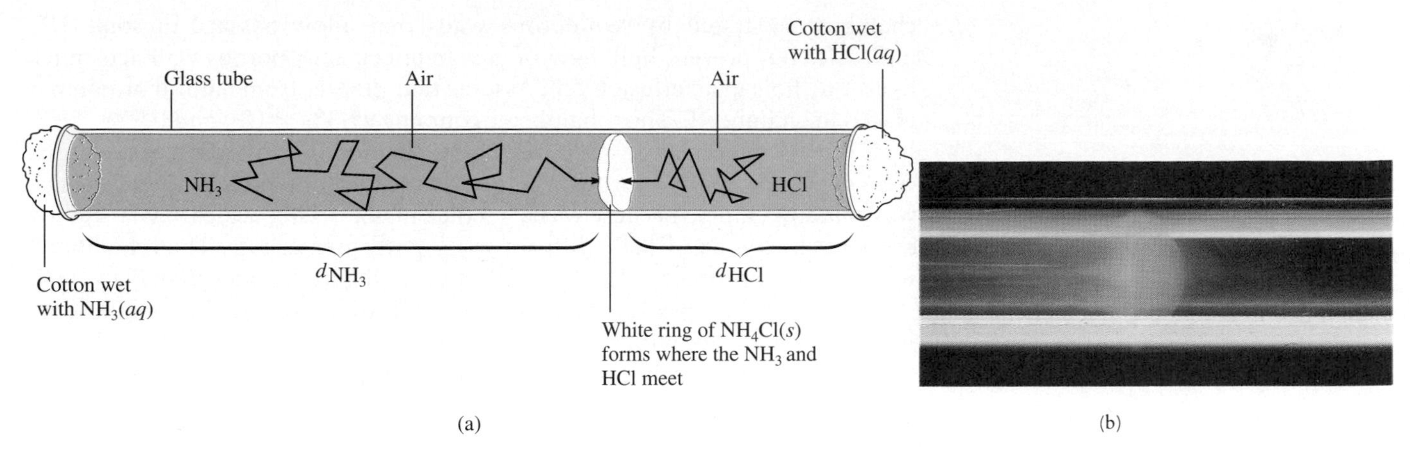

Figure 5.19
(a) A demonstration of the relative diffusion rates of NH_3 and HCl molecules through air. Two cotton plugs, one dipped in HCl(*aq*) and one dipped in NH_3(*aq*), are simultaneously inserted into the ends of the tube. Gaseous NH_3 and HCl vaporizing from the cotton plugs diffuse toward each other and, where they meet, react to form NH_4Cl(*s*). (b) When HCl(*g*) and NH_3(*g*) meet in the tube, a white ring of NH_4Cl(*s*) forms.

However, careful experiments show that this prediction is not borne out—the observed ratio of distances is 1.3, not 1.5 as predicted by Graham's law. This discrepancy is not due to a failure of the kinetic molecular theory or of Graham's law; it exists because this "diffusion" experiment does not involve a simple diffusion process. Rather, it involves a *flow* of ammonia and hydrogen chloride gases through the air in the tube. Because the NH_3 and HCl molecules suffer many collisions with the N_2 and O_2 molecules in the tube, the flow rates of NH_3 and HCl are not directly proportional to their molecular velocities. Higher velocities lead to a higher number of intermolecular collisions, which in turn impedes the flow of the gas. Because of its smaller mass (and thus higher average velocity) the flow of the ammonia gas is impeded more than the flow of the hydrogen chloride gas. Therefore, the NH_3(*g*) travels a smaller distance to meet the HCl(*g*) than is expected from Graham's law (the distance ratio is smaller than 1.5).

While we have given only a qualitative treatment here, the phenomena accompanying the mixing of gases are well understood, and the results of this experiment can be described very accurately by quantitative theories.

Although other technologies are now coming into use for this purpose, gaseous diffusion has played an important role in the enrichment of uranium for use in nuclear reactors (Fig. 5.20). Natural uranium is mostly $^{238}_{92}U$, which cannot be fissioned to produce energy. It contains only about 0.7% of the fissionable nuclide $^{235}_{92}U$. For uranium to be useful as a nuclear fuel, the relative amount of $^{235}_{92}U$ must be increased to about 3%. In the gas diffusion enrichment process, the natural uranium (containing $^{238}_{92}U$ and a small amount of $^{235}_{92}U$) reacts with fluorine to form a mixture of $^{238}UF_6$ and $^{235}UF_6$. Because these molecules have slightly different masses, they will have slightly different velocities at a given temperature, which allows them to be separated by a multistage diffusion process. To understand how this process works, imagine a series of

Figure 5.20
An aerial view of the gas diffusion uranium-enriching plant at Paducah, Kentucky.

chambers separated by semiporous walls that allow passage of some UF_6 molecules but prevent bulk flow of gas. In effect, each porous wall acts much like a tiny hole in an effusion cell. Assume that the UF_6 from natural uranium is placed in chamber 1. Thus chamber 1 contains 99.3% $^{238}UF_6$ and 0.7% $^{235}UF_6$ (that is, 993 molecules of $^{238}UF_6$ for every 7 molecules of $^{235}UF_6$). Some molecules of this UF_6 diffuse through the semiporous barrier into chamber 2, which was initially empty. Because of its smaller mass, $^{235}UF_6$, which has a slightly greater velocity than $^{238}UF_6$, diffuses at a slightly greater rate. Thus chamber 2 will contain a ratio of $^{235}UF_6$ to $^{238}UF_6$ that is slightly greater than 7 to 993.

Although the process is called **gaseous diffusion,** because the chambers are separated by screens with holes just large enough for individual UF_6 molecules to pass through, it behaves like an effusion process. Thus, we can find the actual ratio of the two types of UF_6 in chamber 2 from Graham's law:

$$\frac{\text{Diffusion rate for } ^{235}UF_6}{\text{Diffusion rate for } ^{238}UF_6} = \sqrt{\frac{M^{238}UF_6}{M^{235}UF_6}} = \sqrt{\frac{352.05 \text{ g/mol}}{349.03 \text{ g/mol}}} = 1.0043$$

We can use this factor to calculate the ratio of $^{235}UF_6/^{238}UF_6$ in chamber 2:

$$\underset{\text{Chamber 2}}{\underset{\uparrow}{\frac{^{235}UF_6}{^{238}UF_6}}} = 1.0043 \times \underset{\text{Chamber 1}}{\underset{\uparrow}{\frac{^{235}UF_6}{^{238}UF_6}}} = 1.0043\left(\frac{7}{993}\right)$$

$$= 1.0043(7.0493 \times 10^{-3})$$
$$= 7.0797 \times 10^{-3}$$

This very slight increase represents a change from the ratio of 70,493 $^{235}UF_6$ molecules per 10,000,000 $^{238}UF_6$ molecules in chamber 1 to the ratio of 70,797 $^{235}UF_6$ molecules per 10,000,000 $^{238}UF_6$ molecules in chamber 2.

This enrichment process (in $^{235}UF_6$) continues as the slightly enriched gas in chamber 2 diffuses into chamber 3 and is again enriched by a factor of 1.0043. The same process is repeated until sufficient enrichment occurs. Obviously, this process will take many stages. For example, to calculate the number of steps required to enrich from 0.700% ^{235}U to 3.00% ^{235}U, we have the following equation:

$$\underset{\text{Original ratio}}{\underset{\uparrow}{\frac{0.700\ ^{235}UF_6}{99.3\ ^{235}UF_6}}} \times (1.0043)^N = \underset{\text{Desired ratio}}{\underset{\uparrow}{\frac{3.00\ ^{235}UF_6}{97.0\ ^{238}UF_6}}}$$

where N represents the number of stages. This equation follows from the fact that each stage produces an enrichment by the factor 1.0043. Thus

$$(\text{Original ratio}) \times \underset{\substack{\text{First}\\\text{stage}}}{\underset{\uparrow}{1.0043}} \times \underset{\substack{\text{Second}\\\text{stage}}}{\underset{\uparrow}{1.0043}} \times \underset{\substack{\text{Third}\\\text{stage}}}{\underset{\uparrow}{1.0043}} \times \cdots = \text{final ratio}$$

Solving the above equation for N yields 345. Thus we predict that 345 stages are required to obtain the desired enrichment.

Although we have greatly oversimplified* the actual enrichment process here, this discussion gives you an idea of how it is accomplished. A photo of actual diffusion cells is shown in Fig. 5.21.

Figure 5.21
Uranium-enrichment converters from the Paducah gaseous diffusion plant in Kentucky.

5.8 Collisions of Gas Particles with the Container Walls

In the analysis of the kinetic molecular model that led to the ideal gas equation, we assumed that the pressure a gas exerts is due to the collisions of its particles with the walls of its container. In this section we will consider the details of that phenomenon.

Our goal is to obtain an equation that describes the number of particles that collide per second with a given area of the wall. Although a rigorous derivation of such an equation can be carried out from the details of the kinetic molecular theory, we will not do that. Instead, we will pursue a qualitative strategy, trying to obtain the fundamental relationships from our conceptual understanding of how an ideal gas is expected to behave. We will define the quantity we are looking for as Z_A, the collision rate (per second) of the gas particles with a section of wall that has an area A (in m^2). We expect Z_A to depend on the following factors:

1. The average velocity of the gas particles
2. The size of the area being considered
3. The number of particles in the container

How is Z_A expected to depend on the average velocity of the gas particles? For example, if we double the average velocity, we double the number of wall impacts, so Z_A should double. Thus Z_A depends directly on u_{avg}:

$$Z_A \propto u_{\text{avg}}$$

Similarly, Z_A depends directly on A, the area of the wall under consideration. That is, if we double the area being considered, we will double the number of impacts per second that occur within that section of the wall. Thus $Z_A \propto A$.

Likewise, if the number of particles in the container is doubled, the impacts with the wall will double. For a general case, we need to consider not the absolute number of particles but the number of particles per unit volume (the number density of particles), which can be represented by N/V, the number of particles N divided by the volume V (in m^3). Thus Z_A is expected to depend directly on N/V. That is, $Z_A \propto N/V$.

In summary, Z_A should be directly proportional to u_{avg}, A, and N/V:

$$Z_A \propto u_{\text{avg}} \times A \times \frac{N}{V}$$

*For a more detailed description, *see* W. Spindel and T. Ishida, Isotope Separation, *J. Chem. Ed.*, **68** (1991): 312.

EXAMPLE 5.9

Calculate the impact rate on a 1.00-cm^2 section of a vessel containing oxygen gas at a pressure of 1.00 atm and 27°C.

Solution

To calculate Z_A, we must identify the values of the variables in the equation

$$Z_A = A\frac{N}{V}\sqrt{\frac{RT}{2\pi M}}$$

In this case A is given as 1.00 cm^2. However, to be inserted into the expression for Z_A, A must have the units m^2. The appropriate conversion gives $A = 1.00 \times 10^{-4}$ m^2.

The quantity N/V can be obtained from the ideal gas law by solving for n/V and then converting to the appropriate units:

$$\frac{n}{V} = \frac{P}{RT} = \frac{1.00 \text{ atm}}{\left(0.08206 \frac{\text{L atm}}{\text{K mol}}\right)(300. \text{ K})} = 4.06 \times 10^{-2} \text{ mol/L}$$

To obtain N/V, which has the units (molecules)/m^3, from n/V, we make the following conversion:

$$\frac{N}{V} = 4.06 \times 10^{-2} \frac{\text{mol}}{\text{L}} \times 6.022 \times 10^{23} \frac{\text{(molecules)}}{\text{mol}} \times \frac{1000 \text{ L}}{\text{m}^3}$$
$$= 2.44 \times 10^{25} \text{ (molecules)/m}^3$$

The quantity M represents the molar mass of O_2 in kg. Thus

$$M = 32.0 \frac{\text{g}}{\text{mol}} \times \frac{1 \text{ kg}}{1000 \text{ g}} = 3.20 \times 10^{-2} \text{ kg/mol}$$

Next, we insert these quantities into the expression for Z_A:

$$Z_A = A\frac{N}{V}\sqrt{\frac{RT}{2\pi M}} = (1.00 \times 10^{-4} \text{ m}^2)(2.44 \times 10^{25} \text{ m}^{-3})$$
$$\times \sqrt{\frac{\left(8.3145 \frac{\text{J}}{\text{K mol}}\right)(300. \text{ K})}{(2)(3.14)\left(3.20 \times 10^{-2} \frac{\text{kg}}{\text{mol}}\right)}} = 2.72 \times 10^{23} \text{ s}^{-1}$$

That is, in this gas 2.72×10^{23} collisions per second occur on each 1.00-cm^2 area of the container.

Note that the units for Z_A expected from this relationship are

$$\frac{\text{m}}{\text{s}} \times \text{m}^2 \times \frac{\text{(particles)}}{\text{m}^3} \longrightarrow \frac{\text{(particles)}}{\text{s}} \text{ or } \frac{\text{(collisions)}}{\text{s}}$$

The parentheses are used here because particles and collisions are understood and are not actual units. The correct units for Z_A are 1/s, or s^{-1}. The fact that the product $u_{avg} \times A \times N/V$ gives the units expected for Z_A indicates that we are considering all of the gas properties that influence Z_A. Substituting the expression for u_{avg} gives

$$Z_A \propto \frac{N}{V} A \sqrt{\frac{8RT}{\pi M}}$$

A more detailed analysis of the situation shows that the proportionality constant is $\frac{1}{4}$. Thus the exact equation for Z_A is

$$Z_A = \frac{1}{4}\frac{N}{V} A \sqrt{\frac{8RT}{\pi M}} = A \frac{N}{V} \sqrt{\frac{RT}{2\pi M}}$$

5.9 Intermolecular Collisions

Recall that the postulates of the kinetic molecular model do not take into account collisions between gas particles. Since this model correctly fits ideal gas behavior (that is, the behavior approached by real gases at high T and low P), our conclusion is that intermolecular collisions apparently do not have an important influence on the pressure, volume, or temperature of a gas behaving ideally. That is, the effects of the collisions must somehow "cancel out" relative to the properties P, V, and T of an ideal gas. However, there is much evidence to suggest that collisions do occur among the gas particles in a real gas. For example, a gas that is somehow disturbed from a Maxwell-Boltzmann distribution of velocities will rapidly change until it again reaches an M-B distribution. This behavior must be due to energy exchanges through collisions.

In this section we will consider the collision frequency of the particles in a gas. We will start by considering a single spherical gas particle with diameter d (in meters) that is moving with velocity u_{avg}. As this particle moves through the gas in a straight line, it will collide with another particle only if the other particle has its center in a cylinder with radius d, as shown in Fig. 5.22.

Any particle with its center outside this cylinder will not be hit by our particle. Thus our particle "sweeps out" a cylinder of radius d and length $u_{avg} \times 1$ second during every second of its flight. Therefore, the volume of the cylinder swept out per second is

$$V = \text{volume} = \underbrace{(\pi d^2)}_{\text{Area of cylinder slice}}\underbrace{(u_{avg})(1\ \text{s})}_{\text{Length of cylinder}}$$

As the particle travels through this cylinder, the number of collisions depends on the number of gas particles in that volume. To specify the number of gas particles, we use the number density of the gas, N/V, which indicates the number of gas particles per unit volume. Thus we can write

$$\text{Number of collisions per second} = \left\{\begin{matrix}\text{volume} \\ \text{swept out}\end{matrix}\right\} \times \frac{N}{V} = \pi d^2(u_{avg})\left(\frac{N}{V}\right)$$

$$= \pi d^2\left(\sqrt{\frac{8RT}{\pi M}}\right)\left(\frac{N}{V}\right) = \frac{N}{V} d^2 \sqrt{\frac{8\pi RT}{M}}$$

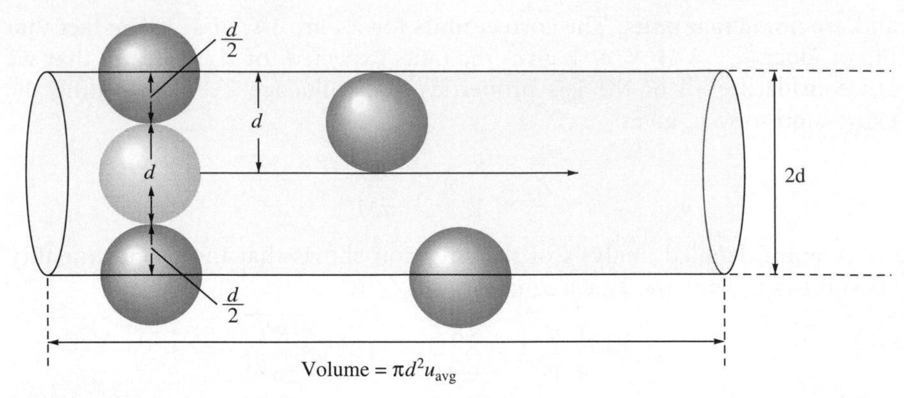

Figure 5.22
The cylinder swept out by a gas particle of diameter *d*.

This equation is not quite correct. If you are thinking carefully about this situation, you may be asking yourself the question, "What about the motions of the other particles?" That is, we have said that the primary particle has velocity u_{avg}, but we have assumed that the other particles are stationary. Of course, they are not really stationary. They are moving in various directions with various velocities. When the motions of the other particles are accounted for (a derivation we will not show here), the *relative velocity* of the primary particle becomes $\sqrt{2}\, u_{avg}$ rather than the value u_{avg} that we have been using. Thus the expression for the collision rate becomes

$$\text{Collision rate (per second)} = Z = \sqrt{2}\frac{N}{V}d^2\sqrt{\frac{8\pi RT}{M}} = 4\frac{N}{V}d^2\sqrt{\frac{\pi RT}{M}}$$

EXAMPLE 5.10

Calculate the collision frequency for an oxygen molecule in a sample of pure oxygen gas at 27°C and 1.0 atm. Assume that the diameter of an O_2 molecule is 300 pm.

Solution

To obtain the collision frequency, we must identify the quantities in the expression

$$Z = 4\frac{N}{V}d^2\sqrt{\frac{\pi RT}{M}}$$

that are appropriate to this case. We can obtain the value of N/V for this sample of oxygen by assuming ideal behavior. From the ideal gas law

$$\frac{n}{V} = \frac{P}{RT} = \frac{1.0 \text{ atm}}{\left(0.08206 \dfrac{\text{L atm}}{\text{K mol}}\right)(300.\text{ K})} = 4.1 \times 10^{-2} \text{ mol/L}$$

Thus

$$\frac{N}{V} = \left(4.1 \times 10^{-2} \frac{\text{mol}}{\text{L}}\right)\left(6.022 \times 10^{23} \frac{\text{molecules}}{\text{mol}}\right)\left(\frac{1000 \text{ L}}{\text{m}^3}\right)$$
$$= 2.5 \times 10^{25} \text{ (molecules)/m}^3$$

From the given information we know that

$$d = 300 \text{ pm} = 300 \times 10^{-12} \text{ m or } 3 \times 10^{-10} \text{ m}$$

Also, for O_2, $M = 3.20 \times 10^{-2}$ kg/mol. Thus

$$Z = 4(2.5 \times 10^{25} \text{ m}^{-3})(3 \times 10^{-10} \text{ m})^2 \times \sqrt{\frac{\pi(8.3145 \text{ J K}^{-1} \text{ mol}^{-1})(300 \text{ K})}{3.20 \times 10^{-2} \text{ kg/mol}}}$$
$$= 4 \times 10^9 \text{ (collisions)}/s = 4 \times 10^9 \; s^{-1}$$

Notice how large this number is. Each O_2 molecule undergoes approximately 4 billion collisions per second in this gas sample.

Mean Free Path

As we saw above, the collision frequency Z represents the number of collisions per second that occur in a given gas sample. On the other hand, the reciprocal of Z gives the time (in seconds) between collisions. Thus if $Z = 4 \times 10^9$ (collisions) per second, then $1/Z = 2.5 \times 10^{-10}$ seconds between collisions. Now if we multiply $1/Z$ by the average velocity, we obtain the **mean free path λ:**

$$\lambda = \frac{1}{Z} \times u_{\text{avg}} = \text{distance between collisions}$$

Time between collisions (s) ↗ ; ↑ Distance traveled per second

Substituting the expressions for $1/Z$ and u_{avg} gives

$$\lambda = \left(\frac{1}{4(N/V)(d^2)\sqrt{\pi RT/M}}\right)\left(\sqrt{\frac{8RT}{\pi M}}\right) = \frac{1}{\sqrt{2}(N/V)(\pi d^2)}$$

EXAMPLE 5.11

Calculate the mean free path in a sample of oxygen gas at 27°C and 1.0 atm.

Solution

Using data from the previous example, we have

$$\lambda = \frac{1}{\sqrt{2}(2.5 \times 10^{25} \text{ m}^{-3})(\pi)(3 \times 10^{-10} \text{ m})^2} = 1 \times 10^{-7} \text{ m}$$

Note that an O_2 molecule travels only a very short distance before it collides with another O_2 molecule. This produces a path for a given O_2 molecule like the one represented in Fig. 5.14, where the length of each straight line is $\sim 10^{-7}$ m.

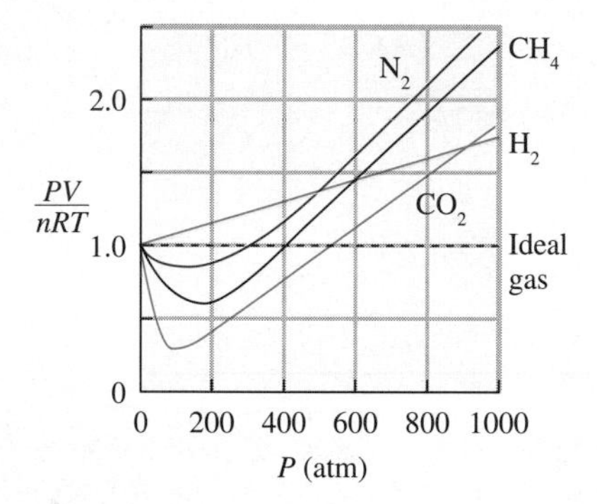

Figure 5.23
Plots of PV/nRT versus P for several gases (200 K). Note the significant deviations from ideal behavior ($PV/nRT = 1$). The behavior is close to ideal only at low pressures (less than 1 atm).

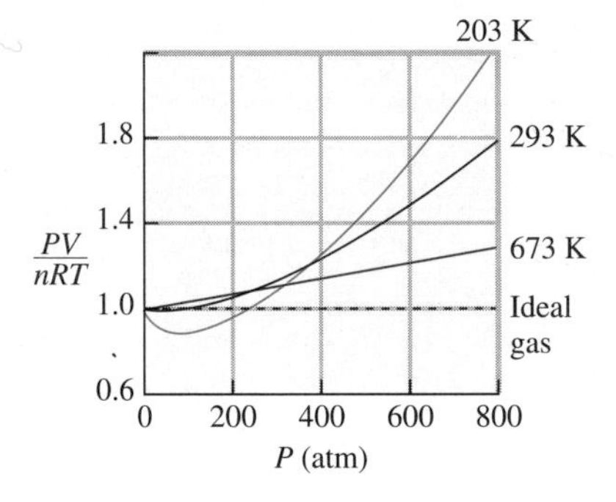

Figure 5.24
Plots of PV/nRT versus P for nitrogen gas at three temperatures. Note that, although nonideal behavior is evident in each case, the deviations are smaller at the higher temperatures.

The fact that PV/nRT is also 1 at high pressures for many gases is due to a canceling of nonideal effects.

5.10 Real Gases

An ideal gas is a hypothetical concept. No gas *exactly* follows the ideal gas law, although many gases come very close at low pressures and/or high temperatures. Thus ideal gas behavior can best be thought of as the behavior *approached by **real gases** under certain conditions.*

We have seen that a very simple model, the kinetic molecular theory, by making some rather drastic assumptions (no interparticle interactions and zero volume for the gas particles), successfully explains ideal behavior. However, it is important that we examine real gas behavior to see how it differs from that predicted by the ideal gas law and to determine what modifications of the kinetic molecular theory are needed to explain the observed behavior. Since a model is an approximation and will inevitably fail, we must be ready to learn from such failures. In fact, we often learn more about nature from the failures of our models than from their successes.

We will examine the experimentally observed behavior of real gases by measuring the pressure, volume, temperature, and number of moles for a gas and noting how the quantity PV/nRT depends on pressure. Plots of PV/nRT versus P are shown for several gases in Fig. 5.23. For an ideal gas PV/nRT equals 1 under all conditions, but notice that for real gases PV/nRT approaches 1 only at low pressures (typically ≤ 1 atm). To illustrate the effect of temperature, we have plotted PV/nRT versus P for nitrogen gas at several temperatures in Fig. 5.24. Notice that the behavior of the gas appears to become more nearly ideal as the temperature is increased. The most important conclusion to be drawn from these plots is that a real gas typically exhibits behavior that is closest to ideal behavior at *low pressures* and *high temperatures.*

How can we modify the assumptions of the kinetic molecular theory to fit the behavior of real gases? An equation for real gases was developed in 1873 by Johannes van der Waals, a physics professor at the University of Amsterdam who in 1910 received a Nobel prize for his work. To follow his analyses, we start with the ideal gas law,

$$P = \frac{nRT}{V}$$

Remember that this equation describes the behavior of a hypothetical gas consisting of volumeless entities that do not interact with each other. In contrast, a real gas consists of atoms or molecules that have finite volumes. Thus the volume available to a given particle in a real gas is less than the volume of the container, because the gas particles themselves take up some of the space. To account for this discrepancy, van der Waals represented the actual volume as the volume of the container, V, minus a correction factor for the volume of the molecules, nb, where n is the number of moles of gas and b is an empirical constant (one determined by fitting the equation to the experimental results).

Thus the volume *actually available* to a given gas molecule is given by the difference $V - nb$.

This modification of the ideal gas equation leads to the expression

$$P' = \frac{nRT}{(V - nb)}$$

P′ is corrected for the finite volume of the particles. The attractive forces have not yet been taken into account.

The volume of the gas particles has now been taken into account.

The next step is to account for the attractions that occur among the particles in a real gas. The effect of these attractions is to make the observed pressure P_{obs} smaller than it would be if the gas particles did not interact:

$$P_{obs} = (P' - \text{correction factor}) = \left(\frac{nRT}{V - nb} - \text{correction factor}\right)$$

This effect can be understood by using the following model. When gas particles come close together, attractive forces occur, which cause the particles to hit the wall with slightly less force than they would in the absence of these interactions (see Fig. 5.25).

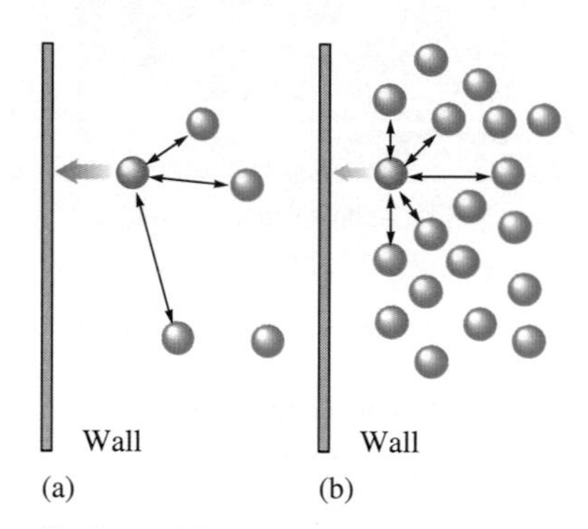

Figure 5.25
(a) Gas at low concentration—relatively few interactions between particles. The indicated gas particle exerts a pressure on the wall close to that predicted for an ideal gas. (b) Gas at high concentration—many more interactions between particles. The indicated gas particle exerts a much lower pressure on the wall than would be expected in the absence of interactions.

The size of the correction factor depends on the concentration of gas molecules defined in terms of moles of gas particles per liter (n/V). The higher the concentration, the more likely a pair of gas particles will be close enough to attract each other. For large numbers of particles, the number of interacting *pairs* of particles depends on the square of the number of particles and thus on the square of the concentration, or $(n/V)^2$. This reasoning can be justified as follows. In a gas sample containing N particles, there are $N - 1$ partners available for each particle, as shown in Fig. 5.26. Since the $1 \cdots 2$ pair is the same as the $2 \cdots 1$ pair, this analysis counts each pair twice. Thus for N particles there are $N(N - 1)/2$ pairs. If N is a very large number, $N - 1$ approximately equals N, giving $N^2/2$ possible pairs. Thus the correction to the ideal pressure for the attractions of the particles has the form

$$P_{obs} = P' - a\left(\frac{n}{V}\right)^2$$

We have now corrected for both the finite volume and the attractive forces of the particles.

where a is a proportionality constant (which includes the factor of $\frac{1}{2}$ from $N^2/2$). The value of a for a given real gas can be determined from observing the actual behavior of that gas. Inserting the corrections for both the volume of the particles and the attractions of the particles gives the equation

P_{obs} is usually called just P.

$$P_{obs} = \frac{nRT}{V - nb} - a\left(\frac{n}{V}\right)^2$$

Observed pressure (↑ P_{obs}); Volume of the container (↑ V); Volume correction (↑ nb); Pressure correction (↑ $a(n/V)^2$)

This equation can be rearranged to give the **van der Waals equation:**

$$\underbrace{\underbrace{\left[P_{obs} + a\left(\frac{n}{V}\right)^2\right]}_{\text{Corrected pressure}}}_{P_{ideal}} \underbrace{\underbrace{(V - nb)}_{\text{Corrected volume}}}_{V_{ideal}} = nRT$$

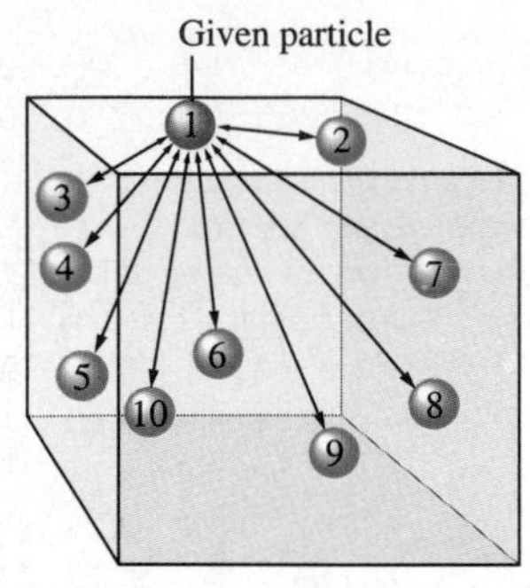

Figure 5.26
Illustration of pairwise interactions among gas particles. In a sample with 10 particles, each particle has 9 possible partners, to give 10(9)/2 = 45 distinct pairs. The factor of $\frac{1}{2}$ arises, because when particle 1 is the particle of interest, we count the ①···② pair; and when particle ② is the particle of interest, we count the ②···① pair. However, ①···② and ②···① are the same pair, which we thus have counted twice. Therefore, we must divide by 2 to get the correct number of pairs.

TABLE 5.3
Values of van der Waals Constants for Some Common Gases

Gas	$a\left(\frac{\text{atm L}^2}{\text{mol}^2}\right)$	$b\left(\frac{\text{L}}{\text{mol}}\right)$
He	0.034	0.0237
Ne	0.211	0.0171
Ar	1.35	0.0322
Kr	2.32	0.0398
Xe	4.19	0.0511
H_2	0.244	0.0266
N_2	1.39	0.0391
O_2	1.36	0.0318
Cl_2	6.49	0.0562
CO_2	3.59	0.0427
CH_4	2.25	0.0428
NH_3	4.17	0.0371
H_2O	5.46	0.0305

The values of the weighting factors, a and b, are determined for a given gas by fitting experimental behavior. That is, a and b are varied until the best fit of the observed pressure is obtained under all conditions. The values of a and b for various gases are given in Table 5.3.

Experimental studies indicate that the changes van der Waals made in the basic assumptions of the kinetic molecular theory corrected the major flaws in the model. First, consider the effects of volume. For a gas at low pressure (large volume), the volume of the container is very large compared with the volumes of the gas particles. That is, the volume available to the gas is essentially equal to the volume of the container, so the gas behaves ideally. On the other hand, for a gas at high pressure (small volume), the volume of the particles becomes significant, so that the volume available to the gas is significantly less than the container volume. These observations are illustrated in Fig. 5.27. Note that the volume-correction constant b generally increases with the size of the gas molecule, which gives further support to these arguments.

The fact that a real gas tends to behave more ideally at high temperatures can also be explained in terms of the van der Waals model. At high temperatures the particles are moving so rapidly that the effects of interparticle interactions are not very important.

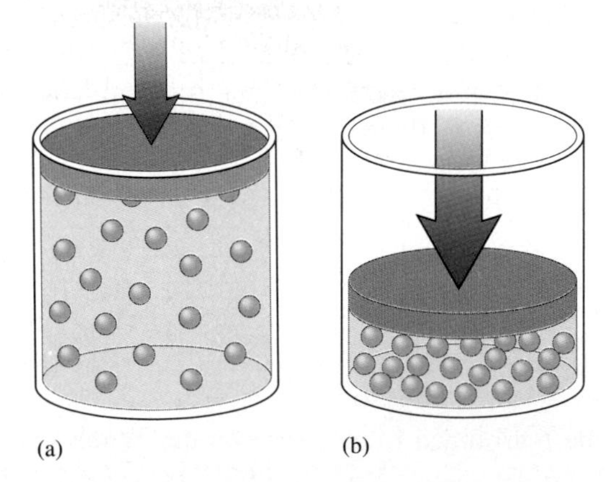

Figure 5.27
The volume occupied by the gas particles themselves is less important at (a) large container volumes (low pressure) than at (b) small container volumes (high pressure).

COOL SOUNDS

As the 1996 prohibition on the production of the ozone-damaging chlorofluorocarbons (CFCs) used in many air conditioners and refrigerators fast approaches, innovative methods for cooling without CFCs are being considered.

Because of the forces that exist among its molecules, a real gas becomes hot when it is compressed and cools when it expands. This phenomenon is at the heart of air conditioning and refrigeration systems that employ gases such as CFCs. The gaseous compound is first compressed using an electrically powered compressor (with the heat being vented to the surroundings) and is then allowed to expand. The expansion lowers the gas's temperature and provides a source of cooling.

A novel idea being explored by Steven L. Garrett and his colleagues at the Naval Postgraduate School in Monterey, CA, is to use sound waves to power a refrigerator. The apparatus is fairly simple: loudspeakers are mounted at each end of a U-shaped tube filled with an inert gas. The sound waves produced by the speakers cause pressure variations in the tube. In low-pressure areas the gas cools; in high-pressure areas the gas becomes hot. The hot areas are cooled by a heat exchanger and over time the gas in the tube becomes very cold. Garrett's acoustic refrigerator uses 205 watts of power to cool its main compartment to 4°C and the freezing compartment to −22°C. These temperatures are accomplished by blasting a sustained 160-decibel tone—a sound 10,000 times louder than the loudest rock concert—into a mixture of inert gases. But because the device is so well insulated, you need a stethoscope to tell whether it is running.

Acoustic refrigeration units have some real advantages: they have no moving parts (except the speaker drivers) and they are environmentally safe. Presently the "sound-fridge" is neither as cheap nor as efficient as conventional cooling devices, but Garrett believes that the unit can be made competitive within 2 or 3 years.

The corrections made by van der Waals to the kinetic molecular theory make physical sense, which makes us confident that we understand the fundamentals of gas behavior at the particle level. This is significant because so much important chemistry takes place in the gas phase. In fact, the mixture of gases called the atmosphere is vital to our existence. In the next section we consider some of the important reactions that occur in the atmosphere.

5.11 Chemistry in the Atmosphere

The gases that are most important to us are located in the **atmosphere** that surrounds the earth's surface. The principal components are N_2 and O_2, but many other important gases, such as H_2O and CO_2 are also present. The average composition of the earth's atmosphere near sea level, with the water vapor removed, is shown in Table 5.4. Because of gravitational effects, the composition of the earth's atmosphere is not constant: heavier molecules tend to be near the earth's surface, and light molecules tend to migrate to higher altitudes and eventually to escape into space. The atmosphere is a highly complex and dynamic system, but for convenience we divide it into several layers based on the way the temperature changes with altitude. (The lowest layer, called the troposphere, is shown in Fig. 5.28.) Note that in contrast to the complex

TABLE 5.4
Atmospheric Composition near Sea Level (dry air)*

Component	Mole Fraction
N_2	0.78084
O_2	0.20948
Ar	0.00934
CO_2	0.000345
Ne	0.00001818
He	0.00000524
CH_4	0.00000168
Kr	0.00000114
H_2	0.0000005
NO	0.0000005
Xe	0.000000087

*The atmosphere contains various amounts of water vapor, depending on conditions.

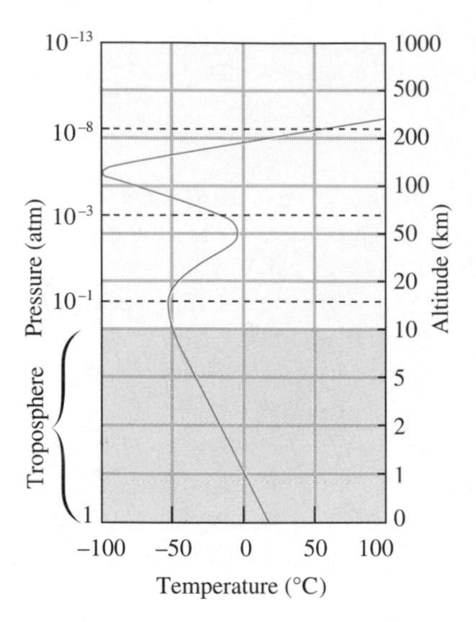

Figure 5.28
The variation of temperature and pressure with altitude. Note that the pressure steadily decreases with increasing altitude but that the temperature does not change monotonically.

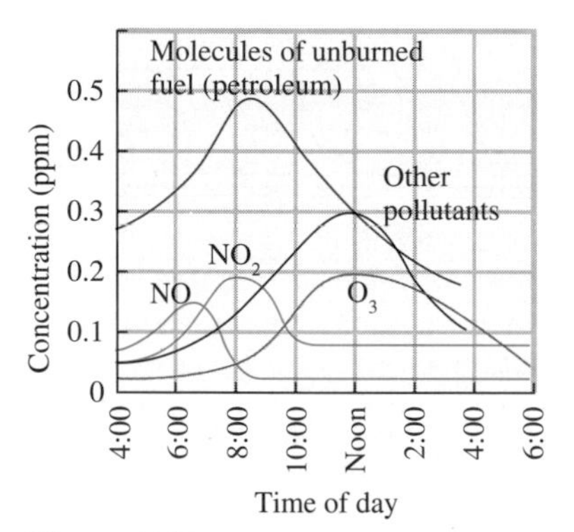

Figure 5.29
Concentration (in molecules per million molecules of "air") of some smog components versus time of day. After P. A. Leighton's classic experiment, "Photochemistry of Air Pollution," in *Physical Chemistry: A Series of Monographs*, ed. Eric Hutchinson and P. Van Rysselberghe, Vol. IX, New York: Academic Press, 1961.

Although represented here as O_2, the actual oxidant is an organic peroxide such as CH_3COO, formed by reaction of O_2 with organic pollutants.

temperature profile of the atmosphere in general, the pressure decreases in a regular way with increasing altitude in the troposphere.

The chemistry occurring in the higher levels of the atmosphere is mostly determined by the effects of high-energy radiation and particles from the sun and other sources in space. In fact, the upper atmosphere serves as an important shield to prevent this high-energy radiation from reaching the earth, where it would damage the relatively fragile molecules sustaining life. In particular, the ozone in the upper atmosphere helps prevent high-energy ultraviolet radiation from penetrating to the earth. Intensive research is in progress to determine the natural factors that control the ozone concentration and to understand how it is affected by chemicals released into the atmosphere.

The chemistry occurring in the troposphere is strongly influenced by human activities. Millions of tons of gases and particulates are released into the troposphere by our highly industrial civilization. Actually, it is amazing that the atmosphere can absorb so much material with relatively small permanent changes.

Significant changes, however, are occurring. Severe **air pollution** is found around many large cities, and it is probable that long-range changes in the planet's weather are taking place. We will deal only with the short-term, localized effects of pollution.

The two main sources of pollution are transportation and the production of electricity. The combustion of petroleum in vehicles produces CO, CO_2, NO, and NO_2, along with unburned molecules from petroleum. When this mixture is trapped close to the ground in stagnant air, reactions occur, producing chemicals that are potentially irritating and harmful to living systems.

The complex chemistry of polluted air appears to center around ozone and the nitrogen oxides (NO_x). At the high temperatures in the gasoline and diesel engines of cars and trucks, N_2 and O_2 react to form a small quantity of NO, which is emitted into the air with the exhaust gases (see Fig. 5.29). This NO is oxidized in air to NO_2, which in turn absorbs energy from sunlight and breaks up into nitric oxide and free oxygen atoms:

$$NO_2(g) \xrightarrow[\text{energy}]{\text{Radiant}} NO(g) + O(g)$$

Oxygen atoms are very reactive and can combine with O_2 to form *ozone:*

$$O(g) + O_2(g) \longrightarrow O_3(g)$$

Ozone is also very reactive. It can react with the unburned hydrocarbons in the polluted air to produce chemicals that cause the eyes to water and burn and are harmful to the respiratory system.

The end-product of this whole process is often referred to as **photochemical smog,** so called because light is required to initiate some of the reactions. The production of photochemical smog can be more clearly understood by examining as a group the reactions discussed above:

$$\begin{aligned} NO_2(g) &\longrightarrow NO(g) + O(g) \\ O(g) + O_2(g) &\longrightarrow O_3(g) \\ NO(g) + \tfrac{1}{2}O_2(g) &\longrightarrow NO_2(g) \\ \hline \text{Net reaction:} \quad \tfrac{3}{2}O_2(g) &\longrightarrow O_3(g) \end{aligned}$$

Note that the NO_2 molecules assist in the formation of ozone without being consumed themselves. The ozone then produces other pollutants.

We can observe this process by analyzing polluted air at various times during a day (see Fig. 5.29). As people drive to work between 6 and 8 A.M., the amounts of NO, NO_2, and unburned molecules from petroleum increase. Later, as the decomposition of NO_2 occurs, the concentration of ozone and other pollutants builds up. Current efforts to combat the formation of photochemical smog are focused on cutting down the amounts of molecules from unburned fuel in automobile exhaust and designing engines that produce less nitric oxide (see Fig. 5.30).

Figure 5.30
Our various modes of transportation produce large amounts of nitrogen oxides, which facilitate the formation of photochemical smog.

The other major source of pollution results from burning coal to produce electricity. Much of the coal found in the Midwest contains significant quantities of sulfur, which, when burned, produces sulfur dioxide:

$$\text{S (In coal)} + O_2(g) \longrightarrow SO_2(g)$$

A further oxidation reaction occurs when sulfur dioxide is changed to sulfur trioxide in the air:

$$2SO_2(g) + O_2(g) \longrightarrow 2SO_3(g)$$

This equation only describes the overall stoichiometry of the process; many different oxidants actually participate in the oxidation of sulfur dioxide (see Chapter 15 for a further discussion). The production of sulfur trioxide is significant because it can combine with droplets of water in the air to form sulfuric acid:

$$SO_3(g) + H_2O(l) \longrightarrow H_2SO_4(aq)$$

Sulfuric acid is very corrosive to both living things and building materials. Another result of this type of pollution is **acid rain** (see Fig. 5.31). In many parts

Figure 5.31
Fir trees damaged by pollution and acid rain.

of the northeastern United States and southeastern Canada, the acid rain has caused some freshwater lakes to become too acidic for fish to live.

The problem of sulfur dioxide pollution is further complicated by the energy crisis. As petroleum supplies dwindle and the price increases, our dependence on coal will grow. As supplies of low-sulfur coal are used up, high-sulfur coal will be utilized. One way to use high-sulfur coal without further harming the air quality is to remove the sulfur dioxide from the exhaust gas by means of a system called a *scrubber* before it is emitted from the power plant stack. A common method of *scrubbing* involves blowing powdered limestone ($CaCO_3$) into the combustion chamber, where it is decomposed to lime and carbon dioxide:

$$CaCO_3(s) \longrightarrow CaO(s) + CO_2(g)$$

The lime then combines with the sulfur dioxide to form calcium sulfite:

$$CaO(s) + SO_2(g) \longrightarrow CaSO_3(s)$$

The calcium sulfite and any remaining unreacted sulfur dioxide is removed by injecting an aqueous suspension of lime into the combustion chamber and the stack, producing a *slurry* (a thick suspension), as shown in Fig. 5.32.

Unfortunately, there are many problems associated with scrubbing. The systems are complicated and expensive and consume a great deal of energy. The large quantities of calcium sulfite produced in the process present a disposal problem. With a typical scrubber approximately 1 ton of calcium sulfite per year is produced per person served by the power plant. Since no use has yet been found for this calcium sulfite, it is usually buried in a landfill. As a result of these difficulties, air pollution by sulfur dioxide continues to be a major problem, one that is expensive in terms of damage to the environment and to human health, as well as in monetary terms.

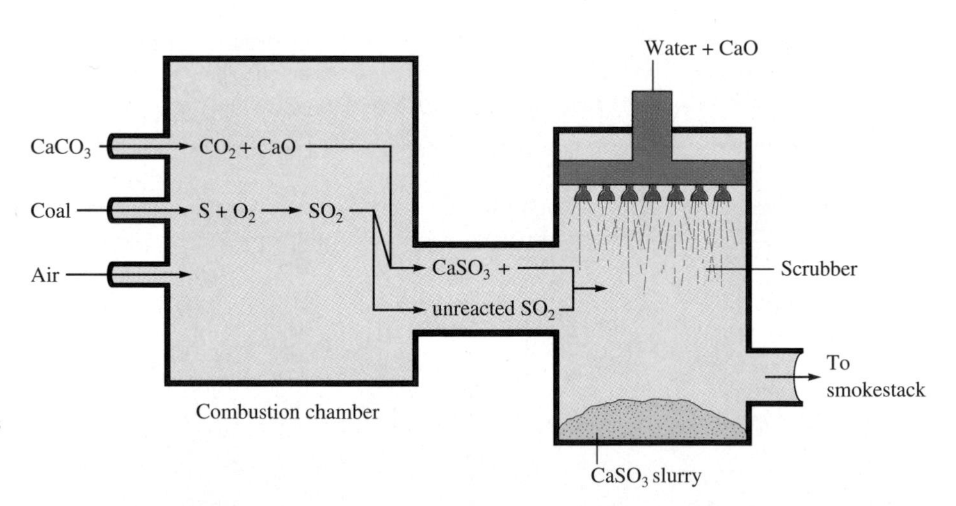

Figure 5.32
A schematic diagram of the process for scrubbing sulfur dioxide from stack gases in power plants.

ACID RAIN: A GROWING PROBLEM

Rainwater, even in pristine, wilderness areas, is slightly acidic, because some of the carbon dioxide present in the atmosphere dissolves in the raindrops to produce H^+ ions by the following reaction:

$$H_2O(l) + CO_2(g) \longrightarrow H^+(aq) + HCO_3^-(aq)$$

This process produces only very small concentrations of H^+ ions in the rainwater. However, gases such as NO_2 and SO_2, which are by-products of energy use, can produce significantly higher H^+ concentrations. Nitrogen dioxide reacts with water to give a mixture of nitrous acid and nitric acid:

$$2NO_2(g) + H_2O(l) \longrightarrow HNO_2(aq) + HNO_3(aq)$$

Sulfur dioxide is oxidized to sulfur trioxide, which then reacts with water to form sulfuric acid:

$$2SO_2(g) + O_2(g) \longrightarrow 2SO_3(g)$$
$$SO_3(g) + H_2O(l) \longrightarrow H_2SO_4(aq)$$

The damage caused by the acid formed in polluted air is a growing worldwide problem. Lakes are dying in Norway, the forests are sick in Germany, and buildings and statues are deteriorating all over the world.

For example, the Field Museum in Chicago contains more white Georgia marble than any other structure in the world. But nearly seventy years of exposure to the elements has taken such a toll on it that the building has just undergone a multimillion-dollar renovation to replace the damaged marble with freshly quarried material.

(left) Sculpture on the Field Museum of Chicago marred by acid rain and soot. (right) Sculpture after a recent restoration.

What is the chemistry of the deterioration of marble by sulfuric acid? Marble is produced by geological processes at high temperatures and pressures from limestone, a sedimentary rock formed by slow deposition of calcium carbonate from the shells of marine organisms. Limestone and marble are chemically identical ($CaCO_3$) but differ in physical properties because limestone is composed of smaller particles of calcium carbonate and is thus more porous and more workable. Although both limestone and marble are used for buildings, marble can be polished to a higher sheen and is often preferred for decorative purposes.

Both marble and limestone react with sulfuric acid to form calcium sulfate. The process can be represented most simply as

$$CaCO_3(s) + H_2SO_4(aq) \longrightarrow Ca^{2+}(aq) + SO_4^{2-}(aq) + H_2O(l) + CO_2(g)$$

In this equation the calcium sulfate is represented by separate hydrated ions because calcium sulfate is quite water-soluble and dissolves in rainwater. Thus in areas bathed by rainwater the marble slowly dissolves away.

In areas of the building protected from the rain, the calcium sulfate can form the mineral gypsum, $CaSO_4 \cdot 2H_2O$. The $\cdot 2H_2O$ in the formula of gypsum indicates the presence of two water molecules (called waters of hydration) for each $CaSO_4$ formula unit in the solid. The smooth surface of the marble is thus replaced by a thin layer of gypsum, a more porous material that binds soot and dust.

What can be done to protect limestone and marble structures from this kind of damage? Of course, one approach is to lower sulfur dioxide and nitrogen oxide emissions from power plants (Fig. 5.32). In addition, scientists are experimenting with coatings to protect marble from the acidic atmosphere. However, a coating can do more harm than good unless it "breathes." If moisture trapped beneath the coating freezes, the expanding ice can fracture the marble. Needless to say, it is difficult to find a coating that will allow water to pass but not allow acid to pass, so the search continues.

Suggested Reading

A. Elena Charola, "Acid Rain Effects on Stone Monuments," *J. Chem. Ed.* 64 (1987), p. 436.

EXERCISES

A blue exercise number indicates that the answer to that exercise appears at the back of this book.

Pressure

1. A sealed-tube manometer as shown below can be used to measure pressures below atmospheric pressure. The tube above the mercury is evacuated. When there is a vacuum in the flask, the mercury levels in both arms of the U-tube are equal. If a gaseous sample is introduced into the flask, the mercury levels are different. The difference h is a measure of the pressure of the gas inside the flask. If h is equal to 4.75 cm, calculate the pressure in the flask in torr, pascals, and atmospheres.

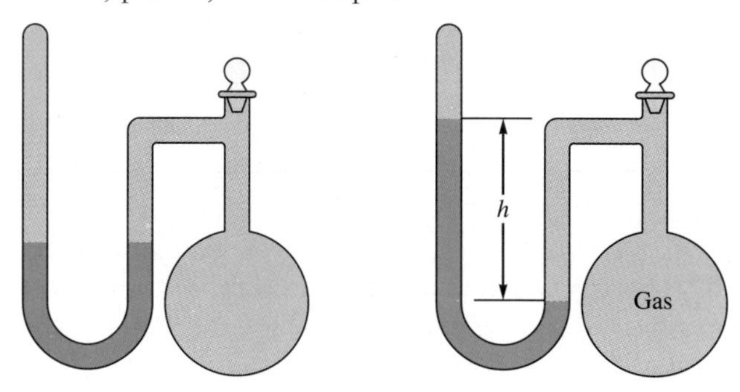

2. A diagram for an open-tube manometer is shown below.

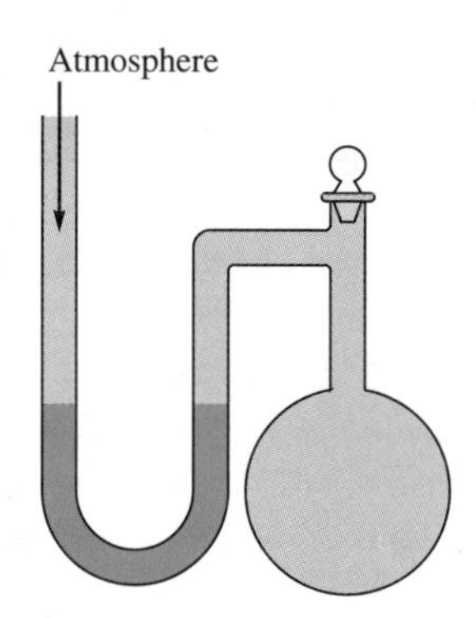

If the flask is open to the atmosphere, the mercury levels are equal. For each of the following situations where a gas is contained in the flask, calculate the pressure in the flask in torr, atmospheres, and pascals.

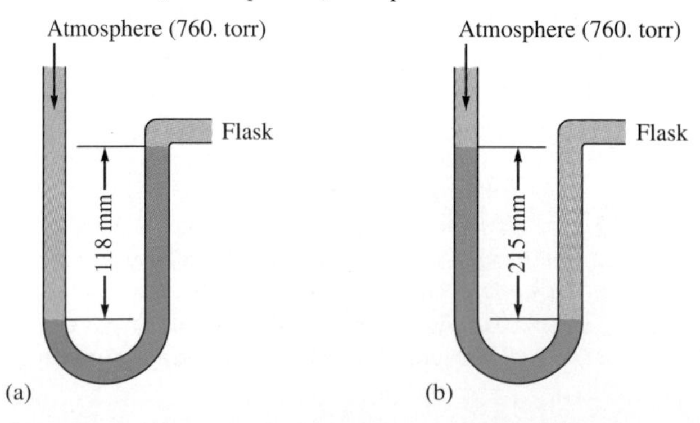

c. Calculate the pressures in the flask in parts a and b (in torr) if the atmospheric pressure is 635 torr.

3. The gravitational force exerted by an object is given by

$$F = mg$$

where F is the force in newtons, m is the mass in kilograms, and g is the acceleration due to gravity, 9.81 m/s^2. Calculate the force exerted per unit of area by a column of mercury (density = 13.59 g/cm^3) 76.0 cm high. How high would a column of water (density = 1.00 g/cm^3) have to be to exert the same force?

4. Assuming the open-tube manometer in Exercise 2 contains a nonvolatile silicone oil (density = 1.30 g/cm^3) instead of mercury (density = 13.6 g/cm^3), what are the pressures in the flask as shown in parts a and b in torr, atmospheres, and pascals?

5. What advantage would there be in using a less dense fluid than mercury in a manometer used to measure relatively small differences in pressure? (See Exercise 4.)

Gas Laws

6. Draw a qualitative graph to show how the first property varies with the second in each of the following (assume 1 mol of an ideal gas and T in Kelvin).
   a. PV versus V with constant T
   b. P versus T with constant V
   c. T versus V with constant P
   d. P versus V with constant T
   e. P versus $1/V$ with constant T
   f. PV/T versus P

7. Consider the flask diagrammed below. What are the final partial pressures of H_2 and N_2 after the stopcock between the two flasks is opened? (Assume the final volume is 3.00 L.) What is the total pressure (in torr)?

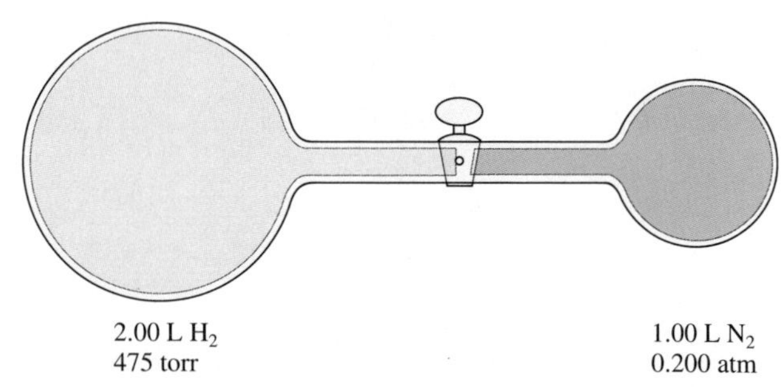

8. A steel cylinder contains 150 mol of argon gas at a temperature of 25°C and a pressure of 7.5 MPa. After some of the argon has been used, the pressure is 1.2 MPa at a temperature of 17°C. What mass of argon remains in the cylinder?

9. A balloon is filled with helium gas at 25°C. The balloon expands until the pressure is equal to the barometric pressure of 720. torr. The balloon rises to an altitude of 6000 ft, where the pressure is 605 torr and the tempera-

ture is 15°C. What is the change in the volume of the balloon as it ascends to 6000 ft?

10. A piece of solid carbon dioxide, with a mass of 22.0 g, is placed in an otherwise empty 4.00-L container at 27°C. What is the pressure in the container after all of the carbon dioxide vaporizes? If 22.0 g of solid carbon dioxide were placed in a similar container already containing air at 740. torr, what would be the partial pressure of carbon dioxide and the total pressure in the container after the carbon dioxide had vaporized?

11. An ideal gas is in a cylinder with a volume of 5.0×10^2 mL at a temperature of 30.°C and a pressure of 710 torr. The gas is compressed to a volume of 25 mL, and the temperature is raised to 820°C. What is the new pressure?

12. The steel bomb of a bomb calorimeter, which has a volume of 75.0 mL, is charged with oxygen to a pressure of 145 atm at 22°C. Calculate the moles of oxygen in the bomb.

13. A container is filled with an ideal gas to a pressure of 40.0 atm at 0°C.
    a. What will be the pressure in the container if it is heated to 125°C?
    b. At what temperature would the pressure be 1.50×10^2 atm?
    c. At what temperature would the pressure be 25.0 atm?

14. A 2.50-L container is filled with 175 g of argon.
    a. If the pressure is 5.00 atm, what is the temperature?
    b. If the temperature is 195 K, what is the pressure?

15. A compressed gas cylinder at 13.7 MPa and 23°C is in a room where a fire raises the temperature to 950°C. What is the new pressure in the cylinder?

16. A bicycle tire is filled with air to a pressure of 100. psi, at a temperature of 19°C. Riding the bike on asphalt on a hot day increases the temperature of the tire to 58°C. The volume of the tire increases by 4.0%. What is the new pressure in the bicycle tire?

17. A hot-air balloon is filled with air to a volume of 4.00×10^3 m^3 at 745 torr and 21°C. The air in the balloon is then heated to 62°C, causing the balloon to expand to a volume of 4.20×10^3 m^3. What is the ratio of the number of moles of air in the heated balloon to the original number of moles of air in the balloon? (*Hint:* openings in the balloon allow air to flow in and out. Thus the pressure in the balloon is always the same as that of the atmosphere.)

18. A sample of oxygen gas is collected over water at 25°C and a total pressure of 745 torr. The volume of gas collected is 150.0 mL. What is the mass of the oxygen collected? (At 25°C the vapor pressure of water is 23.8 torr.)

19. At 0°C a 1.00-L flask contains 2.00×10^{-2} mol of N_2, 1.80×10^2 mg of O_2, and NO at a concentration of 9.00×10^{18} molecules/cm^3. What is the partial pressure of each gas, and what is the total pressure in the flask?

20. At STP, 1.0 L of Br_2 reacts completely with 3.0 L of F_2, producing 2.0 L of a product. What is the formula of the product? (All substances are gases.)

21. A sample of nitrogen gas was collected over water at 20.°C and a total pressure of 1.00 atm. A total volume of 2.50×10^2 mL was collected. What mass of nitrogen was collected? (At 20.°C the vapor pressure of water is 17.5 torr.)

22. Helium is collected over water at 25°C and 1.00 atm total pressure. What total volume of "wet" gas must be collected to obtain 0.586 g of helium? (At 25°C the vapor pressure of water is 23.8 torr.)

23. A mixture of 1.00 g of H_2 and 1.00 g of He exerts a pressure of 0.480 atm. What is the partial pressure of each gas present in the mixture?

24. In a mixture of the two gases, the partial pressures of $CH_4(g)$ and $O_2(g)$ are 0.175 atm and 0.250 atm, respectively.
    a. What is the mole fraction of each gas in the mixture?
    b. If the mixture occupies a volume of 2.00 L at 65°C, calculate the total number of moles of gas in the mixture.
    c. Calculate the number of grams of each gas in the mixture.

Gas Density, Molar Mass, and Reaction Stoichiometry

25. Silicon tetrachloride, $SiCl_4$, and trichlorosilane, $SiHCl_3$, are both starting materials for the production of electronics-grade silicon. Calculate the densities of pure $SiCl_4$ and pure $SiHCl_3$ vapor at 85°C and 758 torr.

26. A compound contains only nitrogen and hydrogen and is 87.4% nitrogen by mass. A gaseous sample of the compound has a density of 0.977 g/L at 710. torr and 100.°C. What is the molecular formula of the compound?

27. A compound has the empirical formula CHCl. A 256-mL flask at 373 K and 750. torr contains 0.800 g of the gaseous compound. Give the molecular formula.

28. One of the chemical controversies of the nineteenth century concerned the element beryllium (Be). Berzelius originally claimed that beryllium was a trivalent element (forming Be^{3+} ions) and that it formed an oxide with the formula Be_2O_3. This assumption resulted in a calculated atomic weight of 13.5 for beryllium. In formulating his periodic table, Mendeleev proposed that beryllium was divalent (forming Be^{2+} ions) and that it gave an oxide with the formula BeO. This assumption gives an atomic

weight of 9.0. In 1894 A. Combes (*Comptes Rendes* 1894, p. 1221) reacted beryllium with the anion $C_5H_7O_2^-$ and measured the density of the gaseous product. Combes's data for two different experiments are as follows:

	I	II
Mass	0.2022 g	0.2224 g
Volume	22.6 cm^3	26.0 cm^3
Temperature	13°C	17°C
Pressure	765.2 torr	764.6 torr

If beryllium is a divalent metal, the molecular formula of the product will be $Be(C_5H_7O_2)_2$; if it is trivalent, the formula will be $Be(C_5H_7O_2)_3$. Show how Combes's data help to confirm that beryllium is a divalent metal.

29. Discrepancies in the experimental values of the molar mass of nitrogen provided some of the first evidence for the existence of the noble gases. If pure nitrogen is collected from the decomposition of ammonium nitrite,

$$NH_4NO_2(s) \xrightarrow{\text{Heat}} N_2(g) + 2H_2O(g)$$

its measured molar mass is 28.01. If O_2, CO_2, and H_2O are removed from air, the remaining gas has an average molar mass of 28.15. Assuming this discrepancy is solely a result of contamination with argon (atomic mass = 39.95), calculate the ratio of moles of Ar to moles of N_2 in air.

30. A sample of methane (CH_4) gas contains a small amount of helium. Calculate the volume percentage of helium if the density of the sample is 0.70902 g/L at 0.0°C and 1.000 atm.

31. Metallic molybdenum can be produced from the mineral molybdenite, MoS_2. The mineral is first oxidized in air to molybdenum trioxide and sulfur dioxide. Molybdenum trioxide is then reduced to metallic molybdenum using hydrogen gas. The balanced equations are

$$MoS_2(s) + 7/2O_2(g) \rightarrow MoO_3(s) + 2SO_2(g)$$

$$MoO_3(s) + 3H_2(g) \rightarrow Mo(s) + 3H_2O(l)$$

Calculate the volumes of air and hydrogen gas at 17°C and 1.00 atm that are necessary to produce 1.00×10^3 kg of pure molybdenum from MoS_2. Assume air contains 21% oxygen by volume and assume 100% yield for each reaction.

32. In 1897 the Swedish explorer Andreé tried to reach the North Pole in a balloon. The balloon was filled with hydrogen gas. The hydrogen gas was prepared from iron splints and diluted sulfuric acid. The reaction is

$$Fe(s) + H_2SO_4(aq) \rightarrow FeSO_4(aq) + H_2(g)$$

The volume of the balloon was 4800 m^3 and the loss of hydrogen gas during filling was estimated at 20%. What mass of iron splints and 98% (by mass) H_2SO_4 were needed to ensure the complete filling of the balloon? Assume a temperature at 0°C, a pressure of 1.0 atm during filling, and 100% yield.

33. Urea (H_2NCONH_2) is used extensively as a nitrogen source in fertilizers. It is produced commercially from the reaction of ammonia and carbon dioxide:

$$2NH_3(g) + CO_2(g) \xrightarrow[\text{Pressure}]{\text{Heat}} H_2NCONH_2(s) + H_2O(g)$$

Ammonia gas at 223°C and 90. atm flows into a reactor at a rate of 500. L/min. Carbon dioxide at 223°C and 45 atm flows into the reactor at a rate of 600. L/min. What mass of urea is produced per minute by this reaction assuming 100% yield?

34. Calculate the volume of O_2, at STP, required for the complete combustion of 125 g of octane (C_8H_{18}) to CO_2 and H_2O.

35. The method used by Joseph Priestley to obtain oxygen made use of the thermal decomposition of mercuric oxide:

$$2HgO(s) \xrightarrow{\text{Heat}} 2Hg(l) + O_2(g)$$

What volume of oxygen gas, measured at 30.°C and 725 torr, can be produced from the complete decomposition of 4.10 g of mercuric oxide?

36. Xenon and fluorine will react to form binary compounds when a mixture of the two gases is heated to 400°C in a nickel reaction vessel. A 100.0-mL nickel container is filled with xenon and fluorine giving partial pressures of 1.24 atm and 10.10 atm, respectively, at a temperature of 25°C. The reaction vessel is heated to 400°C to cause a reaction to occur and then cooled to a temperature at which F_2 is a gas and the xenon fluoride is a nonvolatile solid. The remaining F_2 gas is transferred to another 100.0-mL nickel container where the pressure of F_2 at 25°C is 7.62 atm. Assuming all of the xenon has reacted, what is the formula of the product?

37. The nitrogen content of organic compounds can be determined by the Dumas method. The compound in question is first reacted by passage over hot CuO(*s*):

$$\text{Compound} \xrightarrow[\text{CuO}(s)]{\text{Hot}} N_2(g) + CO_2(g) + H_2O(g)$$

The gaseous products are then passed through a concentrated solution of KOH to remove the CO_2. After passage through the KOH solution the gas contains N_2 and is saturated with water vapor. In a given experiment a 0.253-g sample of a compound produced 31.8 mL of N_2 saturated with water vapor at 25°C and 726 torr. What is the mass percent of nitrogen in the compound? (The vapor pressure of water at 25°C is 23.8 torr.)

38. An organic compound contains C, H, N, and O. Combustion of 0.1023 g of the compound in excess oxygen yielded 0.2766 g of CO_2 and 0.0991 g of H_2O. A sample of 0.4831 g of the compound was analyzed for nitrogen by the Dumas method. At STP, 27.6 mL of dry N_2 was obtained. In a third experiment the density of the compound as a gas was found to be 4.02 g/L at 127°C and 256 torr. What are the empirical formula and the molecular formula of the compound?

39. Consider the first step in the industrial production of nitric acid:

$$4NH_3(g) + 5O_2(g) \rightarrow 4NO(g) + 6H_2O(g)$$

a. What volume of NO, measured at 1.00 atm and 1000°C can be produced from 10.0 L of NH_3 and excess O_2 measured at the same temperature and pressure?
b. What volume of O_2 measured at STP is consumed in reacting with 10.0 kg of NH_3?
c. What mass of NO is produced from the reaction of 5.00×10^2 L of NH_3, measured at 250.°C and 3.00 atm, with excess O_2?
d. What mass of H_2O is produced from the reaction of 65.0 L of NH_3 with 75.0 L of O_2, both measured at STP?
e. How many moles of NO are produced from the mixture in part d?

Kinetic Molecular Theory and Real Gases

40. Using the postulates of the kinetic molecular theory, give a molecular interpretation of Boyle's law, Charles's law, and Dalton's law of partial pressures.

41. Calculate the average kinetic energies of the CH_4 and N_2 molecules at 273 K and 546 K.

42. Calculate the root mean square velocities of the CH_4 and N_2 molecules at 273 K and 546 K.

43. Do all of the molecules in a 1-mol sample of $CH_4(g)$ have the same kinetic energy at 273 K?

44. Consider three identical flasks filled with different gases.

Flask A: CO at 760 torr and 0°C
Flask B: N_2 at 250 torr and 0°C
Flask C: H_2 at 100 torr and 0°C

a. In which flask will the molecules have the greatest average kinetic energy?
b. In which flask will the molecules have the greatest root mean square velocity?
c. Which flask will have the greatest number of collisions per second with the walls of the container?

45. One way of separating oxygen isotopes is by gaseous diffusion of carbon monoxide. Calculate the relative rates of diffusion of $^{12}C^{16}O$, $^{12}C^{17}O$, and $^{12}C^{18}O$. Name some advantages and disadvantages of separating oxygen isotopes by gaseous diffusion of carbon dioxide instead of carbon monoxide.

46. Consider a 1.0-L container of neon gas at STP. Will the average kinetic energy, root mean square velocity, frequency of collisions of gas molecules with each other, frequency of collisions of gas molecules with the walls of the container, and energy of impact of gas molecules with the container increase, decrease, or remain the same under each of the following conditions?
a. The temperature is increased to 100°C.
b. The temperature is decreased to 50°C.
c. The volume is decreased to 0.5 L.
d. The number of moles of neon is doubled.

47. The diffusion rate of an unknown gas is measured and found to be 31.50 mL/min. Under identical experimental conditions the diffusion rate of O_2 is found to be 30.50 mL/min. If the choices are CH_4, CO, NO, CO_2, and NO_2, what is the identity of the unknown gas?

48. The rate of diffusion of a particular gas was measured to be 24.0 mL/min. Under the same conditions the rate of diffusion of pure methane gas, CH_4, is 47.8 mL/min. What is the molar mass of the unknown gas?

49. It took 4.5 min for 1.0 L of helium to effuse through a porous barrier. How long will it take for 1.0 L of NF_3 gas to effuse under identical conditions?

50. Calculate the pressure exerted by 0.5000 mol of N_2 in a 1.0000-L container at 25.0°C. (See Table 5.3.)
a. Use the ideal gas law.
b. Use the van der Waals equation.
c. Compare the results from parts a and b.

51. Calculate the pressure exerted by 0.5000 mol of N_2 in a 10.000-L container at 25.0°C. (See Table 5.3.)
a. Use the ideal gas law.
b. Use the van der Waals equation.
c. Compare the results from parts a and b.
d. Compare the results with those in the previous exercise.

52. Use values of the van der Waals constants for NH_3 (Table 5.3) and show how much of its deviation from ideal behavior can be corrected by the van der Waals equation. Use the data for pressure and volume given in Example 5.1.

53. We state that the ideal gas law tends to hold best at low pressures and high temperatures. Show how the van der Waals equation simplifies to the ideal gas law under these conditions.

Atmosphere Chemistry

54. Use the data in Table 5.4 to calculate the partial pressure of NO in dry air assuming the total pressure is 1.0 atm. Assuming a temperature of 0°C, calculate the number of molecules of NO per cubic centimeter.

55. Atmospheric scientists often use mixing ratios to express the concentrations of trace compounds in air. Mixing ratios are often expressed as ppmv (parts per million volume):

$$\text{ppmv of } X = \frac{\text{vol of } X \text{ at STP}}{\text{total vol of air at STP}} \times 10^6$$

In November 1983 the concentration of carbon monoxide in the air in downtown Denver, Colorado, reached 3.0×10^2 ppmv. The atmospheric pressure at that time was 628 torr and the temperature was 0°C.
a. What was the partial pressure of CO?
b. What was the concentration of CO in molecules per cubic meter?
c. What was the concentration of CO in molecules per cubic centimeter?

56. A 10.0-L sample of air was collected at an altitude of 1.00×10^2 km. What would be the volume of the sample at STP? (See Fig. 5.28.)

57. Write reactions to show how the nitric and sulfuric acids in acid rain react with marble and limestone. Both marble and limestone are primarily calcium carbonate.

58. In a particular urban area the NO_2 concentration reached a maximum value of 0.20 ppm by volume. (See Exercise 55.) Assume this concentration exists to an altitude of 5.0×10^3 ft over an area of 5.0 mi by 10.0 mi. Assuming 1.0 atm pressure and 0°C temperature, calculate the total mass of NO_2 in the atmosphere over this city.

59. Trace organic compounds in the atmosphere are first concentrated and then measured by gas chromatography. In the concentration step, several liters of air are pumped through a tube containing a porous substance that traps organic compounds. The tube is then connected to a gas chromatograph and heated to release the trapped compounds. The organic compounds are separated in the column and the amounts are measured. In an analysis for benzene and toluene in air, a 3.00 L sample of air at 748 torr and 23°C was passed through the trap. The gas chromatography analysis showed that this air sample contained 89.6 ng of benzene (C_6H_6) and 153 ng of toluene (C_7H_8). Calculate the mixing ratio (see Exercise 55) and number of molecules per cubic centimeter for both benzene and toluene.

Additional Exercises

60. Rationalize the following observations.
a. Aerosol cans will explode if heated.
b. You can drink through a soda straw.
c. A thin-walled can will collapse when the air inside is removed by a vacuum pump.
d. Manufacturers produce different types of tennis balls for high and low altitudes.

61. In the "Méthode Champenoise," grape juice is fermented in a wine bottle to produce sparkling wine. The reaction is

$$C_6H_{12}O_6(aq) \longrightarrow 2C_2H_5OH(aq) + 2CO_2(g)$$

Fermentation of 750. mL of grape juice (density = 1.0 g/cm^3) is allowed to take place in a bottle with a total volume of 825 mL until 12% by volume is ethanol (C_2H_5OH). Assuming that the CO_2 is insoluble in H_2O (actually a wrong assumption), what would be the pressure of CO_2 inside the wine bottle at 25°C? (The density of ethanol is 0.79 g/cm^3.)

62. A compressed gas cylinder contains 1.00×10^3 g of argon gas. The pressure inside the cylinder is 2050. psi (pounds per square inch) at a temperature of 18°C. How much gas remains in the cylinder if the pressure is decreased to 650. psi at a temperature of 26°C?

63. The total mass that can be lifted by a balloon is given by the difference between the mass of air displaced by the balloon and the mass of the gas inside the balloon. Consider a hot-air balloon that approximates a sphere 5.00 m in diameter and contains air heated to 65°C. The surrounding air temperature is 21°C. The pressure in the balloon is equal to the atmospheric pressure, which is 745 torr.
a. What total mass can the balloon lift? Assume the average molar mass of air is 29.0. (*Hint:* heated air is less dense than cool air.)
b. If the balloon is filled with enough helium at 21°C and 745 torr to achieve the same volume as in part a, what total mass can the balloon lift?
c. What mass could the hot-air balloon (from part a) lift if it were on the ground in Denver, Colorado, where a typical atmospheric pressure is 630. torr?
d. What mass could the hot air balloon (from part a) lift if it were a cold day with a temperature of −8°C?

64. An important process for the production of acrylonitrile (C_3H_3N) (U.S. production is greater than 10^9 lb) is given by the following reaction:

$$2C_3H_6(g) + 2NH_3(g) + 3O_2(g) \longrightarrow 2C_3H_3N(g) + 6H_2O(g)$$

A 150.-L reactor is charged to the following partial pressures at 25°C:

$$P_{C_3H_6} = 0.500 \text{ MPa}$$
$$P_{NH_3} = 0.800 \text{ MPa}$$
$$P_{O_2} = 1.500 \text{ MPa}$$

What mass of acrylonitrile can be produced from this mixture (MPa = 10^6 Pa)?

65. The oxides of Group 2A metals (symbolized by M here) react with carbon dioxide according to the following reaction:

$$MO(s) + CO_2(g) \longrightarrow MCO_3(s)$$

A 2.85-g sample containing only MgO and CuO is placed in a 3.00-L container. The container is filled with CO_2 to a pressure of 740. torr at 20.°C. After the reaction has gone to completion, the pressure inside the flask is 390. torr at 20.°C. What is the mass percent of MgO in the mixture? Assume that only the MgO reacts with CO_2.

66. Nitrous oxide (N_2O) can be produced by the thermal decomposition of ammonium nitrate:

$$NH_4NO_3(s) \xrightarrow{\text{Heat}} N_2O(g) + 2H_2O(l)$$

What volume of N_2O, collected over water at a total pressure of 97.0 kPa and 18°C, can be produced from the thermal decomposition of 1.55 g of NH_4NO_3? (The vapor pressure of water at 18°C is 15 torr.)

67. Formaldehyde (CH_2O) is sometimes released from foamed insulation used in homes. The federal standard for the allowable amount of CH_2O in air is 1.0 ppbv (parts per billion volume; similar to ppmv as defined in Exercise 55). How many molecules per cubic centimeter is this at STP? If the concentration of formaldehyde in a room is 1.0 ppbv, what total mass of formaldehyde is present at STP if the room measures 18.0 ft × 24.0 ft × 8.0 ft?

68. Acetylene gas, $C_2H_2(g)$, can be produced by reacting solid calcium carbide, CaC_2, with water. The products are acetylene and calcium hydroxide. What volume of wet acetylene is collected at 25°C and 715 torr when 2.50 g of calcium carbide is reacted with an excess of water? (At 25°C the vapor pressure of water is 23.8 torr.)

69. Consider the three flasks in the diagram below. Assuming the connecting tubes have negligible volume, what is the partial pressure of each gas and the total pressure after all of the stopcocks are opened?

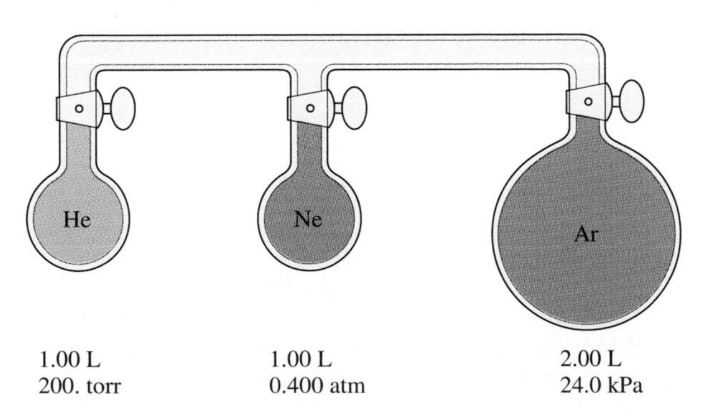

70. A 1.00-L container is evacuated until the pressure is 1.00×10^{-6} torr at 22°C. How many molecules of gas remain in the container? How many molecules are there per cubic centimeter?

71. A compound contains only C, H, and N. It is 58.51% C and 7.37% H by mass. Helium effuses through a porous frit 3.20 times faster than the compound does. Determine the empirical and molecular formulas of this compound.

72. A compound containing only C, H, and N yields the following data:
   i. Complete combustion of 35.0 mg of the compound produced 33.5 mg of CO_2 and 41.1 mg of H_2O.
   ii. A 65.2-mg sample of the compound was analyzed for nitrogen by the Dumas method, giving 35.6 mL of N_2 at 740. torr and 25°C.
   iii. The effusion rate of the compound as a gas was measured and found to be 24.6 mL/min. The effusion rate of argon gas, under identical conditions, is 26.4 mL/min.

   What is the formula of the compound?

73. A 15.0-L tank is filled with H_2 to a pressure of 2.00×10^2 atm. How many balloons (each 2.00 L) can be inflated to a pressure of 1.00 atm from the tank? Assume that there is no temperature change and that the tank cannot be emptied below 1.00 atm pressure.

74. Consider the following diagram:

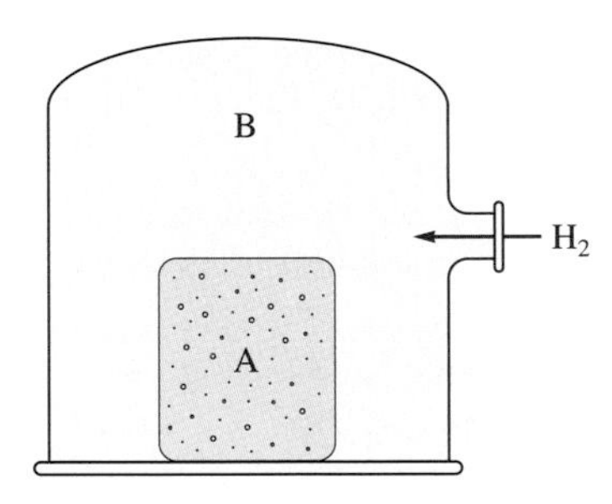

A porous container (A), filled with air at STP, is contained in a large enclosed container (B), which is flushed with $H_2(g)$. What will happen to the pressure inside container (A)? Explain your answer.

75. Without looking at tables of values, which of the following gases would you expect to have the largest value of the van der Waals constant *b*: H_2, N_2, CH_4, C_2H_6, or C_3H_8?

76. From the values in Table 5.3 for the van der Waals constant *a* for the gases H_2, CO_2, N_2, and CH_4, predict which molecule shows the strongest intermolecular attractions.

77. Derive a linear relationship between gas density and temperature and use it to estimate the value of absolute zero temperature (in °C to the nearest 0.1°C) from an air

sample whose density is 1.2930 g/L at 0.0°C and 0.9460 g/L at 100.0°C. Assume air obeys the ideal gas law and that the pressure is held constant.

78. A sample of nitrogen gas weighing 9.30 g at a pressure of 750. torr occupies a volume of 12.3 L when its temperature is 450. K. What is its volume (at constant *P*) when its temperature is 300. K?

79. A 1.00-L gas sample at 100.°C and 600. torr contains 50.0% helium and 50.0% xenon by mass. What are the partial pressures of the individual gases?

80. The stopcock between a 3.00-L bulb containing N_2 at 0.250 atm and a 7.00-L bulb containing argon at 0.500 atm is opened. When equilibrium has been reached, what is the pressure of the gaseous mixture (assume the temperature remains constant)?

81. Is the corrected (ideal) volume of a real gas greater or less than the actual volume of the container? Why?

82. A flask contains $\frac{1}{3}$ mol H_2 and $\frac{2}{3}$ mol He. Compare the force on the wall per impact of H_2 relative to that for He.

83. Consider separate gaseous samples of He, H_2, CH_4, Ne, and Ar. In which sample do the molecules have the largest average velocity?

84. A chemist weighed out 5.14 g of a mixture containing unknown amounts of BaO(*s*) and CaO(*s*) and placed the sample in a 1.50-L flask containing $CO_2(g)$ at 30.0°C and 750. torr. After the reaction to form $BaCO_3(s)$ and $CaCO_3(s)$ was completed, the pressure of $CO_2(g)$ remaining was 230. torr. Calculate the mass percentages of CaO(*s*) and BaO(*s*) in the mixture.

85. The density of a pure gaseous compound was measured at 0.00°C as a function of pressure to give the following results:

Density (g/L)	Pressure (atm)
0.17893	0.2500
0.35808	0.5000
0.53745	0.7500
0.71707	1.000

Calculate the molar mass of this compound, corrected for any nonideal behavior of the gas. Assume the nonideal gas obeys the equation $PV/nRT = 1 + \beta P$. (*Hint:* derive an equation for P/d and plot P/d versus *P*.)

86. The Maxwell-Boltzmann distribution function $f(u)$ increases at small values of *u* and decreases at large values of *u*. Identify the parts of the function responsible for this behavior.

87. A person accidentally swallows a drop of liquid oxygen, $O_2(l)$, which has a density of 1.149 g/mL. Assuming the drop has a volume of 0.050 mL, what volume of gas will be produced in the person's stomach at body temperature (37°C) and a pressure of 1.0 atm?

88. Calculate the root mean square, the most probable, and the average velocities for $N_2(g)$ at 227°C.

89. Consider separate 1.0-L samples of $O_2(g)$ and He(*g*), both at 25°C and the same pressure. Compare the change in momentum per impact and the number of impacts per second in the two samples.

90. Consider separate 1.00-L samples of Ar(*g*), both containing the same number of moles, one at 27°C and the other at 77°C. Compare the change in momentum per impact and the number of impacts per second in the two samples.

91. For a 1.00-L sample containing 1.00 mol of CO_2 at 37°C, calculate the pressure using:
   a. the ideal gas equation
   b. the van der Waals equation (see Table 5.3)

92. A 100.-L flask contains a mixture of methane, CH_4, and argon at 25°C. The mass of argon present is 228 g and the mole fraction of methane in the mixture is 0.650. Calculate the total kinetic energy of the gaseous mixture.

93. Consider separate 1.0-L samples of He(*g*) and $UF_6(g)$, both at 1.00 atm and containing the same number of moles. What ratio of temperatures for the two samples would produce the same collision frequency with the vessel walls?

94. Represent the following plots.
   a. PV/n (*y* axis) versus *P* (*x* axis) for a real gas that obeys the equation $PV/n = \alpha + \beta P$
   b. change in momentum per impact versus mass of an individual gas particle for a series of ideal gases all at the same temperature
   c. *P* versus *V* for an ideal gas where *n* and *T* are constant
   d. *P* versus *T*(K) for an ideal gas where *n* and *V* are constant

95. Consider the reaction between 50.0 mL of liquid methyl alcohol, CH_3OH (density = 0.850 g/mL), and 22.8 L of O_2 at 27°C and a pressure of 2.00 atm. The products of the reaction are $CO_2(g)$ and $H_2O(g)$. Calculate the number of moles of H_2O formed if the reaction goes to completion.

96. Ammonia reacts with O_2 to form either NO(*g*) or $NO_2(g)$ according to these unbalanced equations:

$$NH_3(g) + O_2(g) \longrightarrow NO(g) + H_2O(g)$$
$$NH_3(g) + O_2(g) \longrightarrow NO_2(g) + H_2O(g)$$

In a certain experiment 2.00 mol of $NH_3(g)$ and 10.00 mol of $O_2(g)$ are contained in a closed flask. After the reaction is complete, 6.75 mol of $O_2(g)$ remains. Calculate the number of moles of NO(*g*) in the product mixture. (*Hint:* you cannot do this problem by adding the balanced equations, since you cannot assume that the two reactions will occur with equal probability.)

97. A liquid consists of hexane (C_6H_{14}) and propane (C_3H_8). When 0.2759 g of this liquid is combusted with excess oxygen, 0.8339 g of CO_2 is formed. Calculate the composition (by mass) of the original mixture of hexane and propane.

98. A steel cylinder contains 5.00 mol of graphite (pure carbon) and 5.00 mol of O_2. The mixture is ignited and all of the graphite reacts. Combustion produces a mixture of CO gas and CO_2 gas. After the cylinder has cooled to its original temperature, it is found that the pressure of the cylinder has increased by 17.0%. Calculate the mole fractions of CO, CO_2, and O_2 in the final gaseous mixture.

99. Ethene is converted to ethane by the reaction:

$$C_2H_4(g) + H_2(g) \xrightarrow{\text{catalyst}} C_2H_6(g)$$

C_2H_4 flows into a catalytic reactor at 25.0 atm and 300.°C with a flow rate of 1000. L/min. Hydrogen at 25.0 atm and 300.°C flows into the reactor at a flow rate of 1500. L/min. If 15.0 kg of C_2H_6 are collected per minute, what is the percent yield of the reaction?

100. Methane (CH_4) gas flows into a combustion chamber at a rate of 200. L/min at 1.50 atm and ambient temperature. Air is added to the chamber at 1.00 atm and the same temperature, and the gases are ignited.
   a. To ensure complete combustion of CH_4 to $CO_2(g)$ and $H_2O(g)$, three times as much oxygen as is necessary is reacted. Assuming air is 21 mole percent O_2 and 79 mole percent N_2, calculate the flow rate of air necessary to deliver the required amount of oxygen.
   b. Under the conditions in part a, combustion of methane was not complete as a mixture of $CO_2(g)$ and $CO(g)$ were produced. It was determined that 95.0% of the carbon in the exhaust gas was present in CO_2. The remainder was present as carbon in CO. Calculate the composition of the exhaust gas in terms of mole fraction of CO, CO_2, O_2, N_2, and H_2O. Assume CH_4 is completely reacted and N_2 is unreacted.
   c. Assuming a total pressure of the exhaust gas of 1.00 atm, calculate the partial pressures of the gases in part b.

101. Consider a sample of $Cl_2(g)$ with a volume of 4.0 L, a pressure of 5.0 atm, and a temperature of 127°C. Calculate the average translational kinetic energy per Cl_2 molecule in this sample.

102. A spherical vessel with a volume of 1.00 L was evacuated and sealed. Twenty-four hours later the pressure of air in the vessel was found to be 1.20×10^{-6} atm. During this 24-h period the vessel had been surrounded by air at 27°C and 1.00 atm. Assuming that air is 78 mole percent nitrogen and that the remainder is oxygen, calculate the diameter of the tiny circular hole in the vessel that allowed the air to leak in.

103. Calculate the number of stages needed to change a mixture of $^{13}CO_2$ and $^{12}CO_2$ that is originally 0.10% (by moles) $^{13}CO_2$ to a mixture that is 0.010% $^{13}CO_2$ by a gaseous diffusion process. (The mass of ^{13}C is 13.003355 amu.)

104. Calculate the collision frequency and the mean free path in a sample of helium gas with a volume of 5.0 L at 27°C and 3.0 atm. Assume that the diameter of a helium atom is 50. pm.

105. Two samples of gas are separated in two rectangular 1.00-L chambers by a thin metal wall. One sample is pure helium and the other is pure radon. Both samples are at 27°C and show a pressure of 2.00×10^{-6} atm. Assuming that the metal wall separating the gases suddenly develops a circular hole of radius 1.00×10^{-6} m, calculate the pressure in each chamber after 10.0 h have passed.

106. A certain sample of uranium is reacted with fluorine to form a mixture of $^{235}UF_6(g)$ and $^{238}UF_6(g)$. After 100 diffusion steps the gas contains 1526 $^{235}UF_6$ molecules per 1.000×10^5 total number of molecules in the gas ($^{235}UF_6$ + $^{238}UF_6$). What is the ratio of ^{235}U to ^{238}U atoms in the original sample of uranium?

107. A spherical glass container of unknown volume contains helium gas at 25°C and 1.960 atm. When a portion of the helium is withdrawn and adjusted to 1.00 atm at 25°C, it is found to have a volume of 1.75 cm^3. The gas remaining in the first container shows a pressure of 1.710 atm. Calculate the volume of the spherical container.

108. A compound Z is known to have the composition 34.38% Ni, 28.13% C, and 37.48% O. In an experiment 1.00 L of gaseous Z is mixed with 1.00 L of argon, where each gas is at P = 2.00 atm and T = 25°C. When this mixture of gases is put in an effusion chamber, the ratio of Z molecules to Ar molecules in the effused mixture is 0.4837. Using these data, calculate the following.
   a. the empirical formula for Z
   b. the molar mass for Z
   c. the molecular formula for Z
   d. the mole ratio of Z to argon in a sample of gas obtained by five effusion steps (starting with the original mixture)

109. A 2.00-L sample of $O_2(g)$ was collected over water at a total pressure of 785 torr and 25°C. When the $O_2(g)$ was dried (water vapor removed), the gas had a volume of 1.94 L at 25°C and 785 torr. Calculate the vapor pressure of water at 25°C.

110. Calculate the kinetic energy possessed by 1.00×10^{20} molecules of methane gas (CH_4) at $T = 27°C$, assuming ideal behavior.

111. Hydrogen cyanide gas is commercially prepared by the reaction of methane, $CH_4(g)$, ammonia, $NH_3(g)$, and oxygen, $O_2(g)$, at a high temperature. The other product is gaseous water.

a. Write a balanced chemical equation for the reaction.
b. Methane and ammonia gases flow into a reactor at a rate of 20.0 L/s. Oxygen gas is introduced at a flow rate of 40.0 L/s. All of the reactant gases are at 1.00 atm and 150.°C. What mass of HCN is produced per second by this reaction assuming 100% yield?

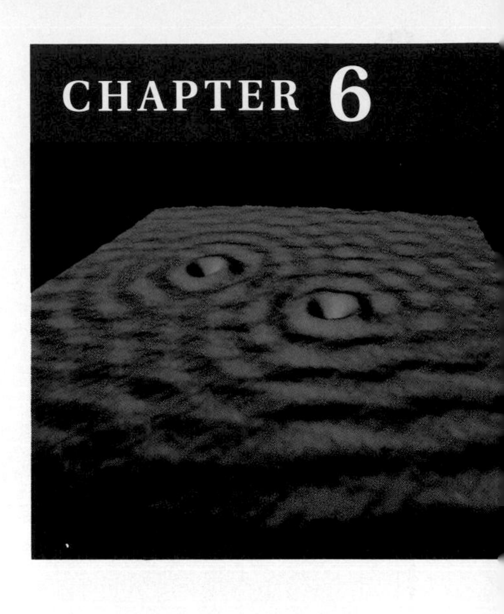

Chemical Equilibrium

In doing stoichiometry calculations, we assume that reactions proceed to completion, that is, until one of the reactants is consumed. Many reactions *do* proceed essentially to completion. For such reactions it can be assumed that the reactants are quantitatively converted to products and that the amount of limiting reactant that remains is negligible. On the other hand, there are many chemical reactions that stop far short of completion. An example is the dimerization of nitrogen dioxide:

$$2NO_2(g) \longrightarrow N_2O_4(g)$$

The reactant NO_2 is a reddish brown gas, and the product N_2O_4 is a colorless gas. When NO_2 is placed in an evacuated, sealed glass vessel at 25°C, the initial dark brown color decreases in intensity as the NO_2 is converted to colorless N_2O_4. However, even over a long period of time, the contents of the reaction vessel do not become colorless. Instead, the intensity of the brown color eventually becomes constant, which means that the concentration of NO_2 is no longer changing. This observation is a clear indication that the reaction has stopped short of completion. In fact, the system has reached **chemical equilibrium,** *the state where the concentrations of all reactants and products remain constant with time.*

Any chemical reaction carried out in a closed vessel will reach equilibrium. For some reactions the equilibrium position so favors the products that the reaction appears to have gone to completion. We say that the equilibrium position for such a reaction lies *far to the right,* in the direction of the products.

Reddish-brown nitrogen gas streaming from a flask where copper is reacting with concentrated nitric acid.

For example, when gaseous hydrogen and oxygen are mixed in stoichiometric quantities and react to form water vapor, the reaction proceeds essentially to completion. The amounts of the reactants that remain when the system reaches equilibrium are so tiny as to be negligible. In contrast, some reactions occur only to a slight extent. For example, when solid CaO is placed in a closed vessel at 25°C, the decomposition to solid Ca and gaseous O_2 is virtually undetectable. In cases like this, the equilibrium position is said to lie *far to the left*, in the direction of the reactants.

In this chapter we will discuss how and why a chemical system comes to equilibrium and the characteristics of a system at equilibrium. In particular, we will discuss how to calculate the concentrations of the reactants and products present for a given system at equilibrium.

6.1 The Equilibrium Condition

Since no changes occur in the concentrations of reactants or products in a reaction system at equilibrium, it may appear that everything has stopped. However, this is not the case. On the molecular level there is frenetic activity. Equilibrium is not static; it is a highly *dynamic* situation. The concept of chemical equilibrium is analogous to two island cities connected by a bridge. Suppose the traffic flow on the bridge is the same in both directions. It is obvious that there is motion, since one can see the cars traveling across the bridge; but the number of cars in each city is not changing because equal numbers are entering and leaving. The result is no *net* change in the car population.

Equilibrium is a dynamic situation.

To see how this concept applies to chemical reactions, consider the reaction between steam and carbon monoxide in a closed vessel at a high temperature, where the reaction takes place rapidly:

$$H_2O(g) + CO(g) \rightleftharpoons H_2(g) + CO_2(g)$$

Assume that the same number of moles of gaseous CO and gaseous H_2O are placed in a closed vessel and allowed to react. The plots of the concentrations of reactants and products versus time are shown in Fig. 6.1. Note that since CO and H_2O were originally present in equal molar quantities, and since they react in a 1 : 1 ratio, the concentrations of the two gases are always equal. Also, since H_2 and CO_2 are formed in equal amounts, they are always present at the same concentrations.

Figure 6.1 is a profile of the progress of the reaction. When CO and H_2O are mixed, they immediately begin reacting to form H_2 and CO_2. This leads to a decrease in the concentrations of the reactants, but the concentrations of the products, which were initially at zero, are increasing. Beyond a certain time, indicated by the dashed line in Fig. 6.1, the concentrations of reactants and products no longer change—equilibrium has been reached. Unless the system is somehow disturbed, no further changes in concentrations will occur. Note that although the equilibrium position lies far to the right, the concentrations of reactants never reach zero; the reactants will always be present in small but constant concentrations.

Why does equilibrium occur? As we will see in much more detail in Chapter 15, chemical reactions occur via collisions of the reacting molecules. The energy associated with a collision can break bonds in the reactant molecules, allowing

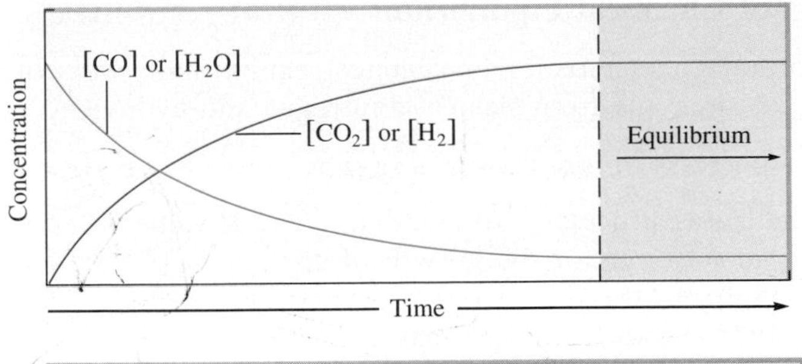

Figure 6.1
The changes in concentrations with time for the reaction $H_2O(g) + CO(g) \rightleftharpoons H_2(g) + CO_2(g)$ when equimolar quantities of $H_2O(g)$ and $CO(g)$ are mixed.

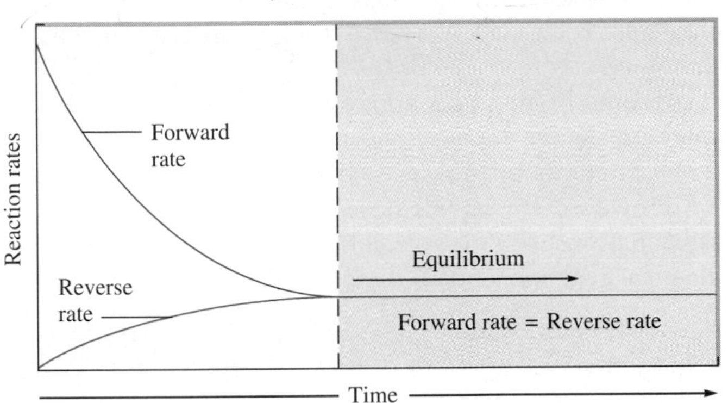

Figure 6.2
The changes with time in the rates of forward and reverse reactions for $H_2O(g) + CO(g) \rightleftharpoons H_2(g) + CO_2(g)$ when equimolar quantities of $H_2O(g)$ and $CO(g)$ are mixed. Note that the rates for the forward and reverse reactions do not change in the same way with time. We will not be concerned with the reasons for this difference at this point.

them to rearrange to form the products. Since the collision rate of the molecules in a gas depends on the concentration of molecules present, the rate of a reaction depends on the concentrations of the reactants. Thus as reactants collide and react to form products

$$H_2O(g) + CO(g) \longrightarrow H_2(g) + CO_2(g)$$

the concentrations of the reactants decrease, causing the rate of this reaction (the *forward* reaction) to decrease—that is, the reaction slows down. (See Fig. 6.2.)

As in the bridge traffic analogy, there is also a reverse direction:

$$H_2(g) + CO_2(g) \longrightarrow H_2O(g) + CO(g)$$

Initially in this experiment no H_2 and CO_2 were present, and this reverse reaction could not occur. However, as the forward reaction proceeds, the concentrations of H_2 and CO_2 build up, and the rate of the reverse reaction increases (Fig. 6.2) as the forward reaction slows down. Eventually, the concentrations reach levels where the rate of the forward reaction equals the rate of the reverse reaction. That is, the concentrations of the reactants and products achieve values such that H_2O and CO are being consumed by the forward reaction at exactly the same rate as they are being produced by the reverse reaction. The system has reached equilibrium.

The equilibrium position of a reaction—left, right, or somewhere in between—is determined by many factors: the initial concentrations, the relative energies of the reactants and products, and the relative "degree of organization" of the reactants and products. Energy and organization come into play because nature tries to achieve minimum energy and maximum disorder. For now, we will simply view the equilibrium phenomenon as an experimentally verified fact. The theoretical origins of equilibrium will be explored in detail in Chapter 10.

The relationship between equilibrium and thermodynamics is explored in Section 10.11.

The Characteristics of Chemical Equilibrium

The United States produces almost 20 million tons of ammonia annually.

To explore the important characteristics of chemical equilibrium, we will consider the synthesis of ammonia from elemental nitrogen and hydrogen:

$$N_2(g) + 3H_2(g) \rightleftharpoons 2NH_3(g)$$

This process (called the **Haber process**) is of great commercial value because ammonia is an important fertilizer for the growth of corn and other crops. Ironically, this beneficial process was discovered in Germany just before World War I in a search for ways to produce nitrogen-based explosives. In the course of this work, German chemist Fritz Haber (1868–1934) pioneered the large-scale production of ammonia.

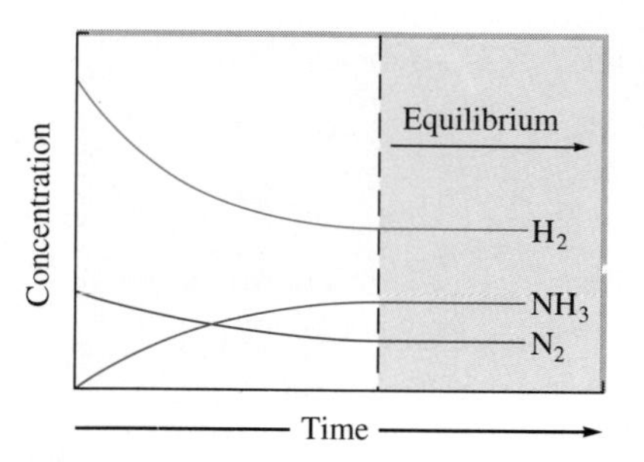

Figure 6.3
A concentration profile for the reaction $N_2(g) + 3H_2(g) \rightleftharpoons 2NH_3(g)$ when only $N_2(g)$ and $H_2(g)$ are mixed initially.

When gaseous nitrogen, hydrogen, and ammonia are mixed in a closed vessel at 25°C, no apparent change in the concentrations occur over time, irrespective of the original amounts of the gases. It would seem that equilibrium has been attained. However, this is not necessarily true.

There are two possible reasons why the concentrations of the reactants and products of a given chemical reaction remain unchanged when mixed:

1. The system is at chemical equilibrium.
2. The forward and reverse reactions are so slow that the system moves toward equilibrium at an undetectable rate.

The second reason applies to the nitrogen, hydrogen, and ammonia mixture at 25°C. Because the molecules involved have strong chemical bonds, mixtures of N_2, H_2, and NH_3 at 25°C can exist with no apparent change over long periods of time. However, under appropriate conditions the system does reach equilibrium, as shown in Fig. 6.3. Note that because of the reaction stoichiometry, H_2 disappears three times as fast as N_2 does, and NH_3 forms twice as fast as N_2 disappears.

6.2 The Equilibrium Constant

The law of mass action is based on experimental observations.

Science is fundamentally empirical—it is based on experiment. The development of the equilibrium concept is typical. From observations of many chemical reactions, two Norwegian chemists, Cato Maximilian Guldberg (1836–1902) and Peter Waage (1833–1900), proposed the **law of mass action** in 1864 as a general description of the equilibrium condition. For a reaction of the type

$$j\text{A} + k\text{B} \rightleftharpoons l\text{C} + m\text{D}$$

where A, B, C, and D represent chemical species and j, k, l, and m are their coefficients in the balanced equation, the law of mass action is represented by the following **equilibrium expression:**

$$K = \frac{[\text{C}]^l[\text{D}]^m}{[\text{A}]^j[\text{B}]^k}$$

The square brackets indicate the concentrations of the chemical species *at equilibrium,* and K is a constant called the **equilibrium constant.**

Guldberg and Waage found that the equilibrium concentrations for every reaction system they studied obeyed this relationship. That is, when the observed equilibrium concentrations are inserted into the equilibrium expression constructed from the law of mass action for a given reaction, the result is a constant (at a given temperature and assuming ideal behavior). Thus the value of the equilibrium constant for a given reaction system can be calculated from the measured concentrations of reactants and products present at equilibrium, a procedure illustrated in Example 6.1.

EXAMPLE 6.1

The following equilibrium concentrations were observed for the Haber process at 127°C:

$$[NH_3] = 3.1 \times 10^{-2} \text{ mol/L}$$

$$[N_2] = 8.5 \times 10^{-1} \text{ mol/L}$$

$$[H_2] = 3.1 \times 10^{-3} \text{ mol/L}$$

a. Calculate the value of K at 127°C for this reaction.

b. Calculate the value of the equilibrium constant at 127°C for the reaction

$$2NH_3(g) \rightleftharpoons N_2(g) + 3H_2(g)$$

c. Calculate the value of the equilibrium constant at 127°C for the reaction given by the equation

$$\tfrac{1}{2}N_2(g) + \tfrac{3}{2}H_2(g) \rightleftharpoons NH_3(g)$$

Solution

a. The balanced equation for the Haber process is

$$N_2(g) + 3H_2(g) \rightleftharpoons 2NH_3(g)$$

Thus using the law of mass action to construct the expression for K, we have

$$K = \frac{[NH_3]^2}{[N_2][H_2]^3} = \frac{(3.1 \times 10^{-2} \text{ mol/L})^2}{(8.5 \times 10^{-1} \text{ mol/L})(3.1 \times 10^{-3} \text{ mol/L})^3}$$

$$= 3.8 \times 10^4 \text{ L}^2/\text{mol}^2$$

b. This reaction is written in the reverse order of the equation given in part a. This leads to the equilibrium expression

$$K' = \frac{[N_2][H_2]^3}{[NH_3]^2}$$

which is the reciprocal of the expression used in part a. So

$$K' = \frac{[N_2][H_2]^3}{[NH_3]^2} = \frac{1}{K} = \frac{1}{3.8 \times 10^4 \text{ L}^2/\text{mol}^2} = 2.6 \times 10^{-5} \text{ mol}^2/\text{L}^2$$

c. We use the law of mass action: $K'' = \dfrac{[NH_3]}{[N_2]^{1/2}[H_2]^{3/2}}$

If we compare this expression with the one obtained in part a, we see that since

$$\frac{[NH_3]}{[N_2]^{1/2}[H_2]^{3/2}} = \left(\frac{[NH_3]^2}{[N_2][H_2]^3}\right)^{1/2}$$

then $$K'' = K^{1/2}$$

Thus $$K'' = K^{1/2} = (3.8 \times 10^4\ L^2/mol^2)^{1/2} = 1.9 \times 10^2\ L/mol$$

We can draw some important conclusions from the results of Example 6.1. For a reaction of the form

$$jA + kB \rightleftharpoons lC + mD$$

the equilibrium expression is

$$K = \frac{[C]^l[D]^m}{[A]^j[B]^k}$$

If this reaction is reversed, the new equilibrium expression is

$$K' = \frac{[A]^j[B]^k}{[C]^l[D]^m} = \frac{1}{K}$$

If the original reaction is multiplied by some factor n to give

$$njA + nkB \rightleftharpoons nlC + nmD$$

the equilibrium expression becomes

$$K'' = \frac{[C]^{nl}[D]^{nm}}{[A]^{nj}[B]^{nk}} = K^n$$

Summary: **Some Characteristics of the Equilibrium Expression**

- **The equilibrium expression for a reaction written in reverse is the reciprocal of that for the original reaction.**
- **When the balanced equation for a reaction is multiplied by a factor n, the equilibrium expression for the new reaction is the original expression raised to the nth power. Thus $K_{new} = (K_{original})^n$.**
- **The apparent units for K are determined by the powers of the various concentration terms. The (apparent) units for K therefore depend on the reaction being considered. We will have more to say about the units for K in Section 6.4.**

The law of mass action applies to solution and gaseous equilibria.

The law of mass action is widely applicable. It correctly describes the equilibrium behavior of all chemical reaction systems whether they occur in solution or in the gas phase. Although, as we will see later, corrections for nonideal behavior must be applied in certain cases, such as for concentrated aqueous solutions and for gases at high pressures, the law of mass action provides a remarkably accurate description of all types of chemical equilibria. For example, consider again the ammonia synthesis reaction. The equilibrium

TABLE 6.1 **Results of Three Experiments for the Reaction $N_2(g) + 3H_2(g) \rightleftharpoons 2NH_3(g)$**

Experiment	Initial Concentrations	Equilibrium Concentrations	$K = \frac{[NH_3]^2}{[N_2][H_2]^3}$
I	$[N_2]_0 = 1.000\ M$ $[H_2]_0 = 1.000\ M$ $[NH_3]_0 = 0$	$[N_2] = 0.921\ M$ $[H_2] = 0.763\ M$ $[NH_3] = 0.157\ M$	$K = 6.02 \times 10^{-2}\ L^2/mol^2$
II	$[N_2]_0 = 0$ $[H_2]_0 = 0$ $[NH_3]_0 = 1.000\ M$	$[N_2] = 0.399\ M$ $[H_2] = 1.197\ M$ $[NH_3] = 0.203\ M$	$K = 6.02 \times 10^{-2}\ L^2/mol^2$
III	$[N_2]_0 = 2.00\ M$ $[H_2]_0 = 1.00\ M$ $[NH_3]_0 = 3.00\ M$	$[N_2] = 2.59\ M$ $[H_2] = 2.77\ M$ $[NH_3] = 1.82\ M$	$K = 6.02 \times 10^{-2}\ L^2/mol^2$

constant K always has the same value at a given temperature. At 500°C the value of K is $6.0 \times 10^{-2}\ L^2/mol^2$. Whenever N_2, H_2, and NH_3 are mixed together at this temperature, the system will always come to an equilibrium position such that

$$\frac{[NH_3]^2}{[N_2][H_2]^3} = 6.0 \times 10^{-2}\ L^2/mol^2$$

This expression has the same value at 500°C, *regardless of the amounts of the gases that are mixed together initially.*

Although the special ratio of products to reactants defined by the equilibrium expression is constant for a given reaction system at a given temperature, the *equilibrium concentrations will not always be the same.* Table 6.1 gives three sets of data for the synthesis of ammonia, showing that even though the individual sets of equilibrium concentrations are quite different for the different situations, the *equilibrium constant, which depends on the ratio of the concentrations, remains the same* (within experimental error). Note that subscripts of zero indicate initial concentrations.

Each *set of equilibrium concentrations* is called an **equilibrium position.** It is essential to distinguish between the equilibrium constant and the equilibrium positions for a given reaction system. There is only *one* equilibrium constant for a particular system at a particular temperature, but there are an *infinite* number of equilibrium positions. The specific equilibrium position adopted by a system depends on the initial concentrations, but the equilibrium constant does not.

For a reaction at a given temperature, there are many equilibrium positions but only one value for K.

6.3 Equilibrium Expressions Involving Pressures

So far we have been describing equilibria involving gases in terms of concentrations. Equilibria involving gases can also be described in terms of pressures. The relationship between the pressure and the concentration of a gas can be seen from the ideal gas equation:

$$PV = nRT \quad \text{or} \quad P = \left(\frac{n}{V}\right)RT = CRT$$

where C equals n/V, or the number of moles of gas (n) per unit volume (V). Thus C represents the *molar concentration of the gas.*

For the ammonia synthesis reaction, the equilibrium expression can be written in terms of concentrations,

$$K = \frac{[NH_3]^2}{[N_2][H_2]^3} = \frac{C_{NH_3}{}^2}{(C_{N_2})(C_{H_2}{}^3)}$$

or in terms of the *equilibrium partial pressures of the gases,*

$$K_p = \frac{P_{NH_3}{}^2}{(P_{N_2})(P_{H_2}{}^3)}$$

K involves concentrations; K_p involves pressures.

In this book K denotes an equilibrium constant in terms of concentrations, and K_p represents an equilibrium constant in terms of partial pressures.

The relationship between K and K_p for a particular reaction follows from the fact that for an ideal gas, $C = P/RT$. For example, for the ammonia synthesis reaction,

$$K = \frac{[NH_3]^2}{[N_2][H_2]^3} = \frac{C_{NH_3}{}^2}{(C_{N_2})(C_{H_2}{}^3)}$$

$$= \frac{\left(\frac{P_{NH_3}}{RT}\right)^2}{\left(\frac{P_{N_2}}{RT}\right)\left(\frac{P_{H_2}}{RT}\right)^3} = \frac{P_{NH_3}{}^2}{(P_{N_2})(P_{H_2}{}^3)} \times \frac{\left(\frac{1}{RT}\right)^2}{\left(\frac{1}{RT}\right)^4}$$

$$= \frac{P_{NH_3}{}^2}{(P_{N_2})(P_{H_2}{}^3)}(RT)^2 = K_p(RT)^2$$

However, for the synthesis of hydrogen fluoride from its elements,

$$H_2(g) + F_2(g) \rightleftharpoons 2HF(g)$$

the relationship between K and K_p is

$$K = \frac{[HF]^2}{[H_2][F_2]} = \frac{C_{HF}{}^2}{(C_{H_2})(C_{F_2})} = \frac{\left(\frac{P_{HF}}{RT}\right)^2}{\left(\frac{P_{H_2}}{RT}\right)\left(\frac{P_{F_2}}{RT}\right)} = \frac{P_{HF}{}^2}{(P_{H_2})(P_{F_2})} = K_p$$

Thus for this reaction K is equal to K_p. This equality occurs because the sum of the coefficients on either side of the balanced equation is identical, so the terms in RT cancel. In the equilibrium expression for the ammonia synthesis reaction, the sum of the powers in the numerator is different from that in the denominator, so K does not equal K_p.

For the general reaction

$$jA + kB \rightleftharpoons lC + mD$$

the relationship between K and K_p is

$$K_p = K(RT)^{\Delta n}$$

where Δn is the sum of the coefficients of the *gaseous* products minus the sum of the coefficients of the *gaseous* reactants. This equation is easy to derive from the definitions of K and K_p and the relationship between pressure and concentration. For the above general reaction,

$$K_p = \frac{(P_C{}^l)(P_D{}^m)}{(P_A{}^j)(P_B{}^k)} = \frac{(C_C \times RT)^l(C_D \times RT)^m}{(C_A \times RT)^j(C_B \times RT)^k}$$

$$= \frac{(C_C{}^l)(C_D{}^m)}{(C_A{}^j)(C_B{}^k)} \times \frac{(RT)^{l+m}}{(RT)^{j+k}} = K(RT)^{(l+m)-(j+k)} = K(RT)^{\Delta n}$$

where $\Delta n = (l + m) - (j + k)$, the difference in the sums of the coefficients for the gaseous products and reactants.

Δn always involves products minus reactants.

We have seen that the (apparent) units of the equilibrium constant depend on the specific reaction. For example, for the reaction

$$H_2(g) + F_2(g) \rightleftharpoons 2HF(g)$$

the units for K and K_p can be found as follows:

$$K = \frac{C_{HF}{}^2}{(C_{H_2})(C_{F_2})} = \frac{(\text{mol/L})^2}{(\text{mol/L})(\text{mol/L})} \Rightarrow \text{no units}$$

$$K_p = \frac{P_{HF}{}^2}{(P_{H_2})(P_{F_2})} = \frac{(\text{atm})^2}{(\text{atm})(\text{atm})} \Rightarrow \text{no units}$$

For this reaction neither K nor K_p has units, and K is equal to K_p. The identical powers in the numerator and denominator cause the units to cancel. For equilibrium expressions in which the powers in the numerator and denominator are not the same, the equilibrium constants will have (apparent) units and K will not equal K_p.

Note that in the above discussion we used the term "apparent units" when referring to equilibrium constants. This term was used because the theoretical foundation for the concept of equilibrium based on thermodynamics includes a *reference state* for each substance, which always causes the units of concentration or pressure to cancel. We will explore this situation thoroughly in Chapter 10, but we will introduce this concept in Section 6.4.

6.4 The Concept of Activity

As we will see in Chapter 10, the "true" equilibrium constant expression does not simply involve the observed equilibrium pressure or concentration for a substance but involves the *ratio* of the equilibrium pressure (or concentration) for a given substance to a *reference* pressure (or concentration) for that substance. This ratio is defined as the **activity** of the substance, which in terms of pressures is defined as

$$\text{Activity } (i\text{th component}) = a_i = \frac{P_i}{P_{\text{reference}}}$$

where P_i = partial pressure of the ith gaseous component

$P_{\text{reference}}$ = 1 atm (exactly)

and where ideal behavior is assumed.

Using the concept of activities, the equilibrium expression for the reaction

$$jA(g) + kB(g) \rightleftharpoons lC(g) + mD(g)$$

is written as

$$K = \frac{(a_C)^l(a_D)^m}{(a_A)^j(a_B)^k} = \frac{\left(\frac{P_C}{P_{ref}}\right)^l\left(\frac{P_D}{P_{ref}}\right)^m}{\left(\frac{P_A}{P_{ref}}\right)^j\left(\frac{P_B}{P_{ref}}\right)^k}$$

With all pressures expressed in atmospheres, we have

$$K_p = \frac{\left(\frac{P_C\ (\cancel{atm})}{1\ \cancel{atm}}\right)^l\left(\frac{P_D\ (\cancel{atm})}{1\ \cancel{atm}}\right)^m}{\left(\frac{P_A\ (\cancel{atm})}{1\ \cancel{atm}}\right)^j\left(\frac{P_B\ (\cancel{atm})}{1\ \cancel{atm}}\right)^k} = \frac{P_C{}^l P_D{}^m}{P_A{}^j P_B{}^k}$$

where K_p is unitless as shown.

When the equilibrium composition of a system is expressed in units of moles per liter, the reference state is (exactly) 1 mol/L.

Because of the difference in reference states, in general,

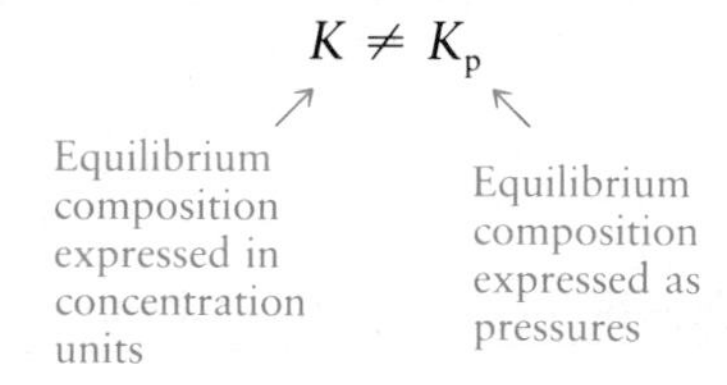

The only exception to this principle occurs when the sum of the powers in the numerator and denominator are the same (as discussed previously for $H_2 + F_2 \rightleftharpoons 2HF$). In such a case $K = K_p$.

In this chapter we will often retain the (apparent) units for K_p or K to remind ourselves which reference state is being used and to facilitate conversions between K and K_p. However, in the remainder of the text, the value of equilibrium constants usually will be given without units.

6.5 Heterogeneous Equilibria

So far, we have discussed equilibria only for systems in the gas phase, where all reactants and products are gases. These situations represent **homogeneous equilibria.** However, many equilibria involve more than one phase and are called **heterogeneous equilibria.** For example, the thermal decomposition of calcium carbonate in the commercial preparation of lime occurs by a reaction involving both solid and gas phases:

Lime is among the top six chemicals manufactured in the United States in terms of amount produced.

$$CaCO_3(s) \rightleftharpoons \underset{\text{Lime}}{\underset{\uparrow}{CaO(s)}} + CO_2(g)$$

Lime, used here as a fertilizer, is produced in the thermal decomposition of calcium carbonate.

Straightforward application of the law of mass action leads to the equilibrium expression

$$K' = \frac{[CO_2][CaO]}{[CaCO_3]}$$

However, experimental results show that the *position of a heterogeneous equilibrium does not depend on the amounts of pure solids or liquids present.* This result makes sense when the meaning of an activity for a pure liquid or solid is understood. For a pure liquid or solid the reference state is the pure liquid or solid. Thus for the decomposition of $CaCO_3$ considered above, we do not insert $[CaCO_3]$ or $[CaO]$ into the equilibrium expression but rather into the activity of each:

Pure solid

$$a_{CaCO_3} = \frac{[CaCO_3]}{[CaCO_3]} = 1$$

Pure solid (reference state)

and

$$a_{CaO} = \frac{[CaO]}{[CaO]} = 1$$

Thus the equilibrium expressions for the decomposition of solid $CaCO_3$ are

$$K = \frac{[CO_2]\ (1)}{1} = [CO_2] \quad \text{and} \quad K_p = \frac{P_{CO_2}\ (1)}{1} = P_{CO_2}$$

In summary, we can make the following general statement: *the activity of a pure solid or liquid is always 1.*

Note that the net effect of inserting an activity of 1 into the equilibrium expression for each pure solid or liquid in the reaction has the same effect as simply disregarding them. If pure solids or pure liquids are involved in a chemical reaction, their concentrations *are not included in the equilibrium expression* for the reaction. This simplification occurs *only* with pure solids or liquids, not with solutions or gases, because in these last two cases the activity cannot be assumed to be 1.

For example, in the decomposition of liquid water to gaseous hydrogen and oxygen,

$$2H_2O(l) \rightleftharpoons 2H_2(g) + O_2(g)$$

where

$$K = [H_2]^2[O_2] \quad \text{and} \quad K_p = (P_{H_2}{}^2)(P_{O_2})$$

water is not included in either equilibrium expression because it is present as a pure liquid ($a_{H_2O(l)} = 1$). However, if the reaction were carried out under conditions in which the water is a gas rather than a liquid,

$$2H_2O(g) \rightleftharpoons 2H_2(g) + O_2(g)$$

then

$$K = \frac{[H_2]^2[O_2]}{[H_2O]^2} \quad \text{and} \quad K_p = \frac{(P_{H_2}{}^2)(P_{O_2})}{P_{H_2O}{}^2}$$

because the concentration or pressure of water vapor can assume different values, depending on the conditions. That is, we cannot assume an activity of 1 in such a case.

6.6 Applications of the Equilibrium Constant

Knowing the equilibrium constant for a reaction allows us to predict several important features of the reaction: the tendency of the reaction to occur (but not the speed of the reaction), whether a given set of concentrations represents an equilibrium condition, and the equilibrium position that will be achieved from a given set of initial concentrations.

The Extent of a Reaction

The inherent tendency for a reaction to occur is indicated by the magnitude of the equilibrium constant. A value of K much larger than 1 means that at equilibrium the reaction system will consist of mostly products—the equilibrium lies to the right. That is, reactions with very large equilibrium constants go essentially to completion. On the other hand, a very small value of K means that the system at equilibrium will consist of mostly reactants—the equilibrium position is far to the left. The given reaction does not occur to any significant extent.

It is important to understand that *the size of K and the time required to reach equilibrium are not directly related.* The time required to achieve equilibrium depends on the reaction rate. The size of K is determined by factors such as the difference in energy between products and reactants, which will be discussed in detail in Chapter 10.

Reaction Quotient

When the reactants and products of a given chemical reaction are mixed, it is useful to know whether the mixture is at equilibrium, and if it is not, in which direction the system will shift to reach equilibrium. If the concentration of one of the reactants or products is zero, the system will shift in the direction that produces the missing component. However, if all the initial concentrations are nonzero, it is more difficult to determine the direction of the move toward equilibrium. To determine the shift in such cases, we use the **reaction quotient** (**Q**). The reaction quotient is obtained by applying the law of mass action, but using *initial concentrations* instead of equilibrium concentrations. For example, for the synthesis of ammonia,

$$N_2(g) + 3H_2(g) \rightleftharpoons 2NH_3(g)$$

the expression for the reaction quotient is

$$Q = \frac{[NH_3]_0^2}{[N_2]_0[H_2]_0^3}$$

where the subscripts of zero indicate initial concentrations.

To determine in which direction a system will shift to reach equilibrium, we compare the values of Q and K. There are three possible cases:

1. Q *is equal to* K. The system is at equilibrium; no shift will occur.
2. Q *is greater than* K. In this case the ratio of initial concentrations of products to initial concentrations of reactants is too large. For the system to

EXAMPLE 6.2

For the synthesis of ammonia at 500°C, the equilibrium constant is 6.0×10^{-2} L²/mol². Predict the direction in which the system will shift to reach equilibrium in each of the following cases.

a. $[NH_3]_0 = 1.0 \times 10^{-3}\,M$; $[N_2]_0 = 1.0 \times 10^{-5}\,M$; $[H_2]_0 = 2.0 \times 10^{-3}\,M$

b. $[NH_3]_0 = 2.00 \times 10^{-4}\,M$; $[N_2]_0 = 1.50 \times 10^{-5}\,M$; $[H_2]_0 = 3.54 \times 10^{-1}\,M$

c. $[NH_3]_0 = 1.0 \times 10^{-4}\,M$; $[N_2]_0 = 5.0\,M$; $[H_2]_0 = 1.0 \times 10^{-2}\,M$

Solution

a. First we calculate the value of Q:

$$Q = \frac{[NH_3]_0^2}{[N_2]_0[H_2]_0^3}$$

$$= \frac{(1.0 \times 10^{-3}\text{ mol/L})^2}{(1.0 \times 10^{-5}\text{ mol/L})(2.0 \times 10^{-3}\text{ mol/L})^3}$$

$$= 1.3 \times 10^{7}\text{ L}^2/\text{mol}^2$$

Since $K = 6.0 \times 10^{-2}$ L²/mol², Q is much greater than K. For the system to attain equilibrium, the concentrations of the products must be decreased and the concentrations of the reactants increased. The system will shift to the left:

$$N_2 + 3H_2 \longleftarrow 2NH_3$$

b. We calculate the value of Q:

$$Q = \frac{[NH_3]_0^2}{[N_2]_0[H_2]_0^3}$$

$$= \frac{(2.00 \times 10^{-4}\text{ mol/L})^2}{(1.50 \times 10^{-5}\text{ mol/L})(3.54 \times 10^{-1}\text{ mol/L})^3}$$

$$= 6.01 \times 10^{-2}\text{ L}^2/\text{mol}^2$$

In this case $Q = K$, so the system is at equilibrium. No shift will occur.

c. The value of Q is

$$Q = \frac{[NH_3]_0^2}{[N_2]_0[H_2]_0^3} = \frac{(1.0 \times 10^{-4}\text{ mol/L})^2}{(5.0\text{ mol/L})(1.0 \times 10^{-2}\text{ mol/L})^3}$$

$$= 2.0 \times 10^{-3}\text{ L}^2/\text{mol}^2$$

Here Q is less than K, so the system will shift to the right, attaining equilibrium by increasing the concentration of the product and decreasing the concentrations of the reactants:

$$N_2 + 3H_2 \longrightarrow 2NH_3$$

reach equilibrium, a net change of products to reactants must occur. The system *shifts to the left,* consuming products and forming reactants, until equilibrium is achieved.

3. Q *is less than* K. In this case the ratio of initial concentrations of products to initial concentrations of reactants is too small. The *system must shift to the right,* consuming reactants and forming products, to attain equilibrium.

Calculating Equilibrium Pressures and Concentrations

A typical equilibrium problem involves finding the equilibrium concentrations (or pressures) of reactants and products, given the value of the equilibrium constant and the initial concentrations (or pressures).

EXAMPLE 6.3

Assume that the reaction for the formation of gaseous hydrogen fluoride from hydrogen and fluorine has an equilibrium constant of 1.15×10^2 at a certain temperature. In a particular experiment at this temperature 3.000 mol of each component were added to a 1.500-L flask. Calculate the equilibrium concentrations of all species.

Solution

The balanced equation for the reaction is

$$H_2(g) + F_2(g) \rightleftharpoons 2HF(g)$$

The equilibrium expression is

$$K = 1.15 \times 10^2 = \frac{[HF]^2}{[H_2][F_2]}$$

We first calculate the initial concentrations:

$$[HF]_0 = [H_2]_0 = [F_2]_0 = \frac{3.000 \text{ mol}}{1.500 \text{ L}} = 2.000\ M$$

From the value of Q,

$$Q = \frac{[HF]_0{}^2}{[H_2]_0[F_2]_0} = \frac{(2.000)^2}{(2.000)(2.000)} = 1.000$$

which is much less than K, we know that the system must shift to the right to reach equilibrium.

What change in the concentrations is necessary? Since the answer to this question is presently unknown, we will define the change needed in terms of x. Let x equal the number of moles per liter of H_2 consumed to reach equilibrium. The stoichiometry of the reaction shows that x mol/L of F_2 will also be consumed and that $2x$ mol/L of HF will be formed:

$$\begin{array}{ccccc} H_2(g) & + & F_2(g) & \longrightarrow & 2HF(g) \\ x \text{ mol/L} & + & x \text{ mol/L} & \longrightarrow & 2x \text{ mol/L} \end{array}$$

Now the equilibrium concentrations can be expressed in terms of x:

Initial Concentration (mol/L)	Change (mol/L)	Equilibrium Concentration (mol/L)
$[H_2]_0 = 2.000$	$-x$	$[H_2] = 2.000 - x$
$[F_2]_0 = 2.000$	$-x$	$[F_2] = 2.000 - x$
$[HF]_0 = 2.000$	$+2x$	$[HF] = 2.000 + 2x$

To solve for x, we substitute the equilibrium concentrations into the equilibrium expression:

$$K = 1.15 \times 10^2 = \frac{[HF]^2}{[H_2][F_2]} = \frac{(2.000 + 2x)^2}{(2.000 - x)^2}$$

The right side of this equation is a perfect square, so taking the square root of both sides gives

$$\sqrt{1.15 \times 10^2} = \frac{2.000 + 2x}{2.000 - x}$$

which yields $x = 1.528$. The equilibrium concentrations can now be calculated:

$$[H_2] = [F_2] = 2.000\ M - x = 0.472\ M$$

$$[HF] = 2.000\ M + 2x = 5.056\ M$$

CHECK: Checking these values by substituting them into the equilibrium expression gives

$$\frac{[HF]^2}{[H_2][F_2]} = \frac{(5.056)^2}{(0.472)^2} = 1.15 \times 10^2$$

which agrees with the given value of K.

6.7 Solving Equilibrium Problems

We have already considered most of the strategies needed to solve equilibrium problems. The typical procedure for analyzing a chemical equilibrium problem can be summarized as shown below.

Summary: **Solving Equilibrium Problems**

- **Write the balanced equation for the reaction.**
- **Write the equilibrium expression using the law of mass action.**
- **List the initial concentrations.**
- **Calculate Q and determine the direction of the shift to equilibrium.**

- **Define the change needed to reach equilibrium, and define the equilibrium concentrations by applying the change to the initial concentrations.**
- **Substitute the equilibrium concentrations into the equilibrium expression, and solve for the unknown.**
- **Check your calculated equilibrium concentrations by making sure they give the correct value of *K*.**

In Example 6.3 we were able to solve for the unknown by taking the square root of both sides of the equation. However, this situation is not very common, so we must now consider a more typical problem. Suppose that for a synthesis of hydrogen fluoride from hydrogen and fluorine, 3.000 moles of H_2 and 6.000 moles of F_2 are mixed in a 3.000-liter flask. The equilibrium constant for the synthesis reaction at this temperature is 1.15×10^2. We calculate the equilibrium concentration of each component as follows:

- We begin as usual by writing the balanced equation for the reaction:

$$H_2(g) + F_2(g) \rightleftharpoons 2HF(g)$$

- The equilibrium expression is

$$K = 1.15 \times 10^2 = \frac{[HF]^2}{[H_2][F_2]}$$

- The initial concentrations are

$$[H_2]_0 = \frac{3.000 \text{ mol}}{3.000 \text{ L}} = 1.000\ M$$

$$[F_2]_0 = \frac{6.000 \text{ mol}}{3.000 \text{ L}} = 2.000\ M$$

$$[HF]_0 = 0$$

- There is no need to calculate Q; since no HF is initially present, we know that the system must shift to the right to reach equilibrium.
- If we let x represent the number of moles per liter of H_2 consumed to reach equilibrium, we can represent the equilibrium concentrations as follows:

Initial Concentration (mol/L)	Change (mol/L)	Equilibrium Concentration (mol/L)
$[H_2]_0 = 1.000$	$-x$	$[H_2] = 1.000 - x$
$[F_2]_0 = 2.000$	$-x$	$[F_2] = 2.000 - x$
$[HF]_0 = 0$	$+2x$	$[HF] = 0 + 2x$

- Substituting the equilibrium concentrations into the equilibrium expression gives

$$K = 1.15 \times 10^2 = \frac{[HF]^2}{[H_2][F_2]} = \frac{(2x)^2}{(1.000 - x)(2.000 - x)}$$

To solve for x, we perform the indicated multiplication,

$$(1.000 - x)(2.000 - x)(1.15 \times 10^2) = (2x)^2$$

to give

$$(1.15 \times 10^2)x^2 - 3.000(1.15 \times 10^2)x + 2.000(1.15 \times 10^2) = 4x^2$$

and collect terms,

$$(1.11 \times 10^2)x^2 - (3.45 \times 10^2)x + 2.30 \times 10^2 = 0$$

This expression is a quadratic equation of the general form

$$ax^2 + bx + c = 0$$

where the roots can be obtained from the quadratic formula:

$$x = \frac{-b \pm \sqrt{b^2 - 4ac}}{2a}$$

In this example $a = 1.11 \times 10^2$, $b = -3.45 \times 10^2$, and $c = 2.30 \times 10^2$. Substituting these values into the quadratic formula gives two values for x:

$$x = 2.14 \text{ mol/L} \qquad \text{and} \qquad x = 0.968 \text{ mol/L}$$

Both of these results cannot be valid (since a *given* set of initial concentrations leads to only *one* equilibrium position). How can we choose between them? Since the expression for the equilibrium concentration of H_2 is

$$[H_2] = 1.000\ M - x$$

the value of x cannot be 2.14 mol/L (because subtracting 2.14 M from 1.000 M gives a negative concentration for H_2, which is physically impossible). Thus the correct value for x is 0.968 mol/L, and the equilibrium concentrations are as follows:

$$[H_2] = 1.000\ M - 0.968\ M = 3.2 \times 10^{-2}\ M$$

$$[F_2] = 2.000\ M - 0.968\ M = 1.032\ M$$

$$[HF] = 2(0.968\ M) = 1.936\ M$$

- We can check these concentrations by substituting them into the equilibrium expression

$$\frac{[HF]^2}{[H_2][F_2]} = \frac{(1.936)^2}{(3.2 \times 10^{-2})(1.032)} = 1.13 \times 10^2$$

This value is in close agreement with the given value for K (1.15×10^2), so the calculated equilibrium concentrations are correct.

Note that although we used the quadratic formula to solve for x in this problem, other methods are also available. For example, trial and error is always a possibility. However, use of successive approximations (see Appendix Section A1.4) is often preferable. For example, in this case successive approximations can be carried out conveniently by starting with the quadratic equation

$$(1.11 \times 10^2)x^2 - (3.45 \times 10^2)x + 2.30 \times 10^2 = 0$$

and dividing it by 1.11×10^2 to give

$$x^2 - 3.11x + 2.07 = 0$$

which can then be rearranged to

$$x^2 = 3.11x - 2.07$$

or

$$x = \sqrt{3.11x - 2.07}$$

Now we can proceed by guessing a value of x, which is then inserted into the square root expression. Next, we calculate a "new" value of x from the expression

$$x = \sqrt{3.11x - 2.07}$$

"New" value calculated ↗ (left-hand x); ↖ Guessed value of x inserted (the x under the root)

When the calculated value (the new value) of x agrees with the guessed value, the equation has been solved.

To solve the algebraic equations you encounter when doing chemistry problems, use whatever method is convenient and comfortable for you.

Treating Systems That Have Small Equilibrium Constants

We have seen that fairly complicated calculations are often necessary to solve equilibrium problems. However, under certain conditions, simplifications can be made that greatly reduce the mathematical difficulties. For example, consider gaseous NOCl, which decomposes to form the gases NO and Cl_2. At 35°C the equilibrium constant is 1.6×10^{-5} mol/L. In an experiment in which 1.0 mole of NOCl is placed in a 2.0-liter flask, what are the equilibrium concentrations?

The balanced equation is

$$2NOCl(g) \rightleftharpoons 2NO(g) + Cl_2(g)$$

and

$$K = \frac{[NO]^2[Cl_2]}{[NOCl]^2} = 1.6 \times 10^{-5} \text{ mol/L}$$

The initial concentrations are

$$[NOCl]_0 = \frac{1.0 \text{ mol}}{2.0 \text{ L}} = 0.50\ M, \qquad [NO]_0 = 0, \qquad [Cl_2]_0 = 0$$

Since there are no products initially, the system will move to the right to reach equilibrium. We will define x as the change in concentration of Cl_2 needed to reach equilibrium. The changes in the concentrations of NOCl and NO can then be obtained from the balanced equation:

$$2NOCl(g) \longrightarrow 2NO(g) + Cl_2(g)$$
$$2x \longrightarrow 2x + x$$

The concentrations can be summarized as follows:

Initial Concentration (mol/L)	Change (mol/L)	Equilibrium Concentration (mol/L)
$[NOCl]_0 = 0.50$	$-2x$	$[NOCl] = 0.50 - 2x$
$[NO]_0 = 0$	$+2x$	$[NO] = 0 + 2x = 2x$
$[Cl_2]_0 = 0$	$+x$	$[Cl_2] = 0 + x = x$

The equilibrium concentrations must satisfy the equilibrium expression:

$$K = 1.6 \times 10^{-5} = \frac{[NO]^2[Cl_2]}{[NOCl]^2} = \frac{(2x)^2(x)}{(0.50 - 2x)^2}$$

Multiplying and collecting terms results in an equation that requires complicated methods to solve directly. However, we can avoid this situation by recognizing that since K is so small (1.6×10^{-5} mol/L), the system will not proceed far to the right to reach equilibrium. That is, x *represents a relatively small number.* Consequently, the term $(0.50 - 2x)$ can be approximated by 0.50. That is, when x is small,

$$0.50 - 2x \approx 0.50$$

Making this approximation allows us to simplify the equilibrium expression:

Approximations can simplify complicated math, but their validity should be carefully checked.

$$1.6 \times 10^{-5} = \frac{(2x)^2(x)}{(0.50 - 2x)^2} \approx \frac{(2x)^2(x)}{(0.50)^2} = \frac{4x^3}{(0.50)^2}$$

Solving for x^3 gives

$$x^3 = \frac{(1.6 \times 10^{-5})(0.50)^2}{4} = 1.0 \times 10^{-6}$$

and $x = 1.0 \times 10^{-2}$ (mol/L).

Next, we must check the validity of the approximation. If $x = 1.0 \times 10^{-2}$, then

$$0.50 - 2x = 0.50 - 2(1.0 \times 10^{-2}) = 0.48$$

The difference between 0.50 and 0.48 is 0.02, or 4% of the initial concentration of NOCl, a relatively small discrepancy that will have little effect on the outcome. That is, since $2x$ is very small compared with 0.50, the value of x obtained in the approximate solution should be very close to the exact value. We use this approximate value of x to calculate the equilibrium concentrations:

A good way to assess whether a 4% error is acceptable here is to examine the precision of the data given. For example, note that the value of K is 1.6×10^{-5}, which can be interpreted as $(1.6 \pm 0.1) \times 10^{-5}$. Thus the uncertainty in K is at least 1 part in 16, or about 6%. Therefore, a 4% error in [NOCl] is acceptable.

$$[NOCl] = 0.50 - 2x = 0.48\ M \approx 0.50\ M$$

$$[NO] = 2x = 2(1.0 \times 10^{-2}\ M) = 2.0 \times 10^{-2}\ M$$

$$[Cl_2] = x = 1.0 \times 10^{-2}\ M$$

CHECK: $$\frac{[NO]^2[Cl_2]}{[NOCl]^2} = \frac{(2.0 \times 10^{-2})^2(1.0 \times 10^{-2})}{(0.50)^2} = 1.6 \times 10^{-5}$$

Since the given value of K is 1.6×10^{-5}, these calculated concentrations are correct.

This problem turned out to be relatively easy to solve because the *small value of* K *and the resulting small shift to the right to reach equilibrium allowed simplification.*

6.8 Le Châtelier's Principle

It is important to understand the factors that control the *position* of a chemical equilibrium. For example, when a chemical is manufactured, the chemists and chemical engineers in charge of production want to choose conditions that favor the desired product as much as possible. In other words, they want the equilibrium to lie far to the right. When Fritz Haber was developing the process for the synthesis of ammonia, he did extensive studies on how the temperature and pressure affect the equilibrium concentration of ammonia. Some of his results are given in Table 6.2. Note that the amount of NH_3 at equilibrium increases with an increase in pressure but decreases with an increase in temperature. Thus the amount of NH_3 present at equilibrium is favored by conditions of low temperature and high pressure.

However, this is not the whole story. Carrying out the process at low temperatures is not feasible, because then the reaction is too slow. Even though the equilibrium tends to shift to the right as the temperature is lowered, the attainment of equilibrium is much too slow at low temperatures to be practical. This observation emphasizes once again that we must study both the thermodynamics (Chapter 10) and the kinetics (Chapter 15) of a reaction before we really understand the factors that control it.

We can qualitatively predict the effects of changes in concentration, pressure, and temperature on a system at equilibrium by using **Le Châtelier's principle,** which states that *if a change in conditions (a "stress") is imposed on a system at equilibrium, the equilibrium position will shift in a direction that tends to reduce that change in conditions.* Although this rule, put forth by Henri Le Châtelier in 1884, sometimes oversimplifies the situation, it works remarkably well.

The Effect of a Change in Concentration

To see how we can predict the effects of a change in concentration on a system at equilibrium, we will consider the ammonia synthesis reaction. Suppose there is an equilibrium position described by these concentrations:

$$[N_2] = 0.399\ M, \qquad [H_2] = 1.197\ M, \qquad [NH_3] = 0.202\ M$$

TABLE 6.2 **The Percent by Mass of NH_3 at Equilibrium in a Mixture of N_2, H_2, and NH_3 as a Function of Temperature and Total Pressure***

	Total Pressure		
Temperature (°C)	300 atm	400 atm	500 atm
400	48% NH_3	55% NH_3	61% NH_3
500	26% NH_3	32% NH_3	38% NH_3
600	13% NH_3	17% NH_3	21% NH_3

*Each experiment was begun with a 3:1 mixture of H_2 and N_2.

What will happen if 1.000 mole per liter of N_2 is suddenly injected into the system? We can answer this question by calculating the value of Q. The concentrations before the system adjusts are

$$[N_2]_0 = 0.399\ M + \underset{\substack{\uparrow \\ \text{Added } N_2}}{1.000\ M} = 1.399\ M$$

$$[H_2]_0 = 1.197\ M$$
$$[NH_3]_0 = 0.202\ M$$

Note that we label these as "initial concentrations" because the system is no longer at equilibrium. Then

$$Q = \frac{[NH_3]_0^2}{[N_2]_0[H_2]_0^3} = \frac{(0.202)^2}{(1.399)(1.197)^3} = 1.70 \times 10^{-2}$$

Since we are not given the value of K, we must calculate it from the first set of equilibrium concentrations:

$$K = \frac{[NH_3]^2}{[N_2][H_2]^3} = \frac{(0.202)^2}{(0.399)(1.197)^3} = 5.96 \times 10^{-2}$$

As we might have expected, because the concentration of N_2 was increased, Q is less than K.

The system will shift to the right to arrive at the new equilibrium position. Rather than do the calculations, we simply summarize the results:

Equilibrium Position I		Equilibrium Position II
$[N_2] = 0.399\ M$	1.000 mol/L	$[N_2] = 1.348\ M$
$[H_2] = 1.197\ M$	$\Longrightarrow$	$[H_2] = 1.044\ M$
$[NH_3] = 0.202\ M$	of N_2 added	$[NH_3] = 0.304\ M$

These data reveal that the equilibrium position does in fact shift to the right: the concentration of H_2 decreases; the concentration of NH_3 increases; and, of

Blue anhydrous cobalt(II) chloride and pink hydrated cobalt(II) chloride. Since the reaction $CoCl_2(s) + 6H_2O(g) \rightarrow CoCl_2 \cdot 6H_2O(s)$ is shifted to the right by water vapor, $CoCl_2$ is often used in novelty devices to detect humidity.

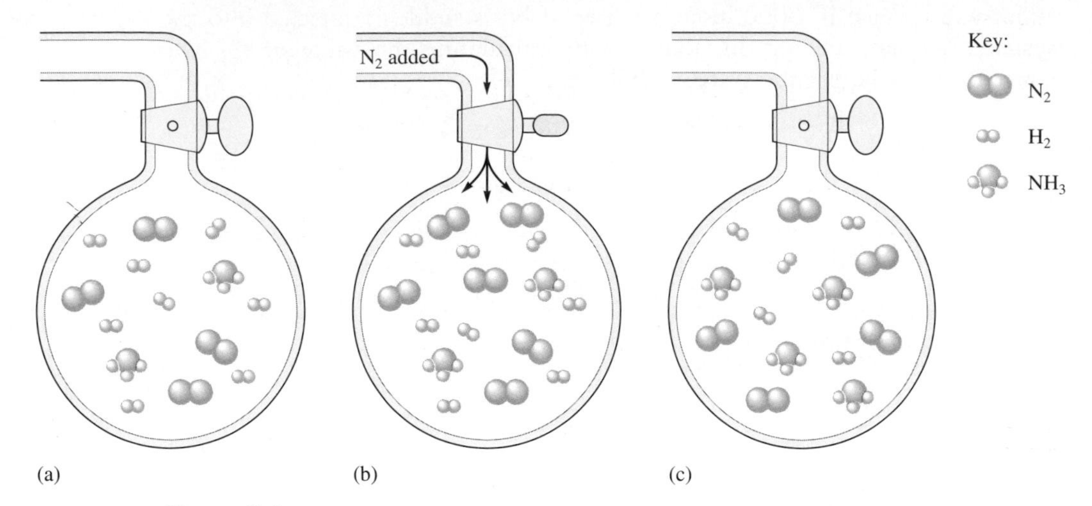

Figure 6.4
(a) The initial equilibrium mixture of N_2, H_2, and NH_3. (b) Addition of N_2. (c) The new equilibrium position for the system containing more N_2 (due to addition of N_2), less H_2, and more NH_3 than the mixture in (a).

course, since nitrogen is added, the concentration of N_2 shows an increase relative to the amount present at the original equilibrium position. (However, note that the nitrogen decreased from the amount present immediately after addition of the 1.000 mol of N_2, because the reaction shifted to the right.)

We can predict this shift qualitatively by using Le Châtelier's principle. Since the stress imposed is the addition of nitrogen, Le Châtelier's principle predicts that the system will shift in a direction that consumes nitrogen. This reduces the effect of the addition. Thus Le Châtelier's principle correctly predicts that adding nitrogen causes the equilibrium to shift to the right (see Fig. 6.4).

The system shifts in the direction that compensates for the imposed change in conditions.

If ammonia had been added instead of nitrogen, the system would have shifted to the left to consume ammonia. So we can paraphrase Le Châtelier's principle for this case as follows: *if a reactant or product is added to a system at equilibrium, the system will shift away from the added component. If a reactant or product is removed, the system will shift toward the removed component.*

EXAMPLE 6.4

Arsenic can be extracted from its ores by first reacting the ore with oxygen (called *roasting*) to form solid As_4O_6, which is then reduced with carbon:

$$As_4O_6(s) + 6C(s) \rightleftharpoons As_4(g) + 6CO(g)$$

Predict the direction of the shift of the equilibrium position for this reaction in response to each of the following changes in conditions.

a. addition of CO

b. addition or removal of C or As_4O_6

c. removal of As_4

Solution

a. Le Châtelier's principle predicts that the shift will be away from the substance whose concentration is increased. The equilibrium position will shift to the left when CO is added.

b. Since the amount of a pure solid has no effect on the equilibrium position, changing the amount of C or As_4O_6 will have no effect.

c. If gaseous As_4 is removed, the equilibrium position will shift to the right to form more products. In industrial processes the desired product is often continuously removed from the reaction system to increase the yield.

The Effect of a Change in Pressure

Basically, there are three ways to change the pressure of a reaction system involving gaseous components at a given temperature:

1. Add or remove a gaseous reactant or product.
2. Add an inert gas (one not involved in the reaction).
3. Change the volume of the container.

We have already considered the addition or removal of a reactant or product. When an inert gas is added, there is no effect on the equilibrium position. *The addition of an inert gas increases the total pressure but has no effect on the concentrations or partial pressures of the reactants or products (assuming ideal gas behavior).* Thus the system remains at the original equilibrium position.

When the volume of the container is changed, the concentrations (and thus the partial pressures) of both reactants and products are changed. We could calculate Q and predict the direction of the shift. However, for systems involving gaseous components there is an easier way: we focus on the volume. The central idea is that *when the volume of the container holding a gaseous system is reduced, the system responds by reducing its own volume. This is done by decreasing the total number of gaseous molecules in the system.*

To see how this works, we can rearrange the ideal gas law to give

$$V = \left(\frac{RT}{P}\right)n$$

or at constant T and P

$$V \propto n$$

That is, at constant temperature and pressure, the volume of a gas is directly proportional to the number of moles of gas present.

Suppose we have a mixture of gaseous nitrogen, hydrogen, and ammonia at equilibrium (Fig. 6.5). If we suddenly reduce the volume, what will happen to the equilibrium position? The reaction system can reduce its volume by reducing the number of molecules present. Consequently, the reaction

$$N_2(g) + 3H_2(g) \rightleftharpoons 2NH_3(g)$$

will shift to the right, since in this direction four molecules (one of nitrogen and

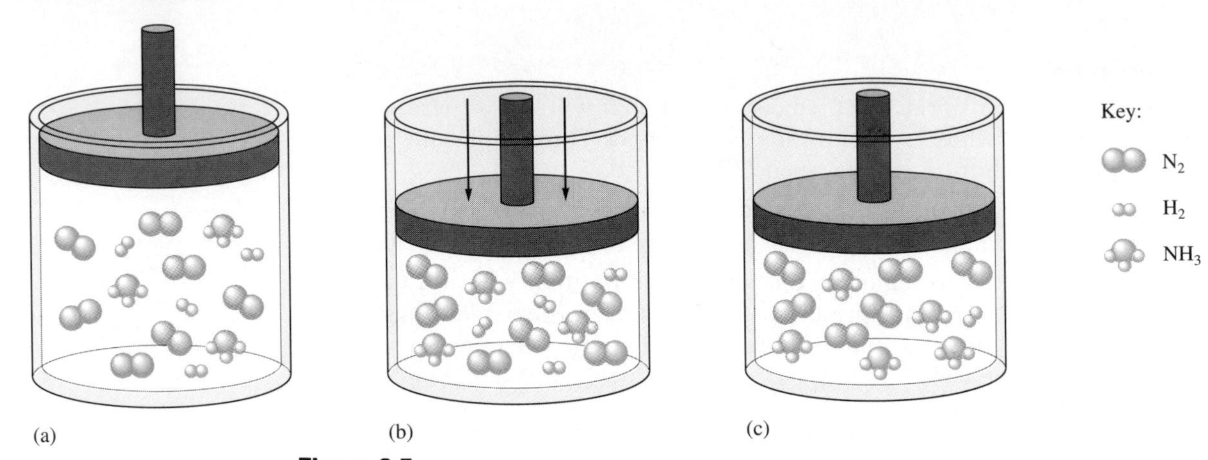

Figure 6.5
(a) A mixture of $NH_3(g)$, $N_2(g)$, and $H_2(g)$ at equilibrium. (b) The volume is suddenly decreased. (c) The new equilibrium position for the system containing more NH_3, less N_2, and less H_2. The reaction $N_2(g) + 3H_2(g) \rightleftharpoons 2NH_3(g)$ shifts to the right (toward the side with fewer molecules) when the container volume is decreased.

three of hydrogen) react to produce two molecules (of ammonia), thus *reducing the total number of gaseous molecules present.* The new equilibrium position will be further to the right than the original one. That is, the equilibrium position will shift toward the side of the reaction involving the smaller number of gaseous molecules in the balanced equation. This phenomenon is illustrated in Fig. 6.6.

The opposite is also true. When the container volume is increased, the system will shift in the direction that increases its volume. An increase in volume in the ammonia synthesis system will produce a shift to the left to increase the total number of gaseous molecules present.

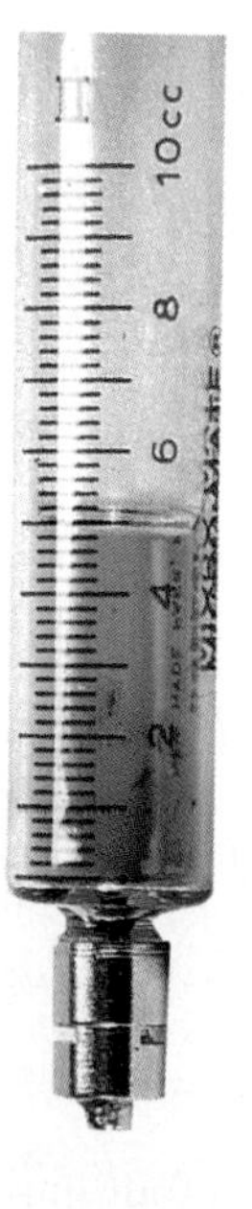

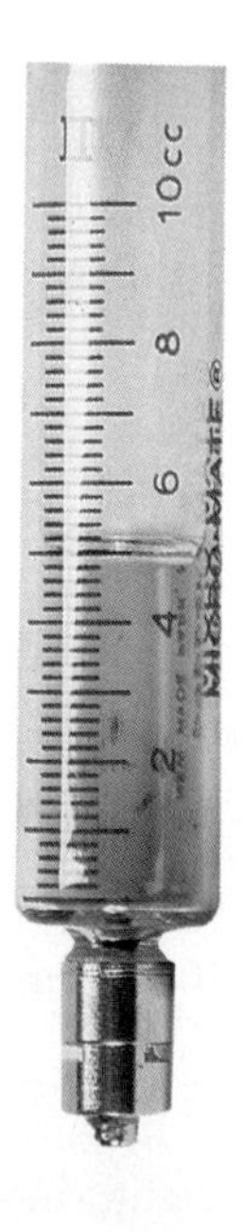

Figure 6.6
(left) Brown $NO_2(g)$ and colorless $N_2O_4(g)$ at equilibrium in a syringe. (center) The volume is suddenly decreased, giving a greater concentration of both N_2O_4 and NO_2 (indicated by the darker brown color). (right) A few seconds after the sudden volume decrease, the color becomes much lighter brown as the equilibrium shifts from brown $NO_2(g)$ to colorless $N_2O_4(g)$. This is predicted by Le Châtelier's principle, since in the equilibrium

$$2NO_2(g) \rightleftharpoons N_2O_4(g)$$

the product side has the smaller number of molecules.

EXAMPLE 6.5

Predict the shift in equilibrium position that will occur for each of the following processes when the volume is reduced.

a. the preparation of liquid phosphorus trichloride:

$$P_4(s) + 6Cl_2(g) \rightleftharpoons 4PCl_3(l)$$

b. the preparation of gaseous phosphorus pentachloride:

$$PCl_3(g) + Cl_2(g) \rightleftharpoons PCl_5(g)$$

c. the reaction of phosphorus trichloride with ammonia:

$$PCl_3(g) + 3NH_3(g) \rightleftharpoons P(NH_2)_3(g) + 3HCl(g)$$

Solution

a. Since P_4 and PCl_3 are a pure solid and a pure liquid, respectively, we need to consider only the effect of the decrease in volume on Cl_2. The position of the equilibrium will shift to the right, since the reactant side contains six gaseous molecules and the product side has none.

b. Decreasing the volume will shift the given reaction to the right, since the product side contains only one gaseous molecule and the reactant side has two.

c. Both sides of the balanced reaction equation have four gaseous molecules. A change in volume will have no effect on the equilibrium position. There is no shift in this case.

The Effect of a Change in Temperature

It is important to recognize that although the changes we have just discussed may alter the equilibrium *position*, they do not alter the equilibrium *constant* (assuming ideal behavior). For example, the addition of a reactant shifts the equilibrium position to the right but has no effect on the value of the equilibrium constant; the new equilibrium concentrations satisfy the original equilibrium constant.

The effect of temperature on equilibrium is different, however, because *the value of* K *changes with temperature.* We can use Le Châtelier's principle to predict the direction of the change.

The synthesis of ammonia from nitrogen and hydrogen releases energy (is *exothermic*). We can represent this situation by treating energy as a product:

$$N_2(g) + 3H_2(g) \rightleftharpoons 2NH_3(g) + \text{energy}$$

If energy in the form of heat is added to this system at equilibrium, Le Châtelier's principle predicts that the shift will be in the direction that consumes energy, in this case to the left. Note that this shift decreases the concentration of NH_3 and increases the concentrations of N_2 and H_2, thus *decreasing the value of K*. The experimentally observed change in K with temperature for this reaction is indicated in Table 6.3. The value of K decreases with increased temperature, as predicted.

(top) At a higher temperature, brown $NO_2(g)$ is favored. (bottom) As the temperature decreases, a shift in equilibrium from brown $NO_2(g)$ to colorless $N_2O_4(g)$ occurs.

TABLE 6.3 **Observed Value of *K* for the Ammonia Synthesis Reaction as a Function of Temperature***

Temperature (K)	K (L^2/mol^2)
500	90
600	3
700	0.3
800	0.04

* For this exothermic reaction the value of K decreases as the temperature increases, as predicted by Le Châtelier's principle.

TABLE 6.4 **Shifts in the Equilibrium Position for the Reaction $N_2O_4(g) \rightleftharpoons 2NO_2(g)$**

Change	Shift
Addition of $N_2O_4(g)$	Right
Addition of $NO_2(g)$	Left
Removal of $N_2O_4(g)$	Left
Removal of $NO_2(g)$	Right
Addition of He(g)	None
Decrease in container volume	Left
Increase in container volume	Right
Increase in temperature	Right
Decrease in temperature	Left

On the other hand, for a reaction that consumes energy (an *endothermic* reaction), such as the decomposition of calcium carbonate,

$$\text{Energy} + CaCO_3(s) \rightleftharpoons CaO(s) + CO_2(g)$$

an increase in temperature will cause the equilibrium to shift to the right and the value of K to increase.

In summary, to use Le Châtelier's principle to describe the effect of a temperature change on a system at equilibrium, treat energy as a reactant (in an endothermic process) or as a product (in an exothermic process), and predict the direction of the shift as if an actual reactant or product is added or removed. Although Le Châtelier's principle cannot predict the size of the change in K, it can correctly predict the direction of the change.

We have seen how Le Châtelier's principle can be used to predict the effect of several types of changes on a system at equilibrium. As a summary of these ideas, Table 6.4 shows how various changes affect the equilibrium position of the endothermic reaction

$$N_2O_4(g) \rightleftharpoons 2NO_2(g)$$

6.9 Equilibria Involving Real Gases

To this point in our discussion of the equilibrium phenomenon, we have assumed ideal behavior for all substances. In fact, the value of K calculated from the law of mass action is the true value of the equilibrium constant for a given reaction system only if the observed pressures (concentrations) are corrected for any nonideal behavior.

To gain some appreciation for the effect of nonideal behavior on the calculation of equilibrium constants, consider the data in Table 6.5, which show the values of K_p (at 723 K) for the reaction

$$N_2(g) + 3H_2(g) \rightleftharpoons 2NH_3(g)$$

calculated from the (uncorrected) observed equilibrium pressures (P^{obs}) at various total pressures. Note that K_p^{obs}, defined as

$$K_p^{obs} = \frac{(P_{NH_3}^{obs})^2}{(P_{N_2}^{obs})(P_{H_2}^{obs})^3}$$

increases significantly with total pressure. This result makes sense in view of the fact that, as we discussed in Section 5.10, $P^{obs} < P^{ideal}$ for a real gas at pressures above 1 atm. Recall that the discrepancy between P^{obs} and P^{ideal} increases with increasing pressure. Thus for this case, one expects K_p^{obs} to increase with increasing total pressure, because the excess of powers in the denominator magnifies the error in pressures there as compared with the numerator.

One common method for finding the limiting value (the "true" value) of K_p is to measure K_p at various values of total pressure (constant temperature) and then to extrapolate the results to zero pressure. Another way to obtain the true value of K_p is to correct the observed equilibrium pressures for any nonideal behavior. For example, we might represent the activity of the ith gaseous component as

$$a_i = \frac{\gamma_i P_i^{obs}}{P_{ref}}$$

where γ_i is called the activity coefficient for correcting P_i^{obs} to the ideal value. Obtaining the values of the activity coefficients is a complex process, which will not be treated here.

In general, for equilibrium pressures of 1 atm or less, the value of K_p calculated from the observed equilibrium pressures is expected to be within about 1% of the true value. However, at high pressures the deviations can be quite severe, as illustrated by the data in Table 6.5.

TABLE 6.5 Values of K_p^{obs} at 723 K for the Reaction $N_2(g) + 3H_2(g) \rightleftharpoons 2NH_3(g)$ as a Function of Total Pressure (at equilibrium)

Total Pressure (atm)	K_p^{obs} (atm^{-2})
10	4.4×10^{-5}
50	4.6×10^{-5}
100	5.2×10^{-5}
300	7.7×10^{-5}
600	1.7×10^{-4}
1000	5.3×10^{-4}

EXERCISES

A blue exercise number indicates that the answer to that exercise appears at the back of this book.

Characteristics of Chemical Equilibrium

1. Characterize a system at chemical equilibrium with respect to each of the following.
   a. the rates of the forward and reverse reactions
   b. the overall composition of the reaction mixture
2. Is the following statement true or false? "Reactions with large equilibrium constants are very fast." Explain your answer.
3. Consider the following reaction:

$$H_2O(g) + CO(g) \rightleftharpoons H_2(g) + CO_2(g)$$

Amounts of H_2O, CO, H_2, and CO_2 are put into a flask so that the composition corresponds to an equilibrium position. If the CO placed in the flask is labeled with radioactive ^{14}C, will ^{14}C be found only in CO molecules for an indefinite period of time? Why or why not?

4. Consider the same reaction as in Exercise 3. In a particular experiment 1.0 mol of $H_2O(g)$ and 1.0 mol of $CO(g)$ are put into a flask and heated to 350°C. In another experiment 1.0 mol of $H_2(g)$ and 1.0 mol of $CO_2(g)$ are put into a different flask with the same volume as the first. This mixture is also heated to 350°C. After equilibrium is reached, will there be any difference in the composition of the mixtures in the two flasks?

The Equilibrium Constant

5. Distinguish between the terms *equilibrium constant* and *equilibrium position.*
6. Distinguish between the terms *equilibrium constant* and *reaction quotient.*
7. In atmospheric chemistry the concentrations of trace substances are usually expressed in units of molecules/cm^3. Below are data for some gas-phase reactions at 300. K (from the NASA publication "Chemical Kinetics and Photochemical Data for Use in Atmospheric Modeling, Evaluation Number 5"). The term K^* indicates the use of molecules/cm^3 for concentration. Calculate K and K_p for each reaction.

   a. $HO_2(g) + NO_2(g) \rightleftharpoons HO_2NO_2(g)$ $K^* = 1.26 \times 10^{-11}$
   b. $CH_3O_2(g) + NO_2(g) \rightleftharpoons CH_3O_2NO_2(g)$ $K^* = 2.09 \times 10^{-12}$
   c. What concentration (in molecules/cm^3) of HO_2NO_2 is in equilibrium with 1.65×10^{10} molecules/cm^3 of HO_2 and 6.00×10^{12} molecules/cm^3 of NO_2?

8. At 127°C, $K = 2.6 \times 10^{-5}$ mol^2/L^2 for the reaction

$$2NH_3(g) \rightleftharpoons N_2(g) + 3H_2(g)$$

Calculate K_p at this temperature.

9. At a given temperature $K = 278$ for the reaction

$$2SO_2(g) + O_2(g) \rightleftharpoons 2SO_3(g)$$

Calculate values of K for the following reactions at this temperature.
   a. $SO_2(g) + \frac{1}{2}O_2(g) \rightleftharpoons SO_3(g)$
   b. $2SO_3(g) \rightleftharpoons 2SO_2(g) + O_2(g)$
   c. $SO_3(g) \rightleftharpoons SO_2(g) + \frac{1}{2}O_2(g)$
   d. $4SO_2(g) + 2O_2(g) \rightleftharpoons 4SO_3(g)$

10. At 427°C a 1.0-L flask contains 20.0 mol of H_2, 18.0 mol of CO_2, 12.0 mol of H_2O, and 5.9 mol of CO at equilibrium. Calculate K for the reaction

$$CO_2(g) + H_2(g) \rightleftharpoons CO(g) + H_2O(g)$$

11. At a particular temperature a 3.00-L flask contains 3.50 mol of HI, 4.10 mol of H_2, and 0.30 mol of I_2 at equilibrium. Calculate K at this temperature for the reaction

$$H_2(g) + I_2(g) \rightleftharpoons 2HI(g)$$

12. An equilibrium mixture contains 0.60 g of solid carbon and gaseous carbon dioxide and carbon monoxide at partial pressures of 2.9 atm and 2.6 atm, respectively. Calculate K_p for the reaction

$$C(s) + CO_2(g) \rightleftharpoons 2CO(g)$$

13. A sample of solid ammonium chloride was placed in an evacuated container and then heated so that it decomposed to ammonia gas and hydrogen chloride gas. After heating, the total pressure in the container was found to be 4.4 atm. Calculate K_p at this temperature for the decomposition reaction

$$NH_4Cl(s) \rightleftharpoons NH_3(g) + HCl(g)$$

14. A sample of gaseous PCl_5 was introduced into an evacuated flask so that the pressure of pure PCl_5 would be 0.50 atm at 523 K. However, PCl_5 decomposes to gaseous PCl_3 and Cl_2, and the actual pressure in the flask was found to be 0.84 atm. Calculate K_p for the decomposition reaction

$$PCl_5(g) \rightleftharpoons PCl_3(g) + Cl_2(g)$$

at 523 K. Also calculate K at this temperature.

15. A flask was filled with 2.00 mol of gaseous SO_2 and 2.00 mol of gaseous NO_2 and then heated. After equilibrium was reached 1.30 mol of gaseous NO was present.

Assume that the reaction

$$SO_2(g) + NO_2(g) \rightleftharpoons SO_3(g) + NO(g)$$

occurs under these conditions. Calculate the value of the equilibrium constant for this reaction.

Equilibrium Calculations

16. The equilibrium constant is 0.0900 at 25°C for the reaction

$$H_2O(g) + Cl_2O(g) \rightleftharpoons 2HOCl(g)$$

For which of the following sets of conditions is the system at equilibrium? For those that are not at equilibrium, in which direction will the system shift?
a. P_{H_2O} = 200. torr, P_{Cl_2O} = 49.8 torr, P_{HOCl} = 21.0 torr
b. P_{H_2O} = 296 torr, P_{Cl_2O} = 15.0 torr, P_{HOCl} = 20.0 torr
c. A 2.0-L flask contains 0.084 mol of HOCl, 0.080 mol of Cl_2O, and 0.98 mol of H_2O.
d. A 3.0-L flask contains 0.25 mol of HOCl, 0.0010 mol of Cl_2O, and 0.56 mol of H_2O.

17. Ethyl acetate is synthesized in a nonreacting solvent (not water) according to the following equation:

$$CH_3CO_2H + C_2H_5OH \rightleftharpoons CH_3CO_2C_2H_5 + H_2O$$

Acetic acid Ethanol Ethyl acetate

for which K = 2.2.

Which of the following mixtures are at equilibrium?
a. $[CH_3CO_2C_2H_5]$ = 0.22 M, $[H_2O]$ = 0.10 M, $[CH_3CO_2H]$ = 0.010 M, $[C_2H_5OH]$ = 0.010 M
b. $[CH_3CO_2C_2H_5]$ = 0.22 M, $[H_2O]$ = 0.0020 M, $[CH_3CO_2H]$ = 0.0020 M, $[C_2H_5OH]$ = 0.10 M
c. $[CH_3CO_2C_2H_5]$ = 0.88 M, $[H_2O]$ = 0.12 M, $[CH_3CO_2H]$ = 0.044 M, $[C_2H_5OH]$ = 6.0 M
d. $[CH_3CO_2C_2H_5]$ = 4.4 M, $[H_2O]$ = 4.4 M, $[CH_3CO_2H]$ = 0.88 M, $[C_2H_5OH]$ = 10.0 M

18. For the reaction in Exercise 17, what must the concentration of water be for a mixture with $[CH_3CO_2C_2H_5]$ = 2.0 M, $[CH_3CO_2H]$ = 0.10 M, and $[C_2H_5OH]$ = 5.0 M to be at equilibrium? Why is water included in the equilibrium expression for this reaction?

19. At 900.°C K_p = 1.04 atm for the reaction

$$CaCO_3(s) \rightleftharpoons CaO(s) + CO_2(g)$$

At a low temperature dry ice (solid CO_2), calcium oxide, and calcium carbonate are introduced into a 50.0-L reaction chamber. The temperature is raised to 900.°C. For the following mixtures, will the initial amount of calcium oxide increase, decrease, or remain the same as the system moves toward equilibrium?
a. 655 g of $CaCO_3$, 95.0 g of CaO, 58.4 g of CO_2
b. 780 g of $CaCO_3$, 1.00 g of CaO, 23.76 g of CO_2
c. 0.14 g of $CaCO_3$, 5000 g of CaO, 23.76 g of CO_2
d. 715 g of $CaCO_3$, 813 g of CaO, 4.82 g of CO_2

20. At 25°C, K = 0.090 for the reaction

$$H_2O(g) + Cl_2O(g) \rightleftharpoons 2HOCl(g)$$

Calculate the concentrations of all species at equilibrium for each of the following cases.
a. 1.0 g of H_2O and 2.0 g of Cl_2O are mixed in a 1.0-L flask.
b. 1.0 mol of pure HOCl is placed in a 2.0-L flask.

21. Iodine is sparingly soluble in pure water. However, it dissolves in solutions containing excess iodide ion:

$$I^-(aq) + I_2(aq) \rightleftharpoons I_3^-(aq) \qquad K = 710 \text{ L/mol}$$

What is the ratio of $[I_3^-]$ to $[I_2]$ if 0.10 mol of I_2 is added to 1.0 L of KI solution having each of the following concentrations?
a. $[I^-]$ = 0.10 M b. $[I^-]$ = 0.50 M c. $[I^-]$ = 2.00 M

22. At a particular temperature, K_p = 2.5 for the reaction

$$SO_2(g) + NO_2(g) \rightleftharpoons SO_3(g) + NO(g)$$

a. A container initially contains SO_2 and NO_2, each at 1.00 atm partial pressure. Calculate the equilibrium partial pressures of the gases.
b. If all four gases are at an initial partial pressure of 1.00 atm, calculate the equilibrium partial pressures of the gases.

23. At 2200°C, K = 0.050 for the reaction

$$N_2(g) + O_2(g) \rightleftharpoons 2NO(g)$$

What is the partial pressure of NO at equilibrium assuming the N_2 and O_2 had initial pressures of 0.80 atm and 0.20 atm, respectively?

24. At 35°C, $K = 1.6 \times 10^{-5}$ mol/L for the reaction

$$2NOCl(g) \rightleftharpoons 2NO(g) + Cl_2(g)$$

Calculate the concentrations of all species at equilibrium for each of the following original mixtures.
a. 2.0 mol of pure NOCl in a 2.0-L flask
b. 2.0 mol of NO and 1.0 mol of Cl_2 in a 1.0-L flask
c. 1.0 mol of NOCl and 1.0 mol of NO in a 1.0-L flask
d. 3.0 mol of NO and 1.0 mol of Cl_2 in a 1.0-L flask
e. 2.0 mol of NOCl, 2.0 mol of NO, and 1.0 mol of Cl_2 in a 1.0-L flask
f. 1.00 mol/L concentration of all three gases

25. At 1100 K, K_p = 0.25 atm^{-1} for the following reaction:

$$2SO_2(g) + O_2(g) \rightleftharpoons 2SO_3(g)$$

Calculate the equilibrium partial pressures of SO_2, O_2, and SO_3 produced from an initial mixture in which $P_{SO_2} = P_{O_2}$ = 0.50 atm and P_{SO_3} = 0.

26. At a particular temperature, $K = 1.0 \times 10^2$ for the reaction

$$H_2(g) + F_2(g) \rightleftharpoons 2HF(g)$$

a. In an experiment 2.0 mol of H_2 and 2.0 mol of F_2 are introduced into a 1.0-L flask. Calculate the concentrations of all species when equilibrium is reached.
b. To the equilibrium mixture in part a, an additional 0.50 mol of H_2 is added. Calculate the new equilibrium concentrations of H_2, F_2, and HF.

27. Lexan is a plastic used to make compact discs, eyeglass lenses, and bullet-proof glass. One of the compounds used to make Lexan is phosgene ($COCl_2$), a poisonous gas. Phosgene is produced by the reaction

$$CO(g) + Cl_2(g) \rightleftharpoons COCl_2(g)$$

for which $K = 4.5 \times 10^9$ L/mol at 100.°C.
a. Calculate K_p at 100.°C.
b. Equal moles of CO and Cl_2 are reacted at 100.°C. If the total pressure is 5.0 atm, calculate the equilibrium partial pressures of all the gases.

28. At a certain temperature $K = 1.1 \times 10^3$ L/mol for the reaction

$$Fe^{3+}(aq) + SCN^-(aq) \rightleftharpoons FeSCN^{2+}(aq)$$

Calculate the concentrations of Fe^{3+}, SCN^-, and $FeSCN^{2+}$ at equilibrium if 0.020 mol of $Fe(NO_3)_3$ is added to 1.0 L of 0.10 *M* KSCN. (Neglect any volume change.)

Le Châtelier's Principle

29. The fragrances of many naturally occurring substances are due to the presence of organic compounds called *esters*. For example, the ester ethyl butyrate smells like pineapple. Ethyl butyrate can be made by the following reaction:

$$\underset{\text{Butyric acid}}{CH_3CH_2CH_2\overset{\overset{\displaystyle O}{\|}}{C}{-}OH} + \underset{\text{Ethanol}}{CH_3CH_2OH}$$

$$\rightleftharpoons \underset{\text{Ethyl butyrate}}{CH_3CH_2CH_2\overset{\overset{\displaystyle O}{\|}}{C}{-}OCH_2CH_3} + H_2O$$

Butyric acid has an objectionable odor. If you were to choose a solvent for preparing ethyl butyrate, which of the following would be the best choice: water, 95% ethanol (5% water), 100% ethanol, or acetonitrile (CH_3CN, a nonreactive solvent)?

30. Changing the pressure in a reaction vessel by changing the volume may shift the position of a gas-phase equilibrium, but changing the pressure by adding an inert gas will not. Why?

31. How will the equilibrium position of a gas-phase reaction be affected by changing the volume of the reaction vessel? Are there reactions that will not have their equilibria shifted by a change in volume? Explain.

32. Consider the reaction

$$Fe^{3+}(aq) + SCN^-(aq) \rightleftharpoons FeSCN^{2+}(aq)$$

How will the equilibrium position shift if
a. water is added, doubling the volume?
b. $AgNO_3(aq)$ is added? (AgSCN is insoluble.)
c. NaOH(*aq*) is added? [$Fe(OH)_3$ is insoluble.]
d. $Fe(NO_3)_3(aq)$ is added?

33. An important reaction for the commercial production of hydrogen is

$$CO(g) + H_2O(g) \rightleftharpoons H_2(g) + CO_2(g) \qquad \text{(Exothermic)}$$

How will the equilibrium position shift in each of the four following cases?
a. Gaseous carbon dioxide is removed.
b. Water vapor is added.
c. The pressure is increased by adding helium gas.
d. The temperature is increased.

34. Novelty devices for predicting rain contain cobalt(II) chloride and are based on the following equilibrium:

$$\underset{\text{Purple}}{CoCl_2(s)} + 6H_2O(g) \rightleftharpoons \underset{\text{Pink}}{CoCl_2 \cdot 6H_2O(s)}$$

What color will such an indicator be if rain is imminent?

35. How will the amount of SO_3 at equilibrium in the reaction

$$2SO_3(g) \rightleftharpoons 2SO_2(g) + O_2(g) \qquad \text{(Endothermic)}$$

change in each of the following cases?
a. Oxygen gas is added.
b. The pressure is increased by decreasing the volume.
c. The pressure is increased by adding argon gas.
d. The temperature is decreased.
e. Gaseous sulfur dioxide is removed.

36. In which direction will the position of the equilibrium

$$H_2(g) + I_2(g) \rightleftharpoons 2HI(g) \qquad \text{(Endothermic)}$$

be shifted for each of the following changes?
a. $H_2(g)$ is added.
b. $I_2(g)$ is removed.
c. HI(*g*) is removed.
d. Some Ar(*g*) is added.
e. The volume of the container is doubled.
f. The temperature is increased.

Additional Exercises

37. At 25°C, $K_p \approx 1 \times 10^{-31}$ for the reaction

$$N_2(g) + O_2(g) \rightleftharpoons 2NO(g)$$

a. Calculate the concentration of NO (in molecules/cm^3) that can exist in equilibrium in air at 25°C. In air P_{N_2} = 0.8 atm and P_{O_2} = 0.2 atm.
b. Typical concentrations of NO in relatively pristine environments range from 10^8 to 10^{10} molecules/cm^3. Why is there a discrepancy between these values and your answer to part a?
c. Calculate the value of K^* for this reaction. (See Exercise 7.)

38. Given the following equilibrium constants at 427°C,

$$Na_2O(s) \rightleftharpoons 2Na(l) + \tfrac{1}{2}O_2(g) \qquad K_1 = 2 \times 10^{-25}$$
$$NaO(g) \rightleftharpoons Na(l) + \tfrac{1}{2}O_2(g) \qquad K_2 = 2 \times 10^{-5}$$
$$Na_2O_2(s) \rightleftharpoons 2Na(l) + O_2(g) \qquad K_3 = 5 \times 10^{-29}$$
$$NaO_2(s) \rightleftharpoons Na(l) + O_2(g) \qquad K_4 = 3 \times 10^{-14}$$

determine the values for the equilibrium constants for the following reactions.
a. $Na_2O(s) + \tfrac{1}{2}O_2(g) \rightleftharpoons Na_2O_2(s)$
b. $NaO(g) + Na_2O(s) \rightleftharpoons Na_2O_2(s) + Na(l)$
c. $2NaO(g) \rightleftharpoons Na_2O_2(s)$

(*Hint:* when reaction equations are added, the equilibrium expressions are multiplied.)

39. Calculate a value for the equilibrium constant for the reaction

$$O_2(g) + O(g) \rightleftharpoons O_3(g)$$

given that

$$NO_2(g) \xrightarrow{h\nu} NO(g) + O(g) \qquad K = 6.8 \times 10^{-49}$$

$$O_3(g) + NO(g) \rightleftharpoons NO_2(g) + O_2(g) \qquad K = 5.8 \times 10^{-34}$$

(See the hint in Exercise 38.)

40. At 90.°C the equilibrium constant is 6.8×10^{-2} for the reaction

$$H_2(g) + S(s) \rightleftharpoons H_2S(g)$$

If 0.15 mol hydrogen and 1.0 mol sulfur are heated to 90.°C in a 1.0-L container, what will be the partial pressure of H_2S at equilibrium?

41. A system at equilibrium is described by the equation

$$\text{Energy} + SO_2Cl_2(g) \rightleftharpoons SO_2(g) + Cl_2(g)$$

Why does the temperature of the system increase when SO_2 is added to the system at equilibrium?

42. Nitric oxide and bromine at initial partial pressures of 98.4 and 41.3 torr, respectively, were allowed to react at 300. K. At equilibrium the total pressure was 110.5 torr. The reaction is

$$2NO(g) + Br_2(g) \rightleftharpoons 2NOBr(g)$$

a. Calculate the value of K_p.
b. What would be the partial pressures of all species if NO and Br_2, both at an initial partial pressure of 0.30 atm, were allowed to come to equilibrium at this temperature?

43. At 125°C, K_p = 0.25 atm^2 for the reaction

$$2NaHCO_3(s) \rightleftharpoons Na_2CO_3(s) + CO_2(g) + H_2O(g)$$

A 1.00-L flask containing 10.0 g of $NaHCO_3$ is evacuated and heated to 125°C.
a. Calculate the partial pressures of CO_2 and H_2O after equilibrium is established.
b. Calculate the masses of $NaHCO_3$ and Na_2CO_3 present at equilibrium.
c. Calculate the minimum container volume necessary for all of the $NaHCO_3$ to decompose.

44. Hydrogen for use in ammonia production is produced by the endothermic reaction

$$CH_4(g) + H_2O(g) \underset{750°C}{\overset{\text{Ni catalyst}}{\rightleftharpoons}} CO(g) + 3H_2(g)$$

What will happen to a reaction mixture at equilibrium if
a. $H_2O(g)$ is removed?
b. the temperature is increased?
c. an inert gas is added?
d. $CO(g)$ is removed?

45. Consider the reaction

$$P_4(g) \longrightarrow 2P_2(g)$$

where $K_p = 1.00 \times 10^{-1}$ atm at 1325 K. In an experiment where $P_4(g)$ was placed into a container at 1325 K, the equilibrium mixture of $P_4(g)$ and $P_2(g)$ has a total pressure of 1.00 atm. Calculate the equilibrium pressures of $P_4(g)$ and $P_2(g)$. Calculate the fraction (by moles) of $P_4(g)$ that has dissociated to reach equilibrium.

46. Consider the reaction

$$3O_2(g) \rightleftharpoons 2O_3(g)$$

At 175°C and a pressure of 128 torr, an equilibrium mixture of O_2 and O_3 has a density of 0.168 g/L. Calculate K_p for the above reaction at 175°C.

47. A sample of gaseous nitrosyl bromide, NOBr, was placed in a rigid flask, where it decomposed at 25°C according to the following reaction:

$$2NOBr(g) \rightleftharpoons 2NO(g) + Br_2(g)$$

At equilibrium the total pressure and the density of the gaseous mixture were found to be 0.0515 atm and 0.1861 g/L, respectively. Calculate the value of K_p for the above reaction.

48. At 25°C, $K_p = 5.3 \times 10^5$ atm^{-2} for the reaction

$$N_2(g) + 3H_2(g) \rightleftharpoons 2NH_3(g)$$

When a certain partial pressure of $NH_3(g)$ is put into an otherwise empty rigid vessel at 25°C, equilibrium is reached when 50.0% of the original ammonia has decomposed. What was the original partial pressure of ammonia before any decomposition occurred?

49. At 207°C, $K_p = 0.267$ atm for the reaction

$$PCl_5(g) \rightleftharpoons PCl_3(g) + Cl_2(g)$$

a. If 0.100 mol of $PCl_5(g)$ is placed in an otherwise empty 12.0-L vessel at 207°C, calculate the partial pressures of $PCl_5(g)$, $PCl_3(g)$, and $Cl_2(g)$ at equilibrium.
b. In another experiment, the total pressure of an equilibrium mixture is 2.00 atm at 207°C. What mass of PCl_5 was introduced into a 5.00-L vessel to reach this equilibrium position?

50. A 1.604-g sample of methane (CH_4) gas and 6.400 g of oxygen gas are sealed into a 2.50-L vessel at 411°C and are allowed to reach equilibrium. Methane can react with oxygen to form gaseous carbon dioxide and water vapor, or methane can react with oxygen to form gaseous carbon monoxide and water vapor. At equilibrium the pressure of oxygen is 0.326 atm, and the pressure of water vapor is 4.45 atm. Calculate the pressures of carbon monoxide and carbon dioxide present at equilibrium.

51. A 2.4156-g sample of PCl_5 was placed in an empty 2.000-L flask and allowed to decompose to PCl_3 and Cl_2 at 250.0°C:

$$PCl_5(g) \rightleftharpoons PCl_3(g) + Cl_2(g)$$

At equilibrium the total pressure inside the flask was observed to be 358.7 torr.
a. Calculate the partial pressure of each gas at equilibrium and the value of K_p at 250.0°C.
b. What are the new equilibrium pressures if 0.250 mol of Cl_2 gas is added to the flask?

52. At 25°C gaseous SO_2Cl_2 decomposes to $SO_2(g)$ and $Cl_2(g)$ to the extent that 12.5% of the original SO_2Cl_2 (by moles) has decomposed to reach equilibrium. The total pressure (at equilibrium) is 0.900 atm. Calculate the value of K_p for this system.

53. At 1000 K the $N_2(g)$ and $O_2(g)$ in air (78% N_2, 21% O_2, by moles) react to form a mixture of $NO(g)$ and $NO_2(g)$. The values of the equilibrium constants are 1.5×10^{-4} and 1.0×10^{-5} atm^{-1} for the formation of $NO(g)$ and $NO_2(g)$, respectively. At what total pressure will the partial pressures of $NO(g)$ and $NO_2(g)$ be equal in an equilibrium mixture of $N_2(g)$, $O_2(g)$, $NO(g)$, and $NO_2(g)$?

54. The partial pressures of an equilibrium mixture of $N_2O_4(g)$ and $NO_2(g)$ are $P_{N_2O_4} = 0.33$ atm and $P_{NO_2} = 1.2$ atm at a certain temperature. The volume of the container is doubled. Find the partial pressures of the two gases when a new equilibrium is established.

55. For the reaction

$$PCl_5(g) \rightleftharpoons PCl_3(g) + Cl_2(g)$$

at 600. K, the equilibrium constant is 11.5 atm. Suppose that 2.450 g of PCl_5 is placed in an evacuated 500.-mL bulb, which is then heated to 600. K.
a. What would the pressure of PCl_5 be if it did not dissociate?
b. What is the partial pressure of PCl_5 at equilibrium?
c. What is the total pressure in the bulb at equilibrium?
d. What is the degree of dissociation of PCl_5 at equilibrium?

56. The equilibrium constant K_p for the reaction

$$CCl_4(g) \rightleftharpoons C(s) + 2Cl_2(g)$$

at 700°C is 0.76 atm. Determine the initial pressure of carbon tetrachloride that will produce a total equilibrium pressure of 1.20 atm at 700°C.

57. At a particular temperature, $K_p = 0.25$ for the reaction

$$N_2O_4(g) \rightleftharpoons 2NO_2(g)$$

a. A flask containing only N_2O_4 at an initial pressure of 4.5 atm is allowed to reach equilibrium. Calculate the equilibrium partial pressures of the gases.
b. A flask containing only NO_2 at an initial pressure of 9.0 atm is allowed to reach equilibrium. Calculate the equilibrium partial pressures of the gases.
c. From your answers to parts a and b, does it matter from which direction an equilibrium position is reached?
d. The volume of the container in part a is decreased to one-half the original volume. Calculate the new equilibrium partial pressures.
e. The volume of the container in part a is tripled. Calculate the new equilibrium partial pressures.

58. An 8.00-g sample of SO_3 was placed in an evacuated container, where it decomposed at 600.°C according to the following reaction:

$$SO_3(g) \rightleftharpoons SO_2(g) + 1/2O_2(g)$$

At equilibrium the total pressure and the density of the gaseous mixture were 1.80 atm and 1.60 g/L, respectively. Calculate K_p for this reaction.

59. A sample of iron(II) sulfate was heated in an evacuated container to 929 K, where the following reactions occurred:

$$2FeSO_4(s) \rightleftharpoons Fe_2O_3(s) + SO_3(g) + SO_2(g)$$

$$SO_3(g) \rightleftharpoons SO_2(g) + 1/2O_2(g)$$

After equilibrium was reached, the total pressure was 0.836 atm and the partial pressure of oxygen was 0.0275 atm. Calculate K_p for each of the above reactions.

60. The gas arsine, AsH_3, decomposes as follows:

$$2AsH_3(g) \rightleftharpoons 2As(s) + 3H_2(g)$$

In an experiment pure $AsH_3(g)$ was placed in an empty, rigid, sealed flask at a pressure of 392.0 torr. After 48 h the pressure in the flask was observed to be constant at 488.0 torr.
   a. Calculate the equilibrium pressure of $H_2(g)$.
   b. Calculate K_p for this reaction.

61. At 450°C, $K_p = 6.5 \times 10^{-3}$ atm^{-2} for the ammonia synthesis reaction. Assume that a reaction vessel with a movable piston initially contains 3.0 mol $H_2(g)$ and 1.0 mol $N_2(g)$. Make a plot to show how the partial pressure of $NH_3(g)$ present at equilibrium varies for the total pressures 1.0 atm, 10.0 atm, 100. atm, and 1000. atm (assuming that K_p remains constant). (*Note:* assume these total pressures represent the initial total pressure of $H_2(g)$ plus $N_2(g)$, where $P^\circ_{NH_3} = 0$.)

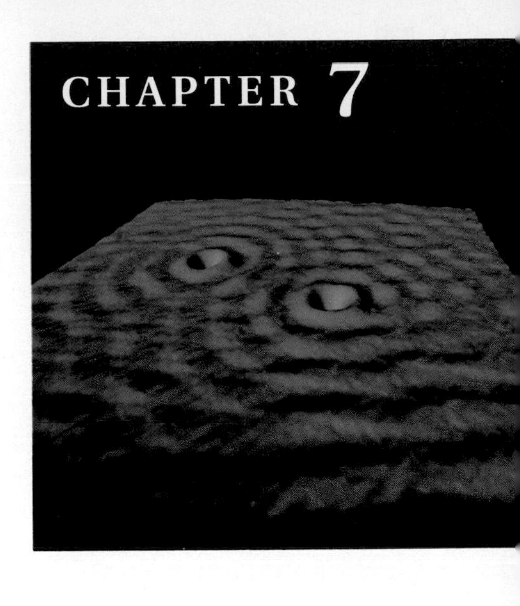

Acids and Bases

In this chapter we reencounter two very important classes of compounds, acids and bases. We will explore their interactions and apply the fundamentals of chemical equilibria discussed in Chapter 6 to systems involving proton transfer reactions.

Acid-base chemistry is important in a wide variety of everyday applications. There are complex systems in our bodies that carefully control the acidity of our blood, since even small deviations may lead to serious illness and death. The same sensitivity exists in other life forms. If you have ever owned tropical fish or goldfish, you know how important it is to monitor and control the acidity of the water in the aquarium.

Acids and bases are also important industrially. For example, the vast quantity of sulfuric acid manufactured in the United States each year is needed to produce fertilizers, polymers, steel, and many other materials.

The influence of acids on living things has assumed special importance in the United States, Canada, and Europe in recent years as a result of the phenomenon of acid rain. This problem is complex, and its diplomatic and economic overtones make it all the more difficult to solve.

7.1 The Nature of Acids and Bases

Acids were first recognized as a class of substances that taste sour. Vinegar tastes sour because it is a dilute solution of acetic acid; citric acid is responsible for the sour taste of a lemon. Bases, sometimes called *alkalis,* are characterized by their bitter taste and slippery feel. Commercial preparations for unclogging drains are highly basic.

The first person to recognize the essential nature of acids and bases was Svante Arrhenius. Based on his experiments with electrolytes, Arrhenius postulated that *acids produce hydrogen ions in aqueous solution, and bases produce hydroxide ions.* At the time of its discovery the **Arrhenius concept** of acids and bases was a major step forward in quantifying acid-base chemistry, but this concept is limited because it applies only to aqueous solutions and allows for only one kind of base—the hydroxide ion. A more general definition of acids and bases was suggested independently by the Danish chemist Johannes N. Brønsted (1879–1947) and the English chemist Thomas M. Lowry (1874–1936) in 1923. In terms of the **Brønsted-Lowry definition,** *an acid is a proton (H^+) donor, and a base is a proton acceptor.* For example, when gaseous HCl dissolves in water, each HCl molecule donates a proton to a water molecule, and so HCl qualifies as a **Brønsted-Lowry acid.** The molecule that accepts the proton, in this case water, is a **Brønsted-Lowry base.**

To understand how water can act as a base, we need to remember that the oxygen of the water molecule has two unshared electron pairs, either of which can form a covalent bond with an H^+ ion. When gaseous HCl dissolves, the following reaction occurs:

$$\mathrm{H{-}\ddot{\underset{\displaystyle\overset{|}{H}}{O}}\!:} + \mathrm{H{-}Cl} \longrightarrow \left[\mathrm{H{-}\ddot{\underset{\displaystyle\overset{|}{H}}{O}}{-}H}\right]^+ + \mathrm{Cl^-}$$

Note that the proton is transferred from the HCl molecule to the water molecule to form H_3O^+, which is called the **hydronium ion.**

The general reaction that occurs when an acid is dissolved in water can best be represented as

Recall that (*aq*) means the substance is hydrated.

$$\underset{\text{Acid}}{HA(aq)} + \underset{\text{Base}}{H_2O(l)} \rightleftharpoons \underset{\text{Conjugate acid}}{H_3O^+(aq)} + \underset{\text{Conjugate base}}{A^-(aq)} \quad (7.1)$$

This representation emphasizes the significant role of the polar water molecule in pulling the proton from the acid. Note that the **conjugate base** is everything that remains of the acid molecule after a proton is lost. The **conjugate acid** is formed when the proton is transferred to the base. A **conjugate acid-base pair** consists of two substances related to each other by the donating and accepting of a single proton. In Equation (7.1) there are two conjugate acid-base pairs: HA and A^-, and H_2O and H_3O^+.

It is important to note that Equation (7.1) really represents *a competition for the proton between the two bases H_2O and A^-*. If H_2O is a much stronger base than A^-, that is, if H_2O has a much greater affinity for H^+ than A^- does, the equilibrium position will be far to the right; most of the acid dissolved will be in the ionized form. Conversely, if A^- is a much stronger base than H_2O, the equilibrium position will lie far to the left. In this case most of the acid dissolved will be present at equilibrium as HA.

The equilibrium expression for the reaction given in Equation (7.1) is

$$K_a = \frac{[H_3O^+][A^-]}{[HA]} = \frac{[H^+][A^-]}{[HA]} \quad (7.2)$$

where K_a is called the **acid dissociation constant.** Both $H_3O^+(aq)$ and $H^+(aq)$ are commonly used to represent the hydrated proton. In this book we will often use simply H^+, but you should remember that it is hydrated in aqueous solutions.

In Chapter 6 we saw that pure solids and liquids are always omitted from the equilibrium expression because they have unit activities. In a dilute solution containing an acid we can assume that the activity of water is 1. Thus the term $[H_2O]$ is not included in Equation (7.2), and the equilibrium expression for K_a has the same form as that for the simple dissociation

$$HA(aq) \rightleftharpoons H^+(aq) + A^-(aq)$$

In this chapter we will always represent an acid as simply dissociating. This does not mean we are using the Arrhenius model for acids. Since water does not affect the equilibrium position, we leave it out of the acid dissociation reaction for simplicity.

You should not forget, however, that water plays an important role in causing the acid to dissociate.

Note that K_a is the equilibrium constant for the reaction in which a proton is removed from HA to form the conjugate base A^-. We use K_a to represent *only* this type of reaction. With this information you can write the K_a expression for any acid, even one that is totally unfamiliar to you.

The Brønsted-Lowry definition is not limited to aqueous solutions; it can be extended to reactions in the gas phase. For example, we discussed the reaction between gaseous hydrogen chloride and ammonia when we studied diffusion (Chapter 5):

$$NH_3(g) + HCl(g) \rightleftharpoons NH_4Cl(s)$$

In this reaction a proton is donated by the hydrogen chloride to the ammonia, as shown by these Lewis structures:

Lewis structures, which represent the electron arrangements in molecules, will be discussed fully in Chapter 13.

$$H{-}\ddot{N}H_2 \;\; H{-}\ddot{\underset{..}{Cl}}: \rightleftharpoons \left[H{-}N(H)_2{-}H\right]^+ + \left[:\ddot{\underset{..}{Cl}}:\right]^-$$

Note that this reaction is not considered an acid-base reaction according to the Arrhenius concept.

7.2 Acid Strength

The strength of an acid is defined by the equilibrium position of its dissociation reaction:

$$HA(aq) + H_2O(l) \rightleftharpoons H_3O^+(aq) + A^-(aq)$$

A **strong acid** is one for which *this equilibrium lies far to the right.* This means that almost all of the original HA is dissociated at equilibrium [see Fig. 7.1 (a)]. There is an important connection between the strength of an acid and that of its conjugate base. *A strong acid yields a weak conjugate base*—one that has a low affinity for a proton. A strong acid can also be described as an acid whose conjugate base is a much weaker base than water (see Fig. 7.2). In this case the water molecules win the competition for the H^+ ions.

Conversely, a **weak acid** is one for which *the equilibrium lies far to the left.* Most of the acid originally placed in the solution is still present as HA at equilibrium. That is, a weak acid dissociates only to a very small extent in aqueous solution [see Fig. 7.1(b)]. In contrast to a strong acid, a weak acid has a

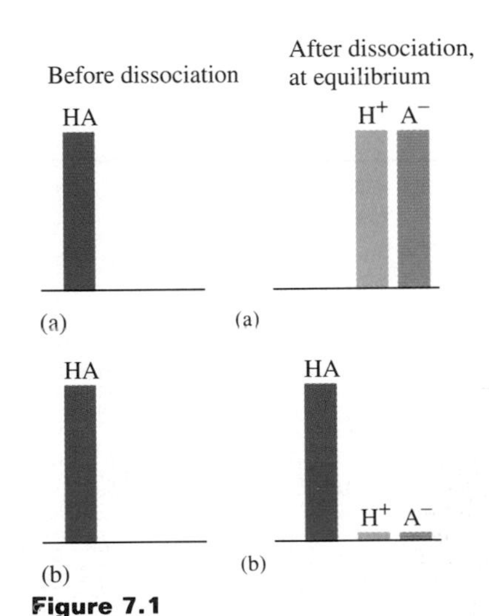

Figure 7.1
Graphical representation of the behavior of acids of different strengths in aqueous solution.
(a) A strong acid. (b) A weak acid.

TABLE 7.1 **Various Ways to Describe Acid Strength**

Property	Strong Acid	Weak Acid
K_a value	K_a is large	K_a is small
Position of the dissociation equilibrium	Far to the right	Far to the left
Equilibrium concentration of H^+ compared with original concentration of HA	$[H^+] \approx [HA]_0$	$[H^+] \ll [HA]_0$
Strength of conjugate base compared with that of water	A^- much weaker base than H_2O	A^- much stronger base than H_2O

$\ll$ means much less than
$\gg$ means much greater than

A strong acid has a weak conjugate base.

Perchloric acid can explode if handled improperly.

conjugate base that is a much stronger base than water. In this case a water molecule is not very successful in pulling an H^+ ion from the conjugate base. *A weak acid yields a relatively strong conjugate base.*

The various ways of describing the strength of an acid are summarized in Table 7.1.

The common strong acids are sulfuric acid [$H_2SO_4(aq)$], hydrochloric acid [$HCl(aq)$], nitric acid [$HNO_3(aq)$], and perchloric acid [$HClO_4(aq)$]. Sulfuric acid is actually a **diprotic acid,** an acid having two acidic protons. The acid H_2SO_4 is a strong acid, virtually 100% dissociated in water:

$$H_2SO_4(aq) \longrightarrow H^+(aq) + HSO_4^-(aq)$$

but the HSO_4^- ion is a weak acid:

$$HSO_4^-(aq) \rightleftharpoons H^+(aq) + SO_4^{2-}(aq)$$

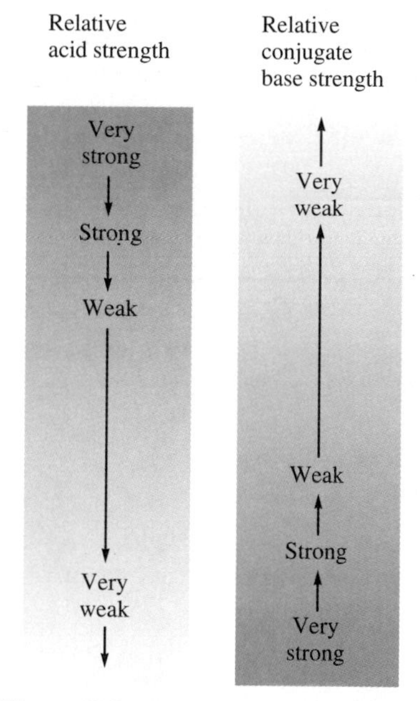

Figure 7.2
The relationship of acid strength and conjugate base strength for the dissociation reaction

$$\underset{\text{Acid}}{HA(aq)} + H_2O(l) \rightleftharpoons H_3O^+(aq) + \underset{\text{Conjugate base}}{A^-(aq)}$$

Most acids are **oxyacids,** in which the acidic proton is attached to an oxygen atom. The strong acids mentioned above, except hydrochloric acid, are typical examples. Many common weak acids, such as phosphoric acid (H_3PO_4), nitrous acid (HNO_2), and hypochlorous acid (HOCl), are also oxyacids. **Organic acids,** those with a carbon atom backbone, commonly contain the **carboxyl group:**

$$-C\begin{matrix}\nearrow\!\!\!\!\!\!= O \\ \searrow O-H\end{matrix}$$

Acids of this type are usually weak. Examples are acetic acid (CH_3COOH), often written $HC_2H_3O_2$, and benzoic acid (C_6H_5COOH).

There are some important acids in which the acidic proton is attached to an atom other than oxygen. The most common of these are the hydrohalic acids HX, where X represents a halogen atom.

Table 7.2 lists common **monoprotic acids** (those having *one* acidic proton) and their K_a values. Note that the strong acids are not listed. When a strong acid molecule such as HCl is placed in water, the position of the dissociation equilibrium

$$HCl(aq) \rightleftharpoons H^+(aq) + Cl^-(aq)$$

TABLE 7.2 Values of K_a for Some Common Monoprotic Acids

Formula	Name	Value of K_a
HSO_4^-	Hydrogen sulfate ion	1.2×10^{-2}
$HClO_2$	Chlorous acid	1.2×10^{-2}
$HC_2H_2ClO_2$	Monochloracetic acid	1.35×10^{-3}
HF	Hydrofluoric acid	7.2×10^{-4}
HNO_2	Nitrous acid	4.0×10^{-4}
$HC_2H_3O_2$	Acetic acid	1.8×10^{-5}
$[Al(H_2O)_6]^{3+}$	Hydrated aluminum(III) ion	1.4×10^{-5}
HOCl	Hypochlorous acid	3.5×10^{-8}
HCN	Hydrocyanic acid	6.2×10^{-10}
NH_4^+	Ammonium ion	5.6×10^{-10}
HOC_6H_5	Phenol	1.6×10^{-10}

Increasing acid strength ↑

Appendix Table A5.1 contains K_a values.

lies so far to the right that [HCl] cannot be measured accurately. This situation prevents an accurate calculation of K_a:

$$K_a = \frac{[H^+][Cl^-]}{[HCl]}$$

↖ Very small and highly uncertain

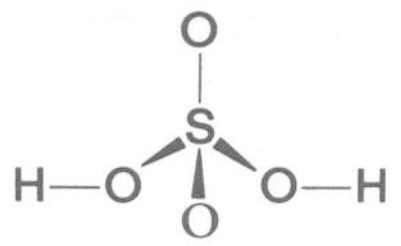

Sulfuric acid

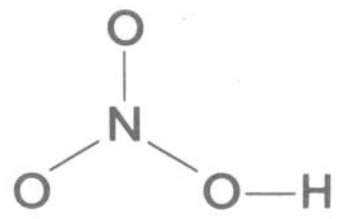

Nitric acid

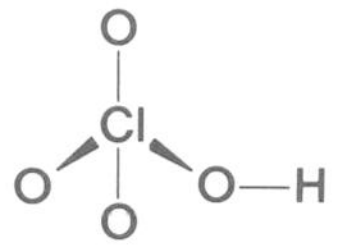

Perchloric acid

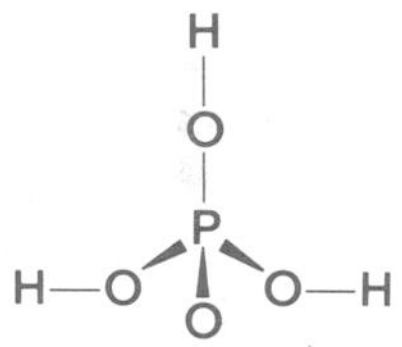

Phosphoric acid

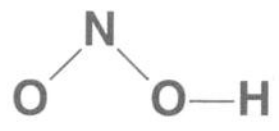

Nitrous acid

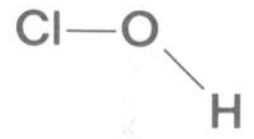

Hypochlorous acid

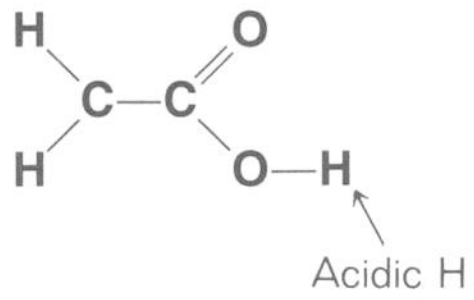

Acetic acid (Acidic H)

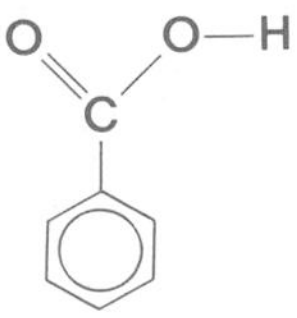

Benzoic acid

EXAMPLE 7.1

Using Table 7.2, arrange the following species according to their strength as bases: H_2O, F^-, Cl^-, NO_2^-, CN^-.

Solution

Remember that water is a stronger base than the conjugate base of a strong acid, but a weaker base than the conjugate base of a weak acid. This rule leads to the following order:

$$Cl^- < H_2O < \text{conjugate bases of weak acids}$$

Weakest bases ⟶ Strongest bases

We can order the remaining conjugate bases by recognizing that the strength of an acid is *inversely related* to the strength of its conjugate base. From Table 7.2 we have

$$K_a \text{ for HF} > K_a \text{ for } HNO_2 > K_a \text{ for HCN}$$

Thus the base strengths increase as follows:

$$F^- < NO_2^- < CN^-$$

The combined order of increasing base strength is

$$Cl^- < H_2O < F^- < NO_2^- < CN^-$$

Water as an Acid and a Base

A substance is said to be *amphoteric* if it can behave either as an acid or as a base. Water is the most common **amphoteric substance.** We see this behavior in the **autoionization** of water, which involves the transfer of a proton from one water molecule to another to produce a hydroxide ion and a hydronium ion:

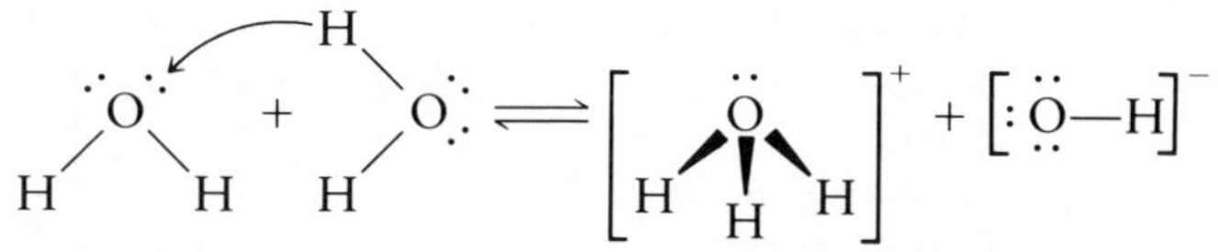

$$H_2O + H_2O \rightleftharpoons [H_3O]^+ + [OH]^-$$

In this reaction one water molecule acts as an acid by furnishing a proton, and the other acts as a base by accepting the proton.

Autoionization can occur in other liquids besides water. For example, in liquid ammonia the autoionization reaction is

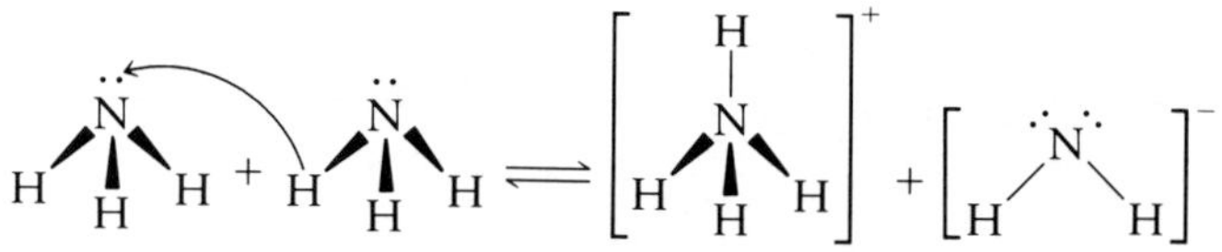

$$NH_3 + NH_3 \rightleftharpoons [NH_4]^+ + [NH_2]^-$$

The autoionization reaction for water

$$2H_2O(l) \rightleftharpoons H_3O^+(aq) + OH^-(aq)$$

leads to the equilibrium expression

$$K_w = [H_3O^+][OH^-] = [H^+][OH^-]$$

where K_w, called the **ion product constant** (or the *dissociation constant*), always refers to the autoionization of water.

Experiment shows that at 25°C

$$[H^+] = [OH^-] = 1.0 \times 10^{-7}\ M$$

which means that at 25°C

$$\begin{aligned} K_w &= [H^+][OH^-] = (1.0 \times 10^{-7}\ \text{mol/L})(1.0 \times 10^{-7}\ \text{mol/L}) \\ &= 1.0 \times 10^{-14}\ \text{mol}^2/\text{L}^2 \end{aligned}$$

The units are customarily omitted, for reasons discussed in Section 6.3.

It is important to recognize the meaning of K_w. In any aqueous solution at 25°C, *no matter what it contains,* the product of $[H^+]$ and $[OH^-]$ must always equal 1.0×10^{-14}. There are three possible situations:

1. A neutral solution, where $[H^+] = [OH^-]$
2. An acidic solution, where $[H^+] > [OH^-]$
3. A basic solution, where $[OH^-] > [H^+]$

In each case, however, at 25°C

$$K_w = [H^+][OH^-] = 1.0 \times 10^{-14}$$

EXAMPLE 7.2

At 60°C the value of K_w is 1×10^{-13}.

a. Using Le Châtelier's principle, predict whether the reaction

$$2H_2O(l) \rightleftharpoons H_3O^+(aq) + OH^-(aq)$$

is exothermic (releases energy) or endothermic (absorbs energy).

b. Calculate $[H^+]$ and $[OH^-]$ in a neutral solution at 60°C.

Solution

a. K_w *increases* from 1×10^{-14} at 25°C to 1×10^{-13} at 60°C. Le Châtelier's principle states that if a system at equilibrium is heated, it will adjust to consume energy. Since the value of K_w increases with temperature, we think of energy as a reactant, and so the process must be endothermic.

b. At 60°C $\qquad [H^+][OH^-] = 1 \times 10^{-13}$

For a neutral solution

$$[H^+] = [OH^-] = \sqrt{1 \times 10^{-13}} = 3 \times 10^{-7}\ M$$

7.3 The pH Scale

The pH scale is a compact way to represent solution acidity. It involves base 10 logs, not natural logs (ln).

Because $[H^+]$ in an aqueous solution is typically quite small, the pH scale provides a convenient way to represent solution acidity. The pH is a log scale based on 10, where

$$pH = -\log[H^+]$$

Thus for a solution where

$$[H^+] = 1.0 \times 10^{-7}\ M$$

then

$$pH = -(-7.00) = 7.00$$

At this point we need to discuss significant figures for logarithms. The rule is that *the number of decimal places in the log is equal to the number of significant figures in the original number*. Thus

2 significant figures

$$[H^+] = 1.0 \times 10^{-9}\ M$$

$$pH = 9.00$$

2 decimal places

Similar log scales are used for representing other quantities. For example,

$$pOH = -\log[OH^-]$$

$$pK = -\log K$$

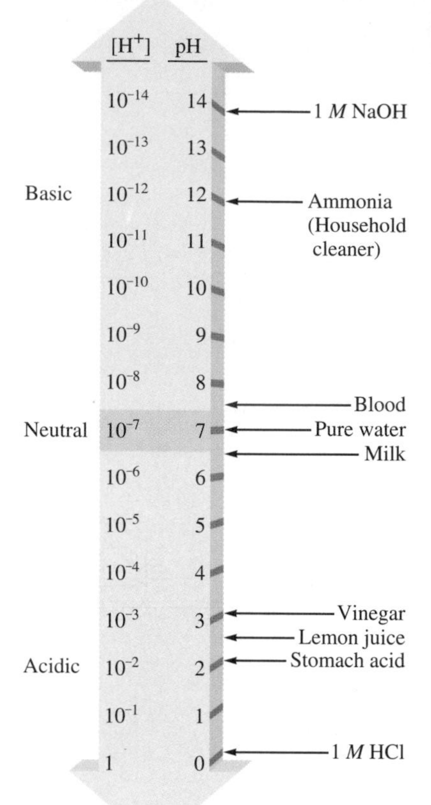

Figure 7.3
The pH scale and pH values of some common substances.

Since pH is a log scale based on 10, *the pH changes by 1 for every power-of-10 change in* $[H^+]$. For example, a solution of pH 3 has an H^+ concentration 10 times that of a solution of pH 4 and 100 times that of a solution of pH 5. Also note that because pH is defined as $-\log[H^+]$, *the pH decreases as* $[H^+]$ *increases*. The pH scale and the pH values for several common substances are shown in Fig. 7.3.

The pH of a solution is usually measured by using a pH meter, an electronic device with a probe that is inserted into a solution of unknown pH. The probe contains an acidic aqueous solution enclosed by a special glass membrane that allows migration of H^+ ions. If the unknown solution has a different pH from the solution in the probe, an electrical potential results, which is registered on the meter (see Fig. 7.4).

Since we have considered all of the fundamental definitions relevant to acid-base solutions, we can proceed to a quantitative description of the equilibria present in these solutions. The main reason that acid-base problems sometimes seem difficult is that, because a typical aqueous solution contains many components, the problems tend to be complicated. However, you can deal with these problems successfully if you use the following general strategies.

Summary: **General Strategies for Solving Acid-Base Problems**

- ***Think chemistry.*** **Focus on the solution components and their reactions. It will almost always be possible to choose one reaction that is the most important.**
- ***Be systematic.*** **Acid-base problems require a step-by-step approach.**
- ***Be flexible.*** **Although all acid-base problems are similar in many ways, important differences do occur. Treat each problem as a separate entity. Do not try to force a given problem to match any**

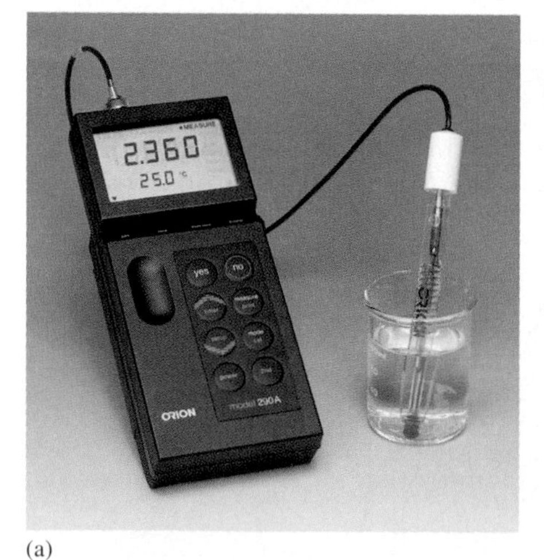

(a)

(b)

Figure 7.4
(a) Measuring the pH of vinegar.
(b) Measuring the pH of aqueous ammonia.

you have solved before. Look for both the similarities and the differences.

- ***Be patient.* The complete solution to a complicated problem cannot be seen immediately in all its detail. Pick the problem apart into its workable steps.**
- ***Be confident.* Look within the problem for the solution, and let the problem guide you. Assume that you can think it out. Do not rely on memorizing solutions to problems. In fact, memorizing solutions is usually detrimental, because you tend to try to force a new problem to be the same as one you have seen before. *Understand and think; don't just memorize.***

7.4 Calculating the pH of Strong Acid Solutions

When we deal with acid-base equilibria, *we must focus on the solution components and their chemistry.* For example, what species are present in a 1.0 *M* solution of HCl? Since hydrochloric acid is a strong acid, we assume that it is completely dissociated. Thus although the label on the bottle says 1.0 *M* HCl, the solution contains virtually no HCl molecules. Typically, container labels indicate the substance(s) used to make up the solution but do not necessarily describe the solution components after dissolution. Thus a 1.0 *M* HCl solution contains H^+ and Cl^- ions rather than HCl molecules.

The next step in dealing with aqueous solutions is to determine which components are significant and which can be ignored. We need to focus on the **major species,** those solution components present in relatively large amounts. In 1.0 *M* HCl, for example, the major species are H^+, Cl^-, and H_2O. Since this solution is very acidic, OH^- is present only in tiny amounts and thus is classed as a minor species. In attacking acid-base problems the importance of *writing the major species in the solution* as the first step cannot be overemphasized. *This single step is the key to solving these problems successfully.*

Always write the major species present in the solution.

To illustrate the main ideas involved, we will calculate the pH of 1.0 *M* HCl. We first list the major species: H^+, Cl^-, and H_2O. Since we want to calculate the pH, we will focus on those major species that can furnish H^+. Obviously, we must consider H^+ from the dissociation of HCl. However, H_2O also furnishes H^+ by autoionization, which is often represented by the simple dissociation reaction

$$H_2O(l) \rightleftharpoons H^+(aq) + OH^-(aq)$$

But is autoionization an important source of H^+ ions? In pure water at 25°C, $[H^+]$ is 10^{-7} *M*. In 1.0 *M* HCl the water will produce even less than 10^{-7} *M* H^+, since by Le Châtelier's principle the H^+ from the dissociated HCl will drive the position of the water equilibrium to the left. Thus the amount of H^+ contributed by water is negligible compared with the 1.0 *M* H^+ from the dissociation of HCl. Therefore we can say that $[H^+]$ in the solution is 1.0 *M*. The pH is then

The H^+ from the strong acid drives the equilibrium $H_2O \rightleftharpoons H^+ + OH^-$ to the left.

$$pH = -\log[H^+] = -\log(1.0) = 0$$

7.5 Calculating the pH of Weak Acid Solutions

Since a weak acid dissolved in water can be viewed as a prototype of almost any equilibrium occurring in aqueous solution, we will proceed carefully and systematically. Although some of the procedures we develop here may seem superfluous, they will become essential as the problems become more complicated. We will develop the necessary strategies by calculating the pH of a 1.00 M solution of HF ($K_a = 7.2 \times 10^{-4}$).

First, as always, write the major species present in the solution.

The first step, as always, is to *write the major species in the solution*. From its small K_a value, we know that hydrofluoric acid is a weak acid and will be dissociated only to a slight extent. Thus when we write the major species, the hydrofluoric acid will be represented in its dominant form, as HF. The major species in solution are HF and H_2O.

The next step is to decide which of the major species can furnish H^+ ions. Actually, both major species can do so:

$$HF(aq) \rightleftharpoons H^+(aq) + F^-(aq) \qquad K_a = 7.2 \times 10^{-4}$$
$$H_2O(l) \rightleftharpoons H^+(aq) + OH^-(aq) \qquad K_w = 1.0 \times 10^{-14}$$

In aqueous solutions typically one source of H^+ can be singled out as dominant. By comparing K_a for HF with K_w for H_2O, we see that hydrofluoric acid, although weak, is still a much stronger acid than water. Thus we will assume that hydrofluoric acid will be the dominant source of H^+. We will ignore the tiny contribution expected from water.

Therefore, it is the dissociation of HF that will determine the equilibrium concentration of H^+ and hence the pH:

$$HF(aq) \rightleftharpoons H^+(aq) + F^-(aq)$$

The equilibrium expression is

$$K_a = 7.2 \times 10^{-4} = \frac{[H^+][F^-]}{[HF]}$$

Just as with gas-phase equilibria, the equilibrium constant expressions for reactions in solution actually require activities. In all of our calculations, we will assume that $a_x = [x]$, where x represents a species in solution.

To solve the equilibrium problem, we follow the procedures developed in Chapter 6 for gas-phase equilibria. First, we list the initial concentrations, the *concentrations before the reaction of interest has proceeded to equilibrium.* Before any HF dissociates, the concentrations of the species in the equilibrium are

$$[HF]_0 = 1.00\ M, \qquad [F^-]_0 = 0, \qquad [H^+]_0 = 10^{-7}\ M \approx 0$$

(Note that the zero value for $[H^+]_0$ is an approximation, since we are neglecting the H^+ ions from the autoionization of water.)

The next step is to determine the change required to reach equilibrium. Since some HF will dissociate to come to equilibrium (but that amount is presently unknown), we let x be the change in the concentration of HF that is required to achieve equilibrium. That is, we assume that x moles per liter of HF will dissociate to produce x moles per liter of H^+ and x moles per liter of F^- as the system adjusts to its equilibrium position. Now the equilibrium concentrations can be defined in terms of x:

$$[HF] = [HF]_0 - x = 1.00 - x$$
$$[F^-] = [F^-]_0 + x = 0 + x = x$$
$$[H^+] = [H^+]_0 + x \approx 0 + x = x$$

Substituting these equilibrium concentrations into the equilibrium expression gives

$$K_a = 7.2 \times 10^{-4} = \frac{[H^+][F^-]}{[HF]} = \frac{(x)(x)}{1.00 - x}$$

This expression produces a quadratic equation that can be solved by using the quadratic formula, as for the gas-phase systems in Chapter 6. However, since K_a for HF is so small, HF will dissociate only slightly; thus x is expected to be small. This will allow us to simplify the calculation. If x is very small compared with 1.00, the term in the denominator can be approximated as follows:

$$1.00 - x \approx 1.00$$

The equilibrium expression then becomes

$$7.2 \times 10^{-4} = \frac{(x)(x)}{1.00 - x} \approx \frac{(x)(x)}{1.00}$$

which yields

$$x^2 \approx (7.2 \times 10^{-4})(1.00) = 7.2 \times 10^{-4}$$
$$x \approx \sqrt{7.2 \times 10^{-4}} = 2.7 \times 10^{-2}$$

How valid is the approximation that $[HF] = 1.00\ M$? Because this question will arise often in connection with acid-base equilibrium calculations, we will consider it carefully. *The validity of the approximation depends on how much accuracy we demand for the calculated value of* $[H^+]$. Typically, the K_a values for acids are known to an accuracy of only about ±5%. Therefore, it is reasonable to apply this figure when determining the validity of the approximation

$$[HA]_0 - x \approx [HA]_0$$

The validity of an approximation should always be checked.

Note that although the 5% rule is an arbitrary choice, it makes sense because of the typical uncertainty in K_a values. You should be aware that the precision of the data for a particular situation should be used to evaluate which approximations are reasonable.

We will use the following test.

First calculate the value of x by making the approximation

$$K_a = \frac{x^2}{[HA]_0 - x} \approx \frac{x^2}{[HA]_0}$$

where

$$x^2 \approx K_a[HA]_0 \quad \text{and} \quad x \approx \sqrt{K_a[HA]_0}$$

Then compare the sizes of x and $[HA]_0$. If the expression

$$\frac{x}{[HA]_0} \times 100$$

is less than or equal to 5%, the value of x is small enough for the approximation

$$[HA]_0 - x \approx [HA]_0$$

to be considered valid.

In our example

$$x = 2.7 \times 10^{-2} \text{ mol/L}$$
$$[HA]_0 = [HF]_0 = 1.00 \text{ mol/L}$$

and

$$\frac{x}{[HA]_0} \times 100 = \frac{2.7 \times 10^{-2}}{1.00} \times 100 = 2.7\%$$

The approximation we made is considered valid, so the value of x calculated by using that approximation is acceptable. Thus

$$x = [H^+] = 2.7 \times 10^{-2}\ M \qquad \text{and} \qquad pH = -\log(2.7 \times 10^{-2}) = 1.57$$

This problem illustrates all the important steps required for solving a typical equilibrium problem involving a weak acid. These steps are summarized below.

A table of K_a values for various weak acids is given in Appendix Table A5.1.

Summary: **Solving Weak Acid Equilibrium Problems**

- **List the major species in the solution.**
- **Choose the species that can produce H^+, and write balanced equations for the reactions producing H^+.**
- **Comparing the values of the equilibrium constants for the reactions you have written, decide which reaction will dominate in the production of H^+.**
- **Write the equilibrium expression for the dominant reaction.**
- **List the initial concentrations of the species participating in the dominant reaction.**
- **Define the change needed to achieve equilibrium; that is, define x.**
- **Write the equilibrium concentrations in terms of x.**
- **Substitute the equilibrium concentrations into the equilibrium expression.**
- **Solve for x the "easy" way—that is, by assuming that $[HA]_0 - x \approx [HA]_0$.**
- **Verify whether the approximation is valid (the 5% rule is the test in this case).**
- **Calculate $[H^+]$ and pH.**

The pH of a Mixture of Weak Acids

Sometimes a solution contains two weak acids of very different strengths. This case is considered in Example 7.3. Note that the usual steps are followed (though not labeled).

EXAMPLE 7.3

Calculate the pH of a solution that contains 1.00 M HCN ($K_a = 6.2 \times 10^{-10}$) and 5.00 M HNO_2 ($K_a = 4.0 \times 10^{-4}$). Also calculate the concentration of cyanide ion (CN^-) in this solution at equilibrium.

Solution

Since HCN and HNO_2 are both weak acids and are thus largely undissociated, the major species in the solution are

$$HCN, \quad HNO_2, \quad H_2O$$

All three of these components produce H^+:

$$HCN(aq) \rightleftharpoons H^+(aq) + CN^-(aq) \qquad K_a = 6.2 \times 10^{-10}$$

$$HNO_2(aq) \rightleftharpoons H^+(aq) + NO_2^-(aq) \qquad K_a = 4.0 \times 10^{-4}$$

$$H_2O(l) \rightleftharpoons H^+(aq) + OH^-(aq) \qquad K_w = 1.0 \times 10^{-14}$$

A mixture of three acids might lead to a very complicated problem. However, the situation is greatly simplified by the fact that even though HNO_2 is a weak acid, it is much stronger than the other two acids present (as revealed by the K values). Thus HNO_2 can be assumed to be the dominant producer of H^+, so we will focus on the equilibrium expression:

$$K_a = 4.0 \times 10^{-4} = \frac{[H^+][NO_2^-]}{[HNO_2]}$$

The initial concentrations, the definition of x, and the equilibrium concentrations are as follows:

Initial Concentration (mol/L)		Equilibrium Concentration (mol/L)
$[HNO_2]_0 = 5.00$ $[NO_2^-]_0 = 0$ $[H^+]_0 \approx 0$	x mol/L HNO_2 dissociates $\longrightarrow$	$[HNO_2] = 5.00 - x$ $[NO_2^-] = x$ $[H^+] = x$

Substituting the equilibrium concentrations into the equilibrium expression and making the approximation that $5.00 - x = 5.00$ gives

$$K_a = 4.0 \times 10^{-4} = \frac{(x)(x)}{5.00 - x} \approx \frac{x^2}{5.00}$$

We solve for x: $\qquad x = 4.5 \times 10^{-2}$

Using the 5% rule we show that the approximation is valid:

$$\frac{x}{[HNO_2]_0} \times 100 = \frac{4.5 \times 10^{-2}}{5.00} \times 100 = 0.90\%$$

Therefore,

$$[H^+] = x = 4.5 \times 10^{-2}\ M \qquad \text{and} \qquad pH = 1.35$$

We also want to calculate the equilibrium concentration of cyanide ion in this solution. The CN^- ions in this solution come from the dissociation of HCN:

$$HCN(aq) \rightleftharpoons H^+(aq) + CN^-(aq)$$

Although the position of this equilibrium lies far to the left and does not contribute *significantly* to $[H^+]$, HCN is the *only source* of CN^-. Thus we must consider the extent of the dissociation of HCN to calculate $[CN^-]$. The equilibrium expression for the above reaction is

$$K_a = 6.2 \times 10^{-10} = \frac{[H^+][CN^-]}{[HCN]}$$

We know $[H^+]$ for this solution from the results for the first part of this problem. Note that *there is only one kind of H^+ in this solution*. It does not matter from which acid the H^+ ions originate. The equilibrium value of $[H^+]$ for the HCN dissociation is 4.5×10^{-2} *M*, even though the H^+ was contributed almost entirely by the dissociation of HNO_2. What is [HCN] at equilibrium? We know $[HCN]_0 = 1.00$ *M*, and since K_a for HCN is so small, a negligible amount of HCN will dissociate.

Thus $$[HCN] = [HCN]_0 - \text{amount of HCN dissociated}$$
$$\approx [HCN]_0 = 1.00\ M$$

Since $[H^+]$ and [HCN] are known, we can find $[CN^-]$ from the equilibrium expression:

$$K_a = 6.2 \times 10^{-10} = \frac{[H^+][CN^-]}{[HCN]} = \frac{(4.5 \times 10^{-2})[CN^-]}{1.00}$$

$$[CN^-] = \frac{(6.2 \times 10^{-10})(1.00)}{4.5 \times 10^{-2}} = 1.4 \times 10^{-8}\ M$$

Note the significance of this result. Since $[CN^-] = 1.4 \times 10^{-8}$ *M*, and since HCN is the only source of CN^-, only 1.4×10^{-8} mol/L of HCN has dissociated. This is a very small amount compared with the initial concentration of HCN, which is exactly what we would expect from its very small K_a value. Thus [HCN] = 1.00 *M* as assumed. Also, this result confirms that HNO_2 is the only significant source of H^+.

Percent Dissociation

It is often useful to specify the amount of weak acid that has dissociated in achieving equilibrium in an aqueous solution. The **percent dissociation** is defined as follows:

$$\text{Percent dissociation} = \frac{\text{amount dissociated (mol/L)}}{\text{initial concentration (mol/L)}} \times 100 \qquad (7.3)$$

For example, we found earlier that in a 1.00 M solution of HF, $[H^+] = 2.7 \times 10^{-2}$ M. For the system to reach equilibrium, 2.7×10^{-2} moles per liter of the original 1.00 M HF dissociates; so

$$\text{Percent dissociation} = \frac{2.7 \times 10^{-2}\ \text{mol/L}}{1.00\ \text{mol/L}} \times 100 = 2.7\%$$

For a given weak acid, the percent dissociation increases as the acid becomes more dilute. For example, the percent dissociation of acetic acid ($HC_2H_3O_2$, $K_a = 1.8 \times 10^{-5}$) is significantly greater in a 0.10 M solution than in a 1.0 M solution.

Demonstrate for yourself (by doing the calculations) that even though the concentration of H^+ ion at equilibrium is smaller in the 0.10 M acetic acid solution than in the 1.0 M acetic acid solution, the percent dissociation is significantly greater in the 0.10 M solution (1.3%) than in the 1.0 M solution (0.42%). This is a general result. *For solutions of any weak acid HA, $[H^+]$ decreases as $[HA]_0$ decreases, but the percent dissociation increases as $[HA]_0$ decreases.*

The more dilute the weak acid solution, the greater the percent dissociation.

This phenomenon can be explained in the following way. Consider the weak acid HA with the initial concentration $[HA]_0$. At equilibrium

$$[HA] = [HA]_0 - x \approx [HA]_0$$

$$[H^+] = [A^-] = x$$

Thus

$$K_a = \frac{[H^+][A^-]}{[HA]} \approx \frac{(x)(x)}{[HA]_0}$$

Now suppose enough water is added to dilute the solution by a factor of 10. The new concentrations before any adjustment occurs are

$$[A^-]_{new} = [H^+]_{new} = \frac{x}{10}$$

$$[HA]_{new} = \frac{[HA]_0}{10}$$

and Q, the reaction quotient, is

$$Q = \frac{(x/10)(x/10)}{[HA]_0/10} = \frac{1(x)(x)}{10[HA]_0} = \frac{1}{10}K_a$$

Since Q is less than K_a, the system must adjust to the right to reach the new equilibrium position. Thus the percent dissociation increases as the acid becomes more dilute. This behavior is summarized in Fig. 7.5. In Example 7.4 we see how the percent dissociation can be used to calculate the K_a value for a weak acid.

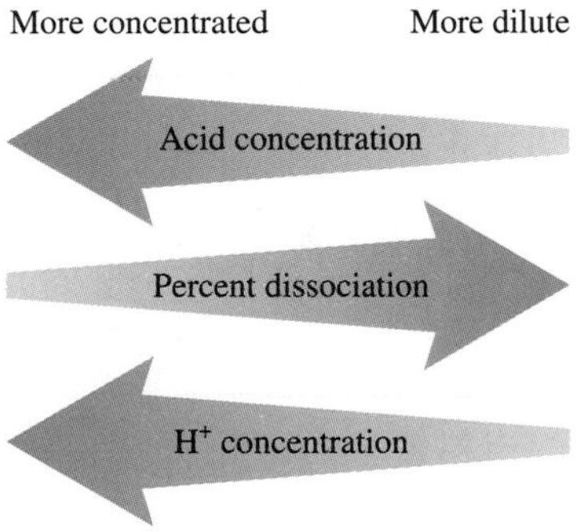

Figure 7.5
The effect of dilution on the percent dissociation and $[H^+]$ of a weak acid solution.

EXAMPLE 7.4

Lactic acid ($HC_3H_5O_3$) is a waste product that accumulates in muscle tissue during exertion, leading to pain and a feeling of fatigue. In a 0.100 M aqueous solution, lactic acid is 3.7% dissociated. Calculate the value of K_a for this acid.

Solution

The small value for the percent dissociation clearly indicates that $HC_3H_5O_3$ is a weak acid. Thus the major species in the solution are the undissociated acid and water:

$$HC_3H_5O_3 \quad \text{and} \quad H_2O$$

Although $HC_3H_5O_3$ is a weak acid, it is much stronger than water and thus will be the dominant source of H^+ in the solution. The dissociation reaction is

$$HC_3H_5O(aq) \rightleftharpoons H^+(aq) + C_3H_5O_3^-(aq)$$

and the equilibrium expression is

$$K_a = \frac{[H^+][C_3H_5O_3^-]}{[HC_3H_5O_3]}$$

The initial and equilibrium concentrations are as follows:

Initial Concentration mol/L		Equilibrium Concentration mol/L
$[HC_3H_5O_3]_0 = 0.10$	x mol/L	$[HC_3H_5O_3] = 0.10 - x$
$[C_3H_5O_3^-]_0 = 0$	$\longrightarrow$ $HC_3H_5O_3$	$[C_3H_5O_3^-] = x$
$[H^+]_0 \approx 0$	dissociates	$[H^+] = x$

The change needed to reach equilibrium can be obtained from the percent dissociation and Equation (7.3). For this acid

$$\text{Percent dissociation} = 3.7\% = \frac{x}{[HC_3H_5O_3]_0} \times 100 = \frac{x}{0.10} \times 100$$

and

$$x = \frac{3.7}{100}(0.10) = 3.7 \times 10^{-3} \text{ mol/L}$$

Now we can calculate the equilibrium concentrations:

$$[HC_3H_5O_3] = 0.10 - x = 0.10\ M \quad \text{(to the correct number of significant figures)}$$

$$[C_3H_5O_3^-] = [H^+] = x = 3.7 \times 10^{-3}\ M$$

These concentrations can now be used to calculate the value of K_a for lactic acid:

$$K_a = \frac{[H^+][[C_3H_5O_3^-]}{[HC_3H_5O_3]} = \frac{(3.7 \times 10^{-3})(3.7 \times 10^{-3})}{0.10} = 1.4 \times 10^{-4}$$

7.6 Bases

In a basic solution pH > 7.

According to the Arrhenius concept, a base is a substance that produces OH^- ions in aqueous solution. According to the Brønsted-Lowry definition, a base is a proton acceptor. The bases sodium hydroxide (NaOH) and potassium hydroxide (KOH) fulfill both criteria. They contain OH^- ions in the solid lattice and behave as strong electrolytes, dissociating completely when dissolving in water:

$$NaOH(s) \xrightarrow{H_2O} Na^+(aq) + OH^-(aq)$$

Thus a 1.0 *M* NaOH solution actually contains 1.0 *M* Na^+ and 1.0 *M* OH^-. Because of their complete dissociation, NaOH and KOH are called **strong bases** in the same sense as we defined strong acids.

All the hydroxides of the Group 1A elements (LiOH, NaOH, KOH, RbOH, and CsOH) are strong bases, but only NaOH and KOH are common laboratory reagents because the lithium, rubidium, and cesium compounds are expensive. The alkaline earth (Group 2A) hydroxides—$Ca(OH)_2$, $Ba(OH)_2$, and $Sr(OH)_2$—are also strong bases. For these compounds 2 moles of hydroxide ion are produced for every mole of metal hydroxide dissolved in aqueous solution.

The alkaline earth hydroxides are not very soluble and are used only when the solubility factor is not important. In fact, the low solubility of these bases can be an advantage. For example, many antacids are suspensions of metal hydroxides such as aluminum hydroxide and magnesium hydroxide. The low solubility of these compounds prevents the formation of a large hydroxide ion concentration that would harm the tissues of the mouth, esophagus, and stomach. Yet these suspensions furnish plenty of hydroxide ion to react with the stomach acid, since the salts dissolve as this reaction proceeds.

Calcium hydroxide, $Ca(OH)_2$, often called **slaked lime,** is widely used in industry because it is inexpensive and plentiful. For example, slaked lime is used in scrubbing stack gases to remove sulfur dioxide from the exhaust of power plants and factories. In the scrubbing process a suspension of slaked lime is sprayed into the stack gases to react with sulfur dioxide gas according to the following equations:

$$SO_2(g) + H_2O(l) \rightleftharpoons H_2SO_3(aq)$$

$$Ca(OH)_2(aq) + H_2SO_3(aq) \rightleftharpoons CaSO_3(s) + 2H_2O(l)$$

Slaked lime is also widely used in water treatment plants for softening hard water, which involves the removal of ions such as Ca^{2+} and Mg^{2+}, which hamper the action of detergents. The softening method most often employed in water treatment plants is the **lime-soda process,** in which *lime* (CaO) and *soda ash* (Na_2CO_3) are added to the water. As we will see in more detail later in this chapter, the CO_3^{2-} ion from soda ash reacts with water to produce the HCO_3^- ion. When the lime is added to the hard water, it forms slaked lime,

$$CaO(s) + H_2O(l) \longrightarrow Ca(OH)_2(aq)$$

which then reacts with the HCO_3^- ion and a Ca^{2+} ion to produce calcium carbonate:

$$Ca(OH)_2(aq) + \underset{\text{From hard water}}{Ca^{2+}(aq)} + 2HCO_3^-(aq) \longrightarrow 2CaCO_3(s) + 2H_2O(l)$$

Thus for every mole of $Ca(OH)_2$ consumed, 1 mole of Ca^{2+} is removed from the hard water, thereby softening it. Some hard water naturally contains bicarbonate ions. In this case no soda ash is needed—simply adding the lime accomplishes the softening.

Hard water contains Ca^{2+} and Mg^{2+} ions, among others, which are detrimental to detergent action.

Calculating the pH of a strong base solution is relatively simple, as illustrated in Example 7.5.

EXAMPLE 7.5

Calculate the pH of a 5.0×10^{-2} *M* NaOH solution.

Solution

The major species in this solution are

$$\underbrace{Na^+, \quad OH^-,}_{\text{From NaOH}} \quad \text{and} \quad H_2O$$

Although the autoionization of water also produces OH^- ions, the pH will be determined by the OH^- ions from the dissolved NaOH. Thus in the solution

$$[OH^-] = 5.0 \times 10^{-2}\ M$$

The concentration of H^+ can be calculated from K_w:

$$[H^+] = \frac{K_w}{[OH^-]} = \frac{1.0 \times 10^{-14}}{5.0 \times 10^{-2}} = 2.0 \times 10^{-13}\ M$$

$$pH = 12.70$$

Note that this solution is basic:

$$[OH^-] > [H^+] \quad \text{and} \quad pH > 7$$

The added OH^- from the salt has shifted the water autoionization equilibrium

$$H_2O(l) \rightleftharpoons H^+(aq) + OH^-(aq)$$

to the left, significantly lowering $[H^+]$ compared with that in pure water.

A base does not have to contain the hydroxide ion.

Many types of proton acceptors (bases) do not contain the hydroxide ion. However, when dissolved in water, these substances increase the concentration of hydroxide ion by reacting with water. For example, ammonia reacts with water as follows:

$$NH_3(aq) + H_2O(l) \rightleftharpoons NH_4^+(aq) + OH^-(aq)$$

The ammonia molecule accepts a proton and thus functions as a base. Water is the acid in this reaction. Note that even though the base ammonia contains no hydroxide ion, it still increases the concentration of hydroxide ion to yield a basic solution.

Bases like ammonia typically have at least one unshared pair of electrons that is capable of forming a bond with a proton. The reaction of an ammonia molecule with a water molecule can be represented as follows:

$$H_3N\!: \; + \; H{-}\ddot{O}{:}{-}H \rightleftharpoons H{-}N^+(H)_2{-}H + {:}\ddot{O}{-}H^-$$

There are many bases like ammonia that produce hydroxide ion by reaction with water. In most of these bases, the lone pair is located on a nitrogen atom. Some examples are:

$H_3C-\ddot{N}H-H$	$H_3C-\ddot{N}H-CH_3$	$H_3C-\ddot{N}(CH_3)-CH_3$	$H-\ddot{N}H-C_2H_5$	$C_5H_5\ddot{N}$
Methylamine	Dimethylamine	Trimethylamine	Ethylamine	Pyridine

Note that the first four bases can be thought of as substituted ammonia molecules where hydrogen atoms are replaced by methyl (CH_3) or ethyl (C_2H_5) groups. The pyridine molecule is like benzene

except that a nitrogen atom has replaced one of the carbon atoms in the ring. The general reaction between a base (B) and water is given by:

$$\underset{\text{Base}}{B(aq)} + \underset{\text{Acid}}{H_2O(l)} \rightleftharpoons \underset{\text{Conjugate acid}}{BH^+(aq)} + \underset{\text{Conjugate base}}{OH^-(aq)} \tag{7.4}$$

Benzene, C_6H_6, is often represented by the symbol

where each vertex represents a carbon atom. The hydrogen atoms are not shown.

The equilibrium reaction for this general solution is

$$K_b = \frac{[BH^+][OH^-]}{[B]}$$

Appendix Table A5.3 contains K_b values.

where K_b *always refers to the reaction of a base with water to form the conjugate acid and the hydroxide ion.*

Bases of the type represented by B in Equation (7.4) compete with OH^-, a very strong base, for the H^+ ion. Thus their K_b values tend to be small (for example, for ammonia $K_b = 1.8 \times 10^{-5}$), and they are called **weak bases.** The values of K_b for some common weak bases are listed in Table 7.3.

Typically, pH calculations for solutions of weak bases are very similar to those for weak acids.

TABLE 7.3 **Values of K_b for Some Common Weak Bases**

Name	Formula	Conjugate Acid	K_b
Ammonia	NH_3	NH_4^+	1.8×10^{-5}
Methylamine	CH_3NH_2	$CH_3NH_3^+$	4.38×10^{-4}
Ethylamine	$C_2H_5NH_2$	$C_2H_5NH_3^+$	5.6×10^{-4}
Aniline	$C_6H_5NH_2$	$C_6H_5NH_3^+$	3.8×10^{-10}
Pyridine	C_5H_5N	$C_5H_5NH^+$	1.7×10^{-9}

Amines

We have seen that many bases have nitrogen atoms with one lone pair of electrons. These bases can be viewed as substituted ammonia molecules, with the general formula $R_xNH_{(3-x)}$. Compounds of this type are called **amines.**

Amines are widely distributed in animals and plants, often serving as messengers or regulators. For example, in the human nervous system there are two amine stimulants, *norepinephrine* and *adrenaline:*

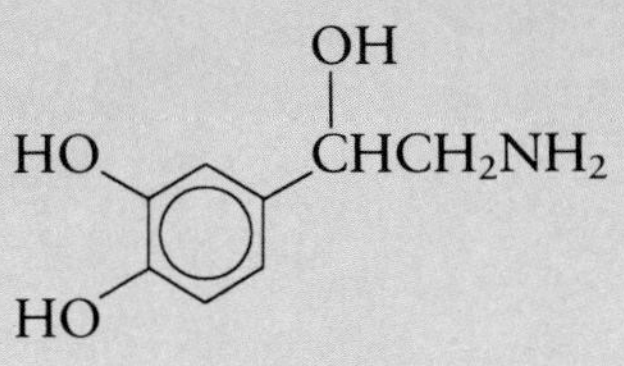

OH

HO $CHCH_2NH_2$

HO

Norepinephrine

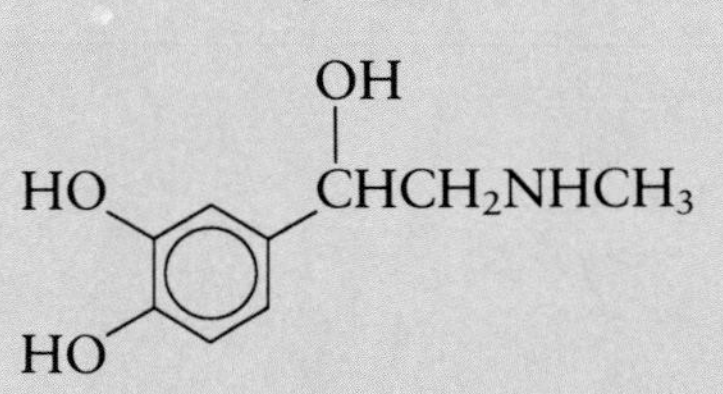

OH

HO $CHCH_2NHCH_3$

HO

Adrenaline

Ephedrine, widely used as a decongestant, was a known drug in China over 2000 years ago. People from cultures in Mexico and the Southwest have used the hallucinogen *mescaline,* extracted from peyote cactus, for centuries.

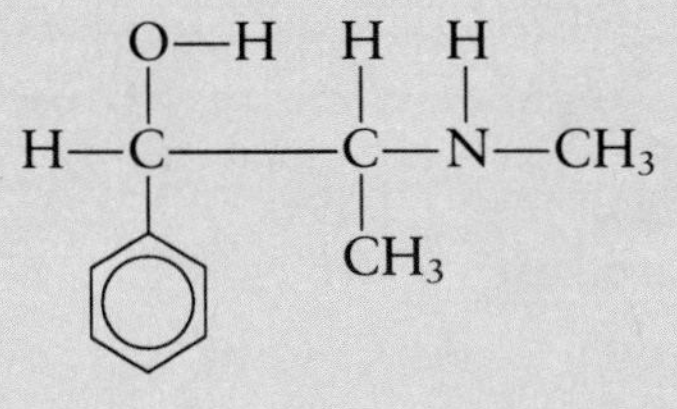

O—H H H

H—C——C—N—CH_3

CH_3

Ephedrine

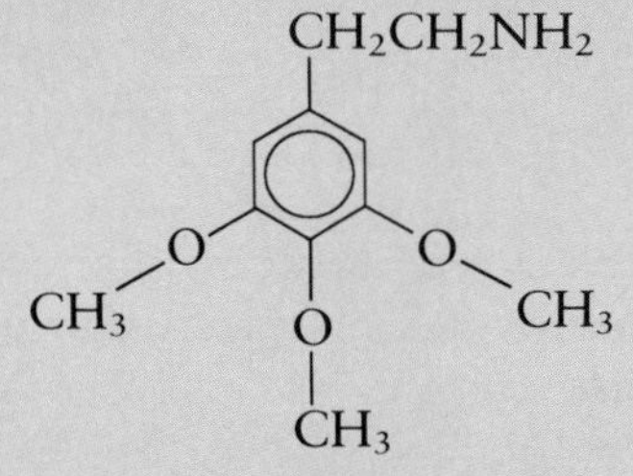

$CH_2CH_2NH_2$

CH_3 O O CH_3

O

CH_3

Mescaline

Many other drugs, such as codeine and quinine, are amines, but they are usually not used in their pure amine forms. Instead, they are treated with an acid to become acid salts. An example of an acid salt is ammonium chloride, obtained by the reaction

$$NH_3 + HCl \longrightarrow NH_4Cl$$

Amines can also be protonated in this way. The resulting acid salt, written as AHCl (where A represents the amine), contains AH^+ and Cl^-. In general, the acid salts are more stable and more soluble in water than the parent amines. For instance, the parent amine of the well-known local anaesthetic *novocaine* is water-insoluble, but the acid salt is much more soluble.

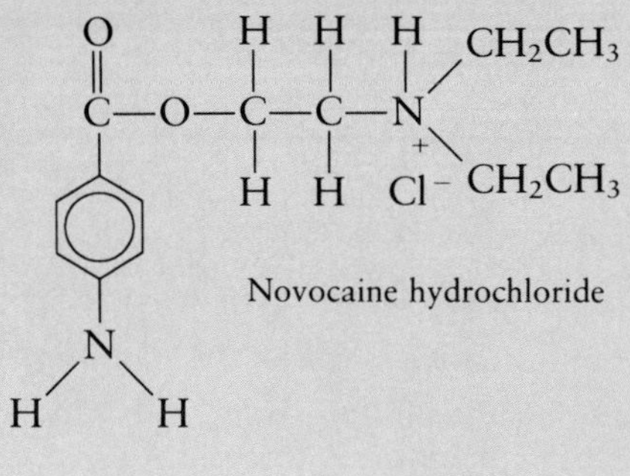

O H H H CH_2CH_3

C—O—C—C—N

H H Cl^- CH_2CH_3

N

H H

Novocaine hydrochloride

EXAMPLE 7.6

Calculate the pH of a 1.0 *M* solution of methylamine ($K_b = 4.38 \times 10^{-4}$).

Solution

Since methylamine (CH_3NH_2) is a weak base, the major species in solution are

$$CH_3NH_2 \quad \text{and} \quad H_2O$$

Both are bases; however, since water can be neglected as a source of OH^-, the dominant equilibrium is

$$CH_3NH_2(aq) + H_2O(l) \rightleftharpoons CH_3NH_3^+(aq) + OH^-(aq)$$

and $$K_b = 4.38 \times 10^{-4} = \frac{[CH_3NH_3^+][OH^-]}{[CH_3NH_2]}$$

The concentrations are as follows:

Initial Concentration (mol/L)		Equilibrium Concentration (mol/L)
$[CH_3NH_2]_0 = 1.0$	x mol/L CH_3NH_2 reacts with H_2O	$[CH_3NH_2] = 1.0 - x$
$[CH_3NH_3^+]_0 = 0$	$\longrightarrow$	$[CH_3NH_3^+] = x$
$[OH^-]_0 \approx 0$	to reach equilibrium	$[OH^-] = x$

Substituting the equilibrium concentrations into the equilibrium expression and making the usual approximation gives

$$K_b = 4.38 \times 10^{-4} = \frac{[CH_3NH_3^+][OH^-]}{[CH_3NH_2]} = \frac{(x)(x)}{1.0 - x} \approx \frac{x^2}{1.0}$$

$$x \approx 2.1 \times 10^{-2}$$

The approximation is valid by the 5% rule, so

$$[OH^-] = x = 2.1 \times 10^{-2}\ M \text{ and pOH} = 1.68$$

Note that since $[H^+][OH^-] = 1.0 \times 10^{-14}$, pH + pOH = 14
Thus, pH = 14.00 − 1.68 = 12.32

7.7 Polyprotic Acids

Some important acids, such as sulfuric acid (H_2SO_4) and phosphoric acid (H_3PO_4), can furnish more than one proton per molecule and are called **polyprotic acids.** A polyprotic acid always dissociates in a *stepwise* manner, one proton at a time. For example, diprotic (two-proton) *carbonic acid* (H_2CO_3), which is vital for maintaining a constant pH in human blood, dissociates in the following steps:

$$H_2CO_3(aq) \rightleftharpoons H^+(aq) + HCO_3^-(aq) \quad K_{a_1} = \frac{[H^+][HCO_3^-]}{[H_2CO_3]} = 4.3 \times 10^{-7}$$

$$HCO_3^-(aq) \rightleftharpoons H^+(aq) + CO_3^{2-}(aq) \quad K_{a_2} = \frac{[H^+][CO_3^{2-}]}{[HCO_3^-]} = 5.6 \times 10^{-11}$$

The successive K_a values for the dissociation equilibria are designated K_{a_1} and K_{a_2}. Note that the conjugate base HCO_3^- of the first dissociation equilibrium becomes the acid in the second step.

Carbonic acid is formed when carbon dioxide gas is dissolved in water. In fact, the first dissociation step for carbonic acid is best represented by the reaction

$$CO_2(aq) + H_2O(l) \rightleftharpoons H^+(aq) + HCO_3^-(aq)$$

since relatively little H_2CO_3 actually exists in solution. However, it is convenient to consider CO_2 in water as H_2CO_3 so that we can treat such solutions by using the familiar dissociation reactions for weak acids.

Phosphoric acid is a **triprotic acid** (three protons) that dissociates in the following steps:

$$H_3PO_4(aq) \rightleftharpoons H^+(aq) + H_2PO_4^-(aq) \qquad K_{a_1} = \frac{[H^+][H_2PO_4^-]}{[H_3PO_4]} = 7.5 \times 10^{-3}$$

$$H_2PO_4^-(aq) \rightleftharpoons H^+(aq) + HPO_4^{2-}(aq) \qquad K_{a_2} = \frac{[H^+][HPO_4^{2-}]}{[H_2PO_4^-]} = 6.2 \times 10^{-8}$$

$$HPO_4^{2-}(aq) \rightleftharpoons H^+(aq) + PO_4^{3-}(aq) \qquad K_{a_3} = \frac{[H^+][PO_4^{3-}]}{[HPO_4^{2-}]} = 4.8 \times 10^{-13}$$

For a typical weak polyprotic acid,

$$K_{a_1} > K_{a_2} > K_{a_3}$$

That is, the acid involved in each successive step of the dissociation is weaker. This is shown by the stepwise dissociation constants given in Table 7.4. These values indicate that the loss of a second or third proton occurs less readily than the loss of the first proton. This result is not surprising; the greater the negative charge on the acid, the more difficult it becomes to remove the positively charged proton.

Although we might expect the pH calculations for solutions of polyprotic acids to be complicated, the most common cases are surprisingly straightforward. To illustrate, we will consider a typical case, phosphoric acid, and a unique case, sulfuric acid.

Phosphoric Acid

Phosphoric acid is typical of most weak polyprotic acids in that its successive K_a values are very different. For H_3PO_4, the ratios of successive K_a values (from Table 7.4) are

$$\frac{K_{a_1}}{K_{a_2}} = \frac{7.5 \times 10^{-3}}{6.2 \times 10^{-8}} = 1.2 \times 10^5$$

$$\frac{K_{a_2}}{K_{a_3}} = \frac{6.2 \times 10^{-8}}{4.8 \times 10^{-13}} = 1.3 \times 10^5$$

So the relative acid strengths are

$$H_3PO_4 \gg H_2PO_4^- \gg HPO_4^{2-}$$

TABLE 7.4 Stepwise Dissociation Constants for Several Common Polyprotic Acids

Name	Formula	K_{a_1}	K_{a_2}	K_{a_3}
Phosphoric acid	H_3PO_4	7.5×10^{-3}	6.2×10^{-8}	4.8×10^{-13}
Arsenic acid	H_3AsO_4	5×10^{-3}	8×10^{-8}	6×10^{-10}
Carbonic acid	H_2CO_3	4.3×10^{-7}	5.6×10^{-11}	
Sulfuric acid	H_2SO_4	Large	1.2×10^{-2}	
Sulfurous acid	H_2SO_3	1.5×10^{-2}	1.0×10^{-7}	
Hydrosulfuric acid*	H_2S	1.0×10^{-7}	$\approx 10^{-19}$	
Oxalic acid	$H_2C_2O_4$	6.5×10^{-2}	6.1×10^{-5}	
Ascorbic acid (vitamin C)	$H_2C_6H_6O_6$	7.9×10^{-5}	1.6×10^{-12}	

* The K_{a_2} value for H_2S is quite uncertain. Its small size makes it very difficult to measure.

This means that in a solution prepared by dissolving H_3PO_4 in water, *only the first dissociation step makes an important contribution to [H^+].* This greatly simplifies the pH calculations for phosphoric acid solutions, as is illustrated in Example 7.7.

For a typical polyprotic acid in water, only the first dissociation step is important in determining the pH.

EXAMPLE 7.7

Calculate the pH of a 5.0 *M* H_3PO_4 solution and determine equilibrium concentrations of the species H_3PO_4, $H_2PO_4^-$, HPO_4^{2-}, and PO_4^{3-}.

Solution

The major species in solution are

$$H_3PO_4 \quad \text{and} \quad H_2O$$

None of the dissociation products of H_3PO_4 is written, since the K_a values are all so small that they will be minor species. The dominant equilibrium will be the dissociation of H_3PO_4:

$$H_3PO_4(aq) \rightleftharpoons H^+(aq) + H_2PO_4^-(aq)$$

where $$K_{a_1} = 7.5 \times 10^{-3} = \frac{[H^+][H_2PO_4^-]}{[H_3PO_4]}$$

The concentrations are as follows:

Initial Concentration (mol/L)		Equilibrium Concentration (mol/L)
$[H_3PO_4]_0 = 5.0$	x mol/L H_3PO_4	$[H_3PO_4] = 5.0 - x$
$[H_2PO_4^-]_0 = 0$	$\xrightarrow{}$	$[H_2PO_4^-] = x$
$[H^+]_0 \approx 0$	dissociates	$[H^+] = x$

Substituting the equilibrium concentrations into the expression for K_{a_1} and making the usual approximation gives

$$K_{a_1} = 7.5 \times 10^{-3} = \frac{[H^+][H_2PO_4^-]}{[H_3PO_4]} = \frac{(x)(x)}{5.0 - x} \approx \frac{x^2}{5.0}$$

Thus

$$x \approx 1.9 \times 10^{-1}$$

Since 1.9×10^{-1} is less than 5% of 5.0, the approximation is acceptable, and

$$[H^+] = x = 0.19\ M, \qquad pH = 0.72$$

So far we have determined that

$$[H^+] = [H_2PO_4^-] = 0.19\ M$$

and $$[H_3PO_4] = 5.0 - x = 4.8\ M$$

The concentration of HPO_4^{2-} can be obtained by using the expression for K_{a_2}:

$$K_{a_2} = 6.2 \times 10^{-8} = \frac{[H^+][HPO_4^{2-}]}{[H_2PO_4^-]}$$

where $$[H^+] = [H_2PO_4^-] = 0.19\ M$$

Thus $$[HPO_4^{2-}] = K_{a_2} = 6.2 \times 10^{-8}\ M$$

To calculate $[PO_4^{3-}]$, we use the expression for K_{a_3} and the values of $[H^+]$ and $[HPO_4^{2-}]$ calculated previously:

$$K_{a_3} = \frac{[H^+][PO_4^{3-}]}{[HPO_4^{2-}]} = 4.8 \times 10^{-13} = \frac{0.19[PO_4^{3-}]}{6.2 \times 10^{-8}}$$

$$[PO_4^{3-}] = \frac{(4.8 \times 10^{-13})(6.2 \times 10^{-8})}{0.19} = 1.6 \times 10^{-19}\ M$$

These results show that the second and third dissociation steps do not make an important contribution to $[H^+]$. This is apparent from the fact that $[HPO_4^{2-}]$ is $6.2 \times 10^{-8}\ M$, indicating that only 6.2×10^{-8} mol/L of $H_2PO_4^-$ has dissociated. The value of $[PO_4^{3-}]$ shows that the dissociation of HPO_4^{2-} is even smaller. We must, however, use the second and third dissociation steps to calculate $[HPO_4^{2-}]$ and $[PO_4^{3-}]$, since these steps are the only sources of these ions.

Sulfuric acid is commonly used to make fertilizers.

Sulfuric Acid

Sulfuric acid is unique among the common acids because it is *a strong acid in its first dissociation step and a weak acid in its second step:*

$$H_2SO_4(aq) \longrightarrow H^+(aq) + HSO_4^-(aq) \qquad K_{a_1} \text{ is very large}$$

$$HSO_4^-(aq) \rightleftharpoons H^+(aq) + SO_4^{2-}(aq) \qquad K_{a_2} = 1.2 \times 10^{-2}$$

Example 7.8 illustrates how to calculate the pH for sulfuric acid solutions.

EXAMPLE 7.8

Calculate the pH of a 1.0 M H_2SO_4 solution.

Solution

The major species in the solution are

$$H^+, \quad HSO_4^-, \quad H_2O$$

where the first two ions are produced by the complete first dissociation step of H_2SO_4. The concentration of H^+ in this solution will be at least 1.0 M, since this amount is produced by the first dissociation step of H_2SO_4. We must now answer this question: "Does the HSO_4^- ion dissociate enough to make a significant contribution to the concentration of H^+?" This question can be answered by calculating the equilibrium concentrations for the dissociation reaction of HSO_4^-:

$$HSO_4^-(aq) \rightleftharpoons H^+(aq) + SO_4^{2-}(aq)$$

where $$K_{a_2} = 1.2 \times 10^{-2} = \frac{[H^+][SO_4^{2-}]}{[HSO_4^-]}$$

The concentrations are as follows:

Initial Concentration (mol/L)		Equilibrium Concentration (mol/L)
$[HSO_4^-]_0 = 1.0$ $[SO_4^{2-}]_0 = 0$ $[H^+]_0 = 1.0$	x mol/L HSO_4^- dissociates → to reach equilibrium	$[HSO_4^-] = 1.0 - x$ $[SO_4^{2-}] = x$ $[H^+] = 1.0 + x$

Note that $[H^+]_0$ is not equal to zero, as is usually the case for a weak acid, because the first dissociation step has already produced some H^+.

Substituting the equilibrium concentrations into the expression for K_{a_2} and making the usual approximation gives

$$K_{a_2} = 1.2 \times 10^{-2} = \frac{[H^+][SO_4^{2-}]}{[HSO_4^-]} = \frac{(1.0 + x)(x)}{1.0 - x} \approx \frac{(1.0)(x)}{(1.0)}$$

Thus $$x \approx 1.2 \times 10^{-2}$$

Since 1.2×10^{-2} is 1.2% of 1.0, the approximation is valid according to the 5% rule. Note that x is not equal to $[H^+]$ in this case. Instead,

$$[H^+] = 1.0\ M + x = 1.0\ M + (1.2 \times 10^{-2})\ M$$
$$= 1.0\ M \quad \text{(to the correct number of significant figures)}$$

Thus since the dissociation of HSO_4^- does not make a significant contribution to the concentration of H^+,

$$[H^+] = 1.0\ M \quad \text{and} \quad pH = 0.00$$

Only in dilute H_2SO_4 solutions does the second dissociation step contribute significantly to $[H^+]$.

Example 7.8 illustrates the most common case for sulfuric acid in which only the first dissociation makes an important contribution to the concentration of H^+. In solutions more dilute than 1.0 M (for example, 0.10 M H_2SO_4), the dissociation of HSO_4^- is important. Solving this type of problem requires use of the quadratic formula, as shown in Example 7.9.

EXAMPLE 7.9

Calculate the pH of a 1.00×10^{-2} M H_2SO_4 solution.

Solution

The major species in solution are

$$H^+, \quad HSO_4^-, \quad H_2O$$

Proceeding as in Example 7.8, we consider the dissociation of HSO_4^-, which leads to the following concentrations:

Initial Concentration (mol/L)		Equilibrium Concentration (mol/L)
$[HSO_4^-]_0 = 0.0100$	x mol/L HSO_4^- dissociates → to reach equilibrium	$[HSO_4^-] = 0.0100 - x$
$[SO_4^{2-}]_0 = 0$		$[SO_4^{2-}] = x$
$[H^+]_0 = 0.0100$ ↑ From dissociation of H_2SO_4		$[H^+] = 0.0100 + x$

Substituting the equilibrium concentrations into the expression for K_{a_2} gives

$$1.2 \times 10^{-2} = K_{a_2} = \frac{[H^+][SO_4^{2-}]}{[HSO_4^-]} = \frac{(0.0100 + x)(x)}{(0.0100 - x)}$$

If we make the usual approximation, then $0.010 + x \approx 0.010$ and $0.010 - x \approx 0.010$; and we have

$$1.2 \times 10^{-2} = \frac{(0.0100 + x)(x)}{(0.0100 - x)} \approx \frac{(0.0100)x}{(0.0100)}$$

The calculated value of x is

$$x = 1.2 \times 10^{-2} = 0.012$$

This value is larger than 0.010, clearly a ridiculous result. Thus we cannot make the usual approximation and must instead solve the quadratic equation. The expression

$$1.2 \times 10^{-2} = \frac{(0.0100 + x)(x)}{(0.0100 - x)}$$

leads to $(1.2 \times 10^{-2})(0.0100 - x) = (0.0100 + x)(x)$

$$(1.2 \times 10^{-4}) - (1.2 \times 10^{-2})x = (1.0 \times 10^{-2})x + x^2$$

$$x^2 + (2.2 \times 10^{-2})x - (1.2 \times 10^{-4}) = 0$$

This equation can be solved by using the quadratic formula,

$$x = \frac{-b \pm \sqrt{b^2 - 4ac}}{2a}$$

where $a = 1$, $b = 2.2 \times 10^{-2}$, and $c = -1.2 \times 10^{-4}$. Use of the quadratic formula gives one negative root (which cannot be correct) and one positive root,

$$x = 4.5 \times 10^{-3}$$

Thus $[H^+] = 0.0100 + x = 0.0100 + 0.0045 = 0.0145$

and $\text{pH} = 1.84$

Note that in this case the second dissociation step produces about half as many H^+ ions as the initial step does.

This problem can also be solved by successive approximations, a method illustrated in Appendix Section A1.4.

Summary: **Characteristics of Weak Polyprotic Acids**

- **Typically, successive K_a values are so much smaller than the first value that only the first dissociation step makes a significant contribution to the equilibrium concentration of H^+. This means that the calculation of the pH for a solution of a weak polyprotic acid is identical to that for a solution of a weak monoprotic acid.**
- **Sulfuric acid is unique in being a strong acid in its first dissociation step and a weak acid in its second step. For relatively concentrated solutions of sulfuric acid (1.0 *M* or higher), the large concentration of H^+ from the first dissociation step represses the second step, which can be neglected as a contributor of H^+ ions. For dilute solutions of sulfuric acid the second step does make a significant contribution, and must be considered in obtaining the total H^+ concentration.**

7.8 Acid-Base Properties of Salts

The term **salt** is often used by chemists as simply another name for *ionic compound*. When a salt dissolves in water, we assume that it breaks up into its ions, which move about independently, at least in dilute solutions. Under certain conditions these ions can behave as acids or bases. In this section we explore such reactions.

Salts That Produce Neutral Solutions

Recall that the conjugate base of a strong acid has virtually no affinity for protons as compared with that of the water molecule. For this reason strong acids completely dissociate in aqueous solution. Thus when anions such as Cl^- and NO_3^- are placed in water, they do not combine with H^+ and therefore have no effect on the pH. Cations such as K^+ and Na^+ from strong bases have no affinity for H^+ and no ability to produce H^+, so they too have no effect on the pH of an aqueous solution. *Salts that consist of the cations of strong bases and the anions of strong acids have no effect on* $[H^+]$ *when dissolved in water*. This means that aqueous solutions of salts such as KCl, NaCl, $NaNO_3$, and KNO_3 are neutral (have a pH of 7).

The salt of a strong acid and a strong base gives a neutral solution.

Salts That Produce Basic Solutions

In an aqueous solution of sodium acetate ($NaC_2H_3O_2$), the major species are

$$Na^+, \quad C_2H_3O_2^-, \quad H_2O$$

What are the acid-base properties of each component? The Na^+ ion has neither acid nor base properties. The $C_2H_3O_2^-$ ion is the conjugate base of acetic acid, a weak acid. This means that $C_2H_3O_2^-$ has a significant affinity for a proton and acts as a base. Finally, water is neither a strong enough acid or base to affect $[H^+]$.

Thus the pH of this solution will be controlled by the $C_2H_3O_2^-$ ion. Since $C_2H_3O_2^-$ is a base, it will react with the best proton donor available. In this case water is the *only* source of protons, and the reaction is:

$$C_2H_3O_2^-(aq) + H_2O(l) \rightleftharpoons HC_2H_3O_2(aq) + OH^-(aq) \qquad (7.5)$$

Note that this reaction, which yields a basic solution, involves a *base reacting with water to produce the hydroxide ion and a conjugate acid*. We have defined K_b as the equilibrium constant for such a reaction. In this case

$$K_b = \frac{[HC_2H_3O_2][OH^-]}{[C_2H_3O_2^-]}$$

The value of K_a for acetic acid is well known (1.8×10^{-5}). But how can we obtain the K_b value for the acetate ion? The answer lies in the relationships among K_a, K_b, and K_w. Note that when the K_a expression for acetic acid is multiplied by the K_b expression for the acetate ion, the result is K_w:

$$K_a \times K_b = \frac{[H^+][C_2H_3O_2^-]}{[HC_2H_3O_2]} \times \frac{[HC_2H_3O_2][OH^-]}{[C_2H_3O_2^-]} = [H^+][OH^-] = K_w$$

This is a very important result. For any weak acid and its conjugate base,

$$K_a \times K_b = K_w$$

Thus when either K_a or K_b is known, the other constant can be calculated. For the acetate ion,

$$K_b = \frac{K_w}{K_a \text{ (for } HC_2H_3O_2)} = \frac{1.0 \times 10^{-14}}{1.8 \times 10^{-5}} = 5.6 \times 10^{-10}$$

This is the K_b value for the reaction described by Equation (7.5). Note that it is obtained from the K_a value of the parent weak acid, in this case acetic acid.

The sodium acetate solution is an example of an important general case. *For any salt whose cation has neutral properties (such as* Na^+ *or* K^+*) and whose anion is the conjugate base of a weak acid, the aqueous solution will be basic.* The K_b value for the anion can be obtained from the relationship $K_b = K_w/K_a$. Equilibrium calculations of this type are illustrated in Example 7.10.

A basic solution is formed if the anion of the salt is the conjugate base of a weak acid.

EXAMPLE 7.10

Calculate the pH of a 0.30 *M* NaF solution. The K_a value for HF is 7.2×10^{-4}.

Solution

The major species in solution are

$$Na^+, \quad F^-, \quad H_2O$$

Since HF is a weak acid, the F^- ion must have a significant affinity for protons. Therefore the dominant reaction will be

$$F^-(aq) + H_2O(l) \rightleftharpoons HF(aq) + OH^-(aq)$$

which yields the K_b expression

$$K_b = \frac{[HF][OH^-]}{[F^-]}$$

The value of K_b can be calculated from K_w and the K_a value for HF:

$$K_b = \frac{K_w}{K_a \text{ (for HF)}} = \frac{1.0 \times 10^{-14}}{7.2 \times 10^{-4}} = 1.4 \times 10^{-11}$$

The concentrations are as follows:

Initial Concentration (mol/L)		Equilibrium Concentration (mol/L)
$[F^-]_0 = 0.30$ $[HF]_0 = 0$ $[OH^-]_0 \approx 0$	x mol/L F^- reacts with H_2O to reach equilibrium $\longrightarrow$	$[F^-] = 0.30 - x$ $[HF] = x$ $[OH^-] = x$

Thus

$$K_b = 1.4 \times 10^{-11} = \frac{[HF][OH^-]}{[F^-]} = \frac{(x)(x)}{0.30 - x} \approx \frac{x^2}{0.30}$$

$$x \approx 2.0 \times 10^{-6}$$

The approximation is valid by the 5% rule, so

$$[OH^-] = x = 2.0 \times 10^{-6}\ M$$

$$pOH = 5.69$$

$$pH = 14.00 - 5.69 = 8.31$$

As expected, the solution is basic.

Base Strength in Aqueous Solution

To emphasize the concept of base strength, let us consider the basic properties of the cyanide ion. One relevant reaction is the dissociation of hydrocyanic acid in water:

$$HCN(aq) + H_2O(l) \rightleftharpoons H_3O^+(aq) + CN^-(aq) \qquad K_a = 6.2 \times 10^{-10}$$

Since HCN is such a weak acid, CN^- appears to be a *strong* base, showing a very high affinity for H^+ *compared with* H_2O, with which it is competing. However, we also need to look at the reaction in which cyanide ion reacts with water:

$$CN^-(aq) + H_2O(l) \rightleftharpoons HCN(aq) + OH^-(aq)$$

where

$$K_b = \frac{K_w}{K_a} = \frac{1.0 \times 10^{-14}}{6.2 \times 10^{-10}} = 1.6 \times 10^{-5}$$

In this reaction CN^- appears to be a weak base; the K_b value is only 1.6×10^{-5}. What accounts for this apparent difference in base strength? The key idea is that in the reaction of CN^- with H_2O, *CN^- is competing with OH^- for H^+, instead of competing with H_2O,* as it does in the HCN dissociation reaction. These equilibria show the following relative base strengths:

$$OH^- > CN^- > H_2O$$

Similar arguments can be made for other "weak" bases, such as ammonia, the acetate ion, and the fluoride ion.

Salts That Produce Acidic Solutions

Some salts produce acidic solutions when dissolved in water. For example, when solid NH_4Cl is dissolved in water, NH_4^+ and Cl^- ions are released, with NH_4^+ behaving as a weak acid:

$$NH_4^+(aq) \rightleftharpoons NH_3(aq) + H^+(aq)$$

The Cl^- ion, having virtually no affinity for H^+ in water, does not affect the pH of the solution.

In general, *a salt whose cation is the conjugate acid of a weak base produces an acidic solution.*

A second type of salt that produces an acidic solution is one that contains a *highly charged metal ion.* For example, when solid aluminum chloride ($AlCl_3$) is dissolved in water, the resulting solution is significantly acidic. Although the Al^{3+} ion is not itself a Brønsted-Lowry acid, the hydrated ion $Al(H_2O)_6^{3+}$ formed in water is a weak acid:

$$Al(H_2O)_6^{3+}(aq) \rightleftharpoons Al(OH)(H_2O)_5^{2+}(aq) + H^+(aq)$$

The high charge on the metal ion polarizes the O—H bonds in the attached water molecules, making the hydrogens in these water molecules more acidic than those in free water molecules. Typically, the higher the charge on the metal ion, the stronger the acidity of the hydrated ion is.

EXAMPLE 7.11

Calculate the pH of a 0.10 *M* NH_4Cl solution. The K_b value for NH_3 is 1.8×10^{-5}.

Solution

The major species in solution are

$$NH_4^+, \quad Cl^-, \quad H_2O$$

Note that both NH_4^+ and H_2O can produce H^+. The dissociation reaction for the NH_4^+ ion is

$$NH_4^+(aq) \rightleftharpoons NH_3(aq) + H^+(aq)$$

for which

$$K_a = \frac{[NH_3][H^+]}{[NH_4^+]}$$

Note that although the K_b value for NH_3 is given, the reaction corresponding to K_b is not appropriate here, since NH_3 is not a major species in the solution. Instead, the given value of K_b is used to calculate K_a for NH_4^+ from the relationship

$$K_a \times K_b = K_w$$

Thus

$$K_a \text{ (for } NH_4^+) = \frac{K_w}{K_b \text{ (for } NH_3)} = \frac{1.0 \times 10^{-14}}{1.8 \times 10^{-5}} = 5.6 \times 10^{-10}$$

Although NH_4^+ is a very weak acid, as indicated by its K_a value, it is stronger than H_2O and thus will dominate in the production of H^+. Thus we will focus on the dissociation reaction of NH_4^+ to calculate the pH of this solution.

We solve the weak acid problem in the usual way:

Initial Concentration (mol/L)		Equilibrium Concentration (mol/L)
$[NH_4^+]_0 = 0.10$ $[NH_3]_0 = 0$ $[H^+]_0 \approx 0$	x mol/L NH_4^+ dissociates → to reach equilibrium	$[NH_4^+] = 0.10 - x$ $[NH_3] = x$ $[H^+] = x$

Thus

$$5.6 \times 10^{-10} = K_a = \frac{[H^+][NH_3]}{[NH_4^+]} = \frac{(x)(x)}{0.10 - x} \approx \frac{x^2}{0.10}$$

$$x \approx 7.5 \times 10^{-6}$$

The approximation is valid by the 5% rule, so

$$[H^+] = x = 7.5 \times 10^{-6}\ M \quad \text{and} \quad pH = 5.12$$

EXAMPLE 7.12

Calculate the pH of a 0.010 M $AlCl_3$ solution. The K_a value for $Al(H_2O)_6^{3+}$ is 1.4×10^{-5}.

Solution

The major species in solution are

$$Al(H_2O)_6^{3+}, \quad Cl^-, \quad H_2O$$

Since the $Al(H_2O)_6^{3+}$ ion is a stronger acid than water, the dominant equilibrium is

$$Al(H_2O)_6^{3+}(aq) \rightleftharpoons Al(OH)(H_2O)_5^{2+}(aq) + H^+(aq)$$

and

$$1.4 \times 10^{-5} = K_a = \frac{[Al(OH)(H_2O)_5^{2+}][H^+]}{[Al(H_2O)_6^{3+}]}$$

This is a typical weak acid problem, which we can solve with the usual procedures.

Initial Concentration (mol/L)		Equilibrium Concentration (mol/L)
$[Al(H_2O)_6^{3+}]_0 = 0.010$	x mol/L $\longrightarrow$	$[Al(H_2O)_6^{3+}] = 0.010 - x$
$[Al(OH)(H_2O)_5^{2+}]_0 = 0$	$Al(H_2O)_6^{3+}$ dissociates	$[Al(OH)(H_2O)_5^{2+}] = x$
$[H^+]_0 \approx 0$	to reach equilibrium	$[H^+] = x$

Thus

$$1.4 \times 10^{-5} = K_a = \frac{[Al(OH)(H_2O)_5^{2+}][H^+]}{[Al(H_2O)_6^{3+}]} = \frac{(x)(x)}{0.010 - x} \approx \frac{x^2}{0.010}$$

$$x \approx 3.7 \times 10^{-4}$$

Since the approximation is valid by the 5% rule,

$$[H^-] = x = 3.7 \times 10^{-4}\ M \quad \text{and} \quad pH = 3.43$$

So far, we have considered salts containing only one ion that has acidic or basic properties. For many salts, such as ammonium acetate ($NH_4C_2H_3O_2$), both ions affect the pH of the aqueous solution. First, we will consider the qualitative aspects of such problems. We can predict whether the solution will be basic, acidic, or neutral by comparing the K_a value for the acidic ion with the K_b value for the basic ion. If the K_a value for the acidic ion is larger than the K_b value for the basic ion, the solution will be acidic. If the K_b value is larger than the K_a value, the solution is basic. Equal K_a and K_b values mean a neutral solution. These facts are summarized in Table 7.5. Table 7.6 summarizes the acid-base properties of aqueous solutions of various salts.

TABLE 7.5
Qualitative Prediction of pH for Solutions of Salts for Which Both Cation and Anion Have Acidic or Basic Properties

$K_a > K_b$	pH < 7 (acidic)
$K_b > K_a$	pH > 7 (basic)
$K_a = K_b$	pH = 7 (neutral)

EXAMPLE 7.13

Predict whether an aqueous solution of each of the following salts will be acidic, basic, or neutral.

a. $NH_4C_2H_3O_2$ b. NH_4CN c. $Al_2(SO_4)_3$

Solution

a. The ions in solution are NH_4^+ and $C_2H_3O_2^-$. As we mentioned previously, K_a for NH_4^+ is 5.6×10^{-10}, and K_b for $C_2H_3O_2^-$ is 5.6×10^{-10}. Thus since K_a for NH_4^+ is equal to K_b for $C_2H_3O_2^-$, the solution will be neutral (pH = 7).

b. The solution will contain NH_4^+ and CN^- ions. The K_a value for NH_4^+ is 5.6×10^{-10}, and

$$K_b \text{ (for } CN^-) = \frac{K_w}{K_a \text{ (for HCN)}} = 1.6 \times 10^{-5}$$

Since K_b for CN^- is much larger than K_a for NH_4^+, this solution will be basic.

c. The solution will contain $Al(H_2O)_6^{3+}$ and SO_4^{2-} ions. The K_a value for $Al(H_2O)_6^{3+}$ is 1.4×10^{-5}, as given in Example 7.12. We must calculate K_b for SO_4^{2-}. The HSO_4^- ion is the conjugate acid of SO_4^{2-}, and its K_a value is K_{a_2} for sulfuric acid, or 1.2×10^{-2}. Therefore,

$$K_b \text{ (for } SO_4^{2-}) = \frac{K_w}{K_{a_2} \text{ (for sulfuric acid)}}$$

$$= \frac{1.0 \times 10^{-14}}{1.2 \times 10^{-2}} = 8.3 \times 10^{-13}$$

This solution will be acidic, since K_a for $Al(H_2O)_6^{3+}$ is much greater than K_b for SO_4^{2-}.

TABLE 7.6 Acid-Base Properties of Aqueous Solutions of Various Types of Salts

Type of Salt	Examples	Comment	pH of Solution
Cation is from strong base; anion is from strong acid	KCl, KNO_3, NaCl, $NaNO_3$	Neither acts as an acid or a base	Neutral
Cation is from strong base; anion is from weak acid	$NaC_2H_3O_2$, KCN, NaF	Anion acts as a base; cation has no effect on pH	Basic
Cation is conjugate acid of weak base; anion is from strong acid	NH_4Cl, NH_4NO_3	Cation acts as acid; anion has no effect on pH	Acidic
Cation is conjugate acid of weak base; anion is conjugate base of weak acid	$NH_4C_2H_3O_2$, NH_4CN	Cation acts as an acid; anion acts as a base	Acidic if $K_a > K_b$, basic if $K_b > K_a$, neutral if $K_a = K_b$
Cation is highly charged metal ion; anion is from strong acid	$Al(NO_3)_3$, $FeCl_3$	Hydrated cation acts as an acid; anion has no effect on pH	Acidic

We have seen that it is possible to make a qualitative prediction of the acidity/basicity of an aqueous solution containing a dissolved salt. We also can give a quantitative description of these solutions by using the procedures we have developed for treating acid-base equilibria. Example 7.14 illustrates this technique.

EXAMPLE 7.14

Calculate the pH of a 0.100 M solution of NH_4CN.

Solution

The major species in solution are

$$NH_4^+, \quad CN^-, \quad H_2O$$

The familiar reactions involving these species are

$$NH_4^+(aq) \rightleftharpoons NH_3(aq) + H^+(aq) \qquad K_a = 5.6 \times 10^{-10}$$

$$CN^-(aq) + H_2O(l) \rightleftharpoons HCN(aq) + OH^-(aq) \qquad K_b = 1.6 \times 10^{-5}$$

$$H_2O(l) + H_2O(l) \rightleftharpoons H_3O^+(aq) + OH^-(aq) \qquad K_w = 1.0 \times 10^{-14}$$

However, noting that NH_4^+ is an acid and CN^- is a base, it is also sensible to consider the reaction

$$NH_4^+(aq) + CN^-(aq) \rightleftharpoons HCN(aq) + NH_3(aq)$$

To evaluate the importance of this reaction, we need the value of its equilibrium constant. Note that

$$\frac{[NH_3][HCN]}{[NH_4^+][CN^-]} = \frac{[H^+][NH_3]}{[NH_4^+]} \times \frac{[HCN]}{[H^+][CN^-]}$$

$$= K_a(NH_4^+) \times \frac{1}{K_a\ (HCN)}$$

Thus the value of the equilibrium constant we need is

$$K = \frac{K_a\ (NH_4^+)}{K_a\ (HCN)} = \frac{5.6 \times 10^{-10}}{6.2 \times 10^{-10}} = 0.90$$

Notice that this equilibrium constant is much larger than those for the other possible reactions. Thus we expect this reaction to be dominant in this solution.

Following the usual procedures, we have the concentrations listed below.

Initial Concentration (mol/L)	Equilibrium Concentration (mol/L)
$[NH_4^+]_0 = 0.100$	$[NH_4^+] = 0.100 - x$
$[CN^-]_0 = 0.100$	$[CN^-] = 0.100 - x$
$[NH_3]_0 = 0$	$[NH_3] = x$
$[HCN]_0 = 0$	$[HCN] = x$

Then $$K = 0.90 = \frac{x^2}{(0.100 - x)^2}$$

Taking the square root of both sides yields

$$0.95 = \frac{x}{0.100 - x}$$

and

$$x = 4.9 \times 10^{-2}\ M = [NH_3] = [HCN]$$

Notice that the reaction under consideration does not involve H^+ or OH^- directly. Thus to obtain the pH, we must consider the position of the HCN or NH_4^+ dissociation equilibrium. For example, for HCN

$$K_a = 6.2 \times 10^{-10} = \frac{[H^+][CN^-]}{[HCN]}$$

From the above calculations

$$[CN^-] = 0.100 - x = 0.100 - 0.049 = 0.051\ M$$
$$[HCN] = x = 4.9 \times 10^{-2}\ M$$

Substituting these values into the K_a expression for HCN gives

$$[H^+] = 6.0 \times 10^{-10}\ M$$

and

$$pH = 9.22$$

Note that this solution is basic, just as we predicted in Example 7.13.

7.9 Acid Solutions in Which Water Contributes to the H^+ Concentration

In most solutions containing an acid, we can assume that the acid dominates in the production of H^+ ions. That is, we typically can assume that the acid produces so much H^+ in comparison with the H^+ produced by water that water can be ignored as a source of H^+. However, in certain cases water must be taken into account when the pH of an aqueous solution is calculated. For example, consider a $1.0 \times 10^{-4}\ M$ solution of a very weak acid HA ($K_a = 1.0 \times 10^{-10}$). A quick calculation shows that if water is ignored, the $[H^+]$ produced by this acid is $1.0 \times 10^{-7}\ M$. This value cannot be the correct $[H^+]$ in this solution at equilibrium, because in pure water $[H^+] = 1.0 \times 10^{-7}\ M$. In this case perhaps the thing to do to get the total $[H^+]$ is to add the two H^+ concentrations:

$$\underset{\substack{\uparrow \\ \text{From } 10^{-4}\ M\ \text{HA}}}{1.0 \times 10^{-7}\ M} + \underset{\substack{\uparrow \\ \text{From } H_2O}}{1.0 \times 10^{-7}\ M} = 2.0 \times 10^{-7}\ M$$

In the typical case involving a weak acid HA in water, the $[H^+]$ produced by the acid is much greater than that produced by water, so that

$$[H^+] = [H^+]_{HA} + [H^+]_{H_2O} = [H^+]_{HA}$$

The HA and H_2O are simultaneously in equilibrium, but in this case we can ignore water as a source of H^+.

However, this procedure is not correct because the two sources of H^+ will affect each other. That is, because both of the reactions

$$H_2O(l) \rightleftharpoons H^+(aq) + OH^-(aq)$$
$$HA(aq) \rightleftharpoons H^+(aq) + A^-(aq)$$

involve H^+, the equilibrium position of each will be affected by the other. Thus we must solve these equilibria simultaneously. This procedure will lead to a concentration of H^+ such that

$$1.0 \times 10^{-7}\ M < [H^+] < 2.0 \times 10^{-7}\ M$$

Note that in these two equilibria there are four unknown concentrations:

$$[H^+], \quad [OH^-], \quad [HA], \quad [A^-]$$

To solve for these concentrations, we need four independent equations that relate them. Two of these equations are provided by the two equilibrium expressions:

$$K_w = [H^+][OH^-] \quad \text{and} \quad K_a = \frac{[H^+][A^-]}{[HA]}$$

A third equation can be derived from the **principle of charge balance:** the positive and negative charges carried by the ions in an aqueous solution must balance. That is, the "concentration of positive charge" must equal the "concentration of negative charge." In this case H^+ is the only positive ion, and the negative ions are A^- and OH^-. Thus the *charge balance* expression is

$$[H^+] = [A^-] + [OH^-]$$

Another relationship can be obtained by recognizing that all of the HA originally dissolved must be present at equilibrium as either A^- or HA. This observation leads to the equation

This expression conserves A.

$$[HA]_0 = [HA] + [A^-]$$

↑ Original concentration of HA dissolved

which is called the **material balance equation.**

These four equations can be used to derive an equation involving only $[H^+]$. We will start with the K_a expression,

$$K_a = \frac{[H^+][A^-]}{[HA]}$$

and use the other relationships to express $[A^-]$ and $[HA]$ in terms of $[H^+]$. Recall that the charge balance equation is

$$[H^+] = [A^-] + [OH^-]$$

Using the K_w expression, we have

$$[OH^-] = \frac{K_w}{[H^+]}$$

and the charge balance equation becomes

$$[H^+] = [A^-] + \frac{K_w}{[H^+]} \quad \text{or} \quad [A^-] = [H^+] - \frac{K_w}{[H^+]}$$

This equation gives $[A^-]$ in terms of $[H^+]$.

The material balance equation is

$$[HA]_0 = [HA] + [A^-] \quad \text{or} \quad [HA] = [HA]_0 - [A^-]$$

Since

$$[A^-] = [H^+] - \frac{K_w}{[H^+]}$$

we have

$$[HA] = [HA]_0 - \left([H^+] - \frac{K_w}{[H^+]}\right)$$

Now we substitute the expressions for $[A^-]$ and $[HA]$ into the K_a expression:

$$K_a = \frac{[H^+][A^-]}{[HA]} = \frac{[H^+]\left([H^+] - \dfrac{K_w}{[H^+]}\right)}{[HA]_0 - \left([H^+] - \dfrac{K_w}{[H^+]}\right)} = \frac{[H^+]^2 - K_w}{[HA]_0 - \dfrac{[H^+]^2 - K_w}{[H^+]}}$$

This expression permits the calculation of the $[H^+]$ in a solution containing a weak acid. That is, it gives the correct $[H^+]$ for any solution made by dissolving a weak acid in pure water.

The equation can be solved by simple trial and error or by the more systematic method of successive approximations. Recall that the usual way of doing successive approximations is to substitute a guessed value of the variable of interest ($[H^+]$ in this case) into the equation everywhere it appears except in one place. The equation is then solved to obtain a new value of the variable, which becomes the "guessed value" in the next round. The process is continued until the calculated value equals the guessed value.

See Appendix 1 (Section A1.4) for information on using successive approximations.

Even though the full equation can be solved in this manner, the process is tedious and time-consuming. We would certainly like to use the simpler method (ignoring the contribution of water to the $[H^+]$) whenever possible. Thus a key question arises: "Under what conditions can problems involving a weak acid be done in the simple way?"

Notice that the term $[H^+]^2 - K_w$ appears twice in the full equation:

$$K_a = \frac{[H^+]^2 - K_w}{[HA]_0 - \dfrac{[H^+]^2 - K_w}{[H^+]}}$$

Now, assume that the condition

$$[H^+]^2 \gg K_w$$

applies, which means that

$$[H^+]^2 - K_w \approx [H^+]^2$$

Under this condition the full equation can be simplified as follows:

$$K_a = \frac{[H^+]^2 - K_w}{[HA]_0 - \dfrac{[H^+]^2 - K_w}{[H^+]}} \approx \frac{[H^+]^2}{[HA]_0 - \dfrac{[H^+]^2}{[H^+]}} = \frac{[H^+]^2}{[HA]_0 - [H^+]}$$

$$= \frac{x^2}{[HA]_0 - x}$$

where $x = [H^+]$ at equilibrium.

This is an important result: if $[H^+]^2 \gg K_w$, the full equation reduces to the typical expression for a weak acid, that originates from ignoring water as a source of H^+. Because uncertainties in K_a values are typically greater than 1%, we can safely assume that "much greater than" means at least 100 times greater. Then since $K_w = 1.0 \times 10^{-14}$, the $[H^+]^2$ must be at least 100×10^{-14}, or 10^{-12}, which corresponds to $[H^+] = 10^{-6}$. Thus if $[H^+]$ is greater than or equal to 10^{-6} *M*, the complicated equation reduces to the simple equation—that is, you get the *same answer* by using either equation.

How do we decide when we must use the complicated equation? The best way to proceed is as follows: calculate the $[H^+]$ in the normal way, ignoring any contribution from H_2O. If $[H^+]$ from this calculation is greater than or equal to 10^{-6} *M*, the answer is correct—that is, the complicated equation will give the same answer. If the $[H^+]$ calculated from the simple equation is less than 10^{-6} *M*, you must use the full equation—that is, water must be considered as a source of H^+.

EXAMPLE 7.15

Calculate the $[H^+]$ in

a. 1.0 *M* HCN ($K_a = 6.2 \times 10^{-10}$)

b. 1.0×10^{-4} *M* HCN ($K_a = 6.2 \times 10^{-10}$)

Solution

a. First, do the weak acid problem the "normal" way. This technique leads to the expression

$$\frac{x^2}{1.0 - x} = 6.2 \times 10^{-10} \approx \frac{x^2}{1.0}$$

$$x = 2.5 \times 10^{-5}\ M = [H^+]$$

Note that the $[H^+]$ from the dissociation of HCN is greater than 10^{-6} *M*, so we are finished. Water makes no important contribution to the $[H^+]$ in this solution.

b. First, do the weak acid problem the "normal" way. This procedure leads to the expression

$$K_a = 6.2 \times 10^{-10} = \frac{x^2}{1.0 \times 10^{-4} - x} \approx \frac{x^2}{1.0 \times 10^{-4}}$$

$$x \approx 2.5 \times 10^{-7}\ M$$

In this very dilute solution of HCN, the $[H^+]$ from HCN alone is less than 10^{-6} *M*, so the full equation must be used to obtain the correct $[H^+]$ in the solution:

$$6.2 \times 10^{-10} = K_a = \frac{[H^+]^2 - 10^{-14}}{1.0 \times 10^{-4} - \dfrac{[H^+]^2 - 10^{-14}}{[H^+]}}$$

We will now solve for $[H^+]$ by use of successive approximations. First we must determine a reasonable guess for $[H^+]$. Note from the above simple calculation that $[H^+]$, ignoring the contribution from water, is 2.5×10^{-7} *M*. Will the actual $[H^+]$ be larger or smaller than this value? It will be a little larger because of the contribution from H_2O. So a reasonable guess for $[H^+]$ is 3.0×10^{-7} *M*.

We now substitute this value for $[H^+]$ into the denominator of the equation, to give

$$K_a = 6.2 \times 10^{-10} = \frac{[H^+]^2 - 1.0 \times 10^{-14}}{1.0 \times 10^{-4} - \dfrac{(3.0 \times 10^{-7})^2 - 1.0 \times 10^{-14}}{3.0 \times 10^{-7}}}$$

$$6.2 \times 10^{-10} = \frac{[H^+]^2 - 1.0 \times 10^{-14}}{1.0 \times 10^{-4} - 2.67 \times 10^{-7}}$$

Now, rearrange this equation so that a value for $[H^+]$ can be calculated:

$$[H^+]^2 = 6.2 \times 10^{-14} - 1.66 \times 10^{-16} + 1.0 + 10^{-14}$$
$$= 7.2 \times 10^{-14}$$
$$[H^+] = \sqrt{7.2 \times 10^{-14}} = 2.68 \times 10^{-7}$$

Recall that the original guessed value of $[H^+]$ was 3.0×10^{-7}. Since the calculated value and the guessed value do not agree, use 2.68×10^{-7} as the new guessed value:

$$K_a = \frac{[H^+]^2 - 1.0 \times 10^{-14}}{1.0 \times 10^{-4} - \dfrac{(2.68 \times 10^{-7})^2 - 1.0 \times 10^{-14}}{2.68 \times 10^{-7}}}$$

Solving for $[H^+]$, we have

$$[H^+] = 2.68 \times 10^{-7} = 2.7 \times 10^{-7}\ M$$

Since the guessed value and the newly calculated value agree, this answer is correct, and it takes into account both contributors to $[H^+]$ (water and HCN). Thus for this solution

$$\text{pH} = -\log(2.7 \times 10^{-7}) = 6.57$$

As you followed the above procedure, you may have noticed that there was an opportunity to simplify the math, but we did not take advantage of it.

Note that the term

$$[HA]_0 - \frac{[H^+]^2 - K_w}{[H^+]}$$

occurs in the denominator of the overall equation. Because $[H^+]$ will be between 10^{-6} and 10^{-7} (otherwise, we would be using the simple equation), we can see that the value of the term

$$\frac{[H^+]^2 - K_w}{[H^+]}$$

will be between zero (if $[H^+] = 10^{-7}\,M$) and 10^{-6} (if $[H^+] = 10^{-6}\,M$). Thus if $[HA]_0 > 2 \times 10^{-5}\,M$ in a given acid solution, then

$$[HA]_0 - \frac{[H^+]^2 - K_w}{[H^+]} \approx [HA]_0$$

within the limits of the 5% rule. Under these conditions the equation

$$K_a = \frac{[H^+]^2 - K_w}{[HA]_0 - \frac{[H^+]^2 - K_w}{[H^+]}}$$

becomes

$$K_a \approx \frac{[H^+]^2 - K_w}{[HA]_0}$$

which can be readily solved for $[H^+]$:

Note that this equation gives the same $[H^+]$ for $1.0 \times 10^{-4}\,M$ HCN as the full equation.

$$[H^+] \approx \sqrt{K_a[HA]_0 + K_w}$$

This simplified equation applies to all cases except for very dilute weak acid solutions.

We will now summarize the conclusions of this section.

Summary: **The pH Calculations for an Aqueous Solution of a Weak Acid HA (major species HA and H_2O)**

- **The full equation for this case is**

$$K_a = \frac{[H^+]^2 - K_w}{[HA]_0 - \frac{[H^+]^2 - K_w}{[H^+]}}$$

- **When the weak acid by itself produces $[H^+] \geq 10^{-6}\,M$, the full equation becomes**

$$K_a = \frac{[H^+]^2}{[HA]_0 - [H^+]}$$

This corresponds to the typical weak acid case.

- **When** $$[HA]_0 \gg \frac{[H^+]^2 - K_w}{[H^+]}$$

the full equation becomes

$$K_a = \frac{[H^+]^2 - K_w}{[HA]_0}$$

which gives $[H^+] = \sqrt{K_a[HA]_0 + K_w}$

7.10 Strong Acid Solutions in Which Water Contributes to the H^+ Concentration

Although in a typical strong acid solution (for example, 0.1 M HCl) the $[H^+]$ is determined by the amount of strong acid present, there are circumstances where the contribution of water must be taken into account. For example, consider a $1.0 \times 10^{-7}\,M$ HNO_3 solution. For this very dilute solution the strong acid and the water make comparable contributions to $[H^+]$ at equilibrium. Because the H^+ from HNO_3 will affect the position of the water equilibrium, the total $[H^+]$ in this solution will not be simply $1.0 \times 10^{-7}\,M + 1.0 \times 10^{-7}\,M = 2.0 \times 10^{-7}\,M$. Rather, $[H^+]$ will be between $1.0 \times 10^{-7}\,M$ and $2.0 \times 10^{-7}\,M$. We can calculate the exact $[H^+]$ by using the principle of charge balance:

$$[\text{Positive charge}] = [\text{negative charge}]$$

In this case we have

$$[H^+] = [NO_3^-] + [OH^-]$$

which, from $K_w = [H^+][OH^-]$, can be written as

$$[H^+] = [NO_3^-] + \frac{K_w}{[H^+]} \quad \text{or} \quad \frac{[H^+]^2 - K_w}{[H^+]} = [NO_3^-]$$

The fact that the solution contains $1.0 \times 10^{-7}\,M$ HNO_3 means that $[NO_3^-] = 1.0 \times 10^{-7}\,M$. Inserting this value, we can solve the above equation to give $[H^+] = 1.6 \times 10^{-7}\,M$.

This approach applies for any strong acid (although it is unnecessary for more typical concentrations). It can also be adapted to calculate the pH of a very dilute strong base solution. (Try your hand at this problem by calculating the pH of a $5.0 \times 10^{-8}\,M$ KOH solution.)

7.11 Strategy for Solving Acid-Base Problems: A Summary

In this chapter we have encountered many different situations involving aqueous solutions of acids and bases, and in the next chapter we will encounter still more. In solving for the equilibrium concentrations in these aqueous solutions, you may be tempted to create a pigeonhole for each possible situation and to memorize the procedures necessary to deal with each particular case. This approach is just not practical and usually leads to frustration: too many pigeonholes are required, because there seem to be an infinite number of cases.

But you can handle any case successfully by taking a systematic, patient, and thoughtful approach. When analyzing an acid-base equilibrium problem, do *not* ask yourself how a memorized solution can be used to solve the problem. Instead, ask yourself this question: "*What are the major species in the solution, and how does each behave chemically?*"

The most important part of doing a complicated acid-base equilibrium problem is the analysis you do at the beginning of a problem:

Which major species are present?

Does a reaction occur that can be assumed to go to completion?

Which equilibrium dominates the solution?

Let the problem guide you. Be patient.

The following steps outline a general strategy for solving problems involving acid-base equilibria.

Summary: **Solving Acid-Base Problems**

- **List the major species in solution.**
- **Look for reactions that can be assumed to go to completion, such as a strong acid dissociating or H^+ reacting with OH^-.**
- **For a reaction that can be assumed to go to completion:**
  a. **Determine the concentrations of the products.**
  b. **Write down the major species in solution after the reaction.**
- **Look at each major component of the solution and decide whether it is an acid or a base.**
- **Pick the equilibrium that will control the pH. Use known values of the dissociation constants for the various species to determine the dominant equilibrium.**
  a. **Write the equation for the reaction and the equilibrium expression.**
  b. **Compute the initial concentrations (assuming that the dominant equilibrium has not yet occurred—for example, there has been no acid dissociation).**
  c. **Define *x*.**
  d. **Compute the equilibrium concentrations in terms of *x*.**
  e. **Substitute the concentrations into the equilibrium expression, and solve for *x*.**
  f. **Check the validity of the approximation.**
  g. **Calculate the pH and other concentrations as required.**

Although these procedures may seem somewhat cumbersome, especially for simpler problems, they will become increasingly helpful as the aqueous solutions become more complicated. If you develop the habit of approaching acid-base problems systematically, the more complex cases will be much easier to manage.

Exercises

A blue exercise number indicates that the answer to that exercise appears at the back of this book.

Nature of Acids and Bases

1. Classify each of the following as a strong acid, weak acid, strong base, or weak base in aqueous solution.
   a. HNO_2
   b. H_3PO_4
   c. CH_3NH_2
   d. NaOH
   e. NH_3
   f. HF
   g. $HC(=O)-OH$
   h. $H_2NCH_2CH_2NH_2$
   i. H_2SO_4
2. Use Table 7.2 to order the following from the strongest to the weakest acid:

   HNO_3, H_2O, HOCl, NH_4^+
3. Use Table 7.2 to order the following from the strongest to the weakest base:

   NO_3^-, H_2O, OCl^-, NH_3
4. You may need Table 7.2 to answer the following questions.
   a. Which is the stronger acid, HI or H_2O?
   b. Which is the stronger acid, H_2O or $HClO_2$?
   c. Which is the stronger acid, HF or HCN?
5. You may need Table 7.2 to answer the following questions.
   a. Which is the stronger base, I^- or H_2O?
   b. Which is the stronger base, H_2O or ClO_2^-?
   c. Which is the stronger base, F^- or CN^-?
6. Write the reaction and the corresponding K_b equilibrium expression for each of the following substances (acting as bases in water).
   a. PO_4^{3-}
   b. HPO_4^{2-}
   c. $H_2PO_4^-$
   d. NH_3
   e. CN^-
   f. pyridine, C_5H_5N
   g. glycine, $NH_2CH_2CO_2H$
   h. ethylamine, $CH_3CH_2NH_2$
   i. aniline, $C_6H_5NH_2$
   j. dimethylamine, $(CH_3)_2NH$
7. The hydride ion (H^-) and methoxide ion (CH_3O^-) have much greater affinities for H^+ than the OH^- ion does. Write equations for the reactions that occur when NaH and $NaOCH_3$ are dissolved in water.
8. Why is H_3O^+ the strongest acid and OH^- the strongest base that can exist in significant amounts in aqueous solutions?
9. Consider the reaction

   $$CH_3CO_2H(aq) + H_2O(l) \rightleftharpoons CH_3CO_2^-(aq) + H_3O^+(aq)$$

   where $K_a = 1.8 \times 10^{-5}$.
   a. Which two bases are competing for the proton?
   b. Which is the stronger base?
   c. In light of your answer to b, why do we classify the acetate ion ($CH_3CO_2^-$) as a weak base? Use an appropriate reaction to justify your answer.
10. In general, as base strength increases, conjugate acid strength decreases. Explain why the conjugate acid of the weak base NH_3 is a weak acid.

Autoionization of Water and pH Scale

11. Values of K_w as a function of temperature are as follows:

Temp (°C)	K_w
0	1.14×10^{-15}
25	1.00×10^{-14}
35	2.09×10^{-14}
40.	2.92×10^{-14}
50.	5.47×10^{-14}

   a. Is the autoionization of water exothermic or endothermic?
   b. What is the pH of pure water at 50.°C?
   c. From a plot of $\ln(K_w)$ versus $1/T$ (using the Kelvin scale), estimate K_w at 37°C, normal physiological temperature.
   d. What is the pH of a neutral solution at 37°C?
12. Supercritical water (water above the critical point of 218 atm and 374°C) has been studied as a medium for the disposal of hazardous wastes.
   a. Assuming pressure has no effect on K_w, use your graph from Exercise 11 to determine K_w for water at 374°C.
   b. How might supercritical water help in the disposal of hazardous wastes?
   c. Why is corrosion a greater problem in high-temperature, high-pressure steam lines than in hot water pipes?

Solutions of Acids

13. List the major species present in 0.250 *M* solutions of each of the following acids, and then calculate the pH for each.
   a. HCl
   b. HBr
   c. $HClO_4$
   d. HNO_3
   e. HNO_2
   f. CH_3CO_2H ($HC_2H_3O_2$)
   g. NH_4Cl
   h. HCN
14. Calculate the pH of 1.0×10^{-12} *M* HCl. Be sure your answer makes sense.
15. A solution is prepared by adding 50.0 mL of concentrated hydrochloric acid and 20.0 mL of concentrated nitric acid to 300 mL of water. More water is added until the final

volume is 1.00 L. Calculate $[H^+]$, $[OH^-]$, and the pH for this solution. [*Hint:* concentrated HCl is 38% HCl (by mass) and has a density of 1.19 g/mL; concentrated HNO_3 is 70.% HNO_3 and has a density of 1.42 g/mL.]

16. Using the K_a values given in Table 7.2, calculate the concentrations of all species present and the pH for each of the following.
    a. 0.20 *M* $HC_2H_3O_2$
    b. 1.5 *M* HNO_2
    c. 0.020 *M* HF
    d. 0.10 *M* lactic acid ($CH_3CH(OH)CO_2H$, $K_a = 1.38 \times 10^{-4}$)

17. Formic acid (HCO_2H) is secreted by ants. Calculate $[H^+]$ and the pH of a 0.025 *M* solution of formic acid ($K_a = 1.8 \times 10^{-4}$).

18. Calculate the pH of a 0.20 *M* solution of iodic acid (HIO_3, $K_a = 0.17$).

19. Boric acid (H_3BO_3) is commonly included in eyewash solutions in chemistry laboratories to neutralize bases splashed in the eye. It acts as a monoprotic acid, but the dissociation reaction is slightly different from that of other acids:

$$B(OH)_3 + H_2O \rightleftharpoons B(OH)_4^- + H^+ \qquad K_a = 5.8 \times 10^{-10}$$

Calculate the pH of a 0.50 *M* solution of boric acid.

20. A solution is prepared by dissolving 0.56 g of benzoic acid ($C_6H_5CO_2H$, $K_a = 6.4 \times 10^{-5}$) in enough water to make 1.0 L of solution. Calculate $[C_6H_5CO_2H]$, $[C_6H_5CO_2^-]$, $[H^+]$, $[OH^-]$, and the pH of this solution.

21. At 25°C a saturated solution of benzoic acid (see Exercise 20) has a pH of 2.80. Calculate the water solubility of benzoic acid in moles per liter and grams per 100. milliliters.

22. A 0.15 *M* solution of a weak acid is 3.0% dissociated. Calculate K_a.

23. A solution with a total volume of 250.0 mL is prepared by diluting 20.0 mL of glacial acetic acid with water. Calculate $[H^+]$ and the pH of this solution. Assume glacial acetic acid is pure liquid acetic acid with a density of 1.05 g/cm³.

24. Calculate the pH of each of the following.
    a. a solution containing 0.10 *M* HCl and 0.10 *M* HOCl
    b. a solution containing 0.050 *M* HNO_3 and 0.50 *M* $HC_2H_3O_2$

25. A solution is prepared by adding 50.0 mL of 0.050 *M* HCl to 150.0 mL of 0.10 *M* HNO_3. Calculate the concentrations of all species in this solution.

26. Calculate the pH of a 0.0010 *M* solution of H_2SO_4.

27. What types of measurements, other than pH measurements, can be made to determine the extent of dissociation of an acid in water?

28. In a 0.100 *M* solution of HF, the percent dissociation is 8.1%. Calculate K_a.

29. Calculate the percent dissociation of the acid in each of the following solutions.
    a. 0.50 *M* acetic acid
    b. 0.050 *M* acetic acid
    c. 0.0050 *M* acetic acid
    d. Use Le Châtelier's principle to explain why percent dissociation increases as the concentration of a weak acid decreases.
    e. Even though the percent dissociation increases from solutions a to c, the $[H^+]$ decreases. Explain.

30. Calculate the percent dissociation of a 0.22 *M* solution of chlorous acid ($HClO_2$, $K_a = 1.2 \times 10^{-2}$).

31. The pH of a 0.063 *M* solution of hypobromous acid (HOBr, but usually written HBrO) is 4.95. Calculate K_a.

32. Trichloroacetic acid (CCl_3CO_2H) is a corrosive acid that is used to precipitate proteins. The pH of a 0.050 *M* solution of trichloroacetic acid is 1.40. Calculate K_a.

33. Calculate $[H^+]$, $[OH^-]$, $[H_3PO_4]$, $[H_2PO_4^-]$, $[HPO_4^{2-}]$, and $[PO_4^{3-}]$ for a 0.100 *M* solution of H_3PO_4.

34. Calculate $[CO_3^{2-}]$ in a 0.010 *M* solution of CO_2 in water (H_2CO_3). If all the CO_3^{2-} in this solution comes from the reaction

$$HCO_3^- \rightleftharpoons H^+ + CO_3^{2-}$$

what percentage of the H^+ ions in the solution is a result of the dissociation of HCO_3^-? When acid is added to a solution of sodium hydrogen carbonate ($NaHCO_3$), vigorous bubbling occurs. How is this reaction related to the existence of carbonic acid (H_2CO_3) molecules in aqueous solution?

Solutions of Bases

35. Using Table 7.3, order the following bases from strongest to weakest:

$$NO_3^-, \quad H_2O, \quad NH_3, \quad CH_3NH_2$$

36. Using Table 7.3, order the following acids from strongest to weakest:

$$HNO_3, \quad H_2O, \quad NH_4^+, \quad CH_3NH_3^+$$

37. Using Table 7.3, answer the following questions.
    a. Which is the stronger base, NO_3^- or NH_3?
    b. Which is the stronger base, H_2O or NH_3?
    c. Which is the stronger base, OH^- or NH_3?
    d. Which is the stronger base, NH_3 or CH_3NH_2?

38. Using Table 7.3, answer the following questions.
    a. Which is the stronger acid, HNO_3 or NH_4^+?
    b. Which is the stronger acid, H_2O or NH_4^+?
    c. Which is the stronger acid, NH_4^+ or $CH_3NH_3^+$?

39. Thallium (I) hydroxide is a strong base used in the synthesis of some organic compounds. Calculate the pH of a solution containing 2.48 g of TlOH per liter.

40. Calculate $[OH^-]$, pOH, and pH for each of the following.
    a. 0.25 M NaOH
    b. 0.00040 M $Ba(OH)_2$
    c. a solution containing 25 g of KOH per liter
    d. a solution containing 150.0 g of NaOH per liter

41. The presence of what element in an organic compound most commonly results in basic properties?

42. Calculate $[OH^-]$, $[H^+]$, and pH of 0.200 M solutions of each of the following amines (the missing K_b values are found in Table 7.3).
    a. ethylamine
    b. diethylamine, $(C_2H_5)_2NH$, $K_b = 1.3 \times 10^{-3}$
    c. triethylamine, $(C_2H_5)_3N$, $K_b = 4.0 \times 10^{-4}$
    d. aniline
    e. pyridine
    f. hydroxylamine, $HONH_2$, $K_b = 1.1 \times 10^{-8}$

43. Draw the structures of the conjugate acids of ephedrine and mescaline. See Section 7.6 for the structures of the amines.

44. Codeine is a derivative of morphine and is used as an analgesic, narcotic, or antitussive. It was once commonly used in cough syrups but is now available only by prescription because of its addictive properties. The formula of codeine is $C_{18}H_{21}NO_3$ and the pK_b is 6.05. Calculate the pH of a 10.0-mL solution containing 5.0 mg of codeine.

45. A codeine-containing cough syrup lists codeine sulfate as a major ingredient instead of codeine. *The Merck Index* gives $C_{36}H_{44}N_2O_{10}S$ as the formula for codeine sulfate. Describe the composition of codeine sulfate. (See Exercise 44.) Why is codeine sulfate used instead of codeine?

46. What is the percent ionization in each of the following solutions?
    a. 0.10 M NH_3
    b. 0.010 M NH_3
    c. 0.10 M CH_3NH_2

47. For the reaction of hydrazine (N_2H_4) with water,

$$H_2NNH_2 + H_2O \rightleftharpoons H_2NNH_3^+ + OH^-$$

K_b is 3.0×10^{-6}. Calculate the pH of a 2.0 M aqueous solution of hydrazine.

48. Quinine ($C_{20}H_{24}N_2O_2$) is the most important alkaloid derived from cinchona bark. It is used as an antimalarial drug.

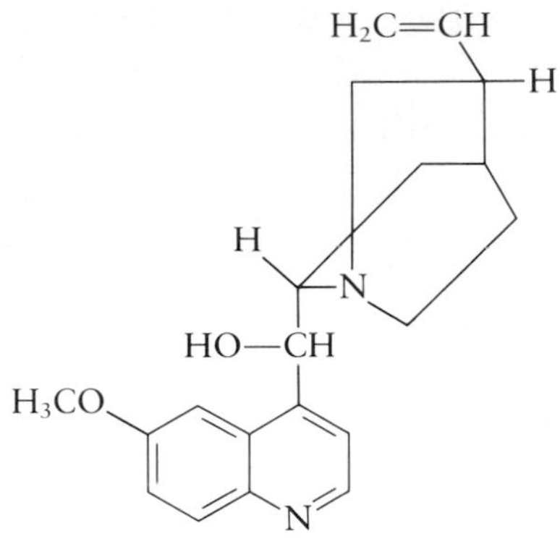

Quinine

For quinine $pK_{b_1} = 5.1$ and $pK_{b_2} = 9.7$. (Recall that $pK_b = -\log K_b$.) One gram of quinine will dissolve in 1900.0 mL of water. Calculate the pH of a saturated aqueous solution of quinine. Consider only the reaction $Q + H_2O \rightleftharpoons QH^+ + OH^-$ described by pK_{b_1}.

49. The pH of a 0.016 M aqueous solution of p-toluidine ($CH_3C_6H_4NH_2$) is 8.60. Calculate K_b.

50. The pH of a 1.00×10^{-3} M solution of pyrrolidine is 10.82. Calculate K_b.

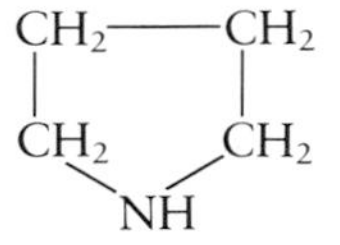

Pyrrolidine

Acid-Base Properties of Salts

51. Derive an expression for the relationship between pK_a and pK_b for a conjugate acid-base pair.

52. Are solutions of the following salts acidic, basic, or neutral? For those that are not neutral, write balanced chemical equations for the reactions causing the solution to be acidic or basic. The relevant K_a and K_b values are found in Tables 7.2, 7.3, and 7.4.
    a. KCl
    b. $NaNO_3$
    c. $NaNO_2$
    d. NH_4NO_3
    e. NH_4NO_2
    f. $NaHCO_3$
    g. $NH_4C_2H_3O_2$
    h. NaF

53. Calculate the pH of each of the following solutions.
    a. 0.10 M CH_3NH_3Cl
    b. 0.050 M NaCN
    c. 0.20 M Na_2CO_3 (consider only the reaction $CO_3^{2-} + H_2O \rightleftharpoons HCO_3^- + OH^-$)
    d. 0.12 M $NaNO_2$
    e. 0.45 M NaOCl

54. Arrange the following 0.10 M solutions in order from most acidic to most basic:

KOH, KBr, KCN, NH_4Br, NH_4CN, HCN

55. Arrange the following 0.10 M solutions in order from most acidic to most basic:

$$H_2O, \quad KNO_2, \quad HNO_3, \quad HNO_2, \quad NH_4NO_3, \quad NH_4NO_2$$

56. Is an aqueous solution of $NaHSO_4$ acidic, basic, or neutral? What reaction occurs with water? Calculate the pH of a 0.10 M solution of $NaHSO_4$. If solid Na_2CO_3 is added to a solution of $NaHSO_4$, what reaction occurs between the CO_3^{2-} and HSO_4^- ions?

57. Sodium azide (NaN_3) is sometimes added to water to kill bacteria. Calculate the concentration of all species in a 0.010 M solution of NaN_3. The K_a value for hydrazoic acid (HN_3) is 1.9×10^{-5}.

58. Given that the K_a value for acetic acid is 1.8×10^{-5} and the K_a value for hypochlorous acid is 3×10^{-8}, which is the stronger base, OCl^- or $C_2H_3O_2^-$?

59. Only 1.0 g of codeine sulfate will dissolve in 30. mL of solution. Calculate the pH of a saturated solution of codeine sulfate. (See Exercises 44 and 45.)

Additional Exercises

60. Liquid ammonia is sometimes used as a solvent for chemical reactions. It undergoes the autoionization reaction:

$$2NH_3 \rightleftharpoons NH_4^+ + NH_2^-$$

a. What species correspond to H^+ and OH^- in liquid ammonia?
b. What is the condition for a neutral solution in liquid ammonia?
c. Sodium metal reacts with liquid ammonia in a manner analogous to the reaction of sodium metal with water that produces $NaOH(aq)$ and $H_2(g)$. Write a balanced chemical equation describing this reaction.
d. What advantages might there be to using liquid ammonia as a solvent?

61. Alka-Seltzer employs the reaction between citric acid and sodium bicarbonate to generate its fizz [$CO_2(g)$]. The structure of citric acid and the values of the acid dissociation constants are

$$\begin{array}{ll} \quad\;\; CH_2—CO_2H & K_{a_1} = 8.4 \times 10^{-4} \\ \quad\;\;\;\; | & \\ HO—C—CO_2H & K_{a_2} = 1.8 \times 10^{-5} \\ \quad\;\;\;\; | & \\ \quad\;\; CH_2—CO_2H & K_{a_3} = 4.0 \times 10^{-6} \end{array}$$

Calculate the value of the equilibrium constant for each of the following reactions (we abbreviate citric acid as H_3Cit), using the K_a values for citric acid and for carbonic acid:

$$H_3Cit + HCO_3^- \rightleftharpoons H_2Cit^- + H_2O + CO_2$$

$$H_3Cit + 3HCO_3^- \rightleftharpoons Cit^{3-} + 3H_2O + 3CO_2$$

(*Hint:* when reactions are added, the corresponding equilibrium constants are multiplied.)

62. The pH of a 1.00×10^{-2} M solution of cyanic acid (HOCN) is 2.77 at 25°C. Calculate K_a for HOCN from this result.

63. a. The principal equilibrium in a solution of $NaHCO_3$ is

$$HCO_3^- + HCO_3^- \rightleftharpoons H_2CO_3 + CO_3^{2-}$$

Calculate the value of the equilibrium constant for this reaction.
b. At equilibrium, what is the relationship between $[H_2CO_3]$ and $[CO_3^{2-}]$?
c. Using the equilibrium

$$H_2CO_3 \rightleftharpoons 2H^+ + CO_3^{2-}$$

and the result from part b, derive an expression for the pH of the solution in terms of K_{a_1} and K_{a_2}.
d. What is the pH of a solution of $NaHCO_3$?

64. Calculate the value for the equilibrium constant for each of the following reactions.

a. $NH_3 + H_3O^+ \rightleftharpoons NH_4^+ + H_2O$
b. $NO_2^- + H_3O^+ \rightleftharpoons HNO_2 + H_2O$
c. $NH_4^+ + CH_3CO_2^- \rightleftharpoons NH_3 + CH_3CO_2H$
d. $H_3O^+ + OH^- \rightleftharpoons 2H_2O$
e. $NH_4^+ + OH^- \rightleftharpoons NH_3 + H_2O$
f. $HNO_2 + OH^- \rightleftharpoons H_2O + NO_2^-$

65. Hemoglobin (abbreviated Hb) is a protein that is responsible for the transport of oxygen in the blood of mammals. Each hemoglobin molecule contains four iron atoms that are the binding sites for O_2 molecules. The oxygen binding is pH-dependent. The relevant equilibrium reaction is as follows:

$$HbH_4^{4+} + 4O_2 \rightleftharpoons Hb(O_2)_4 + 4H^+$$

Use Le Châtelier's principle to answer the following.
a. What form of hemoglobin, HbH_4^{4+} or $Hb(O_2)_4$, is favored in the lungs? What form is favored in the cells?
b. When a person hyperventilates, the concentration of CO_2 in the blood is decreased. How does this result affect the oxygen-binding equilibrium? How does breathing into a paper bag help to counteract this effect?
c. When a person has suffered a cardiac arrest, injection of a sodium bicarbonate solution is given. Why is this injection necessary?

66. Calculate the pH of a 1.0×10^{-7} M solution of NaOH in water.

67. Making use of the assumptions we ordinarily make in calculating the pH of an aqueous solution of a weak acid, calculate the pH of a 1.0×10^{-6} M solution of hypobromous acid (HBrO, $K_a = 2 \times 10^{-9}$). What is wrong with your answer? Why is it wrong? Without trying to solve the problem, tell what has to be included to solve the problem correctly.

68. Calculate the pH of 0.0200 *M* chloroacetic acid (K_a = 1.39×10^{-3}).

69. Calculate the pH of 0.010 *M* HIO_3 ($K_a = 1.7 \times 10^{-1}$).

70. Calculate the pH of a 7.00×10^{-7} *M* HCl solution.

71. A 0.050 *M* solution of the salt NaB has a pH of 9.00. Calculate the pH of a 0.010 *M* solution of HB.

72. Calculate the pH and the concentration of acetate ion in a solution prepared by dissolving 1.0×10^{-3} mol of HCl(*g*) in 1.0 L of 1.0 *M* acetic acid.

73. A certain acid, HA, has a vapor density of 5.11 g/L when in the gas phase at a temperature of 25°C and a pressure of 1.00 atm. When 1.50 g of this acid is dissolved in enough water to make 100.0 mL of solution, the pH is found to be 1.80. Calculate K_a for HA.

74. What mass of NaOH(*s*) must be added to 1.0 L of 0.050 *M* NH_3 to ensure that the percent ionization of NH_3 is no greater than 0.0010%?

75. Acrylic acid ($CH_2{=}CHCO_2H$) is a precursor for many important plastics. K_a for acrylic acid is 5.6×10^{-5}.
    a. Calculate the pH of a 0.10 *M* solution of acrylic acid.
    b. Calculate the percent dissociation of a 0.10 *M* solution of acrylic acid.
    c. Calculate the $[H^+]$ necessary to ensure that the percent dissociation of a 0.10 *M* solution of acrylic acid is less than 0.010%.
    d. Calculate the pH of a 0.050 *M* solution of sodium acrylate ($NaC_3H_3O_2$).

76. The equilibrium constant K_a for the reaction

$$Fe(H_2O)_6^{3+}(aq) + H_2O(l) \rightleftharpoons Fe(H_2O)_5(OH)^{2+}(aq) + H_3O^+(aq)$$

is 6.0×10^{-3}.
    a. Calculate the pH of a 0.10 *M* solution of $Fe(H_2O)_6^{3+}$.
    b. Calculate the pH necessary for 99.90% of the iron(III) to be in the form $Fe(H_2O)_6^{3+}$.
    c. Will a 1.0 *M* solution of iron(II) nitrate have a higher or lower pH than a 1.0 *M* solution of iron(III) nitrate? Explain.

77. A 0.100 *M* solution of the salt BHX has a pH of 8.00, where B is a weak base and X^- is the anion of the weak acid HX. Calculate the K_a value for HX if the K_b value for B is 1.0×10^{-3}.

78. Calculate the pH of a solution containing 1.0 *M* HCl and 1.0 *M* HCN.

79. How many moles of HCl(*g*) must be added to 1.0 L of 2.0 *M* NaOH to achieve a pH of 0.00? (Neglect any volume changes.)

80. An aqueous solution contains a mixture of 0.0500 *M* HCOOH ($K_a = 1.77 \times 10^{-4}$) and 0.150 *M* CH_3CH_2COOH ($K_a = 1.34 \times 10^{-5}$). Calculate the pH of this solution.

81. Consider 50.0 mL of a solution of weak acid HA (K_a = 1.00×10^{-6}), which has a pH of 4.000. What volume of water must be added to make the pH 5.000?

82. Calculate the pH of a solution prepared by mixing equal volumes of 0.10 *M* NH_4Cl and 0.10 *M* NaCN.

83. A chemist dissolves 0.135 mol of $CO_2(g)$ in 2.50 L of 0.105 *M* Na_2CO_3. Calculate the pH of the resulting solution.

84. Calculate the pH of 5.0×10^{-8} *M* HNO_3.

85. Calculate the pH of 5.0×10^{-4} *M* HCN.

86. Calculate the pH of 4.0×10^{-5} *M* phenol (K_a = 1.6×10^{-10}).

87. Calculate the pH of 6.0×10^{-4} *M* $NaNO_2$.

88. Calculate the pH of a solution prepared by mixing equal volumes of 1.0×10^{-4} *M* NH_3 and 1.0×10^{-4} *M* HCl.

89. A solution contains a mixture of acids: 0.50 *M* HA (K_a = 1.0×10^{-3}), 0.20 *M* HB ($K_a = 1.0 \times 10^{-10}$), and 0.10 *M* HC ($K_a = 1.0 \times 10^{-12}$). Calculate the $[H^+]$ in this solution.

90. One mole of a weak acid HA was dissolved in 2.0 L of water. After the system had come to equilibrium, the concentration of HA was found to be 0.45 *M*. Calculate K_a for HA.

91. Calculate $[OH^-]$ in a 3.0×10^{-7} *M* solution of $Ca(OH)_2$.

92. Calculate $[OH^-]$ in a solution obtained by adding 0.0100 mol of solid NaOH to 1.00 L of 15.0 *M* NH_3.

93. Calculate the pH of an aqueous solution containing 1.0×10^{-2} *M* HCl, 1.0×10^{-2} *M* H_2SO_4, and 1.0×10^{-2} *M* HCN.

94. Calculate $[OH^-]$ and the pH for an aqueous solution of 15.0 *M* NH_3 ($K_b = 1.8 \times 10^{-5}$).

95. A solution is made by adding 50.0 mL of 0.200 *M* acetic acid ($K_a = 1.8 \times 10^{-5}$) to 50.0 mL of 1.00×10^{-3} *M* HCl.
    a. Calculate the pH of the solution.
    b. Calculate the acetate ion concentration.

96. Will 0.10 *M* solutions of the following salts be acidic, basic, or neutral?
    a. ammonium bicarbonate
    b. sodium dihydrogen phosphate
    c. sodium hydrogen phosphate
    d. ammonium dihydrogen phosphate
    e. ammonium formate

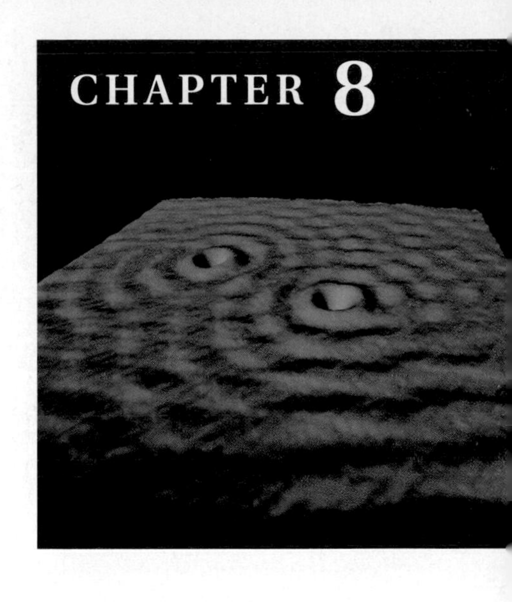

Applications of Aqueous Equilibria

Much important chemistry, including most of the chemistry of the natural world, occurs in aqueous solution. We have already introduced one very significant class of aqueous equilibria, acid-base reactions. In this chapter we consider more applications of acid-base chemistry and introduce two additional types of aqueous equilibria, those involving the solubility of salts and the formation of complex ions.

The interplay of acid-base, solubility, and complex ion equilibria is often important in natural processes, such as the weathering of minerals, the uptake of nutrients by plants, and tooth decay. For example, limestone ($CaCO_3$) will dissolve in water made acidic by dissolved carbon dioxide:

$$CO_2(aq) + H_2O(l) \rightleftharpoons H^+(aq) + HCO_3^-(aq)$$
$$H^+(aq) + CaCO_3(s) \rightleftharpoons Ca^{2+}(aq) + HCO_3^-(aq)$$

This process and its reverse account for the formation of limestone caves and the stalactites and stalagmites found there. The acidic water (containing carbon dioxide) dissolves the underground limestone deposits, thereby forming a cavern. As the water drips from the ceiling of the cave, the carbon dioxide is lost to the air, and solid calcium carbonate forms by the reverse of the above process to produce stalactites on the ceiling and stalagmites where the drops hit the cave floor.

Before we consider the other types of aqueous equilibria, we will deal with acid-base equilibria in more detail.

8.1 Solutions of Acids or Bases Containing a Common Ion

In Chapter 7 we were concerned with calculating the equilibrium concentrations of species (particularly H^+ ions) in solutions containing an acid or a base. In this section we discuss solutions that contain not only the weak acid HA but also its salt NaA. Although this case appears to be a new type of problem, it can be handled rather easily by using the procedures developed in Chapter 7.

Suppose we have a solution containing the weak acid hydrofluoric acid (HF, $K_a = 7.2 \times 10^{-4}$) and its salt sodium fluoride (NaF). Recall that when a salt dissolves in water, it breaks up completely into its ions—it is a strong electrolyte:

$$NaF(s) \xrightarrow{H_2O(l)} Na^+(aq) + F^-(aq)$$

Since hydrofluoric acid is a weak acid and only slightly dissociated, the major species in the solution are HF, Na^+, F^-, and H_2O. The **common ion** in this solution is F^-, since it is produced by both hydrofluoric acid and sodium fluoride. What effect does the presence of the dissolved sodium fluoride have on the dissociation equilibrium of hydrofluoric acid?

To answer this question we compare the extent of dissociation of hydrofluoric acid in two different solutions, the first containing 1.0 M HF and the second containing 1.0 M HF and 1.0 M NaF. According to Le Châtelier's principle the dissociation equilibrium for HF

$$HF(aq) \rightleftharpoons H^+(aq) + F^-(aq)$$

in the second solution will be *driven to the left by the presence of the* F^- *ions from the NaF*. Thus the extent of dissociation of HF will be *less* in the presence of dissolved NaF. The shift in equilibrium position that occurs because of the addition of an ion already involved in the equilibrium reaction is called the **common ion effect.** This effect makes a solution of NaF and HF less acidic than a solution of HF alone.

The common ion effect is quite general. For example, when solid NH_4Cl is dissolved in a 1.0 M NH_3 solution

$$NH_4Cl(s) \xrightarrow{H_2O} NH_4^+(aq) + Cl^-(aq)$$

the added ammonium ions cause the position of the ammonia-water equilibrium

$$NH_3(aq) + H_2O(l) \rightleftharpoons NH_4^+(aq) + OH^-(aq)$$

to shift to the left, reducing the concentration of OH^- ions.

The common ion effect is also important in solutions of polyprotic acids. The production of protons by the first dissociation step greatly inhibits the succeeding dissociation steps, which also produce protons, the common ion in this case. We will see later in this chapter that the common ion effect is also important in dealing with the solubility of salts.

EXAMPLE 8.1

In Section 7.5 we found that the equilibrium concentration of H^+ in a 1.0 *M* HF solution is 2.7×10^{-2} *M* and the percent dissociation of HF is 2.7%. Calculate $[H^+]$ and the percent dissociation of HF in a solution containing both 1.0 *M* HF ($K_a = 7.2 \times 10^{-4}$) and 1.0 *M* NaF.

Solution

As the aqueous solutions we consider become more complex, it becomes increasingly important to be systematic and to *focus on the chemistry* occurring in the solution before thinking about mathematical procedures. *Always* write the major species first and consider the chemical properties of each component.

In a solution containing 1.0 *M* HF and 1.0 *M* NaF, the major species are

$$HF, \quad F^-, \quad Na^+, \quad H_2O$$

Since Na^+ ions have neither acidic nor basic properties, and since water is such a weak acid or base, the important species are HF and F^-; they participate in the acid dissociation equilibrium that control $[H^+]$ in this solution. That is, the position of the equilibrium

$$HF(aq) \rightleftharpoons H^+(aq) + F^-(aq)$$

will determine $[H^+]$ in the solution. The equilibrium expression is

$$K_a = \frac{[H^+][F^-]}{[HF]} = 7.2 \times 10^{-4}$$

The important concentrations are listed in the following table.

Initial Concentration (mol/L)		Equilibrium Concentration (mol/L)
$[HF]_0 = 1.0$ (from dissolved HF)		$[HA] = 1.0 - x$
$[F^-]_0 = 1.0$ (from dissolved NaF)	x mol/L HF dissociates → to reach equilibrium	$[F^-] = 1.0 + x$
$[H^+]_0 = 0$ (neglect contribution from H_2O)		$[H^+] = x$

Note that $[F^-]_0 = 1.0$ *M* from the dissolved sodium fluoride and that the equilibrium $[F^-] > 1.0$ *M* because when the acid dissociates, it produces F^- as well as H^+. Then

$$K_a = 7.2 \times 10^{-4} = \frac{[H^+][F^-]}{[HF]} = \frac{(x)(1.0 + x)}{1.0 - x} \approx \frac{(x)(1.0)}{1.0}$$

(since x is expected to be small).

Solving for x gives

$$x = \frac{1.0}{1.0}(7.2 \times 10^{-4}) = 7.2 \times 10^{-4}$$

Noting that x is small compared with 1.0, we conclude that this result is acceptable. Thus

$$[H^+] = x = 7.2 \times 10^{-4}\ M \qquad \text{(The pH is 3.14.)}$$

The percent dissociation of HF in this solution is

$$\frac{[H^+]}{[HF]_0} \times 100 = \frac{7.2 \times 10^{-4}\ M}{1.0\ M} \times 100 = 0.072\%$$

Compare these values for $[H^+]$ and percent dissociation of HF with those for a 1.0 M HF solution, where $[H^+] = 2.7 \times 10^{-2}\ M$ and the percent dissociation is 2.7%. The large difference clearly shows that the presence of the F^- ions from the dissolved NaF greatly inhibits the dissociation of HF. The position of the acid dissociation equilibrium has been shifted to the left by the presence of F^- ions from NaF.

8.2 Buffered Solutions

The most important buffering system in the blood involves HCO_3^- and H_2CO_3.

The most important application of acid-base solutions containing a common ion is buffering. A **buffered solution** is one that *resists a change in pH* when either hydroxide ions or protons are added. The most important practical example of a buffered solution is human blood, which can absorb the acids and bases produced by biological reactions without changing its pH. A constant pH for blood is vital, because cells can survive only in a very narrow pH range around 7.4.

A buffered solution may contain a weak acid and its salt (for example, HF and NaF) or a weak base and its salt (for example, HN_3 and NH_4Cl). By choosing the appropriate components, a solution can be buffered at virtually any pH.

In treating buffered solutions in this chapter, we will start by considering the equilibrium calculations. We will then use these results to show how buffering works. That is, we will answer the question: "How does a buffered solution resist changes in pH when an acid or base is added?"

As you do the calculations associated with buffered solutions, keep in mind that they are merely solutions containing weak acids or bases and that the procedures required are the same ones we have already developed. Be sure to use the systematic approach introduced in Chapter 7.

EXAMPLE 8.2

A buffered solution contains 0.50 M acetic acid ($HC_2H_3O_2$, $K_a = 1.8 \times 10^{-5}$) and 0.50 M sodium acetate ($NaC_2H_3O_2$).

a. Calculate the pH of this solution.

Solution

The major species in the solution are

$HC_2H_3O_2$,	Na^+,	$C_2H_3O_2^-$,	H_2O
↑	↑	↑	↑
Weak acid	Neither acid nor base	Base (conjugate base of $HC_2H_3O_2$)	Very weak acid or base

Examination of the solution components leads to the conclusion that the acetic acid dissociation equilibrium, which involves both $HC_2H_3O_2$ and $C_2H_3O_2^-$, will control the pH of the solution:

$$HC_2H_3O_2(aq) \rightleftharpoons H^+(aq) + C_2H_3O_2^-(aq)$$

$$K_a = 1.8 \times 10^{-5} = \frac{[H^+][C_2H_3O_2^-]}{[HC_2H_3O_2]}$$

The concentrations are as follows:

Initial Concentration (mol/L)		Equilibrium Concentration (mol/L)
$[HC_2H_3O_2]_0 = 0.50$	x mol/L of $HC_2H_3O_2$ dissociates to reach equilibrium →	$[HC_2H_3O_2] = 0.50 - x$
$[C_2H_3O_2^-]_0 = 0.50$		$[C_2H_3O_2^-] = 0.50 + x$
$[H^+]_0 \approx 0$		$[H^+] = x$

Then

$$K_a = 1.8 \times 10^{-5} = \frac{[H^+][C_2H_3O_2^-]}{[HC_2H_3O_2]} = \frac{(x)(0.50 + x)}{0.50 - x} \approx \frac{(x)(0.50)}{0.50}$$

and $x = 1.8 \times 10^{-5}$

The approximation is valid (by the 5% rule), so

$$[H^+] = x = 1.8 \times 10^{-5}\ M \quad \text{and} \quad pH = 4.74$$

b. Calculate the change in pH that occurs when 0.010 mol of solid NaOH is added to 1.0 L of the buffered solution. Compare this pH change with the change that occurs when 0.010 mol of solid NaOH is added to 1.0 L of water.

Solution

Since the added solid NaOH will completely dissociate, the major species in solution *before any reaction occurs* are $HC_2H_3O_2$, Na^+, $C_2H_3O_2^-$, OH^-, and H_2O. Note that the solution contains a relatively large amount of the very strong base, the hydroxide ion, which has a great affinity for protons. The best source of protons is the acetic acid, so the reaction that will occur is

$$OH^- + HC_2H_3O_2 \longrightarrow H_2O + C_2H_3O_2^-$$

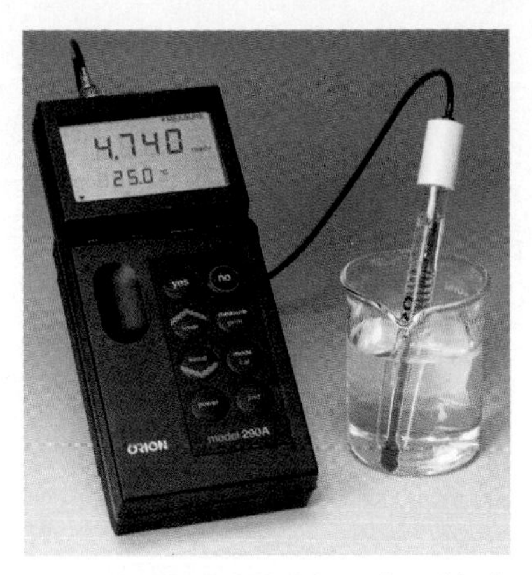

A digital pH meter shows the pH of the buffered solution to be 4.74.

Although acetic acid is a weak acid, the hydroxide ion is such a strong base that the above reaction will *proceed essentially to completion* (until the OH^- ions are consumed).

The best approach to this problem involves two distinct steps: (1) assume the reaction goes to completion and carry out the stoichiometric calculations; then (2) carry out the equilibrium calculations.

1. *The stoichiometry problem.* The reaction occurs as shown below.

	$HC_2H_3O_2$	+	OH^-	$\rightarrow$	$C_2H_3O_2^-$	+ H_2O
Before reaction:	$1.0\ L \times 0.50\ M$ $= 0.50$ mol		0.010 mol		$1.0\ L \times 0.50\ M$ $= 0.50$ mol	
After reaction:	$0.50 - 0.01$ $= 0.49$ mol		$0.010 - 0.010$ $= 0$ mol		$0.50 + 0.01$ $= 0.51$ mol	

Note that 0.01 mol of $HC_2H_3O_2$ has been converted to 0.01 mol of $C_2H_3O_2^-$ by the added OH^-.

2. *The equilibrium problem.* After the reaction between OH^- and $HC_2H_3O_2$ has run to completion, the major species in solution are

$$HC_2H_3O_2, \quad Na^+, \quad C_2H_3O_2^-, \quad H_2O$$

The dominant equilibrium involves the dissociation of acetic acid.

This problem is very similar to that in part a. The only difference is that the addition of 0.01 mol of OH^- has consumed some $HC_2H_3O_2$ and produced some $C_2H_3O_2^-$, yielding the following concentrations:

Initial Concentration (mol/L)		Equilibrium Concentration (mol/L)
$[HC_2H_3O_2]_0 = 0.49$	x mol/L $HC_2H_3O_2$ dissociates to reach equilibrium $\longrightarrow$	$[HC_2H_3O_2] = 0.49 - x$
$[C_2H_3O_2^-]_0 = 0.51$		$[C_2H_3O_2^-] = 0.51 + x$
$[H^+]_0 \approx 0$		$[H^+] = x$

Note that the initial concentrations are defined after the reaction with OH^- is complete but before the system adjusts to equilibrium.

Following the usual procedures gives

$$K_a = 1.8 \times 10^{-5} = \frac{[H^+][C_2H_3O_2^-]}{[HC_2H_3O_2]} = \frac{(x)(0.51 + x)}{0.49 - x} \approx \frac{(x)(0.51)}{0.49}$$

$$x \approx 1.7 \times 10^{-5}$$

The approximations are valid (by the 5% rule), so

$$[H^+] = x = 1.7 \times 10^{-5} \quad \text{and} \quad pH = 4.76$$

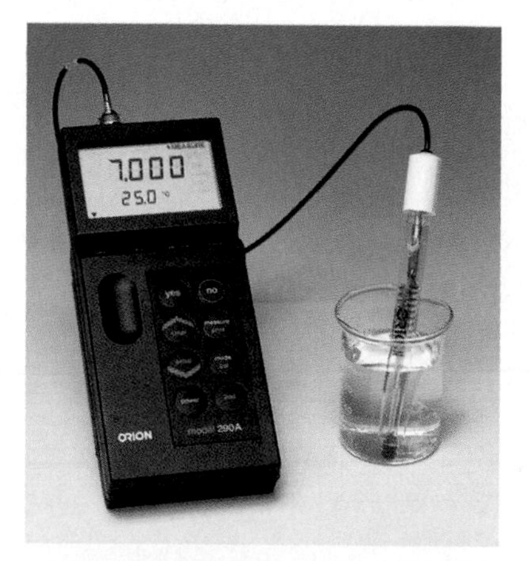

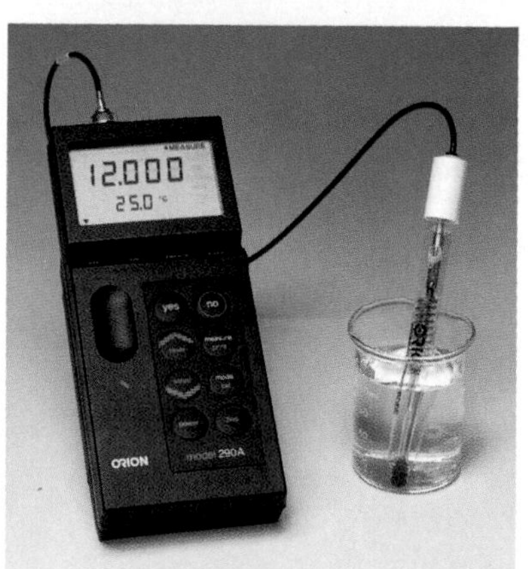

(top) Pure water at pH 7.00. (bottom) When 0.01 mol NaOH is added to 1.0 L of pure water, the pH jumps to 12.00.

The change in pH produced by the addition of 0.01 mol of OH^- to this buffered solution is then

$$\underset{\text{New solution}}{\underset{\uparrow}{4.76}} - \underset{\text{Original solution}}{\underset{\uparrow}{4.74}} = +0.02$$

The pH has increased by 0.02 pH units.

Now compare this result with what happens when 0.01 mol of solid NaOH is added to 1.0 L of water to give 0.01 *M* NaOH. In this case, $[OH^-] = 0.01\ M$ and

$$[H^+] = \frac{K_w}{[OH^-]} = \frac{1.0 \times 10^{-14}}{1.0 \times 10^{-2}} = 1.0 \times 10^{-12}$$

$$pH = 12.00$$

Thus the change in pH is

$$\underset{\text{New solution}}{\underset{\uparrow}{12.00}} - \underset{\text{Pure water}}{\underset{\uparrow}{7.00}} = +5.00$$

The increase is 5.00 pH units. Note how well the buffered solution resists a change in pH, compared with pure water.

Example 8.2 is a typical buffer problem. It contains all of the concepts necessary for handling the calculations for buffered solutions containing weak acids. Pay special attention to the following points:

1. Buffered solutions are simply solutions of weak acids or bases containing a common ion. The pH calculations for buffered solutions require exactly the same procedures previously introduced in Chapter 7. *This is not a new type of problem.*

2. When a strong acid or base is added to a buffered solution, it is best to deal with the stoichiometry of the resulting reaction first. After the stoichiometric calculations are completed, then consider the equilibrium calculations. This procedure can be represented as follows:

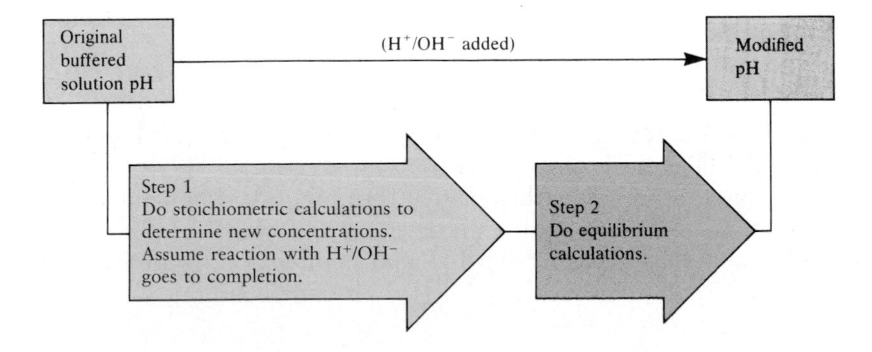

Buffering: How Does It Work?

Example 8.2 demonstrates the ability of a buffered solution to absorb hydroxide ions without a significant change in pH. *But how does a buffer work?* Suppose a buffered solution contains relatively large quantities of a weak acid HA and its conjugate base A^-. Since the weak acid represents the best source of protons, the following reaction occurs when hydroxide ions are added to the solution:

$$OH^- + HA \longrightarrow A^- + H_2O$$

The net result is that OH^- ions are not allowed to accumulate but are replaced by A^- ions.

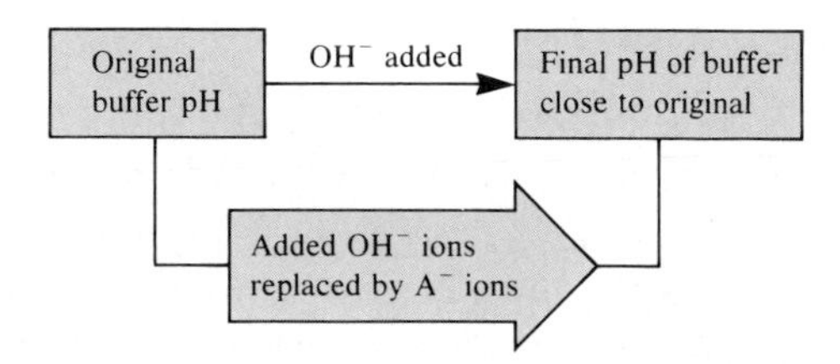

The stability of the pH under these conditions can be understood by examining the equilibrium expression for the dissociation of HA:

$$K_a = \frac{[H^+][A^-]}{[HA]} \qquad \text{or rearranging,} \qquad [H^+] = K_a \frac{[HA]}{[A^-]}$$

In a buffered solution the pH is governed by the ratio $[HA]/[A^-]$.

In other words, the *equilibrium concentration of* H^+ *and thus the pH are determined by the ratio* $[HA]/[A^-]$. When OH^- ions are added, HA is converted to A^-, causing the ratio $[HA]/[A^-]$ to decrease. However, *if the amounts of HA and* A^- *originally present are very large compared with the amount of* OH^- added, the change in the $[HA]/[A^-]$ ratio is small.

In Example 8.2

$$\frac{[HA]}{[A^-]} = \frac{0.50}{0.50} = 1.0 \qquad \text{Initially}$$

$$\frac{[HA]}{[A^-]} = \frac{0.49}{0.51} = 0.96 \qquad \text{After adding 0.01 mol/L of } OH^-$$

The change in the ratio $[HA]/[A^-]$ is very small. Thus the $[H^+]$ and the pH remain essentially constant.

The essence of buffering, then, is that [HA] and $[A^-]$ are large compared with the amount of OH^- added. Thus when the OH^- is added, the concentrations of HA and A^- change, but only by small amounts. Under these conditions the $[HA]/[A^-]$ ratio and thus the $[H^+]$ stay virtually constant.

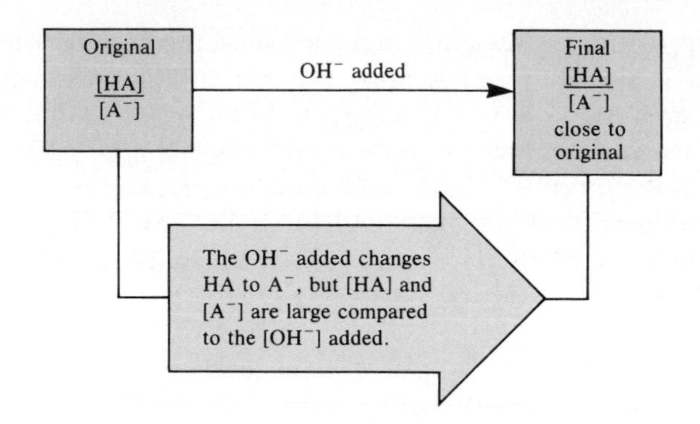

Similar reasoning applies when protons are added to a buffered solution containing a weak acid and a salt of its conjugate base. Because the A^- ion has a high affinity for H^+, the added H^+ ions react with A^- to form the weak acid:

$$H^+ + A^- \longrightarrow HA$$

Thus free H^+ ions do not accumulate. In this case there will be a net change of A^- to HA. However, if $[A^-]$ and [HA] are large compared with the $[H^+]$ added, only a slight change in the pH occurs.

An alternative form of the acid dissociation equilibrium expression,

$$[H^+] = K_a \frac{[HA]}{[A^-]} \tag{8.1}$$

is often useful for calculating $[H^+]$ in buffered solution, since [HA] and $[A^-]$ are usually known. For example, to calculate $[H^+]$ in a buffered solution containing 0.10 *M* HF ($K_a = 7.2 \times 10^{-4}$) and 0.30 *M* NaF, we simply substitute the respective concentrations into Equation (8.1):

$$[H^+] = \underset{K_a \nearrow}{(7.2 \times 10^{-4})} \frac{\overset{[HF] \searrow}{0.10}}{\underset{\nwarrow [F^-]}{0.30}} = 2.4 \times 10^{-4}\ M$$

Another useful form of Equation (8.1) can be obtained by taking the negative log of both sides:

$$-\log[H^+] = -\log(K_a) - \log\left(\frac{[HA]}{[A^-]}\right)$$

That is,

$$pH = pK_a - \log\left(\frac{[HA]}{[A^-]}\right)$$

or inverting the log term and reversing the sign,

$$pH = pK_a + \log\left(\frac{[A^-]}{[HA]}\right) = pK_a + \log\left(\frac{[base]}{[acid]}\right) \tag{8.2}$$

This log form of the expression for K_a is called the **Henderson-Hasselbalch equation** and is useful for calculating the pH of solutions when the ratio $[HA]/[A^-]$ is known.

For a particular buffering system (acid-conjugate base pair), all solutions that have the same ratio $[A^-]/[HA]$ have the same pH. For example, a buffered solution containing 5.0 M $HC_2H_3O_2$ and 3.0 M $NaC_2H_3O_2$ has the same pH as one containing 0.050 M $HC_2H_3O_2$ and 0.030 M $NaC_2H_3O_2$. This result can be shown as follows:

System	$[A^-]/[HA]$
5.0 M $HC_2H_3O_2$ and 3.0 M $NaC_2H_3O_2$	$\frac{3.0\ M}{5.0\ M} = 0.60$
0.050 M $HC_2H_3O_2$ and 0.030 M $NaC_2H_3O_2$	$\frac{0.030\ M}{0.050\ M} = 0.60$

Thus

$$\text{pH} = \text{p}K_a + \log\left(\frac{[C_2H_3O_2^-]}{[HC_2H_3O_2]}\right) = 4.74 + \log(0.60)$$

$$= 4.74 - 0.22 = 4.52$$

Note that in using this equation, we have assumed that the equilibrium concentrations of A^- and HA are equal to their initial concentrations. That is, we are assuming the validity of the approximations

$$[A^-] = [A^-]_0 + x \approx [A^-]_0 \quad \text{and} \quad [HA] = [HA]_0 - x \approx [HA]_0$$

where x represents the amount of acid that dissociates. Since the initial concentrations of HA and A^- are expected to be relatively large in a buffered solution, this assumption is generally acceptable.

EXAMPLE 8.3

A buffered solution contains 0.25 M NH_3 ($K_b = 1.8 \times 10^{-5}$) and 0.40 M NH_4Cl.

a. Calculate the pH of this solution.

Solution

The major species in solution are

$$NH_3, \quad \underbrace{NH_4^+, \quad Cl^-,}_{\text{From the dissolved } NH_4Cl} \quad H_2O$$

Since Cl^- is such a weak base and water is a weak acid or base, the important equilibrium is

$$NH_3(aq) + H_2O(l) \rightleftharpoons NH_4^+(aq) + OH^-(aq)$$

and

$$K_b = 1.8 \times 10^{-5} = \frac{[NH_4^+][OH^-]}{[NH_3]}$$

The appropriate concentrations are as follows:

Initial Concentration (mol/L)		Equilibrium Concentration (mol/L)
$[NH_3]_0 = 0.25$ $[NH_4^+]_0 = 0.40$ $[OH^-] \approx 0$	$\xrightarrow[\text{reacts with } H_2O]{x \text{ mol/L } NH_3}$	$[NH_3] = 0.25 - x$ $[NH_4^+] = 0.40 + x$ $[OH^-] \approx x$

Thus

$$K_b = 1.8 \times 10^{-5} = \frac{[NH_4^+][OH^-]}{[NH_3]} = \frac{(0.40 + x)(x)}{0.25 - x} \approx \frac{(0.40)(x)}{0.25}$$

$$x = 1.1 \times 10^{-5}$$

The approximations are valid (by the 5% rule), so

$$[OH^-] = x = 1.1 \times 10^{-5}$$

$$pOH = 4.95$$

$$pH = 14.00 - 4.95 = 9.05$$

This case is typical of a buffered solution in that the initial and equilibrium concentrations of buffering materials are essentially the same.

ALTERNATIVE SOLUTION: There is another way to solve this problem. Since the solution contains relatively large quantities of *both* NH_4^+ and NH_3, we can use the equilibrium

$$NH_3(aq) + H_2O(l) \rightleftharpoons NH_4^+(aq) + OH^-(aq)$$

to calculate $[OH^-]$ and then calculate $[H^+]$ from K_w, as we have just done. Or we can use the dissociation equilibrium for NH_4^+,

$$NH_4^+(aq) \rightleftharpoons NH_3(aq) + H^+(aq)$$

to calculate $[H^+]$ directly. *Either choice will give the same answer,* since the same equilibrium concentrations of NH_3 and NH_4^+ must satisfy both equilibria.

We can obtain the K_a value for NH_4^+ from the given K_b value for NH_3, since $K_a \times K_b = K_w$:

$$K_a = \frac{K_w}{K_b} = \frac{1.0 \times 10^{-14}}{1.8 \times 10^{-5}} = 5.6 \times 10^{-10}$$

Then using the Henderson-Hasselbalch equation, we have

$$pH = pK_a + \log\left(\frac{[\text{base}]}{[\text{acid}]}\right)$$

$$= 9.25 + \log\left(\frac{0.25\ M}{0.40\ M}\right) = 9.25 - 0.20 = 9.05$$

b. Calculate the pH of the solution that results when 0.10 mol of gaseous HCl is added to 1.0 L of the buffered solution from part a.

Solution

Before any reaction occurs, the solution contains the following major species:

$$NH_3, \quad NH_4^+, \quad Cl^-, \quad H^+, \quad H_2O$$

Remember: Think about the chemistry first. Ask yourself if a reaction will occur among the major species.

What reaction can occur? We know that H^+ will not react with Cl^- to form HCl. In contrast to Cl^-, the NH_3 molecule has a great affinity for protons [this is demonstrated by the fact that NH_4^+ is such a weak acid ($K_a = 5.6 \times 10^{-10}$)]. Thus NH_3 will react with H^+ to form NH_4^+:

$$NH_3(aq) + H^+(aq) \longrightarrow NH_4^+(aq)$$

Since this reaction can be assumed to go essentially to completion forming the very weak acid NH_4^+, we will do the stoichiometric calculations before we consider the equilibrium calculations. That is, we will let the reaction run to completion and then consider the equilibrium.

The stoichiometric calculations for this process are shown below.

	NH_3	+	H^+	→	NH_4^+
Before reaction:	(1.0 L)(0.25 *M*) = 0.25 mol		0.10 mol ↑ Limiting reactant		(1.0 L)(0.40 *M*) = 0.40 mol
After reaction:	0.25 − 0.10 = 0.15 mol		0		0.40 + 0.10 = 0.50 mol

After the reaction goes to completion, the solution contains the major species

$$NH_3, \quad NH_4^+, \quad Cl^-, \quad H_2O$$

and

$$[NH_3]_0 = \frac{0.15 \text{ mol}}{1.0 \text{ L}} = 0.15\ M$$

$$[NH_4^+]_0 = \frac{0.50 \text{ mol}}{1.0 \text{ L}} = 0.50\ M$$

We can use the Henderson-Hasselbalch equation, where

$$[\text{Base}] = [NH_3] \approx [NH_3]_0 = 0.15\ M$$
$$[\text{Acid}] = [NH_4^+] \approx [NH_4^+]_0 = 0.50\ M$$

Then

$$\text{pH} = \text{p}K_a + \log\left(\frac{[\text{NH}_3]}{[\text{NH}_4^+]}\right) = 9.25 + \log\left(\frac{0.15\ M}{0.50\ M}\right)$$
$$= 9.25 - 0.52 = 8.73$$

Note that the addition of HCl only slightly decreases the pH, as we would expect in a buffered solution.

We can now summarize the most important characteristics of buffered solutions.

Summary: **Characteristics of Buffered Solutions**

- **Buffered solutions contain relatively large concentrations of a weak acid and its corresponding weak base. They can involve a weak acid HA and the conjugate base A^- or a weak base B and the conjugate acid BH^+.**
- **When H^+ is added to a buffered solution, it reacts essentially to completion with the weak base present:**

$$H^+ + A^- \longrightarrow HA \quad \text{or} \quad H^+ + B \longrightarrow BH^+$$

- **When OH^- is added to a buffered solution, it reacts essentially to completion with the weak acid present:**

$$OH^- + HA \longrightarrow A^- + H_2O \quad \text{or} \quad OH^- + BH^+ \longrightarrow B + H_2O$$

- **The pH of the buffered solution is determined by the ratio of the concentrations of the weak base and weak acid. As long as this ratio remains virtually constant, the pH will remain virtually constant. This will be the case as long as the concentrations of the buffering materials (HA and A^- or B and BH^+) are large compared with the amounts of H^+ or OH^- added.**

8.3 Buffer Capacity

The **buffering capacity** of a buffered solution is defined in terms of the amount of protons or hydroxide ions it can absorb without a significant change in pH. A buffer with a large capacity contains large concentrations of buffering components and so can absorb a relatively large amount of protons or hydroxide ions and show little pH change. *The pH of a buffered solution is determined by the ratio $[A^-]/[HA]$. The capacity of a buffered solution is determined by the magnitudes of $[HA]$ and $[A^-]$.*

A buffer with a large capacity contains large concentrations of the buffering components.

EXAMPLE 8.4

Calculate the change in pH that occurs when 0.010 mol of gaseous HCl is added to 1.0 L of each of the following solutions.

Solution A: 5.00 *M* $HC_2H_3O_2$ and 5.00 *M* $NaC_2H_3O_2$
Solution B: 0.050 *M* $HC_2H_3O_2$ and 0.050 *M* $NaC_2H_3O_2$

For acetic acid, $K_a = 1.8 \times 10^{-5}$.

Solution

For both solutions the initial pH can be determined from the Henderson-Hasselbalch equation:

$$pH = pK_a + \log\left(\frac{[C_2H_3O_2^-]}{[HC_2H_3O_2]}\right)$$

In each case $[C_2H_3O_2^-] = [HC_2H_3O_2]$. Thus for both A and B,

$$pH = pK_a + \log(1) = pK_a = -\log(1.8 \times 10^{-5}) = 4.74$$

After the addition of HCl to each of these solutions, the major species *before any reaction occurs* are

$$HC_2H_3O_2, \quad Na^+, \quad C_2H_3O_2^-, \quad \underbrace{H^+, \quad Cl^-,}_{\text{From the added HCl}} \quad H_2O$$

Will any reactions occur among these species? Note that we have a relatively large quantity of H^+ that will readily react with any effective base. We know that Cl^- will not react with H^+ to form HCl in water. However, $C_2H_3O_2^-$ will react with H^+ to form the weak acid $HC_2H_3O_2$:

$$H^+(aq) + C_2H_3O_2^-(aq) \longrightarrow HC_2H_3O_2(aq)$$

Because $HC_2H_3O_2$ is a weak acid, we assume that this reaction runs to completion; the 0.010 mol of added H^+ will convert 0.010 mol of $C_2H_3O_2^-$ to 0.010 mol of $HC_2H_3O_2$.

For solution A (since the solution volume is 1.0 L, the number of moles equals the molarity), we have the following concentrations:

	H^+	+	$C_2H_3O_2^-$	→	$HC_2H_3O_2$
Before reaction:	0.010 *M*		5.00 *M*		5.00 *M*
After reaction:	0		4.99 *M*		5.01 *M*

Original solution → New solution

$$\frac{[A^-]}{[HA]} = \frac{5.00}{5.00} = 1.00 \xrightarrow[\text{added}]{H^+} \frac{[A^-]}{[HA]} = \frac{4.99}{5.01} = 0.996$$

The new pH can be obtained by substituting the new concentrations into the Henderson-Hasselbalch equation:

$$pH = pK_a + \log\left(\frac{[C_2H_3O_2^-]}{[HC_2H_3O_2]}\right) = 4.74 + \log\left(\frac{4.99}{5.01}\right)$$
$$= 4.74 - 0.0017 = 4.74$$

There is virtually no change in pH for solution A when 0.010 mol of gaseous HCl is added.

For solution B we have the following calculations:

	H^+	+ $C_2H_3O_2^-$	→ $HC_2H_3O_2$
Before reaction:	0.010 *M*	0.050 *M*	0.050 *M*
After reaction:	0	0.040 *M*	0.060 *M*

The new pH is

$$\text{pH} = 4.74 + \log\left(\frac{0.040}{0.060}\right) = 4.74 - 0.18 = 4.56$$

Original solution | New solution

$$\frac{[A^-]}{[HA]} = \frac{.050}{.050} = 1.0 \xrightarrow[\text{added}]{H^+} \frac{[A^-]}{[HA]} = \frac{0.040}{0.060} = 0.67$$

Although the pH change for solution B is small, a change did occur, which is in contrast to the case for solution A.

These results show that solution A, which contains much larger quantities of buffering components, has a much higher buffering capacity than solution B.

We have seen that the pH of a buffered solution depends on the ratio of the concentrations of buffering components. When this ratio is least affected by added protons or hydroxide ions, the solution is the most resistant to a change in pH. To find the ratio that gives optimum buffering, suppose that we have a buffered solution containing a large concentration of acetate ion and only a small concentration of acetic acid. Addition of protons to form acetic acid will produce a relatively large *percentage* change in the concentration of acetic acid and so will produce a relatively large change in the ratio $[C_2H_3O_2^-]/[HC_2H_3O_2]$ (see Table 8.1). Similarly, if hydroxide ions are added to remove some acetic acid, the percentage change in the concentration of acetic acid is again large. The same effects are seen if the initial concentration of acetic acid is large and that of acetate ion is small.

Because large changes in the ratio $[A^-]/[HA]$ will produce large changes in pH, we want to avoid this situation for the most effective buffering. This type of reasoning leads us to the general conclusion that optimum buffering will occur when [HA] is equal to $[A^-]$. It is under this condition that the ratio $[A^-]/[HA]$ is

TABLE 8.1 Change in $[C_2H_3O_2^-]/[HC_2H_3O_2]$ for Two Solutions When 0.01 mol of H^+ Is Added to 1.0 L of Each Solution

Solution	$\left(\frac{[C_2H_3O_2^-]}{[HC_2H_3O_2]}\right)_{orig}$	$\left(\frac{[C_2H_3O_2^-]}{[HC_2H_3O_2]}\right)_{new}$	Change	Percent Change
A	$\frac{1.00\ M}{1.00\ M} = 1.00$	$\frac{0.99\ M}{1.01\ M} = 0.98$	1.00 → 0.98	2.00%
B	$\frac{1.00\ M}{0.01\ M} = 100$	$\frac{0.99\ M}{0.02\ M} = 49.5$	100 → 49.5	50.5%

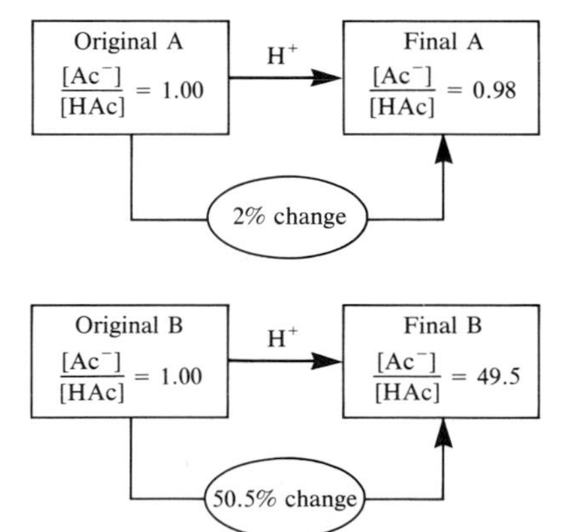

most resistant to change when H^+ or OH^- is added. Thus when choosing the buffering components for a specific application, we want $[A^-]/[HA]$ to equal 1. It follows that since

$$pH = pK_a + \log\left(\frac{[A^-]}{[HA]}\right) = pK_a + \log(1) = pK_a$$

the *p*K_a *of the weak acid selected for the buffer should be as close as possible to the desired pH*. For example, suppose we need a buffered solution with a pH of 4.00. The most effective buffering will occur when [HA] is equal to $[A^-]$. From the Henderson-Hasselbalch equation,

$$\underset{\substack{\uparrow \\ 4.00}}{pH} = pK_a + \log\underset{\substack{\uparrow \\ \text{Ratio = 1 for most effective buffer}}}{\left(\frac{[A^-]}{[HA]}\right)}$$

That is,

$$4.00 = pK_a + \log(1) = pK_a + 0 \qquad \text{and} \qquad pK_a = 4.00$$

Thus in this case the best choice is a weak acid that has $pK_a = 4.00$, or $K_a = 1.0 \times 10^{-4}$.

EXAMPLE 8.5

A chemist needs to prepare a solution buffered at pH 4.30 using one of the following acids (and its sodium salt):

a. chloroacetic acid ($K_a = 1.35 \times 10^{-3}$)
b. propanoic acid ($K_a = 1.3 \times 10^{-5}$)
c. benzoic acid ($K_a = 6.4 \times 10^{-5}$)
d. hypochlorous acid ($K_a = 3.5 \times 10^{-8}$)

Calculate the ratio of $[HA]/[A^-]$ required for each system to yield a pH of 4.30. Which system will work best?

Solution

A pH of 4.30 corresponds to

$$[H^+] = 10^{-4.30} = \text{antilog}(-4.30) = 5.0 \times 10^{-5}\ M$$

Since K_a values rather than pK_a values are given for the various acids, we use Equation (8.1),

$$[H^+] = K_a \frac{[HA]}{[A^-]}$$

instead of the Henderson-Hasselbalch equation. We substitute the required $[H^+]$ and K_a for each acid into Equation (8.1) to calculate each ratio of $[HA]/[A^-]$ needed. The results are as follows:

Acid	$[H^+] = K_a \frac{[HA]}{[A^-]}$	$\frac{[HA]}{[A^-]}$
a. Chloroacetic	$5.0 \times 10^{-5} = 1.35 \times 10^{-3}\left(\frac{[HA]}{[A^-]}\right)$	3.7×10^{-2}
b. Propanoic	$5.0 \times 10^{-5} = 1.3 \times 10^{-5}\left(\frac{[HA]}{[A^-]}\right)$	3.8
c. Benzoic	$5.0 \times 10^{-5} = 6.4 \times 10^{-5}\left(\frac{[HA]}{[A^-]}\right)$	0.78
d. Hypochlorous	$5.0 \times 10^{-5} = 3.5 \times 10^{-8}\left(\frac{[HA]}{[A^-]}\right)$	1.4×10^{3}

Since $[HA]/[A^-]$ for benzoic acid is closest to 1, the system of benzoic acid and its sodium salt is the best choice among those given for buffering a solution at pH 4.3. This example demonstrates the principle that the optimum buffering system has a pK_a value close to the desired pH. The pK_a for benzoic acid is 4.19.

8.4 Titrations and pH Curves

As we saw in Chapter 4, a titration is commonly used to analyze for the amount of acid or base in a solution. This process involves delivering a solution of known concentration (the titrant) from a buret into the unknown solution until the substance being analyzed is just consumed. The stoichiometric or equivalence point is usually signaled by the color change of an indicator. In this section we will discuss the pH changes that occur during an acid-base titration. We will use this information later in this chapter to show how an appropriate indicator can be chosen for a particular titration.

The progress of an acid-base titration is often monitored by plotting the pH of the solution being analyzed as a function of the amount of titrant added. Such a plot is called a **pH curve,** or **titration curve.**

Strong Acid–Strong Base Titrations

The reaction for a strong acid–strong base titration is

$$H^+(aq) + OH^-(aq) \longrightarrow H_2O(l)$$

To compute $[H^+]$ at a given point in the titration, we must determine the moles of H^+ remaining at that point and divide by the total volume of the solution. Before we proceed, we need to consider a new unit, which is especially convenient for titrations. Since titrations usually involve small quantities (burets are typically graduated in milliliters), the mole is inconveniently large. Therefore, we will use the **millimole** (abbreviated **mmol**), where

$$1 \text{ mmol} = \frac{1 \text{ mol}}{1000} = 10^{-3} \text{ mol}$$

1 millimole = 1×10^{-3} mol

1 mL = 1×10^{-3} L

$$\frac{\text{mmol}}{\text{mL}} = \frac{\text{mol}}{\text{L}} = M$$

and

$$\text{Molarity} = \frac{\text{mol of solute}}{\text{L of solution}} = \frac{\dfrac{\text{mol of solute}}{1000}}{\dfrac{\text{L of solution}}{1000}} = \frac{\text{mmol of solute}}{\text{mL of solution}}$$

A 1.0 M solution thus contains 1.0 mole of solute per liter of solution or, *equivalently*, 1.0 millimole of solute per milliliter of solution. Just as we obtain the number of moles of solute from the product of the volume in liters and the molarity, we obtain the number of millimoles of solute from the product of the volume in milliliters and the molarity.

We will illustrate the calculations involved in a strong acid–strong base titration by considering the titration of 50.0 mL of 0.200 M HNO_3 with 0.100 M NaOH. We will calculate the pH of the solution at selected points during the course of the titration where specific volumes of 0.100 M NaOH have been added.

- **No NaOH has been added.**
Since HNO_3 is a strong acid (is completely dissociated), the solution contains the major species

$$H^+, \quad NO_3^-, \quad H_2O$$

The pH is determined by the H^+ from the nitric acid. Since 0.200 M HNO_3 contains 0.20 M H^+,

$$[H^+] = 0.20\ M \quad \text{and} \quad pH = 0.70$$

- **10.0 mL of 0.100 M NaOH have been added.**
In the mixed solution *before any reaction occurs*, the major species are

$$H^+, \quad NO_3^-, \quad Na^+, \quad OH^-, \quad H_2O$$

Note that large quantities of both H^+ and OH^- are present. The 1.0 mmol (10.0 mL × 0.10 M) of added OH^- will react with 1.0 mmol of H^+ to form water:

	H^+	+	OH^-	→	H_2O
Before reaction:	50.0 mL × 0.200 M = 10.0 mmol		10.0 mL × 0.10 M = 1.00 mmol		
After reaction:	10.0 − 1.0 = 9.0 mmol		1.0 − 1.0 = 0		

After the reaction is complete the solution contains

$$H^+, \quad NO_3^-, \quad Na^+, \quad H_2O \quad \text{(The } OH^- \text{ ions have been consumed.)}$$

The pH will be determined by the H^+ remaining:

$$[H^+] = \frac{\text{mmol } H^+ \text{ left}}{\text{volume of solution (mL)}} = \frac{9.0 \text{ mmol}}{(50.0 + 10.0) \text{ mL}} = 0.15\ M$$

↗ Original volume of HNO_3 solution ↖ Volume of NaOH added

The final solution volume is the sum of the original volume of HNO_3 and the volume of added NaOH.

$$pH = -\log(0.15) = 0.82$$

- **20.0 mL (total) of 0.100 *M* NaOH have been added.**
We consider this point from the perspective that a total of 20.0 mL of NaOH has been added to the *original* solution, rather than that 10.0 mL have been added to the solution from the previous point. It is best to go back to the original solution each time a calculation is performed, so a mistake made at an earlier point does not show up in each succeeding calculation. As before, the added OH^- will react with H^+ to form water:

Go back to the original solution each time you perform a calculation to avoid mistakes.

	H^+	+	OH^-	$\rightarrow$	H_2O
Before reaction:	50.0 mL × 0.200 *M* = 10.0 mmol		20.0 mL × 0.10 *M* = 2.0 mmol		
After reaction:	10.0 − 2.0 = 8.0 mmol		2.0 − 2.0 = 0 mmol		

After the reaction

$$[H^+] = \frac{8.0 \text{ mmol} \;(H^+ \text{ remaining})}{(50.0 + 20.0) \text{ mL}} = 0.11\ M$$

$$\text{pH} = 0.94$$

- **50.0 mL (total) of 0.100 *M* NaOH have been added.**
Proceeding exactly as for the previous two points, we find the pH to be 1.30.

- **100.0 mL (total) of 0.100 *M* NaOH have been added.**
At this point the amount of NaOH that has been added is

$$100.0 \text{ mL} \times 0.100\ M = 10.0 \text{ mmol}$$

The original amount of nitric acid was

$$50.0 \text{ mL} \times 0.200\ M = 10.0 \text{ mmol}$$

Enough OH^- has been added to react exactly with all of the H^+ from the nitric acid. This is the **stoichiometric point,** or **equivalence point,** of the titration. At this point the major species in solution are

$$Na^+, \quad NO_3^-, \quad H_2O$$

Since Na^+ has no acid or base properties and NO_3^- is the anion of the strong acid HNO_3 and is therefore a very weak base, neither NO_3^- nor Na^+ affects the pH. Thus the solution is neutral (the pH is 7.00).

- **150.0 mL (total) of 0.100 *M* NaOH have been added.**
The titration reaction is as follows:

	H^+	+	OH^-	$\rightarrow$	H_2O
Before reaction:	50.0 mL × 0.200 *M* = 10.0 mmol		150.00 mL × 0.100 *M* = 15.0 mmol		
After: reaction:	10.0 − 10.0 = 0 mmol		15.0 − 10.0 = 5.0 mmol ↑ Excess OH^- added		

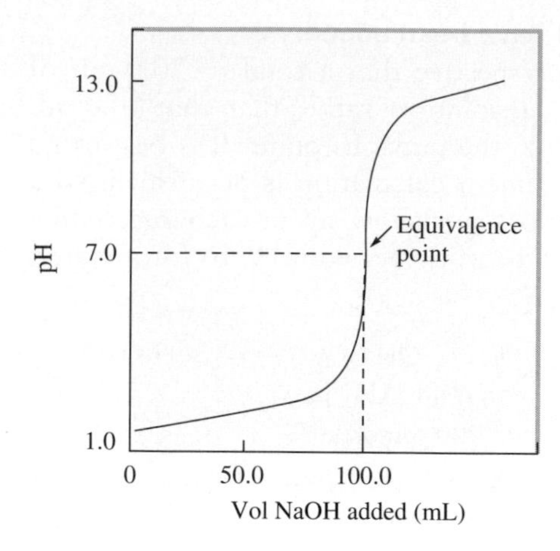

Figure 8.1
The pH curve for the titration of 50.0 mL of 0.200 M HNO_3 with 0.100 M NaOH. Note that the equivalence point occurs when 100.0 mL of NaOH have been added, the point where exactly enough OH^- has been added to react with all of the H^+ originally present. The pH of 7 at the equivalence point is characteristic of a strong acid–strong base titration.

Now the OH^- is *in excess* and thus will determine the pH:

$$[OH^-] = \frac{\text{mmol } OH^- \text{ in excess}}{\text{volume (mL)}} = \frac{5.0 \text{ mmol}}{(50.0 + 150.0)\text{ mL}} = \frac{5.0 \text{ mmol}}{200.0 \text{ mL}}$$
$$= 0.025\ M$$

Since $[H^+][OH^-] = 1.0 \times 10^{-14}$,

$$[H^+] = \frac{1.0 \times 10^{-14}}{2.5 \times 10^{-2}} = 4.0 \times 10^{-13}\ M \quad \text{and} \quad \text{pH} = 12.40$$

- **200.0 mL (total) of 0.100 *M* NaOH have been added.**
Proceeding as for the previous point, we find the pH to be 12.60.

The results of these calculations are summarized by the pH curve shown in Fig. 8.1. Note that the pH changes very gradually until the titration is close to the equivalence point, where a dramatic change occurs. This behavior is due to the fact that early in the titration there is a relatively large amount of H^+ in the solution, and the addition of a given amount of OH^- thus produces only a small change in pH. However, near the equivalence point $[H^+]$ is relatively small, and the addition of a small amount of OH^- produces a large change.

The pH curve in Fig. 8.1, typical of the titration of a strong acid with a strong base, has the following characteristics:

Before the equivalence point, $[H^+]$ (and hence the pH) can be calculated by dividing the number of millimoles of H^+ remaining at that point by the total volume of the solution in milliliters.

At the equivalence point the pH is 7.00.

After the equivalence point $[OH^-]$ can be calculated by dividing the number of millimoles of excess OH^- by the total volume of the solution. Then $[H^+]$ is obtained from K_w.

The titration of a strong base with a strong acid requires reasoning very similar to that used above, except, of course, that OH^- is in excess before the equivalence point and H^+ is in excess after the equivalence point. The pH curve for the titration of 100.0 mL of 0.50 M NaOH with 1.0 M HCl is shown in Fig. 8.2.

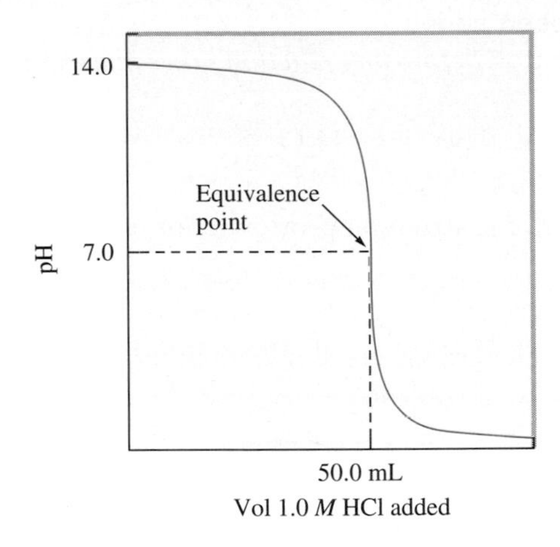

Figure 8.2
The pH curve for the titration of 100.0 mL of 0.50 *M* NaOH with 1.0 *M* HCl. The equivalence point occurs when 50.00 mL of HCl have been added, since at this point 5.0 mmol of H^+ ions have been added to react with the original 5.0 mmol of OH^- ions.

Titrations of Weak Acids with Strong Bases

We have seen that since strong acids and strong bases are completely dissociated, the calculations required to obtain the pH curves for titrations involving the two are quite straightforward. However, when the acid being titrated is a weak acid, there is a major difference: to calculate $[H^+]$ after a certain amount of strong base has been added, we must deal with the weak acid dissociation equilibrium. We dealt with this same type of situation earlier in this chapter when we treated buffered solutions. Calculation of the pH curve for a titration of a weak acid with a strong base really amounts to a series of buffer problems. In performing these calculations, it is very important to remember that even though the acid is weak, it *reacts essentially to completion* with hydroxide ion, a very strong base.

Calculating the pH curve for a weak acid–strong base titration involves the following two-step procedure.

Summary: **Titration Curve Calculations**

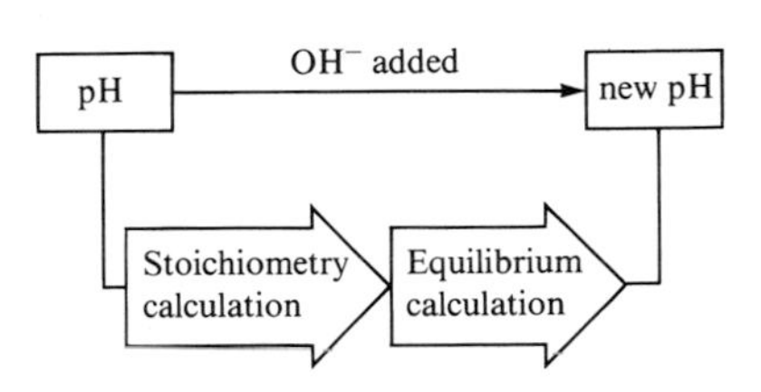

- ***A stoichiometry problem.*** **The reaction of hydroxide ion with the weak acid is assumed to run to completion, and the concentrations of the acid *remaining* and the conjugate base *formed* are determined.**
- ***An equilibrium problem.*** **The position of the weak acid equilibrium is determined, and the pH is calculated.**

It is *essential* to do these steps *separately*. Note that the procedures necessary for solving these problems have all been used before.

Treat the stoichiometry and equilibrium problems separately.

As an illustration, we will consider the titration of 50.0 mL of 0.10 *M* acetic acid ($HC_2H_3O_2$, $K_a = 1.8 \times 10^{-5}$) with 0.10 *M* NaOH. As before, we will calculate the pH at various points representing volumes of added NaOH.

- **No NaOH has been added.**
  This is a typical weak acid calculation of the type introduced in Chapter 7. The pH is 2.87. (Check this value yourself.)

- **10.0 mL of 0.10 *M* NaOH have been added.**
The major species in the mixed solution *before any reaction takes place* are

$$HC_2H_3O_2, \quad OH^-, \quad Na^+, \quad H_2O$$

The strong base OH^- will react with the strongest proton donor, which in this case is $HC_2H_3O_2$.

THE STOICHIOMETRY PROBLEM: The calculations are shown in tabular form.

	OH^-	+	$HC_2H_3O_2$	$\rightarrow$	$C_2H_3O_2^-$	+	H_2O
Before reaction:	$10\ \text{mL} \times 0.10\ M = 1.0\ \text{mmol}$		$50.0\ \text{mL} \times 0.10\ M = 5.0\ \text{mmol}$		0 mmol		
After reaction:	$1.0 - 1.0 = 0\ \text{mmol}$ ↑ Limiting reactant		$5.0 - 1.0 = 4.0\ \text{mmol}$		1.0 mmol		

THE EQUILIBRIUM PROBLEM: We examine the major components left in the solution *after the reaction takes place* to select the dominant equilibrium. The major species are

$$HC_2H_3O_2, \quad C_2H_3O_2^-, \quad Na^+, \quad H_2O$$

Since $HC_2H_3O_2$ is a much stronger acid than H_2O, and since $C_2H_3O_2^-$ is the conjugate base of $HC_2H_3O_2$, the pH will be determined by the position of the acetic acid dissociation equilibrium:

$$HC_2H_3O_2(aq) \rightleftharpoons H^+(aq) + C_2H_3O_2^-(aq)$$

where
$$K_a = \frac{[H^+][C_2H_3O_2^-]}{[HC_2H_3O_2]}$$

We follow the usual steps to complete the equilibrium calculations:

The initial concentrations are defined after the reaction with OH^- has gone to completion but before any dissociation of $HC_2H_3O_2$ has occurred.

Initial Concentration		Equilibrium Concentration
$[HC_2H_3O_2]_0 = \dfrac{4.0\ \text{mmol}}{(50.0 + 10.0)\ \text{mL}} = \dfrac{4.0}{60.0}$	x mmol/mL $HC_2H_3O_2$ dissociates $\longrightarrow$	$[HC_2H_3O_2] = \dfrac{4.0}{60.0} - x$
$[C_2H_3O_2^-]_0 = \dfrac{1.0\ \text{mmol}}{(50.0 + 10.0)\ \text{mL}} = \dfrac{1.0}{60.0}$		$[C_2H_3O_2^-] = \dfrac{1.0}{60.0} + x$
$[H^+]_0 \approx 0$		$[H^+] = x$

Then

$$1.8 \times 10^{-5} = K_a = \frac{[H^+][C_2H_3O_2^-]}{[HC_2H_3O_2]} = \frac{x\left(\frac{1.0}{60.0} + x\right)}{\frac{4.0}{60.0} - x} \approx \frac{x\left(\frac{1.0}{60.0}\right)}{\frac{4.0}{60.0}}$$

$$= \left(\frac{1.0}{4.0}\right)x$$

$$x = \left(\frac{4.0}{1.0}\right)(1.8 \times 10^{-5}) = 7.2 \times 10^{-5} = [H^+]$$

and $\text{pH} = 4.14$

Note that the approximations made are well within 5% uncertainty limits.

- **25.0 mL (total) of 0.10 *M* NaOH have been added.**
  The procedure here is very similar to that used for the previous point and will be summarized briefly. The stoichiometry problem is as follows:

	OH^-	+	$HC_2H_3O_2$	→	$C_2H_3O_2^-$	+	H_2O
Before reaction:	25.0 mL × 0.10 *M* = 2.5 mmol		5.0 mL × 0.10 *M* = 5.0 mmol		0 mmol		
After reaction:	2.5 − 2.5 = 0		5.0 − 2.5 = 2.5 mmol		2.5 mmol		

After the reaction the major species in solution are

$$HC_2H_3O_2, \quad C_2H_3O_2^-, \quad Na^+, \quad H_2O$$

The equilibrium that will control the pH is

$$HC_2H_3O_2(aq) \rightleftharpoons H^+(aq) + C_2H_3O_2^-(aq)$$

Initial Concentration		Equilibrium Concentration
$[HC_2H_3O_2]_0 = \frac{2.5 \text{ mmol}}{(50.0 + 25.0) \text{ mL}}$	x mmol/mL $HC_2H_3O_2$ dissociates →	$[HC_2H_3O_2] = \frac{2.5}{75.0} - x$
$[C_2H_3O_2^-]_0 = \frac{2.5 \text{ mmol}}{(50.0 + 25.0) \text{ mL}}$		$[C_2H_3O_2^-] = \frac{2.5}{75.0} + x$
$[H^+]_0 \approx 0$		$[H^+] = x$

Therefore

$$1.8 \times 10^{-5} = K_a = \frac{[H^+][C_2H_3O_2^-]}{[HC_2H_3O_2]} = \frac{x\left(\frac{2.5}{75.0} + x\right)}{\frac{2.5}{75.0} - x} \approx \frac{x\left(\frac{2.5}{75.0}\right)}{\frac{2.5}{75.0}}$$

$$x = 1.8 \times 10^{-5} = [H^+] \quad \text{and} \quad \text{pH} = 4.74$$

This is a special point in the titration because it is *halfway to the equivalence point*. The original solution, 50.0 mL of 0.10 M $HC_2H_3O_2$, contained 5.0 mmol of $HC_2H_3O_2$. Thus 5.0 mmol of OH^- are required to reach the equivalence point. This corresponds to 50 mL of NaOH, since

$$(50.0 \text{ mL})(0.100\ M) = 5.00 \text{ mmol}$$

At this point, half of the acid has been used up, so $[HC_2H_3O_2] = [C_2H_3O_2^-]$

After 25.0 mL of NaOH have been added, half of the original $HC_2H_3O_2$ has been converted to $C_2H_3O_2^-$. At this point in the titration, $[HC_2H_3O_2]_0$ is equal to $[C_2H_3O_2^-]_0$. We can neglect the effect of dissociation; that is,

$$[HC_2H_3O_2] = [HC_2H_3O_2]_0 - x \approx [HC_2H_3O_2]_0$$
$$[C_2H_3O_2^-] = [C_2H_3O_2^-]_0 + x \approx [C_2H_3O_2^-]_0$$

The expression for K_a at the halfway point is

$$K_a = \frac{[H^+][C_2H_3O_2^-]}{[HC_2H_3O_2]} = \frac{[H^+][C_2H_3O_2^-]_0}{[HC_2H_3O_2]_0} = [H^+]$$

Equal at the halfway point

Thus *at the halfway point* in the titration,

$$[H^+] = K_a \quad \text{and} \quad pH = pK_a$$

- **40.0 mL (total) of 0.10 M NaOH have been added.**
The procedures required here are the same as those used for the previous two points. The pH is 5.35. (Check this value yourself.)

- **50.0 mL (total) of 0.10 M NaOH have been added.**
This point is the equivalence point of the titration; 5.0 mmol of OH^- have been added, which is just enough to react with the 5.0 mmol of $HC_2H_3O_2$ originally present. At this point the solution contains these major species:

$$Na^+, \quad C_2H_3O_2^-, \quad H_2O$$

Note that the solution contains $C_2H_3O_2^-$, which is a base. Remember that a base will combine with a proton, and the only source of protons in this solution is water. Thus the reaction will be

$$C_2H_3O_2^-(aq) + H_2O(l) \rightleftharpoons HC_2H_3O_2(aq) + OH^-(aq)$$

This is a *weak base* reaction characterized by K_b:

$$K_b = \frac{[HC_2H_3O_2][OH^-]}{[C_2H_3O_2^-]} = \frac{K_w}{K_a} = \frac{1.0 \times 10^{-14}}{1.8 \times 10^{-5}} = 5.6 \times 10^{-10}$$

The relevant concentrations are:

Initial Concentration (before any $C_2H_3O_2^-$ reacts with H_2O)		Equilibrium Concentration
$[C_2H_3O_2^-]_0 = \frac{5.0 \text{ mmol}}{(50.0 + 50.0) \text{ mL}} = 0.050\ M$	$\xrightarrow[\text{with } H_2O]{x \text{ mmol/mL } C_2H_3O_2^- \text{ reacts}}$	$[C_2H_3O_2^-] = 0.050 - x$
$[OH^-]_0 \approx 0$		$[OH^-] \approx x$
$[HC_2H_3O_2]_0 = 0$		$[HC_2H_3O_2] = x$

Then

$$5.6 \times 10^{-10} = K_b = \frac{[HC_2H_3O_2][OH^-]}{[C_2H_3O_2^-]} = \frac{(x)(x)}{0.050 - x} \approx \frac{x^2}{0.050}$$

$$x = 5.3 \times 10^{-6}$$

The approximation is valid (by the 5% rule), so

$$[OH^-] = 5.3 \times 10^{-6}\ M$$

and

$$[H^+][OH^-] = K_w = 1.0 \times 10^{-14}$$

$$[H^+] = 1.9 \times 10^{-9}\ M \quad \text{and} \quad \text{pH} = 8.72$$

This is another important result: *the pH at the equivalence point of a titration of a weak acid with a strong base is always greater than 7.* This result occurs because the anion of the acid, which remains in solution at the equivalence point, is a base. In contrast, for the titration of a strong acid with a strong base, the pH at the equivalence point is 7, because the anion remaining in this case is *not* an effective base.

The pH at the equivalence point of a titration of a weak acid with a strong base is always greater than 7.

- **60.0 mL (total) of 0.10 *M* NaOH have been added.**
At this point excess OH^- has been added:

	OH^-	+	$HC_2H_3O_2$	→	$C_2H_3O_2^-$	+	H_2O
Before reaction:	$60.0 \text{ mL} \times 0.10\ M = 6.0 \text{ mmol}$		$50.0 \text{ mL} \times 0.10\ M = 5.0 \text{ mmol}$		0 mmol		
After reaction:	$6.0 - 5.0 = 1.0$ mmol in excess		$5.0 - 5.0 = 0$		5.0 mmol		

After the reaction is complete, the solution contains these major species:

$$Na^+, \quad C_2H_3O_2^-, \quad OH^-, \quad H_2O$$

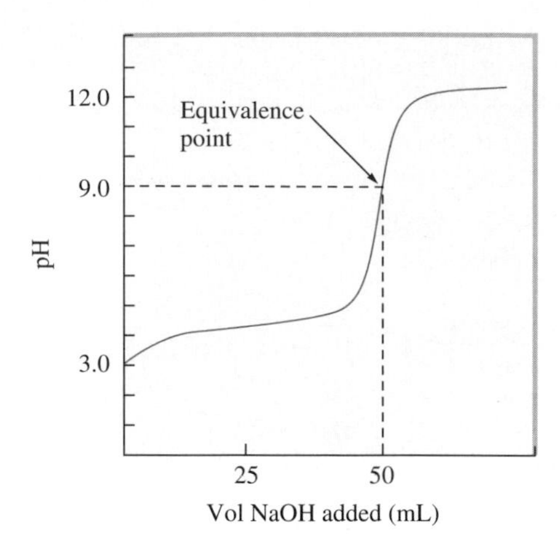

Figure 8.3
The pH curve for the titration of 50.0 mL of 0.100 *M* $HC_2H_3O_2$ with 0.100 *M* NaOH. Note that the equivalence point occurs when 50.0 mL of NaOH have been added, where the amount of added OH^- ions exactly equals the original amount of acid. The pH at the equivalence point is greater than 7 because the $C_2H_3O_2^-$ ion present at this point is a base and reacts with water to produce OH^-.

There are two bases in this solution, OH^- and $C_2H_3O_2^-$. However, $C_2H_3O_2^-$ is a weak base compared with OH^-. Therefore the amount of OH^- produced by the reaction of $C_2H_3O_2^-$ with H_2O will be small compared with the excess OH^- already in solution. You can verify this conclusion by looking at the previous point, where only 5.3×10^{-6} *M* OH^- was produced by $C_2H_3O_2^-$. The amount in this case will be even smaller, since the excess OH^- will push the K_b equilibrium to the left.

Thus the pH is determined by the excess OH^-:

$$[OH^-] = \frac{\text{mmol of } OH^- \text{ in excess}}{\text{volume (in mL)}} = \frac{1.0 \text{ mmol}}{(50.0 + 60.0) \text{ mL}}$$
$$= 9.1 \times 10^{-3}\ M$$

$$\text{and } [H^+] = \frac{1.0 \times 10^{-14}}{9.1 \times 10^{-3}} = 1.1 \times 10^{-12}\ M, \qquad pH = 11.96$$

- **75.0 mL (total) of 0.10 *M* NaOH have been added.**
The procedure needed here is very similar to that for the previous point. The pH is 12.30. (Check this value.)

The pH curve for this titration is shown in Fig. 8.3. Note the differences between this curve and the curve in Fig. 8.1. For example, the shapes of the plots are quite different before the equivalence point, but they are very similar after that point. (The shapes of the strong and weak acid curves are the same after the equivalence points because excess OH^- controls the pH in this region in both cases.) Near the beginning of the titration of the weak acid, the pH increases more rapidly than it does in the strong acid case. It levels off near the halfway point and then increases rapidly again. The leveling off near the halfway point is caused by buffering effects. Earlier in this chapter we saw that optimum buffering occurs when [HA] is equal to $[A^-]$. This is exactly the case at the halfway point of the titration. As we can see from the curve, the pH changes least around this point in the titration.

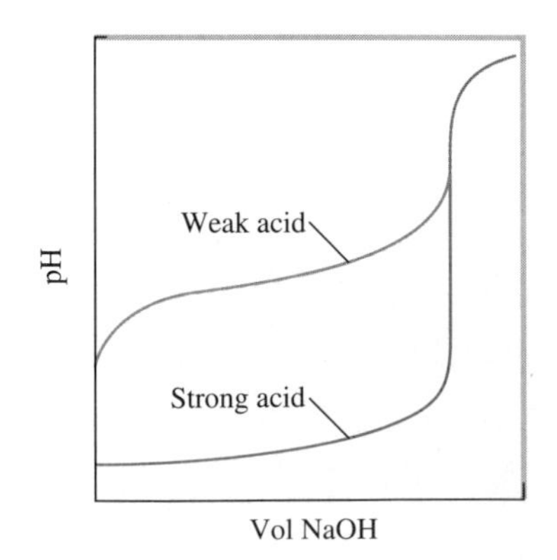

The other notable difference between the curves for strong and weak acids is the value of the pH at the equivalence point. For the titration of a strong acid, the equivalence point occurs at a pH of 7. For the titration of a weak acid, the pH at the equivalence point is greater than 7 because of the basicity of the conjugate base of the weak acid.

It is important to understand that the equivalence point in an acid-base titration is *defined by the stoichiometry, not by the pH*. The equivalence point occurs when enough titrant has been added to react exactly with all of the acid or base being titrated.

The equivalence point is defined by the stoichiometry, not by the pH.

EXAMPLE 8.6

Hydrogen cyanide gas (HCN) is a powerful respiratory inhibitor that is highly toxic. It is a very weak acid ($K_a = 6.2 \times 10^{-10}$) when dissolved in water. If a 50.0-mL sample of 0.100 *M* HCN is titrated with 0.100 *M* NaOH, calculate the pH of the solution at the following points.

a. after 8.00 mL of 0.100 *M* NaOH have been added

Solution

THE STOICHIOMETRY PROBLEM: After 8.00 mL of 0.100 *M* NaOH have been added, we have the following:

	HCN	+	OH^-	→	CN^-	+ H_2O
Before reaction:	50.0 mL × 0.100 *M* = 5.00 mmol		8.00 mL × 0.100 *M* = 0.800 mmol		0 mmol	
After reaction:	5.00 − 0.80 = 4.20 mmol		0		0.800 mmol	

THE EQUILIBRIUM PROBLEM: Since the solution contains the major species

$$HCN, \quad CN^-, \quad Na^+, \quad H_2O$$

the position of the acid dissociation equilibrium

$$HCN(aq) \rightleftharpoons H^+(aq) + CN^-(aq)$$

will determine the pH.

Initial Concentration		Equilibrium Concentration
$[HCN]_0 = \dfrac{4.2 \text{ mmol}}{(50.0 + 8.0) \text{ mL}}$		$[HCN] = \dfrac{4.2}{58.0} - x$
$[CN^-]_0 = \dfrac{0.800 \text{ mmol}}{(50.0 + 8.0) \text{ mL}}$	x mmol/mL HCN dissociates ⟶	$[CN^-] = \dfrac{0.80}{58.0} + x$
$[H^+]_0 \approx 0$		$[H^+] = x$

The approximations made here are well within 5% uncertainty limits.

Substituting into the expression for K_a gives

$$6.2 \times 10^{-10} = K_a = \frac{[H^+][CN^-]}{[HCN]} = \frac{x\left(\frac{0.80}{58.0} + x\right)}{\frac{4.2}{58.0} - x} \approx \frac{x\left(\frac{0.80}{58.0}\right)}{\left(\frac{4.2}{58.0}\right)}$$

$$= x\left(\frac{0.80}{4.2}\right)$$

$$x = 3.3 \times 10^{-9}\ M = [H^+] \quad \text{and} \quad pH = 8.49$$

b. at the halfway point in the titration

Solution

The amount of HCN originally present can be obtained from the original volume and molarity:

$$50.0\ \text{mL} \times 0.100\ M = 5.00\ \text{mmol}$$

Thus the halfway point will occur when 2.50 mmol of OH^- have been added:

$$\text{Volume of NaOH (in mL)} \times 0.100\ M = 2.50\ \text{mmol}\ OH^-$$

or

$$\text{Volume of NaOH} = 25.0\ \text{mL}$$

As was pointed out previously, at the halfway point, [HCN] is equal to $[CN^-]$ and pH is equal to pK_a. Thus after 25.0 mL of 0.100 M NaOH have been added,

$$pH = pK_a = -\log(6.2 \times 10^{-10}) = 9.21$$

c. at the equivalence point

Solution

The equivalence point will occur when a total of 5.00 mmol of OH^- has been added. Since the NaOH solution is 0.100 M, the equivalence point occurs when 50.0 mL of NaOH have been added. This results in the formation of 5.00 mmol of CN^-. The major species in solution at the equivalence point are

$$CN^-, \quad Na^+, \quad H_2O$$

Thus the reaction that will control the pH involves the basic cyanide ion extracting a proton from water:

$$CN^-(aq) + H_2O(l) \rightleftharpoons HCN(aq) + OH^-(aq)$$

and

$$K_b = \frac{K_w}{K_a} = \frac{1.0 \times 10^{-14}}{6.2 \times 10^{-10}} = 1.6 \times 10^{-5} = \frac{[HCN][OH^-]}{[CN^-]}$$

Initial Concentration		Equilibrium Concentration
$[CN^-]_0 = \frac{5.00 \text{ mmol}}{(50.0 + 50.0) \text{ mL}} = 5.00 \times 10^{-2} M$ $[HCN]_0 = 0$ $[OH^-]_0 \approx 0$	$\xrightarrow[\text{with } H_2O]{x \text{ mmol/mL of } CN^- \text{ reacts}}$	$[CN^-] = (5.00 \times 10^{-2}) - x$ $[HCN] = x$ $[OH^-] = x$

Substituting the equilibrium concentrations into the expression for K_b and solving in the usual way gives

$$[OH^-] = x = 8.9 \times 10^{-4}$$

Then from K_w we have

$$[H^+] = 1.1 \times 10^{-11} \quad \text{and} \quad pH = 10.96$$

Two important conclusions can be drawn from a comparison of the titration of 50.0 mL of 0.1 *M* acetic acid covered earlier in this section with that of 50.0 mL of 0.1 *M* hydrocyanic acid (Example 8.6). First, the same amount of 0.1 *M* NaOH is required to reach the equivalence point in both cases. The fact that HCN is a much weaker acid than $HC_2H_3O_2$ has no bearing on the amount of base required. It is the *amount* of acid, not its strength, that determines the equivalence point. Second, the *pH value* at the equivalence point *is* affected by the acid strength. For the titration of acetic acid, the pH at the equivalence point is 8.72; for the titration of hydrocyanic acid, the pH at the equivalence point is 10.96. This difference occurs because the CN^- ion is a much stronger base than the $C_2H_3O_2^-$ ion. Also, the pH at the halfway point of the titration is much higher for HCN than for $HC_2H_3O_2$, again because of the greater base strength of the CN^- ion (or equivalently, the smaller acid strength of HCN).

The amount of acid present, not its strength, determines the equivalence point.

The strength of a weak acid has a significant effect on the shape of its pH curve. Figure 8.4 shows pH curves for 50-mL samples of 0.10 *M* solutions of various acids titrated with 0.10 *M* NaOH. Note that the equivalence point occurs at the same volume of 0.10 *M* NaOH for each case, but that the shapes of the curves are dramatically different. The weaker the acid, the greater the pH value at the equivalence point. In particular, note that the vertical region surrounding the equivalence point becomes shorter as the acid being titrated becomes weaker. We will see in the next section how this affects the choice of an indicator for such a titration.

Besides being used to analyze for the amount of acid or base in a solution, titrations can also be used to determine the values of equilibrium constants, as shown in Example 8.7.

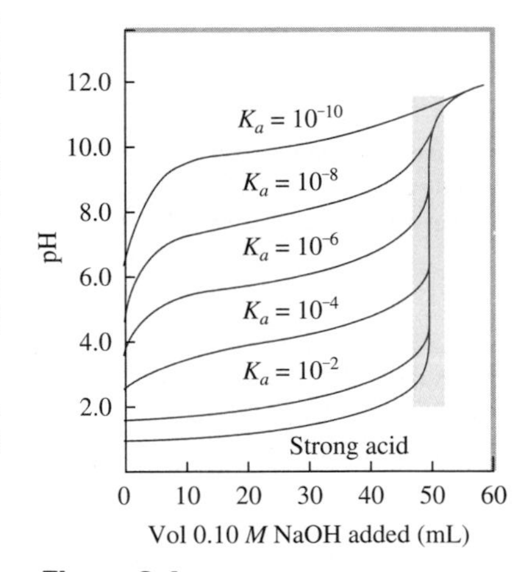

Figure 8.4
The pH curves for the titrations of 50.0-mL samples of 0.10 *M* solutions of various acids with 0.10 *M* NaOH.

EXAMPLE 8.7

A chemist has synthesized a monoprotic weak acid and wants to determine its K_a value. To do so, the chemist dissolves 2.00 mmol of the solid acid in 100.0 mL of water and titrates the resulting solution with 0.0500 M NaOH. After 20.0 mL of NaOH has been added, the pH is 6.00. What is the K_a value for the acid?

Solution

2.00 mmol HA

↓ add OH^-

1.00 mmol HA
1.00 mmol A^-

THE STOICHIOMETRY PROBLEM: We represent the monoprotic acid as HA. The stoichiometry for the titration reaction is shown below.

	HA	+	OH^-	→	A^-	+ H_2O
Before reaction:	2.00 mmol		20.0 mL × 0.0500 M = 1.00 mmol		0 mmol	
After reaction:	2.00 − 1.00 = 1.00 mmol		1.0 − 1.0 = 0		1.00 mmol	

THE EQUILIBRIUM PROBLEM: After the reaction the solution contains the major species

$$HA, \quad A^-, \quad Na^+, \quad H_2O$$

The pH will be determined by the equilibrium

$$HA(aq) \rightleftharpoons H^+(aq) + A^-(aq)$$

$$K_a = \frac{[H^+][A^-]}{[HA]}$$

Initial Concentration		Equilibrium Concentration
$[HA]_0 = \frac{1.00 \text{ mmol}}{(100.0 + 20.0) \text{ mL}} = 8.33 \times 10^{-3} M$		$[HA] = 8.33 \times 10^{-3} - x$
$[A^-]_0 = \frac{1.00 \text{ mmol}}{(100.0 + 20.0) \text{ mL}} = 8.33 \times 10^{-3} M$	x mmol/mL HA dissociates →	$[A^-] = 8.33 \times 10^{-3} + x$
$[H^+]_0 \approx 0$		$[H^+] = x$

Note that x is actually known here because the pH at this point is known to be 6.00. Thus

$$x = [H^+] = \text{antilog}(-\text{pH}) = 1.0 \times 10^{-6} M$$

Substituting the equilibrium concentrations into the expression for K_a allows calculation of the K_a value:

$$K_a = \frac{[H^+][A^-]}{[HA]} = \frac{x(8.33 \times 10^{-3} + x)}{(8.33 \times 10^{-3}) - x}$$

$$= \frac{(1.0 \times 10^{-6})(8.33 \times 10^{-3} + 1.0 \times 10^{-6})}{(8.33 \times 10^{-3}) - (1.0 \times 10^{-6})}$$

$$= \frac{(1.0 \times 10^{-6})(8.33 \times 10^{-3})}{8.33 \times 10^{-3}} = 1.0 \times 10^{-6}$$

There is an easier way to think about this problem. The original solution contained 2.00 mmol of HA, and since 20.0 mL of added 0.0500 *M* NaOH contain 1.00 mmol of OH^-, this is the halfway point in the titration (where [HA] is equal to $[A^-]$). Thus

$$[H^+] = K_a = 1.0 \times 10^{-6}$$

Titrations of Weak Bases with Strong Acids

Titrations of weak bases with strong acids can be treated using the procedures we have introduced previously. As always, you should *first think about the major species in solution* and decide whether a reaction occurs that runs essentially to completion. If such a reaction does occur, let it run to completion and then do the stoichiometric calculations. Finally, choose the dominant equilibrium and calculate the pH.

The calculations involved for the titration of a weak base with a strong acid are illustrated by the following titration of 100.0 mL of 0.050 *M* NH_3 with 0.10 *M* HCl. The strategies needed at several key areas in the titration will be described qualitatively. The actual calculations are summarized in Table 8.2.

- **Before the addition of any HCl.**

1. Major species:

$$NH_3, \quad H_2O$$

NH_3 is a base and will seek a source of protons. In this case H_2O is the only available source of protons.

2. No reactions occur that go to completion, since NH_3 cannot readily take a proton from H_2O. This is evidenced by the small K_b value for NH_3.

TABLE 8.2 Summary of Results for the Titration of 100.0 mL 0.050 *M* NH_3 with 0.10 *M* HCl

Volume of 0.10 *M* HCl added (mL)	$[NH_3]_0$	$[NH_4^+]_0$	$[H^+]$	pH
0	0.05 *M*	0	1.1×10^{-11} *M*	10.96
10.0	$\frac{4.0 \text{ mmol}}{(100 + 10) \text{ mL}}$	$\frac{1.0 \text{ mmol}}{(100 + 10) \text{ mL}}$	1.4×10^{-10} *M*	9.85
25.0*	$\frac{2.5 \text{ mmol}}{(100 + 25) \text{ mL}}$	$\frac{2.5 \text{ mmol}}{(100 + 25) \text{ mL}}$	5.6×10^{-10} *M*	9.25
50.0†	0	$\frac{5.0 \text{ mmol}}{(100 + 50) \text{ mL}}$	4.3×10^{-6} *M*	5.36
60.0‡	0	$\frac{5.0 \text{ mmol}}{(100 + 60) \text{ mL}}$	$\frac{1.0 \text{ mmol}}{160 \text{ mL}} = 6.2 \times 10^{-3}$ *M*	2.21

* Halfway point.
† Equivalence point.
‡ $[H^+]$ determined by the 1.0 mmol of excess H^+.

3. The equilibrium that controls the pH involves the reaction of ammonia with water:

$$NH_3(aq) + H_2O(l) \rightleftharpoons NH_4^+(aq) + OH^-(aq)$$

Use K_b to calculate $[OH^-]$. Although NH_3 is a weak base (compared with OH^-), it produces much more OH^- in this reaction than is produced from the autoionization of H_2O.

- **Before the equivalence point.**

1. Major species (before any reaction occurs):

$$NH_3, \quad \underbrace{H^+, \quad Cl^-,}_{\text{From added HCl}} \quad H_2O$$

2. The NH_3 will react with H^+ from the added HCl:

$$NH_3(aq) + H^+(aq) \rightleftharpoons NH_4^+(aq)$$

This reaction proceeds essentially to completion, because the NH_3 readily reacts with a free proton. This case is much different from the previous case, where H_2O was the only source of protons. The stoichiometric calculations are then carried out using the known volume of 0.10 M HCl added.

3. After the reaction of NH_3 with H^+ is run to completion, the solution contains the following major species:

$$NH_3, \quad NH_4^+, \quad Cl^-, \quad H_2O$$

(NH_4^+: Formed in titration reaction)

The solution contains NH_3 and NH_4^+; thus the equilibria involving these species will determine $[H^+]$. The $[H^+]$ can be calculated using either the dissociation reaction of NH_4^+,

$$NH_4^+(aq) \rightleftharpoons NH_3(aq) + H^+(aq)$$

or the reaction of NH_3 with H_2O,

$$NH_3(aq) + H_2O(l) \rightleftharpoons NH_4^+(aq) + OH^-(aq)$$

to calculate the pH.

- **At the equivalence point.**

1. By definition, the equivalence point occurs when all of the original NH_3 is converted to NH_4^+. Thus the major species in solution are

$$NH_4^+, \quad Cl^-, \quad H_2O$$

2. No reactions occur that go to completion.
3. The equilibrium that controls $[H^+]$ is the dissociation of the weak acid NH_4^+, for which

$$K_a = \frac{K_w}{K_b \text{ (for } NH_3)}$$

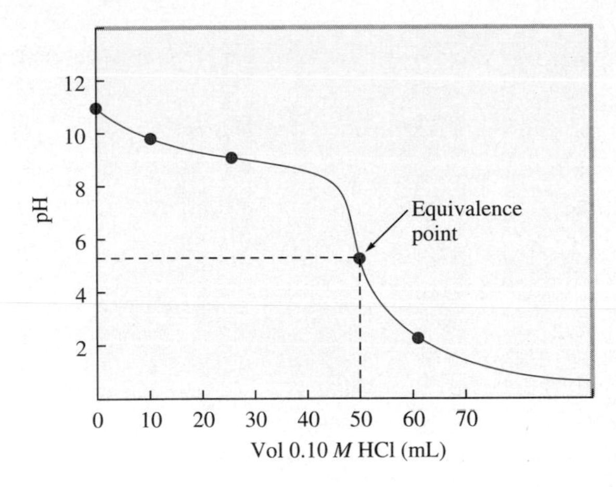

Figure 8.5
The pH curve for the titration of 100.0 mL of 0.050 *M* NH_3 with 0.10 *M* HCl. Note that the pH at the equivalence point is less than 7, since the solution contains the weak acid NH_4^+.

- **Beyond the equivalence point.**
  1. Excess HCl has been added and the major species are

$$H^+, \quad NH_4^+, \quad Cl^-, \quad H_2O$$

  2. No reaction occurs that goes to completion.
  3. Although NH_4^+ will dissociate, it is such a weak acid that $[H^+]$ will be determined simply by the excess H^+:

$$[H^+] = \frac{\text{mmol of } H^+ \text{ in excess}}{\text{mL of solution}}$$

The results of these calculations are shown in Table 8.2. The pH curve is shown in Fig. 8.5.

8.5 Acid-Base Indicators

There are two common methods for determining the equivalence point of an acid-base titration:

1. Use a pH meter to monitor the pH and then plot a titration curve. The center of the vertical region of the pH curve indicates the equivalence point (for example, see Figs. 8.1 through 8.5).
2. Use an **acid-base indicator,** which marks the end point of a titration by changing color. Although the *equivalence point of a titration, defined by the stoichiometry, is not necessarily the same as the end point* (where the indicator changes color), careful selection of the indicator will ensure only negligible error.

The *end point* is defined by the change in color of the indicator. The *equivalence point* is defined by the reaction stoichiometry.

The most common acid-base indicators are complex molecules that are themselves weak acids and are represented by HIn. They exhibit one color when the proton is attached to the molecule and a different color when the proton is absent. For example, **phenolphthalein,** a commonly used indicator, is colorless in its HIn form and pink in its In^-, or basic, form. (See Fig. 8.6.)

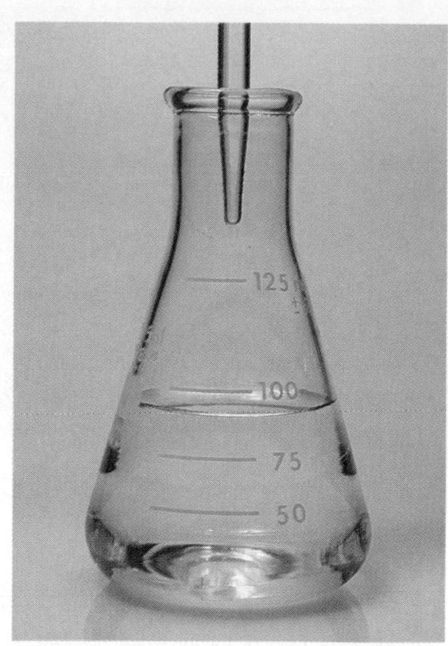

Figure 8.6
The indicator phenolphthalein is pink in basic solution and colorless in acidic solution.

To see how molecules function as indicators, consider the following equilibrium for some hypothetical indicator HIn, a weak acid with $K_a = 1.0 \times 10^{-8}$:

$$\underset{\text{Red}}{\text{HIn}(aq)} \rightleftharpoons \text{H}^+(aq) + \underset{\text{Blue}}{\text{In}^-(aq)}$$

$$K_a = \frac{[\text{H}^+][\text{In}^-]}{[\text{HIn}]}$$

By rearranging, we get

$$\frac{K_a}{[\text{H}^+]} = \frac{[\text{In}^-]}{[\text{HIn}]}$$

Suppose we add a few drops of this indicator to an acidic solution whose pH is 1.0 ($[\text{H}^+] = 1.0 \times 10^{-1}$). Then

$$\frac{K_a}{[\text{H}^+]} = \frac{1. \times 10^{-8}}{1.0 \times 10^{-1}} = 10^{-7} = \frac{[\text{In}^-]}{[\text{HIn}]} = \frac{1}{10{,}000{,}000}$$

This ratio shows that the predominant form of the indicator is HIn, resulting in a red solution. As OH^- is added to this solution in a titration, $[H^+]$ decreases and the equilibrium shifts to the right, changing HIn to In^-. At some point in the titration, enough of the In^- form will be present in the solution so that a purple tint will be noticeable. That is, a color change from red to reddish purple will occur.

How much In^- must be present in the solution for the human eye to detect that the color is different from the original one? For most indicators about one-tenth of the initial form must be converted to the other form before a new color is apparent. We will assume, then, that in the titration of an acid with a base, the color change will be apparent at a pH where

$$\frac{[\mathrm{In}^-]}{[\mathrm{HIn}]} = \frac{1}{10}$$

EXAMPLE 8.8

Bromthymol blue, an indicator with a K_a value of 1.0×10^{-7}, is yellow in its HIn form and blue in its In^- form. Suppose we put a few drops of this indicator in a strongly acidic solution. If the solution is then titrated with NaOH, at what pH will the indicator color change first be visible?

Solution

For bromthymol blue,

$$K_a = 1.0 \times 10^{-7} = \frac{[\mathrm{H}^+][\mathrm{In}^-]}{[\mathrm{HIn}]}$$

We assume the color change is visible when

$$\frac{[\mathrm{In}^-]}{[\mathrm{HIn}]} = \frac{1}{10}$$

That is, we assume we can see the first hint of a greenish tint (yellow plus a little blue) when the solution contains one part blue and ten parts yellow (see Fig. 8.7). Thus

$$K_a = 1.0 \times 10^{-7} = \frac{[\mathrm{H}^+](1)}{10}$$

$$[\mathrm{H}^+] = 1.0 \times 10^{-6} \quad \text{or} \quad \mathrm{pH} = 6.0$$

The color change is first visible at a pH of 6.0.

The Henderson-Hasselbalch equation is very useful in determining the pH at which an indicator changes color. For example, application of Equation (8.2) to the K_a expression for the general indicator HIn yields

$$\mathrm{pH} = \mathrm{p}K_a + \log\left(\frac{[\mathrm{In}^-]}{[\mathrm{HIn}]}\right)$$

where K_a is the dissociation constant for the acid form (HIn). Since we assume that the color change is visible when

$$\frac{[\mathrm{In}^-]}{[\mathrm{HIn}]} = \frac{1}{10}$$

we have the following equation for determining the pH at which the color change occurs:

$$\mathrm{pH} = \mathrm{p}K_a + \log\left(\tfrac{1}{10}\right) = \mathrm{p}K_a - 1$$

(a)

(b)

(c)

Figure 8.7
(a) Yellow acid form of bromthymol blue; (b) a greenish tint is seen when the solution contains 1 part blue and 10 parts yellow; (c) blue basic form.

For bromthymol blue ($K_a = 1 \times 10^{-7}$, or $pK_a = 7$), the pH at the color change is

$$pH = 7 - 1 = 6$$

as we calculated in Example 8.8.

When a basic solution is titrated, the indicator HIn will initially exist as In^- in solution, but as acid is added, more HIn will form. In this case the color change will be visible when there is a mixture of 10 parts In^- to 1 part HIn. That is, a color change from blue to a blue-green color will occur (see Fig. 8.7) owing to the presence of some of the yellow HIn molecules. This color change will be first visible when

$$\frac{[In^-]}{[HIn]} = \frac{10}{1}$$

Note that this expression is the reciprocal of the ratio for the titration of an acid. Substituting this ratio into the Henderson-Hasselbalch equation gives

$$pH = pK_a + \log(\tfrac{10}{1}) = pK_a + 1$$

For bromthymol blue ($pK_a = 7$) we have a color change at

$$pH = 7 + 1 = 8$$

In summary, when bromthymol blue is used for the titration of an acid, the starting form will be HIn (yellow), and the color change occurs at a pH of about 6. When bromthymol blue is used for the titration of a base, the starting form is In^- (blue), and the color change occurs at a pH of about 8. Thus the useful pH range for bromthymol blue is

$$pK_a \text{ (bromthymol blue)} \pm 1 = 7 \pm 1$$

or from 6 to 8. The useful pH ranges for several common indicators are shown in Fig. 8.8.

When we choose an indicator for a titration, we want the indicator end point (where the color changes) and the titration equivalence point to be as close as possible. Choosing an indicator is easier if there is a large change in pH near the equivalence point of the titration. The dramatic change in pH near the equivalence point in a strong acid–strong base titration (Figs. 8.1 and 8.2) produces a sharp end point; that is, the complete color change (from the acid-to-base or the base-to-acid colors) usually occurs over one drop of added titrant.

What indicator should we use for the titration of 100.00 mL of 0.100 *M* HCl with 0.100 *M* NaOH? We know that the equivalence point occurs at pH 7.00. In the initially acidic solution the indicator will exist predominantly in the HIn form. As OH^- ions are added, the pH will increase rather slowly at first (Fig. 8.1) and then will rise rapidly at the equivalence point. This sharp change causes the indicator dissociation equilibrium

$$HIn \rightleftharpoons H^+ + In^-$$

to shift suddenly to the right, producing enough In^- ions to give a color change.

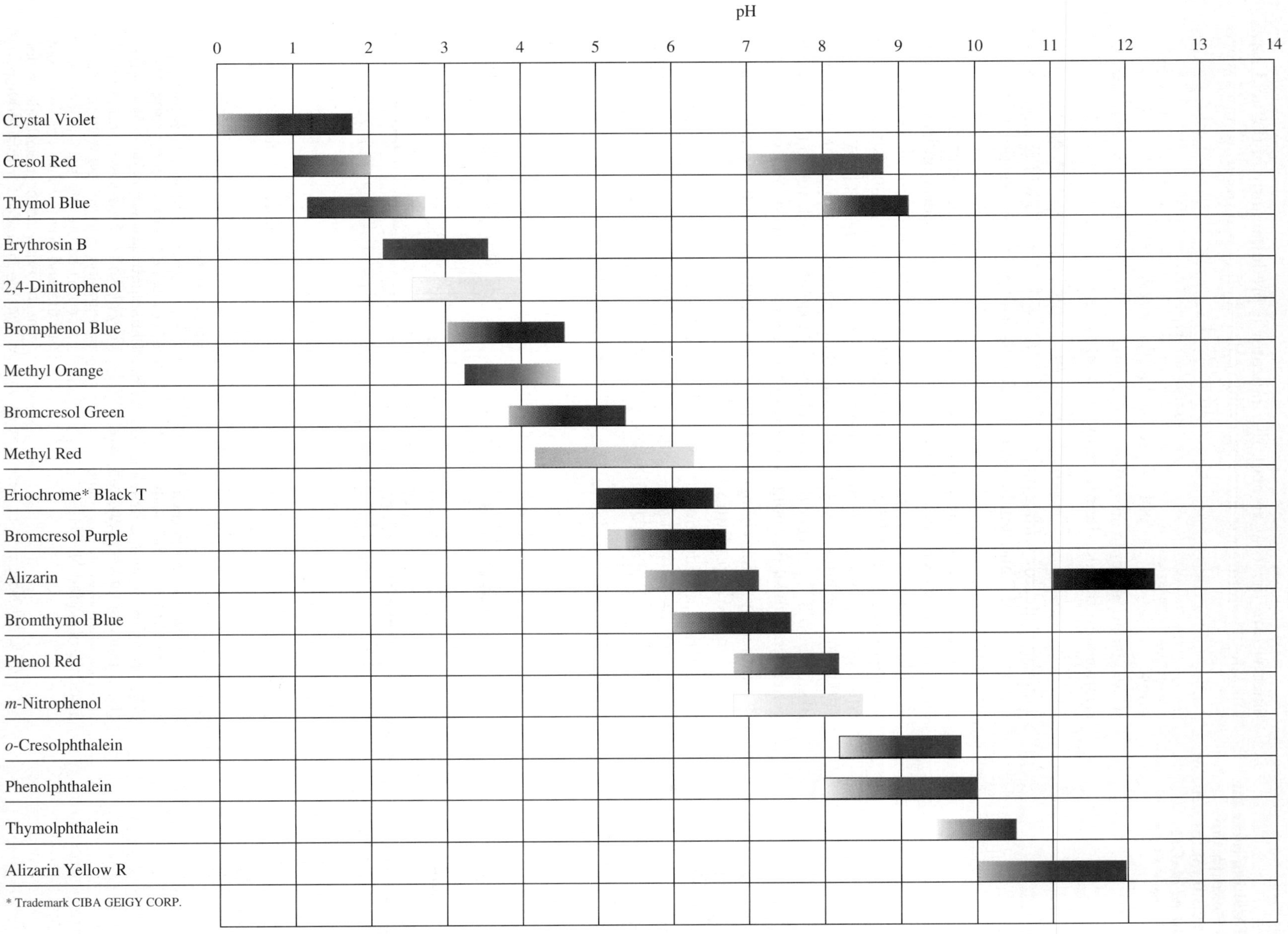

Figure 8.8
The useful pH ranges for several common indicators. Note that most indicators have a useful range of about two pH units, as predicted by the expression $pK_a \pm 1$.

TABLE 8.3
Selected pH Values near the Equivalence Point in the Titration of 100.0 mL of 0.10 *M* HCl with 0.10 *M* NaOH

NaOH added (mL)	pH
99.99	5.3
100.00	7.0
100.01	8.7

Since we are titrating an acid, the indicator is predominantly in the acid form initially. Therefore the first observable color change will occur at a pH where

$$\frac{[\mathrm{In}^-]}{[\mathrm{HIn}]} = \frac{1}{10}$$

Thus $$\mathrm{pH} = \mathrm{p}K_a + \log(\tfrac{1}{10}) = \mathrm{p}K_a - 1$$

If we want an indicator that will change color at pH 7, we can use this relationship to find the pK_a value of a suitable indicator:

$$\mathrm{pH} = 7 = \mathrm{p}K_a - 1 \quad \text{or} \quad \mathrm{p}K_a = 7 + 1 = 8$$

Thus an indicator with a pK_a value of 8 ($K_a = 1 \times 10^{-8}$) will change color at about a pH of 7 and will be ideal to mark the end point for a strong acid–strong base titration.

How crucial is it in a strong acid–strong base titration that the indicator change color exactly at a pH of 7? We can answer this question by examining the pH change near the equivalence point of the titration of 100.0 mL of 0.10 *M* HCl with 0.10 *M* NaOH. The data for a few points at or near the equivalence point are shown in Table 8.3. Note that in going from 99.99 mL to 100.01 mL of added NaOH solution (about half of a drop), the pH changes from 5.3 to 8.7—a very dramatic change. This behavior leads to the following general conclusions about indicators for a strong acid–strong base titration:

Indicator color changes will be sharp, occurring with the addition of a single drop of titrant.

There is a wide choice of suitable indicators. The results will agree within one drop of titrant, using indicators with end points as far apart as pH = 5 and pH = 9 (see Fig. 8.9).

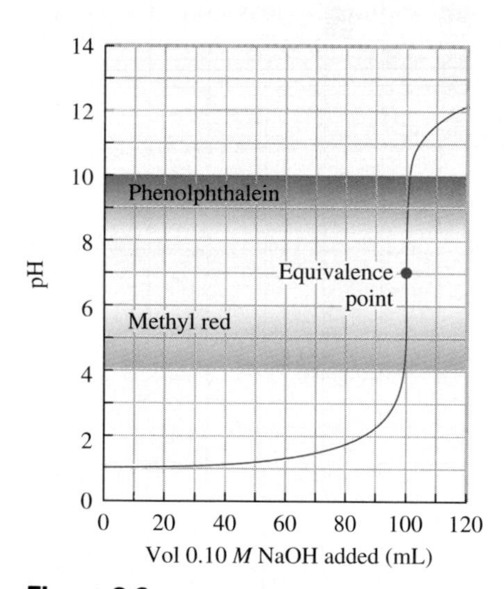

Figure 8.9
The pH curve for the titration of 100.0 mL of 0.10 *M* HCl with 0.10 *M* NaOH. Note that phenolphthalein and methyl red have end points at virtually the same amounts of added NaOH.

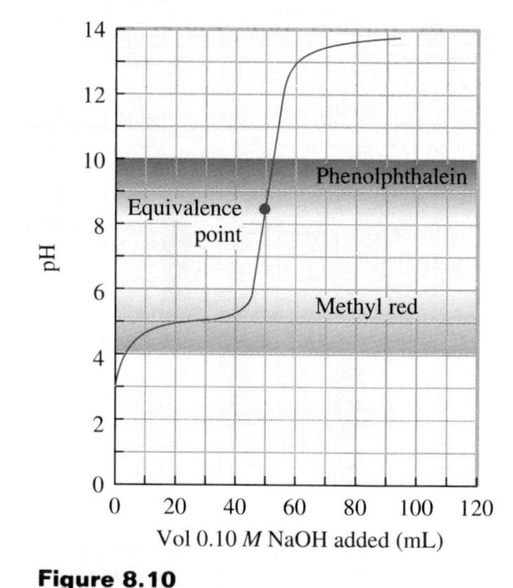

Figure 8.10
The pH curve for the titration of 50 mL of 0.1 *M* $HC_2H_3O_2$ with 0.1 *M* NaOH. Phenolphthalein will give an end point very close to the equivalence point of the titration. Methyl red will change color well before the equivalence point (so the end point will be very different from the equivalence point) and is not a suitable indicator for this titration.

The titration of weak acids is somewhat different. Figure 8.4 shows that the weaker the acid being titrated, the smaller the vertical area is around the equivalence point. This allows much less flexibility in choosing the indicator. We must choose an indicator whose useful pH range has a midpoint as close as possible to the pH at the equivalence point. For example, we saw earlier that in the titration of 0.1 M $HC_2H_3O_2$ with 0.1 M NaOH, the pH at the equivalence point is 8.7 (see Fig. 8.3). A good indicator choice is phenolphthalein, since its useful pH range is 8–10. Thymol blue is also acceptable, but methyl red is not. The choice of an indicator is illustrated graphically in Fig. 8.10.

8.6 Titration of Polyprotic Acids

The acid titrations we have considered so far have involved only monoprotic acids. When a polyprotic acid is titrated, the pH calculations are similar in many ways to those for a monoprotic acid, but enough differences exist to warrant special coverage.

In the titration of a typical polyprotic acid, the various acidic protons are titrated in succession. For example, as sodium hydroxide is used to titrate phosphoric acid, the first reaction that takes place can be represented as

$$H_3PO_4(aq) + OH^-(aq) \longrightarrow H_2PO_4^-(aq) + H_2O(l)$$

This reaction occurs until the H_3PO_4 is consumed (to reach the first equivalence point). Therefore, at the first equivalence point the solution contains the major species Na^+, $H_2PO_4^-$, and H_2O. Then as more sodium hydroxide is added, the reaction

$$H_2PO_4^-(aq) + OH^-(aq) \longrightarrow HPO_4^{2-}(aq) + H_2O(l)$$

occurs to give a solution that contains Na^+, HPO_4^{2-}, and H_2O as the major species at the second equivalence point. As sodium hydroxide is added beyond the second equivalence point, the reaction that occurs can be represented as

$$HPO_4^{2-}(aq) + OH^-(aq) \longrightarrow PO_4^{3-}(aq) + H_2O(l)$$

As mentioned above, the calculations involved in obtaining the pH curve for the titration of a polyprotic acid are closely related to those for a monoprotic acid. The same principles apply, but we must be very careful in identifying which of the various equilibria is appropriate to use in a given case. The secret to success here is, as always, identifying the major species in solution at any given point in the titration. We summarize the various cases in Table 8.4 for a triprotic acid H_3A with dissociation constants K_{a_1}, K_{a_2}, and K_{a_3}.

A point that cannot be overemphasized is that the appropriate equilibrium expression is chosen by knowing what major species are present. Thus if the solution contains HA^{2-} and A^{3-}, we must use K_{a_3} to determine the pH, and so on. Note that in two instances in Table 8.4 we did not specify the equilibrium expression to be used. These two cases need to be considered in more detail; they are discussed following the table.

TABLE 8.4 A Summary of Various Points in the Titration of a Triprotic Acid

Point in the Titration	Major Species Present	Equilibrium Expression Used to Obtain the pH
No base added	H_3A, H_2O	$K_{a_1} = \frac{[H^+][H_2A^-]}{[H_3A]}$
Base added		
Before the first equivalence point	H_3A, H_2A^-, H_2O	$K_{a_1} = \frac{[H^+][H_2A^-]}{[H_3A]}$
At the first equivalence point	H_2A^-, H_2O	See the following discussion
Between the first and second equivalence points	H_2A^-, HA^{2-}, H_2O	$K_{a_2} = \frac{[H^+][HA^{2-}]}{[H_2A^-]}$
At the second equivalence point	HA^{2-}, H_2O	See the following discussion
Between the second and third equivalence points	HA^{2-}, A^{3-}, H_2O	$K_{a_3} = \frac{[H^+][A^{3-}]}{[HA^{2-}]}$
At the third equivalence point	A^{3-}, H_2O	$K_b = \frac{K_w}{K_{a_3}} = \frac{[HA^{2-}][OH^-]}{[A^{3-}]}$
Beyond the third equivalence point	A^{3-}, OH^-, H_2O	pH determined by excess OH^-

Solutions Containing Amphoteric Anions as the Only Acid-Base Major Species

At the first equivalence point in the titration of the acid H_3A with sodium hydroxide, the major species are H_2A^- and H_2O. What equilibrium will control the $[H^+]$ in such a solution?

The key to answering this question is to recognize that H_2A^- is an amphoteric species: it is both an acid,

$$H_2A^-(aq) \rightleftharpoons H^+(aq) + HA^{2-}(aq)$$

and a base,

$$H_2A^-(aq) + H^+(aq) \rightleftharpoons H_3A(aq)$$

What is the best source of protons for H_2A^- behaving as a base? There are two possible sources: H_2O and other H_2A^- ions in the solution. By now we realize that H_2O is a very weak acid. In a typical case H_2A^- will be a much stronger acid than H_2O. Thus in a solution containing H_2A^- and H_2O as the major species, we expect the reaction

$$H_2A^-(aq) + H_2A^-(aq) \rightleftharpoons H_3A(aq) + HA^{2-}(aq)$$

to be the most important acid-base reaction. Notice that this reaction expresses both the acidic and basic properties of H_2A^- (in the above equation one H_2A^-

is behaving as the acid and the other H_2A^- as the base). This reaction leads to the equilibrium expression

$$K = \frac{[H_3A][HA^{2-}]}{[H_2A^-]^2}$$

We can obtain the value for K by recognizing that

$$\frac{[H_3A][HA^{2-}]}{[H_2A^-][H_2A^-]} = \frac{[H_3A]}{[H^+][H_2A^-]} \times \frac{[H^+][HA^{2-}]}{[H_2A^-]} = \frac{1}{K_{a_1}} \times K_{a_2}$$

Thus $$K = \frac{K_{a_2}}{K_{a_1}}$$

The position of this equilibrium will determine the concentrations of H_3A, H_2A^-, and HA^{2-} in the solution and thus will determine the pH. We can obtain an expression for $[H^+]$ from the equation

$$\frac{K_{a_2}}{K_{a_1}} = \frac{[H_3A][HA^{2-}]}{[H_2A^-]^2}$$

by recognizing that if the reaction

$$H_2A^-(aq) + H_2A^-(aq) \rightleftharpoons H_3A(aq) + HA^{2-}(aq)$$

is the only important reaction involving these species, then

$$[H_3A] = [HA^{2-}]$$

This condition allows us to write

$$\frac{K_{a_2}}{K_{a_1}} = \frac{[H_3A][HA^{2-}]}{[H_2A^-]^2} = \frac{[H_3A]^2}{[H_2A^-]^2}$$

From the expression for K_{a_1} we have

$$\frac{[H^+]}{K_{a_1}} = \frac{[H_3A]}{[H_2A^-]}$$

Thus $$\frac{K_{a_2}}{K_{a_1}} = \frac{[H_3A]^2}{[H_2A^-]^2} = \frac{[H^+]^2}{K_{a_1}{}^2}$$

or $$[H^+]^2 = K_{a_1}{}^2 \times \frac{K_{a_2}}{K_{a_1}} = K_{a_1}K_{a_2}$$

and $$[H^+] = \sqrt{K_{a_1}K_{a_2}}$$

In terms of pH this equation becomes

$$pH = \frac{pK_{a_1} + pK_{a_2}}{2}$$

Note that this equation applies only to a solution containing the major species H_2A^- and H_2O. It does not apply, for example, to a solution that contains the major species H_2A^-, HA^{2-}, and H_2O. In the former case H_2A^- is simultaneously the best acid and the best base, and thus the pH is determined by the reaction of H_2A^- with itself. In the latter case the solution contains the acid H_2A^- and its conjugate base HA^{2-}. In this case H_2A^- is the best acid and

EXAMPLE 8.9

Calculate the pH of a 1.0 *M* solution of NaH_2PO_4. (For H_3PO_4, $K_{a_1} = 7.5 \times 10^{-3}$, $K_{a_2} = 6.2 \times 10^{-8}$, and $K_{a_3} = 4.8 \times 10^{-13}$.)

Solution

The major species in solution are

$$Na^+, \quad H_2PO_4^-, \quad H_2O$$

This is an example of a solution containing the amphoteric anion $H_2PO_4^-$, which is, at the same time, the best acid and the best base. Both properties must be considered to calculate the pH correctly. Use the formula involving the average of the p*K*'s:

$$pH = \frac{pK_{a_1} + pK_{a_2}}{2} = \frac{2.12 + 7.21}{2} = 4.67$$

HA^{2-} is the best base, and the equilibrium that controls the pH is that described by K_{a_2}:

$$H_2A^-(aq) \rightleftharpoons H^+(aq) + HA^{2-}(aq)$$

Example 8.9 shows how the pH is calculated at the first equivalence point of the titration of phosphoric acid with sodium hydroxide, where the solution contains $H_2PO_4^-$. Note that the pH of the solution does not depend on the concentration of $H_2PO_4^-$ in the solution. Thus the pH is 4.67 at the first equivalence point in any typical titration of H_3PO_4. Likewise, at the second equivalence point, where the major species are HPO_4^{2-} and H_2O, the pH is calculated from the expression

$$pH = \frac{pK_{a_2} + pK_{a_3}}{2}$$

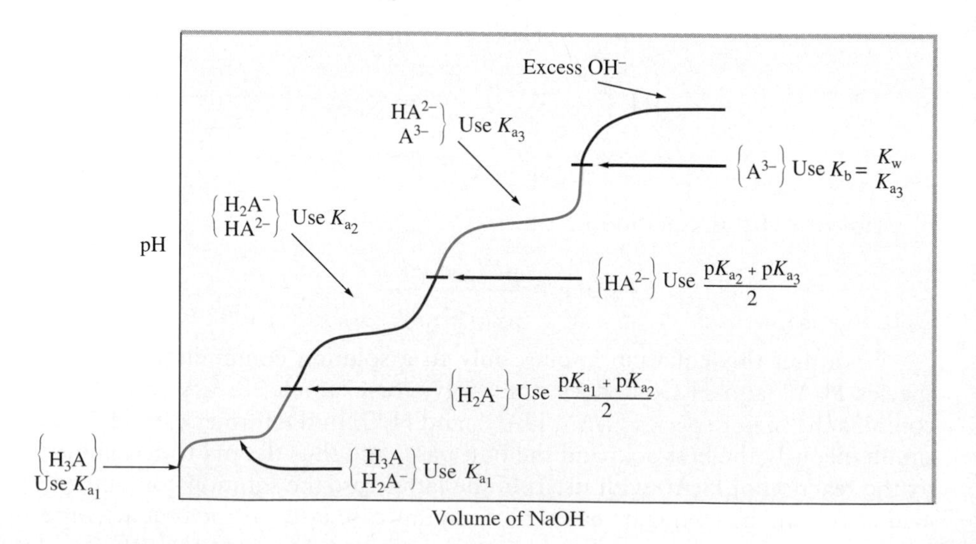

Figure 8.11
A summary of the important equilibria at various points in the titration of a triprotic acid.

Titration of a Polyprotic Acid with Sodium Hydroxide—A Summary

At this point it is useful to summarize the pH calculations associated with the titration of a triprotic acid H_3A. Figure 8.11 shows which expression should be used for the major species in the solution at a given point in the titration.

8.7 Solubility Equilibria and the Solubility Product

Solubility is a very important phenomenon. The fact that substances such as sugar and table salt dissolve in water allows us to flavor foods easily. The fact that calcium sulfate is less soluble in hot water than in cold water causes it to coat tubes in boilers, reducing thermal efficiency. Tooth decay involves solubility: when food lodges between the teeth, acids form that dissolve tooth enamel, which contains a mineral called hydroxyapatite, $Ca_5(PO_4)_3OH$. Tooth decay can be reduced by treating teeth with fluoride.* Fluoride replaces the hydroxide in hydroxyapatite to produce the corresponding fluorapatite, $Ca_5(PO_4)_3F$, and calcium fluoride, CaF_2, both of which are less soluble in acids than the original enamel. Another important consequence of solubility occurs in the use of a suspension of barium sulfate to improve the clarity of X rays of the gastrointestinal tract. The very low solubility of barium sulfate, which contains the toxic ion Ba^{2+}, makes ingestion of the compound safe.

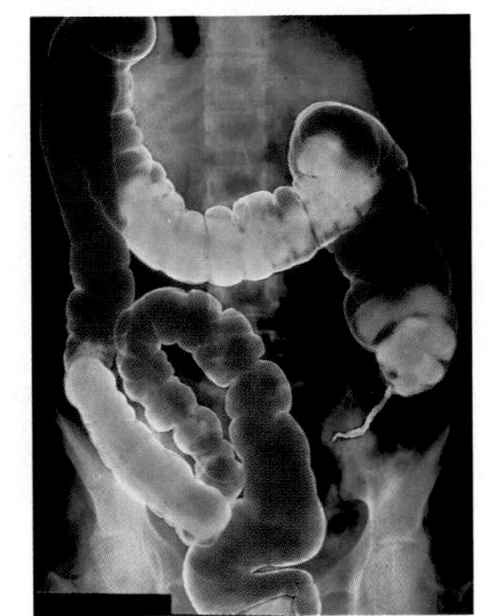

An X ray of the upper gastrointestinal tract clarified by barium sulfate.

In this section we consider the equilibria associated with solids dissolving in water to form aqueous solutions. When a typical ionic solid dissolves in water, it dissociates into separate hydrated cations and anions. For example, calcium fluoride dissolves in water as follows:

$$CaF_2(s) \xrightarrow{H_2O} Ca^{2+}(aq) + 2F^-(aq)$$

When the solid salt is first added to the water, no Ca^{2+} or F^- ions are present. However, as the dissolution proceeds, the concentrations of Ca^{2+} and F^- increase, making it more and more likely that these ions will collide and re-form the solid phase. Thus two competing processes are occurring, the above reaction and the reverse reaction:

$$Ca^{2+}(aq) + 2F^-(aq) \longrightarrow CaF_2(s)$$

Ultimately, dynamic equilibrium is reached:

$$CaF_2(s) \rightleftharpoons Ca^{2+}(aq) + 2F^-(aq)$$

At this point, no more solid can dissolve (the solution is said to be *saturated*).

*Adding F^- to drinking water is controversial. See "Fluoridation of Water" by Bette Hileman, *Chem. and Eng. News*, Aug. 1, 1988, p. 26.

We can write an equilibrium expression for this process according to the law of mass action:

$$K_{sp} = [Ca^{2+}][F^-]^2$$

where $[Ca^{2+}]$ and $[F^-]$ are expressed in mol/L. The constant K_{sp} is called the **solubility product constant,** or simply the **solubility product** for the equilibrium expression.

Pure liquids and pure solids are never included in an equilibrium expression because they have an activity of 1.

Since CaF_2 is a pure solid, it is not included in the equilibrium expression; it has an activity of 1. The fact that the amount of excess solid present does not affect the position of the solubility equilibrium might seem strange at first; more solid means more surface area exposed to the solvent, which would seem to result in greater solubility. This is not the case, however. When the ions in solution re-form the solid, they do so on the surface of the solid. Thus doubling the surface area of the solid not only doubles the rate of dissolving but also doubles the rate of re-formation of the solid. The amount of excess solid present

TABLE 8.5 K_{sp} Values at 25°C for Common Ionic Solids

Ionic solid	K_{sp} (at 25°C)
Fluorides	
BaF_2	2.4×10^{-5}
MgF_2	6.4×10^{-9}
PbF_2	4×10^{-8}
SrF_2	7.9×10^{-10}
CaF_2	4.0×10^{-11}
Chlorides	
$PbCl_2$	1.6×10^{-5}
AgCl	1.6×10^{-10}
Hg_2Cl_2*	1.1×10^{-18}
Bromides	
$PbBr_2$	4.6×10^{-6}
AgBr	5.0×10^{-13}
Hg_2Br_2*	1.3×10^{-22}
Iodides	
PbI_2	1.4×10^{-8}
AgI	1.5×10^{-16}
Hg_2I_2*	4.5×10^{-29}
Sulfates	
$CaSO_4$	6.1×10^{-5}
Ag_2SO_4	1.2×10^{-5}
$SrSO_4$	3.2×10^{-7}
$PbSO_4$	1.3×10^{-8}
$BaSO_4$	1.5×10^{-9}
Chromates	
$SrCrO_4$	3.6×10^{-5}

Ionic solid	K_{sp} (at 25°C)
Hg_2CrO_4*	2×10^{-9}
$BaCrO_4$	8.5×10^{-11}
Ag_2CrO_4	9.0×10^{-12}
$PbCrO_4$	2×10^{-16}
Carbonates	
$NiCO_3$	1.4×10^{-7}
$CaCO_3$	8.7×10^{-9}
$BaCO_3$	1.6×10^{-9}
$SrCO_3$	7×10^{-10}
$CuCO_3$	2.5×10^{-10}
$ZnCO_3$	2×10^{-10}
$MnCO_3$	8.8×10^{-11}
$FeCO_3$	2.1×10^{-11}
Ag_2CO_3	8.1×10^{-12}
$CdCO_3$	5.2×10^{-12}
$PbCO_3$	1.5×10^{-15}
$MgCO_3$	1×10^{-15}
Hg_2CO_3*	9.0×10^{-15}
Hydroxides	
$Ba(OH)_2$	5.0×10^{-3}
$Sr(OH)_2$	3.2×10^{-4}
$Ca(OH)_2$	1.3×10^{-6}
AgOH	2.0×10^{-8}
$Mg(OH)_2$	8.9×10^{-12}
$Mn(OH)_2$	2×10^{-13}
$Cd(OH)_2$	5.9×10^{-15}
$Pb(OH)_2$	1.2×10^{-15}
$Fe(OH)_2$	1.8×10^{-15}

Ionic solid	K_{sp} (at 25°C)
$Co(OH)_2$	2.5×10^{-16}
$Ni(OH)_2$	1.6×10^{-16}
$Zn(OH)_2$	4.5×10^{-17}
$Cu(OH)_2$	1.6×10^{-19}
$Hg(OH)_2$	3×10^{-26}
$Sn(OH)_2$	3×10^{-27}
$Cr(OH)_3$	6.7×10^{-31}
$Al(OH)_3$	2×10^{-32}
$Fe(OH)_3$	4×10^{-38}
$Co(OH)_3$	2.5×10^{-43}
Sulfides	
MnS	2.3×10^{-13}
FeS	3.7×10^{-19}
NiS	3×10^{-21}
CoS	5×10^{-22}
ZnS	2.5×10^{-22}
SnS	1×10^{-26}
CdS	1.0×10^{-28}
PbS	7×10^{-29}
CuS	8.5×10^{-45}
Ag_2S	1.6×10^{-49}
HgS	1.6×10^{-54}
Phosphates	
Ag_3PO_4	1.8×10^{-18}
$Sr_3(PO_4)_2$	1×10^{-31}
$Ca_3(PO_4)_2$	1.3×10^{-32}
$Ba_3(PO_4)_2$	6×10^{-39}
$Pb_3(PO_4)_2$	1×10^{-54}

*Contains Hg_2^{2+} ions. $K_{sp} = [Hg_2^{2+}][X^-]^2$ for Hg_2X_2 salts.

therefore has no effect on the equilibrium position. Similarly, although increasing the surface area by grinding up the solid or stirring the solution speeds up the attainment of equilibrium, neither procedure changes the amount of solid dissolved at equilibrium. Neither the amount of excess solid nor the size of the particles will shift the *position* of the solubility equilibrium.

It is very important to distinguish between the *solubility* of a given solid and its *solubility product*. The solubility product is an *equilibrium constant* and thus has only *one* value for a given solid at a given temperature. Solubility, on the other hand, is an *equilibrium position* and has an *infinite number* of possible values at a given temperature, depending on the other conditions (such as the presence of a common ion). The K_{sp} values at 25°C for many common ionic solids are listed in Table 8.5. The units are customarily omitted.

K_{sp} is an equilibrium constant; solubility is an equilibrium position.

EXAMPLE 8.10

Calculate the K_{sp} value for bismuth sulfide (Bi_2S_3), which has a solubility of 1.0×10^{-15} mol/L at 25°C.

Solution

The system initially contains H_2O and solid Bi_2S_3. The solid dissolves as follows:

$$Bi_2S_3(s) \rightleftharpoons 2Bi^{3+}(aq) + 3S^{2-}(aq)$$

Therefore $$K_{sp} = [Bi^{3+}]^2[S^{2-}]^3$$

Since no Bi^{3+} or S^{2-} ions are present in solution before the Bi_2S_3 dissolves,

$$[Bi^{3+}]_0 = [S^{2-}]_0 = 0$$

Thus the equilibrium concentrations of these ions will be determined by the amount of salt that dissolves to reach equilibrium, which in this case is 1.0×10^{-15} mol/L. Since each Bi_2S_3 unit contains $2Bi^{3+}$ and $3S^{2-}$ ions,

$$1.0 \times 10^{-15} \text{ mol/L } Bi_2S_3(s) \longrightarrow 2(1.0 \times 10^{-15} \text{ mol/L})\ Bi^{3+}(aq) + 3(1.0 \times 10^{-15} \text{ mol/L})\ S^{2-}(aq)$$

The equilibrium concentrations are

$$[Bi^{3+}] = [Bi^{3+}]_0 + \text{change} = 0 + 2.0 \times 10^{-15} \text{ mol/L}$$
$$[S^{2-}] = [S^{2-}]_0 + \text{change} = 0 + 3.0 \times 10^{-15} \text{ mol/L}$$

Then

$$K_{sp} = [Bi^{3+}]^2[S^{2-}]^3 = (2.0 \times 10^{-15})^2(3.0 \times 10^{-15})^3 = 1.1 \times 10^{-73}$$

Precipitation of bismuth sulfide.

In Example 8.10 we used the solubility of an ionic solid to calculate its K_{sp} value. The reverse is also possible: the solubility of an ionic solid can be calculated if its K_{sp} value is known.

EXAMPLE 8.11

The K_{sp} value for copper(II) iodate, $Cu(IO_3)_2$, is 1.4×10^{-7} at 25°C. Calculate its solubility at 25°C.

Solution

The system initially contains H_2O and solid $Cu(IO_3)_2$. The solid dissolves according to the following equilibrium:

$$Cu(IO_3)_2(s) \rightleftharpoons Cu^{2+}(aq) + 2IO_3^-(aq)$$

Therefore $$K_{sp} = [Cu^{2+}][IO_3^-]^2$$

To find the solubility of $Cu(IO_3)_2$, we must find the equilibrium concentrations of the Cu^{2+} and IO_3^- ions. We do this in the usual way by specifying the initial concentrations (before any solid has dissolved) and then defining the change required to reach equilibrium. Since in this case we do not know the solubility, we will assume that x mol/L of the solid dissolves to reach equilibrium. The 1:2 stoichiometry of the salt means that

$$x \text{ mol/L } Cu(IO_3)_2(s) \longrightarrow x \text{ mol/L } Cu^{2+}(aq) + 2x \text{ mol/L } IO_3^-(aq)$$

The concentrations are as follows:

Initial Concentration (mol/L) [before any $Cu(IO_3)_2$ dissolves]		Equilibrium Concentration (mol/L)
$[Cu^{2+}]_0 = 0$	x mol/L dissolves $\longrightarrow$ to reach equilibrium	$[Cu^{2+}] = x$
$[IO_3^-]_0 = 0$		$[IO_3^-] = 2x$

Substituting the equilibrium concentrations into the expression for K_{sp} gives

$$1.4 \times 10^{-7} = K_{sp} = [Cu^{2+}][IO_3^-]^2 = (x)(2x)^2 = 4x^3$$

Then $$x = \sqrt[3]{3.5 \times 10^{-8}} = 3.3 \times 10^{-3} \text{ mol/L}$$

Thus the solubility of solid $Cu(IO_3)_2$ is 3.3×10^{-3} mol/L.

Relative Solubilities

A salt's K_{sp} value provides information about its solubility. However, we must be careful in using K_{sp} values to predict the *relative* solubilities of a group of salts. There are two possible cases:

1. The salts being compared produce the same number of ions. For example, consider

$$AgI(s): \quad K_{sp} = 1.5 \times 10^{-16}$$

$$CuI(s): \quad K_{sp} = 5.0 \times 10^{-12}$$

$$CaSO_4(s): \quad K_{sp} = 6.1 \times 10^{-5}$$

Each of these solids dissolves to produce two ions:

$$\text{Salt} \rightleftharpoons \text{cation} + \text{anion}$$

$$K_{sp} = [\text{cation}][\text{anion}]$$

If x is the solubility in mol/L, then at equilibrium

$$[\text{Cation}] = x$$

$$[\text{Anion}] = x$$

$$K_{sp} = [\text{cation}][\text{anion}] = x^2$$

$$x = \sqrt{K_{sp}} = \text{solubility}$$

Thus in this case we can compare the solubilities of these solids by comparing their K_{sp} values:

$$CaSO_4(s) > CuI(s) > AgI(s)$$

Most soluble; largest K_{sp} — Least soluble; smallest K_{sp}

2. The salts being compared produce different numbers of ions. For example, consider

$$CuS(s): \quad K_{sp} = 8.5 \times 10^{-45}$$
$$Ag_2S(s): \quad K_{sp} = 1.6 \times 10^{-49}$$
$$Bi_2S_3(s): \quad K_{sp} = 1.1 \times 10^{-73}$$

Since these salts produce different numbers of ions when they dissolve, the K_{sp} values cannot be compared *directly* to determine the relative solubilities. In fact, if we calculate the solubilities (using the procedure in Example 8.11), we obtain the results summarized in Table 8.6. The order of solubilities is

$$Bi_2S_3(s) > Ag_2S(s) > CuS(s)$$

Most soluble — Least soluble

which is opposite to the order of the K_{sp} values.

Remember that relative solubilities can be predicted by comparing K_{sp} values *only* for salts that produce the same total number of ions.

TABLE 8.6
Calculated Solubilities for CuS, Ag_2S, and Bi_2S_3 at 25°C

Salt	K_{sp}	Calculated Solubility (mol/L)
CuS	8.5×10^{-45}	9.2×10^{-23}
Ag_2S	1.6×10^{-49}	3.4×10^{-17}
Bi_2S_3	1.1×10^{-73}	1.0×10^{-15}

Common Ion Effect

So far we have considered ionic solids dissolved in pure water. We will now see what happens when the water contains an ion in common with the dissolving salt. For example, consider the solubility of solid silver chromate (Ag_2CrO_4, $K_{sp} = 9.0 \times 10^{-12}$) in a 0.100 M solution of $AgNO_3$. Before any Ag_2CrO_4 dissolves, the solution contains the major species Ag^+, NO_3^-, and H_2O. Since NO_3^- is not found in Ag_2CrO_4, we can ignore it. The relevant initial concentrations (before any Ag_2CrO_4 dissolves) are

$$[Ag^+]_0 = 0.100\ M \quad \text{(from the dissolved } AgNO_3\text{)}$$

$$[CrO_4^{2-}]_0 = 0$$

Silver chromate is a brown solid.

The system comes to equilibrium as Ag_2CrO_4 dissolves according to the reaction

$$Ag_2CrO_4(s) \rightleftharpoons 2Ag^+(aq) + CrO_4^{2-}(aq)$$

for which

$$K_{sp} = [Ag^+]^2[CrO_4^{2-}] = 9.0 \times 10^{-12}$$

We assume that x mol/L of Ag_2CrO_4 dissolves to reach equilibrium, which means that

$$x \text{ mol/L } Ag_2CrO_4(s) \longrightarrow 2x \text{ mol/L } Ag^+(aq) + x \text{ mol/L } CrO_4^{2-}(aq)$$

Now we can specify the equilibrium concentrations in terms of x:

$$[Ag^+] = [Ag^+]_0 + \text{change} = 0.100 + 2x$$
$$[CrO_4^{2-}] = [CrO_4^{2-}]_0 + \text{change} = 0 + x = x$$

Substituting these concentrations into the expression for K_{sp} gives

$$9.0 \times 10^{-12} = [Ag^+]^2[CrO_4^{2-}] = (0.100 + 2x)^2(x)$$

The mathematics required here appears to be complicated, since the right-hand side of the equation produces an expression that contains an x^3 term. However, as is usually the case, we can make simplifying assumptions. Since the K_{sp} value for Ag_2CrO_4 is small (the position of the equilibrium lies far to the left), x is expected to be small compared with 0.100 M. Therefore, $0.100 + 2x \approx 0.100$, which allows simplification of the expression:

$$9.0 \times 10^{-12} = (0.100 + 2x)^2(x) \approx (0.100)^2(x)$$

Then $$x \approx \frac{9.0 \times 10^{-12}}{(0.100)^2} = 9.0 \times 10^{-10} \text{ mol/L}$$

Since x is much less than 0.100 M, the approximation is valid. Thus

$$\text{Solubility of } Ag_2CrO_4 \text{ in } 0.100\ M\ AgNO_3 = x = 9.0 \times 10^{-10} \text{ mol/L}$$

and the equilibrium concentrations are

$$[Ag^+] = 0.100 + 2x = 0.100 + 2(9.0 \times 10^{-10}) = 0.100\ M$$
$$[CrO_4^{2-}] = x = 9.0 \times 10^{-10}\ M$$

Now we compare the solubilities of Ag_2CrO_4 in pure water and in 0.100 M $AgNO_3$:

$$\text{Solubility of } Ag_2CrO_4 \text{ in pure water} = 1.3 \times 10^{-4} \text{ mol/L}$$
$$\text{Solubility of } Ag_2CrO_4 \text{ in } 0.100\ M\ AgNO_3 = 9.0 \times 10^{-10} \text{ mol/L}$$

Note that the solubility of Ag_2CrO_4 is much less in the presence of the Ag^+ ions from $AgNO_3$. This is another example of the common ion effect. The solubility of a solid is lowered if the solution already contains ions common to the solid.

Complications Inherent in Solubility Calculations

So far in this section we have assumed a direct relationship between the observed solubility of a salt and the concentrations of the component ions in the solution. However, this procedure is fraught with difficulties. For example, when we calculated the K_{sp} for Bi_2S_3 earlier, no allowance was made for the fact that S^{2-} is an excellent base, causing a significant amount of the reaction

$$S^{2-}(aq) + H_2O(l) \rightleftharpoons HS^-(aq) + OH^-(aq)$$

to occur in aqueous solution. Thus although we assumed in the earlier calculation that all of the sulfide from dissolving Bi_2S_3 exists as S^{2-} in solution, this is clearly not the case. The solubility of a salt containing a basic anion can be calculated accurately by simultaneously considering the K_{sp} and K_b equilibria. However, we will not show that calculation here.

Another complication that clouds the relationship between the measured solubility of a salt and its K_{sp} value is ion pairing. For example, when $CaSO_4$ dissolves in water, the solution contains

$$Ca^{2+}(aq), \quad SO_4^{2-}(aq), \quad \text{and } CaSO_4(aq)$$

the latter representing an ion pair surrounded by water molecules:

H_2O H_2O H_2O $Ca^{2+} \cdots SO_4^{2-}$ H_2O H_2O H_2O H_2O

Therefore, from a measured solubility of $CaSO_4$ of $\approx 10^{-3}\ M$, one cannot safely assume that

$$[Ca^{2+}] = [SO_4^{2-}] = 10^{-3}\ M$$

since significant numbers of Ca^{2+} and SO_4^{2-} might be present as ion pairs. In very accurate solubility calculations, the activities of the ions are used instead of the stoichiometric concentrations that we have used in this chapter. In obtaining the ion activities, corrections are made for effects such as ion pairing. However, these calculations are beyond the scope of our treatment of solubility.

Yet another complication in K_{sp} calculations involves the formation of complex ions. For example, when AgCl is dissolved in water, some of the ions combine to form $AgCl_2^-$, a complex ion. Thus a saturated solution of AgCl will contain at least the ions Ag^+, Cl^-, and $AgCl_2^-$. In addition, other complex ions such as $AgCl_3^{2-}$ may exist as well as AgCl ion pairs. Again, the assumption that the concentrations of Ag^+ and Cl^- ions can be obtained directly from the measured solubility of AgCl is suspect. These effects can be corrected for by treating the solubility and complex ion equilibria simultaneously.

The point is this: the assumption that the ion concentrations for a particular solid can be obtained directly from its observed solubility cause our results to be, at best, approximations. We will not deal here with the procedures for correcting solubility calculations for these various effects.

pH and Solubility

The pH of a solution can affect a salt's solubility quite significantly. For example, magnesium hydroxide dissolves according to the equilibrium

$$Mg(OH)_2(s) \rightleftharpoons Mg^{2+}(aq) + 2OH^-(aq)$$

Addition of OH^- ions (an increase in pH) will, by the common ion effect, force the equilibrium to the left, decreasing the solubility of $Mg(OH)_2$. On the other hand, an addition of H^+ ions (a decrease in pH) increases the solubility, because OH^- ions are removed from solution by reacting with the added H^+ ions. In response to the lower concentration of OH^-, the equilibrium position moves to the right. This explains how a suspension of solid $Mg(OH)_2$, known as milk of magnesia, dissolves in the stomach to combat excess acidity.

This idea also applies to salts with other types of anions. For example, the solubility of silver phosphate, Ag_3PO_4, is greater in acid than in pure water, because the PO_4^{3-} ion is a strong base that reacts with H^+ to form the HPO_4^{2-} ion. The reaction

$$H^+ + PO_4^{3-} \longrightarrow HPO_4^{2-}$$

occurs in acidic solution, thus lowering the concentration of PO_4^{3-} and shifting the solubility equilibrium

$$Ag_3PO_4(s) \rightleftharpoons 3Ag^+(aq) + PO_4^{3-}(aq)$$

to the right. This in turn increases the solubility of silver phosphate.

Silver chloride (AgCl), however, has the same solubility in acid as in pure water. Why? Since the Cl^- ion is a very weak base, the addition of H^+ to a solution containing Cl^- does not affect $[Cl^-]$ and thus has no effect on the solubility of a chloride salt.

The general rule is that if the anion X^- is an effective base—that is, if HX is a weak acid—the salt MX will show increased solubility in an acidic solution. Examples of common anions that are effective bases are OH^-, S^{2-}, CO_3^{2-}, $C_2O_4^{2-}$, and CrO_4^{2-}. Salts containing these anions are much more soluble in an acidic solution than in pure water.

As mentioned at the beginning of this chapter, one practical result of the increased solubility of carbonates in acid is the formation of huge limestone caves such as Mammoth Cave in Kentucky or Carlsbad Caverns in New Mexico. Carbon dioxide dissolved in groundwater makes it acidic, increasing the solubility of calcium carbonate and eventually producing huge caverns. As the carbon dioxide escapes to the air, the pH of the dripping water goes up and the calcium carbonate precipitates, forming stalactites and stalagmites.

8.8 Precipitation and Qualitative Analysis

So far we have considered solids dissolving in aqueous solutions. Now we will consider the reverse process—the formation of precipitates. When solutions are mixed, various reactions can occur. We have already considered acid-base reactions in some detail. In this section we show how to predict whether a precipitate will form when two solutions are mixed. We will use the **ion product,** which is defined just like the K_{sp} expression for a given solid except that *initial concentrations are used* instead of equilibrium concentrations. For solid CaF_2 the expression for the ion product (Q) is written

$$Q = [Ca^{2+}]_0[F^-]_0^2$$

If we add a solution containing Ca^{2+} ions to a solution containing F^- ions, a precipitate may or may not form, depending on the concentrations of these ions in the mixed solution. To predict whether precipitation will occur, we consider the relationship between Q and K_{sp}:

If *Q is greater than K_{sp}*, precipitation occurs and will continue until the concentrations of ions remaining in solution satisfy K_{sp}.

If *Q is less than K_{sp}*, no precipitation occurs.

Sometimes, we will want to do more than simply predict whether precipitation occurs; we will want to calculate the equilibrium concentrations in the solution after precipitation is complete. For example, we will next calculate the equilibrium concentrations of Pb^{2+} and I^- ions in a solution formed by mixing 100.0 mL of 0.0500 *M* $Pb(NO_3)_2$ and 200.0 mL of 0.100 *M* NaI. First, we must determine whether solid PbI_2 ($K_{sp} = 1.4 \times 10^{-8}$) forms when the solutions are mixed. To do so, we first calculate $[Pb^{2+}]_0$ and $[I^-]_0$ before any reaction occurs:

$$[Pb^{2+}]_0 = \frac{\text{mmol of } Pb^{2+}}{\text{mL of solution}} = \frac{(100.0 \text{ mL})(0.0500 \text{ mmol/mL})}{300.0 \text{ mL}} = 1.67 \times 10^{-2}\ M$$

$$[I^-]_0 = \frac{\text{mmol of } I^-}{\text{mL of solution}} = \frac{(200.0 \text{ mL})(0.100 \text{ mmol/mL})}{300.0 \text{ mL}} = 6.67 \times 10^{-2}\ M$$

The ion product for PbI_2 is

$$Q = [Pb^{2+}]_0[I^-]_0^2 = (1.67 \times 10^{-2})(6.67 \times 10^{-2})^2 = 7.43 \times 10^{-5}$$

Since Q is greater than K_{sp}, a precipitate of PbI_2 will form.

The equilibrium constant for formation of solid PbI_2 is $1/K_{sp}$, or 7×10^7, so this equilibrium lies far to the right.

Note that since the K_{sp} for PbI_2 is quite small (1.4×10^{-8}), only very small quantities of Pb^{2+} and I^- can coexist in aqueous solution. In other words, when Pb^{2+} and I^- are mixed, most of these ions will precipitate out as PbI_2. That is, the reaction

$$Pb^{2+}(aq) + 2I^-(aq) \longrightarrow PbI_2(s)$$

which is the reverse of the dissolution reaction, goes essentially to completion.

If, when two solutions are mixed, a reaction that goes virtually to completion occurs, it is essential to do the stoichiometric calculations before considering the equilibrium calculations. So in this case we let the system go completely in the direction toward which it tends. Then we will let it adjust back to equilibrium. If we let Pb^{2+} and I^- react to completion, we have the following calculations:

	Pb^{2+}	+	$2I^-$	$\rightarrow$	PbI_2
Before reaction:	(100.0 mL)(0.0500 *M*) = 5.00 mmol		(200.0 mL)(0.100 *M*) = 20.0 mmol		The amount of PbI_2 formed does not influence the equilibrium
After reaction:	0 mmol		20.0 − 2(5.00) = 10.0 mmol		

Next we must allow the system to reach equilibrium. At equilibrium $[Pb^{2+}]$ is not zero, because the reaction really does not quite go to completion. The best way to think about this is that once the PbI_2 is formed, a very small amount redissolves to reach equilibrium. Since I^- is in excess, this PbI_2 is dissolving into a solution that contains 10.0 mmol of I^- per 300.0 ml of solution, or 3.33×10^{-2} M I^-.

We can state the resulting problem as follows: what is the solubility of solid PbI_2 in a 3.33×10^{-2} M NaI solution? The lead iodide dissolves according to the equation

$$PbI_2(s) \rightleftharpoons Pb^{2+}(aq) + 2I^-(aq)$$

The concentrations are as follows:

Initial Concentration (mol/L)		Equilibrium Concentration (mol/L)
$[Pb^{2+}]_0 = 0$ $[I^-]_0 = 3.33 \times 10^{-2}$	x mol/L $PbI_2(s)$ dissolves $\longrightarrow$	$[Pb^{2+}] = x$ $[I^-] = 3.33 \times 10^{-2} + 2x$

Substituting into the expression for K_{sp} gives

$$K_{sp} = 1.4 \times 10^{-8} = [Pb^{2+}][I^-]^2 = (x)(3.33 \times 10^{-2} + 2x)^2$$
$$\approx (x)(3.33 \times 10^{-2})^2$$

Then

$$[Pb^{2+}] = x = 1.3 \times 10^{-5}\ M$$
$$[I^-] = 3.33 \times 10^{-2}\ M$$

Note that $3.33 \times 10^{-2} \gg 2x$, so the approximation is valid. These Pb^{2+} and I^- concentrations thus represent the equilibrium concentrations in the solution formed by mixing 100.0 mL of 0.050 M $Pb(NO_3)_2$ and 200.0 mL of 0.100 M NaI.

Selective Precipitation

Mixtures of metal ions in aqueous solution are often separated by **selective precipitation,** that is, by using a reagent whose anion forms a precipitate with only one of the metal ions in the mixture. For example, suppose we have a solution containing both Ba^{2+} and Ag^+ ions. If NaCl is added to the solution, AgCl precipitates as a white solid; but since $BaCl_2$ is soluble, the Ba^{2+} ions remain in solution.

EXAMPLE 8.12

A solution contains 1.0×10^{-4} M Cu^+ and 2.0×10^{-3} M Pb^{2+}. If a source of I^- is added to this solution gradually, will PbI_2 ($K_{sp} = 1.4 \times 10^{-8}$) or CuI ($K_{sp} = 5.3 \times 10^{-12}$) precipitate first? Specify the concentration of I^- necessary to begin precipitation of each salt.

Solution

For PbI_2 the K_{sp} expression is

$$1.4 \times 10^{-8} = K_{sp} = [Pb^{2+}][I^-]^2$$

Since $[Pb^{2+}]$ in this solution is known to be $2.0 \times 10^{-3}\ M$, the greatest concentration of I^- that can be present without causing precipitation of PbI_2 can be calculated from the K_{sp} expression:

$$1.4 \times 10^{-8} = [Pb^{2+}][I^-]^2 = (2.0 \times 10^{-3})[I^-]^2$$
$$[I^-] = 2.6 \times 10^{-3}\ M$$

Any I^- in excess of this concentration will cause solid PbI_2 to form.

Similarly, for CuI the K_{sp} expression is

$$5.3 \times 10^{-12} = K_{sp} = [Cu^+][I^-] = (1.0 \times 10^{-4})[I^-]$$

and

$$[I^-] = 5.3 \times 10^{-8}\ M$$

A concentration of I^- in excess of $5.3 \times 10^{-8}\ M$ will cause formation of solid CuI.

As I^- is added to the mixed solution, CuI will precipitate first, since the $[I^-]$ required is less. Therefore, Cu^+ can be separated from Pb^{2+} by using this reagent.

Since metal sulfide salts differ dramatically in their solubilities, the sulfide ion is often used to separate metal ions by selective precipitation. For example, consider a solution containing a mixture of $10^{-3}\ M\ Fe^{2+}$ and $10^{-3}\ M\ Mn^{2+}$. Since FeS ($K_{sp} = 3.7 \times 10^{-19}$) is much less soluble than MnS ($K_{sp} = 2.3 \times 10^{-13}$), careful addition of S^{2-} to the mixture will precipitate Fe^{2+} as FeS, leaving Mn^{2+} in solution.

We can directly compare K_{sp} values to find relative solubilities because FeS and MnS produce the same number of ions in solution.

One real advantage of using the sulfide ion as a precipitating reagent is that because it is basic, its concentration can be controlled by regulating the pH of the solution. H_2S is a diprotic acid that dissociates in two steps, as shown in the following reactions:

$$H_2S(aq) \rightleftharpoons H^+(aq) + HS^-(aq) \qquad K_{a_1} = 1.0 \times 10^{-7}$$
$$HS^-(aq) \rightleftharpoons H^+(aq) + S^{2-}(aq) \qquad K_{a_2} \approx 10^{-19}$$

Note from the small K_{a_2} value that the S^{2-} ion has a high affinity for protons. In an acidic solution (large $[H^+]$), $[S^{2-}]$ will be relatively small, since under these conditions the dissociation equilibria will lie far to the left. On the other hand, in basic solutions $[S^{2-}]$ will be relatively large, since the very small value of $[H^+]$ will pull both equilibria to the right, producing relatively large amounts of S^{2-}.

Thus the most insoluble sulfide salts, such as CuS ($K_{sp} = 8.5 \times 10^{-45}$) and HgS ($K_{sp} = 1.6 \times 10^{-54}$), can be precipitated from an acidic solution, leaving

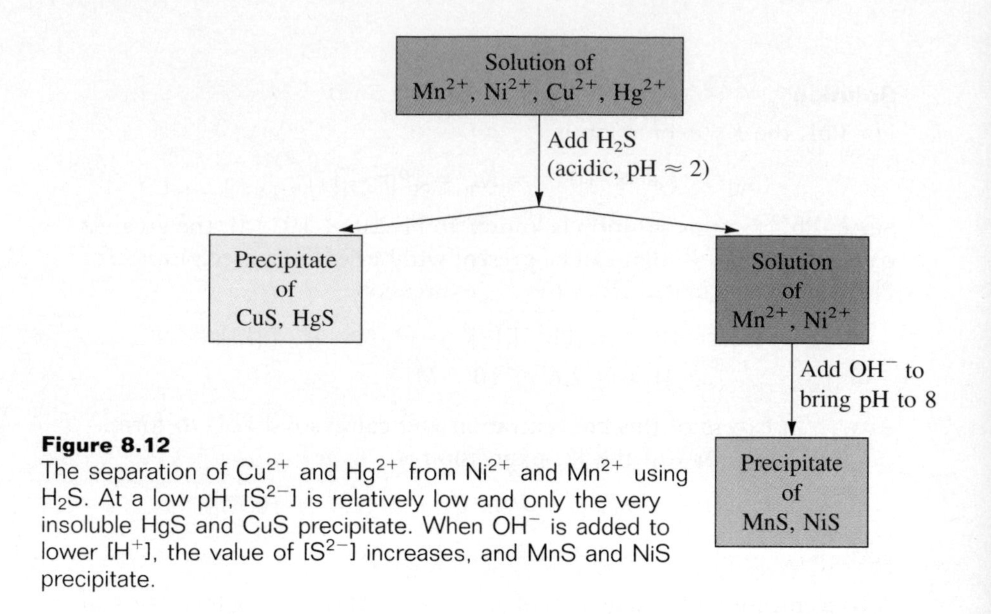

Figure 8.12
The separation of Cu^{2+} and Hg^{2+} from Ni^{2+} and Mn^{2+} using H_2S. At a low pH, $[S^{2-}]$ is relatively low and only the very insoluble HgS and CuS precipitate. When OH^- is added to lower $[H^+]$, the value of $[S^{2-}]$ increases, and MnS and NiS precipitate.

the more soluble ones, such as MnS ($K_{sp} = 2.3 \times 10^{-13}$) and NiS ($K_{sp} = 3 \times 10^{-21}$), still dissolved. The more soluble sulfides can then be precipitated by making the solution slightly basic. This procedure is diagramed in Fig. 8.12.

Qualitative Analysis

The classic scheme for **qualitative analysis** of a mixture containing all of the common cations (listed in Fig. 8.13) involves first separating the cations into five major groups based on solubilities. (These groups are not directly related to the groups of the periodic table.) Each group is then treated further to separate and identify the individual ions. We will be concerned here only with separation of the major groups.

- **Group I—insoluble chlorides.**
  When dilute aqueous HCl is added to a solution containing a mixture of the common cations, only Ag^+, Pb^{2+}, and Hg_2^{2+} will precipitate as insoluble chlorides. All other chlorides are soluble and remain in solution. The Group I precipitate is removed, leaving the other ions in solution for treatment with sulfide ion.

- **Group II—sulfides insoluble in acid solution.**
  After the insoluble chlorides are removed, the solution is still acidic, since HCl was added. If H_2S is added to this solution, only the most insoluble sulfides (those of Hg^{2+}, Cd^{2+}, Bi^{3+}, Cu^{2+}, and Sn^{4+}) will precipitate, since $[S^{2-}]$ is relatively low because of the high concentration of H^+. The more soluble sulfides will remain dissolved under these conditions. The precipitate of the insoluble salts is removed.

- **Group III—sulfides insoluble in basic solution.**
  The solution is made basic at this stage and more H_2S is added. As we saw earlier, a basic solution produces a higher $[S^{2-}]$, which leads to precipitation of the more soluble sulfides. The cations precipitated as sulfides at this stage are Co^{2+}, Zn^{2+}, Mn^{2+}, Ni^{2+}, and Fe^{2+}. If any Cr^{3+} and Al^{3+} ions are present, they will also precipitate, but as insoluble hydroxides (remember that the solution is now basic). The precipitate is separated from the solution containing the rest of the ions.

- **Group IV—insoluble carbonates.**
  At this point all of the cations have been precipitated except those from Groups 1A and 2A of the periodic table. The Group 2A cations form insoluble carbonates and can be precipitated by the addition of CO_3^{2-}. For example, Ba^{2+}, Ca^{2+}, and Mg^{2+} form solid carbonates and can be removed from the solution.

- **Group V—alkali metal and ammonium ions.**
  The only ions remaining in solution at this point are the Group 1A cations and the NH_4^+ ion, all of which form soluble salts with the common anions. The Group 1A cations are usually identified by the characteristic colors they produce when heated in a flame. These colors are due to the emission spectra of these ions.

The qualitative analysis scheme for cations based on the selective precipitation procedure described above is summarized in Fig. 8.13.

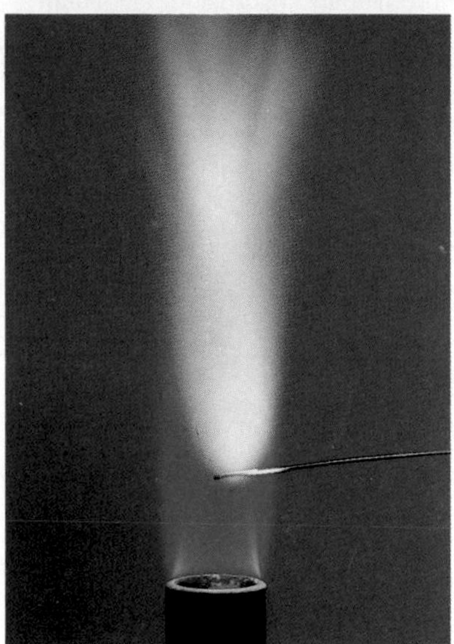

(top) Flame test for potassium. (bottom) Flame test for sodium.

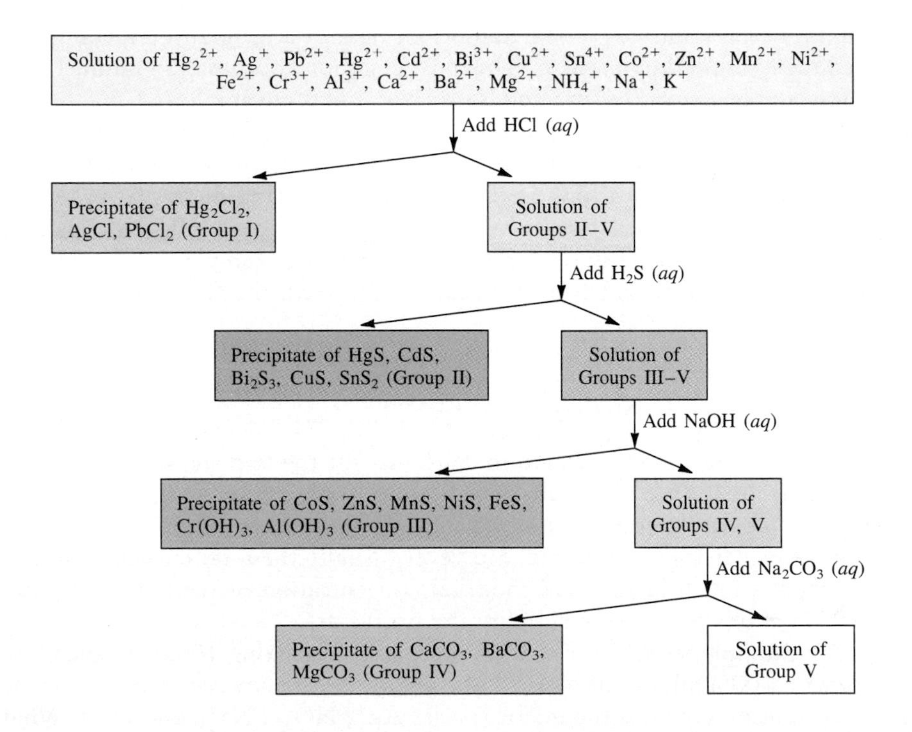

Figure 8.13
A schematic diagram of the classic method for separating the common cations by selective precipitation.

From left to right, cadmium sulfide, chromium(III) hydroxide, aluminum hydroxide, and nickel(II) hydroxide.

8.9 Complex Ion Equilibria

A **complex ion** is a charged species consisting of a metal ion surrounded by *ligands*. A ligand is a molecule or ion having a lone pair of electrons that can be donated to the metal ion to form a covalent bond. Some common ligands are H_2O, NH_3, Cl^-, and CN^-. The number of ligands attached to a metal ion is called the *coordination number*. The most common coordination numbers are 6, for example, in $Co(H_2O)_6{}^{2+}$ and $Ni(NH_3)_6{}^{2+}$; 4, for example, in $CoCl_4{}^{2-}$ and $Cu(NH_3)_4{}^{2+}$; and 2, for example, in $Ag(NH_3)_2{}^+$; but others are known.

The properties of complex ions will be discussed in more detail in Chapter 20. For now we will just look at the equilibria involving these species. Metal ions add ligands one at a time in steps characterized by equilibrium constants called **formation constants, or stability constants.** For example, when solutions containing Ag^+ ions and NH_3 molecules are mixed, the following reactions take place:

$$Ag^+(aq) + NH_3(aq) \rightleftharpoons Ag(NH_3)^+(aq) \qquad K_1 = 2.1 \times 10^3$$
$$Ag(NH_3)^+(aq) + NH_3(aq) \rightleftharpoons Ag(NH_3)_2{}^+(aq) \qquad K_2 = 8.2 \times 10^3$$

where K_1 and K_2 are the formation constants for the two steps. In a solution containing Ag^+ and NH_3, all of the species NH_3, Ag^+, $Ag(NH_3)^+$, and $Ag(NH_3)_2{}^+$ exist at equilibrium. Calculating the concentrations of all these components can be complicated. However, usually the total concentration of the ligand is much larger than the total concentration of the metal ion, and approximations can greatly simplify the problems.

For example, consider a solution prepared by mixing 100.0 mL of 2.0 *M* NH_3 with 100.0 mL of 1.0×10^{-3} *M* $AgNO_3$. *Before any reaction occurs,* the mixed solution contains the major species Ag^+, $NO_3{}^-$, NH_3, and H_2O. What reaction or reactions will occur in this solution? From our discussions of acid-base chemistry, we know that one reaction is

$$NH_3(aq) + H_2O(l) \rightleftharpoons NH_4^+(aq) + OH^-(aq)$$

However, we are interested in the reaction between NH_3 and Ag^+ to form complex ions, and since the position of the above equilibrium lies far to the left (K_b for NH_3 is 1.8×10^{-5}), we can neglect the amount of NH_3 consumed in the reaction with water. So before any complex ion formation occurs, the concentrations in the mixed solution are

$$[Ag^+]_0 = \frac{(100.0\text{ mL})(1.0 \times 10^{-3}\ M)}{200.0\text{ mL}} = 5.0 \times 10^{-4}\ M$$

← Total volume

$$[NH_3]_0 = \frac{(100.0\text{ mL})(2.0\ M)}{200.0\text{ mL}} = 1.0\ M$$

As mentioned already, the Ag^+ ion reacts with NH_3 in a stepwise fashion to form $AgNH_3^+$ and then $Ag(NH_3)_2^+$:

$$Ag^+(aq) + NH_3(aq) \rightleftharpoons Ag(NH_3)^+(aq) \qquad K_1 = 2.1 \times 10^3$$
$$AgNH_3^+(aq) + NH_3(aq) \rightleftharpoons Ag(NH_3)_2^+(aq) \qquad K_2 = 8.2 \times 10^3$$

Since both K_1 and K_2 are large, and since there is a large excess of NH_3, *both reactions can be assumed to go essentially to completion.* This is equivalent to writing the net reaction in the solution as follows:

$$Ag^+ + 2NH_3 \longrightarrow Ag(NH_3)_2^+$$

The stoichiometric calculations are summarized below:

	Ag^+	+	$2NH_3$	→	$Ag(NH_3)_2^+$
Before reaction:	$5.0 \times 10^{-4}\ M$		$1.0\ M$		0
After reaction:	0		$1.0 - 2(5.0 \times 10^{-4}) \approx 1.0\ M$		$5.0 \times 10^{-4}\ M$

↗ Twice as much NH_3 as Ag^+ is required

Note that in this case we have used molarities when performing the calculations and we have assumed this reaction to be complete, using all of the original Ag^+ to form $Ag(NH_3)_2^+$. In reality, a *very small* amount of the $Ag(NH_3)_2^+$ formed will dissociate to produce small amounts of $Ag(NH_3)^+$ and Ag^+. However, since the amount of $Ag(NH_3)_2^+$ dissociating will be so small, we can safely assume that $[Ag(NH_3)_2^+]$ is $5.0 \times 10^{-4}\ M$ at equilibrium. Also, we know that since so little NH_3 has been consumed, $[NH_3]$ is essentially $1.0\ M$ at equilibrium. We can use these concentrations to calculate $[Ag^+]$ and $[Ag(NH_3)^+]$ using the K_1 and K_2 expressions.

To calculate the equilibrium concentration of $Ag(NH_3)^+$, we use

$$K_2 = 8.2 \times 10^3 = \frac{[Ag(NH_3)_2^+]}{[Ag(NH_3)^+][NH_3]}$$

since $[Ag(NH_3)_2^+]$ and $[NH_3]$ are known. Rearranging and solving for $[Ag(NH_3)^+]$ gives

$$[Ag(NH_3)^+] = \frac{[Ag(NH_3)_2^+]}{K_2[NH_3]} = \frac{5.0 \times 10^{-4}}{(8.2 \times 10^3)(1.0)} = 6.1 \times 10^{-8}\ M$$

Now the equilibrium concentration of Ag^+ can be calculated by using K_1:

$$K_1 = 2.1 \times 10^3 = \frac{[Ag(NH_3)^+]}{[Ag^+][NH_3]} = \frac{6.1 \times 10^{-8}}{[Ag^+](1.0)}$$

$$[Ag^+] = \frac{6.1 \times 10^{-8}}{(2.1 \times 10^3)(1.0)} = 2.9 \times 10^{-11}\ M$$

So far, we have assumed that $Ag(NH_3)_2^+$ is the dominant silver-containing species in solution. Is this a valid assumption? The calculated concentrations are

$$[Ag(NH_3)_2^+] = 5.0 \times 10^{-4}\ M$$
$$[AgNH_3^+] = 6.1 \times 10^{-8}\ M$$
$$[Ag^+] = 2.9 \times 10^{-11}\ M$$

Essentially all of the Ag^+ ions originally present end up in $Ag(NH_3)_2^+$ at equilibrium.

These values clearly support the conclusion that

$$[Ag(NH_3)_2^+] \gg [AgNH_3^+] \gg [Ag^+]$$

Thus the assumption that $[Ag(NH_3)_2^+]$ is dominant is valid, and the calculated concentrations are correct.

This analysis shows that although complex ion equilibria have many species present and look complicated, the calculations are actually quite straightforward, especially if the ligand is present in large excess.

EXAMPLE 8.13

Calculate the concentrations of Ag^+, $Ag(S_2O_3)^-$, and $Ag(S_2O_3)_2^{3-}$ in a solution prepared by mixing 150.0 mL of $1.00 \times 10^{-3}\ M$ $AgNO_3$ with 200.0 mL of 5.00 M $Na_2S_2O_3$. The stepwise formation equilibria are

$$Ag^+(aq) + S_2O_3^{2-}(aq) \rightleftharpoons Ag(S_2O_3)^-(aq) \qquad K_1 = 7.4 \times 10^8$$
$$Ag(S_2O_3)^-(aq) + S_2O_3^{2-}(aq) \rightleftharpoons Ag(S_2O_3)_2^{3-}(aq) \qquad K_2 = 3.9 \times 10^4$$

Solution

The concentrations of the ligand and metal ion in the mixed solution *before any reaction occurs* are

$$[Ag^+]_0 = \frac{(150.0\ \text{mL})(1.00 \times 10^{-3}\ M)}{150.0\ \text{mL} + 200.0\ \text{mL}} = 4.29 \times 10^{-4}\ M$$

$$[S_2O_3^{2-}]_0 = \frac{(200.0\ \text{mL})(5.00\ M)}{150.0\ \text{mL} + 200.0\ \text{mL}} = 2.86\ M$$

Since $[S_2O_3^{2-}]_0 \gg [Ag^+]_0$, and since K_1 and K_2 are large, both formation reactions can be assumed to go to completion. The net reaction in the solution is as follows:

	Ag^+	+	$2S_2O_3^{2-}$	$\rightarrow$	$Ag(S_2O_3)_2^{3-}$
Before reaction:	$4.29 \times 10^{-4}\ M$		$2.86\ M$		0
After reaction:	≈ 0		$2.86 - 2(4.29 \times 10^{-4}) \approx 2.86\ M$		$4.29 \times 10^{-4}\ M$

Note that Ag^+ is limiting and that the amount of $S_2O_3^{2-}$ consumed is negligible. Also note that since all of these species are in the same solution, the molarities can be used to do the stoichiometry problem.

Of course, the concentrations calculated above do not represent the equilibrium concentrations. For example, the concentration of Ag^+ is not zero at equilibrium, and there is some $Ag(S_2O_3)^-$ in the solution. To calculate the equilibrium concentrations of these species, we must use the K_1 and K_2 expressions. We can calculate the concentration of $Ag(S_2O_3)^-$ from K_2:

$$3.9 \times 10^4 = K_2 = \frac{[Ag(S_2O_3)_2^{3-}]}{[Ag(S_2O_3)^-][S_2O_3^{2-}]} = \frac{4.29 \times 10^{-4}}{[Ag(S_2O_3)^-](2.86)}$$

$$[Ag(S_2O_3)^-] = 3.8 \times 10^{-9}\ M$$

We can calculate $[Ag^+]$ from K_1:

$$7.4 \times 10^8 = K_1 = \frac{[Ag(S_2O_3)^-]}{[Ag^+][S_2O_3^{2-}]} = \frac{3.8 \times 10^{-9}}{[Ag^+](2.86)}$$

$$[Ag^+] = 1.8 \times 10^{-18}\ M$$

These results show that

$$[Ag(S_2O_3)_2^{3-}] \gg [Ag(S_2O_3)^-] \gg [Ag^+]$$

Thus the assumption that essentially all of the original Ag^+ is converted to $Ag(S_2O_3)_2^{3-}$ at equilibrium is valid.

Complex Ions and Solubility

Often ionic solids that are only slightly soluble in water must be dissolved in aqueous solutions. For example, when the various qualitative analysis groups are precipitated, the precipitates must be redissolved to separate the ions within each group. Consider a solution of cations that contains Ag^+, Pb^{2+}, and Hg_2^{2+}, among others. When dilute aqueous HCl is added to this solution, the Group I ions will form the insoluble chlorides AgCl, $PbCl_2$, and Hg_2Cl_2. Once this mixed precipitate is separated from the solution, it must be redissolved to identify the cations individually. How can this be done? We know that some solids are more soluble in acidic than in neutral solutions. What about chloride salts? For example, can AgCl be dissolved by using a strong acid? The answer is no, because Cl^- ions have virtually no affinity for H^+ ions in aqueous solution. The position of the dissolution equilibrium

$$AgCl(s) \rightleftharpoons Ag^+(aq) + Cl^-(aq)$$

is not affected by the presence of H^+.

(top) Aqueous ammonia is added to silver chloride (white). (bottom) The silver chloride, insoluble in water, dissolves to form $Ag(NH_3)_2^+(aq)$ and $Cl^-(aq)$.

When reactions are added, the equilibrium constant for the overall process is the product of the constants for the individual reactions.

How can we pull the dissolution equilibrium to the right, even though Cl^- is an extremely weak base? The key is to lower the concentration of Ag^+ in solution by forming complex ions. For example, Ag^+ reacts with excess NH_3 to form the stable complex ion $Ag(NH_3)_2^+$. As a result, AgCl is quite soluble in concentrated ammonia solutions. The relevant reactions are

$$AgCl(s) \rightleftharpoons Ag^+(aq) + Cl^-(aq) \qquad K_{sp} = 1.6 \times 10^{-10}$$

$$Ag^+(aq) + NH_3(aq) \rightleftharpoons Ag(NH_3)^+(aq) \qquad K_1 = 2.1 \times 10^3$$

$$Ag(NH_3)^+(aq) + NH_3(aq) \rightleftharpoons Ag(NH_3)_2^+(aq) \qquad K_2 = 8.2 \times 10^3$$

The Ag^+ ion produced by dissolving solid AgCl combines with NH_3 to form $Ag(NH_3)_2^+$. This causes more AgCl to dissolve until the point at which

$$[Ag^+][Cl^-] = K_{sp} = 1.6 \times 10^{-10}$$

Here $[Ag^+]$ refers only to the Ag^+ ion that is present as a separate species in solution. It does *not* represent the total silver content of the solution, which is

$$[Ag]_{\text{total dissolved}} = [Ag^+] + [AgNH_3^+] + [Ag(NH_3)_2^+]$$

As we saw in the previous section, virtually all of the Ag^+ from the dissolved AgCl ends up in the complex ion $Ag(NH_3)_2^+$, so we can represent the dissolving of solid AgCl in excess NH_3 by the equation

$$AgCl(s) + 2NH_3(aq) \rightleftharpoons Ag(NH_3)_2^+(aq) + Cl^-(aq)$$

Since this equation is the *sum of the three stepwise reactions* given above, the equilibrium constant for this reaction is the product of the constants for the above three reactions. (Demonstrate this result to yourself by multiplying the three expressions for K_{sp}, K_1 and K_2.) The equilibrium expression is

$$K = \frac{[Ag(NH_3)_2^+][Cl^-]}{[NH_3]^2}$$

$$= K_{sp} \times K_1 \times K_2$$

$$= (1.6 \times 10^{-10})(2.1 \times 10^3)(8.2 \times 10^3)$$

$$= 2.8 \times 10^{-3}$$

Using this expression, we will now calculate the solubility of solid AgCl in a 10.0 M NH_3 solution. If we let x be the solubility (in mol/L) of AgCl in this solution, we can then write the following expressions for the equilibrium concentrations of the pertinent species:

$$[Cl^-] = x$$

$$[Ag(NH_3)_2^+] = x$$

x mol/L of AgCl dissolves to produce x mol/L of Cl^- and x mol/L of $Ag(NH_3)_2^+$

$$[NH_3] = 10.0 - 2x$$

Formation of x mol/L of $Ag(NH_3)_2^+$ requires $2x$ mol/L of NH_3, since each complex ion contains two NH_3 ligands

Substituting these concentrations into the equilibrium expression gives

$$K = 2.8 \times 10^{-3} = \frac{[Ag(NH_3)_2^+][Cl^-]}{[NH_3]^2} = \frac{(x)(x)}{(10.0 - 2x)^2} = \frac{x^2}{(10.0 - 2x)^2}$$

No approximations are necessary here. Taking the square root of both sides of the equation gives

$$\sqrt{2.8 \times 10^{-3}} = \frac{x}{10.0 - 2x}$$

$$x = 0.48 \text{ mol/L} = \text{solubility of AgCl}(s) \text{ in } 10.0\ M\ NH_3$$

Thus the solubility of AgCl in 10.0 *M* NH_3 is much greater than in pure water, which is

$$\sqrt{K_{sp}} = 1.3 \times 10^{-5} \text{ mol/L}$$

In this chapter, we have considered two strategies for dissolving a water-insoluble ionic solid. If the *anion* of the solid is a good base, the solubility is greatly increased by acidifying the solution. In cases where the anion is not sufficiently basic, the ionic solid can often be dissolved in a solution containing a ligand that forms stable complex ions with its *cation*.

Sometimes, solids are so insoluble that combinations of reactions are needed to dissolve them. For example, to dissolve the extremely insoluble HgS ($K_{sp} = 10^{-54}$), we must use a mixture of concentrated HCl and concentrated HNO_3, called *aqua regia*. The H^+ ions in the aqua regia react with the S^{2-} ions to form H_2S, and Cl^- reacts with Hg^{2+} to form various complex ions such as $HgCl_4^{2-}$. In addition, NO_3^- oxidizes S^{2-} to elemental sulfur. These processes lower the concentrations of Hg^{2+} and S^{2-} and thus promote the solubility of HgS.

Since the solubility of many salts increases with temperature, simple heating is sometimes enough to make a salt sufficiently soluble. For example, earlier in this section we considered the mixed chloride precipitates of the Group I ions—$PbCl_2$, AgCl, and Hg_2Cl_2. The effect of temperature on the solubility of $PbCl_2$ is such that we can precipitate $PbCl_2$ with cold aqueous HCl and then redissolve it by heating the solution to near boiling. In this way, $PbCl_2$ can be separated from the silver and mercury(I) chlorides, since they are not significantly soluble in hot water. Subsequently, solid AgCl can be dissolved and separated from Hg_2Cl_2 by using aqueous ammonia. The solid Hg_2Cl_2 reacts with NH_3 to form a mixture of elemental mercury and $HgNH_2Cl$:

$$Hg_2Cl_2(s) + 2NH_3(aq) \longrightarrow \underset{\text{White}}{HgNH_2Cl(s)} + \underset{\text{Black}}{Hg(l)} + NH_4^+(aq) + Cl^-(aq)$$

The mixed precipitate appears gray. This is an oxidation-reduction reaction, in which one mercury(I) ion in Hg_2Cl_2 is oxidized to Hg^{2+} in $HgNH_2Cl$ and the other mercury(I) ion is reduced to Hg, or elemental mercury.

The treatment of the Group I ions is summarized in Fig. 8.14. Note that the presence of Pb^{2+} is confirmed by adding CrO_4^{2-}, which forms bright yellow lead(II) chromate ($PbCrO_4$). Also note that H^+ added to a solution containing $Ag(NH_3)_2^+$ and Cl^- reacts with the NH_3 to form NH_4^+, thus destroying the $Ag(NH_3)_2^+$ complex. Silver chloride then re-forms:

$$2H^+(aq) + Ag(NH_3)_2^+(aq) + Cl^-(aq) \longrightarrow 2NH_4^+(aq) + AgCl(s)$$

Note that the qualitative analysis of cations by selective precipitation involves all of the types of reactions we have discussed and represents an excellent application of the principles of chemical equilibrium.

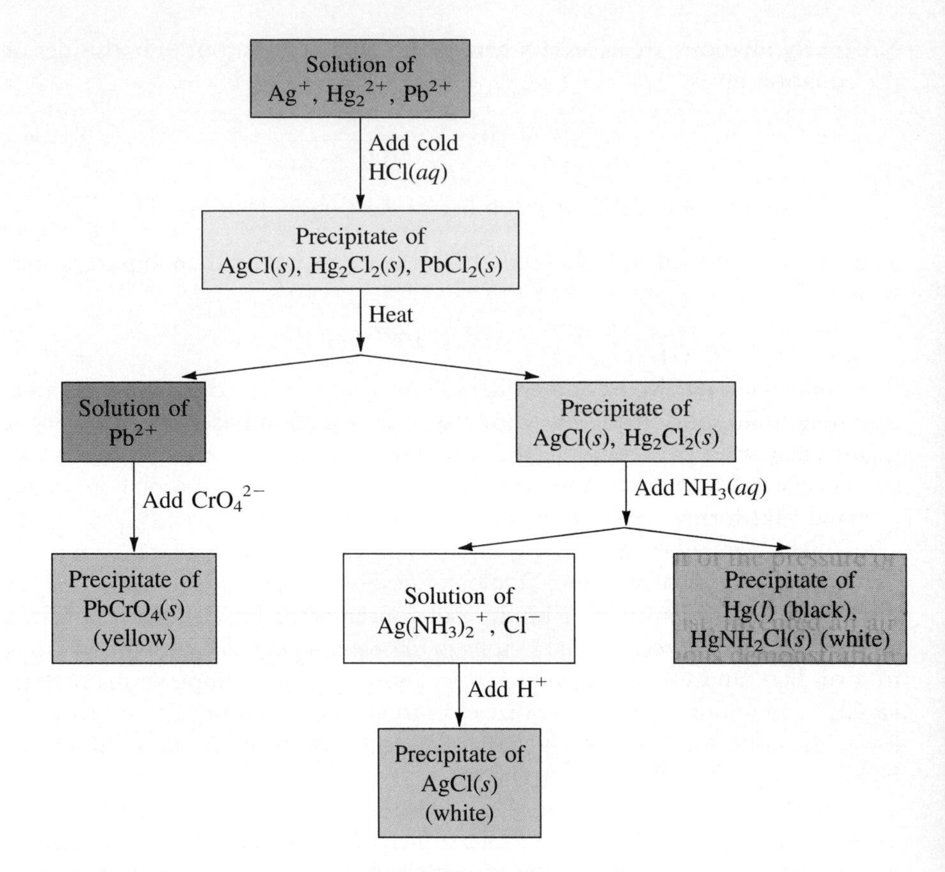

Figure 8.14
The separation of the Group I ions in the classic scheme of qualitative analysis.

Exercises

A blue exercise number indicates that the answer to that exercise appears at the back of the book.

Buffers

1. What components must be present in order to have a buffered solution? For what purpose is a buffer used?
2. A certain buffer is made by dissolving $NaHCO_3$ and Na_2CO_3 in water. Write equations to show how this buffer neutralizes added H^+ and OH^-.
3. What is meant by the capacity of a buffer? How do the following buffers differ in capacity? How do they differ in pH?
   a. 0.01 *M* acetic acid and 0.01 *M* sodium acetate
   b. 0.1 *M* acetic acid and 0.1 *M* sodium acetate
   c. 1.0 *M* acetic acid and 1.0 *M* sodium acetate
4. Derive an equation analogous to the Henderson-Hasselbalch equation that relates pOH and pK_b of a buffered solution composed of a weak base and its conjugate acid, such as NH_3 and NH_4^+.
5. Calculate the pH of each of the following solutions.
   a. 0.100 *M* propanoic acid ($HC_3H_5O_2$, $K_a = 1.3 \times 10^{-5}$)
   b. 0.100 *M* sodium propanoate ($NaC_3H_5O_2$)
   c. pure H_2O
   d. 0.100 *M* $HC_3H_5O_2$ and 0.100 *M* $NaC_3H_5O_2$
6. Calculate the pH after 0.020 mol of HCl is added to 1.00 L of each of the four solutions in Exercise 5.
7. Calculate the pH after 0.020 mol of NaOH is added to 1.00 L of each of the four solutions in Exercise 5.
8. The results of Exercises 5–7 illustrate an important property of buffered solutions. Which solution in Exercise 5 is the buffered solution and what important property is illustrated by the results?
9. Using tabulated K_a and K_b values, calculate the pH of each of the following solutions.
   a. a solution containing 0.10 *M* HNO_2 and 0.15 *M* $NaNO_2$
   b. 25.0 g of pure (glacial) acetic acid and 40.0 g of sodium acetate diluted to 500. mL with water
   c. 50.0 mL of 1.0 *M* HOCl and 30.0 mL of 0.80 *M* NaOH diluted to 250 mL with water
   d. 100.0 g of NH_4Cl and 65.0 g of NaOH diluted to 1.0 L with water
   e. 26.4 g of sodium acetate and 50.0 mL of 6.00 *M* hydrochloric acid diluted to 5.00×10^2 mL with water

10. Calculate the ratio $[NH_3]/[NH_4^+]$ present in each of the following buffered solutions containing ammonia and ammonium chloride.
    a. pH = 9.00 c. pH = 10.00
    b. pH = 8.80 d. pH = 9.60
11. What mass of solid NaOH must be added to 1.0 L of 2.0 *M* $HC_2H_3O_2$ to produce a solution buffered at each pH?
    a. pH = pK_a b. pH = 4.00 c. pH = 5.00
12. A buffered solution is made by adding 75.0 g of sodium acetate to 500.0 mL of a 0.64 *M* solution of acetic acid. What is the pH of the final solution? (Assume no volume change.)
13. Calculate the pH after 0.010 mol of gaseous HCl is added to 250.0 mL of each of the following buffered solutions.
    a. 0.050 *M* NH_3 and 0.15 *M* NH_4Cl
    b. 0.50 *M* NH_3 and 1.50 *M* NH_4Cl
14. The normal pH of blood is 7.41. What is the ratio of CO_2 (usually written H_2CO_3) to HCO_3^- in blood?

$$H_2CO_3 \rightleftharpoons HCO_3^- + H^+ \qquad K_a = 4.3 \times 10^{-7}$$

15. Which of the following result in buffered solutions when equal volumes of the two solutions are mixed?
    a. 0.1 *M* HCl and 0.1 *M* NH_4Cl
    b. 0.1 *M* HCl and 0.1 *M* NH_3
    c. 0.2 *M* HCl and 0.1 *M* NH_3
    d. 0.1 *M* HCl and 0.2 *M* NH_3
16. Which of the following are buffers?
    a. a solution containing 0.1 *M* KNO_3 and 0.1 *M* HNO_3
    b. a solution containing 0.1 *M* $NaNO_2$ and 0.15 *M* HNO_2
    c. 250 mL of a solution of 0.10 *M* acetic acid with 0.5 g KOH added
    d. a solution containing 0.10 *M* Na_2CO_3 and 0.05 *M* Na_3PO_4

Acid-Base Titrations

17. Consider the titration of a generic weak acid HA with a strong base that gives the following titration curve:

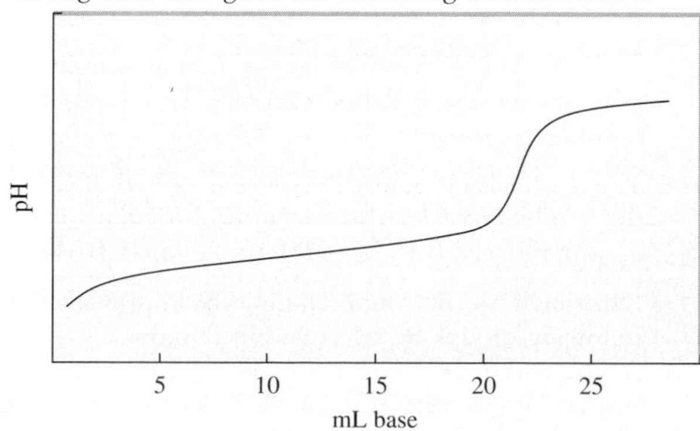

On the curve indicate the points that correspond to the following.
    a. the equivalence point
    b. the maximum buffering region
    c. pH = pK_a
    d. pH depends only on $[HA]_0$
    e. pH depends only on $[A^-]$
    f. pH depends only on the amount of excess strong base added
18. A 25.0-mL sample of 0.100 *M* lactic acid ($HC_3H_5O_3$, pK_a = 3.86) is titrated with a 0.100 *M* NaOH solution. Calculate the pH after the addition of 0.0, 4.0, 8.0, 12.5, 20.0, 24.0, 24.5, 24.9, 25.0, 25.1, 26.0, 28.0, and 30.0 mL of the NaOH. Plot the results of your calculations as pH versus milliliters of NaOH added.
19. Repeat the procedure in Exercise 18, but for the titration of 25.0 mL of 0.100 *M* NH_3 ($K_b = 1.8 \times 10^{-5}$) with 0.100 *M* HCl.
20. Repeat the procedure in Exercise 18, but for the titration of 25.0 mL of 0.100 *M* pyridine with 0.100 *M* hydrochloric acid (K_b for pyridine is 1.7×10^{-9}). Do not do the points at 24.9 and 25.1 mL.
21. Repeat the procedure in Exercise 18, but for the titration of 25.0 mL of 0.100 *M* NaOH with 0.100 *M* HNO_3.
22. Show that the pH at the halfway point of the titration of a weak acid with a strong base is equal to pK_a.
23. Is it possible to determine the K_a of a weak acid from data points other than at the halfway point of a titration?
24. Calculate the pH at the equivalence point for each of the following titrations.
    a. 0.104 g of sodium acetate ($K_b = 5.6 \times 10^{-10}$) is dissolved in 25.0 mL of water and titrated with 0.0996 *M* HCl.
    b. 50.00 mL of 0.0426 *M* HOCl ($K_a = 3.5 \times 10^{-8}$) is titrated with 0.1028 *M* NaOH.
    c. 50.0 mL of 0.205 *M* HBr (a strong acid) is titrated with 0.356 *M* KOH.
25. Estimate the pH of a solution in which crystal violet is yellow and methyl orange is red. (See Fig. 8.8.)
26. Estimate the pH of a solution in which bromcresol green is blue and thymol blue is yellow. (See Fig. 8.8.)
27. A solution has a pH of 9.0. What would be the color of the solution if each of the following indicators were added? (See Fig. 8.8.)
    a. methyl orange c. bromcresol green
    b. alizarin d. thymol blue
28. A solution has a pH of 4.5. What would be the color of the solution if each of the following indicators were added? (See Fig. 8.8.)
    a. methyl orange c. bromcresol green
    b. alizarin d. phenolphthalein

29. Is it possible for thymol blue to contain only a single $—CO_2H$ group and no other acidic or basic functional group? Why or why not?
30. Define the equivalence point and the end point of a titration. Why does one choose an indicator so that the two points coincide? Is it necessary that the pH of the two points be within ±0.01 pH unit of each other? Why or why not?
31. Why does an indicator change from its acid to its base color over a range of pH values?
32. Which of the indicators in Fig. 8.8 should be used for doing the titrations in Exercises 18 and 20?
33. Which of the indicators in Fig. 8.8 should be used for doing the titrations in Exercises 19 and 21?
34. Methyl red (K_a = 5.0×10^{-6}) undergoes a color change from red to yellow as the solution gets more basic. Calculate an approximate pH range for which methyl red is useful. What color change occurs and what is the pH at the color change when a weak acid is titrated with a strong base, using methyl red as an indicator? What color change occurs and what is the pH at the color change when a weak base is titrated with a strong acid, using methyl red as an indicator? For which of the titrations in Exercises 18–21 would methyl red be a suitable indicator?
35. Indicators can be used to estimate the pH values of solutions. To determine the pH of a 0.01 *M* weak acid (HX) solution, a few drops of three different indicators are added to separate portions of 0.01 *M* HX. The resulting colors of the HX solution are summarized in the last column of the accompanying table. What is the approximate pH of the 0.01 *M* HX solution? What is the approximate K_a value for HX?

Indicator (HIn)	Color of HIn	Color of In^-	pK_a of HIn	Color of 0.01 *M* HX
Bromphenol Blue	Yellow	Blue	4.0	Blue
Bromcresol Purple	Yellow	Purple	6.0	Yellow
Bromcresol Green	Yellow	Blue	4.8	Green

Solubility Equilibria

36. Use the information given to calculate values of K_{sp} for the following salts.
    a. A sample of 4.8×10^{-5} mol of calcium oxalate (CaC_2O_4) dissolves in 1.0 L of water to produce a saturated solution.

$$CaC_2O_4(s) \rightleftharpoons Ca^{2+}(aq) + C_2O_4^{2-}(aq)$$

    b. The concentration of Pb^{2+} in a solution saturated with $PbBr_2$ is 2.14×10^{-2} *M*.
    c. The molar solubility of BiI_3 is 1.32×10^{-5} mol/L.
    d. The solubility of iron(II) oxalate (FeC_2O_4) is 65.9 mg/L at 25°C.
    e. A 0.100-L sample of a saturated solution of copper(II) periodate [$Cu(IO_4)_2$] contains 0.146 g of the dissolved salt.
    f. The solubility of lithium carbonate (Li_2CO_3) is 5.48 g/L.
37. Calculate the solubility of each of the following compounds in moles per liter and grams per liter. (Ignore any acid/base properties.)
    a. $Al(OH)_3$, $K_{sp} = 2 \times 10^{-32}$
    b. $Be(IO_4)_2$, $K_{sp} = 1.57 \times 10^{-9}$
    c. $CaSO_4$, $K_{sp} = 6.1 \times 10^{-5}$
    d. $CaCO_3$, $K_{sp} = 8.7 \times 10^{-9}$
    e. $MgNH_4PO_4$, $K_{sp} = 3 \times 10^{-13}$
    f. Hg_2Cl_2, $K_{sp} = 1.1 \times 10^{-18}$ (Hg_2^{2+} is the cation in solution)
    g. $SrSO_4$, $K_{sp} = 3.2 \times 10^{-7}$
    h. Ag_2CO_3, $K_{sp} = 8.1 \times 10^{-12}$
    i. $Ag_2Cl_2O_7$, $K_{sp} = 2 \times 10^{-7}$ ($Cl_2O_7^{2-}$ is the anion in solution)
    j. Cu_2S, $K_{sp} = 2 \times 10^{-47}$
38. Under what circumstances can the relative solubilities of two salts be compared by directly comparing values of the solubility products?
39. Calculate the solubility (in mol/L) of $Fe(OH)_3$ ($K_{sp} \approx 4 \times 10^{-38}$) in each of the following.
    a. water (assume pH is 7.0 and constant)
    b. a solution buffered at pH = 5.0
    c. a solution buffered at pH = 11.0
40. For the substances in Exercise 37, which will show increased solubility as the pH of the solution becomes more acidic? Write equations for the reactions that occur to increase the solubility.
41. The K_{sp} of hydroxyapatite, $Ca_5(PO_4)_3OH$, is 6.8×10^{-37}. Calculate the solubility of hydroxyapatite in pure water in moles per liter. How is the solubility of hydroxyapatite affected by adding acid? When hydroxyapatite is treated with fluoride, the mineral fluorapatite, $Ca_5(PO_4)_3F$, forms. The K_{sp} of this substance is 1×10^{-60}. Calculate the solubility of fluorapatite in water. How do these calculations provide a rationale for the fluoridation of drinking water?
42. The K_{sp} for lead iodide (PbI_2) is 1.4×10^{-8}. Calculate the solubility of lead iodide in each of the following.
    a. water b. 0.10 *M* $Pb(NO_3)_2$ c. 0.010 *M* NaI
43. Describe how you could separate the ions in each of the following groups by selective precipitation.
    A. Ag^+, Mg^{2+}, Cu^{2+}
    b. Pb^{2+}, Ca^{2+}, Fe^{2+}
    c. Cl^-, Br^-, I^-
    d. Pb^{2+}, Bi^{3+}

44. Silica (SiO_2) can undergo the following reactions:

$$SiO_2(s) + 2H_2O(l) \rightleftharpoons H_4SiO_4(aq) \qquad K = 2 \times 10^{-3}$$
Silicic acid

$$H_4SiO_4(aq) \rightleftharpoons H_3SiO_4^-(aq) + H^+(aq) \qquad pK_a = 9.46$$

Will silica be more soluble in an acidic or basic solution? Why?

45. What are the concentrations of all of the ions in solution after 100.0 mL of 0.020 *M* $Pb(NO_3)_2$ and 100.0 mL of 0.020 *M* NaCl are mixed?

46. A solution is prepared by mixing 75.0 mL of 0.020 *M* $BaCl_2$ and 125 mL of 0.040 *M* H_2SO_4. What are the concentrations of barium and sulfate ions in this solution? Assume only SO_4^{2-} ions (no HSO_4^-) are present.

Complex Ion Equilibria

47. Given the following data,

$$Mn^{2+}(aq) + C_2O_4^{2-}(aq) \rightleftharpoons MnC_2O_4(aq) \qquad K_1 = 7.9 \times 10^3$$

$$MnC_2O_4(aq) + C_2O_4^{2-}(aq) \rightleftharpoons Mn(C_2O_4)_2^{2-}(aq) \qquad K_2 = 7.9 \times 10^1$$

calculate the value for the overall formation constant for $Mn(C_2O_4)_2^{2-}$:

$$K = \frac{[Mn(C_2O_4)_2^{2-}]}{[Mn^{2+}][C_2O_4^{\ 2-}]^2}$$

48. Concentrated ammonia is added to an aqueous solution of copper(II) sulfate. Initially, a white precipitate forms. As more ammonia is added, the precipitate dissolves and the solution becomes a deep bluish purple. Write equations describing the reactions that are occurring.

49. The overall formation constant for HgI_4^{2-} is 1.0×10^{30}. That is,

$$1.0 \times 10^{30} = \frac{[HgI_4^{2-}]}{[Hg^{2+}][I^-]^4}$$

What is the concentration of Hg^{2+} in 500.0 mL of a solution that was originally 0.010 *M* Hg^{2+} and had 65 g of KI added to it? The reaction is

$$Hg^{2+}(aq) + 4I^-(aq) \rightleftharpoons HgI_4^{2-}(aq)$$

50. A solution is prepared by adding 0.10 mol of $Ni(NH_3)_6Cl_2$ to 0.50 L of 3.0 *M* NH_3. Calculate $[Ni(NH_3)_6^{2+}]$ and $[Ni^{2+}]$ in this solution [*K* (overall) for $Ni(NH_3)_6^{2+}$ is 5.5×10^8].

51. As a sodium chloride solution is added to a solution of silver nitrate, a white precipitate forms. Ammonia is added to the mixture and the precipitate dissolves. When potassium bromide solution is then added, a pale yellow precipitate appears. When a solution of sodium thiosulfate is added, the yellow precipitate dissolves. Finally, potassium iodide is added to the solution and a yellow precipitate forms. Write reactions for all of the changes mentioned above. What conclusions can you draw concerning the sizes of the K_{sp} values for AgCl, AgBr, and AgI? What can you say about the relative values of the formation constants of $Ag(NH_3)_2^+$ and $Ag(S_2O_3)_2^{3-}$?

52. Solutions of sodium thiosulfate are used to dissolve unexposed AgBr in the developing process for black-and-white film. What mass of AgBr can dissolve in 1.00 L of 0.500 *M* $Na_2S_2O_3$? Assume the overall formation constant for $Ag(S_2O_3)_2^{3-}$ is 2.9×10^{13} and K_{sp} for AgBr is 5.0×10^{-13}. (*Hint:* do not round off any numbers until your final answer.)

53. Will 0.10 mol of AgBr completely dissolve in 1.0 L of 3.0 *M* NH_3? Assume the overall formation constant for $Ag(NH_3)_2^+$ is 1.7×10^7 and K_{sp} for AgBr is 5.0×10^{-13}.

Additional Exercises

54. Will a precipitate of $Cd(OH)_2$ form if 1.0 mL of 1.0 *M* $Cd(NO_3)_2$ is added to 1.0 L of 5.0 *M* NH_3?

$$Cd^{2+}(aq) + 4NH_3(aq) \rightleftharpoons Cd(NH_3)_4^{2+}(aq) \qquad K = 1.0 \times 10^7$$

$$Cd(OH)_2(s) \rightleftharpoons Cd^{2+}(aq) + 2OH^-(aq) \qquad K_{sp} = 5.9 \times 10^{-15}$$

55. Borax ($Na_2B_4O_7 \cdot 10H_2O$) dissolves in water according to the reaction

$$Na_2B_4O_7 \cdot 10H_2O(s) \longrightarrow 2Na^+(aq) + 3H_2O(l) + 2B(OH)_3(aq) + 2B(OH)_4^-(aq)$$

Boric acid reacts with water according to the following reaction:

$$B(OH)_3(aq) + H_2O(l) \rightleftharpoons B(OH)_4^-(aq) + H^+(aq) \qquad K_a = 5.8 \times 10^{-10}$$

a. Calculate the pH of the solution formed when 28.6 g of borax is dissolved to make 1.0 L of solution.
b. If 100. mL of 0.10 *M* NaOH is added to the solution in part a, what is the new pH?

56. Tris(hydroxymethyl)aminomethane, commonly called TRIS or Trizma, is often used as a buffer in biochemical studies. Its buffering range is from pH 7 to 9, and K_b is 1.19×10^{-6} for the reaction

$$\underset{\text{TRIS}}{(HOCH_2)_3CNH_2(aq)} + H_2O(l) \rightleftharpoons \underset{\text{TRISH}}{(HOCH_2)_3CNH_3^+(aq)} + OH^-(aq)$$

a. What is the optimum pH for TRIS buffers?
b. Calculate the ratio [TRIS]/[TRISH$^+$] at pH = 7.00 and at pH = 9.00.
c. A buffer is prepared by diluting 50.0 g of TRIS base and 65.0 g of TRIS hydrochloride (written as TRISHCl) to a total volume of 2.0 L. What is the pH of this buffer? What is the pH after 0.50 mL of 12 *M* HCl is added to a 200.0-mL portion of the buffer?

57. Cacodylic acid, $(CH_3)_2AsO_2H$, is a toxic compound that behaves as a weak acid. It is used to prepare buffered solutions. For the following reaction $pK_a = 6.19$:

$$(CH_3)_2AsO_2H(aq) + H_2O(l) \rightleftharpoons H_3O^+(aq) + (CH_3)_2AsO_2^-(aq)$$

Calculate the masses of cacodylic acid and sodium cacodylate that should be used to prepare 500.0 mL of a buffer at pH = 6.60 that has a total concentration of all arsenic-containing species equal to 0.25 *M*; that is,

$$[(CH_3)_2AsO_2H] + [(CH_3)_2AsO_2^-] = 0.25\ M$$

58. You have the following reagents on hand:

Solids (pK_a of acid form is given)	
$(CH_3)_2AsO_2Na$ (6.19)	Sodium acetate (4.74)
TRISHCl (8.08)	Potassium fluoride (3.14)
Benzoic acid (4.19)	Ammonium chloride (9.26)
Solutions	
5.0 *M* HCl	2.6 *M* NaOH
Glacial acetic acid	

What combinations of these reagents would you use to prepare buffers at the following pH values?

a. 3.0 c. 5.0 e. 9.0
b. 4.0 d. 7.0

59. a. Calculate the pH of a buffered solution that is 0.100 *M* in $C_6H_5CO_2H$ (benzoic acid, $K_a = 6.4 \times 10^{-5}$) and 0.100 *M* in $C_6H_5CO_2Na$.
b. Calculate the pH after 20.0% (by moles) of the benzoic acid is converted to benzoate anion by addition of base. Use the dissociation equilibrium:

$$C_6H_5CO_2H(aq) \rightleftharpoons C_6H_5CO_2^-(aq) + H^+(aq)$$

c. Do the same calculation as in part b, but use the following equilibrium to calculate the pH:

$$C_6H_5CO_2^-(aq) + H_2O(l) \rightleftharpoons C_6H_5CO_2H(aq) + OH^-(aq)$$

d. Do your answers in parts b and c agree? Why or why not?

60. One method for determining the purity of aspirin (empirical formula, $C_9H_8O_4$) is to hydrolyze it with NaOH solution and then to titrate the remaining NaOH. The reaction of aspirin with NaOH is as follows:

$$\underset{\text{Aspirin}}{C_9H_8O_4(s)} + 2OH^-(aq) \xrightarrow[\text{10 min}]{\text{Boil}} \underset{\text{Salicylate ion}}{C_7H_5O_3^-(aq)} + \underset{\text{Acetate ion}}{C_2H_3O_2^-(aq)} + H_2O(l)$$

A sample of aspirin with a mass of 1.427 g was boiled in 50.00 mL of 0.500 *M* NaOH. After the solution was cooled, it took 31.92 mL of 0.289 *M* HCl to titrate the excess NaOH. Calculate the purity of the aspirin. What indicator should be used for this titration? Why?

61. For solutions containing salts of the form NH_4X, the pH is determined by using the equation

$$pH = \frac{pK_a(NH_4^+) + pK_a(HX)}{2}$$

a. Derive this equation. (*Hint:* review Section 8.6 on the pH of solutions containing amphoteric species.)
b. Use this equation to calculate the pH of the following solutions: ammonium formate, ammonium acetate, and ammonium bicarbonate.
c. Solutions of ammonium acetate are commonly used as pH = 7 buffers. Write equations to show how an ammonium acetate solution neutralizes added H^+ and OH^-.

62. Another way to treat data from a pH titration is to graph the absolute value of the change in pH per change in milliliters added versus milliliters added (ΔpH/ΔmL versus mL added). Make this graph using the calculations you did in Exercise 18. What advantage might this method have over the traditional method for treating titration data?

63. Potassium hydrogen phthalate, known as KHP (molar mass = 204.22 g/mol), can be obtained in high purity and is used to determine the concentration of solutions of strong bases by the reaction

$$HP^-(aq) + OH^-(aq) \longrightarrow H_2O(l) + P^{2-}(aq)$$

If a typical titration experiment begins with approximately 0.5 g of KHP and has a final volume of about 100 mL, what is an appropriate indicator to use? The pK_a for HP^- is 5.51.

64. The equilibrium constant for the following reaction is 1.0×10^{23}:

$$Cr^{3+}(aq) + H_2EDTA^{2-}(aq) \rightleftharpoons CrEDTA^-(aq) + 2H^+(aq)$$

$EDTA^{4-}$ = $(^-O_2C{-}CH_2)_2N{-}CH_2{-}CH_2{-}N(CH_2{-}CO_2^-)_2$

Ethylenediaminetetraacetate

EDTA is used as a complexing agent in chemical analysis. Solutions of EDTA, usually containing the disodium salt Na_2H_2EDTA, are used to treat heavy metal poisoning. Calculate $[Cr^{3+}]$ at equilibrium in a solution originally 0.0010 M in Cr^{3+} and 0.050 M in H_2EDTA^{2-} and buffered at pH = 6.00.

65. Calculate the concentration of Pb^{2+} in each of the following.
    a. a saturated solution of $Pb(OH)_2$; $K_{sp} = 1.2 \times 10^{-15}$
    b. a saturated solution of $Pb(OH)_2$ buffered at pH = 13.00
    c. 0.010 mol of $Pb(NO_3)_2$ added to 1.0 L of aqueous solution, buffered at pH = 13.00 and containing 0.050 M Na_4EDTA. Does $Pb(OH)_2$ precipitate from this solution? For the reaction,

$$Pb^{2+}(aq) + EDTA^{4-}(aq) \rightleftharpoons PbEDTA^{2-}(aq) \qquad K = 1.1 \times 10^{18}$$

66. Calculate the concentration of Fe^{3+} in blood (pH = 7.41), using the solubility equilibrium of $Fe(OH)_3$ ($K_{sp} = 4 \times 10^{-38}$). Iron levels in blood serum range from 60 to 150 μg/100 mL. How can you account for the discrepancy between your value and the accepted range?

67. Using the K_{sp} for $Cu(OH)_2$ (1.6×10^{-19}) and the overall formation constant for $Cu(NH_3)_4^{2+}$ (1.0×10^{13}), calculate a value for the equilibrium constant for the reaction

$$Cu(OH)_2(s) + 4NH_3(aq) \rightleftharpoons Cu(NH_3)_4^{2+}(aq) + 2OH^-(aq)$$

68. Use the value of the equilibrium constant you calculated in Exercise 67 to calculate the solubility (in mol/L) of $Cu(OH)_2$ in 5.0 M NH_3. In 5.0 M NH_3 the concentration of OH^- is about 0.010 M.

69. For which salt in each of the following groups will the solubility depend on pH?
    a. AgF, AgCl, AgBr
    b. PbO, $PbCl_2$
    c. $Sr(NO_3)_2$, $Sr(NO_2)_2$
    d. $Ni(NO_3)_2$, $Ni(CN)_2$

70. Use the equilibrium constants from Exercise 44 to calculate the molar solubility (in mol/L) of silica at a pH of 7.0 and at a pH of 10.0.

71. Calculate the final concentrations of $K^+(aq)$, $C_2O_4^{2-}(aq)$, $Ba^{2+}(aq)$, and $Br^-(aq)$ in a solution prepared by adding 0.100 L of 0.200 M $K_2C_2O_4$ to 0.150 L of 0.250 M $BaBr_2$. (For BaC_2O_4, $K_{sp} = 2.3 \times 10^{-8}$.)

72. Calculate the solubility of silver acetate in a buffered solution with pH = 3.00. (K_{sp} for $AgC_2H_3O_2$ is 2.5×10^{-3}.)

73. a. Calculate the molar solubility of SrF_2 in water, ignoring the basic properties of F^-. (For SrF_2, $K_{sp} = 7.9 \times 10^{-10}$.)
    b. Would the measured molar solubility of SrF_2 be greater than or less than the value calculated in part a? Explain.
    c. Calculate the molar solubility of SrF_2 in a solution buffered at pH = 2.00. (K_a for HF is 7.2×10^{-4}.)

74. a. Show that the solubility of $Al(OH)_3$, as a function of $[H^+]$, obeys the equation:

$$S = [H^+]^3 K_{sp}/K_w^3 + KK_w/[H^+]$$

where S = solubility = $[Al^{3+}] + [Al(OH)_4^-]$ and K is the equilibrium constant for:

$$Al(OH)_3(s) + OH^-(aq) \rightleftharpoons Al(OH)_4^-(aq)$$

    b. The value of K is 40.0 and K_{sp} for $Al(OH)_3$ is 2×10^{-32}. Plot the solubility of $Al(OH)_3$ in the pH range 4–12.

75. What volume of 0.0100 M NaOH must be added to 1.00 L of 0.0500 M HOCl to achieve a pH of 8.00?

76. Calculate the pH of a solution prepared by mixing 350.0 mL of 0.200 M HCl with 100.0 mL of 0.400 M NaF.

77. An aqueous solution contains dissolved NH_4Cl and NH_3. The concentration of NH_3 is 0.500 M, and the pH is 8.95.
    a. Calculate the equilibrium concentration of NH_4^+.
    b. Calculate the pH after 4.00 g of NaOH(s) are added to 1.00 L of this solution. (Neglect any volume changes.)

78. Consider 100.0 mL of a solution of 0.200 M Na_2A, where A^{2-} is a base with corresponding acids H_2A ($K_a = 1.0 \times 10^{-3}$) and HA^- ($K_a = 1.0 \times 10^{-8}$).
    a. How much 1.00 M HCl must be added to this solution to reach pH = 8.00?
    b. Calculate the pH at the second stoichiometric point of the titration of 0.200 M Na_2A, with 1.00 M HCl.

79. A 10.00-g sample of the ionic compound NaA, where A^- is the anion of a weak acid, was dissolved in enough water to make 100.0 mL of solution and was then titrated with 0.100 M HCl. After 500.0 mL of HCl was added, the pH was measured and found to be 5.00. The experimenter found that 1.00 L of 0.100 M HCl was required to reach the stoichiometric point of the titration.
    a. What is the molar mass of NaA?
    b. Calculate the pH of the solution at the stoichiometric point of the titration.

80. Consider a solution containing 0.10 M ethylamine ($C_2H_5NH_2$), 0.20 M $C_2H_5NH_3^+$, and 0.20 M Cl^-.
    a. Calculate the pH of this solution.
    b. Calculate the pH after 0.050 mol of KOH(s) is added to 1.00 L of this solution. (Ignore any volume changes.)

81. Consider the titration of 100.0 mL of 0.250 M aniline with 0.500 M HCl. What is the pH of the solution at the stoichiometric point?

82. A certain acetic acid solution has pH = 2.68. Calculate the volume of 0.0975 M KOH required to "neutralize" 25.0 mL of this solution.

83. A 75.0-mL sample of a solution of an acid with a K_a value of 1.5×10^{-4} has a pH of 3.58. How much 9.5×10^{-2} M NaOH is required to just react with the acid present?

84. Consider a solution formed by mixing 50.0 mL of 1.0 M Na_3PO_4, 270.0 mL of 0.50 M HCl, and 100.0 mL of 0.10 M NaCN.
   a. Calculate the pH of this solution.
   b. Calculate the concentration of HCN in this solution.

85. Calculate the volume of 1.50×10^{-2} M NaOH that must be added to 500.0 mL of 0.200 M HCl to give a solution that has pH = 2.15.

86. A 0.400 M solution of ammonia was titrated with hydrochloric acid to the equivalence point, where the total volume was 1.50 times the original volume. At what pH does the equivalence point occur?

87. Consider a solution formed by mixing 50.0 mL of 0.100 M H_2SO_4, 30.0 mL of 0.10 M HOCl, 25.0 mL of 0.20 M NaOH, 25.0 mL of 0.10 M $Ca(OH)_2$, and 10.0 mL of 0.15 M KOH. Calculate the pH of this solution.

88. Consider the titration of 100.0 mL of a 1.00×10^{-4} M solution of an acid HA ($K_a = 5.0 \times 10^{-10}$), with 1.00×10^{-3} M NaOH. Calculate the pH for the following conditions.
   a. before any NaOH has been added
   b. after 5.00 mL of NaOH has been added
   c. at the stoichiometric point

89. A student intends to titrate a solution of a weak acid with a sodium hydroxide solution but reverses the two solutions and places the weak acid solution in the buret. After 23.75 mL of the weak acid solution has been added to 50.0 mL of the 0.100 M NaOH solution, the pH of the resulting solution is 10.50. Calculate the original concentration of the solution of weak acid.

90. In the titration of 100.0 mL of a 0.0500 M solution of acid H_3A ($K_{a_1} = 1.0 \times 10^{-3}$, $K_{a_2} = 5.0 \times 10^{-8}$, $K_{a_3} = 2.0 \times 10^{-12}$), calculate the volume of 1.00 M NaOH required to reach pH values of 9.50 and 4.00.

91. Consider 100.0 mL of a 0.100 M solution of H_3A ($K_{a_1} = 1.5 \times 10^{-4}$, $K_{a_2} = 3.0 \times 10^{-8}$, $K_{a_3} = 5.0 \times 10^{-12}$).
   a. Calculate the pH of this solution.
   b. Calculate the concentration of A^{3-} in this solution.
   c. Calculate the pH of the solution after 10.0 mL of 1.00 M NaOH has been added to the original solution.
   d. Calculate the pH of the solution after 25.0 mL of 1.00 M NaOH has been added to the original solution.

92. Consider the titration of 100.0 mL of a 0.0500 M solution of the hypothetical weak acid H_3X ($K_{a_1} = 1.0 \times 10^{-3}$, $K_{a_2} = 1.0 \times 10^{-7}$, $K_{a_3} = 1.0 \times 10^{-12}$) with 0.100 M KOH. Calculate the pH of the solution under the following conditions.
   a. before any KOH has been added
   b. after 10.0 mL of 0.100 M KOH has been added
   c. after 25.0 mL of 0.100 M KOH has been added
   d. after 50.0 mL of 0.100 M KOH has been added
   e. after 60.0 mL of 0.100 M KOH has been added
   f. after 75.0 mL of 0.100 M KOH has been added
   g. after 100.0 mL of 0.100 M KOH has been added
   h. after 125.0 mL of 0.100 M KOH has been added
   i. after 150.0 mL of 0.100 M KOH has been added
   j. after 200.0 mL of 0.100 M KOH has been added
   k. What volume of 0.100 M KOH is required to reach a pH of 10.00 in this titration?

93. A 50.0-mL sample of 0.00200 M $AgNO_3$ is added to 50.0 mL of 0.0100 M $NaIO_3$. What is the equilibrium concentration of Ag^+ in solution? (K_{sp} for $AgIO_3$ is 3.0×10^{-8}.)

94. The copper(I) ion forms a chloride salt that has $K_{sp} = 1.2 \times 10^{-6}$. Copper(I) also forms a complex ion with Cl^-:

$$Cu^+(aq) + 2Cl^-(aq) \rightleftharpoons CuCl_2^-(aq) \qquad K = 8.7 \times 10^4$$

   a. Calculate the solubility of copper(I) chloride in pure water.
   b. Calculate the solubility of copper(I) chloride in 0.10 M NaCl.

95. The ionic solid MX(s),

$$MX(s) \rightleftharpoons M^+(aq) + X^-(aq)$$

shows a solubility of 3.17×10^{-8} mol/L in a solution that initially contained 1.00 M H^+. Calculate the solubility of MX(s) in water. Assume K_a for HX is 1.00×10^{-15}.

96. Calculate the solubility of silver(I) bromide in a 0.200 M NH_3 solution.

97. Silver(I) chloride dissolves readily in 2 M NH_3 but is quite insoluble in 2 M NH_4NO_3. Explain.

98. For a tripotic acid H_3A, $K_{a_1} = 6.0 \times 10^{-3}$, $K_{a_2} = 1.0 \times 10^{-7}$, and $K_{a_3} = 3.0 \times 10^{-12}$. Which of the following choices will produce a buffered solution with pH = 7.00? (The molar mass of Na_3A is 208 g/mol.)
   a. Dissolve 20.8 g Na_3A in 500 mL of 0.100 M HCl.
   b. Dissolve 20.8 g Na_3A in 1000 mL of 0.100 M HCl.
   c. Dissolve 20.8 g Na_3A in 1500 mL of 0.100 M HCl.
   d. Dissolve 20.8 g Na_3A in 2000 mL of 0.100 M HCl.
   e. Dissolve 20.8 g Na_3A in 2500 mL of 0.100 M HCl.

99. In the titration of 50.0 mL of 1.0 M methylamine, CH_3NH_2 ($K_b = 4.0 \times 10^{-4}$), with 0.50 M HCl, cal-

culate the pH under the following conditions.

a. after 50.0 mL of 0.50 *M* HCl has been added
b. at the stoichiometric point

100. Consider the titration of 100.0 mL of 0.100 *M* H_3A (K_{a_1} = 5.0 × 10^{-4}, K_{a_2} = 1.0 × 10^{-8}, K_{a_3} = 1.0 × 10^{-11}) with 0.0500 *M* NaOH.
    a. Calculate the pH after 100.0 mL of 0.0500 *M* NaOH has been added.
    b. What total volume of 0.0500 *M* NaOH is required to reach a pH of 8.67?

101. Calculate the pH of a solution formed by mixing 100.0 mL of 0.100 *M* NaF and 100.0 mL of 0.025 *M* HCl.

102. A solution is formed by mixing 50.0 mL of 10.0 *M* NaX with 50.0 mL of 2.0 × 10^{-3} *M* $CuNO_3$. Assume that Cu(I) forms complex ions with X^- as follows:

$$Cu^+(aq) + X^-(aq) \rightleftharpoons CuX(aq) \qquad K_1 = 1.0 \times 10^2$$
$$CuX(aq) + X^-(aq) \rightleftharpoons CuX_2^-(aq) \qquad K_2 = 1.0 \times 10^4$$
$$CuX_2^-(aq) + X^-(aq) \rightleftharpoons CuX_3^{2-}(aq) \qquad K_3 = 1.0 \times 10^3$$

Calculate the following concentrations at equilibrium.
a. CuX_3^{2-} b. CuX_2^- c. Cu^+

103. The titration of Na_2CO_3 with HCl has the following qualitative profile:

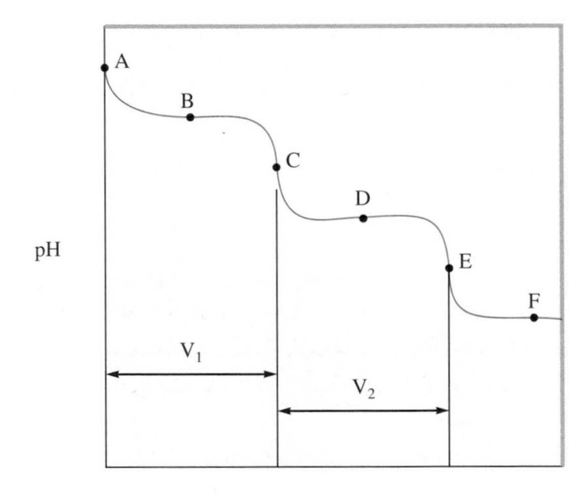

a. Identify the major species in solution at points A–F.
b. For the titration of 25.00 mL of 0.100 *M* Na_2CO_3 with 0.100 *M* HCl, calculate the pH at points A–E. (B and D are half-way points to equivalence.)
c. If a mixture of $NaHCO_3$ and Na_2CO_3 was titrated, what would be the relative sizes of V_1 and V_2?
d. If a mixture of NaOH and Na_2CO_3 was titrated, what would be the relative sizes of V_1 and V_2?
e. A sample contains a mixture of $NaHCO_3$ and Na_2CO_3. When 0.350 g of the sample was titrated with 0.100 *M* HCl, it took 18.9 mL to reach the first stoichiometric point and an additional 36.7 mL to reach the second stoichiometric point. What is the composition in mass percent of the sample?

104. When phosphoric acid is titrated with a NaOH solution, only two stoichiometric points are seen. Why?

105. Consider the following two acids:

Salicylic acid (OH, CO_2H) pK_{a_1} = 2.98; pK_{a_2} = 13.40

$HO_2CCH_2CH_2CH_2CH_2CO_2H$ pK_{a_1} = 4.41; pK_{a_2} = 5.28
Adipic acid

In two separate experiments the pH was measured during the titration of 5.00 mmol of each acid with 0.200 *M* HCl. Each experiment showed only one stoichiometric point when the data were plotted. In one experiment the stoichiometric point was at 25.00 mL of added HCl and in the other experiment the stoichiometric point was at 50.00 mL of HCl. Explain these results.

106. The solubility of calcium benzoate, $Ca(C_6H_5CO_2)_2 \cdot 3H_2O$, in water is 0.080 mol/L.
    a. Calculate the concentrations of Ca^{2+}, $C_6H_5CO_2^-$, $C_6H_5CO_2H$, OH^-, and H^+, and the pH of a saturated solution of calcium benzoate.
    b. Calculate the molar solubility of calcium benzoate in a solution buffered at pH = 4.00.
    c. Why does addition of H^+ increase the solubility of calcium benzoate?

107. A sample of a certain monoprotic weak acid was dissolved in water and titrated with 0.125 *M* NaOH, requiring 16.00 mL to reach the equivalence point. During the titration, the pH after adding 2.00 mL of NaOH was 6.912. Calculate K_a for the weak acid.

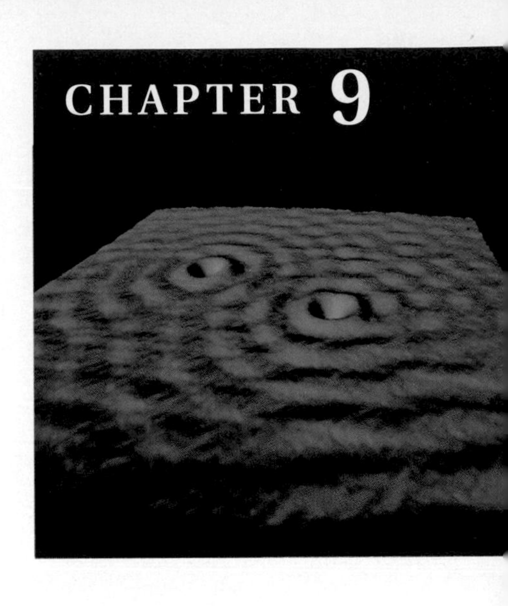

Energy, Enthalpy, and Thermochemistry

Energy is the essence of our very existence as individuals and as a society. The food that we eat furnishes the energy to live, work, and play, just as the coal and oil consumed by manufacturing and transportation systems power our modern industrialized civilization.

Huge quantities of carbon-based fossil fuels have been available for the taking. This abundance of fuels has led to a world society with a voracious appetite for energy, consuming millions of barrels of petroleum every day. We are now dangerously dependent on the dwindling supplies of oil, and this dependence is an important source of tension among nations in today's world. In an incredibly short time we have moved from a period of ample and cheap supplies of petroleum to one of high prices and uncertain supplies. If our present standard of living is to be maintained, we must find alternatives to petroleum. To do so, we need to know the relationship between chemistry and energy, which we explore in this chapter.

There are additional problems with fossil fuels. The waste products from burning fossil fuels significantly affect our environment. For example, when a carbon-based fuel is burned, the carbon reacts with oxygen to form carbon dioxide, which is released into the atmosphere. Although much of this carbon dioxide is consumed in various natural processes such as photosynthesis and the formation of carbonate minerals, the amount of carbon dioxide in the atmosphere is steadily increasing. This increase is significant because atmospheric carbon dioxide absorbs heat radiated from the earth's surface and radiates it back toward the earth. Since this is an important mechanism for controlling the earth's temperature, many scientists fear that an increase in the concentration of carbon dioxide will warm the earth, causing significant changes in climate. In addition, impurities in the fuels react with components of the air to produce air pollution.

In this chapter we will cover the fundamental concepts of energy and take a brief look at the practical aspects of the energy supply and pollution. Additional theoretical aspects of energy will be presented in Chapter 10.

9.1 The Nature of Energy

Although the concept of energy is quite familiar, energy is rather difficult to define precisely. We will define **energy** as the *capacity to do work or to produce heat*. In this chapter we will concentrate on the transfer of energy via heat flow that accompanies chemical processes.

The total energy content of the universe is constant.

One of the most important characteristics of energy is that it is conserved. The **law of conservation of energy** states that *energy can be converted from one form to another but can be neither created nor destroyed*. That is, the energy of the universe is constant. Energy can be classified as either potential energy or kinetic energy. **Potential energy** is energy due to position or composition. For example, water behind a dam has potential energy that can be converted to work when the water flows down through turbines, thereby creating electricity. Attractive and repulsive forces also lead to potential energy. The energy released when gasoline is burned results from differences in the attractive forces between nuclei and electrons in the reactants and products. The **kinetic energy** of an object is due to the motion of the object and depends on the mass of the object (m) and its velocity (v): KE $= \frac{1}{2}mv^2$.

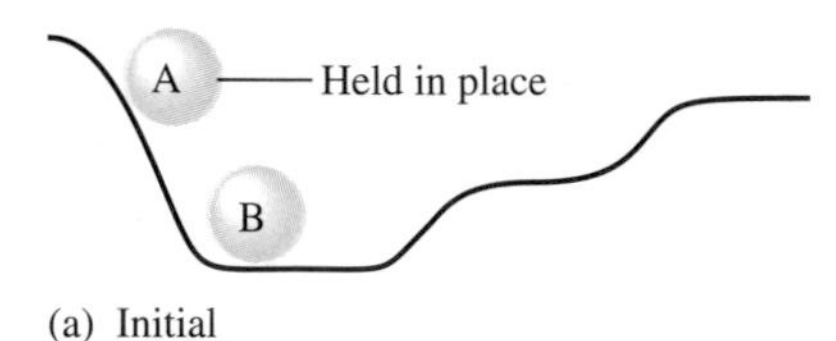

(a) Initial

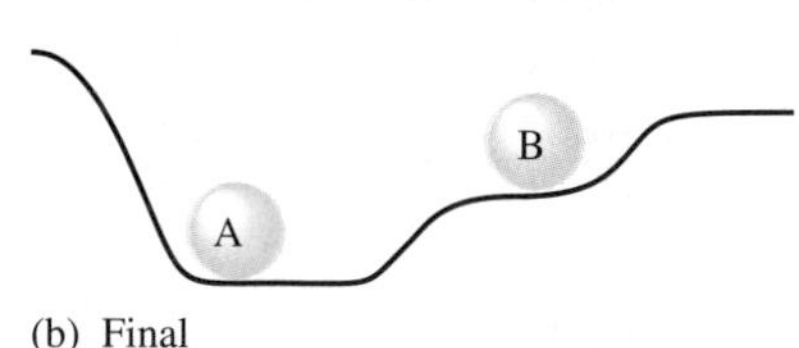

(b) Final

Figure 9.1
(a) In the initial positions ball A has a higher potential energy than ball B. (b) After A has rolled down the hill, the potential energy lost by A has been converted to random motions of the components of the hill (frictional heating) and to the increase in the potential energy of B.

Energy can be converted from one form to another easily. For example, consider the two balls in Fig. 9.1(a). Ball A, because of its higher position, initially has more potential energy than ball B. When A is released, it moves down the hill and strikes B. Eventually the arrangement shown in Fig. 9.1(b) is achieved. What has happened in going from the initial to the final arrangement? The potential energy of A has decreased, but since energy is conserved, all of the energy lost by A must be accounted for. How is this energy distributed?

Initially, the potential energy of A is changed to kinetic energy as the ball rolls down the hill. Part of this kinetic energy has been transferred to B, causing it to be raised to a higher final position. Thus B has increased potential energy. However, since the final position of B is lower than the original position of A, some of the energy is still unaccounted for. Both balls are at rest in their final positions, so the missing energy cannot be due to their motions. What has happened to the remaining energy?

The answer lies in the interaction between the hill's surface and the ball. As A rolls down the hill, some of its kinetic energy is transferred to the surface of the hill as heat. This transfer of energy is called *frictional heating*. The temperature of the hill increases very slightly as the ball rolls down.

Heat involves a *transfer* of energy.

At this point, it is important to recognize that heat and temperature are decidedly different. Recall that temperature is a property that reflects the random motions of the particles in a particular substance. **Heat**, on the other hand, involves the *transfer* of energy between two objects due to a temperature difference. Heat is not a substance contained in an object, although we often talk of heat as if this were true.

Note that in going from the initial to the final arrangements in Fig. 9.1, ball B gains potential energy because ball A has done work on B. **Work** is defined as *a force acting over a distance.* Work is required to raise B from its original position to a higher one. Part of the original energy stored as potential energy in A has been transferred through work to B, thereby increasing B's potential energy. Thus there are two ways to transfer energy: through work and through heat.

In rolling to the bottom of the hill as shown in Fig. 9.1, ball A will always lose the same amount of potential energy. However, the way that this energy transfer is divided between work and heat depends on the specific conditions—the **pathway.** For example, the surface of the hill might be so rough that the energy of A is expended completely through frictional heating; A is moving so slowly when it hits B that it cannot move B to the next level. In this case no work is done. Regardless of the condition of the hill's surface, the *total energy* transferred will be constant. However, the amounts of heat and work will differ. Energy change is independent of the pathway; however, work and heat are both dependent on the pathway.

This brings us to a very important concept: the **state function** or *state property.* A state function refers to a property of the system that depends only on its *present state.* A state function (property) does not depend in any way on the system's past (or future). In other words, the value of a state function does not depend on how the system arrived at the present state; it depends only on the characteristics of the present state.

Stated more precisely, one very important characteristic of a state function is that a change in this function (property) in going from one state to another state is independent of the particular pathway taken between the two states.

Of the functions considered in our present example, energy is a state function, but work and heat are not state functions.

Energy is a state function; work and heat are not.

Chemical Energy

The ideas we have just illustrated using mechanical examples also apply to chemical systems. The combustion of methane, for example, is used to heat many homes in the United States:

$$CH_4(g) + 2O_2(g) \longrightarrow CO_2(g) + 2H_2O(g) + \text{energy (heat)}$$

To discuss this reaction, we divide the universe into two parts: the system and the surroundings. The **system** is the part of the universe on which we wish to focus attention; the **surroundings** include everything else in the universe. In this case we define the system as the reactants and products of the reaction. The surroundings consist of the reaction container, the room, and everything else other than the reactants and products.

When a reaction results in the evolution of heat, it is said to be **exothermic** (*exo-* is a prefix meaning "out of"); that is, energy flows *out of the system.* For example, in the combustion of methane, energy flows out of the system as heat. Reactions that absorb energy from the surroundings are said to be **endothermic.** That is, when the heat flow is *into a system,* the process is endothermic. For example, the formation of nitric oxide from nitrogen and oxygen is endothermic:

$$N_2(g) + O_2(g) + \text{energy (heat)} \longrightarrow 2NO(g)$$

A familiar endothermic physical process is the vaporization of water:

$$H_2O(l) + \text{energy} \longrightarrow H_2O(g)$$

Where does the energy, released as heat, come from in an exothermic reaction? The answer lies in the difference in potential energy between the products and the reactants. In an exothermic reaction, which has lower potential energy, the reactants or the products? We know that total energy is conserved and that energy flows from the system into the surroundings in an exothermic reaction. This means that *the energy gained by the surroundings must be equal to the energy lost by the system*. For methane combustion the energy content of the system decreases, which means that 1 mole of CO_2 and 2 moles of H_2O molecules (the products) possess less potential energy than do 1 mole of CH_4 and 2 moles of O_2 molecules (the reactants). The heat flow into the surroundings results from a lowering of the potential energy of the reaction system. This always holds true. *In any exothermic reaction, the potential energy stored in the chemical bonds is being converted to thermal energy (random kinetic energy) via heat.*

The energy diagram for the combustion of methane is shown in Fig. 9.2, where Δ(PE) represents the *change* in potential energy stored in the bonds of the products as compared with the bonds of the reactants. In other words, this quantity represents the difference between the energy required to break the bonds in the reactants and the energy released when the bonds in the products are formed. In an exothermic process the bonds in the products are stronger (on the average) than those of the reactants. That is, more energy is released by forming the new bonds in the products than is consumed to break the bonds in the reactants. The net result is that the quantity of energy Δ(PE) is transferred to the surroundings through heat.

For an endothermic reaction the situation is reversed, as shown in Fig. 9.3. Energy that flows into the system as heat is used to increase the potential energy of the system. In this case the products have higher potential energy (weaker bonds on average) than the reactants.

The study of energy and its interconversions is called **thermodynamics.** The law of conservation of energy is often called the **first law of thermodynamics** and is stated as follows: *the energy of the universe is constant.*

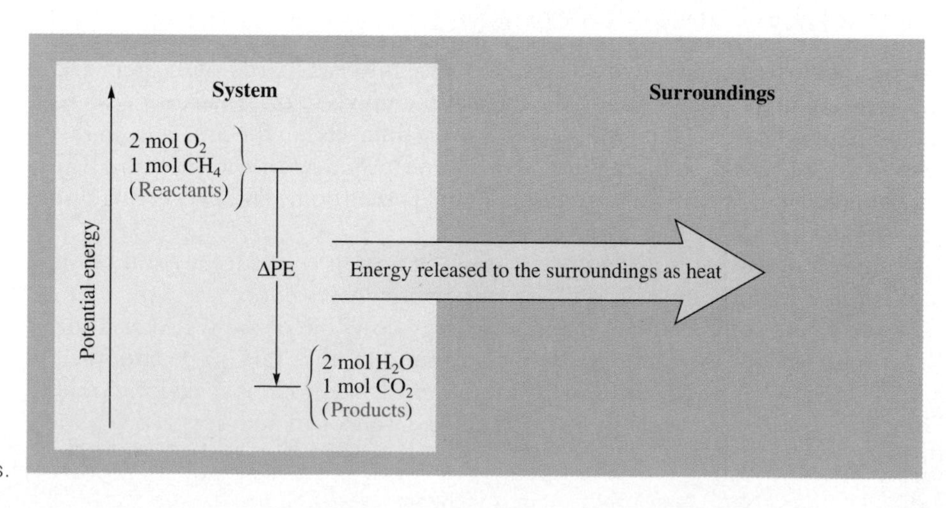

Figure 9.2
The combustion of methane releases the quantity of energy Δ(PE) to the surroundings via heat flow. This is an exothermic process.

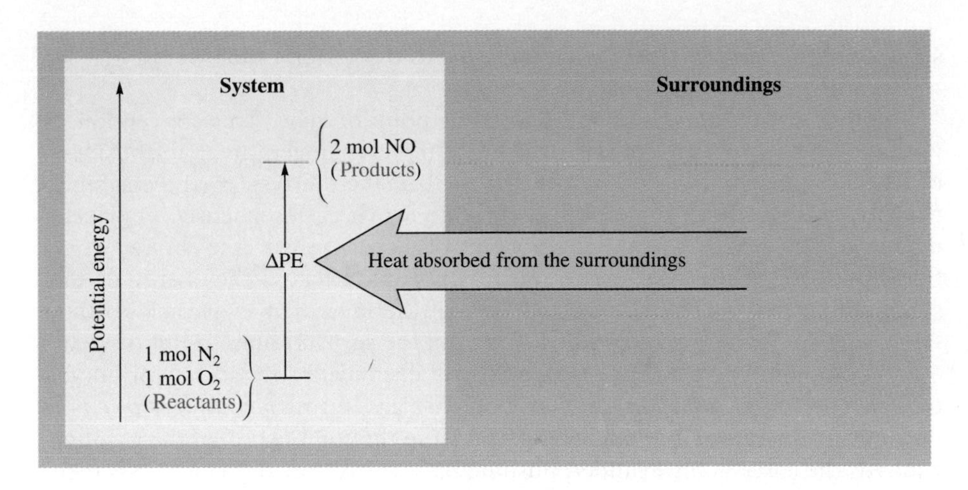

Figure 9.3
The energy diagram for the reaction of nitrogen and oxygen to form nitric oxide. This is an endothermic process.

The **internal energy** (E) of a system can be defined most precisely as the sum of the kinetic and potential energies of all of the "particles" in the system. The internal energy of a system can be changed by a flow of work, heat, or both. That is,

$$\Delta E = q + w$$

where ΔE represents the change in the system's internal energy, q represents heat, and w represents work.

Thermodynamic quantities always consist of two parts: a *number*, giving the magnitude of the change; and a *sign*, indicating the direction of the flow. *The sign reflects the system's point of view*. For example, if a quantity of energy flows *into* the system via heat (an endothermic process), q is equal to $+x$, where the *positive* sign indicates that the *system's energy is increasing*. On the other hand, when energy flows *out of* the system via heat (an exothermic process), q is equal to $-x$, where the *negative* sign indicates that the *system's energy is decreasing*.

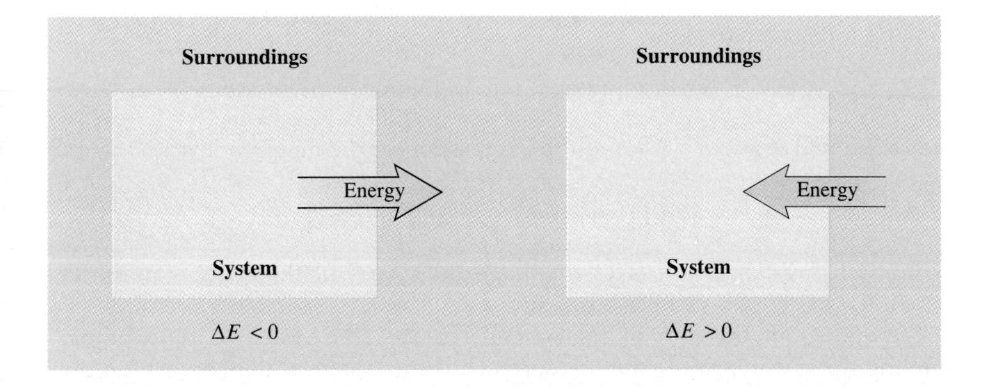

In this text the same conventions also apply to the flow of work. If the system does work on the surroundings (energy flows out of the system), w is negative. If the surroundings do work on the system (energy flows into the system), w is positive. We define work from the system's point of view to be consistent for all thermodynamic quantities. That is, in this convention the signs of both q and w reflect what happens to the system; thus we use $\Delta E = q + w$.

The convention in this text is to take the system's point of view: $q = -x$ denotes an exothermic process, and $q = +x$ denotes an endothermic one.

In this text we *always* take the system's point of view. This convention is not followed in every area of science. For example, engineers are in the business of designing machines to do work, that is, to make the system (the machine) transfer energy to its surroundings through work. Consequently, engineers define work from the surroundings' point of view. In their convention, work that flows out of the system is treated as positive because the energy of the surroundings has increased. The first law of thermodynamics is then written $\Delta E = q - w'$, where w' signifies work from the surroundings' point of view.

A common type of work associated with chemical processes is work done by a gas (through *expansion*) or work done to a gas (through *compression*). For example, in an automobile engine the heat from the combustion of the gasoline expands the gases in the cylinder, pushing back the piston. This motion is then translated into the motion of the car.

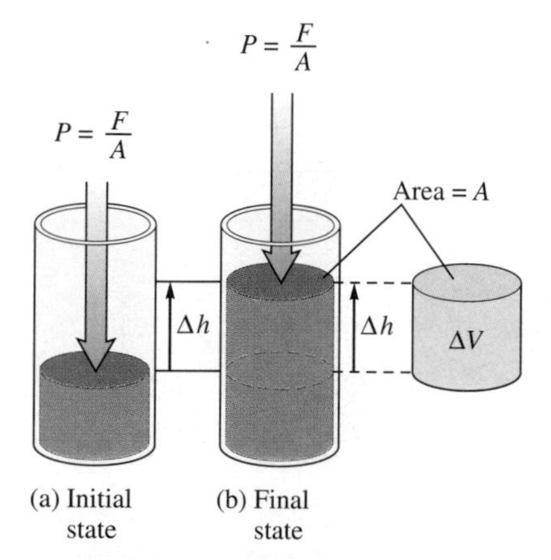

Figure 9.4
(a) The piston, moving a distance Δh against a pressure P, does work on the surroundings. (b) Since the volume of a cylinder is the area of the base times its height, the change in volume of the gas is given by $\Delta h \times A = \Delta V$.

Suppose we have a gas confined to a cylindrical container with a movable piston, as shown in Fig. 9.4, where F is the force acting on a piston of area A. Since pressure is defined as force per unit area, the pressure of the gas is

$$P = \frac{F}{A}$$

Work is defined as a force applied over a given distance, so if the piston moves a distance Δh, as shown in Fig. 9.4, then the magnitude of the work is

$$|\text{Work}| = |\text{force} \times \text{distance}| = |F \times \Delta h|$$

Since $P = F/A$, or $F = P \times A$, then

$$|\text{Work}| = |F \times \Delta h| = |P \times A \times \Delta h|$$

Since the volume of the cylinder equals the area of the piston times the height of the cylinder (Fig. 9.4), the change in volume ΔV resulting from the piston moving a distance Δh is

$$\Delta V = \text{final volume} - \text{initial volume} = A \times \Delta h$$

Substituting $\Delta V = A \times \Delta h$ into the expression for the magnitude of the work gives

$$|\text{Work}| = |P \times A \times \Delta h| = |P\Delta V|$$

Note that this expression gives the *magnitude* (size) of the work required to expand a gas by ΔV against a pressure P.

w and $P\Delta V$ have opposite signs since when the gas expands (ΔV is positive), work flows into the surroundings (w is negative).

What about the sign of the work? The gas (the system) is expanding, moving the piston against the pressure. Thus the system is doing work on the surroundings, so from the system's point of view, the sign of the work should be negative.

For an *expanding gas* ΔV is a positive quantity because the volume is increasing. Thus ΔV and w must have opposite signs, which leads to the equation

$$w = -P\Delta V$$

Note that for a gas expanding against an external pressure P, w is a negative quantity as required, since work flows out of the system. When a gas is *compressed,* ΔV is a negative quantity (the volume decreases), which makes w a positive quantity (work flows into the system).

In dealing with "PV work," keep in mind that the P in $P\Delta V$ always refers to the external pressure —the pressure that causes a compression or that resists an expansion.

The work accompanying a change in volume of a gas is often called "*PV* work."

EXAMPLE 9.1

A balloon is inflated to its full extent by heating the air inside it. In the final stages of this process the volume of the balloon changes from 4.00×10^6 L to 4.50×10^6 L by addition of 1.3×10^8 J of energy as heat. Assuming the balloon expands against a constant pressure of 1.0 atm, calculate ΔE for the process.

Solution

To calculate ΔE we use the equation

$$\Delta E = q + w$$

Since the problem states that 1.3×10^8 J of energy is *added* as heat,

$$q = +1.3 \times 10^8 \text{ J}$$

The work done can be calculated from the expression

$$w = -P\Delta V$$

In this case $P = 1.0$ atm (the external pressure) and

$$\begin{aligned} \Delta V &= V_{\text{final}} - V_{\text{initial}} \\ &= 4.50 \times 10^6 \text{ L} - 4.00 \times 10^6 \text{ L} = 0.50 \times 10^6 \text{ L} \\ &= 5.0 \times 10^5 \text{ L} \end{aligned}$$

Thus

$$w = -1.0 \text{ atm} \times 5.0 \times 10^5 \text{ L} = -5.0 \times 10^5 \text{ L atm}$$

Note that the negative sign for w makes sense, since the gas is expanding and thus doing work on the surroundings.

To calculate ΔE, we must sum q and w. However, since q is given in units of J and w is given in units of L atm, we must change the work to units of joules:

$$w = -5.0 \times 10^5 \text{ L atm} \times \frac{101.3 \text{ J}}{\text{L atm}} = -5.1 \times 10^7 \text{ J}$$

Then

$$\Delta E = q + w = (+1.3 \times 10^8 \text{ J}) + (-5.1 \times 10^7 \text{ J}) = 8 \times 10^7 \text{ J}$$

Since more energy is added through heating than the gas expends doing work, there is a net increase in the energy of the gas in the balloon. Hence ΔE is positive.

The joule (J) is the fundamental SI unit for energy:

$$\text{J} = \frac{\text{kg m}^2}{\text{s}^2}$$

The conversion factor between L atm and J can be obtained from the values of R:

$$0.08206 \frac{\text{L atm}}{\text{K mol}}$$

and

$$8.3145 \frac{\text{J}}{\text{K mol}}$$

9.2 Enthalpy

So far we have discussed the internal energy of a system. A less familiar property of a system is its **enthalpy** (H), which is defined as

$$H = E + PV$$

where E is the internal energy of the system, P is the pressure of the system, and V is the volume of the system.

Enthalpy is a state function. A change in enthalpy does not depend on the pathway between two states.

Since internal energy, pressure, and volume are all state functions, *enthalpy is also a state function*. But what exactly is enthalpy? To help answer this question, consider a process carried out at constant pressure, where the only work allowed is pressure-volume work ($w = -P\Delta V$). Under these conditions the expression

$$\Delta E = q_p + w$$

becomes

$$\Delta E = q_p - P\Delta V$$

or

$$q_p = \Delta E + P\Delta V$$

where q_p is the heat at constant pressure.

We will now relate q_p to the change in enthalpy. The definition of enthalpy is $H = E + PV$. Therefore

$$(\text{change in } H) = (\text{change in } E) + (\text{change in } PV)$$

or

$$\Delta H = \Delta E + \Delta(PV)$$

Since P is constant, the change in PV is due only to a change in volume. Thus

$$\Delta(PV) = P\Delta V$$

and

$$\Delta H = \Delta E + P\Delta V$$

This expression is identical to the one we obtained for q_p:

$$q_p = \Delta E + P\Delta V$$

$\Delta H = q$ at constant pressure, where only "PV work" is allowed.

Thus for a process carried out at constant pressure, where the only work allowed is that from a volume change,

$$\Delta H = q_p$$

At constant pressure (where only PV work is allowed) the change in enthalpy (ΔH) of the system is equal to the energy flow as heat. This means that, for a reaction studied at constant pressure, the flow of heat is a measure of the change in enthalpy for the system. For this reason, the terms *heat of reaction* and *change in enthalpy* are used interchangeably for reactions studied at constant pressure.

The change in enthalpy of a system has no easily interpreted meaning except at constant pressure, where ΔH = heat.

For a chemical reaction the enthalpy change is given by the equation

$$\Delta H = H_{\text{products}} - H_{\text{reactants}}$$

In a case in which the products of a reaction have greater enthalpy than the reactants, ΔH will be positive. Thus heat is absorbed by the system, and the reaction is endothermic. On the other hand, if the enthalpy of the products is less than that of the reactants, ΔH is negative. In this case the overall decrease in enthalpy is achieved by the generation of heat, and the reaction is exothermic.

At constant pressure exothermic means ΔH is negative; endothermic means ΔH is positive.

9.3 Thermodynamics of Ideal Gases

In developing the concepts of thermodynamics, we often find it useful to refer to the properties of matter in the simplest possible context. For this reason we often start with the thermodynamic characteristics of the ideal gas—the hypothetical condition approached by real gases at high temperatures and low pressures such that they obey the relationship $PV = nRT$.

Recall from Chapter 5 that for an ideal gas

$$(\mathrm{KE})_{\mathrm{avg}} = \tfrac{3}{2}RT$$

where $(\mathrm{KE})_{\mathrm{avg}}$ represents the average, random, translational energy for 1 mole of gas at a given temperature T (in Kelvins). The only way to change the kinetic energy of an ideal gas is to change its temperature. The energy ("heat") required to change the temperature of 1 mole of an ideal gas by ΔT is

$$\text{Energy ("heat") required} = \tfrac{3}{2}R\Delta T$$

Note that for a temperature change of 1 K ($\Delta T = 1$) the energy required is $\frac{3}{2}R$.

The **molar heat capacity** of a substance is defined as the energy required to raise the temperature of 1 mole of that substance by 1 K. Thus we might conclude that the molar heat capacity of an ideal gas is $\frac{3}{2}R$. However, we will have to qualify this conclusion when we consider the implications of the PV work that can occur when a gas is heated.

Heating an Ideal Gas at Constant Volume

When an ideal gas is heated in a rigid container where no change in volume occurs, there can be no PV work ($\Delta V = 0$). Under these conditions all of the energy that flows into the gas is used to increase the translational energies of the gas molecules. Thus C_v, the molar heat capacity of an ideal gas *at constant volume,* is $\frac{3}{2}R$, the result anticipated in the above discussion:

For an ideal gas, work occurs only when its volume changes. Thus if a gas is heated at constant volume, the pressure increases but no work occurs.

$$C_v = \tfrac{3}{2}R = \begin{array}{l}\text{"heat" required to change the temperature}\\ \text{of 1 mol of gas by 1 K at constant volume}\end{array}$$

Heating an Ideal Gas at Constant Pressure

When an ideal gas is heated at constant pressure, its volume increases and PV work occurs. So when a gas is heated at constant pressure, energy must be supplied both to change the translational energy of the gas and to provide for the work the gas does as it expands:

$$\text{Energy required} = \text{“heat”} = \begin{matrix}\text{energy needed} \\ \text{to change the} \\ \text{translational energy}\end{matrix} + \begin{matrix}\text{energy needed to} \\ \text{do the } PV \text{ work}\end{matrix}$$

The heat needed to increase the translational energy is $\frac{3}{2}R$, as we concluded above.

The *quantity* of work done as the gas expands by ΔV is $P\Delta V$. Using the ideal gas law, we see that

$$P\Delta V = nR\Delta T = R\Delta T \qquad \text{(per mole)}$$

Thus for a 1-K change in temperature ($\Delta T = 1$ K) the work is R, so

$$\begin{matrix}\text{Heat required to increase the temperature} \\ \text{of 1 mol of gas by 1 K (constant } P)\end{matrix} = \tfrac{3}{2}R + R = \tfrac{5}{2}R$$
$$= C_v + R$$

$C_v = \frac{3}{2}R$
$C_p = \frac{5}{2}R = C_v + R$

Therefore, we have shown that C_p, the molar heat capacity of an ideal gas at constant pressure, is $\frac{5}{2}R$ or $C_v + R$.

Heating a Polyatomic Gas

We have established that for an ideal gas the molar heat capacity at constant volume is $\frac{3}{2}R$. This value of C_v assumes that an ideal gas consists of “particles” that have no structure. That is, we assume that the particles are monatomic (consisting of a single atom). Monatomic real gases, such as helium, have measured values of C_v very close to $\frac{3}{2}R$. However, gases such as SO_2 and $CHCl_3$ that contain polyatomic molecules have observed values for C_v that are significantly greater than $\frac{3}{2}R$. For example, the value of C_v for SO_2 is almost $4R$ at 25°C. This larger value for C_v results because polyatomic molecules absorb energy to excite rotational and vibrational motions in addition to translational motions. That is, at 25°C the molecules in such a gas are rotating, and the atoms in the molecule are vibrating, as if the bonds were springs.

As a polyatomic gas is heated, the gas molecules absorb energy to increase their rotational and vibrational motions as well as to move through space (translate) at higher speeds. Recall from our previous discussions that the temperature of a monatomic ideal gas is an index of the average random *translational* energy of the gas. Thus when a gas is heated, the temperature only increases to the extent that the translational energies of the molecules increase. Any energy that is absorbed to increase the vibrational and rotational energies does not contribute directly to the translational kinetic energy; so for a gas that consists of diatomic or polyatomic molecules, much of the heat absorbed is used in processes that do not directly increase the temperature. Thus its heat capacity (the energy required to change its *temperature* by 1 K) is greater than $\frac{3}{2}R$.

C_v is greater than $\frac{3}{2}R$ for a gas in which the individual particles are molecules, because some of the energy added via heat flow is “stored” in motions that do not directly raise the temperature of the gas.

Note that the elevated value of C_v for a gas whose particles are molecules is not due to nonideal behavior. That is, it does not depend on whether or not the gas obeys the ideal gas law. Rather, it is simply due to the internal structure of

TABLE 9.1 **Molar Heat Capacities of Various Gases at 298 K**

Gas	$C_v \left(\frac{J}{K\ mol}\right)$	$C_p \left(\frac{J}{K\ mol}\right)$	$C_p - C_v$
He, Ne, Ar	12.47	20.80	8.33
H_2	20.54	28.86	8.32
N_2	20.71	29.03	8.32
N_2O	30.38	38.70	8.32
CO_2	28.95	37.27	8.32
C_2H_6	44.60	52.92	8.32

the molecules that enables them to absorb energy for processes other than translational motions.

Recall that C_p is greater than C_v because of the work done by the heated gas as it expands at constant pressure. Thus if we assume that a given polyatomic gas obeys the ideal gas law, the expression

$$C_p = C_v + R$$

can be used to calculate C_p if the value of C_v for the gas is known.

The observed heat capacities of several gases are shown in Table 9.1. Notice that the monatomic gases have values of C_v equal to $\frac{3}{2}R$ (12.47 J K^{-1} mol^{-1}). Note also that as the molecules become more complex (more atoms), C_v increases. This result is expected because the presence of more atoms means that more nontranslational motions are available to absorb energy. Finally, notice that in all cases $C_p - C_v = R$, as expected for gases that closely obey the ideal gas law.

Heating a Gas: Energy and Enthalpy

Recall that the average translational energy of an ideal gas E is given by the expression

$$E = \tfrac{3}{2}RT \qquad \text{(per mole)}$$

for a monatomic ideal gas. The energy can be changed only by changing the temperature:

$$\Delta E = \tfrac{3}{2}R\Delta T \qquad \text{(per mole)}$$

Note that this expression corresponds to

$$\Delta E = C_v\Delta T \qquad \text{(per mole)}$$

or

$$\Delta E = nC_v\Delta T \qquad (n \text{ moles})$$

The constant-volume heat capacity appears in this expression because when a gas is heated at constant volume, all of the input energy (heat) goes toward increasing E (no heat is needed to do work).

On the other hand, when a gas is heated at constant pressure, the volume changes and work occurs. In this case (for n moles of gas),

$$\begin{aligned} \text{“Heat” required} &= q_p = nC_p\Delta T \\ &= n(C_v + R)\Delta T \\ &= \underbrace{nC_v\Delta T}_{\Delta E} + \underbrace{nR\Delta T}_{P\Delta V\,=\,\text{work required}} \end{aligned}$$

Notice that although this process is carried out at constant pressure, ΔE is still given by $nC_v\Delta T$. This result seems contradictory at first glance, but actually, it makes good sense. Because E for an ideal gas depends only on T (it does not depend on pressure or volume, for example), $\Delta E = nC_v\Delta T$ when an ideal gas is heated whether the process occurs at constant volume or constant pressure.

Next, consider the change in enthalpy when a gas is heated. Recall that by definition

$$H = E + PV$$

Thus, in general, a change in enthalpy is given by the expression

$$\Delta H = \Delta E + \Delta(PV)$$

which (using the ideal gas law) becomes

$$\Delta H = \Delta E + \Delta(nRT) = \Delta E + nR\Delta T$$

for a sample of ideal gas containing n moles. Substituting $\Delta E = nC_v\Delta T$, we have

$$\begin{aligned} \Delta H &= nC_v\Delta T + nR\Delta T \\ &= n(C_v + R)\Delta T = nC_p\Delta T \end{aligned}$$

Note that we have shown that

$$\Delta H = nC_p\Delta T$$

The only way to change H and E for an ideal gas is to change the temperature of the gas. Thus for any process involving *an ideal gas at constant temperature*, $\Delta H = 0$ and $\Delta E = 0$.

even though we have not assumed constant pressure (or volume). Thus we have shown that for an ideal gas we can always use the expression $nC_p\Delta T$ to calculate the change in enthalpy when n moles of an ideal gas is heated, regardless of any conditions on pressure or volume.

Again, it may seem contradictory that C_p appears in this expression for ΔH even though the pressure may or may not be constant in the process. However, note that the enthalpy ($H = E + PV$) of an ideal gas depends on E and the product PV. We have seen that E depends directly on T, and from the ideal gas law we can easily show that PV depends directly on T ($PV = nRT$) for a given sample of ideal gas (containing n moles). Thus both the energy E and the enthalpy H of an ideal gas depend only on T, not on P or V (individually):

$$E \propto T \qquad \text{and} \qquad H \propto T$$

For energy (E) the proportionality constant is C_v (per mole), and for enthalpy (H) the proportionality constant is C_p (per mole).

In considering these ideas, we must distinguish among q, ΔH, and ΔE. In the calculation of the heat flow for an ideal gas the equation

$$q = nC\Delta T$$

TABLE 9.2 Thermodynamic Properties of an Ideal Gas

Expression	Application
$C_v = \frac{3}{2}R$	Monatomic ideal gas
$C_v > \frac{3}{2}R$	Polyatomic ideal gas (value must be measured experimentally)
$C_p = C_v + R$	All ideal gases
$C_p = \frac{5}{2}R = \frac{3}{2}R + R$	Monatomic ideal gas
$C_p > \frac{5}{2}R$	Polyatomic ideal gas (specific value depends on the value of C_v)
$\Delta E = nC_v\Delta T$	All ideal gases
$\Delta H = nC_p\Delta T$	All ideal gases

applies, where C_v or C_p is used depending on the conditions. In contrast, $\Delta H = nC_p\Delta T$ and $\Delta E = nC_v\Delta T$ for a temperature change of an ideal gas regardless of whether pressure or volume (or neither) is constant. Also, note that the heat flow equals ΔE at constant volume ($\Delta E = q_v$) but the heat flow equals ΔH at constant pressure ($\Delta H = q_p$). These results are summarized in Table 9.2.

We will illustrate these concepts in Example 9.2.

EXAMPLE 9.2

Consider 2.00 mol of a monatomic ideal gas that is taken from state A ($P_A = 2.00$ atm, $V_A = 10.0$ L) to state B ($P_B = 1.00$ atm, $V_B = 30.0$ L) by two different pathways:

$$\begin{pmatrix} V_C = 30.0\ \text{L} \\ P_C = 2.00\ \text{atm} \end{pmatrix}$$

$$\underset{\text{State A}}{\begin{pmatrix} V_A = 10.0\ \text{L} \\ P_A = 2.00\ \text{atm} \end{pmatrix}} \quad \xrightarrow{1} C \xrightarrow{2} \quad \underset{\text{State B}}{\begin{pmatrix} V_B = 30.0\ \text{L} \\ P_B = 1.00\ \text{atm} \end{pmatrix}}$$

$$A \xrightarrow{3} \begin{pmatrix} V_D = 10.0\ \text{L} \\ P_D = 1.00\ \text{atm} \end{pmatrix} \xrightarrow{4} B$$

Calculate q, w, ΔE, and ΔH for both pathways.

Solution

Before we do any calculations, it is useful to summarize the processes described above using the "PV diagram" shown in Fig. 9.5.

STEP 1

Notice from Fig. 9.5 that this step corresponds to an expansion from 10.0 to 30.0 L at a constant pressure of 2.00 atm. This process must

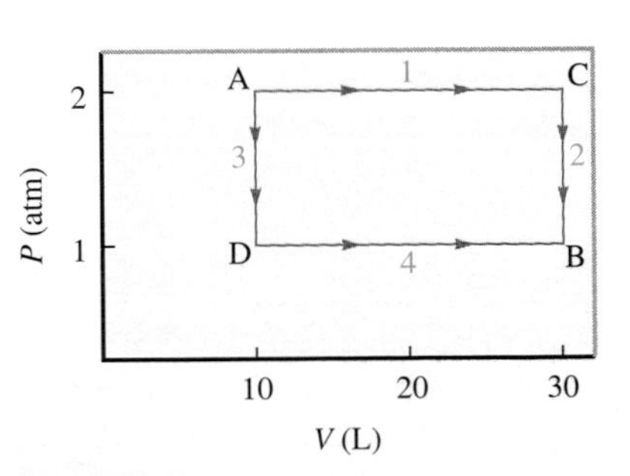

Figure 9.5
Summary of the two pathways discussed in Example 9.2.

occur by heating the gas to produce some temperature change ΔT (not specified in the given data). From the ideal gas law we know that

$$P\Delta V = nR\Delta T$$

In this case $\Delta V = 30.0\text{ L} - 10.0\text{ L}$, so

$$P\Delta V = (2.00\text{ atm})(20.0\text{ L}) = 4.00 \times 10^1\text{ L atm}$$

or if we convert to joules,

$$P\Delta V = 4.00 \times 10^1\text{ L atm} \times \frac{101.3\text{ J}}{\text{L atm}} = 4.05 \times 10^3\text{ J}$$

It follows that

$$\Delta T = \frac{P\Delta V}{nR} = \frac{4.05 \times 10^3\text{ J}}{nR}$$

We know that $w = -P\Delta V$, and because in this case the gas expands against a constant external pressure of 2.00 atm, we have

$$\begin{aligned} w_1 &= -(2.00\text{ atm})(30.0\text{ L} - 10.0\text{ L}) = -4.00 \times 10^1\text{ L atm} \\ &= -4.05 \times 10^3\text{ J} \end{aligned}$$

Also, in this case (constant P)

$$\begin{aligned} q_1 &= q_p = nC_p\Delta T \\ &= n\left(\frac{5}{2}R\right)\left(\frac{4.05 \times 10^3\text{ J}}{nR}\right) = 1.01 \times 10^4\text{ J} \end{aligned}$$

(Note the signs of w and q are as expected: the gas expands, so work flows out of the system; the gas is heated, so heat flows into the system.)

We can calculate ΔE_1 and ΔH_1 as follows:

$$\Delta E_1 = nC_v\Delta T = n\left(\frac{3}{2}R\right)\left(\frac{4.05 \times 10^3\text{ J}}{nR}\right) = 6.08 \times 10^3\text{ J}$$

$$\Delta H_1 = nC_p\Delta T = n\left(\frac{5}{2}R\right)\left(\frac{4.05 \times 10^3\text{ J}}{nR}\right) = 1.01 \times 10^4\text{ J}$$

Note that in this case $q_1(q_p)$ equals ΔH_1, as expected, because this process is carried out at constant pressure.

STEP 2

In this step the gas pressure decreases from 2.00 atm to 1.00 atm at constant volume. This step must correspond to the cooling of the gas by a quantity ΔT, which we can find from the ideal gas law:

$$\begin{aligned} \Delta PV &= nR\Delta T \\ \Delta T &= \frac{\Delta PV}{nR} = \frac{(1.00\text{ atm} - 2.00\text{ atm})(30.0\text{ L})}{nR} \\ &= \frac{-30.0\text{ L atm}}{nR} = \frac{-3.04 \times 10^3\text{ J}}{nR} \end{aligned}$$

Note that ΔT is negative, as expected for a cooling process.

Because in this step $\Delta V = 0$, thus $w_2 = 0$. In this case

$$q_2 = q_v = nC_v\Delta T = n\left(\frac{3}{2}R\right)\left(\frac{-3.04 \times 10^3 \text{ J}}{nR}\right)$$
$$= -4.56 \times 10^3 \text{ J}$$

Also,

$$\Delta E_2 = nC_v\Delta T = n\left(\frac{3}{2}R\right)\left(\frac{-3.04 \times 10^3 \text{ J}}{nR}\right)$$
$$= -4.56 \times 10^3 \text{ J} = q_v$$

and

$$\Delta H_2 = nC_p\Delta T = n\left(\frac{5}{2}R\right)\left(\frac{-3.04 \times 10^3 \text{ J}}{nR}\right)$$
$$= -7.60 \times 10^3 \text{ J}$$

Notice that in this case $q_2 = q_v = \Delta E$, as expected for a constant-volume process.

Using similar reasoning, we can compute the required quantities for Steps 3 and 4.

STEP 3

$$\Delta T = \frac{\Delta PV}{nR} = \frac{(-1.00 \text{ atm})(10.0 \text{ L})}{nR} = \frac{-10.0 \text{ L atm}}{nR}$$
$$= \frac{-1.01 \times 10^3 \text{ J}}{nR}$$

$$w_3 = 0 \qquad (\Delta V = 0)$$

$$q_3 = q_v = nC_v\Delta T = n\left(\frac{3}{2}R\right)\left(\frac{-1.01 \times 10^3 \text{ J}}{nR}\right)$$
$$= -1.52 \times 10^3 \text{ J}$$

$$\Delta E_3 = q_v = -1.52 \times 10^3 \text{ J}$$

$$\Delta H_3 = nC_p\Delta T = n\left(\frac{5}{2}R\right)\left(\frac{-1.01 \times 10^3 \text{ J}}{nR}\right)$$
$$= -2.53 \times 10^3 \text{ J}$$

STEP 4

$$\Delta T = \frac{P\Delta V}{nR} = \frac{(1.00 \text{ atm})(20.0 \text{ L})}{nR} = \frac{20.0 \text{ L atm}}{nR}$$
$$= \frac{2.03 \times 10^3 \text{ J}}{nR}$$

$$w_4 = -P\Delta V = -(1.00 \text{ atm})(20.0 \text{ L}) = -20.0 \text{ L atm}$$
$$= -2.03 \times 10^3 \text{ J}$$

$$q_4 = q_p = nC_p\Delta T = n\left(\frac{5}{2}R\right)\left(\frac{2.03 \times 10^3 \text{ J}}{nR}\right)$$
$$= 5.08 \times 10^3 \text{ J}$$
$$\Delta E_4 = nC_v\Delta T = n\left(\frac{3}{2}R\right)\left(\frac{2.03 \times 10^3 \text{ J}}{nR}\right)$$
$$= 3.05 \times 10^3 \text{ J}$$
$$\Delta H_4 = nC_p\Delta T = n\left(\frac{5}{2}R\right)\left(\frac{2.03 \times 10^3 \text{ J}}{nR}\right)$$
$$= 5.08 \times 10^3 \text{ J} = q_p$$

SUMMARY

- Pathway one (Steps 1 and 2):

$$q_{one} = q_1 + q_2 = 1.01 \times 10^4 \text{ J} - 4.56 \times 10^3 \text{ J}$$
$$= 5.5 \times 10^3 \text{ J}$$
$$w_{one} = w_1 + w_2 = -4.05 \times 10^3 \text{ J}$$
$$q_{one} + w_{one} = 1.5 \times 10^3 \text{ J} = \Delta E_{one}$$
$$\Delta H_{one} = \Delta H_1 + \Delta H_2$$
$$= 1.01 \times 10^4 \text{ J} - 7.60 \times 10^3 \text{ J}$$
$$= 2.5 \times 10^3 \text{ J}$$

- Pathway two (Steps 3 and 4):

$$q_{two} = q_3 + q_4 = -1.52 \times 10^3 \text{ J} + 5.08 \times 10^3 \text{ J}$$
$$= 3.56 \times 10^3 \text{ J}$$
$$w_{two} = w_3 + w_4 = -2.03 \times 10^3 \text{ J}$$
$$q_{two} + w_{two} = 3.55 \times 10^3 \text{ J} - 2.03 \times 10^3 \text{ J}$$
$$= 1.52 \times 10^3 \text{ J} = \Delta E_{two}$$
$$\Delta H_{two} = \Delta H_3 + \Delta H_4$$
$$= -2.53 \times 10^3 \text{ J} + 5.08 \times 10^3 \text{ J}$$
$$= 2.55 \times 10^3 \text{ J}$$

Notice from the results of Example 9.2 that the work and heat are different for the two pathways between states A and B. For example, in pathway one the expansion (by 20.0 L) is carried out at a pressure of 2.00 atm, and in pathway two the expansion (by 20.0 L) is carried out at 1.00 atmosphere. Thus since $|w| = |P\Delta V|$, twice as much work is obtained by way of pathway one as via pathway two. This result is shown graphically in Fig. 9.6, where the colored areas represent $|P\Delta V|$. These results again emphasize that heat and work are both pathway-dependent. On the other hand, note that the sum of q and w is

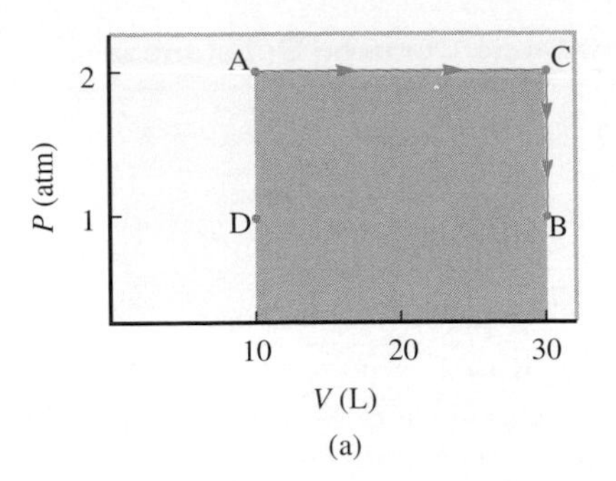

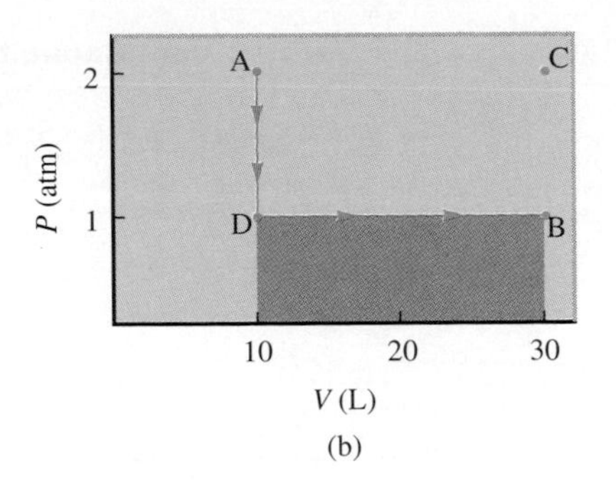

Figure 9.6
The magnitudes of the work for path-way one (a) and pathway two (b) are shown by the colored areas: $|w| = |P\Delta V|$.

the same for both pathways (within round-off differences). This is expected. Recall that

$$\Delta E = q + w$$

and that E is a state function. Note also that the overall ΔH value for pathway one equals that for pathway two (within rounding errors) as expected, because enthalpy is a state function also.

9.4 Calorimetry

We can determine the heat associated with a chemical reaction experimentally by using a device called a **calorimeter. Calorimetry,** the science of measuring heat, is based on observing the temperature change when a body absorbs or discharges energy as heat. Substances respond differently to being heated. We have already discussed the response of ideal gases to heating. Now we expand that discussion to include other substances. In general terms, the **heat capacity** (C) of a substance is defined as

$$C = \frac{\text{heat absorbed}}{\text{increase in temperature}}$$

When an element or a compound is heated, the energy required to reach a certain temperature will depend on the amount of the substance present (for example, it takes twice as much energy to raise the temperature of 2 g of water by 1°C as it takes to raise the temperature of 1 g of water by 1°C). Thus in defining the heat capacity of a substance, the amount of substance must be specified. If the heat capacity is given *per gram* of substance, it is called the **specific heat capacity** with units of $J\ K^{-1}\ g^{-1}$ or $J°C^{-1}\ g^{-1}$. If the heat capacity is given *per mole* of the substance, it is called the **molar heat capacity,** which has the units $J\ K^{-1}\ mol^{-1}$ or $J°C^{-1}\ mol^{-1}$. The specific heat capacities of some common substances are given in Table 9.3.

Specific heat capacity: the energy required to raise the temperature of 1 g of a substance by 1°C.

Molar heat capacity: the energy required to raise the temperature of 1 mol of a substance by 1°C.

Although the calorimeters used for highly accurate work are precision instruments, a very simple calorimeter can be used to examine the fundamentals of calorimetry. All we need are two nested Styrofoam cups with a cover through which a stirrer and thermometer can be inserted, as shown in Fig. 9.7. This device is called a *coffee cup calorimeter*. The outer cup is used to provide extra insulation. The inner cup holds the solution in which the reaction occurs.

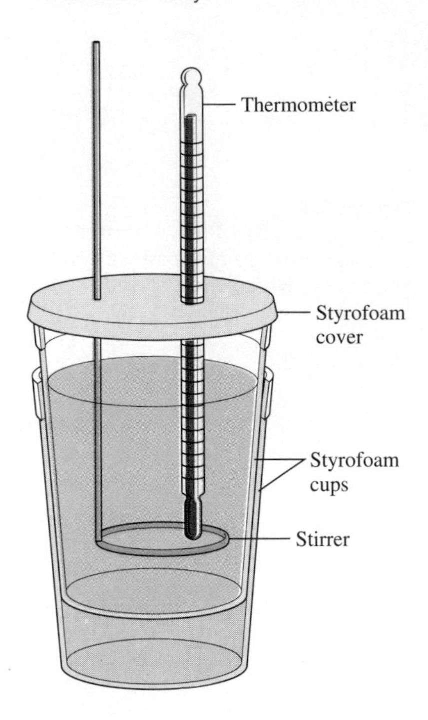

Figure 9.7
A coffee cup calorimeter made of two Styrofoam cups.

TABLE 9.3 The Specific Heat Capacities of Some Common Substances

Substance	Specific Heat Capacity (J °C^{-1} g^{-1})
$H_2O(l)$	4.18
$H_2O(s)$	2.03
$Al(s)$	0.89
$Fe(s)$	0.45
$Hg(l)$	0.14
$C(s)$ (graphite)	0.71

The measurement of heat using a simple calorimeter such as that shown in Fig. 9.7 is an example of **constant-pressure calorimetry,** since the pressure (atmospheric pressure) remains constant during the process. Constant-pressure calorimetry is used in determining the changes in enthalpy occurring in solution. Recall that under these conditions the change in enthalpy equals the heat.

For example, suppose we mix 50.0 mL of 1.0 *M* HCl at 25.0°C with 50.0 mL of 1.0 *M* NaOH also at 25°C in a calorimeter. After the reactants are mixed, the temperature is observed to increase to 31.9°C. As we saw in Chapter 4, the net ionic equation for this reaction is

$$H^+(aq) + OH^-(aq) \longrightarrow H_2O(l)$$

When these reactants (both originally at the same temperature) are mixed, the temperature of the mixed solution is observed to increase. Thus the chemical reaction must be releasing energy as heat. This increases the random motions of the solution components, which in turn increases the temperature. The quantity of energy released can be determined from the temperature increase, the mass of the solution, and the specific heat capacity of the solution. For an approximate result we will assume that the calorimeter does not absorb or leak any heat and that the solution can be treated as if it were pure water with a density of 1.0 g/mL.

If two reactants at the same temperature are mixed and the resulting solution gets warmer, this means the reaction taking place is exothermic. An endothermic reaction cools the solution.

We also need to know the heat required to raise the temperature of a given amount of water by 1°C. Table 9.3 lists the specific heat capacity of water as 4.18 J °C^{-1} g^{-1}. This means that 4.18 J of energy is required to raise the temperature of 1 g of water by 1°C.

From these assumptions and definitions we can calculate the heat (change in enthalpy) for the neutralization reaction:

Energy released by the reaction = energy absorbed by the solution
= specific heat capacity × mass of solution × increase in temperature

Note that the typical units for molar heat capacity ($\Delta E/\Delta T$ per mole) are

$$\frac{J}{K\ mol}$$

Because 1 K equals 1°C, the units

$$\frac{J}{°C\ mol}$$

are used also.

where the increase in temperature = 31.9°C − 25.0°C = 6.9°C, and where

$$\text{mass of solution} = 100.0\ \text{mL} \times 1.0\ \text{g/mL} = 1.0 \times 10^2\ \text{g}$$

Thus

$$\text{Energy released} = \left(4.18 \frac{J}{°C\ g}\right)(1.0 \times 10^2\ g)(6.9°C) = 2.9 \times 10^3\ J$$

Enthalpies of reaction are often expressed in terms of moles of reacting substances. The number of moles of H^+ ions consumed in the above experiment is

$$50.0 \text{ mL} \times \frac{1 \text{ L}}{1000 \text{ mL}} \times \frac{1.0 \text{ mol}}{\text{L}} \text{ H}^+ = 5.0 \times 10^{-2} \text{ mol H}^+$$

Thus 2.9×10^3 J of heat were released when 5.0×10^{-2} mole of H^+ ions reacted. Thus

$$\frac{2.9 \times 10^3 \text{ J}}{5.0 \times 10^{-2} \text{ mol H}^+} = 5.8 \times 10^4 \text{ J}$$

is the heat released per 1.0 mole of H^+ ions neutralized. Thus the *magnitude* of the enthalpy of reaction per mole for

$$\text{H}^+(aq) + \text{OH}^-(aq) \longrightarrow \text{H}_2\text{O}(l)$$

at constant pressure is 58 kJ/mol. Since heat is *evolved*, $\Delta H = -58$ kJ/mol.

Notice that in this example we mentally keep track of the direction of the energy flow and assign the correct sign at the end of the calculation.

EXAMPLE 9.3

When 1.00 L of 1.00 *M* $Ba(NO_3)_2$ at 25.0°C is mixed with 1.00 L of 1.00 *M* Na_2SO_4 at 25°C in a calorimeter, the white solid $BaSO_4$ forms and the temperature of the mixture increases to 28.1°C. Assuming that the calorimeter absorbs only a negligible quantity of heat, that the specific heat capacity of the solution is 4.18 J $°C^{-1}$ g^{-1}, and that the density of the final solution is 1.0 g/mL, calculate the enthalpy change per mole of $BaSO_4$ formed.

Solution

The ions present before any reaction occurs are Ba^{2+}, NO_3^-, Na^+, and SO_4^{2-}. The Na^+ and NO_3^- ions are spectator ions, since $NaNO_3$ is very soluble in water and will not precipitate under these conditions. The net ionic equation of the reaction is therefore

$$\text{Ba}^{2+}(aq) + \text{SO}_4^{2-}(aq) \longrightarrow \text{BaSO}_4(s)$$

Since the temperature increases, the formation of solid $BaSO_4$ must be exothermic; ΔH will be negative.

$$\begin{aligned}\text{Heat evolved by reaction} &= \text{heat absorbed by solution}\\ &= \text{specific heat capacity} \times \text{mass of solution}\\ &\quad \times \text{increase in temperature}\end{aligned}$$

Since 1.00 L of each solution is used, the total solution volume is 2.00 L, and

$$\begin{aligned}\text{Mass of solution} &= 2.00 \text{ L} \times \frac{1000 \text{ mL}}{1 \text{ L}} \times \frac{1.0 \text{ g}}{\text{mL}}\\ &= 2.0 \times 10^3 \text{ g}\\ \text{Temperature increase} &= 28.1°\text{C} - 25.0°\text{C} = 3.1°\text{C}\\ \text{Heat evolved} &= (4.18 \text{ J °C}^{-1} \text{ g}^{-1})(2.0 \times 10^3 \text{ g})(3.1°\text{C})\\ &= 2.6 \times 10^4 \text{ J}\end{aligned}$$

Thus $$q = q_p = \Delta H = -2.6 \times 10^4 \text{ J}$$

Since 1.00 L of 1.00 *M* $Ba(NO_3)_2$ contains 1 mol of Ba^{2+} ions, and 1.00 L of 1.00 *M* Na_2SO_4 contains 1.00 mol of SO_4^{2-} ions, 1.00 mol of solid $BaSO_4$ is formed in this experiment. Thus the enthalpy change per mole of $BaSO_4$ formed is

$$\Delta H = -2.6 \times 10^4 \text{ J/mol} = -26 \text{ kJ/mol}$$

Calculation of ΔH and ΔE for Cases Where *PV* Work Occurs

In the examples of constant-pressure calorimetry we have considered so far, the reactions have occurred in solution, where no appreciable volume changes occur (that is, the total volume of the reactant solution is the sum of the volumes of the solutions that are mixed and remains constant as the reaction proceeds). Under these conditions no work occurs (since $\Delta V = 0$, $P\Delta V = 0$, and $w = 0$). Thus since $\Delta H = q_p$ (constant pressure) and $w = 0$,

$$\Delta E = q_p + w = \Delta H + 0$$

Thus at constant pressure, where $\Delta V = 0$, no work is done and $\Delta E = \Delta H = q_p$.

However, when a reaction involving gases is studied at constant pressure, ΔE may not equal ΔH. The reaction

$$2SO_2(g) + O_2(g) \longrightarrow 2SO_3(g)$$

provides an example of this case. Picture this reaction being carried out at constant temperature and pressure, as shown in Fig. 9.8.

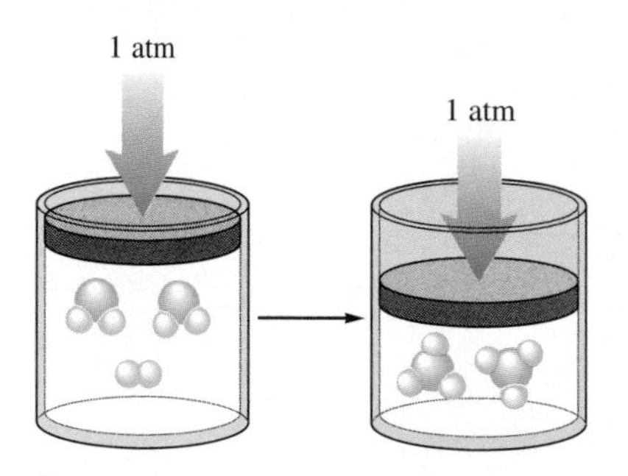

Figure 9.8
A schematic to show the change in volume for the reaction
$2SO_2(g) + O_2(g) \longrightarrow 2SO_3(g)$

In going from reactants to products, the volume of the system decreases, because the number of moles of gas decreases. Since ΔV ($= V_{\text{final}} - V_{\text{initial}}$) is negative, *w* is positive:

$$w = -\underbrace{P\Delta V}_{\text{Negative in this case}}$$

Thus work flows into the system. In addition, because

$$\Delta E = q + w$$

and at constant pressure

$$\Delta H = q_p$$

then

$$\Delta E = q_p + w = \Delta H + w$$

In this case $w \neq 0$, so ΔE and ΔH are different. This case is illustrated in Example 9.4.

EXAMPLE 9.4

When 2.00 mol of $SO_2(g)$ react completely with 1.00 mol of $O_2(g)$ to form 2.00 mol of $SO_3(g)$ at 25°C and a constant pressure of 1.00 atm, 198 kJ of energy are released as heat. Calculate ΔH and ΔE for this process.

Solution

Because the pressure is constant for this process, $\Delta H = q_p$. The description of the experiment states that 198 kJ of heat is *released*. Thus $\Delta H = q_p = -198$ kJ, where the negative sign indicates that energy flows *out* of the system.

The value of ΔE can be calculated from the relationship

$$\Delta E = q + w$$

Since q is known (-198 kJ), we only need the value for w. We know that

$$w = -P\Delta V$$

Solving the ideal gas law for ΔV gives

$$\Delta V = \Delta n\left(\frac{RT}{P}\right)$$

where only n changes (T and P are constant) as the reaction occurs.

In this case

$$\Delta n = n_{final} - n_{initial}$$

$$n_{final} = \underset{\text{Moles of } SO_3}{2 \text{ mol}}$$

$$n_{initial} = \underset{\text{Moles of } SO_2}{2 \text{ mol}} + \underset{\text{Moles of } O_2}{1 \text{ mol}}$$

So

$$\Delta n = 2 \text{ mol} - 3 \text{ mol} = -1 \text{ mol}$$

Now we can calculate w:

$$w = -P\Delta V = -P\underbrace{\left(\Delta n \times \frac{RT}{P}\right)}_{\Delta V} = -\Delta nRT$$

where

$$\Delta n = -1 \text{ mol}$$
$$R = 8.3145 \text{ J K}^{-1} \text{ mol}^{-1}$$
$$T = 25°\text{C} + 273 = 298 \text{ K}$$

Thus

$$w = -(-1 \text{ mol})\left(8.3145 \frac{\text{J}}{\text{K mol}}\right)(298 \text{ K}) = 2.48 \text{ kJ}$$

Using the values of q and w, we can calculate ΔE:

$$\Delta E = q + w = \Delta H + w = -198 \text{ kJ} + 2.48 \text{ kJ} = -196 \text{ kJ}$$

Note that ΔE and ΔH are different for this case because the volume changes (and therefore work occurs).

Calorimetry experiments can also be performed at **constant volume.** For example, when a photographic flashbulb flashes, the bulb becomes very hot, since the reaction of the zirconium or magnesium wire with the oxygen inside the bulb is exothermic. The reaction occurs inside the flashbulb, which is rigid (does not change volume). Under these conditions no work is done, since the volume must change for pressure-volume work to be performed. To study the energy changes in reactions under conditions of constant volume, a **bomb calorimeter** (Fig. 9.9) is often used. Weighed reactants are placed inside a rigid steel container (the "bomb") and ignited. The energy change is determined by measuring the increase in the temperature of the water and other calorimeter parts. For a constant-volume process the change in volume (ΔV) is equal to zero, and so the work is also equal to zero. Therefore,

$$\Delta E = q + w = q = q_v \qquad \text{(constant volume)}$$

Suppose we wish to measure the energy of combustion of octane (C_8H_{18}), a component of gasoline. A 0.5269-g sample of octane is placed in a bomb calorimeter known to have a heat capacity of 11.3 kJ/°C. This means that 11.3 kJ of energy are required to raise the temperature of the water and other parts of the calorimeter by 1°C. The octane is ignited in the presence of excess oxygen, causing the temperature of the calorimeter to increase by 2.25°C. The amount of energy released by the reaction is calculated as follows:

$$\begin{aligned}\text{Energy released by the reaction} &= \text{temperature increase} \times \text{energy required to change the temperature by } 1^\circ\text{C} \\ &= \Delta T \times \text{heat capacity of calorimeter} \\ &= 2.25^\circ\text{C} \times 11.3 \text{ kJ/}^\circ\text{C} = 25.4 \text{ kJ}\end{aligned}$$

This means that 25.4 kJ of energy were released by the combustion of 0.5269 g

Figure 9.9

A commercial bomb calorimeter.

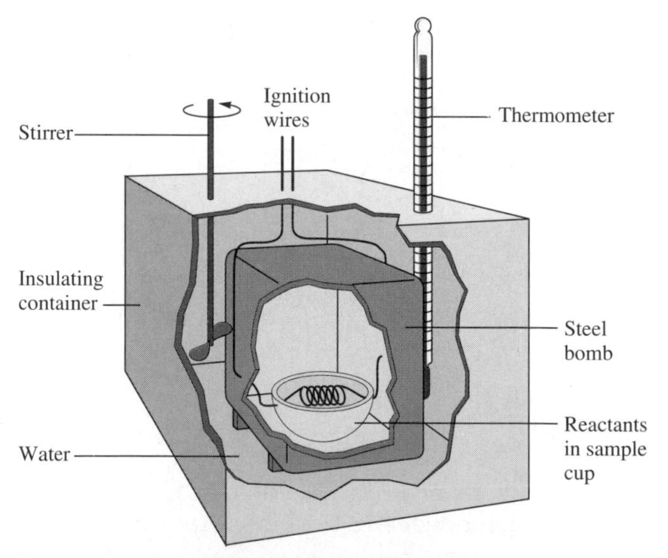

Schematic of a bomb calorimeter. The reaction is carried out inside a rigid steel "bomb," and the heat evolved is absorbed by the surrounding water and the other calorimeter parts. The quantity of energy produced by the reaction can be calculated from the temperature increase.

of octane. The number of moles of octane is

$$0.5269 \text{ g octane} \times \frac{1 \text{ mol octane}}{114.2 \text{ g octane}} = 4.614 \times 10^{-3} \text{ mol octane}$$

Since 25.4 kJ of energy was released for 4.614×10^{-3} mol of octane, the energy released per mole is

$$\frac{25.4 \text{ kJ}}{4.614 \times 10^{-3} \text{ mol}} = 5.50 \times 10^3 \text{ kJ/mol}$$

Since the reaction is exothermic, ΔE is negative:

$$\Delta E_{\text{combustion}} = -5.50 \times 10^3 \text{ kJ/mol}$$

Note that since no work is done in this case, ΔE is equal to the heat:

$$\begin{aligned} \Delta E &= q + w = q \\ &= -5.50 \times 10^3 \text{ kJ/mol} \end{aligned}$$

EXAMPLE 9.5

It has been suggested that hydrogen gas obtained from the decomposition of water might be a substitute for natural gas (principally methane). To compare the energies of combustion of these fuels, the following experiment was carried out, using a bomb calorimeter with a heat capacity of 11.3 kJ/°C. When a 1.50-g sample of methane gas was burned with excess oxygen in the calorimeter, the temperature increased by 7.3°C. When a 1.15-g sample of hydrogen gas was burned with excess oxygen, the temperature increase was 14.3°C. Calculate the energy of combustion (per gram) for hydrogen and methane.

Hydrogen's potential as a fuel is discussed in Section 9.8.

Solution

We calculate the energy of combustion for methane, using the heat capacity of the calorimeter (11.3 kJ/°C) and the observed temperature increase of 7.3°C:

$$\begin{aligned} \text{Energy } \textit{released} \text{ in the combustion of 1.50 g of } CH_4 &= (11.3 \text{ kJ/°C})(7.3\text{°C}) \\ &= 83 \text{ kJ} \end{aligned}$$

$$\text{Energy released in the combustion of 1 g of } CH_4 = \frac{83 \text{ kJ}}{1.50 \text{ g}} = 55 \text{ kJ/g}$$

Similarly, for hydrogen,

$$\begin{aligned} \text{Energy } \textit{released} \text{ in the combustion of 1.15 g of } H_2 &= (11.3 \text{ kJ/°C})(14.3\text{°C}) \\ &= 162 \text{ kJ} \end{aligned}$$

$$\text{Energy released in the combustion of 1 g of } H_2 = \frac{162 \text{ kJ}}{1.15 \text{ g}} = 141 \text{ kJ/g}$$

The energy released by the combustion of 1 g of hydrogen is approximately 2.5 times that for 1 g of methane, indicating that hydrogen gas is a potentially useful fuel.

The direction of energy flow is indicated by words in this example. Using signs, we have

$$\Delta E_{\text{combustion}} = -55 \text{ kJ/g}$$

for methane and

$$\Delta E_{\text{combustion}} = -141 \text{ kJ/g}$$

for hydrogen.

9.5 Hess's Law

ΔH is not dependent on the reaction pathway.

Since enthalpy is a state function, the change in enthalpy in going from some initial state to some final state is independent of the pathway. This means that *in going from a particular set of reactants to a particular set of products, the change in enthalpy is the same whether the reaction takes place in one step or in a series of steps.* This principle is known as **Hess's law** and can be illustrated by examining the oxidation of nitrogen to produce nitrogen dioxide. The overall reaction can be written in one step, where the enthalpy change is represented by ΔH_1:

$$N_2(g) + 2O_2(g) \longrightarrow 2NO_2(g) \qquad \Delta H_1 = 68 \text{ kJ}$$

This reaction can also be carried out in two distinct steps, with enthalpy changes designated by ΔH_2 and ΔH_3:

$$N_2(g) + O_2(g) \longrightarrow 2NO(g) \qquad \Delta H_2 = 180 \text{ kJ}$$
$$2NO(g) + O_2(g) \longrightarrow 2NO_2(g) \qquad \Delta H_3 = -112 \text{ kJ}$$

$$\text{Net reaction: } N_2(g) + 2O_2(g) \longrightarrow 2NO_2(g) \quad \Delta H_2 + \Delta H_3 = 68 \text{ kJ}$$

Note that the sum of these two steps gives the net, or overall, reaction and that

$$\Delta H_1 = \Delta H_2 + \Delta H_3 = 68 \text{ kJ}$$

The principle of Hess's law is shown schematically in Fig. 9.10.

Characteristics of Enthalpy Changes

In order to use Hess's law to compute enthalpy changes for reactions, we must understand two characteristics of ΔH for a reaction:

Reversing a reaction changes the sign of ΔH.

1. If a reaction is reversed, the sign of ΔH is also reversed.
2. The magnitude of ΔH is directly proportional to the quantities of reactants and products in a reaction. If the coefficients in a balanced reaction are multiplied by an integer, the value of ΔH is multiplied by the same integer.

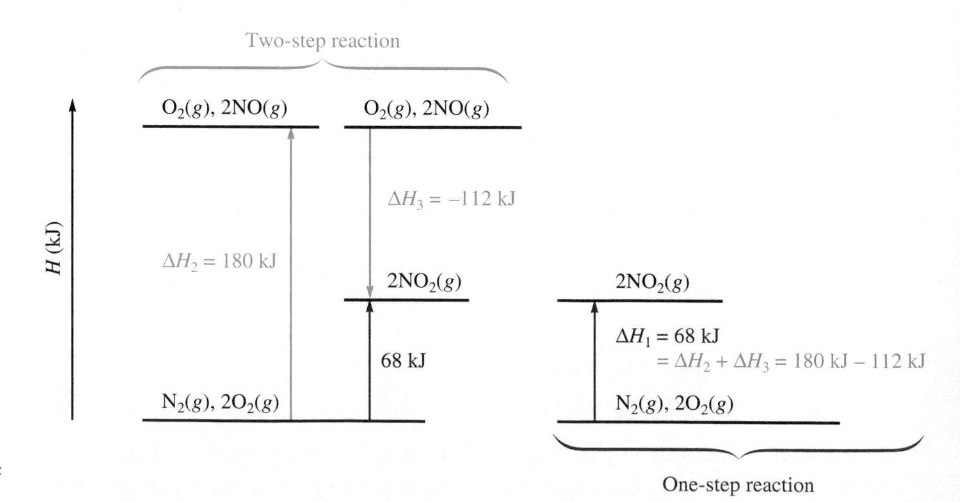

Figure 9.10
The principle of Hess's law. The same change in enthalpy occurs when nitrogen and oxygen react to form nitrogen dioxide, regardless of whether the reaction occurs in one (red) or two (blue) steps.

Both of these rules follow in a straightforward way from the properties of enthalpy changes. The first rule can be explained by recalling that the *sign* of ΔH indicates the *direction* of the heat flow at constant pressure. If the direction of the reaction is reversed, the direction of the heat flow is also reversed. To see this, consider the preparation of xenon tetrafluoride, which was the first binary compound made from a noble gas:

$$\mathrm{Xe}(g) + 2\mathrm{F}_2(g) \longrightarrow \mathrm{XeF}_4(s) \qquad \Delta H = -251\ \mathrm{kJ}$$

This reaction is exothermic, and 251 kJ of energy flows into the surroundings as heat. On the other hand, if the colorless XeF_4 crystals are decomposed into the elements, according to the equation

$$\mathrm{XeF}_4(s) \longrightarrow \mathrm{Xe}(g) + 2\mathrm{F}_2(g)$$

the opposite energy flow occurs because 251 kJ of energy have to be added to the system in this case. Thus for this reaction $\Delta H = +251$ kJ.

The second rule comes from the fact that ΔH is an extensive property, depending on the amount of substances reacting. For example, 251 kJ of energy are evolved for the reaction

$$\mathrm{Xe}(g) + 2\mathrm{F}_2(g) \longrightarrow \mathrm{XeF}_4(s)$$

Thus for a preparation involving twice the quantities of reactants and products,

$$2\mathrm{Xe}(g) + 4\mathrm{F}_2(g) \longrightarrow 2\mathrm{XeF}_4(s)$$

twice as much heat would be evolved:

$$\Delta H = 2(-251\ \mathrm{kJ}) = -502\ \mathrm{kJ}$$

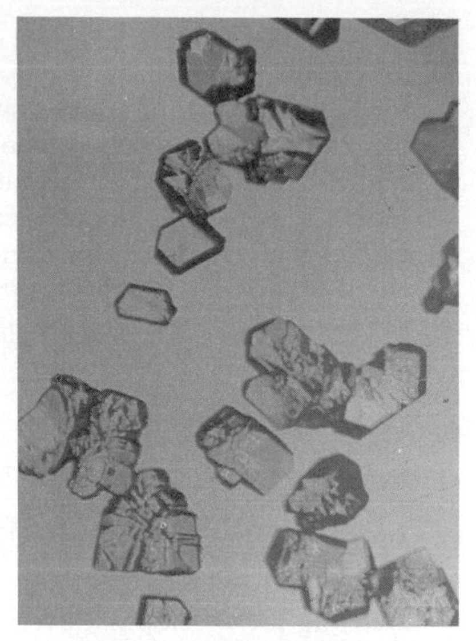

Crystals of xenon tetrafluoride, the first reported binary compound containing a noble gas element.

Hints for Using Hess's Law

Calculations involving Hess's law typically require that several reactions be manipulated and combined to finally give the reaction of interest. In doing this procedure, you should work *backward* from the required reaction, using the reactants and products to decide how to manipulate the other reactions at your disposal. Reverse any reactions as needed to give the required reactants and products, and then multiply the reactions to give the correct numbers of reactants and products. This process involves some trial and error but can be very systematic if you always allow the final reaction to guide you.

EXAMPLE 9.6

Diborane (B_2H_6) is a highly reactive boron hydride, which was once considered as a possible rocket fuel for the U.S. space program. Calculate ΔH for the synthesis of diborane from its elements, according to the equation

$$2\mathrm{B}(s) + 3\mathrm{H}_2(g) \longrightarrow \mathrm{B}_2\mathrm{H}_6(g)$$

using the following data:

Reaction	ΔH
(a) $2\mathrm{B}(s) + \frac{3}{2}\mathrm{O}_2(g) \rightarrow \mathrm{B}_2\mathrm{O}_3(s)$	-1273 kJ
(b) $\mathrm{B}_2\mathrm{H}_6(g) + 3\mathrm{O}_2(g) \rightarrow \mathrm{B}_2\mathrm{O}_3(s) + 3\mathrm{H}_2\mathrm{O}(g)$	-2035 kJ
(c) $\mathrm{H}_2(g) + \frac{1}{2}\mathrm{O}_2(g) \rightarrow \mathrm{H}_2\mathrm{O}(l)$	-286 kJ
(d) $\mathrm{H}_2\mathrm{O}(l) \rightarrow \mathrm{H}_2\mathrm{O}(g)$	44 kJ

Solution

To obtain ΔH for the required reaction, we must somehow combine equations (a), (b), (c), and (d) to produce that reaction, and add the corresponding ΔH values. This procedure can best be done by focusing on the reactants and products of the required reaction. The reactants are $B(s)$ and $H_2(g)$, and the product is $B_2H_6(g)$. How can we obtain the correct equation? Reaction (a) has $B(s)$ as a reactant, as needed in the required equation. Thus reaction (a) will be used as it is. Reaction (b) has $B_2H_6(g)$ as a reactant, but this substance is needed as a product. Thus reaction (b) must be reversed, and the sign of ΔH changed accordingly. Up to this point we have

$$\begin{array}{lll} \text{(a)} & 2B(s) + \tfrac{3}{2}O_2(g) \longrightarrow B_2O_3(s) & \Delta H = -1273 \text{ kJ} \\ -\text{(b)} & B_2O_3(s) + 3H_2O(g) \longrightarrow B_2H_6(g) + 3O_2(g) & \Delta H = -(-2035 \text{ kJ}) \\ \hline \text{Sum:} & B_2O_3(s) + 2B(s) + \tfrac{3}{2}O_2(g) + 3H_2O(g) \longrightarrow B_2O_3(s) + B_2H_6(g) + 3O_2(g) & \Delta H = 762 \text{ kJ} \end{array}$$

Deleting the species that occur on both sides gives

$$2B(s) + 3H_2O(g) \longrightarrow B_2H_6(g) + \tfrac{3}{2}O_2(g) \qquad \Delta H = 762 \text{ kJ}$$

We are closer to the required reaction, but we still need to remove $H_2O(g)$ and $O_2(g)$ and introduce $H_2(g)$ as a reactant. We can do so by using reactions (c) and (d). If we multiply reaction (c) and its ΔH value by 3 and add the result to the above equation, we have

$$\begin{array}{lll} & 2B(s) + 3H_2O(g) \longrightarrow B_2H_6(g) + \tfrac{3}{2}O_2(g) & \Delta H = 762 \text{ kJ} \\ 3 \times \text{(c)} & 3[H_2(g) + \tfrac{1}{2}O_2(g) \longrightarrow H_2O(l)] & \Delta H = 3(-286 \text{ kJ}) \\ \hline \text{Sum:} & 2B(s) + 3H_2(g) + \tfrac{3}{2}O_2(g) + 3H_2O(g) \longrightarrow B_2H_6(g) + \tfrac{3}{2}O_2(g) + 3H_2O(l) & \Delta H = -96 \text{ kJ} \end{array}$$

We can cancel the $\tfrac{3}{2}O_2(g)$ on both sides, but we cannot cancel the H_2O because it is gaseous on one side and liquid on the other. This problem can be solved by adding reaction (d), multiplied by 3:

$$\begin{array}{lll} & 2B(s) + 3H_2(g) + 3H_2O(g) \longrightarrow B_2H_6(g) + 3H_2O(l) & \Delta H = -96 \text{ kJ} \\ 3 \times \text{(d)} & 3[H_2O(l) \longrightarrow H_2O(g)] & \Delta H = 3(44 \text{ kJ}) \\ \hline & 2B(s) + 3H_2(g) + 3H_2O(g) + 3H_2O(l) \longrightarrow B_2H_6(g) + 3H_2O(l) + 3H_2O(g) & \Delta H = +36 \text{ kJ} \end{array}$$

This step gives the reaction required by the problem:

$$2B(s) + 3H_2(g) \longrightarrow B_2H_6(g) \qquad \Delta H = +36 \text{ kJ}$$

Thus ΔH for the synthesis of 1 mol of diborane is +36 kJ.

FIREWALKING: MAGIC OR SCIENCE?

For millennia people have been amazed at the ability of Eastern mystics to walk across beds of glowing coals without any apparent discomfort. Even in the United States thousands of people have performed feats of firewalking as part of motivational seminars. How is this feat possible? Do firewalkers have supernatural powers?

Actually, there are sound scientific explanations of why firewalking is possible. The first important factor concerns the heat capacity of feet. Because human tissue is mainly composed of water, it has a relatively large specific heat capacity. This means that a large amount of energy must be transferred from the coals to significantly change the temperature of the feet. During the brief contact between feet and coals, there is relatively little time for energy flow, so the feet do not reach a high enough temperature to cause damage.

Second, although the surface of the coals has a very high temperature, the red-hot layer is very thin. Therefore, the quantity of energy available to heat the feet is smaller than might be expected. This factor points out the difference between temperature and heat. Temperature reflects the *intensity* of the random kinetic energy in a given sample of matter. The amount of energy available for heat flow, on the other hand, depends on the quantity of matter at a given temperature—10 g of matter at a given temperature contains ten times as much thermal energy as 1 g of the same matter. For example, the tiny spark from a sparkler does not hurt when it hits your hand. The spark has a very high temperature but has so little mass that no significant energy transfer occurs to your hand. This same argument applies to the very thin hot layer on the coals.

A third factor that aids firewalkers is the so-called Leidenfrost effect—the phenomenon that allows water droplets to skate around on a hot griddle for a surprisingly long time. The part of the droplet in contact with the hot surface vaporizes first, providing a gaseous layer that both allows the droplet to skate around and acts as a barrier through which energy does not readily flow to the rest of the droplet. The perspiration on the feet of the presumably tense firewalker would have the same effect. In addition, because firewalking is often done at night, with moist grass surrounding the bed of coals, the firewalker's feet are probably damp before the walk, providing ample moisture for the Leidenfrost effect to occur.

Thus although firewalking is an impressive feat, there are several sound scientific reasons why anyone should be able to do it with the proper training and a properly prepared bed of coals.

Firewalking ritual in India.

9.6 Standard Enthalpies of Formation

For a reaction studied under conditions of constant pressure, we can obtain the enthalpy change by using a calorimeter. However, this process can be very difficult. In fact, in some cases it is impossible, since certain reactions do not lend themselves to such study. An example is the conversion of solid carbon from its graphite form to its diamond form:

$$C_{graphite}(s) \longrightarrow C_{diamond}(s)$$

The value of ΔH for this process cannot be obtained readily by measurement in a calorimeter. We will show next how to *calculate* ΔH for chemical reactions and physical changes by using standard enthalpies of formation.

The **standard enthalpy of formation** (ΔH_f°) of a compound is defined as the *change in enthalpy that accompanies the formation of 1 mole of a compound from its elements with all substances in their standard states.*

The *superscript zero* on a thermodynamic function (for example ΔH°) indicates that the corresponding process has been carried out under standard conditions. The **standard state** for a substance is a precisely defined reference state. Because thermodynamic functions often depend on the concentrations (or pressures) of the substances involved, we must use a common reference state to properly compare the thermodynamic properties of two substances. This is especially important because for most thermodynamic properties, we can measure only *changes* in that property. For example, we have no method for determining absolute values of enthalpy. We can measure only enthalpy changes (ΔH values) by performing heat flow experiments.

Standard states are defined as follows:

Standard state is *not* the same as standard temperature and pressure (STP) for a gas.

The standard state for oxygen is $O_2(g)$ at a pressure of 1 atmosphere; the standard state for sodium is $Na(s)$; the standard state for mercury is $Hg(l)$; and so on.

Definitions of Standard States

- **For a gas the standard state is a pressure of exactly 1 atmosphere.**
- **For a substance present in a solution, the standard state is a concentration of exactly 1 *M* at an applied pressure of 1 atmosphere.**
- **For a pure substance in a condensed state (liquid or solid), the standard state is the pure liquid or solid.**
- **For an element the standard state is the form in which the element exists (is most stable) under conditions of 1 atmosphere and the temperature of interest (usually 25°C).**

Several important characteristics of the definition of the enthalpy of formation will become clearer if we again consider the formation of nitrogen dioxide from the elements in their standard states:

$$\tfrac{1}{2}N_2(g) + O_2(g) \longrightarrow NO_2(g) \qquad \Delta H_f^\circ = 34 \text{ kJ/mol}$$

Note that the reaction is written so that both elements are in their standard states, and 1 mole of product is formed. Enthalpies of formation are *always* given per mole of product with the product in its standard state.

The formation reaction for methanol is written as

$$C(s) + 2H_2(g) + \tfrac{1}{2}O_2(g) \longrightarrow CH_3OH(l) \qquad \Delta H_f^\circ = -239 \text{ kJ/mol}$$

The standard state of carbon is graphite, the standard states for oxygen and hydrogen are the diatomic gases, and the standard state for methanol is the liquid.

The ΔH_f° values for some common substances are shown in Table 9.4. More values are found in Appendix 4. The importance of tabulated ΔH_f° values is that enthalpies for many reactions can be calculated using these numbers. To see how this is done, we will calculate the standard enthalpy change for the combustion of methane:

$$CH_4(g) + 2O_2(g) \longrightarrow CO_2(g) + 2H_2O(l)$$

Enthalpy is a state function, so we can invoke Hess's law and choose *any* convenient pathway from reactants to products and then sum the enthalpy changes. A convenient pathway, shown in Fig. 9.11, involves taking the reactants apart to the respective elements in their standard states in reactions (a) and (b), and then forming the products from these elements in reactions (c) and (d). This general pathway will work for any reaction, since atoms are conserved in a chemical reaction.

Note from Fig. 9.11 that reaction (a), where methane is taken apart into its elements,

$$CH_4(g) \longrightarrow C(s) + 2H_2(g)$$

is just the reverse of the formation reaction for methane:

$$C(s) + 2H_2(g) \longrightarrow CH_4(g) \qquad \Delta H_f^\circ = -75 \text{ kJ/mol}$$

Since reversing a reaction means changing the sign of ΔH but keeping the same magnitude, ΔH for reaction (a) is $-\Delta H_f^\circ$, or 75 kJ. Thus $\Delta H^\circ_{(a)} = 75$ kJ.

Next, we consider reaction (b). Here oxygen is already an element in its standard state, so no change is necessary. Thus $\Delta H^\circ_{(b)} = 0$.

The next steps, reactions (c) and (d), use the elements formed in reactions (a) and (b) to form the products. Note that reaction (c) is simply the formation reaction for carbon dioxide:

$$C(s) + O_2(g) \longrightarrow CO_2(g) \qquad \Delta H_f^\circ = -394 \text{ kJ mol}$$

and $\qquad \Delta H^\circ_{(c)} = \Delta H_f^\circ \text{ [for } CO_2(g)] = -394 \text{ kJ}$

TABLE 9.4
Standard Enthalpies of Formation for Several Compounds at 25°C

Compound	ΔH_f° (kJ/mol)
$NH_3(g)$	−46
$NO_2(g)$	34
$H_2O(l)$	−286
$Al_2O_3(s)$	−1676
$Fe_2O_3(s)$	−826
$CO_2(g)$	−394
$CH_3OH(l)$	−239
$C_8H_{18}(l)$	−269

Note that although the tabulated values of ΔH_f° usually correspond to a temperature of 25°C, values of ΔH_f° can be obtained at any temperature.

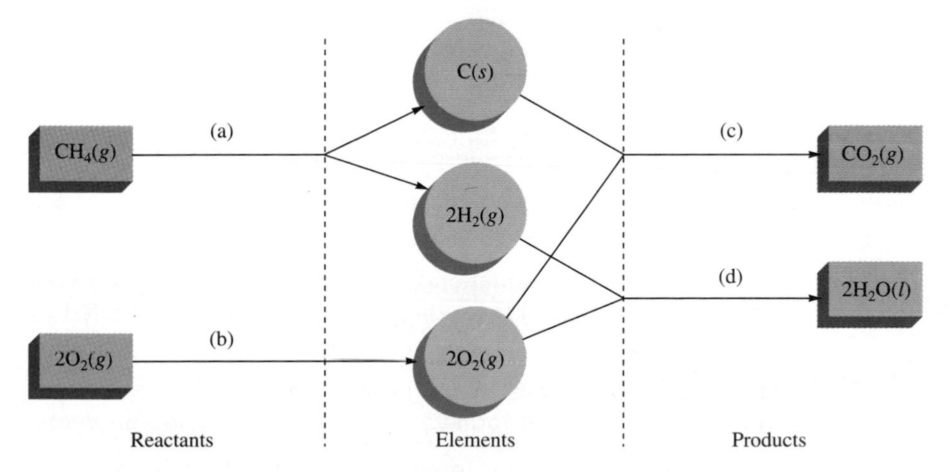

Figure 9.11
In this pathway for the combustion of methane, the reactants are first taken apart in reactions (a) and (b) to form the constituent elements in their standard states, which are then used to assemble the products in reactions (c) and (d).

Reaction (d) is the formation reaction for water:

$$H_2(g) + \tfrac{1}{2}O_2(g) \longrightarrow H_2O(l) \qquad \Delta H_f^\circ = -286 \text{ kJ/mol}$$

However, since 2 moles of water are required in the balanced equation, we must form 2 moles of water from the elements:

$$2H_2(g) + O_2(g) \longrightarrow 2H_2O(l)$$

Thus

$$\Delta H^\circ_{(d)} = 2 \times \Delta H_f^\circ \text{ [for } H_2O(l)] = 2(-286 \text{ kJ}) = -572 \text{ kJ}$$

We have now completed the pathway from the reactants to the products. The change in enthalpy for the overall reaction is the sum of the ΔH values (including their signs) for the steps:

$$\begin{aligned}
\Delta H^\circ_{\text{reaction}} &= \Delta H^\circ_{(a)} + \Delta H^\circ_{(b)} + \Delta H^\circ_{(c)} + \Delta H^\circ_{(d)} \\
&= -\Delta H_f^\circ \text{ [for } CH_4(g)] + 0 + \Delta H_f^\circ \text{ [for } CO_2(g)] \\
&\quad + 2 \times \Delta H_f^\circ \text{ [for } H_2O(l)] \\
&= -(-75 \text{ kJ}) + 0 + (-394 \text{ kJ}) + (-572 \text{ kJ}) \\
&= -891 \text{ kJ}
\end{aligned}$$

Let us examine carefully the pathway we used in this example. First, the reactants were broken down into the elements in their standard states. This step involved reversing the formation reactions and thus switching the signs of the respective enthalpies of formation. The products were then constructed from these elements. This step involved formation reactions and thus enthalpies of formation. We can summarize the entire process as follows: *The enthalpy change for a given reaction can be calculated by subtracting the enthalpies of formation of the reactants from the enthalpies of formation of the products.* Remember to multiply the enthalpies of formation by integers as required by the balanced equation. This procedure can be represented symbolically as follows:

$$\Delta H^\circ_{\text{reaction}} = \Sigma \Delta H_f^\circ \text{ (products)} - \Sigma \Delta H_f^\circ \text{ (reactants)} \qquad (9.1)$$

where the symbol Σ (sigma) means "to take the sum of the terms."

Elements are not included in the calculation since elements require no change in form. We have in effect *defined* the enthalpy of formation of an element in its standard state as zero, since we have chosen this as our reference point for calculating enthalpy changes in reactions.

Summary: **Key Concepts for Doing Enthalpy Calculations**

- **When a reaction is reversed, the magnitude of ΔH remains the same, but the sign changes.**
- **When the balanced equation for a reaction is multiplied by an integer, the value of ΔH for that reaction must be multiplied by the same integer.**
- **The change in enthalpy for a given reaction can be calculated from the enthalpies of formation of the reactants and products:**

$$\Delta H^\circ_{\text{reaction}} = \Sigma \Delta H_f^\circ \text{ (products)} - \Sigma \Delta H_f^\circ \text{ (reactants)}$$

- **Elements in their standard states are not included in the $\Delta H_{reaction}$ calculations. That is, ΔH_f° for an element in its standard state is zero.**

EXAMPLE 9.7

Using the standard enthalpies of formation listed in Table 9.4, calculate the standard enthalpy change for the overall reaction that occurs when ammonia is burned in air to form nitrogen dioxide and water. This is the first step in the manufacture of nitric acid.

$$4NH_3(g) + 7O_2(g) \longrightarrow 4NO_2(g) + 6H_2O(l)$$

Solution

We will use the pathway in which the reactants are broken down into elements in their standard states, and then are used to form the products (see Fig. 9.12).

STEP 1

Decomposition of $NH_3(g)$ into elements [reaction (a) in Fig. 9.12]. The first step involves decomposing 4 mol of NH_3 into N_2 and H_2:

$$4NH_3(g) \longrightarrow 2N_2(g) + 6H_2(g)$$

The above reaction is 4 times the *reverse* of the formation reaction for NH_3:

$$\tfrac{1}{2}N_2(g) + \tfrac{3}{2}H_2(g) \longrightarrow NH_3(g) \qquad \Delta H_f^\circ = -46 \text{ kJ/mol}$$

Thus

$$\Delta H^\circ_{(a)} = 4 \text{ mol}[-(-46 \text{ kJ/mol})] = 184 \text{ kJ}$$

STEP 2

Elemental oxygen [reaction (b) in Fig. 9.12]. Since $O_2(g)$ is an element in its standard state, $\Delta H^\circ_{(b)} = 0$.

We now have the elements $N_2(g)$, $H_2(g)$, and $O_2(g)$, which can be combined to form the products of the overall reaction.

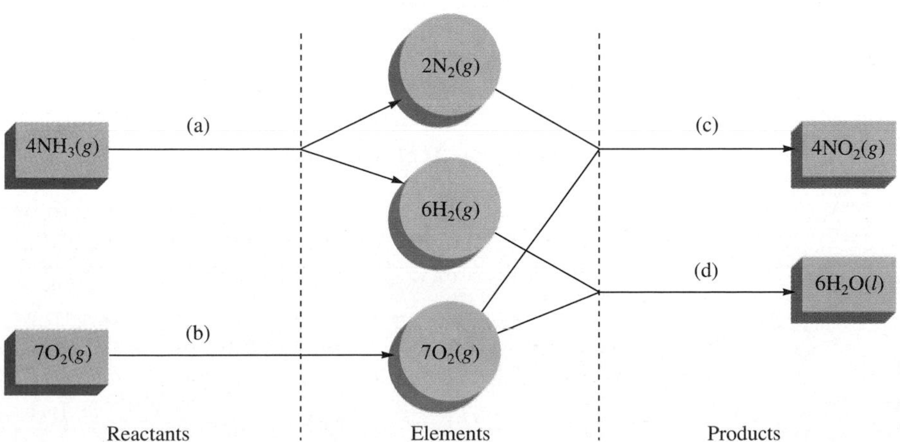

Figure 9.12
A pathway for the combustion of ammonia.

STEP 3

Synthesis of $NO_2(g)$ from the elements [reaction (c) in Fig. 9.12]. The overall reaction equation has 4 mol of NO_2. Thus the required reaction is 4 times the formation reaction for NO_2:

$$4 \times [\tfrac{1}{2}N_2(g) + O_2(g) \longrightarrow NO_2(g)]$$

and

$$\Delta H^\circ_{(c)} = 4 \times \Delta H_f^\circ \text{ [for } NO_2(g)]$$

From Table 9.4, ΔH_f° [for $NO_2(g)$] = 34 kJ/mol, so

$$\Delta H^\circ_{(c)} = 4 \text{ mol} \times 34 \text{ kJ/mol} = 136 \text{ kJ}$$

STEP 4

Synthesis of $H_2O(l)$ from the elements [reaction (d) in Fig. 9.12]. Since the overall reaction equation has 6 mol of $H_2O(l)$, the required reaction is 6 times the formation reaction for $H_2O(l)$:

$$6 \times [H_2(g) + \tfrac{1}{2}O_2(g) \longrightarrow H_2O(l)]$$

and

$$\Delta H^\circ_{(d)} = 6 \times \Delta H_f^\circ \text{ [for } H_2O(l)]$$

From Table 9.4, ΔH_f° [for $H_2O(l)$] = −286 kJ/mol, so

$$\Delta H^\circ_{(d)} = 6 \text{ mol}(-286 \text{ kJ/mol}) = -1716 \text{ kJ}$$

To summarize, we have done the following:

$$\begin{array}{ccccc} 4NH_3(g) & \xrightarrow{\Delta H^\circ_{(a)}} & \left\{ \begin{array}{l} 2N_2(g) + 6H_2(g) \\ 7O_2(g) \end{array} \right\} & \xrightarrow{\Delta H^\circ_{(c)}} & 4NO_2(g) \\ 7O_2(g) & \xrightarrow{\Delta H^\circ_{(b)} = 0} & & \xrightarrow{\Delta H^\circ_{(d)}} & 6H_2O(l) \end{array}$$

Elements in their standard states

We now add the ΔH° values for the steps to obtain ΔH° for the overall reaction:

$$\begin{aligned} \Delta H^\circ_{\text{reaction}} &= \Delta H^\circ_{(a)} + \Delta H^\circ_{(b)} + \Delta H^\circ_{(c)} + \Delta H^\circ_{(d)} \\ &= 4 \times -\Delta H_f^\circ \text{ [for } NH_3(g)] + 0 + 4 \times \Delta H_f^\circ \text{ [for } NO_2(g)] \\ &\quad + 6 \times \Delta H_f^\circ \text{ [for } H_2O(l)] \\ &= 4 \times \Delta H_f^\circ \text{ [for } NO_2(g)] + 6 \times \Delta H_f^\circ \text{ [for } H_2O(l)] \\ &\quad - 4 \times \Delta H_f^\circ \text{ [for } NH_3(g)] \\ &= \Delta H_f^\circ \text{ (products)} - \Delta H_f^\circ \text{ (reactants)} \end{aligned}$$

Remember that elemental reactants and products do not need to be included, since ΔH_f° for an element in its standard state is zero. Note that we have again obtained Equation (9.1). The final solution is

$$\begin{aligned} \Delta H^\circ_{\text{reaction}} &= 6 \times (-286 \text{ kJ}) + 4 \times (34 \text{ kJ}) - 4 \times (-46 \text{ kJ}) \\ &= -1396 \text{ kJ} \end{aligned}$$

Now that we have shown the basis for Equation (9.1), we will make direct use of it to calculate ΔH for reactions in succeeding examples.

EXAMPLE 9.8

Methanol (CH_3OH) is sometimes used as a fuel in high-performance engines. Using the data in Table 9.4, compare the standard enthalpy of combustion per gram of methanol with that of gasoline. Gasoline is actually a mixture of compounds, but assume for this problem that gasoline is pure liquid octane (C_8H_{18}).

Racing cars sometimes use methanol for fuel.

Solution

The combustion reaction for methanol is

$$2CH_3OH(l) + 3O_2(g) \longrightarrow 2CO_2(g) + 4H_2O(l)$$

Using the standard enthalpies of formation from Table 9.4 and Equation (9.1), we have

$$\begin{aligned}\Delta H^\circ_{\text{reaction}} &= 2 \times \Delta H_f^\circ\,[\text{for } CO_2(g)] + 4 \times \Delta H_f^\circ\,[\text{for } H_2O(l)] \\ &\quad - 2 \times \Delta H_f^\circ\,[\text{for } CH_3OH(l)] \\ &= 2 \times (-394\text{ kJ}) + 4 \times (-286\text{ kJ}) - 2 \times (-239\text{ kJ}) \\ &= -1454\text{ kJ}\end{aligned}$$

Thus 1454 kJ of heat is evolved when 2 mol of methanol burns. The molar mass of methanol is 32.0 g/mol. This means that 1454 kJ of energy are produced when 64.0 g of methanol burn. The enthalpy of combustion per gram of methanol is

$$\frac{-1454\text{ kJ}}{64.0\text{ g}} = -22.7\text{ kJ/g}$$

The combustion reaction for octane is

$$2C_8H_{18}(l) + 25O_2(g) \longrightarrow 16CO_2(g) + 18H_2O(l)$$

Using the standard enthalpies of formation from Table 9.4 and Equation (9.1), we have

$$\begin{aligned}\Delta H^\circ_{\text{reaction}} &= 16 \times \Delta H_f^\circ\,[\text{for } CO_2(g)] + 18 \times \Delta H_f^\circ\,[\text{for } H_2O(l)] \\ &\quad - 2 \times \Delta H_f^\circ\,[\text{for } C_8H_{18}(l)] \\ &= 16 \times (-394\text{ kJ}) + 18 \times (-286\text{ kJ}) - 2 \times (-269\text{ kJ}) \\ &= -1.09 \times 10^4\text{ kJ}\end{aligned}$$

This value is the amount of heat evolved when 2 mol of octane burns. Since the molar mass of octane is 114.2 g/mol, the enthalpy of combustion per gram of octane is

$$\frac{-1.09 \times 10^4\text{ kJ}}{2(114.2\text{ g})} = -47.7\text{ kJ/g}$$

The enthalpy of combustion per gram of octane is approximately twice that per gram of methanol. On this basis, gasoline appears to be superior to methanol for use in a racing car, where weight considerations are usually very important. Why, then, is methanol used in racing cars? The answer is that methanol burns much more smoothly than gasoline in high-performance engines, and this advantage more than compensates for its weight disadvantage.

9.7 Present Sources of Energy

Woody plants, coal, petroleum, and natural gas provide a vast resource of energy that originally came from the sun. By the process of photosynthesis, plants store energy that can be claimed by burning the plants themselves or the decay products that have been converted to **fossil fuels.** Although the United States currently depends heavily on petroleum for energy, this dependency is a relatively recent phenomenon, as shown in Fig. 9.13. In this section we discuss some sources of energy and their effects on the environment.

Petroleum and Natural Gas

Although how they were produced is not completely understood, petroleum and natural gas were most likely formed from the remains of marine organisms that lived approximately 500 million years ago. **Petroleum** is a thick, dark liquid composed mostly of compounds called *hydrocarbons* that contain carbon and hydrogen. Table 9.5 lists the formulas and names of several common hydrocarbons. **Natural gas,** usually associated with petroleum deposits, consists mostly of methane but also contains significant amounts of ethane, propane, and butane.

The composition of petroleum varies somewhat, but it consists mostly of hydrocarbons having chains that contain from 5 to more than 25 carbons. To be used efficiently, the petroleum must be separated into fractions by boiling.

TABLE 9.5
Formulas and Names for Some Common Hydrocarbons

Formula	Name
CH_4	Methane
C_2H_6	Ethane
C_3H_8	Propane
C_4H_{10}	Butane
C_5H_{12}	Pentane
C_6H_{14}	Hexane
C_7H_{16}	Heptane
C_8H_{18}	Octane

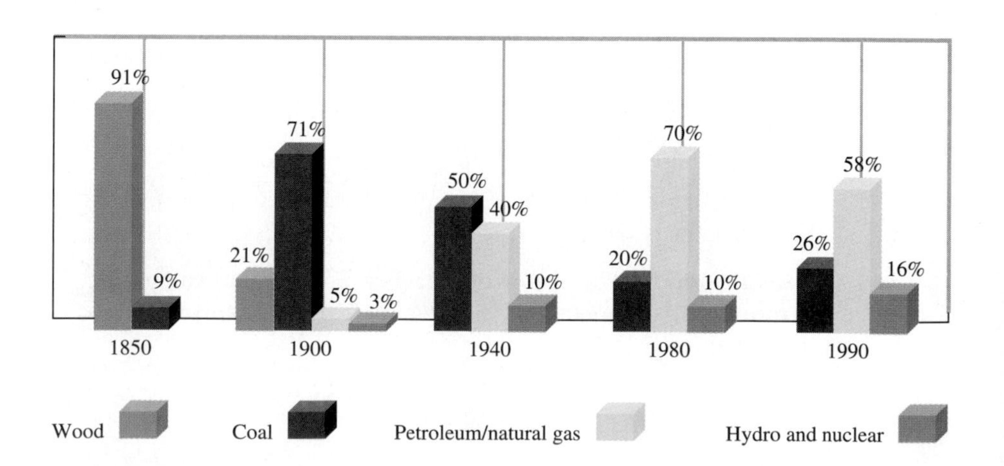

Figure 9.13
Energy sources used in the United States.

The lighter molecules (having the lowest boiling points) can be boiled off, leaving the heavier ones behind. The uses of various petroleum fractions are shown in Table 9.6.

The petroleum era began when the demand for lamp oil during the Industrial Revolution outstripped the traditional sources, animal fats and whale oil. In response to this increased demand, Edwin Drake drilled the first oil well in 1859 at Titusville, Pennsylvania. The petroleum from this well was refined to produce *kerosene* (fraction C_{10}–C_{18}), which served as an excellent lamp oil. *Gasoline* (fraction C_5–C_{10}) had limited use and was often discarded. However, the development of the electric light decreased the need for kerosene, and the advent of the "horseless carriage" with its gasoline-powered engine signaled the birth of the gasoline age.

As gasoline became more important, new ways were sought to increase the yield of gasoline obtained from each barrel of petroleum. William Burton invented a process at Standard Oil of Indiana called *pyrolytic (high-temperature) cracking.* In this process the heavier molecules of the kerosene fraction are heated to about 700°C, causing them to break (crack) into the smaller molecules of hydrocarbons in the gasoline fraction. As cars became larger, more efficient internal combustion engines were designed. Because of the uneven burning of the gasoline then available, these engines "knocked," producing unwanted noise and even engine damage. Intensive research to find additives that would promote smoother burning produced tetraethyl lead $(C_2H_5)_4Pb$, a very effective "antiknock" agent.

The addition of tetraethyl lead to gasoline became a common practice, and by 1960 gasoline contained as much as 3 grams of lead per gallon. As we have discovered so often in recent years, technological advances can produce environmental problems. To prevent air pollution from automobile exhaust, manufacturers have added catalytic converters to car exhaust systems. The effectiveness of these converters, however, is destroyed by lead. The use of leaded gasoline has also greatly increased the amount of lead in the environment, where it can be ingested by animals and humans. For these reasons, the use of lead in gasoline is being phased out, which has required extensive (and expensive) modifications of engines and of the gasoline-refining process.

TABLE 9.6
Uses of the Various Petroleum Fractions

Petroleum Fraction in Terms of Numbers of Carbon Atoms	Major Uses
C_5–C_{10}	Gasoline
C_{10}–C_{18}	Kerosene Jet fuel
C_{15}–C_{25}	Diesel fuel Heating oil Lubricating oil
$>C_{25}$	Asphalt

Coal

Coal was formed from the remains of plants that were buried and subjected to pressure and heat over long periods of time. Plant materials have a high content of cellulose, a complex molecule whose empirical formula is CH_2O but whose molar mass is around 500,000. After the plants and trees that flourished on the earth at various times and places died and were buried, chemical changes gradually lowered the oxygen and hydrogen content of the cellulose molecules. Coal "matures" through four stages: lignite, subbituminous, bituminous, and anthracite. Each stage has higher carbon-to-oxygen and carbon-to-hydrogen ratios; that is, the relative carbon content gradually increases. Typical elemental compositions of the various coals are given in Table 9.7. The energy available from the combustion of a given mass of coal increases as the carbon content increases. Anthracite is the most valuable coal, and lignite the least.

Coal has variable composition depending on both its age and location.

Coal is an important and plentiful fuel in the United States, currently furnishing approximately 20% of our energy. As the supply of petroleum

TABLE 9.7 Elemental Composition of Various Types of Coal

	Mass Percent of Each Element				
Type of Coal	C	H	O	N	S
Lignite	71	4	23	1	1
Subbituminous	77	5	16	1	1
Bituminous	80	6	8	1	5
Anthracite	92	3	3	1	1

dwindles, the share of the energy supply from coal is expected to increase to approximately 30% by the year 2000. However, coal is expensive and dangerous to mine underground, and the strip mining of fertile farmland in the Midwest or of scenic land in the West causes obvious problems. In addition, the burning of coal, especially high-sulfur coal, yields air pollutants such as sulfur dioxide, which in turn can lead to acid rain, as we learned in Chapter 5. However, even if coal were pure carbon, the carbon dioxide produced when it was burned would still have significant effects on the earth's climate.

Effects of Carbon Dioxide on Climate

The earth receives a tremendous quantity of radiant energy from the sun, about 30% of which is reflected into space by the earth's atmosphere. The remaining energy passes through the atmosphere to the earth's surface. Some of this energy is absorbed by plants to drive photosynthesis and some by the oceans to evaporate water, but most of it is absorbed by soil, rock, and water resulting in an increase in the temperature of the earth's surface. This energy is in turn radiated from the heated surface mainly as *infrared radiation*, often called heat radiation.

Sulfur-Eating Bacteria Clean Up Coal

Coal is a very important energy resource in the United States. Although enormous quantities of coal are used to produce electricity in the United States every day, tremendous reserves of this fossil fuel remain underground across the country. One major problem in using coal is that a great percentage of these reserves have a high sulfur content. Burning high-sulfur coal requires expensive treatment of the exhaust gases to remove the SO_2 that would lead to acid rain and other air pollution problems.

One strategy for dealing with high-sulfur coal is pretreatment to remove the sulfur before the combustion process. The best way to remove the sulfur appears to be by use of bacteria. For example, scientists at the Institute of Gas Technology have used genetic engineering to produce bacteria that selectively remove sulfur from coal without degrading the coal itself. By growing the bacteria under conditions where all nutrients except sulfur are furnished in abundance, the scientists are encouraging the growth of those strains of bacteria that metabolize sulfur most efficiently.

Although the bacterial treatment of coal to remove sulfur is currently not economically feasible, this method shows great promise.

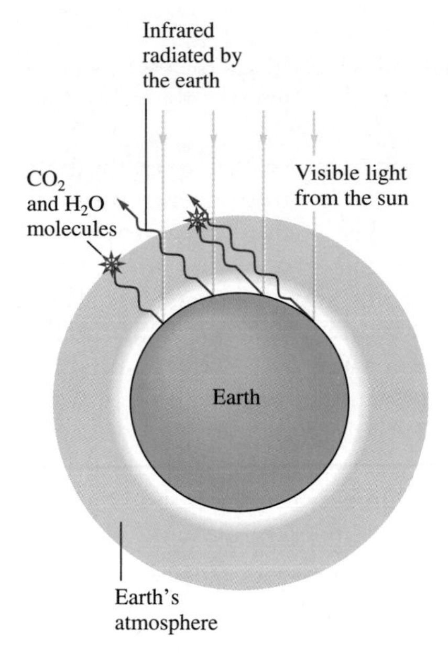

Figure 9.14
The earth's atmosphere is transparent to visible light from the sun. This visible light strikes the earth, where part of it is changed to infrared radiation. This infrared radiation from the earth's surface is strongly absorbed by CO_2, H_2O, and other molecules present in smaller amounts (for example, CH_4 and N_2O) in the atmosphere. In effect, the atmosphere traps some of the energy, acting like the glass in a greenhouse and keeping the earth warmer than it would otherwise be.

The atmosphere, like window glass, is transparent to visible light but does not allow all of the infrared radiation to pass through. Molecules in the atmosphere, principally H_2O and CO_2, strongly absorb infrared radiation and radiate it back toward the earth, as shown in Fig. 9.14. A net amount of thermal energy is thus retained by the earth's atmosphere, which causes the earth to be much warmer than it would be without its atmosphere. In a way the atmosphere acts like the glass of a greenhouse, which is transparent to visible light but absorbs infrared radiation, thus raising the temperature inside the building. This **greenhouse effect** is seen even more spectacularly on Venus, where the dense atmosphere is mainly responsible for the high surface temperature of that planet.

The electromagnetic spectrum including visible and infrared radiation is discussed in Chapter 12.

Thus the temperature of the earth's surface is controlled to a significant extent by the carbon dioxide and water content of the atmosphere. The effect of atmospheric moisture (humidity) is apparent in the Midwest. In summer when the humidity is high, the heat of the sun is retained well into the night, giving very high nighttime temperatures. On the other hand, in winter the coldest temperatures always occur on clear nights, when the low humidity allows efficient radiation of energy back into space.

The average temperature of the earth's surface is 288 K. It would be $\approx$255 K without the "greenhouse gases."

The atmosphere's water content is controlled by the water cycle (evaporation and precipitation), and the average content remains constant over the years. However, as fossil fuels have come into more extensive use, the carbon dioxide concentration has increased significantly. This increase, which was 16% from 1880 to 1980, has rapidly escalated in the past decade (Fig. 9.15). Projections indicate that the carbon dioxide content of the atmosphere may be double in the twenty-first century what it was in 1880. As a result, the earth's average temperature could increase by as much as 3°C, causing dramatic changes in climate and greatly affecting the growth of food crops.

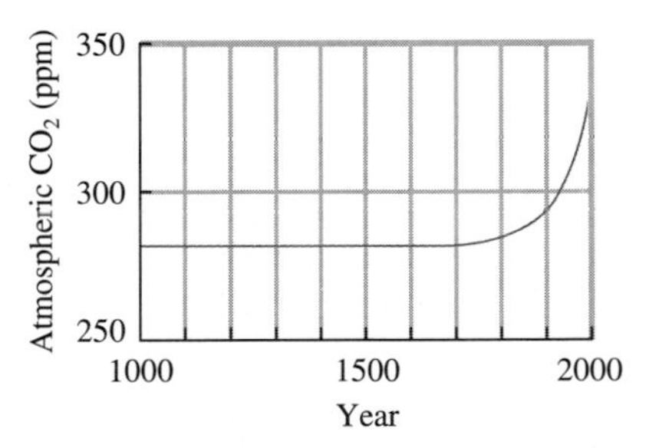

Figure 9.15
The atmospheric CO_2 concentration over the past 1000 years, based on ice core data and direct readings (since 1958). Note the dramatic increase in the past 100 years.

How well can we predict long-term effects? Because weather has been studied for a period of time that is miniscule compared with the age of the earth, the factors that control the earth's climate in the long range are not clearly understood. For example, we do not understand what causes the earth's periodic ice ages. So, indeed, it is difficult to estimate the impact of the increasing carbon dioxide levels.

In fact, the variation in the earth's average temperature over the past century is somewhat confusing. In the northern latitudes during the past century, the average temperature rose by 0.8°C over a period of 60 years, then cooled by 0.5°C during the next 25 years, and finally warmed by 0.2°C in the past 15 years. Such fluctuations do not match the steady increase in carbon dioxide. However, in southern latitudes and in areas near the equator, the average temperature showed a steady increase totaling 0.4°C over the past century. This figure is in reasonable agreement with the predicted effect of the increasing carbon dioxide concentration over that period.

Another significant fact is that the past 10 years appear to be the warmest decade on record. Although the exact relationship between the carbon dioxide concentration in the atmosphere and the earth's temperature is not known at present, one thing is clear: the increase in the atmospheric concentration is quite dramatic (see Fig. 9.15). We must consider the implications of this increase as we consider our future energy needs.

9.8 New Energy Sources

As we search for the energy sources of the future, we need to consider economic, climatic, and supply factors. There are several potential energy sources: the sun (solar), nuclear processes (fission and fusion), biomass (plants), and synthetic fuels. Direct use of the sun's radiant energy to heat our homes and run our factories and transportation systems seems a sensible long-term goal. But what do we do now? Conservation of fossil fuels is one obvious step, but substitutes for fossil fuels must be found eventually. We will discuss some alternative sources of energy here. Nuclear power will be considered in Chapter 21.

Coal Conversion

One alternative energy source involves using a traditional fuel—coal—in new ways. Since transportation costs for solid coal are high, more energy-efficient fuels are being developed from coal. One possibility is to produce a gaseous fuel. Substances like coal that contain large molecules have high boiling points and tend to be solids or thick liquids. To convert coal from a solid to a gas therefore requires reducing the size of the molecules; the coal structure must be broken down in a process called *coal gasification*. This process is carried out by treating the coal with oxygen and steam at high temperatures to break many of the carbon-carbon bonds. These bonds are replaced by carbon-hydrogen and carbon-oxygen bonds as the coal fragments react with the water and oxygen. This process is represented in Fig. 9.16. The desired fuel consists of a mixture of carbon monoxide and hydrogen called *synthetic gas*, or **syngas**, and methane (CH_4) gas. Since all the components of this product can react with oxygen to release heat in a combustion reaction, this gas is a useful fuel.

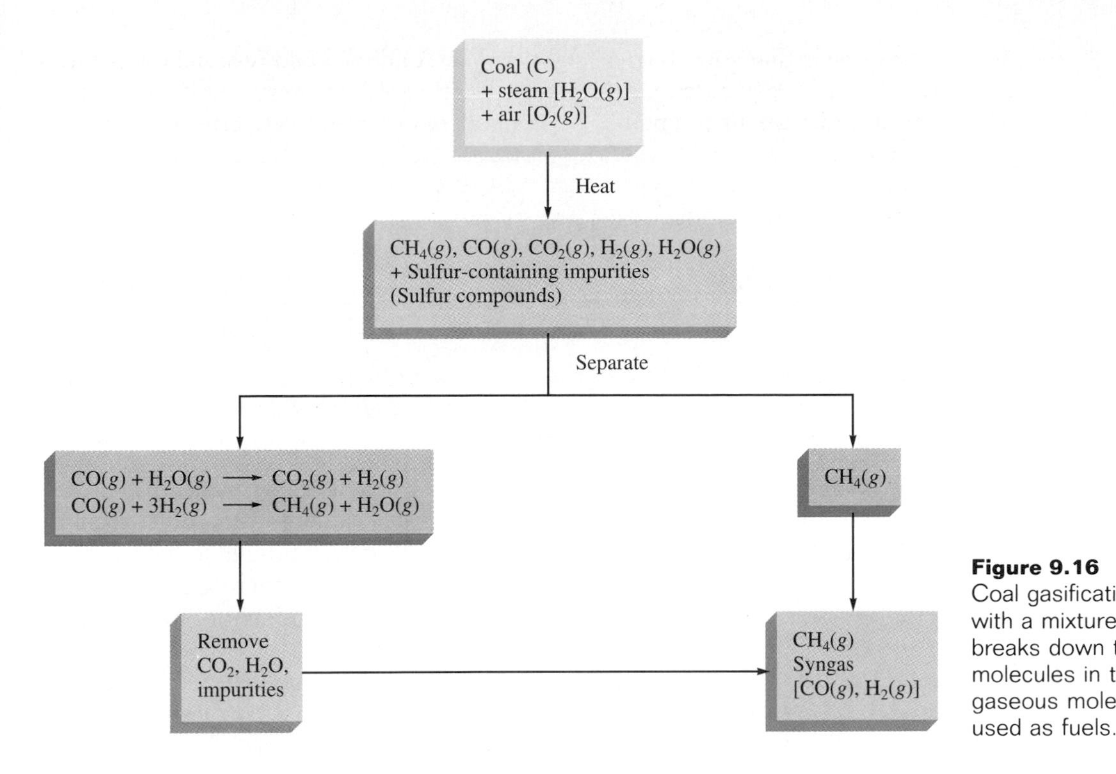

Figure 9.16
Coal gasification. Reaction of coal with a mixture of steam and air breaks down the large hydrocarbon molecules in the coal to smaller gaseous molecules, which can be used as fuels.

One of the most important considerations in designing an industrial process is the efficient use of energy. In coal gasification some of the reactions are exothermic:

An industrial process must be energy-efficient.

$$C(s) + 2H_2(g) \longrightarrow CH_4(g) \qquad \Delta H^\circ = -75 \text{ kJ}$$
$$C(s) + \tfrac{1}{2}O_2(g) \longrightarrow CO(g) \qquad \Delta H^\circ = -111 \text{ kJ}$$
$$C(s) + O_2(g) \longrightarrow CO_2(g) \qquad \Delta H^\circ = -394 \text{ kJ}$$

Other gasification reactions are endothermic, for example:

$$C(s) + H_2O(g) \longrightarrow H_2(g) + CO(g) \qquad \Delta H^\circ = 131 \text{ kJ}$$

If the rate at which the coal, air, and steam are combined is carefully controlled, the correct temperature can be maintained in the process without using any external energy source. That is, an energy balance is attained.

As we stated earlier, syngas can be used directly as a fuel, but it is also important as a raw material for producing other fuels. For example, syngas can be directly converted to methanol:

$$CO(g) + 2H_2(g) \longrightarrow CH_3OH(l)$$

Methanol is used in the production of synthetic fibers and plastics and can also be used as a fuel. In addition, it can be converted directly to gasoline. Approximately half of South Africa's gasoline supply comes from methanol produced from syngas.

In addition to coal gasification, the formation of *coal slurries* is another new use of coal. A slurry is a suspension of fine particles in a liquid. Coal must be pulverized and mixed with water to form a slurry. The resulting slurry can be handled, stored, and burned in ways similar to those used for *residual oil,* a heavy fuel oil from petroleum accounting for 13% of U.S. petroleum imports.

One hope is that coal slurries might replace solid coal and residual oil as fuels for electricity-generated power plants. However, the water needed for slurries might place an unacceptable burden on water resources, especially in the western states.

Hydrogen as a Fuel

If you have ever seen a lecture demonstration where hydrogen-oxygen mixtures were ignited or if you have seen a newsfilm or a photo of the *Hindenburg* disaster, you have witnessed a demonstration of hydrogen's potential as a fuel. The combustion reaction is

$$H_2(g) + \tfrac{1}{2}O_2(g) \longrightarrow H_2O(l) \qquad \Delta H^\circ = -286 \text{ kJ}$$

As we saw in Example 9.5, the heat of combustion of $H_2(g)$ per gram is approximately 2.5 times that of natural gas. In addition, hydrogen has a real advantage over fossil fuels in that the only product of hydrogen combustion is water; fossil fuels also produce carbon dioxide. But even though it appears that hydrogen is a very logical choice for a major future fuel, there are three main problems that are associated with its use: the costs of production, storage, and transport.

First, let's look at the production problem. Although hydrogen is very abundant on earth, virtually none of it exists as the free gas. Currently, the main source of hydrogen gas is from the treatment of natural gas with steam:

$$CH_4(g) + H_2O(g) \longrightarrow 3H_2(g) + CO(g)$$

We can calculate ΔH for this reaction by using Equation (9.1):

$$\begin{aligned}\Delta H^\circ &= \Sigma\Delta H_f^\circ \text{ (products)} - \Sigma\Delta H_f^\circ \text{ (reactants)} \\ &= \Delta H_f^\circ \text{ [for } CO(g)] - \Delta H_f^\circ \text{ [for } CH_4(g)] - \Delta H_f^\circ \text{ [for } H_2O(g)] \\ &= -111 \text{ kJ} - (-75 \text{ kJ}) - (-242 \text{ kJ}) = 206 \text{ kJ}\end{aligned}$$

Note that this reaction is highly endothermic; treating methane with steam is not an efficient way to obtain hydrogen for fuel. It would be much more economical to burn the methane directly.

A virtually inexhaustible supply of hydrogen exists in the waters of the world's oceans. However, the reaction

$$H_2O(l) \longrightarrow H_2(g) + \tfrac{1}{2}O_2(g)$$

requires 286 kJ of energy per mole of liquid water, and under current circumstances large-scale production of hydrogen from water is not economically feasible. However, several methods for such production are currently being studied: electrolysis of water, thermal decomposition of water, and biological decomposition of water.

Electrolysis will be discussed in Chapter 11.

Electrolysis of water involves passing an electrical current through it. The present cost of electricity makes the hydrogen produced by electrolysis too expensive to be competitive as a fuel. However, if in the future we develop more efficient sources of electricity, this situation could change.

Thermal decomposition is another method for producing hydrogen from water. This method involves heating the water to several thousand degrees, where it spontaneously decomposes into hydrogen and oxygen. However, attaining temperatures in this range would be very expensive even if a practical heat source and a suitable reaction container were available.

OLD TIRES—A CLEAN SOURCE OF ENERGY?

Old tires are a big problem—a city of 500,000 people can generate as many as 2000 worn-out tires a day. In the past these unsightly discards have been piled up or sent to landfills. However, that may soon change. Texaco is pioneering a process that converts old tires and used lubricating oil into a gaseous fuel that can be used to generate electricity. It turns out that tires are a very rich source of energy—a 20 lb tire can furnish as much energy as 2.5 gallons of gasoline.

The first step in Texaco's process takes place in a liquefaction reactor where old tires are cooked for half an hour at 700°F in a "solvent" composed of waste lubricating oil, transmission oil or fuel oil. In this process petroleum gases and light oils are removed and the steel reinforcing belts in the tires are reclaimed as scrap steel (see accompanying diagram). The tire oil slurry is then fed to a gasification reactor where it is combined with oxygen to yield a gaseous mixture of hydrogen and carbon monoxide (called synthesis gas or syn-gas). After removal of contaminants such as sulfur dioxide, the syn-gas can be used to fuel a turbine to generate electricity or can be used as a feedstock to manufacture ammonia and methanol.

Texaco has built a demonstration plant in Montebello, Calif., which can process up to 2000 tires and an equal amount (by mass) of waste oil per day. The clean synthesis gas that results can be used as fuel to a "combined-cycle" turbine which includes a gas turbine powered electric generator. The hot exhaust gases from the gas turbine are then used to generate steam, which in turn is used to run a steam turbine powered generator to produce more electric power. Combined-cycle generators achieve efficiencies of about 50% (50% of the energy originally available in the gaseous fuel is converted into electricity) as compared to ~33% efficiency for conventional power generators.

It's time to view old tires as an energy resource rather than an eyesore.

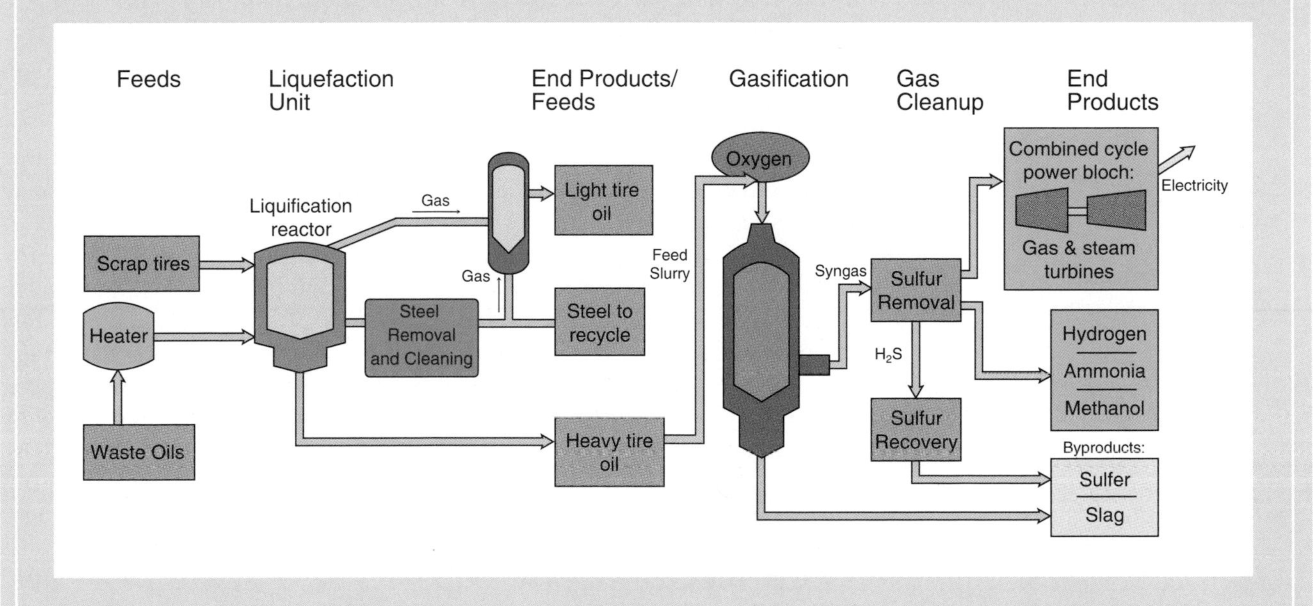

In the thermochemical decomposition of water, chemical reactions, as well as heat, are used to "split" water into its components. One such system involves the following reactions (the temperature required for each is given in parentheses):

$$2HI \longrightarrow I_2 + H_2 \qquad (425°C)$$
$$2H_2O + SO_2 + I_2 \longrightarrow H_2SO_4 + 2HI \qquad (90°C)$$
$$H_2SO_4 \longrightarrow SO_2 + H_2O + \tfrac{1}{2}O_2 \qquad (825°C)$$

$$\text{Net reaction:} \quad H_2O \longrightarrow H_2 + \tfrac{1}{2}O_2$$

Note that the HI is not consumed in this reaction. Note also that the maximum temperature required is 825°C, a temperature that is feasible if a nuclear reactor is used as a heat source. A current research goal is to find a system for which the required temperatures are low enough that sunlight can be used as the energy source.

But what about the systems on earth that biologically decompose water without the aid of electricity or high temperatures? In the process of photosynthesis, green plants absorb carbon dioxide and water and use them along with energy from the sun to produce the substances needed for growth. Scientists have studied photosynthesis for years, hoping to get answers to humanity's food and energy shortages. At present much of this research involves attempts to modify the photosynthetic process so that plants will release hydrogen gas from water instead of using the hydrogen to produce complex compounds. Small-scale experiments have shown that under certain conditions plants do produce hydrogen gas, but the yields are far from being commercially useful. Thus the economical production of hydrogen gas remains unrealized.

The storage and transportation of hydrogen present two problems. First, on metal surfaces the H_2 molecule decomposes to atoms. Since the atoms are so small, they can migrate into the metal, causing structural changes that make it brittle. This might lead to a pipeline failure if hydrogen were pumped under high pressure.

A second problem is the relatively small amount of energy that is available *per unit volume* of hydrogen. Although the energy available per gram of hydrogen is significantly greater than that per gram of methane, the energy available per given volume of hydrogen is about one-third that available from the same volume of methane. Could hydrogen be considered as a potential fuel for automobiles? This is an intriguing question. The internal combustion engines in automobiles can be easily adapted to burn hydrogen. However, the primary difficulty is the storage of enough hydrogen to give an automobile a reasonable range. This is illustrated in Example 9.9.

EXAMPLE 9.9

Assuming that the combustion of hydrogen gas provides three times as much energy per gram as gasoline, calculate the volume of liquid H_2 (density = 0.0710 g/mL) required to furnish the energy contained in 80.0 L (about 20 gal) of gasoline (density = 0.740 g/mL). Calculate also the volume that this hydrogen would occupy as a gas at 1.00 atm and 25°C.

Solution

The mass of 80.0 L of gasoline is

$$80.0 \text{ L} \times \frac{1000 \text{ mL}}{1 \text{ L}} \times \frac{0.740 \text{ g}}{\text{mL}} = 59{,}200 \text{ g}$$

Since H_2 furnishes three times as much energy per gram as gasoline, only a third as much liquid hydrogen is needed to furnish the same energy:

$$\text{Mass of } H_2(l) \text{ needed} = \frac{59{,}200 \text{ g}}{3} = 19{,}700 \text{ g}$$

Since density = mass/volume, volume = mass/density, and the volume of $H_2(l)$ needed is

$$V = \frac{19{,}700 \text{ g}}{0.0710 \text{ g/mL}} = 2.77 \times 10^5 \text{ mL} = 277 \text{ L}$$

Thus 277 L of liquid H_2 are needed to furnish the same energy of combustion as 80.0 L of gasoline.

To calculate the volume that this hydrogen would occupy as a gas at 1.00 atm and 25°C, we use the ideal gas law:

$$PV = nRT$$

In this case P = 1.00 atm, T = 273 + 25°C = 298 K, and R = 0.08206 L atm K^{-1} mol^{-1}. Also,

$$n = 19{,}700 \text{ g } H_2 \times \frac{1 \text{ mol } H_2}{2.02 \text{ g } H_2} = 9.75 \times 10^3 \text{ mol } H_2$$

Thus

$$V = \frac{nRT}{P} = \frac{(9.75 \times 10^3 \text{ mol})(0.08206 \text{ L atm K}^{-1} \text{ mol}^{-1})(298 \text{ K})}{1.00 \text{ atm}}$$

$$= 2.38 \times 10^5 \text{ L} = 238{,}000 \text{ L}$$

At 1 atm and 25°C, the hydrogen gas needed to replace 20 gal of gasoline occupies a volume of 238,000 L.

You can see from Example 9.9 that an automobile would need a huge tank to hold enough hydrogen gas to have a typical mileage range. Clearly, hydrogen must be stored in some other way, possibly as a liquid. Is this feasible? Because of its very low boiling point (20 K), storage of liquid hydrogen requires a superinsulated container that can withstand high pressures. Storage in this manner would be both expensive and hazardous because of the potential for explosion. Thus storage of hydrogen in the individual automobile as a liquid does not seem practical.

A much better alternative seems to be the use of metals that absorb hydrogen to form solid metal hydrides:

Metal hydrides are discussed in Chapter 18.

$$H_2(g) + M(s) \longrightarrow MH_2(s)$$

Heat Packs

A skier is trapped by a sudden snowstorm. After building a snow cave for protection, she realizes her hands and feet are freezing; she is in danger of frostbite. Then she remembers the four small packs in her pocket. She removes the plastic cover from each one to reveal a small paper packet. She places one packet in each boot and one in each mitten. Soon her hands and feet are toasty warm.

These "magic" packets of energy contain a mixture of powdered iron, activated carbon, sodium chloride, cellulose (sawdust), and zeolite, all moistened by a little water. The paper cover is permeable to air.

The exothermic reaction that produces the heat is a very common one—the rusting of iron. The overall reaction can be represented as

$$4Fe(s) + 3O_2(g) \longrightarrow 2Fe_2O_3(s) \quad \Delta H^\circ = -1652 \text{ kJ}$$

although in reality it is somewhat more complicated. When the plastic envelope is removed, O_2 molecules penetrate the paper causing the reaction to begin.

The oxidation of iron by oxygen occurs naturally. Any steel surface exposed to the atmosphere inevitably rusts. But this process is quite slow—much too slow to be useful in hot packs. However, if the iron is ground into a fine powder, the resulting increase in surface area causes the reaction with oxygen to be fast enough to warm hands and feet. The packet can produce heat for up to six hours.

In this method of storage, hydrogen gas would be pumped into a tank containing the solid metal, where it would be absorbed to form a hydride, whose volume would be little more than that of the metal. This hydrogen would then be available for combustion in the engine by release of $H_2(g)$ from the hydride as needed:

$$MH_2(s) \longrightarrow M(s) + H_2(g)$$

Several types of solids that absorb hydrogen to form hydrides are being studied for use in hydrogen-powered vehicles.

Other Energy Alternatives

Many other energy sources are being considered for future use. The western states, especially Colorado, contain huge deposits of *oil shale,* which consists of a complex carbon-based material called kerogen contained in porous rock formations. These deposits have the potential of being a larger energy source than the vast petroleum deposits of the Middle East. However, the main problem with oil shale is that the trapped fuel is not fluid and cannot be pumped. For recovery of the fuel, the rock must be heated to a temperature of 250°C or higher to decompose the kerogen to smaller molecules that produce gaseous and liquid products. This process is expensive and yields large quantities of waste rock, which have a negative environmental impact.

Ethanol (C_2H_5OH) is another fuel with the potential to supplement, if not replace, gasoline. The most common method of producing ethanol is fermentation, a process in which sugar is changed to alcohol by the action of yeast. The sugar can come from virtually any source, including fruits and grains, although

ANAEROBIC ENGINES: ENERGY WITHOUT OXYGEN

A fireman frantically tries to rescue a person trapped in a burning building, but his chain saw dies from a lack of oxygen. This potential tragedy could be averted if the chain saw were powered by an engine whose fuel did not require oxygen—an anaerobic engine. Such an engine has been suggested by two Canadian scientists* who have demonstrated that the compound $(CH_3)_3COOC(CH_3)_3$, called di-tert-butylperoxide or DTBP, decomposes exothermically when compressed in the cylinder of an engine, thus providing energy to run the engine without oxygen. The decomposition reaction is

$$(CH_3)_3COOC(CH_3)_3(g) \longrightarrow C_2H_6(g) + 2(CH_3)_2CO(g)$$

Under conditions where oxygen is plentiful (aerobic conditions) the engine runs normally, with the DTBP fuel reacting with O_2 to produce CO_2 and H_2O. However, when the oxygen supply is severely limited, as in the smoke from a fire, the engine continues to run (although more slowly) from the energy produced by the decomposition of DTBP described by the above equation. Thus the DTBP fuel would allow a chain saw to be used under the conditions encountered at the scene of the fire.

The Canadian scientists have also suggested that large anaerobic engines could be developed to power equipment used in mine rescues and other situations where lack of oxygen would prevent an engine with normal fuel from operating.

*H. O. Pritchard and P. Q. E. Clothier, "Anaerobic Operation of an Internal Combustion Engine," *J. Chem. Soc., Chem. Commun.* (1986), p. 1529.

fuel-grade ethanol would probably come mostly from corn. Car engines can burn pure alcohol or *gasohol,* an alcohol-gasoline mixture (10% ethanol in gasoline), with little modification. Gasohol is now widely available in the United States. The use of pure alcohol as a motor fuel is not feasible in most of the United States because it does not vaporize easily when temperatures are low. However, pure ethanol could be a very practical fuel in warm climates. For example, in Brazil large quantities of ethanol fuel are being produced for cars.

Methanol (CH_3OH), an alcohol similar to ethanol, which has been used successfully for many years in race cars, is now being evaluated as a motor fuel in California. A major gasoline retailer has agreed to install pumps at 25 locations to dispense a fuel that is 85% methanol and 15% gasoline for use in specially prepared automobiles. The California Energy Commission feels that methanol has great potential for providing a secure, long-term energy supply that would help alleviate air quality problems. Arizona and Colorado are also considering methanol as a major source of portable energy.

Another potential source of liquid fuel is oil squeezed from seeds (*seed oil*). For example, some farmers in North Dakota, South Africa, and Australia are now using sunflower oil to replace diesel fuel. Oil seeds, found in a wide variety of plants, can be processed to produce a "biodiesel" oil composed mainly of carbon and hydrogen, which of course reacts with oxygen to produce carbon dioxide, water, and heat. The main advantage of seed oil as a fuel is that it is renewable. It is hoped that oil seed plants can be developed that will thrive under soil and climatic conditions unsuitable for corn and wheat. Ideally, fuel would be grown just like food crops.

Exercises

A blue exercise number indicates that the answer to that exercise appears at the back of this book.

Potential and Kinetic Energy

1. Consider the accompanying diagram. Ball A is allowed to fall and strike ball B. Assume that all of ball A's energy is transferred to ball B at point I, and that there is no loss of energy to other sources. Calculate the kinetic energy and the potential energy of ball B at point II. For a falling object, the potential energy is given by PE = mgz, where m is the mass in kilograms, g is the gravitational constant (9.8 m/s²), and z is the distance in meters.

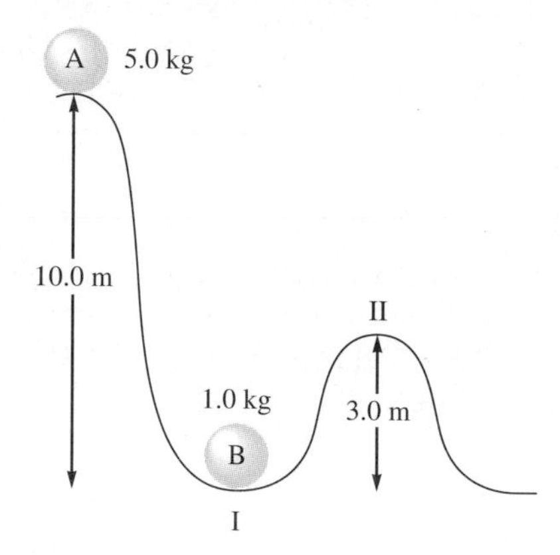

2. Which has greater kinetic energy, an object with a mass of 2.0 kg and a velocity of 1.0 m/s, or an object with a mass of 1.0 kg and a velocity of 2.0 m/s?

3. A balloon filled with 39.1 mol of helium has a volume of 876 L at 0.0°C and 1.00 atm pressure. At constant pressure, the temperature of the balloon is increased to 38.0°C, causing the balloon to expand to a volume of 998 L. Calculate q, w, and ΔE for the helium in the balloon. (The molar heat capacity for helium gas is 20.8 J °C^{-1} mol^{-1}.)

4. Consider a mixture of air and gasoline vapor in a cylinder with a piston. The original volume is 40. cm³. If the combustion of this mixture releases 950. J of energy, to what volume will the gases expand against a constant pressure of 650. torr if all of the energy of combustion is converted into work to push back the piston?

5. A balloon contains 313 g He at a pressure of 1.00 atm. The volume of the balloon is 1910. L. The temperature is decreased by 15°C and the volume is decreased to 1814 L. Assuming a constant pressure of 1.00 atm, calculate q, w, and ΔE for the helium in the balloon. (The molar heat capacity of helium gas is 20.8 J °C^{-1} mol^{-1}.)

Properties of Enthalpy

6. The reaction

$$SO_3(g) + H_2O(l) \longrightarrow H_2SO_4(aq)$$

is the last step in the commercial production of sulfuric acid. The enthalpy change for this reaction is −266 kJ. In the design of a sulfuric acid plant, is it necessary to provide for heating or cooling the reaction mixture? Explain your answer.

7. The enthalpy change for the following reaction is −891 kJ:

$$CH_4(g) + 2O_2(g) \longrightarrow CO_2(g) + 2H_2O(l)$$

Calculate the enthalpy change for each of these cases.
a. 1.00 g of methane is burned in an excess of oxygen.
b. 1.00×10^3 L of methane gas at 740. torr and 25°C is burned in an excess of oxygen.

8. Consider the reaction

$$S(s) + O_2(g) \longrightarrow SO_2(g) \qquad \Delta H° = -296 \text{ kJ/mol}$$

a. How much heat is evolved when 275 g of sulfur is burned in excess O_2?
b. How much heat is evolved when 25 mol of sulfur is burned in excess O_2?
c. How much heat is evolved when 150. g of sulfur dioxide is produced?

9. The enthalpy change for the conversion of white phosphorus to red phosphorus is −17.6 kJ/mol. Will one mole of red phosphorus or one mole of white phosphorus give off more heat when burned in air?

The Thermodynamics of Ideal Gases

10. Calculate the energy required to heat 1.00 kg of ethane gas (C_2H_6) from 25.0°C to 75.0°C first under conditions of constant volume and then at a constant pressure of 2.00 atm. Calculate ΔE, ΔH, and w for these processes also. (See Table 9.1 for relevant data.)

11. Calculate q, w, ΔE, and ΔH for the process in which 88.0 g of nitrous oxide gas (N_2O) is cooled from 165°C to 55°C at a constant pressure of 5.00 atm. (See Table 9.1.)

12. Consider a sample containing 5.00 mol of a monatomic ideal gas that is taken from state A to state B by the following two pathways:

Pathway one: $P_A = 3.00$ atm, $V_A = 15.0$ L $\xrightarrow{1}$ $P_C = 3.00$ atm, $V_C = 55.0$ L $\xrightarrow{2}$ $P_B = 6.00$ atm, $V_B = 20.0$ L

Pathway two:

$$\begin{array}{lcll} P_A = 3.00\text{ atm} & \xrightarrow{3} & P_D = 6.00\text{ atm} & \\ V_A = 15.0\text{ L} & & V_D = 15.0\text{ L} & \\ & & \xrightarrow{4} & P_B = 6.00\text{ atm} \\ & & & V_B = 20.0\text{ L} \end{array}$$

For each step, assume that the external pressure is constant and equals the final pressure of the gas for that step. Calculate q, w, ΔE, and ΔH for each step, and calculate overall values for each pathway.

13. Consider a sample containing 2.00 mol of a monatomic ideal gas that undergoes the following changes:

$$\begin{array}{lclcl} P_A = 10.0\text{ atm} & \xrightarrow{1} & P_B = 10.0\text{ atm} & \xrightarrow{2} & P_C = 20.0\text{ atm} \\ V_A = 10.0\text{ L} & & V_B = 5.0\text{ L} & & V_C = 5.0\text{ L} \\ & & & \xrightarrow{3} & P_D = 20.0\text{ atm} \\ & & & & V_D = 25.0\text{ L} \end{array}$$

For each step, assume that the external pressure is constant and equals the final pressure of the gas for that step. Calculate q, w, ΔE, and ΔH for each step and for the overall change from state A to state D.

Calorimetry and Heat Capacity

14. Why is ΔH obtained directly from the heat flow using a coffee cup calorimeter and ΔE obtained from the heat flow using a bomb calorimeter?

15. The specific heat capacity of aluminum is $0.900\text{ J °C}^{-1}\text{ g}^{-1}$.
    a. How much energy is needed to raise the temperature of an 8.50×10^2 g block of aluminum from 22.8°C to 94.6°C?
    b. What is the molar heat capacity of aluminum?

16. It takes 78.2 J to raise the temperature of 45.6 g of lead by 13.3°C. What is the specific heat capacity of lead? What is the molar heat capacity of lead?

17. A 28.2-g sample of nickel is heated to 99.8°C and placed in a coffee cup calorimeter containing 150.0 g of water at a temperature of 23.5°C. After the metal cools, the final temperature of metal and water is 25.0°C. Calculate the specific heat capacity of nickel, assuming that no heat escapes to the surroundings or is transferred to the calorimeter.

18. A coffee cup calorimeter initially contains 125 g of water at a temperature of 24.2°C. After potassium bromide (10.5 g), also at 24.2°C, is added to the water, the temperature becomes 21.1°C. What is the heat of solution (the heat accompanying the dissolving of the salt) of potassium bromide in J/g and kJ/mol? Assume the specific heat capacity of the solution is $4.18\text{ J °C}^{-1}\text{ g}^{-1}$ and that no heat is transferred to the surroundings or to the calorimeter.

19. In a coffee cup calorimeter 100.0 mL of 1.0 M NaOH and 100.0 mL of 1.0 M HCl are mixed. Both solutions are originally at 24.6°C. After the reaction, the temperature is 31.3°C. Assuming all solutions have a density of 1.0 g/cm^3 and a specific heat capacity of $4.18\text{ J °C}^{-1}\text{ g}^{-1}$, what is the enthalpy change for the neutralization of HCl by NaOH? Assume that no heat is lost to the surroundings or to the calorimeter.

20. In a coffee cup calorimeter 50.0 mL of 0.100 M $AgNO_3$ and 50.0 mL of 0.100 M HCl are mixed. The following reaction occurs:

$$Ag^+(aq) + Cl^-(aq) \longrightarrow AgCl(s)$$

If the two solutions are initially at 22.60°C, and if the final temperature is 23.40°C, calculate ΔH for the above reaction in kJ/mol of AgCl formed. Assume a mass of 100.0 g for the combined solution and a specific heat capacity of $4.18\text{ J °C}^{-1}\text{ g}^{-1}$.

21. Camphor ($C_{10}H_{16}O$) has a heat of combustion of −5903.6 kJ/mol. A sample of camphor with a mass of 0.1204 g is burned in a bomb calorimeter. The temperature in the calorimeter increases by 2.28°C. What is the heat capacity of the calorimeter?

22. A 0.1964-g sample of quinone ($C_6H_4O_2$) is burned in a bomb calorimeter that has a heat capacity of 1.56 kJ/°C. The temperature of the calorimeter increases by 3.2°C. Calculate the energy of combustion of quinone per gram and per mole.

23. The combustion of 0.1584 g of benzoic acid increases the temperature of a bomb calorimeter by 2.54°C. Calculate the heat capacity of the calorimeter. (The energy released by combustion of benzoic acid is 26.42 kJ/g.) A 0.2130-g sample of vanillin ($C_8H_8O_3$) is then burned in the same calorimeter. The temperature increases by 3.25°C. What is the energy of combustion per gram of vanillin and per mole of vanillin?

24. A swimming pool, 10.0 m by 4.0 m, is filled to a depth of 3.0 m with water at a temperature of 20.2°C. How much energy is required to raise the temperature of the water to 30.0°C?

25. In a bomb calorimeter the bomb is surrounded by water that must be added for each experiment. Since the amount of water is not constant from experiment to experiment, mass must be measured in each case. The heat capacity of the calorimeter is broken down into two parts: the water and the calorimeter components. If a calorimeter contains 1.00 kg of water and has a total heat capacity of 10.84 kJ/°C, what is the heat capacity of the calorimeter components?

26. The bomb calorimeter in Exercise 25 is filled with 987 g of water. The initial temperature of the calorimeter contents is 23.32°C. A 1.056-g sample of benzoic acid ($\Delta E_{comb} = -26.42$ kJ/g) is combusted in the calorimeter. What is the final temperature of the calorimeter contents?

Hess's Law

27. Given the following data:

$$S(s) + \tfrac{3}{2}O_2(g) \longrightarrow SO_3(g) \qquad \Delta H° = -395.2 \text{ kJ}$$
$$2SO_2(g) + O_2(g) \longrightarrow 2SO_3(g) \qquad \Delta H° = -198.2 \text{ kJ}$$

calculate $\Delta H°$ for the reaction

$$S(s) + O_2(g) \longrightarrow SO_2(g)$$

28. Given the following data:

$$N_2(g) + 2O_2(g) \longrightarrow 2NO_2(g) \qquad \Delta H° = 67.7 \text{ kJ}$$
$$N_2(g) + 2O_2(g) \longrightarrow N_2O_4(g) \qquad \Delta H° = 9.7 \text{ kJ}$$

calculate $\Delta H°$ for the dimerization of NO_2:

$$2NO_2(g) \longrightarrow N_2O_4(g)$$

29. Given the following data:

$$H_2(g) + \tfrac{1}{2}O_2(g) \longrightarrow H_2O(l) \qquad \Delta H° = -285.8 \text{ kJ}$$
$$N_2O_5(g) + H_2O(l) \longrightarrow 2HNO_3(l) \qquad \Delta H° = -76.6 \text{ kJ}$$
$$\tfrac{1}{2}N_2(g) + \tfrac{3}{2}O_2(g) + \tfrac{1}{2}H_2(g) \longrightarrow HNO_3(l) \qquad \Delta H° = -174.1 \text{ kJ}$$

calculate $\Delta H°$ for the reaction

$$2N_2(g) + 5O_2(g) \longrightarrow 2N_2O_5(g)$$

30. Given the following data:

$$Fe_2O_3(s) + 3CO(g) \longrightarrow 2Fe(s) + 3CO_2(g) \qquad \Delta H° = -23 \text{ kJ}$$
$$3Fe_2O_3(s) + CO(g) \longrightarrow 2Fe_3O_4(s) + CO_2(g) \qquad \Delta H° = -39 \text{ kJ}$$
$$Fe_3O_4(s) + CO(g) \longrightarrow 3FeO(s) + CO_2(g) \qquad \Delta H° = +18 \text{ kJ}$$

calculate $\Delta H°$ for the reaction

$$FeO(s) + CO(g) \longrightarrow Fe(s) + CO_2(g)$$

31. The standard enthalpy for the combustion of solid carbon to form carbon dioxide is −393.7 kJ per mole of carbon. The standard enthalpy for the combustion of carbon monoxide to form carbon dioxide is −283.3 kJ per mole of CO. Use these data to calculate $\Delta H°$ for the reaction

$$2C(s) + O_2(g) \longrightarrow 2CO(g)$$

32. Given the following data:

$$C_2H_2(g) + \tfrac{5}{2}O_2(g) \longrightarrow 2CO_2(g) + H_2O(l) \qquad \Delta H° = -1300. \text{ kJ}$$
$$C(s) + O_2(g) \longrightarrow CO_2(g) \qquad \Delta H° = -394 \text{ kJ}$$
$$H_2(g) + \tfrac{1}{2}O_2(g) \longrightarrow H_2O(l) \qquad \Delta H° = -286 \text{ kJ}$$

calculate $\Delta H°$ for the reaction

$$2C(s) + H_2(g) \longrightarrow C_2H_2(g)$$

33. Given the following data:

$$2O_3(g) \longrightarrow 3O_2(g) \qquad \Delta H° = -427 \text{ kJ}$$
$$O_2(g) \longrightarrow 2O(g) \qquad \Delta H° = +495 \text{ kJ}$$
$$NO(g) + O_3(g) \longrightarrow NO_2(g) + O_2(g) \qquad \Delta H° = -199 \text{ kJ}$$

calculate $\Delta H°$ for the reaction

$$NO(g) + O(g) \longrightarrow NO_2(g)$$

34. The bombardier beetle uses an explosive discharge as a defensive measure. The chemical reaction involved is the oxidation of hydroquinone by hydrogen peroxide to produce quinone and water:

$$C_6H_4(OH)_2(aq) + H_2O_2(aq) \longrightarrow C_6H_4O_2(aq) + 2H_2O(l)$$

Calculate $\Delta H°$ for this reaction from the following data:

$$C_6H_4(OH)_2(aq) \longrightarrow C_6H_4O_2(aq) + H_2(g) \qquad \Delta H° = +177.4 \text{ kJ}$$
$$H_2(g) + O_2(g) \longrightarrow H_2O_2(aq) \qquad \Delta H° = -191.2 \text{ kJ}$$
$$H_2(g) + \tfrac{1}{2}O_2(g) \longrightarrow H_2O(g) \qquad \Delta H° = -241.8 \text{ kJ}$$
$$H_2O(g) \longrightarrow H_2O(l) \qquad \Delta H° = -43.8 \text{ kJ}$$

35. Given the following data:

$$O_2(g) + H_2(g) \longrightarrow 2OH(g) \qquad \Delta H° = +77.9 \text{ kJ}$$
$$O_2(g) \longrightarrow 2O(g) \qquad \Delta H° = +495 \text{ kJ}$$
$$H_2(g) \longrightarrow 2H(g) \qquad \Delta H° = +435.9 \text{ kJ}$$

calculate $\Delta H°$ for the reaction

$$O(g) + H(g) \longrightarrow OH(g)$$

36. Calculate $\Delta H°$ for the reaction

$$N_2H_4(l) + O_2(g) \longrightarrow N_2(g) + 2H_2O(l)$$

given the following data:

$$2NH_3(g) + 3N_2O(g) \longrightarrow 4N_2(g) + 3H_2O(l) \qquad \Delta H° = -1010. \text{ kJ}$$
$$N_2O(g) + 3H_2(g) \longrightarrow N_2H_4(l) + H_2O(l) \qquad \Delta H° = -317 \text{ kJ}$$

$$2NH_3(g) + \tfrac{1}{2}O_2(g) \longrightarrow N_2H_4(l) + H_2O(l) \quad \Delta H° = -143 \text{ kJ}$$
$$H_2(g) + \tfrac{1}{2}O_2(g) \longrightarrow H_2O(l) \quad \Delta H° = -286 \text{ kJ}$$

37. Given the following data:

$$NH_3(g) \rightarrow \tfrac{1}{2}N_2(g) + \tfrac{3}{2}H_2(g) \qquad \Delta H° = 46 \text{ kJ}$$
$$2H_2(g) + O_2(g) \rightarrow 2H_2O(g) \qquad \Delta H° = -484 \text{ kJ}$$

calculate $\Delta H°$ for the reaction

$$2N_2(g) + 6H_2O(g) \rightarrow 3O_2(g) + 4NH_3(g)$$

On the basis of the enthalpy change, is this a useful reaction for the synthesis of ammonia?

Standard Enthalpies of Formation

38. Given the definition of the standard enthalpy of formation for a substance, write separate reactions for the formation of NaCl, H_2O, $C_6H_{12}O_6$, and $PbSO_4$ that have $\Delta H°$ values equal to $\Delta H_f°$ for each compound.

39. Use the values of $\Delta H_f°$ in Appendix 4 to calculate $\Delta H°$ for the following reactions.
   a. $2NH_3(g) + 3O_2(g) + 2CH_4(g) \rightarrow 2HCN(g) + 6H_2O(g)$
   b. $Ca_3(PO_4)_2(s) + 3H_2SO_4(l) \rightarrow 3CaSO_4(s) + 2H_3PO_4(l)$
   c. $NH_3(g) + HCl(g) \rightarrow NH_4Cl(s)$
   d. $SiCl_4(l) + 2H_2O(l) \rightarrow SiO_2(s) + 4HCl(aq)$
   e. $MgO(s) + H_2O(l) \rightarrow Mg(OH)_2(s)$

40. The Ostwald process for the commercial production of nitric acid from ammonia and oxygen involves the following steps:

$$4NH_3(g) + 5O_2(g) \longrightarrow 4NO(g) + 6H_2O(g)$$
$$2NO(g) + O_2(g) \longrightarrow 2NO_2(g)$$
$$3NO_2(g) + H_2O(l) \longrightarrow 2HNO_3(aq) + NO(g)$$

   a. Use the values of $\Delta H_f°$ in Appendix 4 to calculate the value of $\Delta H°$ for each of the above reactions.
   b. Write the overall equation for the production of nitric acid by the Ostwald process by combining the above equations. (Water is also a product.) Is the overall reaction exothermic or endothermic?

41. Calculate $\Delta H°$ for each of the following reactions, using the data in Appendix 4:

$$4Na(s) + O_2(g) \longrightarrow 2Na_2O(s)$$
$$2Na(s) + 2H_2O(l) \longrightarrow 2NaOH(aq) + H_2(g)$$
$$2Na(s) + CO_2(g) \longrightarrow Na_2O(s) + CO(g)$$

Use these values to show why a water or carbon dioxide fire extinguisher might not be effective in putting out a sodium fire.

42. The reusable booster rockets of the space shuttle use a mixture of aluminum and ammonium perchlorate as fuel. A possible reaction is

$$3Al(s) + 3NH_4ClO_4(s) \longrightarrow Al_2O_3(s) + AlCl_3(s) + 3NO(g) + 6H_2O(g)$$

Calculate $\Delta H°$ for this reaction.

43. The space shuttle orbiter utilizes the oxidation of methyl hydrazine by dinitrogentetroxide for propulsion. The balanced reaction is

$$5N_2O_4(l) + 4N_2H_3CH_3(l) \longrightarrow 12H_2O(g) + 9N_2(g) + 4CO_2(g)$$

Calculate $\Delta H°$ for this reaction.

44. Does the reaction in Exercise 42 or that in Exercise 43 produce more energy per kilogram of reactant mixture (stoichiometric amounts)?

45. At 298 K, the standard enthalpies of formation for $C_2H_2(g)$ and $C_6H_6(l)$ are 227 kJ/mol and 49 kJ/mol, respectively.
   a. Calculate $\Delta H°$ for

$$C_6H_6(l) \rightarrow 3C_2H_2(g)$$

   b. Both acetylene (C_2H_2) and benzene (C_6H_6) can be used as fuels. Which compound would liberate more energy per gram when combusted in air?

46. Calculate $\Delta H°$ for each of the following reactions, which occur in the atmosphere.
   a. $C_2H_4(g) + O_3(g) \rightarrow CH_3CHO(g) + O_2(g)$
   b. $O_3(g) + NO(g) \rightarrow NO_2(g) + O_2(g)$
   c. $SO_3(g) + H_2O(l) \rightarrow H_2SO_4(aq)$
   d. $2NO(g) + O_2(g) \rightarrow 2NO_2(g)$

47. Use the reaction

$$2ClF_3(g) + 2NH_3(g) \longrightarrow N_2(g) + 6HF(g) + Cl_2(g) \quad \Delta H° = -1196 \text{ kJ}$$

to calculate $\Delta H_f°$ for $ClF_3(g)$.

48. The enthalpy of combustion of ethene gas, $C_2H_4(g)$, is −1411.1 kJ/mol at 298 K. Given the following enthalpies of formation, calculate $\Delta H_f°$ for $C_2H_4(g)$.

$CO_2(g)$	−393.5 kJ/mol
$H_2O(l)$	−285.9 kJ/mol

Energy Consumption and Sources

49. Assume that the energy in Exercise 24 comes from the combustion of methane (CH_4). What volume of methane, measured at STP, must be burned to produce this amount of energy? [$\Delta H°_{combustion}$ (for CH_4) = −891 kJ/mol CH_4.]

50. Ethanol (C_2H_5OH) has been proposed as an alternative fuel. Calculate the enthalpy of combustion per gram of liquid ethanol.

51. Syngas can be burned directly or converted to methanol. Calculate $\Delta H°$ for the conversion reaction:

$$CO(g) + 2H_2(g) \longrightarrow CH_3OH(l)$$

52. Photosynthetic plants use the following reaction to produce glucose and cellulose:

$$6CO_2(g) + 6H_2O(l) \xrightarrow{\text{Sunlight}} C_6H_{12}O_6(s) + 6O_2(g)$$

How might extensive destruction of forests exacerbate the greenhouse effect?

53. Some automobiles and buses have been equipped to burn propane (C_3H_8) as a fuel. Compare the amount of energy that can be obtained per gram of $C_3H_8(g)$ with that per gram of gasoline, assuming that gasoline is octane, $C_8H_{18}(l)$. (See Example 9.8.) Look up the physical properties of propane. What disadvantages are there to using propane instead of gasoline as a fuel?

54. What safety hazards might be associated with storing hydrogen as a metal hydride?

55. The sun supplies energy at a rate of about 1.0 kilowatt per square meter of surface area (1 watt = 1 J/s). The plants in an agricultural field produce the equivalent of 20. kg of sucrose ($C_{12}H_{22}O_{11}$) per hour per hectare (1 ha = 10,000 m^2). Assuming that sucrose is produced by the reaction

$$12CO_2(g) + 11H_2O(l) \rightarrow C_{12}H_{22}O_{11}(s) + 12O_2(g) \qquad \Delta H = 5640 \text{ kJ}$$

calculate the percentage of sunlight used to produce the sucrose, that is, determine the efficiency of photosynthesis.

56. The best solar panels currently available are about 13% efficient in converting sunlight to electricity. A typical home will use about 40. kWh of electricity a day (1 kWh = 1 kilowatt hour; 1 kW = 1000 J/s). Assuming 8.0 hours of useful sunlight per day, calculate the minimum solar panel surface area necessary to provide all of a typical home's electricity. (See Exercise 55 for the energy rate supplied by the sun.)

Additional Exercises

57. Consider a gas at an initial pressure of 5.0 atm and an initial volume of 1.0 L that undergoes a change to 2.0 atm and 4.0 L by two different pathways.

Pathway one:

$$\begin{matrix} P_i = 5.0 \text{ atm} \\ V_i = 1.0 \text{ L} \end{matrix} \longrightarrow \begin{matrix} P_i = 5.0 \text{ atm} \\ V_f = 4.0 \text{ L} \end{matrix} \longrightarrow \begin{matrix} P_f = 2.0 \text{ atm} \\ V_f = 4.0 \text{ L} \end{matrix}$$

Pathway two:

$$\begin{matrix} P_i = 5.0 \text{ atm} \\ V_i = 1.0 \text{ L} \end{matrix} \longrightarrow \begin{matrix} P_f = 2.0 \text{ atm} \\ V_i = 1.0 \text{ L} \end{matrix} \longrightarrow \begin{matrix} P_f = 2.0 \text{ atm} \\ V_f = 4.0 \text{ L} \end{matrix}$$

These pathways are summarized on the following graph of P versus V.

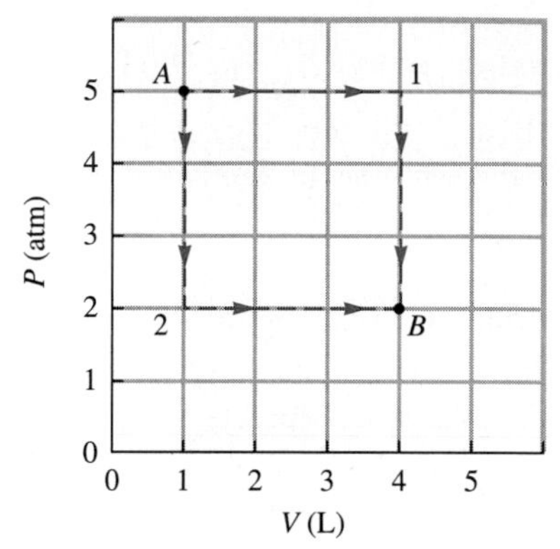

Calculate the work associated with the two pathways. Is work a state function? Explain.

58. A lead bullet with a mass of 7.8 g and a velocity of 5.0 × 10^4 cm/s strikes a 1.00-kg wooden block and becomes embedded in the wood. If both the bullet and the block are initially at a temperature of 25.0°C, what is the final temperature? Assume that there is no heat loss to the surroundings and that all of the kinetic energy of the bullet is converted to heat. Specific heat capacities: wood, 2.1 J $°C^{-1}$ g^{-1}; lead, 0.13 J $°C^{-1}$ g^{-1}.

59. Consider the reaction

$$2HCl(aq) + Ba(OH)_2(aq) \longrightarrow BaCl_2(aq) + 2H_2O(l) \qquad \Delta H = -118 \text{ kJ}$$

How much heat is evolved when 100.0 mL of 0.500 M HCl is reacted with 300.0 mL of 0.500 M $Ba(OH)_2$?

60. A piece of chocolate cake contains about 400 Calories. A nutritional Calorie is equal to 1000 calories (thermochemical calories). How many 8-in-high steps must a 180-lb man climb to expend the 400 Cal from the piece of cake? See Exercise 1 for the formula for potential energy.

61. Hess's law is really just another statement of the first law of thermodynamics. Explain.

62. Hydrazine (N_2H_4) is used as a fuel in liquid-fueled rockets. Either oxygen (assume $O_2(g)$ for this problem) or dinitrogentetroxide [$N_2O_4(l)$] can be used as the oxidizing

agent. In both cases the products are nitrogen gas and gaseous water. Write balanced equations for the two reactions and calculate $\Delta H°$ using standard enthalpies of formation. Compare these values with the result in Exercise 44. Which of the three combinations (per kilogram of the stoichiometric reactant mixture) is the most efficient rocket fuel?

63. Consider the following reaction at 248°C and 1.00 atm:

$$CH_3Cl(g) + H_2(g) \rightarrow CH_4(g) + HCl(g)$$

for which the enthalpy change at 248°C is −83.3 kJ/mol. At constant pressure the molar heat capacities (C_p) for the compounds are: CH_3Cl (48.5 J K^{-1} mol^{-1}), H_2 (28.9 J K^{-1} mol^{-1}), CH_4 (41.3 J K^{-1} mol^{-1}), and HCl (29.1 J K^{-1} mol^{-1}).

a. Assuming the C_p values are independent of temperature, calculate $\Delta H°$ for this reaction at 25°C.
b. Calculate $\Delta H_f°$ for CH_3Cl using data from Appendix 4 and the result from part a.

64. Combustion of table sugar produces $CO_2(g)$ and $H_2O(l)$. When 1.46 g of table sugar is combusted in a constant volume (bomb) calorimeter, 24.00 kJ of heat is liberated.

a. Assuming that table sugar is pure sucrose, $C_{12}H_{22}O_{11}(s)$, write the balanced equation for the combustion reaction.
b. Calculate $\Delta E°$ for the combustion reaction of sucrose.
c. Calculate $\Delta H°$ for the combustion reaction of sucrose.

65. Consider the following changes:

a. $H_2O(g) \rightarrow H_2O(l)$
b. $H_2(g) + Cl_2(g) \rightarrow 2HCl(g)$
c. $2H_2(g) + O_2(g) \rightarrow 2H_2O(g)$
d. $Xe(g) + F_2(g) \rightarrow XeF_2(s)$
e. $NiCl_2 \cdot 6H_2O(s) \rightarrow NiCl_2(s) + 6H_2O(g)$
f. $CO_2(s) \rightarrow CO_2(g)$

At constant pressure, in which of these changes is work done by the system on the surroundings? By the surroundings on the system? In which is no work done?

66. Assume all of the heat in Example 9.1 comes from the combustion of propane (C_3H_8). What mass of propane must be burned to furnish this amount of energy?

67. What is a state function? Enthalpy and internal energy are state functions as a direct consequence of the first law of thermodynamics. Why is this so?

68. Write reactions that correspond to the following enthalpy changes:

a. $\Delta H_f°$ for solid aluminum oxide
b. the standard enthalpy of combustion of liquid ethanol, $C_2H_5OH(l)$
c. the standard enthalpy of neutralization of barium hydroxide solution by hydrochloric acid
d. $\Delta H_f°$ for gaseous vinyl chloride, $C_2H_3Cl(g)$
e. the enthalpy of combustion of liquid benzene, $C_6H_6(l)$
f. the enthalpy of solution of solid ammonium bromide

69. The bomb of a certain bomb calorimeter has a volume of 50.0 mL. It is charged with oxygen to a pressure of 125 atm at 22°C. What is the maximum mass of benzoic acid, $C_7H_6O_2$, that can be burned in this calorimeter if you want to have 25 times as much oxygen in the bomb as will be needed?

70. High-quality audio amplifiers generate large amounts of heat. To dissipate the heat and prevent damage to the electronic devices, manufacturers use heat-radiating metal fins. Would it be better to make these fins out of iron or aluminum? Why? (See Table 9.3 for specific heat capacities.)

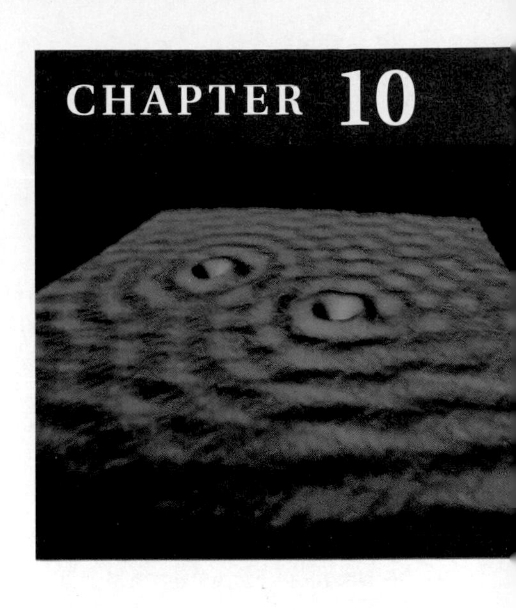

Spontaneity, Entropy, and Free Energy

The *first law of thermodynamics* is a statement of the law of conservation of energy: energy can be neither created nor destroyed. In other words, *the energy of the universe is constant*. Although the total energy remains constant, the various forms of energy can be interchanged through physical and chemical processes. For example, if you drop a book, some of the initial potential energy of the book is changed to kinetic energy, which is transferred to the atoms in the air and the floor as random motion. The net effect of this process is to change a given quantity of potential energy to exactly the same quantity of thermal energy. Energy has been converted from one form to another, but the same quantity of energy exists before and after the process.

Now we will consider a chemical example. When methane is burned in excess oxygen, the major reaction is

$$CH_4(g) + 2O_2(g) \longrightarrow CO_2(g) + 2H_2O(g) + \text{energy}$$

This reaction produces a quantity of energy which is released as heat. This energy flow results from a lowering of the potential energy stored in the bonds of CH_4 and O_2 as they react to form CO_2 and H_2O (see Fig. 10.1). Potential energy is converted to thermal energy, but the energy content of the universe remains constant.

The first law of thermodynamics is used mainly for energy bookkeeping, that is, to answer questions such as the following:

How much energy is involved in the change?
Does energy flow into or out of the system?
What form does the energy finally assume?

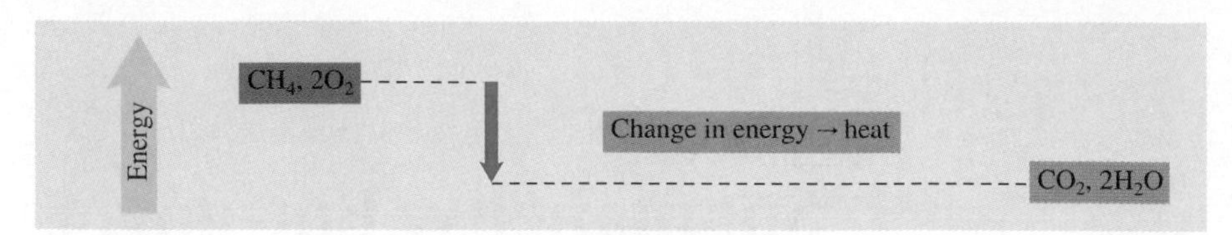

Figure 10.1
When methane and oxygen react to form carbon dioxide and water, the products have lower potential energy than the reactants. This change in potential energy results in energy flow (heat) to the surroundings.

The first law of thermodynamics: the energy of the universe is constant.

Although the first law of thermodynamics provides the means to account for energy changes, it gives no hint as to *why* a particular process occurs in a given direction. This is the main question to be considered in this chapter.

10.1 Spontaneous Processes and Entropy

Spontaneous does not mean fast.

A process is said to be *spontaneous* if it *occurs without outside intervention.* Spontaneous processes may be fast or slow. As we will see in this chapter, thermodynamics can tell us the *direction* in which a process will occur but can say nothing about the *speed* (rate) of the process. As we will explore in detail in Chapter 15, the rate of a reaction depends on many factors, including temperature and concentration. In describing a chemical reaction, the discipline of chemical kinetics (the study of reaction rates) focuses on the pathway between reactants and products; in contrast, thermodynamics considers only the initial and final states and does not require knowledge of the pathway between the reactants and products (see Fig. 10.2).

In summary, thermodynamics lets us predict whether a process will occur but gives no information about the amount of time required. For example, according to the principles of thermodynamics, a diamond should change spontaneously to graphite. The fact that we do not observe this process does not mean the prediction is wrong; it simply means the process is too slow to observe. Thus we need both thermodynamics and kinetics to describe reactions fully.

To explore the idea of spontaneity, consider the following physical and chemical processes:

A ball rolls down a hill but never spontaneously rolls back up the hill.

If exposed to air and moisture, steel rusts spontaneously. However, the iron oxide in rust does not spontaneously change back to iron metal and oxygen gas.

A gas fills its container uniformly. It never spontaneously collects at one end of the container.

Heat flow always occurs from a hot object to a cooler one. The reverse process never occurs spontaneously.

Wood burns spontaneously in an exothermic reaction to form carbon dioxide and water, but wood is not formed when carbon dioxide and water are heated together.

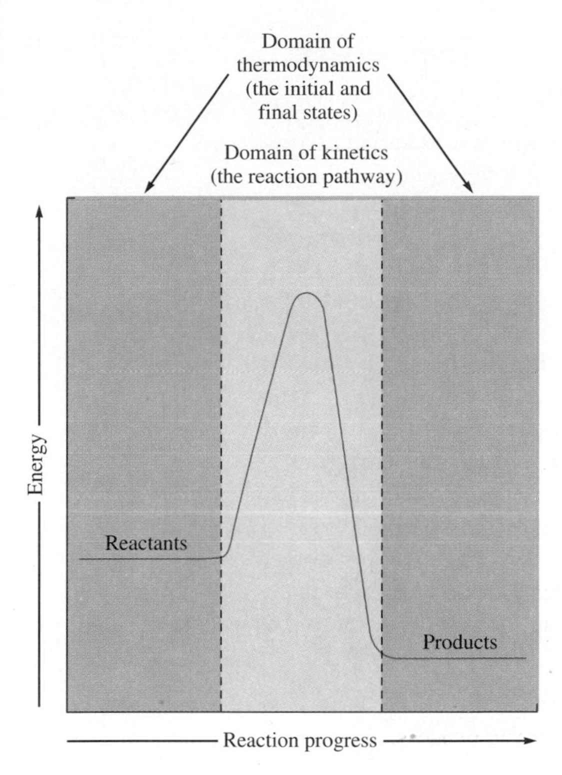

Figure 10.2
The rate of a reaction depends on the pathway from reactants to products; this is the domain of kinetics. Thermodynamics tells us whether or not a reaction is spontaneous based only on the properties of the reactants and products. The predictions of thermodynamics do not require knowledge of the pathway between reactants and products.

At temperatures below 0°C, water spontaneously freezes; and at temperatures above 0°C, ice spontaneously melts.

What thermodynamic principle will provide an explanation why, under a given set of conditions, each of these diverse processes occurs in one direction and never in the reverse? In searching for an answer, we could explain the behavior of a ball on a hill in terms of gravity. But what does gravity have to do with the rusting of a nail or the freezing of water? Early developers of thermodynamics thought that exothermicity might be the key, that a process would be spontaneous if it were exothermic. Although this factor does appear to be important, since many spontaneous processes are exothermic, it is not the only factor. For example, the melting of ice, which occurs spontaneously at temperatures greater than 0°C, is an endothermic process.

What common characteristic causes the processes listed earlier to be spontaneous in one direction only? After many years of observation scientists have concluded that the characteristic common to all spontaneous processes is an increase in a property called **entropy** (S). *The driving force for a spontaneous process is an increase in the entropy of the universe.*

What is entropy? In qualitative terms, *entropy can be viewed as a measure of randomness or disorder*. The natural progression of things is from order to disorder, from lower entropy to higher entropy. You only have to think about the condition of your room to be convinced of this. Your room naturally tends to get messy (disordered), because an ordered room requires everything to be in its place. There are simply many more ways for things to be out of place than to be in place as the following poem by Heather Ryphemi Stregay, written while she was a student in general chemistry, observes.

CHAOS, KEEP IT COMING!

Can you imagine how life would be
If there were no entropy?
Or, making matters even worse,
The laws of entropy were reversed?
Books would get straighter on their shelves,
And children's rooms would clean themselves!
And every rock or stick or tree
Would form a crystal, perfectly.
There'd be no anarchy or war
For everyone would know the score.
Every thing and every face
Would have its certain time and place.
Replacing every beach would pass
An endless stretch of flawless glass.
The sea would be the brightest blue,
And every day the sky would too.
How beautiful would be our world
If order did command it.
If all were straight and never curled:
Perhaps we should demand it.

You'd think a world sans entropy
Would be a lovely place to be.
I said this recently myself,
As all my books fell off their shelf.
Yet pondering this ordered bliss,
I noticed things that I would miss,
Like rolling waves upon the sea,
Or sugar for my morning tea:
The sugar won't dissolve, it's true,
That anti-entropy holds like glue.
And after that, I saw with grief,
There'd be no fractaled maple leaf:
No beauty in the summer wood,
Should chaos disappear for good.
What a bore, to know each day
Would turn out in the same old way.
If entropy would disappear
There'd be no fortune, fate or luck
And even after many years,
Vegas wouldn't make a buck.

Heather Ryphemi Stregay

As another example, suppose you have a deck of playing cards ordered in some particular way. You throw these cards into the air and pick them all up at random. Looking at the new sequence of the cards, you would be very surprised to find that it matched the original order. Such an event would be possible but *very improbable*. There are billions of ways for the deck to be disordered but only one way to be ordered according to your definition. Thus the chances of picking the cards up out of order are much greater than of picking them up in order. It is natural for disorder to increase.

Entropy is a thermodynamic function that describes the *number of arrangements* (positions and/or energy levels) that are *available to a system* existing in a given state. Entropy is closely associated with probability. The key concept is that the more ways a particular state can be achieved, the greater is the likelihood (probability) that that state will occur. In other words, *nature spontaneously proceeds toward the states that have the highest probabilities of existing*. This conclusion is not surprising at all. The difficulty comes in connecting this concept to real-life processes. For example, what does the spontaneous rusting of steel have to do with probability?

Understanding the connection between entropy and spontaneity will allow us to answer such questions. We will begin to explore this connection by considering a very simple process, the expansion of an ideal gas into a vacuum, as represented in Fig. 10.3. Why is this process spontaneous? What causes the gas to expand to a uniform state? The driving force is probability. Because there are more ways of having the gas evenly spread throughout the container than

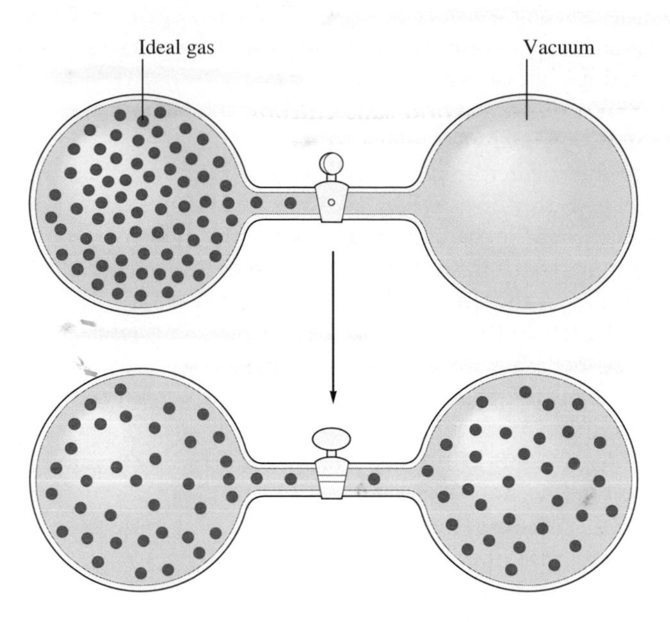

Figure 10.3
The expansion of an ideal gas into an evacuated bulb.

there are ways for it to be in any other possible state, the gas spontaneously attains the uniform distribution.

To understand this conclusion, we will greatly simplify the system and consider some of the possible arrangements of only four gas molecules in a two-bulbed container (Fig. 10.4). How many ways can each arrangement (state) be achieved? Arrangement I can be achieved in only one way—all the molecules must be in one end. Arrangement II can be achieved in four ways, as shown in Table 10.1. Each configuration that gives a particular arrangement is called a *microstate*. Arrangement I has one microstate, and arrangement II has four microstates. Arrangement III can be achieved in six ways (six microstates), as shown in Table 10.1. *Which arrangement is most likely to occur?* The one

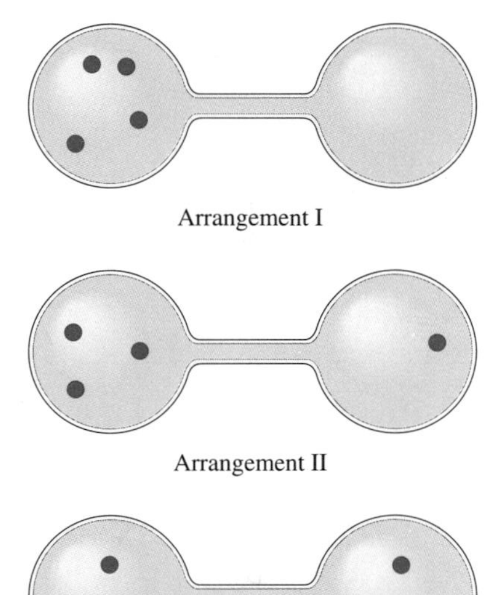

Figure 10.4
Three possible arrangements (states) of four molecules in a two-bulbed flask.

TABLE 10.1 The Microstates That Give a Particular Arrangement (State)

Arrangement	Microstates	
I		A B C D | —
II	B D C | A	A B D | C
	A C D | B	A B C | D
III	A B | C D	B C | A D
	A C | B D	B D | A C
	A D | B C	C D | A B

that can be achieved in the greatest number of ways. Thus arrangement III is most probable, and the relative probabilities of arrangements III, II, and I are 6:4:1. We have discovered an important principle: *the probability of occurrence of a particular arrangement (state) depends on the number of ways (microstates) in which that arrangement can be achieved.*

The consequences of this principle are dramatic for large numbers of molecules. One gas molecule in the flask in Fig. 10.4 has one chance in two of being in the left bulb. We say that the probability of finding the molecule in the left bulb is $\frac{1}{2}$. For two molecules in the flask there is one chance in two of finding each molecule in the left bulb, so there is one chance in four ($\frac{1}{2} \times \frac{1}{2} = \frac{1}{4}$) that *both* molecules will be in the left bulb. As the number of molecules increases, the relative probability of finding all of them in the left bulb decreases, as shown in Table 10.2. For 1 mole of gas the probability of finding all the molecules in the left bulb is so small that this arrangement would "never" occur.

For two molecules in the flask, there are four possible microstates:

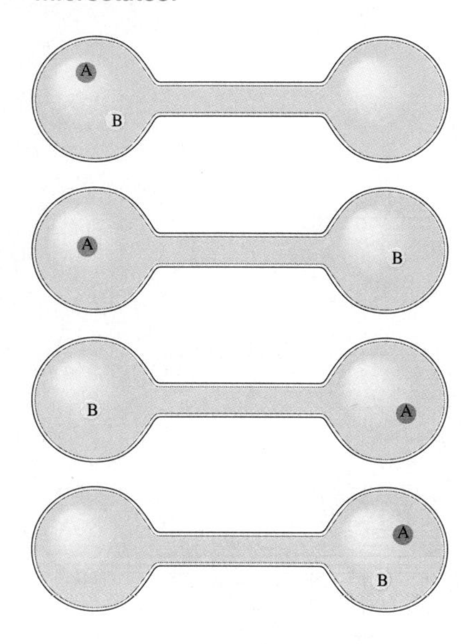

Thus a gas placed in one end of a container will spontaneously expand to fill the entire vessel evenly, because for a large number of gas molecules there is a huge number of microstates corresponding to equal numbers of molecules in both ends. On the other hand, the opposite process,

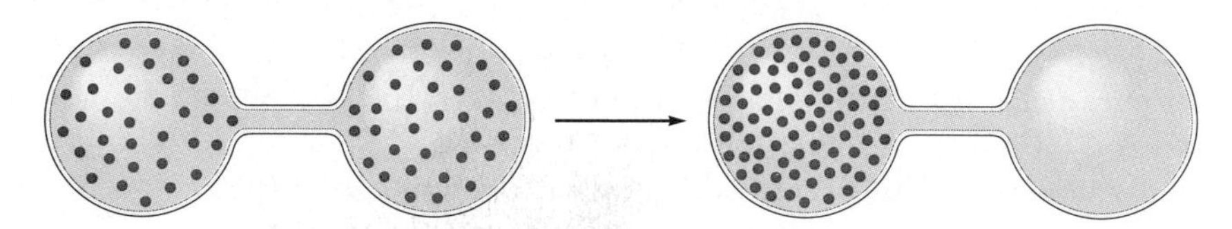

although not impossible, is *highly* improbable since only one microstate leads to this arrangement. Therefore, this process does not occur spontaneously.

Thus there is one chance in four of finding

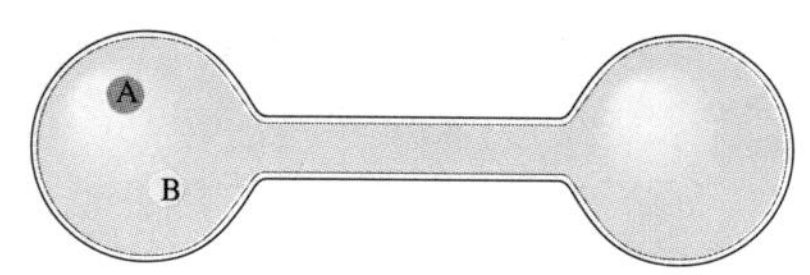

TABLE 10.2 Probability of Finding All the Molecules in the Left Bulb as a Function of the Total Number of Molecules

Number of Molecules	Relative Probability of Finding All Molecules in the Left Bulb
1	$\frac{1}{2}$
2	$\frac{1}{2} \times \frac{1}{2} = \frac{1}{2^2} = \frac{1}{4}$
3	$\frac{1}{2} \times \frac{1}{2} \times \frac{1}{2} = \frac{1}{2^3} = \frac{1}{8}$
5	$\frac{1}{2} \times \frac{1}{2} \times \frac{1}{2} \times \frac{1}{2} \times \frac{1}{2} = \frac{1}{2^5} = \frac{1}{32}$
10	$\frac{1}{2^{10}} = \frac{1}{1024}$
n	$\frac{1}{2^n} = \left(\frac{1}{2}\right)^n$
6×10^{23} (1 mole)	$\left(\frac{1}{2}\right)^{6 \times 10^{23}} = 10^{-(2 \times 10^{23})}$

The type of probability we have been considering in this example is called **positional probability** because it depends on the number of configurations in space (positional microstates) that yield a particular state. A gas expands into a vacuum to give a uniform distribution because the expanded state has the highest positional probability—that is, the largest entropy—of the states available to the system.

Positional probability is also illustrated by changes of state. In general, positional entropy increases in going from solid to liquid to gas. A mole of a substance has a much smaller volume in the solid state than in the gaseous state. In the solid state the molecules are close together, with relatively few positions available to them; in the gaseous state the molecules are far apart, with many more positions available to them. The liquid state is somewhat closer to the solid state than to the gaseous state. We can summarize these comparisons as follows:

$$S_{solid} < S_{liquid} \ll S_{gas}$$

Positional entropy is also important in the formation of solutions. The entropy change associated with the mixing of two pure substances is expected to be positive. An increase in entropy is expected because there are many more microstates for the mixed condition than for the separated condition owing to the increased volume available to the particles of each component of the mixture. For example, when two liquids are mixed, the molecules of each liquid have more available space and thus more available positions. This will be discussed in detail in Chapter 17.

EXAMPLE 10.1

For each of the following pairs, choose the substance with the higher positional entropy (per mole) at a given temperature.

a. Solid CO_2 and gaseous CO_2

b. N_2 gas at 1 atm and N_2 gas at 1.0×10^{-2} atm

Solution

a. Since a mole of gaseous CO_2 has the greater volume, the molecules have many more available positions than in a mole of solid CO_2. Thus gaseous CO_2 has the higher positional entropy.

b. A mole of N_2 gas at 1×10^{-2} atm has a volume 100 times that (at a given temperature) of a mole of N_2 gas at 1 atm. Thus N_2 gas at 1×10^{-2} atm has the higher positional entropy.

EXAMPLE 10.2

Predict the sign of the entropy change for each of the following processes.

a. Solid sugar is added to water to form a solution.

b. Iodine vapor condenses on a cold surface to form crystals.

Iodine subliming and forming crystals on the sides of the flask.

Solution

a. The sugar molecules become randomly dispersed in the water when the solution forms. The sugar molecules have access to a larger volume and thus have more positions available to them. Thus the positional disorder increases. There is an increase in entropy; ΔS is positive.

b. Gaseous iodine is forming a solid. This process involves a change from a relatively large volume to a much smaller volume, which results in lower positional disorder. For this process ΔS is negative (the entropy decreases).

10.2 The Isothermal Expansion and Compression of an Ideal Gas

In this section we will lay the groundwork for several fundamental concepts of thermodynamics by considering the isothermal expansion and compression of an ideal gas. An **isothermal process** is one in which the temperatures of the system and the surroundings remain constant at all times. Recall that the energy of an ideal gas can be changed only by changing its temperature. Therefore, for any isothermal process *involving an ideal gas,*

$$\Delta E = 0$$

and since $$\Delta E = q + w = 0$$

then $$q = -w$$

To illustrate the work and heat effects that accompany the expansion/compression of an ideal gas, consider the apparatus shown in Fig. 10.5. Assume that the pulley is frictionless and that the cable and pan have zero mass.

Initially, assume that the gas occupies a volume V_1 at pressure P_1, where P_1 is just balanced by a mass M_1 on the pan. Thus

$$P_1 = \frac{\text{force}}{\text{area}} = \frac{M_1 g}{A}$$

where A is the area of the piston and g is the gravitational constant. Thus state 1 of the gas is defined by P_1, V_1, n, and T, where n and T remain constant as the expansion occurs.

One-Step Expansion—No Work

If mass M_1 is removed from the pan, the gas will expand, moving the piston to the right end of the cylinder. After expansion the gas occupies a volume $V_2 = 4V_1$ and pressure $P_2 = P_1/4$.

When the process goes from state 1 (P_1, V_1) to state 2 ($P_1/4$, $4V_1$) with no mass on the pan, no heat flows into or out of the gas because T is constant and no work is done (no mass is lifted). Thus work $= w_0 = 0$. This is called a *free expansion*.

P_{external} (the pressure against which the gas expands) is zero in a free expansion.

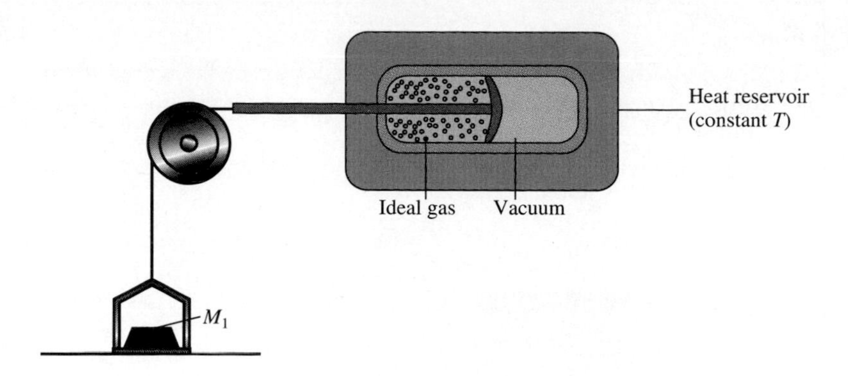

Figure 10.5
A device for the isothermal expansion/compression of an ideal gas.

One-Step Expansion

Now consider an experiment with the gas initially at state 1 where the mass M_1 is replaced by a mass $M_1/4$. The gas will now expand against the pressure (P_{external}):

$$P_{\text{ex}} = \frac{\left(\frac{M_1}{4}\right)g}{A} = \frac{P_1}{4}$$

The mass is lifted and the gas expands until the pressure is $P_1/4$. The new volume is $4V_1$. In this case work is performed:

$$|\text{Work}| = |w_1| = \left(\frac{M_1}{4}\right)gh$$

where h is the *change* in height of the mass.

This work can also be expressed in terms of the external pressure (P_{ex}) on the gas as it expands and the change in volume (ΔV). Recall from Chapter 9 that

$$w = -P_{\text{ex}}\Delta V$$

In this expansion the magnitude (absolute value) of the work is

$$|w_1| = P_{\text{ex}}\Delta V = \frac{P_1}{4}(V_2 - V_1) = \frac{P_1}{4}(4V_1 - V_1) = \frac{3}{4}P_1V_1$$

Since in this case work flows out of the system into the surroundings, the correct sign is

$$w_1 = -P_{\text{ex}}\Delta V = -\tfrac{3}{4}P_1V_1$$

Two-Step Expansion

Next, we will expand the gas in two steps by using two different weights on the pan. First, we put a weight with mass $M_1/2$ on the pan. In this case

$$P_{\text{ex}} = \frac{\left(\frac{M_1}{2}\right)g}{A} = \frac{P_1}{2}$$

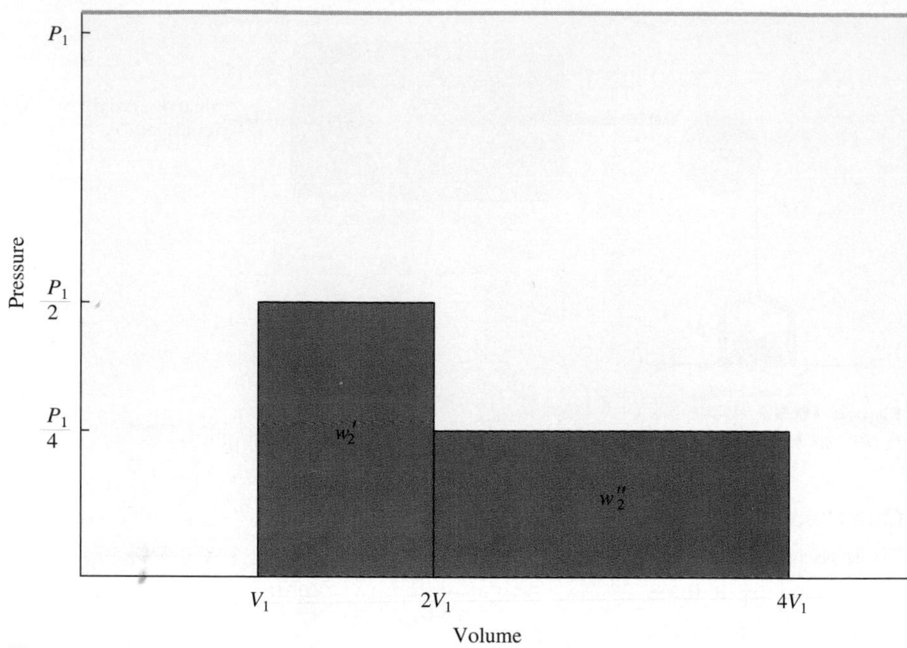

Figure 10.6
The *PV* diagram for a two-step expansion.

and the gas expands until $P_2 = P_1/2$ and $V_2 = 2V_1$. The magnitude of the work is

$$|w_2'| = \frac{P_1}{2}(V_2 - V_1) = \frac{P_1}{2}(2V_1 - V_1) = \frac{P_1V_1}{2}$$

Next, replace the mass $M_1/2$ with a mass $M_1/4$. The gas expands again until $P_3 = P_1/4$ and $V_3 = 4V_1$. The quantity of work in this step is

$$|w_2''| = \frac{P_1}{4}(4V_1 - 2V_1) = \frac{P_1V_1}{2}$$

The total quantity of work in this two-step expansion is

$$|w_2| = \frac{P_1V_1}{2} + \frac{P_1V_1}{2} = P_1V_1$$

With its correct sign $w_2 = -P_1V_1$. This process is diagramed in Fig. 10.6. Note that $|w_2| > |w_1| > |w_0|$, even though in each case the gas is taken from state 1 (P_1, V_1) to state 2 $(P_1/4, 4V_1)$. This result illustrates a property of work we have discussed before: work is pathway-dependent—it is not a state function.

Six-Step Expansion

Next, we will consider the expansion of the gas from state 1 to state 2 in six steps, using several masses between M_1 and $M_1/4$. This process is summarized in Fig. 10.7. In this case $|w_6|$, which is the sum of these six steps, is clearly greater than $|w_2|$.

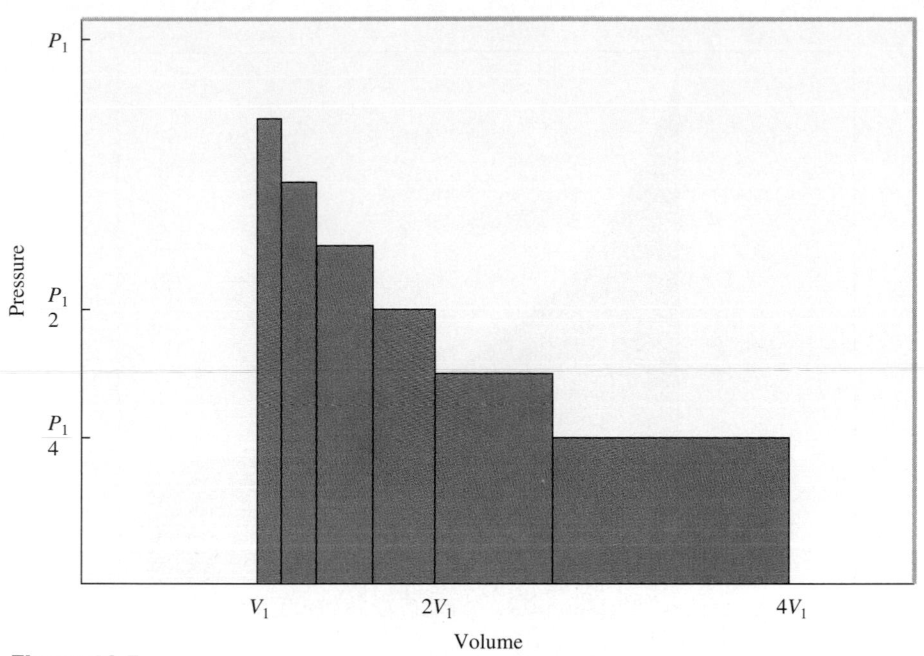

Figure 10.7
The *PV* diagram for a six-step expansion.

Infinite-Step Expansion

If one continues to increase the number of steps, the magnitude of w_n (for an n-step process),

$$|w_n| = \sum_{i=1}^{n} P_i \Delta V_i$$

continues to increase.

Now we consider the limiting case—a process in which P_{ex} is changed by infinitesimally small increments. This case corresponds to the use of an infinite number of weights, each differing from the previous one by an infinitesimally small mass. Under these conditions the successive volume changes become infinitesimally small (dV), and the process requires an infinite number of steps. The mathematical operation needed to sum the steps in this case is the integral

$$|\text{Work}| = \int_{V_1}^{V_2} P_{ex} dV$$

The diagram corresponding to this process is given in Fig. 10.8.

It is important to recognize that when the expansion of the gas is carried out in an infinite number of steps, the *external pressure is always almost exactly equal to the pressure produced by the gas*. That is, at any given time P_{ex} is only less than P_{gas} ($= P$) by an infinitesimally small amount dP. Thus one can assume that at any point in the process $P \approx P_{ex}$.

A process like this one, carried out so that the system is always at equilibrium, is called a *reversible process*. The expansions described earlier that were carried out in a finite number of steps were not done reversibly, because in these processes P_{ex} was smaller than P by a finite amount during each step. We will have more to say presently about what the term *reversible* means in this context.

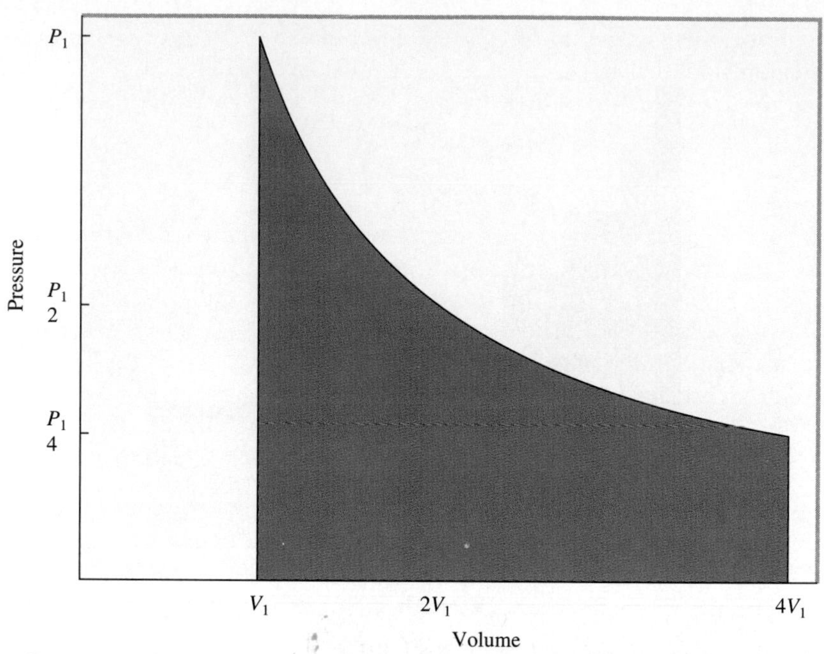

Figure 10.8
The *PV* diagram for the reversible expansion.

Since $P_{ex} \approx P_{gas} = P$ in the reversible expansion, by use of the ideal gas law,

$$P_{ex} \approx P = \frac{nRT}{V}$$

and

$$|\text{Total work}|^* = |w_\infty| = |w_{rev}| = \int_{V_1}^{V_2} \frac{nRT\,dV}{V}$$

Since n and T are held constant in this experiment,

$$|w_{rev}| = nRT \int_{V_1}^{V_2} \frac{dV}{V} = nRT(\ln V_2 - \ln V_1) = nRT \ln\left(\frac{V_2}{V_1}\right)$$

In this specific experiment $V_2 = 4V_1$. Therefore,

$$|w_{rev}| = nRT \ln 4 = 1.4nRT$$

And since $P_1V_1 = nRT$,

$$|w_{rev}| = 1.4P_1V_1$$

for this particular expansion.

Note that as the number of steps in the expansion increases, the amount of work the gas performs also increases. The maximum work that a given amount

---

* In these calculations we deal with the magnitude, for convenience. Actually, for the expansion of a gas the work has a negative sign.

of gas can perform in going from V_1 to V_2 at constant temperature occurs in the reversible expansion. Thus

$$|w_{max}| = |w_{rev}| = nRT \ln\left(\frac{V_2}{V_1}\right)$$

for the isothermal expansion of n moles of an ideal gas.

In doing this hypothetical experiment, we have assumed that the gas behaves ideally. When a given amount of an ideal gas expands at constant temperature, the internal energy of the gas remains constant, that is, $\Delta E = 0$. As mentioned above, this condition means that $q = -w$. Thus a quantity of energy q (equal in magnitude to w) flows into the gas (as heat) as the expansion occurs, and the work (w) is performed. Therefore, the surroundings furnished the energy (through heat flow) necessary to perform the work.

Since work flows out of the system in the reversible expansion, it has a negative sign:

$$w_{rev} = -nRT \ln\left(\frac{V_2}{V_1}\right)$$

and

$$q_{rev} = -w_{rev} = nRT \ln\left(\frac{V_2}{V_1}\right)$$

The Isothermal Compression of an Ideal Gas

Now let's consider the opposite experiment: compressing a gas at pressure $P_1/4$ and volume $4V_1$ to pressure P_1 and volume V_1. This experiment can be done in one step or a number of steps.

One-Step Compression

Initially, the gas is at pressure $P_1/4$ and volume $4V_1$. When mass M_1 is placed on the pan, the gas will be rapidly compressed to the state described by P_1 and V_1. To see how this process works, consider the diagram for the gas at $P_1/4$ and $4V_1$ (Fig. 10.9). We raise the mass M_1 to the pan, which then causes the gas to be compressed to P_1 and V_1 as the pan returns to the original level.

The work performed to recompress the gas is equal to the work performed to lift the weight up to the pan. It also can be expressed in terms of $P_{ex}\Delta V$:

$$|w_1'| = M_1gh = P_1\Delta V = P_1(4V_1 - V_1) = 3P_1V_1$$

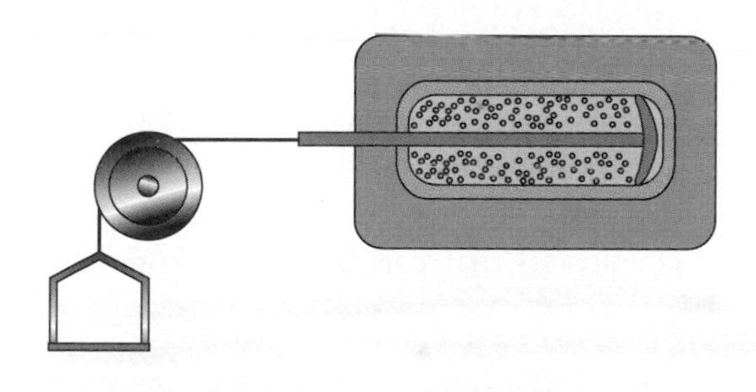

Figure 10.9
The situation before the one-step compression.

Two-Step Compression

We can compress the gas in two steps. For example, a mass $M_1/2$ is lifted onto the pan, and after the gas has been partially compressed, M_1 is put onto the pan in place of $M_1/2$ to finish the compression. In this case the total work required for the compression is

$$|w_2'| = \frac{P_1}{2}(4V_1 - 2V_1) + P_1(2V_1 - V_1) = 2P_1V_1$$

Infinite-Step Compression

Notice that in compressing the gas isothermally, as the number of steps increases, the work required to compress the gas decreases.

If we compress the gas in an infinite number of steps (in which, at all times, $P_{ex} \approx P$), the work required is

$$|w_\infty'| = \int_{4V_1}^{V_2} PdV = nRT \ln\left(\frac{V_2}{V_1}\right)$$

$$= nRT \ln\left(\frac{V_1}{4V_1}\right) = 1.4P_1V_1$$

Because $P_{ex} \approx P$ throughout the process, this is a *reversible* compression. In this case w has a positive sign, because we are performing work on the system. Thus in the reversible, isothermal compression of the gas, $w_\infty' = 1.4P_1V_1$; and since $\Delta E = 0$,

$$q_\infty = -w_\infty' = -1.4P_1V_1$$

As the gas is compressed reversibly and isothermally, it releases $1.4P_1V_1$ (L atm) of energy as heat to the surroundings. In other words, the same quantity of energy flows into the gas as work and flows out of the gas as heat to produce no net change in E as the compression occurs.

In general terms, for a reversible compression from V_1 to V_2,

$$w_{rev} = -q_{rev} = -nRT \ln\left(\frac{V_2}{V_1}\right)$$

or in terms of pressures,

$$w_{rev} = -q_{rev} = -nRT \ln\left(\frac{P_1}{P_2}\right)$$

Note that since this process is a compression ($V_2 < V_1$), $\ln(V_2/V_1)$ has a negative sign, making w_{rev} positive and q_{rev} negative, as expected.

Summary

We summarize the results of these expansion and compression experiments in Table 10.3. The most important conclusion that can be drawn from these results can be stated as follows: only when the expansion and compression *are both done reversibly* (by an infinite number of steps) is the universe the same after the cyclic process (the expansion and the subsequent compression of the gas back to its original state). That is, only for the reversible processes is the heat absorbed during expansion exactly equal to the heat released during compression. In all of the processes carried out using a finite number of steps, more

TABLE 10.3 **Summary of the Isothermal Expansion and Compression Experiments**

	Number of Steps	w	q
Expansion (constant T)	0 (no mass)	0	0
	1	$-0.75P_1V_1$	$0.75P_1V_1$
	2	$-1P_1V_1$	$1P_1V_1$
	4	$-1.16P_1V_1$	$1.16P_1V_1$
	∞	$-1.4P_1V_1$	$1.4P_1V_1$
Compression (constant T)	1	$3P_1V_1$	$-3P_1V_1$
	2	$2P_1V_1$	$-2P_1V_1$
	4	$1.67P_1V_1$	$-1.67P_1V_1$
	∞	$1.4P_1V_1$	$-1.4P_1V_1$

heat is released into the surroundings than is absorbed in the comparable expansion (same number of steps).

For example, in the one-step expansion and compression the net work done is

$$w_{\text{net}} = \underset{\text{Expansion}}{\underset{\uparrow}{-0.75P_1V_1}} + \underset{\text{Compression}}{\underset{\uparrow}{3P_1V_1}} = 2.25P_1V_1$$

and the net heat flow is $-2.25P_1V_1$.

In terms of the *system* this one-step expansion/compression is cyclic. That is, the system starts at state 1 before the expansion process and ends up back at state 1 after the compression process. However, although the system is returned to the same state in this cyclic expansion-compression process, the *surroundings* is not the same. In fact, we might say that in the surroundings "work has been changed to heat." In this one-step cyclic process, $2.25P_1V_1$ (L atm) of work is changed to $2.25P_1V_1$ (L atm) of heat (thermal energy) in the surroundings. This amount represents the net extra work required to lift the weights onto the pan to compress the gas, compared with the work that was obtained as the gas expanded.

This is a general result. In any finite-step, cyclic expansion-compression process "work is always converted to heat":

$$\underset{\text{Ordered energy}}{\text{work}} \longrightarrow \underset{\text{Disordered energy}}{\text{heat}}$$

This result applies whenever the process is carried out in a nonreversible (irreversible) manner. In other words, in an *irreversible* cyclic process more work must be input to the system than the system produces. In all of the finite gas compressions the work required is greater than $1.4P_1V_1$, which is the maximum work available from the expansion.

Now, of course, all real processes are irreversible, because they cannot be carried out in an infinite number of steps without taking an infinite amount of time. In other words, *all real processes are irreversible* (in a thermodynamic sense).

All real processes are irreversible.

The maximum work corresponds to the reversible expansion.

Another important conclusion to be drawn from the above example is that the *maximum work obtainable from the gas occurs when the expansion is carried out reversibly* ($w_{max} = w_{rev}$). This result is always true for PV work, as well as for any other type of work, such as electrical work performed by an electrochemical cell. We will examine this latter example in the next chapter.

The final point that this experiment reemphasizes is that work and heat are pathway-dependent and thus are not state functions. Energy, on the other hand, is a state function. In each of these isothermal expansions and compressions between (P_1, V_1) and ($P_1/4$, $4V_1$), ΔE is always zero, regardless of the number of steps, since T is constant.

At this point we can precisely define the terms *reversible* and *irreversible* in a thermodynamic sense. In a reversible, cyclic process *both the system and the surroundings are returned exactly to their original conditions*. As it turns out, this process is hypothetical. On the other hand, an irreversible process is one in which, even when the system is cycled (state 1 $\rightarrow$ state 2 $\rightarrow$ state 1) and thus returned to its original state, the surroundings are changed in a permanent way. All real processes are irreversible.

10.3 The Definition of Entropy

So far we have discussed entropy and its relationship to disorder in a very qualitative way. Here we want to give a precise, quantitative definition of entropy.

Ludwig Boltzmann's tomb in Vienna. Notice Boltzmann's equation on the monument.

We have said that entropy is related to probability. If a system has several states available to it, the one that can be achieved in the greatest number of ways (has the largest number of microstates) is the one most likely to occur. That is, the state with the greatest probability has the highest entropy.

Now we will connect entropy and probability quantitatively by defining the entropy function S as follows:

$$S = k_B \ln\Omega$$

where

k_B = Boltzmann's constant, the gas constant per molecule (R/N_A)

Ω = the number of microstates corresponding to a given state (including both position and energy)

This definition of entropy shows its exact relationship to probability. However, it is not useful in a practical sense for the typical types of samples used by chemists, because these samples contain so many components. For example, a mole of gas contains 6.022×10^{23} individual particles. And according to one estimate, describing the positions and velocities of this mole of particles would require a stack of paper 10 *light years* tall—and this description would only apply for an instant. Clearly, we cannot deal directly with this definition of entropy for typical-sized samples. We must find a way to connect entropy to the macroscopic properties of matter. To do so, we will consider an ideal gas that expands isothermally from volume V_1 to volume $2V_1$ (see Fig. 10.10).

Focus on an individual particle in this gas. When the gas goes from volume V_1 to $2V_1$, each particle has double the number of positions available to it. That is, $\Omega_2 = 2\Omega_1$.

Now we will use the definition of entropy to calculate ΔS for this expansion. In this case $S_1 = k_B \ln\Omega_1$ and $S_2 = k_B \ln\Omega_2$, so

$$\Delta S = S_2 - S_1 = k_B \ln\Omega_2 - k_B \ln\Omega_1 = k_B \ln\left(\frac{\Omega_2}{\Omega_1}\right)$$

$$= k_B \ln\left(\frac{2\Omega_1}{\Omega_1}\right) = k_B \ln 2$$

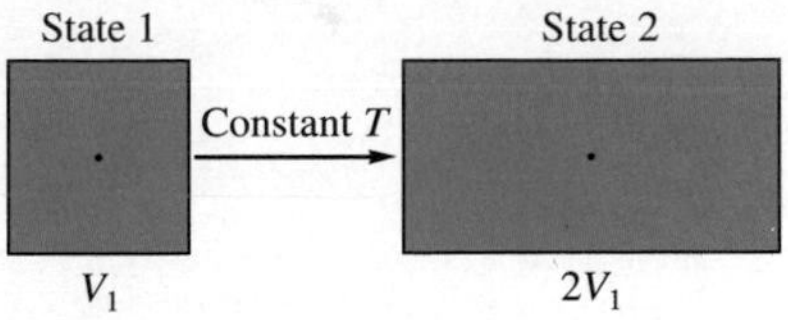

State 1 | Constant T | State 2

V_1 | $2V_1$

Figure 10.10
A particle in a gas that expands from volume V_1 to volume $2V_1$.

This quantity represents ΔS for each particle in the gas. If the gas contains five particles, then the ratio of Ω_2 to Ω_1 is

$$\frac{\Omega_2}{\Omega_1} = 2 \times 2 \times 2 \times 2 \times 2 = 2^5$$

where each 2 represents the fact that each particle has twice as many positions available to it in $2V_1$ as in V_1. This argument can be readily extended to a sample with 1 mole of particles:

$$\frac{\Omega_2}{\Omega_1} = 2^{6 \times 10^{23}}$$

and

$$\Delta S = k_B \ln(2)^{6 \times 10^{23}} = (6 \times 10^{23})(k_B \ln 2)$$
$$= N_A k_B \ln 2 = R \ln 2$$

where R is the gas constant.

Now for a general case in which a gas is expanded from V_1 to V_2, the ratio of Ω_2 to Ω_1 is

$$\frac{\Omega_2}{\Omega_1} = \frac{V_2}{V_1}$$

and for 1 mole of gas

$$\Delta S_{V_1 \rightarrow V_2} = R \ln\left(\frac{V_2}{V_1}\right)$$

For n moles of gas we have

$$\Delta S_{V_1 \rightarrow V_2} = nR \ln\left(\frac{V_2}{V_1}\right)$$

What we have accomplished here is to use the definition of entropy in terms of probability to derive an expression for ΔS that depends on volume, a macroscopic property of the gas. We can now relate the change in entropy to heat flow by noting the striking similarity between the above equation for ΔS and the one derived in Section 10.2 describing q_{rev} for the isothermal expansion-compression of an ideal gas. Compare

$$\Delta S = nR \ln\left(\frac{V_2}{V_1}\right) \quad \text{with} \quad q_{rev} = nRT \ln\left(\frac{V_2}{V_1}\right)$$

Combining these equations gives

$$\Delta S = \frac{q_{rev}}{T}$$

This very important relationship is the macroscopic (thermodynamic) definition of ΔS. In our treatment we started with the definition of entropy based on probability, because that definition better emphasizes the fundamental character of entropy. However, it is also very important to know how entropy changes relate to changes in macroscopic properties, such as volume and heat, because these changes are relatively easy to measure.

10.4 Entropy and Physical Changes

Although chemists deal primarily with the chemical changes of matter, physical changes are also very important. In this section we will consider how the entropy of a substance depends on its temperature and on its physical state.

Temperature Dependence of Entropy

For an isothermal process we have seen that the change in entropy is defined by the relationship

$$\Delta S = \frac{q_{\text{rev}}}{T}$$

We can calculate ΔS for a change in temperature from T_1 to T_2 by summing infinitesimal increments in entropy at each temperature T:

$$dS = \frac{dq_{\text{rev}}}{T}$$

Using integration, we have

$$\Delta S_{T_1 \to T_2} = \int_{T_1}^{T_2} dS = \int_{T_1}^{T_2} \frac{dq_{\text{rev}}}{T}$$

If the process is carried out at constant pressure, then

$$dq_{\text{rev}} = nC_{\text{p}}\, dT$$

for n moles of substance. Thus

$$\Delta S_{T_1 \to T_2} = \int_{T_1}^{T_2} \frac{dq_{\text{rev}}}{T} = nC_{\text{p}} \int_{T_1}^{T_2} \frac{dT}{T}$$

assuming C_{p} is constant between T_1 and T_2. Performing the integration gives

$$\Delta S_{T_1 \to T_2} = nC_{\text{p}} \ln\left(\frac{T_2}{T_1}\right)$$

Similarly, for a process carried out at constant volume

$$\Delta S_{T_1 \to T_2} = nC_{\text{v}} \ln\left(\frac{T_2}{T_1}\right)$$

Entropy Changes Associated with Changes of State

At the normal melting point or boiling point of a substance the two states of matter present at that temperature and at 1 atm pressure are in equilibrium. That is, the two states can coexist indefinitely if the system is isolated (left totally undisturbed). Recall that a reversible process can occur only at equilibrium. Thus since a change of state from solid to liquid at the substance's melting point is a reversible process, we can calculate the change in entropy for this process by using the equation

$$\Delta S = \frac{q_{rev}}{T}$$

where

$$q_{rev} = \Delta H_{fusion} = \begin{array}{l}\text{energy required to melt 1 mol}\\ \text{of solid at the melting point}\end{array}$$

$$T = \text{melting point in K}$$

The same reasoning applies to a change from liquid to gas at the boiling point, except in this case $q_{rev} = \Delta H_{vaporization}$ and T = boiling point.

EXAMPLE 10.3

Calculate the change in entropy that occurs when a sample containing 2.00 mol of water is heated from 50.°C to 150.°C at 1 atm pressure. The molar heat capacities for $H_2O(l)$ and $H_2O(g)$ are 75.3 J K^{-1} mol^{-1} and 36.4 J K^{-1} mol^{-1}, respectively, and the enthalpy of vaporization for water is 40.7 kJ/mol at 100°C.

Solution

Since water changes from liquid to gas at 100°C, we will do this calculation in three (reversible) steps:

1. $\Delta S(l)_{50°C \rightarrow 100°C}$ (heat the liquid)
2. $\Delta S^{100°C}_{(l) \rightarrow (g)}$ (change the liquid to a gas)
3. $\Delta S(g)_{100°C \rightarrow 150°C}$ (heat the gas)

and total the results.

1. The entropy change for heating 2.00 mol of liquid water from 50°C to 100°C is

$$\Delta S_{(1)} = (2.00 \text{ mol})\left(75.3 \frac{\text{J}}{\text{K mol}}\right) \ln\left(\frac{373}{323}\right) = 21.7 \text{ J/K}$$

2. The entropy change for the vaporization of 2.00 mol of water at 100°C can be calculated from the equation

$$\Delta S = \frac{q_{rev}}{T} = \frac{\Delta H_{vap}}{T_{bp}}$$

$$= \frac{4.07 \times 10^4 \text{ J/mol}}{373 \text{ K}} = 1.09 \times 10^2 \text{ J K}^{-1} \text{ mol}^{-1}$$

In this case 2 mol of water is vaporized, so

$$\Delta S = \left(1.09 \times 10^2 \frac{\text{J}}{\text{K mol}}\right)(2 \text{ mol}) = 2.18 \times 10^2 \text{ J/K}$$

3. The entropy change for heating 2.00 mol of gaseous water from 100°C to 150°C is

$$\Delta S_{(3)} = (2.00 \text{ mol})\left(36.4 \frac{\text{J}}{\text{K mol}}\right) \ln\left(\frac{423}{373}\right) = 9.16 \text{ J/K}$$

The total entropy change is

$$\Delta S_{50°\text{C}\rightarrow 150°\text{C}} = \Delta S_{(1)} + \Delta S_{(2)} + \Delta S_{(3)}$$
$$= 21.7 \text{ J/K} + 218 \text{ J/K} + 9.16 \text{ J/K} = 249 \text{ J/K}$$

10.5 Entropy and the Second Law of Thermodynamics

The total energy of the universe is constant, but the entropy is increasing.

We have seen that processes are spontaneous when they result in an increase in disorder. Nature always moves toward the most probable state available to it. We can state this principle in terms of entropy: *in any spontaneous process there is always an increase in the entropy of the universe.* This is the **second law of thermodynamics.** Contrast this law with the first law of thermodynamics, which tells us that the energy of the universe is constant. Energy is conserved in the universe, but entropy is not. In fact, the second law can be paraphrased as follows: *the entropy of the universe is increasing.*

As in Chapter 9, we find it convenient to divide the universe into a system and the surroundings. Thus we can represent the change in the entropy of the universe as

$$\Delta S_{\text{univ}} = \Delta S_{\text{sys}} + \Delta S_{\text{surr}}$$

where ΔS_{sys} and ΔS_{surr} represent the changes in entropy that occur in the system and in the surroundings, respectively.

To predict whether a given process will be spontaneous, we must know the sign of ΔS_{univ}. If ΔS_{univ} is positive, the entropy of the universe increases, and the process is spontaneous in the direction written. If ΔS_{univ} is negative, the process is spontaneous in the *opposite* direction. If ΔS_{univ} is zero, the process has no tendency to occur, indicating that the system is at equilibrium. To predict whether a process is spontaneous, we must consider the entropy changes that occur both in the system and in the surroundings.

EXAMPLE 10.4

In a living cell large molecules are assembled from simple ones. Is this process consistent with the second law of thermodynamics?

Solution

To reconcile the operation of an order-producing cell with the second law of thermodynamics, we must remember that ΔS_{univ}, not ΔS_{sys}, must be positive for a process to be spontaneous. A process for which ΔS_{sys} is negative can be spontaneous if ΔS_{surr} is large and positive. The operation of a cell is such a process.

10.6 The Effect of Temperature on Spontaneity

To explore the interplay of ΔS_{sys} and ΔS_{surr} in determining the sign of ΔS_{univ}, we will first discuss the change of state for 1 mole of water from liquid to gas,

$$H_2O(l) \longrightarrow H_2O(g)$$

considering the water to be the system and everything else the surroundings.

What happens to the entropy of water in this process? A mole of liquid water (18 g) has a volume of approximately 18 mL. A mole of gaseous water at 1 atm and 100°C occupies a volume of approximately 31 L. Clearly, there are many more positions available to the water molecules in a volume of 31 L than in 18 mL; thus the vaporization of water is favored by this increase in positional probability. That is, for this process the entropy of the system increases; ΔS_{sys} has a positive sign.

What about the entropy change in the surroundings? Although we will not prove it here, entropy changes in the surroundings are determined primarily by the flow of energy into or out of the system as heat. To understand this observation, suppose an exothermic process transfers 50 joules of energy as heat to the surroundings, where it becomes thermal energy, that is, kinetic energy associated with the random motions of atoms. Thus this flow of energy into the surroundings increases the random motions of atoms there and so increases the entropy of the surroundings. The sign of ΔS_{surr} is positive. When an endothermic process occurs in the system, it produces the opposite effect. Heat flows from the surroundings to the system, causing the random motions of the atoms in the surroundings to decrease, thus decreasing the entropy of the surroundings. The vaporization of water is an endothermic process. Thus for this change of state, ΔS_{surr} is negative.

In an endothermic process heat flows from the surroundings into the system. In an exothermic process heat flows into the surroundings.

Remember it is the sign of ΔS_{univ} that tells us whether or not the vaporization of water is spontaneous. We have seen that ΔS_{sys} is positive and favors the process and that ΔS_{surr} is negative and unfavorable. Thus the components of ΔS_{univ} are in opposition. Which one controls the situation? The answer *depends on the temperature*. We know that at a pressure of 1 atmosphere water changes spontaneously from liquid to gas at all temperatures above 100°C. Below 100°C the opposite process (condensation) is spontaneous.

Since ΔS_{sys} and ΔS_{surr} are in opposition for the vaporization of water, the temperature must have an effect on the relative importance of these two terms. To understand why, we must discuss in more detail the factors that control the entropy changes in the surroundings. The central idea is that *the entropy changes in the surroundings are primarily determined by heat flow*. An exothermic process in the system increases the entropy of the surroundings, because the energy flow increases the random motions in the surroundings. This means that exothermicity is an important driving force for spontaneity. In earlier chapters we have seen that a system tends to undergo changes that lower its energy. We now understand the reason for that tendency. When a system at constant temperature moves to a lower energy, the energy it gives up is transferred to the surroundings. Some of this energy is transferred as heat, thus leading to an increase in entropy in the surroundings.

The significance of exothermicity as a driving force *depends on the temperature at which the process occurs*. That is, the magnitude of ΔS_{surr} depends on the temperature at which the heat is transferred. We will not attempt to prove this fact here. Instead, we offer an analogy. Suppose that you have $50 to give away. Giving it to a millionaire will not create much of an impression—a millionaire has money to spare. However, to a poor college student, $50 represents a significant sum and will be received with considerable joy. The same principle can be applied to energy transfer via the flow of heat. If 50 joules of energy are transferred to the surroundings, the impact of that event depends greatly on the temperature. If the temperature of the surroundings is very high, the atoms there are in rapid motion. The 50 joules of energy will not make a large percentage change in these motions. On the other hand, if 50 joules of energy are transferred to the surroundings at a very low temperature, where the motions are slow, the energy will cause a large percentage change in the motions. Thus the impact of the transfer of a given quantity of energy as heat to or from the surroundings is greatest at low temperatures.

For our purposes the entropy changes that occur in the surroundings have two important characteristics.

In a process occurring at constant temperature, the tendency for the system to lower its energy is due to the resulting positive ΔS_{surr}.

1. *The sign of ΔS_{surr} depends on the direction of the heat flow.* At constant temperature an exothermic process in the system causes heat to flow into the surroundings, increasing the random motions and thereby increasing the entropy of the surroundings. For this case ΔS_{surr} is positive. The opposite is true for an endothermic process in a system at constant temperature. This principle is often stated in terms of energy. An important driving force in nature results from the tendency of a system to achieve the lowest possible energy.
2. *The magnitude of ΔS_{surr} depends on the temperature.* The transfer of a given quantity of energy as heat produces a much greater percentage change in the randomness of the surroundings at a low temperature than it does at a high temperature. Thus ΔS_{surr} depends directly on the quantity of heat transferred and inversely on temperature. In other words, the tendency for the system to lower its energy becomes a more important driving force at lower temperatures:

$$\begin{array}{c}\text{Driving force provided by} \\ \text{the energy flow (heat)}\end{array} = \begin{array}{c}\text{magnitude of the} \\ \text{entropy change of} \\ \text{the surroundings}\end{array} = \frac{\text{quantity of heat (J)}}{\text{temperature (K)}}$$

These ideas are summarized as follows:

Exothermic process:

ΔS_{surr} = positive

Endothermic process:

ΔS_{surr} = negative

$$\text{Exothermic process:} \qquad \Delta S_{surr} = +\frac{\text{quantity of heat (J)}}{\text{temperature (K)}}$$

$$\text{Endothermic process:} \qquad \Delta S_{surr} = -\frac{\text{quantity of heat (J)}}{\text{temperature (K)}}$$

We can express ΔS_{surr} in terms of the change in enthalpy (ΔH) for a process occurring at constant pressure (where only PV work is allowed), since under these conditions

$$\text{Heat flow (constant } P\text{)} = \text{change in enthalpy} = \Delta H$$

When no subscript is present, the quantity (for example, ΔH) refers to the system.

Recall that ΔH consists of two parts: a sign and a number. The *sign* indicates the direction of flow; a plus sign means into the system (endothermic) and a minus sign means out of the system (exothermic). The *number* indicates the quantity of energy.

Combining all these concepts yields the following definition of ΔS_{surr} for a reaction that takes place under conditions of constant temperature (K) and pressure:

$$\Delta S_{surr} = -\frac{\Delta H}{T}$$

The minus sign is necessary because the sign of ΔH is determined with respect to the reaction system, and this equation expresses a property of the surroundings. This means that if the reaction is exothermic, ΔH has a negative sign; but since heat flows into the surroundings, ΔS_{surr} is positive.

EXAMPLE 10.5

In the metallurgy of antimony the pure metal is recovered by different reactions, depending on the composition of the ore. For example, iron is used to reduce antimony in sulfide ores:

$$Sb_2S_3(s) + 3Fe(s) \longrightarrow 2Sb(s) + 3FeS(s) \qquad \Delta H = -125 \text{ kJ}$$

and carbon is used as the reducing agent in oxide ores:

$$Sb_4O_6(s) + 6C(s) \longrightarrow 4Sb(s) + 6CO(g) \qquad \Delta H = 778 \text{ kJ}$$

Calculate ΔS_{surr} for each of these reactions at 25°C and 1 atm.

Solution

We use

$$\Delta S_{surr} = -\frac{\Delta H}{T}$$

where

$$T = 25 + 273 = 298 \text{ K}$$

For the sulfide ore reaction,

$$\Delta S_{surr} = -\frac{-125 \text{ kJ}}{298 \text{ K}} = 0.419 \text{ kJ/K} = 419 \text{ J/K}$$

Note that ΔS_{surr} is positive, as expected, since this reaction is exothermic; energy flows to the surroundings as heat, increasing the randomness of the surroundings.

For the oxide ore reaction,

$$\Delta S_{surr} = -\frac{778 \text{ kJ}}{298} = -2.61 \text{ kJ/K} = -2.61 \times 10^3 \text{ J/K}$$

In this case ΔS_{surr} is negative because heat flows from the surroundings to the system.

Stibnite contains Sb_2S_3.

We have seen that the spontaneity of a process is determined by the entropy change it produces in the universe. We have also seen that ΔS_{univ} has two components, ΔS_{sys} and ΔS_{surr}. If for some process both ΔS_{sys} and ΔS_{surr} are positive, then ΔS_{univ} is positive, and the process is spontaneous. If, on the other hand, both ΔS_{sys} and ΔS_{surr} are negative, the process does not occur in the direction indicated but is spontaneous in the opposite direction. Finally, if ΔS_{sys}

TABLE 10.4 **Interplay of ΔS_{sys} and ΔS_{surr} in Determining the Sign of ΔS_{univ}**

Signs of Entropy Changes			
ΔS_{sys}	ΔS_{surr}	ΔS_{univ}	Process Spontaneous?
+	+	+	Yes
−	−	−	No (process will occur in opposite direction)
+	−	?	Yes, if ΔS_{sys} has a larger magnitude than ΔS_{surr}
−	+	?	Yes, if ΔS_{surr} has a larger magnitude than ΔS_{sys}

and ΔS_{surr} have opposite signs, the spontaneity of the process depends on the sizes of the opposing terms. These cases are summarized in Table 10.4.

We can now understand why spontaneity is often dependent on temperature and thus why water spontaneously freezes below 0°C and melts above 0°C. The term ΔS_{surr} is temperature-dependent. Since

$$\Delta S_{surr} = -\frac{\Delta H}{T}$$

at constant pressure, the value of ΔS_{surr} changes markedly with temperature. The magnitude of ΔS_{surr} is very small at high temperatures and increases as the temperature decreases. That is, exothermicity is most important as a driving force at low temperatures.

10.7 Free Energy

So far we have used ΔS_{univ} to predict the spontaneity of a process. Now we will define another thermodynamic function that is also related to spontaneity and is especially useful in dealing with the temperature dependence of spontaneity. This function is called **free energy** (G) and is defined as

The symbol G for free energy honors Josiah Willard Gibbs (1839–1903), who was Professor of Mathematical Physics at Yale University from 1871 to 1903. He laid the foundations of many areas of thermodynamics, particularly as they apply to chemistry.

$$G = H - TS$$

where H is the enthalpy, T is the Kelvin temperature, and S is the entropy.

For a process that occurs at constant temperature, the change in free energy (ΔG) is given by the equation

$$\Delta G = \Delta H - T\Delta S$$

Note that all quantities here refer to the system. From this point on we will follow the usual convention that when no subscript is included, the quantity refers to the system.

To see how this equation relates to spontaneity, we divide both sides of the equation by $-T$ to produce

$$-\frac{\Delta G}{T} = -\frac{\Delta H}{T} + \Delta S$$

Remember that at constant temperature and pressure

$$\Delta S_{surr} = -\frac{\Delta H}{T}$$

So we can write

$$-\frac{\Delta G}{T} = -\frac{\Delta H}{T} + \Delta S = \Delta S_{surr} + \Delta S = \Delta S_{univ}$$

We have shown that

$$\Delta S_{univ} = -\frac{\Delta G}{T} \quad \text{at constant } T \text{ and } P$$

This result is very important. It means that a process carried out at constant temperature and pressure will be spontaneous only if ΔG is negative. That is, *a process (at constant* T *and* P*) is spontaneous in the direction in which the free energy decreases* ($-\Delta G$ means $+\Delta S_{univ}$).

Now we have two functions that can be used to predict spontaneity: the entropy of the universe, which applies to all processes; and free energy, which can be used for processes carried out at constant temperature and pressure. Since so many chemical reactions occur under the latter conditions, the free energy function is more useful to chemists.

Let's use the free energy equation to predict the spontaneity of the melting of ice:

$$H_2O(s) \longrightarrow H_2O(l)$$

for which

$$\Delta H° = 6.03 \times 10^3 \text{ J/mol} \quad \text{and} \quad \Delta S° = 22.1 \text{ J K}^{-1} \text{ mol}^{-1}$$

The superscript degree symbol (°) indicates all substances are in their standard states.

Results for the calculations of ΔS_{univ} and $\Delta G°$ at $-10°C$, 0°C, and 10°C are shown in Table 10.5. These data predict that the process is spontaneous at 10°C; that is, ice melts at this temperature since ΔS_{univ} is positive and $\Delta G°$ is negative. The opposite is true at $-10°C$, where water freezes spontaneously.

Why is this so? The answer lies in the fact that ΔS_{sys} ($\Delta S°$) and ΔS_{surr} oppose each other. The term $\Delta S°$ favors the melting of ice because of the increase in positional entropy, and ΔS_{surr} favors the freezing of water because it is an exothermic process. At temperatures below 0°C the change of state occurs in the exothermic direction because ΔS_{surr} is larger in magnitude than ΔS_{sys}. But above 0°C the change occurs in the direction in which ΔS_{sys} is favorable, since in this case ΔS_{sys} is larger in magnitude than ΔS_{surr}. At 0°C the *opposing tendencies*

TABLE 10.5 Results of the Calculation of ΔS_{univ} and $\Delta G°$ for the Process $H_2O(s) \rightarrow H_2O(l)$ at $-10°C$, 0°C, and 10°C*

T (°C)	T (K)	$\Delta H°$ (J/mol)	$\Delta S°$ (J K^{-1} mol^{-1})	$\Delta S_{surr} = -\frac{\Delta H°}{T}$ (J K^{-1} mol^{-1})	$\Delta S_{univ} = \Delta S° + \Delta S_{surr}$ (J K^{-1} mol^{-1})	$T\Delta S°$ (J/mol)	$\Delta G° = \Delta H° - T\Delta S°$ (J/mol)
−10	263	6.03×10^3	22.1	−22.9	−0.8	5.81×10^3	$+2.2 \times 10^2$
0	273	6.03×10^3	22.1	−22.1	0	6.03×10^3	0
10	283	6.03×10^3	22.1	−21.3	+0.8	6.25×10^3	-2.2×10^2

* Note that at 10°C, $\Delta S°$ (ΔS_{sys}) controls, and the process occurs even though it is endothermic. At $-10°C$ the magnitude of ΔS_{surr} is larger than that of $\Delta S°$, so the process is spontaneous in the opposite (exothermic) direction.

TABLE 10.6 Various Possible Combinations of ΔH and ΔS for a Process and the Resulting Dependence of Spontaneity on Temperature

Case	Result
ΔS positive, ΔH negative	Spontaneous at all temperatures
ΔS positive, ΔH positive	Spontaneous at high temperatures (where exothermicity is relatively unimportant)
ΔS negative, ΔH negative	Spontaneous at low temperatures (where exothermicity is dominant)
ΔS negative, ΔH positive	Process not spontaneous at *any* temperature (reverse process is spontaneous at *all* temperatures)

just balance, and the two states coexist; there is no driving force in either direction. An equilibrium exists between the two states of water. Note that ΔS_{univ} is equal to 0 at 0°C.

We can reach the same conclusions by examining $\Delta G°$. At $-10°C$, $\Delta G°$ is positive because the $\Delta H°$ term is larger than the $T\Delta S°$ term. The opposite is true at 10°C. At 0°C, $\Delta H°$ is equal to $T\Delta S°$ and $\Delta G°$ is equal to 0. This means that solid H_2O and liquid H_2O have the same free energy at 0°C ($\Delta G° = G_{(l)} - G_{(s)}$), and the system is at equilibrium.

We can understand the temperature dependence of spontaneity by examining the behavior of ΔG. For a process occurring at constant temperature and pressure,

$$\Delta G = \Delta H - T\Delta S$$

If ΔH and ΔS favor opposite processes, spontaneity will depend on temperature in such a way that the exothermic direction will be favored at low temperatures. For example, for the process

$$H_2O(s) \longrightarrow H_2O(l)$$

ΔH and ΔS are both positive. The natural tendency for this system to lower its energy is in opposition to its natural tendency to increase its positional randomness. At low temperatures ΔH dominates, and at high temperatures ΔS dominates. The various cases are summarized in Table 10.6.

EXAMPLE 10.6

At what temperatures is the following process spontaneous at 1 atm?

$$Br_2(l) \longrightarrow Br_2(g)$$

where $\Delta H° = 31.0$ kJ/mol and $\Delta S° = 93.0 \text{ J K}^{-1} \text{ mol}^{-1}$

What is the normal boiling point of liquid Br_2?

Solution

The vaporization process will be spontaneous at all temperatures where $\Delta G°$ is negative. Note that $\Delta S°$ favors the vaporization process because of the increase in positional entropy, and ΔH° favors the *opposite* process, which is exothermic. These opposite tendencies will exactly balance at the boiling point of liquid Br_2, since at this temperature liquid and gaseous Br_2 are in equilibrium ($\Delta G^\circ = 0$). We can find this temperature by setting $\Delta G^\circ = 0$ in the equation

$$\Delta G^\circ = \Delta H^\circ - T\Delta S^\circ$$

Thus we have

$$0 = \Delta H^\circ - T\Delta S^\circ$$

$$\Delta H^\circ = T\Delta S^\circ$$

and

$$T = \frac{\Delta H^\circ}{\Delta S^\circ} = \frac{3.10 \times 10^4 \text{ J/mol}}{93.0 \text{ J K}^{-1} \text{ mol}^{-1}} = 333 \text{ K}$$

At temperatures above 333 K, $T\Delta S^\circ$ has a larger magnitude than ΔH°, and ΔG° (or $\Delta H^\circ - T\Delta S^\circ$) is negative. Above 333 K the vaporization process is spontaneous; the opposite process occurs spontaneously below this temperature. At 333 K liquid and gaseous Br_2 will coexist in equilibrium. These observations can be summarized as follows (the pressure is 1 atm in each case):

1. $T > 333$ K. The term ΔS° controls. The increase in entropy occurring when liquid Br_2 is vaporized is dominant.
2. $T < 333$ K. The process is spontaneous in the direction in which it is exothermic. The term ΔH° controls.
3. $T = 333$ K. The opposing driving forces are just balanced ($\Delta G^\circ = 0$), and the liquid and gaseous phases of bromine coexist. This is the normal boiling point.

10.8 Entropy Changes in Chemical Reactions

The second law of thermodynamics tells us that a process will be spontaneous if the entropy of the universe increases when the process occurs. We saw in Section 10.7 that for a process at constant temperature and pressure, we can use the change in free energy of the system to predict the sign of ΔS_{univ} and thus the direction in which it is spontaneous. So far we have applied these ideas only to physical processes, such as changes of state and the formation of solutions. However, the main business of chemistry is studying chemical reactions, and therefore, we want to apply the second law to reactions.

First we will consider the entropy changes accompanying chemical reactions that occur under conditions of constant temperature and pressure. As for the other types of processes we have considered, the entropy changes in the *surroundings* are determined by the heat flow that occurs as the reaction takes place. However, the entropy changes in the *system* (the reactants and products of the reaction) are determined primarily by positional probability.

For example, in the ammonia synthesis reaction

$$N_2(g) + 3H_2(g) \longrightarrow 2NH_3(g)$$

four reactant molecules are changed to two product molecules, lowering the number of independent units in the system, thus leading to lower positional disorder. *Fewer molecules mean fewer possible configurations.* To help clarify this result, consider a special container with a million compartments, each large enough to hold a hydrogen molecule. There are a million ways one H_2 molecule can be placed in this container. But suppose we break the H—H bond and place the two independent H atoms in the same container. A little thought will convince you that there are *many* more than a million ways to place the two separate atoms in the container. The number of arrangements possible for the two independent atoms is much greater than the number for the molecule. Thus for the process

$$H_2 \longrightarrow 2H$$

positional entropy increases.

Does positional entropy increase or decrease when the following reaction takes place?

$$4NH_3(g) + 5O_2(g) \longrightarrow 4NO(g) + 6H_2O(g)$$

In this case 9 gaseous molecules are changed to 10 gaseous molecules, and the positional entropy increases. There are more independent units as products than as reactants. In general, when a reaction involves gaseous molecules, *the change in positional entropy is dominated by the relative numbers of molecules of gaseous reactants and products.* If the number of molecules of the gaseous products is greater than the number of molecules of the gaseous reactants, positional entropy typically increases, and ΔS is positive for the reaction.

EXAMPLE 10.7

Predict the sign of $\Delta S°$ for each of the following reactions.

a. the thermal decomposition of solid calcium carbonate:

$$CaCO_3(s) \longrightarrow CaO(s) + CO_2(g)$$

b. the oxidation of SO_2 in air:

$$2SO_2(g) + O_2(g) \longrightarrow 2SO_3(g)$$

Solution

a. Since in this reaction a gas is produced from a solid reactant, the positional entropy increases, and $\Delta S°$ is positive.

b. Here three molecules of gaseous reactants become two molecules of gaseous products. Since the number of gas molecules decreases, positional entropy decreases, and $\Delta S°$ is negative.

Absolute Entropies and the Third Law of Thermodynamics

In thermodynamics it is the *change* in a certain function that is usually important. The change in enthalpy determines whether a reaction is exothermic or endothermic at constant pressure. The change in free energy determines

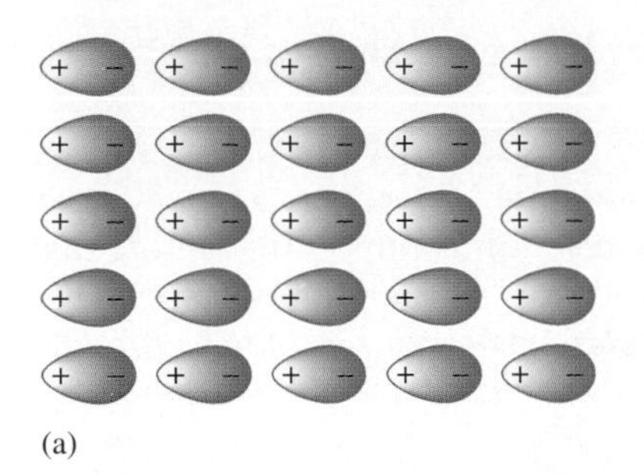

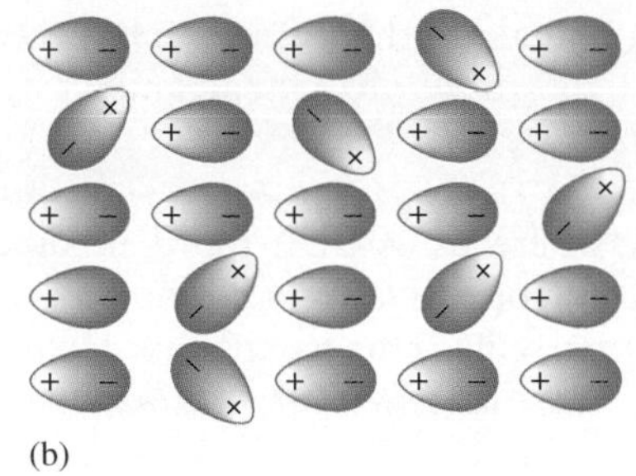

Figure 10.11
(a) A perfect crystal of hydrogen chloride at 0 K; the dipolar HCl molecules are represented by (+|−). The entropy is zero ($S = 0$) for this perfect crystal at 0 K. (b) As the temperature rises above 0 K, lattice vibrations allow some dipoles to change their orientations, producing some disorder and an increase in entropy ($S > 0$).

whether a process is spontaneous at constant temperature and pressure. It is fortunate that changes in thermodynamic functions are sufficient for most purposes, because absolute values for many thermodynamic characteristics of a system (such as enthalpy or free energy) cannot be determined.

However, we can assign *absolute* entropy values. Consider a solid at 0 K, where molecular motion virtually ceases. If it is a perfect crystal, its internal arrangement is absolutely regular [see Fig. 10.11(a)]. There is only *one way* to achieve this perfect order; every particle must be in its place. For example, with N coins there is only one way to achieve the state of all heads. Thus a perfect crystal represents the lowest possible entropy; that is, *the entropy of a perfect crystal at 0 K is zero.* This is a statement of the **third law of thermodynamics.**

As the temperature of a perfect crystal is increased, the random vibrational motions increase, and disorder increases within the crystal [see Fig. 10.11(b)]. Thus the entropy of a substance increases with temperature. Since S is zero for a perfect crystal at 0 K, the entropy value for a substance at a particular temperature can be calculated if we know the temperature dependence of entropy.

We have shown that the change in entropy that accompanies a change in temperature of a substance from T_1 to T_2 can be calculated from the expression

$$\Delta S_{T_1 \rightarrow T_2} = nC \ln\left(\frac{T_2}{T_1}\right)$$

where C is C_p or C_v, depending on the conditions. In using this equation, we are assuming that the heat capacity is independent of temperature. Unfortunately, this assumption is usually not true, especially over a large temperature range. Thus the dependence of C on temperature must be taken into account for accurate calculations. However, we will not be concerned with these procedures here.

When a substance is heated from temperature T_1 to T_2, a change of state may occur. If it does, as we saw in Section 10.4, we use the expression

$$\Delta S = \frac{\Delta H}{T}$$

at the melting point or boiling point to account for the entropy change that accompanies the change of state.

The *standard entropy values* ($S°$) of many common substances at 298 K and 1 atmosphere are listed in Appendix 4. From these values you will see that the entropy of a substance does indeed increase in going from solid to liquid to gas.

The standard entropy values represent the increase in entropy that occurs when a substance is heated from 0 K to 298 K at 1 atm pressure.

Because *entropy is a state function of the system* (it is not pathway-dependent), the entropy change for a given chemical reaction can be calculated by taking the difference between the standard entropy values of the products and those of the reactants:

$$\Delta S^\circ_{\text{reaction}} = \Sigma S^\circ_{\text{products}} - \Sigma S^\circ_{\text{reactants}}$$

where, as usual, Σ represents the sum of the terms. It is important to note that entropy is an extensive property (it depends on the amount of substance present). This means that *the number of moles of a given reactant or product must be taken into account.*

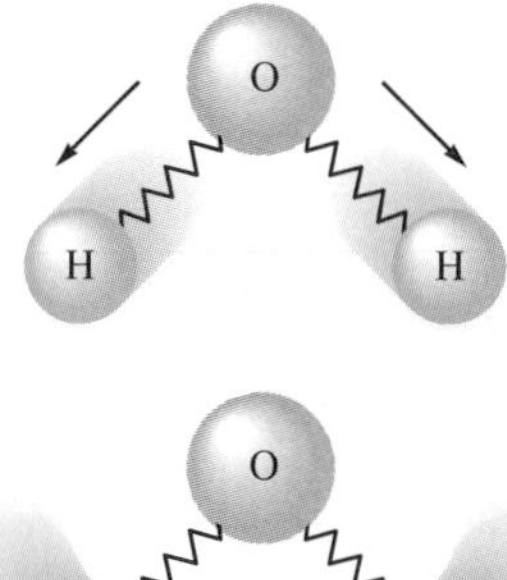

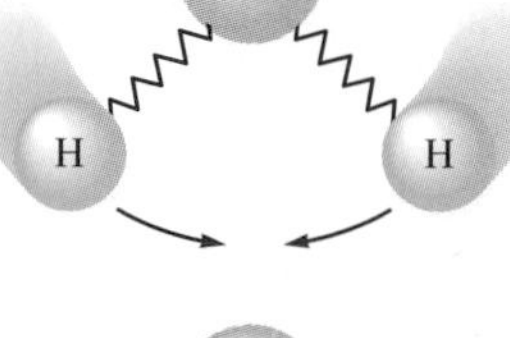

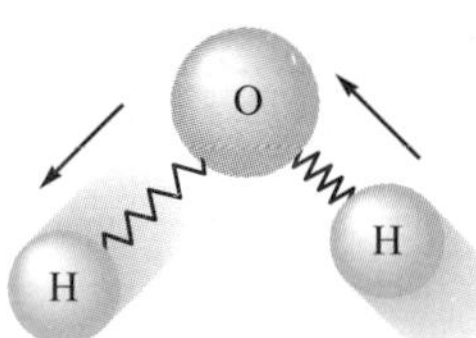

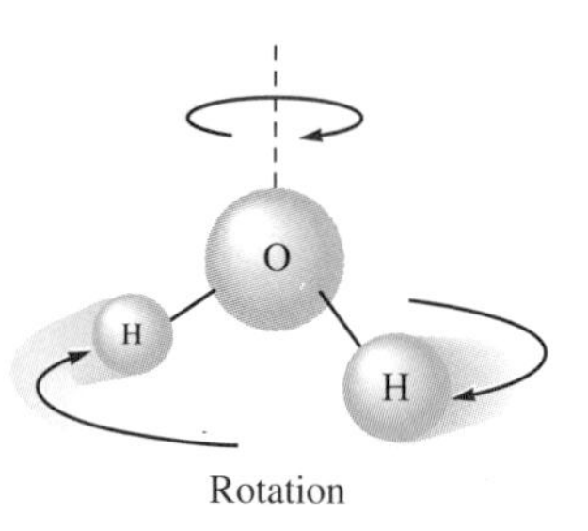

Figure 10.12
The H_2O molecule can vibrate and rotate in several ways, some of which are shown here. This freedom of motion leads to a higher entropy for water than for a substance like hydrogen, which consists of a simple diatomic molecule with fewer possible motions.

EXAMPLE 10.8

Calculate ΔS° for the reduction of aluminum oxide by hydrogen gas:

$$Al_2O_3(s) + 3H_2(g) \longrightarrow 2Al(s) + 3H_2O(g)$$

using the following standard entropy values.

Substance	S° ($J\ K^{-1}\ mol^{-1}$)
$Al_2O_3(s)$	51
$H_2(g)$	131
$Al(s)$	28
$H_2O(g)$	189

Solution

$$\begin{aligned}\Delta S^\circ &= \Sigma S^\circ_{\text{products}} - \Sigma S^\circ_{\text{reactants}} \\ &= 2S^\circ_{Al(s)} + 3S^\circ_{H_2O(g)} - 3S^\circ_{H_2(g)} - S^\circ_{Al_2O_3(s)} \\ &= 2\text{ mol}\left(28\frac{\text{J}}{\text{K mol}}\right) + 3\text{ mol}\left(189\frac{\text{J}}{\text{K mol}}\right) \\ &\quad -3\text{ mol}\left(131\frac{\text{J}}{\text{K mol}}\right) - 1\text{ mol}\left(51\frac{\text{J}}{\text{K mol}}\right) \\ &= 56\text{ J/K} + 567\text{ J/K} - 393\text{ J/K} - 51\text{ J/K} \\ &= 179\text{ J/K}\end{aligned}$$

The reaction considered in Example 10.8 involves 3 moles of hydrogen gas on the reactant side and 3 moles of water vapor on the product side. Would you expect ΔS to be large or small for such a case? We have assumed that ΔS depends on the relative numbers of molecules of gaseous reactants and products. On the basis of that assumption, ΔS should be near zero for the present reaction. However, ΔS is large and positive. Why? The large value for ΔS results from the difference in the entropy values for hydrogen gas and water vapor. The reason for this difference can be traced to the difference in molecular structure. Because it is a nonlinear, triatomic molecule, H_2O has more rotational and vibrational motions (see Fig. 10.12) than does the diatomic H_2 molecule. Thus the standard entropy value for $H_2O(g)$ is greater than that for $H_2(g)$. Generally, *the more complex the molecule, the higher the standard entropy value.*

10.9 Free Energy and Chemical Reactions

For chemical reactions we are often interested in the **standard free energy change** (ΔG°), *the change in free energy that occurs if the reactants in their standard states are converted to the products in their standard states.* For example, for the ammonia synthesis reaction at 25°C,

$$N_2(g) + 3H_2(g) \longrightarrow 2NH_3(g) \qquad \Delta G^\circ = -33.3 \text{ kJ} \qquad (10.1)$$

This ΔG° value represents the change in free energy that occurs when 1 mole of nitrogen gas at 1 atm reacts with 3 moles of hydrogen gas at 1 atm to produce 2 moles of gaseous NH_3 at 1 atm.

It is important to recognize that the standard free energy change for a reaction is not measured directly. For example, we can measure heat flow in a calorimeter to determine ΔH°, but we cannot measure ΔG° this way. The value of ΔG° for the ammonia synthesis in Equation (10.1) was *not* obtained by mixing 1 mole of N_2 with 3 moles of H_2 in a flask and measuring the change in free energy as 2 moles of NH_3 formed. For one thing, if we mixed 1 mole of N_2 and 3 moles of H_2 in a flask, the system would go to equilibrium rather than to completion. Also, we have no instrument that directly measures free energy. Although we cannot directly measure ΔG° for a reaction, we can calculate it from other measured quantities, as we will see later in this section.

Why is it useful to know ΔG° for a reaction? As we will see in more detail later, knowing the ΔG° values for several reactions allows us to compare the relative tendency of these reactions to occur. The more negative the value of ΔG°, the further a reaction will go to the right to reach equilibrium. We must use standard state free energies to make this comparison because free energy depends on pressure or concentration. Thus to get an accurate comparison of reaction tendencies, we must compare all reactions under the same pressure or concentration conditions. We will have more to say about the significance of ΔG° later.

The value of ΔG° tells us nothing about the rate of a reaction, only its eventual equilibrium position.

There are several ways to calculate ΔG°. One common method employs the equation

$$\Delta G^\circ = \Delta H^\circ - T\Delta S^\circ$$

which applies to a reaction carried out at constant temperature. For example, consider the reaction

$$C(s) + O_2(g) \longrightarrow CO_2(g)$$

The values of ΔH° and ΔS° are known to be -393.5 kJ and 3.05 J/K, respectively, and ΔG° can be calculated at 298 K as follows:

$$\begin{aligned}\Delta G^\circ &= \Delta H^\circ - T\Delta S^\circ = -3.935 \times 10^5 \text{ J} - (298 \text{ K})(3.05 \text{ J/K}) \\ &= -3.944 \times 10^5 \text{ J} = -394.4 \text{ kJ} \qquad (\text{per mole of } CO_2)\end{aligned}$$

EXAMPLE 10.9

Consider the reaction

$$2SO_2(g) + O_2(g) \longrightarrow 2SO_3(g)$$

carried out at 25°C and 1 atm. Calculate ΔH°, ΔS°, and ΔG° using the following data.

Substance	ΔH_f° (kJ/mol)	S° (J K^{-1} mol^{-1})
$SO_2(g)$	−297	248
$SO_3(g)$	−396	257
$O_2(g)$	0	205

Solution

The value of ΔH° can be calculated from the enthalpies of formation using the equation we discussed in Section 9.6:

$$\Delta H^\circ = \Sigma \Delta H^\circ_{f\,(\text{products})} - \Sigma \Delta H^\circ_{f\,(\text{reactants})}$$

Then

$$\begin{aligned}\Delta H^\circ &= 2\Delta H^\circ_{f\,(SO_3(g))} - 2\Delta H^\circ_{f\,(SO_2(g))} - \Delta H^\circ_{f\,(O_2(g))}\\ &= 2\text{ mol}(-396\text{ kJ/mol}) - 2\text{ mol}(-297\text{ kJ/mol}) - 0\\ &= -792\text{ kJ} + 594\text{ kJ} = -198\text{ kJ}\end{aligned}$$

The value of ΔS° can be calculated by using the standard entropy values and the equation discussed in Section 10.8:

$$\Delta S^\circ = \Sigma S^\circ_{\text{products}} - \Sigma S^\circ_{\text{reactants}}$$

Thus

$$\begin{aligned}\Delta S^\circ &= 2S^\circ_{SO_3(g)} - 2S^\circ_{SO_2(g)} - S^\circ_{O_2(g)}\\ &= 2\text{ mol}(257\text{ J K}^{-1}\text{ mol}^{-1}) - 2\text{ mol}(248\text{ J K}^{-1}\text{ mol}^{-1})\\ &\quad - 1\text{ mol}(205\text{ J K}^{-1}\text{ mol}^{-1})\\ &= 514\text{ J/K} - 496\text{ J/K} - 205\text{ J/K} = -187\text{ J/K}\end{aligned}$$

We expect ΔS° to be negative since three molecules of gaseous reactants give two molecules of gaseous products.

The value of ΔG° can now be calculated from the equation

$$\Delta G^\circ = \Delta H^\circ - T\Delta S^\circ$$

So

$$\begin{aligned}\Delta G^\circ &= -198\text{ kJ} - (298\text{ K})\left(-187\frac{\text{J}}{\text{K}}\right)\left(\frac{1\text{ kJ}}{1000\text{ J}}\right)\\ &= -198\text{ kJ} + 55.7\text{ kJ} = -142\text{ kJ}\end{aligned}$$

Like enthalpy and entropy, *free energy is a state function.* Thus we can use procedures for finding ΔG similar to those for finding ΔH using Hess's law.

To illustrate this second method for calculating the free energy change, we will obtain ΔG° for the reaction

$$2CO(g) + O_2(g) \longrightarrow 2CO_2(g) \qquad (10.2)$$

from the following data:

$$2CH_4(g) + 3O_2(g) \longrightarrow 2CO(g) + 4H_2O(g) \qquad \Delta G^\circ = -1088\text{ kJ} \qquad (10.3)$$

$$CH_4(g) + 2O_2(g) \longrightarrow CO_2(g) + 2H_2O(g) \qquad \Delta G^\circ = -801\text{ kJ} \qquad (10.4)$$

Note that $CO(g)$ is a reactant in Equation (10.2), so Equation (10.3) must be reversed, since $CO(g)$ is a product in the reaction as written. When a reaction is reversed, the sign of $\Delta G°$ is also reversed. In Equation (10.4), $CO_2(g)$ is a product, as it is in Equation (10.2), but only one molecule of CO_2 is formed. Thus Equation (10.4) must be multiplied by 2, which means the $\Delta G°$ value for Equation (10.4) must also be multiplied by 2. Free energy is an extensive property since it is defined by two extensive properties, H and S.

Reversed Equation (10.3)

$$2CO(g) + 4H_2O(g) \longrightarrow 2CH_4(g) + 3O_2(g) \qquad \Delta G° = -(-1088\ kJ)$$

2 × Equation (10.4)

$$2CH_4(g) + 4O_2(g) \longrightarrow 2CO_2(g) + 4H_2O(g) \qquad \Delta G° = 2(-801\ kJ)$$

$$2CO(g) + O_2(g) \longrightarrow 2CO_2(g) \qquad \Delta G° = -(-1088\ kJ) + 2(-801\ kJ) = -514\ kJ$$

This example shows that ΔG values for reactions are manipulated in exactly the same way as ΔH values.

EXAMPLE 10.10

Using the following data (at 25°C),

$$C_{(s)}^{diamond} + O_2(g) \longrightarrow CO_2(g) \qquad \Delta G° = -397\ kJ \qquad (10.5)$$
$$C_{(s)}^{graphite} + O_2(g) \longrightarrow CO_2(g) \qquad \Delta G° = -394\ kJ \qquad (10.6)$$

calculate $\Delta G°$ for the reaction

$$C_{(s)}^{diamond} \longrightarrow C_{(s)}^{graphite}$$

Solution

We reverse Equation (10.6) to make graphite a product, as required, and then add the new equation to Equation (10.5):

$$C_{(s)}^{diamond} + O_2(g) \longrightarrow CO_2(g) \qquad \Delta G° = -397\ kJ$$

Reversed Equation (10.6)

$$CO_2(g) \longrightarrow C_{(s)}^{graphite} + O_2(g) \qquad \Delta G° = -(-394\ kJ)$$

$$C_{(s)}^{diamond} \longrightarrow C_{(s)}^{graphite} \qquad \Delta G° = -397\ kJ + 394\ kJ = -3\ kJ$$

Since $\Delta G°$ is negative for this process, diamond should spontaneously change to graphite at 25°C and 1 atm. However, the reaction is so slow under these conditions that we do not observe the process. This example shows *kinetic* rather than *thermodynamic* control of a reaction. That is, thermodynamically, diamond should change to graphite, but this spontaneous change is not observed because its rate is so slow. We say that diamond is kinetically stable with respect to graphite, even though it is thermodynamically unstable.

Diamond and graphite.

In Example 10.10 we saw that the process

$$C_{(s)}^{\text{diamond}} \longrightarrow C_{(s)}^{\text{graphite}}$$

is spontaneous but very slow at 25°C and 1 atm. The reverse process can be forced to occur at high temperatures and pressures. Diamond has a more compact structure and thus a higher density than graphite, so the exertion of very high pressure causes it to become thermodynamically favored. If high temperatures are also used to make the process fast enough to be feasible, diamonds can be made from graphite. The conditions usually involve temperatures greater than 1000°C and pressures of about 10^5 atm. About half of all industrial diamonds are made this way. We will discuss this process in more detail in Chapter 16.

The standard state of an element is its most stable state at 25°C and 1 atm.

A third method for calculating the free energy change for a reaction uses standard free energies of formation. The **standard free energy of formation** (ΔG_f°) of a substance is defined as the *change in free energy that accompanies the formation of 1 mole of that substance from its constituent elements with all reactants and products in their standard states.* For example, for the formation of glucose ($C_6H_{12}O_6$) the appropriate reaction is

$$6C(s) + 6H_2(g) + 3O_2(g) \longrightarrow C_6H_{12}O_6(s)$$

The standard free energy associated with this process is called the free energy of formation of glucose. Values of the standard free energy of formation are useful in calculating ΔG° for specific chemical reactions, using the equation

$$\Delta G^\circ = \Sigma \Delta G^\circ_{f\,(\text{products})} - \Sigma \Delta G^\circ_{f\,(\text{reactants})}$$

Calculating ΔG° from free energies of formation is very similar to calculating ΔH°, as shown in Section 9.6.

Values of ΔG_f° for many common substances are listed in Appendix 4. Note that, analogous to the enthalpy of formation, *the standard free energy of formation of an element in its standard state is zero.*

EXAMPLE 10.11

Methanol is a high-octane fuel used in high-performance racing engines. Calculate ΔG° for the reaction

$$2CH_3OH(g) + 3O_2(g) \longrightarrow 2CO_2(g) + 4H_2O(g)$$

given the following free energies of formation:

Substance	ΔG_f° (kJ/mol)
$CH_3OH(g)$	−163
$O_2(g)$	0
$CO_2(g)$	−394
$H_2O(g)$	−229

Solution

We use

$$\Delta G_f^\circ = \Sigma \Delta G^\circ_{f\,(\text{products})} - \Sigma \Delta G^\circ_{f\,(\text{reactants})}$$
$$= 2\Delta G^\circ_{f\,(CO_2(g))} + 4\Delta G^\circ_{f\,(H_2O(g))} - 3\Delta G^\circ_{f\,(O_2(g))} - 2\Delta G^\circ_{f\,(CH_3OH(g))}$$

$$= 2\text{ mol}(-394\text{ kJ/mol}) + 4\text{ mol}(-229\text{ kJ/mol}) - 3(0)$$
$$- 2\text{ mol}(-163\text{ kJ/mol})$$
$$= -1378\text{ kJ}$$

The large magnitude and the negative sign of $\Delta G°$ indicate that this reaction is very favorable thermodynamically.

EXAMPLE 10.12

A chemical engineer wants to determine the feasibility of making ethanol (C_2H_5OH) by reacting water with ethylene (C_2H_4) according to the equation

$$C_2H_4(g) + H_2O(l) \longrightarrow C_2H_5OH(l)$$

Is this reaction spontaneous under standard conditions?

Solution

To determine the spontaneity of this reaction under standard conditions, we must determine $\Delta G°$ for the reaction by using the appropriate standard free energies of formation at 25°C from Appendix 4:

$$\Delta G°_{f\,(C_2H_5OH(l))} = -175\text{ kJ/mol}$$
$$\Delta G°_{f\,(H_2O(l))} = -237\text{ kJ/mol}$$
$$\Delta G°_{f\,(C_2H_4(g))} = 68\text{ kJ/mol}$$

Thus $$\Delta G° = \Delta G°_{f\,(C_2H_5OH(l))} - \Delta G°_{f\,(H_2O(l))} - \Delta G°_{f\,(C_2H_4(g))}$$
$$= -175\text{ kJ} - (-237\text{ kJ}) - 68\text{ kJ} = -6\text{ kJ}$$

and the process is spontaneous under standard conditions at 25°C.

Although the reaction considered in Example 10.12 is spontaneous, other features of the reaction must be studied to see whether the process is feasible. For example, the chemical engineer will need to study the kinetics of the reaction to determine whether it is fast enough to be useful and, if it is not, whether a catalyst can be found to enhance the rate. In doing these studies, the engineer must remember that $\Delta G°$ depends on temperature:

$$\Delta G° = \Delta H° - T\Delta S°$$

Thus if the process must be carried out at high temperatures to be fast enough to be feasible, $\Delta G°$ must be recalculated at that temperature using the $\Delta H°$ and $\Delta S°$ values for the reaction.

10.10 The Dependence of Free Energy on Pressure

In this chapter we have seen that a reaction system at constant temperature and pressure will proceed spontaneously in the direction that lowers its free energy.

For this reason reactions proceed until they reach equilibrium. As we will see later in this section, the equilibrium position represents the lowest free energy value available to a particular reaction system. The free energy of a reaction system changes as the reaction proceeds, because free energy is dependent on the pressure of a gas (or on the concentration of species in solution). We will deal only with the pressure dependence of the free energy of an ideal gas. The dependence of free energy on concentration can be developed using similar reasoning.

To understand the pressure dependence of free energy, we need to know how pressure affects the thermodynamic functions that constitute free energy, that is, enthalpy and entropy (recall that $G = H - TS$). For an ideal gas enthalpy is not pressure-dependent. However, entropy *does* depend on pressure because of its dependence on volume. Consider 1 mole of an ideal gas at a given temperature. At a volume of 10.0 liters, the gas has many more positions available for the molecules than if its volume is 1.0 liter. The positional entropy is greater for the larger volume. In summary, at a given temperature for 1 mole of ideal gas

$$S_{\text{large volume}} > S_{\text{small volume}}$$

or since pressure and volume are inversely related,

$$S_{\text{low pressure}} > S_{\text{high pressure}}$$

We have shown qualitatively that the entropy and therefore the free energy of an ideal gas depends on its pressure. From a more detailed argument, which we will not consider here, one can show that

$$G = G^\circ + RT\ln(P)$$

where G° is the free energy of the gas at a pressure of 1 atmosphere, G is the free energy of the gas at a pressure of P atmospheres, R is the universal gas constant, and T is the Kelvin temperature.

To see how the change in free energy for a reaction depends on pressure, we will consider the ammonia synthesis reaction

$$N_2(g) + 3H_2(g) \longrightarrow 2NH_3(g)$$

The absolute value of the free energy of a substance cannot be obtained. We use it symbolically here to show that it is the change in free energy that is really significant.

In general, $$\Delta G = \Sigma G_{\text{products}} - \Sigma G_{\text{reactants}} \qquad (10.7)$$

For this reaction $$\Delta G = 2G_{NH_3} - G_{N_2} - 3G_{H_2}$$

where
$$G_{NH_3} = G^\circ_{NH_3} + RT\ln(P_{NH_3})$$
$$G_{N_2} = G^\circ_{N_2} + RT\ln(P_{N_2})$$
$$G_{H_2} = G^\circ_{H_2} + RT\ln(P_{H_2})$$

Substituting these values into Equation (10.7) gives

$$\begin{aligned}\Delta G &= 2[G^\circ_{NH_3} + RT\ln(P_{NH_3})] - [G^\circ_{N_2} + RT\ln(P_{N_2})] - 3[G^\circ_{H_2} + RT\ln(P_{H_2})]\\ &= 2G^\circ_{NH_3} - G^\circ_{N_2} - 3G^\circ_{H_2} + 2RT\ln(P_{NH_3}) - RT\ln(P_{N_2}) - 3RT\ln(P_{H_2})\\ &= \underbrace{(2G^\circ_{NH_3} - G^\circ_{N_2} - 3G^\circ_{H_2})}_{\Delta G^\circ_{\text{reaction}}} + RT[2\ln(P_{NH_3}) - \ln(P_{N_2}) - 3\ln(P_{H_2})]\end{aligned}$$

The first term in parentheses is ΔG° for the reaction. Thus we have

$$\Delta G = \Delta G^\circ_{\text{reaction}} + RT[2\ln(P_{NH_3}) - \ln(P_{N_2}) - 3\ln(P_{H_2})]$$

Since $$2 \ln(P_{NH_3}) = \ln(P_{NH_3}{}^2)$$
$$-\ln(P_{N_2}) = \ln\left(\frac{1}{P_{N_2}}\right)$$
and $$-3 \ln(P_{H_2}) = \ln\left(\frac{1}{P_{H_2}{}^3}\right)$$

the equation becomes

$$\Delta G = \Delta G° + RT \ln\left[\frac{P_{NH_3}{}^2}{(P_{N_2})(P_{H_2}{}^3)}\right]$$

But the term $$\frac{P_{NH_3}{}^2}{(P_{N_2})(P_{H_2}{}^3)}$$

is the reaction quotient (Q) discussed in Section 6.6. Therefore, we have

$$\Delta G = \Delta G° + RT \ln(Q)$$

where Q is the reaction quotient (from the law of mass action), T is the temperature (K), R is the gas law constant, $\Delta G°$ is the free energy change for the reaction with all reactants and products at a pressure of 1 atmosphere, and ΔG is the free energy change for the reaction at the specified pressures of reactants and products.

EXAMPLE 10.13

One method for synthesizing methanol (CH_3OH) involves reacting gaseous carbon monoxide and hydrogen:

$$CO(g) + 2H_2(g) \longrightarrow CH_3OH(l)$$

Calculate ΔG at 25°C for this reaction, in which carbon monoxide gas at 5.0 atm and hydrogen gas at 3.0 atm are converted to liquid methanol.

Solution

To calculate ΔG for this process, we use the equation

$$\Delta G = \Delta G° + RT \ln(Q)$$

We must first compute $\Delta G°$ from standard free energies of formation (see Appendix 4). Since

$$\Delta G°_{f\,(CH_3OH(l))} = -166 \text{ kJ}$$
$$\Delta G°_{f\,(H_2(g))} = 0$$
$$\Delta G°_{f\,(CO(g))} = -137 \text{ kJ}$$

then

$$\Delta G° = -166 \text{ kJ} - (-137 \text{ kJ}) - 0 = -29 \text{ kJ} = -2.9 \times 10^4 \text{ J}$$

Note that this is the value of $\Delta G°$ for the reaction of 1 mol CO with 2 mol H_2 to produce 1 mol CH_3OH. We might call this the value of $\Delta G°$ for one "round" of the reaction or for 1 mol of the reaction. Thus the $\Delta G°$ value might be written more accurately as -2.9×10^4 J/mol of reaction, or -2.9×10^4 J/mol rxn.

Note that in this case ΔG is defined for one mole of the reaction, that is, for 1 mol $CO(g)$ reacting with 2 mol $H_2(g)$ to form 1 mol $CH_3OH(l)$. Thus ΔG, $\Delta G°$, and $RT \ln(Q)$ all have units of J/mol of reaction. In this case the units of R are actually J K^{-1} (mol of reaction)$^{-1}$, although they are usually not written this way.

We can now calculate ΔG, where

$$\Delta G° = -2.9 \times 10^4 \text{ J/mol rxn}$$
$$R = 8.3145 \text{ J K}^{-1} \text{ mol}^{-1}$$
$$T = 273 + 25 = 298 \text{ K}$$
$$Q = \frac{1}{(P_{CO})(P_{H_2}{}^2)} = \frac{1}{(5.0)(3.0)^2} = 2.2 \times 10^{-2}$$

Note that the pure liquid methanol is not included in the calculation of Q because a pure liquid has an activity of 1, as discussed in Chapter 6. Thus

$$\begin{aligned}\Delta G &= \Delta G° + RT \ln(Q) \\ &= (-2.9 \times 10^4 \text{ J/mol rxn}) \\ &\quad + [8.3145 \text{ J K}^{-1} (\text{mol rxn})^{-1}](298 \text{ K})[\ln(2.2 \times 10^{-2})] \\ &= (-2.9 \times 10^4 \text{ J/mol rxn}) - (9.4 \times 10^3 \text{ J/mol rxn}) \\ &= -3.8 \times 10^4 \text{ J/mol rxn} \\ &= -38 \text{ kJ/mol rxn}\end{aligned}$$

Note that ΔG is significantly more negative than $\Delta G°$, implying that the reaction is more spontaneous at reactant pressures greater than 1 atm. We might expect this result from Le Châtelier's principle.

The Meaning of ΔG for a Chemical Reaction

In this section we have learned to calculate ΔG for chemical reactions under various conditions. For example, in Example 10.13 the calculations show that the formation of $CH_3OH(l)$ from $CO(g)$ at 5.0 atm reacting with $H_2(g)$ at 3.0 atm is spontaneous. What does this result mean? Does it mean that if we mixed 1.0 mol $CO(g)$ and 2.0 mol $H_2(g)$ together at pressures of 3.0 atm and 5.0 atm, respectively, that 1.0 mol of $CH_3OH(l)$ would form in the reaction flask? The answer is no. This answer may surprise you in view of what has been said in this section. It is true that 1.0 mol of $CH_3OH(l)$ has a lower free energy than 1.0 mol of $CO(g)$ at 5.0 atm plus 2.0 mol of $H_2(g)$ at 3.0 atm. However, when $CO(g)$ and $H_2(g)$ are mixed under these conditions, there is *an even lower free energy available to this system than 1.0 mol pure $CH_3OH(l)$*. For reasons we will discuss shortly, *the system can achieve the lowest possible free energy by going to equilibrium, not by going to completion*. At the equilibrium position some of the $CO(g)$ and $H_2(g)$ will remain in the reaction flask. So even though 1.0 mol of pure $CH_3OH(l)$ is at a lower free energy than 1.0 mol $CO(g)$ and 2.0 mol $H_2(g)$ at 5.0 atm and 3.0 atm, respectively, the reaction system will stop short of forming 1.0 mol of $CH_3OH(l)$. The reaction stops short of completion because the equilibrium mixture of $CH_3OH(l)$, $CO(g)$, and $H_2(g)$ exists at the lowest possible free energy available to the system.

To illustrate this point, we will explore a mechanical example. Consider balls rolling down the two hills shown in Fig. 10.13. Note that in both cases point B has a lower potential energy than point A.

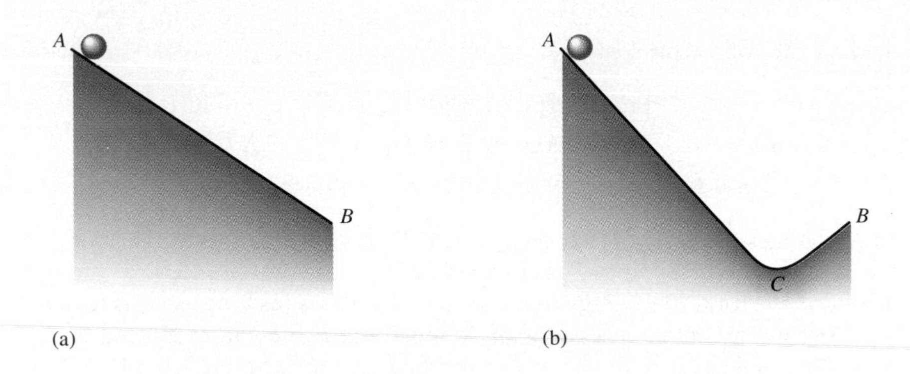

Figure 10.13
Schematic representations of balls rolling down two types of hills.

In Fig. 10.13(a) the ball will roll to point *B*. This diagram is analogous to a phase change. For example, at 25°C ice will spontaneously change completely to liquid water, because the latter has the lowest free energy. In this case liquid water is the only choice. There is no intermediate substance with lower free energy.

The situation is different for a chemical reaction system, as illustrated in Fig. 10.13(b). In Fig. 10.13(b) the ball will not get to point *B*, because there is a lower potential energy at point *C*. Like the ball, a chemical system will seek the *lowest possible* free energy, which, for reasons we will discuss below, is the equilibrium position.

Therefore, although the value of ΔG for a given reaction system tells us whether the products or reactants are favored under a given set of conditions, it does not mean that the system will proceed to pure products (if ΔG is negative) or remain at pure reactants (if ΔG is positive). Instead, the system will spontaneously go to the equilibrium position, the lowest possible free energy available to it. In the next section we will see that the value of $\Delta G°$ for a particular reaction tells us exactly where this position will be.

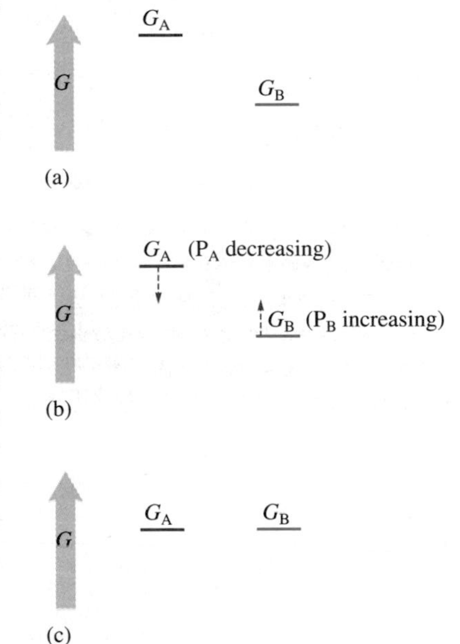

Figure 10.14
(a) The initial free energies of A and B. (b) As A(g) changes to B(g), the free energy of A decreases and that of B increases. (c) Eventually, pressures of A and B are achieved such that $G_A = G_B$, the equilibrium position.

10.11 Free Energy and Equilibrium

When the components of a given chemical reaction are mixed, they will proceed, rapidly or slowly depending on the kinetics of the process, to the equilibrium position. In Chapter 6 we defined the equilibrium position as the point at which the forward and reverse reaction rates are equal. In this chapter we look at equilibrium from a thermodynamic point of view, and we find that *the equilibrium point occurs at the lowest value of free energy available to the reaction system.* As it turns out, the two definitions give the same equilibrium state, which must be the case for both the kinetic and thermodynamic models to be valid.

To understand the relationship between free energy and equilibrium, let's consider the following simple hypothetical reaction:

$$A(g) \rightleftharpoons B(g)$$

where 1.0 mole of gaseous A is initially placed in a reaction vessel at a pressure of 2.0 atm. The free energies for A and B are diagramed as shown in Fig. 10.14(a). As A reacts to form B, the total free energy of the system changes, yielding the following results:

$$\text{Free energy of A} = G_A = G_A^\circ + RT\ln(P_A)$$
$$\text{Free energy of B} = G_B = G_B^\circ + RT\ln(P_B)$$
$$\text{Total free energy of system} = G = G_A + G_B$$

As A changes to B, G_A decreases because P_A is decreasing [Fig. 10.14(b)]. In contrast, G_B increases since P_B is increasing. The reaction proceeds to the right as long as the total free energy of the system decreases (as long as G_B is less than G_A). At some point the pressures of A and B reach the values P_A^e and P_B^e that make G_A equal to G_B. *The system has reached equilibrium* [Fig. 10.14(c)]. Since A at pressure P_A^e and B at pressure P_B^e have the same free energy (G_A equals G_B), ΔG is zero for A at pressure P_A^e changing to B at pressure P_B^e. *The system has reached minimum free energy.* There is no longer any driving force to change A to B or B to A, so the system remains at this position (the pressures of A and B remain constant).

Although we cannot obtain the absolute free energy of a substance, we use it here symbolically to illustrate the relationship between free energy and equilibrium.

Suppose that for the experiment described above the plot of free energy versus the mole fraction of A reacted is defined as shown in Fig. 10.15(a). In this experiment minimum free energy is reached when 75% of A has been changed to B. At this point the pressure of A is 0.25 times the original pressure, or

$$(0.25)(2.0\text{ atm}) = 0.50\text{ atm}$$

The pressure of B is

$$(0.75)(2.0\text{ atm}) = 1.5\text{ atm}$$

Since this is the equilibrium position, we can use the equilibrium pressures to calculate a value for K for the reaction in which A is converted to B:

$$K = \frac{P_B^e}{P_A^e} = \frac{1.5\text{ atm}}{0.50\text{ atm}} = 3.0$$

For the reaction $A(g) \rightleftharpoons B(g)$ the pressure is constant during the reaction, since the same number of gas molecules is always present.

Exactly the same equilibrium point will be achieved if we place 1.0 mole of pure B(*g*) in the flask at a pressure of 2.0 atm. In this case B will change to A until equilibrium ($G_B = G_A$) is reached. See Fig. 10.15(b).

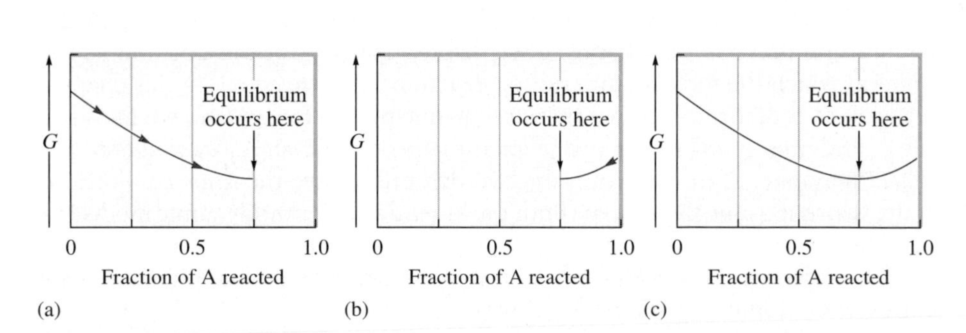

Figure 10.15
(a) The change in free energy to reach equilibrium, beginning with 1.0 mol of A(*g*) at $P_A = 2.0$ atm. (b) The change in free energy to reach equilibrium, beginning with 1.0 mol of B(*g*) at $P_B = 2.0$ atm. (c) The free energy profile for $A(g) \rightleftharpoons B(g)$ in a system containing 1.0 mol (A plus B) at $P_{TOTAL} = 2.0$ atm. Each point on the curve corresponds to the total free energy for a given combination of A and B.

The overall free energy curve for this system is shown in Fig. 10.15(c). Note that any mixture of A(g) and B(g) containing 1.0 mole of A plus B at a total pressure of 2.0 atm will react until it reaches the minimum on the curve.

In summary, when substances undergo a chemical reaction, the reaction proceeds to the minimum free energy (equilibrium), which corresponds to the point where $G_{\text{products}} = G_{\text{reactants}}$, or

$$\Delta G = G_{\text{products}} - G_{\text{reactants}} = 0$$

Here G_{products} and $G_{\text{reactants}}$ represent the sums for all products and all reactants, respectively.

We can now establish a quantitative relationship between free energy and the value of the equilibrium constant. We have seen that

$$\Delta G = \Delta G^\circ + RT\ln(Q)$$

and at equilibrium ΔG equals 0 and Q equals K. So

$$\Delta G = 0 = \Delta G^\circ + RT\ln(K)$$

or

$$\Delta G^\circ = -RT\ln(K)$$

We must note the following characteristics of this very important equation.

Case 1: $\Delta G^\circ = 0$. When ΔG° equals zero for a particular reaction, the free energies of the reactants and products are equal when all components are in the standard states (1 atm for gases). The system is at equilibrium when the pressures of all reactants and products are 1 atm, which means that K equals 1.

Case 2: $\Delta G^\circ < 0$. In this case ΔG° ($G^\circ_{\text{products}} - G^\circ_{\text{reactants}}$) is negative, which means that

$$G^\circ_{\text{products}} < G^\circ_{\text{reactants}}$$

If a flask contains the reactants and products, all at 1 atm, the system is *not* at equilibrium. Since $G^\circ_{\text{products}}$ is less than $G^\circ_{\text{reactants}}$, the system adjusts to the right to reach equilibrium. In this case K is *greater than 1,* since the pressures of the products at equilibrium are greater than 1 atm and the pressures of the reactants at equilibrium are less than 1 atm.

Case 3: $\Delta G^\circ > 0$. Since ΔG° ($G^\circ_{\text{products}} - G^\circ_{\text{reactants}}$) is positive,

$$G^\circ_{\text{reactants}} < G^\circ_{\text{products}}$$

If a flask contains the reactants and products, all at 1 atm, the system is *not* at equilibrium. In this case the system adjusts to the left (toward the reactants, which have a lower free energy) to reach equilibrium. The value of K is *less than 1,* since at equilibrium the pressures of the reactants are greater than 1 atm and the pressures of the products are less than 1 atm.

TABLE 10.7
Qualitative Relationship Between the Change in Standard Free Energy and the Equilibrium Constant for a Given Reaction

ΔG°	K
$\Delta G^\circ = 0$	$K = 1$
$\Delta G^\circ < 0$	$K > 1$
$\Delta G^\circ > 0$	$K < 1$

These results are summarized in Table 10.7. The value of K for a specific reaction can be calculated from the equation

$$\Delta G^\circ = -RT\ln(K)$$

as is shown in Examples 10.14 and 10.15.

EXAMPLE 10.14

Consider the ammonia synthesis reaction

$$N_2(g) + 3H_2(g) \rightleftharpoons 2NH_3(g)$$

where $\Delta G° = -33.3$ kJ per mole of N_2 consumed at 25°C. For each of the following mixtures of reactants and products at 25°C, predict the direction in which the system will shift to reach equilibrium.

a. $P_{NH_3} = 1.00$ atm, $P_{N_2} = 1.47$ atm, $P_{H_2} = 1.00 \times 10^{-2}$ atm
b. $P_{NH_3} = 1.00$ atm, $P_{N_2} = 1.00$ atm, $P_{H_2} = 1.00$ atm

Solution

a. We can predict the direction of the shift to equilibrium by calculating the value of ΔG, using the equation

$$\Delta G = \Delta G° + RT\ln(Q)$$

The units of ΔG, $\Delta G°$, and $RT\ln(Q)$ are all per "mole of reaction," although only the "per mole" is indicated for R (as is customary).

where $$Q = \frac{P_{NH_3}{}^2}{(P_{N_2})(P_{H_2}{}^3)} = \frac{(1.00)^2}{(1.47)[(1.00 \times 10^{-2})^3]} = 6.80 \times 10^5$$

$$T = 25 + 273 = 298 \text{ K}$$

$$R = 8.3145 \text{ J K}^{-1}\text{ mol}^{-1}$$

and $$\Delta G° = -33.3 \text{ kJ/mol} = -3.33 \times 10^4 \text{ J/mol}$$

Thus

$$\Delta G = (-3.33 \times 10^4 \text{ J}) + (8.3145 \text{ J K}^{-1}\text{ mol}^{-1})(298 \text{ K})\ln(6.80 \times 10^5)$$
$$= (-3.33 \times 10^4 \text{ J/mol}) + (3.33 \times 10^4 \text{ J/mol}) = 0$$

Since $\Delta G = 0$, the reactants and products have the same free energies at the given partial pressures. The system is already at equilibrium, and no shift occurs.

b. The partial pressures given here are all 1.00 atm, indicating that the system is in the standard state. That is,

$$\Delta G = \Delta G° + RT\ln(Q) = \Delta G° + RT\ln\frac{(1.00)^2}{(1.00)(1.00)^3}$$
$$= \Delta G° + RT\ln(1.00) = \Delta G° + 0 = \Delta G°$$

For this reaction at 25°C

$$\Delta G° = -33.3 \text{ kJ/mol}$$

The negative value for $\Delta G°$ means that in their standard states the products have a lower free energy than the reactants. Thus the system moves to the right to reach equilibrium. That is, K is greater than 1.

EXAMPLE 10.15

The overall reaction for the corrosion (rusting) of iron by oxygen is

$$4Fe(s) + 3O_2(g) \rightleftharpoons 2Fe_2O_3(s)$$

Using the following data, calculate the equilibrium constant for this reaction at 25°C.

Substance	ΔH_f° (kJ/mol)	S° (J K^{-1} mol^{-1})
$Fe_2O_3(s)$	−826	90
$Fe(s)$	0	27
$O_2(g)$	0	205

Solution

To calculate K for this reaction, we will use the equation

$$\Delta G^\circ = -RT\ln(K)$$

We must first calculate ΔG° from

$$\Delta G^\circ = \Delta H^\circ - T\Delta S^\circ$$

where

$$\begin{aligned}\Delta H^\circ &= 2\Delta H^\circ_{f\,(Fe_2O_3(s))} - 3\Delta H^\circ_{f\,(O_2(g))} - 4\Delta H^\circ_{f\,(Fe(s))}\\ &= 2\text{ mol}(-826\text{ kJ/mol}) - 0 - 0\\ &= -1652\text{ kJ} = -1.652 \times 10^6\text{ J}\\ \Delta S^\circ &= 2S^\circ_{Fe_2O_3} - 3S^\circ_{O_2} - 4S^\circ_{Fe}\\ &= 2\text{ mol}(90\text{ J K}^{-1}\text{ mol}^{-1}) - 3\text{ mol}(205\text{ J K}^{-1}\text{ mol}^{-1})\\ &\quad - 4\text{ mol}(27\text{ J K}^{-1}\text{ mol}^{-1})\\ &= -543\text{ J/K}\end{aligned}$$

and $$T = 273 + 25 = 298\text{ K}$$

Then

$$\begin{aligned}\Delta G^\circ &= \Delta H^\circ - T\Delta S^\circ = (-1.652 \times 10^6\text{ J}) - (298\text{ K})(-543\text{ J/K})\\ &= -1.490 \times 10^6\text{ J}\end{aligned}$$

and $$\begin{aligned}\Delta G^\circ &= -RT\ln(K) = -1.490 \times 10^6\text{ J}\\ &= -(8.3145\text{ J K}^{-1}\text{ mol}^{-1})(298\text{ K})\ln(K)\end{aligned}$$

The units of ΔG, ΔG°, and $RT\ln(Q)$ are all per "mole of reaction," although only the "per mole" is indicated for R (as is customary).

Thus $$\ln(K) = \frac{1.490 \times 10^6}{2.48 \times 10^3} = 601 \quad\text{and}\quad K = e^{601}$$

In terms of base 10, $$K = 10^{261}$$

This is a very large equilibrium constant. The rusting of iron is clearly very favorable from a thermodynamic point of view.

The Temperature Dependence of K

In Chapter 6 we used Le Châtelier's principle to predict qualitatively how the value of K for a given reaction would change with a change in temperature. Now we can specify the quantitative dependence of the equilibrium constant on temperature from the relationship

$$\Delta G^\circ = -RT\ln(K) = \Delta H^\circ - T\Delta S^\circ$$

We can rearrange this equation to give

$$\ln(K) = -\frac{\Delta H^\circ}{RT} + \frac{\Delta S^\circ}{R} = -\frac{\Delta H^\circ}{R}\left(\frac{1}{T}\right) + \frac{\Delta S^\circ}{R}$$

Note that this is a linear equation of the form $y = mx + b$, where $y = \ln(K)$, $m = -\Delta H^\circ/R$ = slope, $x = 1/T$, and $b = \Delta S^\circ/R$ = intercept. This means that if values of K for a given reaction are determined at various temperatures, a plot of $\ln(K)$ versus $1/T$ will be linear, with slope $-\Delta H^\circ/R$ and intercept $\Delta S^\circ/R$. This result assumes that both ΔH° and ΔS° are independent of temperature over the temperature range considered. This assumption is good only over a relatively small temperature range.

An important conclusion that can be drawn from the above equation is that the sign of the slope of the plot of $\ln(K)$ versus $1/T$ depends on the sign of ΔH° for the reaction. Note that an exothermic reaction ($\Delta H^\circ < 0$) will show a positive slope (ΔH° is negative so $-\Delta H^\circ/R$ is positive) for the $\ln(K)$ versus $1/T$ plot. In this case $\ln(K)$ will increase as $1/T$ increases (T decreases). Thus K increases as T is decreased or conversely, K decreases as T is increased. This is exactly the temperature dependence of K predicted for an exothermic reaction by Le Châtelier's principle (see Section 6.8). Of course, the reverse applies for the equilibrium constant for an endothermic reaction: the value of K increases as the temperature is increased.

10.12 Free Energy and Work

One of the main reasons we are interested in physical and chemical processes is because we want to use them to do work for us, and we want this work done as efficiently and economically as possible. We have already seen that at constant temperature and pressure the sign of the change in free energy tells us whether a given process is spontaneous. This information is very useful, because it prevents us from wasting effort on a process that has no inherent tendency to occur. Although a thermodynamically favorable chemical reaction may not occur to any appreciable extent at a given temperature because it is too slow, finding a catalyst to speed up the reaction makes sense in this case. On the other hand, if the reaction is prevented from occurring by its thermodynamic characteristics, we would be wasting our time looking for a catalyst.

In addition to being important qualitatively (telling us whether a process is spontaneous), the change in free energy is important quantitatively because it can tell us how much work can be done through a given process. In fact, as we will show below, the *maximum possible useful work obtainable from a process at constant temperature and pressure is equal to the change in free energy:*

$$w_{\text{useful}}^{\text{max}} = \Delta G$$

This relationship explains why this function is called the *free* energy. Under certain conditions ΔG for a spontaneous process represents the energy that is *free to do useful work*. On the other hand, for a process that is not spontaneous, the value of ΔG tells us the minimum amount of work that must be *expended* to make the process occur.

Recall that the maximum work would only occur along the hypothetical reversible pathway and is thus unattainable (although it can be approached closely in some situations). In any case, knowing the maximum work for a process is still important, because then we can evaluate the efficiency of any machine that might be based on the process.

We will now prove the above relationship between ΔG and $w_{\text{useful}}^{\text{max}}$. First, we define the total work w:

$$w = w_{\text{useful}} + \underbrace{w_{\text{useless}}}_{PV \text{ work}} = w_{\text{useful}} + w_{\text{pv}}$$

Note that PV work is related to the expansion or contraction of the system and is not counted as useful work. From the definition of ΔE, and assuming constant P and T, we have

$$\begin{aligned}\Delta E &= q_{\text{p}} + w = q_{\text{p}} + w_{\text{useful}} + w_{\text{pv}} \\ &= q_{\text{p}} + w_{\text{useful}} - P\Delta V\end{aligned}$$

From the definition of enthalpy,

$$H = E + PV$$

we have

$$\begin{aligned}\Delta H &= \Delta E + P\Delta V \\ &= \underbrace{q_{\text{p}} + w_{\text{useful}} - P\Delta V}_{\Delta E} + P\Delta V \\ &= q_{\text{p}} + w_{\text{useful}}\end{aligned}$$

Next, from the definition of free energy,

$$G = H - TS$$

we have

$$\Delta G = \Delta H - T\Delta S = \underbrace{q_{\text{p}} + w_{\text{useful}}}_{\Delta H} - T\Delta S$$

For the reversible pathway

$$w_{\text{useful}} = w_{\text{useful}}^{\text{max}} \qquad \text{and} \qquad q_{\text{p}} = q_{\text{p}}^{\text{rev}}$$

Thus for the reversible pathway

$$\Delta G = q_{\text{p}}^{\text{rev}} + w_{\text{useful}}^{\text{max}} - T\Delta S$$

and since

$$\Delta S = \frac{q_{\text{p}}^{\text{rev}}}{T}$$

then

$$q_{\text{p}}^{\text{rev}} = T\Delta S$$

So we have

$$\Delta G = T\Delta S + w_{\text{useful}}^{\text{max}} - T\Delta S \qquad \text{or} \qquad \Delta G = w_{\text{useful}}^{\text{max}}$$

Thus we have shown that at constant temperature and pressure the change in free energy for a process gives the maximum useful work available from that process.

Let us consider a few more points in connection with these relationships. If a process is carried out so that $w_{\text{useful}} = 0$, then the expression

$$\Delta G = q_p + w_{\text{useful}} - T\Delta S$$

becomes

$$\Delta G = q_p - T\Delta S$$

And since $\Delta G = \Delta H - T\Delta S$, we have

$$\Delta H - T\Delta S = q_p - T\Delta S$$
$$\Delta H = q_p$$

This relationship between ΔH and q_p is used frequently in thermochemical studies. We bring it up again to emphasize that $\Delta H = q_p$ only at constant pressure *and when no useful work is done* (only PV work is allowed). This last condition is often neglected.

If a process is carried out so that w_{useful} is at a maximum (the hypothetical reversible pathway where $\Delta G = w_{\text{useful}}$), then from the expression

$$\Delta G = q_p + w_{\text{useful}} - T\Delta S$$

we have

$$q_p = T\Delta S$$

Thus q_p, which is pathway-dependent, varies between ΔH (when $w_{\text{useful}} = 0$) and $T\Delta S$ (when $w_{\text{useful}} = w_{\text{useful}}^{\text{max}}$). The quantity $T\Delta S$ represents the minimum heat flow that must accompany the process under consideration. That is, $T\Delta S$ represents the minimum energy that must be "wasted" through heat flow as the process occurs.

In summary, at constant T and P,

$$q_p = \Delta H \quad \text{if} \quad w_{\text{useful}} = 0$$
$$q_p = T\Delta S \quad \text{if} \quad w_{\text{useful}} = w_{\text{useful}}^{\text{max}}$$

10.13 Reversible and Irreversible Processes: A Summary

As we demonstrated in the analysis of the isothermal expansion-compression of an ideal gas in Section 10.2, the amount of work we actually obtain from a spontaneous process is *always* less than the maximum possible amount.

To explore this idea more fully in a more realistic context than that of an ideal gas, let's consider an electric current flowing through the starter motor of a car. The current is generated from a chemical change in a battery. Since we can calculate ΔG for the battery reaction, we can determine the energy available to do work. Can we use all of this energy to do work? No, because a current flowing through a wire causes frictional heating, and the greater the current, the greater the heat. This heat represents wasted energy—it is not useful for running the starter motor. We can minimize this energy waste by running very low currents through the motor circuit. However, zero current flow would be necessary to eliminate frictional heating entirely, and we cannot derive any work from the motor if no current flows. This example shows the

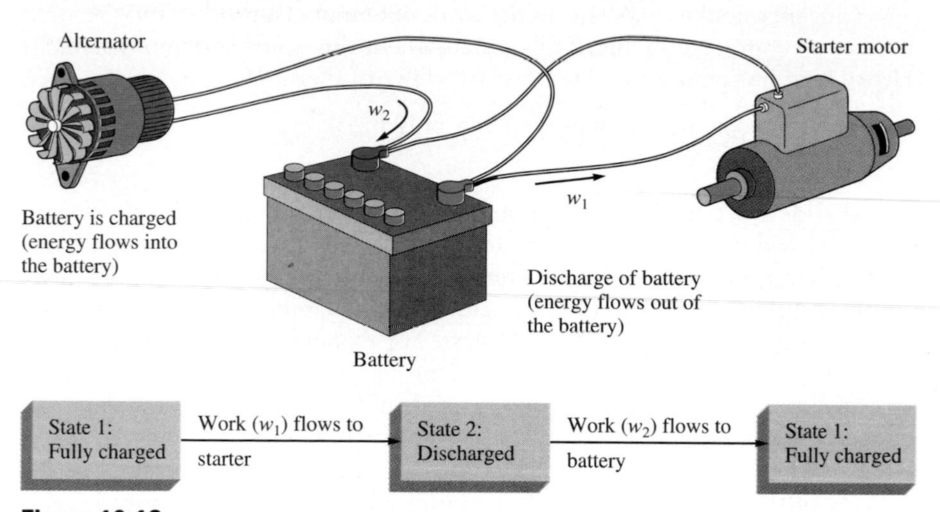

Figure 10.16
A battery can do work by sending current to a starter motor. The battery can then be recharged by forcing current through it. If the current flow in both processes is infinitesimally small, $|w_1| = |w_2|$. This is a *reversible process.* But if the current flow is finite, as it would be in any real case, $|w_2| > |w_1|$. This is an *irreversible process* (the *universe is different* after the cyclic process occurs). All real processes are irreversible.

difficulty nature places us in. Using a process to do work requires that some of the energy be wasted; and usually, the faster we run the process, the more energy we waste.

Achieving the maximum work available from a spontaneous process can only occur via a hypothetical pathway. Any real pathway wastes energy in the sense that the maximum work is not obtained. If we could discharge the battery infinitely slowly by an infinitesimally small current flow, we could achieve the maximum useful work. Also, if we could then recharge the battery by using an infinitesimally small current, exactly the same amount of energy would be used to return the battery to its original state as was obtained in the infinitesimally slow discharge. After we cycle the battery in this way, the universe (the system and surroundings) is exactly the same as it was before the cyclic process. Therefore this is a reversible process (see Fig. 10.16).

However, if the battery is discharged to run the starter motor and then recharged by using a *finite* current flow, as is actually the case, *more* work will always be required to recharge the battery than the battery produces as it discharges. Thus, even though the battery (the system) has returned to its original state, the surroundings has not, because the surroundings had to furnish a net amount of work as the battery was cycled. The *universe is different* after this cyclic process is performed, and this process is irreversible. *All real processes are irreversible.*

Recall that after any real cyclic process is carried out in a system, the surroundings has less ability to do work and contain more thermal energy. In other words, in any real cyclic process work is changed to heat in the surroundings, and the entropy of the universe increases. This is another way of stating the second law of thermodynamics.

Thus thermodynamics tells us the work potential of a process and then tells us that we can never achieve this potential. In this spirit, thermodynamicist Henry Bent has paraphrased the first two laws of thermodynamics as follows:

First law: You can't win, you can only break even.

Second law: You can't break even.

When energy is used to do work, it becomes less organized and less concentrated and thus less useful.

The ideas we have discussed in this section are applicable to the energy crisis that will probably increase in severity over the next 25 years. The crisis is obviously not one of supply; the first law tells us that the universe contains a constant supply of energy. The problem is the availability of *useful* energy. *As we use energy, we degrade its usefulness.* For example, when gasoline reacts with oxygen in the combustion reaction, the change in potential energy results in heat flow. Thus the energy concentrated in the bonds of the gasoline and oxygen molecules ends up *spread* over the surroundings as thermal energy, where it is much more difficult to harness for useful work. In this way the entropy of the universe increases: concentrated energy becomes spread out—more disordered and less useful. Therefore, the crux of the energy problem is that we are rapidly consuming the concentrated energy found in fossil fuels. It took millions of years to concentrate the sun's energy in these fuels, which we will consume in a few hundred years. Thus we must consider carefully our use of these energy sources and use them as wisely as possible.

EXERCISES

A blue exercise number indicates that the answer to that exercise appears at the back of this book.

Spontaneity and Entropy

1. Define what is meant by a *spontaneous process.*
2. Which of the following processes require energy as they occur?
   a. Salt dissolves in H_2O.
   b. A clear solution becomes a uniform color after a few drops of dye are added.
   c. A cell produces proteins from amino acids.
   d. Iron rusts.
   e. A house is built.
   f. A satellite is launched into orbit.
   g. A satellite falls back to earth.
3. Consider the following energy levels, each capable of holding two objects:

$$E = 2\text{ kJ} \quad \underline{\qquad}$$
$$E = 1\text{ kJ} \quad \underline{\qquad}$$
$$E = 0 \quad \underline{\text{XX}}$$

   Draw all of the possible arrangements of the two identical particles (represented by X) in the three energy levels. What total energy is most likely, that is, occurs the greatest number of times? Assume the particles are indistinguishable from each other.
4. Do Exercise 3 with two particles A and B, which can be distinguished from each other.
5. a. Define *entropy.*
   b. Is entropy a form of energy?
6. Which of the following involve an increase in the entropy of the system under consideration?
   a. melting of a solid
   b. evaporation of a liquid
   c. sublimation
   d. freezing
   e. mixing
   f. separation
   g. diffusion
7. In the roll of two dice, what *total* number is the most likely to occur? Is there an energy reason why this number is favored? Would energy have to be spent to increase the probability of getting a particular number (that is, to cheat)?
8. Entropy can be calculated by a relationship proposed by Ludwig Boltzmann:
$$S = k_B \ln \Omega$$
   where $k_B = 1.38 \times 10^{-23}$ J/K and Ω is the number of ways a particular state can be obtained. (This equation is engraved on Boltzmann's tombstone.) Calculate S for the three arrangements of particles in Table 10.1.
9. The standard entropy values ($S°$) for $H_2O(l)$ and $H_2O(g)$ are 70. J K^{-1} mol^{-1} and 189 J K^{-1} mol^{-1}, respectively. Calculate the ratio of Ω_g to Ω_l for water using Boltzmann's equation. (See Exercise 8.)

10. Describe how the following changes affect the positional entropy of a substance.
    a. increase in volume of a gas at constant T
    b. increase in temperature of a gas at constant V
    c. increase in pressure of a gas at constant T

11. Choose the compound with the greatest positional entropy in each case.
    a. 1 mol of H_2 at STP or 1 mol of N_2O at STP
    b. 1 mol of H_2 at STP or 1 mol of H_2 at 100°C and 0.5 atm
    c. 1 mol of N_2 at STP or 1 mol of N_2 at 100 K and 2.0 atm
    d. 1 mol of $H_2O(s)$ at 0°C or 1 mol of $H_2O(l)$ at 20°C

Entropy and the Second Law of Thermodynamics: Free Energy

12. Define each of the following.
    a. system
    b. surroundings
    c. closed system
    d. open system

13. The synthesis of glucose directly from CO_2 and H_2O and the synthesis of proteins directly from amino acids are both nonspontaneous processes under standard conditions. Yet these processes must occur in order for life to exist. In light of the second law of thermodynamics, how can life exist?

14. Could life exist if every organism lived in a closed system? Explain.

15. For each of the following pairs of substances, which substance has the greater value of $S°$?
    a. glucose ($C_6H_{12}O_6$) or sucrose ($C_{12}H_{22}O_{11}$)
    b. H_2O (at 0 K) or H_2O (at 0°C)
    c. $H_2O(l)$ (at 25°C) or $H_2S(g)$ (at 25°C)
    d. N_2O (at 0 K) or He (at 10 K)
    e. $N_2O(g)$ (1 atm, at 25°C) or He(g) (1 atm, at 25°C)
    f. HF(g) (1 atm, at 25°C) or HCl(g) (1 atm, at 25°C)

16. Predict the sign of $\Delta S°$ for each of the following changes.
    a.

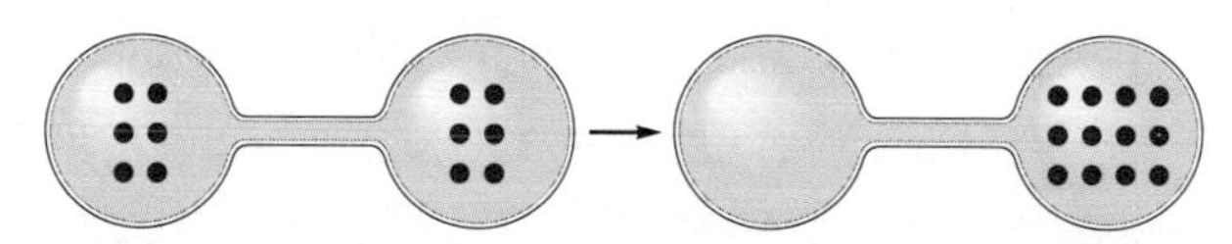

    b. $AgCl(s) \rightarrow Ag^+(aq) + Cl^-(aq)$
    c. $2H_2(g) + O_2(g) \rightarrow 2H_2O(l)$
    d. $Na(s) + \frac{1}{2}Cl_2(g) \rightarrow NaCl(s)$
    e. $HCl(g) \rightarrow H^+(aq) + Cl^-(aq)$
    f. $NaCl(s) \rightarrow Na^+(aq) + Cl^-(aq)$

17. Calculate $\Delta S°$ for each of the following reactions.
    a. $H_2(g) + \frac{1}{2}O_2(g) \rightarrow H_2O(g)$
    b. $3O_2(g) \rightarrow 2O_3(g)$
    c. $N_2(g) + O_2(g) \rightarrow 2NO(g)$

18. For the reaction

$$CS_2(g) + 3O_2(g) \longrightarrow CO_2(g) + 2SO_2(g)$$

$\Delta S°$ is equal to -143 J/K. Use this value and data from Appendix 4 to calculate the value of $S°$ for $CS_2(g)$.

19. For the reaction

$$2Al(s) + 3Br_2(l) \longrightarrow 2AlBr_3(s)$$

$\Delta S°$ is equal to -144 J/K. Use this value and data from Appendix 4 to calculate the value of $S°$ for solid aluminum bromide.

20. The boiling point of chloroform ($CHCl_3$) is 61.7°C. The enthalpy of vaporization is 31.4 kJ/mol. Calculate the entropy of vaporization for $CHCl_3(l)$.

21. For mercury the enthalpy of vaporization is 58.51 kJ/mol and the entropy of vaporization is 92.92 J K^{-1} mol^{-1}. What is the normal boiling point of mercury?

22. The melting point of tungsten is the second highest among the elements (only that of carbon is higher). The melting point of tungsten is 3680 K. The enthalpy of fusion is 35.2 kJ/mol. What is the entropy of fusion?

23. For ammonia (NH_3) the enthalpy of fusion is 5.65 kJ/mol and the entropy of fusion is 28.9 J K^{-1} mol^{-1}.
    a. Will $NH_3(s)$ spontaneously melt at 200. K?
    b. What is the approximate melting point of ammonia?

24. Two crystalline forms of white phosphorus are known. Both forms contain P_4 molecules, but the molecules are packed together in different ways. The α form is always obtained when the liquid freezes. However, below -76.9°C the α form spontaneously converts to the β form:

$$P_4(s, \alpha) \longrightarrow P_4(s, \beta)$$

    a. Predict the signs of ΔH and ΔS for this process.
    b. Predict which form of phosphorus has the more ordered crystalline structure.

Free Energy and Chemical Reactions

25. From data in Appendix 4, calculate $\Delta H°$, $\Delta S°$, and $\Delta G°$ for each of the following reactions at 25°C.
    a. $CH_4(g) + 2O_2(g) \rightarrow CO_2(g) + 2H_2O(g)$
    b. $6CO_2(g) + 6H_2O(l) \rightarrow \underset{\text{Glucose}}{C_6H_{12}O_6(s)} + 6O_2(g)$
    c. $P_4O_{10}(s) + 6H_2O(l) \rightarrow 4H_3PO_4(s)$
    d. $HCl(g) + NH_3(g) \rightarrow NH_4Cl(s)$

26. For the reaction

$$SF_4(g) + F_2(g) \longrightarrow SF_6(g)$$

the value of $\Delta G°$ is -374 kJ. Use this value and data from Appendix 4 to calculate the value of $\Delta G_f°$ for $SF_4(g)$.

27. The value of $\Delta G°$ for the reaction

$$2Al(OH)_3(s) \longrightarrow Al_2O_3(s) + 3H_2O(g)$$

is +7 kJ. Use this value and data from Appendix 4 to calculate the value of the standard free energy of formation of aluminum hydroxide.

28. Acrylonitrile is the starting material used in the manufacture of acrylic fibers (U.S. production capacity is more than 2 million pounds). Three industrial processes for the production of acrylonitrile are given below. Using data from Appendix 4, calculate $\Delta S°$, $\Delta H°$, and $\Delta G°$ for each process. For part a, assume that $T = 25°C$; for part b, $T = 70.° C$; and for part c, $T = 700.° C$.

a. $CH_2{-}CH_2(g)$ (ring closed through O) $+ HCN(g) \rightarrow CH_2{=}CHCN(g) + H_2O(l)$

Ethylene oxide — Acrylonitrile

b. $$HC{\equiv}CH(g) + HCN(g) \xrightarrow[70\text{–}90°C]{CaCl_2 \cdot HCl} CH_2{=}CHCN(g)$$

c. $$4CH_2{=}CHCH_3(g) + 6NO(g) \xrightarrow[Ag]{700°C} 4CH_2{=}CHCN(g) + 6H_2O(g) + N_2(g)$$

29. Consider two reactions for the production of ethanol:

$$C_2H_4(g) + H_2O(g) \longrightarrow CH_3CH_2OH(l)$$
$$C_2H_6(g) + H_2O(g) \longrightarrow CH_3CH_2OH(l) + H_2(g)$$

Which would be more thermodynamically feasible? Why?

30. When most biological enzymes are heated, they lose their catalytic activity. The change

Original enzyme $\longrightarrow$ new form

that occurs upon heating is endothermic and spontaneous. Is the structure of the original enzyme or its new form more ordered? Explain your answer.

31. Consider the reaction

$$H_2(g) \longrightarrow 2H(g)$$

a. Predict the signs of ΔH and ΔS.
b. Will the reaction be more spontaneous at high or low temperatures?

32. Hydrogen cyanide is produced industrially by the following reaction:

$$2NH_3(g) + 3O_2(g) + 2CH_4(g) \xrightarrow[Pt\text{-}Rh]{1000°C} 2HCN(g) + 6H_2O(g)$$

Is the high temperature needed for thermodynamic or kinetic reasons?

33. The autoionization of water at 25°C,

$$H_2O(l) \rightleftharpoons H^+(aq) + OH^-(aq)$$

has the equilibrium constant 1.00×10^{-14}. Calculate $\Delta G°$ for this process at 25°C.

34. The Ostwald process for the commercial production of nitric acid involves three steps:

$$4NH_3(g) + 5O_2(g) \xrightarrow[825°C]{Pt} 4NO(g) + 6H_2O(g)$$
$$2NO(g) + O_2(g) \longrightarrow 2NO_2(g)$$
$$3NO_2(g) + H_2O(l) \longrightarrow 2HNO_3(l) + NO(g)$$

a. Calculate $\Delta H°$, $\Delta S°$, $\Delta G°$, and K (at 298 K) for each of the three steps in the Ostwald process (see Appendix 4).
b. Calculate the equilibrium constant for the first step at 825°C.
c. What is the reason for the high temperature in the first step?

35. One of the reactions that destroys ozone in the upper atmosphere is

$$NO(g) + O_3(g) \rightleftharpoons NO_2(g) + O_2(g)$$

Using data from Appendix 4, calculate $\Delta G°$ and K (at 298 K) for this reaction.

36. Hydrogen sulfide can be removed from natural gas by the reaction

$$2H_2S(g) + SO_2(g) \rightleftharpoons 3S(s) + 2H_2O(g)$$

Calculate $\Delta G°$ and K (at 298 K) for this reaction. Will this reaction be most favored at a high or a low temperature?

37. Cells use the hydrolysis of adenosine triphosphate, abbreviated ATP, as a source of energy. Symbolically, this reaction can be represented as

$$ATP(aq) + H_2O(l) \longrightarrow ADP(aq) + H_2PO_4^-(aq)$$

where ADP represents adenosine diphosphate. For this reaction $\Delta G° = -30.5$ kJ/mol.
a. Calculate K at 25°C.
b. If all of the free energy from the metabolism of glucose

$$C_6H_{12}O_6(s) + 6O_2(g) \longrightarrow 6CO_2(g) + 6H_2O(l)$$

goes into the production of ATP, how many ATP molecules can be produced for every molecule of glucose?

38. Carbon monoxide is toxic because it binds much more strongly to the iron in hemoglobin than does O_2. The equilibrium constant for the binding of CO is about two hundred times that for the binding of O_2. That is, for the reactions

$$(heme)Fe + O_2 \overset{K_{O_2}}{\rightleftharpoons} (heme)Fe - O_2$$
$$(heme)Fe + CO \overset{K_{CO}}{\rightleftharpoons} (heme)Fe - CO$$

$K_{CO}/K_{O_2} = 2.1 \times 10^2$. Calculate the difference in $\Delta G°$ for the binding of CO and O_2 to hemoglobin at 25°C.

39. One reaction that occurs in human metabolism is

$$\underset{\text{Glutamic acid}}{HO_2CCH_2CH_2\underset{\substack{|\\ NH_2}}{C}HCO_2H(aq)} + NH_3(aq) \rightleftharpoons$$

$$\underset{\text{Glutamine}}{H_2N\overset{\substack{O\\ \|}}{C}CH_2CH_2\underset{\substack{|\\ NH_2}}{C}HCO_2H(aq)} + H_2O(l)$$

For this reaction $\Delta G° = 14$ kJ at 25°C.

a. Calculate K for this reaction at 25°C.
b. In a living cell this reaction is coupled with the hydrolysis of ATP. (See Exercise 37.) Calculate $\Delta G°$ and K at 25°C for the following reaction:

$$\text{Glutamic acid}(aq) + \text{ATP}(aq) + NH_3(aq) \rightleftharpoons \text{Glutamine}(aq) + \text{ADP}(aq) + H_2PO_4^-(aq)$$

40. The standard free energies of formation and the standard enthalpies of formation at 298 K for difluoroacetylene (C_2F_2) and hexafluorobenzene (C_6F_6) are

	$\Delta G_f°$ (kJ/mol)	$\Delta H_f°$ (kJ/mol)
$C_2F_2(g)$	191.2	241.3
$C_6F_6(g)$	78.2	132.8

For the following reaction:

$$C_6F_6(g) \rightleftharpoons 3C_2F_2(g)$$

a. calculate $\Delta S°$ at 298 K.
b. calculate K at 298 K.
c. estimate K at 3000. K.

Free Energy and Pressure

41. Calculate ΔG for the reaction

$$NO(g) + O_3(g) \longrightarrow NO_2(g) + O_2(g)$$

for the following conditions:

$T = 298$ K

$P_{NO} = 1.00 \times 10^{-6}$ atm $\quad P_{O_3} = 2.00 \times 10^{-6}$ atm

$P_{NO_2} = 1.00 \times 10^{-7}$ atm $\quad P_{O_2} = 1.00 \times 10^{-3}$ atm

For $\Delta G°$ use your answer from Exercise 35.

42. Calculate ΔG for the reaction

$$2H_2S(g) + SO_2(g) \rightleftharpoons 3S(s) + 2H_2O(g)$$

for the following conditions at 25°C:

$P_{H_2S} = 1.0 \times 10^{-4}$ atm

$P_{SO_2} = 1.0 \times 10^{-2}$ atm

$P_{H_2O} = 3.0 \times 10^{-2}$ atm

See Exercise 36 for $\Delta G°$.

43. Using data from Appendix 4, calculate $\Delta H°$, $\Delta S°$, and K (at 298 K) for the synthesis of ammonia by the Haber process:

$$N_2(g) + 3H_2(g) \rightleftharpoons 2NH_3(g)$$

Calculate ΔG for this reaction under the following conditions (assume an uncertainty of ± 1 in all quantities).

a. $T = 298$ K, $P_{N_2} = P_{H_2} = 200$ atm, $P_{NH_3} = 50$ atm
b. $T = 298$ K, $P_{N_2} = 200$ atm, $P_{H_2} = 600$ atm, $P_{NH_3} = 200$ atm
c. $T = 100$ K, $P_{N_2} = 50$ atm, $P_{H_2} = 200$ atm, $P_{NH_3} = 10$ atm
d. $T = 700$ K, $P_{N_2} = 50$ atm, $P_{H_2} = 200$ atm, $P_{NH_3} = 10$ atm

Additional Exercises

44. If you calculate a value for $\Delta G°$ for a reaction using the values of $\Delta G_f°$ in Appendix 4 and get a negative number, is it correct to say that the reaction is always spontaneous?

45. When the environment is contaminated by a toxic or potentially toxic substance—for example, from a chemical spill or from the use of insecticides—the substance tends to disperse. How is this event consistent with the second law of thermodynamics? In terms of the second law, which requires the least work: cleaning the environment after it has been contaminated, or trying to prevent the contamination before it occurs? Explain your answer.

46. A green plant synthesizes glucose by photosynthesis as shown in the reaction

$$6CO_2(g) + 6H_2O(l) \longrightarrow C_6H_{12}O_6(s) + 6O_2(g)$$

Animals use glucose as a source of energy:

$$C_6H_{12}O_6(s) + 6O_2(g) \longrightarrow 6CO_2(g) + 6H_2O(l)$$

If we were to assume that both of these processes occur to the same extent in a cyclic process, what thermodynamic property must have a nonzero value?

47. Using the relationship

$$\ln(K) = -\frac{\Delta H°}{RT} + \frac{\Delta S°}{R}$$

show that for a system at equilibrium the equilibrium will shift to the right for an endothermic process when the temperature is increased.

48. a. Use the equation in Exercise 47 to determine ΔH° and ΔS° for the autoionization of water:

$$H_2O(l) \rightleftharpoons H^+(aq) + OH^-(aq)$$

T (°C)	K
0	1.14×10^{-15}
25	1.00×10^{-14}
35	2.09×10^{-14}
40.	2.92×10^{-14}
50.	5.47×10^{-14}

b. Estimate the value of ΔG° for the autoionization of water at its critical temperature, 374°C.

c. It has been postulated that at very high temperatures and pressures, water exists as an ionic solid, $H_3O^+OH^-$. Assuming ΔH° and ΔS° are temperature independent, is there a temperature at which $\Delta G^\circ = 0(K = 1)$ for the autoionization of water?

49. Using data from Appendix 4, calculate ΔH°, ΔG° and K_p (at 298 K) for the production of ozone from oxygen:

$$3O_2(g) \rightleftharpoons 2O_3(g)$$

At 30 km above the surface of the earth, the temperature is about 230. K and the partial pressure of oxygen is about 1.0×10^{-3} atm. Estimate the partial pressure of ozone in equilibrium with oxygen at 30 km above the earth's surface. Is it reasonable to assume that the equilibrium between oxygen and ozone is maintained under these conditions? Explain.

50. Elemental sulfur can exist in two crystalline forms, rhombic and monoclinic. Calculate the temperature for the conversion of monoclinic sulfur to rhombic sulfur given the following data:

	ΔH_f° (kJ/mol)	S° (J K^{-1} mol^{-1})
S (rhombic)	0	31.88
S (monoclinic)	0.30	32.55

51. Using data from Appendix 4, calculate ΔH°, ΔS°, and ΔG° for the following reactions that produce acetic acid:

$$CH_4(g) + CO_2(g) \longrightarrow CH_3\overset{\overset{O}{\|}}{C}{-}OH(l)$$

$$CH_3OH(g) + CO(g) \longrightarrow CH_3\overset{\overset{O}{\|}}{C}{-}OH(l)$$

Which reaction would you choose as a commercial method for producing acetic acid, CH_3CO_2H? What temperature conditions would you choose for the reaction?

52. In the text we derived the equation

$$\Delta G = \Delta G^\circ + RT\ln(Q)$$

for gaseous reactions, where the quantities in Q were expressed in units of pressure. We can also use units of mol/L for the quantities in Q. With this in mind, calculate ΔG for the reaction

$$H_2O(l) \rightleftharpoons H^+(aq) + OH^-(aq)$$

under the following conditions at 25°C.

a. $[H^+] = [OH^-] = 1.00 \times 10^{-7}\ M$
b. $[H^+] = 1.00 \times 10^{-5}\ M$, $[OH^-] = 1.00 \times 10^{-9}\ M$
c. $[H^+] = [OH^-] = 1.00 \times 10^{-10}\ M$
d. $[H^+] = 10.0\ M$, $[OH^-] = 1.00 \times 10^{-7}\ M$
e. $[H^+] = 1.00\ M$, $[OH^-] = 1.00\ M$

See Exercise 33 for the ΔG° value. On the basis of the calculated ΔG values, in what direction will the system shift to reach equilibrium in each case? Are these results consistent with Le Châtelier's principle?

53. Many biochemical reactions that occur in cells require relatively high concentrations of potassium ion (K^+). The concentration of K^+ in muscle cells is about 0.15 M. The concentration of K^+ in blood plasma is about 0.0050 M. The high internal concentration in cells is maintained by pumping K^+ from the plasma. How much work must be done to transport 1.0 mol of K^+ from the blood to the inside of a muscle cell at 37°C (normal body temperature)? When 1.0 mol of K^+ is transferred from blood to the cells, do any other ions have to be transported? Why or why not? Much of the ATP (see Exercise 37) formed from metabolic processes is used to provide energy for transport of cellular components. How much ATP must be hydrolyzed to provide the energy for the transport of 1.0 mol of K^+?

54. Entropy has been described as "time's arrow." Interpret this view of entropy.

55. The lines from T. S. Eliot's *The Hollow Men (V)*

> . . this is the way the world ends—
> not with a bang, but a whimper

have been said by some to describe the eventual heat death of the universe. Comment on this view.

(Excerpt from *The Hollow Men* in *Collected Poems 1909–1962* by T. S. Eliot, copyright 1936 by Harcourt Brace Jovanovich, Inc.; copyright © 1963, 1964 by T. S. Eliot. Reprinted by permission of the publisher.)

56. Human DNA contains almost twice as much information as is needed to code for all of the substances produced in the body. Likewise, the digital data sent from *Voyager 2* contains one redundant bit out of every two bits of information. The space telescope transmits three redundant bits for every bit of information. How is entropy related to

the transmission of information? What do you think is accomplished by having so many redundant bits of information in both DNA and the space probes?

57. How much energy must be added at constant volume to 15.0 g He(g) to raise its temperature from 23.0°C to 79.5°C?

58. An electrical heater in a system consisting of a cylinder with a movable piston delivers 4.53 kJ of energy (as heat) to the system. The piston moves against an external pressure such that $P\Delta V = 2.74$ kJ. What is the change in internal energy for the system?

59. A sample of metallic aluminum weighing 50.0 g is heated to 85.0°C. It is then added to an insulated vessel containing 250.0 g of H_2O at 25.0°C. The final temperature of the mixture is 27.5°C. Calculate the approximate value for the specific heat capacity of aluminum.

60. Calculate the energy required to change the temperature of 1.00 kg of ethane (C_2H_6) from 25.0°C to 73.4°C in a rigid vessel. (C_v for C_2H_6 is 44.60 J K^{-1} mol^{-1}.) Calculate the energy required for this same temperature change at constant pressure. Calculate the change in internal energy of the gas in each of these processes.

61. The molar heat capacities for carbon dioxide at 298.0 K are

$$C_v = 28.95 \text{ J K}^{-1} \text{ mol}^{-1}$$

$$C_p = 37.27 \text{ J K}^{-1} \text{ mol}^{-1}$$

The molar entropy of carbon dioxide gas at 298.0 K and 1.000 atm is 213.64 J K^{-1} mol^{-1}.

a. Calculate the energy required to change the temperature of 1.000 mol of carbon dioxide gas from 298.0 K to 350.0 K, both at constant volume and at constant pressure.
b. Calculate the molar entropy of $CO_2(g)$ at 350.0 K and 1.000 atm.
c. Calculate the molar entropy of $CO_2(g)$ at 350.0 K and 1.174 atm.

62. For nitrogen gas the values of C_v and C_p at 25°C are 20.8 J K^{-1} mol^{-1} and 29.1 J K^{-1} mol^{-1}, respectively. When a sample of nitrogen is heated at constant pressure, what fraction of the energy is used to increase the internal energy of the gas? How is the remainder of the energy used? How much energy is required to raise the temperature of 100.0 g N_2 from 25.0°C to 85.0°C in a vessel having a constant volume?

63. At 1500 K the process

$$\underset{10 \text{ atm}}{I_2(g)} \longrightarrow \underset{10 \text{ atm}}{2I(g)}$$

is not spontaneous. However, the process

$$\underset{0.10 \text{ atm}}{I_2(g)} \longrightarrow \underset{0.10 \text{ atm}}{2I(g)}$$

is spontaneous at 1500 K. Explain.

64. The standard enthalpy of vaporization for nitric oxide (nitrogen monoxide) is 13.8 kJ/mol at its normal boiling point (121 K). Calculate $\Delta S°$ for the vaporization of NO(l).

65. The molar entropy of helium gas at 25°C and 1.00 atm is 126.1 J K^{-1} mol^{-1}. Assuming ideal behavior, calculate the entropy of the following.
a. 0.100 mol He(g) at 25°C and a volume of 5.00 L
b. 3.00 mol He(g) at 25°C and a volume of 3000.0 L

66. It is quite common for a solid to change from one structure to another at a temperature below its melting point. For example, sulfur undergoes a phase change from the rhombic crystal structure to the monoclinic form at 95.3°C. What must be the sign of the change in entropy for this phase change? Assuming that ΔH for this phase change is 0.400 kJ/mol, calculate ΔS for this phase change.

67. Calculate the entropy change for the vaporization of liquid methane and hexane, using the following data.

	Boiling Point (1 atm)	ΔH_{vap}
Methane	112 K	8.20 kJ/mol
Hexane	342 K	28.9 kJ/mol

Compare the molar volume of gaseous methane at 112 K with that of gaseous hexane at 342 K. How do the differences in molar volume affect the values of ΔS_{vap} for these liquids?

68. One mole of an ideal gas is contained in a cylinder with a movable piston. The temperature is constant at 77°C. Weights are removed suddenly from the piston to give the following sequence of three pressures:
a. P_1 = 5.00 atm (initial state)
b. P_2 = 2.24 atm
c. P_3 = 1.00 atm (final state)

What is the total work (in joules) in going from the initial to the final states by way of the above two steps? What would be the total work if the process were carried out reversibly?

69. Calculate the entropy change for a process in which 3.00 mol of liquid water at 0°C is mixed with 1.00 mol of water at 100.°C in a perfectly insulated container. (Assume that the molar heat capacity of water is constant at 75.3 J K^{-1} mol^{-1}.)

70. Calculate the change in entropy that occurs when 18.02 g of ice at −10.0°C is placed in 54.05 g of water at 100.0°C

in a perfectly insulated vessel. Assume that the molar heat capacities for $H_2O(s)$ and $H_2O(l)$ are 37.5 J K^{-1} mol^{-1} and 75.3 J K^{-1} mol^{-1}, respectively, and the molar enthalpy of fusion for ice is 6.01 kJ/mol.

71. A sample of ice weighing 18.02 g, initially at −30.0°C, is heated to 140.0°C at a constant pressure of 1.00 atm. Calculate q, w, ΔE, ΔH, and ΔS for this process. The molar heat capacities (C_p) for solid, liquid, and gaseous water, 37.5 J K^{-1} mol^{-1}, 75.3 J K^{-1} mol^{-1}, and 36.4 J K^{-1} mol^{-1}, respectively, are assumed to be temperature-independent. The enthalpies of fusion and vaporization are 6.01 kJ/mol and 40.7 kJ/mol, respectively. Assume ideal gas behavior.

72. A chunk of iron weighing 111.7 g is taken from a beaker of boiling water (at 1.00 atm) and placed in 1000 gal of water at 0°C. Calculate ΔS for the iron (molar heat capacity 25.1 J K^{-1} mol^{-1}) and for the water.

73. Consider the isothermal expansion of 1.00 mol of ideal gas at 27°C. The volume increases from 30.0 L to 40.0 L. Calculate q, w, ΔE, ΔH, ΔS, and ΔG for two situations.
    a. a free expansion
    b. a reversible expansion

74. Consider 1.00 mol of an ideal gas that is expanded isothermally at 25°C from 2.45×10^{-2} atm to 2.45×10^{-3} atm in the following three irreversible steps:

    Step 1: from 2.45×10^{-2} atm to 9.87×10^{-3} atm
    Step 2: from 9.87×10^{-3} atm to 4.93×10^{-3} atm
    Step 3: from 4.93×10^{-3} atm to 2.45×10^{-3} atm

    Calculate q, w, ΔE, ΔS, ΔH, and ΔG for each step and for the overall process.

75. Consider 1.00 mol of an ideal gas at 25°C.
    a. Calculate q, w, ΔE, ΔS, ΔH, and ΔG for the expansion of this gas isothermally and irreversibly from 2.45×10^{-2} atm to 2.45×10^{-3} atm in one step.
    b. Calculate q, w, ΔE, ΔS, ΔH, and ΔG for the same change of pressure as in part a but performed isothermally and reversibly.
    c. Calculate q, w, ΔE, ΔS, ΔH, and ΔG for the one-step isothermal, irreversible compression of 1.00 mol of an ideal gas at 25°C from 2.45×10^{-3} atm to 2.45×10^{-2} atm.
    d. Construct the PV diagrams for the processes described in parts a, b, and c.
    e. Calculate the entropy change in the surroundings for the processes described in parts a, b, and c.

76. For the reaction at 298 K,

$$2NO_2(g) \rightleftharpoons N_2O_4(g)$$

the values of ΔH° and ΔS° are −58.03 kJ/mol and −176.6 J K^{-1} mol^{-1}, respectively. What is the value of ΔG° at 298 K? Assume that ΔH° and ΔS° do not depend upon temperature. At what temperature is $\Delta G^\circ = 0$? Is ΔG° negative above, or below, this temperature?

77. Calculate ΔH° and ΔS° at 25°C for the reaction

$$2SO_2(g) + O_2(g) \longrightarrow 2SO_3(g)$$

at a constant pressure of 1.00 atm, using thermodynamic data in Appendix 4. Also calculate ΔH° and ΔS° at 227°C and 1.00 atm, assuming that the constant-pressure molar heat capacities for $SO_2(g)$, $O_2(g)$, and $SO_3(g)$ are 39.9 J K^{-1} mol^{-1}, 29.4 J K^{-1} mol^{-1}, and 50.7 J K^{-1} mol^{-1}, respectively. (*Hint*: construct a thermodynamic cycle and consider how enthalpy and entropy depend on temperature.)

78. The heat of vaporization of water at the normal boiling point, 373.2 K, is 40.66 kJ/mol. The specific heat capacity of liquid water is 4.184 J K^{-1} g^{-1} and of gaseous water is 2.02 J K^{-1} g^{-1}. Assume that these values are independent of temperature. What is the heat of vaporization of water at 340.2 K?

79. Using thermodynamic data from Appendix 4, calculate ΔG° at 25°C for the process

$$2SO_2(g) + O_2(g) \longrightarrow 2SO_3(g)$$

where all gases are at 1.00 atm pressure. Also calculate ΔG at 25°C for this same reaction but with all gases at 10.0 atm pressure.

80. Calculate the values of ΔS and ΔG for each of the following processes at 298 K:

$$H_2O(l, 298\ K) \longrightarrow H_2O\ (g, V = 1000.\ L/mol)$$
$$H_2O(l, 298\ K) \longrightarrow H_2O(g, V = 100.\ L/mol)$$

The standard enthalpy of vaporization for water at 298 K is 44.02 kJ/mol. Does either of these processes occur spontaneously?

81. Calculate the changes in free energy, enthalpy, and entropy when 1.00 mol Ar(g) at 27°C is compressed isothermally from 100.0 L to 1.00 L.

82. Calculate the difference in free energy between 1.00 M HCl and 0.100 M HCl at 25°C.

83. The equilibrium constant for a certain reaction increases by a factor of 10.0 when the temperature changes from 300.0 K to 350.0 K. Calculate the standard change in enthalpy for this reaction (assuming ΔH° is temperature-independent).

84. Although we often assume that the heat capacity of a substance is not temperature-dependent, this is not strictly true, as shown by the following data for ice.

Temperature (°C)	C_p (J K^{-1} mol^{-1})
−200.	12
−180.	15
−160.	17
−140.	19
−100.	24
−60.	29
−30.	33
−10.	36
0	37

Use these data to calculate graphically the change in entropy for heating ice from −200.°C to 0°C. (*Hint:* recall that

$$\Delta S_{T_1 \rightarrow T_2} = \int_{T_1}^{T_2} \frac{C_p\, dT}{T}$$

and that integration from T_1 to T_2 sums the area under the curve of a plot of C_p/T versus T from T_1 to T_2.)

85. Consider the following C_p values for $N_2(g)$:

C_p (J K^{-1} mol^{-1})	T (K)
28.7262	300.0
29.2937	400.0
29.8545	500.0

Assume that C_p can be expressed in the form

$$C_p = a + bT + CT^2$$

Estimate the value of C_p for $N_2(g)$ at 900. K. Assuming that C_p shows this temperature dependence over the range 100 K to 900 K, calculate ΔS for heating 1.00 mol $N_2(g)$ from 100. K to 900. K.

86. The standard entropy of vaporization of water is greater than that of a "typical" liquid. Give a possible explanation.

87. A 1.00-mol sample of an ideal gas in a vessel with a movable piston initially occupies a volume of 5.00 L at an external pressure of 5.00 atm.
   a. If P_{ex} is suddenly lowered to 2.00 atm and the gas is allowed to expand isothermally, calculate the following quantities for the system: ΔE, ΔH, ΔS, ΔG, w, and q.
   b. Show by the second law that this process will occur spontaneously.

88. Consider the reaction

$$2CO(g) + O_2(g) \longrightarrow 2CO_2(g)$$

Using data from Appendix 4,
   a. Calculate K at 298 K.
   b. Calculate $S°$ for $O_2(g)$ at 298 K.
   c. What is ΔS for this reaction at T = 298 K if the reactants, each at 10.0 atm, are changed to products at 10.0 atm?

89. A cylinder with an initial volume of 10.0 L is fitted with a frictionless piston and is filled with 1.00 mol of an ideal gas at 25°C. Assume that the surroundings is large enough so that if heat is withdrawn from or added to it, the temperature does not change.
   a. The gas expands isothermally and reversibly from 10.0 L to 20.0 L. Calculate the work and the heat.
   b. The gas expands isothermally and irreversibly from 10.0 L to 20.0 L as the external pressure changes instantaneously from 2.46 atm to 1.23 atm. Calculate the work and the heat.

90. Consider the reactions

$$Ni^{2+}(aq) + 6NH_3(aq) \longrightarrow Ni(NH_3)_6^{2+}(aq) \quad (1)$$

$$Ni^{2+}(aq) + 3en(aq) \longrightarrow Ni(en)_3^{2+}(aq) \quad (2)$$

where

$$en = H_2N{-}CH_2{-}CH_2{-}NH_2$$

The ΔH values for the two reactions are quite similar, yet $K_{\text{reaction 2}} > K_{\text{reaction 1}}$. Explain.

91. For the dissociation of the hydrogen halides in aqueous solution,

$$HX(aq) \longrightarrow H^+(aq) + X^-(aq)$$

the following entropy changes occur:

HX	ΔS_{diss} (J K^{-1} mol^{-1})
HF	−88
HCl	−54
HBr	−38
HI	−13

Explain why ΔS_{diss} is negative in all cases, and rationalize the observed order.

92. Consider the system

$$A(g) \longrightarrow B(g)$$

at 25°C.
   a. Assuming that G_A° = 8996 J/mol and G_B° = 11,718 J/mol, calculate the value of the equilibrium constant for this reaction.

b. Calculate the equilibrium pressures that result if 1.00 mol $A(g)$ at 1.00 atm and 1.00 mol $B(g)$ at 1.00 atm are mixed at 25°C.
c. Show by calculations that $\Delta G = 0$ at equilibrium.

93. Consider the reaction

$$H_2(g) + Br_2(g) \rightleftharpoons 2HBr(g)$$

where $\Delta H° = -103.8$ kJ/mol. In a particular experiment 1.00 atm of $H_2(g)$ and 1.00 atm of $Br_2(g)$ were mixed in a 1.00-L flask at 25°C and allowed to reach equilibrium. Then the molecules of H_2 were counted by using a very sensitive technique, and 1.10×10^{13} molecules were found. For this reaction, calculate the values of K, $\Delta G°$, and $\Delta S°$.

94. One mole of an ideal gas with a volume of 6.67 L and a pressure of 1.50 atm is contained in a vessel with a movable piston. The external pressure is suddenly increased to 5.00 atm and the gas compressed isothermally (T = 122 K). Calculate ΔE, ΔH, ΔS, w, q, ΔS_{surr}, ΔS_{univ}, and ΔG.

95. One mole of an ideal gas with a volume of 1.0 L and a pressure of 5.0 atm is allowed to expand isothermally into an evacuated bulb to give a total volume of 2.0 L. Calculate w and q. Also calculate q_{rev} for this change of state.

96. Consider the process

$$\underset{75°C}{A(l)} \longrightarrow \underset{155°C}{A(g)}$$

which is carried out at constant pressure. The total ΔS for this process is known to be 75.0 J K^{-1} mol^{-1}. For $A(l)$ and $A(g)$, the C_p values are 75.0 J K^{-1} mol^{-1} and 29.0 J K^{-1} mol^{-1}, respectively, and are not dependent on temperature. Calculate $\Delta H_{vaporization}$ for $A(l)$ at 125°C (its boiling point).

97. Consider a rigid, insulated box containing 0.400 mol of $He(g)$ at 20.0°C and 1.00 atm in one compartment and 0.600 mol of $N_2(g)$ at 100.0°C and 2.00 atm in the other compartment. These compartments are connected by a partition that transmits heat. What is the final temperature in the box at thermal equilibrium? [For $He(g)$, C_v = 12.5 J K^{-1} mol^{-1}, and for $N_2(g)$, C_v = 20.7 J K^{-1} mol^{-1}.]

98. When a salt dissolves in water, $\Delta H°$ is usually positive. What does this fact suggest about the temperature dependence of the solubility of salts for which $\Delta H°_{solution} > 0$? A significant number of salts for which $\Delta H°_{solution} > 0$ actually show a decrease in solubility with increasing temperature. Give a possible explanation. [See J. A. Campbell, *J. Chem. Ed.*, **62**, 231 (1985), for a discussion of this phenomenon.]

99. If wet silver carbonate is dried in a stream of hot air, the air must have a certain concentration level of carbon dioxide to prevent silver carbonate from decomposing by the reaction

$$Ag_2CO_3(s) \rightleftharpoons Ag_2O(s) + CO_2(g)$$

$\Delta H°$ for this reaction is 79.14 kJ/mol in the temperature range of 25–125°C. Given that the partial pressure of carbon dioxide in equilibrium with pure solid silver carbonate is 6.23×10^{-3} torr at 25°C, calculate the partial pressure of CO_2 necessary to prevent decomposition of Ag_2CO_3 at 110.°C.

100. Benzene (C_6H_6) has a melting point of 5.5°C and an enthalpy of fusion of 10.04 kJ/mol at 25.0°C. The molar heat capacities at constant pressure for solid and liquid benzene are 100.4 J K^{-1} mol^{-1} and 133.0 J K^{-1} mol^{-1}, respectively. For the reaction

$$C_6H_6(l) \rightleftharpoons C_6H_6(s)$$

calculate ΔS_{sys} and ΔS_{surr} at 10.0°C.

101. Impure nickel, refined by smelting sulfide ores in a blast furnace, can be converted into metal from 99.90% to 99.99% purity by the Mond process. The primary reaction involved in the Mond process is

$$Ni(s) + 4CO(g) \rightleftharpoons Ni(CO)_4(g)$$

a. Without referring to Appendix 4, predict the sign of $\Delta S°$ for the above reaction. Explain.
b. The spontaneity of the above reaction is temperature dependent. Predict the sign of ΔS_{surr} for this reaction. Explain.
c. For $Ni(CO)_4(g)$, $\Delta H_f° = -607$ kJ/mol and $S°$ = 417 J K^{-1} mol^{-1} at 298 K. Using these values and data in Appendix 4, calculate $\Delta H°$ and $\Delta S°$ for the above reaction.
d. Calculate the temperature at which $\Delta G° = 0$ (K = 1) for the above reaction, assuming that $\Delta H°$ and $\Delta S°$ do not depend on temperature.
e. The first step of the Mond process involves equilibrating impure nickel with $CO(g)$ and $Ni(CO)_4(g)$ at about 50°C. The purpose of this step is to convert as much nickel as possible into the gas phase. Calculate the equilibrium constant for the above reaction at 50.°C.
f. In the second step of the Mond process, the gaseous $Ni(CO)_4$ is isolated and heated to 227°C. The purpose of this step is to deposit as much nickel as possible as pure solid (the reverse of the above reaction). Calculate the equilibrium constant for the above reaction at 227°C.
g. Why is temperature increased for the second step of the Mond process?

h. The Mond process relies on the volatility of $Ni(CO)_4$ for its success. Only pressures and temperatures at which $Ni(CO)_4$ is a gas are useful. A recently developed variation of the Mond process carries out the first step at higher pressures and a temperature of 152°C. Estimate the maximum pressure of $Ni(CO)_4(g)$ that can be attained before the gas will liquify at 152°C. The boiling point for $Ni(CO)_4$ is 42°C and the enthalpy of vaporization is 29.0 kJ/mol. (*Hint:* the phase change reaction and the corresponding equilibrium expression are

$$Ni(CO)_4(l) \rightleftharpoons Ni(CO)_4(g); \; K_p = P_{Ni(CO)4}$$

$Ni(CO)_4(g)$ will liquify when the pressure of $Ni(CO_4)_4$ is greater than the K_p value.)

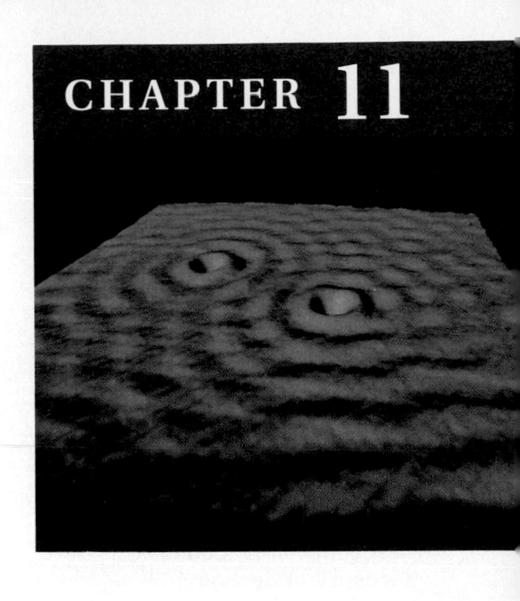

Electrochemistry

Electrochemistry constitutes one of the most important interfaces between chemistry and everyday life. Every time you start your car, turn on your calculator, look at your digital watch, or listen to a radio at the beach, you are depending on electrochemical reactions. Our society sometimes seems to run almost entirely on batteries. Certainly, the advent of small, dependable batteries along with silicon chip technology has made possible the tiny calculators, tape recorders, and clocks that we take for granted.

Electrochemistry is important in other less obvious ways. For example, the corrosion of iron, which has tremendous economic implications, is an electrochemical process. In addition, many important industrial materials such as aluminum, chlorine, and sodium hydroxide are prepared by electrolytic processes. In analytical chemistry, electrochemical techniques employ electrodes that are specific for a given molecule or ion, including H^+ (pH meters), F^-, Cl^-, and many others. These increasingly important methods are used to analyze for trace pollutants in natural waters or for the tiny quantities of chemicals in human blood that may signal the development of a specific disease.

Electrochemistry is best defined as *the study of the interchange of chemical and electrical energy*. It is primarily concerned with two processes that involve oxidation-reduction reactions: the generation of an electric current from a chemical reaction, and the opposite process, the use of a current to produce chemical change.

11.1 Galvanic Cells

Recall from Chapter 4 that an **oxidation-reduction (redox) reaction** involves a transfer of electrons from the **reducing agent** to the **oxidizing agent,** and that **oxidation** involves a *loss of electrons* (an increase in oxidation number) and **reduction** involves a *gain of electrons* (a decrease in oxidation number).

To understand how a redox reaction can be used to generate a current, we will consider the reaction between MnO_4^- and Fe^{2+}:

$$8H^+(aq) + MnO_4^-(aq) + 5Fe^{2+}(aq) \longrightarrow Mn^{2+}(aq) + 5Fe^{3+}(aq) + 4H_2O(l)$$

In this reaction Fe^{2+} is oxidized and MnO_4^- is reduced; electrons are transferred from Fe^{2+} (the reducing agent) to MnO_4^- (the oxidizing agent).

Balancing half-reactions is discussed in Section 4.11.

It is useful to break a redox reaction into two **half-reactions,** one involving oxidation and the other involving reduction. For the above reaction the half-reactions are

$$8H^+ + MnO_4^- + 5e^- \longrightarrow Mn^{2+} + 4H_2O$$
$$5(Fe^{2+} \longrightarrow Fe^{3+} + e^-)$$

Note that the second half-reaction must occur five times for each time the first reaction occurs. The balanced overall reaction is the sum of the half-reactions.

When MnO_4^- and Fe^{2+} are present in the same solution, the electrons are transferred directly as the reactants collide. Under these conditions no useful work is obtained from the chemical energy associated with the reaction, which instead is released as heat. How can we harness this energy? The key is to separate the oxidizing agent from the reducing agent, thus requiring the electron transfer to occur through a wire. The current produced in the wire by the electron flow can then be directed through a device, such as an electric motor, to provide useful work.

For example, consider the system illustrated in Fig. 11.1. If our reasoning has been correct, electrons should flow through the wire from Fe^{2+} to MnO_4^-. However, when we construct the apparatus as shown, no electron flow is apparent. Why? Careful observation would show that when we connect the

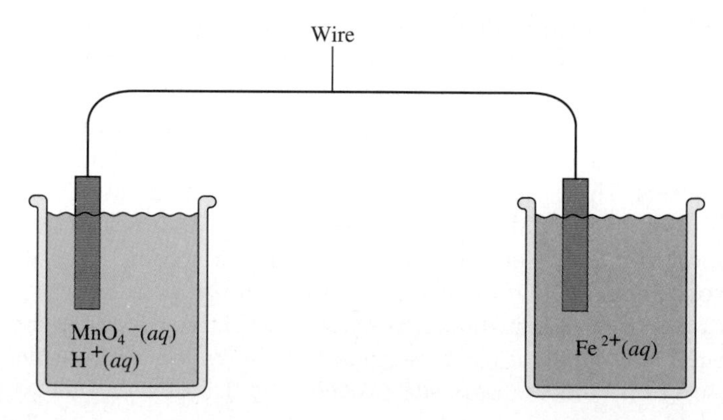

Figure 11.1
Schematic of a method to separate the oxidizing and reducing agents in a redox reaction. (The solutions also contain counter ions to balance the charge.)

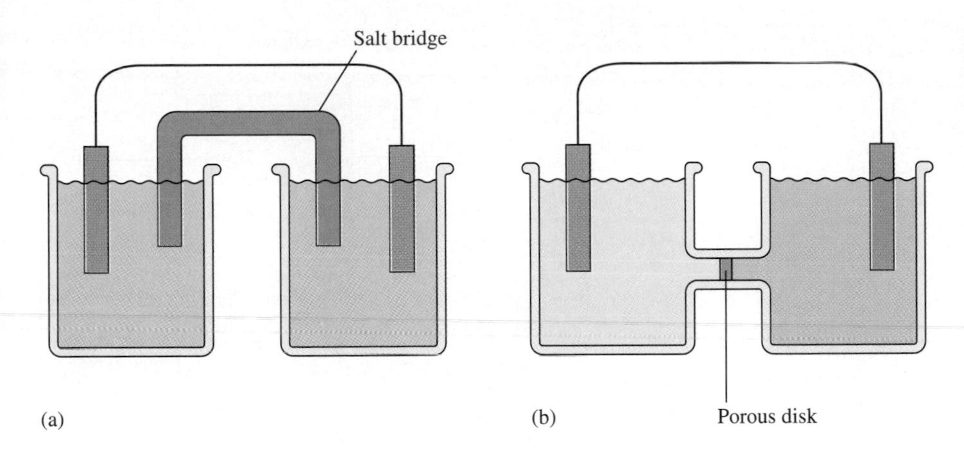

Figure 11.2
Galvanic cells can contain a salt bridge as in (a) or a porous-disk connection as in (b). A salt bridge contains a strong electrolyte held in a Jello-like matrix. A porous disk contains tiny passages that allow hindered flow of ions.

wires from the two compartments, current flows for an instant and then ceases. The current stops flowing because of charge buildups in the two compartments. If electrons flowed from the right to the left compartment in the apparatus as shown, the left compartment (receiving electrons) would become negatively charged, and the right compartment (losing electrons) would become positively charged. Creating a charge separation of this type requires a large amount of energy. Thus sustained electron flow cannot occur under these conditions.

We can, however, solve this problem very simply. The solutions must be connected so that ions can flow to maintain the net charge of zero in each compartment. This connection might involve a **salt bridge** (a U-tube filled with an electrolyte) or a **porous disk** in a tube connecting the two solutions (see Fig. 11.2). Either of these devices allows ion flow without extensive mixing of the solutions. When we make the provision for ion flow, the circuit is complete. Electrons flow through the wire from reducing agent to oxidizing agent, and ions flow between the compartments to keep the net charge zero in each.

A galvanic cell uses a spontaneous redox reaction to produce a current that can be used to do work.

We now have covered all the essential characteristics of a **galvanic cell,** *a device in which chemical energy is changed to electrical energy.* (The opposite process, *electrolysis,* will be considered in Section 11.7.)

The reaction in an electrochemical cell occurs at the interface between an electrode and the solution where the electron transfer occurs. The electrode at which *oxidation* occurs is called the **anode;** the electrode at which *reduction* occurs is called the **cathode** (see Fig. 11.3).

Oxidation occurs at the anode. Reduction occurs at the cathode.

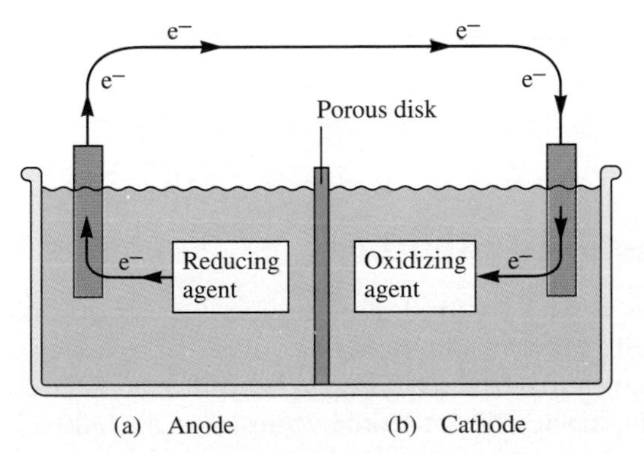

Figure 11.3
An electrochemical process involves electron transfer at the interface between the electrode and the solution. (a) The species in solution acting as the reducing agent supplies electrons to the anode. (b) The species in solution acting as the oxidizing agent receives electrons from the cathode.

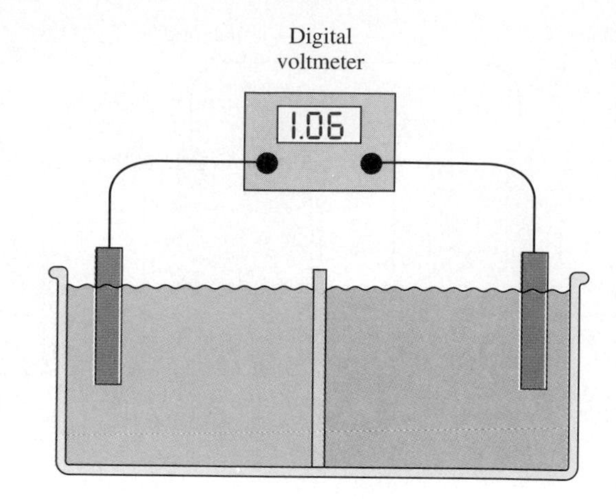

Figure 11.4
Digital voltmeters draw only a negligible current and are convenient to use.

Cell Potential

A galvanic cell consists of an oxidizing agent in one compartment that pulls electrons through a wire from a reducing agent in the other compartment. The "pull," or driving force, on the electrons is called the **cell potential** ($\mathscr{E}_{cell}$), or the **electromotive force** (emf), of the cell. The unit of electrical potential is the **volt** (abbreviated V), which is defined as 1 joule of work per coulomb of charge transferred.

A volt is 1 joule of work per coulomb of charge transferred: 1 V = 1 J/C.

How can we measure the cell potential? One possible instrument is a crude **voltmeter,** which works by drawing current through a known resistance. However, when current flows through a wire, the frictional heating that occurs wastes some of the useful energy of the cell. A traditional voltmeter will therefore measure a potential that is lower than the maximum cell potential. The key to determining the maximum potential is to perform the measurement under conditions of zero current so that no energy is wasted. Traditionally, this measurement has been accomplished by inserting a variable-voltage device (powered from an external source) in *opposition* to the cell potential. The voltage on this instrument, called a **potentiometer,** is adjusted until no current flows in the cell circuit. Under such conditions the cell potential is equal in magnitude and opposite in sign to the voltage setting of the potentiometer. This value represents the *maximum* cell potential, since no energy is wasted heating the wire. Advances in electronic technology have allowed the design of *digital voltmeters* that draw only a negligible amount of current (see Fig. 11.4). Since these instruments are more convenient to use, they have replaced potentiometers in the modern laboratory.

11.2 Standard Reduction Potentials

The name *galvanic cell* honors Luigi Galvani (1737–1798), an Italian scientist generally credited with the discovery of electricity. These cells are sometimes called *voltaic cells* after Alessandro Volta (1745–1827), another Italian, who first constructed cells of this type around 1800.

The reaction in a galvanic cell is always an oxidation-reduction reaction that can be broken down into two half-reactions. It would be convenient to assign a potential to *each* half-reaction so that when we construct a cell from a given pair of half-reactions, we can obtain the cell potential by summing the half-cell

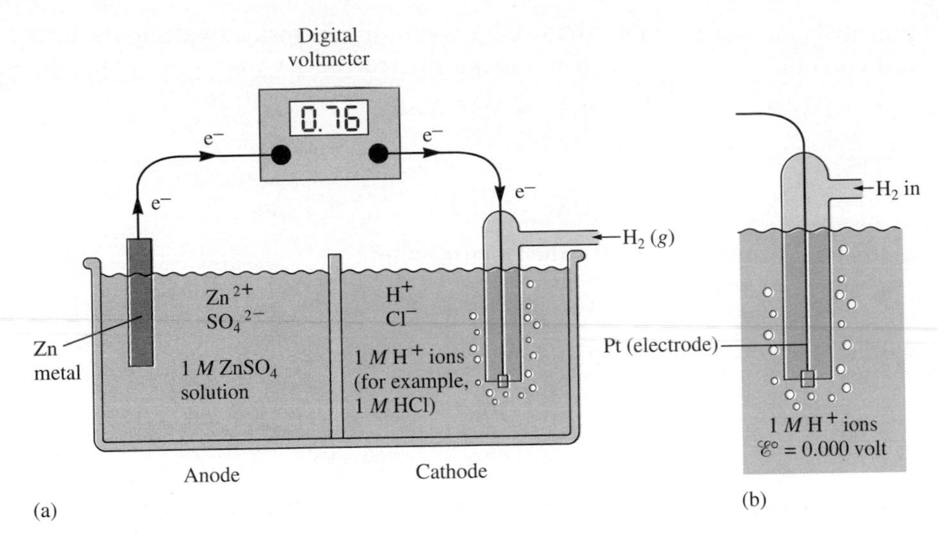

Figure 11.5
(a) A galvanic cell involving the reactions $Zn \rightarrow Zn^{2+} + 2e^-$ (at the anode) and $2H^+ + 2e^- \rightarrow H_2$ (at the cathode) has a potential of 0.76 V. (b) The standard hydrogen electrode, where $H_2(g)$ at 1 atm is passed over a platinum electrode in contact with 1 M H^+ ions. This electrode process (assuming ideal behavior) is arbitrarily assigned a value of exactly 0 V.

potentials. For example, the observed potential for the cell shown in Fig. 11.5(a) is 0.76 volt, and the cell reaction* is

$$2H^+(aq) + Zn(s) \longrightarrow Zn^{2+}(aq) + H_2(g)$$

For this cell the anode compartment contains a zinc metal electrode with Zn^{2+} and SO_4^{2-} ions in an aqueous solution that bathes the electrode. The anode reaction is the oxidation half-reaction:

$$Zn \longrightarrow Zn^{2+} + 2e^-$$

Each zinc atom loses two electrons to produce a Zn^{2+} ion that enters the solution. The electrons flow through the wire. For now we will assume that all cell components are in their standard states, so in this case the solution in the anode compartment will contain 1 M Zn^{2+}. The cathode reaction of this cell is

$$2H^+ + 2e^- \longrightarrow H_2$$

The cathode consists of a platinum electrode (used because it is a chemically inert conductor) in contact with 1 M H^+ ions and bathed by hydrogen gas at 1 atm. Such an electrode, called the **standard hydrogen electrode,** is shown in Fig. 11.5(b).

Although we can measure the *total* potential of this cell (0.76 V), there is no way to measure the potentials of the individual electrodes. Thus if we desire

*In this text we will follow the convention of indicating the physical states of the reactants and products only in the overall redox reaction. For simplicity, half-reactions will *not* include the physical states.

potentials for half-reactions (half-cells), we must arbitrarily divide up the total cell potential. For example, if we assign the reaction

$$2H^+ + 2e^- \longrightarrow H_2$$

where

$$[H^+] = 1\ M \qquad \text{and} \qquad P_{H_2} = 1\ \text{atm}$$

a potential of exactly 0 volts, then the reaction

$$Zn \longrightarrow Zn^{2+} + 2e^-$$

will have a potential of 0.76 volt, since

$$\underset{0.76\ V}{\overset{\uparrow}{\mathscr{E}^\circ_{cell}}} = \underset{0\ V}{\overset{\uparrow}{\mathscr{E}^\circ_{H^+ \to H_2}}} + \underset{0.76\ V}{\overset{\uparrow}{\mathscr{E}^\circ_{Zn \to Zn^{2+}}}}$$

Standard states were discussed in Section 9.6.

The standard hydrogen potential is the reference potential against which all half-reaction potentials are assigned.

Recall that the superscript ° indicates that *standard states* are employed.

By setting the standard potential for the half-reaction $2H^+ + 2e^- \rightarrow H_2$ equal to zero, we can assign values to all other half-reactions. For example, the measured potential for the cell shown in Fig. 11.6 is 1.10 V. The cell reaction is

$$Zn(s) + Cu^{2+}(aq) \longrightarrow Zn^{2+}(aq) + Cu(s)$$

which can be divided into the half-reactions

$$\text{Anode:} \qquad Zn \longrightarrow Zn^{2+} + 2e^-$$
$$\text{Cathode:} \qquad Cu^{2+} + 2e^- \longrightarrow Cu$$

Thus

$$\mathscr{E}^\circ_{cell} = \mathscr{E}^\circ_{Zn \to Zn^{2+}} + \mathscr{E}^\circ_{Cu^{2+} \to Cu}$$

Since $\mathscr{E}^\circ_{Zn \to Zn^{2+}}$ is assigned a value of 0.76 V, the value of $\mathscr{E}^\circ_{Cu^{2+} \to Cu}$ must be 0.34 volt:

$$1.10\ V = 0.76\ V + 0.34\ V$$

The scientific community has universally accepted the values for half-reaction potentials based on the assignment of 0 volts to the process $2H^+ + 2e^- \rightarrow H_2$ (under standard conditions where ideal behavior is assumed). How-

Copper being plated onto the copper metal cathode on the left, and zinc dissolving from the zinc metal anode on the right. Note the salt bridge connecting the two solutions, which allows the ion flow to balance the electron flow through the wire.

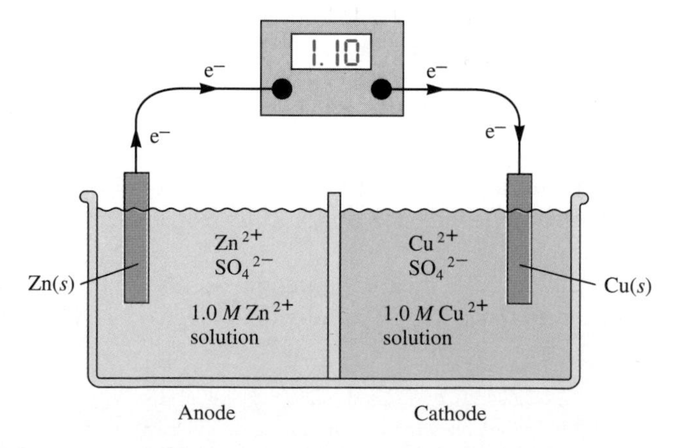

Figure 11.6
A galvanic cell involving the half-reactions $Zn \rightarrow Zn^{2+} + 2e^-$ (anode) and $Cu^{2+} + 2e^- \rightarrow Cu$ (cathode), with $\mathscr{E}^\circ_{cell} = 1.10$ V.

ever, before we can use these values, we need to understand several essential characteristics of half-cell potentials.

The accepted convention is to give the potentials of half-reactions as *reduction* processes. For example,

$$2H^+ + 2e^- \longrightarrow H_2$$
$$Cu^{2+} + 2e^- \longrightarrow Cu$$
$$Zn^{2+} + 2e^- \longrightarrow Zn$$

The $\mathscr{E}°$ values corresponding to these half-reactions are called **standard reduction potentials.** Standard reduction potentials for the most common half-reactions are given in Table 11.1 and Appendix Section 5.5.

All half-reactions are given as reduction processes in standard tables.

Combining two half-reactions to obtain a balanced oxidation-reduction reaction often requires two manipulations:

1. One of the reduction half-reactions must be reversed (since redox reactions must involve a substance being oxidized and a substance being reduced), which means that the *sign* of the potential for this half-reaction must also be *reversed.*

When a half-reaction is reversed, the sign of $\mathscr{E}°$ is reversed.

TABLE 11.1 **Standard Reduction Potentials at 25°C (298 K) for Many Common Half-reactions**

Half-reaction	$\mathscr{E}°$ (V)	Half-reaction	$\mathscr{E}°$ (V)
$F_2 + 2e^- \rightarrow 2F^-$	2.87	$O_2 + 2H_2O + 4e^- \rightarrow 4OH^-$	0.40
$Ag^{2+} + e^- \rightarrow Ag^+$	1.99	$Cu^{2+} + 2e^- \rightarrow Cu$	0.34
$Co^{3+} + e^- \rightarrow Co^{2+}$	1.82	$Hg_2Cl_2 + 2e^- \rightarrow 2Hg + 2Cl^-$	0.34
$H_2O_2 + 2H^+ + 2e^- \rightarrow 2H_2O$	1.78	$AgCl + e^- \rightarrow Ag + Cl^-$	0.22
$Ce^{4+} + e^- \rightarrow Ce^{3+}$	1.70	$SO_4^{2-} + 4H^+ + 2e^- \rightarrow H_2SO_3 + H_2O$	0.20
$PbO_2 + 4H^+ + SO_4^{2-} + 2e^- \rightarrow PbSO_4 + 2H_2O$	1.69	$Cu^{2+} + e^- \rightarrow Cu^+$	0.16
$MnO_4^- + 4H^+ + 3e^- \rightarrow MnO_2 + 2H_2O$	1.68	$2H^+ + 2e^- \rightarrow H_2$	0.00
$2e^- + 2H^+ + IO_4^- \rightarrow IO_3^- + H_2O$	1.60	$Fe^{3+} + 3e^- \rightarrow Fe$	−0.036
$MnO_4^- + 8H^+ + 5e^- \rightarrow Mn^{2+} + 4H_2O$	1.51	$Pb^{2+} + 2e^- \rightarrow Pb$	−0.13
$Au^{3+} + 3e^- \rightarrow Au$	1.50	$Sn^{2+} + 2e^- \rightarrow Sn$	−0.14
$PbO_2 + 4H^+ + 2e^- \rightarrow Pb^{2+} + 2H_2O$	1.46	$Ni^{2+} + 2e^- \rightarrow Ni$	−0.23
$Cl_2 + 2e^- \rightarrow 2Cl^-$	1.36	$PbSO_4 + 2e^- \rightarrow Pb + SO_4^{2-}$	−0.35
$Cr_2O_7^{2-} + 14H^+ + 6e^- \rightarrow 2Cr^{3+} + 7H_2O$	1.33	$Cd^{2+} + 2e^- \rightarrow Cd$	−0.40
$O_2 + 4H^+ + 4e^- \rightarrow 2H_2O$	1.23	$Fe^{2+} + 2e^- \rightarrow Fe$	−0.44
$MnO_2 + 4H^+ + 2e^- \rightarrow Mn^{2+} + 2H_2O$	1.21	$Cr^{3+} + e^- \rightarrow Cr^{2+}$	−0.50
$IO_3^- + 6H^+ + 5e^- \rightarrow \frac{1}{2}I_2 + 3H_2O$	1.20	$Cr^{3+} + 3e^- \rightarrow Cr$	−0.73
$Br_2 + 2e^- \rightarrow 2Br^-$	1.09	$Zn^{2+} + 2e^- \rightarrow Zn$	−0.76
$VO_2^+ + 2H^+ + e^- \rightarrow VO^{2+} + H_2O$	1.00	$2H_2O + 2e^- \rightarrow H_2 + 2OH^-$	−0.83
$AuCl_4^- + 3e^- \rightarrow Au + 4Cl^-$	0.99	$Mn^{2+} + 2e^- \rightarrow Mn$	−1.18
$NO_3^- + 4H^+ + 3e^- \rightarrow NO + 2H_2O$	0.96	$Al^{3+} + 3e^- \rightarrow Al$	−1.66
$ClO_2 + e^- \rightarrow ClO_2^-$	0.954	$H_2 + 2e^- \rightarrow 2H^-$	−2.23
$2Hg^{2+} + 2e^- \rightarrow Hg_2^{2+}$	0.91	$Mg^{2+} + 2e^- \rightarrow Mg$	−2.37
$Ag^+ + e^- \rightarrow Ag$	0.80	$La^{3+} + 3e^- \rightarrow La$	−2.37
$Hg_2^{2+} + 2e^- \rightarrow 2Hg$	0.80	$Na^+ + e^- \rightarrow Na$	−2.71
$Fe^{3+} + e^- \rightarrow Fe^{2+}$	0.77	$Ca^{2+} + 2e^- \rightarrow Ca$	−2.76
$O_2 + 2H^+ + 2e^- \rightarrow H_2O_2$	0.68	$Ba^{2+} + 2e^- \rightarrow Ba$	−2.90
$MnO_4^- + e^- \rightarrow MnO_4^{2-}$	0.56	$K^+ + e^- \rightarrow K$	−2.92
$I_2 + 2e^- \rightarrow 2I^-$	0.54	$Li^+ + e^- \rightarrow Li$	−3.05
$Cu^+ + e^- \rightarrow Cu$	0.52		

When a half-reaction is multiplied by an integer, $\mathscr{E}°$ remains the same.

2. Since the number of electrons lost must equal the number gained, the half-reactions must be multiplied by integers as necessary to achieve electron balance. However, the *value of* $\mathscr{E}°$ *is not changed* when a half-reaction is multiplied by an integer. Since a standard reduction potential is an *intensive property* (it does not depend on how many times the reaction occurs), the potential is *not* multiplied by the integer required to balance the cell reaction.

Consider a galvanic cell based on the redox reaction

$$Fe^{3+}(aq) + Cu(s) \longrightarrow Cu^{2+}(aq) + Fe^{2+}(aq)$$

The pertinent half-reactions are

$$Fe^{3+} + e^- \longrightarrow Fe^{2+} \qquad \mathscr{E}° = 0.77\ V \qquad (1)$$

$$Cu^{2+} + 2e^- \longrightarrow Cu \qquad \mathscr{E}° = 0.34\ V \qquad (2)$$

To balance the cell reaction and calculate the standard cell potential, we must first reverse reaction (2):

$$Cu \longrightarrow Cu^{2+} + 2e^- \qquad -\mathscr{E}° = -0.34\ V$$

Note the change in sign for the $\mathscr{E}°$ value.

Then, since each Cu atom produces two electrons but each Fe^{3+} ion accepts only one electron, reaction (1) must be multiplied by 2:

$$2Fe^{3+} + 2e^- \longrightarrow 2Fe^{2+} \qquad \mathscr{E}° = 0.77\ V$$

Note that the value of $\mathscr{E}°$ is not changed in this case.

Now we can obtain the balanced cell reaction by summing the appropriately modified half-reactions:

$$Cu \longrightarrow Cu^{2+} + 2e^- \qquad -\mathscr{E}° = -0.34\ V$$

$$2Fe^{3+} + 2e^- \longrightarrow 2Fe^{2+} \qquad \mathscr{E}° = 0.77\ V$$

Cell reaction: $Cu(s) + 2Fe^{3+}(aq) \longrightarrow Cu^{2+}(aq) + 2Fe^{2+}(aq) \qquad \mathscr{E}°_{cell} = -0.34\ V + 0.77\ V = 0.43\ V$

Next, we want to consider how to describe a galvanic cell fully, given just its half-reactions. This description will include the cell reaction, the cell potential, and the physical setup of the cell. Let's consider a galvanic cell based on the following half-reactions:

$$Fe^{2+} + 2e^- \longrightarrow Fe \qquad \mathscr{E}° = -0.44\ V$$

$$MnO_4^- + 5e^- + 8H^+ \longrightarrow Mn^{2+} + 4H_2O \qquad \mathscr{E}° = 1.51\ V$$

In a working galvanic cell one of these reactions must run in reverse. Which one?

A galvanic cell runs spontaneously in the direction that gives a positive value for $\mathscr{E}°_{cell}$.

We can answer this question by considering the sign of the potential of a working cell: *a cell will always run spontaneously in the direction that produces a positive cell potential.* Thus in the present case the half-reaction involving iron must be reversed, since this choice leads to a positive cell potential:

$$Fe \longrightarrow Fe^{2+} + 2e^- \qquad -\mathscr{E}° = 0.44\ V$$

$$MnO_4^- + 5e^- + 8H^+ \longrightarrow Mn^{2+} + 4H_2O \qquad \mathscr{E}° = 1.51\ V$$

where

$$\mathscr{E}^\circ_{\text{cell}} = 0.44 \text{ V} + 1.51 \text{ V} = 1.95 \text{ V}$$

The balanced cell reaction is obtained as follows:

$$5(\text{Fe} \longrightarrow \text{Fe}^{2+} + 2\text{e}^-)$$

$$2(\text{MnO}_4^- + 5\text{e}^- + 8\text{H}^+ \longrightarrow \text{Mn}^{2+} + 4\text{H}_2\text{O})$$

$$2\text{MnO}_4^-(aq) + 5\text{Fe}(s) + 16\text{H}^+(aq) \longrightarrow 5\text{Fe}^{2+}(aq) + 2\text{Mn}^{2+}(aq) + 8\text{H}_2\text{O}(l)$$

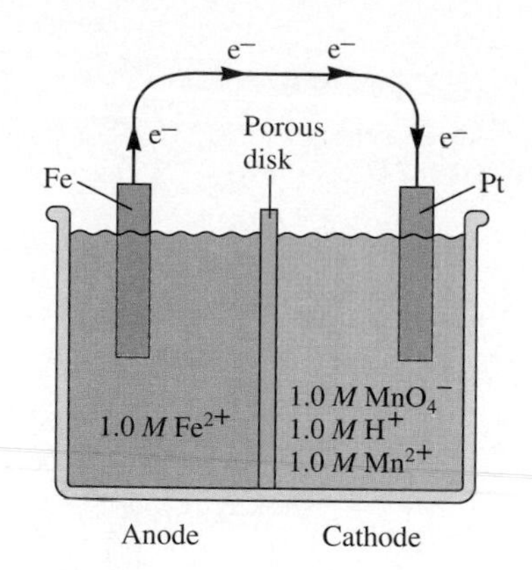

Figure 11.7
The schematic of a galvanic cell based on the half-reactions

$\text{Fe} \rightarrow \text{Fe}^{2+} + 2\text{e}^-$

$\text{MnO}_4^- + 5\text{e}^- + 8\text{H}^+ \rightarrow \text{Mn}^{2+} + 4\text{H}_2\text{O}$

Now consider the physical setup of the cell, shown schematically in Fig. 11.7. In the left compartment the active components in their standard states are pure metallic iron (Fe) and 1.0 *M* Fe^{2+}. The anion present depends on the iron salt used. In this compartment the anion does not participate in the reaction but simply balances the charge. The half-reaction that takes place at this electrode is

$$\text{Fe} \longrightarrow \text{Fe}^{2+} + 2\text{e}^-$$

which is an oxidation reaction, so this compartment is the anode. The electrode consists of pure iron metal.

In the right compartment the active components in their standard states are 1.0 *M* MnO_4^-, 1.0 *M* H^+, and 1.0 *M* Mn^{2+}, with appropriate unreacting ions (often called *counter ions*) to balance the charge. The half-reaction in this compartment is

$$\text{MnO}_4^- + 5\text{e}^- + 8\text{H}^+ \longrightarrow \text{Mn}^{2+} + 4\text{H}_2\text{O}$$

which is a reduction reaction, so this compartment is the cathode. Since neither MnO_4^- nor Mn^{2+} ion can serve as the electrode, a nonreacting conductor must be used. The usual choice is platinum.

The next step is to determine the direction of electron flow. In the left compartment the half-reaction is the oxidation of iron:

$$\text{Fe} \longrightarrow \text{Fe}^{2+} + 2\text{e}^-$$

In the right compartment the half-reaction is the reduction of MnO_4^-:

$$\text{MnO}_4^- + 5\text{e}^- + 8\text{H}^+ \longrightarrow \text{Mn}^{2+} + 4\text{H}_2\text{O}$$

Thus the electrons flow from Fe to MnO_4^- in this cell, or from the anode to the cathode, as is always the case.

Summary: **Items Needed for a Description of a Galvanic Cell**

- **The cell potential (always positive for a galvanic cell) and the balanced cell reaction.**
- **The direction of electron flow, obtained by inspecting the half-reactions and using the directions that give a positive $\mathscr{E}^\circ_{\text{cell}}$.**
- **Designation of the anode and the cathode.**
- **The nature of each electrode and the ions present in each compartment. A chemically inert conductor is required if none of the substances participating in the half-reaction is a conducting solid.**

EXAMPLE 11.1

Describe completely the galvanic cell based on the following half-reactions under standard conditions:

$$Ag^{+} + e^{-} \longrightarrow Ag \qquad \mathscr{E}° = 0.80 \text{ V} \qquad (1)$$

$$Fe^{3+} + e^{-} \longrightarrow Fe^{2+} \qquad \mathscr{E}° = 0.77 \text{ V} \qquad (2)$$

Solution

Since a positive $\mathscr{E}°_{cell}$ value is required, reaction (2) must run in reverse:

$$Ag^{+} + e^{-} \longrightarrow Ag \qquad \mathscr{E}° = 0.80 \text{ V}$$

$$Fe^{2+} \longrightarrow Fe^{3+} + e^{-} \qquad -\mathscr{E}° = -0.77 \text{ V}$$

$$\text{Cell reaction: } Ag^{+}(aq) + Fe^{2+}(aq) \longrightarrow Fe^{3+}(aq) + Ag(s) \qquad \mathscr{E}°_{cell} = 0.03 \text{ V}$$

Since Ag^{+} receives electrons and Fe^{2+} loses electrons in the cell reaction, the electrons flow from the compartment containing Fe^{2+} to the compartment containing Ag^{+}.

Oxidation occurs in the compartment containing Fe^{2+}. Hence this compartment functions as the anode. Reduction occurs in the compartment containing Ag^{+}, so this compartment is the cathode.

The electrode in the Ag/Ag^{+} compartment is silver metal; and an inert conductor, such as platinum, must be used in the Fe^{2+}/Fe^{3+} compartment. Appropriate counter ions are assumed to be present. The diagram for this cell is shown in Fig. 11.8.

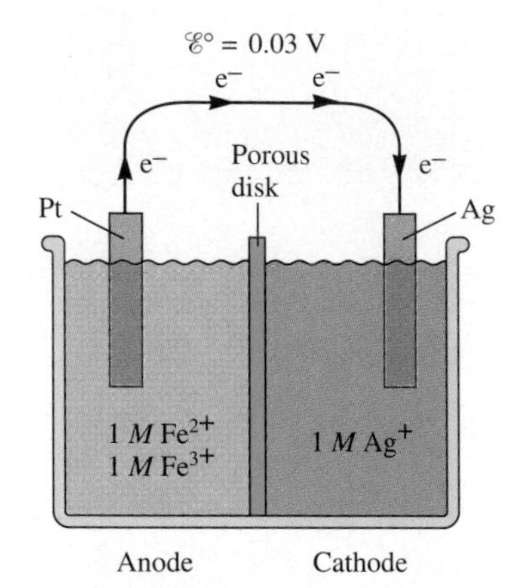

Figure 11.8
Schematic diagram for the galvanic cell based on the half-reactions

$$Ag^{+} + e^{-} \rightarrow Ag$$

$$Fe^{2+} \rightarrow Fe^{3+} + e^{-}$$

11.3 Cell Potential, Electrical Work, and Free Energy

So far we have considered electrochemical cells in a very practical fashion without much theoretical background. The next step will be to explore the relationship between thermodynamics and electrochemistry.

The work that can be accomplished when electrons are transferred through a wire depends on the "push" (the thermodynamic driving force) behind the electrons. This driving force (the emf) is defined in terms of a *potential* difference (in volts) between two points in the circuit. Recall that a volt represents a joule of work per coulomb of charge transferred:

$$\text{emf} = \text{potential difference (V)} = \frac{\text{work (J)}}{\text{charge (C)}}$$

Thus 1 joule of work is produced or required (depending on the direction) when 1 coulomb of charge is transferred between two points in the circuit that differ by a potential of 1 volt.

In this book *work is viewed from the point of view of the system.* Thus work flowing out of the system is indicated by a minus sign. When a cell produces a current, the cell potential is positive, and the current can be used to do work—to run a motor, for instance. Thus the cell potential ($\mathscr{E}$) and the work (w) have opposite signs:

$$\mathscr{E} = \frac{-w}{q}$$

Work ↙ (points to $-w$); Charge ↖ (points to q)

Thus $$-w = q\mathscr{E}$$

From this equation we can see that the maximum work in a cell is obtained at the maximum cell potential:

$$-w_{max} = q\mathscr{E}_{max} \quad \text{or} \quad w_{max} = -q\mathscr{E}_{max}$$

However, there is a problem. For the system to perform electrical work, current must flow. When current flows, some energy is inevitably wasted through frictional heating, so the maximum work is not obtained. This observation reflects the important general principle introduced in Section 10.12: *in any real, spontaneous process some energy is always wasted—the actual work realized is always less than the calculated maximum.* This principle is a consequence of the fact that the entropy of the universe must increase in any spontaneous process. Recall from Section 10.12 that the only process from which maximum work could be realized is the hypothetical reversible process. For a galvanic cell, this would involve an infinitesimally small current flow and thus an infinite amount of time to do the work. Even though we can never achieve the maximum work through the actual discharge of a galvanic cell, we can measure the maximum potential. There is negligible current flow when a cell potential is measured with a potentiometer or an efficient digital voltmeter. No current flow implies no waste of energy, so the measured potential is the maximum.

The work is never the maximum possible if any current is flowing.

Although we can never actually realize the maximum work from a cell reaction, its value is still useful for evaluating the efficiency of a real process based on the cell reaction. For example, suppose a certain galvanic cell has a maximum potential of 2.50 V. In a particular experiment 1.33 moles of electrons pass through this cell at an average actual potential of 2.10 V. The actual work done is

$$w = -q\mathscr{E}$$

where $\mathscr{E}$ represents the actual potential difference at which the current flowed (2.10 V or 2.10 J/C) and q is the quantity of charge transferred (in coulombs). The charge on 1 mole of electrons is called the **faraday** (abbreviated F): *96,485 coulombs of charge per mole of electrons.* Thus q equals the number of moles of electrons times the charge per mole of electrons:

$$q = nF = 1.33 \text{ mol e}^- \times 96{,}485 \text{ C/mol e}^-$$

For the experiment above, the actual work is

$$w = -q\mathscr{E} = -\left(1.33 \text{ mol e}^- \times 96{,}485 \frac{\text{C}}{\text{mol e}^-}\right) \times \left(2.10 \frac{\text{J}}{\text{C}}\right)$$
$$= -2.69 \times 10^5 \text{ J}$$

Michael Faraday lecturing at the Royal Institution before Prince Albert and others (1855).

For the maximum possible work, the calculation is similar, except that the maximum potential is used:

$$w_{\text{max}} = -q\mathscr{E}_{\text{max}} = -\left(1.33 \text{ mol e}^- \times 96{,}485 \frac{\text{C}}{\text{mol e}^-}\right)\left(2.50 \frac{\text{J}}{\text{C}}\right)$$
$$= -3.21 \times 10^5 \text{ J}$$

Thus in its actual operation, the efficiency of this cell is

$$\frac{w}{w_{\text{max}}} \times 100 = \frac{-2.69 \times 10^5 \text{ J}}{-3.21 \times 10^5 \text{ J}} \times 100 = 83.8\%$$

Next we want to relate the potential of a galvanic cell to free energy. In Section 10.12 we saw that for a process carried out at constant temperature and pressure, the change in free energy equals the maximum useful work obtainable from that process:

$$w_{\text{max}} = \Delta G$$

For a galvanic cell

$$w_{\text{max}} = -q\mathscr{E}_{\text{max}} = \Delta G$$

Since

$$q = nF$$

we have

$$\Delta G = -q\mathscr{E}_{\text{max}} = -nF\mathscr{E}_{\text{max}}$$

From now on the subscript on $\mathscr{E}_{\text{max}}$ will be deleted, with the understanding that any potential given in this book is the maximum potential. Thus

$$\Delta G = -nF\mathscr{E}$$

For standard conditions

$$\Delta G^\circ = -nF\mathscr{E}^\circ$$

This equation states that *the maximum cell potential is directly related to the free energy difference between the reactants and the products in the cell.* This relationship is important because it provides an experimental means of obtaining ΔG for a reaction. It also confirms that a galvanic cell runs in the direction that gives a positive value for $\mathscr{E}_{\text{cell}}$; a positive $\mathscr{E}_{\text{cell}}$ value corresponds to a negative ΔG value, which is the condition for spontaneity.

EXAMPLE 11.2

Using the data in Table 11.1, calculate ΔG° for the reaction

$$Cu^{2+}(aq) + Fe(s) \longrightarrow Cu(s) + Fe^{2+}(aq)$$

Is this reaction spontaneous?

Solution

The half-reactions are

$$Cu^{2+} + 2e^- \longrightarrow Cu \qquad \mathscr{E}^\circ = 0.34 \text{ V}$$
$$Fe \longrightarrow Fe^{2+} + 2e^- \qquad -\mathscr{E}^\circ = 0.44 \text{ V}$$
$$Cu^{2+} + Fe \longrightarrow Fe^{2+} + Cu \qquad \mathscr{E}^\circ_{\text{cell}} = 0.78 \text{ V}$$

We can calculate ΔG° from the equation

$$\Delta G^\circ = -nF\mathscr{E}^\circ$$

Since two electrons are transferred in the reaction, 2 moles of electrons are required per mole of reactants and products. Thus $n = 2 \text{ mol e}^-$, $F = 96{,}485 \text{ C/mol e}^-$, and $\mathscr{E}^\circ = 0.78 \text{ V} = 0.78 \text{ J/C}$. Therefore,

$$\Delta G^\circ = -(2 \text{ mol e}^-)\left(96{,}485 \frac{\text{C}}{\text{mol e}^-}\right)\left(0.78 \frac{\text{J}}{\text{C}}\right)$$

$$= -1.5 \times 10^5 \text{ J}$$

The process is spontaneous, as indicated both by the negative sign of ΔG° and the positive sign of $\mathscr{E}^\circ_{\text{cell}}$.

This reaction is used industrially to deposit copper metal from solutions containing dissolved copper ores.

EXAMPLE 11.3

Using the data from Table 11.1, predict whether 1 *M* HNO_3 will dissolve gold metal to form a 1 *M* Au^{3+} solution.

Solution

The half-reaction for HNO_3 acting as an oxidizing agent is

$$NO_3^- + 4H^+ + 3e^- \longrightarrow NO + 2H_2O \qquad \mathscr{E}^\circ = 0.96 \text{ V}$$

The reaction for the oxidation of solid gold to Au^{3+} ions is

$$Au \longrightarrow Au^{3+} + 3e^- \qquad -\mathscr{E}^\circ = -1.50 \text{ V}$$

The sum of these half-reactions gives the required reaction:

$$Au(s) + NO_3^-(aq) + 4H^+(aq) \longrightarrow Au^{3+}(aq) + NO(g) + 2H_2O(l)$$

and $$\mathscr{E}^\circ_{\text{cell}} = 0.96 \text{ V} - 1.50 \text{ V} = -0.54 \text{ V}$$

Since the $\mathscr{E}^\circ$ value is negative, the process will *not* occur under standard conditions. That is, gold will not dissolve in 1 *M* HNO_3 to give 1 *M* Au^{3+}. In fact, a mixture (1:3 by volume) of concentrated nitric and hydrochloric acid, called *aqua regia,* is required to dissolve gold.

A gold ring does not dissolve in nitric acid.

11.4 Dependence of the Cell Potential on Concentration

So far we have described cells under standard conditions. In this section we consider the dependence of the cell potential on concentration. For example, under standard conditions (all concentrations 1 *M*) the cell with the reaction

$$Cu(s) + 2Ce^{4+}(aq) \longrightarrow Cu^{2+}(aq) + 2Ce^{3+}(aq)$$

has a potential of 1.36 V. What will the cell potential be if $[Ce^{4+}]$ is greater than 1.0 *M*? This question can be answered qualitatively in terms of Le Châtelier's principle. An increase in the concentration of Ce^{4+} will favor the forward reaction and thus increase the driving force on the electrons. The cell potential will increase. On the other hand, an increase in the concentration of a product (Cu^{2+} or Ce^{3+}) will oppose the forward reaction, thus decreasing the cell potential.

These ideas are illustrated in Example 11.4.

EXAMPLE 11.4

For the cell reaction

$$2Al(s) + 3Mn^{2+}(aq) \longrightarrow 2Al^{3+}(aq) + 3Mn(s) \qquad \mathscr{E}^\circ_{cell} = 0.48 \text{ V}$$

predict whether $\mathscr{E}_{cell}$ is larger or smaller than $\mathscr{E}^\circ_{cell}$ for the following cases.

a. $[Al^{3+}] = 2.0\ M$, $[Mn^{2+}] = 1.0\ M$

b. $[Al^{3+}] = 1.0\ M$, $[Mn^{2+}] = 3.0\ M$

Solution

a. A product concentration has been raised above 1.0 *M*. This will oppose the cell reaction, causing $\mathscr{E}_{cell}$ to be less than $\mathscr{E}^\circ_{cell}$ ($\mathscr{E}_{cell} < 0.48$ V).

b. A reactant concentration has been increased above 1.0 *M* and $\mathscr{E}_{cell}$ will be greater than $\mathscr{E}^\circ_{cell}$ ($\mathscr{E}_{cell} > 0.48$ V).

The Nernst Equation

The dependence of the cell potential on concentration results directly from the dependence of free energy on concentration. Recall from Chapter 10 that the equation

$$\Delta G = \Delta G^\circ + RT\ln(Q)$$

where Q is the reaction quotient, was used to calculate the effect of concentration on ΔG. Since $\Delta G = -nF\mathscr{E}$ and $\Delta G^\circ = -nF\mathscr{E}^\circ$, the equation becomes

$$-nF\mathscr{E} = -nF\mathscr{E}^\circ + RT\ln(Q)$$

Dividing each side of the equation by $-nF$ gives

$$\mathscr{E} = \mathscr{E}^\circ - \frac{RT}{nF}\ln(Q) \qquad (11.1)$$

Equation (11.1), which gives the relationship between the cell potential and the concentrations of the cell components, is commonly called the **Nernst equation** after the German chemist Hermann Nernst (1864–1941).

Nernst was one of the pioneers in the development of electrochemical theory and is generally given credit for first stating the third law of thermodynamics. He won the Nobel prize in chemistry in 1920 for his contributions to our understanding of thermodynamics.

The Nernst equation is often given in terms of a log(base-10) form that is valid at 25°C:

$$\mathscr{E} = \mathscr{E}^\circ - \frac{0.0592}{n}\log(Q)$$

Using this relationship we can calculate the potential of a cell in which some or all of the components are not in their standard states.

For example, $\mathscr{E}^\circ_{cell}$ is 0.48 volt for the galvanic cell based on the reaction

$$2Al(s) + 3Mn^{2+}(aq) \longrightarrow 2Al^{3+}(aq) + 3Mn(s)$$

Consider a cell in which

$$[Mn^{2+}] = 0.50\ M \qquad \text{and} \qquad [Al^{3+}] = 1.50\ M$$

The cell potential at 25°C for these concentrations must be calculated using the Nernst equation:

$$\mathscr{E}_{cell} = \mathscr{E}^\circ_{cell} - \frac{0.0592}{n}\log(Q)$$

We know that

$$\mathscr{E}^\circ_{cell} = 0.48\ \text{V} \qquad \text{and} \qquad Q = \frac{[Al^{3+}]^2}{[Mn^{2+}]^3} = \frac{(1.50)^2}{(0.50)^3} = 18$$

Since the half-reactions are

$$2Al \longrightarrow 2Al^{3+} + 6e^- \qquad \text{and} \qquad 3Mn^{2+} + 6e^- \longrightarrow 3Mn$$

we know that

$$n = 6$$

Thus

$$\mathscr{E}_{cell} = 0.48 - \frac{0.0592}{6}\log(18) = 0.47\ \text{V}$$

Note that the cell voltage decreases slightly because of the nonstandard concentrations. This change is consistent with the predictions of Le Châtelier's principle (see Example 11.4). In this case, since the reactant concentration is lower than 1.0 M and the product concentration is higher than 1.0 M, $\mathscr{E}_{cell}$ is less than $\mathscr{E}^\circ_{cell}$.

The potential calculated from the Nernst equation is the maximum potential before any current flow occurs. As the cell discharges and current flows from anode to cathode, the concentrations change; as a result, $\mathscr{E}_{cell}$ changes. In fact, *the cell will spontaneously discharge until it reaches equilibrium,* at which point

$$Q = K \ \text{(the equilibrium constant)} \qquad \text{and} \qquad \mathscr{E}_{cell} = 0$$

A "dead" battery is one in which the cell reaction has reached equilibrium; there is no longer any chemical driving force to push electrons through the wire. In other words, *at equilibrium the components in the two cell compartments have the same free energy;* that is, $\Delta G = 0$ for the cell reaction at the equilibrium concentrations. The cell no longer has the ability to do work.

EXAMPLE 11.5

Describe the cell based on the following half-reactions:

$$VO_2^+ + 2H^+ + e^- \longrightarrow VO^{2+} + H_2O \qquad \mathscr{E}° = 1.00 \text{ V} \quad (1)$$

$$Zn^{2+} + 2e^- \longrightarrow Zn \qquad \mathscr{E}° = -0.76 \text{ V} \quad (2)$$

where

$$T = 25°\text{C} \qquad [VO_2^+] = 2.0\ M \qquad [VO^{2+}] = 1.0 \times 10^{-2}\ M$$

$$[H^+] = 0.50\ M \qquad [Zn^{2+}] = 1.0 \times 10^{-1}\ M$$

Solution

The balanced cell reaction is obtained by reversing reaction (2) and multiplying reaction (1) by 2:

2 × reaction (1)

$$2VO_2^+ + 4H^+ + 2e^- \longrightarrow 2VO^{2+} + 2H_2O \qquad \mathscr{E}° = 1.00 \text{ V}$$

Reaction (2) reversed

$$Zn \longrightarrow Zn^{2+} + 2e^- \qquad -\mathscr{E}° = 0.76 \text{ V}$$

Cell reaction:

$$2VO_2^+(aq) + 4H^+(aq) + Zn(s) \longrightarrow 2VO^{2+}(aq) + 2H_2O(l) + Zn^{2+}(aq) \qquad \mathscr{E}°_{\text{cell}} = 1.76 \text{ V}$$

Since the cell contains components at concentrations other than 1 M, we must use the Nernst equation, where $n = 2$ (since two electrons are transferred), to calculate the cell potential. At 25°C we can use the equation in the form

$$\mathscr{E} = \mathscr{E}° - \frac{0.0592}{n}\log(Q)$$

Thus

$$\mathscr{E} = 1.76 - \frac{0.0592}{2}\log\left(\frac{[Zn^{2+}][VO^{2+}]^2}{[VO_2^+]^2[H^+]^4}\right)$$

$$= 1.76 - \frac{0.0592}{2}\log\left[\frac{(1.0 \times 10^{-1})(1.0 \times 10^{-2})^2}{(2.0)^2(0.50)^4}\right]$$

$$= 1.76 - \frac{0.0592}{2}\log(4 \times 10^{-5}) = 1.76 + 0.13 = 1.89 \text{ V}$$

The cell diagram is given in Fig. 11.9.

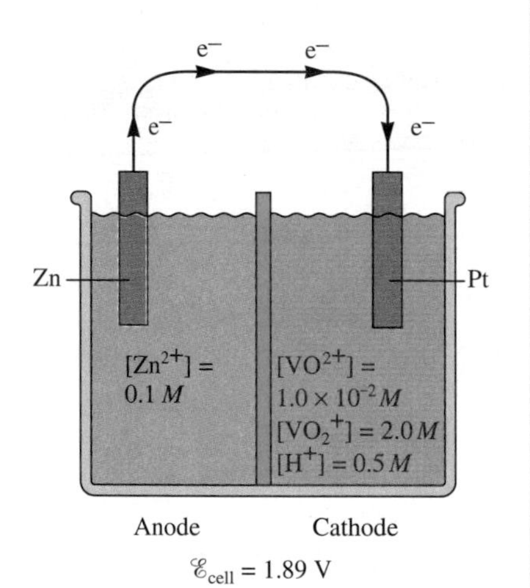

Figure 11.9
Schematic diagram of the cell described in Example 11.5.

Ion-Selective Electrodes

Because the cell potential is sensitive to the concentrations of the reactants and products involved in the cell reaction, measured potentials can be used to determine the concentration of an ion. A pH meter is a familiar example of an instrument that measures concentration from an observed potential. The pH meter has three main components: a standard electrode of known potential, a special **glass electrode** that changes potential depending on the concentration of H^+ ion in the solution into which it is dipped, and a potentiometer that measures the potential between the two electrodes. The potentiometer reading is automatically converted electronically to a direct reading of the pH of the solution being tested.

The glass electrode (see Fig. 11.10) contains a reference solution of dilute hydrochloric acid in contact with a thin glass membrane. The electrical potential of the glass electrode depends on the difference in $[H^+]$ between the reference solution and the solution into which the electrode is dipped. Thus the electrical potential varies with the pH of the solution being tested.

Electrodes that are sensitive to the concentration of a particular ion are called **ion-selective electrodes,** of which the glass electrode for pH measurement is just one example. Glass electrodes can be made sensitive to ions such as Na^+, K^+, or NH_4^+ by changing the composition of the glass. Other ions can be detected if an appropriate crystalline solid replaces the glass membrane. For example, a crystal of lanthanum(III) fluoride (LaF_3) can be used in an electrode to measure $[F^-]$. Solid silver sulfide (Ag_2S) can be used to measure $[Ag^+]$ and $[S^{2-}]$. Some of the ions that can be detected by ion-selective electrodes are listed in Table 11.2.

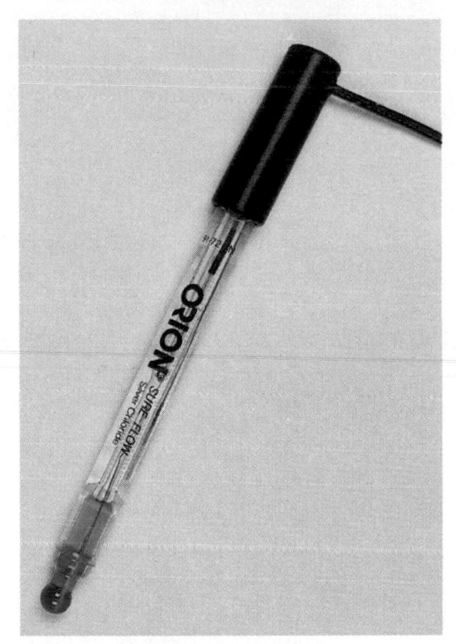

Figure 11.10
A glass electrode contains a reference solution of dilute hydrochloric acid in contact with a thin glass membrane, in which a silver wire coated with silver chloride has been embedded. When the electrode is dipped into a solution containing H^+ ions, the electrode potential is determined by the difference in $[H^+]$ between the two solutions.

Calculation of Equilibrium Constants for Redox Reactions

The quantitative relationship between $\mathscr{E}°$ and $\Delta G°$ allows the calculation of equilibrium constants for redox reactions. For a cell at equilibrium

$$\mathscr{E}_{\text{cell}} = 0 \qquad \text{and} \qquad Q = K$$

Applying these conditions to the form of the Nernst equation valid at 25°C,

$$\mathscr{E} = \mathscr{E}° - \frac{0.0592}{n}\log(Q)$$

gives

$$0 = \mathscr{E}° - \frac{0.0592}{n}\log(K)$$

or

$$\log(K) = \frac{n\mathscr{E}°}{0.0592} \qquad \text{at } 25°\text{C}$$

TABLE 11.2
Some Ions Whose Concentrations Can Be Detected by Ion-Selective Electrodes

Cations	Anions
H^+	Br^-
Cd^{2+}	Cl^-
Ca^{2+}	CN^-
Cu^{2+}	F^-
K^+	NO_3^-
Ag^+	S^{2-}
Na^+	

EXAMPLE 11.6

For the oxidation-reduction reaction

$$S_4O_6^{2-}(aq) + Cr^{2+}(aq) \longrightarrow Cr^{3+}(aq) + S_2O_3^{2-}(aq)$$

the appropriate half-reactions are

$$S_4O_6^{2-} + 2e^- \longrightarrow 2S_2O_3^{2-} \qquad \mathscr{E}° = 0.17\text{ V} \qquad (1)$$

$$Cr^{3+} + e^- \longrightarrow Cr^{2+} \qquad \mathscr{E}° = -0.50\text{ V} \qquad (2)$$

Balance the redox reaction and calculate $\mathscr{E}°$ and K (at 25°C).

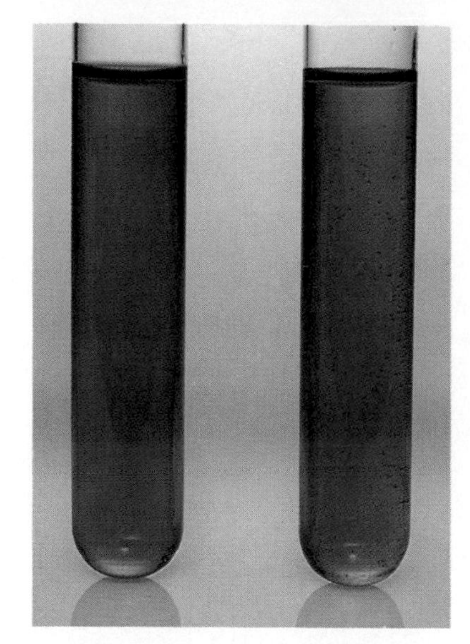

The blue solution on the left contains Cr^{2+} ions and the green solution contains Cr^{3+} ions.

Solution

To obtain the balanced reaction, we must reverse reaction (2), multiply it by 2, and add it to reaction (1):

2 × reaction (2) reversed

$$2(Cr^{2+} \longrightarrow Cr^{3+} + e^-) \qquad -\mathscr{E}° = -(-0.50) \text{ V}$$

Reaction (1)

$$S_4O_6^{2-} + 2e^- \longrightarrow 2S_2O_3^{2-} \qquad \mathscr{E}° = 0.17 \text{ V}$$

Cell reaction:

$$2Cr^{2+}(aq) + S_4O_6^{2-}(aq) \longrightarrow 2Cr^{3+}(aq) + 2S_2O_3^{2-}(aq) \qquad \mathscr{E}° = 0.67 \text{ V}$$

In this reaction 2 mol of electrons are transferred for every unit of reaction; that is, for every 2 mol of Cr^{2+} reacting with 1 mol of $S_4O_6^{2-}$ to form 2 mol of Cr^{3+} and 2 mol of $S_2O_3^{2-}$. Thus $n = 2$. Then

$$\log(K) = \frac{n\mathscr{E}°}{0.0592} = \frac{2(0.67)}{0.0592} = 22.6$$

$$K = 10^{22.6} = 4 \times 10^{22}$$

This very large equilibrium constant is not unusual for a redox reaction.

Concentration Cells

Because cell potentials depend on concentration, we can construct galvanic cells where both compartments contain the same components but at different concentrations. For example, in the cell in Fig. 11.11 both compartments contain aqueous $AgNO_3$, but with different molarities. Let's consider the potential of this cell and the direction of electron flow. The half-reaction relevant to both compartments of this cell is

$$Ag^+ + e^- \longrightarrow Ag \qquad \mathscr{E}° = 0.80 \text{ V}$$

If the cell had 1 M Ag^+ in both compartments,

$$\mathscr{E}°_{\text{cell}} = 0.80 \text{ V} - 0.80 \text{ V} = 0 \text{ V}$$

However, in the cell shown in Fig. 11.11 the concentrations of Ag^+ in the two compartments are 1 M and 0.1 M. Because the concentrations of Ag^+ are unequal, the actual half-cell potentials will not be identical. Thus the cell will exhibit a positive voltage. In which direction will the electrons flow in this cell? The best way to think about this question is to recognize that nature will try to equalize the concentrations of Ag^+ in the two compartments. This can be done by transferring electrons from the compartment containing 0.1 M Ag^+ to the one containing 1 M Ag^+ (left to right in Fig. 11.11). This electron transfer will produce Ag^+ in the left compartment and consume Ag^+ (to form Ag) in the right compartment.

To calculate the potential at 25°C for the cell shown in Fig. 11.11, we use the Nernst equation in the form

$$\mathscr{E} = \mathscr{E}° - \frac{0.0592}{n}\log(Q)$$

e⁻ e⁻ e⁻ e⁻
Porous disk
Ag Ag
0.1 M Ag^+
0.1 M NO_3^-
1 M Ag^+
1 M NO_3^-
Anode Cathode

Figure 11.11
A concentration cell that contains a silver electrode and aqueous silver nitrate in both compartments. Because the right compartment contains 1 M Ag^+ and the left compartment contains 0.1 M Ag^+, there is a driving force to transfer electrons from left to right. Silver is deposited on the right electrode, thus lowering the concentration of Ag^+ in the right compartment. In the left compartment the electrode dissolves to raise the concentration of Ag^+ in solution.

In this case $n = 1$ because the reaction in both compartments is $Ag^+ + e^- \rightarrow Ag$, but running in opposite directions. What is $\mathscr{E}°$ for this cell? Because $\mathscr{E}°$ refers to standard conditions, this corresponds to a cell like the one in Fig. 11.11, *except* that $[Ag^+] = 1\,M$ in both compartments. Such a cell has a potential of zero. That is, $\mathscr{E}° = 0$ for this cell as noted previously. Next, we need to consider the form of Q. Recall that Q is a ratio of product to reactant concentrations. We can represent the process taking place in this cell as

$$Ag^+(1\,M) \longrightarrow Ag^+(0.1\,M)$$

Thus we can write the Nernst equation for this case:

$$\mathscr{E} = \mathscr{E}° - \frac{0.0592}{n}\log(Q) = 0 - \frac{0.0592}{1}\log\left(\frac{0.10}{1.0}\right)$$

$$= -\frac{0.0592}{1}(-1.00) = 0.0592\text{ V}$$

Thus the cell potential is 0.0592 V.

A cell in which both compartments have the same components, but at different concentrations, is called a **concentration cell.** The difference in concentration is the only factor that produces a cell potential in this case, and the voltages are typically small.

EXAMPLE 11.7

Determine the direction of electron flow, designate the anode and cathode, and calculate the potential at 25°C for the cell represented in Fig. 11.12.

Solution

The concentrations of Fe^{2+} ion in the compartments can eventually be equalized by transferring electrons from the left compartment to the right. This will cause Fe^{2+} to be formed in the left compartment, and iron metal will be deposited on the right electrode, thus consuming Fe^{2+} ions. Therefore, electron flow is from left to right, oxidation occurs in the left compartment (the anode), and reduction occurs in the right (the cathode).

To calculate the cell potential, we use the Nernst equation in the form

$$\mathscr{E} = \mathscr{E}° - \frac{0.0592}{n}\log(Q)$$

where $n = 2$ because the cell half-reaction is $Fe^{2+} + 2e^- \rightarrow Fe$ or its opposite. Also, $\mathscr{E}° = 0$ (as always) for a concentration cell, and $Q = 0.01/0.10$, because Fe^{2+} is being formed in the compartment with the lower concentration. Thus

$$\mathscr{E} = 0 - \frac{0.0592}{2}\log\left(\frac{0.01}{0.10}\right) = -0.0296(-1.00) = 0.0296\text{ V}$$

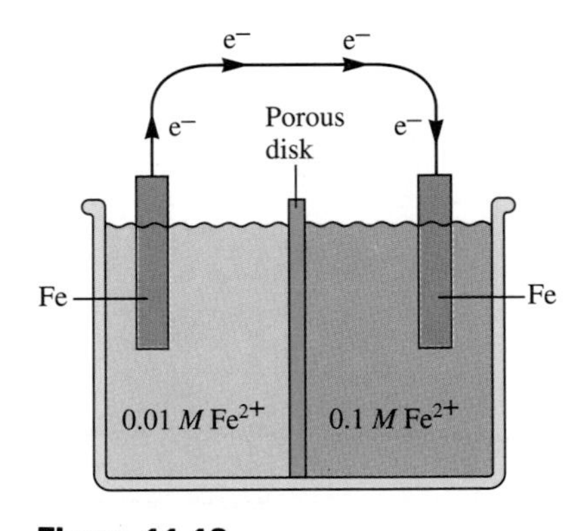

Figure 11.12
A concentration cell containing iron electrodes and different concentrations of Fe^{2+} ion in the two compartments.

Because the potential of an electrochemical cell depends on the concentrations of the participating ions, the observed potential can be used as a sensitive method for measuring ion concentrations in solution. We have already mentioned the ion-selective electrodes that work by this principle. Another application of the relationship between cell potential and concentration is the determination of equilibrium constants for reactions that are not redox reactions. For example, consider a modified version of the silver concentration cell shown in Fig. 11.11. If the 0.10 *M* $AgNO_3$ solution in the left-hand compartment is replaced by 1.0 *M* NaCl and an excess of solid AgCl is added to the cell, the observed cell potential can be used to determine the concentration of Ag^+ in equilibrium with the AgCl(*s*). In other words, at 25°C we can write the Nernst equation as

Ag^+ in equilibrium with AgCl(*s*)
↓

$$\mathscr{E} = 0 - \frac{0.0592}{1}\log\left(\frac{[Ag^+]}{1.0}\right)$$

Note that the measurement of $\mathscr{E}$ for the cell allows the $[Ag^+]$ in equilibrium with AgCl(*s*) to be calculated, which, in turn, allows the K_{sp} for AgCl to be calculated:

$$K_{sp} = [Ag^+][Cl^-]$$

In this case $[Cl^-] = 1.0\,M$ from the 1.0 *M* NaCl, and $[Ag^+]$ can be obtained from the measured value for $\mathscr{E}$. We will illustrate this process in Example 11.8.

EXAMPLE 11.8

A silver concentration cell similar to the one shown in Fig. 11.11 is set up at 25°C with 1.0 *M* $AgNO_3$ in the left compartment and 1.0 *M* NaCl along with excess AgCl(*s*) in the right compartment. The measured cell potential is 0.58 V. Calculate the K_{sp} value for AgCl at 25°C.

Solution

In this case at 25°C

$$\mathscr{E} = 0.58\text{ V} = 0 - \frac{0.0592}{1}\log\left(\frac{[Ag^+]}{1.0}\right)$$

where $[Ag^+]$ represents the equilibrium concentration of Ag^+ in the compartment containing 1.0 *M* NaCl and AgCl(*s*). We calculate $[Ag^+]$ as follows:

$$\log[Ag^+] = -\frac{0.58}{0.0592} = -9.79 \quad\text{and}\quad [Ag^+] = 1.6 \times 10^{-10}\,M$$

Thus

$$K_{sp} = [Ag^+][Cl^-] = (1.6 \times 10^{-10})(1.0) = 1.6 \times 10^{-10}$$

This calculation neglects any complications from complex ions and ion pairs.

Because the measured potential of an electrochemical cell provides a very sensitive method for the experimental determination of equilibrium concentrations, the values of equilibrium constants are often determined from electrochemical measurements.

11.5 Batteries

A **battery** is a galvanic cell or, more commonly, *a group of galvanic cells connected in series,* where the potentials of the individual cells add to give the total battery potential. Batteries are a source of direct current and have become an essential source of portable power in our society. In this section we examine the most common types of batteries. Some new batteries currently being developed are described at the end of the chapter.

Lead Storage Battery

Since about 1915 when self-starters were first used in automobiles, the **lead storage battery** has been a major factor in making the automobile a practical means of transportation. This type of battery can function for several years under temperature extremes from −30°F to 100°F and under incessant punishment from rough roads.

In this battery, lead serves as the anode and lead coated with lead dioxide serves as the cathode. The electrodes dip into an electrolyte solution of sulfuric acid. The electrode reactions are as follows:

Anode reaction: $Pb + HSO_4^- \longrightarrow PbSO_4 + H^+ + 2e^-$

Cathode reaction: $PbO_2 + HSO_4^- + 3H^+ + 2e^- \longrightarrow PbSO_4 + 2H_2O$

Cell reaction: $Pb(s) + PbO_2(s) + 2H^+(aq) + 2HSO_4^-(aq) \longrightarrow 2PbSO_4(s) + 2H_2O(l)$

The typical automobile lead storage battery has six cells connected in series. Each cell contains multiple electrodes in the form of grids (Fig. 11.13) and produces approximately 2 volts, to give a total battery potential of about 12 volts. Note from the cell reaction that sulfuric acid is consumed as the battery discharges, which lowers the density of the electrolyte solution from its initial value of about 1.28 g/cm^3 in the fully charged battery. As a result, the condition of the battery can be monitored by measuring the density of the sulfuric acid solution. The solid lead sulfate formed in the cell reaction during discharge adheres to the grid surfaces of the electrodes. The battery is recharged by forcing current through the battery in the opposite direction to reverse the cell reaction. A car's battery is continuously charged by an alternator driven by the automobile's engine.

An automobile with a dead battery can be jump-started by connecting its battery to the battery in a running automobile. This process can be dangerous, however, because the resulting flow of current causes electrolysis of water in the dead battery, producing hydrogen and oxygen gases (see Section 11.7 for details). Disconnecting the jumper cables after the disabled car starts causes an arc that can ignite the gaseous mixture. If this happens, the battery may explode, ejecting corrosive sulfuric acid. This problem can be avoided by connecting the ground jumper cable to a part of the engine remote from the

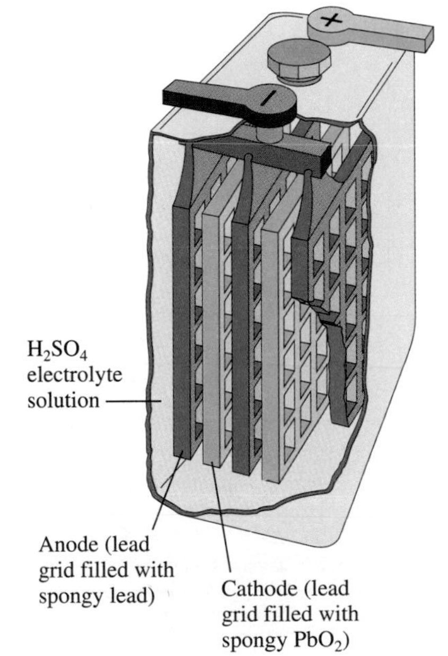

Figure 11.13
One of the six cells in a lead storage battery. The anode consists of a lead grid filled with spongy lead, and the cathode is a lead grid filled with lead dioxide. The cell also contains 38% (by mass) sulfuric acid.

ELECTROCHEMICAL WINDOW SHADES

Everybody likes large windows in their homes and workplaces. However, windows present major problems in making buildings energy efficient—they admit sunlight in the summer increasing air conditioning demands and they allow heat to flow out of the building in the winter. To combat these problems researchers from Electro-Optics Technology Center at Tufts University are developing a "smart" window that uses a short pulse of electricity to vary the transparency of the window in response to changing climatic conditions.

The Tufts smart window consists of seven thin layers as shown on the accompanying illustration. Production of the incredibly thin layers (the seven layers have a total thickness of about 10^{-3} mm) has a lot in common with the manufacture of computer chips, and has proved to be very difficult. The center layer of the window contains an electrolyte which allows ion flow to both of the surrounding layers. One of the sandwiching layers contains tungsten (VI) oxide (WO_3) and the other contains lithium cobalt(III) oxide ($LiCoO_2$). Because the transparency and reflectivity of these compounds change in response to electron and ion flow, they can be used to regulate the passage of light through the window. Depending on the size of the current pulse delivered across the window layers, a continuum of optical properties can be produced ranging from completely transparent to opaque.

Although many practical problems remain to be solved, there is real hope that smart windows will be available for buildings and cars in the near future. This could lead to a significant change in the energy demands of our society—it is estimated that smart windows can reduce a building's energy usage by as much as 50%.

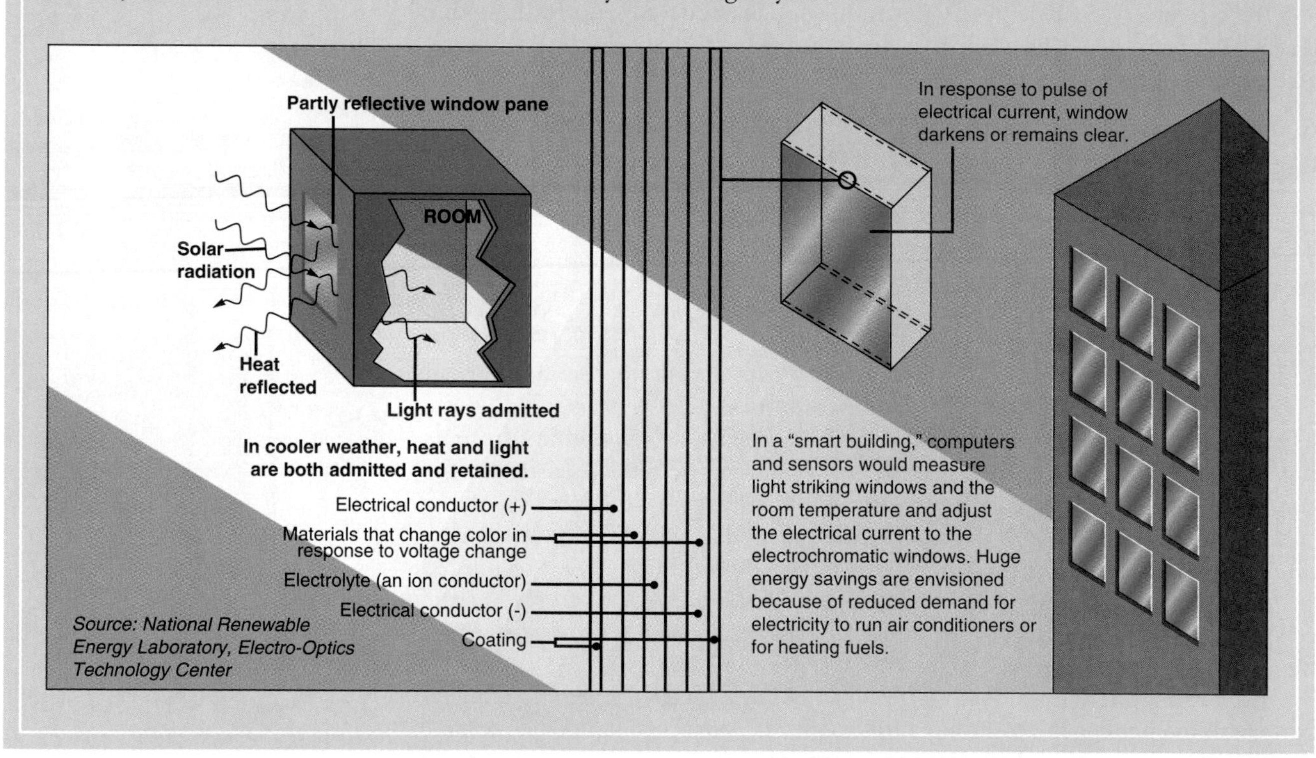

Partly reflective window pane
ROOM
Solar radiation
Heat reflected
Light rays admitted
In cooler weather, heat and light are both admitted and retained.
Electrical conductor (+)
Materials that change color in response to voltage change
Electrolyte (an ion conductor)
Electrical conductor (-)
Coating
In response to pulse of electrical current, window darkens or remains clear.
In a "smart building," computers and sensors would measure light striking windows and the room temperature and adjust the electrical current to the electrochromatic windows. Huge energy savings are envisioned because of reduced demand for electricity to run air conditioners or for heating fuels.

Source: National Renewable Energy Laboratory, Electro-Optics Technology Center

battery. Any arc produced when this cable is disconnected will then be harmless.

Traditional types of storage batteries require periodic "topping off," because the water in the electrolyte solution is depleted by the electrolysis that accompanies the charging process. Recent types of batteries have electrodes made of an alloy of calcium and lead that inhibits the electrolysis of water. These batteries can be sealed since they require no addition of water.

It is rather amazing that in the 75 years lead storage batteries have been used, no better system has been found. Although a lead storage battery does provide excellent service, it has a useful lifetime of only 3–5 years in an automobile. Although it might seem that the battery could undergo an indefinite number of discharge-charge cycles, physical damage from road shock, which tends to shake the solid $PbSO_4$ from the electrodes, and chemical side reactions eventually cause it to fail.

Dry Cell Batteries

The calculators, electronic watches, portable radios, and tape players that are so familiar to us are all powered by small, efficient, **dry cell batteries.** The common dry cell battery was invented more than 100 years ago by George Leclanché (1839–1882), a French chemist. In its *acid version* the dry cell battery contains a zinc inner case that acts as the anode and a carbon rod in contact with a moist paste of solid MnO_2, solid NH_4Cl, and carbon that acts as the cathode (Fig. 11.14). The half-cell reactions are complex but can be approximated as follows:

Anode reaction: $Zn \longrightarrow Zn^{2+} + 2e^-$

Cathode reaction: $2NH_4^+ + 2MnO_2 + 2e^- \longrightarrow Mn_2O_3 + 2NH_3 + H_2O$

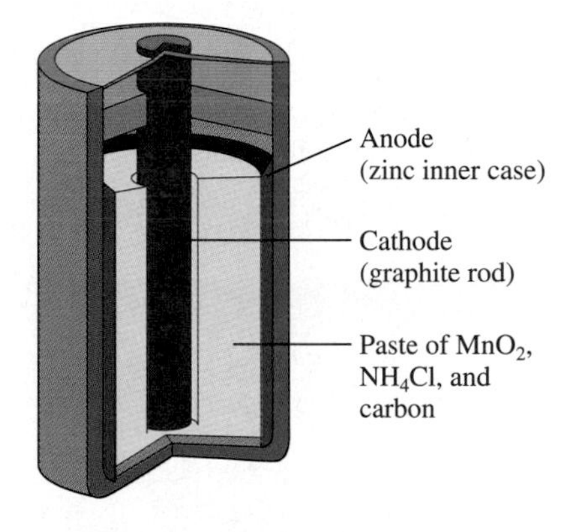

Figure 11.14
A common dry cell battery.

This cell produces a potential of about 1.5 volts.

In the *alkaline version* of the dry cell battery the solid NH_4Cl is replaced with KOH or NaOH. In this case the half-reactions can be approximated as follows:

Anode reaction: $Zn + 2OH^- \longrightarrow ZnO + H_2O + 2e^-$

Cathode reaction: $2MnO_2 + H_2O + 2e^- \longrightarrow Mn_2O_3 + 2OH^-$

The alkaline dry cell lasts longer than the acidic cell mainly because the zinc anode corrodes less rapidly under basic conditions than under acidic conditions.

Other types of dry cell batteries include the *silver cell,* which has a Zn anode. Its cathode employs Ag_2O as the oxidizing agent in a basic environment. *Mercury cells,* often used in calculators, also have a Zn anode. The cathode uses HgO as the oxidizing agent in a basic medium (see Fig. 11.15).

An especially important type of dry cell is the *nickel-cadmium battery,* in which the electrode reactions are as follows:

Anode reaction: $Cd + 2OH^- \longrightarrow Cd(OH)_2 + 2e^-$

Cathode reaction: $NiO_2 + 2H_2O + 2e^- \longrightarrow Ni(OH)_2 + 2OH^-$

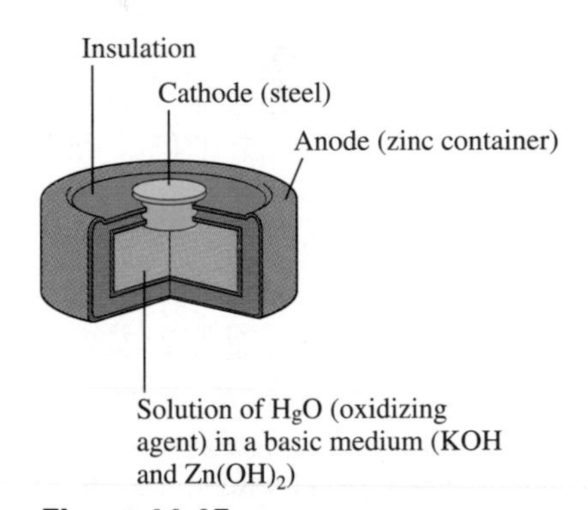

Figure 11.15
A mercury battery of the type used in small calculators.

As in the lead storage battery, the products adhere to the electrodes. Therefore, a nickel-cadmium battery can be recharged an indefinite number of times.

Fuel Cells

A **fuel cell** is *a galvanic cell in which the reactants are continuously supplied.* To illustrate the principles of fuel cells, we will consider the exothermic redox

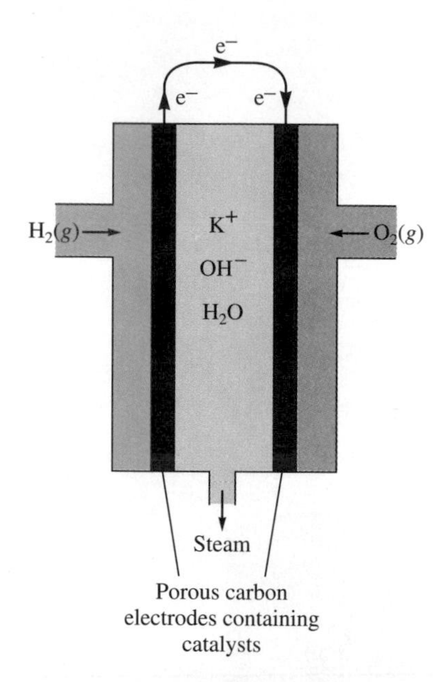

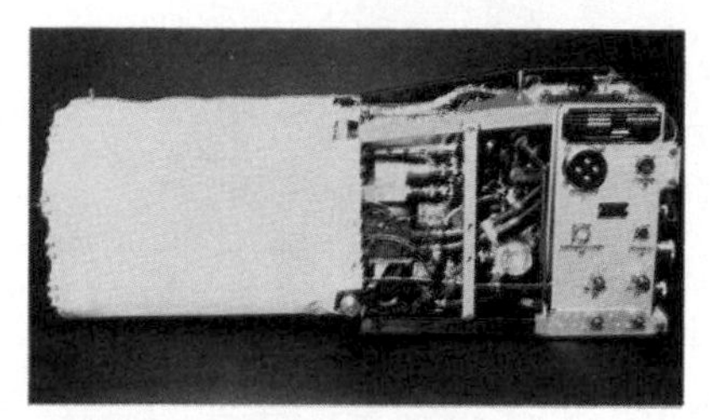

Figure 11.16
(top) Schematic of the hydrogen-oxygen fuel cell. (bottom) One of the three fuel cells that provide electrical power for the Space Shuttle orbiter.

reaction of methane with oxygen:

$$CH_4(g) + 2O_2(g) \longrightarrow CO_2(g) + 2H_2O(g) + \text{energy}$$

Usually, the energy from this reaction is released as heat to warm homes and to run machines. However, in a fuel cell designed to use this reaction, the energy is used to produce an electric current; the electrons flow from the reducing agent (CH_4) to the oxidizing agent (O_2) through a conductor.

The U.S. space program has supported extensive research to develop fuel cells. The shuttle missions use a fuel cell based on the reaction of hydrogen and oxygen to form water:

$$2H_2(g) + O_2(g) \longrightarrow 2H_2O(l)$$

A schematic of a fuel cell that employs this reaction is shown in Fig. 11.16. The half-cell reactions are as follows:

Anode reaction:	$2H_2 + 4OH^- \longrightarrow 4H_2O + 4e^-$
Cathode reaction:	$4e^- + O_2 + 2H_2O \longrightarrow 4OH^-$

A cell of this type weighing about 500 pounds has been designed for space vehicles, but neither this nor any other type of fuel cell is practical enough for general use as a source of portable power. Current research on portable electrochemical power sources seems focused mainly on rechargeable storage batteries with high power-to-weight ratios.

Fuel cells are finding some use, however, as permanent power sources. A power plant built in New York City contains stacks of hydrogen-oxygen fuel cells, which can be rapidly put on-line in response to fluctuating power demands. The hydrogen gas is obtained by decomposing the methane in natural gas. A plant of this type has also been constructed in Tokyo.

11.6 Corrosion

Corrosion can be viewed as the process of returning metals to their natural state—the ores from which they were originally obtained. Corrosion involves oxidation of the metal. Since corroded metal often loses its structural integrity and attractiveness, this spontaneous process has great economic impact. Approximately one-fifth of the iron and steel produced annually is used to replace rusted metal.

Metals corrode because they oxidize easily. Table 11.1 shows that, with the exception of gold, metals commonly used for structural and decorative purposes all have standard reduction potentials less positive than that of oxygen gas. When any one of these half-reactions is reversed (to show oxidation of the metal) and combined with the reduction half-reaction for oxygen, the result is a positive $\mathscr{E}°$ value. Thus the oxidation of most metals by oxygen is spontaneous (although we cannot tell from the potential how fast it will occur).

Some metals, such as copper, gold, silver, and platinum, are relatively difficult to oxidize. They are often called noble metals.

In view of the large differences in the reduction potentials between oxygen and most metals, it is surprising that the problem of corrosion does not prevent the use of metals in air. However, most metals develop a thin oxide coating that tends to protect their internal atoms against further oxidation. The metal that best demonstrates this phenomenon is aluminum. With a reduction potential of

-1.7 volts, aluminum should be easily oxidized by O_2. According to the apparent thermodynamics of the reaction, an aluminum airplane could dissolve in a rainstorm. The fact that this very active metal can be used as a structural material is due to the formation of a thin, adherent layer of aluminum oxide, Al_2O_3, more properly represented as $Al_2(OH)_6$, which greatly inhibits further corrosion. The potential of the "passive," oxide-coated aluminum is -0.6 volt, a value that causes it to behave much like a noble metal.

Iron can also form a protective oxide coating. This coating is not an infallible shield against corrosion, however; when steel is exposed to oxygen in moist air, the oxide that forms tends to scale off, exposing new metal surfaces to corrosion.

The corrosion products of noble metals such as copper and silver are complex and affect the use of these metals as decorative materials. Under normal atmospheric conditions copper forms an external layer of greenish copper carbonate called *patina*. *Silver tarnish* is silver sulfide (Ag_2S), which in thin layers gives the silver surface a richer appearance. Gold with a positive standard reduction potential (1.50 volts), significantly larger than that for oxygen (1.23 volts), shows no appreciable corrosion in air.

Corrosion of Iron

Since steel is the main structural material for bridges, buildings, and automobiles, controlling its corrosion is extremely important. To do so, we must understand the corrosion mechanism. Instead of being a direct oxidation process, as we might expect, the corrosion of iron is an electrochemical reaction, as illustrated in Fig. 11.17.

Steel has a nonuniform surface because its chemical composition is not completely homogeneous. In addition, physical strains leave stress points in the metal. These nonuniformities produce areas where the iron is more easily oxidized (*anodic regions*) than it is at others (*cathodic regions*). In the anodic regions each iron atom gives up two electrons to form the Fe^{2+} ion:

$$Fe \longrightarrow Fe^{2+} + 2e^-$$

The electrons that are released flow through the steel, as they do through the wire of a galvanic cell, to a cathodic region where they react with oxygen:

$$O_2 + 2H_2O + 4e^- \longrightarrow 4OH^-$$

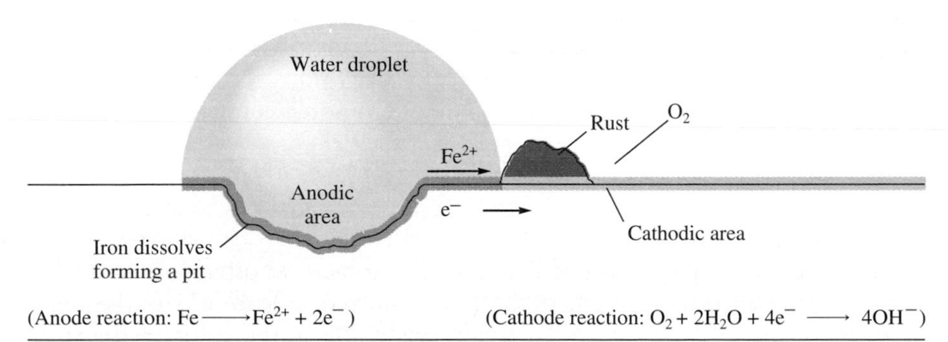

Figure 11.17
The electrochemical corrosion of iron.

Refurbishing the Lady

The restoration of the Statue of Liberty in New York harbor represents a fascinating blend of science, technology, and art. The statue consists of copper sheets attached to a framework of iron, which had become so weakened by corrosion during its 100 years of exposure to the elements that it was in danger of collapsing.

Gustave Eiffel, the French engineer who designed the ingenious support structure, knew from experience that if the copper touched the iron framework, the more active iron would corrode very rapidly. Why does this happen? It is apparent from the reduction potentials

$$Cu^{2+} + 2e^- \longrightarrow Cu \qquad \mathscr{E}° = 0.34 \text{ V}$$

$$Fe^{2+} + 2e^- \longrightarrow Fe \qquad \mathscr{E}° = -0.44 \text{ V}$$

that Cu^{2+} will spontaneously oxidize iron. However, as with many electrochemical processes, the situation is more complex than it first appears. Since we are talking about two metal strips touching each other, the question is: Where do the Cu^{2+} ions (the oxidizing agents) come from? In fact, research on this process suggests that the copper simply acts as a conductor for electrons, and that the oxidizing agent is not Cu^{2+} at all but probably O_2 or oxides of N or S.

Whatever the mechanism, Eiffel attempted to combat the corrosion problem by inserting asbestos pads between the copper sheets and the frame. However, this idea did not work, probably because copper is such a good conductor that *any* contact between the two metals anywhere on the statute totally thwarted the effect of the insulation. In fact, workers found that the iron framework was so corroded it had to be completely replaced with stainless steel, which is much more corrosion-resistant.

Stainless steel has its own problems, however. In being bent to achieve the intricate shapes needed, the steel bars often became brittle. Flexibility was restored by heating each bar to a very high temperature, using a current of 30,000 amperes and then cooling the bar suddenly. Unfortunately, this process also destroyed the corrosion resistance of the stainless steel. That resistance had to be restored by soaking the bars in nitric acid, an oxidizing acid that reforms the protective oxide coating removed by heating.

Another problem faced by the restorers was the removal of layers of coal tar and paint that had been applied in vain attempts to protect the statue's interior. Although the iron could be cleaned by blasting with aluminum oxide powder, the more fragile copper sheets required a gentler treatment. The restorers discovered that liquid nitrogn (77 K) cracked the paint and caused it to peel away. The coal-tar layer under the paint was removed by blasting with baking soda, $NaHCO_3$, a substance sometimes used by museum curators for polishing dinosaur bones. Although this treatment worked very well, it created another chemical problem: where the baking soda seeped between the seams in

The Fe^{2+} ions formed in the anodic regions travel to the cathodic regions through the moisture on the surface of the steel, just as ions travel through a salt bridge in a galvanic cell. In the cathodic regions Fe^{2+} ions react with oxygen to form rust, which is hydrated iron(III) oxide of variable composition:

$$4Fe^{2+}(aq) + O_2(g) + (4 + 2n)H_2O(l) \longrightarrow \underbrace{2Fe_2O_3 \cdot nH_2O(s)}_{\text{Rust}} + 8H^+(aq)$$

Because of the migration of ions and electrons, rust often forms at sites that are remote from those where the iron dissolved to form pits in the steel. Also the degree of hydration of the iron oxide affects the color of the rust, which may vary from black to yellow to the familiar reddish brown.

the copper plates onto the exterior, the statue turned from green to light blue when it rained. After this phenomenon was observed for the first time, workers stationed on the exterior scaffolding immediately cleaned off any leaking $NaHCO_3$.

One of the most interesting aspects of the chemistry of the Statue of Liberty is the green patina on its surface. Copper metal exposed to the atmosphere changes from its bright reddish brown metallic luster, first to an almost black color, then to the familiar green patina. The initial blackening is mainly due to formation of copper oxide and copper sulfide. The green patina that forms consists of thin layers of two types of basic copper sulfates: brochantite, $CuSO_4 \cdot 3Cu(OH)_2$, and antlerite, $CuSO_4 \cdot 2Cu(OH)_2$. Crystals of these compounds seem to be cemented onto the copper surface by organic molecules from the air.

In recent years large areas of the statue's left side have been observed to darken as if the patina were being removed. Scientists speculated that this darkening may be the result of acid rain, which converts brochantite to the more soluble antlerite, which is then washed off by rainwater.

To make any new copper sheets look like the old ones, the restorers transplanted the patina from a weathered piece of copper to the new surface. Applying acetone, an organic solvent, to the weathered surface and scraping with an abrasive cloth caused tiny flakes of patina to fall off. These flakes, applied to the new copper, attached themselves permanently in one to three weeks of exposure to the atmosphere.

The restoration of the Statue of Liberty made use of the latest advances in chemistry as well as facts known to most general chemistry students. It's an example of the fascinating and varied problems faced by chemists as they pursue their profession.

Suggested Reading

Ivars Peterson, "Lessons Learned from a Lady," *Science News* **130** (1986): 392.

The newly restored Statue of Liberty in New York harbor.

The electrochemical nature of the rusting of iron explains the importance of moisture in the corrosion process. Moisture must be present to act as a "salt bridge" between anodic and cathodic regions. Steel does not rust in dry air, a fact that explains why cars last much longer in the arid Southwest than in the relatively humid Midwest. Salt also accelerates rusting, a fact all too easily recognized by car owners in the colder parts of the United States, where salt is used on roads to melt snow and ice. The severity of rusting is greatly increased because the dissolved salt on the moist steel surface increases the conductivity of the aqueous solution formed there and thus accelerates the electrochemical corrosion process. Chloride ions also form very stable complex ions with Fe^{3+}, and this factor tends to encourage the dissolving of the iron, further accelerating the corrosion.

Prevention of Corrosion

Prevention of corrosion is an important way of conserving our natural resources of energy and metals. The primary means of protection is the application of a coating, most commonly paint or metal plating, to protect the metal from oxygen and moisture. Chromium and tin are often used to plate steel (Section 11.8) because they react with oxygen to form a durable, effective oxide coating. Zinc, also used to coat steel in a process called **galvanizing,** does not form an oxide coating. However, since it is a more active metal than iron, as the potentials for the oxidation half-reactions show,

$$Fe \longrightarrow Fe^{2+} + 2e^- \qquad -\mathscr{E}° = 0.44 \text{ V}$$
$$Zn \longrightarrow Zn^{2+} + 2e^- \qquad -\mathscr{E}° = 0.76 \text{ V}$$

any oxidation that occurs dissolves zinc rather than iron. Recall that the reaction with the most positive standard potential has the greatest thermodynamic tendency to occur. Thus zinc acts as a "sacrificial" coating on steel.

Alloying is also used to prevent corrosion. *Stainless steel* contains chromium and nickel, both of which form oxide coatings that change steel's reduction potential to one characteristic of the noble metals. A new technology is now being developed to create surface alloys. Instead of forming a metal alloy such as stainless steel, which has the same composition throughout, a cheaper,

Paint That Stops Rust—Completely

Traditionally, paint has provided the most economical method for protecting steel against corrosion. However, as people who live in the Midwest know well, paint cannot prevent a car from rusting indefinitely. Eventually, flaws develop in the paint that allow the ravages of rusting to take place.

This situation may soon change. Chemists at Glidden Research Center in Ohio have developed a paint called Rustmaster Pro that worked so well to prevent rusting in its initial tests that the scientists did not believe their results. Steel coated with the new paint showed no signs of rusting after an astonishing 10,000 hours of exposure in a salt spray chamber at 38°C.

Rustmaster is a water-based polymer formulation that prevents corrosion in two different ways. First, the polymer layer that cures in air forms a barrier impenetrable to both oxygen and water vapor. Second, the chemicals in the coating react with the steel surface to produce an interlayer between the metal and the polymer coating. This interlayer is a complex mineral called pyroaurite that contains cations of the form $[M_{1-x}Z_x(OH)_2]^{x+}$, where M is a 2+ ion (Mg^{2+}, Fe^{2+}, Zn^{2+}, Co^{2+}, or Ni^{2+}), Z is a 3+ ion (Al^{3+}, Fe^{3+}, Mn^{3+}, Co^{3+}, or Ni^{3+}), and x is a number between 0 and 1. The anions in pyroaurite are typically CO_3^{2-}, Cl^-, and/or SO_4^{2-}.

This pyroaurite interlayer is the real secret of the paint's effectiveness. Because the corrosion of steel has an electrochemical mechanism, motion of ions must be possible between the cathodic and anodic areas on the surface of the steel for rusting to occur. However, the pyroaurite interlayer grows into the neighboring polymer layer, thus preventing this crucial movement of ions. In effect, this layer prevents corrosion in the same way that removing the salt bridge prevents current from flowing in a galvanic cell.

In addition to having an extraordinary corrosion-fighting ability, Rustmaster yields an unusually small quantity of volatile solvents as it dries. A typical paint can produce from 1 to 5 kg of volatiles per gallon; Rustmaster produces only 0.05 kg. This paint may signal a new era in corrosion prevention.

carbon steel is treated by ion bombardment to produce a thin layer of stainless steel or other desirable alloy on the surface. In this process a "plasma" or "ion gas" of the alloying ions is formed at high temperatures and is then directed onto the surface of the metal.

Cathodic protection is a method most often employed to protect steel in buried fuel tanks and pipelines. An active metal, such as magnesium, is connected by a wire to the pipeline or tank to be protected (Fig. 11.18). Because the magnesium is a better reducing agent than iron, electrons are furnished by the magnesium, keeping the iron from being oxidized. As oxidation occurs, the magnesium anode dissolves, and so it must be replaced periodically.

Ground level
Cathode (buried iron pipe)
Connecting insulated wire
Anode (magnesium)
Electrolyte (moist soil)

Figure 11.18
Cathodic protection of an underground pipe.

11.7 Electrolysis

A galvanic cell produces current when an oxidation-reduction reaction proceeds spontaneously. A similar apparatus, an **electrolytic cell,** uses electrical energy to produce chemical change. The process of **electrolysis** involves *forcing a current through a cell to produce a chemical change for which the cell potential is negative;* that is, electrical work causes an otherwise nonspontaneous chemical reaction to occur. Electrolysis has great practical importance; for example, charging a battery, producing aluminum metal, and chrome plating an object are all done electrolytically.

An electrolytic cell uses electrical energy to produce a chemical change that would otherwise not occur spontaneously.

To illustrate the difference between a galvanic cell and an electrolytic cell, consider the cell shown in Fig. 11.19(a) as it runs spontaneously to produce 1.10 volts. In this *galvanic* cell the reaction at the anode is

$$Zn \longrightarrow Zn^{2+} + 2e^-$$

and the cathode reaction is

$$Cu^{2+} + 2e^- \longrightarrow Cu$$

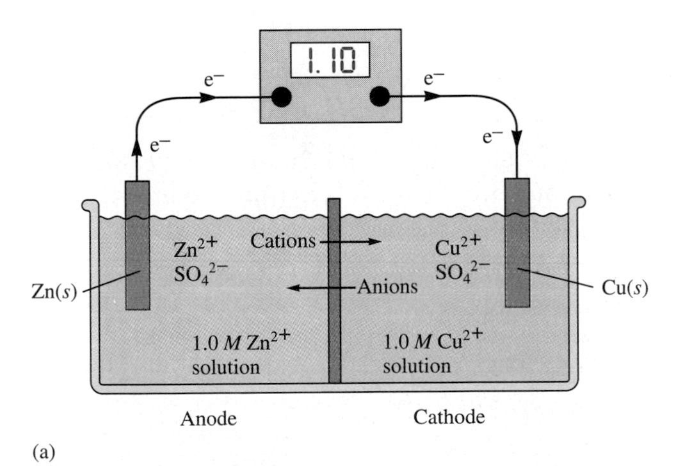

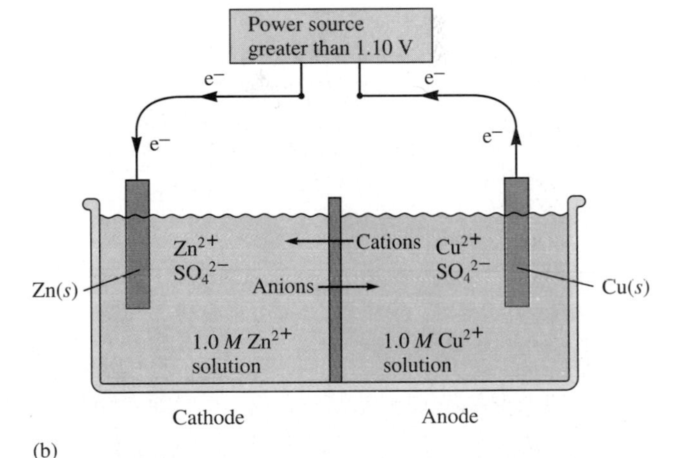

Figure 11.19
(a) A standard galvanic cell based on the spontaneous reaction $Zn + Cu^{2+} \rightarrow Zn^{2+} + Cu$.

(b) A standard electrolytic cell. A power source forces the reaction $Cu + Zn^{2+} \rightarrow Cu^{2+} + Zn$ to occur.

Figure 11.19(b) shows an external power source forcing electrons through the cell in the *opposite* direction to that in (a). This requires an external potential greater than 1.10 V, which must be applied in opposition to the natural cell potential. This device is an *electrolytic cell.* Notice that since electron flow is opposite in the two cases, the anode and cathode are reversed in (a) and (b). Also, ion flow through the salt bridge is opposite in the two cells.

Now we will consider the stoichiometry of electrolytic processes, that is, *how much chemical change occurs with the flow of a given current for a specified time.* Suppose we wish to determine the mass of copper that is plated out when a current of 10.0 amperes [an **ampere** (amp) is *1 coulomb of charge per second*] is passed for 30.0 minutes through a solution containing Cu^{2+}. *Plating* means depositing the neutral metal on the electrode surface by reducing the metal ions in solution. In this case each Cu^{2+} ion requires two electrons to become an atom of copper metal:

1 A = 1 C/s

$$Cu^{2+}(aq) + 2e^- \longrightarrow Cu(s)$$

This reduction process occurs at the cathode of the electrolytic cell.

To solve this stoichiometry problem, we employ the following steps:

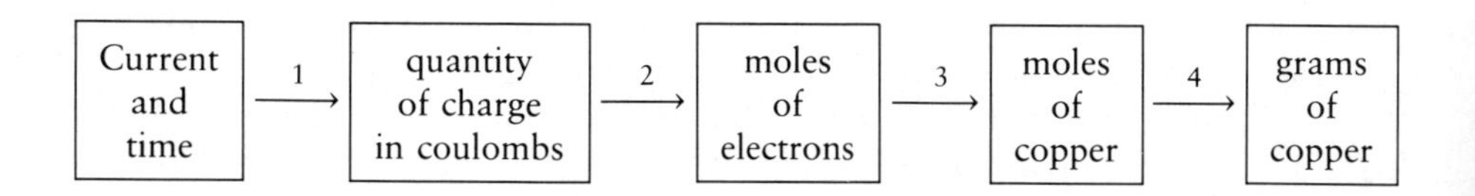

1. Since an amp is a coulomb of charge per second, we multiply the current by the time in seconds to obtain the total coulombs of charge passed into the solution at the cathode:

$$\text{Coulombs of charge} = \text{amps} \times \text{seconds} = \frac{\text{C}}{\text{s}} \times \text{s}$$
$$= 10.0 \frac{\text{C}}{\text{s}} \times 30.0 \text{ min} \times 60.0 \frac{\text{s}}{\text{min}}$$
$$= 1.80 \times 10^4 \text{ C}$$

2. Since 1 mole of electrons carries a charge of 1 faraday, or 96,485 coulombs, we can calculate the number of moles of electrons required to carry 1.80×10^4 coulombs of charge:

$$1.80 \times 10^4 \text{ C} \times \frac{1 \text{ mol e}^-}{96{,}485 \text{ C}} = 1.87 \times 10^{-1} \text{ mol e}^-$$

This means that 0.187 mole of electrons flowed into the solution containing Cu^{2+}.

3. Each Cu^{2+} ion requires two electrons to become a copper atom. Thus each mole of electrons produces $\frac{1}{2}$ mole of copper metal:

$$1.87 \times 10^{-1} \text{ mol e}^- \times \frac{1 \text{ mol Cu}}{2 \text{ mol e}^-} = 9.35 \times 10^{-2} \text{ mol Cu}$$

4. Since we now know the moles of copper metal plated onto the cathode, we can calculate the mass of copper formed:

$$9.35 \times 10^{-2} \text{ mol Cu} \times \frac{63.546 \text{ g}}{\text{mol Cu}} = 5.94 \text{ g Cu}$$

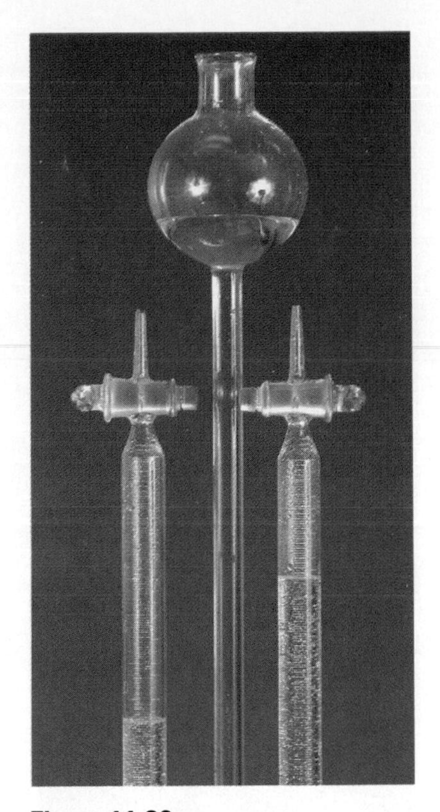

Figure 11.20
The electrolysis of water produces hydrogen gas at the cathode (on the left) and oxygen gas at the anode (on the right). Note that some provision (a salt bridge or porous disk) must be made to allow ion flow.

Electrolysis of Water

We have seen that hydrogen and oxygen combine spontaneously to form water and that the accompanying decrease in free energy can be used to run a fuel cell to produce electricity. The reverse process, which is of course nonspontaneous, can be forced by electrolysis:

Anode reaction:	$2H_2O \longrightarrow O_2 + 4H^+ + 4e^-$	$-\mathscr{E}° = -1.23$ V
Cathode reaction:	$4H_2O + 4e^- \longrightarrow 2H_2 + 4OH^-$	$\mathscr{E}° = -0.83$ V
Net reaction:	$6H_2O \longrightarrow 2H_2 + O_2 + \underbrace{4(H^+ + OH^-)}_{4H_2O}$	$\mathscr{E}° = -2.06$ V
or	$2H_2O \longrightarrow 2H_2 + O_2$	

Note that these potentials assume an anode chamber with 1 *M* H^+ and a cathode chamber with 1 *M* OH^-. In pure water, where $[H^+] = [OH^-] = 10^{-7}$ *M*, the potential for the overall process is -1.23 V.

In practice, however, if platinum electrodes connected to a 6-volt battery are dipped into pure water, no reaction is observed; pure water contains so few ions that only a negligible current can flow. However, addition of even a small amount of a soluble salt causes an immediate evolution of bubbles of hydrogen and oxygen, as illustrated in Fig. 11.20.

Electrolysis of Mixtures of Ions

Suppose a solution in an electrolytic cell contains the ions Cu^{2+}, Ag^+, and Zn^{2+}. If the voltage, which is initially very low, is gradually turned up, in which order will the metals be plated out onto the cathode? This question can be answered by looking at the standard reduction potentials of these ions:

$$Ag^+ + e^- \longrightarrow Ag \qquad \mathscr{E}° = 0.80 \text{ V}$$
$$Cu^{2+} + 2e^- \longrightarrow Cu \qquad \mathscr{E}° = 0.34 \text{ V}$$
$$Zn^{2+} + 2e^- \longrightarrow Zn \qquad \mathscr{E}° = -0.76 \text{ V}$$

Remember that the more *positive* the $\mathscr{E}°$ value, the more the reaction has a tendency to proceed in the direction indicated. Of the three reactions listed, the order of oxidizing ability is

$$Ag^+ > Cu^{2+} > Zn^{2+}$$

This means that silver will plate out first as the potential is increased, followed by copper, and finally zinc.

Figure 11.21
Charles Martin Hall (1863–1914) was a student at Oberlin College in Ohio when he first became interested in aluminum. One of his professors commented that anyone who could manufacture aluminum cheaply would make a fortune, so Hall decided to give it a try. The 21-year-old Hall worked in a wooden shed near his house with an iron frying pan as a container, a blacksmith's forge as a heat source, and galvanic cells constructed from fruit jars. Using these crude galvanic cells, Hall found that he could produce aluminum by passing a current through a molten Al_2O_3/ Na_3AlF_6 mixture. By a strange coincidence, Paul Heroult, a Frenchman who was born and died in the same years as Hall, made the same discovery at about the same time.

The principle described in this section is very useful, but it must be applied with some caution. For example, in the electrolysis of an aqueous solution of sodium chloride, we should be able to use $\mathscr{E}°$ values to predict which products are expected. Of the major species in the solution (Na^+, Cl^-, and H_2O), only Cl^- and H_2O can be readily oxidized. The half-reactions (written as oxidization processes) are:

$$2Cl^- \longrightarrow Cl_2 + 2e^- \qquad -\mathscr{E}° = -1.36 \text{ V}$$
$$2H_2O \longrightarrow O_2 + 4H^+ + 4e^- \qquad -\mathscr{E}° = -1.23 \text{ V}$$

Since water has the more positive potential, we would expect to see O_2 produced at the anode. However, this does not happen. As the voltage is increased in the cell, the Cl^- ion is the first to be oxidized. A much higher potential than expected is required to oxidize water. The voltage required in excess of the expected value (called the *overvoltage*) is much greater for the production of O_2 than for Cl_2, which explains why chlorine is produced at the lower voltage.

The causes of overvoltage are very complex. Basically, the phenomenon is caused by difficulties in transferring electrons from the species in the solution to the atoms on the electrode across the electrode-solution interface. Because of this situation, $\mathscr{E}°$ values must be used cautiously in predicting the actual order of oxidation or reduction of species in an electrolytic cell.

11.8 Commercial Electrolytic Processes

The chemistry of metals is characterized by their ability to donate electrons to form ions. Because metals are typically such good reducing agents, most are found in nature in *ores,* mixtures of ionic compounds often containing oxide, sulfide, and silicate anions. The noble metals, such as gold, silver, and platinum, are more difficult to oxidize and are often found as pure metals.

Production of Aluminum

Aluminum is one of the most abundant elements on earth, ranking third behind oxygen and silicon. Since aluminum is a very active metal, it is found in nature as its oxide in an ore called *bauxite* (named after Les Baux, France, where it was discovered in 1821). Production of aluminum metal from its ore proved to be more difficult than production of most other metals. In 1782 Lavoisier recognized aluminum as a metal "whose affinity for oxygen is so strong that it cannot be overcome by any known reducing agent." As a result, pure aluminum metal remained unknown. Finally, in 1854 a process was found for producing metallic aluminum using sodium, but aluminum remained a very expensive rarity. In fact, it is said that Napoleon III served his most honored guests with aluminum forks and spoons, while the others had to settle for gold and silver utensils.

The breakthrough came in 1886 when two men, Charles M. Hall in the United States and Paul Heroult in France, almost simultaneously discovered a practical electrolytic process for producing aluminum (see Fig. 11.21). The key factor in the *Hall-Heroult process* is the use of molten cryolite (Na_3AlF_6) as the solvent for the aluminum oxide.

The Chemistry of Sunken Treasure

When the galleon *Atocha* was destroyed on a reef by a hurricane in 1622, it was bound for Spain carrying approximately 47 tons of copper, gold, and silver from the New World. The bulk of the treasure was silver bars and coins packed in wooden chests. When treasure hunter Mel Fisher salvaged the silver in 1985, corrosion and marine growth had transformed the shiny metal into something that looked like coral. Restoring the silver to its original condition required an understanding of the chemical changes that had occurred in 350 years of being submerged in the ocean. Much of this chemistry we have already considered at various places in this text.

As the wooden chests containing the silver decayed over the years, the oxygen supply was depleted. This favored the growth of certain bacteria that use the sulfate ion rather than oxygen as an oxidizing agent to generate energy. As these bacteria consume sulfate ions, they release hydrogen sulfide gas that reacts with silver to form black silver sulfide:

$$2Ag(s) + H_2S(aq) \longrightarrow Ag_2S(s) + H_2(g)$$

Thus over the years the surface of the silver became covered with a tightly adhering layer of corrosion, which fortunately protected the silver underneath, thus preventing total conversion of the silver to silver sulfide.

Another change that took place as the wood decomposed was the formation of carbon dioxide.

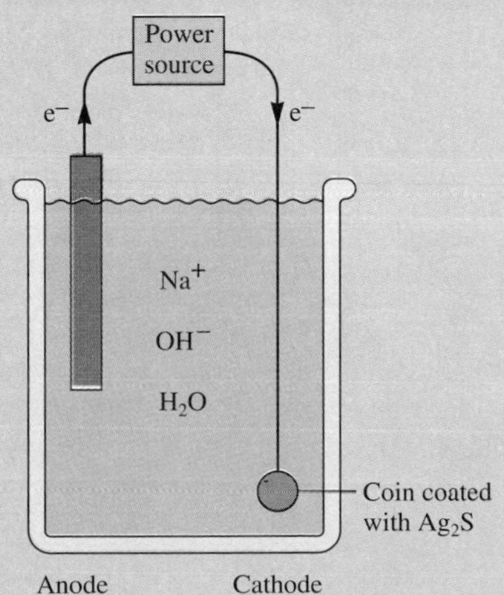

This shifted the equilibrium that is present in the ocean,

$$CO_2(aq) + H_2O(l) \rightleftharpoons HCO_3^-(aq) + H^+(aq)$$

to the right, producing higher concentrations of HCO_3^-. In turn, the HCO_3^- reacted with Ca^{2+} ions present in the seawater to form calcium carbonate:

$$Ca^{2+}(aq) + HCO_3^-(aq) \rightleftharpoons CaCO_3(s) + H^+(aq)$$

Calcium carbonate is the main component of limestone. Thus over time the corroded silver coins and bars became encased in limestone.

Both the limestone formation and the corrosion had to be dealt with. Since $CaCO_3$ contains the basic anion CO_3^{2-}, acid dissolves limestone:

$$2H^+(aq) + CaCO_3(s) \longrightarrow Ca^{2+}(aq) + CO_2(g) + H_2O(l)$$

Soaking the mass of coins in a buffered acidic bath for several hours allowed the individual pieces to be separated, and the black Ag_2S on the surfaces was revealed. An abrasive could not have been used to remove this corrosion; it would have destroyed the details of the engraving—a very valuable feature of the coins to a historian or a collector—and it would have washed away some of the silver. Instead, the corrosion reaction was reversed through electrolytic reduction. The coins were connected to the cathode of an electrolytic cell in a dilute sodium hydroxide solution, as represented in the figure.

As electrons flow, the Ag^+ ions in the silver sulfide are reduced to silver metal,

$$Ag_2S + 2e^- \longrightarrow Ag + S^{2-}$$

As a by-product, bubbles of hydrogen gas from the reduction of water,

$$2H_2O + 2e^- \longrightarrow H_2 + 2OH^-$$

form on the surface of the coins.

The agitation caused by the bubbles loosens the flakes of metal sulfide and helps clean the coins.

Using these procedures, technicians have been able to restore the treasure to very nearly the same condition it was in when the *Atocha* sailed many years ago.

TABLE 11.3
The Price of Aluminum Over the Past Century

Date	Price of Aluminum ($/lb)*
1855	100,000
1885	100
1890	2
1895	0.50
1970	0.30
1980	0.80
1990	0.74

* Note the precipitous drop in price after the discovery of the Hall-Heroult process.

Electrolysis is possible only if ions can move to the electrodes. A common method for producing ion mobility is dissolving the substance to be electrolyzed in water. This method cannot be used for aluminum because water is more easily reduced than Al^{3+}, as the following standard reduction potentials show:

$$Al^{3+} + 3e^- \longrightarrow Al \qquad \mathscr{E}° = -1.66 \text{ V}$$
$$2H_2O + 2e^- \longrightarrow H_2 + 2OH^- \qquad \mathscr{E}° = -0.83 \text{ V}$$

Thus aluminum metal cannot be plated out of an aqueous solution of Al^{3+}.

Ion mobility can also be produced by melting the salt. But the melting point of solid Al_2O_3 is much too high (2050°C) to allow practical electrolysis of the molten oxide. A mixture of Al_2O_3 and Na_3AlF_6, however, has a melting point of 1000°C, and the resulting molten mixture can be used to obtain aluminum metal electrolytically. Because of this discovery by Hall and Heroult, the price of aluminum plunged (see Table 11.3), and its use became economically feasible.

Bauxite is not pure aluminum oxide (called *alumina*) but also contains the oxides of iron, silicon, and titanium, and various silicate materials. The pure hydrated alumina ($Al_2O_3 \cdot nH_2O$) is obtained by treating the crude bauxite with aqueous sodium hydroxide. Being amphoteric, alumina dissolves in the basic solution:

$$Al_2O_3(s) + 2OH^-(aq) \longrightarrow 2AlO_2^-(aq) + H_2O(l)$$

The other metal oxides, which are basic, remain as solids. The solution containing the aluminate ion (AlO_2^-) is separated from the sludge of other oxides and is acidified with carbon dioxide gas, causing the hydrated alumina to reprecipitate:

$$2CO_2(g) + 2AlO_2^-(aq) + (n + 1)H_2O(l) \longrightarrow 2HCO_3^-(aq) + Al_2O_3 \cdot nH_2O(s)$$

The purified alumina is then mixed with cryolite and melted, and the aluminum ion is reduced to aluminum metal in an electrolytic cell of the type shown in Fig. 11.22. Because the electrolyte solution contains a large number of

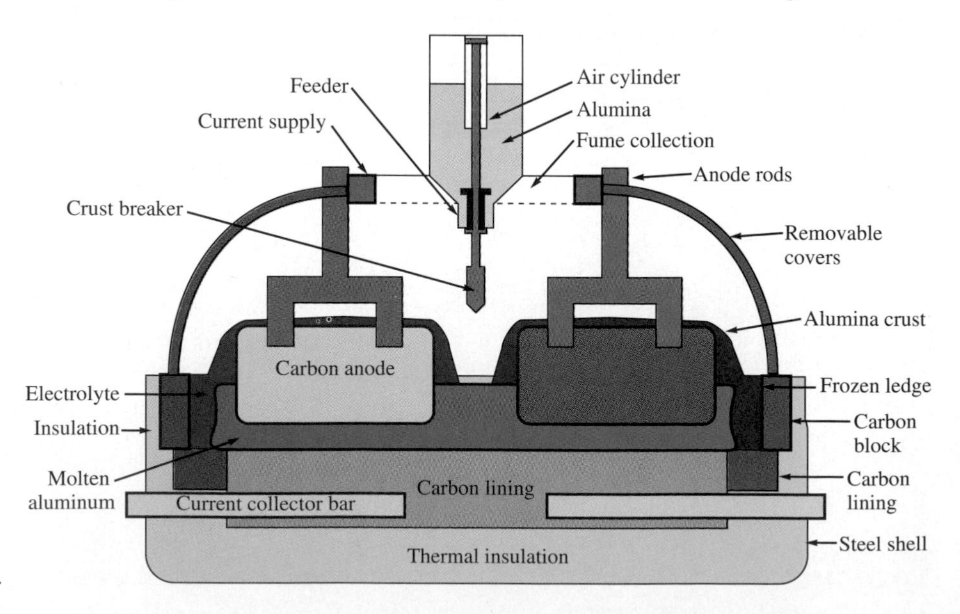

Figure 11.22
A schematic diagram of an electrolytic cell for producing aluminum by the Hall-Heroult process. Because molten aluminum is more dense than the mixture of molten cryolite and alumina, it settles to the bottom of the cell and is drawn off periodically. The graphite electrodes are gradually eaten away and must be replaced from time to time. The cell operates at a current flow of up to 250,000 A.

aluminum-containing ions, the chemistry is not completely understood. However, the alumina probably reacts with the cryolite anion as follows:

$$Al_2O_3 + 4AlF_6^{3-} \longrightarrow 3Al_2OF_6^{2-} + 6F^-$$

The electrode reactions are thought to be the following:

Cathode reaction: $AlF_6^{3-} + 3e^- \longrightarrow Al + 6F^-$

Anode reaction: $2Al_2OF_6^{2-} + 12F^- + C \longrightarrow 4AlF_6^{3-} + CO_2 + 4e^-$

The overall cell reaction can be written as

$$2Al_2O_3 + 3C \longrightarrow 4Al + 3CO_2$$

The aluminum produced in this electrolytic process is 99.5% pure. To be useful as a structural material, aluminum is alloyed with metals such as zinc (used for trailer and aircraft construction) and manganese (used for cooking utensils, storage tanks, and highway signs). The production of aluminum consumes almost 5% of all electricity used in the United States.

Electrorefining of Metals

Purification of metals is another important application of electrolysis. For example, impure copper from the chemical reduction of copper ore is cast into large slabs that serve as the anodes for electrolytic cells. Aqueous copper sulfate is the electrolyte, and thin sheets of ultrapure copper function as the cathodes (see Fig. 11.23).

The main reaction at the anode is

$$Cu \longrightarrow Cu^{2+} + 2e^-$$

Other metals such as iron and zinc are also oxidized from the impure anode:

$$Zn \longrightarrow Zn^{2+} + 2e^-$$
$$Fe \longrightarrow Fe^{2+} + 2e^-$$

Figure 11.23
Ultrapure copper sheets (serving as cathodes) are lowered between slabs of impure copper (serving as anodes) into a tank containing an aqueous solution of copper sulfate ($CuSO_4$). It takes about four weeks for the anodes to dissolve and for the copper to be deposited on the cathodes.

BATTERIES OF THE FUTURE

A possible alternative to the internal combustion engine used in present-day automobiles is an electric motor powered by batteries. This and many other applications have stimulated research into new types of efficient batteries. In fact, some types being developed are capable of producing five to ten times as much power per kilogram as the lead storage battery.

Ion mobility is needed in a battery to prevent charge polarization. Usually this is accomplished with an aqueous electrolyte. But there is now considerable interest in solid ceramic materials that transport ions very efficiently. For example, in the *beta,* or *sodium-sulfur battery,* a solid electrolyte, called beta alumina, composed of a mixture of sodium, aluminum, lithium, and magnesium oxides, separates the electrode compartments. Liquid sodium is the reducing agent (the anode), and liquid sulfur is the oxidizing agent (the cathode). Although the beta battery must be operated at 350°C to keep the reactants liquid, the system shows promise because it can store approximately four times as much energy per kilogram and can undergo approximately three times as many discharge-charge cycles as the lead storage battery. Asea Brown Boveri, a German battery manufacturer, has recently teamed with the automaker BMW to assemble a car powered by a 584-lb liquid sodium-sulfur battery. This car has a top speed of 60 mph, a cruising range of about 100 miles on a full charge, and a battery life projected to be as much as 125,000 miles.

Another promising power source is the *aluminum-air battery,* which is under development by chemist John Cooper at Lawrence Livermore National Laboratory in California. This device is really a combination battery and fuel cell, since it uses a stored quantity of aluminum metal and requires a continuous supply of humid air from the atmosphere. A diagram of the cell is shown in Fig. 11.24. The aluminum-air battery is a flow system, since air and the aqueous sodium hydroxide electrolyte flow past the electrodes. As the cell runs, aluminum and water are consumed, and aluminum hydroxide is produced. Periodically, the aluminum plate must be replaced, water added, and the aluminum hydrox-

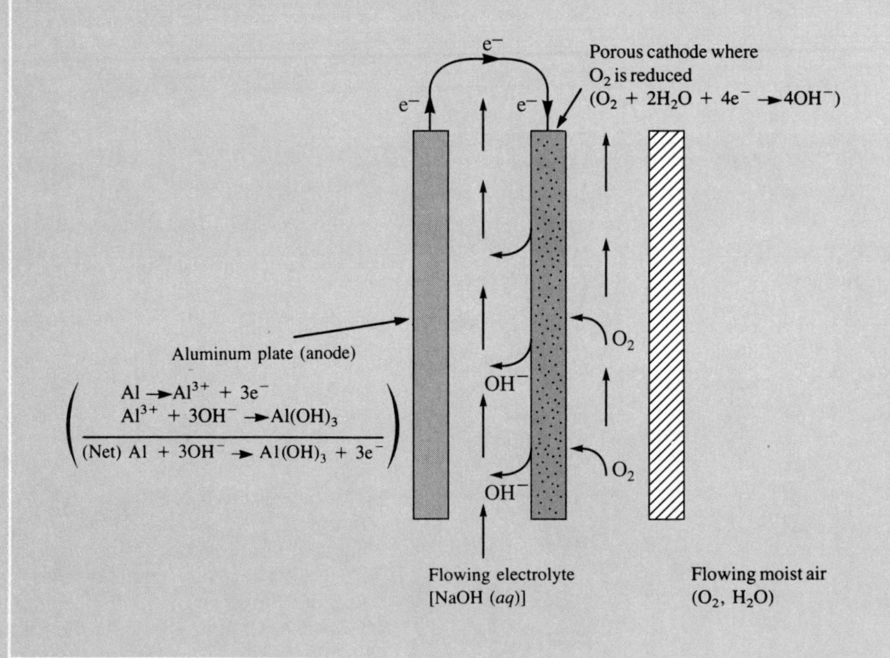

Figure 11.24
The aluminum-air battery. Aluminum is the reducing agent, and oxygen gas from the air is the oxidizing agent. As the aluminum loses electrons and dissolves, it forms aluminum hydroxide in the flowing aqueous sodium hydroxide solution.

ide removed. It is estimated that a small car with an aluminum-air battery weighing approximately 500 lb could travel about 3000 miles before the aluminum needed to be replaced. Every 250 miles during the trip, 6 gal of water would be needed, and the solid aluminum hydroxide would have to be removed.

One new type of battery being studied has no metal electrodes or metal ions [Fig. 11.25(a)]. The electrodes are thin films of giant *polyacetylene* molecules (polymers) of the form shown below:

$$\begin{array}{cccccccccccccccccccccccc} & H & & H & & H & & H & & H & & H & & H & & H & & H & & H & & H & & H \\ & | & & | & & | & & | & & | & & | & & | & & | & & | & & | & & | & & | \\ \sim\sim & C & = & C & - & C & = & C & - & C & = & C & - & C & = & C & - & C & = & C & - & C & = & C\sim\sim \end{array}$$

The battery is charged by attaching it to a power source, which forces electrons from one polyacetylene electrode to the other. As one electrode gains electrons, positive ions from the electrolyte diffuse into the film to balance the charge. The other electrode, which has given up electrons, absorbs anions to balance the charge [Fig. 11.25(b)]. The battery discharges, generating a current to run an electric motor, as the excess electrons flow from the electron-rich anode to the electron-deficient cathode [Fig. 11.25(*c*)].

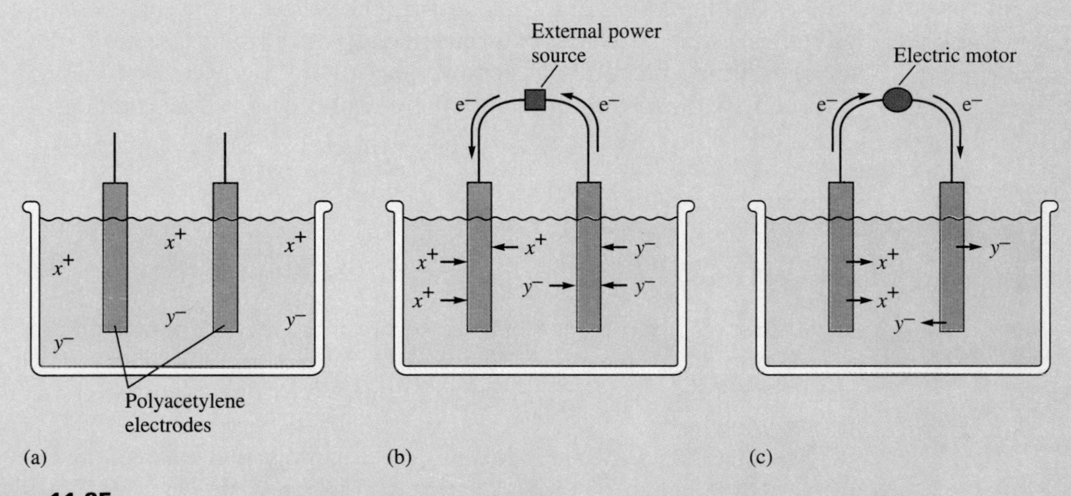

Figure 11.25
A battery with polyacetylene electrodes. (a) The uncharged battery. (b) The charging process, using an external power source. (c) The discharging battery running an electric motor.

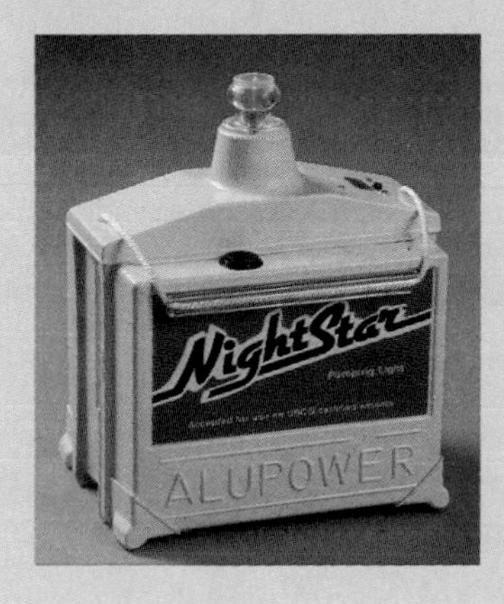

A mooring light powered by an aluminum-air battery. These lights are used to mark the corners of river barges to prevent other boats from running into them.

Suggested Reading

1. Mercouri Kanatzidis, "Conductive Polymers," *Chemical And Engineering News*, December 3, 1990: 36.
2. Richard C. Alkire, "Electrochemical Engineering," *Journal of Chemical Education* 60 (1983): 274.
3. Jacob Jorné, "Flow Batteries," *American Scientist* 71 (1983): 507.
4. Anthony F. Sammells, "Fuel Cells and Electrochemical Energy Storage," *Journal of Chemical Education* 60 (1983): 320.

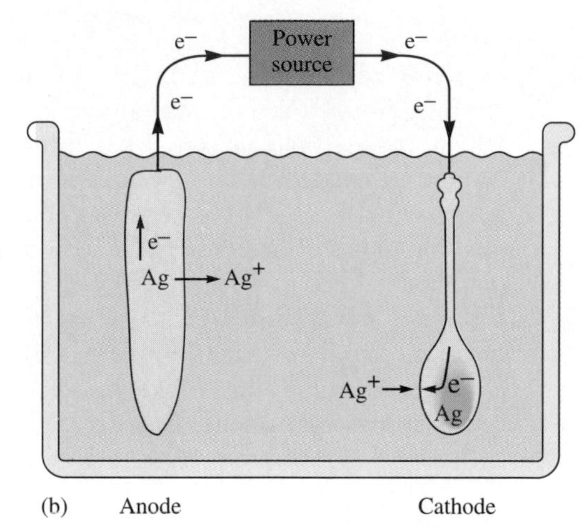

Figure 11.26
Schematic of the electroplating of a spoon. The item to be plated is the cathode, and the anode is a silver bar. Silver is plated out at the cathode:

$$Ag^+ + e^- \rightarrow Ag$$

Note that a salt bridge is not needed here since Ag^+ ions are involved at both electrodes.

Noble metal impurities in the anode are not oxidized at the voltage used; they fall to the bottom of the cell to form a sludge, which is processed to remove the valuable silver, gold, and platinum.

The Cu^{2+} ions from the solution are deposited onto the cathode,

$$Cu^{2+} + 2e^- \longrightarrow Cu$$

producing copper that is 99.95% pure.

Metal Plating

Metals that readily corrode can often be protected by the application of a thin coating of a metal that resists corrosion. Examples are "tin" cans, which are actually steel cans with a thin coating of tin, and chrome-plated steel bumpers for automobiles.

An object can be plated by making it the cathode in a tank containing ions of the plating metal. The silver plating of a spoon is shown schematically in Fig. 11.26. In an actual plating process the solution also contains ligands that form complexes with the silver ion. When the concentration of Ag^+ is lowered in this way, a smooth, even coating of silver is obtained.

Electrolysis of Sodium Chloride

Sodium metal is produced mainly by the electrolysis of molten sodium chloride. Because solid NaCl has a rather high melting point (800°C), it is usually mixed with solid $CaCl_2$ to lower the melting point to about 600°C. The mixture is then electrolyzed in a **Downs cell**, as illustrated in Fig. 11.27, where the reactions are as follows:

Anode reaction: $2Cl^- \longrightarrow Cl_2 + 2e^-$

Cathode reaction: $Na^+ + e^- \longrightarrow Na$

At the temperatures in the Downs cell the sodium is liquid and can be drained off, cooled, and cast into blocks. Because it is so reactive, sodium must be stored in an inert solvent, such as mineral oil, to prevent its oxidation.

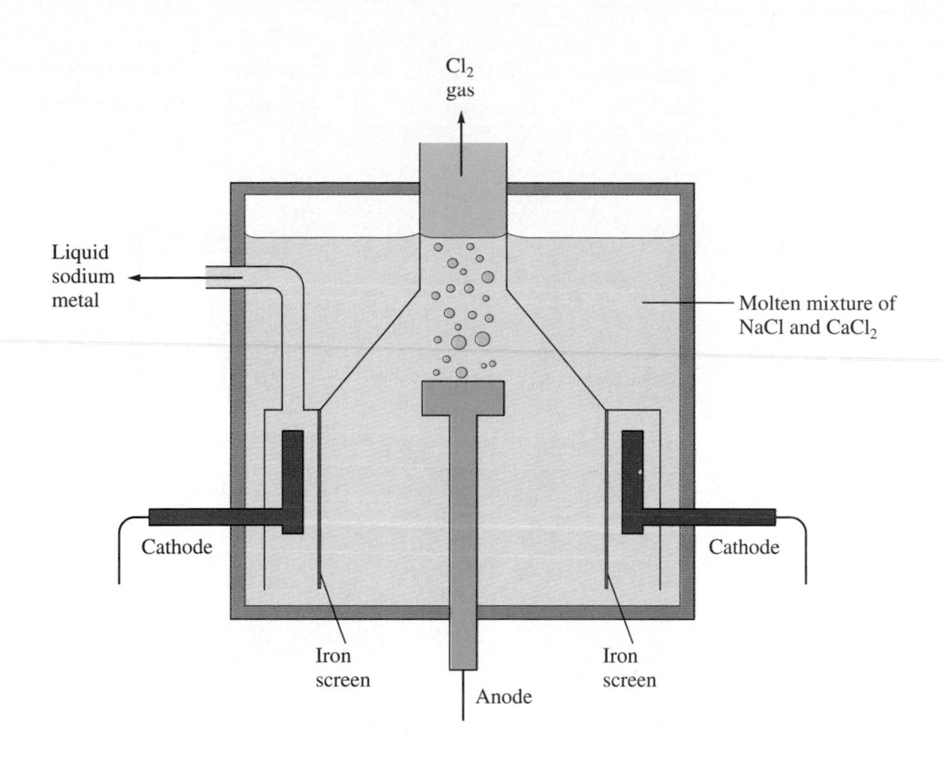

Figure 11.27
The Downs cell for the electrolysis of molten sodium chloride. The cell is designed so that the sodium and chlorine produced cannot come into contact with each other to re-form NaCl.

Electrolysis of aqueous sodium chloride (brine) is an important industrial process for the production of chlorine and sodium hydroxide. In fact, this process is second only to the production of aluminum as a consumer of electricity in the United States. Sodium is not produced in this process under normal circumstances because H_2O is more easily reduced than Na^+, as the standard reduction potentials show:

$$Na^+ + e^- \longrightarrow Na \qquad \mathscr{E}° = -2.71\ V$$
$$2H_2O + 2e^- \longrightarrow H_2 + 2OH^- \qquad \mathscr{E}° = -0.83\ V$$

Hydrogen, not sodium, is produced at the cathode.

For the reasons we discussed in Section 11.7, chlorine gas is produced at the anode. Thus the electrolysis of brine produces hydrogen and chlorine:

$$\text{Anode reaction:} \qquad 2Cl^- \longrightarrow Cl_2 + 2e^-$$
$$\text{Cathode reaction:} \qquad 2H_2O + 2e^- \longrightarrow H_2 + 2OH^-$$

It leaves a solution containing dissolved NaOH and NaCl.

The contamination of the sodium hydroxide by NaCl can be virtually eliminated by using a special **mercury cell** for electrolyzing brine (see Fig. 11.28). In this cell mercury is the conductor at the cathode; and because hydrogen gas has an extremely high overvoltage with a mercury electrode, Na^+ is reduced instead of H_2O. The resulting sodium metal dissolves in the mercury, forming a liquid alloy, which is then pumped to a chamber where the dissolved sodium is reacted with water to produce hydrogen:

$$2Na(s) + 2H_2O(l) \longrightarrow 2Na^+(aq) + 2OH^-(aq) + H_2(g)$$

Relatively pure solid NaOH is recovered from the aqueous solution, and the regenerated mercury is then pumped back to the electrolysis cell. This process,

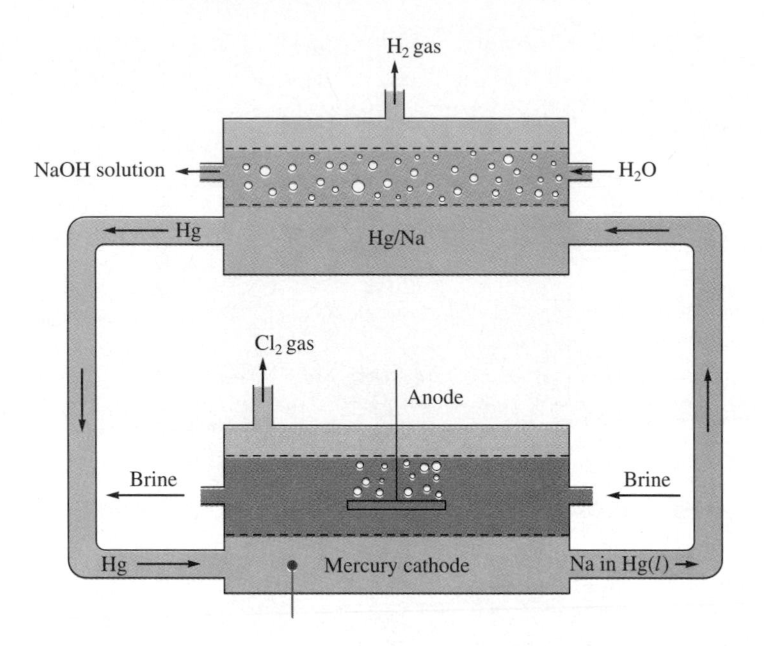

Figure 11.28
The mercury cell for production of chlorine and sodium hydroxide. The large overvoltage required to produce hydrogen at a mercury electrode means that Na^+ ions are reduced rather than water. The sodium formed dissolves in the liquid mercury and is then pumped to a chamber, where it reacts with water.

called the **chlor-alkali process,** has often resulted in significant mercury contamination of the environment; but the waste solutions from this process are now carefully treated to remove mercury.

Because of the environmental problems associated with the mercury cell, it has been largely displaced in the chlor-alkali industry by other technologies. In the United States nearly 75% of chlor-alkali production is now carried out in diaphragm cells. In a diaphragm cell the cathode and the anode are separated by a diaphragm that allows passage of H_2O molecules, Na^+ ions, and, to a limited extent, Cl^- ions. The diaphragm does not allow OH^- ions to pass through it. Thus the H_2 and OH^- formed at the cathode are kept separate from the Cl_2 formed at the anode. The major disadvantage of this process is that the aqueous effluent pumped from the cathode compartment contains a mixture of sodium hydroxide and unreacted sodium chloride, which must be separated if pure sodium hydroxide is a desired product.

In the last twenty-five years a new process has been developed in the chlor-alkali industry that employs a membrane to separate the anode and cathode compartments in brine electrolysis cells. The membrane is superior to the diaphragm used in diaphragm cells because the membrane is impermeable to anions. Only cations can flow through the membrane. Because neither Cl^- nor OH^- ions can pass through the membrane separating the anode and cathode compartments, NaCl contamination of the NaOH formed at the cathode is not a problem. Although membrane technology is only now becoming prominent in the United States, it is already the dominant method for chlor-alkali production in Japan.

EXERCISES

A blue exercise number indicates that the answer to that exercise appears at the back of this book.

Galvanic Cells

1. What is the difference between a galvanic and an electrolytic cell?
2. Why is the use of a salt bridge or porous disk in a galvanic cell necessary?
3. Sketch the galvanic cells based on the following overall reactions. Calculate $\mathscr{E}°$; show the direction of electron flow and the direction of ion migration through the salt bridge; identify the cathode and anode; and give the overall balanced reaction. Assume that all concentrations are 1.0 *M* and that all partial pressures are 1.0 atm.
   a. $Cr^{3+}(aq) + Cl_2(g) \rightleftharpoons Cr_2O_7^{2-}(aq) + Cl^-(aq)$
   b. $Cu^{2+}(aq) + Mg(s) \rightleftharpoons Mg^{2+}(aq) + Cu(s)$
   c. $IO_3^-(aq) + Fe^{2+}(aq) \rightleftharpoons Fe^{3+}(aq) + I_2(s)$
   d. $Zn(s) + Ag^+(aq) \rightleftharpoons Zn^{2+}(aq) + Ag(s)$
4. Draw diagrams for the galvanic cells utilizing the following half-cells. Write balanced equations for the overall cell reactions.

Cathode	Anode
a. Cl_2/Cl^-	Br^-/Br_2
b. IO_4^-/IO_3^-	Mn^{2+}/MnO_4^-
c. Ni^{2+}/Ni	Al/Al^{3+}
d. Co^{3+}/Co^{2+}	Fe^{2+}/Fe^{3+}

Cell Potentials, Standard Reduction Potentials, and Free Energy

5. The saturated calomel electrode, abbreviated SCE, is often used as a reference electrode in making electrochemical measurements. The SCE is composed of mercury in contact with a saturated solution of calomel (Hg_2Cl_2). The electrolyte solution is saturated KCl. $\mathscr{E}_{SCE}$ is +0.242 V relative to the standard hydrogen electrode. Calculate the potential for each of the following galvanic cells containing a saturated calomel electrode and the given half-cell components at standard conditions. In each case, indicate whether the SCE is the cathode or the anode. Standard reduction potentials are found in Table 11.1.
   a. $Cu^{2+} + 2e^- \longrightarrow Cu$
   b. $Fe^{3+} + e^- \longrightarrow Fe^{2+}$
   c. $AgCl + e^- \longrightarrow Ag + Cl^-$
   d. $Al^{3+} + 3e^- \longrightarrow Al$
   e. $Ni^{2+} + 2e^- \longrightarrow Ni$
6. Calculate $\mathscr{E}°$ values for the following cells. Which reactions are spontaneous as written, under standard conditions? Balance the reactions that are not already balanced. Standard reduction potentials are found in Table 11.1.
   a. $2Ag^+(aq) + Cu(s) \rightleftharpoons Cu^{2+}(aq) + 2Ag(s)$
   b. $Zn^{2+}(aq) + Ni(s) \rightleftharpoons Ni^{2+}(aq) + Zn(s)$
   c. $MnO_4^-(aq) + I^-(aq) \rightleftharpoons I_2(aq) + Mn^{2+}(aq)$
   d. $MnO_4^-(aq) + F^-(aq) \rightleftharpoons F_2(g) + Mn^{2+}(aq)$
7. Answer the following questions, using data from Table 11.1 (all under standard conditions).
   a. Is $H^+(aq)$ capable of oxidizing $Cu(s)$ to $Cu^{2+}(aq)$?
   b. Is $H^+(aq)$ capable of oxidizing $Mg(s)$?
   c. Is $Fe^{3+}(aq)$ capable of oxidizing $I^-(aq)$?
   d. Is $Fe^{3+}(aq)$ capable of oxidizing $Br^-(aq)$?
8. Answer the following questions, using data from Table 11.1 (all under standard conditions).
   a. Is $H_2(g)$ capable of reducing $Ag^+(aq)$?
   b. Is $H_2(g)$ capable of reducing $Ni^{2+}(aq)$?
   c. Is $Fe^{2+}(aq)$ capable of reducing $VO_2^+(aq)$?
   d. Is $Fe^{2+}(aq)$ capable of reducing $Cr^{3+}(aq)$ to $Cr^{2+}(aq)$?
   e. Is $Fe^{2+}(aq)$ capable of reducing $Cr^{3+}(aq)$ to $Cr(s)$?
   f. Is $Fe^{2+}(aq)$ capable of reducing $Sn^{2+}(aq)$ to $Sn(s)$?
9. Using data from Table 11.1, place the following in order of increasing strength as oxidizing agents (all under standard conditions):

$$MnO_4^-, Cl_2, Cr_2O_7^{2-}, Mg^{2+}, Fe^{2+}, Fe^{3+}$$

10. Using data from Table 11.1, place the following in order of increasing strength as reducing agents (all under standard conditions):

$$Cr^{3+}, H_2, Zn, Li, F^-, Fe^{2+}$$

11. Consider only the species (standard conditions),

$$Br^-, Br_2, H^+, H_2, La^{3+}, Ca, Cd$$

in answering the following questions. Give reasons for your answers.
   a. Which is the strongest oxidizing agent?
   b. Which is the strongest reducing agent?
   c. Which species can be oxidized by MnO_4^- in acid?
   d. Which species can be reduced by $Zn(s)$?
12. Consider only the species (standard conditions),

$$Ce^{4+}, Ce^{3+}, Fe^{2+}, Fe^{3+}, Fe, Mg^{2+}, Mg, Ni^{2+}, Sn$$

in answering the following questions. Give reasons for your answers. (Use data from Table 11.1.)
   a. Which is the strongest oxidizing agent?
   b. Which is the strongest reducing agent?

c. Will iron dissolve in a 1.0 *M* solution of Ce^{4+}? If so, will there be Fe^{2+} or Fe^{3+} ions in the solution?
d. Which of the species can be oxidized by $H^+(aq)$?
e. Which of the species can be reduced by $H_2(g)$?
f. Would you use Mg or Sn to reduce Fe^{3+} to Fe^{2+}?

13. Use the table of standard reduction potentials (Table 11.1) to pick a reagent that is capable of each of the following oxidations (under standard conditions in acid solution).
a. Oxidizes Hg to Hg_2^{2+} but does not oxidize Hg_2^{2+} to Hg^{2+}.
b. Oxidizes Br^- to Br_2 but does not oxidize Cl^- to Cl_2.
c. Oxidizes Mn to Mn^{2+} but does not oxidize Ni to Ni^{2+}.

14. Use the table of standard reduction potentials (Table 11.1) to pick a reagent that is capable of each of the following reductions (under standard conditions in acidic solution).
a. Reduces Fe^{3+} to Fe^{2+} but does not reduce Fe^{2+} to Fe.
b. Reduces Ag^+ to Ag but does not reduce O_2 to H_2O_2.

15. Is it possible for a reagent to reduce I_2 to I^- but not Cu^{2+} to Cu? What range of standard reduction potentials could such a reagent have?

16. The standard reduction potential for the half-reaction

$$Sn^{4+} + 2e^- \longrightarrow Sn^{2+}$$

is +0.15 V. Use data from Table 11.1 to answer the following questions.
a. Is it possible to find a reducing agent that will reduce Sn^{4+} to Sn^{2+} and not reduce Sn^{2+} to Sn? If so, name a specific reagent.
b. Is it possible to find an oxidizing agent that will oxidize Sn to Sn^{2+} but not oxidize Sn^{2+} to Sn^{4+}? If so, name a specific reagent.

17. Hydrogen peroxide can function either as an oxidizing agent or as a reducing agent. Referring to Table 11.1, write half-reactions for H_2O_2 acting in these roles. Calculate $\mathscr{E}°$ for the reaction

$$2H_2O_2 \longrightarrow 2H_2O + O_2$$

18. A 1.0 *M* solution of Cu^+ ion is not stable even if there is no exposure to oxygen in the air. Use data from Table 11.1 to predict what reaction occurs. Calculate $\mathscr{E}°$, $\Delta G°$, and *K* for this reaction at 25°C.

19. Chlorine dioxide (ClO_2), which is produced by the reaction

$$2NaClO_2(aq) + Cl_2(g) \longrightarrow 2ClO_2(g) + 2NaCl(aq)$$

has been tested as a disinfectant for municipal water treatment.
a. Using data from Table 11.1, calculate $\mathscr{E}°$, $\Delta G°$, and *K* at 25°C for the production of ClO_2.
b. One of the concerns in using ClO_2 as a disinfectant is that the carcinogenic chlorate ion (ClO_3^-) might be a by-product. It can be formed from the reaction

$$ClO_2(g) \rightleftharpoons ClO_3^-(aq) + Cl^-(aq)$$

Balance the equation for the decomposition of ClO_2.

20. A patent attorney has asked for your advice concerning the merits of a patent application claiming the invention of an aqueous single galvanic cell capable of producing a 12-V potential. Comment.

21. Under standard conditions, what reaction occurs, if any, when each of the following operations is performed?
a. Crystals of I_2 are added to a solution of NaCl.
b. Cl_2 gas is bubbled into a solution of NaI.
c. A silver wire is placed in a solution of $CuCl_2$.
d. A lead wire is placed in a solution containing Cu^{2+}.
e. A solution of $FeSO_4$ is allowed to sit exposed to air.

22. For reactions that occur in Exercise 21, write a balanced equation and calculate $\mathscr{E}°$, $\Delta G°$, and *K* at 25°C.

23. Calculate $\Delta G°$ and *K* at 25°C for the reactions in Exercises 3 and 4.

24. The amount of manganese in steel is determined by changing it to permanganate ion. The steel is first dissolved in nitric acid, producing Mn^{2+} ions. These ions are then oxidized to the deeply colored MnO_4^- ions by periodate ion (IO_4^-) in acid solution.
a. Complete and balance an equation describing each of the above reactions.
b. Calculate $\mathscr{E}°$, $\Delta G°$, and *K* at 25°C for each reaction.

25. Combine the equations

$$\Delta G° = -nF\mathscr{E}° \quad \text{and} \quad \Delta G° = \Delta H° - T\Delta S°$$

to derive an expression for $\mathscr{E}°$ as a function of temperature. Describe how one can graphically determine $\Delta H°$ and $\Delta S°$ from measurements of $\mathscr{E}°$ at different temperatures.

26. a. Calculate $\mathscr{E}°$ and $\Delta G°$ for the reaction at 298 K:

$$2H_2O(l) \rightleftharpoons 2H_2(g) + O_2(g)$$

b. Calculate $\Delta H°$ and $\Delta S°$ for the reaction using thermodynamic data in Appendix 4.
c. Calculate the values of $\mathscr{E}°$ and $\Delta G°$ for this reaction at temperatures of 90.°C and 0°C.

27. What property would you look for in designing a reference half-cell that would produce a potential relatively stable with respect to temperature? (See Exercise 25.)

28. Calculate $\mathscr{E}°$ for the reaction

$$CH_3OH(l) + \tfrac{3}{2}O_2(g) \longrightarrow CO_2(g) + 2H_2O(l)$$

using values of $\Delta G_f°$ in Appendix 4. Will $\mathscr{E}°$ increase or decrease with an increase in temperature? (See Exercise 25 for the dependence of $\mathscr{E}°$ on temperature.)

29. The equation $\Delta G° = -nF\mathscr{E}°$ can also be applied to half-reactions. For the half-reaction

$$Ag^+ + e^- \longrightarrow Ag \qquad \mathscr{E}° = 0.80 \text{ V}$$

calculate the value of $\Delta G_f°$ for $Ag^+(aq)$.

30. Use standard reduction potentials to estimate $\Delta G_f°$ for $Fe^{2+}(aq)$ and $Fe^{3+}(aq)$.

31. Estimate $\mathscr{E}°$ for the half-reaction

$$2H_2O + 2e^- \longrightarrow H_2 + 2OH^-$$

given the following values of $\Delta G_f°$:

$$H_2O(l) = -237 \text{ kJ/mol}$$
$$H_2(g) = 0.0$$
$$OH^-(aq) = -157 \text{ kJ/mol}$$

Compare this value of $\mathscr{E}°$ with the value of $\mathscr{E}°$ given in Table 11.1.

32. A disproportionation reaction involves a substance that acts as both an oxidizing and a reducing agent, producing both higher and lower oxidation states of the same element in the products. Which of the following disproportionation reactions are spontaneous under standard conditions?
   a. $2Cu^+(aq) \rightarrow Cu^{2+}(aq) + Cu(s)$
   b. $2Sn^{2+}(aq) \rightarrow Sn^{4+}(aq) + Sn(aq)$
   (See Exercise 16 for the Sn^{4+} reduction potential.)
   c. $3Fe^{2+}(aq) \rightarrow 2Fe^{3+}(aq) + Fe(s)$
   d. $HClO_2(aq) \rightarrow ClO_3^-(aq) + HClO(aq)$ (Unbalanced)
   Use these half-reactions:

$$ClO_3^- + 3H^+ + 2e^- \longrightarrow HClO_2 + H_2O \qquad \mathscr{E}° = +1.21 \text{ V}$$
$$HClO_2 + 2H^+ + 2e^- \longrightarrow HClO + H_2O \qquad \mathscr{E}° = +1.65 \text{ V}$$

33. Calculate $\Delta G°$ and K at 25°C for those reactions in Exercise 32 that are spontaneous under standard conditions.

34. For the following half-reaction, $\mathscr{E}° = +0.017$ V:

$$Ag(S_2O_3)_2^{3-} + e^- \longrightarrow Ag + 2S_2O_3^{2-}$$

Calculate the value of the equilibrium constant at 25°C for the reaction

$$Ag^+(aq) + 2S_2O_3^{2-}(aq) \rightleftharpoons Ag(S_2O_3)_2^{3-}$$

(*Hint:* find a half-reaction that can be added to the one above to give the required reaction.)

35. Cadmium sulfide (CdS) is used in some semiconductor applications. Calculate the value of the solubility product (K_{sp}) for CdS given the following standard reduction potentials:

$$CdS + 2e^- \longrightarrow Cd + S^{2-} \qquad \mathscr{E}° = -1.21 \text{ V}$$
$$Cd^{2+} + 2e^- \longrightarrow Cd \qquad \mathscr{E}° = -0.402 \text{ V}$$

36. Calculate $\mathscr{E}°$ for the following half-reaction:

$$AgI(s) + e^- \longrightarrow Ag(s) + I^-$$

(*Hint:* reference the K_{sp} value for AgI and the standard reduction potential for Ag^+.)

The Nernst Equation

37. The Nernst equation can be applied to half-reactions. Calculate the reduction potential at 25°C of each of the following half-cells.
   a. Cu/Cu^{2+} (0.10 M)
   (The half-reaction is $Cu^{2+} + 2e^- \rightarrow Cu$.)
   b. Cu/Cu^{2+} (2.0 M)
   c. Cu/Cu^{2+} (1.0×10^{-4} M)
   d. MnO_4^- (0.10 M)/Mn^{2+} (0.010 M) at pH = 3.00. (The half-reaction is $MnO_4^- + 8H^+ + 5e^- \rightarrow Mn^{2+} + 4H_2O$.)
   e. MnO_4^- (0.10 M)/Mn^{2+} (0.010 M) at pH = 1.00

38. The overall reaction in the lead storage battery is

$$Pb(s) + PbO_2(s) + 2H^+(aq) + 2HSO_4^-(aq) \longrightarrow 2PbSO_4(s) + 2H_2O(l)$$

   a. Calculate $\mathscr{E}$ at 25°C for this battery when $[H_2SO_4] = 4.5$ M; that is, $[H^+] = [HSO_4^-] = 4.5$ M. At 25°C, $\mathscr{E}° = 2.04$ V for the lead storage battery.
   b. For the cell reaction $\Delta H° = -315.9$ kJ and $\Delta S° = 263.5$ J/K. Calculate $\mathscr{E}°$ at −20.°C. (See Exercise 25.)
   c. Calculate $\mathscr{E}$ at −20.°C when $[H_2SO_4] = 4.5$ M.
   d. Based on your previous answers, why does it seem that batteries fail more often on cold days than on warm days?

39. Consider the concentration cell shown below. Calculate the cell potential at 25°C when the concentration of Ag^+ in the compartment on the right has each of the following values.

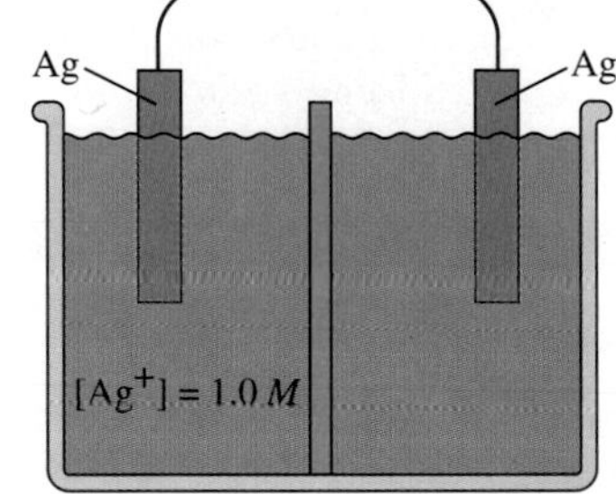

   a. 1.0 M
   b. 2.0 M
   c. 0.10 M
   d. 4.0×10^{-5} M
   e. Calculate the potential when both solutions are 0.10 M in Ag^+.

For each case, identify the cathode, the anode, and the direction in which electrons flow.

40. Can permanganate ion oxidize Fe^{2+} to Fe^{3+} at 25°C under the following conditions (pH = 4.0)?

$$[Mn^{2+}] = 1 \times 10^{-6}\ M \qquad [Fe^{2+}] = 1 \times 10^{-3}\ M$$
$$[MnO_4^-] = 0.01\ M \qquad [Fe^{3+}] = 1 \times 10^{-6}\ M$$

41. An electrochemical cell consists of a standard hydrogen electrode and a copper metal electrode.
 a. What is the potential of the cell at 25°C if the copper electrode is placed in a solution in which $[Cu^{2+}] = 2.5 \times 10^{-4}\ M$?
 b. If the copper electrode is placed in a solution of 0.10 M NaOH that is saturated with $Cu(OH)_2$, what is the cell potential at 25°C? For $Cu(OH)_2$, $K_{sp} = 1.6 \times 10^{-19}$.
 c. The copper electrode is placed in a solution of unknown $[Cu^{2+}]$. The measured potential at 25°C is 0.195 V. What is $[Cu^{2+}]$? (Assume Cu^{2+} is reduced.)
 d. If you wish to construct a calibration curve to show how the cell potential varies with $[Cu^{2+}]$, what should you plot in order to obtain a straight line? What will the slope of this line be?

42. Consider the following half-reactions:

$$Pt^{2+} + 2e^- \longrightarrow Pt \qquad \mathscr{E}° = 1.188\ V$$
$$PtCl_4^{2-} + 2e^- \longrightarrow Pt + 4Cl^- \qquad \mathscr{E}° = 0.755\ V$$
$$NO_3^- + 4H^+ + 3e^- \longrightarrow NO + 2H_2O \qquad \mathscr{E}° = 0.96\ V$$

Explain why platinum metal will dissolve in aqua regia (a mixture of hydrochloric and nitric acids) but not in either concentrated nitric or concentrated hydrochloric acid individually.

43. The physiological properties of chromium depend on its oxidation state. Chromium(III) ion is essential for an enzyme cofactor called the glucose tolerance factor. Chromium deficiency gives symptoms similar to diabetes. The chromium(VI) ion, on the other hand, is toxic and a potent carcinogen, mainly because of the strength of chromium(VI) as an oxidizing agent.
 a. Consider the following half-reaction:

$$CrO_4^{2-} + 4H_2O + 3e^- \longrightarrow Cr(OH)_3 + 5OH^- \qquad \mathscr{E}° = -0.13\ V$$

 [Note that $Cr(OH)_3$ is a solid.] Using the Nernst equation, calculate the potential for this half-reaction at 25°C, where pH = 7.40 and

$$[CrO_4^{2-}] = 1.0 \times 10^{-6}\ M$$

 b. In acidic solution chromium(VI) exists as $Cr_2O_7^{2-}$. Calculate the potential at 25°C for the half-reaction

$$Cr_2O_7^{2-} + 14H^+ + 6e^- \longrightarrow 2Cr^{3+} + 7H_2O \qquad \mathscr{E}° = 1.33\ V$$

 under conditions where pH = 2.00 and

$$[Cr_2O_7^{2-}] = [Cr^{3+}] = 1.0 \times 10^{-6}\ M$$

44. A chemist wishes to determine the concentration of CrO_4^{2-} electrochemically. A cell is constructed consisting of a saturated calomel electrode (SCE, see Exercise 5) and a silver wire coated with Ag_2CrO_4. The $\mathscr{E}°$ value for the following half-reaction is +0.446 V relative to the standard hydrogen electrode:

$$Ag_2CrO_4 + 2e^- \longrightarrow 2Ag + CrO_4^{2-}$$

 a. Calculate $\mathscr{E}_{cell}$ and ΔG at 25°C for the cell reaction when $[CrO_4^{2-}] = 1.00$ mol/L.
 b. Write the Nernst equation for the cell.
 c. If the coated silver wire is placed in a solution (at 25°C) in which $[CrO_4^{2-}] = 1.00 \times 10^{-5}\ M$, what is the expected cell potential?
 d. The measured cell potential at 25°C is 0.504 V when the coated wire is dipped into a solution of unknown $[CrO_4^{2-}]$. What is the $[CrO_4^{2-}]$ for this solution?
 e. Using data from this problem and from Table 11.1, calculate the solubility product (K_{sp}) for Ag_2CrO_4.

45. Write the Nernst equation for the corrosion of iron by oxygen. Will corrosion be a greater problem in an acidic or a basic solution?

46. A cleaning solution for glassware is prepared by dissolving potassium dichromate in concentrated sulfuric acid. (This solution is much less commonly used because of the carcinogenic properties of chromium(VI) compounds.) An unthinking chemist tried to prepare some cleaning solution from some waste potassium dichromate that had been contaminated by sodium chloride. On adding the solid to some sulfuric acid, he was driven from the lab by pungent fumes. What happened?

Electrolysis

47. How long will it take to plate out each of the following with a current of 100.0 A?
 a. 1.0 kg of Al from aqueous Al^{3+}
 b. 1.0 g of Ni from aqueous Ni^{2+}
 c. 5.0 mol of Ag from aqueous Ag^+

48. What mass of each of the following substances can be produced in 1.0 h with a current of 15 A?
 a. Co from aqueous Co^{2+}
 b. Hf from aqueous Hf^{4+}
 c. I_2 from aqueous KI
 d. Cr from molten CrO_3

49. It took 2.30 min with a current of 2.00 A to plate out all of the silver from 0.250 L of a solution containing Ag^+. What was the original concentration of Ag^+ in the solution?

50. What reaction will take place at the cathode and the anode when each of the following is electrolyzed?
 a. molten KF
 b. 0.10 M KF solution
 c. 1.0 M H_2O_2 solution containing 1.0 M H_2SO_4

d. molten $MgCl_2$
e. 1.0 *M* $AgNO_3$

51. A solution at 25°C contains 1.0 *M* Fe^{2+} and 0.010 *M* Ag^+. Which metal will be plated on the cathode first when this solution is electrolyzed? (*Hint:* use the Nernst equation to calculate $\mathscr{E}$ for each half-reaction.)

52. Gold is produced electrochemically from a basic solution of $Au(CN)_4^-$. Gold metal and oxygen gas are produced at the electrodes. Write the overall cell reaction. What volume of pure oxygen gas is produced at 25°C and 740. torr for every kilogram of gold produced?

53. Consider the following half reactions:

$$IrCl_6^{3-} + 3e^- \longrightarrow Ir + 6Cl^- \qquad \mathscr{E}^\circ = 0.77\ V$$

$$PtCl_4^{2-} + 2e^- \longrightarrow Pt + 4Cl^- \qquad \mathscr{E}^\circ = 0.73\ V$$

$$PdCl_4^{2-} + 2e^- \longrightarrow Pd + 4Cl^- \qquad \mathscr{E}^\circ = 0.62\ V$$

A hydrochloric acid solution contains platinum, palladium, and irridium as chloro-complex ions. The solution is 1.0 *M* in chloride ion and 0.020 *M* in each complex ion. Is it feasible to separate the three metals from this solution by electrolysis? (Assume that 99% of a metal must be plated out before another metal begins to plate out.)

54. A solution containing a 3+ metal ion is electrolyzed by a current of 5.00 A for 10.0 min. What is the identity of the metal if 1.19 g of metal is plated out?

55. It takes 74.6 s for a 2.50-A current to plate 0.1086 g of a metal from a solution containing M^{2+} ions. What is the metal?

56. One of the few industrial-scale processes that produces organic compounds electrochemically is used by the Monsanto Company to produce 1,4-dicyanobutane. The reduction reaction is

$$2CH_2{=}CHCN + 2H^+ + 2e^- \longrightarrow NC{-}(CH_2)_4{-}CN$$

The $NC{-}(CH_2)_4{-}CN$ is then chemically reduced by hydrogen to $H_2N{-}(CH_2)_6{-}NH_2$, which is used in the production of nylon. What current must be used to produce 150. kg of $NC{-}(CH_2)_4{-}CN$ per hour?

57. What volume of F_2 gas, at 25°C and 1.00 atm, is produced when molten KF is electrolyzed by a current of 10.0 A for 2.00 h? What mass of potassium metal is produced? At which electrode does each reaction occur?

58. It takes 15 kWh (kilowatt-hours) of electrical energy to produce 1.0 kg of aluminum metal from aluminum oxide by the Hall-Heroult process. Compare this value with the amount of energy necessary to melt 1.0 kg of aluminum metal. Why is it economically feasible to recycle aluminum cans? (The enthalpy of fusion for aluminum metal is 10.7 kJ/mol and 1 watt = 1 J/s.)

59. In the electrolysis of a sodium chloride solution, what volume of Cl_2 is produced in the same time it takes to produce 6.00 L of $H_2(g)$, both volumes measured at 0°C and 1 atm?

60. What volumes of $H_2(g)$ and $O_2(g)$ at STP are produced from the electrolysis of water by a current of 2.50 A in 15.0 min?

61. a. In the electrolysis of an aqueous solution of Na_2SO_4, what reactions occur at the anode and the cathode?

	$\mathscr{E}^\circ$
$S_2O_8^{2-} + 2e^- \longrightarrow 2SO_4^{2-}$	2.01 V
$O_2 + 4H^+ + 4e^- \longrightarrow 2H_2O$	1.23 V
$2H_2O + 2e^- \longrightarrow H_2 + 2OH^-$	−0.83 V
$Na^+ + e^- \longrightarrow Na$	−2.71 V

b. When water containing a small amount (~0.01 *M*) of sodium sulfate is electrolyzed, measurement of the volume of gases generated consistently gives a result that the volume ratio of hydrogen to oxygen is not quite 2:1. To what do you attribute this discrepancy? Predict whether the measured ratio is greater than or less than 2:1.

Additional Exercises

62. Describe each of the following.
a. purification of a metal by electrorefining
b. cathodic protection

63. Give three ways in which metals are protected from corrosion.

64. In 1973 the wreckage of the Civil War ironclad USS *Monitor* was discovered near Cape Hatteras, North Carolina. [The *Monitor* and the CSS *Virginia* (formerly the USS *Merrimack*) fought the first battle between iron-armored ships.] In 1987 investigations were begun to see whether the ship could be salvaged. *Time* reported (June 22, 1987) that scientists were considering adding sacrificial anodes of zinc to the rapidly corroding metal hull of the *Monitor*. Describe how attaching zinc to the hull would protect the *Monitor* from further corrosion.

65. What are the advantages and disadvantages of using fuel cells rather than the corresponding combustion reactions to produce electricity?

66. For the following half-reaction, $\mathscr{E}^\circ = -2.07$ V:

$$AlF_6^{3-} + 3e^- \longrightarrow Al + 6F^-$$

Calculate the equilibrium constant at 25°C for the reaction

$$Al^{3+}(aq) + 6F^-(aq) \rightleftharpoons AlF_6^{3-}(aq)$$

67. What is the maximum work that can be obtained from a hydrogen-oxygen fuel cell that produces 1.00 kg of water at 25°C? Why do we say that this is the maximum work that can be obtained?

68. What happens to $\mathscr{E}_{cell}$ as a battery discharges? Is a battery a system at equilibrium? Explain. What is $\mathscr{E}_{cell}$ when a battery reaches equilibrium?

69. Can cobalt metal be oxidized to Co^{2+} by air at 25°C under each of the following conditions?
 a. under standard conditions in 1.0 M OH^-
 b. under standard conditions in 1.0 M H^+
 c. at pH = 7.00

$$Co^{2+} + 2e^- \longrightarrow Co \qquad \mathscr{E}° = -0.28 \text{ V}$$

70. A crude pH meter is constructed at 25°C by using a saturated calomel electrode (SCE, see Exercise 5) and a standard hydrogen electrode (SHE).
 a. What is $\mathscr{E}°_{cell}$ when the half-cells are the SCE and SHE?
 b. Is the hydrogen electrode the cathode or the anode under these conditions?
 c. Write the Nernst equation showing how this cell is dependent on $[H^+]$.
 d. What is the measured cell potential when the 1.0 M H^+ solution in the hydrogen electrode is replaced by each of the following solutions?
  i. $[H^+] = 1.0 \times 10^{-3}$ M
  ii. $[H^+] = 2.5$ M
  iii. $[H^+] = 1.0 \times 10^{-9}$ M
 e. When an unknown acid solution is measured, the observed potential is 0.285 V. What is the pH of the solution?
 f. Why is the glass electrode more commonly used than the standard hydrogen electrode to measure pH?

71. The measurement of pH using a glass electrode obeys the Nernst equation. The typical response of a pH meter at 25°C is given by the equation

$$\mathscr{E}_{meas} = \mathscr{E}_{ref} + 0.05916 \text{ pH}$$

where $\mathscr{E}_{ref}$ contains the potential of the reference electrode and all other potentials that arise in the cell that are not related to the hydrogen ion concentration. Assume that $\mathscr{E}_{ref} = 0.250$ V and that $\mathscr{E}_{meas} = 0.480$ V.
 a. What is the uncertainty in the values of pH and $[H^+]$ if the uncertainty in the measured potential is ±1 mV (±0.001 V)?
 b. To what accuracy must the potential be measured for the uncertainty in pH to be ±0.02 pH unit?

72. Can NO be produced electrochemically at 25°C by the electrolysis of a 5.0 M aqueous solution of $NaNO_3$?

73. The measurement of F^- ion concentration by ion-selective electrodes at 25°C obeys the equation

$$\mathscr{E}_{meas} = \mathscr{E}_{ref} - 0.05916 \log[F^-]$$

 a. For a given solution $\mathscr{E}_{meas}$ is 0.4462 V. If $\mathscr{E}_{ref}$ is 0.2420 V, what is the concentration of F^- in the solution?
 b. Hydroxide ion interferes with the measurement of F^-. Therefore, the response of a fluoride electrode is

$$\mathscr{E}_{meas} = \mathscr{E}_{ref} - 0.05916 \log([F^-] + k[OH^-])$$

 where $k = 1.00 \times 10^1$ and is called the selectivity factor for the electrode response. Calculate $[F^-]$ for the data in part a if the pH is 9.00. What is the percent error introduced in the $[F^-]$ if the hydroxide interference is ignored?
 c. For the $[F^-]$ in part b, what is the maximum pH such that $[F^-]/k[OH^-] = 50.$?
 d. At low pH, F^- is mostly converted to HF. The fluoride electrode does not respond to HF. What is the minimum pH at which 99% of the fluoride is present as F^- and only 1% is present as HF?
 e. Buffering agents are added to solutions containing fluoride before making measurements with a fluoride selective electrode. Why?

74. Consider a concentration cell that has both electrodes made of some metal M. Solution A in one compartment of the cell contains 1.0 M M^{2+}. Solution B in the other cell compartment has a volume of 1.00 L. At the beginning of the experiment 0.0100 mol of $M(NO_3)_2(s)$ and 0.0100 mol of Na_2SO_4 are dissolved in solution B (ignore volume changes), where the reaction

$$M^{2+}(aq) + SO_4^{2-}(aq) \rightleftharpoons MSO_4(s)$$

occurs. For this reaction equilibrium is rapidly established, whereupon the cell potential is found to be +0.44 V at 25°C. Assume that the process

$$M^{2+} + 2e^- \longrightarrow M$$

has a standard reduction potential of +0.80 V and that no other redox process occurs in the cell. Calculate the value of K_{sp} for $MSO_4(s)$ at 25°C.

75. Consider a cell based on the following half-reactions:

$$Au^{3+} + 3e^- \longrightarrow Au \qquad \mathscr{E}° = 1.50 \text{ V}$$
$$Fe^{3+} + e^- \longrightarrow Fe^{2+} \qquad \mathscr{E}° = 0.77 \text{ V}$$

 a. Draw this cell under standard conditions, labeling the anode, the cathode, the direction of electron flow, and the concentrations, as appropriate.
 b. When enough NaCl(s) is added to the compartment containing gold to make the $[Cl^-] = 0.10$ M, the cell potential is observed to be 0.31 V. Assume that the reaction in the compartment containing gold is

$$Au^{3+}(aq) + 4Cl^-(aq) \rightleftharpoons AuCl_4^-(aq)$$

 Calculate the value of K for this reaction at 25°C.

76. When copper reacts with nitric acid, a mixture of NO(*g*) and NO_2(*g*) is evolved. The volume ratio of the two product gases depends on the concentration of the nitric acid according to the equilibrium

$$2H^+(aq) + 2NO_3^-(aq) + NO(g) \rightleftharpoons 3NO_2(g) + H_2O(l)$$

Consider the following standard reduction potentials at 25°C:

$$3e^- + 4H^+(aq) + NO_3^-(aq) \longrightarrow NO(g) + 2H_2O(l) \quad \mathscr{E}° = 0.957 \text{ V}$$

$$e^- + 2H^+(aq) + NO_3^-(aq) \longrightarrow NO_2(g) + H_2O(l) \quad \mathscr{E}° = 0.775 \text{ V}$$

a. Calculate the equilibrium constant for the above reaction.
b. What concentration of nitric acid will produce a NO and NO_2 mixture with only 0.20% NO_2 (by moles) at 25°C and 1.00 atm? Assume that no other gases are present and that the change in acid concentration can be neglected.

77. The black silver sulfide discoloration of silverware can easily be removed by heating the silver article in a sodium carbonate solution in an aluminum pan. The reaction is

$$3Ag_2S(s) + 2Al(s) \rightleftharpoons 6Ag(s) + 3S^{2-}(aq) + 2Al^{3+}(aq)$$

a. Using data in Appendix 4, calculate $\Delta G°$, K, and $\mathscr{E}°$ for the above reaction. [For $Al^{3+}(aq)$, $\Delta G_f^\circ = -480.$ kJ/mol.]
b. Calculate the value of the standard reduction potential for the following half-reaction:

$$2e^- + Ag_2S(s) \longrightarrow 2Ag(s) + S^{2-}(aq)$$

78. A zinc-copper battery is constructed as follows:

$$\text{Zn} \,|\text{Zn}^{2+}(0.10\ M)|\,|\text{Cu}^{2+}(2.50\ M)|\ \text{Cu}$$

The mass of each electrode is 200. g.
a. Calculate the cell potential when this battery is first connected.
b. Calculate the cell potential after 10.0 A of current has flowed for 10.0 h. (Assume each half-cell contains 1.00 L of solution.)
c. Calculate the mass of each electrode after 10.0 h.
d. How long can this battery deliver a current of 10.0 A before it goes dead?

79. Consider the following reduction potentials:

$$Co^{3+} + 3e^- \longrightarrow Co \quad \mathscr{E}° = 1.26 \text{ V}$$

$$Co^{2+} + 2e^- \longrightarrow Co \quad \mathscr{E}° = -0.28 \text{ V}$$

a. When cobalt metal dissolves in 1.0 *M* nitric acid, will Co^{3+} or Co^{2+} be the primary product?
b. Is it possible to change the concentration of HNO_3 to get a different result in part a?
c. What will be the primary product if cobalt metal is dissolved in hydrogen peroxide?

Quantum Mechanics and Atomic Theory

In the past 200 years a great deal of experimental evidence has accumulated to support the atomic model. This theory has proved to be both extremely useful and physically reasonable. When atoms were first suggested by the Greek philosophers Democritus and Leucippus about 400 B.C., the concept was based mostly on intuition. In fact, for the following 20 centuries, no convincing experimental evidence was available to support the existence of atoms. The first real scientific data were gathered by Lavoisier and others from quantitative measurements of chemical reactions. The results of these stoichiometric experiments led John Dalton to propose the first systematic atomic theory. Dalton's theory, although crude, has stood the test of time extremely well.

Once we came to "believe in" atoms, it was logical to ask: What is the nature of an atom? Does an atom have parts, and, if so, what are they? In Chapter 2 we considered some of the experiments most important for shedding light on the nature of the atom. Now we will see how the atomic theory has evolved to its present state.

One of the most striking things about the chemistry of the elements is the periodic repetition of properties. There are several groups of elements that show great similarities in chemical behavior. As we saw in Chapter 2, these similarities led to the development of the periodic table of the elements. In this chapter we will see that the modern theory of atomic structure accounts for periodicity in terms of the electron arrangements in atoms.

However, before we examine atomic structure, we must consider the revolution that took place in physics in the first 30 years of the twentieth century. During that time experiments were carried out, the results of which could not be explained by the theories of classical physics developed by Isaac Newton and many others who followed him. A radical new theory called quantum

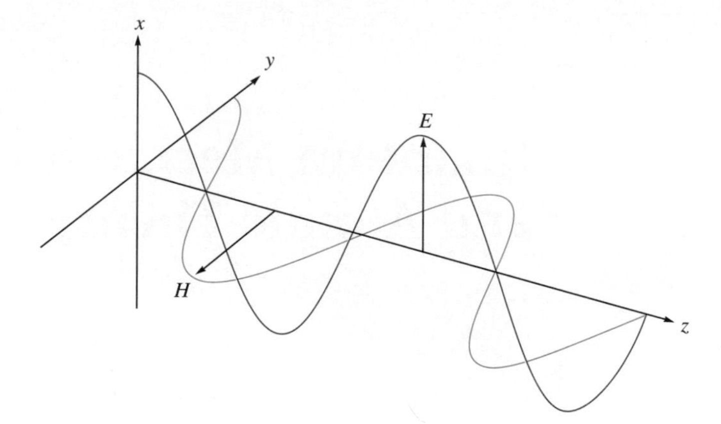

Figure 12.1
Electromagnetic radiation has oscillating electric (*E*) and magnetic (*H*) fields in planes perpendicular to each other and to the direction of propagation.

mechanics was developed to account for the behavior of light and atoms. This "new physics" provides many surprises for people who are used to the macroscopic world but it seems to account flawlessly (within the bounds of necessary approximations) for the behavior of matter.

As the first step in our exploration of this revolution in science, we will consider the properties of light, more properly called electromagnetic radiation.

12.1 Electromagnetic Radiation

One of the ways that energy travels through space is by **electromagnetic radiation.** The light from the sun, the energy used to cook food in a microwave oven, the X rays used by dentists, the radiowaves used by physicians to make MRI maps of body tissues, and the radiant heat from a fireplace are all examples of electromagnetic radiation. Although these forms of radiant energy seem quite different, they all exhibit the same type of wavelike behavior and travel at the speed of light in a vacuum. Electromagnetic radiation is so-named because it has electric and magnetic fields that simultaneously oscillate in planes mutually perpendicular to each other and to the direction of propagation through space (see Fig. 12.1).

Waves are characterized by wavelength, frequency, and speed. **Wavelength** (symbolized by the Greek letter lambda, λ) is the *distance between two consecutive peaks or troughs in a wave,* as shown in Fig. 12.2. The **frequency** (symbolized by the Greek letter nu, ν) is defined as the *number of waves (cycles) per second that pass a given point in space.* Since all types of electromagnetic radiation travel at the speed of light, short-wavelength radiation must have a high frequency. You can see this in Fig. 12.2, where three waves are shown traveling between two points at constant speed. Note that the wave with the shortest wavelength (λ_3) has the highest frequency, and the wave with the longest wavelength (λ_1) has the lowest frequency. This implies an inverse relationship between wavelength and frequency; that is, $\lambda \propto 1/\nu$, or

$$\lambda\nu = c$$

Wavelength (λ) and frequency (ν) are inversely related.

In this equation λ is the wavelength in meters, ν is the frequency in cycles per second, and c is the speed of light, a defined quantity with the exact value of 2.99792458×10^8 m/s. In the SI system, *cycles* is understood, and the unit cycles per second becomes $1/s$, or s^{-1}, which is called the *hertz* (abbreviated Hz).

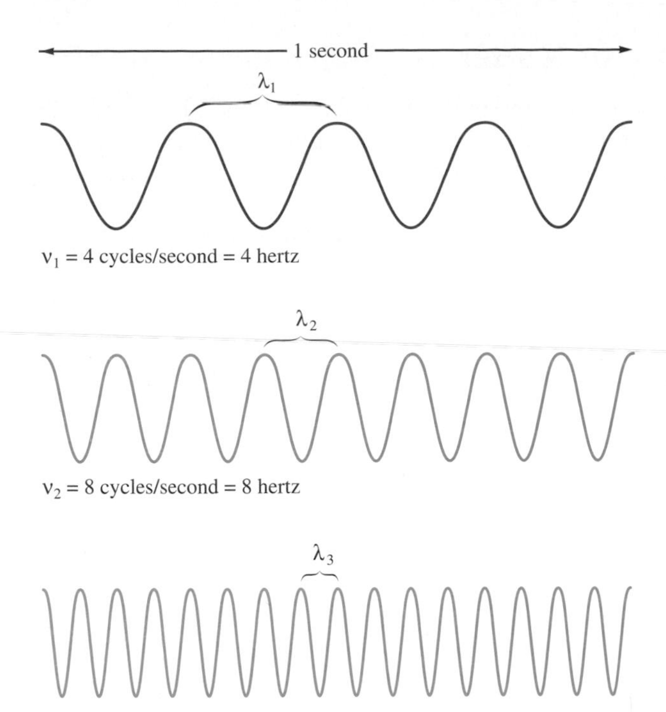

Figure 12.2
The nature of waves. Note that the radiation with the shortest wavelength has the highest frequency. This diagram can represent the oscillating electric or magnetic field of the wave.

Electromagnetic radiation is classified as shown in Fig. 12.3. Radiation provides an important means of energy transfer. For example, the energy from the sun reaches the earth mainly in the forms of visible and ultraviolet radiation, and the glowing coals of a fireplace transmit heat energy by infrared radiation. In a microwave oven the water molecules in food absorb microwave radiation, which increases their motions. This energy is then transferred to other types of molecules via collisions, causing an increase in the food's temperature. As we proceed in the study of chemistry, we will consider many of the classes of electromagnetic radiation and the ways in which they affect matter.

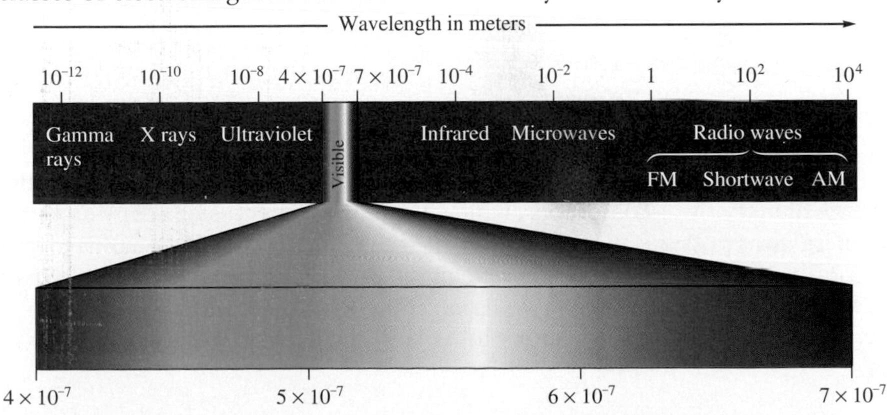

Figure 12.3
Classification of electromagnetic radiation. Spectrum adapted by permission from C. W. Keenan, D. C. Kleinfelter, and J. H. Wood, *General College Chemistry*, 6th edition, Harper & Row Publishers, Inc., 1980.

12.2 The Nature of Matter

It is probably fair to say that at the end of the nineteenth century physicists were feeling rather smug. Available theories could explain phenomena ranging from the motions of the planets to the dispersion of visible light by a prism. Rumor has it that students were being discouraged from pursuing physics as a career because it was felt that all the major problems had been solved or at least described in terms of the current physical theories.

At the end of the nineteenth century the idea prevailed that matter and energy were distinct. Matter was thought to consist of particles, whereas energy in the form of light (electromagnetic radiation) was described as a wave. Particles were things that had mass and whose position in space could be specified. Waves were described as massless and delocalized; that is, their position in space could not be specified. It was also assumed that there was no intermingling of matter and light. Everything known before 1900 seemed to fit neatly into this view.

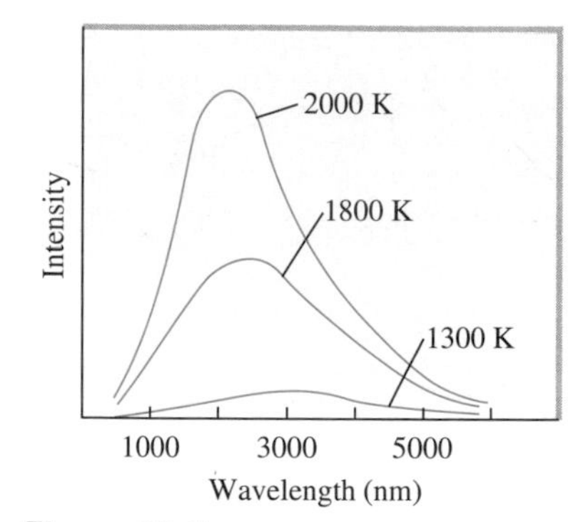

Figure 12.4
The profile of radiation emitted from a blackbody. Note the maximum shifts to shorter wavelengths as the temperature is increased, in agreement with the observed change from a reddish to a white glow as iron is heated to higher temperatures.

At the beginning of the twentieth century, however, certain experimental results suggested that this picture was incorrect. The first important advance came in 1901 from the German physicist Max Planck, who studied the profile (intensity versus wavelength) of the electromagnetic radiation emitted from a solid body heated to incandescence. (An example is a piece of iron that glows red and then white as it is heated to higher and higher temperatures.) The profiles shown in Fig. 12.4 are for so-called blackbody radiation. This may seem like a contradiction in terms, since a blackbody is an idealized object that absorbs all the radiation incident upon it. The term is used in this context to mean radiation that originates from the thermal energy of the body only. It does not include radiation reflected from the object and does not depend on the material composing the object. Blackbody radiation is closely approximated by the radiation emitted through a tiny hole from a cavity inside an object. The main point to be made here is that the radiation profiles shown in Fig. 12.4 are not the ones expected from classical physics. The classical theory of matter, which assumes that matter can absorb or emit any quantity of energy, predicts a radiation profile that has no maximum and goes to infinite intensity at very short wavelengths (an effect often called the **ultraviolet catastrophe**).

Energy can be gained or lost only in integer multiples of $h\nu$.

Planck's constant = 6.626×10^{-34} J s.

Planck found that the observed profiles (with their intensity maxima) could be accounted for by postulating that energy can be gained or lost only in *whole-number multiples* of the quantity $h\nu$, where h is a constant now called **Planck's constant,** determined by experiment to have the value 6.626×10^{-34} J s. That is, the change in energy for a system, ΔE, can be represented by the equation

$$\Delta E = nh\nu$$

where n is an integer (1, 2, 3, . . .), h is Planck's constant, and ν is the frequency of the electromagnetic radiation absorbed or emitted.

Planck's result was a real surprise. Physicists had always assumed that the energy of matter was continuous, which meant that the transfer of any quantity of energy was possible. Now it seemed clear that energy is in fact **quantized** and can only be transferred in discrete units of size $h\nu$. Each of these small "packets" of energy is called a *quantum.* A system can transfer energy only in whole *quanta.* Thus energy seems to have particulate properties.

Blue fireworks over the Brooklyn Bridge.

EXAMPLE 12.1

The blue color in fireworks is often achieved by heating copper(I) chloride (CuCl) to about 1200°C. The hot compound emits blue light having a wavelength of 450 nm. What is the increment of energy (the quantum) that is emitted at 4.50×10^2 nm by CuCl?

Solution

The quantum of energy can be calculated from the equation

$$\Delta E = h\nu$$

The frequency ν for this case can be calculated as follows:

$$\nu = \frac{c}{\lambda} = \frac{2.9979 \times 10^8 \text{ m/s}}{4.50 \times 10^{-7} \text{ m}} = 6.66 \times 10^{14} \text{ s}^{-1}$$

So

$$\Delta E = h\nu = (6.626 \times 10^{-34} \text{ J s})(6.66 \times 10^{14} \text{ s}^{-1}) = 4.41 \times 10^{-19} \text{ J}$$

A sample of CuCl emitting light at 450 nm can lose energy only in increments of 4.41×10^{-19} J, the size of the quantum in this case.

The next important development in the knowledge of atomic structure came when Albert Einstein (see Fig. 12.5) proposed that electromagnetic radiation is itself quantized. Einstein suggested that electromagnetic radiation can be viewed as a stream of "particles" called **photons.** The energy of each photon is given by the expression

$$E_{\text{photon}} = h\nu = \frac{hc}{\lambda}$$

where h is Planck's constant, ν is the frequency of the radiation, and λ is the wavelength of the radiation.

Einstein arrived at this conclusion through his analysis of the **photoelectric effect** (for which he later was awarded the Nobel prize). The photoelectric effect refers to the phenomenon in which electrons are emitted from the surface of a metal when light strikes it. The following observations characterize the photoelectric effect.

1. Studies in which the frequency of the light is varied show that no electrons are emitted by a given metal below a specific threshold frequency ν_0.
2. For light with frequency lower than the threshold frequency, no electrons are emitted regardless of the intensity of the light.
3. For light with frequency greater than the threshold frequency, the number of electrons emitted increases with the intensity of the light.
4. For light with frequency greater than the threshold frequency, the kinetic energy of the emitted electrons increases linearly with the frequency of the light.

Figure 12.5
Albert Einstein (1879–1955) was born in Germany. Nothing in his early development suggested genius; even at the age of 9 he did not speak clearly, and his parents feared that he might be handicapped. When asked what profession Einstein should follow, his school principal replied, "It doesn't matter; he'll never make a success of anything." When he was 10, Einstein entered the Luitpold Gymnasium (high school), which was typical of German schools of that time in being harshly disciplinarian. There he developed a deep suspicion of authority and a skepticism that encouraged him to question and doubt—valuable qualities in a scientist. In 1905, while a patent clerk in Switzerland, Einstein published a paper explaining the photoelectric effect via the quantum theory. For this revolutionary thinking he received a Nobel prize in 1921. Highly regarded by this time, he worked in Germany until 1933, when Hitler's persecution of the Jews forced him to come to the United States. He worked at the Institute for Advanced Studies at Princeton University until his death in 1955.
Einstein was undoubtedly the greatest physicist of our age. Even if someone else had derived the theory of relativity, his other work would have ensured his ranking as the second greatest physicist of his time. Our concepts of space and time were radically changed by ideas he first proposed when he was 26 years old. From then until the end of his life, he attempted unsuccessfully to find a single unifying theory that would explain all physical events.

These observations can be explained by assuming that electromagnetic radiation is quantized (consists of photons), and that the threshold frequency represents the minimum energy required to remove the electron from the metal's surface.

$$\text{Minimum energy required to remove an electron} = E_0 = h\nu_0$$

Because a photon with energy less than E_0 ($\nu < \nu_0$) cannot remove an electron, light with a frequency less than the threshold frequency produces no electrons. On the other hand, for light where $\nu > \nu_0$, the energy in excess of that required to remove the electron is given to the electron as kinetic energy (KE):

$$KE_{\text{electron}} = \tfrac{1}{2}mv^2 = h\nu - h\nu_0$$

Mass of electron (m); Velocity of electron (v); Energy of incident photon ($h\nu$); Energy required to remove electron from metal's surface ($h\nu_0$)

Because in this picture the intensity of light is a measure of the number of photons present in a given part of the beam, a greater intensity means that more photons are available to release electrons (as long as $\nu > \nu_0$ for the radiation).

At about the same time that Einstein was performing his analysis of the photoelectric effect, he was also constructing the theory of special relativity. In connection with this work Einstein derived the famous equation

$$E = mc^2$$

which he published in 1905. The main significance of this equation is that *energy has mass*. This result is more apparent if we rearrange the equation to the following form:

$$m = \frac{E}{c^2}$$

Mass (m); Energy (E); Speed of light (c)

Using this form of the equation, we can calculate the mass associated with a given quantity of energy. For example, we can calculate the apparent mass of a photon. For electromagnetic radiation of wavelength λ the energy of each photon is given by the expression

$$E = \frac{hc}{\lambda}$$

Then the "mass" of a photon of light with wavelength λ is given by

$$m = \frac{E}{c^2} = \frac{hc/\lambda}{c^2} = \frac{h}{\lambda c}$$

Do photons really have mass? The answer *appears* to be yes. In 1922 American physicist Arthur Compton performed experiments involving collisions of X rays with electrons. These experiments showed that photons do exhibit the apparent mass calculated from the above equation. Also, photons do seem to be affected by gravity, as Einstein postulated in his general theory of relativity. However, it is important to recognize that the photon is in no sense a typical particle. A photon has mass only in a relativistic sense—it has no rest mass.

Light as a wave phenomenon

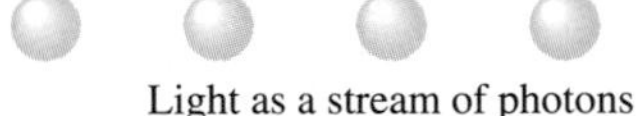

Light as a stream of photons

Figure 12.6
Electromagnetic radiation exhibits wave properties and particulate properties. The energy of each photon of the radiation is related to the wavelength and frequency by the equation $E_{photon} = h\nu = hc/\lambda$.

We can summarize the important conclusions from the work of Planck and Einstein as follows:

Energy is quantized. It can be transferred only in discrete units called quanta.

Electromagnetic radiation, which was previously thought to exhibit only wave properties, seems to show certain characteristics of particulate matter as well. This phenomenon, illustrated in Fig. 12.6, is sometimes referred to as the **dual nature of light**.

Thus light, which was previously thought to be purely wavelike, was found to have certain characteristics of particulate matter. But is the opposite also true? That is, does matter that is normally assumed to be particulate exhibit wave properties? This question was raised in 1923 by a young French physicist named Louis de Broglie (1892–1987). To see how de Broglie supplied the answer to this question, recall that the relationship between mass and wavelength for electromagnetic radiation is $m = h/\lambda c$. For a particle with velocity v the corresponding expression is

Do not confuse ν (frequency) with v (velocity).

$$m = \frac{h}{\lambda v}$$

Rearranging to solve for λ, we have

$$\lambda = \frac{h}{mv}$$

This equation, called de Broglie's equation, allows us to calculate the wavelength for a particle, as shown in Example 12.2.

EXAMPLE 12.2

Compare the wavelength for an electron (mass = 9.11×10^{-31} kg) traveling at a speed of 1.0×10^7 m/s with that for a ball (mass = 0.10 kg) traveling at 35 m/s.

Solution

We use the equation $\lambda = h/mv$, where

$$h = 6.626 \times 10^{-34}\text{ J s} \quad \text{or} \quad 6.626 \times 10^{-34}\text{ kg m}^2/\text{s}$$

since

$$1\text{ J} = 1\text{ kg m}^2/\text{s}^2$$

For the electron

$$\lambda_e = \frac{6.626 \times 10^{-34}\ \dfrac{\text{kg m}^2}{\text{s}}}{(9.11 \times 10^{-31}\text{ kg})(1.0 \times 10^7\text{ m/s})} = 7.3 \times 10^{-11}\text{ m}$$

For the ball

$$\lambda_b = \frac{6.626 \times 10^{-34}\ \dfrac{\text{kg m}^2}{\text{s}}}{(0.10\text{ kg})(35\text{ m/s})} = 1.9 \times 10^{-34}\text{ m}$$

Notice from Example 12.2 that the wavelength associated with the ball is incredibly short. On the other hand, the wavelength of the electron, although quite small, happens to be of the same order as the spacing between the atoms in a typical crystal. This is important because, as we will see presently, it provides a means for testing de Broglie's equation.

Diffraction results when light is scattered from a regular array of points or lines. The diffraction of light from the ridges and grooves of a compact disc is shown in the accompanying photograph. The colors result because the various wavelengths of visible light are not all scattered in the same way. The colors are "separated," giving the same effect as light passing through a prism. Just as a regular arrangement of ridges and grooves produces diffraction, so does a regular array of atoms or ions in a crystal. For example, when X rays are directed onto a crystal of sodium chloride with its regular array of Na^+ and Cl^- ions, the scattered radiation produces a **diffraction pattern** of bright areas and dark spots on a photographic plate, as shown in Fig. 12.7(a). This pattern occurs because the scattered light can interfere constructively (the peaks and troughs of the beams are in phase) to produce a bright area [Fig. 12.7(b)] or destructively (the peaks and troughs are out of phase) to produce a dark spot [Fig. 12.7(c)].

Light diffracted by the closely-spaced grooves on a compact disc shows the visible spectrum.

A diffraction pattern can only be explained in terms of waves. Thus this phenomenon provides a test for the postulate that particles such as electrons have wave properties. As we saw in Example 12.2, an electron with a velocity

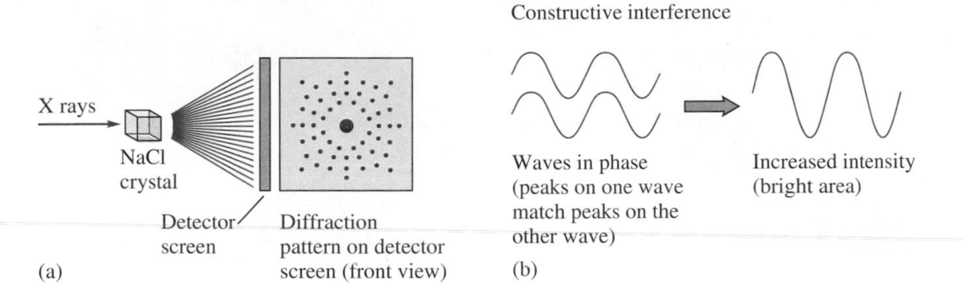

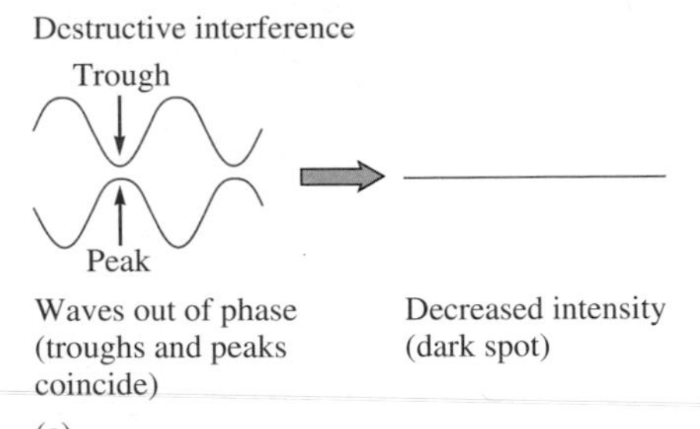

Figure 12.7
(a) Diffraction occurs when electromagnetic radiation is scattered from a regular array of objects, such as the ions in a crystal of sodium chloride. The large spot in the center is from the main incident beam of X rays. (b) Bright areas in the diffraction pattern result from *constructive interference* of waves. The waves are in phase; that is, their peaks match. (c) Dark spots result from *destructive interference* of waves. The waves are out of phase; the peaks of one wave coincide with the troughs of another wave.

of 10^7 m/s (easily achieved by acceleration of the electron in an electric field) has a wavelength of about 10^{-10} m, which is roughly the distance between the components in a typical crystal. This is important because diffraction occurs most efficiently when the spacing between the scattering points is about the same as the wavelength. Thus if electrons actually do have an associated wavelength, a crystal should diffract electrons. An experiment to test this idea was carried out in 1927 by Davisson and Germer at the Bell Laboratories. When they directed a beam of electrons at a nickel crystal, they observed a diffraction pattern similar to that seen from the diffraction of X rays. This result verified de Broglie's relationship, at least for electrons. Larger chunks of matter, such as balls, have wavelengths (Example 12.2) too small to verify experimentally. However, we believe that all matter obeys de Broglie's equation.

Now we have come full circle. Electromagnetic radiation, which at the turn of the twentieth century was thought to be a pure waveform, was found to exhibit particulate properties. Conversely, electrons, which were thought to be particles, were found to have a wavelength associated with them. The significance of these results is that matter and energy are not distinct. Energy is really a form of matter, and all matter shows the same types of properties. That is, *all matter exhibits both particulate and wave properties.* Large "pieces" of matter, like baseballs, exhibit predominantly particulate properties. The associated wavelength is so small that it is not observed. Very small pieces of matter, such as photons, while showing some particulate properties through relativistic effects, exhibit predominantly wave properties. Pieces of matter with intermediate mass, such as electrons, show both the particulate and wave properties of matter.

12.3 The Atomic Spectrum of Hydrogen

Recall from Chapter 2 that key information about the structure of the atom came from several experiments carried out in the early twentieth century. Particularly important were Thomson's discovery of the electron and Rutherford's discovery of the nucleus. Another important experiment concerned the study of the emission of light by excited hydrogen atoms. When a sample of hydrogen gas receives a high-energy spark, the H_2 molecules absorb energy, causing some of the H—H bonds to break. The resulting hydrogen atoms are

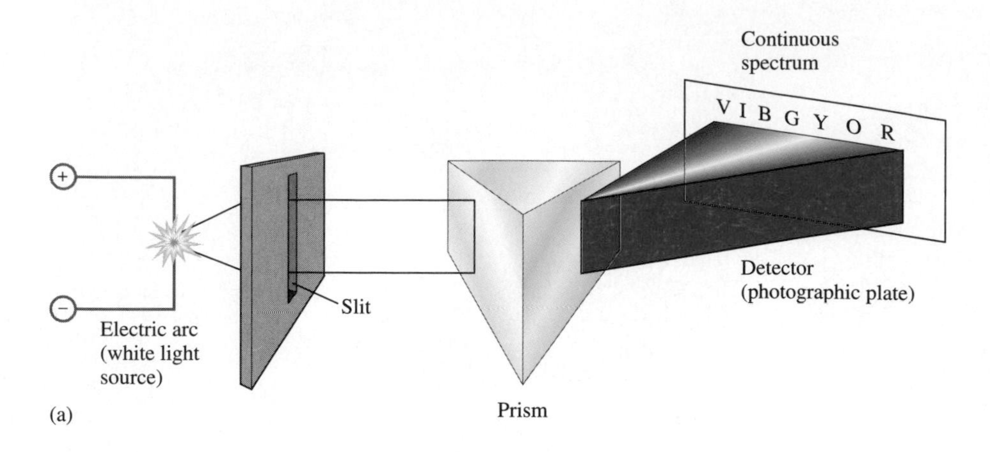

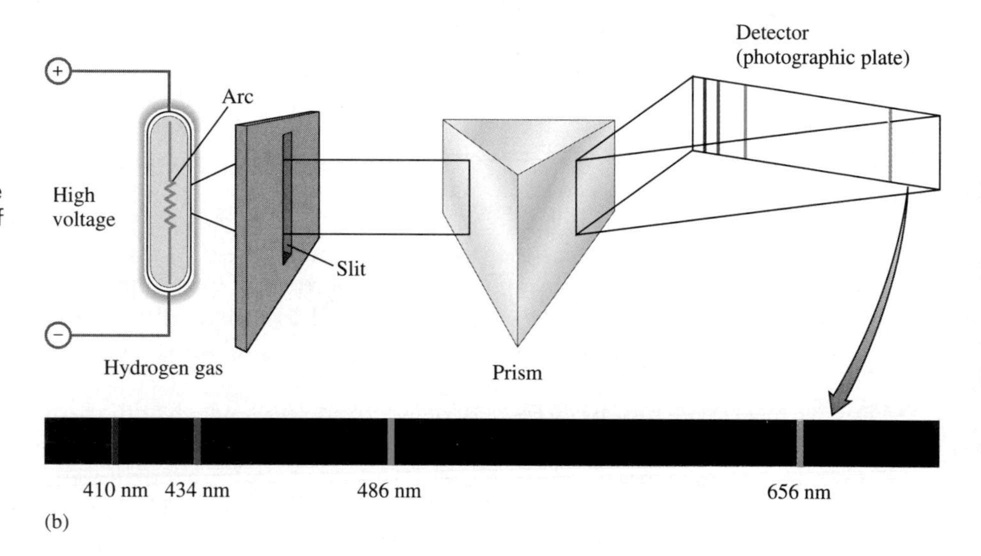

Figure 12.8
(a) A continuous spectrum containing all wavelengths of visible light (indicated by the first letters of the colors of the rainbow).
(b) The hydrogen line spectrum contains only a few discrete wavelengths. Spectrum adapted by permission from C. W. Keenan, D. C. Kleinfelter, and J. H. Wood, *General College Chemistry*, 6th edition, Harper & Row Publishers, Inc., 1980.

excited; that is, they contain excess energy, which they release by emitting light of various wavelengths to produce what is called the *emission spectrum* of the hydrogen atom.

To understand the significance of the hydrogen emission spectrum, we must first describe the **continuous spectrum** that results when white light is passed through a prism, as shown in Fig. 12.8(a). This spectrum, like the rainbow produced when sunlight is dispersed by raindrops, contains *all* the wavelengths of visible light. In contrast, when the hydrogen emission spectrum in the visible region is passed through a prism, as shown in Fig. 12.8(b), we see only a few lines, each corresponding to a discrete wavelength. The hydrogen emission spectrum is called a **line spectrum.**

What is the significance of the line spectrum of hydrogen? It indicates that *only certain energies are allowed for the electron in the hydrogen atom.* In other words, the energy of the electron in the hydrogen atom is *quantized.* Changes in energy between discrete energy levels in hydrogen produce only certain wavelengths of emitted light as shown in Fig. 12.9. For example, a given change in

E

$\Delta E_3 = \frac{hc}{\lambda_3}$

$\Delta E_2 = \frac{hc}{\lambda_2}$

$\Delta E_1 = \frac{hc}{\lambda_1}$

Various energy levels in the hydrogen atom

Figure 12.9
A change between two discrete energy levels emits a photon of light.

SPECTRA AND SPACE

Every element has a characteristic signature: its unique line spectrum produced by the movement of electrons between quantized energy levels. (The spectrum of hydrogen is shown in Fig. 12.8.) The characteristic spectrum of an element is a sure way to verify its presence (or absence) in a given sample. Spectra are particularly useful for the analyses of extraterrestrial material, since collecting actual samples is so difficult. Although we have directly analyzed rocks from the moon and Mars (at great expense), most of our information about the materials in space comes from the light emitted or absorbed by these materials. For example, we have obtained a detailed knowledge of the composition of the sun and other stars by looking for the characteristic spectral lines of the elements in the light emitted by these sources.

Molecules also have characteristic spectral lines that can be used to identify them. The first spectroscopic observations of a comet were carried out by Giovanni Donati in 1864 when he identified species such as C_2, $(CN)_2$, C_3, and CH in the Comet Tempel. Since then, using increasingly sensitive instruments, scientists have found more than 60 different molecules in space, including ethanol (C_2H_5OH), carbon monoxide, hydrogen cyanide (HCN), methyl cyanide (CH_3CN), and water. It is interesting to note that helium was discovered spectroscopically in the sun before it was found on earth.

Electromagnetic radiation is also useful to space scientists in other ways. For example, the microwave radiation that is left over from the postulated "big bang" implies an age for the universe of between 15 and 25 billion years. In addition, infrared radiation is useful for determining the temperature of space objects. The accompanying photo is an infrared picture of the shuttle *Columbia* as it was launched in 1982; the hottest areas appear red and the coolest areas appear violet.

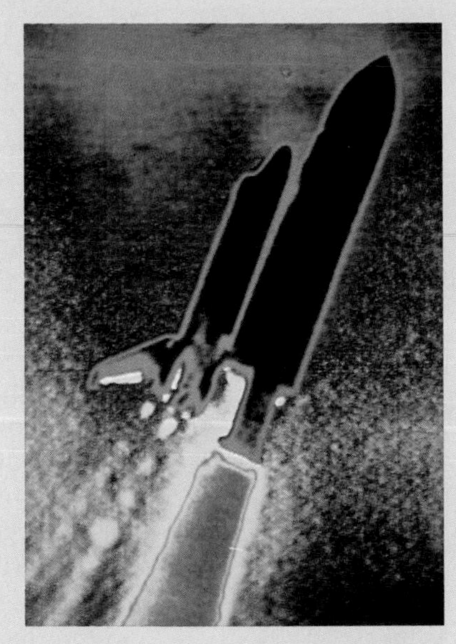

An infrared thermogram of the space shuttle *Columbia*.

Suggested Reading

P. B. Kelter, W. E. Snyder, and C. S. Buchar, "Using NASA and the Space Program to Help High School and College Students Learn Chemistry (Part II—The Current State of Chemistry at NASA)" *J. Chem. Ed.* 64 (1987), p. 228.

energy from a high to a lower level gives a wavelength of light that can be calculated from Einstein's equation:

$$\Delta E = h\nu = \frac{hc}{\lambda}$$

Change in energy (ΔE); Frequency of light emitted (ν); Wavelength of light emitted (λ)

The energy of the electron in the hydrogen atom is quantized.

The discrete line spectrum of hydrogen shows that only certain energies are possible; that is, the electron energy levels are quantized. In contrast, if any energy level were allowed, the emission spectrum would be continuous.

Niels Bohr at 37 years of age, photographed in 1922 when he received the Nobel prize for physics.

12.4 The Bohr Model

In 1913 a Danish physicist named Niels Bohr, aware of the experimental results we have just discussed, developed a **quantum model** for the hydrogen atom. Bohr proposed a model that included the idea that the *electron in a hydrogen atom moves around the nucleus only in certain allowed circular orbits.* He calculated the radii for these allowed orbits by using the theories of classical physics and by making some new assumptions.

From classical physics Bohr knew that a particle in motion tends to move in a straight line and can be made to travel in a circle only by application of a force toward the center of the circle. Thus Bohr reasoned that the tendency of the revolving electron to fly off the atom must be exactly balanced by its attraction for the positively charged nucleus. But classical physics also decreed that a charged particle under acceleration should radiate energy. Since an electron revolving around the nucleus constantly changes its direction, it is constantly accelerating. Therefore, the electron should emit light and lose energy, and thus be drawn into the nucleus. This conclusion, of course, does not correlate with the existence of stable atoms.

Clearly, an atomic model based solely on the theories of classical physics was untenable. Bohr also knew that the correct model had to account for the experimental spectrum of hydrogen, which showed that only certain electron energies were allowed. The experimental data were absolutely clear on this point. Bohr found that his model would fit the experimental results if he assumed that the angular momentum of the electron could occur only in certain increments. It wasn't clear why this should be true, but with this assumption Bohr's model gave hydrogen atom energy levels consistent with the hydrogen emission spectrum. The model is represented pictorially in Fig. 12.10.

Angular momentum equals the product of mass, velocity, and orbital radius.

Although we will not discuss its origins here, the expression for the *energy levels available to the electron in the hydrogen atom is*

$$E = -2.178 \times 10^{-18}\ \text{J}\left(\frac{Z^2}{n^2}\right) \tag{12.1}$$

The "J" in Equation (12.1) stands for joules.

where n is an integer (the larger the value of n, the larger the orbit radius) and Z is the atomic number ($Z = 1$ for hydrogen). The negative sign in Equation (12.1) simply means that the energy of the electron bound to the nucleus is lower than it would be if the electron were at an infinite distance ($n = \infty$) from the nucleus, where there is no interaction and the energy is zero:

Equation (12.1) applies to all one-electron species. In addition to being used to calculate the energy levels in the hydrogen atom, it can also be used for He^+ ($Z = 2$), Li^{2+} ($Z = 3$), Be^{3+} ($Z = 4$), and so on.

$$E = -2.178 \times 10^{-18}\ \text{J}\left(\frac{Z^2}{\infty}\right) = 0$$

The energy of the electron in any orbit is negative relative to this reference state.

Equation (12.1) can be used to calculate the change in energy when the electron changes orbits. For example, suppose the electron in level $n = 6$ of an excited hydrogen atom falls back to level $n = 1$ as the hydrogen atom returns to its lowest possible energy state, its **ground state.** We use Equation (12.1) with $Z = 1$ since the hydrogen nucleus contains a single proton. The energies

The ground state is the lowest possible energy state of an atom.

corresponding to the two states are

$$\text{For } n = 6: \quad E_6 = -2.178 \times 10^{-18}\,\text{J}\left(\frac{1^2}{6^2}\right) = -6.05 \times 10^{-20}\,\text{J}$$

$$\text{For } n = 1: \quad E_1 = -2.178 \times 10^{-18}\,\text{J}\left(\frac{1^2}{1^2}\right) = -2.178 \times 10^{-18}\,\text{J}$$

Note that for $n = 1$ the electron has a more negative energy than it does for $n = 6$, which means that the electron is more tightly bound in the smallest allowed orbit.

The change in energy, ΔE, when the electron falls from $n = 6$ to $n = 1$ is

$$\begin{aligned}\Delta E &= \text{energy of final state} - \text{energy of initial state}\\ &= E_1 - E_6 = (-2.178 \times 10^{-18}\,\text{J}) - (-6.05 \times 10^{-20}\,\text{J})\\ &= -2.118 \times 10^{-18}\,\text{J}\end{aligned}$$

The negative sign for the *change* in energy indicates that the atom has *lost* energy and is now in a more stable state. The energy is carried away from the atom by the production (emission) of a photon.

The wavelength of the emitted photon can be calculated from the equation

$$\Delta E = h\left(\frac{c}{\lambda}\right) \quad \text{or} \quad \lambda = \frac{hc}{\Delta E}$$

where ΔE represents the change in energy of the atom, and thus equals the energy of the emitted photon. We have

$$\lambda = \frac{hc}{\Delta E} = \frac{(6.626 \times 10^{-34}\,\text{J s})(2.9979 \times 10^{8}\,\text{m/s})}{2.118 \times 10^{-18}\,\text{J}} = 9.379 \times 10^{-8}\,\text{m}$$

Note that for this calculation the absolute value of ΔE is used. In this case the direction of energy flow is indicated by saying that a photon of wavelength 9.379×10^{-8} m has been *emitted* from the hydrogen atom. Simply plugging the negative value of ΔE into the equation would produce a negative value for λ, which is physically meaningless.

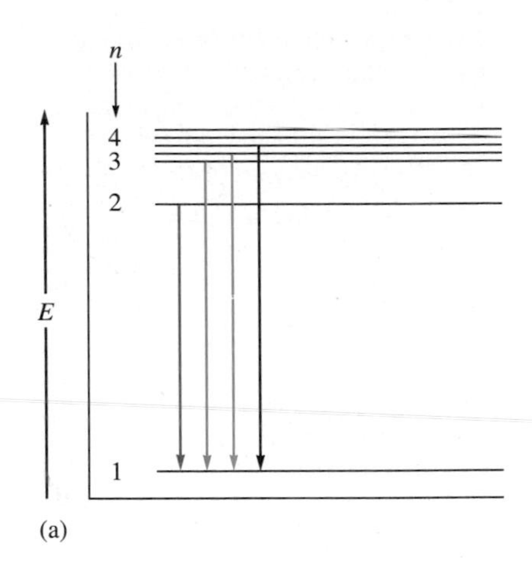

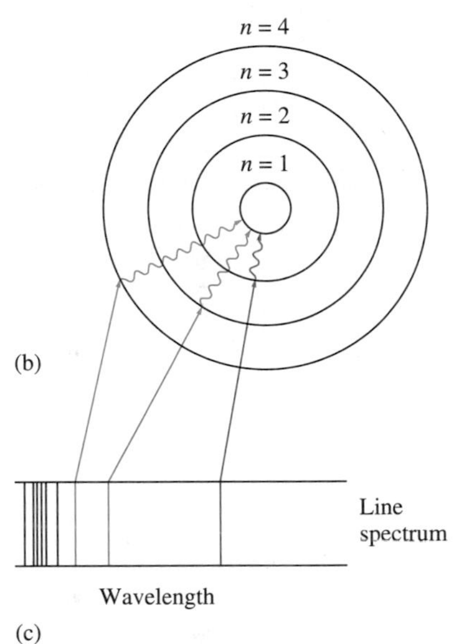

Figure 12.10
Electronic transitions in the Bohr model for the hydrogen atom. (a) An energy-level diagram for electronic transitions. (b) An orbit-transition diagram, which accounts for the experimental spectrum. (c) The resulting line spectrum on a photographic plate.

EXAMPLE 12.3

Calculate the energy required to excite the hydrogen electron from level $n = 1$ to level $n = 2$. Also, calculate the wavelength of light that must be absorbed by a hydrogen atom in its ground state to reach this excited state.

Solution

Using Equation (12.1) with $Z = 1$, we have

$$E_1 = -2.178 \times 10^{-18}\,\text{J}\left(\frac{1^2}{1^2}\right) = -2.178 \times 10^{-18}\,\text{J}$$

$$E_2 = -2.178 \times 10^{-18}\,\text{J}\left(\frac{1^2}{2^2}\right) = -5.445 \times 10^{-19}\,\text{J}$$

$$\begin{aligned}\Delta E &= E_2 - E_1 = (-5.445 \times 10^{-19}\,\text{J}) - (-2.178 \times 10^{-18}\,\text{J})\\ &= 1.634 \times 10^{-18}\,\text{J}\end{aligned}$$

Note from Fig. 12.3 that the light required to produce the transition from the $n = 1$ to $n = 2$ level in hydrogen lies in the ultraviolet region.

The positive value for ΔE indicates that the system has gained energy. The wavelength of light that must be *absorbed* to produce this change is

$$\lambda = \frac{hc}{\Delta E} = \frac{(6.626 \times 10^{-34}\text{ J s})(2.9979 \times 10^{8}\text{ m/s})}{1.634 \times 10^{-18}\text{ J}}$$

$$= 1.216 \times 10^{-7}\text{ m}$$

At this time we must emphasize two important points about the Bohr model:

1. The model correctly fits the quantized energy levels of the hydrogen atom as inferred from its emission spectrum. These energy levels correspond to certain allowed circular orbits for the electrons.
2. As the electron becomes more tightly bound, its energy becomes more negative relative to the zero-energy reference state (corresponding to the electron being an infinite distance from the nucleus). That is, as the electron is brought closer to the nucleus, energy is released from the system.

Using Equation (12.1), we can derive a general equation for the electron moving from one level (n_{initial}) to another level (n_{final}):

$$\Delta E = \text{energy of level } n_{\text{final}} - \text{energy of level } n_{\text{initial}}$$

$$= E_{\text{final}} - E_{\text{initial}}$$

$$= (-2.178 \times 10^{-18}\text{ J})\left(\frac{1^2}{n_{\text{final}}^2}\right) - (-2.178 \times 10^{-18}\text{ J})\left(\frac{1^2}{n_{\text{initial}}^2}\right)$$

$$= -2.178 \times 10^{-18}\text{ J}\left(\frac{1}{n_{\text{final}}^2} - \frac{1}{n_{\text{initial}}^2}\right) \quad (12.2)$$

EXAMPLE 12.4

Calculate the minimum energy required to remove the electron from a hydrogen atom in its ground state.

Solution

Removing the electron from a hydrogen atom in its ground state corresponds to taking the electron from $n_{\text{initial}} = 1$ to $n_{\text{final}} = \infty$. Thus

$$\Delta E = -2.178 \times 10^{-18}\text{ J}\left(\frac{1}{n_{\text{final}}^2} - \frac{1}{n_{\text{initial}}^2}\right)$$

$$= -2.178 \times 10^{-18}\text{ J}\left(\frac{1}{\infty} - \frac{1}{1^2}\right)$$

$$= -2.178 \times 10^{-18}\text{ J}(0 - 1) = 2.178 \times 10^{-18}\text{ J}$$

Thus the energy required to remove the electron from a hydrogen atom in its ground state is 2.178×10^{-18} J.

The New, Improved Atomic Clock

It doesn't have a digital readout like the one found on your trusty clock radio but it certainly keeps better time. Placed in service in April, 1993 at the National Institute of Standards and Technology (NIST), which lies at the foot of the Flatiron Mountains in Boulder, Colorado, the latest atomic clock is accurate within *one second in a million years.* Designated NIST-7, it replaces NBS-6 (NIST used to be called the National Bureau of Standards), the atomic clock which since 1975 had served as the international standard for time.

Under the clock's glistening, 2-meter-long cylindrical case lie several layers of magnetic shielding. At the center of the cylinder is an oven that warms a sample of cesium metal to inject cesium atoms into an evacuated tube. Once in the tube, they are formed into a narrow beam and are bathed in energy from a laser to ensure that all of the cesium atoms are in the same electronic state.

The beam of cesium atoms next enters a chamber 1.6 meters in length where microwaves with the very precise frequency of 9, 192, 631, 770 hertz (cycles per second) are reflecting off the walls. The frequency of these microwaves corresponds exactly to the energy needed to excite a cesium atom from its initial state to an adjacent electronic state. Then, stimulated by a laser, each excited cesium atom emits radiation with a wavelength of 9, 192, 631, 770 hertz, thereby returning to the original electronic state. Electronic circuitry is used to couple this emitted radiation to the microwave generator so the system maintains a constant frequency of 9, 192, 631, 770 hertz. Thus, for the atomic clock one second of time is precisely equal to 9, 192, 631, 770 electromagnetic vibrations. NIST-7 is the world's most precise timepiece.

NIST-7, the seventh generation of atomic clocks at the National Institute of Standards and Technology. Unveiled on April 22, 1993, it keeps time with an uncertainty of less than three parts in 10^{-14} (equivalent to one second in a million years).

Equation (12.2) can be used to calculate the energy change between *any* two energy levels in a hydrogen atom, as shown in Example 12.4.

At first Bohr's model appeared to be very promising. The energy levels calculated by Bohr closely agreed with the values obtained from the hydrogen emission spectrum. However, when Bohr's model was applied to atoms other than hydrogen, it did not work at all. Although some attempts were made to adapt the model using elliptical orbits, it was concluded that Bohr's model is fundamentally incorrect. The model is, however, very important historically, because it shows that the observed quantization of energy in atoms can be explained by making rather simple assumptions. Bohr's model paved the way for later theories. It is important to realize, however, that the current theory of atomic structure is in no way derived from the Bohr model. Electrons do *not* move around the nucleus in circular orbits, as we shall see later in this chapter.

Although Bohr's model fits the energy levels for hydrogen, it is a fundamentally incorrect model for the hydrogen atom.

Fireworks

The art of using mixtures of chemicals to produce explosives is an ancient one. Black powder—a mixture of potassium nitrate, charcoal, and sulfur—was being used in China well before 1000 A.D. and has been used through the centuries in military explosives, in construction blasting, and in fireworks. The DuPont Company, now a major chemical manufacturer, started out as a manufacturer of black powder. In fact, the founder, Eleuthère Du Pont, learned the manufacturing technique from none other than Lavoisier.

Before the nineteenth century fireworks were mainly confined to rockets and loud bangs. Orange and yellow colors came from the presence of charcoal and iron filings. However, with the great advances in chemistry in the nineteenth century, new compounds found their way into fireworks. Salts of copper, strontium, and barium added brilliant colors. Magnesium and aluminum metals gave a dazzling white light. Fireworks, in fact, have changed very little since then.

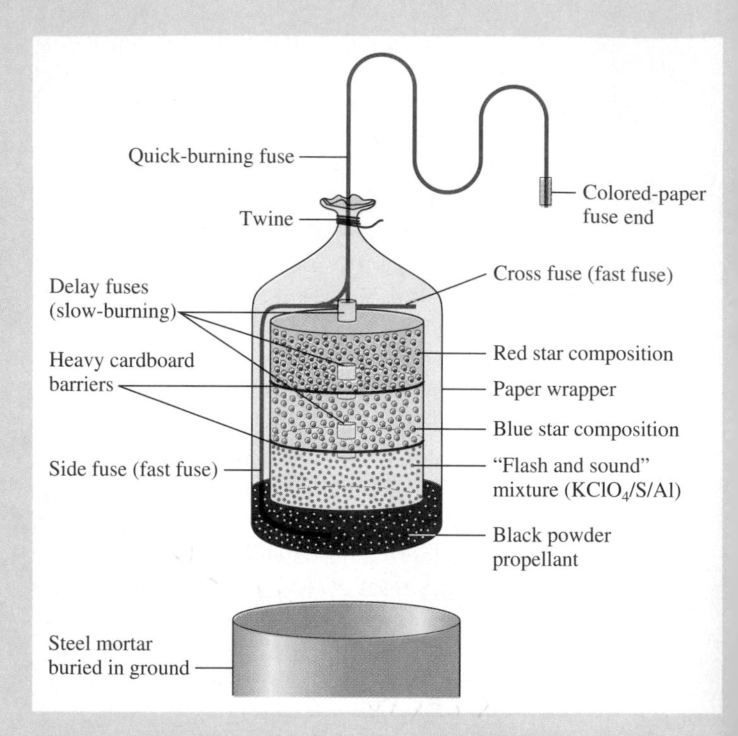

A typical aerial shell used in fireworks displays. Time-delayed fuses cause a shell to explode in stages. In this case a red starburst occurs first, followed by a blue starburst, and finally a flash and loud report. From *Chemical & Engineering News*, June 29, 1981, p. 24. Reprinted with permission. Copyright 1981 American Chemical Society.

Chemicals Commonly Used in the Manufacture of Fireworks

Oxidizers	Fuels	Special Effects
Potassium nitrate	Aluminum	Red flame: strontium nitrate, strontium carbonate
Potassium chlorate	Magnesium	Green flame: barium nitrate, barium chlorate
Potassium perchlorate	Titanium	Blue flame: copper carbonate, copper sulfate, copper oxide
Ammonium perchlorate	Charcoal	Yellow flame: sodium oxalate, cryolite (Na_3AlF_6)
Barium nitrate	Sulfur	White flame: magnesium, aluminum
Barium chlorate	Antimony sulfide	Gold sparks: iron filings, charcoal
Strontium nitrate	Dextrin	White sparks: aluminum, magnesium, aluminum-magnesium alloy, titanium
	Red gum	Whistle effect: potassium benzoate or sodium salicylate
	Polyvinyl chloride	White smoke: mixture of potassium nitrate and sulfur
		Colored smoke: mixture of potassium chlorate, sulfur, and an organic dye

How do fireworks produce their brilliant colors and loud bangs? Actually, only a handful of different chemicals are responsible for most of the spectacular effects. The noise and flashes are produced by an oxidizer (an oxidizing agent) and a fuel (a reducing agent). A common mixture involves potassium perchlorate ($KClO_4$) as the oxidizer and aluminum and sulfur as the fuel. The perchlorate compound oxidizes the fuel in a very exothermic reaction, which produces a brilliant flash, due to the aluminum, and a loud report from the rapidly expanding gases produced. For a color effect an element with a colored emission spectrum is included. Recall that the electrons in atoms can be raised to higher-energy levels when the atoms absorb energy. The excited atoms can then release this excess energy by emitting light of specific wavelengths, often in the visible region. In fireworks the energy to excite the electrons comes from the reaction between the oxidizer and fuel.

Yellow colors in fireworks are due to the 589-nm emission of sodium ions. Red colors come from strontium salts emitting at 606 nm and 636–688 nm. (This red color may be familiar from highway safety flares.) Barium salts give a green color in fireworks, due to a series of emission lines between 505 and 535 nm. A really good blue color, however, is hard to obtain. Copper salts give a blue color, emitting in the 420–460-nm region. But difficulties occur because another commonly used oxidizing agent, potassium chlorate ($KClO_3$), reacts with copper salts to form copper chlorate, a highly explosive compound that is dangerous to store. Paris green, a copper salt containing arsenic, was once extensively used but is now considered to be too toxic.

A typical aerial shell is shown at the top of the next column. The shell is launched from a mortar (a steel cylinder), using black powder as the propellant. Time-delayed fuses are used to fire the shell in stages. A list of chemicals commonly used in fireworks is given in the table.

Although you might think that the chemistry of fireworks is simple, the achievement of the vivid white flashes and the brilliant colors requires com-

An exploding aerial shell.

plex combinations of chemicals. For example, because the white flashes produce high flame temperatures, the colors tend to wash out. Thus oxidizers, such as $KClO_4$, are sometimes used with fuels that produce relatively low flame temperatures. An added difficulty, however, is that perchlorates are very sensitive to accidental ignition and are therefore quite hazardous. Another problem arises from the use of sodium salts. Because sodium produces an extremely bright yellow emission, sodium salts cannot be used when other colors are desired. Carbon-based fuels also give a yellow flame that masks other colors, therefore limiting the use of organic compounds as fuels. You can see that the manufacture of fireworks that produce the desired effects and are also safe to handle requires careful selection of chemicals. And, of course, there is still the dream of that special deep blue flame.

12.5 The Quantum Mechanical Description of the Atom

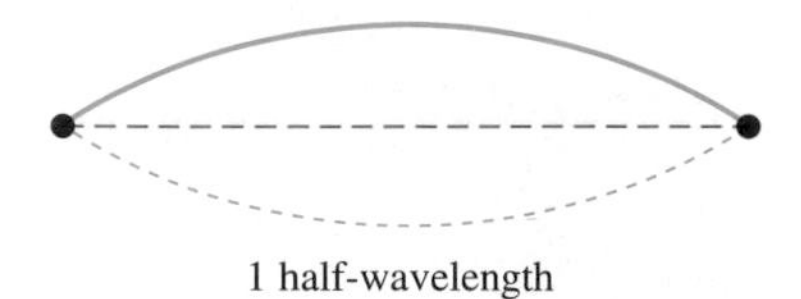

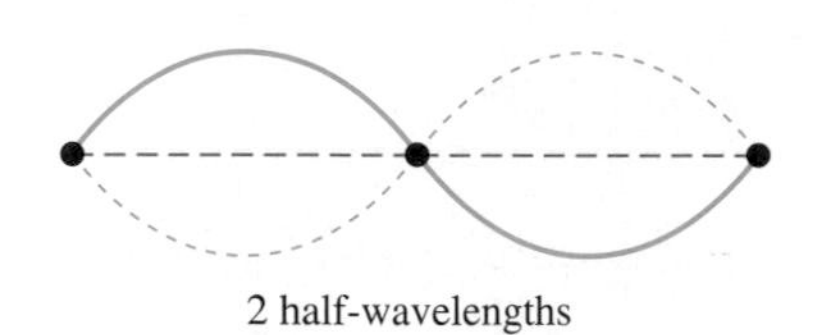

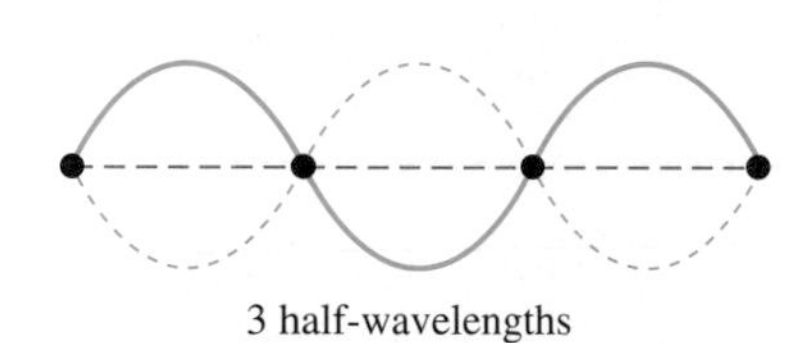

Figure 12.11
The standing wave produced by the vibration of a guitar string fastened at both ends. Each dot represents a node (a point of zero displacement).

By the mid-1920s it had become apparent that the Bohr model could not be made to work. A totally new approach was needed. Three physicists were at the forefront of this effort: Werner Heisenberg, Louis de Broglie, and Erwin Schrödinger. The approach they developed became known as **wave mechanics,** or, more commonly, **quantum mechanics.** As we have already seen, de Broglie originated the idea that the electron, previously considered to be a particle, also shows wave properties. Pursuing this line of reasoning, Schrödinger, an Austrian physicist, decided to attack the problem of atomic structure by giving emphasis to the wave properties of the electron. To Schrödinger and de Broglie, the electron bound to the nucleus seemed similar to a standing wave, and they began research on a wave mechanical description of the atom.

The most familiar example of standing waves occurs in association with musical instruments such as guitars or violins, where a string attached at both ends vibrates to produce the musical tone. The waves are described as standing since they are stationary; the waves do not travel along the length of the string. The motions of the string can be explained as a combination of simple waves of the type shown in Fig. 12.11. The dots in this figure indicate the nodes, or points of zero lateral (sideways) displacement for a given wave. Note that there are limitations on the allowed wavelengths of the standing wave. Since each end of the string is fixed, there is always a node at each end. This means that there must be a whole number of *half*-wavelengths in any of the allowed motions of the string (see Fig. 12.11).

The wave model was applied by de Broglie to the Bohr atom by imagining the electron in the hydrogen atom to be a standing wave. As shown in Fig. 12.12, only certain circular orbits have a circumference into which a whole number of wavelengths of the standing electron wave will "fit." All other orbits would produce destructive interference of the standing electron wave and are not allowed. This seemed like a possible explanation for the observed quantization of the hydrogen atom. The mathematical formalism that Schrödinger developed in 1925 to describe the hydrogen electron as a wave was heavily based on the classical descriptions of wave phenomena. We will first give an overview of this approach before considering it in more detail.

The form of Schrödinger's equation is

$$\hat{H}\psi = E\psi$$

where ψ, called the **wave function,** is a function of the coordinates (x, y, and z) of the electron's position in three-dimensional space, and where $\hat{H}$ represents a

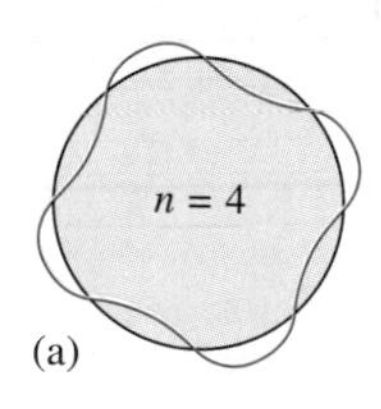

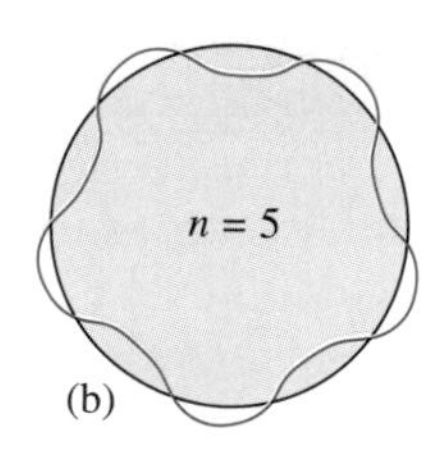

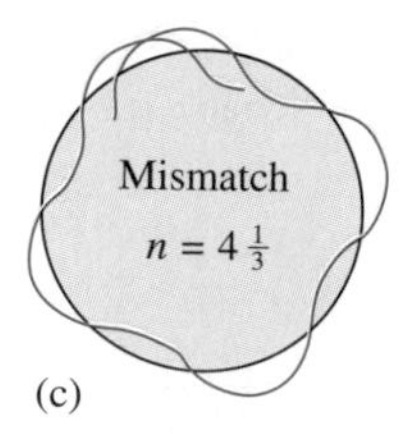

Figure 12.12
The hydrogen electron visualized as a standing wave around the nucleus. The circumference of a particular circular orbit has to correspond to a whole number of wavelengths, as shown in (a) and (b), or else destructive interference occurs, as shown in (c). This model is consistent with the fact that only certain electron energies are allowed; the atom is quantized. (Although this idea has encouraged scientists to use a wave theory, it does not mean that the electron really travels in circular orbits.)

Electrons as Waves

While scientists talk about the dual wave and particle properties of electrons, many non-scientists still believe that electrons are only tiny particles. Rooted as we are in the macroscopic world, it can be difficult for some to picture a particle as also being a wave. One look at the accompanying picture, however, should help change that. What looks like ripples surrounding two barely submerged pebbles in a pool of water is really the surface of a copper crystal.

Although they are true believers in the wave nature of electrons, the physicists at the IBM Almaden Research Center in San Jose, CA were genuinely surprised when their scanning tunneling microscope (STM) produced this image of the copper surface. "We looked at the surface with all these waves and thought, 'Is our machine broken?'" says Michael Crommie, one of the IBM physicists. But the researchers soon realized that the waves were produced by electrons confined to the metal's surface that bounced off impurities (the two pits). Because the electrons are waves, they form interference patterns after reflecting off the impurities, producing standing waves.

To further explore this behavior, the IBM scientists constructed a "quantum corral" by using their STM to place 48 iron atoms on a copper surface in a circle ~14 nm in diameter. Then using the STM to study electron behavior on the copper surface inside the corral they observed the standing electron waves shown in the photo. This image provides a unique visual confirmation of what the Schrödinger equation predicts. Electrons are wavelike. Seeing is believing!

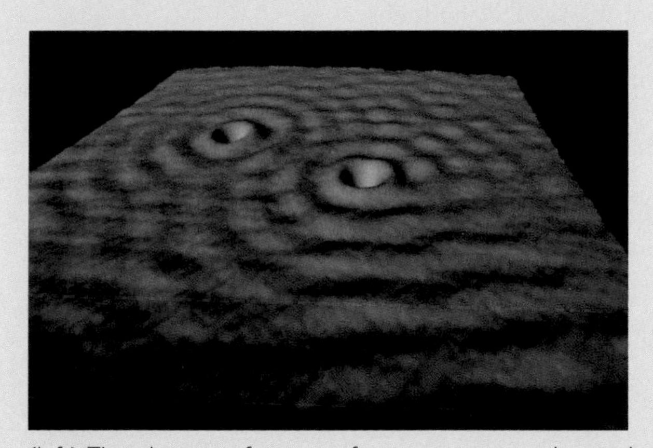

(left) The electrons form interference patterns, the ripples shown here, and produce standing waves. (right) Iron atoms in a circular "corral" cause electron standing waves.

set of mathematical instructions called an *operator*. An operator is a mathematical tool that acts on a function to produce another function. In some special cases the operator gives back the original function simply multiplied by a constant. Note that the Schrödinger equation corresponds to such a special case. In the Schrödinger equation the operator $\hat{H}$, called the Hamiltonian, acts to give back the wave function multiplied by the constant E, which represents the total energy of the atom (the sum of the potential energy due to the attraction between the proton and the electron, and the kinetic energy of the moving electron). When this equation is analyzed, many solutions are found. Each solution consists of a wave function ψ that is characterized by a particular value of E. A specific wave function for a given electron is often called an **orbital.** It is important to recognize that Schrödinger could not be sure that treating an electron as a wave makes any sense: The test would be whether or not the model could correctly fit the experimental data for hydrogen and other atoms.

To illustrate the most important ideas of the wave mechanical (quantum mechanical) model of the atom, we will first concentrate on the wave function corresponding to the lowest energy for the hydrogen atom. This wave function is called the 1*s* orbital. The first point of interest is the meaning of the word *orbital.* One thing is clear: an orbital is *not* a Bohr orbit. The electron in the hydrogen 1*s* orbital is not moving around the nucleus in a circular orbit. How, then, is the electron moving? The answer is somewhat surprising: *we do not know.* The wave function gives us no information about the movements of the electron. This observation is somewhat disturbing. When we solve problems involving the motions of particles in the macroscopic world, we are able to predict their trajectories. For example, when two billiard balls with known velocities collide, we can predict their motions after the collision. However, we cannot predict the electron's motion using the 1*s* orbital function. Does this mean that the theory is useless? Not necessarily: we have already learned that an electron does not behave much like a billiard ball, so we must examine the situation closely before we discard the theory.

The uncertainty principle is sometimes called the indeterminacy principle.

Werner Heisenberg, who was also involved in the development of the quantum mechanical model for the atom, discovered a very important principle in 1927 that helps us to understand the meaning of orbitals—the **Heisenberg uncertainty principle.** Heisenberg's mathematical analysis led him to a surprising conclusion: *there is a fundamental limitation to just how precisely we can know both the position and the momentum of a particle at a given time.* Stated mathematically, the uncertainty principle is

$$\Delta x \cdot \Delta p \geq \frac{\hbar}{2}$$

where Δx is the uncertainty in a particle's position, Δp is the uncertainty in a particle's momentum, and $\hbar$ is Planck's constant divided by 2π ($\hbar = h/2\pi$). Thus the minimum uncertainty in the product $\Delta x \cdot \Delta p$ is $h/4\pi$. This relationship means that the more precisely we know a particle's position, the less precisely we can know its momentum, and vice versa. This limitation is so small for large particles such as baseballs or billiard balls that it is unnoticed. However, for a small particle such as the electron, the limitation becomes quite important. Applied to the electron, the uncertainty principle implies that we cannot know the exact path of the electron as it moves around the nucleus. It is therefore not appropriate to assume that the electron is moving around the nucleus in a well-defined orbit as in the Bohr model.

EXAMPLE 12.5

The hydrogen atom has a radius on the order of 0.05 nm. Assuming that we know the position of an electron to an accuracy of 1% of the hydrogen radius, calculate the uncertainty in the velocity of the electron,

Electrons as Waves

While scientists talk about the dual wave and particle properties of electrons, many non-scientists still believe that electrons are only tiny particles. Rooted as we are in the macroscopic world, it can be difficult for some to picture a particle as also being a wave. One look at the accompanying picture, however, should help change that. What looks like ripples surrounding two barely submerged pebbles in a pool of water is really the surface of a copper crystal.

Although they are true believers in the wave nature of electrons, the physicists at the IBM Almaden Research Center in San Jose, CA were genuinely surprised when their scanning tunneling microscope (STM) produced this image of the copper surface. "We looked at the surface with all these waves and thought, 'Is our machine broken?'" says Michael Crommie, one of the IBM physicists. But the researchers soon realized that the waves were produced by electrons confined to the metal's surface that bounced off impurities (the two pits). Because the electrons are waves, they form interference patterns after reflecting off the impurities, producing standing waves.

To further explore this behavior, the IBM scientists constructed a "quantum corral" by using their STM to place 48 iron atoms on a copper surface in a circle ~14 nm in diameter. Then using the STM to study electron behavior on the copper surface inside the corral they observed the standing electron waves shown in the photo. This image provides a unique visual confirmation of what the Schrödinger equation predicts. Electrons are wavelike. Seeing is believing!

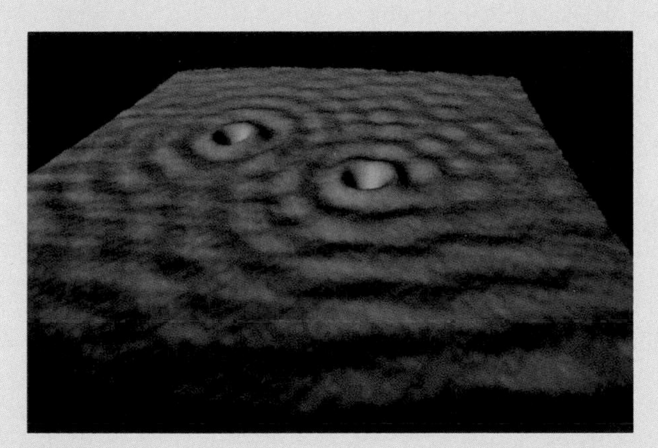

(left) The electrons form interference patterns, the ripples shown here, and produce standing waves. (right) Iron atoms in a circular "corral" cause electron standing waves.

set of mathematical instructions called an *operator*. An operator is a mathematical tool that acts on a function to produce another function. In some special cases the operator gives back the original function simply multiplied by a constant. Note that the Schrödinger equation corresponds to such a special case. In the Schrödinger equation the operator $\hat{H}$, called the Hamiltonian, acts to give back the wave function multiplied by the constant E, which represents the total energy of the atom (the sum of the potential energy due to the attraction between the proton and the electron, and the kinetic energy of the moving electron). When this equation is analyzed, many solutions are found. Each solution consists of a wave function ψ that is characterized by a particular value of E. A specific wave function for a given electron is often called an **orbital.** It is important to recognize that Schrödinger could not be sure that treating an electron as a wave makes any sense: The test would be whether or not the model could correctly fit the experimental data for hydrogen and other atoms.

To illustrate the most important ideas of the wave mechanical (quantum mechanical) model of the atom, we will first concentrate on the wave function corresponding to the lowest energy for the hydrogen atom. This wave function is called the 1*s* orbital. The first point of interest is the meaning of the word *orbital.* One thing is clear: an orbital is *not* a Bohr orbit. The electron in the hydrogen 1*s* orbital is not moving around the nucleus in a circular orbit. How, then, is the electron moving? The answer is somewhat surprising: *we do not know.* The wave function gives us no information about the movements of the electron. This observation is somewhat disturbing. When we solve problems involving the motions of particles in the macroscopic world, we are able to predict their trajectories. For example, when two billiard balls with known velocities collide, we can predict their motions after the collision. However, we cannot predict the electron's motion using the 1*s* orbital function. Does this mean that the theory is useless? Not necessarily: we have already learned that an electron does not behave much like a billiard ball, so we must examine the situation closely before we discard the theory.

The uncertainty principle is sometimes called the indeterminacy principle.

Werner Heisenberg, who was also involved in the development of the quantum mechanical model for the atom, discovered a very important principle in 1927 that helps us to understand the meaning of orbitals—the **Heisenberg uncertainty principle.** Heisenberg's mathematical analysis led him to a surprising conclusion: *there is a fundamental limitation to just how precisely we can know both the position and the momentum of a particle at a given time.* Stated mathematically, the uncertainty principle is

$$\Delta x \cdot \Delta p \geq \frac{\hbar}{2}$$

where Δx is the uncertainty in a particle's position, Δp is the uncertainty in a particle's momentum, and $\hbar$ is Planck's constant divided by 2π ($\hbar = h/2\pi$). Thus the minimum uncertainty in the product $\Delta x \cdot \Delta p$ is $h/4\pi$. This relationship means that the more precisely we know a particle's position, the less precisely we can know its momentum, and vice versa. This limitation is so small for large particles such as baseballs or billiard balls that it is unnoticed. However, for a small particle such as the electron, the limitation becomes quite important. Applied to the electron, the uncertainty principle implies that we cannot know the exact path of the electron as it moves around the nucleus. It is therefore not appropriate to assume that the electron is moving around the nucleus in a well-defined orbit as in the Bohr model.

EXAMPLE 12.5

The hydrogen atom has a radius on the order of 0.05 nm. Assuming that we know the position of an electron to an accuracy of 1% of the hydrogen radius, calculate the uncertainty in the velocity of the electron, using the Heisenberg uncertainty principle. Then compare this value with the uncertainty in the velocity of a ball of mass 0.2 kg and radius 0.05 m whose position is known to an accuracy of 1% of its radius.

Solution

From Heisenberg's uncertainty principle the smallest possible uncertainty in the product $\Delta x \cdot \Delta p$ is $\hbar/2$; that is,

$$\Delta x \cdot \Delta p = \frac{\hbar}{2} = \frac{h}{4\pi}$$

For the electron the uncertainty in position (Δx) is 1% of 0.05 nm, or

$$\Delta x = (0.01)(0.05 \text{ nm}) = 5 \times 10^{-4} \text{ nm}$$

Converting to meters gives

$$5 \times 10^{-4} \text{ nm} \times \frac{10^{-9} \text{ m}}{1 \text{ nm}} = 5 \times 10^{-13} \text{ m}$$

The values of the constants are

$$m = \text{mass of the electron} = 9.11 \times 10^{-31} \text{ kg}$$

$$h = 6.626 \times 10^{-34} \text{ J s} = 6.626 \times 10^{-34} \frac{\text{kg m}^2}{\text{s}}$$

$$\pi = 3.14$$

We can now solve for the uncertainty in momentum:

$$\Delta p = \frac{\hbar}{2 \cdot \Delta x} = \frac{h}{4\pi \cdot \Delta x} = \frac{6.626 \times 10^{-34} \frac{\text{kg m}^2}{\text{s}}}{4(3.14)(5 \times 10^{-13} \text{ m})}$$

$$= 1.05 \times 10^{-22} \text{ kg m/s} \quad \text{(keeping extra significant figures)}$$

Recalling that $p = mv$ and assuming that the electron mass is constant (ignoring any relativistic corrections), we have

$$\Delta p = \Delta(mv) = m\Delta v$$

and the uncertainty in velocity is

$$\Delta v = \frac{\Delta p}{m} = \frac{1.05 \times 10^{-22} \text{ kg m/s}}{9.11 \times 10^{-31} \text{ kg}} = 1.15 \times 10^{8} \text{ m/s} = 1 \times 10^{8} \text{ m/s}$$

Thus if we know the electron's position with a minimum uncertainty of 5×10^{-13} m, the uncertainty in the electron's velocity is at least 1×10^{8} m/s. This is a very large number; in fact, it is the same magnitude as the speed of light (3×10^{8} m/s). At this level of uncertainty we have virtually no idea of the velocity of the electron.

For the ball the uncertainty in position (Δx) is 1% of 0.05 m, or 5×10^{-4} m. Thus the minimum uncertainty in velocity is

$$\Delta v = \frac{\Delta p}{m} = \frac{h}{\Delta x \cdot m \cdot 4\pi} = \frac{6.626 \times 10^{-34} \frac{\text{kg m}^2}{\text{s}}}{(5 \times 10^{-4}\ \text{m})(0.2\ \text{kg})(4)(3.14)}$$
$$= 5 \times 10^{-31}\ \text{m/s}$$

This means there is a very small (undetectable) uncertainty in our measurements of the speed of a ball. Note that this uncertainty is not due to the limitations of measuring instruments; Δv is an *inherent* uncertainty.

Thus the uncertainty principle is negligible in the world of macroscopic objects but is very important for objects with small masses, such as the electron.

12.6 The Particle in a Box

Overview

The Schrödinger wave equation, $\hat{H}\psi = E\psi$, lies at the heart of the quantum mechanical description of atoms. Recall from the above discussion that $\hat{H}$ represents an operator (the Hamiltonian) that "extracts" the total energy E (the sum of the potential and kinetic energies) from the wave function. The wave function ψ depends on the x, y, and z coordinates of the electron's position in space.

Note that the Schrödinger equation requires that when ψ is operated on by $\hat{H}$, the result is ψ multiplied by a constant, E, which represents the total energy of the particular state described by ψ. As we will see, there are many possible solutions to the Schrödinger equation for a given system. For example, for the hydrogen atom there are many functions that satisfy the Schrödinger equation, each one corresponding to a particular energy for hydrogen's electron. Each of these specific wave functions for the hydrogen atom is called an orbital.

Although the detailed solution of the Schrödinger equation for the hydrogen atom is not appropriate in this text, we will illustrate some of the properties of wave mechanics and wave functions by using the wave equation to describe a very simple, hypothetical system commonly called "the particle in a box," a situation where a particle is trapped in a one-dimensional box that has infinitely high "sides." It is important to recognize that this situation is not an accurate physical model for the hydrogen atom. That is, the hydrogen atom is really not much like this particle in a box. The reasons for treating the particle in a box are that (1) it illustrates the mathematics of wave mechanics, (2) it gives an indication of the characteristics of wave functions, and (3) it shows how energy quantization arises. Thus this treatment of a particle in a box illustrates the "flavor" of the wave mechanical description of the hydrogen atom, but it should not be taken to be an accurate representation of the hydrogen atom itself.

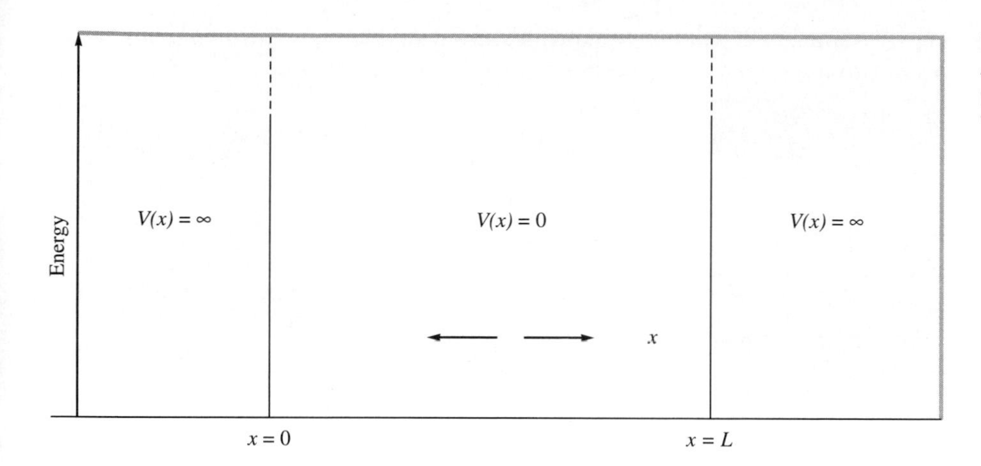

Figure 12.13
A schematic representation of a particle in a one-dimensional box with infinitely high potential walls.

The Particle in a Box as a Model

Consider a particle with mass m that is free to move back and forth along one dimension (we arbitrarily choose x) between the values $x = 0$ and $x = L$ (that is, we are considering a one-dimensional "box" of size L meters). We will assume that the potential energy $V(x)$ of the particle is zero at all points along its path, except at the end points $x = 0$ and $x = L$, where $V(x)$ is infinitely large. In effect, we have a repulsive barrier of infinite strength at each end of the box. Thus the particle is trapped in a one-dimensional box with impenetrable walls (see Fig. 12.13).

As we mentioned before, the Schrödinger equation contains the energy operator $\hat{H}$. In this case, since the potential energy is zero inside the box, the only energy possible is the kinetic energy of the particle as it moves back and forth along the x axis. The operator for this kinetic energy is

$$-\frac{\hbar^2}{2m}\frac{d^2}{dx^2}$$

where $\hbar$ is Planck's constant divided by 2π, m is the mass of the particle, and d^2/dx^2 is the second derivative with respect to x. The form of this operator comes from the description of waves in classical physics. Inserting this operator into the Schrödinger equation $\hat{H}\psi = E\psi$ gives

$$-\frac{\hbar^2}{2m}\frac{d^2\psi}{dx^2} = E\psi$$

where ψ is a function of x ($\psi(x)$). We can rearrange this equation to give

$$\frac{d^2\psi}{dx^2} = -\frac{2mE}{\hbar^2}\psi$$

Our goal is to find specific functions $\psi(x)$ that satisfy this equation. Notice that the solutions to this equation are functions such that $d^2\psi/dx^2 =$ (constant)ψ. That is, each solution must be a function whose second derivative has the same form as the original function. One function that behaves this way is the sine function. For example, consider the function $A \sin(kx)$, where A and k

are constants. We will now take the second derivative of this function with respect to x:

$$\frac{d^2}{dx^2}(A \sin kx) = A\frac{d}{dx}\left(\frac{d \sin kx}{dx}\right) = A\frac{d}{dx}(k \cos kx)$$
$$= Ak\left(\frac{d \cos kx}{dx}\right) = Ak(-k \sin kx)$$
$$= -Ak^2 \sin kx = -k^2 A \sin kx$$

Thus we have shown that

$$\frac{d^2(A \sin kx)}{dx^2} = -k^2(A \sin kx)$$

This is just the type of function that will satisfy the Schrödinger equation for the particle in a box. In fact, when we compare the general form of the Schrödinger equation

$$\frac{d^2\psi}{dx^2} = -\frac{2mE}{\hbar^2}\psi$$

with

$$\frac{d^2(A \sin kx)}{dx^2} = -k^2(A \sin kx)$$

we see that

$$-k^2 = -\frac{2mE}{\hbar^2}$$

which can be rearranged to give an expression for energy:

$$E = \frac{\hbar^2 k^2}{2m}$$

What does this equation mean? We have simply specified that A and k are constants. What values can these constants have? Note that if they could assume any values, this equation would lead to an infinite number of possible energies—that is, a continuous distribution of energy levels. However, this is not correct. For reasons we will discuss presently, we find that only certain energies are allowed. That is, this system is quantized. In fact, the ability of wave mechanics to account for the observed (but initially unexpected) quantization of energy in nature is one of the most important factors in convincing us that it may be a correct description of the properties of matter.

Quantization enters the wave mechanical description of the particle in a box via the boundary conditions. Boundary conditions arise from the physical requirements of natural systems. That is, we must insist that our descriptions of natural systems make physical sense. For example, assume that in describing an aqueous solution containing an acid, we arrive at the expression $[H^+]^2 = 4.0 \times 10^{-8} M^2$. The solutions to this expression are

$$[H^+] = 2.0 \times 10^{-4} M \quad \text{and} \quad [H^+] = -2.0 \times 10^{-4} M$$

In doing such a problem, we automatically reject the second possibility because there is no physical meaning for a negative concentration. What we have done here is apply a type of boundary condition to this situation.

The boundary conditions for the particle in a box enforce the following facts:

1. The particle cannot be outside the box—it is bound inside the box.
2. In a given state the total probability of finding the particle in the box must be 1.
3. The wave function must be continuous.

We have seen that the function $\psi = A\sin(kx)$ satisfies the Schrödinger equation $\hat{H}\psi = E\psi$. We will now define the constants k and A so that this function also satisfies the boundary conditions based on the three constraints listed above. Because the particle must stay inside the box and because the wave function must be continuous, the value of $\psi(x)$ must be zero at each wall. That is,

$$\psi(0) = 0 \qquad \text{and} \qquad \psi(L) = 0$$

Recall that the sine function is zero at angles of 0°, 180° (π radians), 360° (2π radians), and so on. Thus the function $A\sin kx$ is automatically zero when $x = 0$.

The requirement that the wave function must also be zero at the other wall, which can be stated as $\psi(L) = A\sin(kL) = 0$, means that k is limited to the values of $n\pi/L$, where n is an integer (1, 2, 3, . . .). That is,

$$\psi(x) = A\sin\left(\frac{n\pi}{L}x\right)$$

then

$$\psi(L) = A\sin\left(\frac{n\pi}{L}\cdot L\right) = A\sin(n\pi) = 0$$

To assign the value of the constant A, we need to introduce a new idea. In the application of wave mechanics to the description of matter, scientists have learned to associate the square of the wave function with probability. As we will discuss in more detail below, this means that the square of the wave function evaluated at a given point gives the relative probability of finding a particle near that point. This concept is relevant to the boundary conditions for the particle in a box because the total probability in a given state must be 1. To be more precise, the probability of finding the particle on a segment of the x axis of length dx surrounding point x is $\psi^2(x)\,dx$. Because there is one particle in the box the sum of all of these probabilities along the x axis from $x = 0$ to $x = L$ must be 1. We sum these probabilities over the length of the box (from $x = 0$ to $x = L$) by integration from $x = 0$ to $x = L$:

$$\begin{array}{l}\text{Total probability of finding}\\ \text{the particle in the box}\end{array} = \int_0^L \psi^2(x)\,dx = 1$$

Substituting $\psi(x) = A\sin[(n\pi/L)x]$, we have

$$\int_0^L \psi^2(x)\,dx = \int_0^L A^2\sin^2\left(\frac{n\pi}{L}x\right)dx = 1$$

or

$$\int_0^L \sin^2\left(\frac{n\pi}{L}x\right)dx = \frac{1}{A^2}$$

The value of the integral is $L/2$, which means that

$$\frac{L}{2} = \frac{1}{A^2} \quad \text{and} \quad A = \sqrt{\frac{2}{L}}$$

Now that we know the allowed values of k and A, we can specify the wave function for the particle in a one-dimensional box as

$$\psi(x) = \sqrt{\frac{2}{L}} \sin\left(\frac{n\pi}{L} x\right)$$

We can also substitute the value of k into the expression for energy:

$$E = \frac{\hbar^2 k^2}{2m} = \frac{\hbar^2 (n\pi/L)^2}{2m}$$

Substituting $\hbar = h/2\pi$ gives

$$E = \frac{n^2 h^2}{8mL^2} \quad \text{where} \quad n = 1, 2, 3, 4, \ldots$$

Note that this analysis leads to a series of solutions to the Schrödinger equation, where each function corresponds to a given energy state:

n	Function	Energy
1	$\psi_1 = \sqrt{\frac{2}{L}} \sin\left(\frac{\pi}{L} x\right)$	$E_1 = \frac{h^2}{8mL^2}$
2	$\psi_2 = \sqrt{\frac{2}{L}} \sin\left(\frac{2\pi}{L} x\right)$	$E_2 = \frac{4h^2}{8mL^2} = \frac{h^2}{2mL^2}$
3	$\psi_3 = \sqrt{\frac{2}{L}} \sin\left(\frac{3\pi}{L} x\right)$	$E_3 = \frac{9h^2}{8mL^2}$
4	$\psi_4 = \sqrt{\frac{2}{L}} \sin\left(\frac{4\pi}{L} x\right)$	$E_4 = \frac{16h^2}{8mL^2} = \frac{2h^2}{mL^2}$
⋮	⋮	⋮

Notice something very important about these results. The application of the boundary conditions has led to a series of *quantized* energy levels. That is, only certain energies are allowed for the particle bound in the box. This result fits very nicely with the experimental evidence, such as the hydrogen emission spectrum, that nature does not allow continuous energy levels for *bound* systems, as classical physics had led us to expect. Note that the energies are quantized, because the boundary conditions require that n assume only integer values. Consequently, we call n the quantum number for this system.

We can diagram the solutions to the particle-in-a-box problem conveniently by showing a plot of the wave function that corresponds to each energy level. The energy level, wave function, and probability distribution are shown in Fig. 12.14 for the first three levels.

Note that each wave function goes to zero at the edges of the box, as required by the boundary conditions. Another way to say this is that the stand-

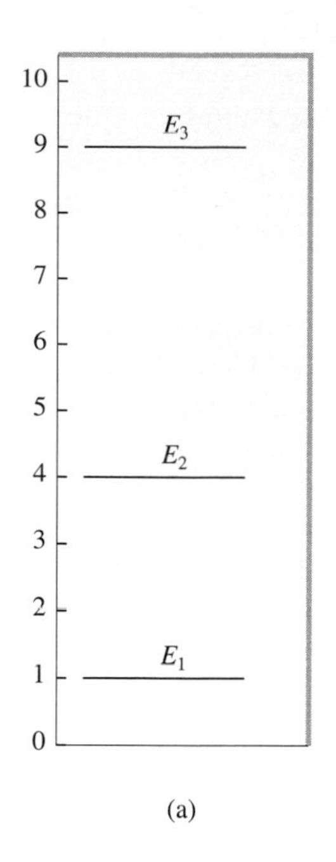

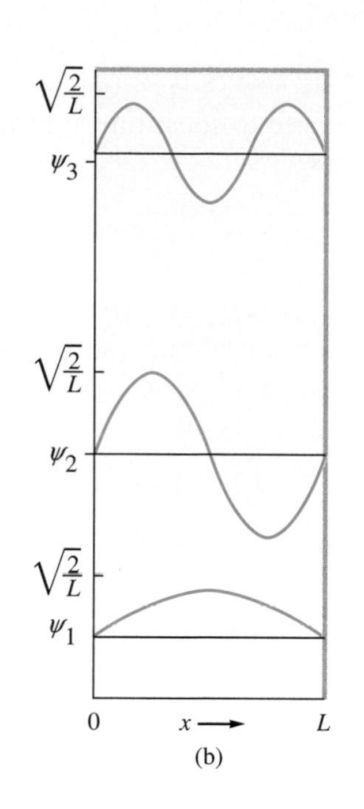

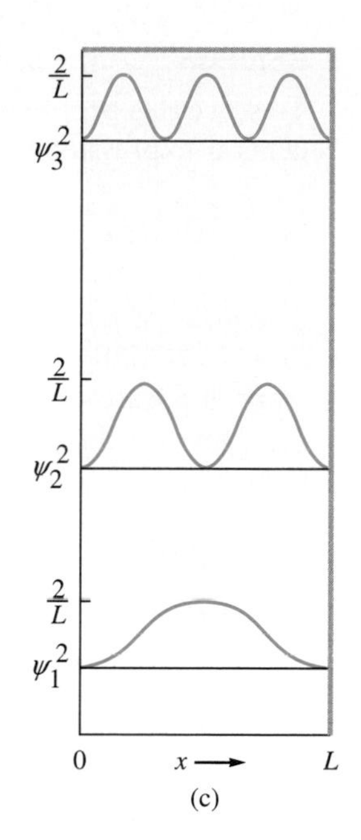

Figure 12.14
(a) The first three energy levels for a particle in a one-dimensional box in increments of $h^2/(8mL^2)$. (b) The wave functions for the first three levels plotted as a function of x. Note that the maximum value is $\sqrt{2/L}$ in each case. (c) The square of the wave functions for the first three levels plotted as a function of x. Note that the maximum value is $2/L$ in each case.

ing waves that represent the particle must have wavelengths such that an *integral number of half-wavelengths exactly equals the size of the box*. Waves with any other wavelengths could not exist because they would destructively interfere over time. Also, note from Figure 12.14 that the probability distribution is significantly different for the three levels. For $n = 1$ (the lowest energy or ground state) the particle is most likely to be found near the center of the box. In contrast, for $n = 2$ the particle has zero probability of being found in the center of the box. This zero point is called a node. Notice that the number of nodes increases with n.

Another interesting characteristic of the particle in a box is that the particle cannot have zero energy (that is, n cannot equal zero). For example, if n were equal to zero, ψ_0 would be zero everywhere in the box ($\sin 0 = 0$). This would mean that ψ_0^2 would also be zero. In this case there could be no particle in the box, which contradicts the boundary conditions. This fact that the particle must have a nonzero energy in its ground state is a characteristic of all particles with quantized energies. In addition, for the particle in a box a value of zero for the energy would mean that the particle was sitting still (zero kinetic energy). This condition would violate the uncertainty principle, because we would simultaneously know the exact values of the momentum (zero) and the position of the particle. For similar reasons all quantized particles must possess a minimum energy, often called the *zero-point energy*.

EXAMPLE 12.6

Assume that an electron is confined to a one-dimensional box 1.50 nm in length. Calculate the lowest three energy levels for this electron, and calculate the wavelength of light necessary to promote the electron from the ground state to the first excited state.

Solution

To solve this problem, we need to substitute appropriate values into the general expression for energy:

$$E = \frac{n^2h^2}{8mL^2}$$

The mass of an electron (m) is 9.11×10^{-31} kg; the dimension of the box (L) is 1.50 nm, or 1.50×10^{-9} m; and the value of Planck's constant is 6.626×10^{-34} J s.

For $n = 1$ we get

$$E_1 = \frac{(1)^2(6.626 \times 10^{-34}\ \text{J s})^2}{(8)(9.11 \times 10^{-31}\ \text{kg})(1.50 \times 10^{-9}\ \text{m})^2} = 2.68 \times 10^{-20}\ \text{J}$$

Similarly, for $n = 2$ we get

$$E_2 = 1.07 \times 10^{-19}\ \text{J}$$

And for $n = 3$ we get

$$E_3 = 2.41 \times 10^{-19}\ \text{J}$$

Note that since

$$E_n = n^2 \frac{h^2}{8mL^2} = n^2E_1$$

then
$$E_2 = (2)^2 \frac{h^2}{8mL^2} = 4E_1 \quad \text{and} \quad E_3 = 9E_1$$

To calculate the wavelength of light necessary to excite the electron from level 1 to level 2 (the first *excited* state), we first need to obtain the energy difference between the two levels:

$$\Delta E = E_2 - E_1 = (n_2^2 - n_1^2) \frac{h^2}{8mL^2}$$

$$= (3)(2.68 \times 10^{-20}\ \text{J}) = 8.04 \times 10^{-20}\ \text{J}$$

Then we find the wavelength required from the equation

$$\Delta E = \frac{hc}{\lambda}$$

Inserting the appropriate values gives

$$\lambda = \frac{hc}{\Delta E} = \frac{(6.626 \times 10^{-34}\ \text{J s})(2.9979 \times 10^8\ \text{m/s})}{8.04 \times 10^{-20}\ \text{J}}$$

$$= 2.47 \times 10^{-6}\ \text{m} = 2470\ \text{nm}$$

12.7 The Wave Equation for the Hydrogen Atom

Unlike the particle in a one-dimensional box, the electron of the hydrogen atom moves in three dimensions and has potential energy, due to its attraction to the positive nucleus at the atom's center. These differences can be easily accounted for by including the second derivatives with respect to all three of the Cartesian coordinates and by inserting a term that specifies the dependence of the electron's potential energy on its position in space.

Because it is more convenient mathematically, the coordinate system is changed from Cartesian to spherical polar coordinates (see Fig. 12.15) before the Schrödinger equation is solved. In spherical polar coordinates a given point in space, specified by values of the Cartesian coordinates x, y, and z, is described by specific values of r, θ, and ϕ.

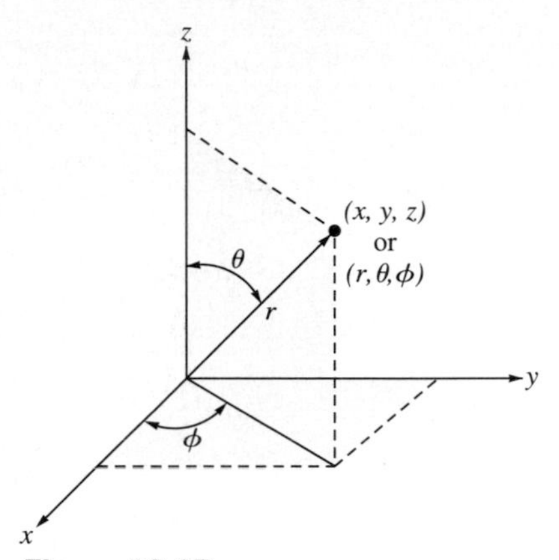

Figure 12.15
The spherical polar coordinate system.

In the spherical polar coordinate system the wave function $\psi(r, \theta, \phi)$ can be written as a product of a function depending on r, one depending on θ, and one depending on ϕ:

$$\psi(r, \theta, \phi) = R(r)\Theta(\theta)\Phi(\phi)$$

This separation of variables allows an exact solution to the Schrödinger equation

$$\hat{H}\psi = E\psi$$

for the hydrogen atom.

In spherical polar coordinates the potential energy (in cgs units) of the electron is

$$V(r) = -\frac{(Ze)(e)}{r}$$

To convert the potential energy from cgs units to SI units (joules), the expression shown must be multiplied $\frac{1}{4}\pi\epsilon_0$, where ϵ_0 (the permittivity of the vacuum) is 8.854×10^{-12} C^2/J m.

where Ze represents the nuclear charge ($Z = 1$ for the hydrogen atom). As with the particle in a box, when the Schrödinger equation for the hydrogen atom is solved and the boundary conditions are applied, a series of wave functions is obtained, each function corresponding to a particular energy. In contrast to the particle in a one-dimensional box, where one quantum number emerges from the mathematics, the three-dimensional hydrogen atom gives rise to three quantum numbers.

The boundary conditions, which differ in some important aspects from those of the particle in the box because of the very different nature of the physical system, will not be discussed here.

The conventional symbols for these quantum numbers are as follows:

n	the *principal quantum number*
ℓ	the *angular momentum quantum number*
m_ℓ	the *magnetic quantum number*

We will have more to say in succeeding sections about what values these quantum numbers can assume and their physical meanings.

The mathematics of wave mechanics leads to the following expression for the allowed energies of hydrogen's electron:

$$E_n = -\frac{Z^2}{n^2}\left(\frac{me^4}{8\epsilon_0^2h^2}\right) = -2.178 \times 10^{-18}\ \mathrm{J}\left(\frac{Z^2}{n^2}\right)$$

where $Z = 1$ for hydrogen and where n can assume only integer values (1, 2, 3, . . .). Several characteristics of this equation are worth emphasizing. First, note that the energy of the electron depends only on the principal quantum number (this is true only for one-electron species). Also note that because n is restricted to integer values, hydrogen's electron can assume only discrete energy values—the energy levels are quantized. Finally, note that this is exactly the same equation for energy as obtained in the Bohr model.

So that you have an idea of what they look like, the first few wave functions for hydrogen are shown in Table 12.1, along with the three quantum numbers n, ℓ, and m_ℓ.

When we solve the Schrödinger equation for the hydrogen atom, some of the solutions contain complex numbers (that is, they contain $i = \sqrt{-1}$).

TABLE 12.1 **Solutions of the Schrödinger Wave Equation for a One-Electron Atom**

n	ℓ	m_ℓ	Orbital	Solution
1	0	0	$1s$	$\psi_{1s} = \frac{1}{\sqrt{\pi}}\left(\frac{Z}{a_0}\right)^{3/2} e^{-\sigma}$
2	0	0	$2s$	$\psi_{2s} = \frac{1}{4\sqrt{2\pi}}\left(\frac{Z}{a_0}\right)^{3/2} (2 - \sigma)e^{-\sigma/2}$
2	1	0	$2p_z$	$\psi_{2p_z} = \frac{1}{4\sqrt{2\pi}}\left(\frac{Z}{a_0}\right)^{3/2} \sigma e^{-\sigma/2} \cos\theta$
2	1	± 1	$2p_x$	$\psi_{2p_x} = \frac{1}{4\sqrt{2\pi}}\left(\frac{Z}{a_0}\right)^{3/2} \sigma e^{-\sigma/2} \sin\theta \cos\phi$
			$2p_y$	$\psi_{2p_y} = \frac{1}{4\sqrt{2\pi}}\left(\frac{Z}{a_0}\right)^{3/2} \sigma e^{-\sigma/2} \sin\theta \sin\phi$
3	0	0	$3s$	$\psi_{3s} = \frac{1}{81\sqrt{3\pi}}\left(\frac{Z}{a_0}\right)^{3/2} (27 - 18\sigma + 2\sigma^2)e^{-\sigma/3}$
3	1	0	$3p_z$	$\psi_{3p_z} = \frac{\sqrt{2}}{81\sqrt{\pi}}\left(\frac{Z}{a_0}\right)^{3/2} (6\sigma - \sigma^2)e^{-\sigma/3} \cos\theta$
3	1	± 1	$3p_x$	$\psi_{3p_x} = \frac{\sqrt{2}}{81\sqrt{\pi}}\left(\frac{Z}{a_0}\right)^{3/2} (6\sigma - \sigma^2)e^{-\sigma/3} \sin\theta \cos\phi$
			$3p_y$	$\psi_{3p_y} = \frac{\sqrt{2}}{81\sqrt{\pi}}\left(\frac{Z}{a_0}\right)^{3/2} (6\sigma - \sigma^2)e^{-\sigma/3} \sin\theta \sin\phi$
3	2	0	$3d_{z^2}$	$\psi_{3d_z^2} = \frac{1}{81\sqrt{6\pi}}\left(\frac{Z}{a_0}\right)^{3/2} \sigma^2 e^{-\sigma/3} (3\cos^2\theta - 1)$
3	2	± 1	$3d_{xz}$	$\psi_{3d_{xz}} = \frac{\sqrt{2}}{81\sqrt{\pi}}\left(\frac{Z}{a_0}\right)^{3/2} \sigma^2 e^{-\sigma/3} \sin\theta \cos\theta \cos\phi$
			$3d_{yz}$	$\psi_{3d_{yz}} = \frac{\sqrt{2}}{81\sqrt{\pi}}\left(\frac{Z}{a_0}\right)^{3/2} \sigma^2 e^{-\sigma/3} \sin\theta \cos\theta \sin\phi$
3	2	± 2	$3d_{xy}$	$\psi_{3d_{xy}} = \frac{1}{81\sqrt{2\pi}}\left(\frac{Z}{a_0}\right)^{3/2} \sigma^2 e^{-\sigma/3} \sin^2\theta \sin 2\phi$
			$3d_{x^2-y^2}$	$\psi_{3d_{x^2-y^2}} = \frac{1}{81\sqrt{2\pi}}\left(\frac{Z}{a_0}\right)^{3/2} \sigma^2 e^{-\sigma/3} \sin^2\theta \cos 2\phi$

Note: $\sigma = Zr/a_0$ where $Z = 1$ for hydrogen; $a_0 = \epsilon_0 h^2/\pi m e^2 = 5.29 \times 10^{-11}$ m.

Because it is more convenient physically to deal with orbitals that contain only real numbers, the complex orbitals are usually combined (added and subtracted) to remove the complex portions. For example, the p_x and p_y orbitals shown in Table 12.1 are combinations of the complex orbitals that correspond to values of m_ℓ of $+1$ and -1. These orbitals are indicated with a brace in Table 12.1. The last four d orbitals listed are also obtained by combination of complex orbitals, as indicated by braces in Table 12.1.

(a)

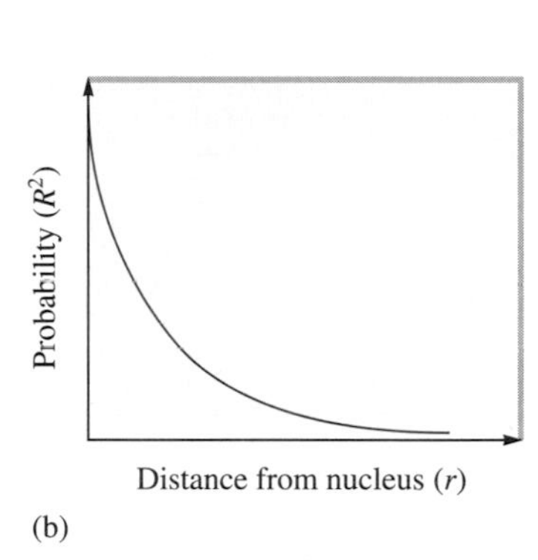

Figure 12.16
(a) The probability distribution for the hydrogen 1s orbital in three-dimensional space. (b) The probability of finding the electron at points along a line drawn outward from the nucleus in any direction for the hydrogen 1s orbital.

12.8 The Physical Meaning of a Wave Function

Now that we have examined some of the mathematical details of the quantum mechanical treatment of the hydrogen atom, we need to consider what it all means. What is a wave function, and what does it tell us about the electron to which it applies? First, a warning: there is always danger in taking a mathematical description of nature and using our human experiences to interpret it. Although our attempts to attach physical significance to mathematical descriptions are quite useful to us as we try to understand how nature operates, they must be viewed with caution. Simple pictorial models of a particular natural phenomenon always oversimplify the phenomenon and should not be taken too literally. With that caveat we will proceed to try to picture what the "quantum mechanical atom" is like.

Recall that the uncertainty principle indicates that there is no way of knowing the detailed movements of the electron in a hydrogen atom. Given this severe limitation, what then is the physical meaning of a wave function for an electron? Although the function itself has no easily visualized meaning, as we mentioned in the treatment of the particle in a box, the square of the wave function does have a physical significance. *The square of the function evaluated at a particular point in space indicates the probability of finding an electron near that point.* For example, suppose we have two positions in space, one defined by the coordinates r_1, θ_1, and ϕ_1, and the other by the coordinates r_2, θ_2, and ϕ_2. The relative probability of finding the electron near positions 1 and 2 is determined by substituting the values of r, θ, and ϕ for the two positions into the wave function, squaring the function value, and computing the following ratio:

The square of the function here means the square of the magnitude, $|\psi|^2$. This distinction is important when orbitals with complex numbers are being considered: $|\psi|^2$ = (real part)^2 + (imaginary part)^2.

$$\frac{[\psi(r_1, \theta_1, \phi_1)]^2\, dv}{[\psi(r_2, \theta_2, \phi_2)]^2\, dv} = \frac{N_1}{N_2}$$

The quotient N_1/N_2 is the ratio of the probabilities of finding the electron in the infinitesimally small volume elements dv around points 1 and 2. For example, if the value of the ratio N_1/N_2 is 100, the electron is 100 times more likely to be found at position 1 than at position 2. The model gives no information concerning when the electron will be at either position or how it moves between the positions. This vagueness is consistent with the concept of the Heisenberg uncertainty principle.

The square of the wave function is most conveniently represented as a **probability distribution,** in which the intensity of color is used to indicate the probability value at a given point in space. The probability distribution for the hydrogen 1s orbital is shown in Fig. 12.16(a). The best way to think about this

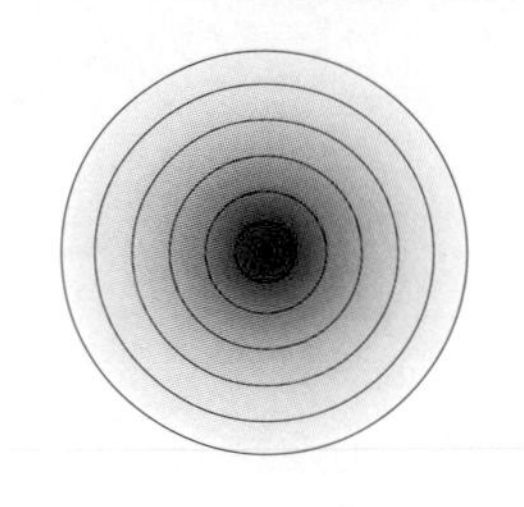

(a)

Radial probability ($4\pi r^2 R^2$)

Distance from nucleus (r)

(b)

Figure 12.17
(a) Cross section of the hydrogen 1s orbital probability distribution divided into successive thin spherical shells. (b) The radial probability distribution. A plot of the total probability of finding the electron in each thin spherical shell as a function of distance from the nucleus.

diagram is as a three-dimensional time exposure, with the electron as a tiny moving light. The more times the electron visits a particular point, the darker the negative becomes. Thus the darkness (intensity) of a point indicates the probability of finding an electron at that position. This diagram is sometimes known as an *electron density map;* electron density and electron probability mean the same thing.

Another way of representing the electron probability distribution for the 1*s* orbital is to calculate the probability at points along a line drawn outward in any direction from the nucleus. The result is shown in Fig. 12.16(b), where R^2 (the square of the radial part—the part that depends on r—of the 1*s* orbital) is plotted versus r. Note that the probability of finding the electron at a particular position is greatest close to the nucleus and that it drops off rapidly as the distance from the nucleus increases.

We are also interested in knowing the *total* probability of finding the electron in the hydrogen atom at a particular *distance* from the nucleus. Imagine that the space around the hydrogen nucleus is made up of a series of thin spherical shells (rather like layers in an onion), as shown in Fig. 12.17(a). When the total probability of finding the electron in each spherical shell is plotted versus the distance from the nucleus, the plot in Fig. 12.17(b) is obtained. This graph is called the **radial probability distribution,** which is a plot of $4\pi r^2 R^2$ versus r, where R represents the radial part of the wave function.

The maximum in the curve occurs because of two opposing effects. The probability of finding an electron at a particular position is greatest near the nucleus, but the volume of the spherical shell increases with the distance from the nucleus. Therefore, as we move away from the nucleus, the probability of finding the electron at a given position decreases. However, we are summing more positions. Thus the total probability increases to a certain radius and then decreases as the electron probability at each position becomes very small. Mathematically, the maximum occurs because in the function $4\pi r^2 R^2$, r^2 increases with r while R^2 decreases with r [see Fig. 12.16(b)]. For the hydrogen 1*s* orbital the maximum radial probability (the distance at which the electron is most likely to be found) occurs at a distance of 5.29×10^{-2} nm, or 0.529 Å (angstrom), from the nucleus. Interestingly, this distance is exactly the radius of the innermost orbit in the Bohr model, and thus is called the Bohr radius, denoted by a_0. Note that in Bohr's model the electron is assumed to have a circular path and so is *always* found at this distance. In the wave mechanical model the specific electron motions are unknown; therefore this is the *most probable* distance at which the electron is found.

1 Å $= 10^{-10}$ m; the angstrom is often used as the unit for atomic radius because of its convenient size. Another convenient unit is the picometer (1 pm $= 10^{-12}$ m).

One more characteristic of the hydrogen 1*s* orbital that we must consider is its size. As we can see from Fig. 12.16, the size of this orbital cannot be precisely defined, since the probability never becomes zero (although it drops to an extremely small value at large values of r). Therefore the hydrogen 1*s* orbital has no distinct size. However, it is useful to have a definition of relative orbital size. *The normally accepted arbitrary definition of the size of the hydrogen 1s orbital is the radius of the sphere that encloses 90% of the total electron probability.* That is, 90% of the time the electron is found inside this sphere. Application of this rule to the hydrogen atom 1*s* orbital gives a sphere with radius 2.6 a_0, or 1.4×10^{-10} *m* (140 pm).

So far we have described only the lowest-energy wave function in the hydrogen atom, the 1*s* orbital. Hydrogen has many other orbitals, which are described in the next section.

12.9 The Characteristics of Hydrogen Orbitals

Quantum Numbers

As we have seen, when we solve the Schrödinger equation for the hydrogen atom, we find many wave functions (orbitals) that satisfy it. Each of these orbitals is characterized by a set of quantum numbers that arise when the boundary conditions are applied. Now we will systematically describe these quantum numbers in terms of the values they can assume and their physical meanings.

The **principal quantum number** (n), which can have integral values (1, 2, 3, . . .), is related to the size and energy of the orbital. As n increases, the orbital becomes larger and the electron spends more time farther from the nucleus. An increase in n also means higher energy, because the electron is less tightly bound to the nucleus, and the energy is less negative. The **angular momentum quantum number** (ℓ) can have integral values from 0 to $n - 1$ for each value of n. This quantum number relates to the angular momentum of an electron in a given orbital. The dependence of the wave functions on ℓ determines the shapes of the atomic orbitals. The value of ℓ for a particular orbital is commonly assigned a letter: $\ell = 0$ is called *s*; $\ell = 1$ is called *p*; $\ell = 2$ is called *d*; and $\ell = 3$ is called *f*. (See Table 12.2.) The **magnetic quantum number** (m_ℓ) can have integral values between ℓ and $-\ell$, including zero. The value of m_ℓ relates to the orientation in space of the angular momentum associated with the orbital. As we mentioned earlier, many of the familiar atomic orbitals are actually a combination of a complex orbital characterized by m_ℓ and one characterized by $-m_\ell$.

TABLE 12.2
The Angular Momentum Quantum Numbers and Corresponding Letter Symbols

Value	Letter Used
0	*s*
1	*p*
2	*d*
3	*f*
4	*g*

$$n = 1, 2, 3, \ldots$$
$$\ell = 0, 1, \ldots, (n-1)$$
$$m_\ell = -\ell, \ldots, 0, \ldots, +\ell$$

The labels *s*, *p*, *d*, and *f* are used for historical reasons. They originally referred to characteristics of lines observed in the atomic spectra: *s* (sharp), *p* (principal), *d* (diffuse), and *f* (fundamental). Beyond *f* the letters become alphabetic: *g*, *h*, ...

Number of Orbitals per Subshell
s = 1
p = 3
d = 5
f = 7
g = 9

EXAMPLE 12.7

For principal quantum level $n = 5$, determine the number of subshells (different values of ℓ) and give the designation of each.

Solution

For $n = 5$ the allowed values of ℓ run from 0 to 4 ($n - 1 = 5 - 1$). Thus the subshells and their designations are

$\ell = 0$	$\ell = 1$	$\ell = 2$	$\ell = 3$	$\ell = 4$
5*s*	5*p*	5*d*	5*f*	5*g*

TABLE 12.3 **Quantum Numbers for the First Four Levels of Orbitals in the Hydrogen Atom**

n	ℓ	Orbital Designation	m_ℓ	Number of Orbitals
1	0	1*s*	0	1
2	0	2*s*	0	1
	1	2*p*	−1, 0, +1	3
3	0	3*s*	0	1
	1	3*p*	−1, 0, 1	3
	2	3*d*	−2, −1, 0, 1, 2	5
4	0	4*s*	0	1
	1	4*p*	−1, 0, 1	3
	2	4*d*	−2, −1, 0, 1, 2	5
	3	4*f*	−3, −2, −1, 0, 1, 2, 3	7

The first four levels of orbitals in the hydrogen atom are listed with their quantum numbers in Table 12.3. Note that each set of orbitals with a given value of ℓ (sometimes called a **subshell**) is designated by giving the value of n and the letter for ℓ. Thus an orbital where $n = 2$ and $\ell = 1$ is symbolized as $2p$. There are three $2p$ orbitals, which have different orientations in space. We will describe these orbitals in the next section.

Orbital Shapes and Energies

We have seen that the meaning of an orbital is illustrated most clearly by a probability distribution. Each orbital in the hydrogen atom has a unique probability distribution. We also have seen that another means of representing an orbital is by the surface that surrounds 90% of the total electron probability. These three types of representations for the hydrogen 1*s*, 2*s*, and 3*s* orbitals are shown in Fig. 12.18. Note the characteristic spherical shape of each of the *s* orbitals. Note also that the 2*s* and 3*s* orbitals contain areas of high probability separated by areas of zero probability. These latter areas are called **nodal surfaces,** or simply **nodes.** The number of nodes increases as n increases. For *s* orbitals the number of nodes is given by $n - 1$. For our purposes, however, we will think of *s* orbitals only in terms of their overall spherical shape, which becomes larger as the value of n increases.

n value
↓
$2p_x$ ←orientation in space
↑
ℓ value

Two types of representations for the 2*p* orbitals (there are no 1*p* orbitals) are shown in Fig. 12.19. Note that the *p* orbitals are not spherical, like *s* orbitals, but have two **lobes** separated by a node at the nucleus. The *p* orbitals are labeled according to the axis of the Cartesian coordinate system along which the lobes lie. For example, the 2*p* orbital with lobes along the x axis is called the $2p_x$ orbital.

As you might expect from our discussion of the *s* orbitals, the 3*p* orbitals have a more complex probability distribution than that of the 2*p* orbitals (see

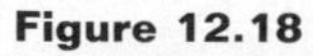

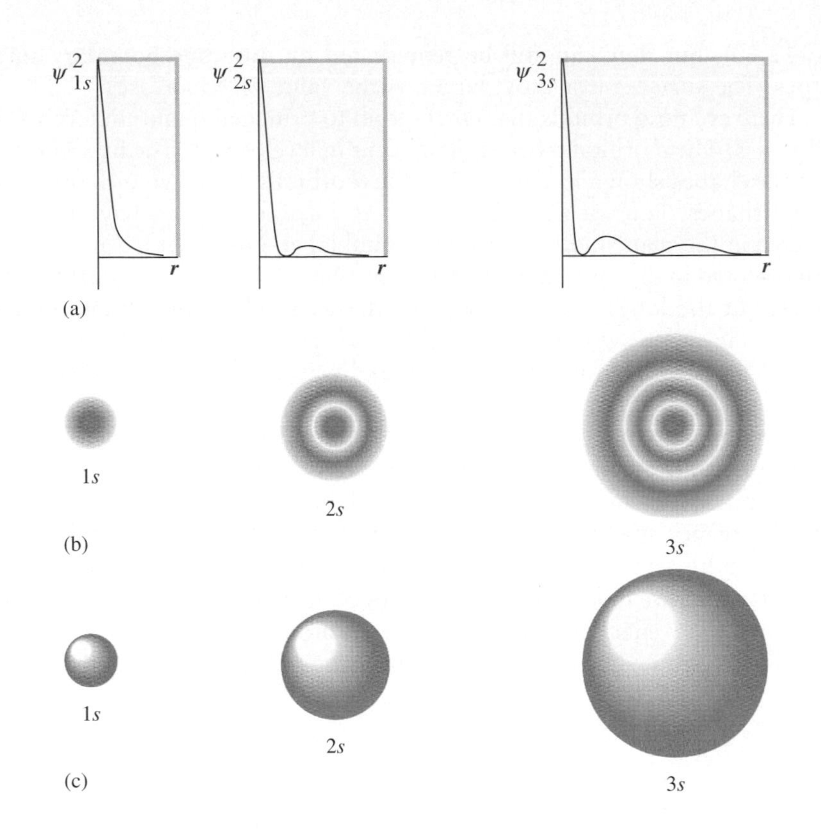

Figure 12.18
Three representations of the hydrogen 1*s*, 2*s*, and 3*s* orbitals. (a) The square of the wave function. (b) "Slices" of the three-dimensional electron density. (c) The surfaces that contain 90% of the total electron probability (the "sizes" of the orbitals).

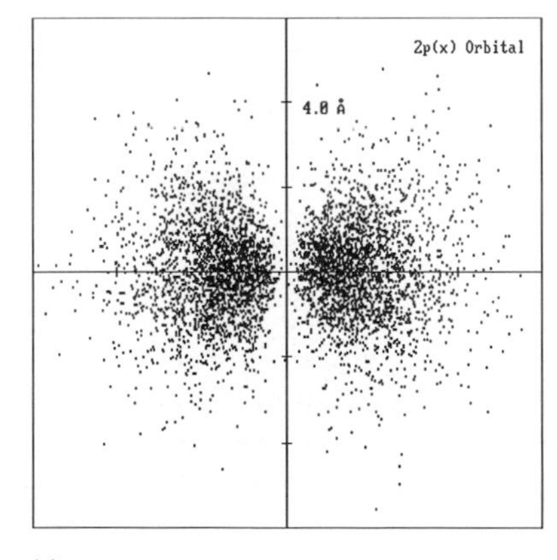

(a)

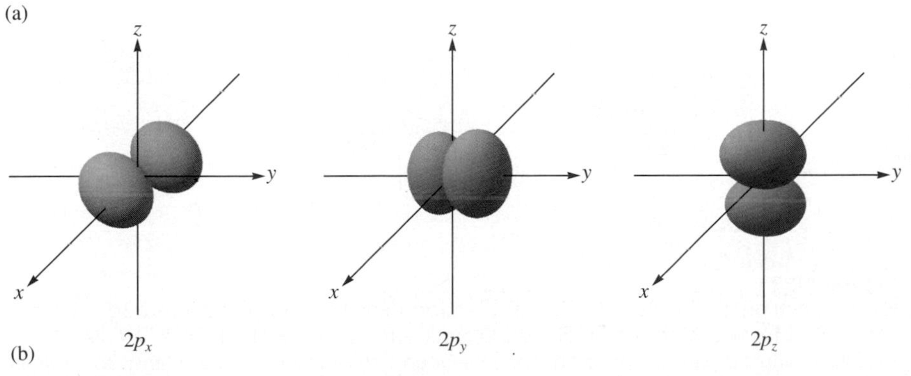

(b)

Figure 12.19
Representation of the 2*p* orbitals. (a) The electron probability distribution for a 2*p* orbital. Generated from a program by Robert Allendoerfer on Project SERAPHIM disk PC2402; printed with permission. (b) The boundary surface representations of all three 2*p* orbitals.

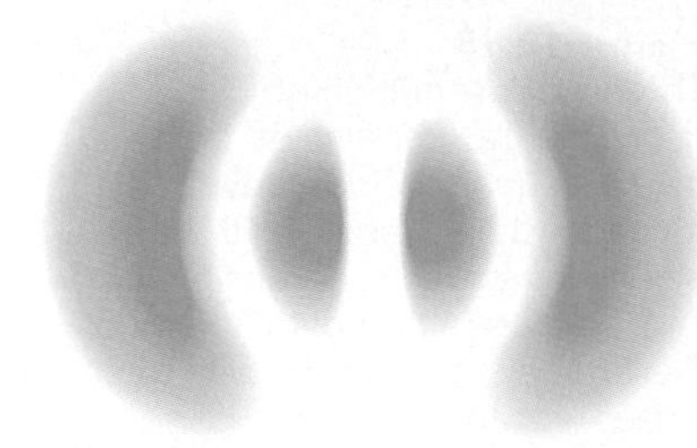

Figure 12.20
A cross section of the electron probability distribution for a 3p orbital.

Fig. 12.20), but they can still be represented by the same boundary surface shapes. The surfaces just grow larger as the value of n increases.

There are no d orbitals that correspond to principal quantum levels $n = 1$ and $n = 2$. The d orbitals ($\ell = 2$) first occur in level $n = 3$. The five 3d orbitals have the shapes shown in Fig. 12.21. The d orbitals have two different fundamental shapes. Four of the orbitals (d_{xz}, d_{yz}, d_{xy}, and $d_{x^2-y^2}$) have four lobes centered in the plane indicated in the orbital label. Note that d_{xy} and $d_{x^2-y^2}$ are both centered in the xy plane; however, the lobes of the $d_{x^2-y^2}$ lie *along* the x and y axes, but the lobes of d_{xy} lie *between* the axes. The fifth orbital, d_{z^2}, has a unique shape with two lobes along the z axis and a "belt" centered in the xy plane. The d orbitals for levels $n > 3$ look like the 3d orbitals but have larger lobes.

The f orbitals first occur in level $n = 4$ and, as might be expected, they have shapes even more complex than those of the d orbitals. Figure 12.22 shows representations of the 4f orbitals ($\ell = 3$) along with their designations. These orbitals are not involved in the bonding in any of the compounds we will consider in this text. Their shapes and labels are included here for completeness.

So far we have talked about the shapes of the hydrogen atomic orbitals but not about their energies. For the hydrogen atom the energy of a particular orbital is determined by its value of n. Thus *all* orbitals with the same value of n

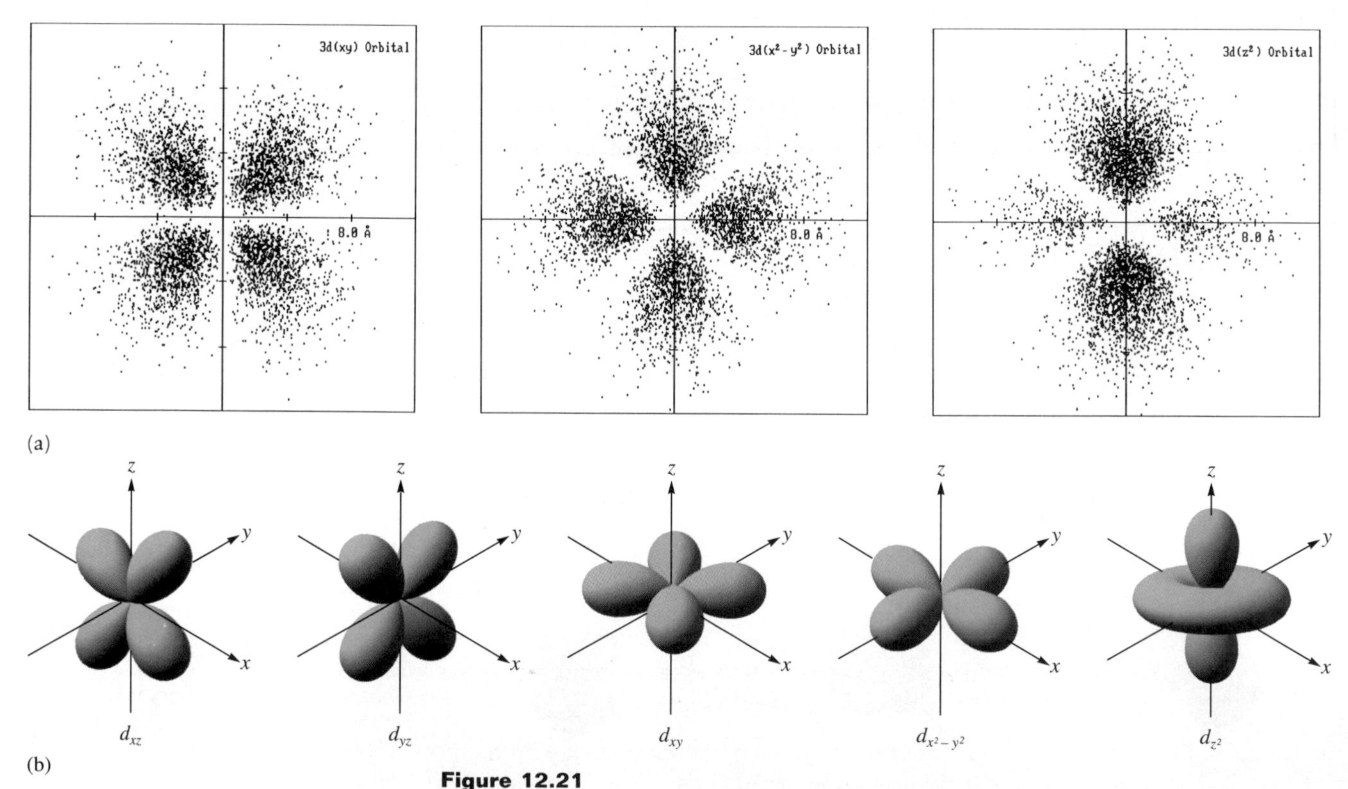

Figure 12.21
Representation of the 3d orbitals. (a) Electron density plots of selected 3d orbitals. Generated from a program by Robert Allendoerfer on Project SERAPHIM disk PC2402; printed with permission. (b) The boundary surfaces of all of the 3d orbitals.

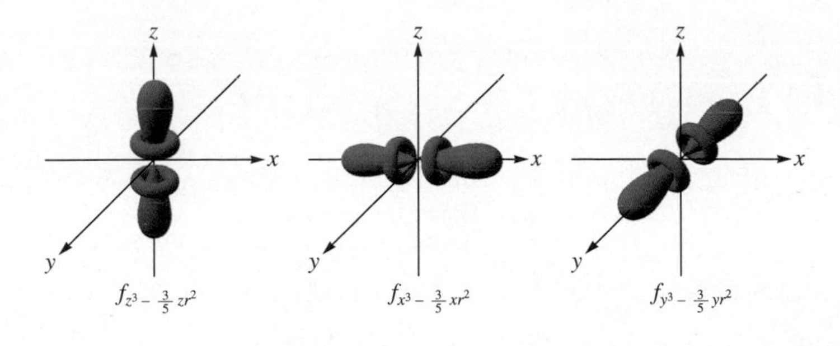

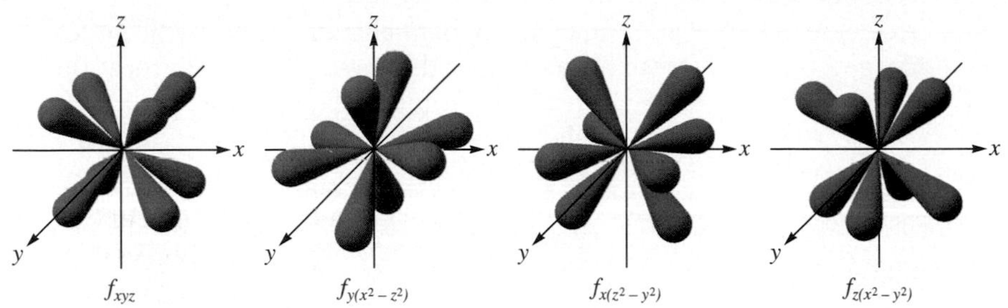

Figure 12.22
Representation of the 4f orbitals in terms of their boundary surfaces.

have the *same energy*—they are said to be **degenerate.** This feature is shown in Fig. 12.23, where the energies for the orbitals in the first three quantum levels for hydrogen are shown.

Hydrogen's single electron can occupy any of its atomic orbitals. However, in the lowest energy state, the ground state, the electron resides in the 1s orbital. If energy is put into the atom, the electron can be transferred to a higher-energy orbital, producing an excited state.

E
3s — 3p — — — 3d — — — — —
2s — 2p — — —
1s —

Figure 12.23
Orbital energy levels for the hydrogen atom.

Summary: **The Hydrogen Atom**

- **In the quantum mechanical model the electron is described as a wave. This representation leads to a series of wave functions (orbitals) that describe the possible energies and spatial distributions available to the electron.**
- **In agreement with the Heisenberg uncertainty principle, the model cannot specify the detailed electron motions. Instead, the square of the wave function represents the probability distribution of the electron in that orbital. This approach allows us to picture orbitals in terms of probability distributions, or electron density maps.**
- **The size of an orbital is arbitrarily defined as the surface that contains 90% of the total electron probability.**
- **The hydrogen atom has many types of orbitals. In the ground state the single electron resides in the 1*s* orbital. The electron can be excited to higher-energy orbitals if the atom absorbs energy.**

How Does the Electron Cross a Node?

A question that often arises when students confront the quantum mechanical picture of the atom is: "If a node in a wave function signifies zero electron probability, how can an electron get from one part of an orbital to another if the two parts are separated by a node?" For example, how does an electron get from one lobe of a $2p$ orbital to the other?

This turns out to be a very subtle question, which brings up the great difficulty in connecting the quantum theory to the behavior of objects familiar to us. For example, it's not at all clear how one should picture an electron. The best advice is probably to not picture it at all, but we find that pictures (analogies to features we observe in the macroscopic world) often help us in understanding and using our theories. Thus we try to form pictures as free of distortion as possible.

So how do we answer the node question? Many ideas have been advanced, and none are entirely satisfactory. One suggestion focuses on the fact that a node is a point. Because probability is evaluated in a finite volume element dv around a given point, the nonzero probability on both sides of a node makes the probability inside dv nonzero. Another suggestion is that because of the uncertainty principle, we cannot say that the electron has a definite path. Therefore, we cannot assume that to get from one point to another, it must pass through the points in between, as is true for a classical particle. A third suggestion is that since the electron is neither a wave nor a particle but behaves like both, the nodes are simply a result of the wave part of its nature and are not a problem. Another explanation states that although the quantum theory allows us to describe mathematically the behavior of atoms, the quantities in the theory have no physical significance. A fifth explanation is that since the Schrödinger equation does not take relativity into account, it is only an approximation and the nodes disappear when this correction is made.

The above list is only a sampling of the "explanations" that have been put forth to try to account for the "node problem." There are many others. The point of this discussion is not to evaluate the merits of the various suggestions but to show how complicated this question turns out to be and how difficult it is to form physical interpretations of the quantum mechanical description of matter.

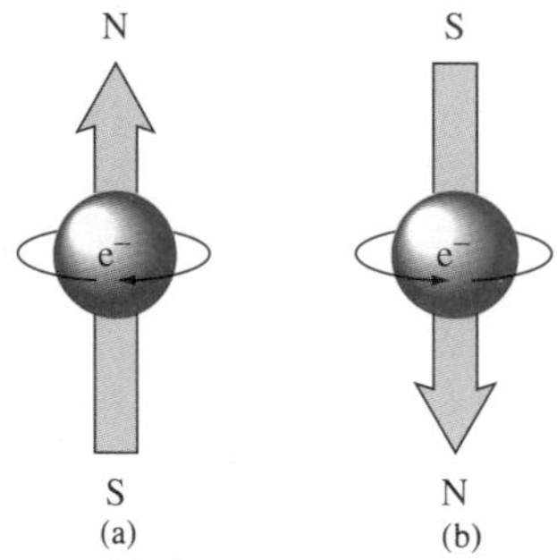

Figure 12.24
A picture of the spinning electron. Spinning in one direction, the electron produces the magnetic field oriented as shown in (a). Spinning in the opposite direction, it gives a magnetic field of the opposite orientation, as shown in (b). This picture of the spinning electron should be viewed with caution. There is considerable dispute about the correct physical interpretation of the spin quantum number.

12.10 Electron Spin and the Pauli Principle

The concept of **electron spin** was developed by Samuel Goudsmit and George Uhlenbeck in 1925 while they were graduate students at the University of Leyden in the Netherlands. They found that a fourth quantum number (in addition to n, ℓ, and m_ℓ) was necessary to account for the details of the emission spectra of atoms. These data indicated that the electron has a magnetic moment, which has two possible orientations when the atom is placed in an external magnetic field. Since they knew from classical physics that a spinning charge produces a magnetic moment, it seemed reasonable to assume that the electron could have two spin states, thus producing the two oppositely directed magnetic moments (see Fig. 12.24). The new quantum number adopted to describe this phenomenon, called the **electron spin quantum number** (m_s), can have only one of two values, $+\frac{1}{2}$ and $-\frac{1}{2}$. We can interpret this to mean that the electron can spin in one of two opposite directions, although other interpretations have also been suggested.

For our purposes the main significance of the electron spin quantum number is connected with the postulate of Austrian physicist Wolfgang Pauli

(1900–1958), which is often stated as follows: *in a given atom no two electrons can have the same set of four quantum numbers* (n, ℓ, m_ℓ, *and* m_s). This is called the **Pauli exclusion principle.** Since electrons in the same orbital have the same values of n, ℓ, and m_ℓ, this postulate requires that they have different values of m_s. Since only two values of m_s are allowed, we might paraphrase the Pauli principle as follows: *an orbital can hold only two electrons, and they must have opposite spins.* This principle will have important consequences when we use the atomic model to relate the electron arrangement of an atom to its position in the periodic table.

$m_s = +\frac{1}{2}$ or $-\frac{1}{2}$

Each orbital can hold a maximum of two electrons.

12.11 Polyelectronic Atoms

The quantum mechanical model provides a description of the hydrogen atom that agrees very well with experimental data. However, the model would not be very useful if it did not account for the properties of the other atoms as well.

To see how the model applies to **polyelectronic atoms**—that is, atoms with more than one electron—let's consider helium, which has two protons in its nucleus and two electrons:

(2+) e^- e^-

There are three energy contributions that must be considered in the description of the helium atom: (1) the kinetic energy of the electrons as they move around the nucleus, (2) the potential energy of attraction between the nucleus and the electrons, and (3) the potential energy of repulsion between the two electrons.

Although this atom can be readily described in terms of the quantum mechanical model, the Schrödinger equation that results cannot be solved exactly. The difficulty arises in dealing with the repulsion between the electrons. This so-called *electron correlation problem* refers to the fact that we cannot rigorously account for the effect a given electron has on the motions of the other electrons in an atom.

The electron correlation problem occurs with all polyelectronic atoms. To treat these systems using the quantum mechanical model, we must make approximations. The simplest approximation involves treating each electron as if it were moving in a *field of charge that is the net result of the nuclear attraction and the average repulsions of all the other electrons.* To see how this is done, let's compare the neutral helium atom and the He^+ ion:

(2+) e^- e^- (2+) e^-

He He^+

What energy is required to remove an electron from each of these species? Experiments show that 2372 kJ of energy are required to remove one electron from all of the atoms in a mole of helium. Removing the one electron from each ion in a mole of He^+ ions requires 5248 kJ of energy. Thus it takes more than twice as much energy to remove an electron from a He^+ ion than from a He atom.

Why such a large difference? In both cases the nucleus has a 2+ charge. However, in the helium atom there are two electrons that repel each other, but in the He^+ ion there is only one electron and thus no electron-electron repulsion. That is, the large difference in the energies required to remove one electron must be due to the electron-electron repulsions in the neutral atom. Each electron in the He atom is much less tightly bound to the nucleus than the electron in the He^+ ion. In other words, the effectiveness of the positively charged nucleus in binding the electrons has been decreased by the repulsions between the electrons. Thus the *effect of the electron repulsions can be thought of as reducing the nuclear charge* to an apparent value of less than 2+ toward a particular electron, as shown below:

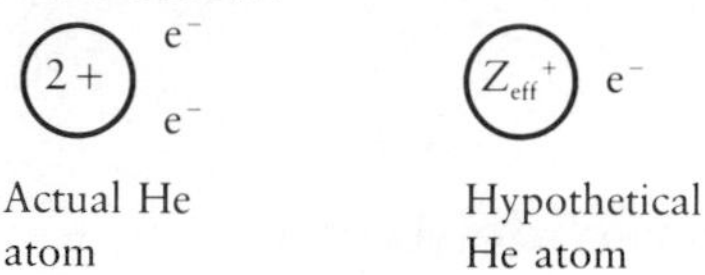

The *apparent* nuclear charge, or the **effective nuclear charge,** is designated Z_{eff}. For a helium atom Z_{eff}, the charge "experienced" by each electron, is less than 2. In general,

$$\text{Effective nuclear charge} = Z_{eff} = Z_{actual} - (\text{effect of electron repulsions})$$

$Z_{eff} = Z -$ effect of electron repulsion.

where $Z_{actual} = Z$, the atomic number (number of protons).

This simplification allows us to treat each electron individually, where each electron is viewed as moving under the influence of a positive nuclear charge Z_{eff}. This simplified atom has one electron like hydrogen, but with a positive nuclear charge of Z_{eff} instead of 1. We therefore can find the energy and wave function for each helium electron by substituting Z_{eff} in place of $Z = 1$ in the hydrogen wave mechanical equations. When we do this, we find that both helium electrons reside in a modified 1*s* orbital that is spherical, like that for the hydrogen atom, but smaller, because Z_{eff} is greater than 1. The larger nuclear charge draws each of the electrons closer to the nucleus, therefore binding each more tightly than the electron in hydrogen is bound. The increased nuclear charge of the helium atom is more important than the repulsions between the two electrons, so that each of the electrons in helium is bound more tightly than the electron in the hydrogen atom.

The model we have just described so greatly oversimplifies the structure of polyelectronic atoms that, although it produces some qualitatively useful ideas about polyelectronic atoms, it is not satisfactory for the description of quantitative atomic properties. To get an accurate description of polyelectronic atoms, we must take into account the electron-electron interactions in a much more detailed manner than simply assuming that they reduce the nuclear charge.

Nothing we do will allow us to solve the problem exactly, because the electron motions are correlated. That is, because electrons repel each other, the movement of a given electron will affect the movements of all of the others. This correlation problem is reflected in the Schrödinger equation for polyelectronic atoms in the following way. Because the equation contains energy terms that simultaneously involve two different electrons, it cannot be separated rigorously into equations that involve only one electron. Thus the Schrödinger equation for polyelectronic atoms cannot be solved exactly.

One approach for dealing with this problem is to solve the equation numerically. That is, a computer is used to find the numerical values of the wave functions at each point in space that produce the lowest overall energy for the atom. Although this approach allows accurate calculation of atomic properties, it suffers from two major disadvantages: it is prohibitively time-consuming for any but the simplest of atoms, and the results are very difficult to interpret physically.

A more practical approach, the **self-consistent field** (SCF) **method,** is now used almost universally to treat polyelectronic atoms. In this method a given electron is assumed to be moving in a potential energy field that is due both to the nucleus and to the average "electron density" of all of the other electrons in the atom (residing in their various orbitals). This approximation allows the many-electron Schrödinger equation to be separated into a set of one-electron equations that can be solved by computers. The orbitals (one-electron functions) that result from this approach have angular properties exactly the same as those of the hydrogen orbitals but have radial characteristics somewhat different from those of the hydrogen orbitals. Although the quantum numbers obtained in the description of the hydrogen atom do not apply exactly to the orbitals obtained from the self-consistent field approach, we still use them as convenient labels for the atomic orbitals in polyelectronic atoms.

We will have more to say later about the self-consistent field approach, but first we will see how the atomic orbitals for polyelectronic atoms can be used to account for the form of the periodic table of the elements.

12.12 The History of the Periodic Table

The modern periodic table contains a tremendous amount of useful information. In this section we will discuss the origin of this valuable tool; later, we will see how the quantum mechanical model for the atom explains the periodicity of chemical properties. Certainly one of the greatest successes of the quantum mechanical model is its ability to account for the arrangement of the elements in the periodic table.

The periodic table was originally constructed to represent the patterns observed in the chemical properties of the elements. As chemistry progressed during the eighteenth and nineteenth centuries, it became evident that the earth is composed of a great many elements with very different properties. Things are much more complicated than the simple model of earth, air, fire, and water suggested by the ancients. At first, the array of elements and properties was bewildering. Gradually, however, patterns were noticed.

The first chemist to recognize patterns was Johann Dobereiner, who found several groups of three elements with similar properties, for example, chlorine, bromine, and iodine. However, as Dobereiner attempted to expand this model of **triads** (as he called them) to the rest of the known elements, it became clear that this concept was severely limited.

The next notable attempt was made by the English chemist John Newlands, who in 1864 suggested that elements should be arranged in **octaves.** He noticed that certain properties seemed to repeat for every eighth element in a way similar to the musical scale, which repeats for every eighth tone. Although this model managed to group several elements with similar properties, it was not generally successful.

The present form of the periodic table was conceived independently by two chemists in 1869: the German Julius Lothar Meyer and the Russian Dmitri Ivanovich Mendeleev (Fig. 12.25). Usually, Mendeleev is given most of the

Figure 12.25
Dmitri Ivanovich Mendeleev (1834–1907), born in Siberia as the youngest of 17 children, taught chemistry at the University of St. Petersburg. In 1860 Mendeleev heard the Italian chemist Cannizzaro lecture on a reliable method for determining the correct atomic masses of the elements. This important development paved the way for Mendeleev's own brilliant contribution to chemistry—the periodic table. In 1861 Mendeleev returned to St. Petersburg, where he wrote a book on organic chemistry. Later Mendeleev also wrote a book on inorganic chemistry, and he was struck by the fact that the systematic approach characterizing organic chemistry was lacking in inorganic chemistry. In attempting to systematize inorganic chemistry, he eventually arranged the elements in the form of the periodic table.

Mendeleev was a versatile genius who was interested in many fields of science. He worked on many problems associated with Russia's natural resources, such as coal, salt, and various metals. Being particularly interested in the petroleum industry, he visited the United States in 1876 to study the Pennsylvania oil fields. His interests also included meteorology and hot-air balloons. In 1887 he made an ascent in a balloon to study a total eclipse of the sun.

credit, because it was he who showed how useful the table could be in predicting the existence and properties of yet unknown elements. For example, in 1872 when Mendeleev first published his table (see Fig. 12.26), the elements gallium, scandium, and germanium were unknown. Mendeleev correctly predicted the existence and properties of these elements from gaps in his periodic table. The data for germanium (which Mendeleev called *ekasilicon*) are shown in Table 12.4. Note the excellent agreement between the actual values and Mendeleev's predictions, which were based on the properties of other members in the group of elements similar to germanium.

TABELLE II

REIHEN	GRUPPE I. — R^2O	GRUPPE II. — RO	GRUPPE III. — R^2O^3	GRUPPE IV. RH^4 RO^2	GRUPPE V. RH^3 R^2O^5	GRUPPE VI. RH^2 RO^3	GRUPPE VII. RH R^2O^7	GRUPPE VIII. — RO^4
1	H = 1							
2	Li = 7	Be = 9,4	B = 11	C = 12	N = 14	O = 16	F = 19	
3	Na = 23	Mg = 24	Al = 27,3	Si = 28	P = 31	S = 32	Cl = 35,5	
4	K = 39	Ca = 40	— = 44	Ti = 48	V = 51	Cr = 52	Mn = 55	Fe = 56, Co = 59, Ni = 59, Cu = 63.
5	(Cu = 63)	Zn = 65	— = 68	— = 72	As = 75	Se = 78	Br = 80	
6	Rb = 85	Sr = 87	?Yt = 88	Zr = 90	Nb = 94	Mo = 96	— = 100	Ru = 104, Rh = 104, Pd = 106, Ag = 108.
7	(Ag = 108)	Cd = 112	In = 113	Sn = 118	Sb = 122	Te = 125	J = 127	
8	Cs = 133	Ba = 137	?Di = 138	?Ce = 140	—	—	—	— — — —
9	(—)	—	—	—	—	—	—	
10	—	—	?Er = 178	?La = 180	Ta = 182	W = 184	—	Os = 195, Ir = 197, Pt = 198, Au = 199.
11	(Au = 199)	Hg = 200	Tl = 204	Pb = 207	Bi = 208	—	—	
12	—	—	—	Th = 231	—	U = 240	—	— — — —

Figure 12.26
Mendeleev's early periodic table, published in 1872. Note the spaces left for missing elements with atomic weights 44, 68, 72, and 100. From *Annalen der Chemie und Pharmacie*, VIII, Supplementary Volume for 1872, page 511.

TABLE 12.4 Comparison of the Properties of Germanium as Predicted by Mendeleev and as Actually Observed

Properties of Germanium	Predicted in 1871	Observed in 1886
Atomic mass	72	72.3
Density	5.5 g/cm^3	5.47 g/cm^3
Specific heat	0.31 J $°C^{-1}$ g^{-1}	0.32 J $°C^{-1}$ g^{-1}
Melting point	Very high	960°C
Oxide formula	RO_2	GeO_2
Oxide density	4.7 g/cm^3	4.70 g/cm^3
Chloride formula	RCl_4	$GeCl_4$
bp of chloride	100°C	86°C

TABLE 12.5 Predicted Properties of Elements 113 and 114

Property	Element 113	Element 114
Chemically like	Thallium	Lead
Atomic mass	297	298
Density	16 g/mL	14 g/mL
Melting point	430°C	70°C
Boiling point	1100°C	150°C

Using his table, Mendeleev was also able to correct several values of atomic masses. For example, the original atomic mass of 76 for indium was based on the assumption that indium oxide had the formula InO. This atomic mass placed indium, which has metallic properties, among the nonmetals. Mendeleev assumed that the atomic mass was probably incorrect and proposed that the formula of indium oxide was really In_2O_3. On the basis of this (correct) formula, indium has an atomic mass of approximately 113, placing the element among the metals. Mendeleev also corrected the atomic masses of beryllium and uranium.

Because of its obvious usefulness Mendeleev's periodic table was almost universally adopted, and it remains one of the most valuable tools at the chemist's disposal. For example, it is still used to predict the properties of elements yet to be discovered, as shown in Table 12.5.

A current version of the periodic table is shown inside the front cover of this book. The only fundamental difference between this table and that of Mendeleev is that the current table lists the elements in order by atomic number rather than by atomic mass. The reason for this will become clear later in this chapter as we explore the electron arrangements of the atom.

12.13 The Aufbau Principle and the Periodic Table

We can use the quantum mechanical model of the atom to show how the electron arrangements in the atomic orbitals of the various atoms account for the organization of the periodic table. Our main assumption here is that all atoms have orbitals similar to those that have been described for the hydrogen atom. *As protons are added one by one to the nucleus to build up the elements, electrons are similarly added to these atomic orbitals.* This is called the **Aufbau principle.**

Aufbau is German for "building up."

Hydrogen has one electron, which occupies the 1*s* orbital in its ground state. The configuration for hydrogen is written as $1s^1$, which can be represented by the following *orbital diagram:*

		1*s*	2*s*	2*p*
H:	$1s^1$	↑					

The arrow represents an electron spinning in a particular direction.

The next element, *helium,* has two electrons. Since two electrons with opposite spins can occupy an orbital, according to the Pauli exclusion principle, the electrons for helium are in the 1*s* orbital with opposite spins. This yields a $1s^2$ configuration:

		1*s*	2*s*	2*p*
He:	$1s^2$	↑↓					

Lithium has three electrons, two of which can go into the 1*s* orbital before the orbital is filled. Since the 1*s* orbital is the only orbital with $n = 1$, the third electron will occupy the lowest-energy orbital with $n = 2$, or the 2*s* orbital, giving a $1s^2 2s^1$ configuration:

We will see in Section 12.14 why the 2*s* orbital is lower in energy than the 2*p* orbital.

		1*s*	2*s*	2*p*
Li:	$1s^2 2s^1$	↑↓	↑				

The next element, *beryllium,* has four electrons, which occupy the 1*s* and 2*s* orbitals:

		1*s*	2*s*	2*p*
Be:	$1s^2 2s^2$	↑↓	↑↓				

Boron has five electrons, four of which occupy the 1*s* and 2*s* orbitals. The fifth electron goes into the second type of orbital with $n = 2$, the 2*p* orbitals:

		1*s*	2*s*	2*p*
B:	$1s^2 2s^2 2p^1$	↑↓	↑↓	↑			

Since all of the 2*p* orbitals have the same energy (are degenerate), it does not matter which 2*p* orbital the electron occupies.

Carbon has six electrons: two electrons occupy the 1*s* orbital, two occupy the 2*s* orbital, and two occupy 2*p* orbitals. Since there are three 2*p* orbitals with the same energy, the mutually repulsive electrons will occupy *separate* 2*p* orbitals.

This behavior is summarized by **Hund's rule** (named for the German physicist F. H. Hund), which states that *the lowest-energy configuration for an atom is the one having the maximum number of unpaired electrons in a particular set of degenerate orbitals allowed by the Pauli principle.*

For an atom with unfilled subshells, the lowest energy is achieved by electrons occupying separate orbitals, as allowed by the Pauli exclusion principle.

The configuration for carbon could be written $1s^2 2s^2 2p^1 2p^1$ to indicate that the electrons occupy separate 2*p* orbitals. However, the configuration is usually given as $1s^22s^22p^2$, and it is understood that the electrons are in different $2p$ orbitals. The orbital diagram for carbon is

		$1s$	$2s$	$2p$
C:	$1s^22s^22p^2$	[↑↓]	[↑↓]	[↑][↑][ ]

Note the unpaired electrons in the $2p$ orbitals, as required by Hund's rule.

The configuration for *nitrogen,* which has seven electrons, is $1s^22s^22p^3$. The three electrons in $2p$ orbitals occupy separate orbitals:

		$1s$	$2s$	$2p$
N:	$1s^22s^22p^3$	[↑↓]	[↑↓]	[↑][↑][↑]

The configuration for *oxygen,* which has eight electrons, is $1s^22s^22p^4$. One of the $2p$ orbitals is now occupied by a pair of electrons with opposite spins, as required by the Pauli exclusion principle:

		$1s$	$2s$	$2p$
O:	$1s^22s^22p^4$	[↑↓]	[↑↓]	[↑↓][↑][↑]

The orbital diagrams and electron configurations for *fluorine* (nine electrons) and *neon* (ten electrons) are given below:

		$1s$	$2s$	$2p$
F:	$1s^22s^22p^5$	[↑↓]	[↑↓]	[↑↓][↑↓][↑]
Ne:	$1s^22s^22p^6$	[↑↓]	[↑↓]	[↑↓][↑↓][↑↓]

With neon, the orbitals with $n = 1$ and $n = 2$ are now completely filled.

For *sodium* the first ten electrons occupy the $1s$, $2s$, and $2p$ orbitals, and the eleventh electron must occupy the first orbital with $n = 3$, the $3s$ orbital. The electron configuration for sodium is $1s^22s^22p^63s^1$. To avoid writing the inner-level electrons, we often abbreviate this configuration as $[Ne]3s^1$, where [Ne] represents the electron configuration of neon, $1s^22s^22p^6$.

[Ne] is shorthand for $1s^22s^22p^6$.

The next element, *magnesium,* has the configuration $1s^22s^22p^63s^2$, or $[Ne]3s^2$. Then the next six elements, *aluminum* through *argon,* have configurations obtained by filling the $3p$ orbitals one electron at a time. Figure 12.27 summarizes the electron configurations of the first 18 elements by giving the number of electrons in the type of orbital occupied last.

At this point it is useful to introduce the concept of **valence electrons,** that is, *the electrons in the outermost principal quantum level of an atom.* The valence electrons of the nitrogen atom, for example, are the $2s$ and $2p$ electrons. For the sodium atom the valence electron is the electron in the $3s$ orbital, and so on. Valence electrons are the most important electrons to chemists, because they are involved in bonding, as we will see in the next two chapters. The inner electrons are known as **core electrons.**

Note in Fig. 12.27 that a very important pattern is developing: *the elements in the same group (vertical column of the periodic table) have the same valence electron configuration.* Remember that Mendeleev originally placed the elements in groups based on similarities in chemical properties. Now we understand the reason behind these groupings. Elements with the same valence electron configuration often show similar chemical behavior.

The element after argon is *potassium.* Since the $3p$ orbitals are fully occupied in argon, we might expect the next electron to go into a $3d$ orbital (recall

H $1s^1$							He $1s^2$
Li $2s^1$	Be $2s^2$	B $2p^1$	C $2p^2$	N $2p^3$	O $2p^4$	F $2p^5$	Ne $2p^6$
Na $3s^1$	Mg $3s^2$	Al $3p^1$	Si $3p^2$	P $3p^3$	S $3p^4$	Cl $3p^5$	Ar $3p^6$

Figure 12.27
The electron configurations in the type of orbital occupied last for the first 18 elements.

that for $n = 3$ the orbitals are $3s$, $3p$, and $3d$). However, the chemistry of potassium is clearly very similar to that of lithium and sodium, indicating that the last electron in potassium occupies the $4s$ orbital instead of one of the $3d$ orbitals, a conclusion confirmed by many types of experiments. The electron configuration of potassium is

[Ar] is shorthand for $1s^22s^22p^63s^23p^6$.

$$\text{K:} \quad 1s^22s^22p^63s^23p^64s^1 \quad \text{or} \quad [\text{Ar}]4s^1$$

The next element is *calcium:*

$$\text{Ca:} \quad [\text{Ar}]4s^2$$

The next element, *scandium,* begins a series of ten elements (scandium through zinc) called the **transition metals,** whose configurations are obtained by adding electrons to the five $3d$ orbitals. The configuration of scandium is

When an electron configuration is given in this text, the orbitals are listed in the order in which they fill.

$$\text{Sc:} \quad [\text{Ar}]4s^23d^1$$

That of *titanium* is $\quad$ Ti: $\quad [\text{Ar}]4s^23d^2$

And that of *vanadium* is $\quad$ V: $\quad [\text{Ar}]4s^23d^3$

Chromium is the next element. The expected configuration is $[\text{Ar}]4s^23d^4$. However, the observed configuration is

$$\text{Cr:} \quad [\text{Ar}]4s^13d^5$$

The explanation for this configuration of chromium is beyond the scope of this book. In fact, chemists are still disagreeing over the exact cause of this anomaly. Note, however, that the observed configuration has both the $4s$ and $3d$ orbitals half-filled. This is a good way to remember the correct configuration.

The next four elements, *manganese* through *nickel,* have the expected configurations:

$$\text{Mn:} \quad [\text{Ar}]4s^23d^5 \qquad \text{Co:} \quad [\text{Ar}]4s^23d^7$$
$$\text{Fe:} \quad [\text{Ar}]4s^23d^6 \qquad \text{Ni:} \quad [\text{Ar}]4s^23d^8$$

The configuration for *copper* is expected to be $[\text{Ar}]4s^23d^9$. However, the observed configuration is

$$\text{Cu:} \quad [\text{Ar}]4s^13d^{10}$$

In this case a half-filled $4s$ orbital and a filled set of $3d$ orbitals characterize the actual configuration.

Zinc has the expected configuration:

$$\text{Zn:} \quad [\text{Ar}]4s^23d^{10}$$

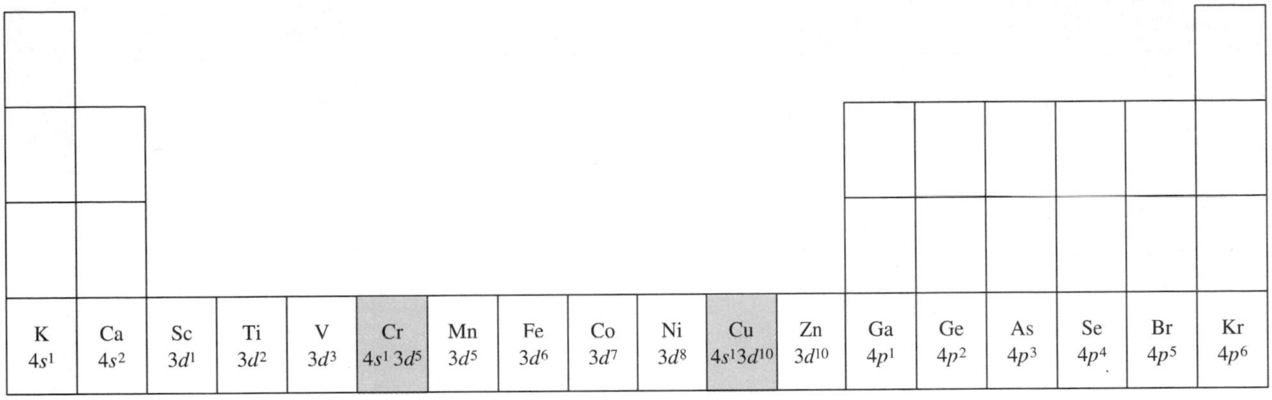

K	Ca	Sc	Ti	V	Cr	Mn	Fe	Co	Ni	Cu	Zn	Ga	Ge	As	Se	Br	Kr
$4s^1$	$4s^2$	$3d^1$	$3d^2$	$3d^3$	$4s^1 3d^5$	$3d^5$	$3d^6$	$3d^7$	$3d^8$	$4s^1 3d^{10}$	$3d^{10}$	$4p^1$	$4p^2$	$4p^3$	$4p^4$	$4p^5$	$4p^6$

Figure 12.28
Electron configurations for potassium through krypton. The transition metals (scandium through zinc) have the general configuration [Ar]$4s^2 3d^n$, except for chromium and copper.

The configurations of the transition metals are shown in Fig. 12.28. The next six elements, *gallium* through *krypton*, have configurations that correspond to filling the $4p$ orbitals (see Fig. 12.28).

The entire periodic table is represented in Fig. 12.29 in terms of which orbitals are being filled. The valence electron configurations are given in Fig. 12.30.

From these two figures, note the following additional points:

1. The $(n + 1)s$ orbitals always fill before the nd orbitals. For example, the $5s$ orbitals fill in rubidium and strontium before the $4d$ orbitals fill in the second row of transition metals (yttrium through cadmium).

The $(n + 1)s$ orbital fills before the nd orbitals.

2. After lanthanum, which has the configuration [Xe]$6s^2 5d^1$, a group of 14 elements called the **lanthanide series,** or the lanthanides, occurs. This series

Lanthanides are elements in which the $4f$ orbitals are being filled.

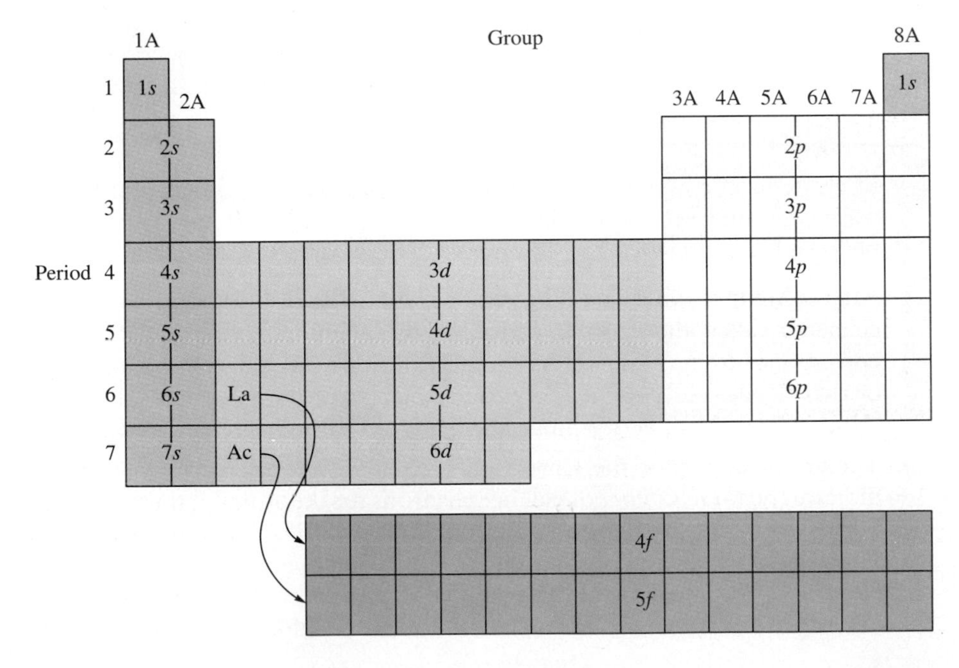

Figure 12.29
The orbitals being filled for elements in various parts of the periodic table. Note that when we move along a horizontal row (a period), the $(n + 1)s$ orbital fills before the nd orbital. The group labels indicate the number of valence electrons (ns plus np electrons) for the elements in each group.

Period number, highest occupied electron level	Representative Elements: 1A ns^1	2A ns^2	d-Transition Elements										Representative Elements: 3A ns^2np^1	4A ns^2np^2	5A ns^2np^3	6A ns^2np^4	7A ns^2np^5	Noble gases: 8A ns^2np^6
1	1 H $1s^1$																	2 He $1s^2$
2	3 Li $2s^1$	4 Be $2s^2$											5 B $2s^22p^1$	6 C $2s^22p^2$	7 N $2s^22p^3$	8 O $2s^22p^4$	9 F $2s^22p^5$	10 Ne $2s^22p^6$
3	11 Na $3s^1$	12 Mg $3s^2$											13 Al $3s^23p^1$	14 Si $3s^23p^2$	15 P $3s^23p^3$	16 S $3s^23p^4$	17 Cl $3s^23p^5$	18 Ar $3s^23p^6$
4	19 K $4s^1$	20 Ca $4s^2$	21 Sc $4s^23d^1$	22 Ti $4s^23d^2$	23 V $4s^23d^3$	24 Cr $4s^13d^5$	25 Mn $4s^23d^5$	26 Fe $4s^23d^6$	27 Co $4s^23d^7$	28 Ni $4s^23d^8$	29 Cu $4s^13d^{10}$	30 Zn $4s^23d^{10}$	31 Ga $4s^24p^1$	32 Ge $4s^24p^2$	33 As $4s^24p^3$	34 Se $4s^24p^4$	35 Br $4s^24p^5$	36 Kr $4s^24p^6$
5	37 Rb $5s^1$	38 Sr $5s^2$	39 Y $5s^24d^1$	40 Zr $5s^24d^2$	41 Nb $5s^14d^4$	42 Mo $5s^14d^5$	43 Tc $5s^14d^6$	44 Ru $5s^14d^7$	45 Rh $5s^14d^8$	46 Pd $4d^{10}$	47 Ag $5s^14d^{10}$	48 Cd $5s^24d^{10}$	49 In $5s^25p^1$	50 Sn $5s^25p^2$	51 Sb $5s^25p^3$	52 Te $5s^25p^4$	53 I $5s^25p^5$	54 Xe $5s^25p^6$
6	55 Cs $6s^1$	56 Ba $6s^2$	57 La* $6s^25d^1$	72 Hf $4f^{14}6s^25d^2$	73 Ta $6s^25d^3$	74 W $6s^25d^4$	75 Re $6s^25d^5$	76 Os $6s^25d^6$	77 Ir $6s^25d^7$	78 Pt $6s^15d^9$	79 Au $6s^15d^{10}$	80 Hg $6s^25d^{10}$	81 Tl $6s^26p^1$	82 Pb $6s^26p^2$	83 Bi $6s^26p^3$	84 Po $6s^26p^4$	85 At $6s^26p^5$	86 Rn $6s^26p^6$
7	87 Fr $7s^1$	88 Ra $7s^2$	89 Ac** $7s^26d^1$	104 Unq $7s^26d^2$	105 Unp $7s^26d^3$	106 Unh $7s^26d^4$	107 Uns $7s^26d^5$	108 Uno $7s^26d^6$	109 Une $7s^26d^7$									

f-Transition Elements

*Lanthanides	58 Ce $6s^24f^15d^1$	59 Pr $6s^24f^35d^0$	60 Nd $6s^24f^45d^0$	61 Pm $6s^24f^55d^0$	62 Sm $6s^24f^65d^0$	63 Eu $6s^24f^75d^0$	64 Gd $6s^24f^75d^1$	65 Tb $6s^24f^95d^0$	66 Dy $6s^24f^{10}5d^0$	67 Ho $6s^24f^{11}5d^0$	68 Er $6s^24f^{12}5d^0$	69 Tm $6s^24f^{13}5d^0$	70 Yb $6s^24f^{14}5d^0$	71 Lu $6s^24f^{14}5d^1$
**Actinides	90 Th $7s^25f^06d^2$	91 Pa $7s^25f^26d^1$	92 U $7s^25f^36d^1$	93 Np $7s^25f^46d^1$	94 Pu $7s^25f^66d^0$	95 Am $7s^25f^76d^0$	96 Cm $7s^25f^76d^1$	97 Bk $7s^25f^96d^0$	98 Cf $7s^25f^{10}6d^0$	99 Es $7s^25f^{11}6d^0$	100 Fm $7s^25f^{12}6d^0$	101 Md $7s^25f^{13}6d^0$	102 No $7s^25f^{14}6d^0$	103 Lr $7s^25f^{14}6d^1$

Figure 12.30
The periodic table with atomic symbols, atomic numbers, and partial electron configurations.

of elements corresponds to the filling of the seven $4f$ orbitals. Note that sometimes one electron occupies a $5d$ instead of a $4f$ orbital. This occurs because the energies of the $4f$ and $5d$ orbitals are very similar.

Actinides are elements in which the $5f$ orbitals are being filled.

3. After actinium, which has the configuration $[Rn]7s^26d^1$, a group of 14 elements called the **actinide series,** or the actinides, occurs. This series corresponds to the filling of the seven $5f$ orbitals. Note that sometimes one or two electrons occupy the $6d$ orbitals instead of the $5f$ orbitals, because these orbitals have very similar energies.

The group label tells the total number of valence electrons for that group.

4. The group labels for the Groups 1A, 2A, 3A, 4A, 5A, 6A, 7A, and 8A indicate the *total number* of valence electrons for the atoms in these groups. For example, all the elements in Group 5A have the configuration ns^2np^3. (The d electrons fill one period late and are usually not counted as valence electrons.) The meaning of the group labels for the transition metals is not as clear as for the A group elements, so these will not be used in this text.

5. The groups labeled 1A, 2A, 3A, 4A, 5A, 6A, 7A, and 8A are often called the **main-group, or representative, elements.** Remember that every member of these groups has the same valence electron configuration.

In 1985 the International Union of Pure and Applied Chemistry (IUPAC), a body of scientists organized to standardize scientific conventions, recommended a new form for the periodic table, which the American Chemical Society has adopted (see Fig. 12.31). In this new version the group number indicates the number of *s*, *p*, and *d* electrons added since the last noble gas. We will not use the new format in this book, but you should be aware that the familiar periodic table may soon be replaced by this or a similar format.

The American Chemical Society has endorsed the following names for elements 104–109:
104 rutherfordium (Rf)
105 hahnium (Ha)
106 seaborgium (Sg)
107 nielsbohrium (Ns)
108 hassium (Hs)
109 meitnerium (Mt)
These names have not yet been officially approved.

The results considered in this section are very important. We have seen that the wave mechanical model can be used to explain the arrangement of elements in the periodic table. This model allows us to understand that the similar chemistry exhibited by the members of a given group arises from the fact that they all have the same valence electron configuration. Only the principal quantum number of the occupied orbitals changes in going down a particular group.

It is important to be able to give the electron configuration for each of the main-group elements. This is most easily done by using the periodic table. If you understand how the table is organized, it is not necessary to memorize the order in which the orbitals fill. Review Fig. 12.29 and Fig. 12.30 to make sure that you understand the correspondence between the orbitals and the periods and groups.

Predicting the configurations of the transition metals (3*d*, 4*d*, and 5*d* elements), the lanthanides (4*f* elements), and the actinides (5*f* elements) is somewhat more difficult, because there are many exceptions of the type encountered in the first-row transition metals (the 3*d* elements). You should

Figure 12.31
A form of the periodic table recommended by IUPAC.

1	2	3	4	5	6	7	8	9	10	11	12	13	14	15	16	17	18
1 H																	2 He
3 Li	4 Be											5 B	6 C	7 N	8 O	9 F	10 Ne
11 Na	12 Mg											13 Al	14 Si	15 P	16 S	17 Cl	18 Ar
19 K	20 Ca	21 Sc	22 Ti	23 V	24 Cr	25 Mn	26 Fe	27 Co	28 Ni	29 Cu	30 Zn	31 Ga	32 Ge	33 As	34 Se	35 Br	36 Kr
37 Rb	38 Sr	39 Y	40 Zr	41 Nb	42 Mo	43 Tc	44 Ru	45 Rh	46 Pd	47 Ag	48 Cd	49 In	50 Sn	51 Sb	52 Te	53 I	54 Xe
55 Cs	56 Ba	57 La	72 Hf	73 Ta	74 W	75 Re	76 Os	77 Ir	78 Pt	79 Au	80 Hg	81 Tl	82 Pb	83 Bi	84 Po	85 At	86 Rn
87 Fr	88 Ra	89 Ac	104 Unq	105 Unp	106 Unh	107 Uns	108 Uno	109 Une									

Lanthanide series	58 Ce	59 Pr	60 Nd	61 Pm	62 Sm	63 Eu	64 Gd	65 Tb	66 Dy	67 Ho	68 Er	69 Tm	70 Yb	71 Lu
Actinide series	90 Th	91 Pa	92 U	93 Np	94 Pu	95 Am	96 Cm	97 Bk	98 Cf	99 Es	100 Fm	101 Md	102 No	103 Lr

Cr: $[Ar]4s^13d^5$
Cu: $[Ar]4s^13d^{10}$

memorize the configurations of chromium and copper, the two exceptions in the first-row transition metals, since these elements are often encountered.

EXAMPLE 12.8

Give the electron configurations for sulfur (S), cadmium (Cd), hafnium (Hf), and radium (Ra), using the periodic table inside the front cover of this book.

Solution

Sulfur, element 16, resides in Period 3, where the $3p$ orbitals are being filled (see Fig. 12.32). Since sulfur is the fourth among the $3p$ elements, it must have four $3p$ electrons. Its configuration is

$$\text{S:} \quad 1s^22s^22p^63s^23p^4 \quad \text{or} \quad [\text{Ne}]3s^23p^4$$

Cadmium, element 48, is located in Period 5 at the end of the $4d$ transition metals, as shown in Fig. 12.32. It is the tenth element in the series and thus has ten electrons in the $4d$ orbitals (in addition to the two electrons in the $5s$ orbital). The configuration is

$$\text{Cd:} \quad 1s^22s^22p^63s^23p^64s^23d^{10}4p^65s^24d^{10} \quad \text{or} \quad [\text{Kr}]5s^24d^{10}$$

Hafnium, element 72, is found in Period 6, as shown in Fig. 12.32. Note that it occurs just after the lanthanide series. Thus the $4f$ orbitals are already filled. Hafnium is the second member of the $5d$ transition series and has two $5d$ electrons. The configuration is

$$\text{Hf:} \quad 1s^22s^22p^63s^23p^64s^23d^{10}4p^65s^24d^{10}5p^66s^24f^{14}5d^2$$
$$\text{or} \quad [\text{Xe}]6s^24f^{14}5d^2$$

Radium, element 88, is in Period 7 (and Group 2A), as shown in Fig. 12.32. Thus radium has two electrons in the $7s$ orbital, and the configuration is

$$\text{Ra:} \quad 1s^22s^22p^63s^23p^64s^23d^{10}4p^65s^24d^{10}5p^66s^24f^{14}5d^{10}6p^67s^2$$
$$\text{or} \quad [\text{Rn}]7s^2$$

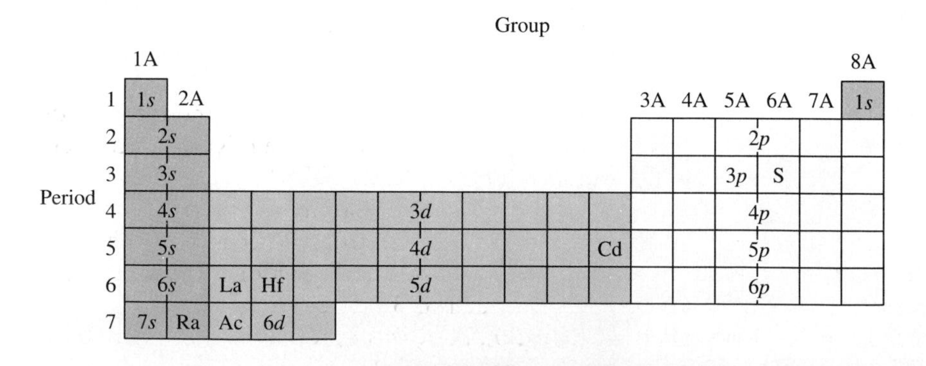

Figure 12.32
The positions of the elements considered in Example 12.8.

12.14 Further Development of the Polyelectronic Model

Before we proceed with further discussion of polyelectronic atoms, we should summarize some of the most important things that have been said about the quantum mechanical description of atoms to this point. Most important, there is a fundamental difference between the solution of the Schrödinger equation for the hydrogen atom and the solutions for all polyelectronic atoms. The Schrödinger equation for the hydrogen atom can be solved exactly to yield the now-familiar hydrogen orbitals. These orbitals are characterized by the quantum numbers n, ℓ, m_ℓ, and m_s, and the energy levels corresponding to these orbitals depend only on n (all orbitals with the same value of n are degenerate). Recall that some of the orbitals directly obtained from the solution to the Schrödinger equation are complex (contain $\sqrt{-1}$). For example, of the three orbitals corresponding to $n = 2$ and $\ell = 1$ ($2p$ orbitals), the orbital corresponding to $m_\ell = 0$ is real (the $2p_z$ orbital), but the orbitals corresponding to the values of m_ℓ of $+1$ and -1 are complex. For ease of physical interpretation these latter two orbitals are combined to produce two real orbitals ($2p_x$ and $2p_y$). These same procedures apply to all of the p orbitals (corresponding to higher values of n). Similarly, for the $3d$ orbitals ($n = 3$, $\ell = 2$) the orbital corresponding to $m_\ell = 0$ is real (d_{z^2}), whereas the orbitals corresponding to m_ℓ values of ± 1 and ± 2 are complex and are used to construct the familiar real orbitals.

In contrast to the Schrödinger equation for the hydrogen atom, the Schrödinger equation for a polyelectronic atom cannot be solved exactly. For example although the hydrogen and helium atoms are similar in many respects, the mathematical descriptions of these atoms are fundamentally different. Because electrons repel each other, the motions of the two helium electrons are correlated (coupled), and this fact prevents the exact separation of the Schrödinger equation for helium into independent, solvable equations for each electron. Thus solving the Schrödinger equation for helium (or any other polyelectronic atom) requires approximations. The approach most commonly used, the self-consistent field (SCF) method, was developed by Hartree and is applied as follows. For an atom containing N electrons, a wave function (orbital) is guessed for each electron except one. For example, assume that orbitals are guessed for electrons 2, 3, 4, . . . , N: $\psi_2, \psi_3, \psi_4, \ldots, \psi_N$. The next step involves solving the Schrödinger equation for electron 1, which is moving in a potential field created by the nucleus and the electrons in orbitals $\psi_2, \psi_3, \ldots, \psi_N$. The repulsions between electron 1 and the other electrons are computed at each point in space from the sum of the average electron densities (probabilities) corresponding to $|\psi_2|^2, |\psi_3|^2, \ldots, |\psi_N|^2$ in volume element dv around that point. With the aid of a computer the problem is solved to yield the wave function for electron 1, which we will label ψ_1'.

The next step is to do the same type of calculation to obtain a new wave function for electron 2 moving in a field of electrons described by the wave functions $\psi_1', \psi_3, \psi_4, \ldots, \psi_N$. This step leads to a new function ψ_2' for electron 2. Now the process is carried out for electron 3 interacting with electrons described by the wave functions $\psi_1', \psi_2', \psi_4, \ldots, \psi_N$ to produce a new function ψ_3'. This procedure continues until all electrons have been covered to yield the wave functions $\psi_1', \psi_2', \psi_3', \ldots, \psi_N'$. Then the entire process starts again with electron

1 and continues through electron N to give the new functions $\psi_1'', \psi_2'', \psi_3'', \ldots, \psi_N''$. The procedure is diagramed in Fig. 12.33. When a given cycle produces a set of wave functions that are virtually identical to the previous set, a self-consistent field is achieved and the procedure is terminated.

The orbitals that arise from the SCF method are quite similar to hydrogen orbitals. They have the same angular characteristics (same type of boundary surfaces) as do the orbitals of hydrogen. However, the radial parts of the orbitals are different from those of the hydrogen orbitals. Although the n quantum number from the treatment of hydrogen does not apply exactly to the SCF orbitals, it is still convenient to retain it as a label. It is important to note that the energies of the SCF orbitals for polyelectronic atoms depend on both n and ℓ, not just n, as for hydrogen.

Finally, although it is not precisely correct to assume that the N electrons in an atom occupy N independent one-electron orbitals, this remains a very useful idea for understanding many atomic properties, including the organization of the periodic table. Recall that in order for us to account for the arrangement of the atoms on the periodic table, the orbitals that correspond to a given value of n must fill in the order ns, then np, then nd, and, finally, nf. From this observation we would expect the energies of the one-electron SCF orbitals to vary in the order

$$E_{ns} < E_{np} < E_{nd} < E_{nf}$$

and this ordering is borne out by the calculations.

We can understand the observed order of orbital energies in a qualitative sense by considering the so-called **penetration effect.** To get an appreciation for this concept, consider the radial probability plots for the 3s, 3p, and 3d orbitals in a sodium atom, as shown in Fig. 12.34. Note that although an electron in the 3s orbital spends most of its time far from the nucleus and outside the core electrons (the electrons in the 1s, 2s, and 2p orbitals), it has a small but significant probability of being quite close to the nucleus. We say it significantly

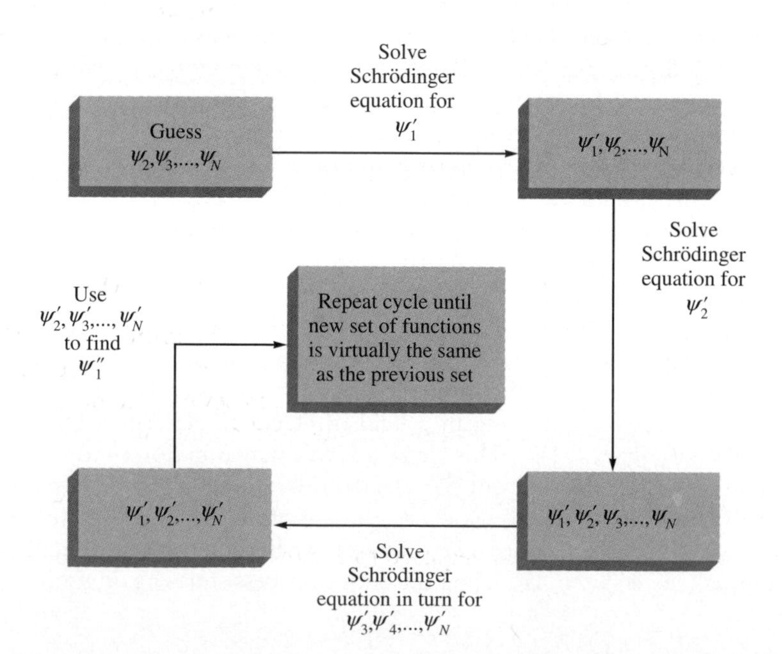

Figure 12.33
A schematic of the self-consistent field method for obtaining the orbitals of a polyelectronic atom.

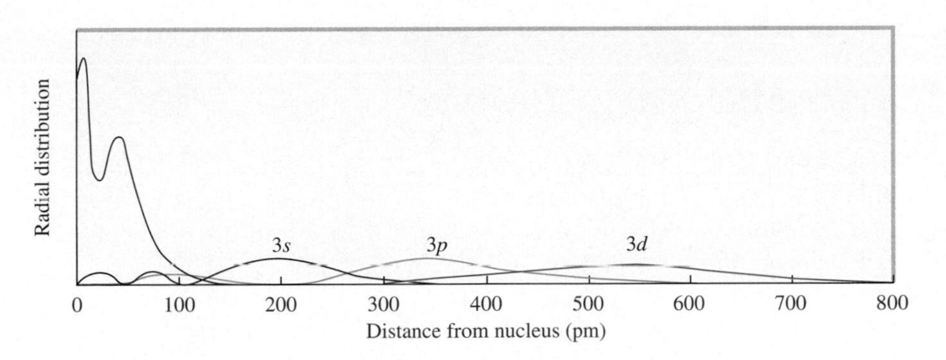

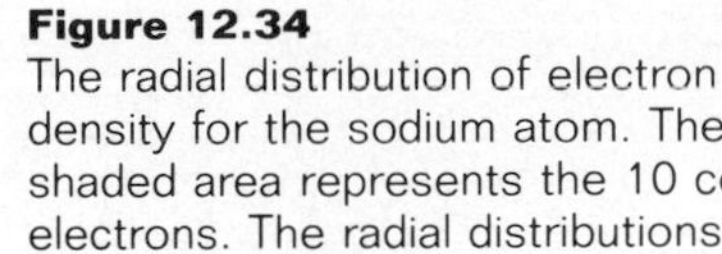
Figure 12.34
The radial distribution of electron density for the sodium atom. The shaded area represents the 10 core electrons. The radial distributions of the 3*s*, 3*p*, and 3*d* orbitals are also shown. Note the difference in the penetration effects of an electron in these three orbitals.

penetrates the core electron "cloud." On the other hand, an electron in a 3*p* orbital does not have a probability maximum close to the nucleus. Thus we can say that an electron in a 3*p* orbital penetrates the core electrons to a lesser extent than an electron in the 3*s* orbital. Similarly, an electron in a 3*d* orbital (see Fig. 12.34) shows much less penetration than a 3*p* electron does. These ideas help us to understand why an electron "prefers" the 3*s* orbital to the 3*p* or 3*d* and why, after the 3*s* orbital is filled, the next electron occupies the 3*p* rather than the 3*d* orbital. That is, the penetration effect helps us to understand qualitatively the order

$$E_{3s} < E_{3p} < E_{3d}$$

The penetration effect also helps to explain why the 4*s* orbital fills before the 3*d* orbital. Recall that potassium has the electron configuration $1s^22s^22p^63s^23p^64s^1$ rather than the expected $1s^22s^22p^63s^23p^63d^1$. We can explain this result by observing that an electron in a 4*s* orbital penetrates much more than an electron in a 3*d* orbital, as shown graphically in Fig. 12.35. Note that although the most probable distance from the nucleus for a 3*d* electron is less than that for a 4*s* electron, the 4*s* electron has a significant probability of penetrating close to the nucleus. This explains why the potassium atom in its lowest-energy state has its last electron in the 4*s* orbital rather than in the 3*d* orbital.

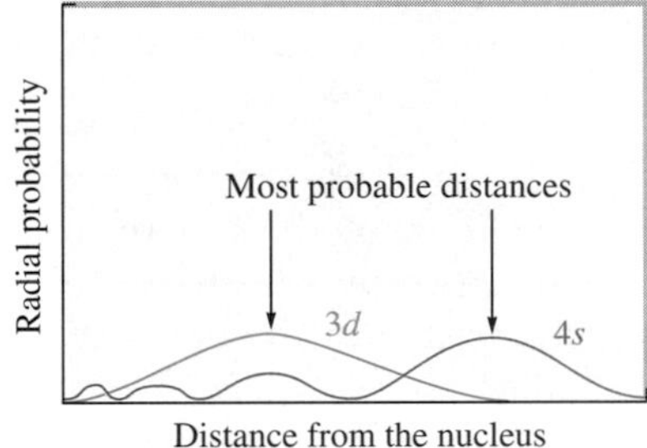

Figure 12.35
Radial probability distributions for the 3*d* and 4*s* orbitals. Note that the most probable distance of the electron from the nucleus for the 3*d* orbital is less than that for the 4*s* orbital. However, the 4*s* orbital allows more electron penetration close to the nucleus and thus is preferred over the 3*d* orbital.

Although the rigorous description of polyelectronic atoms is quite complicated, our simple qualitative ideas about electrons in independent orbitals are often very useful when we try to understand why atoms behave the way they do. We will consider some specific atomic properties in the next section.

12.15 Periodic Trends in Atomic Properties

We have developed a fairly complete picture of polyelectronic atoms that is quite successful in accounting for the periodic table of elements. We will next use the model to account for the observed trends in several important atomic properties: ionization energy, electron affinity, and atomic size.

Ionization Energy

Ionization energy is the energy required to remove an electron from a gaseous atom or ion,

$$X(g) \longrightarrow X^+(g) + e^-$$

Why Is Mercury a Liquid?

The silver liquid called mercury has been known since ancient times. In fact the symbol for mercury (Hg) comes from its Greek name *Hydrargyrum*, which means watery silver. Although elements in the liquid state at ambient temperature and pressure are quite rare (Br_2 is another example), the liquid nature of mercury is especially confounding. For example, compare the properties of mercury and gold:

	Mercury	Gold
Melting point	−39°C	1064°C
Density	13.6g cm^{-3}	19.3 g cm^{-3}
Enthalpy of fusion	2.30 kJ mol^{-1}	12.8 kJ mol^{-1}
Conductivity	10.4 kS m^{-1}	426 kS m^{-1}

It is quite apparent that these metals, which are neighbors on the periodic table, have strikingly different properties. Why? The answer is not at all straightforward—but very interesting. It seems to hinge on relativity.

Recall that Einstein postulated in his theory of special relativity in 1905 that the mass (m) of a moving object increases with its velocity (v):

$$m_{\text{relativistic}} = m_{\text{rest}}/\sqrt{1 - (v/c)^2}$$

where c is the speed of light. In the simple models for the atom we ignore relativistic effects on the electron mass. Although these effects are negligible for light atoms (the mass change is ~0.003% for the hydrogen electron), they become important for heavy elements such as gold and mercury. For example, the relativistic mass for a 1s electron in mercury is ~1.23 times its rest mass and this effect leads to a very significant contraction in the radius of the 1s orbital.

It turns out that relativity has an even more profound impact on atomic theory than the above calculations suggest. A relativistic treatment of the atom fundamentally changes the way we view the electrons in atoms. In fact, as shown by British physicist Paul Dirac, the concept of electron spin is unnecessary in a relativistic treatment of the atom. The point here is not to explain these very complex ideas, but to alert you to concepts you will be learning more about in higher level courses.

How does relativity explain why mercury has a melting point of −39°C while that of neighboring gold is 1064°C? The first step in answering this question involves considering the electron configurations of these atoms:

Au: [Xe] $4f^{14}5d^{10}6s^1$
Hg: [Xe] $4f^{14}5d^{10}6s^2$

where the atom or ion is assumed to be in its ground state. Although the values for ionization energy will be given in this text in terms kilojoules per mole of atoms, it is quite common in other chemical literature to see values given per atom. In that context the term *ionization potential* is used, and the units are electron-volts (eV) per atom (1 eV = 1.602 × 10^{-19} J).

The ionization energy for a particular electron in an atom is a source of information about the energy of the orbital it occupies in the atom. In fact, Koopman's theorem states: *the ionization energy of an electron is equal to the energy of the orbital from which it came.* This rule is an approximation because, among other things, it assumes that the electrons left behind in the resulting ion will not reorganize in response to the removal of an electron. However, ionization energies do provide information that is quite useful in testing the orbital model of the atom.

To introduce some of the characteristics of ionization energy, we will consider the energy required to remove several electrons in succession from

Notice that gold has an unfilled 6*s* subshell but the 6*s* level is filled in mercury. Because of its configuration a gold atom can use its half-filled 6*s* orbital to form a bond to another gold atom. In fact the metal-metal bond in the Au_2 molecule is an astonishingly strong 221 kJ mol^{-1}, a value very close to the bond energy of the Cl_2 molecule (239 kJ mol^{-1}) and greater than the bond energy of I_2 (149 kJ mol^{-1}). In addition, gold has an electron affinity (−220 kJ mol^{-1}) that is higher than that of oxygen and sulfur. Further, gold forms a compound with cesium (CsAu) that exhibits the CsCl crystal structure (see Figure 16.41 and Section 16.8) in which gold atoms take the place of Cl^- ions. Thus, a gold atom seems to behave a lot like a halogen atom.

What causes gold to emulate many properties of the nonmetallic halogens? The apparent answer lies in the dramatic contraction of the gold 6*s* orbital due to relativistic effects. The unexpectedly small radius of the 6*s* orbital of a gold atom results in a much lower energy than is predicted in the absence of relativistic effects. It is this very low energy unfilled 6*s* orbital that causes gold atoms to form very stable Au_2 molecules in the gas phase and to bind strongly to each other in the solid state producing its high melting point. This same low energy 6*s* orbital also leads to gold's unexpectedly high electron affinity and to its unusual color. The fact that gold is not the silvery color exhibited by most metals is due to the absorption of blue light to transfer an electron between the 5*d* and 6*s* orbitals in gold atoms.

So why are gold and mercury so different? The answer lies in the different electron configurations of the two atoms. Unlike gold, the low energy 6*s* orbital in mercury is filled and these two electrons are very tightly bound to the mercury atom. In fact, one can think of Hg as being analogous to He. That is, the low energy pair of 6*s* electrons on mercury causes it to behave like a noble gas atom—it cannot bond to another mercury atom. This explains why mercury is unique among metals in that it is almost entirely monomeric in its gas phase. In contrast, the species Hg_2^{2+} is extremely stable even in aqueous solution. This fact is not surprising once it is realized that Hg^+ is isoelectronic with Au.

Thus it is the unusually low energy of the 6*s* orbitals apparently due by relativistic effects that cause gold to behave like a halogen atom and mercury to behave like a noble gas.

Suggested Reading

"Why Is Mercury Liquid" by Lars J. Norrby, J. Chem. Ed., *68*, 110 (1991).

aluminum atoms in the gaseous state. The ionization energies are

$$Al(g) \longrightarrow Al^+(g) + e^- \qquad I_1 = 580 \text{ kJ/mol}$$
$$Al^+(g) \longrightarrow Al^{2+}(g) + e^- \qquad I_2 = 1815 \text{ kJ/mol}$$
$$Al^{2+}(g) \longrightarrow Al^{3+}(g) + e^- \qquad I_3 = 2740 \text{ kJ/mol}$$
$$Al^{3+}(g) \longrightarrow Al^{4+}(g) + e^- \qquad I_4 = 11{,}600 \text{ kJ/mol}$$

Several important points can be illustrated from these results. In a stepwise ionization process, it is always the highest-energy electron (the one bound least tightly) that is removed first. The energy required to remove the highest-energy electron of an atom is called the **first ionization energy** (I_1). The first electron removed from the aluminum atom comes from the 3*p* orbital (Al has the electron configuration $[Ne]3s^23p^1$). The second electron comes from the 3*s* orbital (since Al^+ has the configuration $[Ne]3s^2$).

Note that the value of I_1 is considerably smaller than the value of I_2, the **second ionization energy.** This result makes sense for several reasons. The primary factor is simply charge. Note that the first electron is removed from a neutral atom (Al), whereas the second electron is removed from a 1 + ion (Al^+). The increase in positive charge binds the electrons more firmly and the ionization energy increases. The same trend shows up in the third (I_3) and fourth (I_4) ionization energies, where the electron is removed from the Al^{2+} and Al^{3+} ions, respectively.

The increase in successive ionization energies for an atom also makes sense in terms of relative orbital energies. The increase from I_1 to I_2 is expected because the first electron is removed from a $3p$ orbital that is higher in energy than the $3s$ orbital from which the second electron is removed. The largest jump in ionization energy by far occurs in going from the third ionization energy (I_3) to the fourth (I_4). This large jump occurs because I_4 corresponds to removing a core electron (Al^{3+} has the configuration $1s^22s^22p^6$), and core electrons are bound much more tightly than valence electrons.

Table 12.6 gives the values of ionization energies for all of the Period 3 elements. Note the large jump in energy in each case in going from the removal of valence electrons to the removal of core electrons.

The values of the first ionization energy for the elements in the first five periods of the periodic table are graphed in Fig. 12.36. Note that, in general, as we go *across a period from left to right, the first ionization energy increases.* To account for this trend qualitatively, we need to discuss the concept of electrons **shielding** each other from the nuclear charge. Shielding occurs because electrons repel each other. Our simple pictures of the atom lead us to expect that core electrons should be quite effective in shielding outer electrons from the nuclear charge, since the core electrons are between the nucleus and the outer electrons. On the other hand, electrons in the same principal quantum level, which on the average are all at about the same distance from the nucleus, are not expected to shield each other very well. Thus as we move from left to right

TABLE 12.6 Successive Ionization Energies in Kilojoules per Mole for the Elements in Period 3

General increase →; General decrease ↑ (down the Element column)

Element	I_1	I_2	I_3	I_4	I_5	I_6	I_7
Na	495	4560					
Mg	735	1445	7730	Core electrons*			
Al	580	1815	2740	11,600			
Si	780	1575	3220	4350	16,100		
P	1060	1890	2905	4950	6270	21,200	
S	1005	2260	3375	4565	6950	8490	27,000
Cl	1255	2295	3850	5160	6560	9360	11,000
Ar	1527	2665	3945	5770	7230	8780	12,000

* Note the large jump in ionization energy in going from removal of valence electrons to removal of core electrons.

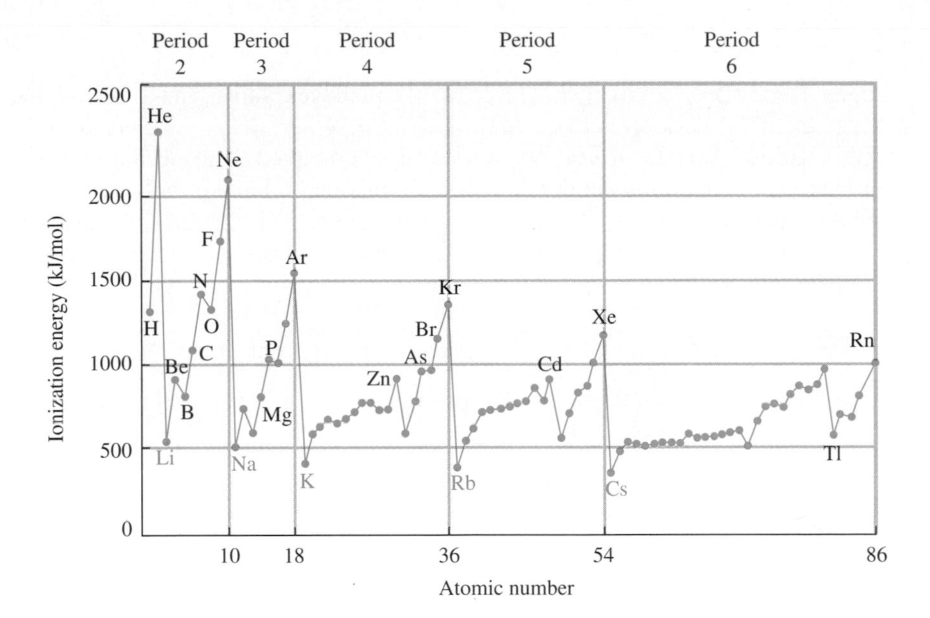

Figure 12.36
The values of first ionization energy for the elements in the first five periods. In general, ionization energy decreases in going down a group. For example, note the decrease in values for Group 1A and Group 8A. In general, ionization energy increases in going left to right across a period. For example, note the sharp increase going across Period 2 from lithium through neon.

in a given period of the periodic table, we do not expect the electrons to completely shield each other from the increasing nuclear charge as the number of protons in the nucleus increases. Thus the electrons are expected to be bound more firmly in going from left to right across a given period, which means that the ionization energy should increase.

Electrons in the same principal quantum level do not shield each other as well as core electrons shield outer electrons.

On the other hand, *the first ionization energy values decrease in going down a group*. This can be seen most clearly by focusing on the Group 1A elements (the alkali metals) and the Group 8A elements (the noble gases), as shown in Table 12.7. The main reason for the decrease in going down a group is that the electrons being removed are, on the average, farther from the nucleus. As n increases, the size of the orbital increases, and the electron is easier to remove.

First ionization energy increases across a period and decreases down a group.

In Fig. 12.36 we see that there are some discontinuities in ionization energy in going across a period. For example, discontinuities occur in Period 2, in going from beryllium to boron and from nitrogen to oxygen. These exceptions to the normal trend can be explained in terms of how electron repulsions depend on the electron configuration. We will discuss the elements in Period 2 individually to further develop the concept of shielding.

The increase in the ionization energy in going from lithium ($1s^22s^1$) to beryllium ($1s^22s^2$) is expected, since the $2s$ electrons do not shield each other completely. The decrease in the ionization energy in going from beryllium ($1s^22s^2$) to boron ($1s^22s^22p^1$) suggests that the electrons in the $2s$ orbital effectively shield the $2p$ electron. This is sensible in view of the greater penetration by the $2s$ electrons as compared with the $2p$ electron. The $2s$ electrons spend more time closer to the nucleus than the $2p$ electrons, where they provide effective shielding. The steady increase in the ionization energy in going from boron ($1s^22s^22p^1$) to carbon ($1s^22s^22p^2$) to nitrogen ($1s^22s^22p^3$) is expected, because the $2p$ electrons are not very effective in shielding each other from the increasing nuclear charge. The drop in the ionization energy in going from

TABLE 12.7
First Ionization Energies for the Alkali Metals and Noble Gases

Atom	I_1 (kJ/mol)
Group 1A	
Li	520.
Na	495
K	419
Rb	409
Cs	382
Group 8A	
He	2377
Ne	2088
Ar	1527
Kr	1356
Xe	1176
Rn	1042

nitrogen ($1s^22s^22p^3$) to oxygen ($1s^22s^22p^4$) can be explained as follows. In nitrogen each $2p$ electron is in a separate orbital. When the extra electron for oxygen is added, one $2p$ orbital becomes doubly occupied. The electron repulsions between the electrons in the doubly occupied orbital makes either of these electrons easier to remove. As we move from oxygen ($1s^22s^22p^4$) to fluorine ($1s^22s^22p^5$) to neon ($1s^22s^22p^6$), the ionization energy increases, because in each

EXAMPLE 12.9

The first ionization energy for phosphorus is 1060 kJ/mol, and that for sulfur is 1005 kJ/mol. Why?

Solution

Phosphorus and sulfur are neighboring elements in Period 3 of the periodic table and have the following valence electron configurations: phosphorus is $3s^23p^3$, and sulfur is $3s^23p^4$.

Ordinarily, the first ionization energy increases as we go across a period, so we might expect sulfur to have a greater ionization energy than phosphorus. However, in this case the fourth p electron in sulfur must be placed in an already occupied orbital. The especially large electron-electron repulsions that result cause this electron to be more easily removed than might be expected.

EXAMPLE 12.10

Consider atoms with the following electron configurations:

$$1s^22s^22p^6$$
$$1s^22s^22p^63s^1$$
$$1s^22s^22p^63s^2$$

Which atom has the largest first ionization energy, and which one has the smallest second ionization energy? Explain your choices.

Solution

The atom with the largest value of I_1 is the one with the configuration $1s^22s^22p^6$ (this is the neon atom), because this element is found at the right end of Period 2. Since the $2p$ electrons do not shield each other very effectively, I_1 will be large. The other configurations given include $3s$ electrons. These electrons are effectively shielded by the core electrons and are farther from the nucleus than the $2p$ electrons in neon. Thus I_1 for these atoms is smaller than I_1 for neon.

The atom with the smallest value of I_2 is the one with the configuration $1s^22s^22p^63s^2$ (the magnesium atom). For magnesium both I_1 and I_2 involve valence electrons. For the atom with the configuration $1s^22s^22p^63s^1$ (sodium), the second electron lost (corresponding to I_2) is a core electron (from a $2p$ orbital).

case the electron being considered is in a doubly occupied orbital. Note that the ionization energy values for oxygen, fluorine, and neon are different from those of boron, carbon, and nitrogen by an amount that represents the extra electron repulsions in doubly occupied orbitals, as shown in Fig. 12.36.

Electron Affinity

Electron affinity *is the energy change associated with the addition of an electron to a gaseous atom:*

$$X(g) + e^- \longrightarrow X^-(g)$$

The sign convention for electron affinity values follows the convention for energy changes used in Chapters 9 and 10.

Because two different conventions have been used, there is a good deal of confusion in the chemical literature about the signs for electron affinity values. Electron affinity has been defined in many textbooks as the energy *released* when an electron is added to a gaseous atom. This convention requires that a positive sign be attached to an exothermic addition of an electron to an atom, which opposes normal thermodynamic conventions. Therefore, in this book we define electron affinity as a *change* in energy. This means that if the addition of the electron is exothermic, the corresponding value for electron affinity will carry a negative sign.

Figure 12.37 shows the electron affinity values for the atoms among the first 20 elements that form stable, isolated, negative ions—that is, the atoms that undergo the addition of an electron as shown above. As expected, all of these elements have negative (exothermic) electron affinities. Note that the *more negative* the energy, the greater the quantity of energy released. Although electron affinities generally become more negative from left to right across a period, there are several exceptions to this rule in each period. The dependence of electron affinity on atomic number can be explained by considering the changes in electron repulsions as a function of electron configurations. For example, the fact that the nitrogen atom does not form a stable, isolated $N^-(g)$ ion, whereas carbon forms $C^-(g)$, reflects the difference in the electron configurations of these atoms. An electron added to nitrogen ($1s^22s^22p^3$) to form the $N^-(g)$ ion ($1s^22s^22p^4$) would have to occupy a $2p$ orbital that already contains one electron. The extra repulsion between the electrons in this doubly occupied orbital causes $N^-(g)$ to be unstable. When an electron is added to carbon ($1s^22s^22p^2$) to form the $C^-(g)$ ion ($1s^22s^22p^3$), no such extra repulsions occur.

In contrast to the nitrogen atom, the oxygen atom can add one electron to form the stable $O^-(g)$ ion. Presumably, oxygen's greater nuclear charge, compared with that of nitrogen, is sufficient to overcome the repulsion associated

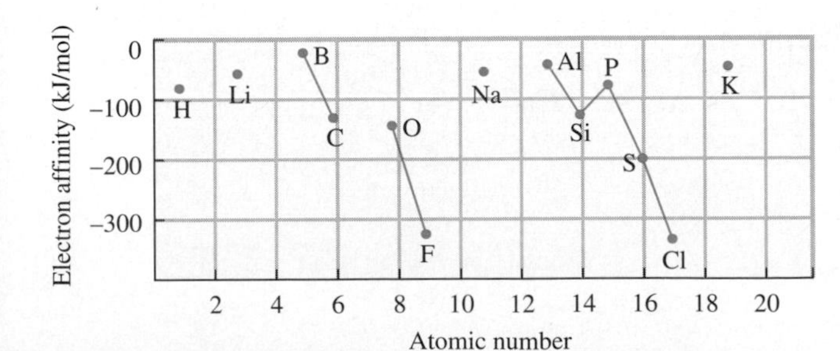

Figure 12.37
The electron affinity values for atoms among the first 20 elements that form stable, isolated X^- ions. The lines shown connect adjacent elements. The absence of a line indicates missing elements (He, Be, N, Ne, Mg, Ar, and Ca), whose atoms do not add an electron exothermically and thus do not form stable, isolated X^- ions.

with putting a second electron into an already occupied $2p$ orbital. However, it should be noted that a second electron *cannot* be added to an oxygen atom [$O^-(g) + e^- \nrightarrow O^{2-}(g)$] to form an isolated oxide ion. This outcome seems strange in view of the many stable oxide compounds (MgO, Fe_2O_3, and so on) that are known. As we will discuss in detail in Chapter 13, the O^{2-} ion is stabilized in ionic compounds by the large attractions that occur among the positive ions and the oxide ions.

When we go down a group, electron affinity should become more positive (less energy released), since the electron is added at increasing distances from the nucleus. Although this is generally the case, the changes in electron affinity in going down most groups are relatively small and numerous exceptions occur. This behavior is demonstrated by the electron affinities of the Group 7A elements (the halogens) shown in Table 12.8. Note that the range of values is quite small compared with the changes that typically occur across a period. Also note that although chlorine, bromine, and iodine show the expected trend, the energy released when an electron is added to fluorine is smaller than might be expected. This smaller energy release has been attributed to the small size of the $2p$ orbitals. Because the electrons must be very close together in these orbitals, there are unusually large electron-electron repulsions. In the other halogens with their larger orbitals, the repulsions are not as severe.

TABLE 12.8
Electron Affinities of the Halogens

Atom	Electron Affinity (kJ/mol)
F	−327.8
Cl	−348.7
Br	−324.5
I	−295.2

Atomic Radius

Just as the size of an orbital cannot be specified exactly, neither can the size of an atom. We must make some arbitrary choices to obtain values for **atomic radii**. These values can be obtained by measuring the distances between atoms in chemical compounds. For example, in the bromine molecule the distance between the two nuclei is known to be 228 pm. The bromine atomic radius is assumed to be half this distance, or 114 pm, as shown in Fig. 12.38. Measurements of this type have led to the values of the atomic radii for the elements shown in Fig. 12.39. These radii are often called *covalent atomic radii* because of the way they are determined. These values are significantly smaller than might be expected from the 90% electron density volumes of isolated atoms, because when atoms form bonds, their electron "clouds" interpenetrate. However, these values form a self-consistent data set that can be used to discuss the

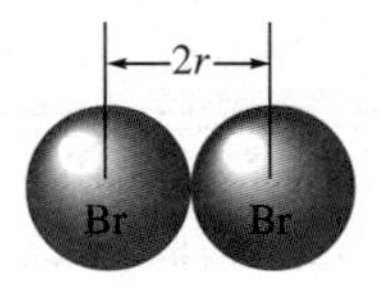

Figure 12.38
The radius of an atom (r) is defined as half the distance between the nuclei in a molecule consisting of identical atoms.

EXAMPLE 12.11

Predict the trend in radius of the following ions: Be^{2+}, Mg^{2+}, Ca^{2+}, and Sr^{2+}.

Solution

All these ions are formed by removing two electrons from an atom of a Group 2A element. In going from beryllium to strontium, we are going down the group, so the sizes increase.

$$Be^{2+} < Mg^{2+} < Ca^{2+} < Sr^{2+}$$

Smallest radius ↑ (Be²⁺)   Largest radius ↑ (Sr²⁺)

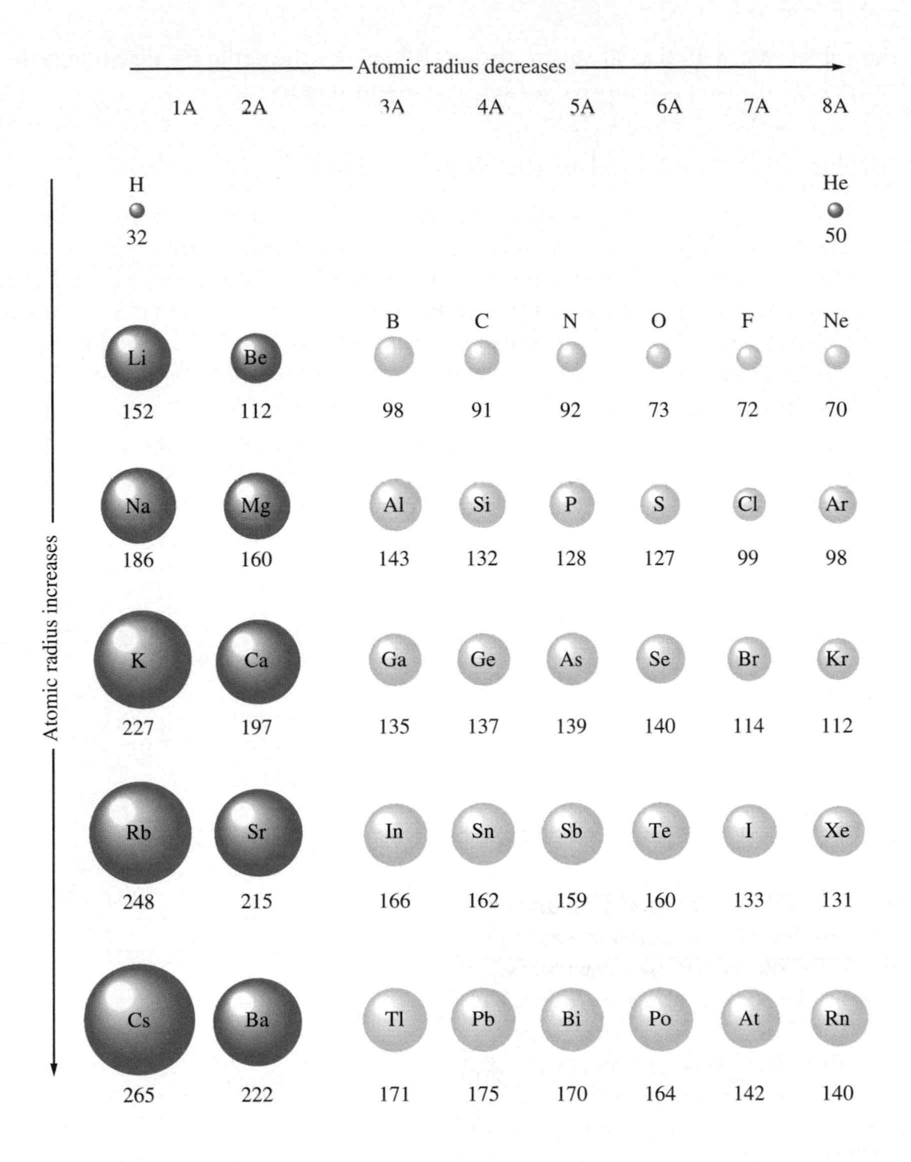

Figure 12.39
Atomic radii (in picometers) for selected atoms. Note that atomic radius decreases going across a period and increases going down a group. The values for the noble gases are estimated, because data from bonded atoms are lacking.

trends in atomic radii. Note from Fig. 12.39 that the atomic radius decreases in going from left to right across a period. This decrease can be explained in terms of the increasing effective nuclear charge (decreasing shielding) in going from left to right. This means that the valence electrons are drawn closer to the nucleus, decreasing the size of the atom.

Atomic radius increases down a group, because of the increases in the orbital sizes in successive principal quantum levels.

Atomic radius decreases across a period and increases down a group.

12.16 The Properties of a Group: The Alkali Metals

We have seen that the periodic table originated as a way to portray the systematic properties of the elements. Mendeleev was primarily responsible for first showing its usefulness in correlating and predicting the elemental properties. In this section we will summarize much of the information available from the table. We will also illustrate the usefulness of the table by discussing the properties of a representative group, the alkali metals.

Information Contained in the Periodic Table

1. The essence of the periodic table is that the groups of representative elements exhibit similar chemical properties that change in a regular way. The quantum mechanical model has allowed us to understand that the similarity of properties of the atoms in a group arises from the identical valence electron configurations shared by group members. *It is the number and type of valence electrons that primarily determine an atom's chemistry.*
2. One of the most valuable types of information available from the periodic table is the electron configuration of any representative element. If you

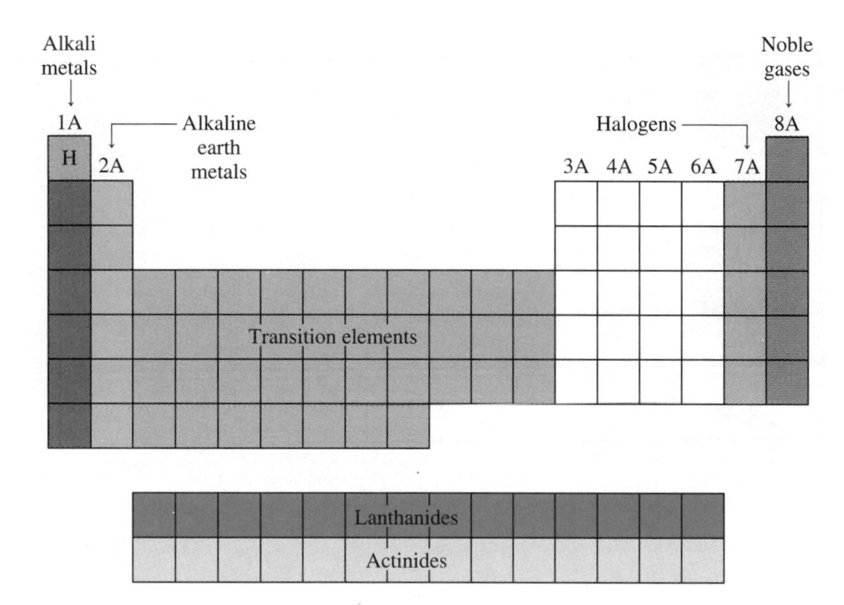

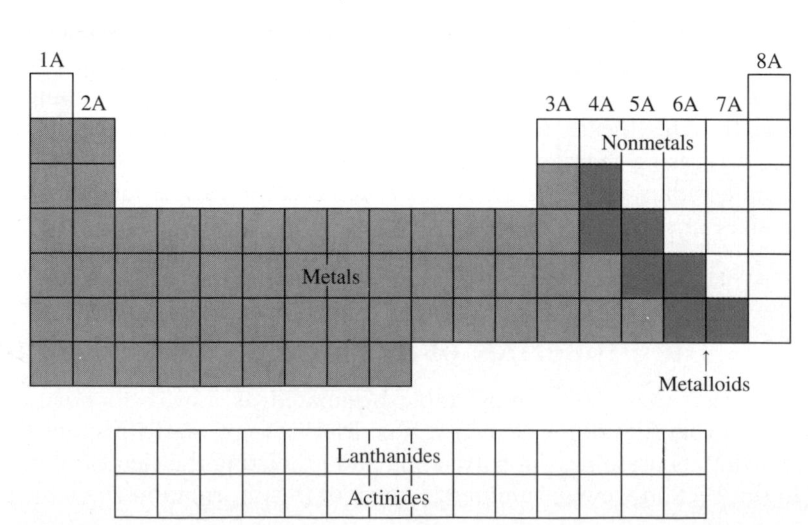

Figure 12.40
Special names for groups in the periodic table.

understand the organization of the table, you do not need to memorize electron configurations for the elements. Although the predicted electron configurations for transition metals are sometimes incorrect, this is not a serious problem. You should, however, memorize the configuration of two exceptions, chromium and copper, since these 3*d* transition elements are found in many important compounds.

3. As we mentioned in Chapter 2, certain groups in the periodic table have special names. These names are summarized in Fig. 12.40.
4. The most fundamental classification of the elements is into metals and nonmetals. The essential chemical property of a metal is the tendency to give up electrons to form a positive ion; metals tend to have low ionization energies. The metallic elements are found on the left side of the table, as shown in Fig. 12.40. The most reactive metals are found on the lower left-hand portion of the table where the ionization energies are smallest. The distinctive chemical property of a nonmetal is the ability to gain electrons to form an anion when reacting with a metal. The nonmetals have large ionization energies and most have negative electron affinities. The nonmetals are found on the right side of the table. The most reactive ones are located in the upper right-hand corner, excluding the noble gas elements, which are quite unreactive. The division between metals and nonmetals shown in Fig. 12.40 is only approximate. Many elements along the division line exhibit both metallic and nonmetallic properties under certain circumstances. These elements are called **metalloids,** or *semimetals.*

The Alkali Metals

The metals of Group 1A, the alkali metals, illustrate well the relationships among the properties of the elements in a group. Lithium, sodium, potassium, rubidium, cesium, and francium are the most reactive of the metals. We will not discuss francium here since it occurs in nature in only very small quantities. Although hydrogen is found in Group 1A it behaves as a nonmetal, in contrast to the other members of that group. The fundamental reason for hydrogen's nonmetallic character is its very small size (see Fig. 12.39). The electron in the small 1*s* orbital is bound very tightly to the nucleus.

Some important properties of the first five alkali metals are shown in Table 12.9. The data in Table 12.9 show that when we move down the group, the first ionization energy decreases and the atomic radius increases. This agrees with the general trends discussed in Section 12.15.

TABLE 12.9 Properties of Five Alkali Metals

Element	Valence Electron Configuration	Density at 25°C (g/cm³)	mp (°C)	bp (°C)	First Ionization Energy (kJ/mol)	Atomic (covalent) Radius (pm)	Ionic (M^+) Radius (pm)
Li	$2s^1$	0.53	180	1330	520	152	60
Na	$3s^1$	0.97	98	892	495	186	95
K	$4s^1$	0.86	64	760	419	227	133
Rb	$5s^1$	1.53	39	688	409	248	148
Cs	$6s^1$	1.87	29	690	382	265	169

When Hard-Pressed, Hydrogen Becomes a Metal

The most active metals are found in Group 1A. In contrast to the other members of this group, hydrogen is typically a nonmetal. The very small size of the hydrogen atom means that its electron is bound quite firmly, and when hydrogen bonds to other nonmetals, it typically does so by sharing electrons rather than losing its electron to become a cation. However, recent research indicates that under extraordinarily high pressures, hydrogen changes to a metal. Ho-Kwang Mao and R. J. Hemley have performed experiments using a diamond-anvil cell in which mechanical means are used to produce millions of atmospheres of pressure in a tiny 20-μm cell between two beveled, gem-quality diamonds. These researchers found that under *3 million atm* of pressure, hydrogen first assumes a solid phase that is transparent and does not conduct electricity, and then it changes to an opaque solid that conducts electricity like a metal. There is a great deal of interest in this metallic form of hydrogen because theory predicts it should be a superconducting solid (zero resistance to current flow) at temperatures near 0°C, much higher than the temperatures required for any presently known superconducting material. However, so far no conditions have been found that stabilize the metallic form of hydrogen other than extraordinarily high pressure. Researchers are hoping that a way can be found to stabilize the metallic hydrogen phase so that it persists under less severe conditions.

The overall increase in density in going down Group 1A is typical of all groups. It occurs because atomic mass generally increases more rapidly than atomic size.

The smooth decrease in melting point (mp) and boiling point (bp) in going down Group 1A is not typical; in most other groups more complicated behavior occurs. Note that the melting point of cesium is only 29°C. Cesium can be melted readily from the heat of your hand. Cesium's low melting point is very unusual—metals typically have high melting points. For example, tungsten melts at 3410°C. The only other metals with low melting points are mercury (mp = −38°C) and gallium (mp = 30°C).

Recall that the chemical property most characteristic of a metal is the ability to lose its valence electrons. The Group 1A elements are very reactive. They have low ionization energies and react readily with nonmetals to form ionic solids. A typical example involves the reaction of sodium with chlorine to form sodium chloride,

$$2\mathrm{Na}(s) + \mathrm{Cl_2}(g) \longrightarrow 2\mathrm{NaCl}(s)$$

This is an oxidation-reduction reaction in which chlorine oxidizes sodium to form Na^+ and Cl^- ions. In the reactions between metals and nonmetals it is typical for the nonmetal to behave as the oxidizing agent and the metal to behave as the reducing agent, as shown by the following reactions:

$$2\mathrm{Na}(s) + \mathrm{S}(s) \longrightarrow \mathrm{Na_2S}(s)$$
Contains Na^+ and S^{2-} ions

$$6\mathrm{Li}(s) + \mathrm{N_2}(g) \longrightarrow 2\mathrm{Li_3N}(s)$$
Contains Li^+ and N^{3-} ions

$$4\mathrm{Na}(s) + \mathrm{O_2}(g) \longrightarrow 2\mathrm{Na_2O}(s)$$
Contains Na^+ and O^{2-} ions

Lithium: Behavior Medicine

More and more people in our society seem to be suffering from the debilitating effects of mania and depression, but the alkali metal lithium can provide help for many. In fact, over 3 million prescriptions for lithium carbonate are filled annually by retail pharmacies.

Although the details are not well understood, the lithium ion seems to alleviate mood disorders by affecting the way that brain cells respond to neurotransmitters, a class of molecules that facilitate the transmission of nerve impulses.

Specifically, physiologists think that the lithium ion may interfere with a complex cycle of reactions that relays and amplifies messages carried to the cells by neurotransmitters and hormones. They theorize that exaggerated forms of behavior, such as mania or depression, arise from the overactivity of this cycle. Thus the fact that lithium inhibits this cycle may be responsible for its moderating effect on behavior.

There is a growing collection of evidence that violent behavior may result at least partially from the improper regulation of neurotransmitters and hormones. For example, a study in Finland showed that violent criminals, especially arsonists, often had low levels of serotonin, a common neurotransmitter. Studies are now under way to determine whether lithium might also be effective for treating these and other aberrant forms of behavior.

For reactions of this type, the relative reducing powers of the alkali metals can be predicted from the first ionization energies listed in Table 12.9. Since it is much easier to remove an electron from a cesium atom than from a lithium atom, cesium should be the better reducing agent. The expected trend in reducing ability is

$$\mathrm{Cs} > \mathrm{Rb} > \mathrm{K} > \mathrm{Na} > \mathrm{Li}$$

This order is observed experimentally for direct reactions between the solid alkali metals and nonmetals. However, this order of reducing ability is not observed when the alkali metals react in aqueous solution. For example, the reduction of water by an alkali metal is very vigorous and exothermic:

$$2\mathrm{M}(s) + 2\mathrm{H_2O}(l) \longrightarrow \mathrm{H_2}(g) + 2\mathrm{M}^+(aq) + 2\mathrm{OH}^-(aq) + \text{energy}$$

The order of reducing abilities observed for this reaction (for the first three group members) is

$$\mathrm{Li} > \mathrm{K} > \mathrm{Na}$$

which is not the order expected from the relative ionization energies of these metals.

This unexpected order occurs because the formation of the M^+ ions in aqueous solution is strongly influenced by the hydration of these ions by the polar water molecules. The hydration energy of an ion represents the change in energy that occurs when water molecules attach to the M^+ ion. The hydration energies for the Li^+, Na^+, and K^+ ions shown in Table 12.10 indicate that the process is exothermic in each case. However, nearly twice as much energy is released by the hydration of the Li^+ ion as compared with the K^+ ion. This difference is caused by size effects; the Li^+ ion is much smaller than the K^+ ion,

TABLE 12.10
Hydration Energies for Li^+, Na^+, and K^+ Ions

Ion	Hydration Energy (kJ/mol)
Li^+	−500
Na^+	−400
K^+	−300

and thus its *charge density* (charge per unit volume) is much greater. This means that the polar water molecules are more strongly attracted to the small Li^+ ion. Because the Li^+ ion is so strongly hydrated, its formation from the lithium atom occurs more readily than the formation of the K^+ ion from the potassium atom. Although a potassium atom in the gas phase loses its valence electron more readily than a lithium atom in the gas phase, the opposite is true in aqueous solution. This anomaly is an example of the importance of the polarity of the water molecule in aqueous reactions.

There is one more surprise involving the highly exothermic reactions of the alkali metals with water. Experiments show that lithium is the best reducing agent in water, so we might expect lithium to react most violently with water. However, it does not. Sodium and potassium react much more vigorously. Why? The answer lies in the relatively high melting point for lithium. When sodium and potassium react with water, the heat evolved causes them to melt, giving a larger area of contact with water. Lithium, on the other hand, does not melt under these conditions and thus reacts more slowly. This example illustrates the important principle (which we will discuss in detail in Chapter 15) that the energy of a reaction and the rate at which it occurs are not necessarily related.

In this section we have seen that the trends in atomic properties summarized by the periodic table can be a great help in understanding the chemical behavior of the elements. This fact will be emphasized over and over as we proceed in our study of chemistry.

Exercises

A blue exercise number indicates that the answer to that exercise appears at the end of this book.

Light and Matter

1. What experimental evidence supports the quantum theory for light?
2. The laser in an audio compact disc player uses light with a wavelength of 7.80×10^2 nm. What is the frequency of this light? What is the energy of a single photon of this light?
3. Microwave radiation has a wavelength on the order of 1.0 cm. Calculate the frequency and the energy of a single photon of this radiation. Calculate the energy of an Avogadro's number of photons (called an einstein) of this electromagnetic radiation.
4. It takes 7.21×10^{-19} J of energy to remove an electron from an iron atom. What is the maximum wavelength of light that can accomplish this?
5. The work function of an element is the energy required to remove an electron from the surface of the solid. The work function for lithium is 279.7 kJ/mol (that is, it takes 279.7 kJ of energy to remove 1 mol of electrons from 1 mol of Li atoms on the surface of Li metal). What is the maximum wavelength of light that can remove an electron from an atom in lithium metal?
6. The primary visible emissions from mercury are at 404.7 nm and 435.8 nm. Calculate the frequencies for these emissions. Calculate the energy of a single photon and of a mole of photons of light with each of these wavelengths.
7. The ionization energy of gold is 890.1 kJ/mol. Is light with a wavelength of 225 nm capable of ionizing a gold atom in the gas phase?
8. It takes 492 kJ to remove 1 mol of electrons from the surface of solid gold. How much energy does it take to remove a single electron from the surface of gold? What is the maximum wavelength of light capable of doing this?
9. It takes 208.4 kJ of energy to remove one mol of electrons from the atoms on the surface of rubidium metal. If rubidium metal is irradiated with 254-nm light, what is the maximum kinetic energy the released electrons can have?

Hydrogen Atom: The Bohr Model

10. In the Bohr model of the hydrogen atom, precisely what is quantized?
11. Calculate the wavelength of light emitted in each of the following spectral transitions in the hydrogen atom.

a. $n = 3 \rightarrow n = 2$ c. $n = 2 \rightarrow n = 1$
b. $n = 4 \rightarrow n = 2$ d. $n = 4 \rightarrow n = 3$

12. What is the maximum wavelength of light capable of removing an electron from a hydrogen atom in the energy states characterized by $n = 1$ and $n = 3$?

13. An electron is excited from the ground state to the $n = 3$ state in a hydrogen atom. Which of the following statements are true? Correct any false statements.
    a. It takes more energy to ionize the electron from $n = 3$ than from the ground state.
    b. The electron is farther from the nucleus on average in the $n = 3$ state than in the ground state.
    c. The wavelength of light emitted if the electron drops from $n = 3$ to $n = 2$ is shorter than the wavelength of light emitted if the electron falls from $n = 3$ to $n = 1$.
    d. The wavelength of light emitted when the electron returns to the ground state from $n = 3$ is the same as the wavelength of light absorbed to go from $n = 1$ to $n = 3$.
    e. The first excited state corresponds to $n = 3$.

14. Calculate the longest and shortest wavelengths of light emitted by electrons in the hydrogen atoms that begin in the $n = 6$ state and then fall to the state with the appropriate smaller value of n.

15. An excited hydrogen atom emits light with a frequency of 1.141×10^{14} Hz to reach the energy level for which $n = 4$. In what principal quantum level did the electron begin?

Wave Mechanics and Particle in a Box

16. Calculate the de Broglie wavelength for each of the following.
    a. a proton with a velocity 5.0% of the speed of light
    b. an electron with a velocity 15% of the speed of light
    c. the fastest measured fastball (a 5.2-oz baseball with a velocity of 100.8 mph)

17. Neutron diffraction is used in determining the structures of molecules.
    a. Calculate the de Broglie wavelength of a neutron moving at 1.00% of the speed of light.
    b. Calculate the velocity of a neutron with a wavelength of 75 pm (1 pm $= 10^{-12}$ m).

18. Calculate the velocities of electrons with the de Broglie wavelengths of 1.0×10^2 nm and 1.0 nm.

19. What would the value of Planck's constant have to be in order for the wavelength of the baseball in Exercise 16, part c, to be 5.0 cm? 5.0×10^2 nm?

20. Summarize some of the evidence supporting the wave properties of matter.

21. The Heisenberg uncertainty principle can be expressed in the form

$$\Delta E \cdot \Delta t \geq \frac{\hbar}{2}$$

where E represents energy and t represents time. Show that the units for this form are the same as the units for the form used in this chapter:

$$\Delta x \cdot \Delta p \geq \frac{\hbar}{2}$$

22. Using the Heisenberg uncertainty principle, calculate Δx for each of the following.
    a. an electron with $\Delta v = 0.100$ m/s
    b. a baseball (mass $= 145$ g) with $\Delta v = 0.100$ m/s
    c. How does the answer in part a compare with the size of a hydrogen atom?
    d. How does the answer in part b correspond to the size of a baseball?

23. Calculate the wavelength of the electromagnetic radiation required to excite an electron from the ground state to the level with $n = 5$ in a one-dimensional box 40.0 pm in length.

24. Discuss what happens to the energy levels for an electron trapped in a one-dimensional box as the length of the box increases.

25. Which has the lowest (ground-state) energy, an electron trapped in a one-dimensional box of length 10^{-6} m or one with length 10^{-10} m?

26. The treatment of a particle in a one-dimensional box can be extended to a rectangular box of dimensions L_x, L_y, and L_z, yielding the following expression for energy:

$$E = \frac{h^2}{8m}\left(\frac{n_x^2}{L_x^2} + \frac{n_y^2}{L_y^2} + \frac{n_z^2}{L_z^2}\right)$$

The three quantum numbers n_x, n_y, and n_z independently can assume only integer values.
    a. Determine the energies of the three lowest levels, assuming that the box is cubic.
    b. Describe the degeneracies of all the levels that correspond to quantum numbers having values of 1 or 2. How will these degeneracies change in a box where $L_x \neq L_y \neq L_z$?

Orbitals and Quantum Numbers

27. Concerning the hydrogen atom, what information do we get from the values of the quantum numbers n, ℓ, and m_ℓ?

28. Which of the following orbital designations are incorrect: 1*s*, 1*p*, 7*d*, 9*s*, 3*f*, 4*f*, 2*d*?

29. Which of the following sets of quantum numbers are not allowed in the hydrogen atom? For the sets of quantum numbers that are incorrect, state what is wrong.
    a. $n = 2, \ell = 1, m_\ell = -1$
    b. $n = 1, \ell = 1, m_\ell = 0$
    c. $n = 8, \ell = 7, m_\ell = -6$
    d. $n = 1, \ell = 0, m_\ell = 2$
    e. $n = 3, \ell = 2, m_\ell = 2$

f. $n = 4, \ell = 3, m_\ell = 4$
g. $n = 0, \ell = 0, m_\ell = 0$
h. $n = 2, \ell = -1, m_\ell = 1$

30. Are the possible values of ℓ for an electron with $n = 2$ the same as those for an electron with $n = 3$?
31. How many orbitals can have the designation $5p$, $3d_{z^2}$, $4d$, $n = 5$, and $n = 4$?
32. In defining the sizes of orbitals, why must we use an arbitrary value, such as 90% of the total probability?
33. From the diagrams of $2p$ and $3p$ orbitals in Fig. 12.19 and Fig. 12.20, draw a rough graph of the square of the wave function for these orbitals in the direction of one of the lobes.
34. How do the $2p$ orbitals differ from each other?
35. How do the $2p$ and $3p$ orbitals differ from each other?
36. What is a nodal surface in an atomic orbital?
37. What is the physical significance of the value of ψ^2 at a particular point?
38. For hydrogen atoms, the wave function for the state $n = 3, \ell = 0, m_\ell = 0$ is

$$\psi_{300} = \frac{1}{81\sqrt{3\pi}}\left(\frac{1}{a_0}\right)^{3/2}(27 - 18\sigma + 2\sigma^2)e^{-\sigma/3}$$

where $\sigma = r/a_0$ and a_0 is the Bohr radius (5.29×10^{-11} m). Calculate the position of the nodes for this wave function.

Polyelectronic Atoms

39. Why can we not account exactly for the repulsions among electrons in a polyelectronic atom?
40. Calculate the ionization energy (in kJ/mol) for each of the following one-electron species, using the Bohr model.
a. H b. He^+ c. Li^{2+} d. C^{5+} e. Fe^{25+}
41. What are the relative sizes of the $1s$ orbitals in the species in Exercise 40?
42. One of the emission spectral lines for Be^{3+} has a wavelength of 253.4 nm for an electronic transition that begins in the state with $n = 5$. What is the principal quantum number of the lower-energy state corresponding to this emission?
43. We define a spin quantum number. Do we know for a fact that an electron literally spins like a ball?
44. What is the maximum number of electrons in an atom that can have these quantum numbers?
a. $n = 4$
b. $n = 5, m_\ell = +1$
c. $n = 5, m_s = +\frac{1}{2}$
d. $n = 3, \ell = 2$
e. $n = 2, \ell = 1$
f. $n = 0, \ell = 0, m_\ell = 0$
g. $n = 2, \ell = 1, m_\ell = -1, m_s = -\frac{1}{2}$
h. $n = 3$
i. $n = 2, \ell = 2$
j. $n = 1, \ell = 0, m_\ell = 0$

45. The elements Si, Ga, As, Ge, Al, Cd, S, and Se are all used in the manufacture of various semiconductor devices. Write the expected electron configuration for these atoms.
46. The elements Cu, O, La, Y, Ba, Tl, and Bi are all found in high-temperature ceramic superconductors. Write the expected electron configuration for these atoms.
47. Write the expected electron configurations for the following atoms: Sc, Fe, P, Cs, Eu, Pt, Xe, Br.
48. Write the expected electron configurations for the following atoms: K, Rb, Fr, Pu, Sb, Os, Pd, Pb, I.
49. Using Fig. 12.30, list elements (ignore the lanthanides and actinides) that have ground-state electron configurations that differ from those we would expect from their positions in the periodic table.
50. Write the electron configuration for each of the following.
a. the smallest halogen
b. the alkali metal with $2p$ and $3p$ electrons
c. the three lightest alkaline earth metals
d. the Group 3A element in the same period as Sn
e. the nonmetallic elements in Group 4A
f. the (as yet undiscovered) noble gas after radon
51. What do we mean when we say that a $4s$ electron is more penetrating than a $3d$ electron?
52. Why do we emphasize the valence electrons in an atom when discussing atomic properties?
53. A certain oxygen atom has the electron configuration $1s^2 2s^2 2p_x^2 2p_y^2$. How many unpaired electrons are present? Is this an excited state for oxygen? In going from this state to the ground state, would energy be released or absorbed?
54. How many unpaired electrons are there in the ground-state for each of the following elements: Sc, Ti, Al, Sn, Te, and Br?
55. Identify the following elements.
a. An excited state of this element has the electron configuration $1s^2 2s^2 2p^5 3s^1$.
b. The ground-state electron configuration is $[Ne]3s^2 3p^4$.
c. An excited state of this element has the electron configuration $[Kr]5s^2 4d^6 5p^2 6s^1$.
d. The ground-state electron configuration contains three unpaired $6p$ electrons.

The Periodic Table and Periodic Properties

56. Arrange the following groups of atoms in order of increasing size.
a. Be, Mg, Ca
b. Te, I, Xe
c. Ga, Ge, In
d. As, N, F
e. S, Cl, F

57. Arrange the atoms in Exercise 56 in order of increasing first ionization energy.

58. The first ionization energies of Ge, As, and Se are 0.7622, 0.944, and 0.9409 MJ/mol, respectively. Rationalize these values in terms of electron configurations.

59. Many times the claim is made that subshells half-filled with electrons are particularly stable. Can you suggest a physical basis for this claim?

60. We expect the atomic radius to increase down a group in the periodic table. Can you suggest why the atomic radius of hafnium breaks this rule? (See data below.)

Element	Atomic Radius (Å)	Element	Atomic Radius (Å)
Sc	1.57	Ti	1.477
Y	1.693	Zr	1.593
La	1.915	Hf	1.476

61. In 1994 at an American Chemical Society meeting, it was proposed that element 106 be named seaborgium, Sg, in honor of Glenn Seaborg, discoverer of the first transuranium element. This would be the first element named after a living scientist.
    a. Write the expected electron configuration for Sg.
    b. What other element would be most like Sg in its properties?
    c. Write the formula for a possible oxide and a possible oxyanion of Sg.

62. Predict some of the properties of element 117 (the symbol is Uus following conventions proposed by the International Union of Pure and Applied Chemistry, or IUPAC).
    a. What will be its electron configuration?
    b. What element will it most resemble chemically?
    c. What will be the formulas of the neutral binary compounds it forms with sodium, magnesium, carbon, and oxygen?
    d. What oxyanions would you expect Uus to form?

63. Order each of the following sets from the least exothermic electron affinity to the most.
    a. O, S
    b. F, Cl, Br, I
    c. N, O, F

64. The changes in electron affinity as one goes down a group in the periodic table are not nearly as large as the variations in ionization energies. Why?

65. The electron affinity of sodium is −52.9 kJ/mol. In principle, should it be possible to add one electron to Na(*g*) to form the sodide ion, Na^-(*g*)?

66. Which has the more negative electron affinity, the oxygen atom or the O^- ion? Explain your answer.

67. In each of the following sets, which atom or ion has the smallest radius?
    a. Li, Na, K
    b. P, As
    c. O^+, O, O^-
    d. S, Cl, Kr
    e. Pd, Ni, Cu

68. In each of the following sets, which atom or ion has the smallest ionization energy?
    a. Cs, Ba, La
    b. Zn, Ga, Ge
    c. Tl, In, Sn
    d. Tl, Sn, As
    e. O, O^-, O^{2-}

69. The electron affinities of the elements from aluminum to chlorine are −44, −120, −74, −200.4, and −348.7 kJ/mol, respectively. Rationalize the trend in these values.

70. Three elements have the electron configurations $1s^22s^22p^63s^23p^6$, $1s^22s^22p^63s^2$, and $1s^22s^22p^63s^23p^64s^1$. The first ionization energies of the three elements (not in the same order) are 0.419, 0.735, and 1.527 MJ/mol. The atomic radii are 1.60, 0.98, and 2.35 Å. Identify the three elements and match the appropriate values of ionization energy and atomic radius to each configuration.

71. Use data in this chapter to determine the following.
    a. the electron affinity of Mg^{2+}
    b. the electron affinity of Al^+
    c. the ionization energy of Cl^-
    d. the ionization energy of Cl
    e. the electron affinity of Cl^+

72. The first ionization energies of the so-called coinage metals, Cu, Ag, and Au, are 745.5, 731.0, and 890.1 kJ/mol, respectively. Is this the trend you might have expected? In the *CRC Handbook of Chemistry and Physics*, look up the first ionization energies of the following sets of elements: (Sc, Y, La), (Ti, Zr, Hf), (Fe, Ru, Os), (Ga, In, Tl), and (As, Sb, Bi). Which groups follow the same trend as the coinage metals?

73. Calculate the electron affinities for Cu^+(*g*), Ag^+(*g*), and Au^+(*g*) using the data in Exercise 72.

74. Why do the successive ionization energies of an atom always increase?

75. Note the successive ionization energies for silicon given in Table 12.6. Would you expect to see any large jumps between successive ionization energies of silicon as you removed all of the electrons, one by one, beyond those shown in the table?

76. For each of the following pairs of elements,

    (Li and K) (S and Sc) (B and N) (F and Cl)

    pick the one with
    a. the more favorable (exothermic) electron affinity
    b. the higher ionization energy
    c. the larger size

The Alkaline Metals

77. Lithium has many more anhydrous salts that are hygroscopic (they readily absorb water) than the other alkali metals. Explain.

78. An ionic compound of potassium and oxygen has the empirical formula KO. Would you expect this compound to be potassium(II) oxide or potassium peroxide? Explain.

79. Complete and balance the equations for the following reactions.
   a. $Li(s) + O_2(g) \rightarrow$ c. $Cs(s) + H_2O(l) \rightarrow$
   b. $K(s) + S(s) \rightarrow$ d. $Na(s) + Cl_2(g) \rightarrow$

80. Cesium was discovered in natural mineral waters in 1860 by R. W. Bunsen and G. R. Kirchhoff, using the spectroscope they invented in 1859. The name came from the Latin word *caesius*, "sky blue," describing the prominent blue line observed for this element at 455.5 nm. Calculate the frequency and energy of a photon of this light.

81. Small daily doses (1–2 g) of lithium carbonate taken orally are often given to treat manic-depressive psychoses. This dosage maintains the level of lithium ion in the blood at approximately 1×10^{-3} mol/L.
   a. What is the formula of lithium carbonate?
   b. How many grams of lithium are present per liter of blood in these patients?

82. Give the name and formula of the binary compound formed by each of the following pairs of elements.
   a. Li and N d. Li and P
   b. Na and Br e. Rb and H
   c. K and S f. Na and H

83. Predict the atomic number of the next alkali metal after francium, and give its ground-state electron configuration.

Additional Exercises

84. The bright yellow light emitted by a sodium vapor lamp consists of two emission lines at 589.0 nm and 589.6 nm. What are the frequency and the energy of a photon of light at each of these wavelengths? What are the energies in kJ/mol?

85. Spectroscopists use emission spectra to confirm the presence of an element in materials of unknown composition. How is this possible?

86. How many unpaired electrons are in each of the following in the ground state: O, O^+, O^-, Fe, Mn, S, F, Ar?

87. On which quantum number(s) does the energy of an electron depend in each of the following?
   a. a one-electron atom or ion
   b. an atom or ion with more than one electron

88. The wave function for the $2p_z$ orbital in the hydrogen atom is

$$\psi_{2p_z} = \frac{1}{4\sqrt{2\pi}}\left(\frac{Z}{a_0}\right)^{3/2} \sigma e^{-\sigma/2} \cos\theta$$

where a_0 is the value for the radius of the first Bohr orbit in meters (5.29×10^{-11}), σ is $Z(r/a_0)$, r is the value for the distance from the nucleus in meters, and θ is an angle. Calculate the value of $\psi_{2p_z}{}^2$ at $r = a_0$ for $\theta = 0$ (z axis) and for $\theta = 90°$ (xy plane).

89. Elements with very large ionization energies also tend to have highly exothermic electron affinities. Explain. Which group of elements would you expect to be an exception to this statement?

90. Using the element phosphorus as an example, write the equation for a process in which the energy change is equal to each of the following.
   a. the ionization energy b. the electron affinity

91. The work function is the energy required to remove an electron from the surface of a metal. How does this definition differ from that for ionization energy?

92. In the hydrogen atom what is the physical significance of the state for which $n = \infty$ and $E = 0$?

93. Which of the following electron configurations correspond to an excited state? Identify the atoms and write the ground-state electron configuration where appropriate.
   a. $1s^2 2s^2 3p^1$ c. $1s^2 2s^2 2p^4 3s^1$
   b. $1s^2 2s^2 2p^6$ d. $[Ar]4s^2 3d^5 4p^1$

94. Does the minimization of electron-electron repulsions correlate with Hund's rule?

95. The electron affinity for sulfur is more exothermic than that for oxygen. How do you account for this?

96. The ionization energy of sodium is 495 kJ/mol. Will gaseous sodium atoms be ionized by light with a wavelength of 589 nm?

97. What role does electron-electron repulsion play in the trend in ionization energies from lithium to neon?

98. Give possible values for the quantum numbers of the valence electrons in an atom of titanium (Ti).

99. Photogray lenses incorporate small amounts of silver chloride in the glass of the lens. When light hits the AgCl particles, the following reaction occurs:

$$AgCl \xrightarrow{h\nu} Ag + Cl$$

The silver metal formed causes the lenses to darken. The enthalpy change for this reaction is 3.10×10^2 kJ/mol. Assuming that all of this energy must be supplied by light, what is the maximum wavelength of light that can cause this reaction?

100. How many electrons in a given atom can have the following sets of quantum numbers?
 a. $n = 3$
 b. $n = 2, \ell = 0$
 c. $n = 2, \ell = 2$
 d. $n = 2, \ell = 0, m_\ell = 0, m_s = +\frac{1}{2}$

101. Although no currently known elements contain electrons in g orbitals in the ground state, it is possible that these elements will be found or that electrons in excited states of known elements could be in g orbitals. For g orbitals the value of ℓ is 4. What is the lowest value of n for which g orbitals could exist? What are the possible values of m_ℓ? How many electrons could a set of g orbitals hold?

102. What is the most important relationship among elements in the same group of the periodic table?

103. Using the data in this chapter, calculate the change in energy expected for each of the following processes.
 a. $2Cu^+(g) \rightarrow Cu(g) + Cu^{2+}(g)$ (For Cu, $I_1 =$ 746 kJ/mol, $I_2 =$ 1958 kJ/mol.)
 b. $Na^-(g) + Na^+(g) \rightarrow 2Na(g)$ (Electron affinity for Na = −52 kJ/mol.)
 c. $Mg^{2+}(g) + K(g) \rightarrow Mg^+(g) + K^+(g)$
 d. $Na(g) + Cl(g) \rightarrow Na^+(g) + Cl^-(g)$
 e. $Mg(g) + F(g) \rightarrow Mg^+(g) + F^-(g)$
 f. $Mg^+(g) + F(g) \rightarrow Mg^{2+}(g) + F^-(g)$
 g. $Mg(g) + 2F(g) \rightarrow Mg^{2+}(g) + 2F^-(g)$

104. Write equations corresponding to the following energy terms.
 a. the fourth ionization energy of Se
 b. the electron affinity of S^-
 c. the electron affinity of Fe^{3+}
 d. the ionization energy of Mg
 e. the work function of Mg (see Exercise 5)

105. Using only the periodic table inside the front cover of the text, write the expected electron configurations for the following.
 a. the third element in Group 5A
 b. element number 116
 c. an element with three unpaired 5d electrons in the ground state
 d. Ti, Ni, and Os

106. Consider the ground state of antimony (Sb).
 a. How many electrons have $\ell = 1$ as one of their quantum numbers?
 b. How many electrons have $m_\ell = 0$?
 c. How many electrons have $m_\ell = 1$?

107. An ion having a 4+ charge and a mass of 49.9 amu has two electrons with $n = 1$, eight electrons with $n = 2$, and ten electrons with $n = 3$. Supply the following properties for the ion. (*Hint:* in forming ions, the 4s electrons are lost before the 3d electrons.)
 a. the atomic number
 b. total number of s electrons
 c. total number of p electrons
 d. total number of d electrons
 e. the number of neutrons in the nucleus
 f. the mass of 3.01×10^{23} atoms
 g. the ground-state electron configuration of the neutral atom

108. Consider the following ionization energies for aluminum.

$Al(g) \longrightarrow Al^+(g) + e^-$ $\quad I_1 = 580$ kJ/mol
$Al^+(g) \longrightarrow Al^{2+}(g) + e^-$ $\quad I_2 = 1815$ kJ/mol
$Al^{2+}(g) \longrightarrow Al^{3+}(g) + e^-$ $\quad I_3 = 2740$ kJ/mol
$Al^{3+}(g) \longrightarrow Al^{4+}(g) + e^-$ $\quad I_4 = 11{,}600$ kJ/mol

 a. Account for the increasing trend in the values of the ionization energies.
 b. Explain the large increase between I_3 and I_4.
 c. Which one of the four ions has the greatest electron affinity? Explain.
 d. List the four aluminum ions given above in order of increasing size, and explain your ordering. (*Hint:* remember that most of the size of an atom or ion is mainly due to its electrons.)

109. Are the following statements true for the hydrogen atom only, true for all atoms, or not true for any atoms?
 a. The principal quantum number completely determines the energy of a given electron.
 b. The angular momentum quantum number, ℓ, determines the shapes of the atomic orbitals.
 c. The magnetic quantum number, m_ℓ, determines the direction that the atomic orbitals point in space.
 d. The d orbitals are always degenerate.

110. Answer the following questions, assuming that m_s has four values rather than two and that the normal rules apply for n, ℓ, and m_ℓ.
 a. How many electrons could an orbital hold?
 b. How many elements would be contained in the first and second periods of the periodic table?
 c. How many elements would be contained in the first transition metal series?
 d. How many electrons would the set of 4f orbitals be able to hold?

111. Assume that we are in another universe with different physical laws. Electrons in this universe are described by four quantum numbers with meanings similar to those we use. We will call these quantum numbers p, q, r, and s. The rules for these quantum numbers are as follows:

$p = 1, 2, 3, 4, 5, \ldots$

q takes on positive odd integer values and $q \leq p$.

r takes on all even integer values from $-q$ to $+q$. (Zero is considered an even number.)

$s = +\frac{1}{2}$ or $-\frac{1}{2}$.

a. Sketch what the first four periods of the periodic table will look like in this universe.
b. What are the atomic numbers of the first four elements you would expect to be least reactive?
c. Give an example, using elements in the first four rows, of ionic compounds with the formulas XY, XY_2, X_2Y, XY_3, and X_2Y_3.
d. How many electrons can have $p = 3$?
e. How many electrons can have $p = 4, q = 3, r = 2$?
f. How many electrons can have $p = 4, q = 3$?
g. How many electrons can have $p = 3, q = 0, r = 0$?
h. What are the possible values of q and r for $p = 5$?
i. How many electrons can have $p = 6$?

112. While Mendeleev predicted the existence of several undiscovered elements, he did not predict the existence of the noble gases, the lanthanides, or the actinides. Propose reasons why Mendeleev was not able to predict the existence of the noble gases.

113. For a hydrogen atom in its ground state, calculate the relative probability of finding the electron in the area described.
a. in a sphere of volume 1.0×10^{-3} pm^3 centered at the nucleus
b. in a sphere of volume 1.0×10^{-3} pm^3 centered on a point 1.0×10^{-11} m from the nucleus
c. in a sphere of volume 1.0×10^{-3} pm^3 centered on a point 53 pm from the nucleus
d. in a shell between two concentric spheres, one with radius 9.95 pm and the other with radius 10.05 pm
e. in a shell between two concentric spheres, one with radius 52.85 pm and one with radius 52.95 pm

114. What is the total probability of finding a particle in a one-dimensional box in level $n = 3$ between $x = 0$ and $x = L/6$?

115. Using the wave function for the $2p_z$ orbital, show that for this orbital the maximum electron probability lies along the z axis and is zero in the xy plane.

116. For the hydrogen atom, which orbitals have the same energy as the 3s orbital? Is the same thing true for the helium atom?

117. An electron is trapped in a tiny hole in an array of aluminum atoms (assume they behave as uniform hard spheres). In this situation the energy of the electron is quantized, and the lowest-energy transition corresponds to a wavelength of 9.50 nm. Assuming that the hole can be approximated as a cube, what is the radius of a spherical ion that will just fit in this hole? (See Exercise 26 for the appropriate energy equation.)

118. Discuss why a function of the type $A \cos(Lx)$ is not an appropriate solution for the particle in a one-dimensional box.

CHAPTER 13

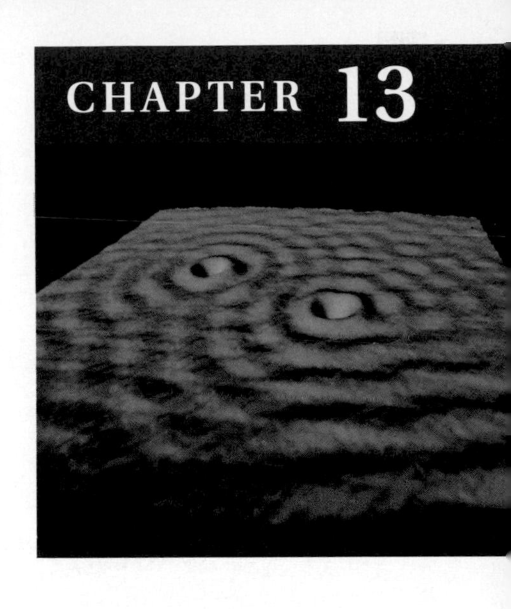

Bonding: General Concepts

The world around us is composed almost entirely of compounds and mixtures of compounds: rocks, coal, soil, petroleum, trees, and human bodies are all complex mixtures of chemical compounds in which different kinds of atoms are bound together. Substances composed of unbound atoms do exist in nature, but they are very rare. Examples are the argon in the atmosphere and the helium mixed with natural gas reserves.

The manner in which atoms are bound together in a given substance has a profound effect on its chemical and physical properties. For example, graphite is a soft, slippery material used as a lubricant in locks, and diamond is one of the hardest materials known, valuable both as a gemstone and in industrial cutting tools. Why do these materials, both composed solely of carbon atoms, have such different properties? The answer, as we will see, lies in the bonding within these substances.

Silicon and carbon are next to each other in Group 4A on the periodic table. From our knowledge of periodic trends, we might expect SiO_2 and CO_2 to be very similar. But SiO_2 is the empirical formula of silica, which is found in sand and quartz, whereas carbon dioxide is a gas, a product of respiration. Why are they so different? We will be able to answer this question after we have developed models for bonding.

Bonding and structure play a central role in determining the course of all chemical reactions, many of which are vital to our survival. Later in this book we will demonstrate the importance of bonding and structure by showing how enzymes facilitate complex chemical reactions, how genetic characteristics are

(left) Quartz grows in beautiful, regular crystals. (right) Two forms of carbon: diamond and graphite.

transferred, and how hemoglobin in the blood carries oxygen throughout the body. All of these fundamental biological reactions hinge on the geometric structures of molecules, sometimes depending on very subtle differences in molecular shape to channel the chemical reaction one way rather than another.

Many of the world's current problems require fundamentally chemical answers: disease and pollution control, the search for new energy sources, the development of new fertilizers to increase crop yields, the improvement of the protein content in various staple grains, and many more. To understand the behavior of natural materials, we must understand the nature of chemical bonding and the factors that control the structures of compounds. In this chapter we will present various classes of compounds that illustrate the different types of bonds and then develop models to describe the structure and bonding that characterize materials found in nature. Later these models will prove useful in understanding chemical reactions.

13.1 Types of Chemical Bonds

What is a chemical bond? There is no simple and yet complete answer to this question. In Chapter 2 we defined bonds as forces that hold groups of atoms together and make the atoms function as a unit.

There are many types of experiments we can perform to determine the fundamental nature of materials. For example, we can study physical properties such as melting point, hardness, and electrical and thermal conductivity. We can also study solubility characteristics and the properties of the resulting solutions. To determine the charge distribution in a molecule, we can study its behavior in an electric field. We can obtain information about the strength of a bonding interaction by measuring the energy required to break the bond, the **bond energy.** Spectroscopy, the study of the interactions of electromagnetic radiation with matter, gives a wealth of information about molecular structure and energy level spacings.

There are several ways atoms can interact with one another to form aggregates. We will consider several specific examples to illustrate the various types of chemical bonds.

When solid sodium chloride is melted, it conducts electricity, a fact that convinces us that sodium chloride contains Na^+ and Cl^- ions. Thus when sodium and chlorine react to form sodium chloride, electrons must be transferred from the sodium atoms to the chlorine atoms to form Na^+ and Cl^- ions, which then aggregate to form solid sodium chloride. Why does this happen? The best simple answer is that *the system can achieve the lowest possible energy by behaving in this way.* Part of the favorable energy change results from the attraction of a chlorine atom for an extra electron. Even more important are the very strong attractions between the oppositely charged ions. The resulting solid sodium chloride is a very sturdy material; it has a melting point of approximately 800°C. The bonding forces that produce this great thermal stability result from the electrostatic attractions of the closely packed, oppositely charged ions. This is an example of *ionic bonding.* Ionic substances are formed when an atom that loses electrons relatively easily reacts with an atom that has a high affinity for electrons. That is, an **ionic compound** results when a metal reacts with a nonmetal.

The energy of interaction between a pair of ions can be calculated by using **Coulomb's law:**

$$V = \frac{Q_1Q_2}{4\pi\epsilon_0 r} = 2.31 \times 10^{-19}\text{ J nm}\left(\frac{Q_1Q_2}{r}\right)$$

where V has units of joules, r is the distance between the ion centers in nanometers, Q_1 and Q_2 are the numerical ion charges, and ϵ_0 is the permittivity of the vacuum. For example, in solid sodium chloride, where the distance between the centers of the Na^+ and Cl^- ions is 276 picometers (0.276 nm), the ionic energy per pair of ions is

$$V = 2.31 \times 10^{-19}\text{ J nm}\left[\frac{(+1)(-1)}{0.276\text{ nm}}\right] = -8.37 \times 10^{-19}\text{ J}$$

The negative sign indicates an attractive force. That is, *the ion pair has lower energy than the separated ions.* For a mole of pairs of Na^+ and Cl^- ions, the energy of interaction is

$$V = \left(-8.37 \times 10^{-19}\frac{\text{J}}{\text{ion pair}}\right)\left(6.022 \times 10^{23}\frac{\text{ion pair}}{\text{mol}}\right)$$
$$= -504\text{ kJ/mol}$$

Note that this energy refers to a mole of $Na^+ \cdot \cdot Cl^-$ ion pairs in the gas phase where a given pair is far from any other pair. In solid sodium chloride, which contains a large array of closely packed Na^+ and Cl^- ions, where a given ion is close to many oppositely charged ions, the energy associated with ionic bonding is much greater than 504 kJ/mol because of the larger numbers of interacting ions.

Coulomb's law can also be used to calculate the repulsive energy when two like-charged ions are brought together. In this case the calculated energy value will have a positive sign.

We have seen that a bonding force develops when two very different atoms react to form oppositely charged ions. But how does a bonding force develop between two identical atoms? Let's explore this situation from a very simple point of view by considering the energy terms that result when two hydrogen atoms are brought close together, as shown in Fig. 13.1(a). For two closely spaced hydrogen atoms, there are two unfavorable energy terms, proton-proton repulsion and electron-electron repulsion, and one favorable term,

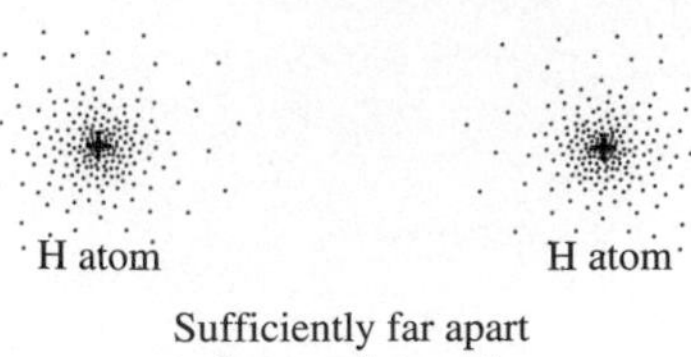

H atom H atom

The atoms begin to interact as they move closer together.

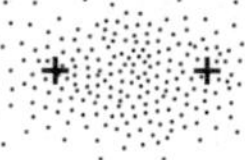

H_2 molecule

Optimum distance to achieve lowest overall energy of system

(a)

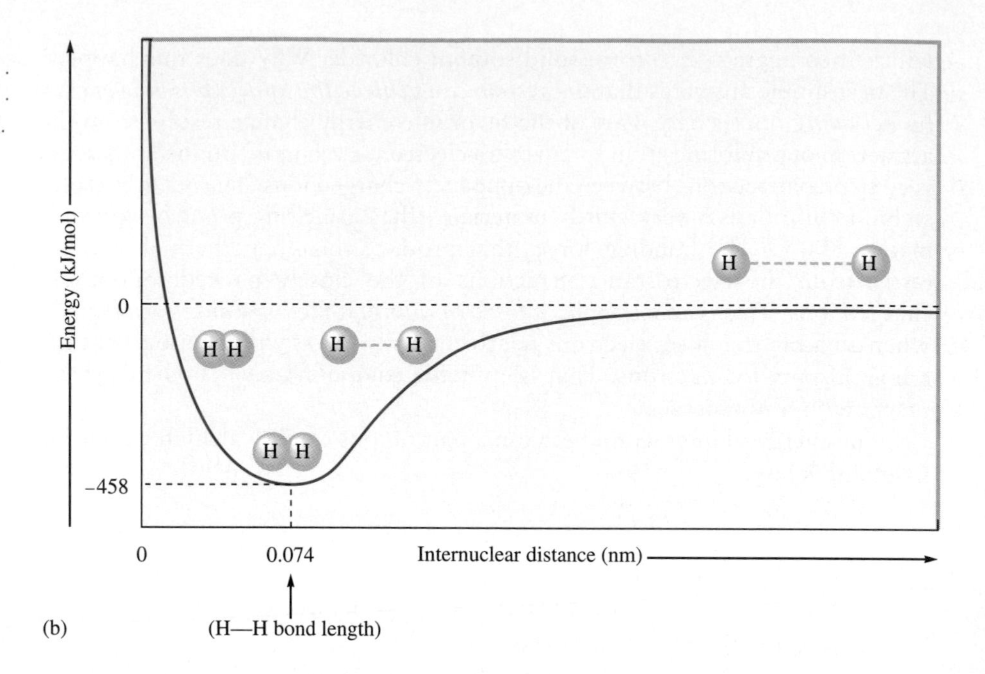

(b)

Figure 13.1
(a) The interaction of two hydrogen atoms. (b) Energy profile as a function of the distance between the nuclei of the hydrogen atoms. As the atoms approach each other, the energy decreases until the distance reaches 0.074 nm (0.74 Å) and then begins to increase again due to repulsions.

Bonding occurs if the energy of the aggregate is lower than that of the separated atoms.

proton-electron attraction. Under what conditions will the H_2 molecule be favored over the separated hydrogen atoms? That is, what conditions will favor bond formation? The answer lies in nature's strong tendency to achieve the lowest possible energy. A bond will form—that is, the two hydrogen atoms will exist as a molecular unit—if the system can lower its total energy in the process.

Therefore, the hydrogen atoms will assume the positions that give the lowest possible energy; the system will act to minimize the sum of the positive (repulsive) energy terms and the negative (attractive) energy term. The distance at which the energy is minimum is called the equilibrium internuclear distance or, more commonly, the **bond length.** The total energy of this system as a function of distance between the hydrogen nuclei is shown in Fig. 13.1(b). Note four important features of this diagram:

1. The energy terms involved are the potential energy that results from the attractions and repulsions among the charged particles and the kinetic energy due to the motions of the electrons.
2. The zero reference point for energy is defined for the atoms at infinite separation.
3. At very short distances the energy rises steeply because of the great importance of the repulsive forces at these distances.
4. The bond length is the distance at which the system has minimum energy.

In the H_2 molecule the electrons reside primarily in the space between the two nuclei, where they are attracted simultaneously by both protons. This positioning is precisely what leads to the stability of the H_2 molecule relative to two separated hydrogen atoms. The potential energy of each electron is lowered because of the increased attractive forces in this area. Although it is not usually discussed in connection with simple models of bonding, the kinetic energy of the electrons also changes when the individual atoms form the molecule. Thus the energy plotted in Fig. 13.1(b) is the total energy of the system, not just the potential energy. When we say that a bond is formed between the hydrogen atoms, we mean that the H_2 molecule is more stable than two separated hydrogen atoms by a certain quantity of energy (the bond energy).

The atoms in H_2 (and all other molecules) actually vibrate back and forth around the equilibrium internuclear distance.

We can also think of a bond in terms of forces. The simultaneous attraction for each electron by the two protons generates a force that pulls the protons toward each other. This attractive force just balances the proton-proton and electron-electron repulsive forces at the distance corresponding to the bond length.

Ionic and covalent bonds are the extreme bond types.

The type of bonding we encounter in the hydrogen molecule and in many other molecules where *electrons are shared by nuclei* is called **covalent bonding.**

So far we have considered two extreme types of bonding. In ionic bonding the participating atoms are so different that one or more electrons are transferred to form oppositely charged ions. The bonding results from electrostatic interactions among the resulting ions. In covalent bonding two identical atoms share electrons equally. The bonding results from the mutual attraction of the two nuclei for the shared electrons. Between these extremes lie intermediate cases in which the atoms are not so different that electrons are completely transferred but are different enough so that unequal sharing results. These are called **polar covalent bonds.** An example of this type of bond occurs in the hydrogen fluoride (HF) molecule. When a sample of hydrogen fluoride gas is placed in an electric field, the molecules tend to orient themselves as shown in Fig. 13.2, with the fluoride end closest to the positive pole and the hydrogen end closest to the negative pole. This result implies that the HF molecule has the following charge distribution:

$$\underset{\delta+}{\mathrm{H}}-\underset{\delta-}{\mathrm{F}}$$

where δ (delta) is used to indicate a fractional charge. This same effect was noted in Chapter 4 where many of water's unusual properties were attributed to the polar O—H bonds in the H_2O molecule.

The most logical explanation for the development of the partial positive and negative charges on the atoms (bond polarity) in such molecules as HF and H_2O is that the electrons in the bonds are not shared equally. For example, we can account for the polarity of the HF molecule by assuming that the fluorine atom has a stronger attraction for the shared electrons than the hydrogen atom. Likewise, in the H_2O molecule the oxygen atom appears to attract the shared electrons more strongly than the hydrogen atoms. Because bond polarity has important chemical implications, we find it useful to quantify the ability of an atom to attract shared electrons. In the next section we show how this is done.

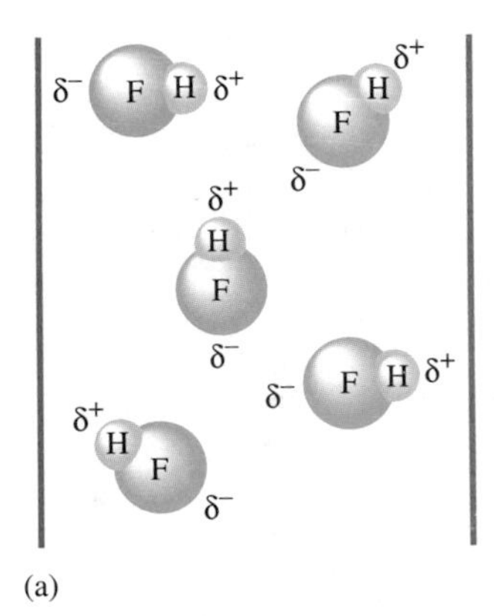

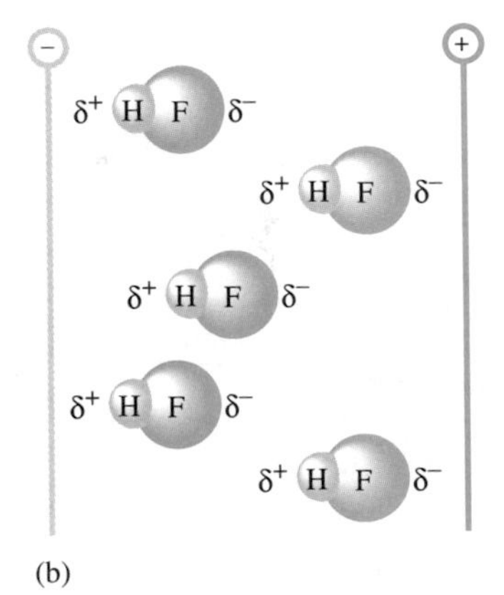

Figure 13.2
The effect of an electric field on hydrogen fluoride molecules. (a) When no electric field is present, the molecules are randomly oriented. (b) When the field is turned on, the molecules tend to line up with their negative ends toward the positive pole and their positive ends toward the negative pole.

13.2 Electronegativity

The different affinities of atoms for the electrons in a bond are described by a property called **electronegativity:** *the ability of an atom in a molecule to attract shared electrons to itself.*

The most widely accepted method for determining electronegativity values is that of Linus Pauling (b. 1901), an American scientist who has won Nobel prizes for both chemistry and peace. To understand Pauling's model, consider a hypothetical molecule HX. The relative electronegativities of the H and X atoms are determined by comparing the measured H—X bond energy with the "expected" H—X bond energy. The expected bond energy is an "average" (actually the geometric mean) of the H—H and X—X bond energies:

$$\text{Expected H—X bond energy} = [(\text{H—H bond energy})(\text{X—X bond energy})]^{1/2}$$

The difference (Δ) between the actual (measured) and expected bond energies is

$$\Delta = (\text{H—X})_{act} - (\text{H—X})_{exp}$$

If H and X have identical electronegativities, $(\text{H—X})_{act}$ and $(\text{H—X})_{exp}$ are the same and Δ is 0. On the other hand, if X has a greater electronegativity than H, the shared electron(s) will tend to be closer to the X atom. The molecule will be polar, with the following charge distribution:

$$\underset{\delta+}{\text{H}}\text{—}\underset{\delta-}{\text{X}}$$

Note that this bond can be viewed as having an ionic, as well as a covalent, component. The electrostatic attraction between the partially charged H and X atoms will lead to a greater bond strength. Thus $(\text{H—X})_{act}$ will be larger than $(\text{H—X})_{exp}$. The greater the differencc in the electronegativities of the atoms, the greater the ionic component of the bond and the greater the value of Δ. Thus the relative electronegativities of H and X can be assigned from the Δ values.

The factor of 0.102 is a conversion factor between kJ and eV (the units originally used by Pauling).

The actual formula Pauling used to calculate electronegativity (EN) differences is

$$\text{EN(X)} - \text{EN(H)} = 0.102\sqrt{\Delta}$$

where all bond energies are in units of kJ/mol. Pauling then obtained absolute electronegativity values for the elements by assigning a value of 4.0 to fluorine (the element with the highest electronegativity).

Electronegativity values have been determined by this process for virtually all of the elements; the results are given in Fig. 13.3. Note that for the representative elements electronegativity generally increases from left to right across a period and decreases down a group. The range of electronegativity values is from 4.0 for fluorine to 0.7 for cesium.

The relationship between electronegativity and bond type is shown in Table 13.1. For identical atoms (electronegativity difference of zero) the electrons in the bond are shared equally and no polarity occurs. When two atoms with widely differing electronegativities interact, electron transfer usually occurs, producing ions—an ionic substance is formed. Intermediate cases give polar covalent bonds with unequal electron sharing.

TABLE 13.1
The Relationship Between Electronegativity and Bond Type

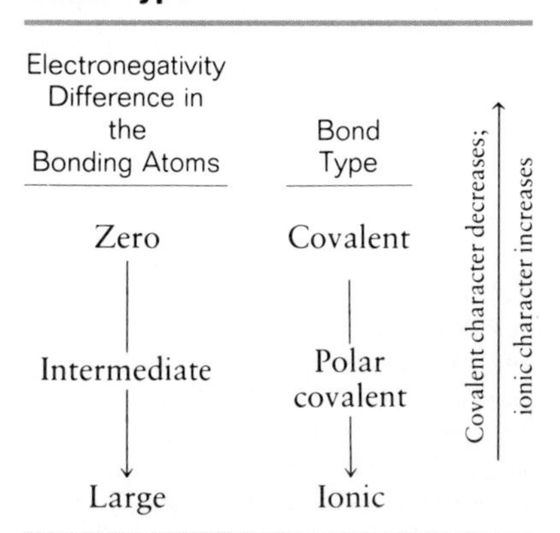

Electronegativity Difference in the Bonding Atoms	Bond Type
Zero	Covalent
↓ Intermediate	Polar covalent
↓ Large	Ionic

Covalent character decreases; ionic character increases

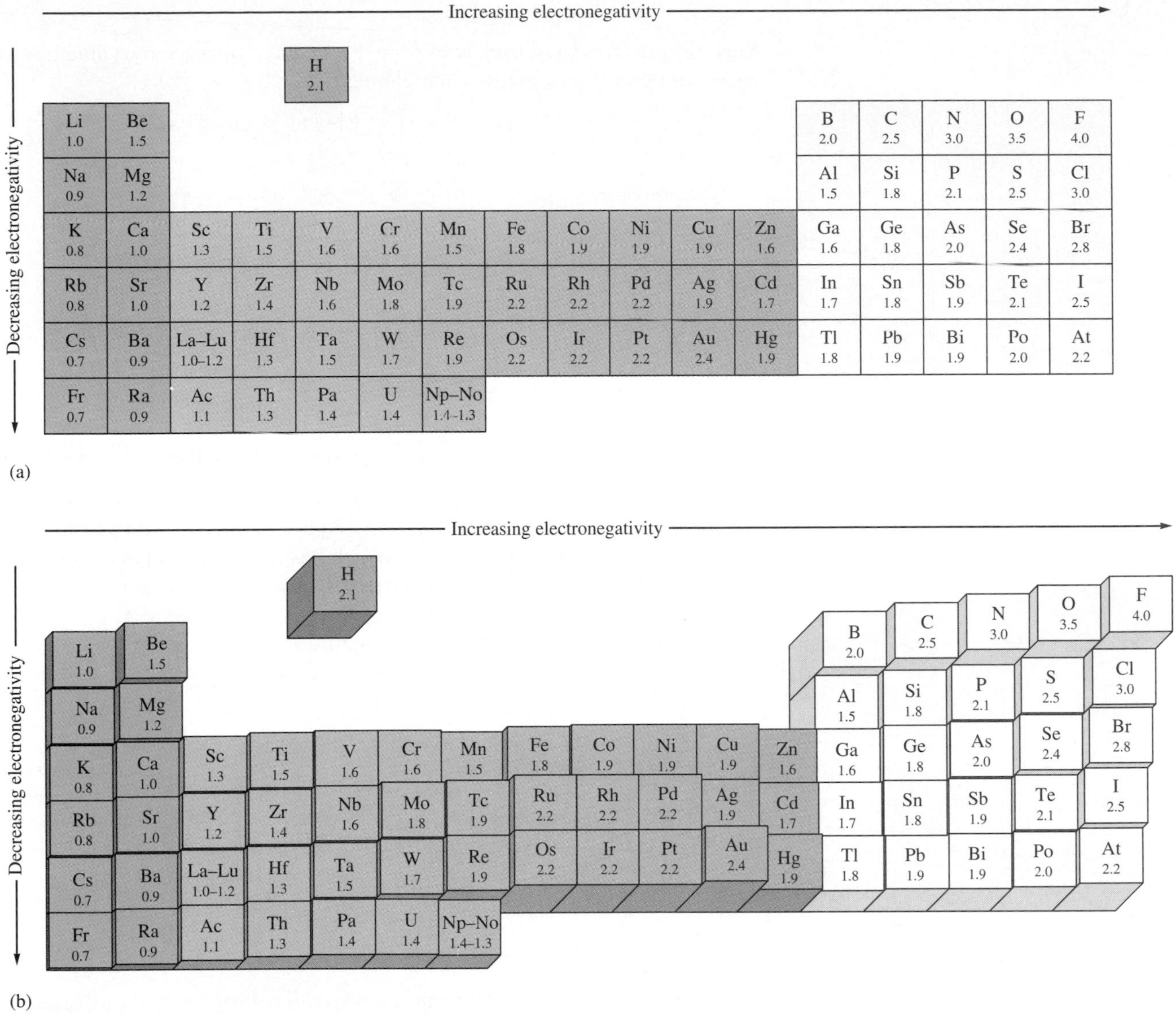

Increasing electronegativity →

Decreasing electronegativity ↓

			H 2.1														
Li 1.0	Be 1.5											B 2.0	C 2.5	N 3.0	O 3.5	F 4.0
Na 0.9	Mg 1.2											Al 1.5	Si 1.8	P 2.1	S 2.5	Cl 3.0
K 0.8	Ca 1.0	Sc 1.3	Ti 1.5	V 1.6	Cr 1.6	Mn 1.5	Fe 1.8	Co 1.9	Ni 1.9	Cu 1.9	Zn 1.6	Ga 1.6	Ge 1.8	As 2.0	Se 2.4	Br 2.8
Rb 0.8	Sr 1.0	Y 1.2	Zr 1.4	Nb 1.6	Mo 1.8	Tc 1.9	Ru 2.2	Rh 2.2	Pd 2.2	Ag 1.9	Cd 1.7	In 1.7	Sn 1.8	Sb 1.9	Te 2.1	I 2.5
Cs 0.7	Ba 0.9	La–Lu 1.0–1.2	Hf 1.3	Ta 1.5	W 1.7	Re 1.9	Os 2.2	Ir 2.2	Pt 2.2	Au 2.4	Hg 1.9	Tl 1.8	Pb 1.9	Bi 1.9	Po 2.0	At 2.2
Fr 0.7	Ra 0.9	Ac 1.1	Th 1.3	Pa 1.4	U 1.4	Np–No 1.4–1.3										

(a)

Increasing electronegativity →

Decreasing electronegativity ↓

(b)

Figure 13.3
(a) The Pauling electronegativity values. Electronegativity generally increases across a period and decreases down a group. (b) This three-dimensional representation emphasizes the periodic trends.

EXAMPLE 13.1

Arrange the following bonds according to increasing polarity: H—H, O—H, Cl—H, S—H, and F—H.

Solution

The polarity of the bond increases as the difference in electronegativity increases. From the electronegativity values in Fig. 13.3, the following variation in bond polarity is expected (the electronegativity value appears in parentheses below each element):

	H—H	< S—H	< Cl—H	< O—H	< F—H
	(2.1)(2.1)	(2.5)(2.1)	(3.0)(2.1)	(3.5)(2.1)	(4.0) (2.1)
Electronegativity difference	0	0.4	0.9	1.4	1.9

Covalent bond $\xrightarrow[\text{Polarity increases}]{}$ polar covalent bond

13.3 Bond Polarity and Dipole Moments

We have seen that when hydrogen fluoride is placed in an electric field, the molecules have a preferential orientation (Fig. 13.2). This follows from the charge distribution in the HF molecule, which has a positive end and a negative end. A molecule like HF that has a center of positive charge and a center of negative charge is said to be *dipolar,* or to have a **dipole moment.** A molecule that has a positive center of charge of magnitude Q and a negative center of charge of magnitude Q separated by a distance R has a dipole moment given by the expression

$$\text{Dipole moment} = \mu = QR$$

The debye is named after Peter Debye, who pioneered in the measurement of dipole moments.

which has SI units of coulomb meter (C m) but is most often given in units of debye [1 debye (D) = 3.336×10^{-30} C m). The dipolar character of a molecule is often represented by an arrow pointing to the negative charge center, with the tail of the arrow indicating the positive center of charge:

$$\underset{\delta+\qquad\quad\delta-}{+\!\!\longrightarrow}$$

TABLE 13.2
The Dipole Moments of Some Diatomic Molecules (gas phase)

Molecule	Dipole Moment (D)
CO	0.112
HF	1.83
HCl	1.11
HBr	0.78
HI	0.38
NaCl	9.00
LiF	6.33
KF	8.60
KBr	10.41

The dipole moment of a molecule gives useful information about its bonding and electron distribution. For example, the observed dipole moment for HF is 1.83 D. If HF were totally ionic (H^+F^-), the expected dipole moment (symbolized by μ) would be

$$\mu = \underset{\substack{\uparrow \\ \text{Electron charge}}}{(1.60 \times 10^{-19}\ \text{C})}\ \underset{\substack{\uparrow \\ \text{H—F bond distance}}}{(9.17 \times 10^{-11}\ \text{m})}$$
$$= 1.47 \times 10^{-29}\ \text{C m} = 4.40\ \text{D}$$

This calculation shows that HF is not fully ionic, since the measured dipole moment is much less than 4.40 D. We can estimate the ionic character of HF by assuming that the hydrogen has a charge $\delta+$ and the fluorine has a charge $\delta-$. Using the measured dipole moment, we have

$$1.83\ \text{D} = (\delta)(9.17 \times 10^{-11}\ \text{m}) \times \frac{1\ \text{D}}{3.336 \times 10^{-30}\ \text{C m}}$$

Solving for δ gives 6.66×10^{-20} C. Since the charge on an electron is 1.60×10^{-19} C, each atom in HF has a fractional charge of

$$\frac{6.66 \times 10^{-20}\ \text{C}}{1.60 \times 10^{-19}\ \text{C}} = 0.416$$

From this argument we might say that HF has 42% ionic bonding. Although this analysis is somewhat oversimplified (it assumes charge distributions can be represented as point charges, for example), it does provide useful information about bonding.

From what has been said so far, we would expect any diatomic molecule with a polar bond (between atoms with different electronegativities) to exhibit a dipole moment. Although this is generally true, the observed dipole moments of diatomic molecules are sometimes smaller than expected. For example, the carbon monoxide molecule CO has a dipole moment of only 0.11 D, much smaller than expected from the polarity of the CO bond. This discrepancy is most likely due to the lone pairs of electrons on the atoms, which make large contributions to the dipole moment in opposition to that from the bond polarity. We will not explore the details of this situation here. The dipole moments of some representative diatomic molecules are listed in Table 13.2.

Polyatomic molecules can also exhibit dipolar behavior. For example, because the oxygen atom in the water molecule has a greater electronegativity than the hydrogen atoms, the molecular charge distribution is that shown in Fig. 13.4(a). This charge distribution causes the water molecule to behave in an electric field as if it had two centers of charge—one positive and one negative—as shown in Fig. 13.4(b). Thus the water molecule has a dipole moment. Similar behavior is observed for the NH_3 molecule (Fig. 13.5). Some molecules have polar bonds but do not have a dipole moment. This occurs when the individual bond polarities are arranged in such a way that they cancel. An example is the CO_2 molecule, a linear molecule that has the charge distribution shown in Fig. 13.6. In this case, since the opposing bond polarities cancel, the carbon dioxide molecule does not have a dipole moment. There is no preferential way for this molecule to line up in an electric field. (Try to find a preferred orientation.)

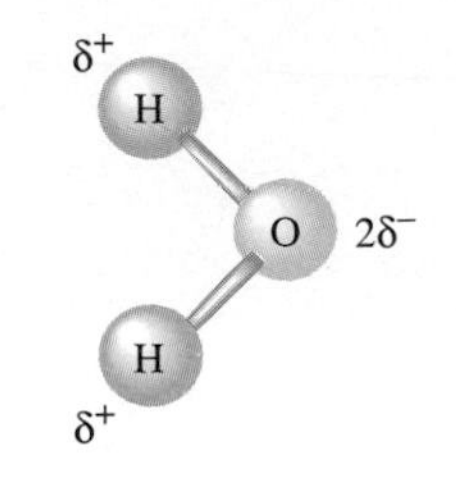

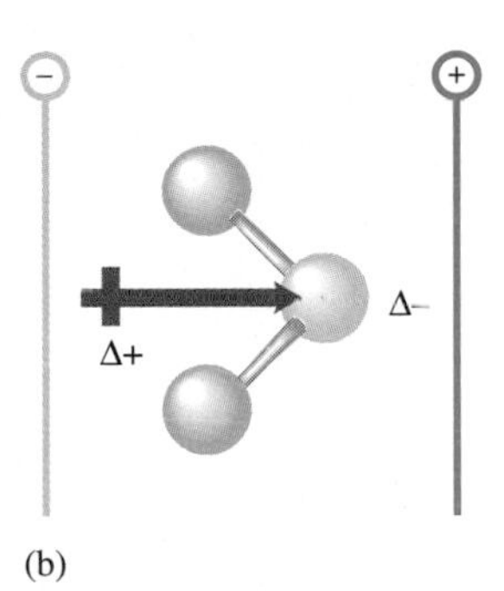

Figure 13.4
(a) The charge distribution in the water molecule. (b) The water molecule in an electric field.

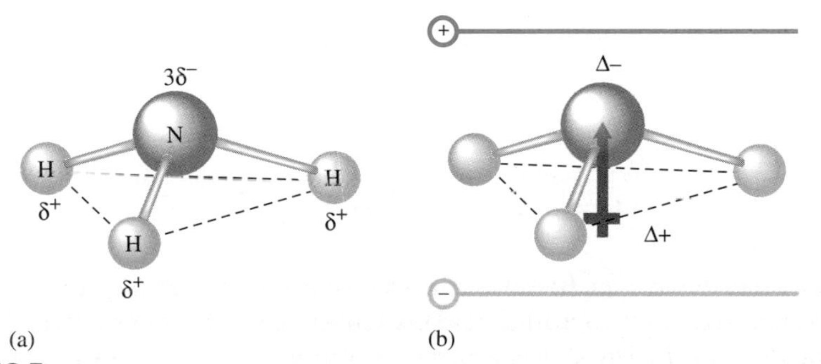

Figure 13.5
(a) The structure and charge distribution of the ammonia molecule. The polarity of the N—H bonds occurs because nitrogen has a greater electronegativity than hydrogen. (b) The dipole moment of the ammonia molecule oriented in an electric field.

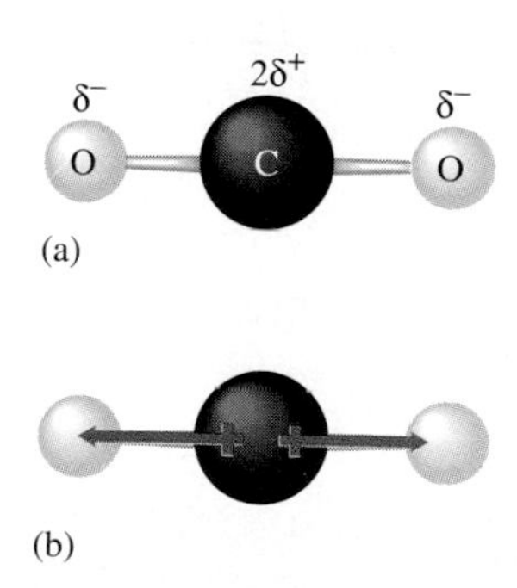

Figure 13.6
(a) The carbon dioxide molecule. (b) The opposed bond polarities cancel, and the carbon dioxide molecule has no dipole moment.

TABLE 13.3 Types of Molecules with Polar Bonds but No Resulting Dipole Moment

Type		Cancellation of Polar Bonds	Example
Linear molecules with two identical bonds	B—A—B	⟵+ +⟶	CO_2
Planar molecules with three identical bonds 120° apart	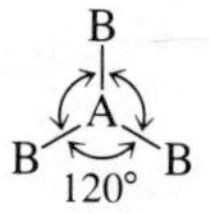 B, A, B, B; 120°	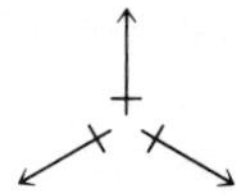	BF_3
Tetrahedral molecules with four identical bonds 109.5° apart	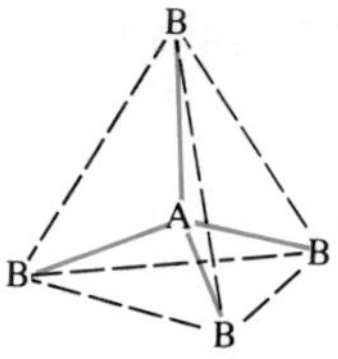 B, A, B, B, B	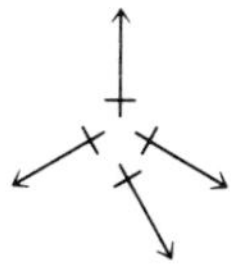	CCl_4

There are many cases where the bond polarities in molecules oppose and exactly cancel each other. Some common types of molecules with polar bonds but without dipole moments are shown in Table 13.3.

EXAMPLE 13.2

For each of the following molecules, show the direction of the bond polarities. Also indicate which ones have dipole moments: HCl, Cl_2, SO_3 (planar), CH_4 (tetrahedral), and H_2S (V-shaped).

Solution

The HCl molecule: Because the electronegativity of chlorine (3.0) is greater than that of hydrogen (2.1), the chlorine is partially negative, and the hydrogen is partially positive. The HCl molecule has a dipole moment oriented as follows:

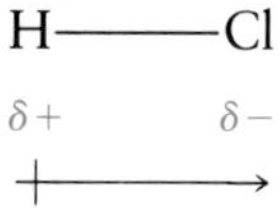

H——Cl
$\delta+$ $\delta-$
+⟶

The Cl_2 molecule: Because the two chlorine atoms share the electrons equally, no bond polarity occurs. The Cl_2 molecule has no dipole moment.

The SO_3 molecule: Because the electronegativity of oxygen (3.5) is greater than that of sulfur (2.5), each oxygen has a partial negative charge, and the sulfur has a partial positive charge:

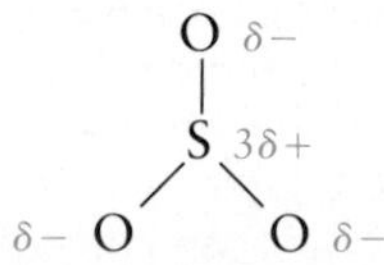

O $\delta-$
|
S $3\delta+$
$\delta-$ O O $\delta-$

However, the bond polarities cancel, and the molecule has no dipole moment.

The CH_4 molecule: Carbon has a slightly higher electronegativity (2.5) than hydrogen (2.1). This leads to small partial positive charges on the hydrogen atoms and a small partial negative charge on the carbon:

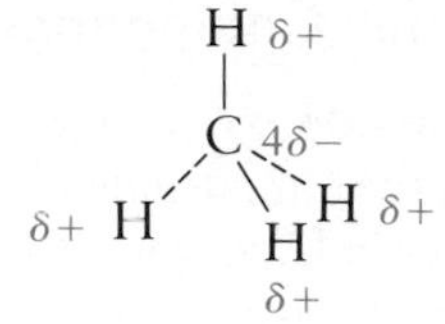

H δ+
C 4δ−
δ+ H H δ+
H
δ+

This case is similar to the third type in Table 13.3. Since the bond polarities cancel, the molecule has no dipole moment.

The H_2S molecule: Since the electronegativity of sulfur (2.5) is greater than that of hydrogen (2.1), the sulfur has a partial negative charge and the hydrogen atoms have a partial positive charge, which can be represented as follows:

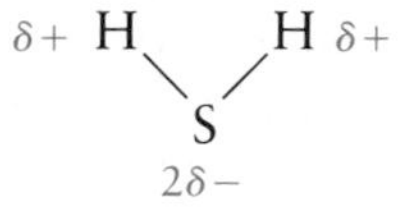

δ+ H H δ+
S
2δ−

This case is analogous to the water molecule. The polar bonds result in a dipole moment oriented as shown:

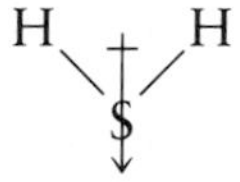

H H
S

The presence of polar bonds does not always yield a polar molecule.

13.4 Ions: Electron Configurations and Sizes

The description of the electron arrangements in atoms that emerged from the quantum mechanical model has helped a great deal in our understanding of what constitutes a stable compound. For example, in a very large number of stable compounds the atoms have noble gas arrangements of electrons. Nonmetallic elements achieve a noble gas electron configuration either by sharing electrons with other nonmetals to form covalent bonds or by taking electrons from metals to form ions. In the latter case the nonmetals form anions and the metals form cations. The following generalizations can be applied to the electron configurations in most stable compounds:

Atoms in stable compounds usually have a noble gas electron configuration.

- When *two nonmetals* react to form a covalent bond, they share electrons in a way that completes the valence electron configurations of both atoms. That is, both nonmetals attain noble gas electron configurations.
- When *a nonmetal and a representative group metal* react to form a binary ionic compound, the ions form so that the valence electron configuration of the nonmetal is completed and the valence orbitals of the metal are emptied. In this way both ions achieve noble gas electron configurations.

As we will see later, there are exceptions to these rules, but they remain a useful place to start.

Although there are some important exceptions, these generalizations apply to the vast majority of compounds and are important to remember. We will deal with covalent bonds more thoroughly later. Next, we will consider what implications these rules hold for ionic compounds.

Predicting Formulas of Ionic Compounds

At the beginning of this discussion we should emphasize that when chemists use the term *ionic compound*, they are usually referring to the solid state of that compound. Solid ionic compounds contain a large collection of positive and negative ions packed together in a way that minimizes the $\ominus \cdot\cdot \ominus$ and $\oplus \cdot\cdot \oplus$ repulsions and maximizes the $\oplus \cdot\cdot \ominus$ attractions. This situation stands in contrast to the gas phase of an ionic substance where discrete ion pairs exist. Thus when we speak in this text of the stability of an ionic compound, we are referring to the solid state, where the large attractive forces present among the oppositely charged ions tend to stabilize (favor the formation of) the ions. For example, as we mentioned in the previous chapter, the O^{2-} ion is not stable as an isolated, gas-phase species but, of course, is very stable in many solid ionic compounds. That is, $MgO(s)$, which contains Mg^{2+} and O^{2-} ions, is very stable, but the isolated, gas-phase ion pair $Mg^{2+} \cdot\cdot O^{2-}$ is not energetically favorable in comparison with the separate neutral gaseous atoms. Thus you should keep in mind that in this section, and in most other cases where we are describing the nature of ionic compounds, the discussion usually refers to the solid state, where many ions are simultaneously interacting.

To illustrate the principles of electron configurations in stable, solid ionic compounds, we will consider the formation of an ionic compound from calcium and oxygen. We can predict what compound will form by considering the valence electron configurations of the two atoms:

$$\text{Ca:} \quad [\text{Ar}]4s^2$$
$$\text{O:} \quad [\text{He}]2s^22p^4$$

From Fig. 13.3 we see that the electronegativity of oxygen (3.5) is much greater than that of calcium (1.0). Because of this large difference, electrons will be transferred from calcium to oxygen to form oxygen anions and calcium cations in the compound. How many electrons are transferred? We can base our prediction on the observation that noble gas configurations are generally the most stable. Note that oxygen needs two electrons to fill its $2s$ and $2p$ valence orbitals and to achieve the configuration of neon ($1s^22s^22p^6$). And by losing two electrons, calcium can achieve the configuration of argon. Two electrons are therefore transferred:

$$\underset{\curvearrowright\; 2e^-}{\text{Ca} + \text{O}} \longrightarrow \text{Ca}^{2+} + \text{O}^{2-}$$

To predict the formula of the ionic compound, we simply recognize that chemical compounds are always electrically neutral—they have the same quantities of positive and negative charges. In this case we must have equal numbers of Ca^{2+} and O^{2-} ions, and the empirical formula of the compound is CaO.

The same principles can be applied to many other cases. For example, consider the compound formed between aluminum and oxygen. Because alumi-

TABLE 13.4 Common Ions with Noble Gas Electron Configurations in Ionic Compounds

Group 1A	Group 2A	Group 3A	Group 6A	Group 7A	Electron Configuration
H^-, Li^+	Be^{2+}				[He]
Na^+	Mg^{2+}	Al^{3+}	O^{2-}	F^-	[Ne]
K^+	Ca^{2+}		S^{2-}	Cl^-	[Ar]
Rb^+	Sr^{2+}		Se^{2-}	Br^-	[Kr]
Cs^+	Ba^{2+}		Te^{2-}	I^-	[Xe]

num has the configuration $[Ne]3s^23p^1$, it must lose three electrons to form the Al^{3+} ion and thus achieve the neon configuration. Therefore, the Al^{3+} and O^{2-} ions form in this case. Since the compound must be electrically neutral, there must be three O^{2-} ions for every two Al^{3+} ions, and the compound has the empirical formula Al_2O_3.

Table 13.4 shows common elements that form ions with noble gas electron configurations in ionic compounds. In losing electrons to form cations, metals in Group 1A lose one electron, those in Group 2A lose two electrons, and those in Group 3A lose three electrons. In gaining electrons to form anions, nonmetals in Group 7A (the halogens) gain one electron, and those in Group 6A gain two electrons. Hydrogen typically behaves as a nonmetal and can gain one electron to form the hydride ion (H^-), which has the electron configuration of helium.

There are some important exceptions to the rules discussed here. For example, tin forms both Sn^{2+} and Sn^{4+} ions, and lead forms both Pb^{2+} and Pb^{4+} ions. Also bismuth forms Bi^{3+} and Bi^{5+} ions, and thallium forms Tl^+ and Tl^{3+} ions. There are no simple explanations for the behavior of these ions. For now, just note them as exceptions to the very useful rule that ions generally adopt noble gas electron configurations in ionic compounds. Our discussion here refers to representative metals. The transition metals exhibit more complicated behavior, forming a variety of ions that will be considered in Chapter 20.

Sizes of Ions

Ion size plays an important role in determining the structure and stability of ionic solids, the properties of ions in aqueous solution, and the biological effects of ions. As with atoms, it is impossible to define precisely the sizes of ions. Most often, ionic radii are determined from the measured distances between ion centers in ionic compounds. This method, of course, involves an assumption about how the distance should be divided up between the two ions. Thus you will note considerable disagreement among ionic sizes given in various sources. Here we are mainly interested in trends and will be less concerned with absolute ion sizes.

Various factors influence ionic size. We will first consider the relative sizes of an ion and its parent atom. Since a positive ion is formed by removing electrons from a neutral atom, the resulting cation is smaller than its parent atom. The opposite is true for negative ions; the addition of electrons to a neutral atom produces an anion significantly larger than its parent atom.

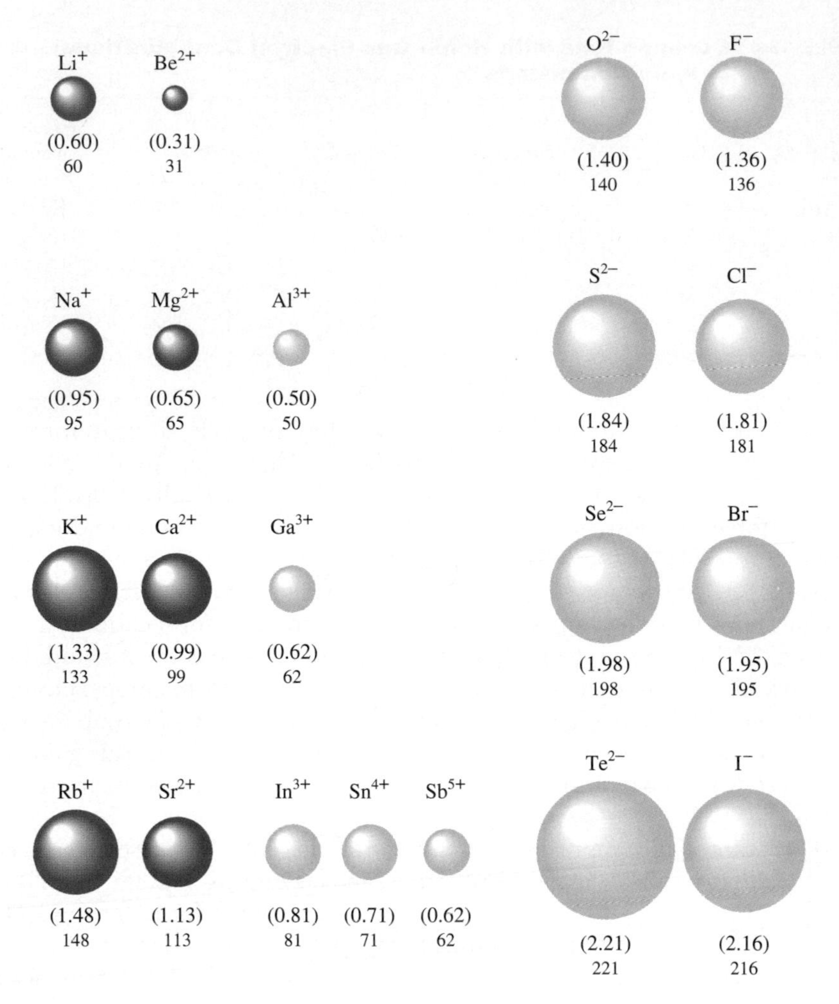

Figure 13.7
Sizes of ions related to positions of elements in the periodic table. Note that size generally increases down a group. Also note that in a series of isoelectronic ions, size decreases with increasing atomic number. The ionic radii are given in units of picometers.

It is also important to know how the sizes of ions vary depending on the positions of the parent elements in the periodic table. Figure 13.7 shows the sizes of the most important ions (each with a noble gas configuration) and their position in the periodic table. Note that ion size increases down a group. The changes that occur horizontally are complicated because of the change from predominantly metals on the left-hand side of the periodic table to nonmetals on the right-hand side. A given period thus contains both elements that give up electrons to form cations and ones that accept electrons to form anions.

One trend worth noting involves the relative sizes of a set of **isoelectronic ions**—*ions containing the same number of electrons*. Consider the ions O^{2-}, F^-, Na^+, Mg^{2+}, and Al^{3+}. Each of these ions has the neon electron configuration. How do the sizes of these ions vary? In general, there are two important facts to consider in predicting the relative sizes of ions: the number of electrons and the number of protons. Since these ions are isoelectronic, the number of electrons is 10 in each case. Electron repulsions should therefore be about the same in all cases. However, the number of protons increases from 8 to 13 as we go from the O^{2-} ion to the Al^{3+} ion. Thus, in going from O^{2-} to Al^{3+}, the 10 electrons experience a greater attraction as the positive charge on the nucleus increases. This causes the ions to become smaller. You can confirm this by looking at Fig. 13.7. In general, for a series of isoelectronic ions, the size decreases as the nuclear charge (Z) increases.

For isoelectronic ions, size generally decreases as Z increases.

EXAMPLE 13.3

Arrange the ions Se^{2-}, Br^-, Rb^+, and Sr^{2+} in order of decreasing size.

Solution

This is an isoelectronic series of ions with the krypton electron configuration. Since these ions all have the same number of electrons, their sizes will depend on nuclear charge. The Z values are 34 for Se^{2-}, 35 for Br^-, 37 for Rb^+, and 38 for Sr^{2+}. Since the nuclear charge is greatest for Sr^{2+}, it is the smallest of these ions. The Se^{2-} ion is largest:

$$\underset{\substack{\uparrow \\ \text{Largest}}}{Se^{2-}} > Br^- > Rb^+ > \underset{\substack{\uparrow \\ \text{Smallest}}}{Sr^{2+}}$$

EXAMPLE 13.4

Choose the largest ion in each of the following groups.

a. Li^+, Na^+, K^+, Rb^+, Cs^+ **b.** Ba^{2+}, Cs^+, I^-, Te^{2-}

Solution

a. The ions are all from Group 1A elements. Since size increases down a group (the ion with the greatest number of electrons is the largest), Cs^+ is the largest ion.

b. This is an isoelectronic series of ions, all of which have the xenon electron configuration. The ion with the smallest nuclear charge is the largest ion:

$$\underset{Z = 52}{Te^{2-}} > \underset{Z = 53}{I^-} > \underset{Z = 55}{Cs^+} > \underset{Z = 56}{Ba^{2+}}$$

Ion size generally increases down a group.

13.5 Formation of Binary Ionic Compounds

In this section we will introduce the factors that influence the stability and the structures of solid binary ionic compounds. We know that metals and nonmetals react by transferring electrons to form cations and anions that are mutually attractive. The resulting ionic solid forms because the aggregated oppositely charged ions have a lower energy than the original elements. Just how strongly the ions attract each other in the solid state is indicated by the **lattice energy**—*the change in energy that takes place when separated gaseous ions are packed together to form an ionic solid:*

$$M^+(g) + X^-(g) \longrightarrow MX(s)$$

The structures of ionic solids will be discussed in detail in Chapter 16.

The lattice energy is often defined as the energy *released* when an ionic solid forms from its ions. However, in this book the sign of an energy term is always determined from the system's point of view: negative if the process is exothermic; positive if endothermic. Thus lattice energy has a negative sign.

We can illustrate the energy changes involved in the formation of an ionic solid by considering the formation of solid lithium fluoride from its elements:

$$Li(s) + \tfrac{1}{2}F_2(g) \longrightarrow LiF(s)$$

To see the energy terms associated with this process, we take advantage of the fact that energy is a state function and break this reaction into steps, the sum of which gives the overall reaction.

STEP 1

Sublimation of solid lithium. Sublimation involves taking a substance from the solid state to the gaseous state:

$$Li(s) \longrightarrow Li(g)$$

The enthalpy of sublimation for Li(s) is 161 kJ/mol.

STEP 2

Ionization of lithium atoms to form Li^+ ions in the gas phase:

$$Li(g) \longrightarrow Li^+(g) + e^-$$

This process corresponds to the first ionization energy for lithium, which is 520 kJ/mol.

STEP 3

Dissociation of fluorine molecules. We need to form 1 mole of fluorine atoms by breaking the F—F bonds in $\frac{1}{2}$ mole of F_2 molecules:

$$\tfrac{1}{2}F_2(g) \longrightarrow F(g)$$

The energy required to break this bond is 154 kJ/mol. In this case we are breaking the bonds in a half mole of fluorine, so the energy required for this step is 154 kJ/2, or 77 kJ.

STEP 4

Formation of F^- ions from fluorine atoms in the gas phase:

$$F(g) + e^- \longrightarrow F^-(g)$$

The energy change for this process corresponds to the electron affinity of fluorine, which is -328 kJ/mol.

STEP 5

Formation of solid lithium fluoride from the gaseous Li^+ and F^- ions:

$$Li^+(g) + F^-(g) \longrightarrow LiF(s)$$

This corresponds to the lattice energy for LiF, which is -1047 kJ/mol.

Since the sum of these five processes yields the desired overall reaction, the sum of the individual energy changes gives the overall energy change:

	Process	Energy Change (kJ)
	$Li(s) \rightarrow Li(g)$	161
	$Li(g) \rightarrow Li^+(g) + e^-$	520
	$\frac{1}{2}F_2(g) \rightarrow F(g)$	77
	$F(g) + e^- \rightarrow F^-(g)$	-328
	$Li^+(g) + F^-(g) \rightarrow LiF(s)$	-1047
Overall:	$Li(s) + \frac{1}{2}F_2(g) \rightarrow LiF(s)$	-617 kJ (per mole of LiF)

This process is summarized by the energy diagram in Fig. 13.8. Note that the formation of solid lithium fluoride from its elements is highly exothermic, mainly because of the very large negative lattice energy. A great deal of energy is released when the ions combine to form the solid. In fact, note that the energy released when an electron is added to a fluorine atom to form the F^- ion (328 kJ/mol) is not enough to remove an electron from lithium (520 kJ/mol). That is, when a metallic lithium atom reacts with a nonmetallic fluorine atom to form *separated* ions,

In doing this calculation, we have ignored the small difference between ΔH_{sub} and ΔE_{sub}.

$$Li(g) + F(g) \longrightarrow Li^+(g) + F^-(g)$$

the process is endothermic and thus unfavorable. Clearly, then, the main impetus for the formation of the ionic compound rather than a covalent

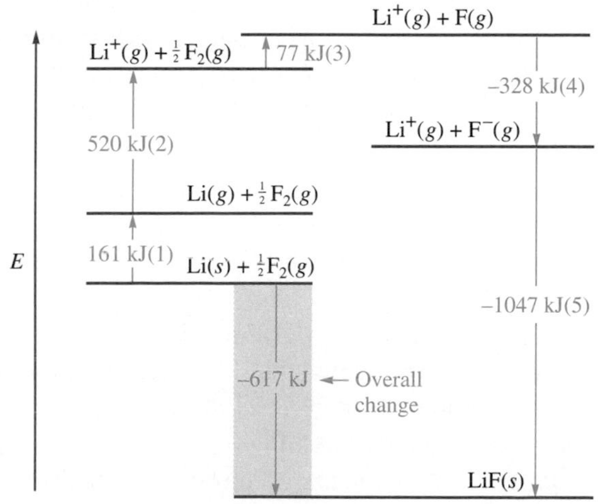

Figure 13.8
The energy changes involved in the formation of solid lithium fluoride from its elements. The numbers in parentheses refer to the reaction steps discussed in the text.

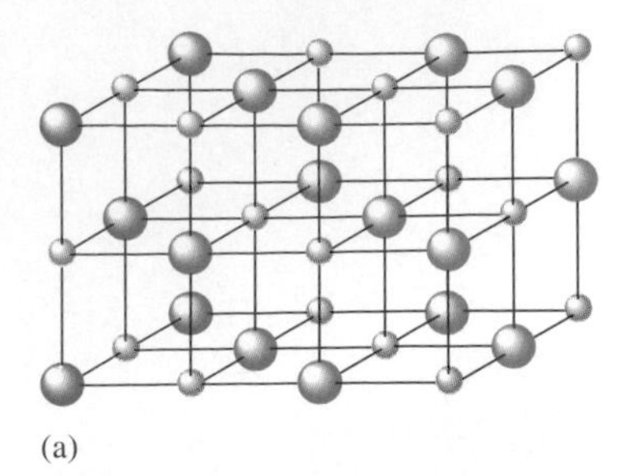
(a)

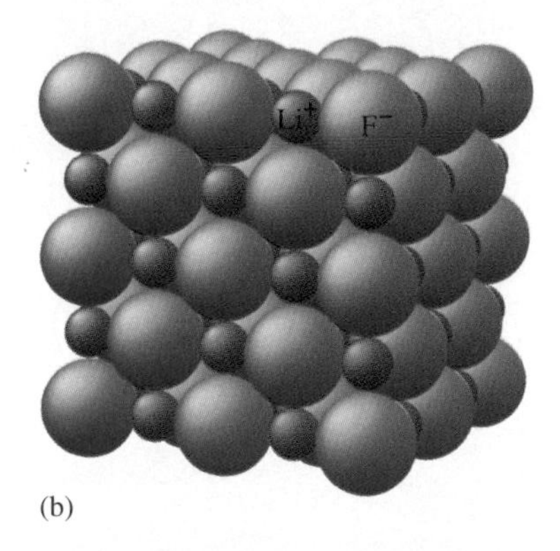

(b)

Figure 13.9
The structure of lithium fluoride. (a) Represented by a ball-and-stick model. Note that each Li^+ ion is surrounded by six F^- ions, and each F^- ion is surrounded by six Li^+ ions. (b) Represented with the ions shown as spheres. The structure is determined by packing the spherical ions in a way that both maximizes the ionic attractions and minimizes the ionic repulsions.

compound results from the strong mutual attractions among the Li^+ and F^- ions in the solid. The lattice energy is the dominant energy term.

The structure of the solid lithium fluoride is represented in Fig. 13.9. Note the alternating arrangement of the Li^+ and F^- ions. Also, note that each Li^+ is surrounded by six F^- ions and each F^- ion is surrounded by six Li^+ ions. This structure can be rationalized by assuming that the ions behave as hard spheres that pack together in a way that both maximizes the attractions among the oppositely charged ions and minimizes the repulsions among the identically charged ions.

All of the binary ionic compounds formed by an alkali metal and a halogen have the structure shown in Fig. 13.9, except for the cesium salts. The arrangement of ions shown in Fig. 13.9 is often called the *sodium chloride structure*, after the most common substance that possesses it.

Lattice Energy Calculations

In discussing the energetics of the formation of solid lithium fluoride, we emphasized the importance of lattice energy in contributing to the stability of the ionic solid. Lattice energy can be represented by a modified form of Coulomb's law,

$$\text{Lattice energy} = k\left(\frac{Q_1Q_2}{r}\right)$$

where k is a proportionality constant that depends on the structure of the solid and the electron configurations of the ions; Q_1 and Q_2 are the charges on the ions, and r is the shortest distance between the centers of the cations and anions. Note that the lattice energy has a negative sign when Q_1 and Q_2 have opposite signs. This result is expected, since bringing cations and anions together is an exothermic process. Also note that the process becomes more exothermic as the ionic charges increase and as the distances between the ions in the solid decrease.

The importance of the charges in ionic solids can be illustrated by comparing the energies involved in the formation of NaF(s) and MgO(s). These solids contain the isoelectronic ions Na^+, F^-, Mg^{2+}, and O^{2-}. The energy diagram for the formation of the two solids is given in Fig. 13.10. Note several important features:

The energy released when the gaseous Mg^{2+} and O^{2-} ions combine to form solid MgO is much greater (more than four times greater) than that released when the gaseous Na^+ and F^- ions combine to form solid NaF.

The energy required to remove two electrons from the magnesium atom (735 kJ/mol for the first and 1445 kJ/mol for the second, yielding a total of 2180 kJ/mol) is much greater than the energy required to remove an electron from a sodium atom (495 kJ/mol).

Energy (737 kJ/mol) is required to add two electrons to the oxygen atom in the gas phase. Addition of the first electron is exothermic (−141 kJ/mol), but addition of the second electron is quite endothermic (878 kJ/mol). This latter energy must be obtained indirectly, since the $O^{2-}(g)$ is not stable.

Since the equation for lattice energy contains the product Q_1Q_2, the lattice energy for a solid with 2+ and 2− ions should be four times that for a solid with 1+ and 1− ions. That is,

$$\frac{(+2)(-2)}{(+1)(-1)} = 4$$

For MgO and NaF the observed ratio of lattice energies (see Fig. 13.10) is

$$\frac{-3925\ \text{kJ}}{-923\ \text{kJ}} = 4.25$$

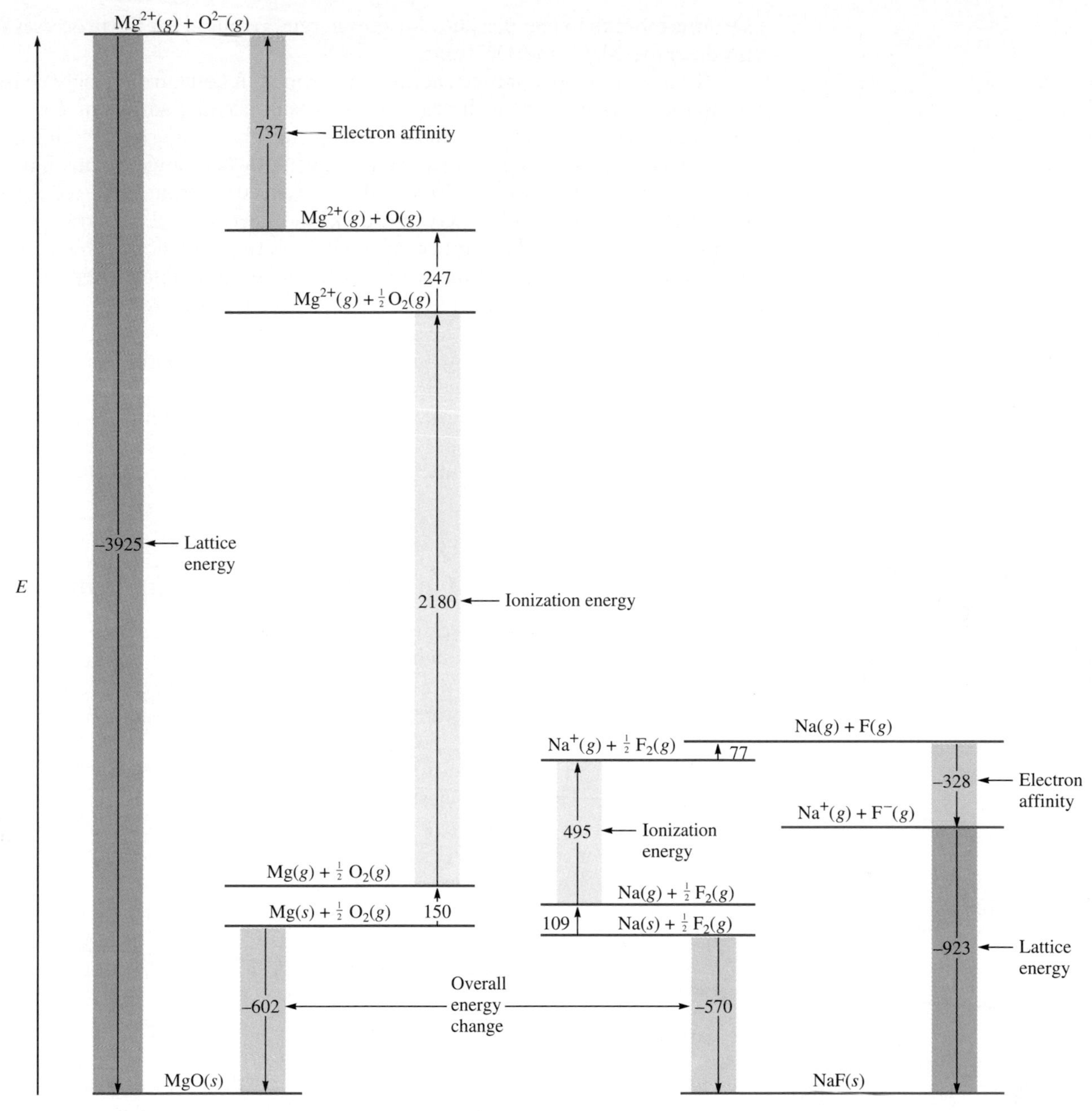

Figure 13.10
Comparison of the energy changes involved in the formation of solid sodium fluoride and solid magnesium oxide. Note the large lattice energy for magnesium oxide (where doubly charged ions are combining), compared with that for sodium fluoride (where singly charged ions are combining).

In view of the facts that twice as much energy is required to remove the second electron from magnesium as to remove the first, and that addition of an electron to the gaseous O^- ion is quite endothermic, it seems puzzling that magnesium oxide contains Mg^{2+} and O^{2-} ions rather than Mg^+ and O^- ions. The answer lies in the lattice energy. Note that the lattice energy for combining gaseous Mg^{2+} and O^{2-} ions to form MgO(*s*) is 3000 kJ/mol more negative than that for combining gaseous Na^+ and F^- ions to form NaF(*s*). Thus the energy released in forming a solid containing Mg^{2+} and O^{2-} ions rather than Mg^+ and

O^- ions more than compensates for the energies required for the processes that produce the Mg^{2+} and O^{2-} ions.

If there is so much lattice energy to be gained in going from singly charged to doubly charged ions in the case of magnesium oxide, why then does solid sodium fluoride contain Na^+ and F^- ions rather than Na^{2+} and F^{2-} ions? We can answer this question by recognizing that both Na^+ and F^- ions have the neon electron configuration. Removal of an electron from Na^+ requires an extremely large quantity of energy (4560 kJ/mol) because a $2p$ electron must be removed. Conversely, the addition of an electron to F^- would require use of the relatively high-energy $3s$ orbital, which is also an unfavorable process. Thus we can say that for sodium fluoride the extra energy required to form the doubly charged ions is greater than the gain in lattice energy that would result.

This discussion of the energies involved in the formation of solid ionic compounds illustrates that a variety of factors operate to determine the composition and structure of these compounds. The most important of these factors involve the balancing of the energies required to form highly charged ions and the energy released when highly charged ions combine to form the solid.

13.6 Partial Ionic Character of Covalent Bonds

(a)

(b)

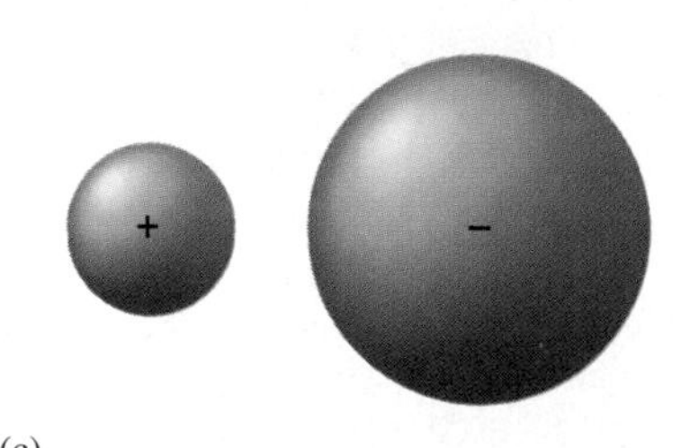

(c)

Figure 13.11
The three possible types of bonds: (a) a covalent bond formed between identical atoms; (b) a polar covalent bond, with both ionic and covalent components; and (c) an ionic bond with no electron sharing.

Recall that when atoms with different electronegativities react to form molecules, the electrons are not shared equally. The possible result is a polar covalent bond or, in the case of a large electronegativity difference, a complete transfer of one or more electrons to form ions. The cases are summarized in Fig. 13.11.

How well can we tell the difference between an ionic bond and a polar covalent bond? The only honest answer to this question is that there are probably no totally ionic bonds between *discrete pairs of atoms*. The evidence for this statement comes from calculations of the percent ionic character for the bonds of various binary compounds in the gas phase. These calculations are based on comparisons of the measured dipole moments for molecules of the type X—Y with the calculated dipole moments for the completely ionic case, X^+Y^-. We performed a calculation of this type for HF in Section 13.3. The percent ionic character of a bond can be defined as

$$\text{Percent ionic character of a bond} = \left(\frac{\text{measured dipole moment of X—Y}}{\text{calculated dipole moment of } X^+Y^-}\right) \times 100$$

Application of this definition to various compounds (in the gas phase) gives the results shown in Fig. 13.12, where percent ionic character is plotted versus the difference in the electronegativity values of X and Y. Note from this plot that ionic character increases with electronegativity difference, as expected. However, none of the bonds reaches 100% ionic character, even though compounds with the maximum possible electronegativity differences are considered. Thus according to this definition, no individual bonds are completely ionic. This conclusion is in contrast to the usual classification of many of these compounds (as solids). All of the compounds shown in Fig. 13.12 with more than 50% ionic character are normally considered to be ionic solids. Recall,

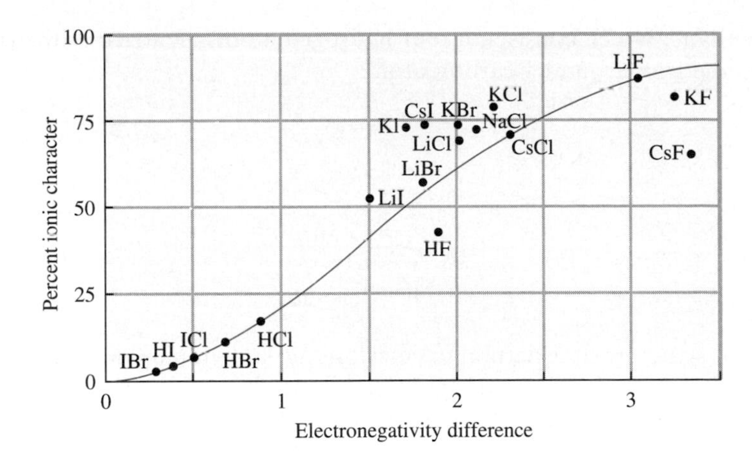

Figure 13.12
The relationship between the ionic character of a covalent bond and the electronegativity difference of the bonded atoms.

however, that the results in Fig. 13.12 are for the gas phase, where individual XY molecules exist. These results cannot necessarily be assumed to apply to the solid state, where the existence of ions is favored by the multiple ion interactions.

Another complication in identifying ionic compounds is that many substances contain polyatomic ions. For example, NH_4Cl contains NH_4^+ and Cl^- ions, and Na_2SO_4 contains Na^+ and SO_4^{2-} ions. The ammonium and sulfate ions are held together by covalent bonds. Thus calling NH_4Cl and Na_2SO_4 ionic compounds is somewhat ambiguous.

We will avoid these problems by adopting an operational definition of ionic compounds: *any compound that conducts an electric current when melted will be classified as ionic*. Also, the generic term *salt* will be used interchangeably with *ionic compound* in this book.

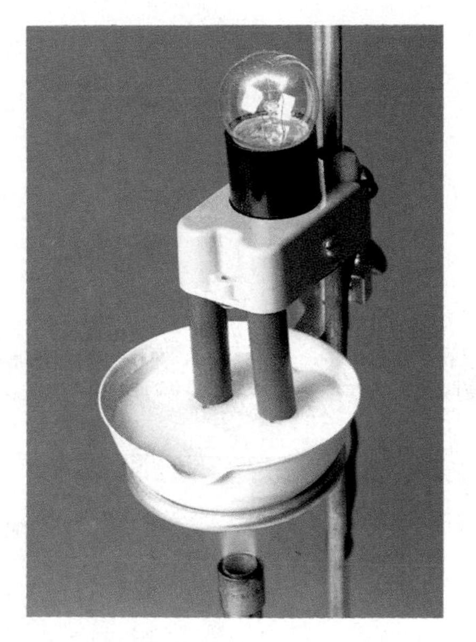

Molten NaCl conducts an electric current, indicating the presence of mobile Na^+ and Cl^- ions.

13.7 The Covalent Chemical Bond: A Model

Before we develop specific models for covalent chemical bonding, it will be helpful to summarize some of the concepts introduced in this chapter.

What is a chemical bond? Chemical bonds can be viewed as forces that cause a group of atoms to behave as a unit.

Why do chemical bonds occur? There is no principle of nature that states that bonds are favored or disfavored. Bonds are neither inherently "good" nor inherently "bad" as far as nature is concerned; they result from the tendency of a system to seek its lowest possible energy. From a simplistic point of view, bonds occur when collections of atoms are more stable (lower in energy) than the separate atoms. For example, approximately 1652 kJ of energy are required to break a mole of methane (CH_4) molecules into separate C and H atoms. Or, taking the opposite view, 1652 kJ of energy are released when 1 mole of methane is formed from 1 mole of gaseous C atoms and 4 moles of gaseous H atoms. Thus we can say that 1 mole of CH_4 molecules in the gas phase is 1652 kJ lower in energy than 1 mole of carbon atoms plus 4 moles of hydrogen atoms. Methane is therefore a stable molecule relative to its separated atoms.

We find it useful to interpret molecular stability in terms of a model called a chemical bond. To help understand why this model was invented, let's continue

A tetrahedron has four equal triangular faces.

with methane, which consists of four hydrogen atoms arranged at the corners of a tetrahedron around a carbon atom:

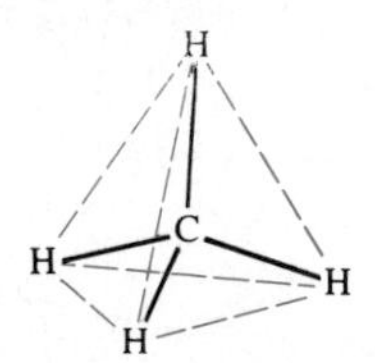

Given this structure, it is natural to envision four individual C—H interactions (we call them bonds). The energy of stabilization of CH_4 is divided equally among the four bonds to give an average C—H bond energy per mole of C—H bonds:

$$\frac{1652 \text{ kJ}}{4} = 413 \text{ kJ}$$

Next, consider methyl chloride, which consists of CH_3Cl molecules having the structure

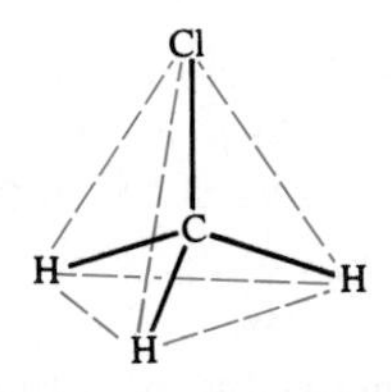

Experiments have shown that approximately 1578 kJ of energy are required to break down 1 mole of gaseous CH_3Cl molecules into gaseous carbon, chlorine, and hydrogen atoms. The reverse process can be represented as

$$C(g) + Cl(g) + 3H(g) \longrightarrow CH_3Cl(g) + 1578 \text{ kJ/mol}$$

A mole of gaseous methyl chloride is lower in energy by 1578 kJ than its separate gaseous atoms. Thus a mole of methyl chloride is held together by 1578 kJ of energy. Again, it is very useful to divide this energy into individual bonds. Methyl chloride can be visualized as containing one C—Cl bond and three C—H bonds. If we assume arbitrarily that a C—H interaction represents the same quantity of energy in any situation (that is, that the strength of a C—H bond is independent of its molecular environment), we can do the following bookkeeping:

$$\begin{aligned}
\text{1 mol of C—Cl bonds plus 3 mol of C—H bonds} &= 1578 \text{ kJ} \\
\text{C—Cl bond energy} + 3(\text{average C—H bond energy}) &= 1578 \text{ kJ} \\
\text{C—Cl bond energy} + 3(413 \text{ kJ/mol}) &= 1578 \text{ kJ} \\
\text{C—Cl bond energy} = 1578 - 1239 &= 339 \text{ kJ/mol}
\end{aligned}$$

These assumptions allow us to associate given quantities of energy with C—H and C—Cl bonds.

It is important to note that the bond concept is a human invention. Bonds provide a method for dividing up the energy evolved when a stable molecule is formed from its component atoms. Thus in this context *a bond represents a quantity of energy* obtained from the molecular energy of stabilization in a rather arbitrary way. This is not to say that the concept of individual bonds is a bad idea. In fact, the modern concept of the chemical bond, conceived by the American chemists G. N. Lewis and Linus Pauling, is one of the most useful ideas chemists have ever developed.

Models: An Overview

The framework of chemistry, like that of any science, consists of models—attempts to explain how nature operates on the microscopic level, based on experiences in the macroscopic world. To understand chemistry, one must understand its models and how they are used. We will use the concept of bonding to reemphasize the important characteristics of models, including their origin, structure, and uses.

Bonding is a model proposed to explain molecular stability.

Models originate from our observations of the properties of nature. For example, the concept of bonds arose from the observations that most chemical processes involve collections of atoms and that chemical reactions involve rearrangements of the ways the atoms are grouped. So to understand reactions, we must understand the forces that bind atoms together.

In natural processes there is a tendency toward lower energy. Collections of atoms therefore occur because the aggregated state has lower energy than the separated atoms. Why? As we have seen earlier in this chapter, the best explanations for the energy change involve atoms sharing electrons or atoms transferring electrons to become ions. In the case of electron sharing we find it convenient to assume that individual bonds occur between pairs of atoms. Let's explore the validity of this assumption and see how it is useful.

In a diatomic molecule such as H_2, it is natural to assume that a bond exists between the atoms, holding them together. It is also useful to assume that individual bonds are present in polyatomic molecules such as CH_4. So instead of thinking of CH_4 as a unit with a stabilization energy of 1652 kJ per mole, we choose to think of CH_4 as containing four C—H bonds, each worth 413 kJ of energy per mole of bonds. Without this concept of individual bonds in molecules, chemistry would be hopelessly complicated. There are millions of different chemical compounds, and if each of these compounds had to be considered as an entirely new entity, the task of understanding chemical behavior would be overwhelming.

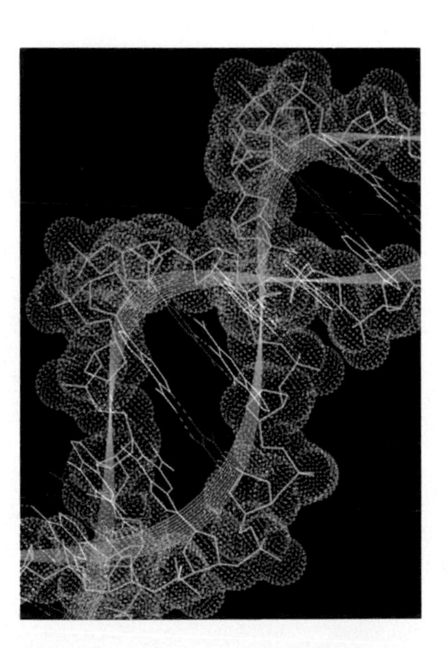

The concept of individual bonds makes it much easier to deal with complex molecules such as DNA. A small segment of a DNA molecule is shown here, using a computer-generated model.

The bonding model provides a framework to systematize chemical behavior by enabling us to think of molecules as collections of common fundamental components. For example, a typical biomolecule, such as a protein, contains hundreds of atoms and might seem discouragingly complex. However, if we think of a protein as constructed of individual bonds, C—C, C—H, C—N, C—O, N—H, and so on, it helps tremendously in predicting and understanding the protein's behavior. The essential idea is that we expect a given bond to behave about the same in any molecular environment. Used in this way, the model of the chemical bond has helped chemists to systematize the reactions of the millions of existing compounds.

In addition to being very useful, the bonding model is also physically sensible. It makes sense that atoms can form stable groups by sharing electrons; shared electrons give a lower energy state because they are simultaneously attracted by two nuclei.

Also, as we will see in the next section, bond energy data support the existence of discrete bonds that are relatively independent of the molecular environment. It is very important to remember, however, that the chemical bond is only a model. Although our concept of discrete bonds in molecules agrees with many of our observations, some molecular properties require that we think of a molecule as a whole, with the electrons free to move through the entire molecule. This is called *delocalization* of the electrons, a concept that will be discussed more completely in the next chapter.

Summary: **Fundamental Properties of Models**

- Models are human inventions, always based on an incomplete understanding of how nature works. *A model does not equal reality.*

- Models are often wrong. This property derives from the first property. Models are based on speculation and are always oversimplifications.

- Models tend to become more complicated as they age. As flaws are discovered in our models, we "patch" them and thus add more detail.

- It is very important to understand the assumptions inherent in a particular model before you use it to interpret observations or to make predictions. Simple models usually involve very restrictive assumptions and can only be expected to yield qualitative information. Asking for a sophisticated explanation from a simple model is like expecting to get an accurate mass for a diamond by using a bathroom scale.

  For a model to be used effectively, we must understand its strengths and weaknesses and ask only appropriate questions. An illustration of this point is the simple Aufbau principle used to explain the electron configurations of the elements. Although this model correctly predicts the configuration for most atoms, chromium and copper do not agree with the predictions. Detailed studies show that the configurations of chromium and copper result from complex electron interactions that are not taken into account in the simple model. However, this does not mean that we should discard the simple model that is so useful for most atoms. Instead, we must apply it with caution and not expect it to be correct in every case.

- When a model is wrong, we often learn much more than when it is right. If a model makes a wrong prediction, it usually means we do not understand some fundamental characteristics of nature. We often learn by making mistakes. (Try to remember that when you get back your next chemistry test.)

13.8 Covalent Bond Energies and Chemical Reactions

In this section we will consider the energies associated with various types of bonds and see how the bonding concept is useful in dealing with the energies of chemical reactions. One important consideration is to establish the sensitivity of a particular type of bond to its molecular environment. For example, consider the stepwise decomposition of methane:

Process	*Energy Required (kJ/mol)*
$CH_4(g) \rightarrow CH_3(g) + H(g)$	435
$CH_3(g) \rightarrow CH_2(g) + H(g)$	453
$CH_2(g) \rightarrow CH(g) + H(g)$	425
$CH(g) \rightarrow C(g) + H(g)$	339
	Total = 1652

$$\text{Average} = \frac{1652}{4} = 413$$

Although a C—H bond is broken in each case, the energy required varies in a nonsystematic way. This example shows that the C—H bond is somewhat sensitive to its environment. We use the *average* of these individual bond dissociation energies even though this quantity only approximates the energy associated with a C—H bond in a particular molecule. The degree of sensitivity of a bond to its environment can also be seen from experimental measurements of the energy required to break the C—H bond in the following molecules:

Molecule	Measured C—H Bond Energy (kJ/mol)
$HCBr_3$	380
$HCCl_3$	380
HCF_3	430
C_2H_6	410

These data show that the C—H bond strength varies significantly with its environment, but the concept of an average C—H bond strength remains useful to chemists. The average values of bond energies for various types of bonds are listed in Table 13.5.

So far, we have discussed bonds in which one pair of electrons is shared. This type of bond is called a **single bond.** As we will see in more detail later, atoms sometimes share two pairs of electrons, forming a **double bond,** or share three pairs of electrons, forming a **triple bond.** The bond energies for these *multiple bonds* are also given in Table 13.5.

A relationship also exists between the number of shared electron pairs and the bond length. As the number of shared electrons increases, the bond length shortens. This relationship is shown for selected bonds in Table 13.6.

TABLE 13.5 Average Bond Energies (kJ/mol)

Single Bonds						Multiple Bonds	
H—H	432	N—H	391	I—I	149	C=C	614
H—F	565	N—N	160	I—Cl	208	C≡C	839
H—Cl	427	N—F	272	I—Br	175	O=O	495
H—Br	363	N—Cl	200			C=O	799
H—I	295	N—Br	243	S—H	347	C≡O	1072
		N—O	201	S—F	327	N=O	607
C—H	413	O—H	467	S—Cl	253	N=N	418
C—C	347	O—O	146	S—Br	218	N≡N	941
C—N	305	O—F	190	S—S	266	C=N	615
C—O	358	O—Cl	203			C≡N	891
C—F	485	O—I	234	Si—Si	340		
C—Cl	339			Si—H	393		
C—Br	276	F—F	154	Si—C	360		
C—I	240	F—Cl	253	Si—O	452		
C—S	259	F—Br	237				
		Cl—Cl	239				
		Cl—Br	218				
		Br—Br	193				

TABLE 13.6 Bond Lengths for Selected Bonds

Bond	Bond Type	Bond Length (Å)	Bond Energy (kJ/mol)
C—C	Single	1.54	347
C=C	Double	1.34	614
C≡C	Triple	1.20	839
C—O	Single	1.43	358
C=O	Double	1.23	799
C—N	Single	1.43	305
C=N	Double	1.38	615
C≡N	Triple	1.16	891

Bond Energy and Enthalpy

Bond energy values can be used to calculate approximate energies for reactions. To illustrate how this is done, we will calculate the change in energy that accompanies the following reaction:

$$H_2(g) + F_2(g) \longrightarrow 2HF(g)$$

This reaction involves breaking one H—H and one F—F bond and forming two H—F bonds. For bonds to be broken, energy must be *added* to the system—an endothermic process. Consequently, the energy terms associated with bond breaking have *positive* signs. The formation of a bond *releases* energy, an exothermic process, and the energy terms associated with bond making carry a *negative* sign. We can write the enthalpy change for a reaction as follows:

ΔH = sum of the energies required to break old bonds (positive signs) plus the sum of the energies released in the formation of new bonds (negative signs)

This leads to the expression

$$\Delta H = \underbrace{\Sigma D \text{ (bonds broken)}}_{\text{Energy required}} - \underbrace{\Sigma D \text{ (bonds formed)}}_{\text{Energy released}}$$

where Σ represents the sum of terms and D represents the bond energy per mole of bonds. (D *always* has a positive sign.)

Since bond energies are typically averages taken from several compounds, the ΔH calculated from bond energies is not expected to agree exactly with that calculated from enthalpies of formation.

In the case of the formation of HF,

$$\begin{aligned} \Delta H &= D_{H—H} + D_{F—F} - 2D_{H—F} \\ &= 1 \text{ mol} \times \frac{432 \text{ kJ}}{\text{mol}} + 1 \text{ mol} \times \frac{154 \text{ kJ}}{\text{mol}} - 2 \text{ mol} \times \frac{565 \text{ kJ}}{\text{mol}} \\ &= -544 \text{ kJ} \end{aligned}$$

Thus when 1 mole of $H_2(g)$ and 1 mole of $F_2(g)$ react to form 2 moles of $HF(g)$, 544 kJ of energy should be released.

This result can be compared with the calculation of ΔH for this reaction from the standard enthalpy of formation for HF (−271 kJ/mol):

$$\Delta H = 2 \times (-271 \text{ kJ/mol}) = -542 \text{ kJ}$$

Thus the use of bond energies to calculate ΔH works quite well in this case.

EXAMPLE 13.5

Using the bond energies listed in Table 13.5, calculate ΔH for the reaction of methane with chlorine and fluorine to give Freon-12, CF_2Cl_2.

$$CH_4(g) + 2Cl_2(g) + 2F_2(g) \longrightarrow CF_2Cl_2(g) + 2HF(g) + 2HCl(g)$$

Solution

The idea here is to break the bonds in the reactants to give individual atoms and then assemble these atoms into the products by forming new bonds:

$$\text{Reactants} \xrightarrow[\text{required}]{\text{Energy}} \text{atoms} \xrightarrow[\text{released}]{\text{Energy}} \text{products}$$

We then combine the energy changes to calculate ΔH:

$$\Delta H = \begin{matrix}\text{energy required} \\ \text{to break bonds}\end{matrix} - \begin{matrix}\text{energy released} \\ \text{when bonds form}\end{matrix}$$

where the minus sign gives the correct sign to the energy terms for the exothermic processes.

Reactant bonds broken:

CH_4	4 mol C—H	$4 \text{ mol} \times \dfrac{413 \text{ kJ}}{\text{mol}}$	= 1652 kJ
$2Cl_2$	2 mol Cl—Cl	$2 \text{ mol} \times \dfrac{239 \text{ kJ}}{\text{mol}}$	= 478 kJ
$2F_2$	2 mol F—F	$2 \text{ mol} \times \dfrac{154 \text{ kJ}}{\text{mol}}$	= 308 kJ
		Total energy required	= 2438 kJ

Product bonds formed:

CF_2Cl_2	2 mol C—F	$2 \text{ mol} \times \dfrac{485 \text{ kJ}}{\text{mol}}$	= 970 kJ
	and 2 mol C—Cl	$2 \text{ mol} \times \dfrac{339 \text{ kJ}}{\text{mol}}$	= 678 kJ
HF	2 mol H—F	$2 \text{ mol} \times \dfrac{565 \text{ kJ}}{\text{mol}}$	= 1130 kJ
HCl	2 mol H—Cl	$2 \text{ mol} \times \dfrac{427 \text{ kJ}}{\text{mol}}$	= 854 kJ
		Total energy released	= 3632 kJ

We now can calculate ΔH:

$$\Delta H = \text{energy required to break bonds} - \text{energy released when bonds form}$$

$$= 2438 \text{ kJ} - 3632 \text{ kJ} = -1194 \text{ kJ}$$

Since the sign of the value for the enthalpy change is negative, this means that 1194 kJ of energy is released per mole of CF_2Cl_2 formed. The value of ΔH calculated for this reaction using enthalpies of formation is -1126 kJ.

In performing the calculations in this section, we have made several approximations. Of course, we made the usual assumption that average bond energies apply regardless of the specific molecular environment. We also ignored the difference between enthalpy and internal energy in these calculations. Recall from Chapter 9 that at constant pressure $\Delta E = \Delta H - P\Delta V$. Thus if a reaction involves a change in volume, a correction should be applied to the calculations of reaction enthalpies from bond energies. However, this correction is usually small compared with the uncertainties inherent in this method.

It should also be noted that the use of bond energies to estimate enthalpies of reaction should be limited to cases where all the reactants and products are in the gas phase where intermolecular interactions are expected to be minimal. Bond energies do not take these interactions into account.

13.9 The Localized Electron Bonding Model

So far we have discussed the general characteristics of the chemical bonding model and have seen that properties such as bond strength and polarity can be assigned to individual bonds. In this section we introduce a specific model used to describe covalent bonds. We need a simple model that can be easily applied even to very complicated molecules and that can be used routinely by chemists to interpret and organize the wide variety of chemical phenomena. The model that serves this purpose is often called the **localized electron (LE) model.** This model assumes that *a molecule is composed of atoms that are bound together by using atomic orbitals to share electron pairs.* The electron pairs in the molecule are assumed to be localized on a particular atom or in the space between two atoms. Those pairs of electrons localized on an atom are called **lone pairs,** and those found in the space between the atoms are called **bonding pairs.**

As we will apply it, the LE model has three parts:

1. Description of the valence electron arrangement in the molecule using Lewis structures (will be discussed in the next section).
2. Prediction of the geometry of the molecule, using the valence shell electron pair repulsion (VSEPR) model (will be discussed in Section 13.13).
3. Description of the types of atomic orbitals used by the atoms to share electrons or hold lone pairs (will be discussed in Chapter 14).

13.10 Lewis Structures

The **Lewis structure** of a molecule represents the arrangement of valence electrons among the atoms in the molecule. These representations are named after G. N. Lewis (Fig. 13.13). The rules for writing Lewis structures are based on the observations of thousands of molecules, which show that *in most stable compounds the atoms achieve noble gas electron configurations.* Although this is not always the case, it is so common that it provides a very useful place to start.

We have already seen that when metals and nonmetals react to form solid binary ionic compounds, electrons are transferred and that the resulting ions typically have noble gas electron configurations. An example is the formation of KBr, where the K^+ ion has the [Ar] electron configuration and the Br^- ion has the [Kr] electron configuration. In writing Lewis structures, the rule is that *only the valence electrons are included.* Using dots to represent electrons, the Lewis structure for KBr is

Lewis structures show only valence electrons.

K   :B̤̈r:

1+ charge   1− charge

No dots are shown on the K^+ ion since it has no valence electrons. The Br^- ion is shown with eight electrons since it has a filled valence shell.

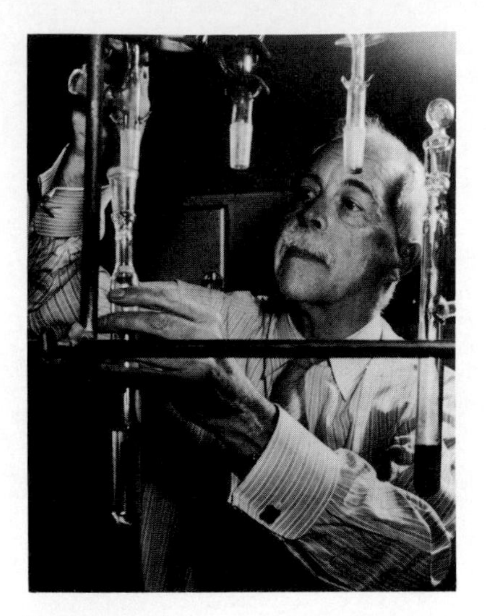

Figure 13.13
G. N. Lewis (above) conceived the octet rule while lecturing to a class of general chemistry students in 1902. He was also one of the two authors of a now classic work on Thermodynamics. Lewis and Randall *Thermodynamics and the Free Energy of Chemical Substances* (1923). (right) This is his original sketch. From G. N. Lewis, *Valence,* Dover Publications, Inc., New York, 1966.

Next, we will consider Lewis structures for molecules with covalent bonds involving elements in the first and second periods. The principle of achieving a noble gas electron configuration applies to these elements as follows:

Hydrogen forms stable molecules where it shares two electrons. That is, it follows a **duet rule.** For example, when two hydrogen atoms, each with one electron, combine to form the H_2 molecule, we have

$$
\begin{array}{ccc} H\cdot & & \cdot H \\ & \searrow \quad \swarrow & \\ & H:H & \end{array}
$$

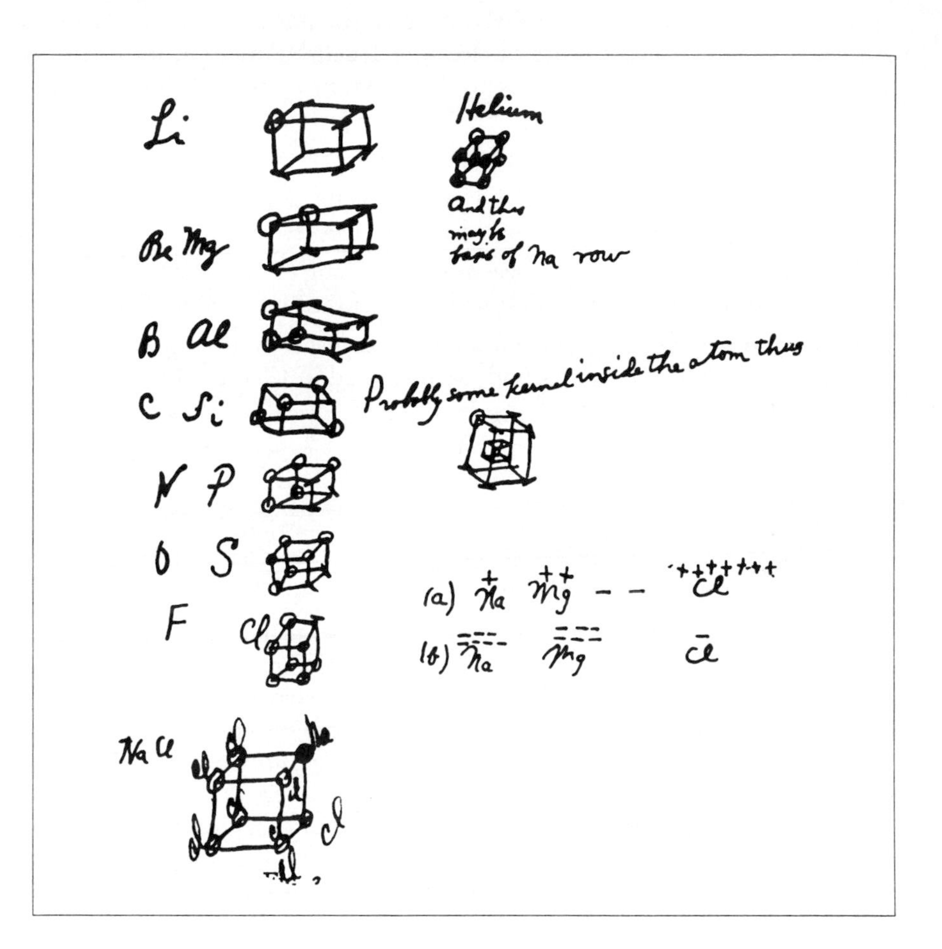

By sharing electrons, each hydrogen in H_2 has two electrons. This gives each hydrogen a filled valence shell.

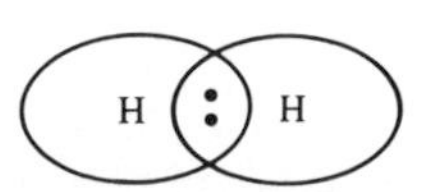

Helium does not form bonds because its valence orbitals are already filled; it is a noble gas. Helium has the electron configuration $1s^2$ and can be represented by the Lewis structure

$$\text{He}:$$

The second-row nonmetals (carbon through fluorine) form stable molecules when they are surrounded by enough electrons to fill the valence orbitals—the 2*s* and the three 2*p* orbitals. Since eight electrons are required to fill the 2*s* and 2*p* orbitals, these elements typically obey the **octet rule**; they are surrounded by eight electrons. An example is the F_2 molecule, which has the following Lewis structure:

Carbon, nitrogen, oxygen, and fluorine almost always obey the octet rule in stable molecules.

$$:\ddot{\underset{\cdot\cdot}{F}}\cdot \longrightarrow :\ddot{\underset{\cdot\cdot}{F}}:\ddot{\underset{\cdot\cdot}{F}}: \longleftarrow \cdot\ddot{\underset{\cdot\cdot}{F}}:$$

F atom with seven valence electrons — F_2 molecule — F atom with seven valence electrons

Note that each fluorine atom in F_2 is, in effect, surrounded by eight electrons, two of which are shared with the other atom. Recall that the shared pair of electrons is called a *bonding pair*. Each fluorine atom also has three pairs of electrons not involved in bonding. These are the *lone pairs*.

Neon does not form bonds since it already has an octet of valence electrons (it is a noble gas). The Lewis structure is

$$:\ddot{\underset{\cdot\cdot}{Ne}}:$$

Note that only the valence electrons of the neon atom ($2s^2 2p^6$) are represented by the Lewis structure. The $1s^2$ electrons are core electrons and take no part in chemical reactions.

From the discussion above we can formulate the following rules for writing Lewis structures of molecules containing atoms from the first two periods.

Summary: **Writing Lewis Structures**

- **Sum the valence electrons from all the atoms. Do not worry about keeping track of which electrons come from which atoms. It is the *total* number of electrons that is important.**
- **Use a pair of electrons to form a bond between each pair of bound atoms.**
- **Arrange the remaining electrons to satisfy the duet rule for hydrogen and the octet rule for the second-row elements.**

To see how these rules are applied, we will construct the Lewis structures for a few molecules. We will first consider the water molecule and follow the rules above.

STEP 1

We sum the *valence* electrons for H_2O as shown:

$$\underset{\text{H}}{1} + \underset{\text{H}}{1} + \underset{\text{O}}{6} = 8 \text{ valence electrons}$$

STEP 2

Using one pair of electrons per bond, we draw in the two O—H single bonds:

H—O—H

Note that *a line instead of a pair of dots is used to indicate each pair of bonding electrons.* This is the standard notation.

STEP 3

We distribute the remaining electrons around the atoms to achieve a noble gas electron configuration for each atom. Since four electrons have been used in forming the two bonds, four electrons (8 − 4) remain to be distributed. Hydrogen is satisfied with two electrons (duet rule), but oxygen needs eight electrons to achieve a noble gas configuration. Thus the remaining four electrons are added to oxygen as two lone pairs. Dots are used to represent the lone pairs:

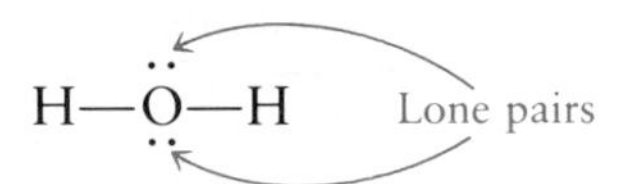

This is the correct Lewis structure for the water molecule. Each hydrogen has two electrons and the oxygen has eight:

H—O—H represents H:O:H.

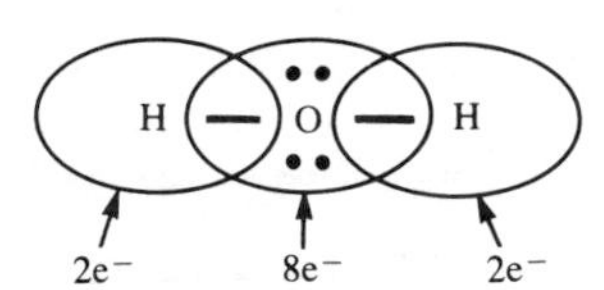

As a second example, we will write the Lewis structure for carbon dioxide. Summing the valence electrons gives

$$4 + 6 + 6 = 16$$

C O O

After forming a bond between the carbon and each oxygen,

O—C—O

the remaining electrons are distributed to achieve noble gas configurations on each atom. In this case we have 12 electrons (16 − 4) remaining after the bonds are drawn. The distribution of these electrons is determined by a trial-and-error process. We have six pairs of electrons to distribute. Suppose we try three pairs on each oxygen to give

:Ö—C—Ö:

To make sure that this structure is correct, we need to check two things:

1. The total number of electrons. There are 16 valence electrons in this structure, which is the correct number.
2. The octet rule for each atom. Each oxygen has 8 electrons, but the carbon only has 4. This cannot be the correct Lewis structure.

EXAMPLE 13.6

Give the Lewis structure for each of the following.

a. HF **b.** N_2 **c.** NH_3 **d.** CH_4 **e.** CF_4 **f.** NO^+

Solution

In each case we apply the three rules for writing Lewis structures. Recall that lines are used to indicate shared electron pairs and that dots are used to indicate nonbonding pairs (lone pairs). We have the following tabulated results:

	Total Valence Electrons	Draw Single Bonds	Calculate Number of Electrons Remaining	Use Remaining Electrons to Achieve Noble Gas Configurations
a. HF	1 + 7 = 8	H—F	6	·· H—F: ··
b. N_2	5 + 5 = 10	N—N	8	:N≡N:
c. NH_3	5 + 3(1) = 8	H—N—H | H	2	·· H—N—H | H
d. CH_4	4 + 4(1) = 8	H | H—C—H | H	0	H | H—C—H | H
e. CF_4	4 + 4(7) = 32	F | F—C—F | F	24	·· :F: ·· | ·· :F—C—F: ·· | ·· :F: ··
f. NO^+	5 + 6 − 1 = 10	N—O	8	$[:N{\equiv}O:]^+$

When writing Lewis structures, don't worry about which electrons came from which atoms. The best way to look at a molecule is to regard it as a new entity that uses all of the available valence electrons of the atoms to achieve the lowest possible energy.* The valence electrons belong to the molecule rather than to the individual atoms. Simply distribute all valence electrons so that the various rules are satisfied, without regard to the origin of each particular electron.

*In a sense this approach corrects for the fact that the localized electron model overemphasizes that a molecule is simply a sum of its parts, that is, that the atoms retain their individual identities in the molecule.

How can we arrange the 16 available electrons to achieve an octet for each atom? Suppose there are two shared pairs between the carbon and each oxygen:

O=C=O represents O::C::O.

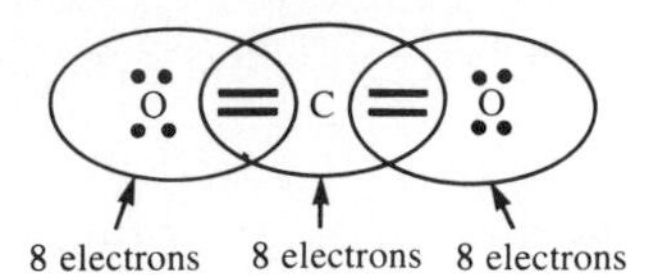

Now each atom is surrounded by 8 electrons, and the total number of electrons is 16, as required. Thus the correct Lewis structure for carbon dioxide has two double bonds.

Finally, let us consider the Lewis structure of the CN^- (cyanide) ion. Summing the valence electrons, we have

$$CN^-$$
$$4 + 5 + 1 = 10$$

Note that the negative charge requires that an extra electron must be added. After drawing a single bond (C—N), we distribute the remaining electrons to achieve a noble gas configuration for each atom. Eight electrons remain to be distributed. We can try various possibilities, for example:

C—N

This structure is incorrect, because C and N have only six electrons each instead of eight. The correct arrangement is

[:C≡N:]⁻

(Satisfy yourself that both carbon and nitrogen have eight electrons.)

13.11 Exceptions to the Octet Rule

The localized electron model is a simple but very successful model, and the rules we have used for Lewis structures apply to most molecules. However, with such a simple model some exceptions are inevitable. Boron, for example, tends to form compounds where the boron atom has fewer than eight electrons around it—it does not have a complete octet. Boron trifluoride (BF_3), a gas at normal temperatures and pressures, reacts very energetically with molecules such as water and ammonia that have available lone pairs. The violent reactivity of BF_3 with electron-rich molecules arises because the boron atom is electron-deficient. Boron trifluoride has 24 valence electrons. The Lewis structure that seems most consistent with the properties of BF_3 is

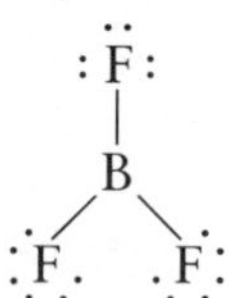

Note that in this structure boron has only six electrons around it. The octet rule for boron can be satisfied by drawing a structure with a double bond, such as

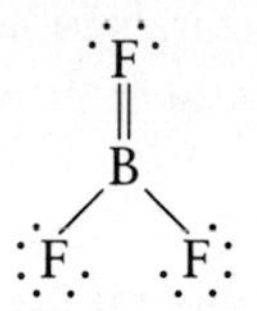

However, since fluorine is so much more electronegative than boron, this structure is questionable. In fact, experiments indicate that each B—F bond is probably best described by the first Lewis structure, which is also consistent with the reactivity of BF_3 toward electron-rich molecules—for example, toward NH_3 to form H_3NBF_3:

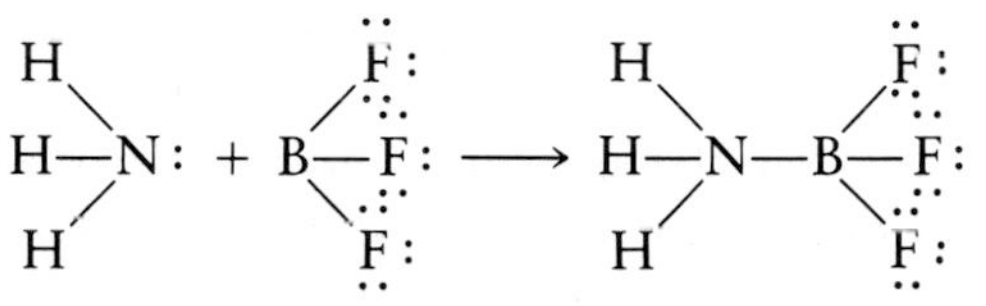

In this stable compound boron has an octet of electrons.

It is characteristic of boron to form molecules where the boron atom is electron-deficient. On the other hand, carbon, nitrogen, oxygen, and fluorine can be counted on to obey the octet rule.

Some atoms exceed the octet rule. This behavior is observed only for those elements in Period 3 of the periodic table and beyond. To see how this arises, we will consider the Lewis structure for sulfur hexafluoride (SF_6). The sum of the valence electrons for SF_6 is

$$6 + 6(7) = 48 \text{ electrons}$$

Indicating the single bonds gives the structure on the left below:

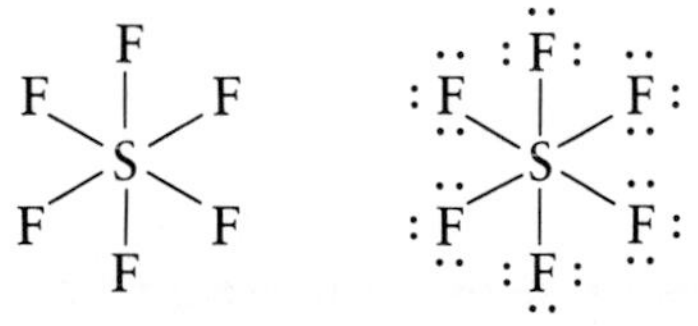

We have used 12 electrons to form the S—F bonds, which leaves 36 electrons. Since fluorine always follows the octet rule, we complete the six fluorine octets to give the structure on the right above. This structure uses all 48 valence electrons for SF_6, but sulfur has 12 electrons around it; that is, sulfur *exceeds* the octet rule. How can this happen?

To answer this question, we need to consider the different types of valence orbitals characteristic of second- and third-period elements. The second-row elements have 2*s* and 2*p* valence orbitals, and the third-row elements have 3*s*, 3*p*, and 3*d* orbitals. The 3*s* and 3*p* orbitals fill with electrons in going from sodium to argon, but the 3*d* orbitals remain empty. For example, the valence orbital diagram for a sulfur atom is

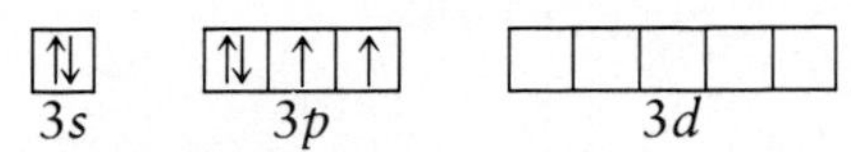

Third-row elements can exceed the octet rule.

The localized electron model assumes that the empty 3*d* orbitals can be used to accommodate extra electrons. Thus the sulfur atom in SF_6 can have 12 electrons around it by using the 3*s* and 3*p* orbitals to hold 8 electrons, and by placing the extra 4 electrons in the formerly empty 3*d* orbitals.

Summary: Lewis Structures and the Octet Rule

- **The second-row elements C, N, O, and F should always be assumed to obey the octet rule.**
- **The second-row elements B and Be often have fewer than eight electrons around them in their compounds. These electron-deficient compounds are very reactive.**
- **The second-row elements never exceed the octet rule, since their valence orbitals (2*s* and 2*p*) can accommodate only eight electrons.**
- **Third-row and heavier elements often satisfy the octet rule but can exceed the octet rule by using their empty valence *d* orbitals.**
- **When writing the Lewis structure for a molecule, first draw single bonds between all bonded atoms, and then satisfy the octet rule for all of the atoms. If electrons remain after the octet rule has been satisfied, place them on the elements having available *d* orbitals (elements in the third period or beyond).**

EXAMPLE 13.7

Write the Lewis structure for PCl_5.

Solution

We can follow the same stepwise procedure we used previously for sulfur hexafluoride.

STEP 1

Sum the valence electrons.

$$\underset{\text{P}}{\underset{\uparrow}{5}} + \underset{\text{Cl}}{\underset{\uparrow}{5}}(7) = 40 \text{ electrons}$$

STEP 2

Indicate single bonds between bound atoms.

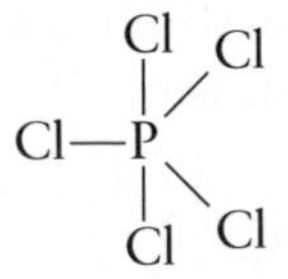

```
  Cl  Cl
   | /
Cl—P
   | \
  Cl  Cl
```

STEP 3

Distribute the remaining electrons. In this case, 30 electrons (40 − 10) remain. These are used to satisfy the octet rule for each chlorine atom. The final Lewis structure is

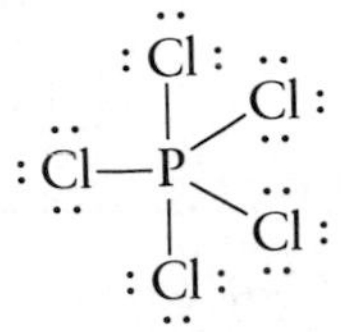

```
        ..
       :Cl:  ..
        |  /Cl:
   ..   | /   ..
  :Cl—P<      ..
   ..   | \   Cl:
        |      ..
       :Cl:
        ..
```

Note that phosphorus, which is a third-row element, exceeds the octet rule by two electrons.

In the PCl_5 and SF_6 molecules, the central atoms (P and S, respectively) must have the extra electrons. However, in molecules having more than one atom that can exceed the octet rule, it is not always clear which atom should have the extra electrons. Consider the Lewis structure for the triiodide ion (I_3^-), which has

$$\underset{\substack{\uparrow \\ \text{I}}}{3(7)} + \underset{\substack{\uparrow \\ 1-\ \text{charge}}}{1} = 22 \text{ valence electrons}$$

Indicating the single bonds gives I—I—I. At this point 18 electrons (22 − 4) remain. Trial and error will convince you that one of the iodine atoms must exceed the octet rule, but *which* one?

The rule we will follow is that *when it is necessary to exceed the octet rule for one of several third-row (or higher) elements, assume that the extra electrons are placed on the central atom.*

Thus for I_3^- the Lewis structure is

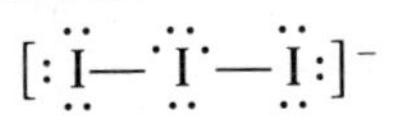

```
  ..   ..   ..
[:I—:I:—I:]⁻
  ..   ..   ..
```

where the central iodine exceeds the octet rule. This structure agrees with known properties of I_3^-.

EXAMPLE 13.8

Write the Lewis structure for each molecule or ion.

a. ClF_3 b. XeO_3 c. $RnCl_2$ d. $BeCl_2$ e. ICl_4^-

Solution

a. The chlorine atom (third row) accepts the extra electrons.

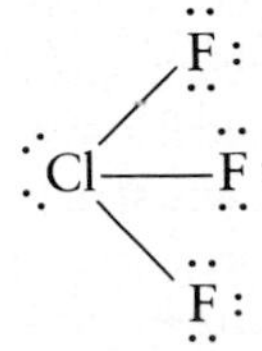

```
          ..
         /F:
        /  ..
  ..   /    ..
  :Cl——F:
  ..   \    ..
        \   ..
         F:
          ..
```

b. All atoms obey the octet rule.

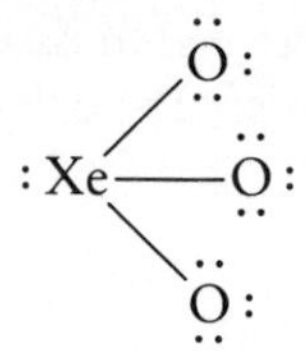

(Lewis structure: Xe with one lone pair, bonded to three O atoms, each with three lone pairs)

c. Radon, a noble gas in Period 6, accepts the extra electrons.

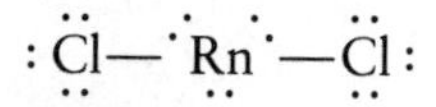

:Cl—Rn—Cl: (Rn with three lone pairs; each Cl with three lone pairs)

d. Beryllium is electron-deficient.

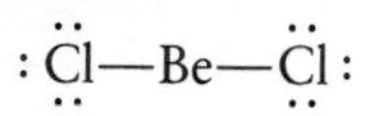

:Cl—Be—Cl: (each Cl with three lone pairs)

e. Iodine exceeds the octet rule.

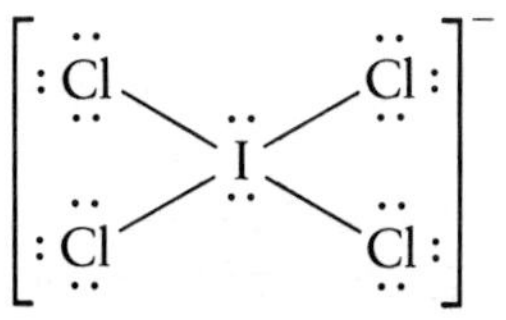

$[ICl_4]^-$ (I with two lone pairs, bonded to four Cl atoms, each with three lone pairs)

13.12 Resonance

A valid Lewis structure is one that obeys the rules we have outlined.

Sometimes more than one valid Lewis structure is possible for a given molecule. For example, consider the Lewis structure for the nitrate ion (NO_3^-), which has 24 valence electrons. So that an octet of electrons surrounds each atom, a structure like the following is required:

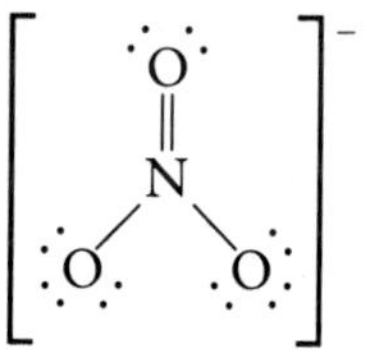

$[NO_3]^-$ (N double-bonded to one O with two lone pairs, single-bonded to two O atoms each with three lone pairs)

If this structure accurately represents the bonding in NO_3^-, there should be two types of N—O bonds observed in the molecule: one shorter bond (the double bond) and two identical longer ones (the two single bonds). However, experiments clearly show that NO_3^- exhibits only *one* type of N—O bond with a length and strength between those expected for a single bond and a double bond. Thus although the structure we have shown above is a valid Lewis structure, it does *not* correctly represent the bonding in NO_3^-. This is a serious problem, and it means that the model must be modified.

Look again at the proposed Lewis structure for NO_3^-. Because there is no reason for choosing a particular oxygen atom to have the double bond, there are really three valid Lewis structures:

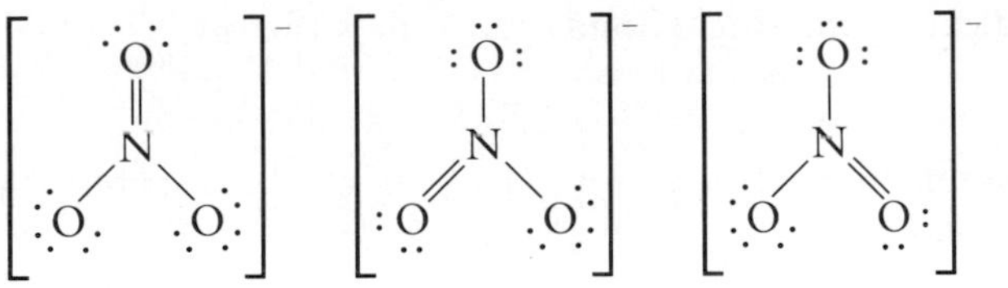

Is any of these structures a correct description of the bonding in NO_3^-? No, because NO_3^- does not have one double and two single bonds—it has three equivalent bonds. We can solve this problem by making the following assumption: the correct description of NO_3^- is *not given by any one* of the three Lewis structures individually but is given only by the *superposition of all three.*

The nitrate ion does not exist as any of the three extreme forms indicated by the individual Lewis structures but instead exists as an average of all three. **Resonance** *occurs when more than one valid Lewis structure can be written for a particular molecule.* The resulting electron structure of the molecule is given by the average of these **resonance structures.** This situation is usually represented by double-headed arrows as follows:

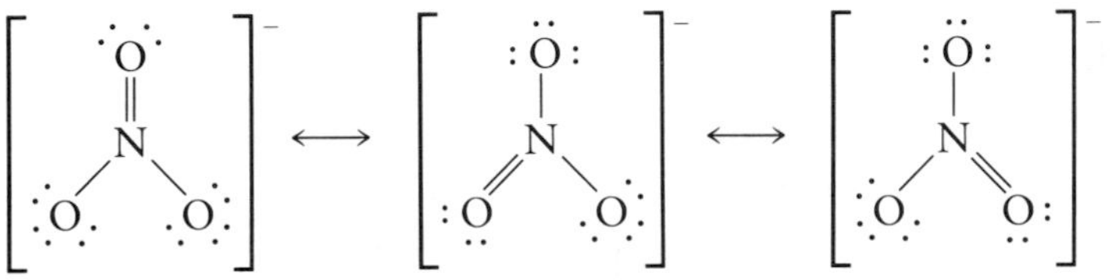

$$\longleftrightarrow \qquad \longleftrightarrow$$

Note that in all of these resonance structures the arrangement of the nuclei is the same. Only the placement of the electrons differs. The arrows do not indicate that the molecule "flips" from one resonance structure to another. They simply show that the *actual structure is an average of the three resonance structures.*

The concept of resonance is necessary because the localized electron model postulates that electrons are localized between a given pair of atoms. However, nature doesn't really operate this way. Electrons are actually delocalized—they can move around the entire molecule. The valence electrons in the NO_3^- molecule distribute themselves to provide equivalent N—O bonds. Resonance is necessary to compensate for this defective assumption of the localized electron model. However, because this model is so useful, we retain the concept of localized electrons and add resonance to accommodate species like NO_3^-.

EXAMPLE 13.9

Describe the electron arrangement in the nitrite anion (NO_2^-), using the localized electron model.

Solution

We will follow the usual procedure for obtaining the Lewis structure for the NO_2^- ion.

In NO_2^- there are $5 + 2(6) + 1 = 18$ valence electrons.

Indicating the single bonds gives the structure

O—N—O

The remaining 14 electrons (18 − 4) can be distributed to produce these structures:

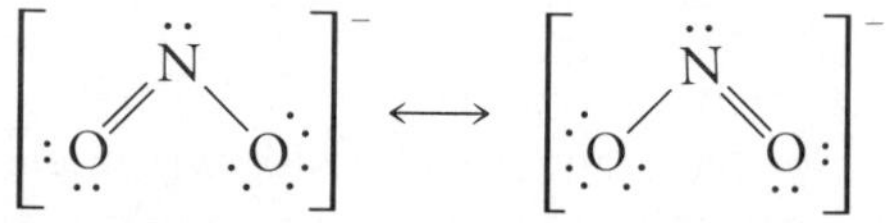

This is a resonance situation. Two equivalent Lewis structures can be drawn. *The electronic structure of the molecule is not correctly represented by either resonance structure but by the average of the two.* There are two equivalent N—O bonds, each one intermediate between a single and double bond.

Odd-Electron Molecules

Relatively few molecules formed from nonmetals contain odd numbers of electrons. One common example is nitric oxide (NO), which is formed when nitrogen and oxygen gases react at the high temperatures present in automobile engines. Nitric oxide is emitted into the air, where it reacts with oxygen to form gaseous nitrogen dioxide (NO_2), another odd-electron molecule.

Since the localized electron model is based on pairs of electrons, it does not handle odd-electron cases in a natural way, although Lewis structures are sometimes written for these species. To treat odd-electron molecules accurately, we need a more sophisticated model.

Formal Charge

Equivalent Lewis structures contain the same numbers of single and multiple bonds. For example, the resonance structures for O_3,

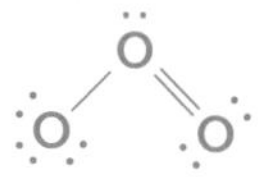

and

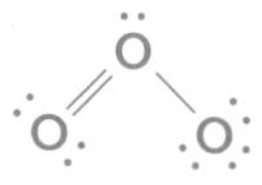

are equivalent Lewis structures. They are equally important in describing the bonding in O_3. Nonequivalent Lewis structures contain different numbers of single and multiple bonds.

Molecules or polyatomic ions containing atoms that can exceed the octet rule often have many nonequivalent Lewis structures, all of which obey the rules for writing Lewis structures. For example, as we will see in detail below, the sulfate ion has a Lewis structure with all single bonds and several Lewis structures that contain double bonds. How do we decide which of the many possible Lewis structures best describes the actual bonding in sulfate? One method involves estimating the charge on each atom in the various possible Lewis structures and using these charges to select the most appropriate structure(s). We will see below how this is done, but first we must decide on a method to evaluate atomic charges in molecules.

In Chapter 4 we discussed one system for obtaining charges for atoms in molecules—oxidation states. However, in assigning oxidation states we always count *both* of the shared electrons as belonging to the more electronegative atom in a bond. This practice leads to highly exaggerated estimates of charge. In other words, although oxidation states are useful for bookkeeping electrons in redox reactions, they are not realistic estimates of the actual charges on individual atoms in a molecule, and so they are not suitable for judging the appropriateness of Lewis structures. A second definition of atomic charges in a molecule, the **formal charge,** is more suitable for evaluating Lewis structures.

The concept of formal charge requires that we compare:

1. The number of valence electrons on the free neutral atom (which has a charge of zero because the number of electrons equals the number of protons) and
2. the number of valence electrons "belonging" to a given atom in a molecule

If an atom in a molecule has the same number of valence electrons as it does in the free state, the positive and negative charges just balance, and the atom has a formal charge of 0. If an atom has one more valence electron in a molecule than it has as a free atom, it has a formal charge of -1 and so on. Thus the formal charge on an atom in a molecule is defined as

$$\text{Formal charge} = \begin{pmatrix}\text{number of valence} \\ \text{electrons on a free atom}\end{pmatrix} - \begin{pmatrix}\text{number of valence electrons} \\ \text{assigned to the atom in the} \\ \text{molecule}\end{pmatrix}$$

To compute the formal charge of an atom in a molecule, we assign the valence electrons to the various atoms, by making the following assumptions:

1. Lone pair electrons belong entirely to the atom in question.
2. Shared electrons are *divided equally* between the two sharing atoms.

Thus the number of valence electrons assigned to a given atom is calculated as follows:

$$(\text{Valence electrons})_{\text{assigned}} = \begin{pmatrix}\text{number of lone} \\ \text{pair electrons}\end{pmatrix} + \tfrac{1}{2}\begin{pmatrix}\text{number of} \\ \text{shared electrons}\end{pmatrix}$$

We will illustrate the procedures for calculating formal charges by considering two of the possible Lewis structures for the sulfate ion, which has 32 valence electrons. For the Lewis structure

$$\left[\begin{array}{ccccc} & & :\ddot{\text{O}}: & & \\ & & | & & \\ :\ddot{\text{O}} & — & \text{S} & — & \ddot{\text{O}}: \\ \ \ \cdot\cdot & & | & & \cdot\cdot \\ & & :\text{O}: & & \\ & & \cdot\cdot & & \end{array}\right]^{2-}$$

each oxygen atom has 6 lone pair electrons and shares 2 electrons with the sulfur atom. Thus according to the above assumptions, each oxygen is assigned 7 valence electrons:

$$\text{Valence electrons assigned to each oxygen} = \underset{\substack{\uparrow \\ \text{Lone pair electrons}}}{6} \text{ plus } \underset{\substack{\uparrow \\ \text{Shared electrons}}}{\tfrac{1}{2}(2)} = 7$$

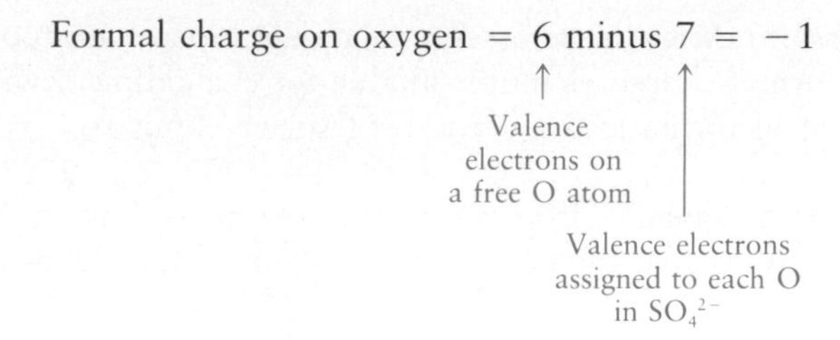

The formal charge on each oxygen is -1.

For the sulfur atom there are no lone pair electrons and eight electrons are shared with the oxygen atoms. Thus for sulfur,

$$\text{Valence electrons assigned to sulfur} = \underset{\text{Lone pair electrons}}{0} \text{ plus } \underset{\text{Shared electrons}}{\tfrac{1}{2}(8)} = 4$$

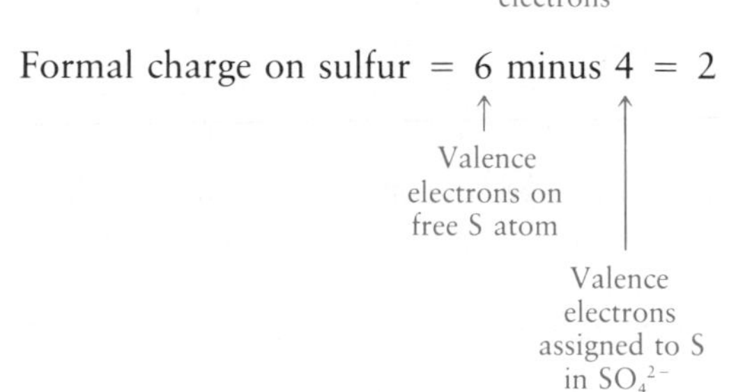

A second possible Lewis structure is

```
       ..
      :O:
       ‖
   ..  |   ..
  :O—S—O:
   ..  ‖   ..
      :O:
       ..        2−
```

In this case the formal charges are as calculated below.

For oxygen atoms with single bonds:

$$\text{Valence electrons assigned} = 6 + \tfrac{1}{2}(2) = 7$$

$$\text{Formal charge} = 6 - 7 = -1$$

For oxygen atoms with double bonds:

$$\text{Valence electrons assigned} = 4 + \underset{\text{Each double bond has 4 electrons}}{\tfrac{1}{2}(4)} = 6$$

$$\text{Formal charge} = 6 - 6 = 0$$

For the sulfur atom:

$$\text{Valence electrons assigned} = 0 + \tfrac{1}{2}(12) = 6$$

$$\text{Formal charge} = 6 - 6 = 0$$

We will use two fundamental assumptions about formal charges to evaluate Lewis structures:

1. Atoms in molecules try to achieve formal charges as close to zero as possible.
2. Any negative formal charges are expected to reside on the most electronegative atoms.

We can use these principles to evaluate the two Lewis structures for the sulfate ion discussed previously. Notice that in the structure with only single bonds, each oxygen has a formal charge of -1, while the sulfur has a formal charge of $+2$. In contrast, in the structure with two double bonds and two single bonds, the sulfur and two oxygen atoms have a formal charge of 0, while two oxygens have a formal charge of -1. From the assumptions given above, the structure with two double bonds is preferred—it has lower average formal charges, and the -1 formal charges are on electronegative oxygen atoms. Thus for the sulfate ion we might expect resonance structures such as

$$\left[\mathrm{O{=}S({-}O)(O{=})}{-}\mathrm{O}\right]^{2-} \longleftrightarrow \left[\mathrm{O{=}S(O)({-}O){=}O}\right]^{2-}$$

to more accurately describe the bonding than the Lewis structure with only single bonds.

Summary: Rules Governing Formal Charge

- **To calculate the formal charge on an atom:**
  1. **Take the sum of the lone pair electrons and one-half of the shared electrons. This is the number of valence electrons assigned to a given atom in the molecule.**
  2. **Subtract the number of assigned electrons from the number of valence electrons on the free, neutral atom to obtain the formal charge.**
- **The sum of the formal charges of all atoms in a given molecule or ion must equal the overall charge on that species.**
- **If nonequivalent Lewis structures exist for a species, those with formal charges closest to zero and with any negative formal charges on the most electronegative atoms are considered to best describe the bonding in the molecule or ion.**

EXAMPLE 13.10

Give possible Lewis structures for XeO_3, an explosive compound of xenon. Which Lewis structure or structures are most appropriate according to the formal charges?

Solution

For XeO_3 (26 valence electrons) we can draw the following possible Lewis structures (formal charges are indicated in parentheses):

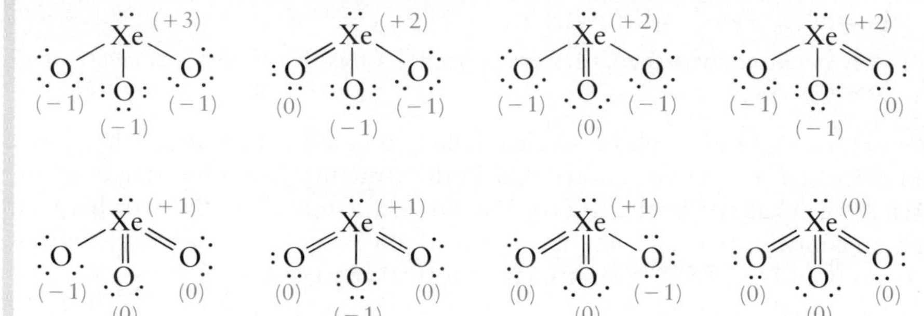

On the basis of the formal charges, we would predict that the Lewis structures with the lower values of formal charge would be most appropriate for describing the bonding in XeO_3.

The concept of formal charge is most often used to evaluate the importance of various Lewis structures for molecules that exhibit resonance. However, formal charge arguments also can be helpful in predicting which, among a given group of atoms, is the central atom in a simple molecule. For example, why is carbon dioxide O—C—O rather than C—O—O? Although this question can be pursued at many different levels of sophistication, the simplest approach involves considering the formal charges in the two possible structures. Note that in the Lewis structure given previously for carbon dioxide

$$:\ddot{O}=C=\ddot{O}:$$

all atoms have formal charges of 0. However, if the atoms are arranged as follows,

C—O—O

all of the Lewis structures give unreasonable formal charges. Consider the following possibilities, where the formal charges are listed below each atom:

$$\underset{-1}{:C}\equiv\underset{+2}{O}-\underset{-1}{\ddot{\underset{..}{O}}:} \qquad \underset{-2}{\ddot{C}}=\underset{+2}{O}=\underset{0}{\ddot{O}} \qquad \underset{-3}{:\ddot{\underset{..}{C}}}-\underset{+2}{O}\equiv\underset{+1}{O:}$$

None of these Lewis structures (with their resulting formal charges) agrees with our observation that oxygen has a significantly greater electronegativity than carbon. That is, it doesn't make sense that a compound would contain a negatively charged carbon atom next to a positively charged oxygen atom.

As a final note, there are several cautions about formal charge to keep in mind. First, although formal charges are closer to actual atomic charges in molecules than are oxidation states, formal charges still are only estimates of charge—they should not be taken as actual atomic charges. Second, the evaluation of Lewis structures using formal charge ideas can lead to erroneous predictions.

In this same vein, note the difference between a "correct" or valid Lewis structure and an electronic structure that accurately accounts for a molecule's observed properties. A valid Lewis structure is one that obeys the rules we have established for Lewis structures. However, this Lewis structure may or may not give an accurate picture of the molecule and its properties. Experiments must be carried out to make the final decisions on the correct description of bonding in a molecule or polyatomic ion.

13.13 Molecular Structure: The VSEPR Model

The structures of molecules play a very important role in determining their chemical properties. As we will see later, structure is particularly important for biological molecules; a slight change in the structure of a large biomolecule can completely destroy its usefulness to a cell or may even change the cell from normal to cancerous.

Many accurate methods now exist for determining **molecular structure,** the three-dimensional arrangement of the atoms in a molecule. These methods must be used if precise information about structure is required. However, it is often useful to predict the approximate molecular structure of a molecule. In this section we consider a simple model that allows us to do this. This model, called the **valence shell electron pair repulsion** (VSEPR) **model,** is useful in predicting the geometries of molecules formed from nonmetals. The main postulate of this model is that *the structure around a given atom is determined principally by minimizing electron pair repulsions.* The idea is that the bonding and nonbonding pairs around a given atom should be positioned as far apart as possible. To see how this model works, we will first consider the molecule $BeCl_2$, which has the Lewis structure

The origin of the "repulsions" among electron pairs probably results more from the operation of the Pauli exclusion principle than from electrostatic effects, but we will not be concerned with that in this text.

$$:\ddot{\underset{\cdot\cdot}{Cl}}—Be—\ddot{\underset{\cdot\cdot}{Cl}}:$$

Note that there are two pairs of electrons around the beryllium atom. What arrangement of these electron pairs allows them to be as far apart as possible to minimize the repulsions? Clearly, the best arrangement places the pairs on opposite sides of the beryllium atom at 180°:

$$—Be—$$
180°

This is the maximum possible separation for two electron pairs. Once we have determined the optimum arrangement of the electron pairs around the central atom, we can specify the molecular structure of $BeCl_2$, that is, the positions of the atoms. Since each electron pair on beryllium is shared with a chlorine atom, the molecule has a **linear structure** with a bond angle of 180°:

$$Cl—Be—Cl$$
180°

Next, let's consider BF_3, which has the Lewis structure

$$\begin{array}{c} :\ddot{F}: \\ | \\ :\ddot{\underset{\cdot\cdot}{F}}—B—\ddot{\underset{\cdot\cdot}{F}}: \end{array}$$

Here the boron atom is surrounded by three pairs of electrons. What arrangement will minimize the repulsions? The electron pairs are farthest apart at angles of 120°:

B 120°

Since each electron pair is shared with a fluorine atom, the molecular structure is

```
     F
     |  120°
     B
   /   \
  F     F
```

This is a planar (flat) and triangular molecule, which is commonly described as **trigonal planar.**

Next, let us consider the methane molecule, which has the Lewis structure

```
     H
     |
  H—C—H
     |
     H
```

There are four pairs of electrons around the central carbon atom. What arrangement of these electron pairs best minimizes the repulsions? First, let's try a square planar arrangement:

```
   |  90°
 —C—
   |
```

The carbon atom and the electron pairs are centered in the plane of the paper, and the angles between the pairs are all 90°.

Is there another arrangement with angles greater than 90° that would put the electron pairs even further away from each other? The answer is yes. The **tetrahedral arrangement** has angles of 109.5°:

C ~109.5°

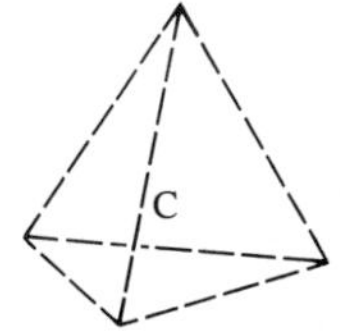

It can be shown that this is the maximum possible separation of four pairs around a given atom. This means that *whenever four pairs of electrons are present around an atom, they should always be arranged tetrahedrally.*

Now that we have obtained the electron pair arrangement that gives the least repulsion, we can determine the positions of the atoms and thus the molecular structure of CH_4. In methane each of the four electron pairs is shared between the carbon atom and a hydrogen atom. Thus the hydrogen atoms are placed as shown in Fig. 13.14, giving the molecule a tetrahedral structure with the carbon atom at the center.

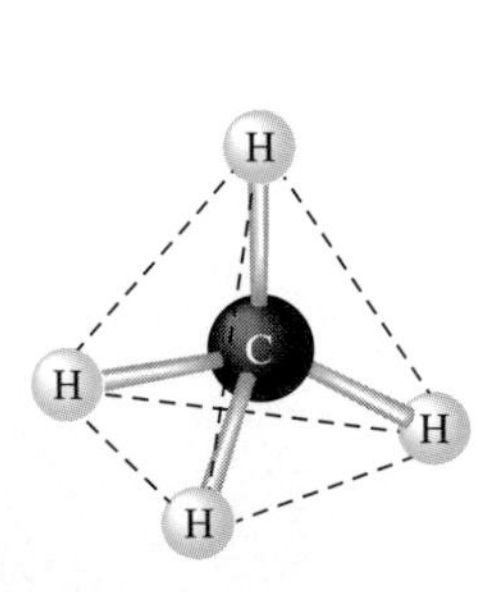

Figure 13.14
The molecular structure of methane. The tetrahedral arrangement of electron pairs produces a tetrahedral arrangement of hydrogen atoms.

Recall that the fundamental idea of the VSEPR model is to find the arrangement of electron pairs around the central atom that minimizes the electron repulsions. Then we can determine the molecular structure from knowing how the electron pairs are shared with the peripheral atoms.

Summary: **Steps for Using the VSEPR Model**

- **Draw the Lewis structure for the molecule.**
- **Count the electron pairs and arrange them in the way that minimizes repulsion (that is, put the pairs as far apart as possible).**
- **Determine the positions of the atoms from the ways the electron pairs are shared.**
- **Name the molecular structure from the positions of the *atoms*.**

We will now predict the structure of ammonia (NH_3) using this approach.

- **Draw the Lewis structure.**

$$\begin{array}{c} \quad \cdot\cdot \\ H-N-H \\ | \\ H \end{array}$$

- **Count the pairs of electrons and arrange them to minimize repulsions.** The NH_3 molecule has four pairs of electrons: three bonding pairs and one nonbonding pair. From the discussion of the methane molecule, we know that the best arrangement of four electron pairs is a tetrahedral array, as shown in Fig. 13.15(a).
- **Determine the positions of the atoms.** The three H atoms share electron pairs, as shown in Fig. 13.15(b).

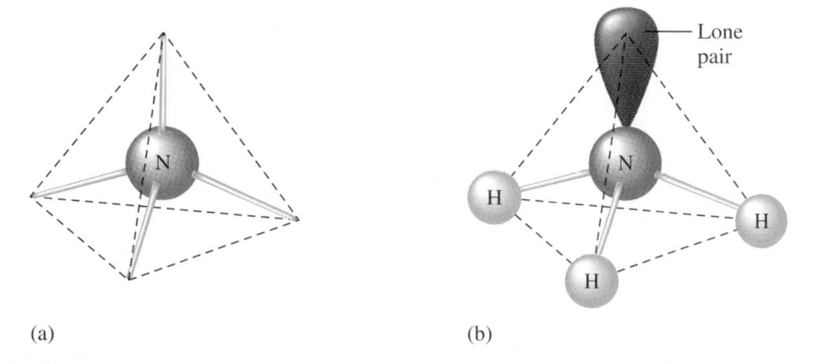

Figure 13.15
(a) The tetrahedral arrangement of electron pairs around the nitrogen atom in the ammonia molecule. (b) Three of the electron pairs around nitrogen are shared with hydrogen atoms, as shown, and the fourth is a lone pair. Although the arrangement of *electron pairs* is tetrahedral, as in the methane molecule, the hydrogen atoms in the ammonia molecule occupy only three corners of the tetrahedron. A lone pair occupies the fourth corner.

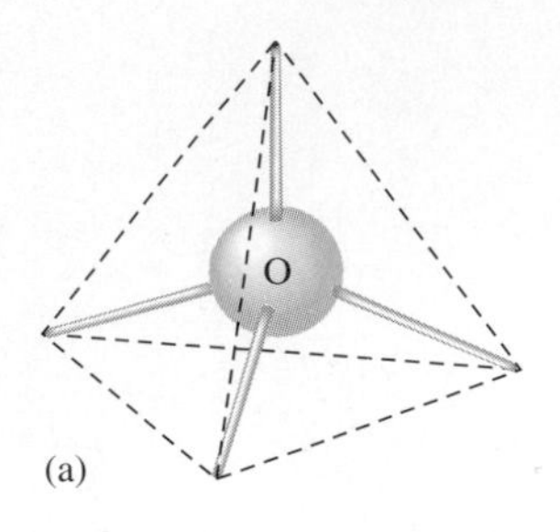

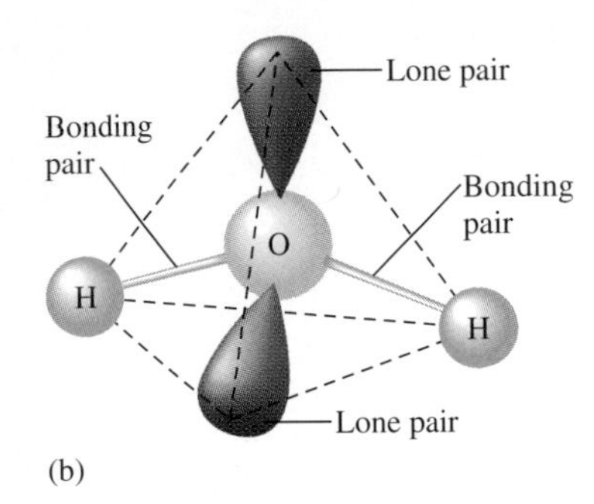

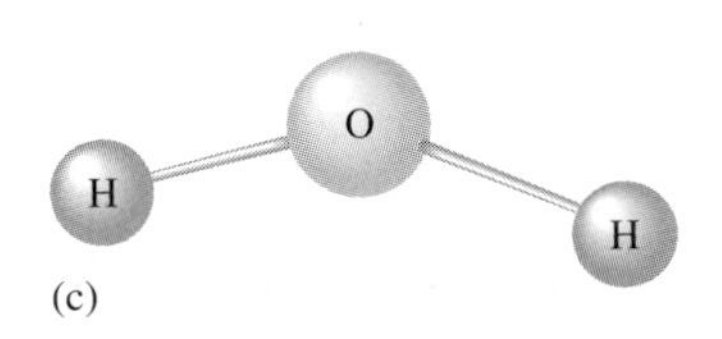

Figure 13.16
(a) The tetrahedral arrangement of the four electron pairs around oxygen in the water molecule. (b) Two of the electron pairs are shared between oxygen and the hydrogen atoms, and two are lone pairs. (c) The V-shaped molecular structure of the water molecule.

- **Name the molecular structure.**
  It is very important to recognize that the *name* of the molecular structure is always based on the *positions of the atoms.* The placement of the electron pairs determines the structure, but the name is based on the positions of the atoms. Thus it is incorrect to say that the NH_3 molecule is tetrahedral. It has a tetrahedral arrangement of electron pairs but not a tetrahedral arrangement of atoms. The molecular structure of ammonia is a **trigonal pyramid** (one triangular side is different from the other three), rather than a tetrahedron.

EXAMPLE 13.11

Describe the molecular structure of the water molecule.

Solution

The Lewis structure for water is

$$H-\ddot{\underset{\cdot\cdot}{O}}-H$$

There are four pairs of electrons: two bonding pairs and two nonbonding pairs. To minimize repulsions, these are best arranged in a tetrahedral array, as shown in Fig. 13.16(a). Although H_2O has a tetrahedral arrangement of electron pairs, it is not a tetrahedral molecule. The atoms in the H_2O molecule form a V shape, as shown in Fig. 13.16(b) and (c).

From Example 13.11 we see that the H_2O molecule is V-shaped, or bent, because of the presence of the lone pairs. If no lone pairs were present, the molecule would be linear, the polar bonds would cancel, and the molecule would have no dipole moment. This would make water very different from the polar substance so familiar to us.

From the previous discussion we would predict that the H—X—H bond angle (where X is the central atom) in CH_4, NH_3, and H_2O should be the tetrahedral angle (109.5°). Experiments, however, show that the actual bond angles are those given in Fig. 13.17. What significance do these results have for the VSEPR model? One possible point of view is that the observed angles are close enough to the tetrahedral angle to be satisfactory. The opposite view is that the deviations are significant enough to require modification of the simple model so that it can more accurately handle similar cases. We will take the latter view.

Methane: 109.5° — Ammonia: 107° — Water: 104.5°

Figure 13.17
The bond angles in the CH_4, NH_3, and H_2O molecules. Note that the bond angle between bonding pairs decreases as the number of lone pairs increases.

Let us examine the following data:

	CH_4	NH_3	H_2O
Number of lone pairs	0	1	2
Bond angle	109.5°	107°	104.5°

One interpretation of the trend observed here is that lone pairs require more space than bonding pairs; in other words, as the number of lone pairs increases, the bonding pairs are increasingly squeezed together.

This interpretation seems to make physical sense if we think in the following terms. A bonding pair is shared between two nuclei, and the electrons can be close to either nucleus. Therefore they are relatively confined between the two nuclei. A lone pair is localized on only one nucleus, so both electrons are close to that nucleus only, as shown schematically in Fig. 13.18. These pictures help us understand why a lone pair may require more space near an atom than a bonding pair.

As a result of these observations, we make the following addition to the original postulate of the VSEPR model: *lone pairs require more room than bonding pairs and tend to compress the angles between the bonding pairs.*

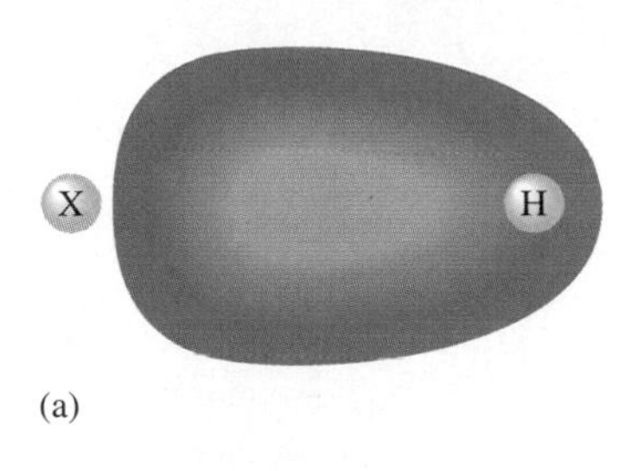

(a)

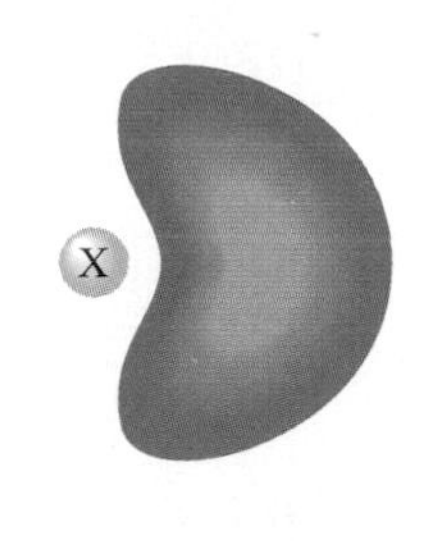

(b)

Figure 13.18
(a) In a bonding pair of electrons the electrons are shared by two nuclei. (b) In a lone pair, since both electrons must be close to a single nucleus, they tend to take up more of the space around that atom.

EXAMPLE 13.12

When phosphorus reacts with excess chlorine gas, the compound phosphorus pentachloride (PCl_5) is formed. In the gaseous and liquid states this substance consists of PCl_5 molecules, but in the solid state it consists of a 1:1 mixture of PCl_4^+ and PCl_6^- ions. Predict the geometric structures of PCl_5, PCl_4^+, and PCl_6^-.

Solution

The Lewis structure for PCl_5 is shown in the margin. Five pairs of electrons around the phosphorus atom require a trigonal bipyramidal arrangement (see Table 13.7). When the chlorine atoms are included, a trigonal bipyramidal molecule results, as shown on the next page.

Lewis structure for PCl_5

The Lewis structure for the PCl_4^+ ion [$5 + 4(7) - 1 = 32$ valence electrons] is shown in the margin. There are four pairs of electrons surrounding the phosphorus atom in the PCl_4^+ ion. This requires a tetrahedral arrangement of the pairs, as shown in the figure in the margin. Since each pair is shared with a chlorine atom, a tetrahedral PCl_4^+ cation results.

Lewis structure for PCl_4^+

The Lewis structure for PCl_6^- [$5 + 6(7) + 1 = 48$ valence electrons] is

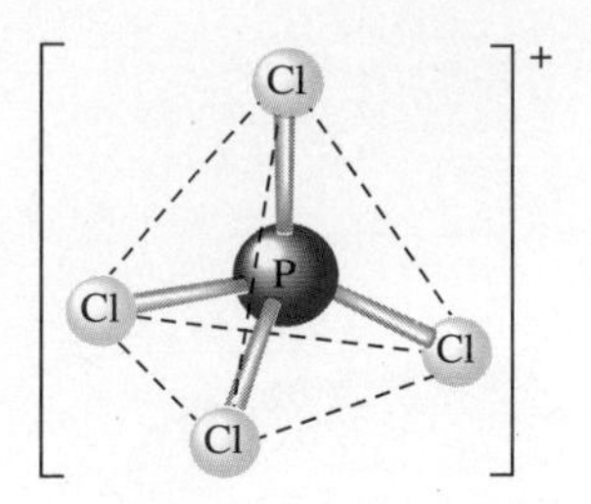

Tetrahedral PCl_4^+ cation

Since phosphorus is surrounded by six pairs of electrons, an octahedral arrangement is required to minimize repulsions, as shown below on the left. Since each electron pair is shared with a chlorine atom, an octahedral PCl_6^- anion is predicted.

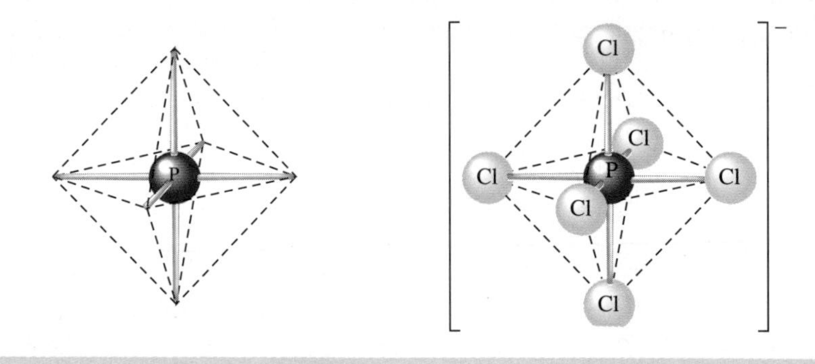

TABLE 13.7 **Arrangements of Electron Pairs Around an Atom Yielding Minimum Repulsion**

Number of Electron Pairs	Arrangement of Electron Pairs		
2	Linear		:—*A*—:
3	Trigonal planar		
4	Tetrahedral		
5	Trigonal bipyramidal		90° 120°
6	Octahedral		

So far we have considered cases with two, three, and four electron pairs around the central atom. These are summarized in Table 13.7. For five pairs of electrons there are several possible electron pair arrangements. The one that produces minimum repulsion is a **trigonal bipyramid.** Note from Table 13.7 that this arrangement has two different angles, 90° and 120°. As the name suggests, the structure formed by this arrangement of pairs consists of two trigonal-based pyramids that share a common base. Six pairs of electrons can best be arranged around a given atom to form an **octahedral structure** with 90° angles, as shown in Table 13.7.

In order to use the VSEPR model to determine the geometric structures of molecules, you should memorize the relationships between the number of electron pairs and their best arrangement.

EXAMPLE 13.13

Because the noble gases have filled *s* and *p* valence orbitals, they are not expected to be chemically reactive. In fact, for many years these elements were called *inert gases* because of this supposed inability to form any compounds. However, in the early 1960s several compounds of krypton, xenon, and radon were synthesized. For example, a team at the Argonne National Laboratory produced the stable colorless compound xenon tetrafluoride (XeF_4). Predict its structure and determine whether it has a dipole moment.

Solution

The Lewis structure for XeF_4 is shown below.

```
 :F̤      F̤:
    \ ‥ /
     Xe
    / ‥ \
 :F̤      F̤:
```

The xenon atom in this molecule is surrounded by six pairs of electrons, requiring an octahedral arrangement:

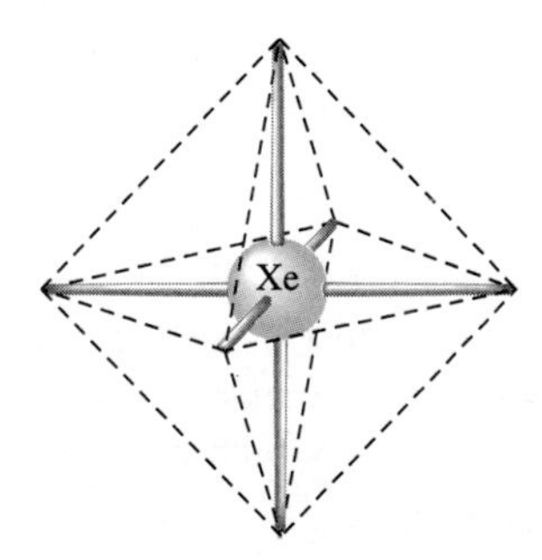

The structure predicted for this molecule depends on how the lone pairs and bonding pairs are arranged. Consider the two possibilities shown in Fig. 13.19. The bonding pairs are indicated by the presence of fluorine atoms. Since the structure predicted differs in the two cases, we must decide which of these arrangements is preferable. The key is to look at the lone pairs. In the structure in part (a) the lone pair–lone pair

Xenon tetrafluoride crystals.

angle is 90°; in the structure in part (b) the lone pairs are separated by 180°. Since lone pairs require more room than bonding pairs, a structure with two lone pairs at 90° is unfavorable. Thus the arrangement in Fig. 13.19(b) is preferred, and the molecular structure is predicted to be square planar. Note that this molecule is *not* described as being octahedral. There is an *octahedral arrangement of electron pairs,* but the *atoms* form a **square planar** structure.

Although each Xe—F bond is polar (fluorine has a greater electronegativity than xenon), the square planar arrangement of these bonds causes the polarities to cancel.

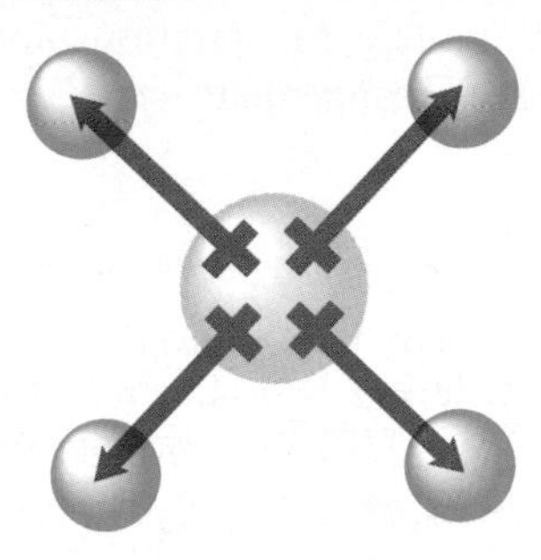

Thus XeF_4 has no dipole moment.

We can further illustrate the use of the VSEPR model for molecules or ions with lone pairs by considering the triiodide ion (I_3^-). The Lewis structure for I_3^- is

$$[:\ddot{\underset{..}{I}}—\ddot{\underset{..}{I}}—\ddot{\underset{..}{I}}:]^-$$

The central iodine atom has five pairs around it, requiring a trigonal bipyramidal arrangement. Several possible arrangements of the lone pairs are shown in Fig. 13.20. Note that structures (a) and (b) have lone pairs at 90°, whereas in (c) all lone pairs are at 120°. Thus structure (c) is preferred. The resulting molecular structure for I_3^- is linear.

$$[I—I—I]^-$$

The VSEPR Model and Multiple Bonds

So far in our treatment of the VSEPR model, we have not considered any molecules with multiple bonds. To see how these molecules are handled by this model, let's consider the NO_3^- ion, which requires three resonance structures to describe its electronic structure:

$$\left[\begin{array}{c} \ddot{O} \\ \| \\ N \\ \diagup \quad \diagdown \\ :\ddot{\underset{..}{O}}: \quad :\ddot{\underset{..}{O}}: \end{array}\right]^- \longleftrightarrow \left[\begin{array}{c} :\ddot{O}: \\ | \\ N \\ /\!/ \quad \diagdown \\ :\underset{..}{O} \quad :\ddot{\underset{..}{O}}: \end{array}\right]^- \longleftrightarrow \left[\begin{array}{c} :\ddot{O}: \\ | \\ N \\ \diagup \quad \backslash\!\backslash \\ :\ddot{\underset{..}{O}}: \quad \underset{..}{O}: \end{array}\right]^-$$

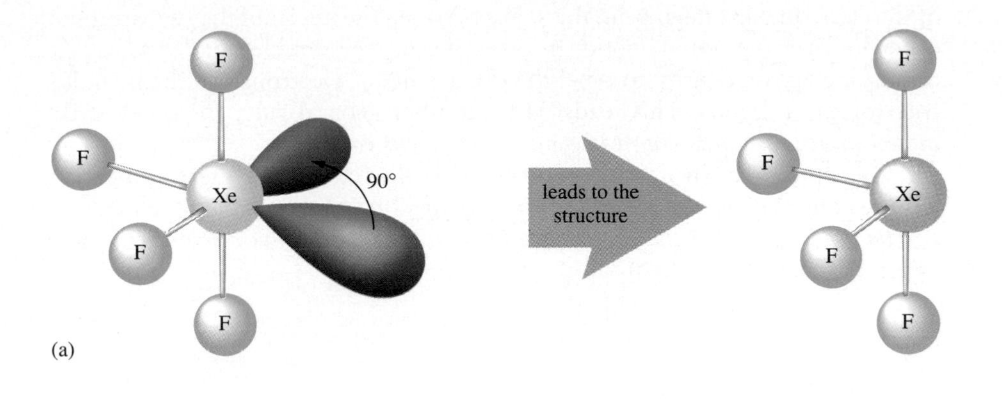

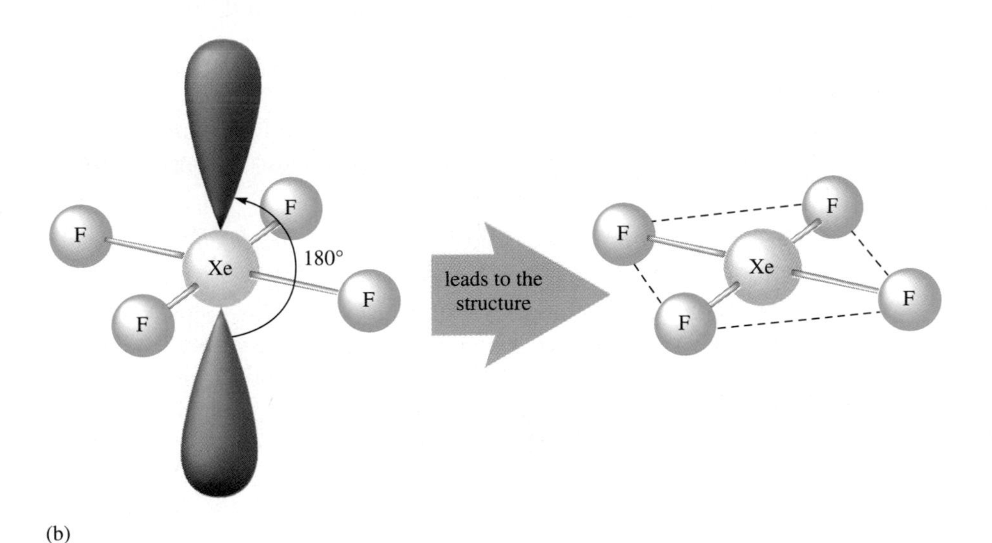

Figure 13.19
Possible electron pair arrangements for XeF_4. Since arrangement (a) has lone pairs 90° apart, it is less favorable than arrangement (b), where the lone pairs are 180° apart.

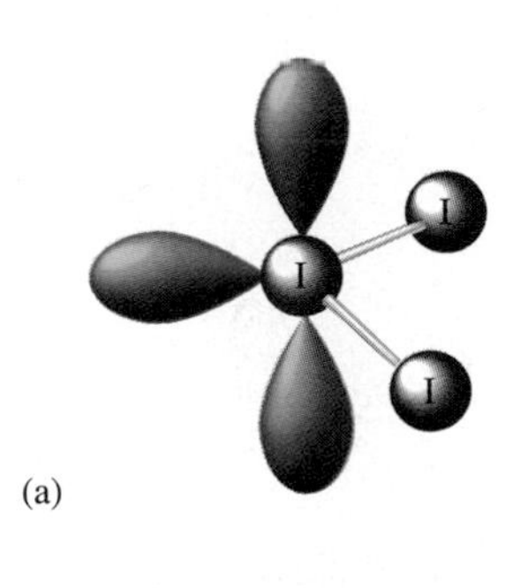

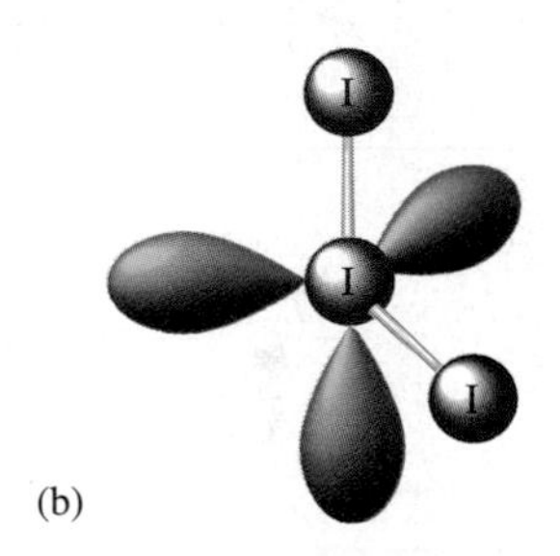

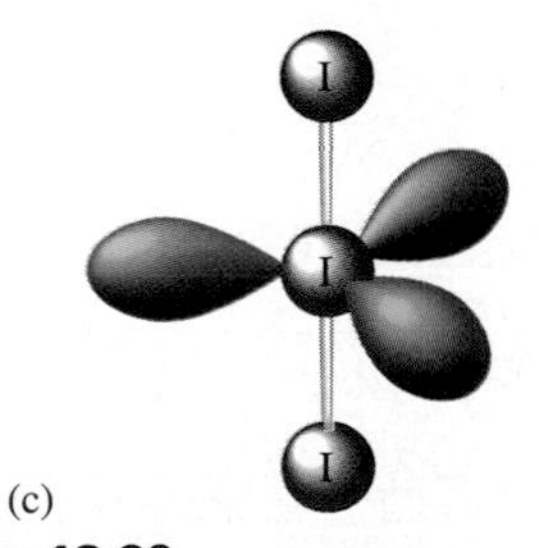

Figure 13.20
Three possible arrangements of the electron pairs in the I_3^- ion. Arrangement (c) is preferred because there are no 90° lone pair–lone pair interactions.

The NO_3^- ion is known to be planar with 120° bond angles:

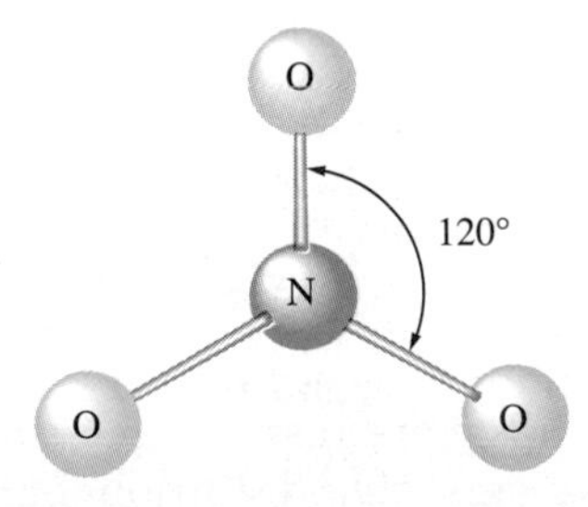

This planar structure is the one expected for three pairs of electrons around a central atom, which means that *a double bond should be counted as one effective pair* in using the VSEPR model. This makes sense because the two pairs of electrons involved in the double bond are *not* independent pairs. Both of the electron pairs must be in the space between the nuclei of the two atoms in order to form the double bond. In other words, the double bond acts as one center of electron density to repel the other pairs of electrons. The same holds true for triple bonds. This leads us to another general rule: *for the VSEPR model multiple bonds count as one effective electron pair.*

The molecular structure of nitrate also illustrates one more important point: *when a molecule exhibits resonance, any one of the resonance structures can be used to predict the molecular structure using the VSEPR model.* These rules are illustrated in Example 13.14.

EXAMPLE 13.14

Predict the molecular structure of the sulfur dioxide molecule. Is this molecule expected to have a dipole moment?

Solution

First, we must determine the Lewis structure for the SO_2 molecule, which has 18 valence electrons. The expected resonance structures are

$$:\ddot{O}=\ddot{S}-\ddot{O}: \longleftrightarrow :\ddot{O}-\ddot{S}=\ddot{O}:$$

To determine the molecular structure, we must count the electron pairs around the sulfur atom. In each resonance structure the sulfur has one lone pair, one pair in a single bond, and one double bond. Counting the double bond as one pair yields three effective pairs around the sulfur. According to Table 13.7, a trigonal planar arrangement is required, yielding a V-shaped molecule:

O–S–O (120°)

Thus the structure of the SO_2 molecule is expected to be V-shaped with a 120° bond angle. The molecule has a dipole moment directed as shown:

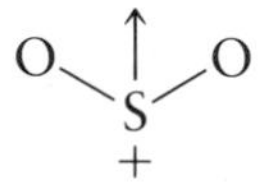

O–S–O (↑, +)

Since the molecule is V-shaped, the polar bonds do not cancel.

It should be noted at this point that lone pairs oriented at least 120° from other pairs do not produce significant distortions of bond angles. For example, the angle in the SO_2 molecule is actually quite close to 120°. We will follow the general principle that *an angle of 120° provides lone pairs with enough space so that distortions do not occur. Angles less than 120° are distorted when lone pairs are present.*

Molecules Containing No Single Central Atom

So far we have considered molecules consisting of one central atom surrounded by other atoms. The VSEPR model can be readily extended to more complicated molecules, such as methanol (CH_3OH). This molecule is represented by the following Lewis structure:

```
      H
      |   ..
  H—C—O—H
      |   ..
      H
```

The molecular structure can be predicted from the arrangement of pairs around the carbon and oxygen atoms. Note that there are four pairs of electrons around the carbon, which calls for a tetrahedral arrangement, as shown in Fig. 13.21(a). The oxygen also has four pairs, requiring a tetrahedral arrangement. However, in this case the tetrahedron will be slightly distorted by the space requirements of the lone pairs [Fig. 13.21(b)]. The overall geometric arrangement for the molecule is shown in Fig. 13.21(c).

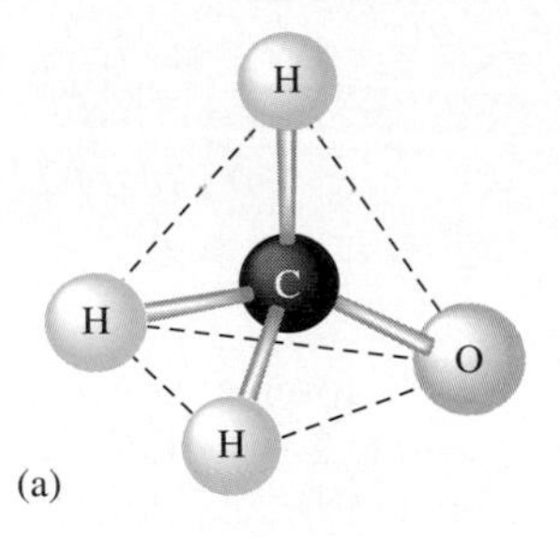

(a)

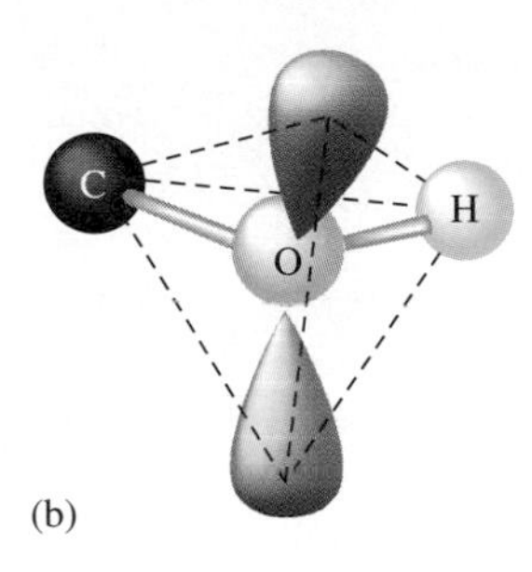

(b)

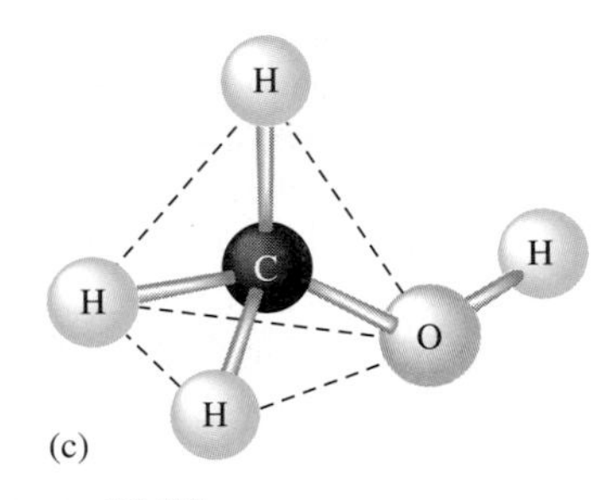

(c)

Figure 13.21 The molecular structure of methanol. (a) The arrangement of electron pairs and atoms around the carbon atom. (b) The arrangement of bonding and lone pairs around the oxygen atom. (c) The molecular structure.

Summary: **The VSEPR Model**

The following rules are helpful in using the VSEPR model to predict molecular structure.

- **Determine the Lewis structure(s) for the molecule.**
- **For molecules with resonance structures, use any of the structures to predict the molecular structure.**
- **Sum the electron pairs around the central atom.**
- **When counting pairs, count each multiple bond as a single effective pair.**
- **Determine the arrangement of the pairs that minimizes electron pair repulsions. These arrangements are shown in Table 13.7.**
- **Lone pairs require more space than bonding pairs. Choose an arrangement that gives the lone pairs as much room as possible, although it appears that an angle of at least 120° between lone pairs provides enough space. Recognize that lone pairs at angles less than 120° may produce distortions from the idealized structure.**

The VSEPR Model—How Well Does It Work?

The VSEPR model is very simple. There are only a few rules to remember, yet the model correctly predicts the molecular structures of most molecules formed from nonmetallic elements. Molecules of any size can be treated by applying the VSEPR model to each appropriate atom (those bonded to at least two other atoms) in the molecule. Thus we can use this model to predict the structures of molecules with hundreds of atoms. It does, however, fail in a few instances. For example, phosphine (PH_3), which has a Lewis structure analogous to that of ammonia,

Chemical Structure and Communication: Semiochemicals

In this chapter we have stressed the importance of being able to predict the three-dimensional structure of a molecule. Molecular structure is important because of its effect on chemical reactivity. This is especially true in biological systems, where reactions must be efficient and highly specific. Among the hundreds of types of molecules in the fluids of a typical biological system, the appropriate reactants must find and react only with each other—they must be very discriminating. This specificity depends largely on structure. The molecules are constructed so that only the appropriate partners can approach each other in a way that allows reaction.

Molecular structure is also central for those molecules used as a means of communication. Examples of chemical communication occurring in humans are the conduction of nerve impulses across synapses, the control of the manufacture and storage of key chemicals in cells, and the senses of smell and taste. Plants and animals also use chemical communication. For example, ants lay down a chemical trail so that other ants can find a certain food supply. Ants also warn their fellow workers of approaching danger by emitting certain chemicals.

Molecules convey messages by fitting into appropriate receptor sites in a very specific way, which is determined by their structure. When a molecule occupies a receptor site, chemical processes are stimulated that produce the appropriate response. Sometimes, receptors can be fooled, as in the use of artificial sweeteners—molecules fit the sites on the taste buds that stimulate a "sweet" response in the brain, but they are not metabolized in the same way as natural sugars. Similar deception is useful in insect control. If an area is sprayed with synthetic female sex attractant molecules, the males of that species become so confused that mating does not occur.

A *semiochemical* is a molecule that delivers a message between members of the same or different species of plant or animal. There are three groups of these chemical messengers: allomones, kairomones, and pheromones. Each is of great ecological importance.

An *allomone* is defined as a chemical that gives adaptive advantage to the producer. For example, leaves of the black walnut tree contain a herbicide, juglone, that appears after the leaves fall to the ground. Juglone is not toxic to grass or certain grains, but it is effective against plants such as apple trees that would compete for the available water and food supplies.

```
  ..            ..
H—P—H         H—N—H
  |             |
  H             H
```

would be predicted to have a molecular structure similar to that for NH_3 with bond angles of approximately 107°. However, the bond angles of phosphine are actually 94°. There are ways of explaining this structure, but more rules have to be added to the model.

This example again illustrates the point that simple models are bound to have exceptions. In introductory chemistry we want to use simple models that fit the majority of cases; we are willing to accept a few failures rather than complicate the model. The amazing thing about the VSEPR model is that such a simple model correctly predicts the structures of so many molecules.

Antibiotics are also allomones, since the microorganisms produce them to inhibit other species from growing near them.

Many plants produce bad-tasting chemicals to protect themselves from plant-eating insects and animals. The familiar compound nicotine deters animals from eating the tobacco plant. The millipede sends an unmistakable "back off" message by squirting a predator with benzaldehyde and hydrogen cyanide.

Defense is not the only use of allomones, however. Flowers use scent to attract pollinating insects. Honeybees, for instance, are guided to alfalfa flowers by a series of sweet-scented compounds.

Kairomones are chemical messengers that bring advantageous news to the receiver. For example, the floral scents are kairomones from the honeybees' viewpoint. Many predators are guided by kairomones emitted by their food. For example, apple skins exude a chemical that attracts the codling moth larva. In some cases kairomones help the underdog. Certain marine mollusks can pick up the "scent" of their predators, the sea stars, and make their escape.

Pheromones are chemicals that affect receptors of the same species as the donor. That is, they are specific within a species. *Releaser pheromones* cause an immediate reaction in the receptor, while *primer pheromones* cause long-term effects. Examples of releaser pheromones are the sex attractants of insects, generated in some species by the males and in others by the females. Sex pheromones have also been found in plants and mammals.

Alarm pheromones are highly volatile compounds (ones easily changed to a gas) released to warn of danger. Honeybees produce isoamyl acetate ($C_7H_{14}O_2$) in their sting glands. Because of its high volatility, this compound does not linger after the state of alert is over. Social behavior in insects is characterized by the use of *trail pheromones,* which are used to indicate a food source. Social insects such as bees, ants, wasps, and termites use these substances. Since trail pheromones are less volatile compounds, the indicators persist for some time.

Primer pheromones, which cause long-term behavioral changes, are harder to isolate and identify. One example, however, is the "queen substance" produced by queen honeybees. All the eggs in a colony are laid by one queen bee. If she is removed from the hive or dies, the worker bees are activated by the absence of the queen substance and begin to feed royal jelly to bee larvae in order to raise a new queen. The queen substance also prevents the development of the workers' ovaries, so that only the queen herself can produce eggs.

Many studies of insect pheromones are now under way in the hope that they will provide a method of controlling insects that is more efficient and safer than the current chemical pesticides.

Exercises

A blue exercise number indicates that the answer to that exercise appears at the back of this book.

Chemical Bonds and Electronegativity

1. Explain the difference between the following pairs of terms.
   a. electronegativity and electron affinity
   b. covalent bond and polar covalent bond
   c. polar covalent bond and ionic bond
2. Use Coulomb's law,

$$V = \frac{Q_1Q_2}{4\pi\epsilon_0 r} = 2.31 \times 10^{-19}\ \text{J nm}\left(\frac{Q_1Q_2}{r}\right)$$

to calculate the energy of interaction for the following two arrangements of charges, each having a magnitude equal to the electron charge.

a.

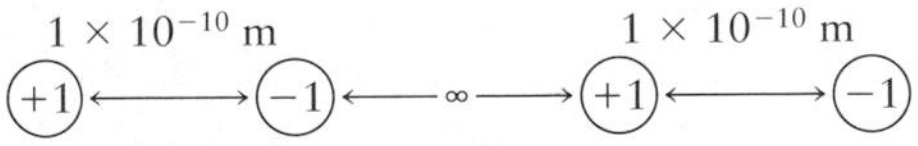

b.

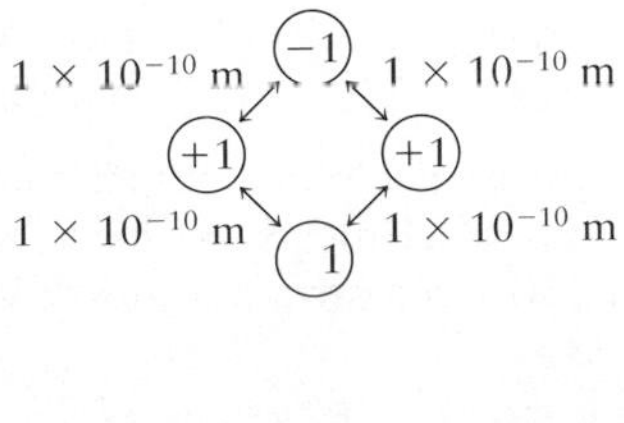

3. Without using Fig. 13.3, predict the order of increasing electronegativity in each of the following groups of elements.
   a. C, N, O
   b. S, Se, Cl
   c. Si, Ge, Sn
   d. Tl, S, Ge
   e. Na, K, Rb
   f. B, O, Ga
4. Without using Fig. 13.3, predict which bond in each of the following groups is the most polar.
   a. C—F, Si—F, Ge—F
   b. P—Cl, S—Cl
   c. S—F, S—Cl, S—Br
   d. Ti—Cl, Si—Cl, Ge—Cl
   e. C—H, Si—H, Sn—H
   f. Al—Br, Ga—Br, In—Br, Tl—Br
5. Repeat Exercises 3 and 4. This time use the values of the electronegativities of the elements given in Fig. 13.3. Are there any differences among your answers?
6. An alternative definition of electronegativity is

$$\text{Electronegativity} = \text{constant (I.E.} - \text{E.A.)}$$

where I.E. is the ionization energy and E.A. is the electron affinity using the sign conventions of this book. Use data in Chapter 12 to calculate the (I.E. − E.A.) term for F, Cl, Br, and I. Do these values show the same trend as the electronegativity values given in this chapter? The first ionization energies of the halogens are 1678, 1255, 1138, and 1007 kJ/mol, respectively. (*Hint:* choose a constant so that the electronegativity of fluorine equals 4.0. Using this constant, calculate relative electronegativities for the other halogens and compare to values given in the text.)

Ionic Compounds

7. For each of the following groups, place the atoms and ions in order of decreasing size.
   a. Cu, Cu^+, Cu^{2+}
   b. Ni^{2+}, Pd^{2+}, Pt^{2+}
   c. O^{2-}, S^{2-}, Se^{2-}
   d. La^{3+}, Eu^{3+}, Gd^{3+}, Yb^{3+}
   e. Te^{2-}, I^-, Xe, Cs^+, Ba^{2+}, La^{3+}
8. Write electron configurations for each of the following.
   a. the cations: Mg^{2+}, Sn^{2+}, K^+, Al^{3+}, Tl^+, As^{3+}
   b. the anions: N^{3-}, O^{2-}, F^-, Te^{2-}
   c. the most stable ion formed by: Be, Rb, Ba, Se, I
9. Which of the following ions have noble gas electron configurations?
   a. Fe^{2+}, Fe^{3+}, Sc^{3+}, Co^{3+}
   b. Tl^+, Te^{2-}, Cr^{3+}
   c. Pu^{4+}, Ce^{4+}, Ti^{4+}
   d. Ba^{2+}, Pt^{2+}, Mn^{2+}
10. Define the term *isoelectronic*. When comparing sizes of monatomic ions of elements in the same period of the periodic table, why is it advantageous to compare isoelectronic species?
11. List three ions that are isoelectronic with the krypton atom.
12. Which compound in each of the following pairs of ionic substances has the most exothermic lattice energy? Justify your answers.
   a. NaCl, KCl
   b. LiF, LiCl
   c. $Mg(OH)_2$, MgO
   d. $Fe(OH)_2$, $Fe(OH)_3$
   e. NaCl, Na_2O
   f. MgO, BaS
13. Following are some important properties of ionic compounds:
   i. low electrical conductivity as solids, and high conductivity in solution or when molten
   ii. relatively high melting and boiling points
   iii. brittleness

   How does the concept of ionic bonding discussed in this chapter account for these properties?
14. Use the following data to estimate ΔH_f° for sodium chloride.

$$Na(s) + \tfrac{1}{2}Cl_2(g) \longrightarrow NaCl(s)$$

Lattice energy for NaCl	−786 kJ/mol
Ionization energy for Na	495 kJ/mol
Electron affinity of Cl	−349 kJ/mol
Bond energy of Cl_2	239 kJ/mol
Enthalpy of sublimation for Na	109 kJ/mol

15. Use the following data to estimate ΔH_f° for barium chloride.

$$Ba(s) + Cl_2(g) \longrightarrow BaCl_2(s)$$

Lattice energy for $BaCl_2$	−2056 kJ/mol
First ionization energy of Ba	503 kJ/mol
Second ionization energy of Ba	965 kJ/mol
Electron affinity of Cl	−349 kJ/mol
Bond energy of Cl_2	239 kJ/mol
Enthalpy of sublimation of Ba	178 kJ/mol

16. Consider the following energy changes:

	ΔE (kJ/mol)
$Mg(g) \rightarrow Mg^+(g) + e^-$	735
$Mg^+(g) \rightarrow Mg^{2+}(g) + e^-$	1445
$O(g) + e^- \rightarrow O^-(g)$	−141
$O^-(g) + e^- \rightarrow O^{2-}(g)$	+878

   a. Magnesium oxide exists as $Mg^{2+}O^{2-}$, not as Mg^+O^-. Explain.

b. What simple experiment could be done to confirm that magnesium oxide does not exist as Mg^+O^-?

17. Use the following data to estimate ΔH for the reaction $S^-(g) + e^- \rightarrow S^{2-}(g)$. Include an estimate of uncertainty.

	ΔH_f°	Lattice Energy	I.E. of M	ΔH_{sub} of M
Li_2S	−500	−2472	520	161
Na_2S	−365	−2203	495	109
K_2S	−381	−2052	419	90
Rb_2S	−361	−1949	409	82
Cs_2S	−360	−1850	382	78

$$S(s) \longrightarrow S(g) \qquad \Delta H = 277 \text{ kJ/mol}$$

$$S(g) + e^- \longrightarrow S^-(g) \qquad \Delta H = -200 \text{ kJ/mol}$$

Assume all values are known to ±1 kJ/mol.

18. Using data from Exercise 16, calculate ΔE for the reaction $O(g) + 2e^- \rightarrow O^{2-}(g)$.

19. The second electron affinity values for both oxygen and sulfur are unfavorable (endothermic). Explain.

20. Rationalize the following lattice energy values:

Compound	Lattice Energy (kJ/mol)
CaSe	−2862
Na_2Se	−2130
CaTe	−2721
Na_2Te	−2095

21. The lattice energies of the oxides and chlorides of iron(II) and iron(III) are −2631, −3865, −5359, and −14,774 kJ/mol. Match the appropriate formula with each lattice energy. Explain your answer.

22. Predict the empirical formulas of the ionic compounds formed from the following pairs of elements. Name each compound.
a. Li and N
b. Ga and O
c. Rb and Cl
d. Ba and S

Bond Energies

23. Use bond energies to predict ΔH for the isomerization of methyl isocyanide to acetonitrile.

$$CH_3NC(g) \longrightarrow CH_3CN(g)$$

24. Use bond energy values in Table 13.5 to estimate ΔH for each of the following reactions in the gas phase.
a. $H_2 + Cl_2 \rightarrow 2HCl$
b. $N_2 + 3H_2 \rightarrow 2NH_3$
c. $H_2C{=}CH_2 + Br_2 \rightarrow CH_2BrCH_2Br$
d. $C_2H_4 + H_2O_2 \rightarrow CH_2(OH){-}CH_2(OH)$

25. Compare your answers from parts a and b of Exercise 24 with ΔH values calculated for each reaction from standard enthalpies of formation in Appendix 4. Do enthalpy changes calculated from bond energies give a reasonable estimate of the actual values?

26. The compound, hexaazaisowurtzitane, is the highest-energy explosive known (*C & E News*, p. 26, Jan. 17, 1994). The compound, also known as CL-20, was first synthesized in 1987. The method of synthesis and detailed performance data are still classified because of CL-20's potential military application in rocket boosters and in warheads of "smart" weapons. The structure of CL-20 is

CL-20

Three possible reactions for the explosive decomposition of CL-20 are

(i) $C_6H_6N_{12}O_{12}(s) \longrightarrow 6CO(g) + 6N_2(g) + 3H_2O(g) + 3/2O_2(g)$

(ii) $C_6H_6N_{12}O_{12}(s) \longrightarrow 3CO(g) + 3CO_2(g) + 6N_2(g) + 3H_2O(g)$

(iii) $C_6H_6N_{12}O_{12}(s) \longrightarrow 6CO_2(g) + 6N_2(g) + 3H_2(g)$

a. Use bond energies to calculate ΔH for these three reactions.
b. Which of the above reactions releases the largest amount of energy per kilogram of CL-20?

27. Calculate ΔH for the following reaction, using bond energies.

$$C_2H_4(g) + O_3(g) \longrightarrow CH_3CHO(g) + O_2(g)$$

Compare this result with that obtained using standard enthalpies of formation in Exercise 46 of Chapter 9.

28. Three processes that have been used for the industrial manufacture of acrylonitrile—an important chemical used in the manufacture of plastics, synthetic rubber, and

fibers—are shown below. Use bond energy values (Tables 13.5 and 13.6) to estimate ΔH for each of the reactions.

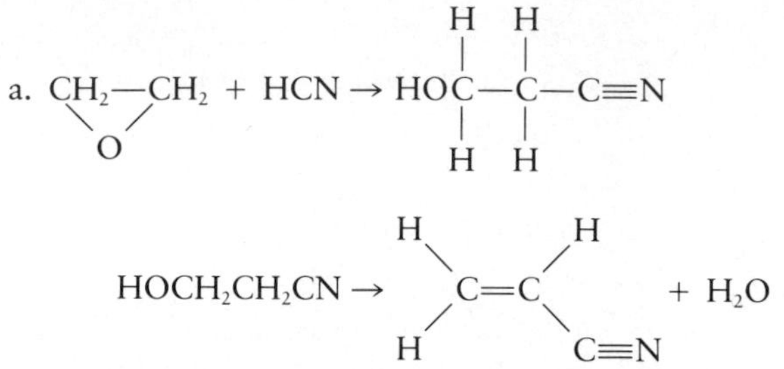

```
                              H   H
                              |   |
a. CH2—CH2 + HCN → HOC—C—C≡N
     \  /                     |   |
      O                       H   H

                    H           H
                     \         /
HOCH2CH2CN →          C=C            + H2O
                     /         \
                    H           C≡N
```

b. $4CH_2{=}CHCH_3 + 6NO \xrightarrow[Ag]{700°C}$

$4CH_2{=}CHCN + 6H_2O + N_2$

The nitrogen-oxygen bond energy in nitric oxide, NO, is 630. kJ/mol.

c. $2CH_2{=}CHCH_3 + 2NH_3 + 3O_2 \xrightarrow[425\text{–}510°C]{Catalyst}$

$2CH_2{=}CHCN + 6H_2O$

29. Is the elevated temperature noted in parts b and c of Exercise 28 needed to provide energy to endothermic reactions?

30. Acetic acid is responsible for the sour taste of vinegar. It can be manufactured using the following reaction:

```
                        O
                        ‖
CH3OH + CO ⟶ CH3C—OH
```

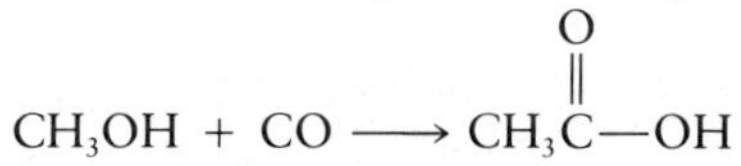

Use tabulated values of bond energies (Table 13.5) to estimate ΔH for this reaction.

31. Use bond energies (Table 13.5), values of electron affinities (Table 12.8), and the ionization energy of hydrogen (1312 kJ/mol) to estimate ΔH for each of the following reactions.
   a. $HF(g) \rightarrow H^+(g) + F^-(g)$
   b. $HCl(g) \rightarrow H^+(g) + Cl^-(g)$
   c. $HI(g) \rightarrow H^+(g) + I^-(g)$
   d. $H_2O(g){:}H^+(g) + OH^-(g)$
   (Electron affinity of OH(g) = −180. kJ/mol.)

32. The standard enthalpies of formation of $S(g)$, $F(g)$, $SF_4(g)$, and $SF_6(g)$ are +278.8, +79.0, −775, and −1209 kJ/mol, respectively.
   a. Use these data to estimate the energy of an S—F bond.
   b. Compare the value that you calculated in part a with the value given in Table 13.5. What conclusions can you draw?
   c. Why are the ΔH_f° values for S(g) and F(g) not equal to zero, even though sulfur and fluorine are elements?

33. Use the following standard enthalpies of formation to estimate the N—H bond energy in ammonia. Compare this with the value in Table 13.5.

N(g)	472.7 kJ/mol
H(g)	216.0 kJ/mol
$NH_3(g)$	−46.1 kJ/mol

34. The standard enthalpy of formation of $NH_3(g)$ is −46 kJ/mol. Use this and values for the N≡N and H—H bond energies to estimate the N—H bond energy. Compare this result to your result from Exercise 33.

35. The standard enthalpy of formation of $N_2H_4(g)$ is 95.4 kJ/mol. Use this information and the data in Exercise 33 to estimate the N—N single-bond energy. Compare your result with the value in Table 13.5.

36. What is the relationship between ΔH_f° for H(g) given in Exercise 33 and the H—H bond energy listed in Table 13.5?

Lewis Structures and Resonance

37. Draw Lewis structures that obey the octet rule for each of the following.
   a. HCN c. $CHCl_3$ e. BF_4^-
   b. PH_3 d. NH_4^+ f. SeF_2

38. Draw a Lewis structure that obeys the octet rule for each of the following molecules and ions. In each case the first atom listed is the central atom.
   a. $POCl_3$, SO_4^{2-}, XeO_4, PO_4^{3-}, ClO_4^-
   b. NF_3, SO_3^{2-}, PO_3^{3-}, ClO_3^-
   c. ClO_2^-, SCl_2, PCl_2^-

39. Considering your answers to Exercise 38, what conclusions can you draw concerning the structures of species containing the same number of atoms and the same number of valence electrons?

40. Draw Lewis structures for the following. Show all resonance structures, where applicable.
   a. NO_2^-, HNO_2, NO_3^-, HNO_3
   b. SO_4^{2-}, HSO_4^-, H_2SO_4
   c. CN^-, HCN
   d. OCN^-, SCN^-, N_3^-
   e. C_2N_2 (atomic arrangement is NCCN)

41. Some of the pollutants in the atmosphere are ozone, sulfur dioxide, and sulfur trioxide. Draw Lewis structures for these three molecules.

42. Peroxyacetyl nitrate, or PAN, is present in photochemical smog. Draw Lewis structures (including resonance) for PAN. The skeletal arrangement is

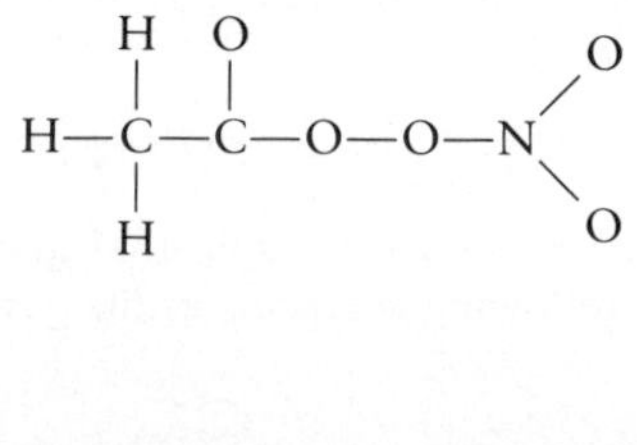

```
   H  O            O
   |  |           /
H—C—C—O—O—N
   |              \
   H               O
```

43. A toxic cloud covered Bhopal, India, in December 1984 when water leaked into a tank of methyl isocyanate, and the product escaped into the atmosphere. Methyl isocyanate is used in the production of many pesticides. Draw the Lewis structures for methyl isocyanate, CH_3NCO, including resonance forms.

44. S_2Cl_2 and SCl_2 are important industrial chemicals used primarily in the vapor-phase vulcanization of certain rubbers. Draw Lewis structures for S_2Cl_2 and SCl_2.

45. Benzene (C_6H_6) consists of a six-membered ring of carbon atoms with one hydrogen bonded to each carbon. Draw Lewis structures for benzene, including resonance structures.

46. An important observation supporting the need for resonance in the localized electron model is that there are only three different structures of dichlorobenzene ($C_6H_4Cl_2$). How does this fact support the need for the concept of resonance?

47. Borazine ($B_3N_3H_6$) has often been called "inorganic" benzene. Draw Lewis structures for borazine. Borazine is a six-membered ring of alternating boron and nitrogen atoms.

48. Draw all the possible Lewis structures for dimethylborazine, $(CH_3)_2B_3N_3H_4$. (See Exercise 47.) Would there be a different number of structures if there was no resonance?

49. Write Lewis structures for the following molecules, which have central atoms that do not obey the octet rule: PF_5, BrF_3, $Be(CH_3)_2$, BCl_3, $XeOF_4$ (Xe is the central atom), XeF_6, SeF_4.

50. ClF_3 and BrF_3 are both used to fluorinate uranium to produce UF_6 in the processing and reprocessing of nuclear fuel. Draw Lewis structures for ClF_3 and BrF_3.

51. Draw Lewis structures for CO_3^{2-}, HCO_3^-, and H_2CO_3. When acid is added to an aqueous solution containing carbonate or bicarbonate ions, carbon dioxide gas is formed. We generally say that carbonic acid (H_2CO_3) is unstable. Use bond energies to estimate ΔH for the reaction

$$H_2CO_3 \longrightarrow CO_2 + H_2O$$

Specify a possible cause for the instability of carbonic acid.

52. Nitrous oxide (N_2O) has three possible Lewis structures:

$$:\ddot{N}{=}N{=}\ddot{O}: \longleftrightarrow :N{\equiv}N{-}\ddot{\underset{..}{O}}: \longleftrightarrow :\ddot{\underset{..}{N}}{-}N{\equiv}O:$$

Given the following bond lengths (1 Å = 10^{-10} m),

N—N	1.67 Å	N=O	1.15 Å
N=N	1.20 Å	N—O	1.47 Å
N≡N	1.10 Å		

rationalize the observations that the N—N bond length in N_2O is 1.12 Å and that the N—O bond length is 1.19 Å.

53. Consider the following bond lengths:

C—O 1.43 Å   C=O 1.23 Å   C≡O 1.09 Å

In the CO_3^{2-} ion, all three C—O bonds have identical bond lengths of 1.36 Å. Why?

54. Order the following species with respect to the carbon-oxygen bond length (longest to shortest):

CO, CO_2, CO_3^{2-}, CH_3OH

What is the order from the weakest to the strongest carbon-oxygen bond?

55. Place the species below in order from the shortest to the longest nitrogen-oxygen bond.

H_2NOH, N_2O, NO^+, NO_2^-, NO_3^-

Formal Charge

56. Assign formal charges to the resonance structures for N_2O in Exercise 52. Can you eliminate any of the resonance structures on the basis of formal charge? Is this consistent with observation?

57. Use formal charge to rationalize why BF_3 does not follow the octet rule.

58. Assign formal charges to the atoms in carbon monoxide. Use these to explain why CO has a much smaller dipole moment than is expected on the basis of electronegativity.

59. Draw Lewis structures that obey the octet rule for the following species. Assign the formal charge to each central atom.

a. $OPCl_3$ c. ClO_4^- e. SO_2Cl_2 g. ClO_3^-
b. SO_4^{2-} d. PO_4^{3-} f. XeO_4 h. NO_4^{3-}

60. Draw the Lewis structures that involve minimum formal charges for the species in Exercise 59.

Molecular Structure and Polarity

61. Predict the molecular structure and the bond angles for each molecule or ion in Exercises 37, 38, and 40.

62. Predict the molecular structure and the bond angles for each of the following ions.

a. I_3^- b. ClF_3 c. IF_4^+ d. SF_5^+

63. Predict the molecular structure and the bond angles for each of the following.

a. SeO_3^{2-} b. SeH_2 c. SeO_4^{2-}

64. Predict the molecular structure and the bond angles for each of the following.

a. BrF_5 b. KrF_4 c. IF_6^+

65. The addition of antimony pentafluoride (SbF_5) to liquid hydrogen produces a solution classified as a *superacid*. Superacids are capable of acting as acids toward many compounds that we normally expect not to act as bases.

For example, in the following reaction HF acts as a base:

$$SbF_5 + 2HF \longrightarrow SbF_6^- + H_2F^+$$

Draw Lewis structures for and predict the molecular structures of the reactants and products of this reaction.

66. Which of the following molecules have dipole moments? For the molecules that are polar, indicate the polarity of each bond and the direction of the net dipole moment of the molecule.
   a. CH_2Cl_2, $CHCl_3$, CCl_4
   b. CO_2, N_2O
   c. PH_3, NH_3, AsH_3

67. Draw Lewis structures and predict the molecular structures of the following.
   a. chromate ion
   b. dichromate ion
   c. thiosulfate ion ($S_2O_3^{2-}$)
   d. peroxydisulfate ion ($S_2O_8^{2-}$)

68. What two requirements must be satisfied for a molecule to be polar?

69. Draw Lewis structures and predict the molecular structures of the following:
   a. OCl_2, Br_3^-, BeH_2, BH_2^-
   b. BCl_3, NF_3, IF_3
   c. CF_4, SeF_4, XeF_4
   d. IF_5, AsF_5

   Which of the above compounds have dipole moments?

70. Draw a Lewis structure and predict the molecular structure and polarity of each of the follow'ng sulfur fluorides: SF_2, SF_4, SF_6, and S_2F_4 (exists as $F_3S—SF$). Predict the F—S—F bond angles in each molecule.

71. The molecules BF_3, CF_4, CO_2, PF_5, and SF_6 are all nonpolar, even though they contain polar bonds. Why?

72. Two molecules exist with the formula N_2F_2. The Lewis structures are

   F—N=N—F (trans) and F—N=N—F (cis)

   a. What are the N—N—F bond angles in the two molecules?
   b. What is the polarity of each molecule?

73. Draw Lewis structures for the following and predict whether each is polar or nonpolar.
   a. HOCN
   b. BeF_2
   c. KrF_4
   d. CF_2Cl_2
   e. H_2NNH_2 (hydrazine)
   f. H_2CO

Additional Exercises

74. Although both the Br_3^- and I_3^- ions are known, the F_3^- ion does not exist. Explain.

75. There are three possible structures of $PF_3(CH_3)_2$, where P is the central atom. Draw them and describe how measurements of dipole moments might be used to distinguish among them.

76. Look up the energies for the bonds in CO and N_2. Although the bond in CO is stronger, CO is considerably more reactive than N_2. Give a possible explanation.

77. Which member of the following pairs would you expect to be more stable? Justify each choice.
   a. SO_4 or SO_4^{2-}
   b. NF_5 or PF_5
   c. OF_6 or SF_6
   d. BH_3 or BH_4^-
   e. MgF or MgO
   f. CsCl or $CsCl_2$
   g. KBr or K_2Br

78. Many times, extra stability is characteristic of a molecule or ion in which resonance is possible. How could this feature be used to explain the acidities of the following compounds? (The acidic hydrogen is marked by an asterisk.) Part c shows resonance in the phenyl ring, C_6H_5.

   a. H—C(=O)—OH*
   b. CH_3—C(=O)—CH=C(OH*)—CH_3
   c. C_6H_5—OH* (phenyl ring with OH*)

79. Would you expect the electronegativity of titanium to be the same in the species Ti, Ti^{2+}, Ti^{3+}, and Ti^{4+}? Explain your answer.

80. Draw a Lewis structure and predict the molecular structure of each of the following.
   a. SiF_4 b. SeF_4 c. KrF_4

   Why do these three molecules have different molecular structures? Which, if any, are polar?

81. Write a Lewis structure and predict the molecular structure of each of the following.
   a. BF_3 b. PF_3 c. BrF_3

   Why are the molecular structures different? Which, if any, molecules are polar?

82. Give a rationalization for the octet rule in terms of orbitals.

83. Draw a Lewis structure for the *N,N*-dimethylformamide molecule. The skeletal structure is

   H—C(O)—N(CH$_3$)—CH_3

Various types of evidence lead to the conclusion that there is some double-bond character to the CN bond. Draw one or more resonance structures that support this observation.

84. Disulfurdinitrogen, S_2N_2, exists as a ring of alternating sulfur and nitrogen atoms. S_2N_2 polymerizes to polythiazyl, which acts as a metallic conductor of electricity along the polymer chain. Draw a Lewis structure for S_2N_2.

85. For each of the following, write an equation that corresponds to the energy given.
a. lattice energy of NaCl
b. lattice energy of NH_4Br
c. lattice energy of MgS
d. ΔH_f° of O(*g*)
e. O—O double-bond energy, beginning with $O_2(g)$ as a reactant

86. Aluminum chloride exists as discrete Al_2Cl_6 molecules in the gas phase. The skeletal structure is

```
Cl       Cl       Cl
   \    /   \    /
    Al        Al
   /    \   /    \
Cl       Cl       Cl
```

a. Complete the Lewis structure.
b. Predict the bond angles around each Al atom.
c. Is the molecule polar or nonpolar?

87. Do the Lewis structures obtained in Exercises 59 and 60 predict the same molecular structure for each case?

88. The xenon-oxygen bond energy in XeO_3 is 84 kJ/mol. Which of the Lewis structures in Example 13.10 is most consistent with this fact?

89. A promising new material with great potential as a fuel in solid rocket motors is ammonium dinitramide, $NH_4N(NO_2)_2$.
a. Draw Lewis structures (including resonance forms) for the dinitramide ion, $N(NO_2)_2^-$.
b. Predict the bond angles about each nitrogen in the dinitramide ion.
c. Ammonium dinitramide can decompose explosively to nitrogen, water, and oxygen. Write a balanced equation for this reaction and use bond energies to estimate ΔH for the explosive decomposition of this compound.

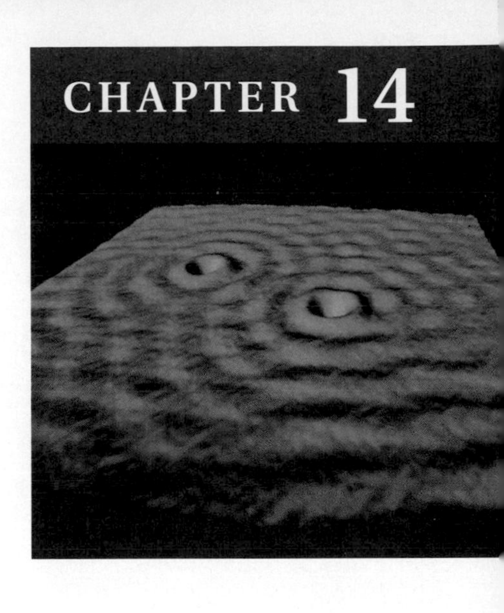

Covalent Bonding: Orbitals

In Chapter 13 we discussed the fundamental concepts of bonding and introduced the most widely used simple model for covalent bonding: the localized electron model. We saw the usefulness of a bonding model as a means for systematizing chemistry by allowing us to look at molecules in terms of individual bonds. We also saw that the approximate molecular structure can be predicted using a model that focuses on minimizing electron pair repulsions. In this chapter we will examine bonding models in more detail, particularly focusing on the role of orbitals.

14.1 Hybridization and the Localized Electron Model

Recall that the localized electron model views a molecule as a collection of atoms bound together by sharing electrons between atomic orbitals. The arrangement of valence electrons is represented by the Lewis structure (or structures, where resonance occurs), and the approximate molecular geometry can be predicted using the VSEPR model. In this section we will describe what types of atomic orbitals are employed by this model to share the electrons and hence to form the bonds.

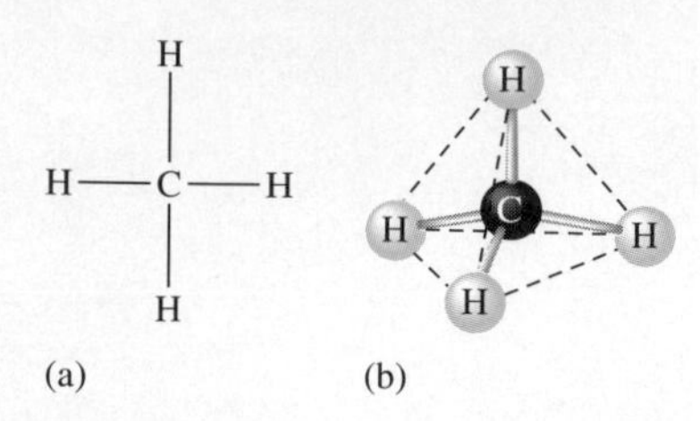

Figure 14.1
(a) The Lewis structure of the methane molecule. (b) The tetrahedral molecular geometry of the methane molecule.

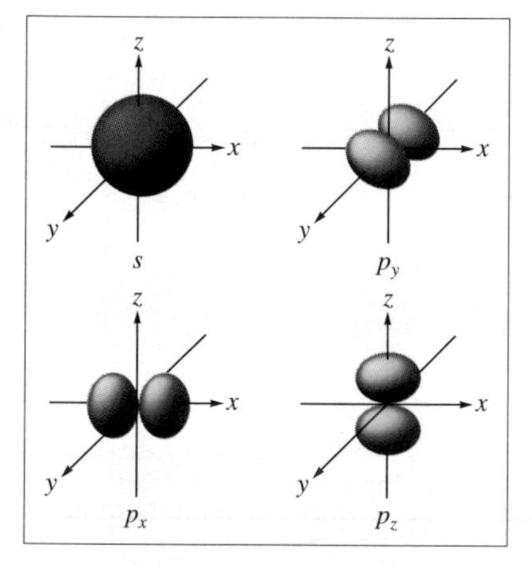

Figure 14.2
The valence orbitals on a free carbon atom: $2s$, $2p_x$, $2p_y$, and $2p_z$.

Hybridization is a modification of the localized electron model to account for the observation that atoms often seem to use special atomic orbitals in forming molecules.

sp^3 hybridization gives a tetrahedral set of orbitals.

sp^3 Hybridization

Let us reconsider the bonding in methane, which has the Lewis structure and molecular geometry shown in Fig. 14.1. In general, we assume that bonding involves only the valence orbitals. This means that the hydrogen atoms in methane use $1s$ orbitals. The valence orbitals of a carbon atom are the $2s$ and $2p$ orbitals shown in Fig. 14.2. Thinking about how carbon can use these orbitals to bond to the hydrogen atoms reveals two related problems:

1. Using the $2p$ and $2s$ atomic orbitals will lead to two different types of C—H bonds: (a) those from the overlap of a $2p$ orbital of carbon and a $1s$ orbital of hydrogen (there will be three of these) and (b) those from the overlap of a $2s$ orbital of carbon and a $1s$ orbital of hydrogen (there will be one of these). This presents a problem because the experimental evidence indicates that methane has four identical C—H bonds.
2. Since the carbon $2p$ orbitals are mutually perpendicular, we might expect the three C—H bonds formed with these orbitals to be oriented at 90° angles:

   However, the methane molecule is known by experiment to be tetrahedral with bond angles of 109.5°.

This analysis suggests that carbon adopts a set of atomic orbitals other than its "native" $2s$ and $2p$ orbitals to bond to the hydrogen atoms in forming the methane molecule. In fact, it is not surprising that the $2s$ and $2p$ orbitals present on an *isolated* carbon atom might not be the best set of orbitals for bonding. That is, a different set of atomic orbitals might better serve the carbon atom to form the most stable CH_4 molecule.

To account for the known structure of methane in terms of this model, we need a set of four equivalent atomic orbitals, arranged tetrahedrally. In fact, such a set of orbitals can be obtained quite readily by combining the carbon $2s$ and $2p$ orbitals, as shown schematically in Fig. 14.3. This mixing of the native atomic orbitals to form special orbitals for bonding is called **hybridization.** The four new orbitals are called sp^3 orbitals since they are formed from one $2s$ and three $2p$ orbitals (s^1p^3). Similarly, we say that the carbon atom undergoes sp^3 **hybridization,** or is sp^3 hybridized. The four sp^3 orbitals are identical in shape, each one having a large lobe and a small lobe (see Fig. 14.4). The four orbitals are oriented in space so that the large lobes form a tetrahedral arrangement, as shown in Fig. 14.3.

The linear combinations of the $2s$ and $2p$ orbitals that give the four sp^3 hybrid orbitals are listed below:

$$\phi_1 = \tfrac{1}{2}[(s) + (p_x) + (p_y) + (p_z)]$$
$$\phi_2 = \tfrac{1}{2}[(s) + (p_x) - (p_y) - (p_z)]$$
$$\phi_3 = \tfrac{1}{2}[(s) - (p_x) + (p_y) - (p_z)]$$
$$\phi_4 = \tfrac{1}{2}[(s) - (p_x) - (p_y) + (p_z)]$$

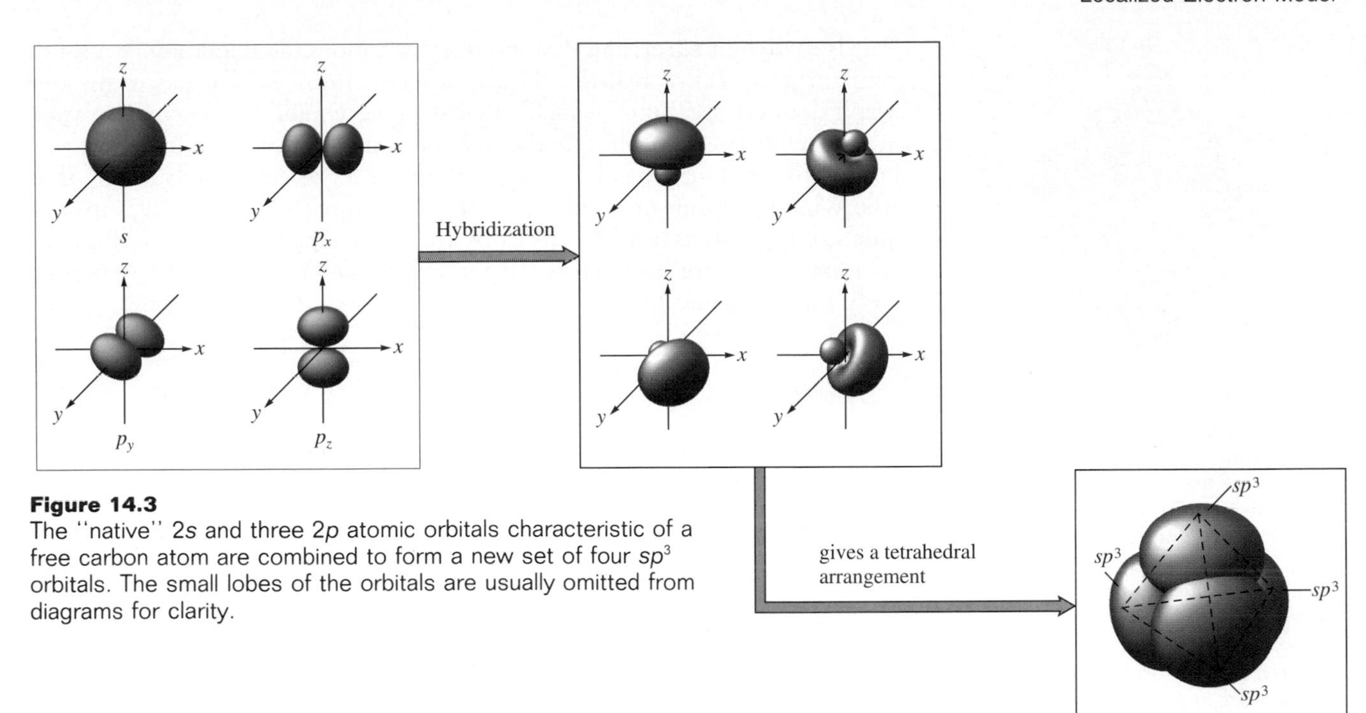

Figure 14.3
The "native" $2s$ and three $2p$ atomic orbitals characteristic of a free carbon atom are combined to form a new set of four sp^3 orbitals. The small lobes of the orbitals are usually omitted from diagrams for clarity.

where (s) and (p) represent $2s$ and $2p$ atomic orbital functions, and where the factor of $\frac{1}{2}$ is present to satisfy the boundary condition that the total probability is one for each orbital. Each of the functions ϕ_1, ϕ_2, ϕ_3, and ϕ_4 represents a separate sp^3 hybrid orbital.

The hybridization of the carbon $2s$ and $2p$ orbitals can also be represented by an orbital energy-level diagram, as shown in Fig. 14.5. Note that electrons have been omitted because we are not concerned with the electron arrangements on the individual atoms—it is the total number of valence electrons and the arrangement of these electrons in the *molecule* that are important.

In summary, the experimentally known structure of methane can be explained by the localized electron model if we assume that carbon adopts a special set of sp^3 atomic orbitals oriented toward the corners of a tetrahedron, which are then used to bond to the hydrogen atoms, as shown in Fig. 14.6.

We can extend this analysis to any molecule that forms a tetrahedral set of bonds: *whenever a set of equivalent, tetrahedral atomic orbitals is required by an atom, this model assumes that the atom forms a set of* sp^3 *orbitals;* the atom becomes sp^3 hybridized.

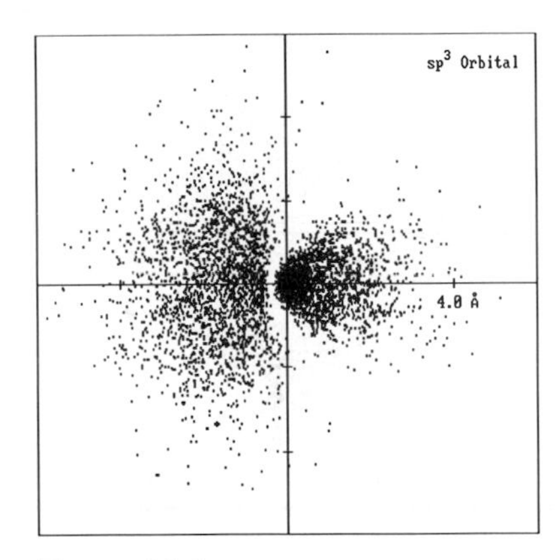

Figure 14.4
Cross section of an sp^3 orbital. Generated from a program by Robert Allendoerfer on Project SERAPHIM disk PC2402; printed with permission.

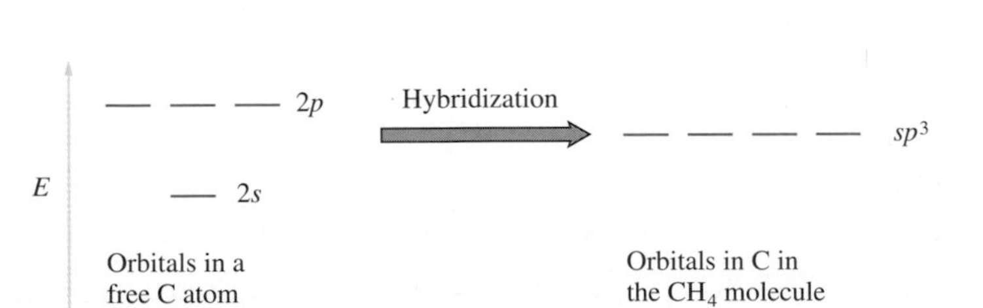

Figure 14.5
An energy-level diagram showing the formation of four sp^3 orbitals.

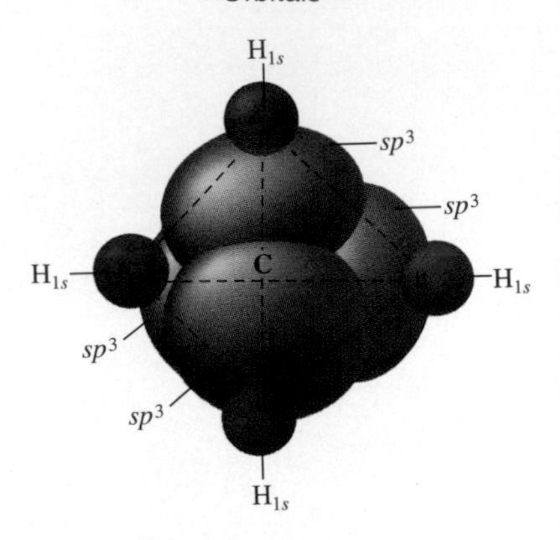

Figure 14.6
The tetrahedral set of four sp^3 orbitals on the carbon atom is used to share electron pairs with the four 1*s* orbitals of the hydrogen atoms to form the four equivalent C—H bonds. This accounts for the known tetrahedral structure of the CH_4 molecule. Note that in this figure and those that follow, the hybrid orbitals are drawn with narrowed lobes to show their orientations more clearly.

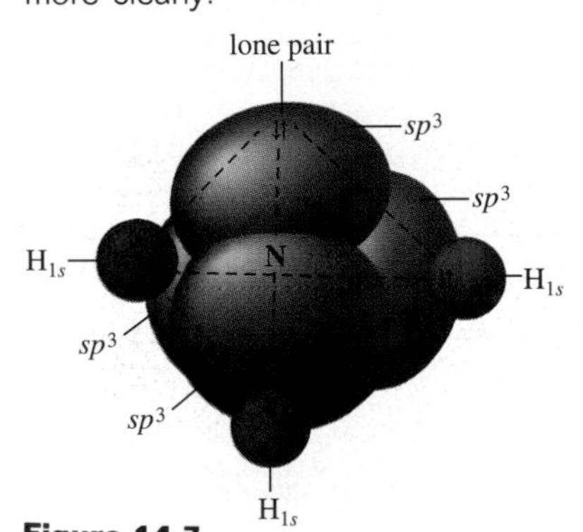

Figure 14.7
The nitrogen atom in ammonia is sp^3 hybridized.

It is really not surprising that an atom in a molecule might adopt a set of atomic orbitals, called **hybrid orbitals,** different from those it has in the free state. It does not seem unreasonable that to achieve minimum energy, an atom uses one set of atomic orbitals in the free state and a different set in a molecule. This is consistent with the idea that a molecule is more than simply a sum of its parts. What the atoms in a molecule were like before the molecule was formed is not as important as how the electrons are best arranged in the molecule. The individual atoms are assumed to respond as needed to achieve the minimum energy for the molecule.

EXAMPLE 14.1

Describe the bonding in the ammonia molecule using the localized electron model.

Solution

A complete description of the bonding involves three steps:

1. Write the Lewis structure.
2. Determine the arrangement of electron pairs using the VSEPR model.
3. Determine the hybrid atomic orbitals used for bonding in the molecule.

The Lewis structure for NH_3 is

$$\begin{array}{c} \ddot{\text{N}} \\ \text{H}-\text{N}-\text{H} \\ | \\ \text{H} \end{array}$$

The four electron pairs around the nitrogen atom require a tetrahedral arrangement. We have seen that a tetrahedral set of sp^3 hybrid orbitals is obtained by combining the 2*s* and three 2*p* orbitals. In the NH_3 molecule three of the sp^3 orbitals are used to form bonds to the three hydrogen atoms, and the fourth sp^3 orbital is used to hold the lone pair, as shown in Fig. 14.7.

sp^2 Hybridization

The plastics shown here were manufactured with ethylene.

Ethylene (C_2H_4) is an important starting material in the manufacture of plastics. The C_2H_4 molecule has 12 valence electrons and the following Lewis structure:

$$\begin{array}{ccc} \text{H} & & \text{H} \\ & \text{C}=\text{C} & \\ \text{H} & & \text{H} \end{array}$$

Recall that in the VSEPR model a double bond acts as one effective pair. Thus in the ethylene molecule each carbon is surrounded by three effective pairs. This model requires a trigonal planar arrangement with bond angles of 120°. What orbitals do the carbon atoms employ in this molecule? The molecular geometry

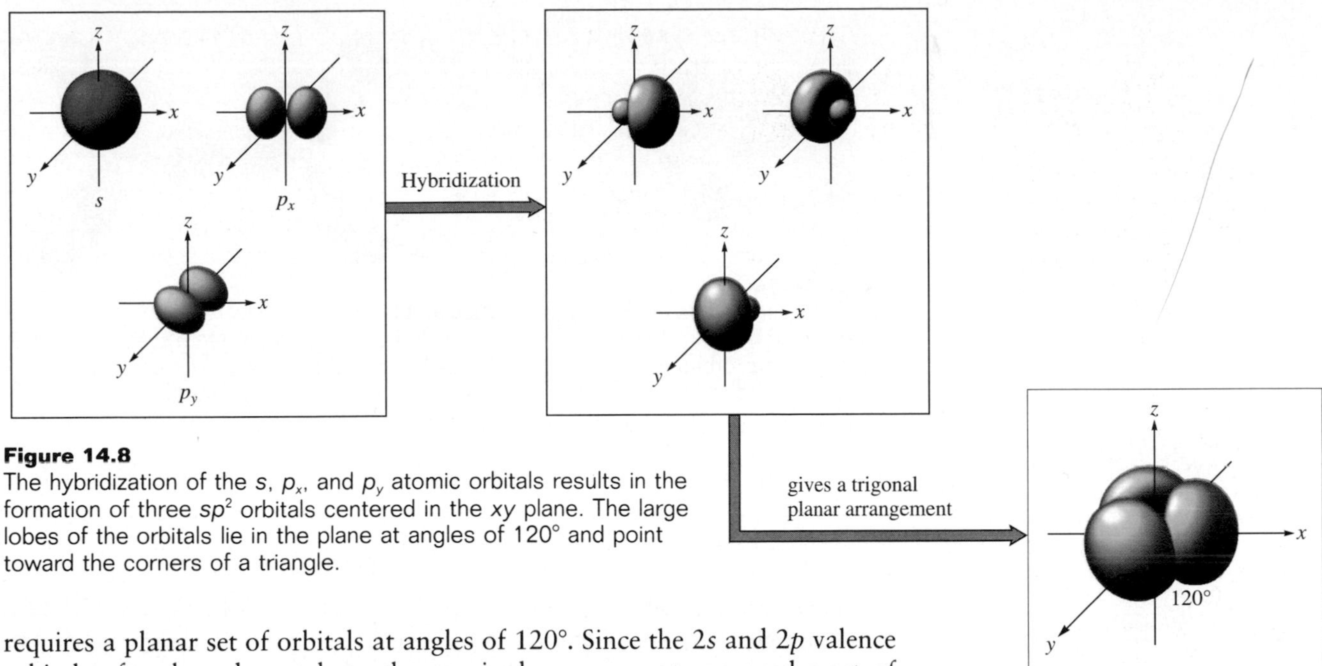

Figure 14.8
The hybridization of the s, p_x, and p_y atomic orbitals results in the formation of three sp^2 orbitals centered in the xy plane. The large lobes of the orbitals lie in the plane at angles of 120° and point toward the corners of a triangle.

requires a planar set of orbitals at angles of 120°. Since the 2s and 2p valence orbitals of carbon do not have the required arrangement, we need a set of hybrid orbitals.

The sp^3 orbitals we have just considered will not work, because they exhibit angles of 109.5° rather than the required 120°. Therefore, in ethylene the carbon atom must hybridize in a different manner. A set of three orbitals arranged at 120° angles in the same plane can be obtained by combining one s orbital and two p orbitals, as shown schematically in Fig. 14.8. The orbital energy-level diagram for this arrangement is shown in Fig. 14.9. Since one 2s and two 2p orbitals are used to form these hybrid orbitals, this is called ***sp*² hybridization.** Note from Fig. 14.8 that the plane of the sp^2 hybridized orbitals is determined by which p orbitals are used. Since in this case we have arbitrarily decided to use the p_x and p_y orbitals, the hybrid orbitals are centered in the xy plane. We will not show the detailed forms of these functions.

The assumption that a double bond acts as one effective electron pair, equivalent to a single bonding pair, works well to give the *approximate* molecular structure.

sp^2 hybridization gives a trigonal planar arrangement of atomic orbitals.

In the formation of the sp^2 orbitals, one 2p orbital on carbon has not been used. This remaining p orbital (p_z) is oriented perpendicular to the plane of the sp^2 orbitals, as shown in Fig. 14.10.

Now we will see how these orbitals can be used to account for the Lewis structure of ethylene. The three sp^2 orbitals on each carbon are used to share electrons, as shown in Fig. 14.11. In each of these bonds the electron pair is

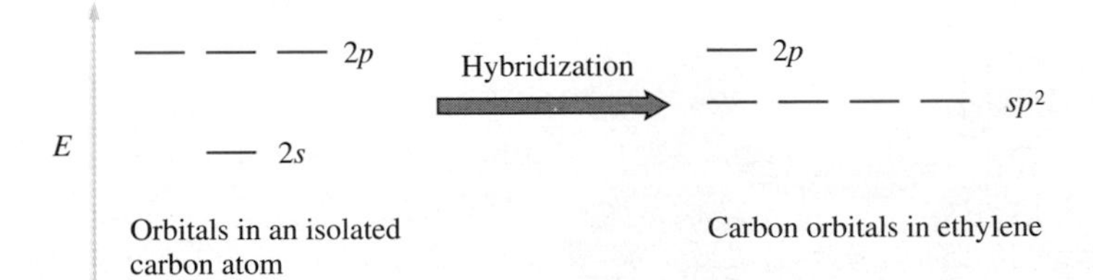

Figure 14.9
An orbital energy-level diagram for sp^2 hybridization. Note that one p orbital remains unchanged.

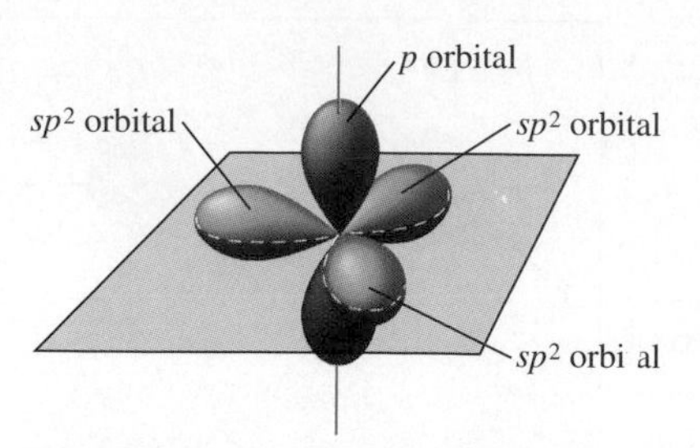

Figure 14.10
When one s and two p orbitals are mixed to form a set of three sp^2 orbitals, one p orbital remains unchanged and is perpendicular to the plane of the hybrid orbitals.

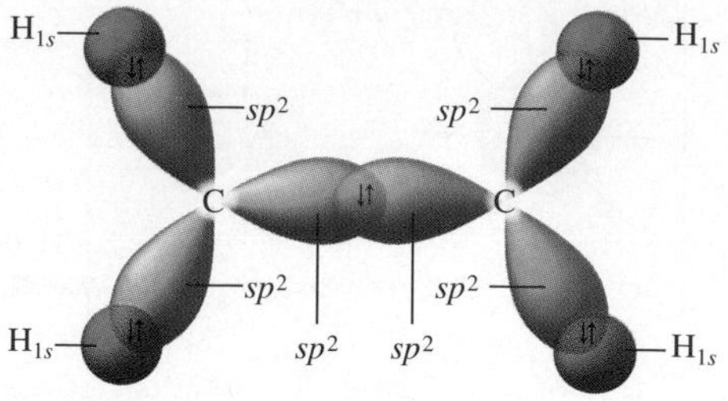

Figure 14.11
The σ bonds in ethylene. Note that for each bond the shared electron pair occupies the region directly between the atoms.

shared in an area centered on a line running between the atoms. This type of covalent bond is called a **sigma (σ) bond.** In the ethylene molecule the σ bonds are formed using sp^2 orbitals on each carbon atom and the $1s$ orbital on each hydrogen atom.

How can we explain the double bond between the carbon atoms? In the σ bond the electron pair occupies the space between the carbon atoms. The second bond must therefore result from sharing an electron pair in the space *above and below* the σ bond. This type of bond can be formed using the $2p$ orbital perpendicular to the sp^2 hybrid orbitals on each carbon atom (refer to Fig. 14.10). These parallel p orbitals can share an electron pair, which occupies the space above and below a line joining the atoms, to form a **pi (π) bond,** as shown in Fig. 14.12.

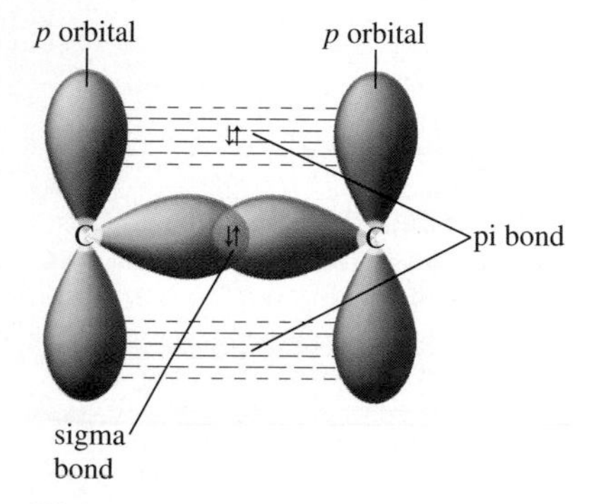

Figure 14.12
A carbon-carbon double bond consists of a σ bond and a π bond. In the σ bond the shared electrons occupy the space directly between the atoms. The π bond is formed from the unhybridized p orbitals on the two carbon atoms. In a π bond the shared electron pair occupies the space above and below a line joining the atoms.

Note that σ bonds are formed from orbitals whose lobes point toward each other, but π bonds result from parallel orbitals. A *double bond consists of one σ bond,* where the electron pair is located directly between the atoms, *and one π bond,* where the shared pair occupies the space above and below the σ bond.

We can now completely specify the orbitals used to form the bonds in the ethylene molecule. As shown in Fig. 14.13, the carbon atoms use sp^2 hybrid orbitals to form the σ bonds to the hydrogen atoms and to each other, and use p orbitals to form the π bond with each other. Note that we have accounted fully for the Lewis structure of ethylene with its carbon-carbon double bond and carbon-hydrogen single bonds.

This example illustrates an important general principle of this model: *whenever an atom is surrounded by three effective pairs, a set of* sp^2 *hybrid orbitals is required.*

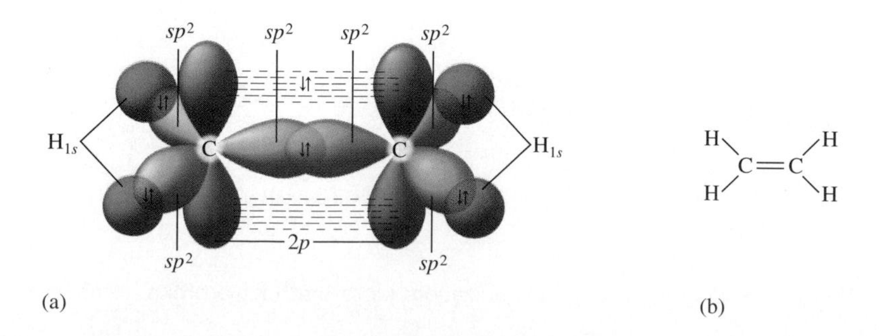

Figure 14.13
(a) The orbitals used to form the bonds in ethylene. (b) The Lewis structure for ethylene.

sp Hybridization

Another type of hybridization occurs in carbon dioxide, which has the following Lewis structure:

$$:\dot{\underset{\cdot}{O}}{=}C{=}\dot{\underset{\cdot}{O}}:$$

In the CO_2 molecule the carbon atom has two effective pairs that are arranged at an angle of 180°. We therefore need a pair of atomic orbitals oriented in opposite directions. This requires a new type of hybridization, since neither sp^3 nor sp^2 hybrid orbitals fit this case. Obtaining two hybrid orbitals arranged at 180° requires ***sp* hybridization,** involving one *s* orbital and one *p* orbital, as shown schematically in Fig. 14.14.

In terms of this model, *two effective pairs around an atom will always require* sp *hybridization of that atom.* The *sp* orbitals of carbon in carbon dioxide are shown in Fig. 14.15, and the corresponding orbital energy-level diagram for their formation is given in Fig. 14.16. The *sp* hybrid orbitals are used to form the σ bonds between carbon and the oxygen atoms. Note that two $2p$ orbitals remain unchanged on the *sp* hybridized carbon. They are used to form the π bonds to the oxygen atoms.

In the CO_2 molecule each oxygen atom has three effective pairs around it, requiring a trigonal planar arrangement of the pairs. Since a trigonal set of hybrid orbitals corresponds to sp^2 hybridization, each oxygen atom can be assumed to be sp^2 hybridized. The orbital on each oxygen left unchanged by the hybridization process is used for the π bond to the carbon atom.

Now we are ready to account for the Lewis structure of carbon dioxide. The *sp* orbitals on carbon form σ bonds with the sp^2 orbitals on the two oxygen

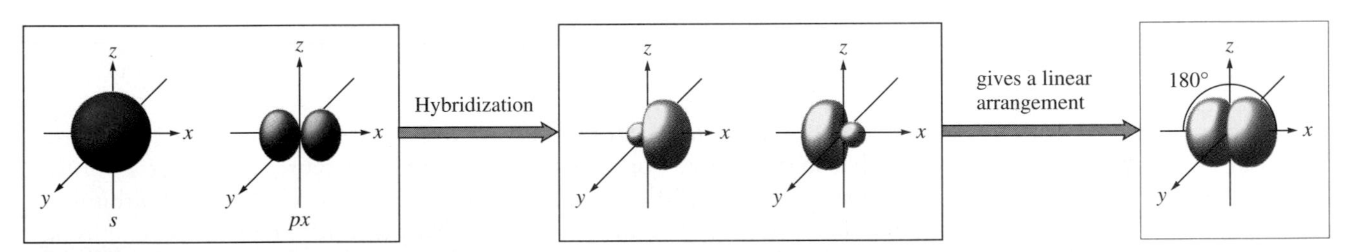

Figure 14.14
When one *s* orbital and one *p* orbital are hybridized, a set of two *sp* orbitals oriented at 180° results.

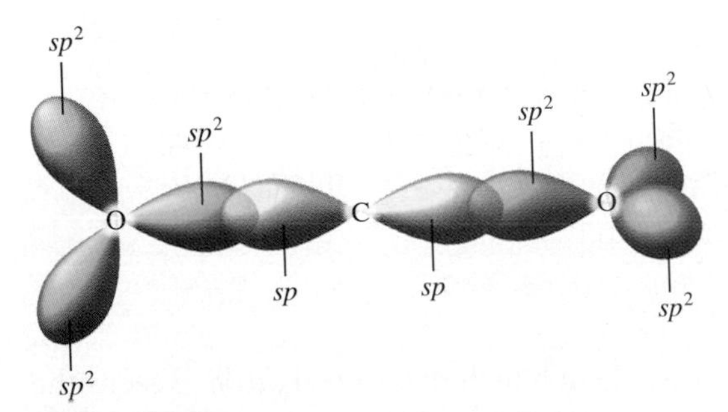

Figure 15.15

The hybrid orbitals in the CO_2 molecule.

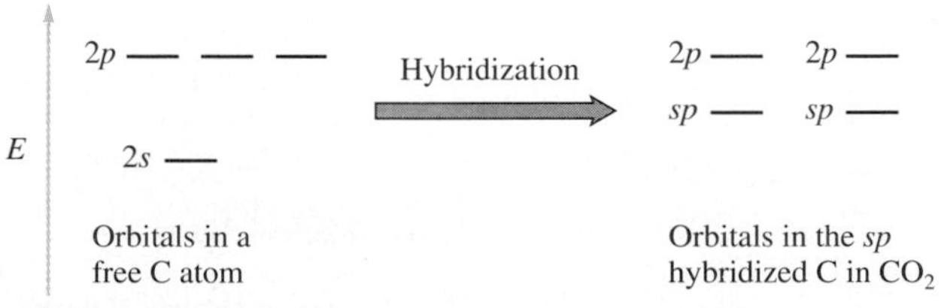

E
2p — — —
2s —
Hybridization
2p — 2p —
sp — sp —
Orbitals in a free C atom
Orbitals in the *sp* hybridized C in CO_2

Figure 14.16
The orbital energy-level diagram for the formation of *sp* hybrid orbitals of carbon.

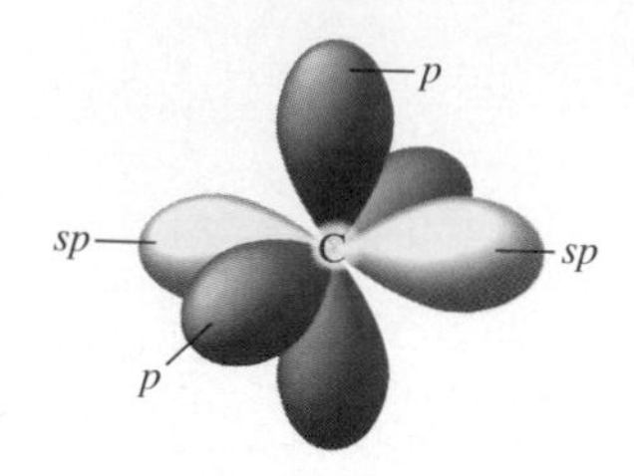

Figure 14.17
The orbitals of an *sp* hybridized carbon atom.

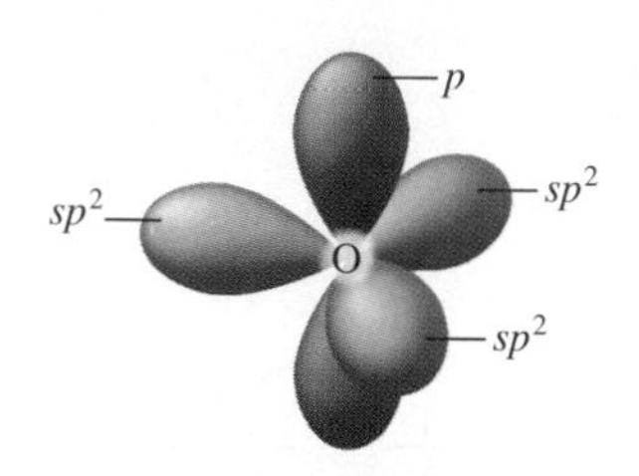

Figure 14.18
The orbital arrangement for an sp^2 hybridized oxygen atom.

atoms (Fig. 14.15). The remaining sp^2 orbitals on the oxygen atoms hold lone pairs. The π bonds between the carbon atom and each oxygen atom are formed by the overlap of parallel $2p$ orbitals. The *sp* hybridized carbon atom has two unhybridized *p* orbitals, pictured in Fig. 14.17. Each of these *p* orbitals is used to form a π bond with an oxygen atom (see Fig. 14.18). The total bonding picture predicted by the localized electron model for the CO_2 molecule is shown in Fig. 14.19. Note that this picture of the bonding neatly explains the arrangement of electrons predicted by the Lewis structure.

It is useful to remind ourselves at this point that because we are using a very simple bonding model, the picture of a molecule we get from this model is an approximate one. The carbon dioxide molecule provides a good case in point. Various types of evidence suggest that the electron density around the two C—O bonds in CO_2 is actually cylindrically symmetric—that is, the electron density is homogeneous all around the O—C—O molecular axis. This result is not consistent with the simple picture given in Fig. 14.19(a) in which the two C—O bonds have their π electron densities centered in two perpendicular planes. The model can be corrected in a straightforward manner to remove this difficulty, but we will not pursue this correction here. The point is that a very simple model is quite useful to us, because it is so easy to apply. Simple models help us organize our observations and often provide a place to start as we pursue the answer to a problem. However, we must be wary of simple models—they often provide a significantly distorted picture.

EXAMPLE 14.2

Describe the bonding in the N_2 molecule.

Solution

The Lewis structure for the nitrogen molecule is

$$:N{\equiv}N:$$

where each nitrogen atom is surrounded by two effective pairs. (Remember that multiple bonds count as one effective pair.) This gives a linear arrangement (180°) requiring a pair of oppositely directed orbitals. Therefore, this situation requires *sp* hybridization. Each nitrogen atom in the nitrogen molecule has two *sp* hybrid orbitals and two unchanged *p* orbitals, as shown in Fig. 14.20(a). The *sp* orbitals are used to form the σ bond between the nitrogen atoms and to hold lone pairs, as shown in Fig. 14.20(b). The *p* orbitals are used to form the two π bonds [see Fig. 14.20(c)]; each pair of overlapping, parallel *p* orbitals holds one electron pair. Such bonding accounts for the electron arrangement given by the Lewis structure. The triple bond consists of a σ bond (overlap of two *sp* orbitals) and two π bonds (each from an overlap of two *p* orbitals). In addition, a lone pair occupies an *sp* orbital on each nitrogen atom.

dsp^3 Hybridization

To illustrate the treatment of a molecule in which the central atom exceeds the octet rule, we will consider the bonding in the phosphorus pentachloride molecule (PCl_5). The Lewis structure

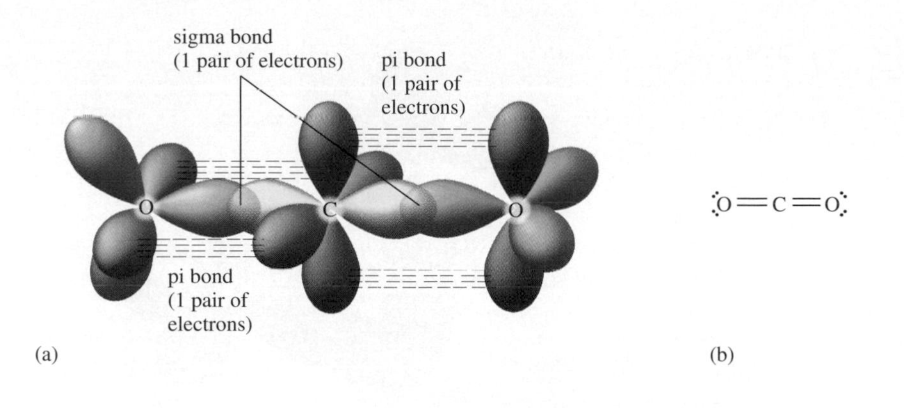

Figure 14.19
(a) The orbitals predicted by the localized electron model to describe the bonds in carbon dioxide. Note that the carbon-oxygen double bonds each consist of one σ bond and one π bond. (b) The Lewis structure for carbon dioxide.

$$\mathrm{PCl_5}\ \text{(Lewis structure: P bonded to five Cl atoms, each Cl with three lone pairs)}$$

shows that the phosphorus atom is surrounded by five electron pairs. Since in the VSEPR model five pairs require a trigonal bipyramidal arrangement, we need a trigonal bipyramidal set of atomic orbitals on phosphorus. Such a set of orbitals is formed by ***dsp*³ hybridization** of one *d* orbital, one *s* orbital, and three *p* orbitals, as shown in Fig. 14.21.

Although it will not be important for our purposes, the dsp^3 hybrid orbital set is different from the hybrids we have considered so far in that the hybrid orbitals pointing to the vertices of the triangle (often called the three equatorial hybrid orbitals) are slightly different in shape than the other two (the *axial* orbitals). This situation stands in contrast to the *sp*, sp^2, and sp^3 hybrid sets in which each orbital in a particular set is identical in shape to the others.

The dsp^3 hybridized phosphorus atom in the PCl_5 molecule uses its five dsp^3 orbitals to share electrons with the five chlorine atoms. Note that according to this model *a set of five effective pairs around a given atom always requires a trigonal bipyramidal arrangement, which in turn requires* dsp³ *hybridization of that atom.*

The Lewis structure for PCl_5 shows that each chlorine atom is surrounded by four electron pairs. This requires a tetrahedral arrangement, which in turn requires a set of four sp^3 orbitals on each chlorine atom.

Now we can describe the bonding in the PCl_5 molecule. The five P—Cl σ bonds are formed by sharing electrons between a dsp^3 orbital on the phospho-

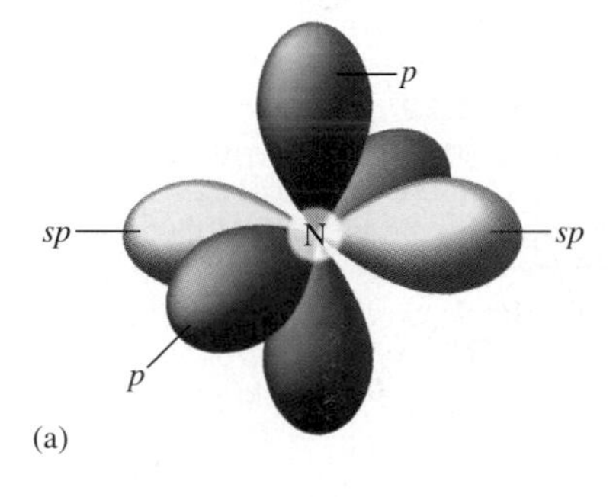

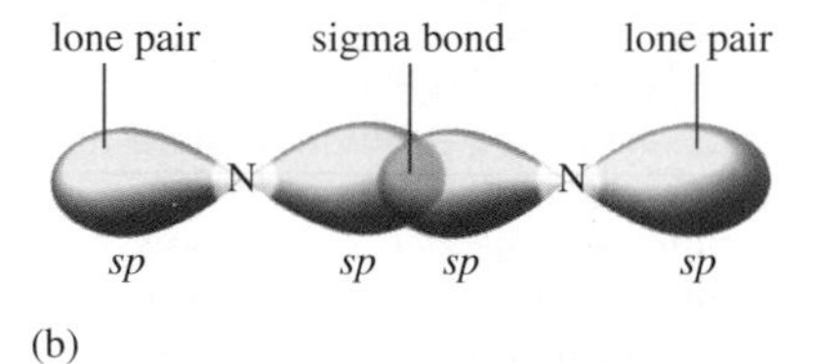

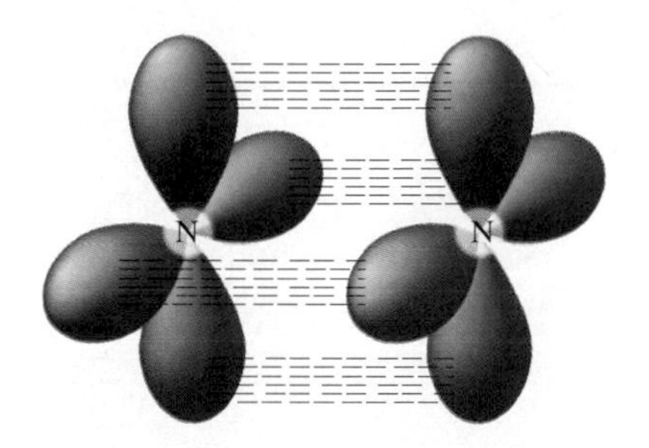

Figure 14.20
(a) An *sp* hybridized nitrogen atom. There are two *sp* hybrid orbitals and two unhybridized *p* orbitals. (b) The σ bond in the N_2 molecule. (c) The two π bonds in N_2 are formed when electron pairs are shared between two sets of parallel *p* orbitals.

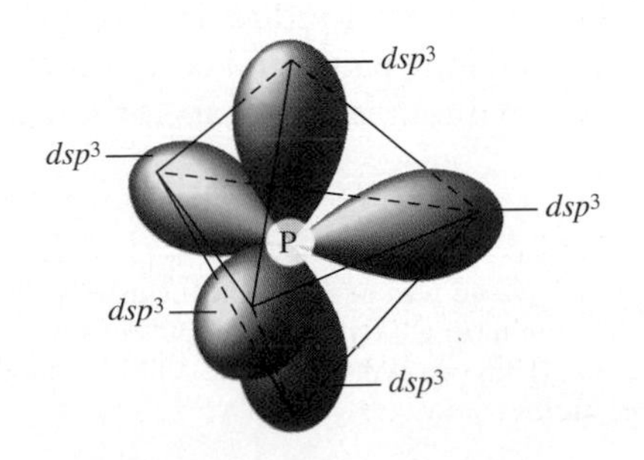

Figure 14.21
A set of dsp^3 hybrid orbitals on a phosphorus atom. Note that the set of five dsp^3 orbitals has a trigonal bipyramidal arrangement. (Each dsp^3 orbital also has a small lobe that is not shown in this diagram.)

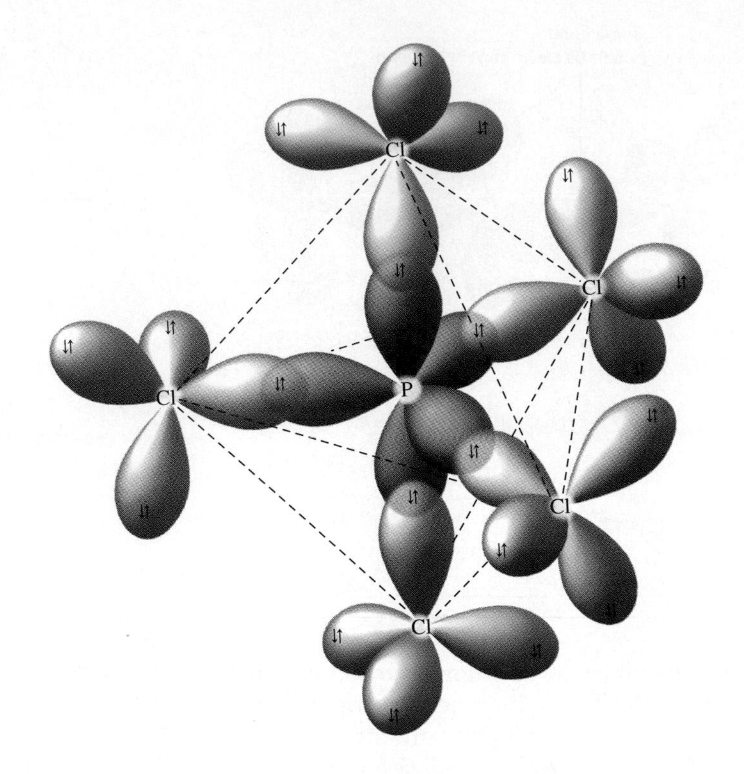

Figure 14.22
The orbitals used to form the bonds in the PCl_5 molecule. The phosphorus uses a set of five dsp^3 orbitals to share electron pairs with sp^3 orbitals on the five chlorine atoms. The other sp^3 orbitals on each chlorine atom hold lone pairs.

Although in its simplest form, the localized electron model of bonding invokes the use of hybrids involving *d* orbitals for trigonal bipyramidal and octahedral structures, there is considerable disagreement about the actual participation of *d* orbitals in such molecules. For a discussion, see A. E. Reed and F. Weinhold, *J. Am. Chem. Soc.,* 1986, 108, 3586, and references therein.

rus atom and an sp^3 orbital on each chlorine.* The other sp^3 orbitals on each chlorine hold lone pairs. This arrangement is shown in Fig. 14.22.

EXAMPLE 14.3

Describe the bonding in the triiodide ion (I_3^-).

Solution

The Lewis structure for I_3^-,

$$\left[:\ddot{\underset{..}{I}}-\dot{\ddot{\underset{..}{I}}}-\ddot{\underset{..}{I}}:\right]^-$$

shows that the central iodine atom has five pairs of electrons. A set of five pairs requires a trigonal bipyramidal arrangement, which in turn requires a set of dsp^3 orbitals. The outer iodine atoms have four pairs of electrons, which calls for a tetrahedral arrangement and sp^3 hybridization.

Thus the central iodine is dsp^3 hybridized. Three of these hybrid orbitals hold lone pairs, and two of them overlap with sp^3 orbitals from the other two iodine atoms to form σ bonds.

*Although we have no way of proving conclusively that each chlorine atom is sp^3 hybridized, we assume that minimizing electron pair repulsions is as important for peripheral atoms as for the central atom. Thus we will apply the VSEPR model and hybridization to both central and peripheral atoms.

d^2sp^3 Hybridization

Some molecules have six pairs of electrons around a central atom; an example is sulfur hexafluoride (SF_6), which has the Lewis structure

```
  ..  ..  ..
 :F  :F: F:
  ..\ | /..
      S
  ../ | \..
 :F  :F: F:
  ..  ..  ..
```

This requires an octahedral arrangement of pairs and, in turn, an octahedral set of six hybrid orbitals. This leads to **d^2sp^3 hybridization,** in which two d orbitals, one s orbital, and three p orbitals are combined (see Fig. 14.23). Note that *six electron pairs around an atom are always arranged octahedrally, requiring* d^2sp^3 *hybridization of the atom.* Each d^2sp^3 orbital on the sulfur atom is used to bond to a fluorine atom. Since there are four pairs on each fluorine atom, the fluorine atoms are assumed to be sp^3 hybridized.

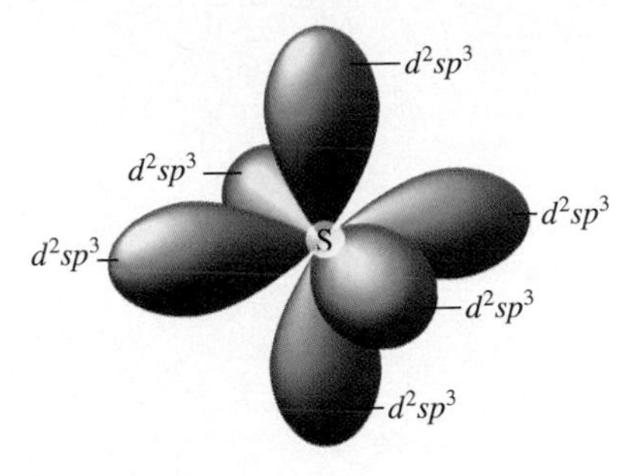

Figure 14.23
An octahedral set of d^2sp^3 orbitals on a sulfur atom. The small lobe of each hybrid orbital has been omitted for clarity.

EXAMPLE 14.4

How is the xenon atom in XeF_4 hybridized?

Solution

The Lewis structure for XeF_4 has six pairs of electrons around xenon that are arranged octahedrally to minimize repulsions. An octahedral set of six atomic orbitals is required to hold these electrons, and the xenon atom is assumed to be d^2sp^3 hybridized.

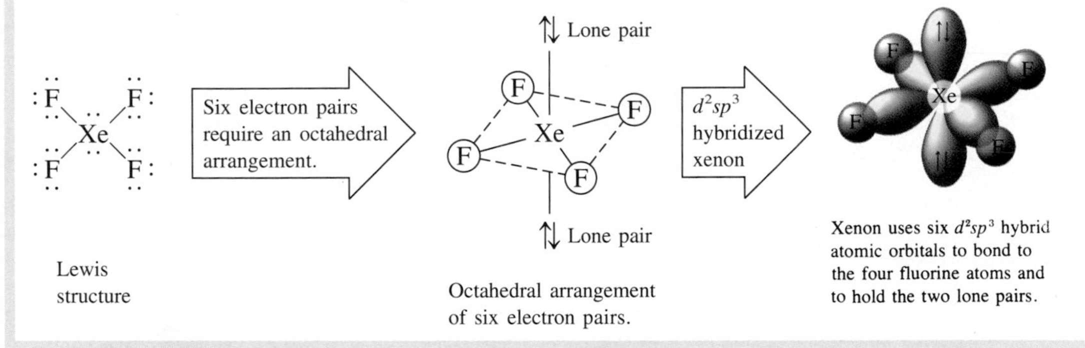

Lewis structure

Octahedral arrangement of six electron pairs.

Xenon uses six d^2sp^3 hybrid atomic orbitals to bond to the four fluorine atoms and to hold the two lone pairs.

The description of a molecule using the localized electron model involves three distinct steps.

Summary: Describing a Molecule with the Localized Electron Model

- **Draw the Lewis structure(s).**
- **Determine the arrangement of electron pairs, using the VSEPR model.**
- **Specify the hybrid orbitals needed to accommodate the electron pairs.**

It is important to perform the steps in this order. For a model to be successful, it must follow nature's priorities. In the case of bonding, it seems clear that the tendency for a molecule to minimize its energy is more important than the maintenance of the characteristics of atoms as they exist in the free state. The atoms adjust to meet the "needs" of the molecule. When considering the bonding in a particular molecule, therefore, we always start with the molecule rather than the component atoms. In the molecule the electrons are arranged to give each atom a noble gas configuration where possible and to minimize electron pair repulsions. We then assume that the atoms adjust their orbitals by hybridization to allow the molecule to adopt the structure that gives the minimum energy.

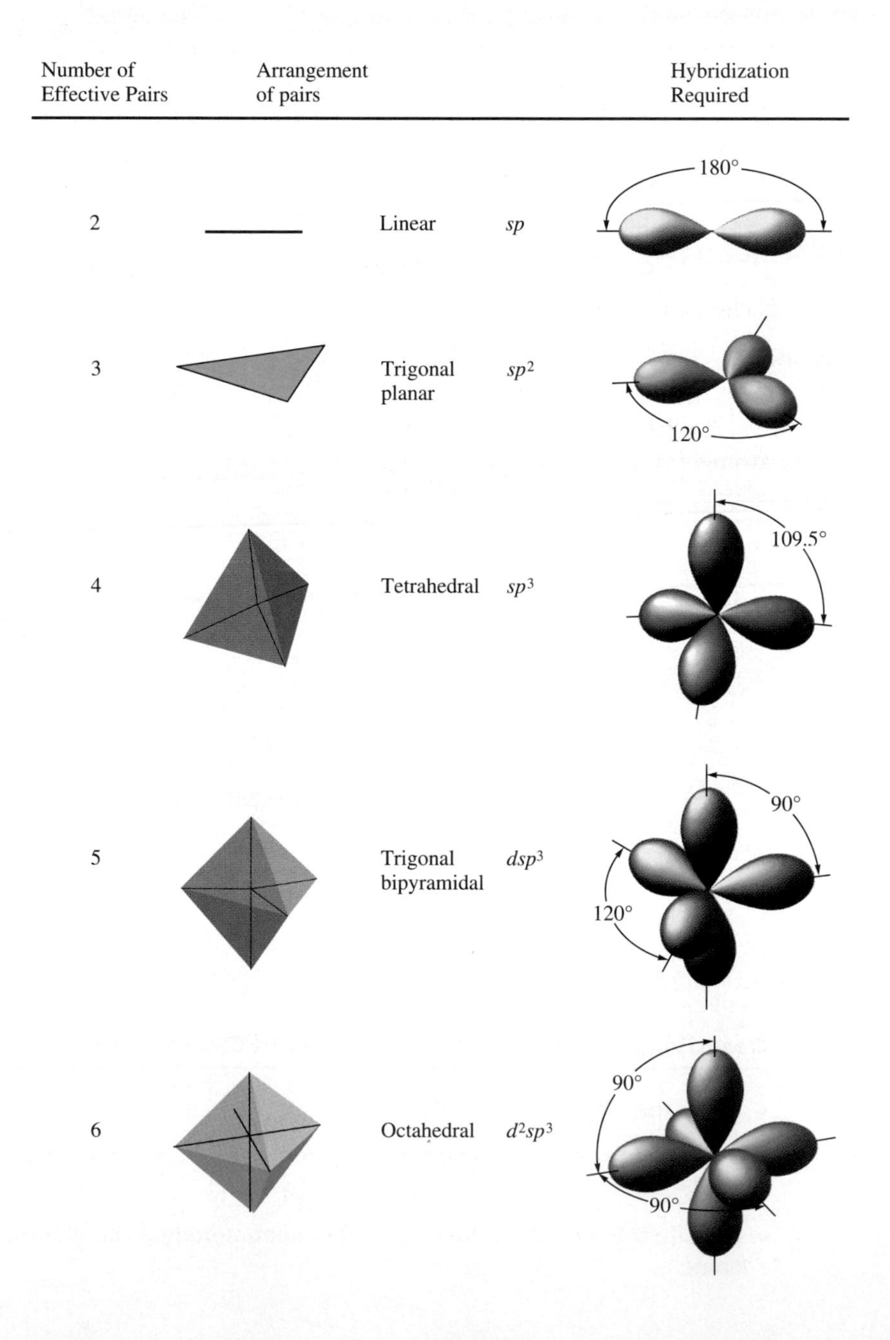

Number of Effective Pairs
Arrangement of pairs
Hybridization Required
2
Linear
sp
180°
3
Trigonal planar
sp2
120°
4
Tetrahedral
sp3
109.5°
5
Trigonal bipyramidal
dsp3
90°
120°
6
Octahedral
d2sp3
90°
90°

Figure 14.24
The relationship among the number of effective pairs, their spatial arrangement, and the hybrid orbital set required.

In applying the localized electron model, we must remember not to overemphasize the characteristics of the separate atoms. The particular atom on which the valence electrons originated is not important; what is important is where they are needed in the molecule to achieve maximum stability. In the same vein, the forms of the orbitals on the isolated atoms are relatively unimportant; what really matters are the forms needed by the molecule to achieve minimum energy.

The requirements for the various types of hybridization are summarized in Fig. 14.24 on the opposite page.

EXAMPLE 14.5

For each of the following molecules or ions, describe the molecular structure and predict the hybridization of each atom.

a. CO b. BF_4^- c. XeF_2

Solution

a. The CO molecule has 10 valence electrons, and its Lewis structure is

$$:C{\equiv}O:$$

Each atom has two effective pairs, which means that both are *sp* hybridized. The triple bond consists of a σ bond produced by the overlap of an *sp* orbital from each atom and two π bonds produced by the overlap of 2*p* orbitals from each atom. The lone pairs are in *sp* orbitals. Since the CO molecule has only two atoms, it must be linear.

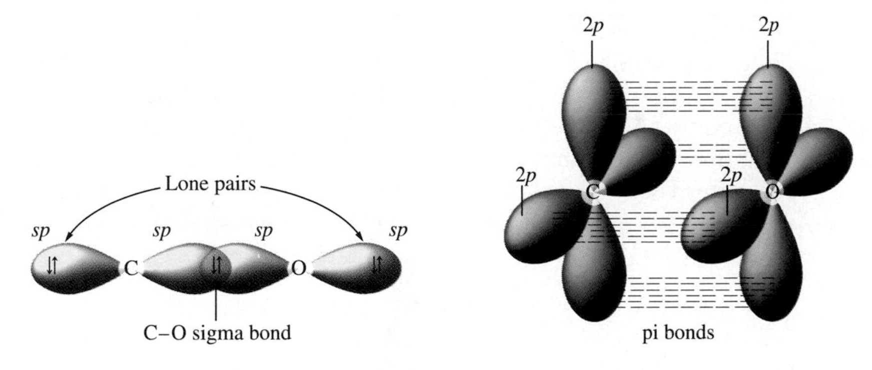

b. The BF_4^- ion has 32 valence electrons. The Lewis structure shows four pairs of electrons around the boron atom, requiring a tetrahedral arrangement:

$$\left[\begin{array}{c} :\ddot{F}: \\ | \\ :\ddot{F}-B-\ddot{F}: \\ | \\ :\ddot{F}: \end{array}\right]^-$$

4 electron pairs around B →

F, B, F, F, F

This requires sp^3 hybridization of the boron atom. Each fluorine atom also has four electron pairs and can be assumed to be sp^3 hybridized (only one sp^3 orbital is shown for each fluorine atom). The BF_4^- ion's molecular structure is tetrahedral.

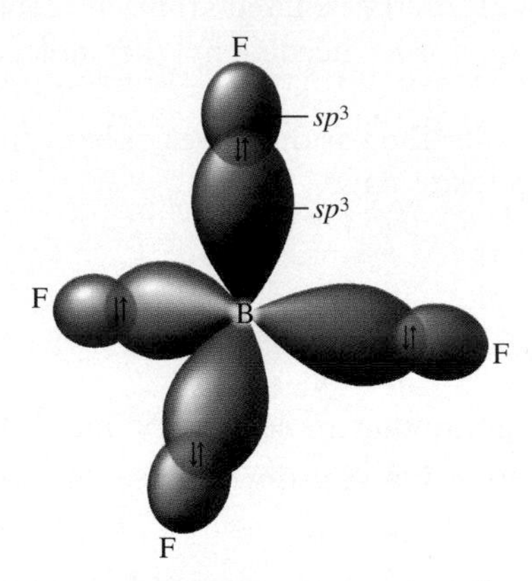

c. The XeF_2 molecule has 22 valence electrons. The Lewis structure shows five electron pairs on the xenon atom, which requires a trigonal bipyramidal arrangement:

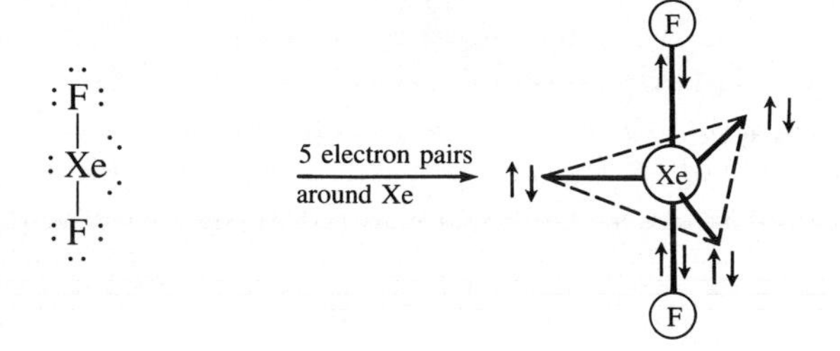

Note that the lone pairs are placed in the plane where they are 120° apart. Accommodating five pairs at the vertices of a trigonal bipyramid requires that the xenon atom adopt a set of five dsp^3 orbitals. Each fluorine atom has four electron pairs and can be assumed to be sp^3 hybridized. The XeF_2 molecule has a linear arrangement of atoms.

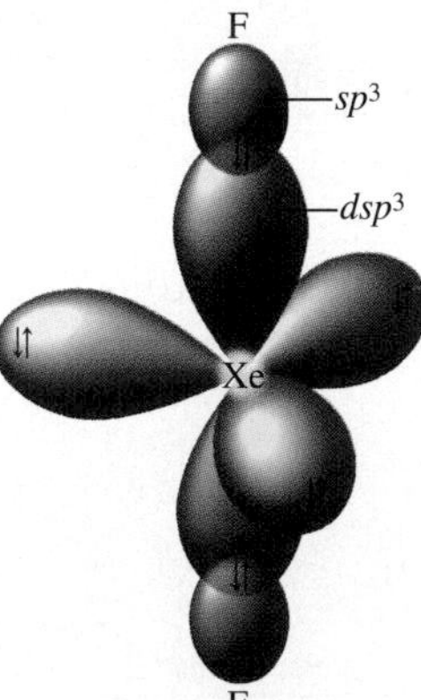

14.2 The Molecular Orbital Model

It should be clear by now that the localized electron model is of great value in interpreting the structures and bonding of molecules. However, there are some problems with this model at this level of approximation. For example, since it incorrectly assumes that electrons are localized, the concept of resonance must be added. Also, the model does not deal easily with molecules containing unpaired electrons. And finally, the model in this form gives no direct information about bond energies.

Another model often used to describe bonding is the **molecular orbital model.** To introduce the assumptions, methods, and results of this model, we will consider the simplest of all molecules, H_2, which consists of two protons and two electrons. A very stable molecule, H_2 is lower in energy than the separated hydrogen atoms by 432 kJ/mol.

Since the hydrogen molecule consists of protons and electrons, the same components found in separated hydrogen atoms, it seems reasonable to use a theory similar to the atomic theory discussed in Chapter 12, which assumes that the electrons in an atom exist in orbitals of a given energy. Can we apply this same type of model to the hydrogen molecule? Yes; in fact, describing the H_2 molecule in terms of quantum mechanics is quite straightforward.

However, even though it is formulated rather easily, this problem cannot be solved exactly. The difficulty is the same as that encountered in dealing with polyelectronic atoms—the electron correlation problem. Since we cannot account for the details of the electron movements, we cannot deal with the electron-electron interactions in a specific way. We need to make approximations that allow the solution of the problem but that do not destroy the model's physical integrity. The success of these approximations can be measured only by comparing predictions from the theory with experimental observations. In this case we will see that the simplified model works well.

Just as atomic orbitals are solutions to the quantum mechanical treatment of atoms, **molecular orbitals (MOs)** are solutions to the molecular problem. Molecular orbitals have many of the same characteristics as atomic orbitals. Two of the most important are: (1) they can hold two electrons with opposite spins and (2) the square of the molecular orbital wave function indicates the electron probability.

As in the application of quantum mechanics to isolated atoms, the molecular orbital treatment can be carried out at various levels of sophistication. In our description of the model we will assume that the molecular orbitals for H_2 are constructed using hydrogen 1*s* orbitals. We say that the 1*s* orbitals form the "basis set" for the molecular orbitals. A more detailed treatment would use a different basis set—one in which the radial part of the atomic orbitals would be allowed to vary to achieve the lowest-energy molecular orbitals for the hydrogen molecule. However, to avoid as many complications as possible, as we discuss the fundamental ideas of the molecular orbital description of molecules, we will use the simplest version of this model.

We will now describe the bonding in the hydrogen molecule using the MO model. The first step is to obtain the hydrogen molecule's orbitals, a process that is greatly simplified if we assume that the molecular orbitals can be constructed from the hydrogen 1*s* orbitals.

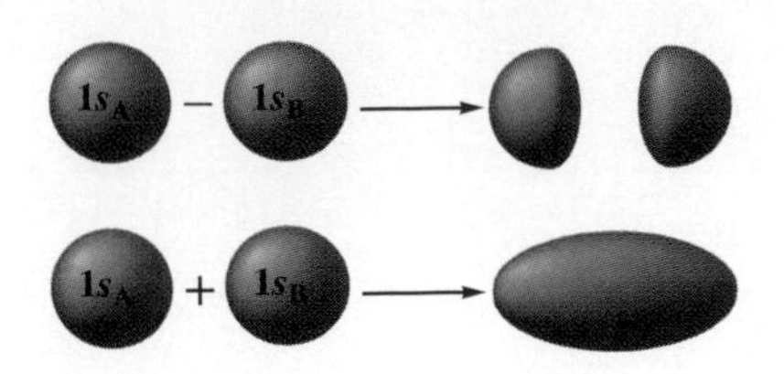

Figure 14.25
The combination of hydrogen 1*s* atomic orbitals to form molecular orbitals.

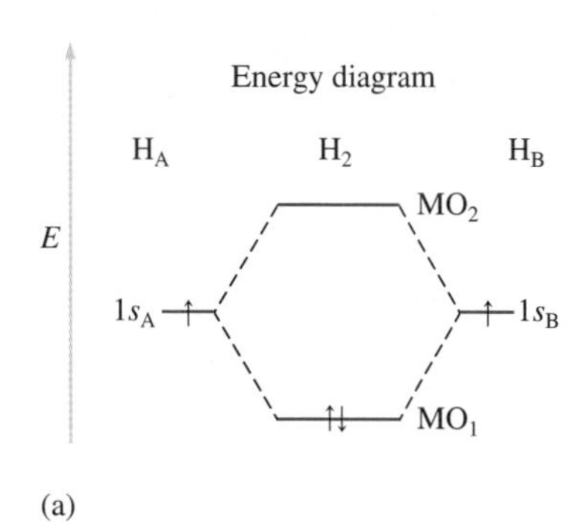

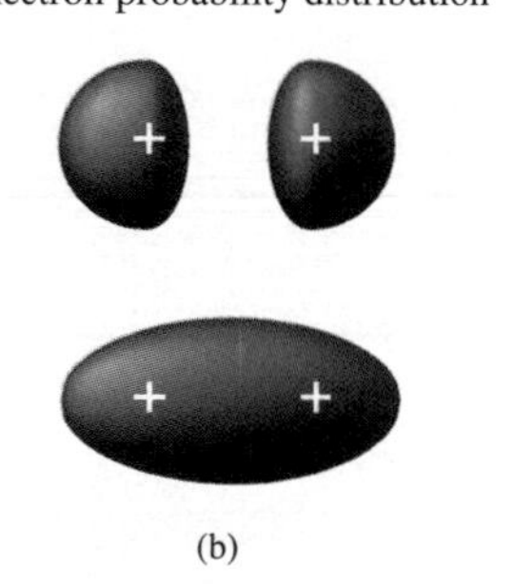

Figure 14.26
(a) The molecular orbital energy-level diagram for the H_2 molecule. (b) The shapes of the molecular orbitals are obtained by squaring the wave functions for MO_1 and MO_2.

Bonding will result if the molecule has lower energy than the separated atoms.

In this approximate treatment of the hydrogen molecule, two molecular orbitals result:

$$MO_1 = 1s_A + 1s_B$$
$$MO_2 = 1s_A - 1s_B$$

where $1s_A$ and $1s_B$ represent the 1*s* orbitals from the two separated hydrogen atoms. This process is shown schematically in Fig. 14.25.

The orbital properties of greatest interest are size, shape (described by the electron probability distribution), and energy. These properties for the hydrogen molecular orbitals are represented in Fig. 14.26. From Fig. 14.26 we can note several important points:

1. The electron probability of both molecular orbitals is centered along the line passing through the two nuclei. For MO_1 the greatest electron probability is *between* the nuclei. For MO_2 it is centered along the molecular axis, but outside the area between the two nuclei. In both molecular orbitals the electron density has cylindrical symmetry with respect to the molecular axis. That is, the electron probability is the same along any line drawn perpendicular to the bond axis at a given point on the axis. This type of cylindrically symmetric electron distribution is described as *sigma* (σ), as in the localized electron model. Accordingly, we refer to MO_1 and MO_2 as **sigma (σ) molecular orbitals.**
2. In the molecule only the molecular orbitals are available for occupation by electrons. The 1*s* atomic orbitals of the hydrogen atoms no longer exist, because the H_2 molecule—a new entity—has its own set of new orbitals.
3. MO_1 is lower in energy than the 1*s* orbitals of free hydrogen atoms, but MO_2 is higher in energy than the 1*s* orbitals. This fact has very important implications for the stability of the H_2 molecule: if the two electrons (one from each hydrogen atom) occupy the lower-energy MO, they will have lower energy than they have in the two separate hydrogen atoms. This situation favors molecule formation, because nature tends to seek the lowest energy state. That is, the driving force for molecule formation here is that the molecular orbital available to the two electrons has lower energy than the atomic orbitals these electrons occupy in the separated atoms. This situation is "*pro*"-*bonding.*

   On the other hand, if the two electrons were forced to occupy the higher-energy MO, they would be definitely "*anti*"-*bonding.* In this case these electrons would have lower energy in the separated atoms than in the molecule; thus the separated state would be favored. Of course, since the lower-energy MO_1 *is* available, the two electrons occupy that MO and the resulting molecule is stable.

   We have seen that the molecular orbitals of the hydrogen molecule fall into two classes: bonding and antibonding. A **bonding molecular orbital** is *lower in energy than the atomic orbitals of which it is composed.* Electrons in this type of orbital favor the molecule; that is, they will favor bonding. An **antibonding molecular orbital** is *higher in energy than the atomic orbitals of which it is composed.* Electrons in this type of orbital will favor the separated atoms (they are antibonding). Figure 14.27 illustrates these ideas.
4. Figure 14.26 shows that for the bonding molecular orbital in the H_2 molecule, the electrons have the greatest probability of being between the

nuclei. This is exactly what we would expect, since the electrons can lower their energies by being simultaneously attracted by both nuclei. On the other hand, the electron distribution for the antibonding molecular orbital is such that the electrons are mainly outside the space between the nuclei. This type of distribution is not expected to provide any bonding force. In fact, it causes the electrons to be higher in energy than in the separated atoms. Thus the molecular orbital model produces electron distributions and energies that agree with our basic ideas of bonding. This fact reassures us that the model is physically reasonable.

5. The labels on molecular orbitals indicate their symmetries (shapes), their parent atomic orbitals, and whether they are bonding or antibonding. Antibonding character is indicated by an asterisk. For the H_2 molecule both MOs have σ symmetry and both are constructed from hydrogen 1*s* atomic orbitals. The molecular orbitals for H_2 are therefore labeled as follows:

$$MO_1 = \sigma_{1s}$$
$$MO_2 = \sigma_{1s}^*$$

6. Molecular electron configurations can be written in much the same way as atomic configurations. Since the H_2 molecule has two electrons in the σ_{1s} molecular orbital, the electron configuration is σ_{1s}^2.
7. Each molecular orbital can hold two electrons, but the spins must be opposite.
8. Orbitals are conserved. The number of molecular orbitals is always the same as the number of atomic orbitals used to construct them.

Many of the above points are summarized in Fig. 14.28.

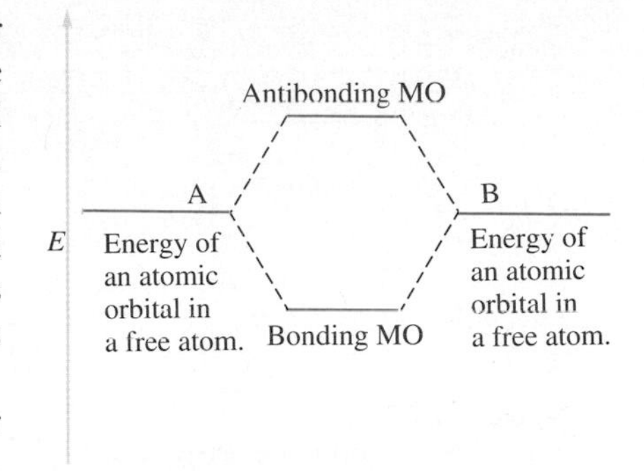

Figure 14.27
Bonding and antibonding molecular orbitals (MOs).

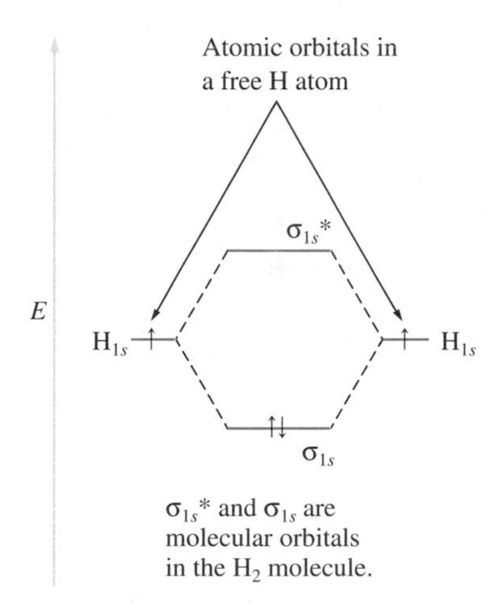

Figure 14.28
A molecular orbital energy-level diagram for the H_2 molecule.

Bond Order

Fundamentally, a molecule forms because it has lower energy than the separated atoms. In the simple molecular orbital model this is reflected by the number of bonding electrons (those that achieve lower energy in going from the free atoms to the molecule) versus the number of antibonding electrons (those that are higher in energy in the molecule than in the free atoms). If the number of bonding electrons is greater than the number of antibonding electrons in a given molecule, the molecule is predicted to be stable.

The quantitative indicator of molecular stability (bond strength) for a diatomic molecule is the **bond order:** *the difference between the number of bonding electrons and the number of antibonding electrons, divided by 2.*

$$\text{Bond order} = \frac{\text{number of bonding electrons} - \text{number of antibonding electrons}}{2}$$

We divide by 2, because we are used to thinking of bonds in terms of *pairs* of electrons.

Bond order is an indication of bond strength, because it reflects the difference between the number of bonding electrons and the number of antibonding electrons, which in turn reflects the quantity of energy released when the

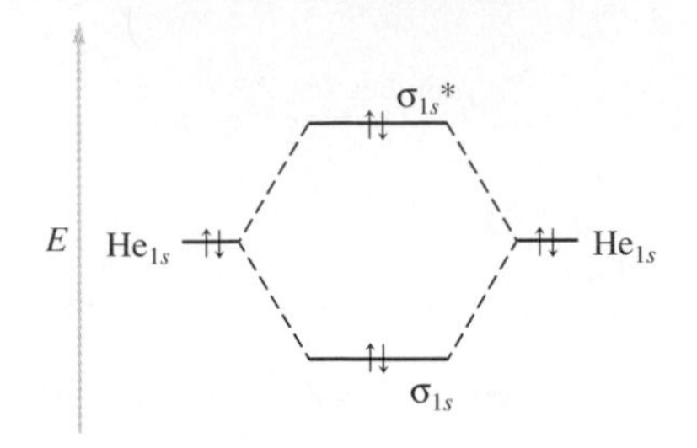

Figure 14.29
The molecular orbital energy-level diagram for the He_2 molecule.

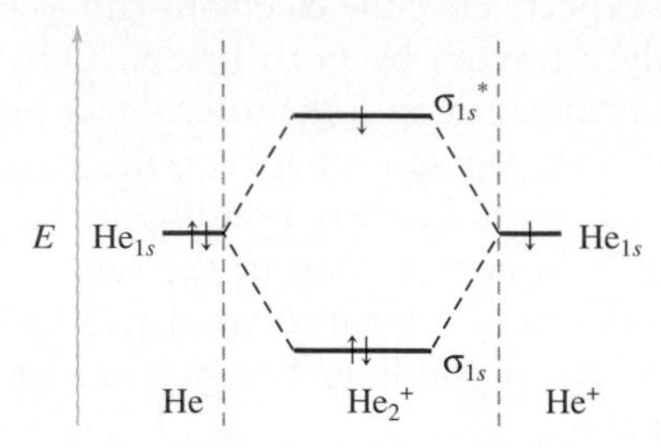

Figure 14.30
The molecular orbital energy-level diagram for the He_2^+ ion. Note that in this case the components from which He_2^+ is "constructed" are He and He^+.

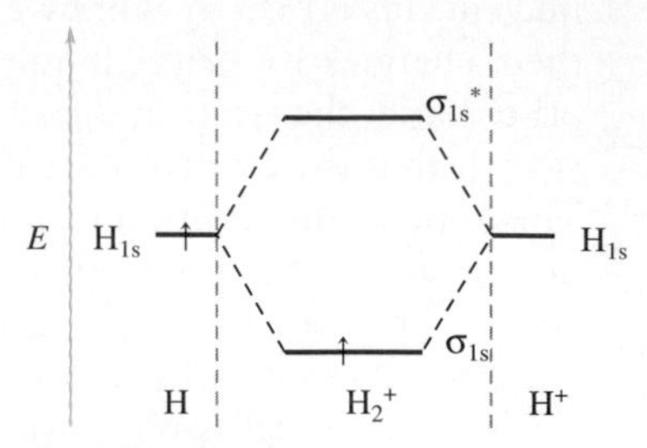

Figure 14.31
The molecular orbital energy-level diagram for the H_2^+ ion.

molecule is formed from its atoms. *Therefore, larger bond order indicates greater bond strength.*

Since the H_2 molecule has two bonding electrons and no antibonding electrons, the bond order is one:

$$\text{Bond order} = \frac{2 - 0}{2} = 1$$

We will now apply the molecular orbital model to the helium molecule (He_2). Does this model predict that this molecule is stable? Since the He atom has a $1s^2$ configuration, $1s$ orbitals are used to construct the molecular orbitals. Therefore the molecule will have four electrons. From the diagram shown in Fig. 14.29, it is apparent that two electrons are raised in energy and two are lowered in energy. Thus the bond order is zero:

$$\frac{2 - 2}{2} = 0$$

This implies that the He_2 molecule is *not* stable with respect to the two free He atoms, which agrees with the observation that helium gas consists of individual He atoms.

Next, we will apply the simple MO model to diatomic ions, starting with He_2^+. Is this ion expected to be stable? Examination of Fig. 14.30 shows that when He is combined with He^+ to form He_2^+, two electrons are lowered in energy and only one is raised in energy. That is, the bond order is $(2 - 1)/2 = \frac{1}{2}$, and He_2^+ is predicted to be stable. Experiments have shown that He_2^+ does indeed exist with a bond energy of 250 kJ/mol.

Likewise, the H_2^+ ion is known to exist with a bond energy of 255 kJ/mol. This result is in agreement with the simple MO model (see Fig. 14.31), which predicts a bond order of $(1 - 0)/2 = \frac{1}{2}$.

Figure 14.32 shows the MO diagram for the H_2^- ion, which is predicted to have a bond order of $(2 - 1)/2 = \frac{1}{2}$. However, in this case the prediction of the model does not agree with experimental results. The H_2^- ion is not known, and all indications are that it is unstable, immediately decomposing to H_2 and a free electron. Thus, again, we are reminded about the perils of simple models. The reasons for the instability of the H_2^- ion are beyond the scope of this treatment.

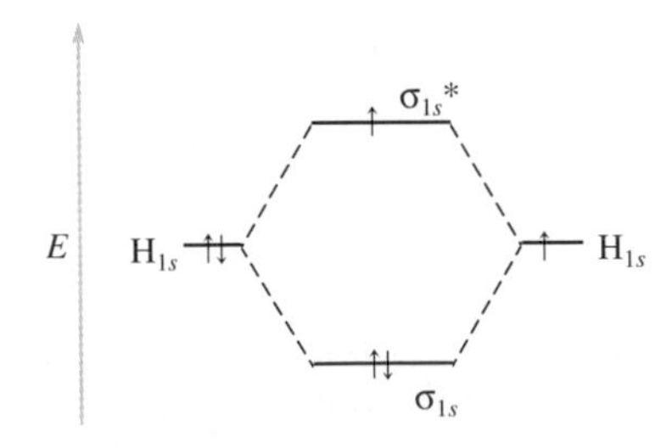

Figure 14.32
The molecular orbital energy-level diagram for the H_2^- ion.

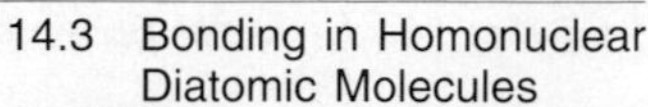

14.3 Bonding in Homonuclear Diatomic Molecules

In this section we consider *homonuclear diatomic molecules* (those composed of two identical atoms) formed by elements in Period 2 of the periodic table. The lithium atom has a $1s^2 2s^1$ electron configuration, and from our discussion in the previous section, it would seem logical to use the Li 1*s* and 2*s* orbitals to form the molecular orbitals of the Li_2 molecule. However, the 1*s* orbitals on the lithium atoms are much smaller than the 2*s* orbitals and therefore do not overlap in space to any appreciable extent (see Fig. 14.33). Thus the two electrons in each 1*s* orbital can be assumed to be localized on a given atom, which means that they do not participate in the bonding. The following general principle applies: *in order to participate in molecular orbitals, atomic orbitals must overlap in space*. This means that only the valence orbitals of atoms contribute significantly to the molecular orbitals of a particular molecule.

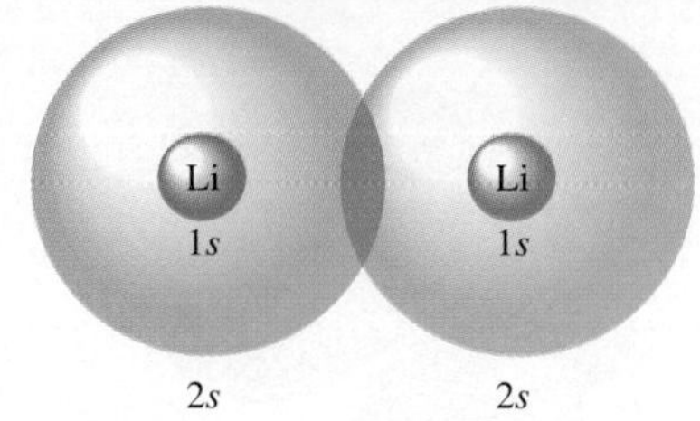

Figure 14.33
The relative sizes of the lithium 1*s* and 2*s* atomic orbitals.

Only valence atomic orbitals contribute significantly to MOs.

The molecular orbital diagram of the Li_2 molecule and the shapes of its bonding and antibonding MOs are shown in Fig. 14.34. The electron configuration for Li_2 (valence electrons only) is $\sigma_{2s}{}^2$, and the bond order is one:

$$\frac{2 - 0}{2} = 1$$

Thus Li_2 is expected to be a stable molecule (has lower energy than two separated lithium atoms). However, this does not mean that Li_2 is the most stable form of elemental lithium. In fact, at normal temperature and pressure, lithium exists as a solid containing many lithium atoms bound together.

For the beryllium molecule (Be_2) the bonding and antibonding orbitals both contain two electrons. In this case the bond order is $(2 - 2)/2 = 0$. Thus the model predicts that since Be_2 is not more stable than two separated Be atoms, no molecule should form. However, experiments indicate that $Be_2(g)$ does exist, although it has a *very* weak bond (bond energy $\approx$ 10 kJ/mol). In contrast, beryllium metal contains many beryllium atoms bonded to each other and is stable for reasons we will discuss in Chapter 16.

Since the boron atom has a $1s^2 2s^2 2p^1$ configuration, we describe the B_2 molecule by considering how *p* atomic orbitals combine to form molecular orbitals. Recall that *p* orbitals have two lobes and that they occur in sets of three mutually perpendicular orbitals [Fig. 14.35(a)]. When two B atoms approach each other, two pairs of *p* orbitals can overlap in a parallel fashion [Figs. 14.35(b) and (c)] and one pair can overlap head-on [Fig. 14.35(d)].

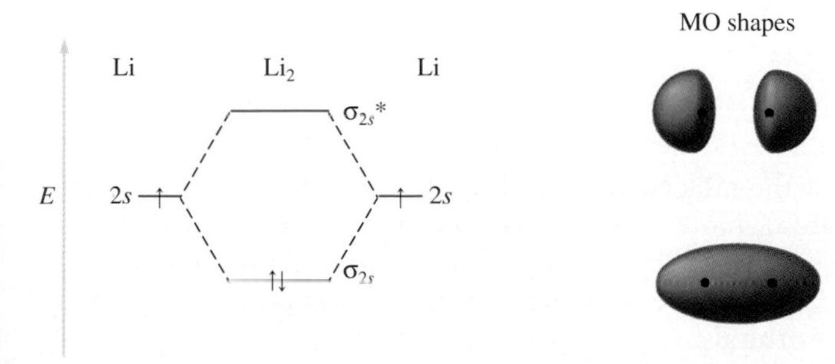

Figure 14.34
The molecular orbital energy-level diagram for the Li_2 molecule.

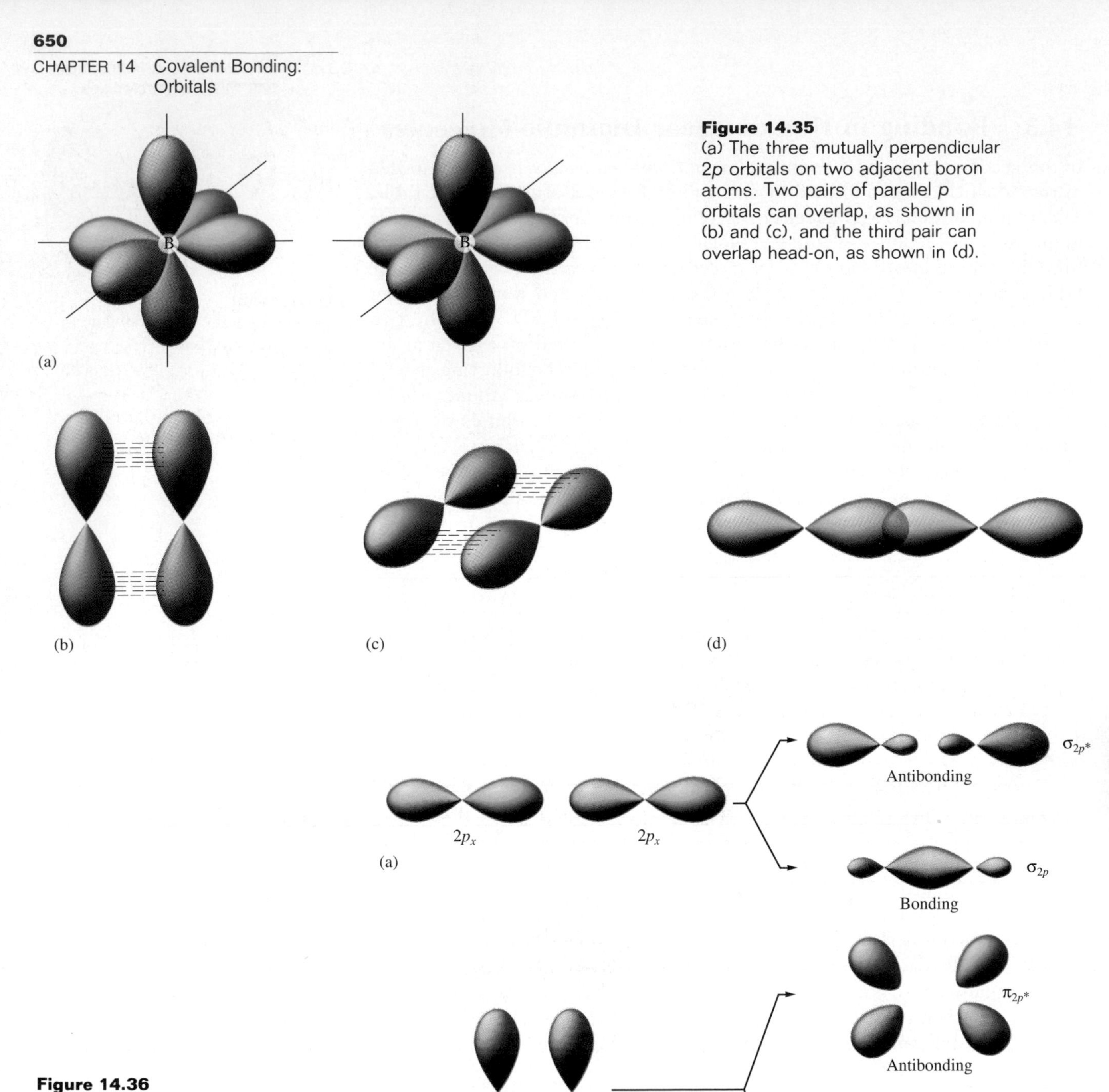

Figure 14.35
(a) The three mutually perpendicular 2*p* orbitals on two adjacent boron atoms. Two pairs of parallel *p* orbitals can overlap, as shown in (b) and (c), and the third pair can overlap head-on, as shown in (d).

Figure 14.36
(a) The two *p* orbitals on the boron atom that overlap head-on produce two σ molecular orbitals, one bonding and one antibonding.
(b) Two *p* orbitals that lie parallel overlap to produce two π molecular orbitals, one bonding and one antibonding.

First, let's consider the molecular orbitals formed by the head-on overlap, as shown in Fig. 14.36(a). Note that the electrons in the bonding MO are, as expected, concentrated between the nuclei, and the electrons in the antibonding MO are concentrated outside the area between the two nuclei. Both of these MOs are σ molecular orbitals.

The p orbitals that overlap in a parallel fashion also produce bonding and antibonding orbitals [Fig. 14.36(b)]. Since the electron probability lies above and below the line between the nuclei, both the orbitals are **pi (π) molecular orbitals.** They are designated as π_{2p} for the bonding MO and $\pi_{2p}{}^*$ for the antibonding MO.

Let's try to make an educated guess about the relative energies of the σ and π molecular orbitals formed from the $2p$ atomic orbitals. Would we expect the electrons to prefer the σ bonding orbital (where the electron probability is concentrated in the area between the nuclei) or the π bonding orbital? The σ orbital would seem to have the lower energy since the electrons are closest to the two nuclei. This agrees with the observation that σ interactions are typically stronger than π interactions.

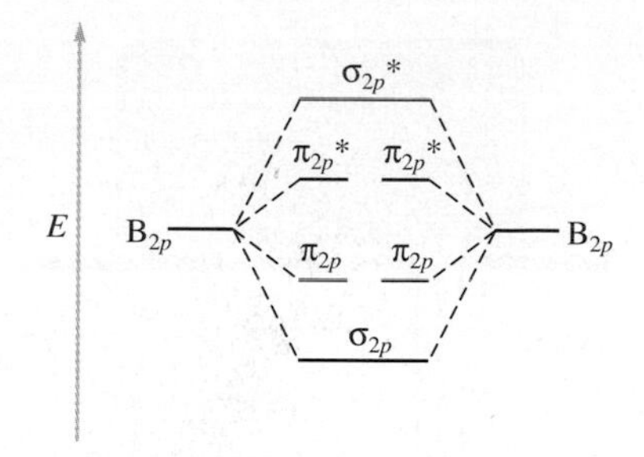

Figure 14.37
The *expected* molecular orbital energy-level diagram for the combination of the $2p$ orbitals on two boron atoms.

Figure 14.37 gives the molecular orbital energy-level diagram *expected* when the two sets of $2p$ orbitals on the two boron atoms combine to form molecular orbitals. Note that there are two π bonding orbitals at the same energy (degenerate orbitals) formed from the two pairs of parallel p orbitals, and there are two degenerate π antibonding orbitals. The energy of the π_{2p} orbitals is expected to be higher than that of the σ_{2p} orbital, because σ interactions are generally stronger than π interactions.

To construct the total molecular orbital diagram for the B_2 molecule, we make the assumption that the $2s$ and $2p$ orbitals combine separately (in other words, there is no $2s$-$2p$ mixing). The resulting diagram is shown in Fig. 14.38. Note that B_2 has six *valence* electrons. (Remember that the $1s$ orbitals and electrons are assumed not to participate in the bonding.) This diagram predicts a bond order of one:

$$\frac{4 - 2}{2} = 1$$

Therefore, B_2 should be a stable molecule.

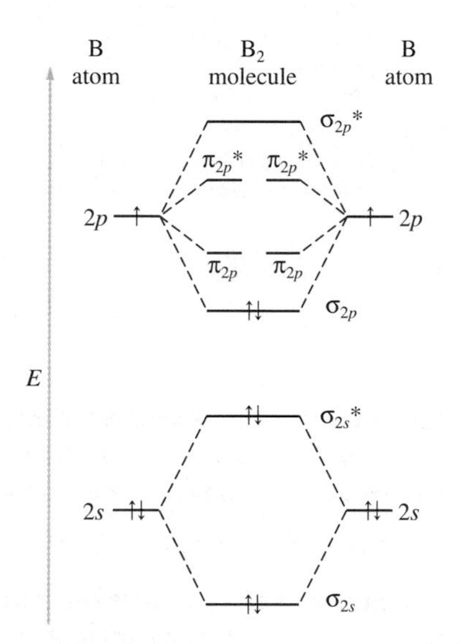

Figure 14.38
The *expected* molecular orbital energy-level diagram for the B_2 molecule.

Paramagnetism

At this point we need to discuss an additional molecular property—magnetism. Most materials have no magnetism until they are placed in a magnetic field. However, in the presence of such a field, magnetism of two types can be induced. **Paramagnetism** causes the substance to be attracted toward the inducing magnetic field. **Diamagnetism** causes the substance to be repelled from the inducing magnetic field. Figure 14.39 illustrates how paramagnetism is measured. The sample is weighed with the electromagnet turned off and then weighed again with the electromagnet turned on. An increase in weight when the field is turned on indicates the sample is paramagnetic. Studies have shown that *paramagnetism is associated with unpaired electrons,** and *diamagnetism is associated with paired electrons.* Any substance that has both paired and unpaired electrons will exhibit net paramagnetism, since the effect of paramagnetism is much stronger than that of diamagnetism. Thus the phenomenon of paramagnetism provides a ready means for testing whether or not a substance contains unpaired electrons.

*This is an oversimplification but one that will not concern us here.

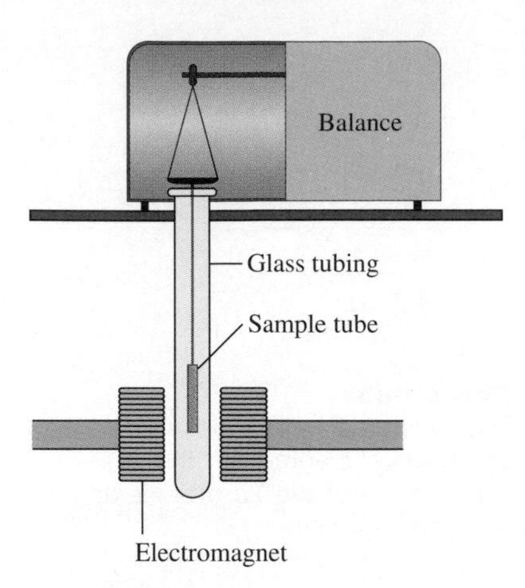

Figure 14.39
An apparatus used to measure the paramagnetism of a sample. A paramagnetic sample will appear heavier when the electromagnet is turned on because the sample is attracted into the inducing magnetic field.

The molecular orbital energy-level diagram represented in Fig. 14.38 predicts that the B_2 molecule will be diamagnetic, since the MOs contain only paired electrons. However, experiments show that B_2 is actually paramagnetic with two unpaired electrons. Why does the model yield the wrong prediction? This is yet another illustration of how models are developed and used. In general, we try to use the simplest possible model that accounts for all of the important observations. In this case, although the simplest model has been relatively successful in describing all of the diatomic molecules up to B_2, it certainly is suspect if it cannot describe the B_2 molecule correctly.

Let's reconsider one of our previous assumptions. In our treatment of B_2 we have assumed that the *s* and *p* orbitals combine separately to form molecular orbitals. Calculations show that when the *s* and *p* orbitals are allowed to mix (participate) in the same molecular orbital, a different energy-level diagram results for B_2 (see Fig. 14.40). Note that even though the *s* and *p* contributions to the MOs are no longer separate, we retain the simple orbital designations. The mixing of *p* and *s* atomic orbitals occurs only in the σ molecular orbitals (σ_{2s}, σ_{2s}^*, σ_{2p}, and σ_{2p}^*). Because the energy of the σ_{2p} orbital is changed by *p-s* mixing, the energies of π_{2p} and σ_{2p} orbitals are reversed. Also, the *p-s* mixing changes the energies of the σ_{2s} and σ_{2s}^* such that they are no longer equally spaced relative to the energy of the free 2*s* orbital.

When the six valence electrons of the B_2 molecule are placed in the modified energy-level diagram, each of the last two electrons goes into one of the degenerate π_{2p} orbitals. This produces a paramagnetic molecule in agreement with experimental results. Thus when the model is extended to allow *p-s* mixing in molecular orbitals, it predicts the correct magnetism. Note that the bond order is $(4 - 2)/2 = 1$, as before.

The remaining homonuclear diatomic molecules of the Period 2 elements can be described using the same molecular orbitals as for B_2 (see Fig. 14.40) and inserting the correct number of electrons. These are summarized in Fig. 14.41, together with experimentally obtained bond strengths and lengths. Several significant points arise from these results.

Bond energy increases with bond order; bond length decreases with increasing bond order.

1. There are definite correlations between bond order, bond energy, and bond length. As the bond order predicted by the molecular orbital model increases, the bond energy increases and the bond length decreases. This is a clear indication that the bond order predicted by the model accurately reflects bond strength, and it strongly supports the reasonableness of the MO model.
2. Comparison of the bond energies of the B_2 and F_2 molecules indicates that bond order cannot automatically be associated with a particular bond energy. Although both molecules have a bond order of 1, the bond in B_2 appears to be about twice as strong as the bond in F_2. As we will see in our later discussion of the halogens, F_2 has an unusually weak single bond due to larger-than-usual electron-electron repulsions (there are 14 valence electrons on the small F_2 molecule).
3. Note the very large bond energy associated with the N_2 molecule, which the molecular orbital model predicts will have a bond order of 3, a triple bond. The very strong bond in N_2 is the principal reason that many nitrogen-containing compounds are used as explosives. The reactions involving these explosives give the very stable N_2 molecule as a product, thus releasing large quantities of energy.

4. The O_2 molecule is known to be paramagnetic. This can be very convincingly demonstrated by pouring liquid oxygen between the poles of a strong magnet, as shown in Fig. 14.42. Because of its paramagnetism, the oxygen is attracted to the magnet gap, where it remains until it evaporates. Significantly, the molecular orbital model correctly predicts oxygen's paramagnetism, but the simplest form of the localized electron model predicts a diamagnetic molecule.

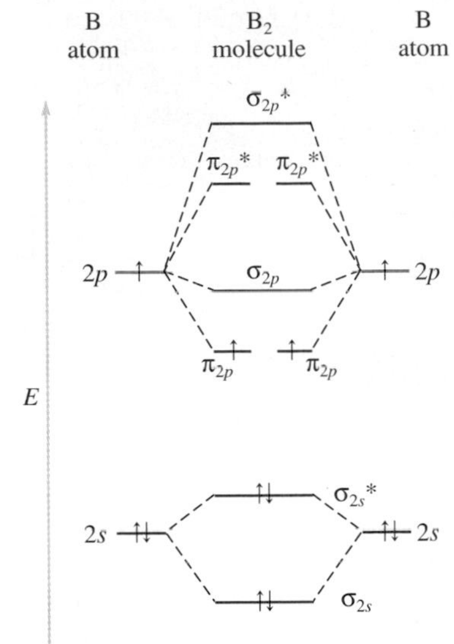

Figure 14.40
The *correct* molecular orbital energy-level diagram for the B_2 molecule. When *p-s* mixing is allowed, the energies of the σ_{2p} and π_{2p} orbitals are reversed. The two electrons from the B 2*p* orbitals now occupy separate, degenerate π_{2p} molecular orbitals and thus have parallel spins. Therefore, this diagram explains the observed paramagnetism of B_2.

EXAMPLE 14.6

For the species O_2, O_2^+, and O_2^-, give the electron configuration and the bond order for each. Which has the strongest bond?

Solution

The O_2 molecule has 12 valence electrons (6 + 6); O_2^+ has 11 valence electrons (6 + 6 − 1); and O_2^- has 13 valence electrons (6 + 6 + 1).

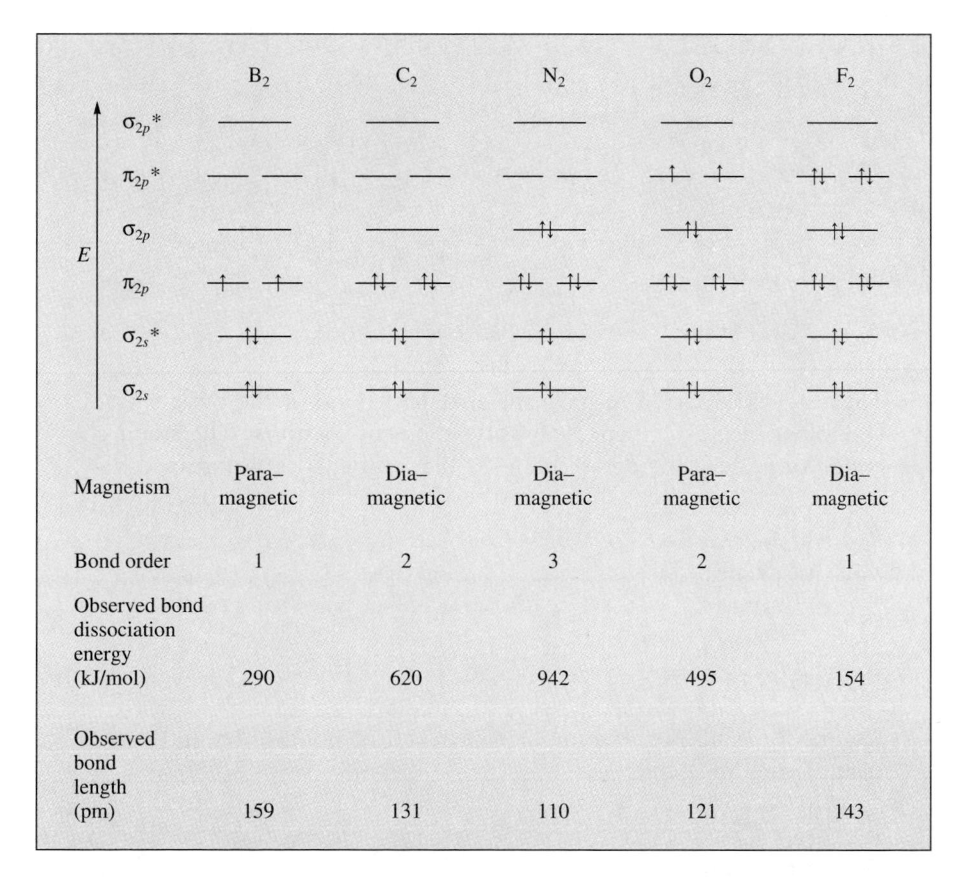

B_2 C_2 N_2 O_2 F_2
σ_{2p}^*
π_{2p}^*
σ_{2p}
E
π_{2p}
σ_{2s}^*
σ_{2s}
Magnetism: Para–magnetic, Dia–magnetic, Dia–magnetic, Para–magnetic, Dia–magnetic
Bond order: 1, 2, 3, 2, 1
Observed bond dissociation energy (kJ/mol): 290, 620, 942, 495, 154
Observed bond length (pm): 159, 131, 110, 121, 143

Figure 14.41
The molecular orbital energy-level diagrams, bond orders, bond energies, and bond lengths for the diatomic molecules B_2 through F_2. There is evidence that for O_2 and F_2 the σ_{2p} orbital is lower in energy than the π_{2p} orbital. However, since this does not affect the predicted magnetism or bond order for these molecules, we will use the same order of MOs for all of the diatomic molecules of the period 2 elements.

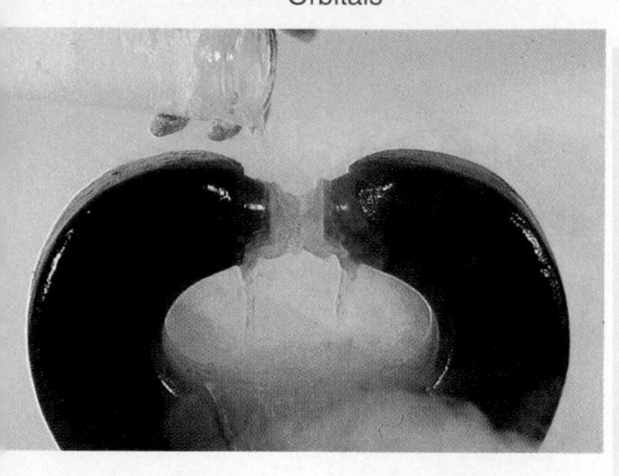

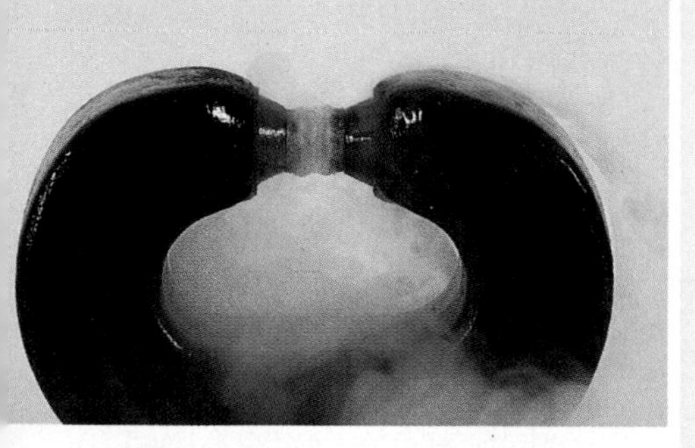

Figure 14.42
When liquid oxygen is poured into the space between the poles of a strong magnet, it remains there until it boils away. This attraction of liquid oxygen for the magnetic field demonstrates the paramagnetism of the O_2 molecule.

We will assume that the ions can be treated using the same molecular orbital diagram used for the neutral diatomic molecule:

	O_2	O_2^+	O_2^-
σ_{2p}^*	—	—	—
π_{2p}^*	↑ ↑	↑ —	↑↓ ↑
σ_{2p}	↑↓	↑↓	↑↓
π_{2p}	↑↓ ↑↓	↑↓ ↑↓	↑↓ ↑↓
σ_{2s}^*	↑↓	↑↓	↑↓
σ_{2s}	↑↓	↑↓	↑↓

The electron configuration for each species can then be taken from the diagram:

$$O_2: \quad (\sigma_{2s})^2(\sigma_{2s}{}^*)^2(\pi_{2p})^4(\sigma_{2p})^2(\pi_{2p}{}^*)^2$$
$$O_2{}^+: \quad (\sigma_{2s})^2(\sigma_{2s}{}^*)^2(\pi_{2p})^4(\sigma_{2p})^2(\pi_{2p}{}^*)^1$$
$$O_2{}^-: \quad (\sigma_{2s})^2(\sigma_{2s}{}^*)^2(\pi_{2p})^4(\sigma_{2p})^2(\pi_{2p}{}^*)^3$$

The bond orders are as follows:

$$\text{For } O_2: \quad \frac{8-4}{2} = 2$$

$$\text{For } O_2{}^+: \quad \frac{8-3}{2} = 2.5$$

$$\text{For } O_2{}^-: \quad \frac{8-5}{2} = 1.5$$

Thus $O_2{}^+$ is expected to have the strongest bond of the three species.

Experimental evidence supports these predictions. The bond energies for $O_2{}^+$, O_2, and $O_2{}^-$ are 643, 495, and 395 kJ/mol, respectively.

EXAMPLE 14.7

Use the molecular orbital model to predict the bond order and magnetism of each molecule.

a. Ne_2 **b.** P_2

Solution

a. The valence orbitals for Ne are $2s$ and $2p$. Thus we can use the molecular orbitals we have already constructed for the diatomic molecules of the Period 2 elements. The Ne_2 molecule has 16 valence electrons (8 from each atom). Placing these electrons in the appropriate molecular orbitals produces the following diagram:

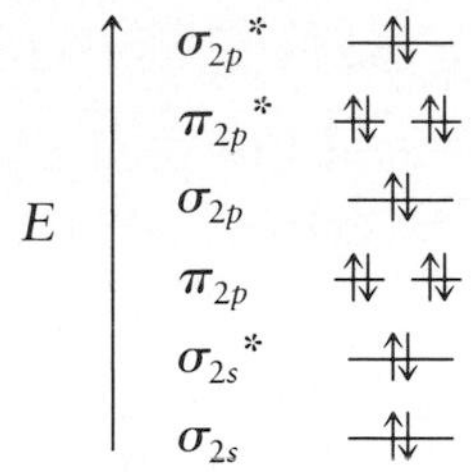

E ↑	
σ_{2p}^*	⇅
π_{2p}^*	⇅ ⇅
σ_{2p}	⇅
π_{2p}	⇅ ⇅
σ_{2s}^*	⇅
σ_{2s}	⇅

The bond order is $(8 - 8)/2 = 0$, so Ne_2 should not exist.

b. The P_2 molecule contains phosphorus atoms from the third row of the periodic table. We will assume that diatomic molecules of the Period 3 elements can be treated very similarly to those from Period 2. The only change will be that the molecular orbitals are formed from 3*s* and 3*p* atomic orbitals. The P_2 molecule has 10 valence electrons (5 from each phosphorus atom). The resulting molecular orbital diagram follows.

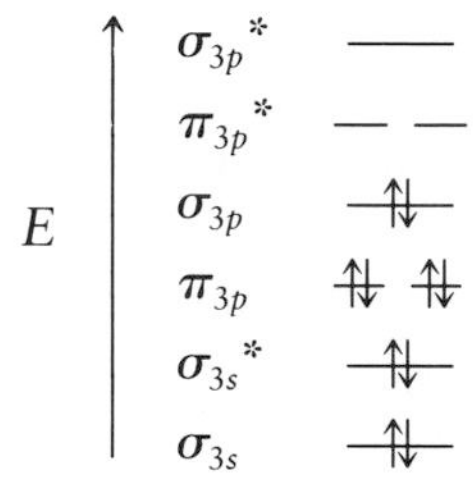

E ↑	
σ_{3p}^*	—
π_{3p}^*	— —
σ_{3p}	⇅
π_{3p}	⇅ ⇅
σ_{3s}^*	⇅
σ_{3s}	⇅

The molecule has a bond order of 3 and is expected to be diamagnetic.

It should be noted that phosphorus exists in nature as P_4 molecules (and as other, more complex, forms).

14.4 Bonding in Heteronuclear Diatomic Molecules

In this section we will deal with selected examples of **heteronuclear diatomic molecules**—those containing two different atoms. A special case involves molecules containing atoms adjacent to each other in the periodic table. Since the atoms involved in such a molecule are so similar, we can use the molecular orbital diagram for homonuclear molecules. For example, we can predict the bond order and magnetism of nitric oxide (NO) by placing its 11 valence electrons (5 from nitrogen and 6 from oxygen) in the molecular orbital energy-level diagram shown in Fig. 14.43. The molecule should be paramagnetic with a bond order of

$$\frac{8 - 3}{2} = 2.5$$

Experimentally, nitric oxide is indeed found to be paramagnetic. Note that this odd-electron molecule is described very naturally by the MO model. In contrast, the localized electron model, in the simple form used in this text, does not readily describe such molecules.

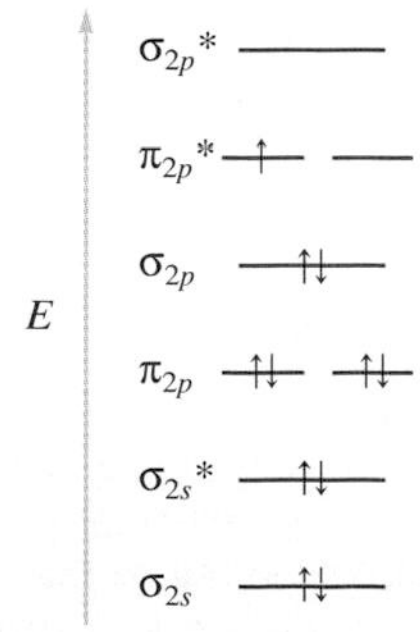

E ↑	
σ_{2p}^*	—
π_{2p}^*	↑ —
σ_{2p}	⇅
π_{2p}	⇅ ⇅
σ_{2s}^*	⇅
σ_{2s}	⇅

Figure 14.43
The molecular orbital energy-level diagram for the NO molecule. The bond order is 2.5.

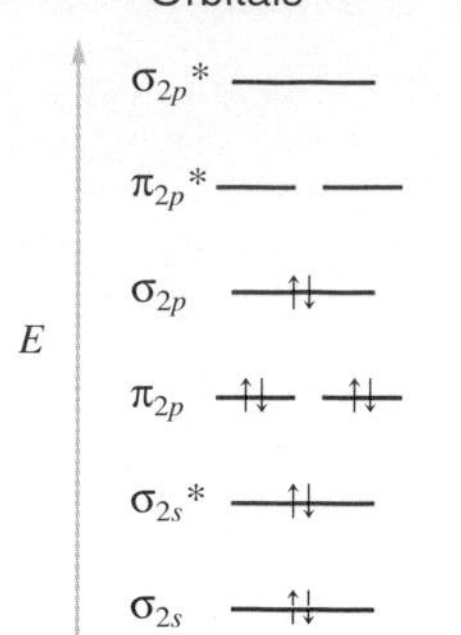

Figure 14.44
The molecular orbital energy-level diagram for both the NO^+ and CN^- ions.

EXAMPLE 14.8

Use the molecular orbital model to predict the magnetism and bond order of the NO^+ and CN^- ions.

Solution

The NO^+ ion has 10 valence electrons ($5 + 6 - 1$). The CN^- ion also has 10 valence electrons ($4 + 5 + 1$). Both ions are therefore diamagnetic and have a bond order of three:

$$\frac{8 - 2}{2} = 3$$

The molecular orbital diagram for these two ions is the same (see Fig. 14.44).

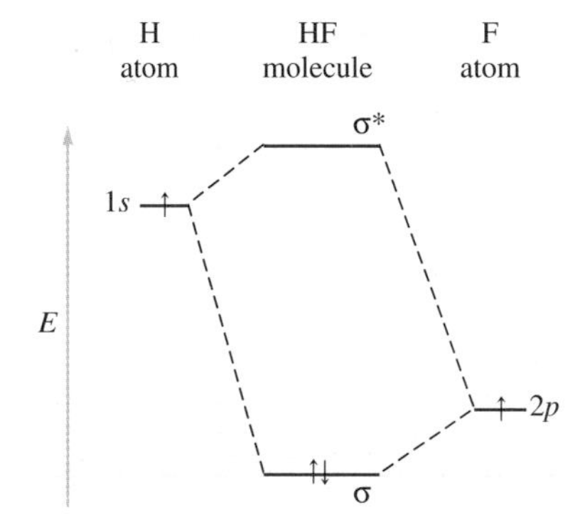

Figure 14.45
A partial molecular orbital energy-level diagram for the HF molecule.

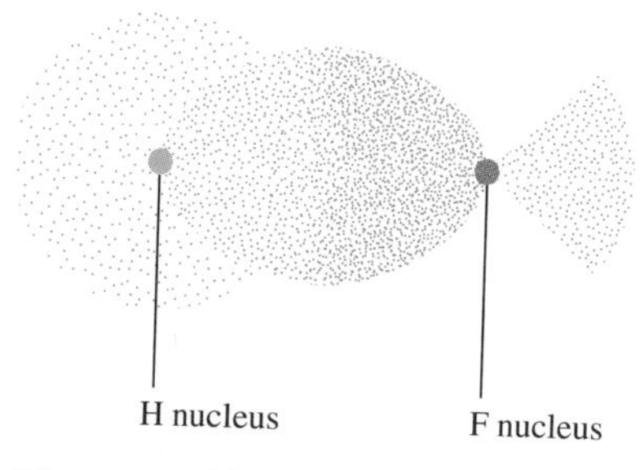

Figure 14.46
The electron probability distribution in the bonding molecular orbital of the HF molecule. Note the greater electron density close to the fluorine atom.

When the two atoms of a diatomic molecule are very different, the energy-level diagram for homonuclear molecules can no longer be used. A new diagram must be devised for each molecule. We will illustrate such a case by considering the hydrogen fluoride (HF) molecule. The electron configurations of the hydrogen and fluorine atoms are $1s^1$ and $1s^2 2s^2 2p^5$, respectively. To keep things as simple as possible, we will assume that fluorine uses only one of its $2p$ orbitals to bond to hydrogen. Thus the molecular orbitals for HF will be composed of fluorine $2p$ and hydrogen $1s$ orbitals. Figure 14.45 gives the partial molecular orbital energy-level diagram for HF, focusing only on the orbitals involved in the bonding. We are assuming that fluorine's other valence electrons remain localized on the fluorine atom. The $2p$ orbital of fluorine is shown at a lower energy than the $1s$ orbital of hydrogen on the diagram because fluorine binds its valence electrons more tightly. Thus the $2p$ electron on a free fluorine atom is at lower energy than the $1s$ electron on a free hydrogen atom. The diagram predicts that the HF molecule should be stable since both electrons are lowered in energy relative to their energies in the free hydrogen and fluorine atoms, and this is the driving force for bond formation.

Because the fluorine $2p$ orbital is lower in energy than the hydrogen $1s$ orbital, the electrons prefer to be closer to the fluorine atom. That is, the σ molecular orbital containing the bonding electron pair shows greater electron probability close to the fluorine (see Fig. 14.46). The electron pair is not shared equally. This causes the fluorine atom to have a slight excess of negative charge and leaves the hydrogen atom partially positive. This is exactly the bond polarity observed for HF. Thus the molecular orbital model accounts in a straightforward way for the different electronegativities of hydrogen and fluorine and the resulting unequal charge distribution.

14.5 Combining the Localized Electron and Molecular Orbital Models

In this text we have treated bonding in terms of simple models not only because we are presenting the material at an introductory level, but also because a lot of first-order thinking by chemists employs these simple, "back-of-the-envelope"

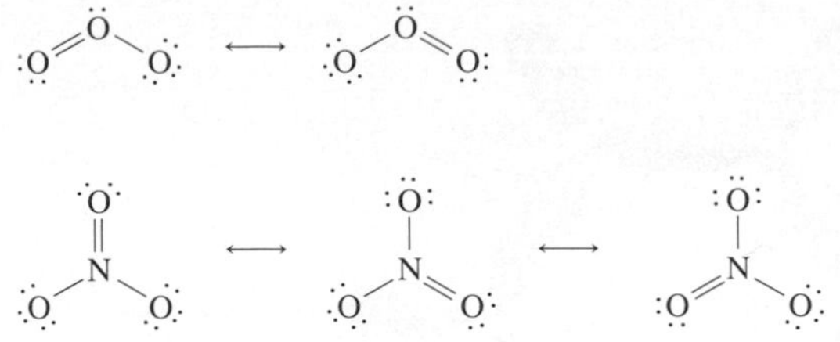

Figure 14.47
The resonance structures for O_3 and NO_3^-. Note that it is the double bond that occupies various positions in the resonance structures.

models. Thus as long as we are aware of the pitfalls of oversimplified models, we can use them to our benefit.

In this section we will attempt to make these simple models even more useful by addressing a particular shortcoming of the localized electron model—its assumption that electrons are localized (restricted to the space between a given pair of atoms). This problem is most apparent for molecules where several valid Lewis structures can be drawn. Recall that none of the resonance structures taken alone adequately describes the electronic structure of the molecule. The concept of resonance was invented to solve this problem. However, even with resonance included, the localized electron model does not describe molecules and ions such as O_3 and NO_3^- in a very satisfying way.

It would seem that the ideal bonding model would be one with the simplicity of the localized electron model but with the delocalization characteristics of the molecular orbital model. We can achieve this by combining the two models to describe molecules that require resonance. Note that for species such as O_3 and NO_3^- the double bond changes position in the resonance structures (see Fig. 14.47). Since a double bond involves one σ and one π bond, there is a σ bond between all bound atoms in each resonance structure. It is really the π bond that has different locations in the various resonance structures.

In molecules that require resonance, it is the π bonding that is most clearly delocalized.

We conclude that the σ electrons in a molecule can be described as being localized with no apparent problems. It is the π electrons that must be treated as being delocalized. Thus for molecules that require resonance, we will use the localized electron model to describe the σ bonding and the molecular orbital model to describe the π bonding. This allows us to keep the bonding model as simple as possible and yet give a more physically accurate description of such molecules.

We will illustrate this procedure by considering the bonding in benzene, an important industrial chemical that must be handled carefully because it is a known carcinogen. The benzene molecule (C_6H_6) consists of a planar hexagon of carbon atoms with one hydrogen atom bound to each carbon atom, as shown in Fig. 14.48(a). In the molecule all six C—C bonds are known to be equivalent. To explain this fact, the localized electron model must invoke resonance [see Fig. 14.48(b)].

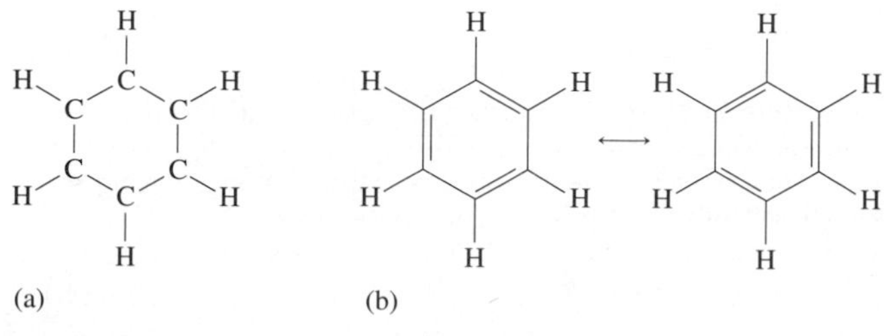

(a)

(b)

Figure 14.48
(a) The benzene molecule consists of a ring of six carbon atoms with one hydrogen atom bound to each carbon; all atoms are in the same plane. All of the C—C bonds are known to be equivalent. (b) Two of the resonance structures for the benzene molecule. The localized electron model must invoke resonance to account for the six equal C—C bonds.

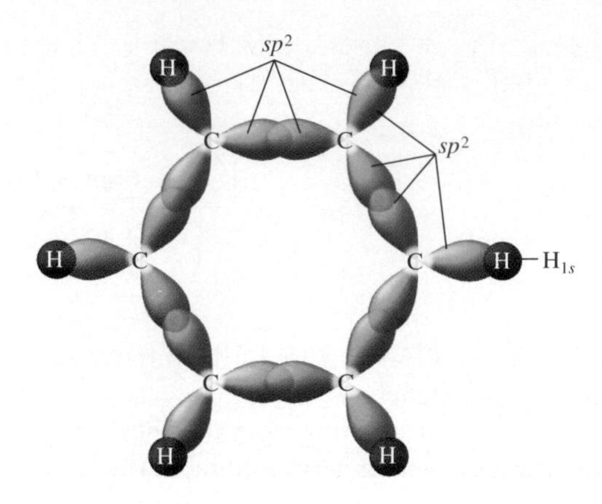

Figure 14.49
The σ bonding system in the benzene molecule.

A better description of the bonding in benzene results when we use a combination of the models, as described above. In this description it is assumed that the σ bonds to carbon involve sp^2 orbitals, as shown in Fig. 14.49. These σ bonds are all centered in the plane of the molecule.

Since each carbon atom is sp^2 hybridized, a p orbital perpendicular to the plane of the ring remains on each carbon atom. These six p orbitals can be used to form π molecular orbitals, as shown in Fig. 14.50(a). The electrons in the resulting π molecular orbitals are delocalized above and below the plane of the ring, as shown in Fig. 14.50(b). This gives six equivalent C—C bonds, as required by the known structure of the benzene molecule. The benzene structure is often written as

to indicate the **delocalized π bonding** in the molecule.

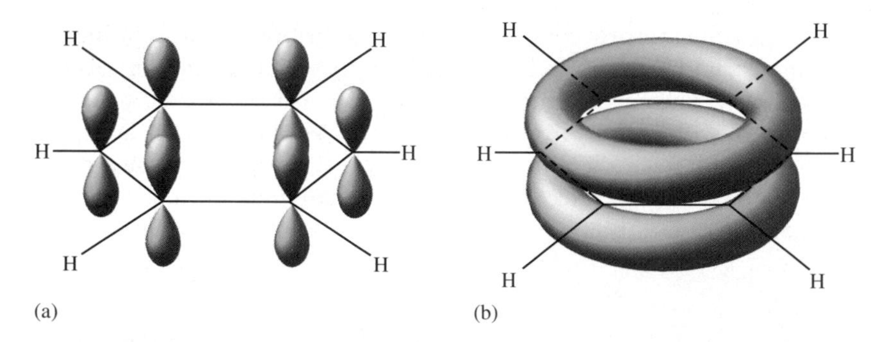

Figure 14.50
(a) The π molecular orbital system in benzene is formed by combining the six p orbitals from the six sp^2 hybridized carbon atoms. (b) The electrons in the resulting π molecular orbitals are delocalized over the entire ring of carbon atoms, giving six equivalent bonds. A composite of these orbitals is represented here.

Very similar treatments can be applied to other planar molecules for which resonance is required by the localized electron model. For example, the NO_3^- ion can be described using the π molecular orbital system shown in Fig. 14.51. In this molecule each atom is assumed to be sp^2 hybridized. This leaves one p orbital on each atom perpendicular to the plane of the ion. These p orbitals can combine to form the π molecular orbital system.

14.6 Orbitals: Human Inventions

In the treatment of the bonding and structure in molecules we have stressed that the descriptions we have given are models—and simple ones, at that. These models are extraordinarily valuable because they provide pictures in our minds of what molecules might look like and how they might behave. This helps us decide what questions to ask (what experiments to do) as we try to understand more completely the complicated behavior of matter. However, as valuable as these pictures are, we must remember that they greatly oversimplify a very complex situation. We must realize that these models can give us a rather superficial view that is often highly misleading.

For example, after all of the talk about orbitals in the last several chapters, you probably have the distinct impression that orbitals actually exist. The truth is, they do not—at least not in the physical sense. Orbitals are mathematical functions—solutions to a modified Schrödinger equation (the changes made to allow separation into independent electron equations).

Thus orbitals are mathematical tools we use to describe atoms and molecules. However, they have no physical reality in the sense that if you could look

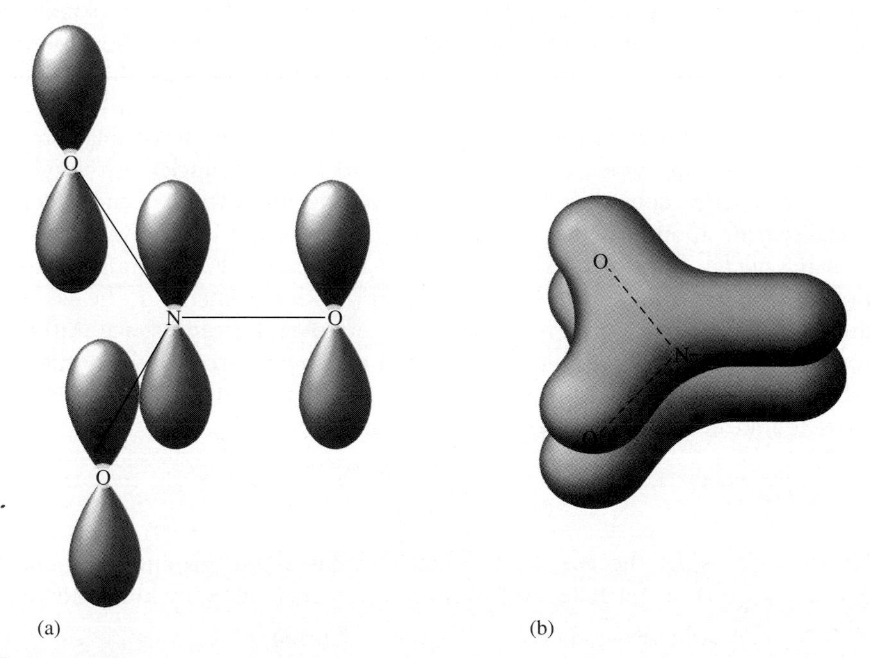

Figure 14.51
(a) The p orbitals used to form the π bonding system in the NO_3^- ion. (b) A representation of the delocalization of the electrons in the π molecular orbital system of the NO_3^- ion.

at a water molecule, you would probably not "see" anything like the pictures we have shown in this book. These pictures only help us visualize the theoretical information we have accumulated for water.

We should also make clear that much more sophisticated treatments of atoms and molecules than we have considered here do exist and are carried out quite routinely by chemists who specialize in this area. These quantum mechanical calculations involve many fewer approximations than are made to obtain the models we have discussed here. However, although these treatments produce accurate mathematical descriptions of atomic and molecular properties, they are usually very difficult to interpret physically.

In contrast, the simple, highly pictorial models we have considered in this text are extremely useful because they help us do first-order chemical thinking and organize the information about matter that we gather from experiments. For example, there is much experimental evidence to suggest that methane has a tetrahedral structure. We have seen that the VSEPR model predicts this structure and that sp^3 hybrid orbitals on the carbon atom are consistent with this structure. Of course, this does not mean that methane really has four localized pairs of electrons, all with the same energy. In fact, the electrons in methane are delocalized and probably do not all have the same energy. However, the fact that methane is more complicated than these models indicate does not destroy the usefulness of the models. They still help us greatly as we organize the vast amount of information we collect about molecules. For example, chemists have learned to expect certain behaviors of "sp^3 carbon atoms" no matter what molecule contains them. A similar statement also applies to "sp^2 carbon atoms," "*sp* carbon atoms," and so on.

There is another point that should be made in connection with the theories of atoms and molecules. Even the experts in the field disagree (sometimes violently) about what the mathematical descriptions of atoms and molecules mean. And even the most fundamental ideas about bonding that we have discussed in this text are subject to controversy. For example, we have stated repeatedly that bonding results primarily from the lowering of the potential energy that occurs when electrons are shared between nuclei. Although the potential energy does decrease when a bonding interaction occurs, there is disagreement about whether the decrease occurs in the "bonding region" or near the nuclei. It has been suggested* that an even more important contribution to bonding results from the lowering of the kinetic energy of the electrons due to delocalization—that is, being spread out over a greater space. Although we cannot explore this idea in detail here, it may help to review the particle-in-a-box problem described in Section 12.6. Recall that the energy levels for a particle in a box are described by the expression

$$E_n = \frac{n^2h^2}{8mL^2}$$

Notice that as L (the size of the box) increases, the ground-state energy decreases. Thus the particle has lower kinetic energy (the only kind allowed in

*Sture Nordholm, "Delocalization—The Key Concept of Covalent Bonding," *J. Chem. Ed.* 65 (1988), 581.

the box) as the size of the box increases. This same effect seems to occur when an electron is delocalized (spread out) when two or more atoms join to form a molecule. Thus it has been suggested that delocalization (with the accompanying lowering of the electrons' kinetic energy) may be a more fundamental reason for bond formation than the lowering of the electrons' potential energy.

The point here is not to present these more complex concepts in detail but to make sure that you appreciate the nature of models and that you recognize the difference between the information learned about matter from experiments and the information learned from theories.* Because matter is so complex, we often tailor our theories to explain a limited area of nature so as to keep the theories simple enough to understand and use. However, it is very dangerous to assume that such a limited theory can then be used in a wider sense. We must be very careful to ask proper questions of our theories or we face the danger of being greatly misled.

14.7 Molecular Spectroscopy

Spectroscopy can be defined as the study of the interaction of electromagnetic radiation with matter. We have already discussed the emission spectrum of the hydrogen atom (Chapter 12) and the importance of the information it provides about hydrogen's quantized energy levels. Spectroscopy is also very useful for the study of molecules. Molecules can absorb electromagnetic radiation to furnish energy for many different processes, all of which have quantized energy levels. For example, a molecule can absorb or emit a photon and go from a lower electronic energy state to a higher electronic energy state or vice versa. This so-called *electronic transition* can be described approximately as a change from one electron arrangement (a particular configuration of electrons in the molecular orbitals) to another. Typically, electronic transitions require photons in the ultraviolet or visible regions of the spectrum. In addition to undergoing electronic transitions, molecules can also undergo vibrational and rotational energy transitions. For example, the atoms in a molecule vibrate around their equilibrium positions, giving rise to quantized vibrational energy levels whose spacings correspond to the energies of photons in the infrared region of the electromagnetic spectrum. Molecules also rotate and have quantized rotational energy levels whose spacings correspond to the energies of photons in the microwave region. In fact, it is the rotational excitation of the water molecules in food and the transfer of this energy to other molecules that form the basis for microwave cooking.

---

*References to some interesting and provocative papers on this subject follow. These articles are probably too sophisticated to be read with great understanding now, but you can try to read them and then put copies of them in your files to be reread as your knowledge of chemistry increases. They raise some interesting questions.

Linus Pauling, "The Nature of the Chemical Bond—1992," *J. Chem. Ed.* **69** (1992), p. 519.

J. F. Ogilvie, "The Nature of the Chemical Bond—1990," *J. Chem. Ed.* **67** (1990), p. 280.

Michael Laing, "No Rabbit Ears on Water," *J. Chem. Ed.* **64** (1987), p. 124.

R. Bruce Martin, "Localized and Spectroscopic Orbitals: Squirrel Ears on Water," *J. Chem. Ed.* **65** (1988), p. 688.

A schematic representation of two molecular electronic states with the accompanying vibrational and rotational energy levels is given in Fig. 14.52. Note that each electronic energy level has a set of quantized vibrational states and that each vibrational state has a set of rotational states.

Depending on the energy of the photons interacting with it, a molecule can undergo a pure rotational transition (change rotational levels but remain in the same vibrational and electronic states), a vibrational transition that may also involve a simultaneous rotational transition (see Fig. 14.52), or an electronic transition that may involve a simultaneous vibrational and/or rotational change. The details of the geometric structure and electronic structure of the molecule determine exactly what types of transitions can occur. We cannot deal with these issues here because they are beyond the scope of this course.

Rotational Spectroscopy

The spacings of the quantized rotational energy levels for a molecule depend on its molecular structure. Therefore, the experimental determination of a molecule's rotational energy-level spacings provides information about its structure. That is, determination of the specific photons of microwave radiation that a given molecule absorbs yields direct information about the rotational energy spacings. This in turn gives information about the details of the molecule's structure. For example, for a linear molecule the energies of the rotational states are given by the formula

$$E_J = \frac{\hbar^2}{2I} J(J + 1)$$

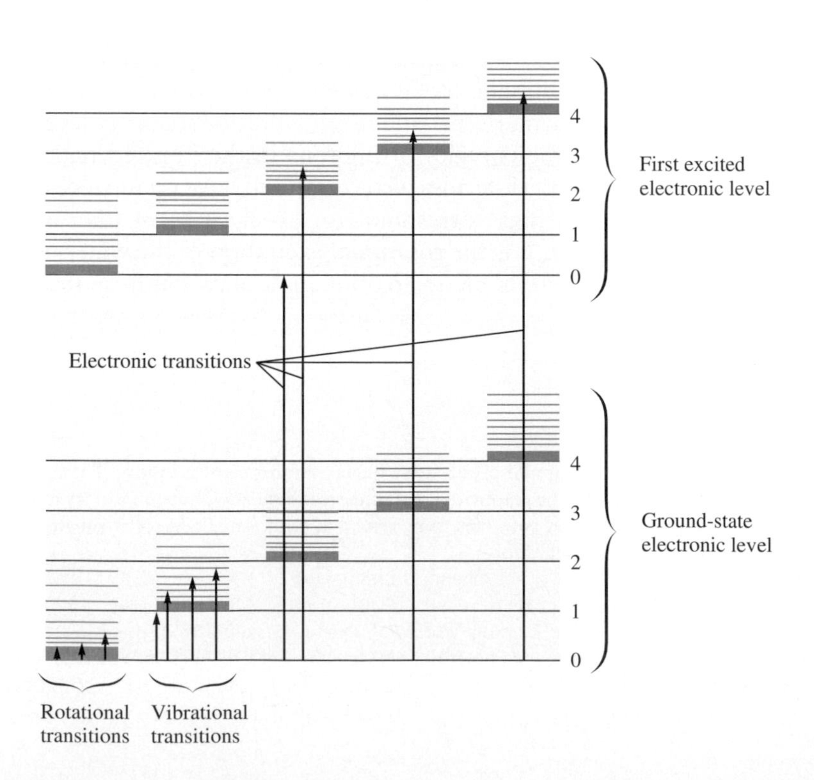

Figure 14.52
A schematic representation of two electronic energy levels in a molecule, with the vibrational (in red) and rotational (in blue) energy levels shown for each electronic state. Note that rotational changes are lowest in energy, followed by vibrational changes and then electronic changes, which require the highest-energy photons.

where J = the rotational quantum number, which can assume only integer values and zero ($J = 0, 1, 2, 3, \ldots$)

$\hbar$ = Planck's constant divided by 2π

I = the moment of inertia of the molecule

For a diatomic molecule containing atoms with masses m_1 and m_2, the moment of inertia is given by the relationship

$$I = \mu R_e^2$$

where μ = reduced mass = $\dfrac{m_1 m_2}{m_1 + m_2}$

R_e = average (equilibrium) bond length

If the energy of the photon necessary to promote a diatomic molecule from E_0 ($J = 0$) to E_1 ($J = 1$) is determined, the value of I for the molecule can be calculated, which, in turn, allows the calculation of R_e. Thus the rotational spectrum of a diatomic molecule provides an accurate method for measuring its average bond length.

The analysis of the rotational spectra for polyatomic molecules is more complex but also can provide accurate details of molecular structure.

Vibrational Spectroscopy

As we have seen, a molecule can be approximated as a collection of atoms held together by bonds. In describing the vibrations in a molecule, we can compare the bond between a given pair of atoms to a spring attached to two masses. As the atoms move apart in a vibrational motion, the bond, like a spring, provides a restoring force that pulls the atoms back toward each other. The atoms vibrate back and forth about the average bond distance R_e which leads to quantized energy levels given by the expression

$$E_v = h\nu_0 \left(v + \tfrac{1}{2}\right)$$

where ν_0 = the characteristic frequency of the vibration

v = the vibrational quantum number, which can assume only the values 0, 1, 2, 3, . . .

Notice that the vibrational energy is not zero when $v = 0$. The energy $h\nu_0/2$ corresponding to $v = 0$ is the zero-point energy of the molecule.

One very important use of vibrational spectroscopy (often called *infrared spectroscopy*) is to assist in the identification of an unknown molecule. The infrared (IR) spectrum of a molecule is typically represented as a plot of the energy transmitted versus the *wave number* of the radiation. The wave number for electromagnetic radiation is the reciprocal of the wavelength in centimeters. This provides a conveniently sized unit commonly used in plotting infrared spectra. A typical spectrum is shown in Fig. 14.53. Note that regions of the spectrum where the energy transmitted is small correspond to regions where the molecule has absorbed a large quantity of this radiation.

A particular bonded pair of atoms has a characteristic vibrational frequency (wave number) that is relatively insensitive to its molecular environment. Thus a vibration that appears in the IR spectrum at that characteristic wave number provides good evidence that this particular atom pair is present in

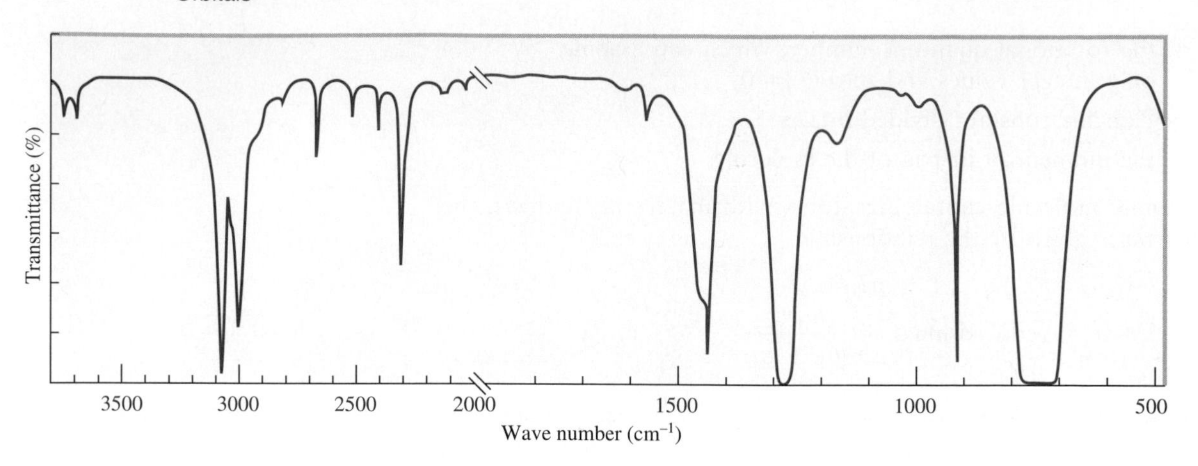

Figure 14.53
The infrared spectrum of CH_2Cl_2. (Note that the wave number scale changes on this spectrum at 2000 cm^{-1}.)

the molecule. For example, a —C—H pair in a molecule will always show a vibration at approximately 3000 cm^{-1} (the range of wave numbers is actually about 2880–3030 cm^{-1}, depending on the specific molecular environment). On the other hand, the —O—H group in a molecule will show a vibration at about 3600 cm^{-1}. Likewise, a carbonyl group

$$\gt C=O$$

always shows a distinctively shaped vibrational band at about 1700 cm^{-1}, a carbon-carbon double bond

$$\gt C=C \lt$$

shows a vibration at about 1650 cm^{-1}, and so on. Therefore, the IR spectrum of a molecule can be a great aid in identifying what atom groupings are present in a molecule and thus can provide valuable information for identifying a specific molecule.

Electronic Spectroscopy

The electronic spectrum of a molecule, which typically occurs in the ultraviolet or visible region of the electromagnetic spectrum, provides information about the spacings of the electronic energy levels in the molecule. The electronic spectrum, a plot of the quantity of radiation absorbed versus the wavelength of the radiation, shows peaks (maxima) at wavelengths where the photons have an energy that matches an energy gap in the molecule.

Often when a molecule absorbs a photon, the resulting excited state is much more chemically reactive than is the molecule in its ground state. The study of the chemistry of these electronically excited molecules is called *photochemistry*. A particularly important area of photochemistry concerns the study of the reactions that occur in the earth's atmosphere. For example, photochemical smog depends to an important extent on the photodecomposition of nitrogen dioxide to form very reactive oxygen atoms that combine with oxygen molecules in the air to form ozone, which then goes on to react with other pollutants. Ironically, the problems of ozone *depletion* in the upper atmosphere also result from a photochemical reaction—in this case, the photodecomposition of chlorofluorocarbons such as CF_2Cl_2 to produce reactive chlorine atoms, which catalyze ozone decomposition. We will discuss this situation in more detail in Chapter 15.

Lasers, which will be discussed in more detail in Chapter 16, have assumed a very important role in electronic spectroscopy in recent years, both because they can deliver large quantities of energy at very precise wavelengths and because the power can be delivered in very short bursts—on the order of only 10^{-12} second (one picosecond) in duration. This allows the excitation of molecules with an ultrashort pulse of energy, after which the molecule can be observed as it decays back to the ground state by various pathways, many of which take place on the picosecond time scale.

In this section we have touched on only three types of spectroscopy. There are many other types, some dealing with the properties of the nucleus, some dealing with the surface properties of solids, some dealing with the various types of emissions by excited molecules, and so on. Spectroscopy represents one of the most powerful tools available to the scientist.

Exercises

A blue exercise number indicates that the answer to that exercise appears at the back of this book.

The Localized Electron Model and Hybrid Orbitals

1. For each of the following molecules, write the Lewis structure, predict the molecular structure (including bond angles), give the expected hybrid orbitals on the central atom, and predict the overall polarity.
   a. CF_4 b. NF_3 c. OF_2 d. BF_3 e. BeH_2 f. TeF_4 g. AsF_5 h. KrF_2 i. KrF_4 j. SeF_6 k. $XeOF_4$ l. $XeOF_2$ m. XeO_4

2. Predict the hybrid orbitals used by the sulfur atom(s) in each of the following.
   a. SO_2
   b. SO_3
   c. $S_2O_3^{2-}$ $\left[S\text{—}S(\text{=}O)(\text{—}O)\text{—}O \right]^{2-}$ (structure: S—S—O with O above and O below the central S)
   d. $S_2O_8^{2-}$ $\left[O\text{—}S\text{—}O\text{—}O\text{—}S\text{—}O \right]^{2-}$ (structure: each S bearing O above and O below)
   e. SO_3^{2-}
   f. SO_4^{2-}
   g. SF_2
   h. SF_4
   i. SF_6
   j. $F_3S\text{—}SF$

3. Phosphorus forms many compounds with three-dimensional cage structures. Some are shown below. Complete the Lewis structures, showing all unshared electron pairs. What is the hybridization of the phosphorus atom in each of the molecules? Predict approximate values of the O—P—O and P—O—P bond angles.

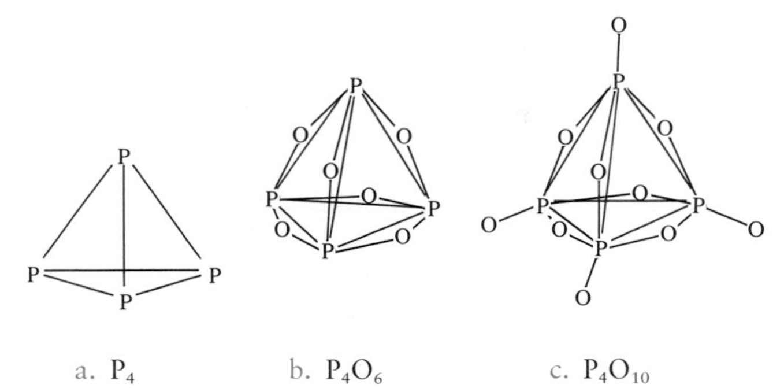

a. P_4 b. P_4O_6 c. P_4O_{10}

4. Why must all six atoms in C_2H_4 be in the same plane?

5. The allene molecule has the following Lewis structure:

   (H)(H)C=C=C(H)(H)

   Are all four hydrogen atoms in the same plane? If not, what is the spatial relationship among them? Why?

6. Biacetyl and acetoin are added to margarine to make it taste more like butter.

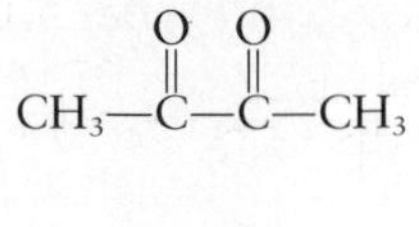

$CH_3-C(=O)-C(=O)-CH_3$ Biacetyl

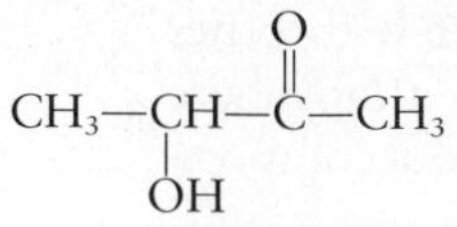

$CH_3-CH(OH)-C(=O)-CH_3$ Acetoin

Complete the Lewis structures, predict values for all C—C—O bond angles, and give the hybridization of all of the carbon atoms in these two compounds. Are the four carbons and two oxygens in biacetyl in the same plane?

7. How many σ bonds and how many π bonds are there in biacetyl and acetoin (see Exercise 6)?

8. Many important compounds in the chemical industry are derivatives of ethylene, C_2H_4. Two of them are methyl methacrylate and acrylonitrile.

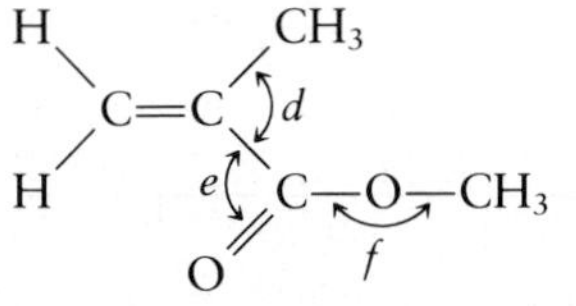

H, H, C=C, CH$_3$, C, O, CH$_3$, O, *d*, *e*, *f*

Methyl methacrylate

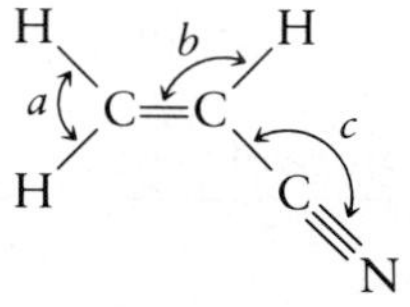

H, H, C=C, H, C≡N, *a*, *b*, *c*

Acrylonitrile

Complete the Lewis structures for these molecules, showing all lone pairs. Give approximate values for bond angles *a* through *f*, and give the hybridization of all carbon atoms. In methyl methacrylate, how many of the atoms in the molecule lie in the same plane?

9. How many σ bonds and how many π bonds are there in methyl methacrylate and acrylonitrile (see Exercise 8)?

10. One of the first drugs to be approved for use in treatment of acquired immune deficiency syndrome (AIDS) is azidothymidine (AZT). Complete the Lewis structure of AZT.

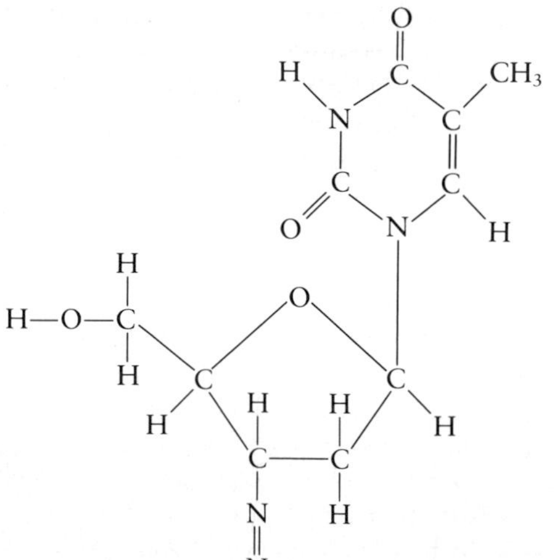

O, H, N, C, C, CH$_3$, C, C, O, N, H, H, H—O—C, H, O, C, C, H, H, H, H, C—C, H, N, N, N

a. How many carbon atoms use sp^3 hybridization?
b. How many carbon atoms use sp^2 hybridization?
c. Which atom is *sp* hybridized?
d. How many σ bonds are in the molecule?
e. How many π bonds are in the molecule?
f. Which ring(s) is/are planar?
g. What is the N—N—N bond angle in the azide ($-N_3$) group?
h. What is the H—O—C bond angle in the side group attached to the five-membered ring?
i. What is the hybridization of the oxygen atom in the $-CH_2OH$ group?

11. Cyanamide (H_2NCN), an important industrial chemical, is produced by the following steps:

$$CaC_2 + N_2 \longrightarrow CaNCN + C$$

$$CaNCN \xrightarrow{\text{Acid}} \underset{\text{Cyanamide}}{H_2NCN}$$

Calcium cyanamide (CaNCN) is used as a direct-application fertilizer, weed killer, and cotton defoliant. It is also used to make cyanamide, dicyandiamide, and melamine plastics:

$$H_2NCN \xrightarrow{\text{Acid}} \underset{\text{Dicyandiamide}}{NCNC(NH_2)_2}$$

$$NCNC(NH_2)_2 \xrightarrow[NH_3]{\text{Heat}}$$ H$_2$N, N, NH$_2$, C, C, N, N, C, NH$_2$

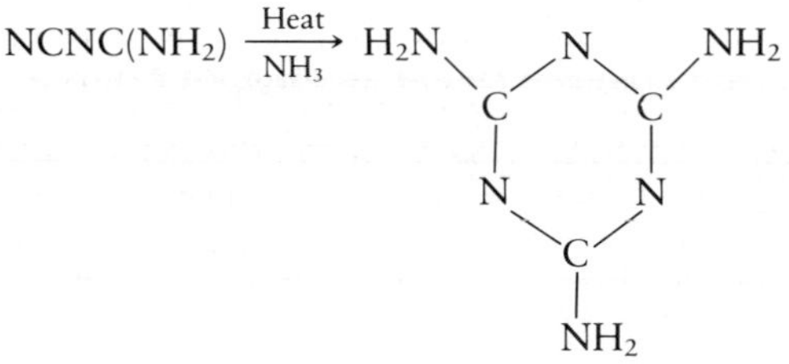

Melamine
(π bonds not shown)

a. Draw Lewis structures for NCN^{2-}, H_2NCN, dicyandiamide, and melamine, including resonance structures where appropriate.
b. Give the hybridization of the C and N atoms in each species.
c. How many σ bonds and how many π bonds are in each species?
d. Is the ring in melamine planar?
e. There are three different types of C—N bonds in dicyandiamide, $NCNC(NH_2)_2$, and the molecule is nonlinear. Of all the resonance structures you drew for this molecule, predict which should be the most important.

12. Hot and spicy foods contain molecules that stimulate pain-detecting nerve endings. Two such molecules are piperine and capsaicin:

Piperine

Capsaicin

Piperine is the active compound in white and black pepper and capsaicin is the active compound in chili peppers.

a. Complete the Lewis structure for piperine and capsaicin showing all lone pairs of electrons.
b. How many carbon atoms are *sp*, sp^2, and sp^3 hybridized in each molecule?
c. Which hybrid orbitals are used by the nitrogen atoms in each molecule?
d. Give approximate values for the bond angles marked *a* through *l* in the above structures.

The Molecular Orbital Model

13. Describe the differences between bonding and antibonding molecular orbitals in terms of electron distribution and energy.

14. Which of the following are predicted by the molecular orbital model to be stable diatomic species?
    a. H_2^+, H_2, H_2^-, H_2^{2-}
    b. He_2^{2+}, He_2^+, He_2
    c. Be_2, B_2, Li_2

15. Using the molecular orbital model to describe the bonding in O_2^+, O_2, O_2^-, and O_2^{2-}, predict the bond orders and the relative bond lengths for these four species. How many unpaired electrons are present in each?

16. Does the molecular orbital model or the localized electron model better account for the bonding in nitric oxide (NO)? Explain.

17. What are the relationships among bond order, bond energy, and bond length? Which of these quantities can be measured?

18. Show how two 2*p* atomic orbitals can combine to form a σ and a π molecular orbital.

19. Show how a H 1*s* atomic orbital and a F 2*p* atomic orbital overlap to form bonding and antibonding molecular orbitals in the hydrogen fluoride molecule. Are these molecular orbitals σ or π molecular orbitals?

20. Using the molecular orbital model, write electron configurations for the following diatomic species and calculate the bond orders. Which ones are paramagnetic?
    a. H_2
    b. B_2
    c. F_2
    d. CN^+
    e. CN
    f. CN^-
    g. N_2
    h. N_2^+
    i. N_2^-

21. Place the species CN^+, CN, and CN^- in order of increasing bond length and increasing bond energy. Use your results from Exercise 20.

22. Place the species N_2, N_2^+, and N_2^- in order of increasing bond length and increasing bond energy. Use your results from Exercise 20.

23. The molecules N_2 and CO are isoelectronic but their properties are quite different. Although as a first approximation we often use the same molecular orbital diagram for both, suggest how the molecular orbitals in N_2 and CO might be different.

24. As compared with CO and O_2, CS and S_2 are very unstable molecules. Give an explanation based on the relative abilities of the sulfur and oxygen atoms to form π bonds.

25. Construct a molecular orbital energy level diagram for the P_2 molecule.

26. Acetylene (C_2H_2) can be produced from the reaction of calcium carbide (CaC_2) with water. Use both the localized electron and molecular orbital models to describe the bonding in the acetylide anion (C_2^{2-}). What neutral, heteronuclear diatomic molecule is isoelectronic with the acetylide anion?

27. Use Figs. 14.45 and 14.46 to answer the following questions.
    a. Would the bonding molecular orbital in HF place greater electron density near the H or the F atom? Why?
    b. Would the bonding molecular orbital have greater fluorine 2*p* character, greater hydrogen 1*s* character, or an equal contribution from both? Why?
    c. Answer the previous two questions for the antibonding molecular orbital in HF.

28. Use orbital energy diagrams like that shown for B_2 in Fig. 14.40 to answer the following questions.
    a. The first ionization energy of N_2 (1501 kJ/mol) is greater than the first ionization energy of atomic nitrogen (1402 kJ/mol). Explain.
    b. Would you expect F_2 to have a lower or higher first ionization energy than atomic fluorine? Why?

29. The diatomic molecule OH exists in the gas phase. OH plays an important part in combustion reactions and is a reactive oxidizing agent in polluted air. The bond length and bond energy have been measured to be 97.06 pm and 424.7 kJ/mol, respectively. Assume that the OH molecule is analogous to the HF molecule discussed in the chapter and that the molecular orbitals result from the overlap of

a p_z orbital from oxygen and the $1s$ orbital of hydrogen (the O—H bond lies along the z axis).

a. Draw pictures of the sigma bonding and antibonding molecular orbitals in OH.
b. Which of the two molecular orbitals has the greater hydrogen $1s$ character?
c. Can the $2p_x$ orbital of oxygen form molecular orbitals with the $1s$ orbital of hydrogen? Explain.
d. Knowing that only the $2p$ orbitals of oxygen interact significantly with the $1s$ orbital of hydrogen, complete the molecular orbital energy-level diagram for OH. Place the correct number of electrons in the energy levels.
e. Estimate the bond order for OH.
f. Predict whether the bond order of OH^+ is greater than, less than, or the same as that of OH. Explain.

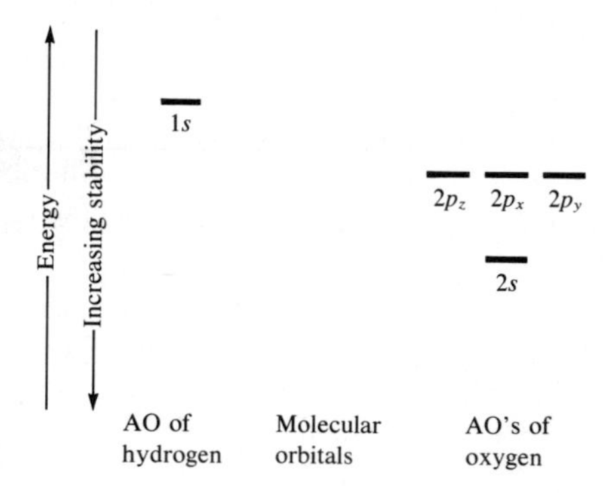

Additional Exercises

30. A variety of chlorine oxide fluorides and related cations and anions are known. They tend to be powerful oxidizing and fluorinating agents. $FClO_3$, the most stable of this group of compounds, has been studied as an oxidizing component of rocket propellents. Draw Lewis structures, predict molecular structures, and describe the bonding (in terms of hybrid orbitals) for the following.
    a. FClO
    b. $FClO_2$
    c. $FClO_3$
    d. F_3ClO
    e. F_3ClO_2

31. $FClO_2$ and F_3ClO can both gain a fluoride ion to form stable anions. F_3ClO and F_3ClO_2 can also lose a fluoride ion to form stable cations. Draw Lewis structures and describe the hybrid orbitals used by chlorine in these four ions.

32. Antimony pentafluoride exists in the liquid phase not as discrete SbF_5 molecules but as a polymer (a large aggregate of small molecules). Describe the molecular structure and the hybrid orbitals in discrete SbF_5 molecules. What hybrid orbitals does antimony employ in the polymer?

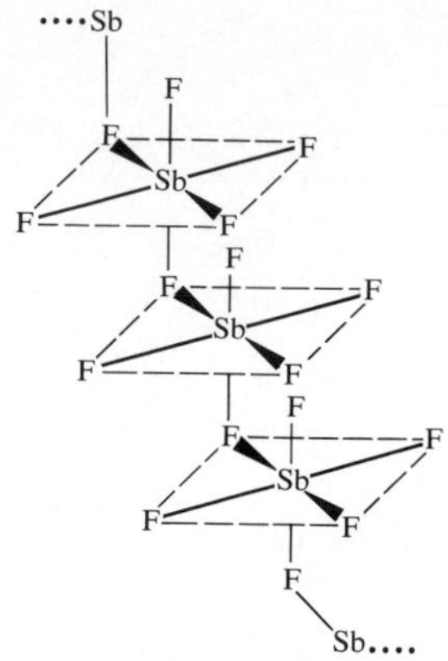

Experimental observations indicate that the fluorine atoms exist in three distinct types of chemical environments in liquid antimony pentafluoride. Indicate the different types of fluorine atoms in the polymer.

33. The antibiotic, thiarubin-A, was discovered by studying the feeding habits of wild chimpanzees in Tanzania. The structure for thiarubin-A is

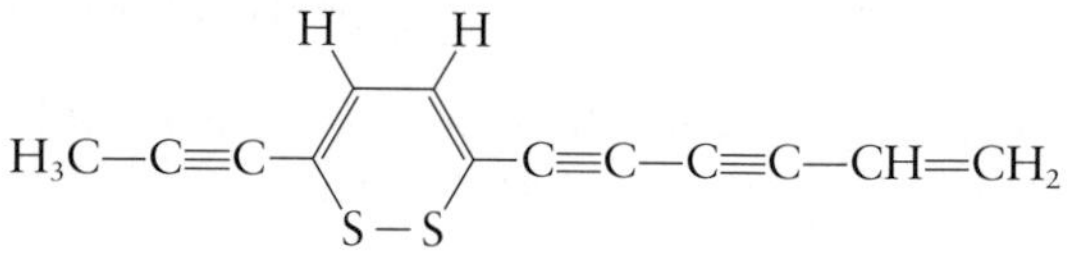

    a. Complete the Lewis structure showing all lone pairs of electrons.
    b. Indicate the hybrid orbitals used by the carbon and sulfur atoms in thiarubin-A.
    c. How many σ and π bonds are present in this molecule?

34. Two structures can be drawn for cyanuric acid,

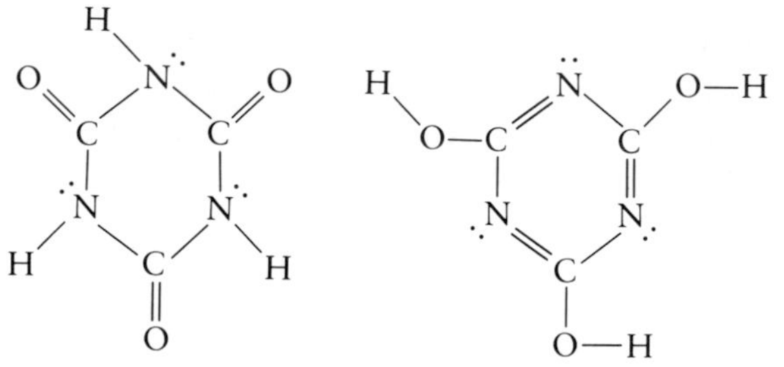

    a. Are these two resonance structures of the same molecule? Why or why not?
    b. Give the hybridization of the carbon and nitrogen atoms in each structure.
    c. Use bond energies (Table 13.5) to predict which form is more stable; i.e., which contains the strongest bonds?

35. Using bond energies from Table 13.5, estimate the barrier to rotation about a carbon-carbon double bond. To do this, consider what must happen to go from

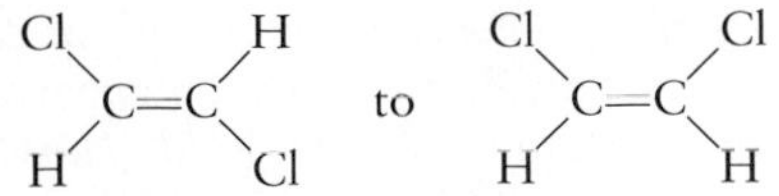

Cl H
C=C to C=C
H Cl

Cl Cl
C=C
H H

in terms of making and breaking chemical bonds; i.e., what happens to the π bond?

36. Describe the bonding in the O_3 molecule and the NO_2^- ion, using the localized electron model. How would the molecular orbital model describe the π bonding in these two species?

37. Two molecules used in the polymer industry are azidocarbonamide and methyl cyanoacrylate. Their structures are

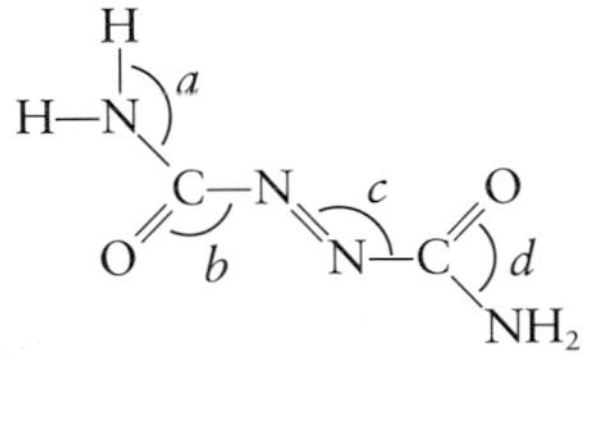

H
H—N (a)
C—N (c) O
O (b) N—C (d)
NH$_2$

e N
C
H_2C=C (f) g
C—O—CH_3 (h)
O

Azidocarbonamide 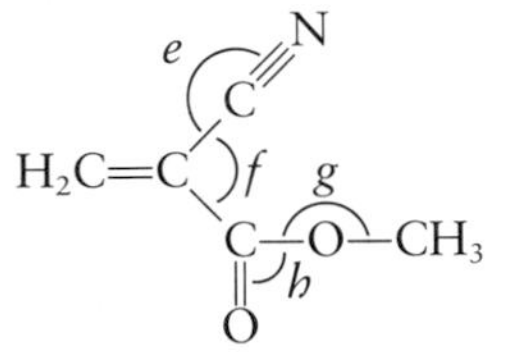 Methyl cyanoacrylate

Azidocarbonamide is used in forming polystyrene. When added to the molten plastic, it decomposes to nitrogen, carbon monoxide, and ammonia gases, which are captured as bubbles in the molten polymer. Methyl cyanoacrylate is the main ingredient in super glue. As the glue sets, methyl cyanoacrylate polymerizes across the carbon-carbon double bond. (See Chapter 22.)

a. Complete the Lewis structures showing all lone pairs of electrons.
b. Which hybrid orbitals are used by the carbon atoms in each molecule and the nitrogen atoms in azidocarbonamide?
c. How many π bonds are present in each molecule?
d. Give approximate values for the bond angles marked *a* through *h* in the above structures.

38. Cholesterol ($C_{27}H_{46}O$) has the following structure:

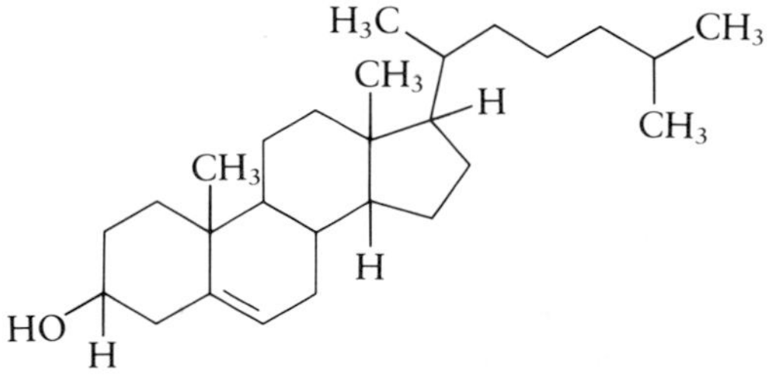

H_3C CH_3
CH_3
H CH_3
CH_3
H
HO
H

In such shorthand structures, each point where lines meet represents a carbon atom, and most H atoms are not shown. Draw the complete structure, showing all carbon and hydrogen atoms. (There will be four bonds to each carbon atom.) Indicate which carbon atoms use sp^2 or sp^3 hybrid orbitals. Are all carbon atoms in the same plane, as implied by the structure?

39. Describe the bonding in NO^+, NO^-, and NO, using both the localized electron and molecular orbital models. From the molecular orbital model, predict the order of bond energies and bond lengths for the nitrogen-oxygen bond in the three species.

40. Describe the bonding in the first excited state of N_2 (the one closest in energy to the ground state), using the molecular orbital model. What differences do you expect in the properties of the molecule in the ground state and in the first excited state? (An excited state of a molecule corresponds to an electron arrangement other than that giving the lowest possible energy.)

41. Describe the bonding in the Be_2 molecule, using the localized electron model. How does this compare with the description of the bonding in Be_2 using the molecular orbital model?

42. What type of experiment can be done to determine whether a material is paramagnetic?

43. A Lewis structure obeying the octet rule can be drawn for O_2 as follows:

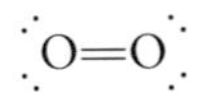

:O=O:

Use the molecular orbital energy-level diagram for O_2 to show that this Lewis structure for O_2 corresponds to an excited state.

44. Complete the Lewis structures of the following molecules. Predict the molecular structure and polarity, bond angles, and the hybrid orbitals used by the atoms marked by asterisks for each molecule.

a. $COCl_2$

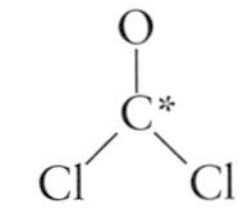

O
C*
Cl Cl

b. N_2F_2 F—N*—N*—F

c. N_2O_3 O—*N—*N(O)(O)

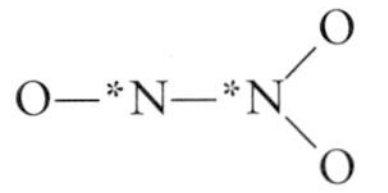

d. COS O—C*—S

e. ICl_3 Cl—I*—Cl
Cl

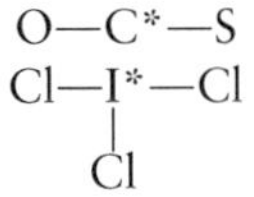

45. Which of the species in Exercise 44 show resonance? Draw all resonance structures for each. Include formal charges in all structures.

46. What is misleading about the following statement: "The methane molecule (CH_4) is a tetrahedral molecule because the carbon atom is sp^3 hybridized."

47. Complete the following resonance structures for $OPCl_3$:

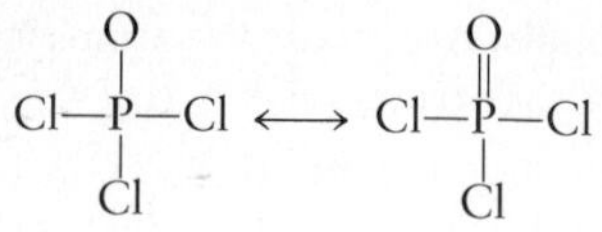

$$\begin{array}{c} O \\ | \\ Cl-P-Cl \\ | \\ Cl \end{array} \longleftrightarrow \begin{array}{c} O \\ \| \\ Cl-P-Cl \\ | \\ Cl \end{array}$$

(A) (B)

a. Would you predict the same molecular structure from each resonance structure?
b. What is the hybridization of P in each structure?
c. What orbitals can the P atom use to form the π bond in structure B?
d. Which resonance structure would be favored on the basis of formal charges?

48. The N_2O molecule is linear and polar.
a. On the basis of this experimental evidence, which arrangement is correct, NNO or NON? Explain your answer.
b. On the basis of your answer in part a, write the Lewis structure of N_2O (including resonance forms). Give the formal charge on each atom and the hybridization of the central atom.
c. How would the multiple bonding in: $:N{\equiv}N{-}\ddot{\underset{\cdot\cdot}{O}}:$ be described in terms of orbitals?

49. Values of measured bond energies may vary greatly depending on the molecule studied. Consider the following reactions:

$$NCl_3(g) \longrightarrow NCl_2(g) + Cl(g) \quad \Delta H = 375 \text{ kJ/mol}$$
$$ONCl(g) \longrightarrow NO(g) + Cl(g) \quad \Delta H = 158 \text{ kJ/mol}$$

Rationalize the difference in the values of ΔH for these reactions, even though each reaction appears to only involve the breaking of one N—Cl bond. (*Hint:* consider the bond order of the NO bond in ONCl and in NO.)

50. Carbon monoxide (CO) forms bonds to a variety of metals and metal ions. Its ability to bond to iron in hemoglobin is the reason CO is so toxic. The bond carbon monoxide forms to metals is through the carbon atom:

$$M{-}C{\equiv}O$$

a. On the basis of electronegativities, would you expect the carbon atom or the oxygen atom to form bonds to metals?
b. Assign formal charges to the atoms in CO. Which atom would you expect to bond to a metal on this basis?
c. In the molecular orbital model, bonding molecular orbitals place more electron density near the more electronegative atom. (See the HF molecule, Figs. 14.45 and 14.46.) Antibonding molecular orbitals place more electron density near the less electronegative atom in the diatomic molecule. Use the molecular orbital model to predict which atom of carbon monoxide should form bonds to metals.

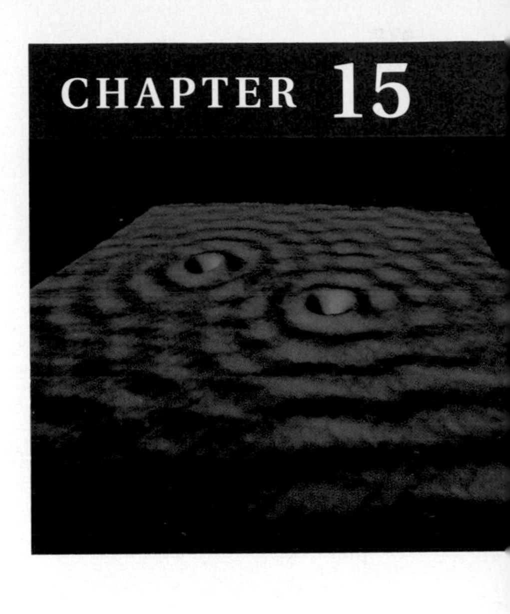

Chemical Kinetics

The applications of chemistry focus largely on chemical reactions, and the commercial use of a reaction requires knowledge of several of its characteristics. A reaction is defined by its reactants and products, whose identities must be learned by experiment. Once the reactants and products are known, the equation for the reaction can be written and balanced and stoichiometric calculations can be carried out. Another very important characteristic of a reaction is its spontaneity. Spontaneity refers to the *inherent tendency* for the process to occur; however, it implies nothing about speed. *Spontaneous does not mean fast.* There are many spontaneous reactions that are so slow that no apparent reaction occurs over a period of weeks or years at normal temperatures. For example, there is a strong inherent tendency for gaseous hydrogen and oxygen to combine to form water,

$$2H_2(g) + O_2(g) \longrightarrow 2H_2O(l)$$

but in fact the two gases can coexist indefinitely at 25°C. Similarly, the gaseous reactions

$$H_2(g) + Cl_2(g) \longrightarrow 2HCl(g)$$
$$N_2(g) + 3H_2(g) \longrightarrow 2NH_3(g)$$

are both highly likely to occur from a thermodynamic standpoint, but we observe no product formation under normal conditions. In addition, the process of changing diamond to graphite is spontaneous, but is so slow that it is not detectable.

To be useful, reactions must occur at a reasonable rate. To produce the 20 million tons of ammonia needed each year for fertilizer, we cannot simply mix nitrogen and hydrogen gases at 25°C and wait for them to react. It is not enough to understand the stoichiometry and thermodynamics of a reaction; we must also understand the factors that govern the rate of the reaction. The area of chemistry that concerns reaction rates is called **chemical kinetics.**

One of the main goals of chemical kinetics is to understand the steps by which a reaction takes place. This series of steps is called the *reaction mechanism.* Understanding the mechanism allows us to find ways to facilitate the reaction. For example, the Haber process for the production of ammonia requires high temperatures to achieve commercially feasible reaction rates. However, even higher temperatures (and more cost) would be required without the use of iron oxide, which speeds up the reaction.

In this chapter we will consider the fundamental ideas of chemical kinetics. We will explore rate laws, reaction mechanisms, and simple models for chemical reactions.

15.1 Reaction Rates

The kinetics of air pollution is discussed in Section 15.9.

To introduce the concept of reaction rate, we will consider the decomposition of nitrogen dioxide, a gas that causes air pollution. Nitrogen dioxide decomposes to nitric oxide and oxygen as follows:

$$2NO_2(g) \longrightarrow 2NO(g) + O_2(g)$$

Suppose in a particular experiment we start with a flask of nitrogen dioxide at 300°C and measure the concentrations of nitrogen dioxide, nitric oxide, and oxygen over time as the nitrogen dioxide decomposes. The results of this experiment are summarized in Table 15.1 and the data are plotted in Fig. 15.1.

Note from these results that the concentration of the reactant (NO_2) decreases with time and that the concentrations of the products (NO and O_2) increase with time. Chemical kinetics deals with the speed at which these

TABLE 15.1 **Concentrations of Reactant and Products as a Function of Time for the Reaction $2NO_2(g) \longrightarrow 2NO(g) + O_2(g)$ (at 300°C)**

	Concentration (mol/L)		
Time (±1 s)	NO_2	NO	O_2
0	0.0100	0	0
50	0.0079	0.0021	0.0011
100	0.0065	0.0035	0.0018
150	0.0055	0.0045	0.0023
200	0.0048	0.0052	0.0026
250	0.0043	0.0057	0.0029
300	0.0038	0.0062	0.0031
350	0.0034	0.0066	0.0033
400	0.0031	0.0069	0.0035

changes occur. The speed, or *rate*, of a process is defined as the change in a given quantity over a specific period of time. For chemical reactions the quantity that changes is the amount or concentration of a reactant or product. So the **reaction rate** of a chemical reaction is defined as the *change in concentration of a reactant or product per unit time:*

$$\text{Rate} = \frac{\text{concentration of A at time } t_2 - \text{concentration of A at time } t_1}{t_2 - t_1}$$

$$= \frac{\Delta[\text{A}]}{\Delta t}$$

where A represents a specific reactant or product and where the square brackets indicate concentration in mol/L. As usual, the symbol Δ indicates a *change* in a given quantity.

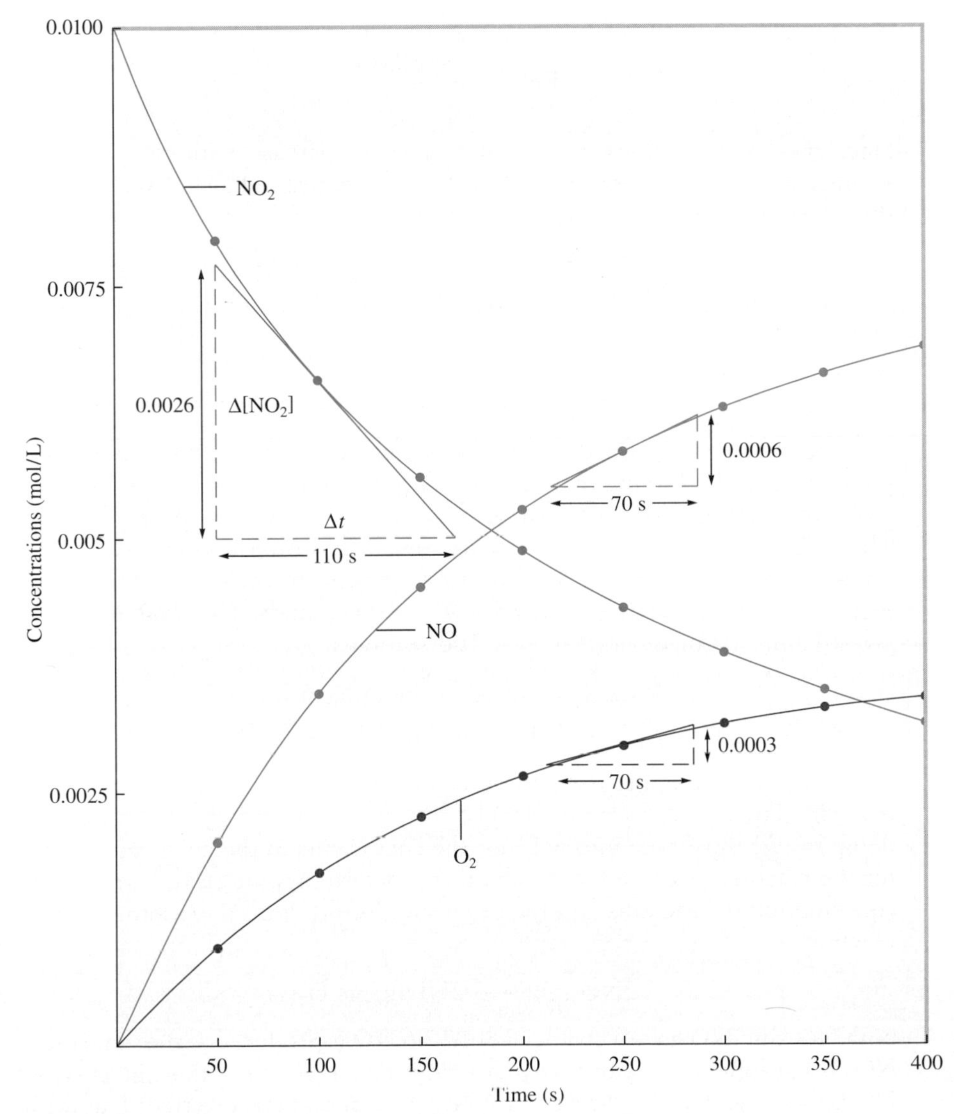

Figure 15.1
Starting with pure nitrogen dioxide at 300°C, the concentrations of nitrogen dioxide, nitric oxide, and oxygen are plotted versus time.

Now let us calculate the average rate at which the concentration of NO_2 changes over the first 50 seconds of the reaction, using the data given in Table 15.1.

$$\text{Rate} = \frac{\Delta[NO_2]}{\Delta t}$$

$$= \frac{[NO_2]_{t=50} - [NO_2]_{t=0}}{50\text{ s} - 0\text{ s}}$$

$$= \frac{0.0079\text{ mol/L} - 0.0100\text{ mol/L}}{50\text{ s}}$$

$$= -4.2 \times 10^{-5}\text{ mol L}^{-1}\text{ s}^{-1}$$

Note that since the concentration of NO_2 decreases with time, $\Delta[NO_2]$ is a negative quantity. Because it is customary to work with *positive* reaction rates, we define the rate of this particular reaction as

$$\text{Rate} = -\frac{\Delta[NO_2]}{\Delta t}$$

Since the concentrations of reactants always decrease with time, any rate expression involving a reactant will include a negative sign. The average rate of this reaction from 0 to 50 seconds is then

$$\text{Rate} = -\frac{\Delta[NO_2]}{\Delta t}$$

$$= -(-4.2 \times 10^{-5}\text{ mol L}^{-1}\text{ s}^{-1})$$

$$= 4.2 \times 10^{-5}\text{ mol L}^{-1}\text{ s}^{-1}$$

TABLE 15.2
Average Rate (in mol L^{-1} s^{-1}) of Decomposition of Nitrogen Dioxide as a Function of Time

$-\frac{\Delta[NO_2]}{\Delta t}$	Time Period (s)
4.2×10^{-5}	$0 \rightarrow 50$
2.8×10^{-5}	$50 \rightarrow 100$
2.0×10^{-5}	$100 \rightarrow 150$
1.4×10^{-5}	$150 \rightarrow 200$
1.0×10^{-5}	$200 \rightarrow 250$

Note: The *rate* decreases with time.

The average rates for this reaction during several other time intervals are given in Table 15.2. Note that the rate is not constant but decreases with time. The rates given in Table 15.2 are *average* rates over 50-second time intervals. The value of the rate at a particular time (the **instantaneous rate**) can be obtained by computing the slope of a line tangent to the curve at that point. Figure 15.1 shows a tangent drawn at $t = 100$ seconds. The *slope* of this line gives the instantaneous rate at $t = 100$ seconds:

$$\text{Rate} = -(\text{slope of the tangent line})$$

$$= 2.4 \times 10^{-5}\text{ mol L}^{-1}\text{ s}^{-1}$$

So far we have discussed the rate of this reaction only in terms of the reactant. The rate can also be defined in terms of the products. However, in doing so, we must take into account the coefficients in the balanced equation for the reaction, because the stoichiometry determines the relative rates of the consumption of reactants and the generation of products. For example, in the reaction

$$2NO_2(g) \longrightarrow 2NO(g) + O_2(g)$$

both NO_2 and NO have a coefficient of 2, so NO is produced at the same rate as NO_2 is consumed. We can verify this from Fig. 15.1. Note that the curve for NO is the same shape as the curve for NO_2 except that it is inverted. This means that at any point in time the slope of the tangent to the curve for NO will be the negative of the slope to the curve for NO_2. (Verify this at the point $t = 100$ seconds on both curves.) In contrast, O_2 has a coefficient of 1, which means it is produced half as fast as NO, which has a coefficient of 2. That is, the rate of NO production is twice the rate of O_2 production.

We can also verify this fact from Fig. 15.1. For example, at $t = 250$ seconds,

$$\text{Slope of the tangent to the NO curve} = 8.6 \times 10^{-6} \text{ mol L}^{-1}\text{ s}^{-1}$$
$$\text{Slope of the tangent to the } O_2 \text{ curve} = 4.3 \times 10^{-6} \text{ mol L}^{-1}\text{ s}^{-1}$$

The slope at $t = 250$ seconds on the NO curve is twice the slope of that point on the O_2 curve, showing that the rate of production of NO is twice that of O_2.

The rate information can be summarized as follows:

Rate of consumption of NO_2	=	rate of production of NO	=	2(rate of production of O_2)

Because the reaction rate changes with time, and because the rate may be different for the various reactants and products (by factors that depend on the coefficients in the balanced equation), we must be very specific when we describe a rate for a chemical reaction.

15.2 Rate Laws: An Introduction

Note that the term *reversible* has different meanings in kinetics and thermodynamics.

Chemical reactions are *reversible*. In our discussion of the decomposition of nitrogen dioxide, we have so far considered only the forward reaction:

$$2NO_2(g) \longrightarrow 2NO(g) + O_2(g)$$

However, the reverse reaction can also occur. As NO and O_2 accumulate, they can react to reform NO_2:

$$O_2(g) + 2NO(g) \longrightarrow 2NO_2(g)$$

When gaseous NO_2 is placed in an otherwise empty container, the dominant reaction initially is

$$2NO_2(g) \longrightarrow 2NO(g) + O_2(g)$$

and the change in the concentration of NO_2 ($\Delta[NO_2]$) depends only on the forward reaction. However, after a period of time enough products accumulate so that the reverse reaction becomes important. Now $\Delta[NO_2]$ depends on the *difference in the rates of the forward and reverse reactions.* This complication can be avoided if we study the rate of a reaction under conditions where the reverse reaction makes only a negligible contribution. Typically, this means that we study a reaction at a point soon after the reactants are mixed, before the products have had time to build up to significant levels.

Under conditions such that the reverse reaction can be neglected, the *reaction rate depends only on the concentrations of the reactants.* For the decomposition of nitrogen dioxide, we can write

$$\text{Rate} = k[NO_2]^n \qquad (15.1)$$

Such an expression, which shows how the rate depends on the concentrations of reactants, is called a **rate law.** The proportionality constant k, called the **rate constant,** and n, called the **order** of the reactant, must both be determined by experiment. The order of a reactant can be positive or negative and can be an integer or a fraction. For the relatively simple reactions we will consider in this book, the orders will generally be positive integers.

Note two important points about Equation (15.1):

1. The concentrations of the products do not appear in the rate law because the reaction rate is being studied under conditions where the reverse reaction does not contribute to the overall rate.
2. The value of the exponent n must be determined by experiment; it cannot be written from the balanced equation.

Before we go further, we must define exactly what we mean by the term *rate* in Equation (15.1). In Section 15.1 we saw that reaction rate means a change in concentration per unit time. However, which reactant or product concentration do we choose in defining the rate? For example, for the decomposition of NO_2 to produce O_2 and NO considered in Section 15.1, we could define the rate in terms of any of these three species. However, since O_2 is produced only half as fast as NO_2, we must be careful to specify which species we are talking about in a given case. For instance, we might choose to define the reaction rate in terms of the consumption of NO_2:

$$\text{Rate} = -\frac{\Delta[NO_2]}{\Delta t} = -\frac{d[NO_2]}{dt} = k[NO_2]^n$$

where d indicates an infinitesimally small change. On the other hand, we could define the rate in terms of the production of O_2:

$$\text{Rate}' = \frac{d[O_2]}{dt} = k'[NO_2]^n$$

Note that because $2NO_2$ molecules are consumed for every O_2 molecule produced,

$$\text{Rate} = 2 \times \text{Rate}'$$

or

$$k[NO_2]^n = 2k'[NO_2]^n$$

and

$$k = 2 \times k'$$

Thus the value of the rate constant depends on how the rate is defined.

In this text we will always be careful to define exactly what is meant by the rate for a given reaction so there will be no confusion about which specific rate constant is being used.

Types of Rate Laws

Notice that the rate law we have used to this point expresses rate as a function of concentration. For example, for the decomposition of NO_2 we have defined the rate as

$$\text{Rate} = -\frac{d[NO_2]}{dt} = k[NO_2]^n$$

which tells us (once we have determined the value of n) exactly how the rate depends on the concentration of the reactant, NO_2. A rate law that expresses how the *rate depends on concentration* is called the **differential rate law,** but it is often simply called the **rate law.** Thus when we use the term *the rate law* in this text, we mean the expression that gives the rate as a function of concentration.

The terms *differential rate law* and *rate law* will be used interchangeably in this text.

A second kind of rate law, the **integrated rate law,** will also be important in our study of kinetics. The integrated rate law expresses how the *concentrations depend on time.* As we will see, a given differential rate law is always related to a certain type of integrated rate law and vice versa. That is, if we determine the differential rate law for a given reaction, we automatically know the form of the integrated rate law for the reaction. This means that once we determine either type of rate law for a reaction, we also know the other one.

Which rate law we choose to determine by experiment often depends on what types of data are easiest to collect. If we can conveniently measure how the rate changes as the concentrations are changed, we can readily determine the differential (rate/concentration) rate law. On the other hand, if it is more convenient to measure the concentration as a function of time, we can determine the form of the integrated (concentration/time) rate law. We will discuss how rate laws are actually determined in the next several sections.

Why are we interested in determining the rate law for a reaction? How does it help us? It helps us because we can work backward from the rate law to find the steps by which the reaction occurs. Most chemical reactions do not take place in a single step but result from a series of sequential steps. To understand a chemical reaction, we must learn what these steps are. For example, a chemist who is designing an insecticide may study the reactions involved in the process of insect growth to see what type of molecule may interrupt this series of reactions. Or an industrial chemist may be trying to make a reaction occur faster, using less expensive conditions. To accomplish this, he or she must know which step is slowest, because it is that step that must be speeded up. Thus a chemist is usually not interested in a rate law for its own sake but for what it tells about the steps by which a reaction occurs. We will develop a process for finding the reaction steps later in this chapter.

Summary: **Rate Laws**

- **There are two types of rate laws.**
  1. **The differential rate law (often called simply the rate law) shows how the rate of a reaction depends on concentrations.**
  2. **The integrated rate law shows how the concentrations of species in the reaction depend on time.**
- **Because we will typically consider reactions under conditions where the reverse reaction is unimportant, our rate laws will involve concentrations of reactants.**

- **Because the differential and integrated rate laws for a given reaction are related in a well-defined way, the experimental determination of *either* of the rate laws is sufficient.**
- **Experimental convenience usually dictates which type of rate law is determined experimentally.**
- **Knowing the rate law for a reaction is important, mainly because we can usually infer the individual steps involved in the reaction from the specific form of the rate law.**

TABLE 15.3
Concentration/Time Data for the Reaction $2N_2O_5(soln) \longrightarrow 4NO_2(soln) + O_2(g)$ (at 45°C)

$[N_2O_5]$ (mol/L)	Time (s)
1.00	0
0.88	200
0.78	400
0.69	600
0.61	800
0.54	1000
0.48	1200
0.43	1400
0.38	1600
0.34	1800
0.30	2000

15.3 Determining the Form of the Rate Law

The first step in understanding how a given chemical reaction occurs is to determine the *form* of the rate law. In this section we will explore ways to obtain the differential rate law for a reaction. First, we will consider the decomposition of dinitrogen pentoxide in carbon tetrachloride solution:

$$2N_2O_5(soln) \longrightarrow 4NO_2(soln) + O_2(g)$$

Data for this reaction at 45°C are listed in Table 15.3 and plotted in Fig. 15.2. In this reaction the oxygen gas escapes from the solution and thus does not react with the nitrogen dioxide, so we do not have to be concerned about the effects of the reverse reaction at any time over the life of the reaction. In other words, the reverse reaction is negligible at all times over the course of this reaction.

Evaluation of the reaction rates at N_2O_5 concentrations of 0.90 M and 0.45 M, by taking the slopes of the tangents to the curve at these points (see Fig. 15.2), yields the following data:

$[N_2O_5]$	Rate (mol L^{-1} s^{-1})
0.90 M	5.4×10^{-4}
0.45 M	2.7×10^{-4}

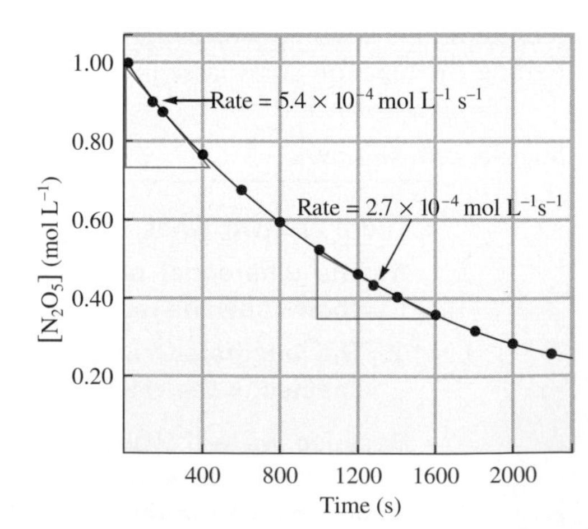

Figure 15.2
A plot of the concentration of N_2O_5 as a function of time for the reaction $2N_2O_5(soln) \longrightarrow 4NO_2(soln) + O_2(g)$ (at 45°C). Note that the reaction rate at $[N_2O_5] = 0.90\ M$ is twice that at $[N_2O_5] = 0.45\ M$.

Note that when $[N_2O_5]$ is halved, the rate is also halved. This means that the rate of this reaction depends on the concentration of N_2O_5 to the *first power*. In other words, the (differential) rate law for this reaction is

$$\text{Rate} = -\frac{d[N_2O_5]}{dt} = k[N_2O_5]$$

Thus the reaction is *first order* in N_2O_5. Note that for this reaction the order is *not* the same as the coefficient of N_2O_5 in the balanced equation for the reaction. This reemphasizes the fact that the order of a particular reactant must be obtained by *observing* how the reaction rate depends on the concentration of that reactant.

First order: rate $= k[A]$. Doubling the concentration of A doubles the reaction rate.

We have seen that determining the instantaneous rate at two different reactant concentrations gives the following rate law for the decomposition of N_2O_5:

$$\text{Rate} = -\frac{d[A]}{dt} = k[A]$$

where A represents N_2O_5.

Method of Initial Rates

The most common method for directly determining the form of the differential rate law for a reaction is the **method of initial rates.** The **initial rate** of a reaction is the instantaneous rate determined just after the reaction begins (just after $t = 0$). The idea is to determine the instantaneous rate before the initial concentrations of reactants have changed significantly. Several experiments are carried out using different initial concentrations, and the initial rate is determined for each run. The results are then compared to see how the initial rate depends on the initial concentrations. This procedure allows the form of the rate law to be determined. We will illustrate the method of initial rates by using the following reaction:

The value of the initial rate is determined for each experiment at the same value of t as close to $t = 0$ as possible.

$$NH_4^+(aq) + NO_2^-(aq) \longrightarrow N_2(g) + 2H_2O(l)$$

Table 15.4 gives initial rates obtained from three experiments involving different initial concentrations of reactants.

TABLE 15.4 **Initial Rates from Three Experiments for the Reaction $NH_4^+(aq) + NO_2^-(aq) \longrightarrow N_2(g) + 2H_2O(l)$**

Experiment	Initial Concentration of NH_4^+	Initial Concentration of NO_2^-	Initial Rate (mol L^{-1} s^{-1})
1	0.100 M	0.0050 M	1.35×10^{-7}
2	0.100 M	0.010 M	2.70×10^{-7}
3	0.200 M	0.010 M	5.40×10^{-7}

The general form of the rate law for this reaction is

$$\text{Rate} = -\frac{d[NH_4^+]}{dt} = k[NH_4^+]^n[NO_2^-]^m$$

We can determine the values of n and m by observing how the initial rate depends on the initial concentrations of NH_4^+ and NO_2^-. In Experiments 1 and 2, where the initial concentration of NH_4^+ remains the same but where the initial concentration of NO_2^- doubles, the observed initial rate also doubles. Since

$$\text{Rate} = k[NH_4^+]^n[NO_2^-]^m$$

we have, for Experiment 1,

$$\text{Rate} = 1.35 \times 10^{-7} \text{ mol L}^{-1}\text{ s}^{-1} = k(0.100 \text{ mol/L})^n(0.0050 \text{ mol/L})^m$$

and for Experiment 2,

$$\text{Rate} = 2.70 \times 10^{-7} \text{ mol L}^{-1}\text{ s}^{-1} = k(0.100 \text{ mol/L})^n(0.010 \text{ mol/L})^m$$

Rates 1, 2, and 3 were determined at the same value of t (very close to $t = 0$).

The ratio of these rates is

$$\frac{\text{Rate 2}}{\text{Rate 1}} = \underbrace{\frac{2.70 \times 10^{-7} \text{ mol L}^{-1}\text{ s}^{-1}}{1.35 \times 10^{-7} \text{ mol L}^{-1}\text{ s}^{-1}}}_{2.00} = \frac{\cancel{k(0.100 \text{ mol/L})^n}(0.010 \text{ mol/L})^m}{\cancel{k(0.100 \text{ mol/L})^n}(0.0050 \text{ mol/L})^m}$$

$$= \underbrace{\frac{(0.010 \text{ mol/L})^m}{(0.0050 \text{ mol/L})^m}}_{(2.0)^m}$$

Thus

$$\frac{\text{Rate 2}}{\text{Rate 1}} = 2.00 = (2.0)^m$$

which means the value of m is 1. The rate law for this reaction is first order in the reactant NO_2^-.

A similar analysis of the results for Experiments 2 and 3 yields the following ratio:

$$\frac{\text{Rate 3}}{\text{Rate 2}} = \frac{5.40 \times 10^{-7} \text{ mol L}^{-1}\text{ s}^{-1}}{2.70 \times 10^{-7} \text{ mol L}^{-1}\text{ s}^{-1}} = \frac{(0.200 \text{ mol/L})^n}{(0.100 \text{ mol/L})^n}$$

$$= 2.00 = \left(\frac{0.200}{0.100}\right)^n = (2.00)^n$$

The value of n is also 1.

We have shown that the values of n and m are both 1. Therefore, the rate law is

$$\text{Rate} = k[NH_4^+][NO_2^-]$$

Overall reaction order is the sum of the orders for each reactant. For a discussion of how this term can be misleading, see "Some Provocative Opinions on the Terminology of Chemical Kinetics" by John C. Reeve, J. Chem. Ed., vol. 68, p. 728, 1991.

This rate law is first order in both NO_2^- and NH_4^+. Note that it is merely a coincidence that n and m have the same values as the coefficients of NH_4^+ and NO_2^- in the balanced equation for the reaction.

The **overall reaction order** is the sum of n and m. For this reaction, $n + m = 2$. The reaction is second order overall.

The value of the rate constant (k) can now be calculated by using the results of *any* of the three experiments shown in Table 15.4. From the data for Experiment 1 we know that

$$\text{Rate} = k[NH_4^+][NO_2^-]$$
$$1.35 \times 10^{-7} \text{ mol L}^{-1} \text{ s}^{-1} = k(0.100 \text{ mol/L})(0.0050 \text{ mol/L})$$

Then

$$k = \frac{1.35 \times 10^{-7} \text{ mol L}^{-1} \text{ s}^{-1}}{(0.100 \text{ mol/L})(0.0050 \text{ mol/L})} = 2.7 \times 10^{-4} \text{ L mol}^{-1} \text{ s}^{-1}$$

EXAMPLE 15.1

The reaction between bromate ions and bromide ions in acidic aqueous solution is given by the following equation:

$$BrO_3^-(aq) + 5Br^-(aq) + 6H^+(aq) \longrightarrow 3Br_2(l) + 3H_2O(l)$$

Table 15.5 gives the results of four experiments involving this reaction. Using these data, determine the orders for all three reactants, the overall reaction order, and the value of the rate constant.

TABLE 15.5 The Results from Four Experiments to Study the Reaction $BrO_3^-(aq) + 5Br^-(aq) + 6H^+(aq) \longrightarrow 3Br_2(l) + 3H_2O(l)$

Experiment	Initial Concentration of BrO_3^- (mol/L)	Initial Concentration of Br^- (mol/L)	Initial Concentration of H^+ (mol/L)	Measured Initial Rate (mol L^{-1} s^{-1})
1	0.10	0.10	0.10	8.00×10^{-4}
2	0.20	0.10	0.10	1.60×10^{-3}
3	0.20	0.20	0.10	3.20×10^{-3}
4	0.10	0.10	0.20	3.20×10^{-3}

Solution

The general form of the rate law for this reaction is

$$\text{Rate} = k[BrO_3^-]^n[Br^-]^m[H^+]^p$$

We can determine the values of n, m, and p by comparing the rates from the various experiments. To determine the value of n, we use the results from Experiments 1 and 2, in which only $[BrO_3^-]$ changes:

$$\frac{\text{Rate 2}}{\text{Rate 1}} = \frac{1.60 \times 10^{-3} \text{ mol L}^{-1} \text{ s}^{-1}}{8.00 \times 10^{-4} \text{ mol L}^{-1} \text{ s}^{-1}}$$

$$= \frac{k(0.20 \text{ mol/L})^n(0.10 \text{ mol/L})^m(0.10 \text{ mol/L})^p}{k(0.10 \text{ mol/L})^n(0.10 \text{ mol/L})^m(0.10 \text{ mol/L})^p}$$

$$2.0 = \left(\frac{0.20 \text{ mol/L}}{0.10 \text{ mol/L}}\right)^n = (2.0)^n$$

Thus n is equal to 1.

To determine the value of m, we use the results from Experiments 2 and 3, in which only $[Br^-]$ changes:

$$\frac{\text{Rate 3}}{\text{Rate 2}} = \frac{3.20 \times 10^{-3}\ \text{mol L}^{-1}\ \text{s}^{-1}}{1.60 \times 10^{-3}\ \text{mol L}^{-1}\ \text{s}^{-1}}$$

$$= \frac{k(0.20\ \text{mol/L})^n(0.20\ \text{mol/L})^m(0.10\ \text{mol/L})^p}{k(0.20\ \text{mol/L})^n(0.10\ \text{mol/L})^m(0.10\ \text{mol/L})^p}$$

$$2.0 = \left(\frac{0.20\ \text{mol/L}}{0.10\ \text{mol/L}}\right)^m = (2.0)^m$$

Thus m is equal to 1.

To determine the value of p, we use the results from Experiments 1 and 4, in which $[BrO_3^-]$ and $[Br^-]$ are constant but $[H^+]$ changes:

$$\frac{\text{Rate 4}}{\text{Rate 1}} = \frac{3.20 \times 10^{-3}\ \text{mol L}^{-1}\ \text{s}^{-1}}{8.00 \times 10^{-4}\ \text{mol L}^{-1}\ \text{s}^{-1}}$$

$$= \frac{k(0.10\ \text{mol/L})^n(0.10\ \text{mol/L})^m(0.20\ \text{mol/L})^p}{k(0.10\ \text{mol/L})^n(0.10\ \text{mol/L})^m(0.10\ \text{mol/L})^p}$$

$$4.0 = \left(\frac{0.20\ \text{mol/L}}{0.10\ \text{mol/L}}\right)^p$$

$$4.0 = (2.0)^p = (2.0)^2$$

Thus p is equal to 2.

The rate of this reaction is first order in BrO_3^- and Br^- and second order in H^+. The overall reaction order is $n + m + p = 4$.

The rate law can now be written:

$$\text{Rate} = k[BrO_3^-][Br^-][H^+]^2$$

The value of the rate constant k can be calculated from the results of any of the four experiments. For Experiment 1 the initial rate is 8.0×10^{-4} mol L^{-1} s^{-1}, and $[BrO_3^-] = 0.100\ M$, $[Br^-] = 0.100\ M$, and $[H^+] = 0.100\ M$. Using these values in the rate law gives

$$8.00 \times 10^{-4}\ \text{mol L}^{-1}\ \text{s}^{-1} = k(0.10\ \text{mol/L})(0.10\ \text{mol/L})(0.10\ \text{mol/L})^2$$

$$8.00 \times 10^{-4}\ \text{mol L}^{-1}\ \text{s}^{-1} = k(1.0 \times 10^{-4}\ \text{mol}^4/\text{L}^4)$$

$$k = \frac{8.00 \times 10^{-4}\ \text{mol L}^{-1}\ \text{s}^{-1}}{1.0 \times 10^{-4}\ \text{mol}^4/\text{L}^4}$$

$$= 8.00\ \text{L}^3\ \text{mol}^{-3}\ \text{s}^{-1}$$

CHECK: Verify that the same value of k can be obtained from the results of the other experiments.

15.4 The Integrated Rate Law

The rate laws we have considered so far express the rate as a function of the reactant concentrations. It is also useful to be able to express the reactant concentrations as a function of time, given the (differential) rate law for the reaction. In this section we will show how this is done.

We will proceed by first looking at reactions involving a single reactant:

$$a\text{A} \longrightarrow \text{products}$$

all of which have a rate law of the form

$$\text{Rate} = -\frac{d[\text{A}]}{dt} = k[\text{A}]^n$$

We will develop the integrated rate laws individually for the cases $n = 1$ (first order), $n = 2$ (second order), and $n = 0$ (zero order).

In the field of kinetics the rate for this type of reaction is usually defined as

$$\text{Rate} = -\frac{1}{a}\frac{d[\text{A}]}{dt}$$

where a is the coefficient of A in the balanced equation. However, to avoid complications, we will leave out the factor of $1/a$, which simply changes the value of the rate constant by a factor of a.

First-Order Rate Laws

For the reaction

$$2N_2O_5(soln) \longrightarrow 4NO_2(soln) + O_2(g)$$

experiments show that the rate law is

$$\text{Rate} = -\frac{d[N_2O_5]}{dt} = k[N_2O_5]$$

Since the rate of this reaction depends on the concentration of N_2O_5 to the first power, it is a **first-order reaction.** This means that if the concentration of N_2O_5 in a flask were suddenly doubled, the rate of production of NO_2 and O_2 would also double. Using calculus, this differential rate law can be integrated, which yields the expression

$$\ln[N_2O_5] = -kt + \ln[N_2O_5]_0$$

where ln indicates the natural logarithm, t is the time, $[N_2O_5]$ is the concentration of N_2O_5 at time t, and $[N_2O_5]_0$ is the initial concentration of N_2O_5 (at $t = 0$, the start of the experiment). Note that such an equation, called the integrated rate law, expresses the *concentration of the reactant as a function of time.*

For a chemical reaction of the form

$$a\text{A} \longrightarrow \text{products}$$

where the kinetics are first order in [A], the differential rate law is of the form

$$\text{Rate} = -\frac{d[\text{A}]}{dt} = k[\text{A}]$$

and the **integrated first-order rate law** is

$$\ln[\text{A}] = -kt + \ln[\text{A}]_0 \qquad (15.2)$$

There are three important things to note about Equation (15.2):

An integrated rate law relates concentration to reaction time.

1. The equation shows how the concentration of A depends on time. If the initial concentration of A and the value of the rate constant k are known, the concentration of A at any time can be calculated.
2. Equation (15.2) is of the form $y = mx + b$, where a plot of y versus x is a straight line with slope m and intercept b. In this case

$$y = \ln[A] \qquad x = t \qquad m = -k \qquad b = \ln[A]_0$$

For a first-order reaction, a plot of ln[A] versus t is a straight line.

Thus for a first-order reaction, plotting the natural logarithm of concentration versus time always gives a straight line. This fact is often used to test whether or not a reaction is first order. For the reaction of the type

$$aA \longrightarrow \text{products}$$

the *reaction is first order in A if a plot of ln[A] versus* t *is a straight line.* Conversely, if the plot is not a straight line, the reaction is not first order.

3. This integrated rate law for a first-order reaction can also be expressed in terms of the *ratio* of [A] and $[A]_0$, as follows:

$$\ln\left(\frac{[A]_0}{[A]}\right) = kt$$

EXAMPLE 15.2

The decomposition of N_2O_5 in the gas phase was studied at constant temperature:

$$2N_2O_5(g) \longrightarrow 4NO_2(g) + O_2(g)$$

The following results were collected:

$[N_2O_5]$ (mol/L)	Time (s)
0.1000	0
0.0707	50
0.0500	100
0.0250	200
0.0125	300
0.00625	400

Using these data, verify that the rate law is first order in $[N_2O_5]$, and calculate the value of the rate constant, where the rate $= -d[N_2O_5]/dt$.

Solution

We can verify that the rate law is first order in $[N_2O_5]$ by constructing a plot of $\ln[N_2O_5]$ versus time. The values of $\ln[N_2O_5]$ at various times are given below, and the plot of $\ln[N_2O_5]$ versus time is shown in Fig. 15.3.

$\ln[N_2O_5]$	Time (s)
−2.303	0
−2.649	50
−2.996	100
−3.689	200
−4.382	300
−5.075	400

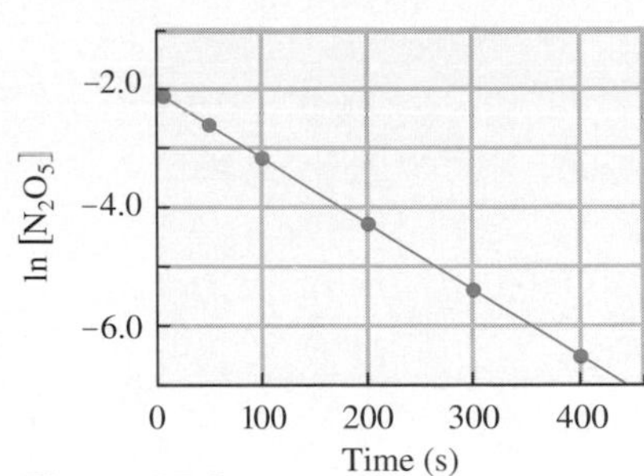

Figure 15.3
A plot of $\ln[N_2O_5]$ versus time.

The plot is a straight line, confirming that the reaction is first order in N_2O_5, since it follows the equation $\ln[N_2O_5] = -kt + \ln[N_2O_5]_0$.

Since the reaction is first order, the slope of the line equals $-k$. In this case

$$k = -(\text{slope}) = 6.93 \times 10^{-3}\ \text{s}^{-1}$$

EXAMPLE 15.3

Using the data given in Example 15.2, calculate $[N_2O_5]$ 150. s after the start of the reaction.

Solution

We know from Example 15.2 that $[N_2O_5] = 0.0500$ mol/L at 100 s and $[N_2O_5] = 0.0250$ mol/L at 200 s. Since 150 s is halfway between 100 s and 200 s, it is tempting to assume that we can use a simple average to obtain $[N_2O_5]$ at that time. This is incorrect, because it is $\ln[N_2O_5]$, not $[N_2O_5]$, that depends directly on t. To calculate $[N_2O_5]$ after 150 s, we must use Equation (15.2):

$$\ln[N_2O_5] = -kt + \ln[N_2O_5]_0$$

where $t = 150.$ s, $k = 6.93 \times 10^{-3}\ \text{s}^{-1}$ (as determined in Example 15.2), and $[N_2O_5]_0 = 0.100$ mol/L.

$$\ln([N_2O_5])_{t=150} = -(6.93 \times 10^{-3}\ \text{s}^{-1})(150.\ \text{s}) + \ln(0.100)$$
$$= -1.040 - 2.303 = -3.343$$
$$[N_2O_5]_{t=150} = \text{antilog}(-3.343) = 0.0353\ \text{mol/L}$$

Note that this value of $[N_2O_5]$ is *not* halfway between 0.0500 mol/L and 0.0250 mol/L.

Half-Life of a First-Order Reaction

The time required for a reactant to reach half of its original concentration is called the **half-life of a reaction** and is designated by the symbol $t_{1/2}$. To illustrate this idea, we can calculate the half-life of the decomposition reaction discussed

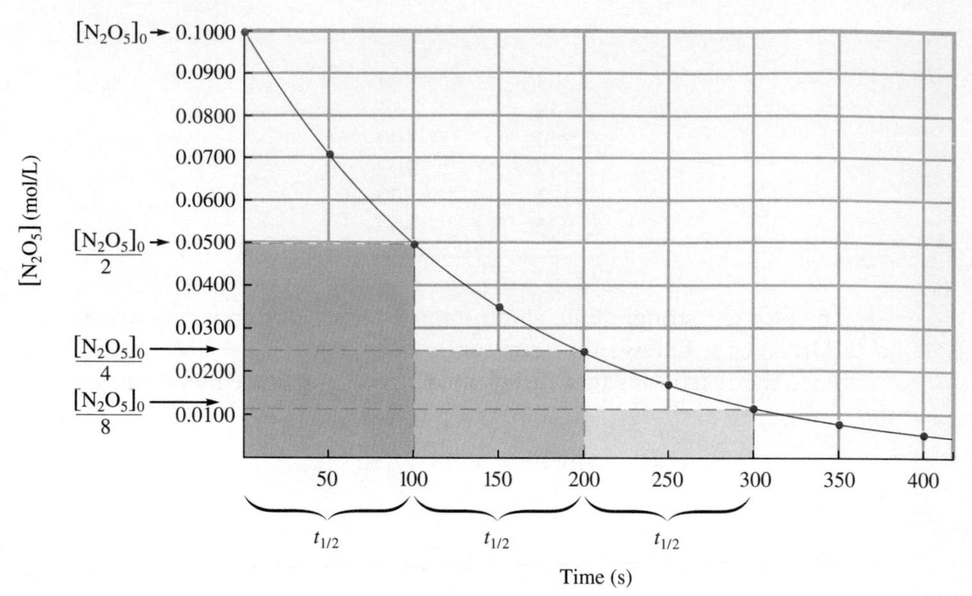

Figure 15.4
A plot of $[N_2O_5]$ versus time for the decomposition reaction of N_2O_5.

in Example 15.2. The data plotted in Fig. 15.4 show that the half-life for this reaction is 100 seconds. We can see this by considering the following numbers:

$[N_2O_5]$(mol/L)	t (s)		
0.100	0		
		$\Delta t = 100$ s;	$\frac{[N_2O_5]_{t=100}}{[N_2O_5]_{t=0}} = \frac{0.050}{0.100} = \frac{1}{2}$
0.0500	100		
		$\Delta t = 100$ s;	$\frac{[N_2O_5]_{t=200}}{[N_2O_5]_{t=100}} = \frac{0.025}{0.050} = \frac{1}{2}$
0.0250	200		
		$\Delta t = 100$ s;	$\frac{[N_2O_5]_{t=300}}{[N_2O_5]_{t=200}} = \frac{0.0125}{0.0250} = \frac{1}{2}$
0.0125	300		

Note that it *always* takes 100 seconds for $[N_2O_5]$ to be halved in this reaction.

A general formula for the half-life of a first-order reaction can be derived from the integrated rate law for the general reaction,

$$a\text{A} \longrightarrow \text{products}$$

If the reaction is first order in [A],

$$\ln\left(\frac{[\text{A}]_0}{[\text{A}]}\right) = kt$$

By definition, when $t = t_{1/2}$, $\qquad [\text{A}] = \frac{[\text{A}]_0}{2}$

Then for $t = t_{1/2}$, the integrated rate law becomes

$$\ln\left(\frac{[A]_0}{[A]_0/2}\right) = kt_{1/2}$$

or

$$\ln(2) = kt_{1/2}$$

Substituting the value of ln(2) and solving for $t_{1/2}$ gives

$$t_{1/2} = \frac{0.693}{k} \qquad (15.3)$$

This is the *general equation for the half-life of a first-order reaction*. Equation (15.3) can be used to calculate $t_{1/2}$ if k is known or k if $t_{1/2}$ is known. Note that for a first-order reaction *the half-life does not depend on concentration.*

For a first-order reaction $t_{1/2}$ is independent of the initial concentration.

EXAMPLE 15.4

A certain first-order reaction has a half-life of 20.0 min.

a. Calculate the rate constant for this reaction.

b. How much time is required for this reaction to be 75% complete?

Solution

a. Solving Equation (15.3) for k gives

$$k = \frac{0.693}{t_{1/2}} = \frac{0.693}{20.0 \text{ min}} = 3.47 \times 10^{-2} \text{ min}^{-1}$$

b. We use the integrated rate law in the form

$$\ln\left(\frac{[A]_0}{[A]}\right) = kt$$

If the reaction is 75% complete, 75% of the reactant has been consumed. This leaves 25% in the original form:

$$\frac{[A]}{[A]_0} \times 100 = 25$$

This means that

$$\frac{[A]}{[A]_0} = 0.25 \quad \text{and} \quad \frac{[A]_0}{[A]} = 4.0$$

Therefore

$$\ln\left(\frac{[A]_0}{[A]}\right) = \ln(4.0) = kt = \left(\frac{3.47 \times 10^{-2}}{\text{min}}\right)t$$

and

$$t = 40. \text{ min}$$

Thus it takes 40. min for this particular reaction to reach 75% completion.

Let's consider another way of solving this problem by using the definition of half-life. After one half-life the reaction has gone 50% to completion. If the initial concentration were 1.0 mol/L, after one half-life the concentration would be 0.50 mol/L. One more half-life would produce a concentration of 0.25 mol/L. Comparing 0.25 mol/L with the original 1.0 mol/L shows that 25% of the reactant is left after two half-lives. This is a general result. (What percentage of reactant remains after three half-lives?) Two half-lives for this reaction is 2(20.0 min), or 40.0 min, which agrees with the above answer.

Second-Order Rate Laws

For a general reaction involving a single reactant,

$$aA \longrightarrow \text{products}$$

Second order: rate $= k[A]^2$. Doubling the concentration of A quadruples the reaction rate; tripling the concentration of A increases the rate by nine times.

which is second order in A, the rate law can be defined as

$$\text{Rate} = -\frac{d[A]}{dt} = k[A]^2 \qquad (15.4)$$

Integration of this differential rate law yields the **integrated second-order rate law:**

$$\frac{1}{[A]} = kt + \frac{1}{[A]_0} \qquad (15.5)$$

Note the following characteristics of Equation (15.5):

For a second-order reaction a plot of $1/[A]$ versus t is linear.

1. A plot of $1/[A]$ versus t will produce a straight line with a slope equal to k.
2. Equation (15.5) shows how [A] depends on time and can be used to calculate [A] at any time t, provided k and $[A]_0$ are known.

When one half-life of a second-order reaction has elapsed ($t = t_{1/2}$), by definition, $[A] = [A]_0/2$. Equation (15.5) then becomes

$$\frac{1}{\frac{[A]_0}{2}} = kt_{1/2} + \frac{1}{[A]_0}$$

and

$$\frac{1}{[A]_0} = kt_{1/2}$$

Solving for $t_{1/2}$ gives *the expression for the half-life of a second-order reaction:*

$$t_{1/2} = \frac{1}{k[A]_0} \qquad (15.6)$$

EXAMPLE 15.5

When two identical molecules combine, the resulting molecule is called a *dimer.*

Butadiene reacts to form its dimer according to the equation

$$2C_4H_6(g) \longrightarrow C_8H_{12}(g)$$

The following data were collected for this reaction at a given temperature:

$[C_4H_6]$ (mol/L)	Time (±1 s)
0.01000	0
0.00625	1000
0.00476	1800
0.00370	2800
0.00313	3600
0.00270	4400
0.00241	5200
0.00208	6200

a. Is this reaction first order or second order?

b. What is the value of the rate constant for the reaction?

c. What is the half-life for the reaction under the conditions of this experiment?

Solution

a. To decide whether the rate law for this reaction is first order or second order, we must see whether the plot of $\ln[C_4H_6]$ versus time is a straight line (first order) or the plot of $1/[C_4H_6]$ versus time is a straight line (second order). The data necessary to make these plots are as follows:

t (s)	$\frac{1}{[C_4H_6]}$	$\ln[C_4H_6]$
0	100	−4.605
1000	160	−5.075
1800	210	−5.348
2800	270	−5.599
3600	319	−5.767
4400	370	−5.915
5200	415	−6.028
6200	481	−6.175

The resulting plots are shown in Fig. 15.5. Since the $\ln[C_4H_6]$ versus t plot is not a straight line, the reaction is *not* first order. The reaction is, however, second order, as shown by the linearity of the $1/[C_4H_6]$ versus t plot. Thus we can now write the rate law for this second-order reaction:

$$\text{Rate} = -\frac{d[C_4H_6]}{dt} = k[C_4H_6]^2$$

b. For a second-order reaction a plot of $1/[C_4H_6]$ versus t produces a straight line with slope k. In terms of the standard equation for a straight line, $y = mx + b$, we have $y = 1/[C_4H_6]$ and $x = t$. In this case,

$$k = \text{slope} = 6.14 \times 10^{-2}\ \text{L mol}^{-1}\ \text{s}^{-1}$$

c. The expression for the half-life of a second-order reaction is

$$t_{1/2} = \frac{1}{k[A]_0}$$

In this case $k = 6.14 \times 10^{-2}$ L mol^{-1} s^{-1} (from part b) and $[A]_0 = [C_4H_6]_0 = 0.01000\ M$ (the concentration at $t = 0$). Thus

$$t_{1/2} = \frac{1}{(6.14 \times 10^{-2}\ \text{L mol}^{-1}\ \text{s}^{-1})(1.000 \times 10^{-2}\ \text{mol/L})}$$
$$= 1.63 \times 10^3\ \text{s}$$

The initial concentration of C_4H_6 is halved in 1630 s.

Figure 15.5
(a) A plot of $\ln[C_4H_6]$ versus t. (b) A plot of $1/[C_4H_6]$ versus t.

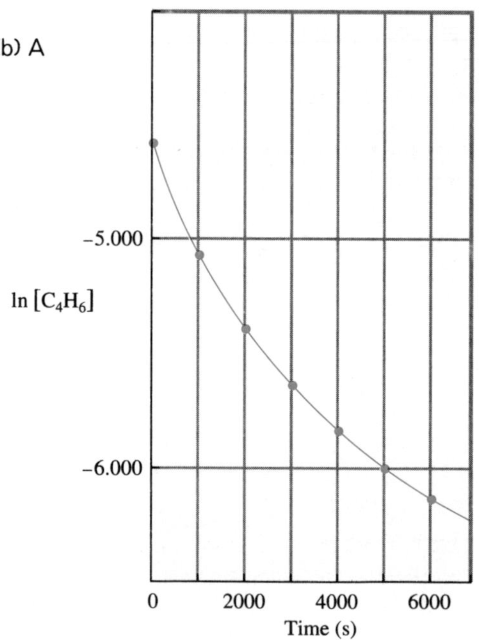

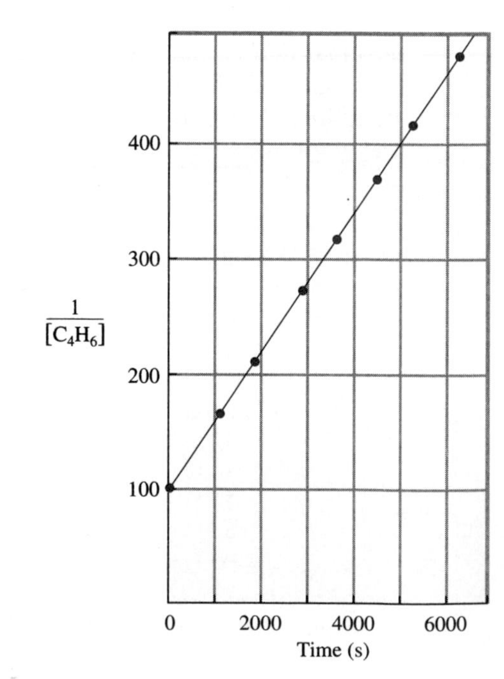

It is important to recognize the difference between the half-life for a first-order reaction and the half-life for a second-order reaction. For a second-order reaction $t_{1/2}$ depends on both k and $[A]_0$; for a first-order reaction $t_{1/2}$ depends only on k. For a first-order reaction a constant time is required to reduce the concentration of the reactant by half, and then by half again, and so on, as the reaction proceeds. In Example 15.5 we saw that this is *not* true for a second-order reaction. For that second-order reaction we found that the first half-life (the time required to go from $[C_4H_6] = 0.010\ M$ to $[C_4H_6] = 0.0050\ M$) is 1630 seconds. We can estimate the second half-life from the concentration data as a function of time. Note that to reach $0.0024\ M\ C_4H_6$ (approximately 0.0050/2) requires 5200 seconds of reaction time. Thus to get from $0.0050\ M\ C_4H_6$ to $0.0024\ M\ C_4H_6$ takes 3570 seconds (5200 − 1630). The second half-life is much longer than the first. This pattern is characteristic of second-order reactions. In fact, *for a second-order reaction each successive half-life is double the preceding one* (provided the effects of the reverse reaction can be ignored, as we are assuming here). Prove this to yourself by examining the equation $t_{1/2} = 1/(k[A]_0)$.

For a second-order reaction $t_{1/2}$ is dependent on $[A]_0$. For a first-order reaction $t_{1/2}$ is independent of $[A]_0$.

For each successive half-life $[A]_0$ is halved. Since $t_{1/2} = 1/k[A]_0$, $t_{1/2}$ doubles.

Zero-Order Rate Laws

Most reactions involving a single reactant show either first-order or second-order kinetics. However, sometimes such a reaction can be a **zero-order reaction.** The rate law for a zero-order reaction is

$$\text{Rate} = k[A]^0 = k(1) = k$$

For a zero-order reaction the rate is constant. It does not change with concentration as it does for first-order or second-order reactions.

A zero-order reaction has a constant rate.

The **integrated rate law for a zero-order reaction** is

$$[A] = -kt + [A]_0 \qquad (15.7)$$

In this case a plot of [A] versus t gives a straight line of slope $-k$, as shown in Fig. 15.6.

The expression for the half-life of a zero-order reaction can be obtained from the integrated rate law. By definition, $[A] = [A]_0/2$ when $t = t_{1/2}$, so

$$\frac{[A]_0}{2} = -kt_{1/2} + [A]_0$$

or

$$kt_{1/2} = \frac{[A]_0}{2}$$

Solving for $t_{1/2}$ gives

$$t_{1/2} = \frac{[A]_0}{2k} \qquad (15.8)$$

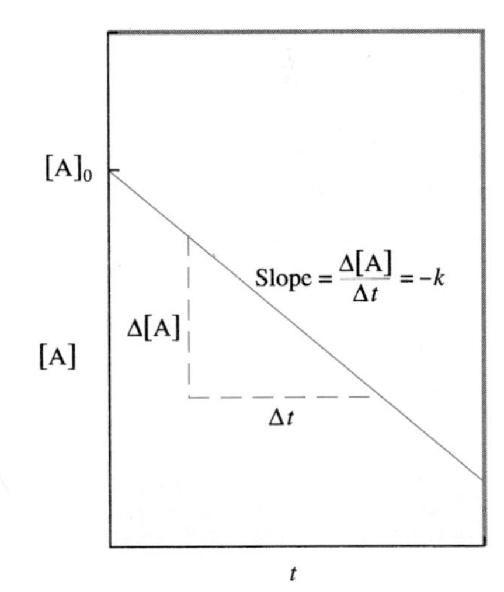

Figure 15.6
A plot of [A] versus t for a zero-order reaction.

Zero-order reactions are most often encountered when a substance such as a metal surface or an enzyme is required for the reaction to occur. For example, the decomposition reaction

$$2N_2O(g) \longrightarrow 2N_2(g) + O_2(g)$$

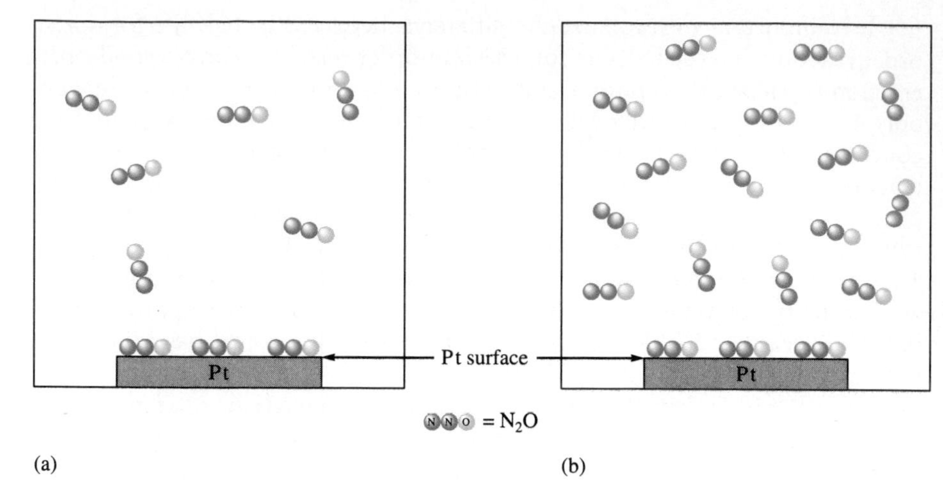

Figure 15.7
The decomposition reaction $2N_2O(g) \longrightarrow 2N_2(g) + O_2(g)$ takes place on a platinum surface. Although $[N_2O]$ is twice as great in (b) as in (a), the rate of decomposition of N_2O is the same in both cases since the platinum surface can only accommodate a certain number of molecules. As a result, this reaction is zero order.

occurs on a hot platinum surface. When the platinum surface is completely covered with N_2O molecules, an increase in the concentration of N_2O has no effect on the rate, since only those N_2O molecules on the surface can react. Under these conditions *the rate is a constant* because it is controlled by what happens on the platinum surface rather than by the total concentration of N_2O, as illustrated in Fig. 15.7. This reaction can also occur at high temperatures with no platinum surface present, but under these conditions it is not zero order.

Integrated Rate Laws for Reactions with More Than One Reactant

So far we have considered the integrated rate laws for simple reactions with only one reactant. Special techniques are required to deal with more complicated reactions. For example, consider the reaction

$$BrO_3^-(aq) + 5Br^-(aq) + 6H^+(aq) \longrightarrow 3Br_2(l) + 3H_2O(l)$$

From experimental evidence we know that the rate law is

$$\text{Rate} = -\frac{d[BrO_3^-]}{dt} = k[BrO_3^-][Br^-][H^+]^2$$

Suppose we run this reaction under conditions where $[BrO_3^-]_0 = 1.0 \times 10^{-3}\ M$, $[Br^-]_0 = 1.0\ M$, and $[H^+]_0 = 1.0\ M$. As the reaction proceeds, $[BrO_3^-]$ decreases significantly; but because the Br^- ion and H^+ ion concentrations are so large initially, relatively little of either of these two reactants is consumed.

Thus $[Br^-]$ and $[H^+]$ remain *approximately constant*. In other words, under the conditions where the Br^- ion and H^+ ion concentrations are much larger than the BrO_3^- ion concentration, we can assume that throughout the reaction

$$[Br^-] = [Br^-]_0 \qquad \text{and} \qquad [H^+] = [H^+]_0$$

This means that the rate law can be written as

$$\text{Rate} = k[Br^-]_0[H^+]_0^2[BrO_3^-] = k'[BrO_3^-]$$

where, since $[Br^-]_0$ and $[H^+]_0$ are constant,

$$k' = k[Br^-]_0[H^+]_0^2$$

The rate law

$$\text{Rate} = k'[BrO_3^-]$$

is first order. However, since this rate law was obtained by simplifying a more complicated one, it is called a **pseudo-first-order rate law.** Under the conditions of this experiment, a plot of $\ln[BrO_3^-]$ versus t gives a straight line with a slope equal to $-k'$. Since $[Br^-]_0$ and $[H^+]_0$ are known, the value of k can be calculated from the equation

$$k' = k[Br^-]_0[H^+]_0^2$$

which can be rearranged to give

$$k = \frac{k'}{[Br^-]_0[H^+]_0^2}$$

Note that the kinetics of complicated reactions can be studied by observing the behavior of one reactant at a time. If the concentration of one reactant is much smaller than the concentrations of the others, then the amounts of those reactants present in large concentrations will not change significantly and can be regarded as constant. The change in concentration with time of the reactant present in a relatively small amount can then be used to determine the order of the reaction in that component. This technique allows us to determine rate laws for complex reactions.

15.5 Rate Laws: A Summary

In the last several sections we have developed the following important points:

1. To simplify the rate laws for reactions, we have always assumed that the rate is being studied under conditions where only the forward reaction is important. This produces rate laws that only contain reactant concentrations.
2. There are two types of rate laws.
   a. The differential rate law (often called *the rate law*) shows how the rate depends on the concentrations. The forms of the rate laws for zero-order, first-order, and second-order kinetics of reactions with single reactants are shown in Table 15.6.

TABLE 15.6 Summary of the Kinetics for Reactions of the Type $aA \longrightarrow$ Products That Are Zero, First, or Second Order in [A]

	Order		
	Zero	First	Second
Rate law	Rate = k	Rate = k[A]	Rate = $k[A]^2$
Integrated rate law	$[A] = -kt + [A]_0$	$\ln[A] = -kt + \ln[A]_0$	$\frac{1}{[A]} = kt + \frac{1}{[A]_0}$
Plot needed to give a straight line	[A] versus t	ln[A] versus t	$\frac{1}{[A]}$ versus t
Relationship of rate constant to the slope of straight line	Slope = $-k$	Slope = $-k$	Slope = k
Half-life	$t_{1/2} = \frac{[A]_0}{2k}$	$t_{1/2} = \frac{0.693}{k}$	$t_{1/2} = \frac{1}{k[A]_0}$

b. The integrated rate law shows how concentration depends on time. The integrated rate laws corresponding to zero-order, first-order, and second-order kinetics of reactions with a single reactant are given in Table 15.6.

3. Whether we determine the differential rate law or the integrated rate law depends on the type of data that can be collected conveniently and accurately. Once we have experimentally determined either type of rate law for a given reaction, we can write the other rate law.

4. The most common method for experimentally determining the differential rate law is the method of initial rates. In this method several experiments are run at different initial concentrations, and the instantaneous rates are determined for each at the same value of t as close to $t = 0$ as possible. The point is to evaluate the rate before the concentrations change significantly from the initial values. From a comparison of the initial rates and the initial concentrations, the dependence of the rate on the concentrations of various reactants can be obtained—that is, the order in each reactant can be determined.

5. To experimentally determine the integrated rate law for a reaction, we measure concentrations at various values of t as the reaction proceeds. Then we see which integrated rate law correctly fits the data. Typically, this is done by ascertaining which type of plot gives a straight line. This information is described for reactions with a single reactant in Table 15.6. Once the correct straight-line plot is found, the correct integrated rate law can be chosen and the value of k obtained from the slope. Also, the (differential) rate law for the reaction can then be written.

6. The integrated rate law for a reaction that involves several reactants can be treated by choosing conditions such that the concentration of only one reactant varies in a given experiment. This is done by having the concentration of one reactant small compared with the concentrations of all of the

others, causing a rate law such as

$$\text{Rate} = k[A]^n[B]^m[C]^p$$

to reduce to

$$\text{Rate} = k'[A]^n$$

where $k' = k[B]_0^m[C]_0^p$ and $[B]_0 \gg [A]_0$ and $[C]_0 \gg [A]_0$. The value of n is obtained by determining whether a plot of [A] versus t is linear ($n = 0$), a plot of ln[A] versus t is linear ($n = 1$), or a plot of 1/[A] versus t is linear ($n = 2$). The value of k' is determined from the slope of the appropriate plot. The values of m, p, and k are found by determining the value of k' at several different concentrations of B and C.

15.6 Reaction Mechanisms

Most chemical reactions occur by a *series of steps* called the **reaction mechanism.** To understand a reaction, we must know its mechanism, and one of the main purposes for studying kinetics is to learn as much as possible about the steps involved in a reaction. In this section we explore some of the fundamental characteristics of reaction mechanisms.

Consider the reaction between nitrogen dioxide and carbon monoxide:

$$NO_2(g) + CO(g) \longrightarrow NO(g) + CO_2(g)$$

The rate law for this reaction is known from experiment to be

$$\text{Rate} = k[NO_2]^2$$

As we will see below, this reaction is more complicated than it appears from the balanced equation. This is quite typical; the balanced equation for a reaction tells us the reactants, the products, and the stoichiometry but gives no direct information about the reaction mechanism.

A balanced equation does not tell us *how* the reactants become products.

For the reaction between nitrogen dioxide and carbon monoxide, the mechanism is thought to involve the following steps:

$$NO_2(g) + NO_2(g) \xrightarrow{k_1} NO_3(g) + NO(g)$$

$$NO_3(g) + CO(g) \xrightarrow{k_2} NO_2(g) + CO_2(g)$$

where k_1 and k_2 are the rate constants of the individual reactions. In this mechanism gaseous NO_3 is an **intermediate,** a species that is neither a reactant nor a product but that is formed and consumed during the reaction sequence.

An intermediate is formed in one step and consumed in a subsequent step and so is never seen as a product.

Each of these reactions is called an **elementary step,** *a reaction whose rate law can be written from its molecularity.* **Molecularity** is defined as the number of species that must collide to produce the reaction indicated by that step. A reaction involving one molecule is called a **unimolecular step.** Reactions involving the collision of two and three species are termed **bimolecular** and **termolecular,** respectively. Termolecular steps are quite rare, because the probability of three molecules colliding simultaneously is very small. Examples of these three types of elementary steps and the corresponding rate laws are shown in Table 15.7. Note from Table 15.7 that the rate law for an elementary step follows *directly* from the molecularity of that step. For example, for a bimolecular step the rate law is always second order, either of the form $k[A]^2$ for a step

The prefix *uni-* means one, *bi-* means two, and *ter-* means three.

A unimolecular elementary step is always first order, a bimolecular step is always second order, and so on.

TABLE 15.7 **Examples of Elementary Steps and Corresponding Rate Laws**

Elementary Step	Molecularity	Rate Law
$A \longrightarrow$ products	*Uni*molecular	Rate $= k[A]$
$A + A \longrightarrow$ products $(2A \longrightarrow$ products$)$	*Bi*molecular	Rate $= k[A]^2$
$A + B \longrightarrow$ products	*Bi*molecular	Rate $= k[A][B]$
$A + A + B \longrightarrow$ products $(2A + B \longrightarrow$ products$)$	*Ter*molecular	Rate $= k[A]^2[B]$
$A + B + C \longrightarrow$ products	*Ter*molecular	Rate $= k[A][B][C]$

with a single reactant or of the form $k[A][B]$ for a step involving two reactants.

We can now define a reaction mechanism more precisely. It is a *series of elementary steps that must satisfy two requirements:*

1. The sum of the elementary steps must give the overall balanced equation for the reaction.
2. The mechanism must agree with the experimentally determined rate law.

To see how these requirements are applied, we will consider the mechanism given above for the reaction of nitrogen dioxide with carbon monoxide. First, note that the sum of the two steps gives the overall balanced equation:

$$NO_2(g) + NO_2(g) \longrightarrow NO_3(g) + NO(g)$$
$$NO_3(g) + CO(g) \longrightarrow NO_2(g) + CO_2(g)$$

$$\cancel{NO_2}(g) + NO_2(g) + \cancel{NO_3}(g) + CO(g) \longrightarrow \cancel{NO_3}(g) + NO(g) + \cancel{NO_2}(g) + CO_2(g)$$

$$\text{Overall reaction:} \quad NO_2(g) + CO(g) \longrightarrow NO(g) + CO_2(g)$$

The first requirement for a correct mechanism is met. To see whether the mechanism meets the second requirement, we need to introduce a new concept: the **rate-determining step.** Multistep reactions often have one step that is much slower than all the others. Reactants can become products only as fast as they can complete this slowest step. That is, the overall reaction can be no faster than the slowest or rate-determining step in the sequence. An analogy for this situation is the rapid pouring of water through a funnel into a container. The water collects in the container at a rate that is essentially determined by the size of the funnel opening and not by the rate of pouring.

A reaction is only as fast as its slowest step.

Which is the rate-determining step in the reaction of nitrogen dioxide with carbon monoxide? Let's *assume* that the first step is rate determining and the second step is relatively fast:

$$NO_2(g) + NO_2(g) \longrightarrow NO_3(g) + NO(g) \quad \text{Slow (rate determining)}$$
$$NO_3(g) + CO(g) \longrightarrow NO_2(g) + CO_2(g) \quad \text{Fast}$$

What we have really assumed here is that the formation of NO_3 occurs much more slowly than its reaction with CO. The rate of CO_2 production is then controlled by the rate of formation of NO_3 in the first step. Since this is an elementary step, we can write the rate law from the molecularity. The bimolecular first step has the rate law

$$\text{Rate of formation of } NO_3 = \frac{d[NO_3]}{dt} = k_1[NO_2]^2$$

Since the overall reaction rate can be no faster than the slowest step,

$$\text{Overall rate} = k_1[NO_2]^2$$

Note that this rate law agrees with the experimentally determined rate law given earlier. Since the mechanism we assumed above satisfies the two requirements stated earlier, it *may* be the correct mechanism for the reaction.

EXAMPLE 15.6

The balanced equation for the reaction of the gaseous nitrogen dioxide and fluorine is

$$2NO_2(g) + F_2(g) \longrightarrow 2NO_2F(g)$$

The experimentally determined rate law is

$$\text{Rate} = k[NO_2][F_2]$$

A suggested mechanism for this reaction is

$$NO_2 + F_2 \xrightarrow{k_1} NO_2F + F \qquad \text{Slow}$$
$$F + NO_2 \xrightarrow{k_2} NO_2F \qquad \text{Fast}$$

Is this an acceptable mechanism? That is, does it satisfy the two requirements?

Solution

The first requirement for an acceptable mechanism is that the sum of the steps should give the balanced equation:

$$NO_2 + F_2 \longrightarrow NO_2F + F$$
$$F + NO_2 \longrightarrow NO_2F$$
$$\overline{2NO_2 + F_2 + F \longrightarrow 2NO_2F + F}$$
$$\text{Overall reaction:} \quad 2NO_2 + F_2 \longrightarrow 2NO_2F$$

The first requirement is met.

The second requirement is that the mechanism must agree with the experimentally determined rate law. Since the proposed mechanism states that the first step is rate determining, the overall reaction rate must be that of the first step. The first step is bimolecular, so the rate law is

$$\text{Rate} = k_1[NO_2][F_2]$$

This has the same form as the experimentally determined rate law. The proposed mechanism is acceptable because it satisfies both requirements. (Note that we have not proved it is *the correct* mechanism.)

How does a chemist deduce the mechanism for a given reaction? The rate law is always determined first. Then using chemical intuition and following the two rules given above, the chemist constructs a possible mechanism. *A mechanism can never be proved absolutely.* We can only say that a mechanism that satisfies the two requirements is *possibly* correct. Deducing the mechanism for a chemical reaction can be difficult; it requires skill and experience. We will only touch on this process in this text.

Mechanisms with Fast Forward and Reverse First Steps

A common type of reaction mechanism is one involving a first step in which *both* the forward and reverse reactions are very fast compared with the reactions in the second step. An example of this type of mechanism is that for the decomposition of ozone to oxygen. The balanced reaction is

$$2O_3(g) \longrightarrow 3O_2(g)$$

The observed rate law is

$$\text{Rate} = k\frac{[O_3]^2}{[O_2]}$$

Note that this rate law is unusual in that it contains the concentration of a *product*. The mechanism proposed for this process is

$$O_3 \underset{k_{-1}}{\overset{k_1}{\rightleftharpoons}} O_2 + O$$

$$O + O_3 \xrightarrow{k_2} 2O_2$$

The double arrows in the first step indicate that both the forward and reverse reactions are important. They have the rate constants k_1 and k_{-1}, respectively.

For this mechanism we will assume that *both* the forward and reverse reactions of the first step are very fast compared with the reactions in the second step. This means that the second step is rate determining. Therefore, the rate for the overall reaction is equal to the rate of the second step:

$$\text{Rate} = k_2[O][O_3]$$

This rate law does not have the same form as the experimentally determined rate law. For one thing, it contains the concentration of the intermediate, an oxygen atom. We can remove [O] and obtain a rate law that agrees with experiment by making an additional assumption. We assume that the rates of the forward and reverse reactions in the first step are equal. That is, we assume that the initial reversible fast step is at equilibrium. This makes sense because the rates of both the forward and reverse reactions for the first step are so much faster than the rate of the second step. For the first step

$$\text{Rate of forward reaction} = k_1[O_3]$$

and

$$\text{Rate of reverse reaction} = k_{-1}[O_2][O]$$

At equilibrium we have

$$k_1[O_3] = k_{-1}[O_2][O]$$

We solve for [O]:

$$[O] = \frac{k_1[O_3]}{k_{-1}[O_2]}$$

Now we substitute the expression for [O] into the rate law for the second step:

$$\text{Rate} = k_2[O][O_3] = k_2\left(\frac{k_1[O_3]}{k_{-1}[O_2]}\right)[O_3] = \frac{k_2k_1[O_3]^2}{k_{-1}[O_2]}$$

$$= k\,\frac{[O_3]^2}{[O_2]}$$

where k is a composite constant representing k_2k_1/k_{-1}.

This rate law, *derived* by postulating the two elementary steps and making assumptions about the relative rates of these steps, agrees with the experimental rate law. Since this mechanism (the elementary steps *plus* the assumptions) also gives the correct overall stoichiometry, it is an acceptable mechanism for the decomposition of ozone to oxygen.

EXAMPLE 15.7

The gas-phase reaction of chlorine with chloroform is described by the equation

$$Cl_2(g) + CHCl_3(g) \longrightarrow HCl(g) + CCl_4(g)$$

The rate law determined from experiment has a noninteger order:

$$\text{Rate} = k[Cl_2]^{1/2}[CHCl_3]$$

A proposed mechanism for this reaction follows:

$$Cl_2(g) \underset{k_{-1}}{\overset{k_1}{\rightleftharpoons}} 2Cl(g)$$ Both fast with equal rates (fast equilibrium)

$$Cl(g) + CHCl_3(g) \xrightarrow{k_2} HCl(g) + CCl_3(g)$$ Slow

$$CCl_3(g) + Cl(g) \xrightarrow{k_3} CCl_4(g)$$ Fast

Is this an acceptable mechanism for the reaction?

Solution

Two questions must be answered. First, does the mechanism give the correct overall stoichiometry? Adding the three steps does yield the correct balanced equation:

$$Cl_2(g) \rightleftharpoons 2Cl(g)$$

$$Cl(g) + CHCl_3(g) \longrightarrow HCl(g) + CCl_3(g)$$

$$CCl_3(g) + Cl(g) \longrightarrow CCl_4(g)$$

$$Cl_2(g) + \cancel{Cl}(g) + CHCl_3(g) + \cancel{CCl_3}(g) + \cancel{Cl}(g)$$
$$\longrightarrow \cancel{2Cl}(g) + HCl(g) + \cancel{CCl_3}(g) + CCl_4(g)$$

Overall reaction: $Cl_2(g) + CHCl_3(g) \longrightarrow HCl(g) + CCl_4(g)$

Second, does the mechanism agree with the observed rate law? Since the overall reaction rate is determined by the rate of the slowest step,

$$\text{Overall rate} = \text{rate of second step} = k_2[Cl][CHCl_3]$$

Since the chlorine atom is an intermediate, we must find a way to eliminate [Cl] in the rate law. This can be done by recognizing that since the first step is at equilibrium, its forward and reverse rates are equal:

$$k_1[Cl_2] = k_{-1}[Cl]^2$$

Solving for $[Cl]^2$ gives

$$[Cl]^2 = \frac{k_1[Cl_2]}{k_{-1}}$$

Taking the square root of both sides yields

$$[Cl] = \left(\frac{k_1}{k_{-1}}\right)^{1/2} [Cl_2]^{1/2}$$

and

$$\text{Rate} = k_2[Cl][CHCl_3] = k_2\left(\frac{k_1}{k_{-1}}\right)^{1/2} [Cl_2]^{1/2}[CHCl_3] = k[Cl_2]^{1/2}[CHCl_3]$$

where

$$k = k_2\left(\frac{k_1}{k_{-1}}\right)^{1/2}$$

The rate law derived from the mechanism agrees with the experimentally observed rate law. This mechanism satisfies the two requirements and thus is an acceptable mechanism.

15.7 The Steady-State Approximation

In the simplest reaction mechanisms one particular step is usually rate determining. However, it is not unusual in complex, multistep reaction mechanisms for different steps to be rate determining under different sets of conditions.

In cases where a specific rate-determining step cannot be chosen, an analysis called the **steady-state approximation** is often used. The central feature of this method is the assumption that the concentration of any intermediate remains constant as the reaction proceeds. An intermediate is neither a reactant nor a product but something that is formed and then consumed as the reaction proceeds.

For example, the reaction between nitric oxide and hydrogen,

$$2NO(g) + H_2(g) \longrightarrow N_2O(g) + H_2O(g)$$

may proceed via the following mechanism:

1. $2NO \underset{k_{-1}}{\overset{k_1}{\rightleftharpoons}} N_2O_2$
2. $N_2O_2 + H_2 \xrightarrow{k_2} N_2O + H_2O$

In this mechanism the intermediate is N_2O_2. To apply the steady-state approximation to this mechanism, we assume the concentration of N_2O_2 remains constant. That is,

$$\frac{d[N_2O_2]}{dt} = 0$$

Next, we will identify the steps that produce N_2O_2 and those that consume N_2O_2 and write the rate law for each. Then we will apply the condition that the concentration of N_2O_2 is constant by setting the total rate of production of N_2O_2 equal to the total rate of consumption of N_2O_2. That is, if

$$\text{Rate of production of } N_2O_2 = \text{rate of consumption of } N_2O_2$$

then
$$\frac{d[N_2O_2]}{dt} = 0$$

Rate of Production of N_2O_2

In this mechanism N_2O_2 is produced only in the forward part of the first elementary step,

$$2NO \xrightarrow{k_1} N_2O_2$$

and the rate law for this step is

$$\frac{d[N_2O_2]}{dt} = k_1[NO]^2$$

Rate of Consumption of N_2O_2

In this mechanism N_2O_2 is consumed in the reverse part of the first step,

$$2NO \xleftarrow{k_{-1}} N_2O_2$$

and in the second step,

$$N_2O_2 + H_2 \xrightarrow{k_2} N_2O + H_2O$$

The rate laws for these steps are

$$-\frac{d[N_2O_2]}{dt} = k_{-1}[N_2O_2] \quad \text{and} \quad -\frac{d[N_2O_2]}{dt} = k_2[N_2O_2][H_2]$$

The Steady-State Condition

Now we equate the rates of production and consumption of N_2O_2:

$$\underbrace{k_1[NO]^2}_{\text{Rate of production}} = \underbrace{k_{-1}[N_2O_2] + k_2[N_2O_2][H_2]}_{\text{Total rate of consumption}}$$

The Rate Law for the Overall Reaction

Next, we will write the rate law for the overall reaction,

$$2NO + H_2 \longrightarrow N_2O + H_2O$$

We can do this in several ways, depending on which reactant or product we use to represent the rate. In this case we will choose the decomposition of H_2 to define the rate:

$$\text{Rate of reaction} = -\frac{d[H_2]}{dt}$$

Note that H_2 is consumed only in the second step of the mechanism:

$$N_2O_2 + H_2 \xrightarrow{k_2} N_2O + H_2O$$

Thus the rate law is

$$-\frac{d[H_2]}{dt} = k_2[N_2O_2][H_2]$$

However, this is not the final form of the rate law for the overall reaction because it contains the concentration of an intermediate. We can remove this concentration from the rate law by solving the steady-state expression

$$k_1[NO]^2 = k_{-1}[N_2O_2] + k_2[N_2O_2][H_2]$$

for $[N_2O_2]$:

$$[N_2O_2] = \frac{k_1[NO]^2}{k_{-1} + k_2[H_2]}$$

We substitute this expression into the rate law,

$$\text{Rate} = -\frac{d[H_2]}{dt} = k_2[N_2O_2][H_2]$$

to give

$$\text{Rate} = -\frac{d[H_2]}{dt} = k_2[H_2]\left(\frac{k_1[NO]^2}{k_{-1} + k_2[H_2]}\right)$$

or

$$\text{Rate} = -\frac{d[H_2]}{dt} = \frac{k_2k_1[H_2][NO]^2}{k_{-1} + k_2[H_2]}$$

This is the overall rate law for the proposed mechanism based on the steady-state analysis. Note that this rate law is quite complicated, which is common for rate laws obtained by assuming steady-state conditions. The usual practice for testing the validity of a complicated rate law involves choosing concentration conditions that produce a simpler form of the rate law. For example, if the reaction between NO and H_2 is studied under conditions where the concentration of H_2 is large enough so that

$$k_2[H_2] \gg k_{-1}$$

then the full rate law

$$\frac{k_2k_1[H_2][NO]^2}{k_{-1} + k_2[H_2]}$$

reduces to

$$\frac{k_2k_1[H_2][NO]^2}{k_2[H_2]} = k_1[NO]^2$$

Thus at sufficiently high concentrations of H_2, the reaction should show second-order dependence on the concentration of nitric oxide, if the suggested mechanism is valid.

On the other hand, at low concentrations of H_2, such that

$$k_{-1} \gg k_2[H_2]$$

the rate law reduces to the form

$$\text{Rate} = \frac{k_2k_1}{k_{-1}}[H_2][NO]^2 = k[H_2][NO]^2$$

Studies of the reaction under these conditions should show first-order dependence on $[H_2]$, as well as second-order dependence on [NO], if the suggested mechanism is valid.

The ideas we have developed above for a specific mechanism can be generalized as follows.

Summary: **Analyzing a Mechanism Using the Steady-State Approximation**

1. **Write the proposed mechanism (the elementary steps).**
2. **Construct a steady-state expression for each intermediate I by applying the criterion**

   $$\frac{d[I]}{dt} = 0$$

   which means that

   Rate of production of I = rate of consumption of I

   This condition is implemented by identifying each step that produces or consumes I and writing the appropriate rate law for each. The sum of the rate laws that produce I are then set equal to the sum of the rate laws that consume I.
3. **From the steady-state approximation for each intermediate I_1, I_2, . . . , solve for $[I_1]$, $[I_2]$, ...**

4. **Construct the rate law for the overall reaction in terms of one of the reactants or products. The decision about which reactant or product to use in constructing the rate law is based on convenience.**

5. **Use the expressions from Step 3 for $[I_1]$, $[I_2]$, . . . to substitute for concentrations of intermediates found in the rate law from Step 4. The goal is to obtain an overall rate law that contains only reactant and/or product concentrations.**

15.8 A Model for Chemical Kinetics

How do chemical reactions occur? We already have given some indications. For example, we have seen that the rates of chemical reactions depend on the concentrations of the reacting species. The initial rate for the reaction

$$a\text{A} + b\text{B} \longrightarrow \text{products}$$

can be described by the rate law

$$\text{Rate} = k[\text{A}]^n[\text{B}]^m$$

where the order of each reactant depends on the detailed reaction mechanism. This explains why reaction rates depend on concentration. But what about some of the other factors affecting reaction rates? For example, how does temperature affect the speed of a reaction?

We can answer this question qualitatively from our experience. We use refrigerators because food spoilage is retarded at low temperatures. The combustion of wood occurs at a measurable rate only at high temperatures. An egg cooks in boiling water much faster at sea level than in Leadville, Colorado (elevation 10,000 feet), where the boiling point of water is approximately 90°C. These observations and others lead us to conclude that *chemical reactions speed up when the temperature is increased.* Experiments have shown that virtually all rate constants show an exponential increase with absolute temperature, as represented in Fig. 15.8.

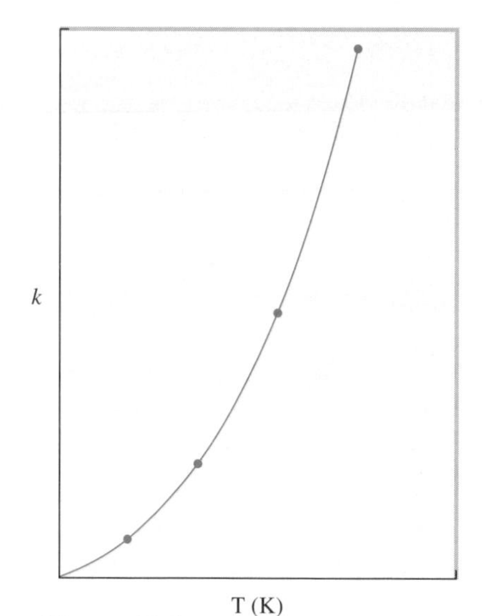

Figure 15.8
A plot showing the exponential dependence of the rate constant on absolute temperature. The exact temperature dependence of k varies with the reaction in question. This plot represents the behavior of a rate constant that doubles for every 10 K increase in temperature.

In this section we will introduce a model that can be used to account for the observed characteristics of reaction rates. This model, the **collision model,** is built around the central idea that *molecules must collide to react.* We have already seen that this assumption can explain the concentration dependence of reaction rates. Now we need to consider whether this model can also account for the observed temperature dependence of reaction rates.

The kinetic molecular theory of gases predicts that an increase in temperature increases molecular velocities and so increases the frequency of intermolecular collisions. This agrees with the observation that reaction rates are greater at higher temperatures. Thus there is qualitative agreement between the collision model and experimental observations. However, it is found that the rate of reaction is much smaller than the calculated collision frequency in a given collection of gas particles. This must mean that *only a small fraction of the collisions produces a reaction.* Why?

This question was first addressed in the 1880s by Svante Arrhenius. He proposed the existence of a *threshold energy,* called the **activation energy,** that

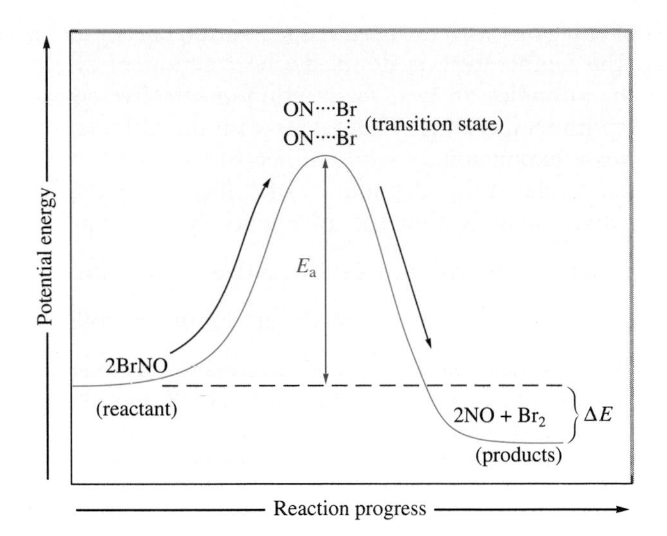

Figure 15.9
The change in potential energy as a function of reaction progress for the reaction $2BrNO \longrightarrow 2NO + Br_2$. The activation energy (E_a) represents the energy needed to disrupt the BrNO molecules, so that they can form products. The quantity ΔE represents the net change in energy in going from reactant to products.

must be overcome to produce a chemical reaction. We can see that this proposal makes sense by considering the decomposition of BrNO in the gas phase:

$$2BrNO(g) \longrightarrow 2NO(g) + Br_2(g)$$

In this reaction two Br—N bonds must be broken and one Br—Br bond must be formed. Breaking a Br—N bond requires considerable energy (243 kJ/mol), which must come from somewhere. The collision model postulates that the energy required to break the bonds comes from the kinetic energies possessed by the reacting molecules before the collision. This kinetic energy is changed into potential energy as the molecules are distorted during a collision, breaking bonds and rearranging the atoms into the product molecules.

We can envision the reaction progress as shown in Fig. 15.9. The arrangement of atoms found at the top of the potential energy "hill," or barrier, is called the **activated complex,** or **transition state.** The conversion of BrNO to NO and Br_2 is exothermic, as indicated by the fact that the products have lower energy than the reactant. However, ΔE has no effect on the rate of the reaction. Rather, the rate depends on the size of the activation energy E_a.

The higher the activation energy, the slower the reaction at a given temperature.

The main point here is that a certain minimum energy is required for two BrNO molecules to "get over the hill" so that products can form. This is furnished by the collision energy. A collision between two BrNO molecules with small kinetic energies will not have enough energy to get over the barrier and no reaction occurs. Thus, at a given temperature only a certain fraction of the collisions possess enough energy to be effective, and thus to result in product formation.

We can be more precise by recalling from Chapter 5 that a distribution of velocities occurs in a sample of gas molecules. Therefore, a distribution of collision energies also occurs, as shown in Fig. 15.10 for two different temperatures. Figure 15.10 also shows the activation energy for the reaction in question. Only collisions with energy greater than the activation energy are able to react (get over the barrier). At the lower temperature T_1, the fraction of effective collisions is quite small. However, as the temperature is increased to T_2,

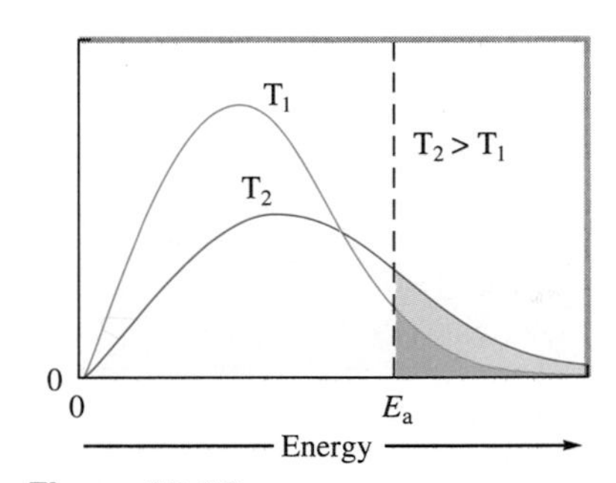

Figure 15.10
Plot showing the number of collisions with a particular energy at (a) T_1 and (b) T_2, where $T_2 > T_1$.

the fraction of collisions with the required activation energy increases dramatically. When the temperature is doubled, the fraction of effective collisions much more than doubles. In fact, the fraction of effective collisions increases *exponentially* with temperature. This agrees with the observation that rates of reactions increase exponentially with temperature.

Arrhenius postulated that the number of collisions having an energy equal to or greater than the activation energy is given by the expression

$$\text{Number of collisions with at least the activation energy} = (\text{total number of collisions})e^{-E_a/RT}$$

where E_a is the activation energy, R is the universal gas constant, and T is the Kelvin temperature. The factor $e^{-E_a/RT}$ represents the fraction of collisions with energy E_a or greater at temperature T.

We have seen that not all molecular collisions are effective in producing chemical reactions because a minimum energy is required for the reaction to occur. There is, however, another complication. Experiments show that the *observed reaction rate is considerably smaller than the rate of collisions with enough energy to surmount the barrier*. This means that many collisions, even though they have the required energy, still do not produce a reaction. Why not?

The answer lies in the **molecular orientations** during collisions. We can illustrate this effect by using the reaction between two BrNO molecules, as shown in Fig. 15.11. Some collision orientations can lead to reaction, and others cannot. Therefore, we must include a correction factor to allow for collisions with nonproductive molecular orientations.

To summarize, two requirements must be satisfied for reactants to collide successfully (to rearrange to form products):

1. The collision must involve enough energy to produce the reaction; that is, the collision energy must equal or exceed the activation energy.
2. The relative orientations of the reactants must allow formation of any new bonds necessary to produce products.

Taking these factors into account, we can represent the rate constant as

$$k = zpe^{-E_a/RT}$$

where z is the collision frequency (the total number of collisions per second). The factor p in this expression, called the **steric factor,** reflects the fraction of collisions with effective orientations. Recall that the factor $e^{-E_a/RT}$ represents the fraction of collisions with sufficient energy to produce a reaction. This expression is most often written in the form

$$k = Ae^{-E_a/RT} \tag{15.9}$$

which is called the **Arrhenius equation.** In this equation A, which replaces zp, is called the **pre-exponential factor** or **frequency factor** for the reaction.

Taking the natural logarithm of each side of the Arrhenius equation yields

$$\ln(k) = -\frac{E_a}{R}\left(\frac{1}{T}\right) + \ln(A) \tag{15.10}$$

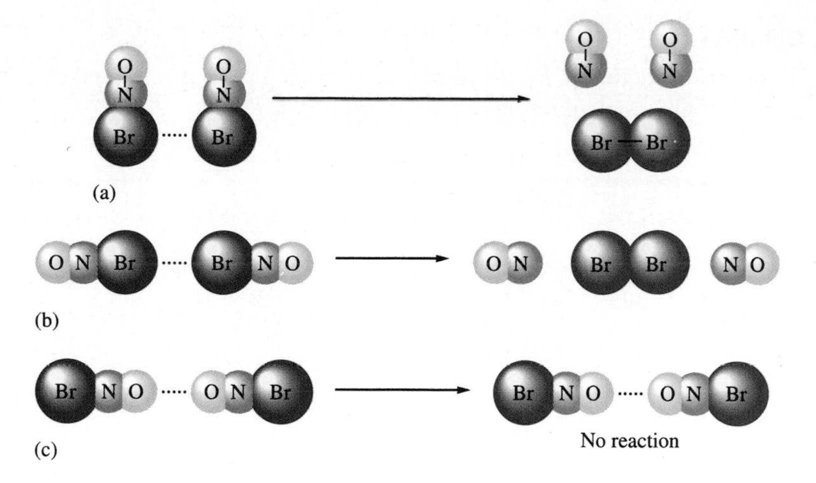

Figure 15.11
Several possible orientations for a collision between two BrNO molecules. Orientations (a) and (b) can lead to a reaction, but orientation (c) cannot.

Equation (15.10) is a linear equation of the type $y = mx + b$, where $y = \ln(k)$, $m = -E_a/R$ = slope, $x = 1/T$, and $b = \ln(A)$ = intercept. Thus for a reaction where the rate constant obeys the Arrhenius equation, a plot of $\ln(k)$ versus $1/T$ gives a straight line. The slope and intercept can be used to determine the values of E_a and A characteristic of that reaction. The fact that most rate constants obey the Arrhenius equation to a good approximation indicates that the collision model for chemical reactions is physically reasonable.

EXAMPLE 15.8

The reaction

$$2N_2O_5(g) \longrightarrow 4NO_2(g) + O_2(g)$$

was studied at several temperatures and the following values of k were obtained:

k (s^{-1})	T (°C)
2.0×10^{-5}	20
7.3×10^{-5}	30
2.7×10^{-4}	40
9.1×10^{-4}	50
2.9×10^{-3}	60

Calculate the value of E_a for this reaction.

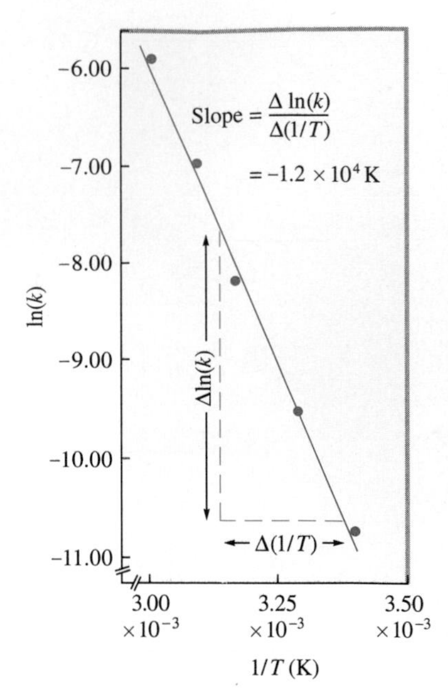

Figure 15.12
Plot of $\ln(k)$ versus $1/T$ for the reaction $2N_2O_5(g) \longrightarrow 4NO_2(g) + O_2(g)$. The value of the activation energy for this reaction can be obtained from the slope of the line, which equals $-E_a/R$.

Solution

To obtain the value of E_a, we need to construct a plot of $\ln(k)$ versus $1/T$. First, we must calculate values of $\ln(k)$ and $1/T$:

T (°C)	T (K)	$1/T$ (K)	k (s^{-1})	$\ln(k)$
20	293	3.41×10^{-3}	2.0×10^{-5}	−10.82
30	303	3.30×10^{-3}	7.3×10^{-5}	−9.53
40	313	3.19×10^{-3}	2.7×10^{-4}	−8.22
50	323	3.10×10^{-3}	9.1×10^{-4}	−7.00
60	333	3.00×10^{-3}	2.9×10^{-3}	−5.84

The plot of $\ln(k)$ versus $1/T$ is shown in Fig. 15.12. The slope is found to be -1.2×10^4 K. Since

$$\text{Slope} = -\frac{E_a}{R}$$

then

$$E_a = -R(\text{slope}) = -(8.3145 \text{ J K}^{-1} \text{ mol}^{-1})(-1.2 \times 10^4 \text{ K})$$
$$= 1.0 \times 10^5 \text{ J/mol}$$

Thus the value of the activation energy for this reaction is 1.0×10^5 J/mol.

The most common procedure for finding E_a for a reaction involves measuring the rate constant k at several temperatures and then plotting $\ln(k)$ versus $1/T$, as shown in Example 15.8. However, E_a can also be calculated from the values of k at only two temperatures using a formula that can be derived as follows from Equation (15.10).

At temperature T_1, the rate constant is k_1; thus

$$\ln(k_1) = -\frac{E_a}{RT_1} + \ln(A)$$

At temperature T_2, the rate constant is k_2; thus

$$\ln(k_2) = -\frac{E_a}{RT_2} + \ln(A)$$

Subtracting the first equation from the second gives

$$\ln\left(\frac{k_2}{k_1}\right) = \frac{E_a}{R}\left(\frac{1}{T_1} - \frac{1}{T_2}\right) \quad (15.11)$$

Therefore the values of k_1 and k_2 measured at temperatures T_1 and T_2 can be used to calculate E_a.

15.9 Catalysis

We have seen that the rate of a reaction increases dramatically with temperature. If a particular reaction does not occur fast enough at normal temperatures, we can speed it up by raising the temperature. However, sometimes this is not feasible. For example, since living cells can survive only in a rather narrow temperature range, the human body is designed to operate at an almost constant temperature of 98.6°F. But many of the complicated biochemical reactions keeping us alive would be much too slow at this temperature without intervention. We survive only because the body contains many substances called **enzymes,** which increase the rates of these reactions even though body temperature remains constant. In fact, almost every biologically important reaction is assisted by a specific enzyme. An important example involves the enzyme carbonic anhydrase, which catalyzes the reaction of carbon dioxide with water:

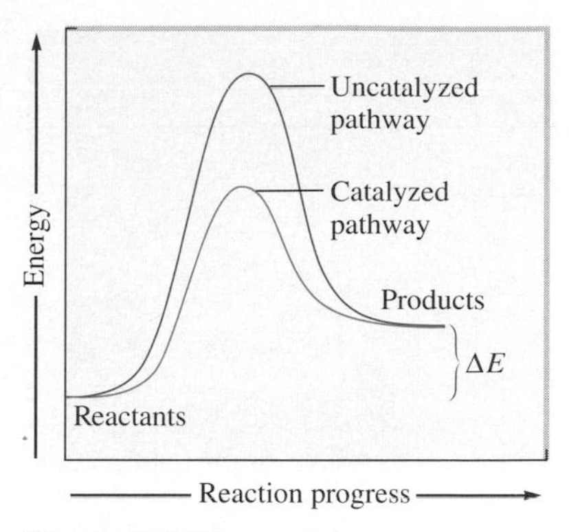

Figure 15.13
Energy plots for a catalyzed and an uncatalyzed pathway for a given reaction.

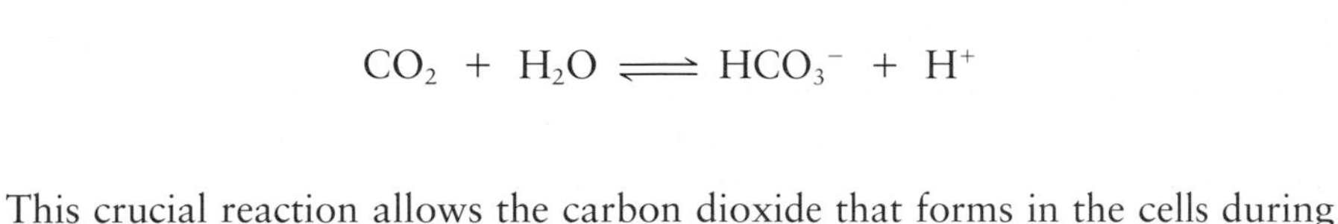

$$CO_2 + H_2O \rightleftharpoons HCO_3^- + H^+$$

This crucial reaction allows the carbon dioxide that forms in the cells during metabolism to be removed. If the carbon dioxide were allowed to accumulate, it would poison the cell. Carbonic anhydrase is so efficient that one molecule of enzyme can catalyze the reaction of over *600,000 carbon dioxide molecules in one second!*

Although it is possible to use higher temperatures to speed up commercially important reactions, such as the Haber process for synthesizing ammonia, this is very expensive. In a chemical plant an increase in temperature means significantly increased energy costs. The use of an appropriate catalyst allows a reaction to proceed rapidly at a relatively low temperature and therefore can hold down production costs.

A **catalyst** is *a substance that speeds up a reaction without being consumed itself.* Just as virtually all vital biological reactions are assisted by enzymes (biological catalysts), almost all industrial processes also benefit from the use of catalysts. For example, the production of sulfuric acid uses vanadium(V) oxide, and the Haber process uses a mixture of iron and iron oxide.

How does a catalyst work? Remember that for each reaction a certain energy barrier must be surmounted. How can we make a reaction occur faster without raising the temperature to increase the molecular kinetic energies? The solution is to provide a new pathway for the reaction, one with a *lower activation energy.* That is what a catalyst does, as shown in Fig. 15.13. Because the catalyst allows the reaction to occur along a pathway with a lower activation energy, a much larger fraction of collisions is effective at a given temperature. Thus the reaction rate is increased. This effect is illustrated in Fig. 15.14. Note from this diagram that although a catalyst lowers the activation energy (E_a) for a reaction, it does not affect the energy difference (ΔE) between products and reactants.

Catalysts are classified as homogeneous or heterogeneous. A **homogeneous catalyst** is one that is *present in the same phase (physical state) as the reacting molecules.* A **heterogeneous catalyst** exists *in a different phase,* usually as a solid.

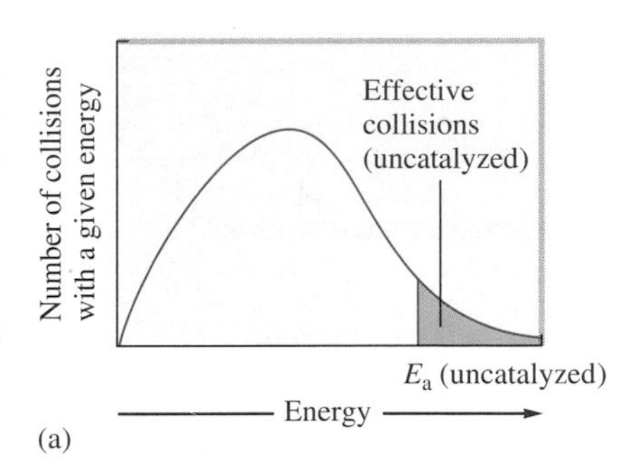

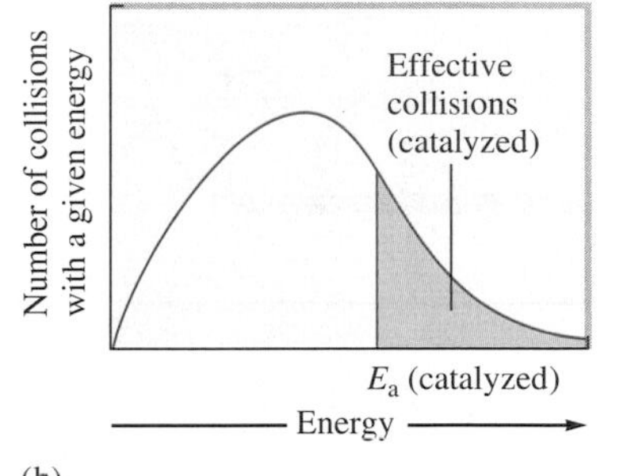

Figure 15.14
Effect of a catalyst on the number of reaction-producing collisions. Because a catalyst provides a reaction pathway with a lower activation energy, a much greater fraction of the collisions is effective for the catalyzed pathway than for the uncatalyzed pathway (at a given temperature). This allows reactants to become products at a much higher rate, even though there is no temperature increase.

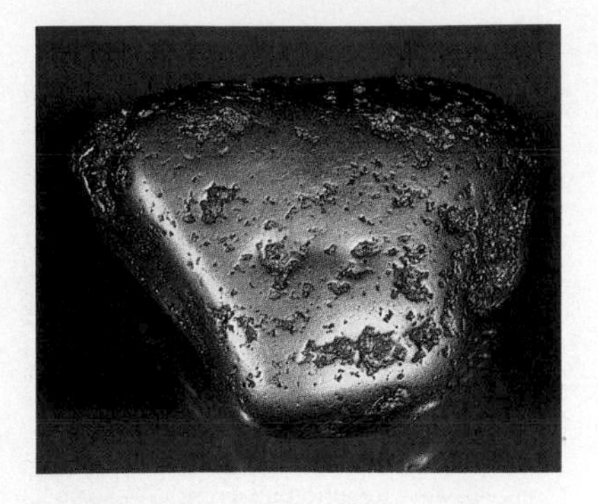

A nugget of platinum. A solid catalyst such as platinum can greatly increase the rate of a hydrogenation reaction.

Heterogeneous Catalysis

Heterogeneous catalysis most often involves gaseous reactants being adsorbed on the surface of a solid catalyst. **Adsorption** refers to the collection of one substance on the surface of another substance; **absorption** refers to the penetration of one substance into another. Water is *absorbed* by a sponge.

One of the earliest examples of heterogeneous catalysis involves the synthesis of ammonia from nitrogen and hydrogen. This process was developed in 1909 by the German chemist Fritz Haber—who tested more than 1000 possible catalysts before settling on iron as the best choice. (Today ammonia manufacturers use a solid catalyst consisting of a mixture of iron, potassium, and calcium that performs better than iron alone.) Although Haber had no means for determining why iron was a good catalyst, it is now understood that the key to iron's effectiveness is that the strong nitrogen-nitrogen and hydrogen-hydrogen bonds are weakened when the H_2 and N_2 molecules are bound to iron atoms on the surface of the metal. Iron turns out to be an ideal catalyst because the iron-nitrogen bond is sufficiently strong so that the nitrogen atoms on the surface do not recombine to form N_2 but is weak enough to allow nitrogen and hydrogen atoms on the surface to combine to form ammonia.

An important example of heterogeneous catalysis occurs in the hydrogenation of unsaturated hydrocarbons, compounds composed mainly of carbon and hydrogen and containing some carbon-carbon double bonds. Hydrogenation is an important industrial process used to change unsaturated fats, occurring as oils, to saturated fats (solid shortenings such as Crisco). In this process the C=C bonds are converted to C—C bonds, through the addition of hydrogen.

A simple example of hydrogenation involves ethylene:

$$\underset{\text{Ethylene}}{H_2C{=}CH_2}(g) + H_2(g) \longrightarrow \underset{\text{Ethane}}{H_3C{-}CH_3}(g)$$

This reaction is quite slow at normal temperatures, mainly because the strong bond in the hydrogen molecule results in a large activation energy for the reaction. However, the reaction rate can be greatly increased by using a solid catalyst of platinum, palladium, or nickel. The hydrogen and ethylene adsorb on the catalyst surface, where the reaction occurs. The main function of the catalyst apparently is to allow formation of metal-hydrogen interactions that weaken the H—H bonds and facilitate the reaction. The mechanism is illustrated in Fig. 15.15.

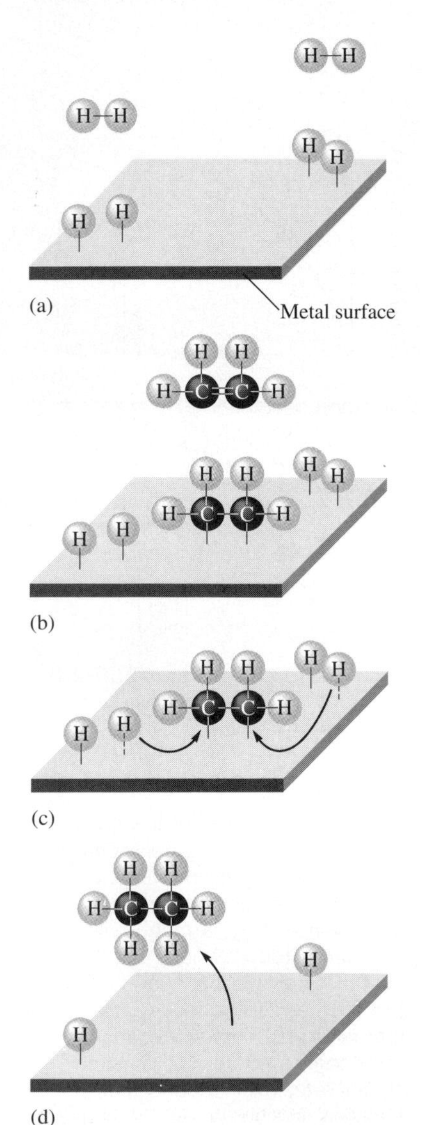

Figure 15.15
Heterogeneous catalysis of the hydrogenation of ethylene. (a) Hydrogen is adsorbed on the metal surface, forming metal-hydrogen bonds and breaking the H—H bonds. (b) During adsorption the C—C π bond in ethylene is broken and metal-carbon bonds are formed. (c) The adsorbed molecules and atoms migrate toward each other on the metal surface, forming new C—H bonds. (d) The C atoms in ethane (C_2H_6) have completely saturated bonding capacities and so cannot bind strongly to the metal surfaces. The C_2H_6 molecule thus escapes.

Typically, heterogeneous catalysis involves four steps:

1. Adsorption and activation of the reactants
2. Migration of the adsorbed reactants on the surface
3. Reaction among the adsorbed substances
4. Escape, or *desorption,* of the products

A variety of carbon and aluminum-supported precious metal catalysts.

Heterogeneous catalysis also occurs in the oxidation of gaseous sulfur dioxide to gaseous sulfur trioxide. This process is especially interesting because it illustrates both positive and negative consequences of chemical catalysis.

The negative side is the formation of damaging air pollutants. Recall that sulfur dioxide, a toxic gas with a choking odor, is formed when sulfur-containing fuels are burned. However, it is sulfur trioxide that causes most of the environmental damage, mainly through the production of acid rain. When sulfur trioxide combines with a droplet of water, sulfuric acid is formed:

$$H_2O(l) + SO_3(g) \longrightarrow H_2SO_4(aq)$$

This sulfuric acid can cause considerable damage to vegetation, buildings and statues, and fish populations.

Sulfur dioxide is *not* rapidly oxidized to sulfur trioxide in clean, dry air. Why, then, is there a problem? The answer is catalysis. Dust particles and water droplets catalyze the reaction between SO_2 and O_2 in the air.

On the positive side, the heterogeneous catalysis of the oxidation of SO_2 is used to advantage in the manufacture of sulfuric acid, where the reaction of O_2 and SO_2 to form SO_3 is catalyzed by a solid mixture of platinum and vanadium(V) oxide.

Heterogeneous catalysis is also utilized in the catalytic converters of automobile exhaust systems. The exhaust gases, containing compounds such as nitric oxide, carbon monoxide, and unburned hydrocarbons, are passed through a converter containing beads of solid catalyst (see Fig. 15.16). The catalyst promotes the conversion of carbon monoxide to carbon dioxide, hydrocarbons to carbon dioxide and water, and nitric oxide to nitrogen gas, to lessen the environmental impact of the exhaust gases. However, this beneficial catalysis can, unfortunately, be accompanied by the unwanted catalysis of the oxidation of SO_2 to SO_3, the latter with the moisture present to form sulfuric acid.

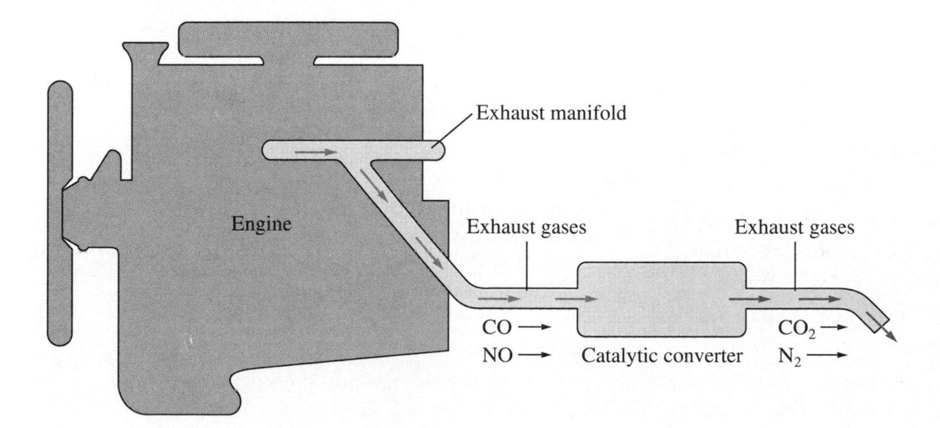

Figure 15.16
The exhaust gases from an automobile engine are passed through a catalytic converter to minimize environmental damage.

Enzymes: Nature's Catalysts

The most impressive examples of homogeneous catalysis occur in nature, where the complex reactions necessary for plant and animal life are made possible by enzymes. Enzymes are large molecules specifically tailored to facilitate a given type of reaction. Usually enzymes are proteins, an important class of biomolecules constructed from α-amino acids that have the general structure

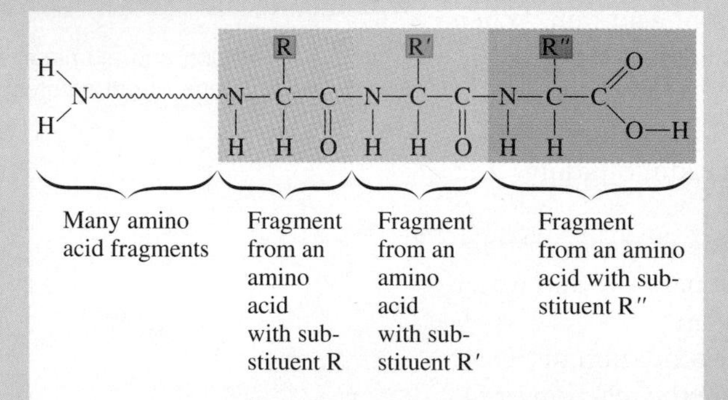

where R represents any one of 20 different substituents. These amino acid molecules can be "hooked together" to form a polymer (a word meaning "many parts") called a *protein*. The general structure of a protein can be represented as follows:

H R R' R" O
N~~~C—C—N—C—C—N—C—C
H H O H H O H H O—H

Many amino acid fragments | Fragment from an amino acid with substituent R | Fragment from an amino acid with substituent R' | Fragment from an amino acid with substituent R"

Since specific proteins are needed by the human body, the proteins in food must be broken into their constituent amino acids, which are then used to construct new proteins in the body's cells. The reaction in which a protein is broken down one amino acid at a time is shown in Fig. 15.17. Note that in this reaction a water molecule reacts with a protein molecule to produce an amino acid and a new protein containing one less amino acid. Without the help of the enzymes found in human cells, this reaction would be much too slow to be useful. One of these enzymes is *carboxypeptidase-A,* a zinc-containing protein (Fig. 15.18).

Carboxypeptidase-A captures the protein to be acted on (called the *substrate*) in a special groove and positions the substrate so that the end is in the *active site,* where the catalysis occurs (Fig. 15.19). Note that the Zn^{2+} ion bonds to the oxygen of the C=O (carbonyl) group. This polarizes the electron density in the carbonyl group, allowing the neighboring C—N bond to be broken much more easily. When the reaction is completed, the remaining portion of the substrate protein and the newly formed amino acid are released by the enzyme.

Water molecule | Protein | Amino acid | New protein

Figure 15.17
The removal of the end amino acid from a protein by reaction with a molecule of water. The products are an amino acid and a new, smaller protein.

The process just described for carboxypeptidase-A is characteristic of the behavior of other enzymes. Enzyme catalysis can be represented by the series of reactions shown below:

$$\mathrm{E} + \mathrm{S} \underset{k_{-1}}{\overset{k_1}{\rightleftharpoons}} \mathrm{E} \cdot \mathrm{S}$$

$$\mathrm{E} \cdot \mathrm{S} \overset{k_2}{\rightleftharpoons} \mathrm{E} + \mathrm{P}$$

where E represents the enzyme, S represents the substrate, E · S represents the enzyme-substrate complex, and P represents the products. The enzyme and substrate form a complex where the reaction occurs. The enzyme then releases the products and is ready to repeat the process. The most amazing thing about enzymes is their efficiency. Because an enzyme plays its catalytic role over and over and very rapidly, only a tiny amount of enzyme is required. This makes the isolation of enzymes for study quite difficult.

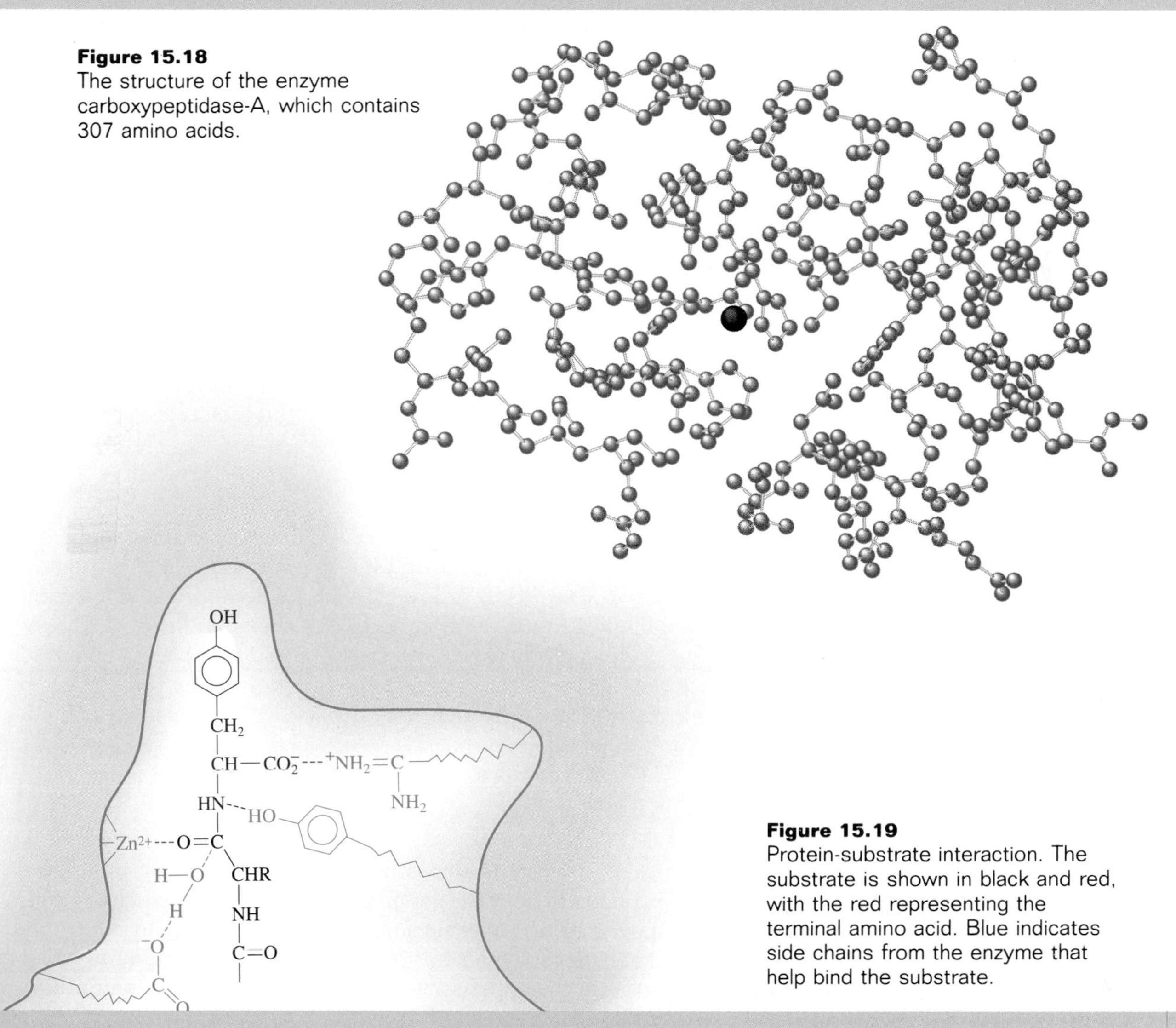

Figure 15.18
The structure of the enzyme carboxypeptidase-A, which contains 307 amino acids.

Figure 15.19
Protein-substrate interaction. The substrate is shown in black and red, with the red representing the terminal amino acid. Blue indicates side chains from the enzyme that help bind the substrate.

Because of the complex nature of the reactions that take place in the converter, a mixture of catalysts is used. The most effective catalytic materials are transition metal oxides and noble metals such as palladium and platinum. A catalytic converter typically consists of platinum and rhodium particles deposited on a ceramic honeycomb, a configuration that maximizes the contact between the metal particles and the exhaust gases. In studies performed during the last ten years researchers at General Motors have shown that rhodium promotes the dissociation of NO molecules adsorbed on its surface, thereby enhancing the conversion of NO, a serious air pollutant, to N_2, a natural component of pure air.

One consequence of the widespread use of catalytic converters has been the need to remove lead from gasoline. Tetraethyl lead was used for more than 50 years as a very effective "octane booster" due to its antiknocking characteristics. However, lead quickly destroys much of a converter's catalytic efficiency. This poisoning effect, along with health concerns about the toxicity of lead, has necessitated the removal of lead from gasoline, which has caused a search for other antiknock additives and the redesigning of engines to run on lower-octane gasoline.

Yet another application of solid-state catalysis occurs in the desulfurization of petroleum. Natural petroleum includes various molecules that contain sulfur atoms. Combustion of this petroleum produces SO_2, which must be removed from the exhaust to prevent air pollution. One way to prevent pollution by SO_2 is to remove the sulfur from the petroleum before it is used for fuel—the desulfurization of petroleum. One type of sulfur containing molecules found in petroleum are thiols, which can be written R—SH, where R represents a molecular fragment containing a long chain of carbon atoms. In desulfurization the goal is to remove the sulfur from this molecule to produce a hydrocarbon (R—H):

$$R{-}SH \longrightarrow R{-}H + S$$

Oil chemists have found that this process is catalyzed by a mixture of molybdenum, cobalt, and sulfur. Because this catalyst is a very complicated substance, we are not certain how it works. It is thought that the desulfurization reaction involves the thiol binding to the catalytic surface as shown in Figure 15.16. In this process the S—H bond is broken, with the R—S and the H being bound to metal atoms on the surface of the catalyst. The H atom is then transferred to the R fragment to form R—H, which migrates away, leaving the S behind.

Homogeneous Catalysis

A homogeneous catalyst exists in the same phase as the reacting molecules. There are many examples in both the gas and liquid phases. One such example is the unusual catalytic behavior of nitric oxide toward ozone. In the troposphere, the part of the atmosphere closest to earth, nitric oxide catalyzes ozone production. However, in the upper atmosphere it catalyzes the decomposition of ozone. Both of these effects are unfortunate environmentally.

In the lower atmosphere NO is produced in any high-temperature combustion process where N_2 is present. The reaction

$$N_2(g) + O_2(g) \longrightarrow 2NO(g)$$

is very slow at normal temperatures because of the very strong N≡N and O=O bonds. However, at elevated temperatures, such as those found in the internal combustion engines of automobiles, significant quantities of NO form. Some of this NO is converted back to N_2 in the catalytic converter, but significant amounts escape into the atmosphere to react with oxygen:

$$2NO(g) + O_2(g) \longrightarrow 2NO_2(g)$$

In the atmosphere NO_2 can absorb light and decompose as follows:

$$NO_2(g) \xrightarrow{\text{Light}} NO(g) + O(g)$$

The oxygen atom formed in this process is very reactive and can combine with oxygen molecules to form ozone:

$$O_2(g) + O(g) \longrightarrow O_3(g)$$

Ozone is a powerful oxidizing agent that can react with other air pollutants to form substances irritating to the eyes and lungs.

Although O_2 is represented here as the oxidizing agent for NO, the actual oxidizing agent is probably some type of peroxide compound produced by the reaction of oxygen with pollutants. The direct reaction of NO with O_2 is very slow.

In this series of reactions nitric oxide is acting as a true catalyst because it assists the production of ozone without being consumed itself. This can be seen by summing the reactions:

$$\begin{aligned} NO(g) + \tfrac{1}{2}O_2(g) &\longrightarrow NO_2(g) \\ NO_2(g) &\xrightarrow{\text{Light}} NO(g) + O(g) \\ O_2(g) + O(g) &\longrightarrow O_3(g) \\ \hline \tfrac{3}{2}O_2(g) &\longrightarrow O_3(g) \end{aligned}$$

In the upper atmosphere the presence of nitric oxide has the opposite effect—the depletion of ozone. The series of reactions involved is

$$\begin{aligned} NO(g) + O_3(g) &\longrightarrow NO_2(g) + O_2(g) \\ O(g) + NO_2(g) &\longrightarrow NO(g) + O_2(g) \\ \hline O(g) + O_3(g) &\longrightarrow 2O_2(g) \end{aligned}$$

Nitric oxide is again catalytic, but here its effect is to change O_3 to O_2. This is a potential problem because O_3, which absorbs ultraviolet light, is necessary to protect us from the harmful effects of this high-energy radiation. That is, we want O_3 in the upper atmosphere to block ultraviolet radiation from the sun. However, we do not want it in the lower atmosphere where we have to breathe it and its oxidation products.

The ozone layer in the upper atmosphere is also threatened by *freons,** a group of stable, noncorrosive compounds used as refrigerants and as propellants in aerosol cans. The most commonly used substance of this type is Freon-12, CCl_2F_2. The chemical inertness of freons makes them more useful but also creates a problem, since they remain for a long time in the environment.

* For more information, see S. Elliott and F. S. Rowland, "Chlorofluorocarbons and Stratospheric Ozone," *J. Chem. Ed.* **64** (1987), p. 387, and P. S. Zurer, "Ozone Depletion's Recurring Surprises Challenge Atmospheric Scientists," *Chemical and Engineering News,* May 24, 1993, p. 8.

CHEMICALS TO PROTECT THE OZONE

Chlorofluorocarbons (CFCs) are ideal compounds for refrigeration and air conditioning applications because they are nontoxic and chemically inert (and thus noncorrosive). However, the unreactivity of these compounds (once thought to be their major virtue) causes them to persist for long periods in the atmosphere (estimated lifetimes are hundreds of years). Eventually, these molecules reach altitudes where ultraviolet light causes them to decompose producing chlorine atoms that catalyze the decomposition of the ozone in the stratosphere. Because of this problem, the world's industrialized nations have signed the Montreal Protocol which bans CFCs by 1996 (with a 10-year grace period for developing countries). So we must find substitutes for the CFCs—and fast.

In fact, the search for substitutes is now well underway. Worldwide production of CFCs has already decreased to half of the 1986 level of 1.13 million metric tons. One strategy for replacing CFCs has been to switch to similar compounds that contain carbon and hydrogen atoms substituted for chlorine atoms. For example, the United States appliance industry has switched from freon 12 (CF_2Cl_2) to the compound CH_2FCH_3 (called HFC-134a) for home refrigerators, and most of the new cars and trucks sold in the United States have air conditioners that employ HFC-134a. Converting the 140 million autos currently on the road in the United States that use CF_2Cl_2 will pose a major headache, but experience suggests that replacement of freon 12 with HFC-134a is less expensive than was originally feared. For example, Volvo Cars of North America estimates that a Volvo can be converted from freon 12 to HFC-134a for around $300.

The electronics industry, which formerly used large quantities of CFCs as precision cleaning agents, is also rapidly switching to other chemicals. For example, IBM's San Jose manufacturing facility, which was the largest industrial emitter of CFCs, has now switched completely to water-based cleaning technologies.

Among the most difficult challenges in making the switch away from CFCs are the centrifugal chillers used in the cooling systems of many commercial buildings. Most of these systems use freon-11 ($CFCl_3$) and will eventually be converted to HCFC-123 ($CHCl_2CF_3$) at a cost estimated to be nearly 2 billion dollars. The hydrogen atoms on the HCFCs make them more reactive in the lower atmosphere (the troposphere) so that most of them disappear before they can reach the stratosphere. However, because a small fraction of these molecules will reach the upper atmosphere, the HCFCs are also regulated under the Montreal Protocol. HCFC molecules, such as HCFC-123, which have relatively low ozone depletion potentials probably will be produced far into the 21st century, although environmental groups are applying pressure to ban HCFCs earlier.

A related environmental issue involves replacing the halons for firefighting applications. In particular, scientists are seeking an effective replacement for CF_3Br (halon-1301), the nontoxic "magic gas" used to flood enclosed spaces such as offices, aircraft, race cars and military tanks in case of fire. The compound CF_3I, which appears to have a lifetime in the atmosphere of only a few days, looks like a promising candidate but much more research on the toxicology and ozone-depleting properties of CF_3I will be required before it receives government approval as a halon substitute.

The chemical industry has responded amazingly fast to the ozone depletion emergency. It is encouraging that we can act rapidly when an environmental crisis occurs. Now we need to get better at keeping the environment at a higher priority as we plan for the future.

Eventually, they migrate into the upper atmosphere to be decomposed by high-energy light. Among the decomposition products are chlorine atoms:

$$CCl_2F_2(g) \xrightarrow{\text{Light}} CClF_2(g) + Cl(g)$$

These chlorine atoms can catalyze the decomposition of ozone:

$$Cl(g) + O_3(g) \longrightarrow ClO(g) + O_2(g)$$
$$O(g) + ClO(g) \longrightarrow Cl(g) + O_2(g)$$
$$\overline{O(g) + O_3(g) \longrightarrow 2O_2(g)}$$

The problem of freons has been brought strongly into focus by the recent discovery of a mysterious "hole" in the ozone layer in the stratosphere over Antarctica. Studies to find the reason for the hole have found unusually high levels of chlorinemonoxide (ClO) in the atmosphere over Antarctica. This strongly implicates the freons in the atmosphere as being at least partially responsible for the ozone destruction in the area.

Because they pose environmental problems, freons were banned years ago by the U.S. government for use in aerosol cans, but they are still widely used in air conditioners and refrigerators. In an attempt to totally phase out the use of freons, usually called CFCs (*c*hloro*f*luoro*c*arbons) in industry, the industrial nations of the world have agreed to a treaty called the Montreal Protocol on Substances that Deplete the Ozone Layer. This treaty calls for the end of production of CFCs by January 1, 1996. The ban on the production of halons (bromofluorocarbons used for firefighting) is already in place. Although the ban on CFCs is not yet in effect, production of these chemicals is already declining, and monitoring stations around the globe have detected a significant slowdown in the atmospheric buildup of Freon-11 ($CFCl_3$) and Freon-12. These measurements suggest that the atmospheric concentrations of these CFCs will peak by the turn of the century and then start to decline, allowing the ozone to replenish itself.

Exercises

A blue exercise number indicates that the answer to that exercise appears at the back of this book.

Reaction Rates

1. Thiosulfate ion is oxidized by iodine according to the following reaction:

$$2S_2O_3^{3-}(aq) + I_2(aq) \longrightarrow S_4O_6^{2-}(aq) + 2I^-(aq)$$

If 0.0080 mol of $S_2O_3^{2-}$ is consumed in 1.0 L of solution per second, what is the rate of consumption of I_2? At what rates are $S_4O_6^{2-}$ and I^- produced?

2. In the Haber process for the production of ammonia,

$$N_2(g) + 3H_2(g) \longrightarrow 2NH_3(g)$$

what is the relationship between the rate of production of ammonia and the rate of consumption of hydrogen?

3. What are the units for each of the following if concentrations are expressed in moles per liter?
   a. rate of a chemical reaction
   b. rate constant for a zero-order rate law*
   c. rate constant for a first-order rate law
   d. rate constant for a second-order rate law
   e. rate constant for a third-order rate law

4. The rate law* for the reaction

$$Cl_2(g) + CHCl_3(g) \longrightarrow HCl(g) + CCl_4(g)$$

is

$$\text{Rate} = k[Cl_2]^{1/2}[CHCl_3]$$

What are the units for k?

5. Concentrations of trace substances in the atmosphere are often expressed in molecules per cubic centimeter. What are the units for the quantities in Exercise 3 if concentrations are expressed this way?

* In the Exercises the term *rate law* always means differential rate law.

6. The hydroxyl radical (OH) is an important oxidizing agent in the atmosphere. At 298 K the rate constant for the reaction of OH with benzene is 1.24×10^{-12} cm^3 molecule^{-1} s^{-1}. Calculate the value of the rate constant in L mol^{-1} s^{-1}.

7. The decomposition of hydrogen iodide on finely divided gold at 150°C is zero order with respect to HI:

$$2HI(g) \xrightarrow{Au} H_2(g) + I_2(g)$$

$$\text{Rate} = -\frac{d[HI]}{dt} = k = 1.20 \times 10^{-4} \text{ mol L}^{-1} \text{ s}^{-1}$$

a. If an experiment has an initial HI concentration of 0.250 mol/L, what is the concentration of HI after 25 min?
b. How long will it take for all of the HI to decompose?
c. Give the rates of formation for H_2 and I_2.

Rate Laws from Experimental Data: Initial-Rates Method

8. The reaction

$$2NO(g) + Cl_2(g) \longrightarrow 2NOCl(g)$$

was studied at −10°C. The following results were obtained, where

$$\text{Rate} = -\frac{d[Cl_2]}{dt}$$

$[NO]_0$ (mol/L)	$[Cl_2]_0$ (mol/L)	Initial Rate (mol L^{-1} min^{-1})
0.10	0.10	0.18
0.10	0.20	0.35
0.20	0.20	1.45

a. What is the rate law*?
b. What is the value of the rate constant?

9. The reaction

$$2I^-(aq) + S_2O_8^{2-}(aq) \longrightarrow I_2(aq) + 2SO_4^{2-}(aq)$$

was studied at 25°C. The following results were obtained, where

$$\text{Rate} = -\frac{d[S_2O_8^{2-}]}{dt}$$

$[I^-]_0$ (mol/L)	$[S_2O_8^{2-}]_0$ (mol/L)	Initial Rate (mol L^{-1} s^{-1})
0.080	0.040	12.50×10^{-6}
0.040	0.040	6.250×10^{-6}
0.080	0.020	5.560×10^{-6}
0.032	0.040	4.350×10^{-6}
0.060	0.030	6.410×10^{-6}

a. Determine the rate law*.
b. Calculate the value of the rate constant for each experiment and an average for the rate constant.

10. The reaction

$$I^-(aq) + OCl^-(aq) \longrightarrow IO^-(aq) + Cl^-(aq)$$

was studied and the following data were obtained:

$[I^-]_0$ (mol/L)	$[OCl^-]_0$ (mol/L)	Initial Rate (mol L^{-1} s^{-1})
0.12	0.18	7.91×10^{-2}
0.060	0.18	3.95×10^{-2}
0.030	0.090	9.88×10^{-3}
0.24	0.090	7.91×10^{-2}

a. What is the rate law*?
b. Calculate the rate constant.

11. The rate of the reaction between hemoglobin (Hb) and carbon monoxide (CO) was studied at 20°C. The following data were collected with all concentration units in μmol/L. (A hemoglobin concentration of 2.21 μmol/L is equal to 2.21×10^{-6} mol/L.)

$[Hb]_0$ (μmol/L)	$[CO]_0$ (μmol/L)	Initial Rate (μmol L^{-1} s^{-1})
2.21	1.00	0.619
4.42	1.00	1.24
4.42	3.00	3.71

a. Determine the orders of this reaction with respect to Hb and CO.
b. Determine the rate law*.
c. Calculate the value of the rate constant.
d. What would be the initial rate for an experiment with $[Hb]_0 = 3.36$ μmol/L and $[CO]_0 = 2.40$ μmol/L?

12. The following data were obtained for the reaction

$$2ClO_2(aq) + 2OH^-(aq) \longrightarrow ClO_3^-(aq) + ClO_2^-(aq) + H_2O(l)$$

where $\text{Rate} = -\frac{d[ClO_2]}{dt}$

$[ClO_2]_0$ (mol/L)	$[OH^-]_0$ (mol/L)	Initial Rate (mol L^{-1} s^{-1})
0.0500	0.100	5.75×10^{-2}
0.100	0.100	2.30×10^{-1}
0.100	0.050	1.15×10^{-1}

* In the Exercises the term *rate law* always means differential rate law.

a. Determine the rate law* and the value of the rate constant.
b. What would be the initial rate for an experiment with $[ClO_2]_0 = 0.175$ mol/L and $[OH^-]_0 = 0.0844$ mol/L?

Integrated Rate Laws from Experimental Data

13. The dimerization of butadiene was studied at 500. K:

$$2C_4H_6(g) \longrightarrow C_8H_{12}(g)$$

The following data were obtained, where

$$\text{Rate} = -\frac{d[C_4H_6]}{dt}$$

Time (s)	$[C_4H_6]$ (mol/L)
195	1.6×10^{-2}
604	1.5×10^{-2}
1246	1.3×10^{-2}
2180	1.1×10^{-2}
6210	0.68×10^{-2}

Determine the forms of the integrated rate law, the differential rate law, and the rate constant for this reaction. (These are actual experimental data, so they may not give a perfectly straight line.)

14. Determine the forms of the integrated and the differential rate laws for the decomposition of benzene diazonium chloride,

$$C_6H_5N_2Cl(aq) \longrightarrow C_6H_5Cl(l) + N_2(g)$$

from the following data, which were collected at 50.°C and 1.00 atm:

Time (s)	N_2 Evolved (mL)
6	19.3
9	26.0
14	36.0
22	45.0
30.	50.4
∞	58.3

The total solution volume was 40.0 mL.

15. The decomposition of hydrogen peroxide was studied at a particular temperature. The following data were obtained, where

$$\text{Rate} = -\frac{d[H_2O_2]}{dt}$$

Time (s)	$[H_2O_2]$ (mol/L)
0	1.0
120 ± 1	0.91
300 ± 1	0.78
600 ± 1	0.59
1200 ± 1	0.37
1800 ± 1	0.22
2400 ± 1	0.13
3000 ± 1	0.082
3600 ± 1	0.050

Determine the integrated rate law, the differential rate law, and the value of the rate constant.

16. The rate of the reaction

$$NO_2(g) + CO(g) \longrightarrow NO(g) + CO_2(g)$$

depends only on the concentration of nitrogen dioxide at temperatures below 225°C. At a temperature below 225°C the following data were collected:

Time (s)	$[NO_2]$ (mol/L)
0	0.500
1.20×10^3	0.444
3.00×10^3	0.381
4.50×10^3	0.340
9.00×10^3	0.250
1.80×10^4	0.174

Determine the integrated rate law, the differential rate law, and the value of the rate constant at this temperature.

17. The rate of the reaction

$$O(g) + NO_2(g) \longrightarrow NO(g) + O_2(g)$$

was studied. This reaction is one step of the nitric oxide–catalyzed destruction of ozone in the upper atmosphere.
a. In the first set of experiments, NO_2 was in large excess, at a concentration of 1.0×10^{13} molecules/cm³ with the following data collected:

Time (s)	[O] (atoms/cm³)
0	5.0×10^9
1.0×10^{-2}	1.9×10^9
2.0×10^{-2}	6.8×10^8
3.0×10^{-2}	2.5×10^8

What is the order of the reaction with respect to oxygen atoms?
b. The reaction is known to be first order with respect to NO_2. Determine the overall rate law* and the value of the rate constant.

* In the Exercises the term *rate law* always means differential rate law.

18. The reaction

$$NO(g) + O_3(g) \longrightarrow NO_2(g) + O_2(g)$$

was studied by performing two experiments. In the first experiment (results shown in following table) the rate of disappearance of NO was followed in a large excess of O_3. (The $[O_3]$ remains effectively constant at 1.0×10^{14} molecules/cm^3):

Time (ms)	[NO] (molecules/cm^3)
0	6.0×10^8
100 ± 1	5.0×10^8
500 ± 1	2.4×10^8
700 ± 1	1.7×10^8
1000 ± 1	9.9×10^7

In the second experiment [NO] was held constant at 2.0×10^{14} molecules/cm^3. The data for the disappearance of O_3 are as follows:

Time (ms)	$[O_3]$ (molecules/cm^3)
0	1.0×10^{10}
50 ± 1	8.4×10^9
100 ± 1	7.0×10^9
200 ± 1	4.9×10^9
300 ± 1	3.4×10^9

a. What is the order with respect to each reactant?
b. What is the overall rate law*?
c. What is the value of the rate constant obtained from each set of experiments?

$$\text{Rate} = k'[NO]^x \qquad \text{Rate} = k''[O_3]^y$$

d. What is the value of the rate constant for the overall rate law?

$$\text{Rate} = k[NO]^x[O_3]^y$$

Half-Life

19. Radioactive decay is a process that follows a first-order rate law: Rate $= kN$, where N is the number of nuclei and the rate is the nuclear disintegrations per unit of time. The half-life for the decay of ^{239}Pu is 24,360 yr. The half-life for the decay of ^{241}Pu is 13 yr.
a. Calculate the rate constants for the decay of ^{239}Pu and ^{241}Pu.
b. Which of the two species decays more rapidly?
c. If one starts with 5.0 g of pure ^{241}Pu, how many disintegrations per second occur initially? What mass of ^{241}Pu is left after 1.0 yr, 10. yr, 100. yr?

20. A certain first-order reaction is 45.0% complete in 65 s. What are the rate constant and the half-life of this reaction?

21. The radioactive isotope ^{32}P decays by first-order kinetics and has a half-life of 14.3 days. How long does it take for 95.0% of a given sample of ^{32}P to decay?

22. It took 143 s for 50.0% of a particular substance to decompose. If the initial concentration is 0.060 M and the decomposition reaction follows second-order kinetics, what is the value of the rate constant?

23. A first-order reaction is 38.5% complete in 480. s.
a. Calculate the rate constant.
b. What is the value of the half-life?
c. How long will it take for the reaction to go to 25%, 75%, and 95% completion?

24. Radioactive copper-64 decays by first-order kinetics and has a half-life of 12.8 days.
a. What is the value of k in s^{-1}?
b. A sample contains 28.0 mg of ^{64}Cu. How many decays will be produced in the first second?
c. A chemist obtains a fresh sample of ^{64}Cu to perform a particular experiment. During this experiment, the radioactivity cannot fall below 3% of the initial measured value. How long does she have to do the experiment?

Reaction Mechanisms

25. Define each of the following.
a. elementary step
b. reaction mechanism
c. rate-determining step

26. Write the rate laws for the following elementary reactions.
a. $CH_3NC(g) \longrightarrow CH_3CN(g)$
b. $O_3(g) + NO(g) \longrightarrow O_2(g) + NO_2(g)$
c. $O_3(g) \longrightarrow O_2(g) + O(g)$
d. $O_3(g) + O(g) \longrightarrow 2O_2(g)$
e. $^{14}_{6}C \longrightarrow ^{14}_{7}N + \beta$ particle (nuclear decay)

27. Is the mechanism

$$NO + Cl_2 \xrightarrow{k_1} NOCl_2$$
$$NOCl_2 + NO \xrightarrow{k_2} 2NOCl$$

consistent with the results you obtained in Exercise 8? If so, which step is the rate-determining step?

28. The mechanism for the decomposition of hydrogen peroxide is

$$H_2O_2 \longrightarrow 2OH$$
$$H_2O_2 + OH \longrightarrow H_2O + HO_2$$
$$HO_2 + OH \longrightarrow H_2O + O_2$$

* In the Exercises the term *rate law* always means differential rate law.

Using your results from Exercise 15, specify which step is the rate-determining step.

29. The reaction

$$2NO(g) + O_2(g) \longrightarrow 2NO_2(g)$$

exhibits the rate law

$$\text{Rate} = k[NO]^2[O_2]$$

Which of the following mechanisms is consistent with this rate law?

a. $NO + O_2 \longrightarrow NO_2 + O$ Slow
$O + NO \longrightarrow NO_2$ Fast

b. $NO + O_2 \rightleftharpoons NO_3$ Fast equilibrium
$NO_3 + NO \longrightarrow 2NO_2$ Slow

c. $2NO \longrightarrow N_2O_2$ Slow
$N_2O_2 + O_2 \longrightarrow N_2O_4$ Fast
$N_2O_4 \longrightarrow 2NO_2$ Fast

d. $2NO \rightleftharpoons N_2O_2$ Fast equilibrium
$N_2O_2 \longrightarrow NO_2 + O$ Slow
$O + NO \longrightarrow NO_2$ Fast

30. The reaction

$$2NO_2Cl \longrightarrow 2NO_2 + Cl_2$$

follows the rate law

$$\text{Rate} = k[NO_2Cl]$$

The mechanism is

$$NO_2Cl \xrightarrow{k_1} NO_2 + Cl$$

$$NO_2Cl + Cl \xrightarrow{k_2} NO_2 + Cl_2$$

Which is the rate-determining step?

31. The reaction

$$5Br^-(aq) + BrO_3^-(aq) + 6H^+(aq) \longrightarrow 3Br_2(l) + 3H_2O(l)$$

is expected to obey the mechanism

$$BrO_3^-(aq) + H^+(aq) \underset{k_{-1}}{\overset{k_1}{\rightleftharpoons}} HBrO_3(aq) \quad \text{Fast equilibrium}$$

$$HBrO_3(aq) + H^+(aq) \underset{k_{-2}}{\overset{k_2}{\rightleftharpoons}} H_2BrO_3^+ \quad \text{Fast equilibrium}$$

$$Br^-(aq) + H_2BrO_3^+(aq) \xrightarrow{k_3} (Br{-}BrO_2)(aq) + H_2O(l) \quad \text{Slow}$$

$$(Br{-}BrO_2)(aq) + 4H^+(aq) + 4Br^-(aq) \longrightarrow \text{products} \quad \text{Fast}$$

Write the rate law for this reaction.

32. The gas phase reaction between Br_2 and H_2 to form HBr proceeds by the following mechanism:

$$Br_2 \underset{k_{-1}}{\overset{k_1}{\rightleftharpoons}} 2Br$$

$$Br + H_2 \underset{k_{-2}}{\overset{k_2}{\rightleftharpoons}} HBr + H$$

$$H + Br_2 \xrightarrow{k_3} HBr + Br$$

$$2Br \xrightarrow{k_4} Br_2$$

a. Under what conditions does the rate law have the form $\frac{d[HBr]}{dt} = k'[Br_2]$?
b. Under what conditions does the rate law have the form $\frac{d[HBr]}{dt} = k''[H_2][Br_2]^{1/2}$?
c. Give expressions for k' and k'' in terms of the rate constants used to define the mechanism.

Temperature Dependence of Rate Constants and the Collision Model

33. Why is the rate of a reaction affected by each of the following?
a. frequency of collisions
b. kinetic energy of collisions
c. orientation of collisions

34. Define what is meant by unimolecular and bimolecular steps.

35. Why are termolecular steps infrequently seen in chemical reactions?

36. Which of the following reactions would you expect to have the larger rate at room temperature? Why? (*Hint:* think of which would have the lower activation energy.)

$$2Ce^{4+}(aq) + Hg_2^{2+}(aq) \longrightarrow 2Ce^{3+}(aq) + 2Hg^{2+}(aq)$$

$$H_3O^+(aq) + OH^-(aq) \longrightarrow 2H_2O(l)$$

37. At 25°C the first-order rate constant for a reaction is $2.0 \times 10^3\ s^{-1}$. The activation energy is 15.0 kJ/mol. What is the value of the rate constant at 75°C?

38. A first-order reaction has rate constants of $4.6 \times 10^{-2}\ s^{-1}$ and $8.1 \times 10^{-2}\ s^{-1}$ at 0°C and 20.°C, respectively. What is the value of the activation energy?

39. For the reaction

$$H_2(g) + I_2(g) \longrightarrow 2HI(g)$$

where the rate law is

$$-\frac{d[H_2]}{dt} = k[H_2][I_2]$$

the rate constants are 2.45×10^{-4} and 0.950 L mol^{-1} s^{-1} at 302°C and 508°C, respectively.
a. Calculate the values of E_a and A for this reaction.
b. Calculate the value of k at 375°C.

40. Experimental values for the temperature dependence of the rate constant for the gas-phase reaction

$$NO(g) + O_3(g) \longrightarrow NO_2(g) + O_2(g)$$

are as follows:

T (K)	k (L mol^{-1} s^{-1})
195	1.08×10^9
230	2.95×10^9
260	5.42×10^9
298	12.0×10^9
369	35.5×10^9

Make the appropriate graph using these data, and determine the activation energy for this reaction.

41. Chemists commonly use a rule of thumb that an increase of 10 K in temperature doubles the rate of a reaction. What must the activation energy be for this statement to be true for a temperature increase from 25°C to 35°C?

42. The rate constant for the gas-phase decomposition of N_2O_5,

$$N_2O_5 \longrightarrow 2NO_2 + \tfrac{1}{2}O_2$$

has the following temperature dependence:

T (K)	k (s^{-1})
338	4.9×10^{-3}
318	5.0×10^{-4}
298	3.5×10^{-5}

Make the appropriate graph using these data, and determine the activation energy for this reaction.

43. For the following reaction profiles, indicate:
a. the positions of reactants and products
b. the activation energy
c. ΔE for the reaction

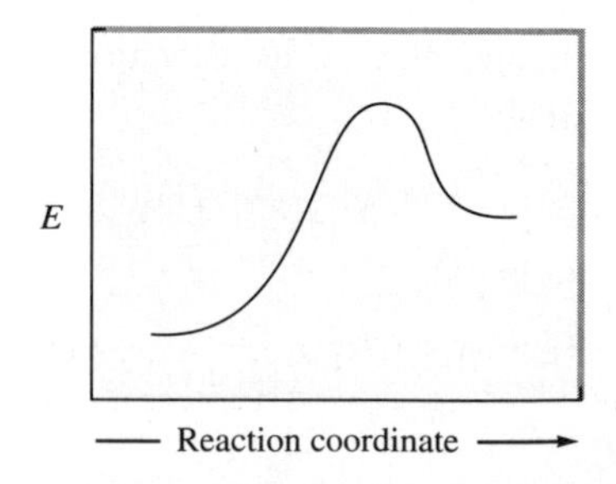

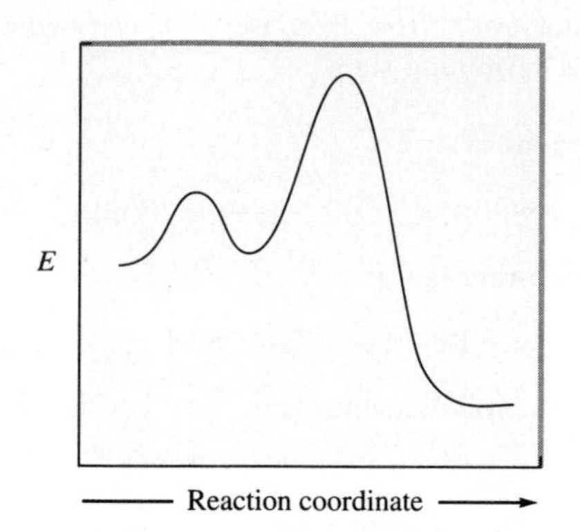

44. Draw a rough sketch of the energy profile for each of the following cases.
a. $\Delta E = +10$ kJ/mol, $E_a = 25$ kJ/mol
b. $\Delta E = -10$ kJ/mol, $E_a = 50$ kJ/mol
c. $\Delta E = -50$ kJ/mol, $E_a = 50$ kJ/mol
Which reaction will have the greatest rate at 298 K? Assume the frequency factor A is the same for all three reactions.

45. The activation energy for the reaction

$$H_2(g) + I_2(g) \longrightarrow 2HI(g)$$

is 167 kJ/mol, and ΔE for the reaction is +28 kJ/mol. What is the activation energy necessary for the decomposition of HI?

Catalysis

46. Define each of the following.
a. homogeneous catalyst
b. heterogeneous catalyst

47. How does a catalyst increase the rate of a chemical reaction?

48. Should the concentration of a homogeneous catalyst appear in the rate law?

49. Would a given reaction have the same rate law for both a catalyzed and an uncatalyzed pathway? Explain.

50. One pathway for the destruction of ozone in the upper atmosphere is

$$O_3(g) + NO(g) \longrightarrow NO_2(g) + O_2(g) \quad \text{Slow}$$
$$NO_2(g) + O(g) \longrightarrow NO(g) + O_2(g) \quad \text{Fast}$$

Overall reaction: $O_3(g) + O(g) \longrightarrow 2O_2(g)$

a. Which species is a catalyst?
b. Which species is an intermediate?

c. E_a for the uncatalyzed reaction

$$O_3(g) + O(g) \longrightarrow 2O_2$$

is 14.0 kJ. E_a for the same reaction when catalyzed is 11.9 kJ. What is the ratio of the rate constant for the catalyzed reaction to that for the uncatalyzed reaction at 25°C? Assume the frequency factor A is the same for each reaction.

51. One of the concerns about the use of freons is that they will migrate to the upper atmosphere, where chlorine atoms can be generated by the reaction

$$\underset{\text{Freon-12}}{CCl_2F_2} \xrightarrow{h\nu} CF_2Cl + Cl$$

Chlorine atoms can also act as a catalyst for the destruction of ozone. The activation energy for the reaction

$$Cl + O_3 \longrightarrow ClO + O_2$$

is 2.1 kJ/mol. Which is the more effective catalyst for the destruction of ozone, Cl or NO? (See Exercise 50.)

52. Assuming the mechanism for the hydrogenation of C_2H_4 given in Section 15.9 is correct, would you predict that the product of the reaction of C_2H_4 with D_2 would be $CH_2D—CH_2D$ or $CHD_2—CH_3$?

53. For enzyme-catalyzed reactions that follow the mechanism

$$E + S \rightleftharpoons E \cdot S$$

$$E \cdot S \rightleftharpoons E + P$$

a graph of the rate as a function of [S], the concentration of the substrate, has the following general appearance:

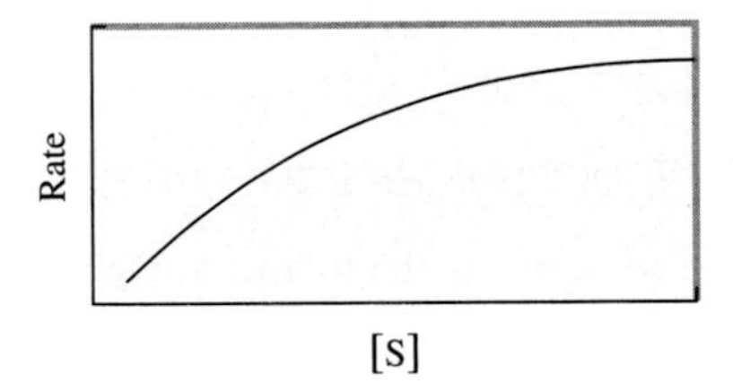

Note that at high substrate concentrations the rate no longer changes with [S]. Suggest a reason for this.

Additional Exercises

54. The reaction

$$H_2SeO_3(aq) + 6I^-(aq) + 4H^+(aq) \longrightarrow Se(s) + 2I_3^-(aq) + 3H_2O(l)$$

was studied at 0°C. The following data were obtained:

$[H_2SeO_3]_0$ (mol/L)	$[H^+]_0$ (mol/L)	$[I^-]_0$ (mol/L)	Initial Rate (mol L^{-1} s^{-1})
1.0×10^{-4}	2.0×10^{-2}	2.0×10^{-2}	1.66×10^{-7}
2.0×10^{-4}	2.0×10^{-2}	2.0×10^{-2}	3.33×10^{-7}
3.0×10^{-4}	2.0×10^{-2}	2.0×10^{-2}	4.99×10^{-7}
1.0×10^{-4}	4.0×10^{-2}	2.0×10^{-2}	6.66×10^{-7}
1.0×10^{-4}	1.0×10^{-2}	2.0×10^{-2}	0.42×10^{-7}
1.0×10^{-4}	2.0×10^{-2}	4.0×10^{-2}	13.2×10^{-7}
1.0×10^{-4}	1.0×10^{-2}	4.0×10^{-2}	3.36×10^{-7}

These relationships hold only if there is an insignificant amount of I_3^- present. Determine the rate law* and the value of the rate constant. (Assume that rate = $-d[H_2SeO_3]/dt$.)

55. The effect of the hydroxide concentration on the rate of the reaction

$$I^-(aq) + OCl^-(aq) \longrightarrow IO^-(aq) + Cl^-(aq)$$

was studied and the following data were obtained:

$[I^-]_0$ (mol/L)	$[OCl^-]_0$ (mol/L)	$[OH^-]_0$ (mol/L)	Initial Rate (mol L^{-1} s^{-1})
0.0013	0.012	0.10	9.4×10^{-3}
0.0026	0.012	0.10	18.7×10^{-3}
0.0013	0.006	0.10	4.7×10^{-3}
0.0013	0.018	0.10	14.0×10^{-3}
0.0013	0.012	0.05	18.7×10^{-3}
0.0013	0.012	0.20	4.7×10^{-3}
0.0013	0.018	0.20	7.0×10^{-3}

Determine the rate law and the value of the rate constant for this reaction.

56. The following data were obtained for the gas-phase decomposition of dinitrogen pentoxide,

$$2N_2O_5(g) \longrightarrow 4NO_2(g) + O_2(g)$$

$[N_2O_5]_0$ (mol/L)	Initial Rate (mol L^{-1} s^{-1})
0.075	8.90×10^{-4}
0.190	2.26×10^{-3}
0.275	3.26×10^{-3}
0.410	4.85×10^{-3}

Defining the rate as $-d[N_2O_5]/dt$, write the rate law and calculate the value of the rate constant.

57. The thermal degradation of silk was studied by Kuruppillai, Hersh, and Tucker ("Historic Textile and Paper Materials," *ACS Advances in Chemistry Series,* No. 212, 1986) by measuring the tensile strength of silk fibers at various times of exposure to elevated temperature. The loss of tensile strength follows first-order kinetics,

$$-\frac{ds}{dt} = ks$$

where s is the strength of the fiber retained after heating and k is the first-order rate constant. The effects of adding a deacidifying agent and an antioxidant to the silk were studied, and the following data were obtained.

	Strength Retained (%)		
Heating Time (days)	Untreated	Deacidifying Agent	Antioxidant
0.00	100.0	100.1	114.6
1.00	67.9	60.8	65.2
2.00	38.9	26.8	28.1
3.00	16.1	—	11.3
6.00	6.8	—	—

a. Determine the first-order rate constants for the thermal degradation of silk for each of the three experiments.

b. Does either of the two additives appear to retard the degradation of silk?

58. Calculate the half-life for the thermal degradation of silk for each of the three experiments in Exercise 57.

59. Sulfuryl chloride, SO_2Cl_2, decomposes to sulfur dioxide, SO_2, and chlorine, Cl_2, by a first-order reaction in the gas phase. The following data were obtained when a sample containing 5.00×10^{-2} mol of sulfuryl chloride was heated to 600K $\pm$ 1 K in a 5.00×10^{-1} L container.

Time (h)	0.00	1.00	2.00	4.00	8.00	16.00
Pressure (atm)	4.93	5.60	6.34	7.33	8.56	9.52

Define the rate as $-d[SO_2Cl_2]/dt$.

a. Determine the value of the rate constant for the decomposition of sulfuryl chloride at 600 K.

b. What is the half-life of the reaction?

c. What would be the pressure in the vessel after 0.500 h and after 12.0 h?

d. What fraction of the sulfuryl chloride remains after 20.0 h?

60. The decomposition of many substances on the surface of a heterogeneous catalyst shows the following behavior:

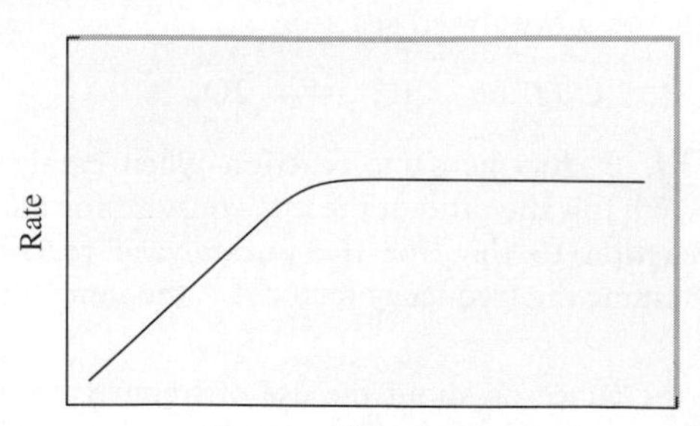

How do you account for the rate law changing from first order to zero order in the concentration of reactant?

61. The decomposition of NH_3 to N_2 and H_2 was studied on two surfaces.

Surface	E_a (kJ/mol)
W	163
Os	197

Without a catalyst the activation energy is 335 kJ/mol.

a. Which surface is the better heterogeneous catalyst for the decomposition of NH_3? Explain.

b. How many times faster is the reaction at 298 K on the W surface compared to the reaction with no catalyst present?

c. The decomposition reaction on the two surfaces obeys a rate law of the form

$$\text{Rate} = k\frac{[NH_3]}{[H_2]}$$

How can you explain the inverse dependence of the rate on the H_2 concentration?

62. The reaction

$$I^-(aq) + OCl^-(aq) \longrightarrow IO^-(aq) + Cl^-(aq)$$

is believed to occur by the following mechanism:

$$OCl^- + H_2O \underset{k_{-1}}{\overset{k_1}{\rightleftharpoons}} HOCl + OH^- \quad \text{Fast equilibrium}$$

$$I^- + HOCl \xrightarrow{k_2} HOI + Cl^- \quad \text{Slow}$$

$$HOI + OH^- \xrightarrow{k_3} H_2O + IO^- \quad \text{Fast}$$

Write the rate law for this reaction. *Note:* Since the reaction is in aqueous solution, the effective concentration of water remains constant. Thus the rate of the forward reaction in the first step can be written as

$$\text{Rate} = k[H_2O][OCl^-] = k_1[OCl^-]$$

63. The rate law for the reaction

$$BrO_3^-(aq) + 3SO_3^{2-}(aq) \longrightarrow Br^-(aq) + 3SO_4^{2-}(aq)$$

is $\quad \text{Rate} = k[BrO_3^-][SO_3^{2-}][H^+]$

The first step in the mechanism is

$$SO_3^{2-}(aq) + H^+(aq) \xrightarrow{k_1} HSO_3^-(aq) \quad \text{Fast}$$

The second step is rate determining. Write a possible second step for the mechanism.

64. In the gas phase the production of phosgene from chlorine and carbon monoxide proceeds by the following mechanism:

$$Cl_2 \underset{k_{-1}}{\overset{k_1}{\rightleftharpoons}} 2Cl \quad \text{Fast equilibrium}$$

$$Cl + CO \underset{k_{-2}}{\overset{k_2}{\rightleftharpoons}} COCl \quad \text{Fast equilibrium}$$

$$COCl + Cl_2 \xrightarrow{k_3} COCl_2 + Cl \quad \text{Slow}$$

$$2Cl \xrightarrow{k_4} Cl_2 \quad \text{Fast}$$

Overall reaction: $CO + Cl_2 \longrightarrow COCl_2$

a. Write the rate law for this reaction.
b. Which species are intermediates?

65. Consider two reaction vessels, one containing A and the other containing B, with equal concentrations at $t = 0$. If both substances decompose by first-order kinetics, where

$$k_A = 4.50 \times 10^{-4}\ s^{-1}$$

$$k_B = 3.70 \times 10^{-3}\ s^{-1}$$

how much time must pass to reach a condition such that $[A] = 4.00[B]$?

66. Evaluate the orders of A and B for the reaction

$$A + B \longrightarrow \text{products}$$

from the following data:

$[A]_0$ (mol/L)	$[B]_0$ (mol/L)	Initial Rate (mol $L^{-1}\ s^{-1}$)
0.2	1.0	3
0.5	1.0	11.8
1.0	1.0	33.5
2.0	1.0	94.7
2.0	2.0	189.5

67. Changing the temperature of a reaction from 275 to 300. K causes the rate constant to increase by a factor of 7.00. Calculate E_a for this reaction.

68. The reaction

$$A \longrightarrow B + C$$

is known to be zero order in A, and to have a rate constant of 5.0×10^{-2} mol $L^{-1}\ s^{-1}$ at 25°C. An experiment was run at 25°C where $[A]_0 = 1.0 \times 10^{-3}$ M.
a. Write the integrated rate law for this reaction.
b. Calculate the half-life for the reaction.
c. Calculate the concentration of B after 5.0×10^{-3} s has elapsed.

69. At 500 K in the presence of a copper surface, ethanol decomposes according to the equation

$$C_2H_5OH(g) \longrightarrow CH_3CHO(g) + H_2(g)$$

The total pressure was measured for this reaction in a given experiment, and the following data were obtained:

t (s)	P_{total} (torr)
0	250.
100.	263
200.	276
300.	289
400.	302
500.	315

a. Predict the total pressure at $t = 900.$ s.
b. With pressures in atmospheres and time in seconds, what is the value of the rate constant, and what are its units?
c. What is the order of the reaction with respect to C_2H_5OH?

70. The following results were obtained at 600 K for the decomposition of ethanol on an alumina (Al_2O_3) surface,

$$C_2H_5OH(g) \longrightarrow C_2H_4(g) + H_2O(g)$$

t (s)	P_{total} (torr)
0	250.
10.	265
20.	280.
30.	295
40.	310.
50.	325

a. Predict P_{total} in torr at $t = 80.$ s.
b. What is the value of the rate constant, and what are its units?

c. What is the order of the reaction?
d. Calculate P_{total} at t = 300. s.

71. At 620. K butadiene dimerizes at a moderate rate. The following data were obtained in an experiment involving this reaction:

t (s)	$[C_4H_6]$ (M)
0	0.01000
1000.	0.00629
2000.	0.00459
3000.	0.00361

a. Determine the order of the reaction in butadiene.
b. In how many seconds is the dimerization 1.0% complete?
c. In how many seconds is the dimerization 2.0% complete?
d. What is the half-life for the reaction if the initial concentration of butadiene is 0.020 M?
e. Use the results from this problem and Exercise 15.13 to calculate the activation energy for the dimerization of butadiene.

72. Two isomers (A and B) of a given compound dimerize as follows:

$$2A \xrightarrow{k_1} A_2$$

$$2B \xrightarrow{k_2} B_2$$

Both processes are known to be second order in reactant, and k_1 is known to be 0.250 L mol^{-1} s^{-1} at 25°C. In a particular experiment A and B were placed in separate containers at 25°C, where $[A]_0 = 1.00 \times 10^{-2}\ M$ and $[B]_0 = 2.50 \times 10^{-2}\ M$. It was found that after each reaction had progressed for 3.00 min, [A] = 3.00[B]. In this case the rate laws are defined as

$$\text{Rate} = -\frac{d[A]}{dt} = k_1[A]^2$$

$$\text{Rate} = -\frac{d[B]}{dt} = k_2[B]^2$$

a. Calculate the concentration of A_2 after 3.00 min.
b. Calculate the value of k_2.
c. Calculate the half-life for the experiment involving A.

73. In 6 M HCl the complex ion $Ru(NH_3)_6^{3+}$ has a half-life of 14 h at 25°C. Under these conditions, how long will it take for the $[Ru(NH_3)_6^{3+}]$ to decrease to 12.5% of its initial value? (Assume first-order kinetics.)

74. For the reaction A → products, successive half-lives are observed to be 10.0, 20.0, and 40.0 min for an experiment in which $[A]_0 = 0.10\ M$. Calculate the concentration of A at the following times.
a. 80.0 min
b. 30.0 min.

75. Consider the hypothetical reaction

$$A + B + 2C \longrightarrow 2D + 3E$$

In a study of this reaction three experiments were run at the same temperature. The rate is defined as $-d[B]/dt$.

Experiment 1:

$[A]_0 = 2.0\ M$ $[B]_0 = 1.0 \times 10^{-3}\ M$ $[C]_0 = 1.0\ M$

[B] (mol/L)	Time (s)
2.7×10^{-4}	1.0×10^5
1.6×10^{-4}	2.0×10^5
1.1×10^{-4}	3.0×10^5
8.5×10^{-5}	4.0×10^5
6.9×10^{-5}	5.0×10^5
5.8×10^{-5}	6.0×10^5

Experiment 2:

$[A]_0 = 1.0 \times 10^{-2}\ M$ $[B]_0 = 3.0\ M$ $[C]_0 = 1.0\ M$

[A] (mol/L)	Time (s)
8.9×10^{-3}	1.0
7.1×10^{-3}	3.0
5.5×10^{-3}	5.0
3.8×10^{-3}	8.0
2.9×10^{-3}	10.0
2.0×10^{-3}	13.0

Experiment 3:

$[A]_0 = 10.0\ M$ $[B]_0 = 5.0\ M$ $[C]_0 = 5.0 \times 10^{-1}\ M$

[C] (mol/L)	Time (s)
0.43	1.0×10^{-2}
0.36	2.0×10^{-2}
0.29	3.0×10^{-2}
0.22	4.0×10^{-2}
0.15	5.0×10^{-2}
0.08	6.0×10^{-2}

Write the rate law for this reaction, and calculate the rate constant.

76. For the reaction

$$2N_2O_5(g) \longrightarrow 4NO_2(g) + O_2(g)$$

the following data were collected, where

$$\text{Rate} = -\frac{d[N_2O_5]}{dt}$$

t (s)	$T = 338$ K $[N_2O_5]$	$T = 318$ K $[N_2O_5]$
0	$1.00 \times 10^{-1}\ M$	$1.00 \times 10^{-1}\ M$
100.	$6.14 \times 10^{-2}\ M$	$9.54 \times 10^{-2}\ M$
300.	$2.33 \times 10^{-2}\ M$	$8.63 \times 10^{-2}\ M$
600.	$5.41 \times 10^{-3}\ M$	$7.43 \times 10^{-2}\ M$
900.	$1.26 \times 10^{-3}\ M$	$6.39 \times 10^{-2}\ M$

Calculate E_a for this reaction.

77. Consider the following reaction:

$$CH_3X + Y \longrightarrow CH_3Y + X$$

At 25°C the following two experiments were run, yielding the following data:

Experiment 1: $[Y]_0 = 3.0\ M$

$[CH_3X]$	Time (h)
$7.08 \times 10^{-3}\ M$	1.0
$4.52 \times 10^{-3}\ M$	1.5
$2.23 \times 10^{-3}\ M$	2.3
$4.76 \times 10^{-4}\ M$	4.0
$8.44 \times 10^{-5}\ M$	5.7
$2.75 \times 10^{-5}\ M$	7.0

Experiment 2: $[Y]_0 = 4.5\ M$

$[CH_3X]$	Time (h)
$4.50 \times 10^{-3}\ M$	0
$1.70 \times 10^{-3}\ M$	1.0
$4.19 \times 10^{-4}\ M$	2.5
$1.11 \times 10^{-4}\ M$	4.0
$2.81 \times 10^{-5}\ M$	5.5

Experiments were also run at 85°C. The value of the rate constant at 85°C was found to be 7.88×10^8 (with the time in units of hours), where $[CH_3X]_0 = 1.0 \times 10^{-2}\ M$ and $[Y]_0 = 3.0\ M$.

a. Determine the rate law and the value of k for this reaction at 25°C.
b. Determine the half-life at 85°C.
c. Determine E_a for the reaction.
d. Given that the C—X bond energy is known to be about 325 kJ/mol, suggest a mechanism that explains the results in parts a and c.

78. Experiments have shown the average frequency of chirping of individual snowy tree crickets (*Oecanthus fultoni*) to be 178 min^{-1} at 25.0°C, 126 min^{-1} at 20.3°C, and 100. min^{-1} at 17.3°C.

a. What is the apparent activation energy of the reaction that controls the chirping?
b. What chirping rate would be expected at 15.0°C?
c. Compare the observed rates and your calculated rate from part b to the rule of thumb that the Fahrenheit temperature is 42 plus 0.80 times the number of chirps in 15 s.

79. Experiments during a recent summer on a number of fireflies (small beetles, *Lampyridae photinus*) showed that the average interval between flashes of individual insects was 16.3 s at 21.0°C and 13.0 s at 27.8°C.

a. What is the apparent activation energy of the reaction that controls the flashing?
b. What would be the average interval between flashes of an individual firefly at 30.0°C?
c. Compare the observed intervals and the one you calculated in part b to the rule of thumb that the Celsius temperature is 54 minus twice the interval between flashes.

80. The following mechanism is proposed for the reduction of NO_3^- by $MoCl_6^{2-}$:

$$MoCl_6^{2-} \underset{k_{-1}}{\overset{k_1}{\rightleftharpoons}} MoCl_5^- + Cl^-$$

$$NO_3^- + MoCl_5^- \xrightarrow{k_2} OMoCl_5^- + NO_2^-$$

a. What is the intermediate?
b. Derive an expression for the rate law (rate $= d[NO_2^-]/dt$) for the overall reaction, using the steady-state approximation.

81. The following mechanism has been proposed to account for the rate law of the decomposition of ozone to $O_2(g)$:

$$O_3 + M \underset{k_{-1}}{\overset{k_1}{\rightleftharpoons}} O_2 + O + M$$

$$O + O_3 \xrightarrow{k_2} 2O_2$$

Apply the steady-state hypothesis to the concentration of atomic oxygen, and derive the rate law for the decomposition of ozone. (M stands for an atom or molecule that can exchange kinetic energy with the particles undergoing the chemical reaction.)

82. The gas-phase decomposition $2N_2O_5 \rightarrow 4NO_2 + O_2$ is first order but not unimolecular. A possible mechanism is

$$M + N_2O_5 \underset{k_{-1}}{\overset{k_1}{\rightleftharpoons}} NO_3 + NO_2 + M$$

$$NO_3 + NO_2 \xrightarrow{k_2} NO + O_2 + NO_2$$

$$NO_3 + NO \xrightarrow{k_3} 2NO_2$$

Apply the steady-state hypothesis to the concentrations of the intermediates NO_3 and NO and derive the rate law for the decomposition of N_2O_5.

83. The compound NO_2Cl is thought to decompose to NO_2 and Cl_2 by the following mechanism:

$$NO_2Cl \underset{k_{-1}}{\overset{k_1}{\rightleftharpoons}} NO_2 + Cl$$

$$NO_2Cl + Cl \xrightarrow{k_2} NO_2 + Cl_2$$

Derive the rate law for the production of Cl_2 using the steady-state approximation.

84. A first-order reaction has an activation energy E_a. When a certain catalyst is added, the activation energy of the new (catalyzed) pathway is $E_a/2$. Calculate the ratio of $k_{\text{uncatalyzed}}/k_{\text{catalyzed}}$ at 25°C and at 250.°C, assuming that $E_a = 50.0$ kJ/mol.

85. Consider the hypothetical reaction

$$B \longrightarrow E + F$$

which is assumed to occur by the mechanism

$$B + B \underset{k_{-1}}{\overset{k_1}{\rightleftharpoons}} B^* + B$$

$$B^* \xrightarrow{k_2} E + F$$

where B* represents a B molecule with enough energy to surmount the reaction energy barrier.

a. Derive the rate law for the production of E using the steady-state approximation.
b. Assume that this reaction is known to be first order. Under what conditions does your derived rate law (from part a) agree with this observation?
c. Explain how a chemical reaction can be first order, since even in a simple case ($B \rightarrow E + F$) molecules must collide to build up enough energy to get over the energy barrier. Why aren't all reactions at least second order? In other words, explain the physical significance of the result from part b.

86. The reaction

$$H_2O_2(aq) + 3I^-(aq) + 2H^+(aq) \longrightarrow I_3^-(aq) + 2H_2O(l)$$

was studied at 25°C and $[H^+] = 1.0\ M$. The following results were obtained, where

$$\text{Rate} = \frac{-d[H_2O_2]}{dt}$$

$[H_2O_2]_0$ (mol/L)	$[I^-]_0$ (mol/L)	Initial Rate (mol L^{-1} min^{-1})
0.10	0.10	7.0×10^{-4}
0.050	0.10	3.5×10^{-4}
0.10	0.20	1.4×10^{-3}

The rate of this reaction is independent of $[H^+]$ at very acidic pHs.

a. Determine the rate law.
b. Calculate the value of the rate constant.
c. What is the initial rate of formation of I_3^- in units of mol/min for a solution with a volume of 0.50 L and $[H_2O_2]_0 = 2[I^-]_0 = 0.50$ mol/L?

87. Hydrogen peroxide and the iodide ion react in acidic solution as follows:

$$H_2O_2(aq) + 3I^-(aq) + 2H^+(aq) \longrightarrow I_3^-(aq) + 2H_2O(l)$$

The kinetics of this reaction were studied by following the decay of the concentration of H_2O_2 and constructing plots of $\ln[H_2O_2]$ versus time. All of the plots were linear and all solutions had $[H_2O_2]_0 = 8.0 \times 10^{-4}$ mol/L. The slopes of these straight lines depended on the initial concentrations of I^- and H^+. The results are shown below:

$[I^-]$ (mol/L)	$[H^+]$ (mol/L)	Slope (min^{-1})
0.1000	0.0400	−0.120
0.3000	0.0400	−0.360
0.4000	0.0400	−0.480
0.0750	0.0200	−0.0760
0.0750	0.0800	−0.118
0.0750	0.1600	−0.174

The rate law for this reaction has the form

$$\text{Rate} = \frac{-d[H_2O_2]}{dt} = (k_1 + k_2[H^+])[I^-]^m[H_2O_2]^n$$

a. Specify the orders of this reaction with respect to $[H_2O_2]$ and $[I^-]$.

b. Calculate the values of the rate constants, k_1 and k_2.
c. What reason could there be for the two-term dependence of the rate on $[H^+]$?

88. Many biochemical reactions are catalyzed by large protein molecules called enzymes. A typical mechanism for the conversion of a biochemical substrate (S) to product (P) catalyzed by an enzyme (E) involves the following steps:

$$E + S \underset{k_{-1}}{\overset{k_1}{\rightleftharpoons}} ES$$

$$ES \xrightarrow{k_2} P$$

The rate-determining step is the decomposition of the intermediate enzyme-substrate complex (ES) to products (P). Under these conditions, show that the overall rate of product formation is

$$\text{Rate} = \frac{d[P]}{dt} = \frac{k_1 k_2 [E]_T [S]}{k_{-1} + k_2 + k_1[S]}$$

where $[E]_T$ equals the total enzyme concentration:

$$[E]_T = [E] + [ES]$$

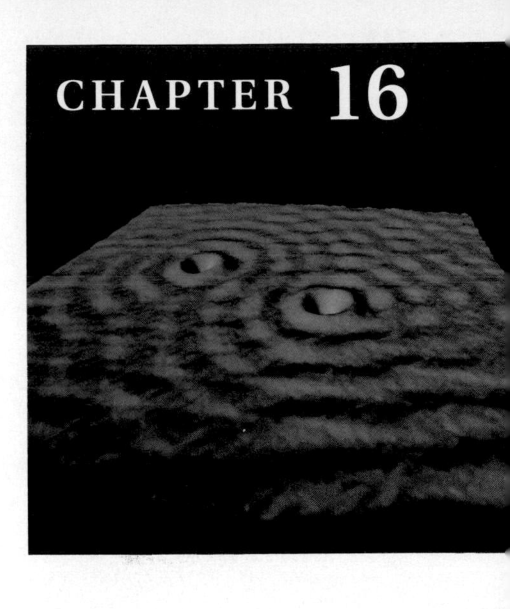

CHAPTER 16

Liquids and Solids

You have only to think about water to appreciate how different the three states of matter are. Flying, swimming, and ice skating are all done in contact with water in its various forms. Clearly, the arrangements of the water molecules must be significantly different in its gas, liquid, and solid forms.

Recall that a gas can be pictured as a substance whose component particles are far apart and are in rapid, random motion, exerting relatively small forces on each other. The kinetic molecular model was constructed to account for the ideal behavior that real gases approach at high temperatures and low pressures.

Solids are obviously very different from gases. Gases have low density, high compressibility, and completely fill a container. Solids have much greater densities, are compressible only to a very slight extent, and are rigid—a solid maintains its shape irrespective of its container. These properties indicate that the components of a solid are close together and exert large attractive forces on each other.

The properties of liquids lie somewhere between those of solids and of gases, but not midway between, as can be seen from some of the properties of the three states of water. For example, compare the enthalpy change for the melting of ice at 0°C (the heat of fusion) with that for vaporizing liquid water at 100°C (the heat of vaporization):

$$H_2O(s) \longrightarrow H_2O(l) \qquad \Delta H^\circ_{\text{fus}} = 6.02 \text{ kJ/mol}$$
$$H_2O(l) \longrightarrow H_2O(g) \qquad \Delta H^\circ_{\text{vap}} = 40.7 \text{ kJ/mol}$$

These values show a much greater change in structure in going from the liquid

TABLE 16.1
Densities of the Three States of Water

State	Density (g/cm^3)
Solid (0°C, 1 atm)	0.9168
Liquid (25°C, 1 atm)	0.9971
Gas (400°C, 1 atm)	3.26×10^{-4}

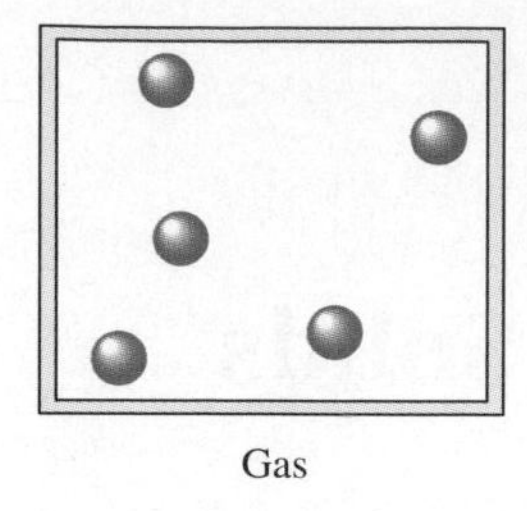

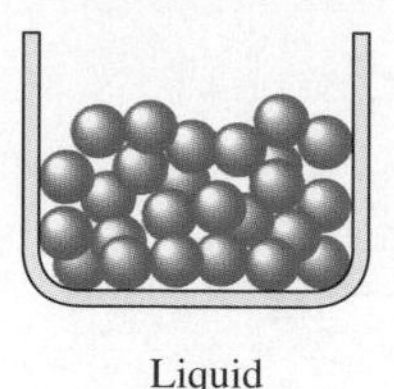

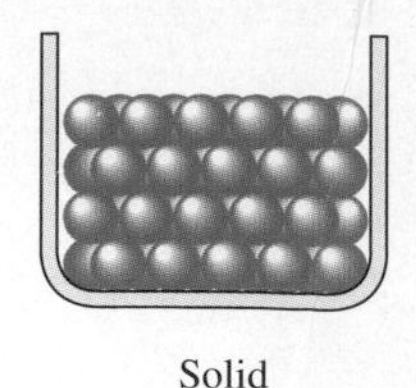

Figure 16.1
The three states of matter.

to the gas than in going from the solid to the liquid. This suggests that there are extensive attractive forces among the molecules in liquid water, similar to but not as strong as those in the solid state.

The relative similarity of the liquid and solid states can also be seen in the densities of the three states of water. As shown in Table 16.1, the densities for liquid and solid water are quite close. Compressibilities can also be used to explore the relationship among water's states. At 25°C the density of liquid water changes from 0.99707 g/cm^3 at a pressure of 1 atm to 1.046 g/cm^3 at 1065 atm. Given the large change in pressure, this is a very small variation in the density. Ice also shows little variation in density with increased pressure. On the other hand, at 400°C the density of gaseous water changes from 3.26×10^{-4} g/cm^3 at 1 atm pressure to 0.157 g/cm^3 at 242 atm—a huge variation.

The conclusion is clear. The liquid and solid states show many similarities and are strikingly different from the gaseous state (see Fig. 16.1). We must bear this in mind as we develop models for the structures of solids and liquids.

16.1 Intermolecular Forces

Recall that atoms can form stable units called molecules by sharing electrons. This is called *intramolecular* (within the molecule) bonding. In this chapter we will consider the properties of the **condensed states** of matter (liquids and solids) and the forces that cause the aggregation of the components of a substance to form a liquid or a solid. These forces may involve covalent or ionic bonding, or they may involve weaker interactions usually called **intermolecular forces** (because they occur between, rather than within, molecules).

Remember that temperature is a measure of the random motions of the particles in a substance.

It is important to recognize that when a substance like water changes from solid to liquid to gas, *the molecules remain intact.* The changes of state are due to changes in the forces *among* the molecules rather than those *within* the molecules. In ice, as we will see later in this chapter, the molecules are virtually locked in place, although they can vibrate about their positions. If energy is added, the motions of the molecules increase, and they eventually achieve the greater movement and disorder characteristic of liquid water; the ice has melted. As more energy is added, the gaseous state is eventually reached, where the individual molecules are far apart and thus interacting relatively little. However, the gas still consists of water molecules. It would take much more energy to overcome the covalent bonds and decompose the water molecules into their component atoms. This can be seen by comparing the energy needed to vaporize 1 mole of liquid water (40.7 kJ) with that needed to break the O—H bonds in 1 mole of water molecules (934 kJ).

Dipole-Dipole Forces

Recall that molecules with polar bonds often behave in an electric field as if they had a center of positive charge and a center of negative charge; that is, they exhibit a dipole moment. Molecules with dipole moments can attract each other electrostatically by lining up so that the positive and negative ends are close to each other, as shown in Fig. 16.2(a). This is called a **dipole-dipole attraction.** In a condensed state such as a liquid, the dipoles find the best compromise between attraction and repulsion, as shown in Fig. 16.2(b).

Dipole-dipole forces are typically only about 1% as strong as covalent or ionic bonds, and they rapidly become weaker as the distance between the dipoles increases. At low pressures in the gas phase, where the molecules are far apart, these forces are relatively unimportant.

Particularly strong dipole-dipole forces are seen among molecules in which hydrogen is bound to a highly electronegative atom, such as nitrogen, oxygen, or fluorine. Two factors account for the strengths of these interactions: the great polarity of the bond and the close approach of the dipoles, allowed by the very small size of the hydrogen atom. Because dipole-dipole attractions of this type are so unusually strong, they are given a special name—**hydrogen bonding.** Figure 16.3 shows hydrogen bonding among water molecules.

Hydrogen bonding has a very important effect on various physical properties. For example, the boiling points for the covalent hydrides of the elements in Groups 4A, 5A, 6A, and 7A are shown in Fig. 16.4. Note that the nonpolar tetrahedral hydrides of Group 4A show a steady increase in boiling point with molar mass (that is, in going down the group), while for the other groups, the lightest member has an unexpectedly high boiling point. Why? The answer lies in the especially large hydrogen-bonding interactions that exist among the smallest molecules with the most polar X—H bonds. These unusually strong hydrogen-bonding forces are due primarily to two factors. One factor is the relatively large electronegativity values of the lightest elements in each group, which leads to especially polar X—H bonds. The second factor is the small size of the first element of each group. This allows for the close approach of the dipoles, further strengthening the intermolecular forces. Because the interactions among the molecules containing the lightest elements in Groups 4A, 5A,

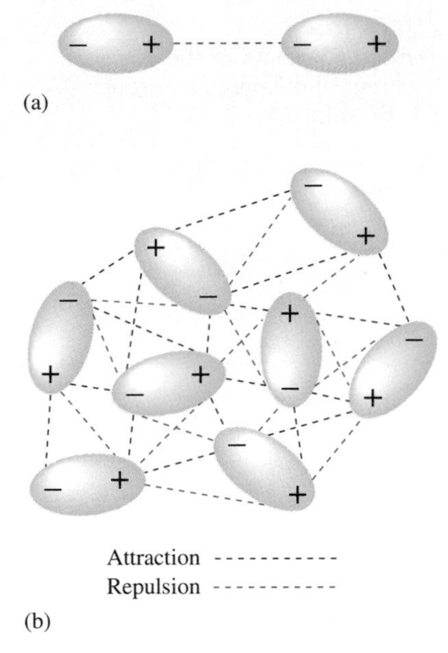

Figure 16.2
(a) The electrostatic interaction of two polar molecules. (b) The interaction of many dipoles in a condensed state.

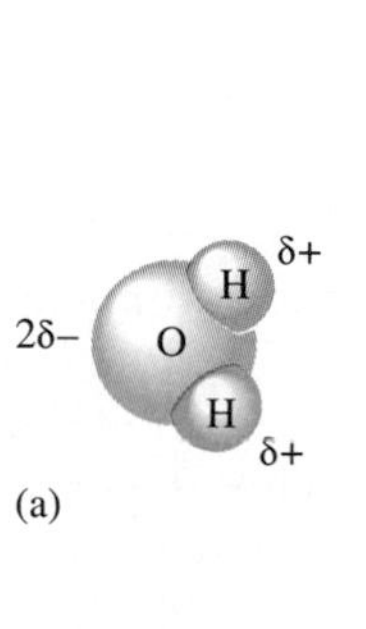

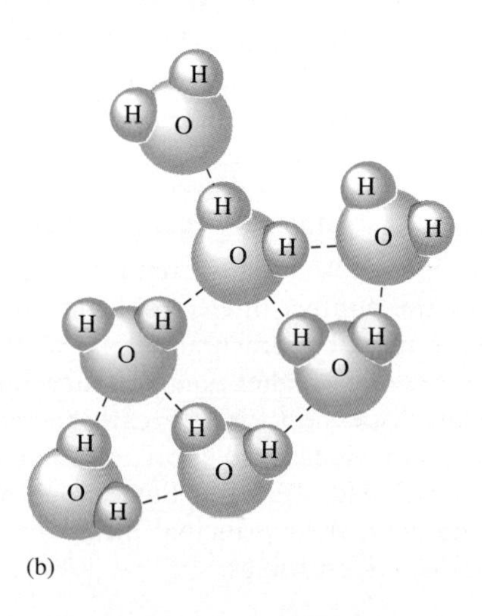

Figure 16.3
(a) The polar water molecule.
(b) Hydrogen bonding among water molecules. Note that the small size of the hydrogen atom allows for close interactions.

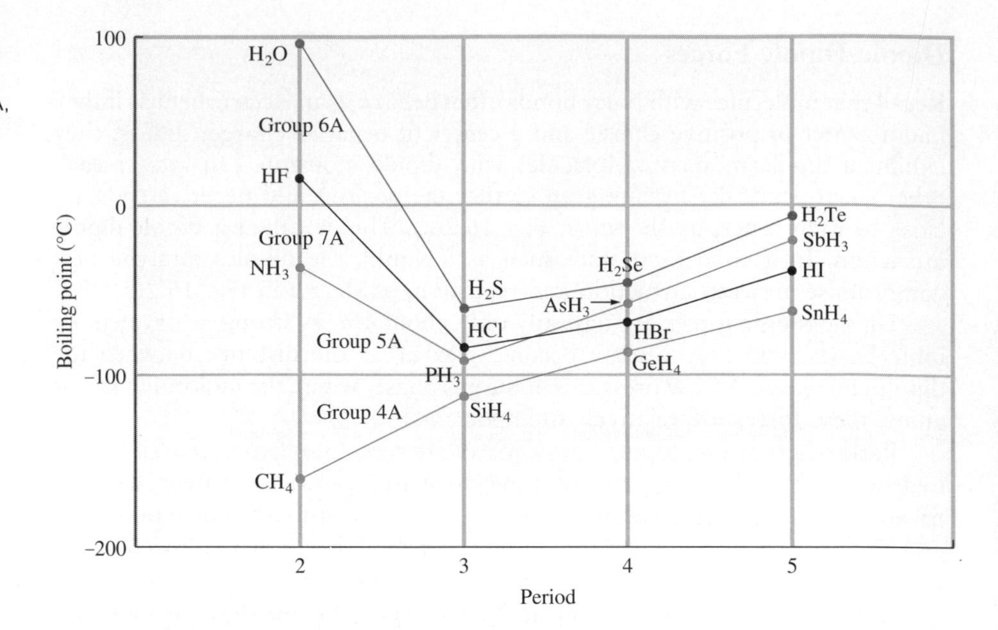

Figure 16.4
The boiling points of the covalent hydrides of elements in Groups 4A, 5A, 6A, and 7A.

and 6A are so strong, an unusually large quantity of energy must be supplied to overcome these interactions and separate the molecules to produce the gaseous state—hence the very high boiling points.

Boiling point will be defined precisely in Section 16.10.

London Dispersion Forces

The dipole-dipole and London dispension forces are sometimes called van der Waals forces after Johannes van der Waals. See Section 5.10.

Even molecules without dipole moments must exert forces on each other. We know they do because all substances—even the noble gases—exist in the liquid and solid states under certain conditions. The relatively weak forces that exist among noble gas atoms and nonpolar molecules are called **London dispersion forces.** To understand the origin of these forces, let us consider a pair of noble gas atoms. Although we usually assume that the electrons of an atom are uniformly distributed about the nucleus, this is apparently not true at every instant. Atoms can develop a momentary nonsymmetrical electron distribution that produces a temporary dipolar arrangement of charge. This *instantaneous dipole* can then *induce* a similar dipole in a neighboring atom, as shown in Fig. 16.5(a). This phenomenon leads to an interatomic attraction that is both weak and short-lived but that can be very significant for large atoms as we will discuss later. For these interactions to become strong enough to produce a solid, the motions of the atoms must be greatly reduced. This explains, for instance, why the noble gas elements have such low freezing points (see Table 16.2).

TABLE 16.2
The Freezing Points of the Group 8A Elements

Element	Freezing Point (°C)
Helium*	−269.7
Neon	−248.6
Argon	−189.4
Krypton	−157.3
Xenon	−111.9

* Helium is the only liquid that does not freeze when the temperature is lowered at 1 atm. It will freeze if pressure is applied.

Note from Table 16.2 that the freezing point rises going down the group. The principal cause for this trend is that as the mass (and the atomic number) increases, the number of electrons increases, so there is an increased chance of the occurrence of momentary dipoles. We say that large atoms with many electrons exhibit a higher *polarizability* than small atoms. Thus the importance of London dispersion forces greatly increases as atomic size increases.

The dispersion forces in molecules with large atoms are quite significant and are often actually more important than dipole-dipole forces.

These same ideas also apply to nonpolar molecules such as H_2, CH_4, CCl_4, and CO_2 [see Fig. 16.5(b)]. Since none of these molecules has a permanent dipole moment, their principal means of attraction for each other is through London dispersion forces.

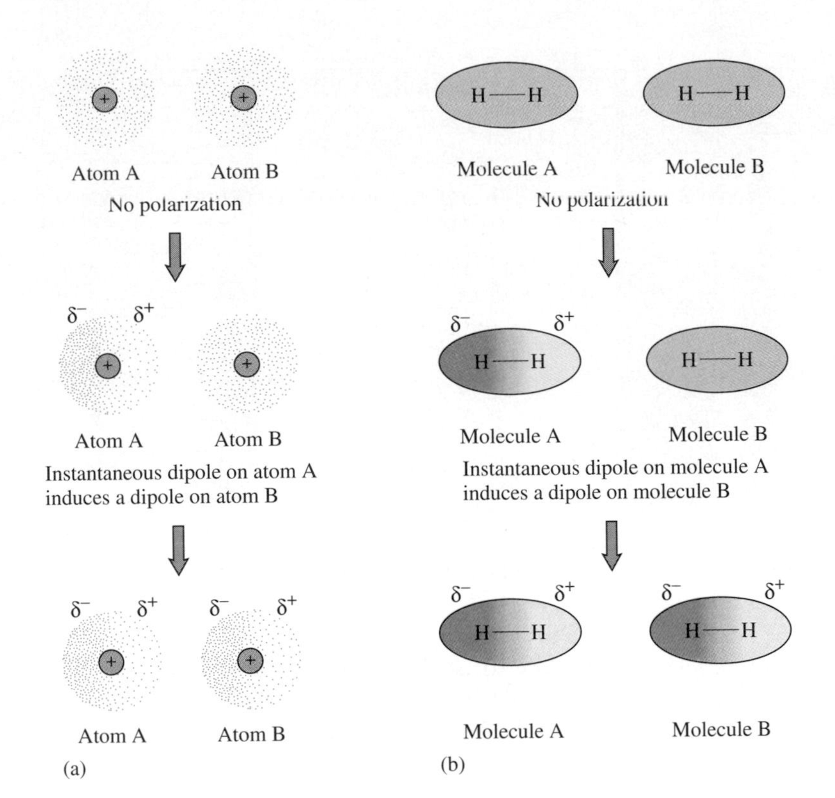

Figure 16.5
(a) An instantaneous polarization can occur on atom A, creating an instantaneous dipole. This dipole creates an induced dipole on neighboring atom B. (b) Nonpolar molecules such as H_2 can develop instantaneous and induced dipoles.

16.2 The Liquid State

Liquids and liquid solutions are vital to our lives. Of course, water is the most important liquid. Besides being essential to life, it provides a medium for food preparation, for transportation, for cooling in many types of machines and industrial processes, for recreation, for cleaning, and for a myriad of other uses.

Liquids exhibit many characteristics that help us understand their natures. We have already mentioned their low compressibility, lack of rigidity, and high density compared with gases. Many of the properties of liquids give us direct information about the forces that exist among the particles. For example, when a liquid is poured onto a solid surface, it tends to bead as droplets, a phenomenon that depends on the intermolecular forces. Although molecules in the interior of the liquid are completely surrounded by other molecules, those at the liquid's surface are subject to attractions only from the side and from below (Fig. 16.6). The effect of this uneven pull on the surface molecules tends to draw them into the body of the liquid and causes a droplet of liquid to assume the shape that has the minimum surface area—a sphere.

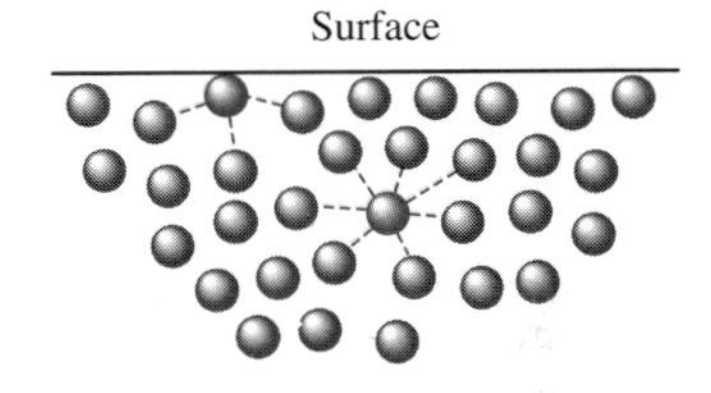

Figure 16.6
A molecule in the interior of a liquid is attracted to the molecules surrounding it, whereas a molecule at the surface of a liquid is attracted only by molecules below it and on each side of it.

For a given volume a sphere has a smaller surface area than any other shape.

To increase a liquid's surface area, molecules must move from the interior of the liquid to the surface. This requires energy, since some intermolecular forces must be overcome. The resistance of a liquid to an increase in its surface area is called the **surface tension** of the liquid. As we would expect, liquids with relatively large intermolecular forces tend to have relatively high surface tensions.

Polar liquids also exhibit **capillary action,** the spontaneous rising of a liquid in a narrow tube. Two different types of forces are responsible for this property: *cohesive forces,* the intermolecular forces among the molecules of the

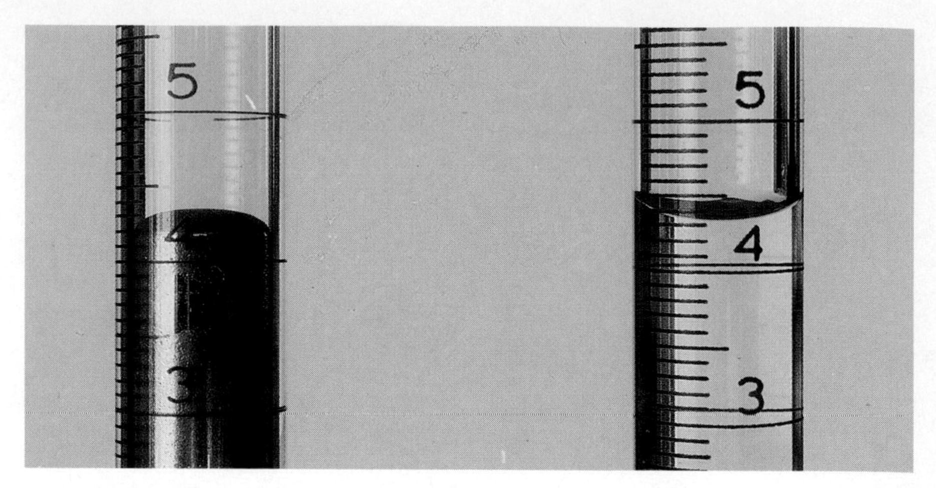

Figure 16.7
Nonpolar liquid mercury forms a convex meniscus in a glass tube; polar water forms a concave meniscus.

The composition of glass is discussed in Section 16.5.

liquid; and *adhesive forces,* the forces between the liquid molecules and their container. We have already seen how cohesive forces operate among polar molecules. Adhesive forces between a polar liquid and a given surface are strongest when the surface is made of a substance that has polar bonds. For example, glass contains many oxygen atoms with partial negative charges that are attractive to the positive end of a polar molecule such as water. This ability of water to "wet" glass makes it creep up the walls of the tube where the water surface touches the glass. This, however, tends to increase the surface area of the water, which is opposed by the cohesive forces that try to minimize the surface area. Thus because water has both strong cohesive (intermolecular) forces and strong adhesive forces to glass, it "pulls itself" up a glass capillary tube (a tube with a small diameter) to a height where the weight of the column of water just balances the water's tendency to be attracted to the glass surface. The concave shape of the meniscus (Fig. 16.7) shows that water's adhesive forces toward the glass are stronger than its cohesive forces. A nonpolar liquid such as mercury shows a convex meniscus in a glass tube. This behavior is characteristic of a liquid in which the cohesive forces are stronger than the adhesive forces toward the glass.

Another property of liquids that is strongly dependent on intermolecular forces is **viscosity,** a measure of a liquid's resistance to flow. As might be expected, liquids with large intermolecular forces tend to be highly viscous. For example, glycerol, whose structure is

```
      H
      |
  H—C—O—H
      |
  H—C—O—H
      |
  H—C—O—H
      |
      H
```

has an unusually high viscosity, mainly due to its high capacity to form hydrogen bonds.

Molecular complexity also leads to higher viscosity because very large molecules can become entangled with each other. For example, nonviscous gasoline contains molecules of the type $CH_3—(CH_2)_n—CH_3$, where n varies from about 3 to 8. However, grease, which is very viscous, contains much larger molecules in which n varies from 20 to 25.

In many respects the development of a structural model for liquids presents greater challenges than the development of such a model for the other two states of matter. In the gaseous state the particles are so far apart and are moving so rapidly that intermolecular forces are negligible under most circumstances. This means we can use a relatively simple model for gases. In the solid state, although the intermolecular forces are large, the molecular motions are minimal, and fairly simple models are again possible. The liquid state, however, has both strong intermolecular forces *and* significant molecular motions. Such a situation precludes the use of really simple models for liquids. Recent advances in spectroscopy—the study of the manner in which substances interact with electromagnetic radiation—make it possible to follow the very rapid changes that occur in liquids. As a result, our models of liquids are becoming more accurate. As a starting point, a typical liquid might best be viewed as containing a large number of regions where the arrangements of the components are similar to those found in the solid, but with more disorder. In addition, a smaller number of regions exist in the liquid where gaps ("holes") occur among the molecules. The situation is highly dynamic, with rapid fluctuations occurring in both types of regions.

16.3 An Introduction to Structures and Types of Solids

There are many ways to classify solids, but the broadest categories are **crystalline solids,** those with a highly regular arrangement of their components, and **amorphous solids,** those with considerable disorder in their structures.

The regular arrangement of the components of a crystalline solid at the microscopic level produces the beautiful, characteristic shapes of crystals, such as those shown in Fig. 16.8. The positions of the components in a crystalline solid are usually represented by a **lattice,** a three-dimensional array of points designating the centers of the components (atoms, ions, or molecules) that shows the repetitious pattern of the components. The *smallest repeating unit* of the lattice is called the **unit cell.** Thus a particular lattice can be generated by repeating (translating) the unit cell in all three dimensions to form the extended structure. Three common unit cells and their lattices are shown in Fig. 16.9.

Although we will concentrate on crystalline solids in this book, there are many important noncrystalline (amorphous) materials. An example is common glass, which is best pictured as a solution whose components are "frozen in place" before they can achieve an ordered arrangement. Although glass is a solid (it has a rigid shape), a great deal of disorder exists in its structure.

X-ray Analysis of Solids

The structures of crystalline solids are most commonly determined by **X-ray diffraction.** Diffraction occurs when beams of light are scattered from a regular

Figure 16.8
Several crystalline solids. (clockwise from upper left) Rhodochrosite; calcite with malachite; pyrite; amethyst.

array of points or lines where the spacings between the components are comparable to the wavelength of the light. Diffraction is due to constructive interference when the waves of parallel beams are in phase and to destructive interference when the waves are out of phase.

When X rays of a single wavelength are directed at a crystal, a diffraction pattern is obtained, as we saw in Fig. 12.7. The light and dark areas on the photographic plate occur because the waves scattered from various atoms may reinforce or cancel each other (see Fig. 16.10). The key to whether the waves reinforce or cancel is the difference in distance traveled by the waves after they strike the atoms. The waves are in phase before they are reflected; so if the difference in distance traveled after reflection is an *integral number of wavelengths,* the waves will still be in phase when they meet again.

Since the distance traveled after reflection depends on the distance between the atoms, the diffraction pattern can be used to determine the interatomic

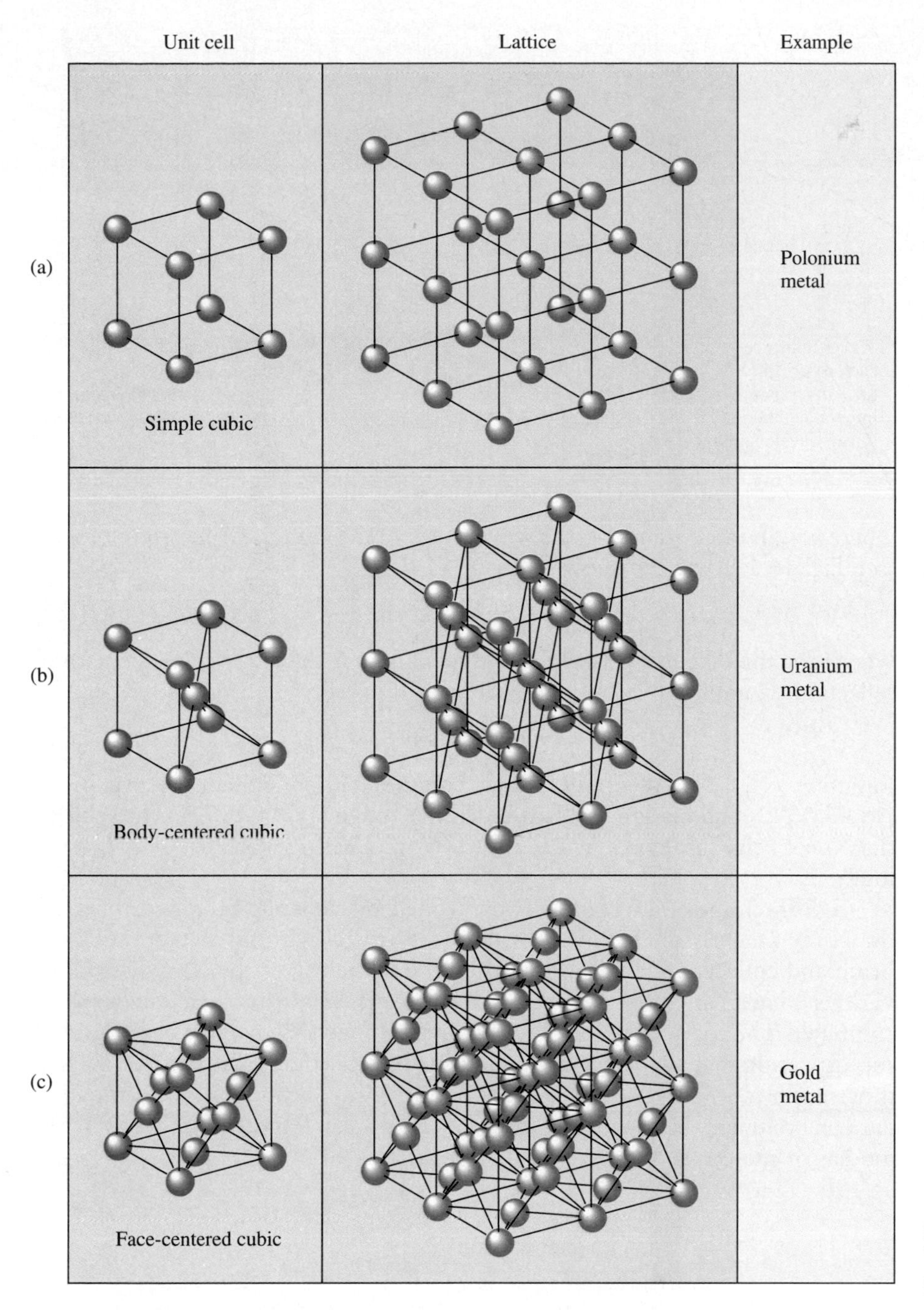

Figure 16.9
Three cubic unit cells and the corresponding lattices.

spacings. The exact relationship can be formulated using the diagram in Fig. 16.11, which shows two in-phase waves being reflected by atoms in two different layers of a crystal. The extra distance traveled by the lower wave is the sum of the distances xy and yz. The waves will be in phase after reflection if

$$xy + yz = n\lambda \tag{16.1}$$

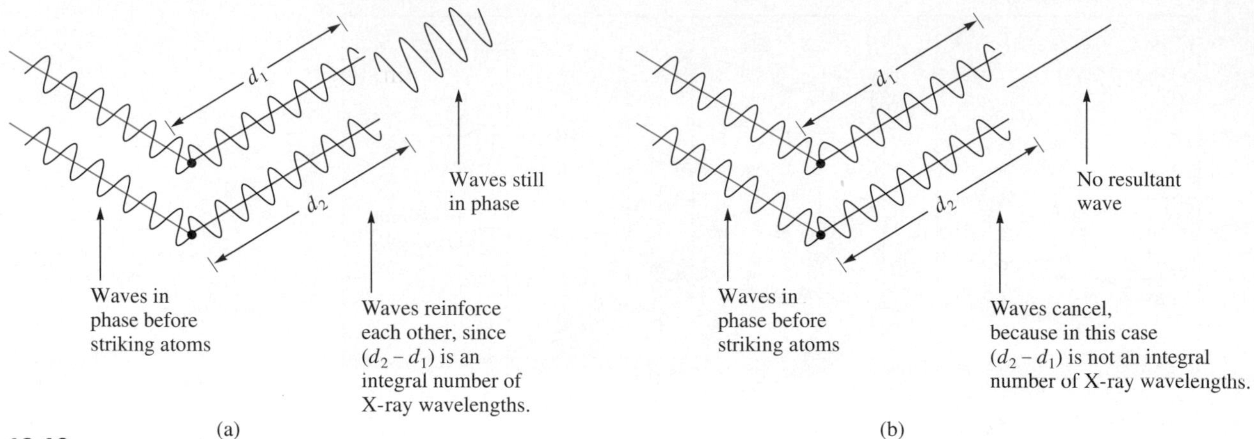

Figure 16.10
X rays that are scattered from two different atoms may either (a) reinforce or (b) cancel each other, depending on whether they are in phase or out of phase.

where n is an integer and λ is the wavelength of the X rays. Using trigonometry (see Fig. 16.11), we can show that

$$xy + yz = 2d \sin \theta \tag{16.2}$$

where d is the distance between the atoms and θ is the angle of incidence and reflection. Combining Equation (16.1) and Equation (16.2) gives

$$n\lambda = 2d \sin \theta \tag{16.3}$$

Equation (16.3) is called the **Bragg equation,** after William Henry Bragg (1862–1942) and his son William Lawrence Bragg (1890–1972). They shared the Nobel prize in physics in 1915 for their pioneering work in X-ray crystallography.

A diffractometer is a computer-controlled instrument used for carrying out the X-ray analysis of crystals. It rotates the crystal with respect to the X-ray beam and collects the data produced by the scattering of the X rays from the various planes of atoms in the crystal. The results are then analyzed by computer. The techniques for crystal structure analysis have reached a level of sophistication that allows the determination of very complex structures, such as those important in biological systems. Using X-ray diffraction, we can gather data on bond lengths and angles and, in doing so, can test the predictions of our models of molecular geometry.

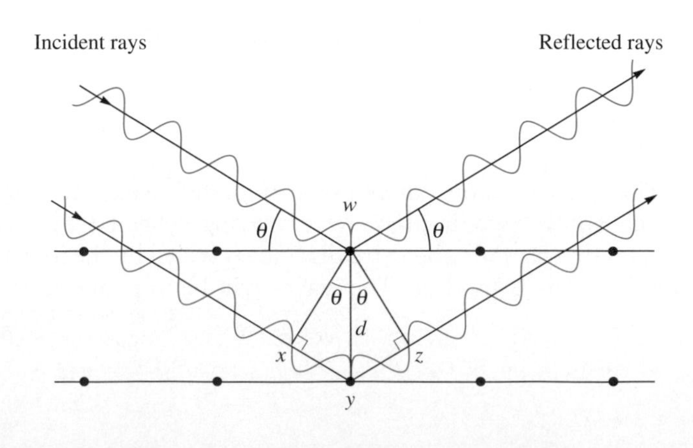

Figure 16.11
Reflection of X rays of wavelength λ from a pair of atoms in two different layers of a crystal. The lower wave travels an extra distance equal to the sum of xy and yz. If this distance is an integral number of wavelengths (n = 1, 2, 3, . . .), the waves reinforce each other when they exit the crystal.

Types of Crystalline Solids

There are many different types of crystalline solids. For example, although both sugar and salt dissolve readily in water, the properties of the resulting solutions are quite different. The salt solution readily conducts an electric current, but the sugar solution does not. This behavior arises from the nature of the components of these two solids. Common salt (NaCl) is an ionic solid; it contains Na^+ and Cl^- ions. When solid sodium chloride dissolves in the polar water, sodium and chloride ions are distributed throughout the resulting solution and are free to conduct electric current. Table sugar (sucrose), on the other hand, is composed of neutral molecules that are dispersed throughout the water when the solid dissolves. Since no ions are present in the solid, the resulting solution does not conduct electricity. These examples illustrate two important types of solids: **ionic solids,** represented by sodium chloride; and **molecular solids,** represented by sucrose.

Diamond, the hardest natural substance known, is used to grind metals.

A third type of solid is illustrated by elements such as graphite, diamond, and buckminsterfullerene (all pure carbon), boron, silicon, and all metals. These substances all have atoms occupying the lattice points; we will call them **atomic solids.** Examples of the three types of solids are shown in Fig. 16.12.

The properties of a solid are determined primarily by the nature of the forces that hold the solid together. For example, although argon, copper, and diamond all form atomic solids, they have strikingly different properties. Argon has a very low melting point ($-189°C$), while diamond and copper melt at high temperatures (about 3500°C and 1083°C, respectively). Copper is an excellent conductor of electricity, but argon and diamond are both insulators. Copper can be easily changed in shape; it is both malleable (can form thin sheets) and ductile (can be pulled into a wire). Diamond, on the other hand, is the hardest natural substance known. The marked differences in properties among these

The internal forces in a solid determine the properties of the solid.

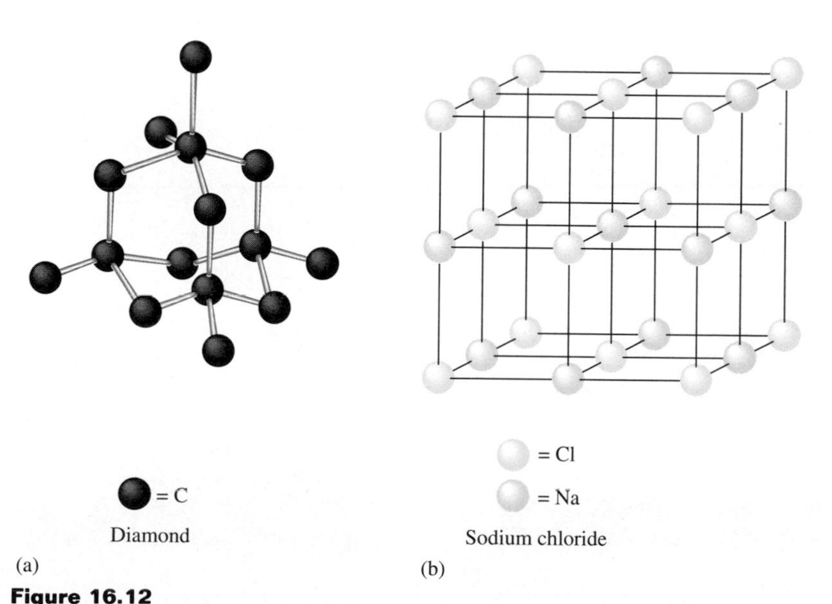

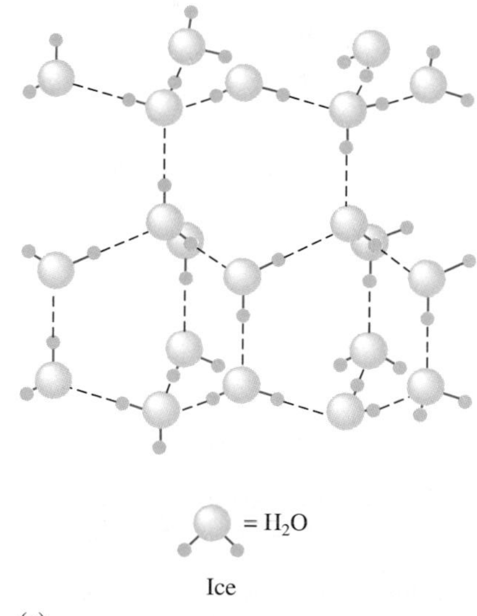

Figure 16.12
Examples of three types of crystalline solids. Only part of the structure is shown in each case. (a) An atomic solid. (b) An ionic solid. (c) A molecular solid. The dashed lines show the hydrogen bonding among the polar water molecules.

three atomic solids are due to bonding differences. We will explore the bonding in solids in the next two sections.

16.4 Structure and Bonding in Metals

The closest packing model for metallic crystals assumes that metal atoms are uniform, hard spheres.

Metals are characterized by high thermal and electrical conductivity, malleability, and ductility. As we will see, these properties can be traced to the nondirectional covalent bonding found in metallic crystals.

A metallic crystal can be pictured as containing spherical atoms packed together and bonded to each other equally in all directions. We can model such a structure by packing uniform, hard spheres in a manner that most efficiently uses the available space. Such an arrangement is called **closest packing.** The spheres are packed in layers, as shown in Fig. 16.13(a), where each sphere is surrounded by six others. In the second layer the spheres do not lie directly over those in the first layer. Instead, each one occupies an indentation (or dimple) formed by three spheres in the first layer [see Fig. 16.13(b)]. In the third layer the spheres can occupy the dimples of the second layer in two possible ways. They can occupy positions so that each sphere in the third layer lies directly over a sphere in the first layer (the *aba* arrangement), or they can occupy positions so that no sphere in the third layer lies over one in the first layer (the *abc* arrangement).

Figure 16.13
The closest packing arrangement of uniform spheres. (a) A typical layer where each sphere is surrounded by six others. (b) The second layer is like the first, but it is displaced so that each sphere in the second layer occupies a dimple in the first layer. (c) The spheres in the third layer can occupy dimples in the second layer so that the spheres in the third layer lie directly over those in the first layer (*aba*), or they can occupy dimples in the second layer so that no spheres in the third layer lie above any in the first layer (*abc*).

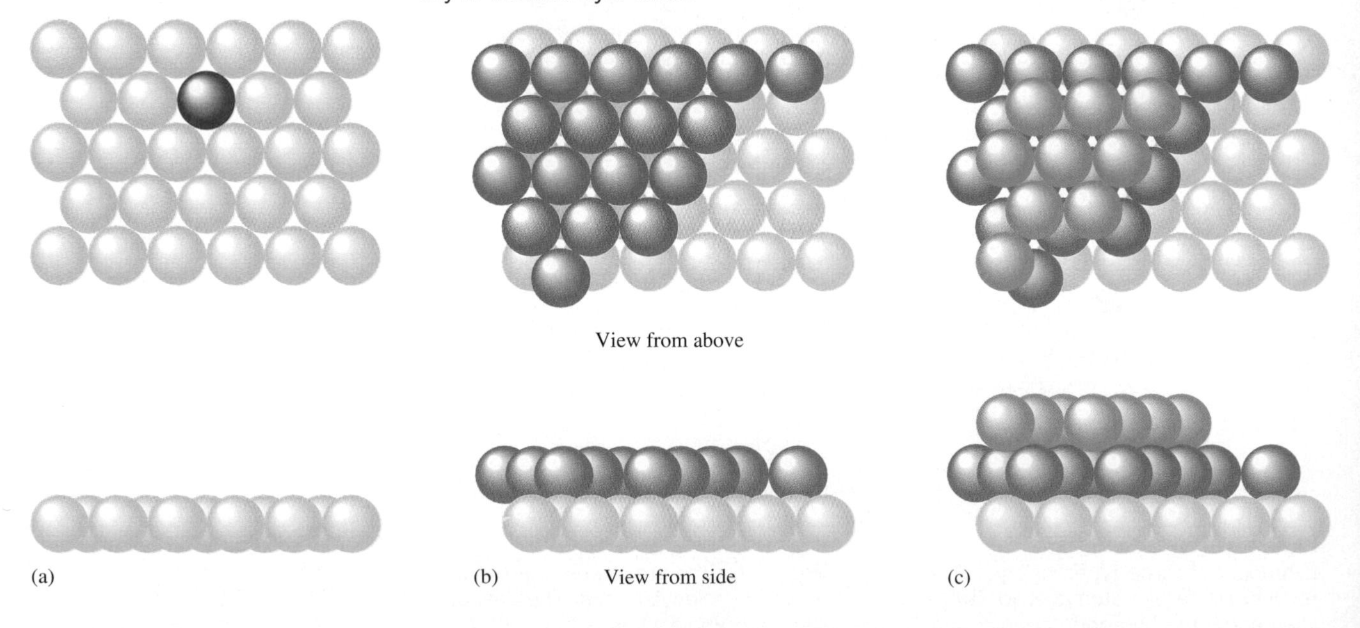

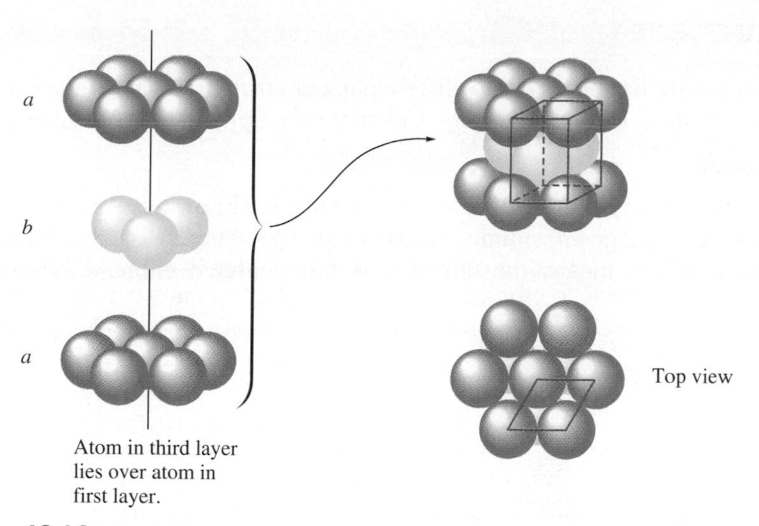

Figure 16.14
When spheres are closest packed so that the spheres in the third layer lie directly over those in the first layer (*aba*), the unit cell forms the hexagonal prism illustrated here in red.

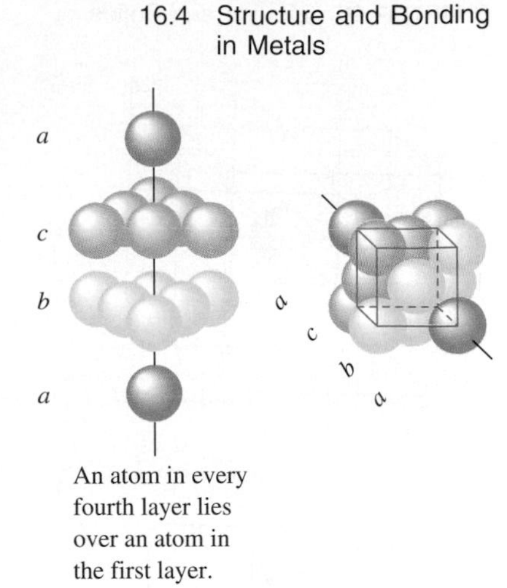

Figure 16.15
When spheres are packed in the *abc* arrangement, the unit cell is face-centered cubic. So that the cubic arrangement is easier to see, the vertical axis has been tilted as shown.

The *aba* arrangement has the *hexagonal* unit cell shown in Fig. 16.14, and the resulting structure is called the **hexagonal closest packed (hcp) structure.** The *abc* arrangement has a *face-centered cubic* unit cell, as shown in Fig. 16.15, and the resulting structure is called the **cubic closest packed (ccp) structure.** Note that in the hcp structure the spheres in every other layer occupy the same vertical position (*ababab* . . .), while in the ccp structure the spheres in every fourth layer occupy the same vertical position (*abcabca* . . .). A characteristic of both structures is that each sphere has 12 equivalent nearest neighbors: 6 in the same layer, 3 in the layer above, and 3 in the layer below (that form the dimples). This is illustrated for the hcp structure in Fig. 16.16.

Knowing the *net* number of spheres (atoms) in a particular unit cell is important for many applications involving solids. To illustrate the procedure for finding the net number of spheres in a unit cell, let us consider a face-centered cubic unit cell (Fig. 16.17). Note that this unit cell is defined by the *centers* of the spheres on the cube's corners. Thus 8 cubes share a given sphere, so $\frac{1}{8}$ of this sphere lies inside each unit cell. Since a cube has 8 corners, there are $8 \times \frac{1}{8}$ pieces, or enough to put together 1 whole sphere. The spheres at the center of each face are shared by 2 unit cells, so $\frac{1}{2}$ of each lies inside a particular unit cell. Since the cube has 6 faces, we have $6 \times \frac{1}{2}$ pieces, or enough to construct 3 whole spheres. Thus the net number of spheres in a face-centered cubic unit cell is

$$\left(8 \times \frac{1}{8}\right) + \left(6 \times \frac{1}{2}\right) = 4$$

In Example 16.1 we will see how the density of a closest packed solid can be calculated.

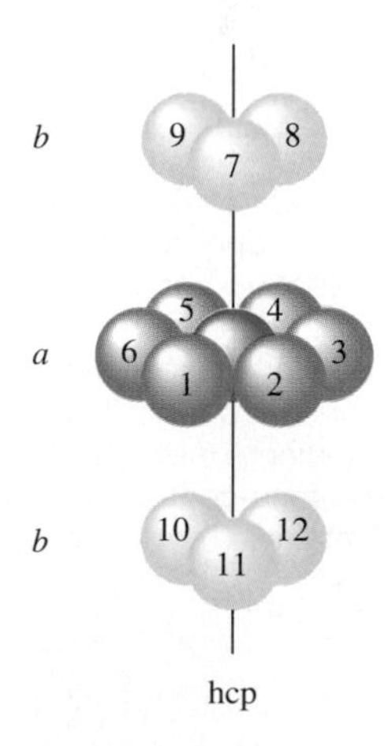

Figure 16.16
The indicated sphere has 12 equivalent nearest neighbors.

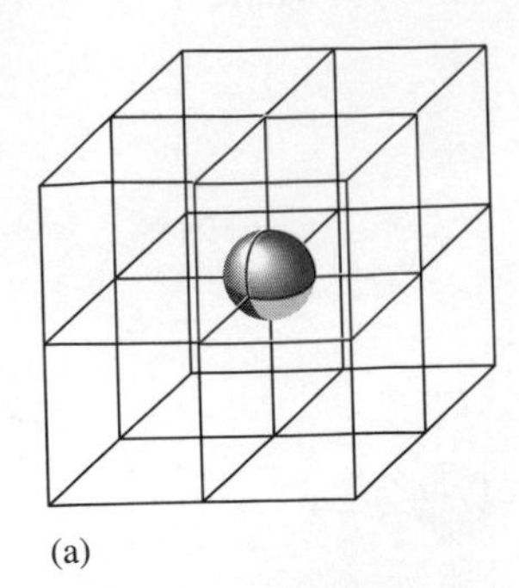

(a)

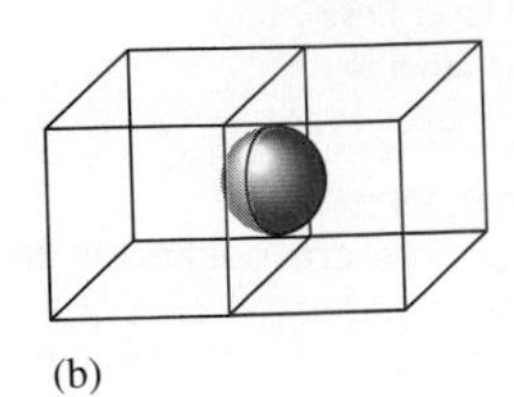

(b)

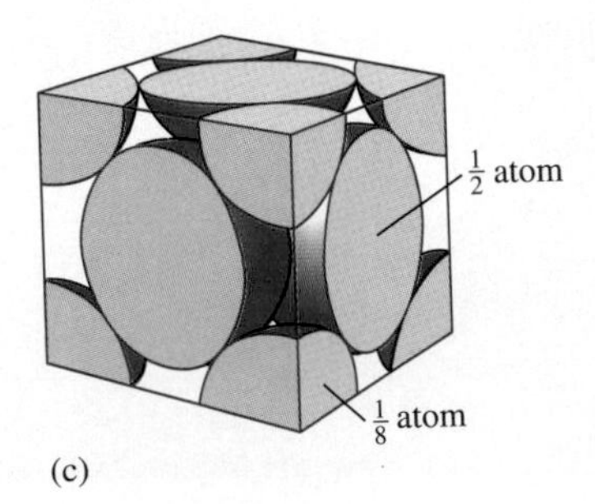

(c)

Figure 16.17
The net number of spheres in a face-centered cubic unit cell. (a) Note that the sphere on a corner of the colored cell is shared with 7 other unit cells. Thus $\frac{1}{8}$ of such a sphere lies within a given unit cell. Since there are 8 corners in a cube, there are 8 of these $\frac{1}{8}$ pieces, the equivalent of 1 net sphere. (b) The sphere on the center of each face is shared by two unit cells, and thus each unit cell has $\frac{1}{2}$ of each of these types of spheres. There are 6 of these $\frac{1}{2}$ spheres, yielding 3 net spheres. (c) Thus the face-centered cubic unit cell contains 4 net spheres.

EXAMPLE 16.1

Silver crystallizes in a cubic closest packed structure. The radius of a silver atom is 1.44 Å (144 pm). Calculate the density of solid silver.

Solution

Density is mass per unit volume. Thus we need to know how many silver atoms occupy a given volume in the crystal. The structure is cubic closest packed, which means the unit cell is face-centered cubic, as shown below.

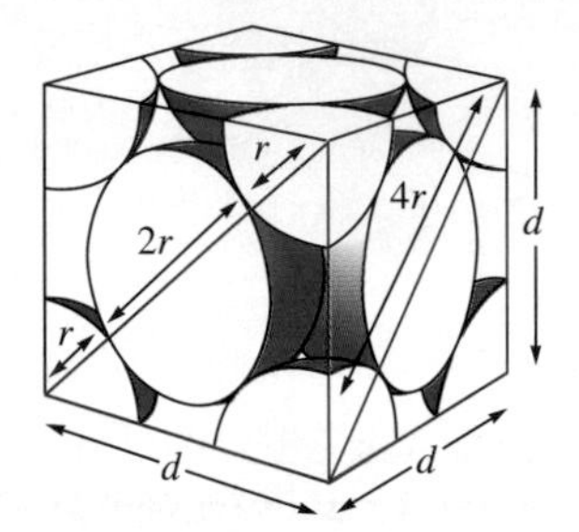

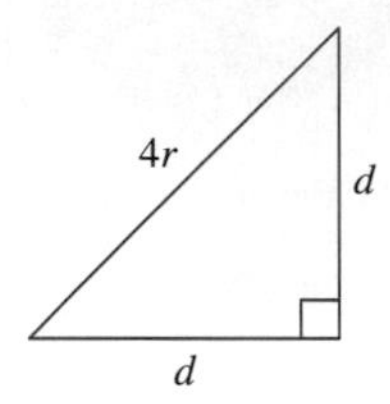

We must find the volume of this unit cell and the net number of atoms it contains. Note that in this structure the atoms touch along the diagonals of each face and not along the edges of the cube. Thus the length of the diagonal is $R + 2R + R$, or $4R$, where R represents the radius of a silver atom. We next find the length of the edge of the cube by the Pythagorean theorem:

$$e^2 + e^2 = (4R)^2$$
$$2e^2 = 16R^2$$
$$e^2 = 8R^2$$
$$e = \sqrt{8R^2} = R\sqrt{8}$$

Since $R = 1.44$ Å for a silver atom,

$$e = (1.44\text{ Å})(\sqrt{8}) = 4.07\text{ Å}$$

The volume of the unit cell is e^3, which is $(4.07\text{ Å})^3$ or 67.4 Å^3. We convert this to cubic centimeters as follows:

$$67.4\text{ Å}^3 \times \left(\frac{1.00 \times 10^{-8}\text{ cm}}{\text{Å}}\right)^3 = 6.74 \times 10^{-23}\text{ cm}^3$$

Since the net number of atoms in a face-centered cubic unit cell is 4, there are 4 silver atoms in a volume of $6.74 \times 10^{-23}\text{ cm}^3$. Therefore

$$\text{Density} = \frac{\text{mass}}{\text{volume}}$$
$$= \frac{(4\text{ atoms})(107.9\text{ g/mol})(1\text{ mol}/6.022 \times 10^{23}\text{ atoms})}{6.74 \times 10^{-23}\text{ cm}^3}$$
$$= 10.6\text{ g/cm}^3$$

Closest packing describes the most efficient method for arranging uniform spheres. We can calculate the fraction of the space actually occupied by the spheres (f_v),

$$f_v = \frac{\text{volume occupied by spheres in the unit cell}}{\text{volume of the unit cell}}$$

by using ideas we have developed in the previous discussion.

Recall that in the cubic closest packing arrangement, the unit cell is face-centered cubic and contains four net spheres. Thus the volume occupied by spheres in the unit cell is four times the volume of each sphere:

$$4 \times \tfrac{4}{3}\pi R^3$$

In Example 16.1 we found that the edge of the unit cell (e) is related to the radius of the packed spheres (R) as follows:

$$e = R\sqrt{8}$$

Therefore the volume of the unit cell is e^3, where

$$e^3 = (R\sqrt{8})^3$$

and the fraction of space occupied by the spheres is

$$f_v = \frac{\overbrace{4 \times \tfrac{4}{3}\pi R^3}^{\text{Volume of spheres}}}{\underbrace{(R\sqrt{8})^3}_{\text{Volume of unit cell}}} = 0.740$$

This result means that in a cubic closest packed solid 74.0% of the space is occupied by spheres (26.0% is open space). This result also applies to hexagonal closest packing, although we will not prove it here.

In contrast, for simple cubic packing (spheres stacked on top of each other in successive layers) the spheres occupy only 52.4% of the space (verify this for yourself).

Examples of metals that are cubic closest packed are aluminum, iron, copper, cobalt, and nickel. Magnesium and zinc exhibit hexagonal closest packing. Calcium and certain other metals can crystallize in either structure.

Body-Centered Cubic Packing

Although most metals assume one of the closest packed structures in the solid state, some metallic elements have structures that are not closest packed. For example, the structures of the alkali metals are characterized by a **body-centered cubic (bcc) unit cell** (see Fig. 16.9). In this structure each sphere has 8 nearest neighbors (count the number of atoms around the atom at the center of the unit cell), as compared with 12 in the closest packed structures.

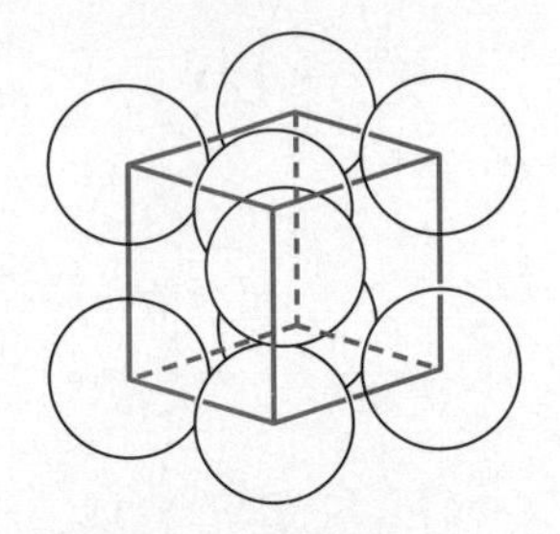

Figure 16.18
In the body-centered cubic unit cell the spheres touch along the body diagonal.

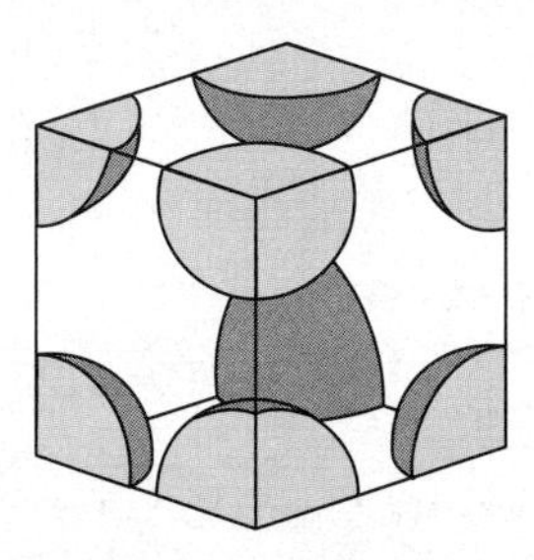

Figure 16.19
The body-centered cubic unit cell with the center sphere deleted.

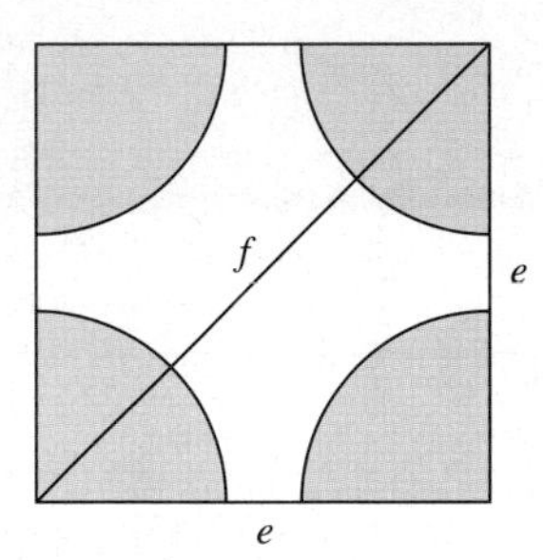

Figure 16.20
One face of the body-centered cubic unit cell. By the Pythagorean theorem $f^2 = e^2 + e^2$.

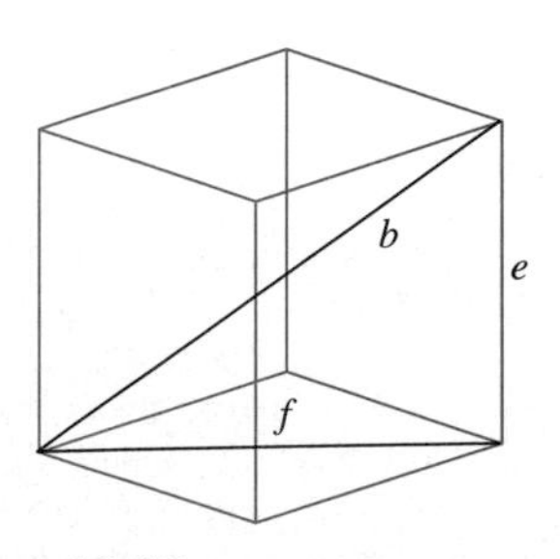

Figure 16.21
The relationship of the body diagonal (b) to the face diagonal (f) and edge (e) for the body-centered cubic unit cell.

Note that in body-centered cubic packing the unit cell is a cube with one sphere at its center. In this structure the spheres touch along the body diagonal (see Fig. 16.18). For clarity the cube for this unit cell is shown in Fig. 16.19 with the center sphere deleted.

The body-centered arrangement of spheres is not a closest packed structure, as we can show by calculating the fraction of space occupied by the spheres. To express the volume of the cube in terms of the radius of the packed spheres, we must use the Pythagorean theorem twice. First, we express the face diagonal f in terms of the edge e (see Fig. 16.20):

$$f^2 = e^2 + e^2 = 2e^2$$

Because the spheres touch along the body diagonal, the length of the body diagonal b is $4R$. We can relate the body diagonal, the face diagonal, and the cube edge as follows (see Fig. 16.21):

$$b^2 = (4R)^2 = e^2 + f^2$$

and

$$(4R)^2 = e^2 + 2e^2 = 3e^2$$

Thus

$$e = \frac{4R}{\sqrt{3}}$$

The body-centered cubic unit cell contains 2 net spheres:

$$\underset{\text{Corners}}{8(\tfrac{1}{8})} + \underset{\text{Center}}{1} = 2$$

Thus we can compute the fraction of space occupied by the spheres:

$$\frac{2 \times \frac{4}{3}\pi R^3 \;(\text{Volume occupied by spheres})}{\left(\frac{4R}{\sqrt{3}}\right)^3 \;(\text{Cube volume})} = \frac{\sqrt{3}\,\pi}{8} = 0.680$$

So in the body-centered cubic arrangement 68.0% of the space is actually occupied by spheres. This is somewhat less than the space occupied in the closest packed structures (74.0%). That is, in contrast to cubic closest packing and hexagonal closest packing, the body-centered cubic method of packing spheres does not represent a closest packed structure.

The reasons why some metals adopt a body-centered cubic structure rather than a closest packed structure are complex and are often not apparent in particular cases.

Bonding in Metals

Any successful bonding model for metals must account for the typical physical properties of metals: malleability, ductility, and the efficient and uniform conduction of heat and electricity in all directions. Although the shapes of most pure metals can be changed relatively easily, most metals are durable and have high melting points. These facts indicate that the bonding in most metals is both *strong* and *nondirectional.* That is, although it is difficult to separate metal atoms, it is relatively easy to move them, provided the atoms stay in contact with each other.

The simplest picture that explains these observations is the **electron sea model,** which envisions a regular array of metal cations in a "sea" of valence electrons (see Fig. 16.22). The mobile electrons conduct heat and electricity, and the cations are easily moved around as the metal is hammered into a sheet or pulled into a wire.

A related model that gives a more detailed view of the electron energies and motions is the **band model,** or the molecular orbital (MO) model, for metals. In this model the electrons are assumed to travel around the metal crystal in molecular orbitals formed from the valence atomic orbitals of the metal atoms (Fig. 16.23).

Recall that in the MO model for the gaseous Li_2 molecule (Section 14.3), two widely spaced molecular orbital energy levels (bonding and antibonding) result when two identical atomic orbitals interact. However, when many metal atoms interact, as in a metal crystal, the large number of resulting molecular orbitals become closely spaced, forming a virtual continuum of levels, called **bands** (see Fig. 16.23).

As an illustration, picture a magnesium metal crystal, which has an hcp structure. Since each magnesium atom has one 3*s* and three 3*p* valence atomic orbitals, a crystal with *n* magnesium atoms has available *n*(3*s*) and 3*n*(3*p*) orbitals to form the molecular orbitals, as illustrated in Fig. 16.24. Note that the core electrons are localized, as shown by their presence in the energy "well" around each magnesium atom produced by its nuclear charge. However, the

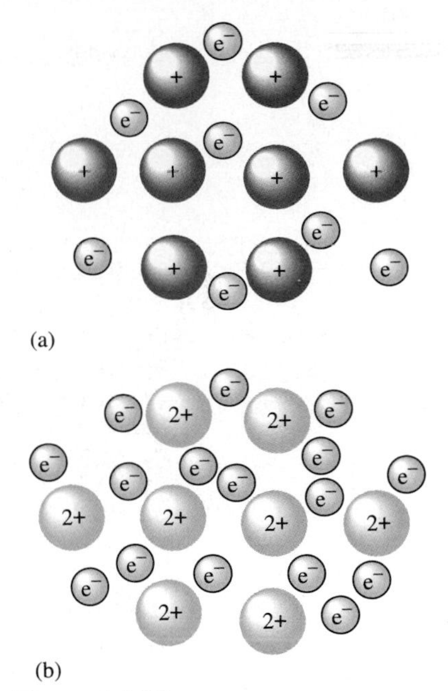

Figure 16.22
The electron sea model for metals postulates a regular array of cations in a "sea" of valence electrons. (a) Representation of an alkali metal (Group 1A) with one valence electron. (b) Representation of an alkaline earth metal (Group 2A) with two valence electrons.

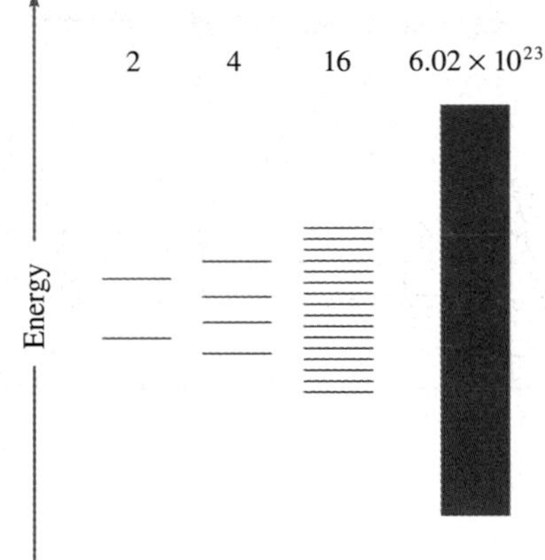

Figure 16.23
The molecular orbital energy levels produced when various numbers of atomic orbitals interact. Note that for two atomic orbitals two rather widely spaced energy levels result. (Recall the description of H_2 in Section 14.2.) As more atomic orbitals become available to form molecular orbitals, the resulting energy levels become more closely spaced, finally producing a band of very closely spaced orbitals.

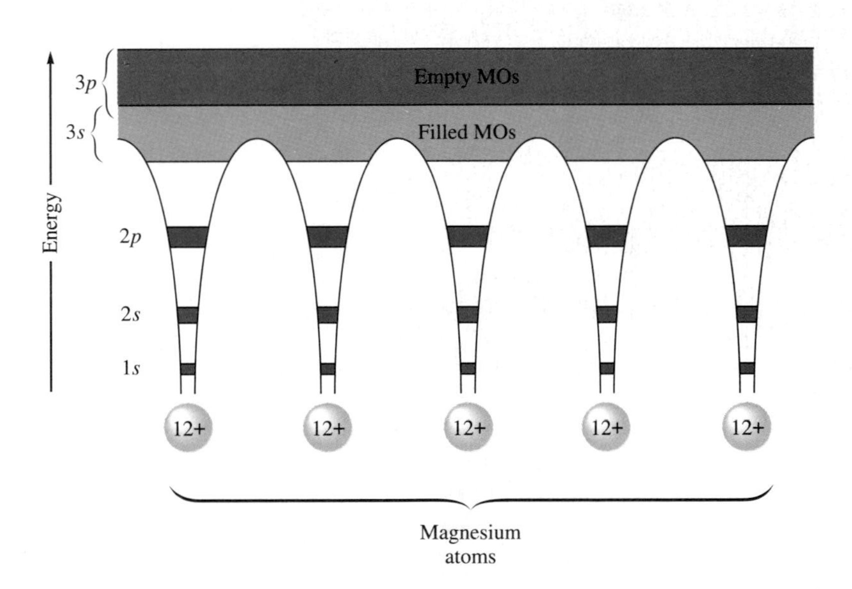

Figure 16.24
A representation of the energy levels (bands) in a magnesium crystal. The electrons in the 1*s*, 2*s*, and 2*p* orbitals are close to the nuclei and thus are localized on each magnesium atom as shown. However, the 3*s* and 3*p* valence orbitals overlap and mix to form molecular orbitals. Electrons in these energy levels can travel throughout the crystal.

SUPERCONDUCTIVITY

Although metals such as copper and aluminum are good conductors of electricity, up to 20% of the total energy is wasted in the transmission of a current by resistance heating of the wires. However, this loss of energy may be avoidable. Certain kinds of materials undergo a remarkable transition as they are cooled; their electrical resistance changes virtually to zero. These substances, called **superconductors,** conduct electricity with no wasted heat energy.

There is, of course, a catch. In the past superconductivity has been observed only at very low temperatures and is thus very expensive to maintain. In 1911 mercury was observed to exhibit superconductivity, but only at ≈4 K, the boiling point of liquid helium. In the next 75 years several niobium alloys showed superconductivity at temperatures as high as 23 K. However, in 1986 researchers realized that certain metallic oxides are superconductors at temperatures well above 23 K. The graph shows how rapidly our knowledge of superconductors is increasing.

The current class of high-temperature superconductors is called *perovskites,* a well-known class of compounds that contain copper, an alkaline earth metal, a lanthanide metal, and oxygen. In the unit cell for the compounds with formula $YBa_2Cu_3O_x$, in which $x = 6.527$, the small open circles indicate the expected oxygen positions, but some of these positions are vacant in the known materials—thus the variable oxygen content.

In addition to exhibiting superconductivity at temperatures above the boiling point of liquid nitrogen (77 K), these ceramic materials can sustain very high currents, an important characteristic necessary for any large-scale application. More recent

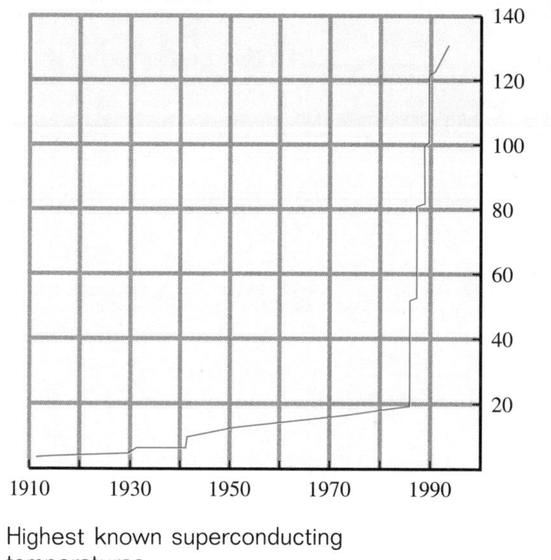

Highest known superconducting temperatures.

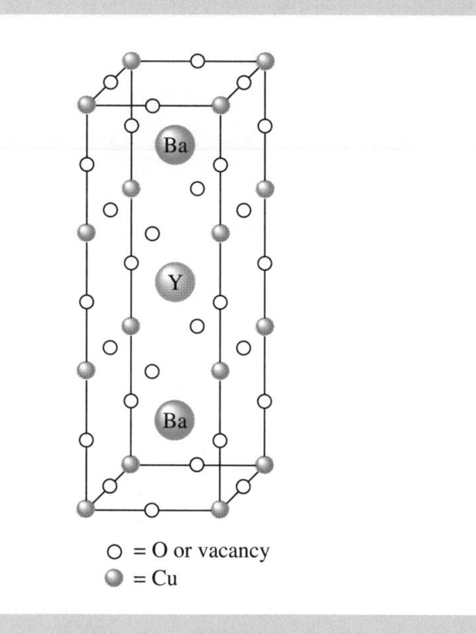

results have shown that materials containing thallium and calcium ions in place of the lanthanide ions superconduct at 125 K. In early 1993 researchers in Zurich, Switzerland discovered a mercury-containing compound that is superconducting at 133 K.

What does all this mean for the future? Superconducting materials could easily cause a revolution similar to that brought about by solid-state electronics. For example, one can envision electrical power transmitted with virtually no loss of energy. Also, new generations of superconducting electromagnets could lead to the operation of particle accelerators at unprecedented energies, the development of nuclear fusion power, the ability to produce incredibly accurate three-dimensional images of the human body for diagnosis of disease, and the development of superfast, magnetically levitated trains.

After a great surge of progress in the late 1980s, the pace slowed greatly in the early 1990s, as theory tried to catch up with the results from experiments. We need to develop an understanding of why these ceramic materials are superconducting so that we can design better materials that superconduct at even higher temperatures. And there are many practical problems. For example, metal conductors can be easily bent into coils to allow construction of electromagnets, but anyone who has ever dropped a plate knows how brittle ceramic materials are. Another question concerns how current can be conducted over long distances by superconducting materials. However, there seems to be little doubt that ingenious solutions will be found to the problems. The answers are already beginning to appear.

For example, American Superconductor (ASC) of Westborough, Massachusetts, in conjunction with Pirelli Cable of Milan, Italy, has produced a prototype, high-temperature superconducting "wire" that exceeds the current-carrying threshold required for commercial underground power transmission cables. The one-meter-long conductor, based on a bismuth-strontium-calcium-copper oxide superconductor, carries over 2300 amps of direct current at liquid nitrogen temperatures (77 K). This is more than twice the current-carrying capability of conventional copper cables.

ASC manufactures superconducting cables by pouring powdered reactants into a silver tube, drawing the tube into a wire, and then heating to produce the superconducting ceramic. Tens to hundreds of these fine conductors are then combined to form superconducting cable. Although ASC is now capable of producing individual superconducting wire filaments that are 1 km in length, much work remains to find a way to automate the multifilament cable manufacturing process.

A magnet is levitated over a superconducting ceramic immersed in liquid nitrogen.

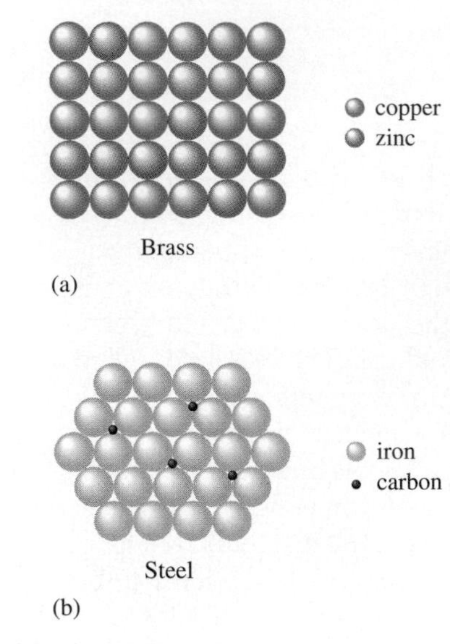

Figure 16.25
Two types of alloys.

valence electrons occupy closely spaced molecular orbitals, which are only partially filled.

The existence of empty molecular orbitals close in energy to filled molecular orbitals explains the thermal and electrical conductivity of metal crystals. Metals conduct electricity and heat very efficiently because of the availability of highly mobile electrons. For example, when an electrical potential is placed across a strip of metal, for current to flow electrons must be free to move from the negative to the positive areas of the metal. In the band model for metals, mobile electrons are furnished when electrons in filled molecular orbitals are excited into empty ones. The conduction electrons are free to travel throughout the metal crystal as dictated by the potential imposed on the metal. The molecular orbitals occupied by these conducting electrons are called **conduction bands.** These mobile electrons also account for the efficiency of the conduction of heat through metals. When one end of a metal rod is heated, the mobile electrons can rapidly transmit the thermal energy to the other end.

Metal Alloys

Because of the nature of the structure and bonding of metals, other elements can be introduced into a metallic crystal relatively easily to produce substances called alloys. An **alloy** is best defined as *a substance that contains a mixture of elements and has metallic properties.* Alloys can be conveniently classified into two types.

In a **substitutional alloy** some of the host metal atoms are *replaced* by other metal atoms of similar size. For example, in brass approximately one-third of the atoms in the host copper metal are replaced by zinc atoms, as shown in Fig. 16.25(a). Sterling silver (93% silver and 7% copper), pewter (85% tin, 7% copper, 6% bismuth, and 2% antimony), and plumber's solder (67% lead and 33% tin) are other examples of substitutional alloys.

An **interstitial alloy** is formed when some of the interstices (holes) in the closest packed metal structure are occupied by small atoms, as shown in Fig. 16.25(b). Steel, the best known interstitial alloy, contains carbon atoms in the holes of an iron crystal. The presence of the interstitial atoms changes the properties of the host metal. Pure iron is relatively soft, ductile, and malleable due to the absence of strong directional bonding. The spherical metal atoms can be rather easily moved with respect to each other. However, when carbon, which forms strong directional bonds, is introduced into an iron crystal, the presence of the directional carbon-iron bonds makes the resulting alloy harder, stronger, and less ductile than pure iron. The amount of carbon present directly affects the properties of steel. *Mild steels,* containing less than 0.2% carbon, are relatively ductile and malleable. These steels are used for nails, cables, and chains. *Medium steels,* containing 0.2–0.6% carbon, are harder than mild steels and are used in rails and structural steel beams. *High-carbon steels,* containing 0.6–1.5% carbon, are tough and hard and are used for springs, tools, and cutlery.

Many types of steel also contain elements in addition to iron and carbon. Such steels are often called *alloy steels* and can be viewed as being mixed interstitial (carbon) and substitutional (other metals) alloys. Bicycle frames, for example, are constructed from a wide variety of alloy steels. The compositions of the two brands of steel tubing used in high-quality racing bicycles are given in Table 16.3.

TABLE 16.3 **The Composition of the Two Brands of Steel Tubing Most Commonly Used to Make Lightweight Racing Bicycles**

Brand of Tubing	% C	% Si	% Mn	% Mo	% Cr
Reynolds	0.25	0.25	1.3	0.20	—
Columbus	0.25	0.30	0.65	0.20	1.0

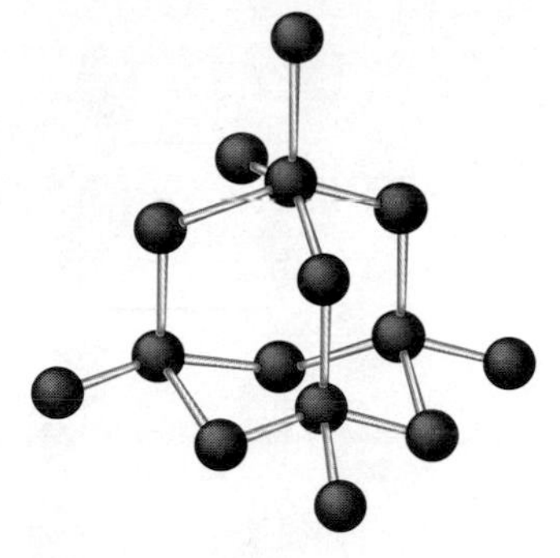
(a) Diamond

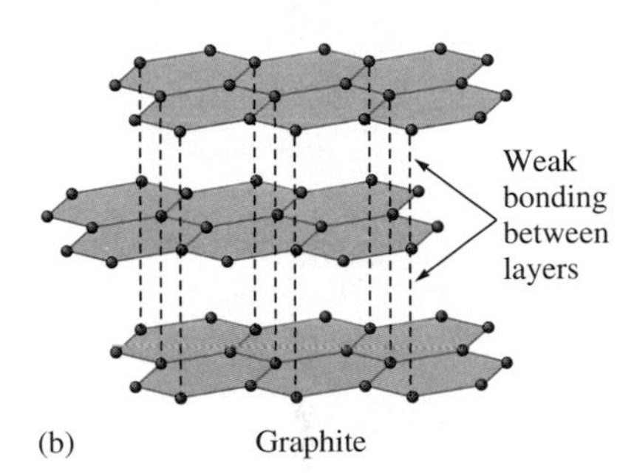

(b) Graphite

Figure 16.26
The structures of diamond and graphite. In each case only a small part of the entire structure is shown.

16.5 Carbon and Silicon: Network Atomic Solids

Many atomic solids contain strong directional covalent bonds. We will call these substances **network solids.** In contrast to metals, these materials are typically brittle and do not efficiently conduct heat or electricity. To illustrate network solids, in this section we will discuss two very important elements, carbon and silicon, and some of their compounds.

Carbon occurs in the *allotropes* (different forms) diamond, graphite, and the fullerenes. The fullerenes are molecular solids (see Section 16.6), but diamond and graphite are typically network solids. In diamond, the hardest naturally occurring substance, each carbon atom is surrounded by a tetrahedral arrangement of other carbon atoms, as shown in Fig. 16.26(a). This structure is stabilized by covalent bonds, which, in terms of the localized electron model, are formed by the overlap of sp^3 hybridized atomic orbitals on each carbon atom.

It is also useful to consider the bonding among the carbon atoms in diamond in terms of the molecular orbital model. Energy-level diagrams for diamond and a typical metal are given in Fig. 16.27. Recall that the conductivity of metals can be explained by postulating that electrons are excited from filled levels into the very near empty levels—the conduction bands. However, note that in the energy-level diagram for diamond there is *a large gap between the filled and the empty levels.* This means that electrons cannot be easily transferred to the empty conduction bands. As a result, diamond is not expected to

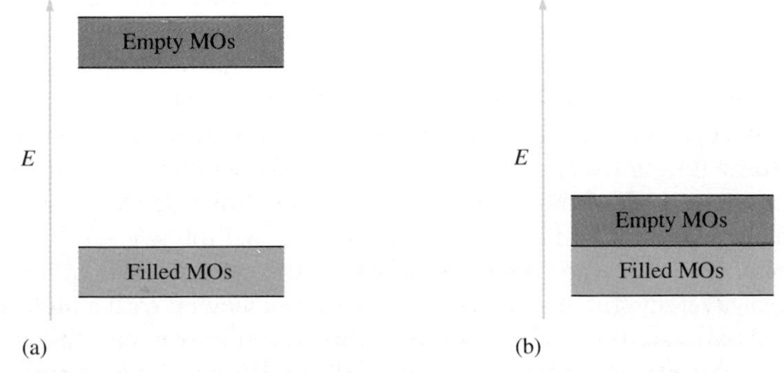

Figure 16.27
Partial representation of the molecular orbital energies in (a) diamond and (b) a typical metal.

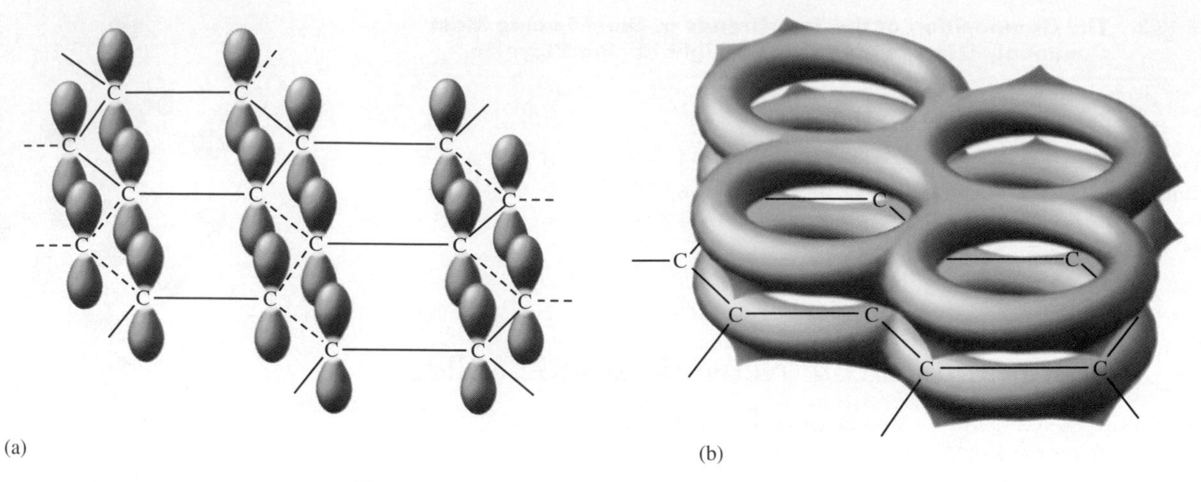

Figure 16.28
The p orbitals (a) perpendicular to the plane of the carbon ring system in graphite can combine to form (b) an extensive π bonding network.

be a good electrical conductor. In fact, this prediction of the model agrees exactly with the observed behavior of diamond, which is known to be an electrical *insulator*—it does not conduct an electric current.

Graphite is very different from diamond. Diamond is hard, basically colorless, and an insulator; but graphite is slippery, black, and a conductor. These differences, of course, arise from the differences in bonding in the two types of solids. In contrast to the tetrahedral arrangement of carbon atoms in diamond, the structure of graphite is based on layers of carbon atoms arranged in fused six-membered rings, as shown in Fig. 16.26(b). Each carbon atom in a particular layer of graphite is surrounded by three other carbon atoms in a trigonal planar arrangement with 120° bond angles. The localized electron model predicts sp^2 hybridization in this case. The three sp^2 orbitals on each carbon are used to form σ bonds to three other carbon atoms. One $2p$ orbital remains unhybridized on each carbon and is perpendicular to the plane of carbon atoms, as shown in Fig. 16.28. These orbitals combine to form a group of closely spaced π molecular orbitals that are important in two ways. First, they contribute significantly to the stability of the graphite layers because of the π bonding. Second, the π molecular orbitals with their delocalized electrons account for the electrical conductivity of graphite. These closely spaced orbitals are exactly analogous to the conduction bands found in metal crystals.

A Raytheon scientist examines individual diamond crystals with a scanning electron microscope.

Graphite is often used as a lubricant in locks (where oil is undesirable because it collects dirt). The characteristic slipperiness of graphite can be explained by noting that graphite has very strong bonding *within* the layers of carbon atoms but little bonding *between* the layers (the valence electrons are all used to form σ and π bonds among carbons within a given layer). This arrangement allows the layers to slide past one another quite readily. Graphite's layered structure is quite obvious when viewed with a high-magnification electron microscope (Fig. 16.29). This structure is in contrast to that of diamond, which has uniform bonding in all directions in the crystal.

Because of their extreme hardness, diamonds are extensively used in industrial cutting implements. Thus it is desirable to convert inexpensive graphite to diamond. As we might expect from the high density of diamond (3.5 g/cm^3) compared with that of graphite (2.2 g/cm^3), this transformation can be accomplished by applying very high pressures to graphite. The application of 150,000 atm of pressure at 2800°C converts graphite virtually completely to diamond. (The high temperature is required to break the strong bonds in graphite so that the rearrangement can occur.)

Figure 16.29
A high-resolution image of crystalline graphite (×2,500,000) taken by an electron microscope.

Silicon is an important constituent of the compounds that make up the earth's crust. In fact, silicon is to geology what carbon is to biology. Just as carbon compounds are the basis for most biologically significant systems, silicon compounds are fundamental to most of the rocks, sands, and soils found in the earth's crust. Although carbon and silicon are next to each other in Group 4A of the periodic table, the carbon-based compounds of biology and the silicon-based compounds of geology have markedly different structures. Carbon compounds typically contain long strings of carbon-carbon bonds, but the most stable silicon compounds involve chains with silicon-oxygen bonds. *The most important silicon compounds contain silicon and oxygen.*

The fundamental silicon-oxygen compound is **silica,** which has the empirical formula SiO_2. Knowing the properties of the similar compound carbon dioxide, one might expect silica to be a gas that contains discrete SiO_2 molecules. In fact, nothing could be further from the truth—quartz and some types of sand are typical of the materials composed of silica. What accounts for this difference? The answer lies in the bonding.

Artificial diamonds are discussed in more detail in Section 16.11.

Recall that the Lewis structure for CO_2 is

$$:\ddot{O}=C=\ddot{O}:$$

and that each C=O bond is described as a combination of a σ bond, involving a carbon *sp* hybrid orbital, and a π bond, involving a carbon 2*p* orbital. On the contrary, silicon cannot use its valence 3*p* orbitals to form strong π bonds to oxygen; the larger size of the silicon atom and its orbitals result in a less effective overlap with the smaller oxygen orbitals. Therefore, instead of forming π bonds, the silicon atom satisfies the octet rule by forming single bonds to four oxygen atoms, as shown in the representation of the structure of quartz in Fig. 16.30. Note that in this structure each silicon atom is at the center of a tetrahedral arrangement of oxygen atoms, which are shared with other silicon atoms. Although the empirical formula for quartz is SiO_2, the structure is based on a *network* of SiO_4 tetrahedra with shared oxygen atoms rather than discrete SiO_2 molecules. Thus the differing abilities of carbon and silicon to form π bonds with oxygen have profound effects on the structures and properties of CO_2 and SiO_2.

The bonding in the CO_2 molecule was described in Section 14.1.

Compounds closely related to silica that are found in most rocks, soils, and clays are the **silicates.** Like silica, the silicates are based on interconnected SiO_4 tetrahedra. However, in contrast to silica, where the O:Si ratio is 2:1, silicates have O:Si ratios greater than 2:1 and contain silicon-oxygen *anions.* This means that for the formation of the neutral solid silicates, cations are needed to balance the excess negative charge. In other words, silicates are salts containing metal cations and polyatomic silicon-oxygen anions. Examples of important silicate anions are shown in Fig. 16.31.

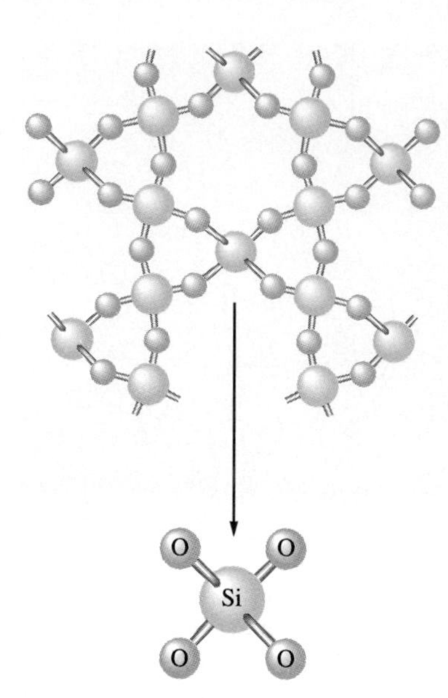

Figure 16.30
The structure of quartz (empirical formula SiO_2). Quartz contains chains of SiO_4 tetrahedra that share oxygen atoms.

When silica is heated above its melting point (about 1600°C) and then cooled rapidly, an amorphous solid called a **glass** results (see Fig. 16.32). Note that a glass contains a good deal of disorder, in contrast to the crystalline nature of quartz. Glass more closely resembles a very viscous solution than a crystalline solid. Common glass results when substances such as Na_2CO_3 are added to the silica melt, which is then cooled. The properties of glass can be varied greatly by varying the additives. For example, addition of B_2O_3 produces a glass (called borosilicate glass) that expands and contracts little under large temperature changes. Thus it is useful for labware and cooking utensils. The most common brand name for this glass is Pyrex. The addition of K_2O produces an especially hard glass that can be ground to the precise shapes needed for eyeglasses and contact lenses. The compositions of several types of glass are shown in Table 16.4.

Ceramics

Ceramics are typically made from clays (which contain silicates) that are hardened by firing at high temperatures. Ceramics are a class of nonmetallic materials that are strong, brittle, and resistant to heat and attack by chemicals.

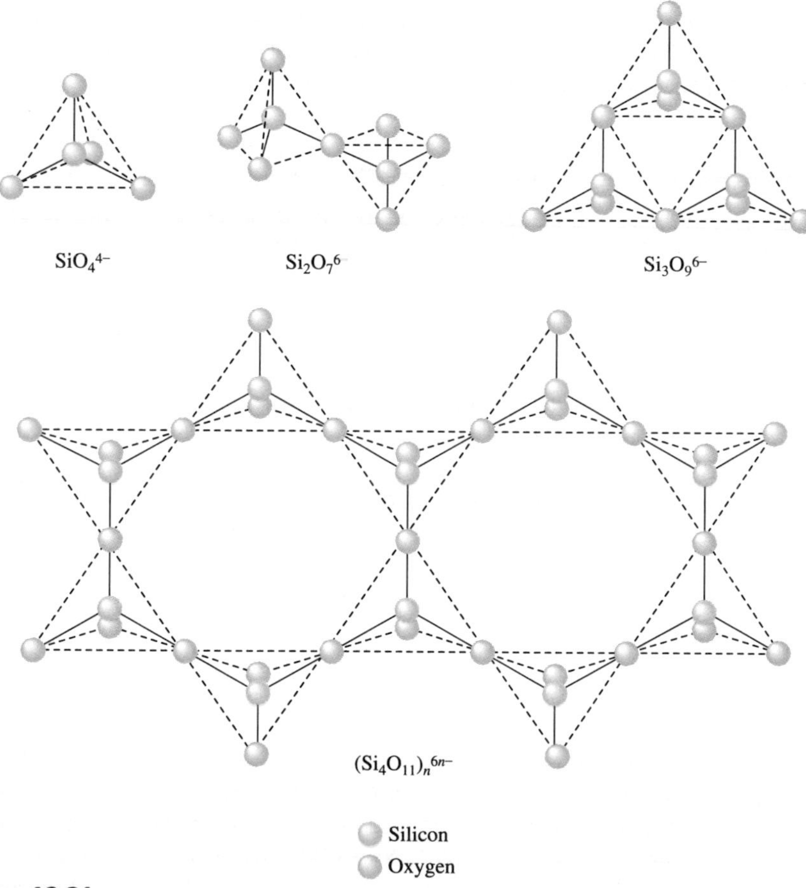

Figure 16.31
Examples of silicate anions, all of which are based on SiO_4^{4-} tetrahedra.

TABLE 16.4 Compositions of Some Common Types of Glass

Type of Glass	Percentages of Various Components						
	SiO_2	CaO	Na_2O	B_2O_3	Al_2O_3	K_2O	MgO
Window (soda-lime glass)	72	11	13	—	0.3	3.8	—
Cookware (aluminosilicate glass)	55	15	—	—	20	—	10
Heat-resistant (borosilicate glass)	76	3	5	13	2	0.5	—
Optical	69	12	6	0.3	—	12	—

(a)

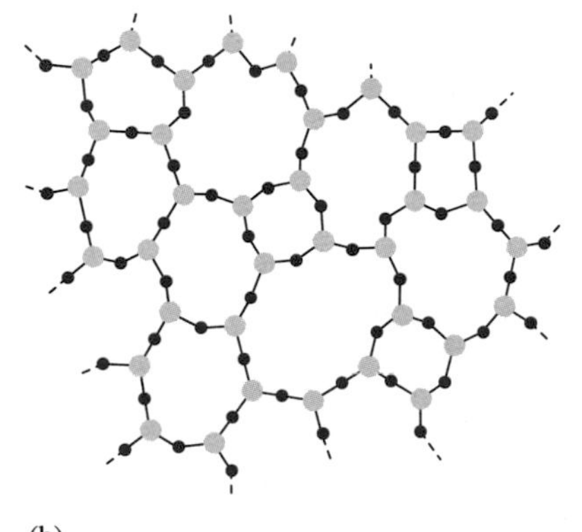

(b)

Figure 16.32 Two-dimensional representations of (a) a quartz crystal and (b) a quartz glass.

Like glass, ceramics are based on silicates, but with that the resemblance ends. Glass can be melted and remelted as often as desired; but once a ceramic has been hardened, it is resistant to extremely high temperatures. This behavior results from the very different structures of glasses and ceramics. A glass is a *homogeneous,* noncrystalline "frozen solution," while a ceramic is *heterogeneous.* A ceramic contains two phases: minute crystals of silicates that are suspended in a glassy cement.

To understand how ceramics harden, one must know something about the structure of clays. Clays are formed by the weathering action of water and carbon dioxide on the mineral feldspar, which is a mixture of silicates with empirical formulas such as $K_2O \cdot Al_2O_3 \cdot 6SiO_2$ and $Na_2O \cdot Al_2O_3 \cdot 6SiO_2$. Feldspar is really an **aluminosilicate** in which aluminum as well as silicon atoms are part of the oxygen-bridged polyanion. The weathering of feldspar produces kaolinite, consisting of tiny thin platelets with the empirical formula $Al_2Si_2O_5(OH)_4$. When dry, the platelets cling together; but when water is present, they can slide over one another, giving clay its plasticity. As clay dries, the platelets begin to interlock again. When the remaining water is driven off during firing, the silicates and cations form a "glass" that binds the tiny crystals of kaolinite.

Ceramics have a very long history. Rocks, which are natural ceramic materials, served as the earliest tools. Later, clay vessels dried in the sun or baked in fires served as containers for food and water. These early vessels were crude and quite porous. With the discovery of glazing, which probably occurred about 3000 B.C. in Egypt, pottery became more serviceable as well as more beautiful. Prized porcelain is essentially the same material as crude earthenware, but specially selected clays and glazings are used for porcelain, which is also fired at a very high temperature.

Semiconductors

Elemental silicon has the same structure as diamond, as might be expected from its position in the periodic table (in Group 4A directly under carbon). Recall that in diamond there is a large energy gap between the filled and empty molecular orbitals (Fig. 16.27). This gap prevents excitation of electrons to the empty molecular orbitals (conduction bands) and makes diamond an insulator. In silicon the situation is similar but the energy gap is smaller. A few electrons

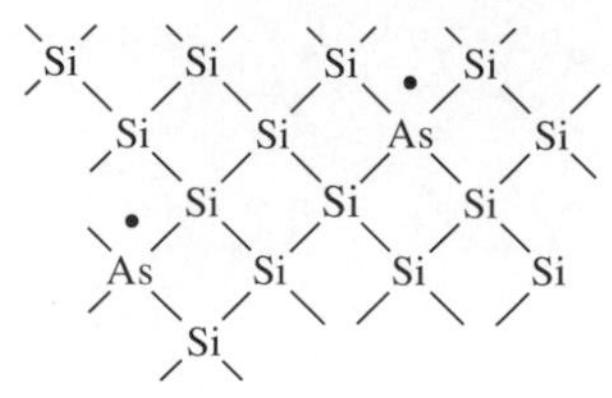

(a)

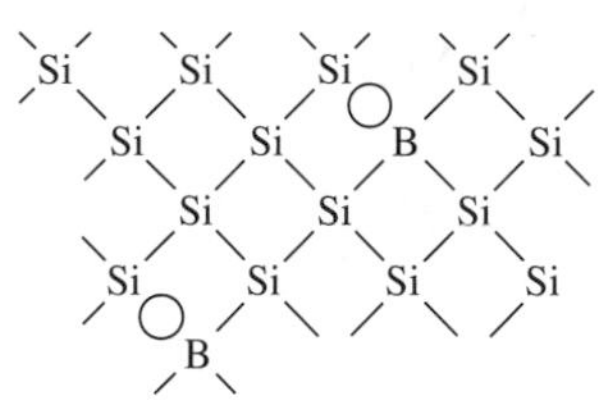

(b)

Figure 16.33
(a) A silicon crystal doped with arsenic, which has one extra valence electron. (b) A silicon crystal doped with boron, which has one less electron than silicon.

Electrons must be in singly occupied molecular orbitals to conduct a current.

can cross the gap at 25°C, making silicon a *semiconducting element,* or **semiconductor.** In addition, at higher temperatures, where more energy is available to excite electrons into the conduction bands, the conductivity of silicon increases. This is typical behavior for a semiconducting element and is in contrast to the behavior of metals. In metals, the conductivity decreases with increasing temperature.

The small conductivity of silicon can be enhanced at normal temperatures if the silicon crystal is *doped* with certain other elements. For example, when a small fraction of silicon atoms is replaced by arsenic atoms, each having *one more* valence electron than silicon, extra electrons become available for conduction, as shown in Fig. 16.33(a). This produces an **n-type semiconductor,** a substance whose conductivity is increased by doping it with atoms having more valence electrons than the atoms in the host crystal. These extra electrons lie close in energy to the conduction bands and can easily be excited into these levels, where they can conduct an electric current [see Fig. 16.34(a)].

We can also enhance the conductivity of silicon by doping the crystal with an element such as boron, which has only three valence electrons, one *less* than silicon. Because boron has one less electron than is required to form the bonds to the surrounding silicon atoms, an electron vacancy, or *hole,* is created, as shown in Fig. 16.33(b). As an electron moves to fill this hole, it leaves a new hole. As this process is repeated, the hole advances through the crystal in a direction opposite to movement of the electrons jumping to fill the hole. Another way of thinking about this phenomenon is that in pure silicon each atom has four valence electrons, so the low-energy molecular orbitals are exactly filled. Replacing silicon atoms with boron atoms leaves vacancies in these molecular orbitals, as shown in Fig. 16.34(b). This means that there is only one electron in some of the molecular orbitals, and these unpaired electrons can function as conducting electrons. Thus the substance becomes a better conductor. When semiconductors are doped with atoms having fewer valence electrons than the atoms of the host crystal, they are called **p-type**

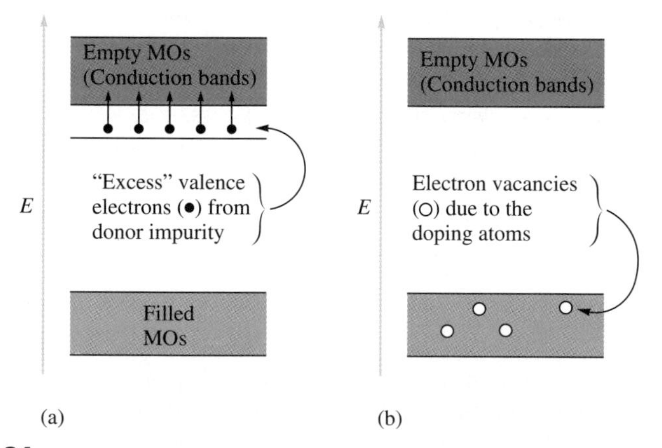

Figure 16.34
Energy-level diagrams for (a) an n-type semiconductor and (b) a p-type semiconductor.

semiconductors, so named because the positive holes can be viewed as the charge carriers.

Most important applications of semiconductors involve the connection of a p-type and an n-type to form a **p-n junction.** Figure 16.35(a) shows a typical junction; the dark circles represent excess electrons in the n-type semiconductor, and the white circles represent holes (electron vacancies) in the p-type semiconductor. At the junction a small number of electrons migrate from the n-type region into the p-type region, where there are vacancies in the low-energy molecular orbitals. The effect of these migrations is to place a negative charge on the p-type region (since it now has a surplus of electrons) and a positive charge on the n-type region (since it has lost electrons, leaving holes in its low-energy molecular orbitals). This charge buildup, called the *contact potential,* or *junction potential,* prevents further migration of electrons.

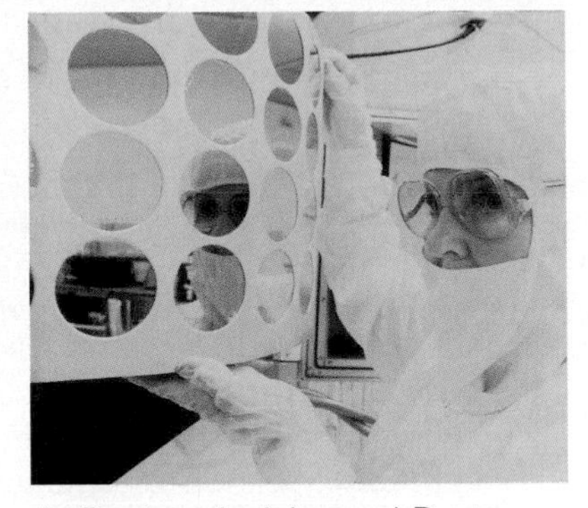

At Raytheon's Advanced Device Center, wafers made from gallium arsenide, a semiconductor, are coated with gold and other metals.

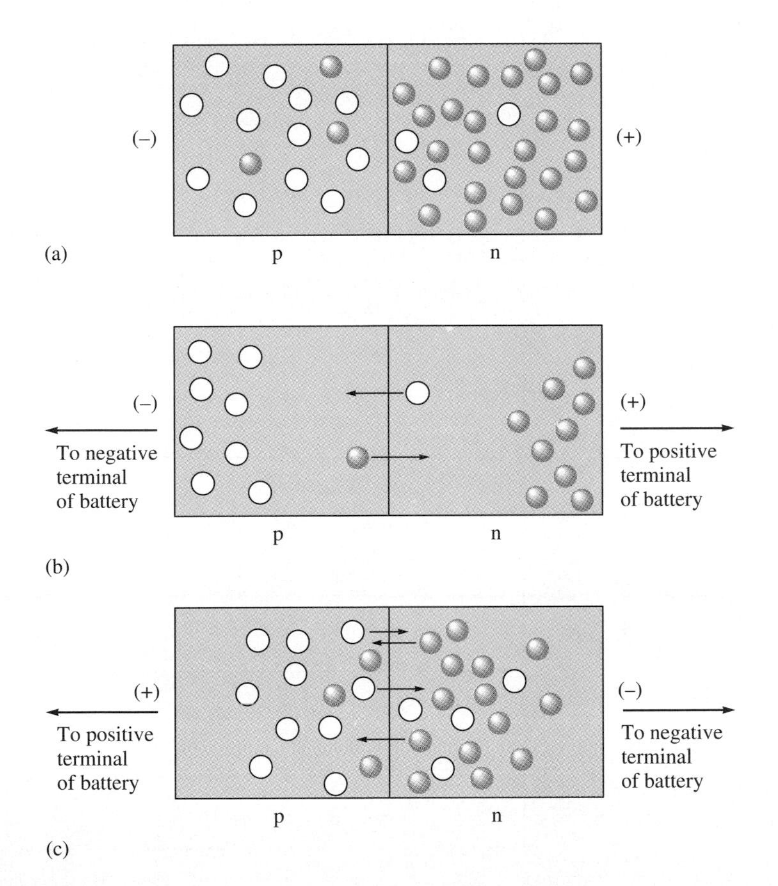

Figure 16.35
The p-n junction involves the contact of a p-type and an n-type semiconductor. (a) The charge carriers of the p-type region are holes (○). In the n-type region the charge carriers are electrons (●). (b) No current flows (reverse bias). (c) Current readily flows (forward bias). Note that each electron that crosses the boundary leaves a hole behind. Thus the electrons and the holes move in opposite directions.

GALLIUM ARSENIDE LASERS

Lasers are an essential part of modern life, providing the most efficient way to carry telephone communications by cable, the best quality music using compact disc players, and the most accurate method for reading grocery prices. Lasers have also become invaluable for precision surgery on the eyes and for removing arterial blockages in persons with heart disease.

The laser (the word *lase* is an acronym for "light amplification by stimulated emission") is a device for producing an intense, coherent beam of light when excited electrons release photons as they change to a lower energy state. Excited electrons can return to a lower energy level by spontaneous, random emission of light. However, there is another mechanism in which the emission process is actually stimulated by a passing photon that has the same energy as the emitted photon. This is the phenomenon on which the laser is based. We will have more to say about this presently.

Gallium arsenide is a prominent member of the so-called 3–5 class of semiconductors, those containing elements from Groups 3A and 5A of the periodic table. An n-type semiconductor results when the arsenic-to-gallium ratio is greater than 1. A p-type semiconductor results when gallium is in excess. By carefully controlling the conditions under which a Ga-As semiconductor is made, one can design p-n junctions into the material. At each p-n junction electrons from the n-type region are available to "fall into" holes in the p-type region, producing photons (see the accompanying figure). As the emitted photons travel through the solid, they *stimulate* the emission of additional photons. A mirror at the edge of the solid reflects these photons back through the substance, where they stimulate even more emission events. All of these stimulated emissions produce photons with their waves in phase. Thus this process leads to a large number of coherent (in-phase) photons, which proceed out the other end of the solid where a "half-mirror" is placed. This mirror reflects some of the photons and allows the rest to leave as an intense, coherent "laser beam" of light. A current driven through the substance by an applied potential difference provides a continuous supply of electrons (and removes electrons to make new holes) so that the process occurs continuously while the power is on.

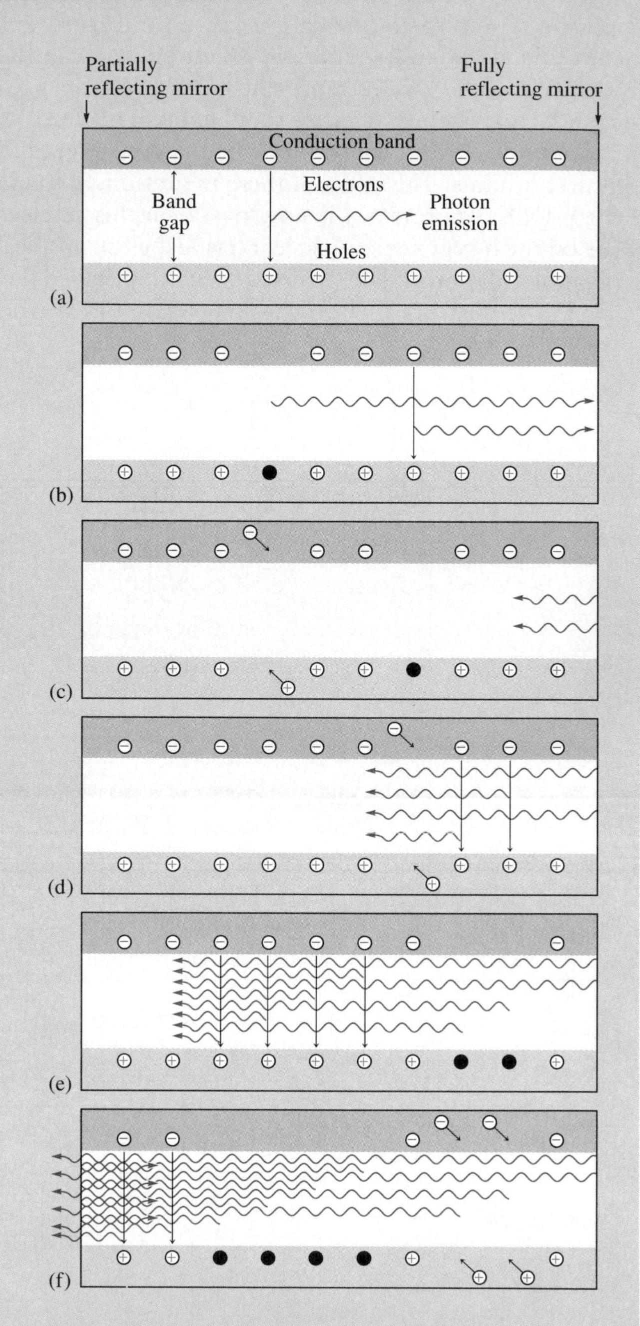

Now suppose an external electrical potential is applied by connecting the negative terminal of a battery to the p-type region and the positive terminal to the n-type region. The situation represented in Fig. 16.35(b) results. Electrons are drawn toward the positive terminal, and the resulting holes move toward the negative terminal—exactly opposite to the natural flow of electrons at the p-n junction. The junction resists the imposed current flow in this direction and is said to be under *reverse bias*. No current flows through the system.

On the other hand, if the battery is connected so that the negative terminal is connected to the n-type region and the positive terminal is connected to the p-type region [Fig. 16.35(c)], the movement of electrons (and holes) is in the favored direction. The junction has low resistance, and a current flows easily. The junction is said to be under *forward bias*.

A p-n junction makes an excellent *rectifier*, a device that produces direct current (flows in one direction) from an alternating current (flows in both directions alternately). When placed in a circuit where the potential is constantly reversing, a p-n junction transmits current only under forward bias, thus converting the alternating current to a direct current. Radios, computers, and other electrical devices formerly used bulky, unreliable vacuum tubes as rectifiers. The p-n junction has revolutionized electronics; modern solid-state components contain p-n junctions in printed circuits.

16.6 Molecular Solids

So far we have considered solids in which atoms occupy the lattice positions. In most cases such crystals can be considered to consist of one giant molecule. However, there are many types of solids that contain discrete molecular units at each lattice position. A common example is ice, where the lattice positions are occupied by water molecules [see Fig. 16.12(c)]. Other examples are dry ice (solid carbon dioxide), some forms of sulfur that contain S_8 molecules [Fig. 16.36(left)], certain forms of phosphorus that contain P_4 molecules [Fig. 16.36(right)], and the fullerenes, which are all of carbon-containing molecules such as C_{60}, C_{70} and others (see Buckminsterfullerene: A New Form of Carbon, p. 42). These substances are characterized by strong covalent bonding *within* the molecules but relatively weak forces *between* the molecules. For example, it takes only 6 kj of energy to melt 1 mole of solid water (ice) because only intermolecular (H_2O—H_2O) interactions must be overcome. However, 470 kJ of energy is required to break a mole of covalent O—H bonds. The differences between the covalent bonds within the molecules and the forces between the molecules are apparent from the comparison of the interatomic and intermolecular distances in solids shown in Table 16.5.

The forces that exist among the molecules in a molecular solid depend on the nature of the molecules. Many molecules such as CO_2, I_2, P_4, and S_8 have zero dipole moments, and the intermolecular forces are London dispersion forces. Because these forces are usually small, we might expect all of these substances to be gaseous at 25°C, as is the case for carbon dioxide. However, as the size of the molecules increases, the London forces become quite large, causing many of these substances to be solids at 25°C.

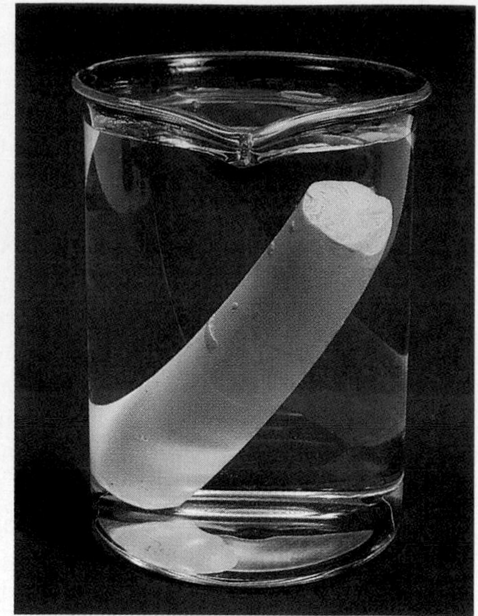

Figure 16.36
(left) Sulfur crystals (yellow) contain S_8 molecules. (right) White phosphorus contains P_4 molecules. It is so reactive with the oxygen in air that it must be stored under water.

TABLE 16.5 **Comparison of Atomic Separations Within Molecules (Covalent Bonds) and Between Molecules (Intermolecular Interactions)**

Solid	Distance Between Atoms in Molecule*	Closest Distance Between Molecules in the Solid
P_4	2.20 Å	3.8 Å
S_8	2.06 Å	3.7 Å
Cl_2	1.99 Å	3.6 Å

* The shorter distances within the molecules indicate stronger bonding.

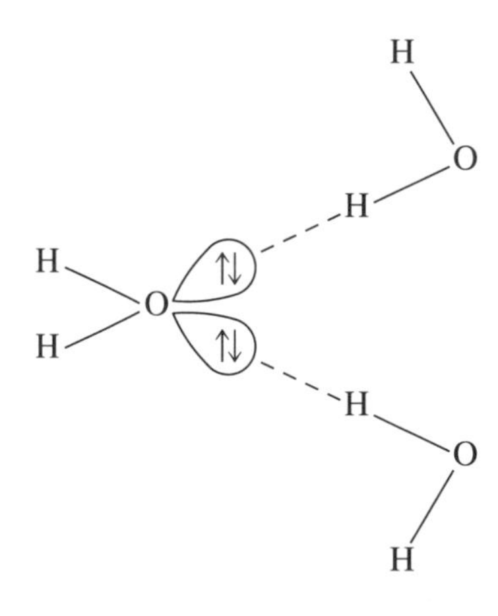

When molecules do have dipole moments, their intermolecular forces are typically greater, especially when hydrogen bonding is possible. Water molecules are particularly well suited to interact with each other, because each molecule has two polar O—H bonds and two lone electron pairs on the oxygen atom. This can lead to the association of four hydrogen atoms with each oxygen: two by covalent bonds and two by dipole forces, as shown in the figure in the margin. Note the two relatively short covalent oxygen-hydrogen bonds and the two longer oxygen-hydrogen dipole interactions that can be seen in the ice structure in Fig. 16.12(c).

16.7 Ionic Solids

Ionic solids are stable, high-melting substances held together by the strong electrostatic forces that exist between oppositely charged ions. The principles governing the structures of ionic solids were introduced in Section 13.5. In this section we will review and extend these principles.

The structures of most binary ionic solids, such as sodium chloride, can be explained by the closest packing of spheres. Typically, the large ions, which are usually the anions, are packed in one of the closest packing arrangements (hcp or ccp), while the smaller cations fit into holes among the close packed anions. The packing is done in a way that maximizes the electrostatic attractions among oppositely charged ions while minimizing the repulsions among ions with like charges.

There are three types of holes in closest packed structures:

1. Trigonal holes are formed by three spheres in the same layer [Fig. 16.37(a)].
2. Tetrahedral holes are formed when a sphere sits in the dimple of three spheres in an adjacent layer [Fig. 16.37(b)].
3. Octahedral holes are formed between two sets of three spheres in adjoining layers of the closest packed structures [Fig. 16.37(c)].

For spheres of a given diameter, the holes increase in size as follows:

$$\text{trigonal} < \text{tetrahedral} < \text{octahedral}$$

In fact, trigonal holes are so small that they are never occupied in binary ionic compounds. Whether the tetrahedral or octahedral holes in a given binary ionic solid are occupied depends mainly on the *relative* sizes of the anion and cation. Next, we will determine the sizes of the octahedral and tetrahedral holes and consider guidelines for their occupation by ions.

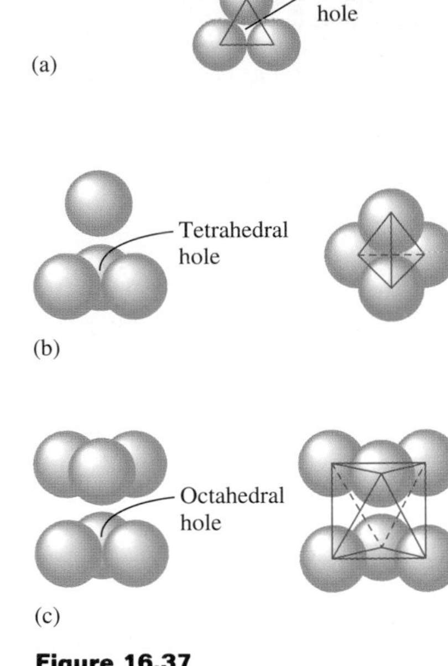

Figure 16.37
The holes that exist among closest packed uniform spheres. (a) The trigonal hole formed by three spheres in a given plane. (b) The tetrahedral hole formed when a sphere occupies a dimple in an adjacent layer. (c) The octahedral hole formed by six spheres in two adjacent layers.

Octahedral Holes

An octahedral hole lies at the center of six equidistant spheres whose centers define an octahedron. Three of these six spheres lie in one closest packed layer and three lie in the adjacent layer [Fig. 16.37(c)]. These six spheres can be rotated to show the octahedron more clearly (Fig. 16.38).

Since the edges of all six spheres are exactly the same distance from the center of the octahedral hole, we can calculate the radius of this hole most easily

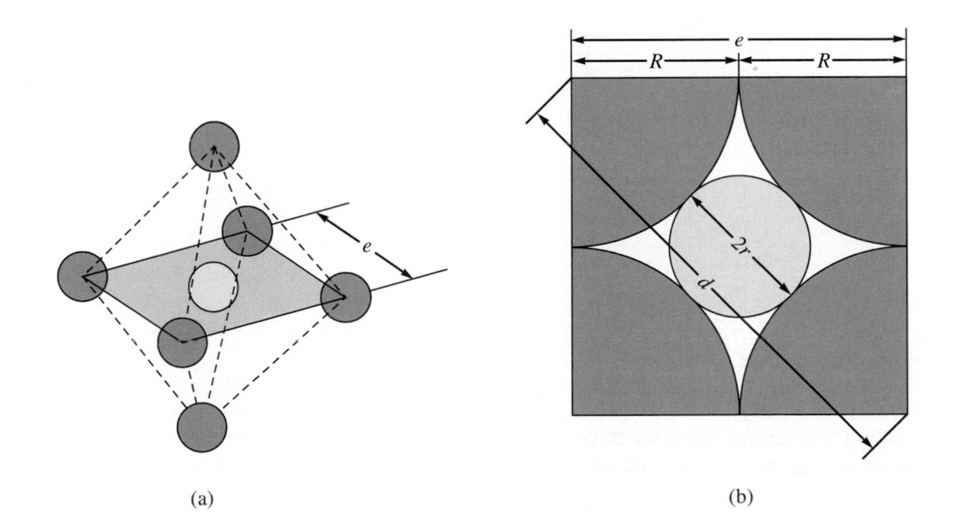

Figure 16.38
(a) The octahedral hole (shown in yellow) lies at the center of six spheres that touch along the edge (e) of the square. The packed spheres (in blue) are shown smaller than actual size for clarity. (b) The diagonal of the square (d) equals $R + 2r + R$, where r is the radius of the octahedral hole and R is the radius of the packed spheres.

by focusing on the four spheres whose centers form a square (Fig. 16.38). Note from Fig. 16.38 that R is the radius of the packed spheres, r is the radius of the octahedral hole, and d is the length of the diagonal of the square. From the Pythagorean theorem

$$(2R)^2 + (2R)^2 = d^2$$

where $$d = R + 2r + R = 2R + 2r = 2(R + r)$$

So we have $$8R^2 = d^2$$

and $$d = \sqrt{8}\,R = 2\sqrt{2}\,R = 2(R + r)$$

Solving for r yields

$$r = \sqrt{2}\,R - R = 1.414R - R = 0.414R$$

This result shows that an octahedral hole in a closest packed structure has a radius that is 0.414 times the radius of the packed spheres (ions).

Tetrahedral Holes

A tetrahedral hole lies at the center of four spheres whose centers form a tetrahedron [Fig. 16.37(b)]. Three of the spheres are in a given closest packed layer, and the fourth is in the next layer, nestling into the dimple formed by the other three.

The most convenient way to do geometric calculations on a tetrahedron is to inscribe it in a cube, as shown in Fig. 16.39. In this cube two spheres touch along the face diagonal of any cube face. This means that the face diagonal of the cube has length $2R$, where R is the radius of the packed spheres. The center of the tetrahedral hole is at the center of the body diagonal of the cube (Fig. 16.40). Now we will use the Pythagorean theorem twice to express the length of

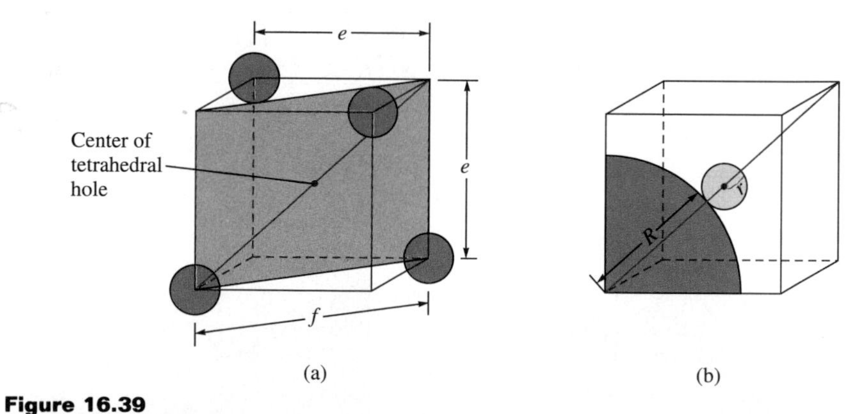

Figure 16.39
The tetrahedral hole. (a) The four spheres around a tetrahedral hole are shown inscribed in a cube. The spheres are shown much smaller than actual size. They actually touch along the face diagonal f. (b) The center of the tetrahedral hole (shown in red) is at the center of the body diagonal b (shown in green).

the body diagonal in terms of R. First, we express the face diagonal f in terms of the edge of the cube e:

$$f^2 = e^2 + e^2 = (2R)^2$$

Thus
$$e = \sqrt{2}\, R$$

Now we express the body diagonal b in terms of f and e:

$$b^2 = f^2 + e^2 = (2R)^2 + (\sqrt{2}\, R)^2$$

which leads to
$$b = \sqrt{6}\, R$$

Now the distance from the center of the body diagonal to the corner of the cube ($b/2$) is

$$r + R = \frac{b}{2} = \frac{\sqrt{6}\, R}{2} = \sqrt{\frac{3}{2}}\, R$$

↑ Radius of tetrahedral hole ↖ Radius of packed spheres

So
$$r = \sqrt{\frac{3}{2}}\, R - R = 1.225R - R = 0.225R$$

Thus we have shown that in a closest packed structure a tetrahedral hole has a radius that is 0.225 times the radius of the packed spheres.

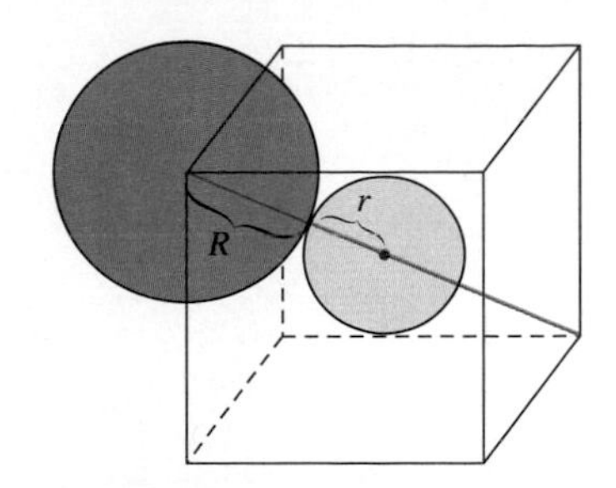

Figure 16.40
One packed sphere and its relationship to the tetrahedral hole. Note that (body diagonal)/2 = $R + r$.

Guidelines for Filling Octahedral and Tetrahedral Holes

The above calculations show that the tetrahedral holes in a given closest packed structure are about half the size of the octahedral holes. That is, for packed spheres with radius R,

$$r^{\text{Tet}} = 0.225R \quad \text{and} \quad r^{\text{Oct}} = 0.414R$$

Now picture an ionic solid MX containing M^+ cations and X^- anions where the X^- ions form a closest packed array. Assuming that the X^- ions have radius $R-$, how do we decide where to put the smaller cations? That is, do they occupy tetrahedral or octahedral holes? To obtain the most stable solid, we want to maximize the $M^+ \cdots X^-$ interactions and minimize the $X^- \cdots X^-$ interactions. We can do this by putting the M^+ ions into holes that are slightly smaller than the size of the M^+ ions. This causes the closest packed X^- ions to be pushed apart slightly. Assuming the ions behave as hard spheres, this will result in a structure where the M^+ and X^- ions touch but where the X^- ions will no longer be touching.

In the closest packed arrangement, the X^- ions are touching. However, if M^+ ions are forced into holes in the structure that are smaller than the M^+ ions, the X^- ions are forced apart, decreasing the $X^- \cdots X^-$ repulsions. On the other hand, M^+ and X^- will be in direct contact, maximizing the $M^+ \cdots X^-$ interactions.

With this idea in mind, we can now establish guidelines for filling the tetrahedral and octahedral holes. In terms of $r+$ (radius of M^+) and $R-$ (radius of X^-), if

$$0.225R- < r+ < 0.414R-$$

↑ Size of tetrahedral holes ↑ Size of octahedral holes

then the M^+ ions are placed in the tetrahedral holes. That is, this condition ensures that the M^+ ions are larger than the tetrahedral holes in the closest packed array of X^- ions. This is what we want.

As we consider MX solids with larger and larger M^+ ions, at some point the M^+ ions become large enough to fill (and exceed the size of) the octahedral holes. Specifically, a cation with radius larger than $0.414R-$ should be placed in the octahedral holes rather than in the tetrahedral holes in the closest packed structure. As we continue to consider larger and larger M^+ ions, what happens? Is there an upper limit to the size of M^+ ions that can be forced into the octahedral holes? The answer is yes. For very large cations the solid switches from a closest packed array of X^- ions to a simple cubic arrangement of X^- ions with an M^+ ion in the center of each cube (Fig. 16.41).

This structure can be described as M^+ occupying a *cubic hole*. Use of geometric arguments similar to those we have used for the tetrahedral and octahedral holes shows that for a cubic hole,

$$r+^{\text{cubic}} = 0.732R-$$

This means that for the solid MX, the M^+ ions occupy the octahedral holes in the range where

$$0.414R- < r+ < 0.732R-$$

For large M^+ ions ($r+ > 0.732R-$), the solid switches to the simple cubic arrangement just described.

The guidelines for filling the tetrahedral, octahedral, and cubic holes are summarized in Table 16.6.

TABLE 16.6
Guidelines for Filling Various Types of Holes for the Ionic Solid *MX*

Size of M^+	Type of Hole Filled
$0.225R- < r+ < 0.414R-$	Tetrahedral
$0.414R- < r+ < 0.732R-$	Octahedral
$0.732R- < r+$	Cubic

A few more points need to be considered. The guidelines just given assume that the ions behave as hard spheres and that only ionic forces occur. Therefore, it is not surprising that these guidelines *are not always obeyed*. They just give us a starting point for describing the structures of ionic solids. In addition, although we have considered only solids of the type MX containing a 1:1 ratio of cations and anions, these guidelines also apply to other types of binary ionic solids. We will consider some specific solids in the next section.

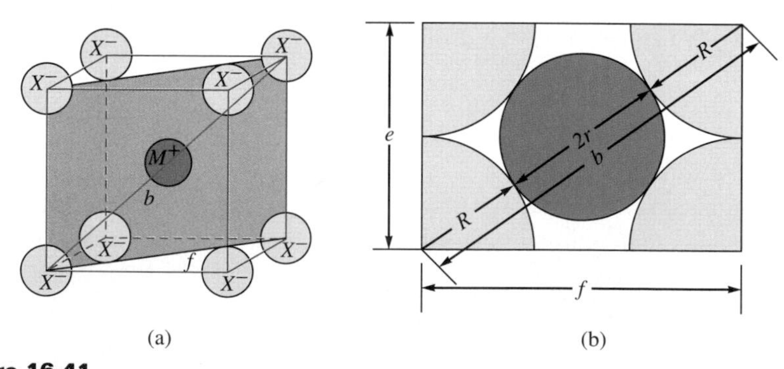

Figure 16.41
(a) A simple cubic array with X^- ions, with an M^+ ion in the center (in the cubic hole). (b) The body diagonal b equals $R + 2r + R$, since X^- and M^+ touch along this body diagonal.

16.8 Structures of Actual Ionic Solids

In this section we will consider some specific binary ionic solids to show how these solids illustrate the ideas of ion packing. Because an ionic solid must be neutral overall, the stoichiometry of the compound (the ratio of the numbers of anions to cations) is determined by the ion charges. On the other hand, the structure of the compound (the placement of the ions in the solid) is determined, at least to a first approximation, by the relative sizes of the ions.

Before we consider specific compounds, we need to consider the locations and relative numbers of tetrahedral and octahedral holes in the closest packed structures. The location of the tetrahedral holes in the face-centered cubic unit cell of the ccp structure is shown in Fig. 16.42(a). Note from this figure that there are eight tetrahedral holes in the unit cell. Recall from the discussion in Section 16.4 that there are four net spheres in the face-centered cubic unit cell. Thus there are *twice as many tetrahedral holes as there are packed spheres* in the closest packed structure.

The location of the octahedral holes in the face-centered cubic unit cell is shown in Fig. 16.43(a). The easiest octahedral hole to find in this structure is the one at the center of the cube. Note that this hold is surrounded by six spheres, as is required to form an octahedran. Since the remaining octahedral holes are shared with other unit cells, they are more difficult to visualize. However, it can be shown that the number of octahedral holes in the ccp structure is the *same* as the number of packed spheres.

Closest packed structures contain twice as many tetrahedral holes as packed spheres. Closest packed structures contain the same number of octahedral holes as packed spheres.

Using these ideas, we will now consider the structures of some specific ionic solids.

The Structures of the Alkali Halides

The structure of sodium chloride, which is the prototype for most of the alkali halides, is best described as a cubic closest packed array of Cl^- ions with the Na^+ ions in all of the octahedral holes [see Fig. 16.43(b)]. The relative sizes of these ions is such that $r_{Na^+} = 0.66R_{Cl^-}$, so this solid obeys the guidelines given previously. Note that the Cl^- ions are forced apart by the Na^+ ions, which are too large for the octahedral holes in the closest packed array of Cl^- ions. Since the number of octahedral holes is the same as the number of packed spheres, all of the octahedral holes must be filled with Na^+ ions to achieve the required 1 : 1 stoichiometry. Most other alkali halides also have the sodium chloride structure. In fact, all of the halides of lithium, sodium, potassium, and rubidium

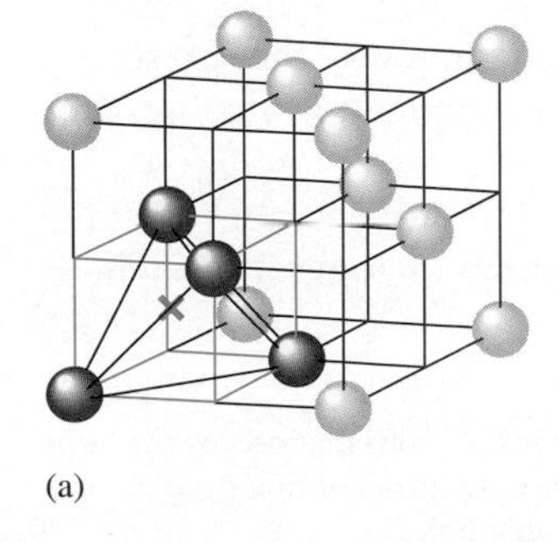

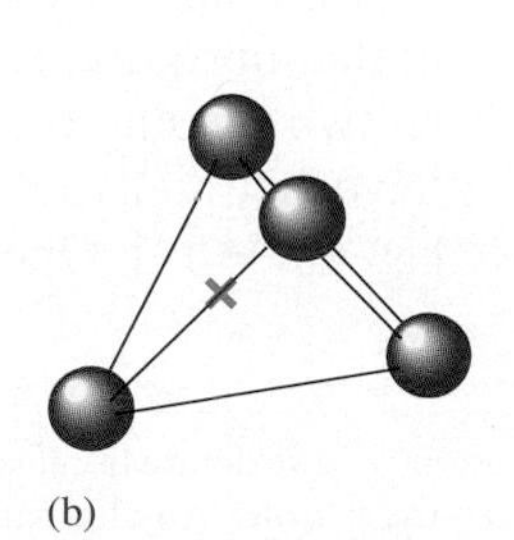

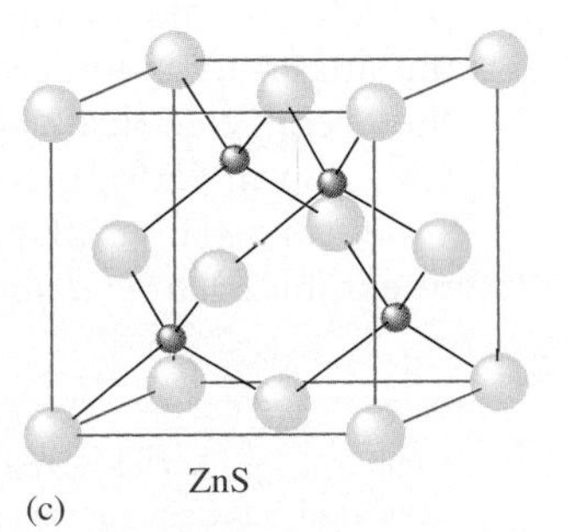

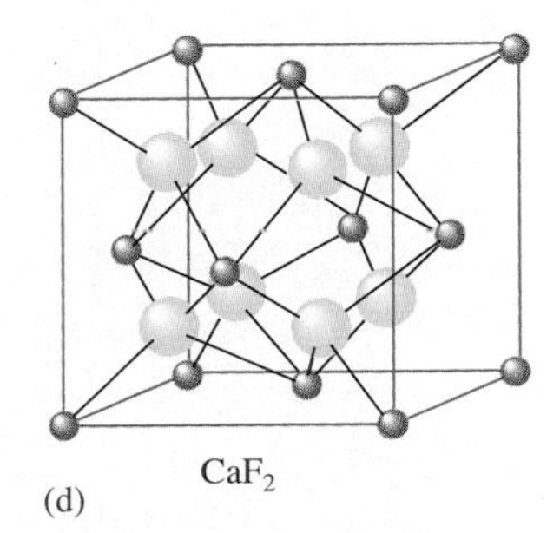

Figure 16.42
(a) The location (×) of a tetrahedral hole in the face-centered cubic unit cell. (b) One of the tetrahedral holes. (c) The unit cell for ZnS where the S^{2-} ions (yellow) are closest packed with the Zn^{2+} ions (red) filling alternate tetrahedral holes. (d) The unit cell for CaF_2, where the Ca^{2+} ions (red) form a face-centered cubic arrangement with the F^- ions (yellow) in all of the tetrahedral holes.

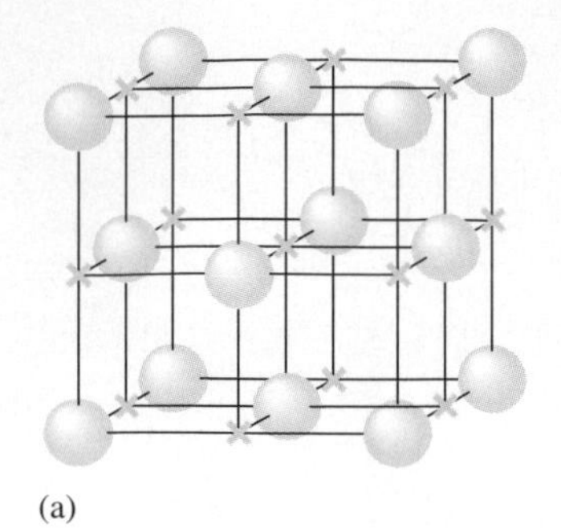

(a)

(b)

Figure 16.43
(a) The locations (×) of the octahedral holes in the face-centered cubic unit cell. (b) Representation of the unit cell for solid NaCl. The Cl^- ions (blue spheres) have a ccp arrangement with Na^+ ions (yellow spheres) filling all of the octahedral holes.

have this structure. Cesium fluoride has the sodium chloride structure; but because of the large size of Cs^+ ions, in this case the Cs^+ ions form a cubic closest packed arrangement with the F^- ions in all of the octahedral holes. On the other hand, cesium chloride, in which the Cs^+ and Cl^- ions are almost the same size, has a simple cubic structure of Cl^- ions, with each Cs^+ ion in the cubic hole in the center of each cube. The compounds cesium bromide and cesium iodide also have this latter structure.

The Structure of Zinc Sulfide

The compound ZnS illustrates another important general type of structure. In ZnS the ion sizes are such that $r_{Zn^{2+}} \approx 0.35R_{S^{2-}}$. According to the guidelines, this should mean that the Zn^{2+} ions occupy the tetrahedral holes among the closest packed S^{2-} ions. This is exactly what is found in ZnS. However, because there are twice as many tetrahedral holes as closest packed ions, only one-half of the tetrahedral holes are occupied in zinc sulfide, producing the required 1:1 ratio of Zn^{2+} and S^{2-} ions [see Fig. 16.42(c)]. Zinc sulfide has two forms: zinc blende, where the S^{2-} ions are cubic closest packed; and wurtzite, where the S^{2-} ions are hexagonal closest packed. In both cases the Zn^{2+} ions occupy half of the tetrahedral holes. The substances ZnO and CdS also show both the zinc blende and wurtzite structures.

The Structure of Calcium Fluoride

The structure of the compound CaF_2 can be described as a face-centered cubic array of Ca^{2+} ions with the F^- ions in all of the tetrahedral holes.* This gives the required 1:2 ratio of Ca^{2+} and F^- ions. This structure is called the fluorite structure [see Fig. 16.42(d)] and is also observed in the compounds SrF_2, $BaCl_2$, PbF_2, and CdF_2, among others.

16.9 Lattice Defects

So far we have assumed that crystalline compounds are perfect, that is, that all of the atoms, ions, or molecules are present and occupy the correct sites. Although crystalline materials are highly ordered and most of the components are where they are expected to be, all real crystals have imperfections called *lattice defects*.

Point defects refer to totally missing particles (atoms, ions, or molecules) or to cases where the particle is in a nonstandard location. A crystal with missing particles is said to have *Schottky defects* [see Fig. 16.44(a)]. When ions are missing from an ionic compound, they must be missing in a way that preserves the overall electrical neutrality of the substance. For example, for every missing Ca^{2+} ion in CaF_2, there must be two missing F^- ions.

Crystals in which particles have migrated to nonstandard positions are said to exhibit *Frenkel defects* [see Fig. 16.44(b)]. One group of compounds where

*The structure of CaF_2 is not a closest packed structure because the Ca^{2+} ions that occupy the face-centered cubic sites are smaller than the F^- ions. An alternative description of this structure is a simple cubic array of F^- ions, with Ca^{2+} ions in half of the cubic holes.

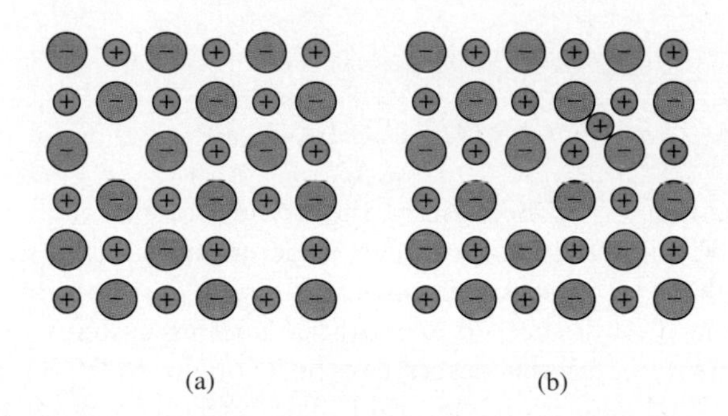

Figure 16.44
Defects in crystalline ionic solids. (a) Schottky defects, in which there are vacant sites. (b) Frenkel defects, in which an atom or an ion of either sign is present at an inappropriate site.

Frenkel defects are present to the extreme is the silver halides, AgCl, AgBr, and AgI. In these compounds the anion positions are mostly those expected from closest packing ideas; however the silver ions are distributed almost randomly in the various holes and can easily travel within the solid structure. This property is a major reason the silver halides are so useful in photographic films.

In some cases the crystal defects involve impurities that lead to *nonstoichiometric compounds*. For example, wüstite, which has the idealized formula FeO, actually varies from $Fe_{1.0}O_{1.0}$ to $Fe_{0.95}O_{1.0}$. These variations occur when some of the Fe^{2+} ions in the compound are replaced by Fe^{3+} ions in a way that preserves the electrical neutrality of the solid. There are many other examples of nonstoichiometric compounds. One extreme case involves TiO, which can vary from $Ti_{1.0}O_{0.7}$ to $Ti_{0.8}O_{1.0}$.

16.10 Vapor Pressure and Changes of State

Now that we have considered the general properties of the three states of matter, we can explore the processes by which matter changes state. One very familiar example of a change in state occurs when a liquid evaporates from an open container. This is clear evidence that the molecules of a liquid can escape from the liquid's surface and form a gas. Called **vaporization,** or *evaporation,* this process is endothermic, because energy is required to overcome the relatively strong intermolecular forces in the liquid. The energy required to vaporize 1 mole of a liquid at a pressure of 1 atm is called the standard **heat of vaporization,** or the standard **enthalpy of vaporization,** and is usually symbolized as ΔH°_{vap}.

Vapor is the usual term for the gas phase of a substance that exists as a solid or liquid at 25°C and 1 atm.

The endothermic nature of vaporization has great practical significance; in fact, one of the most important roles that water plays in our world is to act as a coolant. Because of the strong hydrogen bonding among its molecules in the liquid state, water has an unusually large heat of vaporization (40.7 kJ/mol). A significant portion of the sun's energy that reaches earth is spent evaporating water from the oceans, lakes, and rivers rather than warming the earth. The vaporization of water is also crucial to the body's temperature control system through evaporation of perspiration.

Vapor Pressure

When a liquid is placed in a closed container, the amount of liquid at first decreases but eventually becomes constant. The decrease occurs because there is an initial net transfer of molecules from the liquid to the vapor phase (Fig. 16.45). However, as the number of vapor molecules increases, so does the rate

OZONE-SAFE AIR CONDITIONING

Most commercial buildings have air conditioning systems that use chlorofluorocarbons (CFCs), chemicals that inevitably leak into the atmosphere and eventually contribute to the decomposition of stratospheric ozone (See Section 15.9). One strategy for solving this problem is to replace the CFC refrigerants with other chemicals less likely to harm the ozone layer.

Another approach is to use the endothermicity of the evaporation of water to cool air for buildings. This cooling method does not require a refrigerant. Evaporative coolers, or "swamp coolers" as they are often called, have been around for many years, especially in dry climates. The process involves the same mechanism as cooling the human body by the endothermic evaporation of perspiration. Hot air is drawn through a spray of water or wet porous pads, becoming cooler and more humid as it absorbs water vapor. Currently about 5% of the 86 million households in the U.S. use evaporative coolers, while nearly 70% of U.S. dwellings have CFC-containing air conditioners.

Because of the dryness of the air, evaporative coolers work best in desert areas such as the Southwest. However, by adding a desiccant (a water absorber) to evaporative cooling systems, the dry air of the desert can be brought to all areas of the country. Several U.S. industries are now developing systems that incorporate a rotating wheel containing a solid desiccant which dries the air before it passes through the evaporative cooler. The desiccant is reactivated by air from the outside that is heated by a gas burner (see figure below). The desiccant evaporative cooling system developed by ICC Technologies of Philadelphia employs a patented titanium silicate molecular sieve as the desiccant. Molecular sieves are network solids with an aluminosilicate backbone that have cavities and tunnels which are just the right size to trap small molecules such as water.

The major advantage of the desiccant/evaporative cooling systems is that they allow independent control of temperature and humidity—without the use of CFCs.

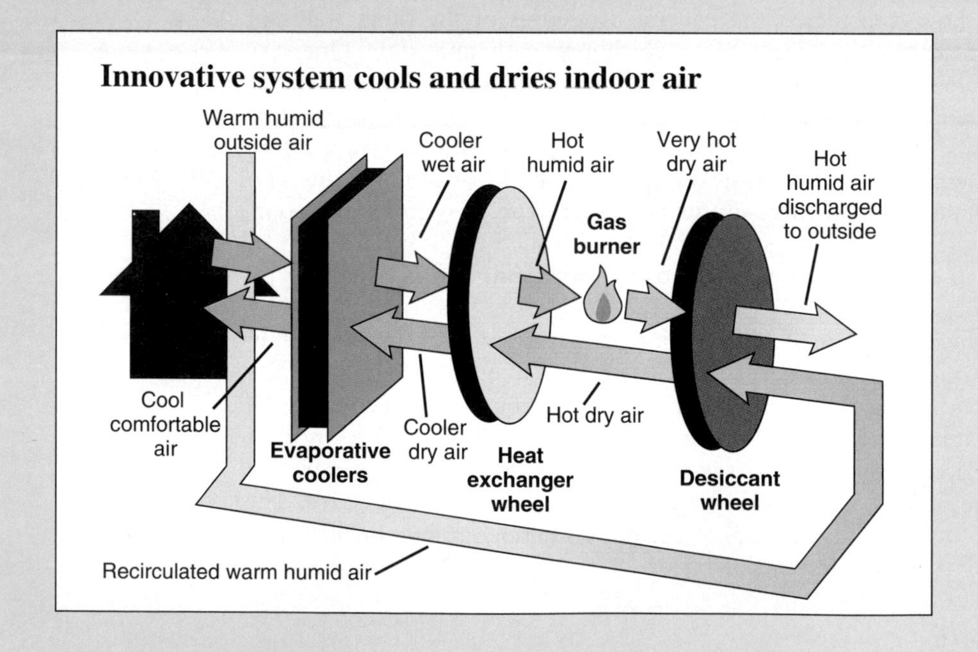

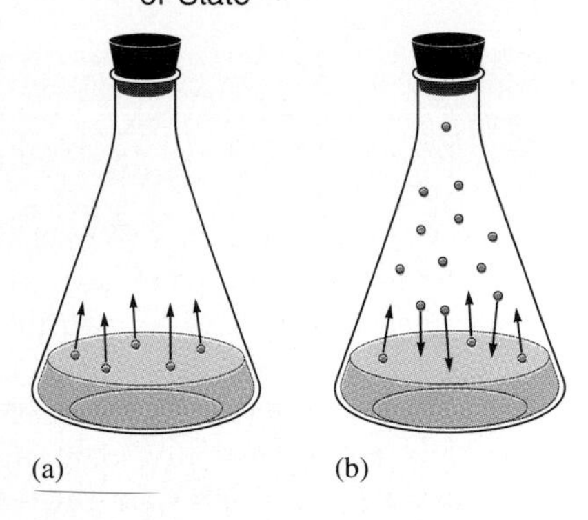

Figure 16.45
Behavior of a liquid in a closed container. (a) Initially, net evaporation occurs as molecules are transferred from the liquid to the vapor phase, so the amount of liquid decreases. (b) As the number of vapor molecules increases, the rate of return to the liquid (condensation) increases, until finally the rate of condensation equals the rate of evaporation. The system is at equilibrium, and no further changes occur in the amounts of vapor or liquid.

of return of these molecules to the liquid. The process by which vapor molecules re-form a liquid is called **condensation.** Eventually, enough vapor molecules are present above the liquid so that the rate of condensation equals the rate of evaporation (see Fig. 16.46). At this point no further net change occurs in the amount of liquid or vapor because the two opposite processes exactly balance each other; the system is at equilibrium. Note that this system is highly dynamic on the molecular level—molecules are constantly escaping from and entering the liquid at a high rate. However, there is no net change, because the two opposite processes just balance each other.

The pressure of the vapor present at equilibrium is called the *equilibrium vapor pressure,* or, more commonly, the **vapor pressure** of the liquid. A simple barometer can measure the vapor pressure of a liquid, as shown in Fig. 16.47. The liquid is injected at the bottom of the tube of mercury and floats to the surface because the mercury is so dense. A portion of the liquid evaporates at the top of the column, producing a vapor whose pressure pushes some mercury out of the tube. When the system reaches equilibrium, the vapor pressure can be determined from the change in the height of the mercury column, since

$$P_{\text{atmosphere}} = P_{\text{vapor}} + P_{\text{Hg column}}$$

Thus

$$P_{\text{vapor}} = P_{\text{atmosphere}} - P_{\text{Hg column}}$$

The vapor pressures of liquids vary widely [see Fig. 16.47(b)]. Liquids with high vapor pressures are said to be *volatile*—they evaporate readily from an open dish.

The vapor pressure of a liquid is principally determined by the size of the *intermolecular forces* in the liquid. Liquids in which the intermolecular forces are large have relatively low vapor pressures, because the molecules need high energies to escape to the vapor phase. For example, although water has a much lower molar mass than diethyl ether, the strong hydrogen bonding forces that exist among water molecules in the liquid cause water's vapor pressure to be much lower than that of diethyl ether [see Fig. 16.47(b)]. In general, substances with large molar masses have relatively low vapor pressures, mainly because of the large dispersion forces. The more electrons a substance has, the more polarizable it is and the greater the dispersion forces are.

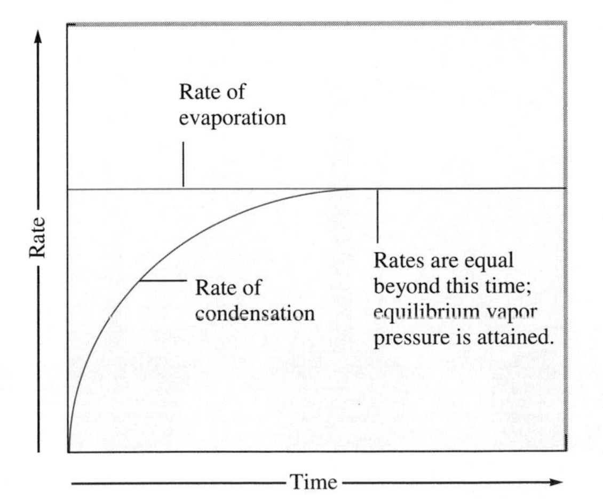

Figure 16.46
The rates of condensation and evaporation over time for a liquid sealed in a closed container. The rate of evaporation remains constant, while the rate of condensation increases as the number of molecules in the vapor phase increases, until the two rates become equal. At this point the equilibrium vapor pressure is attained.

COOL CARS

You've spent a day at the beach and come back to your car to drive home. When you open the door, the heat about knocks you down.

Solar heating of a car on a hot day can drive interior temperatures to 130°F or higher. One choice is to grit your teeth, climb into the car, roll down the windows and put the air conditioner on max. But there's another way if you have an aerosol can of Instant Car Kooler™. In that case you spray the interior of the car for a minute or so and *voilà*—in a few seconds it's 75°F in your car.

This may sound too good to be true, but it's based on a simple phenomenon—the endothermic nature of evaporation. The spray can contains a mixture of 90% water and 10% ethyl alcohol. When the contents of the can are sprayed into the car interior, very fine droplets of the water/alcohol mixture are dispersed in the air. Since the droplets are so small, the mist has a very high surface area and thus evaporates very rapidly. Because the enthalpy of evaporation of water (42.5 kJ/mol at 55°C) is so large, the evaporation process dramatically decreases the temperature of the air inside the car.

Instant Car Kooler was invented by physicist Domingo K. L. Tan who patented the product in 1989 and now markets it in 16-ounce aerosol cans.

Measurements of the vapor pressure for a given liquid at several temperatures show that *vapor pressure increases significantly with temperature.* Figure 16.48 illustrates the distribution of molecular kinetic energies present in a liquid at two different temperatures. To overcome the intermolecular forces in a liquid, a molecule must have a minimum kinetic energy. As the temperature of the liquid is increased, the fraction of molecules having sufficient energy to overcome these forces and escape to the vapor phase increases markedly. Thus the vapor pressure of a liquid increases dramatically with temperature. The vapor pressure values for water at several temperatures are given in Table 16.7.

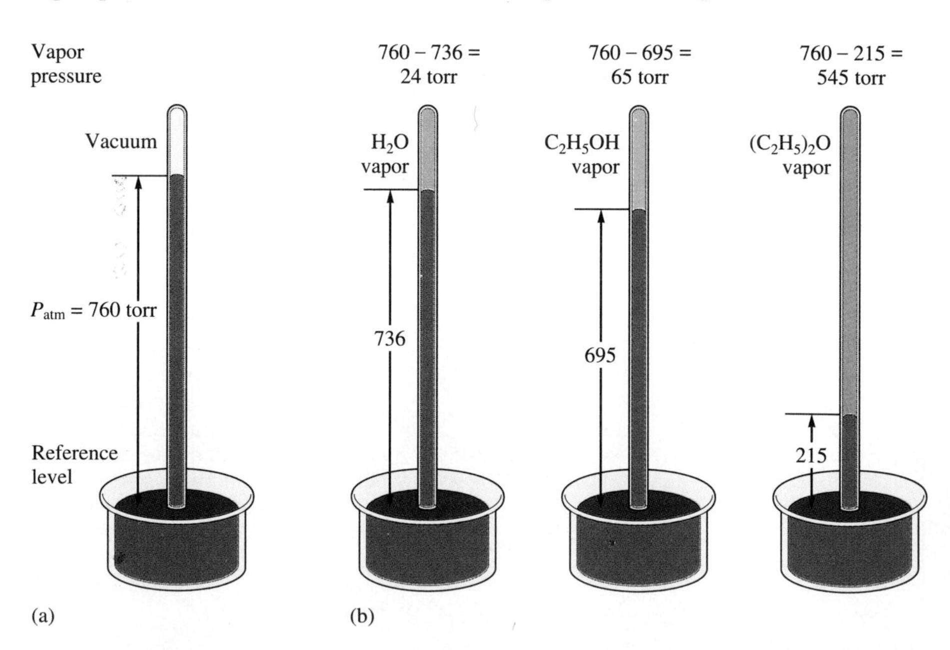

Figure 16.47
(a) The vapor pressure of a liquid can be easily measured by using a simple barometer of the type shown here. (b) The three liquids, water, ethanol (C_2H_5OH), and diethyl ether [$(C_2H_5)_2O$], have quite different vapor pressures. Ether is by far the most volatile of the three. Note that in each case some of the liquid remains (floating on the mercury).

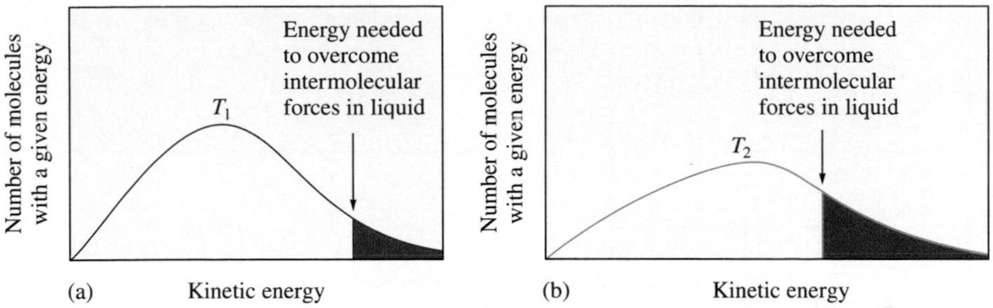

Figure 16.48
The number of molecules in a liquid with a given energy versus kinetic energy at two temperatures. Part (a) shows a lower temperature than that in part (b). Note that the proportion of molecules with enough energy to escape the liquid to the vapor phase (indicated by shaded areas) increases dramatically with temperature. This causes vapor pressure to increase markedly with temperature.

The quantitative nature of the temperature dependence of vapor pressure can be illustrated graphically. Plots of vapor pressure versus temperature for water, ethanol, and diethyl ether are shown in Fig. 16.49(a). Note the nonlinear increase in vapor pressure for all the liquids as the temperature is increased. A straight line can be obtained from these data by plotting $\ln(P_{vap})$ versus $1/T$, where T is the Kelvin temperature [see Fig. 16.49(b)]. We can represent this behavior by the equation

$$\ln(P_{vap}) = -\frac{\Delta H_{vap}}{R}\left(\frac{1}{T}\right) + C \qquad (16.4)$$

where ΔH_{vap} is the enthalpy of vaporization, R is the universal gas constant, and C is a constant characteristic of a given liquid. Note that this equation is a simplified version of the equation obtained in Section 10.11 for the temperature dependence of an equilibrium constant:

$$\ln K = -\frac{\Delta H^\circ}{R}\left(\frac{1}{T}\right) + \frac{\Delta S^\circ}{R}$$

The similarity is not unexpected, since vapor pressure is an equilibrium phenomenon. Note that we can now interpret the constant C in terms of ΔS, the entropy of vaporization for a given liquid.

TABLE 16.7
The Vapor Pressure of Water as a Function of Temperature

T (°C)	P (torr)
0.0	4.579
10.0	9.209
20.0	17.535
25.0	23.756
30.0	31.824
40.0	55.324
60.0	149.4
70.0	233.7
90.0	525.8

EXAMPLE 16.2

Using the plots in Fig. 16.49(b), determine whether water or diethyl ether has the smaller enthalpy of vaporization.

Solution

When $\ln(P_{vap})$ is plotted versus $1/T$, the slope of the resulting straight line is

$$-\frac{\Delta H_{vap}}{R}$$

Note from Fig. 16.49(b) that the slopes of the lines for water and diethyl ether are both negative, as expected. Also note that the line for ether has the smaller slope. Thus ether has the smaller value of ΔH_{vap}. This makes sense because the hydrogen bonding in water causes it to have a relatively large enthalpy of vaporization.

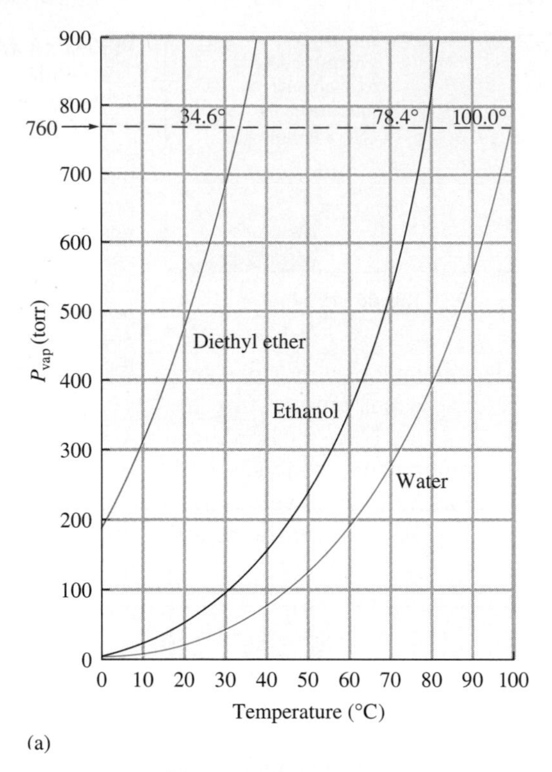

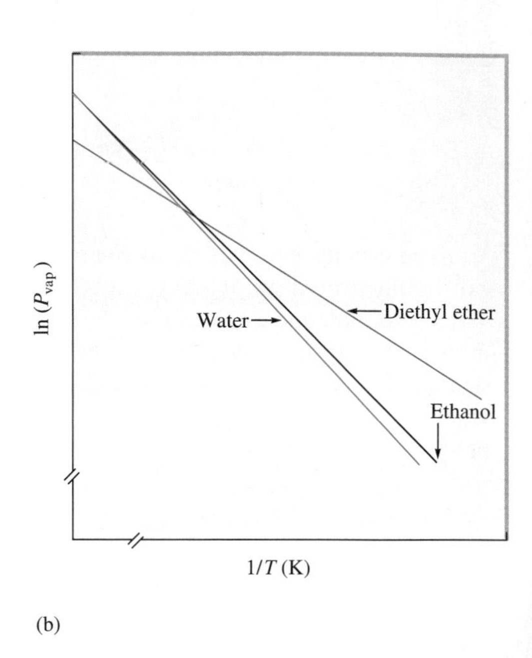

Figure 16.49
(a) The vapor pressure of water, ethanol, and diethyl ether as a function of temperature. (b) Plots of ln(P_{vap}) versus 1/T (Kelvin temperature) of water, ethanol, and diethyl ether.

Equation (16.5) is called the Clausius-Clapeyron equation.

Equation (16.4) is important for several reasons. For example, we can determine the heat of vaporization for a liquid by measuring P_{vap} at several temperatures and then evaluating the slope of a plot of $\ln(P_{vap})$ versus $1/T$. On the other hand, if we know the values of ΔH_{vap} and P_{vap} at one temperature, we can use Equation (16.4) to calculate P_{vap} at another temperature. This can be done by recognizing that the constant C does not depend on temperature. Thus at two temperatures, T_1 and T_2, we can solve Equation (16.4) for C and then write the equality

$$\ln(P_{vap}^{T_1}) + \frac{\Delta H_{vap}}{RT_1} = C = \ln(P_{vap}^{T_2}) + \frac{\Delta H_{vap}}{RT_2}$$

which can be rearranged to

$$\ln(P_{vap}^{T_1}) - \ln(P_{vap}^{T_2}) = \frac{\Delta H_{vap}}{R}\left(\frac{1}{T_2} - \frac{1}{T_1}\right)$$

or

$$\ln\left(\frac{P_{vap}^{T_1}}{P_{vap}^{T_2}}\right) = \frac{\Delta H_{vap}}{R}\left(\frac{1}{T_2} - \frac{1}{T_1}\right) \tag{16.5}$$

Figure 16.50
Sublimation of iodine.

Like liquids, solids have vapor pressures. Figure 16.50 shows iodine vapor in equilibrium with solid iodine in a closed flask. At 25°C and 1 atm iodine *sublimes;* that is, it goes directly from the solid to the gaseous state without passing through the liquid state. **Sublimation** also occurs with dry ice (solid carbon dioxide) under these conditions.

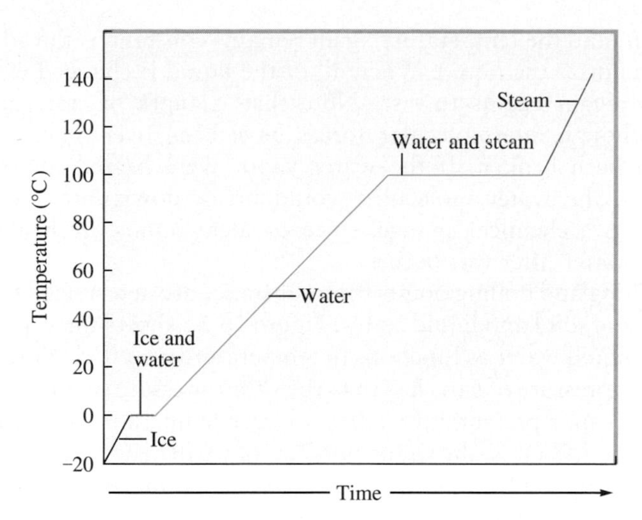

Figure 16.51
The heating curve for a given quantity of water where energy is added at a constant rate. The plateau at the boiling point is longer than the plateau at the melting point because it takes almost seven times more energy (and thus seven times the heating time) to vaporize liquid water than to melt ice. The slopes of the other lines are different because the different states of water have different molar heat capacities (the energy required to raise the temperature of 1 mole of a substance by 1°C).

Changes of State

What happens when a solid is heated? Typically, it melts to form a liquid. If the heating continues, the liquid at some point boils and forms the vapor phase. This process can be represented by a **heating curve:** a plot of temperature versus time for a process where energy is added at a constant rate.

The heating curve for water is given in Fig. 16.51. As energy flows into the ice, the random vibrations of the water molecules increase as the temperature rises. Eventually, the molecules become so energetic that they break loose from their lattice positions, and the change from solid to liquid occurs. This is indicated by a plateau at 0°C on the heating curve. At this temperature, called the **melting point,** all of the added energy is used to disrupt the ice structure by breaking the hydrogen bonds, thus increasing the potential energy of the water molecules. The enthalpy change that occurs at the melting point when a solid melts is called the **heat of fusion** (or more accurately, the **enthalpy of fusion**), ΔH_{fus}. The melting points and enthalpies of fusion for several representative solids are listed in Table 16.8.

Ionic solids such as NaCl and NaF have very high melting points and very high enthalpies of fusion because of the strong ionic forces in these solids. At the other extreme is $O_2(s)$, a molecular solid containing nonpolar molecules with weak intermolecular forces.

The temperature remains constant until all of the solid has changed to liquid; then it begins to increase again. At 100°C the liquid water reaches its

TABLE 16.8 **Melting Points and Enthalpies of Fusion for Several Representative Solids**

Compound	Melting Point (°C)	Enthalpy of Fusion (kJ/mol)
O_2	−218	0.45
HCl	−114	1.99
HI	−51	2.87
CCl_4	−23	2.51
$CHCl_3$	−64	9.20
H_2O	0	6.01
NaF	992	29.3
NaCl	801	30.2

boiling point, and the temperature again remains constant as the added energy is used to vaporize the liquid. When all of the liquid is changed to vapor, the temperature again begins to rise. Note that changes of state are physical changes; although intermolecular forces have been overcome, no chemical bonds have been broken. If the water vapor were heated to much higher temperatures, the water molecules would break down into the individual atoms. This is a chemical change, since covalent bonds are broken. We no longer have water after this occurs.

The melting and boiling points will be defined more precisely later in this section.

The melting and boiling points for a substance are determined by the vapor pressures of the solid and liquid states. Figure 16.52 shows the vapor pressures of solid and liquid water as functions of temperature near 0°C. Note that below 0°C the vapor pressure of ice is less than the vapor pressure of liquid water. Also note that the vapor pressure of ice has a larger temperature dependence than that of the liquid. That is, the vapor pressure of ice increases more rapidly for a given rise in temperature than does the vapor pressure of water. Thus as the temperature of the solid is increased, a point is eventually reached where the *liquid and solid have identical vapor pressures.* This is the melting point.

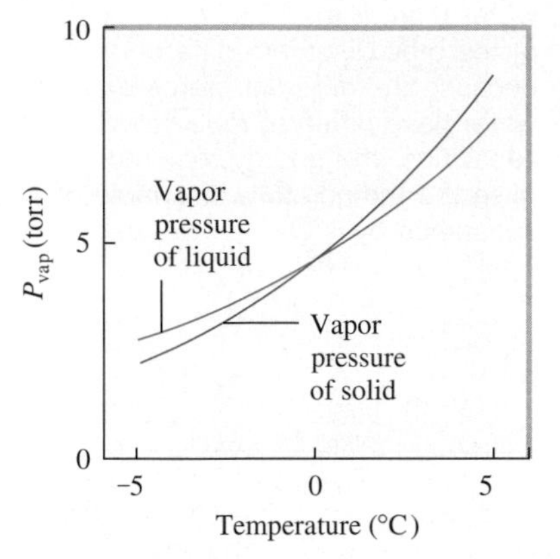

Figure 16.52
The vapor pressures of solid and liquid water as a function of temperature. The data for liquid water below 0°C are obtained from supercooled water. The data for solid water above 0°C are estimated by extrapolation of vapor pressure from below 0°C.

These concepts can be demonstrated experimentally using the apparatus illustrated in Fig. 16.53, where ice occupies one compartment and liquid water the other. Consider the following cases.

Case 1: a temperature at which the vapor pressure of the solid is greater than that of the liquid. At this temperature the solid requires a higher pressure of vapor than the liquid to be in equilibrium with the vapor. Thus, as vapor is released from the solid to try to achieve equilibrium, the liquid absorbs vapor in an attempt to reduce the vapor pressure to its equilibrium value. The net effect is a conversion from solid to liquid through the vapor phase. In fact, no solid can exist under these conditions. The amount of solid steadily decreases and the volume of liquid increases. Finally, there is only liquid in the right compartment, which comes to equilibrium with the water vapor, and no further changes occur in the system. The temperature for Case 1 must be *above the melting point* of ice, since only the liquid state can exist.

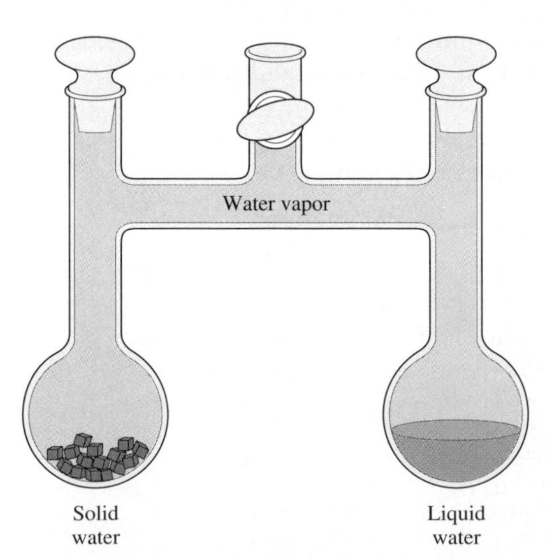

Figure 16.53
An apparatus that allows solid and liquid water to interact only through the vapor state.

Case 2: a temperature at which the vapor pressure of the solid is less than that of the liquid. This is the opposite of the situation in Case 1. In this case the liquid requires a higher pressure of vapor than the solid to be in equilibrium with the vapor. So the liquid gradually disappears, and the amount of ice increases. Finally, only the solid remains, which achieves equilibrium with the vapor. This temperature must be *below the melting point* of ice, since only the solid state can exist.

Case 3: a temperature at which the vapor pressures of the solid and liquid are identical. In this case the solid and liquid states have the same vapor pressure, so they can coexist in the apparatus at equilibrium simultaneously with the vapor. This temperature represents the *melting point* where both the solid and liquid states can exist.

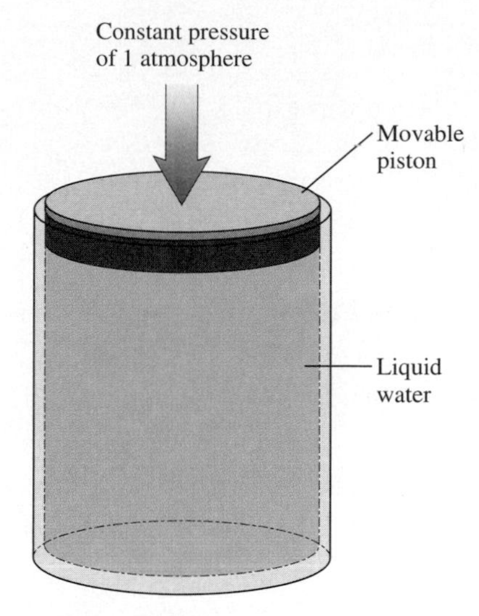

Figure 16.54
Water in a closed system with a pressure of 1 atm exerted on the piston. No bubbles can form within the liquid as long as the vapor pressure is less than 1 atm.

We can now describe the melting point of a substance more precisely. The **normal melting point** is defined as *the temperature at which the solid and liquid states have the same vapor pressure under conditions where the total pressure is 1 atm.*

Boiling occurs when the vapor pressure of a liquid becomes equal to the pressure of its environment. The **normal boiling point** of a liquid is *the temperature at which the vapor pressure of the liquid is exactly 1 atm.* This concept can be understood by reference to Fig. 16.54. At temperatures where the vapor pressure of the liquid is less than 1 atm, no bubbles of vapor can form because the pressure on the surface of the liquid is greater than the pressure in any spaces in the liquid where bubbles are trying to form. Only when the liquid reaches a temperature at which the pressure of vapor in the spaces in the liquid is 1 atm can bubbles form and boiling occur.*

However, observed changes of state do not always occur exactly at the boiling point or melting point. For example, water can be readily **supercooled;** that is, it can be cooled below 0°C at 1 atm pressure and still remain in the liquid state. Supercooling occurs because, as it is cooled, the water may not achieve the degree of organization necessary to form ice at 0°C; thus it continues to exist as the liquid. At some point the correct ordering occurs and ice rapidly forms, releasing energy in the exothermic process and bringing the temperature back up to the melting point, where the remainder of the water freezes (see Fig. 16.55).

A liquid can also be **superheated,** or raised to temperatures above its boiling point, especially if it is heated rapidly. Superheating can occur because bubble formation in the interior of the liquid requires that many high-energy molecules gather in the same vicinity. This may not happen at the boiling point, especially if the liquid is heated rapidly. If the liquid becomes superheated, the vapor pressure in the liquid is greater than the atmospheric pressure. Once a bubble does form, since its internal pressure is greater than that of the atmosphere, it can burst before rising to the surface, blowing the surrounding liquid out of the container. This phenomenon is called *bumping* and has ruined many experiments. It can be avoided by adding boiling chips to the flask containing

*Note that in real life the bubbles seen forming in water as it is being heated to boiling arise from the expulsion of dissolved air.

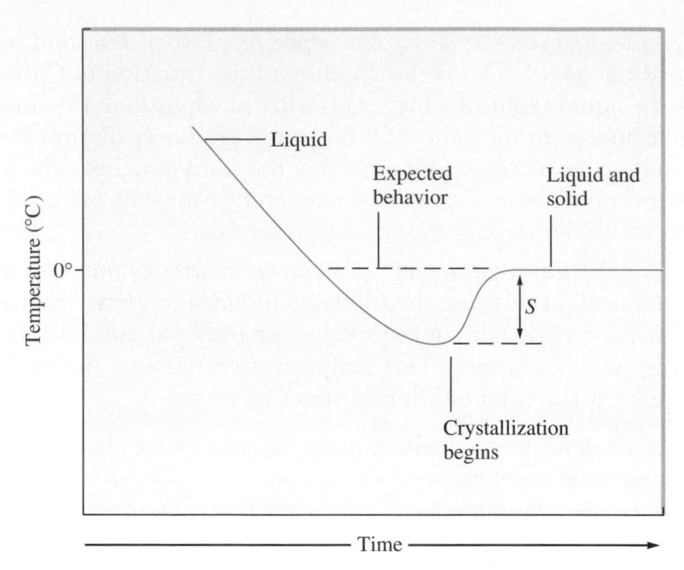

Figure 16.55
The supercooling of water. The extent of supercooling is given by S.

the liquid. Boiling chips are bits of porous ceramic material containing trapped air that escapes on heating, forming tiny bubbles that act as "starters" for vapor bubble formation. These starter bubbles allow a smooth onset of boiling as the boiling point is reached.

16.11 Phase Diagrams

A **phase diagram** is a convenient way of representing the phases of a substance as a function of temperature and pressure. For example, the phase diagram for water (Fig. 16.56) shows which state exists at a given temperature and pressure. It is important to recognize that a phase diagram describes conditions and events for a pure substance in a *closed* system of the type represented in Fig. 16.54, where no material can escape into the surroundings and no air is present.

To show how to interpret the phase diagram for water, we will consider heating experiments at several pressures, shown by the dashed lines in Fig. 16.57.

Experiment 1: pressure is 1 atm. This experiment begins with the cylinder shown in Fig. 16.54 completely filled with ice at a temperature of −20°C, and with the piston exerting a pressure of 1 atm directly on the ice (there is no air space). Since at temperatures below 0°C the vapor pressure of ice is less than 1 atm—which is the constant external pressure on the piston—no vapor is present in the cylinder. As the cylinder is heated, ice remains the only component until the temperature reaches 0°C, where the ice changes to liquid water as energy is added. This is the normal melting point of water. When all of the solid has changed to liquid, the temperature again rises. At this point the cylinder contains only liquid water. *No vapor is present* because the vapor pressure of

Below the critical temperature and pressure the gaseous state of a substance is often referred to as a vapor.

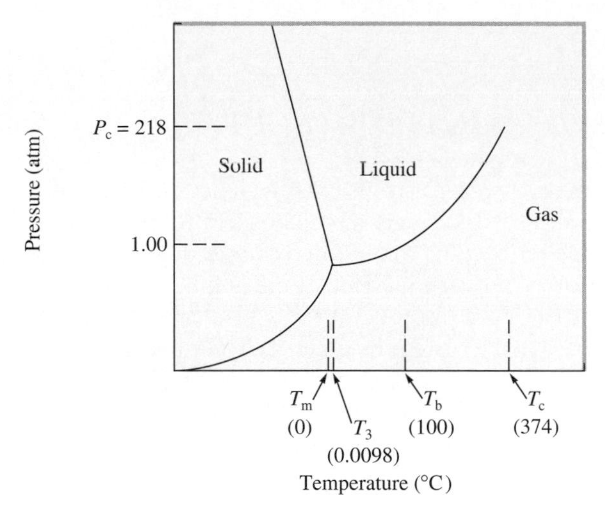

Figure 16.56
The phase diagram for water. T_m represents the normal melting point; T_3 and P_3 denote the triple point; T_b represents the normal boiling point; T_c represents the critical temperature; P_c represents the critical pressure. The negative slope of the solid/liquid line reflects the fact that the density of ice is less than that of liquid water.

liquid water under these conditions is less than 1 atm, the constant external pressure on the piston. Heating continues until the temperature of the liquid water reaches 100°C. At this point the vapor pressure of liquid water is 1 atm, and boiling occurs; the liquid changes to vapor. This is the normal boiling point of water. After all of the liquid has been converted to steam, the temperature again rises as the heating continues. The cylinder now contains only water vapor.

Experiment 2: pressure is 2.0 torr. Again we start with ice as the only component in the cylinder at −20°C. The pressure exerted by the piston in this case is only 2.0 torr. As heating proceeds the temperature rises to −10°C, where the ice changes directly to vapor through the process of

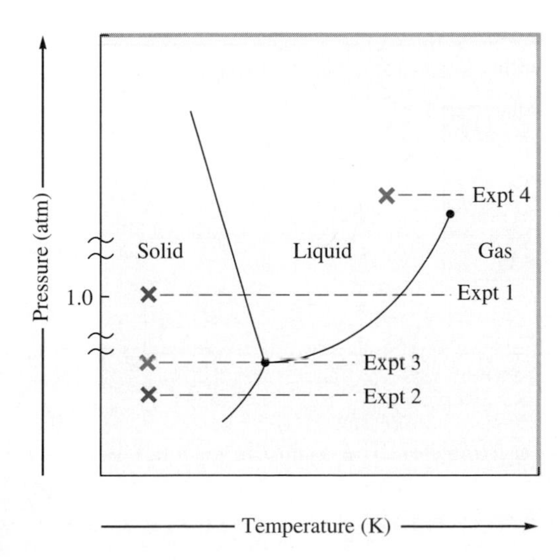

Figure 16.57
Diagrams of various heating experiments on samples of water in closed systems.

Transistors and Printed Circuits

Transistors have had an immense impact on the technology of electronic devices for which signal amplification is needed, such as communications equipment and computers. Before the invention of the transistor at Bell Laboratories in 1947, amplification was provided exclusively by vacuum tubes, which were both bulky and unreliable. The first electronic digital computer, ENIAC, built at the University of Pennsylvania, had 19,000 vacuum tubes and consumed 150,000 W of electricity. Because of the discovery and development of the transistor and the printed circuit, a hand-held calculator run by a small battery has the same computing power as ENIAC.

A *junction transistor* is made by joining n-type and p-type semiconductors to form an n-p-n or a p-n-p junction. The former type is shown in the diagram below. The input signal (to be amplified) occurs in circuit 1, which has a small resistance and a forward-biased n-p junction (junction 1). As the voltage of the input signal to this circuit varies, the current in the circuit varies. This means there is a change in the number of electrons crossing the n-p junction. Circuit 2 has a relatively large resistance and is under reverse bias. The key to the operation of the transistor is that current flows in circuit 2 only when electrons crossing junction 1 also cross junction 2 and travel to the positive terminal. Since the current in circuit 1 determines the number of electrons crossing junction 1, the number of electrons available to cross junction 2 is also directly proportional to the current in circuit 1. The current in circuit 2 therefore varies depending on the current in circuit 1.

The voltage (V), current (I), and resistance (R) in a circuit are related by the equation

$$V = IR$$

Since circuit 2 has a large resistance, a given current in circuit 2 produces a larger voltage than the same current in circuit 1, which has a small resistance. Thus a signal of variable voltage in circuit 1, such as might be produced by a human voice on a telephone, is reproduced in circuit 2, but with much greater voltage changes. That is, the input signal has been *amplified* by the junction transistor. This device replaces the large vacuum tube and is a tiny component of a printed circuit on a silicon chip.

Silicon chips are really "planar" transistors constructed from thin layers of n-type and p-type regions connected by conductors. A chip less than 1 cm wide can contain hundreds of printed circuits and be used in computers, radios, and televisions.

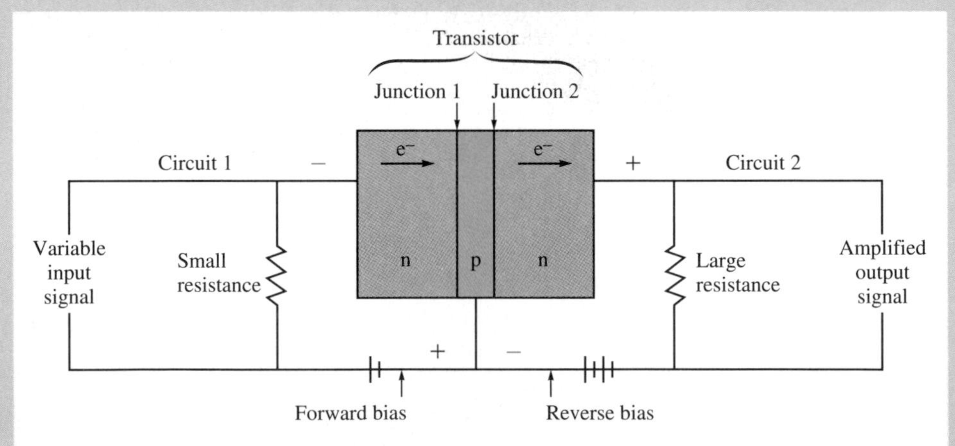

A schematic of two circuits connected by a transistor. The signal in circuit 1 is amplified in circuit 2.

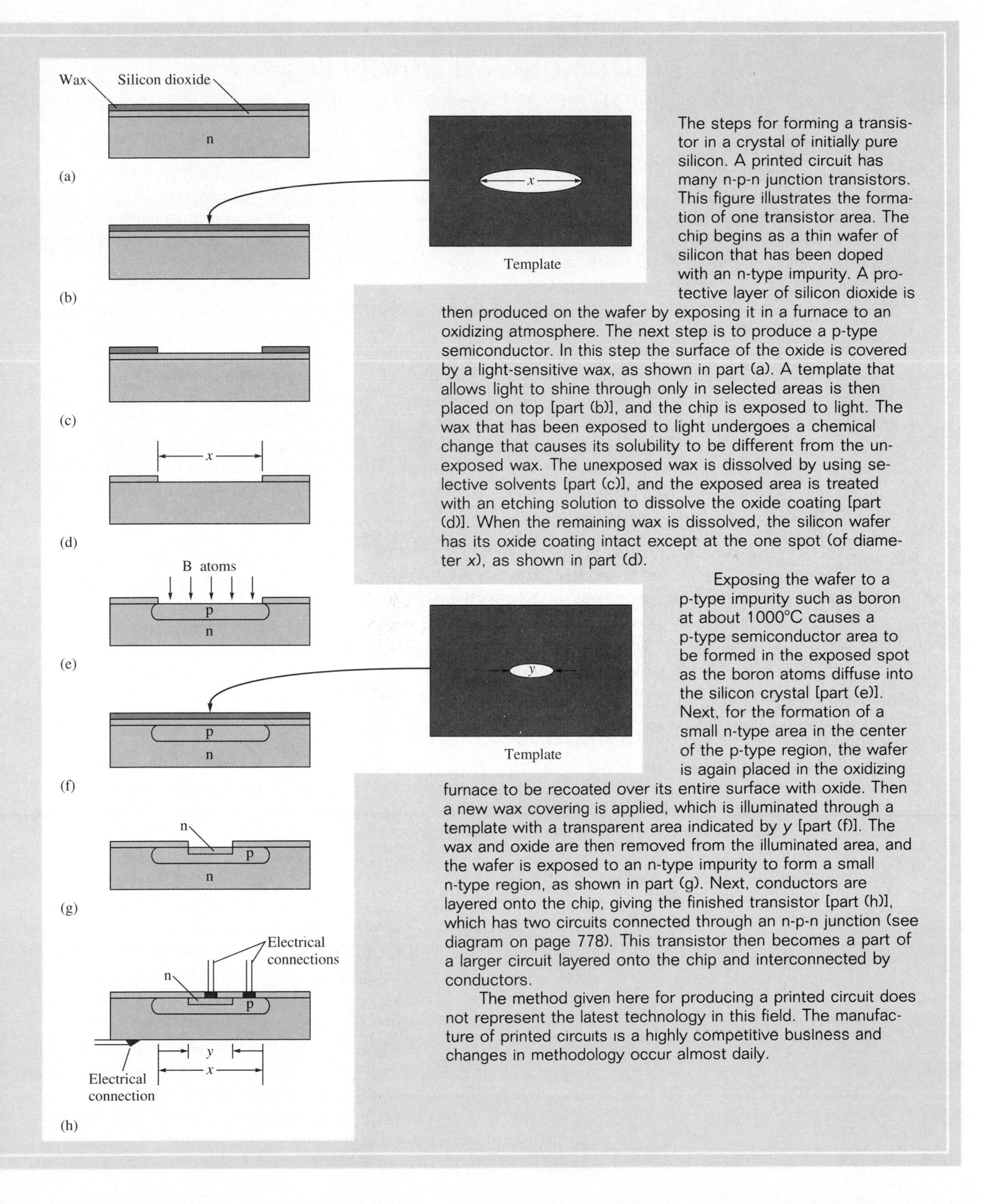

The steps for forming a transistor in a crystal of initially pure silicon. A printed circuit has many n-p-n junction transistors. This figure illustrates the formation of one transistor area. The chip begins as a thin wafer of silicon that has been doped with an n-type impurity. A protective layer of silicon dioxide is then produced on the wafer by exposing it in a furnace to an oxidizing atmosphere. The next step is to produce a p-type semiconductor. In this step the surface of the oxide is covered by a light-sensitive wax, as shown in part (a). A template that allows light to shine through only in selected areas is then placed on top [part (b)], and the chip is exposed to light. The wax that has been exposed to light undergoes a chemical change that causes its solubility to be different from the unexposed wax. The unexposed wax is dissolved by using selective solvents [part (c)], and the exposed area is treated with an etching solution to dissolve the oxide coating [part (d)]. When the remaining wax is dissolved, the silicon wafer has its oxide coating intact except at the one spot (of diameter x), as shown in part (d).

Exposing the wafer to a p-type impurity such as boron at about 1000°C causes a p-type semiconductor area to be formed in the exposed spot as the boron atoms diffuse into the silicon crystal [part (e)]. Next, for the formation of a small n-type area in the center of the p-type region, the wafer is again placed in the oxidizing furnace to be recoated over its entire surface with oxide. Then a new wax covering is applied, which is illuminated through a template with a transparent area indicated by y [part (f)]. The wax and oxide are then removed from the illuminated area, and the wafer is exposed to an n-type impurity to form a small n-type region, as shown in part (g). Next, conductors are layered onto the chip, giving the finished transistor [part (h)], which has two circuits connected through an n-p-n junction (see diagram on page 778). This transistor then becomes a part of a larger circuit layered onto the chip and interconnected by conductors.

The method given here for producing a printed circuit does not represent the latest technology in this field. The manufacture of printed circuits is a highly competitive business and changes in methodology occur almost daily.

MAKING DIAMONDS AT LOW PRESSURES: FOOLING MOTHER NATURE

In 1955 Robert H. Wentorf, Jr., accomplished something that borders on alchemy—he turned peanut butter into diamonds. He and his coworkers at the General Electric Research and Development Center also changed roofing pitch, wood, coal, and many other carbon-containing materials into diamonds, using a process involving temperatures of ≈2000°C and pressures of ≈100,000 atm. Although the first diamonds made by this process looked like black sand because of the impurities present, the process has now been developed to a point such that beautiful, clear, gem-quality diamonds can be produced. General Electric now has the capacity to produce 150 million carats (30,000 kg) of diamonds, virtually all of which is "diamond grit" used for industrial purposes such as abrasive coatings on cutting tools. The production of large, gem-quality diamonds by this process is still too expensive to compete with the natural sources of these stones. However, this may change as methods are developed for making diamonds at low pressures.

The high temperatures and pressures used in the GE process for making diamonds make sense if one looks at the accompanying phase diagram for carbon. Note that graphite—not diamond—is the most stable form of carbon under ordinary conditions of temperature and pressure. However, diamond becomes more stable than graphite at very high pressures (as one would expect from the greater density of diamond). The high temperature used in the GE process is necessary to disrupt the bonds in graphite so that diamond (the most stable form of carbon at the high pressures used in the process) can form.

This brings us again to the difference between kinetic and thermodynamic stability. Under normal conditions of temperature and pressure, graphite has a lower free energy than diamond (by ≈2 kJ/mol). However, to get from diamond to graphite requires a large expenditure of energy. This is the energy of activation that determines the rate of a reaction at a particular temperature. As a result of this high activation energy, diamonds formed at the high pressures found deep in the earth's crust can be brought to the earth's surface by natural geological processes and continue to exist for millions of years. Although these diamonds are thermodynamically unstable (relative to graphite), they are kinetically stable. That is, the reaction to change diamond to graphite is favored by thermodynamics but is disfavored by kinetics—it is so slow that diamonds *almost* last forever.* Diamond is said to be metastable (thermodynamically unstable but kinetically stable) under normal conditions of temperature and pressure.

We have seen that diamond formed at high pressures is "trapped" in this form by slow kinetics,

sublimation. Sublimation occurs when the vapor pressure of ice is equal to the external pressure, which in this case is only 2.0 torr. No liquid water appears under these conditions because the vapor pressure of liquid water is always greater than 2.0 torr; thus it cannot exist at this pressure. If liquid water were placed in a cylinder at such a low pressure, it would vaporize immediately.

Experiment 3: pressure is 4.588 torr. Again we start with ice as the only component in the cylinder at −20°C. In this case the pressure exerted on the ice by the piston is 4.588 torr. As the cylinder is heated, no new phase appears until the temperature reaches 0.0098°C. At this point, called the **triple point,** solid and liquid water have identical vapor pressures of 4.588 torr. Thus *at 0.0098°C and 4.588 torr all three states of*

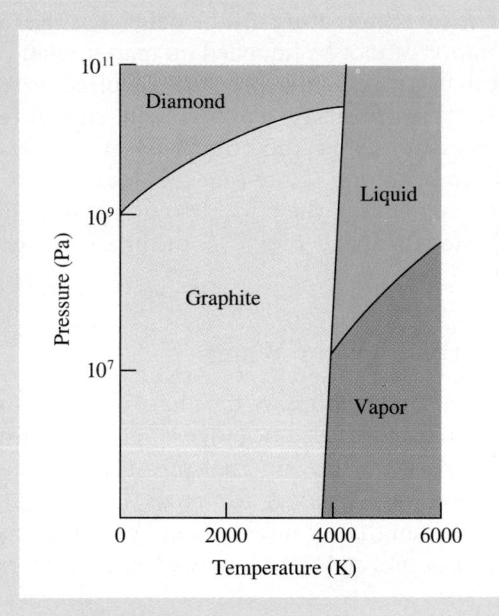

The phase diagram for carbon.

but this process is very expensive. Can diamond be formed at low pressures? The phase diagram for carbon says no. However, researchers have found that under the right conditions diamonds can be "grown" at low pressures. The process used is called chemical vapor deposition (CVD). CVD uses an energy source to release carbon atoms from a compound such as methane into a steady flow of hydrogen gas (some of which is catalytically dissociated to produce hydrogen atoms). The carbon atoms then deposit as a diamond film on a surface maintained at a temperature between 600°C and 900°C. Why does diamond form on this surface rather than the thermodynamically favored graphite? Nobody is sure, but it has been suggested that at these relatively high temperatures the diamond structure grows faster than the graphite structure and so diamond is favored kinetically under these conditions. It has also been suggested that the hydrogen atoms present react much faster with graphite fragments than with diamond fragments, effectively removing any graphite from the growing film. Once it forms, of course, diamond is kinetically trapped. The major advantage of CVD is that there is no need for the extraordinarily high pressures used in the traditional process for synthesizing diamonds.

The first products with diamond films are already on the market. Audiophiles can buy tweeters that have diaphragms coated with a thin diamond film that limits sound distortion. Watches with diamond-coated crystals are planned, as are diamond-coated windows in infrared scanning devices used in analytical instruments and missile guidance systems. These applications represent only the beginning for diamond-coated products.

*In Morocco a 50-km-long slab called Beni Bousera contains chunks of graphite that were probably once diamonds formed in the deposit when it was buried 150 km underground. As this slab slowly rose to the surface over millions of years, the very slow reaction changing diamond to graphite had time to occur. That is, on this time scale thermodynamic control could exert itself. In the diamond-rich kimberlite deposits in South Africa, which rise to the surface much faster, kinetic control exists.

water are present. In fact, *only* under these conditions can all three states of water coexist.

Experiment 4: pressure is 225 atm. In this experiment we start with liquid water in the cylinder at 300°C; the pressure exerted by the piston on the water is 225 atm. Liquid water can be present at this temperature because of the high external pressure. As the temperature increases, something happens that we did not observe in the first three experiments: the liquid gradually changes into a vapor but goes through an intermediate "fluid" region, which is neither true liquid nor vapor. This is quite unlike the behavior at lower temperatures and pressures, say at 100°C and 1 atm, where the temperature remains constant while a definite phase change from liquid to vapor occurs. The unusual behavior at

300°C and 225 atm occurs because the conditions are beyond the critical point for water. The **critical temperature** can be defined as the temperature above which the vapor cannot be liquefied no matter what pressure is applied. The **critical pressure** is the pressure required to produce liquefaction *at* the critical temperature. Together, the critical temperature and the critical pressure define the **critical point.** For water the critical point is 374°C and 218 atm. Note that the liquid/vapor line on the phase diagram for water ends at the critical point. Beyond this point the transition from one state to another involves the intermediate "fluid" region just described.

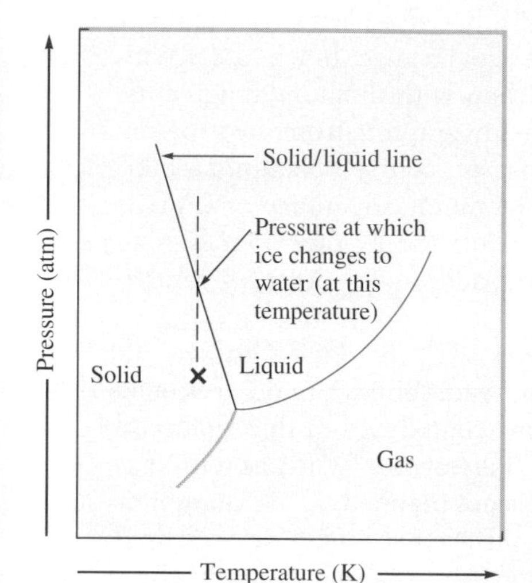

Figure 16.58
The phase diagram for water. At point × on the phase diagram, water is a solid. However, as the external pressure is increased while the temperature remains constant (indicated by the vertical dotted line), the solid/liquid line is crossed and the ice melts.

The motion of an ice skate blade over ice is a complex phenomenon, and other factors, such as frictional heating, may be important.

Applications of the Phase Diagram for Water

There are several additional interesting features of the phase diagram for water. Note that the solid/liquid boundary line has a negative slope. This means that the melting point of water *decreases* as the external pressure *increases*. This behavior, which is opposite to that observed for most substances, occurs because the density of ice is *less* than that of liquid water at the melting point. (The maximum density of water occurs at 4°C.) Thus when liquid water freezes, its volume increases.

We can account for the effect of pressure on the melting point of water by using the following reasoning. At the melting point, liquid and solid water coexist—they are in dynamic equilibrium. What happens if we apply pressure to this system? When subjected to increased pressure, matter reduces its volume. This behavior is most dramatic for gases but it also occurs for condensed states. Since a given mass of ice at 0°C has a larger volume than the same mass of liquid water, the system can reduce its volume in response to the increased pressure by changing to liquid. Thus at 0°C and an external pressure greater than 1 atm, water is liquid. In other words, the freezing point of water is less than 0°C when the pressure is greater than 1 atm.

Figure 16.58 illustrates the effect of pressure on ice. At the point X on the phase diagram, ice is subjected to increased pressure at constant temperature. Note that as the pressure is increased, the solid/liquid line is crossed, indicating that the ice melts. This seems to be what happens in ice skating. The narrow blade of the skate exerts a large pressure, since the skater's weight is supported by the small area of the blade. The ice under the blade melts because of the pressure, providing lubrication. After the blade passes, the liquid refreezes, as normal pressure returns. Without this lubrication effect due to the thawing ice, ice skating would not be the smooth, graceful activity that it can be.

Ice's lower density has other implications. When water freezes in a pipe or an engine block, it expands and breaks the container. For this reason water pipes are insulated in cold climates and antifreeze is used in water-cooled engines. The lower density of ice also means that ice formed on rivers and lakes floats, providing a layer of insulation that helps prevent bodies of water from freezing solid in the winter. Aquatic life can therefore continue to live through periods of freezing temperatures.

A liquid boils at the temperature where the vapor pressure of the liquid equals the external pressure. Thus the boiling point of a substance, like the melting point, depends on the external pressure. This is why water boils at different temperatures at different elevations (see Table 16.9), and any cooking carried out in boiling water will be affected by this variation. For example, it

TABLE 16.9 **Boiling Point of Water at Various Locations**

Location	Feet Above Sea Level	P_{atm} (torr)	Boiling Point (°C)
Top of Mt. Everest, Tibet	29,028	240	70
Top of Mt. McKinley, Alaska	20,320	340	79
Top of Mt. Whitney, Calif.	14,494	430	85
Leadville, Colo.	10,150	510	89
Top of Mt. Washington, N.H.	6,293	590	93
Boulder, Colo.	5,430	610	94
Madison, Wis.	900	730	99
New York City, N.Y.	10	760	100
Death Valley, Calif.	−282	770	100.3

takes longer to hard-boil an egg in Leadville, Colorado (elevation: 10,150 ft), than in San Diego (sea level), since water boils at a lower temperature, 89°C, in Leadville.

As we mentioned earlier, the phase diagram for water describes a closed system. Therefore, we must be very cautious in using the phase diagram to explain the behavior of water in a natural setting, such as on the earth's surface. For example, in dry climates (low humidity), snow and ice seem to sublime—a minimum amount of slush is produced. Wet clothes put on an outside line at temperatures below 0°C freeze and then dry while frozen. However, the phase diagram (Fig. 16.56) shows that ice should *not* be able to sublime at normal atmospheric pressures. What is happening in these cases? Ice in the natural environment is not in a closed system. The pressure is provided by the atmosphere rather than by a solid piston. This means that the vapor produced over the ice can escape from the immediate region as soon as it is formed. The vapor does not come to equilibrium with the solid, and the ice slowly disappears. Sublimation, which seems forbidden by the phase diagram, does seem to occur under these conditions, but it is not sublimation under equilibrium conditions.

The Phase Diagram for Carbon Dioxide

The phase diagram for carbon dioxide (Fig. 16.59) differs significantly from that for water. The solid/liquid line has a positive slope, since solid carbon dioxide is more dense than liquid carbon dioxide. The triple point for carbon dioxide occurs at 5.1 atm and −56.6°C, and the critical point occurs at 72.8 atm and 31°C. At a pressure of 1 atm, solid carbon dioxide sublimes at −78°C, a property that leads to its common name, *dry ice*. No liquid phase occurs under normal atmospheric conditions, making dry ice a convenient coolant.

Carbon dioxide is often used in fire extinguishers, where it exists as a liquid at 25°C under high pressures. Liquid carbon dioxide released from the extinguisher into the environment at 1 atm immediately changes to a vapor. Being heavier than air, this vapor smothers the fire by keeping oxygen away from the flame. The liquid/vapor transition is highly endothermic, so cooling also results, which helps to put out the fire. The "fog" produced by a carbon dioxide extinguisher is not solid carbon dioxide but rather moisture frozen from the air.

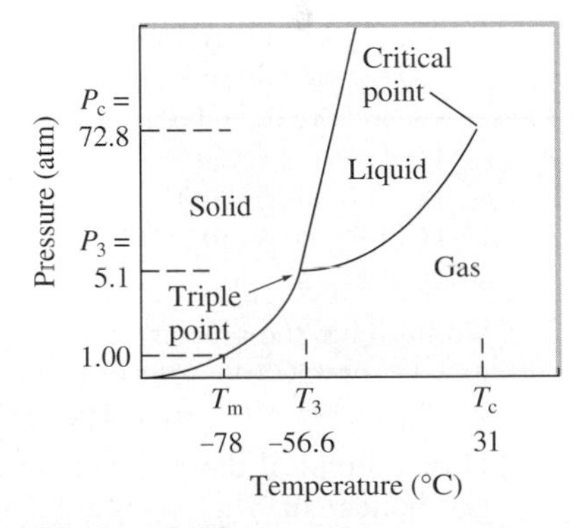

Figure 16.59
The phase diagram for carbon dioxide. The liquid state does not exist at a pressure of 1 atm. The solid/liquid line has a positive slope, since the density of solid carbon dioxide is greater than that of liquid carbon dioxide.

Exercises

A blue exercise number indicates that the answer to that exercise appears at the back of this book.

Intermolecular Forces and Physical Properties

1. Describe the relationship between the polarity of individual molecules and the nature and strength of intermolecular forces.
2. List the major types of intermolecular forces in order of increasing strength. Is there some overlap? That is, can the strongest London dispersion forces be greater than some dipole-dipole forces? Give examples of such instances.
3. Describe the relationship between molecular size and the strength of London dispersion forces.
4. Is it possible for the dispersion forces in a particular substance to be stronger than the hydrogen-bonding forces in another substance? Explain your answer.
5. How do the following physical properties depend on the strength of intermolecular forces?
   a. surface tension
   b. viscosity
   c. melting point
   d. boiling point
   e. vapor pressure
6. Does the nature of intermolecular forces change when a substance changes from a solid to a liquid, or from a liquid to a gas? Explain what causes a substance to undergo a phase change.
7. Identify the most important types of interparticle forces present in the solids of each of the following substances.
   a. $BaSO_4$
   b. H_2S
   c. Xe
   d. C_2H_6
   e. Cs
   f. Hg
   g. P_4
   h. H_2O
   i. NH_4Cl
   j. steel
   k. Teflon, $CF_3(CF_2CF_2)_nCF_3$
   l. polyethylene, $CH_3(CH_2CH_2)_nCH_3$
   m. $CHCl_3$
   n. Ge
   o. NO
   p. BF_3
8. Predict which substance in each of the following pairs would have the greater dipole-dipole forces.
   a. CO_2 or OCS
   b. PF_3 or PF_5
   c. SF_2 or SF_6
   d. SO_3 or SO_2
9. How do typical dipole-dipole forces differ from hydrogen bonds? In what ways are they similar?
10. For which molecule in each of the following pairs would you expect the hydrogen bonding to other molecules of the same type to be greater?
    a. $CH_3CH_2CH_2NH_2$ or $H_2NCH_2CH_2NH_2$
    b. $B(OH)_3$ or BH_3
    c. CH_3OH or H_2CO
    d. HF or HI
11. Rationalize the difference in boiling points for each of the following pairs of substances.
    a. *n*-pentane $CH_3CH_2CH_2CH_2CH_3$ 36.2°C
       neopentane $H_3C{-}C(CH_3)_2{-}CH_3$ 9.5°C
    b. dimethyl ether CH_3OCH_3 −25°C
       ethanol CH_3CH_2OH 79°C
    c. HF 20°C
       HCl −85°C
    d. $TiCl_4$ 136°C
       LiCl 1360°C
    e. HCl −85°C
       LiCl 1360°C
    f. LiCl 1360°C
       CsCl 1290°C
12. Rationalize the following differences in physical properties in terms of intermolecular forces. Compare the first three substances with each other, compare the last three with each other, and then compare all six. Can you account for any anomalies?

Substance	bp (°C)	mp (°C)	ΔH_{vap} (kJ/mol)
Benzene, C_6H_6	80	6	33.9
Naphthalene, $C_{10}H_8$	218	80	51.5
Carbon tetrachloride	76	−23	31.8
Acetone, CH_3COCH_3	56	−95	31.8
Acetic acid, CH_3CO_2H	118	17	39.7
Benzoic acid, $C_6H_5CO_2H$	249	122	68.2

13. Consider the following enthalpy changes:

$$F^-(g) + HF(g) \longrightarrow FHF^-(g) \qquad \Delta H = -155 \text{ kJ/mol}$$

$$(CH_3)_2C{=}O(g) + HF(g) \longrightarrow (CH_3)_2C{=}O{-}{-}{-}HF(g) \qquad \Delta H = -46 \text{ kJ/mol}$$

$$H_2O(g) + HOH(g) \longrightarrow H_2O{-}{-}{-}HOH \text{ (in ice)} \qquad \Delta H = -21 \text{ kJ/mol}$$

How do the strengths of hydrogen bonds vary with the electronegativity of the element to which hydrogen is bonded? Where in the above series would you expect hydrogen bonds of the following type to fall?

$$>N{-}{-}{-}H{-}O{-} \quad \text{and} \quad >N{-}{-}{-}H{-}N<$$

14. The strongest known hydrogen bond is formed in potassium bifluoride, KHF_2. Potassium bifluoride is produced by reacting KF with HF in water. What structure would you predict for the bifluoride ion (FHF^-)? Why?

15. Why is ΔH_{vap} for water much greater than ΔH_{fus}? What does this reveal concerning changes in intermolecular forces in going from solid to liquid to vapor?

16. Using the heats of fusion and vaporization for water, calculate the change in enthalpy for the sublimation of water:

$$H_2O(s) \longrightarrow H_2O(g)$$

Using the ΔH value given in Exercise 13 and the number of hydrogen bonds formed to each water molecule, estimate what portion of the intermolecular forces in ice can be accounted for by hydrogen bonding.

17. Oil of wintergreen, or methyl salicylate, has the following structure:

O
‖
C—OCH_3
OH

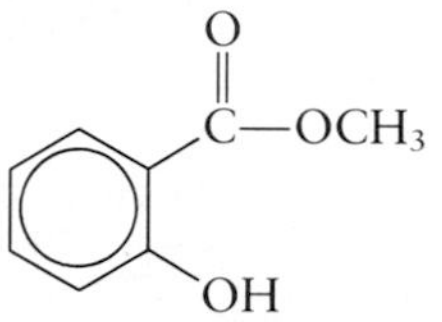

mp = −8°C

Methyl-4-hydroxybenzoate is another molecule with exactly the same molecular formula; it has the following structure:

O
‖
C—OCH_3
OH

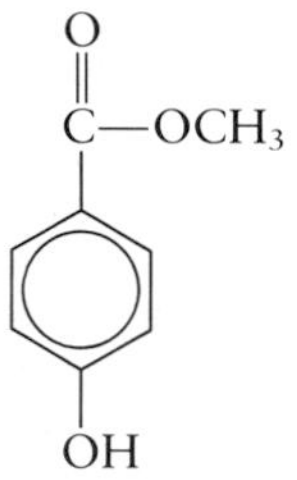

mp = 127°C

Account for the large difference in the melting points of the two substances.

18. Consider the following melting point data:

Compound	NaCl	$MgCl_2$	$AlCl_3$	$SiCl_4$	PCl_3	SCl_2	Cl_2
mp (°C)	801	708	190	−70	−91	−78	−101
Compound	NaF	MgF_2	AlF_3	SiF_4	PF_5	SF_6	F_2
mp (°C)	997	1396	1040	−90	−94	−56	−220

Account for the trends in melting points for the two series of compounds in terms of interparticle forces.

19. In each of the following groups of substances, pick the one that has the given property. Justify each answer.
    a. Highest boiling point: Hg, NaCl, or N_2
    b. Lowest freezing point: H_2, CH_4, or CO
    c. Smallest vapor pressure at 25°C; SiO_2, CO_2, or H_2O
    d. Greatest viscosity: $CH_3CH_2CH_2CH_3$, CH_3CH_2OH, or $HOCH_2CH_2OH$
    e. Strongest hydrogen bonding: NH_3, PH_3, or SbH_3
    f. Greatest heat of vaporization: H_2O, H_2S, H_2Se, or H_2Te
    g. Smallest enthalpy of fusion: H_2O, CO_2, MgO, or Li_2O

20. The nonpolar hydrocarbon, $C_{25}H_{52}$, is a solid at room temperature. Its boiling point is greater than 400°C. Which has the stronger intermolecular forces, $C_{25}H_{52}$ or H_2O? Explain your answer.

21. How could you experimentally determine whether TiO_2 is an ionic solid or a network solid? What would you predict on the basis of electronegativity differences?

22. Distinguish between each of the following.
    a. polarizability and polarity
    b. London dispersion forces and dipole-dipole forces
    c. intermolecular forces and intramolecular forces

23. Titanium(IV) chloride is a liquid that boils at 136°C. What might explain why $TiCl_4$ exists as discrete covalent molecules rather than as an ionic substance?

Properties of Liquids

24. In what ways are liquids similar to solids? In what ways are liquids similar to gases?

25. Define *critical temperature* and *critical pressure*. In terms of the kinetic molecular theory, why is it impossible for a substance to exist as a liquid above its critical temperature?

26. What is the relationship between critical temperature and intermolecular forces?

27. Use the kinetic molecular theory to explain why a liquid in an insulated vessel gets cooler as it evaporates.

28. The shape of the meniscus of water in a glass tube is different from that of mercury in a glass tube. Why?

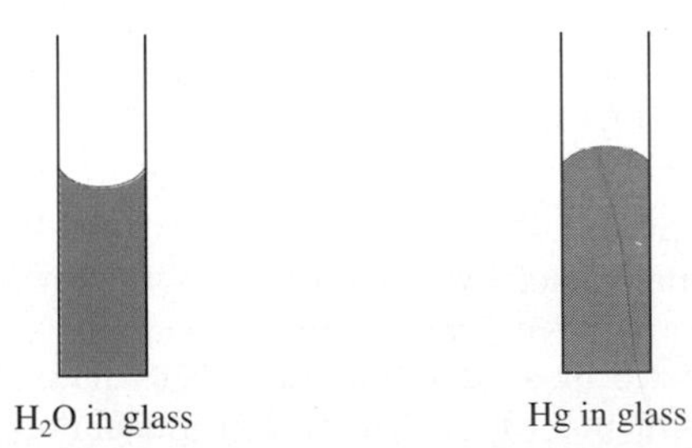

Predict the shape of the meniscus of water in a polyethylene tube. (Polyethylene can be represented as $CH_3(CH_2)_nCH_3$, where n is a large number on the order of 1000.)

29. Will water rise to a greater height by capillary action in a glass tube or in a polyethylene tube of the same diameter?

30. Some of the physical properties of H_2O and D_2O are as follows:

Property	H_2O	D_2O
Density at 20°C (g/ml)	0.997	1.108
Boiling point (°C)	100.00	101.41
Melting point (°C)	0.00	3.79
ΔH°_{vap} (kJ/mol)	40.7	41.61
ΔH°_{fus} (kJ/mol)	6.01	6.3

Account for the differences. (*Note:* D is a symbol often used for ^2H, the deuterium isotope of hydrogen.)

31. Hydrogen peroxide (H_2O_2) is a syrupy liquid with a relatively low vapor pressure and a normal boiling point of 152.2°C. Rationalize the differences between these physical properties and those of water.

32. Carbon diselenide, CSe_2, is a liquid at room temperature. The normal boiling point is 125°C, and the melting point is −45.5°. Carbon disulfide, CS_2, is also a liquid at room temperature, with normal boiling and melting points of 46.5°C and −111.6°C, respectively. How do the strengths of the intermolecular forces vary from CO_2 to CS_2 to CSe_2? Explain your answer.

Structures and Properties of Solids

33. Distinguish between the solids in the following pairs.
    a. crystalline solid and amorphous solid
    b. ionic solid and molecular solid
    c. molecular solid and network solid
    d. metallic solid and network solid

34. Will a crystalline solid or an amorphous solid give a simpler X-ray diffraction pattern? Why?

35. Is it possible to generalize that amorphous solids always have weaker or stronger interparticle forces than crystalline solids? Explain your answer.

36. What type of solid will each of the following substances form?

a. CO_2	e. Ru	i. NaOH	m. GaAs
b. SiO_2	f. I_2	j. U	n. BaO
c. Si	g. KBr	k. $CaCO_3$	o. NO
d. CH_4	h. H_2O	l. PH_3	p. GeO_2

37. When a metal was exposed to X rays, it emitted X rays of a different wavelength. The emitted X rays were diffracted by a LiF crystal (d = 201 pm), and first-order diffraction (n = 1 in the Bragg equation) was detected at an angle of 34.68°. Calcualte the wavelength of the X ray emitted by the metal.

38. The value of $2d$ (see Section 16.3) for mica, a silicate mineral, is 19.93 Å. Calculate the angle for first-order diffraction (n = 1 in the Bragg equation) of X rays from a molybdenum X-ray source (λ = 0.712 Å).

39. X rays from a copper X-ray tube (λ = 1.54 Å) were diffracted at an angle of 14.22° by a crystal of silicon. Assuming first-order diffraction (n = 1 in the Bragg equation), what is the interplanar spacing in silicon?

40. A metallic solid with atoms in a face-centered cubic unit cell with an edge length of 3.92 Å (392 pm) has a density of 21.45 g/cm³. Calculate the atomic mass and the atomic radius of the metal. Identify the metal.

41. Cobalt exists in two crystalline forms. Below 417°C cobalt is in the α-form, which has a hexagonal closest packed structure and a density of 8.90 g/cm³. Above 417°C cobalt is in the β-form, a cubic closest packed arrangement. The atomic radius of cobalt is 1.25 Å. Is there a change in the density of cobalt in going from the α-form to the β-form?

42. Iridium (Ir) has a face-centered cubic unit cell with an edge length of 3.833 Å (383.3 pm). The density of iridium is 22.61 g/cm³. Use these data to calculate a value for Avogadro's number.

43. Titanium metal has a body-centered cubic unit cell. The density of titanium is 4.50 g/cm³. Calculate the edge length of the unit cell and a value for the atomic radius of titanium. (*Hint:* in a body-centered arrangement of spheres, the spheres touch along the body diagonal.)

44. Tungsten metal, which has the highest melting point of all the elements except for carbon, exists in a body-centered cubic structure. The atomic radius of tungsten is 139 pm. Calculate the density of solid tungsten.

45. Nickel has a face-centered cubic unit cell. The density of nickel is 6.84 g/cm³. Calculate a value for the atomic radius of nickel.

46. The ionic radius of gold is 144 pm and the density is 19.32 g/cm³. Does elemental gold have a face-centered cubic structure or body-centered cubic structure?

47. In solid KCl the smallest distance between the center of a potassium ion and the center of a chloride ion is 314 pm. Calculate the length of the edge of the unit cell and determine the density of KCl, assuming it has the same structure as sodium chloride.

48. Magnesium oxide has the same structure as sodium chloride and has a density of 3.58 g/cm³. Calculate the edge length of the unit cell using this information. The ionic radii of Mg^{2+} and O^{2-} are 65 and 140. pm, respectively. Calculate the edge length of the unit cell from these data and compare it with your first answer.

49. The unit cell for a pure xenon fluoride compound is shown in the following diagram.

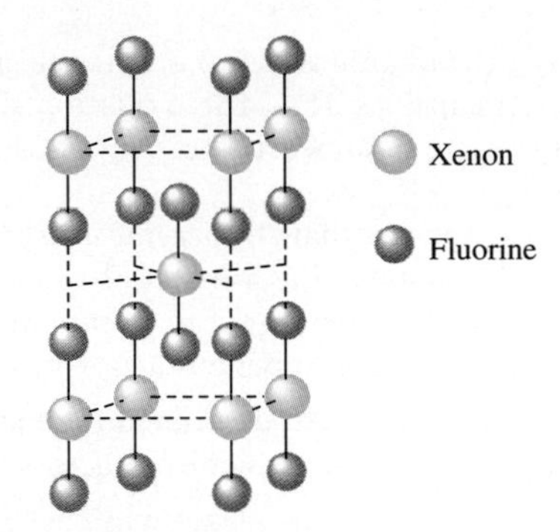

a. What is the formula of the compound?
b. The unit cell of XeF_2 has a height of 702 pm, and the edge of the square base has a length of 432 pm. Calculate the density of XeF_2.

50. Superalloys have been made of nickel and aluminum. (See *Scientific American,* October 1986, p. 159.) The alloy owes its strength to the formation of an ordered phase, called the *gamma-prime phase,* in which Al atoms are at the corners of a cubic unit cell, while Ni atoms are at the face centers. What is the composition (relative numbers of atoms) for this phase of the nickel-aluminum superalloy?

51. The memory metal, nitinol, is an alloy of nickel and titanium. It is called a memory metal because after being deformed, a piece of nitinol wire will return to its original shape. (See *Chem Matters,* October 1993, pp. 4–7.) The structure of nitinol consists of a simple cubic array of Ni atoms and an inner penetrating simple cubic array of Ti atoms. In the extended lattice, a Ti atom is found at the center of a cube of Ni atoms; the reverse is also true.
a. Describe the unit cell for nitinol.
b. What is the empirical formula of nitinol?
c. What are the coordination numbers of Ni and Ti in nitinol?

52. Perovskite is a mineral containing calcium, titanium, and oxygen. The following diagram represents the unit cell.

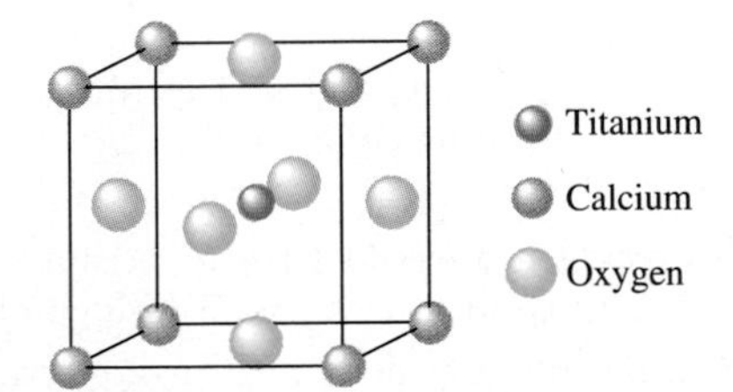

a. What is the formula of perovskite?
b. An alternative way of drawing the unit cell of perovskite has calcium at the center of each cubic unit cell. What are the positions of the titanium and oxygen atoms in this representation of the unit cell? Show that the formula for perovskite is the same for both unit cell representations.
c. How many oxygen atoms surround a given Ti atom in each representation of the unit cell?

53. Materials containing the elements Y, Ba, Cu, and O, that are superconductors (electrical resistance equals zero) at temperatures above that of liquid nitrogen, were recently discovered. The structures of these materials are based on the perovskite structure. Were they to have the ideal perovskite structure, the superconductors would have the structure shown in part (a) of the accompanying figure.

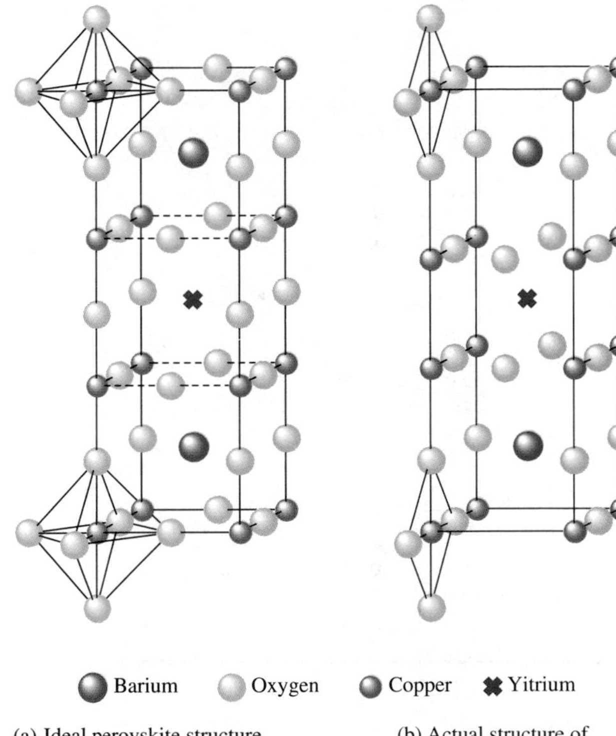

(a) Ideal perovskite structure

(b) Actual structure of superconductor

a. What is the formula of this ideal perovskite material?
b. How is this structure related to the perovskite structure shown in Exercise 52?

These materials, however, do not act as superconductors unless they are deficient in oxygen. The structure of the actual superconducting phase appears to be that shown in part (b) of the figure.

c. What is the formula of this material?

54. The structures of another class of ceramic, high-temperature superconductors are shown below.

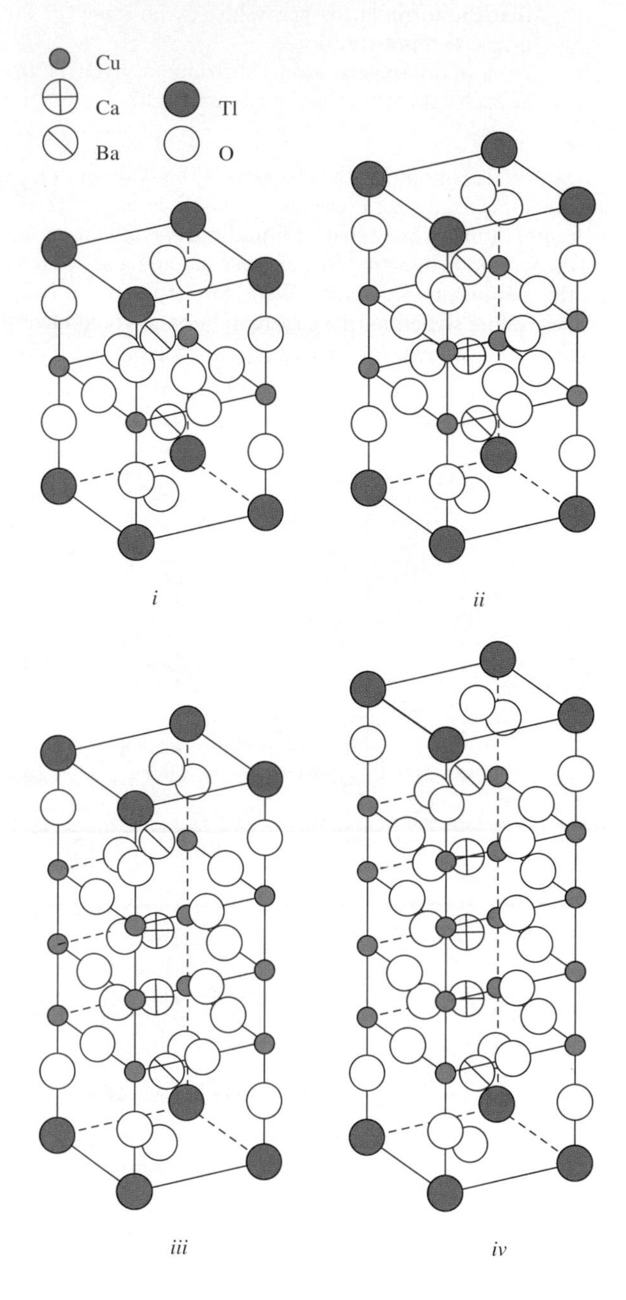

a. Determine the formula of each of these four superconductors.
b. One of the structural features that appears to be essential for high-temperature superconductivity is the presence of planar sheets of copper and oxygen atoms. As the number of sheets in each unit cell increases, the temperature for the onset of superconductivity increases. Order the four structures from lowest to the highest superconducting temperature.
c. Assign oxidation states to Cu in each structure assuming Tl exists as Tl^{3+}. The oxidation states of Ca, Ba, and O are assumed to be +2, +2, and −2, respectively.
d. It also appears that copper must display a mixture of oxidation states for a material to exhibit superconductivity. Explain how this occurs in these materials as well as in the superconductor in Exercise 53.

55. Use the band model to describe differences among insulators, conductors, and semiconductors.

56. Use the band model to explain why each of the following increases the conductivity of a semiconductor.
a. increasing the temperature
b. irradiating with light
c. adding an impurity

57. Selenium is a semiconductor used in photocopying machines. What type of semiconductor would be formed if a small amount of indium impurity were added to pure selenium?

58. The Group 3/Group 5 semiconductors are composed of equal amounts of atoms from Group 3A and Group 5A, for example, InP and GaAs. These types of semiconductors are used in light-emitting diodes and solid-state lasers. What would you add to make a p-type semiconductor from pure GaAs? How would you dope pure GaAs to make an n-type semiconductor?

59. The band gap in aluminum phosphide, AlP, is 2.5 electron-volts ($1\text{ eV} = 1.6 \times 10^{-19}\text{ J}$). What wavelength of light is emitted by an AlP diode?

60. An aluminum antimonide solid-state laser emits light with a wavelength of 730. nm. Calculate the band gap in joules.

Phase Changes and Phase Diagrams

61. Define each of the following:
a. condensation
b. evaporation
c. sublimation
d. supercooled liquid

62. Describe what is meant by dynamic equilibrium in terms of the vapor pressure of a volatile liquid.

63. The temperature inside a pressure cooker is 115°C. Use Equation (16.5) to calculate the vapor pressure of water inside the pressure cooker. Determine the temperature inside the pressure cooker if the vapor pressure of water was 3.50 atm.

64. What pressure would have to be applied to steam at 350.°C to condense the steam to liquid water?

65. How much energy does it take to convert 0.500 kg of ice at −20.°C to steam at 250.°C? Specific heat capacities: ice, 2.1 J g^{-1} °C; liquid, 4.2 J g^{-1} $°C^{-1}$; steam, 2.0 J g^{-1} $°C^{-1}$. $\Delta H^\circ_{vap} = 40.7$ kJ/mol; $\Delta H^\circ_{fus} = 6.01$ kJ/mol.

66. What is the final temperature when 0.850 kJ of energy is added to 10.0 g ice at 0°C? (See Exercise 65.)

67. In regions with dry climates evaporative coolers are used to cool air. A typical electric air conditioner is rated at 1.00×10^4 Btu/h (1 Btu, or British thermal unit, equals the amount of energy needed to raise the temperature of 1 lb of water by 1°F). How much water must be evaporated each hour to dissipate as much heat as a typical electric air conditioner?

68. Plot the following data and from the graph determine ΔH_{vap} for magnesium and lithium. In which metal is the bonding stronger?

Vapor Pressure (mm Hg)	Temperature (°C) Li	Mg
1.	750.	620.
10.	890.	740.
100.	1080.	900.
400.	1240.	1040.
760.	1310.	1110.

69. The enthalpy of vaporization of acetone is 32.0 kJ/mol. The normal boiling point of acetone is 56.5°C. What is the vapor pressure of acetone at 25.0°C?

70. At an elevation of 5300 ft the atmospheric pressure is about 630. torr. Determine the boiling point of acetone (ΔH°_{vap} = 32.0 kJ/mol) at this elevation. Calculate the vapor pressure at 25°C at this elevation. (See Exercise 69.)

71. The enthalpy of vaporization of mercury is 59.1 kJ/mol. The normal boiling point of mercury is 357°C. What is the vapor pressure of mercury at 25°C?

72. How does each of the following affect the rate of evaporation of a liquid in an open dish?
    a. intermolecular forces
    b. temperature
    c. surface area

73. Some water is placed in a sealed glass container connected to a vacuum pump (a device used to pump gases from a container), and the pump is turned on. The water appears to boil and then freezes. Explain these changes by using the phase diagram for water. What would happen to the ice if the vacuum pump was left on indefinitely?

74. Consider the following phase diagram. What phases are present at points A through H? Identify the triple point, normal boiling point, normal freezing point, and critical point. Which phase is denser, solid or liquid?

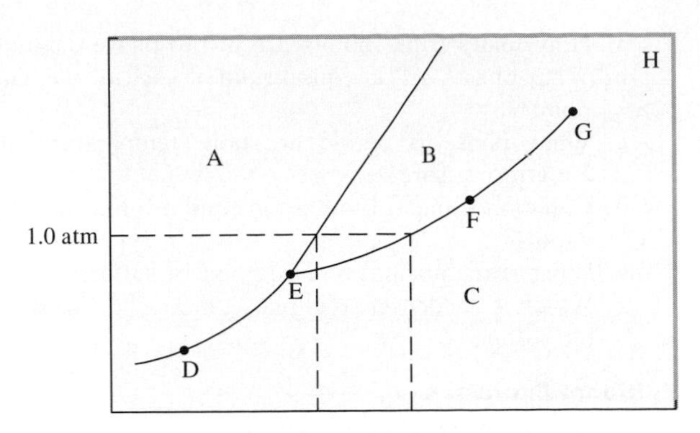

75. Describe how a phase diagram can be constructed from the heating curve for a substance.

76. A substance has the following properties:

		Specific Heat Capacities	
ΔH_{vap}	20 kJ/mol	$C_{(s)}$	3.0 J g^{-1} °C^{-1}
ΔH_{fus}	5 kJ/mol	$C_{(l)}$	2.5 J g^{-1} °C^{-1}
bp	75°C	$C_{(g)}$	1.0 J g^{-1} °C^{-1}
mp	−15°C		

Sketch a heating curve for the substance, starting at −50°C.

77. Why is a burn from steam typically much more severe than a burn from boiling water?

78. A 0.250-g chunk of sodium metal is cautiously dropped into a mixture of 50.0 g of water and 50.0 g of ice, both at 0°C. The reaction is

$$2Na(s) + 2H_2O(l) \longrightarrow 2NaOH(aq) + H_2(g) \qquad \Delta H = -368 \text{ kJ}$$

Will the ice melt? Assuming the final mixture has a specific heat capacity of 4.18 J g^{-1} °C^{-1}, calculate the final temperature.

79. Use the accompanying phase diagram for sulfur to answer the following questions.

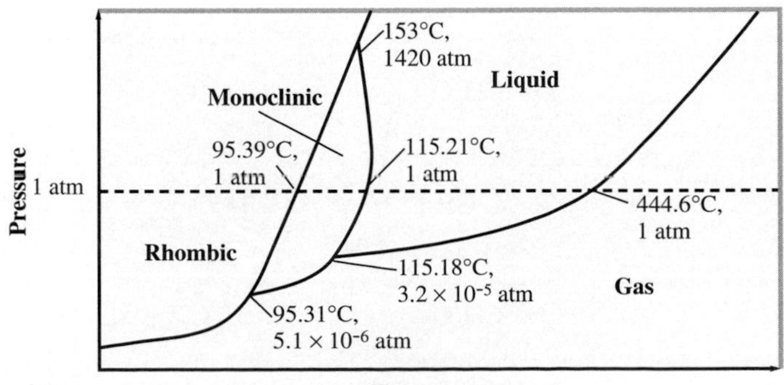

a. How many triple points are in the phase diagram?
b. What phases are in equilibrium at each of the triple points?
c. What phase is stable at room temperature and 1.0 atm pressure?
d. Can monoclinic sulfur exist in equilibrium with sulfur vapor?
e. What is the normal boiling point of sulfur?
f. Which is the denser solid phase, monoclinic or rhombic sulfur?

Additional Exercises

80. Boron nitride (BN) exists in two forms. The first is a slippery solid formed from the reaction of BCl_3 with NH_3, followed by heating in an ammonia atmosphere at 750°C. Subjecting the first form of BN to a pressure of 85,000 atm at 1800°C produces a second form that is the second hardest substance known. Both forms of BN remain solids to 3000°C. Suggest structures for the two forms of BN.

81. From the following data for liquid nitric acid, determine its heat of vaporization and normal boiling point.

Temperature (°C)	Vapor Pressure (mm Hg)
0	14.4
10.	26.6
20.	47.9
30.	81.3
40.	133
50.	208
80.	670

82. When wet laundry is hung on a clothesline on a cold winter day, it freezes, but eventually dries. Explain.

83. What fraction of the total volume of a simple cubic structure is occupied by hard spheres?

84. Many organic acids, such as acetic acid (CH_3CO_2H), whose structure is shown in Exercise 85, exist in the gas phase as hydrogen-bonded dimers (two molecule units). Draw a reasonable structure for such dimers.

85. Rationalize the following boiling points.

$CH_3C(=O)OH$	118°C
$ClCH_2C(=O)OH$	189°C
$CH_3C(=O)OCH_3$	57°C

86. Use the diagram of the unit cell for the hexagonal closest packed structure in Fig. 16.14 to calculate the net number of atoms in the hcp unit cell.

87. Some ice cubes with a total mass of 475 g are placed in a microwave oven and subjected to 750. W (750. J/s) of energy for 5.00 min. Does the water boil? If not, what is the final temperature of the water? (Assume all of the energy of the microwaves is absorbed by the water and that no heat is lost from the water.)

88. An ice cube tray contains enough water at 22.0°C to make 18 ice cubes that each have a mass of 30.0 g. The tray is placed in a freezer that uses CF_2Cl_2 as a refrigerant. The heat of vaporization of CF_2Cl_2 is 158 J/g. What mass of CF_2Cl_2 must be vaporized in the refrigeration cycle to convert all of the water at 22.0°C to ice at −5.0°C? The heat capacities for $H_2O(s)$ and $H_2O(l)$ are 2.08 J g^{-1} °C^{-1} and 4.18 J g^{-1} °C^{-1}, respectively, and the enthalpy of fusion for ice is 6.01 kJ/mol.

89. Calculate the enthalpy of vaporization and the normal boiling point of methanol using the following data.

Temperature (°C)	Vapor Pressure (mm Hg)
−6.0°C	20.0
5.0°C	40.0
12.1°C	60.0
21.2°C	100.0
49.9°C	400.0

90. Use the phase diagram for carbon given in Section 16.11 to answer the following questions.
a. How many triple points are in the phase diagram?
b. What phases can coexist at each triple point?
c. What happens if graphite is subjected to very high pressures at room temperature?
d. If we assume that the density increases with an increase in pressure, which is more dense, graphite or diamond?

91. From the density of cesium chloride (3.97 g/cm^3), calculate the distance between the centers of adjacent Cs^+ and Cl^- ions in the solid. Compare this value with the expected distance based on the sizes of the ions (Fig. 13.7).

92. Consider a cation in a trigonal hole. What size ion will just fit in the hole if the packed spheres have radius R?

93. The compounds K_2O, CuI, and ZrI_4 all can be described as cubic closest packed anions with the cations in tetrahedral holes. What fraction of the tetrahedral holes is occupied for each case?

94. The edge of the LiCl unit cell is 514 pm in length. Assuming that the Li^+ ions just fit in the octahedral holes of the closest packed Cl^- ions, calculate the ionic radii for the Li^+ and Cl^- ions. Compare them with the radii given in Fig. 13.7, and discuss the significance of any discrepancies.

95. Use the relative ionic radii in Fig. 13.7 to predict the structures expected for CsBr and KF. Do these predictions agree with observed structures?

96. Argon has a cubic closest packed structure as a solid. Assuming that argon has a radius of 190. pm, calculate the density of solid argon.

97. Rubidium chloride has the sodium chloride structure at normal pressures but assumes the cesium chloride structure at high pressures. What ratio of densities is expected for these two forms? Does this change in structure make sense on the basis of simple models?

98. Rhenium oxide does not have a closest packed structure. Its structure can best be described as a simple cubic array of rhenium ions with the oxide ions at the center of each edge of the cubic unit cell. What is the charge of the rhenium ion in this compound?

99. Spinel is a mineral that contains 37.9% aluminum, 17.1% magnesium, and 45.0% oxygen, by mass, and has a density of 3.57 g/cm^3. The edge of the cubic unit cell measures 809 pm. How many of each type of atom are present in the unit cell?

100. A given sample of wüstite has the formula $Fe_{0.950}O_{1.00}$. Calculate the fraction of iron ions present as Fe^{3+}. What fraction of the sites normally occupied by Fe^{2+} must be vacant in this solid?

101. A certain oxide of titanium is 28.31% oxygen by mass and contains a mixture of Ti^{2+} and Ti^{3+} ions. Determine the formula of the compound and the relative numbers of Ti^{2+} and Ti^{3+} ions.

102. The vapor pressure of water at 30.0°C is 31.824 torr; at this temperature the density of liquid water is 0.99567 g/cm^3. What is the ratio of the average distance between water molecules in the liquid and in the saturated vapor at this temperature?

103. Dry nitrogen gas is bubbled through liquid benzene (C_6H_6) at 20.0°C. From 100.0 L of the gaseous mixture of nitrogen and benzene, 24.7 g of benzene are condensed by passing the mixture through a trap at a temperature where nitrogen is gaseous and the vapor pressure of benzene is negligible. What is the vapor pressure of benzene at 20.0°C?

104. A sample of dry nitrogen gas weighing 100.0 g is bubbled through liquid water at 25.0°C. The gaseous mixture of nitrogen and water vapor escapes at a total pressure of 700. torr. What mass of water has vaporized? (The vapor pressure of water at 25°C is 23.8 torr.)

105. The molar enthalpy of vaporization of water at 373 K is 41.16 kJ/mol. What fraction of this energy is used to change the internal energy of the water, and what fraction is used to do work against the atmosphere? (Assume that water vapor is an ideal gas.)

106. A special vessel (see Fig. 16.53) contains ice and supercooled water (both at −10°C) connected by vapor space. Describe what happens to the amounts of ice and water as time passes.

107. Manganese crystallizes in a structure that has a body-centered cubic lattice. What is the coordination number (number of nearest neighbors) for Mn in this structure, and how many atoms are there in each unit cell?

108. Consider the ionic solid A_xB_y, which has the following unit cell. The B ions are packed in a cubic arrangement, where each face has the following structure:

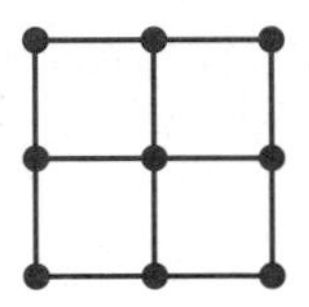

There is one B in the center of the cube. The structure can also be described in terms of three parallel planes of B's of the type shown above. The resulting structure thus contains eight intersecting cubes of B's. The A ions are found in the centers of alternate intersecting cubes (that is, four of every eight cubes have A's in the center). What is the formula of A_xB_y? In the extended structure, how many B's surround each A? What structure do the B's form?

109. You are given a small bar of an unknown metal X. You find the density of the metal to be 10.5 g/cm^3. An X-ray diffraction experiment measures the edge of the face-centered cubic unit cell as 4.09 Å (1 Å = 10^{-10} m). Identify X.

110. You are asked to help set up a historical display in the park by stacking some cannonballs next to a Revolutionary War cannon. You are told to stack them by starting with a triangle in which each side is composed of four touching cannonballs. You are to continue stacking them until you have a single ball on the top centered over the middle of the triangular base.
   a. How many cannonballs do you need?
   b. What type of closest packing is displayed by the cannonballs?
   c. The four corners of the pyramid of cannonballs form the corners of what type of regular geometric solid?

CHAPTER 17

Properties of Solutions

Most of the substances we encounter in daily life are mixtures: wood, milk, gasoline, shampoo, steel, and air are all well-known examples. When the components of a mixture are uniformly intermingled—that is, when a mixture is homogeneous—it is called a **solution.** Solutions can be gases, liquids, or solids, as shown in Table 17.1. However, we will be concerned in this chapter with the properties of liquid solutions, particularly those containing water. Many essential chemical reactions occur in aqueous solutions, since water is capable of dissolving so many substances.

TABLE 17.1 **Various Types of Solutions**

Example	State of Solution	State of Solute	State of Solvent
Air, natural gas	Gas	Gas	Gas
Vodka in water, antifreeze	Liquid	Liquid	Liquid
Brass, steel	Solid	Solid	Solid
Carbonated water (soda)	Liquid	Gas	Liquid
Seawater, sugar solution	Liquid	Solid	Liquid
Hydrogen in platinum	Solid	Gas	Solid

AN ENERGY SOLUTION

In today's society we face many serious problems related to energy: turmoil in the Middle East, the greenhouse effect, and polluted city air. Clearly the search for readily available, economical, and clean energy sources is of crucial importance. The solution may be in our backyards—literally.

When the sun shines on a swimming pool, the water is warmed by the solar energy. Because water is colorless (does not absorb visible light), the visible light from the sun passes through the water and strikes the bottom of the pool, where some of the light is absorbed and transformed to heat energy. This energy causes the water near the pool bottom to become warmer than the water above it. This warmer water, which has a lower density than the cooler water above it, then tends to rise toward the surface. This sets up convection currents, which circulate the warmer water to the surface where it is cooled (mainly by evaporation). The convection currents keep the temperature of the water in the pool relatively uniform throughout.

How can this convective cooling process be stopped, thus allowing a pool of water to be an effective heat sink? The answer lies in adding common salt, NaCl. When sufficient salt is added to a pool of water 2 to 3 m deep, a salinity gradient is established. That is, the salt concentration is not uniform in the pool. In fact, three distinct layers can be identified. The top layer, called the convective layer, has a salt concentration of about 2% by mass and is only about half a meter thick. The bottom layer has a very high salt concentration (about 27% by mass) and serves as the heat storage layer. The middle, so-called nonconvective layer is about 1.5 m thick. Being intermediate in density between the top and bottom layers, the nonconvective layer is trapped between these layers and acts as an insulator.

Because the bottom layer of the solution in the pool is so much more dense than the other layers, it can get very hot without setting up significant convection. With a black, light-absorbing coating on the pool bottom, the water in the deepest layer of the pool can easily reach temperatures in the range of 90–100°C. In fact, a temperature as high as 107°C has been reported. (The boiling point of this very concentrated solution is much higher than 100°C.) Energy from the pool can be extracted if

17.1 Solution Composition

A solute is the substance being dissolved. The solvent is the dissolving medium.

Because a mixture, unlike a chemical compound, has a variable composition, the relative amounts of substances in a solution must be specified. The qualitative terms *dilute* (relatively little solute present) and *concentrated* (relatively large amount of solute) are often used to describe solution content. However, we need to define solution composition more precisely in order to perform calculations. For example, in dealing with the stoichiometry of solution reactions in Chapter 4, we found it useful to describe solution composition in terms of **molarity,** or the number of moles of solute per liter of solution.

When liquids are mixed, the liquid present in the largest amount is called the solvent.

Other ways of describing solution composition are also useful. **Mass percent** (sometimes called *weight percent*) is the percent solute by mass in the solution:

$$\text{Mass percent} = \left(\frac{\text{grams of solute}}{\text{grams of solution}}\right) \times 100$$

Another way of describing solution composition is the **mole fraction** (symbolized by the Greek letter chi, χ), the ratio of the number of moles of a given the hot brine from the storage layer is pumped through a heat exchanger to heat water or some other fluid.

A solar pond can be used to generate electricity by employing special turbines that use a working fluid other than water. Typical steam turbines cannot be used, because the solar pond cannot produce temperatures high enough to generate steam efficiently. However, lower-boiling liquids can be used.

The solar pond in En Bokek, Israel, near the Dead Sea.

Solar ponds are now in use. A 52-acre pond at En Bokek near the Dead Sea produces 2.5 MW of electricity at peak power and could produce as much as 5 MW if more heat exchangers and turbines were installed in the powerhouse. This is a useful amount of electricity, but 200 such ponds would be required to provide the electricity produced by a typical conventional power plant. In the United States such installations probably would be most feasible in the Southwest, with its abundant land and sunshine. However, the concept may be more widely applicable. A study financed by the U.S. Department of Energy indicates that a 27,000-acre solar pond producing 600 MW of electricity would be competitive with conventional power plants in all states except Alaska.

Solar ponds are relatively inexpensive to build, easy to maintain, and nonpolluting if the pond bottom is sealed to prevent any leakage of the salt water. It has been estimated that by 2010 as much as 7% of the energy requirements of the United States could be supplied by solar ponds with a combined surface area about four times that of the Great Salt Lake in Utah.

component to the total number of moles of solution. For a two-component solution, where n_A and n_B represent the number of moles of the two components,

$$\text{Mole fraction of component A} = \chi_A = \frac{n_A}{n_A + n_B}$$

In very dilute aqueous solutions the molality and the molarity are nearly the same.

Still another way of describing solution composition is **molality** (symbolized by m), the number of moles of solute per *kilogram of solvent:*

$$\text{Molality} = \frac{\text{moles of solute}}{\text{kilograms of solvent}}$$

Since molarity depends on the volume of the solution, it changes slightly with temperature. Molality is independent of temperature, since it depends on mass.

17.2 The Thermodynamics of Solution Formation

Dissolving solutes in liquids is very common. We dissolve salt in the water used to cook vegetables, sugar in iced tea, stains in cleaning fluid, gaseous carbon dioxide in water to make soda water, ethanol in gasoline to make gasohol, and so on.

Solubility is important in other ways as well. For example, because the pesticide DDT is fat-soluble, it is retained and concentrated in animal tissues, where it causes detrimental effects. This is why DDT, even though it is effective for killing mosquitos, has been banned in the United States. Also, the solubility of various vitamins is important in determining correct dosages. The insolubility of barium sulfate means it can be safely used to improve X-ray images of the gastrointestinal tract, even though Ba^{2+} ions are quite toxic.

Polar solvents dissolve polar solutes; nonpolar solvents dissolve nonpolar solutes.

What factors affect solubility? The cardinal rule of solubility is *like dissolves like.* We find that we must use a polar solvent to dissolve a polar or ionic solute and a nonpolar solvent to dissolve a nonpolar solute. Now we will attempt to see why this behavior occurs from a thermodynamic point of view. As we will see, solubility is an extraordinarily complex phenomenon, especially when water is the solvent. However, it is useful to explore some of the fundamental aspects of solubility because it has such important consequences. To simplify the discussion, we will assume that the formation of a liquid solution takes place in three distinct steps.

STEP 1

Breaking up the solute into individual components (expanding the solute).

STEP 2

Overcoming intermolecular forces in the solvent to make room for the solute (expanding the solvent).

STEP 3

Allowing the solute and solvent to interact to form the solution.

These steps are illustrated in Fig. 17.1. Steps 1 and 2 require energy, since forces must be overcome to expand the solute and the solvent. Step 3 usually

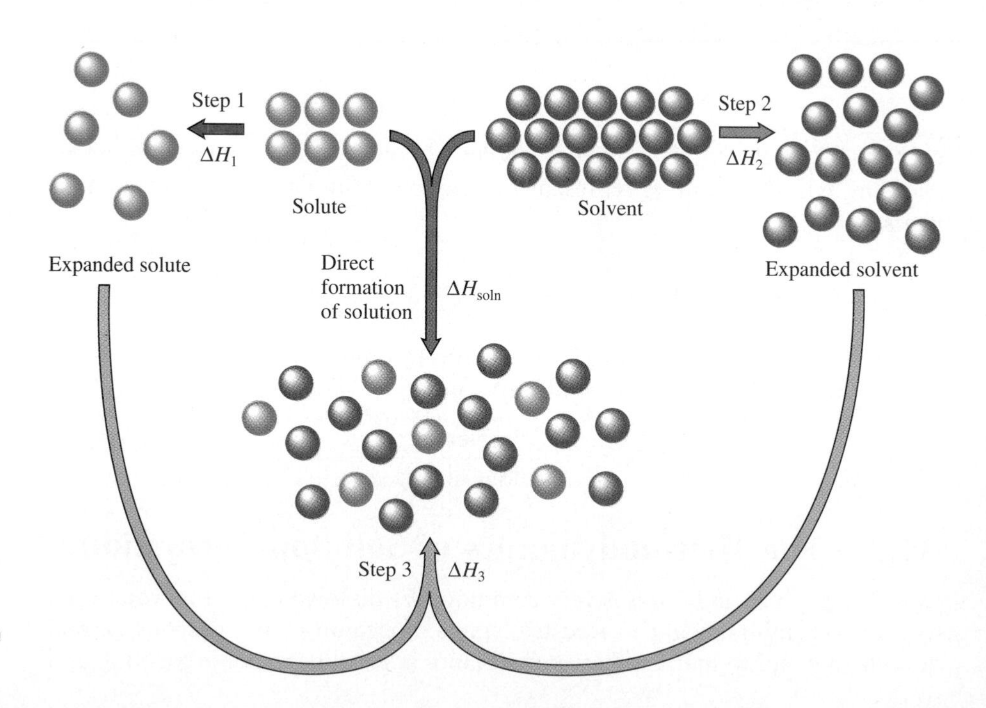

Figure 17.1
The formation of a liquid solution can be divided into three steps: (1) expanding the solute, (2) expanding the solvent, and (3) combining the expanded solute and solvent to form the solution.

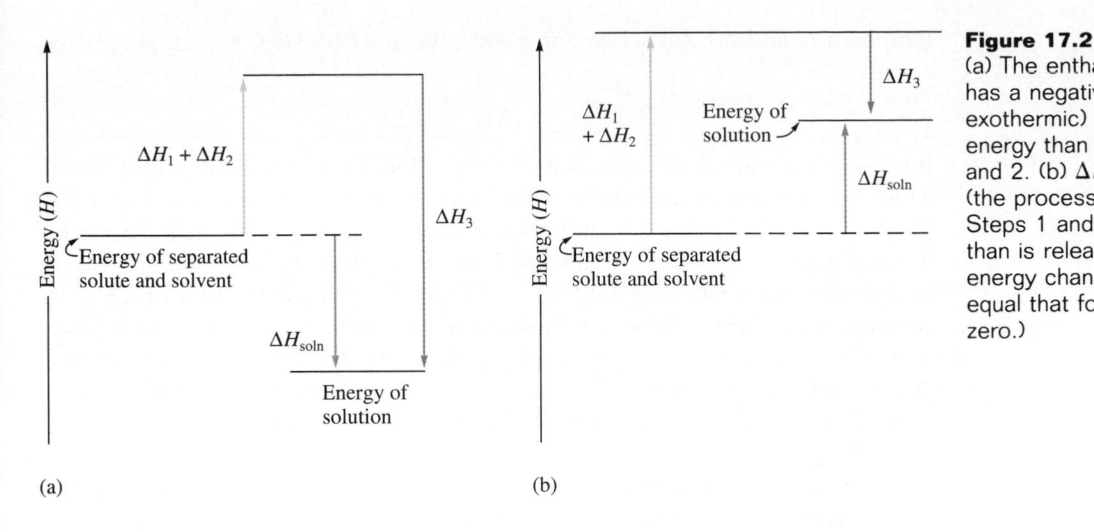

Figure 17.2
(a) The enthalpy of solution ΔH_{soln} has a negative sign (the process is exothermic) if Step 3 releases more energy than is required by Steps 1 and 2. (b) ΔH_{soln} has a positive sign (the process is endothermic) if Steps 1 and 2 require more energy than is released in Step 3. (If the energy changes for Steps 1 and 2 equal that for Step 3, then ΔH_{soln} is zero.)

releases energy. In other words, Steps 1 and 2 are endothermic, while Step 3 is usually exothermic. The overall enthalpy change associated with the formation of the solution, called the **enthalpy (heat) of solution** (ΔH_{soln}), can be viewed as the sum of the ΔH values for the steps:

The enthalpy of solution is the sum of energies used in expanding both the solvent and solute and the energy of solvent-solute interaction.

$$\Delta H_{soln} = \Delta H_1 + \Delta H_2 + \Delta H_3$$

where ΔH_{soln} may have a positive sign (energy absorbed) or a negative sign (energy released), as shown in Fig. 17.2.

To illustrate the importance of the various energy terms in the equation for ΔH_{soln}, we will consider a familiar case: the solubility of sodium chloride in water. It is clear that for NaCl(s) the term ΔH_1 is large and positive because of the strong ionic forces in the crystal that must be overcome. Also, ΔH_2 is expected to be large and positive because of the hydrogen bonds that must be broken in water. Finally, ΔH_3 is expected to be large and negative because of the strong interactions between the ions and the water molecules. In fact, the exothermic and endothermic terms essentially cancel in this case, as shown from the known values:

$$NaCl(s) \longrightarrow Na^+(g) + Cl^-(g) \qquad \Delta H_1^\circ = 786 \text{ kJ/mol}$$

$$H_2O(l) + Na^+(g) + Cl^-(g) \longrightarrow Na^+(aq) + Cl^-(aq)$$

$$\Delta H_{hyd}^\circ = \Delta H_2^\circ + \Delta H_3^\circ = -783 \text{ kJ/mol}$$

The second step shown here combines the terms ΔH_2 (for expanding the solvent) and ΔH_3 (for solvent-solute interactions) and is called the **enthalpy (heat) of hydration** (ΔH_{hyd}). This term represents the enthalpy change associated with the dispersal of a gaseous solute in water. Thus the standard enthalpy of solution for dissolving sodium chloride is the sum of ΔH_1° and ΔH_{hyd}°:

$$\Delta H_{soln}^\circ = 786 \text{ kJ/mol} - 783 \text{ kJ/mol} = 3 \text{ kJ/mol}$$

Note that ΔH_{soln}° is small but positive; the dissolving process requires a small amount of energy. Then why is NaCl so soluble in water? The answer to this question must involve the entropy change for the dissolving process. Recall from Chapter 10 that to predict whether or not a given process (at constant temperature and pressure) is spontaneous, we must consider the change in free energy:

$$\Delta G = \Delta H - T\Delta S$$

It is an experimental fact that NaCl(s) dissolves in water to form 1.0 M NaCl. Thus $\Delta G°$ for this process must be negative. However, the above calculations show that $\Delta H°_{soln}$ is positive and thus unfavorable. Therefore, $\Delta S°_{soln}$ must be positive and large enough to make $\Delta G°$ negative (through the $-T\Delta S°$ term). It is certainly not surprising that $\Delta S°_{soln}$ would be positive for this process. In considering the three steps for dissolution of a solute mentioned above, we would expect ΔS_1 and ΔS_2 to be positive since the solute and solvent are "expanded" in these steps. Also, ΔS_3 would be expected to be positive in a general case as a solute is randomly dispersed in the relatively large volume of solvent.

Thus we might generalize for an ionic (or polar) solute dissolving in a polar solvent as follows. Because $\Delta H°_{soln}$ contains large positive and negative contributions, it is difficult to predict the sign of $\Delta H°_{soln}$. However, even if $\Delta H°_{soln}$ is positive, it is not expected to be so large that it would overwhelm the expected positive value of $\Delta S°_{soln}$ for this process. The overall effect is to make $\Delta G°$ negative; thus the solution forms spontaneously.

Similar arguments can be made to explain why nonpolar solutes dissolve in nonpolar solvents. In this case $\Delta H°_{soln}$ is expected to be small, because the endothermic and exothermic terms in the solution process are expected to be similar in size. Thus the expected positive value of $\Delta S°_{soln}$ again would furnish the driving force for the solution process. These arguments suggest that $\Delta S°_{soln}$ provides the principal driving force for the behavior summarized by the rule "like dissolves like."

Now let's consider a process that is not spontaneous: a nonpolar solute does not dissolve in large quantities in a polar solvent. The floating oil slick that results whenever a major oil spill occurs in the ocean provides graphic evidence of this. What causes $\Delta G°$ to be positive for the dispersal of a nonpolar material in a polar solvent, such as water? For this case $\Delta H°_{soln}$ is expected to be positive (unfavorable) because $\Delta H°_3$, the only exothermic component of $\Delta H°_{soln}$, is not expected to have a very large magnitude. On the other hand, for reasons given above, we expect $\Delta S°_{soln}$ to be positive (favorable). Thus, we can explain the incompatibility of nonpolar and polar substances on the basis of the expected large, positive heat of solution that overwhelms the positive entropy change, thus giving a positive value for $\Delta G°$. Therefore, the solution does not form in this case.

TABLE 17.2
Values of $\Delta S°_{soln}$ for Several Salts Dissolving in Water

Process	$\Delta S°$ (J K^{-1} mol^{-1})
$KCl(s) \rightarrow K^+(aq) + Cl^-(aq)$	75
$LiF(s) \rightarrow Li^+(aq) + F^-(aq)$	−36
$CaS(s) \rightarrow Ca^{2+}(aq) + S^{2-}(aq)$	−138

Water as a Solvent

Because water is the most significant solvent in our world, it is especially important that we understand the solvent properties of water. As we noted in Chapter 16, water is not a typical liquid, with most of its unusual properties arising from the extensive hydrogen bonding present among the molecules. Because of its unique nature, we must be very cautious in using simple arguments to account for the solvent properties of water.

To illustrate the unusual nature of water as a solvent, consider the values of $\Delta S°_{soln}$ listed in Table 17.2 for KCl(s), LiF(s), and CaS(s) forming aqueous solutions. Note that when KCl(s) is dissolved in water to form a 1.0 M solution, the value of ΔS°_{soln} is positive, as expected from the previous discussion. However, note that ΔS°_{soln} is *negative* for the other two salts. Why? How could the random dispersal in water of ions formerly present in a highly ordered solid produce a negative entropy change?

Obviously something must be occurring in the solution process that leads to increased order, which in some cases is large enough to dominate ΔS°_{soln}. There is little doubt that this ordering effect arises from the hydration of the ions. In describing aqueous solutions containing ionic solutes in Chapter 4, we discussed the fact that the polar water molecules are attracted to the ions to form hydrated species. The assembling of a group of water molecules around the ions is an order-producing phenomenon and would be expected to make a negative contribution to ΔS°_{soln}. Studies show that the more charge density an ion possesses, the greater this hydration effect will be. This idea is borne out by the data in Table 17.2. For example, note that ΔS°_{soln} for KCl(*s*) is positive, but the value for LiF(*s*) is negative. This probably results from the smaller sizes (and thus larger charge densities) of Li^+ and F^- as compared with K^+ and Cl^-. The smaller ions presumably are able to bind the hydrating water molecules more firmly and thus show a more negative value for ΔS°_{soln}. The changes on the ions are also important. Note that CaS(*s*) exhibits a value of ΔS°_{soln} that is more negative than that for LiF(*s*), as might be expected for the more highly charged Ca^{2+} and S^{2-} ions.

We have seen that ionic solutes dissolving in water can lead to negative values of ΔS°_{soln}, presumably because of ion hydration effects. In certain cases the dispersal of nonpolar solute particles in water can also produce negative values of ΔS°_{soln}. For example, for benzene [a nonpolar liquid containing C_6H_6 molecules (see Section 14.5)] dissolving in water, ΔS°_{soln} has been estimated to be -58 J K^{-1} mol^{-1}. This very negative value for ΔS°_{soln} suggests that a great deal of ordering occurs when benzene is dispersed in water. This is a surprising result. What is the origin of this ordering? Clearly, the situation here is quite different from that for ionic solutes. The polar water molecules will not strongly hydrate the nonpolar benzene molecules, as they do ions.

Benzene does not dissolve in water to the extent that a 1 *M* solution is possible. The value of ΔS°_{soln} given here is extrapolated from values measured at much lower concentrations.

What could cause ΔS°_{soln} to be negative for this nonpolar solute? The negative value for ΔS°_{soln} in this case seems to arise not from the hydration of the solute, but from the opposite behavior. Instead of hydrating the nonpolar molecule, water seems to form a cage to isolate the nonpolar solute from the bulk water structure. This cage formation probably occurs as follows. So that a nonpolar molecule can be introduced into the water structure, a "hole" must be formed, which requires the breaking of some water-water hydrogen bonds. In response, in an effort to recover the lost hydrogen bonding, the water molecules around the edge of the hole apparently form even more hydrogen bonds than are normally found in bulk water. That is, a highly ordered cage of hydrogen-bonded water molecules forms around the nonpolar solute molecule. This ordering of the water structure makes a negative contribution to the entropy of solution and in certain cases leads to a negative value of ΔS°_{soln}. In fact, this unfavorable entropy contribution resulting from cage formation could be an important reason why nonpolar solutes are insoluble in water.

Summary

The point of this discussion has been to consider some of the thermodynamic aspects of the process that occurs when a solute is dispersed in a solvent to form

Miracle Solvents

When a substance is heated beyond its critical temperature (T_c) and then placed under extremely high pressure, it forms a **supercritical fluid.** Although such fluids resemble liquids, they technically are not liquids. A liquid cannot exist above the critical temperature for a substance. For example, if a gaseous substance is maintained above T_c and pressure is gradually applied, there is no distinct point where the gas changes to a liquid, as would occur below T_c, but the supercritical fluid gradually forms as the pressure is increased.

In recent experiments at the Agriculture Department's Northern Regional Research Center in Peoria, Illinois, scientists have found that supercritical carbon dioxide behaves as a very useful nonpolar solvent for removing fat from meat. At temperatures above 31°C (T_c for CO_2) and several hundred atmospheres of pressure, the carbon dioxide fluid can dissolve virtually all the fat from samples of meat. Even more important, the fluid also will dissolve any pesticide or drug residues that may be present in the meat. When the carbon dioxide fluid is returned to normal pressures, it immediately vaporizes, and the fat, drug, and pesticide molecules come "raining" out to allow easy analysis of the types and amounts of contaminants present in the meat. Therefore, this method provides an efficient way to test meat for trace pesticide and drug residues.

Like supercritical carbon dioxide, supercritical water is also a very interesting substance that has strikingly different properties from those of liquid water. For example, recent experiments have shown that supercritical (superfluid) water can behave simultaneously as both a polar and a nonpolar solvent. While the reasons for this unusual behavior remain unclear, the practical value of this behavior is very clear: it makes superfluid water a very useful reaction medium for a wide variety of substances. One extremely important application of this idea involves the environmentally sound destruction of industrial wastes. Most hazardous organic (nonpolar) substances can be dissolved in supercritical water and oxidized by dissolved O_2 in a matter of minutes. The products of these reactions are water, carbon dioxide, and possibly simple acids (which result when halogen-containing compounds are reacted). Therefore, the aqueous mixture that results from the reaction often can be disposed of with little further treatment. In contrast

a solution. Our observations tell us that "like dissolves like." However, the dissolution process is so complex that predicting whether a particular solute will dissolve in a given solvent is risky. Solubility is difficult to explain and even more difficult to predict, especially when water is the solvent. The only way to be certain about the compatibility of a given solute and solvent is to do the experiment.

17.3 Factors Affecting Solubility

Structure Effects

In the previous section we saw that solubility is favored if the solute and solvent have similar polarities. Since it is the molecular structure that determines polarity, there should be a definite connection between structure and solubility.

to the incinerators used to destroy organic waste products, a supercritical water reactor is a closed system (has no emissions).

Supercritical water has strange and wonderful properties, one of the most astonishing being that a flame can burn within the supercritical fluid during the reaction of O_2 with an organic substance. These properties promise to make supercritical water a versatile medium for chemical reactions.

Suggested Reading

R. W. Shaw, T. B. Brill, A. A. Clifford, C. A. Eckert, and E. U. Franck, "Supercritical Water," *Chem. Eng. News,* Dec. 23, 1991: 26.

(*left*) A multistage supercritical fluid extraction apparatus. The scientist draws off extracted material into a receiver flask. (*above*) Chicken fat obtained by using supercritical carbon dioxide as the extracting agent.

Vitamins provide an excellent example of the relationship among molecular structure, polarity, and solubility.

For the last several years, there has been considerable publicity about the pros and cons of consuming large quantities of vitamins. For example, large doses of vitamin C have been advocated to combat various illnesses, including the common cold. Vitamin E has been extolled as a youth-preserving elixir and a protector against the carcinogenic (cancer-causing) effects of certain chemicals. However, there are possible detrimental effects from taking large amounts of some vitamins, depending on their solubilities.

Vitamins can be divided into two classes: *fat-soluble* (vitamins A, D, E, and K) and *water-soluble* (vitamins B and C). The reason for the differing solubility characteristics can be seen by comparing the structures of vitamins A and C (Fig. 17.3). Vitamin A, composed mostly of carbon and hydrogen atoms that have similar electronegativities, is virtually nonpolar. This causes it to be soluble in nonpolar materials such as body fat, which is also largely composed

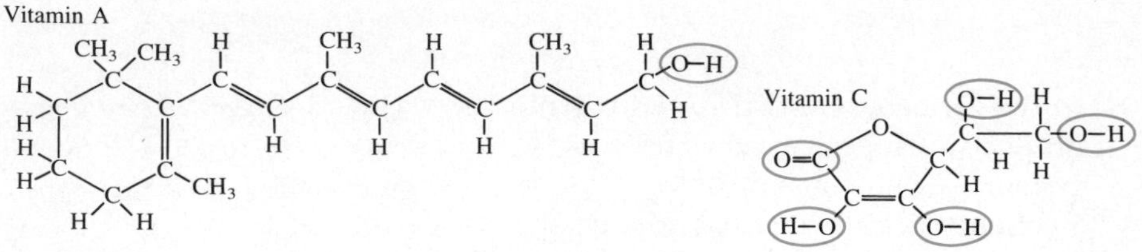

Figure 17.3
The molecular structures of (*left*) vitamin A (nonpolar, fat-soluble) and (*right*) vitamin C (polar, water-soluble). The circles indicate polar bonds. Note that vitamin C contains far more polar bonds than vitamin A.

of carbon and hydrogen, but not soluble in polar solvents such as water. On the other hand, vitamin C has many polar O—H and C—O bonds, making the molecule polar and thus water-soluble. We often describe nonpolar materials like vitamin A as *hydrophobic* (water-fearing) and polar substances like vitamin C as *hydrophilic* (water-loving).

Because of their solubility characteristics, the fat-soluble vitamins can build up in the fatty tissues of the body. This buildup has both positive and negative effects. Since these vitamins can be stored, the body can tolerate for a time a diet deficient in vitamins A, D, E, or K. Conversely, if excessive amounts of these vitamins are consumed, their buildup can lead to the illness *hypervitaminosis*.

In contrast, the water-soluble vitamins are excreted by the body and therefore must be consumed regularly. This fact was first recognized when the British navy discovered that scurvy, a disease often suffered by sailors, could be prevented if the sailors regularly ate fresh limes (which are a good source of vitamin C) when aboard ship (hence the name "limey" for the British sailor).

Pressure Effects

Although pressure has little effect on the solubilities of solids or liquids, it does significantly increase the solubility of a gas. Carbonated beverages, for example, are always bottled at high pressures of carbon dioxide to ensure a high concentration of carbon dioxide in the liquid. The fizzing that occurs when you open a can of soda results from the escape of gaseous carbon dioxide, because the atmospheric pressure of CO_2 is much lower than that used in the bottling process.

The increase in gas solubility with pressure can be understood from Fig. 17.4. Figure 17.4(a) shows a gas in equilibrium with a solution; that is, the gas molecules are entering and leaving the solution at the same rate. If the pressure is suddenly increased [Fig. 17.4(b)], the number of gas molecules per unit volume increases; thus the gas enters the solution at a higher rate than it leaves. As the concentration of dissolved gas increases, the rate of escape of the gas also increases until a new equilibrium is reached [Fig. 17.4(c)]. At this point the solution contains more dissolved gas than before.

The relationship between gas pressure and the concentration of a dissolved gas is given by **Henry's law:**

$$P = k_{\mathrm{H}}\chi$$

Henry's law is often expressed in the form $P = kC$, where C is the concentration of the gas in mol/L. In this case k has the units L atm/mol.

William Henry (1774–1836), a close friend of John Dalton, formulated his law in 1801.

Henry's law holds only when there is no chemical reaction between solute and solvent.

where P represents the partial pressure of the gaseous solute above the solution, χ represents the mole fraction of the dissolved gas, and k_{H} is a constant (the Henry's law constant) characteristic of a particular solution. Henry's law states that *the amount of gas dissolved in a solution is directly proportional to the pressure of the gas above the solution.*

Henry's law is obeyed most accurately for dilute solutions of gases that do not dissociate in or react with the solvent. For example, Henry's law is obeyed

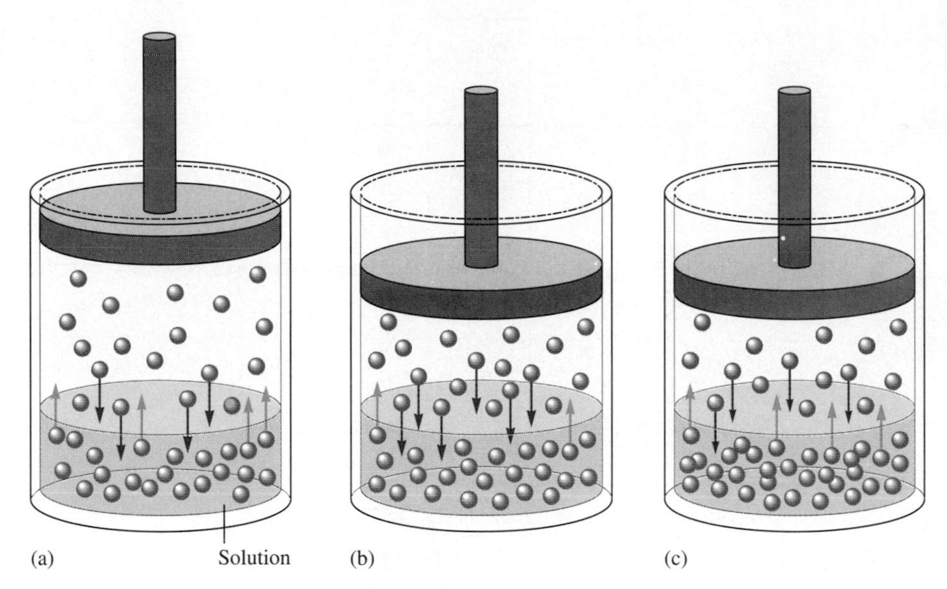

Figure 17.4
(a) A gaseous solute in equilibrium with a solution. (b) The piston is pushed in, increasing the pressure of the gas and the number of gas molecules per unit volume. This increases the rate at which the gas enters the solution, so the concentration of dissolved gas also increases. (c) The greater gas concentration in the solution causes an increase in the rate of escape. A new equilibrium is reached.

by oxygen gas in water, but it does *not* correctly represent the behavior of gaseous hydrogen chloride in water because of the dissociation reaction:

$$HCl(g) \xrightarrow{H_2O} H^+(aq) + Cl^-(aq)$$

The Henry's law constants for the aqueous solutions of several gases are given in Table 17.3.

TABLE 17.3
The Values of Henry's Law Constants for Several Gases Dissolved in Water at 298 K

Gas	k_H (atm)
CH_4	4.13×10^2
CO_2	1.64×10^3
O_2	4.34×10^4
CO	5.71×10^4
H_2	7.03×10^4
N_2	8.57×10^4

Temperature Effects for Aqueous Solutions

Everyday experiences of dissolving substances like sugar may lead you to think that solubility always increases with temperature. This is not the case. The dissolving of a solid occurs *more rapidly* at higher temperatures, but the amount of solid that can be dissolved may increase or decrease with increasing temperature. The effect of temperature on the solubility in water of several common solids is shown in Fig. 17.5. Note that although the solubility of most solids increases with temperature, the solubilities of some substances (such as sodium sulfate and cerium sulfate) decrease with increasing temperature.

Predicting the temperature dependence of solubility is very difficult. For example, although there is some correlation between the sign of ΔH°_{soln} and the variation of solubility with temperature, important exceptions exist.* The only sure way to determine the temperature dependence of a solid's solubility is by experiment.

Because ΔH°_{soln} refers to dissolving a small amount of solute in a large amount of a 1.0 *M* ideal solution, it is not necessarily relevant to the process of dissolving a solid in a saturated solution. Thus ΔH°_{soln} is of limited use in predicting the variation of solubility with temperature.

The behavior of gases dissolving in water appears to be less complex. The solubility of a gas in water typically decreases with increasing temperature, as shown for several cases in Fig. 17.6. This temperature effect has important environmental implications because of the widespread use of water from lakes

In contrast to the behavior of gases in water, gases typically become more soluble in most nonaqueous solvents as temperature increases.

*For more information, see R. S. Treptow, "Le Châtelier's Principle Applied to the Temperature Dependence of Solubility," *J. Chem. Ed.* **61** (1984), p. 499.

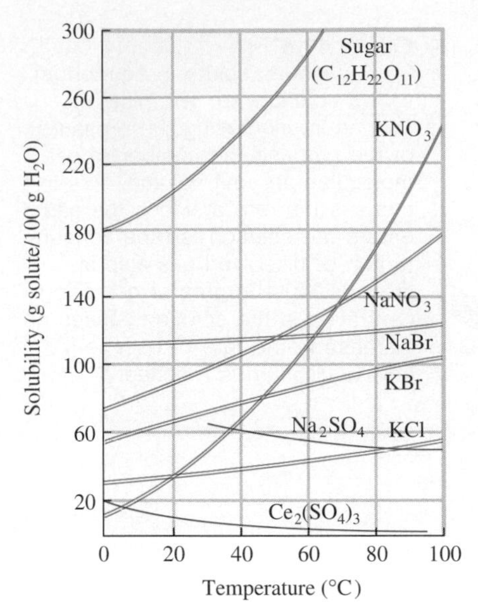

Figure 17.5
The solubilities of several solids as a function of temperature. Note that although most substances become more soluble in water with increasing temperature, sodium sulfate and cerium sulfate become less soluble.

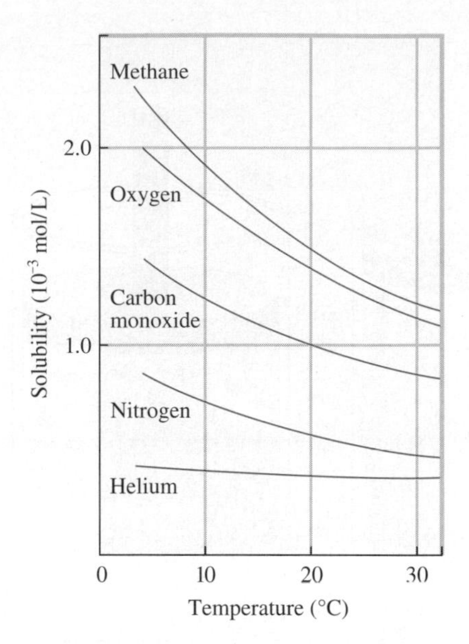

Figure 17.6
The solubilities of several gases in water as a function of temperature at a constant pressure of 1 atm of gas above the solution.

and rivers for industrial cooling. After being used as a coolant, the water is returned to its natural source at a higher-than-ambient temperature (**thermal pollution** has occurred). Because it is warmer, this water contains less than the normal concentration of oxygen and is also less dense; it tends to "float" on the colder water below, thus blocking normal oxygen absorption. This effect can be especially important in deep lakes. The warm upper layer can seriously decrease the amount of oxygen available to aquatic life in the deeper layers of the lake.

The decreasing water solubility of gases with increased temperature is also responsible for the formation of *boiler scale*. As we discussed in Chapter 7, the bicarbonate ion is formed when carbon dioxide is dissolved in water containing the carbonate ion:

$$CO_3^{2-}(aq) + CO_2(aq) + H_2O(l) \longrightarrow 2HCO_3^-(aq)$$

When the water also contains Ca^{2+} ions, this reaction is especially important—calcium bicarbonate is soluble in water, but calcium carbonate is insoluble. When the water is heated, the carbon dioxide is driven off. In order for the system to replace the lost carbon dioxide, the reverse reaction must occur:

$$2HCO_3^-(aq) \longrightarrow H_2O(l) + CO_2(aq) + CO_3^{2-}(aq)$$

This reaction, however, also increases the concentration of carbonate ions, causing solid calcium carbonate to form. This solid is the boiler scale that coats the walls of containers such as industrial boilers and tea kettles. Boiler scale reduces the efficiency of heat transfer and can lead to blockage of pipes (see Fig. 17.7).

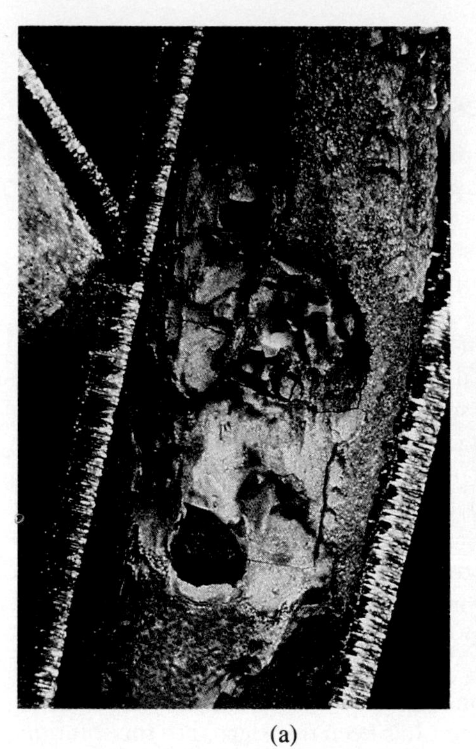
(a)

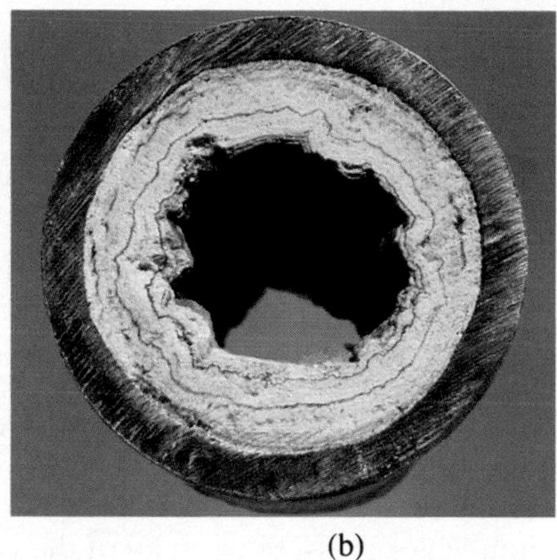
(b)

Figure 17.7
A hot-water pipe cut in half to show collected mineral deposits almost clogging the pipe.

THE LAKE NYOS TRAGEDY

On August 21, 1986, a cloud of gas suddenly boiled from Lake Nyos in Cameroon, killing nearly 2000 people. Although at first it was speculated that the gas was hydrogen sulfide, it now seems clear it was carbon dioxide. What would cause Lake Nyos to emit this huge, suffocating cloud of CO_2? Although the answer may never be known for certain, many scientists believe that the lake suddenly "turned over," bringing to the surface water that contained huge quantities of dissolved carbon dioxide. Lake Nyos is a deep lake that is thermally stratified: layers of warm, less dense water near the surface float on the colder, denser water layers nearer the lake's bottom. Under normal conditions the lake remains this way: there is little mixing among the different layers. Scientists believe that over hundreds or thousands of years, carbon dioxide gas had seeped into the cold water at the lake's bottom and dissolved in great amounts because of the large pressure of CO_2 present (in accordance with Henry's law). For some reason on August 21, 1986, the lake apparently suffered an overturn, possibly due to wind or to unusual cooling of the lake's surface by monsoon clouds. This caused water that was greatly supersaturated with CO_2 to reach the surface and release tremendous quantities of gaseous CO_2 that suffocated humans and animals before they knew what hit them—a tragic, monumental illustration of Henry's law.

Lake Nyos in Cameroon.

17.4 The Vapor Pressures of Solutions

Liquid solutions have physical properties significantly different from those of the pure solvent, a fact that has great practical importance. For example, we add antifreeze to the water in a car's cooling system to prevent freezing in winter and boiling in summer. We also melt ice on sidewalks and streets by spreading salt. These preventatives work because of the solute's effect on the solvent's properties.

A nonvolatile solute has no tendency to escape from solution into the vapor phase.

To explore how a nonvolatile solute affects a solvent, we will consider the experiment represented in Fig. 17.8, in which a sealed container encloses a beaker containing an aqueous sugar solution and a beaker containing pure water. Gradually, the volume of the sugar solution increases and the volume of the pure water decreases. Why? We can explain this observation if the vapor pressure of the pure solvent is greater than that of the solution. Under these conditions the pressure of vapor necessary to achieve equilibrium with the pure solvent is greater than that required to reach equilibrium with the solution. Thus as the pure solvent emits vapor in an attempt to reach equilibrium, the solution absorbs vapor to try to lower the vapor pressure toward its equilibrium value. This process results in a net transfer of water from the pure liquid through the vapor phase to the solution. The system can only reach an equilibrium vapor pressure when all of the water has been transferred to the solution. This experiment is just one of many observations indicating that the presence of a *nonvolatile solute lowers the vapor pressure of a solvent.*

The presence of a nonvolatile solute reduces the tendency of solvent molecules to escape.

We can account for this behavior in terms of the simple model shown in Fig. 17.9. The dissolved nonvolatile solute decreases the number of solvent molecules per unit volume. Thus it lowers the number of solvent molecules at the surface, which proportionately lowers the escaping tendency of the solvent molecules. For example, in a solution consisting of half nonvolatile solute molecules and half solvent molecules, we expect the observed vapor pressure to

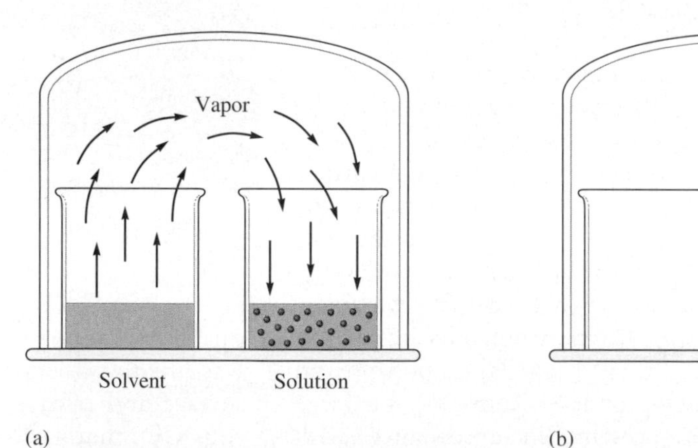

Figure 17.8
A solution and a pure solvent in a closed environment. (a) Initial stage. (b) After a period of time, the solvent is completely transferred to the solution.

be half that of the pure solvent, since only half as many molecules can escape. In fact, this agrees with our observations.

Detailed studies of the vapor pressures of solutions containing nonvolatile solutes were carried out by François M. Raoult. His results are described by the equation known as **Raoult's law:**

$$P_{soln} = \chi_{solvent} P^{\circ}_{solvent}$$

where P_{soln} is the observed vapor pressure of the solution, $\chi_{solvent}$ is the mole fraction of solvent, and $P^{\circ}_{solvent}$ is the vapor pressure of the pure solvent. Note that for a solution containing half solute and half solvent molecules, $\chi_{solvent}$ is 0.5, so the vapor pressure of the solution is half that of the pure solvent. On the other hand, for a solution where three-fourths of the solution molecules are solvent, $\chi_{solvent} = \frac{3}{4} = 0.75$, and $P_{soln} = 0.75P^{\circ}_{solvent}$. The idea is that the nonvolatile solute lowers the vapor pressure simply by diluting the solvent.

Raoult's law is a linear equation of the form $y = mx + b$, where $y = P_{soln}$, $x = \chi_{solvent}$, $m = P^{\circ}_{solvent}$, and $b = 0$. Thus a plot of P_{soln} versus $\chi_{solvent}$ gives a straight line with a slope equal to $P^{\circ}_{solvent}$, as shown in Fig. 17.10.

The effect of the solute on the vapor pressure of a solution gives us a convenient way to "count" molecules and thus provides a means for experimentally determining molar masses. Suppose a certain mass of a compound is dissolved in a solvent, and the vapor pressure of the resulting solution is measured. Using Raoult's law, we can determine the number of moles of solute present. Since the mass of this number of moles is known, we can calculate the molar mass.

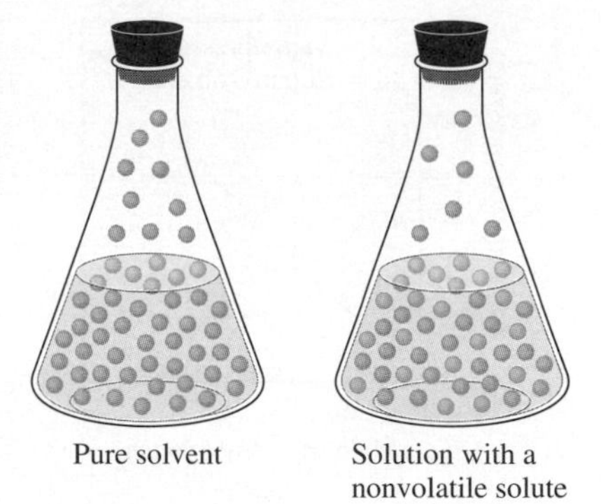

Figure 17.9
The presence of a nonvolatile solute inhibits the escape of solvent molecules from the liquid, thus lowering the vapor pressure of the solvent.

Raoult's law states that the vapor pressure of a solution is directly proportional to the mole fraction of solvent present.

Recyclable Heat

The current rise in athletic activities has led to many injuries and has necessitated new products to treat those injuries. Of the many instant hot and cold packs available, one of the most interesting types produces heat by the exothermic precipitation of a solid from a solution. This heat pack looks like a tiny air mattress with two compartments. The larger one contains a viscous supersaturated solution of sodium thiosulfate ($Na_2S_2O_3$) dissolved in water, and the smaller one contains crystals of solid sodium thiosulfate.

Because sodium thiosulfate crystals do not readily precipitate from aqueous solution, water can be greatly supersaturated with the solid. That is, much more solid $Na_2S_2O_3$ can be dissolved in water than should be possible at a given temperature. However, as soon as the supersaturated solution is exposed to a seed crystal of $Na_2S_2O_3$, which provides a pattern for crystal growth, the solid forms, releasing a considerable quantity of energy as heat over several hours. The pack is activated by squeezing out a seed crystal so that it comes into contact with the solution. The resulting formation of the solid gives a pleasantly warm temperature of 48°C (118°F).

This heat pack can be recycled by simply placing it in boiling water to redissolve the sodium thiosulfate. As the pack cools, the solid remains dissolved—it again becomes supersaturated. The pack can be reused until the supply of seed crystals is depleted.

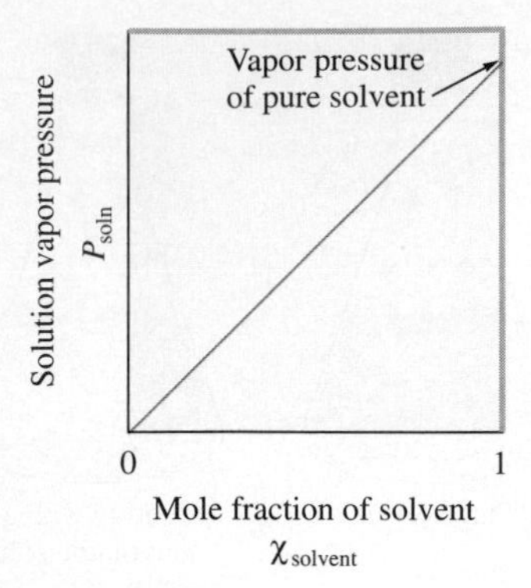

Figure 17.10
For a solution that obeys Raoult's law, a plot of P_{soln} versus $\chi_{solvent}$ yields a straight line.

EXAMPLE 17.1

A solution was prepared by adding 20.0 g of urea to 125 g of water at 25°C, a temperature at which pure water has a vapor pressure of 23.76 torr. The observed vapor pressure of the solution was found to be 22.67 torr. Calculate the molar mass of urea.

Solution

Raoult's law can be rearranged to give

$$\chi_{H_2O} = \frac{P_{soln}}{P^{\circ}_{H_2O}}$$

This form allows us to determine the mole fraction of water in the solution:

$$\chi_{H_2O} = \frac{22.67 \text{ torr}}{23.76 \text{ torr}} = 0.9541$$

However, we are interested in the urea. To find its molar mass, we must find the number of moles represented by 20.0 g. We can calculate the number of moles of urea from the definition of χ_{H_2O}:

$$\chi_{H_2O} = \frac{\text{mol } H_2O}{\text{mol } H_2O + \text{mol urea}} = \frac{n_{H_2O}}{n_{H_2O} + n_{urea}}$$

From the mass of water used to prepare the solution, we have

$$n_{H_2O} = 125 \text{ g } H_2O \times \frac{1 \text{ mol } H_2O}{18.0 \text{ g } H_2O} = 6.94 \text{ mol } H_2O$$

Since $\chi_{H_2O} = 0.9541$,

$$\chi_{H_2O} = 0.9541 = \frac{n_{H_2O}}{n_{H_2O} + n_{urea}} = \frac{6.94}{6.94 + n_{urea}}$$

or

$$0.9541(6.94 + n_{urea}) = 6.94$$

Solving for the number of moles of urea gives

$$n_{urea} = \frac{6.94 - 6.62}{0.9541} = 0.335 \text{ mol}$$

Since 20.0 g of urea was originally dissolved, 0.335 mol of urea weighs 20.0 g. Thus

$$\frac{20.0 \text{ g}}{0.335 \text{ mol}} = 59.7 \text{ g/mol}$$

The value for the molar mass of urea determined in this experiment is thus 59.7 g/mol. Urea has the formula $(NH_2)_2CO$ and a molar mass of 60.0, so the result obtained in this experiment agrees fairly well with the known value.

We can also use vapor pressure measurements to characterize solutions. For example, 1 mole of sodium chloride dissolved in water lowers the vapor pressure approximately twice as much as expected because the ions separate when it dissolves. Thus vapor pressure measurements can give valuable information about the nature of the solute after it dissolves. We will discuss this in more detail in Section 17.7.

The lowering of vapor pressure depends on the number of solute particles present in the solution.

Nonideal Solutions

Any solution that obeys Raoult's law is called an **ideal solution.** One might say that Raoult's law is to solutions what the ideal gas law is to gases. As with gases, ideal behavior for solutions is never perfectly achieved, but is sometimes closely approached. Nearly ideal behavior is often observed when the solute-solute, solvent-solvent, and solute-solvent interactions are very similar. That is, in solutions where the solute and solvent are very much alike, the solute simply acts to dilute the solvent. However, if the solvent has a special affinity for the solute, such as if hydrogen bonding occurs, the tendency of the solvent molecules to escape will be lowered more than expected. In such cases the observed vapor pressure will be *lower* than the value predicted by Raoult's law; there is a *negative deviation from Raoult's law.*

Strong solute-solvent interaction gives a vapor pressure lower than that predicted by Raoult's law.

As mentioned above, for a solution to behave ideally the solute-solute, solvent-solvent, and solute-solvent interactions would have to be identical. This would correspond to a situation where $\Delta H_{soln} = 0$. On the other hand, when a solute and solvent release large quantities of energy in the formation of a solution—that is, when ΔH_{soln} is large and negative—we can assume that strong interactions exist between the solute and solvent. In this case we expect a negative deviation from Raoult's law.

So far we have assumed that the solute is nonvolatile and so does not contribute to the vapor pressure over the solution. However, for liquid-liquid solutions where both components are volatile, a modified form of Raoult's law applies:

$$P_{TOTAL} = P_A + P_B = \chi_A P_A^\circ + \chi_B P_B^\circ$$

where P_{TOTAL} represents the total vapor pressure of a solution containing A and B, χ_A and χ_B are the mole fractions of A and B, P_A° and P_B° are the vapor pressures of pure A and pure B, and P_A and P_B are the partial pressures resulting from molecules of A and B in the vapor above the solution.

Liquid-liquid solutions obeying this form of Raoult's law are said to be *ideal.* However, as with solutions containing nonvolatile solutes, deviations from Raoult's law are often observed. These can be positive or negative, as shown in Fig. 17.11.

Again, large negative heats of solution indicate especially strong solute-solvent interactions, and such solutions are expected to show *negative* deviations from Raoult's law. Both components have a lower escaping tendency in the solution than in the pure liquids. This behavior is illustrated by an acetone-water solution where the molecules can hydrogen bond effectively:

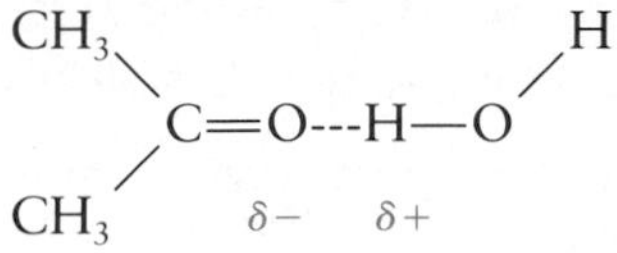

```
CH3                 H
   \               /
    C=O---H—O
   /
CH3      δ−    δ+
```

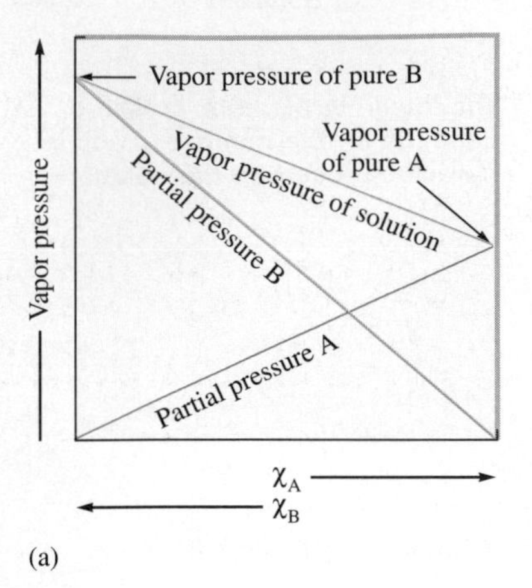

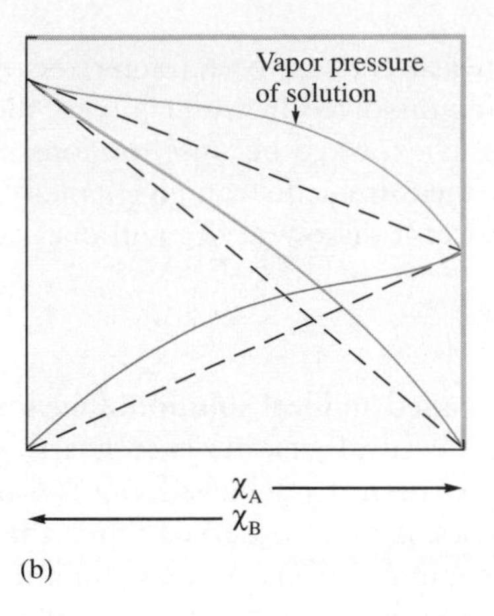

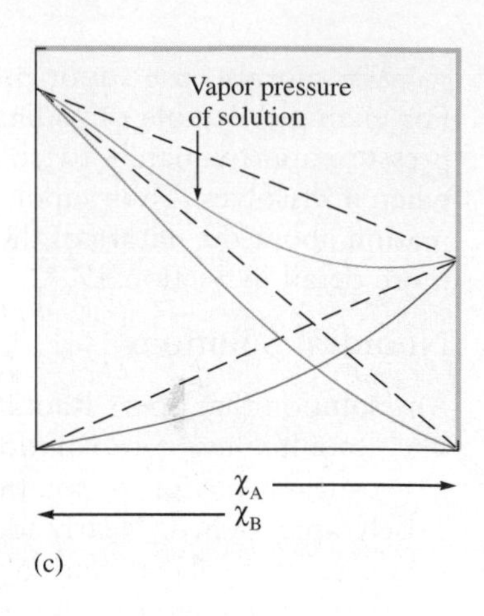

Figure 17.11
Vapor pressure for a solution of two volatile liquids. (a) The behavior predicted for an ideal liquid-liquid solution by Raoult's law. (b) A solution for which P_{TOTAL} is larger than the value calculated from Raoult's law. This solution shows a positive deviation from Raoult's law. (c) A solution for which P_{TOTAL} is smaller than the value calculated from Raoult's law. This solution shows a negative deviation from Raoult's law.

In contrast, if two liquids mix endothermically, this indicates that the solute-solvent interactions are weaker than the interactions among the molecules in the pure liquids. More energy is required to expand the liquids than is released when the liquids are mixed. In this case the molecules in the solution have a higher tendency to escape than expected, and *positive* deviations from Raoult's law are observed. An example of this case is provided by a solution of ethanol and hexane, whose Lewis structures are as follows:

```
    H  H                  H  H  H  H  H  H
    |  |  ..              |  |  |  |  |  |
 H—C—C—O—H            H—C—C—C—C—C—C—H
    |  |  ..              |  |  |  |  |  |
    H  H                  H  H  H  H  H  H

   Ethanol                    Hexane
```

The polar ethanol and the nonpolar hexane molecules are not able to interact effectively. Thus the enthalpy of solution is positive, as is the deviation from Raoult's law.

Finally, for a solution of very similar liquids, such as benzene and toluene,

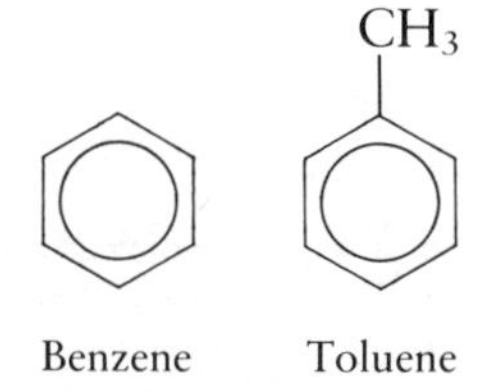

the enthalpy of solution is very close to zero; thus the solution closely obeys Raoult's law (ideal behavior).

A summary of the behavior of various types of solutions is given in Table 17.4.

TABLE 17.4 **Summary of the Behavior of Various Types of Solutions**

Interactive Forces Between Solute (A) and Solvent (B) Particles	ΔH_{soln}	ΔT for Solution Formation	Deviation from Raoult's Law	Example
A⟷A, B⟷B ≡ A⟷B	Zero	Zero	None (ideal solution)	Benzene-toluene
A⟷A, B⟷B < A⟷B	Negative (exothermic)	Positive	Negative	Acetone-water
A⟷A, B⟷B > A⟷B	Positive (endothermic)	Negative	Positive	Ethanol-hexane

17.5 Boiling-Point Elevation and Freezing-Point Depression

In the preceding section we saw how a solute affects the vapor pressure of a liquid solvent. Because changes of state depend on vapor pressure, the presence of a solute also affects the freezing point and boiling point of a solvent. Freezing-point depression, boiling-point elevation, and osmotic pressure (discussed in Section 17.6) are called **colligative properties.** As we will see, they are grouped together because they depend only on the number, and not on the identity, of the solute particles in an ideal solution. Because of their direct relationship to the number of solute particles, the colligative properties are very useful for characterizing the nature of a solute after it is dissolved in a solvent, and for determining the molar masses of substances.

The relationships between vapor pressure and changes of state were discussed in Section 16.10.

Boiling-Point Elevation

The normal boiling point of a liquid is the temperature at which the vapor pressure is equal to 1 atm. We have seen that a nonvolatile solute lowers the vapor pressure of the solvent. Therefore, such a solution must be heated to a higher temperature than the boiling point of the pure solvent to reach a vapor pressure of 1 atm. This means that *a nonvolatile solute elevates the boiling point of the solvent.* Figure 17.12 shows the phase diagram for an aqueous solution containing a nonvolatile solute. Note that the liquid/vapor line is shifted to higher temperatures than those for pure water.

Normal boiling point was defined in Section 16.10.

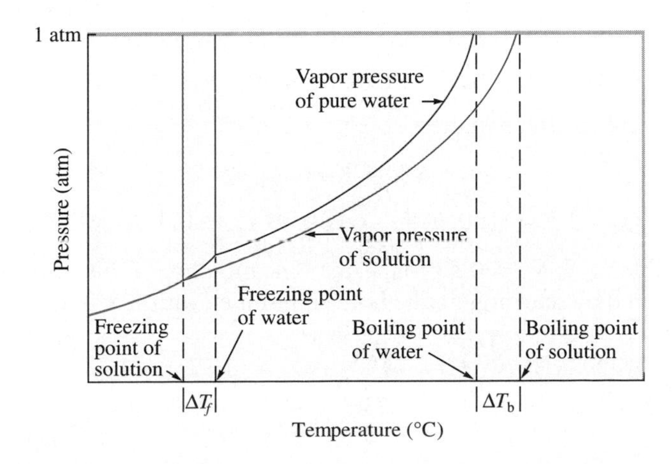

Figure 17.12
Phase diagrams for pure water (red lines) and for an aqueous solution containing a nonvolatile solute (blue lines). Note that the boiling point of the solution is higher than that of pure water. Conversely, the freezing point of the solution is lower than that of pure water. The effect of a nonvolatile solute is to extend the liquid range of a solvent.

TABLE 17.5 Molal Boiling-Point Elevation Constants (K_b) and Freezing-Point Depression Constants (K_f) for Several Solvents

Solvent	Boiling Point (°C)	K_b (°C kg/mol)	Freezing Point (°C)	K_f (°C kg/mol)
Water (H_2O)	100.0	0.51	0.	1.86
Carbon tetrachloride (CCl_4)	76.5	5.03	−22.99	30.
Chloroform ($CHCl_3$)	61.2	3.63	−63.5	4.70
Benzene (C_6H_6)	80.1	2.53	5.5	5.12
Carbon disulfide (CS_2)	46.2	2.34	−111.5	3.83
Ethyl ether ($C_4H_{10}O$)	34.5	2.02	−116.2	1.79
Camphor ($C_{10}H_{16}O$)	208.0	5.95	179.8	40.

As you might expect, the magnitude of the boiling-point elevation depends on the concentration of the solute. The change in boiling point can be represented by the equation

$$\Delta T = K_b m_{\text{solute}}$$

where ΔT is the boiling-point elevation, or the difference between the boiling point of the solution and that of the pure solvent; K_b is a constant that is characteristic of the solvent and is called the **molal boiling-point elevation constant;** and m_{solute} is the *molality* of the solute in the solution. Values of K_b for some common solvents are given in Table 17.5.

The molar mass of a solute can be determined from the observed boiling-point elevation, as shown in Example 17.2.

EXAMPLE 17.2

A solution was prepared by dissolving 18.00 g of glucose in 150.0 g of water. The resulting solution was found to have a boiling point of 100.34°C. Calculate the molar mass of glucose. Glucose is a molecular solid that is present as individual molecules in solution.

Solution

We make use of the equation

$$\Delta T = K_b m_{\text{solute}}$$

where $\Delta T = 100.34°\text{C} - 100.00°\text{C} = 0.34°\text{C}$

From Table 17.5, $K_b = 0.51$ for water. The molality of this solution can be calculated by rearranging the boiling-point elevation equation to give

$$m_{\text{solute}} = \frac{\Delta T}{K_b} = \frac{0.34°\text{C}}{0.51 \dfrac{°\text{C kg}}{\text{mol}}} = 0.67 \text{ mol/kg}$$

The solution was prepared using 0.1500 kg of water. Using the definition of molality, we can find the number of moles of glucose in the solution.

$$m_{\text{solute}} = 0.67 \text{ mol/kg} = \frac{\text{mol solute}}{\text{kg solvent}} = \frac{n_{\text{glucose}}}{0.1500 \text{ kg}}$$

$$n_{\text{glucose}} = (0.670 \text{ mol/kg})(0.1500 \text{ kg}) = 0.10 \text{ mol}$$

Thus 0.10 mol of glucose weighs 18.00 g, and 1.0 mol of glucose weighs 180 g (10 × 18.00 g). The molar mass of glucose is 180 g/mol.

Freezing-Point Depression

Melting point and freezing point both refer to the temperature at which the solid and liquid coexist.

When a solute is dissolved in a solvent, the freezing point of the solution is lower than that of the pure solvent. Why? Recall that the vapor pressures of ice and liquid water are the same at 0°C. Suppose a solute is dissolved in water. The resulting solution does not freeze at 0°C because *the water in the solution has a lower vapor pressure than that of pure ice.* No ice forms under these conditions. However, the vapor pressure of ice decreases more rapidly than that of liquid water as the temperature decreases. Therefore, as the solution is cooled, the vapor pressure of the ice and that of the liquid water in the solution will eventually become equal. The temperature at which this occurs is the new freezing point of the solution and is below 0°C. The freezing point has been *depressed.*

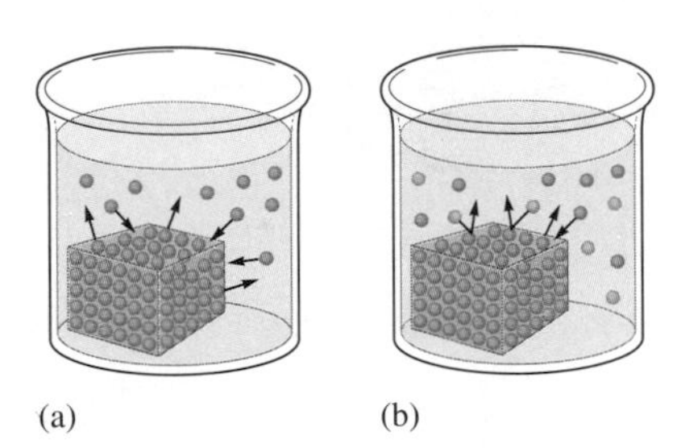

Figure 17.13
(a) Ice in equilibrium with liquid water. (b) Ice in equilibrium with liquid water containing a dissolved solute (shown in orange).

We can account for this behavior in terms of the simple model shown in Fig. 17.13. The presence of the solute lowers the rate at which molecules in the liquid return to the solid state. Thus for an aqueous solution only the liquid state is found at 0°C. As the solution is cooled, the rate at which water molecules leave the solid ice decreases until this rate and the rate of formation of ice become equal and equilibrium is reached. This is the freezing point of the water in the solution.

Because a solute lowers the freezing point of water, compounds such as sodium chloride and calcium chloride are often spread on streets and sidewalks to prevent ice from forming in freezing weather. Of course, if the outside temperature is lower than the freezing point of the resulting salt solution, ice forms anyway. So this procedure is not effective at extremely cold temperatures.

The solid/liquid line for an aqueous solution is shown on the phase diagram for water in Fig. 17.12. Since the presence of a solute elevates the boiling point and depresses the freezing point of the solvent, adding a solute has the effect of extending the liquid range.

The equation for freezing-point depression is analogous to that for boiling-point elevation:

$$\Delta T = K_f m_{\text{solute}}$$

where ΔT is the freezing-point depression, or the difference between the freezing point of the pure solvent and that of the solution; and K_f is a constant that is characteristic of a particular solvent and is called the **molal freezing-point depression constant.** Values of K_f for common solvents are listed in Table 17.5.

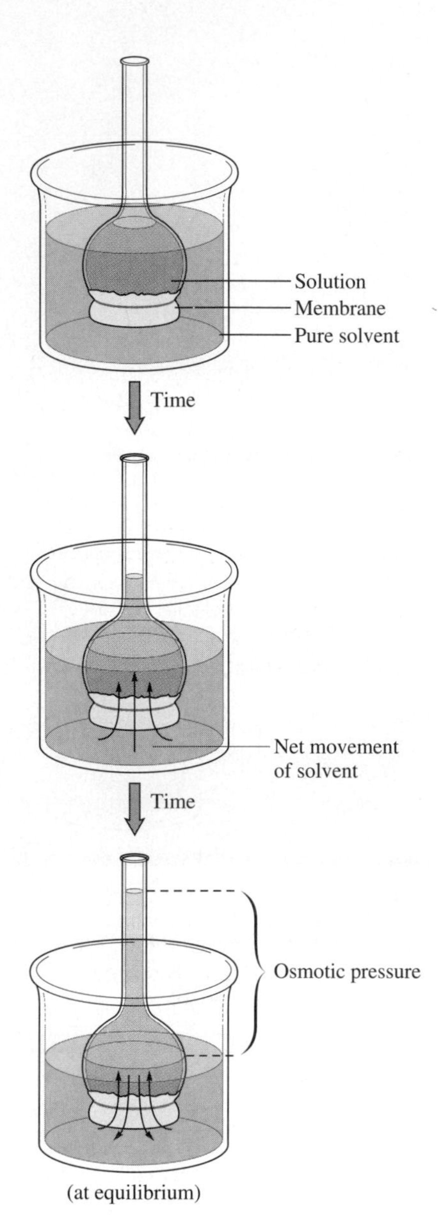

Figure 17.14
A tube with a bulb on the end is covered by a semipermeable membrane. The solution inside the tube is bathed in the pure solvent. There is a net transfer of solvent molecules into the solution until the hydrostatic pressure equalizes the solvent flow in both directions. The solvent level shows little change, because the narrow stem of the bulb has a very small volume.

EXAMPLE 17.3

What mass of ethylene glycol, $C_2H_6O_2$, the main component of antifreeze, must be added to 10.0 L of water to produce a solution for use in a car's radiator that freezes at -10.0°F (-23.3°C)? Assume the density of water is exactly 1 g/mL.

Solution

The freezing point must be lowered from 0°C to -23.3°C. To determine the molality of ethylene glycol needed to accomplish this, we can use the equation

$$\Delta T = K_f m_{\text{solute}}$$

where $\Delta T = 23.3$°C and $K_f = 1.86$ (from Table 17.5). Solving for the molality gives

$$m_{\text{solute}} = \frac{\Delta T}{K_f} = \frac{23.3\text{°C}}{1.86\ \frac{\text{°C kg}}{\text{mol}}} = 12.5 \text{ mol/kg}$$

This means that 12.5 mol of ethylene glycol must be added per kilogram of water. We have 10.0 L, or 10.0 kg, of water. Therefore, the total number of moles of ethylene glycol needed is

$$\frac{12.5 \text{ mol}}{\text{kg}} \times 10.0 \text{ kg} = 1.25 \times 10^2 \text{ mol}$$

The mass of ethylene glycol needed is

$$1.25 \times 10^2 \text{ mol} \times \frac{62.1 \text{ g}}{\text{mol}} = 7.76 \times 10^3 \text{ g (or 7.76 kg)}$$

Like the boiling-point elevation, the observed freezing-point depression can be used to determine molar masses and to characterize solutions.

17.6 Osmotic Pressure

Osmotic pressure, another of the colligative properties, can be understood from Fig. 17.14. A solution and pure solvent are separated by a **semipermeable membrane,** which allows *solvent but not solute* molecules to pass through. As time passes, the volume of the solution increases while that of the solvent decreases. This flow of solvent into the solution through the semipermeable membrane is called **osmosis.** Eventually the liquid levels stop changing, indicating that the system has reached equilibrium. Because the liquid levels are different at this point, there is a greater hydrostatic pressure on the solution than on the pure solvent. This excess pressure is called the **osmotic pressure.** We can take another view of this phenomenon, as illustrated in Fig. 17.15. Osmosis can be prevented by applying a pressure to thc solution. *The pressure that just stops the osmosis is equal to the osmotic pressure of the solution.*

A simple model to explain osmotic pressure can be constructed as shown in Fig. 17.16. The membrane allows only solvent molecules to pass through. However, the initial rates of solvent transfer to and from the solution are not the same. The solute particles interfere with the passage of solvent from the solution, so the rate of transfer is slower from the solution to the solvent than in the reverse direction. Thus there is a net transfer of solvent molecules into the solution, causing the solution volume to increase. As the solution level rises in the tube, the resulting pressure exerts an extra "push" on the solvent molecules in the solution, forcing them back through the membrane. Eventually, enough pressure develops so that the solvent transfer becomes equal in both directions. At this point, equilibrium is achieved and the levels stop changing.

Osmotic pressure can be used to characterize solutions and determine molar masses as can the other colligative properties; however, osmotic pressure is particularly useful because a small concentration of solute produces a relatively large osmotic pressure.

Experiments show that the dependence of the osmotic pressure on solution concentration is represented by the equation

$$\pi = MRT$$

where π is the osmotic pressure in atmospheres, M is the molarity of the solute, R is the gas law constant, and T is the Kelvin temperature.

A molar mass determination using osmotic pressure is illustrated in Example 17.4.

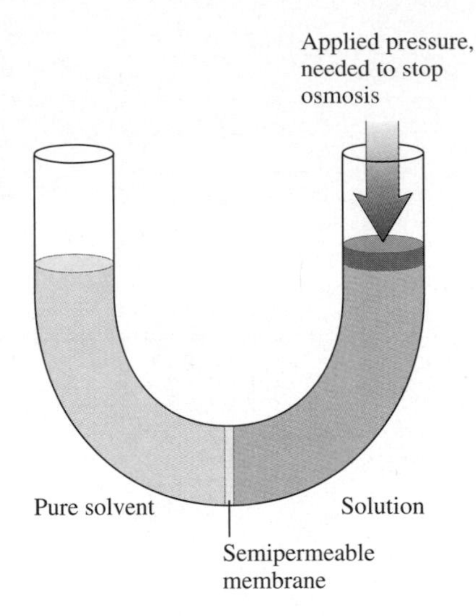

Figure 17.15
The normal flow of solvent into the solution (osmosis) can be prevented by applying an external pressure to the solution. The pressure required to stop the osmosis is equal to the osmotic pressure of the solution.

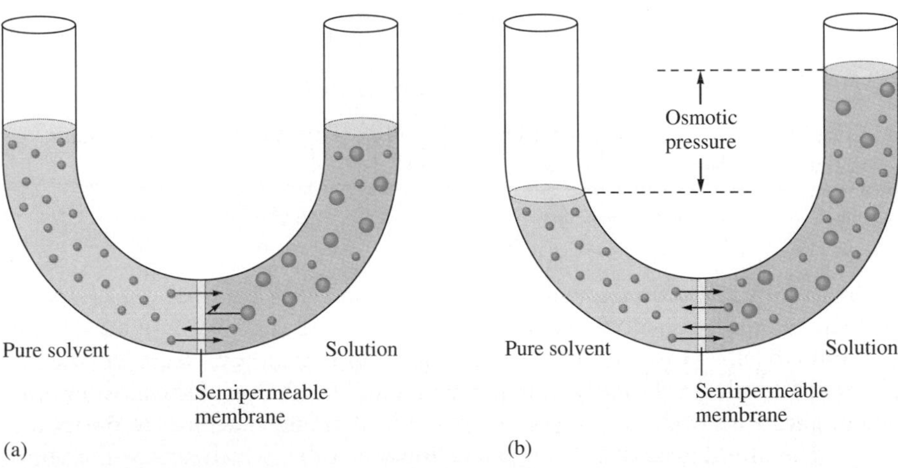

Figure 17.16
(a) A pure solvent and its solution with a nonvolatile solute are separated by a semipermeable membrane through which solvent molecules (blue) can pass but solute molecules (green) cannot. The rate of solvent transfer is greater from solvent to solution than from solution to solvent. (b) The system at equilibrium, where the rate of solvent transfer is the same in both directions.

EXAMPLE 17.4

To determine the molar mass of a certain protein, 1.00×10^{-3} g of the protein was dissolved in enough water to make 1.00 mL of solution. The osmotic pressure of this solution was found to be 1.12 torr at 25.0°C. Calculate the molar mass of the protein.

Solution

We use the equation

$$\pi = MRT$$

In this case

$$\pi = 1.12 \text{ torr} \times \frac{1 \text{ atm}}{760 \text{ torr}} = 1.47 \times 10^{-3} \text{ atm}$$

$$R = 0.08206 \text{ L atm K}^{-1} \text{ mol}^{-1}$$

$$T = 25.0 + 273 = 298 \text{ K}$$

Note that the osmotic pressure must be converted to atmospheres because of the units of R.

Solving for M gives

$$M = \frac{1.47 \times 10^{-3} \text{ atm}}{(0.08206 \text{ L atm K}^{-1} \text{ mol}^{-1})(298 \text{ K})} = 6.01 \times 10^{-5} \text{ mol/L}$$

Since 1.00×10^{-3} g of protein was dissolved in 1 mL of solution, the mass of protein per liter of solution is 1.00 g. The solution's concentration is 6.01×10^{-5} mol/L. This concentration is produced from 1.00×10^{-3} g of protein per milliliter, or 1.00 g/L. Thus 6.01×10^{-5} mol of protein has a mass of 1.00 g, and

$$\frac{1.00 \text{ g}}{6.01 \times 10^{-5} \text{ mol}} = \frac{x \text{ g}}{1.00 \text{ mol}}$$

$$x = 1.66 \times 10^{4} \text{ g/mol}$$

The molar mass of the protein is 1.66×10^4. This molar mass may seem very large, but it is relatively small for a protein.

Measurements of osmotic pressure generally give much more accurate molar mass values than those from freezing-point or boiling-point changes.

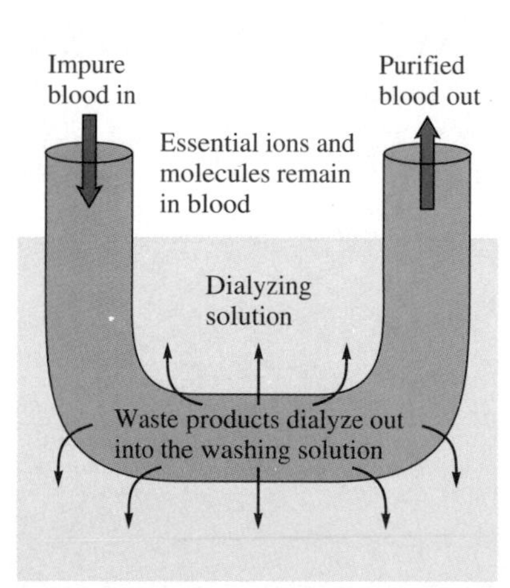

Figure 17.17
Representation of the functioning of an artificial kidney.

In osmosis a semipermeable membrane prevents transfer of *all* solute particles. A similar phenomenon called **dialysis** occurs at the walls of most plant and animal cells. However, in this case the membrane allows transfer of both solvent molecules and *small* solute molecules and ions. One of the most important applications of dialysis is the use of artificial kidney machines to purify the blood. The blood is passed through a cellophane tube, which acts as the semipermeable membrane. The tube is immersed in a dialyzing solution (see Fig. 17.17). This "washing" solution contains the same concentrations of ions and small molecules as blood but has none of the waste products normally removed by the kidneys. The resulting dialysis of waste products cleanses the blood.

Solutions that have identical osmotic pressures are said to be **isotonic solutions.** Fluids administered intravenously must be isotonic with body fluids.

For example, if cells are bathed in a hypertonic solution, which is a solution having an osmotic pressure higher than that of the cell fluids, the cells will shrivel because of a net transfer of water out of the cells. This phenomenon is called *crenation*. The opposite phenomenon, called *lysis,* occurs when cells are bathed in a hypotonic solution, a solution with an osmotic pressure lower than that of the cell fluids. In this case the cells rupture because of the flow of water into the cells.

The brine used in pickling causes the cucumbers to shrivel.

We can use the phenomenon of crenation to our advantage. Food can be preserved by treating its surface with a solute that forms a solution hypertonic to bacteria cells. Bacteria on the food then tend to shrivel and die. This is why salt can be used to protect meat and sugar can be used to protect fruit.

EXAMPLE 17.5

What concentration of sodium chloride in water is needed to produce an aqueous solution isotonic with blood (π = 7.70 atm at 25°C)?

Solution

We can calculate the molarity of the solute from the equation

$$\pi = MRT \qquad \text{or} \qquad M = \frac{\pi}{RT}$$

$$M = \frac{7.70 \text{ atm}}{(0.08206 \text{ L atm K}^{-1} \text{ mol}^{-1})(298 \text{ K})} = 0.315 \text{ mol/L}$$

This represents the total molarity of solute particles. But NaCl gives two ions per formula unit. Therefore, the concentration of NaCl needed is 0.315/2, or 0.158 *M*. That is,

$$\underset{0.1575\,M}{\text{NaCl}} \longrightarrow \underbrace{\underset{0.1575\,M}{\text{Na}^+} + \underset{0.1575\,M}{\text{Cl}^-}}_{0.315\,M}$$

Reverse Osmosis

If a solution in contact with pure solvent across a semipermeable membrane is subjected to an external pressure larger than its osmotic pressure, **reverse osmosis** occurs. The pressure will cause a net flow of solvent from the solution to the solvent, as shown in Fig. 17.18. In reverse osmosis, the semipermeable membrane acts as a "molecular filter" to remove solute particles. This fact is potentially applicable to the **desalination** (removal of dissolved salts) of seawater, which is highly hypertonic to body fluids and thus is not drinkable.

As the population of the sunbelt areas of the United States increases, more demand will be placed on the limited supplies of fresh water there. One obvious source of fresh water is from the desalination of seawater. Various schemes have been suggested, including solar evaporation, reverse osmosis, and even a plan for towing icebergs from Antarctica (remember that pure water freezes out of an aqueous solution). The problem, of course, is that all the available

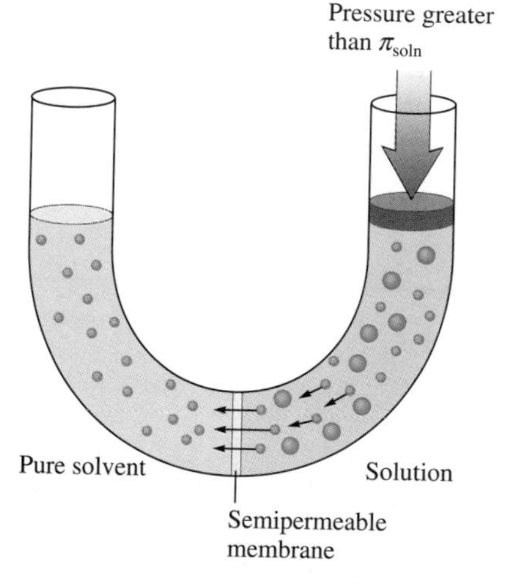

Figure 17.18
Reverse osmosis. A pressure greater than the osmotic pressure of the solution is applied, causing a net flow of solvent molecules (blue) from the solution to the pure solvent. The solute molecules (green) remain behind.

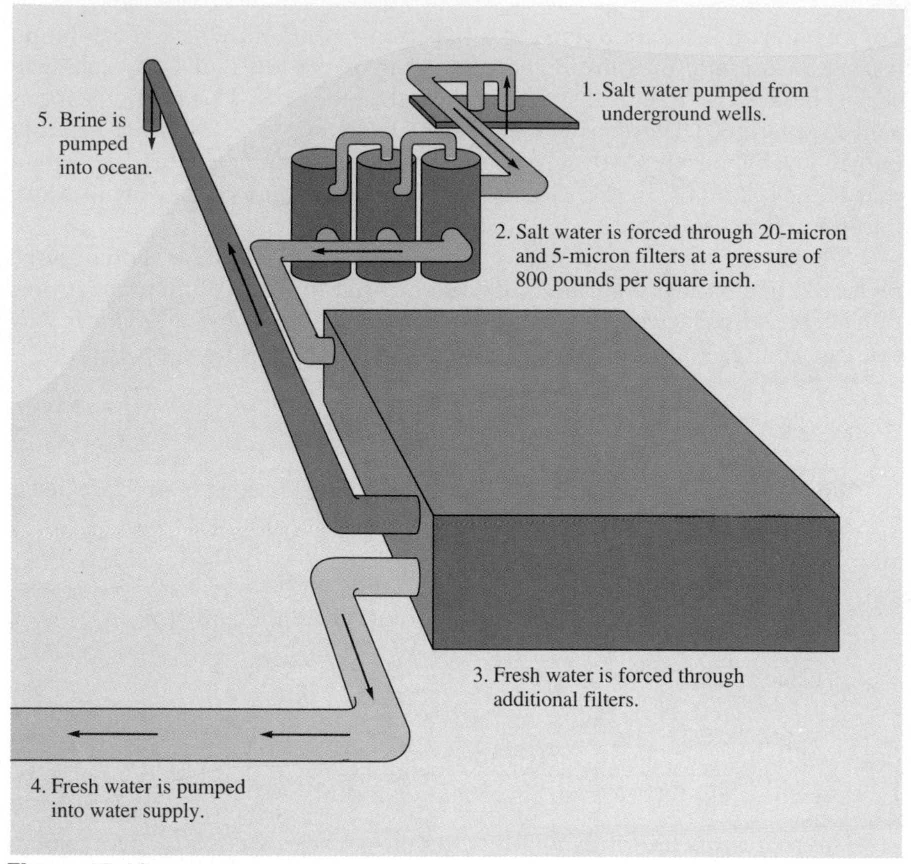
1. Salt water pumped from underground wells.
2. Salt water is forced through 20-micron and 5-micron filters at a pressure of 800 pounds per square inch.
3. Fresh water is forced through additional filters.
4. Fresh water is pumped into water supply.
5. Brine is pumped into ocean.

Figure 17.19
Residents of Catalina Island off the coast of southern California are benefiting from a desalination plant that can supply 132,000 gallons per day, one-third of the island's daily needs.

processes are expensive. However, as water shortages increase, desalination is becoming necessary. For example, the first full-time public desalination plant has just started operations on Catalina Island, just off the coast of California (see Fig. 17.19). This plant, which can produce 132,000 gallons of drinkable water from the Pacific Ocean every day, operates by reverse osmosis. Powerful pumps, developing over 800 lb/in^2 of pressure, are employed to force seawater through synthetic semipermeable membranes.

Catalina Island's plant seems to be just the beginning. The city of Santa Barbara opened a $40 million desalination plant in 1992 that can produce 8 million gallons of drinkable water a day, and other plants are in the planning stages.

A small-scale, manually operated reverse osmosis desalinator has been developed by the U.S. Navy to provide fresh water on life rafts (Fig. 17.20). Potable water can be supplied by this desalinator at the rate of 1.25 gallons of water per hour—enough to keep 25 people alive. This compact desalinator weighs only 10 pounds and will replace the bulky cases of fresh water now stored in Navy life rafts.

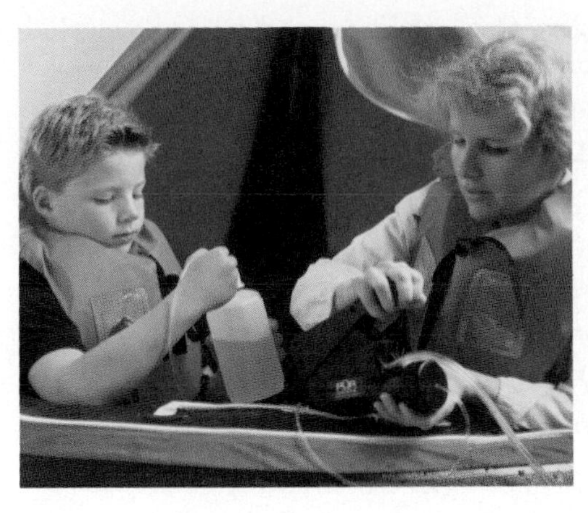

Figure 17.20
A family uses a commerically available desalinator, similar to those developed by the Navy for life rafts. The essential part of these desalinators is a cellophane-like membrane wrapped around a tube with holes in it. Seawater is forced through the tube by a hand-operated pump at pressures of about 70 atm to produce water with a salt content only 40% higher than that from a typical tap.

17.7 Colligative Properties of Electrolyte Solutions

As we have seen, the colligative properties of solutions depend on the total concentration of solute particles. For example, a 0.10 *m* glucose solution shows a freezing-point depression of 0.186°C:

$$\Delta T = K_f m = (1.86°\text{C kg/mol})(0.10 \text{ mol/kg}) = 0.186°\text{C}$$

On the other hand, a 0.10 *m* sodium chloride solution should show a freezing-point depression of 0.37°C, since the solution is 0.10 *m* Na^+ ions and 0.10 *m* Cl^- ions. Therefore, because the solution is 0.20 *m* in solute particles, $\Delta T = (1.86°\text{C kg/mol})(0.20 \text{ mol/kg}) = 0.37°\text{C}$.

The relationship between the moles of solute dissolved and the moles of particles in solution is usually expressed by the **van't Hoff factor** (*i*):

$$i = \frac{\text{moles of particles in solution}}{\text{moles of solute dissolved}}$$

The *expected* value for *i* can be calculated for a salt by noting the number of ions per formula unit. For example, for NaCl, *i* is 2; for K_2SO_4, *i* is 3; and for $Fe_3(PO_4)_2$, *i* is 5. These calculated values assume that when a salt dissolves, it completely dissociates into its component ions, which then move around independently. However, this assumption is not always true. For example, the freezing-point depression observed for 0.10 *m* NaCl is 1.87 times that observed for 0.10 *m* glucose, rather than twice as great. That is, for a 0.10 *m* NaCl solution the observed value for *i* is 1.87 rather than 2. Why? The best explanation is that **ion pairing** occurs in solution (see Fig. 17.21). At a given instant a small percentage of the sodium and chloride ions are paired and thus count as a single particle. In general, ion pairing is most important in concentrated solutions. As the solution becomes more dilute, the ions are farther apart and less ion pairing occurs. For example, in a 0.0010 *m* NaCl solution, the observed value of *i* is 1.97, which is very close to the expected value.

Ion pairing occurs to some extent in all electrolyte solutions. Table 17.6 shows expected and observed values of *i* for a given concentration of various electrolytes. Note that the deviation of *i* from the expected value tends to be greatest when the ions have multiple charges. This is expected because ion pairing ought to be most important for highly charged ions.

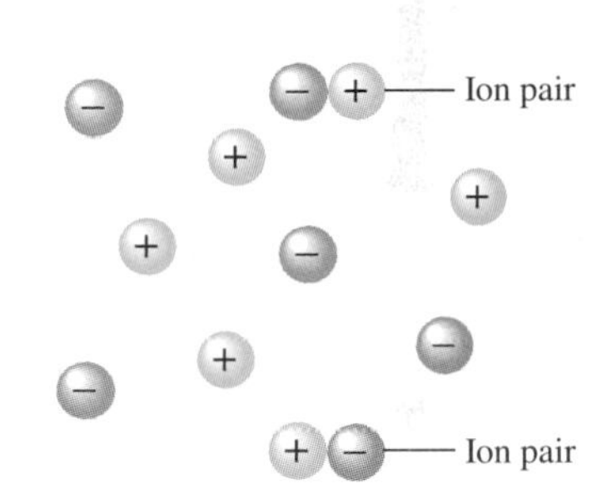

Figure 17.21
In an aqueous solution a few ions aggregate, forming ion pairs that behave as a unit.

TABLE 17.6
Expected and Observed Values of the van't Hoff Factor for 0.05 *m* Solutions of Several Electrolytes

Electrolyte	*i* (expected)	*i* (observed)
NaCl	2.0	1.9
$MgCl_2$	3.0	2.7
$MgSO_4$	2.0	1.3
$FeCl_3$	4.0	3.4
HCl	2.0	1.9
Glucose*	1.0	1.0

* A nonelectrolyte shown for comparison.

Dutch chemist J. H. van't Hoff (1852–1911) received the first Nobel prize in chemistry in 1901.

The colligative properties of electrolyte solutions are described by including the van't Hoff factor in the appropriate equation. For example, for changes in freezing and boiling points the modified equation is

$$\Delta T = imK$$

where K represents the freezing-point depression constant or the boiling-point elevation constant for the solvent.

For the osmotic pressure of electrolyte solutions, the equation is

$$\pi = iMRT$$

EXAMPLE 17.6

The observed osmotic pressure for a 0.10 M solution of $Fe(NH_4)_2(SO_4)_2$ at 25°C is 10.8 atm. Compare the expected and experimental values for i.

Solution

The ionic solid $Fe(NH_4)_2(SO_4)_2$ dissociates in water to produce 5 ions:

$$Fe(NH_4)_2(SO_4)_2 \xrightarrow{H_2O} Fe^{2+} + 2NH_4^{+} + 2SO_4^{2-}$$

Thus the expected value for i is 5. We can obtain the experimental value for i by using the equation for osmotic pressure:

$$\pi = iMRT \qquad \text{or} \qquad i = \frac{\pi}{MRT}$$

where π = 10.8 atm, M = 0.10 mol/L, R = 0.08206 L atm K^{-1} mol^{-1}, and T = 25 + 273 = 298 K. Substituting these values into the equation gives

$$i = \frac{\pi}{MRT} = \frac{10.8\ \text{atm}}{\left(0.10\ \frac{\text{mol}}{\text{L}}\right)\left(0.08206\ \frac{\text{L atm}}{\text{K mol}}\right)(298\ \text{K})} = 4.42$$

The experimental value for i is less than the expected value, presumably because of ion pairing.

17.8 Colloids

Mud can be suspended in water by vigorous stirring. When the stirring stops, most of the particles rapidly settle out; but even after several days some of the smallest particles remain suspended. Although undetected in normal lighting, their presence can be demonstrated by shining a beam of intense light through the suspension. The beam is visible from the side because the light is scattered by the suspended particles. In a true solution, on the other hand, the beam is invisible from the side because the individual ions and molecules dispersed in the solution are too small to scatter visible light.

Ferrofluids: Magnetic Magic

Many of the liquids important in everyday life, such as coffee, wine and gasoline are true solutions but many others are not. For example, liquid soap and most shampoos consist of aqueous suspensions of detergents, and milk is a suspension of protein and fat globules in water. A new type of fluid has been developed for use in industry consisting of tiny particles of iron oxide suspended in water or some other liquid. This "ferrofluid" has unique properties because the iron oxide particles are magnetic and thus the fluid can be controlled by a magnetic field. One very promising application of ferrofluids is for seals on rotating shafts. For example, these "liquid O-ring" seals are used in the manufacture of computer chips. During fabrication the chips are subjected to high temperatures in vacuum chambers. Rotating shafts entering the chambers must be tightly sealed to preserve the vacuum and to eliminate contaminants. Although mechanical seals are available, they produce friction and drag. The liquid seals are constructed by using magnets to hold the ferrofluid in grooves cut into the shaft, as shown on the accompanying diagram. Because the sealing agent is a fluid rather than a solid, it compensates completely for imperfections in the surface of the shaft and thus forms a better seal. This fluid seal also does not wear in the conventional sense. In fact, some seals have been operating in the semiconductor industry for more than ten years. Fluid seals may also be useful in pumps used in refineries and chemical plants. The conventional seals on these pumps, most of which run twenty-four hours a day, may allow enough leakage to contribute significantly to air pollution. Ferrofluidics Corp. has shown that their fluid seals can reduce leakage by a factor of 1000.

Future uses of magnetic fluids may include biomedical applications. For example, it has been proposed that drugs be attached to magnetic particles so that magnets can be used to concentrate the medicine in the part of the body affected by the disease being treated.

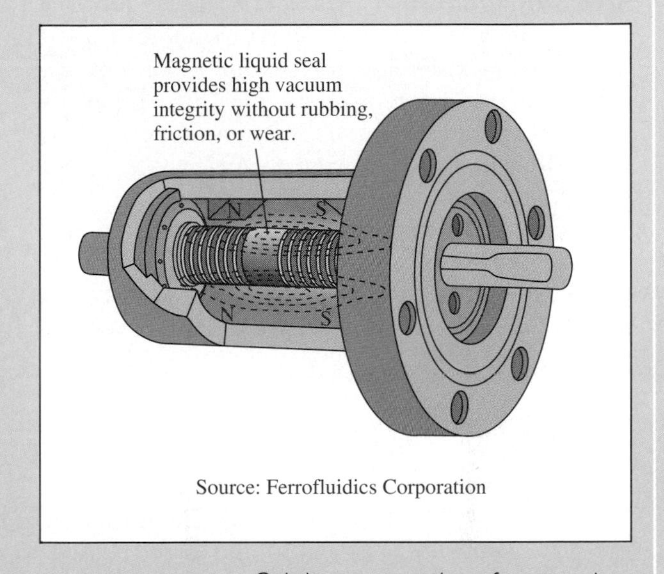
Magnetic liquid seal provides high vacuum integrity without rubbing, friction, or wear.
N
S
N
S

Source: Ferrofluidics Corporation

Subdomain particles of iron oxide suspended in fluid make a highly effective seal. In this mechanism, which is used in chip manufacturing, the magnetic seal prevents contaminants from entering the area where semiconductors are being made.

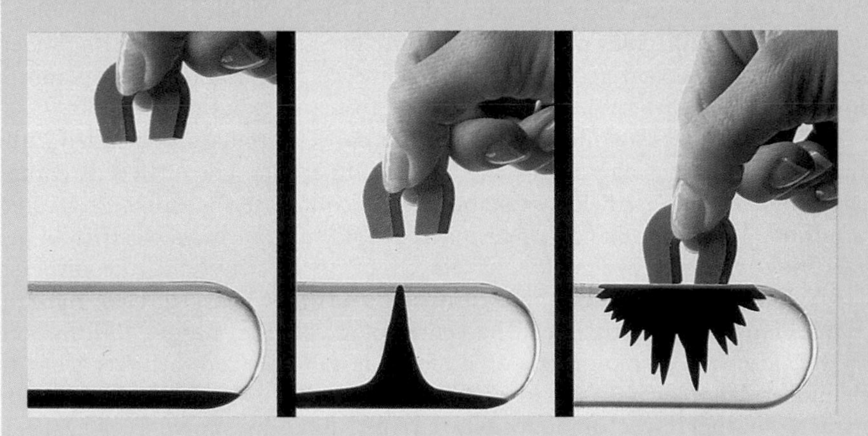

Ferrofluid being attracted by a magnet.

ORGANISMS AND ICE FORMATION

The ice-cold waters of the polar oceans are teeming with fish that seem immune to freezing. One might think that these fish have some kind of antifreeze in their blood. However, studies show that they are protected from freezing in a way that is very different from the way antifreeze protects our cars. As we have seen in this chapter, solutes such as sugars, salts, and ethylene glycol lower the temperature at which the solid and liquid phases of water can coexist. However, the fish could not tolerate high concentrations of solutes in their blood because of the osmotic pressure effects. Instead, they are protected by proteins in their blood. These proteins allow the water in the bloodstream to be supercooled—exist below 0°C—without forming ice. The proteins apparently coat the surface of each tiny ice crystal as soon as it begins to form, thus preventing it from growing to a size that would cause biological damage.

Although it might at first seem surprising, this research on polar fish has attracted the attention of ice cream manufacturers. Premium quality ice cream is smooth; it does not have large ice crystals in it. The makers of ice cream would like to incorporate these polar fish proteins, or at least molecules that behave similarly, into ice cream to prevent growth of ice crystals during storage.

Fruit and vegetable growers have a similar interest: they also want to prevent ice formation that could damage their crops during an unusual cold wave. However, this is a very different kind of problem than keeping polar fish from freezing. Many types of fruits and vegetables are colonized by bacteria that manufacture a protein that *encourages* freezing by acting as a nucleating agent for an ice crystal. Chemists have identified the offending protein in the bacteria as well as the gene that is responsible for making it. They have learned to modify the genetic material of these bacteria in such a way that removes their ability to make the protein that encourages ice crystal formation. If testing shows that these modified bacteria have no harmful effects on the crop or the environment, the original bacteria strain will be replaced with the new form so that ice crystals will not form so readily when a cold snap occurs.

The scattering of light by particles is called the **Tyndall effect** (Fig. 17.22) and is often used to distinguish between a suspension and a true solution.

A suspension of tiny particles in some medium is called a *colloidal dispersion,* or a **colloid.** The suspended particles can be single, large molecules or aggregates of molecules or ions ranging in size from 1 to 1000 nanometers. Colloids are classified according to the states of the dispersed phase and the dispersing medium. Table 17.7 summarizes various types of colloids.

What stabilizes a colloid? Why do the particles remain suspended rather than forming larger aggregates and precipitating out? The answer is complicated, but the main effect that stabilizes the colloid seems to be *electrostatic repulsion.* A colloid, like all other macroscopic substances, is electrically neutral. However, when a colloid is placed in an electric field, the dispersed particles all migrate to the same electrode; thus they must all have the same charge. How is this possible? The center of a colloidal particle (a tiny ionic crystal, a group of molecules, or a single large molecule) attracts from the medium a layer of ions, all of the same charge. This group of ions in turn attracts another layer of oppositely charged ions, as shown in Fig. 17.23. Because the colloidal particles all have an outer layer of ions with the same

Figure 17.22
The Tyndall effect. The yellow solution does not show the path of the light beam whereas the suspension of Iron (III) hydroxide clearly shows the light path.

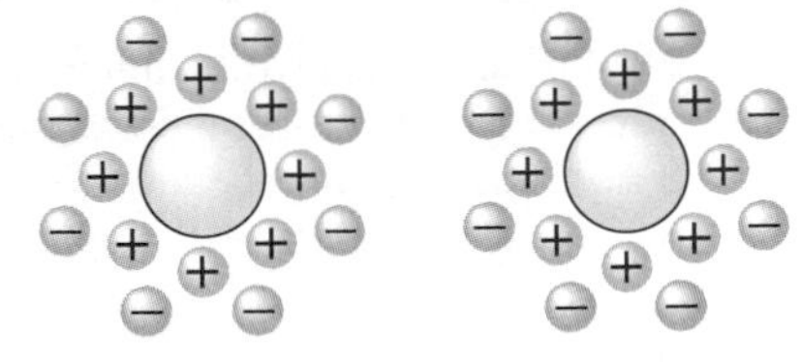

Figure 17.23
A representation of two colloidal particles. In each the center particle is surrounded by a layer of positive ions, with negative ions in the outer layer. Thus although the particles are electrically neutral, they still repel each other because of their outer negative layer of ions.

charge, they repel each other; thus they do not easily aggregate to form particles that are large enough to precipitate.

The destruction of a colloid, called **coagulation,** can usually be accomplished either by heating or by adding an electrolyte. Heating increases the velocities of the colloidal particles, causing them to collide with enough energy so that the ion barriers are penetrated. This allows the particles to aggregate. As this is repeated many times, the particle grows to a point where it settles out. Adding an electrolyte neutralizes the adsorbed ion layers. This is why clay suspended in rivers is deposited where the river reaches the ocean, forming the deltas characteristic of large rivers like the Mississippi. The high salt content of the seawater causes the colloidal clay particles to coagulate.

The removal of soot from smoke is another example of the coagulation of a colloid. When smoke is passed through an electrostatic precipitator (Fig. 17.24), the suspended solids are removed. The use of precipitators has produced an immense improvement in the air quality of heavily industrialized cities.

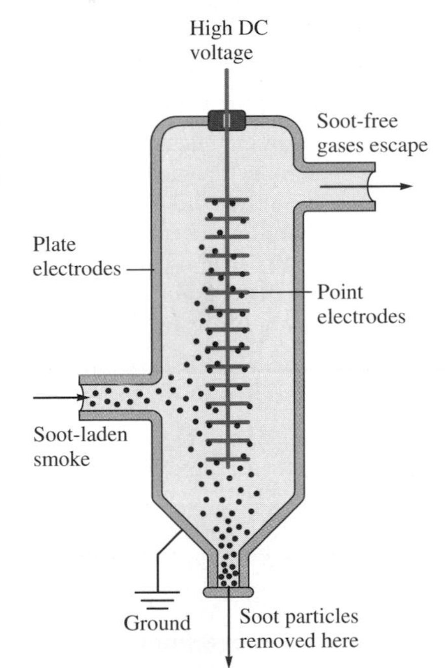

Figure 17.24
The Cottrell precipitator installed in a smokestack. The charged plates attract the colloidal particles because of their ion layers and thus remove them from the smoke.

TABLE 17.7 **Types of Colloids**

Examples	Dispersing Medium	Dispersed Substance	Colloid Type
Fog, aerosol sprays	Gas	Liquid	Aerosol
Smoke, airborne bacteria	Gas	Solid	Aerosol
Whipped cream, soap suds	Liquid	Gas	Foam
Milk, mayonnaise	Liquid	Liquid	Emulsion
Paint, clays, gelatin	Liquid	Solid	Sol
Marshmallow, polystyrene foam	Solid	Gas	Solid foam
Butter, cheese	Solid	Liquid	Solid emulsion
Ruby glass	Solid	Solid	Solid sol

EXERCISES

A blue exercise number indicates that the answer to that exercise appears at the back of this book.

1. A solution is prepared by dissolving 125 g of sucrose, $C_{12}H_{22}O_{11}$, in enough water to produce 1.00 L of solution. What is the molarity of this solution?

2. What mass of sodium oxalate, $Na_2C_2O_4$, is needed to prepare 0.250 L of a 0.100 *M* solution?

3. A solution is prepared by diluting 25.00 mL of a 0.308 *M* solution of $NiCl_2$ to a final volume of 0.500 L. What is the concentration of nickel chloride in this solution? What are the concentrations of nickel ions and chloride ions in this solution?

4. A 158.5-mg sample of pure Cu metal is dissolved in a small amount of concentrated nitric acid, giving Cu^{2+} ions in solution. This solution is then diluted to a final volume of 1.00 L.
   a. What is the molar concentration of Cu^{2+} ions in the final solution?
   b. Assuming the density of the solution is 1.0 g/cm^3, what is the concentration of Cu^{2+} ions in parts per million? (See Exercise 9 in Chapter 4.)

5. An aqueous solution contains 2.8 mg of Cd^{2+} ions per mL of solution.
   a. What is the concentration of Cd^{2+} ions in ppm and ppb? (See Exercise 9 in Chapter 4.)
   b. What is the concentration of Cd^{2+} ions in mol/L?

6. A solution contains 1.06×10^{-3} *M* $Ca(NO_3)_2$.
   a. Determine the molar concentrations of calcium cations and nitrate anions in this solution.
   b. Determine the mass (in grams) of the Ca^{2+} ions in 1.0 mL of this solution.
   c. How many nitrate ions are there in 1.0×10^{-6} L (1.0 μL) of this solution?

7. What volume of 0.25 *M* HCl solution must be diluted to prepare 1.00 L of 0.040 *M* HCl?

8. Write equations representing the dissociation of the following strong electrolytes in water.
   a. HNO_3
   b. Na_2SO_4
   c. $AlCl_3$
   d. $SrBr_2$
   e. $KClO_4$
   f. NH_4Br
   g. NH_4NO_3
   h. $CuSO_4$
   i. NaOH

Concentration of Solutions

9. A commonly purchased disinfectant is a 3.0% (by mass) solution of hydrogen peroxide (H_2O_2) in water. Assuming the density of the solution is 1.0 g/cm^3, calculate the molarity, molality, and mole fraction of H_2O_2.

10. Common commercial acids and bases are aqueous solutions with the following properties:

	Density (g/cm^3)	Mass Percent of Solute
Hydrochloric acid	1.19	38
Nitric acid	1.42	70.
Sulfuric acid	1.84	95
Acetic acid	1.05	99
Ammonia	0.90	28

Calculate the molarity, molality, and mole fraction of each of the above reagents.

11. A solution is prepared by mixing 50.0 mL of toluene ($C_6H_5CH_3$, d = 0.867 g/cm^3) with 125 mL of benzene (C_6H_6, d = 0.874 g/cm^3). Assuming that the volumes add upon mixing, what are the mass percent, mole fraction, molality, and molarity of the toluene in the resulting solution?

12. An aqueous antifreeze solution is 40.0% ethylene glycol ($C_2H_6O_2$) by mass. The density of the solution is 1.05 g/cm^3. Calculate the molality, molarity, and mole fraction of the ethylene glycol.

13. A 1.37 *M* aqueous solution of citric acid ($H_3C_6H_5O_7$) has a density of 1.10 g/cm^3. What are the mass percent, molality, and mole fraction of the citric acid?

14. Does molarity or molality depend on temperature? Explain your answer. Why is molality, not molarity, used in the equations describing freezing-point depression and boiling-point elevation?

15. Determine the molarity and mole fraction of a 1.00 *m* solution of acetone (CH_3COCH_3) dissolved in ethanol (C_2H_5OH). (Density of acetone = 0.788 g/cm^3; density of ethanol = 0.789 g/cm^3.) Assume that the final volume equals the sum of the volumes of acetone and ethanol.

16. In a solution of acetone and ethanol the mole fraction of acetone is 0.40. What is the concentration of acetone in terms of mass percent?

17. A solution is made by dissolving 25 g of NaCl in enough water to make 1.0 L of solution. Assume the density of the solution is 1.0 g/cm^3. Calculate the mass percent, molarity, molality, and mole fraction of NaCl.

18. A solution is prepared by dissolving 50.0 g of cesium chloride (CsCl) in 50.0 g of water. The density of the solution is 1.58 g/cm^3. What are the mass percent, molarity, molality, and mole fraction of the cesium chloride?

Thermodynamics of Solutions and Solubility

19. What specific feature of chemical substances do we refer to when we say that "like dissolves like"?

20. Rationalize the trend in water solubility for the following simple alcohols.

Alcohol	Solubility (g/100 g H_2O at 20°C)
Methanol, CH_3OH	Soluble in all proportions
Ethanol, CH_3CH_2OH	Soluble in all proportions
Propanol, $CH_3CH_2CH_2OH$	Soluble in all proportions
Butanol, $CH_3(CH_2)_2CH_2OH$	8.14
Pentanol, $CH_3(CH_2)_3CH_2OH$	2.62
Hexanol, $CH_3(CH_2)_4CH_2OH$	0.59
Heptanol, $CH_3(CH_2)_5CH_2OH$	0.09

21. Which ion in each of the following pairs would you expect to be more strongly hydrated? Why?
   a. Na^+ or Mg^{2+}
   b. Mg^{2+} or Be^{2+}
   c. Fe^{2+} or Fe^{3+}
   d. F^- or Br^-
   e. Cl^- or ClO_4^-
   f. ClO_4^- or SO_4^{2-}

22. The lattice energy* of KCl is −715 kJ/mol, and the enthalpy of hydration is −684 kJ/mol. Calculate the enthalpy of solution per mole of solid KCl. Describe the process to which this enthalpy change applies.

23. Use the following data to calculate the enthalpy of hydration for cesium iodide and cesium hydroxide.

	Lattice Energy*	ΔH_{soln}
CsI(*s*)	−604 kJ/mol	33 kJ/mol
CsOH(*s*)	−724 kJ/mol	−72 kJ/mol

24. From your answers to Exercise 23, which ion, OH^- or I^-, is more strongly hydrated?

25. What factors cause one solute to be more strongly hydrated than another? For each of the following pairs, predict which substance would be more soluble in water.
   a. CH_3CH_2OH or $CH_3CH_2CH_3$
   b. $CHCl_3$ or CCl_4
   c. CH_3CO_2H or $CH_3(CH_2)_{14}CO_2H$
   d. Na_2S or CuS
   e. $AlCl_3$ or Al_2O_3
   f. CO_2 or SiO_2

*Lattice energy was defined in Chapter 13 as the energy change for the process $M^+(g) + X^-(g) \rightarrow MX(s)$.

26. Which solvent, water or carbon tetrachloride, would you choose to dissolve each of the following?
   a. $Cu(NO_3)_2$
   b. CS_2
   c. CH_3CO_2H
   d. $CH_3(CH_2)_{16}CH_2OH$
   e. HCl
   f. C_6H_6

27. Although $Al(OH)_3$ is insoluble in water, NaOH is very soluble. Explain this difference in terms of lattice energies.

28. In the flushing and cleaning of columns used in liquid chromatography, a series of solvents is utilized. Hexane, chloroform, methanol, and water are passed through the column in that order. Rationalize the order in terms of intermolecular forces and the mutual solubility (miscibility) of the solvents.

29. A typical detergent is sodium dodecylsulfate, or SDS, $CH_3(CH_2)_{10}CH_2SO_4^-Na^+$. In aqueous solution small aggregates of detergent anions called *micelles* form. Propose a structure of the micelles.

30. Define *hydrophobic* and *hydrophilic*.

31. Is Henry's law valid for the solubility of CO_2 in a basic solution? (*Hint:* $CO_2 + OH^- \longrightarrow HCO_3^-$.)

32. Calculate the solubility of O_2 in water at 25°C at a partial pressure of O_2 of 120 torr. The Henry's law constant for O_2 is 7.8×10^2 atm L/mol for Henry's law in the form $P = kC$, where C is the gas concentration (mol/L).

33. Rationalize the temperature dependence of the solubility of a gas in terms of the kinetic molecular theory.

Vapor Pressures of Solutions

34. A solution is prepared by mixing 50.0 g of glucose ($C_6H_{12}O_6$) with 600.0 g of water. What is the vapor pressure of this solution at 25°C? (At 25°C the vapor pressure of pure water is 23.8 torr.)

35. A solution is made by mixing 50.0 g of acetone (CH_3COCH_3) with 50.0 g of methanol (CH_3OH). What is the vapor pressure of this solution at 25°C? What is the composition of the vapor expressed as a mole fraction? Assume ideal solution and gas behavior. (At 25°C the vapor pressures of pure acetone and pure methanol are 271 and 143 torr, respectively.) Is it reasonable to assume that this is an ideal solution? Why or why not?

36. Which of the following will have the lowest total vapor pressure at 25°C?
   a. pure water
   b. a solution of glucose in water with $\chi_{glucose} = 0.01$
   c. a solution of sodium chloride in water with $\chi_{NaCl} = 0.01$
   d. a solution of methanol in water with $\chi_{CH_3OH} = 0.2$ [Consider the vapor pressure of both methanol (see Exercise 35) and water.]

37. In terms of intermolecular forces, what gives rise to positive and negative deviations from Raoult's law?

38. If a solution shows positive deviations from Raoult's law, would you expect it to have a higher or a lower boiling point than if it were ideal? Why?

39. For each case, choose the pair of substances you would expect to give the more nearly ideal solution. Explain your choices.
    a. CF_3CF_3 and $CF_3CF_2CF_3$ or H_2O and $CH_3\overset{\overset{O}{\|}}{C}CH_3$
    b. $CH_3(CH_2)_5CH_3$ and $CH_3(CH_2)_4CH_3$ or CHF_3 and $CH_3{-}O{-}CH_3$
    c. H_3PO_4 and H_2O or CCl_4 and CF_4

40. A solution is made by dissolving 25.8 g of urea, (CH_4N_2O) a nonelectrolyte, in 275 g of water. Calculate the vapor pressures of this solution at 25°C and 45°C. (The vapor pressure of pure water is 23.8 torr at 25°C and 71.9 torr at 45°C.)

41. Benzene and toluene form ideal solutions. Consider a solution of benzene and toluene prepared at 25°C. Assuming the mole fractions of benzene and toluene in the vapor phase are equal, calculate the composition of the solution. At 25°C the vapor pressures of benzene and toluene are 95 and 28 torr, respectively.

42. Pentane, C_5H_{12}, and hexane, C_6H_{14}, combine to form an ideal solution. At 25°C the vapor pressures of pentane and hexane are 511 and 150. torr, respectively. A solution is prepared by mixing 25 mL of pentane (density = 0.63 g/mL) with 45 mL of hexane (density = 0.66 g/mL).
    a. What is the vapor pressure of this solution?
    b. What is the mole fraction of pentane in the vapor that is in equilibrium with this solution?

43. What is the composition of a pentane-hexane solution that has a vapor pressure of 350. torr at 25°C? What is the composition of the vapor in equilibrium with this solution? (See Exercise 42 for vapor pressures of pure hexane and pentane.)

Colligative Properties

44. A solution is prepared by dissolving 4.9 g of sucrose ($C_{12}H_{22}O_{11}$) in 175 g of water. Calculate the boiling point, freezing point, and osmotic pressure of this solution at 25°C. Sucrose is a nonelectrolyte. Assume that the molality and molarity of this solution are equal.

45. Calculate the freezing-point depression and osmotic pressure at 25°C of an aqueous solution of 1.0 g/L of a protein (molar mass = 9.0×10^4 g/mol) if the density of the solution is 1.0 g/cm^3.

46. Considering your answer to Exercise 45, which colligative property, freezing-point depression or osmotic pressure, would be better for determining the molar masses of large molecules. Explain your answer.

47. A 20.0-mg sample of a protein is dissolved in water to make 25.0 mL of solution. The osmotic pressure of the solution is 0.56 torr at 25°C. What is the molar mass of the protein?

48. Reserpine is a natural product isolated from the roots of the shrub *Rauwolfia serpentina*. It was first synthesized in 1956 by Nobel prize winner R. B. Woodward. It is used as a tranquilizer and sedative. When 1.00 g of reserpine is dissolved in 25.0 g of camphor, the freezing-point depression is 2.63°C (K_f for camphor is 40.°C kg/mol). Calculate the molality of the solution and the molar mass of reserpine.

49. Cellophane is a form of cellulose that has undergone the viscose process. The structure of the cellophane polymer is

```
        CH2OH
        |
        CH—O
       /    \
  —CH        CHO—
     \      /
      CH—CH
      |   |
      OH  OH     ]n
```

What structural features allow cellophane to act as a semipermeable membrane for water? (*Hint:* what types of intermolecular forces are possible between cellophane and water?)

50. A 1.60-g sample of a mixture of naphthalene ($C_{10}H_8$) and anthracene ($C_{14}H_{10}$) is dissolved in 20.0 g of benzene (C_6H_6). The freezing point of the solution is 2.81°C. What is the composition of the sample mixture in terms of mass percent? The freezing point of benzene is 5.51°C, and K_f is 5.12°C kg/mol.

51. If the fluid inside a tree is approximately 0.1 *M* more concentrated in solute than the groundwater that bathes the roots, how high will a column of fluid rise in the tree? Assume the density of the fluid is 1.0 g/cm^3. (The density of mercury is 13.6 g/cm^3.)

52. Before refrigeration was common, many foods were preserved by salting them heavily. Fruits were preserved by mixing them with a large amount of sugar (fruit preserves). How do salt and sugar act as preservatives?

53. What volume of ethylene glycol ($C_2H_6O_2$) must be added to 15.0 L of water to produce an antifreeze solution with a freezing point of −30.0°C? What is the boiling point of this solution? (The density of ethylene glycol is 1.11 g/cm^3.)

Properties of Electrolyte Solutions

54. Seawater is approximately 0.6 *M* NaCl. What is the minimum pressure that must be applied at 25°C to purify seawater by reverse osmosis?

55. Consider the following:

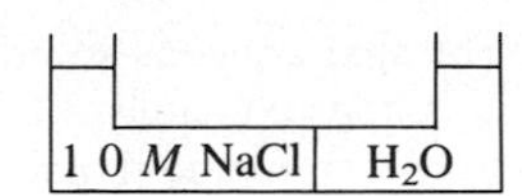

What would happen to the level of liquid in the two arms if the semipermeable membrane was permeable to the following?
a. H_2O only
b. H_2O, Na^+, and Cl^-

56. Place the following solutions in order of increasing freezing-point depression.
a. 1.0 *m* glucose in water
b. 1.0 *m* NaCl in water
c. 1.0 *m* HOCl in water (HOCl is a weak acid.)
d. 1.0 *m* $MgCl_2$ in water

57. From these liquids,

pure water
solution of sucrose (χ = 0.01) in water
solution of NaCl (χ = 0.01) in water
solution of $CaCl_2$(χ = 0.01) in water

choose the one with the following:
a. highest freezing point
b. lowest freezing point
c. highest boiling point
d. lowest boiling point
e. highest osmotic pressure

58. Calculate the freezing point of a 0.5 *m* solution of $Ca(NO_3)_2$ in water. (Assume i = 3.0.)

59. Would you expect the measured freezing point of 0.5 *m* $Ca(NO_3)_2$ to be higher or lower than the value you calculated in Exercise 58? Explain.

60. In some regions of the southwest United States, the water is very hard. For example, in Las Cruces, New Mexico, the tapwater contains about 560 μg of dissolved solids per milliliter. Reverse osmosis units are marketed in this area to soften water. A typical unit exerts a pressure of 8.0 atm and can produce 45 L of water per day.
a. Assuming all of the dissolved solids are $MgCO_3$ and assuming a temperature of 27°C, what total volume of water must be processed to produce 45 L of pure water?
b. Would the same system work for purifying seawater? (Assume seawater is approximately 0.60 *M* NaCl.)

61. Why is the observed freezing-point depression for electrolyte solutions sometimes less than the calculated value? Is the error greater for concentrated or dilute solutions?

62. Use the following data for three different solutions of $CaCl_2$ to calculate the apparent value of the van't Hoff factor.

Molality	Freezing-Point Depression (°C)
0.091	0.440
0.279	1.330
0.475	2.345

63. The freezing-point depression of a 0.091 *m* solution of CsCl is 0.302°C. The freezing-point depression of a 0.091 *m* solution of $CaCl_2$ is 0.440°C. In which solution does ion association (ion-pairing) appear to be greater? Explain your answer.

64. In the winter of 1994, record low temperatures were registered throughout the United States. For example, in Champaign, Illinois, a record low of −29°F was registered. At this temperature can salting icy roads with $CaCl_2$ be effective in melting the ice?
a. Assume i = 3 for $CaCl_2$.
b. Assume the average value of i from Exercise 62.
(The solubility of $CaCl_2$ in cold water is 74.5 g per 100.0 mL of water.)

Additional Exercises

65. The solubility of benzoic acid,

C—O—H
‖
O

is 0.34 g/100 mL in water at 25°C and 10.0 g/100 mL in benzene (C_6H_6) at 25°C. Rationalize this solubility behavior. For a 1.0 molal solution of benzoic acid in benzene, would the measured freezing-point depression be equal to, greater than, or less than 5.12°C? (K_f = 5.12°C kg/mol for benzene.)

66. Would benzoic acid be more or less soluble in a basic aqueous solution than it is in water?

67. Specifications for lactated Ringer's solution, which is used for intravenous (IV) injections, are as follows for each 100. mL of solution:

285–315 mg Na^+
14.1–17.3 mg K^+
4.9–6.0 mg Ca^{2+}
368–408 mg Cl^-
231–261 mg lactate, $C_3H_5O_3^-$

a. Specify the amounts of NaCl, KCl, $CaCl_2 \cdot 2H_2O$, and $NaC_3H_5O_3$ needed to prepare 100. mL of lactated Ringer's solution.
b. What is the range of the osmotic pressure of the solution at 37°C, given the above specifications?

68. When pure methanol is mixed with water, the solution gets warmer to the touch. Would you expect this solution to be ideal? Explain.

69. In the vapor over a pentane-hexane solution at 25°C, the mole fraction of pentane is equal to 0.15. What is the mole fraction of pentane in the solution? (See Exercise 42 for the vapor pressures of the pure liquids.)

70. The normal boiling point of methanol is 64.7°C. A solution containing a nonvolatile solute in methanol has a vapor pressure of 710.0 torr at 64.7°C. What is the mole fraction of methanol in this solution?

71. A solid consists of a mixture of $NaNO_3$ and $Mg(NO_3)_2$. When 6.50 g of this solid is dissolved in 50.0 g of water, the freezing point is lowered by 5.40°C. What is the composition of the solid (by mass)?

72. Anthraquinone contains only carbon, hydrogen, and oxygen. When 4.80 mg of anthraquinone is burned, 14.22 mg of CO_2 and 1.66 mg of H_2O are produced. The freezing point of camphor is lowered by 22.3°C when 1.32 g of anthraquinone is dissolved in 11.4 g of camphor. Calculate the empirical and molecular formulas of anthraquinone.

73. A forensic chemist is given a white solid that is suspected of being pure cocaine ($C_{17}H_{21}NO_4$, molar mass = 303.35). She dissolves 1.22 ± 0.01 g of the solid in 15.60 ± 0.01 g of benzene. The freezing point is lowered by 1.32 ± 0.04°C.
   a. What is the molar mass of the substance? Assuming that the percent of uncertainty in the calculated molar mass is the same as the percent of uncertainty in the temperature change, calculate the uncertainty in the molar mass.
   b. Could the chemist unequivocally state that the substance is cocaine? For example, is the uncertainty small enough to distinguish cocaine from codeine ($C_{18}H_{21}NO_3$, molar mass = 299.36)?
   c. Assuming the absolute uncertainties in the measurements of temperature and mass remain unchanged, how could the chemist improve the precision of her results?

74. A 0.15-g sample of a purified protein is dissolved in water to give 2.0 mL of solution. The osmotic pressure is found to be 18.6 torr at 25°C. Calculate the protein's molar mass.

75. An unknown compound contains only carbon, hydrogen, and oxygen. Combustion analysis of the compound gives mass percents of 31.57% C and 5.30% H. The molar mass is determined by measuring the freezing-point depression of an aqueous solution. A freezing point of −5.20°C is recorded for a solution made by dissolving 10.56 g of the compound in 25.0 mL of water. Determine the empirical formula, molar mass, and molecular formula of the compound. Assume the compound is a nonelectrolyte.

76. The most concentrated aqueous solution of NaOH that can be prepared is approximately 50% NaOH by mass. Calculate the mole fraction and molality of NaOH in this solution.

77. A bottle of wine contains 12.5% ethanol by volume. The density of ethanol (C_2H_5OH) is 0.79 g/cm^3. Calculate the concentration of ethanol in wine in terms of mass percent and molality.

78. In a lab you need 100 mL of each of the following solutions. Explain how you would proceed by using the given information.
   a. 2.0 *m* KCl in water (density unknown to you)
   b. 15% NaOH by mass in water (density unknown to you)
   c. 25% NaOH by mass in CH_3OH (d = 0.79 g/cm^3)
   d. 30% C_2H_5OH by volume in water

79. Calculate the freezing point and boiling point of an antifreeze solution that is 40.0% by mass of ethylene glycol ($HOCH_2CH_2OH$) in water.

80. How would you prepare 1.0 L of an aqueous solution of sucrose ($C_{12}H_{22}O_{11}$) having an osmotic pressure of 15 atm at 25°C?

81. The freezing point of *t*-butanol is 25.50°C and K_f is 9.1°C kg/mol. Usually *t*-butanol absorbs water on exposure to air. If the freezing point of a 10.0-g sample of *t*-butanol is 24.59°C, how many grams of water are present in the sample?

82. Calculate the freezing point of each of the following solutions. (For parts b and c, assume complete dissociation.)
   a. 5.0 g of glucose ($C_6H_{12}O_6$) in 25 g of H_2O
   b. 5.0 g of NaCl in 25 g of H_2O
   c. 2.0 g of $Al(NO_3)_3$ in 15 g of H_2O
   d. 1.0 g of benzoic acid ($C_6H_5CO_2H$) in 10.0 g benzene

83. In Exercise 61 in Chapter 5 you calculated the pressure of CO_2 in a bottle of sparkling wine, assuming that the CO_2 was insoluble in water. This was a bad assumption. Redo this problem by assuming that CO_2 obeys Henry's law. Use the data given in that problem to calculate the partial pressure of CO_2 in the gas phase and the solubility of CO_2 in the wine at 25°C. The Henry's law constant for CO_2 is 32 L atm/mol at 25°C with Henry's law in the form $P = kC$, where C is the concentration of the gas in mol/L.

84. The molar mass of a nonelectrolyte is 58.0 g/mol. Compute the boiling point of a solution containing 24.0 g of this compound and 600. g of water if the barometric pressure is such that pure water boils at 99.725°C.

85. A solution contains 3.75 g of a nonvolatile pure hydrocarbon in 95 g of acetone. The boiling points of pure acetone and the solution are 55.95°C and 56.50°C, respectively.

The molal boiling-point constant of acetone is 1.71°C kg/mol. What is the approximate molar mass of the hydrocarbon?

86. The vapor pressure of water at 28.0°C is 28.35 torr. Compute the vapor pressure at 28.0°C of a solution containing 68.0 g of sucrose, $C_{12}H_{22}O_{11}$, and 1000. g of water.

87. Calculate the osmotic pressure at 17.0°C of an aqueous solution containing 1.75 g of sucrose, $C_{12}H_{22}O_{11}$, per 150. mL of solution.

88. An aqueous solution of urea had a freezing point of −0.52°C. Predict the osmotic pressure of the same solution at 37°C. Assume that the molarity and the molality are the same.

89. Explain the following on the basis of the behavior of atoms and/or ions.
   a. Cooking with hot water is faster in a pressure cooker than in an open pan.
   b. Salt is used on icy roads.
   c. Melted sea ice from the Arctic Ocean produces fresh water.
   d. $CO_2(s)$ (dry ice) does not have a normal boiling point under normal atmospheric conditions, even though CO_2 is a liquid in fire extinguishers.

90. A 0.100-g sample of the weak acid HA (molar mass = 100.0 g/mol) is dissolved in 500.0 g of water. The freezing point of the resulting solution is −0.0056°C. Calculate the value of K_a for this acid.

91. Consider an aqueous solution containing sodium chloride that has a density of 1.01 g/mL. Assume the solution behaves ideally. The freezing point of this solution at 1.0 atm is −1.28°C. Calculate the percent composition of this solution (by mass).

92. A sample containing 0.0500 mol of $Fe_2(SO_4)_3$ is dissolved in enough water to make 1.00 L of solution. This solution contains hydrated SO_4^{2-} and $Fe(H_2O)_6^{3+}$ ions. The latter behave as an acid:

$$Fe(H_2O)_6^{3+} \rightleftharpoons Fe(OH)(H_2O)_5^{2+} + H^+$$

   a. Calculate the expected osmotic pressure of this solution at 25°C if the above dissociation is negligible.
   b. The actual osmotic pressure of the solution is 6.73 atm at 25°C. Calculate K_a for the dissociation reaction of $Fe(H_2O)_6^{3+}$. (To do this calculation, you must assume that none of the ions goes through the semipermeable membrane. Actually, this is not a good assumption for the tiny H^+ ion.)

93. At 25°C the vapor in equilibrium with a solution containing carbon disulfide and acetonitrile has a total pressure of 263 torr and is 85.5 mole percent carbon disulfide. What is the mole fraction of carbon disulfide in the solution? At 25°C the vapor pressure of carbon disulfide is 375 torr. Assume the solution and the vapor exhibit ideal behavior.

94. An unknown solid contains $MgCl_2$ (molar mass = 95.218 g/mol) and NaCl (molar mass = 58.443 g/mol). When 0.5000 g of this solid is dissolved in enough water to form 1.000 L of solution, the osmotic pressure at 25.0°C is observed to be 0.3950 atm. What is the mass percent of $MgCl_2$ in the solid? (Assume ideal behavior for the solution.)

95. The compound VCl_4 undergoes dimerization in solution:

$$2VCl_4 \rightleftharpoons V_2Cl_8$$

When 6.6834 g of VCl_4 is dissolved in 100.0 g of carbon tetrachloride, the freezing point is lowered by 5.97°C. Calculate the value of the equilibrium constant for the dimerization of VCl_4 at this temperature. (The density of the equilibrium mixture is 1.696 g/cm³ and K_f = 29.8°C kg/mol for CCl_4.)

CHAPTER 18

The Representative Elements: Groups 1A Through 4A

So far in this book we have covered the major principles and explored the most important models of chemistry. In particular, we have seen that the chemical properties of the elements can be explained very successfully by the quantum mechanical model of the atom. In fact, the most convincing evidence of that model's validity is its ability to relate the observed periodic properties of the elements to the number of valence electrons in their atoms.

We have learned many properties of the elements and their compounds, but we have not discussed extensively the relationship between the chemical properties of a specific element and its position on the periodic table. In this chapter and the next, we will explore the chemical similarities and differences among the elements in the several groups of the periodic table and will try to interpret these data using the wave mechanical model of the atom. In the process we will illustrate a great variety of chemical properties and further demonstrate the practical importance of chemistry.

18.1 A Survey of the Representative Elements

The traditional form of the periodic table is shown in Fig. 18.1. Recall that the **representative elements,** whose chemical properties are determined by the valence-level *s* and *p* electrons, are designated Groups 1A through 8A. The **transition metals,** in the center of the table, result from the filling of *d* orbitals. The elements that correspond to the filling of the 4*f* and 5*f* orbitals are listed separately as the **lanthanides** and **actinides,** respectively.

1A	2A											3A	4A	5A	6A	7A	8A
H																	He
Li	Be											B	C	N	O	F	Ne
Na	Mg											Al	Si	P	S	Cl	Ar
K	Ca	Sc	Ti	V	Cr	Mn	Fe	Co	Ni	Cu	Zn	Ga	Ge	As	Se	Br	Kr
Rb	Sr	Y	Zr	Nb	Mo	Tc	Ru	Rh	Pd	Ag	Cd	In	Sn	Sb	Te	I	Xe
Cs	Ba	La	Hf	Ta	W	Re	Os	Ir	Pt	Au	Hg	Tl	Pb	Bi	Po	At	Rn
Fr	Ra	Ac	Rf	Ha	Unh	Uns		Une									

Lanthanides	Ce	Pr	Nd	Pm	Sm	Eu	Gd	Tb	Dy	Ho	Er	Tm	Yb	Lu
Actinides	Th	Pa	U	Np	Pu	Am	Cm	Bk	Cf	Es	Fm	Md	No	Lr

Figure 18.1 The periodic table. The elements in the A groups are the representative elements. The elements shown in red are called transition metals. The dark line approximately separates the nonmetals from the metals. The elements that have both metallic and nonmetallic properties (semimetals) are shaded in blue.

The heavy black line in Fig. 18.1 separates the metals from the nonmetals, except for one case. Hydrogen, which appears on the metal side, is a nonmetal. Some elements just on either side of this line, such as silicon and germanium, exhibit both metallic and nonmetallic properties. These elements are often called **metalloids,** or **semimetals.** The fundamental chemical difference between metals and nonmetals is that metals tend to lose their valence electrons to form *cations,* which usually have the valence electron configuration of the noble gas from the preceding period. On the other hand, nonmetals tend to gain electrons to form *anions* that exhibit the electron configuration of the noble gas in the same period. Metallic character is observed to increase in going down a given group, which is consistent with the trends in ionization energy, electron affinity, and electronegativity discussed earlier (see Sections 12.15 and 13.2).

Metallic character increases going down a group in the periodic table.

Atomic Size and Group Anomalies

Although the chemical properties of the members of a group have many similarities, there are also important differences. The most dramatic differences usually occur between the first and second member. For example, hydrogen in Group 1A is a nonmetal, while lithium is a very active metal. This extreme difference results primarily from the very large difference in the atomic radii of hydrogen and lithium, as shown in Fig. 18.2. Since the small hydrogen atom has a much greater attraction for electrons than do the larger members of Group 1A, it forms covalent bonds with nonmetals. In contrast, the other members of Group 1A lose their valence electrons to nonmetals to form 1+ cations in ionic compounds.

The effect of size is also evident in other groups. For example, the oxides of the metals in Group 2A are all quite basic except for the first member of the series; beryllium oxide (BeO) is amphoteric. The basicity of an oxide depends on its ionic character. Ionic oxides contain the O^{2-} ion, which reacts with water to form two OH^- ions. All the oxides of the Group 2A metals are highly ionic except for beryllium oxide, which has considerable covalent character. The

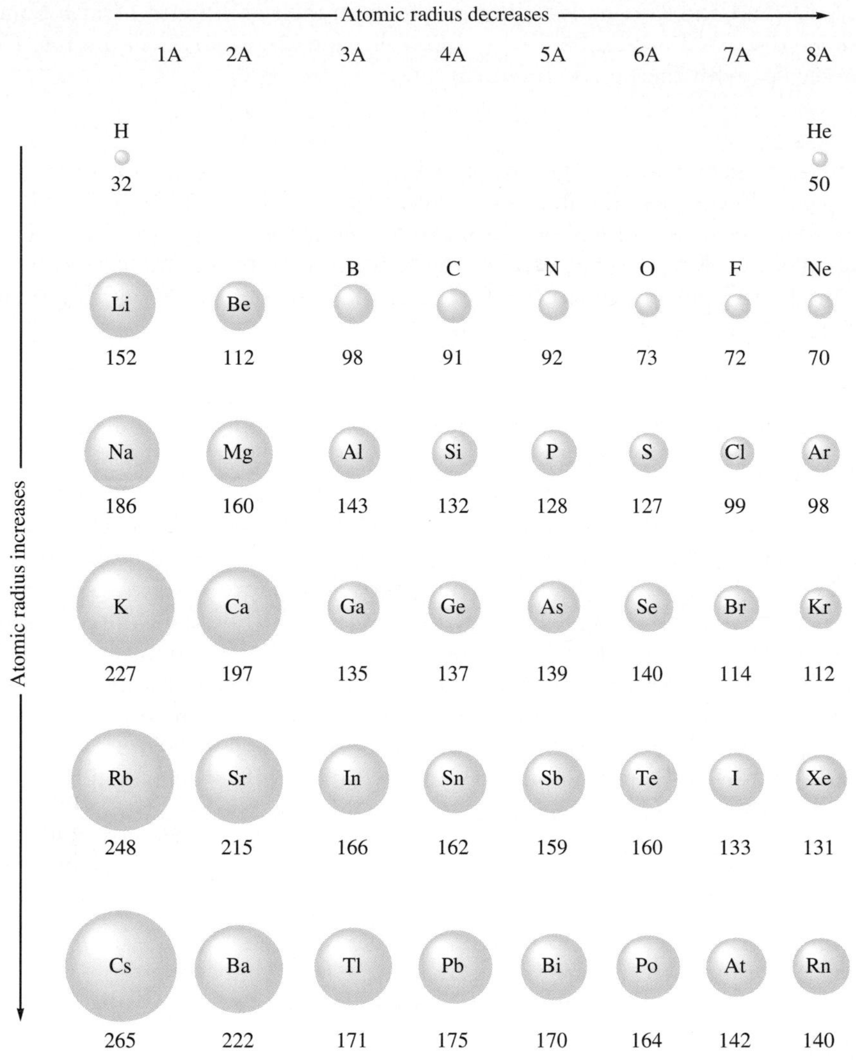

Atomic radius decreases →

Atomic radius increases ↓

1A	2A	3A	4A	5A	6A	7A	8A
H 32							He 50
Li 152	Be 112	B 98	C 91	N 92	O 73	F 72	Ne 70
Na 186	Mg 160	Al 143	Si 132	P 128	S 127	Cl 99	Ar 98
K 227	Ca 197	Ga 135	Ge 137	As 139	Se 140	Br 114	Kr 112
Rb 248	Sr 215	In 166	Sn 162	Sb 159	Te 160	I 133	Xe 131
Cs 265	Ba 222	Tl 171	Pb 175	Bi 170	Po 164	At 142	Rn 140

Figure 18.2
Some atomic radii (in picometers).

small Be^{2+} ion can effectively polarize the electron "cloud" of the O^{2-} ion, thereby producing significant electron sharing. We see the same pattern in Group 3A, where only the small boron atom behaves as a nonmetal, or sometimes as a semimetal, while aluminum and the other members are active metals.

In Group 4A the effect of size is reflected in the dramatic differences between the chemical properties of carbon and silicon. The chemistry of carbon is dominated by molecules containing chains of C—C bonds, but silicon compounds mainly contain Si—O bonds rather than Si—Si bonds. Silicon does form compounds with chains of Si—Si bonds, but these compounds are much more reactive than the corresponding carbon compounds. The reasons for the difference in reactivity between the carbon and silicon compounds are quite complex but are likely related to the differences in the sizes of the carbon and silicon atoms.

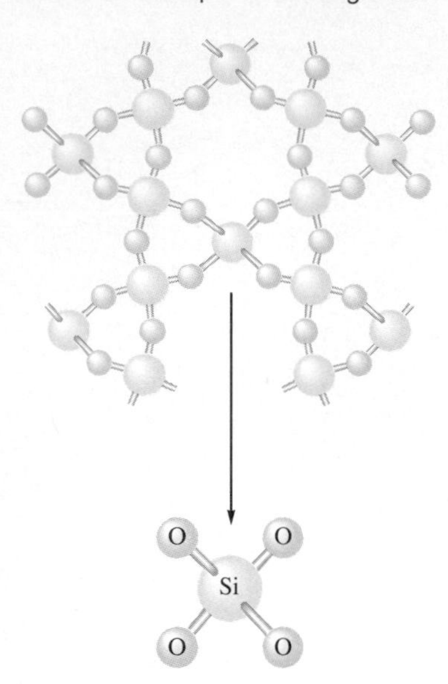

Figure 18.3
The structure of quartz, which has the empirical formula SiO_2. Note that the structure is based on interlocking SiO_4 tetrahedra, in which each oxygen atom is shared by two silicon atoms.

Carbon and silicon also differ markedly in their abilities to form π bonds. As we discussed in Section 14.1, carbon dioxide is composed of discrete CO_2 molecules with the Lewis structure

$$:\ddot{O}=C=\ddot{O}:$$

where the carbon and oxygen atoms achieve the [Ne] configuration by forming π bonds. In contrast, the structure of silica (empirical formula of SiO_2) is based on SiO_4 tetrahedra with Si—O—Si bridges, as shown in Fig. 18.3. The silicon $3p$ valence orbitals do not overlap very effectively with the smaller oxygen $2p$ orbitals to form π bonds; therefore, discrete SiO_2 molecules with the Lewis structure

$$:\ddot{O}=Si=\ddot{O}:$$

are not stable. Instead, the silicon atoms achieve a noble gas configuration by forming four Si—O single bonds.

The importance of π bonding for the relatively small elements of the second period also explains the different elemental forms of the members of Groups 5A and 6A. For example, elemental nitrogen exists as very stable N_2 molecules with the Lewis structure $:N\equiv N:$. Elemental phosphorus forms larger aggregates of atoms, the simplest being the tetrahedral P_4 molecules found in white phosphorus (see Fig. 19.12). Like silicon atoms, the relatively large phosphorus atoms do not form strong π bonds, but prefer to achieve a noble gas configuration by forming single bonds to several other phosphorus atoms. In contrast, its very strong π bonds make the N_2 molecule the most stable form of elemental nitrogen. Similarly, in Group 6A the most stable form of elemental oxygen is the O_2 molecule with a double bond. However, the larger sulfur atom forms bigger aggregates, such as the cyclic S_8 molecule (see Fig. 19.17), which contain only single bonds.

The relatively large change in size in going from the first to the second member of a group also has important consequences for the Group 7A elements. For example, fluorine has a smaller electron affinity than chlorine. This violation of the expected trend can be attributed to the fact that the small size of the fluorine $2p$ orbitals causes unusually large electron-electron repulsions. The relative weakness of the bond in the F_2 molecule can be explained in terms of the repulsions among the lone pairs, shown in the Lewis structure:

$$:\ddot{F}-\ddot{F}:$$

The small size of the fluorine atoms allows close approach of the lone pairs, which leads to much greater repulsions than those found in the Cl_2 molecule with its much larger atoms.

Thus the relatively large increase in atomic radius in going from the first to the second member of a group causes the first element to exhibit properties quite different from the others.

Abundance and Preparation

Table 18.1 shows the distribution of elements in the earth's crust, oceans, and atmosphere. The major element is, of course, oxygen, which is found in the atmosphere as O_2, in the oceans as H_2O, and in the earth's crust primarily in

TABLE 18.1 **Distribution (Mass Percent) of the 18 Most Abundant Elements in the Earth's Crust, Oceans, and Atmosphere**

Element	Mass Percent	Element	Mass Percent
Oxygen	49.2	Titanium	0.58
Silicon	25.7	Chlorine	0.19
Aluminum	7.50	Phosphorus	0.11
Iron	4.71	Manganese	0.09
Calcium	3.39	Carbon	0.08
Sodium	2.63	Sulfur	0.06
Potassium	2.40	Barium	0.04
Magnesium	1.93	Nitrogen	0.03
Hydrogen	0.87	Fluorine	0.03
		All others	0.49

silicate and carbonate minerals. The second most abundant element, silicon, is found throughout the earth's crust in the silica and silicate minerals that form the basis of most sand, rocks, and soil. The most abundant metals, aluminum and iron, are found in ores, in which they are combined with nonmetals, most commonly oxygen. One notable fact revealed by Table 18.1 is the small incidence of most transition metals. Since many of these relatively rare elements are assuming increasing importance in our high-technology society, it is possible that the control of transition metal ores may ultimately have more significance in world politics than control of petroleum supplies will.

Sand such as that found at the Great Sand Dunes National Monument in Colorado is composed of silicon and oxygen.

The distribution of elements in living materials is very different from that found in the earth's crust. Table 18.2 shows the distribution of elements in the human body. Oxygen, carbon, hydrogen, and nitrogen form the basis for all biologically important molecules. The other elements, even though they are found in relatively small amounts, are often crucial for life. For example, zinc is found in over one hundred fifty different biomolecules in the human body.

Only about one-fourth of the elements occur naturally in the free state. Most are found in a combined state. The *process of obtaining a metal from its ore* is called **metallurgy.** Since the metals in ores are found in the form of cations, the chemistry of *metallurgy always involves reduction of the ions to the elemental metal (with an oxidation state of zero).* A variety of reducing agents can be used, but carbon is the usual choice because of its wide availability and relatively low cost. For example, carbon is the primary reducing agent in the production of steel. Carbon can also be used to produce tin and lead from their oxides:

Carbon is the cheapest and most readily available industrial reducing agent for metallic ions.

$$2SnO(s) + C(s) \xrightarrow{\text{Heat}} 2Sn(s) + CO_2(g)$$

$$2PbO(s) + C(s) \xrightarrow{\text{Heat}} 2Pb(s) + CO_2(g)$$

Hydrogen gas is sometimes used as a reducing agent for metal oxides, as in the production of tin:

$$SnO(s) + H_2(g) \xrightarrow{\text{Heat}} Sn(s) + H_2O(g)$$

TABLE 18.2 **Abundance of Elements in the Human Body**

Major Elements	Mass Percent	Trace Elements (in alphabetical order)
Oxygen	65.0	Arsenic
Carbon	18.0	Chromium
Hydrogen	10.0	Cobalt
Nitrogen	3.0	Copper
Calcium	1.4	Fluorine
Phosphorus	1.0	Iodine
Magnesium	0.50	Manganese
Potassium	0.34	Molybdenum
Sulfur	0.26	Nickel
Sodium	0.14	Selenium
Chlorine	0.14	Silicon
Iron	0.004	Vanadium
Zinc	0.003	

Electrolysis is often used to reduce the most active metals. In Chapter 11 we considered the electrolytic production of aluminum metal. The alkali metals are also produced by electrolysis, usually of their molten halide salts.

The preparation of nonmetals varies widely. Elemental nitrogen and oxygen are usually obtained from the **liquefaction** of air, which is based on the principle that a gas cools as it expands. After each expansion, part of the cooler gas is compressed, while the rest is used to carry away the heat of the compression. The compressed gas is then allowed to expand again. This cycle is repeated many times. Eventually, the remaining gas becomes cold enough to form the liquid state. Because liquid nitrogen and liquid oxygen have different boiling points, they can be separated by the distillation of liquid air. Both substances are important industrial chemicals, with nitrogen ranking second in terms of amount manufactured in the United States ($\approx$60 billion pounds per year) and oxygen ranking third (over 40 billion pounds per year). Hydrogen can be obtained from the electrolysis of water, but more commonly, it is obtained from the decomposition of the methane in natural gas. Sulfur is found underground in its elemental form and is recovered by the Frasch process (see Section 19.6). The halogens are obtained by oxidation of the anions from halide salts (see Section 19.7).

The preparation of sulfur and the halogens is discussed in Chapter 19.

18.2 The Group 1A Elements

The Group 1A elements with their ns^1 valence electron configurations are all very active metals (they lose their valence electrons very readily), except for hydrogen, which behaves as a nonmetal. We will discuss the chemistry of hydrogen in the next section. Many of the properties of the **alkali metals** have been given previously (Section 12.16). The sources and methods of preparation of pure alkali metals are given in Table 18.3. The ionization energies, standard

TABLE 18.3 Sources and Methods of Preparation of the Pure Alkali Metals

Element	Source	Method of Preparation
Lithium	Silicate minerals such as spodumene, $LiAl(Si_2O_6)$	Electrolysis of molten LiCl
Sodium	NaCl	Electrolysis of molten NaCl
Potassium	KCl	Electrolysis of molten KCl
Rubidium	Impurity in lepidolite, $Li_2(F,OH)_2Al_2(SiO_3)_3$	Reduction of RbOH with Mg and H_2
Cesium	Pollucite ($Cs_4Al_4Si_9O_{26} \cdot H_2O$) and an impurity in lepidolite (Fig. 18.4)	Reduction of CsOH with Mg and H_2

Figure 18.4
Lepidolite is mainly composed of lithium, aluminum, silicon, and oxygen, but it also contains significant amounts of rubidium and cesium.

TABLE 18.4 Selected Physical Properties of the Alkali Metals

Element	Ionization Energy (kJ/mol)	Standard Reduction Potential (V) for $M^+ + e^- \rightarrow M$	Radius of M^+ (Å)	Melting Point (°C)
Lithium	520	−3.05	0.60	180
Sodium	495	−2.71	0.95	98
Potassium	419	−2.92	1.33	64
Rubidium	409	−2.99	1.48	39
Cesium	382	−3.02	1.69	29

reduction potentials, ionic radii, and melting points for the alkali metals are listed in Table 18.4.

In Section 12.16 we saw that the alkali metals all react vigorously with water to release hydrogen gas:

$$2M(s) + 2H_2O(l) \longrightarrow 2M^+(aq) + 2OH^-(aq) + H_2(g)$$

We will reconsider this process briefly because it illustrates several important concepts. From the ionization energies we might expect lithium to be the weakest of the alkali metals as a reducing agent in water. However, the standard reduction potentials indicate that it is the strongest. This reversal results mainly from the very large energy of hydration of the small Li^+ ion. Because of its relatively high charge density, the Li^+ ion very effectively attracts water molecules. A large quantity of energy is released in the process, favoring the formation of the Li^+ ion and making lithium a strong reducing agent in aqueous solution.

Several properties of the alkali metals are given in Table 12.9.

We also saw in Section 12.16 that lithium, although it is the strongest reducing agent, reacts more slowly with water than sodium or potassium. From the discussions in Chapters 10 and 15, we know that the *equilibrium position* for a reaction (in this case indicated by the $\mathscr{E}°$ values) is controlled by thermodynamic factors, but that the *rate* of a reaction is controlled by kinetic factors. There is *no* direct connection between these factors. Lithium reacts more slowly with water than sodium or potassium because as a solid lithium has a higher melting point than either of the other elements. Since lithium does

Sodium reacts violently with water.

not become molten from the heat of reaction with water as sodium and potassium do, it has a smaller area of contact with the water.

The relative ease with which the alkali metals lose electrons to form M^+ cations means that they react with nonmetals to form ionic compounds. Although we might expect the alkali metals to react with oxygen to form regular oxides of the general formula M_2O, lithium is the only one that does so in the presence of excess oxygen gas:

$$4Li(s) + O_2(g) \longrightarrow 2Li_2O(s)$$

Sodium forms solid Na_2O if the oxygen supply is limited, but in excess oxygen it forms *sodium peroxide:*

$$2Na(s) + O_2(g) \longrightarrow Na_2O_2(s)$$

Sodium peroxide containing the basic O_2^{2-} anion reacts with water to form **hydrogen peroxide** and hydroxide ions:

$$Na_2O_2(s) + 2H_2O(l) \longrightarrow 2Na^+(aq) + H_2O_2(aq) + 2OH^-(aq)$$

Hydrogen peroxide has the structure

H
:Ö—Ö:
H

Hydrogen peroxide is a strong oxidizing agent often used as a bleach for hair and as a disinfectant.

Potassium, rubidium, and cesium react with oxygen to produce **superoxides** of the general formula MO_2, which contain the O_2^- anion. For example, potassium reacts with oxygen as follows:

$$K(s) + O_2(g) \longrightarrow KO_2(s)$$

The superoxides release oxygen gas in reactions with water or carbon dioxide:

$$2MO_2(s) + 2H_2O(l) \longrightarrow 2M^+(aq) + 2OH^-(aq) + O_2(g) + H_2O_2(aq)$$
$$4MO_2(s) + 2CO_2(g) \longrightarrow 2M_2CO_3(s) + 3O_2(g)$$

This chemistry makes superoxides very useful in the self-contained breathing apparatus used by fire fighters. These "air packs" are also used as emergency equipment in labs or production facilities in case toxic fumes are released.

The types of compounds formed by the alkali metals with oxygen are summarized in Table 18.5.

TABLE 18.5
Types of Compounds Formed by the Alkali Metals with Oxygen

General Formula	Name	Examples
M_2O	Oxide	Li_2O, Na_2O
M_2O_2	Peroxide	Na_2O_2
MO_2	Superoxide	KO_2, RbO_2, CsO_2

Lithium is the only alkali metal that reacts with gaseous nitrogen to form a **nitride salt** containing the N^{3-} anion:

$$6Li(s) + N_2(g) \longrightarrow 2Li_3N(s)$$

Table 18.6 summarizes some important reactions of the alkali metals.

The alkali metal ions are very important for the proper functioning of biological systems, such as nerves and muscles; Na^+ and K^+ ions are present in all body cells and fluids. In human blood plasma the concentrations are

$$[Na^+] \approx 0.15\ M \quad \text{and} \quad [K^+] \approx 0.005\ M$$

In the fluids *inside* the cells the concentrations are reversed:

$$[Na^+] \approx 0.005\ M \quad \text{and} \quad [K^+] \approx 0.16\ M$$

Since the concentrations are so different inside and outside the cells, an elaborate mechanism involving selective ligands is needed to transport Na^+ and K^+ ions through the cell membranes.

TABLE 18.6 **Selected Reactions of the Alkali Metals**

Reaction	Comment
$2M + X_2 \longrightarrow 2MX$	X_2 = any halogen molecule
$4Li + O_2 \longrightarrow 2Li_2O$	Excess oxygen
$2Na + O_2 \longrightarrow Na_2O_2$	
$M + O_2 \longrightarrow MO_2$	M = K, Rb, or Cs
$2M + S \longrightarrow M_2S$	
$6Li + N_2 \longrightarrow 2Li_3N$	Li only
$12M + P_4 \longrightarrow 4M_3P$	
$2M + H_2 \longrightarrow 2MH$	
$2M + 2H_2O \longrightarrow 2MOH + H_2$	
$2M + 2H^+ \longrightarrow 2M^+ + H_2$	Violent reaction!

Recently, studies have been carried out concerning the role of the Li^+ ion in the human brain, and lithium carbonate has been used extensively in the treatment of manic-depressive patients. The Li^+ ion apparently affects the levels of neurotransmitters, molecules that assist the transmission of messages along the nerve networks. Incorrect concentrations of these molecules can lead to depression or mania. (See Section 12.16 for a brief discussion of lithium's role in biological systems.)

18.3 The Chemistry of Hydrogen

Under ordinary conditions of temperature and pressure, hydrogen is a colorless, odorless gas composed of H_2 molecules. Because of its low molar mass and nonpolarity, hydrogen has a very low boiling point (−253°C) and melting point (−260°C). Hydrogen gas is highly flammable; mixtures of air containing from 18% to 60% hydrogen by volume are explosive. In a common lecture demonstration hydrogen and oxygen gases are bubbled into soapy water. The resulting bubbles are then ignited with a candle on a long stick, producing a loud explosion.

The major industrial source of hydrogen gas is the reaction of methane with water at high temperatures (800–1000°C) and pressures (10–50 atm) in the presence of a metallic catalyst (often nickel):

$$CH_4(g) + H_2O(g) \xrightarrow[\text{Catalyst}]{\text{Heat, pressure}} CO(g) + 3H_2(g)$$

Large quantities of hydrogen are also formed as a by-product of gasoline production, when hydrocarbons with high molecular weights are broken down (or *cracked*) to produce smaller molecules more suitable for use as a motor fuel.

Very pure hydrogen can be produced by the electrolysis of water (see Section 11.7), but this method is currently not economically feasible for large-scale production because of the relatively high cost of electricity.

The major industrial use of hydrogen is in the production of ammonia by the Haber process. Large quantities of hydrogen are also used for hydrogenating unsaturated vegetable oils (those containing carbon-carbon double

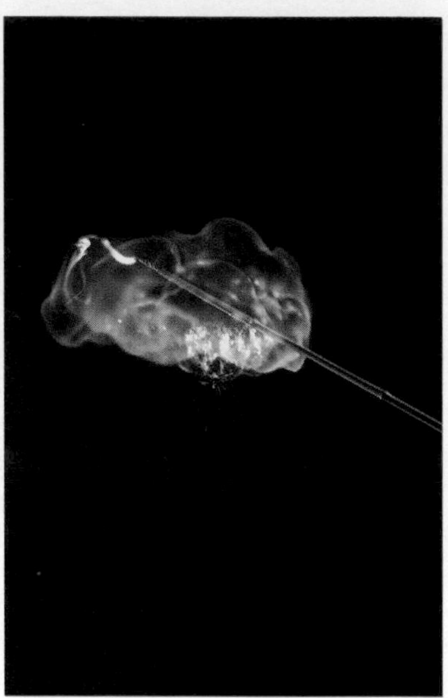

(left) Hydrogen gas being used to blow soap bubbles. (right) As the bubbles float upward, they are lighted by using a candle on a long pole results from the heat of the reaction between hydrogen and oxygen which excites sodium atoms in the soap bubbles.

bonds) to produce solid shortenings that are saturated (containing carbon-carbon single bonds):

$$\sim\sim\overset{\displaystyle H}{\overset{|}{C}}=\overset{\displaystyle H}{\overset{|}{C}}\sim\sim \xrightarrow{H_2} \sim\sim\underset{\displaystyle H}{\underset{|}{\overset{\displaystyle H}{\overset{|}{C}}}}-\underset{\displaystyle H}{\underset{|}{\overset{\displaystyle H}{\overset{|}{C}}}}\sim\sim$$

The catalysis of this process was discussed in Section 15.9.

Chemically, hydrogen behaves as a typical nonmetal, forming covalent compounds with other nonmetals and forming salts with very active metals. Binary compounds containing hydrogen are called **hydrides,** of which there are three classes. The **ionic** (or saltlike) **hydrides** are formed when hydrogen combines with the most active metals, those from Groups 1A and 2A. Examples are LiH and CaH_2, which can best be characterized as containing hydride ions (H^-) and metal cations. Because the presence of two electrons in the small 1*s* orbital produces large electron-electron repulsions and because the nucleus only has a 1^+ charge, the hydride ion is a strong reducing agent. For example, when ionic hydrides are placed in water, a violent reaction takes place. This reaction results in the formation of hydrogen gas, as seen in the equation

$$LiH(s) + H_2O(l) \longrightarrow H_2(g) + Li^+(aq) + OH^-(aq)$$

Boiling points of covalent hydrides are discussed in Section 16.1.

Covalent hydrides are formed when hydrogen combines with other nonmetals. We have encountered many of these compounds already: HCl, CH_4, NH_3, H_2O, and so on. The most important covalent hydride is water. The polarity of the H_2O molecule leads to many of water's unusual properties. Water has a much higher boiling point than is expected from its molar mass. It has a large heat of vaporization and a large heat capacity, both of which make it a very useful coolant. Water has a higher density as a liquid than as a solid, because of the open structure of ice, which results from maximizing the hydro-

Light Produces Hydrogen from Seawater

As it becomes increasingly clear that the greenhouse effect (the warming of the earth due to the accumulation of CO_2 and other gases in the atmosphere) is a serious environmental problem, the search for noncarbon-based fuels is being intensified. One such fuel is hydrogen gas, which produces only water as a combustion product. One of the drawbacks of hydrogen as a fuel is its high cost. In recent years, considerable research has been performed on ways to use solar energy to liberate hydrogen from water. Pioneering work to develop a semiconductor for the solar electrolysis of water was carried out at the University of Texas by A. J. Bard and coworkers, who found that a bipolar photoelectrode consisting of CdSe deposited on titanium foil could produce hydrogen from water using visible light. Another approach to this problem was recently described by H. Ti Tien of Michigan State University. The heart of Tien's hydrogen-producing solar cell is a cadmium selenide *n*-type semiconductor deposited on nickel foil, which separates an electrolyte solution containing Fe(II) and Fe(III) complex ions from seawater. When visible light shines on the semiconductor, electrons are excited into conduction bands, leaving behind holes (low-energy electron vacancies). The electrons in the conduction bands move to the metal foil/seawater interface, where they reduce water to form H_2 gas. The holes, on the other hand, travel to the semiconductor/electrolyte interface, where they are filled with electrons from Fe^{2+} ions (see the accompanying figure), which are regenerated by an external energy source (not shown).

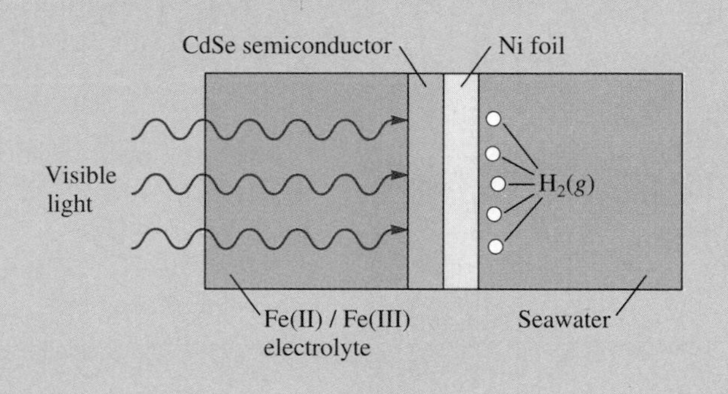

gen bonding (see Fig. 18.5). Because water is an excellent solvent for ionic and polar substances, it provides an effective medium for life processes. In fact, water is one of the few covalent hydrides that is nontoxic to organisms.

The third class of hydrides is the **metallic,** or **interstitial, hydrides,** which are formed when transition metal crystals are treated with hydrogen gas. The hydrogen molecules dissociate at the metal's surface, and the small hydrogen atoms migrate into the crystal structure to occupy holes, or *interstices.* These metal-hydrogen mixtures are more like solid solutions than true compounds. Palladium can absorb about *900 times* its own volume of hydrogen gas. In fact, hydrogen can be purified by placing it under slight pressure in a vessel containing a thin wall of palladium. The hydrogen diffuses into and through the metal wall, leaving the impurities behind.

Although hydrogen can react with transition metals to form compounds such as UH_3 and FeH_6, most of the interstitial hydrides have variable compositions (often called *nonstoichiometric* compositions) with formulas such as $LaH_{2.76}$ and $VH_{0.56}$. The compositions of the nonstoichiometric hydrides vary with the length of exposure of the metal to hydrogen gas.

When interstitial hydrides are heated, much of the absorbed hydrogen is lost as hydrogen gas. Because of this behavior, these materials offer possibilities for storing hydrogen for use as a portable fuel. The internal combustion

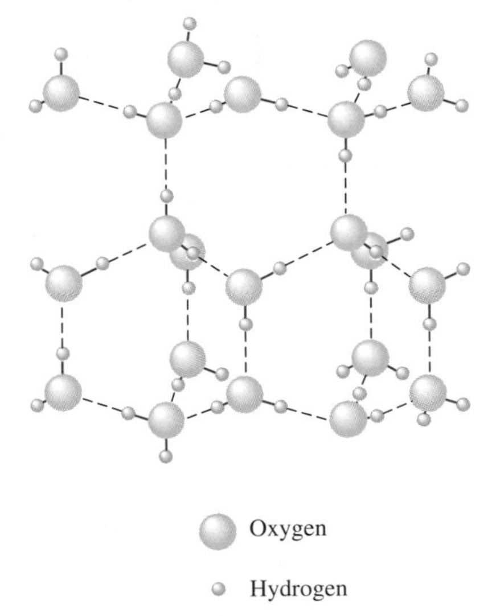

Figure 18.5
The structure of ice, showing the hydrogen bonding.

See Section 9.8 for a discussion of the feasibility of using hydrogen gas as a fuel.

engines in current automobiles can burn hydrogen gas with little modification, but storage of enough hydrogen to provide an acceptable mileage range remains a problem. One possible solution might be to use a fuel tank containing a porous solid that includes a transition metal. The hydrogen gas could be pumped into the solid to form the interstitial hydride. The hydrogen gas could then be released when the engine requires additional energy. This system is now being tested by several automobile companies.

2A
Be
Mg
Ca
Sr
Ba
Ra

18.4 The Group 2A Elements

The Group 2A elements (with the valence electron configuration ns^2) are very reactive, losing their two valence electrons to form ionic compounds that contain M^{2+} cations. These elements are commonly called the **alkaline earth metals** because of the basicity of their oxides:

An amphoteric oxide displays both acidic and basic properties.

$$MO(s) + H_2O(l) \longrightarrow M^{2+}(aq) + 2OH^-(aq)$$

Only the amphoteric beryllium oxide (BeO) also shows some acidic properties, such as dissolving in aqueous solutions containing hydroxide ions:

$$BeO(s) + 2OH^-(aq) + H_2O(l) \longrightarrow Be(OH)_4{}^{2-}(aq)$$

The more active alkaline earth metals react with water as the alkali metals do, producing hydrogen gas:

$$M(s) + 2H_2O(l) \longrightarrow M^{2+}(aq) + 2OH^-(aq) + H_2(g)$$

Calcium metal reacting with water to form bubbles of hydrogen gas.

Calcium, strontium, and barium react vigorously at 25°C. The less easily oxidized beryllium and magnesium show no observable reaction with water at 25°C, although magnesium reacts with boiling water. Table 18.7 summarizes various properties, sources, and preparations of the alkaline earth metals.

The heavier alkaline earth metals react with nitrogen or hydrogen at high temperatures to produce ionic nitride or hydride salts. For example:

$$3Ca(s) + N_2(g) \longrightarrow Ca_3N_2(s)$$
$$Ca(s) + H_2(g) \longrightarrow CaH_2(s)$$

Magnesium, strontium, and barium form similar compounds. Beryllium hydride cannot be formed by direct combination of the elements but can be prepared by the following reaction:

$$BeCl_2 + 2LiH \longrightarrow BeH_2 + 2LiCl$$

Hydrogen bridges between the beryllium atoms produce a polymeric structure for BeH_2, as shown in Fig. 18.6. The localized electron model describes this bonding by assuming that only one electron pair is available to bind each Be—H—Be cluster. This is called a *three-center bond*, since one electron pair is shared among three atoms. Three-center bonds have also been postulated to explain the bonding in other electron-deficient compounds (compounds where there are fewer electron pairs than bonds), such as the boron hydrides (see Section 18.5).

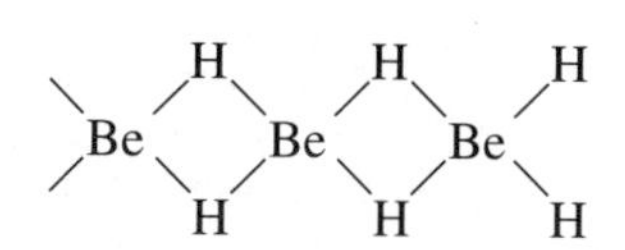

Figure 18.6
The structure of solid BeH_2.

TABLE 18.7 Selected Physical Properties, Sources, and Methods of Preparation for the Group 2A Elements

Element	Radius of M^{2+} (Å)	Ionization Energy (kJ/mol) First	Ionization Energy (kJ/mol) Second	$\mathscr{E}°$ (V) for $M^{2+} + 2e^- \longrightarrow M$	Source	Method of Preparation
Beryllium	≈0.3	900	1760	−1.70	Beryl ($Be_3Al_2Si_6O_{18}$)	Electrolysis of molten $BeCl_2$
Magnesium	0.65	738	1450	−2.37	Magnesite ($MgCO_3$), dolomite ($MgCO_3 \cdot CaCO_3$), carnallite ($MgCl_2 \cdot KCl \cdot 6H_2O$)	Electrolysis of molten $MgCl_2$
Calcium	0.99	590	1146	−2.76	Various minerals containing $CaCO_3$	Electrolysis of molten $CaCl_2$
Strontium	1.13	549	1064	−2.89	Celestite ($SrSO_4$), strontianite ($SrCO_3$)	Electrolysis of molten $SrCl_2$
Barium	1.35	503	965	−2.90	Baryte ($BaSO_4$), witherite ($BaCO_3$)	Electrolysis of molten $BaCl_2$
Radium	1.40	509	979	−2.92	Pitchblende (1 g of Ra/7 tons of ore)	Electrolysis of molten $RaCl_2$

As we saw in Section 18.1, the small size and relatively high electronegativity of the beryllium atom causes its bonds to be more covalent than is usual for a metal. For example, beryllium chloride with the Lewis structure

$$:\ddot{\underset{..}{Cl}}—Be—\ddot{\underset{..}{Cl}}:$$

exists as a linear molecule, as predicted by the VSEPR model. The Be—Cl bonds are covalent, and beryllium is best described as being *sp* hybridized. Note that since the beryllium atom in $BeCl_2$ is very electron-deficient (only four valence electrons surround it), we are not surprised that it is very reactive toward electron pair donors (Lewis bases) such as ammonia:

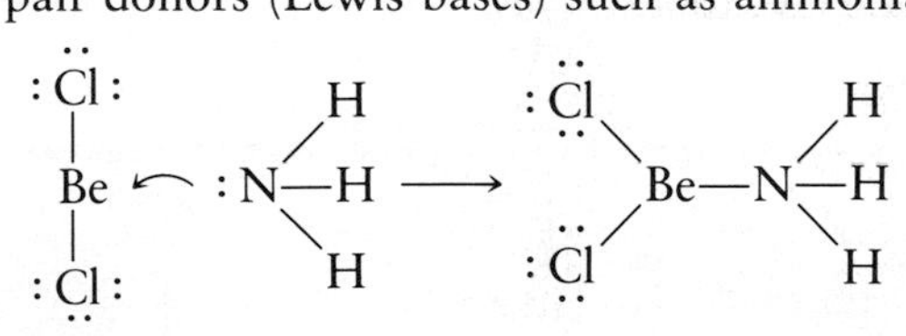

$$Cl_2Be + :NH_3 \longrightarrow Cl_2Be—NH_3$$

As a solid, $BeCl_2$ achieves an octet of electrons around the beryllium atom by forming an extended structure containing Be in a tetrahedral environment, as shown in Fig. 18.7. The lone pairs on the chlorine atoms are used to form Be—Cl bonds.

The alkaline earth metals have great practical importance. Calcium and magnesium ions are essential for human life. Calcium is found primarily in the structural minerals composing bones and teeth. Magnesium (as the Mg^{2+} ion) plays a vital role in metabolism and in muscle functions. Also, magnesium is commonly used to produce the bright light from photographic flash bulbs from its reaction with oxygen:

$$2Mg(s) + O_2(g) \longrightarrow 2MgO(s) + \text{light}$$

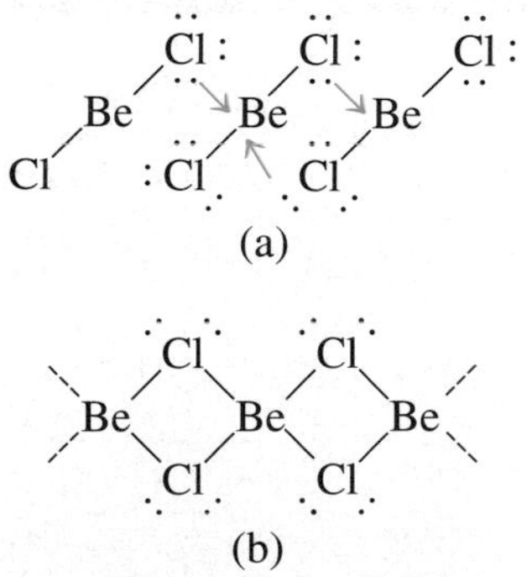

(a)

(b)

Figure 18.7
(a) Solid $BeCl_2$ can be visualized as consisting of many $BeCl_2$ molecules, in which lone pairs on the chlorine atoms are used to bond to the beryllium atoms in adjacent $BeCl_2$ molecules. (b) The extended structure of solid $BeCl_2$.

Chrystalline beryl from Colombia.

TABLE 18.8 Selected Reactions of the Group 2A Elements

Reaction	Comment
$M + X_2 \longrightarrow MX_2$	X_2 = any halogen molecule
$2M + O_2 \longrightarrow 2MO$	Ba gives BaO_2 as well
$M + S \longrightarrow MS$	
$3M + N_2 \longrightarrow M_3N_2$	High temperatures
$6M + P_4 \longrightarrow 2M_3P_2$	High temperatures
$M + H_2 \longrightarrow MH_2$	M = Ca, Sr or Ba; high temperatures; Mg at high pressure
$M + 2H_2O \longrightarrow M(OH)_2 + H_2$	M = Ca, Sr, or Ba
$M + 2H^+ \longrightarrow M^{2+} + H_2$	
$Be + 2OH^- + 2H_2O \longrightarrow Be(OH)_4^{2-} + H_2$	

Because magnesium metal has a relatively low density and displays moderate strength, it is a useful structural material, especially if alloyed with aluminum.

Table 18.8 summarizes some important reactions involving the alkaline earth metals.

Relatively large concentrations of Ca^{2+} and Mg^{2+} ions are often found in natural water supplies. These ions in this so-called **hard water** interfere with the action of detergents and form precipitates with soap. In Section 7.6 we saw that Ca^{2+} is often removed by precipitation as $CaCO_3$ in large-scale water softening. In individual homes Ca^{2+}, Mg^{2+}, and other cations are removed by **ion exchange.** An **ion exchange resin** consists of large molecules (polymers) that have many ionic sites. A cation exchange resin is represented schematically in Fig. 18.8(a), showing Na^+ ions bound ionically to the SO_3^- groups that are covalently attached to the resin polymer. When hard water is passed over the resin, Ca^{2+} and Mg^{2+} bind to the resin in place of Na^+, which is released into the solution [Fig. 18.8(b)]. Replacing Mg^{2+} and Ca^{2+} by Na^+ [Fig. 18.8(c)] "softens" the water, because the sodium salts of soap are soluble.

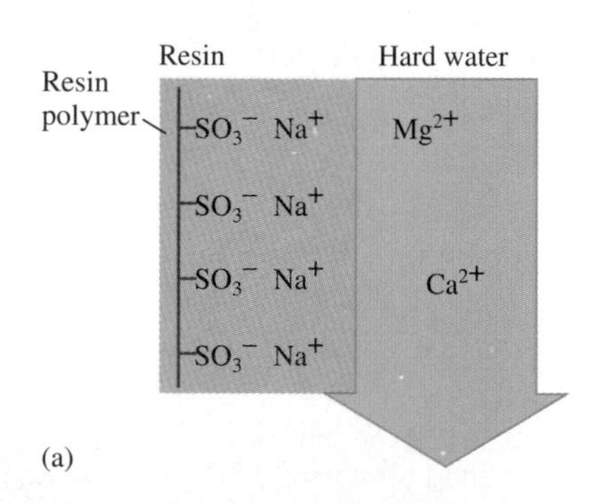

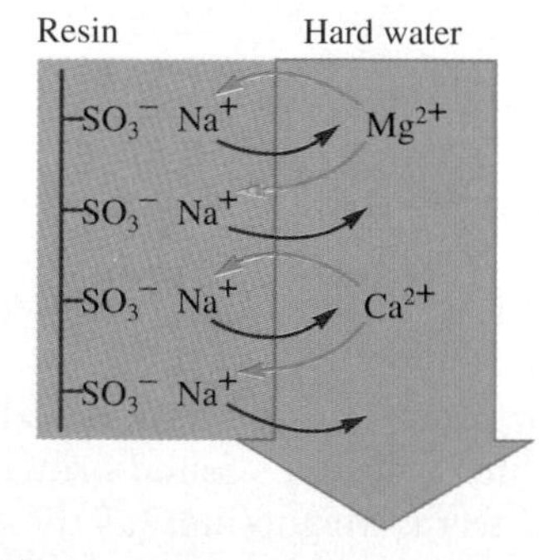

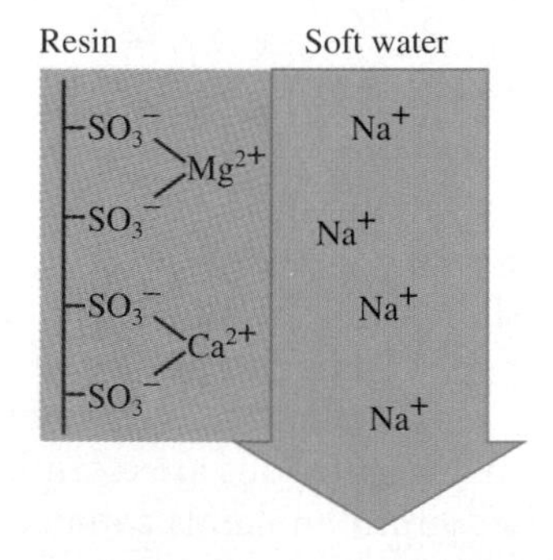

Figure 18.8
(a) A schematic representation of a typical cation exchange resin. (b) and (c) When hard water is passed over the cation exchange resin, the Ca^{2+} and Mg^{2+} bind to the resin.

TABLE 18.9 **Selected Physical Properties, Sources, and Methods of Preparation for the Group 3A Elements**

Element	Radius of M^{3+} (Å)	Ionization Energy (kJ/mol)	$\mathscr{E}°$ (V) for $M^{3+} + 3e^- \longrightarrow M$	Sources	Method of Preparation
Boron	0.2	798	—	Kernite, a form of borax ($Na_2B_4O_7 \cdot 4H_2O$)	Reduction by Mg or H_2
Aluminum	0.51	581	−1.71	Bauxite (Al_2O_3)	Electrolysis of Al_2O_3 in molten Na_3AlF_6
Gallium	0.62	577	−0.53	Traces in various minerals	Reduction with H_2 or electrolysis
Indium	0.81	556	−0.34	Traces in various minerals	Reduction with H_2 or electrolysis
Thallium	0.95	589	0.72	Traces in various minerals	Electrolysis

18.5 The Group 3A Elements

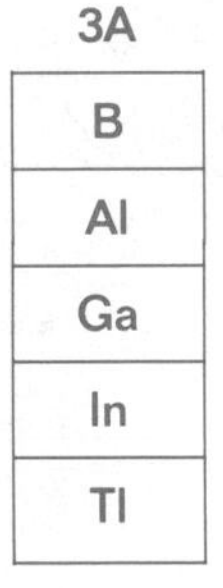

The Group 3A elements (valence electron configuration ns^2np^1) generally show the increase in metallic character in going down the group that is characteristic of the representative elements. Some physical properties, sources, and methods of preparation for the Group 3A elements are summarized in Table 18.9.

Boron is a typical nonmetal and most of its compounds are covalent. The most interesting compounds of boron are the covalent hydrides called **boranes.** We might expect BH_3 to be the simplest hydride, since boron has three valence electrons to share with three hydrogen atoms. However, this compound is unstable, and the simplest known member of the series is diborane (B_2H_6), with the structure shown in Fig. 18.9(a). In this molecule the terminal B—H bonds are normal covalent bonds, each involving one electron pair. The bridging bonds are three-center bonds similar to those in solid BeH_2. Another interesting borane contains the square pyramidal B_5H_9 molecule [Fig. 18.9(b)], which has four three-center bonds situated around the base of the pyramid. Because the boranes are extremely electron-deficient, they are highly reactive. The boranes react very exothermically with oxygen and were once evaluated as potential fuels for rockets in the U.S. space program.

Aluminum, the most abundant metal on earth, has metallic physical properties, such as high thermal and electrical conductivities and a lustrous appearance; however its bonds to nonmetals are significantly covalent. This covalency

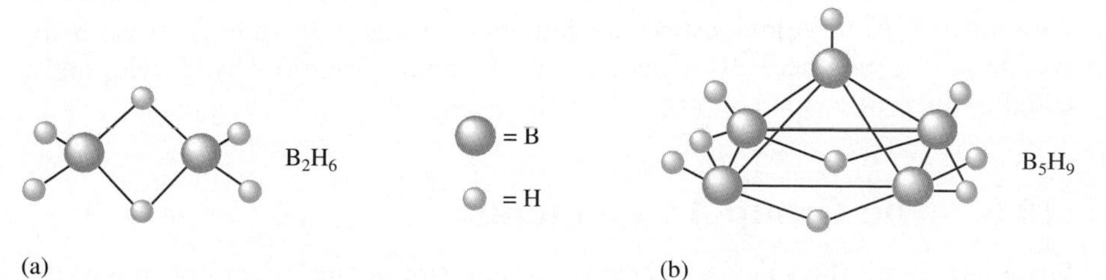

Figure 18.9
(a) The structure of B_2H_6 with its two three-center B—H—B bridging bonds and four "normal" B—H bonds. (b) The structure of B_5H_9. There are five "normal" B—H bonds to terminal hydrogens and four three-center bridging bonds around the base.

TABLE 18.10 Selected Reactions of the Group 3A Elements

Reaction	Comment
$2M + 3X_2 \longrightarrow 2MX_3$	X_2 = any halogen molecule; Tl gives TlX as well, but no TlI_3
$4M + 3O_2 \longrightarrow 2M_2O_3$	High temperatures; Tl gives Tl_2O as well
$2M + 3S \longrightarrow M_2S_3$	High temperatures; Tl gives Tl_2S as well
$2M + N_2 \longrightarrow 2MN$	M = Al only
$2M + 6H^+ \longrightarrow 2M^{3+} + 3H_2$	M = Al, Ga, or In; Tl gives Tl^+
$2M + 2OH^- + 6H_2O \longrightarrow 2M(OH)_4^- + 3H_2$	M = Al or Ga

is responsible for the amphoteric nature of Al_2O_3, which dissolves in acidic or basic solution, and for the acidity of $Al(H_2O)_6^{3+}$ (see Section 7.8):

$$Al(H_2O)_6^{3+}(aq) \rightleftharpoons Al(OH)(H_2O)_5^{2+}(aq) + H^+(aq)$$

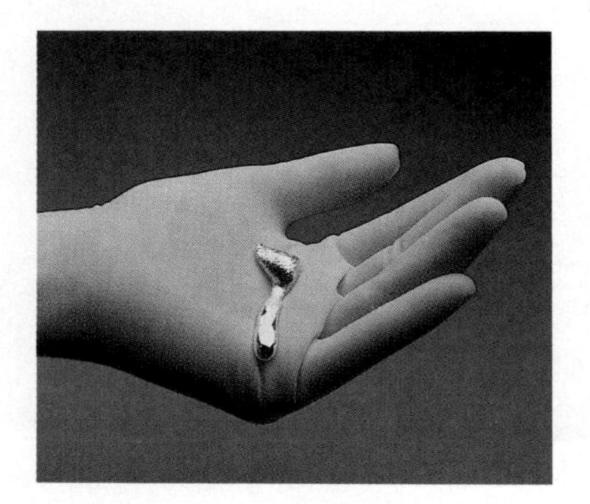

Gallium melts in the hand.

One especially interesting property of *gallium* is its unusually low melting point of 29.8°C, which is in contrast to the 660°C melting point of aluminum. Also, since gallium's boiling point is approximately 2400°C, it has the largest liquid range of any metal. This makes it useful for thermometers, especially to measure high temperatures. Gallium, like water, expands when it freezes. The chemistry of gallium is quite similar to that of aluminum. For example, Ga_2O_3 is amphoteric.

The chemistry of *indium* is similar to that of aluminum and gallium except that compounds containing the 1+ ion are known, such as InCl and In_2O, in addition to those with the more common 3+ ion.

The chemistry of *thallium* is completely metallic. For example, Tl_2O_3 is a basic oxide. Both the +1 and +3 oxidation states are quite common for thallium; the oxides Tl_2O_3 and Tl_2O, and the chlorides $TlCl_3$ and TlCl are all well-known compounds. The tendency for the heavier members of Group 3A to exhibit the +1 as well as the expected +3 oxidation state is often called the **inert pair effect.** This effect is also found in Group 4A, where lead and tin exhibit both +4 and +2 oxidation states.

The explanation for the inert pair effect is very complex, involving a relativistic treatment of the atom, and will not be considered here.

Table 18.10 summarizes some important reactions of the Group 3A elements.

The practical importance of the Group 3A elements mostly centers on aluminum. Since the discovery of the electrolytic production process by Hall and Heroult (Section 11.8), aluminum has become a highly important structural material in a wide variety of applications from aircraft bodies to bicycle components. Aluminum is especially valuable, because it has a high strength-to-weight ratio, and because it protects itself from corrosion by developing a tough, adherent oxide coating.

18.6 The Group 4A Elements

4A
C
Si
Ge
Sn
Pb

Group 4A (with the valence electron configuration ns^2np^2) contains two of the most important elements on earth: carbon, the fundamental constituent of the molecules necessary for life; and silicon, which forms the basis of the geological

TABLE 18.11 Selected Physical Properties, Sources, and Methods of Preparation for the Group 4A Elements

Element	Electronegativity	Melting Point (°C)	Boiling Point (°C)	Sources	Method of Preparation
Carbon	2.5	3727 (sublimes)	—	Graphite, diamond, petroleum, coal	—
Silicon	1.8	1410	2355	Silicate minerals, silica	Reduction of K_2SiF_6 with Al, or reduction of SiO_2 with Mg
Germanium	1.8	937	2830	Germanite (mixture of copper, iron, and germanium sulfides)	Reduction of GeO_2 with H_2 or C
Tin	1.8	232	2270	Cassiterite (SnO_2)	Reduction of SnO_2 with C
Lead	1.9	327	1740	Galena (PbS)	Roasting of PbS with O_2 to form PbO_2 and then reduction with C

world. The change from nonmetallic to metallic properties seen in Group 3A is also apparent in going down Group 4A from carbon, a typical nonmetal, to silicon and germanium, usually considered semimetals, to the metals tin and lead. Table 18.11 summarizes some physical properties, sources, and methods of preparation for the elements in this group.

All of the Group 4A elements can form four covalent bonds to nonmetals, for example, CH_4, SiF_4, $GeBr_4$, $SnCl_4$, and $PbCl_4$. In each of these tetrahedral molecules the central atom is described as sp^3 hybridized by the localized electron model. All of these compounds, except those of carbon, can react with Lewis bases to form two additional covalent bonds. For example, $SnCl_4$, which is a fuming liquid (bp = 114°C), can add two chloride ions:

$$SnCl_4 + 2Cl^- \longrightarrow SnCl_6^{2-}$$

Carbon compounds cannot react in this way because of the small atomic size of carbon, and because there are no *d* orbitals on carbon to accommodate the extra electrons, as there are on the other elements in the group.

TABLE 18.12 Strengths of C—C, Si—Si, and Si—O Bonds

Bond	Bond Energy (kJ/mol)
C—C	347
Si—Si	340
Si—O	368

We have seen that carbon also differs markedly from the other members of Group 4A in its ability to form π bonds. This accounts for the completely different structures and properties of CO_2 and SiO_2. Note from Table 18.12 that C—C bonds and Si—O bonds are stronger than Si—Si bonds. This partly explains why the chemistry of carbon is dominated by C—C bonds, while that of silicon is dominated by Si—O bonds.

The organic chemistry of carbon is discussed in Chapter 22.

Carbon occurs in the allotropic forms graphite and diamond, whose structures were given in Section 16.5.

Although graphite is thermodynamically more stable than diamond, the transformation of diamond to graphite is not observed under normal conditions.

Carbon monoxide (CO), one of three oxides of carbon, is an odorless, colorless, toxic gas at 25°C and 1 atm. It is a by-product of the combustion of carbon-containing compounds when there is a limited oxygen supply. Incidents of carbon monoxide poisoning are especially common in the winter in cold areas of the world when blocked furnace vents limit the availability of oxygen. The bonding in carbon monoxide, which has the Lewis structure :C≡O:, is described in terms of *sp*-hybridized carbon and oxygen atoms that interact to form one σ and two π bonds.

A new form of elemental carbon, the fullerenes, was discussed in Chapter 2.

Carbon dioxide, a linear molecule with the Lewis structure

$$:\ddot{O}=C=\ddot{O}:$$

and an *sp*-hybridized carbon atom, is a product of human and animal respiration and of the combustion of fossil fuels. It is also produced by fermentation, a process by which the sugar in fruits and grains is changed to ethanol (C_2H_5OH) and carbon dioxide:

$$\underset{\text{Glucose}}{C_6H_{12}O_6(aq)} \xrightarrow{\text{Enzymes}} 2C_2H_5OH(aq) + 2CO_2(g)$$

Carbon dioxide dissolves in water to produce an acidic solution:

$$CO_2(aq) + H_2O(l) \rightleftharpoons H^+(aq) + HCO_3^-(aq)$$

Carbon suboxide, the third carbon oxide, is a linear molecule with the Lewis structure

$$:\ddot{O}=C=C=C=\ddot{O}:$$

which contains *sp*-hybridized carbon atoms.

Silicon, the second most abundant element in the earth's crust, is a semimetal found widely distributed in silica and silicates (see Section 16.5). Approximately 85% of the earth's crust is composed of these substances. Although silicon is found in some steel and aluminum alloys, its major use is in semiconductors for electronic devices (see Chapter 16).

Germanium, a relatively rare element, is a semimetal used mainly in the manufacture of semiconductors for transistors and similar electronic devices.

Tin is a soft, silvery metal that can be rolled into thin sheets (tin foil) and has been used for centuries in various alloys such as bronze (20% Sn and 80% Cu), solder (33% Sn and 67% Pb), and pewter (85% Sn, 7% Cu, 6% Bi, and 2% Sb). Tin exists as three allotropes: *white tin,* stable at normal temperatures; *gray tin,* stable at temperatures below 13.2°C; and *brittle tin,* found at temperatures above 161°C. When tin is exposed to low temperatures, it gradually changes to powdery gray tin and crumbles away; this is known as *tin disease.*

Currently, tin is used mainly as a protective coating for steel, especially for cans used as food containers. The thin layer of tin, applied electrolytically, forms a protective oxide coating that prevents further corrosion.

Tin forms compounds in the +2 and +4 oxidation states. For example, all of the possible tin(II) and tin(IV) halides are known. The tin(IV) halides, except for SnF_4, are all relatively volatile (see Table 18.13) and generally behave much more like molecular than ionic compounds. This is no doubt due to the high charge-to-radius ratio that would be characteristic of Sn^{4+} if it existed. Thus instead of containing Sn^{4+} and X^- ions, the tin(IV) halides contain covalent Sn—X bonds. The Sn^{4+} ion almost certainly does not exist in these or any other known compounds. The compound SnF_4 is relatively nonvolatile, not because it contains the Sn^{4+} and F^- ions, but because it consists of alternating SnF_6^{2-} and SnF_2^{2+} ions.

The tin(II) halides are much less volatile than the corresponding tin(IV) compounds; in fact, they are probably ionic, containing Sn^{2+} and X^- ions. Tin(II) chloride in aqueous solution is commonly used as a reducing agent. Tin(II) fluoride (stannous fluoride) was for many years added to toothpaste to help prevent tooth decay. It has been replaced by sodium fluoride.

Lead is easily obtained from its ore, galena (PbS). Because lead melts at such a low temperature, it may have been the first pure metal obtained from its ore. We know that lead was used as early as 3000 B.C. by the Egyptians. It was later used by the Romans to make eating utensils, glazes on pottery, and even

TABLE 18.13
Boiling Points of the Tin(IV) Halides

Compound	Boiling Point (°C)
SnF_4	705 (sublimes)
$SnCl_4$	114
$SnBr_4$	202
SnI_4	364

CONCRETE LEARNING

Concrete has literally paved the way for civilization over the past 5000 years, tracing its roots to the ancient Egyptians. At a cost of about a penny per pound, concrete is ubiquitous in today's world—used in houses, factories, roads, dams, cooling towers, pipes, skyscrapers, and countless other places. In the United States alone there are an estimated $6 trillion worth of concrete-based structures.

Most concretes are based on Portland cement, patented in 1824 by an English bricklayer named J. Aspdin and so named because it forms a product that resembles the natural limestone on the Isle of Portland in England. Portland cement is a powder containing a mixture of calcium silicates [Ca_2SiO_4 (26%) and Ca_3SiO_5 (51%)], calcium aluminate [$Ca_3Al_2O_6$ (11%)], and calcium iron aluminate [$Ca_4Al_2Fe_2O_{10}$ (1%)]. Portland cement is made from a mixture of limestone, sand, shale, clay, and gypsum ($CaSO_4 \cdot 2H_2O$). When the cement is mixed with sand, gravel, and water, it turns into a muddy substance that eventually hardens into the familiar concrete that finds so many uses in our world. The hardening of concrete occurs not through drying but through hydration. The material becomes dry and hard as the water is consumed in building the complex silicate structures present in cured concrete. Although many of the details of this process are poorly understood, the main "glue" that holds the components of concrete together is calcium silicate hydrate, which forms a three-dimensional network mainly responsible for concrete's strength.

Despite its strength when newly produced, concrete contains pockets of air and water dispersed throughout, making it porous and subject to deterioration. Thus despite all of the advantages of concrete, it cracks and deteriorates seriously over time.

Much research is now being carried out to improve the durability of concrete. Most of these efforts are directed toward lowering the porosity of concrete and making it less brittle. One group of additives aimed at solving this problem consists of molecules with carbon atom backbones that have sulfate groups attached. These so-called superplasticizers allow the formation of concrete using much less water, and these chemicals have doubled the strength of ready-mix concrete over the past 20 years.

Researchers have also found that the properties of concrete can be greatly improved by adding fibers of various kinds, including those made of steel, glass, and carbon-based polymers. One type of fiber concrete—called slurry infiltrated fiber concrete (SIFCON), which is tough enough to be used to make missile silos, and can be formed into complex shapes—may be especially useful for structures in earthquake-prone areas.

Other efforts to improve concrete center on replacing Portland cement with other binders such as carbon-based polymers. Although these polymer-based concretes will burn and do lose their shapes at high temperatures, they are much more resistant to the effects of water, acids, and salts than those made with Portland cement.

Despite the fact that most concrete now used is still very similar to that used by the Romans to build the Pantheon, progress is being made, and revolutionary improvements may be just around the corner.

Suggested Reading

For additional information see Gary Stix, "Concrete Solutions," *Scientific American,* April 1993. p. 102–112.

intricate plumbing systems. The Romans also prepared a sweetener called sapa by boiling down grape juice in lead-lined vessels. The sweetness of this syrup was due partly to the formation of lead(II) acetate (formerly called sugar of lead), a very sweet-tasting compound. The problem with these practices is that lead is very toxic. In fact, the Romans had so much contact with lead that it may

(left) Galena from Galena, Illinois. (right) Roman baths such as these in Bath, England, used lead pipes for water.

have contributed to the demise of their civilization. Analysis of bones from that era shows significant levels of lead.

Although lead poisoning has been known since at least the second century B.C., lead continues to be a problem. For example, many children have been exposed to lead by eating chips of lead-based paint. Because of this problem lead-based paints are no longer used for children's furniture and many states ban lead-based paint for interior use. Lead poisoning can also occur when acidic foods and drinks leach the lead from lead-glazed pottery dishes that were improperly fired and when liquor is stored in leaded crystal decanters, producing toxic levels of lead in the drink in a relatively short time. In addition, the widespread use of tetraethyl lead, $(C_2H_5)_4Pb$, as an antiknock agent in gasoline has increased the lead levels in our environment during this century. Concern about the effects of this lead pollution has caused the U.S. government to require the gradual replacement of the lead in gasoline with other antiknock agents. The largest commercial use of lead (over one million tons annually) is for electrodes in the lead storage batteries used in automobiles (Section 11.5).

Lead(II) oxide, known as litharge.

Lead forms compounds in the +2 and +4 oxidation states. The lead(II) halides, all of which are known, exhibit ionic properties and thus can be assumed to contain Pb^{2+} ions. Only PbF_4 and $PbCl_4$ are known among the possible lead(IV) halides, presumably because lead(IV) oxidizes bromide and iodide ions, producing the lead(II) halide and the free halogen:

$$PbX_4 \longrightarrow PbX_2 + X_2$$

Lead(IV) chloride decomposes in this fashion at temperatures above 100°C.

Yellow lead(II) oxide, known as *litharge*, is widely used to glaze ceramic ware. Lead(IV) oxide does not exist in nature, but a substance with the formula $PbO_{1.9}$ can be produced in the laboratory by oxidation of lead(II) compounds in basic solution. The nonstoichiometric nature of this compound is caused by defects in the crystal structure. The crystal has some vacancies in positions where there should be oxide ions. These imperfections in the crystal (called

TABLE 18.14 **Selected Reactions of the Group 4A Elements**

Reaction	Comment
$M + 2X_2 \longrightarrow MX_4$	X_2 = any halogen molecule; M = Ge or Sn; Pb gives PbX_2
$M + O_2 \longrightarrow MO_2$	M = Ge or Sn; high temperatures; Pb gives PbO or Pb_3O_4
$M + 2H^+ \longrightarrow M^{2+} + H_2$	M = Sn or Pb

lattice defects) make lead(IV) oxide an electrical conductor, since the oxide ions jump from hole to hole. This makes possible the use of lead(IV) oxide as an electrode (the cathode) in the lead storage battery.

Lattice defects are discussed in Section 16.9.

Table 18.14 summarizes some important reactions of the Group 4A elements.

EXERCISES

A blue exercise number indicates that the answer to that exercise appears at the back of this book.

Group 1A Elements

1. Although the earth was formed from the same interstellar material as the sun, there is little hydrogen in the earth's atmosphere. How can you explain this?
2. Hydrogen is produced commercially by the reaction of methane with steam:

$$CH_4(g) + H_2O(g) \rightleftharpoons CO(g) + 3H_2(g)$$

   a. Calculate $\Delta H°$ and $\Delta S°$ for this reaction (use the data in Appendix 4).
   b. What temperatures will favor product formation assuming standard pressures?
3. Hydrogen is also produced commercially by the following reactions:

$$3Fe(s) + 4H_2O(g) \rightleftharpoons Fe_3O_4(s) + 4H_2(g)$$
$$C(s) + H_2O(g) \rightleftharpoons CO(g) + H_2(g)$$

   a. Using the data in Appendix 4, calculate $\Delta H°$ and $\Delta S°$ for the above reactions.
   b. At what temperatures are each of the above reactions spontaneous under standard conditions?
4. List two major industrial uses of hydrogen.
5. What are the three types of hydrides? How do they differ?
6. Many lithium salts are hygroscopic (absorb water), but the corresponding salts of the other alkali metals are not. Why are lithium salts different from the others?
7. Complete and balance the following reactions.
   a. $Li_3N(s) + HCl(aq) \longrightarrow$
   b. $Rb_2O(s) + H_2O(l) \longrightarrow$
   c. $Cs_2O_2(s) + H_2O(l) \longrightarrow$
   d. $NaH(s) + H_2O(l) \longrightarrow$
8. Use the Nernst equation (Section 11.4) to calculate the amount of work that must be done to transport 1.0 mol of K^+ ions from the outside of a cell to the inside. Inside muscle cells $[K^+]$ is about 0.15 mol/L, and in blood plasma $[K^+]$ is about 5.0×10^{-3} mol/L.
9. Graph the melting points (Table 18.4) of the alkali metals versus atomic number. Predict the melting point for the element francium. Would you predict francium to be a solid or a liquid at room temperature?
10. Lithium reacts with acetylene in liquid ammonia to produce lithium acetylide ($LiC{\equiv}CH$) and hydrogen gas. Write a balanced equation for this reaction. What type of reaction is this?
11. What evidence supports putting hydrogen in Group 1A of the periodic table? In some periodic tables hydrogen is listed separately from all of the groups. In what ways is hydrogen unlike a Group 1A element?
12. Show how the reaction of NaH with water,

$$NaH(s) + H_2O(l) \longrightarrow Na^+(aq) + OH^-(aq) + H_2(g)$$

   can be described as both an oxidation-reduction and an acid-base reaction.
13. Write balanced equations describing the reaction of potassium metal with each of the following:
   a. O_2 b. S_8 c. P_4 d. H_2 e. H_2O

14. Write formulas for each of the following:
   a. sodium oxide
   b. sodium superoxide
   c. sodium peroxide
15. Describe how potassium superoxide can be used in a self-contained breathing apparatus.

Group 2A Elements

16. All of the Group 1A and 2A metals are produced by electrolysis of molten salts. Why?
17. Predict the geometry around beryllium in the compound Cl_2BeNH_3. What hybrid orbitals are used by beryllium and nitrogen in this compound? What type of acid is $BeCl_2$?
18. How does the acidity of the aqueous solutions of the alkaline earth metal ions (M^{2+}) change in going down the group?
19. Would you expect $BeCl_2NH_3$ or another compound to form from the reaction of $BeCl_2$ with an excess of ammonia? Draw Lewis structures of other products that might be produced.
20. Predict a structure of BeF_2 in the gas phase. What structure would you predict for $BeF_2(s)$?
21. Write balanced equations describing the reactions of Ca with each of the following:
   a. O_2 d. P_4
   b. S_8 e. H_2
   c. N_2 f. H_2O
22. Write formulas for barium oxide and barium peroxide.
23. Magnesium nitride and magnesium phosphide are expected to react with water in a fashion similar to lithium nitride. Write equations describing the reactions of magnesium nitride and magnesium phosphide with water.
24. What current is needed to produce 1.00×10^3 kg of Ca metal in 8.00 h from the electrolysis of molten $CaCl_2$? What mass of Cl_2 is produced?
25. Calculate the solubility of $Mg(OH)_2$ in an aqueous solution buffered at pH = 8.00 (see Table 8.5).

Group 3A Elements

26. Boron hydrides were once evaluated for possible use as rocket fuels. Complete and balance the following reaction:

$$B_2H_6 + O_2 \longrightarrow B(OH)_3$$

27. Name each of the following.
   a. TlOH
   b. In_2S_3
   c. Ga_2O_3
28. The compound $AlCl_3$ is quite volatile and appears to exist as a dimer in the gas phase. Propose a structure for this dimer.
29. Assume that element 113 is produced. Predict the radius of its 3+ ion and the first ionization energy, given the information in Table 18.9.
30. Lithium aluminum hydride ($LiAlH_4$) is a powerful reducing agent used in the synthesis of organic compounds. Assign oxidation states to all atoms in $LiAlH_4$.
31. Ga_2O_3 is an amphoteric oxide and In_2O_3 is a basic oxide. Write equations describing reactions that illustrate these properties.
32. What type of semiconductor is formed when a Group 3A element is added as an impurity to Si or Ge?
33. Tricalcium aluminate is an important component of Portland cement. It is 44.4% calcium and 20.0% aluminum by mass. The remainder is oxygen.
   a. Calculate the empirical formula of tricalcium aluminate.
   b. The structure of tricalcium aluminate was not determined until 1975. The $Al_6O_{18}^{18-}$ anion has the following structure:

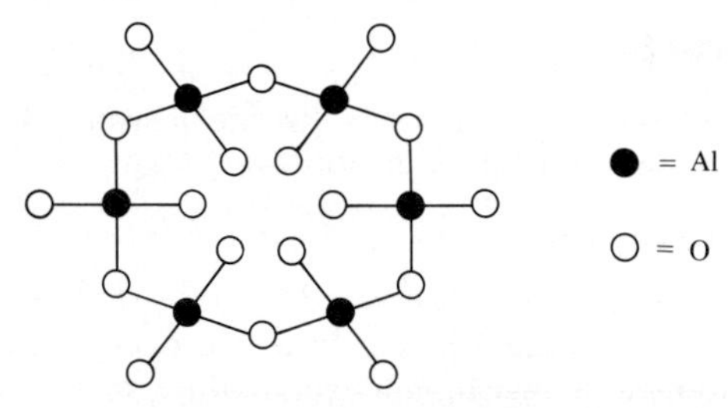

   What is the molecular formula of tricalcium aluminate?
   c. How would you describe the bonding in the $Al_6O_{18}^{18-}$ anion?
34. Write balanced equations describing the reactions of In with each of the following:
   a. F_2 c. O_2
   b. Cl_2 d. HCl
35. Write a balanced equation describing the reaction of aluminum metal with concentrated aqueous sodium hydroxide.
36. Is Al_2O_3 an acidic, basic, or amphoteric oxide? Write balanced chemical equations to support your answer.

Group 4A Elements

37. Discuss the importance of the C—C and Si—Si bond strengths and of π bonding to the properties of carbon and silicon.
38. Which bond would you expect to be more reactive, a Si—H bond or a C—H bond? Why?
39. Carbon suboxide (C_3O_2) has the following Lewis structure:

$$:\ddot{O}=C=C=C=\ddot{O}:$$

a. What is the molecular geometry of C_3O_2?
b. What hybrid orbitals are used by the carbon atoms in C_3O_2?

40. Use bond energies (see Table 13.5) to estimate ΔH for the following reaction.

$$\text{HO}-\overset{\overset{\text{O}}{\|}}{\text{C}}-\text{OH} \longrightarrow CO_2 + H_2O$$

Why is it better to write carbonic acid as $CO_2(aq)$ rather than as H_2CO_3?

41. Carbon and sulfur form compounds with the formulas CS_2 and C_3S_2. Draw Lewis structures for these compounds and predict the shapes of the molecules.

42. Why are the tin(IV) halides more volatile than the tin(II) halides?

43. From the information on the temperature stability of white and gray tin given in this chapter, which form would you expect to have the more ordered structure?

44. The compounds CCl_4 and H_2O do not react with each other. On the other hand, silicon tetrachloride reacts with water according to the equation

$$SiCl_4(l) + 2H_2O(l) \longrightarrow SiO_2(s) + 4HCl(aq)$$

Discuss the importance of thermodynamics and kinetics in the reactivity of water with $SiCl_4$ as compared with water's lack of reactivity with CCl_4.

45. Stannous fluoride (SnF_2) was formerly used to introduce F^- ions into toothpaste. Give another name for stannous fluoride.

46. Silicon carbide (SiC) is an extremely hard substance. Propose a structure for SiC.

47. Hydrofluoric acid (HF) cannot be stored in glass because it reacts with silica (SiO_2) to form the volatile SiF_4. Write a balanced equation for the reaction of HF with glass.

48. What is the proportion of lead(II) to lead(IV) in red lead (Pb_3O_4)?

49. Calculate $\mathscr{E}°$ for the reaction

$$2H_2O(l) + 2Pb(s) + O_2(g) \longrightarrow 2Pb(OH)_2(s)$$

which is analogous to the corrosion of iron by O_2 (see Chapter 11). Are lead pipes more easily corroded by oxygen than iron pipes? For $Pb(OH)_2 + 2e^- \rightarrow Pb + 2OH^-$, $\mathscr{E}° = -0.57$ V and the standard cell potential for the analagous corrosion of iron process is 1.3 V.

50. Calculate the solubility of $Pb(OH)_2$ in water (in mol/L), given that $K_{sp} = 1.2 \times 10^{-15}$.

51. Which Group 4A elements are capable of reducing H^+ to H_2? Write balanced equations describing the reactions.

Additional Exercises

52. Compounds containing the sodide ion (Na^-) have recently been made:

$$2Na + \text{Crypt} \longrightarrow [Na(\text{Crypt})]^+Na^-$$

where Crypt represents a cryptand $N[(C_2H_4O)_2C_2H_4]_3N$:

The cryptand encapsulates the Na^+ ion. Why is it necessary to encapsulate Na^+? (See Exercise 65 in Chapter 12.)

53. The Group 1A metals are extremely soluble in liquid ammonia, releasing hydrogen gas on dissolution. As the metal begins to dissolve, a deep blue solution forms. The color is attributed to the presence of solvated electrons, or $e(NH_3)_x^-$. The metal amide (MNH_2) can be isolated by allowing the ammonia to evaporate.
a. Write a balanced equation for the reaction for the formation of $NaNH_2$.
b. The solubility of sodium in liquid ammonia at $-33.5°C$ is 251.4 g/kg. Calculate the mass percent, mole fraction, and molality of sodium in this solution.

54. Ionic compounds, such as $KMnO_4$, can be dissolved in nonpolar solvents by adding *crown ethers*. The structure of a typical crown ether is as follows:

Suggest how crown ethers make $KMnO_4$ soluble in nonpolar solvents.

55. Beryllium is amphoteric, in contrast to its fellow Group 2A metals. Beryllium metal reacts with aqueous NaOH to produce hydrogen gas and $Be(OH)_4^{2-}$. Write a balanced equation for this reaction. What is the oxidizing agent? What is the reducing agent?

56. Give a formula for each binary compound.
a. beryllium nitride
b. strontium peroxide

57. The compound $BeSO_4 \cdot 4H_2O$ cannot be easily dehydrated by heating. It dissolves in water to produce an acidic solution. How do you account for these observations?

58. Elemental boron is produced by the reduction of boron oxide by magnesium to give boron and magnesium oxide. Write a balanced equation for this reaction.

59. The compound with the formula TlI_3 is a black solid. Given the following standard reduction potentials,

$$Tl^{3+} + 2e^- \longrightarrow Tl^+ \qquad \mathscr{E}° = +1.25 \text{ V}$$

$$I_3^- + 2e^- \longrightarrow 3I^- \qquad \mathscr{E}° = +0.55 \text{ V}$$

would you formulate this compound as thallium(III) iodide or thallium(I) triodide?

60. The compound Ga_2Cl_4 is brown. How could you determine experimentally whether this compound contains two gallium(II) ions or one gallium(I) and one gallium(III) ion? (*Hint:* consider the electron configurations of the three possible ions.)

61. The resistivity (a measure of electrical resistance) of graphite is $(0.4 \text{ to } 5.0) \times 10^{-4}$ ohm-cm in the basal plane. (The basal plane is the plane of the six-membered rings of carbon atoms.) The resistivity is 0.2 to 1.0 ohm-cm along the axis perpendicular to the plane. The resistivity of diamond, 10^{14} to 10^{16} ohm-cm, is independent of direction. How can you account for this behavior in terms of the structures of graphite and diamond?

62. Why is graphite a good lubricant? What advantages does it have over grease- or oil-based lubricants?

63. Elemental silicon was not isolated until 1823 when J. J. Berzelius succeeded in reducing K_2SiF_6 with molten potassium.
   a. Write a balanced equation for this reaction.
   b. Describe the bonding in K_2SiF_6.

64. Dioctyltin compounds are used as stabilizers for polyvinyl chloride polymers. They can be produced by the following reaction:

$$2CH_3CH_2CH_2CH_2CH_2CH_2CH_2CH_2X + Sn \longrightarrow (C_8H_{17})_2SnX_2$$

where X is a halogen. Predict the structure of $(C_8H_{17})_2SnCl_2$.

65. One reason suggested for the instability of long chains of silicon atoms is that the decomposition involves the transition state shown below:

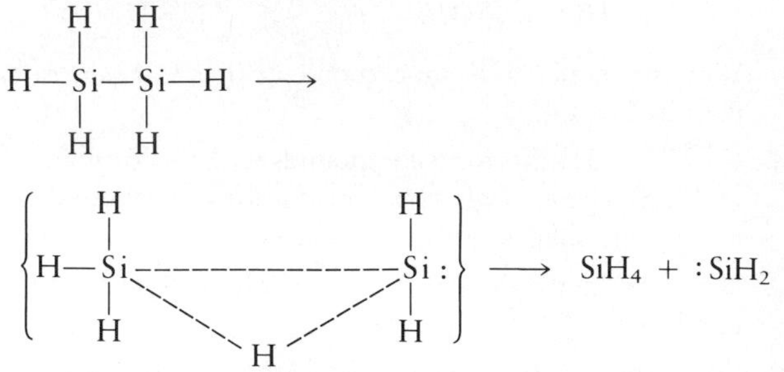

$$H_3Si{-}SiH_3 \longrightarrow \left\{ H_3Si\cdots H\cdots SiH_2{:} \right\} \longrightarrow SiH_4 + {:}SiH_2$$

The activation energy for such a process is 210 kJ/mol, which is less than either the Si—Si or the Si—H bond energy. Why would a similar mechanism not be expected to be very important in the decomposition of long carbon chains?

66. What are some of the structural differences between quartz and amorphous SiO_2?

67. Diagonal relationships in the periodic table exist in addition to the vertical relationships. For example, Be and Al are similar in some of their properties, as are B and Si. Rationalize why these diagonal relationships hold for properties such as size, ionization energy, and electron affinity.

68. What is the inert pair effect? How is it important in the properties of thallium and lead?

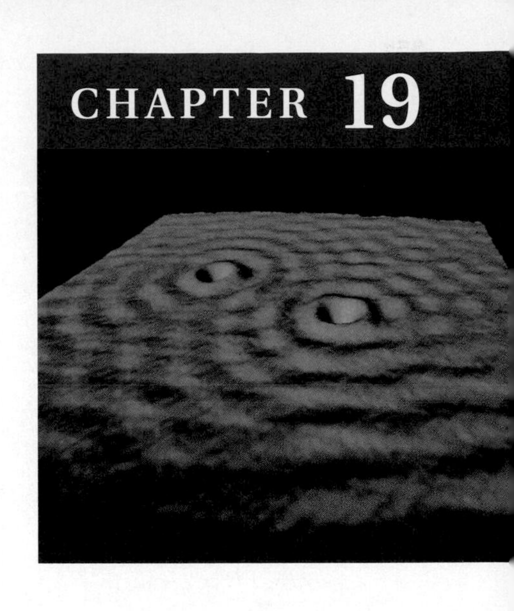

CHAPTER 19

The Representative Elements: Groups 5A Through 8A

In Chapter 18 we saw that vertical groups of elements tend to show similar chemical characteristics because they have identical valence electron configurations. Generally, metallic character increases going down a group, since the electrons are increasingly further from the nucleus. The most dramatic change occurs after the first group member and reflects the large percent increase in atomic size.

As we proceed from Group 1A to Group 7A, the elements change from active metals (electron donors) to strong nonmetals (electron acceptors). Thus it is not surprising that the middle groups show the most varied chemistry: some group members behave principally as metals, others behave mainly as nonmetals, and some show both tendencies. The elements in Groups 5A and 6A show great chemical variety and form many compounds of considerable practical value. The halogens (Group 7A) are nonmetals that are also found in many everyday substances. The elements in Group 8A (the noble gases) are most useful in their elemental forms, but their ability to form compounds, discovered only within the past 35 years, has provided important tests for the theories of chemical bonding.

In this chapter we give an overview of the elements in Groups 5A through 8A, concentrating on the chemistry of the most important elements in these groups: nitrogen, phosphorus, oxygen, sulfur, and the halogens.

5A
N
P
As
Sb
Bi

19.1 The Group 5A Elements

The Group 5A elements (with the valence electron configuration ns^2np^3), which are prepared as shown in Table 19.1, exhibit remarkably varied chemical properties. As usual, metallic character increases going down the group, as is apparent from the electronegativity values (Table 19.1). Nitrogen and phosphorus are nonmetals that can gain three electrons to form 3 − anions in salts with active metals; examples are magnesium nitride (Mg_3N_2) and beryllium phosphide (Be_3P_2). The chemistry of these two important elements is discussed in the next two sections.

Bismuth and *antimony* tend to be metallic, readily losing electrons to form cations. Although these elements have five valence electrons, so much energy is required to remove all five that no ionic compounds containing Bi^{5+} or Sb^{5+} ions are known. Three pentahalides (BiF_5, $SbCl_5$, and SbF_5) are known, but these are molecular rather than ionic compounds. The fact that no other pentahalides of these elements are known no doubt results from the strong oxidizing ability of bismuth and antimony in the +5 oxidation state. In fact, BiF_5 is an excellent fluorinating agent, readily decomposing to fluorine and bismuth trifluoride:

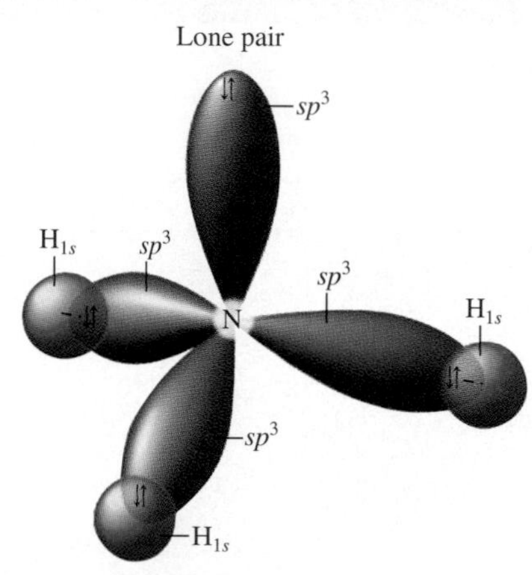

Figure 19.1
The pyramidal shape of the Group 5A MX_3 molecules.

$$BiF_5 \longrightarrow BiF_3 + F_2$$

Salts containing Bi^{3+} or Sb^{3+} ions, such as $Sb_2(SO_4)_3$ and $Bi(NO_3)_3$, are quite common. When these salts are dissolved in water, the resulting hydrated cations are very acidic. For example, the reaction of the Bi^{3+} ion with water can be represented as

$$Bi^{3+}(aq) + H_2O(l) \longrightarrow BiO^+(aq) + 2H^+(aq)$$

where BiO^+ is called the *bismuthyl ion*. If chloride ion is added to this solution, a white salt called bismuthyl chloride (BiOCl) precipitates. Antimony exhibits similar chemistry in aqueous solution.

The Group 5A elements can form molecules or ions that involve three, five, or six covalent bonds to the Group 5A atom. Examples involving three single bonds are NH_3, PH_3, NF_3, and $AsCl_3$. Each of these molecules has a lone pair of electrons (and thus can behave as a Lewis base) and a pyramidal shape as predicted by the VSEPR model (see Fig. 19.1).

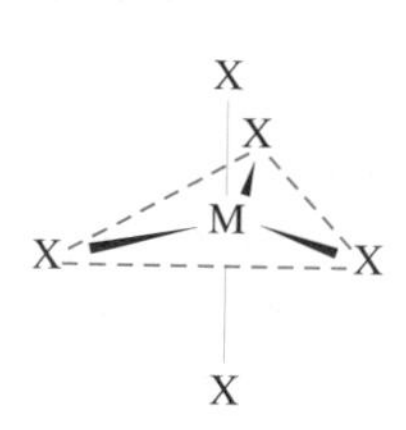

Figure 19.2
The trigonal bipyramidal shape of the MX_5 molecules.

All of the Group 5A elements except nitrogen can form molecules with five covalent bonds (of general formula MX_5). Nitrogen cannot form such molecules because of its small size and lack of available *d* orbitals. The MX_5 molecules have a trigonal bipyramidal shape (see Fig. 19.2) as predicted by the VSEPR model, and the central atom is described as dsp^3 hybridized. The MX_5 molecules can accept an additional electron pair to form ionic species containing six covalent bonds. An example is

$$PF_5 + F^- \longrightarrow PF_6^-$$

where the PF_6^- anion has an octahedral shape (see Fig. 19.3) and the phosphorus atom is described as d^2sp^3 hybridized.

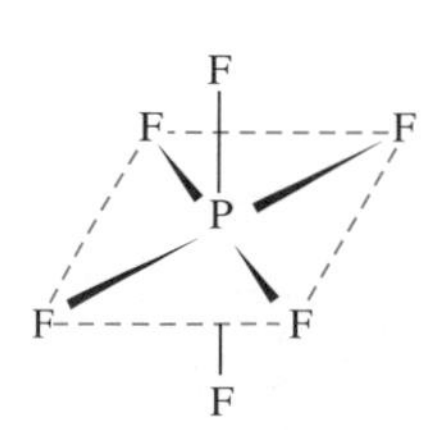

Figure 19.3
The octahedral PF_6^- ion.

Although the MX_5 molecules have a trigonal bipyramidal structure in the gas phase, the solids of many of these compounds contain a 1 : 1 mixture of the ions MX_4^+ and MX_6^- (Fig. 19.4). The MX_4^+ cation is tetrahedral (the atom represented by M is sp^3 hybridized), and the MX_6^- anion is octahedral (the

TABLE 19.1 **Selected Physical Properties, Sources, and Methods of Preparation for the Group 5A Elements**

Element	Electronegativity	Sources	Method of Preparation
Nitrogen	3.0	Air	Liquefaction of air
Phosphorus	2.1	Phosphate rock ($Ca_3(PO_4)_2$), fluorapatite ($Ca_5(PO_4)_3F$)	$2Ca_3(PO_4)_2 + 6SiO_2 \longrightarrow 6CaSiO_3 + P_4O_{10}$ $P_4O_{10} + 10C \longrightarrow 4P + 10CO$
Arsenic	2.0	Arsenopyrite (Fe_3As_2, FeS)	Heating arsenopyrite in the absence of air
Antimony	1.9	Stibnite (Sb_2S_3)	Roasting Sb_2S_3 in air to form Sb_2O_3 and then reduction with carbon
Bismuth	1.9	Bismite (Bi_2O_3), bismuth glance (Bi_2S_3)	Roasting Bi_2S_3 in air to form Bi_2O_3 and then reduction with carbon

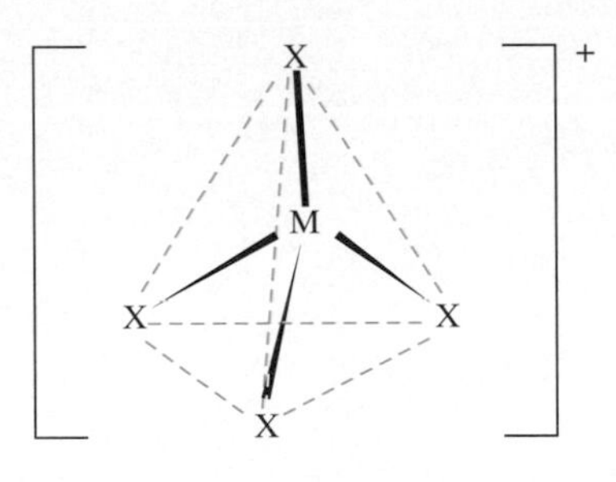

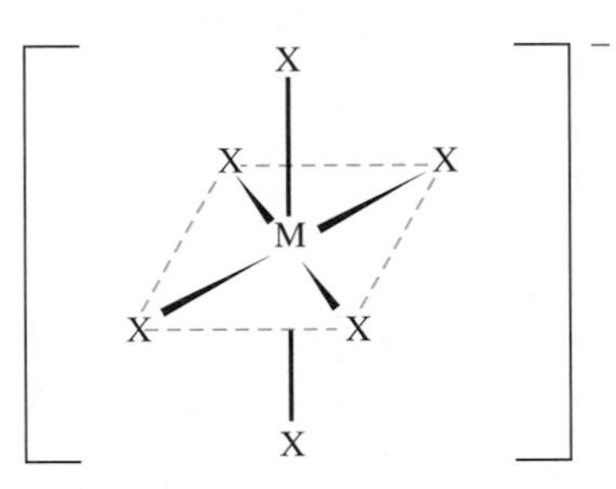

Figure 19.4
The structures of the tetrahedral MX_4^+ and the octahedral MX_6^- ions.

atom represented by M is d^2sp^3 hybridized). Examples are PCl_5 (which in the solid state contains PCl_4^+ and PCl_6^-) and AsF_3Cl_2 (which in the solid state contains $AsCl_4^+$ and AsF_6^-).

As discussed in Section 18.1, the ability of the Group 5A elements to form π bonds decreases dramatically after nitrogen. This explains why elemental nitrogen exists as N_2 molecules containing two π bonds, whereas the other elements in the group exist as larger aggregates containing single bonds. For example, in the gas phase the elements phosphorus, arsenic, and antimony consist of P_4, As_4, and Sb_4 molecules, respectively.

19.2 The Chemistry of Nitrogen

At the earth's surface virtually all elemental nitrogen exists as the N_2 molecule with its very strong triple bond (941 kJ/mol). Because of this large bond strength, the N_2 molecule is so unreactive that it can coexist with most other elements under normal conditions without undergoing any appreciable reaction. This property makes nitrogen gas very useful as a medium for experiments involving substances that react with oxygen or water. Such experiments can be done using an inert atmosphere box of the type shown in Fig. 19.5.

The strength of the triple bond in the N_2 molecule is important both thermodynamically and kinetically. Thermodynamically, the great stability of the N≡N bond means that most binary compounds containing nitrogen decompose exothermically to the elements, for example:

$$N_2O(g) \longrightarrow N_2(g) + \tfrac{1}{2}O_2(g) \qquad \Delta H^\circ = -82 \text{ kJ}$$
$$NO(g) \longrightarrow \tfrac{1}{2}N_2(g) + \tfrac{1}{2}O_2(g) \qquad \Delta H^\circ = -90 \text{ kJ}$$
$$NO_2(g) \longrightarrow \tfrac{1}{2}N_2(g) + O_2(g) \qquad \Delta H^\circ = -34 \text{ kJ}$$
$$N_2H_4(g) \longrightarrow N_2(g) + 2H_2(g) \qquad \Delta H^\circ = -95 \text{ kJ}$$
$$NH_3(g) \longrightarrow \tfrac{1}{2}N_2(g) + \tfrac{3}{2}H_2(g) \qquad \Delta H^\circ = +46 \text{ kJ}$$

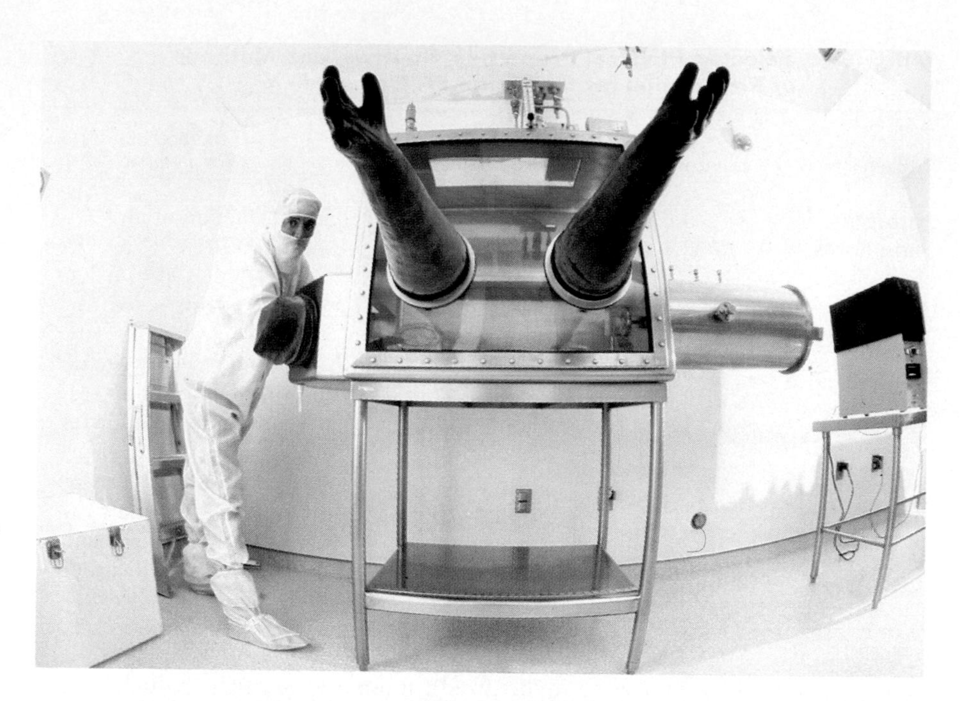

Figure 19.5
An inert atmosphere box used when working with oxygen- or water-sensitive materials. The box is filled with an inert gas such as nitrogen, and work is done through the ports fitted with large rubber gloves.

Of these compounds only ammonia is thermodynamically more stable than its component elements. That is, only for ammonia is energy required to decompose the molecule to its elements. For the remaining molecules energy is released when decomposition to the elements occurs, as a result of the great stability of N_2.

The importance of the thermodynamic stability of N_2 can be clearly seen in the power of nitrogen-based explosives, such as nitroglycerin ($C_3H_5N_3O_9$), which has the following structure:

```
         H          H          H
         |          |          |
      H—C————————C————————C—H
         |          |          |
         O          O          O
         |          |          |
         N          N          N
        / \        / \        / \
       O   O      O   O      O   O
```

When ignited or subjected to sudden impact, nitroglycerin decomposes very rapidly and exothermically:

$$4C_3H_5N_3O_9(l) \longrightarrow 6N_2(g) + 12CO_2(g) + 10H_2O(g) + O_2(g) + \text{energy}$$

An explosion occurs; that is, large volumes of gas are produced in a fast, highly exothermic reaction. Note that 4 moles of liquid nitroglycerin produce 29 (6 + 12 + 10 + 1) moles of gaseous products. This alone produces a large increase in volume. However, also note that the products, which include N_2, are very stable molecules with strong bonds. Their formation is therefore accompanied by the release of large quantities of energy as heat, which increases the gaseous volume. The hot, rapidly expanding gases produce a pressure surge and damaging shock wave.

Most high explosives are organic compounds that, like nitroglycerin, contain nitro ($—NO_2$) groups and produce nitrogen and other gases as products. Another example is *trinitrotoluene* (TNT), a solid at normal temperatures, which decomposes as follows:

$$2C_7H_5N_3O_6(s) \longrightarrow 12CO(g) + 5H_2(g) + 3N_2(g) + 2C(s) + \text{energy}$$

CH_3
NO_2 NO_2
NO_2
TNT

Note that 2 moles of solid TNT produce 20 moles of gaseous products plus energy.

The effect of bond strength on the kinetics of reactions involving the N_2 molecule is illustrated by the synthesis of ammonia from nitrogen and hydrogen, a reaction we have discussed many times before. Because a large quantity of energy is required to disrupt the $N\equiv N$ bond, the ammonia synthesis reaction occurs at a negligible rate at room temperature, even though the equilibrium constant is very large ($K \approx 10^6$). Of course, the most direct way to increase the rate of a reaction is to raise the temperature. However, since this reaction is very exothermic,

$$N_2(g) + 3H_2(g) \longrightarrow 2NH_3(g) \qquad \Delta H° = -92 \text{ kJ}$$

the value of K decreases significantly with a temperature increase (at 500°C, $K \approx 10^{-2}$).

Obviously, the kinetics and the thermodynamics of this reaction are in opposition. A compromise must be reached, involving high pressure to force the equilibrium to the right and high temperature to produce a reasonable rate. The **Haber process** for manufacturing ammonia represents such a compromise (see Fig. 19.6). The process is carried out at a pressure of about 250 atm and a temperature of approximately 400°C. Even higher temperatures would be required if a catalyst consisting of a solid iron oxide mixed with small amounts of potassium oxide and aluminum oxide were not used to facilitate the reaction.

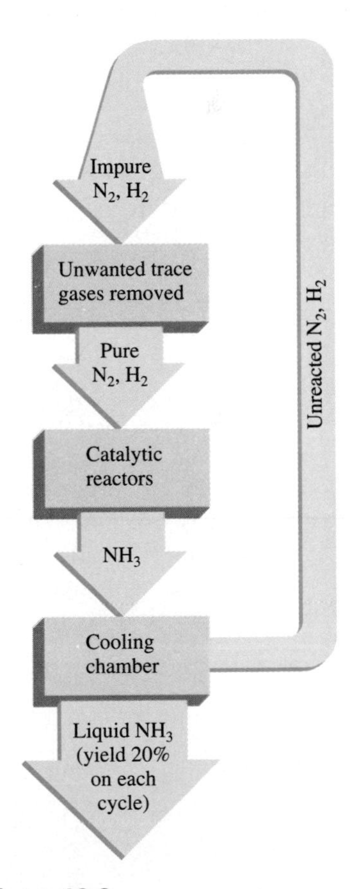

Figure 19.6
A schematic diagram of the Haber process for the manufacture of ammonia.

Nitrogen is essential to living systems. The problem with nitrogen is not one of supply—we are surrounded by it—but rather of changing it from the inert N_2 molecule to a form usable by plants and animals. The process of transforming N_2 to other nitrogen-containing compounds is called **nitrogen fixation.** The Haber process is one example of nitrogen fixation. The ammonia produced can be applied to the soil as a fertilizer, since plants can readily employ the nitrogen in ammonia to make the nitrogen-containing biomolecules essential for their growth.

Nitrogen fixation also results from the high-temperature combustion process in automobile engines. The nitrogen in the air is drawn into the engine and reacts at a significant rate with oxygen to form nitric oxide (NO), which further reacts with oxygen from the air to form nitrogen dioxide (NO_2). This nitrogen dioxide, which contributes to photochemical smog in many urban areas (see Section 15.9), reacts with moisture in the air and eventually reaches the soil to form nitrate salts, which are plant nutrients.

Nitrogen fixation also occurs naturally. For example, lightning provides the energy to disrupt N_2 and O_2 molecules in the air, producing highly reactive nitrogen and oxygen atoms. These atoms in turn attack other N_2 and O_2 molecules to form nitrogen oxides that eventually become nitrates. Although lightning has traditionally been credited with forming about 10% of the total fixed nitrogen, recent studies indicate that lightning may account for as much as half of the fixed nitrogen available on earth. Another natural nitrogen fixation

A BLANKET OF NITROGEN

Inert gases play a very important role as blanketing agents to prevent unwanted chemical reactions. Nitrogen gas, because of its ready availability and low reactivity at normal temperatures, has been the leader in this market. For example, nitrogen is finding increasing use for blanketing fruit after it has been picked to protect it from deterioration until the consumer buys it. Apples that receive no protective treatment deteriorate to the point of being unmarketable within about six weeks after picking (of which four weeks are required for ripening). However, the marketable range of apples can be extended to nearly thirty months by storing them at temperatures just above freezing and by controlling the storage atmosphere. As apples ripen, they consume oxygen and emit carbon dioxide and ethylene (C_2H_4). Ethylene is a ripening hormone that rapidly accelerates the ripening, and thus the deterioration, of the fruit.

State-of-the-art cold-storage facilities for apples have purging and recycling systems to lower the oxygen content of the gas in the room to below 5% within two days. Optimum long-term storage conditions involve oxygen concentrations of $\approx$1.5% by volume, carbon dioxide concentrations of $\approx$3% by volume, and ethylene concentrations below 1 ppm.

One relatively expensive method for producing the nitrogen gas used in these storage facilities is the vaporization of liquid nitrogen stored in cryogenic tanks. A less expensive source of nitrogen gas is the exhaust gas from the combustion reaction of propane and air. The carbon dioxide in the exhaust is removed with activated charcoal. In Europe nitrogen gas is commonly produced by catalytically decomposing ammonia to nitrogen and hydrogen and then removing the hydrogen gas by reaction with oxygen.

However, recent advances in high-volume gas separation technology in the United States have made direct separation of nitrogen from air economically feasible. One separation method is called the pressure-swing adsorption process, in which carbon molecular sieves (substances with precisely sized passages) are used to selectively adsorb the oxygen from air while allowing the nitrogen to pass through. When the molecular sieves become saturated with oxygen, the O_2 molecules are dislodged by a sudden jump in pressure—thus the name of the process. Using more than one bed of molecular sieves in parallel permits a steady flow of nitrogen for controlled-atmosphere applications.

Another gas separation process employs special hollow fiber membranes designed to allow the oxygen and water vapor in a stream of air to "leak" through the walls of the fibers as the air flows through, producing a stream of nitrogen at the end of the tube.

One major advantage of these latter two processes is that they can be used to generate a controlled atmosphere for a product such as fruit while it is in transport by ship, rail, or truck. These gas separation processes have also been adapted for use in other industries. For example, the Fetzer Winery in Redwood Valley, California, uses a membrane separator to produce a nitrogen-rich atmosphere for its fermentation tanks to prevent oxygen-driven decomposition of the wine.

The future of the gas-blanketing market appears anything but inert.

process involves bacteria that reside in the root nodules of plants such as beans, peas, and alfalfa. These **nitrogen-fixing bacteria** readily allow the conversion of nitrogen to ammonia and to other nitrogen-containing compounds useful to plants. The efficiency of these bacteria is intriguing: they produce ammonia at soil temperatures and 1 atm pressure, whereas the Haber process requires severe conditions of 400°C and 250 atm. For obvious reasons, researchers are studying these bacteria intensively.

When plants and animals die and decompose, the elements they consist of are returned to the environment. In the case of nitrogen, the return of the element to the atmosphere as nitrogen gas, called **denitrification,** is carried out by bacteria that change nitrates to nitrogen. The complex **nitrogen cycle** is summarized in Fig. 19.7. It has been estimated that as much as 10 million tons more nitrogen per year is currently being fixed by natural and human processes than is being returned to the atmosphere. This fixed nitrogen is accumulating in the soil, lakes, rivers, and oceans, where it promotes the growth of algae and other undesirable organisms.

Nodules containing nitrogen-fixing bacteria on the roots of peas.

Nitrogen Hydrides

By far the most important hydride of nitrogen is **ammonia.** A toxic, colorless gas with a pungent odor, ammonia is manufactured in huge quantities (≈40 billion pounds per year), mainly for use in fertilizers.

The pyramidal ammonia molecule has a lone pair of electrons on its nitrogen atom (see Fig. 19.1) and polar N—H bonds. This structure leads to a high degree of intermolecular interaction by hydrogen bonding in the liquid state, thereby producing an unusually high boiling point (−33.4°C) for a substance with such a low molar mass. Note, however, that the hydrogen bonding in liquid ammonia is clearly not as important as that in liquid water, which has about the same molar mass but a much higher boiling point. The water molecule has two polar bonds involving hydrogen and two lone pairs—the right combination for optimum hydrogen bonding—in contrast to the one lone pair and three polar bonds of the ammonia molecule.

As we saw in Chapter 7, ammonia behaves as a base, reacting with acids to produce ammonium salts. For example,

$$NH_3(g) + HCl(g) \longrightarrow NH_4Cl(s)$$

A second nitrogen hydride of major importance is **hydrazine** (N_2H_4). The Lewis structure of hydrazine

```
H         H
 \ ..  .. /
  N — N
 /        \
H          H
```

indicates that each nitrogen atom should be sp^3 hybridized with bond angles close to 109.5° (the tetrahedral angle), since the nitrogen atom is surrounded by four electron pairs. The observed structure with bond angles of 112° (see Fig. 19.8) agrees reasonably well with these predictions. Hydrazine, a colorless liquid with an ammoniacal odor, freezes at 2°C and boils at 113.5°C. This boiling point is quite high for a compound with a molar mass of 32; this suggests that considerable hydrogen bonding occurs among the polar hydrazine molecules.

Hydrazine is a powerful reducing agent and has been widely used as a rocket propellant. For example, its reaction with oxygen is highly exothermic:

$$N_2H_4(l) + O_2(g) \longrightarrow N_2(g) + 2H_2O(g) \qquad \Delta H^\circ = -622 \text{ kJ}$$

Since hydrazine also reacts vigorously with the halogens, fluorine is often used instead of oxygen as the oxidizer in rocket engines. Substituted hydrazines,

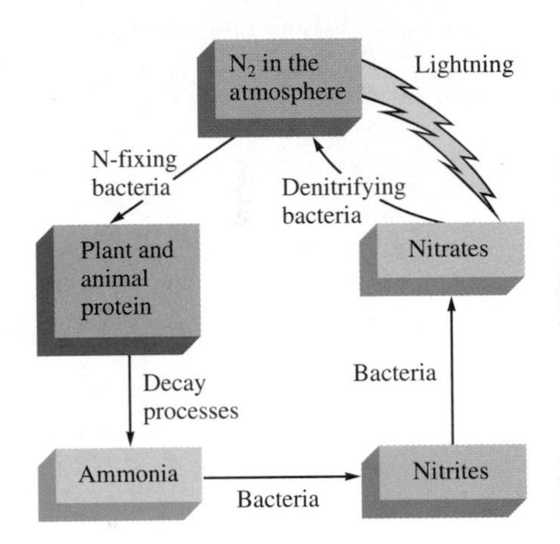

Figure 19.7
The nitrogen cycle. To be used by plants and animals, nitrogen must be converted from N_2 to nitrogen-containing compounds, such as nitrates, ammonia, and proteins. The nitrogen is returned to the atmosphere by natural decay processes.

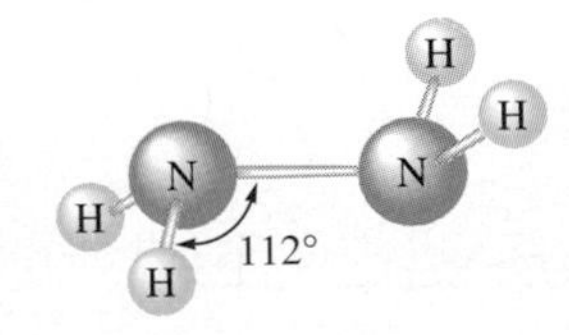

Figure 19.8
The molecular structure of hydrazine (N_2H_4). This arrangement minimizes the repulsion between the lone pairs on the nitrogen atoms by placing them on opposite sides.

The space shuttle orbiter uses monomethylhydrazine mixed with an oxidizer as fuel.

where one or more of the hydrogen atoms are replaced by other groups, are also useful rocket fuels. For example, monomethylhydrazine,

$$\begin{array}{ccccc} CH_3 & & & & H \\ & \diagdown & & \diagup & \\ & & N{-}N & & \\ & \diagup & & \diagdown & \\ H & & & & H \end{array}$$

is used with the oxidizer dinitrogen tetroxide (N_2O_4) to power the U.S. space shuttle orbiter. The reaction is

$$5N_2O_4(l) + 4N_2H_3(CH_3)(l) \longrightarrow 12H_2O(g) + 9N_2(g) + 4CO_2(g)$$

Because of the large number of gaseous molecules produced and the exothermic nature of this reaction, a very high thrust per mass of fuel is achieved. The reaction is also self-starting—it begins immediately when the fuels are mixed—which is a useful property for rocket engines that must be started and stopped frequently.

The use of hydrazine as a rocket propellant is a rather specialized application. The main industrial use of hydrazine is as a "blowing" agent in the manufacture of plastics. Hydrazine decomposes to form nitrogen gas, which causes foaming in the liquid plastic and results in a porous texture. Another major use of hydrazine is in the production of agricultural pesticides. Of the many hundreds of hydrazine derivatives (substituted hydrazines) that have been tested, 40 are used as fungicides, herbicides, insecticides, or plant growth regulators.

The manufacture of hydrazine involves the oxidation of ammonia by the hypochlorite ion in basic solution:

$$2NH_3(aq) + OCl^-(aq) \longrightarrow N_2H_4(aq) + Cl^-(aq) + H_2O(l)$$

Although this reaction looks straightforward, the actual process involves many steps. This reaction also requires high pressure, high temperature, and catalysis to optimize the yield of hydrazine in the face of many competing reactions.

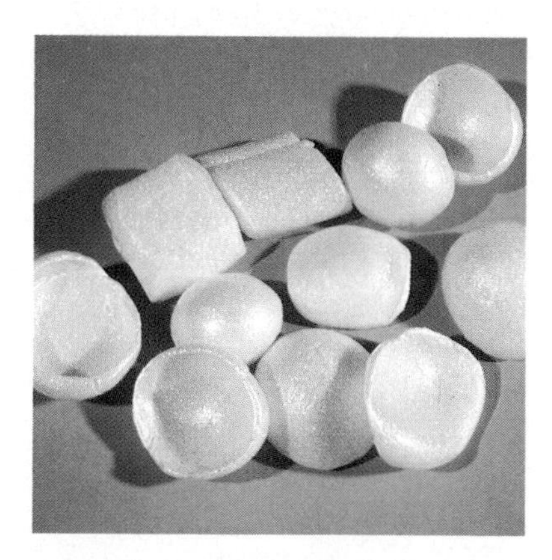

Blowing agents—such as hydrazine, which forms nitrogen gas on decomposition—are used to produce porous plastics like these Styrofoam chips.

Nitrogen Oxides

Nitrogen forms a series of oxides in which its oxidation state ranges from $+1$ to $+5$, as shown in Table 19.2.

Dinitrogen monoxide (N_2O), more commonly called *nitrous oxide* or laughing gas, has an inebriating effect and is often used as a mild anesthetic by dentists. Because of its high solubility in fats, nitrous oxide is widely used as a propellant in aerosol cans of whipped cream. It is dissolved in the liquid inside the can at high pressure and forms bubbles that produce foaming as the liquid is released from the can. A significant amount of N_2O exists in the atmosphere, mostly produced by soil microorganisms, and its concentration appears to be gradually increasing. Because it can strongly absorb infrared radiation, nitrous oxide plays a small but probably significant role in controlling the earth's temperature in the same way that atmospheric carbon dioxide and water vapor do (see the discussion of the greenhouse effect in Section 9.7). Some scientists fear that the rapid decrease of tropical rain forests resulting from the development of countries such as Brazil will significantly affect the rate of production

TABLE 19.2 Some Common Nitrogen Compounds

Oxidation State of Nitrogen	Compound	Formula	Lewis Structure*
−3	Ammonia	NH_3	H—N—N | H
−2	Hydrazine	N_2H_4	H—N—N—H | | H H
−1	Hydroxylamine	NH_2OH	H—N—O—H | H
0	Nitrogen	N_2	:N≡N:
+1	Dinitrogen monoxide (nitrous oxide)	N_2O	N=N=O
+2	Nitrogen monoxide (nitric oxide)	NO	:N=O
+3	Dinitrogen trioxide	N_2O_3	O—N(=O)—N=O
+4	Nitrogen dioxide	NO_2	:O—N=O
+5	Nitric acid	HNO_3	:O—N—O—H ‖ O

* In some cases additional resonance structures are needed to fully describe the electron distribution.

A copper penny reacts with nitric acid to produce NO gas, which is immediately oxidized in air to give brown NO_2.

of N_2O by soil organisms and thus will have important effects on the earth's temperature.

In the laboratory nitrous oxide is prepared by the thermal decomposition of ammonium nitrate:

$$NH_4NO_3(s) \xrightarrow{\text{Heat}} N_2O(g) + 2H_2O(g)$$

This experiment must be done carefully because ammonium nitrate can explode. In fact, one of the greatest industrial disasters in U.S. history occurred in 1947 in Texas when a ship loaded with ammonium nitrate for use as fertilizer exploded and killed nearly 600 people.

Do not attempt this experiment unless you have the proper safety equipment.

Nitrogen monoxide (NO), commonly called nitric oxide, is a colorless gas under normal conditions and can be produced in the laboratory by reacting 6 *M* nitric acid with copper metal:

$$8H^+(aq) + 2NO_3^-(aq) + 3Cu(s) \longrightarrow 3Cu^{2+}(aq) + 4H_2O(l) + 2NO(g)$$

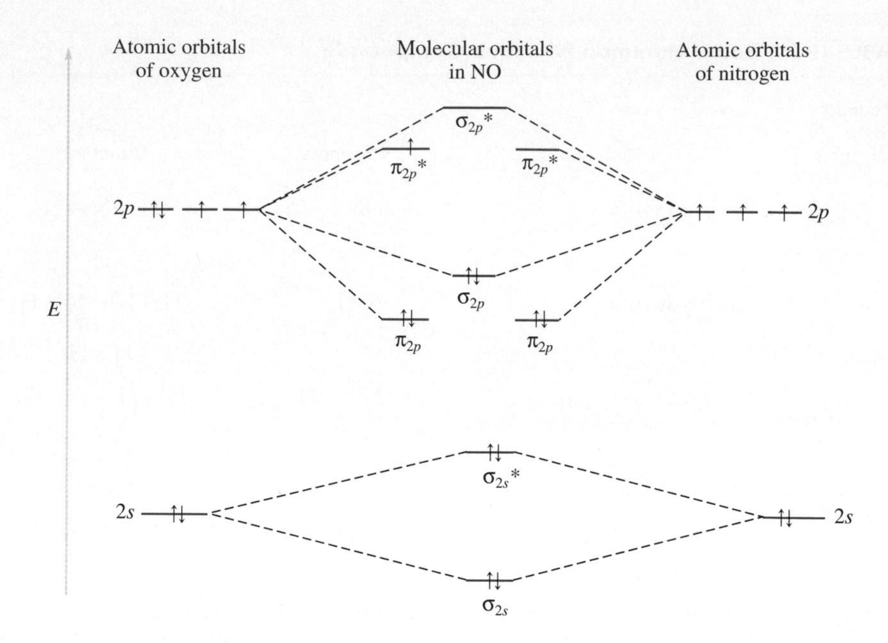

Figure 19.9
The molecular orbital energy-level diagram for nitric oxide (NO). The bond order is 2.5, or (8 − 3)/2.

When this reaction is carried out in the air, the nitric oxide is immediately oxidized by O_2 to brown nitrogen dioxide (NO_2).

Since the NO molecule has an odd number of electrons, it is most conveniently described in terms of the molecular orbital model. The molecular orbital energy-level diagram is shown in Fig. 19.9. Note that the NO molecule should be paramagnetic and have a bond order of 2.5, predictions that are supported by experimental observations. Since the NO molecule has one high-energy electron, it is not surprising that it can be rather easily oxidized to form NO^+, the *nitrosyl ion*. Because an antibonding electron is removed in going from NO to NO^+, the resulting ion should have a stronger bond (the predicted bond order is 3) than the molecule. This is borne out by experiment. The bond lengths and bond energies for nitric oxide and the nitrosyl ion are shown in Table 19.3. The nitrosyl ion is formed when nitric oxide and nitrogen dioxide are dissolved in concentrated sulfuric acid:

$$NO(g) + NO_2(g) + 3H_2SO_4(aq) \longrightarrow 2NO^+(aq) + 3HSO_4^-(aq) + H_3O^+(aq)$$

The ionic compound $NO^+HSO_4^-$ can be isolated from this solution.

Nitric oxide is thermodynamically unstable and decomposes to nitrous oxide and nitrogen dioxide:

$$3NO(g) \longrightarrow N_2O(g) + NO_2(g)$$

Nitrogen dioxide (NO_2), which is also an odd-electron molecule, has a V-shaped structure. The brown, paramagnetic NO_2 molecule readily dimerizes to form dinitrogen tetroxide,

$$2NO_2(g) \rightleftharpoons N_2O_4(g)$$

which is diamagnetic and colorless. The value of the equilibrium constant is ≈1 for this process at 55°C and, since the dimerization is exothermic, K decreases as the temperature increases.

TABLE 19.3
Comparison of the Bond Lengths and Bond Energies for Nitric Oxide and the Nitrosyl Ion

	NO	NO^+
Bond length (Å)	1.15	1.09
Bond energy (kJ/mol)	630	1020
Bond order (predicted by MO model)	2.5	3

Nitrous Oxide: Laughing Gas That Propels Whipped Cream and Cars

Nitrous oxide (N_2O), more properly called dinitrogen monoxide, is a compound with many interesting uses. It was discovered in 1772 by Joseph Priestley (who is also given credit for discovering oxygen gas), and its intoxicating effects were noted almost immediately. In 1798 the 20-year-old Humphry Davy became director of the Pneumatic Institute, which was set up to investigate the medical effects of various gases. Davy tested the effects of N_2O on himself, reporting that after inhaling 16 quarts of the gas in 7 minutes, he became "absolutely intoxicated."

Over the next century "laughing gas," as nitrous oxide became known, was developed as an anesthetic, particularly for dental procedures. Nitrous oxide is still used as an anesthetic, although it has been largely replaced by more modern drugs.

One major use of nitrous oxide today is as the propellant in cans of "instant" whipped cream. The high solubility of N_2O in the whipped cream mixture makes it an excellent candidate for pressurizing the cans of whipping cream.

Another current use of nitrous oxide is to produce "instant horsepower" for street racers. Because the reaction of N_2O with O_2 to form NO actually absorbs heat, this reaction has a cooling effect when placed in the fuel mixture in an automobile engine. This cooling effect lowers combustion temperatures, thus allowing the fuel/air mixture to be significantly more dense (the density of a gas is inversely proportional to temperature). This effect can produce a burst of additional power in excess of 200 horsepower. Because engines are not designed to run steadily at such high power levels, the nitrous oxide is injected from a tank when extra power is desired.

The least common of the nitrogen oxides are *dinitrogen trioxide* (N_2O_3), a blue liquid that readily dissociates to gaseous nitric oxide and nitrogen dioxide, and *dinitrogen pentoxide* (N_2O_5), which under normal conditions is a solid that is best viewed as a mixture of NO_2^+ and NO_3^- ions. Although N_2O_5 molecules can exist in the gas phase, they readily dissociate to nitrogen dioxide and oxygen:

$$2N_2O_5(g) \rightleftharpoons 4NO_2(g) + O_2(g)$$

This reaction follows first-order kinetics, as discussed in Section 15.4.

Oxyacids of Nitrogen

Nitric acid is an important industrial chemical (≈8 million tons produced annually) used in the manufacture of many products, such as nitrogen-based explosives and ammonium nitrate for use as a fertilizer.

Nitric acid is produced commercially by the oxidation of ammonia in the **Ostwald process** (see Fig. 19.10). In the first step of this process ammonia is oxidized to nitric oxide:

$$4NH_3(g) + 5O_2(g) \longrightarrow 4NO(g) + 6H_2O(g) \qquad \Delta H^\circ = -905 \text{ kJ}$$

Although this reaction is highly exothermic, it is very slow at 25°C. A side reaction occurs between nitric oxide and ammonia:

$$4NH_3(g) + 6NO(g) \longrightarrow 5N_2(g) + 6H_2O(g)$$

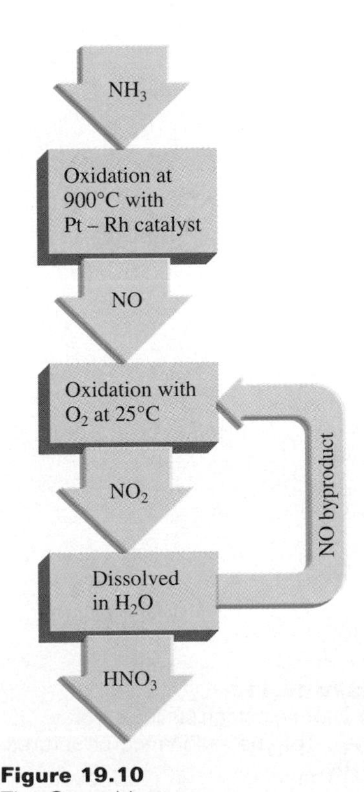

Figure 19.10
The Ostwald process.

which is particularly undesirable because it traps the nitrogen in the very unreactive N_2 molecules. The desired reaction can be accelerated and the effects of the competing reaction can be minimized if the ammonia oxidation is carried out by using a catalyst of a platinum-rhodium alloy heated to 900°C. Under these conditions there is a 97% conversion of the ammonia to nitric oxide.

In the second step nitric oxide is reacted with oxygen to produce nitrogen dioxide:

$$2NO(g) + O_2(g) \longrightarrow 2NO_2(g) \qquad \Delta H° = -113 \text{ kJ}$$

This oxidation reaction has a rate that *decreases* with increasing temperature. Because of this very unusual behavior, the reaction is carried out at ≈25°C and is kept at this temperature by cooling with water.

The third step in the Ostwald process is the absorption of nitrogen dioxide by water:

$$3NO_2(g) + H_2O(l) \longrightarrow 2HNO_3(aq) + NO(g) \qquad \Delta H° = -139 \text{ kJ}$$

The gaseous NO produced in the reaction is recycled so it can be oxidized to NO_2. The aqueous nitric acid from this process is about 50% HNO_3 by mass, which can be increased to 68% by distillation to remove some of the water. The maximum concentration attainable by this method is 68%, because nitric acid and water form an *azeotrope* at this concentration. The solution can be further concentrated to 95% HNO_3 by treatment with concentrated sulfuric acid, which strongly absorbs water; H_2SO_4 is often used as a *dehydrating (water-removing) agent*.

An *azeotrope* is a solution that, like a pure liquid, distills at a constant temperature without a change in composition.

Nitric acid is a colorless, fuming liquid (bp = 83°C) with a pungent odor; it decomposes in sunlight by the following reaction:

$$4HNO_3(l) \xrightarrow{h\nu} 4NO_2(g) + 2H_2O(l) + O_2(g)$$

As a result, nitric acid turns yellow as it ages because of the dissolved nitrogen dioxide. The common laboratory reagent called concentrated nitric acid is 15.9 *M* HNO_3 (70.4% HNO_3 by mass) and is a very strong oxidizing agent. The resonance structures and molecular structure of HNO_3 are shown in Fig. 19.11. Note that the hydrogen is bound to an oxygen atom, rather than to nitrogen as the formula might suggest.

Nitric acid reacts with oxides, hydroxides, carbonates, and with other ionic compounds containing basic anions to form nitrate salts. For example,

$$Ca(OH)_2(s) + 2HNO_3(aq) \longrightarrow Ca(NO_3)_2(aq) + 2H_2O(l)$$

Nitrate salts are generally very soluble in water.

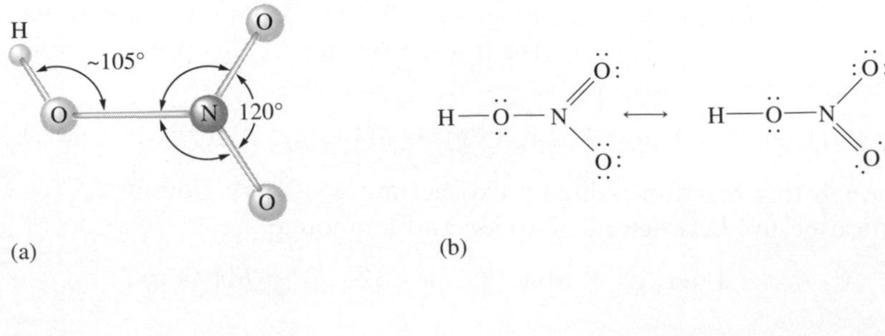

H ~105° O N 120° O O

(a)

H—Ö—N(=O:)(—O:) ⟷ H—Ö—N(—O:)(=O:)

(b)

Figure 19.11
(a) The molecular structure of HNO_3. (b) The resonance structures of HNO_3.

Nitrous acid (HNO_2) is a weak acid,

$$HNO_2(aq) \rightleftharpoons H^+(aq) + NO_2^-(aq) \qquad K_a = 4.0 \times 10^{-4}$$

that forms pale yellow nitrite (NO_2^-) salts. In contrast to nitrates, which are often used as explosives, nitrites are quite stable even at high temperatures. Nitrites are usually prepared by bubbling equal numbers of moles of nitric oxide and nitrogen dioxide into the appropriate aqueous solution of a metal hydroxide. For example,

$$NO(g) + NO_2(g) + 2NaOH(aq) \longrightarrow 2NaNO_2(aq) + H_2O(l)$$

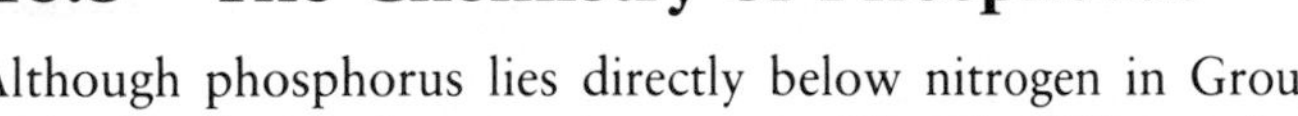

19.3 The Chemistry of Phosphorus

White phosphorus reacts vigorously with the oxygen in air and must be stored under water. Red phosphorus is stable in air.

Although phosphorus lies directly below nitrogen in Group 5A of the periodic table, its chemical properties are significantly different from those of nitrogen. The differences arise mainly from four factors: nitrogen's ability to form much stronger π bonds, the greater electronegativity of nitrogen, the larger size of phosphorus atoms, and the availability of empty valence d orbitals on phosphorus.

The chemical differences between nitrogen and phosphorus are apparent in their elemental forms. In contrast to the diatomic form of elemental nitrogen, which is stabilized by strong π bonds, there are several solid forms of phosphorus that all contain aggregates of atoms. *White phosphorus,* which contains discrete tetrahedral P_4 molecules [see Fig. 19.12(a)], is very reactive; it bursts into flames on contact with air (it is said to be *pyrophoric*). Consequently, white phosphorus is commonly stored under water. White phosphorus is quite toxic; the P_4 molecules are very damaging to tissue, particularly the cartilage and bones of the nose and jaw. The much less reactive forms called *black phosphorus* and *red phosphorus* are network solids (see Section 16.5). Black phosphorus has a regular crystalline structure [Fig. 19.12(b)], but red phosphorus is amorphous and is thought to consist of chains of P_4 units [Fig. 19.12(c)]. Red phosphorus can be obtained by heating white phosphorus in the absence of air at 1 atm. Black phosphorus is obtained from either white or red phosphorus by heating at high pressures.

Even though phosphorus has a lower electronegativity than nitrogen, it will form phosphides (ionic substances containing the P^{3-} anion) such as Na_3P and Ca_3P_2. Phosphide salts react vigorously with water to produce *phosphine* (PH_3), a toxic, colorless gas:

$$2Na_3P(s) + 6H_2O(l) \longrightarrow 2PH_3(g) + 6Na^+(aq) + 6OH^-(aq)$$

Phosphine is analogous to ammonia, although it is a much weaker base ($K_b \approx 10^{-26}$) and is much less soluble in water. Because phosphine has a relatively

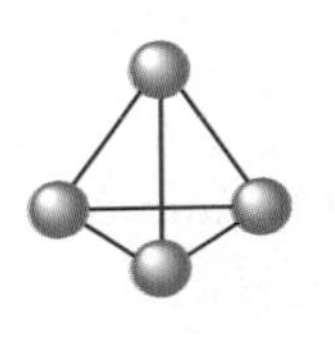

(a)

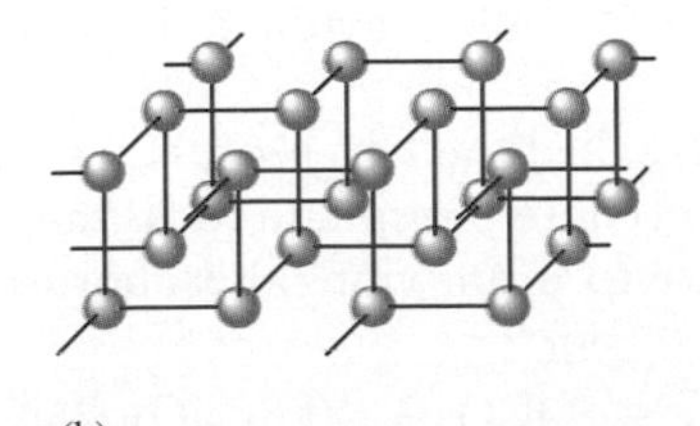

(b)

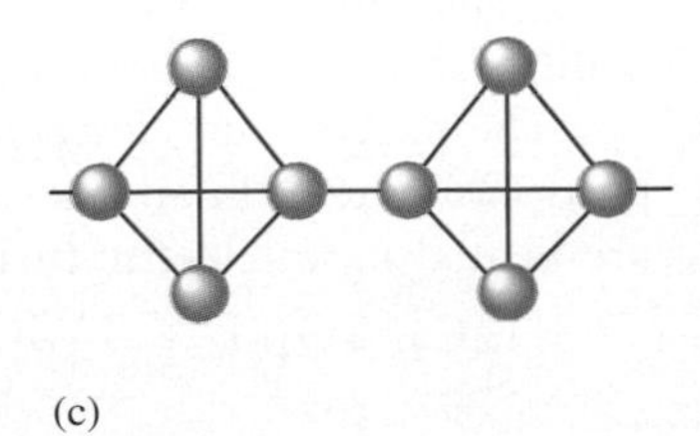

(c)

Figure 19.12
(a) The P_4 molecule found in white phosphorus. (b) The crystalline network structure of black phosphorus. (c) The chain structure of red phosphorus.

small affinity for protons, phosphonium (PH_4^+) salts are very uncommon and not very stable—only PH_4I, PH_4Cl, and PH_4Br are known.

Phosphine has the Lewis structure

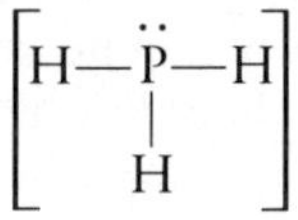

$$\left[\begin{array}{c} H-\ddot{P}-H \\ | \\ H \end{array}\right]$$

and a pyramidal molecular structure as we would predict from the VSEPR model. However, it has bond angles of 94°, rather than 107° as found in the ammonia molecule. The reasons for this are complex; therefore we will simply regard phosphine as an exception to the simple version of the VSEPR model that we use.

Phosphorus Oxides and Oxyacids

Phosphorus reacts with oxygen to form oxides in which its oxidation states are +5 and +3. The oxide P_4O_6 is formed when elemental phosphorus is burned in a limited supply of oxygen, and P_4O_{10} is produced when the oxygen is in excess. Picture these oxides (shown in Fig. 19.13) as being constructed by adding oxygen atoms to the fundamental P_4 structure. The intermediate states, P_4O_7, P_4O_8, and P_4O_9, which contain one, two, and three terminal oxygen atoms, respectively, are also known.

The terminal oxygens are the nonbridging oxygen atoms.

Tetraphosphorus decoxide (P_4O_{10}), which was formerly represented as P_2O_5 and called phosphorus pentoxide, has a great affinity for water and thus is a powerful dehydrating agent. For example, it can be used to convert HNO_3 and H_2SO_4 to their parent oxides, N_2O_5 and SO_3, respectively.

When tetraphosphorus decoxide dissolves in water, **phosphoric acid** (H_3PO_4), also called orthophosphoric acid, is produced:

$$P_4O_{10}(s) + 6H_2O(l) \longrightarrow 4H_3PO_4(aq)$$

Pure phosphoric acid is a white solid that melts at 42°C. Aqueous phosphoric acid is a much weaker acid ($K_{a_1} \approx 10^{-2}$) than nitric acid or sulfuric acid and is a poor oxidizing agent.

Phosphate minerals are the main source of phosphoric acid. Unlike nitrogen, phosphorus is found in nature exclusively in a combined state, principally as the PO_4^{3-} ion in phosphate rock, which is mainly calcium phosphate, $Ca_3(PO_4)_2$, and fluorapatite, $Ca_5(PO_4)_3F$. Fluorapatite can be converted to phosphoric acid by grinding up the phosphate rock and forming a slurry with sulfuric acid:

$$Ca_5(PO_4)_3F(s) + 5H_2SO_4(aq) + 10H_2O(l) \longrightarrow HF(aq) + 5CaSO_4 \cdot 2H_2O(s) + 3H_3PO_4(aq)$$

(A similar reaction can be written for the conversion of calcium phosphate.) The solid product $CaSO_4 \cdot 2H_2O$, called *gypsum,* is used to manufacture wallboard for construction.

The process just described, called the *wet process,* produces only impure phosphoric acid. In another procedure phosphate rock, sand (SiO_2), and coke are heated in an electric furnace to form white phosphorus:

$$12Ca_5(PO_4)_3F + 43SiO_2 + 90C \longrightarrow 9P_4 + 90CO + 20(3CaO \cdot 2SiO_2) + 3SiF_4$$

The white phosphorus is then burned in air to form tetraphosphorus decoxide,

P P P P
Limited O_2
Excess O_2
P_4O_6
P_4O_{10}

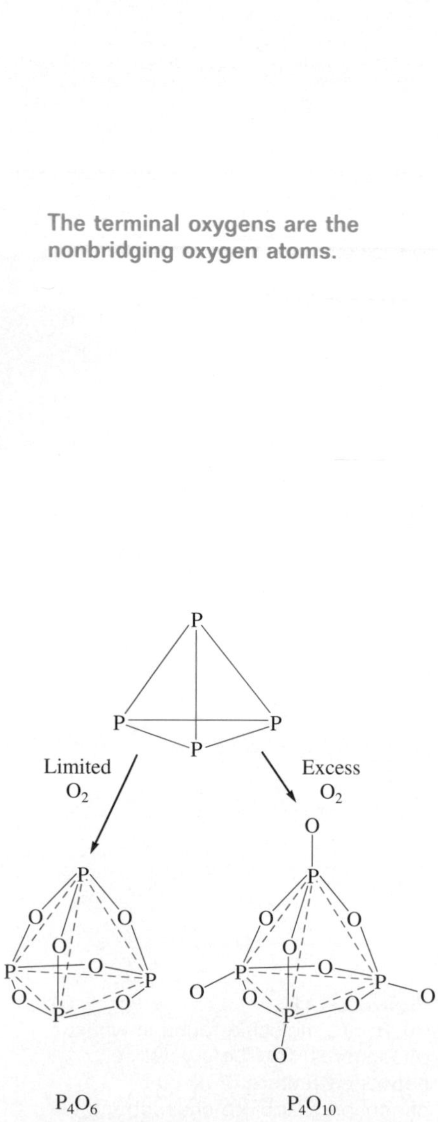

Figure 19.13
The structures of P_4O_6 and P_4O_{10}.

which is then combined with water to give phosphoric acid.

Phosphoric acid easily undergoes **condensation reactions,** in which a molecule of water is eliminated in the joining of two acid molecules:

```
      O                O                      O        O
      |                |                      |        |
H—O—P—O—[H    H—O]—P—O—H  ⟶  H—O—P—O—P—O—H + H₂O
      |                |                      |        |
      O                O                      O        O
      |                |                      |        |
      H                H                      H        H
```

The product ($H_4P_2O_7$) is called *pyrophosphoric acid.* Further heating produces polymers, such as *tripolyphosphoric acid* ($H_5P_3O_{10}$), which has the structure

```
      O     O     O
      |     |     |
H—O—P—O—P—O—P—O—H
      |     |     |
      O     O     O
      |     |     |
      H     H     H
```

The sodium salt of tripolyphosphoric acid is widely used in detergents because the $P_3O_{10}^{5-}$ anion can form complexes with metal ions such as Mg^{2+} and Ca^{2+}, which would otherwise interfere with detergent action.

EXAMPLE 19.1

Determine the molecular structure and the hybridization of the central atom of the phosphoric acid molecule.

Solution

In the phosphoric acid molecule the hydrogen atoms are attached to the oxygen atoms. The Lewis structure is shown below. Note that the phosphorus atom is surrounded by four effective pairs, which are arranged tetrahedrally. The atom is sp^3 hybridized.

```
       :O:
       ‖
   ..  ‖  ..
H—O—P—O—H
   ..  |  ..
       :O:
       |
       H
```

Lewis structure

Molecular structure

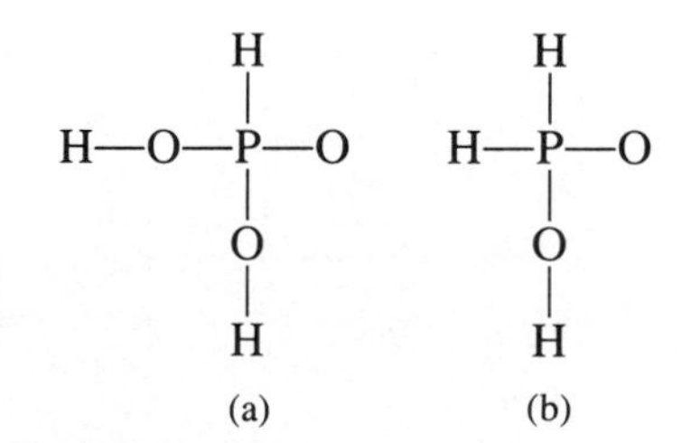

Figure 19.14
(a) The structure of phosphorous acid (H_3PO_3). (b) The structure of hypophosphorous acid (H_3PO_2).

When the oxide P_4O_6 is placed in water, **phosphorous acid** (H_3PO_3) is formed [Fig. 19.14(a)]. Although the formula suggests a triprotic acid, phosphorous acid is a *diprotic* acid. The hydrogen atom bonded directly to the phosphorus atom is not acidic in aqueous solution; only those hydrogen atoms bonded to the oxygen atoms in H_3PO_3 can be released as protons.

A third oxyacid of phosphorus is *hypophosphorous acid* (H_3PO_2) [Fig. 19.14(b)], which is a monoprotic acid.

Phosphorus in Fertilizers

Phosphorus is essential for plant growth. Although most soil contains large amounts of phosphorus, it is often present in insoluble minerals, making it inaccessible to the plants. Soluble phosphate fertilizers are manufactured by treating phosphate rock with sulfuric acid to make **superphosphate of lime,** a mixture of $CaSO_4 \cdot 2H_2O$ and $Ca(H_2PO_4)_2 \cdot H_2O$. If phosphate rock is treated with phosphoric acid, $Ca(H_2PO_4)_2$, or *triple phosphate,* is produced. The reaction of ammonia with phosphoric acid gives *ammonium dihydrogenphosphate* ($NH_4H_2PO_4$), a very efficient fertilizer since it furnishes both phosphorus and nitrogen.

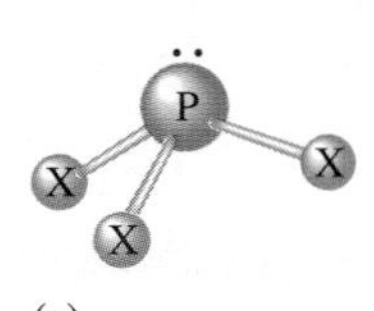

(a)

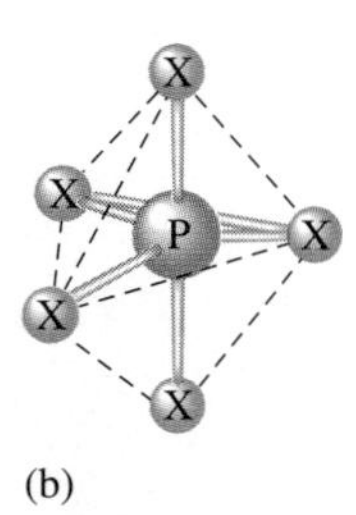

(b)

Figure 19.15
Structures of the phosphorus halides. (a) The PX_3 compounds have pyramidal molecules. (b) The gaseous and liquid phases of the PX_5 compounds are composed of trigonal bipyramidal molecules.

Phosphorus Halides

Phosphorus forms all possible halides of the general formulas PX_3 and PX_5, with the exception of PI_5. The PX_3 molecule has the expected pyramidal structure [Fig. 19.15(a)]. Under normal conditions of temperature and pressure, PF_3 is a colorless gas, PCl_3 is a liquid (bp = 74°C), PBr_3 is a liquid (bp = 175°C), and PI_3 is an unstable red solid (mp = 61°C). All of the PX_3 compounds react with water to produce phosphorous acid:

$$PX_3 + 3H_2O(l) \longrightarrow H_3PO_3(aq) + 3HX(aq)$$

In the gaseous and liquid states the PX_5 compounds have molecules with a trigonal bipyramidal structure [Fig. 19.15(b)]. However, PCl_5 and PBr_5 form ionic solids: solid PCl_5 contains a 1:1 mixture of octahedral PCl_6^- ions and tetrahedral PCl_4^+ ions, and solid PBr_5 appears to consist of PBr_4^+ and Br^- ions.

The PX_5 compounds react with water to form phosphoric acid:

$$PX_5 + 4H_2O(l) \longrightarrow H_3PO_4(aq) + 5HX(aq)$$

19.4 The Group 6A Elements

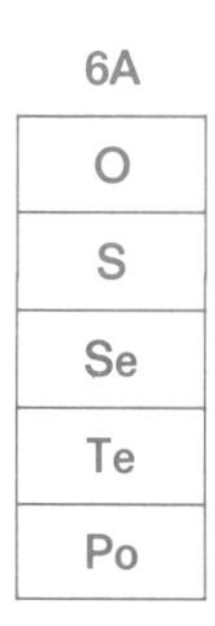

Although in Group 6A (Table 19.4) there is the usual tendency for metallic properties to increase going down the group, none of the Group 6A elements behaves as a typical metal. The most common chemical behavior of a Group 6A element involves reacting with a metal to achieve a noble gas electron configuration by adding two electrons to become a 2− anion in ionic compounds. In fact, for most metals the oxides and sulfides constitute the most common minerals.

The Group 6A elements can form covalent bonds with other nonmetals. For example, they combine with hydrogen to form a series of covalent hydrides of the general formula H_2X. Those members of the group that have valence *d* orbitals available (all except oxygen) commonly form molecules in which they are surrounded by more than eight electrons. Examples are SF_4, SF_6, TeI_4, and $SeBr_4$.

The two heaviest members of Group 6A can lose electrons to form cations. Although they do not lose all six valence electrons because of the high energies that would be required, tellurium and polonium appear to exhibit some chemis-

TABLE 19.4 **Selected Physical Properties, Sources, and Methods of Preparation for the Group 6A Elements**

Element	Electronegativity	Radius of X^{2-} (pm)	Source	Method of Preparation
Oxygen	3.5	140	Air	Distillation from liquid air
Sulfur	2.5	184	Sulfur deposits	Melted with hot water and pumped to the surface
Selenium	2.4	198	Impurity in sulfide ores	Reduction of H_2SeO_4 with SO_2
Tellurium	2.1	221	Nagyagite (mixed sulfide and telluride)	Reduction of ore with SO_2
Polonium	2.0	230	Pitchblende	

try involving their 4+ cations. However, the chemistry of these Group 6A cations is much more limited than that of the Group 5A elements bismuth and antimony.

In recent years there has been a growing interest in the chemistry of selenium, an element found throughout the environment in trace amounts. Selenium's toxicity has long been known, but some medical studies have shown an *inverse* relationship between the incidence of cancer and the selenium levels in soil. It has been suggested that the greater dietary intake of selenium by people living in areas with relatively high selenium levels somehow furnishes protection from cancer. These studies are only preliminary, but selenium is definitely known to be physiologically important (it is involved in the activity of vitamin E and certain enzymes). Selenium is also a semiconductor and therefore finds some application in the electronics industry.

Polonium was discovered in 1898 by Marie and Pierre Curie in their search for the sources of radioactivity in pitchblende. Polonium has 27 isotopes and is highly toxic and very radioactive. It has been suggested that the isotope ^{210}Po, a natural contaminant of tobacco and an α-particle producer (see Section 21.1), might be at least partly responsible for the incidence of cancer in smokers.

19.5 The Chemistry of Oxygen

It is hard to overstate the importance of oxygen, the most abundant element in and near the earth's crust. Oxygen is present in the atmosphere as oxygen gas and ozone; in soil and rocks in oxide, silicate, and carbonate minerals; in the oceans in water; and in our bodies in water and in a myriad of molecules. In addition, most of the energy we need to live and run our civilization comes from the exothermic reactions of oxygen with carbon-containing molecules.

The most common elemental form of oxygen (O_2) constitutes 21% of the volume of the earth's atmosphere. Since nitrogen has a lower boiling point than oxygen, nitrogen can be boiled away from liquid air, leaving oxygen and small amounts of argon, another component of air. Liquid oxygen is a pale blue liquid that freezes at −219°C and boils at −183°C. The paramagnetism of the O_2 molecule can be demonstrated by pouring liquid oxygen between the poles of a strong magnet, where it "sticks" until it boils away (see Fig. 14.42). The paramagnetism of the O_2 molecule can be accounted for by the molecular orbital model (Fig. 14.41), which also explains its bond strength.

The other form of elemental oxygen is **ozone** (O_3), a molecule that can be represented by the resonance structures

$$\ddot{\text{O}}{=}\ddot{\text{O}}{-}\ddot{\text{O}}{:} \longleftrightarrow {:}\ddot{\text{O}}{-}\ddot{\text{O}}{=}\ddot{\text{O}}$$

The bond angle in the O_3 molecule is 117°, in reasonable agreement with the prediction of the VSEPR model (three effective pairs require a trigonal planar arrangement). That the bond angle is slightly less than 120° can be explained by concluding that more space is required for the lone pair than for the bonding pairs.

Ozone can be prepared by passing an electrical discharge through pure oxygen gas. The electrical energy disrupts the bonds in some O_2 molecules, thereby producing oxygen atoms, which react with other O_2 molecules to form O_3. Ozone is much less stable than oxygen at 25°C and 1 atm. For example, $K \approx 10^{-57}$ for the equilibrium

$$3O_2(g) \rightleftharpoons 2O_3(g)$$

A pale blue, highly toxic gas, ozone is a much more powerful oxidizing agent than oxygen. Because of its oxidizing ability, ozone is being considered as a replacement for chlorine in municipal water purification. Chlorine leaves residues of chloro compounds, such as chloroform ($CHCl_3$), which may cause cancer after long-term exposure. Although ozone effectively kills the bacteria in water, one problem with **ozonolysis** is that the water supply is not protected against recontamination, since virtually no ozone remains after the initial treatment. In contrast, for chlorination significant residual chlorine remains after treatment.

The oxidizing ability of ozone can be highly detrimental, especially when it is present in the pollution from automobile exhausts (see Section 5.11).

Scientists have become concerned that freons and nitrogen dioxide are promoting the destruction of the ozone layer (Section 15.9).

Ozone exists naturally in the upper atmosphere of the earth. The *ozone layer* is especially important because it absorbs ultraviolet light and thus acts as a screen to prevent this radiation, which can cause skin cancer, from penetrating to the earth's surface. When an ozone molecule absorbs this energy, it splits into an oxygen molecule and an oxygen atom:

$$O_3 \xrightarrow{h\nu} O_2 + O$$

If the oxygen molecule and atom collide, they will not stay together as ozone unless a "third body," such as a nitrogen molecule, is present to help absorb the energy released by bond formation. The third body absorbs this energy as kinetic energy; its temperature is increased. Therefore, the energy originally absorbed as ultraviolet radiation is eventually changed to thermal energy. Thus the ozone prevents the harmful high-energy ultraviolet light from reaching the earth.

Cinnabar from New Almaden, California.

19.6 The Chemistry of Sulfur

Sulfur is found in nature both in large deposits of the free element and in widely distributed ores, such as galena (PbS), cinnabar (HgS), pyrite (FeS_2), gypsum ($CaSO_4 \cdot 2H_2O$), epsomite ($MgSO_4 \cdot 7H_2O$), and glauberite ($Na_2SO_4 \cdot CaSO_4$).

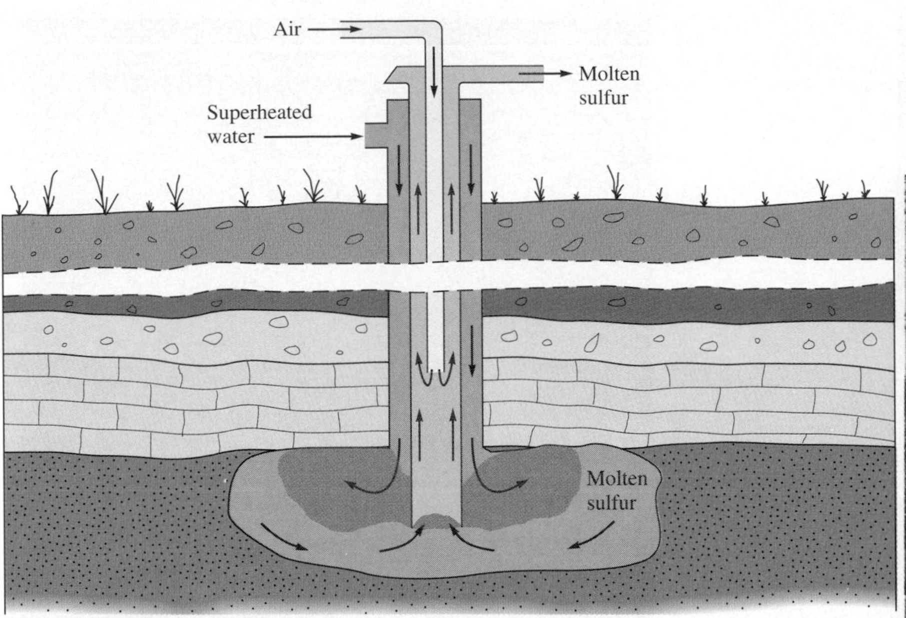

Figure 19.16
The Frasch method for recovering sulfur from underground deposits.

Melted sulfur obtained from underground deposits by the Frasch process.

About 60% of the sulfur produced in the United States comes from the underground deposits of elemental sulfur found in Texas and Louisiana. This sulfur is recovered by using the **Frasch process** developed by Herman Frasch in the 1890s. Superheated water is pumped into the deposit to melt the sulfur (mp = 113°C), which is then forced to the surface by air pressure (see Fig. 19.16). The remaining 40% of sulfur produced in the United States is either a by-product of the purification of fossil fuels before combustion to prevent pollution or from the sulfur dioxide (SO_2) scrubbed from the exhaust gases when sulfur-containing fuels are burned.

The scrubbing of sulfur dioxide from exhaust gases was discussed in Section 5.11.

In contrast to oxygen, elemental sulfur exists as S_2 molecules only in the gas phase at high temperatures. Because sulfur atoms form much stronger σ bonds than π bonds, S_2 is more unstable at 25°C than larger aggregates such as S_6 and S_8 rings and S_n chains (Fig. 19.17). The most stable form of sulfur at 25°C and 1 atm is called *rhombic sulfur* [see Fig. 19.18(a)], which contains stacked S_8 rings. If rhombic sulfur is melted and heated to 120°C, it forms *monoclinic sulfur* as it slowly cools [Fig. 19.18(b)]. The monoclinic form also contains S_8 rings, but the rings are stacked differently than in rhombic sulfur.

As sulfur is heated beyond its melting point, a relatively nonviscous liquid containing S_8 rings forms initially. With continued heating the liquid becomes highly viscous, as the rings first break and then the broken rings link up to form long chains. Further heating lowers the viscosity, because the long chains are broken down as energetic single sulfur atoms break loose. If this liquid is suddenly cooled, a substance called *plastic sulfur,* which contains S_n chains and has rubberlike qualities, is formed. Eventually, this form reverts back to the more stable S_8 rings.

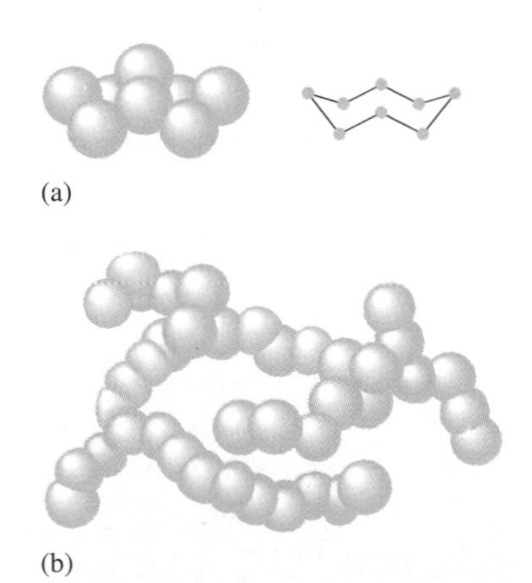

Figure 19.17
(a) The S_8 molecule. (b) Chains of sulfur atoms in viscous liquid sulfur. The chains may contain as many as 10,000 atoms.

Pouring liquid sulfur in water to produce plastic sulfur.

(a) Crystals of rhombic sulfur.

Figure 19.18 (b) Crystals of monoclinic sulfur.

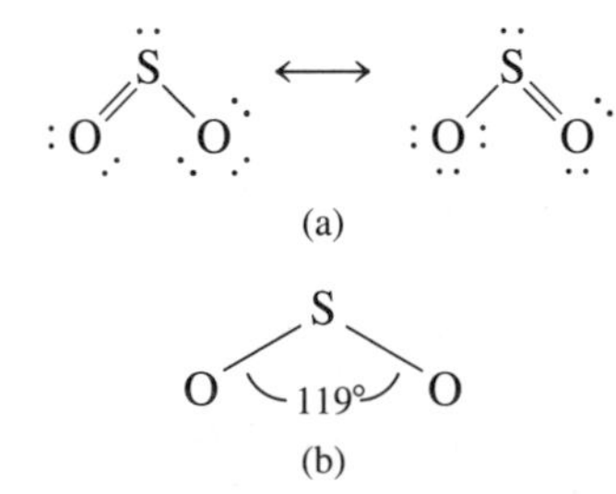

Figure 19.19
(a) The resonance structures for SO_2. (b) SO_2 is a bent molecule with a 119° bond angle, as predicted by the VSEPR model.

Sulfur Oxides

From its position below oxygen in the periodic table, we might expect the simplest stable oxide of sulfur to have the formula SO. However, *sulfur monoxide,* which can be produced in small amounts when gaseous sulfur dioxide (SO_2) is subjected to an electrical discharge, is very unstable. The difference in the stabilities of the O_2 and SO molecules probably reflects the much stronger π bonding between oxygen atoms than between a sulfur and an oxygen atom.

Sulfur burns in air with a bright blue flame to give *sulfur dioxide* (SO_2), a colorless gas with a pungent odor, which condenses to a liquid at $-10°C$ and 1 atm. Sulfur dioxide, a very effective antibacterial agent, is often used to preserve stored fruit. Its structure is given in Fig. 19.19.

Figure 19.20
(a) The resonance structures most commonly given for SO_3. (b) A resonance structure with three double bonds. (c) SO_3 is a planar molecule with 120° bond angles.

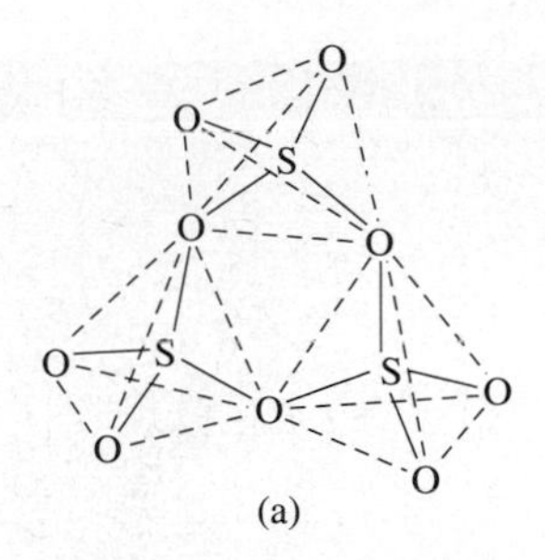

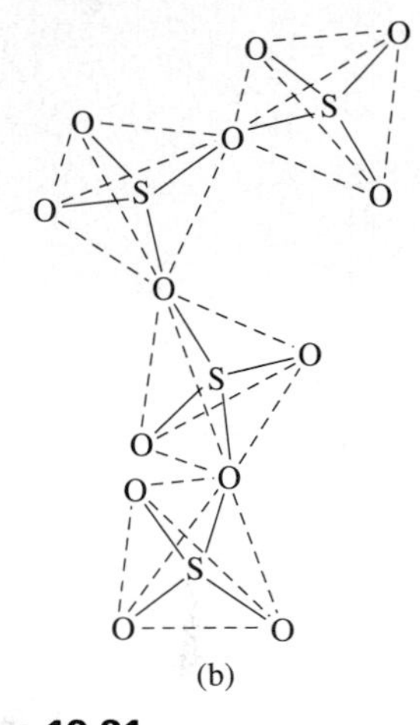

Figure 19.21
Different structures for solid SO_3. (a) S_3O_9 rings. (b) $(SO_3)_x$ chains. In both cases the sulfur atoms are surrounded by a tetrahedral arrangement of oxygen atoms.

Sulfur dioxide reacts with oxygen to produce *sulfur trioxide* (SO_3):

$$2SO_2(g) + O_2(g) \longrightarrow 2SO_3(g)$$

However, this reaction is very slow in the absence of a catalyst. One of the mysteries during early research on air pollution was how the sulfur dioxide produced from the combustion of sulfur-containing fuels is so rapidly converted to sulfur trioxide in the atmosphere. It is now known that dust and other particles can act as heterogeneous catalysts for this process (see Section 15.9). In the preparation of sulfur trioxide for the manufacture of sulfuric acid, either platinum metal or vanadium(V) oxide (V_2O_5) is used as a catalyst, and the reaction is carried out at $\approx$500°C, even though this temperature decreases the value of the equilibrium constant for this exothermic reaction.

The preparation of sulfur trioxide provides an example of the compromise that often must be made between thermodynamics and kinetics.

The bonding in the SO_3 molecule is usually described in terms of the resonance structures shown in Fig. 19.20(a). However, there is some evidence to suggest that the resonance structure given in Fig. 19.20(b) may give the most accurate description of the bonding in SO_3. This molecule is trigonal planar, as predicted by the VSEPR model. Sulfur trioxide is a corrosive gas with a choking odor and forms white fumes of sulfuric acid when it reacts with moisture in the air. Thus sulfur trioxide and nitrogen dioxide (which reacts with water to form a mixture of nitrous and nitric acids) are the major culprits in the formation of acid rain.

Sulfur trioxide condenses to a colorless liquid at 44.5°C and freezes at 16.8°C to give three solid forms, one containing S_3O_9 rings and the other two containing $(SO_3)_x$ chains (Fig. 19.21).

Oxyacids of Sulfur

Sulfur dioxide dissolves in water to form an acidic solution. The reaction is often represented as

$$SO_2(g) + H_2O(l) \longrightarrow H_2SO_3(aq)$$

where H_2SO_3 is called *sulfurous acid*. However, very little H_2SO_3 actually exists in the solution. The major form of sulfur dioxide in water is SO_2, and the acid dissociation equilibria are best represented as

$$SO_2(aq) + H_2O(l) \rightleftharpoons H^+(aq) + HSO_3^-(aq) \qquad K_{a_1} = 1.5 \times 10^{-2}$$
$$HSO_3^-(aq) \rightleftharpoons H^+(aq) + SO_3^{2-}(aq) \qquad K_{a_2} = 1.0 \times 10^{-7}$$

Figure 19.22
The reaction of H_2SO_4 with sucrose to produce a blackened column of carbon.

This situation is analogous to the behavior of carbon dioxide in water (Section 7.7). Although H_2SO_3 cannot be isolated, salts of SO_3^{2-} (*sulfites*) and HSO_3^- (*hydrogen sulfites*) are well known.

Sulfur trioxide reacts violently with water to produce the diprotic acid **sulfuric acid:**

$$SO_3(g) + H_2O(l) \longrightarrow H_2SO_4(aq)$$

Manufactured in greater amounts than any other chemical, sulfuric acid is usually produced by the *contact process,* which is described in Chapter 3. About 60% of the sulfuric acid manufactured in the United States is used to produce fertilizers from phosphate rock (see Section 19.3). The other 40% is used in lead storage batteries and in petroleum refining, steel manufacturing, and for various other purposes in the chemical industry.

Because sulfuric acid has a high affinity for water, it is often used as a dehydrating agent. Gases that do not react with sulfuric acid, such as oxygen, nitrogen, and carbon dioxide, are often dried by bubbling them through concentrated solutions of the acid. Sulfuric acid is such a powerful dehydrating agent that it will remove hydrogen and oxygen from a substance in a 2:1 ratio even when the substance contains no molecular water. For example, concentrated sulfuric acid reacts vigorously with common table sugar (sucrose), leaving a charred mass of carbon (see Fig. 19.22):

$$\underset{\text{Sucrose}}{C_{12}H_{22}O_{11}(s)} + 11H_2SO_4(conc) \longrightarrow 12C(s) + 11H_2SO_4 \cdot H_2O(l)$$

Sulfuric acid is a moderately strong oxidizing agent, especially at high temperatures. Hot concentrated sulfuric acid oxidizes bromide or iodide ions to elemental bromine or iodine; for example,

$$2I^-(aq) + 3H_2SO_4(aq) \longrightarrow I_2(aq) + SO_2(aq) + 2H_2O(l) + 2HSO_4^-(aq)$$

Hot sulfuric acid attacks copper metal:

$$Cu(s) + 2H_2SO_4(aq) \longrightarrow CuSO_4(aq) + 2H_2O(l) + SO_2(aq)$$

The cold acid does not react with copper.

The acidic properties of sulfuric acid solutions are discussed in Section 7.7.

Other Compounds of Sulfur

Sulfur reacts with both metals and nonmetals to form a wide variety of compounds in which it has a +6, +4, +2, 0, or −2 oxidation state (see Table 19.5). The −2 oxidation state occurs in the metal sulfides and in *hydrogen sulfide* (H_2S), a toxic, foul-smelling gas that acts as a diprotic acid when dissolved in water. Hydrogen sulfide is a strong reducing agent in aqueous solution, producing a milk-like suspension of finely divided sulfur as one of the products. For example, hydrogen sulfide reacts with chlorine in aqueous solution as follows:

$$H_2S(g) + Cl_2(aq) \longrightarrow 2H^+(aq) + 2Cl^-(aq) + \underset{\text{Milky suspension of sulfur}}{S(s)}$$

TABLE 19.5 Common Compounds of Sulfur with Various Oxidation States

Oxidation State of Sulfur	Compounds
+6	SO_3, H_2SO_4, SO_4^{2-}, SF_6
+4	SO_2, HSO_3^-, SO_3^{2-}, SF_4
+2	SCl_2
0	S_8 and all other forms of elemental sulfur
−2	H_2S, S^{2-}

Sulfur also forms the **thiosulfate ion** ($S_2O_3^{2-}$), which has the Lewis structure

The prefix *thio* means sulfur.

$$\left[\begin{array}{ccccc} & & :\ddot{O}: & & \\ & & | & & \\ :\ddot{O} & - & S & - & \ddot{S}: \\ & & | & & \\ & & :\ddot{O}: & & \end{array}\right]^{2-}$$

Note that this anion can be viewed as a sulfate ion in which one of the oxygen atoms has been replaced by sulfur, which is reflected in the name *thio*sulfate. The thiosulfate ion can be formed by heating sulfur with a sulfite salt in aqueous solution:

$$S(s) + SO_3^{2-}(aq) \longrightarrow S_2O_3^{2-}(aq)$$

One important use of thiosulfate ion is in photography, where $S_2O_3^{2-}$ dissolves solid silver bromide by forming a complex with the Ag^+ ion (see Section 19.7). Since the thiosulfate ion is also a good reducing agent, it is often used to analyze for iodine:

$$S_2O_3^{2-}(aq) + I_2(aq) \longrightarrow S_4O_6^{2-}(aq) + 2I^-(aq)$$

where $S_4O_6^{2-}$ is called the *tetrathionate ion*.

Sulfur reacts with the halogens to form a variety of compounds, such as S_2Cl_2, SF_4, SF_6, and S_2F_{10}. The structures of these molecules are shown in Fig. 19.23.

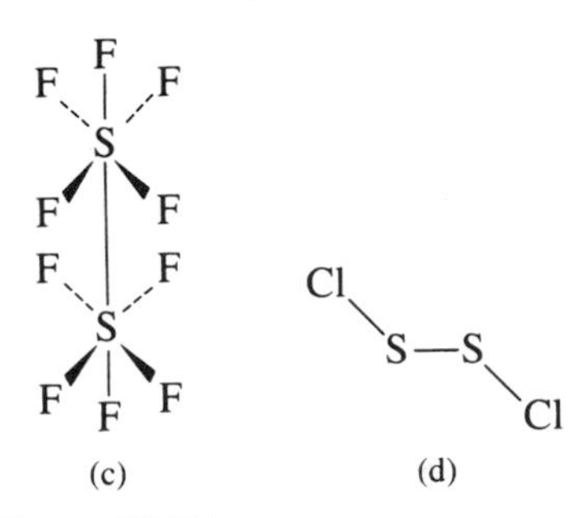

Figure 19.23
The structures of (a) SF_4, (b) SF_6, (c) S_2F_{10}, and (d) S_2Cl_2.

19.7 The Group 7A Elements

7A
F
Cl
Br
I
At

In our coverage of the representative elements we have progressed from the groups of metallic elements (Groups 1A and 2A), through groups in which the lighter members are nonmetals and the heavier members are metals (Groups 3A, 4A, and 5A), to a group containing all nonmetals (Group 6A—although some might prefer to call polonium a metal). The Group 7A elements, the **halogens** (with the valence electron configuration ns^2np^5), are all nonmetals whose properties generally vary smoothly going down the group. The only notable exceptions are the unexpectedly low value for the electron affinity of fluorine and the unexpectedly small bond energy of the F_2 molecule (see Section 18.1). Table 19.6 summarizes the trends in some physical properties of the halogens.

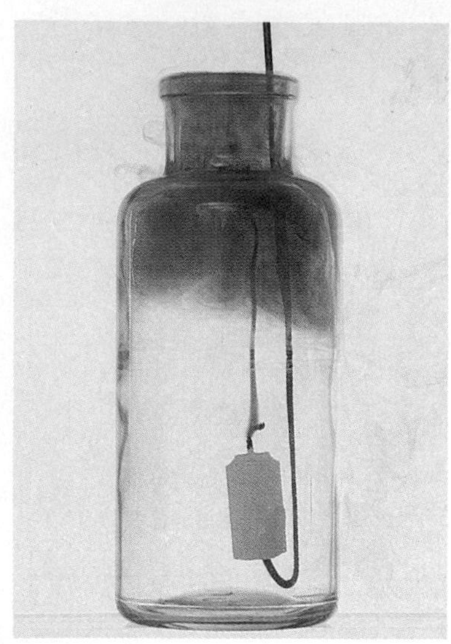

A candle burning in an atmosphere of $Cl_2(g)$. The exothermic reaction, which involves breaking C—C and C—H bonds in the wax and forming C—Cl bonds in their places, produces enough heat to make the gases in the region incandescent (a flame results).

TABLE 19.6 **Trends in Selected Physical Properties of the Group 7A Elements**

Element	Electronegativity	Radius of X^- (pm)	$\mathscr{E}°$ (V) for $X_2 + 2e \rightarrow 2X^-$	Bond Energy of X_2 (kJ/mol)
Fluorine	4.0	136	2.87	154
Chlorine	3.0	181	1.36	239
Bromine	2.8	185	1.09	193
Iodine	2.5	216	0.54	149
Astatine	2.2	—	—	—

Because of their high reactivities, the halogens are not found as free elements in nature. Instead, they are found as halide ions (X^-) in various minerals and in seawater (see Table 19.7).

Although astatine is a member of Group 7A, its chemistry is of no practical importance because all of its known isotopes are radioactive. The longest-lived isotope, ^{210}At, has a half-life of only 8.3 hours.

The halogens, particularly fluorine, have very high electronegativity values (Table 19.6). They tend to form polar covalent bonds with other nonmetals and ionic bonds with metals in their lower oxidation states. When a metal ion is in a higher oxidation state, such as +3 or +4, the metal-halogen bonds are polar and covalent. For example, $TiCl_4$ and $SnCl_4$ are both covalent compounds that are liquids under normal conditions.

Chlorine, bromine, and iodine.

TABLE 19.7 Some Physical Properties, Sources, and Methods of Preparation for the Group 7A Elements

Element	Color and State	Percentage of Earth's Crust	Melting Point (°C)	Boiling Point (°C)	Sources	Method of Preparation
Fluorine	Pale yellow gas	0.07	−220	−188	Fluorospar (CaF_2), cryolite (Na_3AlF_6), fluorapatite ($Ca_5(PO_4)_3F$)	Electrolysis of molten KHF_2
Chlorine	Yellow-green gas	0.14	−101	−34	Rock salt (NaCl), halite (NaCl), sylvite (KCl)	Electrolysis of aqueous NaCl
Bromine	Red-brown liquid	2.5×10^{-4}	−7.3	59	Seawater, brine wells	Oxidation of Br^- by Cl_2
Iodine	Violet-black solid	3×10^{-5}	113	184	Seaweed, brine wells	Oxidation of I^- by electrolysis or MnO_2

PHOTOGRAPHY

In black-and-white photography, light from an object is focused onto a special paper containing an emulsion of solid silver bromide. Silver salts turn dark when exposed to light, because the radiant energy stimulates the transfer of an electron to the Ag^+ ion, forming an atom of elemental silver. When photographic paper (*film*) is exposed to light, the areas exposed to the brightest light form the most silver atoms. The next step in forming an image is the application of a chemical reducing agent to the film, a process called *developing*. The real advantage of silver-based films is that the silver atoms already formed from the exposure to light catalyze the reduction of millions of Ag^+ ions in the immediate vicinity in the developing process. Thus the effect of exposure to light is greatly intensified in this chemical reduction process. Once the image has been developed, the unchanged solid silver bromide must be removed so that the film is no longer light-sensitive and the image is fixed. A solution of sodium thiosulfate (called *hypo*) is used in this *fixing process:*

$$AgBr(s) + 2S_2O_3^{2-}(aq) \longrightarrow Ag(S_2O_3)_2^{3-}(aq) + Br^-(aq)$$

After the excess silver bromide is dissolved and washed away, the fixed image (the *negative*) is ready to produce the positive print. By shining light through the negative onto a fresh sheet of film and repeating the developing and fixing processes, a black-and-white photograph can be produced.

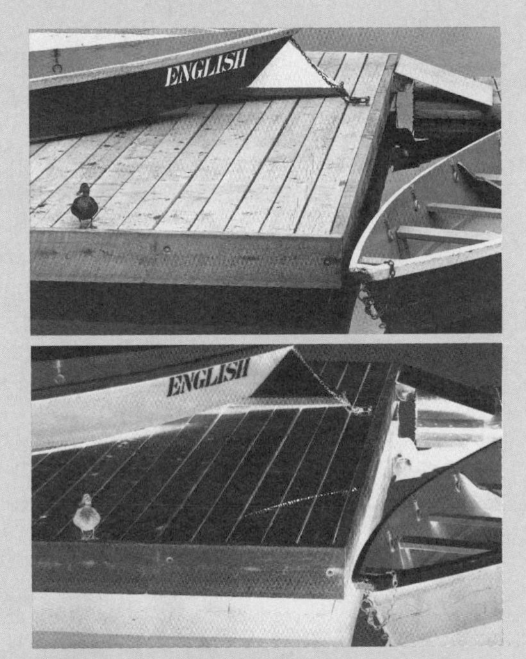

Positive and negative images.

Hydrogen Halides

The hydrogen halides can be prepared by a reaction of the elements:

$$H_2(g) + X_2(g) \longrightarrow 2HX(g)$$

This reaction occurs with explosive vigor when fluorine and hydrogen are mixed. On the other hand, hydrogen and chlorine can coexist with little apparent reaction for relatively long periods in the dark. However, ultraviolet light causes an explosively fast reaction, and this is the basis of a popular lecture demonstration, the "hydrogen-chlorine cannon." Bromine and iodine also react with hydrogen, but more slowly.

The hydrogen halides can also be prepared by treating halide salts with acid. For example, hydrogen fluoride and hydrogen chloride can be prepared as follows:

$$CaF_2(s) + H_2SO_4(aq) \longrightarrow CaSO_4(s) + 2HF(g)$$
$$2NaCl(s) + H_2SO_4(aq) \longrightarrow Na_2SO_4(s) + 2HCl(g)$$

Sulfuric acid is capable of oxidizing Br^- to Br_2 and I^- to I_2. Therefore, it cannot be used to prepare hydrogen bromide and hydrogen iodide. However, phosphoric acid, a nonoxidizing acid, can be used to treat bromides and iodides to form the corresponding hydrogen halides.

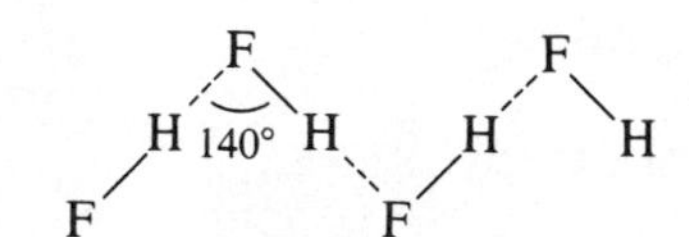

Figure 19.24
The hydrogen bonding among HF molecules in liquid hydrogen fluoride.

Some physical properties of the hydrogen halides are listed in Table 19.8. Note the very high boiling point for hydrogen fluoride, which results from extensive hydrogen bonding among the very polar HF molecules (Fig. 19.24). Fluoride ion has such a high affinity for protons that in concentrated aqueous solutions of hydrogen fluoride, the ion $[F\text{---}H\text{---}F]^-$ exists, in which an H^+ ion is centered between two F^- ions.

When dissolved in water, the hydrogen halides behave as acids, and all except hydrogen fluoride are completely dissociated. Because water is a much stronger base than Cl^-, Br^-, or I^- ion, the acid strengths of HCl, HBr, and HI cannot be differentiated in water. However, in a less basic solvent, such as glacial (pure) acetic acid, the acids show different strengths:

$$\underset{\text{Strongest acid}}{H{-}I} > H{-}Br > H{-}Cl \gg \underset{\text{Weakest acid}}{H{-}F}$$

To see why hydrogen fluoride is the only weak acid in water among the HX molecules, let's consider the dissociation equilibrium,

$$HX(aq) \rightleftharpoons H^+(aq) + X^-(aq) \qquad \text{where} \qquad K_a = \frac{[H^+][X^-]}{[HX]}$$

TABLE 19.8 Some Physical Properties of the Hydrogen Halides

HX	Melting Point (°C)	Boiling Point (°C)	H—X Bond Energy (kJ/mol)
HF	−83	20	565
HCl	−114	−85	427
HBr	−87	−67	363
HI	−51	−35	295

from a thermodynamic point of view. Recall that acid strength is reflected by the magnitude of K_a—a small K_a value means a weak acid. Also recall that the value of an equilibrium constant is related to the standard free energy change for the reaction

$$\Delta G^\circ = -RT\ln(K)$$

As ΔG° becomes more negative, K becomes larger; a *decrease* in free energy favors a given reaction. As we saw in Chapter 10, free energy depends on enthalpy, entropy, and temperature. For a process at constant temperature,

$$\Delta G^\circ = \Delta H^\circ - T\Delta S^\circ$$

Thus to explain the various acid strengths of the hydrogen halides, we must focus on the factors that determine ΔH° and ΔS° for the acid dissociation reaction.

What energy terms are important in determining ΔH° for the dissociation of HX in water? (Keep in mind that large, positive contributions to the value of ΔH° will tend to make ΔG° more highly positive, K_a smaller, and the acid weaker.) One important factor is certainly the H—X bond strength. Note from Table 19.8 that the H—F bond is much stronger than the other H—X bonds. This factor tends to make HF a weaker acid than the others.

Another important contribution to ΔH° is the enthalpy of hydration (see Section 17.2) of X^- (see Table 19.9). As we would expect, the smallest of the halide ions, F^-, has the most negative value—its hydration is the most exothermic. This term favors the dissociation of HF into its ions, more so than it does for the other HX molecules.

So far we have two conflicting factors: the large HF bond energy tends to make HF a weaker acid than the other hydrogen halides, but the enthalpy of hydration favors the dissociation of HF more than that of the others. When we compare data for HF and HCl, the difference in bond energy (138 kJ/mol) is slightly smaller than the difference in the enthalpies of hydration for the anions (144 kJ/mol). If these were the *only* important factors, HF should be a stronger acid than HCl because the large enthalpy of hydration of F^- more than compensates for the large HF bond strength.

As it turns out, the *deciding factor appears to be entropy.* Note from Table 19.9 that the entropy of hydration for F^- is much more negative than the entropy of hydration for the other halides because of the high degree of ordering that occurs as the water molecules associate with the small F^- ion. Remember that a negative change in entropy is unfavorable. Thus although the enthalpy of hydration favors dissociation of HF, the *entropy* of hydration strongly opposes it.

When all of these factors are taken into account, ΔG° for the dissociation of HF in water is positive; that is, K_a is small. In contrast, ΔG° for dissociation of the other HX molecules in water is negative (K_a is large). This example illustrates the complexity of the processes that occur in aqueous solutions and the importance of entropy effects in that medium.

In practical terms, **hydrochloric acid** is the most important of the **hydrohalic acids,** the aqueous solutions of the hydrogen halides. About 3 million tons of hydrochloric acid are produced annually for use in cleaning steel before galvanizing and in the manufacture of many other chemicals.

TABLE 19.9
The Enthalpies and Entropies of Hydration for the Halide Ions

$$X^-(g) \xrightarrow{H_2O} X^-(aq)$$

X^-	ΔH° (kJ/mol)	ΔS° (J/K mol)
F^-	−510	−159
Cl^-	−366	−96
Br^-	−334	−81
I^-	−291	−64

Hydration becomes more exothermic as the charge density of an ion increases. Thus for ions of a given charge the smallest is most strongly hydrated.

When H_2O molecules cluster around an ion, an ordering effect occurs; thus ΔS°_{hyd} is negative.

Stomach acid is 0.1 *M* HCl.

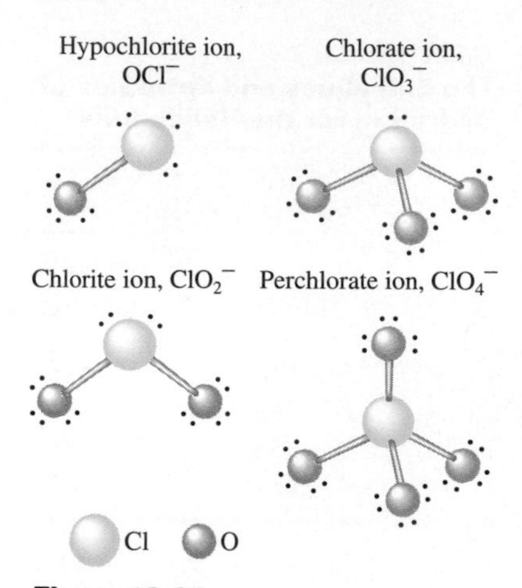

Figure 19.25
The structures of the oxychloro anions.

Hydrofluoric acid is used to etch glass by reacting with the silica in glass to form the volatile gas SiF_4:

$$SiO_2(s) + 4HF(aq) \longrightarrow SiF_4(g) + 2H_2O(l)$$

Oxyacids and Oxyanions

All of the halogens except fluorine combine with various numbers of oxygen atoms to form a series of oxyacids, as shown in Table 19.10. The strengths of these acids vary in direct proportion to the number of oxygen atoms attached to the halogen, the acid strength increasing as more oxygens are added.

The only member of the chlorine series that has been obtained in the pure state is *perchloric acid* ($HOClO_3$), a strong acid and a powerful oxidizing agent. Because perchloric acid reacts explosively with many organic materials, it must be handled with great caution. The other oxyacids of chlorine are known only in solution, although salts containing their anions are well known (Fig. 19.25).

Hypochlorous acid (HOCl) is formed when chlorine gas is dissolved in cold water:

$$Cl_2(aq) + H_2O(l) \rightleftharpoons HOCl(aq) + H^+(aq) + Cl^-(aq)$$

Note that in this reaction chlorine is both oxidized (from 0 in Cl_2 to +1 in HOCl) and reduced (from 0 in Cl_2 to -1 in Cl^-). Such a reaction, *in which a given element is both oxidized and reduced,* is called a **disproportionation reaction.** Hypochlorous acid and its salts are strong oxidizing agents; their solutions are widely used as household bleaches and disinfectants.

Chlorate salts, such as $KClO_3$, are also strong oxidizing agents and are used as weed killers and as oxidizers in fireworks (see Chapter 12) and explosives.

Fluorine forms only one oxyacid, hypofluorous acid (HOF), but it forms at least two oxides. When fluorine gas is bubbled into a dilute solution of sodium hydroxide, the compound *oxygen difluoride* (OF_2) is formed:

The name for OF_2 is oxygen difluoride rather than difluorine oxide because fluorine has a higher electronegativity than oxygen and thus is named as the anion.

$$4F_2(g) + 3H_2O(l) \longrightarrow 6HF(aq) + OF_2(g) + O_2(g)$$

Oxygen difluoride is a pale yellow gas (bp = $-145°C$) that is a strong oxidizing agent. The oxide *dioxygen difluoride* (O_2F_2) is an orange solid that can be prepared by an electric discharge in an equimolar mixture of fluorine and oxygen gases:

$$F_2(g) + O_2(g) \xrightarrow[]{\text{Electric discharge}} O_2F_2(s)$$

TABLE 19.10 **The Known Oxyacids of the Halogens**

Oxidation State of Halogen	Fluorine	Chlorine	Bromine	Iodine*	General Name of Acids	General Name of Salts
+1	HOF	HOCl	HOBr	HOI	Hypohalous acid	Hypohalites, MOX
+3	†	HOClO	†	†	Halous acid	Halites, MXO_2
+5	†	$HOClO_2$	$HOBrO_2$	$HOIO_2$	Halic acid	Halates, MXO_3
+7	†	$HOClO_3$	$HOBrO_3$	$HOIO_3$	Perhalic acid	Perhalates, MXO_4

*Iodine also forms $H_4I_2O_9$ (mesodiperiodic acid) and H_5IO_6 (paraperiodic acid).
†Compound is unknown.

Other Halogen Compounds

The halogens react readily with most nonmetals to form a variety of compounds, some of which are shown in Table 19.11.

Halogens react with each other to form **interhalogen compounds**. These compounds have the general formula AB_n, where n is typically 1, 3, 5, or 7 and A is the larger of the two halogens. The structures of these compounds (see Fig. 19.26) are predicted accurately by the VSEPR model. The interhalogens are volatile, highly reactive substances that act as strong oxidizing agents. They react readily with water, forming the halide ion of the more electronegative halogen and the hypohalous acid of the less electronegative halogen. For example,

$$ICl(s) + H_2O(l) \longrightarrow H^+(aq) + Cl^-(aq) + HOI(aq)$$

The halogens react with carbon to form many commercially important compounds. Especially significant are a group of polymers, or long-chain compounds (see Section 22.5), made from ethylene molecules:

$$\begin{array}{ccc} H & & H \\ & C{=}C & \\ H & & H \end{array}$$

TABLE 19.11 Some Compounds of the Halogens with Nonmetals

Compounds with Group 3A Nonmetals	Compounds with Group 4A Nonmetals	Compounds with Group 5A Nonmetals	Compounds with Group 6A Nonmetals	Compounds with Group 7A Nonmetals
BX_3 (X = F,Cl,Br,I)	CX_4 (X = F,Cl,Br,I)	NX_3 (X = F,Cl,Br,I)	OF_2	ICl
BF_4^-		N_2F_4	O_2F_2	IBr
	SiF_4		OCl_2	BrF
	SiF_6^{2-}	PX_3 (X = F,Cl,Br,I)	OBr_2	BrCl
	$SiCl_4$	PF_5		ClF
		PCl_5	SF_2	
	GeF_4	PBr_5	SCl_2	ClF_3
	GeF_6^{2-}		S_2F_2	BrF_3
	$GeCl_4$	AsF_3	S_2Cl_2	ICl_3
		AsF_5	SF_4	IF_3
			SCl_4	
		SbF_3	SF_6	ClF_5
		SbF_5		BrF_5
			SeF_4	IF_5
			SeF_6	
			$SeCl_2$	IF_7
			$SeCl_4$	
			$SeBr_4$	
			TeF_4	
			TeF_6	
			$TeCl_4$	
			$TeBr_4$	
			TeI_4	

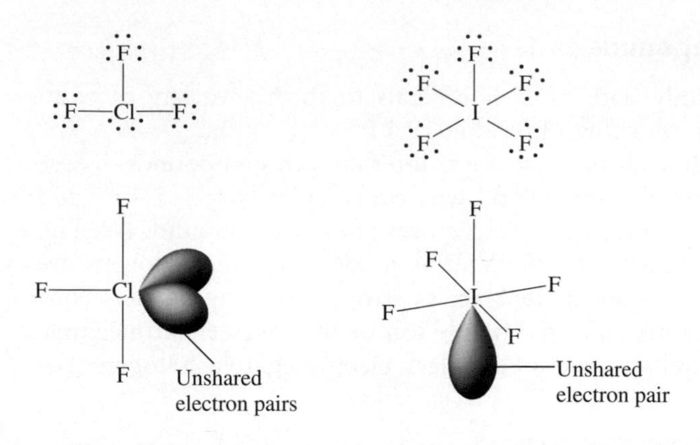

(a) ClF_3 is T-shaped (b) IF_5 is square pyramidal

Figure 19.26
The idealized structures of the interhalogens ClF_3 and IF_5. In reality, the lone pairs cause the bond angles to be slightly less than 90°.

For example, the compound perfluoroethylene,

$$F_2C{=}CF_2$$

can react with itself to give the polymer

$$\sim\!\!\sim CF_2{-}CF_2{-}CF_2{-}CF_2 \sim\!\!\sim$$

called *Teflon*, a substance with high thermal stability and resistance to chemical attack, used as a coating in many different applications.

The **freons,** compounds of the general formula CCl_xF_{4-x}, are very stable compounds that are used as coolant fluids in refrigerators and air conditioners.

Chlorine is found in many other useful carbon compounds. For example, the molecule

$$H_2C{=}CHCl$$

called vinyl chloride, can be polymerized to form *polyvinyl chloride,* or PVC:

$$\sim\!\!\sim CH_2{-}CHCl{-}CH_2{-}CHCl{-}CH_2 \sim\!\!\sim$$

Chlorine is also found in many pesticides.

Bromine is used in silver bromide-based photographic films. It is also used to make NaBr and KBr, two chemicals used in sedatives and soporifics (sleeping aids). Bromine-containing compounds are also used as flame retardants for cloth (see Section 22.5).

8A
He
Ne
Ar
Kr
Xe
Rn

19.8 The Group 8A Elements

The Group 8A elements, the **noble gases,** are characterized by filled s and p valence orbitals (electron configuration of $2s^2$ for helium and ns^2np^6 for the others). Because of their completed valence shells, these elements are very unreactive. In fact, no noble gas compounds were known 35 years ago. Selected properties of the Group 8A elements are summarized in Table 19.12.

Automatic Sunglasses

Sunglasses can be troublesome. It seems they are always getting lost or sat on. One solution to this problem for people who wear glasses is photochromic glass—glass that darkens in response to intense light. Recall that glass is a complex, noncrystalline material that is composed of polymeric silicates (see Chapter 16). Of course, glass transmits visible light—its transparency is its most useful property.

Glass can be made photochromic by adding tiny silver chloride crystals which get trapped in the glass matrix as the glass solidifies. Silver chloride has the unusual property of darkening when struck by light—the property that makes the silver halide salts so useful for photographic films. This darkening occurs because light causes an electron transfer from Cl^- to Ag^+ in the silver chloride crystal, forming a silver atom and a chlorine atom. The silver atoms formed in this way tend to migrate to the surface of the silver chloride crystal, where they aggregate to form a tiny crystal of silver metal, which is opaque to light.

In photography the image defined by the grains of silver is fixed by chemical treatment so that it remains permanent. However, in photochromic glass this process must be reversible—the glass must become fully transparent again when the person goes back indoors. The secret to the reversibility of photochromic glass is the presence of Cu^+ ions. The added Cu^+ ions serve two important functions. First, they reduce the Cl atoms formed in the light-induced reaction. This prevents them from escaping from the crystal:

$$Ag^+ + Cl^- \xrightarrow{h\nu} Ag + Cl$$

$$Cl + Cu^+ \longrightarrow Cu^{2+} + Cl^-$$

Second, when the exposure to intense light ends (the person goes indoors), the Cu^{2+} ions migrate to the surface of the silver chloride crystal, where they accept electrons from silver atoms as the tiny crystal of silver atoms disintegrates:

$$Cu^{2+} + Ag \longrightarrow Cu^+ + Ag^+$$

The Ag^+ ions that are re-formed in this way then return to their places in the silver chloride crystal, making the glass transparent once again.

Typical photochromic glass decreases to about 20% transmittance (transmits 20% of the light that strikes it) in strong sunlight, and then over a period of a few minutes returns to about 80% transmittance indoors (normal glass has 92% transmittance).

TABLE 19.12 **Selected Properties of Group 8A Elements**

Element	Melting Point (°C)	Boiling Point (°C)	Atmospheric Abundance (% by volume)	Examples of Compounds
Helium	−272	−269	5×10^{-4}	None
Neon	−249	−246	1×10^{-3}	None
Argon	−189	−186	9×10^{-1}	None
Krypton	−157	−153	1×10^{-4}	KrF_2
Xenon	−112	−107	9×10^{-6}	XeF_4, XeO_3, XeF_6

Helium was identified by its characteristic emission spectrum as a component of the sun before it was found on earth. The major sources of helium on earth are natural gas deposits, where helium was formed from the α-particle decay of radioactive elements. The α particle is a helium nucleus that can easily pick up electrons from the environment to form a helium atom. Although helium forms no compounds, it is an important substance that is used as a coolant, as a pressurizing gas for rocket fuels, as a diluent in the gases used for deep-sea diving and spaceship atmospheres, and as the gas in lighter-than-air airships (blimps).

Like helium, *neon* and *argon* form no compounds. However, they are used extensively. For example, neon is employed in luminescent lighting (neon signs), and argon is used to provide the noncorrosive atmosphere in incandescent light bulbs, which prolongs the life of the tungsten filament.

Of the Group 8A elements, only *krypton* and *xenon* have been observed to form stable chemical compounds. The first of these was prepared in 1962 by Neil Bartlett (b. 1932), an English chemist who made an ionic compound that he thought had the formula $XePtF_6$. Subsequent studies have indicated the compound might be better represented as $XeFPtF_6$ and contains the XeF^+ and PtF_6^- ions.

Less than a year after Bartlett's report, a group at Argonne National Laboratory near Chicago prepared xenon tetrafluoride by reacting xenon and fluorine gases in a nickel reaction vessel at 400°C and 6 atm:

$$Xe(g) + 2F_2(g) \longrightarrow XeF_4(s)$$

Xenon tetrafluoride forms stable colorless crystals. Two other xenon fluorides, XeF_2 and XeF_6, were synthesized by the group at Argonne, and a highly explosive xenon oxide (XeO_3) was also found. The xenon fluorides react with water to form hydrogen fluoride and oxycompounds; for example:

$$XeF_6(s) + 3H_2O(l) \longrightarrow XeO_3(aq) + 6HF(aq)$$
$$XeF_6(s) + H_2O(l) \longrightarrow XeOF_4(aq) + 2HF(g)$$

In the past 25 years other xenon compounds have been prepared. Examples are XeO_4 (explosive), $XeOF_4$, $XeOF_2$, and, XeO_3F_2. These compounds contain discrete molecules with covalent bonds between the xenon atom and the other atoms. A few compounds of krypton, such as KrF_2 and KrF_4, have also been observed. There is evidence that radon also reacts with fluorine, but the radioactivity of radon makes its chemistry very difficult to study.

EXAMPLE 19.2

Use the VSEPR model to predict whether or not XeF_6 has an octahedral structure.

Solution

The XeF_6 molecule contains 50 [8 + 6(7)] valence electrons, and the Lewis structure is

```
        ..
  ..   :F:   ..
 :F\    |   /F:
  ..  \ Xe /  ..
  ..  / ..|\  ..
 :F/    |   \F:
  ..   :F:   ..
        ..
```

The xenon atom has seven effective pairs of electrons surrounding it (one lone pair and six bonding pairs), one more pair than can be accommodated in an octahedral arrangement. Thus XeF_6 should not have an octahedral structure but instead should be distorted from this geometry by the extra electron pair. There is experimental evidence that the structure of XeF_6 is not octahedral.

EXERCISES

A blue exercise number indicates that the answer to that exercise appears at the back of this book.

Group 5A Elements

1. The oxyanion of nitrogen in which it has the highest oxidation state is the nitrate ion (NO_3^-). The corresponding oxyanion of phosphorus is PO_4^{3-}. The NO_4^{3-} ion is known but is not very stable. The PO_3^- ion is not known. Account for these differences in terms of the bonding in the four anions.

2. Using data from Appendix 4, calculate $\Delta H°$, $\Delta S°$, and $\Delta G°$ for the reaction

$$N_2(g) + O_2(g) \longrightarrow 2NO(g)$$

Why does the NO formed in an automobile engine not readily decompose back to N_2 and O_2 in the atmosphere?

3. What is the hybridization of the nitrogen atoms in the compounds listed in Table 19.2?

4. Which of the compounds in Table 19.2 exhibit resonance? Draw the corresponding resonance structures for those that show resonance.

5. Use bond energies (Table 13.5) to show that the preferred products for the decomposition of N_2O_3 are NO_2 and NO, rather than O_2 and N_2O. (The N—O single-bond energy is 201 kJ/mol.) (*Hint:* consider the reaction kinetics.)

6. Considering the information given about N_2O_5 in Section 19.2, draw a possible structure for this molecule (in the gas phase).

7. What trade-offs must be made between kinetics and thermodynamics in the Haber process for the production of ammonia? How did the discovery of an appropriate catalyst make the process feasible?

8. Oxidation of the cyanide ion produces the stable cynate ion, OCN^-. The fulminate ion, CNO^-, on the other hand, is very unstable. Fulminate salts explode upon being struck; $Hg(CNO)_2$ is used in blasting caps. Draw Lewis structures and assign formal charges for the cynate and fulminate ions. Why is the fulminate ion so unstable?

9. Write an equation for the reaction of hydrazine with fluorine gas to produce nitrogen gas and hydrogen fluoride gas. Estimate $\Delta H°$ for this reaction, using bond energies from Table 13.5.

10. Write balanced equations that describe how the following can be produced.
    a. $NO(g)$
    b. $N_2O(g)$
    c. $KNO_2(aq)$

11. In each of the following pairs of substances, one is stable and known, while the other is unstable. For each pair, choose the stable substance and explain why the other compound is unstable.
    a. NF_5 or PF_5
    b. AsF_5 or AsI_5
    c. NF_3 or NBr_3

12. Write balanced equations for the reactions described in Table 19.1 for the production of Bi and Sb.

13. Predict the relative acid strengths of the following.
    a. H_3PO_4 and H_3PO_3
    b. H_3PO_4, $H_2PO_4^-$, and HPO_4^{2-}
14. Isohypophosphonic acid, $H_4P_2O_6$, and diphosphonic acid, $H_4P_2O_5$, are tri- and diprotic acids, respectively. Draw Lewis structures for these acids that are consistent with their behavior as acids.
15. Account for the reactivity with oxygen of white phosphorus, red phosphorus, and black phosphorus in terms of their structures.
16. What experiments can be carried out to verify that bismuthyl chloride (BiOCl) is composed of BiO^+ and Cl^- ions and not Bi^+ and OCl^- ions?
17. Arsenic reacts with oxygen to form oxides analogous to the phosphorus oxides. These arsenic oxides react with water similarly to the phosphorus oxides. Write balanced chemical equations describing the reaction of arsenic with oxygen and the reaction of the oxides with water.
18. Compare the description of the localized electron model (Lewis structures) to the molecular orbital model for the bonding in NO, NO^+, and NO^-. Account for any discrepancies between the two models.
19. Sodium tripolyphosphate ($Na_5P_3O_{10}$) is used in many synthetic detergents. Its major effect is to soften the water by complexing Mg^{2+} and Ca^{2+} ions. It also increases the efficiency of surfactants, or wetting agents that lower a liquid's surface tension. The pK value for the formation of $MgP_3O_{10}^{3-}$ is -8.60. The reaction is $Mg^{2+} + P_3O_{10}^{5-} \rightleftharpoons MgP_3O_{10}^{3-}$. Calculate the concentration of Mg^{2+} in a solution that was originally 50. ppm Mg^{2+} (50. mg/L of solution) after 40. g of $Na_5P_3O_{10}$ is added to 1.0 L of the solution.
20. Sodium bismuthate ($NaBiO_3$) is used to test for the presence of Mn^{2+} in solution by the reaction

$$Mn^{2+}(aq) + NaBiO_3(s) \longrightarrow MnO_4^-(aq) + BiO_3^{3-}(aq)$$

    a. Balance the above equation.
    b. Given that bismuth does not form double bonds with oxygen in BiO_3^- and that $NaBiO_3$ is relatively insoluble in water, what type of structure must $NaBiO_3$ have to account for this behavior?
21. One of the most strongly acidic solutions known is a mixture of antimony pentafluoride (SbF_5) and fluorosulfonic acid (HSO_3F). The dominant equilibria are

$$SbF_5 + HSO_3F \rightleftharpoons F_5SbOSO_2FH$$
$$F_5SbOSO_2FH + HSO_3F \rightleftharpoons H_2SO_3F^+ + F_5SbOSO_2F^-$$

    a. Draw Lewis structures for all of the species shown in the preceding reactions.
    b. This *superacid* solution is capable of protonating (adding H^+) to virtually every known organic compound. What is the active protonating agent in the superacid solution?
22. Draw Lewis structures for the $AsCl_4^+$ and $AsCl_6^-$ ions. What type of reaction (acid-base, oxidation-reduction, or the like) is the following?

$$2AsCl_5(g) \longrightarrow AsCl_4AsCl_6(s)$$

23. Complete and balance each of the following.
    a. $P_4O_6(s) + O_2(g) \longrightarrow$
    b. $P_4O_{10}(s) + H_2O(l) \longrightarrow$
    c. $PCl_5(l) + H_2O(l) \longrightarrow$

Group 6A Elements

24. Use bond energies to estimate the maximum wavelength of light that will cause the reaction

$$O_3 \xrightarrow{h\nu} O_2 + O$$

25. Ozone is desirable in the upper atmosphere but undesirable in the lower atmosphere. A dictionary states that ozone has the scent of a fresh spring day. How can these seemingly conflicting statements be reconciled in terms of the chemical properties of ozone?
26. The teflate anion ($OTeF_5^-$) often acts like a halide ion. For example, $P(OTeF_5)_3$ has a structure similar to that of PF_3. Draw Lewis structures for $OTeF_5^-$ and $P(OTeF_5)_3$.
27. The first acid dissociation constants of H_2S, H_2Se, and H_2Te are 1.03×10^{-7}, 1.3×10^{-4}, and 2.3×10^{-3}, respectively.
    a. To what do you attribute this trend? (See the data in the table following part b.)
    b. To the nearest power of 10, predict the K_a value for H_2Po.

Bond Energy (kJ/mol)		Electron Affinity (kJ/mol)	
H—S	363	S	−200.4
H—Se	276	Se	−195
H—Te	238	Te	−190

28. The structure of TeF_5^- is

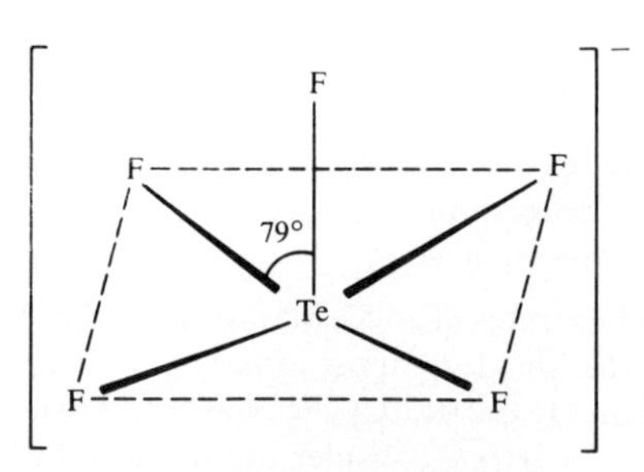

Draw a complete Lewis structure for TeF_5^-. How do you account for the distortion from the ideal square pyramidal geometry?

29. Sulfur will dissolve in an aqueous solution that contains some sulfide ion. Sulfur will then precipitate from this solution if nitric acid is added. Explain this behavior.

30. When nitric acid is added to a solution of sodium sulfide, elemental sulfur forms.
    a. What type of reaction is this?
    b. Use the table of standard reduction potentials (Table 11.1) to predict the reduction product of HNO_3 in this reaction.
    c. Write a balanced equation for this process.

31. Write a balanced equation describing the reduction of H_2SeO_4 by SO_2 to produce selenium.

32. The xerographic (dry-writing) process was invented in 1938 by C. Carlson. In xerography an image is produced on a photoconductor by exposing it to light. Selenium is commonly used, since its conductivity increases three orders of magnitude upon exposure to light in the range from 400 to 500 nm. What color light should be used to cause selenium to become conductive? (See Fig. 20.24.)

33. Cadmium sulfide, which has recently been tested as a photoconductor for xerography, is a semiconducting material with a band gap of 2.42 eV (1 eV, or electron-volt, equals 96.5 kJ/mol). What is the minimum wavelength of light needed to promote electrons from the valence band to the conduction band in CdS? What color of light is this? (See Fig. 20.24.)

34. The SF_5^- ion can be formed by the reaction

$$CsF + SF_4 \longrightarrow Cs^+ + SF_5^-$$

Predict the structure of the SF_5^- ion.

Group 7A Elements

35. Draw Lewis structures and predict geometries for each of the following compounds:
    a. ClF_5
    b. IF_3
    c. Cl_2O_7

```
 O             O
  \           /
O—Cl—O—Cl—O
  /           \
 O             O
```

    d. $FBrO_2$

36. Hypofluorous acid is the most recently prepared of the halogen oxyacids. Weighable amounts were first obtained in 1971 by M. H. Studies and E. N. Appelman, using the fluorination of ice. Hypofluorous acid is exceedingly unstable, decomposing spontaneously (with a half-life of 30 min) to HF and O_2 in a Teflon container at room temperature. It reacts rapidly with water to produce HF, H_2O_2, and O_2. In dilute acid H_2O_2 is the major product; in dilute base O_2 is the major product.
    a. Write balanced chemical equations for the reactions described above.
    b. Assign oxidation states to the elements in hypofluorous acid. Do these oxidation states suggest why hypofluorous acid is so unstable? Explain.

37. Draw the Lewis structure of O_2F_2. Assign oxidation numbers and formal charges to the atoms in O_2F_2. The compound O_2F_2 is a vigorous and potent oxidizing and fluorinating agent. Are oxidation numbers or formal charges more useful in accounting for these properties of O_2F_2?

38. Give two reasons why F_2 is the most reactive of the halogens.

39. Chlorine is used as a disinfectant in the treatment of water supplies.
    a. Calculate $\mathscr{E}°$ for the reaction

$$Cl_2(aq) \longrightarrow OCl^-(aq) + Cl^-(aq) \qquad \text{(not balanced)}$$

    b. Is the reaction spontaneous under standard conditions?
    c. What will be the effect of pH on $\mathscr{E}$ for this reaction? (*Hint:* HOCl is a weak acid.)

(*Note:* for the half-reaction $2OCl^- + 4H^+ + 2e^- \longrightarrow 2H_2O + Cl_2$, $\mathscr{E}° = 1.63$ V.)

40. What is a disproportionation reaction? Use the following reduction potentials,

$$ClO_3^- + 3H^+ + 2e^- \longrightarrow HClO_2 + H_2O \qquad \mathscr{E}° = 1.21 \text{ V}$$

$$HClO_2 + 2H^+ + 2e^- \longrightarrow HClO + H_2O \qquad \mathscr{E}° = 1.65 \text{ V}$$

to predict whether $HClO_2$ will disproportionate.

41. Sodium perchlorate is produced by the electrolytic oxidation of aqueous $NaClO_3$:

$$ClO_3^- + H_2O \longrightarrow ClO_4^- + H_2$$

More than half of the annual U.S. production of $NaClO_4$ is converted to NH_4ClO_4 for use as a rocket fuel. The two booster rockets on the space shuttle use NH_4ClO_4 (70% by mass) as an oxidizer and powdered aluminum (30% by mass) as fuel. About 7×10^5 kg of NH_4ClO_4 is required for each launch.
    a. What is the minimum electrical potential that must be applied to produce $NaClO_4$? For the half-reaction $ClO_4^- + 2H^+ + 2e^- \rightarrow ClO_3^- + H_2O$, $\mathscr{E}° = +1.19$ V.
    b. A possible reaction in the booster rocket is

$$3Al(s) + 3NH_4ClO_4(s) \longrightarrow Al_2O_3(s) + AlCl_3(s) + 3NO(g) + 6H_2O(g)$$

Estimate the total heat generated by the booster rocket during a space shuttle launch.

42. Photogray lenses contain small embedded crystals of solid silver chloride. Silver chloride is light-sensitive because of the reaction

$$AgCl(s) \xrightarrow{h\nu} Ag(s) + Cl$$

Small particles of metallic silver cause the lenses to darken. In the lenses this process is reversible. When the light is removed, the reverse reaction occurs. However, when pure white silver chloride is exposed to sunlight, it darkens; the reverse reaction does not occur in the dark.

a. How do you explain this difference?

b. Photogray lenses do become permanently dark in time. How do you account for this?

Group 8A Elements

43. The noble gases were among the latest elements discovered; their existence was not predicted by Mendeleev when he published his first periodic table. Why do you think this was so?

44. Although He is the second most abundant element in the universe, it is very rare on earth. Why?

45. In chemistry textbooks written before 1962, the noble gases were referred to as the inert gases. Why do we no longer use the term *inert gases?*

46. Using the data in Table 19.12, calculate the mass of xenon at 25°C and 1.0 atm in a room 10.0 m × 5.0 m × 3.0 m. How many Xe atoms are in this room? How many Xe atoms do you inhale in one breath (approximately 2 L) of air?

47. Draw Lewis structures for and predict the shapes of the following.

a. XeO_3
b. XeO_4
c. $XeOF_4$
d. $XeOF_2$
e. XeO_3F_2

48. Xenon difluoride has proved to be a versatile fluorinating agent. For example, in the reaction

$$C_6H_6(l) + XeF_2(g) \longrightarrow C_6H_5F(l) + Xe(g) + HF(g)$$

the by-products Xe and HF are easily removed, leaving pure C_6H_5F. Xenon difluoride is stored in an inert atmosphere free from oxygen and water. Why is this necessary?

Additional Exercises

49. The compound NF_3 is quite stable, but NCl_3 is very unstable (NCl_3 was first synthesized in 1811 by P. L. Dulong, who lost three fingers and an eye studying its properties). The compounds NBr_3 and NI_3 are unknown, although the explosive $NI_3 \cdot NH_3$ is known. How do you account for the instability of these halides of nitrogen?

50. Many structures of phosphorus- and sulfur-containing compounds are drawn with some P=O and P=S bonds. These bonds are not the typical π bonds we've considered, which involve the overlap of two p orbitals. They result instead from the overlap of a d orbital on the phosphorus or sulfur atom with a p orbital on oxygen. This type of π bonding is sometimes used as an explanation for why H_3PO_3 has the first structure below rather than the second:

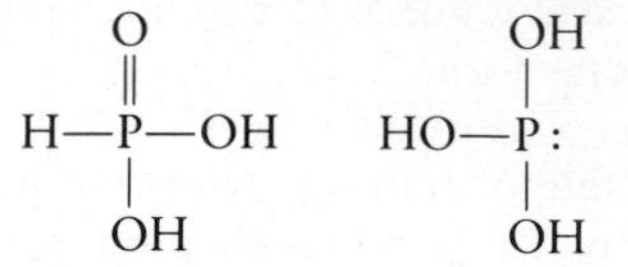

```
     O                OH
     ‖                |
 H—P—OH         HO—P:
     |                |
     OH               OH
```

Draw a picture showing how a d orbital and a p orbital overlap to form a π bond.

51. The xenon halides, oxyhalides, and oxides are isoelectronic with many other compounds and ions containing halogens. Give a compound or ion, in which iodine is the central atom, that is isoelectronic with each of the following.

a. XeO_4
b. XeO_3
c. XeF_2
d. XeF_4
e. XeF_6
f. $XeOF_3$

52. The energy required to break a particular bond is not always constant. Compare the N—Cl bond energies in NOCl and NCl_3:

$$NOCl \longrightarrow NO + Cl \qquad \Delta H° = 158 \text{ kJ/mol}$$

$$NCl_3 \longrightarrow NCl_2 + Cl \qquad \Delta H° \approx 375 \text{ kJ/mol}$$

Why is there such a great discrepancy in the apparent N—Cl bond energies? (*Hint:* consider what happens to the nitrogen-oxygen bond in the first reaction.)

53. Use data in Appendix 4 to calculate $\Delta H°$, $\Delta S°$, and $\Delta G°$ for the reaction

$$3NO(g) \rightleftharpoons N_2O(g) + NO_2(g)$$

At what temperatures is this reaction spontaneous at standard pressures?

54. Describe the structural changes that occur when plastic sulfur becomes brittle.

55. In many natural waters nitrogen and phosphorus are the least abundant nutrients available for plant life. Some waters that become polluted from agricultural runoff or municipal sewage become infested with algae. The algae consume most of the dissolved oxygen in the water that can kill the fish. Describe how these events are chemically related.

56. Although the perchlorate and periodate ions have been known for a long time, the perbromate ion proved to be quite elusive and was not synthesized until 1965. Perbromate was synthesized by oxidizing bromate ion with aqueous xenon difluoride, producing xenon, hydrofluoric acid, and perbromate ion. Write a balanced chemical equation for this reaction.

Transition Metals and Coordination Chemistry

Transition metals have many uses in our society. Iron is used for steel, copper for electrical wiring and water pipes, titanium for paint, silver for photographic paper, manganese, chromium, vanadium, and cobalt as additives to steel, platinum for industrial and automotive catalysts, and so on.

One indication of the importance of transition metals is the great concern shown by the U.S. government for continuing the supply of these elements. The United States is a net importer of more than 60 "strategic and critical" minerals, including cobalt, manganese, platinum, palladium, and chromium. All of these metals play a vital role in the U.S. economy and defense, but approximately 90% of the required amounts must be imported (see Table 20.1).

In addition to being important in industry, transition metal ions also play a vital role in living organisms. For example, complexes of iron provide for the transport and storage of oxygen, molybdenum and iron compounds are catalysts in nitrogen fixation, zinc is found in more than 150 biomolecules in humans, copper and iron play a crucial role in the respiratory cycle, and cobalt is found in essential biomolecules such as vitamin B_{12}.

In this chapter we explore the general properties of transition metals, paying particular attention to the bonding, structure, and properties of the complex ions of these metals.

TABLE 20.1 Some Transition Metals Important to the U.S. Economy and Defense

Metal	Uses	Percentage Imported
Chromium	Stainless steel (especially for parts exposed to corrosive gases and high temperatures)	~91%
Cobalt	High-temperature alloys in jet engines, magnets, catalysts, drill bits	~93%
Manganese	Steelmaking	~97%
Platinum and palladium	Catalysts	~87%

20.1 The Transition Metals: A Survey

General Properties

One striking characteristic of the representative elements is that their chemistry changes markedly across a given period as the number of valence electrons changes. The chemical similarities occur mainly within the vertical groups. In contrast, *the transition metals show great similarities within a given period as well as within a given vertical group*. This difference occurs because the last electrons added to the transition metal elements are inner electrons: *d* electrons for the *d*-block transition metals and *f* electrons for the lanthanides and actinides. These inner *d* and *f* electrons cannot participate in bonding as readily

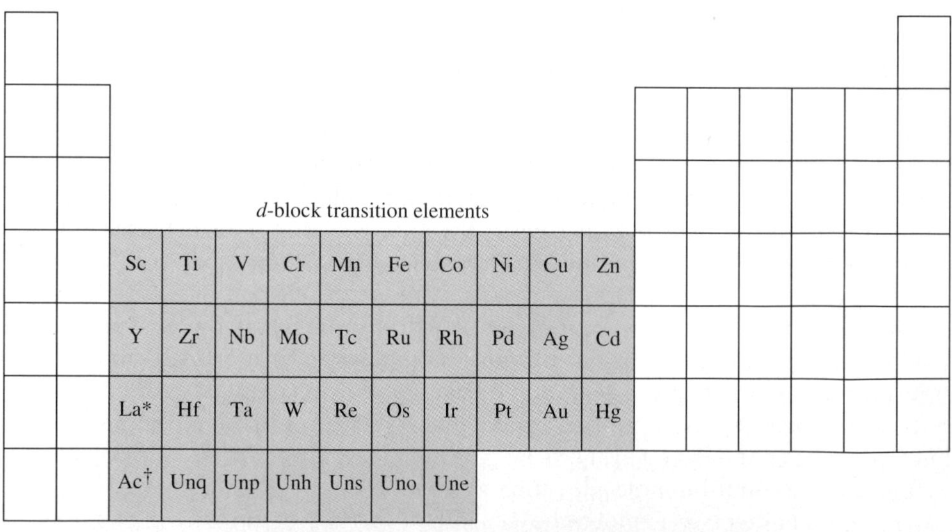

Figure 20.1
The position of the transition elements on the periodic table. The *d*-block elements correspond to filling the 3*d*, 4*d*, 5*d*, or 6*d* orbitals. The inner transition metals correspond to filling the 4*f* (lanthanides) or 5*f* (actinides) orbitals.

(clockwise from upper left) Calcite stalactites colored by traces of iron. Quartz is often colored by the presence of transition metals such as Mn, Fe, and Ni. Wulfenite contains $PbMoO_4$. Rhodochrosite is a mineral containing $MnCO_3$.

as the valence *s* and *p* electrons. Thus the chemistry of the transition elements is not as greatly affected by the gradual change in the number of electrons as is the chemistry of the representative elements.

Group designations are traditionally given on the periodic table for the *d*-block transition metals (Fig. 20.1). However, these designations do not relate as directly to the chemical behavior of these elements as do the designations for the representative elements (the A groups), so we will not use them.

As a class, the transition metals behave as typical metals, exhibiting metallic luster and relatively high electrical and thermal conductivities. Silver is the best conductor of heat and electrical current. However, copper is a close second, which explains copper's wide use in the electrical systems of homes and factories.

In spite of their many similarities, the properties of the transition metals do vary considerably. For example, tungsten with a melting point of 3400°C is used for filaments in light bulbs; in contrast, mercury is a liquid at 25°C. Some transition metals such as iron and titanium are hard and strong and therefore make very useful structural materials; others such as copper, gold, and silver are relatively soft. The chemical reactivity of the transition metals also varies significantly. Some react readily with oxygen to form oxides. Of these metals, some, such as chromium, nickel, and cobalt, form oxides that adhere tightly to the metallic surface, thereby protecting the metal from further oxidation. Others, such as iron, form oxides that scale off, constantly exposing new metal to the corrosion process. On the other hand, the noble metals—primarily gold, silver, platinum, and palladium—do not readily form oxides.

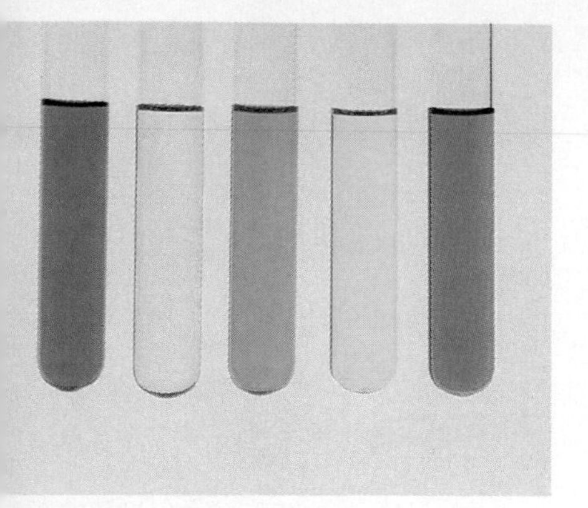

(from left to right) Aqueous solutions containing the metal ions Co^{2+}, Mn^{2+}, Cr^{3+}, Fe^{3+}, and Ni^{2+}.

In forming ionic compounds with nonmetals, the transition metals exhibit several typical characteristics:

More than one oxidation state is often found. For example, iron combines with chlorine to form $FeCl_2$ and $FeCl_3$.

The cations are often **complex ions,** species in which *the transition metal ion is surrounded by a certain number of ligands* (molecules or ions that behave as Lewis bases). For example, the compound $[Co(NH_3)_6]Cl_3$ contains $Co(NH_3)_6{}^{3+}$ cations and Cl^- anions:

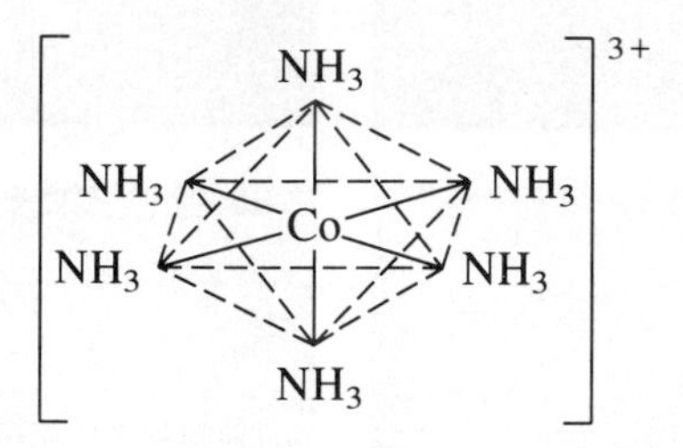

The $Co(NH_3)_6{}^{3+}$ ion

Most compounds are colored, because the typical transition metal ion in a complex ion can absorb visible light of specific wavelengths.

Many compounds are paramagnetic (they contain unpaired electrons).

In this chapter we will concentrate on the **first-row transition metals** (scandium through zinc) because they are representative of the other transition series and because they have great practical significance. Some important properties of these elements are summarized in Table 20.2 and are discussed below.

Electron Configurations

Chromium has the electron configuration $[Ar]4s^13d^5$.

The electron configurations of the first-row transition metals were discussed in Section 12.13. The $3d$ orbitals begin to fill after the $4s$ orbital is complete, that is, after calcium ($[Ar]4s^2$). The first transition metal, *scandium,* has one electron in the $3d$ orbitals; the second, *titanium,* has two; and the third, *vanadium,* has three. We would expect *chromium,* the fourth transition metal, to have the electron configuration $[Ar]4s^23d^4$. However, the actual configuration is $[Ar]4s^13d^5$; it has a half-filled $4s$ orbital and a half-filled set of $3d$ orbitals (one electron in each of the five $3d$ orbitals). It is tempting to say that the configuration results because half-filled "shells" are especially stable. Although there are some reasons to think that this explanation might be valid, it is an oversimplification. For instance, tungsten, which is in the same vertical group as chromium, has the configuration $[Xe]6s^24f^{14}5d^4$, where half-filled s and d shells are not found. There are several similar cases that dispute the importance of half-filled shells.

Basically, the configuration found in chromium occurs because the energies of the $3d$ and $4s$ orbitals are very similar for the first-row transition elements. We saw in Section 12.13 that when electrons are placed in a set of degenerate orbitals, they first occupy each orbital singly to minimize electron repulsions.

TABLE 20.2 Selected Properties of the First-Row Transition Metals

Property	Scandium	Titanium	Vanadium	Chromium	Manganese	Iron	Cobalt	Nickel	Copper	Zinc
Atomic number	21	22	23	24	25	26	27	28	29	30
Electron configuration*	$4s^23d^1$	$4s^23d^2$	$4s^23d^3$	$4s^13d^5$	$4s^23d^5$	$4s^23d^6$	$4s^23d^7$	$4s^23d^8$	$4s^13d^{10}$	$4s^23d^{10}$
Atomic radius† (pm)	162	147	134	130	135	126	125	124	128	138
Ionization energies (eV/atom)										
First	6.54	6.82	6.74	6.77	7.44	7.87	7.86	7.64	7.73	9.39
Second	12.80	13.58	14.65	16.50	15.64	16.18	17.06	18.17	20.29	17.96
Third	24.76	27.49	29.31	30.96	33.67	30.65	33.50	35.17	36.83	39.72
Reduction potential‡ (V)	−2.08	−1.63	−1.2	−0.91	−1.18	−0.44	−0.28	−0.23	+0.34	−0.76
Common oxidation states	+3	+2, +3, +4	+2, +3, +4, +5	+2, +3, +6	+2, +3, +4, +7	+2, +3	+2, +3	+2	+1, +2	+2
Melting point (°C)	1397	1672	1710	1900	1244	1530	1495	1455	1083	419
Density (g/cm^3)	2.99	4.49	5.96	7.20	7.43	7.86	8.9	8.90	8.92	7.14
Electrical conductivity§	—	2	3	10	2	17	24	24	97	27

* Each atom has an argon inner-core configuration.
† Covalent atomic radii.
‡ For the reduction process $M^{2+} + 2e^- \longrightarrow M$ (except for scandium, where the ion is Sc^{3+}).
§ Compared with an arbitrarily assigned value of 100 for silver.

Since the 4*s* and 3*d* orbitals are virtually degenerate in the chromium atom, we would expect the configuration

$$4s\ \underline{\uparrow}\quad 3d\ \underline{\uparrow}\ \underline{\uparrow}\ \underline{\uparrow}\ \underline{\uparrow}\ \underline{\uparrow}$$

rather than

$$4s\ \underline{\uparrow\downarrow}\quad 3d\ \underline{\uparrow}\ \underline{\uparrow}\ \underline{\uparrow}\ \underline{\uparrow}\ \underline{\ \ }$$

The second arrangement has greater electron-electron repulsions and thus has a higher energy.

A set of orbitals with the same energy is said to be degenerate.

The only other unexpected configuration among the first-row transition metals is that of copper, which is $[Ar]4s^13d^{10}$ rather than the expected $[Ar]4s^23d^9$.

Copper has the electron configuration $[Ar]4s^13d^{10}$.

In contrast to the neutral transition metals, where the 3*d* and 4*s* orbitals have very similar energies, the *energy of the* 3d *orbitals in transition metal ions is significantly less than that of the* 4s *orbital.* This means that the electrons remaining after the ion is formed occupy the 3*d* orbitals, since they are lower in energy. *First-row transition metal ions do not have* 4s *electrons.* For example, manganese has the configuration $[Ar]4s^23d^5$, but that of Mn^{2+} is $[Ar]3d^5$. The neutral titanium atom has the configuration $[Ar]4s^23d^2$, but that of Ti^{3+} is $[Ar]3d^1$.

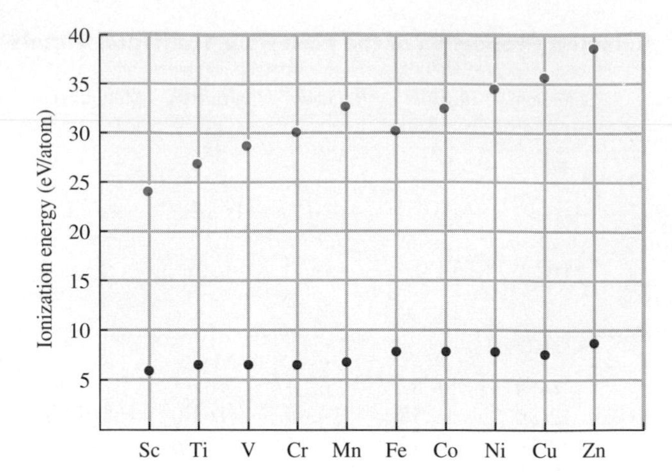

Figure 20.2
Plots of the first (red dots) and third (blue dots) ionization energies for the first-row transition metals.

Oxidation States and Ionization Energies

The transition metals can form a variety of ions by losing one or more electrons. The common oxidation states of these elements are shown in Table 20.2. Note that for the first five metals the maximum possible oxidation state corresponds to the loss of all the 4*s* and 3*d* electrons. For example, the maximum oxidation state of chromium ($[Ar]4s^1 3d^5$) is +6. Toward the right end of the period, the maximum oxidation states are not observed; in fact, the 2+ ions are the most common. The higher oxidation states are not seen for these metals because the 3*d* orbitals become lower in energy as the nuclear charge increases, making the electrons increasingly difficult to remove. From Table 20.2 we see that ionization energy increases gradually from left to right across the period. However, the third ionization energy (corresponding to the removal of an electron from a 3*d* orbital) increases faster than the first ionization energy, clear evidence of the significant decrease in the energy of the 3*d* orbitals in going across the period (see Fig. 20.2).

Standard Reduction Potentials

When a metal acts as a *reducing agent,* the half-reaction is

$$M \longrightarrow M^{n+} + ne^-$$

This is the reverse of the conventional listing of half-reactions in tables. Thus, when we rank the transition metals in order of reducing ability, it is most convenient to reverse the reactions and the signs given in Table 20.2. The metal with the most positive potential is thus the best reducing agent. The transition metals are listed in order of reducing ability in Table 20.3.

Recall that $\mathscr{E}°$ is zero by definition for the process

$$2H^+ + 2e^- \longrightarrow H_2$$

Therefore, all of the metals except copper can reduce H^+ ions in 1 *M* aqueous solutions of strong acid to hydrogen gas:

$$M(s) + 2H^+(aq) \longrightarrow H_2(g) + M^{2+}(aq)$$

TABLE 20.3 **Relative Reducing Abilities of the First-Row Transition Metals in Aqueous Solution**

Reaction	Potential (V)
$Sc \rightarrow Sc^{3+} + 3e^-$	2.08
$Ti \rightarrow Ti^{2+} + 2e^-$	1.63
$V \rightarrow V^{2+} + 2e^-$	1.2
$Mn \rightarrow Mn^{2+} + 2e^-$	1.18
$Cr \rightarrow Cr^{2+} + 2e^-$	0.91
$Zn \rightarrow Zn^{2+} + 2e^-$	0.76
$Fe \rightarrow Fe^{2+} + 2e^-$	0.44
$Co \rightarrow Co^{2+} + 2e^-$	0.28
$Ni \rightarrow Ni^{2+} + 2e^-$	0.23
$Cu \rightarrow Cu^{2+} + 2e^-$	−0.34

Reducing ability ↑

As Table 20.3 shows, the reducing abilities of the first-row transition metals generally decrease going from left to right across the period. Only chromium and zinc do not follow this trend.

The 4*d* and 5*d* Transition Series

In comparing the 3*d*, 4*d*, and 5*d* transition series, it is instructive to consider the atomic radii of these elements (Fig. 20.3). Note that there is a general, although not regular, decrease in size going from left to right across each of the series. Also note that although there is a significant increase in radius in going from the 3*d* to the 4*d* metals, the 4*d* and 5*d* metals are remarkably similar in size. This latter phenomenon is the result of the **lanthanide contraction.** In the **lanthanide series,** which consists of the elements between lanthanum and hafnium (see Fig. 20.1), electrons are filling the 4*f* orbitals. Since the 4*f* orbitals are buried in the interior of these atoms, the additional electrons do not add to the atomic size. In fact, the increasing nuclear charge (remember that a proton is added to the nucleus with each electron) causes the radii of the lanthanide elements to decrease significantly going from left to right. This lanthanide contraction just

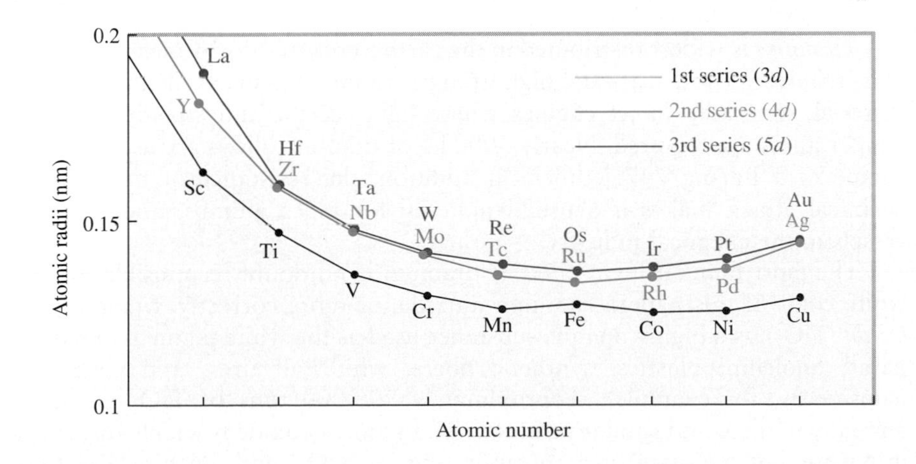

Figure 20.3
Atomic radii of the 3*d*, 4*d*, and 5*d* transition series.

offsets the normal increase in size due to changing from one principal quantum level to another. Thus the 5*d* elements, instead of being significantly larger than the 4*d* elements, are almost identical to them in size. This leads to a great similarity in the chemistry of the 4*d* and 5*d* elements in a given vertical group. For example, the chemical properties of hafnium and zirconium are remarkably similar, and they always occur together in nature. Their separation, which is probably more difficult than the separation of any other pair of elements, often requires fractional distillation of their compounds.

Niobium was originally called columbium and is still occasionally referred to by that name.

In general, the differences between the 4*d* and 5*d* elements in a group increase gradually going from left to right. For example, niobium and tantalum are also quite similar, but less so than zirconium and hafnium.

Although generally less well known than the 3*d* elements, the 4*d* and 5*d* transition metals have certain very useful properties. For example, zirconium and zirconium oxide (ZrO_2) are quite resistant to high temperatures and are used, along with niobium and molybdenum alloys, for space vehicle parts that are exposed to high temperatures during reentry into the earth's atmosphere. Niobium and molybdenum are also important alloying materials for certain types of steel. Tantalum, which has a high resistance to attack by body fluids, is often used for surgical clips. The *platinum group metals*—ruthenium, osmium, rhodium, iridium, palladium, and platinum—are all quite similar and are widely used as catalysts for many types of industrial processes.

20.2 The First-Row Transition Metals

We have seen that while the transition metals are similar in many ways, they also show important differences. We will now explore some of the specific properties of each of the 3*d* transition metals.

Scandium is a rare element that exists in compounds mainly in the +3 oxidation state, for example, in $ScCl_3$, Sc_2O_3, and $Sc_2(SO_4)_3$. The chemistry of scandium strongly resembles that of the lanthanides, with most of its compounds being colorless and diamagnetic. This is not surprising, since Sc^{3+} has no *d* electrons. As we will see in Section 20.6, the color and magnetism of transition metal compounds usually arise from the *d* electrons on the metal ion. Scandium metal, which can be prepared by electrolysis of molten $ScCl_3$, is not widely used because of its rarity. However, it is found in some electronic devices, such as high-intensity lamps.

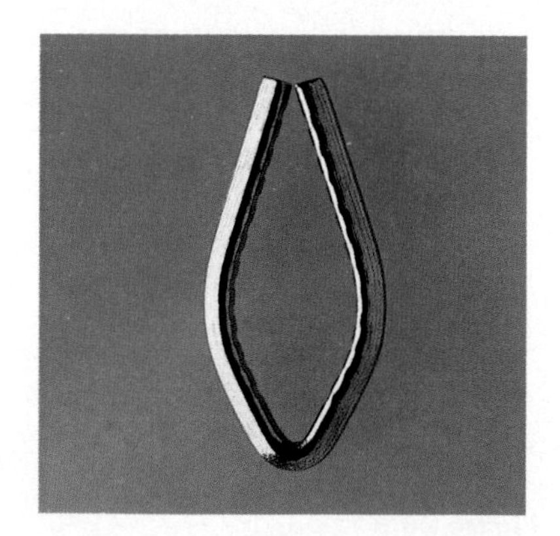

Surgical clip constructed from tantalum metal.

Titanium is widely distributed in the earth's crust (0.6% by mass). Because of its relatively low density and high strength, titanium is an excellent structural material, especially in jet engines where light weight and stability at high temperatures are required. Nearly 5000 kg of titanium alloys are used in each engine of a Boeing 747 jetliner. In addition, the resistance of titanium to chemical attack makes it a useful material for pipes, pumps, and reaction vessels in the chemical industry.

The most familiar compound of titanium is no doubt responsible for the white color of this paper. Titanium dioxide, or more correctly, *titanium(IV) oxide* (TiO_2), is a highly opaque substance used as the white pigment in paper, paint, linoleum, plastics, synthetic fibers, whitewall tires, and cosmetics (sunscreens, for example). Approximately 700,000 tons of TiO_2 are used annually in these and similar products. Titanium(IV) oxide is widely dispersed in nature, but the main ores are rutile (impure TiO_2) and ilmenite ($FeTiO_3$).

Rutile is processed by treatment with chlorine to form volatile $TiCl_4$, which is then separated from the impurities and burned to form TiO_2:

$$TiCl_4(g) + O_2(g) \longrightarrow TiO_2(s) + 2Cl_2(g)$$

Ilmenite is treated with sulfuric acid to form a soluble sulfate,

$$FeTiO_3(s) + 2H_2SO_4(aq) \longrightarrow Fe^{2+}(aq) + TiO^{2+}(aq) + 2SO_4^{2-}(aq) + 2H_2O(l)$$

When this aqueous mixture is allowed to stand under vacuum, solid $FeSO_4 \cdot 7H_2O$ forms and is removed. The mixture remaining is then heated, and the insoluble titanium(IV) oxide hydrate ($TiO_2 \cdot H_2O$) forms. The water of hydration is driven off by heating to form pure TiO_2:

$$TiO_2 \cdot H_2O(s) \xrightarrow{\text{Heat}} TiO_2(s) + H_2O(g)$$

Liquid titanium IV chloride being added to water, forming a cloud of solid titanium oxide and hydrochloric acid.

In its compounds titanium most often exists in the +4 oxidation state. Examples are TiO_2 and $TiCl_4$, the latter a colorless liquid (bp = 137°C) that fumes in moist air to produce TiO_2:

$$TiCl_4(l) + 2H_2O(l) \longrightarrow TiO_2(s) + 4HCl(g)$$

Titanium(III) compounds can be produced by reduction of compounds containing titanium in the +4 state. In aqueous solution Ti^{3+} exists as the purple $Ti(H_2O)_6^{3+}$ ion, which is slowly oxidized to titanium(IV) by air. Titanium(II) is not stable in aqueous solution but does exist in the solid state in compounds such as TiO and dihalides of the type TiX_2.

Vanadium is widely dispersed throughout the earth's crust (0.02% by mass). It is used mostly in alloys with other metals such as iron (80% of vanadium is used in steel) and titanium. Vanadium(V) oxide (V_2O_5) is used as an industrial catalyst in the production of materials such as sulfuric acid.

The manufacture of sulfuric acid was discussed in Section 3.8.

Pure vanadium can be obtained from the electrolytic reduction of fused salts, such as VCl_2, to produce a metal similar to titanium that is steel gray, hard, and corrosion-resistant. Often the pure element is not required for alloying. For example, *ferrovanadium,* produced by reducing a mixture of V_2O_5 and Fe_2O_3 with aluminum, is added to iron to form *vanadium steel,* a hard steel used for engine parts and axles.

The most common oxidation state for vanadium is +5.

The principal oxidation state of vanadium is +5, found in compounds such as the orange V_2O_5 (mp = 650°C) and the colorless VF_5 (mp = 19.5°C). The oxidation states ranging from +5 to +2 all exist in aqueous solution (see Table 20.4). The higher oxidation states, +5 and +4, do not exist as hydrated ions of the type $V^{n+}(aq)$ because these highly charged ions cause the attached water molecules to be very acidic. The H^+ ions are lost to give the oxycations VO_2^+ and VO^{2+}. The hydrated V^{3+} and V^{2+} ions are easily oxidized and thus can function as reducing agents in aqueous solution.

TABLE 20.4
Oxidation States and Species for Vanadium in Aqueous Solution

Oxidation State of Vanadium	Species in Aqueous Solution
+5	VO_2^+ (yellow)
+4	VO^{2+} (blue)
+3	$V^{3+}(aq)$ (blue-green)
+2	$V^{2+}(aq)$ (violet)

Although *chromium* is relatively rare, it is a very important industrial material. The chief ore of chromium, chromite ($FeCr_2O_4$), can be reduced by carbon to give *ferrochrome,*

$$FeCr_2O_4(s) + 4C(s) \longrightarrow \underbrace{Fe(s) + 2Cr(s)}_{\text{Ferrochrome}} + 4CO(g)$$

which can be added directly to iron in the steelmaking process. Chromium

Titanium Makes Great Bicycles

One of the most interesting characteristics of the world of bicycling is the competition among various frame materials. Bicycle frames are now built from steel, aluminum, carbon fiber composites, and titanium, with each material having advantages and disadvantages. Steel is strong, economical, adaptable, and (unfortunately) "rustable." Aluminum is light and stiff but has relatively low fatigue limits (resistance to repeated stresses). Carbon fiber composites have amazing strength-to-mass ratios, and have shock- and vibration-dampening properties superior to any metal; however, they are very expensive. Titanium has a density ≈43% less than that of steel, a yield strength (when alloyed with metals such as aluminum and tin) that is 30% greater than that of steel, an extraordinary resistance to fatigue, and a high resistance to corrosion, but is expensive and difficult to work.

Of all these materials, titanium gives the bicycle that fanatics seem to love the most (Fig. 20.4). After their first ride on a bicycle with a titanium frame, most experienced cyclists find themselves shaking their heads and searching hard for the right words to describe the experience. Typically, the word *magic* is used a great deal in the ensuing description.

The magic of titanium results from its combination of toughness, stretchability, and resilience. A bicycle that is built stiff to resist pedaling loads usually responds by giving a harsh, uncomfortable ride. A titanium bike is very stiff against high pedaling torques, but it seems to transmit much less road shock than bikes made of competitive materials. Why titanium excels in dampening vibrations is not entirely clear. Despite titanium's significantly lower density than steel's, shock waves travel more slowly in titanium than steel. Whatever the explanation for its shock-absorbing abilities, titanium provides three things that cyclists find crucial: light weight, stiffness, and a smooth ride—magic.

Figure 20.4
A titanium bicycle.

Titanium is quite abundant in the earth's crust, ranking ninth of all the elements and second among the transition elements. The metallurgy of titanium presents special challenges. Carbon, the reducing agent most commonly used to obtain metals from their oxide ores, cannot be used because it forms intractable interstitial carbides with titanium. These carbides are extraordinarily hard and have melting points above 3000°C. However, if chlorine gas is used in conjunction with carbon to treat the ore, volatile $TiCl_4$ is formed, which can be distilled off and then reduced with magnesium or sodium at ≈1000°C to form a titanium "sponge." This sponge is then ground up, cleaned with aqua regia (a 1:3 mixture of concentrated HNO_3 and concentrated HCl), melted under a blanket of inert gas (to prevent reaction with oxygen), and cast into ingots. Titanium, a lustrous, silvery metal with a high melting point (1667°C), crystallizes in a hexagonal closest packed structure. Because titanium tends to become quite brittle when trace impurities such as C, N, and O are present, it must be fabricated with great care.

Titanium's unusual ability to stretch makes it hard to machine. It tends to push away even from a very sharp cutting blade, giving a rather unpredictable final dimension. Also, because titanium is embrittled by reaction with oxygen, all welding operations must be carried out under a shielding gas such as argon.

However, the bicycle that results is worth all these difficulties. One woman described a titanium bicycle as "the one God rides on Sunday."

metal, which is often used to plate steel, is hard and brittle. It maintains a bright surface by developing a tough, invisible oxide coating.

Chromium commonly forms compounds in which it has an oxidation state of +2, +3, or +6, as shown in Table 20.5. The Cr^{2+} (chromous) ion is a powerful reducing agent in aqueous solution. In fact, traces of O_2 in other gases can be removed from other gases by bubbling the gaseous mixture through a Cr^{2+} solution:

$$4Cr^{2+}(aq) + O_2(g) + 4H^+(aq) \longrightarrow 4Cr^{3+}(aq) + 2H_2O(l)$$

The chromium(VI) species are excellent oxidizing agents, especially in acidic solution, where chromium(VI) as the dichromate ion ($Cr_2O_7^{2-}$) is reduced to the Cr^{3+} ion:

$$Cr_2O_7^{2-}(aq) + 14H^+(aq) + 6e^- \longrightarrow 2Cr^{3+}(aq) + 7H_2O(l) \quad \mathscr{E}° = 1.33 \text{ V}$$

The oxidizing ability of the dichromate ion is strongly pH-dependent; it increases as $[H^+]$ increases, as predicted by Le Châtelier's principle. In basic solution chromium(VI) exists as the chromate ion, a much less powerful oxidizing agent:

$$CrO_4^{2-}(aq) + 4H_2O(l) + 3e^- \longrightarrow Cr(OH)_3(s) + 5OH^-(aq) \quad \mathscr{E}° = -0.13 \text{ V}$$

The structures of the $Cr_2O_7^{2-}$ and CrO_4^{2-} ions are shown in Fig. 20.5.

Red chromium(VI) oxide (CrO_3) dissolves in water to give a strongly acidic, red-orange solution:

$$2CrO_3(s) + H_2O(l) \longrightarrow 2H^+(aq) + Cr_2O_7^{2-}(aq)$$

It is possible to precipitate bright orange dichromate salts, such as $K_2Cr_2O_7$, from these solutions. When made basic, the solution turns yellow and chromate salts such as Na_2CrO_4 can be obtained. A mixture of chromium(VI) oxide and concentrated sulfuric acid, commonly called *cleaning solution,* is a powerful oxidizing medium that can remove organic materials from analytical glassware, yielding a very clean surface.

Manganese is relatively abundant (0.1% of the earth's crust), although no significant sources are found in the United States. The most common use of manganese is in the production of an especially hard steel used for rock crushers, bank vaults, and armor plate. One interesting source of manganese is *manganese nodules* found on the ocean floor. These roughly spherical "rocks" contain mixtures of manganese and iron oxides as well as smaller amounts of other metals such as cobalt, nickel, and copper. Apparently, the nodules were formed at least partly by the action of marine organisms. Because of the abundance of these nodules, there is much interest in developing economical methods for their recovery and processing.

TABLE 20.5
Typical Chromium Compounds

Oxidation State of Chromium	Examples of Compounds (X = halogen)
+2	CrX_2
+3	CrX_3
	Cr_2O_3 (green)
	$Cr(OH)_3$ (blue-green)
+6	$K_2Cr_2O_7$ (orange)
	Na_2CrO_4 (yellow)
	CrO_3 (red)

Chromium VI oxide dissolving in water.

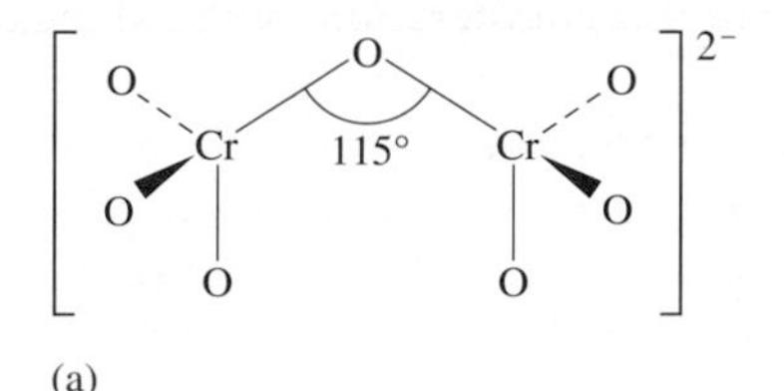

Figure 20.5
The structures of the chromium(VI) anions: (a) $Cr_2O_7^{2-}$, which exists in acidic solution; and (b) CrO_4^{2-} which exists in basic solution.

TABLE 20.6
Some Compounds of Manganese in Its Most Common Oxidation States

Oxidation State of Manganese	Examples of Compounds
+2	$Mn(OH)_2$ (pink)
	MnS (salmon)
	$MnSO_4$ (reddish)
	$MnCl_2$ (pink)
+4	MnO_2 (dark brown)
+7	$KMnO_4$ (purple)

TABLE 20.7
Typical Compounds of Iron

Oxidation State of Iron	Examples of Compounds
+2	FeO (black)
	FeS (brownish black)
	$FeSO_4 \cdot 7H_2O$ (green)
	$K_4Fe(CN)_6$ (yellow)
+3	$FeCl_3$ (brownish black)
	Fe_2O_3 (reddish brown)
	$K_3Fe(CN)_6$ (red)
	$Fe(SCN)_3$ (red)
+2, +3 (mixture)	Fe_3O_4 (black)
	$KFe[Fe(CN)_6]$ (deep blue, Prussian blue)

TABLE 20.8
Typical Compounds of Cobalt

Oxidation State of Cobalt	Examples of Compounds
+2	$CoSO_4$ (dark blue)
	$[Co(H_2O)_6]Cl_2$ (pink)
	$[Co(H_2O)_6](NO_3)_2$ (red)
	CoS (black)
	CoO (greenish brown)
+3	CoF_3 (brown)
	Co_2O_3 (charcoal)
	$K_3[Co(CN)_6]$ (yellow)
	$[Co(NH_3)_6]Cl_3$ (yellow)

Manganese can exist in all oxidation states from +2 to +7, although +2 and +7 are the most common. Manganese(II) forms an extensive series of salts with all of the common anions. In aqueous solution Mn^{2+} forms $Mn(H_2O)_6{}^{2+}$, which has a light pink color. Manganese(VII) is found in the intensely purple permanganate ion ($MnO_4{}^-$). Widely used as an analytical reagent in acidic solution, the $MnO_4{}^-$ ion behaves as a strong oxidizing agent, producing Mn^{2+}:

$$MnO_4{}^-(aq) + 8H^+(aq) + 5e^- \longrightarrow Mn^{2+}(aq) + 4H_2O(l) \qquad \mathscr{E}° = 1.51\text{ V}$$

Several typical compounds of manganese are listed in Table 20.6.

Iron is the most abundant heavy metal (constituting 4.7% of the earth's crust) and the most important to our civilization. It is a white, lustrous, not particularly hard metal that is very reactive toward oxidizing agents. For example, in moist air it is rapidly oxidized by oxygen to form rust, a mixture of iron oxides.

The chemistry of iron mainly involves its +2 and +3 oxidation states. Typical compounds are shown in Table 20.7. In aqueous solutions, iron(II) salts are generally light green because of the presence of $Fe(H_2O)_6{}^{2+}$. Although the $Fe(H_2O)_6{}^{3+}$ ion is colorless, aqueous solutions of iron(III) salts are usually yellow to brown in color, due to the presence of $Fe(OH)(H_2O)_5{}^{2+}$. This latter ion results from the acidity of $Fe(H_2O)_6{}^{3+}$ ($K_a = 6 \times 10^{-3}$):

$$Fe(H_2O)_6{}^{3+}(aq) \rightleftharpoons Fe(OH)(H_2O)_5{}^{2+}(aq) + H^+(aq)$$

Although *cobalt* is relatively rare, it is found in ores such as smaltite ($CoAs_2$) and cobaltite (CoAsS) in large enough concentrations to make its production economically feasible. Cobalt is a hard, bluish white metal mainly used in alloys such as stainless steel and stellite, which is an alloy of iron, copper, and tungsten used in surgical instruments.

The chemistry of cobalt involves mainly its +2 and +3 oxidation states, although compounds containing cobalt in the 0, +1, or +4 oxidation state are known. Aqueous solutions of cobalt(II) salts contain the $Co(H_2O)_6{}^{2+}$ ion, which has a characteristic rose color. Cobalt forms a wide variety of coordination compounds, many of which will be discussed in later sections of this chapter. Some typical cobalt compounds are listed in Table 20.8.

Nickel ranks twenty-fourth in elemental abundance in the earth's crust. It is found in ores, where it is combined mainly with arsenic, antimony, and sulfur. Nickel metal, a silver-white substance with high electrical and thermal conductivities, is quite resistant to corrosion and is often used for plating more active metals. Nickel is also widely used in the production of alloys such as steel.

Nickel is found almost exclusively in the +2 oxidation state in its compounds. Aqueous solutions of nickel(II) salts contain the $Ni(H_2O)_6{}^{2+}$ ion, which has a characteristic emerald green color. Coordination compounds of nickel(II) will be discussed later in this chapter. Some typical nickel compounds are shown in Table 20.9.

Copper, widely distributed in nature in ores containing sulfides, arsenides, chlorides, and carbonates, is valued for its high electrical conductivity and its resistance to corrosion. It is widely used for plumbing, and 50% of all copper produced annually is used for electrical applications. Copper is a major constituent in several well-known alloys (see Table 20.10).

Although copper is not highly reactive (it will not reduce H^+ to H_2, for example), this reddish-colored metal does slowly corrode in air, producing the characteristic green *patina* consisting of basic copper sulfate,

$$3Cu(s) + 2H_2O(l) + SO_2(g) + 2O_2(g) \longrightarrow \underset{\text{Basic copper sulfate}}{Cu_3(OH)_4SO_4}$$

and other similar compounds.

Copper roofs and bronze statues, such as the Statue of Liberty, turn green in air because $Cu_3(OH)_4SO_4$ and $Cu_4(OH)_6SO_4$ form.

The chemistry of copper principally involves the +2 oxidation state, but many compounds containing copper(I) are also known. Aqueous solutions of copper(II) salts are a characteristic bright blue color due to the presence of the $Cu(H_2O)_6^{2+}$ ion. Table 20.11 lists some typical copper compounds.

Although trace amounts of copper are essential for life, it is quite toxic in large amounts; copper salts are used to kill bacteria, fungi, and algae. For example, paints containing copper are used on ship hulls to prevent fouling by marine organisms.

Widely dispersed in the earth's crust, *zinc* is mainly refined from sphalerite (ZnFe)S, which often occurs with galena (PbS). Zinc is a white, lustrous, very active metal that behaves as an excellent reducing agent and that tarnishes rapidly. About 90% of the zinc produced is used for galvanizing steel. Zinc forms colorless salts containing Zn^{2+} ions.

TABLE 20.9
Typical Compounds of Nickel

Oxidation State of Nickel	Examples of Compounds
+2	$NiCl_2$ (yellow)
	$[Ni(H_2O)_6]Cl_2$ (green)
	NiO (greenish black)
	NiS (black)
	$[Ni(H_2O)_6]SO_4$ (green)
	$[Ni(NH_3)_6](NO_3)_2$ (blue)

TABLE 20.10 Alloys Containing Copper

Alloy	Composition (% by mass in parentheses)
Brass	Cu (20–97), Zn (2–80), Sn (0–14), Pb (0–12), Mn (0–25)
Bronze	Cu (50–98), Sn (0–35), Zn (0–29), Pb (0–50), P (0–3)
Sterling silver	Cu (7.5), Ag (92.5)
Gold (18-karat)	Cu (5–14), Au (75), Ag (10–20)
Gold (14-karat)	Cu (12–28), Au (58), Ag (4–30)

TABLE 20.11
Typical Compounds of Copper

Oxidation State of Copper	Examples of Compounds
+1	Cu_2O (red)
	Cu_2S (black)
	CuCl (white)
+2	CuO (black)
	$CuSO_4 \cdot 5H_2O$ (blue)
	$CuCl_2 \cdot 2H_2O$ (green)
	$[Cu(H_2O)_6](NO_3)_2$ (blue)

20.3 Coordination Compounds

Transition metal ions characteristically form **coordination compounds,** which are usually colored and often paramagnetic. A coordination compound typically consists of a *complex ion,* a transition metal ion with its attached ligands (see Section 8.9), and **counter ions,** anions or cations as needed to produce a compound with no net charge. The substance $[Co(NH_3)_5Cl]Cl_2$ is a typical coordination compound. The square brackets indicate the composition of the complex ion, in this case $Co(NH_3)_5Cl^{2+}$, and the two Cl^- counter ions are shown outside the brackets. Note that in this compound one Cl^- acts as a ligand along with the five NH_3 molecules. In the solid state this compound consists of the large $Co(NH_3)_5Cl^{2+}$ cations with twice as many Cl^- anions, all packed together as efficiently as possible. When dissolved in water, the solid behaves like any ionic solid; the cations and anions are assumed to separate and move about independently:

$$[Co(NH_3)_5Cl]Cl_2(s) \xrightarrow{H_2O} Co(NH_3)_5Cl^{2+}(aq) + 2Cl^-(aq)$$

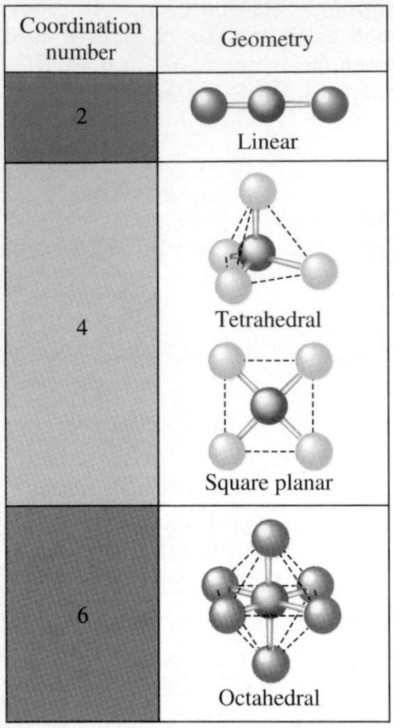

Coordination number	Geometry
2	Linear
4	Tetrahedral
4	Square planar
6	Octahedral

Figure 20.6
The ligand arrangements for coordination numbers 2, 4, and 6.

Coordination compounds have been known since about 1700, but their true nature was not understood until the 1890s when a young Swiss chemist named Alfred Werner proposed that transition metal ions have two types of valence (combining ability). One type of valence, which Werner called the secondary valence, refers to the ability of a metal ion to bind to Lewis bases (ligands) to form complex ions. The other type, the primary valence, refers to the ability of the metal ion to form ionic bonds with oppositely charged ions. Thus Werner explained that the compound, originally written as $CoCl_3 \cdot 5NH_3$, is actually $[Co(NH_3)_5Cl]Cl_2$, in which the Co^{3+} ion has a primary valence of 3, satisfied by the three Cl^- ions, and a secondary valence of 6, satisfied by the six ligands (five NH_3 and one Cl^-). We now call the primary valence the **oxidation state** and the secondary valence the **coordination number.** The latter reflects the number of bonds formed between the metal ion and the ligands in the complex ion.

Coordination Number

The number of bonds formed by metal ions to ligands in complex ions varies from two to eight, depending on the size, charge, and electron configuration of the transition metal ion. As shown in Table 20.12, 6 is the most common coordination number, followed closely by 4, with a few metal ions showing a coordination number of 2. Many metal ions show more than one coordination number, and there is really no simple way to predict what the coordination number is in a particular case. The typical geometries for the various common coordination numbers are shown in Fig. 20.6. Note that six ligands produce an octahedral arrangement around the metal ion. Four ligands can form either a tetrahedral or a square planar arrangement, and two ligands give a linear structure.

Ligands

A **ligand** is a *neutral molecule or ion having a lone pair that can be used to form a bond to a metal ion.* In the 1920s G. N. Lewis suggested a general definition for acid-base behavior in terms of electron pairs. Lewis defined an **acid** as an **electron pair acceptor** and a **base** as an **electron pair donor.** Because a ligand donates an electron pair to an empty orbital on a metal ion, the formation of a metal-ligand bond can be described as the interaction between a Lewis base (the ligand) and a Lewis acid (the metal ion). This is often called a **coordinate covalent bond.**

The Lewis definition of acids and bases also encompasses the Brønsted-Lowry definition. For example,

$$H^+ + :NH_3 \rightarrow NH_4^+$$

TABLE 20.12 **Typical Coordination Numbers for Some Common Metal Ions**

M^+	Coordination Numbers	M^{2+}	Coordination Numbers	M^{3+}	Coordination Numbers
Cu^+	2, 4	Mn^{2+}	4, 6	Sc^{3+}	6
Ag^+	2	Fe^{2+}	6	Cr^{3+}	6
Au^+	2, 4	Co^{2+}	4, 6	Co^{3+}	6
		Ni^{2+}	4, 6		
		Cu^{2+}	4, 6	Au^{3+}	4
		Zn^{2+}	4, 6		

TABLE 20.13 Some Common Ligands

Type	Examples	
Unidentate/monodentate	H_2O	SCN^- (thiocyanate)
	NH_3	OH^-
	CN^-	X^- (halides)
	NO_2^- (nitrite)	
Bidentate	Oxalate	Ethylenediamine (en)
	(−) :O−C(=O)−C(=O)−O: (−), both O bound to M	H_2N−CH_2−CH_2−NH_2, both N bound to M
Polydentate	Diethylenetriamine (dien) $H_2N-(CH_2)_2-NH-(CH_2)_2-NH_2$ Three coordinating atoms	
	Ethylenediaminetetraacetate (EDTA) $((-){:}O-C(=O)-H_2C)_2N-(CH_2)_2-N(CH_2-C(=O)-O{:}(-))_2$ Six coordinating atoms	

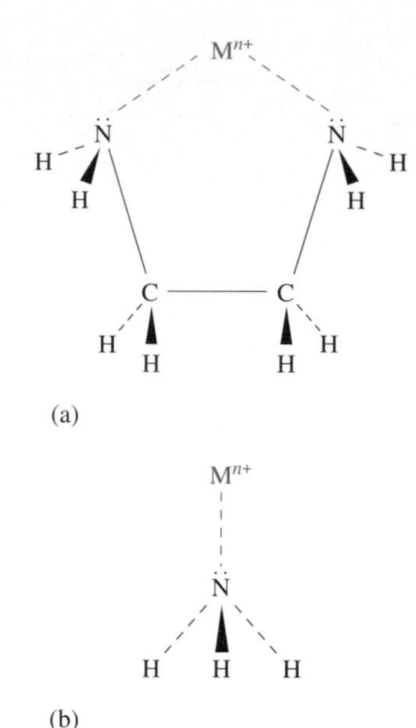

Figure 20.7
(a) The bidentate ligand ethylenediamine can bond to the metal ion through the lone pair on each nitrogen atom, thus forming two coordinate covalent bonds. (b) Ammonia is a monodentate ligand.

A *ligand that can form one bond to a metal ion* is called a **monodentate ligand** or a **unidentate ligand** (from root words meaning "one tooth"). Examples of unidentate ligands are shown in Table 20.13.

Some ligands have more than one atom with a lone pair that can be used to bond to a metal ion. Such ligands are said to be **chelating ligands,** or **chelates** (from the Greek word *chela,* meaning "claw"). A ligand that can form two bonds to a metal ion is called a **bidentate ligand.** A very common bidentate ligand is ethylenediamine (abbreviated en), which is shown coordinating to a metal ion in Fig. 20.7(a). Note the relationship between this ligand and the unidentate ligand ammonia [Fig. 20.7(b)]. Oxalate, another typical bidentate ligand, is shown in Table 20.13.

Ligands that can form more than two bonds to a metal ion are called **polydentate ligands.** Some ligands can form as many as six bonds to a metal ion. The best-known example of such a ligand is ethylenediaminetetraacetate (abbreviated EDTA), which is shown in Table 20.13. This ligand virtually surrounds the metal ion (Fig. 20.8), coordinating through six atoms (a

Figure 20.8
The coordination of EDTA with a 2+ metal ion.

hexadentate ligand). As might be expected from the large number of coordination sites, EDTA forms very stable complex ions with most metal ions. It is therefore useful as a "scavenger" to remove toxic heavy metals such as lead from the human body. It is also used as a reagent to analyze solutions for the metal ion content. EDTA is found in countless consumer products, including soda, beer, salad dressings, bar soaps, and most cleaners. In these products EDTA ties up trace metal ions that would otherwise catalyze decomposition and produce unwanted precipitates.

Even more complicated ligands are found in biological systems, where metal ions play crucial roles in catalyzing reactions, transferring electrons, and transporting and storing oxygen. A discussion of these complex ligands will follow in Section 20.8.

Nomenclature

In Werner's lifetime no system was used to name coordination compounds. Names of the compounds were commonly based on colors and names of discoverers. As the field expanded and more coordination compounds were identified, an orderly system of nomenclature became necessary. A simplified version of this system is summarized by the following rules.

Summary: **Rules for Naming Coordination Compounds**

- **As with any ionic compound, *the cation is named before the anion.***
- **In naming a complex ion, *the ligands are named before the metal ion.***
- **In naming ligands, *an o is added to the root name of an anion.* For example, the halides as ligands are called fluoro, chloro, bromo, and iodo; hydroxide is hydroxo; and cyanide is cyano. *For a neutral ligand the name of the molecule is used,* with the exception of H_2O, NH_3, CO, and NO, as illustrated in Table 20.14.**
- ***The prefixes mono-, di-, tri-, tetra-, penta-, and hexa- are used to denote the number of simple ligands.* The prefixes *bis-*, *tris-*, *tetrakis-*, and the like, are also used, especially for more complicated ligands or ones that already contain *di-*, *tri-*, and so on.**
- ***The oxidation state of the central metal ion is designated by a Roman numeral in parentheses.***
- ***When more than one type of ligand is present, ligands are named in alphabetical order.** Prefixes do not affect the order.**
- ***If the complex ion has a negative charge, the suffix -ate is added to the name of the metal.* Sometimes the Latin name is used to identify the metal (see Table 20.15).**

These rules are applied in Example 20.1.

*In an older system the negatively charged ligands were named first, followed by neutral ligands, with positively charged ligands named last. We will follow the newer convention in this text.

TABLE 20.14 **Names of Some Common Unidentate Ligands**

Neutral Molecules		Anions	
Aqua	H_2O	Fluoro	F^-
Ammine	NH_3	Chloro	Cl^-
Methylamine	CH_3NH_2	Bromo	Br^-
Carbonyl	CO	Iodo	I^-
Nitrosyl	NO	Hydroxo	OH^-
		Cyano	CN^-

TABLE 20.15
Latin Names Used for Some Metal Ions in Anionic Complex Ions

Metal	Anionic Complex Base Name
Iron	Ferrate
Copper	Cuprate
Lead	Plumbate
Silver	Argentate
Gold	Aurate
Tin	Stannate

EXAMPLE 20.1

Give the systematic name for each of the following coordination compounds.

a. $[Co(NH_3)_5Cl]Cl_2$ b. $K_3Fe(CN)_6$ c. $[Fe(en)_2(NO_2)_2]_2SO_4$

Solution

a. To determine the oxidation state of the metal ion, we examine the charges of all ligands and counter ions. The ammonia molecules are neutral and the chloride ions each have a 1− charge, so the cobalt ion must have a 3+ charge to produce a neutral compound. Since cobalt has the oxidation state +3, we use cobalt(III) in the name.

The ligands include one Cl^- ion and five NH_3 molecules. The chloride ion is designated as *chloro,* and each ammonia molecule is designated *ammine.* The prefix *penta-* indicates that there are five NH_3 ligands present. The name of the complex cation is therefore pentaamminechlorocobalt(III). Note that the ligands are named alphabetically, disregarding the prefix. Since the counter ions are chloride ions, the compound is named as a chloride salt:

$$\underbrace{\text{pentaamminechlorocobalt(III)}}_{\text{Cation}}\ \underbrace{\text{chloride}}_{\text{Anion}}$$

b. First, we determine the oxidation state of the iron by considering the other charged species. The compound contains three K^+ ions and six CN^- ions. Therefore, the iron must carry a charge of 3+, giving a total of six positive charges to balance the six negative charges. Thus the complex ion present is $Fe(CN)_6^{3-}$. The cyanide ligands are each designated *cyano,* and the prefix *hexa-* indicates that six are present. Since the complex ion is an anion, we use the Latin name *ferrate.* The oxidation state is indicated by ferrate(III) in the name. The anion name is therefore hexacyanoferrate(III). The cations are K^+ ions, which are simply named potassium. Combining all of this gives the name

$$\underbrace{\text{potassium}}_{\text{Cation}}\ \underbrace{\text{hexacyanoferrate(III)}}_{\text{Anion}}$$

(The common name of this compound is potassium ferricyanide.)

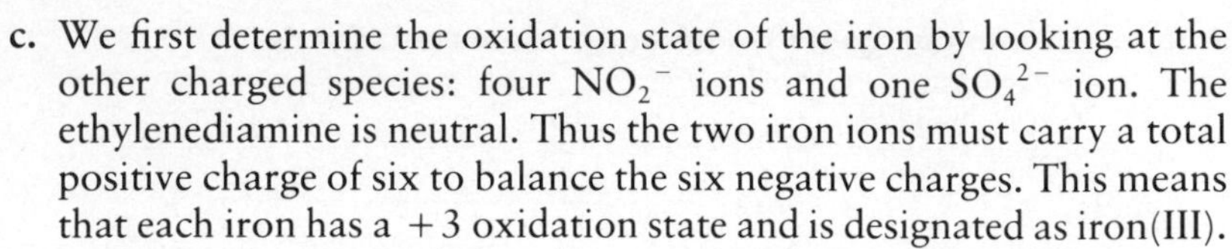

c. We first determine the oxidation state of the iron by looking at the other charged species: four NO_2^- ions and one SO_4^{2-} ion. The ethylenediamine is neutral. Thus the two iron ions must carry a total positive charge of six to balance the six negative charges. This means that each iron has a +3 oxidation state and is designated as iron(III).

Since the name ethylenediamine already contains *di*, we use *bis-* instead of *di-* to indicate the presence of two en ligands. The name for NO_2^- as a ligand is *nitro*, and the prefix *di-* indicates the presence of two NO_2^- ligands. The anion is sulfate. Therefore, the compound's name is

bis(ethylenediamine)dinitroiron(III) sulfate

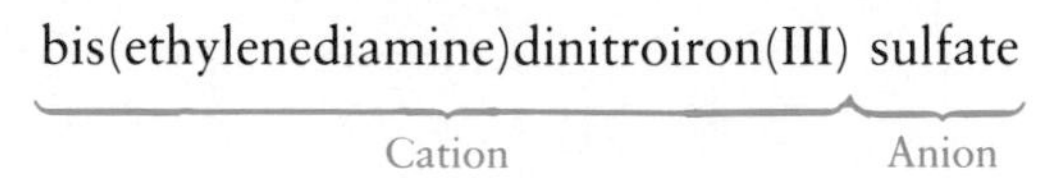

Since the complex ion is a cation in this case, the Latin name for iron is not used.

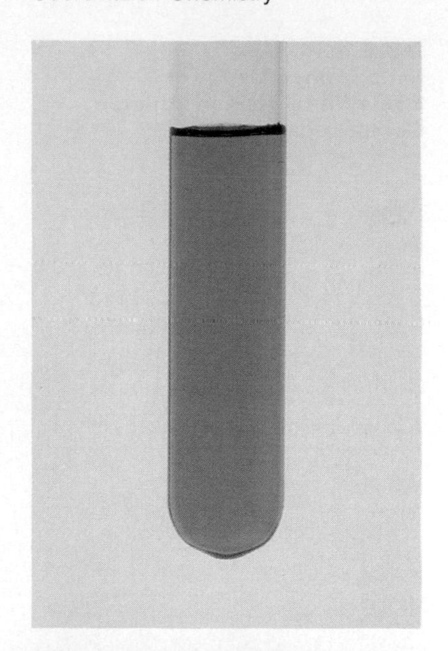

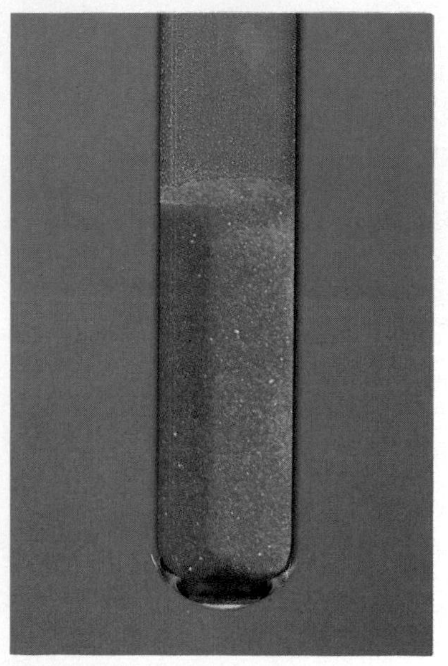

(top) An aqueous solution of $[Co(NH_3)_5Cl]\ Cl_2$. (bottom) Solid $K_3Fe(CN)_6$.

EXAMPLE 20.2

Given the following systematic names, predict the formula of each coordination compound.

a. Triamminebromoplatinum(II) chloride
b. Potassium hexafluorocobaltate(III)

Solution

a. *Triammine* signifies three ammonia ligands, and *bromo* indicates one bromide ion as a ligand. The oxidation state of platinum is +2, as indicated by the Roman numeral II. Thus the complex ion is $[Pt(NH_3)_3Br]^+$. One chloride ion is needed to balance the 1+ charge of this cation. The formula of the compound is $[Pt(NH_3)_3Br]Cl$. Note that square brackets enclose the complex ion.

b. The complex ion contains six fluoride ligands attached to a Co^{3+} ion to give CoF_6^{3-}. Note the *-ate* ending that indicates that the complex ion is an anion. Three K^+ cations are required to balance the 3^- charge on the complex ion. Thus the formula is $K_3[CoF_6]$.

20.4 Isomerism

When two or more species have the same formula but exhibit different properties, they are said to be **isomers.** Although isomers contain exactly the same types and numbers of atoms, the arrangements of their atoms differ, and this leads to different properties. We will consider two main types of isomerism: **structural isomerism,** in which the isomers contain the same atoms but one or more bonds differ; and **stereoisomerism,** in which all of the bonds in the isomers are the same but the spatial arrangements of the atoms are different. Each of these classes also has subclasses (see Fig. 20.9), which we will now consider.

Alfred Werner: Coordination Chemist

During the early and middle parts of the nineteenth century, chemists prepared a large number of colored compounds containing transition metals and other substances such as ammonia, chloride ion, cyanide ion, and water. These compounds were very interesting to chemists who were trying to understand the nature of bonding (Dalton's atomic theory of 1808 was very new at this time), and many theories were suggested to explain these substances. The most widely accepted early theory was the *chain theory*, championed by Sophus Mads Jorgensen (1837–1914), professor of chemistry at the University of Copenhagen. The chain theory got its name from the postulate that metal-ammine* complexes contain chains of NH_3 molecules. For example, Jorgensen proposed the structure

```
   NH3—Cl
  /
Co—NH3—NH3—NH3—NH3—Cl
  \
   NH3—Cl
```

for the compound $Co(NH_3)_6Cl_3$. In the late nineteenth century this theory was used in classrooms around the world to explain the nature of metal-ammine compounds.

However, in 1890 a young Swiss chemist named Alfred Werner, who had just obtained a Ph.D. in the field of organic chemistry, became so interested in these compounds that he apparently even dreamed about them. In the middle of one night Werner awoke realizing that he had discovered the correct explanation for the constitution of these compounds. Writing furiously the rest of that night and into the late afternoon of the following day, he constructed a scientific paper containing his now famous *coordination theory*. This model postulates an octahedral arrangement of ligands around the Co^{3+} ion, leading to the $Co(NH_3)_6{}^{3+}$ complex ion with three Cl^- ions as counter ions. Thus Werner's picture of $Co(NH_3)_6Cl_3$ differed greatly from the chain theory.

In his paper on the coordination theory, Werner explained not only the metal-ammine compounds but most of the other known transition metal compounds, and the importance of his contribution was recognized immediately. He was appointed professor at the University of Zurich, where he spent the rest of his life studying coordination compounds and refining his theory. Alfred Werner was a confident, impulsive man of seemingly boundless energy, who was known for his inspiring lectures, his intolerance of incompetence (he once threw a chair at a student who performed poorly on an oral exam), and his intuitive scientific brilliance. For example, he was the first to show that stereochemistry is a general phenomenon, not one exhibited only by carbon, as was previously thought. He also recognized and named many types of isomerism.

For his work on coordination chemistry and stereochemistry, Werner became the fourteenth Nobel prize winner in chemistry and the first Swiss chemist to be so honored. Werner's work is even more remarkable when one realizes that his ideas preceded any real understanding of the nature of covalent bonds by many years.

*Ammine is the name for NH_3 as a ligand.

Structural Isomerism

The first type of structural isomerism we will consider is **coordination isomerism,** in which the composition of the complex ion varies. For example, $[Cr(NH_3)_5SO_4]Br$ and $[Cr(NH_3)_5Br]SO_4$ are coordination isomers. In the first case $SO_4{}^{2-}$ is coordinated to Cr^{3+}, while Br^- acts as the counter ion; in the second case the roles of these ions are reversed.

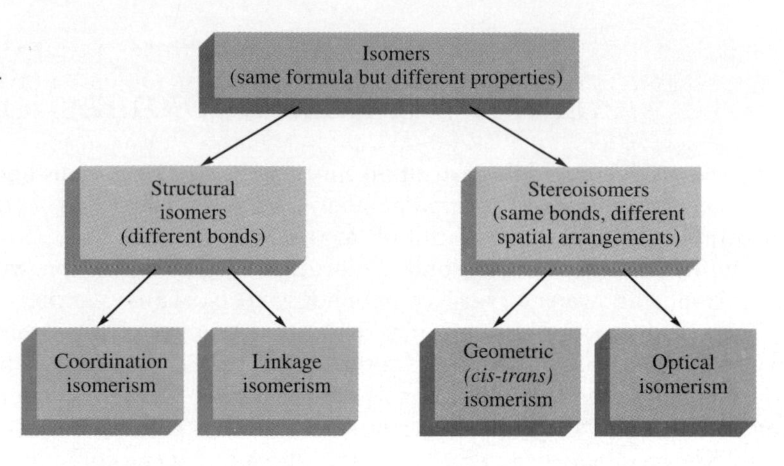

Figure 20.9
Some classes of isomers.

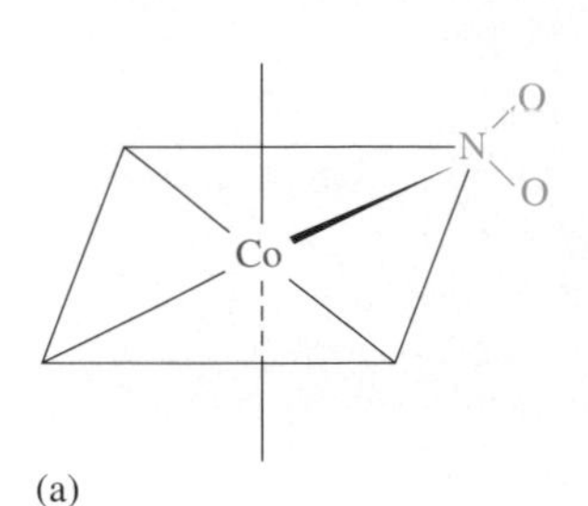

(a)

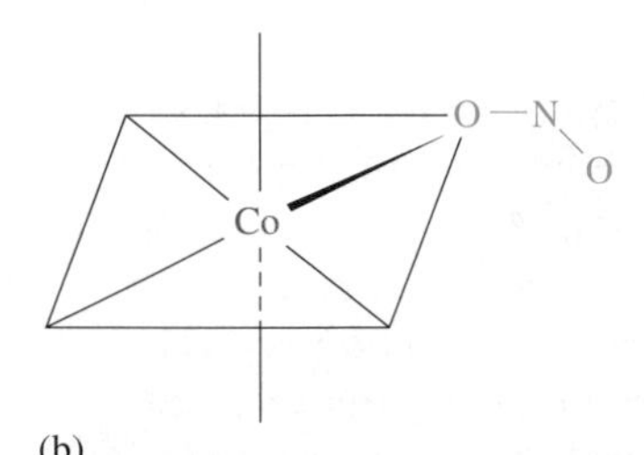

(b)

Figure 20.10
As a ligand, NO_2^- can bond to a metal ion (a) through a lone pair on the nitrogen atom or (b) through a lone pair on one of the oxygen atoms.

Another example of coordination isomerism involves the $[Co(en)_3][Cr(ox)_3]$ and $[Cr(en)_3][Co(ox)_3]$ pair, where ox represents the oxalate ion, a bidentate ligand shown in Table 20.13.

In a second type of structural isomerism, **linkage isomerism,** the composition of the complex ion is the same, but the point of attachment of at least one of the ligands differs. Two ligands that can attach to metal ions in different ways are thiocyanate SCN^- (which can bond through lone pairs on the nitrogen or the sulfur atom), and the nitrite ion NO_2^- (which can bond through lone pairs on the nitrogen or the oxygen atom). For example, the following two compounds are linkage isomers:

$$[Co(NH_3)_4(NO_2)Cl]Cl$$

Tetraamminechloronitrocobalt(III) chloride
(yellow)

$$[Co(NH_3)_4(ONO)Cl]Cl$$

Tetraamminechloronitritocobalt(III) chloride
(red)

In the first case the NO_2^- ligand is called *nitro* and is attached to Co^{3+} through the nitrogen atom; in the second case the NO_2^- ligand is called *nitrito* and is attached to Co^{3+} through an oxygen atom (see Fig. 20.10).

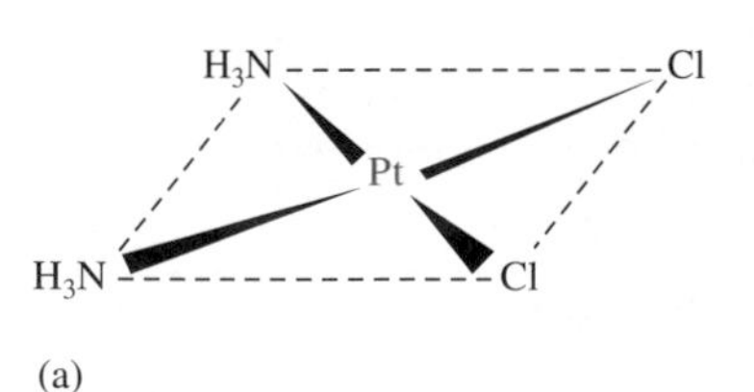

(a)

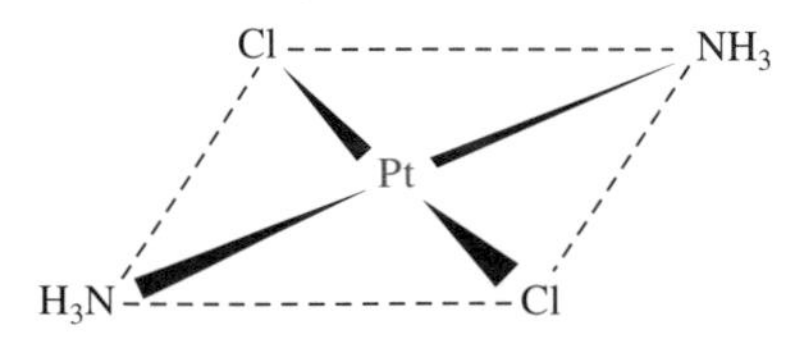

(b)

Figure 20.11
(a) The *cis* isomer of $Pt(NH_3)_2Cl_2$ (yellow). (b) The *trans* isomer of $Pt(NH_3)_2Cl_2$ (pale yellow).

Stereoisomerism

Stereoisomers have the same bonds, but they exhibit different spatial arrangements of their atoms. One type, **geometrical isomerism,** or ***cis-trans*** **isomerism,** occurs when atoms or groups of atoms can assume different positions around a rigid ring or bond. An important example is the compound $Pt(NH_3)_2Cl_2$, which has a square planar structure. The two possible arrangements of the ligands are shown in Fig. 20.11. In the ***cis*** **isomer** the ammonia molecules are next (*cis*) to each other. In the ***trans*** **isomer** the ammonia molecules are across (*trans*) from each other.

Geometrical isomerism also occurs in octahedral complex ions. For example, the compound $[Co(NH_3)_4Cl_2]Cl$ has *cis* and *trans* isomers (Fig. 20.12).

A second type of stereoisomerism is called **optical isomerism,** because the isomers have opposite effects on plane-polarized light. When light is emitted from a source such as a glowing filament, the oscillating electric fields of the photons in the beam are oriented randomly, as shown in Fig. 20.13. If this light is passed through a polarizer, only the photons with electric fields oscillating in a single plane remain, constituting *plane-polarized light.*

In 1815 a French physicist, Jean Biot, discovered that certain crystals could rotate the plane of polarization of light. Later, scientists found that solutions of certain compounds could do the same thing (see Fig. 20.14). Louis Pasteur was the first to understand this behavior. In 1848 he noted that solid sodium ammonium tartrate ($NaNH_4C_4H_4O_4$) existed as a mixture of two types of crystals, which he painstakingly separated with tweezers. Separate solutions of these two types of crystals rotated plane-polarized light in exactly opposite directions. This led to a connection between optical activity and molecular structure.

We now realize that optical activity is exhibited by molecules that have *nonsuperimposable mirror images*. Your hands are nonsuperimposable mirror

Figure 20.12

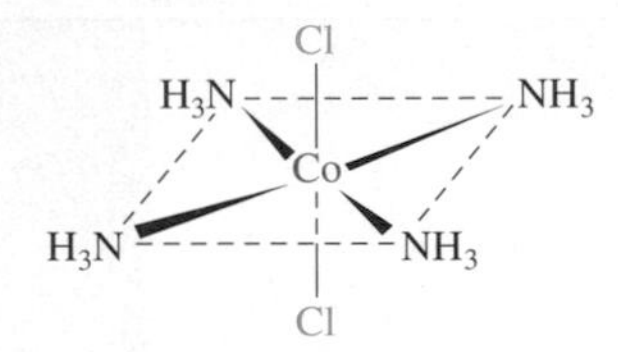

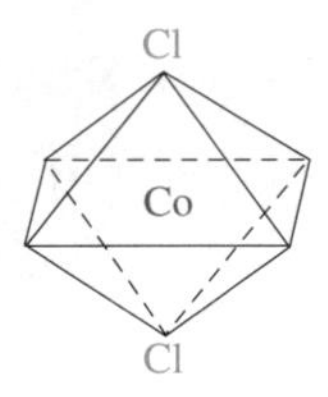

(a)

(a) The *trans* isomer of $[Co(NH_3)_4Cl_2]^+$. The chloride ligands are directly across from each other.

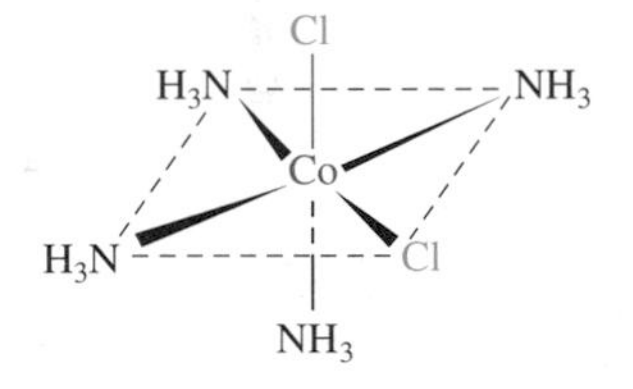

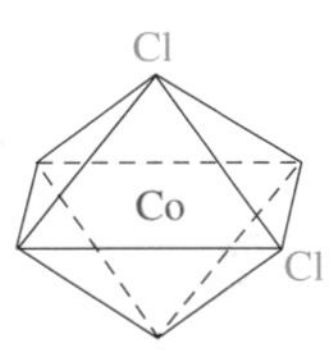

(b)

(b) The *cis* isomer of $[Co(NH_3)_4Cl_2]^+$. The chloride ligands in this case share an edge of the octahedron. Because of their different structures, the *trans* isomer of $[Co(NH_3)_4Cl_2]Cl$ is green, and the *cis* isomer is violet.

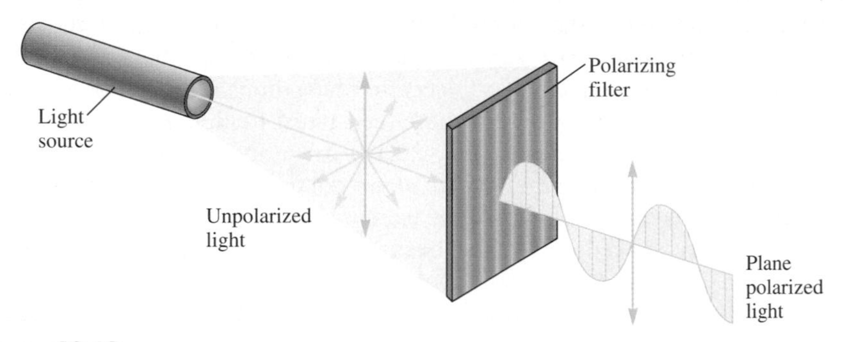

Figure 20.13
Unpolarized light consists of waves vibrating in many different planes (indicated by the arrows). The polarizing filter blocks all waves except those vibrating in a given plane.

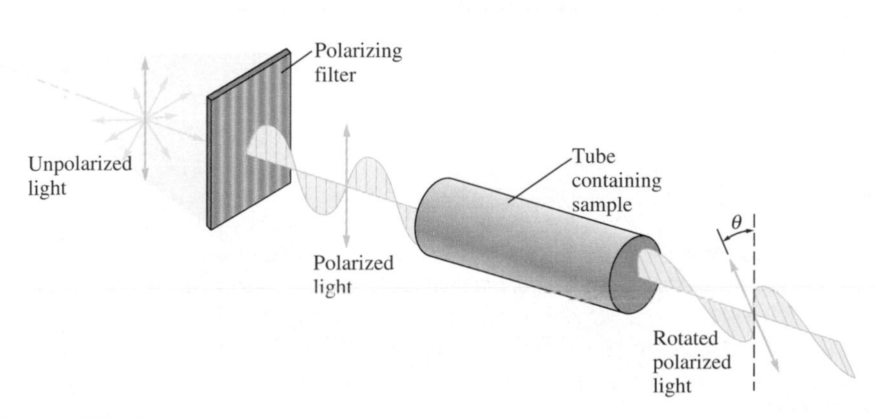

Figure 20.14
The rotation of the plane of polarized light by an optically active substance. The angle of rotation is called theta (θ).

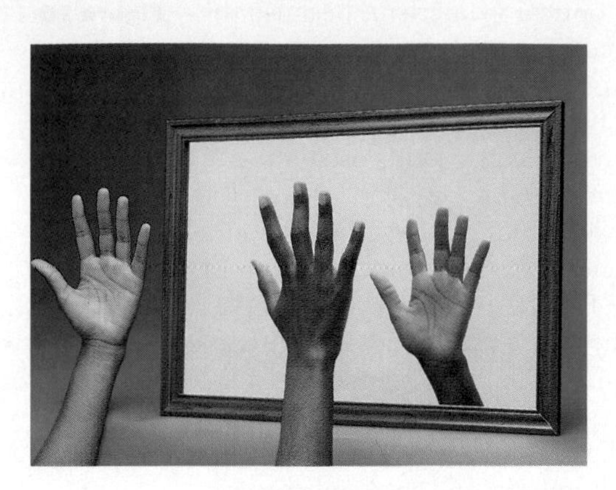

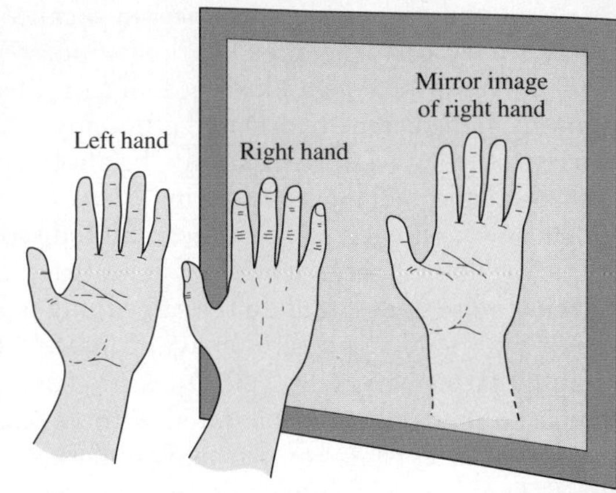

Figure 20.15
A human hand has a nonsuperimposable mirror image. Note that the mirror image of the right hand (while identical to the left hand) cannot be turned in any way to make it identical to (superimposable on) the actual right hand.

images (Fig. 20.15). That is, human hands are related like an object and its mirror image; and one hand cannot be turned to make it identical to the other. Many molecules show this same feature, for example, the complex ion $[Co(en)_3]^{3+}$ shown in Fig. 20.16. Objects that have nonsuperimposable mirror images are said to be **chiral** (from the Greek word for hand, *cheir*).

The isomers of $[Co(en)_3]^{3+}$ (Fig. 20.16) are nonsuperimposable mirror images called **enantiomers;** they rotate plane-polarized light in opposite directions and are thus optical isomers. The isomer that rotates the plane of light to the right (when viewed down the beam of oncoming light) is said to be *dextrorotatory,* designated by *d*. The isomer that rotates the plane of light to the left is *levorotatory* (*l*). An equal mixture of the *d* and *l* isomers in solution, called a *racemic mixture,* does not rotate the plane of the polarized light at all because the two opposite effects cancel.

Geometrical isomers are not necessarily optical isomers. For instance, the *trans* isomer of $[Co(en)_2Cl_2]^+$ shown in Fig. 20.17 is identical to its mirror

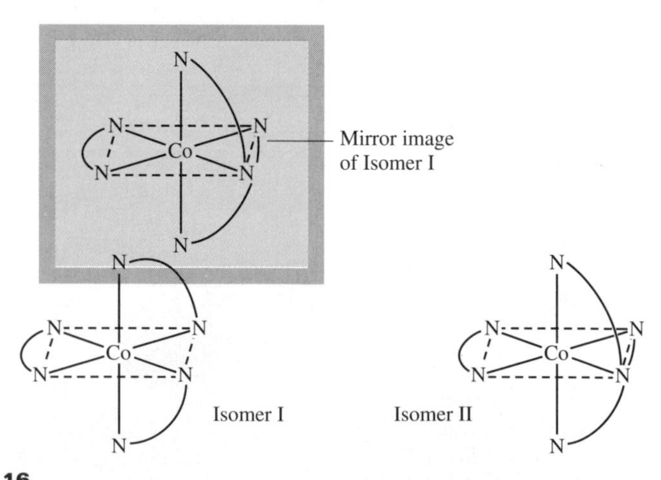

Figure 20.16
Isomers I and II of $Co(en)_3{}^{3+}$ are mirror images (the mirror image of I is identical to II) that cannot be superimposed. That is, there is no way that I can be turned in space so that it is the same as II.

The *trans* isomer and its mirror image are identical. They are not isomers of each other.

trans

(a)

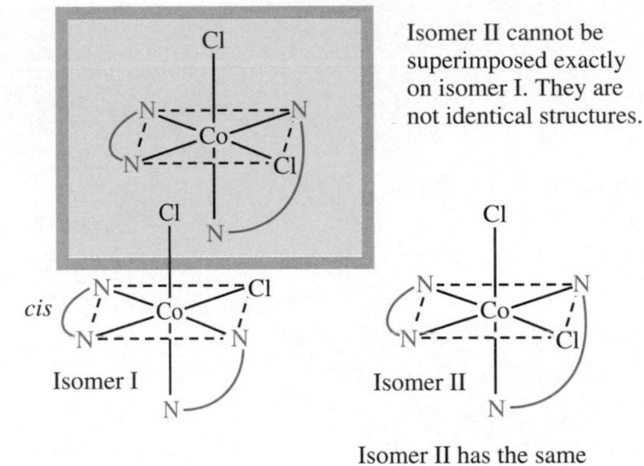
Isomer II cannot be superimposed exactly on isomer I. They are not identical structures.
cis
Isomer I
Isomer II
Isomer II has the same structure as the mirror image of isomer I.
(b)

Figure 20.17
(a) The *trans* isomer of $Co(en)_2Cl_2^+$ and its mirror image are identical (superimposable). (b) The *cis* isomer of $Co(en)_2Cl_2^+$ and its mirror image are not superimposable and are thus a pair of optical isomers.

THE IMPORTANCE OF BEING CIS

Some of the most important advancements of science are the results of accidental discoveries, for example, penicillin, Teflon, and the sugar substitutes cyclamate and aspartame. Another important chance discovery occurred in 1964, when a group of scientists using platinum electrodes to apply an electric field to a colony of *E. coli* bacteria noticed that the bacteria failed to divide but continued to grow, forming long fibrous cells. Further study revealed that cell division was inhibited by small concentrations of the compounds *cis*-$Pt(NH_3)_2Cl_2$ and *cis*-$Pt(NH_3)_2Cl_4$ formed electrolytically in the solution.

Cancerous cells multiply very rapidly because cell division is uncontrolled. Thus these and similar platinum complexes were evaluated as antitumor agents, which inhibit the division of cancer cells. The results showed that *cis*-$Pt(NH_3)_2Cl_2$ was active against a wide variety of tumors, including testicular and ovarian tumors, which are very resistant to treatment by more traditional methods. However, although the *cis* complex showed significant antitumor activity, the corresponding *trans* complex had no effect on tumors. This illustrates the importance of isomerism in biological systems. When drugs are synthesized, great care must be taken to obtain the correct isomer.

Unfortunately, although *cis*-$Pt(NH_3)_2Cl_2$ has proved to be a valuable drug, it has some troublesome side effects, the most serious being kidney damage. As a result, the search continues for even more effective antitumor agents. Promising candidates are shown in Fig. 20.18. Note that they are all *cis* complexes.

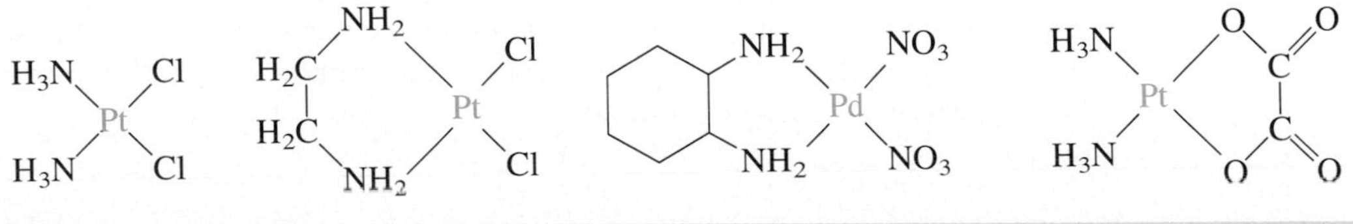

Figure 20.18
Some *cis* complexes of platinum and palladium that show significant antitumor activity. It is thought that the *cis* complexes work by losing two adjacent ligands and then forming coordinate covalent bonds to adjacent bases on a DNA molecule.

Chirality: Why Is It Important?

A molecule is said to be chiral if it can exist as isomers (called enantiomers) that are nonsuperimposable mirror images of each other. We often say these molecules exhibit "handedness," after our nonsuperimposable mirror image left and right hands. Enantiomers rotate plane-polarized light by the same angle but in opposite directions; however, the importance of this type of isomerism goes far beyond this rather curious behavior. In fact, many of the molecules produced by organisms exhibit a specific handedness. This is important because the response of an organism to a particular molecule often depends on how that molecule fits a particular site on a receptor molecule in the organism. Just as a left hand requires a left-handed glove, a left-handed receptor requires a particular enantiomer for a correct fit. Therefore, in designing pharmaceuticals, chemists must be concerned about which enantiomer is the active one—the one that fits the intended receptor.

Ideally, the pharmaceutical should consist of the pure active isomer. One way to obtain the compound as a pure active isomer is to produce the chemical by using organisms, because the production of biomolecules in organisms is stereospecific (yields a specific stereoisomer). For example, amino acids, vitamins, and hormones are naturally produced by yeast in the fermentation of sugar and can be harvested from the ferment. Biotechnology, in which the gene for a particular molecule is inserted into the DNA of a bacterium, provides another approach. Insulin is now produced in this way.

In contrast to the synthesis of biomolecules by organisms where a specific isomer is produced, when chiral molecules are made by "normal" chemical procedures (reactants are mixed and allowed to react), a mixture of the enantiomers is obtained. For example, when one chiral center is present in a molecule, normal chemical synthesis gives an equal mixture of the two mirror image isomers—called a racemic mixture. How does one deal with a pharmaceutical produced as a racemic mixture? One possibility is to administer the drug in its racemic form, assuming that the inactive form (50% of the mixture) will have no effect, positive or negative. In fact, this procedure is being followed for many drugs now on the market. However, it is a procedure that is growing increasingly controversial as evidence mounts that the "inactive" form of the drug may actually produce detrimental effects often totally unrelated to the effect of the active isomer. In effect, a drug administered as a racemic mixture contains a 50% impurity, the impact of which is not well understood.

The alternative to using racemic mixtures is to find a way to produce the substance as a pure isomer or a way to separate the isomers from the racemic mixtures. Both of these options are difficult and thus expensive. However, it is becoming increasingly clear that many pharmaceuticals must be administered as pure isomers to produce the desired results with no side effects. Therefore, a great deal of effort is now being directed toward the synthesis and separation of chiral compounds.

image. Since this isomer is superimposable on its mirror image, it does not exhibit optical isomerism and is therefore not chiral. On the other hand, *cis*-$[Co(en)_2Cl_2]^+$ is *not* superimposable on its mirror image; thus a pair of enantiomers exists for the complex ion, making the *cis* isomer chiral.

EXAMPLE 20.3

Does the complex ion $[Co(NH_3)Br(en)_2]$ exhibit geometrical isomerism? Does it exhibit optical isomerism?

Solution

The complex ion exhibits geometrical isomerism, since the ethylenediamine ligands can be across from or next to each other:

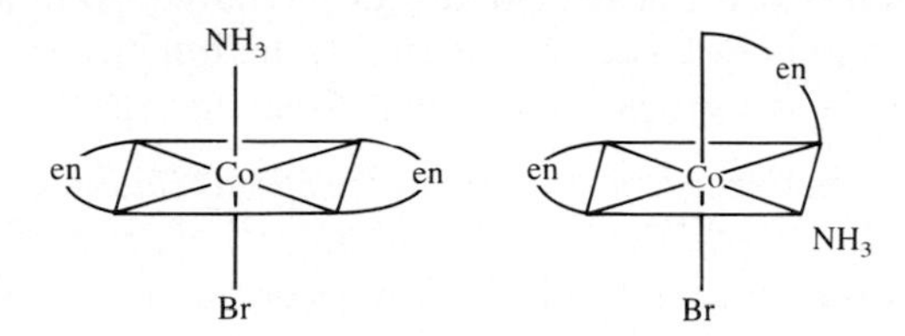

The *cis* isomer of the complex ion also exhibits optical isomerism since its mirror images,

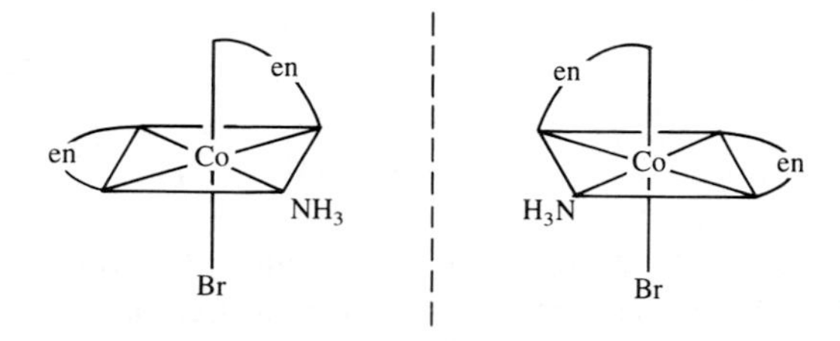

cannot be turned in any way to make them superimposable. Thus these two *cis* isomers are shown to be enantiomers that will rotate plane-polarized light in opposite directions.

Most important, since biomolecules are chiral, their reactions are highly structure-dependent. For example, a drug might have a particular effect because its molecules can bind to chiral molecules in the body. For the binding to be correct, the correct optical isomer of the drug must be administered. Just as the right hand of one person requires the right hand of another to perform a handshake, a given isomer in the body requires a specific isomer of the drug for an interaction to occur. Consequently, the syntheses of drugs, which are usually very complicated molecules, must be carried out in a way that produces the correct "handedness," a requirement that greatly adds to the difficulties in bringing these substances to market.

20.5 Bonding in Complex Ions: The Localized Electron Model

By this point in your study of chemistry, you no doubt recognize that the localized electron model, although very simple, is a very useful model for describing the bonding in molecules. Recall that a central feature of the model is the formation of hybrid atomic orbitals that are used for sharing electron pairs to form σ bonds between atoms. This same model can be used to account for the bonding in complex ions, but there are two important points to keep in mind.

1. The VSEPR model for predicting structure *does not work for complex ions.* However, we can safely assume that a complex ion with a coordination number of 6 has an octahedral arrangement of ligands, and that complexes with two ligands are linear. On the other hand, complex ions with a coordination number of 4 can be either tetrahedral or square planar; there is no reliable way to predict which will occur in a particular case.

2. The interaction between a metal ion and a ligand can be viewed as a Lewis acid-base reaction, with the ligand donating a lone pair of electrons to an *empty* orbital on the metal ion to form a coordinate covalent bond:

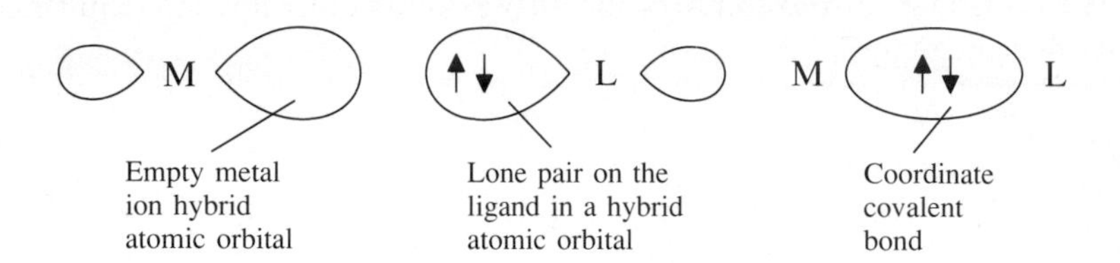

The hybrid orbitals used by the metal ion depend on the number and arrangement of the ligands. For example, accommodating the lone pair from each ammonia molecule in the octahedral $Co(NH_3)_6^{3+}$ ion requires a set of six empty hybrid atomic orbitals in an octahedral arrangement. Recall that an octahedral set of orbitals is formed by the hybridization of two *d*, one *s*, and three *p* orbitals to give six d^2sp^3 orbitals (see Fig. 20.19).

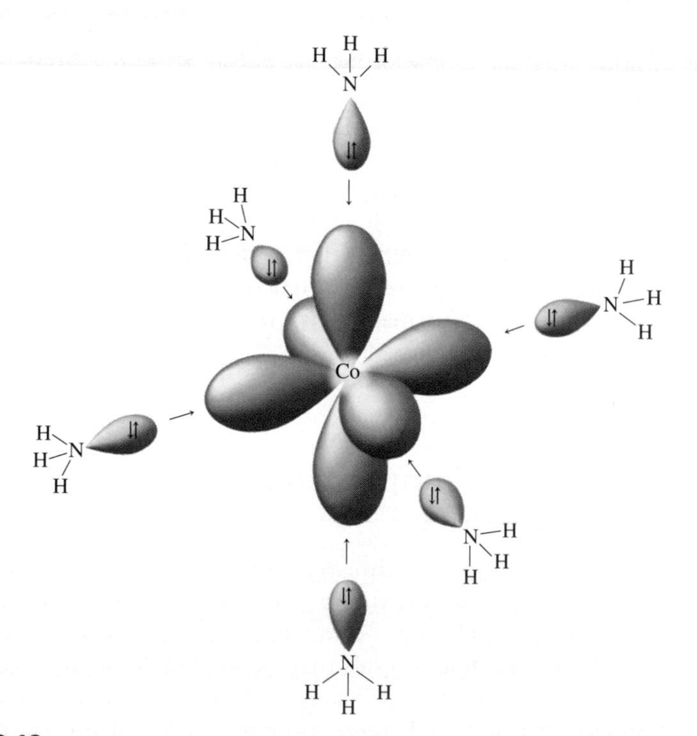

Figure 20.19
A set of six d^2sp^3 hybrid orbitals on Co^{3+} can accept an electron pair from each of six NH_3 ligands to form the $Co(NH_3)_6^{3+}$ ions.

The hybrid orbitals predicted for the metal ion in four-coordinate complexes depend on whether the structure is tetrahedral or square planar. For a tetrahedral arrangement of ligands an sp^3 hybrid set is required (see Fig. 20.20). For example, in the tetrahedral $CoCl_4^{2-}$ ion the Co^{2+} is described as sp^3 hybridized. A square planar arrangement of ligands requires a dsp^2 hybrid orbital set on the metal ion (Fig. 20.20). For example, in the square planar $Ni(CN)_4^{2-}$ the Ni^{2+} is described as dsp^2 hybridized.

A linear complex requires two hybrid orbitals 180° from each other. This arrangement is given by an *sp* hybrid set (Fig. 20.20). Thus in the linear $Ag(NH_3)_2^+$ ion the Ag^+ is described as *sp* hybridized.

Although the localized electron model can account in a general way for metal-ligand bonds, it is rarely used today because it cannot predict important properties of complex ions, such as magnetism and color. Thus we will not pursue the model any further.

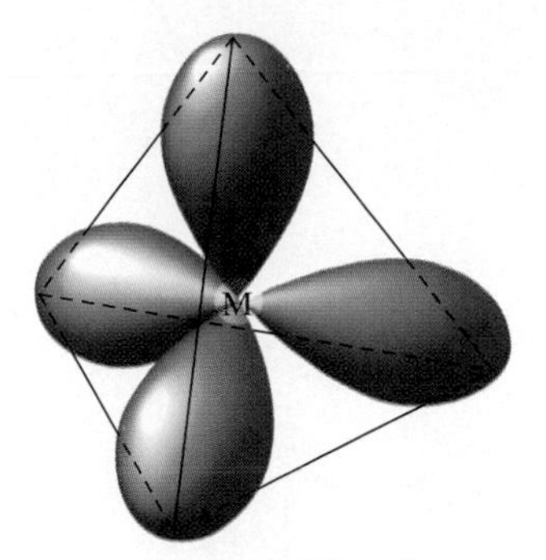

Tetrahedral ligand arrangement; sp^3 hybridization

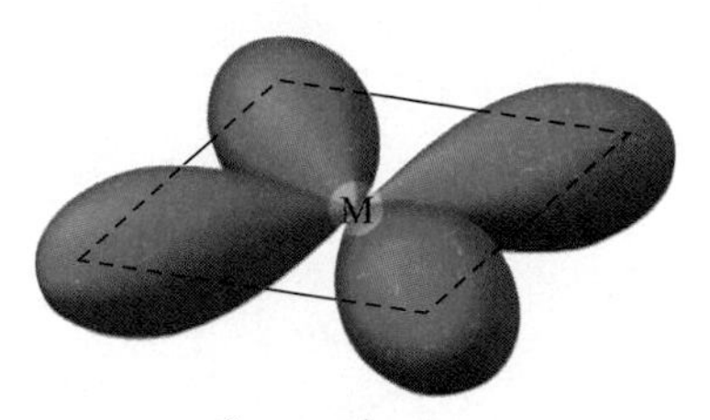

Square planar ligand arrangement; dsp^2 hybridization

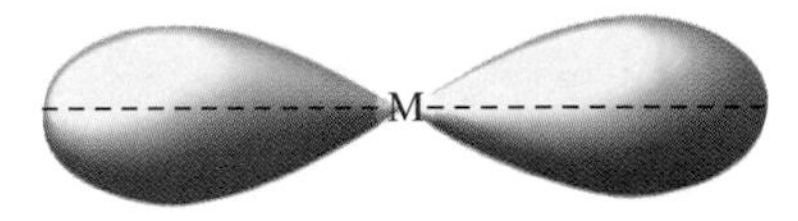

Linear ligand arrangement; *sp*

Figure 20.20
The hybrid orbitals required for tetrahedral, square planar, and linear complex ions. The metal ion hybrid orbitals are empty, so the metal ion bonds to the ligands by accepting lone pairs.

20.6 The Crystal Field Model

The main reason the localized electron model cannot fully account for the properties of complex ions is that in its simplest form it gives no information about how the energies of the *d* orbitals are affected by complex ion formation. This is critical because, as we will see, the color and magnetism of complex ions result from changes in the energies of the metal ion *d* orbitals caused by the metal-ligand interactions.

The **crystal field model** focuses on the energies of the *d* orbitals. In fact, this model is not so much a bonding model as it is an attempt to account for the colors and magnetic properties of complex ions. In its simplest form the crystal field model assumes that the ligands can be approximated by *negative point charges* and that metal-ligand bonding is *entirely ionic*.

Octahedral Complexes

We will illustrate the fundamental principles of the crystal field model by applying it to an octahedral complex. Figure 20.21 shows the orientation of the 3*d* orbitals relative to an octahedral arrangement of point-charge ligands. The important thing to note is that two of the orbitals, d_{z^2} and $d_{x^2-y^2}$, point their lobes *directly at* the point-charge ligands while three of the orbitals, d_{xz}, d_{yz}, and d_{xy}, point their lobes *between* the point charges.

To understand the effect of this difference, we need to consider which type of orbital is lower in energy. Because the negative point-charge ligands repel negatively charged electrons, the electrons will prefer the *d* orbitals farthest from the ligands. In other words, the d_{xz}, d_{yz}, and d_{xy} orbitals (called the t_{2g} set) are at a *lower energy* in the octahedral complex than are the d_{z^2} and $d_{x^2-y^2}$ orbitals (the e_g set). This is shown in Fig. 20.22. The negative point-charge ligands increase the energies of all the *d* orbitals. However, the orbitals that point at the ligands are raised in energy more than those that point between the ligands.

It is this **splitting of the 3*d* orbital energies** (symbolized by Δ) that explains the color and magnetism of complex ions of the first-row transition metal ions. For example, in an octahedral complex of Co^{3+} (a metal ion with six 3*d*

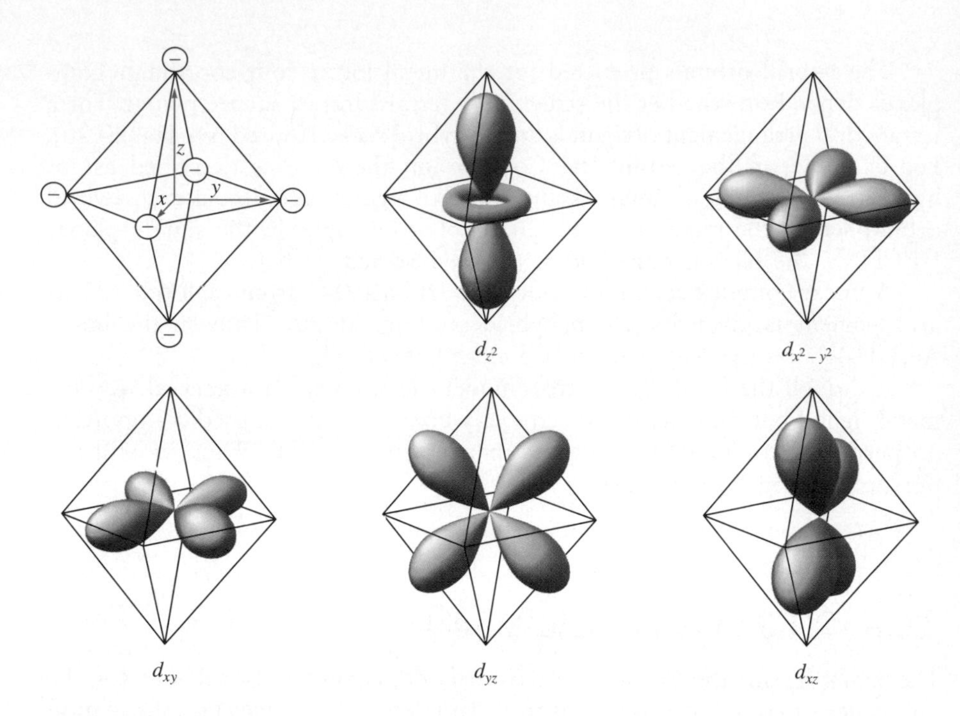

Figure 20.21
An octahedral arrangement of point-charge ligands and the orientation of the 3*d* orbitals.

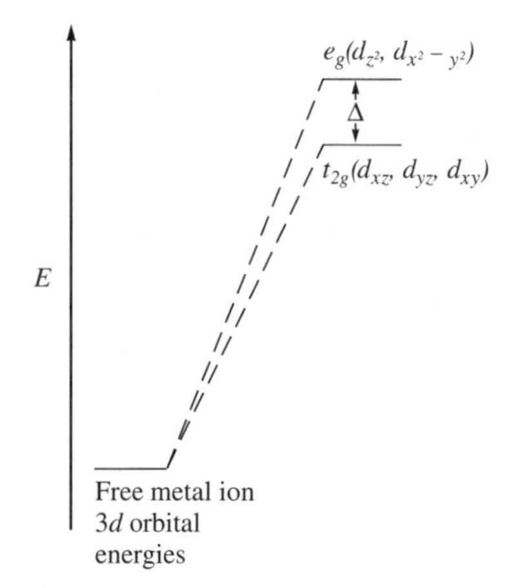

Figure 20.22
The energies of the 3*d* orbitals for a metal ion in an octahedral complex. The 3*d* orbitals are degenerate (all have the same energy) in the free metal ion. In the octahedral complex the orbitals are split into two sets as shown. The difference in energy between the two sets is designated as Δ (delta).

electrons), there are two possible ways to place the electrons in the split 3*d* orbitals (Fig. 20.23). If the splitting produced by the ligands is very large, a situation called the **strong-field case,** the electrons will pair in the lower-energy t_{2g} orbitals. This gives a *diamagnetic* complex in which all electrons are paired. On the other hand, if the splitting is small (the **weak-field case**), the electrons will occupy all five orbitals before pairing occurs. In this case the complex has four unpaired electrons and thus is *paramagnetic.*

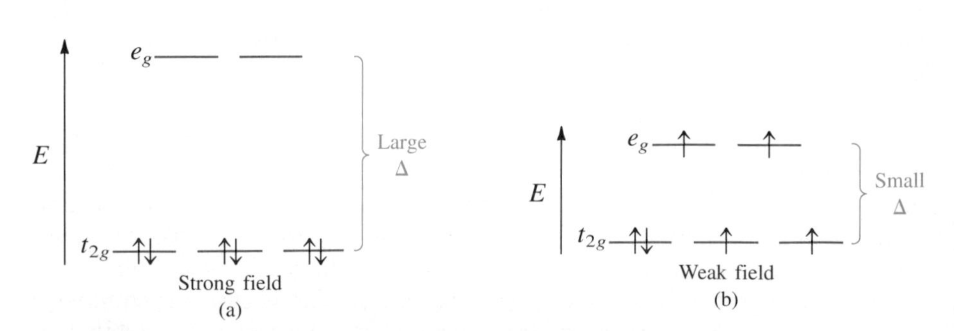

Figure 20.23
Possible electron arrangements in the split 3*d* orbitals of an octahedral complex of Co^{3+} (electron configuration $3d^6$). (a) In a strong field (large Δ value), the electrons fill the t_{2g} set first, giving a diamagnetic complex. (b) In a weak field (small Δ value), the electrons occupy all five orbitals before any pairing occurs.

EXAMPLE 20.4

The $Fe(CN)_6^{3-}$ ion is known to have one unpaired electron. Does the CN^- ligand produce a strong or weak field?

Solution

Since the ligand is CN^- and the overall complex ion charge is 3−, the metal ion must be Fe^{3+}, which has a $3d^5$ electron configuration. The two possible arrangements of the five electrons in the d orbitals split by the octahedrally arranged ligands are

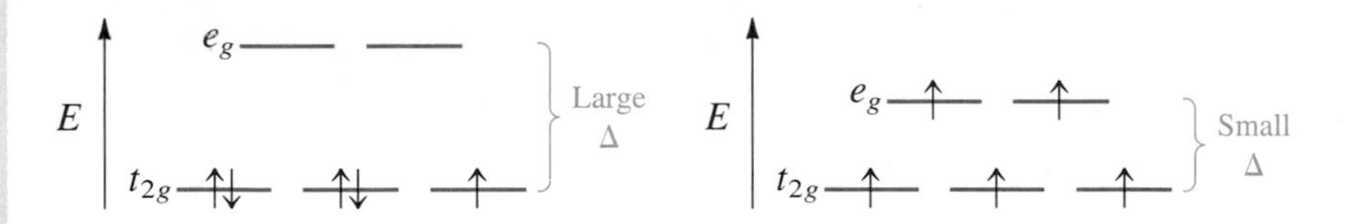

The strong-field case gives one unpaired electron, which agrees with the experimental observation. The CN^- ion is a strong-field ligand toward the Fe^{3+} ion.

The crystal field model allows us to account for the differences in the magnetic properties of $Co(NH_3)_6^{3+}$ and CoF_6^{3-}. The $Co(NH_3)_6^{3+}$ ion is known to be diamagnetic and thus corresponds to the strong-field case, also called the **low-spin case,** since it yields the *minimum* number of unpaired electrons. In contrast, the CoF_6^{3-} ion, which is known to have four unpaired electrons, corresponds to the weak-field case, also known as the **high-spin case,** since it gives the *maximum* number of unpaired electrons.

From studies of many octahedral complexes, we can arrange ligands in order of their ability to produce d-orbital splitting. A partial listing of ligands in this so-called **spectrochemical series** is

$$CN^- > NO_2^- > en > NH_3 > H_2O > OH^- > F^- > Cl^- > Br^- > I^-$$

Strong-field ligands (large Δ) — Weak-field ligands (small Δ)

The ligands are arranged in order of decreasing Δ values toward a given metal ion.

It has also been observed that *the magnitude of* Δ *for a given ligand increases as the charge on the metal ion increases.* For example, NH_3 is a weak-field ligand toward Co^{2+} but acts as a strong-field ligand toward Co^{3+}. This makes sense; as the metal ion charge increases, the ligands are drawn closer to the metal ion because of its increased charge density. As the ligands move closer, they cause greater splitting of the d orbitals, thereby producing a larger Δ value.

We have seen how the crystal field model accounts for the magnetic properties of octahedral complexes. The same model also explains the colors of

EXAMPLE 20.5

Predict the number of unpaired electrons in the complex ion $[Cr(CN)_6]^{4-}$.

Solution

The net charge of 4− means that the metal ion present must be Cr^{2+} $(-6 + 2 = -4)$, which has a $3d^4$ electron configuration. Since CN^- is a strong-field ligand (see the spectrochemical series), the correct crystal field diagram for $[Cr(CN)_6]^{4-}$ is

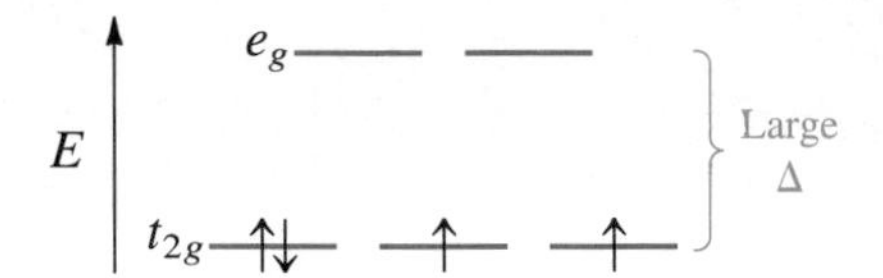

The complex ion will have two unpaired electrons. Note that the CN^- ligand produces such a large splitting that two of the electrons will be paired in the same orbital rather than force one electron up through the energy gap Δ.

these complex ions. For example, consider $Ti(H_2O)_6^{3+}$, an octahedral complex of Ti^{3+}, which has a $3d^1$ electron configuration. This complex ion is violet because it absorbs light in the middle of the visible region of the spectrum (see Fig. 20.24). When a substance absorbs certain wavelengths of light in the visible region, the color of that substance is determined by the wavelengths of visible light that remain. We say that the substance exhibits the color *complementary* to those absorbed. The $Ti(H_2O)_6^{3+}$ ion is violet because it absorbs light in the yellow-green region, thus allowing red light and blue light to pass. The red and blue colors not absorbed produce the observed violet color. This effect is shown schematically in Fig. 20.25. Table 20.16 shows the general relationship between the wavelengths of visible light absorbed and the approximate color observed.

The reason that the $Ti(H_2O)_6^{3+}$ ion absorbs specific wavelengths of visible light can be traced to the transfer of the lone d electron between the split d orbitals (Fig. 20.26). A given photon of light can be absorbed by a molecule

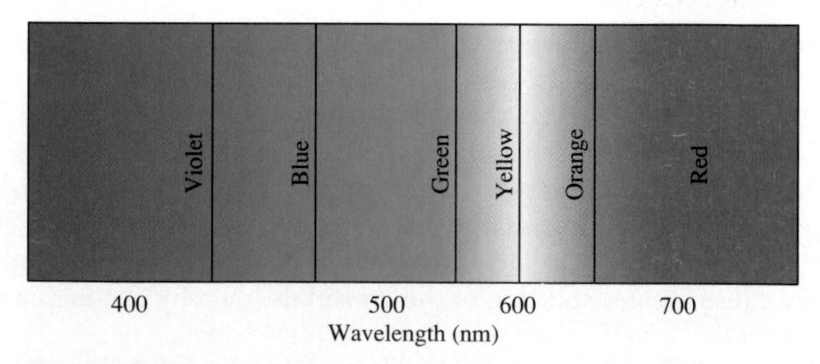

Figure 20.24
The visible spectrum.

TABLE 20.16 Approximate Relationship of Wavelength of Visible Light Absorbed to Color Observed

Absorbed Wavelength in nm (color)	Observed Color
400 (violet)	Greenish yellow
450 (blue)	Yellow
490 (blue-green)	Red
570 (yellow-green)	Violet
580 (yellow)	Dark blue
600 (orange)	Blue
650 (red)	Green

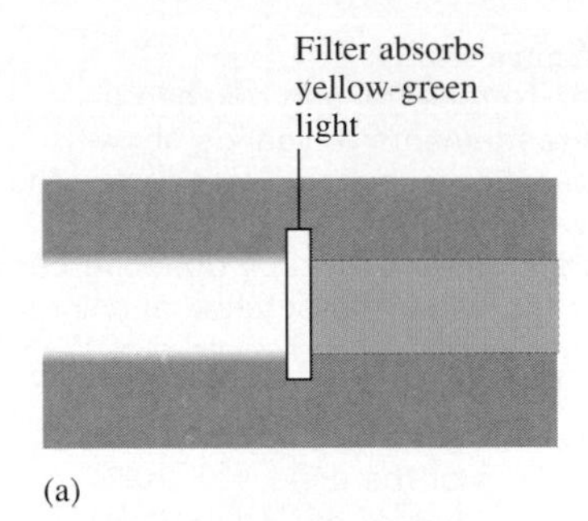

(a)

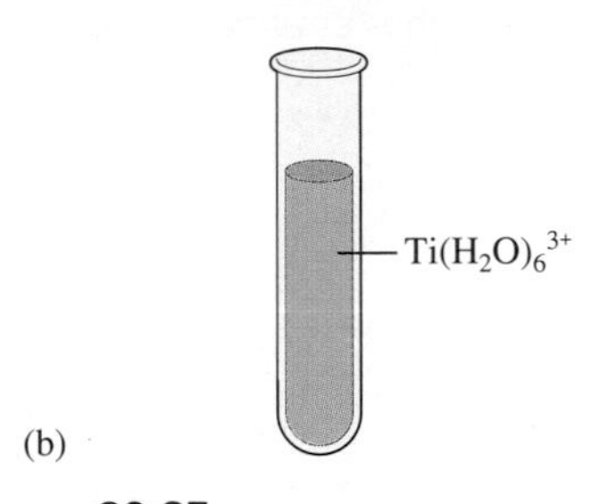

(b)

Figure 20.25
(a) When white light shines on a filter that absorbs wavelengths in the yellow-green region, the emerging light is violet. (b) Because the complex ion $Ti(H_2O)_6^{3+}$ absorbs yellow-green light, a solution of it is violet.

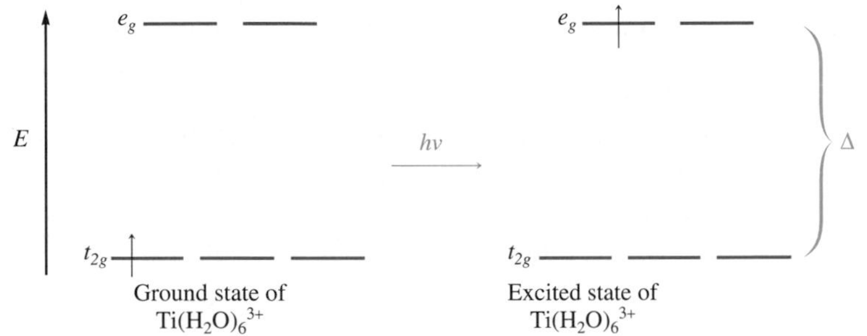

Figure 20.26
The complex ion $Ti(H_2O)_6^{3+}$ can absorb visible light in the yellow-green region to transfer the lone *d* electron from the t_{2g} to the e_g set.

only if the wavelength of the light provides the exact amount of energy needed by the molecule. In other words, whether a wavelength is absorbed is determined by the relationship

$$\Delta E = \frac{hc}{\lambda}$$

where ΔE represents the energy spacing in the molecule (we have used simply Δ in this chapter) and λ represents the wavelength of light needed to provide exactly that amount of energy. Because the *d*-orbital splitting in most octahedral complexes corresponds to the energies of photons in the visible region, octahedral complex ions are usually colored.

Since the ligands coordinated to a given metal ion determine the size of the *d*-orbital splitting, the color changes as the ligands are changed. This occurs because a change in Δ means a change in the wavelength of light is needed to transfer electrons between the t_{2g} and e_g orbitals. Several octahedral complexes of Cr^{3+} and their colors are listed in Table 20.17.

TABLE 20.17 Several Octahedral Complexes of Cr^{3+} and Their Colors

Isomer	Color
$[Cr(H_2O)_6]Cl_3$	Violet
$[Cr(H_2O)_5Cl]Cl_2$	Blue-green
$[Cr(H_2O)_4Cl_2]Cl$	Green
$[Cr(NH_3)_6]Cl_3$	Yellow
$[Cr(NH_3)_5Cl]Cl_2$	Purple
$[Cr(NH_3)_4Cl_2]Cl$	Violet

Other Coordination Geometries

Using the same principles developed for octahedral complexes, we will now consider complexes with other geometries. For example, Fig. 20.27 shows a tetrahedral arrangement of point charges in relation to the 3*d* orbitals of a metal ion. There are two important facts to note.

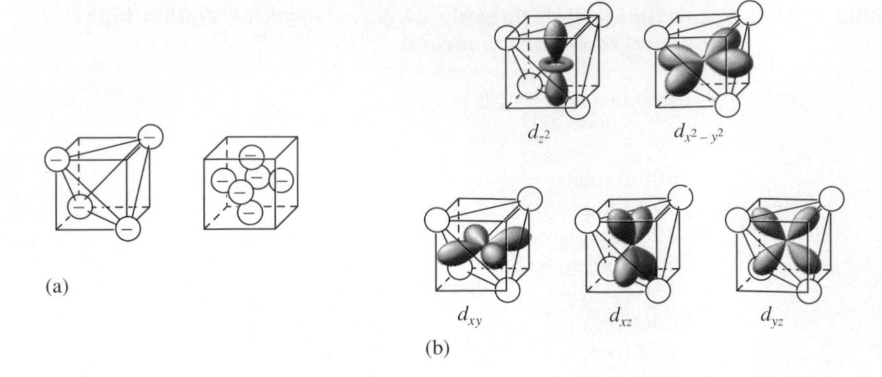

Figure 20.27
(a) Tetrahedral and octahedral arrangements of ligands shown inscribed in cubes. Note that in the two types of arrangements, the point charges occupy opposite parts of the cube: the octahedral point charges are at the centers of the cube faces, while the tetrahedral point charges occupy opposite corners of the cube. (b) The orientations of the 3*d* orbitals relative to the tetrahedral set of point charges.

1. None of the 3*d* orbitals "points at the ligands" in the tetrahedral arrangement, as the $d_{x^2-y^2}$ and d_{z^2} orbitals do in the octahedral case. Thus the tetrahedrally arranged ligands do not differentiate the *d* orbitals as much in the tetrahedral as in the octahedral case. That is, the difference in energy between the split *d* orbitals is significantly less in tetrahedral complexes. Although we will not derive it here, the tetrahedral splitting is $\frac{4}{9}$ that of the octahedral splitting for a given ligand and metal ion:

$$\Delta_{\text{tet}} = \tfrac{4}{9}\Delta_{\text{oct}}$$

2. Although not exactly pointing at the ligands, the d_{xy}, d_{xz}, and d_{yz} orbitals are closer to the point charges than are the d_{z^2} and $d_{x^2-y^2}$ orbitals. This means that the *d*-orbital splitting is opposite to that for the octahedral arrangement. The two arrangements are contrasted in Fig. 20.28. Because the *d*-orbital splitting is relatively small for the tetrahedral case, the weak-field case (high-spin case) *always* applies. There are no known ligands powerful enough to produce the strong-field case in a tetrahedral complex.

EXAMPLE 20.6

Give the crystal field diagram for the tetrahedral complex ion $CoCl_4^{2-}$.

Solution

The complex ion contains Co^{2+}, which has a $3d^7$ electron configuration. The splitting of the *d* orbitals will be small since this is a tetrahedral complex, giving the high-spin case with three unpaired electrons.

E

↑ d_{xy} ↑ d_{xz} ↑ d_{yz}

Small Δ

↑↓ d_{z^2} ↑↓ $d_{x^2-y^2}$

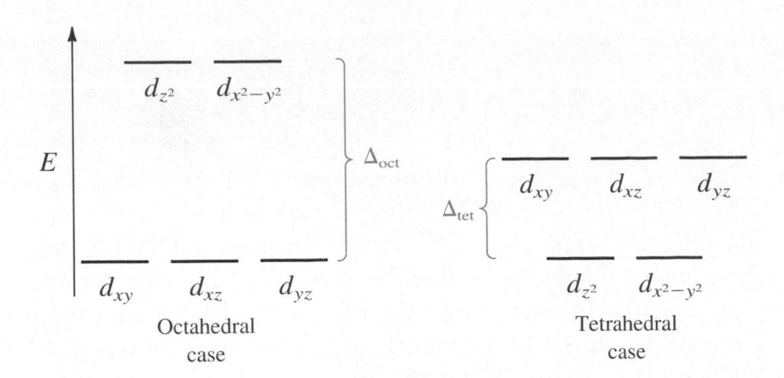

Figure 20.28
The crystal field diagrams for octahedral and tetrahedral complexes. The relative energies of the sets of *d* orbitals are reversed. For a given type of ligand the splitting is much larger for the octahedral complex ($\Delta_{oct} > \Delta_{tet}$). This occurs because in the octahedral arrangement the d_{z^2} and $d_{x^2-y^2}$ point their lobes directly at the point charges and are thus relatively high in energy.

The crystal field model also applies to square planar and linear complexes. The crystal field diagrams for these cases are shown in Fig. 20.29. The ranking of orbitals in these diagrams can be explained by considering the relative orientations of the point charges and the orbitals. The diagram in Fig. 20.28 for the octahedral arrangement can be used to obtain these orientations. We can obtain the square planar complex by starting with an octahedral arrangement of six point charges and then removing the two point charges along the z axis. Removal of the two point charges on the z axis greatly lowers the energy of d_{z^2}. Now only the four lobes of the $d_{x^2-y^2}$ orbital point directly at the four remaining point charges. Therefore, $d_{x^2-y^2}$ is the highest-energy orbital for the square planar case. The relative energies of the remaining orbitals depend on how close their lobes are to the four point charges. The ordering of d_{z^2} and d_{xy} is not entirely obvious, because d_{z^2} has a significant band of electron density centered in the xy plane. We will not deal with this issue here. We can obtain the linear complex from the octahedral arrangement by arbitrarily leaving the two ligands along the z axis and removing the four in the xy plane. This means only the d_{z^2} orbital points at the ligands and is highest in energy.

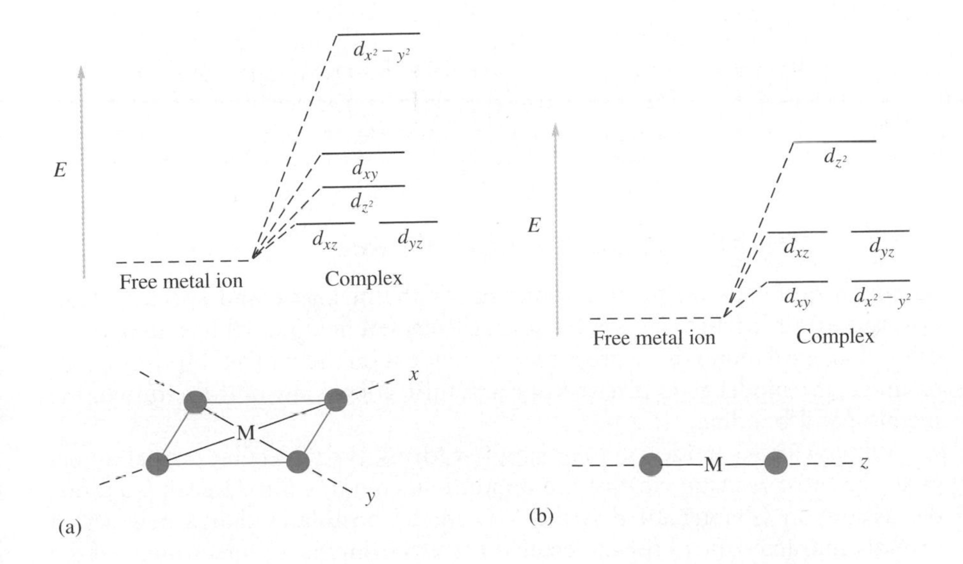

Figure 20.29
(a) The crystal field diagram for a square planar complex oriented in the *xy* plane with ligands along the *x* and *y* axes. The position of the d_{z^2} orbital is higher than those of the d_{xz} and d_{yz} orbitals because of the "doughnut" of electron density in the *xy* plane. The actual position of d_{z^2} is somewhat uncertain and varies in different square planar complexes. (b) The crystal field diagram for a linear complex where the ligands lie along the *z* axis.

Transition Metal Ions Lend Color to Gems

The beautiful pure color of gems, so valued by cultures everywhere, arises from trace transition metal ion impurities in minerals that would otherwise be colorless. For example, the stunning red of a ruby, the most valuable of all gemstones, is caused by Cr^{3+} ions, which replace about 1% of the Al^{3+} ions in the mineral corundum, which is a form of aluminum oxide (Al_2O_3) that is nearly as hard as diamond. In the corundum structure the Cr^{3+} ions are surrounded by six oxide ions at the vertices of an octahedron. This leads to the characteristic octahedral splitting of chromium's $3d$ orbitals, such that the Cr^{3+} ions absorb strongly in the blue-violet and yellow-green regions of the visible spectrum but transmit red light to give the characteristic ruby color. (On the other hand, if some of the Al^{3+} ions in corundum are replaced by a mixture of Fe^{2+}, Fe^{3+}, and Ti^{4+} ions, the gem is a sapphire with its brilliant blue color; or if some of the Al^{3+} ions are replaced by Fe^{3+} ions, the stone is a yellow topaz.)

Emeralds are derived from the mineral beryl, which is a beryllium aluminum silicate (empirical formula: $3BeO \cdot Al_2O_3 \cdot 6SiO_2$). When some of the Al^{3+} ions in beryl are replaced by Cr^{3+} ions, the characteristic green color of emerald results. In this environment the splitting of the Cr^{3+} $3d$ orbitals causes it to strongly absorb yellow and blue-violet light and to transmit green light.

A gem closely related to ruby and emerald is alexandrite, named after Alexander II of Russia. This gem is based on the mineral chrysoberyl, a beryllium aluminate with the empirical formula $BeO \cdot Al_2O_3$ in which $\approx$1% of the Al^{3+} ions are replaced by Cr^{3+} ions. In the chrysoberyl environment Cr^{3+} absorbs strongly in the yellow region of the spectrum. Alexandrite has the interesting property of changing colors depending on the light source. When the first alexandrite stone was discovered deep in a mine in the Russian Ural Mountains in 1831, it appeared to be a deep red color in the firelight of the miners' lamps. However, when the stone was brought to the surface, its color was blue. This seemingly magical color change occurs because the firelight of a miner's helmet is rich in the yellow and red wavelengths of the visible spectrum but does not contain much blue. Absorption of the yellow by the stone produces a reddish color. However, daylight has much more intensity in the blue region than firelight. Thus the extra blue in the light transmitted by the stone gives it a bluish color in daylight.

Once the structure of a natural gem is known, it is usually not very difficult to make the gem artificially. For example, rubies and sapphires are made on a large scale by fusing $Al(OH)_3$ with the appropriate transition metal salts at $\approx$1200°C to make the "doped" corundum. With these techniques gems of astonishing size can be manufactured. Rubies as large as 10 lb and sapphires up to 100 lb have been synthesized. Smaller synthetic stones produced for jewelry are virtually identical to the corresponding natural stones, and it takes great skill for a gemologist to tell the difference.

20.7 The Molecular Orbital Model

Although quite successful in accounting for the magnetic and spectral (light absorption) properties of complex ions, the crystal field model has limited use other than explaining those properties dependent on the d-orbital splitting. For example, the model gives a very crude and misleading view of the nature of the metal-ligand bonding.

Of the various models in their simplest forms, the molecular orbital model gives the most realistic view of the bonding in complex ions. Recall from our discussions in Chapter 14 that the MO model postulates that a new set of orbitals characteristic of the molecule is formed from the atomic orbitals of the component atoms. To illustrate how this model can be applied to complex ions, we will describe the molecular orbitals in an octahedral complex of general formula ML_6^{n+}. To keep things as simple as possible, we will focus only on those ligand orbitals having lone pairs that interact with the metal ion valence orbitals (3*d*, 4*s*, and 4*p*). There are two important considerations in predicting how atomic orbitals will interact to form molecular orbitals:

1. *Extent of orbital overlap*. Atomic orbitals must have a net overlap in space to form molecular orbitals. Figure 20.30 shows an octahedral arrangement of ligands with lone pair orbitals. Recall from our previous discussion that two of the metal ion's 3*d* orbitals (d_{z^2} and $d_{x^2-y^2}$) point at the ligands and thus will form molecular orbitals with the ligand lone pair orbitals. On the other hand, the d_{xy}, d_{xz}, and d_{yz} orbitals point *between* the ligands and thus will not be involved in the σ bonding with the ligands. The spherical 4*s* orbital of the metal ion overlaps with all of the ligand lone pair orbitals, and each of the 4*p* orbitals overlaps with pairs of ligand lone pair orbitals on the three coordinate axes. Thus the d_{z^2}, $d_{x^2-y^2}$, 4*s*, 4_{p_x}, 4_{p_y}, and 4_{p_z} orbitals will be involved in the σ molecular orbitals in the complex ion.
2. *Relative orbital energies*. Atomic orbitals that are close in energy will interact more strongly than those widely separated in energy.

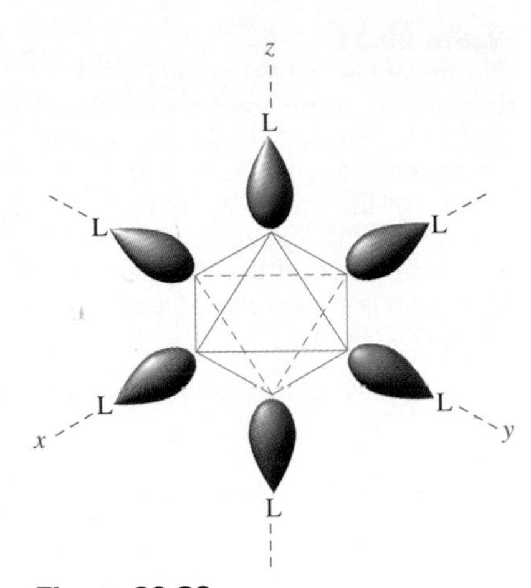

Figure 20.30
An octahedral arrangement of ligands showing their lone pair orbitals.

When we apply these principles to the general complex ML_6^{n+}, we obtain the energy-level diagram shown in Fig. 20.31. Note that the σ_s, σ_p, and σ_d molecular orbitals are bonding MOs; they are *lower* in energy than the ligand orbitals and metal ion orbitals that mix to form them. Electrons in these molecular orbitals have lower energies than they do in either the isolated metal ion or the ligands. These bonding electrons are mainly responsible for the stability of the complex ion.

The molecular orbital model was introduced in Section 14.2.

Because the d_{xz}, d_{yz}, and d_{xy} orbitals (the t_{2g} set) of the metal ion do not overlap with the ligand orbitals, they remain unchanged. Thus they have the same energy in the complex ion as they had in the free metal ion and make no contribution to the stability of the complex. They are called *nonbonding orbitals*.

The e_g* molecular orbitals are antibonding orbitals, since they are higher in energy than the atomic orbitals that mix to form them. Since electrons in these orbitals have higher energies than they do in the free metal ion, they destabilize the complex ion relative to the separated metal ion and ligands. However, *the most important characteristic of the* e_g* *orbitals is that they are primarily composed of* d_{z^2} *and* $d_{x^2-y^2}$ *atomic orbitals,* with relatively little contribution from ligand orbitals. This lack of mixing is due to the large energy difference between the ligand orbitals and the metal ion 3*d* orbitals. Thus we can see that the molecular orbital model predicts the same type of *d*-orbital splitting as the crystal field model, with the added advantage of giving a much more realistic picture of both the metal-ligand bonding interaction and the origin of the splitting. Because it is a more realistic physical model, we can use the molecular orbital model to explain why different ligands produce different magnitudes of splitting. In particular, a ligand with a very electronegative donor atom will have lone pair orbitals of very low energy (the electrons are very firmly bound to the ligand); these orbitals do not mix very thoroughly with the metal ion orbitals. This will result in a small difference between the t_{2g} (nonbonding) and

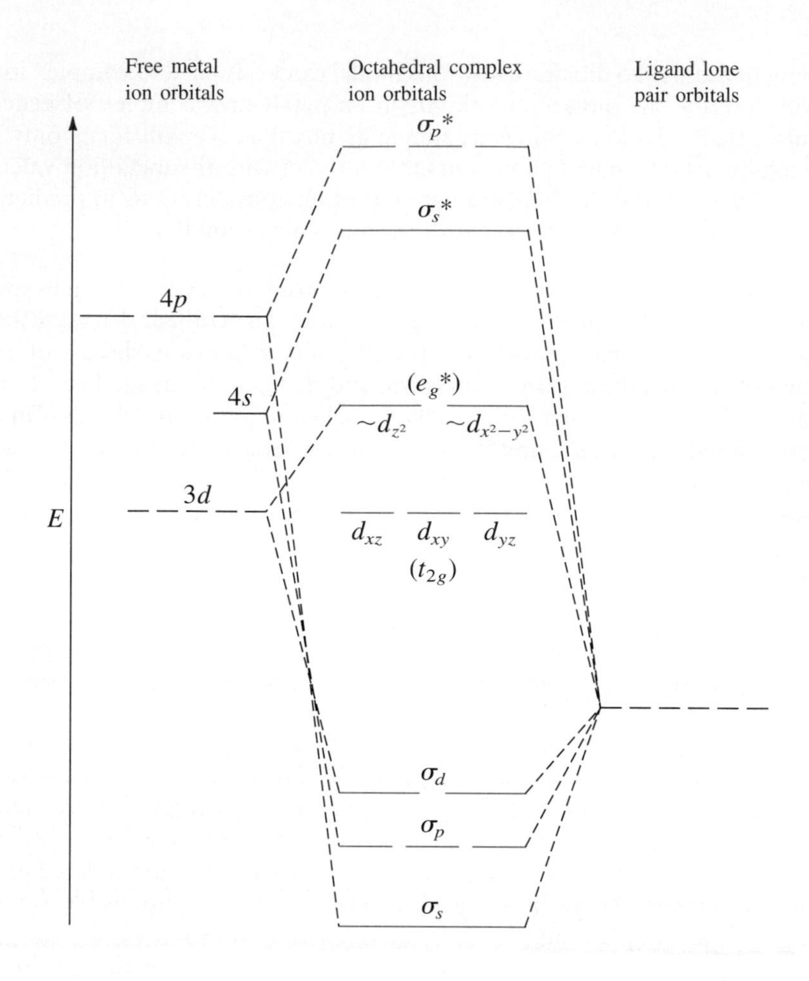

Figure 20.31
The molecular orbital energy-level diagram for an octahedral complex ion ML_6^{n+}. The e_g* molecular orbitals are essentially pure d_{z^2} and $d_{x^2-y^2}$ orbitals of the metal ion. Little mixing with the ligand orbitals occurs because of the large energy difference between the 3*d* and ligand orbitals.

e_g* (antibonding) orbitals. In other words, in this case the e_g orbitals are not much perturbed by the small amount of mixing that occurs. This means that the e_g orbital energies are not changed much by complex ion formation. For example, in CoF_6^{3-} [see Fig. 20.32(a)] the low-energy orbitals of the very electronegative fluoride ion ligands mix only to a small extent with the cobalt ion orbitals. A small amount of *d*-orbital splitting results, giving the high-spin case.

On the other hand, in $Co(NH_3)_6^{3+}$ the ammonia lone pair orbitals are closer in energy to the metal ion orbitals, a situation that produces a larger degree of mixing between the two sets of orbitals. This in turn gives a relatively large amount of *d*-orbital splitting, and the low-spin case results [see Fig. 20.32(b)].

We have seen that the molecular orbital model correctly predicts the *d*-orbital splitting in octahedral complexes, thereby accounting for the magnetic and spectral properties of these species. Moreover, it has a major advantage in that it accounts in a realistic way for metal-ligand bonding. However, it suffers from the disadvantage of being much more complicated to apply than the crystal field model. To take advantage of the relative strengths of the molecular orbital and crystal field models, the two have been combined to give the **ligand field model.** We will not consider the details of this model here.

The molecular orbital model can be applied to all types of complex ions, although we will not extend the model to other cases here. In addition, note

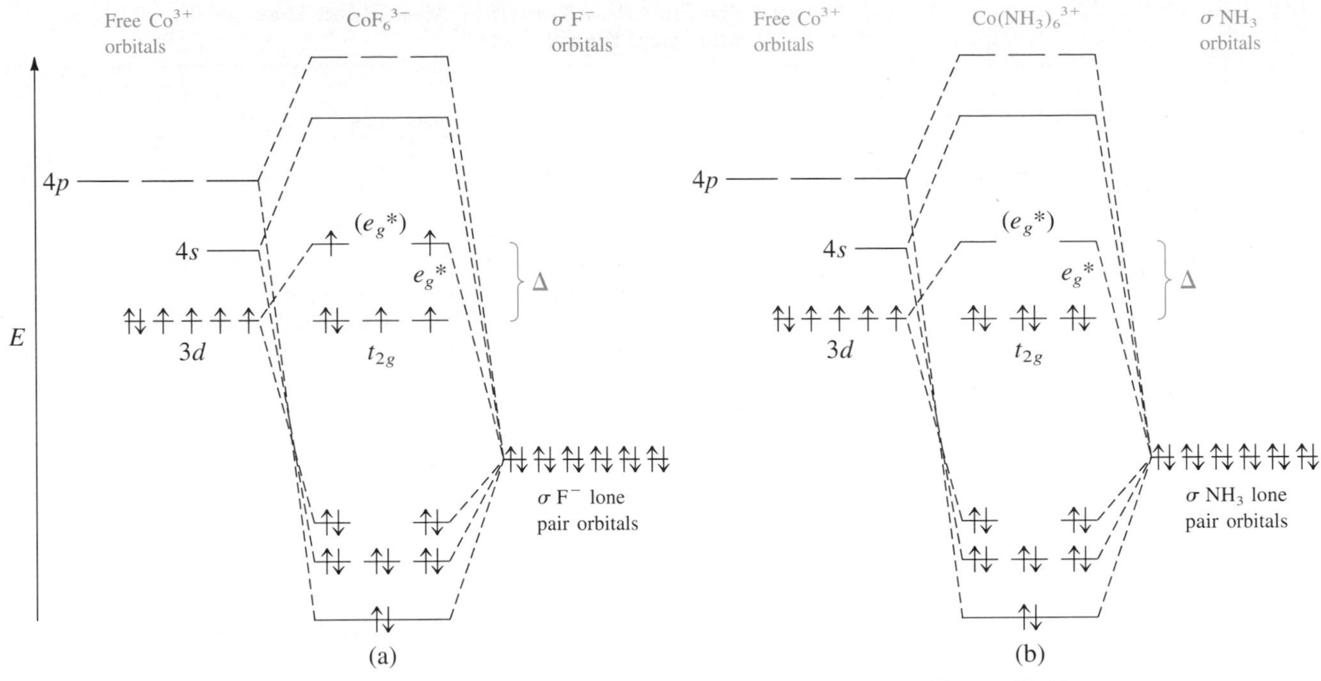

Figure 20.32
(a) The molecular orbital energy-level diagram for CoF_6^{3-}, which yields the high-spin case. (b) The molecular orbital energy-level diagram for $Co(NH_3)_6^{3+}$, which results in the low-spin case.

that in our use of the molecular orbital model to describe octahedral complexes, we considered only σ bonding effects to keep the model as simple as possible. For a complete treatment of complex ions, π bonding interactions also would have to be considered.

20.8 The Biological Importance of Coordination Complexes

The ability of metal ions to coordinate with and then release ligands and to easily undergo oxidation and reduction makes them ideal for use in biological systems. For example, metal ion complexes are used in humans for the transport and storage of oxygen, as electron transfer agents, as catalysts, and as drugs. Most of the first-row transition metals are essential for human health, as is summarized in Table 20.18. We will concentrate on iron's role in biological systems, since several of its coordination complexes have been studied extensively.

Iron plays a central role in almost all living cells. In mammals the principal source of energy comes from the oxidation of carbohydrates, proteins, and fats. Although oxygen is the oxidizing agent for these processes, it does not react directly with the nutrient molecules. Instead, the electrons from the breakdown of these nutrients are passed along a complex chain of molecules, called the *respiratory chain,* eventually reaching the O_2 molecule. The principal electron transfer molecules in the respiratory chain are iron-containing species called **cytochromes,** consisting of two main parts: an iron complex called **heme** and a protein. The structure of the heme complex is shown in Fig. 20.33. Note that it contains an iron ion (it can be either Fe^{2+} or Fe^{3+}) coordinated to a rather complicated planar ligand called a **porphyrin.** As a class, porphyrins all contain

A protein is a large molecule assembled from α-amino acids, which have the general structure

$$H_2N-\underset{H}{\overset{R}{C}}-COOH$$

where R varies.

TABLE 20.18 The First-Row Transition Metals and Their Biological Significance

First-Row Transition Metal	Biological Function(s)
Scandium	None known
Titanium	None known
Vanadium	None known in humans
Chromium	Assists insulin in the control of blood sugar; may also be involved in the control of cholesterol
Manganese	Necessary for a number of enzymatic reactions
Iron	Component of hemoglobin and myoglobin; involved in the electron transport chain
Cobalt	Component of vitamin B_{12}, which is essential for the metabolism of carbohydrates, fats, and proteins
Nickel	Component of the enzymes urease and hydrogenase
Copper	Component of several enzymes; assists in iron storage; involved in the production of color pigments of hair, skin, and eyes
Zinc	Component of insulin and many enzymes

the same central ring structure but have different substituent groups at the edges of the rings. The various porphyrin molecules act as tetradentate ligands for many metal ions, including iron, cobalt, and magnesium. In fact, *chlorophyll,* a substance essential to the process of photosynthesis, is a magnesium-porphyrin complex of the type shown in Fig. 20.34.

In addition to participating in the transfer of electrons from nutrients to oxygen, iron also plays a principal role in the transport and storage of oxygen in mammalian blood and tissues. Oxygen is stored using a molecule called **myoglobin,** which consists of a heme complex and a protein in a structure very similar to that of the cytochromes. In myoglobin the Fe^{2+} ion is coordinated to four nitrogen atoms of the porphyrin ring and to one nitrogen atom of the protein chain, as shown in Fig. 20.35. Since Fe^{2+} is normally six-coordinate, this leaves one position open for attachment of an O_2 molecule.

One especially interesting feature of myoglobin is that it involves an O_2 molecule attaching directly to Fe^{2+}. However, if gaseous O_2 is bubbled into an aqueous solution containing "bare" heme (no protein attached), the Fe^{2+} is

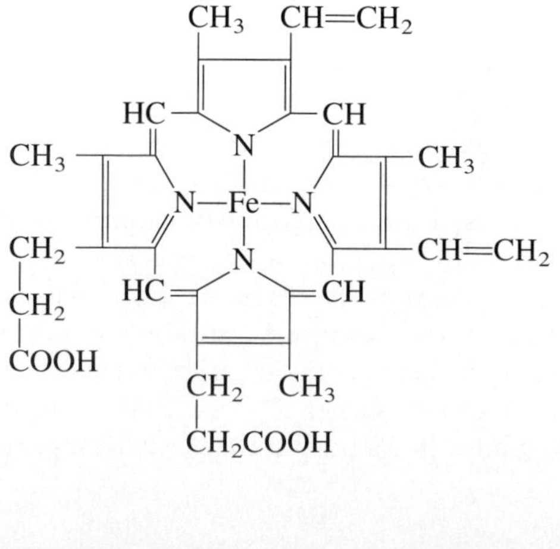

Figure 20.33
The heme complex, in which an Fe^{2+} ion is coordinated to four nitrogen atoms of a planar porphyrin ligand.

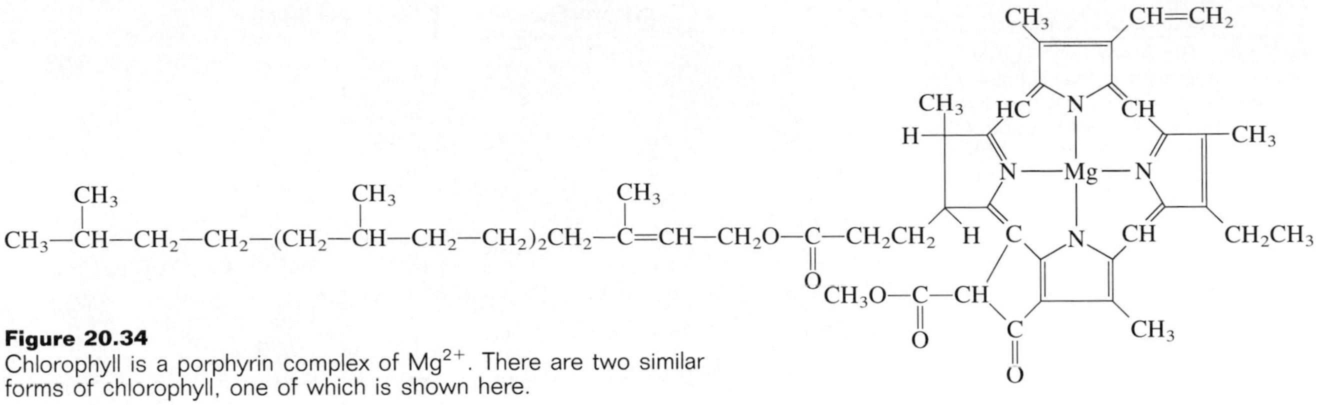

Figure 20.34
Chlorophyll is a porphyrin complex of Mg^{2+}. There are two similar forms of chlorophyll, one of which is shown here.

immediately oxidized to Fe^{3+}. This oxidation of the Fe^{2+} in heme does not happen in myoglobin. This fact is of crucial importance because Fe^{3+} does not form a coordinate covalent bond with O_2. Therefore, myoglobin would not function if the bound Fe^{2+} could be oxidized. Since Fe^{2+} in the "bare" heme complex can be oxidized, it must be the protein that somehow prevents the oxidation. How? Research results indicate that the oxidation of Fe^{2+} to Fe^{3+} involves an oxygen bridge between two iron ions (the circles indicate the ligands):

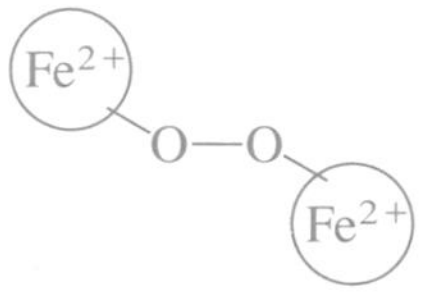

The bulky protein around the heme group in myoglobin prevents two myoglobin molecules from getting close enough to form the oxygen bridge; therefore, oxidation of the Fe^{2+} is prevented.

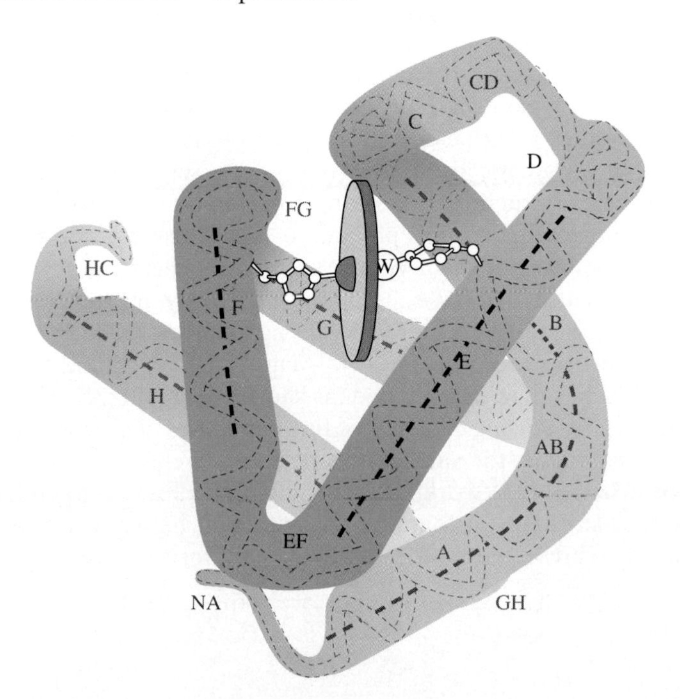

Figure 20.35
A representation of the myoglobin molecule. The Fe^{2+} ion is coordinated to four nitrogen atoms in the porphyrin of the heme (represented by the disk in the figure) and one nitrogen in the protein chain. This leaves a sixth coordination position (indicated by the W) available for an oxygen molecule.

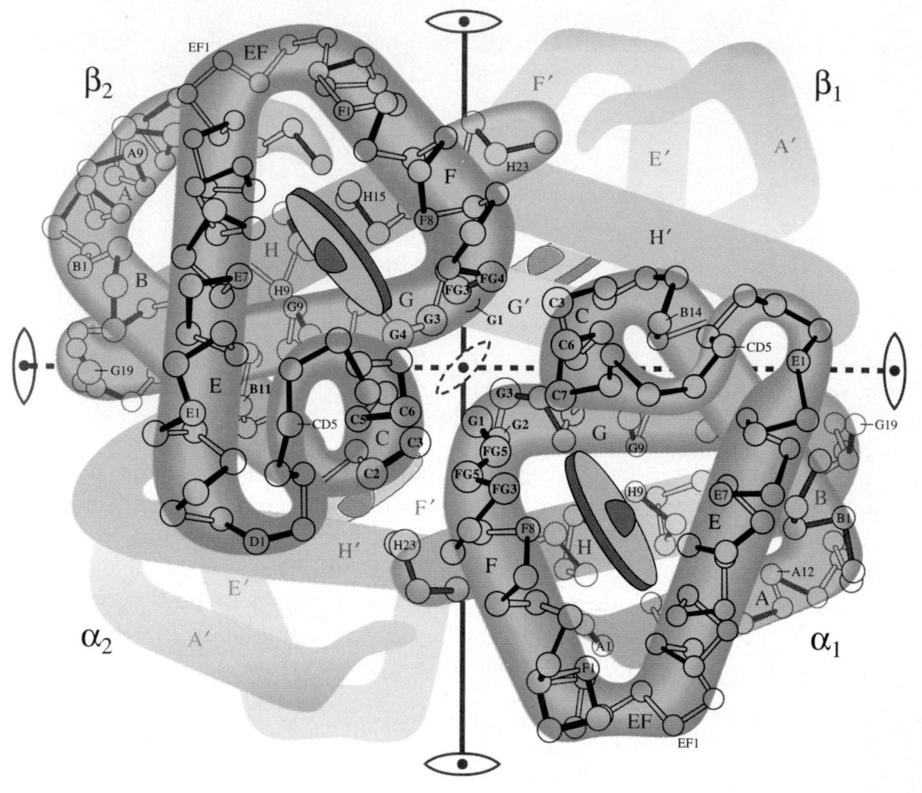

Figure 20.36
A representation of the hemoglobin structure. There are two slightly different types of protein chains (α and β). Each hemoglobin has two α chains and two β chains, each with a heme complex near the center. Thus each hemoglobin molecule can complex with four O_2 molecules.

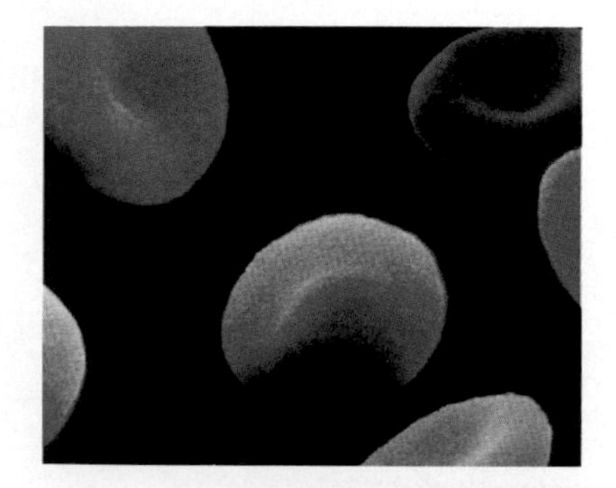

Figure 20.37
Normal red blood cells (top) and sickle-shaped red blood cells (bottom).

The transport of O_2 in the blood is carried out by **hemoglobin,** a molecule consisting of four myoglobin-like units, as shown in Fig. 20.36. Each hemoglobin can therefore bind four O_2 molecules to form a bright red diamagnetic complex. The diamagnetism occurs because oxygen is a strong-field ligand toward Fe^{2+}, which has a $3d^6$ electron configuration. When the oxygen molecule is released, a water molecule occupies the sixth coordination position around each Fe^{2+}, yielding a bluish paramagnetic complex (H_2O is a weak-field ligand toward Fe^{2+}) that gives venous blood its characteristic bluish tint.

Hemoglobin dramatically demonstrates how sensitive the function of a biomolecule is to its structure. In certain people, in the synthesis of the proteins needed for hemoglobin, an improper amino acid is inserted into the protein in two places. This may not seem very serious, since there are several hundred amino acids present. However, because the incorrectly inserted amino acid has a nonpolar substituent instead of the polar one found on the proper amino acid, the hemoglobin drastically changes its shape. The red blood cells are then sickle-shaped rather than disk-shaped, as shown in Fig. 20.37. The misshapen cells can aggregate, causing clogging of tiny capillaries. This condition, known as *sickle cell anemia,* is the subject of intense research.

Our knowledge of the workings of hemoglobin allows us to understand the effects of high altitudes on humans. The reaction between hemoglobin and oxygen can be represented by the following equilibrium:

$$\underset{\text{Hemoglobin}}{Hb(aq)} + 4O_2(g) \rightleftharpoons \underset{\text{Oxyhemoglobin}}{Hb(O_2)_4(aq)}$$

At high altitudes, where the oxygen content of the air is lower than at sea level, the position of this equilibrium will shift to the left, according to Le Châtelier's principle. Because less oxyhemoglobin is formed, fatigue, dizziness, and even a serious illness called *high-altitude sickness* can result. One way to combat this problem is to use supplemental oxygen, as most high-altitude mountain climbers do. However, this is impractical for people who live at high elevations. In fact, the human body adapts to the lower oxygen concentrations by making more hemoglobin, causing the equilibrium to shift back to the right. Someone moving from Chicago to Boulder, Colorado (elevation 5300 feet), would notice the effects of the new altitude for a couple of weeks; but as the hemoglobin level increases, the effects disappear. This change is called *high-altitude acclimatization,* which explains why athletes who want to compete at high elevations should practice under such conditions for several weeks prior to the event.

Our understanding of the biological role of iron also enables us to explain the toxicities of substances such as carbon monoxide and the cyanide ion. Both CO and CN^- are very good ligands toward iron and so can interfere with the normal workings of the iron complexes in the body. For example, carbon monoxide has about 200 times the affinity for the Fe^{2+} in hemoglobin as oxygen does. The resulting stable complex, **carboxyhemoglobin,** prevents the normal uptake of O_2, thus depriving the body of needed oxygen. Asphyxiation can result if enough carbon monoxide is present in the air. The mechanism for the toxicity of the cyanide ion is somewhat different. Cyanide coordinates strongly to cytochrome oxidase, an iron-containing cytochrome enzyme that catalyzes the oxidation-reduction reactions of certain cytochromes. The coordinated cyanide thus prevents the electron transfer process, causing rapid death. Because of its behavior, cyanide is called a *respiratory inhibitor.*

Exercises

A blue exercise number indicates that the answer to that exercise appears at the back of this book.

Transition Metals

1. Write electron configurations for each of the following.
   a. Co, Co^{2+}, Co^{3+}
   b. Pt, Pt^{2+}, Pt^{4+}
   c. Fe, Fe^{2+}, Fe^{3+}
   d. Au, Au^+, Au^{3+}
   e. Cu, Cu^+, Cu^{2+}
2. Use the following ionization energy values to decide whether ilmenite ($FeTiO_3$) is composed of Fe^{2+} and Ti^{4+} ions or Fe^{3+} and Ti^{3+} ions.

Ionization Energies (kJ/mol)	Fe	Ti
First	759	658
Second	1561	1310
Third	2957	2652
Fourth	—	4175

Use the following reduction potentials to answer the same question.

$$TiO^{2+} + 2H^+ + e^- \longrightarrow Ti^{3+} + H_2O \quad \mathscr{E}° = +0.099 \text{ V}$$

$$Fe^{3+} + e^- \longrightarrow Fe^{2+} \quad \mathscr{E}° = +0.77 \text{ V}$$

Which answer is consistent with the information presented in this chapter? Comment on any discrepancies.

3. Molybdenum is obtained as a by-product of copper mining or is mined directly (primary deposits are in the Rocky Mountains in Colorado). In both cases it is obtained as MoS_2, which is then converted to MoO_3. The MoO_3 can be used directly in the production of stainless steel for high-speed tools (this accounts for about 85% of the molybdenum used). Molybdenum can be purified by dissolving MoO_3 in aqueous ammonia and crystallizing ammonium molybdate. Depending on conditions, either $(NH_4)_2Mo_2O_7$ or $(NH_4)_6Mo_7O_{24} \cdot 4H_2O$ is obtained.
   a. Give names for MoS_2 and MoO_3.
   b. What is the oxidation state of Mo in each of the compounds mentioned above?
   c. Complete and balance the following equations:

$$MoS_2(s) + O_2(g) \longrightarrow MoO_3(s)$$

$$NH_3(aq) + MoO_3(s) \longrightarrow (NH_4)_2Mo_2O_7(s)$$

$$NH_3(aq) + MoO_3(s) \longrightarrow (NH_4)_6Mo_7O_{24} \cdot 4H_2O(s)$$

4. Titanium dioxide, the most widely used white pigment, occurs naturally but is often colored by the presence of impurities. The chloride process is often used in purifying rutile, a mineral form of titanium dioxide.
   a. Show that the unit cell for rutile, illustrated below, conforms to the formula TiO_2. (*Hint:* recall the discussion in Section 16.4.)

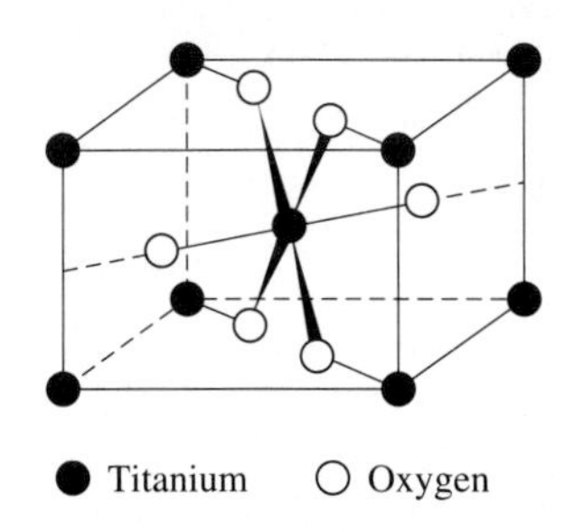

● Titanium ○ Oxygen

   b. The reactions for the chloride process are

$$2TiO_2(s) + 3C(s) + 4Cl_2(g) \xrightarrow{950°C} 2TiCl_4(g) + CO_2(g) + 2CO(g)$$

$$TiCl_4(g) + O_2(g) \xrightarrow{1000\text{–}1400°C} TiO_2(s) + 2Cl_2(g)$$

   Assign oxidation states to the elements in both reactions. Which elements are being reduced, and which are being oxidized? Identify the oxidizing agent and the reducing agent in each reaction.

5. The melting and boiling points of the titanium tetrahalides are given below.

	bp (°C)	mp (°C)
TiF_4	284	—
$TiCl_4$	−24	136.5
$TiBr_4$	38	233.5
TiI_4	155	377

   Rationalize these data in terms of the bonding in and the intermolecular forces among these compounds.

6. Manganese is found in nature in silicate, oxide, and carbonate minerals. Commercially, the most important ores are pyrolusite (MnO_2) and rhodochrosite ($MnCO_3$). Give systematic names for each of these compounds.

7. Ores of cobalt have been used to impart a blue color to glass since 2600 B.C. Impure cobalt metal was first isolated in 1735 by G. Brandt, a Swedish chemist. The name is probably derived from *Kobold,* the German word for "goblin" or "evil spirit." The miners of northern Europe thought that spiteful spirits were responsible for these ores, which on smelting not only failed to yield the cobalt metal but also produced highly toxic fumes (As_4O_6). The principal ores of cobalt are smaltite ($CoAs_2$), cobaltite ($CoAsS$), and linnaeite (Co_3S_4).
   a. Write a reaction for the roasting of smaltite by oxygen, which produces cobalt(II) oxide and As_4O_6. Name As_4O_6. Propose a structure for As_4O_6 on the basis of the discussion of phosphorus chemistry in Chapter 19.
   b. Cobalt is usually obtained as a by-product of the production of copper, nickel, or lead. The ore containing the mixture of metals is first treated with oxygen to form the metal oxides. These oxides are then treated with sulfuric acid, which dissolves the oxides of iron, nickel, and cobalt, but not that of copper, which is left behind. After iron is precipitated as its hydroxide, the basic solution is treated with hypochlorite ion, which oxidizes the Co^{2+} ion to Co^{3+} and leads to precipitation of cobalt(III) hydroxide:

$$Co^{2+}(aq) + OCl^-(aq) + OH^-(aq) \longrightarrow Co(OH)_3(s) + Cl^-(aq)$$

   Balance this equation.
   c. Calculate the concentration of Co^{3+} present in a saturated aqueous solution of $Co(OH)_3$ ($K_{sp} = 2.5 \times 10^{-43}$).
   d. Calculate the concentration of Co^{3+} present in a saturated solution of $Co(OH)_3$ buffered at pH = 10.00.

8. Zinc metal tarnishes quickly in moist air. Write a reaction for this process. Using data from Chapter 11 calculate $\mathscr{E}°$, $\Delta G°$, and K (at 298 K) for this reaction. For $Zn(OH)_2 + 2e^- \rightarrow Zn + 2OH^-$, $\mathscr{E}° = -1.24$ V.

Coordination Compounds

9. Define each of the following.
   a. ligand
   b. chelate
   c. bidentate
   d. complex ion

10. Name the following coordination compounds.
   a. $[CO(NH_3)_6]Cl_2$
   b. $[Co(H_2O)_6]I_3$
   c. $K_2[PtCl_4]$
   d. $K_4[PtCl_6]$
   e. $[Co(NH_3)_5Cl]Cl_2$
   f. $[Co(NH_3)_3(NO_2)_3]$

11. Name the following complex ions.
   a. $Ru(NH_3)_5Cl^{2+}$
   b. $Fe(CN)_6^{4-}$
   c. $Mn(NH_2CH_2CH_2NH_2)_3^{2+}$
   d. $Co(NH_3)_5NO_2^{2+}$

12. Name the following compounds containing complex ions.
   a. $Na_4[Ni(C_2O_4)_3]$
   b. $K_2[CoCl_4]$
   c. $[Cu(NH_3)_4]SO_4$
   d. $[Co(en)_2(SCN)Cl]Cl$

13. Give formulas for the following.
   a. hexakispyridinecobalt(III) chloride
   b. pentaammineiodochromium(III) iodide

c. trisethylenediaminenickel(II) bromide
d. potassium tetracyanonickelate(II)
e. tetraamminedichloroplatinum(IV)tetrachloroplatinate(II)

14. Give formulas for the following complex ions.
a. tetrachloroferrate(III) ion
b. pentaammineaquaruthenium(III) ion
c. iodopentakispyridineplatinum(IV) ion
d. amminetrichloroplatinate(II) ion

15. How many bonds could each of the following chelates form with a metal ion?
a. acetylacetone(acacH)

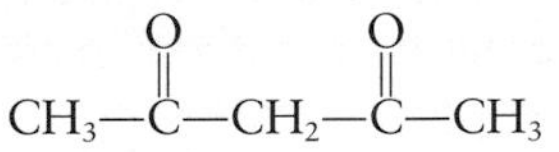

$$CH_3-\overset{\overset{O}{\|}}{C}-CH_2-\overset{\overset{O}{\|}}{C}-CH_3$$

b. diethylenetriamine

$$NH_2-CH_2-CH_2-NH-CH_2-CH_2-NH_2$$

c. salen

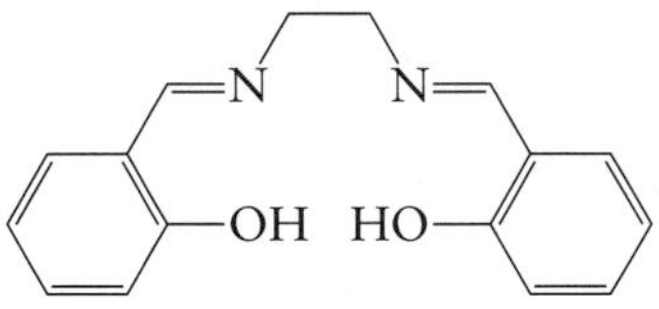

d. porphine

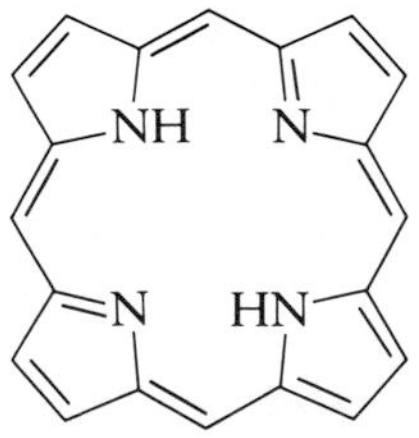

16. Define each of the following and give examples.
a. isomers
b. structural isomers
c. stereoisomers
d. coordination isomers
e. linkage isomers
f. geometric isomers
g. optical isomers

17. Draw geometrical isomers of each of the following complex ions.
a. $[Co(C_2O_4)_2(H_2O)_2]^-$
b. $[Pt(NH_3)_4I_2]^{2+}$
c. $[Ir(NH_3)_3Cl_3]$
d. $[Cr(en)(NH_3)_2I_2]^+$

18. Which of the following ligands are capable of linkage isomerism? Explain your answer.

$$SCN^-, N_3^-, NO_2^-, NH_2CH_2CH_2NH_2, OCN^-, I^-$$

19. Draw all geometrical and linkage isomers of square planar $[Pt(NH_3)_2(SCN)_2]$.

20. BAL is a chelating agent used in treating heavy metal poisoning. It acts as a bidentate ligand. What types of linkage isomers are possible when BAL coordinates to a metal ion?

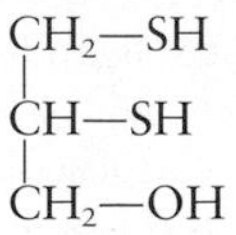

$$\begin{array}{l} CH_2-SH \\ | \\ CH-SH \\ | \\ CH_2-OH \end{array}$$
BAL

21. Draw structures for each of the following.
a. *cis*-dichloroethylenediamineplatinum(II)
b. *trans*-dichlorobisethylenediaminecobalt(II)
c. *cis*-tetraamminechloronitrocobalt(III) ion
d. *trans*-tetraamminechloronitritocobalt(III) ion
e. *trans*-diaquabisethylenediaminecopper(II) ion

22. Amino acids can act as ligands toward transition metal ions. The simplest amino acid is glycine, $NH_2CH_2CO_2H$. Draw a structure of the glycinate anion, $NH_2CH_2CO_2^-$, acting as a bidentate ligand. Draw the structural isomers of the square planar complex $Cu(NH_2CH_2CO_2)_2$.

23. A coordination compound of cobalt(III) contains four ammonia molecules, one sulfate ion, and one chloride ion. Addition of aqueous $BaCl_2$ solution to an aqueous solution of the compound gives no precipitate. Addition of aqueous $AgNO_3$ to an aqueous solution of the compound produces a white precipitate. Propose a structure for this coordination compound.

24. The carbonate ion, CO_3^{2-}, can act as either a monodentate or a bidentate ligand. Draw a picture of CO_3^{2-} coordinating to a metal ion as a bidentate and as a monodentate ligand. The carbonate ion can also act as a bridge between two metal ions. Draw a picture of a CO_3^{2-} ion bridging between two metal ions.

25. Draw the geometrical isomers of $[Cr(en)(NH_3)_2BrCl]^+$. Which of these isomers also has an optical isomer? Draw them.

Bonding, Color, and Magnetism in Coordination Compounds

26. Define each of the following.
a. weak-field ligand
b. strong-field ligand
c. low-spin complex
d. high-spin complex

27. Compounds of copper(II) are generally colored, but compounds of copper(I) are not. Why?

28. Would you expect $Cd(NH_3)_4Cl_2$ to be colored? Why or why not?

29. Henry Taube, the 1983 Nobel prize winner in chemistry, has studied the mechanisms of oxidation-reduction reactions involving transition metal complexes. In one experiment he and his students studied the following reaction:

$$Cr(H_2O)_6^{2+}(aq) + Co(NH_3)_5Cl^{2+}(aq) \longrightarrow Cr(III)\ complexes + Co(II)\ complexes$$

Chromium(III) and cobalt(III) complexes are substitutionally inert (no exchange of ligands) under conditions of the experiment. However, chromium(II) and cobalt(II) complexes can exchange ligands very rapidly. One of the products of the reaction is $Cr(H_2O)_5Cl^{2+}$. Is this consistent with the reaction proceeding through formation of $(H_2O)_5Cr—Cl—Co(NH_3)_5$ as an intermediate? Explain your answer.

30. How many unpaired electrons are in the following complex ions?
    a. $Ru(NH_3)_6^{2+}$ (low-spin case)
    b. $Fe(CN)_6^{3-}$ (low-spin case)
    c. $Ni(H_2O)_6^{2+}$
    d. $V(en)_3^{3+}$
    e. $CoCl_4^{2-}$

31. The complex ion $Ru(phen)_3^{2+}$ has been used as a probe for the structure of DNA.
    a. What type of isomerism is found in $Ru(phen)_3^{2+}$?
    b. $Ru(phen)_3^{2+}$ is diamagnetic (as are all complex ions of Ru^{2+}). Draw the crystal field diagram for the *d* orbitals in this complex ion.

phen = 1,10-phenanthroline =

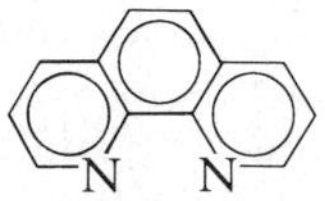

32. The complex ion $NiCl_4^{2-}$ contains two unpaired electrons, but $Ni(CN)_4^{2-}$ is diamagnetic. Propose structures for these two complex ions.

33. Tetrahedral complexes of Co^{2+} are quite common. Use a *d*-orbital splitting diagram to rationalize the stability of Co^{2+} tetrahedral complex ions.

34. Draw the *d*-orbital splitting diagrams for the octahedral complex ions of each of the following.
    a. Fe^{2+} (high and low spin)
    b. Fe^{3+} (high spin)
    c. Ni^{2+}
    d. Zn^{2+}
    e. Co^{2+} (high and low spin)

35. The compound $Ni(H_2O)_6Cl_2$ is green, but $Ni(NH_3)_6Cl_2$ is purple. Predict the predominant color of light absorbed by each compound. Which compound absorbs light with the shorter wavelength? Predict in which compound Δ is greater and whether H_2O or NH_3 is a stronger field ligand. Do your conclusions agree with the spectrochemical series?

36. The complex ion $Fe(CN)_6^{3-}$ is paramagnetic with one unpaired electron. The complex ion $Fe(SCN)_6^{3-}$ has five unpaired electrons. Where does SCN^- lie in the spectrochemical series with respect to CN^-?

37. Would it be better to use octahedral Ni^{2+} complexes or octahedral Cr^{2+} complexes to determine whether a ligand is a high-field or low-field ligand by determining experimentally the number of unpaired electrons? How else could the relative ligand field strengths be determined?

Additional Exercises

38. Nickel can be purified by producing the volatile compound nickel tetracarbonyl, $Ni(CO)_4$. Nickel is the only metal that reacts directly with CO at room temperature. What is the oxidation state of nickel in $Ni(CO)_4$?

39. How would transition metal ions be classified using the Lewis definition of acids and bases?

40. When an aqueous solution of KCN is added to a solution containing Ni^{2+} ions, a precipitate forms, which redissolves upon addition of more KCN solution. No precipitate forms when H_2S is bubbled into this solution. Write reactions describing what happens in this solution. [*Hint:* CN^- is a Brønsted-Lowry base ($K_b \approx 10^{-5}$) and a Lewis base.]

41. Until the discoveries of Werner, it was thought that the presence of carbon in a compound was required for it to be optically active. Werner prepared the following compound containing OH^- ions as bridging groups and then separated the optical isomers.

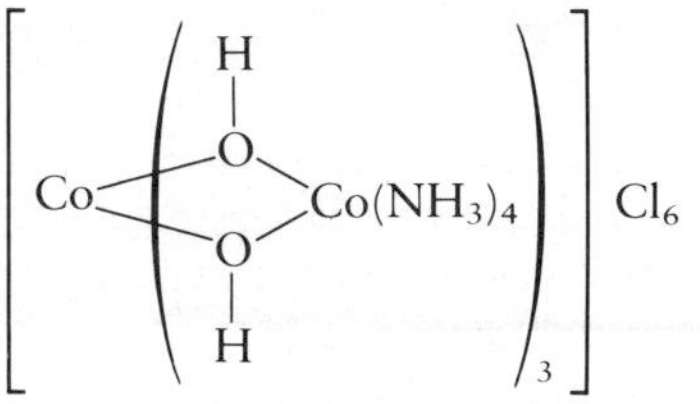

$$\left[Co\left(\begin{matrix} H \\ O \\ O \\ H \end{matrix} Co(NH_3)_4\right)_3\right]Cl_6$$

    a. Draw structures of the two optically active isomers of this compound.
    b. What are the oxidation states of the cobalt ions?
    c. How many unpaired electrons are present if the complex is the low-spin case?

42. Ammonia and potassium iodide solutions were added to an aqueous solution of $Cr(NO_3)_3$. A solid was isolated (compound A) and the following data were collected:
    i. When 0.105 g of compound A was strongly heated in excess O_2, 0.0203 g of CrO_3 was formed.
    ii. In a second experiment it took 32.93 mL of 0.100 *M* HCl to titrate completely the NH_3 present in 0.341 g of compound A.
    iii. Compound A was found to contain 73.53% iodine by mass.
    iv. The freezing point of water was lowered by 0.62°C when 0.601 g of compound A was dissolved in 10.00 g of H_2O (K_f = 1.86°C kg/mol).

What is the formula of the compound? What is the structure of the complex ion present? (*Hints:* Cr^{3+} is expected to be six-coordinate with NH_3 and (possibly) I^- acting as ligands. The I^- ions will be the counter ions if needed.)

43. Why are CN^- and CO toxic to humans?

44. Oxalic acid is often used to remove rust stains. Explain what properties of oxalic acid allow it to accomplish this.

45. Why do transition metal ions often have several oxidation states, but other metals generally have one?

46. Compounds of Sc^{3+} are not colored, but those of Ti^{3+} and V^{3+} are. Why?

47. Chelating ligands often form more stable complex ions than the corresponding monodentate ligands form with the same donor atoms. For example,

$$N^{2+}(aq) + 6NH_3(aq) \rightleftharpoons Ni(NH_3)_6^{2+}(aq) \qquad K_f = 3.2 \times 10^8$$

$$Ni^{2+}(aq) + 3en(aq) \rightleftharpoons Ni(en)_3^{2+}(aq) \qquad K_f = 1.6 \times 10^{18}$$

$$Ni^{2+}(aq) + penten(aq) \rightleftharpoons Ni(penten)^{2+}(aq) \qquad K_f = 2.0 \times 10^{19}$$

where penten is

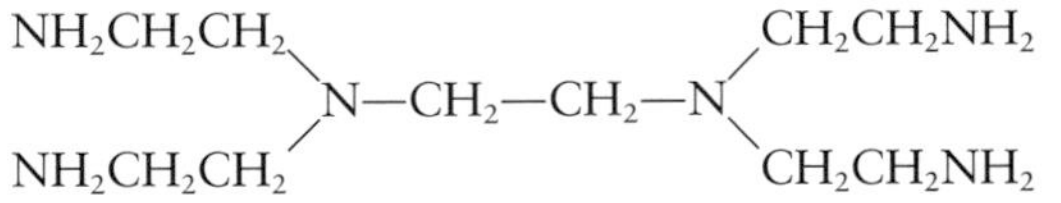

$$(NH_2CH_2CH_2)_2N{-}CH_2{-}CH_2{-}N(CH_2CH_2NH_2)_2$$

This increased stability that results is called the chelate effect. From bond energies, would you expect the enthalpy changes for the above reactions to be very different? What is the order (from least favorable to most favorable) of the entropy changes of the above reactions? How do the values of the formation constants correlate with $\Delta S°$? How can this be used to explain the chelate effect?

48. Consider the following data:

$$Co^{3+} + e^- \longrightarrow Co^{2+} \qquad \mathscr{E}° = 1.82\ V$$

$$Co(en)_3^{2+} \qquad K_f = 1.5 \times 10^{12}$$

$$Co(en)_3^{3+} \qquad K_f = 2.0 \times 10^{47}$$

where en = ethylenediamine.

a. Calculate $\mathscr{E}°$ for the half-reaction

$$Co(en)_3^{3+} + e^- \longrightarrow Co(en)_3^{2+}$$

b. Based on your answer to part a, which is the stronger oxidizing agent, Co^{3+} or $Co(en)_3^{3+}$?

c. Use the crystal field model to rationalize the result in part b.

49. Acetylacetone (see Exercise 15, part a), abbreviated acacH, is a bidentate ligand. It loses a proton and coordinates as $acac^-$, as shown below:

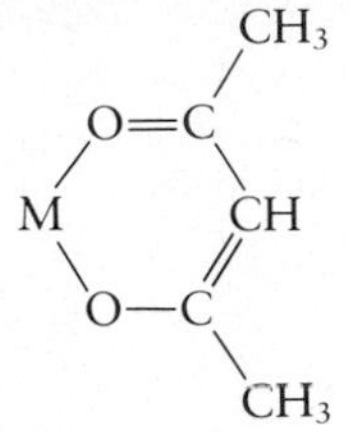

```
           CH3
          /
      O=C
     /    \
   M       CH
     \    //
      O—C
          \
           CH3
```

a. Draw all of the resonance structures for $acac^-$ coordinating to a metal ion.

b. Which of the following complexes are optically active: *cis* $Cr(acac)_2(H_2O)_2$, *trans* $Cr(acac)_2(H_2O)_2$, and $Cr(acac)_3$?

c. Acetylacetone reacts with an ethanol solution of $Eu(NO_3)_3$ to give a compound that is 40.1% C and 4.71% H by mass. Combustion of 0.286 g of the compound gives 0.112 g of Eu_2O_3. What is the molecular formula of the compound formed from the reaction of acetylacetone and europium(III) nitrate? [Assume the compound contains one europium(III) ion.]

50. A compound related to acetylacetone is 1,1,1-trifluoroacetylacetone (abbreviated Htfa):

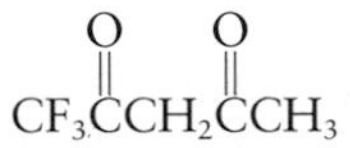

$$CF_3\overset{O}{\overset{\|}{C}}CH_2\overset{O}{\overset{\|}{C}}CH_3$$

Htfa forms complexes in a manner similar to acetylacetone. (See Exercise 49.) Both Be^{2+} and Cu^{2+} form complexes with tfa^- having the formula $M(tfa)_2$. Two isomers are formed for each metal complex.

a. The Be^{2+} complexes are tetrahedral. Draw the two isomers of $Be(tfa)_2$. What type of isomerism is exhibited by $Be(tfa)_2$?

b. The Cu^{2+} complexes are square planar. Draw the two isomers of $Cu(tfa)_2$. What type of isomerism is exhibited by $Cu(tfa)_2$?

51. Draw the structure of the nickel(II) complex of EDTA (see Table 20.13). Is this complex ion optically active?

52. Qualitatively draw the crystal field splitting of the d orbitals in a trigonal planar complex ion. (Let the z axis be perpendicular to the plane of the complex.)

53. Qualitatively draw the crystal field splitting for a trigonal bipyramidal complex ion. (Let the z axis be perpendicular to the trigonal plane.)

54. What is the lanthanide contraction? How does the lanthanide contraction affect the properties of the $4d$ and $5d$ transition metals?

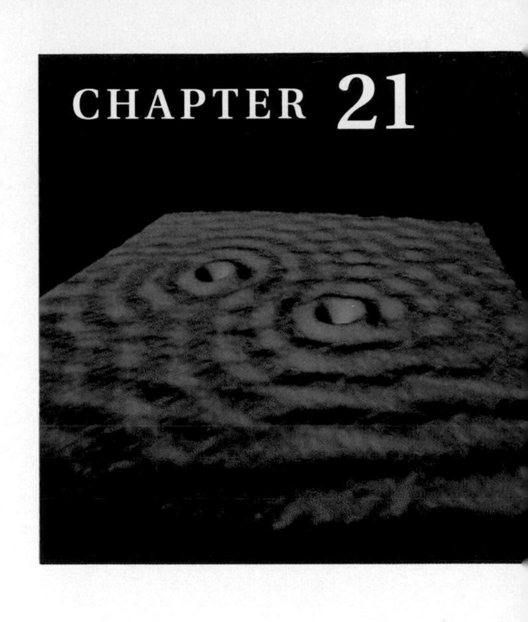

The Nucleus: A Chemist's View

Since the chemistry of an atom is determined by the number and arrangement of its electrons, the properties of the nucleus are not of primary importance to chemists. In the simplest view, the nucleus provides the positive charge to bind the electrons in atoms and molecules. However, a quick reading of any daily newspaper will show you that the nucleus and its properties have an important impact on our society. This chapter considers those aspects of the nucleus about which everyone should have some knowledge.

Several aspects of the nucleus are immediately impressive: its very small size, its very large density, and the magnitude of the energy that holds it together. The radius of a typical nucleus appears to be about 10^{-13} cm. This can be compared with the radius of a typical atom, which is on the order of 10^{-8} cm. A visualization will help you appreciate the small size of the nucleus: if the nucleus of the hydrogen atom were the size of a Ping-Pong ball, the electron in the 1*s* orbital would be, on the average, 0.5 kilometer (0.3 mile) away. The density of the nucleus is equally impressive—approximately 1.6×10^{14} g/cm^3. A sphere of nuclear material the size of a Ping-Pong ball would have a mass of 2.5 *billion tons!* In addition, the energies involved in nuclear processes are typically millions of times larger than those associated with normal chemical reactions. This fact makes nuclear processes very attractive for feeding the voracious energy appetite of our civilization.

Atomos, the Greek root of the word *atom,* means "indivisible." It was originally believed that the atom was the ultimate indivisible particle of which all matter was composed. However, as we discussed in Chapter 2, Lord Rutherford showed in 1911 that the atom is not homogeneous but instead has a dense, positively charged center surrounded by electrons. Subsequently, scientists learned that the nucleus of the atom can be described as containing neutrons and protons. In fact, in the past two decades it has become widely accepted that even the protons and neutrons are composed of smaller particles called *quarks*.

For most purposes the nucleus can be regarded as a collection of nucleons (neutrons and protons), and the internal structures of these particles can be ignored. Recall that the number of protons in a particular nucleus is the atomic number (Z) and that the sum of the neutrons and protons is the mass number (A). Atoms that have identical atomic numbers but different mass number values are called isotopes. The general term *nuclide* is applied to each unique atom and is represented by ${}^{A}_{Z}X$, where X represents the symbol for a particular element.

21.1 Nuclear Stability and Radioactive Decay

Nuclear stability is the central topic of this chapter and forms the basis for all of the important applications related to nuclear processes. Nuclear stability can be considered from both a kinetic and a thermodynamic point of view. Thermodynamic stability, as we use the term here, refers to the potential energy of a particular nucleus as compared with the sum of the potential energies of its component protons and neutrons. We will use the term *kinetic stability* to describe the probability that a nucleus will undergo decomposition to form a different nucleus—a process called radioactive decay. We will consider radioactivity in this section.

Many nuclei are radioactive; that is, they decompose, forming another nucleus and producing one or more particles. An example is carbon-14, which decays as follows:

$$^{14}_{6}C \longrightarrow {}^{14}_{7}N + {}^{0}_{-1}e$$

where ${}^{0}_{-1}e$ represents an electron, which is called a **beta particle, or β particle,** in nuclear terminology. This equation is typical of those representing radioactive decay in that both A and Z must be conserved. That is, the Z values must give the same sum on both sides of the equation ($6 = 7 - 1$), as must the A values ($14 = 14 + 0$).

Of the approximately 2000 known nuclides, only 279 are stable with respect to radioactive decay. Tin has the largest number of stable isotopes—10.

It is instructive to examine how the numbers of neutrons and protons in a nucleus are related to its stability with respect to radioactive decay. Figure 21.1 shows a plot of the positions of the stable nuclei as a function of the number of protons (Z) and the number of neutrons ($A - Z$). The stable nuclides are said to reside in the **zone of stability.**

The following are some important observations concerning radioactive decay:

All nuclides with 84 or more protons are unstable with respect to radioactive decay.

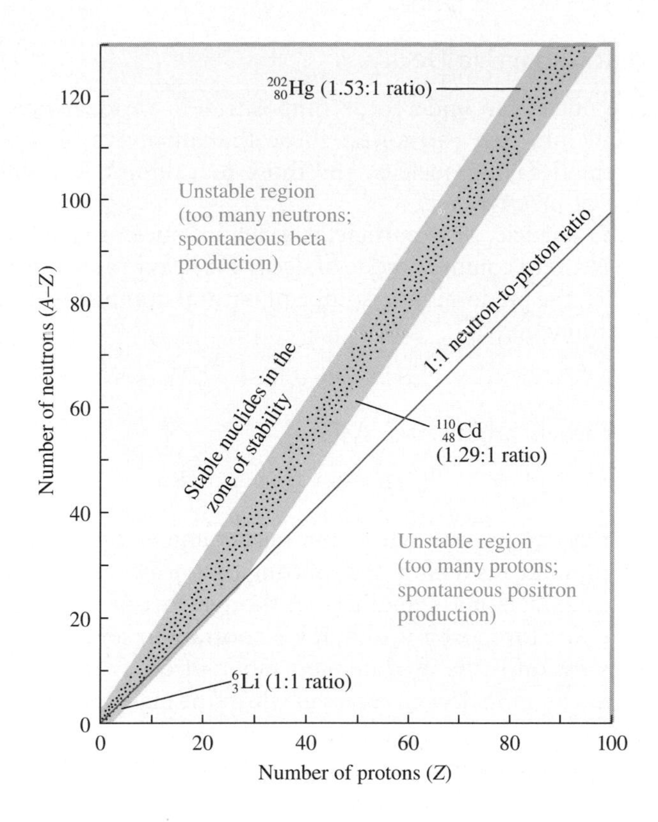

Figure 21.1
The zone of stability. The dots indicate the nuclides that *do not* undergo radioactive decay. Note that as the number of protons in a nuclide increases, the neutron/proton ratio required for stability also increases.

Light nuclides are stable when Z equals $A - Z$, that is, when the neutron/proton ratio is 1. However, for heavier elements the neutron/proton ratio required for stability is greater than 1 and increases with Z.

Certain combinations of protons and neutrons seem to confer special stability. For example, nuclides with even numbers of protons and neutrons are more often stable than those with odd numbers, as shown by the data in Table 21.1.

There are certain specific numbers of protons or neutrons that produce especially stable nuclides. These so-called **magic numbers** are 2, 8, 20, 28, 50, 82, and 126. This behavior is reminiscent of that for atoms, where certain numbers of electrons (2, 10, 18, 36, 54, and 86) produce special chemical stability (the noble gases).

TABLE 21.1 **Number of Stable Nuclides Related to Numbers of Protons and Neutrons**

Number of Protons	Number of Neutrons	Number of Stable Nuclides	Examples
Even	Even	168	$^{12}_{6}C$, $^{16}_{8}O$
Even	Odd	57	$^{13}_{6}C$, $^{47}_{22}Ti$
Odd	Even	50	$^{19}_{9}F$, $^{23}_{11}Na$
Odd	Odd	4	$^{2}_{1}H$, $^{6}_{3}Li$

Note: Even numbers of protons and neutrons seem to favor stability.

Types of Radioactive Decay

Radioactive nuclei can undergo decomposition in various ways. These decay processes fall into two categories: those that involve a change in the mass number of the decaying nucleus, and those that do not. We will consider the former type of process first.

α-particle production involves a change in A for the decaying nucleus; β-particle production has no effect on A.

An **alpha particle, or α particle,** is a helium nucleus (^4_2He). **Alpha-particle production** is a very common mode of decay for heavy radioactive nuclides. For example, $^{238}_{92}U$, the predominant isotope of natural uranium (99.3%), decays by α-particle production:

$$^{238}_{92}U \longrightarrow {}^4_2He + {}^{234}_{90}Th$$

Another α-particle producer is $^{230}_{90}Th$:

$$^{230}_{90}Th \longrightarrow {}^4_2He + {}^{226}_{88}Ra$$

Another decay process where the mass number of the decaying nucleus changes is **spontaneous fission,** the splitting of a heavy nuclide into two lighter nuclides with similar mass numbers. Although this process occurs at an extremely slow rate for most nuclides, it is important in some cases. For instance, spontaneous fission is the predominant mode of decay for $^{254}_{98}Cf$.

The most common decay process in which the mass number of the decaying nucleus remains constant is **β-particle production.** For example, the thorium-234 nuclide produces a β particle and is converted to protactinium-234:

$$^{234}_{90}Th \longrightarrow {}^{234}_{91}Pa + {}^{0}_{-1}e$$

Iodine-131 is also a β-particle producer:

$$^{131}_{53}I \longrightarrow {}^{0}_{-1}e + {}^{131}_{54}Xe$$

The β particle is assigned the mass number 0, since its mass is tiny compared with that of a proton or neutron. Because the value of Z is -1 for the β particle, the atomic number for the new nuclide is greater by 1 than for the original nuclide. Thus *the net effect of β-particle production is to change a neutron to a proton.* We therefore expect nuclides that lie above the zone of stability (those nuclides whose neutron/proton ratios are too high) to be β-particle producers.

It should be pointed out that although the β particle is an electron, the emitting nucleus does not contain electrons. As we shall see later in this chapter, a given quantity of energy (which is best regarded as a form of matter) can become a particle (another form of matter) under certain circumstances. The unstable nuclide creates an electron as it releases energy in the decay process. The electron thus results from the decay process, rather than being present before the decay occurs. Think of this process as being analogous to talking: words are not stored inside us but are formed as we speak. Later in this chapter we will discuss in more detail this very interesting phenomenon in which matter in the form of particles and matter in the form of energy can interchange.

A **gamma ray, or γ ray,** refers to a high-energy photon. Frequently, γ-ray production accompanies nuclear decays and particle reactions, such as in the α-particle decay of $^{238}_{92}U$:

$$^{238}_{92}U \longrightarrow {}^4_2He + {}^{234}_{90}Th + 2\,{}^0_0\gamma$$

in which two γ rays of different energies are sometimes produced in addition to the α particle. The emission of γ rays is one way a nucleus with excess energy (in an excited nuclear state) can relax to its ground state.

Positron production occurs for nuclides below the zone of stability (those nuclides whose neutron/proton ratios are too small). The positron is a particle with the same mass as the electron but opposite charge. An example of a nuclide that decays by positron production is sodium-22:

$$^{22}_{11}\text{Na} \longrightarrow {}^{0}_{1}\text{e} + {}^{22}_{10}\text{Ne}$$

Note that *the net effect of this process is to change a proton to a neutron,* causing the product nuclide to have a higher neutron/proton ratio than the original nuclide.

Besides being oppositely charged, the positron shows an even more fundamental difference from the electron: it is the **antiparticle** of the electron. When a positron collides with an electron, the particulate matter is changed to electromagnetic radiation in the form of high-energy photons:

$$^{0}_{-1}\text{e} + {}^{0}_{1}\text{e} \longrightarrow {}^{0}_{0}\gamma$$

This process, which is characteristic of matter-antimatter collisions, is called *annihilation* and is another example of the interchange of the forms of matter.

Electron capture is a process in which one of the inner-orbital electrons is captured by the nucleus, as illustrated by the process

$$^{201}_{80}\text{Hg} + \underset{\substack{\uparrow \\ \text{Inner-orbital electron}}}{{}^{0}_{-1}\text{e}} \longrightarrow {}^{201}_{79}\text{Au} + {}^{0}_{0}\gamma$$

This reaction would have been of great interest to the alchemists, but unfortunately, it does not occur at a rate that would make it a practical means for changing mercury to gold. The various types of radioactive decay are summarized in Table 21.2.

Often a radioactive nucleus cannot reach a stable state through a single decay process. In such a case a **decay series** occurs until a stable nuclide is formed. A well-known example is the decay series that starts with $^{238}_{92}\text{U}$ and ends with $^{206}_{82}\text{Pb}$, as shown in Fig. 21.2. Similar series exist for $^{235}_{92}\text{U}$:

$$^{235}_{92}\text{U} \xrightarrow[\text{decays}]{\text{Series of}} {}^{207}_{82}\text{Pb}$$

and for $^{232}_{90}\text{Th}$:

$$^{232}_{90}\text{Th} \xrightarrow[\text{decays}]{\text{Series of}} {}^{208}_{82}\text{Pb}$$

TABLE 21.2 Various Types of Radioactive Processes Showing the Changes That Take Place in the Nuclides

Process	Change in A	Change in Z	Change in Neutron/Proton Ratio	Example
β-particle (electron) production	0	+1	Decrease	$^{227}_{89}\text{Ac} \longrightarrow {}^{227}_{90}\text{Th} + {}^{0}_{-1}\text{e}$
Positron production	0	−1	Increase	$^{13}_{7}\text{N} \longrightarrow {}^{13}_{6}\text{C} + {}^{0}_{1}\text{e}$
Electron capture	0	−1	Increase	$^{73}_{33}\text{As} + {}^{0}_{-1}\text{e} \longrightarrow {}^{73}_{32}\text{Ge}$
α-particle production	−4	−2	Increase	$^{210}_{84}\text{Po} \longrightarrow {}^{206}_{82}\text{Pb} + {}^{4}_{2}\text{He}$
γ-ray production	0	0	—	Excited nucleus $\longrightarrow$ ground-state nucleus + $^{0}_{0}\gamma$
Spontaneous fission	—	—	—	$^{254}_{98}\text{Cf} \longrightarrow$ lighter nuclides + neutrons

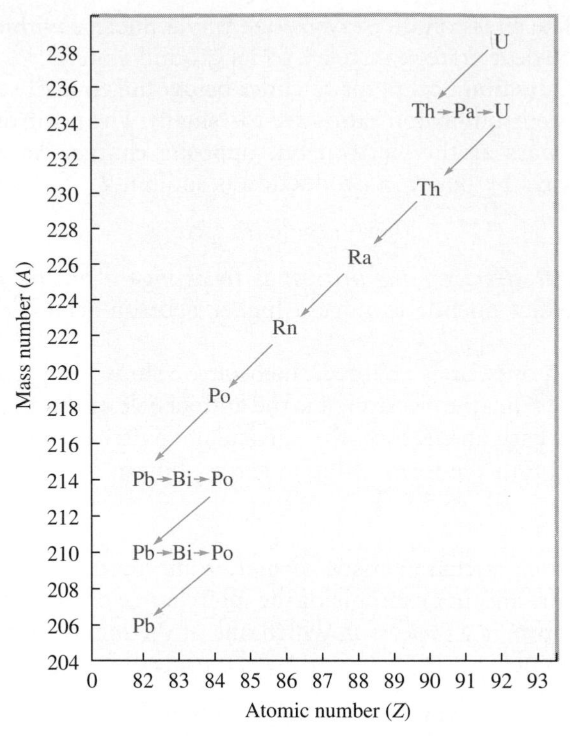

Figure 21.2
The decay series from $^{238}_{92}U$ to $^{206}_{82}Pb$. Each nuclide in the series except $^{206}_{82}Pb$ is unstable. Thus the successive transformations (shown by the arrows) continue until $^{206}_{82}Pb$ is finally formed. Note that horizontal arrows indicate processes in which A is unchanged, and diagonal arrows signify that both A and Z change.

21.2 The Kinetics of Radioactive Decay

Rates of reaction are discussed in Chapter 15.

In a sample consisting of radioactive nuclides of a given type, each nuclide has a certain probability of undergoing decay. Suppose that a sample containing 1000 atoms of a certain nuclide produces 10 decay events per hour. This means that over the span of an hour, 1 out of every 100 nuclides will decay. Given that this probability of decay is characteristic for this type of nuclide, we could predict that a 2000-atom sample would give 20 decay events per hour. Thus for radioactive nuclides the **rate of decay,** which is the negative of the change in the number of nuclides per unit time ($-\Delta N/\Delta t$), is directly proportional to the number of nuclides (N) in a given sample:

$$\text{Rate} = -\frac{\Delta N}{\Delta t} \propto N$$

The negative sign is included because the number of nuclides is decreasing. We now insert a proportionality constant k to give

$$\text{Rate} = -\frac{\Delta N}{\Delta t} = -\frac{dN}{dt} = kN$$

Recall from Chapter 15 that this is the rate law for a first-order process. The integrated first-order rate law is

$$\ln\left(\frac{N}{N_0}\right) = -kt$$

where N_0 represents the original number of nuclides (at $t = 0$) and N represents the number of nuclides *remaining* at time t.

Half-Life

The **half-life** ($t_{1/2}$) of a radioactive sample is defined as the time required for the number of nuclides to reach half of the original value ($N_0/2$). We can use this definition in connection with the integrated first-order rate law (see Section 15.4) to produce the following expression for $t_{1/2}$:

$$t_{1/2} = \frac{\ln(2)}{k} = \frac{0.693}{k}$$

Thus if the half-life of a radioactive nuclide is known, the rate constant can be easily calculated, and vice versa.

As we saw in Section 15.4, the half-life for a first-order process is constant. This is illustrated for the β-particle decay of strontium-90 in Fig. 21.3; it takes 28.8 years for each halving of the amount of $^{90}_{38}Sr$. Contamination of the environment with $^{90}_{38}Sr$ poses serious health hazards because of the similar chemistry of strontium and calcium (both are in Group 2A). Strontium-90 in grass and hay is incorporated into cow's milk along with calcium and is then passed on to humans, where it lodges in the bones. Because of its relatively long half-life, ^{90}Sr persists for years in humans, causing radiation damage that may lead to cancer.

The harmful effects of radiation will be discussed in Section 21.7.

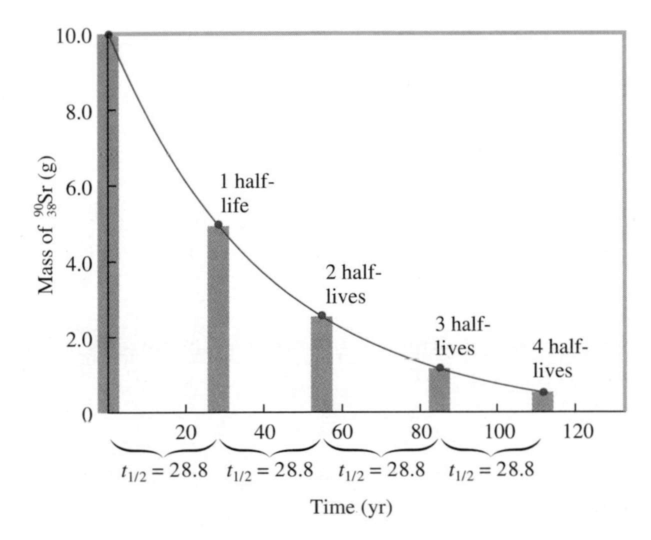

Figure 21.3
The decay of a 10.0-g sample of strontium-90 over time. Note that the half-life is a constant 28.8 years.

Stellar Nucleosynthesis

How did all of the matter around us originate? The scientific answer to this question is a theory called *stellar nucleosynthesis,* or literally, the formation of nuclei in the stars.

Many scientists believe that our universe originated as a cloud of neutrons that became unstable and produced an immense explosion, giving this model its name—*the big bang theory*. The model postulates that following the initial explosion, neutrons decomposed into protons and electrons,

$$^{1}_{0}n \longrightarrow ^{1}_{1}H + ^{0}_{-1}e$$

which eventually recombined to form clouds of hydrogen. Over the eons gravitational forces caused many of these hydrogen clouds to contract and heat up sufficiently to reach temperatures where proton fusion was possible, releasing large quantities of energy. When the tendency to expand due to the heat from fusion and the tendency to contract due to the forces of gravity are balanced, a stable young star such as our sun can be formed.

Eventually, when the supply of hydrogen is exhausted, the core of the star again contracts with further heating until temperatures are reached where fusion of helium nuclei can occur. This leads to the formation of $^{12}_{6}C$ and $^{16}_{8}O$ nuclei. In turn, when the supply of helium nuclei is depleted, further contraction and heating occurs, until the fusion of heavier nuclei takes place. This process occurs repeatedly, forming heavier and heavier nuclei until iron nuclei are formed. Because the iron nucleus is the most stable of all, energy is required to fuse iron nuclei. This endothermic fusion process cannot furnish energy to sustain the star; therefore it cools to a small, dense *white dwarf*.

The evolution just described is characteristic of small and medium-sized stars. Much larger stars, however, become unstable at some time during their evolution and undergo a *supernova explosion*. In this explosion, some medium-mass nuclei are fused to form heavy elements. Also, some light nuclei capture neutrons. These neutron-rich nuclei then produce β particles, increasing their atomic number with each event. This eventually leads to nuclei with large atomic numbers. In fact, almost all nuclei beyond iron are thought to originate from supernova explosions. The debris of a supernova explosion thus contains a large variety of elements and might eventually form a solar system such as our own.

Although other theories for the origin of matter have been suggested, there is much evidence to support the big bang theory. It continues to be widely accepted, though the details continue to be modified. For more information, see "Formation of the Chemical Elements and the Evolution of our Universe" by V. E. Viola [*J. Chem. Ed.* **67** (1990), p. 723].

EXAMPLE 21.1

The half-life of molybdenum-99 is 67.0 h. How much of a 1.000-mg sample of $^{99}_{42}Mo$ remains after 335 h?

Solution

The easiest way to solve this problem is to recognize that 335 h represents five half-lives for $^{99}_{42}Mo$:

$$335 = 5 \times 67.0$$

We can sketch the change that occurs, as is shown in Fig. 21.4. Thus after 335 h, 0.031 mg of $^{99}_{42}Mo$ remains.

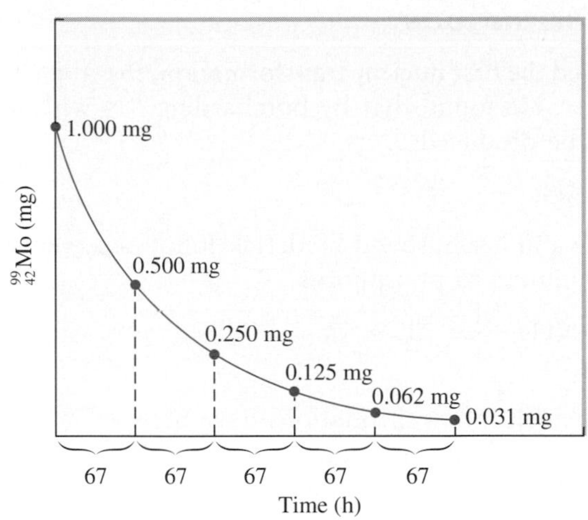

Figure 21.4
The change in the amount of $^{99}_{42}Mo$ with time ($t_{1/2}$ = 67 h).

The half-lives of radioactive nuclides vary over a tremendous range. For example, $^{144}_{60}Nd$ has a half-life of 5×10^{15} years, while $^{214}_{84}Po$ has a half-life of 2×10^{-4} second. To give perspective on this, the half-lives of the nuclides in the $^{238}_{92}U$ decay series are given in Table 21.3.

TABLE 21.3 The Half-Lives of Nuclides in the $^{238}_{92}U$ Decay Series

Nuclide	Particle Produced	Half-Life
Uranium-238 ($^{238}_{92}U$) ↓	α	4.51×10^9 years
Thorium-234 ($^{234}_{90}Th$) ↓	β	24.1 days
Protactinium-234 ($^{234}_{91}Pa$) ↓	β	6.75 hours
Uranium-234 ($^{234}_{92}U$) ↓	α	2.48×10^5 years
Thorium-230 ($^{230}_{90}Th$) ↓	α	8.0×10^4 years
Radium-226 ($^{226}_{88}Ra$) ↓	α	1.62×10^3 years
Radon-222 ($^{222}_{86}Rn$) ↓	α	3.82 days
Polonium-218 ($^{218}_{84}Po$) ↓	α	3.1 minutes
Lead-214 ($^{214}_{82}Pb$) ↓	β	26.8 minutes
Bismuth-214 ($^{214}_{83}Bi$) ↓	β	19.7 minutes
Polonium-214 ($^{214}_{84}Po$) ↓	α	1.6×10^{-4} second
Lead-210 ($^{210}_{82}Pb$) ↓	β	20.4 years
Bismuth-210 ($^{210}_{83}Bi$) ↓	β	5.0 days
Polonium-210 ($^{210}_{84}Po$) ↓	α	138.4 days
Lead-206 ($^{206}_{82}Pb$)	—	Stable

21.3 Nuclear Transformations

In 1919 Lord Rutherford observed the first **nuclear transformation,** *the conversion of one element into another.* He found that by bombarding $^{14}_{7}N$ with α particles, the $^{17}_{8}O$ nuclide could be produced:

$$^{14}_{7}N + ^{4}_{2}He \longrightarrow ^{17}_{8}O + ^{1}_{1}H$$

Fourteen years later, Irene Curie and her husband Frederick Joliot observed a similar transformation from aluminum to phosphorus,

$$^{27}_{13}Al + ^{4}_{2}He \longrightarrow ^{30}_{15}P + ^{1}_{0}n$$

where $^{1}_{0}n$ represents a neutron.

Over the years, many other nuclear transformations have been achieved, mostly using **particle accelerators,** which, as the name reveals, are devices used to give particles very high velocities. Because of the electrostatic repulsion between the target nucleus and a positive ion, accelerators are needed when positive ions are used as bombarding particles. The particle, accelerated to a very high velocity, can overcome the repulsion and penetrate the target nucleus, thus effecting the transformation. A schematic diagram of one type of particle accelerator, the **cyclotron,** is shown in Fig. 21.5. The ion is introduced at the center of the cyclotron and is accelerated in an expanding spiral path by use of alternating electric fields in the presence of a magnetic field. The **linear accelerator** illustrated in Fig. 21.6 employs changing electrical fields to achieve high velocities along a linear pathway.

In addition to positive ions, neutrons are often employed as bombarding particles to effect nuclear transformations. Because neutrons are uncharged and thus not repelled electrostatically by a target nucleus, they are readily absorbed by many nuclei, leading to the formation of new nuclides. The most common source of neutrons for this purpose is a fission reactor (see Section 21.6).

By using neutron and positive-ion bombardment, scientists have been able to extend the periodic table. Prior to 1940 the heaviest known element was uranium ($Z = 92$). However, in 1940 neptunium ($Z = 93$) was produced by neutron bombardment of $^{238}_{92}U$. The process initially gives $^{239}_{92}U$, which then decays to $^{239}_{93}Np$ by β-particle production:

$$^{238}_{92}U + ^{1}_{0}n \longrightarrow ^{239}_{92}U \xrightarrow[t_{1/2} = 23 \text{ min}]{} ^{239}_{93}Np + ^{0}_{-1}e$$

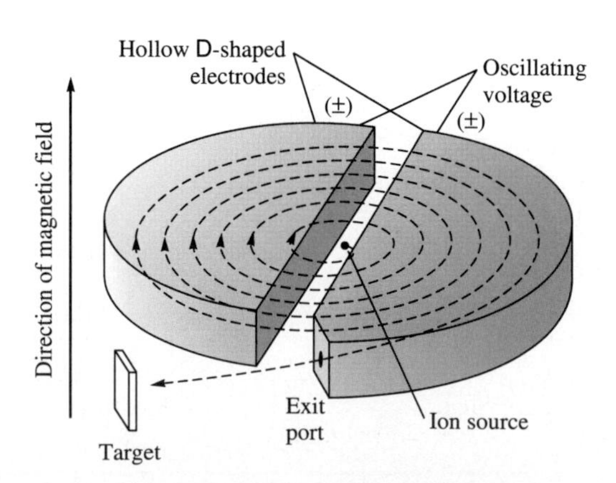

Figure 21.5
A schematic diagram of a cyclotron. The ion is introduced in the center and is then pulled back and forth between the hollow D-shaped electrodes by constant reversals of the electric field. Magnets above and below these electrodes produce a spiral path that expands as the particle velocity increases. When the particle has sufficient speed, it exits the accelerator and is directed at the target nucleus.

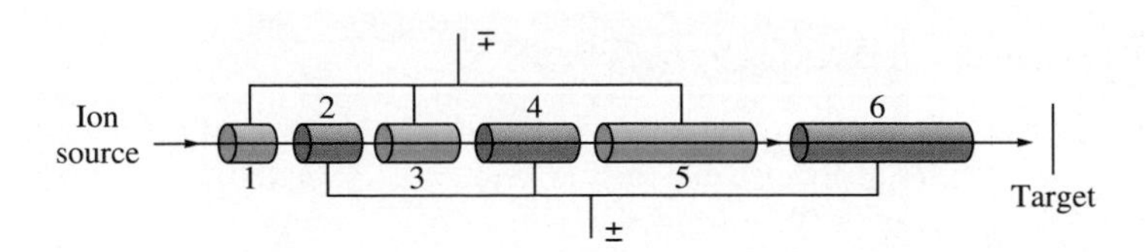

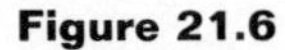
Figure 21.6
Schematic diagram of a linear accelerator, which uses a changing electric field to accelerate a positive ion along a linear path. As the ion leaves the source, the odd-numbered tubes are negatively charged, and the even-numbered tubes are positively charged. The positive ion is thus attracted into tube 1. As the ion leaves tube 1, the tube polarities are reversed. Now tube 1 is positive, repelling the positive ion; and tube 2 is negative, attracting the positive ion. This process continues, eventually producing high particle velocity.

In the years since 1940 the elements with atomic numbers 93 through 106, called the **transuranium elements,*** have been synthesized. Many of these elements have very short half-lives, as shown in Table 21.4. As a result, only a few atoms of some of the transuranium elements have ever been formed. This, of course, makes the chemical characterization of these elements extremely difficult.

21.4 Detection and Uses of Radioactivity

Various instruments measure radioactivity levels, the most familiar being the **Geiger-Müller counter,** or **Geiger counter** (see Fig. 21.7). This instrument takes advantage of the fact that the high-energy particles from radioactive decay processes produce ions when they travel through matter. The probe of the Geiger counter is filled with argon gas, which can be ionized by a rapidly moving particle:

Geiger counters are often called survey meters in industry.

$$\mathrm{Ar}(g) \xrightarrow[\text{particle}]{\text{High-energy}} \mathrm{Ar}^+(g) + \mathrm{e}^-$$

TABLE 21.4 Syntheses of Some of the Transuranium Elements

Element	Neutron Bombardment	Half-Life
Neptunium ($Z = 93$)	${}^{238}_{92}\mathrm{U} + {}^{1}_{0}\mathrm{n} \longrightarrow {}^{239}_{92}\mathrm{U} \longrightarrow {}^{239}_{93}\mathrm{Np} + {}^{0}_{-1}\mathrm{e}$	2.35 days (${}^{239}_{93}\mathrm{Np}$)
Plutonium ($Z = 94$)	${}^{239}_{93}\mathrm{Np} \longrightarrow {}^{239}_{94}\mathrm{Pu} + {}^{0}_{-1}\mathrm{e}$	24,400 years (${}^{239}_{94}\mathrm{Pu}$)
Americium ($Z = 95$)	${}^{239}_{94}\mathrm{Pu} + 2\ {}^{1}_{0}\mathrm{n} \longrightarrow {}^{241}_{94}\mathrm{Pu} \longrightarrow {}^{241}_{95}\mathrm{Am} + {}^{0}_{-1}\mathrm{e}$	458 years (${}^{241}_{95}\mathrm{Am}$)

Element	Positive-Ion Bombardment	Half-Life
Curium ($Z = 96$)	${}^{239}_{94}\mathrm{Pu} + {}^{4}_{2}\mathrm{He} \longrightarrow {}^{242}_{96}\mathrm{Cm} + {}^{1}_{0}\mathrm{n}$	163 days (${}^{242}_{96}\mathrm{Cm}$)
Californium ($Z = 98$)	${}^{242}_{96}\mathrm{Cm} + {}^{4}_{2}\mathrm{He} \longrightarrow {}^{245}_{98}\mathrm{Cf} + {}^{1}_{0}\mathrm{n}$ or ${}^{238}_{92}\mathrm{U} + {}^{12}_{6}\mathrm{C} \longrightarrow {}^{246}_{98}\mathrm{Cf} + 4\ {}^{1}_{0}\mathrm{n}$	44 minutes (${}^{245}_{98}\mathrm{Cf}$)
Element 104	${}^{249}_{98}\mathrm{Cf} + {}^{12}_{6}\mathrm{C} \longrightarrow {}^{257}_{104}\mathrm{X} + 4\ {}^{1}_{0}\mathrm{n}$	
Element 105	${}^{249}_{98}\mathrm{Cf} + {}^{15}_{7}\mathrm{N} \longrightarrow {}^{260}_{105}\mathrm{X} + 4\ {}^{1}_{0}\mathrm{n}$	
Element 106	${}^{249}_{98}\mathrm{Cf} + {}^{18}_{8}\mathrm{O} \longrightarrow {}^{263}_{106}\mathrm{X} + 4\ {}^{1}_{0}\mathrm{n}$	

* For an excellent article on the history of these elements, see G. B. Kauffman, "Beyond Uranium," *Chemical and Engineering News,* November 19, 1990, p. 18.

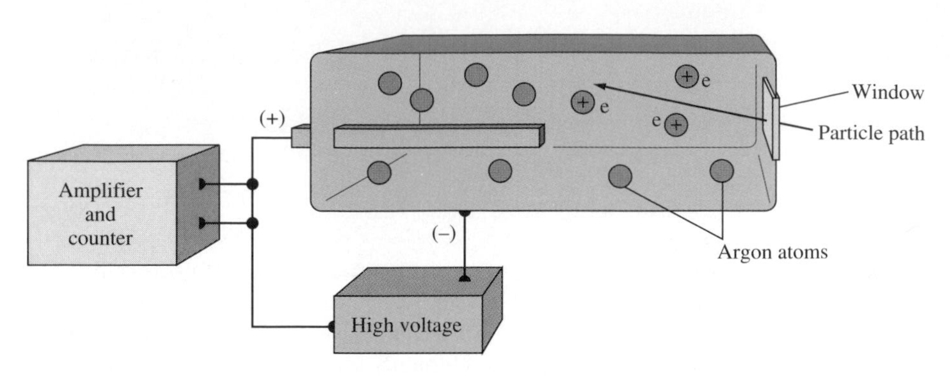

Figure 21.7
A schematic representation of a Geiger-Müller counter. The high-energy radioactive particle enters the window and ionizes argon atoms along its path. The resulting ions and electrons produce a momentary current pulse, which is amplified and counted.

Normally, a sample of argon gas does not conduct a current when an electrical potential is applied. However, the formation of ions and electrons produced by the passage of the high-energy particle allows a momentary current to flow. Electronic devices detect this current flow, so the number of these events can be counted. Thus the decay rate of the radioactive sample can be determined.

Another instrument often used to detect levels of radioactivity is a **scintillation counter,** which takes advantage of the fact that certain substances, such as zinc sulfide, give off light when they are struck by high-energy radiation. A photocell senses the flashes of light that occur as the radiation strikes and thus measures the number of decay events per unit time.

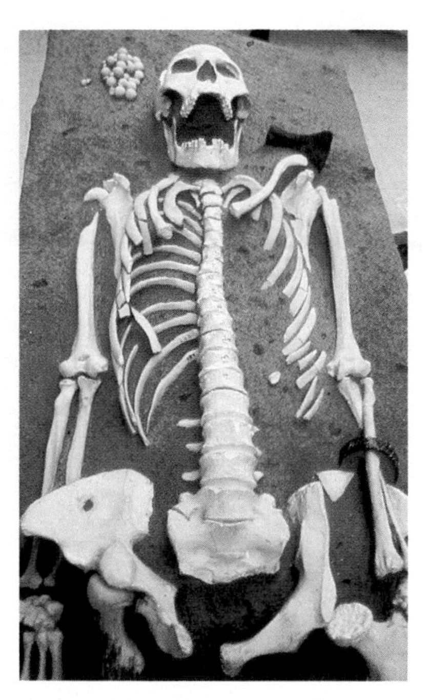

Carbon-14 radioactivity is often used to date human skeletons found at archeological sites.

Dating by Radioactivity

Archaeologists, geologists, and others involved in reconstructing the ancient history of the earth rely heavily on radioactivity to provide accurate dates for artifacts and rocks. A method that has been very important for dating ancient articles made from wood or cloth is **radiocarbon dating,** or **carbon-14 dating.** This technique was originated in the 1940s by Willard Libby, an American chemist who received a Nobel prize for his efforts in this field.

Radiocarbon dating is based on the radioactivity of the nuclide ${}^{14}_{6}C$, which decays via β-particle production:

$$ {}^{14}_{6}C \longrightarrow {}^{0}_{-1}e + {}^{14}_{7}N $$

Carbon-14 is continuously produced in the atmosphere when high-energy neutrons from space collide with nitrogen-14:

$$ {}^{14}_{7}N + {}^{1}_{0}n \longrightarrow {}^{14}_{6}C + {}^{1}_{1}H $$

Thus carbon-14 is continuously produced by this process, and it continuously decomposes through β-particle production. Over the years the rates for these two processes have become equal; and like a participant in a chemical reaction at equilibrium, the amount of ${}^{14}_{6}C$ that is present in the atmosphere remains approximately constant.

Carbon-14 can be used to date wood and cloth artifacts because the ${}^{14}_{6}C$, along with the other carbon isotopes in the atmosphere, reacts with oxygen to form carbon dioxide. A living plant consumes carbon dioxide in the photosynthesis process and incorporates the carbon, including ${}^{14}_{6}C$, into its molecules. As long as the plant lives, the ${}^{14}_{6}C/{}^{12}_{6}C$ ratio in its molecules remains the same as that in the atmosphere because of the continuous uptake of carbon. However, as soon as a tree is cut to make a wooden bowl or a flax plant is harvested to make linen, the $^{14}_{6}C/^{12}_{6}C$ ratio begins to decrease because of the radioactive decay of $^{14}_{6}C$ (the $^{12}_{6}C$ nuclide is stable). Since the half-life of $^{14}_{6}C$ is 5730 years, a wooden bowl found in an archaeological dig showing a $^{14}_{6}C/^{12}_{6}C$ ratio half that found in currently living trees must be approximately 5730 years old. This reasoning assumes that the current $^{14}_{6}C/^{12}_{6}C$ ratio is the same as that found in ancient times.

The $^{14}_{6}C/^{12}_{6}C$ ratio is the basis for carbon-14 dating.

Dendrochronologists, scientists who date trees from annual growth rings, have used data collected from long-lived species of trees, such as bristlecone pines and sequoias, to show that the $^{14}_{6}C$ content of the atmosphere has changed significantly over the ages. These data have been used to derive correction factors that allow very accurate dates to be determined from the observed $^{14}_{6}C/^{12}_{6}C$ ratio in an artifact, especially for artifacts up to 10,000 years old. Recent measurements of uranium/thorium ratios in ancient coral have raised questions about the accuracy of ^{14}C dates extending back 20,000 to 30,000 years, suggesting that errors up to 3000 years may have occurred. Efforts are now being made to recalibrate the ^{14}C dates over this period.

One drawback of radiocarbon dating is that a fairly large piece of the object (up to several grams) must be burned to form carbon dioxide, which is then analyzed for radioactivity. Another method for counting $^{14}_{6}C$ nuclides avoids destruction of a significant portion of a valuable artifact. This technique, requiring only about 10^{-3} gram, uses a mass spectrometer (Chapter 3), in which the carbon atoms are ionized and accelerated through a magnetic field that deflects their paths. Because of their different masses, the various ions are deflected by different amounts and can be counted separately. This allows a very accurate determination of the $^{14}_{6}C/^{12}_{6}C$ ratio in the sample.

Because the half-life of $^{238}_{92}U$ is very long compared with those of the other members of the decay series (Table 21.3) to reach $^{206}_{82}Pb$, the number of nuclides in intermediate stages of decay is negligible. That is, once a $^{238}_{92}U$ nuclide starts to decay, it reaches $^{206}_{82}Pb$ relatively fast.

In their attempts to establish the geological history of the earth, geologists have made extensive use of radioactivity. For example, since $^{238}_{92}U$ decays to the stable $^{206}_{82}Pb$ nuclide, the ratio of $^{206}_{82}Pb$ to $^{238}_{92}U$ in a rock can, under favorable circumstances, be used to estimate the age of the rock.

EXAMPLE 21.2

A rock containing $^{238}_{92}U$ and $^{206}_{82}Pb$ was examined to determine its approximate age. Analysis showed the ratio of $^{206}_{82}Pb$ atoms to $^{238}_{92}U$ atoms to be 0.115. Assuming that no lead was originally present, that all the $^{206}_{82}Pb$ formed over the years has remained in the rock, and that the number of nuclides in the intermediate stages of decay between $^{238}_{92}U$ and $^{206}_{82}Pb$ is negligible, calculate the age of the rock. The half-life of $^{238}_{92}U$ is 4.5×10^9 yr.

Solution

The problem can be solved using the integrated first-order rate law:

$$\ln\left(\frac{N}{N_0}\right) = -kt = -\left(\frac{0.693}{4.5 \times 10^9 \text{ yr}}\right)t$$

where N/N_0 represents the ratio of $^{238}_{92}U$ atoms now found in the rock to the number present when the rock was formed. We are assuming that each $^{206}_{82}Pb$ nuclide present must have come from decay of a $^{238}_{92}U$:

$$^{238}_{92}U \longrightarrow {}^{206}_{82}Pb$$

Thus

$$\boxed{\begin{array}{c}\text{Number of } ^{238}_{92}\text{U atoms}\\ \text{originally present}\end{array}} = \boxed{\begin{array}{c}\text{number of } ^{206}_{82}\text{Pb}\\ \text{atoms now present}\end{array}} + \boxed{\begin{array}{c}\text{number of } ^{238}_{92}\text{U}\\ \text{atoms now present}\end{array}}$$

$$\frac{\text{Atoms of } ^{206}_{82}\text{Pb now present}}{\text{Atoms of } ^{238}_{92}\text{U now present}} = 0.115 = \frac{0.115}{1.000} = \frac{115}{1000}$$

Think carefully about what this means. For every 1115 $^{238}_{92}U$ atoms originally present in the rock, 115 have been changed to $^{206}_{82}Pb$ and 1000 remain as $^{238}_{92}U$. Thus

$$\frac{N}{N_0} = \frac{\overset{\text{Now present}}{^{238}_{92}\text{U}}}{\underbrace{^{206}_{82}\text{Pb} + ^{238}_{92}\text{U}}_{^{238}_{92}\text{U originally present}}} = \frac{1000}{1115} = 0.8969$$

$$\ln\left(\frac{N}{N_0}\right) = \ln(0.8969) = -\left(\frac{0.693}{4.5 \times 10^9 \text{ yr}}\right)t$$

$$t = 7.1 \times 10^8 \text{ yr}$$

This is the approximate age of the rock. It was formed sometime during the Cambrian period.

Medical Applications of Radioactivity

Although the rapid advances of the medical sciences in recent decades are due to many causes, one of the most important has been the discovery and use of **radiotracers,** which are radioactive nuclides that can be introduced into organisms through food or drugs and whose pathways can be *traced* by monitoring their radioactivity (see Fig. 21.8). For example, the incorporation of nuclides such as $^{14}_{6}C$ and $^{32}_{15}P$ into nutrients has produced important information about metabolic pathways.

Iodine-131 has proved very useful in the diagnosis and treatment of illnesses of the thyroid gland. Patients drink a solution containing small amounts of $Na^{131}I$, and the uptake of the iodine by the thyroid gland is monitored with a scanner.

The *m* in Technetium-99m designates an excited nuclear state of ^{99}Tc that decays to the ground state by γ production:

$$^{99m}\text{Tc} \longrightarrow {}^{99}\text{Tc} + \gamma$$

Thallium-201 can be used to assess the damage to the heart muscle in a person who has suffered a heart attack, because thallium is concentrated in healthy muscle tissue. Technetium-99m is also taken up by normal heart tissue and is used for damage assessment in a similar way.

Radiotracers provide sensitive and noninvasive methods for learning about biological systems, for detection of disease, for monitoring the action and effectiveness of drugs, and for early detection of pregnancy; their usefulness should continue to grow. Some useful radiotracers are listed in Table 21.5.

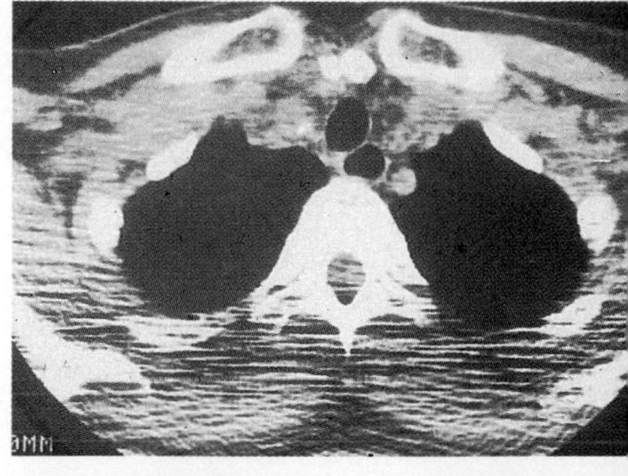

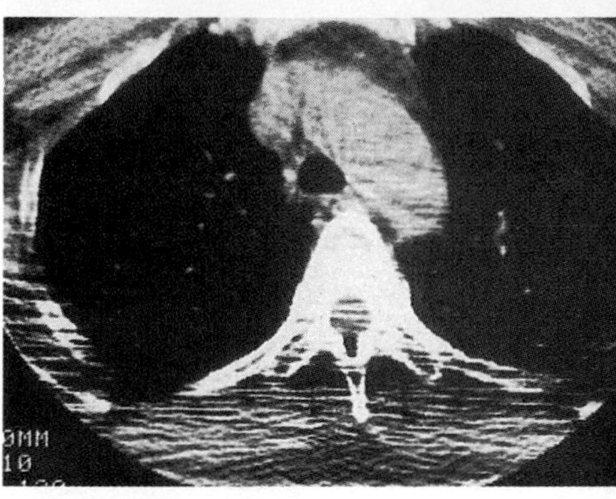

Figure 21.8
A radioisotopic scan showing normal (top) and enlarged (bottom) thyroids.

TABLE 21.5 **Some Radioactive Nuclides, with Half-Lives and Medical Applications as Radiotracers**

Nuclide	Half-Life	Area of the Body Studied
^{131}I	8.1 days	Thyroid
^{59}Fe	45.1 days	Red blood cells
^{99}Mo	67 hours	Metabolism
^{32}P	14.3 days	Eyes, liver, tumors
^{51}Cr	27.8 days	Red blood cells
^{87}Sr	2.8 hours	Bones
^{99}Tc	6.0 hours	Heart, bones, liver, lungs
^{133}Xe	5.3 days	Lungs
^{24}Na	14.8 hours	Circulatory system

21.5 Thermodynamic Stability of the Nucleus

We can determine the thermodynamic stability of a nucleus by calculating the change in potential energy that would occur if that nucleus were formed from its constituent protons and neutrons. For example, let's consider the hypothetical process of forming a $^{16}_{8}O$ nucleus from eight neutrons and eight protons:

$$8\ ^{1}_{0}n + 8\ ^{1}_{1}H \longrightarrow\ ^{16}_{8}O$$

The energy change associated with this process can be calculated by comparing the sum of the masses of eight protons and eight neutrons with the mass of the oxygen nucleus:

$$\text{Mass of } (8\ ^{1}_{0}n + 8\ ^{1}_{1}H) = \underbrace{8(1.67493 \times 10^{-24}\text{ g})}_{\text{Mass of } ^{1}_{0}n} + \underbrace{8(1.67262 \times 10^{-24}\text{ g})}_{\text{Mass of } ^{1}_{1}H}$$

$$= 2.67804 \times 10^{-23}\text{ g}$$

$$\text{Mass of } ^{16}_{8}O \text{ nucleus} = 2.65535 \times 10^{-23}\text{ g}$$

The difference in mass for one nucleus is

$$\text{Mass of } ^{16}_{8}O - \text{mass of } (8\ ^{1}_{0}n + 8\ ^{1}_{1}H) = -2.269 \times 10^{-25}\text{ g}$$

The difference in mass for the formation of 1 mole of $^{16}_{8}O$ nuclei is therefore

$$(-2.269 \times 10^{-25}\text{ g/nucleus})(6.022 \times 10^{23}\text{ nuclei/mol}) = -0.1366\text{ g/mol}$$

Thus 0.1366 gram of mass would be lost if 1 mole of oxygen-16 were formed from the constituent protons and neutrons. Why does the mass change, and how can this information be used to calculate the energy change that accompanies this process?

The answers to these questions can be found in the work of Albert Einstein. As we discussed in Section 12.2, Einstein's theory of relativity showed that energy should be considered a form of matter. His famous equation,

Energy is a form of matter.

$$E = mc^2$$

where c is the speed of light, gives the relationship between a quantity of energy

The energy changes associated with normal chemical reactions are small enough that the corresponding mass changes are not detectable.

and its mass. When a system gains or loses energy, it also gains or loses a quantity of mass, given by E/c^2. Thus the mass of a nucleus is less than that of its component nucleons because the process is so exothermic.

Einstein's equation in the form

$$\text{Energy change} = \Delta E = \Delta mc^2$$

where Δm represents the change in mass (the **mass defect**) that can be used to calculate ΔE for the formation of a nucleus from its component nucleons.

The thermodynamic stability of a particular nucleus is normally represented as energy released per nucleon. To illustrate how this quantity is obtained, we will continue to consider $^{16}_{8}O$. First, we calculate ΔE for one mole of $^{16}_{8}O$ nuclei from the equation

$$\Delta E = \Delta mc^2$$

where $c = 3.00 \times 10^8$ m/s

and $\Delta m = -0.1366$ g/mol $= -1.366 \times 10^{-4}$ kg/mol

as calculated above. Thus

$$\Delta E = (-1.366 \times 10^{-4} \text{ kg/mol})(3.00 \times 10^8 \text{ m/s})^2 = -1.23 \times 10^{13} \text{ J/mol}$$

Next, we calculate ΔE per nucleus by dividing the molar value by Avogadro's number

$$\Delta E \text{ per } ^{16}_{8}\text{O nucleus} = \frac{-1.23 \times 10^{13} \text{ J/mol}}{6.022 \times 10^{23} \text{ nuclei/mol}} = -2.04 \times 10^{-11} \text{ J/nucleus}$$

In terms of a more convenient energy unit, the MeV (million electron-volts), where

$$1 \text{ MeV} = 1.60 \times 10^{-13} \text{ J}$$

we have

$$\Delta E \text{ per } ^{16}_{8}\text{O nucleus} = \left(-2.04 \times 10^{-11} \frac{\text{J}}{\text{nucleus}}\right)\left(\frac{1 \text{ MeV}}{1.60 \times 10^{-13} \text{ J}}\right)$$
$$= -1.28 \times 10^2 \text{ MeV/nucleus}$$

Next, we can calculate the value of ΔE per nucleon by dividing by A, the sum of the numbers of neutrons and protons:

$$\Delta E \text{ per nucleon for } ^{16}_{8}\text{O} = \frac{-1.28 \times 10^2 \text{ MeV/nucleus}}{16 \text{ nucleons/nucleus}}$$
$$= -8.00 \text{ MeV/nucleon}$$

This means that 8.00 MeV of energy per nucleon would be *released* if $^{16}_{8}O$ were formed from free neutrons and protons. The energy required to *decompose* this nucleus into its components has the same numeric value but is positive. This is called the **binding energy** per nucleon for $^{16}_{8}O$.

The values of the binding energy per nucleon for the various nuclides are shown in Fig. 21.9. Note that the most stable nuclei (those requiring the largest energy per nucleon to decompose the nucleus) occur at the top of the curve. The most stable nucleus known is $^{56}_{26}Fe$, which has a binding energy per nucleon of 8.79 MeV.

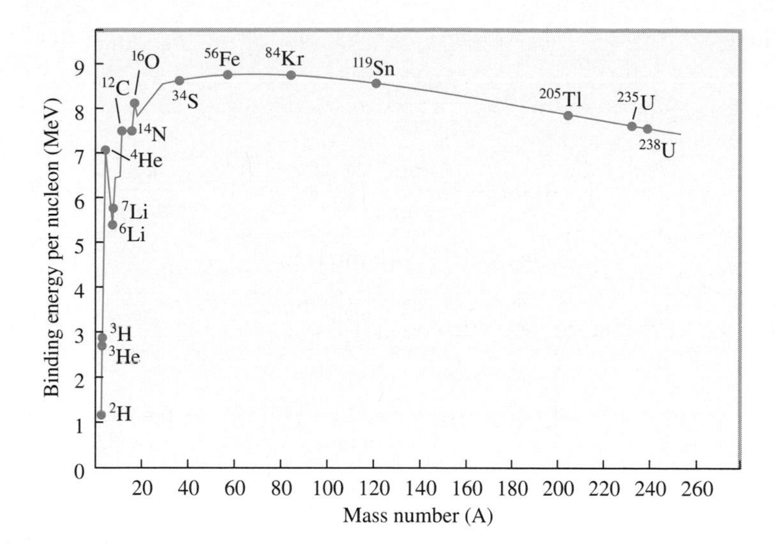

Figure 21.9
The binding energy per nucleon as a function of mass number. The most stable nuclei are at the top of the curve. The most stable nucleus is $^{56}_{26}Fe$.

EXAMPLE 21.3

Calculate the binding energy per nucleon for the $^{4}_{2}He$ nucleus (atomic masses: $^{4}_{2}He$ = 4.0026 amu; $^{1}_{1}H$ = 1.0078 amu).

Solution

First, we must calculate the mass defect (Δm) for $^{4}_{2}He$. Since atomic masses (which include the electrons) are given, we must decide how to account for the electron mass:

$$4.0026 = \text{mass of } ^{4}_{2}\text{He atom} = \text{mass of } ^{4}_{2}\text{He nucleus} + 2m_e$$

Electron mass

$$1.0078 = \text{mass of } ^{1}_{1}\text{H atom} = \text{mass of } ^{1}_{1}\text{H nucleus} + m_e$$

Thus, since a $^{4}_{2}He$ nucleus is "synthesized" from two protons and two neutrons, we see that

$$\Delta m = \underbrace{(4.0026 - 2m_e)}_{\text{Mass of } ^{4}_{2}\text{He nucleus}} - [2\underbrace{(1.0078 - m_e)}_{\text{Mass of } ^{1}_{1}\text{H nucleus (proton)}} + 2\underbrace{(1.0087)}_{\text{Mass of neutron}}]$$

$$= 4.0026 - 2m_e - 2(1.0078) + 2m_e - 2(1.0087)$$
$$= 4.0026 - 2(1.0078) - 2(1.0087)$$
$$= -0.0304 \text{ amu}$$

Note that in this case the electron mass cancels in taking the difference. This will always happen in this type of calculation if the atomic masses are used both for the nuclide of interest and for the $^{1}_{1}H$. Thus 0.0304 amu of mass is *lost* per $^{4}_{2}He$ nucleus formed.

The corresponding energy change can be calculated from

$$\Delta E = \Delta mc^2$$

where

$$\Delta m = -0.0304 \frac{\text{amu}}{\text{nucleus}}$$
$$= \left(-0.0304 \frac{\text{amu}}{\text{nucleus}}\right)\left(1.66 \times 10^{-27} \frac{\text{kg}}{\text{amu}}\right)$$
$$= -5.04 \times 10^{-29} \text{ kg/nucleus}$$

and

$$c = 3.00 \times 10^{8} \text{ m/s}$$

Thus

$$\Delta E = \left(-5.04 \times 10^{-29} \frac{\text{kg}}{\text{nucleus}}\right)\left(3.00 \times 10^{8} \frac{\text{m}}{\text{s}}\right)^2$$
$$= -4.54 \times 10^{-12} \text{ J/nucleus}$$

This means that 4.54×10^{-12} J of energy is *released* per nucleus formed and that 4.54×10^{-12} J would be required to decompose the nucleus into the constituent neutrons and protons. Thus the binding energy (BE) per nucleon is

$$\text{BE per nucleon} = \frac{4.54 \times 10^{-12} \text{ J/nucleus}}{4 \text{ nucleons/nucleus}}$$
$$= 1.14 \times 10^{-12} \text{ J/nucleon}$$
$$= \left(1.14 \times 10^{-12} \frac{\text{J}}{\text{nucleon}}\right)\left(\frac{1 \text{ MeV}}{1.60 \times 10^{-13} \text{ J}}\right)$$
$$= 7.13 \text{ MeV/nucleon}$$

21.6 Nuclear Fission and Nuclear Fusion

The graph shown in Fig. 21.9 has very important implications for the use of nuclear processes as sources of energy. Recall that energy is released, that is, ΔE is negative, when a process goes from a less stable to a more stable state. The higher a nuclide is on the curve, the more stable it is. This means that two types of nuclear processes are exothermic (see Fig. 21.10):

1. Combining two light nuclei to form a heavier, more stable nucleus. This process is called **fusion.**
2. Splitting a heavy nucleus into two nuclei with smaller mass numbers. This process is called **fission.**

Because of the large binding energies involved in holding the nucleus together, both of these processes involve energy changes more than a million times larger than those associated with chemical reactions.

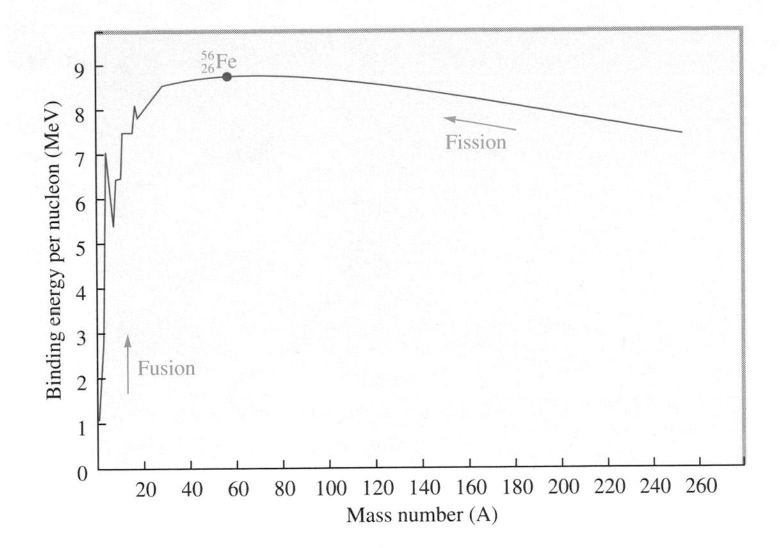

Figure 21.10
Both fission and fusion produce more stable nuclides and are thus exothermic.

Nuclear Fission

Nuclear fission was discovered in the late 1930s when $^{235}_{92}U$ nuclides bombarded with neutrons were observed to split into lighter elements:

$$^{1}_{0}n + ^{235}_{92}U \longrightarrow ^{142}_{56}Ba + ^{91}_{36}Kr + 3\ ^{1}_{0}n$$

This process, shown schematically in Fig. 21.11, releases 3.5×10^{-11} J of energy per event, which translates to 2.1×10^{13} J per mole of $^{235}_{92}U$. Compare this figure with that for the combustion of methane, which releases only 8.0×10^{5} J of energy per mole. The fission of $^{235}_{92}U$ produces about 26 million times as much energy as the combustion of methane.

The process shown above is only one of the many fission processes that $^{235}_{92}U$ can undergo. Another is

$$^{1}_{0}n + ^{235}_{92}U \longrightarrow ^{137}_{52}Te + ^{97}_{40}Zn + 2\ ^{1}_{0}n$$

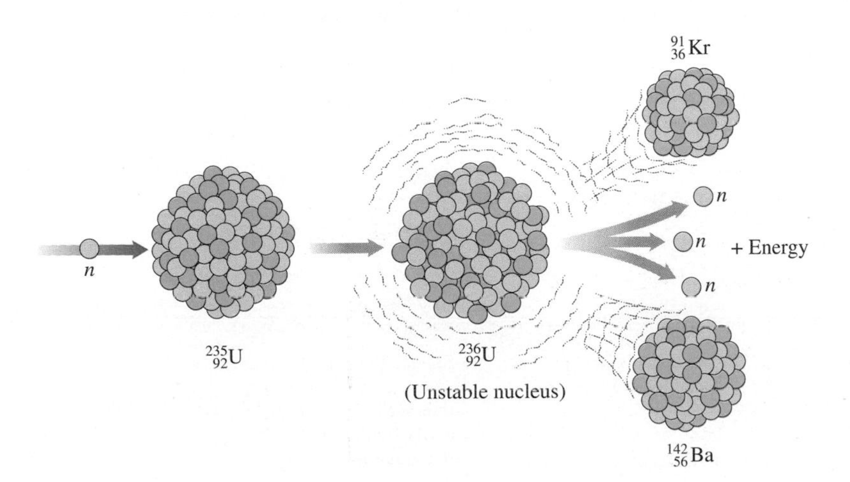

Figure 21.11
Upon capturing a neutron, the $^{235}_{92}U$ nucleus undergoes fission to produce two lighter nuclides, free neutrons (typically three), and a large amount of energy.

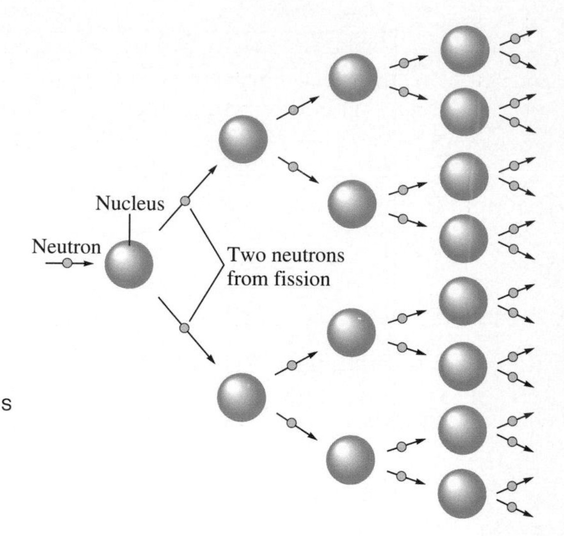

Figure 21.12
Representation of a fission process in which each event produces two neutrons, which can go on to split other nuclei, leading to a self-sustaining chain reaction.

In fact, over 200 different isotopes of 35 different elements have been observed among the fission products of $^{235}_{92}U$.

In addition to the product nuclides, neutrons are also produced in the fission reactions of $^{235}_{92}U$. This makes it possible to produce a self-sustaining fission process—a **chain reaction** (see Fig. 21.12). In order for the fission process to be self-sustaining, at least one neutron from each fission event must go on to split another nucleus. If, on the average, *less than one* neutron causes another fission event, the process dies out; the reaction is said to be **subcritical.** If *exactly one* neutron from each fission event causes another fission event, the process sustains itself at the same level and is said to be **critical.** If *more than one* neutron from each fission event causes another fission event, the process rapidly escalates and the heat buildup causes a violent explosion. This situation is described as **supercritical.**

The critical state requires a certain mass of fissionable material, called the **critical mass.** If the sample is too small, too many neutrons escape before they have a chance to cause a fission event; thus the process stops. This is illustrated in Fig. 21.13.

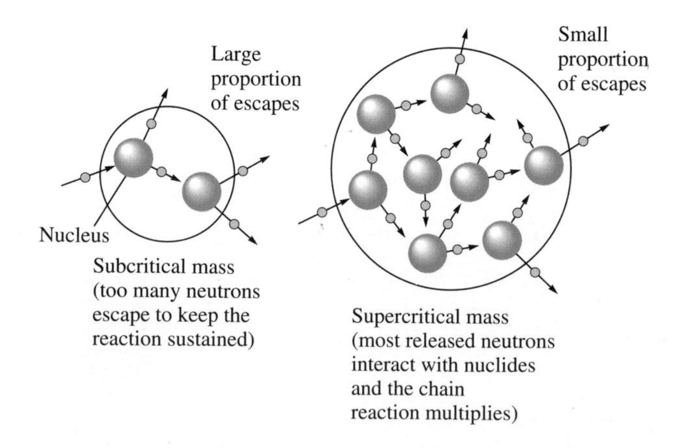

Figure 21.13
If the mass of fissionable material is too small, most of the neutrons escape before causing another fission event; thus the process dies out.

During World War II an intense research effort (the Manhattan Project) was carried out by the United States to build a bomb based on the principles of nuclear fission. This program produced the fission bombs that were used with devastating effects on the cities of Hiroshima and Nagasaki in 1945. Basically, a fission bomb operates by suddenly combining two subcritical masses of fissionable material to form a supercritical mass, thereby producing an explosion of incredible intensity.

Nuclear Reactors

Because of the tremendous energies involved, it seemed desirable after World War II to develop the fission process as an energy source for producing electricity. Therefore, reactors were designed in which controlled fission can occur. The resulting energy is used to heat water to produce steam to run turbine generators, in much the same way that a coal-burning power plant generates energy. A schematic diagram of a nuclear power plant is shown in Fig. 21.14.

In the **reactor core,** shown in Fig. 21.15, uranium that has been enriched to approximately 3% ${}^{235}_{92}U$ (natural uranium contains only 0.7% ${}^{235}_{92}U$) is housed in metal cylinders. A **moderator** surrounds the cylinders to slow down the neutrons so that the uranium fuel can capture them more efficiently. **Control rods,** composed of substances that absorb neutrons, are used to regulate the power level of the reactor. The reactor is designed so that should a malfunction occur, the control rods are automatically inserted into the core to stop the reaction. A liquid (usually water) is circulated through the core to extract the heat generated by the energy of fission; the energy is then passed on via a heat exchanger to water in the turbine system.

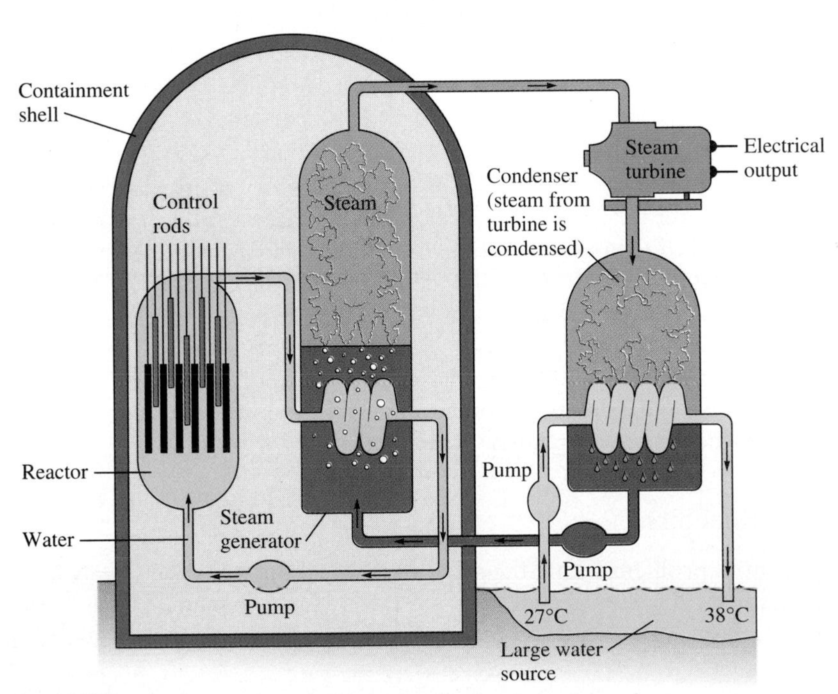

Figure 21.14
A schematic diagram of a nuclear power plant.

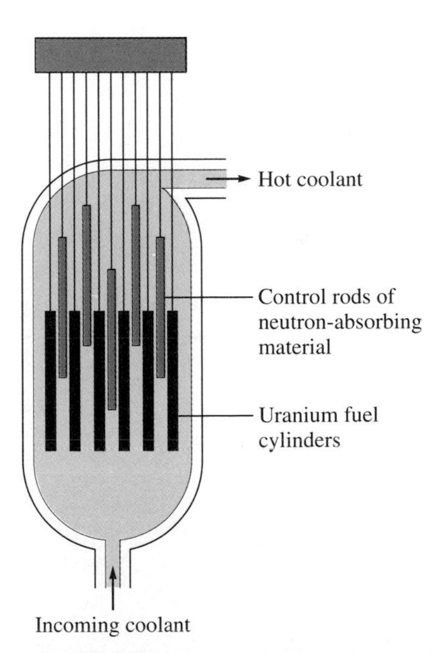

Figure 21.15
A schematic of the reactor core. The position of the control rods determines the level of energy production by regulating the amount of fission taking place.

NUCLEAR POWER: COULD IT STAGE A COMEBACK?

As everyone knows, nuclear power has been a very controversial issue in the United States over the past three decades. Safety issues and cost, along with the attendant political sensitivities, have stalled the growth of nuclear power. However, this situation might change. The intense concern about the warming effects of the carbon dioxide produced by fossil fuels, the uncertain rate of progress in harnessing solar energy, and new reactor designs all may lead to a resurgence of interest in fission-powered nuclear plants.

Indeed, the French have demonstrated that nuclear power can be economically and politically feasible. France, which uses breeder reactors to provide most of its electricity, has an excellent safety record and is pioneering new methods for the storage of radioactive wastes.

In the United States, experiments at Argonne National Laboratory in Illinois indicate that a new generation of breeder reactors can be developed that will be inherently safer than the pressurized boiling-water reactors now in use. This research also shows that these reactors can be cost competitive with coal-fired power plants. The reactor under investigation at Argonne, called an integral fast reactor (IFR), is quite different from the current reactors used in power plants. The key differences are that (1) the IFR uses liquid sodium instead of water as the coolant, (2) the IFR uses its fuel much more efficiently and, in fact, breeds new fissionable fuel as it runs, and (3) the IFR produces only a fraction of the radioactive wastes produced by current reactors. Because the boiling point of sodium is near 900°C, the liquid sodium does not need to be under pressure to prevent boiling, as is required for the water in current power reactors. This greatly lessens the chance of a very rapid loss of coolant should a leak develop. Liquid sodium has other advantages over water, such as not being corrosive to steel. Also, the use of sodium as a coolant allows for much faster neutrons than are possible in present water-cooled reactors. These fast neutrons produce extra power and greatly reduce the radioactive waste from the reactor, because they cause the fission (and thus the destruction) of many radioactive products of the fission process.

^{235}U enrichment is described in Section 5.7.

Although the concentration of $^{235}_{92}U$ in the fuel elements is not great enough to allow a supercritical mass to develop in the core, a failure of the cooling system can lead to temperatures high enough to melt the core. As a result, the building housing the core must be designed to contain the core even if a meltdown occurs. A great deal of controversy now exists about the efficiency of the safety systems in nuclear power plants. Accidents such as the one at the Three Mile Island facility in Pennsylvania in 1979 and the one in Chernobyl,* USSR, in 1986 have led many people to question the wisdom of continuing to build fission-based power plants.

Breeder Reactors

One potential problem facing the nuclear power industry is the supply of $^{235}_{92}U$. Some scientists have suggested that we have nearly depleted those uranium deposits rich enough in $^{235}_{92}U$ to make production of fissionable fuel economi-

*For a detailed account of this incident, see C. A. Atwood, "Chernobyl—What Happened?" *J. Chem. Ed.* **65** (1988), p. 1037.

A breeder reactor at a nuclear power plant in St. Laurent-Des Eaux, France.

In addition, the IFR uses a new type of fuel element, which is an alloy of uranium, plutonium, and zirconium. Because these new metal fuel elements conduct heat so much more efficiently than the ceramic fuel elements now commonly in use, lower core temperatures are possible. Also, these metal fuel elements allow electrochemical reprocessing, which leads to very efficient recycling of usable fuel. This means that virtually all of the initial uranium fuel is eventually consumed. In contrast, present-day reactors consume only a small percentage of the original uranium; the rest is removed with the radioactive fission products. Because the fuel recycling for the IFR does not require a large facility, it can be done at the reactor site, thus eliminating the transportation of large amounts of radioactive materials.

If development goes as smoothly as projected, the integral fast reactor may be providing safe, economical power shortly after the year 2000, while adding no carbon dioxide or other pollutants to the atmosphere.

cally feasible. Because of this possibility, **breeder reactors** have been developed, in which fissionable fuel is actually produced while the reactor runs. In a breeder reactor the major component of natural uranium, nonfissionable $^{238}_{92}U$, is changed to fissionable $^{239}_{94}Pu$. The reaction involves absorption of a neutron, followed by production of two β particles:

$$^{1}_{0}n + ^{238}_{92}U \longrightarrow ^{239}_{92}U$$
$$^{239}_{92}U \longrightarrow ^{239}_{93}Np + ^{0}_{-1}e$$
$$^{239}_{93}Np \longrightarrow ^{239}_{94}Pu + ^{0}_{-1}e$$

The Tokamak Fusion Test Reactor facility at Princeton University.

As the reactor runs and $^{235}_{92}U$ is split, some of the excess neutrons are absorbed by $^{238}_{92}U$ to produce $^{239}_{94}Pu$. The $^{239}_{94}Pu$ is then separated out and used to fuel another reactor. Such a reactor thus "breeds" nuclear fuel as it operates.

Although breeder reactors are now used in France, the United States is proceeding slowly with their development because of their controversial nature. One problem involves the hazards in handling plutonium, which flames upon contact with air and is very toxic.

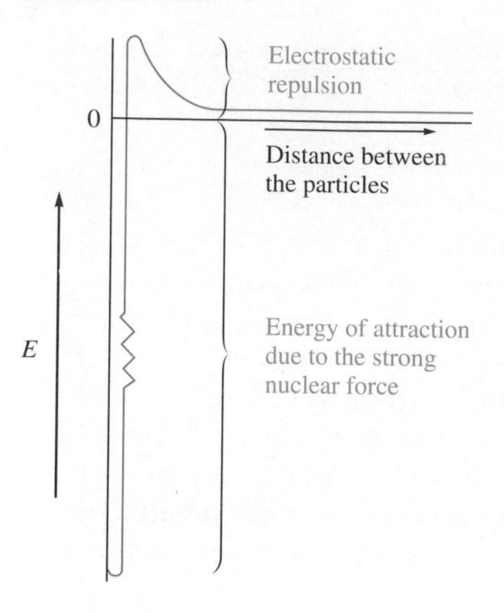

Figure 21.16
A plot of energy versus the separation distance for two ^2_1H nuclei. The nuclei must have sufficient velocities to get over the electrostatic repulsion "hill" and get close enough for the nuclear binding forces to become effective, thus "fusing" the particles into a new nucleus and releasing large quantities of energy. The binding force is at least 100 times the electrostatic repulsion.

Fusion

Large quantities of energy are also produced by the fusion of two light nuclei. In fact, stars produce their energy through nuclear fusion. Our sun, which presently consists of 73% hydrogen, 26% helium, and 1% other elements, gives off vast quantities of energy from the fusion of protons to form helium:

$$^1_1H + ^1_1H \longrightarrow ^2_1H + ^0_1e$$
$$^1_1H + ^2_1H \longrightarrow ^3_2He$$
$$^3_2He + ^3_2He \longrightarrow ^4_2He + 2\ ^1_1H$$
$$^3_2He + ^1_1H \longrightarrow ^4_2He + ^0_1e$$

Intense research is underway to develop a feasible fusion process because of the ready availability of many light nuclides (deuterium, ^2_1H, in seawater, for example) that can serve as fuel in fusion reactors. The major stumbling block is the high temperatures required to initiate fusion. The forces that bind nucleons together to form a nucleus are effective only at *very small* distances ($\approx 10^{-13}$ cm). Thus for two protons to bind together and thereby release energy, they must get very close together. But protons, because they are identically charged, repel each other electrostatically. This means that in order to get two protons (or two deuterons) close enough to bind together (the nuclear binding force is *not* electrostatic), they must be "shot" at each other at speeds high enough to overcome the electrostatic repulsion.

The electrostatic repulsion forces between two ^2_1H nuclei are so great that a temperature of 4×10^7 K is required to give them velocities large enough to cause them to collide with sufficient energy that the nuclear forces can bind the particles together and thus release the binding energy. This situation is represented in Fig. 21.16.

Currently, scientists are studying two types of systems to produce the extremely high temperatures required: high-powered lasers, and heating by electric currents. At present many technical problems remain to be solved, and it is not clear which method will prove more useful or when fusion might become a practical energy source. However, there is still hope that fusion will be a major energy source in the twenty-first century.

21.7 Effects of Radiation

The ozone layer is discussed in Section 19.5.

Everyone knows that being hit by a train is very serious. The problem is the energy transfer involved. In fact, any source of energy is potentially harmful to organisms. Energy transferred to cells can break chemical bonds and cause malfunctioning of the cell systems. This fact is behind the concern about the ozone layer in the earth's upper atmosphere, which screens out high-energy ultraviolet radiation from the sun. Radioactive elements, which are sources of high-energy particles, are also potentially hazardous, although the effects are usually quite subtle. The reason for the subtlety of radiation damage is that even though high-energy particles are involved, the quantity of energy actually deposited in tissues *per event* is quite small. However, the resulting damage is no less real, although the effects may not be apparent for years.

COLD FUSION

In March 1989 two chemists, B. Stanley Pons and Martin Fleischmann, reported to an astonished world that they had succeeded in producing nuclear fusion in an aqueous solution. Instead of the millions of degrees assumed to be required to initiate fusion, these scientists reported what they thought was fusion in connection with the electrolysis of D_2O containing LiOD at 25°C using palladium electrodes. Cold fusion, as the phenomenon was dubbed, was immediately trumpeted by the media to be the solution to the earth's energy problems.

But now, several years after the first announcement of the phenomenon, it is still not clear whether cold fusion actually occurs. Some of the best scientists in the world have been working with feverish intensity on the problem with, at best, uncertain results. Indeed, this episode is an excellent illustration of how difficult and complex scientific research can be. Given the talent of the scientists and the myriad of modern instruments available for the work, one might assume that this issue would have been resolved quickly.

The lab setup used by Drs. Pons and Fleischmann at the University of Utah for their experiments in cold fusion.

The possibility of cold fusion arose when Pons and Fleischmann observed very large bursts of heat accompanying the electrolysis of D_2O/LiOD mixtures using palladium electrodes at high current densities. In addition to the extraordinary amounts of energy, the two scientists reported that tritium and neutrons were also produced. None of these things could be explained by any known chemical reactions. The energy production was, at times, 100 to 1000 times that expected from a chemical reaction, and the presence of tritium and/or neutrons would seem to suggest some sort of nuclear process.

However, there are several problems. The excess heat production is intermittent and unpredictable. Some cells run for weeks with no excess energy production, then suddenly will churn out large quantities of energy for hundreds of hours, and then—just as suddenly—give out. Also, because neutrons are very difficult to detect, great uncertainty remains about how many neutrons, if any, are really produced. Although the detection of tritium is much more straightforward, it is possible that the tritium observed is due to contamination. Also, only about one-fourth of the cells seem to produce tritium, which is very hard to explain. On the positive side, in early 1991 several research groups investigating the electrolysis of D_2O/LiOD using a Pd electrode reported the detection of helium, a product expected in the fusion of deuterium atoms.

So what is going on? Is cold fusion responsible for these effects, or do they arise from a combination of contamination, difficulties in making measurements, and mysterious chemical and/or physical processes? Although hope remains that a process that will prove useful for energy production is occurring, the verdict remains in doubt.

Radiation damage to organisms can be classified as somatic or genetic damage. **Somatic damage** is damage to the organism itself, resulting in sickness or death. The effects may appear almost immediately if a massive dose of radiation is received; for smaller doses damage may appear years later, usually in the form of cancer. **Genetic damage** is damage to the genetic machinery, which produces malfunctions in the offspring of the organism.

The biological effects of a particular source of radiation depend on several factors:

1. *The energy of the radiation.* The higher the energy content of the radiation, the more damage it can cause. Radiation doses are measured in **rads** (which is short for *r*adiation *a*bsorbed *d*ose), where 1 rad corresponds to 10^{-2} J of energy deposited per kilogram of tissue.
2. *The penetrating ability of the radiation.* The particles and rays produced in radioactive processes vary in their abilities to penetrate human tissue: γ rays are highly penetrating; β particles can penetrate approximately 1 cm; and α particles are stopped by the skin.
3. *The ionizing ability of the radiation.* Extraction of electrons from biomolecules to form ions is particularly detrimental to their functions. The ionizing ability of radiation varies dramatically. For example, γ rays penetrate very deeply but cause only occasional ionization. On the other hand, α particles, although not very penetrating, are very effective at causing ionization and producing a dense trail of damage. Thus ingestion of an α-particle producer, such as plutonium, is particularly damaging.
4. *The chemical properties of the radiation source.* When a radioactive nuclide is ingested into the body, its effectiveness in causing damage depends on its residence time. For example, $^{85}_{36}Kr$ and $^{90}_{38}Sr$ are both β-particle producers. However, since krypton is chemically inert, it passes through the body quickly and does not have much time to do damage. Strontium, being chemically similar to calcium, can collect in bones, where it may cause leukemia and bone cancer.

TABLE 21.6
Effects of Short-Term Exposures to Radiation

Dose (rem)	Clinical Effect
0–25	Nondetectable
25–50	Temporary decrease in white blood cell counts
100–200	Strong decrease in white blood cell counts
500	Death of half the exposed population within 30 days after exposure

TABLE 21.7
Typical Radiation Exposures for a Person Living in the United States (1 millirem = 10^{-3} rem)

Source	Exposure (millirems/year)
Cosmic radiation	50
From the earth	47
From building materials	3
In human tissues	21
Inhalation of air	5
Total from natural sources	126
X-ray diagnosis	50
Radiotherapy	10
Internal diagnosis/therapy	1
Nuclear power industry	0.2
TV tubes, industrial wastes, etc.	2
Radioactive fallout	4
Total from human activities	67
Total	193

Because of the differences in the behavior of the particles and rays produced by radioactive decay, both the energy dose of the radiation and its effectiveness in causing biological damage must be taken into account. The **rem** (which is short for *r*oentgen *e*quivalent for *m*an) is defined as follows:

$$\text{Number of rems} = (\text{number of rads}) \times \text{RBE}$$

where RBE represents the relative effectiveness of the radiation in causing biological damage.

Table 21.6 shows the physical effects of short-term exposure to various doses of radiation, and Table 21.7 gives the sources and amounts of radiation exposure for a typical person in the United States. Note that natural sources contribute about twice as much as human activities to the total exposure. However, although the nuclear industry contributes only a small percentage of the total exposure, the major controversy associated with nuclear power plants is the *potential* for radiation hazards. These arise mainly from two sources: accidents allowing the release of radioactive materials, and improper disposal of the radioactive products in spent fuel elements. The radioactive products of the fission of $^{235}_{92}U$, although constituting only a small percentage of the total products, have half-lives of several hundred years and remain dangerous for a

Nuclear Physics: An Introduction

Nuclear physics is concerned with the fundamental nature of matter. The central focuses of this area of study are the relationship between a quantity of energy and its mass, given by $E = mc^2$, and the fact that matter can be converted from one form (energy) to another (particulate) in particle accelerators. Collisions between high-speed particles have produced a dazzling array of new particles—hundreds of them. These events can best be seen as conversions of kinetic energy into particles. For example, a collision of sufficient energy between a proton and a neutron can produce four particles: two protons, one antiproton, and a neutron:

$$^{1}_{1}H + ^{1}_{0}n \longrightarrow 2\,^{1}_{1}H + ^{1}_{-1}H + ^{1}_{0}n$$

where $^{1}_{-1}H$ is the symbol for an *antiproton,* which has the same mass as a proton but the opposite charge. This process is a little like throwing one baseball at a very high speed into another and having the collision produce four baseballs.

The results of such accelerator experiments have led scientists to postulate the existence of three types of forces important in the nucleus: the *strong force,* the *weak force,* and the *electromagnetic force.* Along with the *gravitational force,* these forces are thought to account for all types of interactions found in matter. These forces are believed to be generated by the exchange of particles between the interacting pieces of matter. For example, gravitational force is thought to be carried by particles called *gravitons.* The electromagnetic force (the classical electrostatic force between charged particles) is assumed to be exerted through the exchange of *photons.* The strong force, not charge-related and effective only at very short distances ($\approx 10^{-13}$ cm), is postulated to involve the exchange of particles called *gluons.* The weak force is 100 times weaker than the strong force and seems to be exerted by the exchange of two types of large particles, the W (has a mass 70 times the proton mass) and the Z (has a mass 90 times the proton mass).

The particles discovered have been classified into several categories. Three of the most important classes are as follows:

1. *Hadrons* are particles that respond to the strong force and have internal structure.
2. *Leptons* are particles that do not respond to the strong force and have no internal structure.
3. *Quarks* are particles with no internal structure that are thought to be the fundamental constituents of hadrons. Neutrons and protons are hadrons that are thought to be composed of three quarks each.

The world of particle physics appears mysterious and complicated. For example, particle physicists have discovered new properties of matter they call "color," "charm," and "strangeness" and have postulated conservation laws involving these properties. This area of science is extremely important because it should help us to understand the interactions of matter in a more elegant and unified way. For example, the classification of forces into four categories is probably necessary only because we do not understand the true nature of forces. All forces may be special cases of a single, all-pervading force field that governs all of nature. In fact, Einstein spent the last 30 years of his life looking for a way to unify gravitational and electromagnetic forces—without success. Physicists may now be on the verge of accomplishing what Einstein failed to do.

Although the practical aspects of work in nuclear physics are not yet totally apparent, a more fundamental understanding of the way nature operates could lead to presently undreamed-of devices for energy production, communication, and so on, which could revolutionize our lives.

The accelerator tunnel at Fermilab, a high-energy particle accelerator in Batavia, Illinois.

Nuclear Waste Disposal

Our society has not had a very impressive record for safe disposal of industrial wastes. We have polluted our water and air, and some land areas have become virtually uninhabitable because of our improper burial of chemical wastes. As a result, many people are wary about the radioactive wastes from nuclear reactors. The potential threats of cancer and genetic mutations make these materials especially frightening.

Because of its controversial nature, no nuclear waste generated over the past 40 years has been permanently disposed of. However, in 1982 the U.S. Congress passed the Nuclear Waste Policy Act, which established a timetable for choosing and preparing sites for deep underground disposal of radioactive materials. The program is funded by a tax of 0.1% per kilowatt hour on electricity generated by nuclear power.

The tentative disposal plan calls for the incorporation of the spent nuclear fuel into blocks of borosilicate glass. These blocks will be packed in corrosion-resistant metal containers and then buried in a deep, stable rock formation (see the accompanying figure).

There are indications that this method will isolate the waste until the radioactivity decays to safe levels. One reassuring indication comes from the natural fission "reactor" at Oklo in Gabon, Africa. Initiated about 2 billion years ago when uranium in ore deposits there formed a critical mass, the "reactor" produced fission and fusion products for several thousand years. Although some of these products have migrated away from the site in the intervening 2 billion years, most have stayed in place. Another indication of the possible success of underground isolation is the disposal program carried out at Oak Ridge National Laboratory. In this program the waste has been ground up, mixed with cement, fly ash, and clay, and injected at high pressures into a shale bed 1000 ft underground. This method of injection causes fractures in the rock into which the liquid waste seeps and then solidifies. Monitoring of the wastes has shown little movement of radioactivity. Although this procedure has worked well, the materials involved are not nearly as radioactive as those from power plants, for which this method is not thought suitable. However, this experiment has demonstrated that such wastes can be immobilized.

Choosing the waste repository sites is an especially sensitive issue. Many states have resisted the plan, but Congress has the power to override a state's disapproval. In fact, Congress amended the Nuclear Waste Policy Act in 1987 to make Yucca

long time. Various schemes have been advanced for the disposal of these wastes. The one that seems to hold the most promise is the incorporation of the wastes into ceramic blocks and the burial of these blocks in geologically stable formations. At present, however, no disposal method has been accepted, and nuclear wastes continue to accumulate in temporary storage facilities.

Even if a satisfactory method for permanent disposal of nuclear wastes is found, there will continue to be concern about the effects of exposure to low levels of radiation. Exposure is inevitable from natural sources such as cosmic rays and radioactive minerals, and many people are also exposed to low levels of radiation from reactors, radioactive tracers, and diagnostic X rays. Currently, we have little reliable information on the long-term effects of low-level exposure to radiation.

Two models of radiation damage, illustrated in Fig. 21.17, have been proposed: the *linear model* and the *threshold model*. The linear model postulates that damage from radiation is proportional to the dose, even at low levels

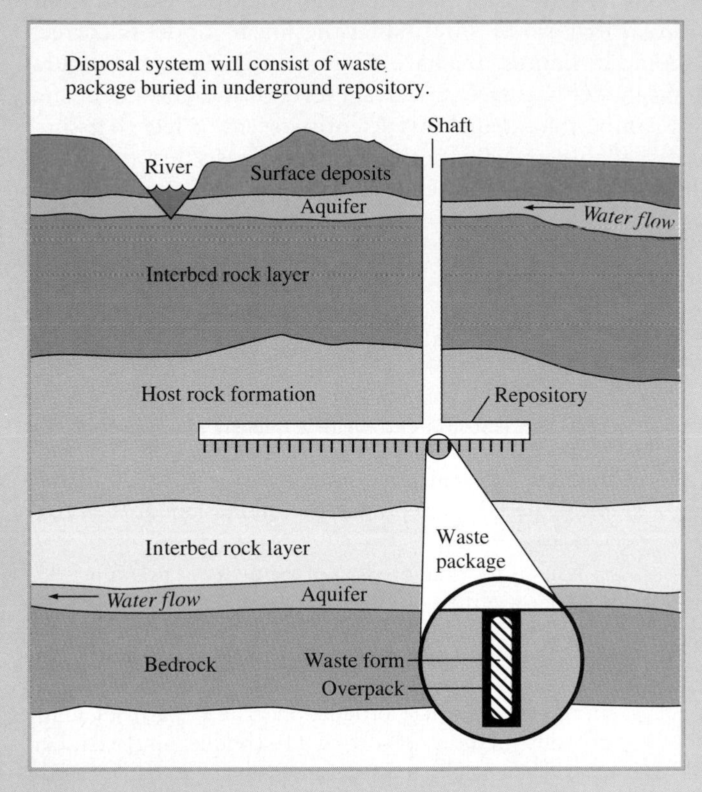

A schematic diagram of the tentative plan for deep underground isolation of nuclear waste. (From *Chemical and Engineering News*, July 18, 1983, pp. 20–38. Reprinted with permission. Copyright 1983 American Chemical Society.)

Mountain in Nevada the primary potential site. Studies are now being carried out to evaluate the feasibility of this site as a safe repository for nuclear waste.

In a related situation the Department of Energy (DOE) is preparing the Waste Isolation Pilot Plant (WIPP) near Carlsbad, New Mexico, for the long-term storage of defense-related nuclear waste. This storage facility is sited 2150 ft below the surface in salt beds that have remained stable and free of ground water for about 225 million years, according to studies by DOE. In the next stages of the testing of WIPP, the Department of Energy will store several thousand drums of waste, which will be monitored for effects such as gas formation.

An alternative to permanent isolation of the spent nuclear fuel is monitored, retrievable storage. This method has an advantage since, if reprocessing to remove potentially useful materials such as plutonium becomes feasible, the wastes are still accessible. The argument for this plan is that the technologies and alternatives available to society a few hundred years from now may be so different from those of the present that these wastes could become valuable resources. At the present time isolation of the wastes seems the most likely event; even so, these underground deposits may become plutonium mines in the future.

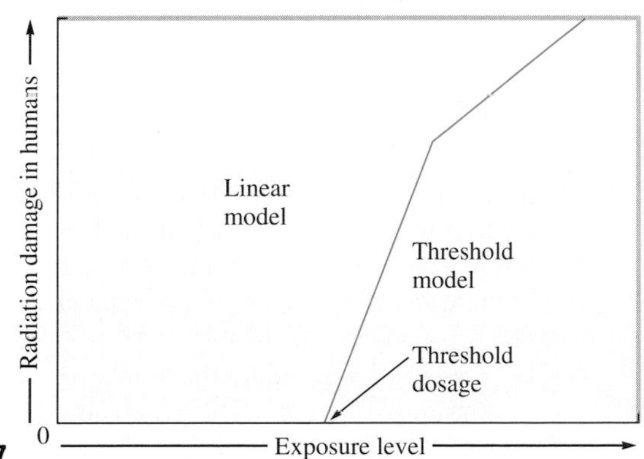

Figure 21.17
The two models for radiation damage. In the linear model even a small dose causes a proportional risk. In the threshold model risk begins only after a certain dose.

of exposure. Thus any exposure is dangerous. The threshold model, on the other hand, assumes that no significant damage occurs below a certain exposure, called the *threshold exposure*. Note that if the linear model is correct, radiation exposure should be limited to a bare minimum (ideally at the natural levels). If the threshold model is correct, a certain level of radiation exposure beyond natural levels can be tolerated. Most scientists seem to feel that since there is little evidence available to evaluate these models, it is safest to assume that the linear hypothesis is correct and to minimize radiation exposure.

Exercises

A blue exercise number indicates that the answer to that exercise appears at the back of this book.

Radioactive Decay and Nuclear Transformations

1. Write an equation describing the radioactive decay of each of the following nuclides. (The particle produced is shown in parentheses.)
   a. $^{3}_{1}H$ (β)
   b. $^{8}_{3}Li$ (β followed by α)
   c. $^{7}_{4}Be$ (electron capture)
   d. $^{8}_{5}B$ (positron)
   e. $^{32}_{15}P$ (β)
2. Write an equation describing the radioactive decay of each of the following nuclides.
   a. ^{60}Co (β)
   b. ^{97}Tc (electron capture)
   c. ^{99}Tc (β)
   d. ^{239}Pu (α)
3. The stable isotopes of boron are boron-10 and boron-11. Four radioactive isotopes with mass numbers 8, 9, 12, and 13 are also known. Predict possible modes of radioactive decay for these four nuclides.
4. Complete the following equations, which describe nuclear reactions that occur in the sun.
   a. $^{1}_{1}H + ^{14}_{7}N \rightarrow ? + ^{4}_{2}He$
   b. Two helium-3 nuclei undergo fusion, forming one helium-4 and two protons.
   c. $^{1}_{1}H + ^{1}_{1}H \rightarrow ^{2}_{1}H + ?$
   d. $^{1}_{1}H + ^{12}_{6}C \rightarrow ?$
5. Many elements have been synthesized by bombarding relatively heavy atoms with high-energy particles in particle accelerators. Complete the following nuclear reactions, which have been used to synthesize elements.
   a. $? + ^{4}_{2}He \rightarrow ^{243}_{97}Bk + ^{1}_{0}n$
   b. $^{238}_{92}U + ^{12}_{6}C \rightarrow ? + 6\,^{1}_{0}n$
   c. $^{249}_{98}Cf + ? \rightarrow ^{263}_{106}Unh + 4\,^{1}_{0}n$
   d. $^{249}_{98}Cf + ^{10}_{5}B \rightarrow ^{257}_{103}Lr + ?$
6. Define *fission* and *fusion*. For what types of nuclei are you likely to see each process?
7. When nuclei undergo nuclear transformations, γ rays of characteristic frequencies are observed. How does this fact, along with other information in the chapter on nuclear stability, suggest that a model similar to the quantum mechanics used for atoms may apply to the nucleus?
8. There are four stable isotopes of iron with mass numbers 54, 56, 57, and 58. There are also two radioactive isotopes: iron-53 and iron-59. Predict modes of decay for these two isotopes and write a nuclear reaction for each.
9. Uranium-235 undergoes a decay series in which the following particles are produced in succession: α, β, α, β, α, α, α, α, β, α, β.
   a. What is the final product of the decay of uranium-235?
   b. What are the ten intermediate nuclides?
10. One type of commercial smoke detector contains a minute amount of radioactive americium-241 (^{241}Am), which decays by α-particle production. The α particles ionize molecules in the air, allowing it to conduct an electric current. When smoke particles enter, the conductivity of the air changes and the alarm buzzes.
    a. Write the equation for the decay of $^{241}_{95}Am$ by α-particle production.
    b. The complete decay of ^{241}Am involves successively α, α, β, α, α, β, α, α, α, β, α, and β production. What is the final stable nucleus produced in this decay series?
    c. Identify the eleven intermediate nuclides.
11. Predict whether each of the following nuclides are stable or unstable (radioactive). If the nuclide is unstable, predict the type of radioactivity you would expect it to exhibit.
    a. $^{45}_{19}K$
    b. $^{56}_{26}Fe$
    c. $^{20}_{11}Na$
    d. $^{194}_{81}Tl$
12. In 1994 it was proposed that element 106 be named seaborgium, Sg, in honor of Glenn T. Seaborg, discoverer of the transuranium elements. This is the first time it has been proposed that an element be named for a living person.
    a. ^{263}Sg was produced by the bombardment of ^{249}Cf with a beam of ^{18}O nuclei. Complete and balance an equation for this reaction.
    b. ^{263}Sg decays by α emission. What is the other product resulting from the α decay of ^{263}Sg?

Kinetics of Radioactive Decay

13. How many disintegrations occur in the first second from 1.00 mol of a radioactive nuclide with each of the following half-lives: 12,000 yr? 12 h? 12 min? 12 s?
14. What fraction of the original carbon-14 remains in a piece of wood that was cut from a tree 2200 years ago?
15. What assumptions are necessary and what problems arise in using carbon-14 dating?
16. A chemist wishes to do an experiment using ^{47}Ca (half-life = 4.5 days). He needs 5.0 μg of the nuclide. What mass of $^{47}CaCO_3$ must he order if it takes 48 h for delivery from the supplier?
17. Cobalt-60 is commonly used as a source of β particles. How long does it take for 10.0% of a sample of cobalt-60 to decay (half-life = 5.26 yr)?
18. A living plant contains approximately the same fraction of carbon-14 as atmospheric carbon dioxide. The observed rate of decay of carbon-14 from a living plant is 15.3 disintegrations per minute per gram of carbon. How many disintegrations per minute per gram of carbon will be measured from a 15,000-yr-old sample? Will radiocarbon dating work well for small samples of 10 mg or less?
19. What is the ratio $^{206}Pb/^{238}U$ by mass in a rock that is 4.5×10^9 yr old? (For ^{238}U, $t_{1/2} = 4.5 \times 10^9$ yr.)
20. The mass ratios of ^{40}Ar to ^{40}K can also be used to date geological materials. Potassium-40 decays by two processes:

$$^{40}_{19}K \longrightarrow {}^{40}_{18}Ar \text{ (electron capture, 10.7\%)} \quad t_{1/2} = 1.27 \times 10^9 \text{ yr}$$

$$^{40}_{19}K \longrightarrow {}^{40}_{20}Ca + {}^{0}_{-1}e \text{ (89.3\%)}$$

a. Why are $^{40}Ar/^{40}K$ ratios used to date materials rather than $^{40}Ca/^{40}K$ ratios?
b. What assumptions must be made in using this technique?
c. A sedimentary rock has a $^{40}Ar/^{40}K$ ratio of 0.95. Calculate the age of the rock.
d. How will the measured age of a rock compare with the actual age if some ^{40}Ar has escaped from the sample?

21. Phosphorus-32 is a commonly used radioactive nuclide in biochemical research, particularly in studies of nucleic acids. The half-life of phosphorus-32 is 14.3 days. What mass of phosphorus-32 is left from an original sample of 175 mg of $Na_3{}^{32}PO_4$ after 35.0 days?
22. The *curie* (Ci) is a commonly used unit for measuring nuclear radioactivity: 1 curie of radiation is equal to 3.7×10^{10} disintegrations per second. This is the number of disintegrations from 1 g of radium in 1 s.
a. What is the activity in mCi (millicuries) of 175 mg of $Na_3{}^{32}PO_4$? (See Exercise 21.)
b. What is the activity in mCi of 1.0 mol of plutonium-239 ($t_{1/2}$ = 24,000 yr)?
23. The half-life of sulfur-38 is 2.87 h.
a. What mass of $Na_2{}^{38}SO_4$ has an activity of 10.0 mCi? (See Exercise 22.)
b. How long does it take for 99.99% of a sample of sulfur-38 to decay?
24. The first atomic explosion was detonated in the desert north of Alamogordo, New Mexico, on July 16, 1945. What fraction of the strontium-90 ($t_{1/2}$ = 28.8 yr) originally produced by that explosion still remains?
25. A bottle of wine, which was analyzed for its tritium (^3H) content, was found to contain 2.0% of the ^3H that was originally present when the wine was bottled. How old is the bottle of wine? (The half-life of ^3H is 12.3 yr.)

Energy Changes in Nuclear Reactions

26. The sun radiates 3.9×10^{23} J of energy into space every second. What is the rate at which mass is lost from the sun?
27. The earth receives 1.8×10^{14} kJ/s of solar energy. What mass of solar material is converted to energy over a 24-h period to provide the daily amount of solar energy to the earth? What mass of coal would have to be burned to provide the same amount of energy? Coal releases 32kJ of energy per gram when burned.
28. Calculate the binding energy per nucleon for ^{24}Mg and ^{27}Mg. The atomic masses are ^{24}Mg, 23.9850 amu; and ^{27}Mg, 26.9843 amu. (Since these are atomic masses, they include the mass of magnesium's 12 electrons. See Example 21.3.)
29. The atomic mass of lithium-6 is 6.015126 amu. What is the mass of a lithium-6 nucleus? Calculate the total binding energy per mole of lithium-6.
30. Calculate the binding energy per nucleon for ^2_1H and ^3_1H. The atomic masses are ^2_1H, 2.01410 amu; and ^3_1H, 3.01605 amu.
31. Consider the reaction discussed in the feature on nuclear physics in Section 21.7:

$$^1_1H + {}^1_0n \longrightarrow 2\,{}^1_1H + {}^1_0n + {}^{1}_{-1}H$$

Calculate the energy change for this reaction. Is energy released or absorbed? What is a possible source for this energy?

32. A positron and an electron annihilate each other upon colliding, thereby producing energy:

$$^{0}_{-1}e + {}^{0}_{+1}e \longrightarrow 2\,{}^0_0\gamma$$

Assuming both γ rays have the same energy, calculate the wavelength of the electromagnetic radiation produced.

33. A small atomic bomb releases an amount of energy equivalent to the detonation of 20,000 tons of TNT; one ton of TNT releases 4×10^9 J of energy upon explosion. Using 2×10^{13} J/mol as the energy released by fission of ^{235}U, estimate a value for the critical mass of ^{235}U. What assumptions must be made in the calculation?

34. Using the kinetic molecular theory (Section 5.6), calculate the average kinetic energy of a ^2_1H nucleus at a temperature of 4×10^7 K. (See Exercise 36 for the appropriate mass values.)

35. Calculate the amount of energy released per gram of hydrogen for the following reaction. The atomic masses are ^1_1H, 1.00782 amu; and ^2_1H, 2.01410 amu. (*Hint:* think carefully about how to account for the electron mass.)

$$^1_1H + ^1_1H \longrightarrow ^2_1H + ^0_{+1}e$$

36. The easiest fusion reaction to initiate is

$$^2_1H + ^3_1H \longrightarrow ^4_2He + ^1_0n$$

Calculate the energy released per nucleus of ^4_2He produced and per mole of ^4_2He produced. The atomic masses are ^2_1H, 2.01410 amu; ^3_1H, 3.01605 amu; and ^4_2He, 4.00260 amu. The masses of the electron and neutron are 5.4858×10^{-4} amu and 1.00866 amu, respectively.

Detection, Uses, and Health Effects of Radiation

37. Why are the chemical properties of a radioactive substance important in assessing its potential health hazards?

38. When using a Geiger-Müller counter to measure radioactivity, one must maintain the same geometrical orientation between the sample and the Geiger-Müller tube in order to compare different measurements. Why?

39. The typical response of a Geiger-Müller tube is shown below. Explain the shape of this curve.

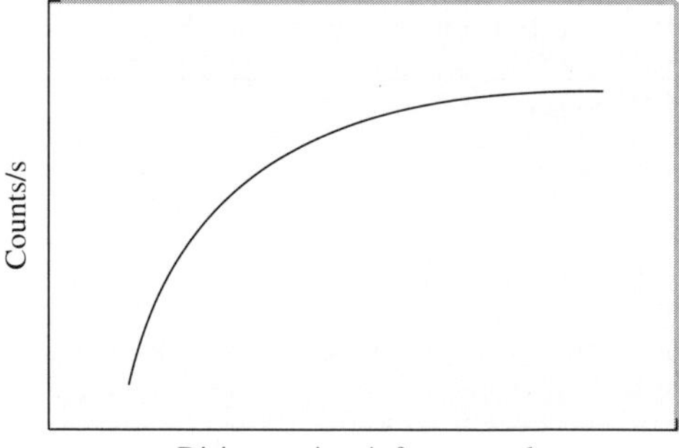

40. Photosynthesis in plants can be represented by the following overall reaction:

$$6CO_2(g) + 6H_2O(l) \xrightarrow{\text{Light}} C_6H_{12}O_6(s) + 6O_2(g)$$

Algae grown in water containing some ^{18}O (in $H_2{}^{18}O$) evolve oxygen with the same isotopic composition as the oxygen in the water. When algae growing in water containing only ^{16}O were furnished with carbon dioxide containing ^{18}O, no ^{18}O was found to be evolved. What conclusions about photosynthesis can be drawn from these experiments?

41. How could a radioactive nuclide be used to demonstrate that chemical equilibrium is a dynamic process?

42. There is a trend in the United States toward using coal-fired power plants to generate electricity rather than building new nuclear fission power plants. Is the use of coal-fired power plants without risk? Make a list of the risks to society from the use of each type of power plant.

43. A 0.10-cm^3 sample of a solution containing a radioactive nuclide (5.0×10^3 disintegrations per minute per milliliter) is injected into a rat. Several minutes later, 1.0 cm^3 of blood is removed. The blood shows 48 disintegrations of radioactivity per minute. What is the volume of blood in the rat? What assumptions must be made in performing this calculation?

44. A chemist studied the reaction mechanism for the reaction

$$2NO(g) + O_2(g) \longrightarrow 2NO_2(g)$$

by reacting $N^{16}O$ with $^{18}O_2$. If the reaction mechanism is

$$NO + O_2 \rightleftharpoons NO_3 \text{ (fast)}$$

$$NO_3 + NO \longrightarrow 2NO_2 \text{ (slow)}$$

what distribution of ^{18}O would you expect in the NO_2?

Additional Exercises

45. Recent attempts to provide a unified field theory for matter have proposed that the fundamental particle is the superstring. A superstring is $\approx 10^{-35}$ m long (10^{-20} times the diameter of a proton). Use the Heisenberg uncertainty principle (Chapter 12),

$$\Delta x \cdot \Delta(mv) \geq \frac{h}{4\pi}$$

to calculate the uncertainty in the mass of a superstring. (Assume that $\Delta x = 1 \times 10^{-35}$ m and that the velocity is 10% the speed of light.) Compare this uncertainty in mass with the electron and proton masses.

46. Consider the following information:
   i. The layer of dead skin on our bodies is sufficient to protect us from most α-particle radiation.
   ii. Plutonium is an α-particle producer.
   iii. The chemistry of Pu^{4+} is similar to that of Fe^{3+}.
   iv. $Pu^{4+} + 4e^- \rightarrow Pu \quad \mathscr{E}° = -1.28$ V

Why is plutonium one of the most toxic substances known?

47. Fusion processes are more likely to occur for lighter elements, but fission processes are more likely to occur for heavier elements. Explain.

48. Why are elevated temperatures necessary to initiate fusion reactions but not fission reactions?

49. Much of the research on controlled fusion focuses on the problem of how to contain the reacting material. Magnetic fields appear to be the most promising mode of containment. Why is containment such a problem? Why must one resort to magnetic fields for containment?

50. How might the discovery of high-temperature superconducting materials affect the feasibility of using fusion as a practical power source?

51. What are the purposes of the moderator and control rods in a fission reactor?

52. Many metals become brittle when subjected to intense radiation from radioactive materials. Explain this observation. How might this fact affect the design of nuclear reactors?

53. Zirconium is one of the few metals that retain their structural integrity upon exposure to radiation. The fuel rods in most nuclear reactors therefore are often made of zirconium. Answer the following questions about the redox properties of zirconium based on the half-reaction

$$ZrO_2 \cdot H_2O + H_2O + 4e^- \longrightarrow Zr + 4OH^- \qquad \mathscr{E}^\circ = -2.36\ \text{V}$$

a. Is zirconium metal capable of reducing water to form hydrogen gas?
b. Write a balanced equation for the reduction of water by zirconium.
c. Calculate $\mathscr{E}^\circ$, ΔG°, and K for the reduction of water by zirconium metal.
d. The reduction of water by zirconium occurred during the accidents at Three Mile Island in 1979. The hydrogen produced was successfully vented and no chemical explosion occurred. If 1.00×10^3 kg of Zr reacts, what mass of H_2 is produced? What volume of H_2 at 1 atm and 1000°C is produced?
e. At Chernobyl in 1986, hydrogen was produced by the reaction of superheated steam with the graphite reactor core:

$$C(s) + H_2O(g) \longrightarrow CO(g) + H_2(g)$$

It was not possible to prevent a chemical explosion at Chernobyl. In light of this, do you think it was a correct decision to vent the hydrogen and other radioactive gases into the atmosphere at Three Mile Island? Explain.

54. In addition to the process described in the text, a second process called the carbon-nitrogen cycle occurs in the sun:

$$^{1}_{1}H + ^{12}_{6}C \longrightarrow ^{13}_{7}N + ^{0}_{0}\gamma$$
$$^{13}_{7}N \longrightarrow ^{13}_{6}C + ^{0}_{+1}e$$
$$^{1}_{1}H + ^{13}_{6}C \longrightarrow ^{14}_{7}N + ^{0}_{0}\gamma$$
$$^{1}_{1}H + ^{14}_{7}N \longrightarrow ^{15}_{8}O + ^{0}_{0}\gamma$$
$$^{15}_{8}O \longrightarrow ^{15}_{7}N + ^{0}_{+1}e$$
$$^{1}_{1}H + ^{15}_{7}N \longrightarrow ^{12}_{6}C + ^{4}_{2}He + ^{0}_{0}\gamma$$

$$\text{Overall reaction:} \quad 4\,^{1}_{1}H \longrightarrow ^{4}_{2}He + 2\,^{0}_{+1}e$$

a. What is the catalyst in the above scheme?
b. What nucleons are intermediates?
c. How much energy is released per mole of hydrogen atoms in the overall reaction? (See Exercises 35 and 36 for the appropriate mass values.)

55. Which do you think would be the greater health hazard: the release of a radioactive nuclide of Sr or a radioactive nuclide of Xe into the environment? Assume the amount of radioactivity is the same in each case. Explain your answer on the basis of the chemical properties of Sr and Xe.

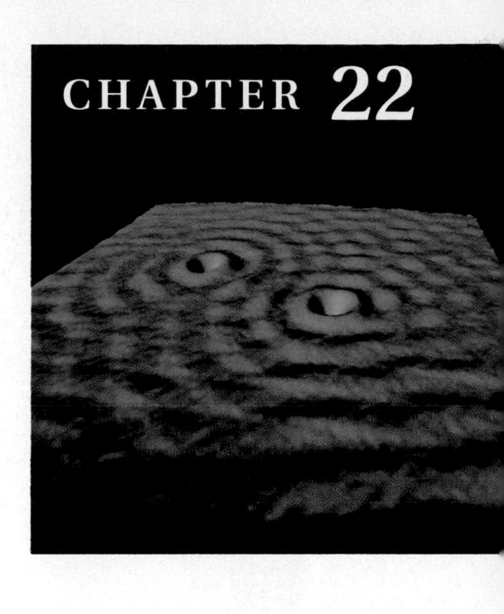

Organic Chemistry

Two Group 4A elements, carbon and silicon, form the basis of most natural substances. Silicon, with its great affinity for oxygen, forms chains and rings containing Si—O—Si bridges to produce the silica and silicates that form the basis for most rocks, sands, and soils. What silicon is to the geological world, carbon is to the biological world. Carbon has the unusual ability of bonding strongly to itself to form long chains or rings of carbon atoms. In addition, carbon forms strong bonds to other nonmetals such as hydrogen, nitrogen, oxygen, sulfur, and the halogens. Because of these bonding properties, there are a myriad of carbon compounds; several million are now known, and the number continues to grow rapidly. Among these many compounds are the **biomolecules,** those responsible for maintaining and reproducing life.

The study of carbon-containing compounds and their properties is called **organic chemistry.** Although a few compounds involving carbon, such as its oxides and carbonates, are considered to be inorganic substances, the vast majority are organic compounds that typically contain chains or rings of carbon atoms.

Originally, the distinction between inorganic and organic substances was based on whether or not a compound was produced by living systems. For example, until the early nineteenth century it was believed that organic compounds had some sort of "life force" and could only be synthesized by living organisms. This view was dispelled in 1828 when the German chemist Friedrich

Wöhler (1800–1882) prepared urea from the inorganic salt ammonium cyanate by simple heating:

$$\underset{\text{Ammonium cyanate}}{NH_4OCN} \xrightarrow{\text{Heat}} \underset{\text{Urea}}{H_2N{-}\underset{\underset{\displaystyle O}{\|}}{C}{-}NH_2}$$

Urea is a component of urine, so it is clearly an organic material; yet here was clear evidence that it could be produced in the laboratory as well as by living things.

Organic chemistry plays a vital role in our quest to understand living systems. Beyond that, the synthetic fibers, plastics, artificial sweeteners, and drugs that are such an accepted part of modern life are products of industrial organic chemistry. In addition, the energy on which we rely so heavily to power our civilization is based mostly on the organic materials found in coal and petroleum.

Because organic chemistry is such a vast subject, we can provide only a brief introduction to it in this book. We will begin with the simplest class of organic compounds, the hydrocarbons, and then show how most other organic compounds can be considered to be derivatives of hydrocarbons.

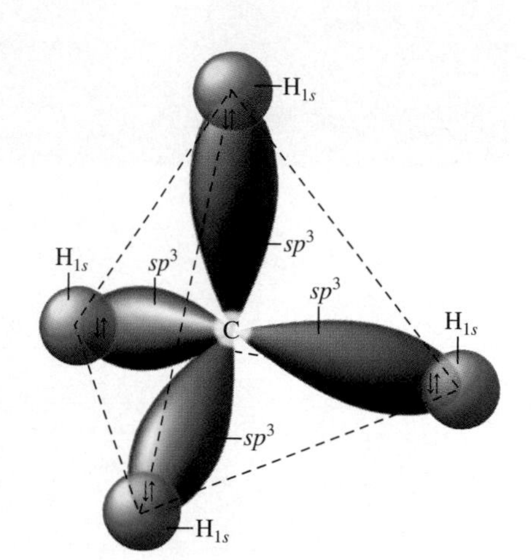

Figure 22.1
The C—H bonds in methane.

$$H{-}\underset{\underset{\displaystyle H}{|}}{\overset{\overset{\displaystyle H}{|}}{C}}{-}\underset{\underset{\displaystyle H}{|}}{\overset{\overset{\displaystyle H}{|}}{C}}{-}H$$

(a)

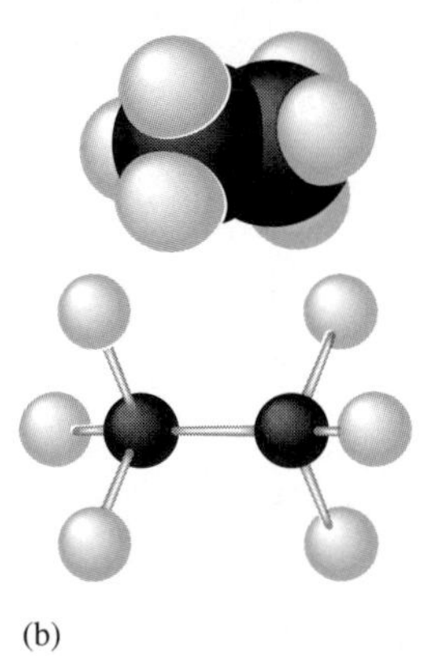

(b)

Figure 22.2
(a) The Lewis structure of ethane (C_2H_6). (b) The molecular structure of ethane represented by ball-and-stick and space-filling models.

22.1 Alkanes: Saturated Hydrocarbons

As the name indicates, **hydrocarbons** are compounds composed of carbon and hydrogen. Those compounds whose carbon-carbon bonds are all single bonds are said to be **saturated,** because each carbon is bound to four atoms, the maximum number. Hydrocarbons containing carbon-carbon multiple bonds are described as being **unsaturated,** since the carbon atoms involved in a multiple bond can react with additional atoms, as shown by the *addition* of hydrogen to ethylene:

$$\underset{\text{Unsaturated}}{H_2C{=}CH_2} + H_2 \longrightarrow \underset{\text{Saturated}}{H{-}\underset{\underset{\displaystyle H}{|}}{\overset{\overset{\displaystyle H}{|}}{C}}{-}\underset{\underset{\displaystyle H}{|}}{\overset{\overset{\displaystyle H}{|}}{C}}{-}H}$$

Note that each carbon in ethylene is bonded to three atoms (one carbon and two hydrogens) but that each can bond to one additional atom if one bond of the carbon-carbon double bond is broken.

The simplest member of the saturated hydrocarbons, which are also called the **alkanes,** is *methane* (CH_4). As discussed in Section 14.1, methane has a tetrahedral structure and can be described in terms of a carbon atom using an sp^3 hybrid set of orbitals to bond to the four hydrogen atoms (see Fig. 22.1). The next alkane, the one containing two carbon atoms, is *ethane* (C_2H_6), as shown in Fig. 22.2. Each carbon in ethane is surrounded by four atoms and thus adopts a tetrahedral arrangement and sp^3 hybridization, as predicted by the localized electron model.

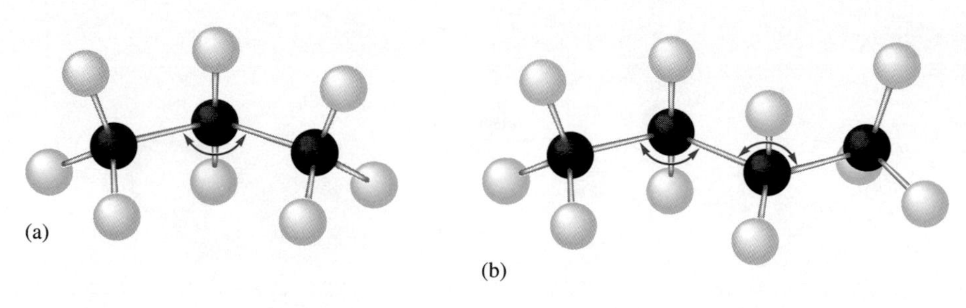

Figure 22.3
The structures of (a) propane ($CH_3CH_2CH_3$) and (b) butane ($CH_3CH_2CH_2CH_3$). Each angle shown in red is 109.5°.

The next two members of the series are *propane* (C_3H_8) and *butane* (C_4H_{10}), shown in Fig. 22.3. Again, each carbon is bonded to four atoms and is described as sp^3 hybridized.

Alkanes in which the carbon atoms form long "strings" or chains are called **normal, straight-chain,** or **unbranched hydrocarbons.** As can be seen from Fig. 22.3, the chains in normal alkanes are not really straight but zig-zag, since the tetrahedral C—C—C angle is 109.5°. The normal alkanes can be represented by the structure

$$H_3C\left(\begin{array}{c}H\\|\\C\\|\\H\end{array}\right)_n CH_3$$

where n is an integer. Note that each member is obtained from the previous one by inserting a *methylene* (CH_2) group. We can condense the structural formulas by omitting some of the C—H bonds. For example, the general formula for normal alkanes shown above can be condensed to

$$CH_3\text{—}(CH_2)_n\text{—}CH_3$$

The first ten normal alkanes and some of their properties are listed in Table 22.1. Note that all alkanes can be represented by the general formula C_nH_{2n+2}. For example, nonane, which has nine carbon atoms, is represented by $C_9H_{(2\times 9)+2}$, or C_9H_{20}. Also note from Table 22.1 that the melting points and boiling points increase as the molar masses increase, as we would expect.

TABLE 22.1 Selected Properties of the First Ten Normal Alkanes

Name	Formula	Molar Mass	Melting Point (°C)	Boiling Point (°C)	Number of Structural Isomers
Methane	CH_4	16	−182	−162	1
Ethane	C_2H_6	30	−183	−89	1
Propane	C_3H_8	44	−187	−42	1
Butane	C_4H_{10}	58	−138	0	2
Pentane	C_5H_{12}	72	−130	36	3
Hexane	C_6H_{14}	86	−95	68	5
Heptane	C_7H_{16}	100	−91	98	9
Octane	C_8H_{18}	114	−57	126	18
Nonane	C_9H_{20}	128	−54	151	35
Decane	$C_{10}H_{22}$	142	−30	174	75

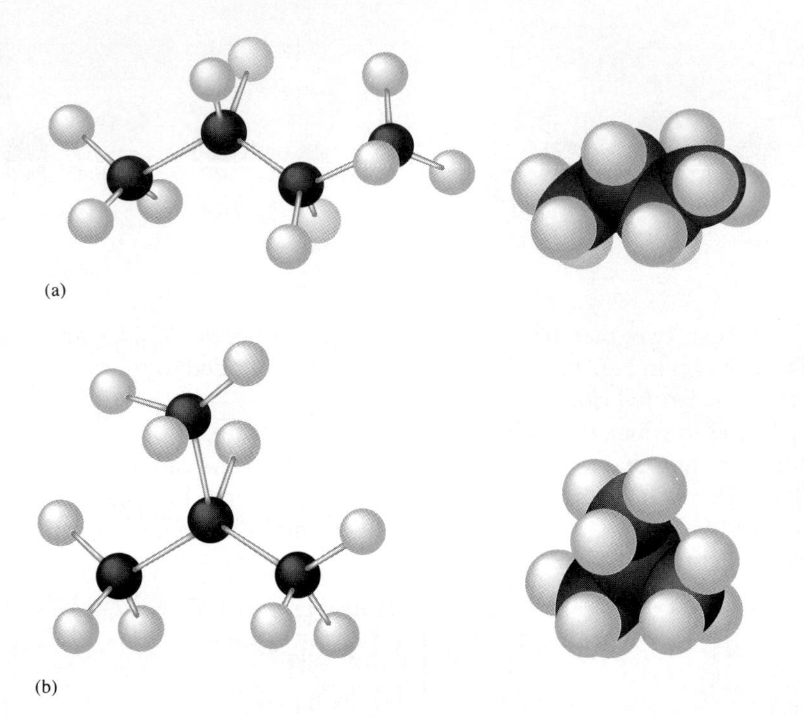

Figure 22.4
(a) Normal butane (abbreviated *n*-butane). (b) The branched isomer of butane (called isobutane).

Isomerism in Alkanes

Butane and all succeeding members of the alkanes exhibit **structural isomerism.** Recall from Section 20.4 that structural isomerism occurs when two molecules have the same atoms but different bonds. For example, butane can exist as a straight-chain molecule (normal butane, or *n*-butane) or with a branched-chain structure (called isobutane), as shown in Fig. 22.4. Because of their different structures, these molecules exhibit different properties. For example, the boiling point of *n*-butane is −0.5°C, whereas that of isobutane is −12°C.

EXAMPLE 22.1

Draw the isomers of pentane.

Solution

Pentane (C_5H_{12}) has the following isomeric structures:

1. $CH_3—CH_2—CH_2—CH_2—CH_3$

n-Pentane

2.

3.

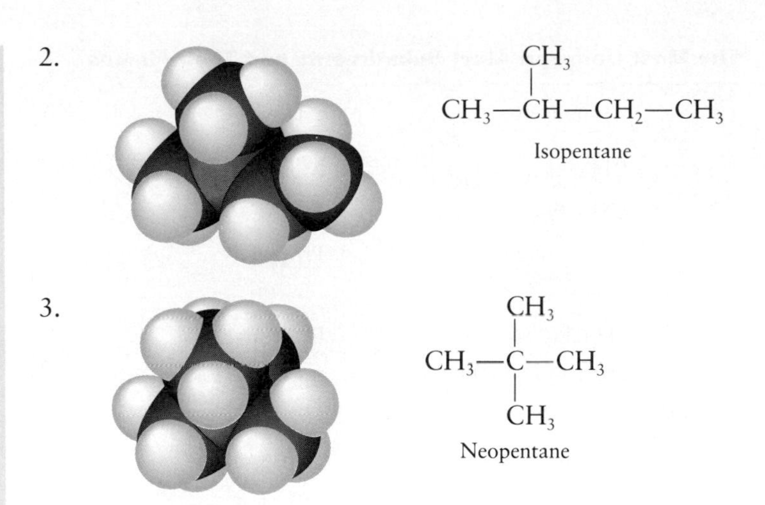

Note that the structures

$$CH_3-CH_2-\underset{}{\overset{\overset{\displaystyle CH_3}{|}}{C}H}-CH_3 \qquad CH_3-\underset{\underset{\displaystyle CH_3}{|}}{C}H-CH_2-CH_3$$

$$CH_3-CH_2-\underset{\underset{\displaystyle CH_3}{|}}{C}H-CH_3$$

which might appear to be other isomers, are actually identical to structure 2.

Nomenclature

Because there are literally millions of organic compounds, it would be impossible to remember common names for all of them. We must have a systematic method for naming them. The following rules are used in naming alkanes.

Rules for Naming Alkanes

1. The names of the alkanes beyond butane are obtained by adding the suffix *-ane* to the Greek root for the number of carbon atoms (*pent-* for five, *hex-* for six, and so on). For a branched hydrocarbon, the longest continuous chain of carbon atoms determines the root name for the hydrocarbon. For example, in the alkane

$$\begin{array}{c} CH_3 \\ | \\ CH_2 \\ | \\ CH_2 \\ | \\ CH_3-CH_2-CH-CH_2-CH_3 \end{array} \quad \text{Six carbons}$$

TABLE 22.2 **The Most Common Alkyl Substituents and Their Names**

Structure*	Name†
$—CH_3$	Methyl
$—CH_2CH_3$	Ethyl
$—CH_2CH_2CH_3$	Propyl
$CH_3\overset{\vert}{C}HCH_3$	Isopropyl
$—CH_2CH_2CH_2CH_3$	Butyl
$CH_3\overset{\vert}{C}HCH_2CH_3$	*sec*-Butyl
$—CH_2—\underset{\underset{CH_3}{\vert}}{\overset{\overset{H}{\vert}}{C}}—CH_3$	Isobutyl
$—\underset{\underset{CH_3}{\vert}}{\overset{\overset{CH_3}{\vert}}{C}}—CH_3$	*tert*-Butyl

* The bond with one end open shows the point of attachment of the substituent to the carbon chain.
† For the butyl groups, *sec-* indicates attachment to the chain through a secondary carbon, a carbon atom attached to *two* other carbon atoms. The designation *tert-* signifies attachment through a tertiary carbon, a carbon attached to *three* other carbon atoms.

the longest chain contains six carbon atoms, and this compound is named as a hexane.

2. When alkane groups appear as substituents, they are named by dropping the *-ane* and adding *-yl.* For example, $—CH_3$ is obtained by removing a hydrogen from methane and is called *methyl,* $—C_2H_5$ is called *ethyl,* $—C_3H_7$ is called *propyl,* and so on. The compound above is therefore an ethylhexane. (See Table 22.2.)

3. The positions of substituent groups are specified by numbering the longest chain of carbon atoms sequentially, starting at the end closest to the branching. For example, the compound

$$CH_3—CH_2—\overset{\overset{CH_3}{\vert}}{C}H—CH_2—CH_2—CH_3$$

1	2	3	4	5	6	Correct numbering
6	5	4	3	2	1	Incorrect numbering

is called 3-methylhexane. Note that the top set of numbers is correct since the left end of the molecule is closest to the branching, and this gives the smallest number for the position of the substituent. Also, note that a hyphen is written between the number and the substituent name.

4. The location and name of each substituent are followed by the root alkane name. The substituents are listed in alphabetical order, and the prefixes *di-*, *tri-*, and so on, are used to indicate multiple, identical substituents.

EXAMPLE 22.2

Draw the structural isomers for the alkane C_6H_{14} and give the systematic name for each one.

Solution

We will proceed systematically, starting with the longest chain and then rearranging the carbons to form the shorter, branched chains.

1. $CH_3CH_2CH_2CH_2CH_2CH_3$ Hexane

   Note that although a structure such as

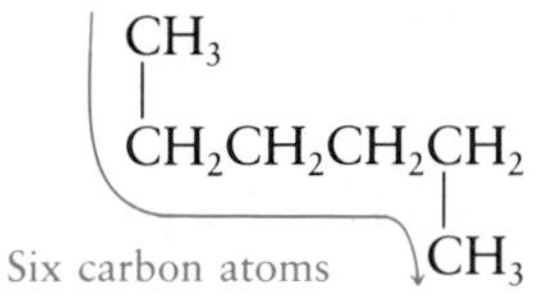

$$\begin{array}{l} CH_3 \\ | \\ CH_2CH_2CH_2CH_2 \\ \qquad\qquad\quad\;\; | \\ \qquad\qquad\quad\; CH_3 \end{array}$$

   Six carbon atoms

   may look different it is still hexane, since the longest carbon chain has six atoms.

2. We now take one carbon out of the chain and make it a methyl substituent.

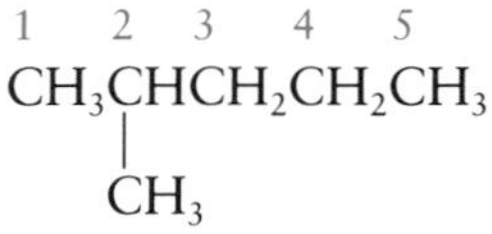

$$\begin{array}{l} \;1 \quad\; 2 \quad 3 \qquad 4 \qquad 5 \\ CH_3CHCH_2CH_2CH_3 \\ \qquad\;\, | \\ \qquad CH_3 \end{array}$$

   2-Methylpentane

   Since the longest chain consists of five carbons, this is a substituted pentane: 2-methylpentane. The 2 indicates the position of the methyl group on the chain. Note that if we numbered the chain from the right end, the methyl group would be on carbon 4. Because we want the smallest possible number, the numbering shown is correct.

3. The methyl substituent can also be on carbon 3 to give

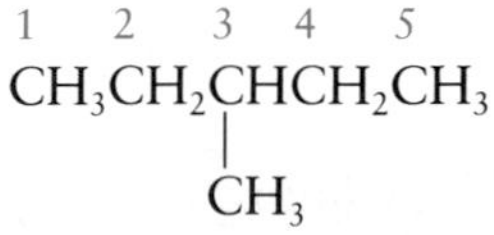

$$\begin{array}{l} \;1 \quad\; 2 \qquad 3 \quad 4 \qquad 5 \\ CH_3CH_2CHCH_2CH_3 \\ \qquad\qquad\;\, | \\ \qquad\qquad CH_3 \end{array}$$

   3-Methylpentane

   Note that we have now exhausted all possibilities for placing a single methyl group on pentane.

4. Next, we can take two carbons out of the original six-member chain:

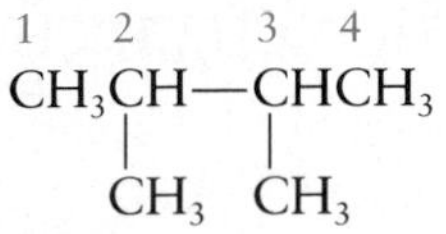

$$\begin{array}{l} \;1 \quad\; 2 \qquad\quad 3 \quad 4 \\ CH_3CH—CHCH_3 \\ \qquad\;\, | \qquad\;\; | \\ \qquad CH_3 \;\; CH_3 \end{array}$$

   2,3-Dimethylbutane

Since the longest chain now has four carbons, the root name is butane. Since there are two methyl groups, we use the prefix *di-*. The numbers denote that the two methyl groups are positioned on the second and third carbons in the butane chain. Note that when two or more numbers are used, they are separated by a comma.

5. The two methyl groups can also be attached to the same carbon atom as shown here:

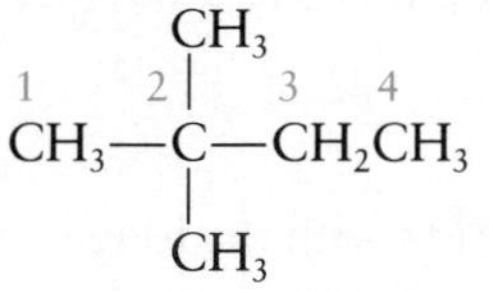

```
        CH3
 1    2 |  3   4
CH3—C—CH2CH3        2,2-Dimethylbutane
        |
        CH3
```

We might also try ethyl-substituted butanes, such as

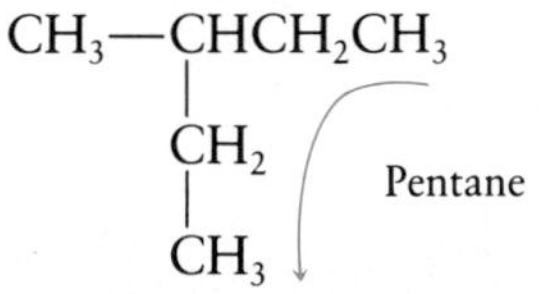

```
CH3—CHCH2CH3
       |
       CH2        Pentane
       |
       CH3
```

However, note that this is instead a pentane (3-methylpentane), since the longest chain has five carbon atoms. Thus this is not a new isomer. Trying to reduce the chain to three atoms provides no further isomers either. For example, the structure

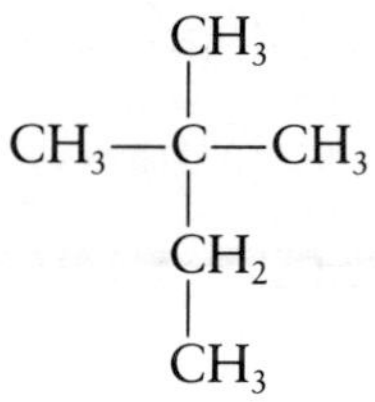

```
          CH3
          |
   CH3—C—CH3
          |
          CH2
          |
          CH3
```

is actually 2,2-dimethylbutane.

Thus there are only five distinct structural isomers of C_6H_{14}: hexane, 2-methylpentane, 3-methylpentane, 2,3-dimethylbutane, and 2,2-dimethylbutane.

EXAMPLE 22.3

Determine the structure for each of the following compounds.

a. 4-ethyl-3,5-dimethylnonane **b.** 4-*tert*-butylheptane

Solution

a. The root name *nonane* signifies a nine-carbon chain. Thus we have

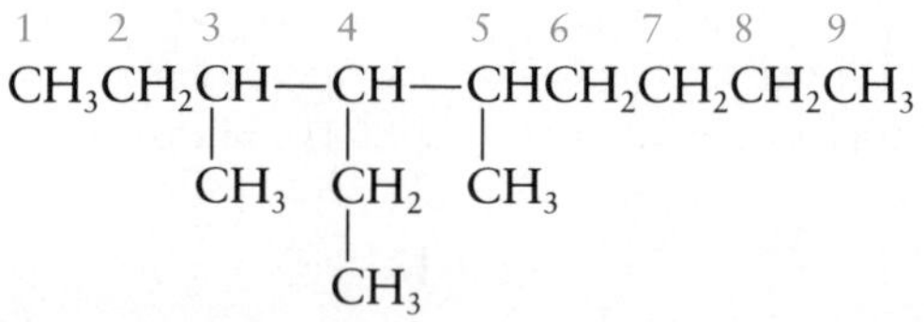

```
 1   2   3     4     5  6   7   8   9
CH3CH2CH—CH—CHCH2CH2CH2CH3
          |     |     |
        CH3  CH2  CH3
                |
               CH3
```

b. Heptane signifies a seven-carbon chain, and the *tert*-butyl group is

$$\begin{array}{c} | \\ H_3C\text{—}C\text{—}CH_3 \\ | \\ CH_3 \end{array}$$

Thus we have

$$\begin{array}{c} \overset{1}{CH_3}\overset{2}{CH_2}\overset{3}{CH_2}\overset{4}{C}H\overset{5}{CH_2}\overset{6}{CH_2}\overset{7}{CH_3} \\ | \\ H_3C\text{—}C\text{—}CH_3 \\ | \\ CH_3 \end{array}$$

Reactions of Alkanes

Because they are saturated compounds and because the C—C and C—H bonds are relatively strong, the alkanes are fairly unreactive. For example, at 25°C they do not react with acids, bases, or strong oxidizing agents. This chemical inertness makes them valuable as lubricating materials and as the backbone for structural materials such as plastics.

At a sufficiently high temperature alkanes do react vigorously and exothermically with oxygen, and these **combustion reactions** are the basis for their widespread use as fuels. For example, the reaction of butane with oxygen is

$$2C_4H_{10}(g) + 13O_2(g) \longrightarrow 8CO_2(g) + 10H_2O(g)$$

The alkanes can also undergo **substitution reactions,** primarily where halogen atoms replace hydrogen atoms. For example, methane can be successively chlorinated as follows:

The *hν* above the arrow represents ultraviolet light.

$$CH_4 + Cl_2 \xrightarrow{h\nu} \underset{\text{Chloromethane}}{CH_3Cl} + HCl$$

$$CH_3Cl + Cl_2 \xrightarrow{h\nu} \underset{\text{Dichloromethane}}{CH_2Cl_2} + HCl$$

$$CH_2Cl_2 + Cl_2 \xrightarrow{h\nu} \underset{\substack{\text{Trichloromethane} \\ \text{(chloroform)}}}{CHCl_3} + HCl$$

$$CHCl_3 + Cl_2 \xrightarrow{h\nu} \underset{\substack{\text{Tetrachloromethane} \\ \text{(carbon tetrachloride)}}}{CCl_4} + HCl$$

Note that the products of the last two reactions have two names; the systematic name is given first, followed by the common name in parentheses. (This format will be used throughout this chapter for compounds that have common names.) Also, note that ultraviolet light (*hν*) furnishes the energy to break the Cl—Cl bond to produce chlorine atoms:

$$Cl_2 \longrightarrow Cl\cdot + Cl\cdot$$

A chlorine atom has an unpaired electron, as indicated by the dot, which makes it very reactive and able to attack the C—H bond.

As we mentioned before, substituted methanes with the general formula CF_xCl_{4-x} containing both chlorine and fluorine as substituents are called chlorofluorocarbons (CFCs) and are also known as *freons*. These substances are very unreactive and are extensively used as coolant fluids in refrigerators and air conditioners. Unfortunately, their chemical inertness allows freons to remain in the atmosphere so long that they eventually reach altitudes where they are a threat to the protective ozone layer (see Section 15.9), and the use of these compounds is being rapidly phased out.

Alkanes can also undergo **dehydrogenation reactions** in which hydrogen atoms are removed and the product is an unsaturated hydrocarbon. For example, in the presence of chromium(III) oxide at high temperatures, ethane can be dehydrogenated, yielding ethylene:

$$CH_3CH_3 \xrightarrow[500°C]{Cr_2O_3} \underset{\text{Ethylene}}{CH_2{=}CH_2} + H_2$$

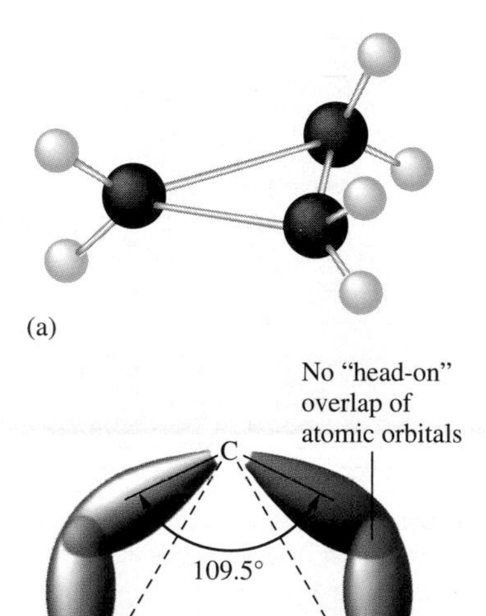

Figure 22.5
(a) The molecular structure of cyclopropane (C_3H_6). (b) The overlap of the sp^3 orbitals that form the C—C bonds in cyclopropane.

Cyclic Alkanes

Besides forming chains, carbon atoms also form rings. The simplest of the **cyclic alkanes** (general formula C_nH_{2n}) is cyclopropane (C_3H_6), shown in Fig. 22.5(a). Since the carbon atoms in cyclopropane form an equilateral triangle with 60° bond angles, their sp^3 hybrid orbitals do not overlap head-on as in normal alkanes [Fig. 22.5(b)]. This results in unusually weak, or *strained*, C—C bonds; thus the cyclopropane molecule is much more reactive than straight-chain propane. The carbon atoms in cyclobutane (C_4H_8) form a square with 90° bond angles, and cyclobutane is also quite reactive.

The next two members of the series, cyclopentane (C_5H_{10}) and cyclohexane (C_6H_{12}), are quite stable, because their rings have bond angles very close to the tetrahedral angles, which allows the sp^3 hybrid orbitals on adjacent carbon atoms to overlap head-on and form normal C—C bonds, which are quite strong. To attain the tetrahedral angles, the cyclohexane ring must "pucker," that is, become nonplanar. Cyclohexane can exist in two forms, the *chair* or the *boat* forms, as shown in Fig. 22.6. The two hydrogen atoms above the ring in the boat form are quite close to each other, and the resulting repulsion between these atoms causes the chair form to be preferred. At 25°C more than 99% of cyclohexane exists in the chair form.

For simplicity, the cyclic alkanes are often represented by the following structures:

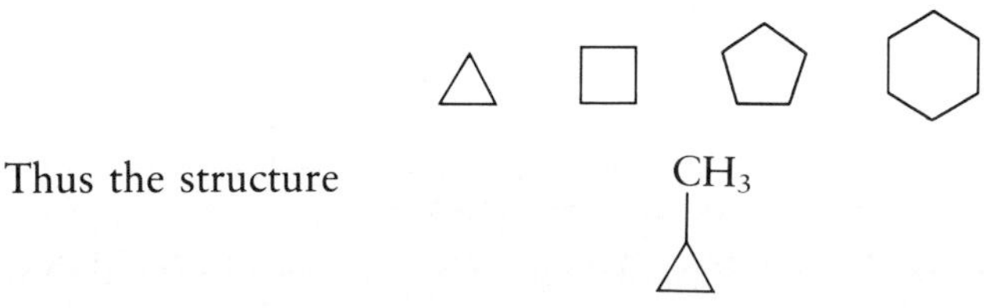

Thus the structure

represents methylcyclopropane.

The nomenclature for cycloalkanes follows the same rules as for the other alkanes except that the root name is preceded by the prefix *cyclo-*. The ring is numbered to yield the smallest substituent numbers possible.

EXAMPLE 22.4

Name each of the following cyclic alkanes.

a. $CH_3—CH—CH_3$ (attached to a cyclohexane ring bearing a CH_3 group)

b. Cyclobutane ring bearing CH_2CH_3 and $CH_2CH_2CH_3$ groups

Solution

a. The six-carbon cyclohexane ring is numbered as follows:

$CH_3—CH—CH_3$ at carbon 1; ring carbons numbered 1, 2, 3, 4, 5, 6; CH_3 at carbon 3

There is an isopropyl group at carbon 1 and a methyl group at carbon 3. The name is 1-isopropyl-3-methylcyclohexane, since the alkyl groups are named in alphabetical order.

b. This is a cyclobutane ring, which is numbered as follows:

CH_2CH_3 at carbon 1; ring carbons numbered 1, 2, 3, 4; $CH_2CH_2CH_3$ at carbon 2

The name is 1-ethyl-2-propylcyclobutane.

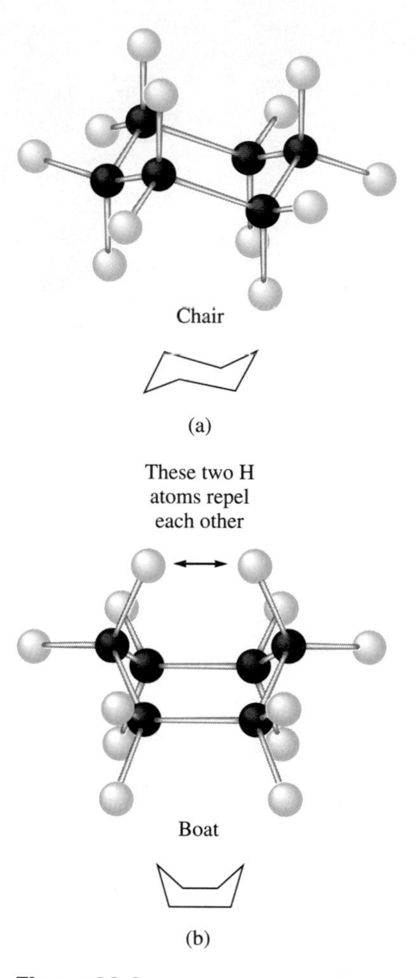

Figure 22.6
The (a) chair and (b) boat forms of cyclohexane.

22.2 Alkenes and Alkynes

Multiple carbon-carbon bonds result when hydrogen atoms are removed from alkanes. Hydrocarbons that contain at least one carbon-carbon double bond are called **alkenes** and have the general formula C_nH_{2n}. The simplest alkene (C_2H_4), commonly known as *ethylene,* has the Lewis structure

$$\begin{matrix} H & & H \\ & C{=}C & \\ H & & H \end{matrix}$$

As discussed in Section 14.1, each carbon in ethylene can be described as sp^2 hybridized. The C—C σ bond is formed by sharing an electron pair between sp^2 orbitals, and the π bond is formed by sharing a pair of electrons between p orbitals (Fig. 22.7).

The systematic nomenclature for alkenes is quite similar to that for alkanes.

1. The root hydrocarbon name ends in *-ene* rather than *-ane*. Thus the systematic name for C_2H_4 is *ethene* and the name for C_3H_6 is propene.

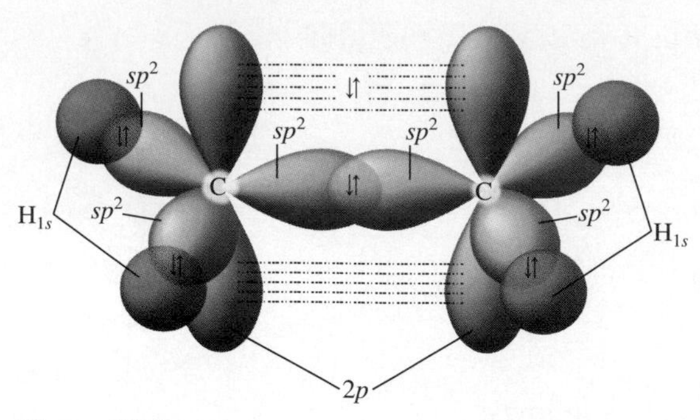

Figure 22.7
The bonding in ethylene.

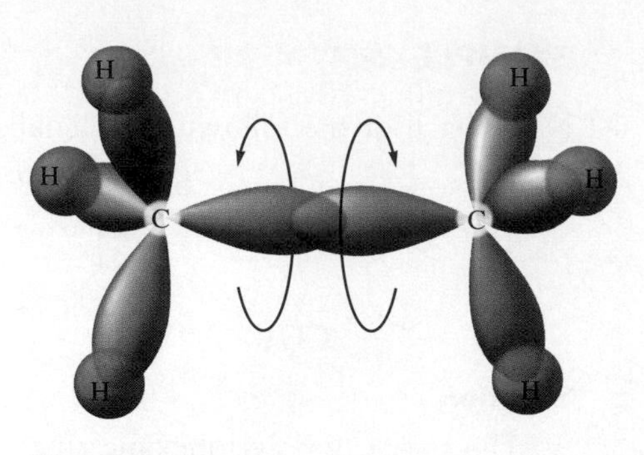

Figure 22.8
The bonding in ethane.

2. In alkenes containing more than three carbon atoms, the location of the double bond is indicated by the lowest numbered carbon atom involved in the bond. Thus $CH_2{=}CHCH_2CH_3$ is called 1-butene, and $CH_3CH{=}CHCH_3$ is called 2-butene.

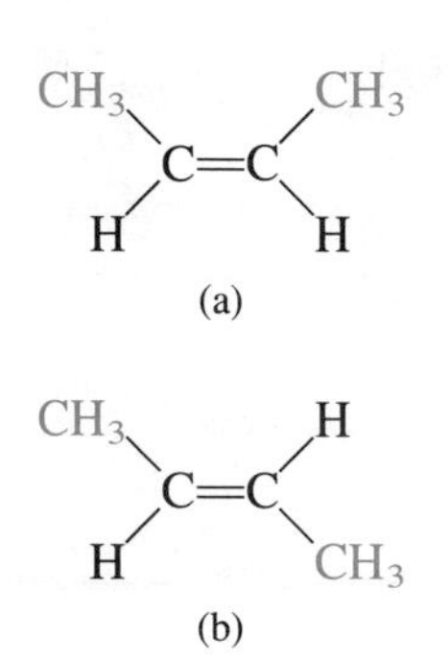

Figure 22.9
The two stereoisomers of 2-butene: (a) *cis*-2-butene and (b) *trans*-2-butene.

Note from Fig. 22.7 that the *p* orbitals on the two carbon atoms in ethylene must be lined up (parallel) to allow formation of the π bond. This prevents rotation of the two CH_2 groups relative to each other at ordinary temperatures, in contrast to alkanes, where free rotation is possible (see Fig. 22.8). The restricted rotation about doubly bonded carbon atoms means that alkenes exhibit ***cis-trans*** **isomerism.** For example, there are two stereoisomers of 2-butene (Fig. 22.9). Identical substituents on the same side of the double bond are designated *cis* and those on opposite sides are labeled *trans.*

Alkynes are unsaturated hydrocarbons containing at least one triple carbon-carbon bond. The simplest alkyne is C_2H_2 (commonly called *acetylene*), which has the systematic name *ethyne.* As discussed in Section 14.1, the triple bond in acetylene can best be described as one σ bond between two *sp* hybrid orbitals on the two carbon atoms and two π bonds involving two 2*p* orbitals on each carbon atom (Fig. 22.10).

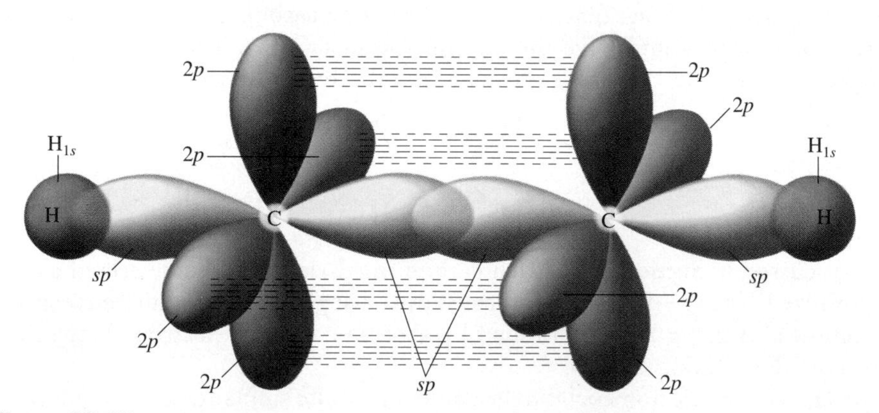

Figure 22.10
The bonding in acetylene.

The nomenclature for alkynes involves the use of *-yne* as a suffix to replace the *-ane* of the parent alkane. Thus the molecule $CH_3CH_2C{\equiv}CCH_3$ has the name 2-pentyne.

Like alkanes, unsaturated hydrocarbons can exist as ringed structures, for example,

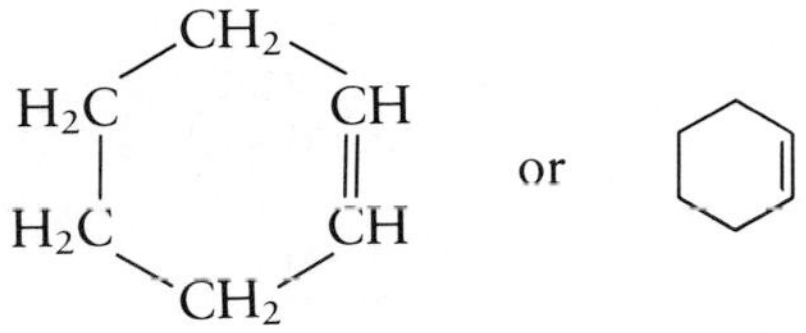

$$\begin{array}{ccc} & CH_2 & \\ H_2C & & CH \\ | & & \| \\ H_2C & & CH \\ & CH_2 & \end{array} \quad \text{or} \quad \text{(hexagon with one double bond)}$$

Cyclohexene

$$\begin{array}{ccc} & CH_3 & \\ & | & \\ & CH & \\ H_2C & & CH_2 \\ & HC{=}CH & \end{array} \quad \text{or} \quad \text{(pentagon with one double bond and } CH_3\text{)}$$

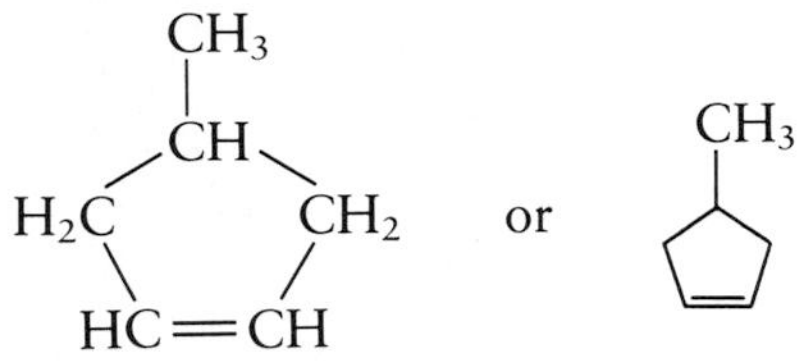

4-Methyl-cyclopentene

For cyclic alkenes, number through the double bond toward the substituent.

EXAMPLE 22.5

Name each of the following molecules.

a.
$$\begin{array}{ccccc} & H & & CH_3 & \\ & & C{=}C & & \\ CH_3CH_2CH & & & H & \\ | & & & & \\ CH_3 & & & & \end{array}$$

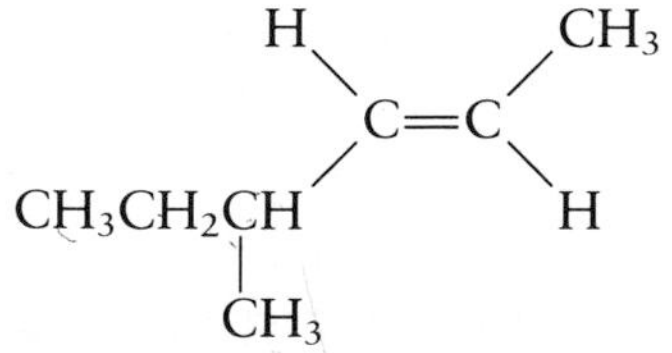

b.
$$\begin{array}{l} CH_3CH_2C{\equiv}CCHCH_2CH_3 \\ \qquad\qquad\quad | \\ \qquad\qquad\quad CH_2 \\ \qquad\qquad\quad | \\ \qquad\qquad\quad CH_3 \end{array}$$

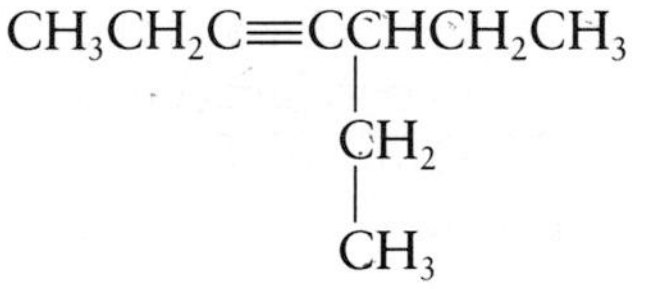

Solution

a. The longest chain, which contains six carbon atoms, is numbered as follows:

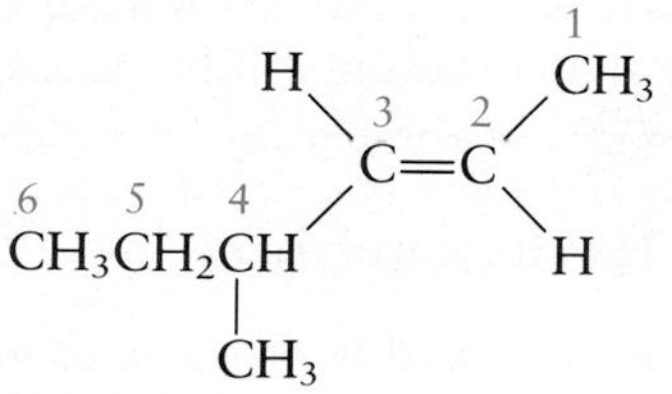

$$\begin{array}{ccccc} & H & & \overset{1}{CH_3} & \\ & & \overset{3}{C}{=}\overset{2}{C} & & \\ \overset{6}{C}H_3\overset{5}{C}H_2\overset{4}{C}H & & & H & \\ | & & & & \\ CH_3 & & & & \end{array}$$

Thus the hydrocarbon is a 2-hexene. Since the hydrogen atoms are located on opposite sides of the double bond, this molecule corresponds to the *trans* isomer. The name is 4-methyl-*trans*-2-hexene.

b. The longest chain, consisting of seven carbon atoms, is numbered as shown (giving the triple bond the lowest possible number):

$$\overset{1}{C}H_3\overset{2}{C}H_2\overset{3}{C}\equiv\overset{4}{C}\overset{5}{C}H\overset{6}{C}H_2\overset{7}{C}H_3 \quad (\text{C-5 bears } CH_2CH_3)$$

The hydrocarbon is a 3-heptyne. The full name is 5-ethyl-3-heptyne, where the position of the triple bond is indicated by the lower-numbered carbon atom involved in this bond.

Reactions of Alkenes and Alkynes

Because alkenes and alkynes are unsaturated, their most important reactions are **addition reactions.** In these reactions π bonds, which are weaker than the C—C σ bonds, are broken, and new σ bonds are formed to the atoms being added. For example, **hydrogenation reactions** involve the addition of hydrogen atoms:

$$\underset{\text{1-Propene}}{CH_2{=}CHCH_3} + H_2 \xrightarrow{\text{Catalyst}} \underset{\text{Propane}}{CH_3CH_2CH_3}$$

In order for this reaction to proceed rapidly at normal temperatures, a catalyst of platinum, palladium, or nickel is employed. The catalyst serves to help break the relatively strong H—H bond, as was discussed in Section 15.9. Hydrogenation of alkenes is an important industrial process, particularly in the manufacture of solid shortenings where unsaturated fats (fats containing double bonds), which are generally liquid, are converted to solid saturated fats.

Halogenation of unsaturated hydrocarbons involves addition of halogen atoms, for example,

$$\underset{\text{1-Pentene}}{CH_2{=}CHCH_2CH_2CH_3} + Br_2 \longrightarrow \underset{\text{1,2-Dibromopentane}}{CH_2BrCHBrCH_2CH_2CH_3}$$

Another important reaction involving certain unsaturated hydrocarbons is **polymerization,** a process in which many small molecules are joined together to form a large molecule. Polymerization will be discussed in Section 22.5.

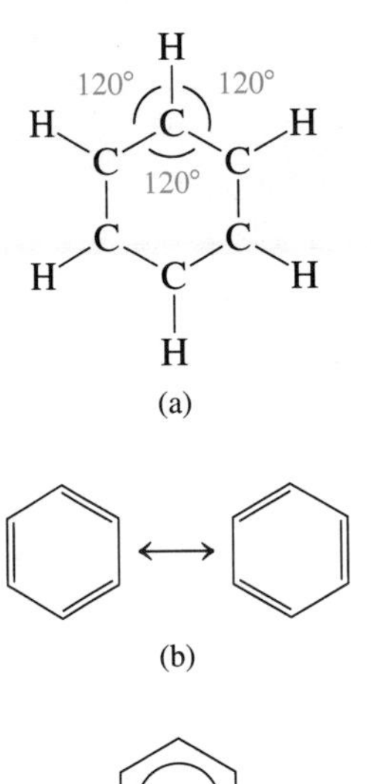

Figure 22.11
(a) The structure of benzene, a planar ring system in which all bond angles are 120°. (b) Two of the resonance structures of benzene. (c) The usual representation of benzene. The circle represents the electrons in the delocalized π system. All C—C bonds in benzene are equivalent.

22.3 Aromatic Hydrocarbons

A special class of cyclic unsaturated hydrocarbons is known as the **aromatic hydrocarbons.** The simplest of these is benzene (C_6H_6), which has a planar ring structure, as shown in Fig. 22.11(a). In the localized electron model of the bonding in benzene, resonance structures of the type shown in Fig. 22.11(b) are used to account for the known equivalence of all of the carbon-carbon bonds. But as we discussed in Section 14.5, the best description of the benzene molecule assumes that sp^2 hybrid orbitals on each carbon are used to form the C—C and C—H σ bonds, while the remaining $2p$ orbital on each carbon is used to form π molecular orbitals. The delocalization of these π electrons is usually indicated by a circle inside the ring [Fig. 22.11(c)].

The delocalization of the π electrons makes the benzene ring behave quite differently from a typical unsaturated hydrocarbon. As we have seen previously, unsaturated hydrocarbons typically undergo rapid addition reactions. However, benzene does not. Instead, it undergoes substitution reactions in which *hydrogen atoms are replaced by other atoms,* for example,

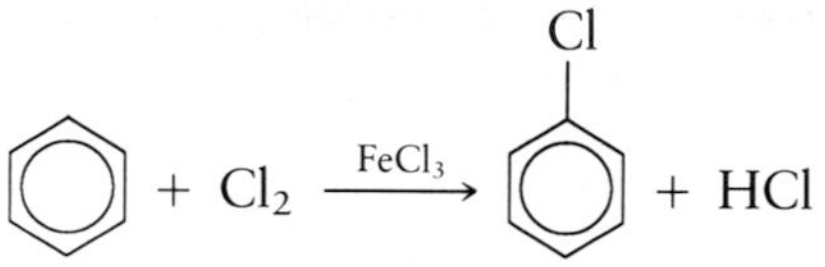

$$C_6H_6 + Cl_2 \xrightarrow{FeCl_3} C_6H_5Cl + HCl$$

Chlorobenzene

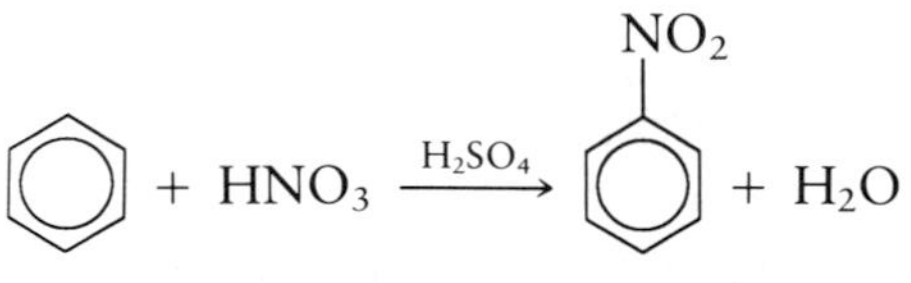

$$C_6H_6 + HNO_3 \xrightarrow{H_2SO_4} C_6H_5NO_2 + H_2O$$

Nitrobenzene

$$C_6H_6 + CH_3Cl \xrightarrow{AlCl_3} C_6H_5CH_3 + HCl$$

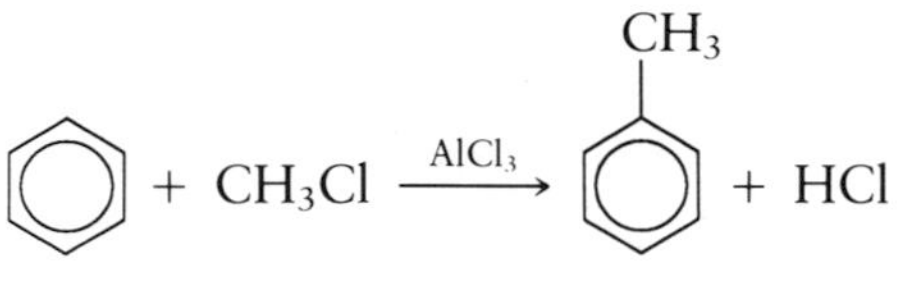

Toluene

In each case the substance shown over the arrow is needed to catalyze these substitution reactions.

Substitution reactions are characteristic of saturated hydrocarbons, and addition reactions are characteristic of unsaturated ones. The fact that benzene reacts more like a saturated hydrocarbon indicates the great stability of the delocalized π electron system.

The nomenclature of benzene derivatives is similar to the nomenclature for saturated ring systems. If there is more than one substituent present, numbers are used to indicate substituent positions. For example, the compound

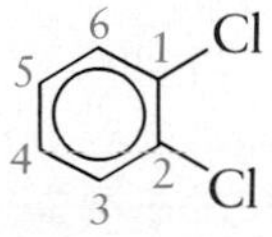

(benzene ring with positions numbered 1–6; Cl at positions 1 and 2)

is named 1,2-dichlorobenzene. Another nomenclature system employs the prefix *ortho-* (*o-*) for two adjacent substituents, *meta-* (*m-*) for two substituents with one carbon between them, and *para-* (*p-*) for two substituents opposite each other. When benzene is used as a substituent, it is called the **phenyl group.** Examples of some aromatic compounds are shown in Fig. 22.12.

Benzene is the simplest aromatic molecule. More complex aromatic systems can be viewed as consisting of a number of "fused" benzene rings. Some examples are given in Table 22.3.

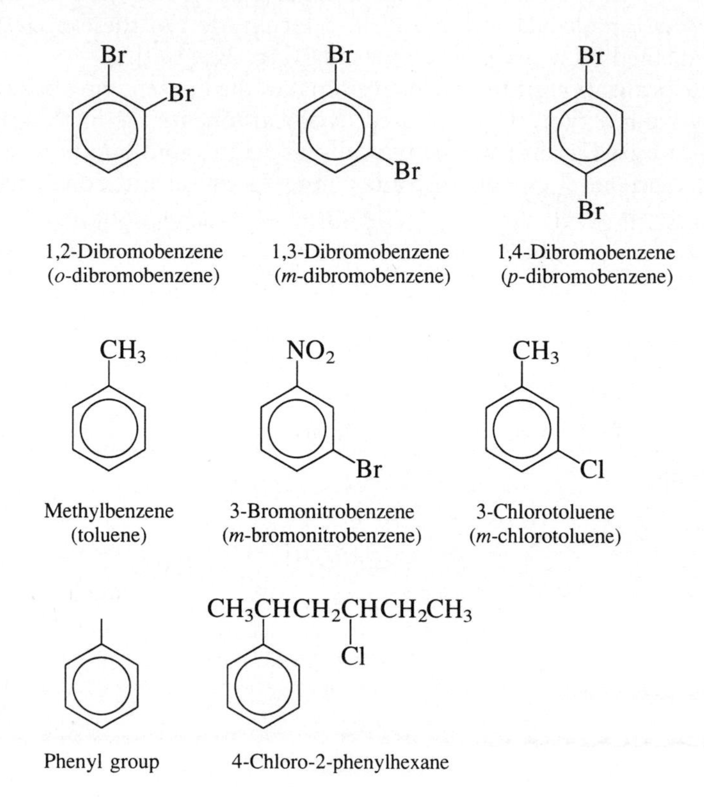

Figure 22.12
Some selected substituted benzenes and their names. Common names are given in parentheses.

TABLE 22.3 More Complex Aromatic Systems

Structural Formula	Name	Use or Effect
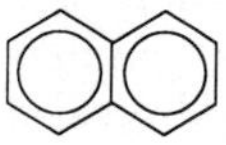	Naphthalene	Mothballs
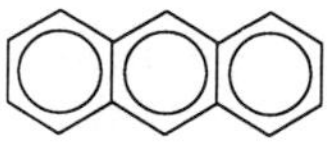	Anthracene	Dyes
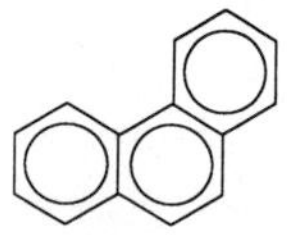	Phenanthrene	Dyes, explosives, and synthesis of drugs
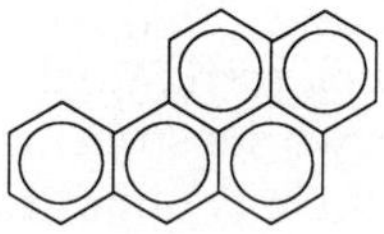	3,4-Benzpyrene	Active carcinogen found in smoke and smog

22.4 Hydrocarbon Derivatives

The vast majority of organic molecules contain elements in addition to carbon and hydrogen. However, most of these substances can be viewed as **hydrocarbon derivatives,** molecules that are fundamentally hydrocarbons but that have additional atoms or groups of atoms called **functional groups.** The common functional groups are listed in Table 22.4. Because each functional group exhibits characteristic chemistry, we will consider them separately.

Alcohols

Alcohols are characterized by the presence of the hydroxyl group (—OH). Some common alcohols are shown in Table 22.5. The systematic name for an alcohol is obtained by replacing the final *-e* of the parent hydrocarbon with *-ol.* The position of the —OH group is specified by a number (where necessary)

TABLE 22.4 The Common Functional Groups

Class	Functional Group	General Formula*	Example
Halohydrocarbons	—X (F,Cl,Br,I)	R—X	CH_3I Iodomethane (methyl iodide)
Alcohols	—OH	R—OH	CH_3OH Methanol (methyl alcohol)
Ethers	—O—	R—O—R′	CH_3OCH_3 Dimethyl ether
Aldehydes	—C(=O)—H	R—C(=O)—H	CH_2O Methanal (formaldehyde)
Ketones	—C(=O)—	R—C(=O)—R′	CH_3COCH_3 Propanone (dimethyl ketone or acetone)
Carboxylic acids	—C(=O)—OH	R—C(=O)—OH	CH_3COOH Ethanoic acid (acetic acid)
Esters	—C(=O)—O—	R—C(=O)—O—R′	$CH_3COOCH_2CH_3$ Ethyl acetate
Amines	$—NH_2$	$R—NH_2$	CH_3NH_2 Aminomethane (methylamine)

* R and R′ represent hydrocarbon fragments.

TABLE 22.5 Some Common Alcohols

Formula	Systematic Name	Common Name
CH_3OH	Methanol	Methyl alcohol
CH_3CH_2OH	Ethanol	Ethyl alcohol
$CH_3CH_2CH_2OH$	1-Propanol	*n*-Propyl alcohol
$CH_3CH(OH)CH_3$	2-Propanol	Isopropyl alcohol

chosen so that it is the smallest of the substituent numbers. Alcohols are classified according to the number of hydrocarbon fragments bonded to the carbon where the —OH group is attached,

$$R{-}CH_2OH \qquad \begin{matrix} R \\ & \diagdown \\ & & CHOH \\ & \diagup \\ R' \end{matrix} \qquad \begin{matrix} & R \\ & | \\ R'{-} & COH \\ & | \\ & R'' \end{matrix}$$

Primary alcohol (one R group) | *Secondary* alcohol (two R groups) | *Tertiary* alcohol (three R groups)

where R, R′, and R″ (which may be the same or different) represent hydrocarbon fragments.

Alcohols usually have much higher boiling points than might be expected from their molar masses. For example, both methanol and ethane have a molar mass of 30, but the boiling point for methanol is 65°C while that for ethane is −89°C. This difference can be understood if we consider the types of intermolecular attractions that occur in these liquids. Ethane molecules are nonpolar and exhibit only weak London dispersion interactions. However, the polar —OH group of methanol produces extensive hydrogen bonding similar to that found in water (see Section 16.1), which results in the relatively high boiling point.

Although there are many important alcohols, the simplest ones, methanol and ethanol, have the greatest commercial value. Methanol, also known as *wood alcohol* because it was formerly obtained by heating wood in the absence of air, is prepared industrially (≈4 million tons annually in the United States) by the hydrogenation of carbon monoxide:

$$CO + 2H_2 \xrightarrow[ZnO/Cr_2O_3]{400°C} CH_3OH$$

Methanol is used as a starting material for the synthesis of acetic acid and for many types of adhesives, fibers, and plastics. It is also used (and such use may increase) as a motor fuel. Methanol is highly toxic to humans and can lead to blindness and death if ingested.

Ethanol is the alcohol found in beverages such as beer, wine, and whiskey; it is produced by the fermentation of glucose in corn, barley, grapes, and so on:

$$\underset{\text{Glucose}}{C_6H_{12}O_6} \xrightarrow{\text{Yeast}} \underset{\text{Ethanol}}{2CH_3CH_2OH} + 2CO_2$$

EXAMPLE 22.6

For each of the following alcohols, give the systematic name and specify whether the alcohol is primary, secondary, or tertiary.

a. 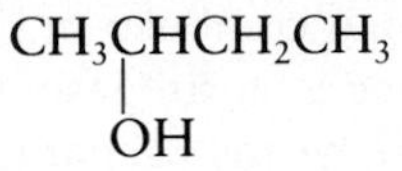 $CH_3CHCH_2CH_3$ (with OH on C-2)

b. $ClCH_2CH_2CH_2OH$

c. 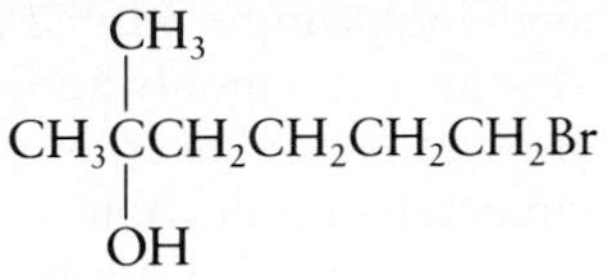 $CH_3CCH_2CH_2CH_2CH_2Br$ (with CH_3 and OH on C-2)

Solution

a. The chain is numbered as follows:

$$\overset{1}{CH_3}\underset{|\atop OH}{\overset{2}{C}H}\overset{3}{C}H_2\overset{4}{C}H_3$$

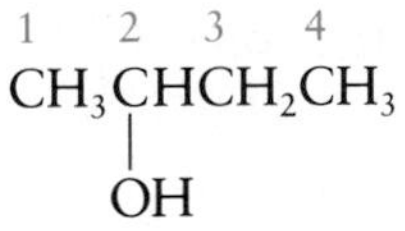

The compound is called 2-butanol, since the —OH group is located at the number 2 position of a four-carbon chain. Note that the carbon to which the —OH is attached also has $—CH_3$ and $—CH_2CH_3$ groups attached:

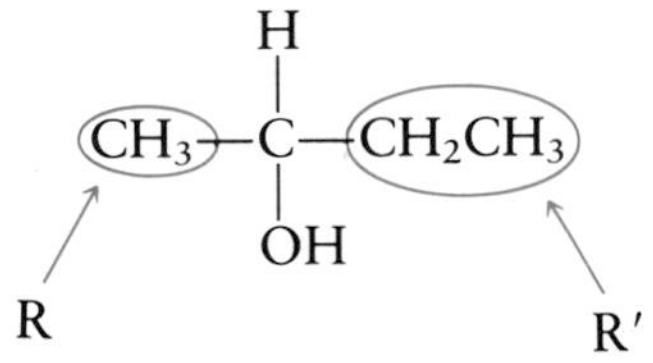

$$\underset{R}{(CH_3)}—\overset{H}{\underset{OH}{C}}—\underset{R'}{(CH_2CH_3)}$$

Therefore, this is a *secondary* alcohol.

b. The chain is numbered as follows:

$$\overset{}{Cl}—\overset{3}{CH_2}—\overset{2}{CH_2}—\overset{1}{CH_2}—OH$$

The name is 3-chloro-1-propanol. This is a *primary* alcohol:

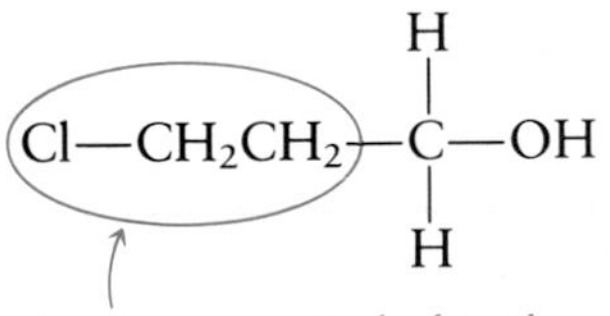

$$(Cl—CH_2CH_2)—\overset{H}{\underset{H}{C}}—OH$$

One R group attached to the carbon with the —OH group

c. The chain is numbered as follows:

$$\overset{1}{(CH_3)}—\overset{(CH_3)}{\underset{OH}{\overset{2}{C}}}—(\overset{3}{C}H_2—\overset{4}{C}H_2—\overset{5}{C}H_2—\overset{6}{C}H_2Br)$$

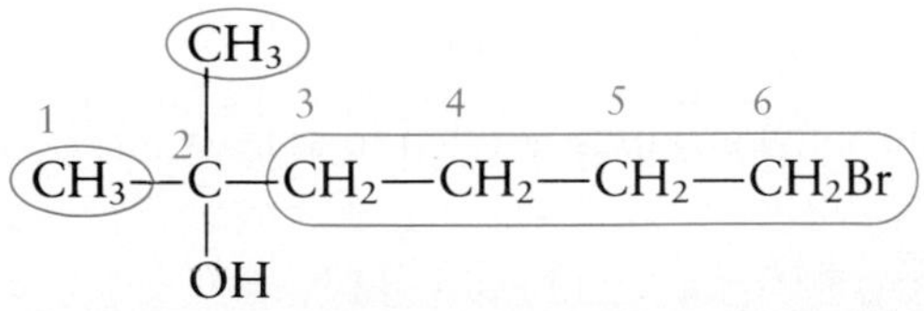

The name is 6-bromo-2-methyl-2-hexanol. This is a *tertiary* alcohol since the carbon where the —OH is attached also has three other R groups attached.

Wine fermentation tanks at Half Moon Bay, California.

The reaction is catalyzed by the enzymes found in yeast. This reaction can proceed only until the alcohol content reaches approximately 13% (that found in most wines), at which point the yeast can no longer survive. Beverages with higher alcohol content are made by distilling the fermentation mixture.

Ethanol, like methanol, can be burned in the internal combustion engines of automobiles and is now commonly added to gasoline to form gasohol (see Section 9.8). It is also used in industry as a solvent and for the preparation of acetic acid. The commercial production of ethanol (one half million tons per year in the United States) is carried out by reaction of water with ethylene:

$$CH_2{=}CH_2 + H_2O \xrightarrow[\text{Catalyst}]{\text{Acid}} CH_3CH_2OH$$

Many polyhydroxyl (more than one —OH group) alcohols are known, the most important being *1,2-ethanediol* (ethylene glycol),

$$\begin{array}{c} H_2C{-}OH \\ | \\ H_2C{-}OH \end{array}$$

a toxic substance that is the major constituent of most automobile antifreeze solutions.

The simplest aromatic alcohol is

$$C_6H_5{-}OH$$

which is commonly called **phenol.** Most of the 1 million tons of phenol produced annually in the United States are used to produce polymers for adhesives and plastics.

Aldehydes and Ketones

Aldehydes and ketones contain the **carbonyl group,**

$$\rangle C{=}O$$

In **ketones** this group is bonded to two carbon atoms, as in acetone,

$$\begin{array}{c} CH_3{-}C{-}CH_3 \\ \| \\ O \end{array}$$

In **aldehydes** the carbonyl group is bonded to at least one hydrogen atom, as in formaldehyde,

$$\begin{array}{c} H{-}C{-}H \\ \| \\ O \end{array}$$

or acetaldehyde,

$$\begin{array}{c} CH_3{-}C{-}H \\ \| \\ O \end{array}$$

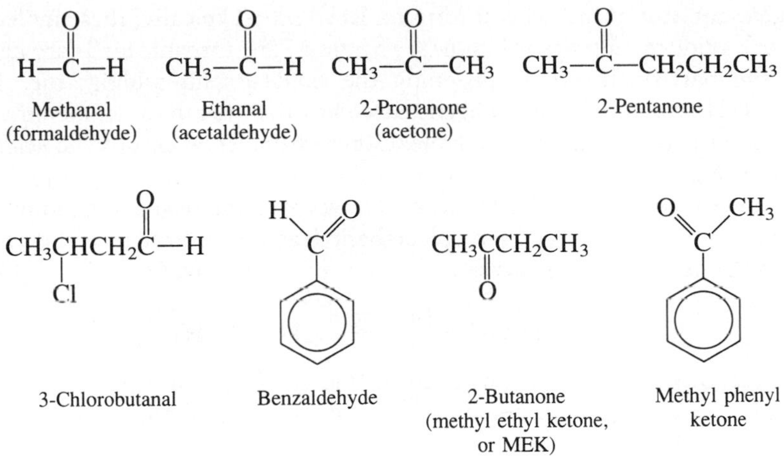

$$\underset{\substack{\text{Methanal}\\ \text{(formaldehyde)}}}{H{-}\overset{O}{\overset{\|}{C}}{-}H}\quad \underset{\substack{\text{Ethanal}\\ \text{(acetaldehyde)}}}{CH_3{-}\overset{O}{\overset{\|}{C}}{-}H}\quad \underset{\substack{\text{2-Propanone}\\ \text{(acetone)}}}{CH_3{-}\overset{O}{\overset{\|}{C}}{-}CH_3}\quad \underset{\text{2-Pentanone}}{CH_3{-}\overset{O}{\overset{\|}{C}}{-}CH_2CH_2CH_3}$$

$$\underset{\text{3-Chlorobutanal}}{CH_3\underset{Cl}{\underset{|}{C}}HCH_2\overset{O}{\overset{\|}{C}}{-}H}\qquad \underset{\text{Benzaldehyde}}{C_6H_5{-}CHO}\qquad \underset{\substack{\text{2-Butanone}\\ \text{(methyl ethyl ketone,}\\ \text{or MEK)}}}{CH_3\underset{O}{\underset{\|}{C}}CH_2CH_3}\qquad \underset{\substack{\text{Methyl phenyl}\\ \text{ketone}}}{C_6H_5{-}C(=O){-}CH_3}$$

Figure 22.13
Some common ketones and aldehydes. Note that since the aldehyde functional group always appears at the end of a carbon chain, carbon is assigned the number 1 when the compound is named.

The systematic name for an aldehyde is obtained from the parent alkane by removing the final *-e* and adding *-al*. For ketones the final *-e* is replaced by *-one*, and a number indicates the position of the carbonyl group where necessary. Examples of common aldehydes and ketones are shown in Fig. 22.13. Note that since the aldehyde functional group always occurs at the end of the carbon chain, the aldehyde carbon is assigned the number one when substituent positions are listed in the name.

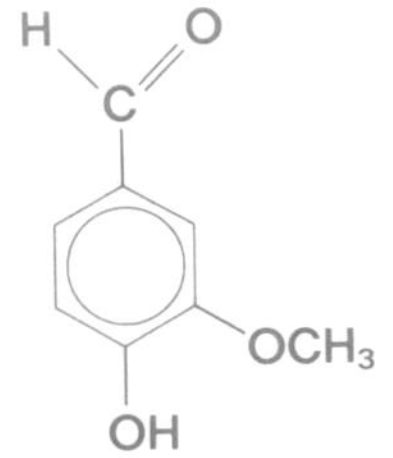

Ketones often have useful solvent properties (acetone is found in nail polish remover, for example) and are frequently used in industry for this purpose. Aldehydes typically have strong odors. Vanillin is responsible for the pleasant odor in vanilla beans; cinnamaldehyde produces the characteristic odor of cinnamon. On the other hand, the unpleasant odor in rancid butter arises from the presence of butyraldehyde.

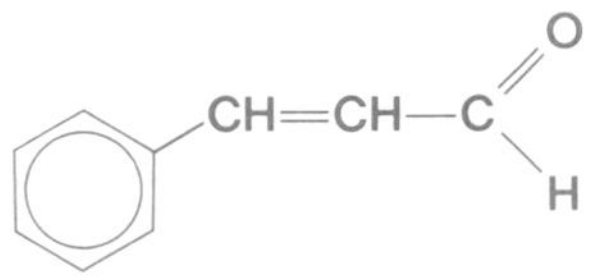

Vanillin: H, O, C, OCH_3, OH

Cinnamaldehyde: $C_6H_5{-}CH{=}CH{-}C(=O){-}H$

Butyraldehyde: $CH_3CH_2CH_2C(=O){-}H$

Aldehydes and ketones are most often produced commercially by the oxidation of alcohols. For example, oxidation of a *primary* alcohol yields the corresponding aldehyde:

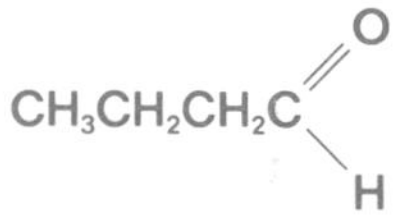

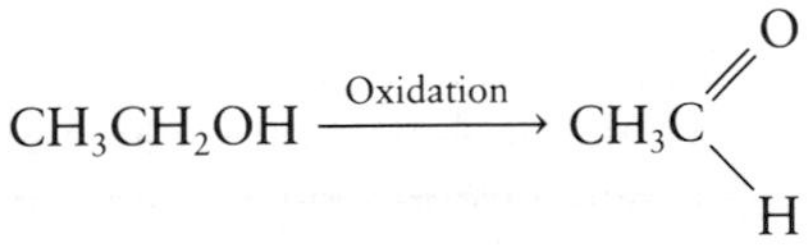

$$CH_3CH_2OH \xrightarrow{\text{Oxidation}} CH_3C\begin{smallmatrix}{}^{\diagup\!\!\diagup O}\\ {}_{\diagdown H}\end{smallmatrix}$$

Oxidation of a *secondary* alcohol results in a ketone:

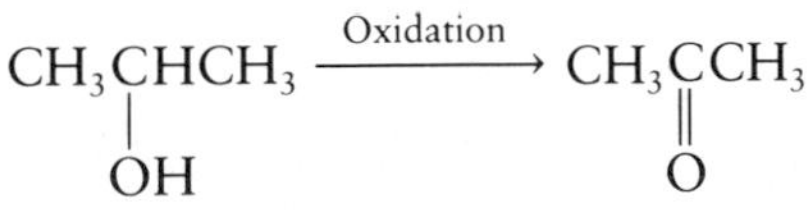

$$CH_3\underset{OH}{\underset{|}{C}}HCH_3 \xrightarrow{\text{Oxidation}} CH_3\underset{O}{\underset{\|}{C}}CH_3$$

Carboxylic Acids and Esters

Carboxylic acids are characterized by the presence of the **carboxyl group**

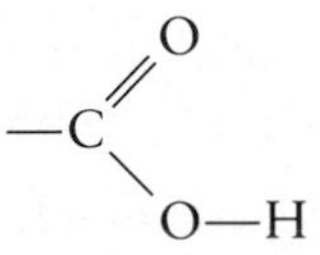

$$-C\begin{smallmatrix}{}^{\diagup\!\!\diagup O}\\ {}_{\diagdown O-H}\end{smallmatrix}$$

$CH_3CH_2CH_2COOH$

Butanoic acid

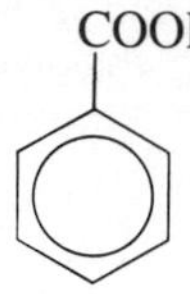

C_6H_5COOH

Benzoic acid

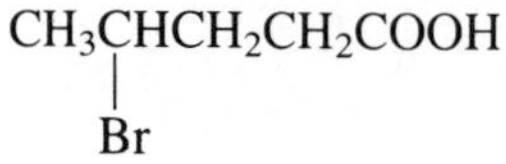

$CH_3CH(Br)CH_2CH_2COOH$

4-Bromopentanoic acid

$Cl_3C{-}COOH$

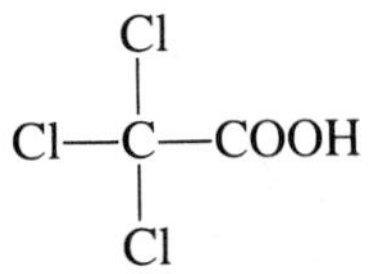

Trichloroethanoic acid (trichloroacetic acid)

Figure 22.14
Some carboxylic acids.

that gives an acid of the general formula RCOOH. Typically, these molecules are weak acids in aqueous solution (see Section 7.5). Organic acids are named from the parent alkane by dropping the final *-e* and adding *-oic*. Thus CH_3COOH, commonly called acetic acid, has the systematic name ethanoic acid, since the parent alkane is ethane. Other examples of carboxylic acids are shown in Fig. 22.14.

Many carboxylic acids are synthesized by oxidizing primary alcohols with a strong oxidizing agent. For example, ethanol can be oxidized to acetic acid by using potassium permanganate:

$$CH_3CH_2OH \xrightarrow{KMnO_4(aq)} CH_3COOH$$

A carboxylic acid reacts with an alcohol to form an **ester** and a water molecule. For example, the reaction of acetic acid with ethanol produces ethyl acetate and water:

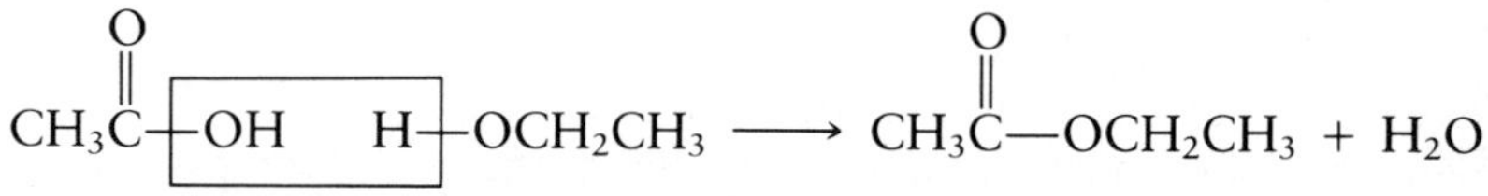

$$CH_3\overset{O}{\overset{\|}{C}}{-}\boxed{OH \quad H}{-}OCH_2CH_3 \longrightarrow CH_3\overset{O}{\overset{\|}{C}}{-}OCH_2CH_3 + H_2O$$

React to form water

Esters often have a sweet, fruity odor that is in contrast to the often pungent odors of the parent carboxylic acids. For example, the odor of bananas is due to *n*-amyl acetate,

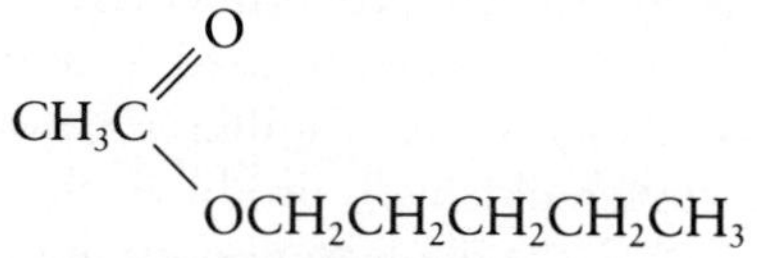

$$CH_3\overset{O}{\overset{\|}{C}}{-}OCH_2CH_2CH_2CH_2CH_3$$

and that of oranges is due to *n*-octyl acetate,

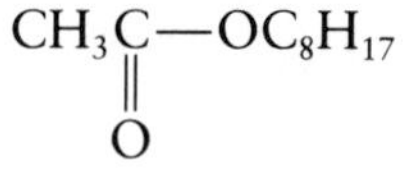

$$CH_3\underset{O}{\underset{\|}{C}}{-}OC_8H_{17}$$

The systematic name for an ester is formed by changing the *-oic* ending of the parent acid to *-oate*. The parent alcohol chain is named first with a *-yl* ending. For example, the systematic name for *n*-octyl acetate is *n*-octylethanoate (from ethanoic acid).

A very important ester is formed from the reaction of salicylic acid and acetic acid:

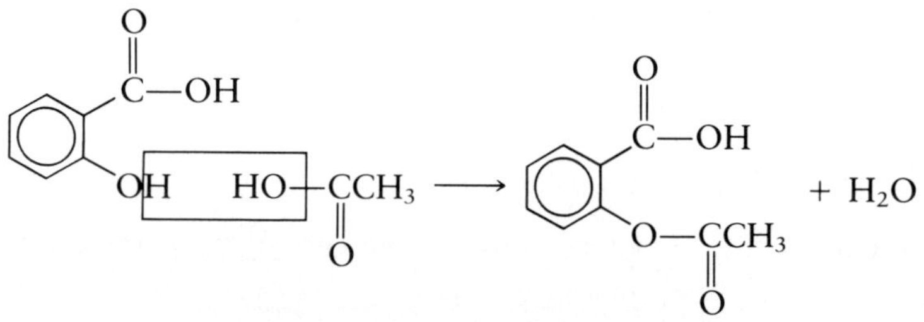

$$C_6H_4(COOH)\boxed{OH \quad HO}{-}\underset{O}{\underset{\|}{C}}CH_3 \longrightarrow C_6H_4(COOH)(O{-}\underset{O}{\underset{\|}{C}}CH_3) + H_2O$$

Salicylic acid | Acetic acid | Acetylsalicylic acid

The product is acetylsalicylic acid, commonly known as *aspirin*, which is used in huge quantities as an analgesic (painkiller).

TABLE 22.6 **Some Common Amines**

Formula	Common Name	Type
CH_3NH_2	Methylamine	Primary
$CH_3CH_2NH_2$	Ethylamine	Primary
$(CH_3)_2NH$	Dimethylamine	Secondary
$(CH_3)_3N$	Trimethylamine	Tertiary
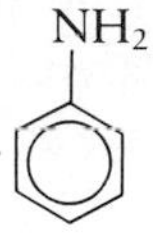 $C_6H_5NH_2$ (NH₂ on benzene ring)	Aniline	Primary
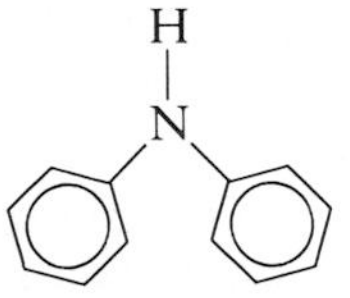 $(C_6H_5)_2NH$ (N—H bonded to two benzene rings)	Diphenylamine	Secondary

Amines

Amines are probably best viewed as derivatives of ammonia in which one or more N—H bonds are replaced by N—C bonds. The resulting amines are classified as *primary* if one N—C bond is present, *secondary* if two N—C bonds are present, and *tertiary* if all three N—H bonds in NH_3 have been replaced by N—C bonds (Fig. 22.15). Examples of some common amines are given in Table 22.6.

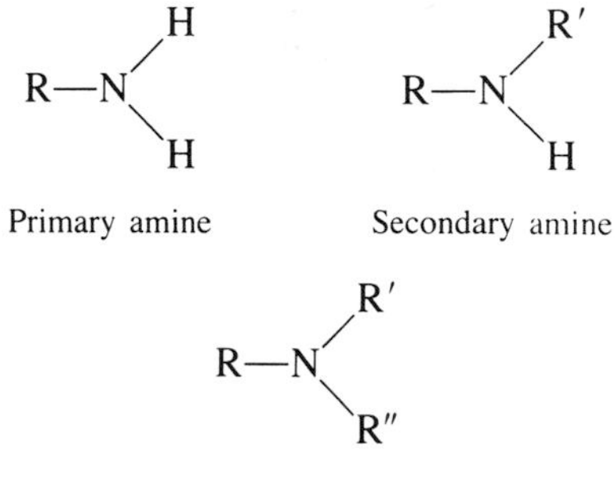

Common names are often used for simple amines; the systematic nomenclature for more complex molecules uses the name *amino-* for the $—NH_2$ functional group. For example, the molecule

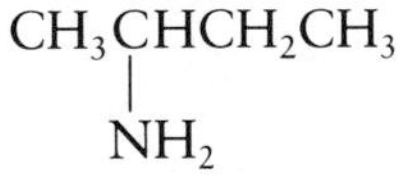

$$CH_3\underset{\substack{|\\NH_2}}{C}HCH_2CH_3$$

is named 2-aminobutane.

Many amines have unpleasant "fishlike" odors. For example, the odors associated with decaying animal and human tissues are due to amines such as putrescine ($H_2NCH_2CH_2CH_2NH_2$) and cadaverine ($H_2NCH_2CH_2CH_2CH_2CH_2NH_2$).

Aromatic amines are primarily used to make dyes. Since many of them are carcinogenic, they must be handled with great care.

R—N(H)(H) Primary amine

R—N(R′)(H) Secondary amine

R—N(R′)(R″) Tertiary amine

Figure 22.15
The general formulas for primary, secondary, and tertiary amines. R, R′, and R″ represent carbon-containing substituents.

22.5 Polymers

Polymers are large, usually chainlike molecules that are built from small molecules called *monomers*. Polymers form the basis for synthetic fibers, rubbers, and plastics and have played a leading role in the revolution that has been brought about in daily life by chemistry during the past 50 years. It has been estimated that approximately 50% of the industrial chemists in the United States work in some area of polymer chemistry, a fact that illustrates just how important polymers are to our economy and standard of living.

The Development and Properties of Polymers

The development of the polymer industry provides a striking example of the importance of serendipity in the progress of science. Many discoveries in polymer chemistry arose from accidental observations that scientists followed up.

The age of plastics might be traced to a day in 1846 when Christian Schoenbein, a chemistry professor at the University of Basel in Switzerland, spilled a flask containing nitric and sulfuric acids. In his hurry to clean up the spill, he grabbed his wife's cotton apron, which he then rinsed out and hung up in front of a hot stove to dry. Instead of drying, the apron flared and burned.

Very interested in this event, Schoenbein repeated the reaction under more controlled conditions and found that the new material, which he correctly concluded to be nitrated cellulose, had some surprising properties. As he had experienced, the nitrated cellulose is extremely flammable and, under certain circumstances, highly explosive. In addition, he found that it could be molded at moderate temperatures to give objects that were, upon cooling, tough but elastic. Predictably, the explosive nature of the substance was initially of more interest than its other properties, and cellulose nitrate rapidly became the basis for smokeless gun powder. Although Schoenbein's discovery cannot be described as a truly synthetic polymer (because he simply found a way to modify the natural polymer cellulose) it formed the basis for a large number of industries that grew up to produce photographic films, artificial fibers, and molded objects of all types.

The first synthetic polymers were produced as by-products of various organic reactions and were regarded as unwanted contaminants. Thus the first preparations of many of the polymers now regarded as essential to our modern lifestyle were thrown away in disgust. One chemist who refused to be defeated by the "tarry" products obtained when he reacted phenol with formaldehyde was the Belgian-American chemist Leo H. Baekeland (1863–1944). Baekeland's work resulted in the first completely synthetic plastic (called Bakelite), a substance that when molded to a certain shape under high pressure and temperature cannot be softened again or dissolved. Bakelite is a **thermoset polymer.** In contrast, cellulose nitrate is a **thermoplastic polymer;** that is, it can be remelted after it has been molded.

The discovery of Bakelite in 1907 spawned a large plastics industry, producing telephones, billiard balls, and insulators for electrical devices. During the early days of polymer chemistry, there was a great deal of controversy over the nature of these materials. Although the German chemist Hermann Staudinger speculated in 1920 that polymers were very large molecules held together by strong chemical bonds, most chemists of the time assumed that these materials were much like colloids, in which small molecules are aggregated into large units by forces weaker than chemical bonds.

One chemist who contributed greatly to the understanding of polymers as giant molecules was Wallace H. Carothers of the DuPont Chemical Company. Among his accomplishments was the preparation of nylon. The nylon story further illustrates the importance of serendipity in scientific research. When nylon is first prepared, the resulting product is a sticky material with little structural integrity. Because of this, it was initially put aside as having no apparently useful characteristics. However, Julian Hill, a chemist in the Carothers research group, one day put a small ball of this nylon on the end of a stirring rod and drew it away from the remaining sticky mass, forming a string.

He noticed the silky appearance and strength of this thread and realized that nylon could be drawn into useful fibers.

The reason for this behavior of nylon is now understood. When nylon is first formed, the individual polymer chains are oriented randomly, like cooked spaghetti, and the substance is highly amorphous. However, when drawn out into a thread, the chains tend to line up (the nylon becomes more crystalline), which leads to increased hydrogen bonding between adjacent chains. This increase in crystallinity, along with the resulting increase in hydrogen-bonding interactions, leads to strong fibers and thus to a highly useful material. Commercially, nylon is produced by forcing the raw material through a *spinneret*, a plate containing small holes, which forces the polymer chains to line up.

Another property that adds strength to polymers is **crosslinking**, the existence of covalent bonds between adjacent chains. The structure of Bakelite is highly crosslinked, which accounts for the strength and toughness of this polymer. Another example of crosslinking occurs in the manufacture of rubber. Raw natural rubber consists of chains of the type

$$\sim\sim CH_2-CH_2-CH=\underset{\displaystyle CH_3}{\underset{|}{C}}-CH_2-CH_2-CH=\underset{\displaystyle CH_3}{\underset{|}{C}}-CH_2\sim\sim$$

and is a soft, sticky material unsuitable for tires. However, in 1839 Charles Goodyear (1800–1860), an American chemist, accidentally found that if sulfur is added to rubber and the resulting mixture is heated (a process called **vulcanization**), the resulting rubber is still elastic (reversibly stretchable) but is much stronger. This change in character occurs because sulfur atoms become bonded between carbon atoms on different chains. These sulfur atoms form bridges between the polymer chains, thus linking the chains together.

Charles Goodyear tried for many years to change natural rubber into a useful product. In 1839 he accidentally dropped some rubber containing sulfur on a hot stove. Noting that the rubber did not melt as expected, Goodyear pursued this lead and developed vulcanization.

Types of Polymers

The simplest and one of the best-known polymers is *polyethylene,* which is constructed from ethylene monomers:

$$nCH_2{=}CH_2 \xrightarrow{\text{Catalyst}} \left(-\overset{\displaystyle H}{\underset{\displaystyle H}{\overset{|}{\underset{|}{C}}}}-\overset{\displaystyle H}{\underset{\displaystyle H}{\overset{|}{\underset{|}{C}}}}- \right)_n$$

where n represents a large number (usually several thousand). Polyethylene is a tough, flexible plastic used for piping, bottles, electrical insulation, packaging films, garbage bags, and many other purposes. Its properties can be varied by using substituted ethylene monomers. For example, when tetrafluoroethylene is the monomer, the polymer Teflon is obtained:

$$n\left(\begin{matrix} F & & F \\ & C{=}C & \\ F & & F \end{matrix} \right) \longrightarrow \left(-\overset{\displaystyle F}{\underset{\displaystyle F}{\overset{|}{\underset{|}{C}}}}-\overset{\displaystyle F}{\underset{\displaystyle F}{\overset{|}{\underset{|}{C}}}}- \right)_n$$

Tetrafluoroethylene — Teflon

Plastic films made from ethylene and derivatives are used to package most consumer products. This photo shows molten plastic being formed into a bag.

The discovery of Teflon, a very important substituted polyethylene, is another illustration of the role of chance in chemical research. In 1938 a DuPont chemist named Roy Plunkett was studying the chemistry of gaseous tetrafluoroethylene. He synthesized about 100 pounds of the chemical and stored it in steel cylinders. When one of the cylinders failed to produce perfluoroethylene gas when the valve was opened, the cylinder was cut open to reveal a white powder. This powder turned out to be a polymer of perfluoroethylene, which was eventually developed into Teflon. Because of the resistance of the strong C—F bonds to chemical attack, Teflon is an inert, tough, and nonflammable material widely used for electrical insulation, nonstick coatings on cooking utensils, and bearings for low-temperature applications.

Other polyethylene-type polymers are made from monomers containing chloro, methyl, cyano, and phenyl substituents, as summarized in Table 22.7. In each case the double carbon-carbon bond in the substituted ethylene monomer becomes a single bond in the polymer. The different substituents lead to a wide variety of properties.

The polyethylene polymers illustrate one of the major types of polymerization reactions, called **addition polymerization,** in which the monomers simply "add together" to produce the polymer. No other products are formed. The polymerization process is initiated by a **free radical** (a species with an unpaired electron) such as the hydroxyl radical (HO·). The free radical attacks and breaks the π bond of an ethylene molecule to form a new free radical,

$$\begin{array}{c} \text{H} \quad\quad \text{H} \\ \diagdown \quad \diagup \\ \text{C=C} \\ \diagup \quad \diagdown \\ \text{H} \;\uparrow\; \quad \text{H} \\ \dot{\text{O}}\text{H} \end{array} \longrightarrow \begin{array}{ccc} & \text{H} & \text{H} \\ & | & | \\ \text{H—} & \text{C—} & \text{C}\cdot \\ & | & | \\ & \text{OH} & \text{H} \end{array}$$

which is then available to attack another ethylene molecule:

$$\begin{array}{c} \text{H} \quad\quad \text{H} \\ \diagdown \quad \diagup \\ \text{C=C} \\ \diagup \quad \diagdown \\ \text{H} \quad\quad \text{H} \end{array} \longrightarrow \begin{array}{cccccc} & \text{H} & \text{H} & \text{H} & \text{H} \\ & | & | & | & | \\ \text{H—} & \text{C—} & \text{C—} & \text{C—} & \text{C}\cdot \\ & | & | & | & | \\ & \text{OH} & \text{H} & \text{H} & \text{H} \end{array}$$

$$\begin{array}{ccc} & \text{H} & \text{H} \\ & | & | \\ \text{H—} & \text{C—} & \text{C}\cdot \\ & | & | \\ & \text{HO} & \text{H} \end{array}$$

Repetition of this process thousands of times creates a long-chain polymer. Termination of the growth of the chain occurs when *two radicals* react to form a bond, a process that consumes two radicals without producing any others.

Another common type of polymerization is **condensation polymerization,** in which a small molecule, such as water, is formed for each extension of the polymer chain. The most familiar polymer produced by condensation is *nylon.* Nylon is a **copolymer,** since two different types of monomers combine to form the chain; a **homopolymer** is the result of polymerizing a single type of monomer. One common form of nylon is produced when hexamethylenediamine and adipic acid react by splitting out a water molecule to form a C—N bond:

TABLE 22.7 Some Common Synthetic Polymers, Their Monomers, and Applications

Monomer		Polymer		
Name	Formula	Name	Formula	Uses
Ethylene	$H_2C{=}CH_2$	Polyethylene	$-\!(CH_2{-}CH_2)_n\!-$	Plastic piping, bottles, electrical insulation, toys
Propylene	$H_2C{=}C(H)(CH_3)$	Polypropylene	$-\!(CH(CH_3){-}CH_2{-}CH(CH_3){-}CH_2)_n\!-$	Film for packaging, carpets, lab wares, toys
Vinyl chloride	$H_2C{=}C(H)(Cl)$	Polyvinyl chloride (PVC)	$-\!(CH_2{-}CH(Cl))_n\!-$	Piping, siding, floor tile, clothing, toys
Acrylonitrile	$H_2C{=}C(H)(CN)$	Polyacrylonitrile (PAN)	$-\!(CH_2{-}CH(CN))_n\!-$	Carpets, fabrics
Tetrafluoroethylene	$F_2C{=}CF_2$	Teflon	$-\!(CF_2{-}CF_2)_n\!-$	Cooking utensils, electrical insulation, bearings
Styrene	$H_2C{=}C(H)(C_6H_5)$	Polystyrene	$-\!(CH_2CH(C_6H_5))_n\!-$	Containers, thermal insulation, toys
Butadiene	$H_2C{=}C(H){-}C(H){=}CH_2$	Polybutadiene	$-\!(CH_2CH{=}CHCH_2)_n\!-$	Tire tread, coating resin
Butadiene and styrene	(See above.)	Styrene-butadiene rubber	$-\!(CH(C_6H_5){-}CH_2{-}CH_2{-}CH{=}CH{-}CH_2)_n\!-$	Synthetic rubber

$$\underset{\text{Hexamethylenediamine}}{H_2N{-}(CH_2)_6{-}N(H){-}H} + \underset{\text{Adipic acid}}{H{-}O{-}C(=O){-}(CH_2)_4{-}C(=O){-}O{-}H} \longrightarrow$$

$$H_2N{-}(CH_2)_6{-}N(H){-}C(=O){-}(CH_2)_4{-}C(=O){-}O{-}H + H_2O$$

Figure 22.16
The reaction to form nylon can be carried out at the interface of two immiscible liquid layers in a beaker. The bottom layer contains adipoyl chloride,

$$\mathrm{Cl{-}\underset{\underset{O}{\|}}{C}{-}(CH_2)_4{-}\underset{\underset{O}{\|}}{C}{-}Cl}$$

dissolved in CCl_4, and the top layer contains hexamethylenediamine,

$$\mathrm{H_2N{-}(CH_2)_6{-}NH_2}$$

dissolved in water. A molecule of HCl is formed as each C—N bond forms.

The molecule formed, called a **dimer** (two monomers joined), can undergo further condensation reactions since it has an amino group at one end and a carboxyl group at the other. Thus both ends are free to react with another monomer. Repetition of this process leads to a long chain of the type

$$\left(\mathrm{\overset{\overset{H}{|}}{N}{-}(CH_2)_6{-}\overset{\overset{H}{|}}{N}{-}\overset{\overset{O}{\|}}{C}{-}(CH_2)_4{-}\overset{\overset{O}{\|}}{C}}\right)_n$$

which is the basic structure of nylon. The reaction to form nylon occurs quite readily and is often used as a lecture demonstration (see Fig. 22.16). The properties of nylon can be varied by changing the number of carbon atoms in the chain of the acid or amine monomer.

More than 1 million tons of nylon are produced annually in the United States for use in clothing, carpets, rope, and so on. Many other types of condensation polymers are also produced. For example, Dacron is a copolymer formed from the condensation reaction of ethylene glycol (a dialcohol) and *p*-terephthalic acid (a dicarboxylic acid):

$$\mathrm{HOCH_2CH_2O\boxed{H\quad HO}{-}\overset{\overset{O}{\|}}{C}{-}C_6H_4{-}\overset{\overset{O}{\|}}{C}{-}O{-}H} \qquad \downarrow\ \mathrm{H_2O}$$

Ethylene glycol — H_2O — *p*-Terephthalic acid

The repeating unit of Dacron is

$$\left(\mathrm{OCH_2CH_2{-}O{-}\overset{\overset{O}{\|}}{C}{-}C_6H_4{-}\overset{\overset{O}{\|}}{C}}\right)_n$$

Wallace Hume Carothers

Wallace H. Carothers, a brilliant organic chemist who was principally responsible for the development of nylon and the first synthetic rubber (Neoprene), was born in 1896 in Burlington, Iowa. As a youth, Carothers was fascinated by tools and mechanical devices and spent many hours experimenting. In 1915 he entered Tarkio College in Missouri. Carothers so excelled in chemistry that even before his graduation, he was made a chemistry instructor.

Carothers eventually moved to the University of Illinois at Urbana-Champaign, where he was appointed to the faculty when he completed his Ph.D. in organic chemistry in 1924. He moved to Harvard University in 1926, and then to DuPont in 1928 to participate in a new program in fundamental research. At DuPont, Carothers headed the organic chemistry division, and during his ten years there played a prominent role in laying the foundations of polymer chemistry.

By the age of 33, Carothers had become a world-famous chemist whose advice was sought by almost everyone working in polymers. He was the first industrial chemist to be elected to the prestigious National Academy of Sciences.

Carothers was an avid reader of poetry and a lover of classical music. Unfortunately, he also suffered from severe bouts of depression that finally led to his suicide in 1937 in a Philadelphia hotel room, where he drank a cyanide solution. He was 41 years old. Despite the brevity of his career, Carothers was truly one of the finest American chemists of all time. His great intellect, his love of chemistry, and his insistence on perfection produced his special genius.

Wallace Carothers

Note that this polymerization involves a carboxylic acid and an alcohol forming an ester group:

$$R-O-\overset{\overset{\large O}{\|}}{C}-R_1$$

Thus Dacron is called a **polyester.** By itself or blended with cotton, Dacron is widely used in fibers for the manufacture of clothing.

Polymers Based on Ethylene

A large section of the polymer industry involves the production of macromolecules from ethylene or substituted ethylenes. As discussed previously, ethylene molecules polymerize by addition after the double bond has been broken by some initiator:

$$X-\underset{H}{\overset{H}{\underset{|}{\overset{|}{C}}}}-\underset{H}{\overset{H}{\underset{|}{\overset{|}{C}}}}\cdot \curvearrowright \underset{H}{\overset{H}{\underset{|}{\overset{|}{C}}}}=\underset{H}{\overset{H}{\underset{|}{\overset{|}{C}}}} \longrightarrow X-\underset{H}{\overset{H}{\underset{|}{\overset{|}{C}}}}-\underset{H}{\overset{H}{\underset{|}{\overset{|}{C}}}}-\underset{H}{\overset{H}{\underset{|}{\overset{|}{C}}}}-\underset{H}{\overset{H}{\underset{|}{\overset{|}{C}}}}\cdot$$

This process continues by adding new ethylene molecules to eventually give polyethylene, a thermoplastic material.

There are two forms of polyethylene: low-density polyethylene (LDPE) and high-density polyethylene (HDPE). The chains in LDPE contain many branches and thus do not pack as tightly as those in HDPE, which consist of mostly straight-chain molecules.

psi is the abbreviation for pounds per square inch; 15 psi $\approx$1 atm.

Traditionally, LDPE has been manufactured under conditions of high pressure ($\approx$20,000 psi) and high temperature (500°C). These severe reaction conditions require specially designed equipment, and for safety reasons the reaction usually has been run behind a reinforced concrete barrier. More recently, lower reaction pressures and temperatures have become possible through the use of catalysts. One catalytic system using triethylaluminum, $Al(C_2H_5)_3$, and titanium(IV) chloride was developed by Karl Ziegler in Germany and Giulio Natta in Italy. Although this is a very efficient catalyst, it catches fire on contact with air and must be handled very carefully. A safer catalytic system was developed at Phillips Petroleum Company. It uses a chromium(III) oxide (Cr_2O_3) and aluminosilicate catalyst and has largely taken over in the United States. The product of the catalyzed reaction is highly linear (unbranched) and is often called *linear low-density polyethylene*. It is very similar to HDPE.

The major use of LDPE is in the manufacture of the tough transparent film that is used in packaging so many consumer goods. Two-thirds of the approximately 10 billion pounds of low-density polyethylene produced annually in the United States are used for this purpose. The major use of HDPE is for blow-molded products, such as bottles for consumer products (see Fig. 22.17).

Molecular weight (not molar mass) is the common terminology in the polymer industry.

The useful properties of polyethylene are due primarily to its high molecular weight (molar mass). Although the strengths of the interactions between specific points on the nonpolar chains are quite small, the chains are so long that these small attractions accumulate to a very significant value, so that the chains stick together very tenaciously. There is also a great deal of physical tangling of the lengthy chains. The combination of these interactions gives the polymer strength and toughness. However, a material like polyethylene can be

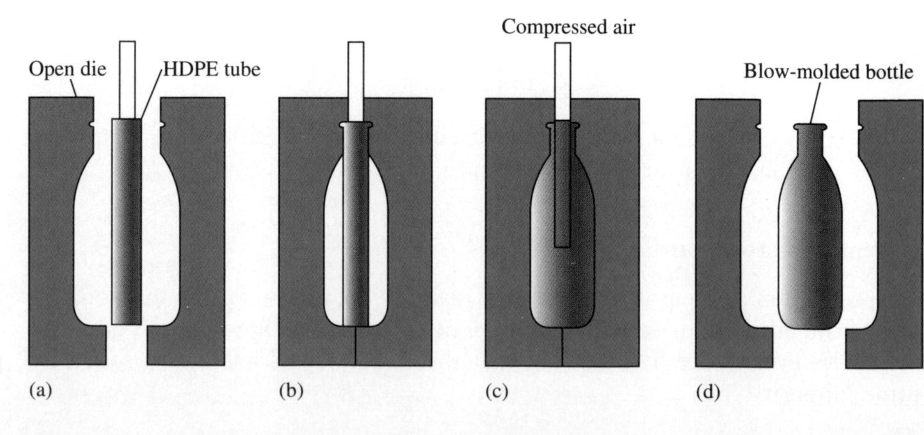

Figure 22.17
A major use of HDPE is for blow-molded objects such as bottles for soft drinks, shampoos, bleaches, and so on. (a) A tube composed of HDPE is inserted into the mold (die). (b) The die closes, sealing the bottom of the tube. (c) Compressed air is forced into the warm HDPE tube, which then expands to take the shape of the die. (d) The molded bottle is removed from the die.

melted and formed into a new shape (thermoplastic behavior), because in the melted state the molecules can readily flow past one another.

Since a high molecular weight gives a polymer useful properties, one might think that the goal would be to produce polymers with chains as long as possible. However, this is not the case—polymers become much harder to process as the molecular weights increase. Most industrial operations require that the polymer flow through pipes as it is processed. But as the chain lengths increase, viscosity also increases. In practice, the upper limit of a polymer's molecular weight is set by the flow requirements of the manufacturing process. Thus the final product often reflects a compromise between the optimum properties for application and those for ease of processing.

Although many polymer properties are greatly influenced by molecular weight, some other important properties are not. For example, chain length does not affect a polymer's resistance to chemical attack. Physical properties such as color, refractive index, hardness, density, and electrical conductivity are also not greatly influenced by molecular weight.

We have already seen that one way of altering the strength of a polymeric material is to vary the chain length. Another method for modifying polymer behavior involves varying the substituents. For example, if we use a monomer of the type

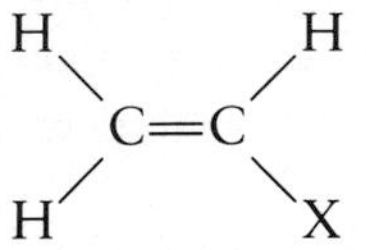

H H
C=C
H X

the properties of the resulting polymer depend on the identity of X. The simplest example is polypropylene, whose monomer is

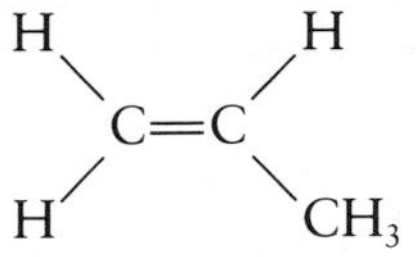

H H
C=C
H CH_3

and that has the form

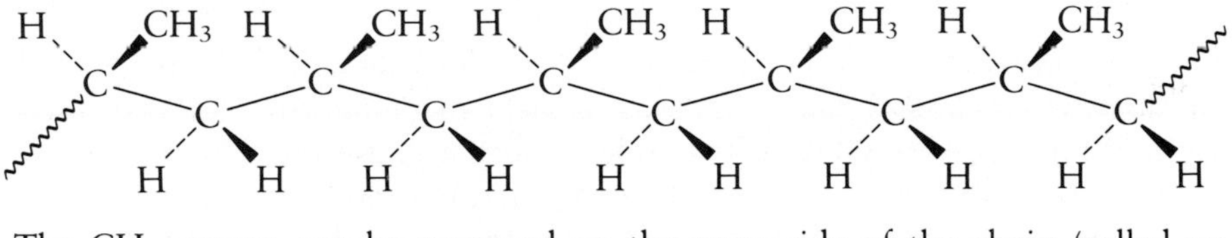

H CH_3 H CH_3 H CH_3 H CH_3 H CH_3
C C C C C C C C C C
H H H H H H H H H H

The CH_3 groups can be arranged on the same side of the chain (called an **isotactic chain**) as shown above, can alternate (called a **syndiotactic chain**) as shown below,

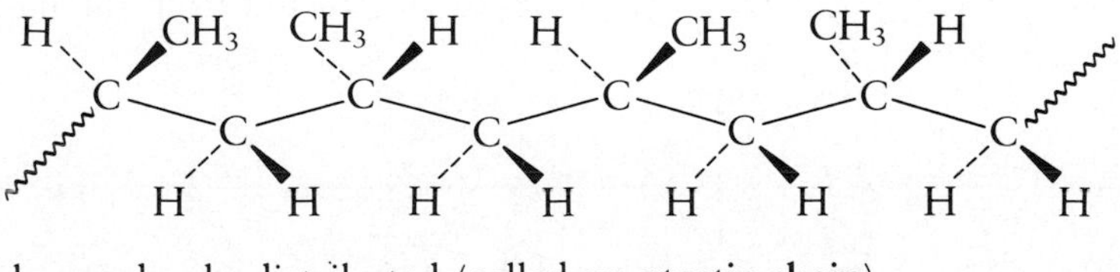

H CH_3 CH_3 H H CH_3 CH_3 H
C C C C C C C C
H H H H H H H H

or can be randomly distributed (called an **atactic chain**).

The chain arrangement has a significant effect on the polymer properties. Most polypropylene is made using the Ziegler-Natta catalyst, $Al(C_2H_5)_3 \cdot TiCl_4$, which produces highly isotactic chains that pack together quite closely. As a result, polypropylene is more crystalline, and therefore

Plastic That Talks and Listens

Imagine a plastic so "smart" that it can be used to sense a baby's breath, measure the force of a karate punch, sense the presence of a person 100 ft away, or make a balloon that sings. There is a plastic film capable of doing all of these things. It's called polyvinylidene difluoride (PVDF), which has the structure

$$\sim\!\sim C(F)_2 - C(H)_2 - C(F)_2 - C(H)_2 \sim\!\sim$$

When this polymer is processed in a particular way, it becomes piezoelectric and pyroelectric. A piezoelectric substance produces an electric current when it is physically deformed or alternately when it undergoes a deformation caused by the application of a current. A pyroelectric material is one that develops an electrical potential in response to a change in its temperature.

Because PVDF is piezoelectric, it can be used to construct a paper-thin microphone; it responds to sound by producing a current proportional to the deformation caused by the sound waves. A ribbon of PVDF plastic one quarter of an inch wide could be strung along a hallway and used to listen to all of the conversations going on as people walk through. On the other hand, electric pulses can be applied to the PVDF film to produce a speaker. A strip of PVDF film glued to the inside of a balloon can play any song stored on a microchip attached to the film—hence a balloon that can sing happy birthday at a party. The PVDF film can also be used to construct a sleep apnea monitor, which, when placed beside the mouth of a sleeping infant, will set off an alarm if the breathing stops, thus helping to prevent sudden infant death syndrome (SIDS). The same type of film is used by the U.S. Olympic karate team to measure the force of kicks and punches as the team trains. Also, gluing two strips of film together gives a material that curls in response to a current, creating an artificial muscle. In addition, because the PVDF film is pyroelectric, it responds to the infrared (heat) radiation emitted by a human as far away as 100 ft, making it useful for burglar alarm systems.

Making the PVDF polymer piezoelectric and pyroelectric requires some very special processing, which makes it costly ($10 per square foot); but this seems a small price to pay for its near-magical properties.

stronger and harder, than polyethylene. The major uses of polypropylene are for molded parts (40%), fibers (35%), and packaging films (10%). Polypropylene fibers are especially useful for athletic wear because they do not absorb water from perspiration, as cotton does. Rather, the moisture is drawn away from the skin to the surface of the polypropylene garment, where it can evaporate. The annual U.S. production of polypropylene is about 7 billion pounds.

Another related polymer, **polystyrene,** is constructed from the monomer styrene,

$$H_2C{=}CH{-}C_6H_5$$

Pure polystyrene is too brittle for many uses, so most polystyrene-based polymers are actually *copolymers* of styrene and butadiene,

$$\mathrm{H_2C{=}CH{-}CH{=}CH_2}$$

thus incorporating bits of butadiene rubber into the polystyrene matrix. The resulting polymer is very tough and is often used as a substitute for wood in furniture.

Another polystyrene-based product is acrylonitrile-butadiene-styrene (ABS), a tough, hard, and chemically resistant plastic used for pipes and for items such as radio housings, telephone cases, and golf club heads, for which shock resistance is an essential property. Originally, ABS was produced by copolymerization of the three monomers:

$$\underset{\text{Acrylonitrile}}{\mathrm{H_2C{=}CH{-}CN}} + \underset{\text{Styrene}}{\mathrm{H_2C{=}CH{-}C_6H_5}} + \underset{\text{Butadiene}}{\mathrm{H_2C{=}CH{-}CH{=}CH_2}} \longrightarrow$$

$$\left(\!\!-\mathrm{CH_2{-}\underset{\displaystyle CN}{\underset{|}{CH}}{-}CH_2{-}\underset{\displaystyle C_6H_5}{\underset{|}{CH}}{-}CH_2{-}\underset{\displaystyle CH_2{=}CH}{\underset{|}{CH}}}-\!\!\right)_n$$

It is now prepared by a special process called *grafting,* in which butadiene is polymerized first, and then the cyanide and phenyl substituents are added chemically.

Another high-volume polymer, **polyvinyl chloride** (PVC), is constructed from the monomer vinyl chloride,

$$\mathrm{H_2C{=}CHCl}$$

We discussed the development of PVC in Chapter 1 as an example of the types of problem-solving situations commonly encountered in the chemical-based industries.

EXERCISES

A blue exercise number indicates that the answer to that exercise appears at the back of this book.

Hydrocarbons

1. Draw the five structural isomers of hexane (C_6H_{14}). Give systematic names for each. Which isomer would you expect to have the highest boiling point? Why?
2. Name each of the following.
   a. 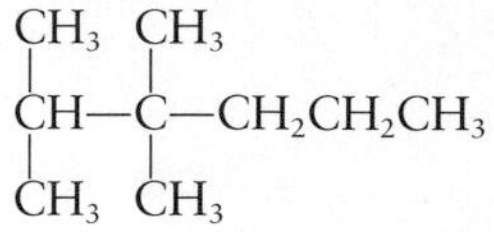 CH_3 CH_3
      $CH—C—CH_2CH_2CH_3$
      CH_3 CH_3
   b. $CH_2CH_2CH(CH_3)_2$ attached to
      $CH_3—C—CH_2—CH_2—CH—CH_2—CH_3$
      with CH_3 on the C and CH_2CH_3 on the CH
   c. $CH_3—CH_2—CH_2—CH—CH_3$
      with CH_2CH_3 on the CH
3. The normal (unbranched) hydrocarbons are often referred to as the straight-chain hydrocarbons. To what does this name refer? Does this mean that all carbon atoms in a straight-chain hydrocarbon actually have a linear arrangement? Explain your answer.
4. Draw the structural formula for each of the following.
   a. 2-methylpentane
   b. 2,2,4-trimethylpentane (Also called isooctane, this substance is the reference, or 100 level, for octane ratings.)
   c. 2-*tert*-butylpentane
   d. The name given in part c is incorrect. Give the correct name for this hydrocarbon.
5. Name each of the following alkenes.
   a. $CH_2{=}CH—CH_2—CH_3$
   b. 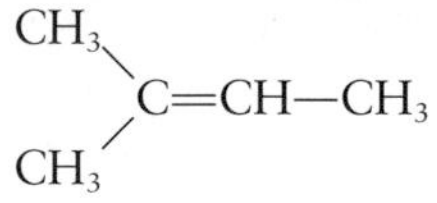 $(CH_3)_2C{=}CH—CH_3$
   c. 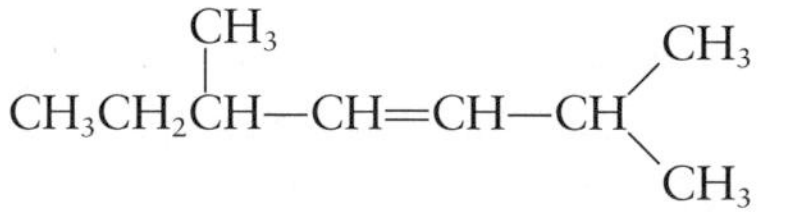 $CH_3CH_2CH(CH_3)—CH{=}CH—CH(CH_3)_2$
6. Give the structure for each of the following.
   a. 3-hexene
   b. 2,4-heptadiene
   c. 2-methyl-3-octene
7. Give the structure for each of the following aromatic hydrocarbons.
   a. *o*-xylene (1,2-dimethylbenzene)
   b. *p*-di-*tert*-butylbenzene
   c. *m*-diethylbenzene
8. Name each of the following.
   a. 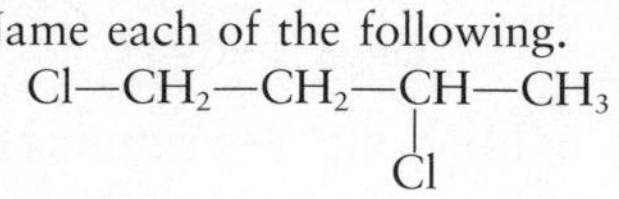 $Cl—CH_2—CH_2—CH—CH_3$
      with Cl on the CH
   b. $CH_3CH_2CH_2CCl_3$
   c. 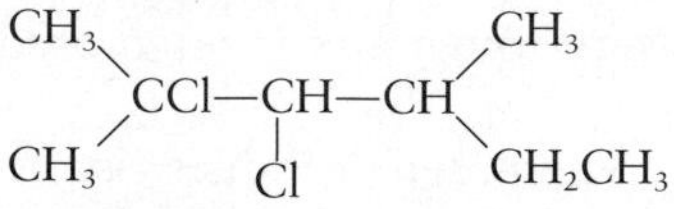 $(CH_3)_2CCl—CH(Cl)—CH(CH_3)(CH_2CH_3)$
   d. CH_2FCH_2F
   e.  Cl on a benzene ring
   f. 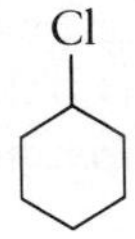 Cl on a cyclohexane ring
   g. 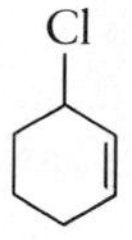 Cl on a cyclohexene ring
9. Name the following compounds.
   a. 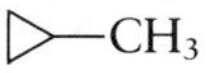 cyclopropane ring—CH_3
   b. 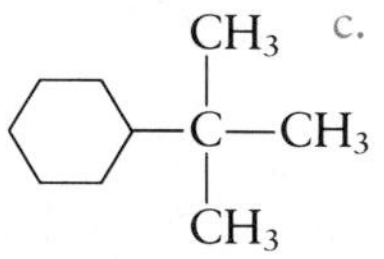 cyclohexane ring—$C(CH_3)_3$
   c. cyclopentene ring with two CH_3 groups
   d. $ClCH{=}CH_2$
   e. 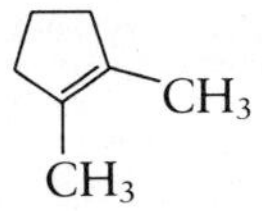 cyclopentene ring with two CH_3 groups
   f.  cyclohexane ring with two Cl on the same carbon
10. Cumene is the starting material for the industrial production of acetone and phenol. The structure of cumene is

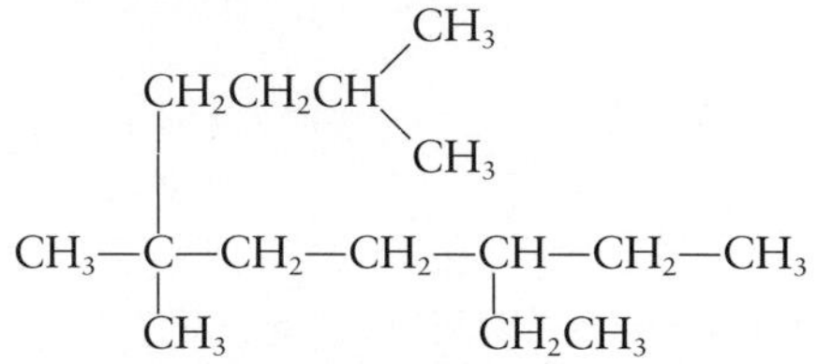

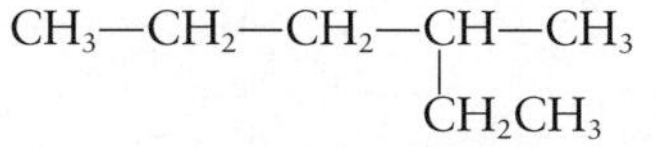

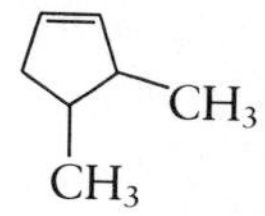

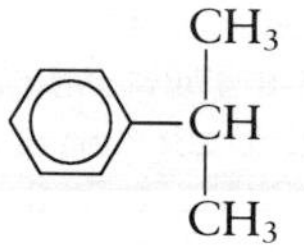

    benzene ring—$CH(CH_3)_2$

    Give the systematic name for cumene.
11. Draw the structure for the hydrocarbon 2-ethyl-3-methyl-5-isopropylhexane. The name given here is incorrect. Supply the correct systematic name.
12. Predict the structures for each of the following.
    a. difluoromethane
    b. 1-bromo-1,2-dichloropropane
    c. *m*-difluorobenzene

Isomerism

13. Distinguish between structural and geometrical isomerism.
14. Distinguish between isomerism and resonance.
15. *Cis-trans* isomerism is also possible in molecules with rings. Draw the *cis* and *trans* isomers of 1,2-dimethylcyclohexane.
16. Draw all of the structural and geometric isomers of $C_3H_4Cl_2$.
17. Draw all of the isomers of dimethylnaphthalene.
18. Draw all of the structural and geometrical isomers of bromochloropropene.

19. Draw all of the isomers of difluoroethene. Which are polar?

20. There is only one compound that is named 1,2-dichloroethane, but there are two distinct compounds that can be named 1,2-dichloroethene. Why?

21. Tautomerism is a property exhibited by molecules that differ in the position of a hydrogen atom. Use bond energies (Table 13.5) to predict which tautomer in each of the following pairs is more stable.

a. 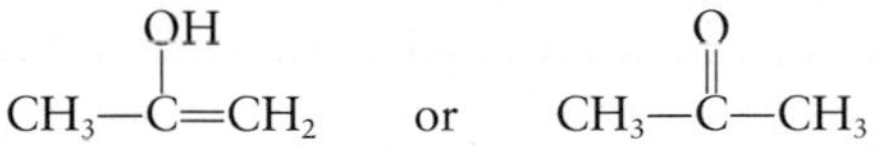 $CH_3—C(OH)=CH_2$ or $CH_3—C(=O)—CH_3$

b. $C_6H_5—C(=O)—N(H)—CH_3$ or $C_6H_5—C(OH)=N—CH_3$

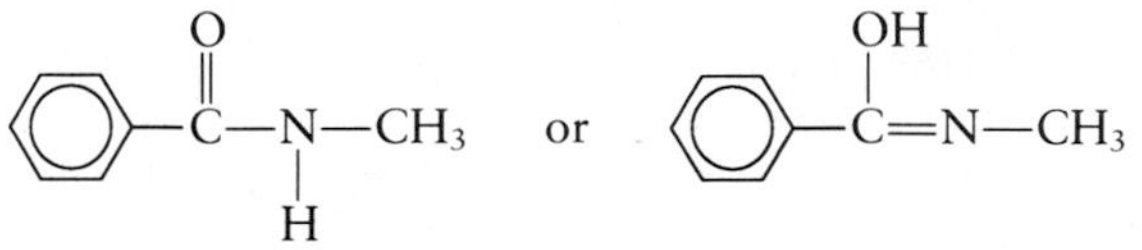

c. $H_3C—C(=O)—CH_2—C(=O)—CH_3$ or $H_3C—C(=O)—CH=C(OH)—CH_3$

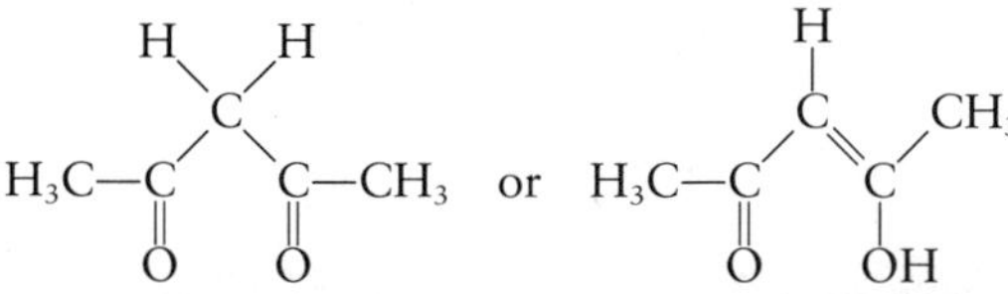

22. Polychlorinated dibenzo-*p*-dioxins, or PCDDs, are highly toxic substances that are present in trace amounts as byproducts of some chemical manufacturing processes. They have been implicated in a number of environmental incidents, for example, the chemical contamination at Love Canal and the herbicide spraying in Vietnam. The structure of dibenzo-*p*-dioxin, along with the customary numbering convention, is

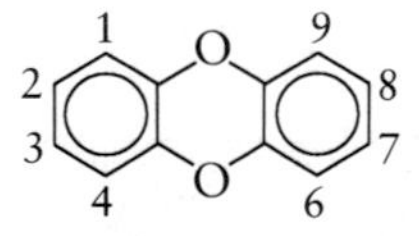

1 9 2 8 3 7 4 6 O O

The most toxic PCDD is 2,3,7,8-tetrachloro-dibenzo-*p*-dioxin. Draw the structure of this compound. Also draw the structures of two other isomers containing four chlorine atoms.

Functional Groups

23. Identify the functional groups present in the following drugs.

a. aspirin

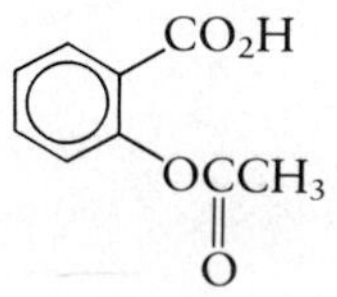

CO_2H $OCCH_3$ O

b. morphine

HO O N—CH_3 HO

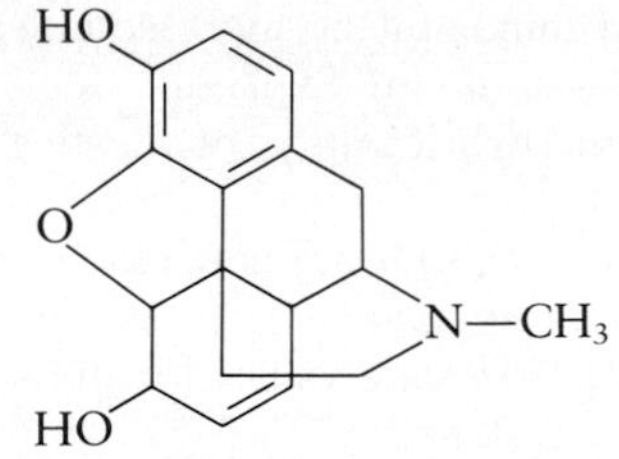

c. naloxone (a narcotic antagonist)

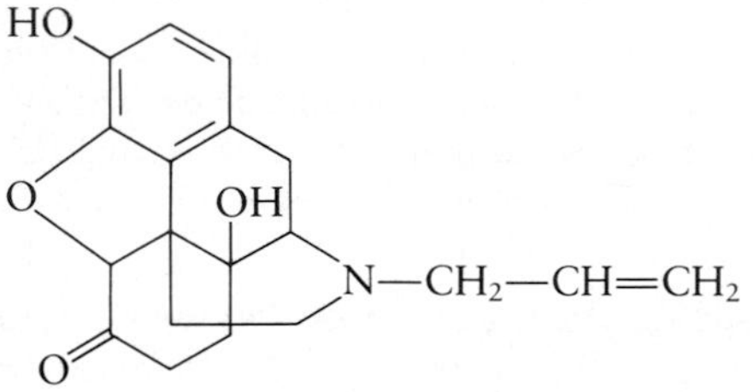

HO O OH N—CH_2—CH=CH_2 O

24. Mycomycin is a naturally occurring antibiotic produced by the fungus *Nocardia acidophilus*. The molecular formula of the substance is $C_{13}H_{10}O_2$, and its systematic name is 3,5,7,8-tridecatetraene-10,12-diynoic acid. Draw the structure of mycomycin.

25. Menthol has the systematic name 2-isopropyl-5-methylcyclohexanol. Draw the structure of menthol.

26. Mimosine is a natural product found in large quantities in the seeds and foliage of some legume plants. It has been shown to cause inhibition of hair growth as well as hair loss in mice.

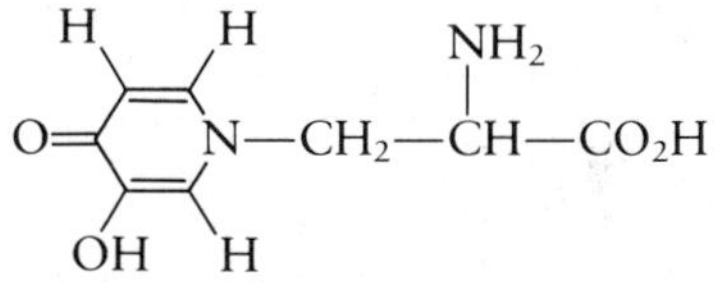

H H NH$_2$ O= N—CH_2—CH—CO_2H OH H

Mimosine, $C_8H_{10}N_2O_4$

a. What functional groups are present in mimosine?
b. Give the hybridization of the eight carbon atoms in mimosine.
c. How many σ and π bonds are found in mimosine?

27. Minoxidil, $C_9H_{15}N_5O$, is a compound produced by the Upjohn Company that has been approved as a treatment for some types of male pattern baldness.

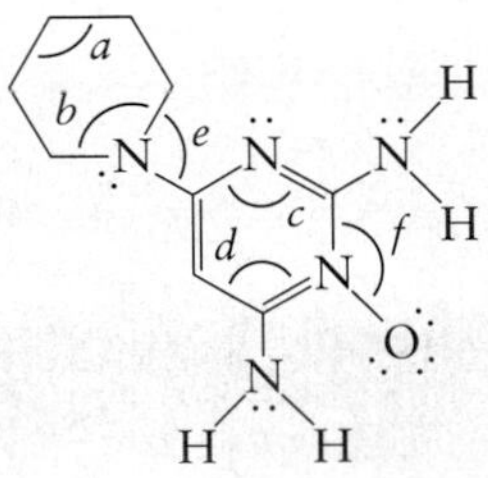

a b N e N N H c d f H N O N H H

a. Would minoxidil be more soluble in acidic or basic aqueous solution? Explain.
b. Give the hybridization of the five nitrogen atoms in minoxidil.
c. Give the hybridization of each of the nine carbon atoms in minoxidil.
d. Give approximate values for the bond angles marked *a*, *b*, *c*, *d*, *e*, and *f*.
e. Including all of the hydrogen atoms, how many σ bonds exist in minoxidil?
f. How many π bonds exist in minoxidil?

28. Many drugs such as morphine (see Exercise 23) are treated with strong inorganic acids. The most commonly used form of morphine is morphine hydrochloride. Draw the structure of morphine hydrochloride. Why are many drugs treated in this way?

29. Ethyl caprate is an ester used in the manufacture of wine bouquets. It is sometimes called "cognac essence." Combustion analysis of a sample of ethyl caprate shows it to be 71.89% carbon, 12.13% hydrogen, and 15.98% oxygen by mass. Hydrolysis (reaction with water) of the ester yields ethanol and an acid. The molar mass of the acid is 172. What is the molecular formula of ethyl caprate?

30. Draw an isomer specified by each of the following.
a. an aldehyde that is an isomer of acetone
b. an ether that is an isomer of 2-propanol
c. a geometrical isomer of *cis*-2-butene
d. a primary amine that is an isomer of trimethylamine
e. a secondary amine that is an isomer of trimethylamine
f. a primary alcohol that is an isomer of 2-propanol

31. Identify each of the following compounds as a carboxylic acid, ester, ketone, aldehyde, or amine.
a. anthraquinone, an important starting material in the manufacture of dyes:

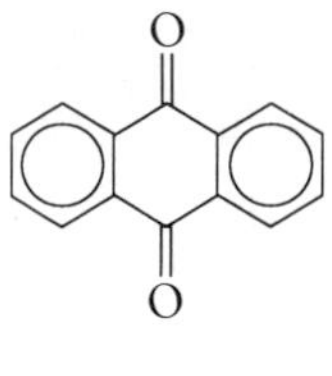

O
‖
(anthraquinone structure)
‖
O

b. 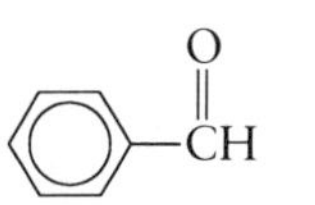 C_6H_5—$\overset{O}{\overset{\|}{C}}$H

c. $CH_3-\overset{O}{\overset{\|}{C}}-CH_2\underset{CH_3}{\underset{|}{C}}HCH_3$

d. C_6H_5—NH—C_6H_5

32. The structure of ephedrine (adrenaline) is

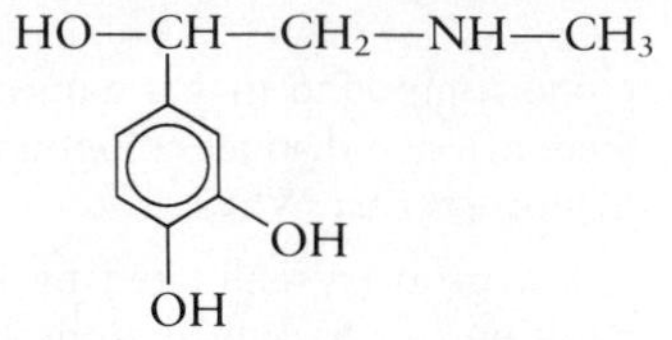

HO—CH—CH_2—NH—CH_3 (CH attached to a benzene ring bearing OH and OH)

What functional groups are present?

Reactions of Organic Compounds

33. Distinguish between substitution and addition reactions. Give an example of each type.

34. Consider the reaction of propane with chlorine.
a. How many monochloro products are formed? Draw their structures.
b. How many dichloro products can be formed? Draw their structures.

35. Complete the following reactions.
a. $CH_2CH_2 + Br_2 \longrightarrow$
b.  (benzene) $+ Br_2 \xrightarrow{Fe}$
c. $CH_3CO_2H + CH_3OH \longrightarrow$

36. Reagents such as HCl, HBr, and HOH can add across carbon-carbon double bonds. Two products are sometimes possible. For the major product, the addition occurs so that the hydrogen atom in the reagent attaches to the carbon atom in the double bond that already has the greater number of hydrogen atoms bonded to it. With this rule in mind, draw the structure of the major product formed in each of the following reactions.
a. $CH_3CH{=}CH_2 + HCl \longrightarrow$
b. $CH_3CH{=}CH_2 + H_2O \longrightarrow$
c. 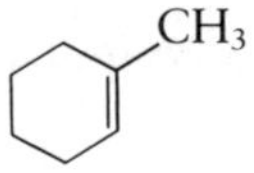 (1-methylcyclohexene, CH_3) $+ HBr \longrightarrow$
d. $H_2C{=}C(CH_3)_2 + H_2O \longrightarrow$
e. (1-methylcyclohexene, CH_3) $+ H_2O \longrightarrow$
f. $CH_3{-}C{\equiv}CH + 2HCl \longrightarrow$

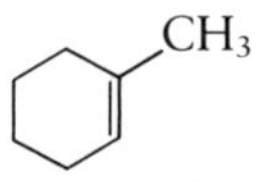

37. Give the structure of the product resulting from the oxidation of each of the following alcohols.
a. CH_3CH_2OH
b. $CH_3\underset{OH}{\underset{|}{C}}H-CH_3$
c. 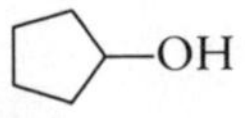 (cyclopentyl)—OH

d. $CH_3-\overset{\displaystyle CH_3}{\underset{\displaystyle CH_3}{C}}-OH$

e. $H_3C-\overset{\displaystyle CH_3}{\underset{\displaystyle CH_3}{C}}-CH_2-OH$

38. Oxidation of an aldehyde yields a carboxylic acid:

$$R-\overset{O}{\overset{\|}{C}}H \xrightarrow{[ox]} R-\overset{O}{\overset{\|}{C}}-OH$$

Thus it is very difficult to obtain only the pure aldehyde from the oxidation of a primary alcohol.

a. Draw structures for the products of the following oxidations.

i. $CH_3\overset{O}{\overset{\|}{C}}H \xrightarrow{[ox]}$

ii. $C_6H_5-\overset{O}{\overset{\|}{C}}-H \xrightarrow{[ox]}$

iii. $(CH_3)_2CH-\overset{O}{\overset{\|}{C}}H \xrightarrow{[ox]}$

b. Which of the reactions in Exercise 37 would result in a mixture of an aldehyde and an acid as products? Draw the structures of the acids formed.

39. How would you make each of the following?

a. 1,2-dibromopropane from propene
b. 1,2-dibromopropane from propyne
c. butyl acetate (butyl ethanoate)
d. ethyl butyrate (ethyl butanoate)
e. acetone from an alcohol

Polymers

40. Define the following and give an example of each.

a. addition polymer
b. condensation polymer
c. copolymer

41. Polyaramid is a term applied to polyamides containing aromatic groups. These polymers were originally made for use as tire cords but have since found many other uses.

a. Kevlar is used in bulletproof vests and many high-strength composites. The structure of Kevlar is

$$\left(\underset{H}{\underset{|}{N}}-C_6H_4-\underset{H}{\underset{|}{N}}\overset{O}{\overset{\|}{C}}-C_6H_4-\overset{O}{\overset{\|}{C}}\underset{H}{\underset{|}{N}}-C_6H_4-\underset{H}{\underset{|}{N}}\overset{O}{\overset{\|}{C}}-C_6H_4-\overset{O}{\overset{\|}{C}} \right)_n$$

Which monomers are used to make Kevlar?

b. Nomex is a polyaramid used in fire-resistant clothing. It is a copolymer of

$H_2N-C_6H_4-NH_2$ (meta) and $HO_2C-C_6H_4-CO_2H$ (meta)

Draw the structure of the Nomex polymer. How do Kevlar and Nomex differ in their structures?

42. The polyester formed from lactic acid,

$$CH_3-\underset{OH}{\underset{|}{C}}H-CO_2H$$

is used for tissue implants and surgical sutures that will dissolve in the body. Draw the structure of a portion of this polymer.

43. Polyimides are polymers that are tough and stable at temperatures up to 400°C. They are used as a protective coating on the quartz fibers in fiber optics. What monomers are used to make the following polyimide?

$$\left(C_6H_4-N\overset{C=O}{\underset{C=O}{<}}C_6H_2\overset{C=O}{\underset{C=O}{>}}N-C_6H_4-N\overset{C=O}{\underset{C=O}{<}}C_6H_2\overset{C=O}{\underset{C=O}{>}}N \right)_n$$

44. "Super glue" contains methyl cyanoacrylate,

$$\underset{O=C-O-CH_3}{NC}{>}C=CH_2$$

which readily polymerizes upon exposure to traces of water or alcohols on the surfaces to be bonded together. The polymer provides a strong bond between the two surfaces. Draw the structure of the polymer formed by methyl cyanoacrylate.

45. Ethylene oxide,

$$\underset{\diagdown O \diagup}{CH_2-CH_2}$$

is an important industrial chemical. Although most ethers are unreactive, ethylene oxide is quite reactive. It resembles C_2H_4 in its reactions in that many addition reactions occur across the C—O bond.

a. Why is ethylene oxide so reactive? (*Hint:* consider the bond angles in ethylene oxide as compared with those predicted by the VSEPR model.)

b. Ethylene oxide undergoes addition polymerization, forming a polymer used in many applications requiring a non-ionic surfactant. Draw the structure of this polymer.

46. Polycarbonates are a class of thermoplastic polymers that are used in the plastic lenses of eyeglasses and in the shells of bicycle helmets. A polycarbonate is made from the reaction of bisphenol A (BPA) with phosgene ($COCl_2$):

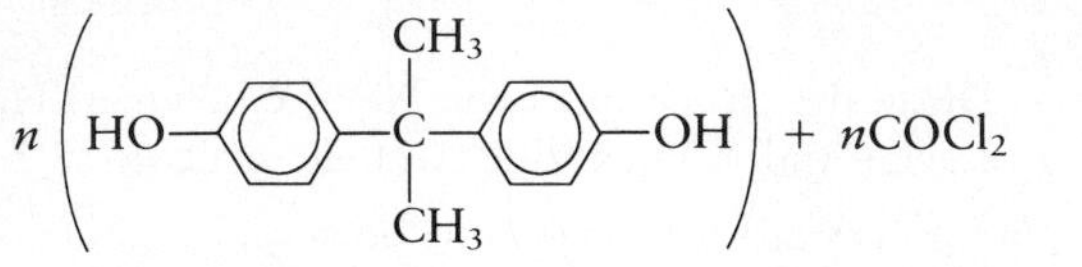

$$n\left(HO-C_6H_4-C(CH_3)_2-C_6H_4-OH\right) + nCOCl_2$$

BPA

$$\xrightarrow[\text{catalyst}]{2n\text{NaOH}} \text{polycarbonate} + 2nNaCl + 2nH_2O$$

Phenol, C_6H_5OH, is used to terminate the polymer (stop its growth).

a. Draw the structure of the polycarbonate chain formed in the above reaction.
b. Is this reaction a condensation or an addition polymerization?

47. Polystyrene can be made more rigid by copolymerizing styrene with divinylbenzene,

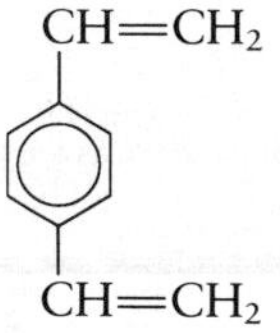

$$CH{=}CH_2-C_6H_4-CH{=}CH_2$$

What purpose does the divinylbenzene serve? Why is the copolymer more rigid?

48. In which polymer, polyethylene or polyvinyl chloride, would you expect to find the stronger intermolecular forces, assuming the average chain lengths are equal?

49. What monomer(s) must be used to produce the following polymers?

a. $\left(CH(F)-CH_2-CH(F)-CH_2-CH(F)-CH_2\right)_n$

b. $\left(O-CH_2-CH_2-\overset{O}{\overset{\|}{C}}-O-CH_2-CH_2-\overset{O}{\overset{\|}{C}}-O-CH_2-CH_2-\overset{O}{\overset{\|}{C}}\right)_n$

c. $\left(O-CH_2-CH_2-O-\overset{O}{\overset{\|}{C}}-CH_2-CH_2-\overset{O}{\overset{\|}{C}}-O-CH_2-CH_2-O-\overset{O}{\overset{\|}{C}}-CH_2-CH_2-\overset{O}{\overset{\|}{C}}\right)_n$

d. 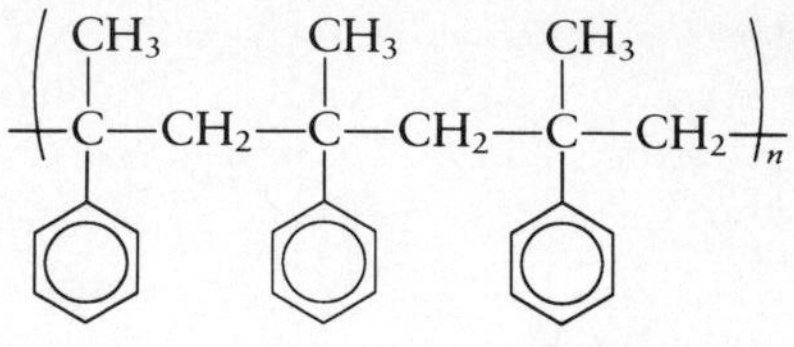 $\left(C(CH_3)(C_6H_5)-CH_2-C(CH_3)(C_6H_5)-CH_2-C(CH_3)(C_6H_5)-CH_2\right)_n$

e. 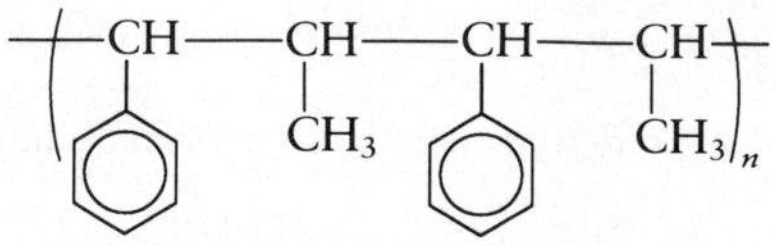 $\left(CH(C_6H_5)-CH(CH_3)-CH(C_6H_5)-CH(CH_3)\right)_n$

f. $+CClFCF_2CClFCF_2CClFCF_2+_n$

g. 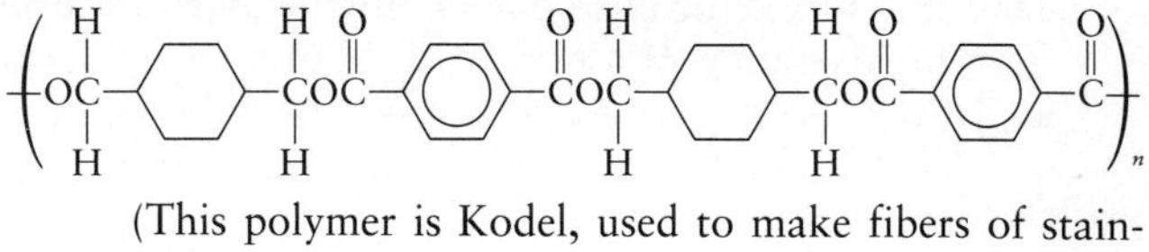 $\left(OCH_2-C_6H_{10}-CH_2OC(=O)-C_6H_4-C(=O)OCH_2-C_6H_{10}-CH_2OC(=O)-C_6H_4-C(=O)\right)_n$

(This polymer is Kodel, used to make fibers of stain-resistant carpeting.)

50. Classify the polymers in Exercise 49 as condensation or addition polymers. Which are copolymers?

Additional Exercises

51. Isoprene is the repeating unit in natural rubber. The structure of isoprene is

$$CH_2{=}C(CH_3)-CH{=}CH_2$$

a. Give a systematic name for isoprene.
b. When isoprene is polymerized, two polymers of the form

$$\left(CH_2-C(CH_3){=}CH-CH_2\right)_n$$

are possible. In natural rubber, the *cis* configuration is found. The polymer with the *trans* configuration about the double bond is called gutta percha and was once used in the manufacture of golf balls. Draw the structure of natural rubber and gutta percha showing three repeating units and the configuration about the carbon-carbon double bonds.

52. Polyesters containing double bonds are often crosslinked by reacting the polymer with styrene. This type of reaction is common in the manufacture of fiberglass.
    a. Draw the structure of the copolymer of

$$HO{-}CH_2CH_2{-}OH \quad \text{and} \quad HO_2C{-}CH{=}CH{-}CO_2H$$

    b. Draw the structure of the crosslinked polymer (after the polyester has been reacted with styrene).

53. Another way of producing highly crosslinked polyesters (see Exercise 52) is to use glycerol. Alkyd resins are a polymer of this type. The polymer forms very tough coatings when baked onto a surface and is used in paints for automobiles and large appliances. Draw the structure of the polymer formed from the condensation of

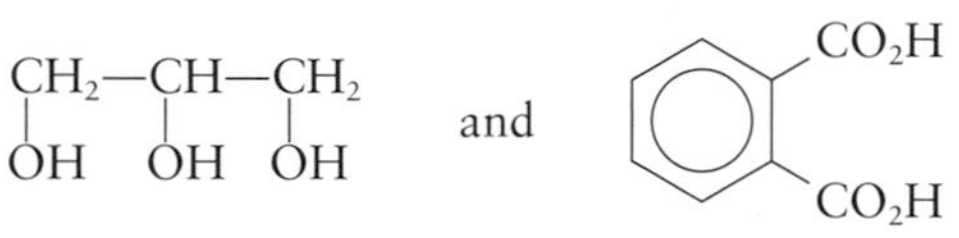

$CH_2{-}CH{-}CH_2$ (each carbon bearing OH) and $C_6H_4(CO_2H)_2$ (ortho)

Glycerol — Phthalic acid

54. Estimate ΔH for the following reactions using bond energies (Table 13.5).

$$3CH_2{=}CH_2(g) + 3H_2(g) \longrightarrow 3CH_3{-}CH_3(g)$$

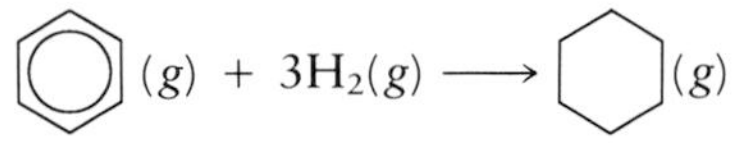

$$C_6H_6 \text{ (benzene)}(g) + 3H_2(g) \longrightarrow C_6H_{12} \text{ (cyclohexane)}(g)$$

The enthalpies of formation of $C_6H_6(g)$ and $C_6H_{12}(g)$ are 82.9 kJ/mol and −90.3 kJ/mol, respectively. Calculate $\Delta H°$ for the two reactions, using standard enthalpies of formation from Appendix 4. How do you account for any discrepancies between results obtained by the two methods?

55. The Amoco Chemical Company has successfully raced a car with a plastic engine. Many of the engine parts, including piston skirts, connecting rods, and valve-train components, are made of a polymer called *Torlon:*

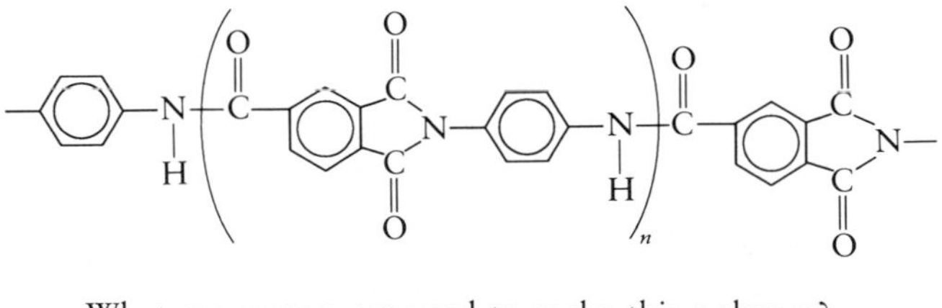

[Structure: repeating unit —C₆H₄—NH—C(=O)—(trimellitimide ring: benzene fused with C(=O)—N—C(=O))—C₆H₄—NH—, in parentheses with subscript n]

What monomers are used to make this polymer?

56. A urethane linkage occurs when an alcohol adds across the carbon-nitrogen double bond in an isocyanate:

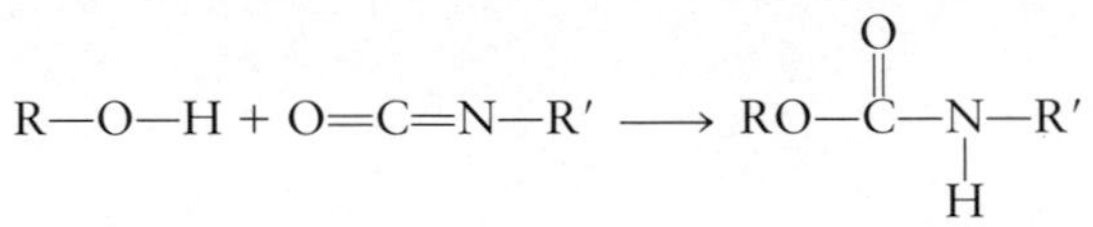

$$R{-}O{-}H + O{=}C{=}N{-}R' \longrightarrow RO{-}\overset{O}{\overset{\|}{C}}{-}\underset{H}{\underset{|}{N}}{-}R'$$

Alcohol — Isocyanate — A urethane

Polyurethanes are formed from the copolymerization of a diol with a diisocyanate. Polyurethanes are used in foamed insulation and a variety of other construction materials. What is the structure of the polyurethane formed by the following reaction?

$$HOCH_2CH_2OH + O{=}C{=}N{-}C_6H_4{-}N{=}C{=}O \longrightarrow$$

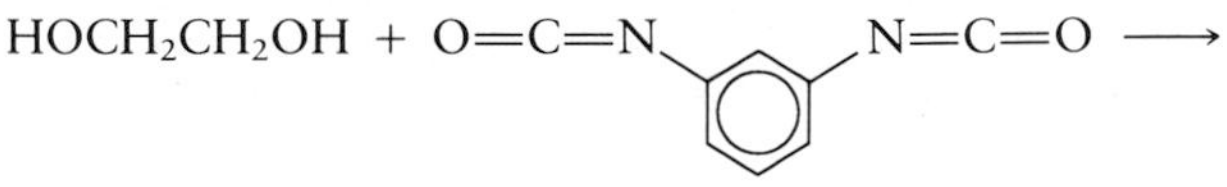

57. Sorbic acid is used to prevent mold and fungus growth in some food products, especially cheeses. The systematic name for sorbic acid is 2,4-hexadienoic acid. Draw structures for the four geometrical isomers of sorbic acid.

58. Poly(lauryl methacrylate) is used as an additive in motor oils to counter the loss of viscosity at high temperature. The structure is

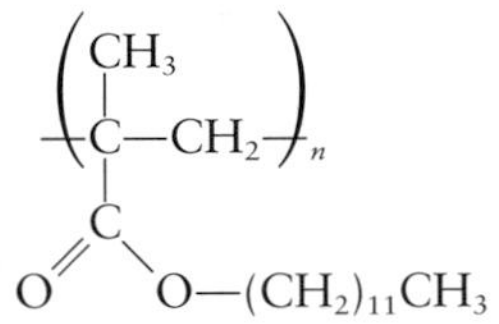

$$\left(\underset{\underset{O{=}C{-}O{-}(CH_2)_{11}CH_3}{|}}{\overset{\overset{CH_3}{|}}{C}}{-}CH_2\right)_n$$

The long hydrocarbon chain of poly(lauryl methacrylate) makes the polymer soluble in oil (a mixture of hydrocarbons with mostly 12 or more carbon atoms). At low temperatures the polymer is coiled into balls. At higher temperatures the balls uncoil and the polymer exists as long chains. Explain how this helps control the viscosity of oil.

CHAPTER 23

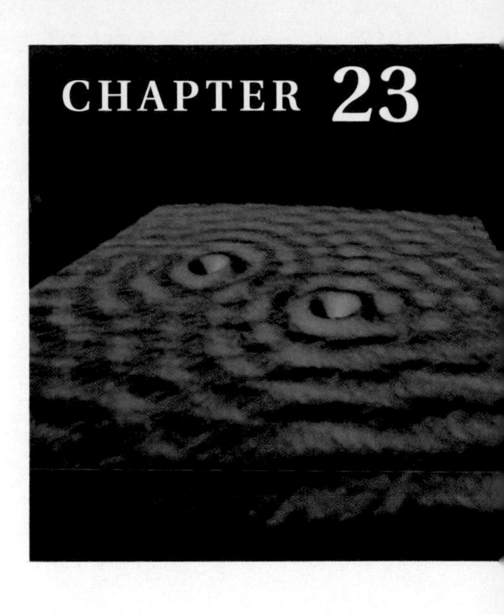

Biochemistry

Biochemistry, which is the study of the chemistry of living systems, is a vast and exciting field in which important discoveries about how life is maintained and how diseases occur are being made every day. In particular, there has been rapid growth in the understanding of how living cells manufacture the molecules necessary for life.

We cannot hope to cover even the majority of the important aspects of biochemistry in this chapter; we will concentrate on the major types of biomolecules that support living systems. First, however, we will survey the elements found in living systems and briefly describe the constitution of a cell.

At present 30 elements are known to be essential or are strongly suspected to be essential to human life. These **essential elements** are shown in Fig. 23.1. The most abundant elements are hydrogen, carbon, nitrogen, and oxygen; but sodium, magnesium, potassium, calcium, phosphorus, sulfur, and chlorine are also present in relatively large amounts. Although present only in trace amounts, the first-row transition metals are essential for the action of many enzymes. For example, zinc, one of the **trace elements,** is present in over 160 biologically important molecules. The functions of the essential elements are summarized in Table 23.1. As more studies are performed, other elements will most certainly be found to be essential.

Life is organized around the functions of the **cell,** the smallest unit in living things that has the properties normally associated with life, such as reproduction, metabolism, mutation, and sensitivity to external stimuli. There are two fundamental types of cells: *prokaryotes,* those without nuclei; and *eukaryotes,*

1A	2A											3A	4A	5A	6A	7A	8A
1 H																	2 He
3 Li	4 Be											5 B	6 C	7 N	8 O	9 F	10 Ne
11 Na	12 Mg											13 Al	14 Si	15 P	16 S	17 Cl	18 Ar
19 K	20 Ca	21 Sc	22 Ti	23 V	24 Cr	25 Mn	26 Fe	27 Co	28 Ni	29 Cu	30 Zn	31 Ga	32 Ge	33 As	34 Se	35 Br	36 Kr
37 Rb	38 Sr	39 Y	40 Zr	41 Nb	42 Mo	43 Tc	44 Ru	45 Rh	46 Pd	47 Ag	48 Cd	49 In	50 Sn	51 Sb	52 Te	53 I	54 Xe
55 Cs	56 Ba	57 La	72 Hf	73 Ta	74 W	75 Re	76 Os	77 Ir	78 Pt	79 Au	80 Hg	81 Tl	82 Pb	83 Bi	84 Po	85 At	86 Rn
87 Fr	88 Ra	89 Ac	104 Unq	105 Unp	106 Unh	107 Uns	108 Uno	109 Une									

58 Ce	59 Pr	60 Nd	61 Pm	62 Sm	63 Eu	64 Gd	65 Tb	66 Dy	67 Ho	68 Er	69 Tm	70 Yb	71 Lu
90 Th	91 Pa	92 U	93 Np	94 Pu	95 Am	96 Cm	97 Bk	98 Cf	99 Es	100 Fm	101 Md	102 No	103 Lr

Figure 23.1
The chemical elements essential for life. Those most abundant in living systems are shown as purple. Nineteen elements, called the trace elements, are shown as green.

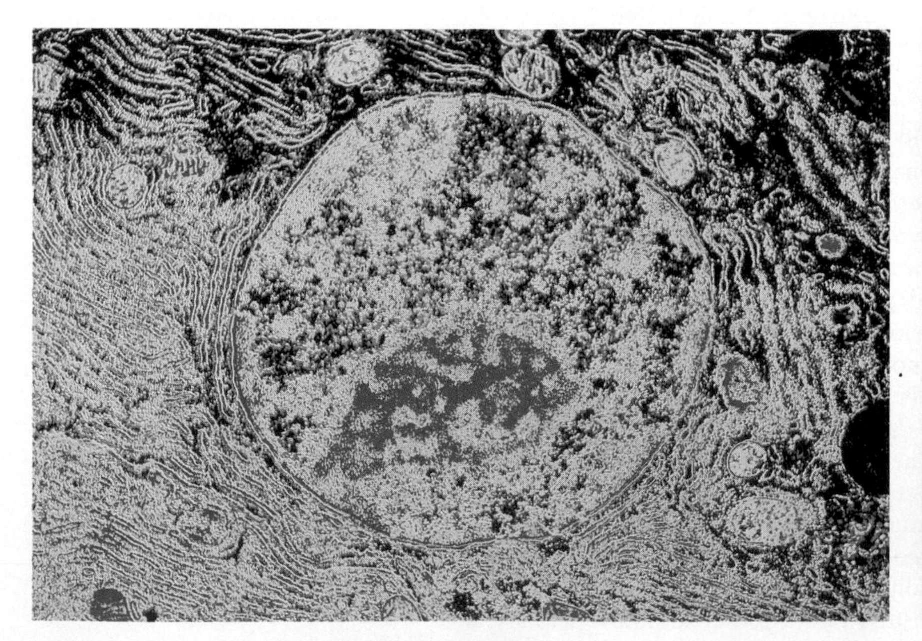

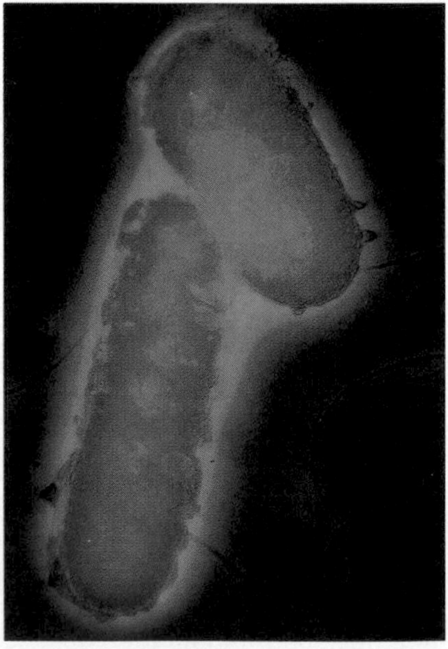

(left) A eukaryotic cell. (right) A prokaryotic cell.

TABLE 23.1 **The Essential Elements and Some of Their Functions**

Element	Percent by Mass in the Human Body	Function
Oxygen	65	Component of water and many organic compounds
Carbon	18	Component of all organic compounds
Hydrogen	10	Component of water and many inorganic and organic compounds
Nitrogen	3	Component of both inorganic and organic compounds
Calcium	1.5	Major component of bone; essential to some enzymes and to muscle action
Phosphorus	1.2	Essential in cellular synthesis and energy transfer
Potassium	0.2	Cation in intracellular fluid
Chlorine	0.2	Anion inside and outside the cells
Sulfur	0.2	Component of proteins and some other organic compounds
Sodium	0.1	Cation in extracellular fluid
Magnesium	0.05	Essential to some enzymes
Iron	<0.05	In hemoglobin, myoglobin, and other proteins (Section 20.8)
Zinc	<0.05	Essential to many enzymes
Cobalt	<0.05	Found in vitamin B_{12}
Copper	<0.05	Essential to several enzymes
Iodine	<0.05	Essential to thyroid hormones
Selenium	<0.01	Essential to some enzymes
Fluorine	<0.01	In teeth and bones
Nickel	<0.01	Essential to some enzymes
Molybdenum	<0.01	Essential to some enzymes
Silicon	<0.01	Found in connective tissue
Chromium	<0.01	Essential in carbohydrate metabolism
Vanadium, tin, manganese, lithium, boron, arsenic, lead, cadmium	<0.01	Exact function not known

those that have nuclei. Bacteria are examples of prokaryotic organisms, while plant and animal cells are examples of eukaryotic cells.

As the fundamental building blocks of all living systems, cells aggregate to form tissues, which in turn are assembled into the organs that make up complex living systems. Thus to understand how life is maintained and reproduced, we must learn how cells operate on the molecular level. This is the main thrust of biochemistry.

The composition and organization of a typical animal cell is shown in Fig. 23.2. The *nucleus,* which contains the *chromosomes,* is separated from the cell fluid (the *cytoplasm*) by a membrane. The chromosomes store both the hereditary information necessary for cell reproduction and the codes needed for the manufacture of essential biomolecules. The cytoplasm contains a variety of

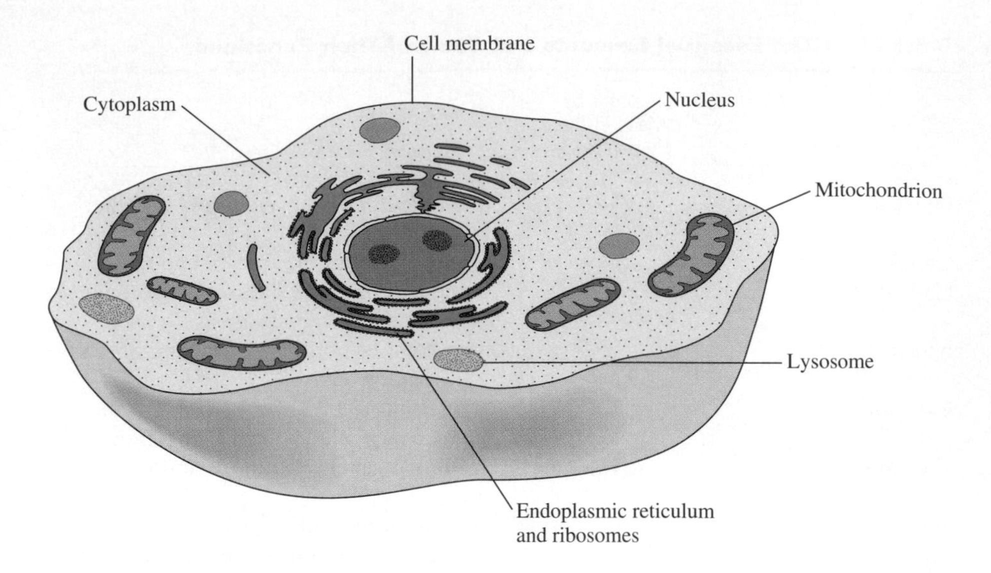

Figure 23.2
A typical animal cell (eukaryotic).

subcellular structures that carry out various cell functions. For example, the *mitochondria* process nutrients and produce energy to be used as required by the cell. The *lysosomes* contain enzymes for "digesting" nutrients such as proteins. The *ribosomes* are the sites for protein synthesis. The cell membrane encloses the cytoplasm, protects the cell components, and allows the passage of nutrients, ions, and wastes.

23.1 Proteins

We have seen that many useful synthetic materials are polymers. Thus it should not be surprising that a great many natural materials are also polymers: starch, hair, silicate chains in soil and rocks, silk and cotton fibers, and the cellulose in woody plants, to name only a few.

In this section we consider a class of natural polymers, the **proteins,** which make up about 15% of our bodies and have molecular weights (molar masses) that range from approximately 6000 to over 1,000,000 grams per mole. Proteins perform many functions in the human body. **Fibrous proteins** provide structural integrity and strength for many types of tissue and are the main components of muscle, hair, and cartilage. Other proteins, usually called **globular proteins** because of their roughly spherical shape, are the "worker" molecules of the body. These proteins transport and store oxygen and nutrients, act as catalysts for the thousands of reactions that make life possible, fight invasion by foreign objects, participate in the body's many regulatory systems, and transport electrons in the complex process of metabolizing nutrients.

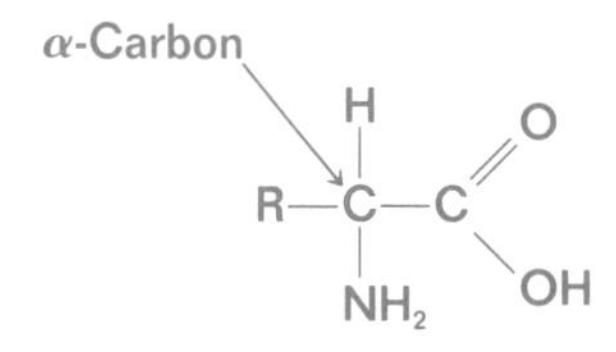

The building blocks of all proteins are the **α-amino acids,** where R may represent H, CH_3, or a more complex substituent. These molecules are called α-amino acids because the amino group ($—NH_2$) is always attached to the α-carbon, the one next to the carboxyl group ($—CO_2H$). The 20 amino acids most commonly found in proteins are shown in Fig. 23.3.

At the pH in biological fluids, the amino acids shown in Fig. 23.3 exist in a different form, with the proton of the —COOH group transferred to the $—NH_2$ group. For example, glycine would be in the form $H_3{}^+NCH_2COO^-$.

Note from Fig. 23.3 that the amino acids are grouped into polar and nonpolar classes, determined by the R groups, or **side chains.** Nonpolar side

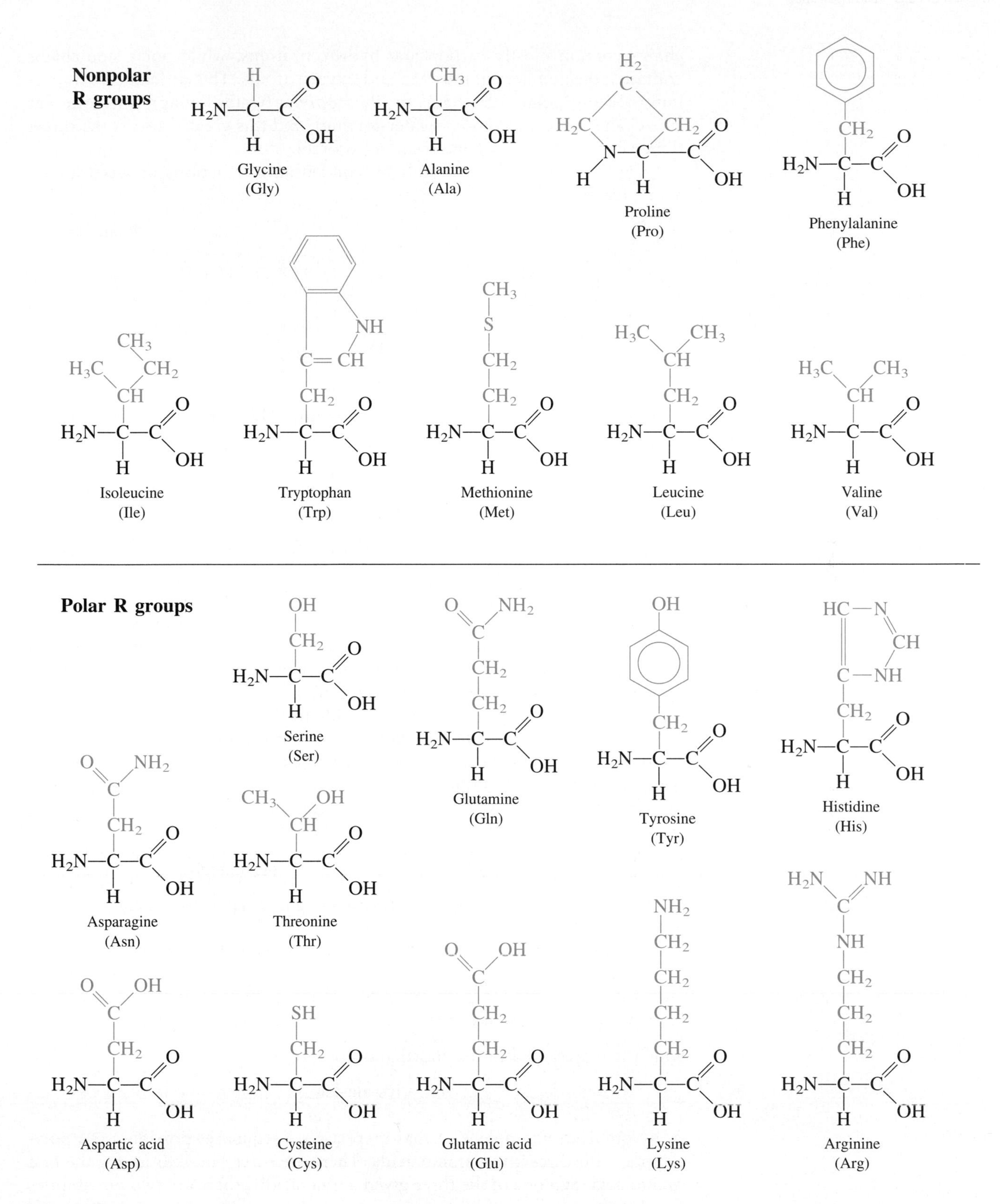

Figure 23.3
The 20 α-amino acids found in most proteins. The R group is shown in color.

chains contain mostly carbon and hydrogen atoms, while polar side chains contain large numbers of nitrogen and oxygen atoms. This difference is important, because polar side chains are *hydrophilic* (water-loving), but nonpolar side chains are *hydrophobic* (water-fearing), and this greatly affects the three-dimensional structure of the resulting protein.

The protein polymer is built by condensation reactions between amino acids, for example,

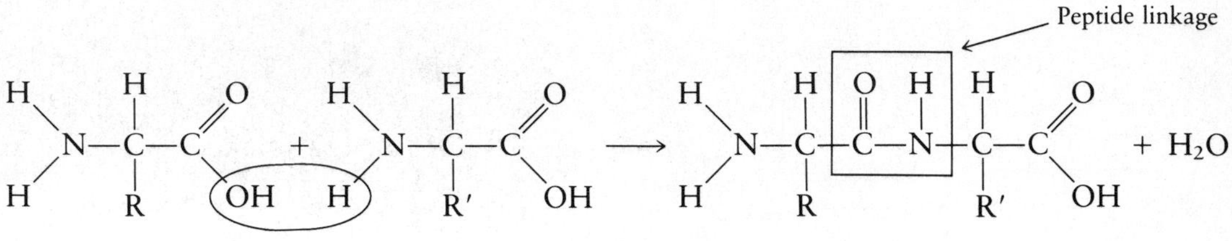

$$H_2N{-}CHR{-}C({=}O){-}OH + H{-}NH{-}CHR'{-}C({=}O){-}OH \longrightarrow H_2N{-}CHR{-}\boxed{C({=}O){-}NH}{-}CHR'{-}C({=}O){-}OH + H_2O$$

Peptide linkage

The product shown above is called a **dipeptide.** This name is used because the structure

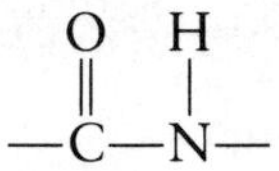

$$-\overset{\overset{\large O}{\|}}{C}-\overset{\overset{\large H}{|}}{N}-$$

The peptide linkage is also found in nylon (see Section 22.5).

is called a **peptide linkage** by biochemists. (The same grouping is called an amide by organic chemists.) Additional condensation reactions lengthen the chain to produce a **polypeptide,** eventually yielding a protein.

You can imagine that with 20 amino acids, which can be assembled in any order, there is essentially an infinite variety possible in the construction of proteins. This flexibility allows an organism to tailor proteins for the many types of functions that must be carried out.

The order or sequence of amino acids in the protein chain is called the **primary structure,** conveniently indicated by using three-letter codes for the amino acids (see Fig. 23.3), where it is understood that the terminal carboxyl group is on the right and the terminal amino group is on the left. For example, one possible sequence for a tripeptide containing the amino acids lysine, alanine, and leucine is

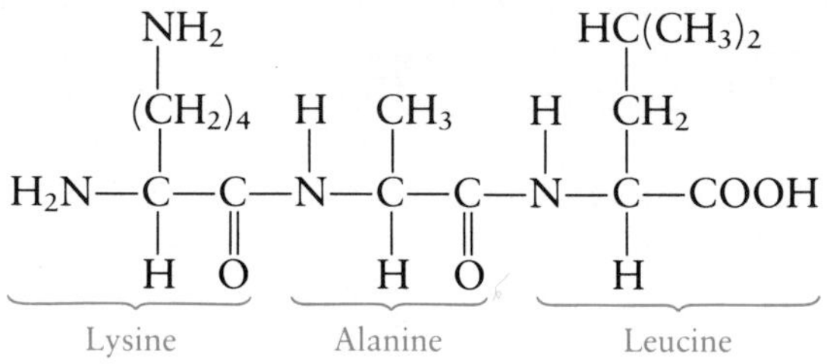

$$\underbrace{H_2N{-}CH[(CH_2)_4NH_2]{-}C({=}O)}_{\text{Lysine}}{-}\underbrace{NH{-}CH(CH_3){-}C({=}O)}_{\text{Alanine}}{-}\underbrace{NH{-}CH[CH_2HC(CH_3)_2]{-}COOH}_{\text{Leucine}}$$

which is represented in the shorthand notation by

lys-ala-leu

Note from Example 23.1 that there are six sequences possible for a polypeptide with three given amino acids. There are three possibilities for the first amino acid (any one of the three given amino acids), there are two possibilities for the second amino acid (one has already been accounted for), but there is

EXAMPLE 23.1

Write the sequences of all possible tripeptides composed of the amino acids tyrosine, histidine, and cysteine.

Solution

There are six possible sequences:

tyr-his-cys	his-tyr-cys	cys-tyr-his
tyr-cys-his	his-cys-tyr	cys-his-tyr

only one possibility left for the third amino acid. Thus the number of sequences is $3 \times 2 \times 1 = 6$. The product $3 \times 2 \times 1$ is often written 3! (and is called 3 factorial). Similar reasoning shows that for a polypeptide with four amino acids, there are 4!, or $4 \times 3 \times 2 \times 1 = 24$, possible sequences.

EXAMPLE 23.2

What number of possible sequences exists for a polypeptide composed of 20 different amino acids?

Solution

The answer is 20!, or

$$20 \times 19 \times 18 \times 17 \times 16 \times \cdots \times 5 \times 4 \times 3 \times 2 \times 1 = 2.43 \times 10^{18}$$

A striking example of the importance of the primary structure of polypeptides can be seen in the differences between *oxytocin* and *vasopressin*. Both of these molecules are nine-unit polypeptides, and they differ by only two amino acids (Fig. 23.4); yet they perform completely different functions in the human body. Oxytocin is a hormone that triggers contraction of the uterus and milk secretion. Vasopressin raises blood pressure levels and regulates kidney function.

cys–tyr–[ile]–gln–asn–cys–pro–[leu]–gly
(a)

cys–tyr–[phe]–gln–asn–cys–pro–[arg]–gly
(b)

Figure 23.4
The amino acid sequences in (a) oxytocin and (b) vasopressin. The differing amino acids are boxed.

A second level of structure in proteins, beyond the sequence of amino acids, is the arrangement of the chain of the long molecule. The **secondary structure** is determined to a large extent by hydrogen bonding between lone pairs on an oxygen in the carbonyl group of an amino acid and a hydrogen atom attached to a nitrogen of another amino acid:

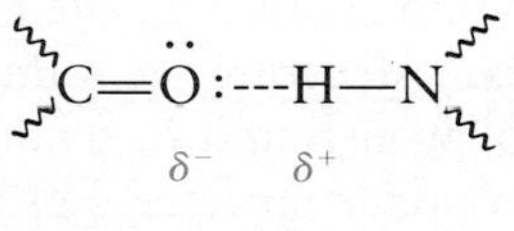

$$\rangle C{=}\ddot{O}{:}\text{---}H{-}N\langle \quad (\delta^- \text{ on O}, \ \delta^+ \text{ on H})$$

Such interactions can occur *within* the chain coils to form a spiral structure called an **α-helix,** as shown in Fig. 23.5 and Fig. 23.6. This type of secondary structure gives the protein elasticity (springiness) and is found in the fibrous

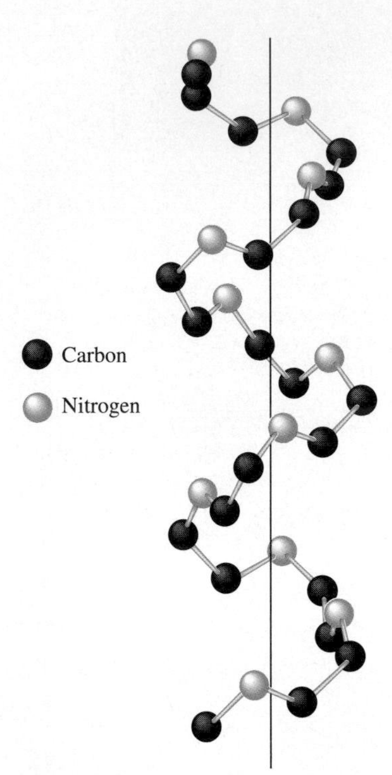

Figure 23.5
Hydrogen bonding within a protein chain causes it to form a stable helical structure called the α-helix. Only the main atoms in the helical backbone are shown here. The hydrogen bonds are not shown.

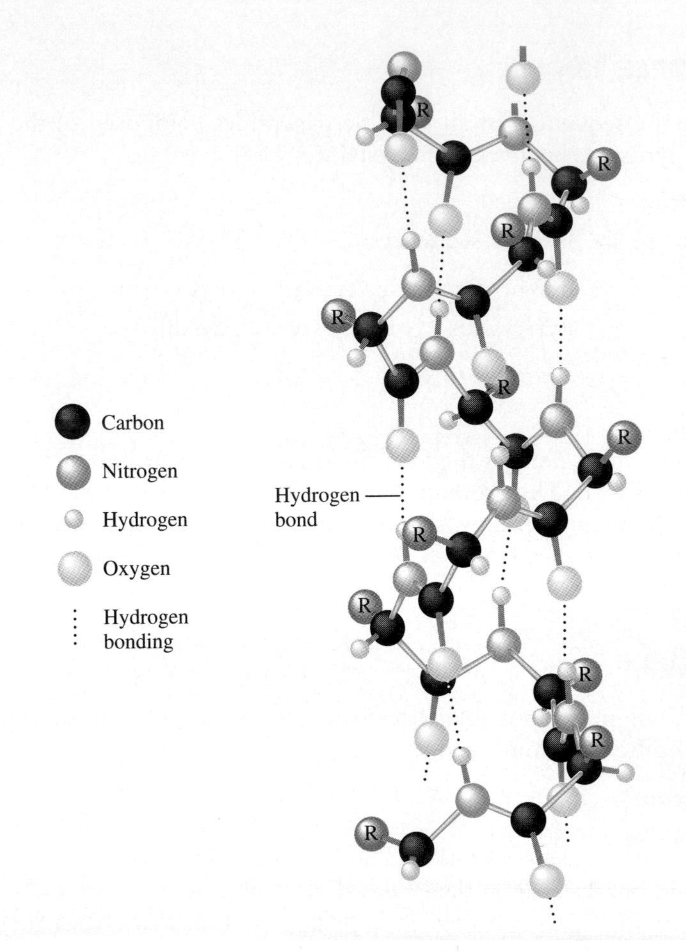

Figure 23.6
Ball-and-stick model of a portion of a protein chain in the α-helical arrangement, showing the hydrogen-bonding interactions.

proteins in wool, hair, and tendons. Hydrogen bonding can also occur *between different* protein chains, joining them together in an arrangement called a **pleated sheet,** as shown in Fig. 23.7. Silk contains this arrangement of proteins, making its fibers flexible yet very strong and resistant to stretching. The pleated sheet is also found in muscle fibers. The hydrogen bonds in the α-helical protein are called *intrachain* (within a given protein chain), and those in the pleated sheet are said to be *interchain* (between protein chains).

As you might imagine, a molecule as large as a protein has a great deal of flexibility and can assume a variety of overall shapes. The specific shape that a protein assumes depends on its function. For long, thin structures, such as hair, wool and silk fibers, and tendons, an elongated shape is required. This may involve an α-helical secondary structure, as found in the protein α-keratin in hair and wool or in the collagen found in tendons [Fig. 23.8(a)]; or it may involve a pleated-sheet secondary structure, as found in silk [Fig. 23.8(b)]. Many of the proteins in the body having nonstructural functions are globular, such as myoglobin (Fig. 23.9). Note that the secondary structure of myoglobin is basically α-helical. However, in the areas where the chain bends to give the

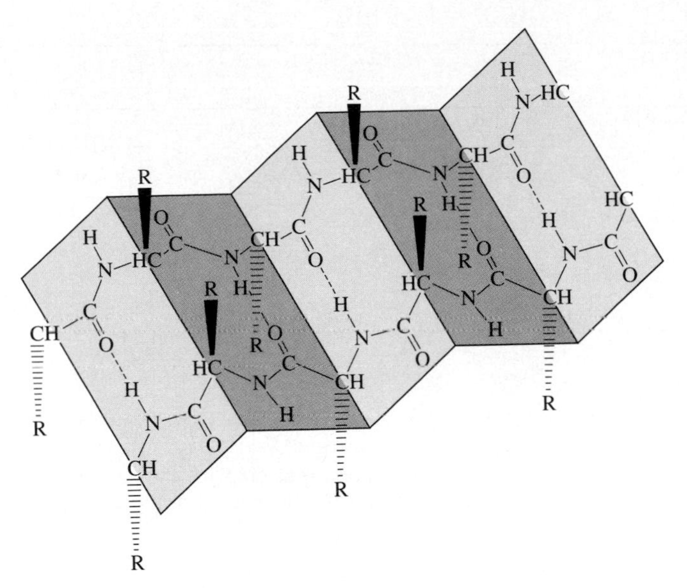

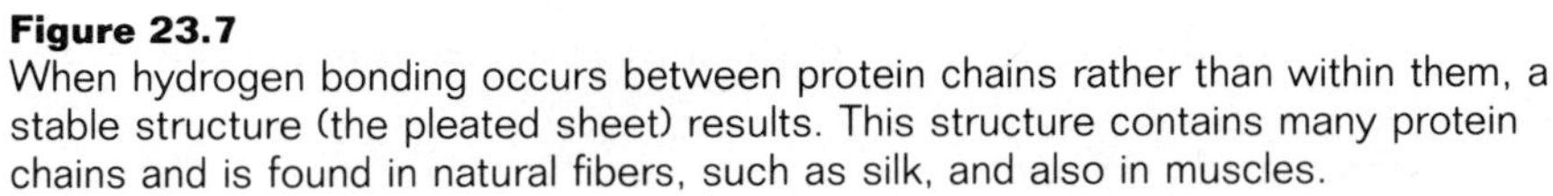
Figure 23.7
When hydrogen bonding occurs between protein chains rather than within them, a stable structure (the pleated sheet) results. This structure contains many protein chains and is found in natural fibers, such as silk, and also in muscles.

(a) (b)

Figure 23.8
(a) Collagen, a protein found in tendons, consists of three protein chains (each with a helical structure) twisted together to form a super helix. The result is a long, relatively narrow protein. (b) The pleated-sheet arrangement of many proteins bound together to form the elongated protein found in silk fibers.

protein its compact globular structure, the α-helix breaks down to give a secondary configuration known as the **random-coil arrangement.**

The overall shape of the protein, long and narrow or globular, is called its **tertiary structure** and is maintained by several different types of interactions: hydrogen bonding, dipole-dipole interactions, ionic bonds, covalent bonds, and London dispersion forces between nonpolar groups. These bonds, which repre-

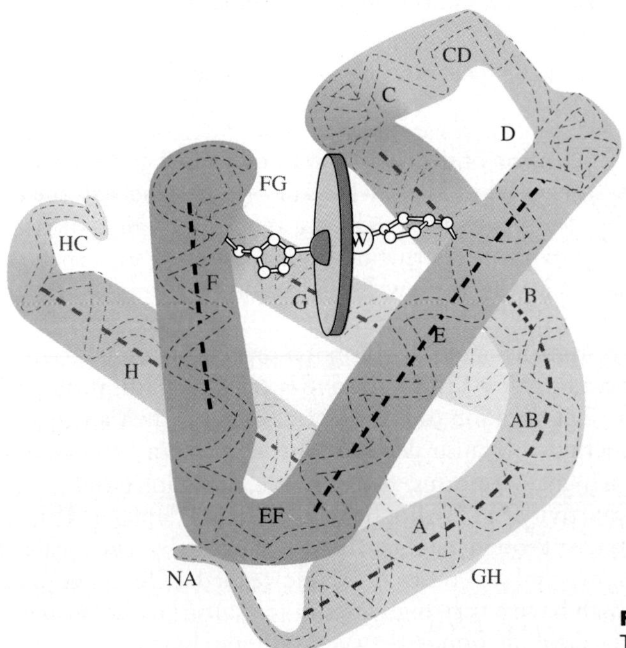

Figure 23.9
The protein myoglobin.

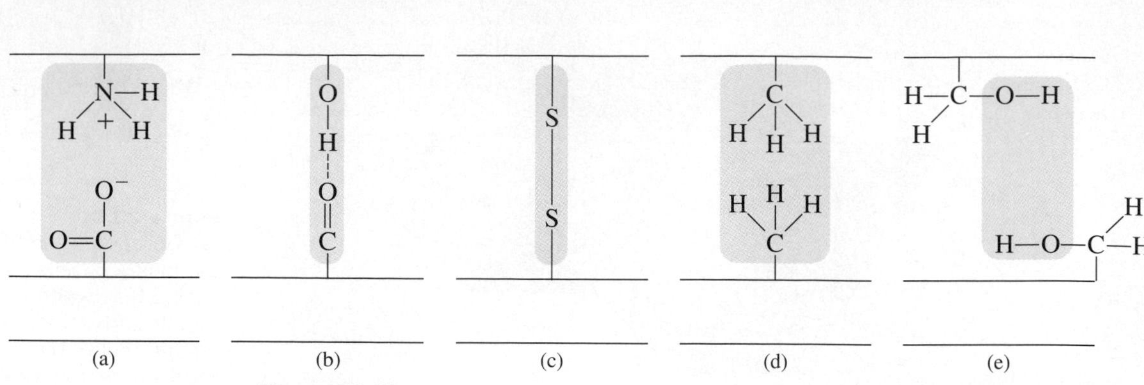

Figure 23.10
Summary of the various types of interactions that stabilize the tertiary structure of a protein: (a) ionic, (b) hydrogen bonding, (c) covalent, (d) London dispersion, and (e) dipole-dipole.

sent all of the bonding types discussed in this text, are summarized in Fig. 23.10.

The amino acid *cysteine*

$$HS-CH_2-\underset{\underset{\displaystyle H\quad H}{\displaystyle N}}{\overset{\displaystyle H}{C}}-\overset{\displaystyle O}{\overset{\|}{C}}-OH$$

plays a special role in stabilizing the tertiary structure of many proteins because the —SH groups on two cysteines can react in the presence of an oxidizing agent to form an S—S bond called a **disulfide linkage:**

$$C-CH_2-S-H + H-S-CH_2-C \longrightarrow C-CH_2-\boxed{S-S}-CH_2-C$$

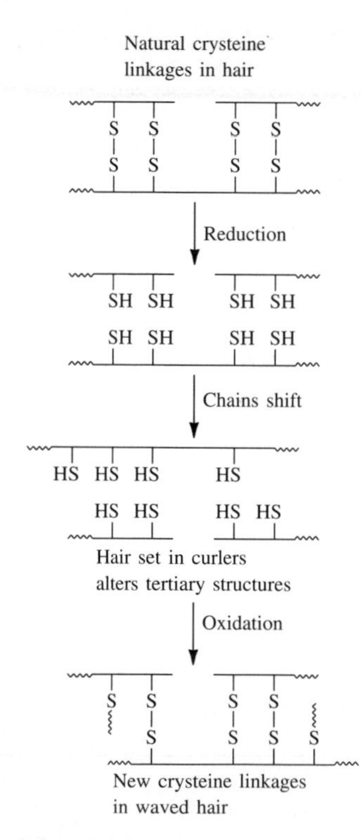

Figure 23.11
The permanent waving of hair.

A practical application of the chemistry of disulfide bonds is permanent waving of hair, as summarized in Fig. 23.11. The S—S linkages in the protein of hair are broken by treatment with a reducing agent. The hair is then set in curlers to change the tertiary protein structure to the desired shape. Then treatment with an oxidizing agent causes new S—S bonds to form, which cause the hair protein to retain the new structure.

The three-dimensional structure of a protein is crucial to its function. The process of breaking down this structure is called **denaturation** (Fig. 23.12). For example, the denaturation of egg proteins occurs when an egg is cooked. Any source of energy can cause denaturation of proteins and is thus potentially dangerous to living organisms. For example, ultraviolet and X-ray radiation or nuclear radioactivity can disrupt protein structure, which may lead to cancer or genetic damage. Protein damage is also caused by chemicals like benzene, trichloroethane, and 1,2-dibromoethane (called EDB). The metals lead and mercury, which have a very high affinity for sulfur, cause protein denaturation by disrupting disulfide bonds between protein chains.

The tremendous flexibility in the various levels of protein structure allows

TABLE 23.2 **Some Functions of Proteins**

1. *Structure.* Proteins provide the high tensile strength of tendons, bones, and skin. Cartilage, hair, wool, fingernails, and claws are mainly protein. Viruses have an outer layer of protein around a nucleic acid core. Proteins called *histones* are bound tightly to DNA in the cells of higher organisms, helping to fold and wrap the DNA in an orderly fashion in chromosomes.
2. *Movement.* Proteins are the major components of muscles and are directly responsible for the ability of muscles to contract. The swimming of sperm is the result of contraction of protein filaments in their tails. The same is true of movements of chromosomes during cell division.
3. *Catalysis.* Nearly all chemical reactions in living organisms are catalyzed by enzymes, which are almost always proteins.
4. *Transport.* Oxygen is carried from the lungs to tissues by the protein hemoglobin in red blood cells. The protein transferrin transports iron in blood plasma from the intestines (where the iron is absorbed) to the spleen (where it is stored) and to the liver and bone marrow (where it is used for synthesis). Proteins in the membranes of cells allow the passage of various molecules and ions.
5. *Storage.* The protein ferritin stores iron in the liver, spleen, and bone marrow.
6. *Energy transformation and storage.* Myosin, a protein in muscle, transforms chemical energy into useful work. Rhodopsin, a protein in the retina of the eye, traps light energy and, working with other membrane components, converts it into the electrical energy of a nerve impulse. Receptor protein molecules that combine with specific small molecules are responsible for the transmission of nerve impulses.
7. *Protection. Antibodies* are special proteins that are synthesized in response to foreign substances and cells, such as bacterial cells. They then bind to those substances or cells, which are called *antigens,* and provide us with immunity to various diseases. We acquire antibodies either from having had the disease or from receiving inactivated viruses in vaccines. Hay fever and food allergies are also caused by the interaction of antibodies with antigens. Interferon, a small protein made and released by cells when they are exposed to a virus, protects other cells against viral infection. Blood-clotting proteins protect against bleeding (hemorrhage). Some antibiotics are polypeptides, which are smaller than proteins.
8. *Control. Hormones* are chemical substances produced in the body that have specific effects on the activity of certain organs. Some hormones are proteins. For example, human growth hormone is a protein. Some hormones such as insulin and glucagon, which are made in the pancreas and control carbohydrate metabolism, are polypeptides. Expression of genetic information is also under the control of proteins.
9. *Buffering.* Because proteins contain both acidic and basic groups on their side chains, they can neutralize both acids and bases. Therefore, proteins provide some buffering for blood and tissues.

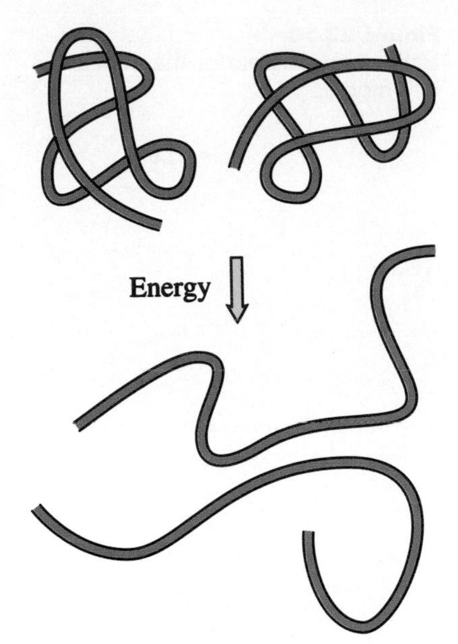

Figure 23.12
A schematic representation of the thermal denaturation of a protein.

the tailoring of proteins for a wide range of specific functions, some of which are given in Table 23.2.

Enzymes

Enzymes are proteins tailored to catalyze specific biological reactions. Without the several hundred enzymes now known, life would be impossible. Enzymes are impressive for their tremendous efficiency (typically 1 to 10 million times as efficient as inorganic catalysts) and their incredible selectivity—they ignore the thousands of molecules in body fluids for which they were not designed.

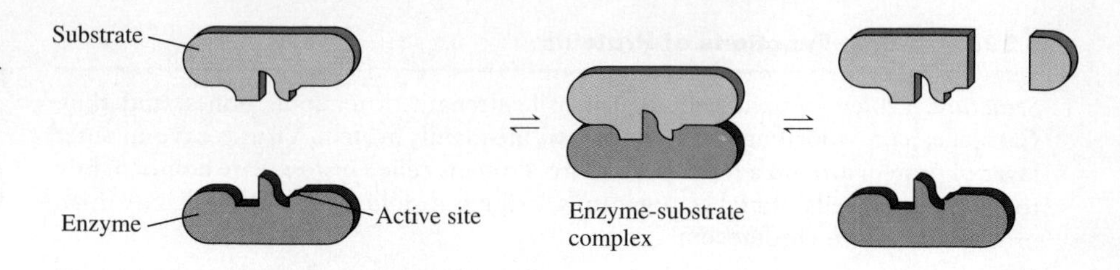

Figure 23.13
Schematic diagram of the lock-and-key model.

Although the mechanisms of catalytic activity are complex and not fully understood in most cases, a simple model called the **lock-and-key model** (see Fig. 23.13) seems to fit many enzymes. This model postulates that the shapes of the reacting molecule, the **substrate,** and the enzyme are such that they fit together much like a key fits a specific lock. The substrate and enzyme attach to each other through hydrogen bonding, ionic bonding, metal ion–ligand bonding, or some combination of these, in such a way that the part of the substrate where the reaction is to occur occupies the active site of the enzyme. This process is summarized schematically in Fig. 23.14 for the enzyme carboxypepti-

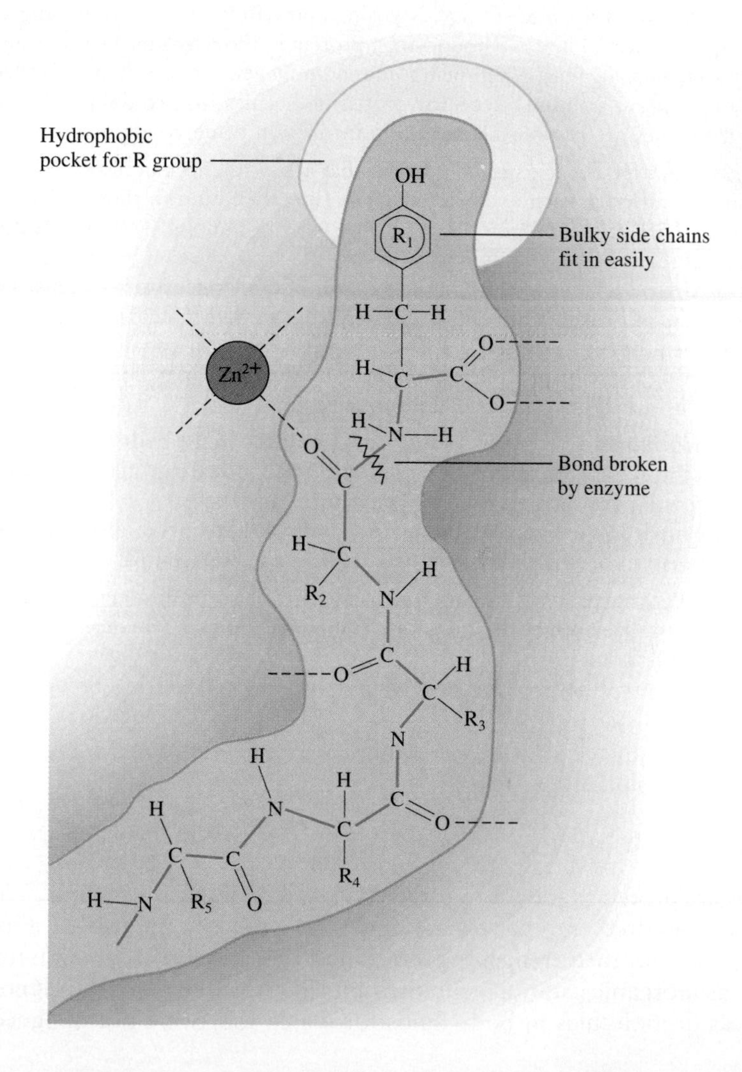

Figure 23.14
Schematic of the binding of a protein by carboxypeptidase.

dase binding a protein. After the reaction occurs, the products are liberated, and the enzyme is ready for a new substrate. Because the enzyme cycles ("turns over") so rapidly, only a tiny amount of enzyme is required. This has made the isolation and study of enzymes quite difficult.

The enzyme carboxypeptidase assists in the digestion of proteins by catalyzing the breaking of the peptide linkage that attaches the end amino acid to the protein:

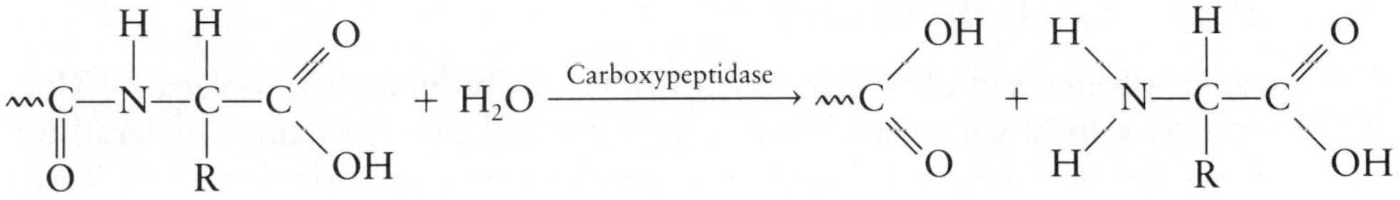

$$\sim C(=O)-N(H)-C(H)(R)-C(=O)OH + H_2O \xrightarrow{\text{Carboxypeptidase}} \sim C(=O)OH + H_2N-C(H)(R)-C(=O)OH$$

The active site contains Zn^{2+}, which coordinates to the carbonyl group of the end amino acid. There is also a hydrophobic pocket nearby to accommodate a large hydrophobic R group on the terminal amino acid. There are other interactions between groups on the enzyme and substrate, which help hold the substrate in place.

We can represent enzyme catalysis by the following mechanism:

STEP 1

The substrate S binds to the active site of the enzyme E:

$$E + S \rightleftharpoons E \cdot S$$

STEP 2

The reaction occurs to give the product P, which is then released from the enzyme:

$$E \cdot S \longrightarrow E + P$$

When a substance other than the substrate is bound to the enzyme's active site, the enzyme is said to be *inhibited.* If the inhibition is permanent, the enzyme is said to be inactivated. Some of the most powerful toxins act by inhibiting or inactivating key enzymes. "Nerve gases," chemicals used as weapons, act in this way. Nerves transfer messages using small molecules called *neurotransmitters.* An example of a neurotransmitter is acetylcholine. Once the nerve impulse has been transmitted—for example, to a muscle—the acetylcholine molecules must be destroyed, because their continued presence would cause overstimulation, eventually leading to convulsions, paralysis, and possibly death. The acetylcholine molecules are removed by reaction with water:

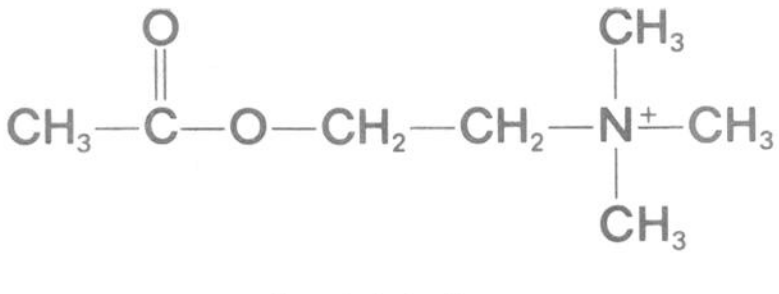

$$CH_3-C(=O)-O-CH_2-CH_2-N^{+}(CH_3)_3$$

Acetylcholine

$$[CH_3COOCH_2CH_2N(CH_3)_3]^+ + H_2O \longrightarrow \underset{\text{Acetic acid}}{CH_3COOH} + \underset{\text{Choline}}{[HOCH_2CH_2N(CH_3)_3]^+}$$

This reaction is catalyzed by the enzyme acetylcholinesterase. The nerve gas diisopropyl phosphorofluoridate (shortened to DIPF) attaches itself permanently to the active site of this enzyme, thus blocking the enzyme's ability to catalyze the removal of acetylcholine. The result is usually death.

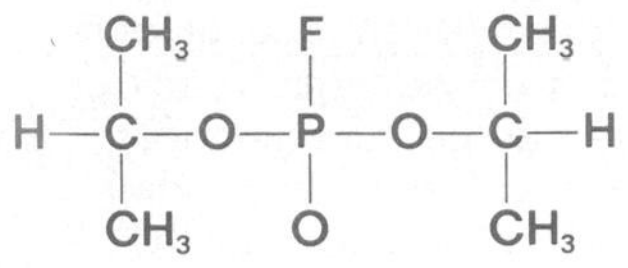

$$H-C(CH_3)_2-O-P(=O)(F)-O-C(CH_3)_2-H$$

DIPF

There are many other ways that enzymes can be deactivated in addition to blockage of the active site by foreign molecules. For example, in **metalloenzymes,** which contain a metal ion at the active site, substitution of a different metal ion for the original ion usually causes the enzyme to malfunction.

And, of course, anything that denatures the enzyme protein destroys its activity.

Because enzymes are so crucial for healthy life and because we hope to learn how to mimic their efficiency in our industrial catalysts, the study of enzymes occupies a prominent role in chemical research.

23.2 Carbohydrates

Carbohydrates form another class of biologically important molecules. They serve as a food source for most organisms and as a structural material for plants. Because many carbohydrates have the empirical formula CH_2O, it was originally believed that these substances were hydrates of carbon, thus accounting for the name.

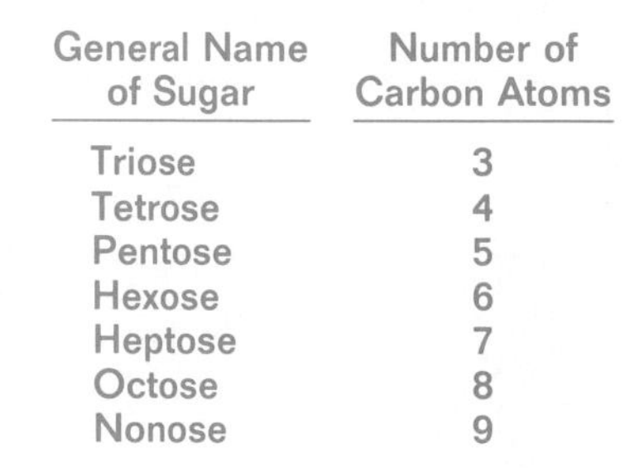

General Name of Sugar	Number of Carbon Atoms
Triose	3
Tetrose	4
Pentose	5
Hexose	6
Heptose	7
Octose	8
Nonose	9

Most important carbohydrates, such as starch and cellulose, are polymers composed of monomers called **monosaccharides,** or **simple sugars.** The monosaccharides are polyhydroxy ketones and aldehydes. The most important contain five carbon atoms (**pentoses**) or six carbon atoms (**hexoses**). One important hexose is *fructose,* a sugar found in honey and fruit. Its structure is

```
        CH2OH
         |
         C=O
         |
   HO—*C—H
         |
    H—*C—OH
         |
    H—*C—OH
         |
        CH2OH
```

Fructose

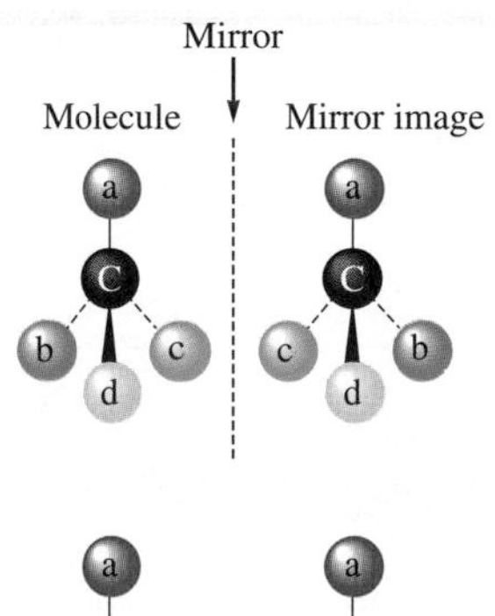

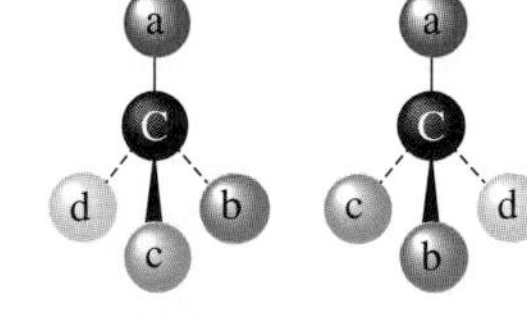

Figure 23.15
When a tetrahedral carbon atom has four different substituents, there is no way that its mirror image can be superimposed. The lower two forms show other possible orientations of the molecule. Compare these with the mirror image and note that they cannot be superimposed.

where the asterisks indicate chiral carbon atoms. In Section 20.4 we saw that molecules with nonsuperimposable mirror images exhibit optical isomerism. A carbon atom with four *different* groups bonded to it in a tetrahedral arrangement *always* has a nonsuperimposable mirror image (see Fig. 23.15), which gives rise to a pair of optical isomers. For example, the simplest sugar, glyceraldehyde,

```
  H     O
   \\   //
     C
     |
 H—*C—OH
     |
    CH2OH
```

which has one chiral carbon, has two optical isomers, as shown in Fig. 23.16.

In fructose each of the three chiral carbon atoms satisfies the requirement of being surrounded by four different groups. This leads to a total of 2^3, or 8, isomers that differ in their ability to rotate polarized light. The particular

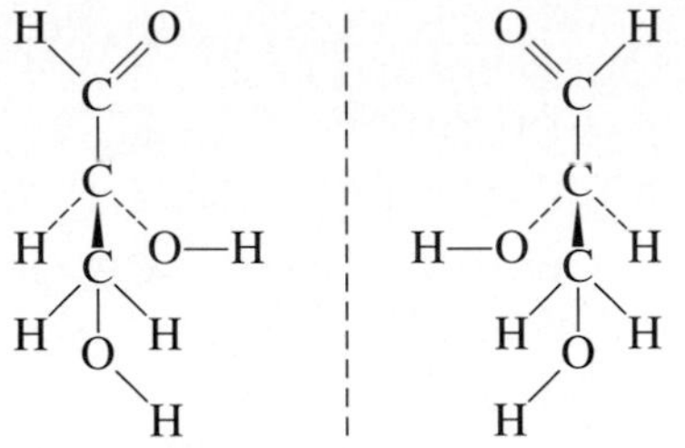

EXAMPLE 23.3

Determine the number of chiral carbon atoms in the following pentose:

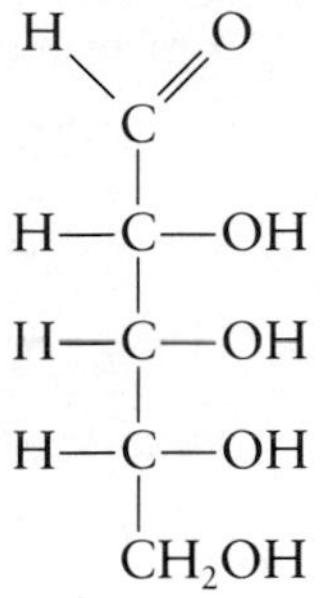

H   O
 \ //
  C
  |
H—C—OH
  |
H—C—OH
  |
H—C—OH
  |
CH_2OH

Solution

We must look for carbon atoms that have four different substituents. The top carbon has only three substituents and thus cannot be chiral. The three carbon atoms shown in blue each have four different groups attached to them:

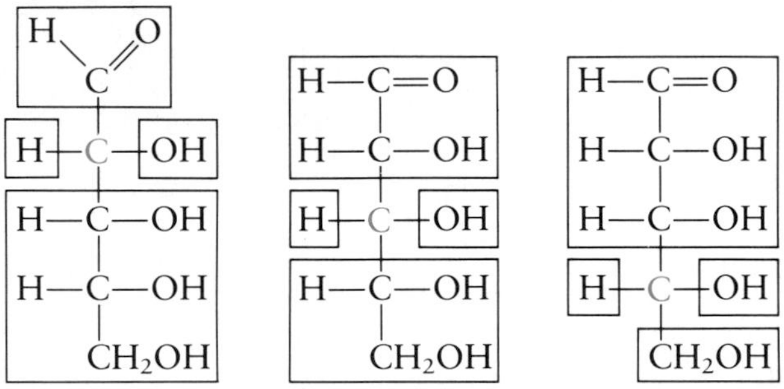

H—C=O | H—C=O | H—C=O
H—C—OH | H—C—OH | H—C—OH
H—C—OH | H—C—OH | H—C—OH
H—C—OH | H—C—OH | H—C—OH
CH_2OH | CH_2OH | CH_2OH

Since the fifth carbon atom has only three types of substituents (it has two hydrogen atoms), it is not chiral.

Thus the three chiral carbon atoms in this pentose are those shown in blue:

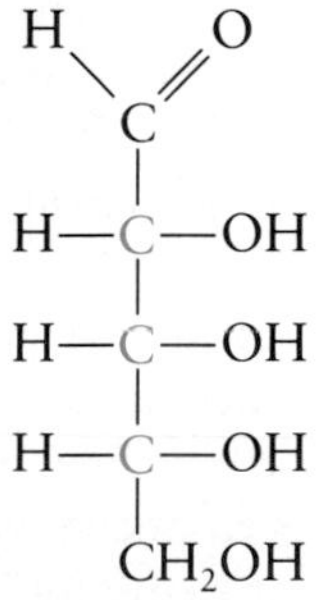

H   O
 \ //
  C
  |
H—C—OH
  |
H—C—OH
  |
H—C—OH
  |
CH_2OH

Note that D-ribose and D-arabinose, shown in Table 23.3, are two of the eight isomers of this pentose.

isomer whose structure is given above is called D-fructose. Generally, monosaccharides have one isomer that is more common in nature than the others. The most important pentoses and hexoses are shown in Table 23.3.

Figure 23.16
The mirror image optical isomers of glyceraldehyde. Note that these mirror images cannot be superimposed.

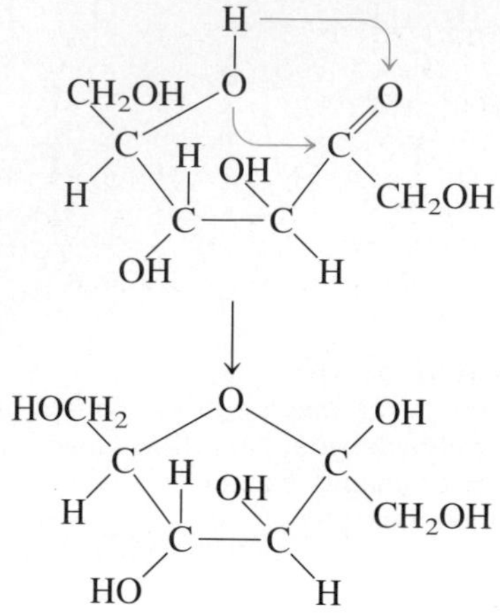

Figure 23.17
The cyclization of D-fructose.

TABLE 23.3 **Some Important Monosaccharides**

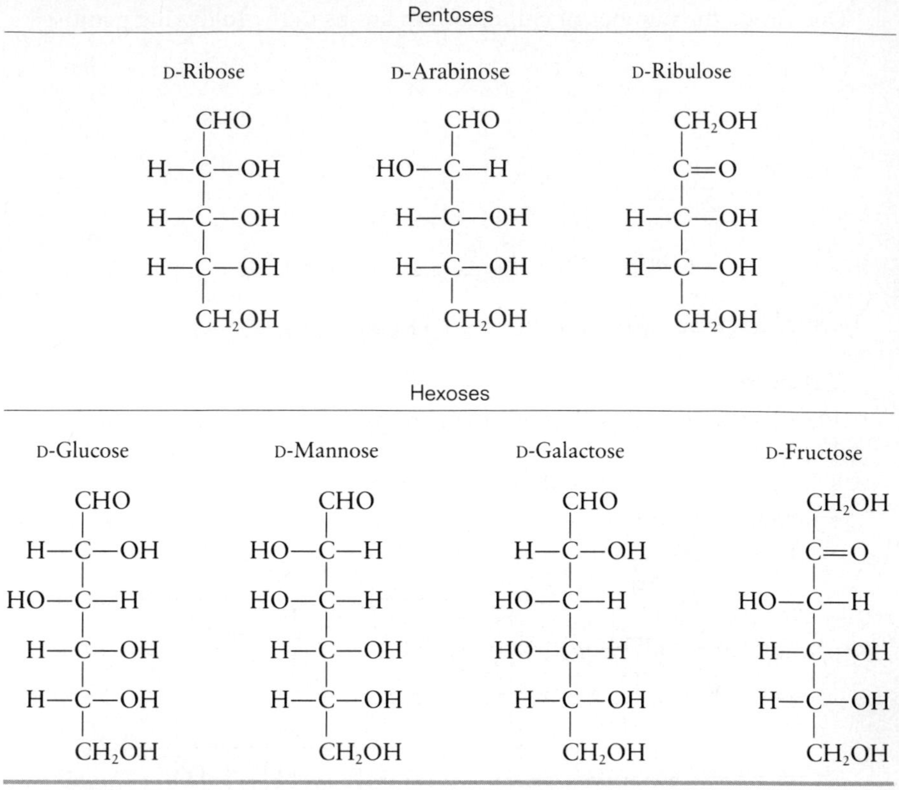

Pentoses

D-Ribose	D-Arabinose	D-Ribulose
CHO	CHO	CH_2OH
H—C—OH	HO—C—H	C=O
H—C—OH	H—C—OH	H—C—OH
H—C—OH	H—C—OH	H—C—OH
CH_2OH	CH_2OH	CH_2OH

Hexoses

D-Glucose	D-Mannose	D-Galactose	D-Fructose
CHO	CHO	CHO	CH_2OH
H—C—OH	HO—C—H	H—C—OH	C=O
HO—C—H	HO—C—H	HO—C—H	HO—C—H
H—C—OH	H—C—OH	HO—C—H	H—C—OH
H—C—OH	H—C—OH	H—C—OH	H—C—OH
CH_2OH	CH_2OH	CH_2OH	CH_2OH

Although we have so far represented the monosaccharides as straight-chain molecules, they usually cyclize, or form a ring structure, in aqueous solution. Figure 23.17 shows this reaction for fructose. Note that a new bond is formed between the oxygen of the terminal hydroxyl group and the carbon of the ketone group. In the cyclic form fructose is a five-membered ring containing a C—O—C bond. The same type of reaction can occur between a hydroxyl group and an aldehyde group, as shown for D-glucose in Fig. 23.18. In this case a six-membered ring is formed.

More complex carbohydrates are formed by combining monosaccharides. For example, **sucrose,** common table sugar, is a **disaccharide** formed from glucose and fructose by elimination of water to form a C—O—C bond between the rings, which is called a **glycoside linkage** (Fig. 23.19). When sucrose is consumed in food, the above reaction is reversed. An enzyme in saliva catalyzes the breakdown of this disaccharide.

Large polymers consisting of many monosaccharide units, called polysaccharides, can form when each ring forms two glycoside linkages, as shown in Fig. 23.19. Three of the most important of these polymers are starch, cellulose, and glycogen. All of these substances are polymers of glucose, differing from each other in the nature of the glycoside linkage, the amount of branching, and molecular weight (molar mass).

Starch, a polymer of α-D-glucose, consists of two parts: *amylose,* a straight-chain polymer of α-glucose [see Fig. 23.20(a)]; and *amylopectin,* a highly

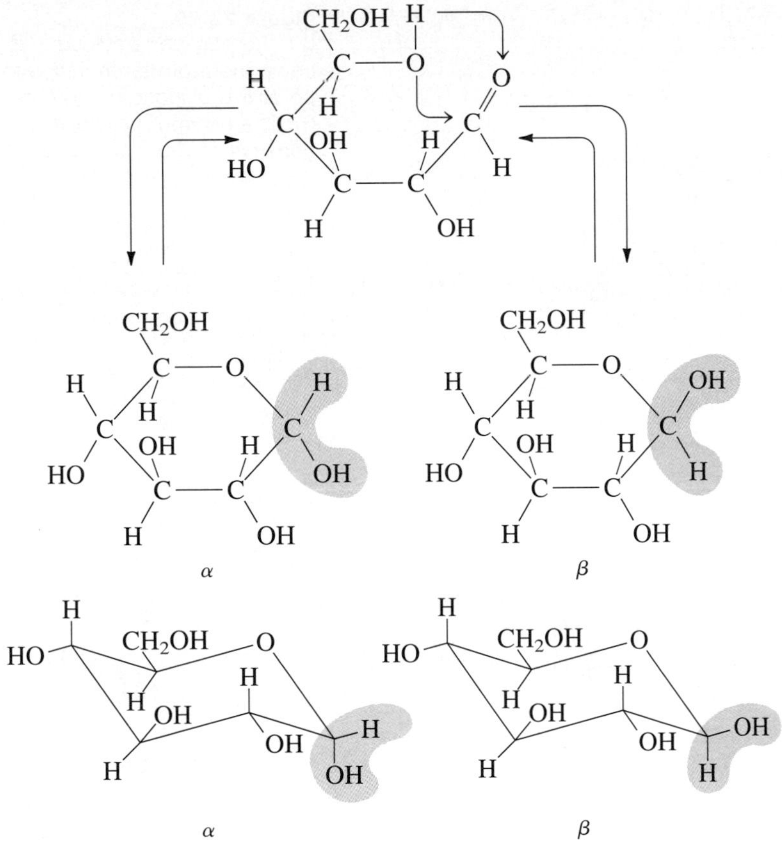

α β α β

Figure 23.18
The cyclization of glucose. Two different rings are possible; they differ in the orientation of the hydroxy group and hydrogen on one carbon, as indicated. The two forms are designated α and β and are shown here in two representations.

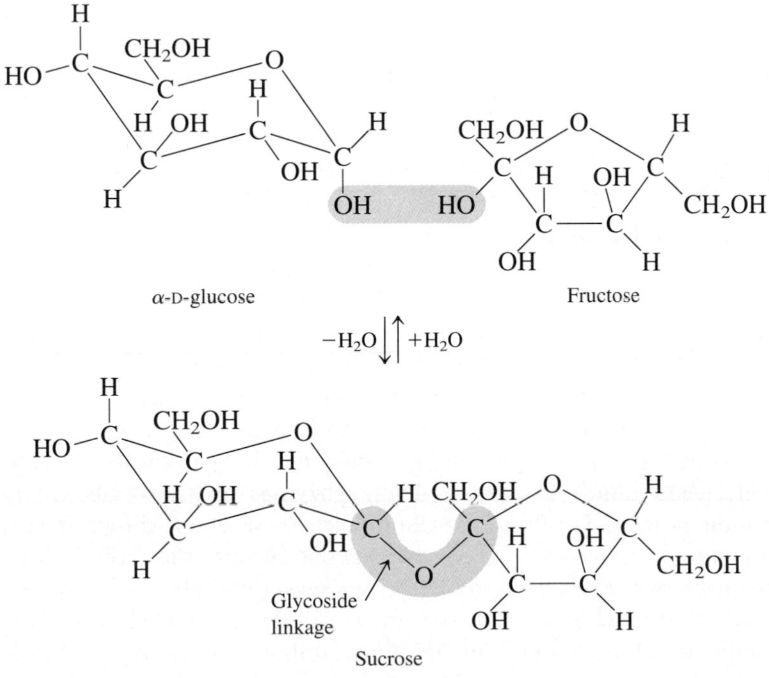

α-D-glucose Fructose

$-H_2O$ $+H_2O$

Glycoside linkage

Sucrose

Figure 23.19
Sucrose is a disaccharide formed from α-D-glucose and fructose.

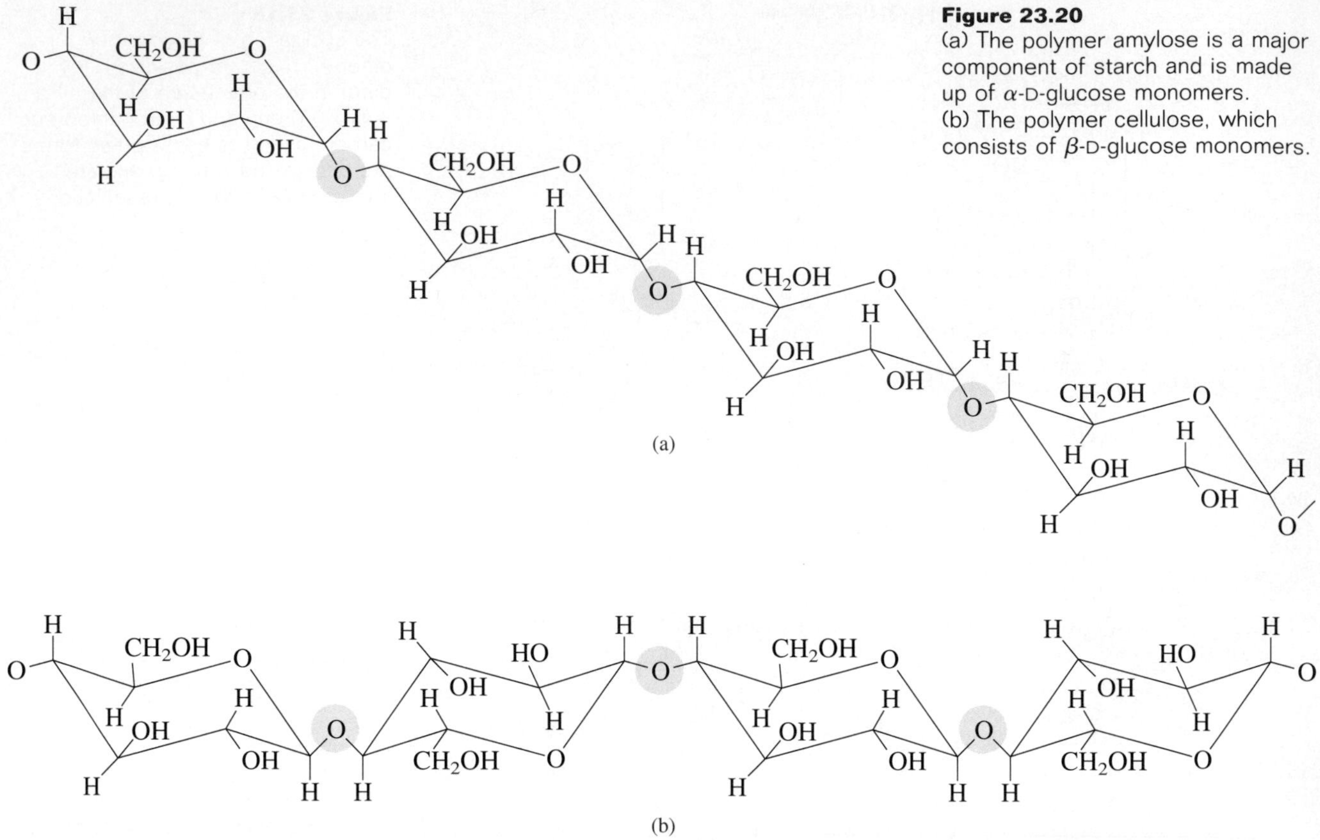

Figure 23.20
(a) The polymer amylose is a major component of starch and is made up of α-D-glucose monomers.
(b) The polymer cellulose, which consists of β-D-glucose monomers.

branched polymer of α-glucose with a molecular weight that is 10–20 times that of amylose. Branching occurs when a third glycoside linkage attaches a branch to the main polymer chain.

Starch, the carbohydrate reservoir in plants, is the form in which glucose is stored by the plant for later use as cellular fuel. Glucose is stored in this high-molecular-weight form because this results in less stress on the plant's internal structure by osmotic pressure. Recall from Section 17.6 that it is the concentration of solute molecules (or ions) that determines the osmotic pressure. Combining the individual glucose molecules into one large chain keeps the concentration of solute molecules relatively low, minimizing the osmotic pressure.

Cellulose, the major structural component of woody plants and natural fibers (such as cotton) is a polymer of β-D-glucose and has the structure shown in Fig. 23.20(b). Note that the β-glycoside linkages in cellulose give the glucose rings a different relative orientation than is found in starch. Although this difference may seem minor, it has very important consequences. The human digestive system contains α-glycosidases, enzymes that can catalyze breakage of the α-glycoside bonds in starch. These enzymes are not effective on the β-glycoside bonds of cellulose, presumably because the different structure results in a poor fit between the enzyme's active site and the carbohydrate. The enzymes necessary to cleave β-glycoside linkages, the β-glycosidases, are found in bacteria that exist in the digestive tracts of termites, cows, deer, and many other animals. Thus, unlike humans, these animals can derive nutrition from cellulose.

Glycogen, the main carbohydrate reservoir in animals, has a structure similar to that of amylopectin but with more branching. It is this branching that is thought to facilitate the rapid breakdown of glycogen into glucose when energy is required.

23.3 Nucleic Acids

Life is possible only because each cell, when it divides, can transmit the vital information about how it works to the next generation. It has been known for a long time that this process involves the chromosomes in the nucleus of the cell. Only since 1953, however, have scientists understood the molecular basis of this intriguing cellular "talent."

The substance that stores and transmits the genetic information is a polymer called **deoxyribonucleic acid (DNA),** a huge molecule with a molecular weight as high as several billion grams per mole. Together with other similar nucleic acids called the **ribonucleic acids (RNA),** DNA is also responsible for the synthesis of the various proteins needed by the cell to carry out its life functions. The RNA molecules, which are found in the cytoplasm outside the nucleus, are much smaller than DNA polymers, with molecular weights of only 20,000 to 40,000.

The monomers of the nucleic acids, called **nucleotides,** are composed of three distinct parts:

1. A *five-carbon sugar,* deoxyribose in DNA and ribose in RNA (Fig. 23.21)
2. A *nitrogen-containing organic base* of the type shown in Fig. 23.22
3. A *phosphoric acid molecule* (H_3PO_4)

The base and the sugar combine as shown in Fig. 23.23(a) to form a unit that in turn reacts with phosphoric acid to create the nucleotide, which is an ester [see

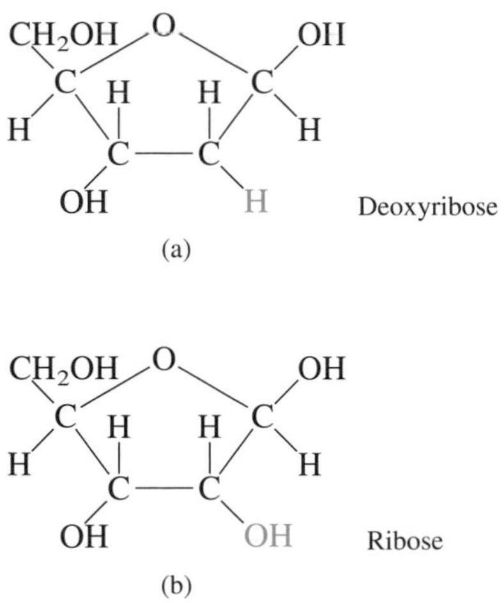

Deoxyribose

(a)

Ribose

(b)

Figure 23.21
The structure of the pentoses (a) deoxyribose and (b) ribose. Deoxyribose is the sugar molecule present in DNA; ribose is found in RNA.

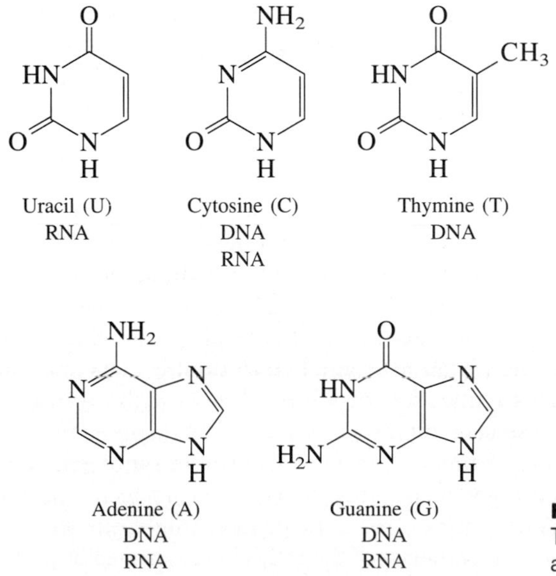

Uracil (U)
RNA

Cytosine (C)
DNA
RNA

Thymine (T)
DNA

Adenine (A)
DNA
RNA

Guanine (G)
DNA
RNA

Figure 23.22
The organic bases found in DNA and RNA.

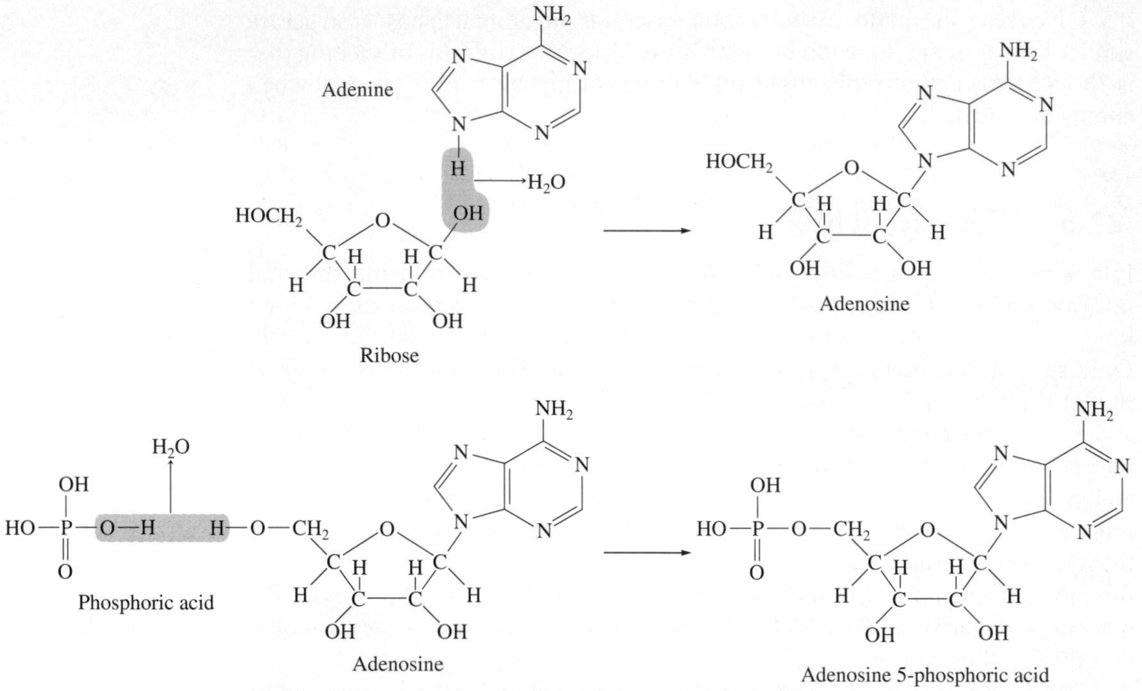

Figure 23.23
(a) Adenosine is formed by the reaction of adenine with ribose. (b) The reaction of phosphoric acid with adenosine to form the ester adenosine 5-phosphoric acid, a nucleotide. (At biological pH, the phosphoric acid would not be fully protonated as is shown here.)

Fig. 23.23(b)]. The nucleotides become connected through condensation reactions that eliminate water to give a polymer of the type represented in Fig. 23.24; such a polymer can contain a *billion* units.

The key to DNA's functioning is its *double-helical structure with complementary bases on the two strands.* The bases form hydrogen bonds to each other, as shown in Fig. 23.25. Note that the structures of cytosine and guanine make them perfect partners for hydrogen bonding, and they are *always* found as pairs on the two strands of DNA. Thymine and adenine form similar hydrogen-bonding pairs.

There is much evidence to suggest that the two strands of DNA unwind during cell division and that new complementary strands are constructed on the unraveled strands (Fig. 23.26). Because the bases on the strands always pair in the same way—cytosine with guanine and thymine with adenine—each unraveled strand serves as a template for attaching the complementary bases (along with the rest of the nucleotide). This process results in two double-helix DNA structures that are identical to the original one. Each new double strand contains one strand from the original DNA double helix and one newly synthesized strand. This replication of DNA allows for the transmission of genetic information as the cells divide.

The other major function of DNA is **protein synthesis.** A given segment of the DNA, called a **gene,** contains the code for a specific protein. These codes transmit the primary structure of the protein (the sequence of amino acids) to the construction "machinery" of the cell. There is a specific code for each amino acid in the protein, which ensures that the correct amino acid will be inserted as the protein chain grows. A code consists of a set of three bases called a **codon.**

DNA stores the genetic information, while RNA molecules are responsible for transmitting this information to the ribosomes, where protein synthesis actually occurs. This complex process involves, first, the construction of a special RNA molecule called **messenger RNA (mRNA).** The mRNA is built in the cell nucleus on the appropriate section of DNA (the gene); the double helix is "unzipped," and the complementarity of the bases is utilized in a process similar to that used in DNA replication. The mRNA then migrates into the cytoplasm of the cell where, with the assistance of the ribosomes, the protein is synthesized.

Small RNA fragments, called **transfer RNA (tRNA),** are tailored to find specific amino acids and then to attach them to the growing protein chain as dictated by the codons in the mRNA. Transfer RNA has a lower molecular weight than messenger RNA. It consists of a chain of 75 to 80 nucleotides, including the bases adenine, cytosine, guanine, and uracil, among others. The chain folds back onto itself in various places as the complementary bases along the chain form hydrogen bonds. The tRNA decodes the genetic message from the mRNA, using a complementary triplet of bases called an **anticodon.** The nature of the anticodon governs which amino acid will be brought to the protein under construction.

The protein is built in several steps. First, a tRNA molecule brings an amino acid to the mRNA (the anticodon of the tRNA must complement the codon of the mRNA (see Fig. 23.27)). Once this amino acid is in place, another tRNA

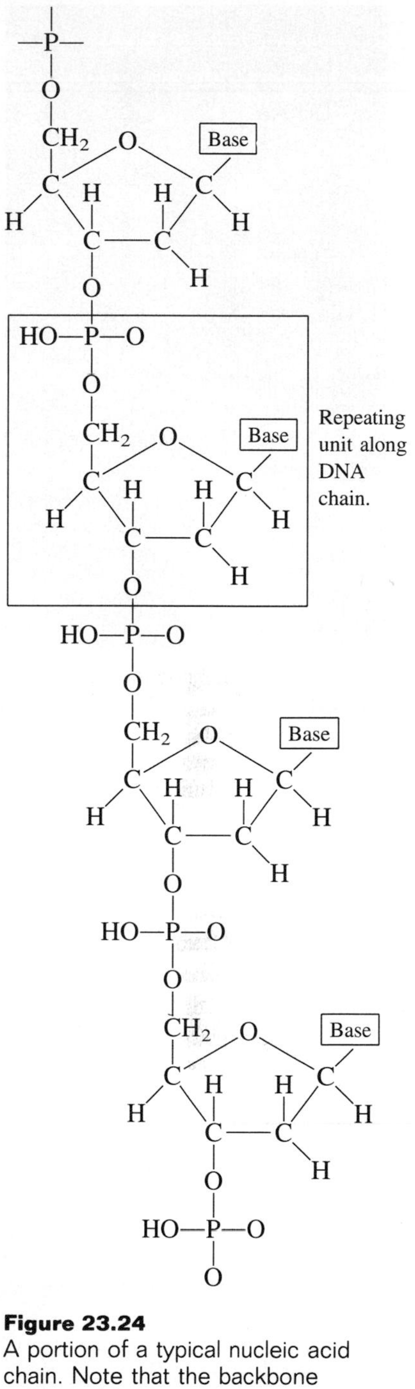

Repeating unit along DNA chain.

Figure 23.24
A portion of a typical nucleic acid chain. Note that the backbone consists of sugar-phosphate esters.

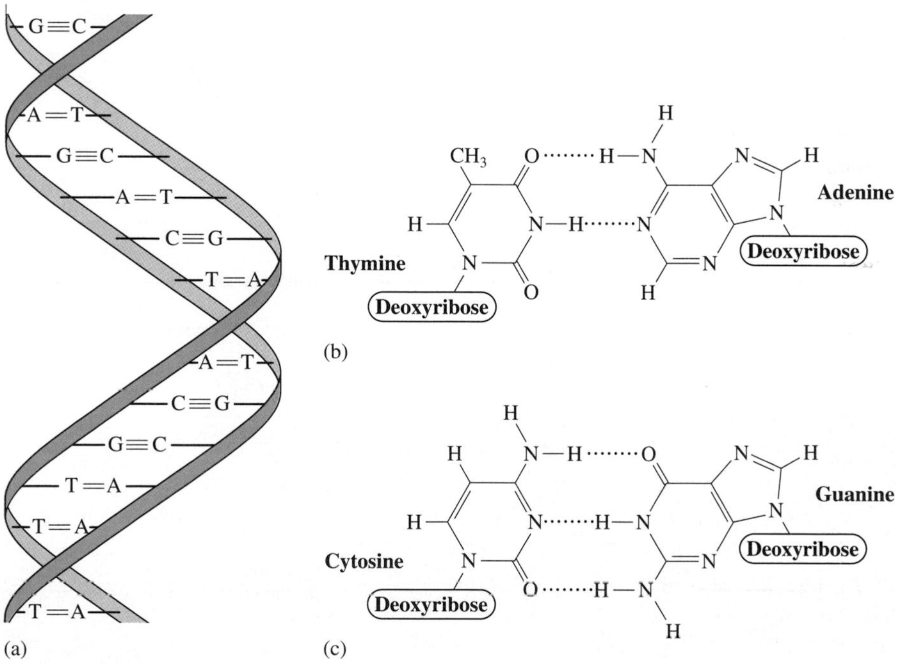

Figure 23.25
(a) The DNA double helix contains two sugar-phosphate backbones, with the bases from the two strands hydrogen-bonded to each other. The complementarity of the (b) thymine-adenine and (c) cytosine-guanine pairs.

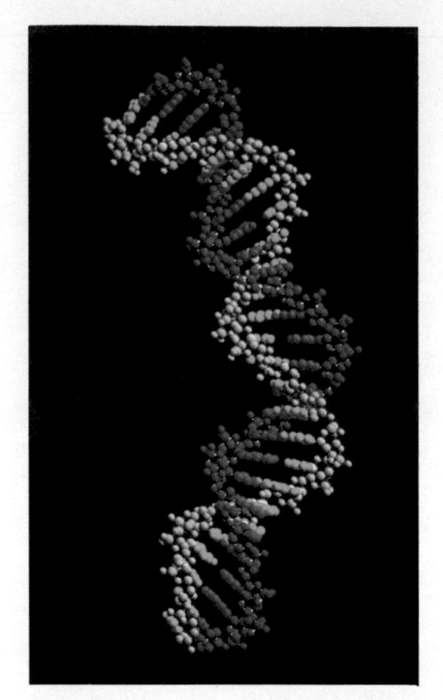
Computer graphic image of the base pairs of DNA.

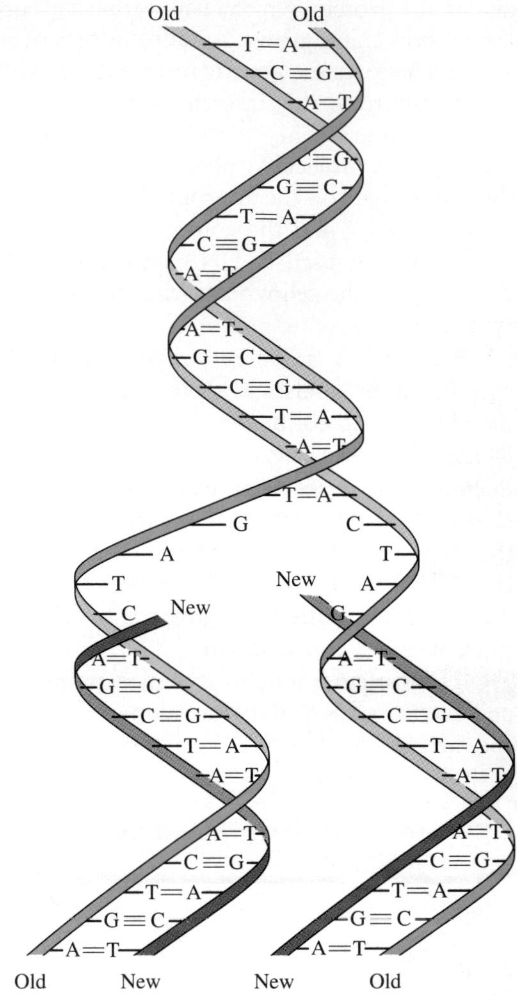

Figure 23.26
During cell division the original DNA double helix unwinds and new complementary strands are constructed on each original strand.

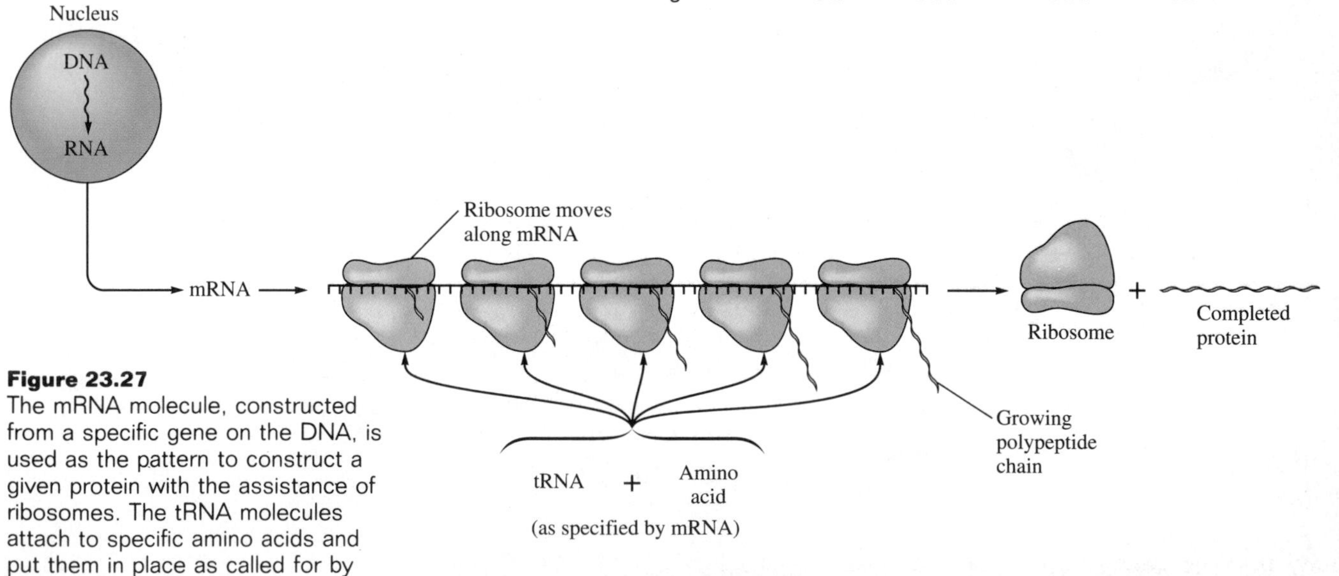

Figure 23.27
The mRNA molecule, constructed from a specific gene on the DNA, is used as the pattern to construct a given protein with the assistance of ribosomes. The tRNA molecules attach to specific amino acids and put them in place as called for by the codons on the mRNA.

moves to the second codon site of the mRNA with its specific amino acid. The two amino acids link via a peptide bond, and the tRNA on the first codon breaks away. The process is repeated down the chain, always matching the tRNA anticodon with the mRNA codon.

23.4 Lipids

The **lipids** are a group of substances classified according to a solubility characteristic. They are defined as water-insoluble substances that can be extracted from cells by nonpolar organic solvents such as ether and benzene. The lipids found in the human body can be divided into four classes according to molecular structure: fats, phospholipids, waxes, and steroids.

The most common **fats** are esters made when glycerol (a polyhydroxy alcohol) reacts with long-chain carboxylic acids called **fatty acids** (see Table 23.4). *Tristearin,* the most common animal fat, is typical of these substances:

$$\begin{array}{l} CH_2{-}OH + H{-}O{-}\overset{\overset{\displaystyle O}{\|}}{C}{-}(CH_2)_{16}{-}CH_3 \\ | \\ CH{-}OH + H{-}O{-}\overset{\overset{\displaystyle O}{\|}}{C}{-}(CH_2)_{16}{-}CH_3 \\ | \\ CH_2{-}OH + H{-}O{-}\overset{\overset{\displaystyle O}{\|}}{C}{-}(CH_2)_{16}{-}CH_3 \end{array} \longrightarrow \begin{array}{l} CH_2{-}O{-}\overset{\overset{\displaystyle O}{\|}}{C}{-}(CH_2)_{16}{-}CH_3 \\ | \\ CH{-}O{-}\overset{\overset{\displaystyle O}{\|}}{C}{-}(CH_2)_{16}{-}CH_3 \\ | \\ CH_2{-}O{-}\overset{\overset{\displaystyle O}{\|}}{C}{-}(CH_2)_{16}{-}CH_3 \end{array} + 3H_2O$$

Glycerol — Three stearic acid molecules — Tristearin

Fats that are esters of glycerol are called **triglycerides** and have the general structure

$$\begin{array}{l} CH_2{-}O{-}\overset{\overset{\displaystyle O}{\|}}{C}{-}R \\ | \\ CH{-}O{-}\overset{\overset{\displaystyle O}{\|}}{C}{-}R' \\ | \\ CH_2{-}O{-}\overset{\overset{\displaystyle O}{\|}}{C}{-}R'' \end{array}$$

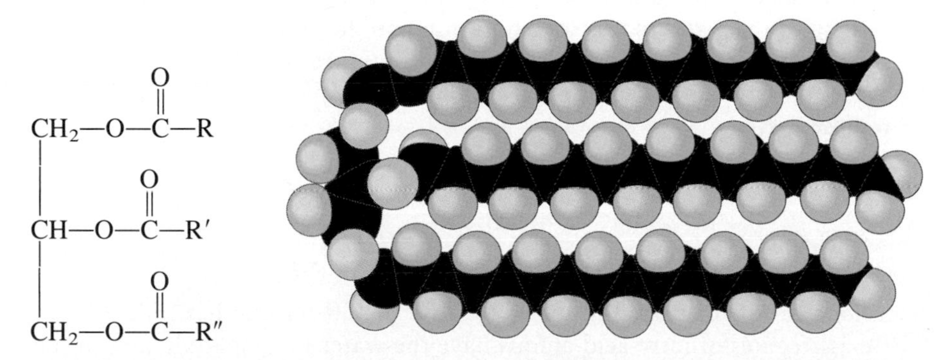

where the three R groups may be the same or different and may be saturated or unsaturated. Vegetable fats tend to be unsaturated and usually occur as oily liquids; animal fats are saturated and occur as solids.

Unsaturated compounds contain one or more C═C bonds.

Triglycerides can be decomposed by treatment with aqueous sodium hydroxide, a process called **saponification.** The products are glycerol and the fatty acid salts; the latter are commonly known as soaps.

TABLE 23.4 **Some Common Fatty Acids and Their Sources**

Name	Formula	Source
Saturated		
Arachidic acid	$CH_3(CH_2)_{18}{-}COOH$	Peanut oil
Butyric acid	$CH_3(CH_2)_2{-}COOH$	Butter
Caproic acid	$CH_3(CH_2)_4{-}COOH$	Butter
Lauric acid	$CH_3(CH_2)_{10}{-}COOH$	Coconut oil
Stearic acid	$CH_3(CH_2)_{16}{-}COOH$	Animal and vegetable fats
Unsaturated		
Oleic acid	$CH_3(CH_2)_7CH{=}CH(CH_2)_7{-}COOH$	Corn oil
Linoleic acid	$CH_3(CH_2)_4CH{=}CH{-}CH_2{-}CH{=}CH(CH_2)_7{-}COOH$	Linseed oil
Linolenic acid	$CH_3CH_2CH{=}CH{-}CH_2CH{=}CH{-}CH_2{-}CH{=}CH{-}(CH_2)_7COOH$	Linseed oil

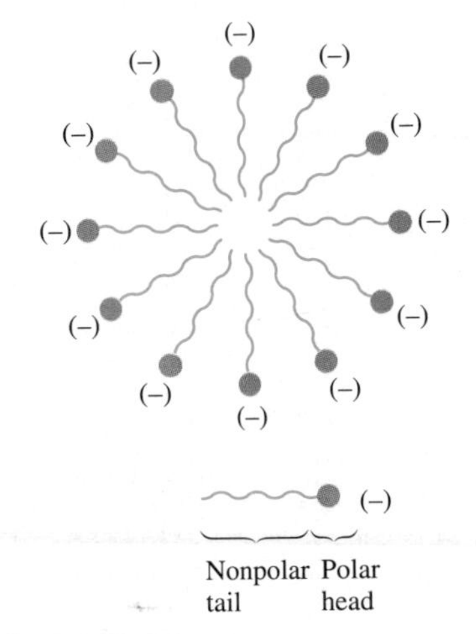

Figure 23.28
The structure of a micelle composed of fatty acid anions. The cations of the salt also surround the negatively charged micelle.

$$\begin{array}{l} CH_2{-}O{-}\overset{\displaystyle O}{\overset{\|}{C}}{-}R \\ | \\ CH{-}O{-}\overset{\displaystyle O}{\overset{\|}{C}}{-}R' \\ | \\ CH_2{-}O{-}\overset{\displaystyle O}{\overset{\|}{C}}{-}R'' \end{array} + 3NaOH \longrightarrow \begin{array}{l} CH_2OH \\ | \\ CHOH \\ | \\ CH_2OH \end{array} + \begin{array}{l} RCOONa \\ \\ R'COONa \\ \\ R''COONa \end{array}$$

Triglyceride Sodium hydroxide Glycerol Soaps

Like dissolves like.

Much of what we call dirt is nonpolar. Grease, for example, consists of long-chain hydrocarbons. However, water, the solvent most commonly available to us, is very polar and will not dissolve "greasy dirt." We need to add something to the water that is somehow compatible with both the polar water and the nonpolar grease. Fatty acid anions are perfect for this role since they have a long nonpolar tail and a polar head. For example, the stearate anion can be represented as

$$CH_3{-}CH_2{-}CH_2{-}CH_2{-}CH_2{-}CH_2{-}CH_2{-}CH_2{-}CH_2{-}CH_2{-}CH_2{-}CH_2{-}CH_2{-}CH_2{-}CH_2{-}CH_2{-}CH_2{-}C(=O){-}O^-$$

Such ions can be dispersed in water because they form **micelles** (Fig. 23.28). These aggregates of fatty acid anions have the water-incompatible tails in the interior, while the anionic parts (the polar heads) point outward to interact with the polar water molecules. A soap "solution" is not a true solution; it does not contain *individual* fatty acid anions dispersed in the water, but rather groups of these ions (micelles). Thus a soap-water mixture is really a suspension of micelles in water. Because the relatively large micelles scatter light, soapy water appears cloudy.

Soap dissolves grease by taking the grease molecules into the nonpolar interior of the micelle (Fig. 23.29), where they can be carried away by the water. Soap can be viewed as an emulsifying agent, since it acts to suspend the normally incompatible grease in the water. Because of this ability to assist water in "wetting" and suspending nonpolar materials, soap is also called a *wetting agent,* or **surfactant.**

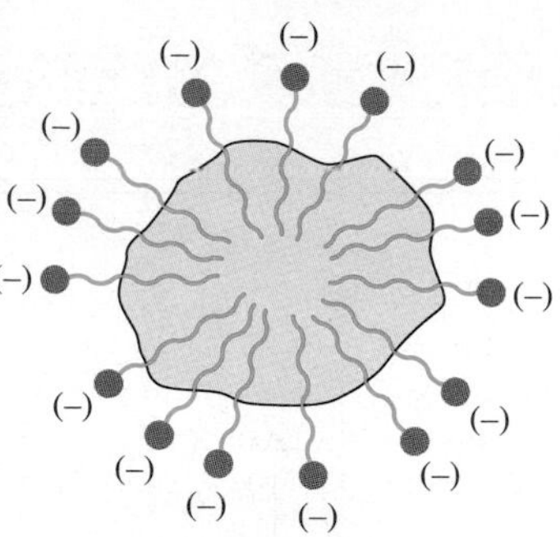

Figure 23.29
Soap micelles absorb grease molecules into their interiors so that the molecules are suspended (emulsified) in the water and can be washed away.

A major disadvantage is that soap anions form precipitates with the cations in hard water, principally Ca^{2+} and Mg^{2+}. This "soap scum" dulls clothes and drastically reduces soap's cleaning efficiency. Water can be softened with slaked lime (Section 7.6) or by ion exchange (Section 18.4). In addition, a huge industry has developed to produce artificial soaps, called detergents. Any molecule that has nonpolar and polar areas similar to those in the fatty acid anions should act to emulsify grease in water. The most widely used class of detergents is the alkylbenzene sulfonates, for example,

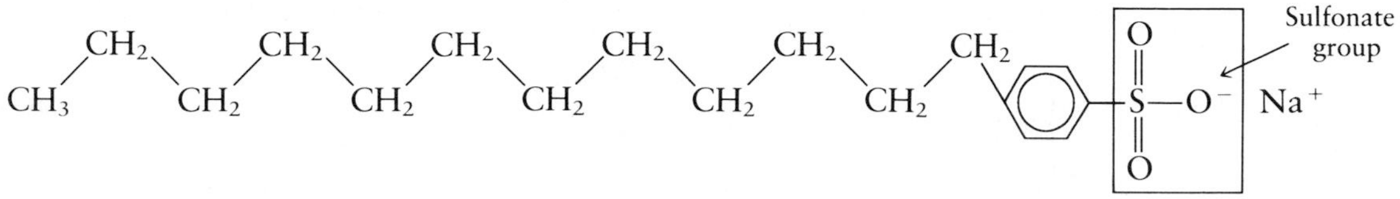

$$CH_3{-}CH_2{-}CH_2{-}CH_2{-}CH_2{-}CH_2{-}CH_2{-}CH_2{-}CH_2{-}CH_2{-}CH_2{-}CH_2{-}C_6H_4{-}\underbrace{\overset{\displaystyle O}{\overset{\|}{\underset{\underset{\displaystyle O}{\|}}{S}}}{-}O^-}_{\text{Sulfonate group}}\ Na^+$$

which can be synthesized from petroleum-based raw materials. Detergent anions have the advantage of not forming insoluble solids with Ca^{2+} and Mg^{2+} ions.

Phospholipids are similar in structure to fats in that they are esters of glycerol. However, unlike fats, they contain only two fatty acids. The third ester linkage involves a phosphate group, which gives phospholipids two distinct parts: the long nonpolar "tail," and the polar, substituted phosphate "head" (Fig. 23.30). Because of this dual nature, phospholipids tend to form **bilayers** in aqueous solution, with the tails in the interior and with the polar heads interfacing with the polar water molecules, as shown in Fig. 23.31(a). Note that this behavior is very similar to that of fatty acid anions. The bilayers of larger phospholipids can close to form *vesicles* [Fig. 23.31(b)].

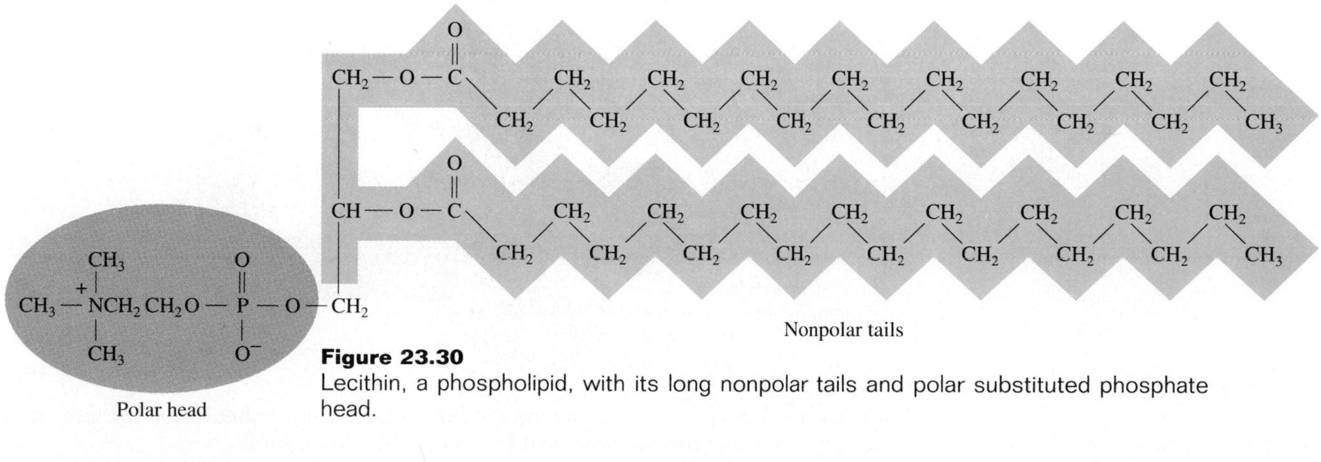

$$\begin{array}{l} CH_2{-}O{-}\overset{\displaystyle O}{\overset{\|}{C}}{-}CH_2{-}CH_2{-}CH_2{-}CH_2{-}CH_2{-}CH_2{-}CH_2{-}CH_2{-}CH_2{-}CH_2{-}CH_2{-}CH_2{-}CH_2{-}CH_2{-}CH_2{-}CH_2{-}CH_3 \\ | \\ CH{-}O{-}\overset{\displaystyle O}{\overset{\|}{C}}{-}CH_2{-}CH_2{-}CH_2{-}CH_2{-}CH_2{-}CH_2{-}CH_2{-}CH_2{-}CH_2{-}CH_2{-}CH_2{-}CH_2{-}CH_2{-}CH_2{-}CH_2{-}CH_2{-}CH_3 \\ | \\ CH_3{-}\overset{+}{N}(CH_3)_2CH_2CH_2O{-}\overset{\displaystyle O}{\overset{\|}{\underset{\underset{\displaystyle O^-}{|}}{P}}}{-}O{-}CH_2 \end{array}$$

Polar head

Nonpolar tails

Figure 23.30
Lecithin, a phospholipid, with its long nonpolar tails and polar substituted phosphate head.

Phospholipids form a significant portion of cell membranes. Figure 23.32 shows a cell membrane in the form of a phospholipid bilayer with proteins distributed in it. The cell membrane must first protect the workings of the cell from the extracellular fluids that bathe it. Its second function is to allow nutrients and other necessary chemicals to enter the cell, while allowing waste products to leave the cell. The detailed operation of the membrane is very complex and not well understood. However, according to the most widely accepted model (called the *fluid mosaic model*) some small uncharged molecules such as water, oxygen, and carbon dioxide diffuse freely through the lipid bilayer, while other substances pass through "gates and passages" provided by specific proteins imbedded in the membrane. The proteins in the membrane also provide the means for communication between cells, via hormones and other messenger molecules.

Like fats and phospholipids, **waxes** are esters. Unlike the former classes, they involve monohydroxy alcohols instead of glycerol. For example, *beeswax*, a substance secreted by the wax glands of bees, is mainly myricyl palmitate,

$$CH_3(CH_2)_{14}-\overset{\overset{\displaystyle O}{\|}}{C}-O-(CH_2)_{29}-CH_3$$

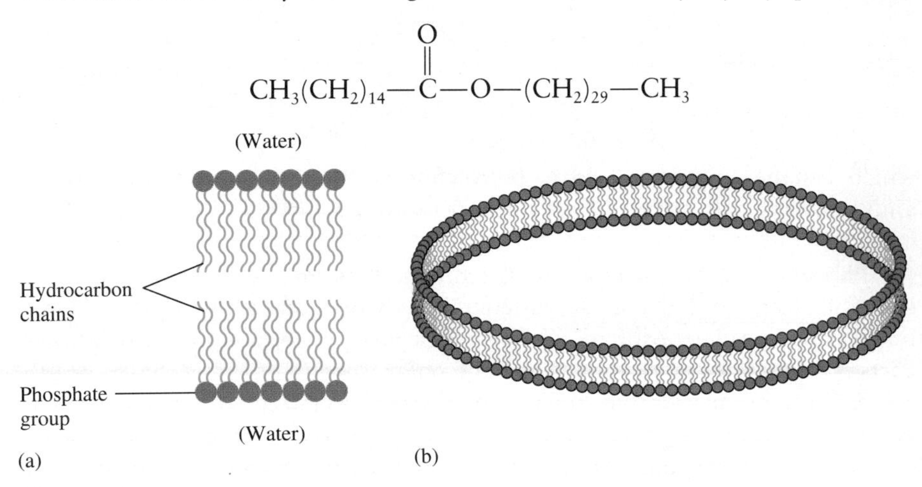

Figure 23.31
(a) A bilayer of phospholipids in aqueous solution. The nonpolar tails remain in the interior of the bilayer, with the polar heads interacting with water molecules. (b) The bilayers can close to form a spherical vesicle, shown here in cross section.

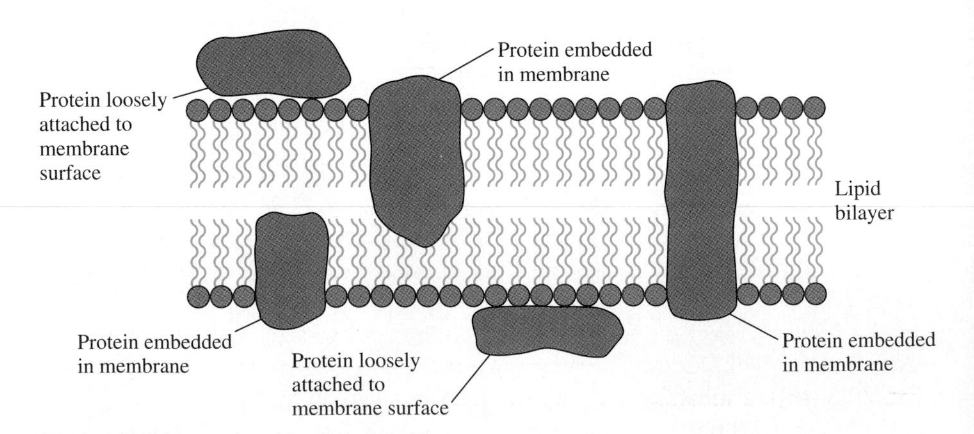

Figure 23.32
A representation of a cell membrane consisting of proteins imbedded in a bilayer of phospholipids that are oriented with their polar heads outward.

formed from palmitic acid,

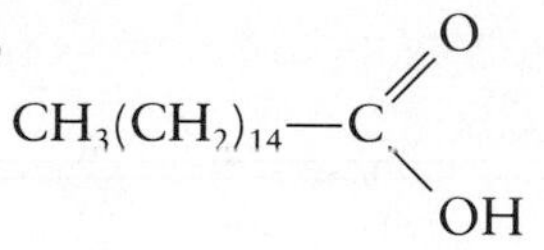

$$CH_3(CH_2)_{14}-C(=O)OH$$

and the alcohol $CH_3(CH_2)_{29}-OH$

Waxes are low-melting solids that furnish waterproof coatings for leaves and fruit and on the skins and feathers of animals. Waxes are also important commercially. For example, whale oil is largely composed of the wax cetyl palmitate. It has been used in so many products, including cosmetics and candles, that the blue whale has been hunted almost to extinction.

Steroids are a class of lipids that have a characteristic fused carbon-ring structure of the type

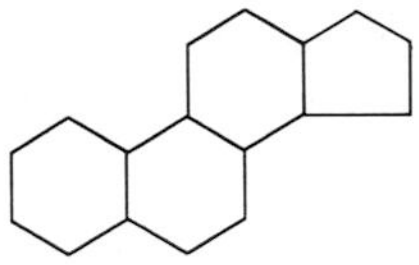

Steroids include four groups: cholesterol, adrenocorticoid hormones, sex hormones, and bile acids.

Cholesterol [Fig. 23.33(a)] is found in virtually all organisms and is the starting material for the formation of the bile acids, steroid hormones, and vitamin D [see Fig. 23.33(b)]. Although cholesterol is essential for human life, it has been implicated in the formation of plaque on the walls of arteries (a process called arteriosclerosis, or hardening of the arteries), which leads to eventual clogging. The phenomenon seems especially important in the arteries that supply blood to the heart. Blockage of these arteries leads to heart damage that often results in death from a heart attack.

The **adrenocorticoid hormones,** synthesized in the adrenal gland, are involved in the regulation of water and the electrolyte balance in body fluids. They also participate in the regulation of the metabolism of proteins and carbohydrates. For example, *cortisol* [Fig. 23.33(c)] slows the construction of proteins so that the amino acids normally used for this purpose can be used by the liver to synthesize extra glucose.

Of the **sex hormones,** the most important male hormone is *testosterone* [Fig. 23.33(d)]. Testosterone is a hormone that controls the growth of the reproductive organs and hair and the development of the muscle structure and deep voice characteristic of males. There are two types of female sex hormones of particular significance: *progesterone* [Fig. 23.33(e)], and a group of estrogens, one of which is *estradiol* [Figure 23.33(f)]. These hormones cause the periodic changes in the ovaries and the uterus responsible for the menstrual cycle. During pregnancy a high level of progesterone is maintained, which prevents ovulation. This effect has led to the use of progesterone-type compounds as birth-control drugs. One of the most common of these drugs is *ethynodiol diacetate* [Figure 23.33(g)].

The **bile acids** are produced from cholesterol found in the liver and are stored in the gallbladder. The primary human bile acid is *cholic acid* [shown in Fig. 23.33(h)], a substance that aids in the digestion of fats by emulsifying them in the intestine. Bile acids can also dissolve cholesterol ingested in food. They are therefore important in limiting cholesterol in the body, since too much cholesterol can be detrimental to human health.

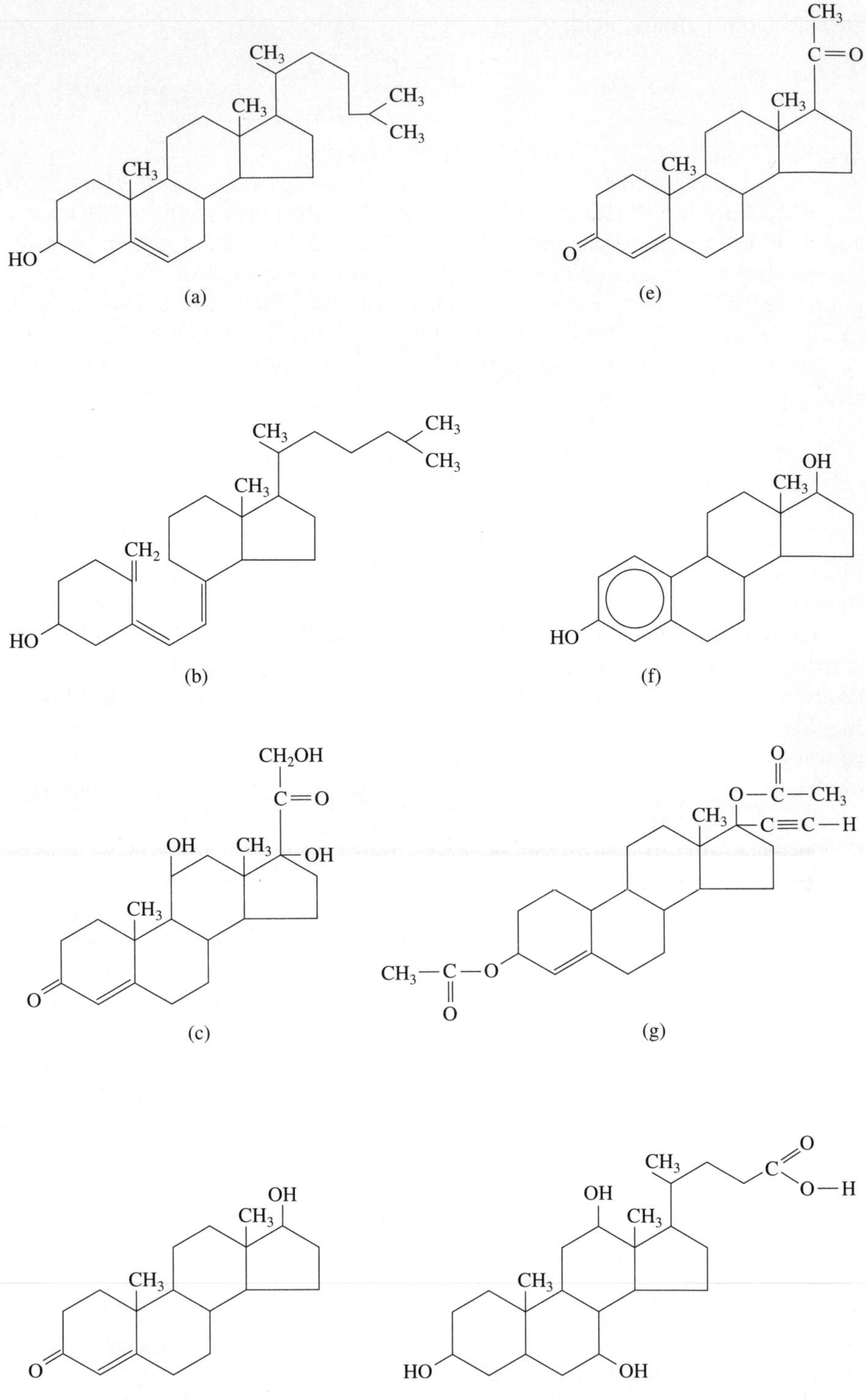

CH_3 CH_3 CH_3 CH_3 CH_3 HO

(a)

CH_3 C=O CH_3 CH_3 O

(e)

CH_3 CH_3 CH_3 CH_3 CH_2 HO

(b)

OH CH_3 HO

(f)

CH_2OH C=O OH CH_3 OH CH_3 O

(c)

O O—C—CH_3 CH_3 C≡C—H CH_3—C—O O

(g)

OH CH_3 CH_3 O

(d)

CH_3 O C O—H OH CH_3 CH_3 HO OH

(h)

Figure 23.33
Several common steroids and steroid derivatives. (a) Cholesterol, (b) vitamin D_3, (c) cortisol, (d) testosterone, (e) progesterone, (f) estradiol, (g) ethynodiol diacetate, and (h) cholic acid.

The Chemistry of Vision

Vision, the most remarkable of our senses, is a complex chemical phenomenon. The human eye is roughly spherical, with an opening in the front to admit light. The light falls on a rear surface lined with cone-shaped and rod-shaped cells. Each eye contains 7 million cones for detecting color and 120 million rods to detect white light and to provide sharpness of visual images. The molecules responsible for vision are attached to the tops of the rods and cones. Here we will consider the function of one of them, rhodopsin.

Rhodopsin has two parts: a protein portion called *opsin,* and a small aldehyde portion called *retinal*. The structure of retinal is shown in its all-*trans* form in Fig. 23.34(a), where all substituents around the double bonds are *trans*. It is interesting to note that vitamin A, a substance known to aid vision (especially night vision) has the same structure as retinal, except that the terminal aldehyde functional group is replaced by a $—CH_2OH$ group to give an alcohol. Retinal can occur in other isomeric forms, one of which is called the 11-*cis* form because of the *cis* arrangement at carbon-11 [see Fig. 23.34(b)]. It is in this *cis* form that retinal is bound to opsin. When rhodopsin absorbs light, the retinal is isomerized to the all-*trans* form, which separates from the opsin.

(a)

(b)

Figure 23.34
The retinal portion of rhodopsin. (a) The *trans* form. (b) The 11-*cis* form. Carbon-11 is shown in red.

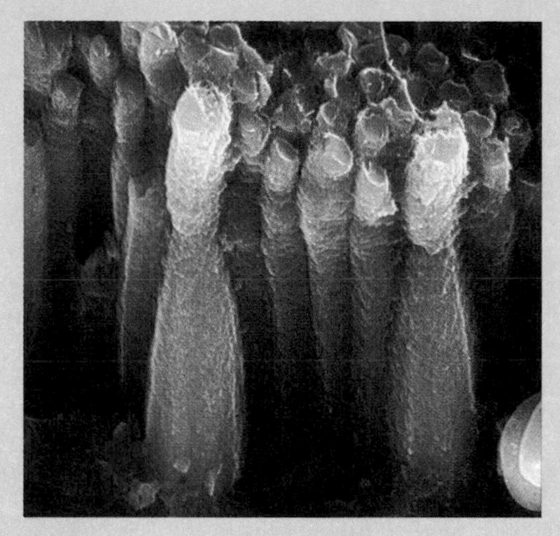

Rods and cones in the human eye (× 2500).

When the two portions separate, the natural reddish purple color of rhodopsin is lost. In addition, the cell to which rhodopsin is attached becomes excited. This receptor cell then excites other cells and sends a message to the brain. Normally, five closely spaced receptor cells must be excited to produce the sensation of vision. This means that only five photons of light are necessary to stimulate the eye, representing a total energy of only 10^{-18} J.

After the activation of rhodopsin and the separation of the *trans* form of retinal, the retinal returns to the 11-*cis* form and reconnects to opsin. This process is relatively slow, which is one reason our eyes need time to adapt to low-light conditions. Also, intensely bright light causes saturation of the light receptors, which causes temporary blindness, since there is no attached retinal to absorb additional photons of light.

Color discrimination is possible because cone cells occur in three groups: those receptive to blue light, those receptive to green light, and those receptive to yellow-red light. Each type can absorb light in a range around its primary color. Thus light in the blue-green region excites both the blue and green receptors.

EXERCISES

A blue exercise number indicates that the answer to that exercise appears at the back of this book.

Proteins and Amino Acids

1. Glycine can exist in water in two forms as shown below; K_a for the carboxylic acid group is 4.3×10^{-3}; K_b for the amino group is 6.0×10^{-3}.

$$H_2N{-}CH_2\overset{O}{\overset{\|}{C}}OH \quad \text{or} \quad {}^{+}H_3N{-}CH_2{-}\overset{O}{\overset{\|}{C}}{-}O^{-}$$

   a. Would you expect the position of the following equilibrium to lie significantly to the right or the left?

$$H_2NCH_2CO_2H \rightleftharpoons {}^{+}H_3NCH_2CO_2{}^{-}$$

   b. Three ions of glycine are possible. Which of these would be predominant in a solution with $[H^+] = 1.0\ M$, or with $[OH^-] = 1.0\ M$?

2. Aspartame, the artificial sweetener marketed under the name NutraSweet, is a methyl ester of a dipeptide. The structure of aspartame is

$$H_2N{-}\underset{CH_2CO_2H}{\underset{|}{CH}}{-}\overset{O}{\overset{\|}{C}}{-}NH{-}\overset{CO_2CH_3}{\overset{|}{CH}}{-}CH_2{-}C_6H_5$$

   a. What two amino acids are used to prepare aspartame?
   b. There is concern that methanol may be produced by the decomposition of aspartame. From what portion of the molecule can methanol be produced? Write an equation for this reaction.

3. When pure crystalline amino acids are heated, decomposition generally occurs before melting. Account for this observation. (*Hint:* see Exercise 1.)

4. Two new amino acids, amino malonic acid and β-carboxy aspartic acid, have recently been isolated from the bacteria *E. coli.* Amino malonic acid has also been found in mammals and seems to be associated with the formation of arteriosclerotic plaque. The structures of these amino acids are

$$HO_2C{-}\underset{NH_2}{\underset{|}{CH}}{-}CO_2H \qquad \begin{matrix} H_2N{-}CH{-}CO_2H \\ | \\ HO_2C{-}CH{-}CO_2H \end{matrix}$$

Amino malonic acid — β-Carboxy aspartic acid

   Would you classify these as hydrophobic or hydrophilic amino acids? Why?

5. Monosodium glutamate (MSG) is commonly used as a flavoring in foods. Draw the structure of MSG.

6. Distinguish between the primary, secondary, and tertiary structures of a protein. Give examples of the types of forces that maintain each type of structure.

7. Sickle-cell anemia is a disease resulting from abnormal hemoglobin molecules. A single glutamic acid in the hemoglobin of a person suffering from this disease is replaced by valine. How might this substitution affect the structure of hemoglobin?

8. Give an example of amino acids that could give rise to the interactions pictured in Fig. 23.10 that maintain the tertiary structures of proteins.

9. What types of interactions can occur between the side chains of phenylalanine and isoleucine? Aspartic acid and lysine?

10. Draw the structures of the two dipeptides that can be formed from serine and alanine.

11. How many tripeptides can be formed from the amino acids ala, ala, and gly?

12. How many polypeptides that contain 25 amino acids can be made from a mixture of 5 different amino acids?

13. Describe how denaturation and inhibition affect the catalytic activity of an enzyme.

14. The rate of enzyme-catalyzed reactions often depends on temperature, as shown in the graph below. How do you account for the shape of this curve?

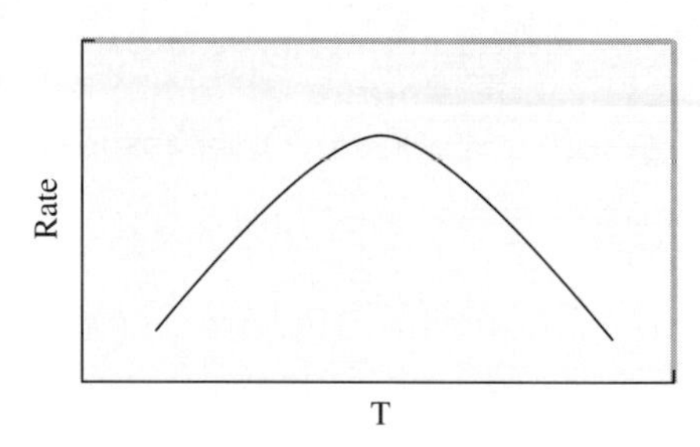

15. Aqueous solutions of amino acids are buffered solutions. Why?

16. Is the primary, secondary, or tertiary structure of an enzyme changed by denaturation?

17. In 1994 chemists at Texas A & M University reported the synthesis of a non-naturally occurring amino acid (*C & E News,* April 18, 1994, pp. 26–27):

$$\begin{matrix} H_2N & & CH_2 & & H \\ & \diagdown & / \; \diagdown & & / \\ & & C{-}{-}C & & \\ & / & & \diagdown & \\ CO_2H & & & & CH_2SCH_3 \end{matrix}$$

   a. To which naturally occurring amino acid is this compound most similar?
   b. Draw the geometrical isomers for this synthetic amino acid.

c. A tetrapeptide, phe-met-arg-phe—NH_2, is synthesized in the brains of rats addicted to morphine and heroin. (The —NH_2 indicates that the peptide ends in $-\overset{\overset{\displaystyle O}{\|}}{C}-NH_2$ instead of —CO_2H.) The TAMU scientists synthesized a similar tetrapeptide, with the synthetic amino acid above replacing one of the original amino acids. Draw a structure for the tetrapeptide containing the synthetic amino acid.

Carbohydrates

18. Draw cyclic structures for D-ribose and D-mannose.
19. Indicate the chiral carbon atoms found in the carbohydrates D-ribose and D-mannose.
20. Sucrose polyester (SPE) is a material being developed by the Procter & Gamble Company as a fat substitute in cooking oil. SPE, which is indigestible and thus adds no calories to food cooked in it, is the polyester formed from sucrose (which has eight hydroxyl groups) and a given fatty acid. The polyesters containing six, seven, or eight fatty acids attached to the hydroxyl groups as esters are not digestible, while those with five or less are digestible.
    a. Draw the structure of the SPE made from sucrose completely esterified with oleic acid.
    b. How many isomers exist when only seven of the hydroxyl groups are esterified (one hydroxyl group remains)?
    c. How many isomers exist when only six of the hydroxyl groups are esterified?
    d. From the discussion of enzymes in Chapters 15 and 23, speculate why the SPE molecules with six, seven, or eight ester functional groups are indigestible while the ones containing fewer fatty acids are digestible.
21. What forces are responsible for the solubility of starch in water?

Optical Isomerism and Chiral Carbon Atoms

22. Distinguish among the following terms: structural isomerism, geometrical isomerism, and optical isomerism.
23. Which of the amino acids in Fig. 23.3 contain more than one chiral carbon atom? Draw the structures of these amino acids and indicate all chiral carbon atoms.
24. Why is glycine not optically active?
25. How many chiral carbon atoms are there in cholesterol? [See Fig. 23.33(a).]
26. Draw the structure of menthol (2-isopropyl-5-methylcyclohexanol), and indicate the chiral carbon atoms.
27. Which of the isomers of bromochloropropene are optically active? (See Exercise 18 in Chapter 22.)
28. The scent of pine trees mainly results from the presence of α-pinene. Oil of turpentine contains roughly 58–65% α-pinene. The structure of α-pinene is

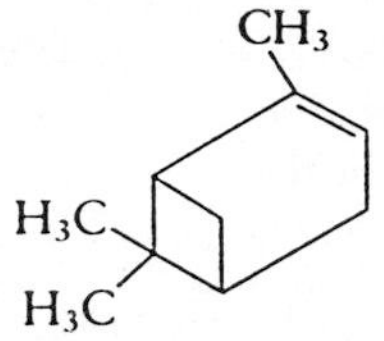

    CH_3

    H_3C

    H_3C

    Is α-pinene optically active? If it is, identify the chiral carbon atoms.
29. Indicate the chiral carbons in the synthetic amino acid illustrated in Exercise 17. Draw the mirror image of the structure shown.

Nucleic Acids

30. The compounds adenine, guanine, cytosine, and thymine are called the nucleic acid bases. What structural features in these compounds make them bases?
31. Each human DNA molecule contains roughly 5×10^9 base pairs. The spacing between base pairs along a given chain is about 340 pm. If a single human DNA molecule were stretched to its full length, how long would it be?
32. Describe the structural differences between DNA and RNA.
33. Part of a DNA sequence is A-T-G-C-G-G-C-A-T. What is its complementary sequence?
34. The codons (words) in DNA that identify which amino acid should be in a protein are three bases long. How many such three-letter words can be made from the four bases adenine, cytosine, guanine, and thymine?
35. Which base will hydrogen-bond with uracil within an RNA molecule? Draw the structure of this base pair.
36. The base sequences in mRNA that code for some of the amino acids are given below.

Glu:	GAA, GAG
Val:	GUU, GUC, GUA, GUG
Met:	AUG
Trp:	UGG
Phe:	UUU, UUC
Asp:	GAU, GAC

These sequences are complementary to the sequences in DNA.
    a. Give the corresponding sequences in DNA for the amino acids listed above.
    b. Give a DNA sequence that would code for the peptide: met-met-phe-asp-trp.
    c. How many different DNA sequences can code for the pentapeptide in part b?

d. What is the peptide that is produced from the DNA sequence C-T-T-A-C-C-A-A-A?

e. What other DNA sequences would yield the same tripeptide as in part d?

37. The change of a single base in the DNA sequence for normal hemoglobin can encode for the abnormal hemoglobin, giving rise to sickle-cell anemia. Which base in the codon for glu in DNA is replaced to give the codon(s) for val? (See Exercises 7 and 36.)

38. The deletion of a single base from a DNA molecule can be a fatal mutation. Substitution of one base for another is often not as serious a mutation. Why?

39. The compound cisplatin (see Chapter 20) appears to kill cancer cells by inhibiting DNA synthesis. Given the following structural information about cisplatin, the information in Exercise 31, and the fact that the chloride ion is easily displaced by other donor molecules in cisplatin, speculate on how cisplatin may interact with DNA.

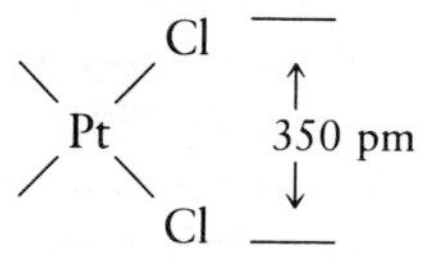

Cl
Pt   350 pm
Cl

40. A tautomeric form (see Exercise 21 in Chapter 22 for a discussion of tautomerism) of thymine has the structure

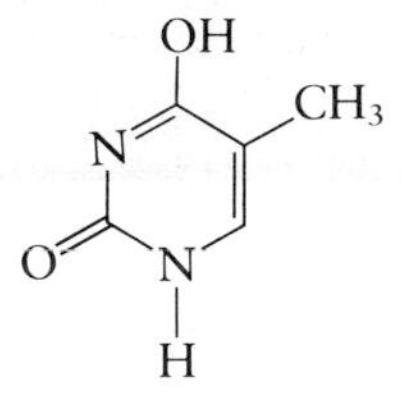

OH
CH_3
N
O
N
H

a. Use bond energies (Table 13.5) to predict whether this tautomer or the one given in Fig. 23.22 is more stable.

b. If the tautomer above, rather than the stable form of thymine, were present in a strand of DNA during replication, what would be the result?

Lipids and Steroids

41. Why can lipids be extracted into organic solvents but carbohydrates cannot?

42. Draw the structure of the triglyceride formed from linoleic acid and glycerol.

43. How many moles of H_2 will it take to completely hydrogenate 1 mol of the triglyceride in Exercise 42? How many products are possible if only 4 mol of H_2 react with this triglyceride? Draw the structures of these products.

44. What is a polyunsaturated fat?

45. The oil of deep-water fish is rich in omega-3 fatty acids. The presence of these fatty acids in the diet appears to lower blood cholesterol levels. The term *omega-3* applies to the location of a carbon-carbon double bond. If we start with the terminal methyl group (the omega carbon), the double bond is located after the third carbon atom. Draw the structures of the omega-3 fatty acids that contain a total of 16 carbon atoms and a total of 18 carbon atoms, respectively.

46. When one spills a solution of sodium hydroxide on one's hands, the skin feels slippery. Why?

47. Lecithin and sphingomyelin are two phospholipids found in amniotic fluid. During gestation, sphingomyelin levels remain relatively constant, while lecithin levels increase. Both substances act as surfactants in the lung, and the ratio of lecithin to sphingomyelin in amniotic fluid is used by physicians to assess whether an infant will be able to breathe unassisted upon birth.

a. Look up and draw the structures of these two substances in a reference such as *The Merck Index*.

b. Indicate the hydrophobic and hydrophilic portions of each molecule.

48. Cetyl palmitate is commonly used as a surfactant in many shampoos and related cosmetic products. Look up cetyl palmitate in an appropriate reference and explain how its structure makes it an effective surfactant.

49. Identify the functional groups present in the steroids shown in Fig. 23.33.

Additional Exercises

50. How many tetrapeptides can be made that contain two phenylalanines (phe) and two glycines (gly)?

51. How many tripeptides can be formed containing one phenylalanine (phe), one glycine (gly), and one alanine (ala)?

52. Use the values of equilibrium constants given in Exercise 1 to calculate equilibrium constants for the following.

a. $^{+}H_3NCH_2CO_2H + H_2O \rightleftharpoons H_2NCH_2CO_2H + H_3O^+$

b. $H_2NCH_2CO_2^- + H_2O \rightleftharpoons H_2NCH_2CO_2H + OH^-$

c. $^{+}H_3NCH_2CO_2H \rightleftharpoons 2H^+ + H_2NCH_2CO_2^-$

53. The isoelectric point of an amino acid is the pH at which the molecule has no net charge. For glycine that would be the pH at which virtually all glycine molecules are in the form $^{+}H_3NCH_2CO_2^-$. If we assume that the principal equilibrium is

$$2\,^{+}H_3NCH_2CO_2^- \rightleftharpoons H_2NCH_2CO_2^- + \,^{+}H_3NCH_2CO_2H \qquad \text{(i)}$$

then at equilibrium

$$[H_2NCH_2CO_2^-] = [^{+}H_3NCH_2CO_2H] \qquad \text{(ii)}$$

Use this result and your answer to part c of Exercise 52 to calculate the pH at which Equation (ii) is true. This will be the isoelectric point of glycine.

54. Look up the structures of the following vitamins in *The Merck Index*. Classify each as water-soluble or fat-soluble.
    a. vitamin A c. vitamin K_5
    b. vitamin E d. vitamin K_6

55. a. Use bond energies (Table 13.5) to estimate ΔH for the reaction of two molecules of glycine to form a peptide linkage.
    b. Would you predict ΔS to favor the formation of peptide linkages between two molecules of glycine?
    c. Would you predict the formation of proteins to be a spontaneous process?

56. The reaction between two nucleotides to form a phosphate ester linkage can be approximated as follows:

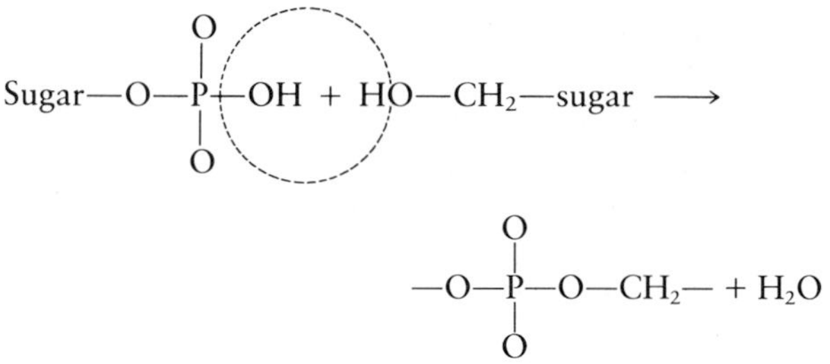

Sugar—O—P(=O)(—O)—OH + HO—CH₂—sugar ⟶

$$\text{Sugar—O—}\overset{\overset{\displaystyle O}{|}}{\underset{\underset{\displaystyle O}{|}}{P}}\text{—OH + HO—CH}_2\text{—sugar} \longrightarrow$$

$$\text{—O—}\overset{\overset{\displaystyle O}{|}}{\underset{\underset{\displaystyle O}{|}}{P}}\text{—O—CH}_2\text{— + H}_2\text{O}$$

Would you predict the formation of a dinucleotide from two nucleotides to be a spontaneous process?

57. Considering your answers to Exercises 55 and 56, how can you justify the existence of proteins and nucleic acids in light of the second law of thermodynamics?

58. The structure of tartaric acid is

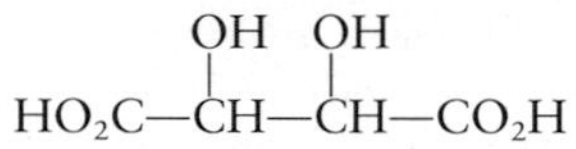

$$\text{HO}_2\text{C—}\overset{\overset{\displaystyle OH}{|}}{\text{CH}}\text{—}\overset{\overset{\displaystyle OH}{|}}{\text{CH}}\text{—CO}_2\text{H}$$

    a. Is the form of tartaric acid pictured below optically active? Why or why not?

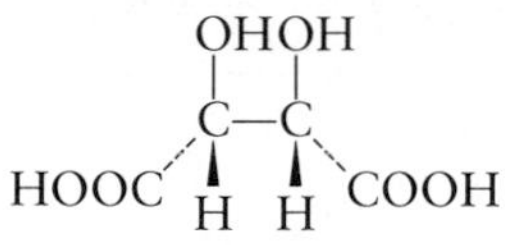

OHOH
| |
C—C
HOOC H H COOH

    b. Draw the optically active forms of tartaric acid.

59. Some organic compounds containing chlorine (e.g., DDT, PCBs, and CCl_4) have been shown to be toxic or carcinogenic. As a result, some environmental groups have proposed a complete ban on the use of chlorine and chlorinated chemicals. Consider the following compounds: Ceclor, Xanax (a top-ten-prescribed pharmaceutical), Claritin, Elocon, Wellbutrin, Femstat, Bonefos, Aclovate, Fareston, Halfan, Sporanox, Selepam, and Melex. Look up the structures and uses for these compounds in *The Merck Index* or the *Physician's Desk Reference*. For *The Merck Index*, use the Eleventh Edition and use the Cross Index of Names. Which features do these drugs have in common? Do you think it is wise to ban the use of all chlorine-containing compounds?

APPENDIXES

APPENDIX ONE Mathematical Procedures

A1.1 Exponential Notation

The numbers characteristic of scientific measurements are often very large or very small; thus it is convenient to express them by using powers of 10. For example, the number 1,300,000 can be expressed as 1.3×10^6, which means multiply 1.3 by 10 six times:

$$1.3 \times 10^6 = 1.3 \times \underbrace{10 \times 10 \times 10 \times 10 \times 10 \times 10}_{10^6 = 1 \text{ million}}$$

Note that each multiplication by 10 moves the decimal point one place to the right, and the easiest way to interpret the notation 1.3×10^6 is that it means move the decimal point in 1.3 to the right six times.

In this notation the number 1985 can be expressed as 1.985×10^3. Note that the usual convention is to write the number that appears before the power of 10 as a number between 1 and 10. Some other examples are given below.

Number	Exponential Notation
5.6	5.6×10^0 or 5.6×1
39	3.9×10^1
943	9.43×10^2
1126	1.126×10^3

To represent a number smaller than 1 in exponential notation, start with a number between 1 and 10 and *divide* by the appropriate power of 10:

$$0.0034 = \frac{3.4}{10 \times 10 \times 10} = \frac{3.4}{10^3} = 3.4 \times 10^{-3}$$

Division by 10 moves the decimal point one place to the *left*. Thus the number 0.00000014 can be written as 1.4×10^{-7}.

To summarize, we can write any number in the form

$$N \times 10^{\pm n}$$

where N is between 1 and 10 and the exponent n is an integer. If the sign preceding n is positive, it means the decimal point in N should be moved n places to the right. If a negative sign precedes n, the decimal point in N should be moved n places to the left.

Multiplication and Division

When two numbers expressed in exponential notation are multiplied, the initial numbers are multiplied and the exponents of 10 are added:

$$(M \times 10^m)(N \times 10^n) = (MN) \times 10^{m+n}$$

For example,

$$(3.2 \times 10^4)(2.8 \times 10^3) = 9.0 \times 10^7$$

When the numbers are multiplied, if a result greater than 10 is obtained for the initial number, the number is adjusted to conventional notation:

$$(5.8 \times 10^2)(4.3 \times 10^8) = 24.9 \times 10^{10} = 2.49 \times 10^{11} = 2.5 \times 10^{11}$$

Division of two numbers expressed in exponential notation involves normal division of the initial numbers and *subtraction* of the exponent of the divisor from that of the dividend. For example,

$$\frac{4.8 \times 10^8}{\underbrace{2.1 \times 10^3}_{\text{Divisor}}} = \frac{4.8}{2.1} \times 10^{(8-3)} = 2.3 \times 10^5$$

Addition and Subtraction

When we add or subtract numbers expressed in exponential notation, *the exponents of the numbers must be the same.* For example, to add 1.31×10^5 and 4.2×10^4, rewrite one number so that the exponents of both are the same:

$$\begin{array}{r} 13.1 \times 10^4 \\ +\ 4.2 \times 10^4 \\ \hline 17.3 \times 10^4 \end{array}$$

In correct exponential notation the result is expressed as 1.73×10^5.

Powers and Roots

When a number expressed in exponential notation is taken to some power, the initial number is taken to the appropriate power and the exponent of 10 is multiplied by that power:

$$(N \times 10^n)^m = N^m \times 10^{m \cdot n}$$

For example,*

$$(7.5 \times 10^2)^3 = 7.5^3 \times 10^{3 \cdot 2} = 422 \times 10^6 = 4.22 \times 10^8$$
$$= 4.2 \times 10^8 \qquad \text{(rounded to 2 significant figures)}$$

When a root is taken of a number expressed in exponential notation, the root of the initial number is taken and the exponent of 10 is divided by the number representing the root:

$$\sqrt{N \times 10^n} = (N \times 10^n)^{1/2} = \sqrt{N} \times 10^{n/2}$$

*Refer to the instruction booklet for your calculator for directions concerning how to take roots and powers of numbers.

For example, $(2.9 \times 10^6)^{1/2} = \sqrt{2.9} \times 10^{6/2} = 1.7 \times 10^3$

Because the exponent of the result must be an integer, we may sometimes have to change the form of the number so that the power divided by the root equals an integer; for example,

$$\begin{aligned}\sqrt{1.9 \times 10^3} &= (1.9 \times 10^3)^{1/2} = (0.19 \times 10^4)^{1/2} \\ &= \sqrt{0.19} \times 10^2 = 0.44 \times 10^2 \\ &= 4.4 \times 10^1\end{aligned}$$

The same procedure is followed for roots other than square roots; for example,

$$\begin{aligned}\sqrt[3]{4.6 \times 10^{10}} &= (4.6 \times 10^{10})^{1/3} = (46 \times 10^9)^{1/3} \\ &= \sqrt[3]{46} \times 10^3 = 3.6 \times 10^3\end{aligned}$$

A1.2 Logarithms

A logarithm is an exponent. Any number N can be expressed as follows:

$$N = 10^x$$

For example,

$$\begin{aligned}1000 &= 10^3 \\ 100 &= 10^2 \\ 10 &= 10^1 \\ 1 &= 10^0\end{aligned}$$

The common, or base 10, logarithm of a number is the power to which 10 must be taken to yield that number. Thus since $1000 = 10^3$,

$$\log 1000 = 3$$

Similarly,

$$\begin{aligned}\log 100 &= 2 \\ \log 10 &= 1 \\ \log 1 &= 0\end{aligned}$$

For a number between 10 and 100, the required exponent of 10 will be between 1 and 2. For example, $65 = 10^{1.8129}$; that is, $\log 65 = 1.8129$. For a number between 100 and 1000, the exponent of 10 will be between 2 and 3. For example, $650 = 10^{2.8129}$ and $\log 650 = 2.8129$.

A number N greater than 0 and less than 1 can be expressed as follows:

$$N = 10^{-x} = \frac{1}{10^x}$$

For example,

$$\begin{aligned}0.001 &= \frac{1}{1000} = \frac{1}{10^3} = 10^{-3} \\ 0.01 &= \frac{1}{100} = \frac{1}{10^2} = 10^{-2} \\ 0.1 &= \frac{1}{10} = \frac{1}{10^1} = 10^{-1}\end{aligned}$$

Thus
$$\log 0.001 = -3$$
$$\log 0.01 = -2$$
$$\log 0.1 = -1$$

Although common logs are often tabulated, the most convenient method for obtaining such logs is to use a calculator.

Since logs are simply exponents, they are manipulated according to the rules for exponents. For example, if $A = 10^x$ and $B = 10^y$, then their product is

$$A \cdot B = 10^x \cdot 10^y = 10^{x+y}$$

and
$$\log AB = x + y = \log A + \log B$$

For division we have
$$\frac{A}{B} = \frac{10^x}{10^y} = 10^{x-y}$$

and
$$\log \frac{A}{B} = x - y = \log A - \log B$$

For a number raised to a power we have

$$A^n = (10^x)^n = 10^{nx}$$

and
$$\log A^n = nx = n \log A$$

It follows that
$$\log \frac{1}{A^n} = \log A^{-n} = -n \log A$$

or for $n = 1$,
$$\log \frac{1}{A} = -\log A$$

When a common log is given, to find the number it represents, we must carry out the process of exponentiation. For example, if the log is 2.673, then $N = 10^{2.673}$. The process of exponentiation is also called taking the antilog, or the inverse logarithm, and is easily carried out by using a calculator.

A second type of logarithm, the natural logarithm, is based on the number 2.7183, which is referred to as e. In this case a number is represented as $N = e^x = 2.7183^x$. For example,

$$N = 7.15 = e^x$$
$$\ln 7.15 = x = 1.967$$

If a natural logarithm is given, to find the number it represents, we must carry out exponentiation to the base e (2.7183) by using a calculator.

A1.3 Graphing Functions

In the interpretation of the results of a scientific experiment, it is often useful to make a graph. It is usually most convenient to graph the function in a form that gives a straight line. The equation for a straight line (a *linear equation*) can be represented by the general form

$$y = mx + b$$

where y is the *dependent variable*, x is the *independent variable*, m is the *slope*, and b is the *intercept* with the y axis.

As an illustration of the characteristics of a linear equation, the function $y = 3x + 4$ is plotted in Fig. A1.1. For this equation $m = 3$ and $b = 4$. Note that the y intercept occurs when $x = 0$. In this case the intercept is 4, as can be seen from the equation ($b = 4$).

The slope of a straight line is defined as the ratio of the rate of change in y to that in x:

$$m = \text{slope} = \frac{\Delta y}{\Delta x}$$

For the equation $y = 3x + 4$, y changes three times as fast as x (since x has a coefficient of 3). Thus the slope in this case is 3. This can be verified from the graph. For the triangle shown in Fig. A1.1:

$$\text{Slope} = \frac{\Delta y}{\Delta x} = \frac{24}{8} = 3$$

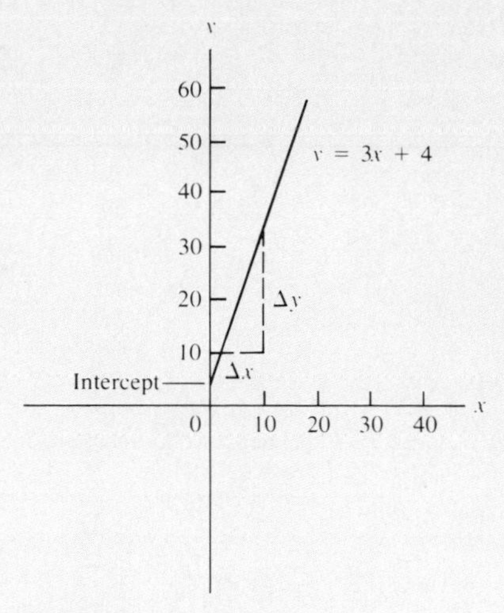

Figure A1.1
Graph of the linear equation $y = 3x + 4$.

Sometimes an equation that is not in standard form can be changed to the form $y = mx + b$ by rearrangement or mathematical manipulation. An example is the equation $k = Ae^{-E_a/RT}$, where A, E_a and R are constants, k is the dependent variable, and $1/T$ is the independent variable. This equation can be changed to standard form by taking the natural logarithm of both sides,

$$\ln k = \ln Ae^{-E_a/RT} = \ln A + \ln e^{-E_a/RT} = \ln A - \frac{E_a}{RT}$$

noting that the log of a product is equal to the sum of the logs of the individual terms and that the natural log of $e^{-E_a/RT}$ is simply the exponent $-E_a/RT$. Thus in standard form the equation $k = Ae^{-E_a/RT}$ is written

$$\underbrace{\ln k}_{y} = \underbrace{-\frac{E_a}{R}}_{m}\underbrace{\left(\frac{1}{T}\right)}_{x} + \underbrace{\ln A}_{b}$$

A plot of $\ln k$ versus $1/T$ (see Fig. A1.2) gives a straight line with slope $-E_a/R$ and intercept $\ln A$.

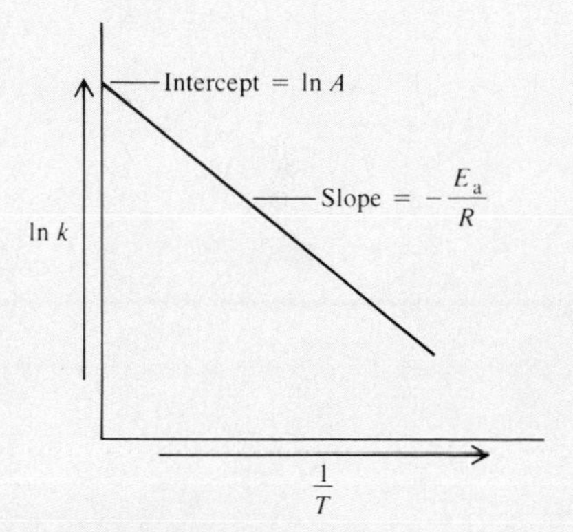

Figure A1.2
Graph of $\ln k$ versus $1/T$.

Of course, many relationships that arise from the description of natural systems are nonlinear, and the "slope" of a curve is continuously changing. In this case the instantaneous slope is given by the tangent to the curve at that point, which is described by a new function obtained by taking the derivative of the original function. For example, for the function in x, $f = ax^2$, the derivative (df/dx) is $2ax$. Thus the slope at each point on the curve defined by the function ax^2 is given by $2ax$.

A1.4 Solving Quadratic Equations

A *quadratic equation*, a polynomial in which the highest power of x is 2, can be written as

$$ax^2 + bx + c = 0$$

One method for finding the two values of x that satisfy a quadratic equation is to use the *quadratic formula:*

$$x = \frac{-b \pm \sqrt{b^2 - 4ac}}{2a}$$

where a, b, and c represent the coefficients of x^2, and x and the constant, respectively. For example, in the determination of $[H^+]$ in a solution of 1.0×10^{-4} *M* acetic acid, the following expression arises:

$$1.8 \times 10^{-5} = \frac{x^2}{1.0 \times 10^{-4} - x}$$

which yields $\quad x^2 + (1.8 \times 10^{-5})x - 1.8 \times 10^{-9} = 0$

where $a = 1$, $b = 1.8 \times 10^{-5}$, and $c = -1.8 \times 10^{-9}$. Using the quadratic formula, we have

$$x = \frac{-b \pm \sqrt{b^2 - 4ac}}{2a}$$

$$= \frac{-1.8 \times 10^{-5} \pm \sqrt{3.24 \times 10^{-10} - (4)(1)(-1.8 \times 10^{-9})}}{2(1)}$$

and $\quad x = \dfrac{6.9 \times 10^{-5}}{2} = 3.5 \times 10^{-5}$

or $\quad x = \dfrac{-10.5 \times 10^{-5}}{2} = -5.2 \times 10^{-5}$

Note that there are two roots, as there always will be for a polynomial in x^2. In this case x represents a concentration of H^+ (see Section 7.5). Thus the positive root is the one that solves the problem, since a concentration cannot be a negative number.

A second method for solving quadratic equations is by *successive approximations,* a systematic method of trial and error. A value of x is guessed and substituted into the equation everywhere x (or x^2) appears, except for one place. For example, for the equation

$$x^2 + (1.8 \times 10^{-5})x - 1.8 \times 10^{-9} = 0$$

we might guess $x = 2 \times 10^{-5}$. Substituting that value into the equation gives

$$x^2 + (1.8 \times 10^{-5})(2 \times 10^{-5}) - 1.8 \times 10^{-9} = 0$$

or $\quad x^2 = 1.8 \times 10^{-9} - 3.6 \times 10^{-10} = 1.4 \times 10^{-9}$

Thus $\quad x = 3.7 \times 10^{-5}$

Note that the guessed value of x (2×10^{-5}) is not the same as the value of x that is calculated (3.7×10^{-5}) after inserting the estimated value. This means that $x = 2 \times 10^{-5}$ is not the correct solution, and we must try another guess.

We take the calculated value (3.7×10^{-5}) as our next guess:

$$x^2 + (1.8 \times 10^{-5})(3.7 \times 10^{-5}) - 1.8 \times 10^{-9} = 0$$

$$x^2 = 1.8 \times 10^{-9} - 6.7 \times 10^{-10} = 1.1 \times 10^{-9}$$

Thus $\quad x = 3.3 \times 10^{-5}$

Now we compare the two values of x again:

Guessed: $x = 3.7 \times 10^{-5}$

Calculated: $x = 3.3 \times 10^{5}$

These values are closer but still not identical.

Next, we try 3.3×10^{-5} as our guess:

$$x^2 + (1.8 \times 10^{-5})(3.3 \times 10^{-5}) - 1.8 \times 10^{-9} = 0$$

$$x^2 = 1.8 \times 10^{-9} - 5.9 \times 10^{-10} = 1.2 \times 10^{-9}$$

Thus $x = 3.5 \times 10^{-5}$

Compare:

Guessed: $x = 3.3 \times 10^{-5}$

Calculated: $x = 3.5 \times 10^{-5}$

Next, we guess $x = 3.5 \times 10^{-5}$, which leads to

$$x^2 + (1.8 \times 10^{-5})(3.5 \times 10^{-5}) - 1.8 \times 10^{-9} = 0$$

$$x^2 = 1.8 \times 10^{-9} - 6.3 \times 10^{-10} = 1.2 \times 10^{-9}$$

Thus $x = 3.5 \times 10^{-5}$

Now the guessed value and the calculated value are the same; we have found the correct solution. Note that this agrees with one of the roots found with the quadratic formula in the first method above.

To further illustrate the method of successive approximations, we will solve Example 7.9 by using this procedure. In solving for $[H^+]$ for 0.010 *M* H_2SO_4, we obtain the following expression:

$$1.2 \times 10^{-2} = \frac{x(0.010 + x)}{0.010 - x}$$

which can be rearranged to give

$$x = (1.2 \times 10^{-2})\left(\frac{0.010 - x}{0.010 + x}\right)$$

We will guess a value for x, substitute it into the right side of the equation, and then calculate a value for x. In guessing a value for x, we know it must be less than 0.010, since a larger value would make the calculated value for x negative and the guessed and calculated values will never match. We start by guessing $x = 0.005$.

The results of the successive approximations are shown in the following table:

Trial	Guessed Value for x	Calculated Value for x
1	0.0050	0.0040
2	0.0040	0.0051
3	0.00450	0.00455
4	0.00452	0.00453

Note that the first guess was close to the actual value and that there was oscillation between 0.004 and 0.005 for the guessed and calculated values. For trial 3 an average of these values was used as the guess, and this led rapidly to the correct value (0.0045 to the correct number of significant figures). Also, note that it is useful to carry extra digits until the correct value is obtained, which is then rounded off to the correct number of significant figures.

The method of successive approximations is especially useful for solving polynomials containing x to a power of 3 or higher. The procedure is the same as for quadratic equations: substitute a guessed value for x into the equation for every x term but one, and then solve for x. Continue this process until the guessed and calculated values agree.

A1.5 Uncertainties in Measurements

The number associated with a measurement is obtained by using some measuring device. For example, consider the measurement of the volume of a liquid in a buret, as shown in Fig. A1.3, where the scale is greatly magnified. The volume is about 22.15 milliliters. Note that the last number must be estimated by interpolating between the 0.1-milliliter marks. Since the last number is estimated, its value may vary depending on who makes the measurement. If five different people read the same volume, the results might be as follows:

Person	Result of Measurement
1	22.15 mL
2	22.14 mL
3	22.16 mL
4	22.17 mL
5	22.16 mL

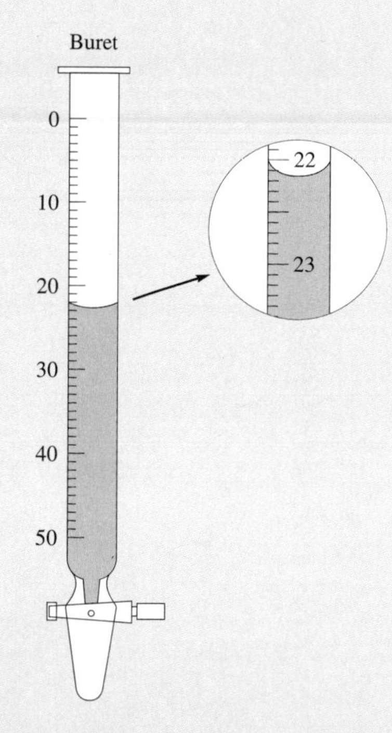

Figure A1.3
Measurement of volume using a buret. The volume is read at the bottom of the liquid curve (called the meniscus).

Note from these results that the first three numbers (22.1) remain the same regardless of who makes the measurement; these are called certain digits. However, the digit to the right of the 1 must be estimated and thus varies; it is called an uncertain digit. We customarily report a measurement by recording all of the certain digits plus the *first* uncertain digit. In our example it would not make any sense to try to record the volume to thousandths of a milliliter, because the value for hundredths of a milliliter must be estimated when using the buret.

It is very important to realize that a *measurement always has some degree of uncertainty.* The uncertainty of a measurement depends on the precision of the measuring device. For example, using a bathroom scale, you might estimate that the mass of a grapefruit is approximately 1.5 pounds. Weighing the same grapefruit on a highly precise balance might produce a result of 1.476 pounds. In the first case the uncertainty occurs in the tenths of a pound place; in the second case the uncertainty occurs in the thousandths of a pound place. Suppose we weigh two similar grapefruit on the two devices and obtain the following results:

	Bathroom Scale	Balance
Grapefruit 1	1.5 lb	1.476 lb
Grapefruit 2	1.5 lb	1.518 lb

Do the two grapefruits have the same mass? The answer depends on which set of results you consider. Thus a conclusion based on a series of measurements depends on the certainty of those measurements. For this reason, it is important to indicate the uncertainty in any measurement. This is done by always recording the certain digits and the first uncertain digit (the estimated number). These numbers are called the **significant figures** of a measurement.

The convention of significant figures automatically gives an indication of the uncertainty in a measurement. The uncertainty in the last number (the estimated number) is usually assumed to be ± 1 unless otherwise indicated. For example, the measurement 1.86 kilograms can be interpreted to mean 1.86 ± 0.01 kilograms.

Precision and Accuracy

Two terms often used to describe uncertainty in measurements are *precision* and *accuracy*. Although these words are frequently used interchangeably in everyday life, they have different meanings in the scientific context. **Accuracy** refers to the agreement of a particular value with the true value. **Precision** refers to the degree of agreement among several measurements of the same quantity. Precision reflects the *reproducibility* of a given type of measurement. The difference between these terms is illustrated by the results of three different target practices shown in Fig. A1.4.

Two different types of errors are also introduced in Fig. A1.4. A **random error** (also called an indeterminate error) means that a measurement has an equal probability of being high or low. This type of error occurs in estimating the value of the last digit of a measurement. The second type of error is called **systematic error** (or determinate error). This type of error occurs in the same direction each time; it is either always high or always low. Figure A1.4(a) indicates large random errors (poor technique). Figure A1.4(b) indicates small random errors but a large systematic error, and Fig. A1.4(c) indicates small random errors and no systematic error.

In quantitative work precision is often used as an indication of accuracy; we assume that the *average* of a series of precise measurements (which should "average out" the random errors because of their equal probability of being high or low) is accurate, or close to the "true" value. However, this assumption is only valid if systematic errors are absent. Suppose we weigh a piece of brass five times on a very precise balance and obtain the following results:

Weighing	Result
1	2.486 g
2	2.487 g
3	2.485 g
4	2.484 g
5	2.488 g

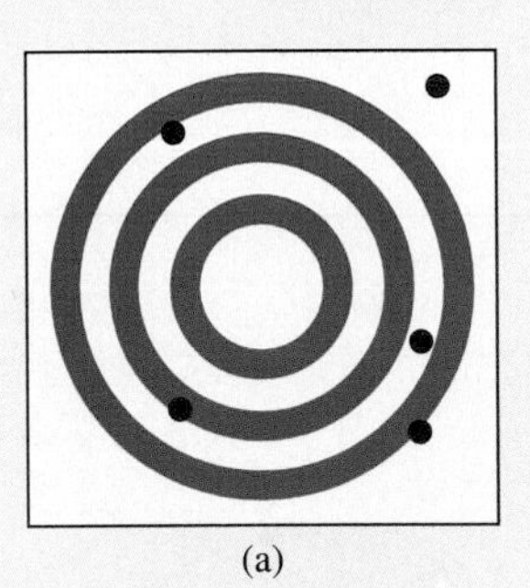

(a)

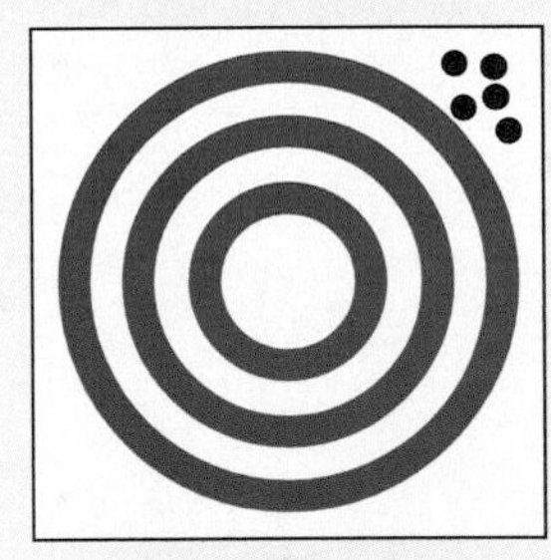

(b)

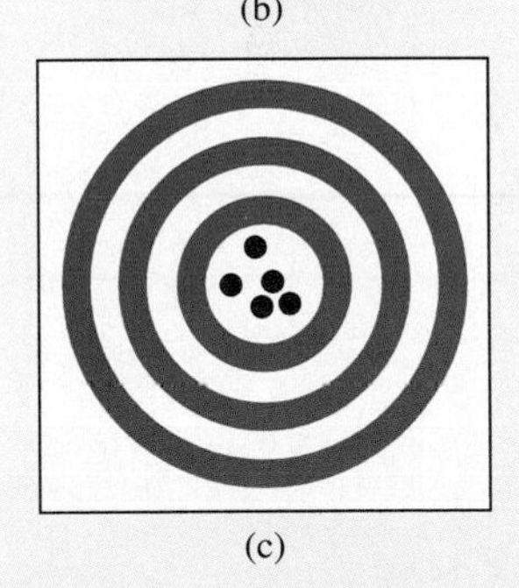

(c)

Figure A1.4
Shooting targets show the difference between *precise* and *accurate*. (a) Neither accurate nor precise (large random errors). (b) Precise but not accurate (small random errors, large systematic error). (c) Bull's-eye! Both precise and accurate (small random errors, no systematic error).

Normally, we would assume that the true mass of the piece of brass is very close to 2.486 grams, which is the average of the five results. However, if the balance has a defect causing it to give a result that is consistently 1.000 gram too high (a systematic error of +1.000 gram), then 2.486 grams would be seriously in error. The point here is that high precision among several measurements is an indication of accuracy *only* if you can be sure that systematic errors are absent.

Expression of Experimental Results

The accuracy of a measurement refers to how close it is to the true value. An inaccurate result occurs as a result of some flaw (systematic error) in the measurement: the presence of an interfering substance, incorrect calibration of an instrument, operator error, and so on. The goal of chemical analysis is to eliminate systematic error, but random errors can only be minimized. In practice, an experiment is almost always done in order to find an unknown value (the true value is not known—someone is trying to obtain that value by doing the experiment). In this case the precision of several replicate determinations is used to assess the accuracy of the result. The results of the replicate experiments are expressed as an average (which we assume is close to the true value) with an error limit that gives some indication of how close the average value may be to the true value. The error limit represents the uncertainty of the experimental result.

To illustrate this procedure, consider a situation that might arise in the pharmaceutical industry. Assume that the specification for a commercial 500-mg acetaminophen (the active painkiller in Tylenol) tablet is that each batch of tablets must contain 450–550 mg of acetaminophen per tablet. Suppose that chemical analysis gave the following results for a batch of acetaminophen tablets: 428, 479, 442, and 435 mg. How can these results be used to decide whether the batch of tablets meets the specification? Although the details of how to draw such conclusions from measured data are beyond the scope of this discussion, we will consider some aspects of this process. We will focus here on the types of experimental uncertainty, the expression of experimental results, and a simplified method for estimating experimental uncertainty when several types of measurements contribute to the final result.

There are two common ways of expressing an average: the mean and the median. The mean ($\bar{x}$) is the arithmetic average of the results, or

$$\text{Mean} = \bar{x} = \sum_{i=1}^{n} \frac{x_i}{n} = \frac{x_1 + x_2 + \cdots + x_n}{n}$$

where Σ means take the sum of the values. The mean is equal to the sum of all the measurements divided by the number of measurements. For the acetaminophen results given previously, the mean is

$$\bar{x} = \frac{428 + 479 + 442 + 435}{4} = 446 \text{ mg}$$

The median is the value that lies in the middle among the results. Half of the measurements are above the median and half are below the median. For results of 465, 485, and 492 mg, the median is 485 mg. When there is an even number of results, the median is the average of the two middle results. For the acetaminophen results the median is

$$\frac{442 + 435}{2} = 439 \text{ mg}$$

There are several advantages to using the median. If a small number of measurements is made, one value can greatly affect the mean. Consider the results for the analysis of acetaminophen: 428, 479, 442, and 435 mg. The mean is 446 mg, which is larger than three of the four results. The median is 439 mg, which lies near the three values that are relatively close to one another.

In addition to expressing an average value for a series of results, we must also express the uncertainty. This usually means expressing either the precision of the measurements or the observed range of the measurements. The range of a series of measurements is defined by the smallest value and the largest value. For the analytical results on the acetaminophen tablets, the range is from 428 to 479 mg. Using this range, we can express the results by saying that the true value lies between 428 and 479 mg. That is, we can express the amount of acetaminophen in a typical tablet as 446 ± 33 mg, where the error limit is chosen to give the observed range (approximately).

The most common way to specify precision is by the standard deviation s, which for a small number of measurements is given by the formula

$$s = \left[\frac{\sum_{i=1}^{n} (x_i - \overline{x})^2}{n - 1}\right]^{1/2}$$

where x_i is an individual result, $\overline{x}$ is the average (either mean or median), and n is the total number of measurements. For the acetaminophen example we have

$$s = \left[\frac{(428 - 446)^2 + (479 - 446)^2 + (442 - 446)^2 + (435 - 446)^2}{4 - 1}\right]^{1/2} = 23$$

Thus we can say that the amount of acetaminophen in a typical tablet in the batch of tablets is 446 mg with a sample standard deviation of 23 mg. Statistically, this means that any additional measurement has a 68% probability (68 chances out of 100) of being between 423 mg (446 − 23) and 469 mg (446 + 23). Thus the standard deviation is a measure of the precision of a given type of measurement.

In scientific calculations it is also useful to be able to estimate the precision of a procedure that involves several measurements by combining the precisions of the individual steps. That is, we want to answer the following question: How do the uncertainties propagate when we combine the results of several different types of measurements? There are many ways to deal with the propagation of uncertainty. We will discuss one simple method below.

Worst-Case Method for Estimating Experimental Uncertainty

To illustrate this method, we will consider the determination of the density of an irregularly shaped solid. In this determination we make three measurements. First, we measure the mass of the object on a balance. Next, we must obtain the volume of the solid. The easiest method for doing this is to partially fill a graduated cylinder with a liquid and record the volume. Then we add the solid and record the volume again. The difference in the measured volumes is the volume of the solid. We can then calculate the density of the solid from the equation

$$D = \frac{M}{V_2 - V_1}$$

where M is the mass of the solid, V_1 is the initial volume of liquid in the graduated cylinder, and V_2 is the volume of liquid plus solid. Suppose we get the following results:

$$M = 23.06 \text{ g}$$
$$V_1 = 10.4 \text{ mL}$$
$$V_2 = 13.5 \text{ mL}$$

The calculated density is

$$\frac{23.06 \text{ g}}{13.5 \text{ mL} - 10.4 \text{ mL}} = 7.44 \text{ g/mL}$$

Now suppose that the precision of the balance used is ± 0.02 g and that the volume measurements are precise to ± 0.05 mL. How do we estimate the uncertainty of the density? We can do this by assuming a worst case. That is, we assume the largest uncertainties in all measurements, and we see what combinations of measurements will give the largest and smallest possible results (the greatest range). Since the density is the mass divided by the volume, the largest value of the density will be that obtained by using the largest possible mass and the smallest possible volume:

Largest possible mass = 23.06 + 0.02

$$D_{max} = \frac{23.08}{13.45 - 10.45} = 7.69 \text{ g/mL}$$

Smallest possible V_2 Largest possible V_1

The smallest value of the density is

Smallest possible mass

$$D_{min} = \frac{23.04}{13.55 - 10.35} = 7.20 \text{ g/mL}$$

Largest possible V_2 Smallest possible V_1

Thus the calculated range is from 7.20 to 7.69, and the average of these values is 7.45. The error limit is the number that gives the high and low range values when added and subtracted from the average. Therefore, we can express the density as 7.45 ± 0.25 g/mL, which is the average value plus or minus the quantity that gives the range calculated by assuming the largest uncertainties.

Analysis of the propagation of uncertainties is useful in drawing qualitative conclusions from the analysis of measurements. For example, suppose that we obtained the above results for the density of an unknown alloy and we want to know if it is one of the following alloys:

Alloy A: $D = 7.58$ g/mL

Alloy B: $D = 7.42$ g/mL

Alloy C: $D = 8.56$ g/mL

We can safely conclude that the alloy is not C. But the values of the densities for alloys A and B are both within the inherent uncertainty of our method. To distinguish between A and B, we need to improve the precision of our determination. The obvious choice is to improve the precision of the volume measurement.

The worst-case method is useful for estimating the maximum uncertainty expected when the results of several measurements are combined to obtain a result. We assume the maximum uncertainty in each measurement and then calculate the minimum and maximum possible results. These extreme values describe the range and thus the maximum error limit associated with a particular determination.

Confidence Limits

A more sophisticated method for estimating the uncertainty of a particular type of determination involves the use of confidence limits. A confidence limit is defined as

$$\text{Confidence limit} = \pm \frac{ts}{\sqrt{n}}$$

where t = a weighting factor based on statistical analysis

s = the standard deviation

n = the number of experiments carried out

In this context an experiment may refer to a single type of measurement (for example, weighing an object) or to a procedure that requires various types of measurements to obtain a given final result (for example, obtaining the percentage of iron in a particular sample of iron ore). Some representative values of t are listed in Table A1.1.

A 95% confidence level means that the true value (the *average* obtained if the experiment were repeated an *infinite* number of times) will lie within $\pm ts/\sqrt{n}$ of the *observed* average (obtained from n experiments) with a 95% probability (95 of 100 times). Thus the factor $\pm ts/\sqrt{n}$ represents an error limit for a given set of results from a particular type of experiment. Thus we might represent the result of n determinations as

$$\bar{x} \pm \frac{ts}{\sqrt{n}}$$

where $\bar{x}$ is the average of the results from the n experiments. This type of error limit is expected to be considerably smaller than that obtained from a worst-case analysis.

TABLE A1.1
Values of *t* for 90% and 95% Confidence Levels

	Values of t for Confidence Intervals	
n	90%	95%
2	6.31	12.7
3	2.92	4.30
4	2.35	3.18
5	2.13	2.78
6	2.02	2.57
7	1.94	2.45
8	1.90	2.36
9	1.86	2.31
10	1.83	2.26

A1.6 Significant Figures

Calculating the final result for an experiment usually involves adding, subtracting, multiplying, or dividing the results of various types of measurements. Thus it is important to be able to estimate the uncertainty in the final result. In the previous section we have considered this process in some detail. A closely related matter concerns the number of digits that should be retained in the result of a given calculation. In other words, how many of the digits in the result are significant (meaningful) relative to the uncertainty expected in the result? From statistical analyses of how uncertainties accumulate when arithmetic operations are carried out, rules have been developed for determining the correct number of significant figures in a final result. First, we must consider how to count the number of significant figures (digits) represented in a particular number.

Rules for Counting Significant Figures (Digits)

1. *Nonzero integers.* Nonzero integers always count as significant figures.
2. *Zeros.* There are three classes of zeros:
   a. *Leading zeros* are zeros that *precede* all of the nonzero digits. They do not count as significant figures. In the number 0.0025 the three zeros simply indicate the position of the decimal point. This number has only two significant figures.
   b. *Captive zeros* are zeros *between* nonzero digits. They always count as significant figures. The number 1.008 has four significant figures.
   c. *Trailing zeros* are zeros at the *right end* of the number. They are significant only if the number contains a decimal point. The number 100 has only one significant figure, whereas the number 1.00×10^2 has

three significant figures. The number one hundred written as 100. also has three significant figures.

3. *Exact numbers.* Many times calculations involve numbers that were not obtained by using measuring devices but were determined by counting: 10 experiments, 3 apples, 8 molecules. Such numbers are called *exact numbers*. They can be assumed to have an infinite number of significant figures. Other examples of exact numbers are the 2 in $2\pi r$ (the circumference of a circle), and the 4 and the 3 in $\frac{4}{3}\pi r^3$ (the volume of a sphere). Exact numbers can also arise from definitions. For example, one inch is defined as exactly 2.54 centimeters. Thus in the statement 1 in = 2.54 cm, neither the 2.54 nor the 1 limits the number of significant figures when used in a calculation.

The following rules apply for determining the number of significant figures in the result of a calculation.

Rules for Significant Figures in Mathematical Operations*

1. *For multiplication or division* the number of significant figures in the result is the same as the number in the least precise measurement used in the calculation. For example, consider this calculation:

$$4.56 \times 1.4 = 6.38 \xrightarrow{\text{Corrected}} 6.4$$

↑ Limiting term has two significant figures (1.4)   ↑ Two significant figures (6.4)

The correct product has only two significant figures, since 1.4 has two significant figures.

2. *For addition or subtraction* the result has the same number of decimal places as the least precise measurement used in the calculation. For example, consider the following sum:

$$\begin{array}{r} 12.11 \\ 18.0 \\ 1.013 \\ \hline 31.123 \end{array} \xrightarrow{\text{Corrected}} 31.1$$

18.0 ←Limiting term has one decimal place

31.1 ↑ One decimal place

The correct result is 31.1, since 18.0 has only one decimal place.

Note that for multiplication and division significant figures are counted. For addition and subtraction the decimal places are counted.

In most calculations you will need to round off numbers to obtain the correct number of significant figures. The following rules should be applied for rounding.

---

*Although these rules work well for most cases, they can give misleading results in certain cases. For a discussion of this, see L. M. Schwartz, "Propagation of Significant Figures," *J. Chem. Ed.* **62** (1985), p. 693.

Rules for Rounding

1. In a series of calculations, carry the extra digits through to the final result, *then* round off.*
2. If the digit to be removed†
   a. is less than 5, the preceding digit stays the same. For example, 1.33 rounds to 1.3.
   b. is equal to or greater than 5, the preceding digit is increased by 1. For example, 1.36 rounds to 1.4.

When rounding, use only the first number to the right of the last significant figure. Do not round off sequentially. For example, the number 4.348 when rounded to two significant figures is 4.3, not 4.4.

APPENDIX TWO Units of Measurement and Conversions Among Units

A2.1 Measurements

Making observations is fundamental to all science. A quantitative observation, or **measurement,** always consists of two parts: a *number* and a scale (a *unit*). Both parts must be present for the measurement to be meaningful.

The two most widely used systems of units are the *English system* used in the United States and the *metric system* used by most of the rest of the industrialized world. This duality obviously causes a good deal of trouble; for example, parts as simple as bolts are not interchangeable between machines built using the different systems. As a result, the United States has begun to adopt the metric system.

For many years, most scientists worldwide have used the metric system. In 1960 an international agreement established a system of units called the *International System (le Système International* in French), abbreviated **SI.** This system is based on the metric system and the units derived from the metric system. The fundamental SI units are listed in Table A2.1.

Because the fundamental units are not always convenient (expressing the mass of a pin in kilograms is awkward), the SI system employs prefixes to change the size of the unit. These prefixes are listed in Table A2.2.

One physical quantity that is very important in chemistry is *volume,* which is not a fundamental SI unit; it is derived from length. A cube with dimensions of 1 m on each edge has a volume of $(1\ m)^3 = 1\ m^3$. Then, recognizing that there are 10 decimeters (dm) in a meter, the volume of the cube is $(10\ dm)^3 =$

---

* This practice will not usually be followed in the examples in this text because we want to show the correct number of significant figures in each step. However, in the answers to the end-of-chapter exercises, only the final answer is rounded.

† This procedure is consistent with the operation of calculators.

TABLE A2.1 The Fundamental SI Units

Physical Quantity	Name of Unit	Abbreviation
Mass	kilogram	kg
Length	meter	m
Time	second	s
Temperature	Kelvin	K
Electric current	ampere	A
Amount of substance	mole	mol
Luminous intensity	candela	cd

1000 dm^3. A cubic decimeter, dm^3, is commonly called a liter (L), which is a unit of volume slightly larger than a quart. Similarly, since 1 dm equals 10 centimeters (cm), the liter $(1\ dm)^3$ contains 1000 cm^3, or 1000 milliliters (mL).

A2.2 Unit Conversions

It is often necessary to convert results from one system of units to another. The most common way of converting units is by the *unit factor method,* more commonly called **dimensional analysis.** To illustrate the use of this method, we will look at a simple unit conversion.

TABLE A2.2 The Prefixes Used in the SI System

Prefix	Symbol	Meaning	Exponential Notation*
exa	E	1,000,000,000,000,000,000	10^{18}
peta	P	1,000,000,000,000,000	10^{15}
tera	T	1,000,000,000,000	10^{12}
giga	G	1,000,000,000	10^{9}
mega	M	1,000,000	10^{6}
kilo	k	1,000	10^{3}
hecto	h	100	10^{2}
deka	da	10	10^{1}
—	—	1	10^{0}
deci	d	0.1	10^{-1}
centi	c	0.01	10^{-2}
milli	m	0.001	10^{-3}
micro	μ	0.000001	10^{-6}
nano	n	0.000000001	10^{-9}
pico	p	0.000000000001	10^{-12}
femto	f	0.000000000000001	10^{-15}
atto	a	0.000000000000000001	10^{-18}

*The most common notations are shown in bold. See Appendix A1.1 if you need a review of exponential notation.

Consider a pin measuring 2.85 cm in length. What is its length in inches? To solve this problem, we must use the equivalence statement

$$2.54 \text{ cm} = 1 \text{ in} \qquad \text{(exactly)}$$

If we divide both sides of this equation by 2.54 cm, we get

$$\frac{2.54 \text{ cm}}{2.54 \text{ cm}} = 1 = \frac{1 \text{ in}}{2.54 \text{ cm}}$$

Note that the expression 1 in/2.54 cm equals 1. This expression is called a **unit factor.** Since 1 in and 2.54 cm are exactly equivalent, multiplying any expression by this unit factor will not change its value.

The pin has a length of 2.85 cm. Multiplying this length by the unit factor gives

$$2.85 \cancel{\text{cm}} \times \frac{1 \text{ in}}{2.54 \cancel{\text{cm}}} = \frac{2.85}{2.54} \text{ in} = 1.12 \text{ in}$$

Note that the centimeter units cancel to give inches for the result. This is exactly what we wanted to accomplish. Note also that the result has three significant figures, as required by the number 2.85. Recall that the 1 and 2.54 in the conversion factor are exact numbers by definition.

Summary: **Converting from One Unit to Another**

- **To convert from one unit to another, use the equivalence statement that relates the two units.**
- **Derive the appropriate unit factor by noting the direction of the required change (to cancel the unwanted units).**
- **Multiply the quantity to be converted by the unit factor to give the quantity with the desired units.**

In dimensional analysis your verification that everything has been done correctly is that the correct units are obtained in the end. *In doing chemistry problems, you should always include the units for the quantities used.* Always check to see that the units cancel to give the correct units for the final result. This provides a very valuable check, especially for complicated problems.

APPENDIX THREE Spectral Analysis

Although volumetric and gravimetric analyses are still commonly used, spectroscopy is the technique most often used for modern chemical analysis. *Spectroscopy* is the study of electromagnetic radiation emitted or absorbed by a given chemical species. Since the quantity of radiation absorbed or emitted can be related to the quantity of the absorbing or emitting species present, this technique can be used for quantitative analysis. There are many spectroscopic techniques, since electromagnetic radiation spans a wide range of energies to include microwaves, X rays, and ultraviolet, infrared, and visible light, to name a few of its familiar forms. However, we will consider here only one procedure, which is based on the absorption of visible light.

If a liquid is colored, it is because some component of the liquid absorbs visible light. In a solution the greater the concentration of the light-absorbing substance, the more light absorbed, and the more intense the color of the solution.

The quantity of light absorbed by a substance can be measured by a *spectrophotometer,* shown schematically in Fig. A3.1. This instrument consists of a source that emits all wavelengths of light in the visible region (wavelengths of ≈400–700 nm); a monochromator, which selects a given wavelength of light; a sample holder for the solution being measured; and a detector, which compares the intensity of incident light, I_0, with the intensity of light after it has passed through the sample, I. The ratio I/I_0, called the *transmittance,* is a measure of the fraction of light that passes through the sample. The amount of light absorbed is given by the *absorbance, A,* where

$$A = -\log \frac{I}{I_0}$$

The absorbance can be expressed by the *Beer-Lambert law:*

$$A = \epsilon l c$$

where ϵ is the molar absorptivity or the molar extinction coefficient (in L mol^{-1} cm^{-1}), l is the distance the light travels through the solution (in cm), and c is the concentration of the absorbing species (in mol/L). The Beer-Lambert law is the basis for using spectroscopy in quantitative analysis. If ϵ and l are known, determining A for a solution allows us to calculate the concentration of the absorbing species in the solution.

Suppose we have a pink solution containing an unknown concentration of $Co^{2+}(aq)$ ions. A sample of this solution is placed in a spectrophotometer, and the absorbance is measured at a wavelength where ϵ for $Co^{2+}(aq)$ is known to be 12 L mol^{-1} cm^{-1}. The absorbance A is found to be 0.60. The width of the sample tube is 1.0 cm. We want to determine the concentration of $Co^{2+}(aq)$ in

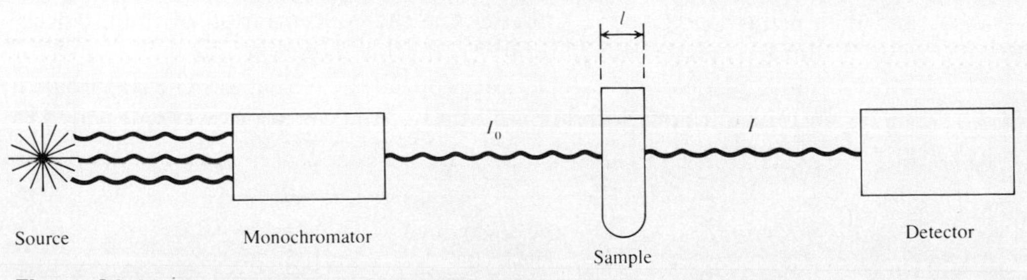

Figure A3.1
A schematic diagram of a simple spectrophotometer. The source emits all wavelengths of visible light, which are dispersed by using a prism or grating and then focused, one wavelength at a time, onto the sample. The detector compares the intensity of the incident light (I_0) with the intensity of the light after it has passed through the sample (I).

the solution. This problem can be solved by a straightforward application of the Beer-Lambert law,

$$A = \epsilon l c$$

where

$$A = 0.60$$

$$\epsilon = \frac{12 \text{ L}}{\text{mol cm}}$$

$$l = \text{light path} = 1.0 \text{ cm}$$

Solving for the concentration gives

$$c = \frac{A}{\epsilon l} = \frac{0.60}{\left(12 \dfrac{\text{L}}{\text{mol cm}}\right)(1.0 \text{ cm})} = 5.0 \times 10^{-2} \text{ mol/L}$$

To obtain the unknown concentration of an absorbing species from the measured absorbance, we must know the product ϵl, since

$$c = \frac{A}{\epsilon l}$$

We can obtain the product ϵl by measuring the absorbance of a solution of *known* concentration, since

$$\epsilon l = \frac{A}{c}$$

(A: Measured using a spectrophotometer; c: Known from making up the solution)

However, a more accurate value of the product ϵl can be obtained by plotting A versus c for a series of solutions. Note that the equation $A = \epsilon l c$ gives a straight line with slope ϵl when A is plotted against c.

For example, consider the following typical spectroscopic analysis. A sample of steel from a bicycle frame is to be analyzed to determine its manganese content. The procedure involves weighing out a sample of the steel, dissolving it in strong acid, treating the resulting solution with a very strong oxidizing agent to convert all the manganese to permanganate ion (MnO_4^-), and then using spectroscopy to determine the concentration of the intensely purple MnO_4^- ions in the solution. To do this, however, the value of ϵl for MnO_4^- must be determined at an appropriate wavelength. The absorbance values for four solutions with known MnO_4^- concentrations were measured to give the following data:

Solution	Concentration of MnO_4^- (mol/L)	Absorbance
1	7.00×10^{-5}	0.175
2	1.00×10^{-4}	0.250
3	2.00×10^{-4}	0.500
4	3.50×10^{-4}	0.875

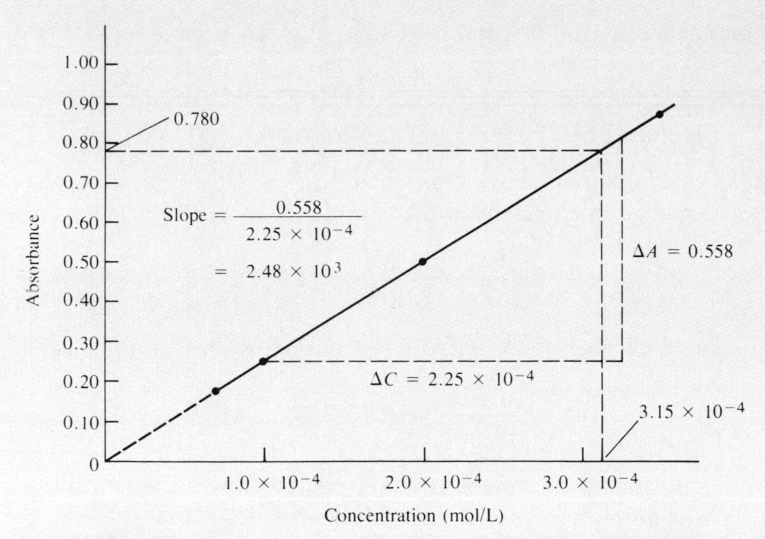

Figure A3.2
A plot of absorbance versus concentration of MnO_4^- in a series of solutions of known concentration.

A plot of absorbance versus concentration for the solutions of known concentration is shown in Fig. A3.2. The slope of this line (change in A/change in c) is 2.48×10^3 L/mol. This quantity represents the product ϵl.

A sample of the steel weighing 0.1523 g was dissolved, and the unknown amount of manganese was converted to MnO_4^- ions. Water was then added to give a solution with a final volume of 100.0 mL. A portion of this solution was placed in a spectrophotometer, and its absorbance was found to be 0.780. We can use these data to calculate the percent manganese in the steel. The MnO_4^- ions from the manganese in the dissolved steel sample show an absorbance of 0.780. Using the Beer-Lambert law, we calculate the concentration of MnO_4^- in this solution:

$$c = \frac{A}{\epsilon l} = \frac{0.780}{2.48 \times 10^3 \text{ L/mol}} = 3.15 \times 10^{-4} \text{ mol/L}$$

However, there is a more direct way for finding c. Using a graph such as that in Fig. A3.2 (often called a Beer's law plot), we can read the concentration that corresponds to $A = 0.780$. This interpolation is shown by dashed lines on the graph. By this method, $c = 3.15 \times 10^{-4}$ mol/L, which agrees with the value obtained above.

Recall that the original 0.1523-g steel sample was dissolved, the manganese was converted to permanganate, and the volume was adjusted to 100.0 mL. We now know that the $[MnO_4^-]$ in that solution is $3.15 \times 10^{-4} M$. Using this concentration, we can calculate the total number of moles of MnO_4^- in that solution:

$$\text{mol of } MnO_4^- = 100.0 \text{ mL} \times \frac{1 \text{ L}}{1000 \text{ mL}} \times 3.15 \times 10^{-4} \frac{\text{mol}}{\text{L}}$$

$$= 3.15 \times 10^{-5} \text{ mol}$$

Each mole of manganese in the original steel sample yields a mole of MnO_4^-. That is,

$$1 \text{ mol of Mn} \xrightarrow{\text{Oxidation}} 1 \text{ mol of } MnO_4^-$$

so the original steel sample must have contained 3.15×10^{-5} mol of manganese. The mass of manganese present in the sample is

$$3.15 \times 10^{-5} \text{ mol of Mn} \times \frac{54.938 \text{ g of Mn}}{1 \text{ mol of Mn}} = 1.73 \times 10^{-3} \text{ g of Mn}$$

Since the steel sample weighed 0.1523 g, the percent manganese in the steel is

$$\frac{1.73 \times 10^{-3} \text{ g of Mn}}{1.523 \times 10^{-1} \text{ g of sample}} \times 100 = 1.14\%$$

This example illustrates a typical use of spectroscopy in quantitative analysis. The steps commonly involved are as follows:

1. Preparation of a calibration plot (a Beer's law plot) from the measured absorbance values of a series of solutions with known concentrations.
2. Measurement of the absorbance of the solution of unknown concentration.
3. Use of the calibration plot to determine the unknown concentration.

APPENDIX FOUR Selected Thermodynamic Data*

Substance and State	ΔH_f° (kJ/mol)	ΔG_f° (kJ/mol)	S° (J K^{-1} mol^{-1})
Aluminum			
$Al(s)$	0	0	28
$Al_2O_3(s)$	−1676	−1582	51
$Al(OH)_3(s)$	−1277		
$AlCl_3(s)$	−704	−629	111
Barium			
$Ba(s)$	0	0	67
$BaCO_3(s)$	−1219	−1139	112
$BaO(s)$	−582	−552	70
$Ba(OH)_2(s)$	−946		
$BaSO_4(s)$	−1465	−1353	132
Beryllium			
$Be(s)$	0	0	10
$BeO(s)$	−599	−569	14
$Be(OH)_2(s)$	−904	−815	47
Bromine			
$Br_2(l)$	0	0	152
$Br_2(g)$	31	3	245
$Br_2(aq)$	−3	4	130
$Br^-(aq)$	−121	−104	82
$HBr(g)$	−36	−53	199
Cadmium			
$Cd(s)$	0	0	52
$CdO(s)$	−258	−228	55
$Cd(OH)_2(s)$	−561	−474	96
$CdS(s)$	−162	−156	65
$CdSO_4(s)$	−935	−823	123
Calcium			
$Ca(s)$	0	0	41
$CaC_2(s)$	−63	−68	70
$CaCO_3(s)$	−1207	−1129	93
$CaO(s)$	−635	−604	40
$Ca(OH)_2(s)$	−987	−899	83
$Ca_3(PO_4)_2(s)$	−4126	−3890	241
$CaSO_4(s)$	−1433	−1320	107
$CaSiO_3(s)$	−1630	−1550	84
Carbon			
$C(s)$ (graphite)	0	0	6

*All values are assumed precise to at least ±1.

Appendix Four (continued)

Substance and State	ΔH_f° (kJ/mol)	ΔG_f° (kJ/mol)	S° (J K^{-1} mol^{-1})
C(*s*) (diamond)	2	3	2
$CO(g)$	−110.5	−137	198
$CO_2(g)$	−393.5	−394	214
$CH_4(g)$	−75	−51	186
$CH_3OH(g)$	−201	−163	240
$CH_3OH(l)$	−239	−166	127
$H_2CO(g)$	−116	−110	219
$HCOOH(g)$	−363	−351	249
$HCN(g)$	135.1	125	202
$C_2H_2(g)$	227	209	201
$C_2H_4(g)$	52	68	219
$CH_3CHO(g)$	−166	−129	250
$C_2H_5OH(l)$	−278	−175	161
$C_2H_6(g)$	−84.7	−32.9	229.5
$C_3H_6(g)$	20.9	62.7	266.9
$C_3H_8(g)$	−104	−24	270
$C_2H_4O(g)$ (ethylene oxide)	−53	−13	242
$CH_2{=}CHCN(g)$	185.0	195.4	274
$CH_3COOH(l)$	−484	−389	160
$C_6H_{12}O_6(s)$	−1275	−911	212
$CCl_4(l)$	−135	−65	216
Chlorine			
$Cl_2(g)$	0	0	223
$Cl_2(aq)$	−23	7	121
$Cl^-(aq)$	−167	−131	57
$HCl(g)$	−92	−95	187
Chromium			
$Cr(s)$	0	0	24
$Cr_2O_3(s)$	−1128	−1047	81
$CrO_3(s)$	−579	−502	72
Copper			
$Cu(s)$	0	0	33
$CuCO_3(s)$	−595	−518	88
$Cu_2O(s)$	−170	−148	93
$CuO(s)$	−156	−128	43
$Cu(OH)_2(s)$	−450	−372	108
$CuS(s)$	−49	−49	67
Fluorine			
$F_2(g)$	0	0	203
$F^-(aq)$	−333	−279	−14
$HF(g)$	−271	−273	174
Hydrogen			
$H_2(g)$	0	0	131
$H(g)$	217	203	115
$H^+(aq)$	0	0	0
$OH^-(aq)$	−230	−157	−11
$H_2O(l)$	−286	−237	70
$H_2O(g)$	−242	−229	189

Substance and State	ΔH_f° (kJ/mol)	ΔG_f° (kJ/mol)	S° (J K^{-1} mol^{-1})
Iodine			
$I_2(s)$	0	0	116
$I_2(g)$	62	19	261
$I_2(aq)$	23	16	137
$I^-(aq)$	−55	−52	106
Iron			
$Fe(s)$	0	0	27
$Fe_3C(s)$	21	15	108
$Fe_{0.95}O(s)$ (wustite)	−264	−240	59
$FeO(s)$	−272	−255	61
$Fe_3O_4(s)$ (magnetite)	−1117	−1013	146
$Fe_2O_3(s)$ (hematite)	−826	−740	90
$FeS(s)$	−95	−97	67
$FeS_2(s)$	−178	−166	53
$FeSO_4(s)$	−929	−825	121
Lead			
$Pb(s)$	0	0	65
$PbO_2(s)$	−277	−217	69
$PbS(s)$	−100	−99	91
$PbSO_4(s)$	−920	−813	149
Magnesium			
$Mg(s)$	0	0	33
$MgCO_3(s)$	−1113	−1029	66
$MgO(s)$	−602	−569	27
$Mg(OH)_2(s)$	−925	−834	64
Manganese			
$Mn(s)$	0	0	32
$MnO(s)$	−385	−363	60
$Mn_3O_4(s)$	−1387	−1280	149
$Mn_2O_3(s)$	−971	−893	110
$MnO_2(s)$	−521	−466	53
$MnO_4^-(aq)$	−543	−449	190
Mercury			
$Hg(l)$	0	0	76
$Hg_2Cl_2(s)$	−265	−211	196
$HgCl_2(s)$	−230	−184	144
$HgO(s)$	−90	−59	70
$HgS(s)$	−58	−49	78
Nickel			
$Ni(s)$	0	0	30
$NiCl_2(s)$	−316	−272	107
$NiO(s)$	−241	−213	38
$Ni(OH)_2(s)$	−538	−453	79
$NiS(s)$	−93	−90	53

Appendix Four (continued)

Substance and State	ΔH_f° (kJ/mol)	ΔG_f° (kJ/mol)	S° (J K^{-1} mol^{-1})
Nitrogen			
$N_2(g)$	0	0	192
$NH_3(g)$	−46	−17	193
$NH_3(aq)$	−80	−27	111
$NH_4^+(aq)$	−132	−79	113
$NO(g)$	90	87	211
$NO_2(g)$	34	52	240
$N_2O(g)$	82	104	220
$N_2O_4(g)$	10	98	304
$N_2O_4(l)$	−20	97	209
$N_2O_5(s)$	−42	134	178
$N_2H_4(l)$	51	149	121
$N_2H_3CH_3(l)$	54	180	166
$HNO_3(aq)$	−207	−111	146
$HNO_3(l)$	−174	−81	156
$NH_4ClO_4(s)$	−295	−89	186
$NH_4Cl(s)$	−314	−203	96
Oxygen			
$O_2(g)$	0	0	205
$O(g)$	249	232	161
$O_3(g)$	143	163	239
Phosphorus			
$P(s)$ (white)	0	0	41
$P(s)$ (red)	−18	−12	23
$P(s)$ (black)	−39	−33	23
$P_4(g)$	59	24	280
$PF_5(g)$	−1578	−1509	296
$PH_3(g)$	5	13	210
$H_3PO_4(s)$	−1279	−1119	110
$H_3PO_4(l)$	−1267	—	—
$H_3PO_4(aq)$	−1288	−1143	158
$P_4O_{10}(s)$	−2984	−2698	229
Potassium			
$K(s)$	0	0	64
$KCl(s)$	−436	−408	83
$KClO_3(s)$	−391	−290	143
$KClO_4(s)$	−433	−304	151
$K_2O(s)$	−361	−322	98
$K_2O_2(s)$	−496	−430	113
$KO_2(s)$	−283	−238	117
$KOH(s)$	−425	−379	79
$KOH(aq)$	−481	−440	9.20
Silicon			
$SiO_2(s)$ (quartz)	−911	−856	42
$SiCl_4(l)$	−687	−620	240
Silver			
$Ag(s)$	0	0	43

Substance and State	ΔH_f° (kJ/mol)	ΔG_f° (kJ/mol)	S° (J K^{-1} mol^{-1})
$Ag^+(aq)$	105	77	73
$AgBr(s)$	−100	−97	107
$AgCN(s)$	146	164	84
$AgCl(s)$	−127	−110	96
$Ag_2CrO_4(s)$	−712	−622	217
$AgI(s)$	−62	−66	115
$Ag_2O(s)$	−31	−11	122
$Ag_2S(s)$	−32	−40	146
Sodium			
$Na(s)$	0	0	51
$Na^+(aq)$	−240	−262	59
$NaBr(s)$	−360	−347	84
$Na_2CO_3(s)$	−1131	−1048	136
$NaHCO_3(s)$	−948	−852	102
$NaCl(s)$	−411	−384	72
$NaH(s)$	−56	−33	40
$NaI(s)$	−288	−282	91
$NaNO_2(s)$	−359	—	—
$NaNO_3(s)$	−467	−366	116
$Na_2O(s)$	−416	−377	73
$Na_2O_2(s)$	−515	−451	95
$NaOH(s)$	−427	−381	64
$NaOH(aq)$	−470	−419	50
Sulfur			
$S(s)$ (rhombic)	0	0	32
$S(s)$ (monoclinic)	0.3	0.1	33
$S^{2-}(aq)$	33	86	−15
$S_8(g)$	102	50	431
$SF_6(g)$	−1209	−1105	292
$H_2S(g)$	−21	−34	206
$SO_2(g)$	−297	−300	248
$SO_3(g)$	−396	−371	257
$SO_4^{2-}(aq)$	−909	−745	20
$H_2SO_4(l)$	−814	−690	157
$H_2SO_4(aq)$	−909	−745	20
Tin			
$Sn(s)$ (white)	0	0	52
$Sn(s)$ (gray)	−2	0.1	44
$SnO(s)$	−285	−257	56
$SnO_2(s)$	−581	−520	52
$Sn(OH)_2(s)$	−561	−492	155
Titanium			
$TiCl_4(g)$	−763	−727	355
$TiO_2(s)$	−945	−890	50
Uranium			
$U(s)$	0	0	50
$UF_6(s)$	−2137	−2008	228

Appendix Four (continued)

Substance and State	ΔH_f° (kJ/mol)	ΔG_f° (kJ/mol)	S° (J K^{-1} mol^{-1})
$UF_6(g)$	−2113	−2029	380
$UO_2(s)$	−1084	−1029	78
$U_3O_8(s)$	−3575	−3393	282
$UO_3(s)$	−1230	−1150	99
Xenon			
$Xe(g)$	0	0	170
$XeF_2(g)$	−108	−48	254
$XeF_4(s)$	−251	−121	146
$XeF_6(g)$	−294	—	—
$XeO_3(s)$	402	—	—
Zinc			
$Zn(s)$	0	0	42
$ZnO(s)$	−348	−318	44
$Zn(OH)_2(s)$	−642	—	—
$ZnS(s)$ (wurtzite)	−193	—	—
$ZnS(s)$ (zinc blende)	−206	−201	58
$ZnSO_4(s)$	−983	−874	120

APPENDIX FIVE Equilibrium Constants and Reduction Potentials

TABLE A5.1 Values of K_a for Some Common Monoprotic Acids

Name	Formula	Value of K_a
Hydrogen sulfate ion	HSO_4^-	1.2×10^{-2}
Chlorous acid	$HClO_2$	1.2×10^{-2}
Monochloracetic acid	$HC_2H_2ClO_2$	1.35×10^{-3}
Hydrofluoric acid	HF	7.2×10^{-4}
Nitrous acid	HNO_2	4.0×10^{-4}
Formic acid	HCO_2H	1.8×10^{-4}
Lactic acid	$HC_3H_5O_3$	1.38×10^{-4}
Benzoic acid	$HC_7H_5O_2$	6.4×10^{-5}
Acetic acid	$HC_2H_3O_2$	1.8×10^{-5}
Hydrated aluminum(III) ion	$[Al(H_2O)_6]^{3+}$	1.4×10^{-5}
Propanoic acid	$HC_3H_5O_2$	1.3×10^{-5}
Hypochlorous acid	$HOCl$	3.5×10^{-8}
Hypobromous acid	$HOBr$	2×10^{-9}
Hydrocyanic acid	HCN	6.2×10^{-10}
Boric acid	H_3BO_3	5.8×10^{-10}
Ammonium ion	NH_4^+	5.6×10^{-10}
Phenol	HOC_6H_5	1.6×10^{-10}
Hypoiodous acid	HOI	2×10^{-11}

TABLE A5.2 Stepwise Dissociation Constants for Several Common Polyprotic Acids

Name	Formula	K_{a_1}	K_{a_2}	K_{a_3}
Phosphoric acid	H_3PO_4	7.5×10^{-3}	6.2×10^{-8}	4.8×10^{-13}
Arsenic acid	H_3AsO_4	5×10^{-3}	8×10^{-8}	6×10^{-10}
Carbonic acid	H_2CO_3	4.3×10^{-7}	5.6×10^{-11}	
Sulfuric acid	H_2SO_4	Large	1.2×10^{-2}	
Sulfurous acid	H_2SO_3	1.5×10^{-2}	1.0×10^{-7}	
Hydrosulfuric acid	H_2S	1.0×10^{-7}	1.3×10^{-13}	
Oxalic acid	$H_2C_2O_4$	6.5×10^{-2}	6.1×10^{-5}	
Ascorbic acid (vitamin C)	$H_2C_6H_6O_6$	7.9×10^{-5}	1.6×10^{-12}	
Citric acid	$H_3C_6H_5O_7$	8.4×10^{-4}	1.8×10^{-5}	4.0×10^{-6}

TABLE A5.3 Values of K_b for Some Common Weak Bases

Name	Formula	Conjugate Acid	K_b
Ammonia	NH_3	NH_4^+	1.8×10^{-5}
Methylamine	CH_3NH_2	$CH_3NH_3^+$	4.38×10^{-4}
Ethylamine	$C_2H_5NH_2$	$C_2H_5NH_3^+$	5.6×10^{-4}
Diethylamine	$(C_2H_5)_2NH$	$(C_2H_5)_2NH_2^+$	1.3×10^{-3}
Triethylamine	$(C_2H_5)_3N$	$(C_2H_5)_3NH^+$	4.0×10^{-4}
Hydroxylamine	$HONH_2$	$HONH_3^+$	1.1×10^{-8}
Hydrazine	H_2NNH_2	$H_2NNH_3^+$	3.0×10^{-6}
Aniline	$C_6H_5NH_2$	$C_6H_5NH_3^+$	3.8×10^{-10}
Pyridine	C_5H_5N	$C_5H_5NH^+$	1.7×10^{-9}

TABLE A5.4 Values of K_{sp} at 25°C for Common Ionic Solids

Ionic Solid	K_{sp} (at 25°C)
Fluorides	
BaF_2	2.4×10^{-5}
MgF_2	6.4×10^{-9}
PbF_2	4×10^{-8}
SrF_2	7.9×10^{-10}
CaF_2	4.0×10^{-11}
Chlorides	
$PbCl_2$	1.6×10^{-5}
$AgCl$	1.6×10^{-10}
Hg_2Cl_2*	1.1×10^{-18}
Bromides	
$PbBr_2$	4.6×10^{-6}
$AgBr$	5.0×10^{-13}
Hg_2Br_2*	1.3×10^{-22}
Iodides	
PbI_2	1.4×10^{-8}
AgI	1.5×10^{-16}
Hg_2I_2*	4.5×10^{-29}
Sulfates	
$CaSO_4$	6.1×10^{-5}
Ag_2SO_4	1.2×10^{-5}
$SrSO_4$	3.2×10^{-7}
$PbSO_4$	1.3×10^{-8}
$BaSO_4$	1.5×10^{-9}
Chromates	
$SrCrO_4$	3.6×10^{-5}
Chromates (*continued*)	
Hg_2CrO_4*	2×10^{-9}
$BaCrO_4$	8.5×10^{-11}
Ag_2CrO_4	9.0×10^{-12}
$PbCrO_4$	2×10^{-16}
Carbonates	
$NiCO_3$	1.4×10^{-7}
$CaCO_3$	8.7×10^{-9}
$BaCO_3$	1.6×10^{-9}
$SrCO_3$	7×10^{-10}
$CuCO_3$	2.5×10^{-10}
$ZnCO_3$	2×10^{-10}
$MnCO_3$	8.8×10^{-11}
$FeCO_3$	2.1×10^{-11}
Ag_2CO_3	8.1×10^{-12}
$CdCO_3$	5.2×10^{-12}
$PbCO_3$	1.5×10^{-15}
$MgCO_3$	1×10^{-15}
Hg_2CO_3*	9.0×10^{-15}
Hydroxides	
$Ba(OH)_2$	5.0×10^{-3}
$Sr(OH)_2$	3.2×10^{-4}
$Ca(OH)_2$	1.3×10^{-6}
$AgOH$	2.0×10^{-8}
$Mg(OH)_2$	8.9×10^{-12}
$Mn(OH)_2$	2×10^{-13}
$Cd(OH)_2$	5.9×10^{-15}
$Pb(OH)_2$	1.2×10^{-15}
$Fe(OH)_2$	1.8×10^{-15}
Hydroxides (*continued*)	
$Co(OH)_2$	2.5×10^{-16}
$Ni(OH)_2$	1.6×10^{-16}
$Zn(OH)_2$	4.5×10^{-17}
$Cu(OH)_2$	1.6×10^{-19}
$Hg(OH)_2$	3×10^{-26}
$Sn(OH)_2$	3×10^{-27}
$Cr(OH)_3$	6.7×10^{-31}
$Al(OH)_3$	2×10^{-32}
$Fe(OH)_3$	4×10^{-38}
$Co(OH)_3$	2.5×10^{-43}
Sulfides	
MnS	2.3×10^{-13}
FeS	3.7×10^{-19}
NiS	3×10^{-21}
CoS	5×10^{-22}
ZnS	2.5×10^{-22}
SnS	1×10^{-26}
CdS	1.0×10^{-28}
PbS	7×10^{-29}
CuS	8.5×10^{-45}
Ag_2S	1.6×10^{-49}
HgS	1.6×10^{-54}
Phosphates	
Ag_3PO_4	1.8×10^{-18}
$Sr_3(PO_4)_2$	1×10^{-31}
$Ca_3(PO_4)_2$	1.3×10^{-32}
$Ba_3(PO_4)_2$	6×10^{-39}
$Pb_3(PO_4)_2$	1×10^{-54}

* Contains Hg_2^{2+} ions. $K_{sp} = [Hg_2^{2+}][X^-]^2$ for Hg_2X_2 salts.

TABLE A5.5 **Standard Reduction Potentials at 25°C (298 K) for Many Common Half-Reactions**

Half-Reaction	$\mathcal{E}^\circ$ (V)	Half-Reaction	$\mathcal{E}^\circ$ (V)
$F_2 + 2e^- \longrightarrow 2F^-$	2.87	$O_2 + 2H_2O + 4e^- \longrightarrow 4OH^-$	0.40
$Ag^{2+} + e^- \longrightarrow Ag^+$	1.99	$Cu^{2+} + 2e^- \longrightarrow Cu$	0.34
$Co^{3+} + e^- \longrightarrow Co^{2+}$	1.95	$Hg_2Cl_2 + 2e^- \longrightarrow 2Hg + 2Cl^-$	0.34
$H_2O_2 + 2H^+ + 2e^- \longrightarrow 2H_2O$	1.78	$AgCl + e^- \longrightarrow Ag + Cl^-$	0.22
$Ce^{4+} + e^- \longrightarrow Ce^{3+}$	1.70	$SO_4^{2-} + 4H^+ + 2e^- \longrightarrow H_2SO_3 + H_2O$	0.20
$PbO_2 + 4H^+ + SO_4^{2-} + 2e^- \longrightarrow PbSO_4 + 2H_2O$	1.69	$Cu^{2+} + e^- \longrightarrow Cu^+$	0.16
$MnO_4^- + 4H^+ + 3e^- \longrightarrow MnO_2 + 2H_2O$	1.68	$2H^+ + 2e^- \longrightarrow H_2$	0.00
$2e^- + 2H^+ + IO_4^- \longrightarrow IO_3^- + H_2O$	1.60	$Fe^{3+} + 3e^- \longrightarrow Fe$	−0.036
$MnO_4^- + 8H^+ + 5e^- \longrightarrow Mn^{2+} + 4H_2O$	1.51	$Pb^{2+} + 2e^- \longrightarrow Pb$	−0.13
$Au^{3+} + 3e^- \longrightarrow Au$	1.50	$Sn^{2+} + 2e^- \longrightarrow Sn$	−0.14
$PbO_2 + 4H^+ + 2e^- \longrightarrow Pb^{2+} + 2H_2O$	1.46	$Ni^{2+} + 2e^- \longrightarrow Ni$	−0.23
$Cl_2 + 2e^- \longrightarrow 2Cl^-$	1.36	$PbSO_4 + 2e^- \longrightarrow Pb + SO_4^{2-}$	−0.35
$Cr_2O_7^{2-} + 14H^+ + 6e^- \longrightarrow 2Cr^{3+} + 7H_2O$	1.33	$Cd^{2+} + 2e^- \longrightarrow Cd$	−0.40
$O_2 + 4H^+ + 4e^- \longrightarrow 2H_2O$	1.23	$Fe^{2+} + 2e^- \longrightarrow Fe$	−0.44
$MnO_2 + 4H^+ + 2e^- \longrightarrow Mn^{2+} + 2H_2O$	1.21	$Cr^{3+} + e^- \longrightarrow Cr^{2+}$	−0.50
$IO_3^- + 6H^+ + 5e^- \longrightarrow \frac{1}{2}I_2 + 3H_2O$	1.20	$Cr^{3+} + 3e^- \longrightarrow Cr$	−0.73
$Br_2 + 2e^- \longrightarrow 2Br^-$	1.09	$Zn^{2+} + 2e^- \longrightarrow Zn$	−0.76
$VO_2^+ + 2H^+ + e^- \longrightarrow VO^{2+} + H_2O$	1.00	$2H_2O + 2e^- \longrightarrow H_2 + 2OH^-$	−0.83
$AuCl_4^- + 3e^- \longrightarrow Au + 4Cl^-$	0.99	$Mn^{2+} + 2e^- \longrightarrow Mn$	−1.18
$NO_3^- + 4H^+ + 3e^- \longrightarrow NO + 2H_2O$	0.96	$Al^{3+} + 3e^- \longrightarrow Al$	−1.66
$ClO_2 + e^- \longrightarrow ClO_2^-$	0.954	$H_2 + 2e^- \longrightarrow 2H^-$	−2.23
$2Hg^{2+} + 2e^- \longrightarrow Hg_2^{2+}$	0.91	$Mg^{2+} + 2e^- \longrightarrow Mg$	−2.37
$Ag^+ + e^- \longrightarrow Ag$	0.80	$La^{3+} + 3e^- \longrightarrow La$	−2.37
$Hg_2^{2+} + 2e^- \longrightarrow 2Hg$	0.80	$Na^+ + e^- \longrightarrow Na$	−2.71
$Fe^{3+} + e^- \longrightarrow Fe^{2+}$	0.77	$Ca^{2+} + 2e^- \longrightarrow Ca$	−2.76
$O_2 + 2H^+ + 2e^- \longrightarrow H_2O_2$	0.68	$Ba^{2+} + 2e^- \longrightarrow Ba$	−2.90
$MnO_4^- + e^- \longrightarrow MnO_4^{2-}$	0.56	$K^+ + e^- \longrightarrow K$	−2.92
$I_2 + 2e^- \longrightarrow 2I^-$	0.54	$Li^+ + e^- \longrightarrow Li$	−3.05
$Cu^+ + e^- \longrightarrow Cu$	0.52		

GLOSSARY

Note to the Student: The Glossary includes brief definitions of some of the fundamental terms used in chemistry. It does not include complex concepts that require detailed explanation for understanding. Please refer to the appropriate sections of the text for complete discussion of particular topics or concepts.

Accuracy the agreement of a particular value with the true value. (A1.5)

Acid a substance that produces hydrogen ions in solution; a proton donor. (4.2)

Acid-base indicator a substance that marks the end point of an acid-base titration by changing color. (8.5)

Acid dissociation constant (K_a) the equilibrium constant for a reaction in which a proton is removed from an acid by H_2O to form the conjugate base and H_3O^+. (7.1)

Acid rain a result of air pollution by sulfur dioxide. (5.11)

Acidic oxide a covalent oxide that dissolves in water to give an acidic solution. (19.3)

Actinide series a group of fourteen elements following actinium in the periodic table, in which the $5f$ orbitals are being filled. (12.13; 18.1)

Activated complex (transition state) the arrangement of atoms found at the top of the potential energy barrier as a reaction proceeds from reactants to products. (15.8)

Activation energy the threshold energy that must be overcome to produce a chemical reaction. (15.8)

Addition polymerization a type of polymerization in which the monomers simply add together to form the polymer, with no other products. (22.5)

Addition reaction a reaction in which atoms add to a carbon-carbon multiple bond. (22.2)

Adsorption the collection of one substance on the surface of another. (15.9)

Air pollution contamination of the atmosphere, mainly by the gaseous products of transportation and production of electricity. (5.11)

Alcohol an organic compound in which the hydroxyl group is a substituent on a hydrocarbon. (22.4)

Aldehyde an organic compound containing the carbonyl group bonded to at least one hydrogen atom. (22.4)

Alkali metal a Group 1A metal. (2.8; 18.2)

Alkaline earth metal a Group 2A metal. (2.8; 18.4)

Alkane a saturated hydrocarbon with the general formula C_nH_{2n+2}. (22.1)

Alkene an unsaturated hydrocarbon containing a carbon-carbon double bond. The general formula is C_nH_{2n}. (22.2)

Alkyne an unsaturated hydrocarbon containing a triple carbon-carbon bond. The general formula is C_nH_{2n-2}. (22.2)

Alloy a substance that contains a mixture of elements and has metallic properties. (16.4)

Alloy steel a form of steel containing carbon plus other metals such as chromium, cobalt, manganese, and molybdenum. (20.2)

Alpha (α) particle a helium nucleus. (21.1)

Alpha particle production a common mode of decay for radioactive nuclides in which the mass number changes. (21.1)

Amine an organic base derived from ammonia in which one or more of the hydrogen atoms are replaced by organic groups. (7.6; 22.4)

α-Amino acid an organic acid in which an amino group and an R group are attached to the carbon atom next to the carboxyl group. (23.1)

Amorphous solid a solid with considerable disorder in its structure. (16.3)

Ampere the unit of electrical current equal to one coulomb of charge per second. (11.7)

Amphoteric substance a substance that can behave either as an acid or as a base. (7.2)

Angular momentum quantum number (ℓ) the quantum number relating to the shape of an atomic orbital, which can assume any integral value from 0 to $n - 1$ for each value of n. (12.7)

Anion a negative ion. (2.9)

Anode the electrode in a galvanic cell at which oxidation occurs. (11.1)

Antibonding molecular orbital an orbital higher in energy than the atomic orbitals of which it is composed. (14.2)

Aqueous solution a solution in which water is the dissolving medium or solvent. (4.0)

Aromatic hydrocarbon one of a special class of cyclic unsaturated hydrocarbons, the simplest of which is benzene. (22.3)

Arrhenius concept a concept postulating that acids produce hydrogen ions in aqueous solution, while bases produce hydroxide ions. (7.1)

Arrhenius equation the equation representing the rate constant as $k = Ae^{-E_a/RT}$ where A represents the product of the collision frequency and the steric factor, and $e^{-E_a/RT}$ is the fraction of collisions with sufficient energy to produce a reaction. (15.8)

Atmosphere the mixture of gases that surrounds the earth's surface. (5.11)

Atomic number the number of protons in the nucleus of an atom. (2.6)

Atomic radius half the distance between the nuclei in a molecule consisting of identical atoms. (12.15)

Atomic solid a solid that contains atoms at the lattice points. (16.3)

Atomic mass (average) the weighted average mass of the atoms in a naturally occurring element. (2.3)

Aufbau principle the principle stating that as protons are added one by one to the nucleus to build up the elements, electrons are similarly added to hydrogenlike orbitals. (12.13)

Autoionization the transfer of a proton from one molecule to another of the same substance. (7.2)

Avogadro's law equal volumes of gases at the same temperature and pressure contain the same number of particles. (5.2)

Avogadro's number the number of atoms in exactly 12 grams of pure ^{12}C, equal to 6.022×10^{23}. (3.2)

Ball-and-stick model a molecular model that distorts the sizes of atoms, but shows bond relationships clearly. (2.7)

Band model a molecular model for metals in which the electrons are assumed to travel around the metal crystal in molecular orbitals formed from the valence atomic orbitals of the metal atoms. (16.4)

Barometer a device for measuring atmospheric pressure. (5.1)

Base a substance that produces hydroxide ions in aqueous solution, a proton acceptor. (7.2)

Base dissociation constant (K_b) the equilibrium constant for the reaction of a base with water to produce the conjugate acid and hydroxide ion. (7.6)

Basic oxide an ionic oxide that dissolves in water to produce a basic solution. (18.4)

Battery a group of galvanic cells connected in series. (11.5)

Beta (β) particle an electron produced in radioactive decay. (21.1)

Beta particle production a decay process for radioactive nuclides in which the mass number remains constant and the atomic number changes. The net effect is to change a neutron to a proton. (21.1)

Bidentate ligand a ligand that can form two bonds to a metal ion. (20.3)

Bilayer a portion of the cell membrane consisting of phospholipids with their nonpolar tails in the interior and their polar heads interfacing with the polar water molecules. (23.4)

Bimolecular step a reaction involving the collision of two molecules. (15.6)

Binary compound a two-element compound. (2.9)

Binding energy (nuclear) the energy required to decompose a nucleus into its component nucleons. (21.5)

Biochemistry the study of the chemistry of living systems. (23)

Biomolecule a molecule responsible for maintaining and/or reproducing life. (22)

Bond energy the energy required to break a given chemical bond. (13.1)

Bond length the distance between the nuclei of the two atoms connected by a bond; the distance where the total energy of a diatomic molecule is minimal. (13.1)

Bond order the difference between the number of bonding electrons and the number of antibonding electrons, divided by two. It is an index of bond strength. (14.2)

Bonding molecular orbital an orbital lower in energy than the atomic orbitals of which it is composed. (14.2)

Bonding pair an electron pair found in the space between two atoms. (13.9)

Borane a covalent hydride of boron. (18.5)

Boyle's law the volume of a given sample of gas at constant temperature varies inversely with the pressure. (5.2)

Breeder reactor a nuclear reactor in which fissionable fuel is produced while the reactor runs. (21.6)

Brønsted-Lowry definition (Model) a model proposing that an acid is a proton donor, and a base is a proton acceptor. (7.1)

Buffer capacity the ability of a buffered solution to absorb protons or hydroxide ions without a significant change in pH; determined by the magnitudes of [HA] and $[A^-]$ in the solution. (8.3)

Buffered solution a solution that resists a change in its pH when either hydroxide ions or protons are added. (8.2)

Calorimetry the science of measuring heat flow. (9.4)

Capillary action the spontaneous rising of a liquid in a narrow tube. (16.2)

Carbohydrate a polyhydroxyl ketone or polyhydroxyl aldehyde or a polymer composed of these. (23.2)

Carboxyhemoglobin a stable complex of hemoglobin and carbon monoxide that prevents normal oxygen uptake in the blood. (20.8)

Carboxyl group the —COOH group in an organic acid. (7.2)

Carboxylic acid an organic compound containing the carboxyl group; an acid with the general formula RCOOH. (22.4)

Catalyst a substance that speeds up a reaction without being consumed. (15.9)

Cathode the electrode in a galvanic cell at which reduction occurs. (11.1)

Cathode rays the "rays" emanating from the negative electrode (cathode) in a partially evacuated tube; a stream of electrons. (2.5)

Cathodic protection a method in which an active metal, such as magnesium, is connected to steel in order to protect it from corrosion. (11.7)

Cation a positive ion. (2.7)

Cell potential (electromotive force) the driving force in a galvanic cell that pulls electrons from the reducing agent in one compartment to the oxidizing agent in the other. (11.1)

Ceramic a nonmetallic material made from clay and hardened by firing at high temperature; it contains minute silicate crystals suspended in a glassy cement. (16.5)

Chain reaction (nuclear) a self-sustaining fission process caused by the production of neutrons that proceed to split other nuclei. (21.6)

Charles's law the volume of a given sample of gas at constant pressure is directly proportional to the temperature in kelvins. (5.2)

Chelating ligand (chelate) a ligand having more than one atom with a lone pair that can be used to bond to a metal ion. (20.3)

Chemical bond the energy that holds two atoms together in a compound. (2.7)

Chemical change the change of substances into other substances through a reorganization of the atoms; a chemical reaction. (2.2)

Chemical equation a representation of a chemical reaction showing the relative numbers of reactant and product molecules. (3.6)

Chemical equilibrium a dynamic reaction system in which the concentrations of all reactants and products remain constant as a function of time. (6)

Chemical formula the representation of a molecule in which the symbols for the elements are used to indicate the types of atoms present and subscripts are used to show the relative numbers of atoms. (2.7)

Chemical kinetics the area of chemistry that concerns reaction rates. (15.1)

Chemical stoichiometry the calculation of the quantities of material consumed and produced in chemical reactions. (3)

Chirality the quality of having nonsuperimposable mirror images. (20.4)

Chlor-alkali process the process for producing chlorine and sodium hydroxide by electrolyzing brine in a mercury cell. (11.8)

Coagulation the destruction of a colloid by causing particles to aggregate and settle out. (17.8)

Codons organic bases in sets of three that form the genetic code. (23.3)

Colligative properties properties of a solution that depend on the number, and not on the identity, of the solute particles. (17.5)

Collision model a model based on the idea that molecules must collide to react; used to account for the observed characteristics of reaction rates. (15.8)

Colloid a suspension of particles in a dispersing medium. (17.8)

Combustion reaction the vigorous and exothermic reaction that takes place between certain substances, particularly organic compounds, and oxygen. (22.1)

Common ion effect the shift in an equilibrium position caused by the addition or presence of an ion involved in the equilibrium reaction. (8.1)

Complete ionic equation an equation that shows all substances that are strong electrolytes as ions. (4.6)

Complex ion a charged species consisting of a metal ion surrounded by ligands. (8.9; 20.1)

Compound a substance with constant composition that can be broken down into elements by chemical processes.

Concentration cell a galvanic cell in which both compartments contain the same components, but at different concentrations. (11.4)

Condensation the process by which vapor molecules reform a liquid. (16.10)

Condensation polymerization a type of polymerization in which the formation of a small molecule, such as water, accompanies the extension of the polymer chain. (22.5)

Condensation reaction a reaction in which two molecules are joined, accompanied by the elimination of a water molecule. (19.3)

Condensed states of matter liquids and solids. (16.1)

Conduction bands the molecular orbitals that can be occupied by mobile electrons, which are free to travel throughout a metal crystal to conduct electricity or heat. (16.4)

Conjugate acid the species formed when a proton is added to a base. (7.1)

Conjugate acid-base pair two species related to each other by the donating and accepting of a single proton. (7.1)

Conjugate base what remains of an acid molecule after a proton is lost. (7.1)

Continuous spectrum a spectrum that exhibits all the wavelengths of visible light. (12.3)

Control rods rods in a nuclear reactor composed of substances that absorb neutrons. These rods regulate the power level of the reactor. (21.6)

Coordinate covalent bond a metal-ligand bond resulting from the interaction of a Lewis base (the ligand) and a Lewis acid (the metal ion). (20.3)

Coordination compound a compound composed of a complex ion and counter ions sufficient to give no net charge. (20.3)

Coordination isomerism isomerism in a coordination compound in which the composition of the coordination sphere of the metal ion varies. (20.4)

Coordination number the number of bonds formed between the metal ion and the ligands in a complex ion. (20.3)

Copolymer a polymer formed from the polymerization of more than one type of monomer. (22.5)

Core electron an inner electron in an atom; one not in the outermost (valence) principal quantum level. (12.13)

Corrosion the process by which metals are oxidized in the atmosphere. (11.6)

Coulomb's law $E = 2.31 \times 10^{-19} (Q_1Q_2/r)$, where E is the energy of interaction between a pair of ions, expressed in joules; r is the distance between the ion centers in nm; and Q_1 and Q_2 are the numerical ion charges. (13.1)

Counter ions anions or cations that balance the charge on the complex ion in a coordination compound. (20.3)

Covalent bonding a type of bonding in which electrons are shared by atoms. (13.1)

Critical mass the mass of fissionable material required to produce a self-sustaining chain reaction. (21.6)

Critical point the point on a phase diagram at which the temperature and pressure have their critical values; the end point of the liquid-vapor line. (16.11)

Critical pressure the minimum pressure required to produce liquefaction of a substance at the critical temperature. (16.11)

Critical reaction (nuclear) a reaction in which exactly one neutron from each fission event causes another fission event, thus sustaining the chain reaction. (21.6)

Critical temperature the temperature above which vapor cannot be liquefied, no matter what pressure is applied. (16.11)

Crosslinking the existence of bonds between adjacent chains in a polymer, thus adding strength to the material. (22.5)

Crystal field model a model used to explain the magnetism and colors of coordination complexes through the splitting of the *d* orbital energies. (20.6)

Crystalline solid a solid with a regular arrangement of its components. (16.3)

Cubic closest packed structure a solid modeled by the closest packing of spheres with an *abcabc* arrangement of layers; the unit cell is face-centered cubic. (16.4)

Cyclotron a type of particle accelerator in which an ion introduced at the center is accelerated in an expanding spiral path by use of alternating electrical fields in the presence of a magnetic field. (21.3)

Cytochromes a series of iron-containing species composed of heme and a protein. Cytochromes are the principal electron-transfer molecules in the respiratory chain. (20.8)

Dalton's law of partial pressures for a mixture of gases in a container, the total pressure exerted is the sum of the pressures that each gas would exert if it were alone. (5.5)

Degenerate orbitals a group of orbitals with the same energy. (12.9)

Dehydrogenation reaction a reaction in which two hydrogen atoms are removed from adjacent carbons of a saturated hydrocarbon, giving an unsaturated hydrocarbon. (22.1)

Delocalization the condition where the electrons in a molecule are not localized between a pair of atoms but can move throughout the molecule. (13.9)

Denaturation the breaking down of the three-dimensional structure of a protein resulting in the loss of its function. (23.1)

Denitrification the return of nitrogen from decomposed matter to the atmosphere by bacteria that change nitrates to nitrogen gas. (19.2)

Density a property of matter representing the mass per unit volume.

Deoxyribonucleic acid (DNA) a huge nucleotide polymer having a double helical structure with complementary bases on the two strands. Its major functions are protein synthesis and the storage and transport of genetic information. (23.3)

Desalination the removal of dissolved salts from an aqueous solution. (17.6)

Dialysis a phenomenon in which a semipermeable membrane allows transfer of both solvent molecules and small solute molecules and ions. (17.6)

Diamagnetism a type of magnetism, associated with paired electrons, that causes a substance to be repelled from the inducing magnetic field. (14.3)

Differential rate law an expression that gives the rate of a reaction as a function of concentrations; often called the rate law. (15.2)

Diffraction the scattering of light from a regular array of points or lines, producing constructive and destructive interference. (12.2)

Diffusion the mixture of gases. (5.7)

Dilution the process of adding solvent to lower the concentration of solute in a solution. (4.3)

Dimer a molecule formed by the joining of two identical monomers. (22.5)

Dipole-dipole attraction the attractive force resulting when polar molecules line up so that the positive and negative ends are close to each other. (16.1)

Dipole moment a property of a molecule whose charge distribution can be represented by a center of positive charge and a center of negative charge. (13.3)

Disaccharide a sugar formed from two monosaccharides joined by a glycoside linkage. (23.2)

Disproportionation reaction a reaction in which a given element is both oxidized and reduced. (19.7)

Disulfide linkage a S—S bond that stabilizes the tertiary structure of many proteins. (23.1)

Double bond a bond in which two pairs of electrons are shared by two atoms. (13.8)

Downs cell a cell used for electrolyzing molten sodium chloride. (11.8)

Dry cell battery a common battery used in calculators, watches, radios, and tape players. (11.5)

Dual nature of light the statement that light exhibits both wave and particulate properties. (12.2)

$E = mc^2$ Einstein's equation proposing that energy has mass; E is energy, m is mass, and c is the speed of light. (12.2)

Effective nuclear charge the apparent nuclear charge exerted on a particular electron, equal to the actual nuclear charge minus the effect of electron repulsions. (12.11)

Effusion the passage of a gas through a tiny orifice into an evacuated chamber. (5.7)

Electrical conductivity the ability to conduct an electric current. (4.2)

Electrochemistry the study of the interchange of chemical and electrical energy. (11)

Electrolysis a process that involves forcing a current through a cell to cause a nonspontaneous chemical reaction to occur. (11.7)

Electrolyte a material that dissolves in water to give a solution that conducts an electric current. (4.2)

Electrolytic cell a cell that uses electrical energy to produce a chemical change that would otherwise not occur spontaneously. (11.7)

Electromagnetic radiation radiant energy that exhibits wavelike behavior and travels through space at the speed of light in a vacuum. (12.1)

Electron a negatively charged particle that moves around the nucleus of an atom. (2.5)

Electron affinity the energy change associated with the addition of an electron to a gaseous atom. (12.15)

Electron capture a process in which one of the inner-orbital electrons in an atom is captured by the nucleus. (21.1)

Electron sea model a model for metals postulating a regular array of cations in a "sea" of electrons. (16.4)

Electron spin quantum number a quantum number representing one of the two possible values for the electron spin; either $+\frac{1}{2}$ or $-\frac{1}{2}$. (12.10)

Electronegativity the tendency of an atom in a molecule to attract shared electrons to itself. (13.2)

Element a substance that cannot be decomposed into simpler substances by chemical or physical means. (2.1)

Elementary step a reaction whose rate law can be written from its molecularity. (15.6)

Empirical formula the simplest whole number ratio of atoms in a compound. (3.5)

Enantiomers isomers that are nonsuperimposable mirror images of each other. (20.4)

End point the point in a titration at which the indicator changes color. (4.9)

Endothermic refers to a reaction where energy (as heat) flows into the system. (9.1)

Energy the capacity to do work or to cause heat flow. (9.1)

Enthalpy a property of a system equal to $E + PV$, where E is the internal energy of the system, P is the pressure of the system, and V is the volume of the system. At constant pressure, where only PV work is allowed, the change in enthalpy equals the energy flow as heat. (9.2)

Enthalpy of fusion the enthalpy change that occurs to melt a solid at its melting point. (16.10)

Entropy a thermodynamic function that measures randomness or disorder. (10.1)

Enzyme a large molecule, usually a protein, that catalyzes biological reactions. (15.9; 23.1)

Equilibrium (thermodynamic definition) the position where the free energy of a reaction system has its lowest possible value. (10.11)

Equilibrium constant the value obtained when equilibrium concentrations of the chemical species are substituted in the equilibrium expression. (6.2)

Equilibrium expression the expression (from the law of mass action) obtained by multiplying the product concentrations and dividing by the multiplied reactant concentrations, with each concentration raised to a power represented by the coefficient in the balanced equation. (6.2)

Equilibrium position a particular set of equilibrium concentrations. (6.2)

Equivalence point (stoichiometric point) the point in a titration when enough titrant has been added to react exactly with the substance in solution being titrated. (4.9; 8.4)

Essential elements the elements definitely known to be essential to human life. (23)

Ester an organic compound produced by the reaction between a carboxylic acid and an alcohol. (22.4)

Exothermic refers to a reaction where energy (as heat) flows out of the system. (9.1)

Exponential notation expresses a number as $N \times 10^M$, a convenient method for representing a very large or very small number and for easily indicating the number of significant figures. (A1.1)

Faraday a constant representing the charge on one mole of electrons; 96,485 coulombs. (11.3)

Fat (glyceride) an ester composed of glycerol and fatty acids. (23.4)

Fatty acid a long-chain carboxylic acid. (23.4)

First law of thermodynamics the energy of the universe is constant; same as the law of conservation of energy. (9.1)

Fission the process of using a neutron to split a heavy nucleus into two nuclei with smaller mass numbers. (21.6)

Formal charge the charge assigned to an atom in a molecule or polyatomic ion derived from a specific set of rules. (13.12)

Formation constant (stability constant) the equilibrium constant for each step of the formation of a complex ion by the addition of an individual ligand to a metal ion or complex ion in aqueous solution. (8.9)

Fossil fuel coal, petroleum, or natural gas; consists of carbon-based molecules derived from decomposition of once-living organisms. (9.7)

Frasch process the recovery of sulfur from underground deposits by melting it with hot water and forcing it to the surface by air pressure. (19.6)

Free energy a thermodynamic function equal to the enthalpy (H) minus the product of the entropy (S) and the Kelvin temperature (T); $G = H - TS$. Under certain conditions the change in free energy for a process is equal to the maximum useful work. (10.7)

Free radical a species with an unpaired electron. (22.5)

Frequency the number of waves (cycles) per second that pass a given point in space. (12.1)

Fuel cell a galvanic cell for which the reactants are continuously supplied. (11.5)

Functional group an atom or group of atoms in hydrocarbon derivatives that contains elements in addition to carbon and hydrogen. (22.4)

Fusion the process of combining two light nuclei to form a heavier, more stable nucleus. (21.6)

Galvanic cell a device in which chemical energy from a spontaneous redox reaction is changed to electrical energy that can be used to do work. (11.1)

Galvanizing a process in which steel is coated with zinc to prevent corrosion. (11.6)

Gamma (γ) ray a high-energy photon. (21.1)

Geiger-Müller counter (Geiger counter) an instrument that measures the rate of radioactive decay based on the ions and electrons produced as a radioactive particle passes through a gas-filled chamber. (21.4)

Gene a given segment of the DNA molecule that contains the code for a specific protein. (23.3)

Geometrical *(cis-trans)* isomerism isomerism in which atoms or groups of atoms can assume different positions around a rigid ring or bond. (20.4; 22.2)

Glass an amorphous solid obtained when silica is mixed with other compounds, heated above its melting point, and then cooled rapidly. (16.5)

Glass electrode an electrode for measuring pH from the potential difference that develops when it is dipped into an aqueous solution containing H^+ ions. (11.4)

Glycosidic linkage a C—O—C bond formed between the rings of two cyclic monosaccharides by the elimination of water. (23.2)

Graham's law of effusion the rate of effusion of a gas is inversely proportional to the square root of the mass of its particles. (5.7)

Gravimetric analysis a method for determining the amount of a given substance in a solution by precipitation, filtration, drying, and weighing. (4.8)

Greenhouse effect a warming effect exerted by the earth's atmosphere (particularly CO_2 and H_2O) due to thermal energy retained by absorption of infrared radiation. (9.7)

Ground state the lowest possible energy state of an atom or molecule. (12.4)

Group (of the periodic table) a vertical column of elements having the same valence electron configuration and showing similar properties. (2.8)

Haber process the manufacture of ammonia from nitrogen and hydrogen, carried out at high pressure and high temperature with the aid of a catalyst. (3.9; 19.2)

Half-life (of a radioactive sample) the time required for the number of nuclides in a radioactive sample to reach half of the original value. (21.2)

Half-life (of a reaction) the time required for a reactant to reach half of its original concentration. (15.4)

Half-reactions the two parts of an oxidation-reduction reaction, one representing oxidation, the other reduction. (4.11; 11.1)

Halogen a Group 7A element. (2.8; 19.7)

Halogenation the addition of halogen atoms to unsaturated hydrocarbons. (22.2)

Hard water water from natural sources that contains relatively large concentrations of calcium and magnesium ions. (18.4)

Heat energy transferred between two objects due to a temperature difference between them. (9.1)

Heat capacity the amount of energy required to raise the temperature of an object by one degree Celsius. (9.4)

Heat of fusion the enthalpy change that occurs to melt a solid at its melting point. (16.10)

Heat of hydration the enthalpy change associated with placing gaseous molecules or ions in water; the sum of the energy needed to expand the solvent and the energy released from the solvent-solute interactions. (17.2)

Heat of solution the enthalpy change associated with dissolving a solute in a solvent; the sum of the energies needed to expand both solvent and solute in a solution and the energy released from the solvent-solute interactions. (17.2)

Heat of vaporization the energy required to vaporize one mole of a liquid at a pressure of one atmosphere. (16.10)

Heating curve a plot of temperature versus time for a substance where energy is added at a constant rate. (16.10)

Heisenberg uncertainty principle a principle stating that there is a fundamental limitation to how precisely both the position and momentum of a particle can be known at a given time. (12.5)

Heme an iron complex. (20.8)

Hemoglobin a biomolecule composed of four myoglobinlike units (proteins plus heme) that can bind and transport four oxygen molecules in the blood. (20.8)

Henderson-Hasselbalch equation an equation giving the relationship between the pH of an acid-base system and the concentrations of base and acid

$$pH = pK_a + \log\left(\frac{[base]}{[acid]}\right). \quad (8.2)$$

Henry's law the amount of a gas dissolved in a solution is directly proportional to the pressure of the gas above the solution. (17.3)

Hess's law in going from a particular set of reactants to a particular set of products, the enthalpy change is the same whether the reaction takes place in one step or in a series of steps; in summary, enthalpy is a state function. (9.5)

Heterogeneous equilibrium an equilibrium involving reactants and/or products in more than one phase. (6.5)

Hexagonal closest packed structure a structure composed of closest packed spheres with an *ababab* arrangement of layers; the unit cell is hexagonal. (16.4)

Homogeneous equilibrium an equilibrium system where all reactants and products are in the same phase. (6.5)

Homopolymer a polymer formed from the polymerization of only one type of monomer. (22.5)

Hund's rule the lowest-energy configuration for an atom is the one having the maximum number of unpaired electrons allowed by the Pauli exclusion principle in a particular set of degenerate orbitals, with all unpaired electrons having parallel spins. (12.13)

Hybrid orbitals a set of atomic orbitals adopted by an atom in a molecule different from those of the atom in the free state. (14.1)

Hybridization a mixing of the native orbitals on a given atom to form special atomic orbitals for bonding. (14.1)

Hydration the interaction between solute particles and water molecules. (4.1)

Hydride a binary compound containing hydrogen. The hydride ion, H^-, exists in ionic hydrides. The three classes of hydrides are covalent, interstitial, and ionic. (18.3)

Hydrocarbon a compound composed of carbon and hydrogen. (22.1)

Hydrocarbon derivative an organic molecule that contains one or more elements in addition to carbon and hydrogen. (22.4)

Hydrogen bonding unusually strong dipole-dipole attractions that occur among molecules in which hydrogen is bonded to a highly electronegative atom. (16.1)

Hydrogenation reaction a reaction in which hydrogen is added, with a catalyst present, to a carbon-carbon multiple bond. (22.2)

Hydrohalic acid an aqueous solution of a hydrogen halide. (19.7)

Hydronium ion the H_3O^+ ion; a hydrated proton. (7.1)

Hypothesis one or more assumptions put forth to explain the observed behavior of nature. (1.3)

Ideal gas a gas that obeys the equation, $PV = nRT$ (5.2)

Ideal gas law an equation of state for a gas, where the state of the gas is its condition at a given time; expressed by $PV = nRT$, where P = pressure, V = volume, n = moles of the gas, R = the universal gas constant, and T = absolute temperature. This equation expresses behavior approached by real gases at high T and low P. (5.3)

Ideal solution a solution whose vapor pressure is directly proportional to the mole fraction of solvent present. (17.4)

Indicator a chemical that changes color and is used to mark the end point of a titration. (4.9; 8.5)

Inert pair effect the tendency for the heavier Group 3A elements to exhibit the +1 as well as the expected +3 oxidation states, and Group 4A elements to exhibit the +2 as well as the +4 oxidation states. (18.5)

Integrated rate law an expression that shows the concentration of a reactant as a function of time. (15.4)

Interhalogen compound a compound formed by the reaction of one halogen with another. (19.7)

Intermediate a species that is neither a reactant nor a product but that is formed and consumed in the reaction sequence. (15.6)

Intermolecular forces relatively weak interactions that occur between molecules. (16.1)

Internal energy a property of a system that can be changed by a flow of work, heat or both; $\Delta E = q + w$, where ΔE is the change in the internal energy of the system, q is heat, and w is work. (9.1)

Ion an atom or a group of atoms that has a net positive or negative charge. (2.7)

Ion exchange (water softening) the process in which an ion-exchange resin removes unwanted ions (for example, Ca^{2+} and Mg^{2+}) and replaces them with Na^+ ions, which do not interfere with soap and detergent action. (18.4)

Ion pairing a phenomenon occurring in solution when oppositely charged ions aggregate and behave as a single particle. (17.7)

Ion-product constant (K_w) the equilibrium constant for the autoionization of water; $K_w = [H^+][OH^-]$. At 25°C, K_w equals 1.0×10^{-14}. (7.2)

Ion-selective electrode an electrode sensitive to the concentration of a particular ion in solution. (11.4)

Ionic bonding the electrostatic attraction between oppositely charged ions. (13.1)

Ionic compound (binary) a compound that results when a metal reacts with a nonmetal to form a cation and an anion. (13.1)

Ionic solid a solid containing cations and anions that dissolves in water to give a solution containing the separated ions, which are mobile and thus free to conduct electrical current. (16.3)

Ionization energy the quantity of energy required to remove an electron from a gaseous atom or ion. (12.15)

Irreversible process any real process. When a system undergoes the changes State 1 → State 2 → State 1 by any real pathway, the universe is different than before the cyclic process took place in the system. (10.2)

Isoelectronic ions ions containing the same number of electrons. (13.4)

Isomers species with the same formula but different properties. (20.4)

Isothermal process a process in which the temperature remains constant. (10.2)

Isotonic solutions solutions having identical osmotic pressures. (17.6)

Isotopes atoms of the same element (the same number of protons) with different numbers of neutrons. They have identical atomic numbers but different mass numbers. (2.5)

Ketone an organic compound containing the carbonyl group

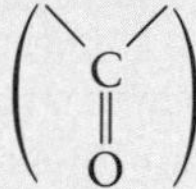

bonded to two carbon atoms. (22.4)

Kinetic energy ($\frac{1}{2}mv^2$) energy due to the motion of an object; dependent on the mass of the object and the square of its velocity. (9.1)

Kinetic molecular theory a model that assumes that an ideal gas is composed of tiny particles (molecules) in constant motion. (5.6)

Lanthanide contraction the decrease in the atomic radii of the lanthanide series elements, going from left to right in the periodic table. (20.1)

Lanthanide series a group of fourteen elements following lanthanum in the periodic table, in which the 4*f* orbitals are being filled. (12.13; 18.1; 20.1)

Lattice a three-dimensional system of points designating the positions of the centers of the components of a solid (atoms, ions, or molecules). (16.3)

Lattice energy the energy change occurring when separated gaseous ions are packed together to form an ionic solid. (13.5)

Law of conservation of energy energy can be converted from one form to another but can be neither created nor destroyed. (9.1)

Law of conservation of mass mass is neither created nor destroyed. (2.2)

Law of definite proportions a given compound always contains exactly the same proportion of elements by mass. (2.2)

Law of mass action a general description of the equilibrium condition; it defines the equilibrium constant expression. (6.2)

Law of multiple proportions when two elements form a series of compounds, the ratios of the masses of the second element that combine with one gram of the first element can always be reduced to small whole numbers. (2.2)

Lead storage battery a battery (used in cars) in which the anode is lead, the cathode is lead coated with lead dioxide, and the electrolyte is a sulfuric acid solution. (11.5)

Le Châtelier's principle if a change is imposed on a system at equilibrium, the position of the equilibrium will shift in a direction that tends to reduce the effect of that change. (6.8)

Lewis acid an electron-pair acceptor. (20.3)

Lewis base an electron-pair donor. (20.3)

Lewis structure a diagram of a molecule showing how the valence electrons are arranged among the atoms in the molecule. (13.10)

Ligand a neutral molecule or ion having a lone pair of electrons that can be used to form a bond to a metal ion; a Lewis base. (20.1)

Lime-soda process a water-softening method in which lime and soda ash are added to water to remove calcium and magnesium ions by precipitation. (7.6)

Limiting reactant (limiting reagent) the reactant that is completely consumed when a reaction is run to completion. (3.9)

Line spectrum a spectrum showing only certain discrete wavelengths. (12.3)

Linear accelerator a type of particle accelerator in which a changing electrical field is used to accelerate a positive ion along a linear path. (21.3)

Linkage isomerism isomerism involving a complex ion where the ligands are all the same but the point of attachment of at least one of the ligands differs. (20.4)

Lipids water-insoluble substances that can be extracted from cells by nonpolar organic solvents. (23.4)

Liquefaction the transformation of a gas into a liquid. (18.1)

Localized electron (LE) model a model that assumes that a molecule is composed of atoms that are bound together by sharing pairs of electrons using the atomic orbitals of the bound atoms. (13.9)

Lock-and-key model a model for the mechanism of enzyme activity postulating that the shapes of the substrate and the enzyme are such that they fit together as a key fits a specific lock. (23.1)

London dispersion forces the forces, existing among noble gas atoms and nonpolar molecules, that involve an accidental dipole that induces a momentary dipole in a neighbor. (16.1)

Lone pair an electron pair that is localized on a given atom; an electron pair not involved in bonding. (13.9)

Magnetic quantum number m_ℓ, the quantum number relating to the orientation of an orbital in space relative to the other orbitals with the same ℓ quantum number. It can have integral values between ℓ and $-\ell$, including zero. (12.9)

Main-group (representative) elements elements in the

groups labeled 1A, 2A, 3A, 4A, 5A, 6A, 7A, and 8A in the periodic table. The group number gives the sum of valence *s* and *p* electrons. (12.13; 18.1)

Major species the components present in relatively large amounts in a solution. (7.4)

Manometer a device for measuring the pressure of a gas in a container. (5.1)

Mass defect the change in mass occurring when a nucleus is formed from its component nucleons. (21.5)

Mass number the total number of protons and neutrons in the atomic nucleus of an atom. (2.6)

Mass percent the percent by mass of a component of a mixture (17.1) or of a given element in a compound. (3.4)

Mass spectrometer an instrument used to determine the relative masses of atoms by the deflection of their ions in a magnetic field. (3.1)

Matter the material of the universe.

Mean free path the average distance a molecule in a given gas sample travels between collisions with other molecules. (5.9)

Messenger RNA (mRNA) a special RNA molecule built in the cell nucleus that migrates into the cytoplasm and participates in protein synthesis. (23.3)

Metal an element that gives up electrons relatively easily and is lustrous, malleable, and a good conductor of heat and electricity. (2.8)

Metalloenzyme an enzyme containing a metal ion at its active site. (23.1)

Metalloids (semimetals) elements along the division line in the periodic table between metals and nonmetals. These elements exhibit both metallic and nonmetallic properties. (12.16; 18.1)

Metallurgy the process of separating a metal from its ore and preparing it for use. (18.1)

Micelles aggregates of fatty acid anions having their hydrophobic tails in the interior and their polar heads pointing outward to interact with the polar water molecules. (23.4)

Millimeters of mercury (mm Hg) a unit of pressure, also called a torr; 760 mm Hg = 760 torr = 101,325 Pa = 1 standard atmosphere. (5.1)

Mixture a material of variable composition that contains two or more substances.

Model (theory) a set of assumptions put forth to explain the observed behavior of matter. The models of chemistry usually involve assumptions about the behavior of individual atoms or molecules. (1.3)

Moderator a substance used in a nuclear reactor to slow down the neutrons. (21.6)

Molal boiling-point elevation constant a constant characteristic of a particular solvent that gives the change in boiling point as a function of solution molality; used in molecular weight determinations. (17.5)

Molal freezing-point depression constant a constant characteristic of a particular solvent that gives the change in freezing point as a function of the solution molality; used in molecular weight determinations. (17.5)

Molality the number of moles of solute per kilogram of solvent in a solution. (17.1)

Molar heat capacity the energy required to raise the temperature of one mole of a substance by one degree Celsius. (9.3)

Molar mass the mass in grams of one mole of molecules or formula units of a substance; also called molecular weight. (3.3)

Molar volume the volume of one mole of an ideal gas; equal to 22.42 liters at STP. (5.4)

Molarity moles of solute per volume of solution in liters. (4.3; 17.1)

Mole (mol) the number equal to the number of carbon atoms in exactly 12 grams of pure ^{12}C; Avogadro's number. One mole represents 6.022×10^{23} units. (3.2)

Mole fraction the ratio of the number of moles of a given component in a mixture to the total number of moles in the mixture. (5.5; 17.1)

Mole ratio (stoichiometry) the ratio of moles of one substance to moles of another substance in a balanced chemical equation. (3.8)

Molecular equation an equation representing a reaction in solution showing the reactants and products in undissociated form, whether they are strong or weak electrolytes. (4.6)

Molecular formula the exact formula of a molecule, giving the types of atoms and the number of each type. (3.5)

Molecular orbital (MO) model a model that regards a molecule as a collection of nuclei and electrons, where the electrons are assumed to occupy orbitals much as they do in atoms, but having the orbitals extend over the entire molecule. In this model the electrons are assumed to be delocalized rather than always located between a given pair of atoms. (14.2)

Molecular orientations (kinetics) orientations of molecules during collisions, some of which can lead to a reaction and some of which cannot. (15.8)

Molecular solid a solid composed of neutral molecules at the lattice points. (16.3)

Molecular structure the three-dimensional arrangement of atoms in a molecule. (13.13)

Molecular weight the mass in grams of one mole of molecules or formula units of a substance; the same as molar mass. (3.3)

Molecularity the number of species that must collide to produce the reaction represented by an elementary step in a reaction mechanism. (15.6)

Molecule a bonded collection of two or more atoms of the same or different elements. (2.7)

Monodentate (unidentate) ligand a ligand that can form one bond to a metal ion. (20.3)

Monoprotic acid an acid with one acidic proton. (7.2)

Monosaccharide (simple sugar) a polyhydroxy ketone or aldehyde containing from three to nine carbon atoms. (23.2)

Myoglobin an oxygen-storing biomolecule consisting of a heme complex and a protein. (20.8)

Natural law a statement that expresses generally observed behavior. (1.3)

Nernst equation an equation relating the potential of an electrochemical cell to the concentrations of the cell components

$$\mathscr{E} = \mathscr{E}^\circ - \frac{0.0592}{n}\log(Q) \text{ at } 25^\circ\text{C. (11.4)}$$

Net ionic equation an equation for a reaction in solution, where strong electrolytes are written as ions, showing only those components that are directly involved in the chemical change. (4.6)

Network solid an atomic solid containing strong directional covalent bonds. (16.5)

Neutralization reaction an acid-base reaction. (4.9)

Neutron a particle in the atomic nucleus with mass virtually equal to the proton's but with no charge. (2.6)

Nitride salt a compound containing the N^{3-} anion. (18.2)

Nitrogen cycle the conversion of N_2 to nitrogen-containing compounds, followed by the return of nitrogen gas to the atmosphere by natural decay processes. (19.2)

Nitrogen fixation the process of transforming N_2 to nitrogen-containing compounds useful to plants. (19.2)

Nitrogen-fixing bacteria bacteria in the root nodules of plants that can convert atmospheric nitrogen to ammonia and other nitrogen-containing compounds useful to plants. (19.2)

Noble gas a Group 8A element. (2.8; 19.8)

Node an area of an orbital having zero electron probability. (12.6)

Nonelectrolyte a substance that, when dissolved in water, gives a nonconducting solution. (4.2)

Nonmetal an element not exhibiting metallic characteristics. Chemically, a typical nonmetal accepts electrons from a metal. (2.8)

Normal boiling point the temperature at which the vapor pressure of a liquid is exactly one atmosphere. (16.10)

Normal melting point the temperature at which the solid and liquid states have the same vapor pressure under conditions where the total pressure on the system is one atmosphere. (16.10)

Normality the number of equivalents of a substance dissolved in a liter of solution. (17.1)

Nuclear atom an atom having a dense center of positive charge (the nucleus) with electrons moving around the outside. (2.5)

Nuclear transformation the change of one element into another. (21.3)

Nucleon a particle in an atomic nucleus, either a neutron or a proton. (2.6)

Nucleotide a monomer of the nucleic acids composed of a five-carbon sugar, a nitrogen-containing base, and phosphoric acid. (23.3)

Nucleus the small, dense center of positive charge in an atom. (2.5)

Nuclide the general term applied to each unique atom; represented by ${}^{A}_{Z}X$, where X is the symbol for a particular element. (21)

Octet rule the observation that atoms of nonmetals tend to form the most stable molecules when they are surrounded by eight electrons (to fill their valence orbitals). (13.10)

Optical isomerism isomerism in which the isomers have opposite effects on plane-polarized light. (20.4)

Orbital a specific wave function for an electron in an atom. The square of this function gives the probability distribution for the electron. (12.7)

***d*-Orbital splitting** a splitting of the *d*-orbitals of the metal ion in a complex such that the orbitals pointing at the ligands have higher energies than those pointing between the ligands. (20.6)

Order (of reactant) the positive or negative exponent, determined by experiment, of the reactant concentration in a rate law. (15.2)

Organic acid an acid with a carbon-atom backbone; often contains the carboxyl group. (7.2)

Organic chemistry the study of carbon-containing compounds (typically chains of carbon atoms) and their properties. (22)

Osmosis the flow of solvent into a solution through a semipermeable membrane. (17.6)

Osmotic pressure (π) the pressure that must be applied to a solution to stop osmosis; $= MRT$. (17.6)

Ostwald process a commercial process for producing nitric acid by the oxidation of ammonia. (19.2)

Oxidation an increase in oxidation state (a loss of electrons). (4.10; 11.1)

Oxidation-reduction (redox) reaction a reaction in which one or more electrons are transferred. (4.10; 11.1)

Oxidation states a concept that provides a way to keep track of electrons in oxidation-reduction reactions according to certain rules. (4.10)

Oxidizing agent (electron acceptor) a reactant that accepts electrons from another reactant. (4.10; 11.1)

Oxyacid an acid in which the acidic proton is attached to an oxygen atom. (7.2)

Ozone O_3, the form of elemental oxygen in addition to the much more common O_2. (19.5)

Paramagnetism a type of induced magnetism, associated with unpaired electrons, that causes a substance to be attracted into the inducing magnetic field. (14.3)

Partial pressures the independent pressures exerted by different gases in a mixture. (5.5)

Particle accelerator a device used to accelerate nuclear particles to very high speeds. (21.3)

Pascal the SI unit of pressure; equal to newtons per meter squared. (5.1)

Pauli exclusion principle in a given atom no two electrons can have the same set of four quantum numbers. (12.10)

Penetration effect the effect whereby a valence electron penetrates the core electrons, thus reducing the shielding effect and increasing the effective nuclear charge. (12.14)

Peptide linkage the bond resulting from the condensation reaction between amino acids; represented by

$$\begin{array}{cc} \text{O} & \text{H} \\ \| & | \\ -\text{C}- & \text{N}- \end{array} \quad (23.1)$$

Percent dissociation the ratio of the amount of a substance that is dissociated at equilibrium to the initial concentration of the substance in a solution, multiplied by 100. (7.5)

Percent yield the actual yield of a product as a percentage of the theoretical yield. (3.9)

Periodic table a chart showing all the elements arranged in columns with similar chemical properties. (2.8)

pH curve (titration curve) a plot showing the pH of a solution being analyzed as a function of the amount of titrant added. (8.4)

pH scale a log scale based on 10 and equal to $-\log[H^+]$; a convenient way to represent solution acidity. (7.3)

Phase diagram a convenient way of representing the phases of a substance in a closed system as a function of temperature and pressure. (16.11)

Phenyl group the benzene molecule minus one hydrogen atom. (22.3)

Phospholipids esters of glycerol containing two fatty acids and a phosphate group. Having nonpolar tails and polar heads, they tend to form bilayers in aqueous solution. (23.4)

Photochemical smog air pollution produced by the action of light on oxygen, nitrogen oxides, and unburned fuel from auto exhaust to form ozone and other pollutants. (5.11)

Photon a quantum of electromagnetic radiation. (12.2)

Physical change a change in the form of a substance, but not in its chemical composition; chemical bonds are not broken in a physical change.

Pi (π) bond a covalent bond in which parallel *p* orbitals share an electron pair occupying the space above and below the line joining the atoms. (14.1)

Planck's constant the constant relating the change in energy for a system to the frequency of the electromagnetic radiation absorbed or emitted; equal to 6.626×10^{-34} J s. (12.2)

Polar covalent bond a covalent bond in which the electrons are not shared equally because one atom attracts them more strongly than the other. (13.1)

Polar molecule a molecule that has a permanent dipole moment. (4.1)

Polyatomic ion an ion containing a number of atoms. (2.7)

Polyelectronic atom an atom with more than one electron. (12.11)

Polymer a large, usually chainlike molecule built from many small molecules (monomers). (22.5)

Polymerization a process in which many small molecules (monomers) are joined together to form a large molecule. (22.2)

Polypeptide a polymer formed from amino acids joined together by peptide linkages. (23.1)

Polyprotic acid an acid with more than one acidic proton. It dissociates in a stepwise manner, one proton at a time. (7.7)

Porous disk a disk in a tube connecting two different solutions in a galvanic cell that allows ion flow without extensive mixing of the solutions. (11.1)

Porphyrin a planar ligand with a central ring structure and various substituent groups at the edges of the ring. (20.8)

Positional probability a type of probability that depends on the number of arrangements in space that yield a particular state. (10.1)

Positron production a mode of nuclear decay in which a particle is formed having the same mass as an electron but opposite charge. The net effect is to change a proton to a neutron. (21.1)

Potential energy energy due to position or composition. (9.1)

Precipitation reaction a reaction in which an insoluble substance forms and separates from the solution. (4.5)

Precision the degree of agreement among several measurements of the same quantity; the reproducibility of a measurement. (A1.5)

Primary structure (of a protein) the order (sequence) of amino acids in the protein chain. (23.1)

Principal quantum number the quantum number relating to the size and energy of an orbital; it can have any positive integer value. (12.9)

Probability distribution the square of the wave function indicating the probability of finding an electron at a particular point in space. (12.8)

Product a substance resulting from a chemical reaction. It is shown to the right of the arrow in a chemical equation. (3.6)

Protein a natural high-molecular-weight polymer formed by condensation reactions between amino acids. (23.1)

Proton a positively charged particle in an atomic nucleus. (2.6; 21)

Pure-substance a substance with constant composition.

Qualitative analysis the separation and identification of individual ions from a mixture. (4.7)

Quantization the concept that energy can occur only in discrete units called quanta. (12.2)

Rad a unit of radiation dosage corresponding to 10^{-2} J of energy deposited per kilogram of tissue (from *r*adiation *a*bsorbed *d*ose). (21.7)

Radioactive decay (radioactivity) the spontaneous decomposition of a nucleus to form a different nucleus. (21.1)

Radiocarbon dating (carbon-14 dating) a method for dating ancient wood or cloth based on the rate of radioactive decay of the nuclide $^{14}_{6}C$. (21.4)

Radiotracer a radioactive nuclide, introduced into an organism for diagnostic purposes, whose pathway can be traced by monitoring its radioactivity. (21.4)

Random error an error that has an equal probability of being high or low. (A1.5)

Raoult's law the vapor pressure of a solution is directly proportional to the mole fraction of solvent present. (17.4)

Rate constant the proportionality constant in the relationship between reaction rate and reactant concentrations. (15.2)

Rate of decay the change in the number of radioactive nuclides in a sample per unit time. (21.2)

Rate-determining step the slowest step in a reaction mechanism, the one determining the overall rate. (15.6)

Rate law (differential rate law) an expression that shows how the rate of reaction depends on the concentration of reactants. (15.2)

Reactant a starting substance in a chemical reaction. It appears to the left of the arrow in a chemical equation. (3.6)

Reaction mechanism the series of elementary steps involved in a chemical reaction. (15.6)

Reaction quotient a quotient obtained by applying the law of mass action to initial concentrations rather than to equilibrium concentrations. (6.6)

Reaction rate the change in concentration of a reactant or product per unit time. (15.1)

Reactor core the part of a nuclear reactor where the fission reaction takes place. (21.6)

Reducing agent (electron donor) a reactant that donates electrons to another substance to reduce the oxidation state of one of its atoms. (4.10; 11.1)

Reduction a decrease in oxidation state (a gain of electrons). (4.10; 11.1)

Rem a unit of radiation dosage that accounts for both the energy of the dose and its effectiveness in causing biological damage (from *r*oentgen *e*quivalent for *m*an). (21.7)

Resonance a condition occurring when more than one valid Lewis structure can be written for a particular molecule. The actual electronic structure is not represented by any one of the Lewis structures but by the average of all of them. (13.12)

Reverse osmosis the process occurring when the external pressure on a solution causes a net flow of solvent through a semipermeable membrane from the solution to the solvent. (17.6)

Reversible process a cyclic process carried out by a hypothetical pathway, which leaves the universe exactly the same as it was before the process. No real process is reversible. (10.2)

Ribonucleic acid (RNA) a nucleotide polymer that transmits the genetic information stored in DNA to the ribosomes for protein synthesis. (23.3)

Root mean square velocity the square root of the average of the squares of the individual velocities of gas particles. (5.6)

Salt an ionic compound. (7.8)

Salt bridge a U-tube containing an electrolyte that connects the two compartments of a galvanic cell, allowing ion flow without extensive mixing of the different solutions. (11.1)

Scientific method the process of studying natural phenomena, involving observations, forming laws and theories, and testing of theories by experimentation. (1.3)

Scintillation counter an instrument that measures radioactive decay by sensing the flashes of light produced in a substance by the radiation. (21.4)

Second law of thermodynamics in any spontaneous process, there is always an increase in the entropy of the universe. (10.5)

Secondary structure (of a protein) the three-dimensional structure of the protein chain (for example, α-helix, random coil, or pleated sheet). (23.1)

Selective precipitation a method of separating metal ions from an aqueous mixture by using a reagent whose anion forms a precipitate with only one or a few of the ions in the mixture. (4.7; 8.8)

Semiconductor a substance conducting only a slight electrical current at room temperature, but showing increased conductivity at higher temperatures. (16.5)

Semipermeable membrane a membrane that allows solvent but not solute molecules to pass through. (17.6)

Shielding the effect by which the other electrons screen, or shield, a given electron from some of the nuclear charge. (12.15)

SI units International System of units based on the metric system and units derived from the metric system. (A2.1)

Side chain (of amino acid) the hydrocarbon group on an amino acid represented by H, CH_3, or a more complex substituent. (23.1)

Sigma (σ) bond a covalent bond in which the electron pair is shared in an area centered on a line running between the atoms. (14.1)

Significant figures the certain digits and the first uncertain digit of a measurement. (A1.5)

Silica the fundamental silicon-oxygen compound, which has the empirical formula SiO_2, and forms the basis of quartz and certain types of sand. (16.5)

Silicates salts that contain metal cations and polyatomic silicon-oxygen anions that are usually polymeric. (16.5)

Single bond a bond in which one pair of electrons is shared by two atoms. (13.8)

Solubility the amount of a substance that dissolves in a given volume of solvent at a given temperature. (4.2)

Solubility product constant the constant for the equilibrium expression representing the dissolving of an ionic solid in water. (8.6)

Solute a substance dissolved in a liquid to form a solution. (4.2; 17.1)

Solution a homogeneous mixture. (17)

Solvent the dissolving medium in a solution. (4.2)

Somatic damage radioactive damage to an organism resulting in its sickness or death. (21.7)

Space-filling model a model of a molecule showing the relative sizes of the atoms and their relative orientations. (2.7)

Specific heat capacity the energy required to raise the temperature of one gram of a substance by one degree Celsius. (9.4)

Spectator ions ions present in solution that do not participate directly in a reaction. (4.6)

Spectrochemical series a listing of ligands in order based on their ability to produce *d*-orbital splitting. (20.6)

Spontaneous fission the spontaneous splitting of a heavy nuclide into two lighter nuclides. (21.1)

Spontaneous process a process that occurs without outside intervention. (10.1)

Standard atmosphere a unit of pressure equal to 760 mm Hg. (5.1)

Standard enthalpy of formation the enthalpy change that accompanies the formation of one mole of a compound at 25°C from its elements, with all substances in their standard states at that temperature. (9.6)

Standard free energy change the change in free energy that will occur for one unit of reaction if the reactants in their standard states are converted to products in their standard states. (10.9)

Standard free energy of formation the change in free energy that accompanies the formation of one mole of a substance from its constituent elements with all reactants and products in their standard states. (10.9)

Standard hydrogen electrode a platinum conductor in contact with 1 M H^+ ions and bathed by hydrogen gas at one atmosphere. (11.2)

Standard reduction potential the potential of a half-reaction under standard state conditions, as measured against the potential of the standard hydrogen electrode. (11.2)

Standard solution a solution whose concentration is accurately known. (4.3)

Standard state a reference state for a specific substance defined according to a set of conventional definitions. (9.6)

Standard temperature and pressure (STP) the condition 0°C and 1 atmosphere of pressure. (5.4)

Standing wave a stationary wave as on a string of a musical instrument; in the wave mechanical model, the electron in the hydrogen atom is considered to be a standing wave. (12.5)

State function a property that is independent of the pathway. (9.1)

States of matter the three different forms in which matter can exist: solid, liquid, and gas.

Stereoisomerism isomerism in which all the bonds in the isomers are the same but the spatial arrangements of the atoms are different. (20.4)

Steric factor the factor (always less than one) that reflects the fraction of collisions with orientations that can produce a chemical reaction. (15.8)

Steroid one of a class of lipids with a characteristic fused carbon-ring structure that includes cholesterol, hormones, and bile acids. (23.4)

Stoichiometric quantities quantities of reactants mixed in exactly the correct amounts so that all are used up at the same time. (3.9)

Strong acid an acid that completely dissociates to produce a H^+ ion and the conjugate base. (4.2; 7.2)

Strong base a metal hydroxide salt that completely dissociates into its ions in water. (4.2; 7.6)

Strong electrolyte a material that, when dissolved in water, gives a solution that conducts an electric current very efficiently. (4.2)

Structural formula the representation of a molecule in which the relative positions of the atoms are shown and the bonds are indicated by lines. (2.7)

Structural isomerism isomerism in which the isomers contain the same atoms but one or more bonds differ. (20.4; 22.1)

Subcritical reaction (nuclear) a reaction in which less than one neutron causes another fission event and the process dies out. (21.6)

Sublimation the process by which a substance goes directly from the solid to the gaseous state without passing through the liquid state. (16.10)

Subshell a set of orbitals with a given angular momentum quantum number. (12.9)

Substitution reaction (hydrocarbons) a reaction in which an atom, usually a halogen, replaces a hydrogen atom in a hydrocarbon. (22.1)

Supercooling the process of cooling a liquid below its freezing point without its changing to a solid. (16.10)

Supercritical reaction (nuclear) a reaction in which more than one neutron from each fission event causes another fission event. The process rapidly escalates to a violent explosion. (21.6)

Superheating the process of heating a liquid above its boiling point without its boiling. (16.10)

Superoxide a compound containing the O_2^- anion. (18.2)

Surface tension the resistance of a liquid to an increase in its surface area. (16.2)

Surfactant a wetting agent, such as soap, that assists water in wetting and suspending nonpolar materials. (23.4)

Surroundings everything in the universe surrounding a thermodynamic system. (9.1)

Syngas synthetic gas, a mixture of carbon monoxide and hydrogen, obtained by coal gasification. (9.8)

System (thermodynamic) that part of the universe on which attention is to be focused. (9.1)

Systematic error an error that always occurs in the same direction. (A1.5)

Termolecular step a reaction involving the simultaneous collision of three molecules. (15.6)

Tertiary structure (of a protein) the overall shape of a protein, long and narrow or globular, maintained by different types of intramolecular interactions. (23.1)

Theoretical yield the maximum amount of a given product that can be formed when the limiting reactant is completely consumed. (3.9)

Theory a set of assumptions put forth to explain some aspect of the observed behavior of matter. (1.3)

Thermal pollution the oxygen-depleting effect on lakes and rivers of using water for industrial cooling and returning it to its natural souce at a higher temperature. (17.3)

Thermodynamic stability (nuclear) the potential energy of a particular nucleus as compared with the sum of the potential energies of its component protons and neutrons. (21.1)

Thermodynamics the study of energy and its interconversions. (9.1)

Third law of thermodynamics the entropy of a perfect crystal at 0 K is zero. (10.8)

Titration a technique in which one solution is used to analyze another. (4.9)

Torr another name for millimeter of mercury (mm Hg). (5.1)

Trace elements metals present only in trace amounts in the human body but essential for the action of many enzymes. (23)

Transfer RNA (tRNA) a small RNA fragment that finds specific amino acids and attaches them to the protein chain as dictated by the codons in mRNA. (23.3)

Transition metals several series of elements in which inner orbitals (*d* or *f* orbitals) are being filled. (12.13; 18.1)

Transuranium elements the elements beyond uranium that are made artifically by particle bombardment. (21.3)

Triple bond a bond in which three pairs of electrons are shared by two atoms. (13.8)

Triple point the point on a phase diagram at which all three states of a substance are present. (16.11)

Tyndall effect the scattering of light by particles in a suspension. (17.8)

Uncertainty (in measurement) the characteristics that any measurement involves estimates and cannot be exactly reproduced. (A1.5)

Unimolecular step a reaction step involving only one molecule. (15.6)

Unit cell the smallest repeating unit of a lattice. (16.3)

Unit factor an equivalence statement between units used for converting from one unit to another. (A2.2)

Universal gas constant the combined proportionality constant in the ideal gas law; 0.08206 L atm/K mol or 8.3145 J/K mol. (5.3)

Valence electrons the electrons in the outermost principal quantum level of an atom. (12.13)

Valence shell electron pair repulsion (VSEPR) model a model whose main postulate is that the structure around a given atom in a molecule is determined principally by minimizing electron-pair repulsions. (13.13)

van der Waals's equation a mathematical expression for describing the behavior of real gases. (5.10)

van't Hoff factor the ratio of moles of particles in solution to moles of solute dissolved. (17.7)

Vapor pressure the pressure of the vapor over a liquid at equilibrium. (16.10)

Vaporization the change in state that occurs when a liquid evaporates to form a gas. (16.10)

Viscosity the resistance of a liquid to flow. (16.2)

Volt the unit of electrical potential defined as one joule of work per coulomb of charge transferred. (11.1)

Voltmeter an instrument that measures cell potential by drawing electric current through a known resistance. (11.1)

Volumetric analysis a process involving titration of one solution with another. (4.9)

Wave function a function of the coordinates of an electron's position in three-dimensional space that describes the properties of the electron. (12.5)

Wave mechanical model a model for the hydrogen atom in which the electron is assumed to behave as a standing wave. (12.7)

Wavelength the distance between two consecutive peaks or troughs in a wave. (12.1)

Wax an ester similar to a fat or a phospholipid but containing a monohydroxy alcohol instead of glycerol. (23.4)

Weak acid an acid that dissociates only slightly in aqueous solution. (4.2; 7.2)

Weak base a base that reacts with water to produce hydroxide ions to only a slight extent in aqueous solution. (4.2; 7.6)

Weak electrolyte a material that, when dissolved in water, gives a solution that conducts only a small electric current. (4.2)

Weight the force exerted on an object by gravity.

Work force acting over a distance. (9.1)

X-ray diffraction a technique for establishing the structures of crystalline solids by directing X rays of a single wavelength at a crystal and obtaining a diffraction pattern from which interatomic spaces can be determined. (16.3)

Zone of nuclear stability the area encompassing the stable nuclides on a plot of their positions as a function of the number of protons and the number of neutrons in the nucleus. (21.1)

ANSWERS TO SELECTED EXERCISES

The answers listed here are from the *Complete Solutions Guide*, in which rounding is carried out at each intermediate step in a calculation in order to show the correct number of significant figures for that step. Therefore, an answer given here may differ in the last digit from the result obtained by carrying extra digits throughout the entire calculation and rounding at the end (the procedure you should follow).

Chapter 2

1. F/S = 1.00 (i), 2.00 (ii), 3.00 (iii); F/S simple whole numbers. **2.** ClF_3 **3.** a. A compound always contains the same number of atoms. b. vol HCL/vol H_2 or Cl_2 = 2 **4.** O, 7.94; Na, 22.8; Mg, 11.9; O and Mg are wrong by factor of ≈ 2; correct formulas are H_2O, Na_2O, and MgO **5.** a. Atoms are conserved in reactions. b. Compounds always contain the same atoms. c. Compounds contain whole numbers of atoms. **6.** Some elements exist as molecules (H_2, Cl_2, O_2, etc.). **7.** Yes. Atomic structure? Atomic masses? Interatomic forces? **8.** Most elements have isotopes. **11.** using $r = 5 \times 10^{-14}$ cm, $d_{nucleus} = 3 \times 10^{15}$ g/cm^3; using $r = 1 \times 10^{-8}$ cm, $d_{atom} = 0.4$ g/cm^3. **15.** Pm, Tc **16.** a. 8; b. 8; c. 18; d. 4; e. 5; f. 3 **17.** a. P; b. I **18.** a. 12p, 12n, 12e; b. 12p, 12n, 10e; c. 27p, 32n, 25e **19.** $^{75}_{33}As^{3+}$, 30e; $^{32}_{16}S^{2-}$, 2−; $^{204}_{81}Tl^{+}$, 80e; $^{197}_{79}Au$, 79p, 118n, 79e, 0; $^{197}_{79}Au^{3+}$, 79p, 118n, 76e, 3+ **21.** Metallic character increases down a group **23.** a. sodium perchlorate; b. magnesium phosphate; c. aluminum sulfate; d. sulfur difluoride **24.** i. dinitrogen tetroxide; j. nitrogen trifluoride; k. dinitrogen tetrafluoride; l. iron(II) sulfate **26.** a. SO_2; b. SO_3; c. Na_2SO_3; d. $KHSO_3$; e. Li_3N; f. $Cr_2(CO_3)_3$ **29.** No difference for pure compounds. **32.** selenate, selenite, tellurate, tellurite **34.** 13.52 **37.** a. Kr; b. Te; c. Ca; d. Ag; e. Pu

Chapter 3

1. 24.31 amu **3.** 185 **5.** 207.2 amu; Pb
8. a. 9.97×10^{-3} g; b. 6.00×10^{-2} g; c. 3.0×10^{3} g; d. 3.323×10^{-23} g; e. 4.00×10^{-14} g; f. 3.60×10^{-10} g; g. 1.0×10^{-7} g **9.** a. 1.661×10^{-22} mol; b. 5.551 mol; c. 8.303×10^{-22} mol; d. 8.953 mol; e. 2.491×10^{-22} mol; f. 0.9393 mol; g. 2.17×10^{-4} mol; h. 1.0×10^{-15} mol; i. 2.5×10^{-8} mol; j. 3.8×10^{-5} mol **12.** 4.4×10^{16} molecules **13.** Y, 13.35%; Ba, 41.22%; Cu, 28.62%; O, 16.81% **14.** InP < $(NPCl_2)_3$ < PF_3 < P_4O_{10} **16.** 1360 g/mol **20.** $C_7H_5N_3O_6$ **22.** $Na_2S_2O_3$ **23.** TiO_2 **25.** C_3H_4; C_9H_{12} **28.** a. $4Fe(s) + 3O_2(g) \rightarrow 2Fe_2O_3(s)$; b. $C_6H_{12}O_6(aq) \rightarrow 2C_2H_5OH(aq) + 2CO_2(g)$; c. $Ca(s) + 2H_2O(l) \rightarrow Ca(OH)_2(aq) + H_2(g)$; d. $Ba(OH)_2(aq) + H_2SO_4(aq) \rightarrow BaSO_4(s) + 2H_2O(l)$ **31.** 4355 g **33.** 97 g **35.** a. $2C_8H_{18}(l) + 25O_2(g) \rightarrow 16CO_2(g) + 18H_2O(g)$; b. 9.6×10^{13} g **38.** 2.51 g (theoretical); 92.0% yield **40.** a. 1.15 g; b. 1.85 g S_8 **43.** a. 795 g; b. 96.2% **45.** Mn; 54.94 **47.** a. Pt, 65.01%; N, 9.335%; H, 2.015%; Cl, 23.63%; b. 72.3 g cisplatin, 35.9 g KCl **49.** Sb_2O_3; Sb_4O_6 **51.** Al_2Se_3 **53.** $C_4H_3O_2$; $C_8H_6O_4$ **55.** 1300 g $CaSO_4$; 630 g H_3PO_4 **57.** a. 18.0 g $HgBr_2$, 2.0 g Br_2 left; b. 35.2 g **59.** $^{12}C^{1}H_2{}^{16}O$ **62.** NH_3, N_2H_4, N_3H; 14.01/1.008 **64.** Pa_3O_8 **66.** 87.8 g/mol **69.** 184 g/mol **70.** X, 14.3%; Z, 85.7% **72.** 32 kg **75.** $SrCO_3$, 54% (53.4%)*; $BaCO_3$, 46% (46.6%) **78.** $C_8H_{18} + 11O_2 \rightarrow 5.7CO_2 + 2.2CO + 0.2CH_4 + 0.74H_2 + 7.9H_2O$

Chapter 4

1. *Solubility* refers to how much dissolves. *Electrolyte* refers to whether ions are produced. **5.** a. 0.2186 *M*; b. 1.876×10^{-3} *M*; c. 8.065×10^{-3} *M* **6.** 0.256 *M*; 1.523×10^{-5} *M* **9.** a. 2.5×10^{-8} *M*; b. 8.4×10^{-9} *M*; c. 1.33×10^{-4} *M*; d. 2.8×10^{-7} *M* **11.** 20.0 ppb; 5.95×10^{-8} *M* **13.** 1 ppm, 5×10^{-5} *M*; 50. ppm, 2.6×10^{-3} *M* **15.** Stock, 2.883×10^{-2} *M*; A, 1.442×10^{-3} *M*; B, 5.768×10^{-5} *M*; C, 1.154×10^{-6} *M* **18.** a. H_2SO_4, then HCl; b. H_2SO_4, then HCl; c. HCl **21.** 0.175 L **24.** 18.8 g **26.** 39.49 mg/tablet; 67.00% **28.** 0.994 g **31.** $C_{22}H_{20}O_{13}$ **34.** 4.7×10^{-2} *M* **37.** a. 0.8393 *M*; b. 5.010% **40.** 0.0776 *M* **43.** a. K, +1; O, −2; Mn, +7; b. Ni, +4; O, −2; c. Fe, +2; d. N, −3; H, +1; P, +5; O, −2; e. P, +3; O, −2 **47.** a. redox; O_2, ox agt; CH_4, red agt; C, ox; O, red; b. redox; HCl, ox agt; Zn, red agt; Zn, ox; H, red; c. not redox; d. redox; O_3, ox agt; NO, red agt; N, ox; O, red **48.** a. $2C_2H_6 + 7O_2 \rightarrow 4CO_2 + 6H_2O$; b. $Mg + 2HCl \rightarrow Mg^{2+} + 2Cl^- + H_2$ **49.** a. $3Cu + 8HNO_3 \rightarrow 3Cu(NO_3)_2 + 2NO + 4H_2O$; b. $14H^+ + Cr_2O_7^{2-} + 6Cl^- \rightarrow 3Cl_2 + 2Cr^{3+} + 7H_2O$; c. $Pb + 2H_2SO_4 + PbO_2 \rightarrow 2PbSO_4 + 2H_2O$ **50.** a. $2H_2O + Al + MnO_4^- \rightarrow Al(OH)_4^- + MnO_2$; b. $2OH^- + Cl_2 \rightarrow Cl^- + ClO^- + H_2O$ **54.** $2H^+ + Mn + 2HNO_3 \rightarrow Mn^{2+} + 2NO_2 + 2H_2O$; $3H_2O + 2Mn^{2+} + 5IO_4^- \rightarrow 2MnO_4^- + 5IO_3^- + 6H^+$ **56.** 1.622×10^{-2} *M* **58.** 34.6% **60.** yes **61.** a. $6H^+ + 8I^- + IO_3^- \rightarrow 3I_3^- + 3H_2O$; b. 3.732 g KI; 5.62×10^{-3} L HCl; c. $2S_2O_3^{2-} + I_3^- \rightarrow 3I^- + S_4O_6^{2-}$; d. 0.0468 *M*; e. requires 1.07 g KIO_3/500.0 mL **62.** 14.63% **66.** KCl, 77.1%; KBr, 22.9% **69.** 9.64×10^{-2} *M* **72.** a. $7H_2O + 2Cr^{3+} + 3S_2O_8^{2-} \rightarrow 6SO_4^{2-} + Cr_2O_7^{2-} + 14H^+$; $14H^+ + Cr_2O_7^{2-} + 6Fe^{2+} \rightarrow 6Fe^{3+} + 2Cr^{3+} + 7H_2O$; b. 3.00×10^{-4} cm **76.** 0.0257 *M* ± 0.0007 *M* **78.** Use a 1-L volumetric flask and weigh to the nearest mg. Weigh between 4.24 g and 4.32 g KIO_3.

Chapter 5

2. a. 642 torr, 0.845 atm, 8.56×10^4 Pa; b. 975 torr, 1.28 atm, 1.30×10^5 Pa; c. 517 torr; 850. torr **4.** a. 749 torr, 0.986 atm, 9.99×10^4 Pa; b. 781 torr, 1.03 atm, 1.04×10^5 Pa **7.** P_{H_2} = 317 torr; P_{N_2} = 50.7 torr; P_{total} = 368 torr **9.** $V_2 = 1.15\ V_1$ **11.** 5.1×10^4 torr **13.** a. 58.3 atm; b. 1.02×10^3 K; c. 171 K **15.** 56.6 MPa **17.** 0.921 **19.** P_{N_2} = 0.448 atm; P_{O_2} = 0.126 atm; P_{NO} = 0.334 atm; P_{total} = 0.907 atm **21.** 0.286 g **23.** P_{H_2} = 0.319 atm; P_{He} = 0.161 atm **25.** 5.77 g/L ($SiCl_4$); 4.60 g/L ($SiHCl_3$) **27.** $C_2H_2Cl_2$ **29.** 1.18×10^{-2} **31.** 4.1×10^6 L air; 7.42×10^5 L H_2 **33.** 3.3×10^4 g/min **35.** 0.247 L **36.** XeF_4 **38.** $C_{12}H_{21}NO$; $C_{24}H_{42}N_2O_2$ **41.** 3.40×10^3 J/mol at 273 K; 6.81×10^3 J/mol at 546 K **44.** a. all the same; b. flask C; c. flask A **47.** NO **49.** 19 min **50.** a. 12.24 atm; b. 12.13 atm; c. The ideal gas law is higher by 0.91%. **52.** At low P and high V, no deviation is apparent. At higher P and smaller V (Experiment 6 in Example 5.1), deviation is ≈1%. **54.** 5×10^{-7} atm; 1×10^{13} molecules/cm^3 **55.** a. 0.19 torr; b. 6.6×10^{21} molecules/m^3 of air; c. 6.6×10^{15} molecules/cm^3 **59.** benzene: 9.47 ppbv, 2.31×10^{11} molecules/cm^3; toluene: 13.7 ppbv, 3.33×10^{11} molecules/cm^3 **61.** 490 atm **63.** a. 1.01×10^4 g; b. 6.65×10^4 g; c. 8.7×10^3 g; d. 1.86×10^4 g **65.** 81.4% **67.** 2.7×10^{10} molecules/cm^3; 1.3×10^{-4} g **69.** P_{He} = 50.0 torr; P_{Ne} = 76.0 torr; P_{Ar} = 90.0 torr; P_{total} = 216.0 torr **71.** C_2H_3N; C_2H_3N **73.** 1490

77. $dT = \dfrac{P\ (\text{molar mass})}{R} = \text{const}$; $-272.6°C$

79. P_{He} = 582 torr; P_{Xe} = 18 torr **82.** 0.7080 **85.** 16.03 g/mol **89.** O_2:He, 2.827:1, 1:2.827 **92.** 60.6 kJ **96.** 0.48 mol **99.** 93.8% **102.** 12 nm **105.** He, 1.94×10^{-6} atm; Rn, 2.06×10^{-6} atm **108.** a. NiC_4O_4; b. 170.8 g/mol; c. NiC_4O_4; d. 2.648×10^{-2} **111.** a. $2CH_4(g) + 2NH_3(g) + 3O_2(g) \rightarrow 2HCN(g) + 6H_2O(g)$; b. 15.6 g/s

Chapter 6

2. false. The size of K is not related to the rate of reaction. **4.** same **7.** a. $K = 7.59 \times 10^9$ L/mol, $K_p = 3.08 \times 10^8$ atm^{-1}; b. $K = 1.26 \times 10^9$ L/mol, $K_p = 5.12 \times 10^7$ atm^{-1}; c. 1.25×10^{12} molecules/cm^3 **9.** a. 16.7; b. 3.60×10^{-3}; c. 6.00×10^{-2}; d. 7.73×10^4 **11.** 10. **13.** 4.8 atm^2 **15.** 3.4 **18.** $[H_2O]$ = 0.55 M; Concentration can change since H_2O is not the solvent. **20.** a. $[HOCl] = 9.2 \times 10^{-3}\ M$, $[H_2O] = 5.1 \times 10^{-2}\ M$, $[Cl_2O] = 1.8 \times 10^{-2}\ M$; b. $[HOCl] = 0.07\ M$, $[H_2O] = [Cl_2O] = 0.22\ M$ **23.** 7.8×10^{-2} atm **25.** P_{SO_2} = 0.38 atm, P_{SO_3} = 0.12 atm, P_{O_2} = 0.44 atm **27.** a. 1.5×10^8 atm^{-1}; b. $P_{CO} = P_{Cl_2} = 1.8 \times 10^{-4}$ atm, P_{COCl_2} = 5.0 atm **29.** 100% ethanol **31.** A change in volume changes concentration (partial pressures). An increase in volume shifts the equilibrium to the side with the greater number of particles in the gas phase. A shift does not occur when reactants and products have the same number of gaseous molecules in the balanced equation. **34.** pink **36.** a. right; b. left; c. right; d. no effect; e. no effect; f. right **38.** a. 4×10^3; b. 8×10^{-2}; c. 8×10^{18} **40.** 0.29 atm **41.** adding SO_2 shifts the reaction position to the left, producing energy as heat. **44.** a. shifts left; b. shifts right; c. no effect; d. shifts right **46.** 1.5 atm^{-1} **48.** 2.1×10^{-3} atm **50.** P_{CO} = 0.58 atm, P_{CO_2} = 1.65 atm **52.** 1.43×10^{-2} atm **54.** P_{NO_2} = 0.71 atm, $P_{N_2O_4}$ = 0.11 atm **56.** 0.88 atm **58.** 0.63 atm$^{1/2}$ **61.** P_{TOTAL} = 1.0 atm: P_{NH_3} = 0.024 atm; P_{TOTAL} = 10.0 atm: P_{NH_3} = 1.4 atm; P_{TOTAL} = 100. atm: P_{NH_3} = 32 atm; P_{TOTAL} = 1000. atm: P_{NH_3} = 440 atm

Chapter 7

3. $NH_3 > OCl^- > H_2O > NO_3^-$ **5.** a. H_2O; b. ClO_2^-; c. CN^- **7.** $H^-(aq) + H_2O(l) \rightarrow H_2(g) + OH^-(aq)$; $OCH_3^-(aq) + H_2O(l) \rightarrow CH_3OH(aq) + OH^-(aq)$ **9.** a. H_2O and $CH_3CO_2^-$; b. $CH_3CO_2^-$; c. Weak bases are stronger bases than water, but are weaker bases than OH^- (K_b values between 10^{-14} and 1). **11.** a. endothermic; b. 6.631; c. 2.35×10^{-14}; d. 6.815 **14.** 7.00 **16.** a. $[H^+] = [C_2H_3O_2^-] = 1.9 \times 10^{-3}\ M$, $[OH^-] = 5.3 \times 10^{-12}\ M$, $[HC_2H_3O_2] = 0.20\ M$, pH = 2.72; b. $[H^+] = [NO_2^-] = 2.4 \times 10^{-2}\ M$, $[OH^-] = 4.2 \times 10^{-13}\ M$, $[HNO_2] = 1.5\ M$, pH = 1.62; c. $[H^+] = [F^-] = 3.5 \times 10^{-3}\ M$, $[OH^-] = 2.9 \times 10^{-12}\ M$, $[HF] = 0.017\ M$, pH = 2.46; d. $[H^+] = [Lac^-] = 3.7 \times 10^{-3}\ M$, $[OH^-] = 2.7 \times 10^{-12}\ M$, $[HLac] = 0.10\ M$, pH = 2.43 **18.** 0.92 **20.** $[H^+] = [Bz^-] = 5.1 \times 10^{-4}\ M$, $[HBz] = 4.1 \times 10^{-3}\ M$, $[OH^-] = 1.9 \times 10^{-11}\ M$, pH = 3.29 **22.** 1.4×10^{-4} **23.** $[H^+] = 5.0 \times 10^{-3}\ M$, pH = 2.30 **25.** $[H^+] = 0.088\ M$, $[Cl^-] = 0.013\ M$, $[NO_3^-] = 0.075\ M$, $[OH^-] = 1.1 \times 10^{-13}\ M$ **27.** conductivity **29.** a. 0.60%; b. 1.9%; c. 5.8%; d. Dilution shifts equilibrium to the side with the greater number of particles (% ionization increases). e. $[H^+]$ also depends on initial concentration of weak acid. **31.** 1.9×10^{-9} **33.** $[H^+] = [H_2PO_4^-] = 2.4 \times 10^{-2}\ M$, $[H_3PO_4] = 0.076\ M$, $[HPO_4^{2-}] = 6.2 \times 10^{-8}\ M$, $[PO_4^{3-}] = 1.2 \times 10^{-18}\ M$, $[OH^-] = 4.2 \times 10^{-13}\ M$ **35.** $CH_3NH_2 > NH_3 > H_2O > NO_3^-$ **37.** a. NH_3; b. NH_3; c. OH^-; d. CH_3NH_2 **39.** 12.049 **41.** nitrogen **43.** In each case H^+ adds to the nitrogen to give a positive ion. **46.** a. 1.3%; b. 4.2%; c. 6.4% **48.** 10.0 **50.** 1.3×10^{-3} **52.** a. neutral; b. neutral; c. basic, NO_2^- is a weak base; d. acidic, NH_4^+ is a weak acid; e. acidic, NH_4^+ is a stronger acid than NO_2^- is as a base; f. basic, HCO_3^- is a stronger base than an acid; g. neutral, acid strength of NH_4^+ is equal to the base strength of $C_2H_3O_2^-$; h. basic, F^- is a weak base **54.** acidic → basic: HCN, NH_4Br, KBr, NH_4CN, KCN, KOH **56.** acidic; $HSO_4^- + H_2O \rightleftharpoons H_3O^+ + SO_4^{2-}$; 1.54; $CO_3^{2-} + HSO_4^- \rightleftharpoons HCO_3^- + SO_4^{2-}$ **58.** OCl^- **61.** 2.0×10^3; 7.6×10^5 **63.** a. 1.3×10^{-4}; b. $[H_2CO_3] = [CO_3^{2-}]$; c. pH = $(pK_{a1} + pK_{a2})/2$; d. 8.31 **65.** a. $Hb(O_2)_4$ in lungs, HbH_4^{4+} in cells; b. Decreasing $[CO_2]$ will decrease $[H^+]$, favoring $Hb(O_2)_4$ formation. Breathing into a bag raises $[CO_2]$. c. $NaHCO_3$ lowers the acidity from accumulated CO_2. **69.** 2.02 **71.** 3.66 **73.** 2.5×10^{-3} **75.** a. 2.62; b. 2.4%; c. 0.56 M; d. 8.48 **77.** 1.0×10^{-5} **79.** 3.0 mol HCl **81.** 4.54 L **83.** 9.92 **85.** 6.24 **87.** 7.20 **89.** $2.2 \times 10^{-2}\ M$ **91.** $6.2 \times 10^{-7}\ M$ **93.** 1.64 **95.** a. 2.80; b. $1.1 \times 10^{-3}\ M$

Chapter 8

1. weak acid, weak base; buffers resist pH change. **3.** The quantity of acid or base that can be absorbed with a negligible pH change. All have the same pH. Capacities: 1.0 M > 0.1 M > 0.01 M. **5.** a. 2.94; b. 8.94; c. 7.00; d. 4.89 **7.** a. 4.29; b. 12.30; c. 12.30; d. 5.07 **8.** solution d; buffers resist pH change. **10.** a. 0.56; b. 0.35; c. 5.6; d. 2.2 **12.** 5.20 **14.** 0.091 **16.** a. no; b. yes; c. yes; d. no **18.** (mL OH^-, pH): (0, 2.43), (4, 3.14), (8, 3.53), (12.5, 3.86), (20, 4.46), (24, 5.24),

(24.5, 5.6), (24.9, 6.3), (25, 8.28), (25.1, 10.3), (26, 11.29), (28, 11.75), (30, 11.96) **20.** (mL H^+, pH): (0, 9.11), (4, 5.95), (8, 5.56), (12.5, 5.23), (20, 4.63), (24, 3.85), (24.5, 3.5), (25, 3.27), (26, 2.71), (28, 2.25), (30, 2.04) **22.** $pH = pK_a + \log\frac{[base]}{[acid]}$; at halfway point, [base] = [acid], so $pH = pK_a$ **24.** a. 3.11; b. 9.97; c. 7.00 **26.** pH > 5, bromcresol green is blue; pH < 8, thymol blue is yellow; pH of solution is between 5 and 8. **28.** a. yellow; b. yellow; c. green; d. colorless **32.** (18) phenolphthalein; (20) 2,4-dintrophenol **35.** pH ≈ 5; $K_a \approx 1 \times 10^{-8}$ **37.** a. 2×10^{-11} mol/L, 2×10^{-9} g/L; b. 7.32×10^{-4} mol/L, 0.286 g/L; c. 7.8×10^{-3} mol/L, 1.1 g/L; d. 9.3×10^{-5} mol/L, 9.3×10^{-3} g/L; e. 7×10^{-5} mol/L, 1×10^{-2} g/L; f. 6.5×10^{-7} mol/L, 3.1×10^{-4} g/L; g. 5.7×10^{-4} mol/L, 0.10 g/L; h. 1.3×10^{-4} mol/L, 3.6×10^{-2} g/L; i. 4×10^{-3} mol/L, 2 g/L; j. 2×10^{-16} mol/L, 3×10^{-14} g/L **39.** a. 4×10^{-17} mol/L; b. 4×10^{-11} mol/L; c. 4×10^{-29} mol/L **42.** a. 1.5×10^{-3} mol/L; b. 1.9×10^{-4} mol/L; c. 1.4×10^{-4} mol/L **45.** $[Pb^{2+}] = [Na^+] = [Cl^-] = 0.010\ M$, $[NO_3^-] = 0.020\ M$ **49.** $3.3 \times 10^{-32}\ M$ **52.** 46.6 g **55.** a. 9.24; b. 9.30 **57.** 4.8 g cacodylic acid, 14.4 ≈ 14 g sodium cacodylate **59.** a. 4.19; b. 4.37; c. 4.37; d. same; both equilibrium expressions are pertinent to this solution and must be satisfied. **61.** a. similar to H_2A^- derivation in Section 8.6; b. 6.50, 7.00, 7.81; c. $NH_4^+ + OH^- \rightarrow NH_3 + H_2O$; $CH_3CO_2^- + H^+ \rightarrow CH_3CO_2H$ **64.** $2.0 \times 10^{-37}\ M$ **66.** $2 \times 10^{-18}\ M$; lowest level of iron in serum = $1 \times 10^{-5}\ M$; complexing agents **68.** 0.056 mol/L **70.** 7.0, 2×10^{-3} mol/L; 10.0, 9×10^{-3} mol/L **72.** 0.38 mol/L **75.** 3.9 L **77.** a. 1.0 *M*; b. 9.1 **79.** a. 100. g/mol; b. 3.02 **81.** 2.68 **83.** 0.56 mL **85.** 4.36 L **87.** 7.46 **89.** 0.210 *M* **91.** a. 2.41; b. $3.6 \times 10^{-17}\ M$; c. 5.67; d. 11.30 **93.** $7.5 \times 10^{-6}\ M$ **95.** 1.00×10^{-11} mol/L **99.** a. 10.60; b. 5.54 **101.** 3.62 **104.** The third stoichiometric point is not seen since K_{a3} for phosphoric acid is so small. **107.** 1.74×10^{-8}

Chapter 9

2. 1.0-kg object with velocity 2.0 m/s **3.** $q = 30.9$ kJ, $w = -12.5$ kJ, $\Delta E = 18.4$ kJ **5.** $q = -24$ kJ, $w = 9.7$ kJ, $\Delta E = -14$ kJ **7.** a. −55.5 kJ; b. -3.55×10^4 kJ **10.** constant *V*: $\Delta E = q_v = 74.3$ kJ, $w = 0$, $\Delta H = 88.1$ kJ; constant *P*: $\Delta H = q_p = 88.1$ kJ, $w = -13.8$ kJ, $\Delta E = 74.3$ kJ **12.** Pathway one: Step 1: $q = 30.4$ kJ, $w = -12.2$ kJ, $\Delta E = 18.2$ kJ, $\Delta H = 30.4$ kJ; Step 2: $q = -28.1$ kJ, $w = 21.3$ kJ, $\Delta E = -6.8$ kJ, $\Delta H = -11$ kJ; Total: $q = 2.3$ kJ, $w = 9.1$ kJ, $\Delta E = 11.4$ kJ, $\Delta H = 19$ kJ; Pathway two: Step 1: $q = 6.84$ kJ, $w = 0$, $\Delta E = 6.84$ kJ, $\Delta H = 11.4$ kJ; Step 2: $q = 7.6$ kJ, $w = -3.0$ kJ, $\Delta E = 4.6$ kJ, $\Delta H = 7.6$ kJ; Total: $q = 14.4$ kJ, $w = -3.0$ kJ, $\Delta E = 11.4$ kJ, $\Delta H = 19.0$ kJ **16.** 0.129 J°C^{-1}g^{-1}, 26.7 J°C^{-1}mol^{-1} **18.** 170 J/g, 20. kJ/mol **20.** −66 kJ/mol **22.** −25 kJ/g, −2700 kJ/mol **24.** 4.9×10^6 kJ **27.** −296.1 kJ **29.** 28.4 kJ **31.** −220.8 kJ **33.** −233 kJ **35.** −427 kJ **37.** 1268 kJ; no, requires too much energy **40.** a. −908 kJ, −112 kJ, −140. kJ; b. $12NH_3(g) + 21O_2(g) \rightarrow 8HNO_3(aq) + 4NO(g) + 14H_2O(g)$; exothermic **42.** −2677 kJ **45.** a. 632 kJ; b. C_2H_2 **47.** −169 kJ/mol **49.** 1.2×10^5 L **51.** −129 kJ **53.** C_3H_8, −50.4 kJ/g; C_8H_{18}, −47.7 kJ/g; Propane is a gas, creating storage and safety hazards. **55.** 0.92% **57.** path 1, $w = -15$ L atm; path 2, $w = -6.0$ L atm; Work is not a state function. Work depends on the path. **59.** 2.95 kJ evolved **60.** 1×10^4 steps **63.** a. −81.8 kJ; b. −85 kJ/mol **65.** Work is done by the system for e and f. Work is done on the system for a, c, and d. No work is done for b. **66.** 0.17 g

Chapter 10

3. 2 kJ is most likely. **7.** Seven; favored by probability, not by energy. Energy must be expended to change the probability. **9.** 1.6×10^6 **11.** a. N_2O; b. H_2 (100°C, 0.5 atm); c. N_2 (STP); d. $H_2O(l)$ **15.** a. $C_{12}H_{22}O_{11}$; b. H_2O (0°C); c. $H_2S(g)$; d. He (10 K); e. N_2O; f. HCl **16.** a. negative; b. positive; c. negative; d. negative; e. negative; f. positive **18.** 238 J K^{-1} mol^{-1} **20.** 93.8 J K^{-1} mol^{-1} **22.** 9.57 J K^{-1} mol^{-1} **25.** a. $\Delta H° = -803$ kJ, $\Delta S° = -4$ J/K, $\Delta G° = -802$ kJ; b. $\Delta H° = 2802$ kJ, $\Delta S° = -262$ J/K, $\Delta G° = 2880.$ kJ; c. $\Delta H° = -416$ kJ, $\Delta S° = -209$ J/K, $\Delta G° = -354$ kJ; d. $\Delta H° = -176$ kJ, $\Delta S° = -284$ J/K, $\Delta G° = -91$ kJ **27.** −1138 kJ/mol **29.** $C_2H_4(g) + H_2O(g) \rightarrow C_2H_5OH(l)$ preferred; other reaction is never spontaneous. **31.** a. $\Delta H(+)$, $\Delta S(+)$; b. high temperatures **33.** 79.9 kJ **35.** $\Delta G° = -198$ kJ, $K = 5.07 \times 10^{34}$ **37.** a. 2.22×10^5; b. 94.3 molecules ATP/molecule glucose **39.** a. 3.5×10^{-3}; b. $\Delta G° = -17$ kJ, $K = 9.5 \times 10^2$ **41.** −188 kJ **43.** $\Delta H° = -92$ kJ, $\Delta S° = -199$ J/K, $\Delta G° = -34$ kJ, $K = 9.1 \times 10^5$; a. $\Delta G = -67$ kJ; b. $\Delta G = -68$ kJ; c. $\Delta G = -86$ kJ; d. $\Delta G = -47$ kJ **46.** ΔS_{univ} **50.** 450 K **53.** at least 8.8 kJ/mol; yes, to preserve electrical neutrality; 0.29 mol ATP **57.** 2.64 kJ **59.** 0.90 J°C^{-1}g^{-1} **61.** a. constant *V*, 1.51 kJ/mol; constant *P*, 1.94 kJ/mol; b. 219.63 J K^{-1} mol^{-1}; c. 218.30 J K^{-1} mol^{-1} **62.** 0.715; remainder of energy used to expand gas; 4.46 kJ **65.** a. 13.2 J/K; b. 471 J/K **67.** Methane, $\Delta S = 73.2$ J K^{-1} mol^{-1}; hexane, $\Delta S = 84.5$ J K^{-1} mol^{-1}. The much greater volume of one mole of hexane gas formed at 342 K compared with the volume of one mole of methane gas formed at 112 K accounts for 82% of the difference in ΔS. **69.** 2.9 J/K **71.** $q = 56.8$ kJ, $w = -3.43$ kJ, $\Delta E = 53.4$ kJ, $\Delta H = 56.8$ kJ, $\Delta S = 163$ J/K **73.** a. $q = 0$, $w = 0$, $\Delta E = 0$, $\Delta H = 0$, $\Delta S = 2.39$ J/K, $\Delta G = -0.717$ kJ; b. $q = 0.717$ kJ, $w = -0.717$ kJ, $\Delta E = 0$, $\Delta H = 0$, $\Delta S = 2.39$ J/K, $\Delta G = -0.717$ kJ **77.** 25°C: $\Delta H° = -198$ kJ, $\Delta S° = -187$ J/K; 227°C: $\Delta H° = -200.$ kJ, $\Delta S° = -191$ J/K **79.** $\Delta G° = -142$ kJ, $\Delta G = -148$ kJ **81.** $\Delta G = 11.5$ kJ, $\Delta H = 0$, $\Delta S = -38.3$ J/K **83.** 40.2 kJ/mol **85.** at 900. K, $C_p = 32.03$ J K^{-1} mol^{-1}; $\Delta S = 63.9$ J/K **87.** a. $\Delta E = 0$, $\Delta H = 0$, $\Delta S = 7.62$ J/K, $\Delta G = -2.32$ kJ, $w = -1.52$ kJ, $q = 1.52$ kJ; b. $\Delta S_{univ} = 2.64$ J/K, spontaneous **89.** a. $w = -1.72$ kJ, $q = 1.72$ kJ; b. $w = -1.25$ kJ, $q = 1.25$ kJ **93.** $K = 2.00 \times 10^{19}$, $\Delta G° = -110.$ kJ, $\Delta S° = 20$ J/K **95.** $w = 0$, $q = 0$; $q_{rev} = 0.35$ kJ **99.** at least 7.4 torr

Chapter 11

5. a. 0.10 V, SCE is anode; b. 0.53 V, SCE is anode; c. 0.02 V, SCE is cathode; d. 1.90 V, SCE is cathode; e. 0.47 V, SCE is cathode **7.** a. no; b. yes; c. yes; d. no **9.** $MnO_4^- > Cl_2 > Cr_2O_7^{2-} > Fe^{3+} > Fe^{2+} > Mg^{2+}$ **11.** a. Br_2, most positive $\mathscr{E}°$; b. Ca, most positive $-\mathscr{E}°$; c. Br^-, H_2, Ca, Cd; all give a positive $\mathscr{E}°_{cell}$; d. Br_2, H^+; both give a positive $\mathscr{E}°_{cell}$ **13.** a. none; b. $Cr_2O_7^{2-}$, O_2, MnO_2, IO_3^-; c. $PbSO_4$, Cd^{2+}, Fe^{2+}, Cr^{3+}, Zn^{2+}, H_2O

15. yes, 0.34 V $< \mathcal{E}^\circ <$ 0.54 V **17.** $H_2O_2 + 2H^+ + 2e^- \rightarrow 2H_2O$ (H_2O_2, oxidizing agent); $O_2 + 2H^+ + 2e^- \rightarrow H_2O_2$ (H_2O_2, reducing agent); $\mathcal{E}^\circ = 1.10$ V **19.** a. $\mathcal{E}^\circ =$ 0.41 V, $\Delta G^\circ = -79$ kJ, $K = 7.1 \times 10^{13}$; b. $3H_2O + 6ClO_2 \rightarrow 5ClO_3^- + Cl^- + 6H^+$ **22.** b. $Cl_2 + 2I^- \rightarrow I_2 + 2Cl^-$, $\mathcal{E}^\circ = 0.82$ V, $\Delta G^\circ = -160$ kJ, $K = 5.0 \times 10^{27}$; d. $Pb + Cu^{2+} \rightarrow Cu + Pb^{2+}$, $\mathcal{E}^\circ = 0.47$ V, $\Delta G^\circ = -91$ kJ, $K = 7.6 \times 10^{15}$; e. $4H^+ + O_2 + 4Fe^{2+} \rightarrow 4Fe^{3+} + 2H_2O$, $\mathcal{E}^\circ = 0.46$ V, $\Delta G^\circ = -180$ kJ, $K = 1.2 \times 10^{31}$ **24.** $3Mn + 8H^+ + 2NO_3^- \rightarrow 2NO + 4H_2O + 3Mn^{2+}$, $\mathcal{E}^\circ = 2.14$ V, $\Delta G^\circ = -1240$ kJ, $K = 10^{217}$; $5IO_4^- + 2Mn^{2+} + 3H_2O \rightarrow 5IO_3^- + 2MnO_4^- + 6H^+$, $\mathcal{E}^\circ = 0.09$ V, $\Delta G^\circ = -90$ kJ, $K = 2 \times 10^{15}$
26. a. $\mathcal{E}^\circ = -1.23$ V, $\Delta G^\circ = 475$ kJ; b. $\Delta H^\circ = 572$ kJ, $\Delta S^\circ = 327$ J/K; c. 90. °C: $\Delta G^\circ = 453$ kJ, $\mathcal{E}^\circ = -1.17$ V; 0°C: $\Delta G^\circ = 483$ kJ, $\mathcal{E}^\circ = -1.25$ V **28.** $\mathcal{E}^\circ = 1.21$ V; $\mathcal{E}^\circ$ will decrease with an increase in T. **29.** 77 kJ/mol **32.** Only a and d are spontaneous. **34.** 1.5×10^{13} **36.** -0.14 V
38. a. 2.12 V; b. 1.98 V; c. 2.05 V; d. Cell potential decreases with decreasing T (also oil becomes more viscous). **40.** yes
43. a. 0.40 V; b. 1.11 V

45. $\mathcal{E} = 0.84\text{ V} - \dfrac{0.0592}{4} \log \left(\dfrac{[OH^-]^4 \, [Fe^{2+}]^2}{P_{O_2}} \right)$; Acidic conditions favor corrosion.

47. a. 30. h; b. 33 s; c. 1.3 h **49.** 1.14×10^{-2} M **51.** Ag
54. In **57.** 9.12 L F_2(anode), 29.2 g K (cathode) **60.** 0.262 L H_2, 0.131 L O_2 **64.** Zinc is preferentially oxidized relative to iron. **66.** 6.0×10^{20} **69.** a. yes; b. yes; c. yes
71. a. pH, ± 0.02; $[H^+]$, $\pm 6 \times 10^{-6}$ M; b. ± 0.001 V **74.** 2.0×10^{-30} **76.** a. 5.98×10^{-10}; b. 1.9 M **79.** a. Co^{2+}; b. no; c. Co^{3+}

Chapter 12

2. 3.84×10^{14} s^{-1}, 2.54×10^{-19} J **4.** 276 nm **7.** no
9. 4.36×10^{-19} J **11.** a. 656.7 nm; b. 486.4 nm; c. 121.6 nm; d. 1886 nm **12.** $n = 1$, 91.20 nm; $n = 3$, 820.8 nm **14.** $n = 6 \rightarrow n = 5$, 7462 nm; $n = 6 \rightarrow n = 1$, 93.79 nm **16.** a. 2.6×10^{-5} nm; b. 1.6×10^{-2} nm; c. 9.8×10^{-26} nm **18.** 1.0×10^2 nm, $v = 7.3 \times 10^3$ m/s; 1.0 nm, $v = 7.3 \times 10^5$ m/s **22.** a. 5.79×10^5 nm; b. 3.64×10^{-33} m; c. much larger than H atom; d. insignificant compared to size of baseball **23.** 0.220 nm **24.** As L increases, E_n will decrease and the energy spacings will decrease. **28.** 1p, 3f, 2d **31.** 5p, 3; 3d_{z^2}, 1; 4d, 5; $n = 5$, 25; $n = 4$, 16
34. orientation in space **38.** The nodes are at a distance from the nucleus of $1.90a_0$ and $7.10a_0$. **40.** a. 1312 kJ/mol; b. 5246 kJ/mol; c. 1.180×10^4 kJ/mol; d. 4.722×10^4 kJ/mol; e. 8.866×10^5 kJ/mol **41.** H: He^+: Li^{2+}: C^{5+}: Fe^{25+}; 1: 1/2: 1/3: 1/6: 1/26 **44.** a. 32; b. 8; c. 25; d. 10; e. 6; f. 0; g. 1; h. 18; i. 0; j. 2 **48.** K: $1s^22s^22p^63s^23p^64s^1$ or $[Ar]4s^1$; Rb: $1s^22s^22p^63s^23p^64s^23d^{10}4p^65s^1$ or $[Kr]5s^1$; Fr: $[Rn]7s^1$; Pu: $[Rn]7s^26d^15f^5$ (expected); Sb: $[Kr]5s^24d^{10}5p^3$; Os: $[Xe]6s^24f^{14}5d^6$; Pd: $[Kr]5s^24d^8$ (expected); Pb: $[Xe]6s^24f^{14}5d^{10}6p^2$; I: $[Kr]5s^24d^{10}5p^5$
50. a. F: $1s^22s^22p^5$; b. K: $[Ar]4s^1$; c. Be: $1s^22s^2$; Mg: $[Ne]3s^2$; Ca: $[Ar]4s^2$; d. In: $[Kr]5s^24d^{10}5p^1$; e. C: $1s^22s^22p^2$; Si: $[Ne]3s^23p^2$; f. element 118: $[Rn]7s^25f^{14}6d^{10}7p^6$ **53.** none; an excited state; energy released **55.** a. Ne; b. S; c. Ag; d. Bi **56.** a. Be $<$ Mg $<$ Ca; b. Xe $<$ I $<$ Te; c. Ge $<$ Ga $<$ In; d. F $<$ N $<$ As; e. F $<$ Cl $<$ S **60.** The 14 lanthanides are between La and Hf. Z(Hf) $>>$ Z(La), which makes Hf significantly smaller than La. **62.** a. $[Rn]7s^25f^{14}6d^{10}7p^5$; b. halogen family, At; c. NaUus, Mg(Uus)$_2$, C(Uus)$_4$, O(Uus)$_2$; d. $UusO^-$, $UusO_2^-$, $UusO_3^-$, $UusO_4^-$ **63.** a. O $<$ S; b. I $<$ Br $<$ F $<$ Cl; c. N $<$ O $<$ F **67.** a. Li; b. P; c. O^+ **68.** a. Cs; b. Ga; c. Tl **71.** a. -1445 kJ/mol; b. -580 kJ/mol; c. 348.7 kJ/mol; d. 1255 kJ/mol; e. -1255 kJ/mol; **74.** Remaining electrons are held more strongly by the nucleus. **76.** a. Li, S, B, Cl; b. Li, S, N, F; c. K, Sc, B, Cl **78.** K_2O_2; K^{2+} is unstable.
80. 6.582×10^{14} s^{-1}, 4.361×10^{-19} J **81.** a. Li_2CO_3; b. 7×10^{-3} g/L **82.** a. Li_3N, lithium nitride; b. NaBr, sodium bromide; c. K_2S, potassium sulfide **83.** element 119: $[Rn]7s^25f^{14}6d^{10}7p^68s^1$ **86.** O, 2; O^+, 3; O^-, 1; Fe, 4; Mn, 5; S, 2; F, 1; Ar, 0 **88.** $r = a_0$, $\theta = 0$: $\psi_{2p_z} = 1.57 \times 10^{14}$, $\psi^2 = 2.46 \times 10^{28}$; $r = a_0$, $\theta = 90°$: $\psi_{2p_z} = 0$, $\psi^2 = 0$
90. a. $P(g) \rightarrow P^+(g) + e^-$; b. $P(g) + e^- \rightarrow P^-(g)$ **92.** The electron is completely separated from the atom. **93.** a. B, excited state; ground state: $1s^22s^22p^1$; b Ne, ground state **96.** no
99. 386 nm **100.** a. 18; b. 2; c. 0; d. 1 **103.** a. 1212 kJ; b. -443 kJ; c. -1026 kJ **105.** a. As: $[Ar]4s^23d^{10}4p^3$; b. $[Rn]7s^25f^{14}6d^{10}7p^4$; c. Ta: $[Xe]6s^24f^{14}5d^3$; Ir: $[Xe]6s^24f^{14}5d^7$; d. Ti: $[Ar]4s^23d^2$; Ni: $[Ar]4s^23d^8$; Os: $[Xe]6s^24f^{14}5d^6$ **107.** a. 24; b. 6; c. 12; d. 2; e. 26; f. 24.9 g; g. $[Ar]4s^13d^5$ **110.** a. 4; b. 4, 16; c. 20; d. 28 **113.** a. 2.2×10^{-9}; b. 1.5×10^{-9}; c. 2.9×10^{-10}; d. 1.9×10^{-4}; e. 1×10^{-3} **114.** 1/6 **117.** 46.5 pm

Chapter 13

1. a. Electronegativity: The ability of an atom in a molecule to attract electrons to itself. Electron affinity: The energy change for the reaction $M(g) + e^- \rightarrow M^-(g)$. The $E.A.$ deals with isolated atoms in the gas phase. b. Covalent bond: sharing of an electron pair. Polar covalent bond: unequal sharing of an electron pair. c. Ionic bond: electrons are no longer shared. Completely unequal sharing (i.e., transfer) of electrons from one atom to another. **3.** a. C $<$ N $<$ O; b. Se $<$ S $<$ Cl; c. Sn $<$ Ge $<$ Si; d. Tl $<$ Ge $<$ S **4.** a. Ge—F; b. P—Cl; c. S—F; d. Ti—Cl **5.** Order of $E.N.$ from Fig. 13.3: a. C (2.5) $<$ N (3.0) $<$ O (3.5), same; b. Se (2.4) $<$ S (2.5) $<$ Cl (3.0), same; c. Si $=$ Ge $=$ Sn (1.8), different; d. Tl (1.8) $=$ Ge (1.8) $<$ S (2.5), different. Most polar bonds using actual $E.N.$ values: a. Si—F and Ge—F equal polarity (Ge—F predicted); b. P—Cl (same as predicted); c. S—F (same as predicted); d. TiCl (same as predicted) **7.** a. Cu $>$ Cu^+ $>$ Cu^{2+}; b. Pt^{2+} $>$ Pd^{2+} $>$ Ni^{2+}; c. Se^{2-} $>$ S^{2-} $>$ O^{2-} **8.** a. Mg^{2+}: $1s^22s^22p^6$; Sn^{2+}: $[Kr]5s^24d^{10}$; K^+: $1s^22s^22p^63s^23p^6$; Al^{3+}: $1s^22s^22p^6$; Tl^+: $[Xe]6s^24f^{14}5d^{10}$; As^{3+}: $[Ar]4s^23d^{10}$; b. N^{3-}, O^{2-}, and F^-: $1s^22s^22p^6$; Te^{2-}: $[Kr]5s^24d^{10}5p^6$ **9.** a. Sc^{3+}; b. Te^{2-} **10.** Isoelectronic: same number of electrons. There are two variables, number of protons and number of electrons, that will determine the size of an ion. Keeping the number of electrons constant, we have to consider only the number of protons to predict trends in size.
12. a. NaCl, Na^+ smaller than K^+; b. LiF, F^- smaller than Cl^-; c. MgO, O^{2-} greater charge than OH^- **14.** $\Delta H_f^\circ = -412$ kJ/mol **16.** a. The lattice energy for $Mg^{2+}O^{2-}$ will be much larger than for Mg^+O^-. b. Mg^+ and O^- both have unpaired electrons. In Mg^{2+} and O^{2-} there are no unpaired electrons. Hence, Mg^+O^- would be paramagnetic; $Mg^{2+}O^{2-}$ would be

diamagnetic. Paramagnetism can be detected by measuring the mass of a sample in the presence and absence of a magnetic field. **18.** $\Delta H = +737$ kJ/mol **19.** The extra electron–electron repulsions are much greater than the attraction of the electron for the nucleus.

21.

Compound	Q_1Q_2	Lattice Energy
$FeCl_2$	(+2)(−1)	−2,631 kJ/mol
FeO	(+2)(−2)	−5,359 kJ/mol
$FeCl_3$	(+3)(−1)	−3,865 kJ/mol
Fe_2O_3	(+3)(−2)	−14,744 kJ/mol

The product of Q_1Q_2 is directly proportional to lattice energy.

22. a. Li_3N, lithium nitride; b. Ga_2O_3, gallium(III) oxide; c. RbCl, rubidium chloride; d. BaS, barium sulfide **23.** −42 kJ **24.** a. −183 kJ; b. −109 kJ **25.** a. −184 kJ/mol (−183 from bond energies); b. −92 kJ/mol (−109 from bond energies); Bond energies seem to give a reasonably good estimate for the enthalpy change of a gas phase reaction. **28.** a. −43 kJ, 37 kJ; b. −1373 kJ; c. −1077 kJ **29.** Since both reactions are highly exothermic, the high temperature is not needed to provide energy. **31.** a. $\Delta H = 1549$; b. $\Delta H = 1390$. **32.** a. Using SF_4 data: $D_{SF} = 343$ kJ/mol; Using SF_6 data: $D_{SF} = 327$ kJ/mol; b. The S—F bond energy in the table is 327 kJ/mol. The value in the table was based on the S—F bond in SF_6. c. S(*g*) and F(*g*) are not the most stable form of the elements at 25°C and 1 atm. The most stable forms are $S_8(s)$ and $F_2(g)$; $\Delta H_f^\circ = 0$ for these two species. **33.** $D_{calc} = 389$ kJ/mol as compared to 391 kJ/mol in the table. **34.** 388 kJ/mol; similar to Exercise 33

37. a. H—C≡N: b. H—P—H (with H on P) c. $CHCl_3$: H—C(—Cl)(—Cl)—Cl

d. $[NH_4]^+$ e. $[BF_4]^-$ f. F—Se—F

38. a. $POCl_3$: O, Cl, Cl, Cl bonded to P

SO_4^{2-}: [O—S(—O)(—O)—O]$^{2-}$

XeO_4: O—Xe(—O)(—O)—O

PO_4^{3-}: [O—P(—O)(—O)—O]$^{3-}$

ClO_4^-: [O—Cl(—O)(—O)—O]$^-$

b. NF_3: F—N(—F)—F

SO_3^{2-}: [O—S(—O)—O]$^{2-}$

PO_3^{3-}: [O—P(—O)—O]$^{3-}$

ClO_3^-: [O—Cl(—O)—O]$^-$

c. ClO_2^-: [O—Cl—O]$^-$

SCl_2: Cl—S—Cl

PCl_2^-: [Cl—P—Cl]$^-$

39. Molecules/ions that have the same number of valence electrons and the same number of atoms will have similar Lewis structures.

40. a. NO_2^-: [O=N—O]$^-$ ⟷ [O—N=O]$^-$

HNO_2: O=N—O—H

NO_3^-: [O=N(—O)—O]$^-$ ⟷ [O—N(=O)—O]$^-$ ⟷ [O—N(—O)=O]$^-$

HNO_3: O=N(—O)—O—H ⟷ O—N(=O)—O—H

b. SO_4^{2-}: [see Exercise 38(a)]

HSO_4^-: [:Ö: above; :Ö—S—Ö—H; :Ö: below]$^-$

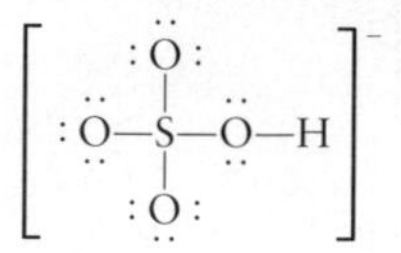

H_2SO_4: :Ö: above; :Ö—S—Ö—H; :Ö—H below

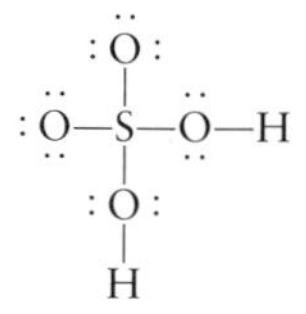

c. CN^-: [:C≡N:]$^-$ HCN: H—C≡N:

41. Ozone O=O—O ⟷ O—O=O

Sulfur dioxide O=S—O ⟷ O—S=O

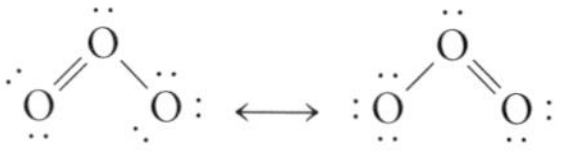

Sulfur trioxide (S bonded to three O; the double bond S=O drawn in each of three positions) ⟷ ⟷

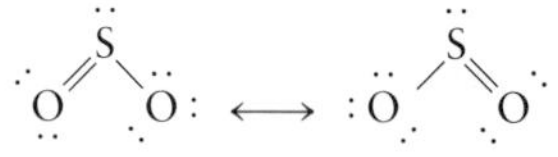

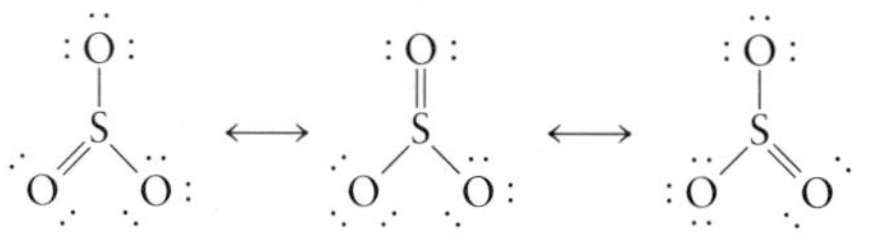

42. PAN, $H_3C_2NO_5$:

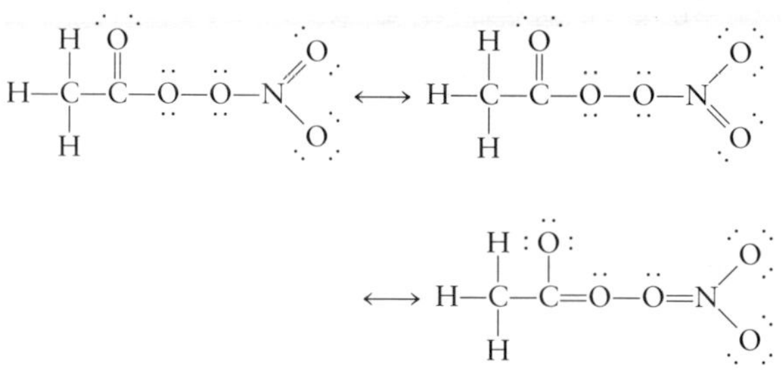

H—C(H)(H)—C(=O)—O—O—N(=O)—O ⟷ H—C(H)(H)—C(=O)—O—O—N(—O)=O

⟷ H—C(H)(H)—C(—O)=O—O=N(—O)—O

45. (two Kekulé structures of benzene, C_6H_6) ⟷

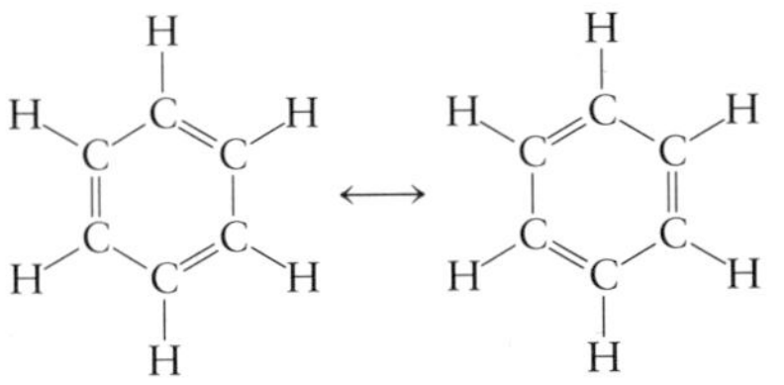

46. With resonance all carbon atoms are equal (we indicate this with a circle in the ring), giving three different structures:

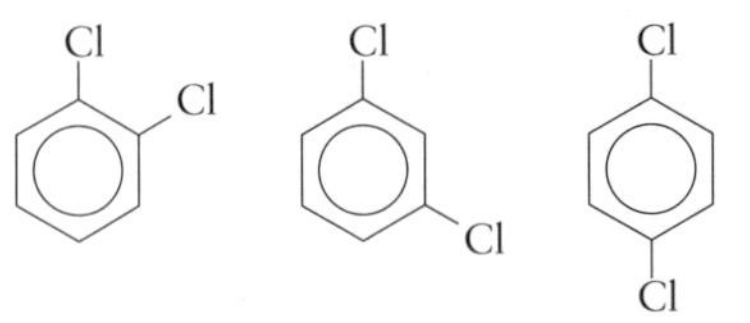

(1,2-, 1,3-, and 1,4-dichlorobenzene; Cl labels on rings)

Localized double bonds give four unique structures.

47. Borazine, $B_3N_3H_6$:

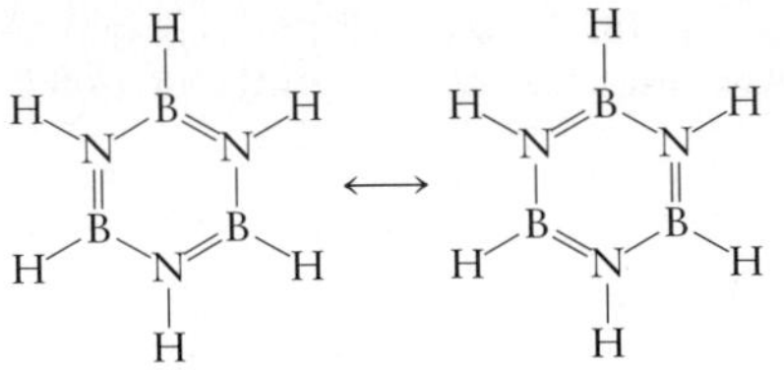

(two resonance structures of the six-membered ring of alternating B and N, each atom bearing H) ⟷

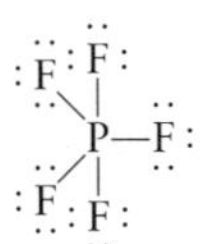

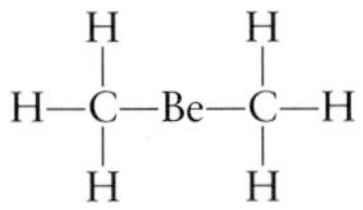

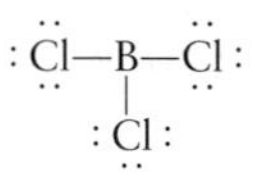

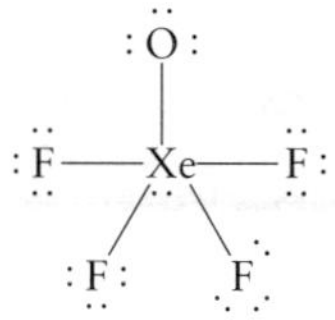

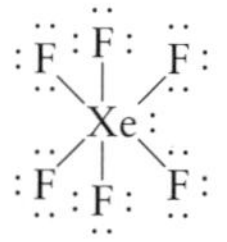

49. PF_5: P bonded to five F (each :F̈:)

BrF_3: :F̈—Br—F̈: with :F̈: below Br (two lone pairs on Br)

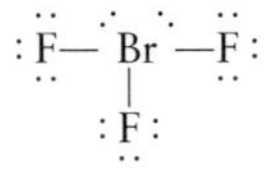

$Be(CH_3)_2$: H—C(H)(H)—Be—C(H)(H)—H

BCl_3: :C̈l—B—C̈l: with :C̈l: below B

$XeOF_4$: :Ö: above Xe; :F̈—Xe—F̈: ; :F̈: and :F̈: below

XeF_6: Xe bonded to six F, with one lone pair on Xe

SeF_4: :F̈—Se—F̈: with :F̈: and :F̈: below; one lone pair on Se

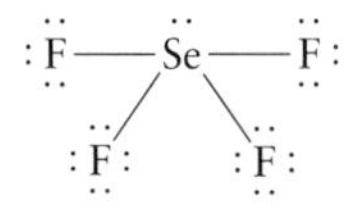

52. The N—N bond is between a double bond and triple bond; the N—O bond is between a single and double bond; the last resonance structure doesn't appear to be as important as the other two.

N=N=O ⟷ :N≡N—Ö:

54. $CH_3OH > CO_3^{2-} > CO_2 > CO$; $CH_3OH < CO_3^{2-} < CO_2 < CO$

56. N=N=O ⟷ :N≡N—Ö: ⟷ :N̈—N≡O:

(−1) (+1) (0) (0) (+1) (−1) (−2) (+1) (+1)

We should eliminate :N̈—N≡O since it has a formal charge of +1 on the most electronegative element. This is consistent

with the observation that the N—N bond is between a double bond and a triple bond and that the N—O bond is between a single and a double bond.

57. The structure

F (+1) = B (−1), with :F: (0) and :F: (0)

obeys the octet rule but has a +1 formal charge on the most electronegative element (fluorine) and a negative formal charge on a much less electronegative element (boron). This is just the opposite of what we expect.

58. :C≡O:

(−1)(+1)

Electronegativity predicts the opposite polarization. The two opposing effects seem to cancel to give a much less polar molecule than expected.

59. See Exercise 38(a) for Lewis structures. a. P, FC = +1; b. S, FC = +2; c. Cl, FC = +3; d. P, FC = +1

60. a. :Cl—P(=O)(—Cl:)—Cl: b. [O=S(=O)(—O:)—O:]$^{2-}$

c. [O=Cl(=O)(=O)—O:]$^{-}$ d. [O=P(—O:)(—O:)—O:]$^{3-}$

61. [37] a. linear, 180°; b. trigonal pyramid, <109.5°; c. tetrahedral, ≈109.5°; d. tetrahedral, 109.5°; e. tetrahedral, 109.5°; f. bent, <109.5°; [38] a. all are tetrahedral, ≈109.5°; b. all are trigonal pyramid, <109.5°; c. all are bent (nonlinear), <109.5°; [40] a. NO_2^-: bent, <120°; HNO_2: bent about N, <120°, bent about N—O—H, <109.5°, NO_3^-: trigonal planar, 120°; HNO_3: trigonal planar about N, 120°, bent about N—O—H, <109.5°; b. tetrahedral about S in all three, 109.5°, bent about S—O—H, <109.5°; c. linear, 180°; d. all are linear, 180°; e. linear about both carbons, 180°

62. a. linear, 180°; b. T-shaped, FClF angles are ≈90°. Since the lone pair will take up more space, the FClF bond angles will probably be slightly less than 90°. c. see saw, ≈90°, ≈120°; d. trigonal bipyramid, 90°, 120°

63. a. trigonal pyramid, <109.5°; b. bent, <109.5°; c. tetrahedral, 109.5°

64. a. square pyramid, ≈90°; b. square planar, 90°; c. octahedral, 90°

66. a. CH_2Cl_2 and $CHCl_3$ have dipole moments. C—Cl bonds are polar (chlorine is the partial negative end). C—H bonds are basically nonpolar. Net dipole points to the middle of the partially negative chlorines. b. N_2O is polar.

$\delta+$ $\delta-$

:N=N=O:

67. a. CrO_4^{2-} is [:O—Cr(—O:)(—O:)—O:]$^{2-}$ tetrahedral;

b. Dichromate ion: $Cr_2O_7^{2-}$ has a roughly tetrahedral arrangement of O atoms about each Cr and the Cr—O—Cr bond is bent.

[(O)(O)(O)Cr—O—Cr(O)(O)(O)]$^{2-}$

68. The two general requirements for a polar molecule are:
1. polar bonds;
2. a structure such that the bond polarities do not cancel.
There are a *few* exceptions to these requirements. O_3, PH_3, and AsH_3 are all slightly polar but have nonpolar bonds. Polarity of these molecules is due to lone pairs on the central atoms.

69. a. OCl_2 is :Cl—O—Cl: bent, polar;

Br_3^- is [:Br—Br—Br:]$^-$ linear, nonpolar;

BeH_2 is H—Be—H linear, nonpolar;

BH_2^- is [H—B—H]$^-$ bent, polar

b. BCl_3 is :Cl—B(—Cl:)—Cl: trigonal planar, nonpolar;

NF_3 is :F—N(—F:)—F: trigonal pyramid, polar;

IF_3 is :F—I(—F:)—F: T-shaped, polar

c. CF_4 is :F—C(—F:)(—F:)—F: tetrahedral, nonpolar;

SeF_4 is :F—Se(—F:)(—F:)—F: see-saw, polar;

XeF_4 is (Lewis structure) square planar, nonpolar

d. IF_5 is (Lewis structure) square pyramid, polar;

AsF_5 is (Lewis structure) trigonal bipyramid, nonpolar

70. :F—S—F: bent, polar, <109.5°;

(Lewis structure of SF_4) see-saw, polar, ≈90°, ≈120°;

(Lewis structure of SF_6) octahedral, nonpolar, 90°;

(Lewis structure of F_3S—SF) nonlinear, polar, ≈90°, <109.5°, ≈120°

73. a. H—O—C≡N: polar;

b. :F—Be—F: nonpolar;

c. (Lewis structure of KrF_4) nonpolar;

d. (Lewis structure of CF_2Cl_2) polar

74. The general structure of a trihalide ion is

$[X—X—X]^-$ (Lewis structure)

Bromine and iodine are large and have empty *d* orbitals to accommodate the expanded octet. Fluorine is small and its valence shell contains only 2*s* and 2*p* orbitals (4 orbitals) and cannot expand its octet.

76. C≡O 1072 kJ/mol, N≡N 941 kJ/mol. CO is polar while N_2 is nonpolar. This may lead to a greater reactivity.
77. a. SO_4^{2-}; SO_4 does not obey octet rule. b. PF_5; N does not expand octet. c. SF_6; O does not expand octet. d. BH_4^-; BH_3 does not obey octet rule.
79. No, we would expect the more highly charged ions to have a greater attraction for an electron. Thus, the electronegativity of an element does depend on its oxidation state.
81. a. BF_3 is (Lewis structure) trigonal planar, nonpolar;

b. PF_3 is (Lewis structure) trigonal pyramid, polar;

c. BrF_3 is (Lewis structure) T-shaped, polar;

Molecular structure depends on the number of atoms *and* the number of lone pairs surrounding the central atom.

83. H—C(=O)—N(CH_3)—CH_3 ⟷ H—C(—O:)=N(CH_3)—CH_3 (resonance structures)

85. a. $Na^+(g) + Cl^-(g) \rightarrow NaCl(s)$; b. $NH_4^+(g) + Br^-(g) \rightarrow NH_4Br(s)$; c. $Mg^{2+}(g) + S^{2-}(g) \rightarrow MgS(s)$; d. $\frac{1}{2}O_2(g) \rightarrow O(g)$; e. $O_2(g) \rightarrow 2O(g)$ **87.** yes

Chapter 14

1. a. tetrahedral, 109.5°, sp^3, nonpolar

(Lewis structure of CF_4)

b. trigonal pyramid, <109.5°, sp^3, polar

(Lewis structure of NF_3)

c. bent, <109.5°, sp^3, polar

(Lewis structure of OF_2)

d. trigonal planar, 120°, sp^2, nonpolar

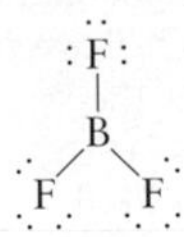

e. linear, 180°, sp, nonpolar

H—Be—H

f. see-saw
$a \approx 120°$,
$b \approx 90°$,
$c \approx 180°$
dsp^3
polar

g. trigonal bipyramid
a 90°, b 120°
dsp^3
nonpolar

h. linear, 180°, dsp^3, nonpolar

:F—Kr—F:

i. square planar, 90°, d^2sp^3, nonpolar

2. a. bent, sp^2

(only one resonance form shown);

b. trigonal planar, sp^2

(only one resonance form shown)

c. tetrahedral, sp^3

d. tetrahedral geometry about each S, sp^3

e. trigonal pyramid, sp^3

f. tetrahedral, sp^3

3. a. sp^3, P—P—P bond angle 60°

b. Lewis structure has one lone pair on each P and two lone pairs on each O. sp^3, POP and OPO angles <109.5°

5. No, the CH_2 planes are mutually perpendicular due to the orientation of the p orbitals.

6. Biacetyl

All CCO angles are 120°. The six atoms are not in the same plane.

Acetoin

angle a = 120°, angle b = 109.5°

7. In biacetyl there are 11 σ and 2 π bonds. In acetoin there are 13 σ and 1 π bonds. **10.** To complete the Lewis structure, add lone pairs to the O and N atoms to satisfy octet. a. 6; b. 4; c. The center N in —N=N=N group; d. 33 σ; e. 5 π; f. The six-membered ring is planar. g. 180°; h. <109.5°; i. sp^3 **12.** a. Each oxygen has two lone pairs and each nitrogen has one lone pair. b. piperine: 0 sp, 11 sp^2, 6 sp^3; capsaicin: 0 sp, 9 sp^2, 9 sp^3; c. sp^3; d. angles a, b, c, d, g, i, and k are ≈120°; angles e, f, h, j, and l are ≈109.5° **14.** a. H_2^+: stable; H_2: stable; H_2^-: stable; H_2^{2-}: not stable; b. He_2^{2+}: stable; He_2^+: stable; He_2: not stable; c. Be_2: not stable; B_2: stable; Li_2: stable **16.** MO model does a better job. NO has an odd number of electrons, which makes it impossible to draw a Lewis structure that obeys the octet rule. **17.** Bond energy is proportional to bond order. Bond length is inversely proportional to bond order. Bond energy and length can be measured. **18.** See Fig. 14.36. **19.** See Figs. 14.45 and 14.46. **20.** a. $(\sigma_{1s})^2$; B.O. = 1; diamagnetic; b. $(\sigma_{2s})^2(\sigma_{2s}^*)^2(\pi_{2p})^2$; B.O. = 1; paramagnetic; c. $(\sigma_{2s})^2(\sigma_{2s}^*)^2(\pi_{2p})^4(\sigma_{2p})^2(\pi_{2p}^*)^4$; B.O. = 1; diamagnetic; d. $(\sigma_{2s})^2(\sigma_{2s}^*)^2(\pi_{2p})^4$; B.O. = 2; diamagnetic; e. $(\sigma_{2s})^2(\sigma_{2s}^*)^2(\pi_{2p})^4(\sigma_{2p})^1$; B.O. = 2.5; paramagnetic; f. $(\sigma_{2s})^2(\sigma_{2s}^*)^2(\pi_{2p})^4(\sigma_{2p})^2$; B.O. = 3; diamagnetic **25.** See Example 14.7 **26.** C_2^{2-} has 10 valence electrons. $[:C\equiv C:]^{2-}$; sp hybrid orbitals used; $(\sigma_{2s})^2(\sigma_{2s}{}^*)^2(\pi_{2p})^4(\sigma_{2p})^2$; B.O. = 3. Both give the same picture, a triple bond composed of one σ and two π bonds. Both predict the ion to be diamagnetic. Lewis structures deal well with diamagnetic (all electrons paired) species. The Lewis model cannot really predict magnetic properties. $C_2{}^{2-}$ is isoelectronic with the CO molecule. **27.** a. The electrons would be closer to F. The F atom is more electronegative than the H atom. b. The $2p$ orbital of F is lower in energy than the $1s$ orbital of H. The bonding MO would have more fluorine $2p$ character since it is closer in energy to the $2p$ atomic orbital. c. The antibonding MO would place more electron density closer to H and would have a greater contribution from the hydrogen $1s$. **28.** a. The electron removed from N_2 is in the σ_{2p} molecular orbital, which is lower in energy than the $2p$ atomic orbital from which the electron in atomic nitrogen is removed. Since the electron removed from N_2 is lower in energy than the electron in N, the ionization energy of N_2 is greater than for N. b. F_2 should have a lower ionization energy than F. The electron removed from F_2 is in a $\pi_{2p}{}^*$ antibonding molecular orbital, which is higher in energy than the $2p$ atomic orbitals.

30. a. :F—Cl—O: bent, sp^3;

b. :F—Cl(—O:)—O: trigonal pyramid, sp^3;

c. :O—Cl(—F:)(—O:)—O: tetrahedral, sp^3

d. ClF_3O (:F:, :F:, :F:, :O: around :Cl:) see-saw, dsp^3;

e. ClF_3O_2 (:O:, :O:, :F:, :F:, :F: around Cl) trigonal bipyramid, dsp^3

31. $F_2ClO_2{}^-$: [:O:, :O:, :F:, :F: around :Cl:]$^-$ see-saw, dsp^3;

F_4ClO^-: [:O:, four :F: around :Cl:]$^-$ square pyramid, d^2sp^3;

F_2ClO^+: [:F—Cl(—F:)—O:]$^+$ trigonal pyramid, sp^3;

$F_2ClO_2{}^+$: [:O:, :F—Cl—O:, :F:]$^+$ tetrahedral, sp^3

35. In order to rotate about the double bond, the π bond must be broken, while the σ bond remains intact. Bond energies are 347 kJ/mol for a C—C and 614 kJ/mol for a C=C. If we take the single bond as the strength of the σ bond, then the strength of the π bond is $(614 - 347) = 267$ kJ/mol. Thus, 267 kJ/mol must be supplied to rotate about a carbon–carbon double bond. **39.** NO^+: Lewis structure and MO model agree. Both predict NO^+ has a triple bond and is diamagnetic. NO: Lewis structures are inadequate for odd electron species. MO model predicts a bond order of 2.5 and predicts NO is paramagnetic. NO^-: Both models predict a double bond. MO model correctly predicts NO^- is paramagnetic while Lewis structure predicts NO^- is diamagnetic (incorrect). Bond energy: $NO^- < NO < NO^+$; Bond length: $NO^+ < NO < NO^-$ **40.** N_2: $(\sigma_{2s})^2(\sigma_{2s}{}^*)^2(\pi_{2p})^4(\sigma_{2p})^2$ in ground state, B.O. = 3, diamagnetic; 1st excited state: $(\sigma_{2s})^2(\sigma_{2s}{}^*)^2(\pi_{2p})^4(\sigma_{2p})^1(\pi_{2p}{}^*)^1$, B.O. = 2, paramagnetic (two unpaired electrons) **41.** We can draw a Lewis structure for Be_2 although it does not obey the octet rule. The MO model predicts a bond order of zero; the MO model predicts that Be_2 shouldn't exist. Be_2 has not yet been observed. **42.** Paramagnetic: Unpaired electrons are present. Measure the mass of a substance in the presence and absence of a magnetic field.

43.

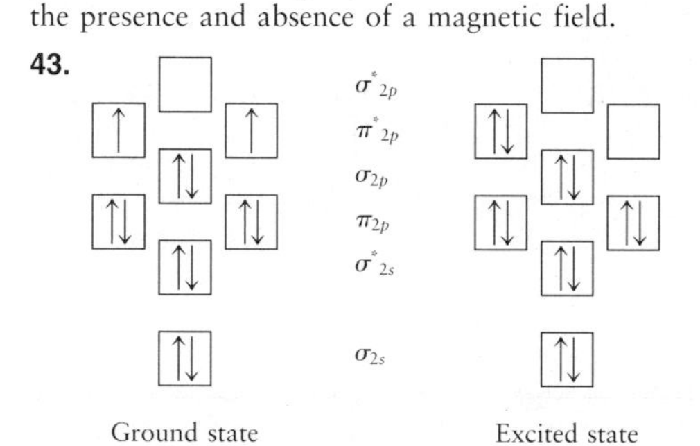

Lewis structure shows no unpaired electrons. The ground state MO diagram for O_2 has two unpaired electrons. It takes energy to pair electrons in the same orbital. The MO diagram with no unpaired electrons is at a higher energy and is an excited state.

47. a. yes, tetrahedral; b. sp^3 in both; c. d orbitals; d. structure B; Formal charges are all zeros.

Chapter 15

1. $-d[I_2]/dt = 0.0040$ mol L^{-1} s^{-1}; $d[S_4O_6{}^{2-}]/dt = 0.0040$ mol L^{-1} s^{-1}; $d[I^-]/dt = 0.0080$ mol L^{-1} s^{-1} **3.** a. mol L^{-1} s^{-1}; b. Rate = k, k has units of mol L^{-1} s^{-1}; c. Rate = k[A], k has units of s^{-1}; d. Rate = $k[A]^2$, k has units of L mol^{-1} s^{-1}; e. Rate = $k[A]^3$, k has units of L^2 mol^{-2} s^{-1} **5.** a. molecules cm^{-3} s^{-1}; b. molecules cm^{-3} s^{-1}; c. s^{-1}; d. cm^3 molecules^{-1} s^{-1}; e. cm^6 molecules^{-2} s^{-1} **6.** $k = 7.47 \times 10^8$ L mol^{-1} s^{-1} **7.** a. 0.07 mol/L; b. $t = 2.08 \times 10^3$ s; c. Both H_2 and I_2 are formed at the rate of 6.00×10^{-5} mol L^{-1} s^{-1}. **8.** a. Rate = $k[NO]^2[Cl_2]$; b. $k = 1.8 \times 10^2$ L^2 mol^{-2} min^{-1} **9.** a. Rate = $k[I^-][S_2O_8{}^{2-}]$; b. k (L mol^{-1} s^{-1}): 3.9×10^{-3}; 3.9×10^{-3}; 3.5×10^{-3}; 3.4×10^{-3}; 3.6×10^{-3}, $k_{mean} = 3.7 \times 10^{-3}$ L mol^{-1} s^{-1} $\pm$ 0.3×10^{-3} L mol^{-1} s^{-1} **11.** a. first order in Hb, first order in CO; b. Rate = k[Hb][CO]; c. 0.280 L μmol^{-1} s^{-1}; d. 2.26 μmol L^{-1} s^{-1} **13.** integrated rate law: $1/[C_4H_6] = kt + 1/[C_4H_6]_0$; differential rate law: Rate = $k[C_4H_6]^2$; $k = 1.4 \times 10^{-2}$ L mol^{-1} s^{-1} **14.** integrated rate

law: $\ln[C_6H_5N_2Cl] = -kt + \ln[C_6H_5N_2Cl]_0$; differential rate law: Rate $= k[C_6H_5N_2Cl]$; $k = 6.9 \times 10^{-2}\ s^{-1}$
15. $\ln[H_2O_2] = -kt + \ln[H_2O_2]_0$; Rate $= k[H_2O_2]$; $k = 8.3 \times 10^{-4}\ s^{-1}$ **17.** a. first order with respect to O; b. Rate $= k[NO_2][O]$; $k = 1.0 \times 10^{-11}$ cm^3 molecules^{-1} s^{-1}
19. a. ^{239}Pu, $k = 2.845 \times 10^{-5}\ yr^{-1} = 9.021 \times 10^{-13}\ s^{-1}$; ^{241}Pu, $k = 0.053\ yr^{-1} = 1.7 \times 10^{-9}\ s^{-1}$; b. ^{241}Pu; c. 2.1×10^{13} disintegrations/s; 4.7 g ^{241}Pu left after 1 yr; 2.9 g left after 10 yr; 0.024 g left after 100 yr **20.** $k = 9.2 \times 10^{-3}\ s^{-1}$; $t_{1/2} = 75$ s **21.** 62 days **22.** $k = 0.12$ L mol^{-1} s^{-1} **25.** a. An elementary reaction (step) is one in which the rate law can be written from the molecularity. b. The mechanism of a reaction is the series of elementary reactions that occur to give the overall reaction. The sum of all the steps in the mechanism gives the balanced equation. c. The rate-determining step is the slowest elementary reaction in any given mechanism. **26.** a. Rate $= k[CH_3NC]$; b. Rate $= k[O_3][NO]$ **27.** This is a possible mechanism with the second step being rate determining. **29.** b
30. The first step is rate determining. **33.** a. The greater the frequency of collisions, the greater the opportunities for molecules to react, and hence, the greater the rate. b. Chemical reactions involve the making and breaking of chemical bonds. The kinetic energy of the collision can be used to break bonds. c. For a reaction to occur, the reactive portion of each molecule must be involved in a collision. Thus, only some collisions have the right orientation. **34.** In a unimolecular reaction, a single reactant molecule decomposes to products. In a bimolecular reaction, two molecules collide to give products. **35.** The probability of the simultaneous collision of three molecules with the correct energy and orientation is exceedingly small. **36.** $H_3O^+(aq) + OH^-(aq) \rightarrow 2H_2O(l)$ should have the faster rate. H_3O^+ and OH^- will be electrostatically attracted to each other; Ce^{4+} and Hg_2^{2+} will repel each other. **37.** $4.8 \times 10^3\ s^{-1}$ **40.** The ln$k$ vs. $1/T$ plot gives a straight line with slope equal to $-E_a/R$. From the slope, E_a equals 11.9 kJ/mol. **41.** 53 kJ/mol

43.

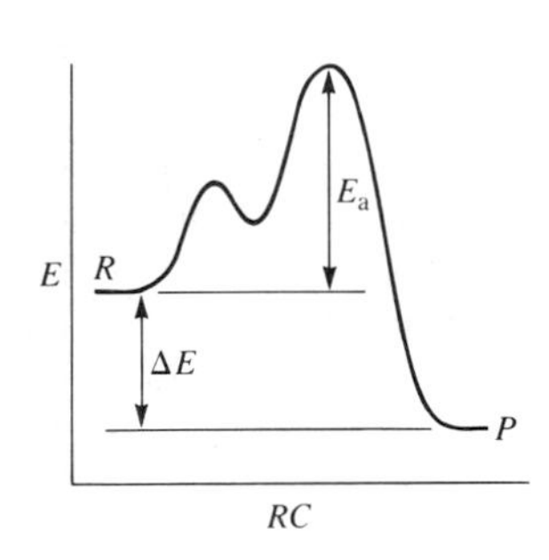

46. a. A homogeneous catalyst is in the same phase as the reactants. b. A heterogeneous catalyst is in a different phase than the reactants. **47.** A catalyst increases the rate of a reaction by providing an alternate pathway (mechanism) with lower activation energy. **48.** Yes, the catalyst takes part in the mechanism and will appear in the rate law. **49.** No, the catalyzed reaction has a different mechanism and hence a different rate law. **50.** a. NO is the catalyst; b. NO_2 is an intermediate; c. 2.3 **53.** At high [S], the enzyme is completely saturated with substrate. No free enzyme is available to catalyze further substrate conversion to product. **54.** Rate $= k[H_2SeO_3][H^+]^2[I^-]^3$; $k_{ave} = 5.2 \times 10^5$ L^5 mol^{-5} s^{-1} **57.** a. untreated: $k = 0.465$ day^{-1}; deacidifying agent: $k = 0.659$ day^{-1}; antioxidant: $k = 0.779$ day^{-1}; b. No, the silk degrades more rapidly with the additives.

58. untreated: $t_{1/2} = 1.49$ day; deacidifying agent: $t_{1/2} = 1.05$ day; antioxidant: $t_{1/2} = 0.890$ day **61.** a. W; lower E_a than Os; b. 1.41×10^{30} times faster; c. H_2 decreases the rate of the reaction. For the decomposition to occur, NH_3 molecules must be adsorbed on the surface of the catalyst. If H_2 that is produced from the reaction is also adsorbed on the catalyst, then there are fewer sites for NH_3 molecules to be adsorbed and the rate is decreased. **62.** Rate $= k_2k_1[I^-][OCl^-]/k_{-1}[OH^-]$ **63.** $BrO_3^- + HSO_3^- \rightarrow SO_4^{2-} + HBrO_2$ (slow) **65.** 4.27×10^2 s
67. 54 kJ/mol **69.** a. 370 torr; b. 1.7×10^{-4} atm/s; c. zero order **71.** a. second order; b. 17.1 s; c. 34.6 s; d. 850 s; e. 31 kJ/mol **74.** a. 1.1×10^{-2} M; b. 2.5×10^{-2} M **75.** Rate $= k[A][B]^2$; $k = 1.4 \times 10^{-2}$ L^2 mol^{-2} s^{-1} **77.** a. Rate $= k[CH_3X]$; $k = 0.93$ h^{-1}; b. 8.80×10^{-10} h; c. 3.0×10^2 kJ/mol; d. $CH_3X \rightarrow CH_3 + X$ (slow), $CH_3 + Y \rightarrow CH_3Y$ (fast) **79.** a. 24 kJ/mol; b. 12 s; c. Rule of thumb is good to 2 significant figures.

80. a. $MoCl_5^-$;

b. $$\frac{d[NO_2^-]}{dt} = \frac{k_1k_2[NO_3^-][MoCl_6^{2-}]}{k_{-1}[Cl^-] + k_2[NO_3^-]}$$

82. $$\frac{-d[N_2O_5]}{dt} = \left[\frac{2k_1k_2}{k_{-1}[M] + 2k_2}\right][N_2O_5][M]$$

86. a. Rate $= k[H_2O_2][I^-]$; b. 7.0×10^{-2} L mol^{-1} min^{-1}; c. 4.4×10^{-3} mol/min

Chapter 16

1. There is an electrostatic attraction between the permanent dipoles of the polar molecules. The greater the polarity, the greater the attraction among molecules. **2.** London dispersion forces (LDF) < dipole-dipole < H-bonding < metallic bonding, covalent network, ionic. Yes, there is considerable overlap. Consider some of the examples in Exercise 12. Benzene (only LDF) has a higher boiling point than acetone (dipole-dipole). Also, there is even more overlap of the stronger forces (metallic, covalent, and ionic). **3.** As the size of the molecule increases, the strength of the London dispersion forces also increases. **4.** yes; Many large, nonpolar compounds are solids at room temperature. The London dispersion forces in these compounds must be stronger than the H-bonding interactions in H_2O, which is a liquid at room temperature. **5.** As the strengths of interparticle forces increase, surface tension, viscosity, melting point, and boiling point increase, while the vapor pressure decreases. **6.** The nature of the forces stays the same. As the temperature increases and the phase changes (solid $\rightarrow$ liquid $\rightarrow$ gas) occur, a greater fraction of the forces are overcome by the increased thermal (kinetic) energy of the particles. **7.** a. ionic; b. LDF, dipole; c. LDF d. LDF; e. metallic; f. metallic; g. LDF; h. H-bonding; i. ionic; j. metallic; k. LDF mostly (C—F bonds are polar, but polymers like Teflon are so large that the LDF are the predominant interparticle forces.) **8.** a. OCS: polar; CO_2: nonpolar; b. PF_3: polar; PF_5: nonpolar **9.** Dipole forces are generally weaker than hydrogen bonds. They are similar in that they arise from an unequal sharing of electrons. We can look at a hydrogen bond as a particularly strong dipole force. **10.** a. $H_2NCH_2CH_2NH_2$; b. $B(OH)_3$ **11.** a. Neopentane is more compact than *n*-pentane resulting in weaker interparticle forces and a lower boiling point. b. Ethanol is capable of H-bonding, dimethyl ether is not. c. HF has stronger H-bonding. d. LiCl is ionic while $TiCl_4$ is a nonpolar molecular substance. Ionic forces are much stronger than the intermolecular forces for covalent compounds. **12.** Benzene, LDF; naphthalene, LDF; carbon tetrachloride, LDF; acetone, LDF, dipole; acetic acid, LDF, dipole, H-bonding; benzoic acid, LDF, dipole, H-bonding; in terms of size and shape: $CCl_4 < C_6H_6 < C_{10}H_8$. The strengths of LDF should be in this order. The physical properties are consistent with this order. We would predict the strength of interparticle forces of the last three molecules to be

acetone < acetic acid < benzoic acid

↑	↑	↑
Polar	H-bond	H-bond, but greater LDF because of greater size

This is consistent with the values given for bp, mp, and H_{vap}. The order of the strengths of interparticle forces based on physical properties is acetone < CCl_4 < C_6H_6 < acetic acid < naphthalene < benzoic acid. Acetone is lowest because, although it is polar, the LDF in acetone must be very small. Naphthalene has very strong LDF.

14. F—H—F⁻ is formed from an HF molecule and an F^- ion

$$F{-}H + F^- \rightarrow [F{-}H{-}F]^-$$

Linear

to give equal F—H distances. The structure is linear as predicted by VSEPR.

15. Only a fraction of the hydrogen bonds are broken in going from solid to liquid. Most of the hydrogen bonds are still present in the liquid and must be broken during the liquid-to-gas phase change.

17. Both molecules are capable of H-bonding. However, in oil of wintergreen the hydrogen bonding is *intra*molecular:

O, C, CH_3, O, H, O

In methyl-4-hydroxybenzoate, the H-bonding is *inter*molecular, resulting in greater intermolecular forces and a higher melting point.

19. a. NaCl: strong ionic bonding in lattice; b. H_2: nonpolar like CH_4, smaller than CH_4, weaker LDF; c. SiO_2: covalent network solid vs. a gas and a liquid; d. $HOCH_2CH_2OH$: greatest amount of H-bonding since two —OH groups are present. **20.** $C_{25}H_{52}$ has stronger intermolecular forces since it boils at a higher temperature than water. The large size of $C_{25}H_{52}$ creates relatively strong LDF. **22.** a. Polarizability of an atom refers to the ease of distorting the electron cloud. It can also refer to distorting the electron clouds in molecules or ions. Polarity refers to the presence of a permanent dipole moment in a molecule. b. London dispersion forces are present in all substances. LDF can be described as accidental dipole-induced dipole forces. Dipole-dipole forces involve the attraction of molecules with permanent dipoles for each other. c. Inter: between; intra: within. For example, in H_2 the covalent bond is an intramolecular force, holding the two H-atoms together in the molecule. The much weaker London dispersion forces are intermolecular forces of attraction.
24. Liquids and solids both have characteristic volume and are not very compressible. Liquids and gases flow and assume the shape of their containers. **25.** *Critical temperature:* The temperature above which a liquid cannot exist, i.e., the vapor cannot be liquefied by increased pressure. *Critical pressure:* The pressure that must be applied to a substance at its critical temperature to produce a liquid. At the critical temperature, all molecules have kinetic energies greater than the interparticle forces and a liquid will not form. **26.** As the interparticle forces increase, the critical temperature increases. **28.** The attraction of H_2O for glass is stronger than the H_2O–H_2O attraction. The meniscus is concave to increase the area of contact between glass and H_2O. The Hg–Hg attraction is greater than the Hg–glass attraction. The meniscus is convex to minimize the Hg–glass contact. Polyethylene is a nonpolar substance. The H_2O–H_2O attraction is stronger than the H_2O–polyethylene attraction; thus, the meniscus will have a convex shape. **31.** The structure of H_2O_2 (H—O—O—H) produces greater hydrogen bonding than water. Long chains of hydrogen-bonded H_2O_2 molecules then become entangled. **33.** a. Crystalline solid: Regular, repeating structure. Amorphous solid: Irregular arrangement of atoms or molecules. b. Ionic solids: Made up of ions held together by ionic bonding. Molecular solid: Made up of discrete covalently bonded molecules held together in the solid by weaker forces (LDF, dipole, or hydrogen bonds). c. Molecular solid: Discrete molecules. Covalent network solid: No discrete molecules. A covalent network is like a large polymer. The interparticle forces are the covalent bonds between atoms.
34. Crystalline solid: A regular, repeating arrangement is necessary to produce planes of atoms that will diffract the X rays.
35. No; Some amorphous solids like glass have stronger interparticle forces than some crystalline solids like water. The reverse is also true. **37.** 0.229 nm **39.** 313 pm = 3.13 Å **40.** 195 g/mol; 139 pm = 1.39 Å; platinum **41.** d = 8.81 g/cm³ for β form; slight change **42.** 6.038×10^{23} **43.** 3.28×10^{-8} cm = 328 pm; radius = 1.42×10^{-8} cm = 142 pm = 1.42 Å **46.** if fcc, d = 19.4 g/cm³; if bcc, d = 17.7 g/cm³; The measured density is consistent with a face-centered cubic unit cell. **49.** a. XeF_2; b. 4.29 g/cm³
51. a. Either a simple cubic array of Ni atoms with a Ti atom in the center or vice versa. b. NiTi; c. Both are 8. **52.** a. $CaTiO_3$; b. Ti atoms are at each corner and O atoms are at the center of each edge in the cubic unit cell. Since the 12 cube edges are each shared by 4 unit cells, both representations give $CaTiO_3$ formula. c. 6 **55.** Conductor: Partially filled valence band; electrons can move through valence band. Insulator: Filled valence band, large band gap; electrons cannot move in valence band and cannot jump to next band. Semiconductor: Filled valence band, small band gap; electrons cannot move in valence band but can jump to next band. **56.** a. As the temperature is increased, more electrons in the valence band have sufficient kinetic energy to jump from the valence band to the conduction band. b. A photon of light is absorbed by an electron, which then has sufficient energy to jump from the valence band to the conduction band. c. An impurity either adds electrons at an energy near the conduction band (n-type) or adds holes (vacant energy levels) at an energy near the valence band (p-type). **58.** To make a p-type semiconductor, we need to dope the material with atoms with fewer valence electrons. The average number of valence electrons is four.

We could dope with more of the Group 3A elements or with atoms of Zn or Cd (most commonly used to produce p-type GaAs). To make n-type GaAs, we could use an excess of As or dope with a Group 6 element such as S. **59.** 5.0×10^2 nm **62.** Equilibrium: There is no change in composition; the vapor pressure is constant. Dynamic: Two processes, vapor $\rightarrow$ liquid and liquid $\rightarrow$ vapor, are occurring at equal rates. **63.** 1.7 atm; 413 K or 140.°C **65.** 1680 kJ **67.** 4.65 kg/h **68.** for Li: 158 kJ/mol; for Mg: 141 kJ/mol; The bonding is stronger in Li. **69.** 221 torr **72.** a. As intermolecular forces increase, the rate of evaporation decreases. b. increase *T*: increase rate; c. increase surface area: increase rate **73.** As *P* is lowered, the water boils. The evaporation of the water is endothermic, and the water is cooled to give some ice. If the pump remained on, all of the ice would sublime. **74.** A: solid; B: liquid; C: vapor; D: solid + vapor; E: solid + liquid + vapor; F: liquid + vapor; G: liquid + vapor; H: vapor; triple point: E; critical point: G; normal boiling point: temperature of liquid-vapor line at 1.0 atm; normal freezing point: temperature of solid-liquid line at 1.0 atm; Solid phase is denser. **78.** Heat released is 2.00 kJ, but to melt 50.0 g of ice requires 16.7 kJ. Not all of the ice will melt (remains at 0°C). **79.** a. 3; b. triple point at 95.31°C; rhombic, monoclinic, vapor; triple point at 115.18°C: monoclinic, liquid, vapor; triple point at 153°C: rhombic, monoclinic, liquid; c. rhombic; d. yes; e. 444.6°C; f. rhombic **80.** The two forms of BN have structures similar to those of graphite and diamond. **83.** 0.524 or 52.4%

```
              O---H—O
             //       \
84. H3C—C             C—CH3
             \        //
              O—H---O
```

86. There are atoms at each of the eight corners (shared by eight unit cells) plus one atom inside the unit cell ($\frac{2}{3}$ of one atom and $\frac{1}{6}$ of two others). Thus, there are two atoms in the unit cell. **88.** 1490 g **91.** From the density, the distance between the centers of adjacent Cs^+ and Cl^- ions is 358 pm. From ionic radii, the distance is 350. pm. **93.** K_2O: 100% (all); CuI: 50% (1/2); ZrI_4: 12.5% (1/8) **96.** 1.71 g/cm^3 **98.** Re^{6+} **100.** 0.11 = fraction of Fe^{3+}; 5.0% (1/20) of Fe sites vacant **102.** distance(*l*)/distance(*g*) = 0.03123 **104.** 2.25 g **107.** coord. no. of Mn is 8; 2 Mn atoms per unit cell **109.** Ag

Chapter 17

1. 0.365 *M* **2.** 3.35 g **3.** 1.54×10^{-2} *M*; $M_{Ni^{2+}} = 1.54 \times 10^{-2}$ mol/L; $M_{Cl^-} = 3.08 \times 10^{-2}$ mol/L **4.** a. 2.49×10^{-3} *M*; b. 160 ppm **5.** a. 2800 ppm; 2.8×10^6 ppb; b. 2.5×10^{-2} *M* **6.** a. $M_{Ca^{2+}} = 1.06 \times 10^{-3}$ mol/L; $M_{NO_3^-} = 2.12 \times 10^{-3}$ mol/L; b. 4.2×10^{-5} g; c. 1.3×10^{15} NO_3^- ions **7.** 0.16 L = 160 mL **9.** 0.88 *M*; 0.91 mol/kg; $\chi = 1.6 \times 10^{-2}$ **10.** Hydrochloric acid: 12 *M*; 17 mol/kg; $\chi = 0.23$; Nitric acid: 16 *M*; 37 mol/kg; $\chi = 0.39$; Sulfuric acid: 18 *M*; 200 mol/kg; $\chi = 0.76$; Acetic acid: 17 *M*; 2000 mol/kg; $\chi = 0.96$; Ammonia: 15 *M*; 23 mol/kg; $\chi = 0.29$ **11.** 28.5%; $\chi = 0.252$; 4.32 mol/kg; 2.69 *M* **14.** Since volume is temperature dependent and mass is not, molarity is temperature dependent. In describing T_f and T_b, we are interested in how a temperature depends on composition. Thus, we do not want our expression of composition to be dependent on temperature also. **15.** 0.746 *M*; $\chi = 0.0441$ **17.** 2.5%; 0.43 *M*; 0.44 molal; $\chi = 7.9 \times 10^{-3}$ **19.** the nature of the interparticle forces **20.** As the hydrocarbon chain gets longer, the alcohol becomes more nonpolar and the solubility decreases. **21.** a. Mg^{2+}; smaller, higher charge; b. Be^{2+}; smaller; d. F^-; smaller; e. Cl^-; smaller **22.** $KCl(s) \rightarrow K^+(aq) + Cl^-(aq)$, ΔH = 31 kJ/mol **23.** CsI: $\Delta H_{hyd} = -571$ kJ/mol; CsOH: $\Delta H_{hyd} = -796$ kJ/mol **25.** polarity, ability to form H-bonding interactions, size and charge of ions in ionic compounds; a. CH_3CH_2OH; b. $CHCl_3$; c. CH_3CO_2H; d. Na_2S **26.** a. H_2O; b. CCl_4; c. H_2O; d. CCl_4 **27.** Both $Al(OH)_3$ and NaOH are ionic. Since the lattice energy is proportional to charge, the lattice energy of aluminum hydroxide is greater than that of sodium hydroxide. The attraction of water molecules for Al^{3+} and OH^- cannot overcome the extra lattice energy, and $Al(OH)_3$ is insoluble. **30.** hydrophobic: water hating; hydrophilic: water loving **31.** $CO_2 + OH^- \rightarrow HCO_3^-$; No, the reaction of CO_2 with OH^- greatly increases the solubility of CO_2 by forming the soluble bicarbonate ion. **32.** 2.0×10^{-4} mol/L **34.** 23.6 torr **35.** 189 torr. In the vapor, $\chi_{acetone} = 0.512$ and $\chi_{methanol} = 0.488$. It is probably not a good assumption that this solution is ideal because of the possibility of methanol-acetone hydrogen bonding. **36.** The vapor pressure of (c) will be the lowest. **37.** If solute-solvent attraction > solvent-solvent and solute-solute attractions, there is a negative deviation from Raoult's law. If solute-solvent < solvent-solvent and solute-solute attractions, there is a positive deviation from Raoult's law. **38.** The boiling point is lower since the vapor pressure of the solution is higher than an ideal solution. **39.** a. CF_3CF_3 and $CF_3CF_2CF_3$; b. C_7H_{16} and C_6H_{14}. Compounds that are similar in size and polarity will form the most ideal solution. **40.** 23.2 torr at 25°C; 70.0 torr at 45°C **42.** a. 290 torr; b. in the vapor phase, $\chi_{pen} = 0.69$ **44.** $T_b = 100.042$°C; $T_f = -0.15$°C; $\pi = 2.0$ atm **45.** $\Delta T_f = 2.0 \times 10^{-5}$ °C; $\pi = 0.20$ torr **46.** osmotic pressure; Freezing-point depression is too small to measure precisely. **48.** 6.6×10^{-2} mol/kg; 610 g/mol **50.** 43% naphthalene; 57% anthracene **53.** 13.5 L; 108.2°C **54.** 30. atm **55.** a. The water would migrate from right to left. The level in the right arm would go down and the level in the left arm would go up. b. The levels would be equal. The concentration of NaCl would be equal in both chambers. **56.** $MgCl_2$ > NaCl > HOCl > glucose **58.** $T_f = -3$°C **59.** The measured freezing point should be higher because of ion association. **61.** Ion pairing can occur, resulting in fewer particles than expected. Ion pairing will increase as the concentration of electrolyte increases. **64.** a. yes; b. yes **65.** Benzoic acid is capable of hydrogen bonding. However, it is more soluble in nonpolar benzene than in water. In benzene, a hydrogen-bonded dimer forms. The dimer is relatively nonpolar and thus more soluble in benzene than in water. The freezing point depression would be less than 5.12°C since dimer formation reduces the effective particle concentration. **66.** Benzoic acid would be more soluble in a basic solution because of the reaction $C_6H_5CO_2H + OH^- \rightarrow C_6H_5CO_2^- + H_2O$. **67.** a. An analytical balance can weigh to the nearest 0.1 mg. We would use 601.2 mg NaCl, 29.9 mg KCl, 20.0 mg $CaCl_2 \cdot 2H_2O$, and 309.5 mg $NaC_3H_5O_3$. b. Osmotic pressure ranges from 6.59 atm to 7.30 atm. **68.** No, ΔH_{soln} for methanol/water is not zero. **72.** empirical formula: C_7H_4O; molecular formula: $C_{14}H_8O_2$ **75.** empirical formula: $C_2H_4O_3$; molecular formula: $C_4H_8O_6$; molar mass = 152.10 g/mol **77.** 10.%; 2.5 molal **80.** Dissolve 209 g sucrose in water and dilute to

1.00 L in a volumetric flask. **81.** 0.018 g H_2O **82.** a. T_f = -2.0°C; b. T_f = -13°C **84.** 100.08°C **87.** 0.812 atm **90.** 1.0×10^{-3} **91.** 1.97% NaCl **93.** 0.600 **94.** 72%

Chapter 18

2. a. $\Delta H° = 207$ kJ; $\Delta S° = 216$ J/K; b. The reaction is spontaneous at $T > 958$ K. **3.** For $3Fe(s) + 4H_2O(g) \rightarrow Fe_3O_4(s) + 4H_2(g)$: a. $\Delta H° = -149$ kJ; $\Delta S° = -167$ J/K; b. spontaneous at $T < 892$ K; For $C(s) + H_2O(g) \rightarrow CO(g) + H_2(g)$: a. $\Delta H° = 132$ kJ; $\Delta S° = 134$ J/K; b. spontaneous at $T > 985$ K **4.** (1) ammonia production; (2) hydrogenation of vegetable oils **7.** a. $Li_3N(s) + 3HCl(aq) \rightarrow 3LiCl(aq) + NH_3(aq)$; b. $Rb_2O(s) + H_2O(l) \rightarrow 2RbOH(aq)$; c. $Cs_2O_2(s) + 2H_2O(l) \rightarrow 2CsOH(aq) + H_2O_2(aq)$; d. $NaH(s) + H_2O(l) \rightarrow NaOH(aq) + H_2(g)$ **8.** 8.4 kJ **10.** $2Li(s) + 2C_2H_2(g) \rightarrow 2LiC_2H(s) + H_2(g)$; oxidation-reduction **13.** a. $K(s) + O_2(g) \rightarrow KO_2(s)$; b. $16K(s) + S_8(s) \rightarrow 8K_2S(s)$; c. $12K(s) + P_4(s) \rightarrow 4K_3P(s)$; d. $2K(s) + H_2(g) \rightarrow 2KH(s)$; e. $2K(s) + 2H_2O(l) \rightarrow H_2(g) + 2KOH(aq)$ **14.** a. sodium oxide: Na_2O; b. sodium superoxide: NaO_2; c. sodium peroxide: Na_2O_2 **17.** trigonal planar; Be uses sp^2 hybrid orbitals; N uses sp^3 hybrid orbitals; $BeCl_2$ is a Lewis acid.

19. $BeCl_2(NH_3)_2$ would form in excess ammonia. A structure for this molecule can be drawn that obeys the octet rule for all atoms.

```
      ..
   H :Cl: H
   |   |   |
H—N—Be—N—H
   |   |   |
   H :Cl: H
      ..
```

23. $Mg_3N_2(s) + 6H_2O(l) \rightarrow 2NH_3(g) + 3Mg^{2+}(aq) + 6OH^-(aq)$; $Mg_3P_2(s) + 6H_2O(l) \rightarrow 2PH_3(g) + 3Mg^{2+}(aq) + 6OH^-(aq)$ **24.** 1.67×10^5 A; 1.77×10^3 kg of Cl_2 **26.** $B_2H_6 + 3O_2 \rightarrow 2B(OH)_3$ **27.** a. thallium(I) hydroxide; b. indium(III) sulfide; c. gallium(III) oxide **29.** The radius of the 3+ ion is about 1.0 Å (100 pm). The ionization energy is close to that of Tl. **31.** $In_2O_3(s) + 6H^+(aq) \rightarrow 2In^{3+}(aq) + 3H_2O(l)$; $In_2O_3(s) + OH^-(aq) \rightarrow$ no reaction; $Ga_2O_3(s) + 2OH^-(aq) + 3H_2O(l) \rightarrow 2Ga(OH)_4^-(aq)$; $Ga_2O_3(s) + 6H^+(aq) \rightarrow 2Ga^{3+}(aq) + 3H_2O(l)$ **32.** p-type **33.** a. $Ca_3Al_2O_6$; b. $Ca_9Al_6O_{18}$; c. There are covalent bonds between Al and O atoms in the $[Al_6O_{18}]^{18-}$ anion. sp^3 hybrid orbitals on Al overlap with sp^3 hybrid orbitals on O to form the σ bonds. **34.** a. $2In(s) + 3F_2(g) \rightarrow 2InF_3(s)$; b. $2In(s) + 3Cl_2(g) \rightarrow 2InCl_3(s)$; c. $4In(s) + 3O_2(g) \rightarrow 2In_2O_3(s)$; d. $2In(s) + 6HCl(aq) \rightarrow 3H_2(g) + 2InCl_3(aq)$ **39.** a. linear; b. sp **40.** $\Delta H = -83$ kJ; ΔH and ΔS are favorable for the decomposition of H_2CO_3 to CO_2 and H_2O. Hence, H_2CO_3 will spontaneously decompose to CO_2 and H_2O. **41.** $\ddot{S}{=}C{=}\ddot{S}$, linear; $\ddot{S}{=}C{=}C{=}C{=}\ddot{S}$, linear **43.** Gray tin has the more ordered structure. **44.** Thermodynamics predicts both should react with water. The answer must lie in kinetics. $SiCl_4$ reacts because an activated complex can easily form by a water molecule attaching to a silicon atom. Carbon cannot expand its coordination number like silicon and a C—Cl bond must be broken to form an activated complex. **46.** SiC would have a covalent network structure similar to diamond. **47.** $SiO_2(s) + 4HF(aq) \rightarrow SiF_4(g) + 2H_2O(l)$ **49.** $\mathscr{E}° = 0.97$ V; Fe pipes corrode more easily than Pb pipes. **50.** 6.7×10^{-6} mol/L **51.** $Sn(s) + 2H^+(aq) \rightarrow Sn^{2+}(aq) + H_2(g)$; $Pb(s) + 2H^+(aq) \rightarrow Pb^{2+}(aq) + H_2(g)$ **52.** The purpose of the cryptand is to encapsulate the Na^+ ion so that it does not come in contact with the Na^- ion and oxidize it to sodium metal. **55.** $Be(s) + 2H_2O(l) + 2OH^-(aq) \rightarrow Be(OH)_4^{2-}(aq) + H_2(g)$. Be is the reducing agent. H_2O is the oxidizing agent. **57.** Be^{2+} ion is a Lewis acid. The ion in solution is $Be(H_2O)_4^{2+}$. The acidic solution results from the reaction $Be(H_2O)_4^{2+}(aq) \rightleftharpoons Be(H_2O)_3(OH)^+(aq) + H^+(aq)$. **59.** In solution, Tl^{3+} can oxidize I^- to I_3^-. Thus, we expect TlI_3 to be thallium(I) triiodide. **60.** If the compound contained Ga(II), it would be paramagnetic. This can easily be determined by measuring the mass of a sample in the presence and absence of a magnetic field. Paramagnetic compounds will have an apparent greater mass in a magnetic field. **63.** a. $K_2SiF_6(s) + 4K(l) \rightarrow 6KF(s) + Si(s)$; b. K_2SiF_6 is an ionic compound, composed of K^+ cations and SiF_6^{2-} anions. The SiF_6^{2-} anion is held together by covalent bonds. **64.** tetrahedral arrangement around Sn **65.** Carbon is much smaller than Si and cannot form a fifth bond in the transition state. **67.** Size decreases from left to right and increases going down. Going one element right and one element down would result in a similar size for the two elements diagonal to each other. The ionization energies will also be similar for the diagonal elements. Electron affinities are harder to predict, but the similar size and ionization energies would lead to similar properties.

Chapter 19

2. $\Delta H° = 180.$ kJ; $\Delta G° = 174$ kJ; $\Delta S° = 25$ J/K; At high temperature the reaction $N_2 + O_2 \rightarrow 2NO$ becomes spontaneous. In the atmosphere, even though $2NO \rightarrow N_2 + O_2$ is spontaneous, it doesn't occur because the rate is too slow. **3.** NH_3: sp^3; N_2H_4: sp^3; NH_2OH: sp^3; N_2: sp; N_2O, central N: sp; NO: sp^2; N_2O_3: both are sp^2; NO_2: sp^2; HNO_3: sp^2 **4.** Resonance is possible for N_2O, NO, N_2O_3, NO_2, and HNO_3.

N_2O

$\ddot{N}{=}N{=}\ddot{O} \longleftrightarrow :N{\equiv}N{-}\ddot{O}: \longleftrightarrow :\ddot{N}{-}N{\equiv}O:$ last form not important

NO

$\dot{N}{=}\ddot{O}: \longleftrightarrow :\dot{N}{=}\ddot{O} \longleftrightarrow :\dot{N}{=}\ddot{O}:$

N_2O_3

```
:O                    :O:
  \\    ..              \    ..
   N—N=O:    ⟷         N—N=O:   ⟷
  /                     //
:O:                    O
 ..                    ..

:O:                    :O:
  \                     \
   N=N=O:    ⟷         N—N≡O:
  /                     /
:O:                    :O:
```

last 2 not important

NO_2

```
   N          ⟷      N          ⟷      N          ⟷
 //  \             /  \\              /  \\
O    :O:        :O:    O           :O:    O
```

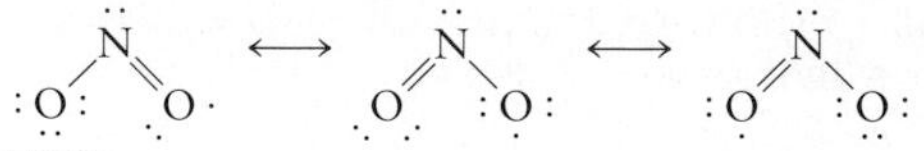

HNO_3 (resonance structures)

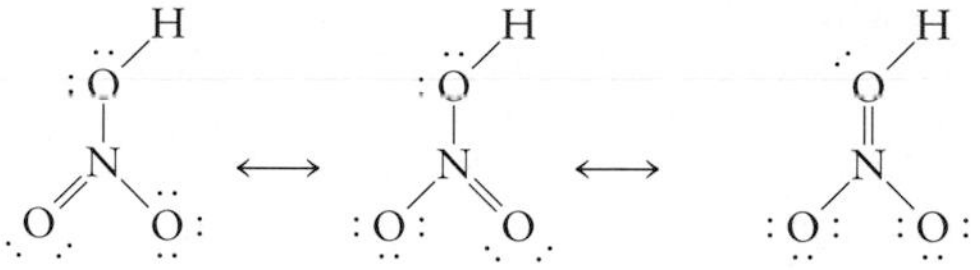

last one not important

5. For

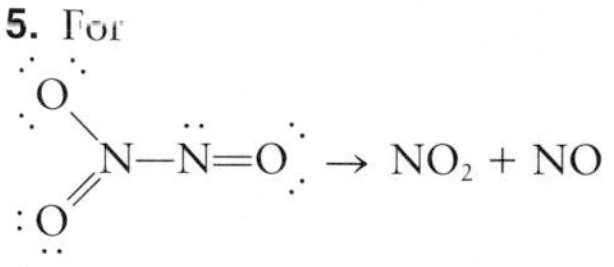

$O_2N{-}N{=}O \rightarrow NO_2 + NO$

the activation energy must in some way involve the breaking of a nitrogen–nitrogen bond. For the reaction

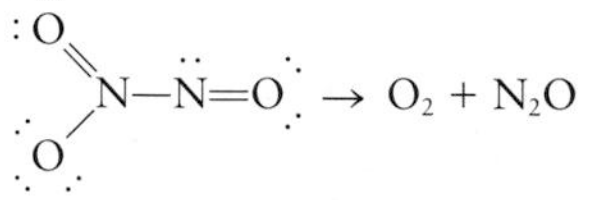

$O_2N{-}N{=}O \rightarrow O_2 + N_2O$

at some point nitrogen–oxygen bonds must be broken. N—N single bonds (160 kJ/mol) are weaker than N—O single bonds (201 kJ/mol). Resonance structures indicate that there is more double bond character to the N—O bonds than the N—N bond. Thus, NO_2 and NO are preferred by kinetics because of the lower activation energy.

8. OCN^-: $[\ddot{O}{=}C{=}\ddot{N}]^- \longleftrightarrow$

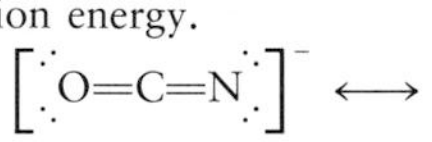

Formal charge 0 0 −1

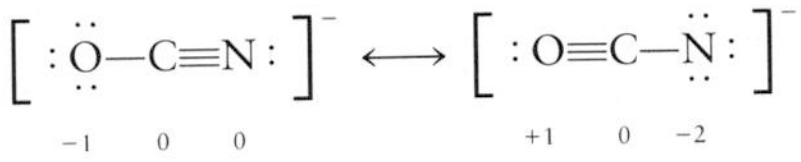

$[:\ddot{O}{-}C{\equiv}N:]^- \longleftrightarrow [:O{\equiv}C{-}\ddot{N}:]^-$

−1 0 0 +1 0 −2

CNO^-: $[\ddot{C}{=}N{=}\ddot{O}]^- \longleftrightarrow$

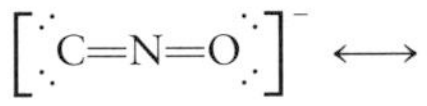

Formal charge −2 +1 0

$[:C{\equiv}N{-}\ddot{O}:]^- \longleftrightarrow [:\ddot{C}{-}N{\equiv}O:]^-$

−1 +1 −1 −3 +1 +1

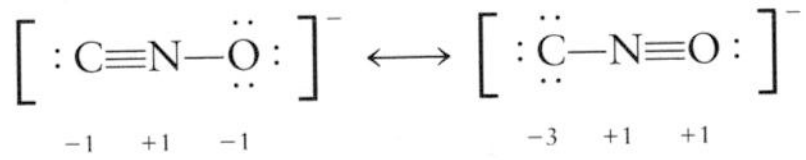

All of the resonance structures for fulminate involve greater formal charges than in cyanate, making fulminate more reactive (less stable). **10.** a. $8H^+(aq) + 2NO_3^-(aq) + 3Cu(s) \rightarrow 3Cu^{2+}(aq) + 4H_2O(l) + 2NO(g)$; b. $NH_4NO_3(s) \xrightarrow{\text{heat}} N_2O(g) + 2H_2O(g)$; c. $NO(g) + NO_2(g) + 2KOH(aq) \rightarrow 2KNO_2(aq) + H_2O(l)$ **12.** $2Sb_2S_3(s) + 9O_2(g) \rightarrow 2Sb_2O_3(s) + 6SO_2(g)$; $2Sb_2O_3(s) + 3C(s) \rightarrow 4Sb(s) + 3CO_2(g)$; $2Bi_2S_3(s) + 9O_2(g) \rightarrow 2Bi_2O_3(s) + 6SO_2(g)$; $2Bi_2O_3(s) + 3C(s) \rightarrow 4Bi(s) + 3CO_2(g)$ **13.** a. $H_3PO_4 > H_3PO_3$; b. $H_3PO_4 > H_2PO_4^- > HPO_4^{2-}$

14. $H_4P_2O_6$: H—P(=O)(—O—H)—O—P(=O)(—O—H)—O—H

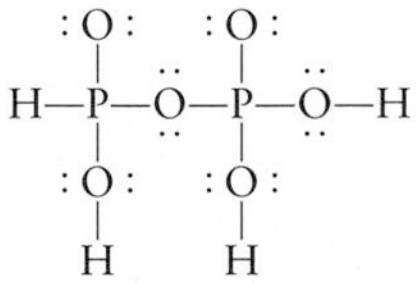

$H_4P_2O_5$: H—O—P(=O)(H)—O—P(=O)(H)—O—H

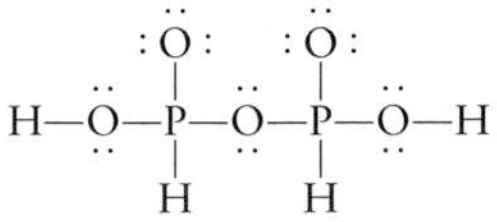

18.

		Bond order	# unpaired e^-
M.O.	NO	2.5	1
	NO^+	3	0
	NO^-	2	2

Lewis NO^+ $[:N{\equiv}O:]^+$

NO $\dot{N}{=}\ddot{O} \longleftrightarrow N{=}\dot{O} \longleftrightarrow N{=}O$

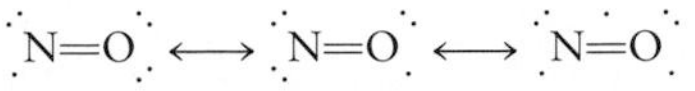

NO^- $[\ddot{N}{=}\ddot{O}]^-$

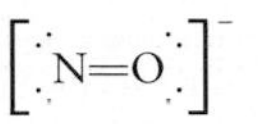

Lewis structures are not adequate for NO and NO^-. The M.O. model gives correct results for all three species. For NO, Lewis structures fail for odd electron species. For NO^-, Lewis structures fail to predict that NO^- is paramagnetic.

19. $4.8 \times 10^{-11} M$ **20.** a. $8H_2O(l) + 2Mn^{2+}(aq) + 5NaBiO_3(s) \rightarrow 2MnO_4^-(aq) + 16H^+(aq) + 5BiO_3^{3-}(aq) + 5Na^+(aq)$; b. Bismuthate exists as a covalent network: $(BiO_3^-)_x$. **23.** a. $P_4O_6(s) + 2O_2(g) \rightarrow P_4O_{10}(s)$; b. $P_4O_{10}(s) + 6H_2O(l) \rightarrow 4H_3PO_4(aq)$; c. $PCl_5(l) + 4H_2O(l) \rightarrow H_3PO_4(aq) + 5HCl(aq)$

26. 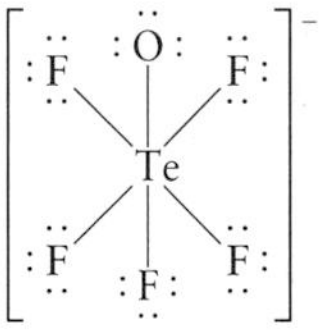 $[TeOF_5]^-$ (octahedral: O and five F about Te) $F_5TeO{-}\ddot{P}(OTeF_5){-}OTeF_5$

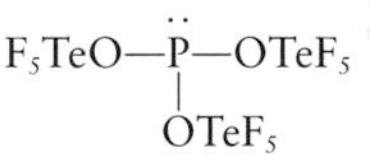

27. a. As we go down the family, K_a increases. This is consistent with the bond to hydrogen getting weaker. b. Po is below Te, so K_a should be larger. The K_a for H_2Po should be on the order of 10^{-2} or 10^{-1}. **29.** Sulfur forms polysulfide ions, S_n^{2-}, which are soluble in water, i.e., $S_8 + S^{2-} \longleftrightarrow S_9^{2-}$. Nitric acid oxidizes S^{2-} to S. **32.** Light from violet to green will work.

35. a. ClF_5:

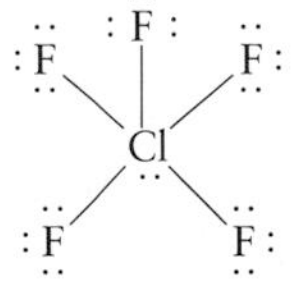

Square pyramid

b. IF_3:

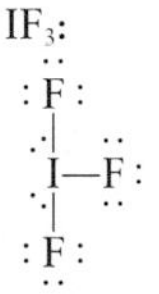

T-shaped

c. Cl_2O_7: $O_3Cl{-}O{-}ClO_3$

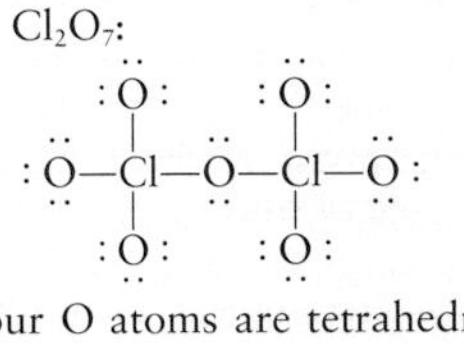

The four O atoms are tetrahedrally arranged about each Cl. The Cl—O—Cl bond angle is close to the tetrahedral bond angle.

36. a. $F_2 + H_2O \rightarrow HOF + HF$; $2HOF \rightarrow 2HF + O_2$; $HOF + H_2O \rightarrow HF + H_2O_2$ (acid); $2OF^- \rightarrow 2F^- + O_2$ (base); b. HOF: assign +1 to H, −1 to F; Oxidation number of O is zero. Oxygen is very electronegative. A zero oxidation state would not be very stable as O is a powerful oxidizing agent.

37. :F—O—O—F:

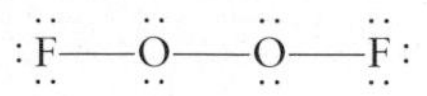

	F	O	O	F
Formal charge	0	0	0	0
Oxidation number	−1	+1	+1	−1

Oxidation numbers are more useful. We are forced to assign +1 as the oxidation number to oxygen. Oxygen is very electronegative and +1 is not a stable oxidation state for this element.

41. a. Assuming standard conditions, a potential greater than 1.19 V must be applied. b. 5×10^9 kJ **42.** a. $AgCl(s) \xrightarrow{h\nu} Ag(s) + Cl$. The reactive chlorine atom is trapped in the crystal. When the light is removed, it reacts with silver atoms to reform AgCl. b. Over time chlorine is lost and the dark silver metal is permanent. **44.** Helium is unreactive and doesn't combine with any other elements. It is a very light gas and would easily escape the earth's gravitational pull as the planet was formed. **45.** The heavier members are not really inert. Xe and Kr have been shown to react and form compounds with other elements. **46.** 5×10^{-2} g; 2×10^{20} atoms; A 2-L breath contains 5×10^{15} atoms of Xe.

47. XeO_3: :O—X—O: (with :O: above) trigonal pyramid;

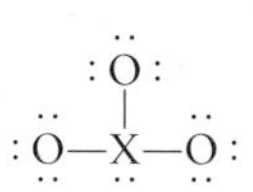

XeO_4: :O—Xe—O: (with :O: above and below) tetrahedral;

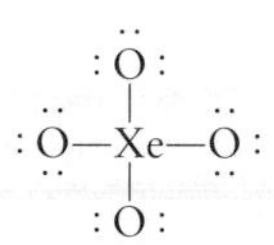

$XeOF_4$: Xe bonded to O and four F square pyramid;

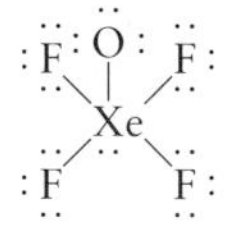

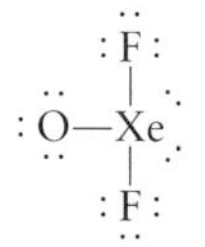

$XeOF_2$: :O—Xe with :F: above and below T-shaped;

XeO_3F_2: Xe bonded to three O and two F trigonal bipyramid

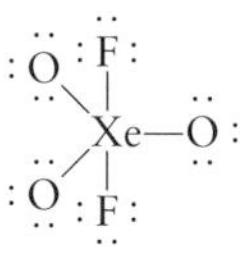

49. As the halogen atoms get larger, it becomes more difficult to fit three halogen atoms around the small N, and the NX_3 molecule becomes less stable. **51.** a. IO_4^-; b. IO_3^-; c. IF_2^-; d. IF_4^-; e. IF_6^-; f. IOF_3^- **52.** For $NCl_3 \rightarrow NCl_2 + Cl$, only the N—Cl bond is broken. For O=N—Cl → NO + Cl, the NO bond gets stronger (bond order increases from 2.0 to 2.5), making ΔH for the reaction smaller than the energy necessary to break the N—Cl bond. **54.** Plastic sulfur consists of long chains of sulfur atoms. It becomes brittle as the long chains break down into S_8 rings. **55.** The pollution provides nitrogen and phosphorus nutrients so that algae can grow. The algae consume oxygen, causing fish to die. **56.** $H_2O(l) + BrO_3^-(aq) + XeF_2(aq) \rightarrow BrO_4^-(aq) + Xe(g) + 2HF(aq)$

Chapter 20

1. a. Co: $[Ar]4s^23d^7$; Co^{2+}: $[Ar]3d^7$; Co^{3+}: $[Ar]3d^6$; b. Pt: $[Xe]6s^14f^{14}5d^9$; Pt^{2+}: $[Xe]4f^{14}5d^8$; Pt^{4+}: $[Xe]4f^{14}5d^6$; c. Fe: $[Ar]4s^23d^6$; Fe^{2+}: $[Ar]3d^6$; Fe^{3+}: $[Ar]3d^5$ **2.** From ionization energy values, we would predict that Fe^{3+} and Ti^{3+} is more stable. From $\mathscr{E}°$ values, we would predict Fe(II) and Ti(IV) to be more stable. The electro-chemical data are consistent with the information in the text. These data are for solutions while ionization energies are for gas-phase reactions. Solution data are more representative of the conditions from which ilmenite formed.
4. a. 4 O on faces × $\frac{1}{2}$ O/face = 2 O atoms; 2 O atoms inside body; total: 4 O atoms; 8 corners × $\frac{1}{8}$ Ti/corner + 1 Ti/body center = 2 Ti; formula of unit cell Ti_2O_4 to give the empirical formula TiO_2.

b. $\overset{+4\,-2}{2TiO_2} + \overset{0}{3C} + \overset{0}{4Cl_2} \longrightarrow \overset{+4\,-1}{2TiCl_4} + \overset{+4\,-2}{CO_2} + \overset{+2\,-2}{2CO}$

C is being oxidized. C is the reducing agent. Cl is being reduced. Cl_2 is the oxidizing agent.

$\overset{+4\,-1}{TiCl_4} + \overset{0}{O_2} \longrightarrow \overset{+4\,-2}{TiO_2} + \overset{0}{2Cl_2}$

Cl is being oxidized. $TiCl_4$ is the reducing agent. O is being reduced. O_2 is the oxidizing agent. **5.** TiF_4: ionic compound containing Ti^{4+} and F^- ions. $TiCl_4$, $TiBr_4$, and TiI_4: covalent compounds composed of discrete, tetrahedral TiX_4 molecules. TiF_4 has the largest freezing point since ionic forces are stronger than covalent intermolecular forces. For the covalent compounds, as size increases, the boiling points and melting points increase since the London dispersion forces increase. **6.** Pyrolusite, manganese(IV) oxide; rhodochrosite, manganese(II) carbonate
7. a. $2CoAs_2(s) + 4O_2(g) \longrightarrow 2CoO(s) + As_4O_6(s)$; As_4O_6: arsenic(III) oxide; As_4O_6 has a cage structure similar to P_4O_6. b. $2Co^{2+}(aq) + 4OH^-(aq) + H_2O(l) + OCl^-(aq) \longrightarrow 2Co(OH)_3(s) + Cl^-(aq)$; c. 2.5×10^{-22} *M* d. 2.5×10^{-31} *M*
9. a. ligand: species that donates a pair of electrons to form a covalent bond to a metal ion (a Lewis base); b. chelate: ligand that can form more than one bond; c. bidentate: ligand that can form two bonds; d. complex ion: metal ion plus ligands
10. a. hexaamminecobalt(II) chloride; b. hexaaquacobalt(III) iodide **11.** a. pentaaminechlororuthenium(III) ion; b. hexacyanoferrate(II) ion **12.** a. sodium *tris*(oxalato)nickelate(II); b. potassium tetrachlorocobaltate(II)
13. a. $[Co(C_5H_5N)_6]Cl_3$; b. $[Cr(NH_3)_5I]I_2$; c. $[Ni(NH_2CH_2CH_2NH_2)_3]Br_2$ **14.** a. $FeCl_4^-$; b. $[Ru(NH_3)_5H_2O]^{3+}$ **15.** a. 2; b. 3; c. 4; d. 4

17. a. $[Co(H_2O)_2(C_2O_4)_2]^-$ structures:

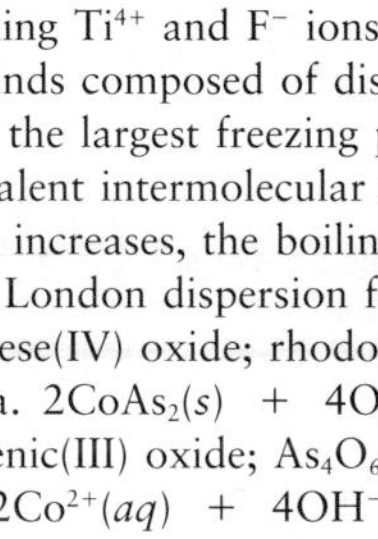

cis

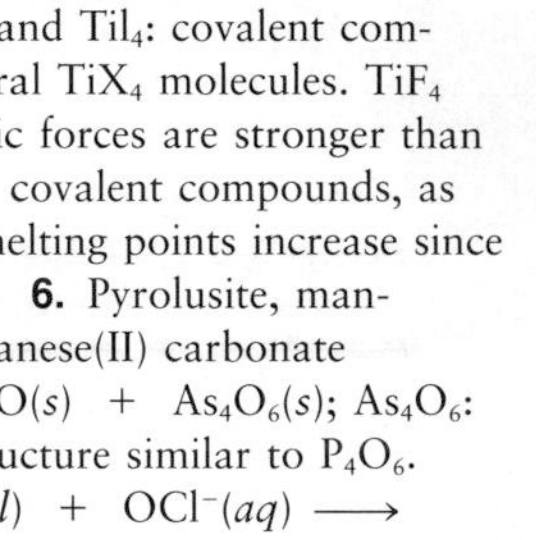

trans

b. $\left[\text{cis-}Pt(NH_3)_4I_2 \right]^{2+}$ — cis (I, I adjacent; H_3N—Pt—NH_3, H_3N, NH_3); $\left[\text{trans-}Pt(NH_3)_4I_2 \right]^{2+}$ — trans (I above, I below; NH_3, H_3N—Pt—NH_3, H_3N)

cis *trans*

18. SCN^-, NO_2^-, and OCN^- can form linkage isomers; all are able to bond to the metal ion in two different ways.

19.

H_3N—Pt—NH_3 with SCN, NCS; H_3N—Pt—NH_3 with SCN, SCN; H_3N—Pt—NH_3 with NCS, SCN

H_3N—Pt—SCN with SCN, H_3N; H_3N—Pt—SCN with NCS, H_3N; H_3N—Pt—NCS with NCS, H_3N

21. a. CH_2—H_2N / CH_2—H_2N chelated to Pt, with Cl, Cl

b. CH_2—H_2N / CH_2—H_2N and NH_2—CH_2 / NH_2—CH_2 chelated to Co, with Cl, Cl

c. $\left[Co(NH_3)_4(Cl)(NO_2) \right]^{+}$ (H_3N, NH_3, Cl, H_3N, NH_3, NO_2 around Co)

23. The octahedral complex ion is $[Co(NH_3)_4SO_4]^+$ where sulfate is acting as a bidentate ligand. The formula for the coordination compound is $[Co(NH_3)_4SO_4]Cl$.

26. a. ligand that will give complex ions with the maximum number of unpaired electrons; b. ligand that will give complex ions with the minimum number of unpaired electrons **30.** a. 0; b. 1; c. 2; d. 2; e. 3 **32.** $NiCl_4^{2-}$ is tetrahedral and $Ni(CN)_4^{2-}$ is square planar.

34. a. Fe^{2+} High spin: ↑ ↑ / ⇅ ↑ ↑; Low spin: — — / ⇅ ⇅ ⇅

b. Fe^{3+} High spin: ↑ ↑ / ↑ ↑ ↑

c. Ni^{2+}: ↑ ↑ / ⇅ ⇅ ⇅

35. $Ni(H_2O)_6Cl_2$ absorbs red light; $Ni(NH_3)_6Cl_2$ absorbs yellow-green light. $Ni(NH_3)_6Cl_2$ absorbs the shorter wavelength light; Δ is larger for $Ni(NH_3)_6Cl_2$. NH_3 is a stronger field ligand than H_2O, consistent with the spectrochemical series. **36.** $Fe(CN)_6^{3-}$: low spin; $Fe(SCN)_6^{3-}$: high spin; CN^- is a stronger field ligand than SCN^-. **38.** 0 **39.** Lewis acids **42.** $[Cr(NH_3)_5I]I_2$; $[Cr(NH_3)_5I]^{2+}$ **44.** $Fe_2O_3(s) + 6H_2C_2O_4(aq) \rightarrow 2Fe(C_2O_4)_3^{3-}(aq) + 3H_2O(l) + 6H^+(aq)$; a soluble complex ion forms **47.** No, since in all three cases, six bonds are formed between Ni^{2+} and nitrogen. $\Delta S°$ for formation of the complex ion is most negative for $6NH_3$ molecules reacting with a metal ion (seven independent species become one). For penten reacting with a metal ion, two independent species form one, so $\Delta S°$ is less negative. Thus, the chelate effect occurs because the more bonds a chelating agent can form to the metal, the more favorable $\Delta S°$ is for the formation of the complex ion, and the larger the formation constant. **51.** See Fig. 20.8. It is optically active.

52. — — $d_{x^2-y^2}, d_{xy}$
— d_{z^2}
— — d_{xz}, d_{yz}

Chapter 21

1. a. ${}^{3}_{1}H \rightarrow {}^{0}_{-1}e + {}^{3}_{2}He$; b. ${}^{8}_{3}Li \rightarrow {}^{8}_{4}Be + {}^{0}_{-1}e$; ${}^{8}_{4}Be \rightarrow 2\,{}^{4}_{2}He$; c. ${}^{7}_{4}Be + {}^{0}_{-1}e \rightarrow {}^{7}_{3}Li$ **2.** ${}^{60}_{27}Co \rightarrow {}^{60}_{28}Ni + {}^{0}_{-1}e$; b. ${}^{97}_{43}Tc + {}^{0}_{-1}e \rightarrow {}^{97}_{42}Mo$ **3.** ${}^{8}B$ and ${}^{9}B$ might decay by either positron emission or electron capture. ${}^{12}B$ and ${}^{13}B$ contain too many neutrons or too few protons, so ${}^{12}B$ and ${}^{13}B$ are expected to be β-emitters. **4.** a. ${}^{1}_{1}H + {}^{14}_{7}N \rightarrow {}^{11}_{6}C + {}^{4}_{2}He$; b. $2\,{}^{3}_{2}He \rightarrow {}^{4}_{2}He + 2\,{}^{1}_{1}H$ **5.** a. ${}^{240}_{95}Am + {}^{4}_{2}He \rightarrow {}^{243}_{97}Bk + {}^{1}_{0}n$; b. ${}^{238}_{92}U + {}^{12}_{6}C \rightarrow {}^{244}_{98}Cf + 6\,{}^{1}_{0}n$ **9.** a. ${}^{207}_{82}Pb$; b. ${}^{231}_{90}Th$; ${}^{231}_{91}Pa$; ${}^{227}_{89}Ac$; ${}^{227}_{90}Th$; ${}^{223}_{88}Ra$; ${}^{219}_{86}Rn$; ${}^{215}_{84}Po$; ${}^{211}_{82}Pb$; ${}^{211}_{83}Bi$; ${}^{207}_{81}Tl$ **11.** a. unstable; beta emission; b. stable; c. unstable; positron emission or electron capture; d. unstable; positron emission or electron capture **13.** $t_{1/2}$ = 12,000 yr, rate = 1.1×10^{12} disintegrations/s; $t_{1/2}$ = 12 h, rate = 9.6×10^{18} disintegrations/s; $t_{1/2}$ = 12 min, rate = 5.8×10^{20} disintegrations/s; $t_{1/2}$ = 12 s, rate = 3.5×10^{22} disintegrations/s **14.** 0.76 **18.** 2.5 disintegrations per minute per g of C; for 10. mg C: 0.025 disintegrations/min; It would take roughly 40 min to see a single disintegration. The background radiation would probably be much greater than the ${}^{14}C$ activity. Thus, ${}^{14}C$ dating is not practical for small samples. **19.** 0.866 **20.** a. The decay of ${}^{40}K$ is not the sole source of ${}^{40}Ca$. b. Decay of ${}^{40}K$ is the sole source of ${}^{40}Ar$ and no ${}^{40}Ar$ is lost over the years. c. 4.2×10^9 years old; d. Measured age would be less than actual age. **22.** a. 9.7×10^6 mCi; b. 1.5×10^4 mCi **26.** 4.3×10^6 kg of mass/s **27.** 1.8×10^5 kg solar material; 5.0×10^{14} kg coal **28.** ${}^{24}_{12}Mg$: 1.324×10^{-12} J/nucleon; ${}^{27}_{12}Mg$: 1.324×10^{-12} J/nucleon **32.** 2.427×10^{-3} nm **35.** 2.0×10^{10} J/g **36.** 2.820×10^{-12} J/atom; 1.698×10^{12} J/mol **37.** The chemical properties may determine where a radioactive material may be concentrated in the body or how easily it may be excreted. **38.** Not all of the emitted radiation enters the Geiger-Müller tube. The fraction of radiation entering the tube must be constant. **39.** The Geiger-Müller tube has a certain response time. After the gas in the tube ionizes to produce a "count," some time must elapse for the gas to return to an electrically neutral state. The response of the tube levels off because at high activities, radioactive particles enter the tube faster than the tube can respond. **40.** All evolved O_2 comes from water. **41.** A nonradioactive substance can be put in equilibrium with a radioactive substance. The two materials can then be checked to see if all the radioactivity remains in the original material or if it has been scrambled by the equilibrium.

43. Assuming that (1) the radionuclide is long lived enough that no significant decay occurs during the time of the experiment, and (2) the total activity is uniformly distributed only in the rat's blood; $V = 10.$ mL. **45.** 2×10^{-7} kg; This is 2×10^{23} times the electron mass and 1×10^{20} times the proton mass. **46.** (i) and (ii) mean that Pu is not a significant threat outside the body. If Pu gets inside the body, it is easily oxidized to Pu^{4+} (iv), which is chemically similar to Fe^{3+} (iii). Thus, Pu^{4+} will concentrate in tissues where Fe^{3+} is found, such as the bone marrow where red blood cells are produced. Here α particles cause considerable damage. **48.** For fusion, a collision of sufficient energy must occur between two positively charged particles to initiate the reaction. This requires high temperatures. In fission, an electrically neutral neutron collides with the positively charged nucleus. This has a much lower activation energy. **49.** The temperatures of fusion reactions are so high that all physical containers would be destroyed. At these high temperatures most of the electrons are stripped from the atoms. A plasma of gaseous ions is formed, which can be controlled by magnetic fields. **51.** moderator: slows the neutrons; control rods: absorb neutrons to slow or halt fission **54.** a. ${}^{12}_{6}C$; b. ${}^{13}N$, ${}^{13}C$, ${}^{14}N$, ${}^{15}O$, and ${}^{15}N$ are intermediates. c. 5.950×10^{11} J/mol H

Chapter 22

1. $CH_3—CH_2—CH_2—CH_2—CH_2—CH_3$, hexane or *n*-hexane (highest bp, least branched);

$CH_3—CH(CH_3)—CH_2—CH_2—CH_3$ 2-methylpentane;

$CH_3—CH_2—CH(CH_3)—CH_2—CH_3$ 3-methylpentane;

$CH_3—C(CH_3)_2—CH_2—CH_3$ 2,2-dimethylbutane;

$CH_3—CH(CH_3)—CH(CH_3)—CH_3$ 2,3-dimethylbutane

2. a. 2,3,3-trimethylhexane; b. 8-ethyl-2,5,5-trimethyldecane; c. 3-methylhexane **3.** There is only one consecutive chain of C atoms. They are not all in a straight line since the bond angle at each carbon is a tetrahedral angle of 109.5°.

4. a. $CH_3—CH(CH_3)—CH_2—CH_2CH_3$

b. $CH_3—C(CH_3)_2—CH_2—CH(CH_3)—CH_3$

c. $CH_3—CH(C(CH_3)_3)—CH_2CH_2CH_3$ [$CH_3—CH—CH_2CH_2CH_3$ with $CH_3—C—CH_3$ / CH_3 on the second carbon]

d. 2,2,3-trimethylhexane

5. a. 1-butene; b. 2-methyl-2-butene; c. 2,5-dimethyl-3-heptene

6. a. $CH_3—CH_2—CH{=}CH—CH_2—CH_3$

b. $CH_3CH{=}CHCH{=}CHCH_2CH_3$

c. $CH_3—CH(CH_3)—CH{=}CHCH_2CH_2CH_2CH_3$

7. a. benzene ring with CH_3 and CH_3 on adjacent carbons

b. $H_3C—C(CH_3)_2—C_6H_4—C(CH_3)_2—CH_3$ (para)

c. benzene ring with CH_2CH_3 and CH_2CH_3 (meta)

8. a. 1,3-dichlorobutane; b. 1,1,1-trichlorobutane; e. chlorobenzene; f. chlorocyclohexane; g. 3-chlorocyclohexene **9.** a. methylcyclopropane; b. *t*-butylcyclohexane; c. 3,4-dimethylcyclopentene **10.** isopropylbenzene or 2-phenylpropane

12. a. H—C(F)(H)—F b. H—C(Br)(Cl)—C(H)(Cl)—C(H)(H)—H

13. Structural isomers: Difference in types of bonds, either the kinds of bonds present or the way in which bonds connect atoms to each other. Geometrical isomers: Same bonds, differ in arrangement in space about rigid bond or ring. **14.** Resonance: All atoms are in the same position. Only the position of π electrons is different. Isomerism: Atoms are in different locations in space.

15.

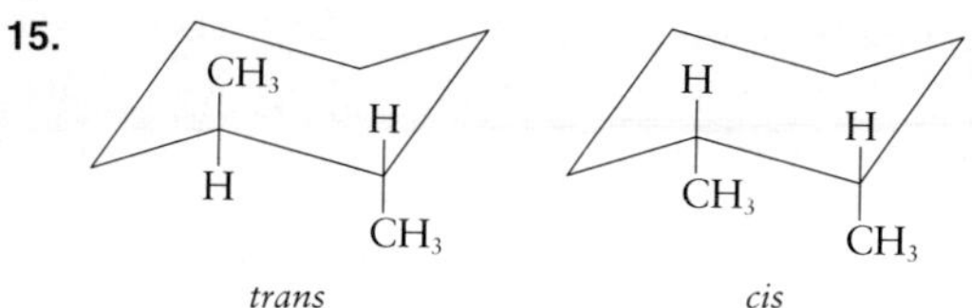

trans *cis*

16. $Cl_2C{=}CH—CH_3$ $CH_2{=}CCl—CH_2Cl$

$CH_2{=}CH—CHCl_2$

(H)(Cl)C=C(Cl)(CH₃); (Cl)(H)C=C(Cl)(CH₃); (Cl)(H)C=C(H)(CH₂Cl); (H)(Cl)C=C(H)(CH₂Cl)

1,1-dichlorocyclopropane; 1,2-dichlorocyclopropane *cis*; 1,2-dichlorocyclopropane *trans*

19. (F)(F)C=C(H)(H) Polar; (F)(H)C=C(F)(H) Polar; (F)(H)C=C(H)(F) Nonpolar

21. a. $CH_3\overset{O}{\overset{\|}{C}}CH_3$ is more stable; it contains the stronger bonds.

b. $C_6H_5{-}\overset{O}{\overset{\|}{C}}{-}\underset{H}{\underset{|}{N}}{-}CH_3$ is more stable; it contains the stronger bonds.

23. a. Carboxylic acid (CO_2H); Ester ($O{-}\overset{O}{\overset{\|}{C}}{-}CH_3$)

b. alcohol (phenol) (HO on ring); ether (O bridge); amine ($N{-}CH_3$); alcohol (HO)

c. alcohol (HO on ring); alcohol (OH); ether (O bridge); amine ($N{-}CH_2{-}CH{=}CH_2$); ketone ($O{=}$)

24.

$$HC{\equiv}C{-}C{\equiv}C{-}HC{=}C{=}CH{-}HC{=}CH{-}HC{=}CH{-}CH_2{-}\overset{O}{\overset{\|}{C}}{-}OH$$

13 12 11 10 9 8 7 6 5 4 3 2 1

26. a. ketone ($O{=}C$); alcohol ($H{-}O$); primary amine (NH_2); tertiary amine (ring N); carboxylic acid ($\overset{O}{\overset{\|}{C}}{-}O{-}H$)

b. 5 carbons in ring and in $-CO_2H$: sp^2; the other two carbons: sp^3; c. 24 σ bonds; 4 π bonds

29. $C_{12}H_{24}O_2$; Ethyl caprate is:

$$CH_3(CH_2)_8\overset{O}{\overset{\|}{C}}{-}OCH_2CH_3$$

30. a. $CH_3{-}CH_2{-}\overset{O}{\overset{\|}{C}}{-}H$ b. $CH_3{-}O{-}CH_2{-}CH_3$

c. $\underset{H}{\overset{H_3C}{}}C{=}C\underset{CH_3}{\overset{H}{}}$ d. $CH_3CH_2CH_2NH_2$

31. a. ketone; b. aldehyde; c. ketone; d. amine

33. Substitution: An atom or group is replaced by another atom or group. H in benzene is replaced by Cl.

$$C_6H_6 + Cl_2 \xrightarrow{\text{catalyst}} C_6H_5{-}Cl + HCl$$

Addition: Atoms or groups are added to a molecule. Cl_2 adds to ethene.

$$CH_2{=}CH_2 + Cl_2 \rightarrow CH_2Cl{-}CH_2Cl$$

34. a. two monochloro products: $CH_2ClCH_2CH_3$ and $CH_3CHClCH_3$; b. four dichloro products: $CHCl_2CH_2CH_3$, $CH_2ClCHClCH_3$, $CH_2ClCH_2CH_2Cl$, and $CH_3CCl_2CH_3$

35. a. $CH_2{=}CH_2 + Br_2 \rightarrow CH_2Br{-}CH_2Br$;
b. $C_6H_6 + Br_2 \xrightarrow{Fe} C_6H_5Br + HBr$;
c. $CH_3CO_2H + CH_3OH \rightarrow CH_3CO_2CH_3 + H_2O$

36. a. $CH_3CH{=}CH_2 + HCl \rightarrow CH_3CHClCH_3$
b. $CH_3CH{=}CH_2 + H_2O \rightarrow CH_3\underset{OH}{\underset{|}{C}}HCH_3$

c. 1-methylcyclohexene (CH_3) + HBr ⟶ 1-bromo-1-methylcyclohexane (CH_3, Br)

37. a. $CH_3{-}\overset{O}{\overset{\|}{C}}{-}H$ b. $CH_3{-}\overset{O}{\overset{\|}{C}}{-}CH_3$

38. a. i. $CH_3{-}\overset{O}{\overset{\|}{C}}{-}OH$ ii. $C_6H_5{-}\overset{O}{\overset{\|}{C}}{-}OH$

iii. $(CH_3)_2CH\overset{O}{\overset{\|}{C}}{-}OH$

b. $CH_3CH_2OH \xrightarrow{[ox]} CH_3\overset{O}{\overset{\|}{C}}H + CH_3\overset{O}{\overset{\|}{C}}OH$

$(CH_3)_3C{-}CH_2OH \xrightarrow{[ox]} (CH_3)_3C{-}\overset{O}{\overset{\|}{C}}H + (CH_3)_3C{-}\overset{O}{\overset{\|}{C}}{-}OH$

39. a. $CH_3CH{=}CH_2 + Br_2 \rightarrow CH_3CHBrCH_2Br$;

b. $CH_3C{\equiv}CH + H_2 \xrightarrow{\text{catalyst}} CH_3CH{=}CH_2 + Br_2 \rightarrow CH_3CHBrCH_2Br$

c. $CH_3CO_2H + CH_3CH_2CH_2CH_2OH \rightarrow$

$$CH_3\overset{O}{\overset{\|}{C}}{-}O{-}CH_2CH_2CH_2CH_3 + H_2O$$

40. a. addition polymer: polymer formed by adding monomer units to a double bond; Teflon; b. condensation polymer: polymer that forms when two monomers combine, eliminating a small molecule; nylon and Dacron; c. copolymer: polymer formed from more than one type of monomer; nylon and Dacron

42. $\left({-}O{-}\underset{}{\overset{CH_3}{\overset{|}{C}}}H{-}\overset{O}{\overset{\|}{C}}{-}O{-}\overset{CH_3}{\overset{|}{C}}H{-}\overset{O}{\overset{\|}{C}}{-}O{-}\overset{CH_3}{\overset{|}{C}}{-}\overset{O}{\overset{\|}{C}}{-}\right)_n$

44. $\left(\begin{array}{c} CN \\ | \\ C \\ | \\ C{-}OCH_3 \\ \| \\ O \end{array} {-}CH_2{-} \begin{array}{c} CN \\ | \\ C \\ | \\ C{-}OCH_3 \\ \| \\ O \end{array} {-}CH_2{-} \right)_n$

46. a.

$+O{-}C_6H_4{-}C(CH_3)_2{-}C_6H_4{-}O{-}\overset{O}{\overset{\|}{C}}{-}O{-}C_6H_4{-}C(CH_3)_2{-}C_6H_4{-}O{-}\overset{O}{\overset{\|}{C}}+_n$

b. condensation

47. Divinylbenzene crosslinks different chains to each other. The chains cannot move past each other because of the crosslinks; thus, the polymer is more rigid. **48.** The stronger interparticle forces would be found in polyvinyl chloride, since there are also dipole–dipole forces in PVC that are not present in polyethylene. **49.** a. $CHF{=}CH_2$; b. $HO{-}CH_2CH_2{-}CO_2H$; c. copolymer of $HOCH_2CH_2OH$ and $HO_2CCH_2CH_2CO_2H$

51. a. 2-methyl-1,3-butadiene;

b.

$\left(H_2C(H_3C)C{=}CH{-}CH_2{-}CH_2{-}C(H_3C){=}CH{-}CH_2{-}CH_2{-}C(H_3C){=}CH{-}CH_2 \right)_n$

cis-polyisoprene (natural rubber)

$\left(CH_2{-}C(H_3C){=}CH{-}CH_2{-}CH_2{-}C(H_3C){=}CH{-}CH_2{-}CH_2{-}C(H_3C){=}CH{-}CH_2 \right)_n$

trans-polyisoprene (gutta percha)

52. a. $\left(OCH_2CH_2O\overset{O}{\overset{\|}{C}}CH{=}CH\overset{O}{\overset{\|}{C}} \right)_n$

b.

$\left(OCH_2CH_2O\overset{O}{\overset{\|}{C}}{-}CH{-}CH\overset{O}{\overset{\|}{C}} \right)_n$ with CH_2 on the first CH, linked through $CH(C_6H_5)$ to

$\left(OCH_2CH_2O\underset{O}{\underset{\|}{C}}{-}CH{-}CH\overset{O}{\overset{\|}{C}} \right)_n$ with CH_2 linked to $CH(C_6H_5)$ ⋮

54. For the reaction $3CH_2{=}CH_2(g) + 3H{-}H(g) \rightarrow 3CH_3{-}CH_3(g)$, $\Delta H = -381$ kJ using bond energies. From enthalpies of formation, $\Delta H^\circ = -410.$ kJ. The two values agree fairly well. For $C_6H_6(g) + 3H_2(g) \rightarrow C_6H_{12}(g)$, we would get the same ΔH from bond energies, -381 kJ. From enthalpies of formation, $\Delta H^\circ = -173.2$ kJ. There is a large discrepancy between these two values of ΔH. The reason is that benzene is more stable (lower in energy) than we expect from bond energies. This extra stability is a result of the delocalized π electrons found in the benzene ring.

56.

$\left(OCH_2CH_2O\overset{O}{\overset{\|}{C}}NH{-}C_6H_4{-}NH\overset{O}{\overset{\|}{C}}OCH_2CH_2O\overset{O}{\overset{\|}{C}}NH{-}C_6H_4{-}NH\overset{O}{\overset{\|}{C}} \right)_n$

Chapter 23

1. a. $K = 2.6 \times 10^9$; equilibrium lies far to the right. b. $^+H_3NCH_2CO_2H$ (1.0 M H^+); $H_2NCH_2CO_2^-$ (1.0 M OH^-) **2.** a. aspartic acid and phenylalanine; b. Aspartame contains the methyl ester of phenylalanine. This ester can hydrolyze to form methanol, $R{-}CO_2CH_3 + H_2O \rightleftharpoons RCO_2H + CH_3OH$.

3. Crystalline amino acids exist as zwitterions, $^+H_3N{-}\underset{}{\overset{R}{\overset{|}{C}}}H{-}CO_2^-$. The interparticle forces are strong. Before the temperature gets high enough to break the ionic bonds, the amino acid decomposes. **4.** They are both hydrophilic amino acids because both contain highly polar R groups. **7.** Glutamic acid has a polar R group and valine has a nonpolar R group. The change in polarity of the R groups could affect the tertiary structure of hemoglobin and affect the ability of hemoglobin to bond to oxygen.

10.

$H_2NCH(CH_2OH)\overset{O}{\overset{\|}{C}}{-}NHCH(CH_3)CO_2H$ — ser-ala

$H_2NCH(CH_3)\overset{O}{\overset{\|}{C}}{-}NHCH(CH_2OH)CO_2H$ — ala-ser

11. three **13.** Both denaturation and inhibition reduce the catalytic activity of an enzyme. Denaturation changes the structure of an enzyme; inhibition involves the attachment of an incorrect molecule at the active site. **14.** The initial increase in rate is a result of the effect of temperature on the rate constant. At higher temperatures the enzyme begins to denature, losing its activity, and the rate decreases. **15.** An amino acid contains both acidic and basic functional groups. Thus, an amino acid can act as both a weak acid and a weak base; this is the requirement for a buffer.

18.

D-ribose: five-membered ring — $HOCH_2$, O, H; H, H, H, OH; OH, OH (chiral carbons marked *)

D-mannose: six-membered ring — CH_2OH, O; H, H, H; HO, OH, HO, OH; H, H (chiral carbons marked *)

19. Chiral carbons are marked with asterisks in Exercise 18. **22.** structural isomers: same formula, different functional groups or chain lengths (different bonds); geometrical isomers: same functional groups (same bonds), but different arrangement of some groups in space; optical isomers: compounds that are nonsuperimposable mirror images of each other.

23.

$H_3C—C^*H(—CH_2CH_3)—C^*H(NH_2)—CO_2H$ (isoleucine)

$H_3C—C^*H(—OH)—C^*H(NH_2)—CO_2H$ (threonine)

isoleucine threonine

25.

CH_3 CH_3 CH_3 CH_3 CH_3 HO

eight chiral carbon atoms (marked with *)

28.

CH_3 H_3C CH_3

two chiral carbon atoms (marked with *)

30. nitrogen atoms with lone pairs **31.** 1.7 m ≈ 2 m
33. T-A-C-G-C-C-G-T-A **34.** 64
35. Uracil will H-bond to adenine.

O HN O N H O---H—NH N NH-----N N N sugar N O sugar

uracil adenine

36. a. glu: CTT, CTC; val: CAA, CAG, CAT, CAC; met: TAC; trp: ACC; phe: AAA, AAG; asp: CTA, CTG;

b. TAC-TAC-AAA-CTA-ACC;
or or
AAG CTG

c. four; d. glu-trp-phe; e. C-T-C-A-C-C-A-A-A; C-T-T-A-C-C-A-A-G; C-T-C-A-C-C-A-A-G **40.** a. The structure with two C=O bonds is more stable. b. It could hydrogen-bond to guanine, forming a G-T base pair instead of A-T. **41.** Lipids are nonpolar and soluble in organic solvents. Carbohydrates contain several —OH groups capable of hydrogen-bonding and are soluble in water.

42.

$CH_2—OC(=O)(CH_2)_7(CH=CHCH_2)_2CH_2CH_2CH_2CH_3$
$CH—OC(=O)(CH_2)_7(CH=CHCH_2)_2CH_2CH_2CH_2CH_3$
$CH_2—OC(=O)(CH_2)_7(CH=CHCH_2)_2CH_2CH_2CH_2CH_3$

43. It will take 6 mol of H_2 to hydrogenate this triglyceride completely. Nine products with 2 double bonds are possible. Shorthand notation for the different products are

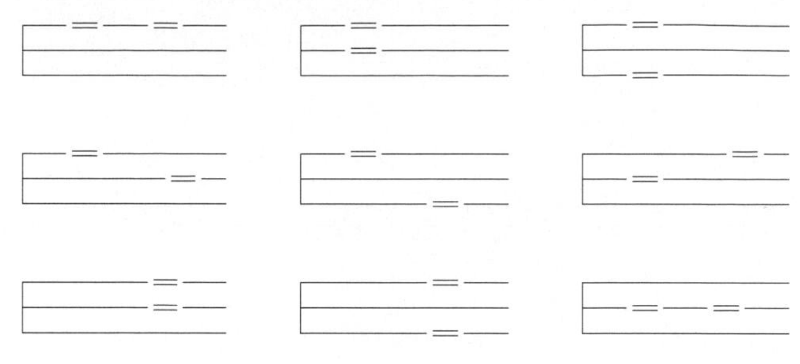

47.

$CH_2—O—C(=O)—R$
$HC—O—C(=O)—R$
$H_2C—O—P(=O)(O^-)—OCH_2CH_2N^+(CH_3)_3$

Hydrophobic
Hydrophilic

Lecithin

$CH_3(CH_2)_{12}—CH=CH$
HOCH
R—CHNCH (C=O)
$CH_2—O—P(=O)(OH)—OCH_2CH_2N^+(CH_3)_3$

Hydrophobic
Hydrophilic

Sphingomyelin

50. six **52.** a. 1.7×10^{-12}; b. 2.3×10^{-12}; c. 7.3×10^{-15} **53.** pH = 7.07 **55.** a. $\Delta H = -23$ kJ; b. For gly + gly ⇌ gly-gly, ΔS is negative (unfavorable); c. ΔG is probably positive because of the negative entropy change. **56.** In the reaction, P—O and O—H bonds are broken and P—O and O—H bonds are formed. Thus $\Delta H \approx 0$. $\Delta S < 0$, since two molecules are going to one molecule. Thus $\Delta G > 0$, not spontaneous. **57.** A cell is not a closed system. There is an external source of energy to drive the reaction. A photosynthetic plant uses sunlight, and animals use the carbohydrates produced by plants as sources of energy. For a cell $\Delta S_{sys} < 0$, but $\Delta S_{surr} > 0$ and

$$|\Delta S_{surr}| > |\Delta S_{sys}|$$

Therefore, ΔS_{univ} increases.

Photo Credits

Title page and chapter openers Courtesy, IBM Corporation, Research Division, Almaden Research Center.

Chapter 1 p. 2, H. Mark Weidman; p. 3, Kristen Brochmann/Fundamental Photographs; p. 11, Larry Lefever/Grant Heilman Photography; p. 15 (top), Courtesy, Kathleen Miller; p. 15 (bottom), Courtesy, Betsy Hovey; p. 16, Courtesy, Keith Reese; p. 17, Courtesy, Kathy Wolfgram; p. 18, Courtesy, Paul Schiller.

Chapter 2 p. 20, Courtesy, Roald Hoffman, Cornell University; p. 21, (detail) *Antoine Laurent Lavoisier and His Wife,* oil on canvas by Jacques Louis David. (The Metropolitan Museum of Art, Purchase, Mr. And Mrs. Charles Wrightsman Gift, in honor of Everett Fahy, 1977. #1977.10); p. 22, The Granger Collection; p. 23, The Granger Collection; p. 26, *Jons Jacob Berzelius,* oil by Olof Johan Sodermark, 1843 (Courtesy, The Royal Swedish Academy of Sciences); p. 27 (top), Roger Du Buisson/The Stock Market; p. 27 (bottom), Michael W. Davidson/Photo Researchers, Inc.; p. 31, *Ernest Rutherford,* oil by Oswald Birley, 1932 (Royal Society, London/Art Resource); p. 34 (both), Tim Alt/Digital Art; p. 35 (top left), E.R. Degginger; p. 35 (top right), Ken O'Donoghue; p. 35 (bottom left), Ken O'Donoghue; p. 35 (bottom right), E.R. Degginger; p. 35 (far right), Tom Pantages; p. 36 (right), E.R. Degginger; p. 38, Tom Pantages; p. 39, E.R. Degginger; p. 41 (left), Tom Pantages; p. 41 (right), E.R. Degginger; p. 42, Ken Eward/Science Source/Photo Researchers, Inc.; p. 43, Ken O'Donoghue.

Chapter 3 p. 55, Ken O'Donoghue; p. 61, CNRI/Science Photo Library/Photo Researchers, Inc.; p. 65, Ken O'Donoghue; p. 67, Yoav/Phototake; p. 68, Ken O'Donoghue; p. 71 (both), Ken O'Donoghue; p. 72, Ken Straiton/First Light; p. 75, Courtesy, Chevron Chemical Company.

Chapter 4 p. 92 (both), Ken O'Donoghue; p. 98, Ken O'Donoghue; p. 100, Ken O'Donoghue; p. 101, Ken O'Donoghue; p. 102, E.R. Degginger; p. 108, Runk/Schoenberger/Grant Heilman Photography; p. 111 (top three), Ken O'Donoghue; p. 111 (bottom), Courtesy, Clean Harbors Environmental Services, Inc.; p. 114, Ken O'Donoghue; p. 124, Tom McHugh/Photo Researchers, Inc.; p. 126, E.R. Degginger.

Chapter 5 p. 145, E.R. Degginger; p. 148, Courtesy, Ford Motor Company; p. 150, Darryl Torckler/Tony Stone Images; p. 156 (all) Ken O'Donoghue; p. 161 (top), Ken O'Donoghue; p. 161 (bottom), Paducah Gaseous Diffusion Plant, managed by Martin Marietta Utility Services, Inc. for the United States Enrichment Corporation under Contract No. USECHQ-93-C-0001; p. 163, Courtesy, Energy Technology Visuals Collection/U.S. Department of Energy; p. 173 (top), John Lawlor/Tony Stone Images; p. 173 (bottom), Oliver Strewe/Tony Stone Images; p. 175 (both), Don & Pat Valenti/Tony Stone Images.

Chapter 6 p. 186, E.R. Degginger; p. 195, Jerry Howard/Positive Images; p. 205, Ken O'Donoghue; p. 208 (all), Ken O'Donoghue; p. 210 (both), Ken O'Donoghue.

Chapter 7 p. 226 (both), Ken O'Donoghue; p. 242, Gene Stein/West Light.

Chapter 8 p. 271, Ken O'Donoghue; p. 272 (both), Ken O'Donoghue; p. 300 (both), E.R. Degginger; p. 302 (all), Ken O'Donoghue; p. 309, CNRI/Science Library/Photo Researchers, Inc.; p. 311, Ken O'Donoghue; p. 314, E.R. Degginger; p. 321 (both), E.R. Degginger; p. 322, E.R. Degginger; p. 326 (both), E.R. Degginger.

Chapter 9 p. 358, Phil Degginger; p. 361, Courtesy, Argonne National Laboratory; p. 363, Al Rubin/Stock, Boston; p. 369, Richard Dole/Duomo.

Chapter 10 p. 396, Tom Pantages; p. 404, Steven S. Zumdahl; p. 411, Brian Parker/Tom Stack & Associates; p. 421, Paul Silverman/Fundamental Photographs.

Chapter 11 p. 452, Ken O'Donoghue; p. 457, *Michael Faraday Lecturing at The Royal Institution in 1855* (Courtesy, The Royal Institution of Great Britain); p. 459, E.R. Degginger; p. 464, E.R. Degginger; p. 470 (bottom), Courtesy, NASA; p. 473, Ron Watts/West Light; p. 477, Yoav/Phototake; p. 478, Courtesy, Oberlin College Archives; p. 481, Courtesy, Kennecott; p. 483, Courtesy, Alupower, Inc.

Chapter 12 p. 499, Grant LeDuc/Stock, Boston; p. 500, The Granger Collection; p. 502, Lightscapes/The Stock Market; p. 505, Howard Sochurek; p. 506, The Granger Collection; p. 509, Courtesy, National Institute of Standards & Technology; p. 511, Richard Megna/Fundamental Photographs; p. 513, (both) Courtesy, IBM Corporation, Research Division, Almaden Research Center; p. 536, The Bettmann Archive.

Chapter 13 p. 568 (left), Manfred Kage/Peter Arnold, Inc.; p. 568 (right), Runk/Schoenberger/Grant Heilman Photography; p. 587, Ken O'Donoghue; p. 589, Will & Deni McIntyre/Photo Researchers, Inc.; p. 596, The Bancroft Library/University of California; p. 617, Courtesy, Argonne National Laboratory.

Chapter 14 p. 634, Bill Pogue/Tony Stone Images; p. 654 (both), Donald Clegg.

Chapter 15 p. 710, E.R. Degginger; p. 711, Courtesy, The Degussa Corporation.

Chapter 16 p. 736, Yoav/Phototake; p. 738 (top left), Brian Parker/Tom Stack & Associates; p. 738 (top right), Gary Milburn/Tom Stack & Associates; p. 738 (bottom left), Brian Parker/Tom Stack & Associates; p. 738 (bottom right), Brian Parker/Tom Stack & Associates; p. 741, Wolfgang Volz/Bilderberg/The Stock Market; p. 749, Yoav/Phototake; p. 752, Courtesy, Raytheon. Photograph by Don Bernstein; p. 753, Professor Williams, LeHigh University; p. 757, Courtesy Raytheon. Photograph by Don Bernstein; p. 760 (left), Paul Silverman/Fundamental Photographs; p. 760 (right), E.R. Degginger; p. 772, E.R. Degginger.

Chapter 17 p. 795, Richard T. Nowitz; p. 801 (both), Courtesy of the U.S. Department of Agriculture; p. 805 (top two), Betz/Visuals Unlimited; p. 805 (bottom), T. Orbar/Sygma; p. 819, Courtesy, Recovery Engineering, Inc.; p. 821, Courtesy Ferrofluidics; p. 823, Stephen Frisch/Stock, Boston.

Chapter 18 p. 835, Mark Heifner/Panographics; p. 837, Barry L. Runk/Grant Heilman; p. 838, Ken O'Donoghue; p. 840 (both), Donald Clegg; p. 842, Tom Pantages; p. 844, Carl Frank/Photo Researchers, Inc.; p. 846, Richard Megna/Fundamental Photographs; p. 850 (top left), E.R. Degginger; p. 850 (top right), Michael Holford; p. 850 (bottom), E.R. Degginger.

Chapter 19 p. 858, © Roger Ressmeyer/Starlight; p. 861, Hugh Spencer/Photo Researchers, Inc.; p. 862 (top), Courtesy, NASA; p. 862 (bottom), E.R. Degginger; p. 863, Richard Megna/Fundamental Photographs; p. 867, Stephen Frisch/Stock, Boston; p. 872, E.R. Degginger; p. 873, Farrell Grehan/Photo Researchers, Inc.; p. 874 (all), E.R. Degginger; p. 876, Larry Stepanowitz/Panographics; p. 878 (top), Yoav/Phototake; p. 878 (bottom), E.R. Degginger; p. 879 (both), Andrew Brilliant/Carol Palmer.

Chapter 20 p. 893 (top left), Paul Silverman/Fundamental Photographs; p. 893 (top right), Paul Silverman/Fundamental Photographs; p. 893 (bottom left) Gary Retherford/Photo Researchers, Inc.; p. 893 (bottom right), Paul Silverman/Fundamental Photographs; p. 894, Ken O'Donoghue; p. 898, Courtesy Pilling Weck; p. 899, Ken O'Donoghue; p. 900, Tony Scarpetta; p. 901, Ken O'Donoghue; p. 908 (both), Ken O'Donoghue; p. 912, Martin Bough/Fundamental Photographs; p. 930 (top), Ken Eward/Science Source/Photo Researchers, Inc.; p. 930 (bottom), Bill Longcore/Science Source/Photo Researchers, Inc.

Chapter 21 p. 948, University of Pennsylvania; p. 951 (both), CNRI/Photo Researchers, Inc.; p. 959 (top), Mike Mazzaschi/Stock, Boston; p. 959 (bottom), Courtesy, Raytheon. Photograph by Don Bernstein; p. 961, News and Information Service, University of Utah; p. 963 Dan McCoy/Rainbow.

Chapter 22 p. 990 (top), Kevin Schafer/Tom Stack & Associates; p. 995, Bob Masini/Phototake; p. 998, Donald Clegg; p. 999, Courtesy, Dupont.

Chapter 23 p. 1012 (left), Dr. Don Faucett/Photo Researchers, Inc.; p. 1012 (right), G. Musil/Visuals Unlimited; p. 1032, Chemical Design Ltd.; p. 1039, Ralph Eagle Jr./Photo Researchers, Inc.

INDEX